W9-BWM-872

ENCYCLOPEDIA OF HUMAN BIOLOGY

VOLUME 3 Di–Gl

ENCYCLOPEDIA OF HUMAN BIOLOGY

VOLUME 3 Di–Gl

Editor–in–Chief

Renato Dulbecco

The Salk Institute

La Jolla, California

ACADEMIC PRESS, INC. *Harcourt Brace Jovanovich, Publishers*

San Diego New York Boston London Sydney Tokyo Toronto

This book is printed on acid-free paper. ∞

Academic Press, Inc.
San Diego, California 92101

United Kingdom Edition published by
Academic Press Limited
24–28 Oval Road, London NW1 7DX

Library of Congress Cataloging-in-Publication Data

Encyclopedia of human biology / [edited by] Renato Dulbecco.
p. cm.
Includes index.
ISBN 0-12-226751-6 (v. 1). -- ISBN 0-12-226752-4 (v. 2). -- ISBN 0-12-226753-2 (v. 3). -- ISBN 0-12-226754-0 (v. 4). -- ISBN 0-12-226755-9 (v. 5). -- ISBN 0-12-226756-7 (v. 6). -- ISBN 0-12-226757-5 (v. 7). -- ISBN 0-12-226758-3 (v. 8)
1. Human biology--Encyclopedias. I. Dulbecco, Renato, 1914-
[DNLM: 1. Biology--encyclopedias. 2. Physiology--Encyclopedias.
QH 302.5 E56]
QP11.E53 1991
612'.003--dc20
DNLM/DLC
for Library of Congress 91-45538
CIP

PRINTED IN THE UNITED STATES OF AMERICA

91 92 93 94 9 8 7 6 5 4 3 2 1

CONTENTS OF VOLUME 3

HOW TO USE THE ENCYCLOPEDIA

We have organized this encyclopedia in a manner that we believe will be the most useful to you and would like to acquaint you with some of its features.

The volumes are organized alphabetically as you would expect to find them in, for example, magazine articles. Thus, "Food Toxicology" is listed as such and would *not* be found under "Toxicology, Food." If the first words in a title are *not* the primary subject matter contained in an article, the main subject of the title is listed first: (e.g., "Sex Differences, Biocultural," "Sex Differences, Psychological," "Aging, Psychiatric Aspects," "Bone, Embryonic Development.") This is also true if the primary word of a title is too general (e.g., "Coenzymes, Biochemistry.") Here, the word "coenzymes" is listed first as "biochemistry" is a very broad topic. Titles are alphabetized letter-by-letter so that "Gangliosides" is followed by "Gangliosides and Neuronal Differentiation" and then "GangliosideTransport."

Each article contains a brief introductory Glossary wherein terms that may be unfamiliar to you are defined *in the context of their use in the article*. Thus, a term may appear in another article defined in a slightly different manner or with a subtle pedagogic nuance that is specific to that particular article. For clarity, we have allowed these differences in definition to remain so that the terms are defined relative to the context of each article.

Articles about closely related subjects are identified in the Index of Related Articles at the end of the last volume (Volume 8.) The article titles that are cross-referenced within each article may be found in this index, along with other articles on related topics.

The Subject Index contains specific, detailed information about any subject discussed in the *Encyclopedia*. Entries appear with the source volume number in boldface followed by a colon and the page number in that volume where the information occurs (e.g., "Diuretics, **3:** 93"). Each article is also indexed by its title (or a shortened version thereof) and the page ranges of the article appear in boldface (e.g., "Abortion, **1: 1–10**" means that the primary coverage of the topic of abortion occurs on pages 1–10 of Volume 1).

If a topic is covered primarily under one heading but is occasionally referred to in a slightly different manner or by a related term, it is indexed under the term that is most commonly used and a cross-reference is given to the minor usage. For example, "B Lymphocytes" would contain all page entries where relevant information occurs, followed by "*see also* B Cells." In addition, "B Cells, *see* B Lymphocytes" would lead the reader to the primary usages of the more general term. Similarly, "*see under*" would mean that the subject is covered under a subheading of the more common term.

An additional feature of the Subject Index is the identification of Glossary terms. These appear in the index where the word "defined" (or the words "definition of") follows an entry. As we noted earlier, there may be more than one definition for a particular term and, as when using a dictionary, you will be able to choose among several different usages to find the particular meaning that is specifically of interest to you.

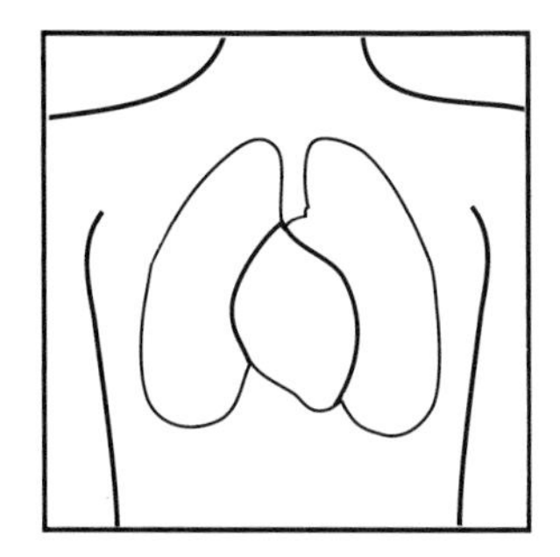

Diagnostic Radiology

ALEXANDER R. MARGULIS AND RUEDI F. THOENI, *University of California–San Francisco*

Glossary

Angiography Radiographic visualization of blood vessels following the introduction of contrast material into the vasculature

Computed tomography Two-dimensional representation of the linear X-ray attenuation coefficient distribution through a narrow planar cross-section of the body

Contrast medium Substance which is introduced into the body because of the difference in absorption of X-rays by the contrast medium and the surrounding tissues and which permits radiographic visualization of that structure

Interventional radiology Roentgenographic technique using the ability to observe and control manipulations within the body via a large variety of catheters and guide wires

Magnetic resonance imaging Imaging of the body which involves the interaction of nuclei of a selected atom with an external magnetic field and an external oscillating (i.e., radiofrequency) electromagnetic field that is changing as a function of time at a particular frequency

Mammography Roentgenography of the breast

Roentgenography, or radiography The making of film records (i.e., radiographs) of internal structures of the body by passage of X-rays or γ-rays through the body to act on specially sensitized film. Roentgen rays are produced when electrons generated by high voltage and moving at high velocity impinge on various substances, particularly heavy metal.

Ultrasonography Visualization of deep structures of the body by recording the reflection of ultrasonic waves directed into the tissues

DIAGNOSTIC RADIOLOGY embodies multiple imaging approaches to render diagnoses on the basis of altered morphology. It consists of conventional radiography and fluoroscopy, ultrasonography, computed X-ray tomography, and magnetic resonance imaging. Nuclear medicine, which involves imaging with radioactive isotopes, is not included here.

I. Plain Film Radiography

X-Rays were discovered in 1895 by Wilhelm Conrad Röntgen, who published his first results in 1896. X-Rays are produced when cathode rays in a vacuum tube strike the anode, producing photons of energy up to 140 kV in modern diagnostic X-ray machines. After filtration the useful range in diagnostic procedures is between 80 and 120 kV.

Fluoroscopic rooms permit direct visualization of structures inside the body, including bones and soft tissue, via image intensifiers and television surveying systems, whereas simple radiographic rooms are used only for obtaining radiographs. Modern fluoroscopic equipment consists of a generator, X-ray tube, image-intensifying tube, and fluoroscopic viewing system. An up-to-date radiographic room does not need fluoroscopic viewing or image-amplifying systems. Some modern radiographic rooms, however, are equipped with digital systems. These digital systems are used to minimize radiation exposure to the patient, to reduce the amount of contrast material administered, and to permit image manipulation (e.g., instant subtraction angiography).

In addition to the standard radiographic and fluoroscopic equipment, specialized rooms for angiography and interventional radiography also have digital subtraction devices. These depict the background before the injection of contrast medium in a positive or negative mode and display the same background after the injection of contrast medium into the vessel in the opposite mode. The subtraction results in deletion of all of the anatomic structures, except the vessels. Angiographic and interventional suites are frequently equipped with biplane fluoroscopic equipment, which has two X-ray tubes and two image intensifiers, providing the simultaneous viewing of images at right angles to each other. The image intensifier and the X-ray tubes are often attached to semicircular mobile devices (i.e., C arms) which hold the X-ray tube mount on one end and the image-intensifier device on the other. This permits rapid movement of the X-ray tube and the image intensifier without the need for additional alignment of the two.

Some angiographic rooms are also equipped with serial film changes, devices which permit multiple X-ray exposures on films that are quickly changed and synchronized to X-ray exposures. Such equipment allows the recording of rapidly recurring events, such as the flow of radiopaque contrast medium through blood vessels or heart chambers. The majority, however, use cine filming devices and/or videotape. Mammographic rooms have equipment that allows the display of fine soft tissue detail of the breast.

Portable mobile X-ray equipment is being used with increasing frequency in hospital rooms and hospital operating rooms. It is flexible and capable of providing superb images, and its use reflects the greater orientation of hospitals toward surgical procedures. Most of the portable machines are equipped with a high-energy rechargeable battery to render the unit independent of electric power fluctuation in the hospital.

II. Plain Film Findings

A. Skull

Films of the skull are today generally used for screening: to rule out skull fractures in emergency rooms and to diagnose scalp abnormalities, including tumors or inflammatory processes of the cranium. Today, the brain itself is best examined with magnetic resonance imaging (MRI) or computed tomography (CT).

B. Spine

Plain film radiography of the spine is used for the screening examination of fractures of the cervical, thoracic, or lumbar spine; for the evaluation of osteoarthritis associated with osteophytes, which represent bone spurs resulting, often, from even mild injuries; for the evaluation of the height of the intervertebral disks; for the diagnosis of primary and metastatic tumors; and for the study of congenital anomalies (e.g., fused vertebrae, hemivertebrae, and defects in the neural arch). In the cervical region traumatic dislocations or fractures are of particular importance, and plain film radiography is essential. In the lumbosacral spine slipping of vertebral bodies on each other, because of congenital defects in the intervertebral articulating processes or because of trauma, is another pathological change in which plain film radiography is helpful.

C. Chest

Plain film evaluation of heart size (Figs. 1 and 2) and shape and the detection of coronary calcifications are still of great clinical value. A diagnosis of enlargement of the atrium and/or ventricle on the left or the right side of the heart can be made in this way. Plain film radiography of the lungs is the primary method of evaluating lung disease and still the most frequent radiographic procedure performed. It can demonstrate diffuse or lobar processes, primary and metastatic neoplasms, and a whole range of alveolar or interstitial disease.

In the examination of the mediastinum (i.e., the space between the lungs), plain film radiography is of less use than methods that provide cross-sectional anatomy (CT or MRI). Nevertheless, widening or change of shape of the mediastinum can be well evaluated, and plain film radiography is often used as a screening method for abnormalities in this region.

D. Abdomen

Plain film radiography of the abdomen is still one of the most important examinations performed in a radiography institution, particularly in emergency sit-

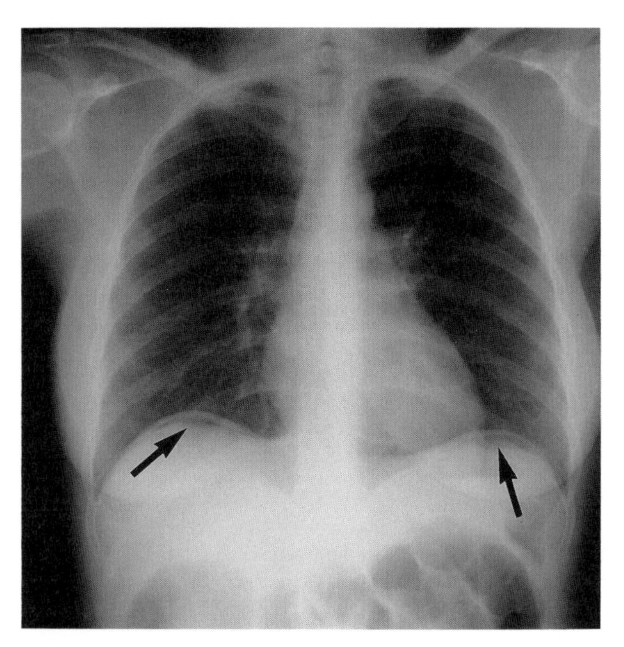

FIGURE 1 Conventional chest radiograph (posterior–anterior projection) of a 24-year-old patient with mild upper abdominal pain. The lungs are clear, and the heart is normal, but a small amount of air (arrows) is seen under each hemidiaphragm.

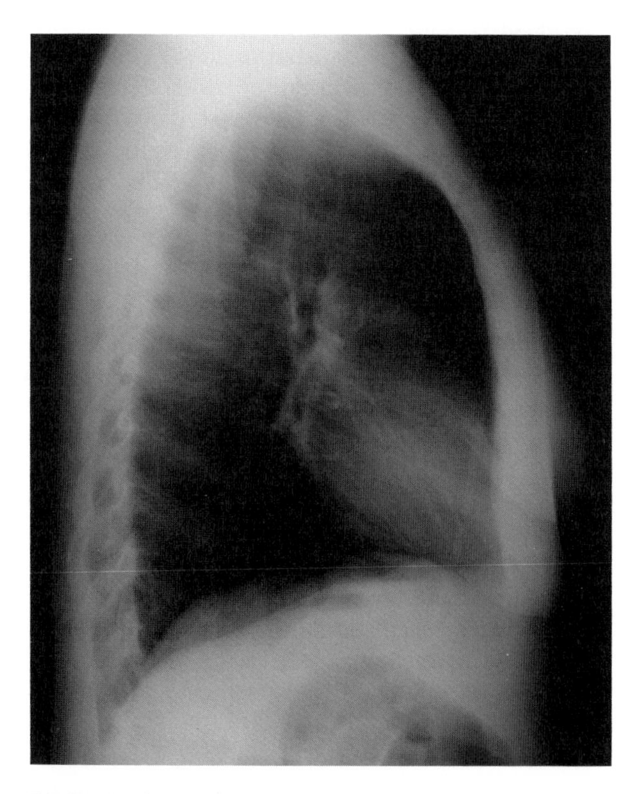

FIGURE 2 Conventional chest radiograph (lateral projection). No definite abnormality is seen on this lateral projection of the chest. A normal-sized heart is well outlined.

uations. Perforations of viscera are diagnosed based on the detection of free air caught under the diaphragm in the upright position or between the edge of the liver and the abdominal wall, with the patient lying on the left side. With this approach even a few milliliters of free air can be accurately detected. Obstructions of the intestinal tube can also be properly visualized by obtaining upright and recumbent films, which demonstrate dilated loops of bowel with air–fluid levels. However, these examinations often must be supplemented by administering radiopaque contrast material either by mouth or by rectum to determine the exact location and course of the obstruction. On radiographs calcium appears white and fat dark, but less dark than air. Adynamic ileus, which shows multiple radiolucent (i.e., dark) bowel loops, represents decreased or absent motility, and dilatation of the small bowel can usually be differentiated from mechanical obstruction by the presence of gas distension throughout the small and large bowels, particularly in the rectum.

This finding is in contradistinction to obstruction which is diagnosed based on distension of the bowel proximal (i.e., toward the mouth) to the obstruction and collapse of the bowel distal (i.e., toward the anus) to it. Masses in the abdomen are best seen on plain radiographs if they contain calcification or fat. In the absence of these features, masses are detected by displacement of the gas-filled bowel. Occasionally, masses are diagnosed because they obliterate fat planes enclosed in the retroperitoneum. Calcifications in tumors, in the peritoneal cavity, or in arteries or veins and stones in the kidneys, gallbladder, or urinary bladder are also well seen on plain films.

E. Extremities

Plain radiographs of extremities—including wrists, hands, ankles, and feet—are obtained with great frequency to detect fractures, dislocations, inflammations, and/or tumors of the soft tissues or bones.

III. Angiographic Principles

A. Contrast Media

Contrast media are used in roentgenography to increase the contrast between specific organs and surrounding tissues. They easily demonstrate vessels injected by either the intravenous or arterial route. Modern contrast media are based on the radiopacity of iodine incorporated in organic compounds. Current ionic intravascular contrast media have been used for about three decades, but have been challenged by nonionic intravascular contrast agents, which apparently are safer and have fewer side effects. All of these contrast agents are excreted by the kidneys and are used for excretory urography, angiography, and CT. If intravenous contrast agents are used for CT, tissue contrast is increased, in addition to the demonstration of vessels. With digital subtraction angiography, smaller amounts of contrast media can be used, particularly if arterial injections are employed.

B. Angiography

Angiography uses rapid film sequences during a bolus of contrast material, which consists of rapid injection of contrast medium (e.g., 60 ml at a rate of 2–5 ml/second). This bolus is administered with a mechanical adjustable injection. For this purpose catheters are placed through arteries or veins, using a femoral, cubital, axillary, jugular, or carotid approach. In recent years digital angiography has permitted the use of smaller doses of contrast material, due to computer manipulation of the data obtained, which includes subtraction techniques and edge enhancement.

IV. Gastrointestinal Tract

Fine-particle barium suspensions are used orally or by enema to examine the gastrointestinal tube. For the upper gastrointestinal examination the best detail is obtained by adding carbon-dioxide-forming crystals together with surface tension-reducing agents (e.g., simethicone) before administering barium. For examination of the colon, air is insufflated into the rectum after barium has been administered by enema. Both of these procedures are called double-contrast examinations, because air and barium combine to permit a "see-through" effect. For the single-contrast barium enema, barium alone is infused.

Double-contrast examination of the small bowel is obtained by enteroclysis (Fig. 3). The examination consists of passing a tube beyond the duodenojejunal junction and injecting barium followed by methylcellulose. This method permits better control of distention and allows the demonstration of superb mucosal detail. In many instances enteroclysis provides more information than that obtainable by

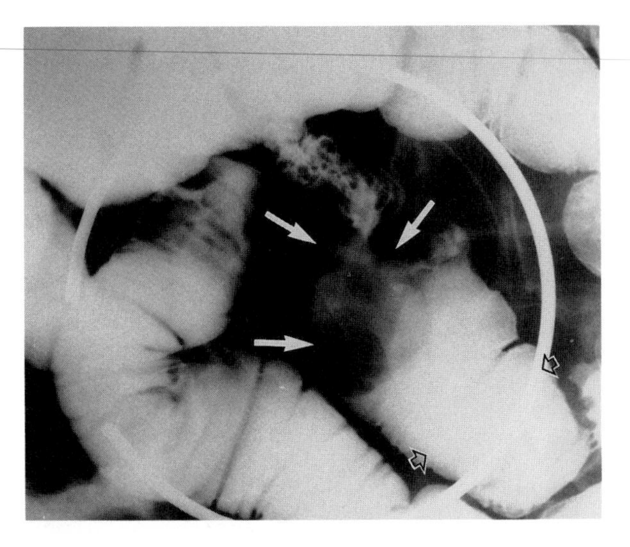

FIGURE 3 Single-contrast enteroclysis. A large mass (white arrows) is seen. Note dilatation of the small bowel (open arrows) proximal to the lesion.

oral administration of barium and assessment of the small bowel by means of spot films and serial overhead films.

For spot films the X-ray tube is below the patient and the film is above. Spot films are obtained during fluoroscopy, with the examiner recording on film any anatomic detail or abnormality seen during screening. Overhead films are obtained by the technologist following fluoroscopy, with the patient lying on the table, the film in the cassette located underneath, and the X-ray tube above the patient.

Water-soluble iodine-containing contrast media are used if perforation of the intestinal tube is suspected, because they are absorbed from the soft tissues and do not produce an inflammatory reaction, resulting in scarring, such as is caused by barium. The same substances are used for cleansing a colon impacted with viscous fecal material, since they act as laxatives, similar to milk of magnesia, due to their high salt concentration (i.e., osmolarity). If an aspiration (i.e., inadvertent passage of food or liquid into the pulmonary system) or esophagopulmonary fistula is suspected, water-soluble iodine-containing contrast media should never be used orally because pulmonary edema could ensue, due to its high osmolarity.

In the performance of gastrointestinal examinations, the esophagus, stomach, colon, and small bowel are all examined first fluoroscopically and then with overhead radiographic films. Fluoroscopy is considered to be important since it is directing the

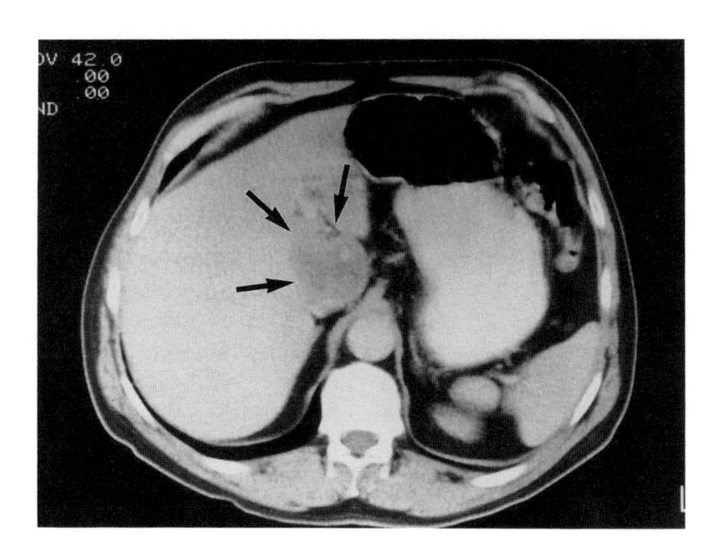

FIGURE 4 CT of the upper abdomen. A low-density lesion (arrows) is seen in the caudate lobe of the liver of a patient with metastatic colon carcinoma.

subsequent conduct of the examination. Barium examinations of the gastrointestinal tract are predominantly for diagnosing ulcers, inflammatory conditions, tumors, congenital abnormalities, and obstruction.

V. Endoscopic Retrograde Cholangiography and Pancreatography

Endoscopic retrograde cholangiography and pancreatography is an examination that combines endoscopic and radiographic techniques and consists of introducing a fiber-optic endoscope by mouth into the duodenum and a catheter via scope and through the papilla of Vater into the biliary and pancreatic ducts. Stones, inflammatory processes, or tumors can thus be detected. By combining the endoscopic diagnostic technique with cauterization for sphincterotomy (i.e., cutting of the Oddi's sphincter, which controls the opening of the common duct into the duodenum), sweeping of the duct with a balloon for stones, or introduction of a stent to bridge an obstruction and permit biliary drainage, surgery can frequently be avoided. A stent is a tube, usually made of plastic, which has a configuration (i.e., curled or flared ends) that prevents it from getting dislodged.

VI. Mammography

After a period of controversy about the dose of radiation administered by radiography to breast tissues, new low-dose techniques have been introduced for the screening of women at risk. High-resolution techniques are capable of demonstrating nonpalpable cancers in addition to inflammatory disease and benign tumors. Screening mammography has significantly decreased mortality from breast cancer in countries where it is widely used.

VII. Computed Tomography

CT is an X-ray-based cross-sectional imaging modality using computer assistance. Images are reconstructed in a process converting detector readings from hundreds of thousands of data samples into an electronic picture that represents the scan section. The picture is composed of a matrix of picture elements (i.e., pixels), each of which has a density value represented by its CT number. Newer CT devices offer high resolution (six to 20) line pairs per centimeter), high speed (50 msec to 2 seconds per scan), fast patient throughput, and a wide range of software options. Ultrafast CT can be used in the evaluations of cardiac function and in pediatric cases in which patient cooperation without sedation is difficult to obtain.

Because the price of CT devices has precipitously declined, and because of their versatility, they have become widespread in the Western world, and their use has encroached on conventional radiography. It

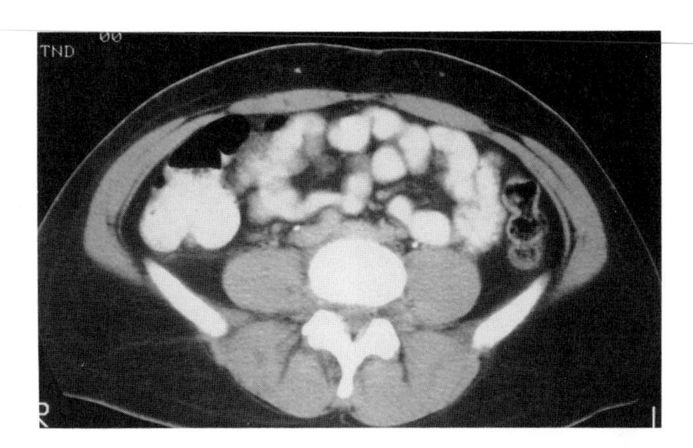

FIGURE 5 CT of the lower abdomen. Multiple small-bowel loops are well demonstrated, due to good filling of the bowel with oral contrast material.

is the examination of choice for the lungs and the abdomen, particularly the peritoneal cavity and the kidneys. Because of its wide field of view, it is an excellent screening method, but does require the use of intravenous and oral iodine-containing contrast media. Although MRI is superior to CT in many areas of the body and the head, CT is still widely used for examinations of the head, spine, heart, lower abdomen, and extremities (Figs. 4 and 5). This is largely due to the relative scarcity of MRI devices.

VIII. Magnetic Resonance Imaging

MRI requires a large magnet into which the patient can fit. Systems with permanent, resistive, or superconducting magnets are available. All of these systems are connected to computer image-displaying monitors and use radiofrequency-emitting and -receiving coils. This method is based on the fact that, in a strong external magnetic field, nuclei with an uneven number of particles have a magnetic moment and become oriented by the field. In MRI protons (i.e., hydrogen nuclei) are used predominantly because of their ubiquity in the body and greatest sensitivity. By bombarding the aligned protons with specific (Larmor) frequency, their direction of spin is changed. They emit signals of the same radiofrequency as the one absorbed, and after cessation of the pulse they return to the original direction of spin. By introducing a known gradient of radiofrequencies and with the use of computer-assisted reconstruction approaches, images are produced in any plane, with superb soft tissue contrast resolution. [*See* MAGNETIC RESONANCE IMAGING.]

MRI is the method of choice for the examination of pathological processes involving the brain, spinal cord, and musculoskeletal system, including the spine and the pelvis. It is also an excellent method for demonstrating abnormalities of the large vessels and is the preferred method for the diagnosis of dissecting aneurysms of the aorta. MRI is as good as CT for examining the neck, mediastinum, heart, retroperitoneum, liver, spleen, and large vessels (Figs. 6 and 7). MRI is superior to CT in the area of the pelvis, particularly in evaluation of the male and female organs. Its use in these areas is getting more frequent as the number of MRI devices and experience with them increase.

IX. Ultrasonography

B-mode ultrasonography uses a transducer that converts a short electric energizing pulse into an ultrasonic pulse, which propagates along the line of the sight of the transducer. Part of the energy in this ultrasonic pulse is reflected. Tissues vary in the echoes reflected to the transducer, where they are

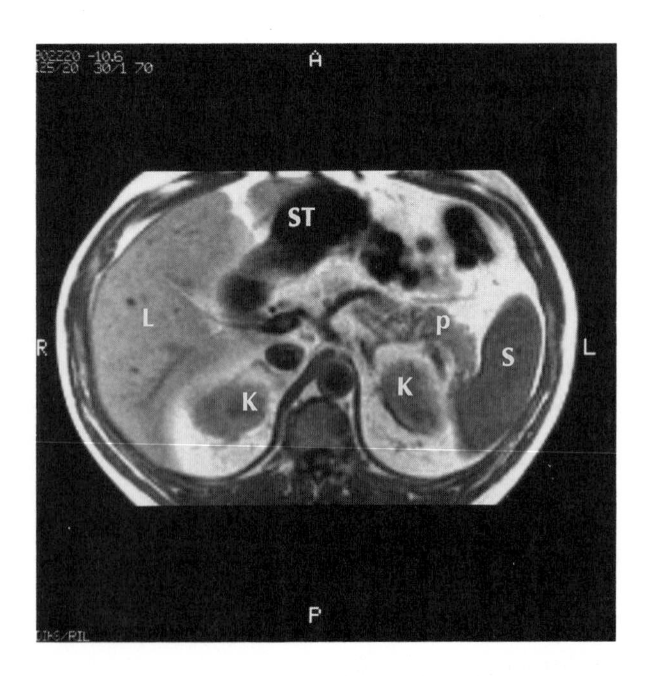

FIGURE 6 Breath-hold MRI of the upper abdomen (transverse plane). Due to the absence of respiratory motion, anatomical detail is well seen. p, Pancreas; S, spleen; L, liver; ST, stomach filled with air; K, kidneys.

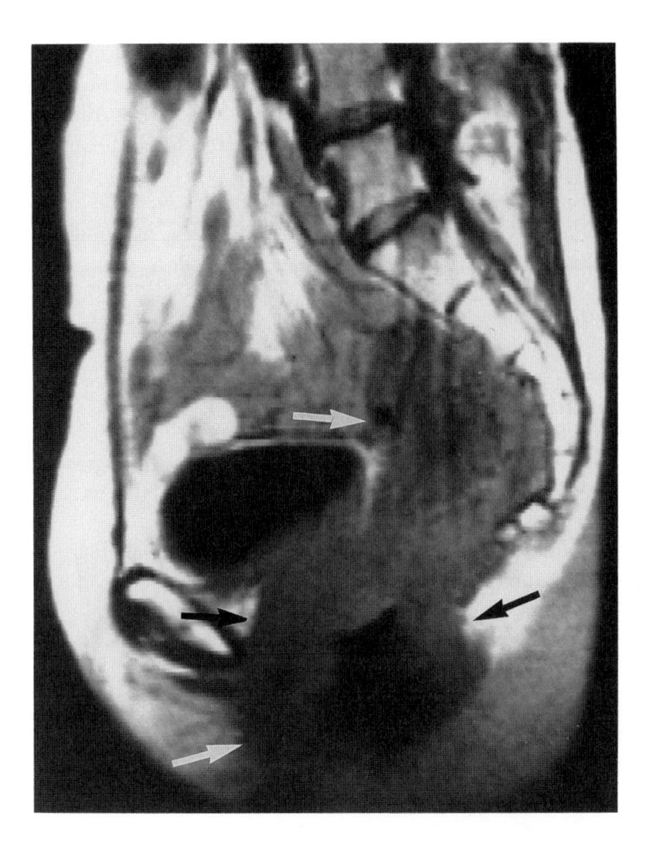

FIGURE 7 MRI of the pelvis (sagittal plane). A large mass (arrows) is seen in the rectum, with invasion of the bladder and a large ulceration near the anus.

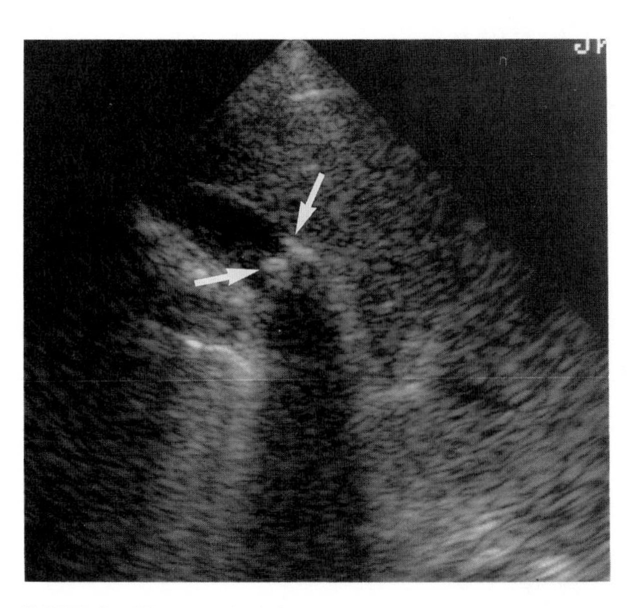

FIGURE 8 Ultrasound of right upper quadrant. Ultrasonography demonstrates well the gallbladder with stones (arrows).

reconverted into electric pulses and amplified. In modern devices a cross-sectional image is obtained in gray scale and is digitized. Ultrasonography is the leading screening method for examination of the abdomen and for the detection and localization of masses. It is also used for screening neonates for suspected congenital anomalies of the brain. It is used intraoperatively with sterile wrapped transducers to gauge the success of tumor removal. Ultrasonography is the standard method for examination of the gallbladder and detection of gallstones (Fig. 8). [*See* ULTRASOUND PROPAGATION IN TISSUE.]

The most frequent use of ultrasonography, and possibly its greatest value, is in obstetrics, in which it is employed for diagnosis as well as for guidance of interventional procedures. High-risk obstetrics depends heavily on the expert use of ultrasonography, which has greatly reduced morbidity and mortality in this branch of medicine. Newer ultrasound devices are also equipped with a Doppler apparatus, which can depict arteries and veins and their patency, direction, amount, and speed of flow. The Doppler apparatus is a noninvasive means of assessing blood flow. This is an excellent method for screening the patency of major blood vessels and is presently heavily used in transplant surgery.

X. Interventional Radiology

Using fluoroscopy, ultrasonography, or CT for guidance, a new approach to therapy has developed, called interventional radiology. This method permits diagnostic biopsies, drainage of abscesses, introduction of stents which bypass obstructions, and balloon dilatations of stenosis of the gastrointestinal tube or blood vessels (i.e., angioplasty). Angioplasty is used with increasing frequency for dilatation of stenoses of coronary and peripheral arteries. Laser techniques as well as "roto-rooter" devices under fluoroscopic guidance have also been used, but are presently experimental.

Interventional procedures are of particular importance in deep areas of the brain and the spinal cord, where surgery can produce adverse results. They are used either instead of surgery or to facilitate surgical approaches in the occlusion of aneurysms (i.e., dilatations of blood vessels which result from weakness of the wall and are prone to rupture) with detachable balloons or in the occlusion of arteriovenous malformations.

Much of the success of neurointerventional techniques is due to the development of microinstruments. Tracking techniques have also been helpful. These are based on the fact that it is possible to freeze an image of the background and the contrast-filled vessels on the screen without further irradiation and superimpose on this background the active image of the advancing catheter. As the instruments become smaller, it is of the utmost importance to improve the resolution of the image intensifiers.

Bibliography

Callen, P. W. (1988). "Ultrasonography in Obstetrics and Gynecology." Saunders, Philadelphia, Pennsylvania.

Feig, S. A., and McLelland, R. (1983). "Breast Carcinoma: Current Diagnosis and Treatment." Masson, New York.

Fraser, R. G., and Pare, J. A. (1988). "Diagnosis of Diseases of the Chest," 3rd ed. Saunders, Philadelphia, Pennsylvania.

Higgins, C. G., and Hricak, H. (1987). "Magnetic Resonance Imaging of the Body." Raven, New York.

Margulis, A. R., and Burhenne, H. I. (1988). "Alimentary Tract Radiology," 4th ed. Mosby, St. Louis, Missouri.

Moss, A. A., Gamsu, G., and Genant, H. K. (1983). "Computed Tomography of the Body." Saunders, Philadelphia, Pennsylvania.

Newton, T. H., and Potts, D. G. (1974). "Radiology of the Skull and Brain," Vol. 2. Mosby, St. Louis, Missouri.

Newton, T. H., Hasso, A. N., and Dillon, W. P. (1988). "Computed Tomography of the Head and Neck," Vol. 3. Raven, New York.

Ney, C., and Friedenberg, R. M. (1981). "Radiographic Atlas of the Genitourinary System," Vols. I and II. Lippincott, Philadelphia, Pennsylvania.

Resnick, D., and Niwayama, G. (1988). "Diagnosis of Bone and Joint Disorders," 2nd ed., Vols. 1–6. Saunders, Philadelphia, Pennsylvania.

Ring, E. J., and McLean, G. K. (1981). "Interventional Radiology: Principles and Techniques." Little, Brown, Boston, Massachusetts.

Stark, D. D., and Bradley, W. G. (1988). "Magnetic Resonance Imaging." Mosby, St. Louis, Missouri.

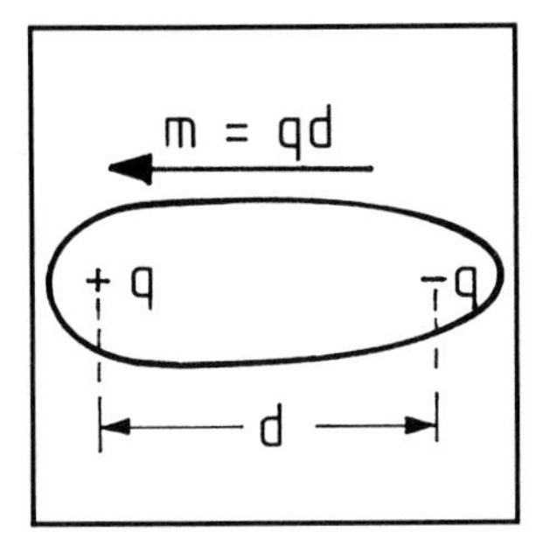

Dielectric Properties of Tissue

RONALD PETHIG, *University of Wales, Bangor*

Glossary

Dielectric Material having a relatively low electrical conductivity and capable of supporting electrostatic stress

Dielectric loss Amount of energy dissipated as heat in a dielectric when it is subjected to a time-varying electric field; this energy loss is associated with the frictional work done in the orientation of polar molecules along the field direction

Dielectric relaxation When an electric field is removed from a dielectric, the orientation of polar molecules is lost by thermal agitation; for a given polar molecule, this relaxation of field-induced orientation occurs at a characteristic relaxation time, the reciprocal of which defines the relaxation frequency of the corresponding dielectric loss

Electrical capacitance Measure of the amount of electrical charge that must be given to a material to raise its electrical potential by 1 unit; capacitance of 1 farad requires 1 coulomb of charge to raise its potential by 1 volt

Electrical conductivity Measure of the ease with which electrical charges can move through a material under the influence of an applied electric field; it is quantified by the amount of charge transferred across 1 unit area per unit voltage gradient per second; reciprocal of resistivity

Impedance Measure of the opposition of a material to the passage of electric current; for direct current this is just the resistance of the material, but for alternating current the effect of the electrical capacitance of the material must also be taken into account

Relative permittivity Principal property of a dielectric, also known as dielectric constant; it is the factor by which, for a given charge distribution, the electric field in a vacuum exceeds that in the dielectric

THE PASSIVE ELECTRICAL PROPERTIES of a material can be completely characterized by measurement of its electrical conductance (G; units Ohm^{-1} or Siemens) and electrical capacitance (C; units farads). Taken together, these two parameters define the dielectric properties of the test material. If measurements are made on the same sample using the same electrode arrangement, then the conductance and capacitance can be defined by two equations:

$$G = k\sigma \text{ and } C = k\varepsilon_0\varepsilon, \quad (1)$$

where k is a geometry factor (units meter) often referred to as the measurement cell constant; and the conductivity (σ; units siemens/meter) is the proportionality factor between the induced electric current density and the applied electric field, and is a measure of the ease with which delocalized electrical charges can move through the material under the influence of the field. For aqueous biological materials, the conductivity is mostly associated with mobile protons and hydrated ions. The factor ε_0 is the dielectric permittivity of free space (of value 8.854×10^{-12} farad/meter), while ε is the permittivity of the material relative to that of free space. For historical reasons, ε is often referred to as the dielectric constant, but because it is a material property that varies (e.g., with electrical frequency and temperature), the term relative permittivity is preferable. The relative permittivity is a measure of the extent to which localized charge distributions can be distorted or polarized by the applied field. For biological materials, such effects are mainly associated with electrical double layers occurring at membrane surfaces or around solvated macromolecules, or with molecules that possess a permanent electric dipole moment.

I. Underlying Theory

A. Background

Dielectric measurements have played a significant role in the biological sciences. For example in 1910, the electrical impedance of suspensions of erythrocytes was measured and found to decrease with increasing frequency, a result that implied that the cells were composed of a poorly conducting membrane surrounding a cytoplasm of relatively low resistivity. In later years, a more detailed analysis of such electrical measurements led to the first indications of the ultra-thin molecular thickness of this membrane. Later observations of a negative capacitance effect in the electrical properties of squid axons led directly to the concept of voltage-gated membrane pores as the vital mechanism of neurotransmission, and dielectric measurements on protein solutions have provided quantitative details of the molecular size, shape, and extent of hydration of protein molecules. More recently, dielectric studies have provided insights into the diffusional motions of proteins and lipids in cell membranes, as well as the influence of hydration on enzyme activity and on protonic and ionic transport processes in protein structures. Studies of the ways that electric fields interact with tissues are also of importance to the development of radiofrequency and microwave diathermy and clinical hyperthermia, in impedance plethysmography and tomography, in the gentle thawing of cryogenically preserved tissue, and in the use of pulsed electric fields to aid tissue and bone regeneration and healing.

B. Theory

Figure 1 shows in a highly schematic form the basic concept of a molecule possessing a dipole moment and examples of electrical double layers around a globular protein and at the surface of a biological membrane. The simplest molecular dipole consists of a pair of opposite electric charges, of magnitude +q and −q, separated by a distance (d), and, in this case, the molecular dipole moment m is given as m = qd and has units of coulomb meter. For a protein molecule, the presence of ionizable acidic and basic

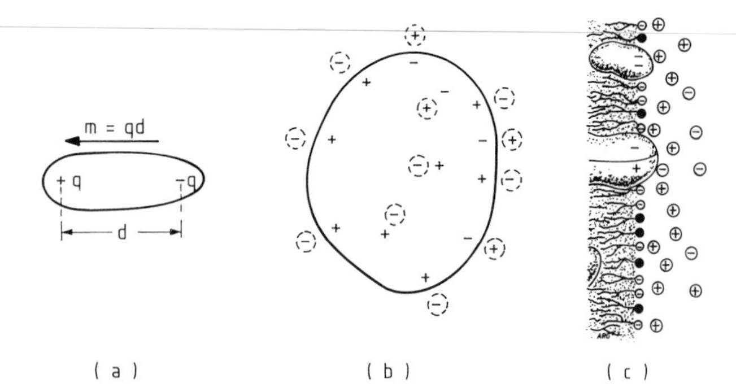

FIGURE 1 (a) The basic form of a molecule possessing a dipole moment (m), consisting of a pair of opposite electrical charges (+q and −q) separated by a distance (d). (b) A schematic illustration of an electrical double layer formed around a charged solvated protein and (c) at the surface of a charged biological membrane.

amino acid residues in its structure gives rise to a comparatively large permanent dipole moment whose value varies with pH and protein conformation. Counterions in the ionic aqueous solution around the protein form an electrical double layer, an effect that also occurs at the surface of all charged biological membranes. When an electric field is applied to such molecular dipoles or electrical double layers, the dipoles tend to align with the field and charge displacements occur in the double layers; i.e., they become polarized. This polarization does not occur instantly on application of the field, and each polarizable entity will respond in its own characteristic way. To describe this, the relative permittivity is written as a complex function of the form

$$\varepsilon(\omega) = \varepsilon_\infty + (\varepsilon_S - \varepsilon_\infty)/(1 + j\omega\tau). \qquad (2)$$

In this equation, ε_∞ is the relative permittivity measured at a frequency so high that the polarizable entity cannot respond quickly enough to the electric field, and ε_S is the limiting low-frequency permittivity where the polarization is fully manifest. The characteristic response time or relaxation time is given as τ, and ω is the angular frequency (radians/sec) of the sinusoidal electric field. The real and imaginary components of the complex permittivity are given as

$$\varepsilon = \varepsilon' - j\varepsilon'', \qquad (3)$$

where, as in equation (2), j is $\sqrt{-1}$. The real part ε', corresponding to the permittivity giving rise to the capacitance described in equation (1), is given by

$$\varepsilon'(\omega) = \varepsilon_\infty + (\varepsilon_S - \varepsilon_\infty)/(1 + \omega^2\tau^2). \qquad (4)$$

The imaginary component ε'' is given by

$$\varepsilon'' = (\varepsilon_S - \varepsilon_\infty)\omega\tau/(1 + \omega^2\tau^2) \quad (5)$$

and corresponds to the dissipative loss associated with the movement of polarizable charges in phase with the electric field. The frequency variations of ε' and ε'' for pure water are shown in Fig. 2. Water is a relatively simple molecule having one effective relaxation time (τ). More complicated polarizable systems have a distribution of relaxation times, and the plots of ε' and ε'' extend over a wider frequency range than is shown in Fig. 2. Because the dielectric loss factor ε'' reflects the extent to which the polarizable charges move in phase with the field, ε contributes to the overall conductivity, which can be written as:

$$\sigma = \sigma_0 + \sigma(\omega) = \sigma_0 + \omega\varepsilon_0\varepsilon''. \quad (6)$$

In this equation, σ_0 is the steady-state conductivity arising from mobile charges, and $\sigma(\omega)$ is the frequency-dependent conductivity arising from dielectric polarization. Also, because the total energy associated with the electric field is constant and is either stored [as reflected in $\varepsilon'(\omega)$] or dissipated [as reflected in $\varepsilon''(\omega)$] by the material, the change in conductivity is directly proportional to the change in permittivity as a dielectric loss process is traversed through its frequency range. In terms of the parameters $\Delta\varepsilon'$ and $\Delta\sigma$ shown in Fig. 2, the relationship is

$$\Delta\sigma = 2\pi f_c \varepsilon_0 \Delta\varepsilon', \quad (7)$$

where f_c is the relaxation frequency ($f_c = 1/2\pi\tau$).

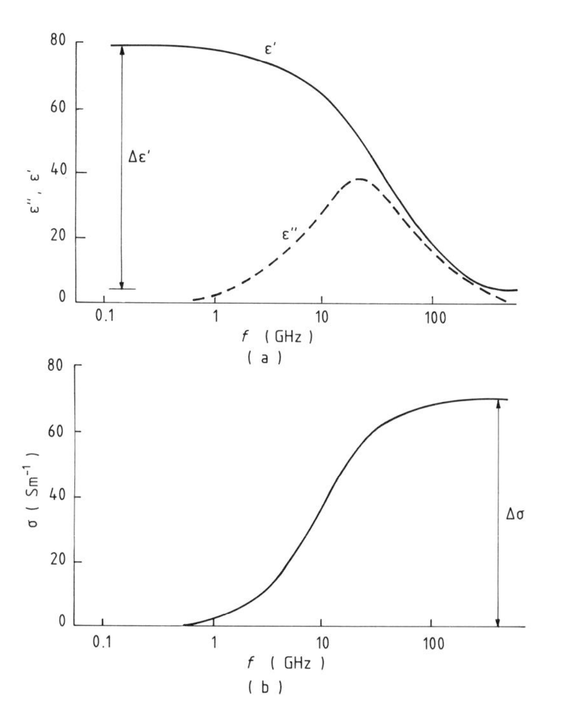

FIGURE 2 The dielectric dispersion exhibited by pure water at 20°C as a function of frequency (f), illustrated in terms of the changes (a) in the dielectric permittivity parameters ε' and ε'' and (b) of the conductivity σ.

II. Tissues

The permittivity of biological tissues typically decreases with increasing frequency in three major steps (i.e., dispersions), which are usually designated the α, β, and γ dispersions. This is shown for muscle tissue in Fig. 3, together with the corresponding increments in the conductivity. For frequencies below around 100 MHz, the conductivity of most tissues is relatively insensitive to the measurement frequency, whereas at higher frequencies, and especially in the GHz range, the conductivity rapidly increases with increasing frequency.

Although the various controlling factors are not fully understood, the α dispersion is generally associated with the tangential and radial displacement of counterions at membrane surfaces. Membranes, largely composed of lipid and protein molecules, are poor conductors. At low frequencies, this high resistance of cell membranes insulates the cell interior (cytoplasm) from external electric fields, and no

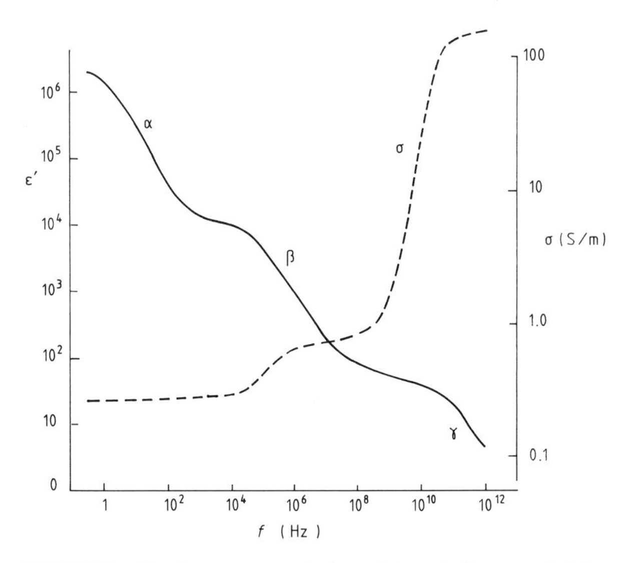

FIGURE 3 The frequency variation of the relative permittivity (ε) and conductivity (σ) for skeletal muscle. Many body tissues exhibit the three distinctive dielectric dispersions α, β, and γ.

current is induced within the cell interior. Effectively, the membrane acts rather like a capacitor and becomes electrically "charged-up." At higher frequencies, the short-circuiting effect of the membrane capacitance allows the electric field to penetrate into the cell until at a sufficiently high frequency the effective membrane resistance becomes vanishingly small and the cell appears dielectrically as a globule of cytoplasm dispersed in an electrolyte. Thus, the effective permittivity and conductivity of cellular structures will fall and rise, respectively, with increasing frequency, leading to the existence of the β dispersion shown in Fig. 3. As the frequency is increased to the GHz range, dielectric relaxation of free-water molecules begins to occur and gives rise to the γ dispersion. [*See* MEMBRANES, BIOLOGICAL.]

As a result of complicating experimental factors such as electrode polarization effects, comparatively little reliable dielectric data exists for tissues in the frequency range up to a few kilohertz. A good example of the α dispersion is given in Fig. 4 for normal and malignant breast tissue, where it may be seen that malignancy appears to markedly influence the dielectric properties. After excision of tissue, the α dispersion steadily decreases with time, and this effect is consistent with the loss of integrity and physiological viability of the cell membranes. Another example of low-frequency data is that shown in Fig. 5 for human knee ligament and human cancellous bone. The dielectric properties of bone are direction-dependent, and for cancellous bone this is considered to be related to the orientation of the trabeculae.

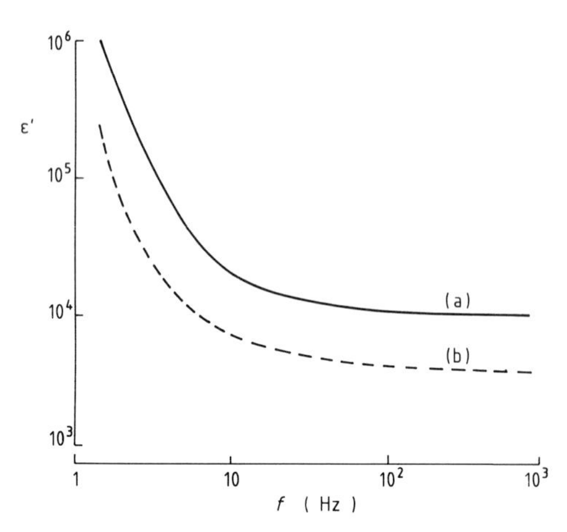

FIGURE 4 Frequency variation of the relative permittivity for (a) normal breast tissue and (b) breast tissue with a malignant tumor.

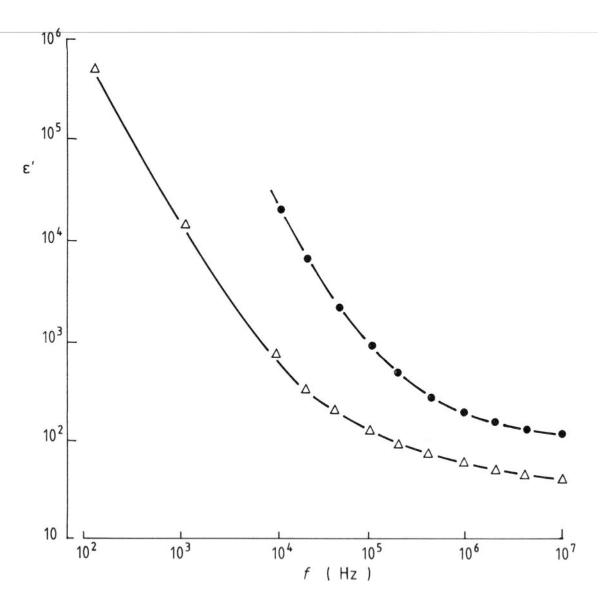

FIGURE 5 Frequency variation of the relative permittivity for human cancellous bone and knee ligament. △, bone; ●, ligament.

A good example of the β dispersion is shown in Fig. 6 for an aqueous suspension of eye lens fibers (bovine). In this figure, the effect of lysing the cell membranes using digitonin is shown, and clearly the β dispersion depends on the integrity of cellular membranes for its existence. After excision of tissue, the β dispersion, like the α dispersion, steadily decreases with time. Therefore, dielectric measurements can be used to monitor the "fresh-

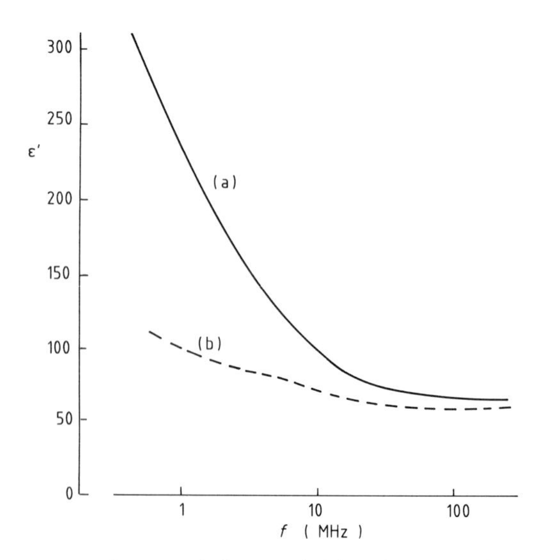

FIGURE 6 The relative permittivity of a suspension of eye lens fibers (a) before and (b) after lysis of the cell membranes using digitonin.

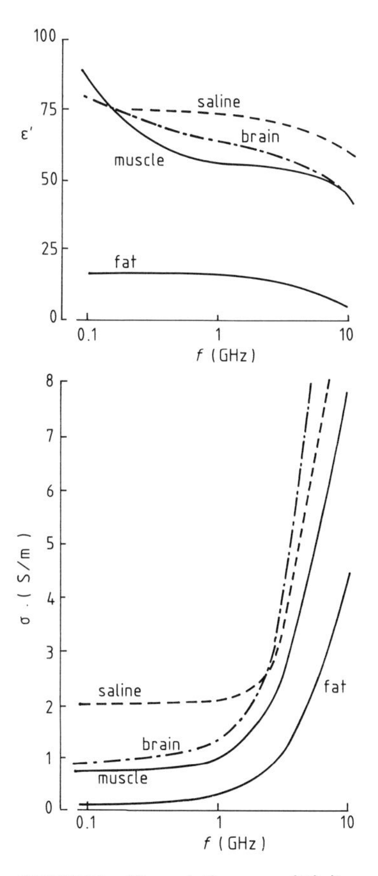

FIGURE 7 The relative permittivity and conductivity at microwave frequencies for muscle, brain tissue, fat, and physiological-strength (0.9%) saline solution.

ness'' of tissue. Sometimes, as for the cases of ligament and bone shown in Fig. 5, the α and β dispersions tend to merge into one another.

For frequencies >100 MHz, where electrical charging effects of the cell membranes become negligible, the dielectric characteristics of tissues can be expected to reflect the properties of the inter- and intracellular electrolytes. As such we will expect to see evidence of the dielectric dispersion associated with the relaxation of water molecules, as can be seen in Fig. 7. Here, the dielectric properties of brain, fat, and muscle >100 MHz are shown alongside that for saline solution of normal physiological strength. Apparently, the microwave dielectric properties of tissues are influenced by their water contents. Muscle, with a water content ranging from 73 to 78 wt%, has a much greater permittivity and conductivity than fat, for example, which has a water content ranging from 5 to 15 wt%.

The dielectric properties of various tissues at 37°C are given in Tables I and II for several frequencies commonly used for clinical therapeutic and diagnostic purposes. Because of the variability of water content values, the data presented in these tables should be regarded as average values. Also, when reviewing the literature, it is important to note that different measurement methods can provide large differences in data. An example of this can be seen by comparing the results for excised brain tissue given in Tables I and II with the *in vivo* data shown in Figure 7. Table III gives the normal ranges of water contents for various tissues and organs; this can be used to estimate the likely spread of dielectric properties expected at microwave frequencies for *in vivo* measurements.

III. Skin

The average *in vivo* dielectric properties of skin over the frequency range 1 Hz to 10 GHz are shown in Fig. 8. These properties exhibit considerable regional variability over the body, with permittivity and conductivity greatest in areas such as the palms, where sweat ducts are in abundance. The dielectric properties of skin can be understood in terms of its inhomogeneous structure and composition and the way in which these features vary from the skin surface into the underlying dermis and subcutaneous tissue. [*See* SKIN.]

Close study of the dielectric dispersion exhibited by normal skin over the frequency range 0.5 Hz to 10 kHz shows that it is characterized by two separate relaxation processes, centered around 80 Hz and 2 kHz, respectively. These relaxation processes are considered to be located within the stratum corneum and to be associated with relaxation of counterions surrounding the corneal cells. Psoriatic skin exhibits significantly different dielectric properties from those of normal skin.

IV. Blood

The relationship between the conductivity of whole blood and hematocrit is shown in Fig. 9. At a frequency of 100 kHz and at 37°C, the numerical relationship between conductivity (σ) and hematocrit (H) is given by

$$\sigma = 1.5 \exp(-0.025 \text{ H\%}) \text{Sm}^{-1}. \quad (8)$$

TABLE I Relative Permittivity of Biological Tissues at 37°C

Material	13.56 MHz	27.12 MHz	433 MHz	915 MHz	2.45 GHz
Artery	—	—	—	—	43
Blood	155	110	66	62	60
Bone					
With marrow	11	9	5.2	4.9	4.8
In Hank's solution	28	24	—	—	—
Bowel (plus contents)	73	49	—	—	—
Brain					
White matter	182	123	48	41	35.5
Grey matter	310	186	57	50	43
Fat	38	22	15	15	12
Kidney	402	229	60	55	50
Liver	288	182	47	46	44
Lung					
Inflated	42	29	—	—	—
Deflated	94	57	35	33	—
Muscle	152	112	57	55.4	49.6
Ocular tissues					
Choroid	240	144	60	55	52
Cornea	132	100	55	51.5	49
Iris	240	150	59	55	52
Lens cortex	175	107	55	52	48
Lens nucleus	50.5	48.5	31.5	30.8	26
Retina	464	250	61	57	56
Skin	120	98	47	45	44
Spleen	269	170	—	—	—

TABLE II Conductivity (Sm^{-1}) of Biological Tissues at 37°C

Material	13.56 MHz	27.12 MHz	433 MH	915 MHz	2.45 GHz
Artery	—	—	—	—	1.85
Blood	1.16	1.19	1.27	1.41	2.04
Bone					
With marrow	0.03	0.04	0.11	0.15	0.21
In Hank's solution	0.021	0.024	—	—	—
Brain					
White matter	0.27	0.33	0.63	0.77	1.04
Grey matter	0.40	0.45	0.83	1.0	1.43
Fat	0.21	0.21	0.26	0.35	0.82
Kidney	0.72	0.83	1.22	1.41	2.63
Liver	0.49	0.58	0.89	1.06	1.79
Lung					
Inflated	0.11	0.13	—	—	—
Deflated	0.29	0.32	0.71	0.78	—
Muscle	0.74	0.76	1.12	1.45	2.56
Ocular tissues					
Choroid	0.97	1.0	1.32	1.40	2.30
Cornea	1.55	1.57	1.73	1.90	2.50
Iris	0.90	0.95	1.18	1.18	2.10
Lens cortex	0.53	0.58	0.80	0.97	1.75
Lens nucleus	0.13	0.15	0.29	0.50	1.40
Retina	0.90	1.0	1.50	1.55	2.50
Skin	0.25	0.40	0.84	0.97	—
Spleen	0.86	0.93	—	—	—

TABLE III Water Content Values for Various Tissues and Organs

Tissue	Wt% (water content)
Bone	44–55
Bone marrow	8–16
Bowel	60–82
Brain	
White matter	68–73
Grey matter	82–85
Fat	5–15
Kidney	78–79
Liver	73–77
Lung	80–83
Muscle	73–78
Ocular tissues	
Choroid	78
Cornea	75
Iris	77
Lens	65
Retina	89
Skin	60–76
Spleen	76–81

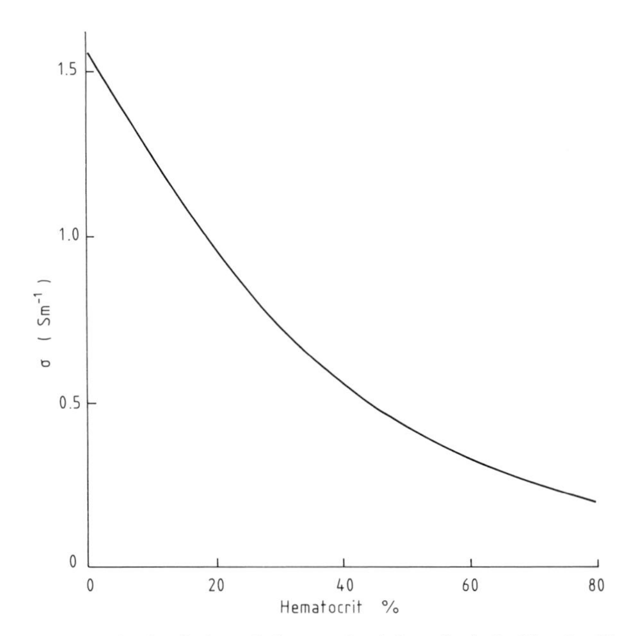

FIGURE 9 Variation of the conductivity of whole blood with hematocrit (H%) at 100 kHz and 37°C (normal H is around 42%).

No difference is found between the conductivity and hematocrit relationship for adult normal, neonatal, and placental blood. Reconstituted, time-expired bank blood exhibits a different relationship of the form

$$\sigma = 1.9 \exp(-0.022\ H\%)\mathrm{Sm}^{-1}. \quad (9)$$

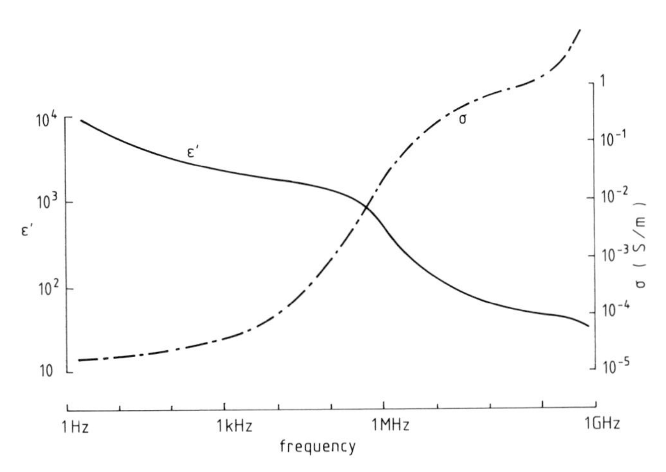

FIGURE 8 Frequency variation of the relative permittivity and conductivity of skin at 37°C.

For patients on hemodialysis, the blood conductivity varies according to the linear relationship

$$\sigma = 4.8 + 8(H\%)^{-1}\mathrm{Sm}^{-1}. \quad (10)$$

This markedly higher conductivity probably is related to the elevated levels of urea, sodium, and potassium found in the blood predialysis. The relative permittivity of whole blood falls markedly in the frequency range from 1 to 100 MHz, typically from around 350 down to 70, whereas the conductivity increases by no more than around 10%. [*See* HEMOGLOBIN.]

Bibliography

Pethig, R. (1987). Dielectric properties of body tissues. *Clin. Phys. Physiol. Meas.* **8A,** 5–12.

Pethig, R., and Kell, D. B. (1987). The passive electrical properties of biological systems: Their significance in physiology, biophysics and biotechnology. *Phys. Med. Biol.* **32,** 933–970.

Schwan, H. P. (1985). Dielectric properties of the cell surface and biological systems. *Stud. Biophys.* **110,** 13–18.

Diet

E. F. PATRICE JELLIFFE and DERRICK B. JELLIFFE, *University of California, Los Angeles*

Glossary

Anorexia nervosa Severe eating disorder, characterized by refusal to eat, commonest in young women, seeking exaggerated thinness
Anthropometry Measurement of the human body, especially weight, height, subcutaneous fat
Bioavailability Digestibility and absorption from the alimentary canal
Bulimia Severe eating disorder with excessive food intake followed by food restriction and "purging" (induced vomiting, laxatives, and/or strenuous exercise), usually in young females, seeking exaggerated thinness
Cartesian Belief that the universe (including human biology) follows mathematical rules (from René Descartes, 1696–1750)
Kwashiorkor Severe form of protein-energy malnutrition (PEM), usually in weanling children (1–2 years of age), with water retention (edema), caused by low-protein diet containing carbohydrate calories, combined with frequent infections
Malting Village-level or commercial treatment of cereal grains to increase nutrient concentration and digestibility (via increased availability of starch-splitting enzyme, amylase)
Marasmus Severe form of PEM, usually in infants, caused by marked lack of calories and protein in the diet, often with infective diarrhea
Multifactorial Condition caused by several interacting factors or causes
Multimix Mixture of foods that are nutritionally complementary
Osteoporosis Condition characterized by poorly calcified bones, commonest in older women, related to long-term low-calcium intake and postmenopausal hormonal changes, characterized by fractures and bent back (kyphosis)
Protein-energy malnutrition (PEM) Various conditions caused by lack of calories and protein in the diet, usually with added infections, commonly in young children in poor communities
Reductionism Analysis or scientific consideration focused on one or very limited factors in what are actually highly complicated circumstances

DIET usually refers to the customary mixture of foods consumed. It is influenced by adaptations dating from gatherer-hunter times in human evolution, modified by historical and technological changes in life-style from agriculture to traditional cities to modern cities.

Present-day diets consumed vary greatly in different ecologies and cultures, but all have to supply nutrients needed for all stages of life and levels of activity. Modification of diet by various means can best be emphasized for two contrasting populations: the often poorly fed majority in less technically developed countries, and the frequently overnourished in more technically developed regions.

I. Customary Diets

A. Selection

The customary or general diet of a population or an individual comprises the usual mixture of foods consumed. This is influenced by availability (agricultural practices, seasonal variation, transport, preservation, and storage facilities), existing cultural attitudes (food classification, culinary methods, meal patterns), and accessibility (cultivable land and income). Customary diets need to contain sufficient amounts of energy (calories), essential nutrients [protein (amino acids), carbohydrates, fats (fatty acids), vitamins, minerals (calcium, phosphorus, magnesium), trace elements (iron, zinc, iodine, copper, fluoride etc.), electrolytes (sodium, potassium), and water] and certain nonnutrients (e.g., fiber) (1) to supply the body's requirements with foods of appropriate consistency for basic metabolic maintenance at different biological stages of life (including the pregnant woman and her fetus, and the elderly), (2) to sustain "good health," avoid nutrient deficiency, and protect against infections, (3) to provide sufficient nutrients for varying levels of physical activity and body build at different phases of life and climatic conditions, (4) to provide for growth, especially during pregnancy, early childhood, and adolescence, and (5) to minimize the risk of some multifactorial chronic, so-called degenerative diseases that can develop in adult life (e.g., obesity, diabetes, coronary heart disease, cirrhosis of the liver, certain types of cancer) in which prolonged overnutrition or excess or imbalance of some nutrients are considered "risk factors." Sometimes an over-simplistic view of single nutrients is made. Although there are individual functions and needs for each, complex interactions occur between different nutrients in the alimentary canal and after absorption into the body. Some are undoubtedly still unrecognized or ill-defined. [*See* NUTRIENTS, ENERGY CONTENT.]

The role of saturated fats and cholesterol have dominated the dietary scene in recent years. In fact, cholesterol is necessary for the formation of various essential body tissues (e.g., the brain in infancy, cell membranes, and some hormones). Normally, it is synthesized in the liver. Excessive dietary intakes may be deposited as plaques in blood vessels, notably the coronary arteries of the heart and brain. This can lead to obstruction and coronary heart disease or stroke. However, the deposition of cholesterol in arteries is increased by the presence in the blood of high density lipoproteins (HDL) found in saturated fat, mostly in animal fats, palm oil, and coconut oil. Conversely, unsaturated fats present in all other plant oils make available low density lipoproteins (LDL), which assist in the removal of arterial cholesterol. The situation is further complicated by variation in individual susceptibility to cholesterol intake—in extreme instances a basic genetic hypercholesteremia occurs independent of dietary intake. [*See* CHOLESTEROL; FATS AND OILS (NUTRITION).]

During some stages of social evolution and in different cultures in the past and present, varied diets, usually centered on the local staple, have been adopted, which successfully supply all nutrient needs. Without this process, these communities could not have survived. Originally, selection must have been the result of long-term trial. This included recognizing and avoiding poisonous items (e.g., polar bear liver by Eskimos [toxic levels of vitamin A]) or making them harmless, as by removing hydrocyanic acid from cassava (in many cultures). Also, some inherent biological drives must have contributed to dietary selection via uncertain mechanisms, using edible items made possible by the climate and geography of the particular area.

Recognizable physiological effects (satiation, [feeling of repletion, largely related to the fat content and retention in the stomach], "mouth-feel," digestibility) and genetic variation in ability to metabolize certain nutrients have influenced the foods included in diets in different communities. More recently, economic affordability, agricultural and food technology, rapid forms of transportation, persuasion by commercial marketing practices, and fast-paced urban life-styles have become increasingly significant.

In all cultures (including so-called Westernized societies), the development of often little appreciated classifications of foods restricts or encourages the use of some of the potentially available edible items for all the population, particularly for vulnerable groups (i.e., pregnant women and young children). Common categories in such classifications include (Table I) foods/nonfoods; "cultural superfoods" (usually the cereal grain or root crop staple); indigenous concepts of body physiology and disease (e.g., "hot–cold" classification of foods and illnesses); prestige/celebration foods; foods desirable or otherwise for some groups (e.g., women, young children); "sympathetic magic" foods; food items forbidden, restricted, or accepted for religious reasons (e.g., kosher foods); and/or for moral concerns, as with vegans (strict vegetarians). Examples of cultural influences include religious fasts (e.g., the Muslim month of Ramadan) or feasts (e.g., Thanksgiving in the United States) and the currently cultural ideal "body-image" (e.g., *gordito* or "little fat one" for infants in some Latin American circumstances; the emaciated, overslim balle-

TABLE I Cultural Classification of Foods Commonly Used in Many Parts of the World

	Definition	Examples: More technically developed countries	Examples: Less technically developed countries
Nonfood*	Edible items not usually eaten	Rats, cats, dogs, insects (Europe, N. America)	Beef (Hindus, India)
"Cultural superfood"	Main food (staple) source of calories and other nutrients, often historical, mythical, and religious significance	Wheat (Europe—now shared with the potato)	Rice (S. China) Maize (corn) (Central America) Potato (Andes)
Body physiology	Related to local concepts of body physiology (e.g., "hot–cold," yin-yang)	Bodily humors (milk = melancholic—not food for young men) (UK)	Animal milk = garam (hot); not for infants with diarrhea ("hot" illness) (parts of India)
Age group	Suitable or unsuitable, often for women (especially pregnant) and young children		Fish not appropriate for young children as can produce worms (some communities in rural Malaysia)
Sympathetic magic	Conveying some similar property to consumer	Underdone steak to give strength to athletes (UK)	Walnut (similar appearance to human brain) (brain power) (Gujerat, India)
Prestige/**celebration	Rare, expensive	"Game" (venison, boar, salmon, honorific, forbidden to population at large in Medieval times)	Special milk-dessert (India)

* Nonfoods: only eaten if severe shortage, as in famines.
** Usually animal product even in vegetarian societies (i.e., milk-based desserts with Hindus in India).

rina–fashion model look for women in some Western countries; or the preferred large majestic bulk in Polynesia).

Also, in all cultures, foods, the types of food, and composition of meals eaten together have psychosocial symbolism and bonding significance, as indicated by sayings and proverbs in many languages. This is especially so on culturally important occasions, including such religious commemorations as Hanukkah and Christmas.

Currently, and increasingly in the future, high-level technology and the vagaries of international economics, competitive forces in trade, and the food industry will influence the range and decisions, positively and negatively, concerning dietary choices all over the world, in both poor and wealthy communities.

B. Evolutionary Perspectives

1. General

During the past 2 million years, humankind has evolved into increasingly complex forms of social organization, levels of technology, and dietary patterns from preindustrial—gatherer–hunters (GH), agriculturalists, pastoralists, traditional city dwellers—to industrial—early and modern techno-cities.

2. Gatherer–Hunters (99.9% of Human Existence)

During this predominant period of time, an extremely mixed diet mainly consisted of wild roots, berries, nuts (including acorns in California), fruits, seeds, grains (including wild rice), insects, birds' eggs, etc., eaten on the move ("grazing") and as available, together with varying amounts of relatively lean game meat, hunted, trapped, or scavenged from predators.

Probably the only communal "meal" that could take place was in the evening, after the discovery of fire and its protection against nocturnal carnivores. Meat and roots would be barbecued, resulting in improved masticability and taste and a limited degree of food preservation. Usually, the diet was mainly plant foods, unless climatic and geographic circumstances made game abundant.

Such diets, currently still found in the few, rapidly diminishing GH peoples (e.g., the !Kung of the

Kalahari and the Hadza of Tanzania) are high in fiber, vitamin C, potassium, and calcium (from vegetable sources and bones), usually adequate in calories (mainly derived from complex plant carbohydrates) even for their highly active way of life. Also, they are low in salt (sodium), fat, and sugars. Vitamin D is both a nutrient and a hormone synthesized in the skin after exposure to the ultraviolet light in sunshine. [*See* VITAMIN D.]

Nevertheless, sweet foods [breast milk (7% lactose), ripe fruits, and especially honey] and fat, calorie-rich portions of animals (abdominal omentum, bone marrow) are prized. Physiologically, this may be correlated with the location of the taste buds on the tongue (anterior "acceptors": sweet, salt; posterior "rejectors": bitter, sour) and with the satiation effect of fatty foods.

In GH peoples, so-called prolonged, on-request breastfeeding for several years is the norm, followed by the introduction in later infancy of various softer foods, including bone marrow, and meat and roots prechewed by the mother.

Dietary studies in present-day GH and paleontological investigations of types of teeth of prehistoric humans (incisors, canines, molars) indicate an omnivorous diet (with the use of all kinds of foods) but suggest a special emphasis on plant foods. However, a basic characteristic is to eat as much as available at any time, especially prized fatty and sweet foods. This helps to develop a reserve for possible less abundant food supplies later.

As a generalization, GH probably had and currently have little overt malnutrition because of their very varied food collection over large areas. A short life span is likely as a result of infections and especially trauma (accidents, animal attacks) of special importance for survival because of the need for rapid mobility to keep up with the group.

3. Agriculturalists and Pastoralists

Revolutionary changes in the diet occurred when "food obtaining" was succeeded by "food production" (early agriculture, domestication of animals). These were dependent on the development of essential agricultural knowledge and village technology (metallurgy, pottery-making), including storage, preservation [as with the complex freeze-dried potato (chũno) in the Andes], and food preparation (as with various techniques for the removal of hydrocyanic acid from cassava) and the recognition of the special properties of the gluten in wheat flour for bread-making. Pastoralist and nomad groups developed, centered on different milk-flesh-producing animals, including sheep, cattle, camels, horses, and reindeer (Lapland). In these communities, the use of animal milk and its products was dominant, as opposed to some other cultures, such as those in much of Polynesia (including Hawaii) where no animal milk was available.

Agriculturalists developed a necessarily more static way of life around the cultivation of a particular cereal grain or root crop staple. The range of different foods consumed was much more limited than with the GH but would be more abundant, provided growing conditions were favorable, without drought, floods, or pests and storage feasible and successful. Often, fatness would be an advantage as a reserve for the annual "lean" or "hungry" season, before the crops were harvested. In some African societies, a degree of obesity was purposely produced in women by overfeeding before the nutritional drain of marriage and repeated reproductive cycles. Sometimes, particularly early in this evolutionary change, a mixed, "transitional" life-style occurred (and still occurs) with limited agriculture and some hunting and food gathering or with a settled village, but with mobile flocks herded by men during the dry season in search of pasture and water.

In all these circumstances, successful traditional diets moved toward nutritionally complementary food mixtures. Food selection and recipes were influenced by cultural and physiological factors including the "mouth-feel," taste, texture, and other sensations perceived during eating (organolepsis). Such diets were normally centered on a staple cereal grain plus a legume [Mexican corn tortillas and red bean frijoles, rice and dhal (lentil) in India] or a root crop and animal product [poi (cocoyam) and shellfish in Polynesian Hawaii]. In some diets, excessive fiber can be consumed, with the phytates present limiting absorption of iron and/or zinc.

Genetic intolerance to certain items (lactose in animal milk) or apparent physiological adaptation (the traditional, almost exclusively sweet potato diet of the New Guinea Highlands) also played a part in evolving dietary patterns, either as cause or effect. Sometimes, such indigenous diets necessitated consuming known potentially harmful items in times of shortage. These include cereal infected with the fungus aflatoxin (with possible liver damage) or use of the grain *Lathyrus sativus* (with potential damage to the spinal cord).

Very extensive food mixtures have developed in

many cultures. For example, Chinese cuisine is characterized by an exceptionally wide range and mixture of foods used, with small quantities of meat and fat, fresh ingredients, and rapid stir-cooking using minimal fuel. All are nutritionally and ecologically desirable practices.

4. Traditional Cities

Traditional cities could only develop when sufficient food was available and transported from the agricultural countryside or grown nearby. Most written historical accounts of diets, including meal patterns and eating utensils, come from the possibly laudatory records of banquets of the aristocracy, where often exotic meat dishes figured prominently and vegetables were little used. All over the world, the least nutritionally desirable, most "white" overmilled cereal grain staple was preferred as being more expensive and exclusive, including white wheat bread and highly polished rice. The usually little mentioned basic diet of the general population would have consisted of less refined staple, fortunately often combined with legumes (such as soy bean preparations in some Asian countries, or other pulses elsewhere) or less expensive animal products, [e.g., "white meat" (cheese)].

In Europe, ill-preserved meat became tainted in the winter, and the search for "masking" spices was one of the motives for the explorations commencing in the 15th century. This trend also resulted in important widespread changes in foods in many areas, with the importation of the potato and corn (maize), with sugar moving from an expensive luxury item to an increasingly common food, and with the limited availability of mainly "flavor foods," such as cocoa and vanilla. Conversely, the diet is what is now Latin America was altered by the introduction of cattle, pork, and wheat.

In most preindustrial societies, problems usually relate to bacterial contamination of food (associated with environmental factors such as poor household hygiene, limited contaminated water), to general food shortage, in the annual "lean" or "hungry" season, especially in drought, and to supplying the high dietary needs of physiologically and culturally vulnerable groups in the weaning period of infancy and in pregnancy, especially if restrictive customs limit foods permitted. In some circumstances, specific nutrients may be generally borderline and sometimes inadequate in the usual diet (e.g., iodine, vitamin A, thiamin). As an unusual example, iron deficiency anemia was common in older children of the Bahima people in Uganda, because of an overemphasis on widely available (but iron-poor) cow's milk as their main food. Geographical nutritional factors may be principally responsible in some multifactorial diseases, as with the severe cardiac illness (Keshan disease) seen in parts of China, where the soil has a low selenium content.

5. Early Industrial Cities

The Industrial Revolution commencing 150 years ago led to a massive urban migration in Europe and elsewhere in search of employment. The diet of the poor, dependent on a cash economy, often deteriorated. This was especially so in infants for whom breastfeeding was often supplanted by inadequate "hand-feeding" (artificial feeding) with paps and dilute contaminated animal milk, while mothers worked in factories for low wages. This practice leads to marasmus and infective diarrhea, as it does now under similar slum circumstances in less technically developed countries. Rickets, caused by deficiency of vitamin D, also increased, as little exposure to sunlight was possible, as did scurvy (lack of vitamin C), especially in young children, as a result of limited availability of fresh fruit and vegetables.

The latter part of the 19th century was characterized by the development of an overconfident (but actually very limited) nutritional science, mainly concerned with the gross composition of foods and the needs of the working poor. It was also paralleled by early examples of mechanical food technology (e.g., canned condensed milk, corned beef) and, as a good early example of "techno-food," (e.g., margarine, originally manufactured in 1869 from beef fat as a cheap butter substitute).

The social impact on the family and its "foodways" was considerable and is currently still increasing. Traditional family meal patterns and foods consumed changed with the advent of earlier versions of store-bought foods. The level and type of food production at village level became changed everywhere, as did the role of family members, especially women. Often in countries under colonial rule, large scale, usually foreign-owned plantation agriculture was developed to produce the raw materials for food products exported and processed in industrialized countries. These included plantations for palm and coconut oil, cane sugar, cocoa, and bananas. Social and nutritional disruptive effects also occurred in the agricultural food-producing countries with major emphasis moving away from indigenous foods for local consumption to cash crops and the use of slaves or imported indentured labor to work the plantations.

6. Modern Techno-Cities

Changes in the food chain, beneficial and otherwise, have accelerated geometrically since World War II, with increasing emphasis on agribusiness. Influences include advances in mass production (chemical agriculture, hybrid seeds, etc.), modern preservation (canning, dehydration, irradiation, freeze-drying), marketing [including rapid transport, large-scale food outlets (supermarkets, fast-food establishments)], and home technology (refrigerator, deep-freeze, microwave oven).

Modern high technology has lead to the availability of the year-round supply of foods needed for increasingly large nonfood-producing urban populations and, in more technically developed wealthy communities, to a bewildering array of highly advertised, blended processed products containing numbers of unfamiliar preservatives, emulsifiers, solvents, stabilizers, and artificial colors. Another innovation has been the technological development of novel food items. These include *mechanically modified* fish protein concentrate and textured meat analogues from spun soy fiber, *surimi* (mock crab meat formulated from trash fish); *underused primary foods* antarctic prawns (krill *Euphasia superba*, yeast, plankton, leaf protein, seaweed, algae (*Spirulina*), fungus; *edible waste products* whey protein (from the cheese industry), and even processed feathers and wool; and *substitute foods* carob bean flour for chocolate. These are being used in blended processed products in industrialized communities in various ways but also have formed part of the food-aid mixtures employed in emergency famine and refugee situations.

Modern processed foods have also often posed problems (1) from thousands of inadequately tested chemicals that can be used in the agribusiness chain, including pesticide residues, antibiotics and hormones, additives, and preservatives of little-known long-term, cumulative significance, and (2) from items with high levels of calories, fat (particularly if saturated animal fat rich in cholesterol) and salt and a low "nutrient density" (amounts of other essential nutrients compared with calories), including an overuse of so-called junk foods. The widespread availability of many high-calorie modern processed foods has coincided with the emergence of mechanical energy-reducing devices, for work, convenience, and pleasure, including automobiles, television, power tools, and central heating. This has resulted in an imbalance between calorie intake from energy-rich foods and energy output from physical activity.

The nutritional results in areas where techno-food is abundant has been the virtual elimination of old-style malnutrition, except among the impoverished and disadvantaged, including some minority communities, unskilled immigrants, the homeless, and the elderly. Instead, the major issues (risk factors) are related in varying degrees directly or partly to dietary excess or to chemical additives in the food chain. These conditions include obesity (also involving adolescents and the elderly) and such multifactorial conditions as coronary heart disease, stroke, diabetes, hypertension, dental caries, some intestinal problems (e.g., constipation), cirrhosis of the liver (overconsumption of alcohol), possibly certain forms of cancer (large intestine, breast, prostate), and osteoporosis.

As a reaction to overprocessed, highly advertised foods, a "health food movement" emerged in more technically developed countries in recent decades concerned with emphasizing the use of more natural, traditional foods produced without chemicals. Indeed, this movement has lead the way for changes in some general and scientific dietary concepts and practices. Examples include an increased emphasis on better soluble and insoluble fiber in the diet, awareness of the unknown risks of chemical agriculture, and the value of exercise. This trend can be seen in products sold in some supermarkets. For example, there has been an awakening of interest in soluble fiber in the diet, as in oat bran or in bran from the Indian grain psyllium, because of its cholesterol-lowering potential, as well as the anti-constipation effect of the insoluble fiber. Nutritious, abandoned "new–old" cereal grains, such as amaranth and Andean quinoa (*Chenopodium quinoa*), have claimed attention. At the same time, the health food movement has become riddled with financially exploitive enterprises ("pseudo-nutrition") with bizarre items promoted as having incredible medieval magical properties. These include, for example, royal bee jelly, pineapple capsules, "negative calorie" meals, and unnecessary high, even dangerous, intakes of vitamins for "stress." A recent example of a harmful alleged panacea has been L-tryptophan, now shown to cause a serious blood disorder in susceptibles. [*See* DIETARY FIBER, CHEMISTRY AND PROPERTIES.]

Scientific research has also continued at a rapid pace but still tends to be overconfident concerning the mechanical mathematical certainty of nutrition. For example, the intake of different nutrients needed for health often have become regarded as precise numerical truths.

Fortunately, it is now more appreciated (1) that there is much inherent genetic interpersonal variation between individuals and communities long accustomed to certain diets, (2) the bioavailability of nutrients, such as iron and calcium, depends on the mixture and interaction of different foods and some medicines when actually present in the intestine, (3) the mathematical tools used for nutritional assessment are all useful approximations, and (4) sensational, alarming scientific reports of limited studies (Cartesian reductionism) and subsequent media publicity give rise to a unwarranted overattention, information overload and public confusion ("national nutritional neurosis"), with a specific "nutrient or food of the month or year." Currently, the important micronutrient selenium has been touted as an unproven preventive against a wide range of maladies ("selenophilia"). Olive oil (justifiably) and garlic (much less certainly) seem to be highly prized foods of the present.

Optimal health and nutrition cannot be defined, and even the widely quoted RDA or RDI (recommended dietary allowances or intakes) are extremely approximate figures, based on imperfect facts derived from limited studies from humans and animals. In the United States, they are intended for use with groups, not individuals ("needs of the majority of the healthy population"), with a "safety factor" of 25% added. The necessity for caution in their use is emphasized by comparison of RDI from one American compilation (1980) to the next (1989) and between various countries. Different figures are sometimes given for certain nutrients; the age groupings selected vary; some countries give two levels, one for individuals and the other for planning for population groups.

II. Dietary Modification

The aim for an individual, a family, or a nation should be to consume an enjoyable, affordable, and culturally acceptable diet, largely based on local foods, with a sufficient intake of all nutrients, leading to the best possible level of health, both currently and in later life. National policy and advice on dietary details will need to vary greatly in different cultures, but everywhere have to be approximations based on presently available knowledge. Much effort has been given by governments, international agencies, and nutritionists as to how to modify those aspects of diets that appear to be dangerous or in need of improvement in particular community. This can be made more difficult by inadequate training in nutrition for physicians and nurses whose advice is often asked. As food habits are learned early in life and often difficult to change, demonstration and motivation are needed, as well as information. Also, the role of nondietary causes (e.g., genetics, infections, parasites, and psychologic and social stresses) in the causation of different forms of malnutrition and in long-term effects of dietary intakes have to be taken into account.

A. Assessment of the Diet, Its Causes and Nutritional Problems

Such an assessment is logically needed but is difficult to carry out. The techniques and tools used in dietary assessment are all approximate, qualitative, yet useful guides. They need, when possible, to comprise several validating methods of nutritional appraisal, such as a combination of clinical signs, anthropometry, laboratory tests and food consumption). Dietary intake may be measured by questionnaires (often 24-hr dietary recall, especially with the limited range of foods in Third World countries), by observation and measuring food consumption and preparation (including, as far as possible, foods eaten out of the home), and, in some literate communities, by dietary records. Errors can result from an atypical, too short period, or from deliberate over- or underreporting of some items, including alcohol consumption.

Biologically, individual (or community) variation in nutrient needs have to be recognized. Also, the bioavailability of nutrients from foods depends on physiological modifications (e.g., increased iron absorption in pregnancy) and on the variable digestion, absorption, and utilization at the cellular level resulting from the actual mixtures of foods in the alimentary canal (e.g., vitamin C–containing foods increase the absorption of iron; phytates interfere with the uptake of calcium and zinc, and tea with iron; a range of factors in breast milk facilitate absorption of many nutrients, such as iron).

However, as in all community studies, investigations into the causes of interrelated dietary inadequacy and infection (or other factors), it is difficult to differentiate those that are "associated" from those that have a predominant or minor "causative" effect. Mathematical analysis (including multiple regression analysis) may be helpful, but a sensible practical view is to regard highly suspect influences as potential "risk factors."

Lastly, dietary analysis in a country is usually complicated by the need to consider many socioeconomic and/or cultural groups, as is the case in America. Also, all over the world, urbanization, the persuasive influence of commercial marketing, and a continually increasing number of technologically modified foods are changing and tending to Westernize dietary habits, especially in cities. This has occurred, for example, in Japan.

B. Nutritional Dietary Modification

This is often termed *nutrition education*. It comprises information, advice, and probably more importantly, persuasion and motivation to change, often related to local cultural considerations (preferred body image; concepts of body physiology, such as "hot–cold" greater attention given to male children). Such approaches include the following overlapping methods: (1) mathematical, (2) food groups, (3) key nutrients, and (4) general advice.

1. Mathematical Method

Mathematical nutrition education is usually self-defeating or unattainable, although appealing to Western Cartesian culture. First, the numbers suggested are often debatable and vary between individuals. Also, no normal person knows—still less customarily calculates—the amount in grams, calories etc., or the percentages of even the most obvious nutrients consumed daily. Indeed, in modern urban circumstances, these calculations are impossible, with the unsuspected "invisible" fat, sugar, and salt in many processed products (e.g., pastries, canned soups, etc . . .). The only way this could be done would be by a trained dietitian calculating the intake of major nutrients (e.g., fat) and advising on reduction or increase in practical domestic measures (e.g., using specified spoonfuls less or more oil) and/or in customary food units (e.g., locally standard tortillas). In some modern Westernized circumstances, some individuals become expert at calorie or cholesterol counting by memorizing (or having lists) of the calorie or cholesterol contents of commonly used foods.

A major role for an approximate mathematical approach to modification is with therapeutic or preventive diets worked out by dietitians for use in institutions (e.g., schools or hospitals) and for nutrition-related diseases (e.g., diabetes). Specialized guidance will be essential for recently developed hospital "nutrition support" services, particularly with parenteral feeding, using the intravenous route in seriously ill persons.

Major dietary issues include (1) obesity, eating disorders (anorexia nervosa and bulimia), diabetes, gout, renal and cardiac illnesses, food allergies, (2) the management of various "inborn errors of metabolism" [such as phenylketonuria (PKU), in which individuals cannot metabolize the amino acid phenylalanine from birth, with serious consequences, including mental retardation], and (3) the role of attention to diet in the supportive treatment of cancer and AIDS.

2. The "Food Group" Method

This approach has been used in the United States and other countries for some decades, usually based on four (or sometimes more) so-called "groups"—meat and meat substitutes, milk and milk products, fruits and vegetables, grains (bread and cereal products)—often directed toward a difficult-to-define "balanced diet" in which all needed nutrients are present and with a mutually complementary effect.

In fact, this concept has only limited usefulness in modern America, especially with the increasing number of commercial processed products. Also, this group method ignores food quality (e.g., nutrient density, fiber content, method of preparation). It gives rise to misconceptions, for example, that protein is only present in animal products. The system is culturally biased; for example, the majority of the world obtains most of its protein from cereal grains and legumes.

In technically less developed countries, the U.S. "food group" classification is still less relevant, and advice and information should be "staple-centered." The diet will always have to be envisaged as being the staple (usually a cereal grain such as rice, or a root crop such as potato or yam) together with other nutritionally complementary foods.

3. Key Nutrient Approach

This approach may be practicable and useful in different circumstances, based on knowledge of foods rich or low in particular nutrients. In parts of

Indonesia, the main nutrient lacking is vitamin A, or its precursor, β-carotene. In such areas, available beta carotene or vitamin A would be key nutrients needing special attention, either in the diet or as a supplement. Foods advised would include yellow-orange pigment vegetables and fruits, as well as red palm oil and dark green leafy vegetables, especially for pregnant and lactating women and young children.

Conversely in overfed communities, the key nutrient is often saturated fat, so that major emphasis should be directed toward discouraging its consumption by substituting other foods and/or (more realistically), limiting or modifying various popular items. For example, in the United States, foods high in animal fat include "hot dogs," regular hamburgers, french fries, ice cream, and fattier bacon.

At the same time, current consensus in pediatric nutrition is that restriction of fat and cholesterol is not indicated in young, rapidly growing children in the first 2 years. After this, moderation seems prudent, partly because of the need to influence lifelong food habits.

4. General Advice

Usually, nutrition education and advice have to be mainly in general terms, at best quasimathematical, within the economic, hygienic, culinary, and cultural restraints indicated earlier, and channeled through schools, the press and mass media, and individual counselling. They must be geared to minimizing locally important nutritional problems ("key nutrient approach"), and be directed to the physiologically and socially at-risk, including pregnant women, young children, teenagers, and the elderly. Three universal generalizations are (1) that the more mixed and varied any diet the better, especially if based on unprocessed foods, (2) that available animal products should be well-cooked, and (3) plant foods should be as well-washed as possible.

Traditional time-periods at which nutrition behavior is implanted or modified is by the family diet and as a result of school meals.

C. Nonnutritional Dietary Modification

For better and worse, urban influences and industrialization increasingly affect the diet all over the world, especially in more technically developed countries. These include marketing (advertising, promotion) (including less desirable foods [e.g., infant formulas, "fast foods"] and processed products containing many ingredients unknown to the consumer), the introduction of artificial substitutes or "fake foods" [e.g., (such as saccharine, aspartarine, etc.), nonabsorbable fat substitutes (such as "Simplesse" and "Olestra"), and the influence of special interest groups (e.g., the milk, meat, egg, and dairy industries)]. These groups can affect the items given emphasis in large-scale programs (e.g., school meals in the United States or as part of international food aid, especially for emergency situations). However, in recent years, in more educated technically developed communities, some marketing practices have changed to increase sales appeal for more educated consumers (i.e., low fat, pesticide-free foods). Food categories particularly likely to expand in sales are those labeled "natural," organic, "diet," "convenience," and "light," despite difficulties in definition.

Dietary choices also for better and worse have also been much influenced by the "health food industry." This movement originated as a reaction to overmechanized, overchemicalized food production, preparation, and consumption. In many ways, it has publicized and headed the present-day moves toward a "prudent diet," with an emphasis on natural foods. At the same time, it has lead to a completely unwarranted quackery, with mystical values attributed to high dosage vitamin supplements and to certain foods, especially many unusual, often exotic, and usually high-priced items, particularly including those directed to psychologically vulnerable groups (e.g., athletes, teenagers, and the elderly).

Dietary frauds are also obvious in the continuing flow of books and regimens offering miraculous "guaranteed" methods of slimming or weight reduction, sometimes combined with "diet pills" and apparatus for "passive exercise." Their value is probably related to their usual high cost and temporary novelty. Their lack of success ("yo-yo weight regain") is indicated by their continual change. The Western obsession with overweight is indicated by the fact that the word *dieting* usually refers to food restriction. This is exemplified by serious eating problems that are probably on the increase, especially in young women. These are anorexia nervosa and bulimia, both rooted in the desire for an ultra-thin, almost emaciated figure. [*See* EATING DISORDERS.]

III. World Perspectives

Satisfactory mixtures of a wide range of foods make up the present-day diet of the world's many cultures, including those in the United States. It is then only possible to highlight two main dietary themes and changes currently occurring in them.

A. Non-Western Diets

This encompasses an extremely wide array of different food combinations and most societies usually include an often small "Westernized" urban minority. On the whole, the basic food mixtures of such traditional diets are usually satisfactory. The main deficiency is often in quantity, particularly the staple. This may itself be changing, as with the increasing spread of the consumption of potato and cassava around the world. In many areas, this problem has been made worse by increasing cost and decreasing production of foods for local use for various reasons. These include overdependence on a limited range of cash crops (e.g., coffee, cane sugar, or cocoa) (whose world price varies yearly depending on yields) and recent financial economic adjustments (1980s) made to try to rectify huge accumulated national debts. Such changes often lead to greatly increased food prices and a move to cheaper, less preferred items [e.g., from imported "salt-fish" (cod) to shark in the Caribbean].

A marked change in non-Western style diets has been in the feeding practices for young children. In many countries, breastfeeding has declined or become accompanied by falsely prestigious apparently "modern" bottle feeding (mixed milk feeding), with locally available animal milk or, increasingly, highly expensive infant formulas, often imported. This practice has spread for a variable mixture of reasons: perceived status, unethical commercial advertising and promotion, unhelpful health services (with professionals untrained in breastfeeding management), and, especially in some urban areas, by women employed in salaried work far away from home. As a result, there has been an increase in avoidable marasmus as a result of over-dilute feeds, often made worse by infective diarrhea.

In addition, dietary problems still occur commonly in the traditionally and physiologically dangerous periods, especially when the young, rapidly growing child is in transition from sufficient breast milk to the full adult diet. At this time many infections occur and add to the nutritional burden in various ways. It is at this stage of life (1–2 or 3 years) that kwashiorkor (and similar less severe forms of PEM) occur more commonly. A major consideration in nutrition in such circumstances is to protect against infections and to institute easy inexpensive treatment (by improved hygiene, immunization, oral rehydration etc.) and to devise home-prepared weaning "multimixes" of local foods, which are nutritious, digestible, culturally acceptable, affordable and culinarily practicable.

Pregnant and lactating women are also especially vulnerable, particularly after repeated reproductive cycles of pregnancy and lactation, often made worse by specific culturally defined food avoidances and limitations. This can lead to various forms of maternal depletion, most commonly anemia due to deficiency of iron or folate. In areas of special risk, other deficiencies may occur, for example, osteomalacia [adult rickets from vitamin D (actually mainly a hormone produced by exposure of the skin to sunlight) and calcium deficiency], where strict purdah (veiling and seclusion of women) is customary, or beriberi (severe thiamin deficiency), when the diet consists almost entirely of polished rice]. Under these special circumstances, the issue of vitamin supplements may be indicated.

B. Westernized Diets

Until recently, Western diets and those of Westernized minorities in the urban elite in many countries have been characterized by (1) usually abundant food supplies, with often hundreds or thousands of new processed items in supermarkets annually, (2) low intakes of fiber and complex carbohydrates (e.g., cereal grains), (3) increasing fat consumption (especially saturated animal fats, such as meat, eggs, and dairy products, including cholesterol), sugar (sucrose), sodium (mainly sodium chloride or salt), and nutritionally unnecessary amounts of protein (sometimes 2–3 times real needs), all of which are present in many purchased, processed food products and fast-food items (e.g., pizza), (4) recognition, diagnosis and overdiagnosis of food allergies (possibly including hyperkinesis in young children), (5) overemphasis on vitamin supplements as being universally needed (over-the-counter vitamin sales in the US ± $1.5 billion per year), (6) increased intake of up to 50,000 chemicals used in agriculture and food processing and preservation (e.g., sulfite in preventing wilting in lettuce, dioxin present in bleached-paper milk cartons, lead and aluminum in canning, and (7) an unexpected rise in illness and unproductive sick, even deaths, caused by food poisoning (microbial infections: salmonella, listeria and campylobacter organisms, viruses, botulism toxin, etc.), most frequently from animal products (e.g., chicken [contaminated disemboweling machines, water-pooling of chicken carcasses], undercooked eggs, "soft" cheese, seafood, and meat infected during mass production, inadequately inspected importation from abroad and distribution, and by unsatisfactory home culinary practices, including storage, hand-washing and undercooking). Especially dangerous items include raw steak *tartare* and *sushi* (particularly the potentially toxic *fugu* fish). Other syndromes sometimes related to modern diet include a range of allergies and possibly hyperkinesis (hyperactive) in young children believed by some to be due to additives in food, especially certain artificial colorings.

Apart from recent mass-production induced infections, the risks of such diets are not in early onset malnutrition, but rather as risk factors in the development of multifactorial so-called degenerative illnesses (mentioned previously) occurring as chronic burdens in later adult life, and thereby limiting *active* longevity (sometimes termed the "epidemiological transition"). Awareness of these ill effects have become increasingly known to the literate public through informed newspapers and magazines. Such educational information seems to have had some effect. It has caused changes (1) in some restaurants (use of vegetable oil, french fries cooked in vegetable oil, rather than a mixture of animal fat (lard) and vegetable oil as before; grilled items rather than fried, opening of salad bars, turkey meat hot dogs, availability of nutritional menu cards emphasizing fish and lean meat); (2) in supermarkets (conspicuous labeling as "pesticide free," etc.); (3) in advice on home food preparation (increased use of vegetable oil and fish, decreasing use of items high in animal fat, e.g., poultry skin, "marbled" beef, eggs), methods of minimizing the chances of "food poisoning" (actual bacterial infection), including avoidance of raw or "underdone" fish and meat, and care with refrigerated foods, and (4) increased emphasis on physical exercise as part of a nutritional/fitness regimen.

As a result of aroused awareness recognition of the dangers of Westernized diets during the past 20 years, at least 15 countries, including Norway, UK, and the United States have produced national nutritional guidelines. All are similar in their general emphasis: decrease calories to suit energy needs (especially animal fat and sugar); limit salt intake; increase fiber (especially soluble fiber) and, possibly, calcium for females from childhood (who are at risk of postmenopausal osteoporosis, largely caused by hormonal changes.) A high emphasis on sugar (especially more sticky preparations) as the main cause of dental caries has been reduced somewhat because of the clearly proven protective effects of adequate intakes of the nutrient fluoride nowadays present in many items, including toothpaste, and in water, where safe fluoridation [1 part per million (ppm)] has been introduced. Recent scientific debate also suggests that advice concerning limiting cholesterol intake remains, but with less *direct* emphasis than before. This can, in fact, be taken care of by limitation of dietary consumption of saturated, mainly animal fat and some tropical vegetable oils (palm, coconut).

TABLE II Dietary Recommendations

- Reduce fat intake to 30% or less calories. Reduce saturated fat intake to less than 10% of calories and the intake of cholesterol to less than 300 mg daily.
- Every day eat five or more one-half cup servings of a combination of vegetables and fruits, especially green and yellow vegetables and citrus fruits. Also increase intake of starches and other complex carbohydrates by eating six or more daily servings of a combination of breads, cereals, and legumes. Carbohydrates should total more than 55% of calories.
- Maintain protein intake at moderate levels, i.e., approximately the currently Recommended Dietary Allowance (RDA) for protein, but not exceeding twice that amount for 1.6 g/kg of body weight for adults.
- Balance food intake and physical activity to maintain appropriate body weight.
- We do not recommend alcoholic beverages. If you do drink, limit yourself to less tahn 1 ounce of pure alcohol daily. This is equivalent of two cans of beer, two small glasses of wine, or two average cocktails. Pregnant women should avoid alcoholic beverages altogether.
- Limit total daily intake of salt to 6 g or less. Limit the use of salt in cooking and avoid adding it to food at the table. Salty, salt preserved, and salt-pickled foods should be consumed sparingly.
- Maintain adequate calcium intake
- Avoid taking dietary supplements in excess of the RDA for 1 day.
- Maintain an optimal intake of fluoride, particularly during the years of primary and secondary tooth formation and growth.

Reproduced with Permission from National Academy of Sciences. "Diet and Health: Implications for Reducing Chronic Disease Risk" (1989). National Academy Press, Washington, D.C.

TABLE III General Guide to an Optimal Diet: Major Points

General	Some specifics	Practical examples
Decrease		
Overall intake of fat	Especially of animal origin, rich in saturated fatty acids and cholesterol	Use vegetable cooking oil and margarine Use lean red meat, fish, poultry
Sodium intake		Use least amount of salt in cooking
Sugar intake	Especially sweet sticky cariogenic candy	
Increase		
Fiber		Whole grain bread
Complex carbohydrates	Vegetables	Particularly potatoes
Vitamin C		Fruits Vegetables (especially fresh)
Exercise		
Variable emphasis		
Vitamin supplements	Usually do not need if mixed diet taken Megadoses to be avoided	

In the United States, "Dietary Recommendations" have been published by the National Research Council (1989). The main general points are given in Table II. However, in their application there are obvious problems in translating mathematical figures for daily intake (e.g., 6 g salt, 1.6 g/kg protein) into practical terms. This is particularly difficult when considering populations of different ages, size, and cultural dietary practices anywhere in the world, including America. Also, no mention is made in these recommendations concerning food safety regarding either chemicals present or bacterial contamination. A simplified guide is given in Table III.

IV. Suggestions for the Future

Essentially, a mismatch has developed in more technically developed countries between diets supplying genetically determined physiological nutritional "*needs*" evolved over thousands of years and the recent availability of abundant, highly advertised, largely processed, often fatty, salty or sugary foods catering to human "*wants*." Unfortunately, an opposite situation has occurred for the majority of populations in many poor, less technically developed countries, so that dietary advice has to be distinctly different—yet with the dangers of advocating elements of a Westernized diet that are currently being warned against. For example, the general population of such countries need more calories and other nutrients in their diet. These programs should be based, when possible, on improved environmental sanitation, on the prevention of contributory infections, and on "appropriate technology" to increase local production, preservation, and storage of foods making up the traditional diet, including small scale urban agriculture. The last include breast milk for infants and the use of homemade, village-level multimix "weaning foods," with special attention to ensuring a high calorie intake, via frequent feeds, added oils or fats, and malting. Usually, the need is to retain the traditional diet but to try to increase its availability and to reinforce when really indicated (e.g., iodine fortification in areas with goiter and fetal iodine-deficiency syndromes; vitamins A or D in regions where these nutrients are lacking). Unfortunately, imitation of "Western" practices and unaffordable status seeking modern advertising is tending increasingly to modify dietary patterns (e.g., infant formulas, calorie dense fatty fast foods), involving cities and both the poor and the relatively well-to-do.

In Westernized societies, the mathematical specifics of nutritional recommendations (Table II) always have to be simplified to useful, practical generalizations, such as lists of foods that should be "limited" or "consumed in greater amounts," unless more definite guidance can be obtained from a dietitian. This will rarely be the case, except for sick individuals or for institutional catering.

Legislation in some countries appears to be moving slowly, usually with delaying tactics from industrial concerns involved, toward limiting the use of chemicals at the many different links in the food chain, including soil pollutants, and making easily understandable, enforceable labeling of products mandatory. A major need is for clearly visible, easily interpretable information on the three commonest dietary excesses (saturated fats, calories, and salt) and on chemical additives or residues, as well as small print details of the percentage of RDA nutrients present. Regulations are needed for false, or more usually misleading, labels giving, for example, information on one item only (e.g., ''cholesterol free'') and ignoring others.

Apart from an universal need to decrease the chemical pollution of foods, future dietary needs in different parts of the world are simple to state in general, although extremely complex, variable, and uncertain in detail. In poor, less technically developed countries, more food is needed, particularly in rural areas, especially for young children and pregnant women, including, paradoxically, ''compact calories'' (oil and fats) during the weaning period. In urbanized, more technically developed countries, a decreased intake is often the main priority, with special reference to foods mentioned earlier. Attention particularly needs to be given to use of smaller portions and to limiting foods rich in animal fats (as concentrated sources of calories, saturated fats, and cholesterol) as well as pesticides and other chemicals, which are mostly lipid-soluble. The consumption of less fatty varieties of fish and new forms of lean pork (50% less fat) and beef is receiving attention, as is the use of nontropical vegetable oils (e.g., corn oil, safflower), which are composed of polyunsaturated fats with no cholesterol, and fiber-rich foods. More recently, mono-unsaturated vegetable oils—olive oil, canola or rape seed oil, and peanut oil—seem preferable as containing no cholesterol (as with all plant foods), having no saturated fat and also not having the risk of being related to some forms of cancer, especially of the colon, for which polyunsaturated vegetable oils are considered by some scientists to be risk factors.

The future of the world's diet is impossible to foresee. Factors that influence it in different directions include the geometric global increase in population size (especially in cities in less technically developed countries, including rising numbers of elderly not looked after by the extended family paralleled by worsening world food reserves; changes in agriculture (use of land for food production vs. urban-style development, ''alternative'' nonchemical biological pest-control, deforestation; mechanized mass production of animal meat (opposed by ''animal rights'' -groups); newer developments in biogenetics (e.g., leaner breeds of cattle and pigs, high milk yielding cows); appropriate technology in small scale food production (aquaculture, urban agriculture); food technology (artificial sweeteners, fat substitutes, soy meat analogues); large-scale man-made and ''natural'' catastrophes (including war and civil strife; massive invasion by pests, such as locusts; widespread climatic changes, such as droughts, and especially the effects of damage to the earth's ozone layer); commercial influences (changes in economics [recession, national debt, altering prices of export cash crops (usually down) and of local basic food commodities (usually increased)]; inappropriate advertising vs. modifying products' profitably to suit real nutritional needs); and national and international political decisions (including the pros and cons of food aid, and national price interventions, food subsidies, rationing in situations of shortage).

In general terms, it seems clear that the healthiest prudent diet individually or in a national policy would have at its core an enjoyable, palatable modernized version of that consumed by ancestral gatherer-hunters, commencing with a full diet in pregnancy (with some nutrient supplements, most commonly iron and/or folate, if indicated), breast-feeding and a nutritionally appropriate transitional or weaning diet. In general, the emphasis would be on a wide mixture of fresh, nonchemicalized, nonprocessed foods, which are high in fiber, low in saturated fats, calories and salt, but high in complex carbohydrates and nutrients considered to be protective (e.g., beta carotene). Animal products should be well-cooked, and plant foods well-washed. This would afford the greatest degree of risk reduction of malnutrition and nutrition-related degenerative maladies, together with improved food safety.

This applies all over the world, but obviously will vary with the abundance and range of foods which are economically and geographically available, and also culturally acceptable in a traditional rural society, or for the urban poor, as well as in mechanized and hectic lifestyle of the present-day prosperous city dweller, whose hunting ground is the confusing abundance of the supermarket.

Bibliography

American Dietetic Association. (1989). ''The Dietary Guidelines: Seven Ways to Help Yourself to Good Health and Nutrition.'' American Dietetic Association, Chicago, Illinois.

Eaton, S. B., Shostak, M., and Konneer, M. (1988). ''The Paleolithic Prescription.'' Harper and Row, New York.

The Hunger Project. (1989). ''Ending Hunger.'' Praeger Publishers, Sparks, Nevada.

James, W. P. T. (1988). ''Healthy Nutrition.'' WHO Regional Office ior Europe, Copenhagen.

Jelliffe, D. B., and Jelliffe, E. F. P. (1990). ''Community Nutritional Assessment.'' Oxford University Press, Oxford.

Latham, M., and Van Veen, M. S., eds. (1989). ''Dietary Guidelines: Proceedings of an International Conference, Toronto, Canada.'' Cornell International Nutrition Monograph Series No. 21, Ithaca, New York.

National Research Council. (1989). ''Diet and Health: Implications for Reducing Chronic Disease Risk.'' U.S. National Academy Press, Washington, D.C.

National Research Council. (1989). ''Alternative Agriculture.'' National Academy of Sciences, Washington, D.C.

National Research Council. (1989). ''Recommended Dietary Allowances.'' National Academy Press, Washington, D.C.

The Surgeon General's Report on Nutrition and Health. (1988). US Department of Health and Human Services Public. Nos. 88-50211, U.S. Government Printing Office, Washington, D.C.

Truswell, A. S., ed. (1983). Recommended dietary intakes around the world. *Nutr Abstr Rev* **53,** 11.

U.S. Department of Agriculture and U.S. Department of Health and Human Services (1990). Nutrition and Your Health: Dietary Guidelines for Americans. 3rd Edition.

World Health Organization. (1989). ''Health Surveillance and Management Procedures for Food-handling Personnel.'' Tech Rept Ser. No. 785, WHO, Geneva.

Dietary Cravings and Aversions during Pregnancy

FORREST D. TIERSON, *University of Colorado at Colorado Springs*

Glossary

Aversion Compelling distaste for a certain food item or items in the diet

Craving Urgent, imperative longing or intense, compulsive desire for one or more of a wide variety of articles of diet

Geophagy Special form of pica in which clay or earth is eaten

Pica Craving for a substance that is not normally considered edible

THE OCCURRENCE OF DIETARY cravings and aversions during pregnancy is well known. Unfortunately, little is known of the etiology or epidemiology of this complex of symptoms. The anthropological and nutritional literatures contain numerous descriptions of cravings and aversions, food taboos, and foods for which either prescriptions or proscriptions exist during pregnancy, in both traditional and modern populations. However, although a wealth of qualitative information exists (usually in the form of narrative examples), little information of a quantitative nature has been published on the subject. In addition, most descriptive studies have been retrospective in nature, reporting observations made relatively late in pregnancy.

I. Background

Special attention has been given to the diets of pregnant women over most of recorded history. Prohibitions or restrictions during pregnancy have long existed for some kinds of foods, while other foods have often been regarded as essential to successful pregnancy outcome.

Many factors are known to affect maternal diet during pregnancy. These factors include cultural practices, beliefs, food availability, economic conditions, and individual food preferences. Not surprisingly, women generally consume more food when they are pregnant than when they are not pregnant. Unexpectedly, we find increased consumption during the early stages of pregnancy, when maternal caloric needs are not as great as during the second and third trimesters. This process of "excess" consumption in early pregnancy is thought to provide the fetus with necessary nutrients without putting undue physiological stress on the mother, although precisely how this process works in humans is not well known. The mechanisms that bring about changes in maternal appetite during pregnancy are also incompletely understood, but these influences on intake levels of nutrients can be substantial.

II. Pica of Pregnancy

Pica is probably the most notorious of the better-known forms of cravings during pregnancy, especially the specific type of pica expressed as geophagy. However, the nutritional and obstetrical literatures contain many references to cravings during pregnancy for items other than clay, earth, and/or laundry starch, including cravings for such diverse items as baking soda, coal, soap, disinfectant, toothpaste, mothballs, gasoline, tar, paraffin, wood, chalk, and even pencil erasers. While certainly not exhaustive, this list does provide an idea of the wide range of items for which cravings during pregnancy have been reported.

By far, cravings for clay, laundry starch, and soil

(earth) outnumber cravings for other items in most areas of the world. Among black women of lower socioeconomic status in the rural southern United States, some studies have found that between 50 and 75% of pregnant women report pica for clay and/or laundry starch. Similar cravings for clay are found among aboriginal women in Australia and among several different groups of women in Africa. In general, explanations for the existence of pica of this type fall into four major categories: (1) superstitious beliefs and/or folklore; (2) physiological explanations claiming that pica exists to compensate for some existing dietary deficiency; (3) psychological factors invoking explanations of insecurity, suggestion, attention-seeking, and other more severely psychotic reasons on the part of the pregnant women; and (4) alterations in the sense of taste and smell during pregnancy. Most likely, the majority of pica can generally be explained by tradition and the folkways of the affected people. For example, black women in the rural southern United States also experience pica for clay and starch when not pregnant. These items are considered to be especially important for successful pregnancy outcome and are supplied by friends and family members to the pregnant woman.

In general, pica of pregnancy is less frequent in Western nations. However, in more educated populations, pica often may not be reported because many individuals consider cravings for odd, nonfood items as aberrant and perverse behavior. Thus, the actual occurrence of pica among more-educated Western populations is not clear. During a recent prospective study of the cravings and aversions of 400 white pregnant women of high socioeconomic status in Albany, New York, no cases of pica were reported.

III. Cravings and Aversions for Food Items

Historically, most studies of dietary cravings and aversions during pregnancy have dealt mainly with pica and secondarily with cravings for food items. Aversions during pregnancy were often not considered at all; however, dietary aversions are at least as common as dietary cravings during pregnancy, and they are most certainly related.

The reasons why cravings and aversions for food items develop are essentially unexplained, although a number of explanations have been offered. Cravings may represent a physiological response to maternal and/or fetal nutritional needs. Aversions may represent a response to low levels of toxins present in the foods for which aversions have developed. Cravings or aversions may also be caused by mediating factors such as maternal metabolic changes and changes in olfactory and taste sensitivity during pregnancy. Cravings and aversions may not be biologically determined at all but, instead, may simply be learned behaviors, or they may be solely idiosyncratic in nature. Despite the cause, studies significantly show that cravings and aversions for food items are not limited to a few isolated individuals, but that they affect a sizable fraction of the pregnant population studied.

The widespread occurrence of dietary cravings and aversions during pregnancy among human populations is demonstrated by the high frequency with which this phenomenon is the focus of folk beliefs concerning pregnancy and health. Cravings and aversions are expressed for a number of common dietary constituents, not just idiosyncratic items such as pickles and ice cream, or clay and starch. Studies have demonstrated that at least 10% of women in specific populations report cravings for several different dietary items, with others also reporting high frequencies of aversions for some food items. Since some aversions are to substances known to be embryotoxic, such as tobacco smoke and alcohol, the occurrence of these aversions may result in reduced fetal exposure to such toxins.

If dietary cravings and aversions affect maternal nutrition by increasing the intake of nutrients or by decreasing the ingestion of embryotoxic agents, then their potential impact on maternal (and, subsequently, fetal) health in human populations might be considerable. Because dietary cravings and aversions during pregnancy exist among many populations, such an impact could be strong and widespread.

A. Occurrence of Dietary Cravings and Aversions

In a relatively recent prospective study of 400 white, well-nourished women in Albany, New York, whose pregnancies were ascertained by the 13th week, 76% of the women reported craving at least one item, while 85% reported at least one aversion during pregnancy. According to information obtained from 7-day diet histories, the greatest changes in frequency of consumption of specific food and beverage items occurred between the last menstrual period and the 12th week of pregnancy. Most cravings and aversions also occurred early in the pregnancy, with aversions occurring earlier, on the average, than cravings. Women reporting cravings increased their consumption of the items craved, and they decreased their consumption of those items for which they reported aversions. At least 15% of the women reported cravings for foods in the categories of ice cream, chocolate candy, cookies, citrus products, and fruits and/or fruit juices (other than citrus). An additional 10% expressed cravings for foods with Italian sauce (essentially tomato sauce). At least 10% of the women reported aversions to foods categorized as fish, beef, foods with Italian sauce, and meat in general. Over 20% reported aversions to alcoholic beverages and fully 34% reported aversions to coffee during their pregnancies. In addition, approximately 13% reported aversions to tobacco smoke.

Less comprehensive results from other studies substantiate many of these results, keeping in mind that most other studies have been retrospective in nature.

B. Cravings and Aversions and Fetal Outcome

Although associations between separate cravings and measures of fetal outcome were not demonstrated in the Albany study, several associations existed between the occurrence of specific dietary aversions and several measures of fetal outcome, especially fetal growth index (FGI), which is a measure of mean size for gestation length adjusted for population means in the specific geographic area. A significant positive association existed between the occurrence of any aversions during pregnancy and increased FGI. Several positive associations were noted between increased FGI and aversions to several different meats, implying that decreased consumption of meat resulted in increased FGI (greater birthweight). Other studies of cravings and aversions have not reported similar findings, although aversions to some meats have been routinely reported in the literature. Additionally, in most populations, there are always certain foods that are thought to be important in terms of their effect on pregnancy outcome—either foods necessary for successful outcome or foods to be carefully avoided during pregnancy.

C. Origins of Cravings and Aversions

The origins of the majority of dietary cravings and aversions reported during pregnancy are defined by the pregnant women as being endogenous in nature. The origins of aversions appear to be more closely related to endogenous reasons than the origins of cravings. Women might express cravings for items that they would prefer to eat anyway (e.g., chocolate), but for which they would feel constrained against eating when not pregnant. Aversions, on the other hand, are usually more abrupt behaviors that are initiated in response to some specific stimulus (e.g., morning sickness). Most of the nausea and vomiting during pregnancy that can be attributed directly to the pregnancy also occurs early in pregnancy. Since the origin of most dietary aversions appears to be more directly related to physiological considerations, perhaps these two symptoms (cravings and aversions) should not be united into a single complex but should be considered as two separate patterns of behavior. [*See* PREGNANCY, NAUSEA AND VOMITING.]

Bibliography

Dickens, G., and Trethowan, W. H. (1971). Cravings and aversions during pregnancy. *J. Psychosom. Res.* **15,** 259–268.

Hook, E. B. (1978). Dietary cravings and aversions during pregnancy. *Am. J. Clin. Nutr.* **31,** 1355–1362.

O'Rourke, D. E., Quinn, J. G., Nicholson, J. O., and Gibson, H. H. (1967). Geophagia during pregnancy. *Obstet. Gynecol.* **29,** 581–584.

Tierson, F. D., and Hook, E. B. (1989). Dietary cravings and aversions during pregnancy and association with pregnancy outcome. *Am. J. Phys. Anthropol.* **78,** 314–315 (abstract).

Tierson, F. D., Olsen, C. L., and Hook, E. B. (1985). Influence of cravings and aversions on diet in pregnancy. *Ecol. Food Nutr.* **17,** 117–129.

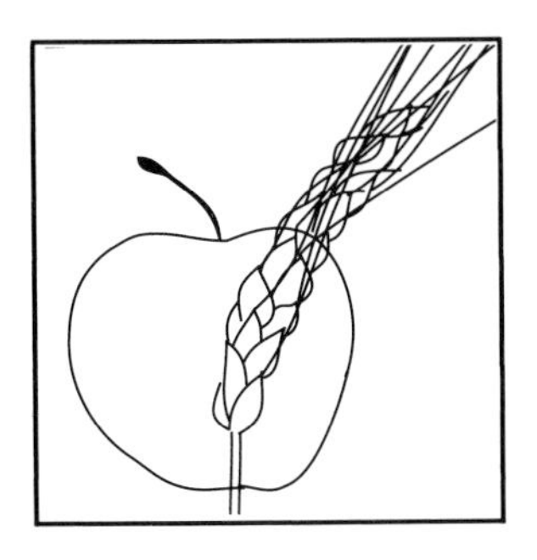

Dietary Fiber, Chemistry and Properties

ROBERT RASIAH SELVENDRAN, *AFRC Institute of Food Research, Norwich Laboratory*

Glossary

α- and β-linkages Aglycone (glycosidically linked group) on carbon-1 is axial when α-linked and equatorial when β-linked; this confers helical structure on starch and linear structure on cellulose; alimentary enzymes are α-glucosidases and, therefore, hydrolyse starch but not cellulose

Cation exchange capacity Ability of a fiber preparation to exchange some of the cations it contains for others in solution

Cellulose Linear polymer of β-(1 $\rightarrow$ 4)-linked D-glucose and the only truly fibrous component of plant cell walls

Dietary fiber Mainly composed of plant structural materials, and resistant to digestion by endogenous enzymes

Hemicelluloses Complex cell wall polysaccharides composed mainly of neutral sugars; they are not water-soluble but are soluble in cold alkali

Lignin High-molecular weight aromatic polymer associated with woody and some specialized tissues

Nonstarch polysaccharides Dietary polysaccharides other than starch, and mainly plant cell wall in origin

Pectin Branched polymer containing a backbone of partially methyl-esterified D-galacturonic acid interspersed with rhamnose residues, some of which carry short side chains of neutral sugars

Resistant starch Starch in cooked and processed foods that is resistant to amylolytic degradation

Starch Plant storage polymer of glucose, which exists in two forms: amylose, an unbranched molecule containing α-(1$\rightarrow$4)-linkages, and amylopectin, a branched molecule containing α-(1$\rightarrow$4)- and α-(1 $\rightarrow$ 6)-linkages

Water-holding capacity Amount of water that can be taken up by 1 g of dry fiber

DIETARY FIBER (DF) may be considered a structural component of plant foods unavailable to digestion by human intestinal enzymes during normal gut transit. The cell walls of vegetables, fruits, and cereals are important as sources of DF. The composition and physicochemical properties of DF polymers influence the rate of digestion and absorption of nutrients, particularly in the small intestine. The DF polymers eventually pass from the small intestine to the colon, which contains vast numbers of anaerobic bacteria that can break down the polymers to release simpler molecules, some of which can be absorbed to provide energy. The fermentable components of DF also contribute to bacterial growth and, therefore, to fecal biomass, whereas some materials resist fermentation and only provide physical bulk to the feces. Both these effects have a laxative effect which influences the long-term health of the large bowel.

I. Underlying Hypothesis and Definition of DF

The role of DF in human nutrition and health has become increasingly topical in recent years. This revival of interest can be traced to the hypothesis that the consumption of low-fiber diets is a common etiological factor in many metabolic and gastrointestinal diseases of the Western world. Although the amount of fiber ingested is small in relation to the total diet, fiber appears to exert a major influence on the metabolism of the gastrointestinal tract, and these effects are dependent on the physicochemical properties of the fiber and its components. The following two statements outline the "DF hypothesis."

1. A diet rich in foods that contain plant cell walls (e.g., high-extraction cereals, fruits, vegetables) is protective against a range of diseases, in particular those prevalent in affluent Western communities (e.g., constipation, diverticular disease, large bowel cancer, coronary heart disease, diabetes, obesity, gallstones).
2. In some instances, a diet providing a low intake of plant cell walls is a causative factor in the etiology of the disease, and in others it provides the conditions under which other etiological factors are more active.

The hypothesis, as stated, implies that the essential difference between protective diets and nonprotective diets is the amount of plant cell wall material they provide, and the protection is, or is derived from, the plant cell walls in the diet.

DF was initially defined as "the skeletal remains of plant cells, in our diet, that are resistant to hydrolysis by the digestive enzymes of man." As this definition did not include the polysaccharides present in some food additives, the definition was later extended to include "all the polysaccharides and lignin in the diet that are not digested by the endogenous secretions of the human digestive tract." Accordingly, for analytical purposes, the term DF refers mainly to the nonstarch polysaccharides (NSP) and lignin in the diet. While the revised definition is generally accepted, it should be borne in mind that the polysaccharides of food additives generally constitute only a very small proportion (<2–3%) of the DF component of most diets. Hence, most of our DF intake comes from the cell walls of vegetables, fruits, cereal products, and other seeds.

The principal components of DF are complex polysaccharides, some of which are associated with polyphenolics (including lignin) and proteins. The other noncarbohydrate components of cell walls, such as cutin, waxes, suberin, phenolic esters, and metal ions (e.g., Ca^{2+}), are quantitatively minor constituents of most plant foods, but some of them have significant effects on the properties and physiological effects of DF. Although most of the DF constituents may survive digestion in the proximal gastrointestinal tract, a significant proportion of them are degraded by the microorganisms of the human colon.

The average intake of DF in the United Kingdom (and in most Western countries) is about 20 g per person per day and, of this, about one-third comes from cereal sources. There is considerable variation between individuals; those with a high consumption tend to obtain more from cereal foods. The British intake of DF is small compared with that of a rural African diet, which might contain 60–150 g DF per day.

II. Main Components of DF

A diversity of plant tissues and organs constitute the vegetables, fruits, and cereals in the diet, but three main types of tissue predominate: (1) ground tissue (parenchyma), (2) supporting and conducting tissue (vascular), and (3) covering or protecting tissue (epidermal). Some supporting and strengthening tissue (sclereids) may sometimes be present. Parenchymatous tissues are the main source of vegetable and fruit fibers because the vascular bundles and sclereids of many vegetables are still relatively immature and only slightly lignified when the vegetables are harvested and eaten, and the lignin content of the cell wall material is usually <5% (w/w). Lignified tissues are of greater importance in cereal sources—such as wheat bran and bran-based and dehulled oat products. Wheat bran contains the lignified outer layers of the grain and the aleurone layer, and the lignin content of the cell wall material is about 12% (w/w). Wheat bran makes a significant contribution to cereal fiber intake. Because the nature of the carbohydrate polymers associated with various types of tissue from dicotyledonous and monocotyledonous plants are different, the polymers present in these species will be discussed separately. Water is also an important component of the cell wall and is present in various amounts—high (~80% w/w) in the parenchyma cell walls of most tissues, except mature dry seeds, but low (~10–15% w/w) in secondary walls.

The NSP content of a plant food, which is a measure of the cell wall polysaccharides, is obtained by removing the starch from the extractive-free residue by enzymatic treatments and determining the component sugars in the starch-depleted residue.

Two hydrolytic procedures are used to release the sugars from the noncellulosic polysaccharides and noncellulosic polysaccharides + cellulose. The liberated sugars are usually derivatized and estimated, as their alditol acetates, by gas–liquid chromatography. The uronic acid content of the preparation is estimated by colorimetric methods or by decarboxylation and measurement of CO_2 evolution. Some processed starch-rich products such as potato powder and cornflakes contain a significant but variable level of retrograded or "resistant" starch. Resistant starch is a term used to describe starch that is resistant to hydrolysis by the amylolytic enzymes that are used in the analysis of DF. It is formed when gelatinized starch retrogrades when cooled and dried, and could arise from a strong hydrogen-bond formation between OH groups on adjacent starch molecules. This starch becomes susceptible to enzymatic hydrolysis only after vigorous chemical treatment, usually with alkali. Most of the resistant starch of processed products would survive small intestinal transit to be degraded by colonic bacteria.

A. Cell Walls of Dicotyledonous Plants

The nonstarch polysaccharides are the major contributors to the DF content of fruits and vegetables and usually comprise between 2 and 2.5% of the plant organs' fresh weight. These organs consist mainly of parenchymatous and growing tissues. Leguminous seeds have a more variable NSP content (e.g., dried peas, 18.6%; haricot beans, 17.1%; chick peas, 9.9%; butter beans, 15.9%) on a dry weight basis.

1. Parenchymatous and Growing Tissues

The parenchyma cell wall is a specialized form of extracellular matrix surrounding the protoplast and exterior to the plasmalemma of the cell. Each cell interacts with its neighbors through the intercellular layers of the middle lamellae, which are rich in pectic substances and are cross-linked by Ca^{2+} to form various tissue types. Removal of the Ca^{2+} leads to cell separation. The dry matter of the cell walls of soft tissues is composed of pectic substances (40–45%), cellulose (30–35%), hemicellulosic polymers (10–15%), proteins (5–10%), and polyphenolics (5%). The proteins can occur as glycoproteins or as proteoglycans, in which the protein component carries polysaccharide substituents, and these appear to serve as cross-linking agents within the walls. Simple- and polyphenolics can also serve to cross-link the wall polymers. By virtue of the various types of cross-links, the primary cell wall components are organized into a complex covalently linked "macromolecular" matrix. Within this matrix are two separate morphological phases; a complexed continuous matrix, containing mainly noncellulosic polysaccharides, which are ester cross-linked and in which distinct microfibrillar structures, composed mainly of cellulose, are dispersed. These appear to be cross-linked to the noncellulosic polysaccharides via glycoproteins and highly branched pectic polysaccharides. The whole structure is bathed in an aqueous medium. It is, therefore, essential to recognize that in food DF should not be regarded as mainly the sum of the component cell wall polymers—the organization of the components into discrete structures confers the properties on DF.

a. The Microfibrillar Component—Cellulose Cellulose is the major structural polysaccharide of the cell walls of all higher plants. Cellulose is a polymer characterized by long chains of β-(1 $\rightarrow$ 4)-linked D-glucopyranosyl residues, whose degree of polymerization is of the order 10,000. Because of the high-molecular chain length and considerable capacity of the free hydroxyl groups to hydrogen-bond, both inter- and intramolecularly, cellulose forms fibers of remarkable strength. The fibers are arranged in an ordered manner within the microfibrils, which have crystalline and amorphous regions. The relative proportions of these regions may account for the differences in properties between cellulose in the primary and secondary walls. The amorphous regions may incorporate other polymers, which probably link the microfibrils to the matrix polysaccharides. This inference is based on the fact that the α-cellulose residues isolated from the cell walls of all parenchymatous tissues of vegetables and fruits contain small but significant amounts of associated pectic polysaccharides, wall glycoproteins, and polyphenolic compounds.

Cellulose is insoluble in water but is soluble in 72% (w/w) sulfuric acid because sulfonation takes place at the hydroxyl group on carbon-6 (C-6), breaking the hydrogen bonds. In primary cell walls, the cellulose microfibrils are "coated" with a layer of xyloglucans (bound by hydrogen bonds), and this enables the insoluble cellulose to be dispersed within the wall matrix. A proportion of the C-6 hydroxyl groups can be chemically replaced with carboxyl groups to give carboxymethyl-cellulose (CMC), which is water-soluble and is used as a food additive. The water-holding capacity (WHC) of cellulose is ~5 g water/g cellulose, whereas that of

CMC is 50–60 g water/g CMC. This shows how the properties of DF polymers can be dramatically altered.

b. The Matrix Components—Noncellulosic Polymers The major *hemicellulosic polymers* of parenchymatous tissues are xyloglucans, glucomannans, and proteoglycan complexes. The xyloglucans and proteoglycans constitute approximately 7–10% and about 5% of the dry weight of the cell walls, respectively. Xyloglucans contain a cellulosic β-D-glucan backbone to which short side chains are attached at C-6 of at least one-half of the glucosyl residues. α-D-xylopyranosyl residues appear to be directly linked to glucosyl residues, but these side chains may be extended by the apposition of β-D-galactopyranosyl, α-L-arabinofuranosyl, or α-L-fucopyranosyl-(1 → 2)-β-D-galactopyranosyl residues. As mentioned before, the xyloglucans are hydrogen-bonded to cellulose microfibrils, which aids cellulose dispersion through the wall.

Growing evidence indicates the occurrence of small but significant amounts of proteoglycans—complexes containing polysaccharide, protein, and polyphenolic moieties—in cell walls of soft tissues. Glycoproteins are also present in small amounts, and both polymers seem to cross-link the wall polysaccharides. The hydroxyproline contents of both types of protein complex vary, the glycoproteins generally containing higher levels. Parenchymatous tissues of most vegetables contain relatively small amounts of wall glycoproteins, although the cell walls of legume parenchyma, such as runner bean pods, contain fairly high levels of hydroxyproline-rich glycoproteins. [*See* PROTEOGLYCANS.]

The *pectic substances* are a complex mixture of colloidal polysaccharides, which can only be partially extracted from the cell walls with hot water, hot dilute mineral acid, or hot aqueous solutions of chelating agents such as ethylenediaminetetra-acetate or ammonium oxalate. The solvent action of the chelating agents depends, in part, on their ability to complex with Ca^{2+} held in the walls by pectins. Recent work has shown that the above conditions of extraction result in considerable breakdown of the pectins by β-eliminative degradation, giving rise to two distinct polysaccharide fractions—one enriched in galacturonic acid with relatively small amounts of neutral sugar residues (usually designated rhamnogalacturonans), and the other low in acidic residues but rich in neutral sugar residues (e.g., pectic arabinans, galactans).

Rhamnogalacturonans are the major constituents of pectic substances in which a proportion of the galacturonic acid residues are present as methyl esters. The rhamnogalacturonan backbone consists of chains of α-(1 → 4)-D-galacturonosyl residues interspersed with (1 → 2)- and (1 → 2,4)-linked L-rhamnopyranosyl residues. Attached to the main chains mostly through C-4 of rhamnosyl residues are side chains consisting primarily of D-galactose and L-arabinose, and also lesser amounts of D-xylose, D-mannose, L-fucose, D-glucuronic acid, and some rare methylated as well as unusual sugars.

Recent studies have shown that most of the "neutral" pectic polysaccharides such as arabinans and galactans that have previously been isolated are in fact degradation products of more complex pectic polysaccharides. Arabinans contain mainly α-(1 → 5)-linked L-arabinofuranosyl residues to some of which terminal arabinofuranosyl residues are attached through C-2 and/or C-3, and galactans contain primarily chains of β-(1 → 4)-linked D-galactopyranosyl residues. Using nondegradative conditions of extraction, only very small amounts of neutral pectic polysaccharides have been solubilized from a variety of vegetables and fruits.

Recent work on pectic polymers has shed new light on the nature and distribution of the pectic polymers within the walls of parenchymatous tissues. The main conclusions are as follows: (1) The pectic polymers of the middle lamellae are rich in galacturonic acid and are highly methyl-esterified, and a significant proportion of them can be solubilized by a dilute solution of cyclohexanediamine-tetra-acetate (CDTA, Na salt) at 20°C. These pectic polymers are only slightly branched and serve to cross-link adjacent cells mainly through Ca^{2+}-pectate bridges, and probably give rise to rhamnogalacturonans under degradative conditions of extraction. (2) The bulk of the pectic polymers of primary cell walls can only be solubilized (after extraction of cell walls with CDTA) under mild alkaline conditions, which suggests that in addition to removing the bridging Ca^{2+}, the ester cross-links between the galacturonic acid residues and hydroxyl groups of sugar residues, and phenolics elsewhere in the wall matrix must be hydrolyzed. (3) A small but significant proportion of the highly branched pectic polymers is associated with the α-cellulose residue. (4) The ratio of rhamnose to galacturonic acid is ~1 : 40–50 for the bulk of the pectic polymers from the middle lamellae and ~1 : 10–20 for the pectic polymers from the primary cell walls; the relative amount of branched rhamnosyl residues is much higher in the latter. These comments serve to emphasize the extent to which the pectic polysaccharides are cross-linked within the parenchyma wall matrix.

2. Lignified Tissues (Secondary Cell Walls)

During maturation, considerable thickening of certain primary cell walls occurs, mainly through the preferential deposition of cellulose, hemicelluloses, and lignins, which constitute about 40%, 25–30%, and 20–25%, respectively, of the dry weight of the secondary walls. The hemicelluloses deposited are mainly acidic xylans and small amounts of glucomannans. Acidic xylans contain primarily β-(1 $\rightarrow$ 4)-linked D-xylopyranosyl residues, ~10–15% of which carry 4-*O*-CH_3-glucuronic acid or glucuronic acid residues on C-2. Lignins are high-molecular weight aromatic polymers and consist, for the most part, of variously linked phenylpropane residues. The relative proportions of these residues vary in the lignins of different plants. Lignin can account for up to 25–30% of the dry weight of the walls of most supporting structures and appears to be covalently linked, mainly through ester links, with the glucuronic acid residues of xylans. This has the effect of cementing together the cell wall polysaccharides to form a rigid matrix and stratifying the wall. In commonly eaten vegetables and fruits, the overall lignin content of the cell walls is usually <5%. When the lignin content is >8–10% the texture becomes unacceptably tough.

3. Seeds

The seeds of dicotyledonous plants can be classified as those that are free of an endosperm (i.e., nonendospermic [e.g., pea, bean]) and those that have an endosperm (i.e., endospermic, as in certain leguminous species [e.g., guar, locust bean]). The former type of seed usually has starch as the main storage polysaccharide, and their cell walls are derived mainly from the tissues of the cotyledons with some contribution from the testa. The cell wall polysaccharides of the cotyledons are similar to those of parenchymatous tissues and are composed mainly of pectic substances, cellulose, hemicelluloses (e.g., xyloglucans), and glycoproteins. The notable difference is that pectic arabinans (in the case of peas) and arabinogalactans (in the case of soy beans), both probably linked to rhamnogalacturonans, tend to predominate in the seeds.

In contrast to nonendospermic seeds, most endospermic seeds contain galactomannans, which are located in the endosperm cell walls. The galactomannans are solubilized during germination of the seeds. The galactomannans are essentially linear molecules but are highly substituted on C-6 of the β-(1 $\rightarrow$ 4)-linked D-mannosyl residues with single galactopyranosyl residues, which renders them hydrophilic and greatly enhances their ability to bind water.

4. Epidermal Protective Coverings of the Wall

The outer walls of epidermal cells of leaves, fruits, and many other aerial organs of plants are covered with a protective layer of waxes and cutin, and these are quantitatively minor constituents of DF. Generally the cutin penetrates and intermingles with the outer polysaccharides of the wall. Underground organs (e.g., tubers) are protected by another type of lipid-derived polymeric material—suberin. Both cutin and suberin are embedded in and overlaid with a complex mixture of relatively nonpolar lipids, which are called waxes. Because cutin and waxes are resistant to bacterial degradation, cutinized tissues may serve an important role in restricting the access of intestinal bacteria (and bacterial enzymes) to the cell wall polysaccharides of some vegetables and fruits, especially when they are not cooked. This phenomenon may be particularly significant in the case of leafy vegetables (e.g., lettuce, cabbage), which have spongy parenchyma cells sandwiched between cutinized layers.

B. Cell Walls of Monocotyledonous Plants (Cereals)

Dietary sources of monocotyledonous plants are mainly cereal grains, and these are an important source of DF. However, some vegetables, notably onions and leeks, are monocotyledons, and these have a NSP content and composition similar to the dicotyledonous fruits and vegetables discussed previously. The NSP content of cereal grains, on a dry weight basis, is variable (e.g., brown rice, 2.1%; pearl barley, 7.8%; whole wheat, 10.4%; porridge oats, 7.1%). The NSP content of cereal products also depends on the extent of refinement; high-extraction wheat products (e.g., whole meal bread, whole meal breakfast cereals) contain more DF than products from low-extraction white flours (e.g., whole wheat flour, 10.4%; whole wheat bread, 10%; wheat bran, 41.7%; 72% extraction white flour, 3.3%). Products derived from grains are also commonly used as breakfast foods in the form of bran-based products, flakes, or porridge. The milling process of wheat is a complex operation, involving separation into three main fractions: bran or pericarp with the attached testa and aleurone layer (in all about 12–17%), the endosperm itself (about 80–85%), and the wheat germ (about 3%). The removal of the bran may result in a loss of up to 75% of the total DF content of the grain. Therefore, a knowledge of the cell walls of the endospermous and lignified tissues of the grain is important, and this will be considered for some important cereals: wheat, barley (which is comparable to oats), and rice.

1. Endospermous Tissues and Aleurone Layer

Unlike parenchymatous tissues of dicots, the endosperm and aleurone cell walls are virtually free of pectic polysaccharides, contain very small amounts of cellulose (~2%), and are rich in hemicelluloses (~80–85%). Protein, containing negligible amounts of hydroxyproline, accounts for ~8–10% of the walls. The endosperm cell walls consist of an amorphous matrix of hemicelluloses in which microfibrillar structures are dispersed. The hemicelluloses of wheat endosperm are mainly highly branched arabinoxylans (~80%), associated with small amounts of phenolics such as ferulic acid, and the mixed linkage β-D-glucan content is only ~1–2%. Arabinoxylans contain primarily β-$(1 \rightarrow 4)$-linked xylopyranosyl residues to some of which terminal L-arabinofuranosyl residues are attached through C-2 and/or C-3. In contrast, barley and oat hemicelluloses are rich in unbranched β-glucans (~75%) and are relatively poor in arabinoxylans (~15–20%). The ratio of $(1 \rightarrow 3)$- to $(1 \rightarrow 4)$-linkages in the barley β-glucans is ~3 : 7, and most of the β-glucans seem to be proteoglycans (polysaccharide–protein complexes). The hot water-soluble β-glucans of oat endosperm have an inherent viscosity not unlike that of pectins and specialist food gums such as guar gum.

Unlike wheat and barley endosperm cell walls, rice endosperm cell walls contain appreciable amounts of cellulose (~48%) and galacturonic acid-containing pectic substances rich in arabinose and xylose (~10%), acidic and neutral arabinoxylans, and some β-glucans. Therefore, the composition of rice endosperm cell walls appears to be intermediate between that of parenchymatous tissues of dicotyledons and that of wheat and barley.

The walls of wheat and barley aleurone cells are comparable and contain mainly arabinoxylans (~65–70%), with some β-glucans (~30%) and proteins. Small amounts of cellulose (~2%) and glucomannans are also present, and the wall polysaccharides are cross-linked by phenolic esters (e.g., ferulic acid). Lignin is absent. The arabinoxylans are neutral and are relatively slightly branched. The absence of glucuronic acid in these arabinoxylans and the absence of lignin indicates that the cross-linking of the wall polymers is not extensive. This scarcity of cross-linking contributes to their relative ease of degradation by bacteria.

2. Lignified Tissues of Wheat Grain (Beeswing Wheat Bran)

Beeswing wheat bran consists mainly of the outer coating of the grain and is about three cells thick, containing the cuticle, epidermis, and hypodermis. The cell walls contain about 53% hemicelluloses, 30% cellulose, 12% phenolics and lignin, and 5% protein. The hemicelluloses are of three types: in decreasing order of abundance these are highly branched acidic arabinoxylans, slightly branched acidic arabinoxylans, and arabinoxylan–xyloglucan complexes. Cross-linkages have been found between the acidic arabinoxylans, xyloglucans, phenolic esters, and lignin; the glucuronic acid component of the arabinoxylans is also involved in cross-link formation. This high degree of cross-linkage limits the solubility of beeswing bran and, hence, accounts for its low degradability by bacteria.

C. Polysaccharide Food Additives: Gums and Mucilages

1. Food Gums

A range of polysaccharides, the majority of which are heteroglycans with branched structures, are used in small amounts in the food industry to give the desired texture to processed products. The food gums are included under the broader definition of the term DF, but their overall contribution to the total DF intake is not significant, except in certain special cases (e.g., guar-based breads for diabetics). Among the leading materials, in decreasing order of use in food, are pectins, gum arabic, alginates, guar gum, CMC, carrageenan, locust-bean gum, and modified starches. With the exception of modified starches and gum arabic, all the polymers listed above are derived from plant cell walls.

2. Seed Mucilage

The mucilage from the seed husk of *Plantago ovata* Forsk. (usually referred to as Ispaghula husk) is obtained by milling the seeds and extracting with water. The main sugars of the mucilage are D-xylose, L-arabinose, and L-rhamnose, which are present in the ratio 10 : 3.2 : 1, and the major polysaccharide has a highly substituted xylan backbone. The mucilage has important physiological effects on large bowel action and is widely used for treating large bowel disorders such as diverticular disease. The mucilage forms a gel in water, retaining many times its own weight of water. This property of the mucilage is probably one of the factors responsible for its laxative action.

III. Properties of DF

A. General Considerations

Diets rich in fiber exhibit characteristic properties depending on (1) the structure of the tissue types, (2) the nature of the intracellular compounds, and (3) the form in which the food is taken—fresh, cooked, or processed. During cooking, a variable degree of disruption of the cellular structure occurs. The cell walls, which act as physical barriers for the diffusion of cell contents and the entry of digestive secretions, therefore determine the fate of the food in the gastrointestinal tract to a significant extent. During boiling of starch-rich materials, (e.g., potatoes, legume seeds), considerable cell separation and sloughing of the cells takes place, making the tissues soft. Cellular disruption is enhanced by mastication and continues in the intestinal tract. The particle size of milled cereal products has an important effect on the properties, such as the WHC and solubility of the constituent DF polymers, by virtue of the increased surface area as the particle size decreases. There is also an increase in the degradability of the starch as particle size decreases. The above considerations serve to show the importance of considering the physical state of the food product when assessing the role of DF. Some of the properties of DF that may be of importance in relation to its gastrointestinal function, and that have definite or plausible links with the etiology of the diseases associated with low-fiber intake, are discussed below.

B. Physicochemical Properties

1. WHC

The WHC of a fiber preparation is taken as a measure of its ability to immobilize water within its matrix. Water outside the matrix can be removed by filtration or by centrifugation and is referred to as "free" water. The water molecules that absorb to the DF polymers hydrogen-bond at positions not otherwise involved in intra- and intermolecular bonding of the polysaccharide molecules. The WHC of pectins and wood cellulose are ~55–80 and 5, respectively; the WHC was determined by a dialysis bag method and is expressed as gram water/gram dry polymer after exposure to water at ambient temperature for 48 hr. The WHC of vegetable and wheat bran cell wall preparations measured by the centrifugation or filtration methods range from 20 to 25 g and 3 to 4 g/g dry cell wall material, respectively. Vegetable fiber contains much pectic material, and the reduced WHC relative to free pectin is due to the association of the pectic polysaccharides with each other through the formation of Ca^{2+} bridges and ester cross-links to form a "compact" structure within the walls. The WHC of wheat bran is low primarily due to (1) the presence of significant amounts of lignified fibers, which bind water poorly, and (2) the fact that the cells of bran are much smaller compared with parenchymatous tissues of vegetables and, therefore, "imbibe" much less water on hydration.

2. Viscosity-Enhancing and Gel-Forming Properties

Viscosity-enhancing and gel-forming properties of certain DF components (e.g., guar gum, oat gum, pectins) are important for two main reasons: (1) viscous components can, under appropriate conditions, delay gastric emptying and (2) viscous components possibly reduce absorption rates in the small intestine. The rate at which food is emptied from the stomach determines the availability of nutrients for absorption in the small intestine. Hence, a delay in gastric emptying contributes to slower transit and absorption. Mixing viscous polymers with food may modify the release of gastrointestinal and pancreatic hormones, gastrointestinal mobility, and morphology. Within the small intestine, viscous polymers may slow absorption by trapping nutrients, digestive enzymes, or steroids within the gel matrix and by slow mixing and diffusion. Recent work has shown that viscous polymers increase the apparent thickness of the "unstirred" layer at the mucosal surface, which slows the rate of transport of water-soluble nutrients and cholesterol across the mucosal wall. These properties are probably responsible for some of the potential therapeutic applications of guar gum, citrus pectin, oat gum, rolled oats, and oat bran in the management of hyperglycemia (in diabetics) and hypercholesterolemia. Fermentation of viscous polysaccharides in the hind gut may also have some role in the hypocholesterolemic effects of viscous polymers through the potential inhibitory effect of propionate on hepatic synthesis.

3. Bile Salt-Binding Characteristics

DF preparations from different sources have been shown to have variable capacity to bind bile salts, and the binding was estimated from the change in bile salt concentration on exposure of the solution (usually 1–5 m*M*) to the adsorbent. These and other related studies suggested that some sources of DF bind bile salts and cause an increase in their excretion in feces, which results in an increased demand for cholesterol conversion to bile salts to maintain bile salt pool sizes. Conversion of cholesterol to bile salts but without a concomitant increase in rate of cholesterol synthesis is a possible contributory factor in the reduction of serum cholesterol levels (hypocholesterolemic effect). Deconjugated bile salts were more strongly bound than conjugated bile salts, and the binding was shown to be dependent on time of exposure and pH of the medium; maximal binding occurred at acidic pH mainly due to the "un-ionized" pectic polysaccharides, and the binding was considerably reduced at alkaline pH. Also, lignified fiber preparations from spent barley grains, wheat bran, and rice bran bound significantly higher levels of some bile salts compared with the delignified preparations. In the case of fiber from barley grains and alfalfa, the binding of bile salts was significantly reduced (~75%) after delignification; somewhat similar results were obtained with cell walls from heavily lignified tissues of mature runner bean pods. These results suggest that lignin may also play a role in the binding of bile salts.

Vegetable and cereal fibers, and some food gums, also have the capacity to interact with and bind lipid components of micelles and partially disrupt them and may consequently inhibit lipid absorption. The ability of "fiber" preparations to bind lipids from micelles decreases in the order guar gum > lignin > alfalfa > wheat bran > cellulose. For a realistic view of the effects of DF on lipid metabolism, the above-mentioned properties of fiber should be considered in conjunction with those discussed in Section III.B.2.

4. Cation-Exchange Capacity and Mineral Absorption

The main functional groups of cell wall polymers that can bind cations are the carboxyl groups of uronic acids. However, the nature of the uronic acid-containing polymers, their mode of occurrence within the cell wall complex, and the form in which the food is taken (fresh, cooked, or processed) are important considerations when assessing the cation-exchange capacity (CEC) of DF. Phytates, which are present in significant amounts in some plant foods, particularly cereal products (e.g., wheat bran, bran-based products), can also bind a range of divalent ions such as Ca, Mg, Fe, and Zn and, thus, influence their bioavailability. A few examples will be given to illustrate the above points.

1. In the case of vegetable and fruit fibers, the nonesterified regions of the pectic polysaccharides of the middle lamellae and primary cell walls, most of which are involved in the formation of "egg box" junction zones with Ca^{2+}, have the potential to bind cations. The CEC of the fibers from immature cabbage leaves and apples is ~1.5 meq/g fiber. This value is less than the total uronic acid content of the fiber preparations, which is ~2.0 meq/g fiber, because some of the galacturonic acid residues of the pectic polysaccharides are esterified. While pectic polysaccharides of vegetable and fruit fibers may bind divalent metallic ions *in vitro*, the effect of such fibers on mineral bioavailability *in vivo* is not significant. In fact, in human studies, isolated pectin was shown to have no effect on Ca, Fe, and Zn balance, and vegetable and fruit fibers had a small variable effect. The latter observation is probably due to degradation of the pectates in the colon by bacteria and subsequent absorption of the associated minerals in the large bowel.

2. Cell walls of endospermous tissues of cereals contain negligible amounts of uronic acids, and the cell walls of lignified tissues (e.g., wheat bran) have

only small amounts of uronic acid associated with the hemicelluloses. The associated glucuronic acids are ester-linked to lignin and phenolics and are not available for cation exchange. However, phytates are present in the aleurone cells of wheat bran, and the acidic phosphate groups of potassium phytate would be available for cation binding provided the molecules are exposed. Processing (e.g., milling) and cooking bran-containing products would be expected to disrupt the cellular structures of the aleurone cells and expose the phytates, which can then cause malabsorption of Ca, Fe, and Zn. Recent work has confirmed that phytates have a marked inhibitory effect on the absorption of nonheme iron, which accounts for 85–90% of the total dietary iron intake. It is of interest that only small amounts of phytates (5–10 mg phytate phosphorus) are needed in a meal to reduce iron absorption by half and that the effect of phytates is much less marked in a meal supplemented with ascorbic acid and/or meat. High intake of phytate-rich foods can also reduce the availability of zinc, while low-phytate fiber-rich foods have no effect on zinc absorption. In a balanced Western diet, where fiber-rich foods account for 20–30% of the daily energy and zinc intake, the effect of phytate is easily counteracted by the animal protein content of the diet. Phytates have also been shown to bind calcium and reduce its absorption. However, for Ca, as for many other minerals, the net retention is determined not only by variations in absorption but also by variations in losses. The effect of high intakes of fiber, from bran-enriched products, on Fe, Zn, and Ca absorption may be undesirable for infants, children, and young adolescents and recommendations for DF intakes in these groups should be different from those of adults.

5. DF and the Human Colon

a. Site of DF Breakdown and Action of Colonic Bacteria Substantial evidence now indicates that the polysaccharides of DF are broken down during passage through the human alimentary tract. The site of breakdown of DF is undoubtedly the large intestine where the polysaccharides are degraded by the anaerobic bacterial flora. In recent studies, using ileostomy subjects as a model, it has been shown that the NSP from oats, wheat, and corn escape digestion in the small intestine; other studies using a similar model have shown complete recovery of NSP from pectin and wheat bran. In the first study, it was also found that the breakdown of starch in unprocessed cereal foods (rolled and steamed oats) was complete but that in processed foods (white bread and cornflakes) some "resistant" starch escaped digestion by the enzymes of the small intestine. In another investigation, the same workers studied the digestibility of starch in unripe and ripe bananas by the ileostomy subjects. Among other things, their results showed that with unripe bananas the amount of starch not hydrolyzed and absorbed from the small intestine and, therefore, passing into the colon, may be up to eight times more than the NSP present. The starch in ripe bananas is degraded to a greater extent in the small intestine.

These studies demonstrate that not only the NSP of DF escape breakdown in the small bowel, but that a proportion of the starch also does. In addition to exogenous polysaccharides, endogenous polysaccharides such as mucins and mucopolysaccharides are also available to bacteria in the colon. The bacteria in the human colon are mostly (99%) anaerobic, saccharolytic organisms, which may be present at levels 10^{10}–10^{11} cells/g in feces making up 25–55% of fecal solids; saccharolytic organisms derive their energy primarily from carbohydrates and their derivatives. the number of bacteria in the colon (10^9/g) is less than in feces. From fecal studies, it appears that the five major genera of the colon are *Bacteroides, Eubacteria, Bifidobacteria, Peptostreptococci,* and *Fusobacteria*. Of these, the *Bacteroides* are thought to account for about one-third of the total organisms.

The main products of bacterial degradation are short-chain fatty acids (SCFA) (acetate, propionate, and butyrate), H_2, CO_2, and methane. In all regions of the large intestine, the SCFA are present in relatively high concentrations (90–130 mmol) and can be absorbed from the colonic lumen and metabolized. The SCFA contribute to normal energy metabolism and represent 60–70% of the energy available had the carbohydrate been absorbed in the small intestine. The gases produced by the bacteria diffuse into the bloodstream for expiration in the breath but are mostly expelled through the rectum as flatus.

b. DF and Fecal Weight The well-established properties of DF are those of increasing fecal bulk and reducing transit time, and many studies have shown that while cereal fiber (enriched with bran) is very effective, vegetable fiber is less so. An increase in digesta bulk may cause colonic propulsion and consequent reduction in transit time, but there are also other nondietary determinants of transit time. In a mixed diet, increase in fecal weight is probably the result of a combination of three main factors: (1) increased mass from undegraded fiber, (2) increase in bacterial mass as a result of fermentation of the fiber, and (3) increase in water retained by the residual fiber–bacteria complex. This component will contain water entrapped by the degraded fiber and bacterial cells. An appreciable proportion of the water will be contained within the bacterial cells, which are ~80% water. With a typical British type of diet, bacteria account for about 55% of the total mass of fecal solids; undigested fiber and water-soluble material account for 17% and 24%, respectively. The water-soluble material consists mainly of minerals, fats, and nitrogen-containing compounds. It has been shown that the fermentability of a DF preparation is a major factor in its effect on fecal bulk. The fiber of vegetables and fruits (e.g., cabbage, carrots, apples) is far more rapidly degraded by colonic bacteria than fiber of cereal products containing lignified tissues (e.g., wheat bran); in the case of wheat bran, the polysaccharides of the aleurone layer are preferentially degraded. This is because vegetable fiber contains much higher levels of soluble cell wall polysaccharides, which allow the bacteria to penetrate the fiber matrix, and even the cellulose, which is not impregnated with lignin, can be readily degraded.

It has been shown in clinical trials that cabbage fiber is extensively degraded (~92%), stimulates bacterial growth, and leads to a modest increase in fecal weight. In contrast, fiber from wheat bran is only ~36% degraded due to the low solubility and extensive cross-linking of the component polysaccharides, which significantly reduce access to bacteria and subsequent breakdown. As a consequence, with bran the incompletely degraded fiber (which has appreciable WHC), associated water, and bacteria gave rise to a significant increase in fecal weight, about twice that of the vegetable fiber. However, the importance of the physical characteristics of the DF is illustrated by the fact that grinding wheat bran to a fine particle size can greatly reduce its fecal bulking ability. This is a result of the increased available surface area and also increased packing density of smaller particles. These and related studies show that the mechanisms by which fiber affects colonic physiology are dependent on the fermentability of the fiber, and this is related to the type of fiber ingested.

IV. Concluding Remarks

DF is derived primarily from the cell walls of edible plant organs; therefore, an attempt has been made in this chapter to describe the chemistry of some of the better characterized cell wall polymers of edible plants with particular emphasis on how this knowledge enhances our understanding of (1) the chemical and physical properties of DF, and the DF content of a range of plant foods, and (2) the mode of action of DF in the human alimentary tract, particularly the human colon, which suggests that most of the established physiological effects of DF are probably protective ones.

As it is becoming increasingly clear that the nature of the cell wall polymers associated with various tissue types are different, an effort has been made to describe the chemistry of cell wall polymers from different tissues. In general, the parenchymatous tissues of vegetables and fruits are rich in pectic polysaccharides and cellulose, whereas those of cereals are rich in hemicelluloses (β-glucans and/or arabinoxylans), and the constituent polymers are not highly cross-linked by phenolics. This is in contrast to the hemicelluloses of lignified tissues (of cereals) in which the polymers are highly cross-linked by phenolics and the cellulose is impregnated with lignin. Disruption of the fibrous structure results in increased access of the pancreatic enzymes to cell contents and, hence, increases the digestibility of starch, etc. Cooking encourages cell separation and increases the amount of soluble DF polysaccharides (e.g., pectin). Pectins and soluble food gums (e.g., guar, β-glucans) have a high viscosity in aqueous solution and can alter gastrointestinal function so as to slow the absorption of sugars, which is beneficial for diabetics.

One of the major roles of DF is the provision of NSP as substrates for the colonic bacteria, and these together with unabsorbed sugars and starch are the main sources of energy for the bacteria. The DF not degraded by bacteria, together with associated bacteria and entrapped water, make up the major components of feces, so that enhanced excretion of fecal matter is a characteristic of high-fiber diets. Several good reviews and books are devoted to various aspects of cell walls of edible plant organs and DF, and the reader is referred to them for further information.

Bibliography

Birch, G. G., and Parker, K. J. (eds.) (1983). "Dietary Fibre." Applied Science Publishers, London and New York.

Leeds, A. R. (ed.) (1985). "Dietary Fibre Perspectives—Reviews and Bibliography," John Libbey and Co. Ltd., London and Paris.

Selvendran, R. R., and O'Neill, M. A. (1987). Isolation and analysis of cell walls from plant material. *In* "Methods of Biochemical Analysis," Vol. 32 (D. Glick, ed.). John Wiley & Sons, Inc., New York.

Selvendran, R. R., Stevens, B. J. H., and DuPont, M. S. (1987). Dietary fiber: Chemistry, analysis, and properties. *In* "Advances in Food Research," Vol. 31 (C. O. Chichester, E. M. Mrak, and B. S. Schweigert, eds.). Academic Press, New York.

Selvendran, R. R., Verne, A. V. F. V., and Faulks, R. M. (1989). Methods for analysis of dietary fibre. *In* "Modern Methods of Plant Analysis—Plant Fibers," Vol. 10 (H. F. Linskens and J. F. Jackson, eds.). Springer-Verlag, Berlin, Heidelberg, New York, and London.

Southgate, D. A. T. (1989). Dietary fibre and the diseases of affluence. *In* "A Balanced Diet?" (J. Deobbing, ed.). Springer-Verlag, Berlin, Heidelberg, New York, and London.

Taylor, T. G., and Jenkins, N. K. (eds.) (1986). "Proceedings of the XIII International Congress of Nutrition." John Libbey & Co. Ltd., London and Paris.

Trowell, H., Burkitt, D., and Heaton, K. (1985). "Dietary Fibre, Fibre-Depleted Foods and Disease." Academic Press, London.

Vahouny, G., and Kritchevsky, D. (1986). "Dietary Fibre—Basic and Clinical Aspects." Plenum Press, New York and London.

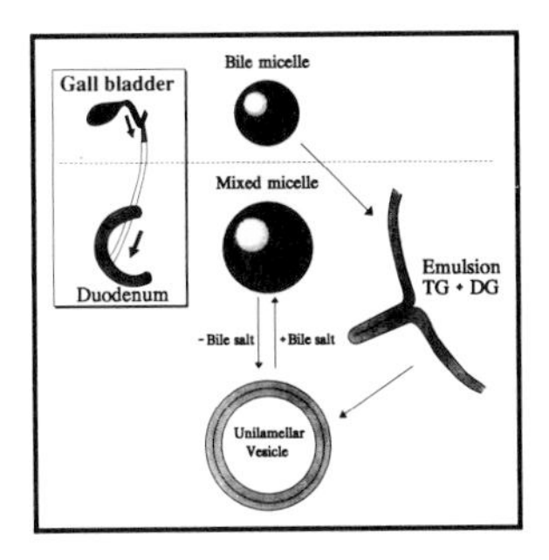

Digestion and Absorption of Human Milk Lipids

OLLE HERNELL AND LARS BLÄCKBERG *University of Umeå, Umeå, Sweden*

Glossary

Amphiphilic substance Substance with both polar and nonpolar properties, often detergents

Bile salt Physiological detergent supplied with the bile into the intestinal lumen [e.g., cholate ($3\alpha,7\alpha,12\alpha$-trihydroxycholanoate)]

Chyme Semifluid mass of partly digested food expelled by the stomach to the duodenum

Emulsion Droplets of oil (mainly triacylglycerol) in water stabilized with a surface coat of amphiphilic substances (detergents)

Lipase Enzyme that hydrolyzes ester bonds of emulsified acylglycerols (triacylglycerol hydrolases, EC 3.1.1.3)

Micelle Aggregate of bile salt spontaneously formed above a certain concentration (i.e., the critical micellar concentration) which solubilizes products of lipolysis (mainly monoacylglycerol and free fatty acid) to form mixed micelles

Milk fat Globules of mainly triacylglycerol (>98%) covered with a surface coat (membrane) of mainly phospholipid and protein

***sn*-2-Monoacylglycerol** Ester between a fatty acid and the hydroxyl group at the 2-position (middle position) of glycerol; *sn*, stereospecifically numbered

Vesicle Bilayered structure of lipolysis products saturated with bile salt that coexist with mixed micelles in the aqueous portion of postprandial upper small intestinal contents

DURING THE LAST decade there has been an increasing awareness that not only is the nutritional composition of human milk ideal to meet the requirements of the newborn infant, but the bioavailability of many nutrients is also better from milk than from formulas made with a composition as similar as possible to milk. For example, it is well known that the coefficient of fat absorption is higher from raw human milk than from such formulas. To achieve the same high coefficient of absorption, manufacturers have used either a high percentage of medium-chain fatty acids or long-chain polyunsaturated fatty acids (e.g., linoleic acid). Recently, there have been calls for caution in using both types of formulas due to observed, or expected, side effects.

For scientific as well as practical reasons it is therefore important to fully understand why the coefficient of fat absorption is so unexpectedly high in breast-fed infants. Even more so since when the accepted model of fat digestion and absorption in healthy adults was applied to the breast-fed infant, it became evident that the model was an oversimplification. Below we review our current understanding of the mechanisms of lipid digestion and absorption in breast-fed infants and why these mechanisms in some respects might be different, and therefore less efficient, in formula-fed infants. In doing so we also exemplify that regarding human milk merely as a source of nutrients and energy, albeit of optimal composition to the newborn, is no longer valid. Rather, breast-feeding could be considered an extension of the umbilical cord, through which immunological, biochemical, and endocrine support and information are communicated.

I. Introduction

An ever-increasing demand for energy is one of the major characteristics of the dynamic and vulnerable perinatal period (i.e., around the time of birth). The energy requirements of the fetus are mainly for growth and tissue maintenance; during the last trimester the weight of the fetus more than triples. Through the placental transport of nutrients, the mother provides the fetus with the required energy substrates. Although carbohydrate (i.e., glucose) is the dominating fetal energy substrate, it seems clear that lipids [i.e., free fatty acids (FFAs) and ketone bodies] may be used as alternative fuels. Via placental transfer and/or synthesis the fetus is also supplied with essential lipid nutrients (i.e., fat-soluble vitamins and essential fatty acids), including long-chain polyenoic derivatives of both the *n*-3 and *n*-6 series, required for membrane synthesis, particularly of the rapidly developing brain and retina. [*See* LIPIDS.]

From a nutritional point of view, birth represents a radical transition. Due to its relative concentration in human milk, and in milk formulas patterned after human milk, lipid now becomes the major energy substrate. In fact, newborn infants ingest three- to fivefold more lipid per kilogram of body weight per day than do adults consuming a typical Western diet. Therefore, the human neonate immediately becomes dependent on effective mechanisms for the digestion and absorption of dietary lipids and on dietary sources of essential nutrients, including lipids.

II. Composition of Human Milk Fat

Energy-rich triacylglycerol (i.e., triglyceride) account for at least 98% of the lipids of human milk. The balance is made up by phospholipids, cholesterol, fat-soluble vitamins, and other minor constituents. The triacylglycerols are synthesized in the endoplasmic reticulum and cytosol of the epithelial cells of the lactating mammary gland. They form the cores of the milk fat globules, which, during synthesis and intracellular storage, are stabilized by a surrounding monolayer of phospholipid. When extruded from the epithelial cell, the globules become enveloped by the apical part of the plasma membrane. Hence, this so-called "milk fat globule membrane" is a trilayer composed mainly of phospholipid, other amphiphilic lipids, and membrane proteins. Some of the proteins are high-molecular-weight glycoproteins that form a remarkable array of filaments, approximately 5×10^2 nm in length, sticking out from the surface. These filaments, absent from bovine milk globules, are extracted by heat treatment (i.e., pasteurization) of the milk. It has been suggested that the filaments might enhance the digestibility of the globules, and therefore could be one reason for the observed reduced fat absorption from pasteurized milk.

Although the sizes of the globules vary, they typically have a diameter of $1–4 \times 10^3$ nm. Thus, composed milk fat globules represent a very stable physiological lipid, or oil-in-water emulsion. However, the stabilizing surface coat, independent of heat treatment, also functions as a barrier with regard to accessibility of the milk triacylglycerol for the necessary gastrointestinal digestion.

III. Lipases of Gastrointestinal Fat Digestion

To be utilized efficiently, the nonabsorbable dietary triacylglycerol must be hydrolyzed to absorbable products. This digestion is accomplished by certain triacylglycerol hydrolases, or lipases, operating in the lumen of the upper gastrointestinal tract. Fig. 1 is a schematic presentation of the main lipases involved in the digestion of human milk triacylglycerol. Also indicated are tissues of secretion for the respective lipases.

A. Colipase-Dependent Pancreatic Lipase

The pancreata of all studied mammalian species secrete a potent lipase. It is an approximately 45-kDa single-chain glycoprotein. The amino acid sequence is known for several species. Recently, the complete gene structure was revealed for the canine lipase. This pancreatic lipase is designed to act preferentially at lipid–water interfaces of the chyme entering the upper small intestine from the stomach. Although it also has some activity on water-soluble ester substrates (i.e., esterase activity), its activity is much higher when acting at the surface of aggre-

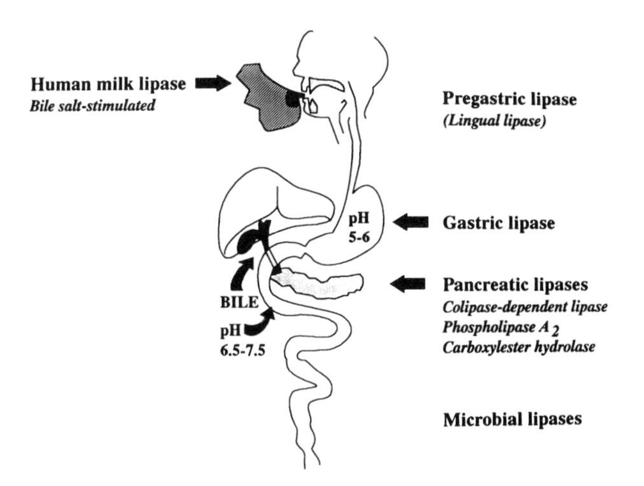

FIGURE 1 The main sources of lipases relevant to milk triacylglycerol digestion.

TABLE I Lipases in the Gastrointestinal Tract of Breast-Fed Infants

Lipase	Origin	Apparent mass (kDa)	Cofactor	Site of action
Colipase-dependent lipase	Pancreas	45	Colipase	Duodenal contents
Gastric lipase	Gastric mucosa	42	None	Gastric contents
BSSL/CEH	Milk/pancreas	107/100	Primary bile salt	Duodenal contents

gated substrates such as emulsion particles. This so-called "surface activation" is a property unique to this lipase when compared to the two other gastrointestinal lipases to be discussed. For optimal function pancreatic lipase depends on a protein cofactor: colipase. This 9-kDa protein is secreted from the pancreas in amounts approximately equimolar to those of the lipase. One important function of colipase is to anchor the lipase to lipid substrates covered with amphiphilic components (e.g., bile salts, phospholipids, and proteins). As mentioned, the milk fat globule is a physiologically relevant representative of such a lipid substrate. Colipase also affects the catalytic efficiency of the bound lipase. Hence, this pancreatic lipase is now often designated "colipase-dependent lipase" (Table I).

B. Gastric Lipase

Gastric contents of several species, including humans, contain lipase activity (Table I). The responsible human lipase is secreted almost exclusively by the gastric mucosal, chief cells of the gastric fundus. In contrast, in the rat the lipase is secreted from the von Ebner's glands localized beneath the circumvallate papillae of the tongue, and in the calf it is secreted from pharyngeal tissue. Due to these differences in the tissue of origin, the lipase(s) has been given several names [e.g., lingual lipase (in rats), pregastric lipase (in calves), or gastric lipase (in humans)]. The properties of the purified enzymes are similar in all cases, implying that, regardless of origin, they have a common physiological function. From the amino acid sequences of the rat and human enzymes, it is clear that they are similar on the molecular level; the sequence homology between the two is more than 75%. Here, we use the name "gastric lipase."

The properties of gastric lipase make it ideally suited to function in gastric contents. It has an acidic pH optimum, is very stable in the acidic environment of the stomach, and is also resistant to digestion by pepsin, which is the gastric proteolytic enzyme. In contrast, when treated with pancreatic proteases (i.e., trypsin and chymotrypsin), the enzyme activity is rapidly lost, particularly in the presence of bile salts. Consequently, it does not remain active in intestinal contents, and hence it is not expected to make a major contribution to intraduodenal lipolysis in healthy human adults.

C. Bile Salt-Stimulated Lipase

The bile salt-stimulated lipase (BSSL) has, during the last decade, become one of the most thoroughly studied of all human milk enzymes. The concentration in milk (0.1 mg/ml) is, compared to that of other milk enzymes, remarkably little influenced by the length of gestation, duration of lactation, or diurnal rhythm. BSSL represents a striking example of how the mother, via her milk, provides the newborn with a biologically active protein important for its growth and development. This interplay is further illustrated by the activation mechanism of the enzyme. The lipase is inactive in the milk, but becomes activated when, after passage through the stomach into the intestinal lumen, the partially digested milk mixes with bile salts. Only free or conjugated primary bile salts (i.e., cholate and chenodeoxycholate) are effective activators. The exact mechanism of activation is unknown. In duodenal contents bile salts not only activate, but also protect the enzyme from inactivation by proteases.

For a long time BSSL was thought to be a constituent of milk only of the highest primates, since a corresponding activity was not found in the milk of cows, horses, pigs, goats, rats, or rhesus monkeys. Recently, however, a BSSL activity was demonstrated in the milk of squirrel monkeys and African Green monkeys. Although, these activities were much lower, they gave evidence that the enzyme is present in Old World as well as New World monkeys. Furthermore, milk from cats and dogs contains a BSSL activity which is 20–50% of that found in human milk. A structural relationship between these lipases and the human lipase was implicated by immunochemical cross-reactivity demonstrated by the use of antibodies to the purified human BSSL. There is no explanation as to why this lipase is secreted into milk in some species, but not in others. A correlation to the milk content of long-chain triacylglycerol was recently suggested; seal milk, which has an extremely high fat content, also has one of the highest BSSL activities found in any species.

The human exocrine pancreas secretes an enzyme, carboxylic ester hydrolase (CEH), which immunochemically and functionally seems to be identical to BSSL, although there is a slight difference in apparent molecular masses: 107 kDa for BSSL and 100 kDa for CEH. They are, however, most likely products of the same gene. In contrast to the enzyme in milk (i.e., BSSL), which is absent in many species, the pancreatic counterpart (i.e., CEH) has been found in all species tested so far. In fact, CEH is phylogenetically older than the colipase-dependent pancreatic lipase. Probably the opposite could not have been possible. CEH, like BSSL, but in contrast to colipase-dependent lipase, is a nonspecific lipase inasmuch as it hydrolyzes not only tri-, di-, and monoacylglycerols (see below) but also cholesteryl esters and esters of fat-soluble vitamins. Recent observations suggest that BSSL and CEH might be of particular importance for the utilization of dietary essential fatty acids, in particular the long-chain polyunsaturated derivatives of linoleic and α-linolenic acids. Furthermore, its activity is not restricted to emulsified substrates. Micellar and soluble substrates are also effectively hydrolyzed.

The apparent molecular masses of CEHs from various species range from around 70 kDa (in rats) to above 100 kDa (in humans). Recently, the complete amino acid sequence for human BSSL was deduced. It consists of more than 700 amino acid residues, and shows very little homology to colipase-dependent lipase or gastric lipase. There is, however, in the amino-terminal half, including the tentative active site, a marked homology to typical esterases. This agrees well with the unusually broad substrate specificity—BSSL is a nonspecific lipase. Furthermore, there is a strong homology to CEH from rat and cow, extending for about 500 residues from the amino terminus. This strongly suggests that CEHs and BSSLs from different species represent the product of one common ancestral gene. The differences in apparent molecular masses reflects differences in both the carboxy-terminal part of the peptide chain and in the degree of glycosylation. From immunocytochemical studies and from studies of protein synthesis in tissue biopsies *in vitro*, it could be clearly shown that BSSL is indeed synthesized by the lactating human mammary gland. Since BSSL accounts for 1% of the total protein in milk, it could be calculated that milk protein accounts for about one-half of the total protein synthesis of the gland.

Measurements of intraduodenal lipase activities in preterm infants 1 hour after consecutive feeds of raw and pasteurized human milk, respectively, showed that about two-thirds of the combined BSSL–CEH activities found was from the milk, indicating the milk as the major source of enzyme in breast-fed infants. Interestingly, in the suckling rat CEH is the main lipolytic enzyme secreted from the pancreas, and it is not until weaning that the colipase-dependent lipase becomes dominating. This further substantiates a physiological relevance of CEH/BSSL during the neonatal period.

D. Relative Physiological Importance of the Lipases

The exact relative quantitative contribution of each of the three lipases to the net fat digestion is difficult to assess. Some conclusions might be drawn from observations of triacylglycerol utilization under pathological conditions represented by low levels of at least one of the lipases. In the healthy adult the capacity to digest and absorb dietary triacylglycerol is high. For example, it has been calculated that, theoretically, the amount of colipase-dependent lipase secreted into the intestinal lumen is in 1000-fold excess of what is required. However, the importance of other lipases is illustrated by the fact that patients suffering from isolated deficiency of colipase-dependent lipase and/or colipase might still absorb 50% or more of dietary triacylglycerol. Likewise, cystic fibrosis patients, suffering from a complete pancreatic insufficiency, but known to have normal or close-to-normal activities of gastric lipase, often digest and absorb at least half of the dietary triacylglycerol, also without supplementation with pancreatic extracts. Isolated gastric lipase or CEH deficiencies have not been described. From indirect evidence, however, it has been argued that when gastric lipolysis is circumvented, fat absorption might be reduced by approximately 30%.

The physiological significance of BSSL in milk triacylglycerol digestion was first suggested from its characteristics and from the frequent observation that fat malabsorption is more common in newborn infants fed cow milk-based formulas (devoid of BSSL activity) than in breast-fed infants. More direct evidence came from fat balance studies comparing the coefficients of absorption from raw and pasteurized human milk. Pasteurization inactivates the milk enzyme and reduces fat absorption by as much as 30% in preterm infants. The recent finding of a BSSL activity in cat milk made it possible to directly study the importance of BSSL in this species. The effect of BSSL on the weight gain of kittens was studied by the use of three groups of kittens fed either raw cat milk, a kitten formula supplemented with purified human BSSL, or the formula alone. Kittens fed formula supplemented with BSSL had a weight gain comparable to milk-fed kittens, but twice that of kittens fed the unsupplemented formula, strongly implying BSSL as the causative factor.

IV. Two-Phase Model of Fat Digestion

Lipid digestion in general, and triacylglycerol digestion in particular, were, until recently, regarded as a two-phase process occurring almost exclusively in the upper part of the small intestinal contents. In short, the triacylglycerol enters the duodenum as an emulsion of oil in water [i.e., large water-insoluble lipid droplets of hydrodynamic radius (R_h = 125–250 nm) dispersed in the aqueous phase of intestinal contents]. At the oil–water interphase colipase-dependent pancreatic lipase hydrolyzes the triacylglycerols. Since this lipase cannot release the fatty acid esterified to the *sn*-2 position of the glycerol molecule, each triacylglycerol gives rise to one *sn*-2-monoacylglycerol and two FFAs. The bile lipids (mainly bile salts and phospholipids) distribute between the aqueous phase and the oil–water interphase, where they displace amphiphilic substances, including lipolysis products, from the interphase into the water. In the aqueous phase bile salts, above their critical micellar concentration, spontaneously form aggregates called micelles, which dramatically increase the solubility of the products of lipolysis in the aqueous phase. Mixed micelles of bile salts and lipolysis products are spherical particles ($R_h \leq 4$ nm) from which absorption occurs.

According to the described two-phase model of fat digestion and absorption, low intraluminal concentrations of colipase-dependent lipase and/or bile salts should be obvious reasons for fat malabsorption. Considering their high lipid intake, one would expect newborn infants to have high intraluminal concentrations of lipase and bile salts. However, because the pancreatic and liver functions at birth are not fully developed, this is not the situation. Intraluminal fasting levels of digestive enzymes secreted from the pancreas are considerably lower than in adults during the first 2–12 months, and adult levels of response to pancreozymin might not be seen until the end of the first year of life. Consistent with this, levels of colipase-dependent lipase in duodenal contents after a meal are, on average, only 4% in preterm infants when compared with adults. Similarly, others have reported a five- to ten-fold lower postprandial intraluminal bile salt concentration in preterm infants than in adults. In fact, such low concentrations (i.e., 1–2 m*M*) are close to or even below that required for micelle formation. Hence, they would not, according to theory, be compatible with efficient solubilization, transport, and ultimate absorption of lipolytic products. Nonetheless, digestion and absorption of triacylglycerol are often remarkably efficient even in preterm breast-fed infants.

In contrast to colipase-dependent lipase, the capacity for gastric lipase secretion is also fully developed in preterm newborns. Since BSSL is supplied with the milk, intraduodenal levels are not dependent on the developmental stage of the infant.

V. Accessibility of Human Milk Triacylglycerol for Hydrolysis

To fully understand the exact function of the three lipases for milk triacylglycerol utilization, it was necessary to make model experiments by using purified enzymes and the natural dietary lipid substrate for the breast-fed newborn (i.e., human milk fat globule triacylglycerol). Table II shows the lipolysis rates obtained with each of the three lipases with human milk fat as substrate, compared with those obtained with a synthetic triacylglycerol emulsion under conditions chosen to be suitable for the respective lipase. Pancreatic colipase-dependent lipase was, itself, even in the presence of colipase, unable to hydrolyze human milk triacylglycerol. The reason is that the lipase does not bind to the milk fat globule membrane, and thus it does not get access to the substrate. Obviously, BSSL is devoid of activity against milk fat in milk as secreted, since no bile salts are present. However, even when bile salts were added in concentrations sufficient for activation with the artificial substrate, no lipolysis of milk fat occurred (Table II). Consequently, neither colipase-dependent lipase nor BSSL activities *in vitro* are, even in the presence of their respective cofactors, by themselves or together, sufficient to account for the efficient hydrolysis of human milk fat found *in vivo*. In sharp contrast gastric lipase hydrolyzed human milk triacylglycerol at the same rate as the synthetic substrate. Thus, gastric lipase is unique among the gastrointestinal lipases in that it can, itself, attack human milk fat globule triacylglycerol. Model experiments have shown that the surface pressure, which depends on the nature and packing of the constituents of the emulsion surface coat (e.g., the milk fat globule membrane) affects gastric lipase and colipase-dependent lipase activities differently. The conclusion to be drawn is that it is an oversimplification to consider the relative contribution of each lipase to gastrointestinal lipolysis in merely quantitative terms.

TABLE II Accessibility of Human Milk Triacylglycerol for Gastrointestinal Lipases[a]

Lipase	Substrate: Triolein/ gum arabic	Substrate: Milk fat
Colipase-dependent lipase[b]	500	<0.5
BSSL[c]	100	<0.5
Gastric lipase[c]	70	80

[a] Conditions were chosen to obtain optimal activity for the respective lipases.
[b] Human pancreatic juice was used as the source of enzyme and cofactor. Values are expressed as micromoles per minute multiplied by milliliters of juice.
[c] Values are expressed as micromoles per minute multiplied by milligrams of enzyme protein.

VI. Sequential and Concerted Action of the Lipases

To reveal the physiological function of the different lipases also from a qualitative viewpoint, it was important to study their concerted effects. If human milk fat globules are pretreated with gastric lipase, resulting in the hydrolysis of only 5–10% of the triacylglycerol ester bonds, this dramatically alters the accessibility of the remaining triacylglycerol for subsequent hydrolysis. Unlike the intact milk fat globules, these partially digested globules are readily cleaved by colipase-dependent lipase, or BSSL. This further emphasizes a unique function of gastric lipase. For hydrolysis by colipase-dependent lipase, this triggering effect is caused by the released FFAs. FFAs enable the lipase–colipase complex to bind efficiently to the substrate surface. Thus, *in vitro,* this effect of gastric lipase can be replaced by the addition of exogenous FFAs, particularly long-chain FFAs. The synergistic action of gastric lipase and colipase-dependent lipase results in the hydrolysis of about two-thirds of the total triacylglycerol ester bonds, giving rise to *sn*-2-monoacylglycerols and FFAs. As discussed in Section IV, these are considered the relevant steps of triacylglycerol digestion in adults.

In breast-fed newborns BSSL also contributes to the net triacylglycerol digestion. The exact nature of the triggering effect of gastric lipase on BSSL-catalyzed hydrolysis is unclear; the essential causing factor is not the released fatty acids. The partial acylglycerols, also generated by gastric lipase, might be of importance. The route of lipid digestion encountering gastric lipase and BSSL, but circumventing colipase-dependent lipase, is unique to the breast-fed infant. It might be of particular importance in infants born prematurely (i.e., with particularly low levels of colipase-dependent lipase and/or bile salts) or in breast-fed infants with pancreatic insufficiency for pathological reasons.

In the physiological situation milk lipid digestion results from the concerted action of all three lipases. Based on *in vitro* experiments with purified human enzymes and cofactors, and using human milk as substrate, a model for triacylglycerol digestion in the breast-fed newborn can be proposed. The results of such experiments are shown in Fig. 2.

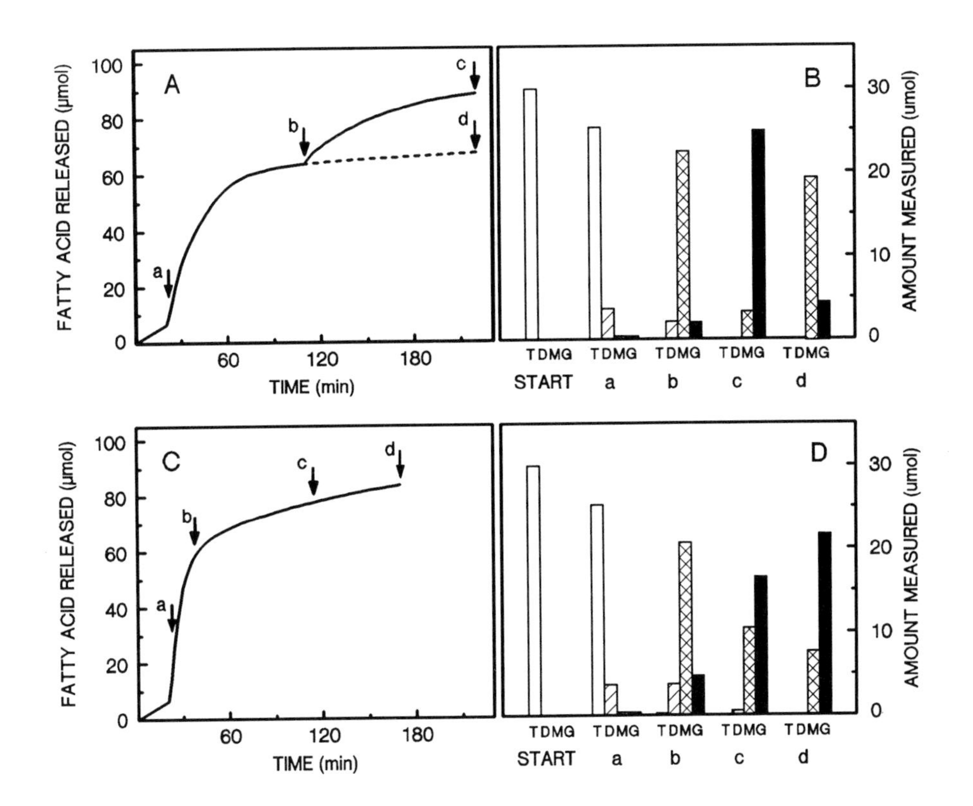

FIGURE 2. Hydrolysis of heat-treated human milk fat *in vitro*. Conditions and concentrations of purified lipases and cofactors were chosen to resemble the physiological situation. A and C shows the continuous recording of released fatty acids. At 0 minutes gastric lipase was added (A and C), after 20 minutes (a) the pH was raised from 6.0 to 7.5 and bile salts, colipase-dependent lipase, and colipase were added (A and C). BSSL was added either after 115 minutes (A, solid line) or after 20 minutes (C). At the start and at the times indicated (a–d) samples were removed and the product pattern was determined. B and D shows the decrease in relative concentration of triacylglycerol (T) and the increase in diacylglycerol (D), monoacylglycerol (M), and glycerol (G) in the incubations of (A) and (C), respectively. [For further details see S. Bernbäck, L. Bläckberg, and O. Hernell, *J. Clin. Invest.* **85** (1990). Reproduced by copyright permission of the American Society for Clinical Investigation.]

Each lipase contributes its specific properties. Under conditions resembling the environment of gastric contents, gastric lipase initiates lipolysis, which is, however, limited. The changes in conditions to resemble the environment of upper small intestinal contents (i.e., raised pH and the addition of colipase-dependent lipase, colipase, and bile salts) (Fig. 2A,a) result in the hydrolysis of more than 60% of the total ester bonds, giving rise to FFAs and monoacylglycerol as end products. BSSL, being a nonspecific lipase, lacks positional specificity. Therefore, it can also hydrolyze *sn*-2-monoacylglycerol. Hence, when BSSL was added to the incubation after the milk fat had been extensively hydrolyzed by gastric lipase and colipase-dependent lipase (Fig. 2A,b), hydrolysis recommenced (Fig. 2A,c) due to hydrolysis of the monoacylglycerol by BSSL. Although the released FFAs were not removed by absorption, as they would have been *in vivo,* the concerted action of all three lipases resulted in the hydrolysis of more than 90% of the total triacylglycerol ester bonds. The hydrolysis of monoacylglycerol was confirmed by product compositional analyses after the various steps of hydrolysis (Fig. 2B). Together with FFAs, free glycerol was the main product formed. In fact, more than 80 mol% of the glycerol initially bound as triacylglycerol was released as free glycerol.

The sequential addition of the lipases, however, is not a true representation of the *in vivo* situation. Obviously, colipase-dependent lipase and BSSL act simultaneously. When this was mimicked *in vitro* (Fig. 2C), BSSL not only affected the product pattern, but also the initial rate of lipolysis. The rate was doubled compared with that obtained with colipase-dependent lipase alone (*cf.* Fig. 2A and C), illustrating its contribution to tri- and diacylglycerol digestion. *In vivo* monoacylglycerol and FFAs are expected to be removed from the site of lipolysis by micellar solubilization and transport. However, since the capacity of micellar transport depends on the intraluminal bile salt concentration, this is a rapid and efficient process only when this concentration is high. Most monoacylglycerol is not available for hydrolysis, and BSSL contributes to lipolysis chiefly by supporting colipase-dependent lipase in the hydrolysis of tri- and diacylglycerol. If, however, the bile salt concentration is low, which might be the normal state in many newborn preterm infants, the monoacylglycerol remains in the lumen for a longer time and BSSL completes lipolysis by also hydrolyzing the monoacylglycerol.

VII. Physicochemical State of Dietary Lipids in Postprandial Intestinal Contents

Because the aqueous solubilities of partially ionized long-chain FFAs and monoacylglycerol are extremely low, but increase considerably in the presence of bile salts above their critical micellar concentrations, mixed micelles have, according to the two-phase model, been considered the sole vehicles for the solubilization and transport of lipolysis products from the emulsion surfaces to the enterocytes for subsequent absorption. Furthermore, lipid uptake is believed to be monomeric, and therefore mixed micelles provide a favorable concentration gradient across unstirred layers and mucosal surfaces. In fact, solubilization of lipolysis products within bile salt micelles has been considered a rate-limiting step in the absorption of dietary lipids. However, in the complete absence of intraluminal bile salts (e.g., in patients with bile fistulas), 50–75% of dietary triacylglycerol is absorbed. Moreover, it has been reported that in breast-fed, in contrast to formula-fed, infants the coefficient of fat absorption does not correlate to the intraluminal bile salt concentration. These observations suggest that nonmicellar mechanisms might be important for the dispersion of lipolysis products in aqueous duodenal contents, and hence for product absorption.

Recently, direct evidence was found for the coexistence of small (i.e., R_h = 20–60 nm) unilamellar vesicles and mixed micelles in the aqueous-rich portion of postprandial duodenal contents in healthy adults. It seems that, during lipolysis, products first locate mainly at the emulsion surface, presumably as multilamellar liquid–crystalline bilayers (i.e., liposomes). As, during the progress of lipolysis, the core of the emulsion droplet shrinks, parts of the excess surface coat, due to forced phase separation resulting from increased surface pressure, pinch off as large liquid–crystalline structures (Fig. 3). Provided sufficient bile salt is present, bile salt catalyzes the formation of small unilamellar vesicles from these multilamellar liposomes and, via the continued presence of bile, a two-phase system of vesicles and mixed micelles forms. Phase separations showed that, at equilibrium, the mixed micelles are composed of bile salt micelles saturated with lipolysis products, while the vesicles are composed of lipolysis products saturated with bile salts. The relative amount of lipolysis products solubi-

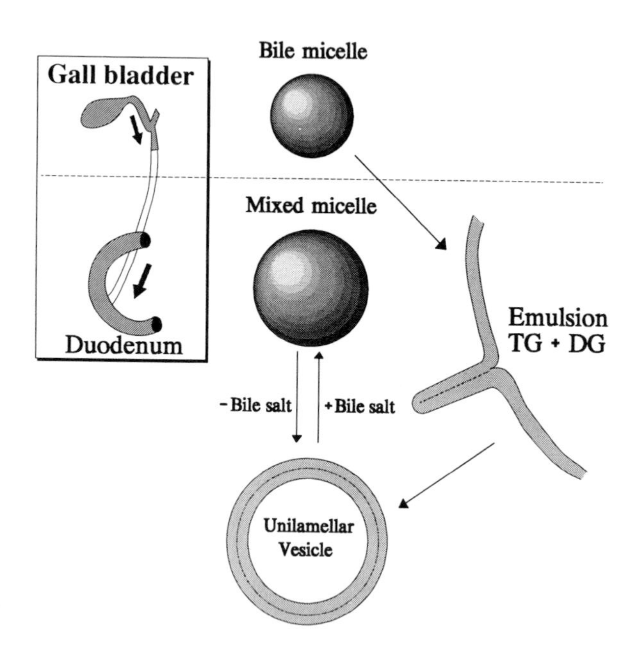

FIGURE 3 Proposed model of physicochemical states of lipids in postprandial upper small intestinal contents. During progress of lipolysis the core of emulsion particles shrinks, while the products formed locate mainly at the interphase. Increased surface pressure force a phase separation and excess material buds off, presumably as multilamellar liquid–crystalline structures. Provided sufficient bile salt is present, bile salt catalyzes the formation of small unilamellar vesicles from these multilamellar liposomes, and via the continued presence of bile salts a physiological two-phase system of vesicles and mixed micelles forms. At low bile salt–lipid ratios vesicles constitute the major product phase from which absorption occurs. However, as long as the micellar phase is not saturated with lipolysis products, there is a spontaneous dissolution of vesicles into mixed micelles. TG, Triacylglycerol; DG, diacylglycerol.

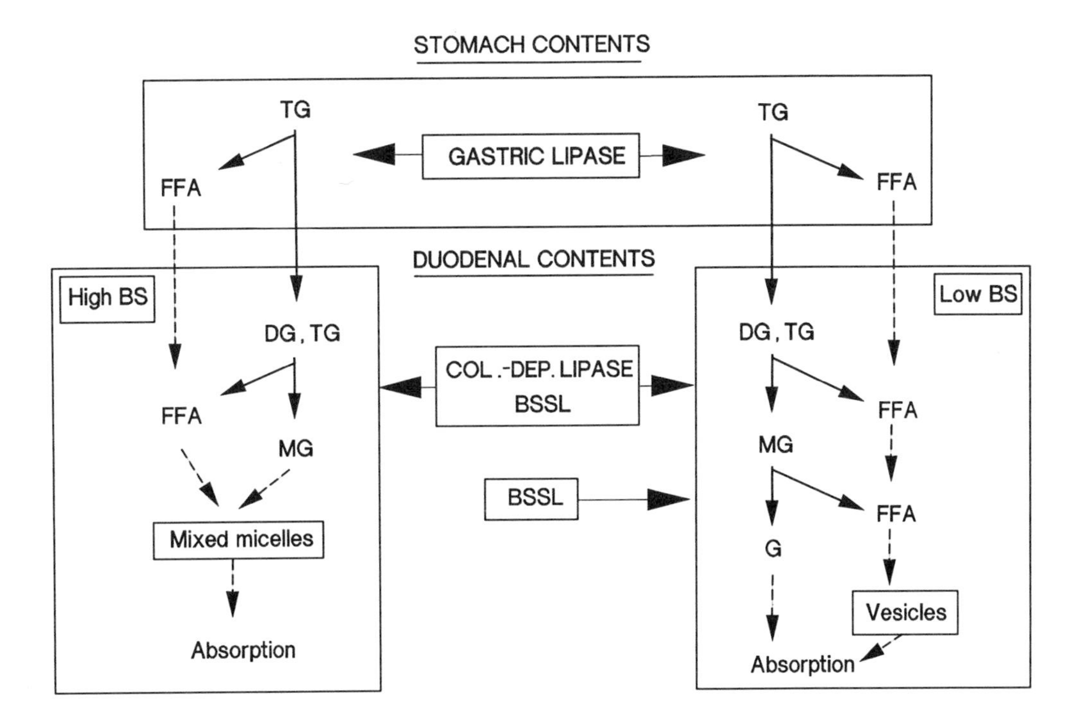

FIGURE 4 The sequential steps involved in digestion of human milk triacylglycerol (TG). FFA, Free fatty acids; BS, intraluminal bile salt concentration; DG, diacylglycerol; MG, monoacylglycerol; G, glycerol. Solid lines represent enzymatic processes; dashed lines, nonenzymatic processes (e.g., transport).

lized in each phase depends on the bile salt concentration relative to the total lipid (i.e., product) concentration (Fig. 3). At low bile salt–lipid ratios the vesicle phase is favored. There is also evidence for the spontaneous dissolution of vesicles into micelles, as long as the micellar phase is unsaturated with lipolysis products.

Saturation of micelles with lipolysis products provides the ideal thermodynamic environment for maximum rates of intestinal absorption. The driving force down a concentration gradient results in lipid product absorption, in healthy adults probably preferentially from micelles, due to their larger number, smaller sizes (i.e., $R_h \leq 4$ nm), and, thereby, their more rapid diffusive access to the mucosal surface. However, absorption could also take place from unilamellar, and perhaps even multilamellar, vesicles. This would explain the relatively unimpaired fat absorption seen at low or absent intraluminal bile salt concentrations (e.g., in patients with bile fistula and infants during the neonatal period). It seems that partially ionized fatty acids have a higher solubility in unilamellar vesicles than have monoacylglycerols. If so, this could be the major explanation as to why a complete hydrolysis of triacylglycerol to water-soluble glycerol and FFAs catalyzed by the milk lipase promotes efficient product absorption in breast-fed infants, and why this absorption is less dependent on the intraluminal bile salt concentration.

VIII. Concluding Remarks

Our present view on human milk triacylglycerol digestion involves the concerted action of three lipolytic enzymes: gastric lipase, pancreatic colipase-dependent lipase, and BSSL of human milk. Figure 4 is a schematic presentation of the sequential steps catalyzed by the respective lipases. Each enzyme has a unique function, but also has overlapping functions with the other two. When the intraluminal bile salt concentration is comparatively high, BSSL aids colipase-dependent lipase in generating monoacylglycerol and FFAs, which, using mixed micelles as vehicles, are absorbed. In infants with low intraluminal bile salt levels, which is normal for many preterm infants, the concerted action of the three lipases results in complete hydrolysis with FFAs and free glycerol as end products. Under these conditions substantial product absorption might occur from unilamellar vesicles.

Bibliography

Abouakil, N., Rogalska, E., Bonicel, J., and Lombardo, D. (1988). Purification of pancreatic carboxylic-ester hydrolase by immunoaffinity and its application to the human milk bile salt-stimulated lipase. *Biochim. Biophys. Acta* **961,** 299.

Bernbäck, S. (1989). "Gastric Lipase and the Sequential Digestion of Human Milk Fat," Ph.D. thesis. Univ. of Umeå, Umeå, Sweden.

Bernbäck, S., Bläckberg, L., and Hernell, O. (1990). Complete digestion of human milk triacylglycerol *in vitro* requires gastric lipase, colipase-dependent lipase, and bile salt-stimulated lipase. *J. Clin. Invest.,* **85,** 221.

Borgström, B., and Brockman, H. L. (1984). "Lipases." Elsevier, Amsterdam.

Hamosh, M. (1989). Enzymes in human milk: Their role in nutrient digestion, gastrointestinal function, and nutrient delivery to the newborn infant. *In* "Textbook of Gastroenterology and Nutrition in Infancy" (E. Lebenthal, ed.), 2nd ed. pp. 121–134. Raven, New York.

Hernell, O., Bläckberg, L., and Bernbäck, S. (1989). Milk lipases and in vivo lipolysis. *In* "Protein and Nonprotein Nitrogen in Human Milk" (S. Atkinson and B. Lönnerdahl, eds.), pp. 221–236. CRC Press, Boca Raton, Florida.

Hernell, O., Bläckberg, L., and Lindberg, T. (1989). Human milk enzymes with emphasis on the lipases. *In* "Textbook of Gastroenterology and Nutrition in Infancy" (E. Lebenthal, ed.), 2nd ed. pp. 209–217. Raven, New York.

Hernell, O., Staggers, J. E., and Carey, M. C. (1990). Physical–chemical behavior of dietary and biliary lipids during intestinal digestion and absorption: II. Phase analysis and aggregation states of luminal lipids during duodenal fat digestion in healthy adult human beings. *Biochemistry,* **209,** 2041.

Nilsson, J., Bläckberg, L., Carlsson, P., Enerbäck, S., Hernell, O., and Bjursell, G. (1990). cDNA cloning of human milk bile salt-stimulated lipase, and evidence for identity to pancreatic carboxylic ester hydrolase. *Eur. J. Biochem.* p. **192,** 543.

Patton, J. S., Vetter, R. D., Hamosh, M., Borgström, B., Lindström, M., and Carey, M. C. (1985). The light microscopy of triglyceride digestion. *Food Microstruct.* **4,** 29.

Wang, C.-S., Martindale, M. E., King, M. M., and Tang, J. (1989). Bile salt-activated lipase: Effect on kitten growth rate. *Am. J. Clin. Nutr.* **49,** 457.

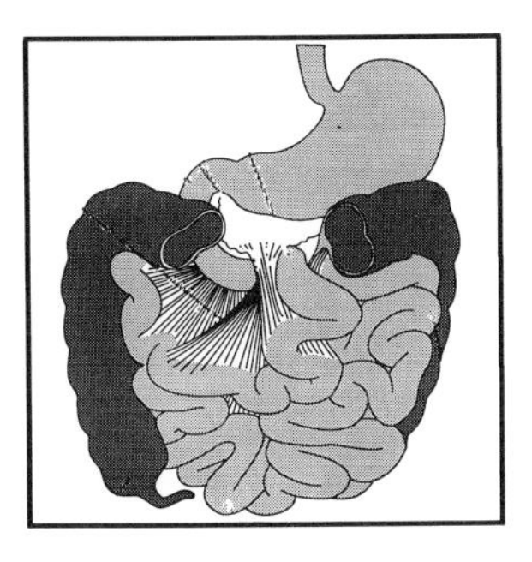

Digestive System, Anatomy

DAVID J. CHIVERS, *University of Cambridge*

Glossary

Attachment, primary Original body wall attachment

Attachment, secondary Subsequent attachment of mesentery of organ, to body wall or to mesentery of another organ

Faunivory Consumption of animal matter, vertebrate or invertebrate

Folivory Consumption of vegetative parts of plants—leaves (blade, midrib, buds, young, mature), stems, exudates (gums/saps), bark

Foregut In abdomen, stomach and first part of duodenum

Frugivory Consumption of reproductive plant parts (primary production)—fruits (pulp and/or seeds, mature/immature), flowers

Hindgut Left colon, rectum

Mesentery Two-layered membrane attaching organ to body wall

Midgut End of duodenum, jejunum, ileum, caecum, right colon, transverse colon

Peritoneum, parietal Membrane lining body cavity

Peritoneum, visceral membrane covering organ

Retroperitoneal Behind parietal peritoneum, organ with no mesentery

ALTHOUGH DIGESTION may start in the oral cavity or mouth, through the secretion of salivary enzymes, which continues to act as the food passes through the pharynx (the crossroads of air and food pathways) and esophagus, it is the gastrointestinal tract in the abdomen and pelvic cavities that is considered here as the digestive system. These biochemical processes of digestion and absorption contrast with the mechanical processes of food degradation that occur in the mouth, through movements of the jaws and occlusion of the teeth in regular cycles. Thus, the anatomy of the human digestive system concerns the structure and arrangement of the stomach, small intestine, caecum, colon, and rectum in relation to function and evolution. In contrast to the gastrointestinal tract of most other mammalian orders, which are specialized either for meat-eating or for leaf-eating, those of primates, especially humans, are relatively unspecialized.

1. Background

A. Development

When, in early development, the embryonic disc, made up of three cellular layers, folds up to assume fetal form, one of the layers overlying ectoderm forms the epidermis of the skin, whereas the underlying layer, the endoderm, forms the mucosa (inner lining) of the digestive system. Initially they are separated cranially and caudally from the degenerating yolk sac (Fig. 1). the intervening layer, the mesoderm, forms the dermis of the skin, the muscles and bones of the body wall, various other organ systems, and the smooth muscle of the digestive system. Thus, the digestive system becomes tubular, divisible into foregut (future pharynx, esophagus, stomach, and start of duodenum), midgut [still in connection with the yolk sac (rest of duodenum, jejunum, ileum, caecum, and right and transverse colon)], and hindgut (left colon, rectum, and urogenital sinus); these three parts of the gut maintain

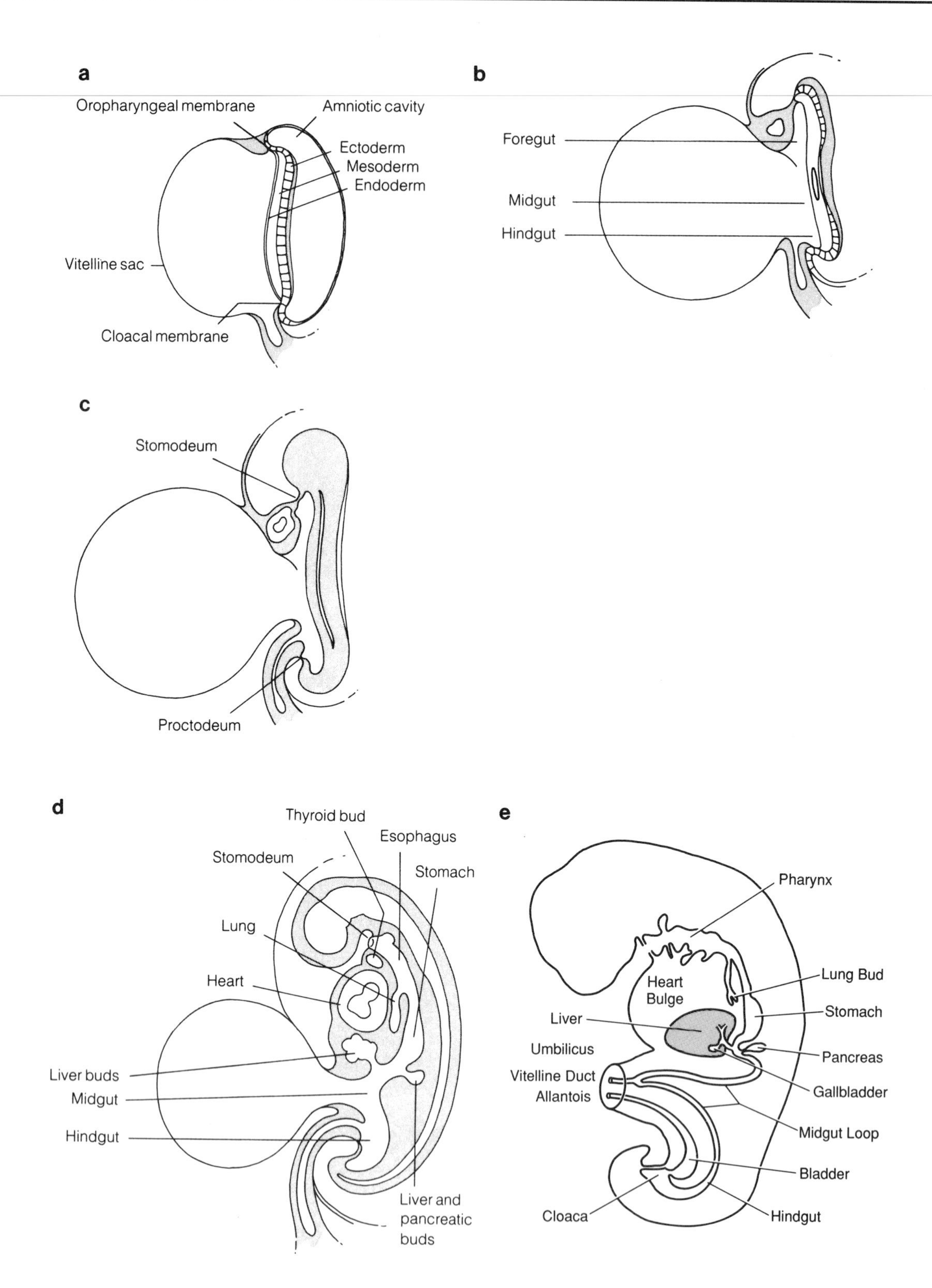

FIGURE 1 Development of gut tube in fetus: (a) beginnings late in third week, (b) elongation of embryo, (c) folding in fourth week, (d) tubular form and organ buds by fifth week, and (e) key features of fetal gut. [Source: Gaudin, A. J., and Jones, K. C. (1989). "Human Anatomy and Physiology." Harcourt Brace Jovanovich, San Diego, p. 442. Reproduced with permission.]

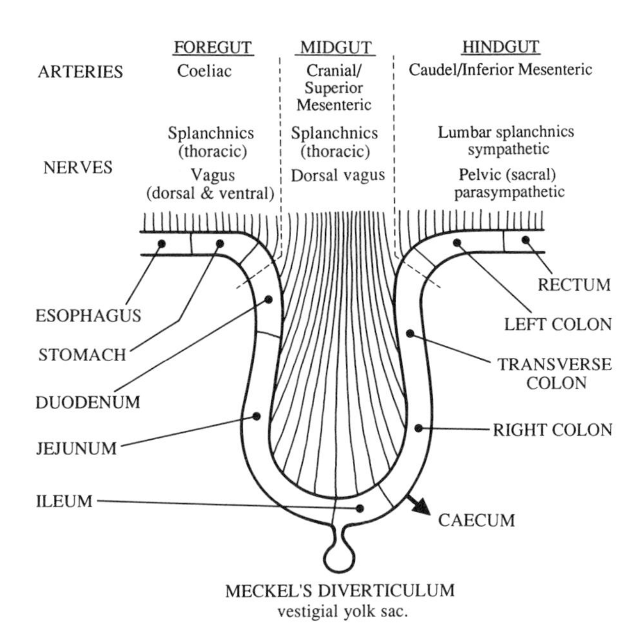

FIGURE 2 Basic design of gastrointestinal tract, with blood supply and innervation.

distinctive blood supplies and innervation through development (See Color Plate 1).

The abdominal foregut expands into the stomach, which rotates to the left acquiring a greater curvature, and becomes U-shaped, in part because of anchorage of the start of the duodenum to the septum transversum [i.e., the mass of mesoderm ventral to the stomach, between the pericardial and peritoneal cavities, within which the liver develops and the cranial vestige of which becomes the main part of the diaphragm (the partition between thoracic and abdominal cavities)]. The anchorage comes about because of a duodenal diverticulum (outgrowth) that develops into the bile duct, gallbladder, and biliary system, ramifying through the liver tissue. Two other diverticula from the foregut part of the duodenum develop into the dorsal and ventral lobes of the pancreas, which subsequently fuse in part.

The midgut loop, retaining a connection with the yolk sac, herniates through the umbilicus as it elongates. It is subsequently withdrawn and laid down within the abdomen more or less simultaneously from both ends in an anti-clockwise direction (viewed from ventrally). Thus, the duodenum ends up to the right and dorsally, and the transverse colon crosses from left to right to the ascending or right colon (and caecum). The coils of small intestine (jejunum and ileum) are bundled in last, and a vestige of the yolk sac (Meckel's diverticulum) may persist in a few individuals. By this time the umbilicus is closed down to allow passage only of the two umbilical arteries and the one umbilical vein (see Fig. 2). The hindgut remains in all species as a simple tube, the left colon leading into the rectum (and thence anus) in the pelvic cavity.

None of these structures are actually in the peritoneal cavity; they are all supported by (and contained within) a double peritoneal membrane (or mesentery) attached initially to the dorsal midline. Subsequent differential growth and movements of the gut tube mean that these original dorsal attachments are often distorted or modified (for the fore- and midgut parts) (Fig. 3). Furthermore, the caudal migration of the liver out of the septum transversum and the cranial migration of parts of the urogenital system and umbilical arteries out of the caudal pelvic wall lead to the acquisition of "ventral" mesenteries. Humans are unique, even among primates, in the secondary reattachment of alternate parts of

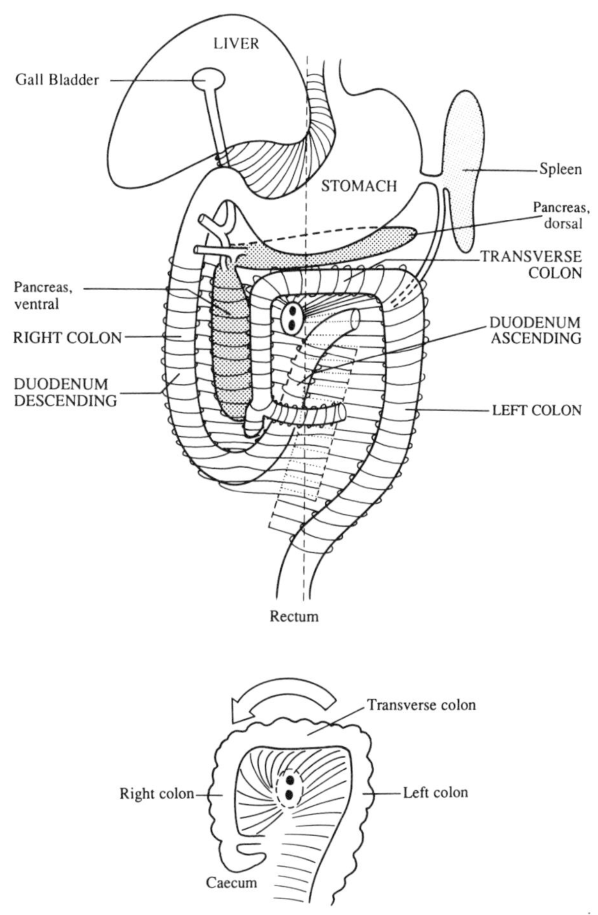

FIGURE 3 Basic topography of mammalian gastrointestinal tract, showing rotations and mesenteries, ventral view.

the gut back behind the peritoneum of the dorsal abdominal wall; this must relate to the acquisition of bipedalism and the need to move back the center of gravity as close as possible to the vertebral column (spine).

B. Gross Structure and Diet

Various measures can be obtained. The functional significance of surface area is for absorption of volume, for fermentation, and, of weight, for muscular activity in digestion.

The stomach and first part of the large intestine are associated with fermentation among mammals in general, whereas the small intestine is the main, but not the only, site of absorption; the amount of muscle reflects the degree of activity in each region, whether it be to pass the food through more rapidly, to mix it more thoroughly, or to reverse its passage to prolong digestion.

Most mammals are either consumers of animal matter [invertebrate or vertebrate (faunivores)] or of the vegetative parts of the primary production [foliage of some kind (folivores)]. It is anatomically and physiologically impossible to digest significant amounts of both kinds of foods, hence the inappropriateness of the term *omnivore*. In faunivores (e.g., insectivores, carnivores, cetaceans, pholidote), where food is rare but readily digested, the gut is simple and dominated by small intestine. Folivores are exploiting a common but indigestible food; they depend on the fermenting action of bacteria and have to expand either the stomach and/or the caecum and right colon (midgut derivatives, hence the error of hindgut fermentation common in the literature). Body size tends to be larger, to house a more voluminous gut. The more specialized foregut fermenters (e.g., ruminants and colobine monkeys) contrast with the caeco-colic fermenters (e.g., rodents, equids, some strepsirhine (Prosiman) primates, and some monkeys and apes) (Table I).

Apart from bats, some rodents, and pigs, only primates have really exploited food of intermediate availability and digestibility (i.e., the reproductive parts of plants, fruit). It is this flexibility of diet, along with the capacity for year-round breeding afforded by the menstrual cycle, that has made primates so successful and led to the evolution of humans. Such frugivores have guts of more even proportions, dominated by small intestine but with some expansion of the stomach and/or, more often, the caecum and first part of the colon. Not all amino acids occur in fruit, hence the smaller frugivores supplement their fruit diet with animal matter and the larger ones with foliage, with the appropriate modifications of gut dimensions (Table I).

From Table I the main contrasts can be seen. In carnivores, the gut is dominated by small intestine, and the stomach is large to accommodate the infrequent meals; the guts are relatively light to facilitate the chase, digestion is aided by prolonged retention time. The guts of ungulates are bulky, especially if they subsist on leaves. Pigs have a comparable proportion of small intestine to carnivores but a smaller stomach and large caecum and colon. Ruminants have a massive stomach for fermentation, whereas the horse has a relatively larger caecum and colon.

The frugivorous (ancestrally) monkeys and apes

TABLE I Gut Dimensions in Various Mammals

Dietary staple	Species	Body weight (kg)	% Gut Surface Area			Total surface area (cm^2)
			Stomach	Small intestine	Caecum colon	
Meat	Dog	9.7	20	67	14	1,320
Meat	Cat	2.5	21	57	22	580
Fruit	Pig	58	3	65	32	18,860
Leaves	Goat	56	67	28	14	35,930
Leaves	Horse	202	2	22	76	48,950
	Monkey					
Leaves	langur	5.8	35	40	25	3,518
Fruit	macaque	3.2	15	48	37	2,129
	Ape					
	Orangutan					
Fruit	Chimpanzee	34–51	7–10	38–49	44–52	10,514
	Gorilla					(7,662–13,373)
Meat	Human	61	8	71	21	6,320

have reduced areas of small intestine and increased areas of caecum and colon (for fermantation of the leaf components); leaf-eating colobine monkeys (e.g., langurs) have much enlarged stomachs and reduction of caecum and colon; in all cases, just under half the gut area is small intestine. Humans have an even greater proportion of small intestine than carnivores, but smaller stomachs and slightly enlarged caecum and colon, reflecting their long meat-eating ancestry. Modern techniques of food preparation and processing allow humans to be omnivorous; data on long-term vegetarians are still needed to confirm the expectation that their gut proportions are more similar to those of apes. The full range of parameters of the human gastrointestinal tract is given in Table II.

C. Basic Anatomy

The gastrointestinal tract is basically a tube of smooth muscle (mesodermal), lined with a mucous membrane (endodermal) and covered by peritoneum (mesodermal). The muscular wall is divided into two layers: an inner circular layer and an outer longitudinal layer [separated into discrete bundles (taenia coli) in the large intestine]; there is an innermost oblique layer reinforcing the body of the stomach. Striated muscle extends some way down the esophagus, and the initial rapid propulsion initially is transformed into the slow, rhythmic contractions of peristalsis lower down, which continue through the rest of the tract.

The mucous membrane is thick but loose, allowing considerable dilation to occur. The lining is divided into two layers (mucosa and submucosa) by a thin sheet of smooth muscle (muscularis mucosae).

TABLE II Human Gut Dimensions[1]

	Stomach		Small Intestine		Caecum + colon			
	x̄	%	x̄	%	x̄	%	x̄	%
Surface area (cm²)	498	8	4,520	7	66	1	1,234	20
Weight (g)	150	14	587	53	22	2	348	31
Volume (cm³)	1,069	24	2,470	56	48	1	854	18
Length (cm)	—		662	81	8	1	146	18

[1] (*n*, 6; body weight, 60.8 kg; body length (crown–rump), 84.6 cm; % gut total)

The mucosa, a single layer of columnar cells, is thrown into folds or crypts, into which cells secrete, forming glands, that vary through the tract. The surface of the small intestine is projected into villi that increase the surface area significantly and contain capillaries and lacteals (lymphatic vessels). The submucosa contains blood vessels, Meissner nerve plexus (Auerbach's plexus is myenteric, between longitudinal and circular muscles), and lymphatic follicles (Fig. 4).

II. Foregut

A. Stomach

This is the most dilated part of the gastrointestinal tract, with a short lesser curvature running from the esophageal orifice (cardia) just to the left of the midline behind the 7th costal cartilage to the pyloric orifice, which is located just to the right of the midline at the level of the 1st/2nd lumbar vertebrae (behind the 9th costal cartilage); thus it faces right, then cranially and finally to the left. The muscular bag is created by the great expansion of the greater curvature (primitive dorsal surface), which faces left, then caudally, and finally, near the pylorus, to the right; to it is attached the equally expanded dorsal mesentery (dorsal mesogastrium, greater omentum) to which it is attached to the dorsal body wall. The cavity created therein (the omental bursa or lesser sac) is a diverticulum of the main peritoneal cavity. This shape has come about by differential growth and rotation of the primitive gastric spindle to create a large sac mainly to the left of the midline (see Color Plate 2).

The fundus is the cranial projection to the left of the esophagus, up against the dome of the diaphragm. As in the body of the stomach, the mucosa is thrown into folds or rugae containing fundic glands, secreting acid and gastric enzymes. These give way to mucous-secreting glands in the pylorus. The lesser curvature is attached to the liver by the lesser omentum (or hepato-gastric ligament); it extends a short way along the duodenum, containing the bile duct from the gallbladder in its free edge. This "ventral" mesentery of the stomach is produced by the caudal migration/expansion of the gastric spindle from over the septum transversum (mesodermal block in embryo between thorax and abdomen) and by the growth of the liver out of the septum transversum.

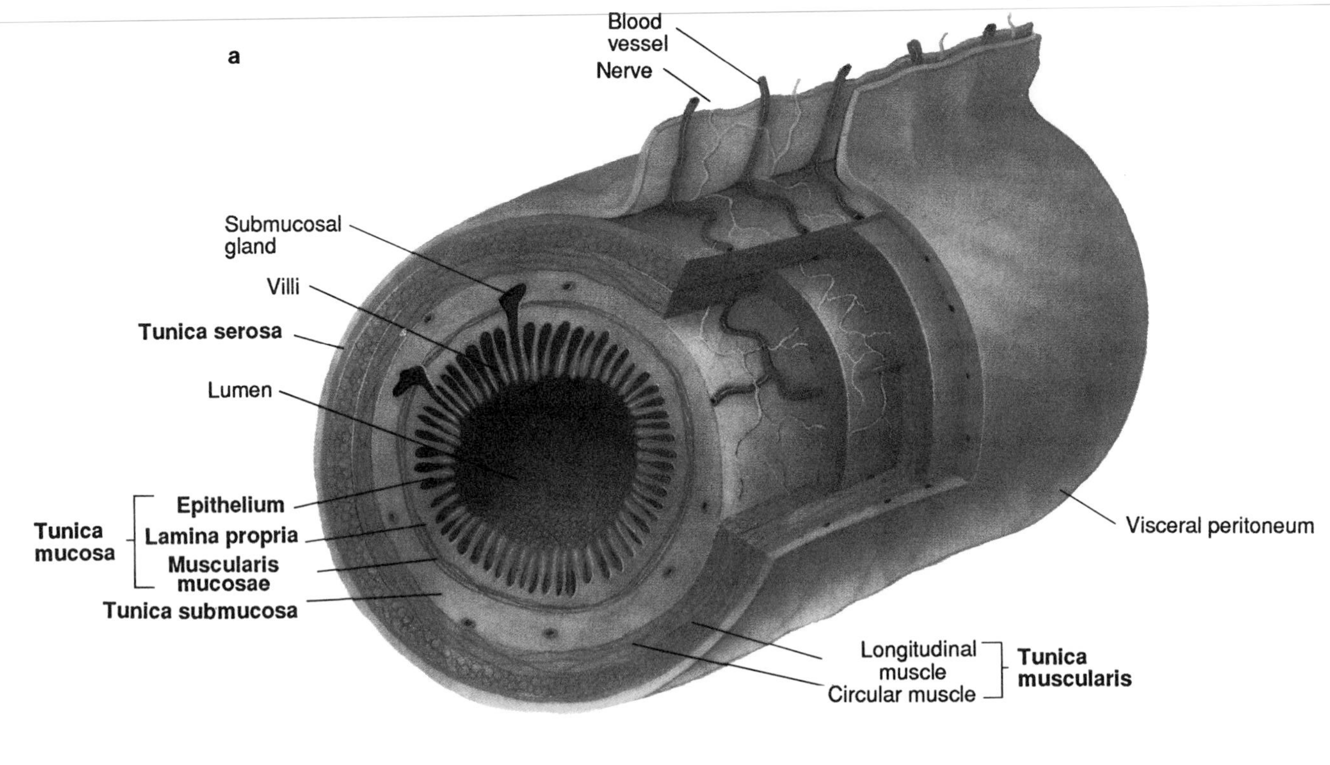

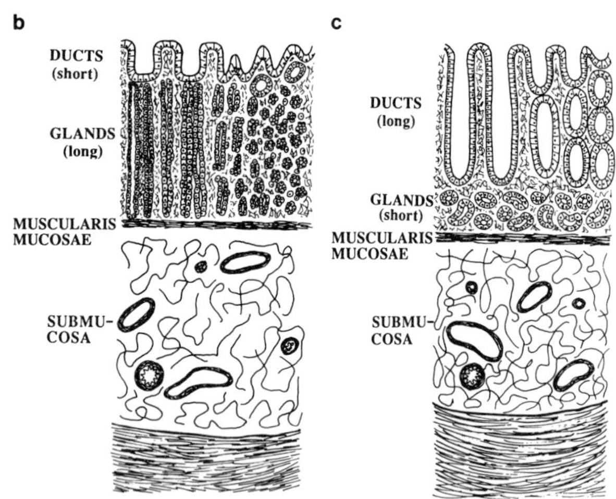

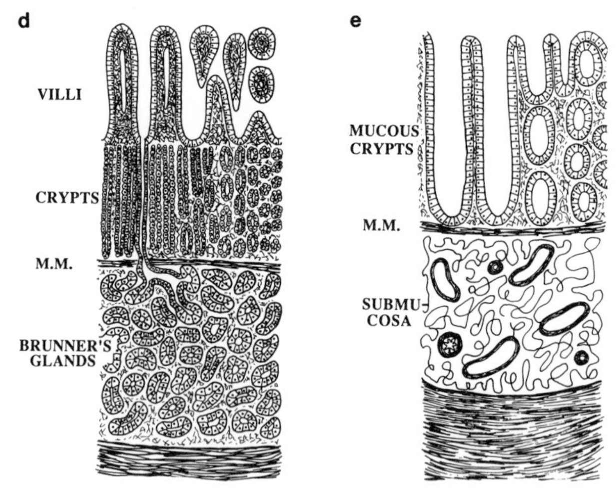

FIGURE 4 Histology of the gastrointestinal tract: (a) general; mucosa and submucosa in (b) body of stomach, (c) pylorus of stomach, (d) duodenum, and (e) colon. [From Last, R. J. (1978). "Anatomy: Regional and Applied," 6th ed. Churchill Livingstone, New York.] [Source: Gaudin, A. J., and Jones, K. C. (1989). "Human Anatomy and Physiology." Harcourt Brace Jovanovich, San Diego, p. 443. Reproduced with permission.]

Hence the stomach is related to the diaphragm and liver cranially, the abdominal wall ventrally, the kidneys and vertebral column dorsally, the transverse colon caudally (and intestinal mass), the spleen to the left, and the liver and duodenum to the right.

B. Duodenum

The forgut extends as far as the entrance of bile and pancreatic ducts, near the end of the second part of the duodenum. The first part, or duodenal cap, loops 5 cm upward and then downward, in front of the right crus of the diaphragm, the right psoas muscle, and the head of the pancreas (Figure 5). The next 7 cm curves caudally over the hilus of the right

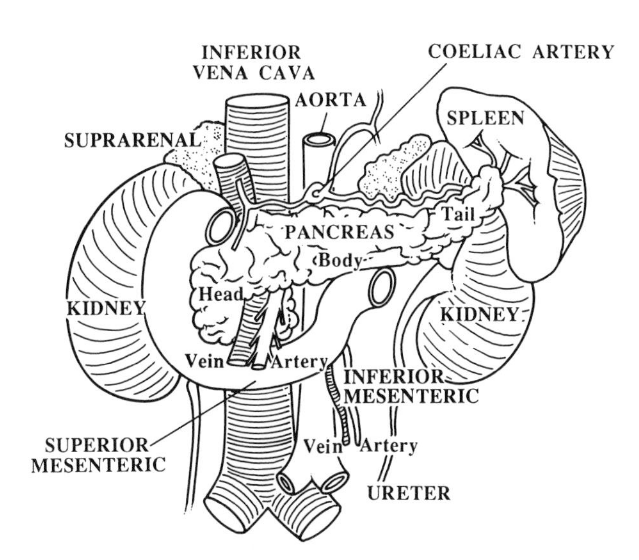

FIGURE 5 Duodenum and pancreas, in relation to major vessels, spleen, kidney, and suprarenal glands. [From Last, R. J. (1978). "Anatomy: Regional and Applied." 6th ed. Churchill Livingstone, New York.]

kidney, behind the right colon and its mesentery. The common bile duct and ventral pancreatic duct open together at the duodenal papilla 2 cm from the end of this part of the duodenum. The duodenum is retroperitoneal; it is not suspended in a mesentery in humans but has been reattached to the dorsal body wall (see Section V,A). Glands secreting digestive enzymes are concentrated at the start of the duodenum, extending through the muscularis mucosa into the submucosa as Brunner's glands, opening into the crypts of Lieberkuhn that extend into the mucosa throughout the intestine. The villi are largest and most numerous at the start of the small intestine; mucous-secreting cells, which are scarce, are represented by the large Paneth cells.

C. Accessory Glands

Reference has been made to the development of the liver in the septum transversum from a meshwork of fetal blood vessels (umbilical veins, increasingly just the left, and vitelline veins from the gut, developing into the hepatic portal vein, bringing nutrients and waste products of digestion) infiltrated by bile canaliculi from the bile/gallbladder diverticulum of the duodenum.

The liver is wedge-shaped, with a small left lobe and large right one (divisible into right, quadrate, and caudate processes) (see Color Plate 3). The parietal surface, covered by peritoneum, except for the left and right triangular and coronary ligaments attaching it to the diaphragm, relates closely to the dome of the diaphragm around the inferior vena cava, which is connected to the heart. The caudally facing visceral surface relates to the stomach on the left and the duodenum, right kidney, and colon on the right, receiving the hepatic portal vein, hepatic artery, and proper bile ducts at the porta at the lower limit of the attachment of the lesser omentum. The left and right lobes are separated by the ligamentum teres, the peritoneal fold from the umbilicus, and abdominal floor, containing the vestige of the left umbilical vein.

The gallbladder is situated between the quadrate and right lobe (see Color Plate 3), receiving bile from the liver lobules, which contain cells in intimate relation with bile canaliculi and with capillaries from hepatic artery and vein and hepatic portal vein. The common bile duct then runs down to the duodenal papilla in the free edge of the lesser omentum.

The pancreas develops as two buds: the ventral bud in common with the bile duct, and the dorsal bud in the dorsal mesentery of the duodenum. The latter enlarges toward the left in relation to the greater curvature of the stomach and transverse colon. The ventral bud escapes from the confines of the ventral mesentery, when the duodenum rotates in response to the rotations of the stomach; it moves under the peritoneal covering of the duodenum into the greater spaces of the dorsal mesentery, extending caudally within the duodenal loop, which is large in most mammals. This rotation brings the ventral bud into contact with the dorsal bud, and they fuse at their bases to form the body of the pancreas, with dorsal (or left) and ventral (or right) lobes around the superior mesenteric arteries and veins (the latter a main branch of the hepatic portal vein).

The pancreas is compact in humans because of the small, C-shaped duodenal loop (Fig. 5), and the body is moulded to the concavity of the duodenum, in front of the inferior vena cava and renal veins. The right lobe is reduced to an uncinate process, in front of the aorta and behind the mesenteric vessels; the left lobe, or tail, extends right across to the left kidney.

The main pancreatic duct (originally the ventral one) opens into the common bile duct; the accessory duct (originally dorsal) opens into the duodenum opposite and nearer the pylorus. These two ducts are only separate in 9% of humans; in 61% the dorsal duct loses contact with the duodenum, so that drainage is through a link into the ventral duct. In the remaining 30% of cases, the dorsal duct is reduced from its original shape and the connecting duct increased.

Mention should finally be made of the spleen, which, although it serves no digestive function, develops from the mesoderm of the stomach wall to migrate out into the greater omentum and into one leaf of the two-layered mesentery, remaining close to the greater curvature to the left of the stomach against the diaphragm at the level of ribs 9–11. The left leaf of greater omentum comes together at the hilus on the visceral side of the spleen through which pass the vessels and nerves of the organ.

D. Blood Vessels and Nerves

Three unpaired arteries leave the aorta [that to the foregut is the coeliac artery (Fig. 4), the most cranial of the three]. It has three main branches: (1) the splenic, to the spleen and greater curvature of the stomach, left gastro-epiploic, proximally; (2) the common hepatic, to the liver, pyloric end of the stomach, greater and lesser curvatures (right gastro-epiploic and right gastric, respectively), and (3) proximal duodenum (superior pancreatico-duodenal, anastomosing with the inferior artery from the midgut artery) (see Section III,F).

As with the rest of the gut, the veins coming from the wall of the gut do not closely follow the arteries; gastric, gastro-epiploic, splenic, and pancreatico-duodenal veins converge on the mesenteric veins to form the hepatic portal vein, which runs into the porta of the liver (Fig. 6).

The motor innervation of the foregut muscular wall comes from the thoracic sympathetic chain and from the vagus (10th cranial) nerve, parasympathetic (Fig. 7). Together they form plexuses along the walls of the arteries and carry sensory fibers from the mucosa back to the central nervous system. The sympathetic fibers, if they have not synapsed in one of the last six thoracic chain ganglia through which they pass, synapse in the coeliac ganglion having passed through the crura of the diaphragm and/or the aortic hiatus at the base of the coeliac artery, thus, the coeliac plexus contains only postganglionic neurons. The parasympathetic neurons synapse in the gut wall and have short post-ganglionic fibers; the vagal fibers enter the abdomen through the esophageal hiatus of the diaphragm, dorsal and ventral to the esophagus (each being a mixture of left and right vagal fibers in the thorax). The ventral vagus, predictably, runs along the lesser curvature of the stomach and right side of the start of the duodenum and up into the liver, (i.e., within the lesser omentum). By contrast, the dorsal vagus, in dorsal mesentery, has access to the midgut, as well as along the coeliac artery to the greater curvature of the stomach and start of duodenum, as well as spleen and part of pancreas. [*See* GASTRIC CIRCULATION.]

III. Midgut

A. Duodenum

The third part of the C-shaped retroperitoneal duodenum passes back toward the midline for 8 cm

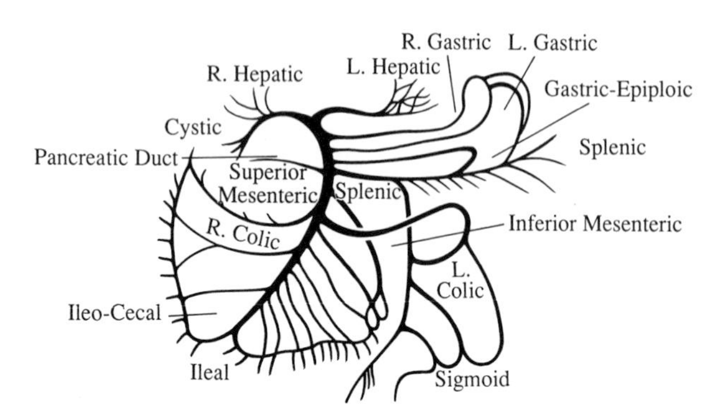

FIGURE 6 Hepatic portal vein and gastrointestinal branches. [From Last, R. J. (1978). "Anatomy: Regional and Applied." 6th ed. Churchill Livingstone, New York.]

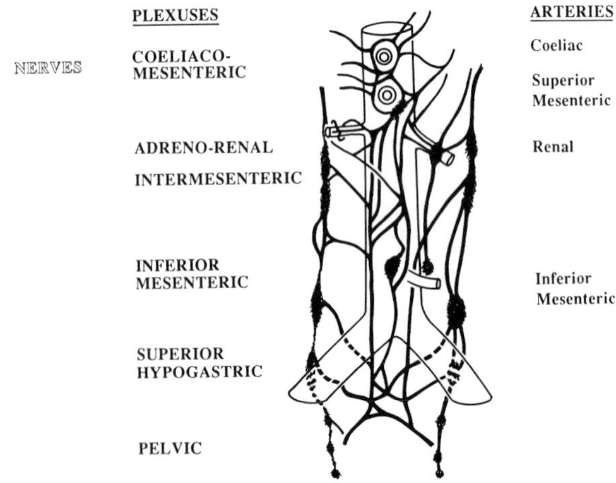

FIGURE 7 Sympathetic and parasympathetic nerves in abdomen, in relation to arterial trunks. [From Last, R. J. (1978). "Anatomy: Regional and Applied." 6th ed. Churchill Livingstone, New York.]

over the psoas muscle, gonadal vessels, and ureter to cross the inferior vena cava and aorta (Fig. 5). The final 5 cm ascend to the left of the aorta to reach the lower border (caudal) of pancreas. This first part of the midgut leads into the midgut loop that is anchored cranially at this duodeno-jejunal flexure by the ligament of Treitz, which descends from the right crus of the diaphragm in front of the aorta and behind the pancreas, to be anchored to the left psoas muscle by connective tissue.

B. Jejunum

The remainder of the small intestine retains it mesentery and is liberally coiled below and in front of the transverse colon. It is the main site of digestion and absorption, with villi getting slowly fewer and smaller, digestive juice secretion decreasing, and the frequency of mucous cells (lubricative) increasing; the wall is thick. Meckel's diverticulum, vestige of the yolk sac (with which the gut lumen was originally confluent), persists in 2% of humans, about 60 cm from the end of the small intestine (Fig. 2).

C. Ileum

This is the terminal part of the small intestine, crossing caudally (inferiorly) from left to right, to the ileo-caeco-colic junction in the lower right part of the abdomen. It is thin-walled and narrower, with Peyer's patches (aggregated lymphatic follicles) sometimes visible externally.

D. Caecum

This blind pouch of the large intestine hangs free in the right iliac fossa, secondarily attached by a mesentery to the ileum; the terminal appendix is a concentration of lymphatic tissue (Fig. 8).

E. Colon

The first part of the colon [right, ascending (15 cm)] is retroperitoneal crossing the iliac crest and ascending in front of the hypaxial muscles; the longitudinal muscle has separated into three bands (taenia coli), which remain distinctive until the rectum. In front of the lower pole of the right kidney, it forms the right colic flexure into the transverse colon, which hangs down in a mesentery as it crosses to the left colic flexure across the descending duodenum and pancreas to the lower pole of the left

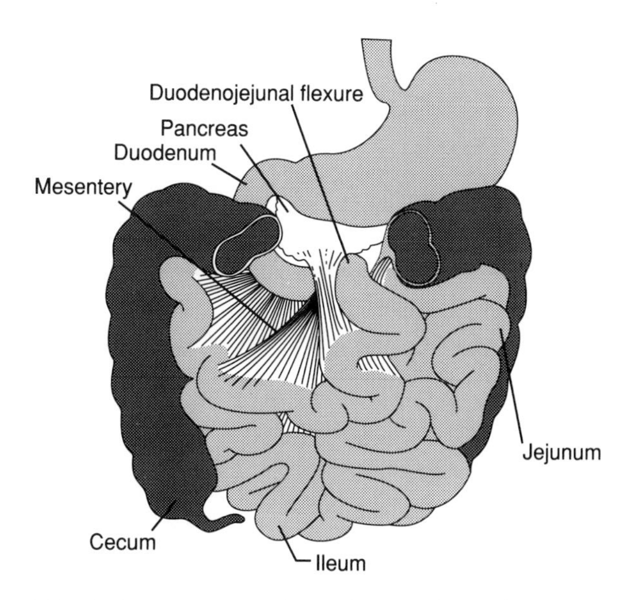

FIGURE 8 Arrangement of small intestine, in relation to stomach and colon. [Source: Gaudin, A. J., and Jones K. C. (1989). "Human Anatomy and Physiology." Harcourt Brace Jovanovich, San Diego, p. 455. Reproduced with permission.]

kidney (see Color Plate 4). This terminal 45 cm of the midgut loop is attached to the diaphragm below the spleen at the level of the 11th rib by a peritoneal fold (the phrenico-colic ligament). The mucosa has a honey-comb appearance under the lower-power microscope, reflecting the profusion of crypts into which mucous cells secrete and through which absorption occurs.

F. Blood Vessels and Nerves

The cranial or superior mesenteric artery is the artery of the midgut. Its first branches (inferior pancreatico-duodenal and common colic) go to the opposite, dorsal ends of the midgut loop; the terminal branches (jejunal) pass to all parts of the main loop, forming anastomosing arcades on the intestinal wall. The common colic artery gives branches to the ileum, caecum, and ascending and transverse colons. Corresponding veins form the superior mesenteric vein, which heads past the pancreas into the hepatic portal vein (Fig. 9).

As mentioned above, parasympathetic innervation comes from the dorsal vagus, with preganglionic fibers that ramify along the artery, being joined by postganglionic sympathetic fibers from the superior mesenteric ganglion at the base of the artery (Fig. 7). [*See* INTESTINAL BLOOD FLOW REGULATION.]

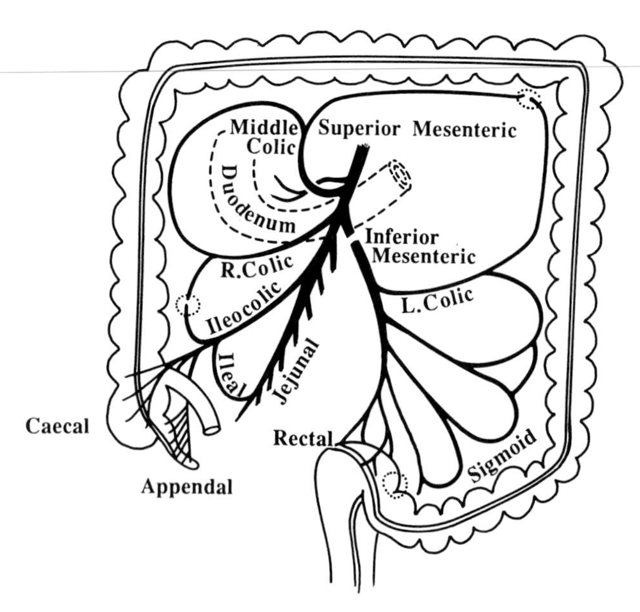

FIGURE 9 Superior and inferior mesenteric veins, in relation to color. [From Last, R. J. (1978). "Anatomy: Regional and Applied." Churchill, Livingstone, New York.]

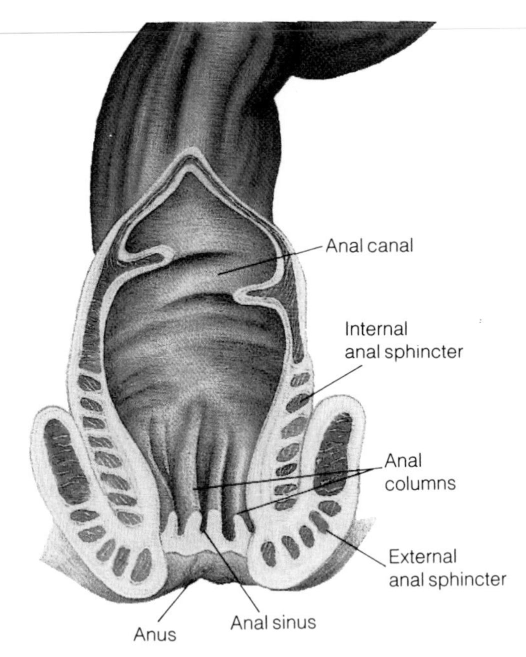

FIGURE 10 Rectum and anal canal. [Source: Gaudin, A. J., and Jones, K. C. (1989). "Human Anatomy and Physiology." Harcourt Brace Jovanovich, San Diego, p. 456. Reproduced with permission.]

IV. Hindgut

A. Colon

The left or descending colon descends for 30 cm behind the peritoneum, crossing the iliac crest and iliac fossa to the pelvic brim, where it runs into the 45 cm of sigmoid colon, which loops up in a mesentery against the psoas muscle and then down in front of the sacrum into the rectum (see Color Plate 4).

B. Rectum

The rectum is retroperitoneal, curving against the sacrum to pass into the anus, with, once again, a continuous sleeve of longitudinal muscle; the upper part is usually empty but the lower part is expanded into the ampulla, containing flatus and feces (Fig. 10).

C. Blood Vessels and Nerves

The inferior mesenteric artery is the artery of the hindgut, giving left colic, sigmoid, and superior rectal branches; caudally the rectum is supplied by branches of the internal iliac artery. Veins form the inferior mesenteric vein, which joins the superior one near its base (Fig. 9).

The sympathetic nerves come from the lumbar outflow, which pass to and have all synapsed by the inferior mesenteric artery at the base of the artery of the same name; postganglionic fibers form a plexus on the left colic artery and run down into the pelvis in the two hypogastric nerves to form a plexus along the visceral branches of the internal iliac artery. Here they are joined by parasympathetic fibers from the sacral outflow to reach the pelvic parts of the gut; this sacral outflow reaches the left colon either by running up the wall of the gut or by running up the hypogastric nerve to the left colic artery. Sensory fibers return to the spinal cord and brain by any of these routes but more often by the sympathetic routes back into the thorac-lumbar spinal cord.

Throughout the gastrointestinal tract, parasympathetic fibers stimulate muscular contraction and glandular secretions, whereas sympathetic fibers are inhibitory to the gut muscle and stimulate the closure of sphincter and blood vessels.

V. Overview

A. Attachments

The anti-clockwise rotations (viewed from ventrally) of the midgut loop (Fig. 11a) pull the original

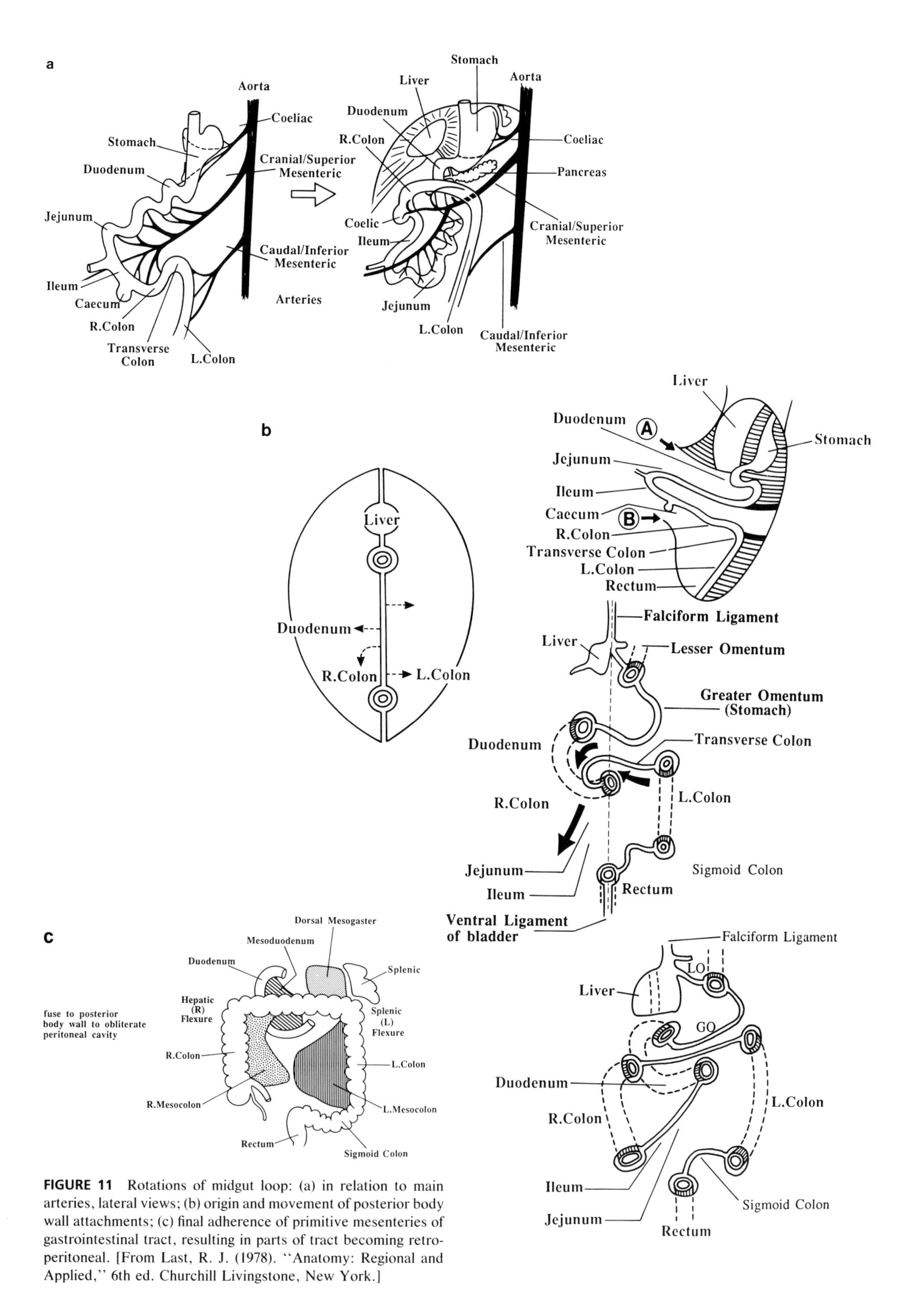

FIGURE 11 Rotations of midgut loop: (a) in relation to main arteries, lateral views; (b) origin and movement of posterior body wall attachments; (c) final adherence of primitive mesenteries of gastrointestinal tract, resulting in parts of tract becoming retroperitoneal. [From Last, R. J. (1978). "Anatomy: Regional and Applied," 6th ed. Churchill Livingstone, New York.]

a

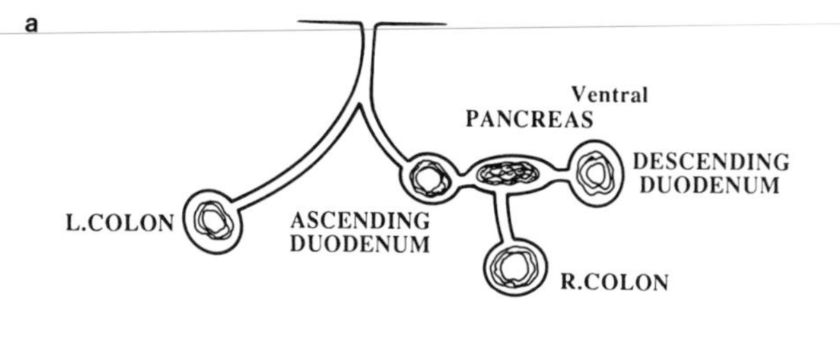

b

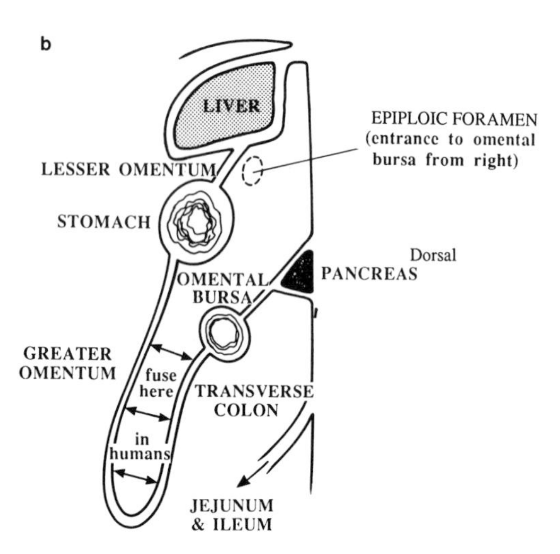

c

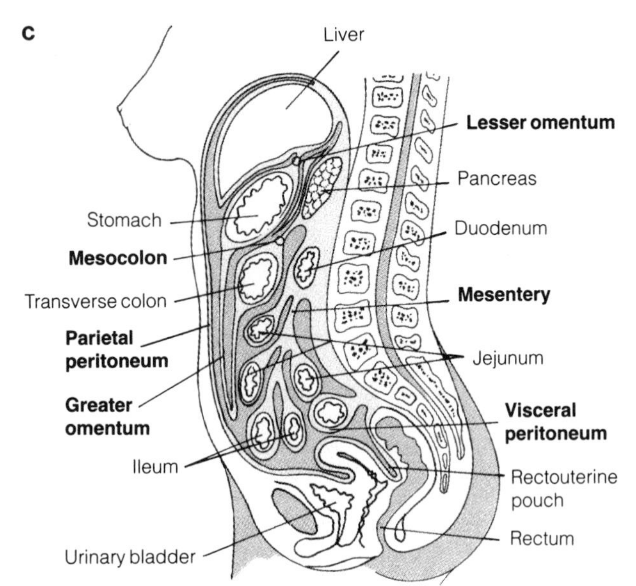

FIGURE 12 Secondary attachments of mesenteries of gastrointestinal tract: (a) transverse section in mammal such as dog, (b) longitudinal section in humans, and (c) sagittal section through human abdomen and pelvis. [Source: Gaudin, A. J., and Jones, K. C. (1989). "Human Anatomy and Physiology." Harcourt Brace Jovanovich, San Diego. p. 472. Reproduced with permission.]

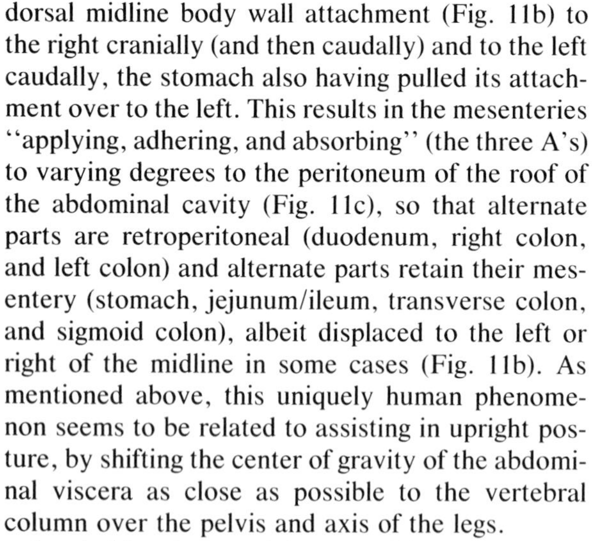

dorsal midline body wall attachment (Fig. 11b) to the right cranially (and then caudally) and to the left caudally, the stomach also having pulled its attachment over to the left. This results in the mesenteries "applying, adhering, and absorbing" (the three A's) to varying degrees to the peritoneum of the roof of the abdominal cavity (Fig. 11c), so that alternate parts are retroperitoneal (duodenum, right colon, and left colon) and alternate parts retain their mesentery (stomach, jejunum/ileum, transverse colon, and sigmoid colon), albeit displaced to the left or right of the midline in some cases (Fig. 11b). As mentioned above, this uniquely human phenomenon seems to be related to assisting in upright posture, by shifting the center of gravity of the abdominal viscera as close as possible to the vertebral column over the pelvis and axis of the legs.

Inevitably, the cranial or caudal movement of parts of the intestine result in the mesentery of one being secondarily attached to another; this is more obvious in other mammals, where no parts of the intestine are retroperitoneal—where the whole tract is suspended in mesentery. Hence, the jejunal mass (most ventral) may be attached to the base of the colon mesentery, which in turn is attached to the mesenteries of stomach and duodenum (most dorsal—retracted from the umbilical hernia first, the jejunal coils last) (Fig. 11a).

B. Relations

So we return to our basic layout (see Color Plate 1 and Fig. 3). In mammals the right colon can be traced back to the dorsal body wall via the mesentery of the duodenum and thence the mesentery of the left colon, and the transverse colon can be traced back via the greater omentum containing the dorsal (left) limb of the pancreas (Fig. 12a). In humans, with the duodenum and dorsal pancreas retroperitoneal, the mesentery of the transverse colon is fused to the peritoneal covering of the peritoneum and the greater omentum (Fig. 12b).

A distinctive feature in all mammals is the voluminous ballooning of the greater omentum from the caudally facing part of the greater curvature of the stomach (the spleen restricts such expansion more proximally (from the left-facing part) (Fig. 12c). Thus, the omental sac comes to lie ventral to the intestinal mass, providing a protective cushion in quadrupeds and acting in all species as a site of fat storage and supposedly having a physical and immunological role in blocking and healing any damage to the intestinal wall.

Thus, we have an enlarged stomach to the left cranially, with the bulk of the liver displaced to the right, a short duodenum in humans to the right, a large loop of colon arching from caudally to the left, across cranially close to the stomach and down on the right, with coils of jejunum centrally and ventrally, all obscured from view (ventrally) by the expansion of the greater omentum caudally.

Although the primate gut is relatively unspecialized for fruit-eating, in contrast to the meat- or leaf-eating specializations of most other mammals, the human gut conforms to this pattern, although assuming more the proportions of a meat-eater, with unique adaptations in attachments for bipedalism.

Bibliography

Chivers, D. J., and Hladik, C. M. (1980). Morphology of the gastro-intestinal tract in primates: Comparisons with other mammals in relation to diet. *J. Morphol* **166,** 337–386.

Hill, W. C. O. (1958). Pharynx, oesophagus, stomach, small and large intestine: Form and position. *Primatologia* **3,** 139–207.

Langer, P. (1987). Evolutionary patterns of Perissodactyla and Artiodactyla (Mammalia) with different types of digestion. *Z. Zool. Syst. Evolut-forsch.* **25,** 212–236.

Last, R. J. (1978). "Anatomy: Regional and Applied," 6th ed. Churchill Livingstone, New York.

MacLarnon, A. M., Chivers, D. J., and Martin, R. D. (1986). Gastro-intestinal allometry in primates and other mammals including new species. *In* "Primate Ecology and Conservation," vol. 2. (J. G. Else and P. C. Lee, eds.), pp. 75–85. Cambridge University Press, Cambridge.

Martin, R. D., Chivers, D. J., MacLarnon, A. M., and Hladik, C. M. (1985). Gstro-intestinal allometry in primates and other mammals. *In* "Size and Scaling in Primate Biology," pp. 61–89. (W. L. Jungers, ed.). Plenum Press, New York.

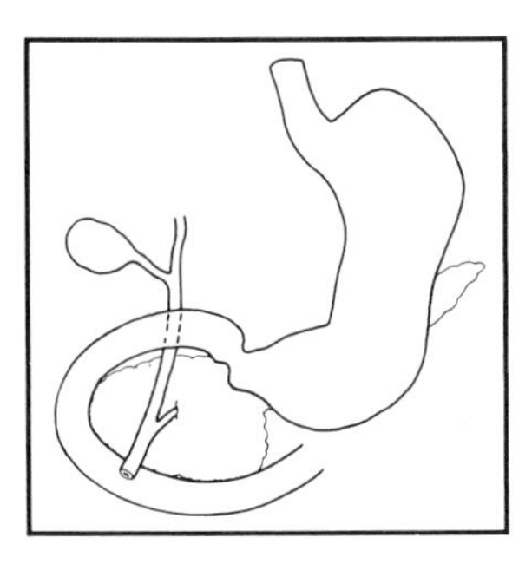

Digestive System, Physiology and Biochemistry

ELDON A. SHAFFER, *Division of Gastroenterology, University of Calgary*

Glossary

Alimentary Relevant to food or nutrition

Assimilate To absorb and incorporate digested food products into the body

Carbohydrates Saccharides or sugars composed of carbon, hydrogen, and oxygen, with the empirical formula $(CH_2O)_n$

Chyme A creamy, semifluid, or gruellike material produced by partial digestion in the stomach

Digestive system System of structures which processes and absorbs food; synonymous with the alimentary tract (or canal) or the gastrointestinal tract

Electrolyte Compound that dissolves in water to form charged ions, such as sodium (Na^+), hydrogen (H^+), bicarbonate (HCO_3^-), or chloride (Cl^-). Such an electrolyte solution conducts an electrical current. Acids, bases, or salts (inorganic or organic) are electrolytes if the substance forms ions in solution.

Endocrine Internal secretions (e.g., hormones) of a gland, usually into the blood. Although typical endocrine glands are the thyroid, adrenal, and reproductive glands, the gastrointestinal tract is the largest endocrine gland in the body.

Enzyme Specialized protein molecules which act as catalysts to enhance the rate of a specific chemical reaction. The substance on which the enzyme works is the substrate; each substrate usually is changed by a specific enzyme. Many enzymes are named by adding the suffix -ase to the name of the substrate on which they act. For example, the enzyme lactase acts on lactose. Other enzymes have less informative names (e.g., pepsin, which means to cook or digest).

Epithelia Cells which cover a surface

Exocrine External, or outward, secretion of a gland, usually via a collecting or duct system (e.g., pancreatic enzymes and bile). Some glands, such as the pancreas, have both exocrine (e.g., enzymes) and endocrine (e.g., the hormone insulin) functions.

Gastrointestinal Referring to both the stomach and the intestine; gastroenteric

Hormone An organic substance, produced in one organ or part of the body, which regulates another organ or cell type. Some hormones reach distant sites of action by circulating in the blood; others are released locally and diffuse toward their target cells. A number of hormones are formed by ductless glands; other hormones, such as cholecystokinin and secretin, are produced by the gastrointestinal tract. Such gut hormones are polypeptides, short protein chains of 10–100 amino acids.

Lumen The cavity in a tubular structure

Mucosa Mucous membrane, composed primarily of epithelial cells, which lines the inner surface of the digestive system

Mucus A clear viscid secretion of mucous membranes

Secretion Production of a substance by a cell or group of cells (a gland), usually in liquid form

Transport To move a substance from one site to another

Triglycerides The major form of fat present in the diet of humans. This lipid is a neutral fat consisting of glycerol, to which are attached three fatty acids.

HUMANS REQUIRE food, water, and electrolytes for sustenance, growth, and survival. The digestive system (also known as the gastrointestinal tract or alimentary canal) is a continuous hollow tube with specialized parts of distinct size, shape, and function. Its purpose is to make nutrients useful. Food enters the mouth, passes down the esophagus to the stomach, and is propelled through the small and large intestines before emptying out the anus (Fig. 1). This long winding canal and its accessory organs (i.e., the salivary glands, pancreas, and liver) are responsible for digestion, the simplification or chemical breakdown of complex molecules. Digestion is facilitated physically by contractions of the stomach and the intestine and biochemically by enzymes and bile secreted from the exocrine organs. Digestive enzymes are secreted by the salivary glands, the pancreas, and the epithelial cells lining the stomach and the small intestine; bile is secreted by the liver, stored in the gallbladder, and emptied into the upper small intestine. Once food has been processed to a sufficiently simple state, absorption occurs across the mucosal lining of the gastrointestinal tract. The absorbed nutrients are then carried by the bloodstream for further metabolic processing in the liver and other body tissues. Fluid and electrolyte absorption also occurs in the intestine. Nondigestible and unabsorbed materials are eliminated via the large intestine or colon.

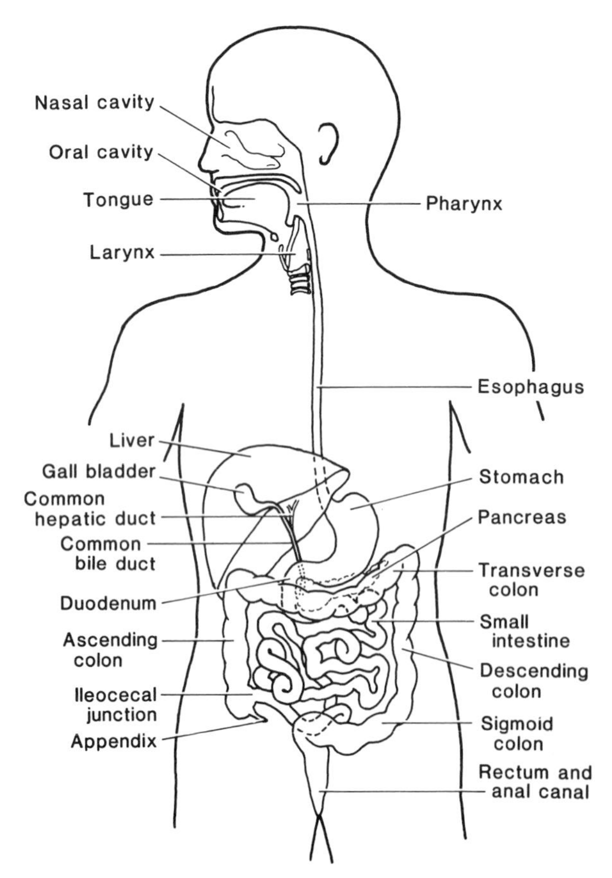

FIGURE 1 Digestive system from the mouth to the anus.

I. Anatomic and Functional Organization of the Digestive System

The digestive system is the normal avenue by which nutrition enters the body and is therefore essential for life. Humans require food (commonly divided into carbohydrates, proteins, and fats), water, inorganic salts, minerals, and vitamins. The food we eat consists of large molecules, too complex for direct use by the body; food must first be digested. These large, sometimes water-insoluble, substances are mechanically and chemically broken down into simpler water-soluble molecules. Digestion takes place within the hollow cavity, or lumen, of the alimentary canal. Absorption is the movement or transport of the digested and dissolved nutrients across the lining of the small intestine. The digestive system is selective, keeping out microorganisms and eliminating waste and toxic material.

The major functions of this system, then, are digestion and absorption, which are accomplished by muscular contractions of the canal and the secretion of digestive juices. Food enters the body from the external environment through the oral cavity and is then propelled through different regions of the digestive system. A number of digestive juices are secreted en route. The resultant smaller molecules can then be absorbed into the transporting fluids of the body: blood and lymph.

The organs of digestion consist of more than the stomach and the intestinal tract. Collectively, the digestive tract includes the mouth, esophagus, stomach, small and large intestines, plus the following glandular outpouchings, whose secretions provide important digestive function: salivary glands, liver and biliary system (bile ducts and gallbladder), and pancreas (Fig. 1). The inner surface of this canal is lined by a continuous sheet of cells (i.e., epithelial cells), some of which secrete digestive juices, while others are involved in absorption. This important lining is termed the mucosa.

The gastrointestinal tract structurally is well suited for its functions: a long coiled tube, into which a variety of glands secrete digestive juices and around which a muscular wall contracts and relaxes to churn and propel the contents. From the upper end of the esophagus to the lower end of the rectum, this wall contains an inner layer of circular muscle (which constricts the lumen) and an outer layer of longitudinal muscle (which shortens the gut).

Two basic types of muscular activity are involved in digestion and absorption: propulsive movements and mixing movements. A repetitive pattern of altered muscle tension either facilitates or impedes progress through the gut. For propulsion, increased pressure in one region squeezes the contents toward a lower region, in which there is reduced pressure within the lumen. Peristaltic contractions are coordinated, sequential, ringlike waves of muscular constriction, which pushes liquids and solids forward.

The rate at which the contents traverse the digestive tract varies between regions, depending on the different functions occurring in that specific organ. Entry of food, for example, through the mouth, pharynx, and esophagus is aided by gravity and is rapid, a conduit to the stomach. The stomach and the small intestine, which are responsible for the majority of digestion and absorption, involve the second process, mixing. Increased tone in the absence of forward progression allows the stomach to grind solids into finer products and mixes the digestive juices with the ingested nutrients. Thus, transit from the stomach through the intestines is slow. Mixing in the small and large intestines increases the exposure of chyme to the surface lining, enhancing absorption.

The circular muscle layer becomes thickened at certain points along the alimentary tract, creating areas of constriction, waistlike sphincters (see the pyloric sphincter depicted later in Fig. 2). Sphincters control the forward and backward fluxes of material; they remain closed until it is appropriate for forward flux to occur, to limit regurgitation (backward fluxes) at all times. They generally function like one-way muscular valves, but are not simple flap valves. Rather, sphincters are high-pressure zones which tighten to retard backward passage, yet relax to allow forward passage when contents arrive from above. A variety of nervous, hormonal, and intrinsic muscular controls regulate the tension and, hence, resistance in these zones. Control of traffic is found in the following sites: at the entrance to the esophagus (the upper esophageal sphincter); at the outlet of the esophagus (the lower esophageal sphincter); between the stomach and the first portion of the small intestine, the duodenum (the pylorus); at the entrance of the pancreatic and biliary ducts into the duodenum (Oddi's sphincter); between the small and large intestines (ileocecal sphincter); and at the outlet (the internal and external anal sphincters). Although sphincters do not necessarily occlude the lumen completely or continuously, they do tend to create "closed" systems.

The musculature of the digestive tract is composed of smooth (involuntary) muscle, except for the entrance and exit of the canal. The striated (voluntary)-muscled organs are the mouth, the pharynx, the upper one-third of the esophagus, and the external anal sphincter. These striated muscles are dependent on nerves for voluntary muscular contraction. The smooth muscle cells differ; they exhibit involuntary spontaneous activity. Nerves and hormones modify, rather than initiate, these contractions. [*See* SMOOTH MUSCLE.]

The gastrointestinal tract is innervated by the autonomic nervous system, which reaches out from many levels of the brain and the spinal cord to connect with local nerves contained in the wall of the gut. The autonomic nervous system consists of two divisions: the sympathetic and the parasympathetic nervous systems. Either can transmit inhibitory or excitatory impulses to smooth-muscled organs or secretory glands. These autonomic nerves also contain sensory fibers involved in the reflex regulation of gut function and nociception (including pain, nausea, and vomiting). The local nerves in the wall are termed the enteric nervous system. These nerves are unique. Although situated outside the central nervous system, they can regulate gut function on their own, even when separated from the central nervous system. That is, the enteric nervous system is truly "autonomous," yet is modulated in turn by impulses from the central nervous system reaching these local nerves via the autonomic nerves. [*See* AUTONOMIC NERVOUS SYSTEM; BRAIN REGULATION OF GASTROINTESTINAL FUNCTION.]

Smooth muscle contraction/relaxation is also controlled by hormones, the most important of which are the gut hormones synthesized by endocrine cells within the gastrointestinal tract. The autonomic nervous system also influences gut hormone production and release. Activation of the vagus nerve of the parasympathetic nervous system promotes the release of hormones such as gastrin, cholecystokinin, and secretin which enhance digestion; the sympathetic nervous system has the opposite effect.

II. Mouth

Although the preparation and cooking of meals may initiate some alteration of complex foods, the digestive process truly begins when food enters the mouth.

A. Mastication

The mechanical chewing of food dramatically reduces the size of food particles, so that swallowing is then possible through the relatively narrow esophagus. Smaller pieces also have relatively more surface area on which digestive processes can work. Chewing is both a voluntary and an involuntary process; food in the mouth stimulates a chewing reflex. The presence of food in the mouth and its smell also trigger nervous reflexes which stimulate digestive function in anticipation of the arrival of food. For example, the vagus nerve increases gastric and pancreatic secretions and initiates gallbladder emptying.

B. Salivation

The salivary glands consist of three paired structures: the large parotid glands near the angle of the jaw, the submandibular glands under the middle of the jaw, and sublingual glands under the tongue. [*See* SALIVARY GLANDS AND SALIVA.]

Saliva has two important functions: digestion and protection. For digestion, saliva lubricates the food as it is chewed into smaller pieces, facilitating swallowing. Saliva also possesses enzymes which initiate the breakdown (i.e., hydrolysis) of starch and fat. As a protective secretion, saliva dissolves and washes away retained food particles, preventing dental caries, possesses some activity against bacteria, and physically buffers ingested substances which could be noxious (from hot beverages to acidic drinks). The alkalinity of saliva when swallowed also protects the lower esophagus from acid refluxing up from the stomach.

III. Esophagus

A. Anatomy

The esophagus in adults is a hollow tube about 25 cm long, extending from the pharynx in the lower part of the neck through the chest (i.e., thorax) and via an opening in the diaphragm into the abdomen for 2 cm before joining the stomach. In its passage from the thoracic cavity to the abdominal cavity, the esophagus moves from a low-pressure region (especially when one inhales) to a high-pressure area. To maintain this pressure differential, the lower esophageal sphincter acts as a one-way "gatekeeper," keeping corrosive gastric acid from rising up the esophagus from below, while allowing ingested food to pass unhindered from above. The esophagus, like the mouth, is lined with a stratified squamous epithelium, layers of cells like the skin but without the superficial coat of keratin (Color Plate 5).

B. Swallowing

Swallowing involves the mouth, pharynx, and esophagus. It begins when a food bolus is pushed backward by the tongue into the pharynx (a funnel-shaped tube between the mouth and the esophagus). This solid ball of food sets up nerve reflexes to the brain, which automatically protects the airways: The soft portion of the roof of the mouth (soft palate) moves upward to close off the communication between the nose and the mouth, the vocal cords in the larynx close together, and a rooflike structure (the epiglottis) swings backward to help seal off the entrance to the lungs (the tracheal tube). Breathing momentarily ceases. Movements of the tongue and the pharynx propel the food back and down to the esophagus.

As food enters the esophagus, involuntary but coordinated contraction waves spread down the esophagus, progessively narrowing the lumen in a propulsive manner. The lower esophageal sphincter relaxes to allow food to be pushed into the stomach. Normally, food passes from the pharynx to the stomach in 5–10 seconds.

IV. Stomach

A. Anatomy

The stomach is the most dilated region of an otherwise narrow digestive tract. It is connected above

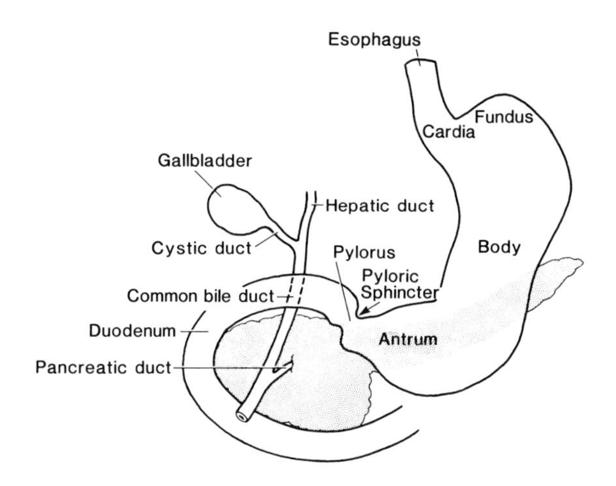

FIGURE 2 Anatomic relationships of the lower esophagus, stomach (including its regions), duodenum, pancreas (stippled gland), and biliary tract.

with the esophagus and ends in the first portion of the duodenum. Its anatomy has been divided into regions (see Fig. 2): (1) the cardia (surrounding the entrance of the esophagus), (2) the fundus (the dome of the stomach to the left and above the cardia), (3) the body (the major portion— the long stem of the J-shaped stomach, (4) the antrum (the more horizontal lower end from the right angulation to the pylorus), and (5) the pylorus (the distal, narrowed, tubular part, whose thickened muscular wall forms the pyloric sphincter).

The surface lining of the human stomach is composed of a simple columnar epithelium (simple meaning a single cell thick; columnar meaning long cells). Numerous pits on the surface lead into branched tubular glands which extend deep into the wall of the stomach (Color Plate 6). Functionally, there are two major gland areas: (1) an acid-secreting portion located in the upper 80% of the stomach and (2) the antrum, which produces the hormone gastrin. Gastrin is released into the bloodstream and circulates to stimulate the cells in the body of the stomach to secrete acid into the lumen of the stomach. Other surface cells produce a slimy protective coat of mucus.

B. Gastric Motility

The motor function of the stomach represents its major contribution to nutrition. Among mammals, its structure and function are tailored to the feeding habits of the animal. Ruminants (such as the cow) have several separate cavities; rodents, an upper stomach clearly defined into a fundus and body; and humans, a single chamber with a less apparent differences between parts.

Gastric motility consists of three functions: (1) accommodating and storing the large volumes of material ingested, (2) mixing this food with gastric secretions and grinding solids into small particles, and (3) emptying the partially digested gastric contents in a controlled manner into the duodenum.

Two anatomic divisions of the stomach coordinate these functions. The upper stomach (the fundus and the upper part of the body) is involved in storage. The antrum grinds particulate matter, while the pylorus sieves and retains solid masses until reduced to an appropriate size (less than about 1 mm in diameter).

1. Storage Function of the Stomach

When fasting, the cavity of the stomach is quite empty, except for small volumes of secretions. Though empty and supposedly at rest, periodic antral contractions occur and may be responsible for the "hunger pains" we normally experience. When food enters, distending the stomach, muscle tone relaxes, enabling the average stomach to accommodate 1 liter with little increase in intragastric pressure.

2. Mixing and Grinding of Food

Weak contractions in the body of the stomach gently mix and blend food with the gastric secretions. Antral contraction waves are propulsive and more vigorous, churning and grinding the solid content of food before the small particles are allowed to exit through the narrowed pylorus. The pyloric sphincter controls delivery of the gastric contents into the duodenum.

3. Gastric Emptying

Emptying occurs concurrently with mixing and grinding and can take several hours. Liquids are squeezed out by the fundus, which contracts down like a piston. Emptying into the duodenum quite rapidly, a liquid meal will leave the stomach in 90 minutes or less, depending on the total volume imbibed. Solids empty only after being ground down to particles less than 1–2 mm in diameter. Nondigestible solids, which cannot be broken down, are eliminated later—by contraction waves which sweep residual contents out of the stomach every 90 minutes—a true "housekeeper," cleaning the gastrointestinal tract during fasting.

C. Gastric Secretion

The important constituents of gastric juice are gastric acid in the form of hydrochloric acid (HCl), a protein-digesting enzyme (pepsinogen), and a factor essential to vitamin B_{12} absorption (intrinsic factor).

The human stomach normally contains about 1 billion specialized cells to produce acid at high concentrations (≈150 m*M*) and large volumes (1–2 liters ≠day). The resultant fluid has a pH of 1.0 during fasting, compared to blood at a pH of 7.4. Under basal or fasting conditions, acid secretion is less than 10% of maximum. Gastric acid secretion is stimulated by three substances which have separate receptors on the back surface (i.e., basolateral membrane) of the acid-secreting cell: (1) gastrin, the hormone synthesized in the antrum; (2) acetylcholine, a chemical substance liberated by nerve endings; and (3) histamine, from cells usually associated with inflammation, mast cells, or some other specialized cells containing histamine. With meals, gastric secretion is stimulated in a classical sequence.

(1) In the cephalic phase, the sight, smell, taste, or chewing of food initiates nervous impulses via the vagus nerve to release acetylcholine. This response can easily be "conditioned" psychologically.

(2) In the gastric phase, food in the stomach further increases gastric secretion through nerve reflexes stimulated by the mechanical distension of the stomach and from an increased release of gastrin into the bloodstream. Different cell types in the stomach wall sense distension, an increase in the pH (less acidity, with food-buffering HCl), and the presence of partially digested protein. This results in increased gastrin.

(3) In the intestinal phase, chyme in the upper small intestine also stimulates acid secretion. The intestine is also a site for powerful inhibitory mechanisms which suppress acid secretion in response to fat, hyperosmolar contents, or acid.

To prevent the stomach from digesting itself under the onslaught of acid and digestive enzymes, several defense mechanisms are available. Mucus is secreted by superficial epithelial cells throughout the stomach (Color Plate 5). This viscous gel adheres to the surface, providing an important lubricating and mechanical protective function. Bicarbonate is also secreted, creating an alkaline environment at the base of the mucus coat, acting as a neutralizing buffer to protect the underlining epithilial cells. Gastric acid is not only important in protein digestion, but also functions to reduce the entry of microorganisms ingested from the environment.

V. Small Intestine

Most digestion and virtually all absorption of the major dietary products take place in the small intestine, making it and its associated structures (i.e., the pancreas and the liver) the most important parts of the alimentary tract.

A. Anatomy

This 300-cm tube begins at the pylorus and ends at the ileocecal junction (Fig. 1). It is divided into three parts: the duodenum, the jejunum, and the ileum.

The duodenum, about 25 cm in length, forms a C-shaped curve, which cradles the head of the pancreas (Fig. 2). The pancreatic and bile ducts open into its central part. The remainder of the small intestine consists of the jejunum (arbitrarily, the upper two-fifths) and the ileum (the distal three-fifths). The jejunum is generally located in the upper left portion of the abdomen; the ileum is lower and to the right.

Three modifications dramatically increase the surface area exposed for absorption: circular folds involving the lining, mucosal villi, and absorptive cells microvilli. First are the transverse folds, which are quite obvious to the naked eye. Next, the mucosal villi, evident at a microscopic level, are numerous, 0.5- to 1-mm, fingerlike projections of the luminal surface (Color Plate 7). Each contains a small artery, vein, and lymph vessel and is covered by columnar epithelial cells. Millions of these villi are present throughout the small bowel, like the surface of a rug. Villi increase the total surface area for absorption to approximately 10 m^2. They are also capable of swaying movements, which stir the food lying adjacent to the intestinal mucosa, aiding absorption. Pitlike crypts encircle the villi around their bases (Fig. 3). The epithelial cells divide in the crypts and mature as they are pushed out of the crypts onto the sides of the villi, to be extruded within 2–3 days at the tips. A third feature increases the surface area of the small intestine 30-fold: multiple, tiny, brushlike projections of the luminal surface of each intestinal epithelial cell, forming mi-

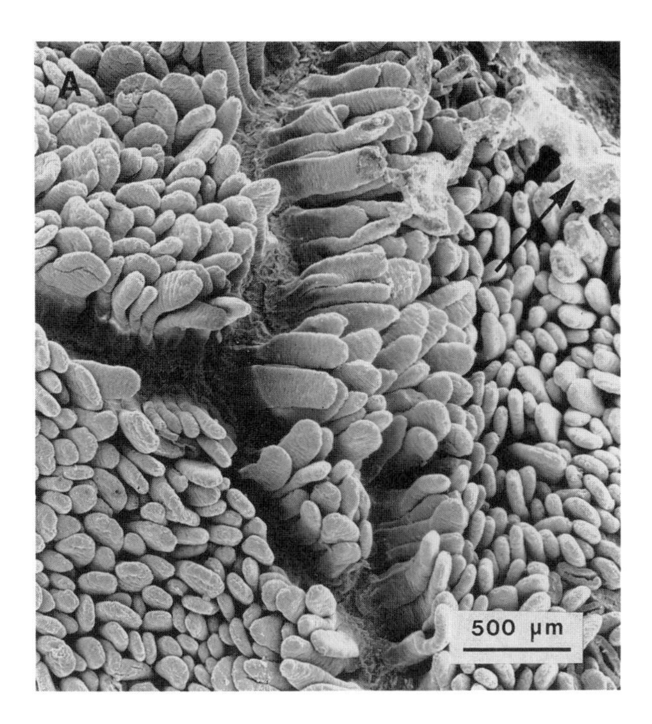

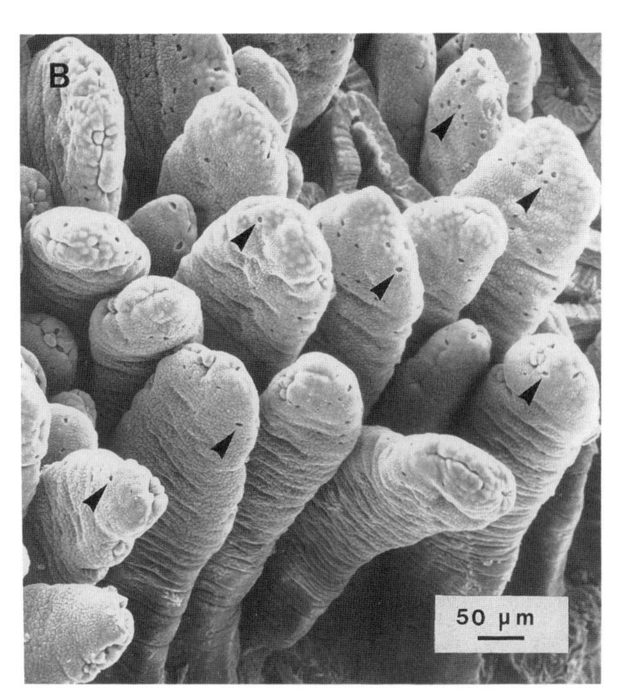

FIGURE 3 Scanning electron micrographs of the small-intestine mucosa, demonstrating the many frondlike villi protruding into the lumen. (A) At the top right corner (arrow) is a patch of mucus. (B) This higher magnification shows the tips of the villi. The tiny holes (arrowheads) are the openings to mucus-secreting cells. (Courtesy of André Buret.)

crovilli, termed the "brush border" membrane because of their appearance under a high-powered microscope. This microvillous membrane also contains digestive enzymes.

B. Motility

The motor action of the small intestine is organized to optimize the processes of digestion and absorption of nutrients. The small intestine receives a stream of chyme from the stomach. The chyme, which is partially digested solids and liquids, almost immediately meets secretions from the pancreas and the biliary tract in the duodenum. These endogenous (i.e., from the body) secretions contain important enzymes and solubilizers to complete the digestive process. Thorough mixing is necessary for chemical digestion and for absorption through contact with the mucosa. Failure of digestion and absorption within the small intestine allows the nutrients to pass into the colon, where bacterial metabolism produces osmotically active end products, gas, and certain fatty acids which can result in diarrhea. Finally, small-bowel contractions must move along any indigestible solids; if left, mechanical obstruction ensues.

Such mixing and movement of contents through the small intestine are controlled by an intrinsic activity of individual smooth muscles modified by the autonomic nervous system and certain gut hormones.

Three general sequences of contractions are recognized: (1) peristalsis, one or more ringed contractions of the circular muscle, propel the contents in an oral-to-anal direction; (2) segmentation, two or more isolated contraction rings, form short occluded segments, which chop the luminal contents into sausagelike portions; and (3) pendular movements, periodic contractions of longitudinal muscle, cause a to-and-fro motion of intestinal contents which contribute to mixing. Peristalsis moves intestinal contents along as if encircling one's fingers around a long thin tube of paste and pulling the tube and squeezing the paste forward. Segmentation and pendular movements provide mixing.

The gastrointestinal tract does not rest during fasting. About every 90 minutes in humans, electrical and motor activity migrates from the lower esophagus or antrum through the small intestine to the terminal ileum. This interdigestive contractile ring complex moves down the small bowel at 6–8 cm/min. Its periodicity is likely governed by the enteric nervous system plexus, but can be modified by other neural or hormonal factors. This motor complex is believed to act as a "housekeeper" during fasting, clearing the small intestine of residual food, secretions, and sloughed epithelial cells into the colon. During the active phase of these complexes, the epithelium switches from absorption to a secretory state, aiding this cleaning function. Ridding the small intestine of such debris prevents any ingested or resident bacteria from proliferating and overgrowing the region. Excessive bacteria in the upper small bowel impairs digestion and absorption.

VI. Exocrine Organs

Two gut-related organs provide important digestive juices to the duodenum. The pancreas and liver are termed exocrine organs because they produce an aqueous solution (containing pancreatic enzymes and bile, respectively) and deliver it via a conduit or ductal system directly to the lumen of the duodenum.

A. Pancreas

1. Anatomy

This large gland, 12–15 cm long, lies transversely across the upper abdomen, behind the stomach, with its head cradled to the right in the C-shaped sweep of the duodenum (Fig. 2). The body and the tail of the pancreas continue from the head and the neck to the left and upward.

The pancreas has a dual function, with two important glandular cell types. Endocrine cells (about 20%) produce important hormones, such as insulin, which circulate in the blood. Exocrine cells (80%) make digestive enzymes, which are secreted via a duct system into the duodenum.

2. Pancreatic Secretions

The pancreas secretes approximately 1 liter of fluid per day, consisting of two components: digestive enzymes and a high sodium bicarbonate fluid. The enzyme-rich juice originates in the exocrine (i.e., glandular) cells, while the bicarbonate component is produced by the duct cells. Two gut hormones regulate pancreatic secretion: cholecystokinin results in a concentrated enzyme output; secretin causes the duct cells to produce a large volume of bicarbonate-rich fluid. Secretin is released from the first few centimeters of the duodenum when the acidic chyme enters from the stomach. Cholecystokinin is released from the duodenum in response to the digested products of dietary fat (i.e., fatty acids) and proteins (i.e., amino acids). Both hormones enter the bloodstream and circulate to act at another site. Secretin got its name because it causes the pancreas to secrete. Cholecystokinin causes the gallbladder (which stores bile) to empty; its name means "to move bile." Other gut hormones and autonomic nerves also regulate pancreatic secretion.

B. Liver and Biliary System

1. Anatomy

The liver, the largest solid organ in the body, weighs 1500 g in adults. Its strategic location in the right upper quadrant of the abdomen (Fig. 1) accommodates a double blood supply: the portal vein brings blood from the intestine (commencing in the villi), and the hepatic artery conveys well-oxygenated blood from the aorta and the heart. Venous drainage is into the inferior vena cava. Bile, the yellow–green fluid produced by the liver, flows along a ductal system and exits by the common bile duct into the duodenum (Fig. 2). The gallbladder, a distensable sac, can accommodate up to 50 ml of bile.

2. Bile Secretion

The liver transports a variety of substances from blood to bile. The liver cell, or hepatocyte, has one surface bathed in plasma and another forming a groove with an adjacent liver cell. Bile is secreted into these tiny canals between liver cells, then empties into small ducts and eventually into the large hepatic ducts. During fasting, most bile enters the gallbladder to be stored, rather than exiting into the duodenum (Fig. 2). Eating causes neural (i.e., vagus nerve) and hormonal (i.e., primarily cholecystokinin) stimulation of the gallbladder smooth muscle, which contracts, expelling its contents. Cholecystokinin also causes the sphincter at the outlet of the common bile duct (i.e., Oddi's sphincter) to relax, facilitating bile output into the duodenum.

Bile, a complex solution of organic and inorganic components, provides the main excretory pathway for toxic metabolites, cholesterol, lipid waste products, and bilirubin (a pigment which is the breakdown product of hemoglobin in red blood cells). The liver is the most important site of cholesterol synthesis (the small bowel being the second most important) and the means by which this lipid is excreted from the body in bile. Bile is 90% water, yet cholesterol is water insoluble. The liver transforms cholesterol into a biological detergent: bile salts. Bile salts solubilize the lipids in bile and also aid in the digestion and the absorption of dietary fat in the upper small intestine. Thus, bile is a fluid "waste" from the liver, yet provides an important function for the assimilation of fat. [*See* BILE ACIDS.]

3. Other Functions

The liver also plays a central role in the handling of carbohydrates, fats, and proteins. It stores glucose as glycogen and mobilizes this sugar when necessary. The liver takes up and stores fat and cholesterol and also synthesizes lipids, all in a well-controlled balance. It also synthesizes most proteins that circulate in the plasma. The liver is also involved in the metabolism of drugs and chemicals, excreting some in bile and altering others into products which can be safely stored or eliminated from the body through the kidneys.

VII. Digestion and Absorption

Digestion is the chemical decomposition of foodstuff into simpler components to facilitate absorption. Although digestion of certain nutrients may begin in the mouth and the stomach, the process is completed in the upper small intestine. Complex molecules are hydrolyzed. Hydrolysis introduces a water molecule to split the bonds joining the two component molecules: Part of the water molecule goes with one component; part, with another. This process occurs with all three major foods—carbohydrates and proteins, which are water soluble, and fat, which is insoluble in water. The resultant products are smaller, more water soluble, and more readily absorbed.

Absorption is the transport across the mucosal lining of the gastrointestinal tract from the lumen into epithelial cells. For substances readily absorbed, transport is a simple matter of diffusion, passive movement down concentration gradients. For others, active transport requiring energy is necessary. The rate and the efficiency of absorption vary in different parts of the digestive tract. For most digested nutrients, this takes place in the upper jejunum, with some notable exceptions: ethyl alcohol (i.e., the drinkable kind) in the stomach, iron in the duodenum, and bile salts and vitamin B_{12} in the terminal ileum.

A. Carbohydrates

1. Digestion

Carbohydrates or saccharides (sugars) are organic compounds consisting of carbon, hydrogen, and oxygen. The ratio of hydrogen to oxygen atoms is always 2 : 1, as can be seen in the formulas for glucose ($C_6H_{12}O_6$) and sucrose ($C_{12}H_{22}O_{11}$). Carbohydrates can be divided into three major groups: (1) monosaccharides, the single units (e.g., glucose); (2) disaccharides, consisting of two joined monosaccharides (e.g., sucrose); and (3) polysaccharides, containing very long chains (eight or more) of monosaccharides (e.g., starch and cellulose) and have molecular weights of several million. Unlike the simpler sugars, polysaccharides are not sweet. Glucose, the most important component, fuels our energy requirements which are "glucose dependent." [*See* CARBOHYDRATES (NUTRITION).]

Our daily intake of carbohydrates is about 350–450 g, consisting of 50% starch, 30% sucrose, and less than 10% lactose (milk sugar). Starch, a complex polymer of many simple sugars, must be hydrolyzed to smaller simpler units before absorption can occur. Both salivary and especially pancreatic amylase attack dietary starch, eventually producing disaccharides. Further hydrolysis depends on disaccharidases, specific enzymes bound to the surface (microvillous) membrane of the intestinal epithelial cell. The resultant single sugar, usually glucose, can then be absorbed.

2. Transport

Active transport occurs against a concentration gradient and requires energy. The active transport of glucose is indirect, being linked to sodium uptake. Glucose and sodium move into the cell together. Once inside, sodium is actively pumped out the other side of the cell into the interstitial space. This basal surface of the epithelial cell possesses an energy-requiring sodium pump. Pumping Na^+ out the bottom of the cell into the high Na^+ concentration of the interstitial fluid is active transport, which maintains a low intracellular sodium concentration. The low concentration of sodium in the cell relative to that in the lumen encourages further uptake of sodium (and with it, glucose). Meanwhile, the entry of glucose raises its intracellular concentration, favoring its passive movement out of the base of the epithelial cell toward the blood. Other sugars, including some glucose, traverse the luminal cell membrane by passive diffusion. This simple transport process is controlled primarily by physicochemical laws, the difference in concentration gradients across cell membranes. The driving force is the random movements of the molecules, which are dependent on the body temperature.

The absorptive process appears not to be an easy one, yet, for sugars, the capacity of the small intestine is enormous. The equivalent of more than 10 kg of table sugar (i.e., sucrose), equivalent to 40,000 kcal, can be absorbed daily. Ironically, sugar substitutes, such as sorbitol, are not well absorbed and, if consumed in excess, reach the colon, where they act as osmotically active particles to retain water and cause diarrhea.

B. Proteins

These complex organic compounds contain carbon, hydrogen, oxygen, and nitrogen. Amino acids are the building units of proteins, linked head to tail by peptide bonds; the carbon end (i.e., the carboxy terminus) of each amino acid is joined to the nitrogen (i.e., the amino terminus) of another amino acid. Such polymers are called polypeptide chains, with molecular weights of 4000 to 1 million or more. [*See* PROTEINS (NUTRITION).]

1. Digestion

Hydrolysis breaks the peptide bond between amino acids. This begins in the lumen of the stomach. The stomach secretes pepsinogen, which is activated in the presence of acid to pepsin. Pepsin cleaves amino acids from large protein molecules, resulting in smaller combinations of amino acids (i.e., polypeptides). Gastric acid also denatures protein. The pancreas, in response to cholecystokinin, secretes proteolytic enzymes, which further reduce the size of the polypeptides inside the lumen of the upper small intestine. Enzymes within the microvillous membrane of the intestinal epithelial cells complete the job, producing free amino acids. Membrane digestion and absorption are closely related.

2. Absorption

Several different systems exist for amino acid absorption. Some require sodium, much like the active transport of glucose; others are transferred by simple diffusion. A small percentage of short-chain peptides (with two or three amino acids) are absorbed intact. Some are further hydrolyzed inside the cell by cellular enzymes; the rest, a small percentage of peptides, enter the blood intact. In some people, these represent a potential source of allergic reactions.

Dietary protein is efficiently absorbed, less than 1% being eliminated through the colon each day. About 60% is absorbed in the jejunum.

C. Fats

Fats or lipids are water-insoluble organic substances. The Western diet contains upward of 80–160 g of fat, 95% of which is triglycerides. This represents half of the daily calorie intake. Triglycerides are also the major storage forms of fat in animal and plant cells. Triglycerides, like carbohydrates, consist of carbon, hydrogen, and oxygen in two components: a glycerol molecule and three fatty acid molecules. Each fatty acid is attached to a separate carbon atom of the glycerol molecule. By definition, fat is hydrophobic, or water hating. This presents a dilemma. The luminal contents of the gastrointestinal tract and its secretions are predominantly in a water medium. Further, the primary enzyme that breaks down these bulky lipids, pancreatic lipase, acts only at surfaces. Like a salad dressing of oil (i.e., fat) and vinegar (i.e., water) left standing, the oil rises to the top, while the water remains a separate phase at the bottom. The surface area between these two phases is relatively small compared to the droplets which can be produced by shaking. This dispersion of fine oil droplets through a water phase produces an emulsion. Thus, within the lumen of the small intestine, the dilemma is resolved by emulsification. Inside the body, any lipid must also be compatible with biological fluids such as blood, which is primarily water. [*See* FATS AND OILS (NUTRITION).]

1. Digestion

The digestion of lipids involves progressive hydrolysis of the bonds linking the three fatty acids to the glycerol backbone of triglycerides. This is carried out by lipase enzymes. Several are present in human breast milk, saliva, the stomach, and, most importantly, pancreatic juice. Lipase from the mouth and (for infants) in breast milk initiate lipid digestion in the stomach. This is aided by the churning action of the antrum, which disperses fat as an emulsion of six triglycerides. As fat is retained in the stomach for up to several hours, significant (10–40%) hydrolysis can occur. Lipolysis in the stomach is critical to the newborn, whose diet is primarily milk, yet it has an immature pancreatic secretory function.

The duodenum is the major site of fat digestion in adults. There is a well-orchestrated entry of bile from the biliary system to further assist in the dispersion of fat into fine oil particles as a suspension and pancreatic juice containing enzymes and bicarbonate. Remember that fatty acids release cholecystokinin from the duodenal mucosa, and this hormone causes the gallbladder to expel its bilious contents of detergents and emulsifiers into the duodenum. Cholecystokinin also stimulates the pancreas to secrete pancreatic lipase. Pancreatic lipase acts at the oil droplet surface, splitting off one fatty acid at a time from the triglycerides, leaving a di- or monoglyceride (with two or one fatty acid still attached, respectively). The products of lipid digestion—fatty acids, monoglycerides, and glycerol—are carried from the surface of the oil droplet by a detergent mixture of bile salts. Bile salts themselves are not absorbed in the upper jejunum, but seem to shuttle back and forth, transferring the lipid digestive products from the oil droplet toward the mucosal surface. Eventually, bile salts are actively absorbed farther on, in the terminal ileum.

2. Absorption

The first barrier through which these lipid products must pass is a fine unstirred water layer adjacent to the epithelial cell lining. Passage across the cell membrane is then comparatively easy. Absorption is passive. For some lipids, the whole process, however, is relatively inefficient: Less than half of ingested cholesterol is absorbed. The inner fluid of the cell (i.e., cytosol) is mainly water, requiring some carrier or mechanism to get these water-hating compounds from the plasma membrane through the cell.

Once inside the intestinal epithelial cell, the components are resynthesized into triglycerides. Covered with a protein coat and some cholesterol, the resulting chylomicrons (i.e., emulsified fat) are assembled and then expelled into the space between the epithelial cells to be transported in lymph. The relatively large chylomicrons, about 1 μm in diameter, cause the lymph (and, in some individuals, the blood) to become milky. With minor changes, chylomicrons soon reach the liver for further processing.

VIII. Fluids and Electrolytes

Over 9–10 liters of fluid enters the gastrointestinal tract each day. We imbibe about 2 or more liters, and another 7–8 liters enters from internal secretions: 1 liter as saliva, 2–3 liters from gastric secretions, 0.5 liter as bile, 1–2 liters from pancreatic juice, plus a smaller contribution by the small intestine. Thus, large volumes of fluid must be absorbed, or we would soon become dehydrated. The absorptive capacity of the small intestine is quite significant, up to and perhaps exceeding 12 liters per day. Of the 1–2 liters presented to the large intestine each day, all but 100–200 ml is absorbed, representing a 90% efficiency.

The gut thus absorbs most of the salt and water that enters. There is no active transport process for water: fluid movement occurs secondary to the transport of either an electrolyte (mainly sodium) or a nonelectrolyte (e.g., glucose). Both active and passive transport mechanisms are responsible for solute movement in the intestine, although their relative contributions differ in the jejunum, ileum, and colon. Passive diffusion involves the movement of solutes from an area of high concentration (i.e., overcrowding) to one of lower concentration. To maintain osmotic equilibrium, water accompanies solute transport. The jejunum, whose epithelium is relatively leaky, has a low resistance to the passage of water and electrolytes and readily allows passive transfer. Active transport, being energy dependent and one way, requires an epithelium more resistant to movement in order to retain the transported particles. In general terms, absorption of sodium and water occurs passively in the jejunum, whereas, in the ileum, the active transport of sodium and glucose draws in water. At a microscopic level, the villi predominantly absorb, whereas the crypts secrete. At any moment, net transport is a balance between these two opposing movements of electrolytes and fluid.

Other important electrolytes, such as potassium, bicarbonate, chloride, and hydrogen, also have one or more transport mechanisms. Iron is primarily absorbed in the duodenum; calcium, in the jejunum.

IX. Colon and Rectum

A. Anatomy

The large intestine is continuous with the ileum (Fig. 1). It begins with the cecum in the lower right quadrant of the abdomen. The appendix, about 10 cm in length, protrudes from the back surface of the cecum. This wormlike tube ends blindly.

The colon consists of the following parts: the ascending colon, which runs from the cecum on the lower righthand side of the abdomen up to below the liver to become the transverse colon; the transverse colon, which crosses the upper abdomen beneath the rib cage to the left upper quadrant; here, it makes another acute turn to become the descending colon, which passes down the left side to enter the pelvis, where, as the sigmoid colon, it is joined to the rectum. The rectum is not completely straight, despite its name, but has a gentle concavity, sloping forward in front of the sacrum and coccyx (i.e., tailbone). The anal canal is the termination of the digestive tract and is surrounded by sphincter muscles. The entire length of the large intestine in adults is about 150 cm in length. Its diameter, though somewhat larger than that of the small intestine, diminishes from the cecum to the anus. The wall of the large intestine consists of layers similar to that of the small intestine, but the arrangement of columnar cells lining the lumen differs (Color Plate 8). There are no true villi in the colon. Instead, the absorptive surface is flat, with many straight tubular crypts.

B. Motility

The colon contains smooth muscle throughout its length, except for the external anal sphincter, which is composed of skeletal (i.e., voluntary) muscle. The gross pattern of flow and motility through the colon is not completely understood. "Mass movement" is a sudden translocation of significant fecal mass over a long segment of the colon, reaching one-third to one-half of the colonic length. Such shifts do not occur regularly, perhaps only a few times per day, and are fleeting, lasting a few seconds. They certainly have a temporal relationship to eating and defecation. The flow of luminal contents through the colon is not steadily progressive, unlike that in the small intestine. Nonpropulsive segmental contractions may move luminal contents in both directions, forward and back. Solid contents in general move much more slowly than does liquid or gas.

Continence is maintained by the internal and external anal sphincters. Large amounts of material in the rectum can initiate reflex relaxation of the tonically contracted internal sphincter. Once this happens, material passes spontaneously into the anal canal. Volitional contractions of the external sphincter may prevent defecation, but once the urge to stool occurs, this skeletal muscle relaxes and elimination follows.

C. Absorption

The liquid residue which remains once the chyme has passed through the small intestine enters the cecum. This residue consists of undigested matter plus fluid taken in with food or secreted as digestive juices, but not resorbed by the small intestine. The colon is presented with about 1 liter of fluid from the ileum each day, but releases less than 200 ml. The absorption of water which accompanies sodium and chloride transport takes place continuously as the residue passes through the large intestine—a journey of 12–14 hours. Some bacterial hydrolysis of colonic matter produces nutrients, such as short-chain fatty acids (with less than 12 carbon atoms); when absorbed, these can provide energy. As water is absorbed, the material in the colon becomes increasingly more solid. The solid (though normally soft) which collects in the rectum is the feces, or stool. No nutrients are absorbed in the colon. Hence, sodium absorption is not linked to that of glucose or other nonelectrolytes, as in the small intestine. The colon must rely on active sodium transport and a small element of passive absorption. Its normal absorptive capacity is 4 liters per day. The net result is feces, consisting of water and solid materials—undigested food residue, microorganisms, and mucus.

X. Conclusions

The digestive system constitutes a simple, yet elegant, winding tube about 5 m in length with specialized regions extending from the mouth to the anus. The food we eat enters from the external environment in a rough and complex form. Propelled and chopped up by the digestive system, this food is mixed with digestive juices, which convert complex molecules to simpler forms. These small digestive products are absorbed into the portal venous blood and lymph for subsequent distribution as vital nutrients to the cells of the body. Unabsorbed residues and other waste products are moved to the end of the tract and are eliminated from the body. The total time from ingestion to elimination is normally 40–60 hours. Nutrient digestion and absorption take 3–6 hours, a relatively short time for processes essential to life.

Bibliography

Aries, I. M., Jakoby, W. B., Popper, H., Schachter, D., and Shafritz, D. A. (1988). "The Liver: Biology and Pathobiology," 2nd ed. Raven, New York.

Davenport, H. W. (1988). "Digest of Digestion." Year Book, Chicago.

Davison, J. S., ed. (1989). "Gastrointestinal Secretion." Wright, London.

Johnson, L. R., ed. (1987). "Physiology of the Gastrointestinal Tract," 2nd ed. Raven, New York.

Digital Dermatoglyphics

ROBERT J. MEIER, *Indiana University*

Glossary

Admixture Evolutionary process by which individuals from one population mate with those from another and thereby transfer genes between groups
Friction skin Epidermal skin surface of the palms and the soles that bears ridge and furrow systems
Mendelian inheritance Inheritance that is governed by a single genetic locus
Microevolution Genetic change that occurs as a result of one or more of the following processes: mutation, random genetic drift, admixture, and natural selection
Natural selection Evolutionary process based on differential survival and/or differential reproduction among individuals who possess variable traits
Polygenic inheritance Inheritance that is dependent on the interaction of several genes
Sexual dimorphism Observable and measurable differences between males and females

DERMATOGLYPHICS is the study of the epidermal ridges as found on the skin surface of the palms, fingers, soles, and toes. This study encompasses the embryological formation of ridges, their minute structure, and their pattern configurations. It can deal at any one of several levels, from individual digits or separate hands to comparisons between the sexes, among different human populations, or even among different primate species. A primary interest is to learn about the development and variation of dermatoglyphic features both in normal persons and in known cases of birth defects. Each of these topics is discussed below.

I. Development of Dermal Ridges

A. Volar Pads

During the first trimester of gestation, at approximately the sixth week, volar pads appear on the fingertips, as well as on other areas of the palm and the foot (see Fig. 1). Volar pads consist of mesenchymal tissue, forming raised moundlike structures which are thought to directly influence the patterning of epidermal ridges. Volar pads are most prominent in the third to fourth fetal months before regressing, and it is during this time that ridges are beginning to form on the pads.

B. Primary and Secondary Ridge Formation

Primary ridge formation begins at the dermal or basal layer of the skin, initially as undulating folds surrounding paired rows of dermal papillae, from which newly formed skin cells arise (see Fig. 2). These folds underlie the primary ridges, and through them pass the ducts from sweat glands that exit on the top of the ridge as a pore. There are also numerous nerve endings exposed along the ridge. Adjacent to the primary ridges are secondary ridges, which are actually furrows extending deeper into the dermis. Commencing about the sixth fetal month the ridge-and-furrow system is visible on the skin surface. It is after this time that no change will occur in the epidermal ridge patterns themselves; they will simply increase in size in response to the growing fetus. These same patterns will persist as well throughout the remainder of a person's lifetime.

Encyclopedia of Human Biology, Volume 3.

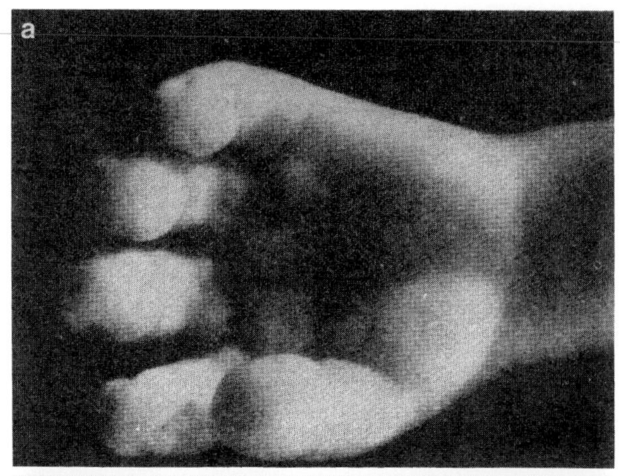

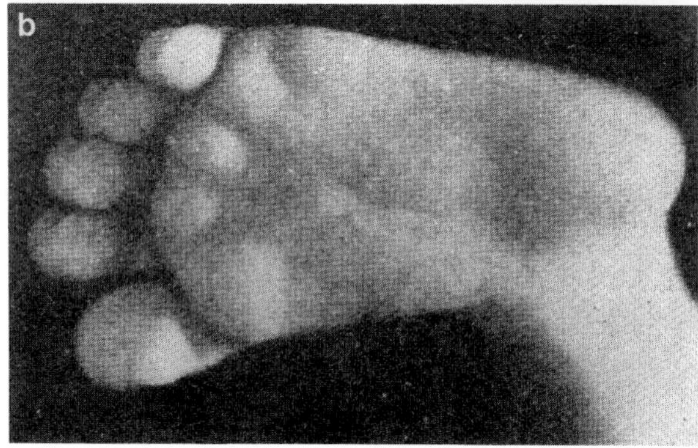

FIGURE 1 Volar pads (a) on the hand of a 10-week human fetus and (b) on the foot of a fetus about 2 weeks older. [From H. Cummins and C. Midlo, "Finger Prints, Palms and Soles. An Introduction to Dermatoglyphics," p. 179. Dover, New York, 1961.]

C. Functions of Epidermal Ridge Systems

Given the rather unique appearance of the epidermal ridged skin surface, there is good reason to expect that it performs some important functions. Indeed, several have been proposed, and all would seem probable to a degree. First, and perhaps foremost, the ridge-and-furrow system serves to retard slipping in a manner similar to automobile tire treads. Accordingly, "friction skin" is sometimes used to denote the antislipping nature of epidermal ridges. Second, since sweat pores open only on the primary ridges and not in the furrows, it is thought that a small amount of sweat bathing the ridges might improve their level of friction and further aid against slipping. Then, since nerve endings are extensive along the ridges, the ridges themselves might well serve to enhance the sense of touch or to improve tactile stimulation. Sir Francis Galton, a pioneer in dermatoglyphic study, performed an experiment which showed that blindfolded people moved their fingertips around in circles when asked to identify objects, as if to actively engage the nerve endings. Finally, there is the possibility that the ridge-and-furrow system provides some added structural stability analogous to corrugated cardboard or metal.

D. Evolutionary and Hereditary Considerations

These many functions probably point to an evolutionary history within primates, all of whom possess "fingerprints," that placed a premium on the ability to carry out secure grasping behaviors, to perceive valuable information about the environment through the sense of touch, and perhaps to withstand extensive use of the volar surfaces in moving about and feeding in trees. In short, primates, in part, adapted to their arboreal niche by means of epidermal ridges and dermatoglyphic patterns. We humans simply continue to reflect some of that ancestral adaptive response, since it is rather unlikely that our current survival strongly depends on functioning epidermal ridge patterns.

Various models of inheritance have been proposed to account for the expression of dermatoglyphic features. Some of these provide evidence of mendelian inheritance or single gene control for the pattern configurations, while others treat the development of the ridges under a polygenic mode of inheritance subject to a certain amount of environmental interaction. Very likely, genic control is effected at various levels, in the formation and regression of the volar pads, as well as in the expression and patterning of the primary and secondary ridges.

II. Pattern Configurations and Classification

A. Pattern Landmarks: Triradius and Core

The two main features found in a pattern are its triradius and core. Figure 3 depicts a pattern on which these landmarks are indicated. The triradius is a place on the pattern from which the ridge system extends along in three directions. The details of the triradius might vary somewhat, but the resultant three radiants are generally easily observed. The core is the centermost ridge or portion of ridge

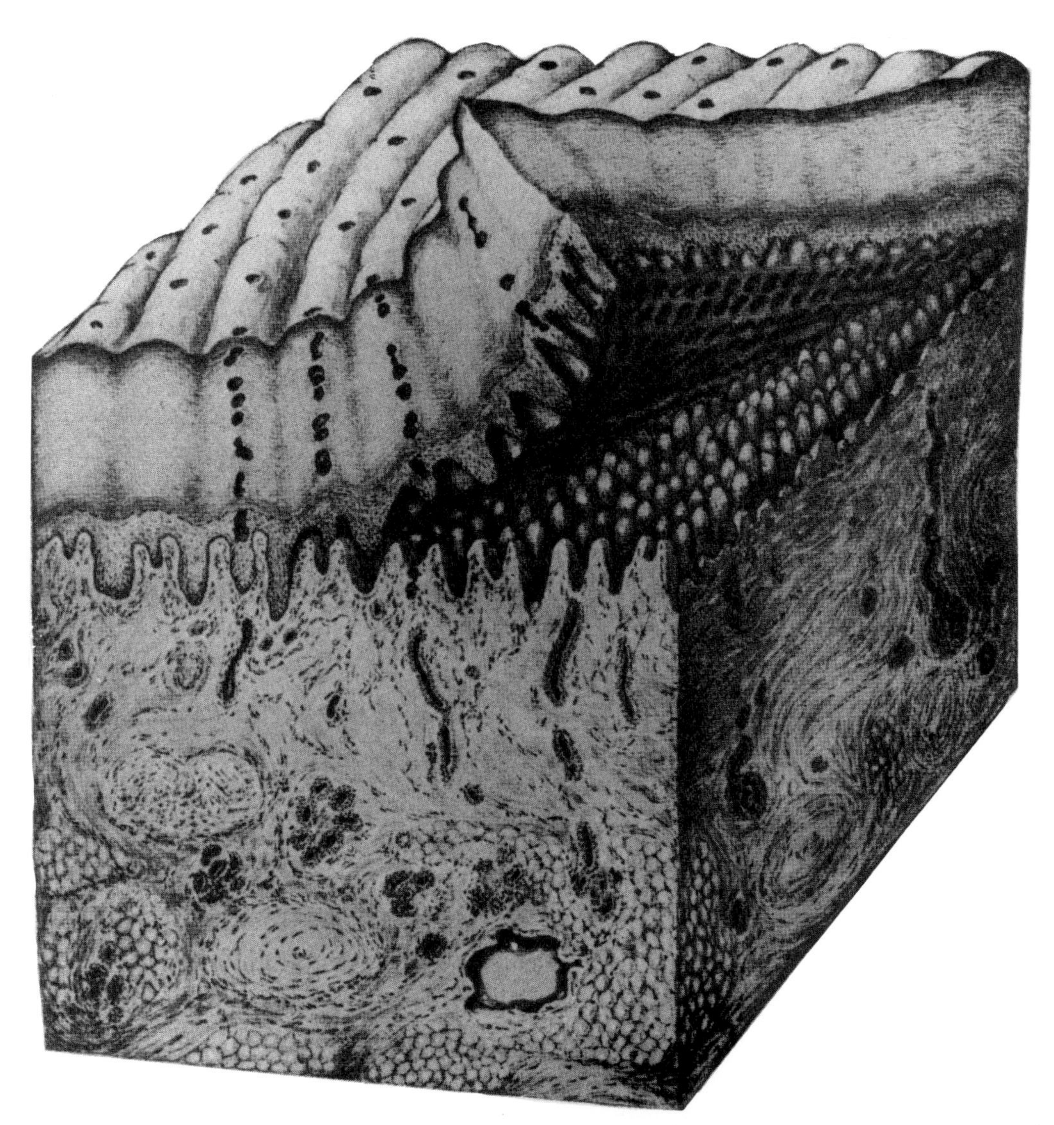

FIGURE 2 The histology of ridged skin. The epidermis is partially lifted away from the dermis to show the underlying dermal papillae. [From H. Cummins and C. Midlo, "Finger Prints, Palms and Soles. An Introduction to Dermatoglyphics," p. 38. Dover, New York, 1961.]

within the pattern. Again, there is variation in this feature, but it is likewise readily determined.

B. Ridge Counting

Many studies utilize the triradius and the core in analyzing fingerprints, initially to classify them into patterns and then to count the number of ridges that

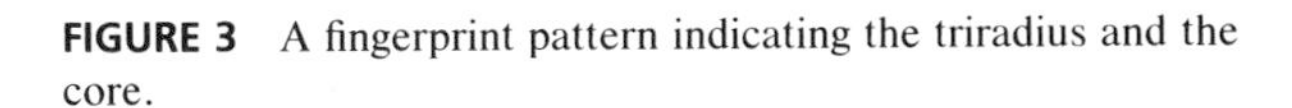

FIGURE 3 A fingerprint pattern indicating the triradius and the core.

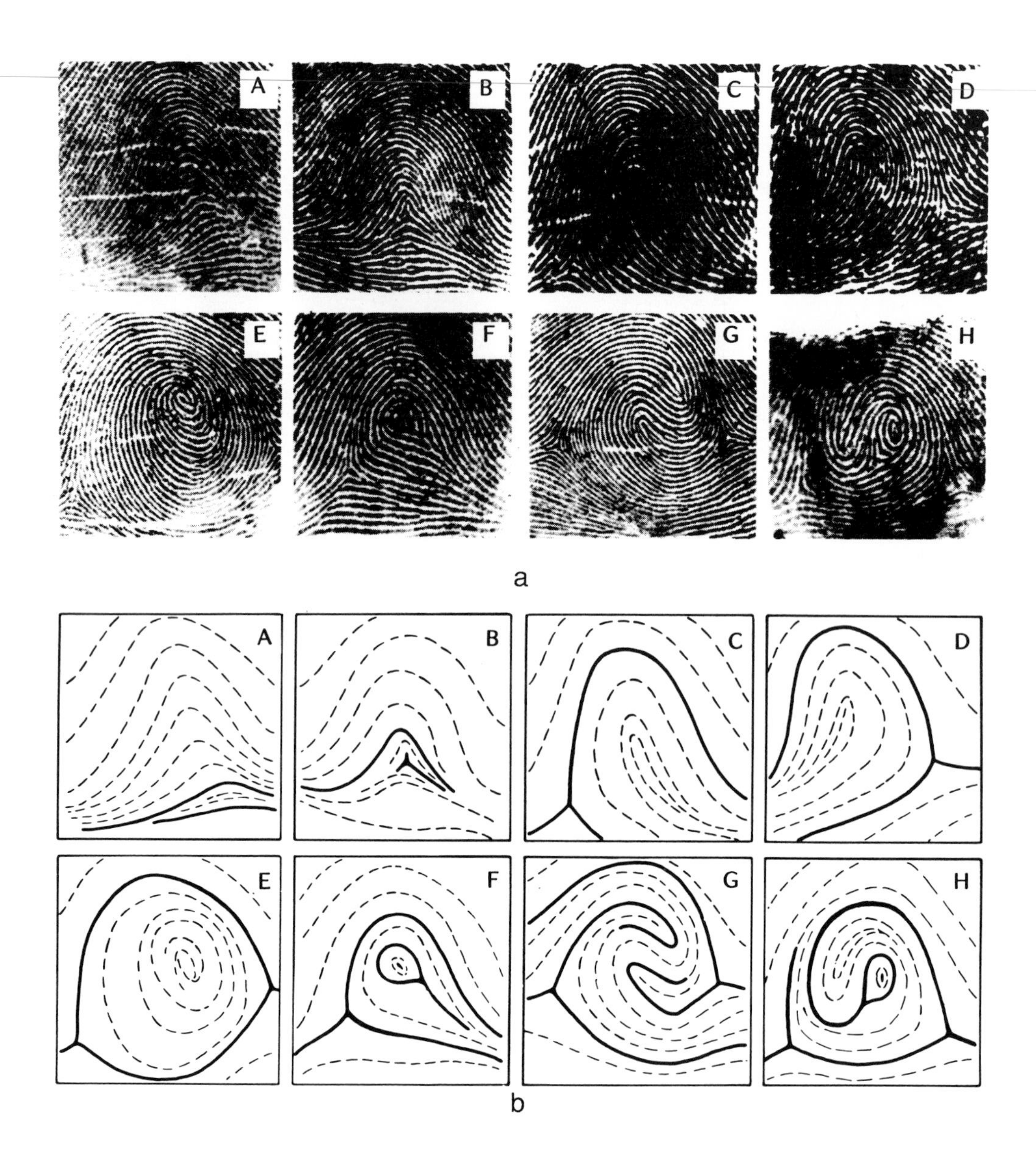

FIGURE 4 Digital pattern types: actual prints (a) with (A) simple arch, (B) tented arch, (C and D) loops, (E) simple whorl, (F and G) variant whorls, (H) accidental pattern. (b) The "type ridges" of each of the patterns. [From M. Alter, Dermatoglyphic analysis as a diagnostic tool. *Medicine* **46,** 35–56, © by Williams & Wilkins, 1966]

make up the patterns. Ridge-counting is done usually with the aid of a magnifier (to 10× at least) and a pointed marker. Ridges are counted between the core and the triradius, but these two landmarks are not included in the count.

C. Arch–Loop–Whorl Classification

Ridge systems form recognizable patterns that can be classified according to the number of triradii and ridge counts they possess. The most basic classification consists of three types—namely, arches, loops, and whorls (see Fig. 4). Arches are the simplest patterns, generally not having a triradius or a core, and hence having a zero ridge count. Loops usually have one triradius, a single core, and one variable ridge count. Loops can be further classified as ulnar if the ridges on the side opposite the triradius open to the ulnar side of the hand (i.e., toward the little finger) or radial if the ridges open to the radial side (i.e., toward the thumb). Whorls have usually two triradii and one or two cores, with two variable ridge counts. There are many subtypes of whorls, which justifies their being considered the

TABLE I Correlations between Total Finger Ridge Count (TFRC) and Pattern Intensity Index (PII)[a]

Population	TFRC No.	TFRC Mean	TFRC SD	PII Mean	PII SD	Correlation[b]
Males						
Point Hope	51	157.7	52.6	16.0	2.8	0.60
Anaktuvuk Pass	35	133.8	40.5	13.5	2.8	0.84
Barter Island	23	134.0	42.0	13.5	2.9	0.74
Barrow	100	133.0	45.9	13.3	3.2	0.77
Wainwright	113	115.9	45.8	12.6	2.8	0.74
Females						
Point Hope	66	136.5	58.6	14.4	3.8	0.69
Anaktuvuk Pass	37	124.8	38.0	13.1	2.7	0.86
Barter Island	16	120.4	52.4	13.0	3.0	0.83
Barrow	118	129.9	47.2	13.4	3.3	0.81
Wainwright	122	104.6	41.9	11.7	3.0	0.81

[a] Modified from R. J. Meier, Dermatoglyphic variation. *In* "Eskimos of Northwestern Alaska. A Biological Perspective" (P. L. Jamison, S. L. Zegura, and F. A. Milan, eds.), p. 83. Dowden, Hutchinson & Ross, Stroudsburg, Pennsylvania, 1978.
[b] All correlations are significant beyond 0.001.

most complex of patterns. Ridge counts might range from one to two ridges in small loops to more than 25 ridges in large whorl patterns. A total finger ridge count (TFRC) is calculated as the sum of the ridges across all fingers, and only the larger of the two counts is used in the case of whorl patterns. An alternative is to record the ridge count of each digit separately in terms of both an ulnar and a radial count. Hence, each person would be represented by a 20-count matrix or array. Another calculated variable is the pattern intensity index (PII), which is the sum of the triradii for all digits. This can range from zero (10 arches) to 20 (10 whorls). There is a strong correlation between TFRC and PII (see Table I).

D. Unusual Dermatoglyphic Features

On the digits, sometimes a pattern configuration is observed that consists of a combination of two patterns, such as a loop and a whorl. These "accidentals" occur in less than 1% of individuals, and then most likely on the index finger (see Fig. 4). Another rare condition involves an unusual appearance of ridges composed of a series of short segments referred to as a "string of pearls". They are more generally classified as ridge dissociations. Frequently, it is possible to see pattern configurations so large that the triradius is not present and is said to be "extralimital," or beyond the ridged-skin surface.

III. Population Comparisons

A. Distribution in Human Groups

It is of considerable interest that dermatoglyphic features express so much variability. There is, of course, the forensic and legal acceptance of the uniqueness of each individual's fingerprints. That is, no two persons have exactly the same fingerprints, and this even applies to identical twins (see Table II). These unique differences actually are found in the detailed makeup of the ridges. Ridges are segmented to form fragments of varying length; they often bifurcate to make forklike structures or rejoin in forming enclosures (see Fig. 5). It is these details (i.e., minutiae) that uniquely identify the individual, but this identification is generally aided by means of pattern configurations.

1. By Digit

Pattern types do not occur randomly across the fingers (see Table III). To illustrate, the thumb (digit I) and the ring finger (digit IV) of each hand tend to have the highest frequency of whorl patterns, while the little finger (digit V) and the middle finger (digit III) bear most of the ulnar loops. The index finger (digit II), has nearly all of the radial loops, and arches are more likely to occur on the index, middle, and little fingers. There might be developmental factors that underlie this differential distribution, which appears to be firmly based, since these ten-

TABLE II Pattern Types (PT) and Ridge Counts (RC) in a Set of Monozygotic (Identical) Twins[a]

Left hand		Digit I	II	III	IV	V
Twin						
PT	A[b]	W	RL	UL	UL	UL
	B[c]	UL	RL	UL	UL	UL
RC	A	21–4	3	17	16	12
	B	22	4	18	18	17
Right hand						
PT	A	UL	UL	UL	UL	UL
	B	UL	UL	UL	UL	UL
RC	A	22	20	16	13	10
	B	23	16	16	12	13

[a] Code, Digit I, Thumb to Digit V, Little Finger. PT, Pattern Type; W, Whorl; UL, Ulnar Loop; RL, Radial Loop; RC, Ridge Count [note that there are two ridge counts for the whorl pattern on the left thumb (digit I) of twin A, while all of the loop patterns have a single ridge count].
[b] TFRC in twin A, 150; PII in twin A, 11.
[c] TFRC in twin B, 159; PII in twin B, 10.

FIGURE 5 Ridge details, or Galton's minutiae. (From H. Cummins and C. Midlo, "Finger Prints, Palms and Soles. An Introduction to Dermatoglyphics," p. 31. Dover, New York, 1961.)

dencies occur throughout nearly all human populations.

2. By Hand

There are also left–right differences in the frequencies of pattern types, usually involving whorls versus arches. Accordingly, whorls are more often found on the right-hand digits, while arches appear more frequently on the left side. Studies of bilateral asymmetry between homologous digits have been done to test notions of canalization of growth. There has also been work in associating fingertip patterns and hand preference, revealing inconsistent findings for right- versus left-handers.

TABLE III Percentages of Frequency of Digital Patterns in 5000 People from the Scotland Yard Files[a]

		Pattern type			
Digit	Side	Whorls	Ulnar loops	Radial loops	Arches
I	R	41.43	55.89	0.22	2.47
	L	29.39	65.90	0.20	4.51
	R+L	35.41	60.89	0.21	3.49
II	R	30.79	32.30	26.03	10.87
	L	28.15	38.10	23.37	10.36
	R+L	29.47	35.20	24.70	10.62
III	R	16.56	74.81	2.53	6.09
	L	16.18	73.32	2.51	7.98
	R+L	16.37	74.07	2.52	7.03
IV	R	41.07	55.61	1.47	1.85
	L	27.82	68.92	0.50	2.75
	R+L	34.44	62.27	0.98	2.30
V	R	13.80	85.46	0.20	0.54
	L	9.03	89.79	0.02	1.17
	R+L	11.41	87.62	0.11	0.85
All	R	28.74	60.83	6.08	4.36
	L	22.12	67.21	5.31	5.35
	R+L	25.42	64.02	5.69	4.85

[a] R, Right; L, left. Modified from H. Cummins and C. Midlo, "Finger Prints, Palms and Soles. An Introduction to Dermatoglyphics," p. 67. Dover, New York, 1961.

3. By Sex

Sexual dimorphism with respect to fingertip pattern frequencies has received considerable attention. The general tendencies show that females have more arches and ulnar loops, whereas males have more whorls and radial loops. Again, there could be minor developmental differences, such as in the timing of embryonic growth and differentiation in males versus females, that might account for the tendencies. Because of the sexual dimorphism in pattern configurations, males generally have higher TFRC and PII values than females, due to their larger percentage of whorls and fewer arches (see Table I).

4. By Population

Strong research interest in population variation has likewise characterized dermatoglyphic study. Indeed, much of the earlier work provided descriptive and comparative analyses of human groups from around the world. While it is not possible to summarize all studies of this nature without obscuring some of the important findings, it might be of interest to bring out once more some general tendencies with regard to the major pattern type distributions (see Table IV). Accordingly, arch patterns have their highest frequencies in sub-Saharan African groups (particularly the Pygmy), while Asian groups show the lowest percentages, and European populations fall in between. With respect to whorls, Asians generally have the highest frequency (along with Pacific Island groups and Australian Aborigines), Africans have the lowest frequency, and Europeans occupy the middle position. Last, Europeans usually have more radial loops on their digits than any of the other populations.

Given the partial genetic control over the expression of dermatoglyphic features, it is to be expected that a certain amount of historical interpopulation differentiation would come about as a result of microevolution, perhaps more through random processes than predominantly through natural selection. Population contacts and any resultant admixture or interbreeding between the groups also can be expected to influence the distribution of digital pattern types and ridge counts. For example, Table V shows the likely consequence of European admixture in Easter Islanders and in Northwestern Alaskan natives by lowering the PII, which is mainly the result of whorl pattern reduction.

A large amount of current research in digital dermatoglyphics is devoted to examining both micro-

TABLE IV The Percentage Ranges and Means of Fingertip Pattern Frequencies in Different Populations[a]

	Fingertip pattern type								
	Whorls			Ulnar L.		Radial L.		Arches	
Population	N[b]	Range	Mean	Range	Mean	Range	Mean	Range	Mean
Caucasians	112	26–49	35.4	50–66	55.6	4–7	4.3	2–9	4.3
Negroes	88	15–42	27.4	53–74	61.4	1–4	2.6	2–18	8.8
Amerindians	76	32–58	42.6	37–58	49.4	1–5	3.1	1–9	5.0
Orientals	55	36–56	46.7	39–57	48.1	1–4	3.0	1–5	1.8
Australasians	60	33–78	52.7	22–64	44.9	0–3	1.1	0–4	1.4
Asian Indians	7	33–56	42.6	33–59	51.8	1–4	2.2	1–7	3.4

[a] Modified from B. Schaumann and M. Alter, "Dermatoglyphics in Medical Disorders," p. 83. Springer-Verlag, New York, 1976; and based on C. Plato, "Variation and Distribution of the Dermatoglyphic Features in Different Populations." Penrose Memorial Symposium, Berlin, 1973.
[b] Number of population samples summarized in table.

evolution and population structure. Dermatoglyphic variables derived from ridge-counting, especially the 20-count array, have been found to be highly suitable for carrying out multivariate analyses. For example, Table VI shows an application of principal component analysis which has reduced the 20 counts to only four components that adequately characterize the nature of dermatoglyphic development according to either the radial or ulnar side of particular sets of digits. Hence, component 1 for males combines the ulnar counts of digits II, III, IV, and V. For females, all five digits fall within this first component. Component 1 obviously relates to a developmental correspondence among ulnar counts. Likewise, the other three components sort out radial counts according to certain digits. For example, component 4 represents the radial count on digit V for both sexes. Very close agreement can be seen between the sexes for the components. This isn't always the case, and in addition, comparable studies have shown population variation in the digits and counts that make up the components. These differences could be reflecting aspects of growth and development along with particular evolutionary histories that characterize human populations.

TABLE V Pattern Intensity Index in Admixed and Unadmixed Groups from Easter Island and Alaska, by Sex

Group	Unadmixed	Admixed
Easter Island		
Males	15.8	14.9
Females	15.0	14.6
Point Hope		
Males	16.0	14.1
Females	14.4	12.0

B. Nonhuman Primate Studies

Although nonhuman primates (e.g., prosimians, monkeys, and apes) have not been studied thoroughly in terms of their dermatoglyphics, enough research has been done to clearly demonstrate a significance in dermatoglyphic variation comparable to that found in humans. However, it appears to be a lesser amount of variation that distinguishes nonhuman primates from humans. Descriptive series have been collected for both New and Old World species, and there is a tentative observation that, with respect to their digital dermatoglyphics, nonhuman primates show less intraindividual variation than is found in humans, with reduced bilateral differences as well as a lesser degree of sexual dimorphism. Additionally, nonhuman primates show

TABLE VI Principal Components Based on a Digital 20-Count Analysis of Northwestern Alaskan Inupiat (Eskimos)[a]

	Component			
Group	1	2	3	4
Males	Ulnar II, III IV, V	Radial II, III IV	Ulnar/radial I	Radial V
Females	Ulnar I, II, III IV, V	Radial II, III IV	Radial I	Radial V

[a] Left and right sides showed identical components.

more uniform expression of the whorl-type pattern across all digits.

At the species and population levels basic comparisons have been made and applied to questions of phylogenetic relationships and taxonomy, with the microevolutionary mechanisms of population differentiation, and with discerning the functional and adaptive significance of dermatoglyphic features. Areas that have received little attention so far, but would be most appropriately examined in nonhuman primates, include investigations of inheritance patterns, the role of the environment in dermatoglyphic expression, and the details surrounding the morphogenesis of volar pads and primary and secondary ridges. Volar pads do persist as prominent structures in some nonhuman primates, which might account for their tendency to express more complex whorl-type patterns. These patterns, in turn, might represent a morphological adaptation involving extensive grasping function during locomotion and other manipulatory activities.

IV. Medical–Clinical Applications

During the prenatal period, when dermatoglyphic features are being formed, there can be disturbances that originate from genetic and/or environmental sources and alter the course of development. An understanding of this interaction assists in applying dermatoglyphic analysis to two medical situations. First, there is the potential of using dermatoglyphics in screening newborns for known birth defects. Furthermore, an analysis of dermatoglyphics of newborns could also be useful in detecting abnormal morphogenesis in general, and thus alert the clinician to closely examine for developmental defects. This application has been moderately successful, as, for example, in profiling Down syndrome patients and patients with aberrant numbers of chromosomes, such as Trisomy 18. Importantly, it is not only the digital dermatoglyphic features that are informative. Very often the palmar and plantar dermatoglyphics are similarly diagnostic and should always be included in the analysis.

The second application involves attempts to discover associations between a given birth defect or medical disorder and unusual dermatoglyphic findings. If such an association is found, this would signify that the defect, or a predisposition toward developing the defect, had possibly originated as early as the end of the first trimester, when dermatoglyphic features are under formation. This kind of study requires an adequate control sample in order to specify just how unusual the patient's dermatoglyphics are, for the differences are rarely any more than statistical peculiarities of already known and described traits. A fairly common finding for digits is that affected cases have an increased frequency of arches or loops and a corresponding reduction in TFRC and PII. Associations have been found between dermatoglyphics and numerous congenital malformations, leukemia, cytomegalic inclusion disease, and rubella embryopathy. Environmental agents having teratogenic effects are also known to alter dermatoglyphic development.

Acknowledgment

Comments from Blanka Schaumann were greatly appreciated.

Bibliography

Cummins, H., and Midlo, C. (1961). "Finger Prints, Palms and Soles. An Introduction to Dermatoglyphics." Dover, New York.

Jantz, R. L. (1987). Anthropological dermatoglyphic research. *Annu. Rev. Anthropol.* **16,** 161–177.

Loesch, D. Z. (1983). "Quantitative Dermatoglyphics. Classification, Genetics, and Pathology." Oxford Univ. Press, Oxford.

Mavalwala, J. (1977). "Dermatoglyphics: An International Bibliography." Mouton, The Hague.

Meier, R. J. (1975). Dermatoglyphics of Easter Islanders analyzed by pattern type, admixture effect, and ridge count variation. *Am. J. Phys. Anthropol.* **42,** 269–276.

Meier, R. J. (1978). Dermatoglyphics variation. *In* "Eskimos of Northwestern Alaska. A Biological Perspective" (P. L. Jamison, S. L. Zegura, and F. Milan, eds.). Dowden, Hutchinson & Ross, Stroudsburg, Pennsylvania.

Meier, R. J. (1980). Anthropological dermatoglyphics: A review. *Yearbook of Phys. Anthropol.* **23,** 147–178.

Meier, R. J. (1981). Sequential developmental components of digital dermatoglyphics. *Hum. Biol.* **53,** 557–573.

Schaumann, B. A., and Alter, M. (1976). "Dermatoglyphics in Medical Disorders." Springer-Verlag, New York.

Wertelecki, W., and Plato, C. C. (1979). "Dermatoglyphics—Fifty Years Later." Alan R. Liss, Inc., New York.

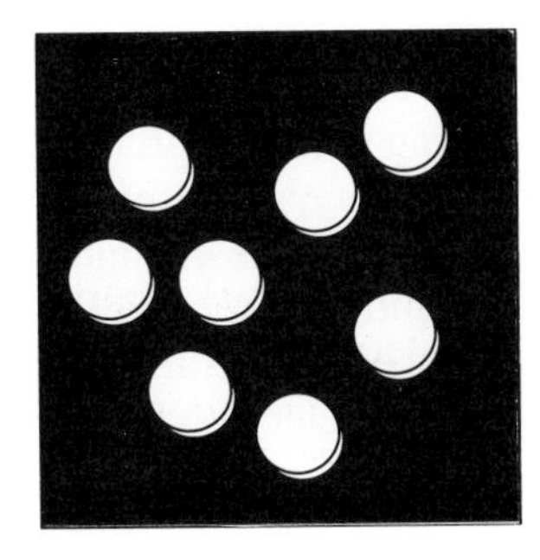

Diuretics

EDWARD J. CAFRUNY, *University of Medicine & Dentistry of New Jersey*

Glossary

Adverse reaction Undesirable or toxic reaction induced by a drug

Edema Surplus of extracellular fluid. The fluid may be confined to a part or spread throughout the body (systemic)

Hypokalemia, hypochloremia, hyponatremia Abnormally low plasma levels of K^+, Cl^-, or Na^+, respectively

Lumen (pl. lumina or lumens) Hollow space or cavity within a duct or tubule. Urine flows through the lumens of the kidney

Natriuretic Substance that increases the renal excretion of sodium salts and fluid

Nephron Functional unit of a vertebrate kidney comprising a glomerulus and attached tubule. More than 1 million nephrons are in each human kidney

Side effect Effect of a drug on an extraneous organ system (i.e., a system the therapist does not intend to influence). Side effects are not necessarily adverse

Tubular reabsorption Movement of any substance from the tubular lumen to the surrounding capillary network

Tubular secretion Movement of any substance from the surrounding capillary network to the tubular lumen

Visceral epithelium Single layer of flattened cells covering glomerular capillaries

Water diuretic Osmotic agent that increases urine flow but has little effect on the excretion of sodium salts

DIURETICS are drugs that act directly on the kidneys to increase the rate of excretion of urine and the sodium salts dissolved in urine. Because the direct source of the fluid and salt excreted is the circulating plasma, blood volume decreases. This leads to the formation of a gradient that favors the passive movement of interstitial fluid (extracellular fluid surrounding cells) into the bloodstream. The volume of the plasma reexpands as the interstitial fluid compartment contracts.

In past years physicians used diuretics primarily to eliminate the disturbing surplus of extracellular fluid and salt (edema) that accumulates in patients with such diseases as chronic congestive heart failure, severe liver damage, or renal insufficiency; use of the drugs has been extended in recent years to ailments in which edema is not present or is not a conspicuous feature. Table I lists some of the common indications for dispensing diuretics. Physicians write more than 100 million prescriptions annually in the United States alone, a large fraction of the total for the treatment of high blood pressure.

I. Physiology of Renal Salt and Water Excretion

Animals that live on land regulate the volume and composition of their body fluids with precision to avoid sizable deviations that can interfere with the normal activity of cells. Because intake is variable, output also must vary. The vertebrate kidney is the primary organ for controlling output, although other organs assist in important ways. The kidney maintains salt and water balance by adjusting the volume and composition of the urine it makes. The renal output of water and dissolved solutes, added to output via other routes (e.g., lung, skin), then matches dietary intake.

TABLE I Some Indications for Diuretic Therapy

Condition	Effect of diuretics	Diuretic class
Congestive heart failure	Mobilize and excrete ECF[a]	All natriuretics except CAI[b]
Cirrhosis of the liver	Mobilize and excrete ECF	All natriuretics except CAI
Hypertension	Lower blood pressure	All natriuretics except CAI
Acute pulmonary edema	Improve lung and heart function	Loop agents
Impending renal failure	Arrest progression	Osmotic or loop agents
Glaucoma	Lower intraocular pressure	CAI only
Renal calcium stones	Decrease urinary calcium levels	Thiazides only
Elevated blood calcium	Increase urinary calcium excretion	Loop diuretics
Chemical intoxication	Increase urine flow to eliminate toxin	Loop diuretics
Hypokalemia	Increase blood potassium	Potassium-sparing diuretics

[a] ECF, extracellular fluid.
[b] CAI, carbonic anhydrase inhibitors.

The complexity of this task is immense. Table II lists the compartments that sequester body fluids. Average total body water content of a 70-kg adult is about 42 liters, 60% of the body weight. More than one-half of this volume is virtually imprisoned within healthy cells, kept there by the osmotic forces of dissolved ions and molecules. Active transport systems control ionic movements into and out of many types of cells (discussed below). Intracellular water comprises about 55% of the total fluid content of the body. Bone and the ducts and lumens of organs contain an additional 10%. These compartments, containing 65% of the total body water, resist rapid changes in volume. The kidney accomplishes its task by communicating with the circulating plasma, the most accessible component of the extracellular fluid compartment. Physico-chemical and hormonal messages transmitted between the kidney and plasma compartments invoke the renal mechanisms that regulate fluid and electrolyte balance.

TABLE II Distribution of Water in an Adult Weighing 70 kg

Compartment	Liters	% of total body water
Extracellular H_2O		
Circulating Plasma	4	9
Interstitial	11	26
Total	15	35
Intracellular H_2O	23	55
Bone and transcellular H_2O	4+	10
Total body H_2O	42+	100

Figure 1 is a diagram of the structure of a single nephron, the basic unit of vertebrate kidneys. It shows the path that fluid filtered out of the bloodstream follows before leaving the body as urine. Some of the tubular segments of the diagram are convoluted; the parts are not drawn to scale. More than a million nephrons are packed in a single human kidney. The first part of each unit, the glomeru-

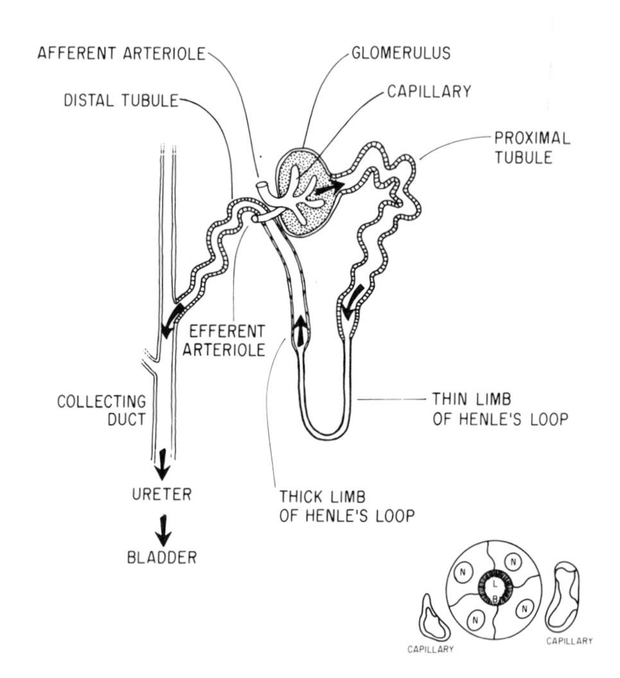

FIGURE 1 Diagram of the structure of a mammalian nephron. The afferent arteriole divides to form a capillary network within the glomerulus. An ultrafiltrate of capillary plasma begins its journey through the tubular lumens (*top arrow*), passing progressively through the proximal tubule, loop of Henle, and distal tubule. Incipient urine then enters the collecting duct and finally exits through the ureter before reaching the bladder. Glomerular capillaries merge and subsequently exit as the efferent arteriole. This arteriole then splits into a tubular capillary network (not shown in the large diagram). The cross section below illustrates the relation between a proximal tubule and the tubular capillary network. L, lumen; B, brush border of proximal cells; N, nucleus of cell. Luminal fluid and solutes traverse the cell to return to the blood via capillaries.

lus, comprises a central region of supporting cells and a capillary network through which blood flows. An outer capsule encloses these structures (Fig. 1). A fraction of plasma water and low molecular weight solutes passes into the tubular lumens. This process is called *glomerular filtration*. The large sizes and shapes of protein molecules hinder their passage through the walls of the glomerular capillaries. Contractions of the heart transmit the energy needed to maintain the filtration process, the first step in the formation of urine. Because glomeruli filter about 180 liters every 24 hr (an amount 45 times greater than the plasma volume), almost all the fluid and solutes, other than waste products, must traverse the tubular cells and subsequently diffuse into tubular capillaries. The rest, usually 1.0–1.5 liters/24 hr, is urine. [*See* URINARY SYSTEM, ANATOMY.]

The crucial second step, tubular reabsorption, is energized by active transport systems in the cellular membranes of transporting segments.

Primary active transport (PAT) describes the movement of an ion or molecule between two compartments in a direction opposite to that predicted from its existing electrochemical gradient, which depends on the salt concentrations in the two compartments. Because the flux is "uphill," from a compartment of low concentration to one of high concentration, the transport is active (i.e., requires expenditure of energy). The sodium pump of the proximal tubule of Fig. 2 illustrates this type of active transport. The energy required to drive sodium out of the cell comes from the enzymatic breakdown of adenosine triphosphate (ATP) in the basolateral membrane. Potassium simultaneously enters the cells through the same pump but exits again because its concentration within the cell is higher than in the surrounding interstitial space. Sodium enters cells passively from the lumen because the pump ejects it so rapidly into the interstitial space that cellular concentration of the ion is always less than luminal concentration. In the illustration, a negatively charged chloride or bicarbonate anion enters passively, thus preventing the electrical charge separation that would supervene if only positively charged sodium ions moved out of the cell. Water follows the ions. The sodium pump is present in all transporting cells of the renal tubule. [*See* ADENOSINE TRIPHOSPHATE (ATP); ION PUMPS.]

Secondary active transport (SAT) is "uphill" movement of an ion across a membrane, together with an ionic partner traveling "downhill." A PAT

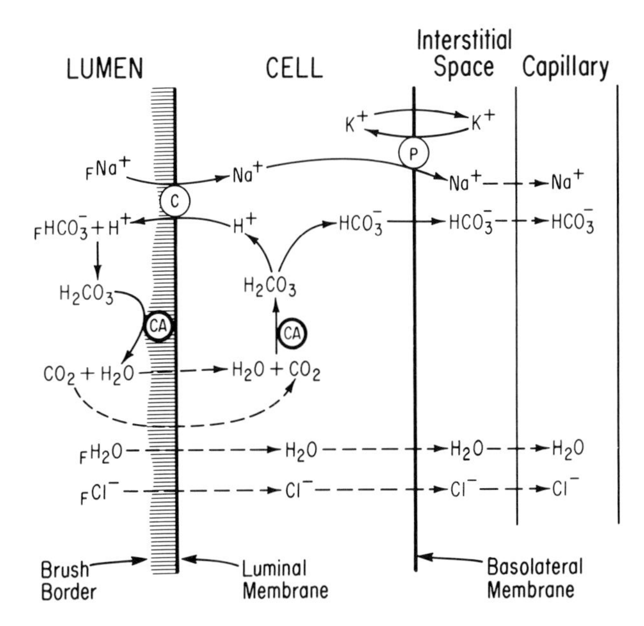

FIGURE 2 Diagram of ion transport systems in the proximal tubule. Filtered sodium (FNa^+) moves into the cell as H^+ moves out. Ions are counter-transported across the membrane (C). Na^+ and K^+ are pumped across the basolateral membrane (P), the first entering the capillary, the second moving back into the cell. Within the cell, carbonic anhydrase (CA) catalyzes the hydration of CO_2 to form H_2CO_3, the source of H^+. Both HCO_3^- and Cl^- can follow Na^+ into the capillary. CA in the brush border catalyzes the dehydration of H_2CO_3 and the products of this reaction enter the cell to reform H_2CO_3. See text.

system drives the partner across a second membrane. The Na-K-2Cl cotransport mechanism located in the thick limb of Henle's loop (see Fig. 1) is an example of SAT. Na^+ and K^+ in the lumens of this tubular segment serve as ionic partners, permitting Cl^- to enter the cell against an opposing electrochemical gradient. The ions are carried by a symporter, a transporter protein that helps an ion (in this instance, Cl^-) pass through a membrane with other ions. PAT across the basolateral membrane (the second membrane) establishes a favorable gradient for the admission of luminal Na^+ into the cell.

The proportions of water and sodium reabsorbed in each segment of the renal tubule are listed in Table III. Values listed are averages for healthy adults. These segments are potential sites of action for diuretics; the efficacy of a given diuretic will depend, in part at least, on the location and reabsorptive capacity of the part(s) of the tubule on which it acts. Note that a 1.0% reduction in tubular reabsorption of water will double the urine flow rate.

TABLE III Reabsorption of Filtered Water and Sodium[a]

Tubular segment	H_2O reabsorbed	Na^+ reabsorbed
	(listed as a percentage of the amount filtered)	
Proximal tubule	70.0	70.0
Loop of Henle	15.0	22.0
Distal tubule and collecting duct	14.0	7.5
Total	99.0	99.5

[a] Glomerular filtration rate = 180 liters/24 hr; urine volume = 1.8 liters/24 hr; Na^+ filtered = 580 g/24 hr; Na^+ excreted = 3 g/24 hr.

II. Classification and Primary Renal Actions of Diuretics

The chemical structures and mechanisms of action of diuretic drugs vary considerably. For this reason, the classification is not rigid or systematic. Table IV illustrates this point. Thiazides are classified according to chemical structure; loop diuretics by site of action in the kidney; the rest by the primary action they exert. Although each group includes many drugs, the table lists prototypes only.

Except for the osmotic group and spironolactone, all the diuretics prescribed in the United States are organic acids or bases. The glomeruli filter and the proximal tubular cells secrete these organic electrolytes. The secretory systems require the expenditure of metabolic energy. Proximal tubular secretion of diuretics enhances both the rate of delivery to and the concentrations achieved in the kidney.

TABLE IV Classification and Primary Action of Diuretics

Class	Primary action	Examples[a]
Carbonic anhydrase inhibitors	Prevent reabsorption of $NaHCO_3$ in proximal and late distal tubules	Diamox
Thiazides and thiazide-related	Inhibit reabsorption of NaCl in early part of the distal tubules	Hydrodiuril Hygroton
Loop agents	Inhibit Na-K-2Cl cotransport in loops of Henle	Lasix Edecrin
Potassium-sparing	Block receptors for Na^+-K^+ exchange in cells of late distal tubules	Aldactone
	Retard the entry of luminal Na^+ into late distal cells	Dyrenium Midamor
Osmotic	Hold fluid in lumens of all tubular segments	Mannitol (USP)

[a] Examples listed are brand names of the most frequently used preparations. Generic names appear in the text.

III. Pharmacology and Therapeutics

A. Carbonic Anhydrase Inhibitors

Carbonic anhydrase (CA) is a zinc-containing enzyme that catalyses the reversible reaction, $CO_2 + H_2O = HCO_3^- + H^+$. The pH of the medium in which the reaction occurs establishes the prevailing direction of the reaction: a rise of pH increases the hydration of CO_2 to form HCO_3^- (bicarbonate), whereas a fall increases the release of CO_2. In the proximal tubule, filtered sodium ions enter the cells passively, exchanging for hydrogen ions that replace luminal sodium. The mechanism of exchange, countertransport, is a form of cotransport and involves a membranal transporter. Figure 2 illustrates the process. Because cellular pH rises when H^+ leaves the cells, CA within the cell accelerates the formation of bicarbonate. The bicarbonate anion then crosses the basolateral membrane with Na^+ and diffuses into surrounding capillaries. CA in brush border membranes, where extruded H^+ ions acidify the environment, converts filtered bicarbonate to CO_2 and H_2O. The intracellular enzyme effectively replaces the filtered bicarbonate ions that are lost by this conversion. The upshot of this seemingly frivolous process is impressive. Replication keeps the plasma concentration of bicarbonate, an important acid-base buffer, from falling after the ion is filtered and subsequently converted to CO_2 and H_2O. In addition, a significant fraction of filtered sodium, accompanied by H_2O and cellular bicarbonate, undergoes reabsorption when H^+ exchanges for Na^+ (Fig. 2).

Drugs that inhibit CA impede the countertransport process in the proximal tubule, and probably in the distal segment as well, by limiting the supply of H^+. The unreabsorbed fraction of Na^+ inevitably expands. Larger quantities are excreted with HCO_3^-. Urine flow rate increases because the ions hold water in what may be described as an "osmotic embrace."

Figure 3 shows the chemical structure of acetazolamide (Diamox), the prototypical CA inhibitor. It is a sulfonamide relative of antibacterial sulfa drugs. Its free sulfamyl group ($-SO_2NH_2$), how-

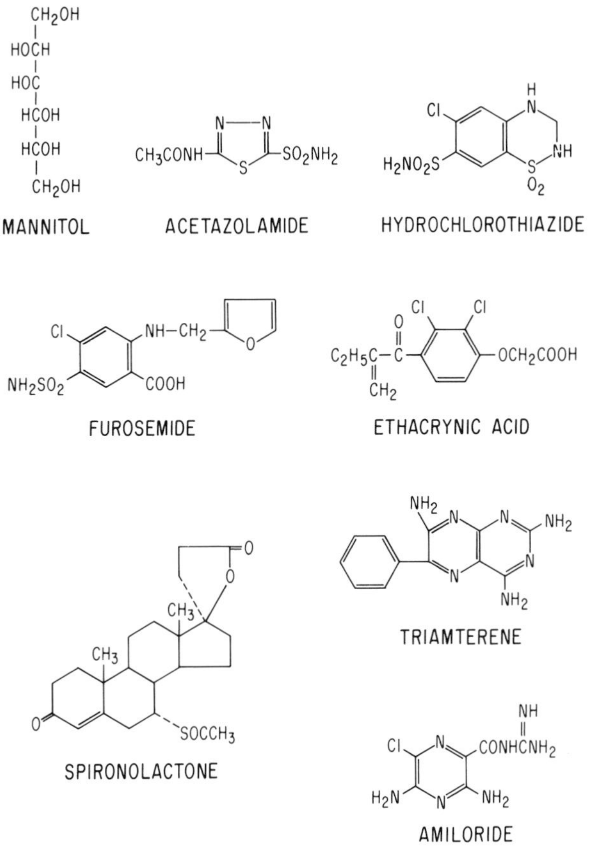

MANNITOL ACETAZOLAMIDE HYDROCHLOROTHIAZIDE FUROSEMIDE ETHACRYNIC ACID TRIAMTERENE SPIRONOLACTONE AMILORIDE

FIGURE 3 Chemical structures of the diuretics discussed in the text.

ever, bestows CA inhibitory activity. Acetazolamide is an effective diuretic but is not often prescribed for several reasons: (1) tolerance to its renal action develops within 3–4 days so that it becomes less effective; (2) acidosis (increased acidity of plasma) occurs when excessive excretion causes plasma bicarbonate levels to fall; and (3) newer diuretics that do not deplete plasma bicarbonate are available. A more frequent indication for CA inhibitors is the treatment of some types of glaucoma (elevated intraocular pressure). Inhibition of CA present in ocular tissues suppresses the production of ocular fluid and thus lowers intraocular pressure. Additional therapeutic indications include the prevention of epileptiform seizures and high-altitude sickness. Acetazolamide alkalinizes the urine, and elevated urinary pH accelerates the excretion of some, but not all, acidic drugs or toxins.

Adverse reactions to CA inhibitors occur less frequently when the dosage is not excessive and drug administration is intermittent or is interrupted periodically. Acidosis and potassium depletion are inevitable consequences of overuse. Depression, malaise, and gastrointestinal intolerance occur less often. Like other sulfonamides, CA inhibitors may rarely elicit signs of hypersensitivity: rash, drug fever, or disordered blood-cell formation.

B. Thiazides and Related Drugs

Successors to CA inhibitors, thiazides, appeared in 1960 and soon became the diuretics of choice because they were continuously effective in the treatment of edema (tolerance did not develop). Use of diuretics increased enormously when clinical researchers reported that thiazides lowered the blood pressure of hypertensive patients without provoking the unpleasant side effects of the poor drugs of the time that were, if anything, only marginally effective in controlling hypertension. Physicians soon discovered that they could reduce the dosage of antihypertensives that were likely to cause side reactions simply by prescribing a thiazide concurrently. Combination therapy not only maintained or improved the response but also reduced the incidence and intensity of adverse reactions.

Chlorothiazide, the progenitor of the thiazide series, promoted the urinary excretion of both bicarbonate and chloride salts. It contained a benzothiadiazide ring flanked by a sulfamyl group ($-SO_2NH_2$), the moiety that imparts CA inhibitory activity. The simple addition of two hydrogen atoms (for the synthesis of hydrochlorothiazide) had an important impact. Hydrochlorothiazide in clinically effective doses increased chloride excretion but had little effect on bicarbonate excretion. The addition of hydrogen atoms had weakened the ability to inhibit the enzyme. The chemical structure of hydrochlorothiazide, the most popular member of the series, appears in Fig. 3. The sulfamyl group is not replaceable but the benzothiadiazide ring structure may be altered without loss of diuretic activity. Absence of significant effect of hydrochlorothiazide and newer members of the thiazide group on bicarbonate excretion solved the problems of acidosis and refractoriness.

The primary site of action of thiazides is the early portion of the distal tubule where they interfere with the electroneutral passage of NaCl across the luminal membrane, retarding entry of the salt into the cells. If the major site of action were in the proximal tubule, a large part of the effect probably would be annulled by a compensatory rise in the activity of the Na-K-2Cl cotransport system in the loop of Henle. Because the entire distal tubule and collecting duct system reabsorbs only about 7–8% of filtered sodium (Table III), an action in the distal segment cannot be expected to be as pronounced as a similar action in the loop of Henle, which controls the reabsorption of as much as 25% of filtered NaCl. However, it is not often desirable or necessary to reduce the reabsorption of sodium ore than 2–3%. A change of this magnitude will usually double or triple the flow of urine.

All the useful thiazides share the same mechanism of action and are equally effective. Differences with respect to oral absorption, distribution and metabolic fate in the body, and rapidity of excretion ("pharmacokinetic properties") account for obvious differences in potency and duration of action. As a general rule, potency (i.e., the dose required to achieve a given response) is not important because efficacy (i.e., the maximal response achievable) is the same for all. [*See* PHARMACOKINETICS.]

Most of the complications of thiazide therapy are manifestations of their pharmacologic effects on the kidney. Depletion of potassium is the chief problem, but other ions also may be depleted. The plasma levels of uric acid and of calcium may rise when the extracellular fluid volume declines. Elevated serum uric acid can cause gouty attacks in susceptible people. Several studies have shown that serum cholesterol may increase during a period of 2–4 months when patients take thiazides daily. The elevation is small, 5–6%, and is not evident in several clinical trials that lasted a year or more. Thiazides also lessen sensitivity to insulin, an effect that prompts readjustment of the latter drug's dosage in diabetics. The clinical relevance of this finding in nondiabetics is not established at this time. [*See* CHOLESTEROL.]

Before 1960, the only drugs that could substantially increase the renal elimination of sodium were the mercurial diuretics (not orally effective) and the CA inhibitors (not continuously effective). The arrival of thiazides resolved these two problems. Similarly, there were no safe and effective drugs for lowering blood pressure. Thiazides filled this void and soon became first-line agents in the pharmacotherapy of hypertension. Although newer drugs can lower pressure more effectively or evoke diuresis in patients who are refractory to the older drugs, thiazides continue to be mainstays. In the United States alone, more than 15 million patients use them.

C. Loop Diuretics

The diuretic response to thiazides is usually hindered or inadequate when the glomerular filtration rate is depressed, heart failure is severe, or the liver is damaged. In such cases, the therapist uses an inhibitor of the Na-K-2Cl cotransport, the SAT system that controls the entry of monovalent ions into cells of the ascending limbs of Henle's loop. These inhibitors are often called "high-ceiling diuretics" because they can induce the renal elimination of enormous quantities of salt and water. Cells of the loop reabsorb 20–25% of the filtered sodium (Table III).

Two chemically unrelated types of loop agent are available in the United States (Fig. 3): furosemide, a sulfamylated derivative of anthranilic acid; and ethacrynic acid, an unsaturated ketone. A third, bumetanide, is a sulfamylated compound pharmacologically akin to but more potent than furosemide.

Most investigators discount, but do not disprove, evidence that ancillary sites of action of the high-ceiling diuretics exist in parts of the nephron other than the loop. The argument against actions in the proximal and distal tubules is based on the assumption that a reduction of NaCl reabsorption proximal to the loop of Henle can be compensated in the loop, whereas a more distal action would be quantitatively trivial in any case. This notion neglects the well-established fact that high-ceiling diuretics limit the capacity of the loop to compensate for an action that takes place upstream. Response to loop diuretics depends on dosage. At low levels, NaCl elimination is equivalent to that elicited by thiazides; at high levels, NaCl and water losses exceed by far any losses that a thiazide can muster. Response, however, depends on the delivery of drug to site of action. When renal function is impaired, less drug gets to the kidney. In such cases, large doses of loop diuretics can be both safe and effective while no quantity of a thiazide will suffice. Furosemide is an agent of choice in acute pulmonary edema. It increases pulmonary and venous compliance, affording rapid relief, and then maintains the beneficial effects by reducing plasma volume.

Adverse reactions to loop diuretics include all the depletion phenomena and metabolic derangements listed for the thiazides. In addition, vertigo and deafness may develop in patients who require large intravenous doses. It is prudent to avoid the concurrent administration of certain (potentially ototoxic) antibiotics.

D. Potassium-Sparing Diuretics

Depletion of body potassium with or without hypokalemia (significant reduction of potassium in the blood) is the most common adverse reaction attributable to diuretics. Depletion can initiate serious disorders in the rhythm of the heart, impair the function of nerves and muscles, and cause intestinal disturbances. Hypokalemia is especially troublesome in patients with congestive heart failure who take both a digitalis preparation (a cardiac glycoside) and a diuretic. Potassium wasting sensitizes the heart to the toxic effects of the cardiac glycoside. Hypokalemia can somtimes be avoided if the patient eats foods that contain large amounts of potassium or, alternatively, takes supplements containing potassium chloride. If these measures fail, oral supplements should be stopped and a potassium-sparing agent started. Because this category of drug influences the tubular transport of only small quantities of Na^+, the diuretic response is weak.

Drugs of this group comprise two pharmacologically distinct groups. The steroidal antagonists of aldosterone resemble the hormone, aldosterone, which is synthesized in the adrenal cortex. Aldosterone attaches to a cytoplasmic receptor in the late portion of the distal tubule. The aldosterone–receptor complex subsequently migrates to the nucleus where it stimulates the production of messenger RNA (mRNA) molecules. The mRNA molecules then move into the cytoplasm. Here, they direct the formation of a protein that increases the permeability of the luminal membrane to Na^+, facilitating the entry of larger quantities of luminal Na^+ into the cell. The sodium pump spurs egress from the cell. Cellular potassium ions move into urine to replace some of the sodium ions that left, thus increasing the excretion of potassium. Because the potassium-sparing drug, spironolactone (Fig. 3), is structurally similar to the natural hormone, it can attach to the same cytoplasmic receptor. However, the two steroids differ enough so that spironolactone cannot initiate formation of the mRNA responsible for generating the permeability-enhancing protein. When the drug is provided, fewer receptor sites are free to react with the natural hormone. For this reason, the drug behaves as a competitive inhibitor.

The second group of potassium-sparing drugs includes two organic bases: (1) triamterene, an aminopteridine chemically related to folic acid, and (2) amiloride, a pyrazinoylguanidine. Both drugs slow the entry of luminal sodium into cells of the late portion of the distal tubule. The molecular mechanism of this action is unknown. Neither drug competes directly with aldosterone. Researchers suggest that both may cause closure of channels through which Na^+ ions gain entry. Like spironolactone, the drugs are weak diuretics but do strongly promote the conservation of potassium.

Triamterene and amiloride are usually prescribed in combination with a thiazide or loop diuretic when maintenance of a normal serum potassium level is essential (e.g., patients with abnormal cardiac rhythms, low levels of serum potassium, or taking a digitalis preparation). They are sometimes used together with another diuretic to amplify the response of individuals who are refractory to a single drug.

All the potassium-sparing diuretics may cause hyperkalemia, even when a potassium-wasting diuretic is prescribed concurrently. This adverse reaction occurs more often when renal function is impaired or renal reserve is limited (e.g., in diabetic or elderly patients). The drugs should not be prescribed together with potassium supplements, when serum potassium exceeds the normal value, or when renal function is poor.

Because of its steroidal structure, spironolactone can also attach to nonrenal hormonal receptors and provoke unpleasant reactions. Examples are menstrual irregularities, hirsutism in females, and gynecomastia in females or males.

E. Osmotic Diuretics

Osmotic agents do not react with receptors or directly inhibit any renal transporting system. Their activity depends, instead, on the osmotic force their molecules exert in solution. When present in interstitial fluid, the molecules attract cellular fluid to the extracellular compartment; in the tubular lumens of the kidney, they oppose the movement of water into renal cells. The amount of water held back depends on the number of molecules of the osmotic diuretic present in the lumens, not on their molecular size. For this reason, the molecular weights of all useful agents are low. Large molecules would increase the weight of the required dose. In addition, the ideal osmotic diuretic should not undergo reabsorption from the tubular fluid or metabolism (chemical alteration) in the body. These events would reduce its concentration in the renal lumens, the sites of action.

Mannitol, a six-carbon sugar (Fig. 3) found in certain plants, approaches this ideal and is the drug of choice. It is not effective when given orally because it does not cross gastrointestinal membranes. Isosorbide and glycerol are effective by the oral route but are less efficient because they penetrate cellular membranes in the kidney (are reabsorbed). When injected intravenously, mannitol is distributed throughout the extracellular fluid compartment but does not penetrate cells. The extracellular fluid volume increases rapidly as the osmotic agent draws fluid from the cells. In the proximal tubule, mannitol slows the passive reabsorption of water that normally follows the active transport of sodium. In effect, the osmotic force of a nonreabsorbable solute opposes the osmotic force of reabsorbable sodium. Urine flow rises in proportion to the dose of the osmotic agent. As a general rule, the fractional excretion of sodium increases but to a much smaller extent than the fractional excretion of water. For this reason, osmotic drugs are not first-line diuretics. Loss of fluid unaccompanied by salt induces thirst. A patient soon reaccumulates fluid. More commonly, mannitol is used to reduce the elevated intracranial pressure of cerebral edema, to prevent the development of renal failure associated with severe trauma or lengthy surgical procedures, or to promote renal "washout" of drugs or toxins.

Initially, the rapid expansion of extracellular fluid brought about by osmotic drugs increases the work of the heart, and some patients suffering from congestive heart failure may develop pulmonary edema. Later, especially when administration of osmotic diuretics is prolonged and the renal response is large, severe volume depletion and electrolyte imbalances supersede.

Bibliography

Cafruny, E. J. (1991). Diuretics: Drugs that increase the excretion of water and electrolytes. *In* "Human Pharmacology" (L. B. Wingard, T. M. Brody, J. Larner, and A. Schwartz, eds.). C. V. Mosby, St. Louis. In press.

Cafruny, E. J., and Itskovitz, H. (1982). Sites of action of loop diuretics. *J. Pharmacol. Exp. Ther.* **223,** 105–109.

Lant, A. (1985). Clinical pharmacology and therapeutic use of diuretics. *Drugs* **29,** 57–87 (Part I), 162–188 (Part II).

Maren, T. H. (1967). Carbonic anhydrase: Chemistry, physiology, and inhibition. *Physiol. Rev.* **47,** 597–781.

Shackleton, R., Wong, N. L. M., and Sutton, R. A. L. (1986). Distal potassium-sparing diuretics. *In* "Diuretics: Physiology, Pharmacology, and Clinical Use" (J. H. Dirks and R. A. L. Sutton, eds.). W. B. Saunders, Philadelphia. pp. 117–134.

Vander, A. J. (1985). "Renal Physiology," 3rd ed. McGraw-Hill, New York.

Velazquez, H. (1988). Thiazide diuretics. *Renal Physiol. (Basel)* **10,** 184–197.

Weiner, I. M. (1990). Drugs affecting renal function and electrolyte metabolism. *In* "The Pharmacological Basis of Therapeutics" (A. G. Gilman, L. S. Goodman, T. W. Rall, and F. Murad, eds.). Macmillan, New York.

DNA Amplification

GEORGE R. STARK, *Imperial Cancer Research Fund*

Glossary

Cellular oncogenes Normally encode proteins that carry out important regulatory functions in cells; when altered by mutation or overexpressed, they can contribute to tumorigenesis

DNA transfer Experimental technique in which exogenous DNA is introduced transiently or stably into recipient cells, with the objective of changing their phenotype

Double minutes Form of amplified DNA visible under the light microscope in metaphase cells; these elements are often paired and have the appearance of normal chromatin; however, they lack centromeres and, therefore, segregate randomly upon cell division

Episomes Submicroscopic extrachromosomal DNA; in mammalian cells, they are probably precursors of double minutes

Inversion and translocation Processes in which the sequence of DNA is rearranged; in inversion, the orientation of part of the DNA within a chromosome is reversed; in a reciprocal translocation, segments of two chromosomes are exchanged

Recombination Reciprocal crossing-over in which two double-stranded DNAs exchange parts of their sequences; if the sequences at the site of the exchange are different, the exchange is nonhomologous (e.g., as in translocation)

Rolling circle replication DNA synthesis proceeds around a circle more than once, forming a linear tandem repeat of the circular template

Sister chromatid exchange Recombination between two newly replicated chromosomes during mitosis; if unequal, there is reciprocal gain and loss of DNA in the two daughter cells

THE PHENOMENON OF DNA amplification concerns an increase in the relative amount of a small part of the genome of a cell, usually less than the amount contained in a full chromosome.

I. Introduction

In some lower organisms, amplification is a normal process involving developmentally programmed increases in DNA copy number in specific cells or tissues at specific times. The purpose is to achieve a rate of gene expression too rapid to be met by maximum expression of a single gene. A well-known example is amplification of the chorion genes in *Drosophila melanogaster*. These genes encode eggshell proteins, and their expression at a high level is needed only in the egg chamber, the only place where this amplification occurs. In humans and other mammals, amplification is not known to occur in normal development; however, it is observed readily in tumor cells, where amplification of cellular oncogenes contributes in important ways to tumorigenesis and metastasis, and in cells in culture, where amplification of genes whose overexpression confers resistance to cytotoxic drugs is seen readily. [*See* GENE AMPLIFICATION.]

II. Amplification of Genes Mediating Drug Resistance

Most information about what happens during DNA amplification comes from studies of drug resistance in cell culture. Any drug whose toxic effect can be overcome by increased expression of a particular protein can be used to select resistant cells that have amplified the gene encoding that protein. Two examples are methotrexate, which selects for amplification of the dihydrofolate reductase gene, and hydroxyurea, which selects for amplification of ribonucleotide reductase. There are now 20–30 different genes whose amplification has been selected in this way. Multidrug resistance is a particularly interesting case. In this phenomenon, a cell selected for resistance to any one of a set of unrelated drugs (e.g., doxorubicin, vinblastine, colchicine) is resistant to each member of the set. The cause is amplification of a single gene, now called *mdr,* encoding a protein that functions as an efflux pump in the plasma membrane. The pump is relatively unspecific and uses the energy of adenosine triphosphate to push a wide spectrum of unwanted compounds out of cells. Because multidrug resistance involves several drugs routinely used in the chemotherapy of cancer, understanding and controlling this phenomenon is particularly relevant to future success in treating cancer.

Although drug-resistant tumors are, unfortunately, commonplace in patients undergoing chemotherapy, it was completely unexpected to find a farmer and his family in which a gene encoding the enzyme butylcholinesterase is amplified about 100-fold and is stably inherited! The amplification event probably took place in a single germ-line cell of an individual exposed to high levels of agricultural organophosphate insecticides. Unless inactivated by this enzyme, these compounds inhibit normal function of germ-line cells, but a cell with a high concentration of butylcholinesterase will survive.

III. Amplification of Cellular Oncogenes in Tumors

Normal mammalian cells amplify their DNA very rarely indeed. It follows that the frequent amplification readily observed in drug-resistant cell lines and in tumors has somehow been activated abnormally in these cells. Furthermore, the probability of amplifying DNA in abnormal cells increases with increasing tumor-forming potential. Since cellular oncogenes are often amplified in tumors, especially in advanced metastatic tumors, it is virtually certain that their overexpression contributes to tumor progession. Specific cellular oncogenes are amplified frequently in specific types of tumors. Some of the best correlations are amplification of *N-myc* in neuroblastoma, *c-myc* in breast and lung cancer, *neu* in breast and stomach cancer, and *int-2* and *hst* in breast cancer and in squamous carcinoma. [*See* ONCOGENE AMPLIFICATION IN HUMAN CANCER.]

IV. Forms of Amplified DNA

When the degree of amplification is high, the extra DNA can often be seen under the light microscope, either as expanded regions of chromosomes or separate from chromosomes as paired, acentric elements called double minutes (Fig. 1). There are also episomes, extrachromosomal forms too small to be seen in the microscope. Double minutes are observed commonly in human tumors analyzed immediately after biopsy; therefore, they are likely to be important in the formation or evolution of amplified DNA in tumors. They have a nucleoprotein structure similar to that of chromosomes and each contains about 5 million bases of DNA. The extrachromosomal forms of amplified DNA replicate only once per cell cycle. Because they lack centromeres, they segregate randomly during cell division and can either accumulate or become lost in a population of cells, depending on whether the selection used requires gain or loss of a particular gene for the cells to survive.

In intrachromosomal amplified DNA, a very large region of flanking sequences, 10 million bases or more, is often coamplified together with each copy of the gene that is the target of selection. This is especially true in the initial event, when a single-copy sequence becomes amplified for the first time (Fig. 2a). If a population of cells already containing an amplification is exposed to a higher concentration of drug, further amplification will allow some cells to survive. In extreme cases, cells containing thousands of copies of a particular amplified gene have been obtained after several steps of selection. If the amplified DNA is extrachromosomal, increased resistance can be achieved by unequal segregation of the element at each mitosis without a new amplification event. However, if the amplified

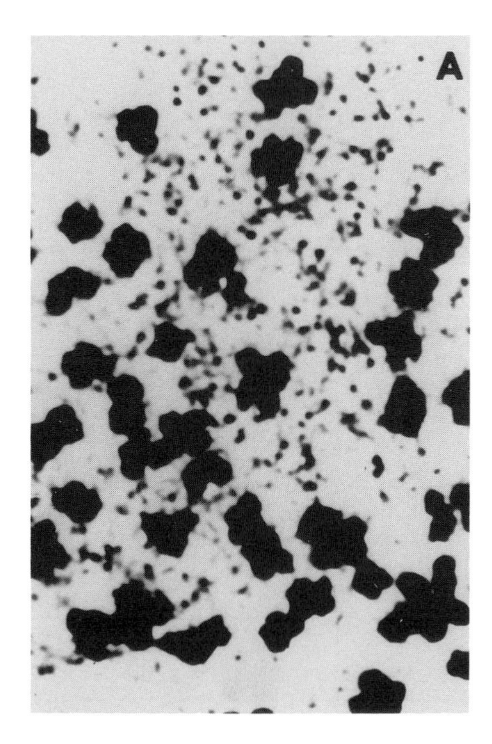

FIGURE 1 Double minute chromosomes. A partial metaphase chromosome spread from a human neuroblastoma cell line is shown.

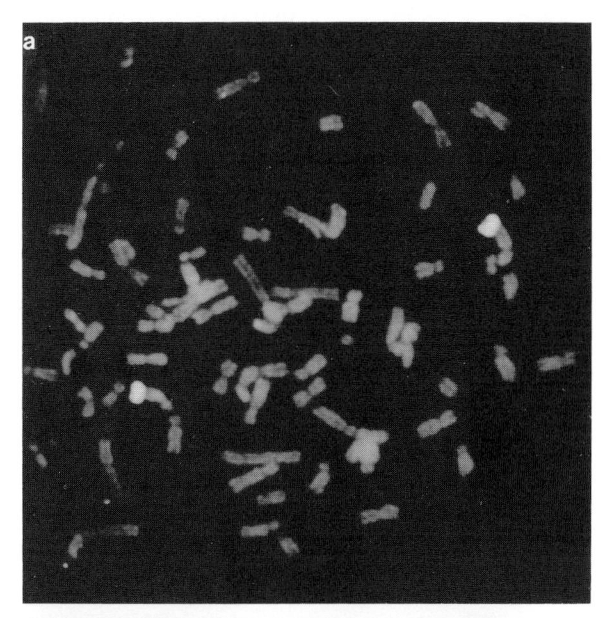

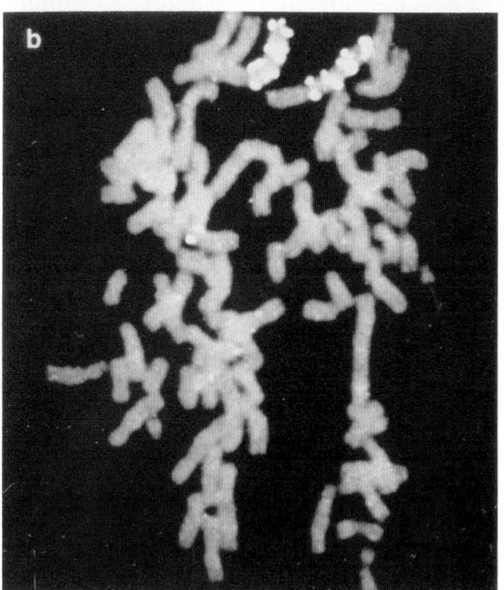

FIGURE 2 *In situ* hybridization to Syrian hamster cells containing amplification of a gene mediating resistance to an inhibitor of uridine monophosphate synthesis. Detection of a biotin-labeled DNA probe employed fluorescein-labeled avidin and antibodies directed against avidin. (a) The first step of amplification includes only a few extra copies of the gene. These are widely spaced and can be seen as pairs of spots on the two sister chromatids. (b) After the third step of amplification, 25 copies of the gene are present in a more condensed form.

DNA is chromosomal, new amplification events are required, in which the genes conferring resistance can become more concentrated (Fig. 2b) and much of the extraneous coamplified DNA can be lost.

V. Stimulation of Amplification

A wide variety of agents or treatments that damage DNA (ultraviolet or ionizing radiation, or carcinogens) or arrest DNA synthesis (hydroxyurea, aphidicolin, or hypoxia) can induce DNA amplification transiently. Severe cellular stress leads to short-term expression of otherwise silent genes whose products, intended to deal with the emergency, may also facilitate DNA amplification. Also, it is possible to isolate stable mutant cell lines, which have a greatly increased probability of amplifying their DNA. In fact, starting with normal cells in which amplification is very rare, one can obtain a series of cell lines with gradually increasing ability to amplify, ending with a line that has a probability as high as 0.001 per cell division per locus! This stable trait can be transferred to low-amplification cells by cell fusion or by DNA transfer, proving that gene expression is responsible for the high-amplification trait and opening the way to eventual isolation of the responsible genes and characterization of their products. It seems likely that similar or identical stimulatory proteins can be induced transiently by stress or stably in the mutant cells.

The ability of a cell to amplify DNA is correlated with its ability to rearrange DNA. It seems likely that a single abnormal process can lead to an unusual transient DNA structure, which can then rearrange in more than one way, sometimes leading to amplification, deletion, other chromosomal abnormalities such as inversion or translocation, or even loss of the affected chromosome. Detailed consideration of mechanisms and structures that might be involved in DNA amplification provides possible explanations for such linked responses. [*See* CHROMOSOME ANOMALIES.]

VI. Mechanisms of Amplification

Studies of developmentally regulated DNA amplification in lower organisms show that there are several different ways to achieve increases in the relative amounts of certain sequences. We know less about the details of abnormal amplifications in mammalian cells, but strongly suspect that several different mechanisms are involved, based on the different structures and locations of the amplified DNA. The mechanisms proposed fall into two major classes. In one, the extra DNA is generated by overreplication within a single cell cycle; in the other, there is no abnormality in replication. Instead, unequal distribution of DNA sequences at division allows one daughter cell to accumulate DNA at the expense of the other.

The best-known example of the first class of mechanisms is shown in Fig. 3. Multiple initiation of bidirectional DNA replication at a single origin within a single cell cycle results in a complex "onionskin" structure that must be resolved before cell division, so that the newly replicated chromosomes can separate. The structure can be resolved in ways that lead to intra- or extrachromosomal amplification, or to deletion if the entire onionskin is removed. A different mechanism within this class requires no abnormality in initiating DNA replication. Instead, following a single initiation event, rolling circle replication copies a DNA segment several times.

Unequal sister chromatid exchange, an example of the second class of mechanisms, is illustrated in Fig. 4. To generate more than one extra copy per cell, more than one unequal exchange must occur in different cell generations. Another mechanism in this class proposes that an extrachromosomal element arises by circularization and deletion of an intrachromosomal segment. This episome then replicates under normal control, once per cell cycle. However, lacking a centromere, it distributes unequally into the daughter cells at each cell division, so that some cells accumulate more copies than others.

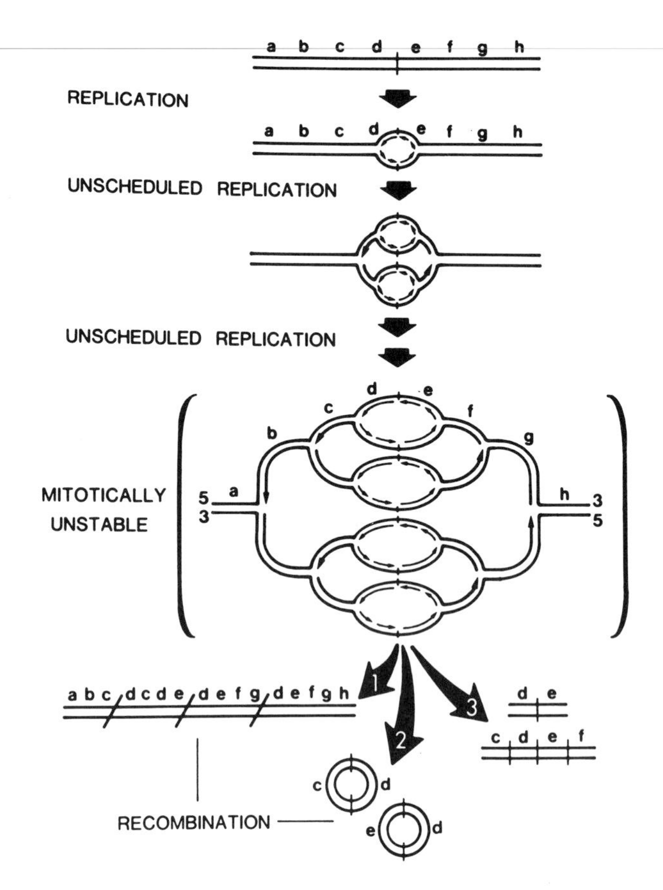

FIGURE 3 The "onionskin" model for amplification involving multiple initiations of replication in a single cell cycle. Unscheduled DNA synthesis and recombination can generate linear intrachromosomal amplified arrays, extrachromosomal circles, or extrachromosomal linear duplexes. Bidirectional replication at an origin generates a bubble that can undergo further rounds of unscheduled DNA replication, resulting in a nested set of partially replicated duplexes. Note that there are only two contiguous chromosomal strands. It is possible for linear duplex DNA to become detached from the structure if two replication forks can approach one another very closely (pathway 3). Recombination within the same duplex could generate extrachromosomal circles (pathway 2), while multiple recombinations among different duplexes could resolve the structure into an intrachromosomal linear array (pathway 1).

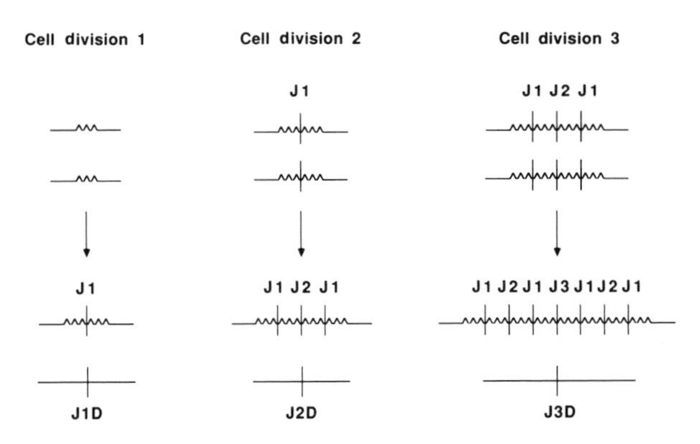

FIGURE 4 A model for gene amplification in which sister chromatids recombine unequally before mitosis. A lineage is shown in which copy number increases maximally at each cell division. Note that only one pair of sister chromatids is shown and that each daughter cell still has an unaltered chromosome. J1–J3, junctions between amplified sequences and; J1D–J3D, junctions between deleted sequences formed at each step. After step 3, there are four copies of J1 and two copies of J2 but only one copy of J3. J2 could also be at single copy number if the increase is not maximal at step 3.

To test these and other mechanistic possibilities critically, we must examine individual cells soon after the initial event. New methods now allow detection of single copies of any DNA sequence in single cells with high resolution, using fluorescence to detect the site of hybridization of a cloned DNA probe (Fig. 2). The structure of newly amplified DNA is thus open to a detailed analysis. Also, as mentioned earlier, it may soon be possible to identify specific genes whose products stimulate amplification. As the mechanisms of DNA amplification become better understood, so too will the connections between amplification and the other chromosomal abnormalities of tumor cells.

Bibliography

Giulotto, E., Knights, C., and Stark, G. R. (1987). Hamster cells with increased rates of DNA amplification, a new phenotype. *Cell* **48,** 837.

Pinkel, D., Straume, T., and Gray, J. W. (1986). Cytogenetic analysis using quantitative high-sensitivity, fluorescence hybridization. *Proc. Natl. Acad. Sci. U.S.A.* **83,** 7790.

Prody, C. A., Dreyfus, P., Zamir, R., Zakut, H., and Soreq, H. (1989). *De novo* amplification within a 'silent' human cholinesterase gene in a family subjected to prolonged exposure to organophosphorus insecticides. *Proc. Natl. Acad. Sci. U.S.A.* **86,** 690.

Saito, I., Groves, R., Giulotto, E., Rolfe, M., and Stark, G. R. (1989). Evolution and stability of chromosomal DNA coamplified with the CAD gene. *Mol. Cell. Biol.* **9,** 2445.

Schimke, R. T. (1988). Gene amplification in cultured cells. *J. Biol. Chem.* **263,** 5989.

Stark, G. R., Debatisse, M., Giulotto, E., and Wahl, G. M. (1989). Recent progress in understanding mechanisms of mammalian DNA amplification. *Cell* **57,** 901.

Stark, G. R., Debatisse, M., Wahl, G. M., and Glover, D. M. (1990). DNA amplification in eucaryotes. *In* "Genome Rearrangement and Amplification: Frontiers in Molecular Biology" (D. Hames and D. M. Glover, eds.). IRL Press, Oxford.

Trask, B. J., and Hamlin, J. L. (1989). Early dihydrofolate gene amplification events in CHO cells usually occur on the same chromosome arm as the original locus. *Genes and Development* **3,** 1913.

Wahl, G. M. (1989). The importance of circular DNA in mammalian gene amplification. *Cancer Res.* **49,** 1333.

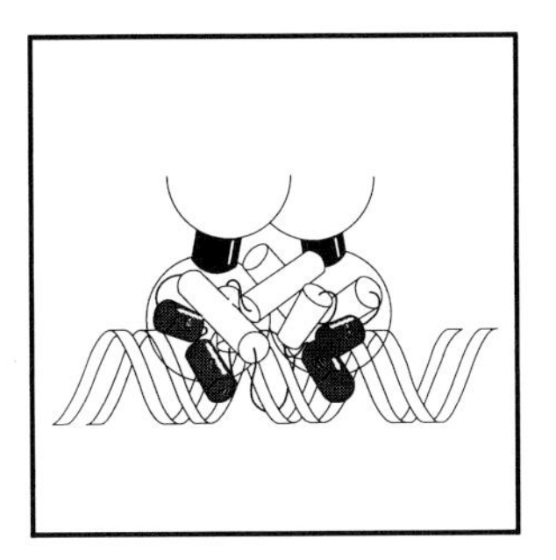

DNA and Gene Transcription

MASAMI MURAMATSU, *The University of Tokyo (Japan)*

Glossary

Amphipathic α-helix α-helical portions of a polypeptide that have affinity to both polar and nonpolar solvents

Cis-acting element When a DNA sequence regulates a gene on the same DNA molecule, the sequence is said to act in cis; in this chapter, refers to any DNA sequence element that participates in the regulation of the gene on the same DNA

Consensus sequence Most common or frequent sequence chosen from available data when variable sequences exist for a cis-acting element (a regulatory signal on DNA)

Domain Portion of a protein molecule having a defined function

Transacting factor Any protein factor that acts on the regulation of a gene but is encoded on a separate DNA molecule; in this article, a regulatory protein of a gene that works by interacting with one of the cis-acting elements

DEOXYRIBONUCLEIC ACID (DNA) is the genetic material and the central molecule of life. Its major function is to express genetic information via messenger RNA to achieve the phenotype of an organism. In this article, the first step of gene expression—transcription—will be detailed together with some fundamental aspects of the molecular structure of DNA.

I. What is DNA?: Structure, Properties, and Function

A. Definition and Structure

DNA is an informational macromolecule contained in virtually all living organisms[1] on earth, and it has all of the genetic information of each species to which it specifically pertains. It is a long, filamentous molecule consisting of a backbone of D-2-deoxyribose moieties connected by 3′-5′-phosphodiester linkages, each having one of the four nitrogenous bases—adenine (A), guanine (G), thymine (T), or cytosine (C)—at the 1′-C atom of the deoxyribose. DNA is actually a polymer of deoxyribonucleotides and is synthesized in the cell from four kinds of deoxyribonucleoside triphosphates (cf. Enzymology of DNA Replication). DNA is made up of two filaments, or strands, connected to each other by hydrogen bonds between A and T, and G and C, forming the Watson–Crick double helix (see Color Plate 9). The two strands are said to be complementary.

The complementary double helix is of prime importance in every aspect of DNA function such as replication, transcription, repair, and recombination. A DNA double helix usually takes one of three different conformations, designated the A-, B-, or Z-form, depending on its environment and sequence. The main features of each form are presented in Table I. The A- and B-forms are right-handed and wind rather smoothly, whereas the Z-form is left-handed and zigzag-shaped (for which it is named). Under physiological conditions, most DNAs appear to exist as the B-form. Some DNAs, which have a special nucleotide sequence such as a stretch of alternate CG or TG, assume the Z-form *in vivo*. This is especially true when some of the C

1. Some viruses have RNA genomes and no DNA.

TABLE I Conformation Comparison of Three Forms of DNA

	A-form	B-form	Z-form
Overall shape	Short and thick	In between	Long and thin
Winding direction	Right-handed	Right-handed	Left-handed
Pitch per turn	24.6 Å	33.2 Å	45.6 Å
Base pairs per turn	10.7	10.0	12
Diameter of helix	25.5 Å	23.7 Å	18.4 Å
Tilt of base to helix axis	+19°	−1°	−9°
Major groove	Narrow and deep	Wide and medium deep	Flat and exposed on surface
Minor groove	Broad and shallow	Narrow and medium deep	Narrow and deep

residues are methylated. Negative supercoiling of DNA that gives a stress of anticlockwise rotation to the long axis of DNA double helix also favors the formation of the Z-form. The specific effects of DNA nucleotide sequence on its higher-order structure are not yet fully understood, particularly when there are DNA–protein interactions, as in chromatin. These aspects of higher-order structure in relation to the interaction with proteins will be one of the most important targets of future research.

B. Properties of DNA

Because double-stranded DNA is a very long, filamentous molecule, DNA solutions have a high viscosity, which disappears by denaturation (separation of the two strands) or shearing. Due to their great length, DNA molecules are rather fragile and are easily broken into smaller fragments by just pipetting through a narrow tube. When DNA molecules are centrifuged in a high-concentration of CsCl, the latter forms a density gradient in the solution and the DNA molecules form a band where the density of the solution is equal to that of the DNA molecule. Because the DNA molecule is much denser than protein molecule, this is an excellent procedure for isolation of DNA from biological materials.

The DNA double helix may be separated by a variety of means, including high temperature and high pH, to break hydrogen bonds, which causes transition from a double helix to a single-strand (denaturation). The temperature at which half of the DNA molecules denature is called the melting temperature (T_m). T_m elevates as the GC content of DNA increases due to the increased amount of hydrogen-bonding; a GC pair has three hydrogen bonds, whereas an AT pair has only two. Denatured single-stranded DNA can reform native double-stranded DNA when incubated at a proper temperature if the two strands have sufficient complementarity (renaturation). Single-stranded RNA having a complementary sequence can also form a (hetero)duplex with a single-stranded DNA by renaturation (hybridization). The ability of a DNA strand to form homo- and heteroduplexes with a complementary nucleic acid (DNA or RNA) strand is the central function of DNA *in vivo* and *in vitro*. These features form the basis of DNA analysis methodology.

Chemically, DNA is much more stable than RNA because of the absence of the 2′-OH group in the ribose and by the protection provided by hydrogen-bonding in the double helix. Nevertheless, DNA may be attacked by a number of enzymes that act in various ways. DNases (endonucleases) randomly cut a DNA molecule within its nucleotide sequence; exonucleases cleave it from either the 3′ or 5′ end. Specific cleavage at a defined nucleotide sequence may be obtained by an enzyme family designated restriction endonuclease (or restriction enzyme). The existence of a large number of restriction enzymes that can cut a variety of possible combinations of nucleotide sequence makes them an important tool of gene engineering. DNA ligases, which can splice DNA; reverse transcriptase, which can synthesize DNA on RNA templates; nuclease S1 (or P1), which digests only single-stranded DNAs; and RNase H, which digests only RNA in DNA–RNA hybrids are among the other DNA-related enzymes frequently used in gene technology.

C. Function of DNA

DNA is the central molecule of life in that it stores genetic information, replicates itself, and then expresses the information as the phenotype of a living organism. The flow of genetic information is drawn schematically as follows:

$$\text{Replication}\circlearrowright \text{DNA} \underset{\text{Reverse transcription}}{\overset{\text{Transcription}}{\rightleftharpoons}} \text{RNA} \xrightarrow{\text{Translation}} \text{Protein}$$

The major functions of DNA are replication and transcription, or, more precisely, gene expression. Other important aspects of DNA function are repair (which is accompanied by DNA synthesis) and recombination (which includes transposition). In this chapter, only one major aspect of DNA function—gene expression (transcription)—will be discussed in molecular detail. The focus will be on mammalian systems, because they are the most relevant to human biology.

II. Mechanism of Transcription

A. Chemistry of Transcription or RNA Synthesis

Cellular RNA, including messenger RNA (mRNA), ribosomal RNA (rRNA), transfer RNA (tRNA), and small nuclear RNA (snRNA), are synthesized on respective DNA templates. The overall stoichiometry may be written as follows:

$$\left\{\begin{matrix} n_1\ \text{ATP} \\ n_2\ \text{GTP} \\ n_3\ \text{CTP} \\ n_4\ \text{UTP} \end{matrix}\right\} + \text{template} \xrightarrow[\text{Mg}^{2+},\ (\text{Mn}^{2+})]{\text{RNA polymerase}}$$
(mononucleotides)

$$\begin{matrix} \text{AMPn}_1 \\ | \\ \text{GMPn}_2 \\ | \\ \text{CMPn}_3 \\ | \\ \text{UMPn}_4 \end{matrix}$$
(a polynucleotide)

$$+ (n_1 + n_2 + n_3 + n_4)\text{PP}_i + \text{template}$$

where ATP and AMP are adenosine tri- and monophosphate, respectively; GTP and GMP are guanosine tri- and monophosphate, respectively; CTP and CMP are cytidine tri- and monophosphate, respectively; UTP and UMP are uridine tri- and monophosphate, respectively; and PP_i is inorganic pyrophosphate n_1–n_4 are variables defined by the base composition of the template DNA. The essential reaction of the polymerization of an RNA chain is the nucleophilic attack of the α-phosphate group of a nucleoside triphosphate by the 3′-hydroxyl group of the ribose moiety at the 3′ end of RNA. Phosphodiester linkages are formed by the energy released from the hydrolysis of the nucleoside triphosphates into component monophosphates. Template DNA assists the alignment of the incoming nucleotide to the proper position by hydrogen-bonding, thus determining the complementary nucleotide sequence. Note that the RNA chain elongates from 5′ to 3′ direction by this mechanism.

B. Initiation of Transcription

For correct readout of DNA, transcription must start from a fixed point, i.e., the top of the gene. In prokaryotes such as *Escherichia coli,* the signal that directs transcription initiation, the promoter, is composed of three regions near and upstream of the transcription start site. These are the Pribnow box at −10 (nucleotide positions are numbered with minus for upstream and with plus for downstream from the transcription start site [+1]), consisting of 5–6 base pairs (consensus: TATAAT), the "−35 sequence," having another specific sequence of 5–6 base pairs (consensus: TTGAC), and a region consisting of several nucleotides around the actual transcription initiation site.

For transcription to begin, RNA polymerase must bind to the initiation site. RNA polymerase is comprised of five subunits; four subunits (two α's, β, β') constitute the core of the enzyme, while the fifth subunit (σ) is required for promoter recognition and transcription initiation. Once transcription begins, the σ factor is released from the transcribing complex and the core polymerase continues RNA chain-elongation. Prokaryotic gene transcription is regulated mainly by a sequence called the operator, located near the promoter. The operator binds repressor molecules, which in turn interact with RNA polymerase and inhibit transcription initiation. Prokaryotes also have positive regulation; for instance, the binding of the catabolite gene activator protein (CAP)–cyclic adenosine monophosphate complex to the promoter of the *lac* operon can stimulate transcription.

III. Transcription Initiation in Eukaryotic Cells

A. Three Polymerase Systems Are in Operation in Eukaryotes

All eukaryotes, from yeast to humans, have three RNA polymerase systems, which transcribe different categories of the gene (Table II). RNA polymerase I transcribes solely ribosomal RNA genes that are repeated several hundred times in most eukaryotic genomes. This polymerase is characterized by complete insensitivity to the fungal toxin α-amanitin but extreme sensitivity to the antibiotic actinomycin D. Ribosomal RNA synthesis is inhibited >95% when 0.05 μg/ml of actinomycin D is added to the culture medium of animal cells. More than 20 times this concentration is required to inhibit the transcription of genes by RNA polymerase II.

RNA polymerase II transcribes protein-coding genes and most of the snRNA genes. It is very sensitive to α-amanitin as shown in Table II. RNA polymerase III transcribes transfer RNA, ribosomal 5S RNA, and some snRNA genes, as well as virus-encoded low-molecular weight RNAs. This enzyme is characterized by a moderate sensitivity to α-amanitin (Table II). All three polymerases consist of two large subunits (resembling *E. coli* β and β') and several smaller subunits. However, all components are different among these three enzyme classes with the possible identity of a couple of subunits.

B. Promoter Sequences and General Transcription Factors

Each category of the eukaryotic gene has its own promoter sequence and is specifically transcribed by the cognate polymerase. For precise initiation and control of gene transcription, eukaryotes have developed multiple transcription factors that recognize and bind the promoter sequence. These protein factors are required for transcription in each polymerase system and are called general transcription factors. Other transcription factors are gene- or category-specific (see Section IV,B). Promoter sequences and the general transcription factors for each class of polymerase are described below.

1. RNA Polymerase I System

There is no consensus promoter sequence for RNA polymerase I common to different species, even among vertebrates. This is reflected in the species-specificity of this system among mammalian species such that a crucial general transcription factor cannot be exchanged between the human and mouse. Apparently, this evolutionary divergence occurred only in the RNA polymerase I system. Because it transcribes only ribosomal RNA, RNA polymerase I and its factors may have coevolved with its cognate promoter sequences much faster than the RNA polymerase II system where the polymerase must interact with factors specific for many different genes.

The promoter region of the eukaryotic ribosomal RNA gene can be divided into two parts: one is the core promoter encompassing the transcription start site and part of the upstream sequence (up to −50 depending on species); the other is the upstream control element that is present near −100 (and varies among species). These sequences are recognized and bound by at least one protein factor (designated TFID or SL-1) with the assistance of UBF1, another factor. RNA polymerase I then recognizes and binds this complex with the possible assistance of another factor, TFIA. When the four kinds of nucleoside triphosphates are supplied to this preinitiation complex, RNA polymerase I begins RNA chain-elongation; the other factors are left behind

TABLE II Classes of Eukaryotic RNA Polymerase

RNA polymerase	Localization	α-amanitin sensitivity	Product
I (A)	Nucleolus	—	rRNA
II (B)	Nucleus (chromatin)	+++ (0.0025 μg/ml)[a]	mRNA, snRNA
III (C)	Nucleus (nucleoplasm)	+ (25 μg/ml)[a]	5SRNA, tRNA, snRNA, VA RNA[b]

[a] ID_{50}.
[b] Virus-associated RNA.

on the promoter. A new polymerase I molecule can now join the promoter complex, allowing transcription to be initiated many times. Transcription by RNA polymerase I stops at a termination site having a specific sequence. In the case of the mouse, it is called the *Sal*I box, to which a termination factor is known to bind.

2. RNA Polymerase II System

There are two types of genes in this category: one has a TATA box (the consensus sequence is $TATA^{A}_{T}$, but it is subject to slight changes) at approximately −30, and the other has no TATA box. In many cases, multiple GC boxes (GGGCGG) are found in the upstream region instead. The latter type of promoter is seen frequently in so-called housekeeping genes, which code for general proteins or enzymes required for basic functions of the cell such as energy production. In genes with a TATA box, the first step of transcription initiation is the binding of TFIID (or TATA-binding factor) to the TATA box. RNA polymerase II, with at least four different factors (TFIIA, TFIIB, TFIIE, and TFIIF), can then interact with this complex and begin transcription. Because of the difficulty of purification of these factors, the precise order of the binding and the function of each factor have not been determined unequivocally. However, TFIIA appears to stabilize the TATA box–TFIID complex, and TFIIF has been shown to bind directly to RNA polymerase II. In contrast to this, TFIID does not seem to be required for the genes without a TATA box. Other factors, such as TFIIB, TFIIE, and TFIIF, appear to be necessary. SP1, a protein that binds to the GC box, has been suggested to be involved in transcription regulation, but this is still speculative. To understand the functions of all these factors, we may also require more knowledge about the function of each subunit of RNA polymerase II. Termination of transcription of RNA polymerase II is not well defined but appears to occur far downstream from the poly(A) addition site and may occur at multiple sites, causing transcription to cease gradually.

3. RNA Polymerase III System

A remarkable feature of the promoters for RNA polymerase III is that they are usually present inside of the gene, i.e., in the transcribed region of the gene. In the 5S ribosomal RNA gene, the internal promoter is divided into three regions known as Box A, the I region, and Box C. The sequences and the spacing of these regions are rigorously determined. Transfer RNA genes also have two regions—the A- and B-blocks—as the internal promoter. The A-block has a sequence similar to Box A of the 5S RNA gene. Both the 5S ribosomal RNA gene as well as the rRNA gene, having the internal promoter in common, appear to have an upstream sequence, which enhances the promoter activity.

RNA polymerase III utilizes at least four transcription factors: TFIIIA, TFIIIB, TFIIIC, and TFIIID. In 5S RNA gene transcription, TFIIIA first binds with the internal promoter followed by TFIIIC binding and complex stabilization. RNA polymerase III can then bind and transcribe the gene. TFIIIC, but not TFIIIA, is required for tRNA gene transcription. TFIIIC first binds with the gene recognizing both A- and B-blocks, followed by TFIIIB and RNA polymerase III binding, resulting in transcription. Termination occurs either at a stretch of more than four straight T's (TTTT) or at a sufficiently long repeat of AT's.

IV. How is Gene Transcription Regulated?

A. Cis-Acting Elements

Eukaryotic genes are exquisitely regulated to achieve the developmental program of the genome and to maintain homeostasis of the cell and the whole body against external stimuli and perturbation. Regulation is carried out by a number of sequences that are engaged in the control of gene transcription (Fig. 1). These can be divided into several categories. The first category, the promoter, specifies the transcription start site of every gene. It is composed of the TATA box and, in many cases, some sequence at the transcription start site. In

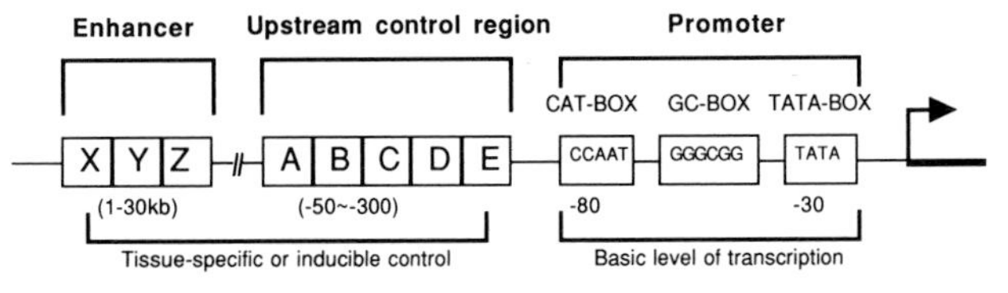

FIGURE 1 Cis-acting elements in gene control region. Not all promoters have all these elements. Both the upstream control region and the enhancer consist of more than one element and may share some of the elements (A ~ E, X ~ Z). Only the enhancer can work at a distance.

genes without a TATA box, GC boxes with certain unidentified sequences near the transcription start site appear to determine transcription initiation. General transcription factors previously described also bind to these structures forming basic transcription apparatus.

The second category is the upstream regulatory region, or the enhancer. Enhancers, when first found in some viruses such as SV40, were defined as activating DNA elements that exerted influence from distant places in an orientation-independent manner, either upstream or downstream of the promoter. Some enhancers, such as the immunoglobulin heavy-chain enhancer, were further found to be tissue-specific. (Tissue specificity may provide a basis for specific gene activation during differentiation). However, it was later found that some other cell- and/or induction-specific regulatory sequences work only at a certain upstream proximity to the promoter and in an orientation-dependent manner. These seem to be intermediate forms of promoters and enhancers and are regarded either as a part of the promoter (together with the core promoter represented by TATA box) or a type of enhancer working at short range. They are sometimes referred to as upstream control regions (or sequences).

Some sequences with positive effects are frequently found near and upstream of the TATA box (e.g., CAT box [CCAATG] and GC-box [GGGCGG]). They are thought to strengthen the promoter activity and are sometimes included in the promoter structure. Most upstream control sequences are found within several hundred base pairs of the promoter; however, other enhancers, many of which are tissue-specific, are located at several kilo-bases (kb) and sometimes as far as 30 kb upstream. Both upstream control sequences and enhancers are generally composed of a few elements clustered within several dozen base pairs. Because many of these control regions share some of the elements with homologous sequence, the cell- or gene-specificity of an enhancer may be determined by the combinatorial interaction of these elements with different transacting factors of a particular cell type. Alternatively, one cis-acting element sometimes partially overlaps or superimposes another element, allowing competition as well as synergism between different transacting factors. Negative control elements, called silencers, or dehancers, which cancel out enhancer effects, have also been described. Some of the cis-acting elements that have been relatively well studied are shown in Table III together with their cognate transacting factors.

B. Transacting Factors

The above-mentioned cis-acting elements exert their gene-activating or -repressing effects by binding specific protein factors. Several examples of this are presented in Table III. These protein factors, generally called transacting factors, are also transcription factors. They are generally low in abundance in the cell. By using gene engineering methodology, various deletion and substitution mutants of these factors have been synthesized and functionally tested in cultured cells. The results indicate that these transacting molecules have separable "domain" structures, each domain having a different function, such as DNA-binding, protein–protein interaction, and transactivating capacities.

1. DNA-Binding Domains

For most of the transacting factors, a DNA-binding domain consisting of several dozen amino acids is required, although not sufficient, for activity. There are at least five, and probably more, categories by which a transacting factor could interact with DNA (Table III; Fig. 2). The first one, known as a zinc finger, was originally found in a general transcription factor TFIIIA (see Section III.B.3) and in the steroid hormone receptor family. Zinc fingers are characterized by about 30 amino acids that include two cysteine and two histidine (or again cysteine) residues forming a fingertiplike structure surrounding a Zn^{+2} ion. The transactivator SP1, which is found rather ubiquitously in mammalian cells and binds with the GC box, has three zinc fingers near its carboxy-terminus. These proteins bind with DNA by inserting the finger structure into the major groove of DNA. The sequence specificity appears to be determined by some amino acids near the finger (e.g., at the base of the stem of the finger).

The second category by which a transacting factor could bind DNA is the helix-turn-helix-type structure. This structure has been well analyzed in prokaryotic repressor molecules and, in eukaryotes, is represented by a so-called homeodomain. This domain, containing about 60 amino acids, was first identified in *Drosophila* homeotic genes that regulate different aspects of embyronic development. It has since been found in a number of mammalian transcription factors, including OCT-1 (or OTF-1, NF-A1, and NFIII), OCT-2 (or OTF-2 and

TABLE III Mammalian Transacting Factors and Binding Sequences[a]

Transacting factor	DNA-binding site (5′→3′)	Class	Some characteristics
SP1	GGGCGG[b]	zinc finger	glutamine-rich domain, O-glycosylated
GR[c]	GGTACAN$_3$TGTTCT[d]	zinc finger	act as dimer
C/EBP	TGTGGAAAG	leucine zipper	rat liver nuclear protein, dimerization
C-Jun	TGA$^{G}_{C}$TCA[e]	leucine zipper	form heterodimer with c-fos
CREB	TGACGTCA[f]	leucine zipper	weakly cross-reactive with c-Jun-binding site
OCT-1	ATTTGCAT[g]	helix-turn-helix[h]	ubiquitous
OCT-2	ATTTGCAT[g]	helix-turn-helix[h]	lymphoid cell-specific
OCT-3	ATTTGCAT[g]	helix-turn-helix[h]	embryonic cell-specific, proline-rich domain
Pit-1 (GHF1)	$^{TT}_{AA}$TATNCAT	helix-turn-helix	pituitary specific, serine, threonine (tyrosin)-rich domain
CTF/NF-1	GCCAAT		proline-rich domain by alternative splicing
SRF	GATGTCCATA TTAGGACATC[i]		dimerization
E12/E47	CANNTG[j]	helix-loop-helix	ubiquitous
MyoD	CANNTG[j]	helix-loop-helix	directs muscle differentiation, regulated by Id having a similar sequence

[a] Only a part of the known transcription factors of mammals.
[b] GC box.
[c] Glucocorticoid receptor.
[d] Glucocorticoid-responsive element (GRE).
[e] TPA-responsive element (TRE).
[f] cAMP-responsive element (CRE).
[g] Octamer.
[h] This is more specifically a "POU domain," which consists of a homeodomain and a POU box.
[i] Serum-responsive element (SRE).
[j] Core consensus sequence.

NFA2), Pit-1 (or GHF1), and OCT-3 (see Table III). Some of these proteins have a POU domain, which consists of a homeodomain and a POU box comprising about 160 amino acids. OCT-1 is found almost ubiquitously and can activate histone H2B gene *in vitro*. In contrast, OCT-2 is lymphoid cell-specific, and OCT-3 is confined to undifferentiated embryonic cells.

The third category of transacting factors consists of proteins that have a so-called leucine zipper structure. The leucine zipper consists of four to five consecutive leucines that appear every seven amino acids. In this way, they are all aligned when the polypeptide chain forms an α-helix. The series of leucines are thought to interact with another similar series on another molecule, facilitating the dimerization of these proteins to form a hetero- or homodimer. In these proteins, a DNA-binding domain of about 30 amino acids with a high net basic charge is present immediately upstream (N-terminal side) of the leucine zipper. Dimer formation is a prerequisite for the specific binding of these proteins to the cognate DNA sequences and is thought to have regulatory significance. The palindromic nature of the target sequences suggests the interaction of each monomer with the half-sequence of the dyad-symmetry. These proteins include C/EBP, Jun, Fos, and CREB. An oncogene c-jun product is known to bind to the AP-1 site (or TRE) as a heterodimer with the oncogene c-fos product and activates the transcription of a gene having this site (Table III).

The fourth category of the DNA-binding domain is found in CTF/NF-1. It has a high content of basic amino acids and can form an α-helix; no similarity is apparent with any of the above-described motifs. The fifth and most recently identified category of the DNA-binding domain is the helix-loop-helix structure. It is characterized by two amphipathic helices separated by an intervening loop. Several hydrophobic amino acid residues in the helices are strictly conserved. Many of these proteins have a basic region just N-terminal side of the first helix. The helices are supposed to interact to form a heterodimer of different classes of these proteins, and the basic domain is required for the interaction with DNA. This category of transacting proteins include

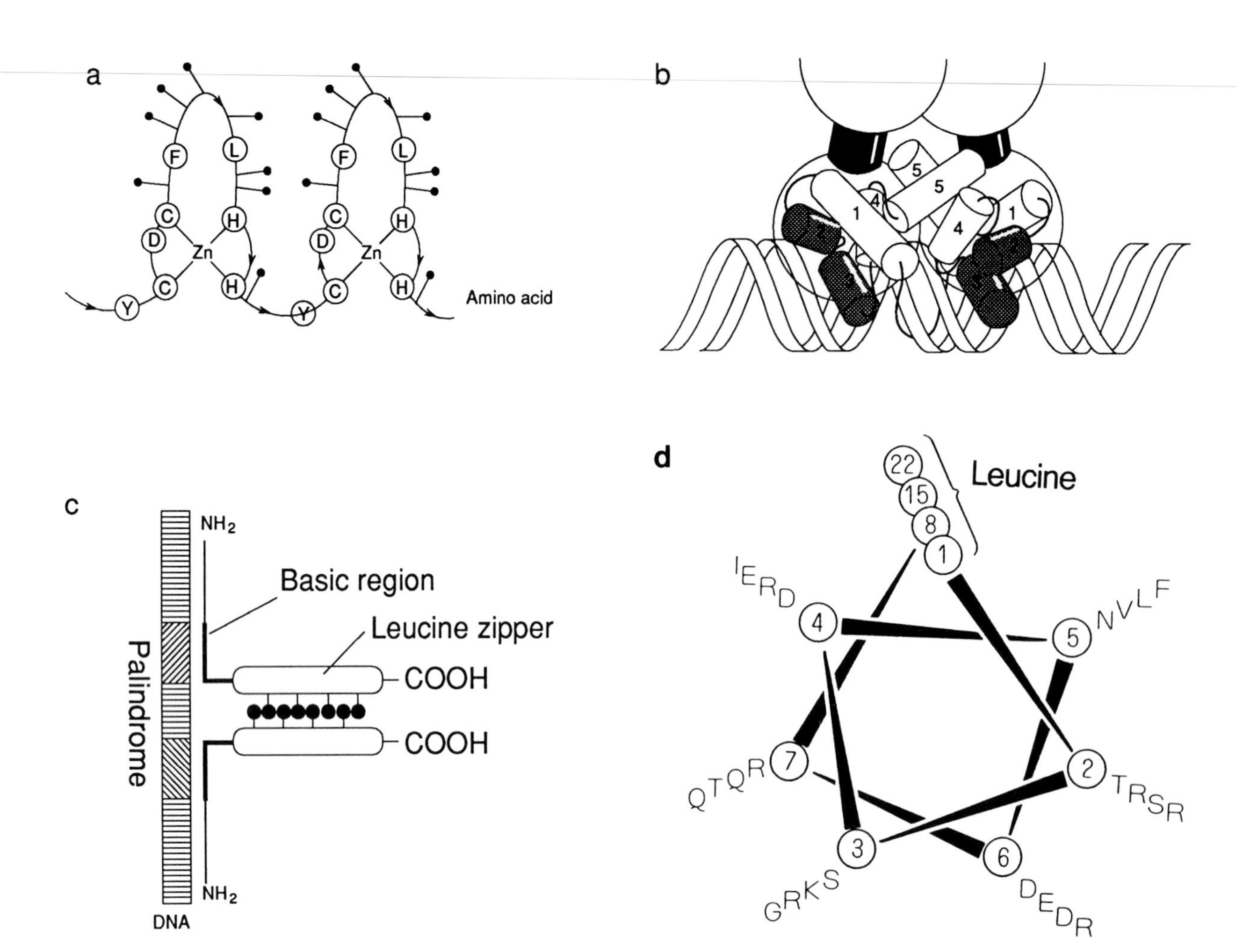

FIGURE 2 Three modes of DNA binding of transacting factor. (a) Zinc finger. Amino acids in the circles are shown by one-letter symbols. (b) Helix-turn-helix. Lambda repressor molecules (as a dimer) bound to an operator site. Helices are shown as rods. Two helices (Nos. 2 and 3) are shown to interact with DNA in the major groove. [Reproduced, with permission, from M. Ptashne, 1980, "A Genetic Switch. Gene Control and Phage," Cell Press, Cambridge Massachusetts, and Blackwell Scientific Publications, Palo Alto, California.] (c) Leucine zipper. Two proteins with a leucine zipper such as Jun and Fos form a dimer and interact with a palindromic sequence such as TRE of DNA via the basic region adjacent to the leucine zipper. (d) Formation of a zipper. An α-helix with four leucines at every seven amino acids aligns the leucine residues in a row, which can interact with a similar structure as shown in (c). Amino acid residues are represented by one-letter symbols.

immunoglobulin enhancer-binding proteins E12/E47, Myo-D, c-Myc, *da* (daughterless) protein, etc. Intriguingly, some of them may have specific regulatory proteins that have a similar helix-loop-helix structure but without the basic DNA-binding domain. They appear to form a superfamily of proteins that are engaged in the regulation of gene expression during development and differentiation. DNA-binding proteins lacking any of the known domain structures are increasing as the number of structurally defined transacting factors increases. Therefore, more different families of transactivating proteins with different modes of interaction will be found in the future.

2. Transactivating Domains

The transactivating domain of a transcription factor has the size of 30–100 amino acids and apparently works independently of its DNA-binding domain. (This has been demonstrated by making chimeric molecules in which a transactivating domain is exchanged between unrelated transcription factors and hooked to the DNA-binding domain of another factor.) Some factors have more than one transactivation domain. These domains are located at different regions of the molecule and work independently of each other. Several different types of transactivating domains are known. One transactivating domain, an acidic domain, is found in yeast GAL4 and GCN4 and is characterized by containing a substantial negative charge (a high content of glutamic and aspartic acids) and the ability to form amphipathic α-helix. The α-helix is sometimes called an "acidic noodle," based on the postulated shape of this domain. In this structure, there appears to be no definite amino acid sequence requirements, only a net negative charge requirement. Jun proteins are one class of proteins that have negatively charged α-helical regions, which may act as a transactivating domain. These acidic noodles are postulated to interact with one of the general transcription factors, possibly TFIID, or with a subunit of RNA polymerase II.

A second type of transacting domain is the glutamin-rich domain, which is found in SP1 and characterized by a high content (about 25%) of glutamine and very few charged amino acids. Regions with a high glutamine content are found in some of the *Drosophila* homeotic genes, yeast genes, mammalian OCT-1, OCT-2, Jun, and other genes, although no sequence homology is apparent among the different groups. A third type of transcriptional activating domain is the proline-rich domain found in the carboxyl-terminus of CTF/NF-1. Proline-rich regions are also noted in a number of mammalian transactivating proteins including Jun, OCT-2, OCT-3, AP-2, and serum responsive factor (SRF). How these different motifs in the transactivating domain interact with the basic promoter apparatus, however, is not yet known. It is tempting to speculate that different motifs interact with different general transcription factors and/or different subunits of RNA polymerase II.

3. Modification of Transacting Factors

Transcription factors are subject to modifications such as phosphorylation and glycosylation. When signal transduction-dependent gene activation is insensitive to protein synthesis inhibitors such as cycloheximide, it is deduced that activation does not involve new transcription factor synthesis but rather is regulated by some modification of pre-existing factors. Although relatively few instances are known at present, yeast heat-shock transcription factor and mammalian CREB factor appear to be positively regulated by phosphorylation. The latter factor is known to dimerize as a result of phosphorylation. Glycosylation may also be a posttranslational regulatory mechanism of transacting factors since SP1, which is highly glycosylated, could be inhibited *in vitro* by wheat germ agglutinin, which binds to the sugar components.

V. Concluding Remarks

The mechanism of regulation of gene transcription is not yet completely clarified. Recent advances in molecular biology with combined gene, protein, and cell technology have established a fundamental approach to this problem. A number of genes are now being analyzed with respect to their DNA signals and the protein factors interacting with them. As a result, some useful pictures are emerging as testable models of macromolecular interaction between DNA and protein as well as protein and protein. One important point that is already apparent is that both the cis-acting elements and the transacting factors are probably limited in number as compared with the complexity of the genome and its regulation. Not only are enhancers and other regulatory regions composed of a limited number of cis-acting elements by combinatorial arrangements, but the transacting factors also appear to be composed of a limited number of DNA-binding and transactivating domains, which are combined during the evolution of the gene. Nevertheless, many more genes must be studied before a comprehensive understanding of the cis-acting elements and the transacting factors is obtained. When sufficient numbers of these components are identified and analyzed, we will then be able to define the network of multicomponent and interdependent gene regulation, which is the basis for development, differentiation, and homeostasis of organisms such as humans. Other closely related subjects not discussed in this article include chromatin structure, nuclear scaffold or matrix, and DNA methylation, for which separate chapters are available in this book. [*See* CHROMATIN FOLDING; DNA METHYLATION AND GENE ACTIVITY; DNA IN THE NUCLEOSOME.]

Bibliography

Evans, R. M. (1988). The steroid and thyroid horome receptor superfamily. *Science* **240,** 889.

Gehring, W. J. (1987). Homeo boxes in the study of development. *Science* **236,** 1245.

Johnson, P. F., and McKnight, S. L. (1989). Eukaryotic transcriptional regulatory proteins. *Annu. Rev. Biochem.* **58,** 799.

Maniatis, T., Goodbourn, S., and Fischer, J. A. (1987). Regulation of inducible and tissue-specific gene expression. *Science* **236,** 1237.

McKnight, S., and Tjian, R. (1986). Transcriptional selectivity of viral genes in mammalian cells. *Cell* **46,** 795.

Mitchell, P. J., and Tjian, R. (1989). Transcriptional regulation in mammalian cells by sequence-specific DNA binding proteins. *Science* **245,** 371.

Ptashne, M. (1986). "A Genetic Switch. Gene Control and Phage." Cell Press, Cambridge, Massachusetts.

Ptashne, M. (1988). How eucaryotic transcriptional enhancers work. *Nature* **336,** 683.

Saltzman, A. G., and Weinmann, R. (1989). Promoter specificity and modulation of RNA polymerase II transcription. *FASEB J.* **3,** 1723.

Watson, J. D., Hopkins, N. H., Roberts, J. W., Steitz, J. A., and Weiner, A. M. (1987). "Molecular Biology of the Gene," 4th ed. Benjamin/Cummings, Menlo Park, California.

TTACCACCGGC
AATGGTGGCCG
CTATCACCGCA
GATAGTGGCGT
CTAACACCGTG
GATTGTGGCAC
TTACCTCTGGC
TTACCACCGGC

DNA Binding Sites

OTTO G. BERG, *Uppsala University Biomedical Center*

Glossary

Nonspecific Refers to properties shared by all DNA sequences

Protein–DNA specificity Ability of a protein molecule to bind preferentially to one particular sequence of DNA base pairs, but not to others

Pseudosites DNA sequences in the genome that resemble specific sites, but have appeared by random chance and do not serve any function

IN ALL ORGANISMS a large fraction of the DNA is covered by bound protein of various kinds. Most of these proteins serve to organize—or roll up—the linear DNA molecule into a more compact structure. These structural proteins are bound mostly in a nonspecific way; that is, they bind anywhere along the DNA. Other regulatory proteins require for their proper function that they bind predominantly to specific sites on the DNA. Such specific sites are defined by a certain sequence of DNA base pairs that the protein can recognize and bind to. The biological function and molecular design of these specific DNA binding sites are the subjects of this article.

I. Purpose of DNA Binding

The genetic information is stored in the huge linear DNA molecule. For this information to be disseminated in a systematic and regulated way, regulatory proteins must be directed to the appropriate regions of the DNA and retained there while their action is needed. Thus, the main purpose of DNA binding sites is to serve as attachment points for gene-regulatory proteins. As a first step in the process of reading the genetic information, genes are transcribed into RNA by the enzyme RNA polymerase. Starting from its attachment site—the promoter—RNA polymerase follows the linear DNA and sequentially copies a gene—or a cluster of genes—into RNA. The promotor not only signals the starting point, but can also determine how often a gene is transcribed. [*See* DNA AND GENE TRANSCRIPTION.]

The need for different gene products depends strongly on the developmental stage of the cell and on its environment. Therefore, a sensitive control is required so that genes can be turned on and off by internal or external signals. This is achieved by a whole range of regulatory proteins that can bind to specific sites in or around the promoters of their particular target genes and interact with the polymerase to either block or stimulate its activity at the promoter site. Genes can be turned off by repressor proteins that bind to operator sites in or near the promoter and thereby block the access of RNA polymerase. Sites for the binding of activator proteins, which stimulate RNA polymerase, are sometimes also located in the promoter region, but often such sites (e.g., the enhancer sites) can be found at a considerable distance.

Almost all of the known specific DNA binding sites for protein are involved in gene regulation. Other such sites are found, for example, at the origin of DNA replication. Some structural proteins could also require specific DNA sequences to direct their binding. It can be expected that the molecular design of these other types of binding sites follows the same principles as discerned for the gene-regulatory sites discussed here.

Gene regulation and its dependence on the structure and arrangements of DNA binding sites are best understood in prokaryotes, and it is mostly on these systems that the present discussion is based. The basic physical principles for protein–DNA recognition are the same also in eukaryotes and higher organisms, but there are important differences in the functional arrangements of binding sequences. For example, prokaryotic genes have distinct promoter sites, while the promoter sites in eukaryotic genes are much more complex in terms of DNA sequence. These differences in molecular design might be connected with the much greater complexity and needs of development for eukaryotes.

The restriction sites offer an example of recognition sites of a somewhat different nature. These sites, where the DNA chain can be cut by special enzymes (i.e., restriction enzymes) are spread throughout the genome, rather than situated at specific locations. These sites, therefore, do not have the same functional requirements as gene-regulatory sites, although in the two cases the recognition mechanism is based on the same kind of physical protein–DNA interactions.

II. Protein–DNA Recognition Is Based on Physical Interactions

The target for a DNA-binding protein—the DNA molecule—constitutes a long regular helical structure. The outer surfaces of this helix look much the same along the entire length. A specific DNA sequence is defined from a certain combination of the four base pairs (i.e., A · T, T · A, C · G, and G · C). The genome of a bacterial cell consists typically of a DNA molecule with a few million base pairs ordered in a linear array; the genome of a higher organism can be 1000-fold larger. Thus, a protein molecule that should bind specifically to only a few sites must be able to distinguish a specific DNA sequence among a vast excess of all other structurally similar parts of the DNA molecule. The resolution of this problem requires both adequate physical interactions and proper base pair coding of individual sites. [*See* HUMAN GENOME AND ITS EVOLUTIONARY ORIGIN.]

A. Structural Fit

The primary requirement for binding is a structural fit among the molecules that allows a sufficient number of weak interactions to be established. Such a structural fit often involves a cleft on the protein where the DNA helix can be lodged; also, chemical groups on the protein can be fitted into the helical grooves of the DNA for more intimate contacts. To achieve sufficient binding strength, this structural complementarity—called "lock-and-key" fit—must be combined with an interactional complementarity such that the groups that are brought into contact also will be held together by physical interactions.

B. Electrostatic Interactions

The binding of protein to DNA has a strong electrostatic component mediated by the attraction between the negative charges on the phosphate groups along the backbone on the outside of the DNA helix and a collection of positive charges on the DNA-binding surface of the protein. This binding is predominantly nonspecific, since the backbone is essentially the same along the DNA molecule and independent of the particular sequence of DNA base pairs from which it is built. The electrostatic interactions provide the main contribution to binding for nonspecific proteins. The binding of proteins to specific sites also has a strong electrostatic component; as a consequence these proteins bind not only to their specific sites, but also, though much more weakly, to all other sequences in the genome.

C. Hydrogen Bonds

A hydrogen bond is formed when a hydrogen atom is shared by two other atoms. It occurs when a hydrogen that is covalently linked to a particular atom (i.e., the donor, most often oxygen or nitrogen) is brought into contact with another atom (i.e., the acceptor, also most often oxygen or nitrogen) with which it can be shared.

Both protein and DNA molecules carry many molecular groups that can function as hydrogen bond donors or acceptors. Of particular importance are those that are exposed in the grooves of the DNA helix; these are sequences specific (i.e., the pattern of hydrogen bond donors and acceptors in the grooves is determined by the base pair sequence of the DNA). Specific binding of a protein to DNA is achieved when the pattern of hydrogen bond donors and acceptors on the protein is complementary to that of a particular DNA sequence, such that the juxtaposition of the molecules can correctly align individual donors with acceptors and vice versa. Even if the juxtaposition leaves only a few donors or acceptors unpaired, the loss of potential binding energy is too large and the complex cannot hold together. Thus, the hydrogen bond complementarity constitutes an extremely sensitive determinant for the specificity of protein–DNA binding and is the main criterion for the molecular design of DNA binding sites.

D. Other Interactions

There are other effects that influence both binding and specificity. For example, hydrophobic interactions, particularly with the methyl group of thymine, can also contribute to specific binding.

Furthermore, the binding of a protein is likely to distort the structure of the DNA somewhat in order to achieve the best fit among interacting groups. Such structural distortions are energetically costly and therefore reduce the binding energy. Since the flexibility of DNA depends on the base pair sequence, the energetic cost is also sequence dependent. Thus, while they can only decrease the overall binding strength, structural DNA distortions can nevertheless contribute to specificity by decreasing the binding to different extents on different sequences.

These effects should be considered mostly as a modulating influence on the basic contributions from electrostatic and hydrogen bond interactions, since they are not sufficiently limiting by themselves to define a specific binding site for a protein.

Some particular sequences of DNA base pairs can take on structures that differ greatly from the usual DNA helix (e.g., "left-handed" DNA or cruciform DNA), and proteins might exist that, through a complementary fit, specifically recognize such alternate structures. This could provide an effective way of defining specific binding sites, but no such case has been identified.

III. Functional Requirements

The most important requirements for the biological function of DNA binding sites are that the protein be able to find its target site(s) within a reasonable time, and, once there, be able to stay long enough to execute its function. The physical protein–DNA interactions and the base pair coding of the specific sites are arranged to meet these requirements.

A. Binding Strength

For many gene-regulatory proteins (e.g., repressors and activators) the biological activity is determined primarily by the probability of occupancy at individual binding sites; the extent to which a gene is repressed, for example, is proportional to the fraction of time that a repressor is bound at the operator. The occupancy, in turn, is determined by the binding strength and the availability of the required protein. A weak binding can to some extent be compensated for by an increased amount of protein to provide a sufficient level of occupancy.

Function is not always determined by binding alone, and in these cases specific activity would be a more important property; for example, RNA polymerase requires a series of activation events in which the DNA helix is opened before transcription initiation can take place. However, binding to the specific recognition sequence is the first step in function. To the extent that signals for the activity are carried by the DNA sequence, for simplicity of discussion in this article, they are lumped together with the binding specificity.

B. Functional Specificity

Apart from their strong binding to specific sites, most gene-regulatory proteins also have a weak—mostly electrostatic—nonspecific affinity for DNA in general. Because of the large amounts of DNA in the cell, even a weak affinity can lead to a strong competitive effect when large fractions of the gene-regulatory proteins are bound at nonspecific and nonfunctional DNA sites. While the specificity of a particular protein is large, exhibiting a large difference in binding strength between a specific and a nonspecific site, in the cell the specificity is effectively much smaller.

In the living cell a large fraction of the genome is covered by structural proteins. Even if this fraction is not available and therefore not contributing to the competitive binding of the gene-regulatory protein, nonspecific competition is expected to be appreciable. Although the exact organization of the DNA in the living cell is not sufficiently well established for any definite conclusion, it can be surmised that an increased availability of the genetic control regions relative to other parts of the DNA could serve to increase effective specificity.

Apart from the purely nonspecific competition, also expected are a whole range of randomly occurring pseudosites that more or less resemble the specific ones in sequence (see Section IV,A). This follows from the limited site sizes used and the sometimes slow decrease in binding strength as more "wrong" base pairs are allowed in a site.

Both a weak binding strength and the binding competition from nonspecific and pseudospecific sites can be compensated for by increasing the amount of protein in the cell. The functional specificity can be defined from the amount of protein that is "lost" on nonfunctional sites. Thus, specificity must be balanced against the physiological cost of investing in more protein.

C. Functional Variability

Specific sites for the same protein require different binding strengths, depending on the physiological need for the particular gene(s) they control. To achieve sufficient fine tuning of the binding strength, some of the base pairs that contribute to specific binding do so only to a smaller degree. In this way a precise adjustment of binding at individual sites can be achieved by replacing weakly interacting base pairs. However, to achieve sufficient specificity within the limited binding site size, other base pairs must be discriminatory and therefore contribute strongly to binding. Thus, the specificity and variability requirements lead to a whole range of base pair interaction levels. As discussed in Section IV,C, these different interaction levels are to some extent reflected in the statistics of base pair choice.

D. Kinetics

Molecular complexes can be formed after random diffusional motions have brought the molecules into contact. The specific binding of protein to DNA requires a precise alignment of the molecules before the interacting groups can form their bonds. This precise alignment has a low probability of occurring by chance when the molecules are tossed around by the random thermal motions. As a consequence the time it would take to form such a specific complex might become exceedingly long.

A nonspecific complex requires much less precision and can therefore be formed much more rapidly; it is sufficient that the positive charges on the DNA-binding surface of the protein are brought into the neighborhood of the negative charges on the outside of the DNA. Once nonspecifically bound, the protein can explore the grooves of the DNA helix for specific interaction possibilities. Such a two-step binding can significantly increase the rate of specific complex formation.

It has been found that some gene-regulatory proteins can make even more efficient use of the nonspecific binding intermediate in that they can actually "slide" in a one-dimensional diffusion along the nonspecific DNA in search of specific binding interactions. While this sliding mechanism can lead to a fast association in experimental systems, its efficiency is expected to be much smaller in the living cell; even a weak nonspecific binding to the large amounts of DNA present leads to competition effects that decrease the efficiency. Consequently, there is a delicate balance between the rate-enhancing and rate-decreasing effects of the nonspecific DNA binding. Nevertheless, the kinetic requirements for a sufficiently fast association could be a major reason that the gene-regulatory protein has a nonspecific binding at all. This could therefore provide and important limitation on the usefulness of too large a specificity.

E. Protein–Protein Interactions

Specific binding sites are sometimes arranged in tandem (i.e., next to each other) on the DNA, such that a regulatory protein bound at one site can interact with proteins bound at neighboring sites. In this way the specific DNA binding of a particular protein can require the presence of several other protein molecules on nearby sites. This use of protein–protein interactions is one way of effectively increasing specificity without unduly increasing the size and specificity requirements of the individual proteins. This arrangement also affords more sensitive and intricate control of gene expression by requiring the simultaneous presence of sometimes many different protein factors.

As only a limited number of sites can be arranged in the regions surrounding a promoter, packing problems can arise when the control mechanisms require more binding sites. One way of avoiding such problems is to make use of the flexibility of the linear DNA molecule, since bending and looping can bring distant DNA segments into close contact. This seems to be the strategy used by the enhancer sites, which can be placed at distances of 1000 base pairs or more away from the promoter site. Proteins bound at such sites can be brought into contact with RNA polymerase at the promoter by looping of the DNA. This kind of arrangement allows much more freedom in the specification and location of specific binding sites when the linear constraints of the DNA molecule are effectively removed.

IV. Coding Constraints and the Statistics of Base Pair Choice

The pattern of interaction possibilities exposed on the DNA-binding surface of the protein has evolved to interact more favorably with some DNA sequences than others. Simultaneously, the specific DNA sequences have been chosen to provide a physiologically adequate binding at the required sites in the DNA molecule. The result of this mutual evolutionary adaptation can be seen in the DNA sequences used to define the functional sites.

A. Site Size

The size of the protein imposes a limit on how many DNA base pairs it can interact with simultaneously; for a typical gene regulatory protein this would be about 15–30 base pairs. This physical limitation on the site size provides an upper bound on the possible number specific interactions, although not all of the base pairs in a site contribute to specificity.

There is also a statistical limitation on the site size: A specific site cannot be too small either. If the number of base pairs that contribute specific interactions to the binding of a gene-regulatory protein is too small, by random chance there could appear a large number of sites that are identical to a specific one. The probability that a specified sequence of n base pairs will appear by random chance is $(\frac{1}{4})^n$, if each of the four possible base pairs is equally likely to be chosen. For $n = 11$, this corresponds approximately to one chance in 4 million. Thus, in a bacterial genome with a few million base pairs, unique specific sites would have to be defined by sequences of about 11 base pairs or more. In higher organisms, with their much larger genomes, this limitation could be much more severe.

This statistical minimal site size estimate is based on an absolute discrimination such that sites with one wrong base pair will not bind at all. In actuality in real sites many of the specific base pairs contribute relatively weakly to binding discrimination, and a site can carry a number of wrong base pairs and still be a fairly effective binding site. This is necessary for the functional variability of different sites. As a consequence, however, the site size must be larger than the minimal estimate to avoid the competitive presence of too many nonfunctional random sites in the genome. In fact, the site sizes actually used seem barely sufficient to keep the expected number of such randomly occurring pseudosites within reasonable limits.

B. Sequence Variability

Most gene-regulatory proteins can bind to several functional sites in the genome. On one extreme there is RNA polymerase, which recognizes 1000 or more promoter sites in the bacterial genome, while some repressor proteins bind only a few operator sites. Although similar in sequence, the different binding sites for the same protein are not identical.

Sometimes this sequence variability is functional: Genes that are regulated by the same protein are needed to different extents and therefore require different binding of the protein. There is also a possibility that a binding site overlaps with other distinct sequences—either a binding site for a different protein factor or part of a gene—such that other base pair requirements must be simultaneously satisfied.

If the same binding and function can be achieved by several somewhat different sequence combinations, there is also a statistical variability. In this case the actual base pair sequence used among the functionally equivalent possibilities might be a matter of evolutionary history and random choice.

The pattern of interaction possibilities on the DNA-binding surface of the protein is expected to define a distinct DNA sequence to which it has the best complementarity and binding. Due to the large variability, however, this best binding sequence is not likely to be in wide use as a functional site in the genome. Nevertheless, binding sites for a particular protein often share sufficient similarities in the sequence choice that potential sites in a stretch of sequenced DNA can be identified by the human eye.

C. Consensus Sequences

By examining the base pair sequences of a set of functional specific sites for a particular protein, one can derive information on the importance of individual base pair choices at the various positions in a site. Some base pairs are conserved and occur almost always at particular positions in all sites of the set. Such base pairs are likely to contribute strong binding interactions for the protein. Others are only weakly preferred and are therefore expected to contribute less to the binding or activity.

By listing the most commonly occurring base pair at each position, one finds the consensus sequence. This can be expected to be similar to the sequence that would exhibit the best complementary fit with the binding surface of the protein. In this way the statistics of base pair choice carry information about the molecular design of the protein.

One can also use the variability in the base pair choices to quantitatively correlate the degree of preference for a certain base pair with the strength of its contribution to protein binding. This makes it possible to predict the relative binding strengths for arbitrary sequences. Although such predictions agree fairly well with experimental data (when available), they are fraught with uncertainties, both conceptual and statistical. Among other concerns care must be taken that conserved base pairs are not required for other purposes, perhaps contributing to the binding of some other factor at overlapping binding sites. When only a few binding sites are known, there is also a large statistical uncertainty in the estimate of the degree of preference for an individual base pair. Consequently, such a prediction can only provide a rough idea of the strength of binding and activity before these quantities have been experimentally determined.

V. Examples

The molecular design of DNA binding sites is based on the physical protein–DNA interactions and on an adequate base pair coding to satisfy the functional requirements discussed above. To see how real systems have developed to cope with these demands, it is helpful to look at some examples.

A. Restriction Sites

As a defense against invading foreign DNA in a cell, restriction enzymes can cut DNA at well-defined sites. Over 100 enzymes with different DNA specificities are known. The recognition sequences are defined by four to 10 base pairs. In contrast to the other types of sites discussed here, the function does not require specified location of the sites, and they are found anywhere. This is one reason that their sequence length can be so much smaller than that of the gene-regulatory sites. A specific sequence of only four base pairs is expected to occur by random chance every few hundred base pairs, while one with ten occurs once in every million base pairs.

Furthermore, the function does not require a graded or varied response at different sites; a site is either cut or not. Thus, there is no requirement for a functional sequence variability and consequently no need for base pairs that contribute only weakly to specific recognition.

This example shows that the physical protein–DNA interactions can be sufficiently discriminatory to distinguish DNA sequences of only a few base pairs. However, the gene-regulatory sites must involve a large number of less discriminatory base pair interactions to satisfy their requirements for a unique location and a functional variability.

B. Operators

The design of bacterial DNA binding sites is best understood for operator sites, which were the first to be studied in systematic detail, both functionally and by physicochemical means. More recently, some repressor–operator complexes have also been examined by X-ray crystallography.

Most repressor proteins are built up from two or more identical subunits, providing a twofold symmetry. Similarly, the operator sites have a matching symmetry in their sequence. This is one way of increasing specificity (i.e., twice the interactions with DNA) without requiring an expanded protein design, since the same protein unit is used twice. Furthermore, the twofold symmetry allows the protein to bind in either orientation, thereby doubling its likelihood of finding a specific site.

Table I lists a set of six recognition sites for two repressors, the cro protein and the λ repressor in phage λ. The first entry is the consensus sequence for which the twofold symmetry is most evident, as

TABLE I Recognition Sequences for cro Protein and λ Repressor

Consensus[a]	TTACCACCGGCGGTGATAA AATGGTGGCCGCCACTATT
O_{R3}	CTATCACCGCAAGGGATAA GATAGTGGCGTTCCCTATT
O_{R2}	CTAACACCGTGCGTGTTGA GATTGTGGCACGCACAACT
O_{R1}	TTACCTCTGGCGGTGATAA AATGGAGACCGCCACTATT
O_{L1}	ATACCACTGGCGGTGATAC TATGGTGACCGCCACTATG
O_{L2}	TTATCTCTGGCGGTGTTGA AATAGAGACCGCCACAACT
O_{L3}	TAACCATCTGCGGTGATAA ATTGGTAGACGCCACTATT

[a] The consensus sequence is formed by taking the most commonly occurring base pair at each position in the six recognition sequences listed.

the lower strand read from right to left is almost identical to the upper one read from left to right. The other entries are the six real sites, labeled with their names. The elements of conservation, variability, and symmetry are obvious in these sequences. Similar properties are found in most, if not all, other lists of both operator and activator sites in prokaryotes.

C. Prokaryotic Promoters

The recognition sites for RNA polymerase (i.e., promoters) are placed at the beginning of a gene and serve to set the polymerase at the correct position for transcription initiation. Since the gene is to be transcribed on one strand only and in one direction, the promoter sequence must also bind the polymerase in a correct orientation. Therefore, in contrast to the operator sites, promoters cannot be symmetrical in sequence.

RNA polymerase is a large protein that can interact with a long stretch of the DNA. The bacterial promoters are approximately 50 base pairs or more in length, but not all of these base pairs are important for recognition and activity. Two regions of six base pairs each seem to contribute most to specificity. These are located at around 10 and 35 base pairs upstream from the point where the transcription is initiated. In the -10 region the consensus sequence is TAtaaT (small letters symbolizing less conserved base pairs), and in the -35 region it is TTGaca.

The importance of these two regions for RNA polymerase activity have been demonstrated in several ways: They contain the most conserved base pairs, most of the deleterious mutations identified are localized there, and chemical probes show that RNA polymerase make important physical contacts with the DNA in and around them.

Clearly, 12 base pairs—of which perhaps only six or fewer are really strongly required—are not sufficient to define unique sites. In fact, based on these criteria, promoterlike elements can be found scattered everywhere in the genome. Although there are other regions with weakly conserved base pairs in the promoter sequences that also contribute to specificity, it has proved difficult to define unequivocally the necessary and sufficient sequence requirements for a functioning promoter. Possibly, also regions outside the promoter contribute to the specificity of transcription initiation by RNA polymerase, as seems to be the case in eukaryotes.

D. Promoters in Higher Eukaryotes

In eukaryotes there are three different forms of RNA polymerase that recognize different kinds of promoters, with varied and sometimes confusing DNA sequence requirements.

RNA polymerase II transcribes those genes that code for protein, and it is the polymerase that must recognize the largest number of sites. Its recognition sequences share some regions that are partially conserved also among different species. One of the most important ones is the so-called TATA box, which is located approximately 25 base pairs upstream from the start site. This seems to serve a role similar to the -10 region in the bacterial promoter: to position the polymerase accurately for transcription initiation. Further upstream from the start site is quite often a CCAAT sequence, but its exact location does not seem to be crucial, and it can even occur on the opposite DNA strand. Also, other upstream sequences seem to be required; some of these are present only for certain kinds of genes and might serve as specific signals to identify such genes.

These upstream sequences seem to be binding sites for the auxiliary protein factors needed before RNA polymerase II can recognize the start site. Thus, the physical recognition of the DNA sequence by the polymerase is partially indirect through other bound protein factors. This could create a much more efficient recognition surface for the polymerase, by removing the requirement for a linear search along the DNA. Exactly what factors are needed and where they bind is currently under intense experimental study.

The logic of this design could be connected to the large size and complexity of the eukaryotic genome; the probability that these various binding elements would occur at random in the same region is quite small, even in a large amount of DNA. At the same time the seemingly lax sequence requirements allow a large measure of freedom in both the construction and control of promoter sites.

E. Enhancer Elements

The activity of polymerase II at promoter sites can also be strongly influenced by DNA sequences far outside, sometimes several thousand base pairs away from, the start site, either upstream or downstream inside the gene. Of particular interest are the enhancer elements, first identified in some DNA tumor viruses. These consist of individual sequences—enhancer motifs—that are organized into fairly long stretches of DNA of up to a few hundred base pairs. Their function is largely independent of position and orientation with respect to the gene itself. The individual sequences are thought to be binding sites for activator (or sometimes inhibitor) proteins, which can be brought into direct contact with the RNA polymerase at the promoter by bending of the DNA into a loop. Other mechanisms are also possible, however; for example, protein binding at the enhancer could change the DNA structure and thereby make the promoter more accessible.

One example of enhancer motif is the hormone-responsive elements present in hormone-sensitive enhancers. These sequences are about 15 base pairs in length and serve as binding sites for hormone-receptor proteins. Steroid hormones work as signal molecules that, by binding to their receptor proteins, change the expression of particular genes. Binding of the hormone activates the receptor protein, and from its recognition sites in the hormone-sensitive enhancers, it can stimulate polymerase activity.

Due to the complex structure of the enhancer elements comprising recognition sequences for a large number of different proteins, their exact function has been difficult to resolve. Clearly, the modular structure of the enhancers, where different binding sequences can be introduced and shuffled around, is similar to the structure of the promoters themselves and allows for complex control mechanisms. The modular structure also allows for a relatively facile reorganization of such control circuits; it is likely that such changes constitute one of the major evolutionary pathways for higher organisms.

VI. Concluding Remarks

There has been explosive development in the study of gene control and the role of DNA binding sites due to recent advances in experimental techniques, notably genetic engineering and DNA sequencing methods. However, this is still the beginning, and many important questions remain. Although the basic principles of protein–DNA binding and the fundamental design of binding sites are fairly well understood, the organization and function of these binding sites in their biological and physiological contexts remains to be elucidated, particularly for higher organisms. This is of immense importance for further advances in both biology and medicine.

Bibliography

Berg, O. G., and von Hippel, P. H. (1988). Selection of DNA binding sites by regulatory proteins. *Trends Biochem. Sci.* **13,** 207.

Brennan, R. D., and Matthews, B. W. (1989). Structural basis of DNA–protein recognition. *Trends Biochem. Sci.* **14,** 286.

Ptashne, M. (1986). "A Genetic Switch: Gene Control and Phage λ." Cell Press, Cambridge, Massachusetts.

Ptashne, M. (1988). How eucaryotic transcriptional enhancers work. *Nature (London)* **335,** 683.

von Hippel, P. H., and Berg, O. G. (1986). On the specificity of DNA–protein interactions. *Proc. Natl. Acad. Sci. U.S.A.* **83,** 1608.

Watson, J. D., Hopkins, N. H., Roberts, J. W., Steitz, J. A., and Weiner, A. M. (1987). "Molecular Biology of the Gene," 4th ed. Benjamin/Cummings, Menlo Park, California.

DNA in the Nucleosome

RANDALL H. MORSE, *National Institutes of Health*

Glossary

Chromatin Ordered complex of protein and DNA, including both nucleosomal and nonnucleosomal proteins, found in eukaryotic cells

Gel electrophoresis Method for separating DNA molecules according to size by allowing them to migrate through a gel matrix under the influence of an electric field

Major and minor grooves of DNA Larger and smaller of the two grooves that follow the path of the helix in double-stranded DNA (see Fig. 1, bottom)

Nuclease Enzyme that cleaves DNA or RNA

IN EUKARYOTIC ORGANISMS (organisms whose cells have nuclei, such as yeast, plants, and animals), DNA is complexed with proteins in an orderly array. The first level of packaging occurs in the nucleosome, a structure consisting of approximately 160 base pairs of DNA wrapped around a disk-shaped complex of proteins called histones. Nucleosomes have been crystallized and, using physical and biochemical techniques, much has been learned about their structure. The packaging of DNA into nucleosomes presents a potential problem inside the cell, where vital processes such as transcription and replication depend on the recognition of particular DNA sequences by protein factors. Whether and how such factors can recognize DNA sequences, which are both structurally deformed and sterically obscured by their incorporation into nucleosomes, is currently under investigation.

I. Introduction

The DNA content of a cell of the bacterium *Escherichia coli* is about 5 million base pairs (bp), whereas human cells have a content of about 3 billion bp, of which there are two copies. These amounts of DNA, if stretched out in the form of solitary double helixes, would measure 1.4 mm and 1 m in length, respectively. Because these lengths are many times greater than the diameter of the vessels that contain them (bacterial cells or eukaryotic cell nuclei), some form of compaction or packaging of the DNA must take place to allow it to fit within the container. Although it is conceivable that the DNA might simply be squashed into the cell like string, this is not the case. In both prokaryotes (bacteria) and eukaryotes (higher organisms), proteins associate with and compact the DNA. In prokaryotes, this function is fulfilled by the histone-like, or HU, proteins, which are not discussed here. In eukaryotes, the packaging of DNA takes place at several levels, giving an overall compaction in length relative to the extended double helix of about 10^4 (Fig. 1). The highest degree of compaction can be visualized as the metaphase chromosome, a complicated structure that allows the orderly segregation of the genetic material to progeny cells during cell division (mitosis). This structure arises from the ordered folding of what is already an organized structure, the 30-nm fiber (a nm is 10^{-9} m, or a millionth of a millimeter). The 30-nm fiber can be further unfolded to yield a 10-nm fiber of chromatin. The 10-nm fiber, seen under the electron microscope, has the appearance of beads on a string (Fig. 2). The string is

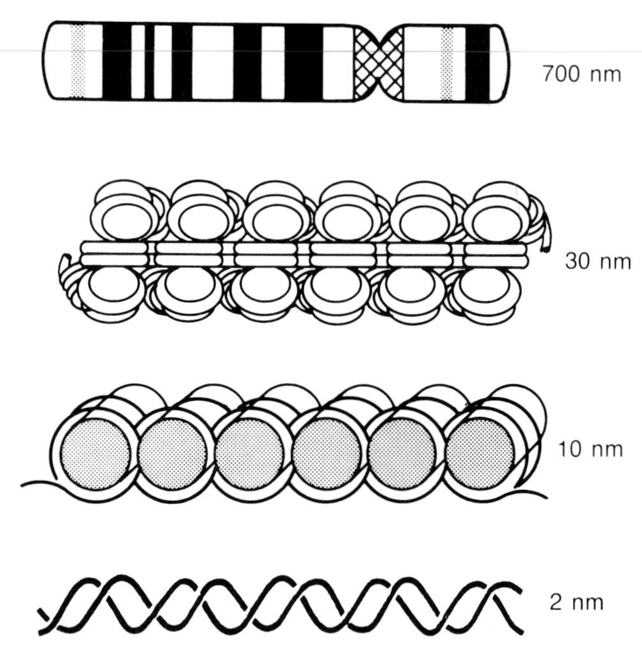

FIGURE 1 Hierarchies of packaging of DNA in eukaryotic cell nuclei. From top: metaphase chromosome; 30-nm chromatin fiber; 10-nm chromatin fiber; naked duplex DNA. The diameters of the various structures are given at the right. Various structures have been proposed for the folding of nucleosomes in the 30-nm fiber; for simplicity, only one is shown. (Redrawn, with permission, from Alberts *et al.*, 1989, "Molecular Biology of the Cell," Garland Publishing, Inc., New York.)

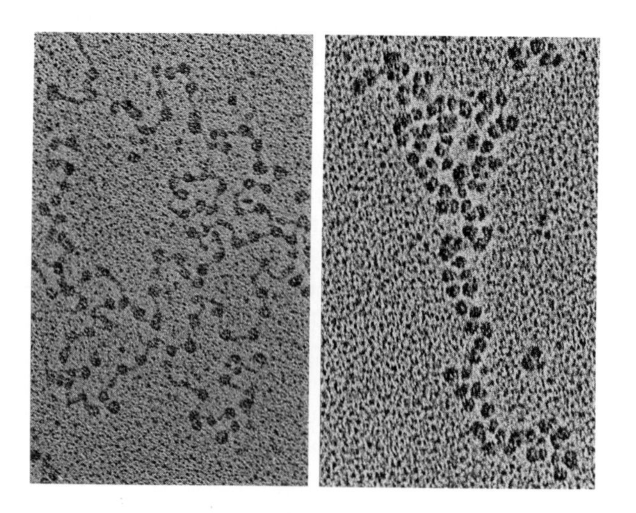

FIGURE 2 Electron micrographs of chromatin. Chromatin as viewed by electron microscopy ($10^5\times$ enlargement) reveals "beads-on-a-string" structure of nucleosomes on the DNA fiber. In the left panel, histone H1 is absent; in the right panel, histone H1 is present, and the chromatin fiber is condensed (see Fig. 1). (Reproduced from Thoma, Koller, and Klug, 1979, *J. Cell. Biol.* **83,** 403, courtesy of Dr. Fritz Thoma; and with permission from The Rockefeller University Press, New York.)

DNA; the beads represent nucleosomes, the basic units of eukaryotic chromatin. [*See* CHROMOSOMES; DNA AND GENE TRANSCRIPTION; HISTONES AND HISTONE GENES.]

II. Structure and Properties of the Nucleosome

A. The General Picture

The nucleosome diagrammed in the 10-nm fiber in Fig. 1 (see also Fig. 3) is a simple structure: roughly two turns of DNA are wrapped around the outside of a disk-shaped object like a length of hose around a hockey puck. DNA, however, is in some respects a more complex structure than a length of hose, and the hockey puck of Fig. 1 is not so featureless either. What the disk is meant to represent is the protein core of the nucleosome. This core is made up of two copies each of the four core histone proteins, H2A, H2B, H3, and H4.

The details of how DNA and histones are arranged in the nucleosome have been deduced primarily from X-ray diffraction and chemical cross-linking studies. The cross-linking work, by establishing particular sites of contact between the histone proteins and DNA, helped in deducing the orientation of the polypeptide chains of the histones when the crystal structure was being solved. A rough picture of the accepted structure is shown in Fig. 3. It consists of 146 bp of DNA wrapped around a disk-shaped core, which is 7.3 nm in diameter and 4.0 nm in height. The entire structure is 11.0 nm in diameter and 5.7 nm along the axis of the disk and possesses near dyad symmetry. (This means that the structure is largely symmetric with respect to a 180-degree rotation around the central axis; as a consequence, if a person could sit on the exact center of the DNA sequence wrapped around the outside, he or she could travel in either direction along the DNA, and the view of the protein core would be nearly the same). The DNA wraps around the outside for about 1.8 turns, with about 80 bp of DNA in a single turn. The histones make contacts all along the inside of the DNA helix but have little if any contact with the outside surface of the DNA. Histones H3 and H4 are assembled in a tetrameric (four-part) structure, which forms the most central part of the nucleosome. Two pairs formed by one histone H2A and one H2B sit on opposite sides of this tetramer, but still inside the DNA loop. These dimers do not sit exactly symmetrically in the crystallized nucleosome. It is not known if this reflects asymmetry also present in nucleosomes in solution (or in the cell) or if it is due to crystal-packing forces.

Finally, an additional protein, histone H1, can bind to the nucleosome at the exit and entry points of the DNA (Fig. 3) to yield a structure termed the *chromatosome*. In most cell types, approximately one molecule of histone H1 is present for each nucleosome. The interaction of histone H1 with the nucleosome core stabilizes the interaction of an additional 20 bp of DNA, so that two full turns, comprising 166 bp of DNA, are stabilized in the chromatosome. Histone H1 (and related forms) are important in higher-order folding of nucleosomes in chromatin. In spite of this important role for H1, the precise nature of its interactions with the nucleosome are largely unknown, and the chromatosome has yet to be crystallized.

B. The Histones

The core histones are small proteins, ranging from 11 to 16 kilodaltons in molecular weight in most eukaryotes (Table I). All four are rich in basic amino acid residues, particularly lysine and arginine, which account for 20–25% of the amino acid content of the histones. This is a property one might expect of proteins designed to bind tightly to an acidic polymer such as DNA. All of the histones, as they exist in the nucleosome, can be divided into a central globular region and a more extended, amino-terminal tail. In some cases, (histones H2A and H3), the carboxyl terminus is also present in an extended configuration. This is evident because the histone tails in the nucleosome are more accessible to proteases (enzymes that cleave the polypeptide chain of proteins) than are the central domains, and from nuclear magnetic resonance studies, which show the amino acid residues in these same regions to be much more mobile than those in the central domains. The histone tails are the site of a number of chemical modifications (acetylation, phosphorylation, and methylation, for example) that take place within the cell. The central globular regions, on the other hand, are subject to virtually no modifications, probably reflecting their relative inaccessibility in the nucleosome.

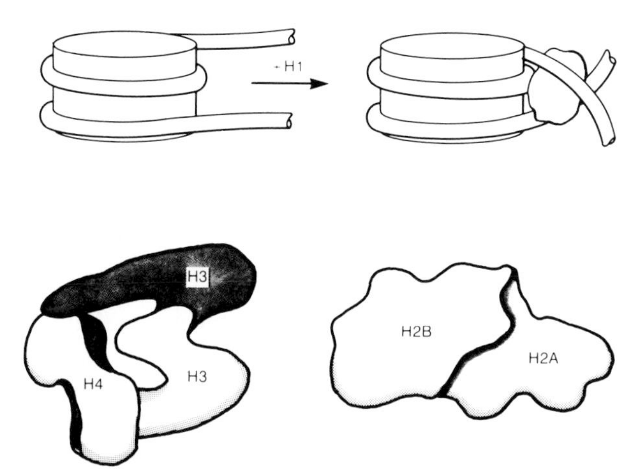

FIGURE 3 The nucleosome. The DNA helix is shown narrower than it would be in proper scale, so that the disk-shaped histone octamer is not obscured. The addition to H1 to the nucleosome core particle gives rise to the chromatosome. The arrangement of the histone proteins in the octamer core is schematized at the bottom, as they would be viewed from the top of the disk in the upper part of the figure. Histones H3 and H4 combine in a tetrameric structure, as shown; the second copy of H4 is represented by the shaded region below the upper copy and is mostly obscured. This tetramer is sandwiched in the nucleosome between two symmetrically disposed dimers of histones H2A and H2B.

TABLE I Properties of the Histone Proteins

	Histone				
	H1	H2A	H2B	H3	H4
Molecular weight (kilodaltons)	23	14–15	14–16	15	11
Lysine content	29%	11%	16–18%	10%	11%
Arginine content	1.5%	7–9%	2–5%	14%	14%

The core histones are remarkably well conserved throughout eukaryotes. For example, the major form of H2A from humans differs from that of chickens in only 5 out of 129 amino acid residues, from that of trout in 10 residues, and from that of yeast in 31 residues. Histones H3 and H4 are even more highly conserved: differences between human and yeast H4 occur in 8 out of 102 residues, and the trout and chicken proteins are identical to human H4. Some sort of evolutionary pressure seems to be operating to keep the histone proteins so highly conserved in their primary structures, but the nature of the damage that an organism would suffer if, for example, some of the amino acids in histone H3 were altered is unknown. (Surprisingly, yeast cells lacking all or part of the N-terminal regions from histones H2A, H2B, or H4—including some highly conserved amino acids—are able to survive quite well.)

Compared with the core histones, histone H1 shows substantial variation among species. Moreover, the types of H1 found in a single organism may vary according to cell type and stage of development and within the cell cycle. Variants of the core histones also can be found, but these represent smaller contributions to the total pool of core histones than the variants of H1 do to its total pool. Some progress has been made toward understanding the role that certain specific histone subtypes play in the cell, but the function of most histone variants, like the function of many of the chemical modifications that the histones undergo, remains enigmatic.

C. Properties of the Nucleosome

Some of the most widely exploited properties of the nucleosome concern the way in which the histone proteins mask nucleosomal DNA from particular chemical and enzymatic agents. Enzymes known as nucleases, for example, will cut naked DNA in solution into very small pieces, so that if the exposure is sufficiently long, the DNA will be degraded to pieces less than 10 bp (or nucleotides) long. One such enzyme is micrococcal nuclease. When chromatin is treated with this enzyme, only the DNA between nucleosomes—the *linker* DNA—is rapidly cut. This results in about 180–200 bp of DNA per nucleosome being protected from digestion under mild conditions and 146 bp showing strong resistance to nuclease cutting even during protracted digestion. Because not all of the linker DNA is immediately cut by the enzyme under mild conditions, when the DNA digestion products are analyzed by gel electrophoresis, a ladder of fragments is seen corresponding to DNA associated with 1, 2, 3, . . . nucleosomes. This allows the length of the linker DNA to be measured. The core DNA—the 146 bp that are strongly protected against micrococcal nuclease digestion in the nucleosome—is virtually invariant in all species studied, again demonstrating the extreme evolutionary conservation of nucleosome structure. The length of the linker DNA, on the other hand, varies among species, among tissues, and even along the DNA within a single cell. There is evidence that histone H1—the linker histone—plays a part in determining the length of the linker DNA, but how this is accomplished, and the significance of variations in linker DNA length, remain to be discovered.

Another nuclease used to probe nucleosome structure is DNase I. This enzyme interacts with chromatin in a very different way from micrococcal nuclease. It is capable of cutting DNA in the nucleosome core, but only every 10 bp, so that the digestion products when visualized by gel electrophoresis form a ladder of fragments of 10, 20, 30, . . . , 140 (and possibly as large as 160 or 170) bp in length; This is because DNase I cuts DNA only in its minor groove and can apparently only gain access to the minor groove where it faces outward on the nucleosomal surface, which occurs once every helical repeat, or about every 10 bp.

Why does DNase I cut DNA in the nucleosome, whereas micrococcal nuclease is much less able to do so? A clue may lie in the crystal structure of DNase I complexed with a small piece of DNA. The structure shows that the DNA fragment is bent away from the enzyme. Nucleosomal DNA bends the same way relative to the enzyme (i.e., when DNase I or any other enzyme approaches DNA in a nucleosome, the DNA is bent away from the enzyme), so the enzyme may have no difficulty in inducing the correct DNA conformation to cut the phosphodiester bond (the bond connecting adjacent nucleotides in one strand of the helix). Indeed, the correct conformation is already present. Micrococcal nuclease, on the other hand, may only cut DNA well when it is straight, or it may even need to bend the DNA toward itself, either of which would be difficult with nucleosomal DNA. This is only a conjecture, however, because the structure of micrococcal nuclease complexed with DNA has not been determined.

In addition to protecting DNA in different ways from digestion by micrococcal nuclease and DNase I, the nucleosome also protects DNA against digestion by restriction endonucleases (enzymes that cut at specific DNA sequences) and hydroxyl radicals. Nucleosomes also stabilize DNA so that it denatures (i.e., separates into the two component strands) at higher temperatures than does naked DNA and prevents DNA from undergoing thermal untwisting (a change corresponding to a slight increase in the separation between adjacent bases, which DNA undergoes with increasing temperature). Other characteristics of the nucleosome include a well-defined sedimentation coefficient, which is greatly different from that of naked DNA, and a change in certain physical characteristics of the DNA, such as the circular dichroic spectrum.

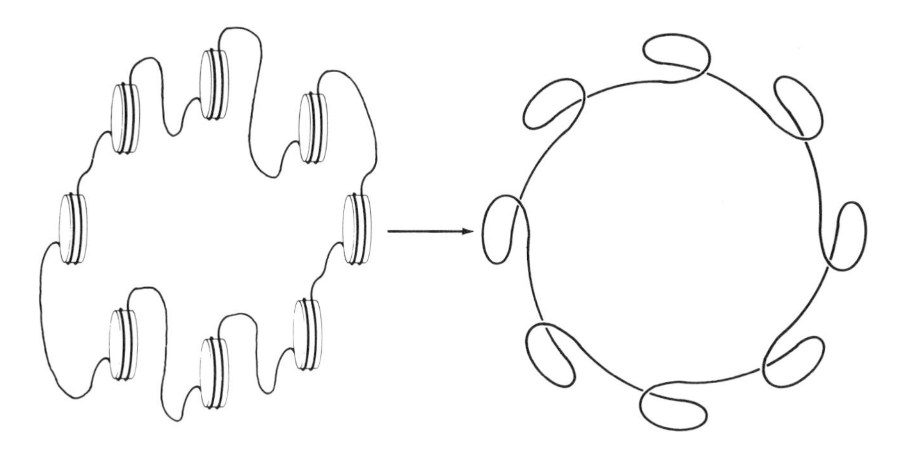

FIGURE 4 Disruption of nucleosomes on a closed, circular DNA molecule leads to the formation of loops, or supercoils, in the DNA. One supercoil arises from each nucleosome. These supercoils are said to be left-handed, or negative. If the strands crossed in the opposite sense the supercoils would be right-handed, or positive.

D. DNA in the Nucleosome

Although the packaging of DNA into nucleosomes is only the first level of compaction in the cell, the double helix must undergo substantial deformation to be wrapped around the histone core. If the 80 bp of DNA that form one turn were smoothly bent, the edge-to-edge separation of adjacent base pairs would increase from 3.4 A (on average) to 3.9 A at the outside edge (a 14% increase), and decrease to 2.9 A at the inside. Some of the details of the crystal structure confirm this kind of deformation: the minor groove of the DNA double helix varies from 7 A on the inner face of the helix (facing the histone core) to 13 A on the outside, and the major groove from 11 A to 20 A. These changes in the conformation of DNA when it is incorporated into nucleosomes may be important in the cell and will be discussed later.

Two other kinds of perturbations that DNA structure undergoes when it is incorporated into the nucleosome are *kinks,* which are sharp bends in the helix, and a change in the *twist,* or helical period. Sharp bends or kinks are found in the DNA helix at one and four helical turns (one turn is about 10 bp) on either side of the dyad (or center) of the DNA in the nucleosome. These bends are apparent in the crystal structure and have also been detected by chemical means. The twist of the DNA helix, which can be defined as the number of base pairs to complete one turn of the helix, changes from an average value in solution of 10.5–10.6 to about 10.0–10.2 in the nucleosome. This may be due to the altered ionic environment seen by DNA in the immediate neighborhood of the nucleosome, caused by the high local concentration of positive charges contributed by the histones.

The wrapping of DNA around the outside of the histone octamer results in another profound change in a fundamental property of the DNA: its writhe, or *supercoiling*. These terms refer to the three-dimensional shape of DNA in space. Supercoiling might be described as how twisted up the helix is on itself. This is a different kind of twisting from that described above; the two strands of the helix can maintain their same helical period, or twist, but the helix as a whole can still be twisted, like a rubber band or the cord to a telephone receiver. Each nucleosome introduces a single supercoil into the DNA, as depicted in Fig. 4. Both the path that the DNA follows around the outside of the histone octamer and the slight change in twist that occurs upon incorporation of DNA into the nucleosome contribute to the exact magnitude of supercoiling that is observed. Supercoils can be put into DNA in the absence of nucleosome formation (e.g., by the prokaryotic enzyme DNA gyrase), but this process requires energy. It has been speculated that removal of a nucleosome or nucleosomes in a local domain of DNA *in vivo* (for instance, in the region of a gene's promoter) could produce a locally supercoiled region of DNA, which might be recognized by factors involved in transcriptional activation. However, to date no direct evidence has been found for this kind of mechanism being involved in transcriptional regulation in eukaryotes.

E. Nucleosome Assembly

Nucleosomes can be assembled *in vitro* from DNA and purified histones. This may seem fantastic—that such a complex structure, involving four different proteins and an intertwined pair of DNA strands, can be faithfully put together from its solubilized constituents—but in fact, the nucleosome is virtually a self-assembling structure. The most commonly used method for reconstituting nucleosomes is simplicity itself: purified histones and DNA are mixed in a buffered solution having sodium chloride at a concentration of 1–2 *M*, and dialyzed against solutions of successively lower ionic strength. The assembly process is strongly aided by the affinities the component histones have for each other. Histones H3 and H4 readily bind together to form a tetramer—$(H3 \cdot H4)_2$—resembling the inner core of the histone ocatmer, and H2A and H2B can form dimers with a structure similar to that in the nucleosome (Fig. 3). Nucleosomes resulting from *in vitro* reconstitution are identical to those isolated from cells in histone content, sedimentation coefficient, supercoiling of DNA, appearance under the electron microscope, protecting DNA against nuclease digestion, and so on.

One important property of chromatin that is not faithfully mimicked by *in vitro* reconstitution is the spacing between nucleosomes. This spacing, as mentioned earlier, is a regular feature in living cells that depends on species, cell type, and so forth. In contrast, when nucleosomes are assembled *in vitro* by salt dialysis onto DNA sufficiently long to accommodate many nucleosomes, the spacing between particles is fairly random. At low nucleosome densities, individual particles are usually spread far apart, with occasional pairs being close-packed (meaning that there is almost no linker DNA between them). At high densities, most of the nucleosomes become close-packed. Certain cell extracts, however (e.g., from frog eggs or oocytes), are capable of assembling chromatin in which the individual nucleosomes are faithfully reconstituted and correct spacing is generated as well. These extracts contain histones as well as assembly factors, such as the protein nucleoplasmin and a factor called N1, which appear to maintain histones in a form that allows for their efficient assembly into chromatin. Investigators are employing such extracts to create templates that can be used in studies of transcription and replication of chromatin. One extract, made from cultured mammalian cells, has the interesting property of causing exogenously added DNA to be replicated and then preferentially assembling the replicated DNA into nucleosomes. Use of this extract should eventually allow dissection of the process by which newly replicated DNA is packaged into chromatin *in vivo*. [*See* CHROMATIN FOLDING; CHROMATIN STRUCTURE AND GENE EXPRESSION IN THE MAMMALIAN BRAIN.]

F. Nucleosome Positioning

Some DNA sequences are incorporated into nucleosomes, *in vivo* and/or *in vitro*, with a preferred rotational and translational orientation with respect to the histone octamer. In such *positioned nucleosomes*, if the DNA sequence that is protected against micrococcal nuclease digestion in the nucleosome were numbered from base pair 1 to 146, base pair 5 (for example) would always be found at the same place in the nucleosome—5 base pairs from one end—base pair 73 in the center, and so forth. But of course the DNA sequence is effectively numbered by its precise nucleotide sequence, so that investigators have been able to show that some sequences do indeed give rise to positioned nucleosomes. Other sequences, however, when reconstituted *in vitro* into nucleosomes, are found to be situated randomly with respect to the histone octamer. Similarly, individual DNA sequences may or may not be associated with positioned nucleosomes *in vivo*.

What determines whether a particular DNA sequence is or is not capable of forming a positioned nucleosome? One determinant is the ease with which a sequence can be bent. Certain DNA sequences form bends in the helix, so that if they are repeated with the same periodicity as the double helix, the DNA is curved overall. Such molecules can even form small circles visible as such under the electron microscope. Given the tight turning of DNA in the nucleosome, it is not surprising that such a sequence would be incorporated in a nucleosome with a preferred rotational orientation. What structural features lead to an exact translational orientation (i.e., why the DNA sequence at, say, +30 from one end is not located at +40, +50, etc.) is not yet clear, although the kinks found near the center of nucleosomal DNA may serve as some kind of positioning signal.

Bending (or perhaps bendability) can allow nucleosome positioning both *in vitro* and *in vivo*. Another mechanism that can position nucleosomes *in vivo* is the presence of other, non-nucleosomal proteins that associate with particular DNA sequences. When this happens (as with some DNA-binding proteins in yeast), the adjacent nucleosomes are restricted in the way the histone octamer can be positioned on the DNA, and this effect can in turn (apparently) ripple out for several nucleosomes. This kind of mechanism for nucleosome positioning may, in fact, have a functional role, as discussed below.

III. Nucleosome Function

A. General Considerations

Processes that use DNA as substrate in eukaryotic cells (e.g., transcription, replication, and repair) depend on the recognition of specific DNA sequences by various regulatory factors. However, for the DNA to be packaged inside the nucleus, it must become tightly associated with histone proteins in the form of nucleosomes, and these in turn are arranged into higher-order structures. The structure of the nucleosome discussed above is likely to hinder the association of DNA with regulatory proteins, which recognize specific DNA sequences. These proteins must recognize either the molecular surface and charge presented by the base pairs (as seen in the minor or major groove), the exact conformation of the phosphate backbone of the double helix, or both. When DNA is organized into nucleosomes, these features are greatly perturbed. Kinks are introduced into the helix; the DNA is strongly bent; the twist is altered, albeit subtly. These changes seem likely to impose strong constraints on any sequence-recognizing protein, requiring that the protein either possess inherent conformational flexibility, or be capable of specifically recognizing those features (bending, kinking) that are specific to nucleosomal DNA. One example of a protein that prefers bent DNA as a substrate and is capable of acting on nucleosomal DNA is DNase I, as discussed earlier.

In addition, the intimate association of the histone proteins with the DNA, as well as the close proximity of the (nearly) two turns of the DNA helix to each other (Fig. 3), may provide steric and ionic barriers sufficient to render the DNA inaccessible to many proteins. Thus, restriction endonucleases are inhibited from cutting DNA that is incorporated into nucleosomes, although the mechanism is unknown. However, restriction endonucleases vary somewhat in their ability to cleave nucleosomal DNA; for instance, two enzymes (HpaII and MspI), which recognize the same four base-pair sequence, were found to differ in their ability to cut DNA within a nucleosome. It may turn out that regulatory proteins cover a whole spectrum, from recognizing only protein-free DNA to recognizing nucleosomal DNA in a particular orientation to being completely indifferent as to whether DNA is packaged into nucleosomes. In other words, from the perspective of proteins that act on DNA, chromatin may appear transparent, completely opaque, or something in between.

B. Transcription, Replication, and Nucleosomes

Nucleosomes could conceivably affect transcription at any of three principal stages: binding of factors required for RNA polymerase recognition and binding; initiation; and transcriptional elongation. *In vitro* studies indicate that eukaryotic RNA polymerase II, which is responsible for production of mRNA, appears incapable of initiating transcription if certain of its promoter sequences are buried within a nucleosome.

Binding of RNA polymerase II to promoter DNA and initiation of transcription require the association of several proteins with the promoter. One of these proteins, transcription factor IID (TFIID), binds to the TATA box, a sequence motif found 25–30 bp upstream of the transcription start site of most eukaryotic genes transcribed by RNA polymerase II. If nucleosomes are assembled *in vitro* onto a gene with a TATA box, transcription by RNA polymerase II is greatly inhibited. If, however, TFIID is allowed to associate with the TATA box prior to nucleosome assembly, transcription is able to proceed when RNA polymerase II and other appropriate factors are added. Similarly, if nucleosomes are assembled onto a gene (the 5S RNA gene) transcribed by RNA polymerase III, another eukaryotic polymerase, transcription is greatly inhibited, but if the transcription factors TFIIIA, TFIIIB, and TFIIIC are first allowed to associate with the gene, nucleosome assembly is not inhibitory. These experiments suggest that either the proteins of the nucleosome or the given transcription factor, but not both, can associate with the relevant DNA sequences and that once TFIID or TFIIIA, TFIIIB, and TFIIIC are bound to the DNA, subsequent nucleosome assembly does not block binding of polymerase or other transcription factors and does not prevent initiation. However, it is not certain that binding of TFIID is all that is required to prevent nucleosomes from inhibiting transcription of TATA-box-containing genes.

Some transcription factors seem to be capable of recognizing DNA in nucleosomes or even capable of displacing nucleosomes. An example is the transcriptional activation from a promoter sequence termed the glucocorticoid response element (GRE). This promoter element activates transcription when it has been recognized by a protein receptor that has bound a hormone molecule. In mouse cells, the free, unactivated GRE appears to be incorporated into a nucleosome; after hormone activation, the nucleosome appears to be lost, and the receptor–hormone complex bound in its place. Moreover, when the GRE is incorporated into a nucleosome *in vitro,* it adopts a particular position within the nucleosome (see the section on nucleosome positioning above), and the receptor–hormone complex is capable of recognizing and binding to the GRE without substantially disrupting nucleosome structure. In the cell, once the active receptor has bound to the GRE, the associated nucleosome may be sufficiently destabilized that other factors (perhaps nucleosome assembly factors) catalyze its dismantling. It is not known whether the particular positioning of the GRE within its nucleosome is important for this process to take place.

A series of experiments that bear on the function of nucleosomes in transcription *in vivo* has been performed in yeast cells. Yeast (*Saccharomyces cerevisi*) have only two copies of the genes for each histone protein, and gene replacement strategies have allowed workers to eliminate one of these (one H3 gene, for example) and mutate or inactivate its partner. The effect of these manipulations on both overall cell viability and transcription of particular genes was studied and correlated with effects on nucleosome or chromatin structure. Little effect was seen upon deletion of one copy of any of the individual histone genes. After the second copies of H3 and H4 were deleted, however, the cells stopped growing because they were unable to progress through part of the cell cycle (G2). Transcription of a number of genes was monitored in these cells. One gene in particular, the PHO5 gene, substantially increased its rate of transcription. Under conditions in which this gene is normally repressed (high inorganic phosphate concentration), the PHO5 promoter is found associated with positioned nucleosomes in normal cells. Some disruption in the chromatin structure of the PHO5 promoter was evident in the cells that had their H3 and H4 genes inactivated, suggesting that the positioned nucleosomes normally found there may have the effect of repressing transcription. Another example in yeast is a study of nucleosome positioning in the presence and absence of the $\alpha 2$ protein. This protein has several functions; one is that upon binding to a sequence in the promoter of the STE6 gene, it represses transcription. When this sequence is placed, by DNA recombination *in vitro,* on a small circular plasmid, which is then introduced into yeast cells, the nucleosomes are strongly positioned in the presence of the $\alpha 2$ factor, and transcription of the nearby gene is low. When $\alpha 2$ is absent, nucleosome positioning is weaker (i.e., nucleosomes are found to adopt different positions on different plasmid molecules), and transcription is increased. It seems that the $\alpha 2$ factor functions as a transcriptional repressor by organizing nucleosomes to incorporate critical promoter sequences, masking them from transcription factors and thereby preventing transcription.

Nucleosomes can also inhibit transcription by creating or allowing formation of higher-order structures. Histone H1, for example, which has been implicated in the folding of chromatin into higher-order structures, has long been known to be associated with inactive genes. Moreover, the 5S RNA gene isolated as chromatin is not transcribed; if histone H1 is removed, the gene is transcribed, and readdition of H1 again abolishes transcription.

Nucleosomes or higher-order chromatin structures can also affect elongation once polymerase has recognized its promoter sequences. *In vitro* studies show that eukaryotic RNA poymerase II can transcribe through nucleosomes, but RNA polymerase III does not. The short genes transcribed *in vivo* by RNA polymerase III, however, have not been shown to be incorporated into nucleosomes, whereas those transcribed by RNA polymerase II are known to be associated with nucleosomes. As to how transcription proceeds through nucleosomes, two principal models have been proposed. One model says that nucleosomes unfold to allow passage of RNA polymerase. This would entail disrupting histone–histone contacts, changing the topology of the DNA in the nucleosome, but maintaining the contacts of at least the noncoding strand with the histone proteins. Following polymerase passage, the nucleosome would refold. The second model postulates that histone–histone contacts are maintained in the core octamer, while the octamer is transiently displaced from transcribed sequences, perhaps by association with neighboring nucleosomes. Studies of chromatin structure over transcribed sequences suggest some transient disruption or modification, but the precise nature of these changes remains obscure.

Less is known about the relation between nucleosomes and DNA replication, partly because the nature and properties of DNA sequences responsible for replication in eukaryotes are still poorly known. However, a replication origin in yeast, which is normally found outside a nucleosome, is inactivated after being incorporated into a nucleosome. The most likely explanation for this result is that the histone proteins prevent recognition of specific DNA sequences by proteins required for replication.

A final question is whether nucleosomes serve any specific function with respect to transcription and replication. Nucleosomes and chromatin may only be important for packaging DNA, and when things are set up normally, they simply stay out of the way of factors needed for transcription and replication. In fact, numerous studies in which genes introduced into cells function as predicted according to their promoter type suggest that transcriptional promoters and their associated binding factors generally are capable of overriding any ability that chromatin may have to mask DNA sequences. However, nature is usually economical, and it is likely to use the capability of nucleosomes to affect cellular processes.

Bibliography

Drlica, K., and Rouviere-Yaniv, J. (1987). Histonelike proteins of bacteria. *Microbiol. Rev.* **51,** 301.

Grunstein, M. (1991). Histones and nucleosomes: Repressors of transcription initiation. *Annu. Rev. Cell Biol.* **6** (in press).

Pederson, D. S., and Simpson, R. T. (1988). Structural and regulatory hierarchies in transcriptionally active chromatin. *ISI Atlas Sci.: Biochem.* **1,** 155.

Pederson, D. S., Thoma, F., and Simpson, R. T. (1986). Core particle, fiber, and transcriptionally active chromatin structure. *Annu. Rev. Cell Biol.* **2,** 117.

Richmond, T. J., Finch, J. T., Rushton, B., Rhodes, D., and Klug, A. (1984). Structure of the nucleosome core particle at 7A solution. *Nature* **311,** 532.

Simpson, R. T. (1991). Nucleosome positioning: Occurrence, mechanisms, and functional consequences. *Prog. Nucleic Acid Res. Mol. Biol.* (in press).

Van Holde, K. E. (1989). "Chromatin." Springer-Verlag, New York.

Wolffe, A. P. (1990). New approaches to chromatin function. *New Biologist* **2,** 9–16.

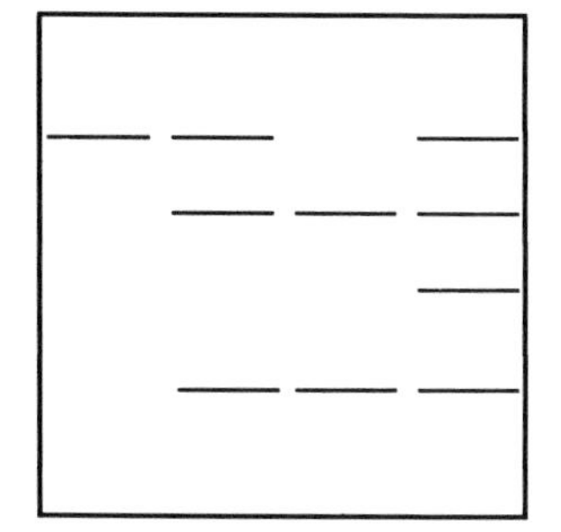

DNA Markers as Diagnostic Tools

ANNE BOWCOCK, *Stanford University and The University of Texas Southwestern Medical Center at Dallas*

Glossary

Allele One of two or more alternate forms of a DNA sequence or a gene occupying the same locus on a particular chomosome; an individual with two similar alleles is said to be homozygous; one with two unlike alleles is said to be heterozygous

Autosomal Chromosome other than a sex chromosome

Carrier Individual heterozygous for a single recessive gene

Dominant Allele manifesting its phenotypic effect also in the heterozygotes; a trait determined by a dominant allele

Genotype Sum total of the genetic information of an organism

Haplotype Combination of individual alleles at linked loci

Karyotype The somatic chromosomal complement of an individual, analyzed according to size and banding patterns of each chromosome

kb Kilobases (1,000 bases)

Locus Position that a gene or a specific sequence of DNA, occupies on a genetic map; alleles are situated at identical loci in homologous chromosomes

Phenotype Observable properties (structural and functional) of an organism; for many DNA polymorphic systems, both alleles are observed and the system is said to be codominant; when only one allele is observed it is considered a dominant trait; the other allele is a recessive trait because it can only be observed when it is on both chromosomes

Polymorphism Existence of two or more different genes or nucleotide sequences in a population; a locus is considered to be polymorphic when the second most common allele is present at a frequency of $>1\%$ in the population

Probe Laboratory definition for a recombinant DNA clone that can be used to detect its specific sequence within complex (e.g., genomic) DNA; can be inserts (of approximately 0.5–5 kb) in plasmid (an extrachromosomal element) vectors, inserts (of approximately 15 kb) in bacteriophage vectors, or inserts of approximately 40 kb in cosmid vectors

Recombination Can occur during reduction division (meiosis), which results in the generation of sperm and eggs that are haploid; refers to the process that gives rise to new combinations of linked genes due to the physical exchange of material between homologous chromosomes; the probability that recombination will occur between two loci is proportional to the physical distance between them

Restriction endonuclease Bacterial enzyme that cuts double-stranded DNA at or near a specific nucleotide sequence

Restriction fragment-length polymorphism Polymorphism in DNA sequence that is recognized by a restriction endonuclease; the loss or gain of the restriction endonuclease site results in a DNA fragment (allele) of different length; the DNA fragment is detected with a specific DNA probe; because both alleles can be observed, the system is said to be codominant; alleles are inherited as Mendelian traits

Vector DNA molecule that can accept a piece of foreign DNA; the resultant recombinant DNA molecule can be taken up by and amplified in a host cell such as a bacterium

RESTRICTION FRAGMENT-LENGTH POLYMORPHISM (RFLPs) are dispersed throughout human chromosomes. The specificity of the restriction endonucleases combined with DNA probes can provide markers that are genetically linked to disease loci. If a marker locus is closely linked to a disease locus, one can follow the segregation of a particular allele in a family that is also segregating a disease allele. The alleles at a marker locus are usually independent of the disease mutation, and in different families different alleles may cosegregate with the disease locus; however, within one family, a particular allele will track a disease gene. For a number of human genetic diseases, prenatal and preclinical diagnoses are now possible by determining which marker alleles have been inherited in the individual at risk. In addition, it is often possible to predict the genotype of relatives. This allows the detection of asymptomatic affected individuals in the case of dominant diseases, carrier females in the case of sex-linked diseases, and the discrimination of carriers versus normals in the case of autosomal recessive diseases. The goal in the detection of disease is to directly detect the mutation giving rise to the disease without relying on linked RFLPs where a low probability of recombination exists between the marker locus and the disease locus. For several diseases, direct detection is now possible.

I. Introduction

The first example of RFLPs used as a tool for antenatal diagnosis, but using a well-characterized gene, was with sickle-cell anemia, which is due to a mutation at codon 6 of the the β-globin gene. A restriction endonuclease, *Hpa*I usually cuts the DNA bearing the β-globin gene into a 7.6-kb fragment in American Blacks. It was found that the β-globin gene with the sickle-cell mutation (β^S) was localized on a 13-kb *Hpa*I fragment >60% of the time. Loss of the first HpaI site and the sickle-cell mutation are unrelated base-pair substitutions, but recombination between the two mutations is very unlikely because they are so closely linked. Thus, in this population, the presence of the 13-kb *Hpa*I fragment also indicated the presence of the sickle-cell mutation, and it was possible to track the inheritance of the β^S allele by tracking the inheritance of the 13-kb *Hpa*I fragment. This development allowed fetal blood sampling to be replaced by DNA analysis of cells obtained from amniocentesis in fetuses at risk of having sickle-cell disease. Many RFLPs within the β-globin gene cluster were later found, resulting in a high probability that genetic diseases at this locus could be diagnosed. It is now possible to directly detect many mutations at this locus, avoiding the use of linked RFLPs. [*See* SICKLE CELL HEMOGLOBIN.]

Because DNA polymorphic sites are scattered throughout the genome, it follows that some are physically close to genes that can be mutated and result in disease. If the DNA polymorphic site and the disease gene region are so close that they are often inherited together, they are said to be linked; i.e., very little recombination occurs between them. The advantage of using linked RFLPs for the diagnosis of genetic diseases is that it is not necessary to have isolated the disease gene. In many genetic diseases, the disease gene is not known or isolated, and prenatal and preclinical diagnoses are only possible with linked RFLPs. [*See* GENETIC DISEASES.]

The alleles generated by restriction endonuclease site polymorphisms can be detected as fragments of different lengths on autoradiographs after Southern blotting (see below). The development of the polymerase chain reaction (PCR; see below) subsequently allowed the direct detection of site polymorphisms. In addition, PCR allows the direct detection of mutations giving rise to disease, once they are identified.

There are now over 2,000 probes detecting RFLPs in the human genome. Some of these have been shown to be linked to disease loci and are used in the diagnosis of their respective genetic diseases.

II. Characterization of RFLPs

A. Detection by Southern Blotting

The detection of RFLPs was first achieved by cleavage of total human DNA with a restriction endonuclease, agarose gel electrophoresis, Southern blotting, hybridization with a radioactive probe, and autoradiography to detect the region of interest (Fig. 1). In the simplest case, where an RFLP is due to the presence or absence of a restriction endonuclease site contained within a region detected by a DNA probe, there will be three possible fragment patterns on the autoradiograph (Fig. 2).

B. Types of RFLPs

RFLPs can be due to an alteration of 1 base pair, which abolishes the restriction endonuclease site;

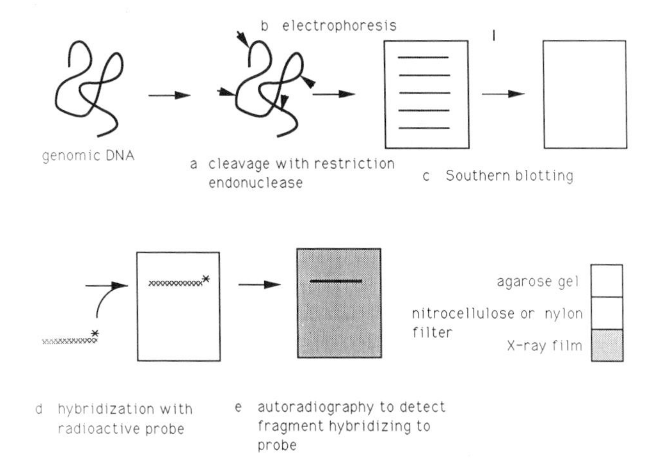

FIGURE 1 Detection of RFLPs. a. Cleavage of human DNA by a 6-base pair cutter (a restriction enzyme that recognizes 6 base pairs and cleaves the DNA at that sequence) generates approximately 1 million DNA fragments. b. DNA fragments are fractionated according to their size by being subejcted to agarose gel electrophoresis in an electric field. This forces the DNA, which is negatively charged, to migrate to the positive pole. For a particular DNA fragment, its rate of migration is inversely proportional to its size. c. Southern blotting involves the denaturation of DNA fragments and their transfer to a solid support (such as nitrocellulose, or a more resilient nylon membrane). d. The addition of a radioactively labeled DNA probe in the appropriate buffer results in hybridization, or annealing, of the probe and the specific DNA fragment to which it is homologous, or complementary. Unbound probe is washed off the membrane. e. The fragment, which is bound to the probe, is vizualized by autoradiography.

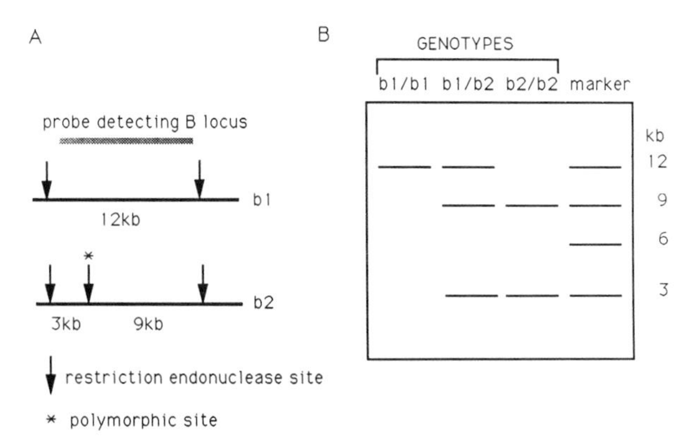

FIGURE 2 Detection of an RFLP. A. The generation of a restriction site in human DNA yields fragments of 3 and 9 kb (b2 allele) instead of 12 kb (b1 allele). B. The different fragment lengths can be visualized at different positions on the autoradiograph, according to where they migrated in the original gel before Southern blotting. DNA markers included in the electrophoresis step allow one to size them accurately. The patterns illustrated are for a homozygote for the b1 allele (b1/b1), a heterozygote for the b1 and b2 alleles (b1/b2), and a homozygote for the b2 allele (b2/b2).

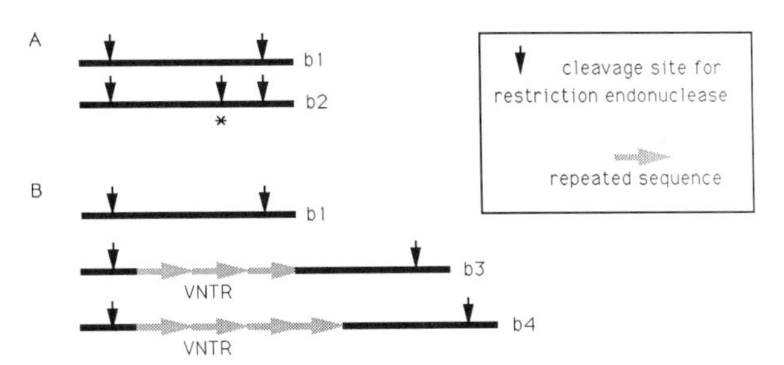

FIGURE 3 Different types of RFLPs. A. Base-pair substitutions at restriction endonuclease sites generate two alleles: those with the site (b2) and those without the site (b1). B. DNA rearrangements such as those due to insertions of variable numbers of tandem repeats (VNTRs) can generate several alleles. Wherease RFLPs due to base-pair substitution can usually only be detected with one restriction endonuclease, RFLPs due to DNA rearrangements can be detected with several enzymes.

alternatively, a DNA rearrangement such as an insertion or deletion can occur, which results in alleles of different fragment lengths that can be detected with a variety of enzymes (Fig. 3).

For RFLPs due to loss of a restriction site, there are usually only two alleles. The frequency of these alleles in the population determines how useful the RFLP is in tracking a disease gene. Some of the most useful RFLPs for diagnostic purposes have many alleles at a single locus. Several workers have identified DNA probes that identify such highly polymorphic loci. Many of these are due to variable numbers of a tandemly repeated sequence (VNTRs). The repeated sequence can be a single nucleotide, dinucleotide (e.g., CA) tri or tetranucleotide, or a "minisatellite" sequence or longer. The minisatellite regions were originally detected because they shared a 10–15-base pair "core" sequence. Some minisatellite probes detect several loci at once, and, thus, typing an individual with only a few of these multilocus probes is likely to yield a genetic fingerprint specific to that individual. This has been particularly useful in paternity testing, immigration disputes, and forensic testing.

VNTR probes are likely to be extremely useful in genetic fingerprinting samples, which enter the laboratory to be analyzed, protecting the laboratory from errors that can occur from sample mix-ups.

C. Types of Probes

Probes need to detect unique restriction fragments in human DNA; therefore, a prerequisite that they do not contain DNA sequences that are repeated elsewhere in the genome is required. One of the most common forms of repeat is the "Alu" sequence, a 300-base pair sequence, occurring in approximately 300,000 copies in the haploid genome. They exist singly, and in clusters, and on average it is estimated that an Alu sequence occurs every 1,800 base pairs.

Hybridizing a Southern blot of human DNA with repeated sequences results in a smear instead of discrete fragments. The larger the insert, the greater the probability that it contains repetitive sequences. To eliminate this problem, it is necessary either to include "cold" (not radioactivity labeled) repeated sequence in the hybridization mix to competitively hybridize with the repetitive DNA in the probe or to subclone the probe (so that only a portion of the insert, lacking the repeated DNA, is contained within the probe). Exceptions to this are the multilocus probes, which hybridize to many different unlinked fragments; however, these probes still allow for discrete fragments to be typed.

Probes can be derived from genomic DNA, or from complementary DNA (cDNA; cDNA is complementary to messenger RNA [mRNA] and is synthesized with an enzyme known as reverse transcriptase). A hypothetical region of the genome showing the origin of different types of probes is shown in Fig. 4. [*See* GENOME, HUMAN.]

D. Applications of Linked RFLPs

Routinely, DNA samples from a family at risk and requesting diagnosis are digested with restriction enzymes that produce RFLPs revealed by DNA

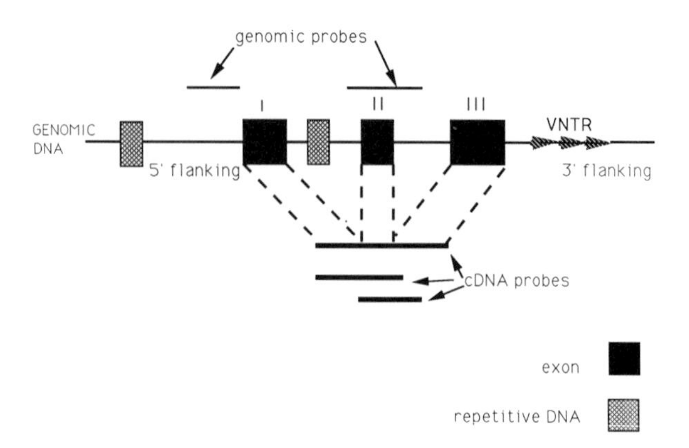

FIGURE 4 Hypothetical region of the genome containing a gene with three exons (I, II, and III), repetitive DNA, and a VNTR in the 3′ flanking region of the gene. The relationship of cDNA probes (obtained by reverse transcriptase off the mRNA template) to genomic probes is shown.

probes identifying loci close to the disease locus. The restriction fragments (alleles) are then scored for each family member (e.g., the affected individual, fetus, sibs of unknown genotype that wish to know if they are carriers or are asymptomatic but affected, parents, grandparents). Because the disease and the alleles are known to cosegretate, the inheritance of the alleles will track the inheritance of the disease. When polymorphic loci, defined by RFLPs, are a distance from the disease locus so that recombination can occur between the two loci, the reliability of the test is lowered, and this is included in the final risk assessment.

The RFLP probes themselves have also allowed the mapping of a number of human diseases to human chromosomes. The first was Huntington's disease (HD) to chromosome 4p with a probe known as G8. The localization of cystic fibrosis (CF) to chromosome 7 was achieved with the demonstration that a random probe, pLAM-917, segregated 85% of the time with the CF defect. The localization of these and other genetic diseases allowed the further identification of other closer DNA markers. These markers can be used in prenatal and preclinical diagnoses, and their reliability depends on the distance they are from the disease locus. The information gained from each marker depends on the frequency of the alleles at the loci in question. When many alleles exist at one locus, such as in the case of the VNTR-containing loci, far more information can be gained than with simple two-allele systems, especially when the frequency of the one allele is low. In a fully informative mating, the inheritance of the parental chromosomes can be determined in the children. Figures 5 and 6 demonstrate the possible diagnoses that can be obtained in sibs of an individual with an autosomal dominant versus an autosomal recessive disease when a two-allele marker locus is linked to a disease locus.

The identification of markers extremely close to a disease locus eventually aids in the identification of the gene and the mutation(s) giving rise to the disease. The aim at this stage is to develop means of detecting the mutation(s) directly. However, for many genetic diseases, the identification of the mutation(s) has not yet been achieved, or the mutations are so numerous that the use of linked RFLPs in diagnosis is more reliable and efficient.

DNA diagnosis is now available for a number of inherited diseases; however, in most cases, because the diagnosis relies on knowing which allele is linked to the disease locus, and either allele may be

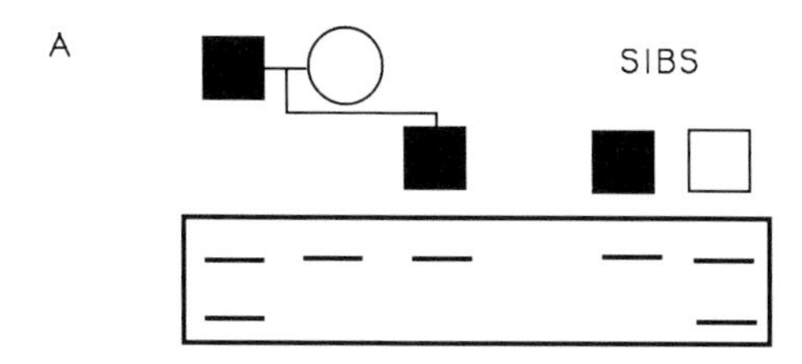

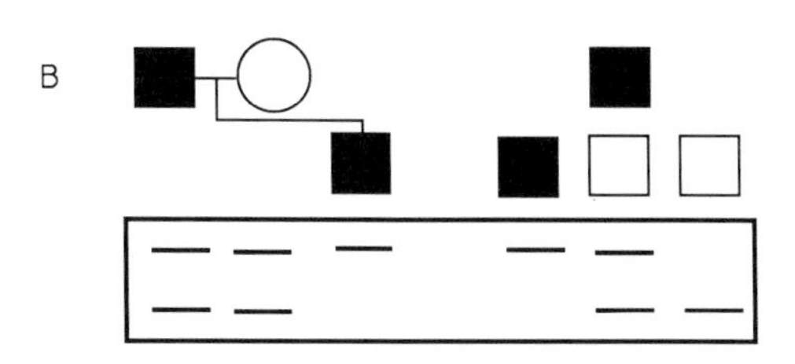

FIGURE 5 Informative (A) and partially informative (B) linked markers in a family segregating for an autosomal dominant trait. Open symbols are unaffected individuals; closed symbols are affected individuals. The fragments (alleles) seen in sibs with and without the trait are shown. In B, sibs that are heterozygotes for the marker alleles have a 50% chance of having the trait.

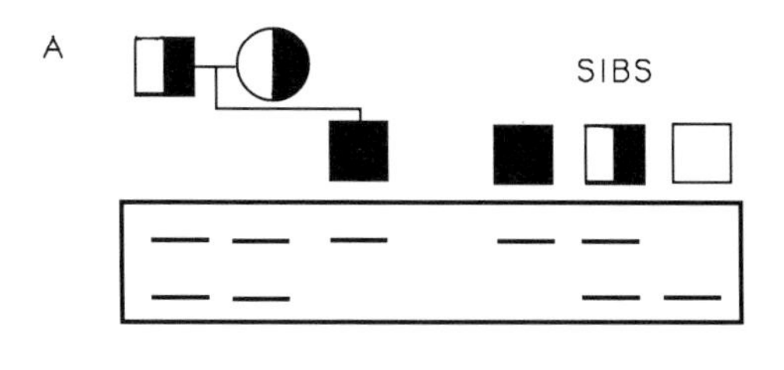

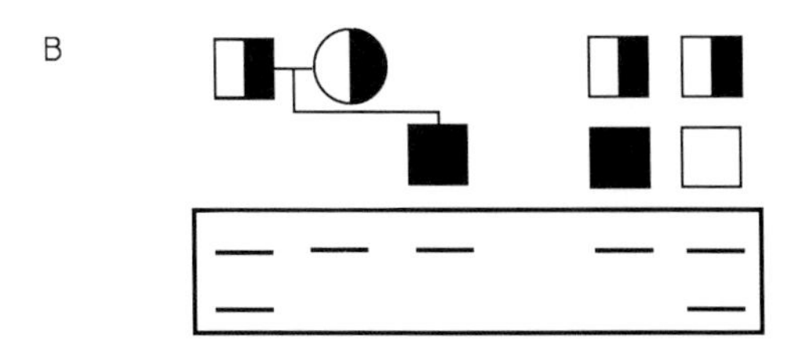

FIGURE 6 Informative (A) and partially informative (B) linked markers in a family segregating for an autosomal recessive trait. Open symbols are unaffected individuals; closed symbols are affected individuals; half-filled symbols are carriers. In B, sibs homozygous for the upper fragment (allele) have a 50% chance of being affected and a 50% chance of being a carrier. Sibs heterozygous for the marker alleles have a 50% chance of being normal and a 50% chance of being carriers.

linked, depending on the family, this type of diagnosis can only be performed in families where at least one child is already affected. Before 1984, DNA analysis was carried out on amniocytes obtained at 15–17 wk gestation. Since then, chorionic villus sampling at 9–11 wk has been available as an alternative way of obtaining fetal cells.

III. The Polymerase Chain Reaction and Allele-Specific Oligonucleotide Probes

With PCR technology RFLPs can be detected in <1 day, in contrast to the time taken to obtain results from Southern blotting, which is usually several weeks. The principle is to use small stretches of single-stranded DNA (oligonucleotide primers) that flank a region of double-stranded DNA of interest (one from one strand, the other from the other strand). The complex (genomic) DNA template, containing the region to be amplified, is denatured to single strands by heating at 94°C, the oligonucleotides are annealed to the DNA substrate at an appropriate temperature (usually between 50 and 60°C), and DNA polymerase extends the oligonucleotides with the DNA substrate as a template. With successive cycles of denaturation, annealing, and extension, it is possible to amplify the stretch of DNA flanked by the oligonucleotide primers. This permits a 220,000-fold amplification of the target sequence with as few as 100 cells as starting material (Fig. 7). More recently, the use of a polymerase derived from a thermostable bacterium has facilitated this procedure because it is active at 72°C and is not denatured at 94°C when the DNA substrate is being denatured. Thus, the same *Taq* polymerase can be cycled repeatedly through the two or three temperatures required for amplification.

Such rapid amplification of a region of interest allows rapid typing of that region for a restriction endonuclease site polymorphism (Fig. 8B1 and 8B2). It is also possible to ask directly if the restriction endonuclease site or a particular DNA sequence is present. This is achieved by hybridization with short probes specific for a particular DNA sequence termed allele-specific oligonucleotides (ASOs). The DNA to be tested is spotted onto a membrane and hybridized with an ASO probe. The ASO only binds if there is a perfect match between it and the amplified DNA sequence (Fig. 8A1 and

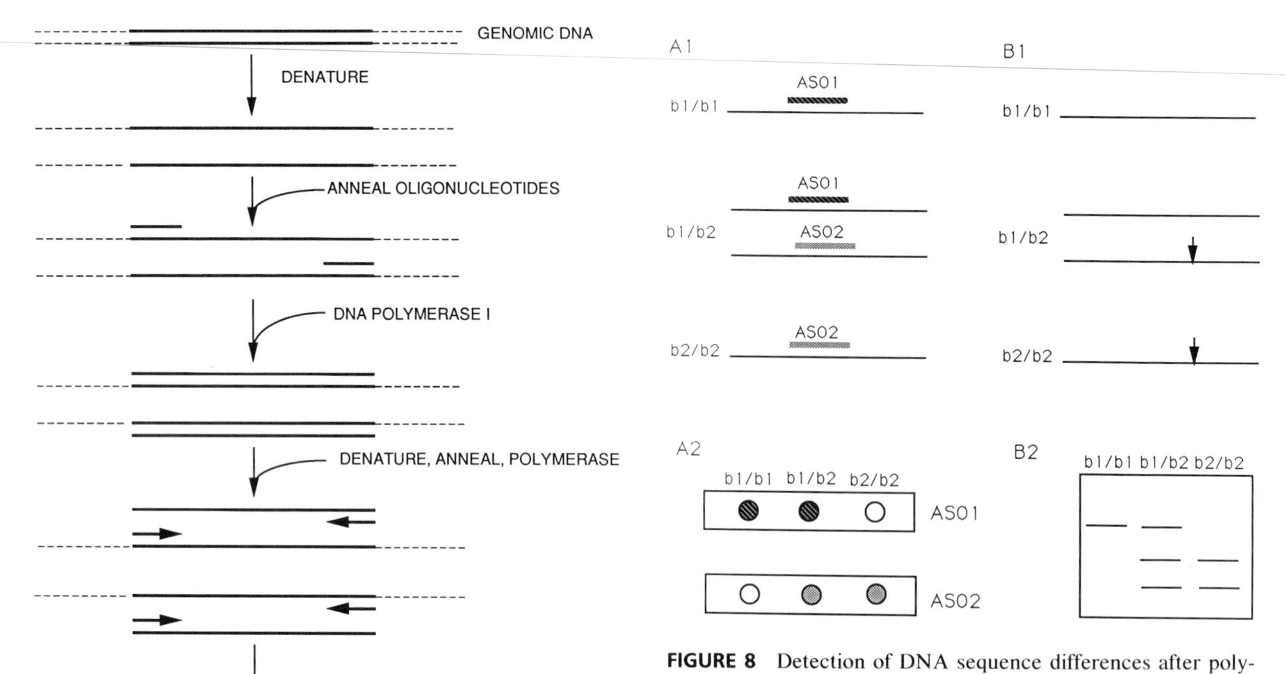

FIGURE 7 The polymerase chain reaction, consisting of successive rounds of denaturation of a complex DNA template, annealing with oligonucleotide primer sequences, and extension with DNA polymerase I.

FIGURE 8 Detection of DNA sequence differences after polymerase chain reaction. A1. The amplified product can be hybridized with allele-specific oligonucleotide (ASOs); in this case ASO1 binds to DNA from the b1 allele and ASO2 binds to DNA from the b2 allele. A2. Amplified product is dot-blotted onto a solid support such as nitrocellulose, and hybridization with a specific ASO determines which allele is present. B1. The alleles can sometimes be discriminated by cleavage with a specific restriction endonuclease. B2. Agarose gel electrophoresis of amplified DNA digested with the appropriate enzyme will indicate which alleles are present.

8A2). The presence of the ASO is detected by labeling the ASO with radioactivity or a substance that can be detected with a chemical reaction that converts a colorless substrate to a colored precipitate. More recently, the ASO has been immobilized on a solid support, and the amplified product has been hybridized to it.

The first application of polymerase chain reaction (PCR) in DNA diagnosis was to enhance the sensitivity of the prenatal diagnosis of sickle-cell anemia. PCR now is being used for prenatal diagnosis and carrier detection of sickle-cell anemia, CF, the thalassemias, Duchenne muscular dystrophy (DMD), hemophilia and a number of other genetic diseases. In conjunction with linked RFLPs, this is a highly reliable, informative diagnostic procedure. In certain cases, when the gene giving rise to the disease is known, but where different mutations are known to give rise to the disease, it is sometimes desirable to determine the mutation in a particular disease family. This can be carried out efficiently by amplification with PCR followed by direct sequencing of the amplified product directly (this determines the base-pair composition of the amplified DNA). This is sometimes carried out in the case of the thalassemias when the battery of ASOs do not detect the mutation. PCR can also be used to detect deletions (as in the case of multiplex amplification of the DMD gene; see below). These direct approaches do not rely on the availability of linked RFLP probes.

The great advantage of PCR is that it can be performed on partially degraded DNA, or where the supply of DNA is limiting. This is illustrated by the demonstration that it is possible to diagnose both CF and phenylketonuria (PKU) from old Guthrie cards (newborn blood spots stored on filters).

IV. DNA Markers for Common Human Genetic Diseases

A. Autosomal Dominant

Table I lists a few autosomal dominant diseases where diagnosis with DNA markers is offered. In

TABLE I Description of Some Autosomal Dominant Diseases Detectable at the DNA Level

Disease (location; incidence in Caucasians)	Mode of detection
Huntington's disease (4p16.2–4p16.3; 1/40,000)	Closely linked RFLPs
Von Recklinghausen neurofibromatosis (17q11.2; 1/4,000)	Flanking RFLPs
Adult polycystic kidney disease (16p13.11–16pter; 1/1,000)	Flanking RFLPs
Retinoblastoma (RB1) (13q14.2; 1/3,400–1/15,000)	RFLPs within and flanking the RB1 gene. PCR to detect DNA sequence polymorphisms
Myotonic dystrophy (19q13.2; 1/5,000–1/15,000)	Closely linked RFLPs

addition to prenatal diagnosis, it is often possible to detect heterozygotes who will develop the disease but who are presymptomatic at the time of the analysis.

1. HD

HD develops late in life, usually after the affected individuals have had children and the opportunity to transmit the HD gene to them. It is a progressive neurodegenerative disorder and results in numerous psychiatric symptoms and a characteristic gait. The age of onset of the disease is usually between 30 and 50 yr, and the clinical symptoms vary.

This disease was mapped to the terminal cytogenetic band of chromosome 4p with a probe known as G8, which identifies the locus D4S10. The genetic distance between D4S10 and HD is 3–4 centiMorgans (cM). A number of closer probes have now been found. Some of these are now used for first trimester exclusion diagnosis. It is now also possible to offer predictive testing for individuals with a family history of HD; in other words, in a large proportion of cases, one can distinguish between carriers without symptoms (because they are too young to have yet developed the disease) and their siblings who have not inherited the mutant gene. Before this, children of individuals affected by HD, who were at 50% risk of the disease, had to wait until they were 65 yr or older to be sure they had escaped HD. A successful predictive test can modify the risk from 50% to as low as 2% or as high as 98% in individuals at risk. Several centers now offer predictive testing to those who spontaneously request such tests.

Figure 9 demonstrates the segregation of alleles at two loci (D4S10 and D4S95) that are closely linked to the HD locus in a family also segregating this disease. In these families rare recombination events have occurred between HD and D4S10, but the D4S95 alleles still segregate with the HD allele.

Although no therapy can be offered at present, knowledge that an individual is likely to have HD allows them to plan their life-style and future with this in mind.

Predictive testing for HD is likely to be a model for the presymptomatic testing of other disorders

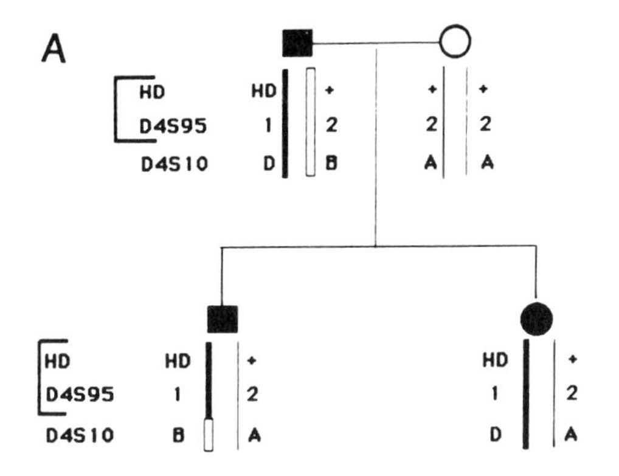

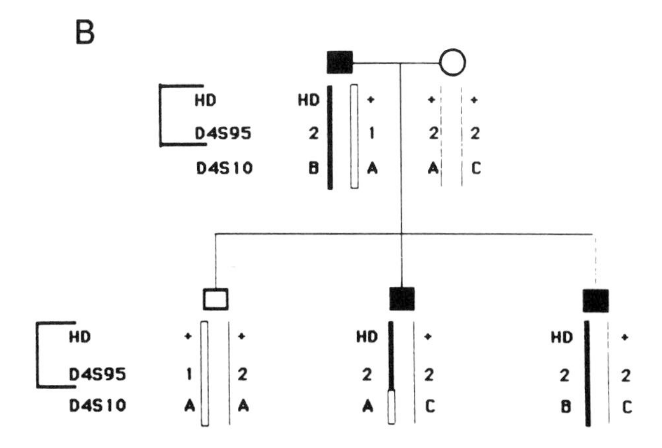

FIGURE 9 Segregation of D4S95 in Huntington's disease (HD) families with recombinants between D4S10 and HD. Open symbols are unaffected individuals; closed symbols are individuals diagnosed as having HD. Squares represent male individuals, circles represent female individuals. The letters listed for D4S10 represent haplotypes A–D. The D haplotype for D4S10 is associated with the disease gene for the family shown in *a*; the B haplotype for D4S10 travels with the disease gene for the family shown in *b*. The first sib in family *a* and the second sib in family *b* each inherited the nondisease haplotype for D4S10, yet they both have Huntington's disease. In neither family has a crossover occurred between D4S95 and HD. [Reproduced, with permission, from Wasmuth *et al.*, 1988, A highly polymorphic locus very tightly linked to the Huntington's disease gene, *Nature* **332,** 734–736.]

such as Alzheimer's disease, schizophrenia, manic depression, and familial cancer once the genetics for these diseases is elucidated and markers are isolated. [*See* HUNTINGTON'S DISEASE.]

2. Neurofibromatosis

Von Recklinghausen neurofibromatosis (NF1) is an autosomal dominant disease, characterized by cafe-au-lait spots, multiple neurofibromas (benign tumors of peripheral nerves) that increase in size and number with age and an increased risk of malignancy especially glioma and neurofibrosarcoma. The disease occurs with variable manifestations and severity. Approximately 30–40% are new mutations. The mutation giving rise to NF1 was shown to be located near the centromere on chromosome 17 with the probe pA10-41. Figure 10 shows the DNA fragments seen with this probe in a family with NF. NF has now been mapped to the long arm of chromosome 17 at 17q11.2. At present, presymptomatic and prenatal diagnoses are performed with many probes closely linked to this locus.

3. Polycystic Kidney Disease

Adult polycystic kidney disease (APCKD) is a common and often lethal disease affecting many organs. The most significant defect is the development and enlargement of cysts in several organs including the liver, pancreas, spleen, and kidneys. Cyst development cannot be prevented by any known therapy and leads to irreversible renal failure at an average age of 51 yr unless dialysis or transplantation is used. The first symptoms tend to occur after the age of 40 yr after individuals have reproduced; however, there is marked variability in onset. In 1985, researchers showed that the APCKD locus (PKD1) is closely linked to the α-globin locus on the short arm of chromosome 16. They used a 3′HVR probe that consists of a tandem repeat sequence from the 3′ end of the α-globin cluster. Subsequently, many probes have been isolated that detect loci flanking the disease locus. These allow an improved method of presymptomatic diagnosis of the disease.

One problem with the diagnosis of APCKD that is likely to be a paradigm for the diagnosis of some other genetic disorders is that in 2–3% of cases, the disease is caused by a mutation at another locus (i.e., APCKD is genetically heterogeneous). This form of the disease, which is also autosomal dominant, has been described for approximately six families and is clinically indistinguishable from the linked form of APCKD.

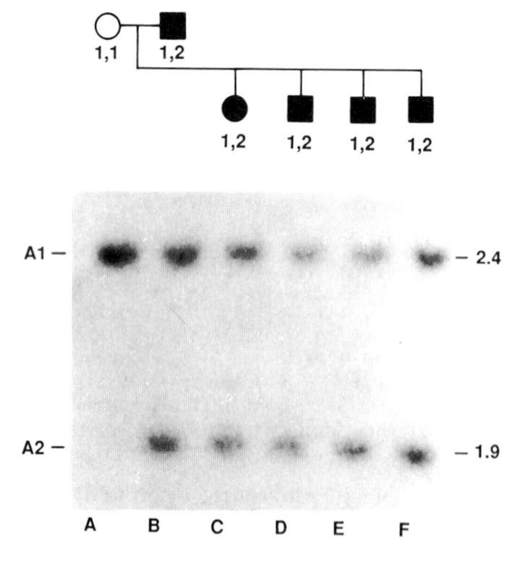

FIGURE 10 Hybridization of probe pA10-41 to MspI-digested DNAs from individuals in a neurofibromatosis kindred. The alleles are a 2.4 kb (A1) or 1.9 kb (A2) band. Squares represent male individuals circles represent females. Open pedigree symbols are unaffected individuals, closed symbols are affected individuals. [Reproduced, with permission, from Barker *et al.*, 1987, Hybridization of probe pA10-41, *Science* **236,** 1100. Copyright 1987 by the AAAS.]

4. Retinoblastoma

Retinoblastoma is a malignant tumor that arises in the eyes of children, usually before the age of 4 yr. Thirty to 40% of patients with retinoblastoma have inherited the predisposition to the tumor, in addition to several other cancers, particularly osteosarcoma. The cancer develops in 80–90% of individuals who carry at this locus any one of a variety of mutant alleles associated with a predisposition to the tumor. Much evidence supports the hypothesis that retinoblastomas that develop in patients without the hereditary form of retinoblastoma arise from retinal cells that have acquired somatic mutations at the same genetic locus.

Retinoblastoma and osteosarcoma arise from cells that have lost both functional copies of the retinoblastoma gene. This can be due to loss of one copy of the gene in somatic tissue in an individual who has inherited an initial mutation on the other allele. Alternatively, the mutations on both chromosomes can occur somatically.

Numerous advances have been made in the study of retinoblastoma at the molecular level, and it is one of the few familial cancers where prenatal and preclinical diagnoses are available.

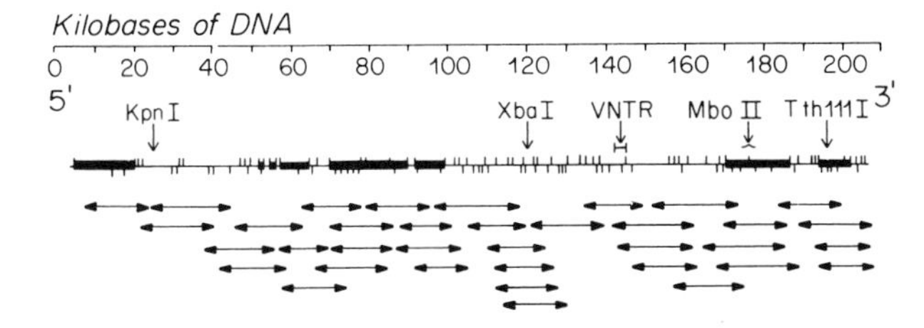

FIGURE 11 Restriction enzyme map of the genomic locus of the retinoblastoma gene. Vertical markers above the map represent the locations of *Hin*dIII sites; those below the map represent the locations of *Eco*RI sites. The boxed areas represent *Hin*dIII fragments that contain sequences found in the cDNA (exons). Each double-headed arrow beneath the map represents a distinct recombinant bacteriophage clone. The positions of the polymorphic DNA sites are indicated with arrows. The location of the polymorphic *Mbo*II site or sites has not been determined precisely but is shown at approximately 175 kb on this map. [Reprinted, by permission of The New England Journal of Medicine, **318,** 152, 1988.]

The human retinoblastoma gene (RB1) contains 27 exons (protein-coding regions) and spans a region of approximately 200 kb. The gene has been cloned, and several RFLPs have been described. The locations of some of these are shown in Figure 11. Approximately 90% of families referred for DNA-based diagnoses are informative for these RFLPs; however, RFLP screening may have detected only 5% of the potentially useful polymorphic sites. Recently, it has been shown that additional DNA sequence polymorphisms can be detected with PCR.

In 10–20% of all families, it is sometimes possible with Southern blotting to determine that a parent has an altered copy of the gene (deleted or rearranged). The identification of children who also have that altered chromosome predicts which ones have a tendency to develop retinoblastoma.

The diagnostic test identifies individuals who have inherited a tumor-predisposing mutation at the retinoblastoma locus and allows eye examinations to be focused on children who carry a mutant allele, reducing the need for such examinations in children thought to have a normal genotype. [*See* RETINOBLASTOMA, MOLECULAR GENETICS.]

B. Autosomal Recessive Diseases

Table II contains a description of some of the more common recessive disorders detected with linked RFLPs. In the case of autosomal recessive disorders where the appropriate DNA markers are available, prenatal diagnosis and the detection of heterozygotes (carriers) is possible.

TABLE II Description of Some Autosomal Recessive Diseases Detectable at the DNA Level

Disease (location; incidence in Caucasians)	Mode of detection
Cystic fibrosis (CF) (7q31; 1/2000)	Closely linked RFLPs PCR and ASOs
Phenylketonuria (12q22–q24.2; 1/4,500–1/16,000)	RFLPs detected with PAH cDNA probe
Sickle-cell anemia (16p13.3; 1/3000 in United States)	Direct detection with restriction enzyme (*Mst*II, *Cvn*I, *Oxa*NI) or PCR and ASOs
β-thalassemia (16p13.3)	Direct detection—battery of ASOs for different ethnic groups
	RFLPs within β-globin cluster
α-1-antitrypsin deficiency (14q32.1)	RFLPs within and flanking the gene
	PCR and ASOs for the Z allele
Friedreich's ataxia (9p22-cen; 1/50,000)	Linked RFLPs
Wilson's disease (13q14.3–q21; 1/33,500)	Flanking RFLPs

PAH, phenylalanine hydroxylase.

1. CF

CF, the most common autosomal recessive disease in Caucasions, was mapped to chromosome 7 in 1985. This was with a probe known as LAM917, which was shown to recombine with the CF locus at an incidence of 15%. Soon after that two other probes, the oncogene MET, and a probe identifying a piece of DNA of unknown function, J3.11 (locus D7S8), were shown to be extremely close to the CF locus and to probably flank it. At this stage, it was possible to offer prenatal diagnosis for CF with a reasonable amount of confidence. Many other RFLP probes closer to CF have been isolated subsequently. These have increased the probability of a fully informative and reliable prenatal diagnosis. It is possible to directly type many of the restriction site polymorphisms giving rise to the closely linked RFLPs with PCR.

In 1989, the CF gene was isolated and shown to span a distance of 250 kb. Twenty-four exons code for an mRNA 6.5 kb long. Analysis of CF versus

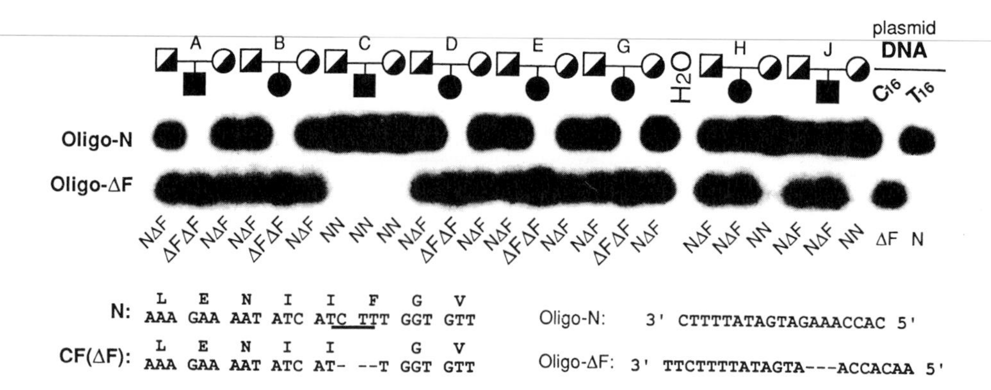

FIGURE 12 Detection of the ΔF_{508} mutation in cystic fibrosis (CF) by oligonucleotide hybridization. Autoradiographs show the hybridization results of genomic DNA from representative CF families with the two specific oligonucleotide probes as indicated. Oligo-N detects the normal DNA sequence, and oligo-ΔF detects the mutant sequence. Genomic DNA samples from each family member were separated by electrophoresis and transferred to a nylon membrane. The membrane was hybridized with radioactively labelled oligonucleotide probes, washed and exposed to X-ray film. Samples without DNA (H_2O) and plasmid DNA, T16 (N cDNA), and C16 (cDNA with the ΔF_{508} deletion) were included as controls. The illustration is based on the assumption that the triplet CTT was deleted; the sequencing data do not allow distinction between deletion of these nucleotides or other combinations. [Reproduced, with permission, from Kerem *et al.*, 1989, Identification of the cystic fibrosis gene: Genetic analysis, *Science* **245,** 1078. Copyright 1989 by the AAAS.]

normal DNA has shown that on 68% of CF chromosomes the defect is due to a deletion of three base pairs within exon 10. This results in deletion of the amino acid phenylalanine. This mutation occurs most commonly on haplotypes that could be considered part of a related group; however, this deletion has been observed on at least one other haplotype, suggesting that this mutation has arisen independently at least twice. The phenylalanine deletion is associated with pancreatic insufficiency, and it is suggested that other mutations may be less severe and sometimes result in the pancreatic-sufficient form of CF. Direct detection of this deletion with PCR and ASOs is shown in Fig. 12.

At least 40 other mutations have been found that can give rise to CF. A large number such as this is not ideal for the nationwide carrier screening of this disease. [*See* Cystic Fibrosis, Molecular Genetics.]

2. PKU

PKU is the most common inborn error of amino acid metabolism in Caucasoid populations and is due to a deficiency of hepatic phenylalanine hydroxylase (PAH). In the normal human liver, PAH catalyzes the rate-limiting step in the hydroxylation of phenylalanine to tyrosine. If PAH is defective, serum phenylalanine accumulates, which results in hyperphenylalaninemia and abnormalities in the metabolism of many compounds derived from phenylalanine. Approximately 1 in 50 individuals is a carrier of the disease. If not detected early, affected children develop severe mental retardation. The consequences can be avoided with a rigid implementation of a restricted diet low in phenylalanine during the first decade of life. Neonatal screening for PKU has been routinely carried out in Western countries, but until recently there was no way of detecting affected fetuses. This is now possible with RFLPs detected with a cDNA probe for the PAH gene. Although genetic analysis has defined two mutations, both associated with different haplotypes, until the identification of all PKU mutations, direct detection is not practical. [*See* Phenylketonuria, Molecular Genetics.]

3. Sickle-Cell Anemia and β-Thalassemia

The detection of sickle-cell anemia with the linked *Hpa*I RFLP has evolved into the direct analysis of the single base-pair change within the β-globin gene causing sickle-cell disease. The specific mutation within the β-globin gene, which is due to an amino acid substitution of glutamic acid for valine, also abolishes the site for the restriction enzyme DdeI and MstII or its isoschizomers CvnI or OxaNI (isoschizomers are restriction enzymes that recognize the same nucleotide sequence). This meant that prenatal diagnosis could be done without family studies and more reliably than with a linked RFLP. With the advent of PCR, and rapid amplification of the region in question, it was possible to ask directly if the restriction site was present, and

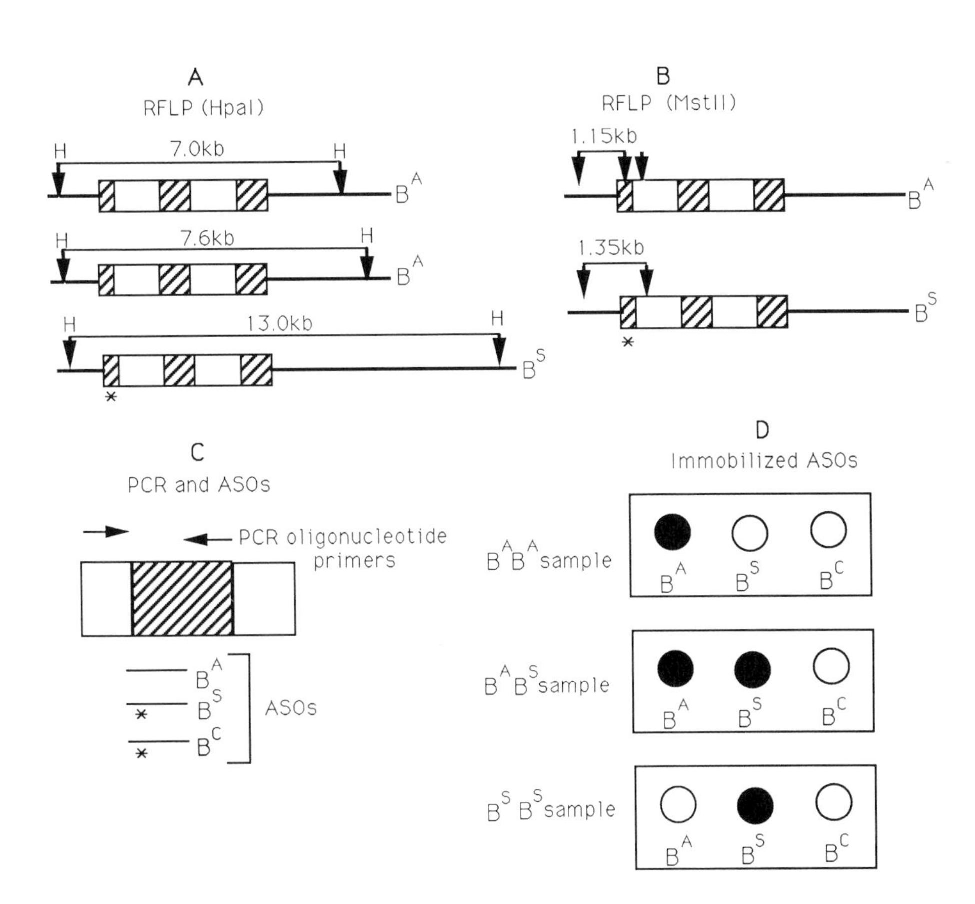

FIGURE 13 Progression of sickle-cell anemia diagnosis. A. Linkage association of the 13.0-kb *Hpa*I fragment with the β-globin gene containing the sickle-cell mutation (B^S). The 7.0-kb and 7.6-kb fragments are associated with the normal β-globin gene (B^A). The cleavage sites for *Hpa*I (H) are shown. The position of the sickle-cell mutation is marked with an asterisk. Hatched boxes are β-globin exons; open boxes are introns. B. The restriction endonuclease *Mst*II cleaves at codon 6 of B^A but not B^S, allowing direct detection of the mutation. The arrows show the *Mst*II sites. C. Allele-specific oligonucleotide (ASO) recognition of B^A, B^S, and B^C. In conjunction with amplification of the DNA containing this sequence with polymerase chain reactions (PCR), this provides a highly specific approach to allele typing. D. Detection of the B^A, B^S, and B^C alleles with ASOs immobilized on a solid support. The amplified DNA, which has incorporated a biotinylated primer, hybridizes to the ASO probes and is detected with a reaction with converts a colorless dye into a colored precipitate. Results are shown after hybridizing DNA from individuals with the genotypes B^A/B^A, B^A/B^S, and B^S/B^S. [Adapted, with permission, from C. Thomas Caskey, 1987, Disease diagnosis by recombinant DNA methods, *Science* **236,** 1223–1229. Copyright 1987 by the AAAS.]

Southern blotting could be avoided entirely. Subsequently, the application of allele-specific oligonucleotide probes allows one to do without a restriction endonuclease, since the presence or absence of the mutant site can be determined directly by dot blotting and hybridization with ASO probes. Developments such as hybridization of the PCR product to an immobilized array of oligonucleotide probes further improve the efficiency of the test. The progression of the diagnosis of sickle-cell anemia is shown in Fig. 13. Direct detection with ASOs is preferable to cleavage by a restriction enzyme such as *Mst*II since a second common mutant allele at codon 6 exists known as beta-C (B^C). This mutation does not alter the endonuclease recognition site of either MstII or DdeI.

The thalassemias are a heterogeneous group of hereditary anemias characterized by defective β-globin synthesis. Ethnic groups at risk for β-thalassemia are Mediterranean, North African, Middle Eastern, Asian Indian, Chinese, Southeast Asian, and Black African. At the end of 1988, >60 β-thalassemia mutations had been discovered, and it is estimated that alleles giving rise to 99% of β-thalassemia genes in the world have been identified. Direct detection of β-thalassemia mutations with PCR has only been possible since these different lesions have been characterized. Before that, indirect detection with RFLPs was carried out. This carried a theoretical error rate of between 1 in 300 and 1 in

500 due to meiotic recombination between a site used in tracking the β-thalassemia mutation and the mutation itself.

Each ethnic group has been found to have its own set of β-thalassemia mutations, and in any particular group, four to six alleles make up 90% of the β-thalassemia genes. This has simplified the detection of the disease-producing mutations. Prenatal diagnosis is now usually done by direct detection of the mutation, which involves a combination of dot-blot hybridization and restriction analysis. Approximately half of the β-thalassemia mutations can be detected with restriction enzyme analysis of the PCR product. PCR can also be used to detect deletions in these genes.

If difficulty is encountered in the direct detection of sickle-cell anemia or β-thalassemia, haplotype analysis of various family members by PCR will allow rapid prenatal diagnosis by indirect detection. This demonstrates the use that linked RFLPs play in prenatal and preclinical diagnoses, even when methods for the direct detection of the mutations are available. Linked RFLPs also provide a way of checking conclusions drawn from direct detection.

4. α-1-Antitrypsin Deficiency

This disease results in predisposition to childhood liver cirrhosis and pulmonary emphysema in early adult life. This occurs earlier in smokers than in nonsmokers (the third decade for smokers, and the fifth decade for nonsmokers).

A number of alleles exist for α-1-antitrypsin. M is the most common, normal allele. There are also a number of alleles that give rise to deficiency, the most common of which is the Z allele, which occurs in 1 out of 7,000 North Americans. This allele can be distinguished by isoelectric focusing and is due to a base-pair substitution in codon 342 of exon V, which results in conversion of glu to lys. Many different null alleles also exist. RFLPs exist within and flanking the AAT gene. The haplotypes derived from these RFLPs have been shown to be strongly associated with different α-1-antitrypsin alleles. Thus, the determination of haplotypes present in diseased individuals can give an indication of the mutant alleles that are present.

It is now possible to detect the Z allele with PCR and ASO typing, and direct detection systems have also been developed for a number of other alleles. However, RFLPs still play an important role in the diagnosis of this disease, due to the number of other mutant alleles that exist that are not detectable with ASOs.

C. Sex-Linked Diseases

Table III describes some sex-linked diseases detectable at the DNA level. In families affected with sex-linked diseases, where the appropriate DNA markers are available, prenatal diagnosis is offered to determine the affected males. The detection of female carriers is also often possible.

1. DMD

Affected males are often confined to a wheelchair by the age of 12 yr and die in their late teens. One third of all cases appear to arise via a new mutation. This disease has been shown to be due to defects in the X-linked dystrophin gene, which normally codes for an mRNA of 14 kb. Becker muscular dystrophy (BMD), which is a clinically milder form of the disease, has been shown to also be due to mutations in the dystrophin gene. No cure or effective therapy exists; however, it is now possible to perform highly accurate prenatal diagnosis and carrier detection of DMD and BMD via Southern analysis using cDNA and genomic clones for dystrophin.

The DMD gene spans more than 2 Mb of DNA and is split into at least 60 exons. Partial gene deletions have been shown to cause 50–60% of DMD lesions. These deletions usually span several hundred kilobases of the DMD locus and generally

TABLE III Description of Some Sex-Linked Diseases Detectable at the DNA Level

Disease (location; incidence in Caucasian males)	Method of detection
Duchenne and Becker muscular dystrophy (xp21.3–p21.1)	cDNA probes to detect deletions
	Multiplex amplification to detect deletions
	Linked RFLPs within and flanking the dystrophin gene
Hemophilia A (Xq28; 1/10,000)	RFLPs within and flanking the factor VIII gene
Hemophilia B (Xq26.3–q27.1; 1/50,000)	RFLPs linked to factor IX gene
X-linked hypohidrotic ectodermal dysplasia (Xq11–q21.1)	Linked RFLPs

overlap two specific regions located approximately 0.5 and 1.2 Mb from the promoter. Fifteen percent of mutations are due to duplications. This gene is too complex to be analyzed simultaneously with the complete cDNA probe: a *Hind*III digest generates at least 65 restriction fragments, and eight cDNA subcloned probes are required to detect all deletions in this gene. For the detection of carrier females, because of the presence of one normal gene, a dosage analysis is necessary in the region of the deletion. Alternatively, pulsed-field gel electrophoresis (which resolves large DNA fragments, i.e., between 50,000 and 5 million bp) can be used to detect an abnormal breakpoint fragment. In the one-third of cases that have no visible deletion, analysis has to be performed with intragenic RFLP markers and several informative DNA sequence polymorphisms have been identified within the dystrophin gene. These allow diagnosis in 37% of samples. One problem with this approach is that the accuracy of diagnosis with linked RFLPs is reduced. This may be due to the large size of the dystrophin gene, because recombination between a linked marker and the mutation occurs in 5% of meioses.

In some laboratories, PCR is now being used for the rapid detection of 80–90% of all dystrophin gene deletions. Currently, 18 separate oligonucleotide primers are being combined in a single reaction to simultaneously amplify nine deletion-prone exons of the gene. This approach, known as multiplex amplification, can be used in both the prenatal and postnatal diagnoses of DMD (Fig. 14); however, it is not applicable to nondeletion cases or for carrier diagnosis.

If one of the bands is missing after multiplex amplification, the extent of the deletion is determined with cDNA probes. Even with multiplex amplification, RFLP analysis with intragenic or extragenic probes is useful to determine if nonpaternity or sample switching has occurred.

Due to the high rate of new mutations in the dystrophin gene, genetic counseling and prenatal diagnosis will only have a limited effect on the incidence of the disease. [*See* MUSCULAR DYSTROPHY, MOLECULAR GENETICS.]

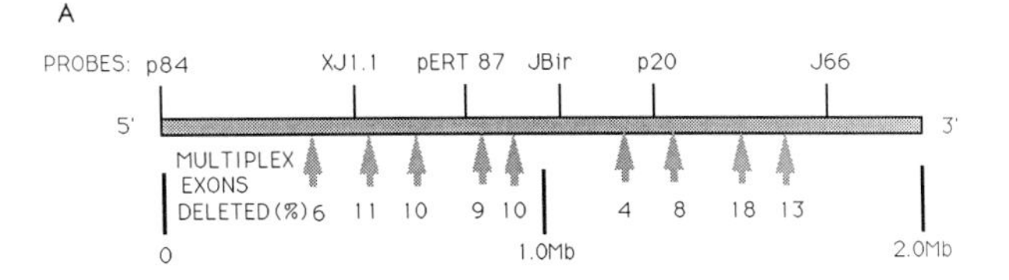

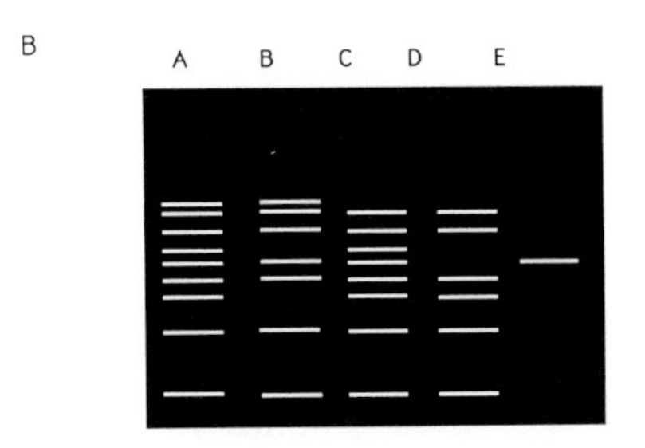

FIGURE 14 DMD mutations. A. Schematic illustration of the DMD gene illustrating the relative locations of the nine currently amplified regions (arrows) relative to six genomic probe markers. The percentage of DMD patients carrying deletions of each of the nine regions is also shown. B. Detection of DNA deletions at the DMD locus by multiplex amplification. Lane A shows the typical pattern of amplification products seen in normal DNA. Lanes B–E show the patterns that can be seen in DMD males with deletions of the dystrophin gene. [Adapted, with permission, from R. A. Gibbs, J. S. Chamberlain and C. Thomas Caskey, 1989, Diagnosis of new mutation diseases using the polymerase chain reaction, *in* "PCR Technology" (H. A. Erlich, ed.), Stockton Press, New York.]

2. Hemophilias

Hemophilias are disorders of blood coagulation due to a deficiency of clotting factors VIII and IX. Hemophilia A, the classic type of hemophilia, is one of the most common severe, inherited bleeding disorders in humans. It is due to a deficiency in a protein involved in coagulation. This protein, known as factor VIII, which normally circulates in the plasma with von Willebrand factor, is an essential protein cofactor in the intrinsic coagulation pathway. The lesion giving rise to hemophilia A is due to a Glu-Gly substitution in exon 7 of factor VIII. Like DMD, the incidence of sporadic (noninherited) cases is high. The human gene for factor VIII has been cloned and sequenced and shown to be 186 kb with 26 exons and 25 introns. Until the identification of the gene, prenatal diagnosis was only possible with fetal blood sampling that was performed in the second trimester. This presented a 5% risk to the fetus. In addition, carrier detection was extremely unreliable due to the wide range of clotting found in normal and heterozygous females.

Since the discovery of several RFLPs within and linked to the factor VIII gene, carrier identification and prenatal diagnosis has been achieved with chorionic villus sampling in the first trimester.

Many mutations have been detected in the factor VIII gene. These mutations are diverse; in a total of 240 patients, 28 molecular defects have been found: 5.8% are deletions, 5.0% base-pair substitutions, and 0.8% are insertions. The existence of many unique mutations underscores the necessity of using linked RFLPs for DNA-based diagnosis where the gene can be tracked in families at risk, while remaining independent of the point mutation causing the hemophilia.

Hemophilia B, or Christmas disease, is due to a defect in the factor IX gene. Diagnosis has previously relied on the familial segregation of linked RFLPs. This is successful in 65% of Caucasoid cases but fails where the marker is uninformative (15% of cases) of where there is homozygosity for closely linked alleles in key family members (20% of cases). Effort is being made to identify the mutations causing the disease, so that they can be detected directly. [*See* HEMOPHILIA, MOLECULAR GENETICS.]

V. Other Uses of DNA Markers and Future Projections

A. Polygenic Diseases

One of the most challenging applications of RFLP research and association of genetic diseases is the study of polygenic diseases such as cardiovascular disease and atherosclerosis. Coronary heart disease is the leading cause of morbidity and mortality in the United States. While it is evident that many different genes are likely to contribute to these diseases, the role and interactions of the different genes are not understood. By using RFLPs to study the inheritance of candidate genes in high-risk families, it is likely that their interaction will slowly be unravelled. [*See* CORONARY HEART DISEASE, MOLECULAR GENETICS.]

B. Autoimmune Diseases

Similarly, RFLPs linked to genes predisposing to familial cancers and autoimmune diseases are likely to be determined, allowing individual risk assessment. A variety of autoimmune diseases such as insulin-dependent diabetes mellitus, rheumatoid arthritis, multiple sclerosis, and myasthenia gravis already have been associated with different serological types of the human leukocyte antigen (HLA) class II antigens; however, it is not known if the HLA association with human autoimmune diseases is due to a specific immune response or is due to linkage disequilibrium with disease-susceptible alleles. Nevertheless, genes at other loci are also likely involved in these autoimmune diseases. [*See* AUTOIMMUNE DISEASE; MULTIPLE SCLEROSIS; MYASTHENIA GRAVIS.]

C. Psychiatric Diseases

There is also hope that RFLP probes will be useful in the study of the inheritance of psychiatric diseases. The extent to which genetic heterogeneity exists is still not known (how many independent genes at different loci give rise to these diseases). Familial Alzheimer's disease (FAD) is a neurodegenerative disorder that occurs late in life; most cases occur after the age of 70 yr. There are several large families where the disease appears to be inherited as an autosomal dominant trait. In these families, FAD has been shown to be linked to loci on the long arm of chromosome 21 in four large FAD families where the age at onset was younger than usually observed (40+ yr). This suggests that early onset and late onset may have different genetic causes. Studies of RFLPs linked to manic depression, schizophrenia, and autism are also likely to provide information and, possibly, the potential for prenatal and preclinical diagnoses. [*See* ALZHEIMER'S DISEASE.]

D. DNA Markers in Tissue Typing

The major histocompatibility complex (MHC) of mammals encodes many different cell-surface glycoproteins and serum complement components that mediate a variety of immunological functions. The human MHC, known as HLA, region lies on the short arm of chromosome 6 and is highly polymorphic. The class I molecules are the classical transplantation antigens and are encoded by HLA A, B, and C loci. The value of HLA typing for transplantation is demonstrated by the increased graft survival observed in HLA matched donor–recipient pairs. Polymorphisms within class I have been defined serologically. These can be used for tissue typing for transplantation. RFLPs are also detected with the HLA genes, although entirely correlating specific serologic types with specific RFLP patterns has not been possible. [*See* IMMUNOBIOLOGY OF TRANSPLANTATION.]

PCR followed by hybridization with ASOs has been used with great success to type HLA polymorphisms, in particular HLA-DQA, HLA-DQB, and HLA-DRB. This provides the opportunity for much more refined HLA typing than that afforded by current serologic methods and should prove useful in tissue typing for transplantation.

E. Paternity Determination and Forensics

Since the combination of alleles an individual possesses serves as a "genetic fingerprint," DNA markers are being used in paternity testing and forensic analysis. Currently, DNA markers detected with PCR, and highly polymorphic DNA probes such as the minisatellite probes and other VNTR probes, are being used for this purpose.

F. Uses of RFLP Probes for Karyotype Detection

As the genome becomes saturated with DNA markers, they likely will be used to determine molecular karyotypes. In particular, subtle chromosomal rearrangements, which are not visible cytogenetically, would be detectable. This would also involve the identification of microdeletions in certain congenital disorders. This approach has been applied in the case of cri-du chat, which is commonly due to a deletion of 5p15.1–pter. A translocation involving this region of chromosome 5, yet undetectable cytogenetically, could be detected with DNA probes, and prenatal diagnosis and carrier testing could be provided.

Conclusion

By providing direct or indirect determination of disease state or individual identity, DNA diagnosis is having a major impact on the diagnosis of genetic disease, tissue typing, paternity testing and possibly karyotype detection in the future.

Acknowledgments

I would like to thank A. J. Cooper (The University of Texas Southwestern Medical Center) and R. T. Taggart (Wayne State University), for helpful comments on this article.

Bibliography

Antonarakis, S. E., and Kazazian, H. H. (1988). The molecular basis of hemophilia A in man. *Trends Genet.* **4,** 233.

Barker, D., Wright, E., Nguyen, K., Cannon, L., Fain, P., Goldgar, D., Bishop, D. T., Carey, J., Baty, B., Kivlin, J., Willard, H., Waye, J. S., Greig, G., Leinwand, L., Nakamura, Y., O'Connell, P., Leppert, M., Lalouel, J.-M., White, R., and Skolnik, M. (1987). Gene for von Recklinghausen neurofibromatosis is in the pericentric region of chromosome 17. *Science* **236,** 1100.

Beaudet, A., Bowcock, A., Buchwald, M., Cavalli-Sforza, L., Farrall, M., King, M.-C., Klinger, K., Lalouel, J.-M., Lathrop, G., Naylor, S., Ott, J., Tsui, L.-C., Wainwright, B., Watkins, P., White, R., and Williamson, R. (1986). Linkage of cystic fibrosis to two tightly linked DNA markers: Joint report from a collaborative study. *Am. J. Hum. Genet.* **39,** 581.

Botstein, D., White, R. L., Skolnik, M., and Davis, R. W. (1980). Construction of a genetic linkage map in man using restriction fragment length polymorphisms. *Am. J. Hum. Genet.* **32,** 314.

Bowcock, A. M., Farrer, L. A., Hebert, J. M., Agger, M., Sternlieb, I., Scheinberg, I. H., Buys, C. H. C. M., Scheffer, H., Frydman, M., Chajek-Saul, T., Bonne-Tamir, B., and Cavalli-Sforza, L. L. (1988). Eight closely linked loci place the Wilson disease locus within 13q14–q21. *Am. J. Hum. Genet.* **43,** 664.

Caskey, C. T. (1987). Disease diagnosis by recombinant DNA methods. *Science* **236,** 1223.

Chamberlain, J. S., Gibbs, R. A., Ranier, J. E., Nguyen, P. N., and Caskey, C. T. (1988). Deletion screening of the Duchenne muscular dystrophy locus via multiplex DNA amplification. *Nucleic Acids Res.* **16,** 11141.

Chamberlain, S., Shaw, J., Rowland, A., Wallis, J., South, S., Nakamura, Y., von Gabain, A., Farrall, M., and Williamson, R. (1988). Mapping of mutation causing Friedreich's ataxia to human chromosome 9. *Nature* **334,** 248.

DiLella, A. G., Marvit, J., Brayton, K., and Woo, S. L. C. (1987). An amino-acid substitution involved in phenylketonuria is in linkage disequilibrium with DNA haplotype 2. *Nature* **327,** 333.

DiLella, A. G., Marvit, J., Lidsky, A. S., Guttler, F., and Woo, S. L. C. (1986). Tight linkage between a splicing mutation and a specific DNA haplotype in phenylketonuria. *Nature* **322,** 799.

Erlich, H. A., Sheldon, E. L., and Horn, G. (1986). HLA typing using DNA probes. *Biotechnology* **4,** 975.

Gibbs, R. A., Chamberlain, J. S., and Caskey, C. T. (1989). Diagnosis of new mutation diseases using the polymerase chain reaction. *In* "PCR Technology" (Henry A. Erlich, ed.), p. 171. Stockton Press, New York.

Gitschier, J., Drayna, D., Tuddenham, E. G. D., White, R. L., and Lawn, R. M. (1985). Genetic mapping and diagnosis of haemophilia A achieved through a BclI polymorphism in the factor VII gene. *Nature* **314,** 738.

Goldgar, D. E., Green, P., Parry, D. M., and Mulvihill, J. J. (1989). Multipoint linkage analysis in neurofibromatosis type I: An international collaboration. *Am. J. Hum. Genet.* **44,** 6.

Jeffreys, A. J., Wilson, V., and Thein, S. L. (1985). Hypervariable 'minisatellite' regions in human DNA. *Nature* **314,** 6006.

Kan, Y. W., and Dozy, A. M. (1978). Polymorphism of DNA sequence adjacent to human beta-globin structural gene: Relationship to sickle mutation. *Proc. Natl. Acad. Sci. U.S.A.* **75,** 5631.

Kazazian, H. H. (1989). Use of PCR in the diagnosis of monogenic disease. *In* "PCR Technology" (Henry A. Erlich, ed.), p. 153. Stockton Press, New York.

Kazazian, H. H., and Boehm, C. D. (1988). Molecular basis and prenatal diagnosis of beta-thalassemia. *Blood* **72,** 1107.

Kerem, N., Rommens, J. M., Buchanan, J. A., Markiewicz, D., Cox, T. K., Chakravarti, A., Buchwald, M., Tsui, L.-C. (1989). Identification of the cystic fibrosis gene: Genetic analysis. *Science* **245,** 1073.

Kogan, S. C., Doherty, M., and Gitschier, J. (1987). An improved method for prenatal diagnosis of genetic diseases by analysis of amplified DNA sequences. Application to hemophilia A. *N. Eng. J. Med.* **317,** 985.

Montandon, A. J., Green, P. M., Gianelli, F., and Bentley, D. R. (1989). Direct detection of point mutations by mismatch analysis: Application to haemophilia B. *Nucleic Acids Res.* **17,** 3347.

Nakamura, Y., Leppert, M., O'Connell, P., Wolff, R., Holm, T., Culver, M., Martin, C., Fujimoto, E., Joff, M., Kumlin, E., and White, R. (1987). Variable number of tandem repeat (VNTR) markers for human gene mapping. *Science* **235,** 1616.

Overhauser, J., Bengtsson, U., McMahon, J., Ulm, J., Butler, M. G., Santiago, L., and Wasmuth, J. J. (1989). Prenatal diagnosis and carrier detection of a cryptic translocation by using DNA markers from the short arm of chromosome 5. *Am. J. Hum. Genet.* **45,** 296.

Reeders, S. T., Keith, T., Green, P., Germino, G. G., Barton, N. J., Lehmann, O. J., Brown, V. A., Phipps, P., Morgan, J., Bear, J. C., and Parfrey, P. (1988). Regional localization of the autosomal dominant polycystic kidney disease locus. *Genomics* **3,** 150.

Rieger, R., Michaelis, A., and Green, M. M. (1976). "Glossary of Genetics and Cytogenetics," 4th ed. Springer Verlag, Berlin.

Riordan, J. R., Rommens, J. M., Kerem, B., Alon, N., Rozmahel, R., Grzelczak, Z., Zielenski, J., Lok, S., Plavsic, N., Chou, J.-L., Drumm, M. L., Iannuzzi, M. C., Collins, F. S., and Tsui, L.-C. (1989). Identification of the cystic fibrosis gene: Cloning and characterization of complementary DNA. *Science* **245,** 1066.

Rommens, J. M., Iannuzzi, M. C., Kerem, B., Drumm, M. L., Melmer, G., Dean, M., Rozmahel, R., Cole, J. L., Kennedy, D., Hidaka, N., Zsiga, M., Buchwald, M., Riordan, J. R., Tsui, L.-C., and Collins, F. S. (1989). Identification of the cystic fibrosis gene: chromosome walking and jumping. *Science* **245,** 1059.

Rubin, C. M., Houck, C. M., Deininger, P. L., Friedman, T., and Schmid, C. W. (1980). Partial nucleotide sequence of the 300-nucleotide interspersed repeated human DNA sequences. *Nature* **284,** 372.

Saiki, R. K., Walsh, P. S., Levenson, C. H., and Erlich, H. A. (1989). Genetic analysis of amplified DNA with immobilized sequence-specific oligonucleotide probes. *Proc. Natl. Acad. Sci. U.S.A.* **86,** 6230.

Shaw, D. J., and Harper, P. S. (1989). Myotonic dystrophy: Developments in molecular genetics. *Brit. Med. Bull.* **45,** 745.

St. George-Hyslop, P. H., Tanzi, R. E., Polinsky, R. J., Haines, J. L., Nee, L., Watkins, P. C., Myers, R. H., Feldman, R. G., Pollen, D., Drachman, D., Growdon, J., Bruni, A., Foncin, J.-F., Salmon, D., Frommelt, P., Amaducci, L., Sorbi, S., Piacentini, S., Stewart, G. D., Hobbs, W. J., Conneally, P. M., and Gusella, J. (1987). The genetic defect causing familial Alzheimer's disease maps on chromosome 21. *Science* **235,** 885.

Tsui, L.-C., Buchwald, M., Barker, D., *et al.* (1985). Cystic fibrosis locus defined by a genetically linked polymorphic DNA marker. *Science* **230,** 104.

Wasmuth, J. J., Hewitt, J., Smith, B., Allard, D., Haines, J., Skarecky, D., Partlow, E., and Hayden, M. R. (1988). A highly polymorphic locus very tightly linked to the Huntington's disease gene. *Nature* **332,** 734.

Weber, J., and May, P. E. (1989). Abundant class of human DNA polymorphisms which can be typed using the polymerase chain reaction. *Am. J. Hum. Genet.* **44,** 388.

Wiggs, J., Nordenskjold, M., Yandell, D., Rapaport, J., Grondin, V., Janson, M., Werelius, B., Petersen, R., Craft, A., Riedel, K., Liberfarb, R., Walton, D., Wilson, W., and Dryja, T. (1988). Prediction of the risk of hereditary retinoblastoma, using DNA polymorphisms within the retinoblastoma gene. *N. Engl. J. Med.* **318,** 151.

Yandell, D. W., and Dryja, T. P. (1989). Detection of DNA sequence polymorphisms by enzymatic amplification and direct genomic sequencing. *Am. J. Hum. Genet.* **45,** 547.

Zonana, J., Sarfarazi, M., Thomas, N. S. T., Clarke, A., Marymee, K., and Harper, P. S. (1989). Improved definition of carrier status in X-linked hypohidrotic ectodermal dysplasia by use of restriction fragment length polymorphism–based linkage analysis. *Jour. Ped.* **114,** 392.

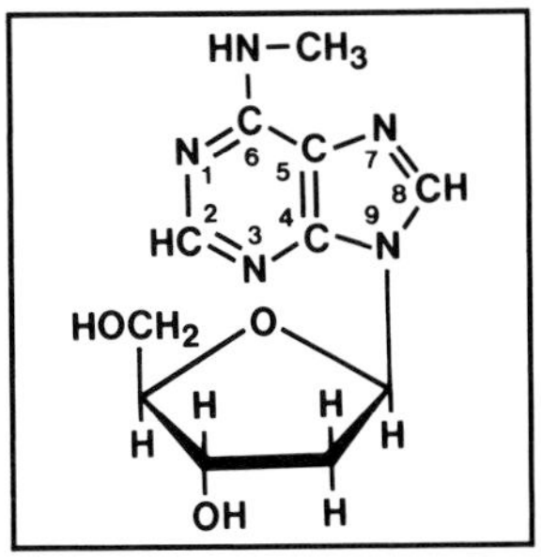

DNA Methylation and Gene Activity

WALTER DOERFLER, *University of Cologne*

Glossary

Demethylation Removal of a 5-methyldeoxycytidine residue, possibly by DNA replication and concomitant inhibition of maintenance methylation at a given sequence; it has been postulated, but not yet proven, that a 5-methyldeoxycytidine residue can also be actively excised, and that the ensuing defect can be repaired by the insertion of a deoxycytidine residue leading to a net demethylation

DNA methylation Postreplicational attachment of a -CH_3 (methyl) group by a DNA methyltransferase to a deoxycytidine or a deoxyadenosine residue in the double-stranded DNA chain

DNA methyltransferase (Cellular) enzyme that transfers a methyl group from *S*-adenosylmethionine as the methyl donor to DNA

DNA–protein interactions Binding of a specific protein or of a protein complex to a specific deoxyribonucleotide sequence in DNA, a so-called sequence motif; unknown rules of DNA–protein interactions can be considered to constitute one of the unresolved, more complex genetic codes

Enhancer Deoxyribonucleotide sequence, some 15–20 deoxynucleotides in length, whose presence increases the activity of the promoter(s) irrespective of location and orientation

Methylation pattern Specific distribution of 5-methyldeoxycytidine residues in a DNA segment, in a promoter, or in an entire eukaryotic genome

Promoter Regulatory region (deoxyribonucleotide sequence) usually in the 5′ flanking sequence of a gene; contains numerous sequence motifs for the interaction of transcription factors; it is probably the combination of several transcription factors binding to a promoter that bestows specificity upon the system. Cofactors bind to transcription factors and via protein–protein interactions, convert the promoter into a complex three-dimensional structure.

Restriction endonucleases Set of enzymes from prokaryotic organisms or encoded by chlorella viruses that cleave DNA endonucleolytically at specific deoxyribonucleotide sequences (e.g., *Hpa*II [from *Haemophilus parainfluenzae*] at

5′-C↓CGG-3′
3′-GGC↑C-5′

sequences [arrows indicate cleavage sites]).

Sequence-specific methylation Introduction of a methyl group to a deoxycytidine residue in a specific deoxynucleotide sequence (e.g., the DNA methyltransferase *Hpa*II modifies the sequence 5′-C*CGG-3′ at the 3′-located C, designated by an *)

Transactivation Activation of a promoter-gene assembly by a transacting function, at a distance, usually by a specific protein—the transactivator; its precise function is not yet known, but it is thought to modify the functions of transcriptional factors and, in that way, to contribute to the activation of a promoter

IN MAMMALIAN DNA, a significant portion (<10%) of all deoxycytidine residues have been converted to 5-methyldeoxycytidine (5-mC) residues. These methylated nucleotides are distributed in specific patterns, which are probably genetically determined. Evidence indicates that 5-mC residues frequently, but not exclusively, occur in 5′-CG-3′ dinucleotide sequences, and that there is a clustering of 5′-CG-3′ sequences in the regulatory (promoter) region of many functional genes. Specific patterns of DNA methylation can be maintained by the cellular DNA methyltransferase(s). Foreign DNA artificially fixed in the mammalian genome by integration becomes specifically methylated in distinct patterns. The *de novo* establishment of these patterns can require many cell generations, and it is not yet known what mechanisms determine these *de novo* methylation events. It has been demonstrated that in the establishment of *de novo* patterns the gradual spreading of DNA methylation, which starts from focal sites, plays an important role. Sequence-specific promoter methylations can lead to the long-term inactivation or inhibition of promoter function. In that way, foreign genes, artificially added to a genome, are often permanently silenced. For a number of model promoter-gene assemblies, this inactivating effect of sequence-specific promoter methylations has been documented both *in vivo* and *in vitro*. Moreover, inverse correlations between promoter methylations and gene activity have been reported for a large number of genes in many different eukaryotic systems. In that sense, promoter methylation can affect many biological functions. The gene-inactivating effects of sequence-specific promoter methylations can be partly abrogated by cis- (enhancers) or transacting (transactivators) functions, apparently without the loss of the methyl group in both DNA complements. It is likely that sequence-specific promoter methylations exert their inhibitory effect on the promoter via the modulation of specific DNA motif-transcription factor interactions, by structural changes in promoter motifs, or by a combination of both effects. Proteins specifically binding to methylated sequences may also be important.

I. Introduction

All genetic information in living organisms is stored and functionally realized in the sequence of the four nucleosides deoxyadenosine (A), deoxycytidine (C), deoxyguanosine (G), and thymidine (T) in deoxyribonucleic acid (DNA). Some viral genomes use RNA as the genetic storage molecule. The genetic codes encompass the primary code for the synthesis of all protein chains in an organism; the human genome is estimated to accommodate some 5×10^4–10^5 different polypeptides. However, more complex genetic codes are ensconced in the nucleotide sequence of DNA (e.g., regulatory sequence motifs whose function is premised on complex DNA–protein interactions, programs for differentiation and development). Other functions facilitated by DNA sequence have not been clearly defined yet. Presently, investigations on whether or not DNA sequence is somehow related to the >5,000 human languages seem to be gaining momentum. [*See* DNA AND GENE TRANSCRIPTION.]

In structural terms, DNA is very flexible in that it can assume a host of different conformations, each of which may have important functional implications; however, there is a more subtle modifier of DNA structure, the addition of methyl ($-CH_3$) groups to two of the deoxyribonucleotides—A or C in DNA. DNA methylation can also influence the structure of DNA (e.g., facilitate the transition from the B [the normal right-handed double helix] to the Z form [the left-handed double helix]). The modified nucleosides 5-mC, 4-methyldeoxycytidine, 5-hydroxymethyldeoxycytidine, and N^6-methyldeoxyadenosine (N^6-mA) have been found in the DNA of prokaryotic organisms. The DNA of the higher eukaryotes, particularly the DNA of humans, is thought to contain 5-mC as the only modified nucleotide. Since the discovery that DNA methylation affects the activity of many restriction endonucleases, the biological role of DNA methylation has been sought in the restriction-modification system. During the last decade, a more all-encompassing function for DNA methylation as a genetic signal has been recognized. A methyl group in a highly specialized position in the major groove of the DNA double helix can theoretically be recognized by proteins or protein complexes that exert their functions by binding to specific sequence motifs in DNA. In that sense, the presence of a methyl group can potentially modulate the function of many DNA-binding proteins, but the functional consequences of this modulation cannot be predicted *a priori*. These sequelae could be positive or negative, depending on the nature of the DNA–protein interaction. The latter statement can be exemplified by the specificity of restriction endonucleases: Many of these enzymes are inhibited by the presence of 5-mC or of

N^6-mA in their recognition sequences. On the other hand, the restriction endonucleases *Dpn*I, *Nan*I, *Nmu*DI, *Nmu*EI, and probably others require the occurrence of N^6-mA in the recognition sequence 5′-GATC-3′ for their activity.

With a 5-mC residue in mammalian DNA affecting the interaction of several regulatory proteins, a wide gamut of biological functions can presumably be regulated by this genetic signal. Evidence indicates that DNA methylation might influence DNA replication or recombination. The most intensely studied function afflicted by sequence-specific methylations is the regulation of gene expression in higher eukaryotic cells, in animal and plant cells alike. Aside from the availability of a very large number of variations in nucleotide sequences as regulatory motifs (e.g., in a promoter sequence), the methylation of specific nucleotides is the only modification of DNA known to date with far-reaching functional consequences. As one or few 5-mC residues in specific promoter sequences can lead to long-term promoter inactivations by modulating DNA–protein interactions, a large number of biological processes can be dominated by site-specific methylations (see Section VII). In organic chemistry, the -CH_3 group is considered the basic and simplest residue. Placed in a decisive nucleotide sequence, this seemingly innocuous chemical asset can be transformed into a powerful genetic signal. The nucleotide 5-mC has, therefore, been considered as the fifth nucleotide in DNA.

II. Patterns of DNA Methylation in Mammalian DNA

At a given time in the development of an organism, the DNA in the genome of each cell or of each organ system in complex organisms has its specific pattern of methylation, (i.e., a specific distribution of 5-mC residues). Due to technical limitations, these patterns cannot be determined in their entirety at present. These patterns appear to be subject to change, again dependent on the state of development or growth. Currently, two techniques that allow the detection of 5-mC residues in a complex genome are available.

1. A considerable number of *restriction endonucleases* are sensitive to sequence-specific methylation. The principle of this type of analysis will be explained for the isoschizomeric restriction endonuclease pair *Hpa*II (*Haemophilus parainfluenzae*) and *Msp*I (*Moraxella* species), which recognize and cleave the same target sequence 5′-CCGG-3′. Methylation of the 3′-located C residue, even in one strand, prevents cleavage at the sequence by *Hpa*II but allows cleavage by *Msp*I. Methylation of the 5′-located C inactivates *Msp*I at this site. The distribution of 5-mC in the 5′-CCGG-3′ sequences in the segment chosen for analysis can be determined by cleaving DNA with *Hpa*II or *Msp*I and by comparing the sizes of the fragments produced by using separation by gel electrophoresis (Southern blotting). This analysis necessarily restricts the investigation to the recognition sequence of the enzyme so that many C residues in a sequence, which could carry a methyl group, are excluded from the analysis. Despite this limitation, this method has proven useful in obtaining a quick survey of methylation patterns. There is evidence that the methylation of 5′-CCGG-3′ sequences serves as a reliable indicator of the true methylation pattern.

2. The method of *genomic sequencing* allows all deoxycytidine residues of a sequence to be analyzed. The technique is based on the limited reaction of hydrazine with 5-mC, whereas at C residues the DNA chain does react and is cleaved by the subsequent piperidine treatment. Thus, in a nucleotide sequencing reaction, 5-mC residues appear as gaps in a conventional sequencing ladder. This method gives a precise account of all 5-mC residues present in a sequence; however, the technique is still difficult and limited to relatively short segments of a few hundred nucleotides of DNA.

As discussed below, the extent of promoter methylation is a significant indicator of gene activity, and more complex patterns of methylation may have important functional significance for the activity states of large genome segments. Methylation patterns in that context provide valuable activity indicators. It will be desirable to determine how patterns of DNA methylation are *de novo* established, how they are then maintained, and under what conditions they can be altered. These questions pertain mainly, but not exclusively, to DNA that has been extraneously added to cells and has become integrated into a mammalian genome (e.g., integrated adenovirus genomes or isolated adenovirus promoters). This viral genome has also served a useful role as a model system for studies on the function of DNA methylation. Because the DNA in the isolated adenovirion is not detectably methylated, the gradual establishment of *de novo* patterns of DNA methylation in adenovirus DNA, which has recently been integrated into cellular DNA, can be studied in considerable detail. The data available demonstrate that the establishment of a particular methylation pattern in the integrated adenovirus genome requires many cell generations, and a pattern seems to arise by the spreading of DNA methylation from one or a few initial foci. The mechanism of spreading, however, is not understood.

The genes for the tumor necrosis factors (TNF) α and β, in particular their 5′-upstream and promoter regions, have been investigated in DNA isolated from human lymphocytes, granulocytes, and sperm. The results are characterized by a very remarkable interindividual concordance of DNA methylation in specific human cell types. The patterns are identical in the DNA from one cell type for different individuals even of different genetic origins, but different in the DNA from different cell types. For example, in the DNA from human granulocytes of 15 different individuals, 5-mC residues have been localized by the genomic sequencing technique in three identical sequence positions in the 5′-upstream region and in one downstream position of the gene for TNF-α. It appears that in human DNA, highly specific patterns of methylation exist in specific regions of the genome and in specific cell types. These patterns may be identical or very similar in the DNA of different individuals (even of very different ethnic origins).

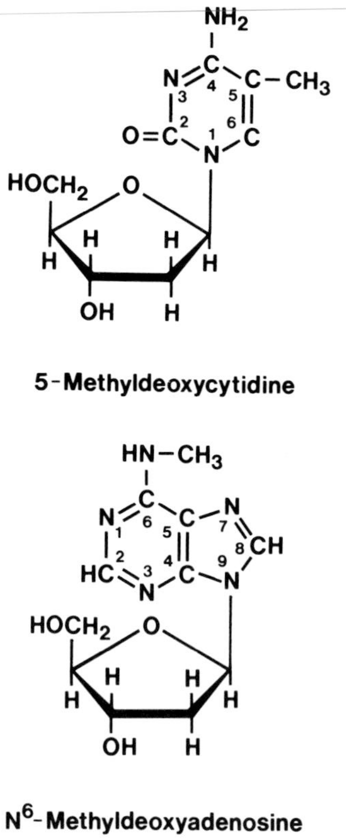

5-Methyldeoxycytidine

N^6-Methyldeoxyadenosine

FIGURE 1 Structural formulas of 5-methyldeoxycytidine and N^6-methyldeoxyadenosine.

III. Basic Information on DNA Methylation

A. Chemistry

The two main modified nucleosides in DNA are 5-mC and N^6-mA (Fig. 1). The nucleosides 4-methyldeoxycytidine and hydroxy-5-methyldeoxycytidine have also been observed. The nucleoside 5-mC probably constitutes the only modified nucleotide in mammalian DNA; however, the occurrence of minute amounts, perhaps at very special sites, of other modifications cannot be excluded. Of the four dinucleotide combinations with 5-mC, 5′-5-mCG-3′, 5′-5-mCC-3′, 5′-5-mCA-3′, and 5′-5-mCT-3′ all have been found in mammalian DNA. The most frequently occurring methylated dinucleotide is 5′-5-mCG-3′. The 5′-CG-3′ dinucleotides are statistically underrepresented in mammalian DNA. Their scarcity has functional implications in that they are clustered at the 5′ flanks of many genes in their promoter. It has been argued that this clustering, and the corresponding paucity of 5′-CG-3′ groups in many other nonpromoter-associated parts of the genome, might be a consequence of frequent deamination to which 5-mC can be prone, being then converted to thymidine. While this reasoning is supported by the facts of elementary nucleotide chemistry, it lacks—when applied exclusively—evolutionary plausibility. Upon spontaneous hydrolytic deamination, a 5-mC residue is converted to thymidine, hence a C/G nucleotide pair could be changed to a T/A pair, unless it was repaired to the previously existing C/G nucleotide pair, which might be equally likely. Any nucleotide sequences occurring in any organism in nature, including sequences containing 5-mC, are probably the products of long-standing selection during evolution. If a 5′-5-mCG-3′ dinucleotide pair was functionally required, it certainly would have been selected for. Moreover, the underrepresentation of 5′-CG-3′ dinucleotide pairs is by no means universal; it is peculiar to higher eukariotic genomes. Secondly, a deamination event involving an original 5′-5-mCG-3′

dinucleotide should lead to a 5′-CG-3′ or 5′-TA-3′ dinucleotide constellation with equal frequency. It is likely, however, that there are important functional reasons for the selection of clusters of 5′-CG-3′ dinucleotide combinations at the promoter flank of many operative genes. While the final explanation for the 5′-CG-3′ clustering is not yet available, the elucidation of the phenomenon may be sought in the code governing DNA–protein interactions and in the structural requirements for functional promoter sequences.

B. DNA Methyltransferase(s)

The 5-mC residues in DNA are thought to be generated after DNA is replicated, when DNA methyltransferases add an *S*-adenosylmethionine (SAM)-derived methyl group to the carbon in the 5 position of the pyrimidine cytidine in a new DNA chain. Very little is known about the specificity of mammalian DNA methyltransferases. The gene for one of these enzymes has been molecularly cloned, and its nucleotide sequence has been determined. The gene consists of a DNA-binding domain and a DNA-methyltransferase domain with an amazingly high degree of homology to the amino acid sequence of prokaryotic DNA methyltransferases. Although the mode of action of DNA methyltransferases is incompletely understood, an interesting experiment sheds light on the complexity of the selection mechanism by which the enzyme chooses its substrates in an immense array of nucleotides. Highly purified preparations of a mammalian DNA methyltransferase were incubated with a segment of the globin gene promoter, which carries a cluster of three 5′-CG-3′ dinucleotide sequences. The DNA methyltransferase critically distinguished between the three sites according to their location in the 5′ to 3′ polarity on the promoter. The most 5′ site was modified the least, the most 3′ site the most extensively. The DNA methyltransferase apparently recognizes specific positions in a nucleotide sequence.

C. Maintenance and *de novo* Methylation

In formal terms, two different types of DNA methylation can be distinguished; maintenance and *de novo* methylation. The former involves the remethylation of a newly synthesized DNA strand after DNA replication in a hemimethylated DNA molecule (Fig. 2), where the old strand is methylated but the new one is not. Hemimethylated sequences

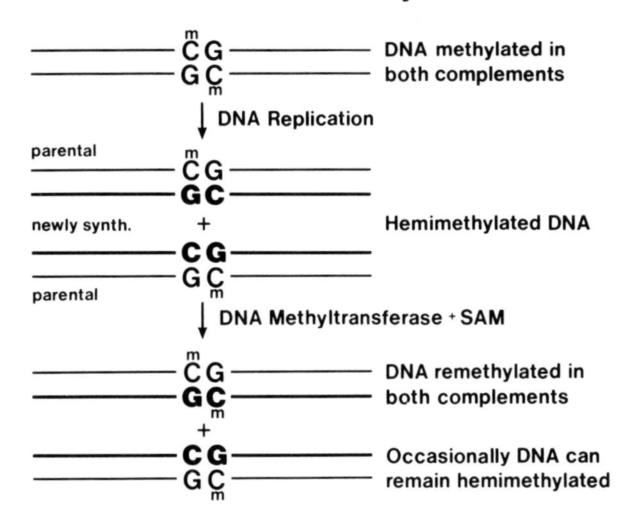

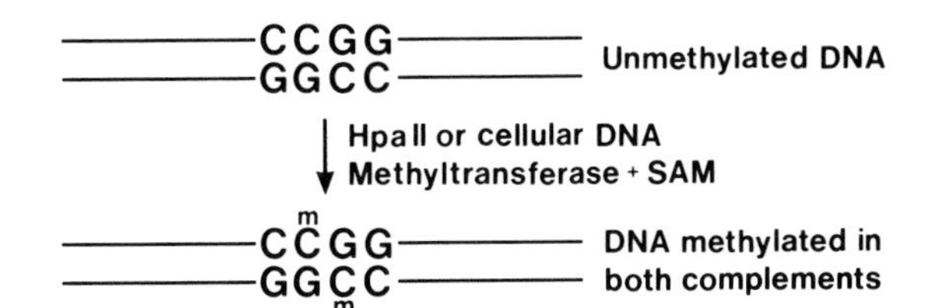

FIGURE 2 Schematic explanation of maintenance and *de novo* DNA methylation. SAM, *S*-adenosylmethionine. For details see text.

in the double-stranded DNA molecule are thought to represent an efficient substrate for the DNA methyltransferase. Evidence indicates that hemimethylated sequences can escape methylation for several cell generations. Thus, there must be more complex regulatory mechanisms that decide about maintenance methylation, in addition to the existence of a hemimethylated site. In *de novo* methylation, a previously completely unmethylated DNA sequence becomes methylated, as has been documented for adenovirus DNA integrated into the DNA of transformed hamster cells. But it is not known how sites of *de novo* methylation are chosen, how methylation spreads to adjacent sequences, or whether or not the same DNA methyltransferase(s) serves both types of DNA methylation.

From a functional point of view, 5-mC can be viewed as a true fifth nucleoside in DNA, serving as an important genetic signal. This genetic signal is almost ubiquitous in nature, although *Drosophila* might be capable of doing without it. For technical reasons, this is still an unresolved issue. The amount of 5-mC in the DNA of different eukaryotes varies considerably. Plant DNA can contain a very high proportion of 5-mC as compared with deoxycytidine.

D. Demethylation

It would be important to know whether a 5-mC residue in a critical promoter or in another DNA sequence can be actively removed. A regulatory genetic signal must be flexibly applied and reversible. There are at least two major possibilities to eliminate 5-mC from a sequence. (1) During DNA replication, demethylated sequences may be produced by the specific inhibition of the DNA methyltransferase responsible for maintenance DNA methylation. However, the regulation of the activity state of this enzyme is not understood, nor is its sequence-specific inhibition. (2) A 5-mC residue could be actively removed, perhaps by a repairlike excision and its replacement by a nonmethylated C, but this mechanism has not been proven. However, good evidence indicates the occurrence of DNA demethylations without concomitant DNA replication. The problem of active demethylation and/or selective demethylations as a consequence of highly specific inhibitions of maintenance DNA methylations remains one of the many gaps in our understanding of how patterns of DNA methylation are introduced, altered, or preserved.

DNA can be artificially demethylated by growing cells in the presence of the cytidine analogue 5-azacytidine, which is presumably incorporated into newly synthesized DNA and can bind the DNA methyltransferase very tightly, probably covalently. As a consequence, the available cellular levels of DNA methyltransferase are significantly decreased, and cellular DNA sequences become progressively demethylated as cell replication proceeds. With the loss of methylated cytidines, a large number of previously long-term inactivated cellular genes become reactivated. Even entire differentiation programs of an organism can be turned on in this way. Although 5-azacytidine, a toxic compound, may have several side effects that could potentially be involved in gene activation, the reactivation of genes by this drug is attributed to unspecific demethylation events (Section IV.C.2).

E. Inheritance of Patterns

Patterns of DNA methylation are inherited, but they may be changeable under certain conditions, with far-reaching functional consequences. The stability of a genomic pattern may entail a certain distribution of activities among the ensemble of genes that are transcribed in each individual cell type in an organism. The effects of these patterns on cellular functions other than transcription have so far not been studied in sufficient detail.

IV. Sequence-Specific Promoter Methylations and Gene Inactivation

A. Inverse Correlations

In adenovirus-transformed hamster cell lines, some of the viral genes (early genes) are expressed, whereas the late viral genes encoding the structural proteins of the virion are often permanently silenced. These cell lines carry parts of or the entire viral genome integrated into the host cell genome. An inverse correlation between the extent of DNA methylation in the various genes and their levels of expression has been observed. Similar observations have been reported for a very large number of viral and nonviral eukaryotic genes. The finding of inverse correlations between promoter methylation and gene activity has started an interesting discussion on whether promoter methylation may be the cause or the consequence of promoter inactivation or inhibition. This problem can be approached by the *in vitro* methylation of promoters.

B. *In vitro* Methylation of Promoter Sequences and Effect on Activity

This topic has been studied in detail with adenoviral promoters as model systems. The E1A genes of the human adenoviruses are the earliest viral genes to be expressed after infection of human cells. The promoter of the E1A gene of adenovirus type 12 has been fused to the prokaryotic gene for chloramphenicol acetyltransferase (CAT) as an activity indicator. Upon transfection of this construct into mammalian cells, the determination of CAT activity in extracts of the transfected cells provides a convenient reporter for promoter function. When the E1A promoter is methylated in two 5′-CCGG-3′ or in eight 5′-GCGC-3′ sequences, it is inactivated. These results already point to the promoter sequence as being decisively sensitive to DNA methylation. Surprisingly, the E1A promoter of adenovirus type 12 DNA can also be inactivated when one N^6-mA is introduced about 280 nucleotides 5′ upstream or about 60 nucleotides downstream from the TATA signal in this promoter. Even this unusual modification can be recognized by the transcriptional control apparatus in mammalian cells.

Another adenovirus promoter, which has been studied extensively, is the late E2A promoter of adenovirus type 2 DNA. The E2A gene codes for a DNA-binding protein essential in viral DNA replication and has two promoter systems, one used early and the second late in infection. Here, only the late E2A promoter will be considered. The functional analysis on the methylation of this promoter starts again with an inverse correlation. In the adenovirus type 2-transformed hamster cell line HE1, the integrated E2A gene is active (i.e., it is transcribed and translated into protein), and the 14 5′-CCGG-3′ sequences in this gene are all unmethylated. Conversely, in cell lines HE2 and HE3, the E2A gene is permanently silenced, and all 14 5′-CCGG-3′ sequences in the E2A region are methylated. The same correlation can be observed after the *in vitro* methylation of all 14 5′-CCGG-3′ sequences in the E2A gene. Inactivation of this gene can be demonstrated by microinjection into nuclei of *Xenopus laevis* oocytes. In this assay, the unmethylated gene exhibits full transcriptional activity. It suffices to methylate the three 5′-CCGG-3′ sequences in the late E2A promoter region to achieve inactivation or strong inhibition of this promoter, whereas the methylation of the 11 *Hpa*II sites in the remainder of the gene has no effect on its activity. In these experiments, only the three 5′-CCGG-3′ sequences in the late E2A promoter at positions −215, +6, and +24, counting from the site of transcription initiation at +1, have been methylated. None of the 5′-CCGG-3′ sequences in the vector and the CAT gene parts of the construct have been modified. These data have provided strong evidence for the notion that promoter methylations are decisive for promoter inactivation. Very similar results have been obtained with a number of other eukaryotic genes.

In *Spodoptera frugiperda* insect cells, which are thought to lack 5-mC in their DNA, an insect baculovirus (*Autographa californica* nuclear polyhedrosis virus) promoter (p10) can be inactivated by *in vitro* 5′-CCGG-3′ methylation. This finding suggests that insect cells can also respond to this genetic signal.

These results were complemented by experiments in which a methylated or an unmethylated promoter-gene construct has been permanently fixed in the cellular genome of permanent cell lines by integration. In these studies, DNA methylation and gene transcription levels have been followed.

Cell lines were generated introducing both the late E2A promoter-CAT gene construct in the methylated or the nonmethylated form and the selectable marker neomycin phosphotransferase, which confers neomycin resistance on the transformed cells. Three types of cell lines with the nonrearranged construct integrated in the genome of hamster cells were obtained. (i) Lines, which contain the nonmethylated construct, express the CAT gene, and the integrated construct remains nonmethylated. Cell lines generated using the premethylated construct have been found (ii) to retain a methylated promoter and to exhibit silenced CAT genes or (iii) to become partly demethylated at the 5′-CCGG-3′ sequences and show expression of the CAT gene. These data confirm the results of previous experiments and show that the effect of methylation is stable over many rounds of cell divisions.

The finding of late E2A promoter sequences that have lost methylation in some of the cell lines poses the additional question of why the promoter became demethylated and reactivated in some of the cell lines. It is conceivable that the site of integration of the late E2A promoter-CAT gene construct into the hamster genome plays an important role in determining the stability of a preimposed methylation pattern; however, other more complicated explanations also exist.

The promoter-silencing effect of sequence-specific methylations in the late E2A promoter has also been shown in a cell-free *in vitro* transcription system. The late E2A promoter of adenovirus type 2 DNA has again served as the model template, either nonmethylated or 5′-CCGG-3′ methylated at nucleotides −215, +6, and +24, relative to the transcriptional initiation site at +1. In control experiments, the methylated late E2A promoter and the unmethylated major late promoter of adenovirus type 2 DNA have been jointly added as templates to the same *in vitro* transcription reaction. The methylated E2A promoter is inactivated under these conditions; the unmethylated major late promoter exhibits full activity.

All data obtained with the late E2A promoter

TABLE I Inactivation of the Late E2A Promoter of Adenovirus Type 2 DNA by Sequence-Specific 5′-CCGG-3′ Methylations: Survey of Test Systems

1. Inverse correlations in adenovirus type 2-transformed hamster cell lines
   HE1: The late E2A promoter is unmethylated at 5′-CCGG-3′ sequences and active
   HE2/HE3: The late E2A promoter is methylated at 5′-CCGG-3′ sequences and inactive
2. *In vitro* 5′-CCGG-3′ methylation of the late E2A promoter
   Transient expression: i) Microinjection into *Xenopus laevis* oocytes
   ii) Transfection into mammalian cells
3. Cell-free transcription system
4. Genomic integration of the late E2A promoter-CAT gene construct
5. Integration of methylated construct in transgenic mice
   Methylation remains stable in all organs investigated.

model system have been summarized in Table I. These data were obtained by the same promoter and the same specific sequences for *in vitro* methylation in a gamut of experimental systems in which promoter inactivation due to methylation has been documented. However, the decisive sequences, which upon methylation cause promoter inactivation, still must be identified experimentally for each promoter.

C. Reversal of Promoter Inactivation

A regulatory signal for gene activity can attain its full potential only when it is pliable (i.e., when an inactivating event like promoter methylation can be rendered functionally reversible). Several possibilities have been realized and will be discussed briefly.

1. DNA replication leads to hemimethylated and eventually to nonmethylated double-stranded DNA when maintenance methylation (which can, but does not have to, coincide with replication) is somehow inhibited. It is not known how maintenance methylation is regulated nor how effectively this mode of demethylation works.

2. As outlined above, 5-azacytidine causes demethylation of replicating DNA, presumably by the specific inhibition of the DNA methyltransferase(s). Consequently, previously shut-down genes can be transcriptionally reactivated, sometimes with a concomitant change in chromatin structure as evidenced by the gene sequence being rendered sensitive to DNase I.

3. However, a promoter that had been inhibited by sequence-specific methylations can be also reactivated in more subtle ways. Trans- and cis-acting factors can abrogate the inhibition of the methylated late E2A promoter of adenovirus type 2 DNA without demethylation occurring, at least not in both DNA complements. Thus, the transactivator of adenovirus DNA encoded by the E1A gene can turn on the previously silenced late E2A promoter with an activity somewhat lower than that of the unmethylated promoter. This transeffect can be elicited with the methylated late E2A promoter either chromosomally or episomally (plasmid) located and with the transactivating gene itself being localized, respectively, episomally or on cellular chromosomes. Transcription from the thus transactivated, methylated E2A promoter initiated normally at the authentic cap site. Similar results were obtained with an enhancer of human cytomegalovirus placed in the vicinity of the methylated late E2A promoter. These results show that the inhibiting effect of sequence-specific methylations on promoter activity is not absolute but must be viewed in the context of other factors influencing promoter function.

D. Sequence-Specific Methylations and DNA–Protein Interactions in Eukaryotic Promoters

This topic is presently under active investigation. Apparently, the interaction of some but not all of the numerous transcription factors with specific promoter motifs can be inhibited by DNA methylation. The effects of these DNA–protein interactions were evaluated in some cases by determining either the protection of specific DNA sequences from DNaseI cleavage by specific proteins bound to these DNA sequences (DNaseI protection assay). In other instances, protein binding was monitored by the delayed migration of DNA fragment–protein complexes in an electrophoretic field (gel shift assay). Genomic footprinting techniques and the reaction with specific chemicals can enhance the sensitivities of these techniques. It was observed that the spreading of methylation in cell lines, which carry the late E2A promoter with three *in vitro* premethylated 5′-CCGG-3′ sequences, initially involves a DNA domain of this promoter devoid of bound proteins. Subsequently, methylation spreads to neighboring regions irrespective of the presence of proteins, and the patterns of transcription factors bound to DNA are altered. DNA methylation at sequences, with which certain transcription factors interact, interferes with protein binding. In contrast, methylation of sequences in the vicinity of sites binding to other factors seems to permit binding. [*See* DNA Binding Sites.]

The inhibitory effect of three 5-mC residues in the late E2A promoter sequence of adenovirus type 2 DNA might be attributed to interference with the binding of proteins. This possibility was confirmed experimentally: The results demonstrate that protein binding (possibly involving the AP2 factor) is abolished by methylation of the same sequences in the late E2A promoter whose methylation inhibits promoter function. Further understanding of how promoter methylation functions in gene inactivation at a biochemical level requires new experimental approaches. Moreover, it will be interesting to study what, if any, effect DNA methylation has on the DNA structure in an active or inactive promoter. In particular, it has been discussed to what extent DNA methylation influences DNA kinks, twists, and other subtle structural alterations in DNA.

A protein isolated from human placenta can bind specifically to certain methylated DNA sequences. This finding raises the question of whether or not large segments of the permanently inactivated mammalian genome are highly methylated and complexed with specific methyl-recognizing proteins. Perhaps the activity of the DNA methyltransferase system is directed by proteins intimately associated with DNA.

V. Spreading of DNA Methylation

Newly arising patterns of DNA methylation in mammalian DNA, particularly in DNA sequences recently integrated into that genome, are characterized by the spreading of DNA methylation from an initial point to adjacent DNA sequences. In several hamster cell lines, an E2A promoter-CAT gene construct, *in vitro* methylated at three 5′-CCGG-3′ sequences, has been genomically fixed (see Section IV.B). With increasing numbers of generations of these cells, methylation extends progressively to neighboring 5′-CG-3′ nucleotide pairs. Eventually, all 5′-CG-3′ sequences are completely methylated, and methylation also involves a 5′-CA-3′ and a 5′-CT-3′ dinucleotide in one of the cell lines, as shown by the genomic sequencing technique.

An adenovirus genome of some 30–35 kilobase pairs, which has been integrated into the hamster cell genome, also becomes methylated progressively with an increasing number of cell generations. This spreading of methylation over an entire integrated, previously nonmethylated viral genome proceeds in discrete steps in a nonrandom manner and is initiated between 30 and 50 map units of the viral genome. It is unlikely that the pre-existing patterns of methylation in the neighboring cellular sequences around the site of integration of the viral DNA codetermine the spreading of methylation. The findings about spreading of methylation lead to the speculation of whether or not the progressive inactivation of an entire X chromosome in mammalian females may be caused by the gradual extension of DNA methylation. Such extensive spreading of DNA methylation could be envisaged by assuming the coalescence of initial foci of methylation spaced in certain intervals on that chromosome. Extending these speculations to the somatic chromosomes, one can pose the question of how huge chromosomal sections are silenced permanently during differentiation and development, and how the decision to express either the maternally or the paternally inherited allele in the genome can also be related to selectively imprinted patterns.

VI. Interdigitating Patterns of DNA Methylation?

Conceivably, although not proven, a complicated pattern of DNA methylation in the mammalian genome is composed of several patterns with functionally different meanings, and perhaps these different patterns interdigitate with each other, giving rise to the sum total of a highly complex pattern involving the entire mammalian genome. The notion of interdigitating, mutually overlapping patterns has arisen on the basis of the following observations. For the silencing of the late E2A promoter of adenovirus type 2 DNA, the methylation of three 5′-CCGG-3′ sequences in this promoter seems to suffice. In the virus-transformed hamster cell lines HE2 or HE3, all of the 5′-CG-3′ sequences and one 5′-CA-3′ and one 5′-CT-3′ dinucleotide in the permanently inactivated E2A promoter are methylated. Is this apparent excess of methylation in the late E2A promoter a safeguard for permanent inactivation, or do some of these 5′-5-mCG-3′ sequences carry a different functional signal value?

In mammalian cells, very little is known about the function of methylated sequences in the regulation of elementary processes other than that of gene transcription. Amazingly, in green algae (chlorella), chlorella-specific viruses encode their own restriction-modification system.

The discussion on whether promoter and/or gene methylation is the cause or the consequence of long-term gene inactivation or inhibition may turn out to be a pseudoargument. On the one hand, the methylation of a few decisive 5′-CG-3′ sequences inactivates or inhibits a eukaryotic promoter; on the other hand, evidence indicates that DNA methylation spreads progressively from an initial focus of methylation involving only a few 5′-CG-3′ sequences to the adjacent 5′-CG-3′ and other C-containing sequences. Thus, the few methyl groups that initially cause promoter inactivation probably serve simultaneously as the origins of extensive spreading for the methyltransfer reaction. Therefore, specifically methylated sequences can be both the cause and the consequence of gene inactivation in that the modification that inhibits a promoter at the same time gives rise to spreading and to a more complex methylation pattern. In particular, by using the insensitive tool of restriction endonucleases for the analyses of methylation patterns, the few initial methylations that caused promoter inhibition would have been overlooked if they were not located in sequences recognized by the restriction endonucleases. Of course, the problem remains of how the DNA methyltransferase system identifies the decisive promoter sequences that need to be initially methylated.

VII. Promoter Inactivation by DNA Methylation: The Basis for More Complex Phenomena in Biology

Gene activity is regulated by many factors; one important contribution to the long-term inactivation or inhibition of promoter function is made by sequence-specific methylations. The functionality of this genetic signal is not absolute but is at least partly reversible (e.g., by transacting proteins, by enhancers). By regulating gene activity, DNA methylation can influence several more complex biological phenomena (Table II).

TABLE II Promoter Inactivation by DNA Methylation: The Basis for More Complex Phenomena in Biology

1. Development and differentiation
2. Inactivation of the X chromosome
3. Maternal and paternal imprinting
4. Inactivation of immunoglobulin genes
5. Persisting viral infections
6. Gene activities in tumor cells
7. Protracted occurrence of clinical symptoms in hereditary disease (not yet investigated?)
8. Plant molecular biology

1. Development and differentiation are thought to encompass a temporally and spatially precisely programmed sequence of gene activations and gene inactivations. The genetic basis for these programs is not understood. Apparently, patterns of methylation are inversely correlated with changing activity levels of individual genes and groups of genes.

2. The inactivation of genes on one of the X chromosomes has been related to DNA methylation. Some of the inactive, X-localized genes can be reactivated by 5-azacytidine treatment of cells growing in culture.

3. Evidence indicates that maternal and paternal imprinting of the mammalian chromosome is related to DNA methylation. Patterns of methylation are inherited in different ways depending on their maternal or paternal origin. Most of this evidence has been deduced from work with transgenic animals. To what extent the methylation pattern of the transgene is codetermined by the flanking sequences into which it has been integrated has not yet been clarified.

4. The activation of the immunoglobulin genes entails complex rearrangement and activation events. The activation of some of these genes appears to be accompanied by specific changes in DNA methylation patterns.

5. Specific DNA methylation patterns have been observed in persisting viral genes in adenovirus-transformed cells, and DNA methylation has been shown to play a role in viral persistence, particularly in the persistence of the herpes virus genome.

6. A characteristic difference between tumor and nontumor cells in any organ system is the differential pattern of gene expression. Genes that have been inactivated in the nontumor cells can become reactivated in tumor cells and vice versa. Concomitant with these changes, alterations in the patterns of DNA methylation in specific genes have been observed in virus-induced tumor cells and also in human tumors.

7. It is interesting that for some human (genetic) defects the onset of clinically symptomatic disease

is delayed until relatively late in the afflicted individual's life (e.g., in Huntington's disease). Studies on the specific changes in methylation patterns of the human genes affected in these individuals might shed light on this medically relevant problem. Perhaps the defect in some of these diseases lies in the proper regulation of DNA methylation in (several) segments of the human genome.

8. In plant molecular biology, the role that DNA methylation plays in gene regulation has also been recognized. Plant DNA sequences can be very highly methylated, involving 30% and even higher proportions of all C residues. At the level of individual genes, evidence has been adduced that the transposase activity of the Ac-controlling element in maize is regulated by DNA methylation. The Ac element encodes a function that is directly or indirectly involved in the transposition event in maize. An active transposase promoter is hypomethylated; the inactive promoter is strongly methylated.

VIII. Access to Complex Patterns of DNA Methylation: Direction of Future Research

In a living cell of a certain developmental state, the specific functional conditioning is determined by the sum of the activities of all transcribed genes and of the silenced parts of the cellular genome. DNA methylation levels in the promoter parts of the genes may serve as a first indicator of gene activities. Thus, access to the analyses of patterns of methylation in promoter segments of genes might also provide a survey of gene activities. Presently, only limited technical facilities are available to solve this problem. The total genome of a cell could be compared with the net of road intersections in a city with each gene or important genetic element corresponding to one intersection. "Traffic lights," 5′-CG-3′ groups, control these focal points of activity. Methylation of such a sequence would silence a center like a red traffic light stops the flow of traffic. Absence of methylation at the decisive sequence releases the block like switching the traffic light to green. Dysregulations of gene activities are probably at the core of many diseases, particularly of tumor diseases. The balance of lights (5′-CG-3′ sequences) turned to red (methylated) compared with those switched to green (unmethylated) decides about the efficient flow of traffic in a city, much like that of the functional state of a cell.

Future research will be directed toward the elucidation of many of the complex mechanisms briefly outlined in the preceding sections. At another fundamental level, the ways in which certain patterns of methylation in DNA are established, maintained, and probably altered according to the functional necessities in a cell will have to be studied in detail. The mechanism of action and the regulation of the activity of the DNA methyltransferase(s) will assume an important place in that work. We will also have to understand the biochemical mechanisms by which promoter methylations affect the function of transcription factors and other proteins that can bind specifically to promoter sequence motifs and can modulate promoter activity. A topic that has so far received little attention is the mechanism by which a promoter can be tuned to higher or lower activity levels. Whether or not there is a parallel in the gradation of methylation of sequences in a promoter is not known.

Acknowledgments

I am indebted to Petra Böhm for excellent editorial work. Research in the author's laboratory has been supported by the Deutsche Forschungsgemeinschaft through SFB74 (to the end of 1988) and SFB 274 (beginning in 1989).

Bibliography

Arber, W. (1974). DNA modification and restriction. *Prog. Nucleic Acid Res. Mol. Biol.* **14,** 1.

Bestor, T., Laudano, A., Mattaliano, R., and Ingram, V. (1988). Cloning and sequencing of a cDNA encoding DNA methyltransferase of mouse cells. The carboxyl-terminal domain of the mammalian enzymes is related to bacterial restriction methyltransferases. *J. Mol. Biol.* **203,** 971.

Bird, A. P. (1986). CpG-rich islands and the function of DNA methylation. *Nature* **321,** 209.

Cantoni, G. L., and Razin, A. (eds.) (1985). "Biochemistry and Biology of DNA Methylation." Alan R. Liss, Inc., New York.

Church, G. M., and Gilbert, W. (1984). Genomic sequencing. *Proc. Natl. Acad. Sci. U.S.A.* **81,** 1991.

Doerfler, W. (1981). DNA methylation—A regulatory signal in eukaryotic gene expression. *J. Gen. Virol.* **57,** 1.

Doerfler, W. (1982). In search of more complex genetic codes—Can linguistics be a guide? *Med. Hypoth.* **9,** 563.

Doerfler, W. (1983). DNA methylation and gene activity. *Annu. Rev. Biochem.* **52,** 93.

Doerfler, W. (1989). Complexities in gene regulation by promoter methylation. *In* "Nucleic Acids and Molecular Biology," Vol. 3 (F. Eckstein and D. M. J. Lilley, eds.), pp. 92–119. Springer Verlag, Berlin and Heidelberg.

Doerfler, W., Toth, M., Kochanek, S., Achten, S., Freisem-Rabien, U., Behn-Krappa, A., and Orend, G. (1990). Eukaryotic DNA methylation: facts and problems. *FEBS Letters* **268,** 329.

Feinberg, A. P., and Vogelstein, B. (1983). Hypomethylation distinguishes genes of some human cancers from their normal counterparts. *Nature* **301,** 89.

Gartler, S. M., and Riggs, A. D. (1983). Mammalian X-chromosome inactivation. *Annu. Rev. Genet.* **17,** 155.

Groudine, M., and Conkin, K. F. (1985). Chromatin structure and *de novo* methylation of sperm DNA: Implications for activation of the paternal genome. *Science* **228,** 1061.

Hermann, R., Hoeveler, A., and Doerfler, W. (1989). Sequence-specific methylation in a downstream region of the late E2A promoter of adenovirus type 2 DNA prevents protein binding. *J. Mol. Biol.* **210,** 411.

Holliday, R. (1987). The inheritance of epigenetic defects. *Science* **238,** 163.

Jones, P. A., and Taylor, S. M. (1980). Cellular differentiation, cytidine analogs and DNA methylation. *Cell* **20,** 85.

Klein, G. (1989). Viral latency and transformation: The strategy of Epstein–Barr virus. *Cell* **58,** 5.

Kochanek, S., Toth, M., Dehmel, A., Renz, D., and Doerfler, W. (1990). Interindividual concordance of methylation profiles in the human genes for tumor necrosis factors α and β. *Proc. Natl. Acad. Sci. U.S.A.* **87,** 8830.

Kruczek, I., and Doerfler, W. (1983). Expression of the chloramphenicol acetyltransferase gene in mammalian cells under the control of adenovirus type 12 promoters: Effect of promoter methylation on gene expression. *Proc. Natl. Acad. Sci. U.S.A.* **80,** 7586.

Langner, K.-D., Weyer, U., and Doerfler, W. (1986). Trans effect of the E1 region of adenoviruses on the expression of a prokaryotic gene in mammalian cells: Resistance to 5′-CCGG-3′ methylation. *Proc. Natl. Acad. Sci. U.S.A.* **83,** 1598.

McClelland, M., and Nelson, M. (1988). The effect of site-specific DNA methylation on restriction endonucleases and DNA modification methyltransferases—A review. *Gene* **74,** 291.

Razin, A., Szyf, M., Kafri, T., Roll, M., Giloh, H., Scarpa, S., Carotti, D., and Cantoni, G. L. (1986). Replacement of 5-methylcytosine by cytosine: A possible mechanism for transient DNA demethylation during differentiation. *Proc. Natl. Acad. Sci. U.S.A.* **83,** 2827.

Saluz, H. P., and Jost, J. P. (1987). "A Laboratory Guide to Genomic Sequencing." Birkhäuser, Basel and Boston.

Sutter, D., and Doerfler, W. (1980). Methylation of integrated adenovirus type 12 DNA sequences in transformed cells is inversely correlated with viral gene expression. *Proc. Natl. Acad. Sci. U.S.A.* **77,** 253.

Swain, J. L., Stewart, T. A., and Leder, P. (1987). Parental legacy determines methylation and expression of an autosomal transgene: A molecular mechanism for parental imprinting. *Cell* **50,** 719.

Toth, M., Lichtenberg, U., and Doerfler, W. (1989). Genomic sequencing reveals a 5-methylcytosine-free domain in active promoters and the spreading of preimposed methylation patterns. *Proc. Natl. Acad. Sci. U.S.A.* **86,** 3728.

Van Etten, J. L., Xia, Y., Burbank, D. E., and Narva, K. E. (1988). Chlorella viruses code for restriction and modification enzymes. *Gene* **74,** 113.

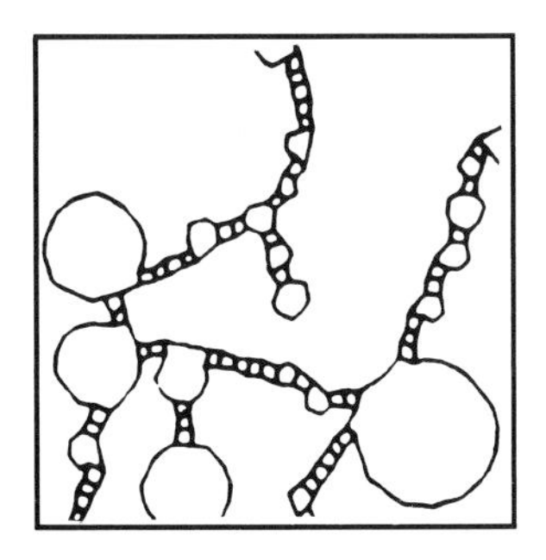

DNA, RNA, and Protein Sequence Analysis by Computer

GUNNAR VON HEIJNE, *Department of Molecular Biology, Karolinska Institute Center for Biotechnology*

Glossary

Consensus pattern A string of semiconserved nucleotides or amino acids that defines a functionally important region, such as a promoter site or a glycosylation site

EMBL data library The European equivalent of GenBank, run by the European Molecular Biology Laboratory in Heidelberg, Federal Republic of Germany

GenBank A central computerized repository of DNA sequences run by Los Alamos National Laboratories

Molecular graphics A technique by which models of biological macromolecules can be displayed and manipulated on a high-resolution computer screen

Restriction site A specific sequence of bases in DNA that is cut by a restriction enzyme

MOLECULAR BIOLOGY is heavily dependent on computers for the storage, handling, and analysis of sequence data. The trend toward automation of data collection and analysis will be a dominant feature of molecular biology in the future.

I. DNA Sequence Data Banks

The sequencing of new genes and other regions of interest on DNA represents a major effort in current molecular biology. The amount of sequence data available has necessitated the establishment of centralized computerized data banks, freely available to both academic and nonacademic scientists. The growth of the GenBank nucleotide sequence data library is shown in Fig. 1. It is estimated that the annual rate of growth in the mid-1990s will be somewhere between 5 and 50 million bases; the present size of GenBank is near 40 million bases.

A typical GenBank entry is shown in Fig. 2. The annotations include taxonomic information, bibliographic data, and a list of sites, such as regulatory signals, splice sites, and start and stop codons for protein coding regions.

GenBank and its European counterpart, the EMBL Data Library, are distributed periodically on computer tape, and they can also be accessed via international computer networks. Many commercial software companies offer versions of these data banks as parts of complete DNA and protein sequence analysis packages for microcomputers. Due to their size, the data banks are now often put on CD-ROM disks (compact disks) and are provided with accessory software that allows efficient information retrieval.

II. Sequence Analysis on Computer

In the modern molecular biology laboratory, the computer is an integral part of the necessary equipment. It is used not only for searching data banks, but for much of the day-to-day routine. Researchers doing DNA sequencing projects need the computer to keep track of partial sequences not yet assembled into a final full sequence, restriction site analysis is next to impossible without the help of a computer, and many complicated cloning projects can be simulated on the computer before they are actually carried out.

A conceptually more challenging use of computers is sequence analysis (i.e., one's attempts to make sense of a new polynucleotide or protein sequence). Where are the regulatory signals that de-

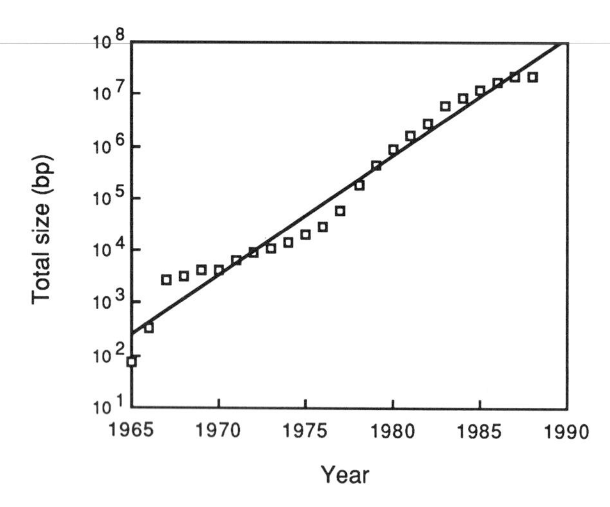

FIGURE 1 Growth of the GenBank DNA sequence data bank. The total size is now near 40 million bases. (Courtesy of Christian Burks, Los Alamos.)

fine the beginning and the end of a gene? Where are the introns and exons? Where are the protein coding regions? What kind of protein will the cell make from this gene? Is the gene or protein similar to other known sequences? Such questions can be profitably approached with the aid of a computer.

A. Finding Related Sequences

An obvious question to ask when a new sequence appears is whether it is related to something already known. To check this, one normally starts with a fast scan of the full data bank. There are algorithms that will find any reasonably similar sequence within a matter of minutes. Such "hits" are then subjected to more refined analyses, which ultimately produce one or more alignments between the new sequence and the related ones (Fig. 3). These alignments can be used to pinpoint conserved regions of possible functional significance, and they can serve as a basis for estimating the evolutionary relationship between the two sequences.

B. DNA

Sequence analysis of DNA usually aims at identifying various signals in the sequence (i.e., local patterns of nucleotides that may serve as recognition sites for regulatory proteins, RNA polymerases, or splicing enzymes). This is normally done by searching for "consensus patterns": short stretches that match most or all positions in the relevant query sequences (Fig. 4). In this way, one can highlight likely promoters and terminators for RNA synthesis, splicing donor and acceptor sites, and other sequences that might serve to regulate gene expression.

C. RNA

mRNA molecules contain signals for the initiation and the termination of protein synthesis. Candidate signals can again be found by consensus pattern searches. However, these searches rarely yield unambiguous results; hence, other methods to locate likely coding regions are also much in use. The latter methods rely on the fact that the relative use of codons coding for the same amino acid is highly nonrandom in most genes. Thus, real coding regions are expected to conform to the particular bias in codon usage of the organism in question, whereas "false" putative coding regions should deviate from this norm (Fig. 5). In practice, codon bias analysis is often a more reliable means of identifying coding regions than the simple signal searches.

Since RNA molecules are single stranded, they can fold into very complicated stem–loop structures by internal base pairing (Fig. 6). There are an almost infinite number of ways to fold all but the very shortest RNA molecules, but algorithms have been developed that will find one or a few of the energetically most stable structures in reasonable computer time. To be reliable, RNA folding analysis should be supplemented by experimental data on, for example, RNase sensitivity of as many regions of the molecule as possible.

D. Proteins

The main use for computer analysis of a protein sequence is getting an idea of what the molecule's three-dimensional structure might be like. This is the classical "protein folding problem," and it has plagued molecular biology since the first crystal structures of proteins were solved in the 1950s. Unfortunately, we are still far from a solution to this problem. [*See* PROTEINS.]

In practice, all one can hope for is to get a vague idea of the secondary structure of a new protein, unless the crystal structure of a related molecule is already known. In the latter case, the new chain can often be modeled after the old one, using molecular graphics techniques.

The less ambitious task of finding sorting signals for protein targeting or candidate sites for post-

```
LOCUS       KAEPULA      4133 bp ds-DNA            BCT       31-DEC-1987
DEFINITION  K.aerogenes pullulanase (pulA) gene, complete cds.
ACCESSION   M16187
KEYWORDS    pullulanase.
SOURCE      K.aerogenes (strain W70) DNA, clone pMP1.
  ORGANISM  Klebsiella aerogenes
            Prokaryota; Bacteria; Gram-negative facultatively anaerobic rods;
            Enterobacteriaceae.
REFERENCE   1  (bases 1 to 4133)
  AUTHORS   Katsuragi,N., Takizawa,N. and Murooka,Y.
  TITLE     Entire nucleotide sequence of the pullulanase gene of Klebsiella
            aerogenes W70
  JOURNAL   J Bacteriol 169, 2301-2306 (1987)
  STANDARD  full staff_review
COMMENT     Draft entry and computer-readable copy of sequence in [1] kindly
            provided by Y.Murooka, 23-JUN-1987.
FEATURES       from  to/span     description
    pept         23  <     1 (c) malX protein
    pept        330     3620     pullulanase precursor (EC 3.2.1.41)
    sigp        330      386     pullulanase signal peptide
    matp        387     3617     pullulanase mature peptide
    refnumbr      1        1     numbered 1 in [1]
BASE COUNT      880 a   1183 c   1223 g    847 t
ORIGIN      566 bp upstream of EcoRV site.

BACTERIA:KAEPULA  Length: 4133  October 18, 1988  13:02  Check: 3972


       1  AACCCCATCA CAGAAACTGG CATTTTAATT CCTTTGCTAA TTTATTTTTC

      51  TGCACGTTGG CGTAATACCC TACCGAAACA GGTAGGTGAT TATTGGCTTT

          .......

    4051  GTGAAAAGCA TCCGCCTCTG GTCGCATCAG CCGCCGTCCT CGCTGGATGA

    4101  TCGCCGGTGA GCCGCGGTCG GATGGTGCAG ATC
```

FIGURE 2 A typical GenBank entry.

```
  1 VLSPADKTNVKAAWGKVGAHAGEYGAEALERMFLSFPTTKTYFPHF.... 46
     || ||   |   ||||||   | | | | ||| | | |   |  |
  1 GLSDGEWQLVLNVWGKVEADIPGHGQEVLIRLFKGHPETLEKFDKFKHLK 50
             .         .         .         .         .
 47 ..DLSHGSAQVKGHGKKVADALTNAVAHVDDMPNALSALSDLHAHKLRVD 94
      |   || ||| ||  |  ||   |   |     | || | || | ||
 51 SEDEMKASEDLKKHGATVLTALGGILKKKGHHEAEIKPLAQSHATKHKIP 100
             .         .         .         .         .
 95 PVNFKLLSHCLLVTLAAHLPAEFTPAVHASLDKFLASVSTVLTSKYR... 141
       | ||| |||  |    |||| |  || ||| |      | | ||
101 VKYLEFISECIIQVLQSKHPGDFGADAQGAMNKALELFRKDMASNYKELG 150
```

FIGURE 3 An alignment of two related proteins (human α-chain hemoglobin and human myoglobin). Note that a gap has been inserted to optimize the alignment.

```
tcTTGACat ..................TAtAaT.....⇒
 -35                          -10      +1
```

FIGURE 4 Consensus sequences for prokaryotic promoter sites, the so-called −35 and −10 boxes. Bases in upper case are more conserved than those in lower case.

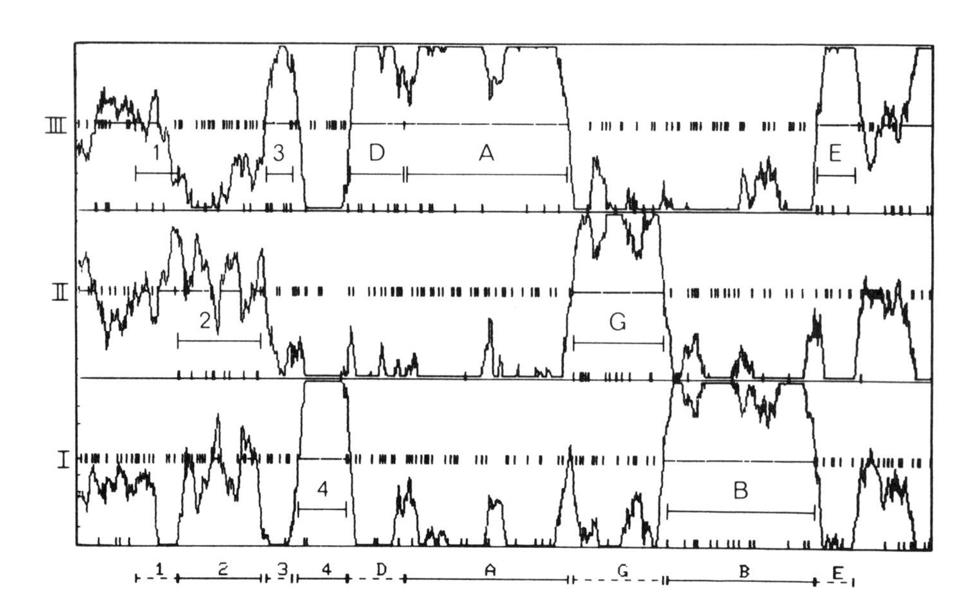

FIGURE 5 A codon bias plot for all three reading frames of the *unc* operon of *Escherichia coli*. The degree of "fit" to the average codon use of *E. coli* is displayed on the *y* axis. Note that the actual coding regions (1–4, D, A, G, B, and E) stand out clearly from the noncoding regions in their respective reading frames. The small vertical bars are stop codons. [Reprinted from R. Staden, *Nucleic Acids Res.* **12,** 551 (1984) by permission.]

translational modifications, such as N- or O-linked glycosylation, phosphorylation, or myristylation, is much easier. In essence, one searches for consensus patterns using methods as in the case of DNA signals, discussed in Section II,B.

III. Future Prospects of Sequence Analysis

To provide the necessary support for the human genome project (i.e., the sequencing of the full 3×10^9 bases of DNA sequence in the human chromosomes), techniques for automated data storage, handling, and analysis must be carried to much higher levels. This is not simply a problem of hardware—supercomputers and parallel processing may speed up many critically important algorithms by orders of magnitude—but, rather, a problem of understanding more about the structures and the interactions of the biological macromolecules themselves. As long as we are unable to predict the structure of a protein directly from its amino acid sequence, and as long as we cannot formulate precise models for DNA-binding protein interaction with their target signals, we will not be able to make full use of the information in our sequence data banks. Computer power is no panacea for a lack of knowledge, although computers are absolutely essential for gaining that knowledge. [*See* HUMAN GENOME AND ITS EVOLUTIONARY ORIGIN.]

Bibliography

Bishop, M. J., and Rawlings, C. J., eds. (1987). "Nucleic Acid and Protein Sequence Analysis: A Practical Approach." IRL Press, Oxford, England.

Doolittle, R. F. (1986). "Of URFs and ORFs: A Primer on How to Analyze Derived Amino Acid Sequences." University Science Books. Mill Valley, California

von Heijne, G. (1987). "Sequence Analysis in Molecular Biology: Treasure Trove or Trivial Pursuit?" Academic Press, Orlando, Florida.

von Heijne, G. (1988). Getting sense out of sequence data. *Nature (London)* **333,** 605–608.

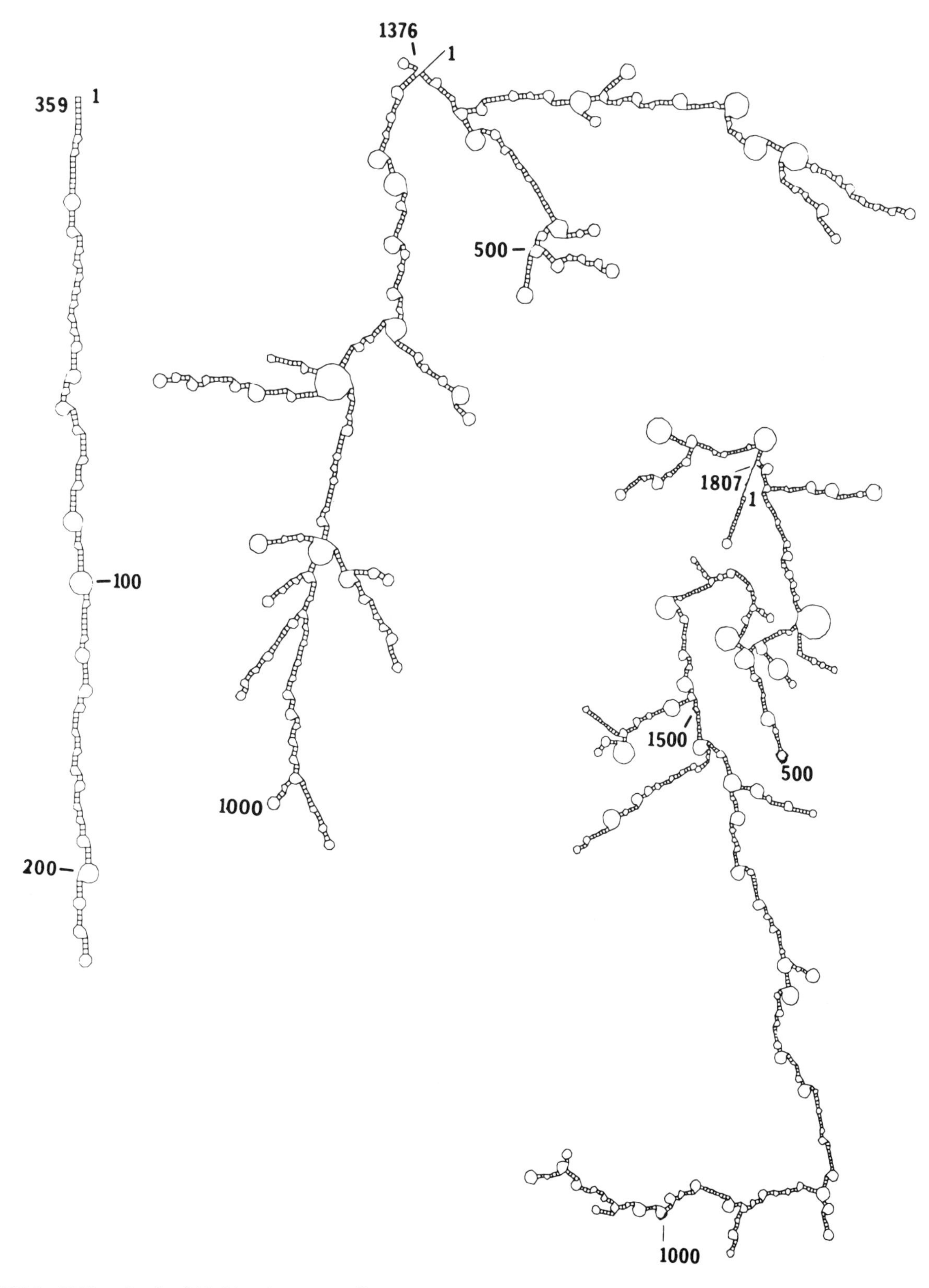

FIGURE 6 RNA molecules folded into low-energy (but not necessarily correct) secondary structures "by computer." [Reprinted from D. W. Mount, *Bio/Technology* p. 791. (September 1984) by copyright © permission.]

DNA Synthesis

SARAN A. NARANG, *National Research Council of Canada*

Glossary

Linkers Self-complementary oligonucleotides that contain endonuclease restriction enzyme site

Nucleosides *N*-glycosides of heterocyclic nitrogenous bases. The *N*-glycosides of purine and pyrimidine with pentoses is either D-ribose or D-2-deoxyribose, both in the furanose form

Plasmid or vectors Extrachromosomal genetic element consisting of circular DNA

Restriction endonuclease Enzymes that cleave DNA only at specific short deoxyribonucleotide sequences

HIDDEN IN the human genome is the coded knowledge for how to build a human and a human brain, and this by far exceeds all the knowledge stored in all the libraries of the world. Recently, it has been possible to synthesize chemically this molecule of life and to recreate and manipulate the information content of a gene. With the availability of automatic DNA synthesizers, synthetic oligonucleotides are being widely used. To cite some examples: the use as primers for initiating the synthesis of longer DNAs in many research applications, for DNA sequence analysis, as probes for detecting genetic defects, and for diagnosis of viral infections. The methodology of the chemical synthesis of oligonucleotides will be outlined in this chapter.

I. Synthetic Objective

DNA is an organic macromolecule consisting of a pair of long polynucleotide chains. Each chain consists of 2-deoxy-D-ribose rings linked by $3' \rightarrow 5'$ phosphodiester linkages carrying one of four possible bases at each $1'$ sugar position. The two chains are held together by hydrogen bonding between pairs of bases in which guanine (purine) is always joined to cytosine (pyrimidine) and adenine (purine) is always joined to thymine (pyrimidine) to form a double helix structure. [*See* DNA AND GENE TRANSCRIPTION.]

The fundamental objective in oligonucleotide synthesis is the specific and sequential formation of internucleotide phosphodiester linkages. Because a deoxyribonucleoside monomer contains two hydroxyl groups ($3'$ and $5'$), one must be protected while the other is specifically phosphorylated and then coupled to the second deoxyribonucleoside unit. Meanwhile, other reactive moieties (e.g., amino groups) and two of the three oxygen atoms of phosphate must also be protected.

II. Protecting Groups

A. Deoxyribose Moiety

The 4,4′-dimethoxytrityl (DMTr) group seems to be the most popular and labile protecting group for $5'$ primary hydroxyl function of deoxyribose. It has the advantage of being introduced selectively and renders the nucleoside lipophilic enough to allow use of acetonitrile or methylene chloride as organic solvents. Dichloro- or trichloroacetic acid in methylene dichloride and aromatic sulfonic acid are the milder reagents for the selective deblocking step, which must be carried out after synthesis is completed. The protection of 3′-secondary hydroxyl group must be in general complementary to the protecting group on 5′-hydroxyl and phosphate moieties. The most common base labile groups are acetyl and benzoyl esters.

B. Base

The amino group of a base requires a permanent protection, and this is achieved by acylation. Although several alternatives have been proposed, the most popular protecting groups remain benzoyl for adenine, anisoyl for cytosine, and isobutyryl for guanine (thymine usually requires no protecting group). These groups are stable to all the normal reactions used in oligonucleotide synthesis but are cleaved at approximately equal rates by concentrated ammonia at the end of synthesis.

C. Phosphorus

Useful phosphate protecting groups are generally the chlorophenyl derivatives when phosphotriester chemistry is used and methyl group when phosphite-triester chemistry is used. In addition, β-cyanoethyl is extensively used because of its ease in selective removal on treatment with organic amines. The chlorophenyl group is cleaved at the end of synthesis by treatment with concentrated ammonia or an oximate anion and the methyl group by the thiophenolate anion.

III. Methods in Oligonucleotide Synthesis

A. Phosphotriester Method

The basic principle of the phosphotriester method is to mask each internucleotidic phosphodiester function by a suitable protecting group during the course of synthesis. As uncharged molecules, the resulting phosphotriester intermediates are soluble in organic solvents and thus amenable to conventional methods of purification (e.g., silica-gel chromatography). After building a defined sequence, all the protecting groups are removed at the final step to generate an oligonucleotide containing natural 3′ → 5′ phosphodiester bonds. The main advantages of this method are large-scale synthesis (10–20 g), significant shorter period in purification steps, and higher yield.

The basic strategy of the phosphotriester method is to start the synthesis from a fully protected deoxymononucleotide containing a 3′-phosphotriester group (Fig. 1). Because the intermediate oligonucleotide contains a fully masked 3′-phosphate group, the necessity for a phosphorylation step at each condensation stage is eliminated. Such a starting material is a protected monomer prepared by treating a 5′-dimethoxyltrityl *N*-acyl deoxyribonucleoside with *p*-chloro-phenylphosphoryl *bis*-triazolide followed by a cyanoethylation reaction. *Bis*-triazolide, a bifunctional phosphorylating reagent, do change the reactivity of the phosphorylating species and avoids the formation of a symmetrical nucleoside phosphoric triester. The fully protected monomer is purified by silica gel and can be used to elongate chain from either end. On acid treatment, 5′-hydroxyl is made free, and in triethylamine, the β-cyanoethyl group is removed selectively to generate a phosphodiester function. The coupling of these two components generate a fully protected dinucleotide containing 3′-phosphotriester group. Using this approach, block condensation leads to the synthesis of longer oligonucleotides.

The development of arylsulfonyl tetrazolides has played a significant role in the development of phosphotriester method. Because the rate of the reaction very much depends on the catalyst present, most notably *N*-methylimidazole, 4-dimethylaminopyridine, and pyridine-*N*-oxide are favored. With these catalysts the reactions can be driven to completion in minutes.

B. Phosphite-Triester Method

The basic feature of the phosphite-triester method is the linking of nucleosides through an internucleotide trivalent phosphorous bond, which on subsequent oxidation generates a phosphotriester bond (Fig. 2). The main advantage of this approach is the speed of reaction with high yield. This method is particularly suitable for DNA synthesis on solid support using appropriately protected deoxynucloside 3′-phosphoramidites. Each synthesis cycle begins by treating dimethoxytrityl deoxynucleoside 3′-linked to solid support with dichloroacetic acid to remove the trityl ether followed by washes with dichloromethane and acetonitrile. The next step proceeds by converting the protected deoxynucleoside 3′-*N*,*N*-diisopropylaminophosphoramidite

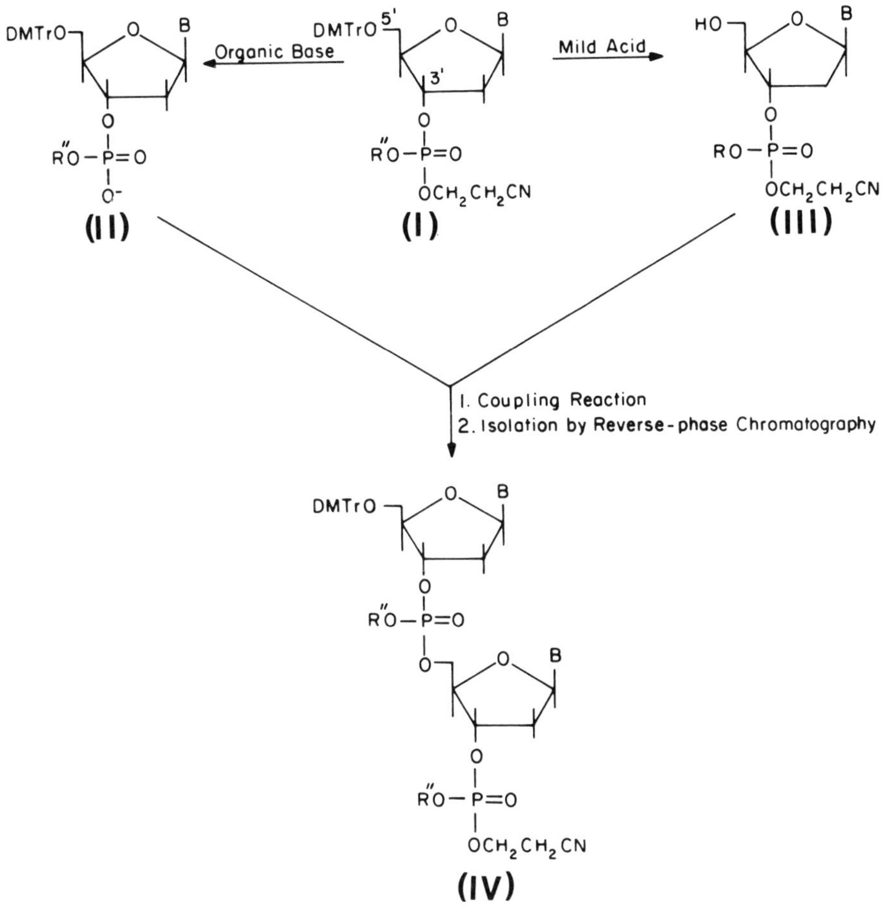

FIGURE 1 Phosphotriester method of oligonucleotide synthesis. (I) represents fully protected deoxymononucleotide having dimethoxytrityl (DMTr) at the 5′ position and chlorphenyl, cyanoethyl protecting group for the phosphate at the 3′ position. (II) represents monomer containing phosphodiester moiety at 3′ position. (III) represents monomer containing free 5′-hydroxyl position. (IV) is the resulting dinucleotide on coupling (II) and (III). B represent the protected base A, C, G, T.

to the tetrazolide and then adding this activated nucleotide in excess to the support. The reaction is completed within a minute. After washing with acetonitrile, the support containing phosphite is sequentially treated with capping reagent, an aqueous hydrolysis solution, and finally an aqueous iodine solution to oxidize to the phosphate. The synthesis proceeds stepwise in 3′ → 5′ direction by the addition of one nucleotide per cycle.

C. Deprotection and Purification

At the end of synthesis, it is essential to remove all the protecting groups in the correct order. On heating with concentrated ammonia, the phosphate protecting groups from the phosphate and base are cleaved first to form the corresponding phosphodiesters and free base followed by acid treatment to remove the terminal DMTr group. The linkage between oligonucleotide and support is cleaved at the same time. For the phosphite-triester approach when using methyl, a good nucleophile such as

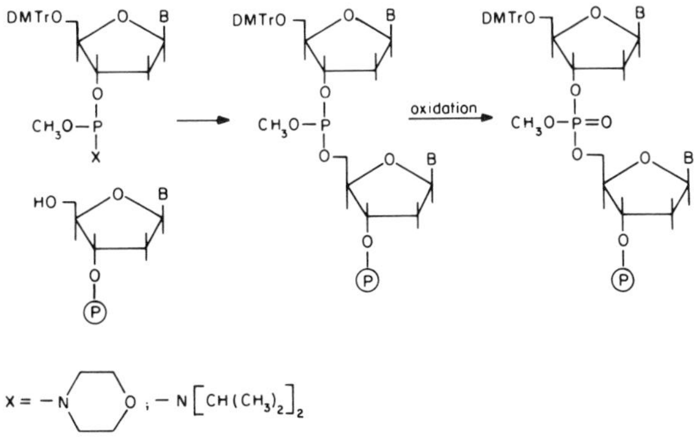

FIGURE 2 Phosphite-triester method of oligonucleotide synthesis. (P), polymer support; B, protected base A, C, G, T.

thiophenolate has to be used. Methyl group can also be replaced by the β-cyanoethyl group.

Purification of the unprotected oligonucleotides can be routinely carried out on 12% polyacrylamide gel electrophoresis (PAGE) in the presence of 7 M urea. The expected oligonucleotide is visualized by ultraviolet shadowing with a fluorescent thin-layer chromatography sheet at 254 nm. The expected band is cut out and extracted with buffer and desalted on a reverse phase (RP) column.

D. Sequence Analysis

The sequence of synthetic oligomers is generally determined by the mobility-shift method. It involves the labeling of the 5′-hydroxyl group with T_4-polynucleotide kinase and [γ-^{32}P]-ATP. The labeled compound is then partially digested with snake venom phosphodiesterase to produce labeled, sequentially degraded products down to mononucleotides. These are then separated by two-dimensional chromatography, first by cellulose acetate electrophoresis at pH 3.5 and then homochromatography on DEAE-cellulose thin-layer plate. The sequence of an oligomer is determined by the characteristic mobility shifts of the labeled degradation product on two-dimensional chromatography.

IV. Applications

A. Gene Synthesis

The natural ability of complementary polynucleotides to form bihelical structures can be exploited to build up double-stranded DNA starting from appropriate chemically synthesized oligonucleotide sequences. This is possible, owing to the existence of an enzyme DNA ligase that joins oligonucleotides provided they are in close apposition. This enzyme catalyzes the joining by transmutating the high-energy pyrophosphate linkage of ATP cofactor into a phosphodiester bond between the 5′-phosphoryl and 3′-hydroxyl termini. Two approaches are in common use for building long DNA sequences from short oligonucleotides. For this purpose complementary oligonucleotides must have overlapping regions, so that when they are annealed together, they form a helix with protruding strands.

1. T_4-DNA Ligation

Synthetic complementary duplex fragments with protruding strands at either end are joined together sequentially or in parallel several at a time using T_4-DNA ligase enzyme. There are two approaches. The sequential method is time-consuming because a purification step is required after each ligation reaction. In case of longer genes, the block synthesis is more popular. In this, the gene is synthesized in blocks with unique restriction endonuclease sites at their ends, inserted into a vector, cloned, and confirmed by DNA sequence analysis. Finally the blocks are removed from the parent vectors, using the appropriate restriction endonuclease, brought together in the right order in terms of unique restriction sites in the target linearized vector, ligated, cloned, and finally sequenced. Further development in this area has been the so-called one-pot ligation procedure. In this case, all the component DNA duplex fragments with significant overlap of four to six bases on either side are mixed, annealed, ligated with a linearized plasmid, and used for transformation of competent *Escherichia coli* cells. Colonies are screened by colony hybridization with radioactive probes and confirmed by DNA sequence analysis.

2. Enzymatic Repair

Escherichia coli DNA polymerase catalyzes a template-directed "repair" synthesis that serves to add specific nucleotides to a primer DNA annealed to the template strand. At 5–9°C, the incorporation reaction stops when the single-stranded section of the template is completely repaired (i.e., made into a double helix), and no initiation of new chains occurs. Partial repair syntheses are also possible with one, two, or three of the four deoxynucleoside triphosphates added to the enzymatic reaction. This method is sometimes used to modify the ends of a DNA piece to be cloned. The large tryptic fragment (Klenow) of DNA polymerase is generally used to avoid 5′-3′ exonucleolysis during synthesis.

3. Hybrid-Gene Synthesis

A novel synthetic strategy, hybrid-gene synthesis, which produces both wild-type and mutant genes simultaneously, has been recently devised. Several overlapping synthetic oligonucleotides containing specific regions of mismatched bases in the middle are annealed and ligated to a linearized plasmid. This heteroduplex plasmid contains several mismatched regions with the wild-type on one strand and the mutant gene on the other strand. On transformation of a bacterial host, each strand acts as a template yielding two plasmids having two related genes: one wild-type, the other mutant. Simultaneous synthesis of DNA sequences encoding for

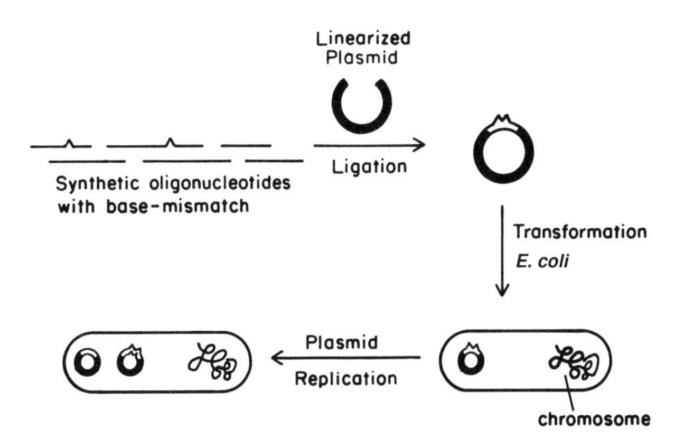

FIGURE 3 Strategy for hybrid-gene synthesis. Overlapping synthetic oligonucleotides with regions of base mismatch were phosphorylated, annealed, and ligated to linearized plasmid vector to yield heteroduplex.

human and mouse epidermal growth factors (EGF) has been achieved (Fig. 3). The two homologous polypeptides differ in 16 of 53 amino acids, a 30% difference. Plasmids bearing either the human or mouse EGF-coding DNA sequence were identified by colony hybridization using the appropriate probe and confirmed by DNA sequence analysis.

B. Gene Probing

Under appropriate conditions, synthetic oligonucleotides form stable complexes when hybridized to DNA or RNA molecules containing complementary sequence. It thus enables one to discriminate DNA sequences that differ as little as a single nucleotide. This sequence specificity has led to the development of oligonucleotides as probe for cloned genes based solely on protein sequence information. Oligonucleotides have also been used to detect single base pair differences in sequence as complex as a human genomic restriction digest. This technique has proven extremely powerful in monitoring human genetic diseases.

C. DNA-Sequencing Analysis

The use of synthetic oligonucleotides as primer for the DNA sequence analysis has become routine by the dideoxy method. The primer is incubated with a template DNA in the presence of *Escherichia coli* polymerase I and four dNTPs (one of them ^{32}P-labeled plus a ddNTP (dideoxy). During the primer extension reaction, at any position where a dideoxy residue is incorporated, the DNA chain is terminated. By using a different ddNTP in four different samples, four families of DNA fragments that share the same 5′ end are generated. The four samples are denatured and fractionated by electrophoresis in a denaturing gel. The sequence of the extended primer, which is complementary to the template, can be read directly from the autoradiogram of the gel. [*See* DNA, RNA, AND PROTEIN SEQUENCE ANALYSIS BY COMPUTER.]

D. Polymerase Chain Reaction

Using two synthetic oligonucleotides as primer and Taq DNA polymerase, polymerase chain reaction (PCR) is used to amplify the segment of DNA that lies between two regions for known sequence. These oligonucleotides typically have a difference sequence and are complementary to sequences that (1) lie on opposite strands of the template DNA and (2) flank the segment of DNA that is to be amplified. The template DNA is first denatured by heating in the presence of large molar excess of each of the two oligonucleotides. The reaction mixture is then cooled to a temperature that allows the primers to anneal to their target sequences, after which annealed primers are extended with DNA polymerase. The cycle of denaturation, annealing, and DNA synthesis is then repeated many times. Because the product of one round of amplification serves as templates for the next, each successive cycle essentially doubles the amount of the desired DNA product. It has extensive application in the diagnosis of genetic disorders.

Acknowledgment

NRC Publication No. 31905.

Bibliography

Engels, J., and Uhlmann, E. (1988). Gene synthesis. *In* "Advances in Biochemical Engineering/Biotechnology," Vol. 37. (A. Feichter, ed.), pp. 73–127. Springer-Verlag, Berlin, Heidelberg.

Gait, M. J., ed. (1984). "Oligonucleotide Synthesis." IRL Press. Oxford, England.

Narang, S. A., ed. (1987). "Synthesis and Applications of DNA and RNA." Academic Press, Orlando, Florida.

Sung, W. L., Zahab, D. M., Yao, F. -L., Wu, R., and Narang, S. A. (1986). Simultaneous synthesis of human-, mouse- and chimeric epidermal growth factor genes via "hybrid gene synthesis" approach. *Nucleic Acids Res* **14,** 6159–6168.

Down's Syndrome, Molecular Genetics

YORAM GRONER and ARI ELSON, *The Weizmann Institute of Science*

GLOSSARY

Cloned DNA Large number of DNA molecules identical to each other and to an ancestral molecule
Genomic clone DNA sequence derived from the genome and carried by a cloning vector
Karyotype Entire chromosome complement of a cell or an individual as determined by the spreading and staining of mitotic metaphase chromosomes
Nod-disjunction Failure of duplicated chromosomes to migrate to opposite poles during cell division
Phenotype Appearance and other characteristics of an individual, resulting mainly from its genetic constitution
Restriction fragment length polymorphism (RFLP) Subtle variations in the precise DNA sequence exist between individuals, some of which create or destroy sites of restriction endonuclease cleavage, thereby changing the length of restriction fragments
Robertsonian translocation Also called centric fusion, a special type of reciprocal translocation occurring between two acrocentric chromosomes that are joined in such a way that the two long arms are essentially preserved
Somatic cell hybrids Cell lines that are the result of fusion of human cells in culture to cells obtained from rodents. Cells of this type gradually lose their human chromosomes and thus serve as an invaluable tool for correlating phenomena with the presence of specific human chromosomes
Transfection of eukaryotic cells Introduction of exogenous cloned DNA into the cells
Transgenic mice Created by introduction of exogenous cloned DNA into the germ line of the animal

DOWN'S SYNDROME (DS or trisomy 21) is a severe genetic disorder caused by the triplication of the distal part of the long arm of chromosome 21. It is assumed that the presence of extra copies of genes that reside in the triplicated region results in the synthesis of added amounts of gene products, which upset the normal biochemical pattern of existence; this in turn causes the physiological symptoms of the disease. One of the main thrusts in research into the molecular biology of DS is the precise definition of the minimal length of the triplicated fragment that causes the disease and the identification of the genes that reside in it. Other efforts are directed toward cloning of these genes and linking the results of their enhanced expression to the physiological symptoms of the syndrome through the use of several types of model systems.

I. Introduction

Down's syndrome is the most common human genetic disease occurring approximately once in every 700–1,000 live births. It is thought to be the single most frequent cause of mental retardation in the industrialized countries. DS patients suffer from a wide range of symptoms. Most obvious among these are morphological defects (e.g., muscular flaccidity, short stature, and the epicanthic eye folds, which give rise to the eye shape characteristic to the syndrome). Patients are mentally retarded, and those who survive past their mid-30s usually develop Alzheimer's disease. Premature aging and an increased incidence of leukemia and other hematological disorders are common in affected individuals, as are cardiac defects, a high susceptibility to infections, and several types of endocrinological disorders. The risk of a child being born with trisomy 21 sharply increases as maternal age progresses into the fourth decade of life, and because presently many couples in Western societies postpone parenthood until these age groups, the incidence of DS is expected to increase. [*See* ALZHEIMER'S DISEASE; GENETIC DISEASES.]

The presently available techniques for prenatal screening for DS (amniocentesis and chorion villi biopsy) are costly and not without risk; they are therefore applied routinely only to at-risk pregnancies, with the result being that most pregnancies are not screened at all for DS. Moreover, because of continuous improvements in all aspects of clinical treatment, the life expectancy of DS patients has tripled during the past two decades; middle-aged DS patients are no longer a rare occurrence. Thus, despite the medical and technological advances of recent years, the prevalence of DS individuals in society is not likely to be significantly decreased in the near future.

The disease was first described by the English physician John Langdon Down in 1866. Early descriptions of the patients used the term *mongol* in reference to their facial appearance; hence, the disease is sometimes referred to as *mongolism.* In 1959, Jerome Lejeune, Marthe Gautier, and Raymond Turpin of the Institute de Progenese in Paris recognized that a strong linkage exists between the presence of an extra copy of chromosome 21 and the occurrence of DS. It has since been found that this is true in about 95% of the cases of DS. The source of the extra chromosome in the great majority of such cases is a maternal nondisjunction event that occurs during the formation of germ cells at the time of the first meiotic division. In the remaining 5% of the cases, only part of chromosome 21 is present in triplicate, translocated onto another chromosome. Analysis of such cases gradually led to the conclusion in the mid-1970s that the presence of an extra copy of the entire chromosome was not an absolute prerequisite for DS; rather, trisomy of the distal half of the long arm, cytologically known as band 21q22, was sufficient. Band 21q22 comprises approximately a third of chromosome 21 and is estimated to contain several hundred genes.

Reports of DS patients whose chromosomes appear to be normal exist in the literature. Such cases are suggested to have arisen by an aberration too fine to be detected by karyotyping, the technique commonly used to verify a clinical diagnosis of DS. The normal appearance of the chromosomes in a DS patient can be explained by assuming that a duplication of a small segment of band 21q22 has taken place; alternatively, one or more recombination events may have occurred, resulting in a very limited translocation of part of band 21q22 to another chromosome. Elucidation of such cases may serve to further narrow down the size of the pathological segment. However, the number of cases of this type that have been reported is small, and it is not clear whether these patients exhibit all the symptoms of "classical" DS.

Moving to the molecular level, the mechanism by which the presence of the extra chromosomal material causes the symptoms of DS is unknown. However, there is thought to exist a direct proportionality between the number of gene copies in the genomes of DS patients and the amount of gene products synthesized; stated somewhat bluntly, 50% more gene copies results in 50% more gene products. These products are thought to be perfectly normal, but their increased amounts are believed to upset the balance of biochemical reactions that ensure normal existence.

Down's syndrome is set apart from most other genetic diseases. The latter are for the most part the result of a problem affecting a single gene. Usually, the resulting gene product is damaged; its capabilities to carry out its functions are greatly reduced or even eliminated. In contrast, in DS the wide range of symptoms is caused by small (~50%) increases in the amounts of a large number of normal gene products, the exact number and identity of which are presently unknown.

However, as will become evident from the description of the nature of gene dosage effects, it is safe to assume that not all the genes that reside in band 21q22 contribute to the same extent in causing the DS phenotype; some should bear a more prominent role in this than others. Elucidating the molecular basis of DS is then an effort to identify the gene products whose augmentation by 50% or more causes the symptoms of DS; to explain the precise mechanisms by which the suspected gene products cause the symptoms of the disease; and to explain the possible interrelations between the gene products that might serve to amplify or ameliorate the severity of the symptoms.

II. Theoretical Aspects of Gene Dosage

A primary gene dosage effect (GDE) is defined as the existence of a direct relation between the number of copies of a gene and the amounts of its product. The hypothesis that DS is caused by GDEs encompasses two levels of reasoning: that GDEs exist in DS and that they have physiological manifestations. As the main emphasis in DS is on GDEs of added DNA, there will be no mention of GDEs of decreased DNA content, as in monosomies or deletions.

The process by which a gene gives rise to its product includes many steps and is regulated by a complex system of checks and balances. In general, not all DNA encoding for products is used at all times or in all tissues; the expression of many gene products is specific for certain tissues or developmental stages in the life of an organism. An extra copy of a gene may exist without being expressed in some cells because of mechanisms that prevent the expression of the gene in question in inappropriate contexts. As the extra DNA includes the normal regulatory signals of the genes, the pattern of expression of the added DNA should be similar to that of the normal DNA complement.

A GDE resulting from the presence of an extra copy of a chromosome might be expected to result in increased synthesis of a large number of gene products. However, attempts to demonstrate such an increase by the technique of two-dimensional electrophoresis of total protein extracts from cells known to be trisomic have not generally succeeded. Some studies have shown that some gene products are present in enhanced amounts and in a manner that can be related to the existence of added chromosomal material. However, the extent of this phenomenon was far below what was expected, and the amount of some of the other gene products was, in fact, reduced. In contrast, an examination of messenger RNA in skin fibroblasts known to be either monosomic, disomic, or trisomic for chromosome 21 revealed that transcription of chromosome-21 specific sequences was proportional to the copy number of chromosome 21. In addition, there have been several successful attempts to demonstrate enhanced synthesis of the products of several specific genes from band 21q22 in cells from DS patients (Table I) as well as in other types of cells. The failure to demonstrate gross changes in trisomic material by two-dimensional gel electrophoresis might stem from the relative insensitivity of this technique. An additional parameter that hinders detection of global GDEs is the timing of expression. It is possible that some genes (e.g., those responsible for defects already evident at birth) are expressed at high levels during embryonal development; analysis of postnatal material might not reveal any abnormalities in their expression.

At the physiological level, the presence of a gene dosage effect does not automatically mean that the normal biochemical pattern of the organism will be disturbed, because the extra amount of a gene product can still be compensated for at the level of protein function. In certain cases the organism can adjust the amounts and activities of other parts of its metabolism to counteract these effects, creating what are known as *secondary GDEs*. An often-quoted example of a secondary GDE in the context of DS is the case of glutathione peroxidase, an enzyme known to be part of the mechanisms that defend cells from the damages of oxidative stress. The gene for this enzyme is located on chromosome 3; yet, its activity is increased in erythrocytes of DS patients, presumably as part of the defense mechanisms against the damage caused by the overproduction of hydrogen peroxide by copper-zinc superoxide dismutase (CuZnSOD), a chromosome-21-encoded enzyme for which a primary GDE has been demonstrated in DS. The reverse is also true: The physiological result of a GDE might be amplified, producing an end result much larger than that expected from an increase by a mere 50% in product synthesis. Again, using an example from chromosome 21, it has been demonstrated that some of the functions triggered in cells by the interferon-alpha receptor, for which a gene dosage effect of 50% has been demonstrated, are increased three- to eightfold when compared with cells devoid of such a GDE.

The physiological effects a GDE can have are intimately linked to the identity of the gene product itself, irrespective of whether the GDE is of primary or of secondary nature.

A. Enzymes

Despite the large body of information that has been accumulated on genes that code for enzymes, there is relatively little experimental information regarding the physiological effects of small changes in enzymatic activity. A theoretical explanation backed

TABLE I Genes Assigned to Chromosome 21

Gene symbol	Name	E.C. number	Chromosomal location	Remarks
AABT	Beta-Amino Acid Transport System		Chromosome 21	Tentative
AD1	Familial Alzheimer's Disease		21q21-21q22.1	
APP	Beta-Amyloid Precursor Protein		21q21	
ASNSL2	Asparagine Synthetase (?)		Chromosome 21	Tentative
BCEI	Estrogen-Inducible Protein from Breast Cancer Cell Line		21q22.3	
CBS	Cystathionine Beta Synthetase	4.2.1.21	21q22	GDE in fibroblasts trisomic for chromosome 21
CD 18	Beta Subunit of LFA Antigen and Related Proteins		21q22.3, distal to ETS-2	
COL6A1 & COL6A2	Alpha-1 and Alpha-2 Chains of Type 6 Collagen		21q22.3	
CRYA1	Alpha-1 Crystallin		21q22.3	
ERG	ETS-Related Gene		21q22.3	
ETS-2	Cellular Homologue of the E26 Avian Tansforming Retrovirus		21q22.3	
GPXP1	Glutathione Peroxidase (?)		Chromosome 21	Tentative
HSPA3	HSP70 Heat Shock Protein		Chromosome 21	Tentative
HTOR	Hydroxytryptophan Oxygenase Regulator (?)		Chromosome 21	Tentative
IFNAR & IFNBR	Receptor(s) for Interferon Type Alpha/Beta		21q21-21qter	GDE in fibroblasts aneploid for various parts of chromosome 21
IFNGT1	Transducer (?) of Interferon Gamma Receptor		Chromosome 21	
MX1, MX2	Myxovirus (Influenza) Resistance Loci		Chromosome 21	
PAIS	Phosphoribosyl Aminoimidazole Synthetase	6.3.3.1	Chromosome 21	
PFKL	Liver-Type Subunit of Phosphofructokinase	2.7.1.11	21q22.3	GDE in erythrocytes and fibroblasts
PGFT	Phosphoribosyl Glycinamide Formyl Transferase	2.1.2.2	21q11.2-21q22.2	
PNY1	Anonymous Polypeptides		Chromosome 21	PNY1, MW = 82KD, PI = 7.0
PNY2				PNY2, MW = 65KD, PI = 6.0
PNY3				PNY3, MW = 33KD, PI = 5.1
PNY4				PNY4, MW = 72KD, PI = 6.3
PNY5				PNY5, MW = 40KD, PI = 5.2
PRGS	Phosphoribosyl Glycinamide Synthetase	6.3.4.13	21q22.1	GDE in fibroblasts aneuploid for regions of chromosome 21
RNR4	Ribosomal RNA		21p12	
S14	Unidentified Surface Antigen		Chromosome 21	
S100-B	Beta Subunit of S100 Protein		21q22	
SOD1	Cu/Zn Superoxide Dismutase	1.15.1.1	21q22.1	GDE in fibroblasts, various types of blood cells

[a] GDE, gene dosage effect; MW, molecular weight; PI, isoelectric point; E.C., enzyme commission classification number.

by some experimental evidence suggests that this is due to the very nature of enzymatic systems. Enzymes do not usually function alone, rather they are part of metabolic pathways and are linked to other enzymes through common substrates, products, or effectors. Multienzymatic pathways in which control of the pathway's flux is not concentrated at a single step have an inherent capacity to absorb small changes in the activity of one of their constituent enzymes so that the total flux through the pathway remains more or less unaffected. One result of this is that the sizes of pools of intermediate metabolites in the pathway are changed. As the number of enzymes in a pathway of the type described increases, the relative share of each of its constituent enzymes in controlling the flux rate decreases. A pathway's buffering capacity is then expected to increase with the number of enzymes that make it up. One can therefore predict that enzymatic GDEs will succeed in provoking a physiological response only in special cases. A prime example of such a case are enzymes that do not abide by the above generalization (e.g., those that catalyze rate-limiting reactions). As the control of flux through the pathway is determined in such cases by the activity of the extra-active enzyme, the degree of freedom left for the remaining enzymes to compensate for this is limited and cannot prevent the flux through the pathway from changing. Another exception to the rule is when changing an enzyme's activity results in the accumulation of a harmful metabolite. In such cases, the physiological effect can be noticed even when the total flux through the pathway is unchanged.

B. Receptors and Ligands

Many cellular processes are initiated by the process of a ligand binding to a receptor molecule located on the outside of the cell. The rules that describe receptor–ligand interactions can be used to show that the fraction of the receptors that bind ligand molecules at a given ligand concentration is independent of the amount of receptors available. Therefore, any increase in the number of receptor molecules on the cell as a result of a GDE will not change the fraction of receptors that actively bind ligand molecules. The absolute number of bound receptors will, however, increase by definition under such circumstances, and it is precisely this parameter that determines the strength of the signal that the cell receives. A cell with more receptors on its surface will receive a stronger signal at a given ligand concentration than a similar cell with less surface receptors; the two types of cells might then be expected to react differently to similar environmental signals. As the degree of binding of receptors to ligands is dependent on the concentration of ligand molecules, GDEs that increase ligand concentrations are expected to have physiological effects similar to those of increased receptor concentrations.

C. Assembly of Macromolecules

Some of the enzymes and structural components of cells are multimers constructed from several smaller protein molecules. Usually the identity and amounts of these subunits are strictly defined. In such cases, a GDE leading to an excess of a subunit could be expected to change the total amount of the multimer only if its amount was the limiting factor for the formation of the multimer in the first place. Cases exist, however, in which the precise composition of the multimer is not strictly defined; rather, it is the result of the random association of whatever types of subunits are present. In such a case, increasing the amount of one type of subunit could result in its increased incorporation into the multimer with possible impacts on its properties and function. An example of the latter case is the glycolytic enzyme phosphofructokinase (PFK). PFK functions as an association of four subunits of which three different types exist. The subunits come together to form active PFK molecules in a random manner; the precise composition and properties of PFK tetramers in different tissues have been shown to reflect the levels of expression of the three types of subunits in these tissues. Because each of the subunit types is known to possess different biochemical properties, enhanced expression of any type of subunit could be expected to sway the composition and properties of the PFK tetramers in its direction. We should add at this point that the gene for the liver-type subunit of PFK has been mapped to the pathological segment of chromosome 21, and the composition of PFK tetramers in erythrocytes of DS patients has been shown to include a higher than normal fraction of L-type subunits. Similar effects are expected to occur in organized structural components of cells. Disruption of the equilibrium of subunit concentrations might change the overall geometry of the multimer and influence the spatial organization of cells and the organs in which they reside.

D. Regulatory Systems

Some of the functions that regulate gene expression are encoded by the genome itself. It is therefore not inconceivable that a GDE in a gene whose product regulates the expression of other genes could result in their incorrect expression. The expression of these genes could be either enhanced or repressed, as determined by the function of the amplified regulatory gene product. The finding mentioned earlier that the amounts of some proteins from total protein extracts from trisomic sources were reduced in comparison to normal controls might loosely be interpreted as indirect demonstration of the increased activity of a regulatory gene product that decreases the expression of these proteins. No direct evidence suggests that the activity of a regulatory gene product is affected by trisomy 21. However, the implications of such a finding would be wide as it could suggest a mechanism by which seemingly unrelated and unexplained secondary GDEs could be justified.

From all that has been described, it is clear that the results of the presence of extra copies of genes may not be noticeable. However, in many of these cases, although the organism might seem normal, its capabilities to maintain its biochemical balance under unfavorable conditions might be decreased. Stressful situations can therefore result in the realization of the potential harm that GDEs can have.

III. Mapping of Chromosome 21

A. Theory

The human genome is estimated to contain 3×10^9 base pairs of DNA and in the order of magnitude of 100,000 genes. Chromosome 21 contains about 1.5% of this (i.e., around 45–50 million base pairs). Assuming that genes are equally distributed within the genome, chromosome 21 is expected to contain some 1,500 genes. Band 21q22, the so-called pathological segment of the chromosome, comprises about a third of the chromosome; its 15 million base pairs are therefore estimated to harbor several hundred genes. The task of identifying the genes that reside on chromosome 21 and mapping their precise location and order is formidable. At the present time, relatively few (i.e., only about 25) of the expected number of genes have been shown to reside on the chromosome (Fig. 1); most of these have been regionally mapped to band 21q22. However, the fine mapping of chromosome 21 is of prime importance to enable a better understanding of the causes of DS and of other diseases whose loci have been mapped to this chromosome (e.g., Alzheimer's disease). A precise map of the chromosome would allow a reduction in what is perceived to be the size of the fragment whose presence in triplicate causes DS; determining which genes reside in this segment is absolutely vital to the process of understanding the molecular mechanisms of the disease. [*See* Genes; Genome, Human.]

A prerequisite for using methods of molecular biology to map particular areas of chromosome 21 (and chromosomes in general) is the existence of specific, unique probes (i.e., DNA fragments that can be used to hybridize to, in this case, chromosome 21). Such a probe must be unique (i.e., it must hybridize only to the gene it was meant to locate and possibly also to closely related sequences and not to repetitive sequences that exist in the genome). For the purpose of mapping, it is not necessary to know the biological function of the DNA used as a probe. Hence, probes can be either parts of known genes or they can be anonymous, in the sense that their identity, functions, or even their precise nucleotide sequence are unknown. Of the several dozen unique DNA probes for chromosome 21 that exist today, the vast majority are anonymous. This bias reflects the fact that anonymous probes are generally easier to isolate because probes that are fragments of known genes are usually the byproduct of the time-consuming task of the cloning and identification of genes in which they reside. It is conceivable that some of the anonymous probes will be recognized in the future as parts of genes that have yet to be identified. [*See* DNA Markers as Diagnostic Tools.]

Once a gene is assigned to a chromosome, attempts are usually made to localize it to a particular subregion of the chromosome, thus producing a map of the chromosome. Two basic types of chromosome maps exist. Physical maps depict the actual physical distance between points on chromosomes; distances are measured in physical maps in base pairs of DNA or in multiples thereof. Genetic linkage maps, on the other hand, depict the degree of recombination between loci; the greater the degree of recombination between two points the farther apart they will be placed on this type of map. Distances in genetic linkage maps are measured in crossing-over units, which are essentially a measure of the probability of a recombination event occurring between points on the map. When comparing a genetic linkage map with a physical map of the same region, we find that the order of the points on both maps is similar, although the distances between points are usually not. Because the recombination rate along a chromosome is not constant, there is no constant linear factor for comparing distances measured in both types of maps. [*See* Genetic Maps.]

A probe used for genetic linkage mapping must, aside from being unique and located on chromosome 21, recognize a site of restriction fragment length polymorphism (RFLP); some 40% of probes used for physical mapping fulfill this additional criterion. In such cases, the probe, in conjunction with the specific restriction endonuclease that generated the RFLP, can be used to mark the fragment and to

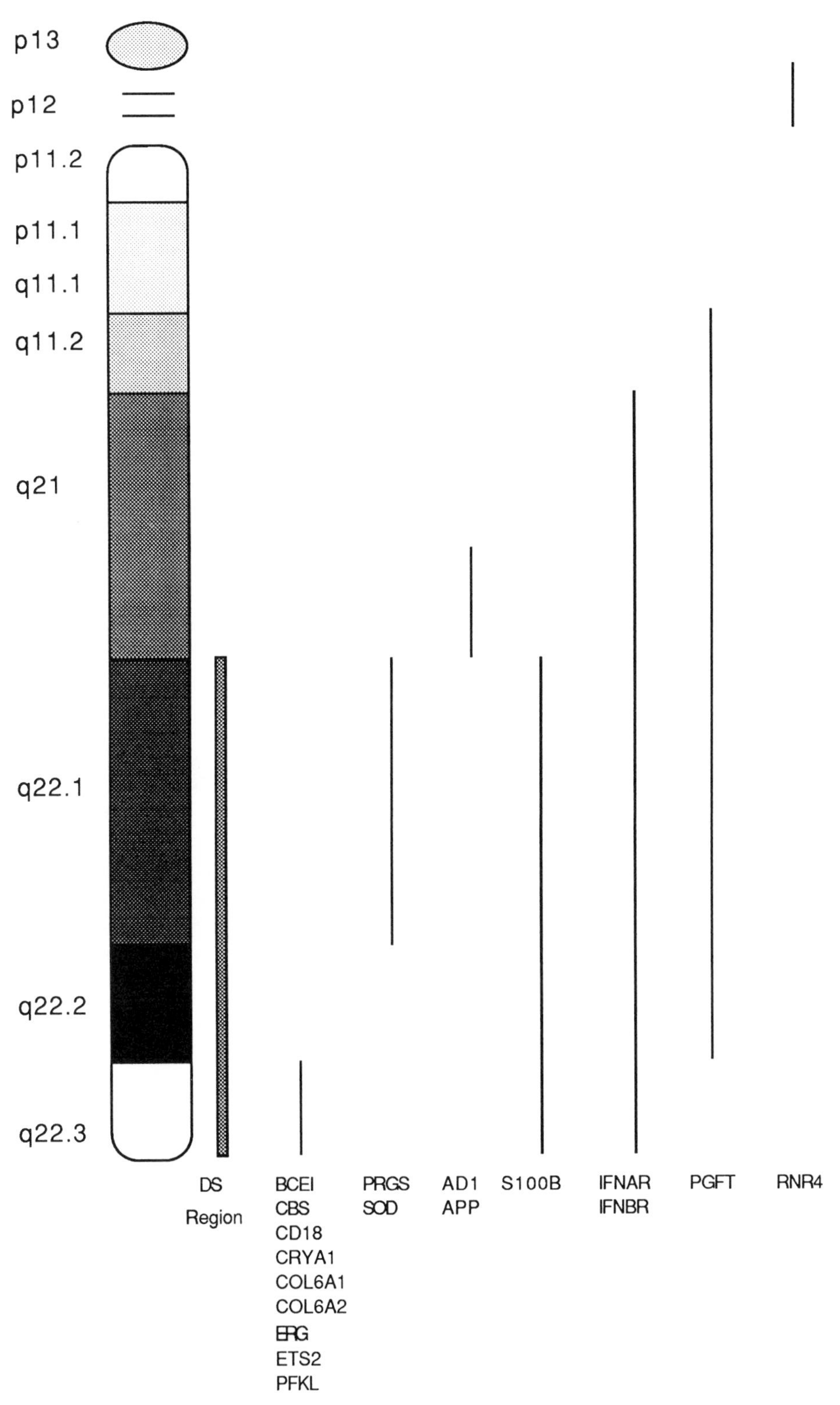

FIGURE 1 Map of human chromosome 21. Loci, mentioned in the text which have been assigned to but not mapped on the chromosome are HSPA3, IFNGT1, MX1, PAIS, PNY1-PNY5, and S14.

measure its size. Each of the different lengths of fragments represents a different sequence version at a certain point in the genome and is essentially an allele at this locus. When used to screen the RFLP patterns of populations, probes can be used to measure the frequency of each allele and to determine which alleles exist in each individual. Following the identity of alleles along a pedigree can reveal the frequency of recombinations, which is the raw data from which genetic maps are constructed.

B. Brief Overview of Genes Mapped to Chromosome 21

Speculations abound as to the possible connections between the genes known to reside on chromosome 21 and particular symptoms of DS. Some of the suggested connections are reasonable; most, however, remain unproven. Moreover, ideas relating to one gene may not necessarily be applicable in a system beset by many GDEs whose combination may produce, cancel, or change the intensity of symptoms in an as yet unpredictable manner. Theoretical linkages between overexpression of specific genes and symptoms of DS are therefore generally not discussed here; connections demonstrated experimentally are described in following sections. Table I lists the genes that have been assigned to chromosome 21; here we present a brief overview of some of them.

1. AD1—The Locus for Early-Onset Familial Alzheimer's Disease

Not all cases of Alzheimer's disease are thought to be caused by a genetic defect. However, in some large families the disease is transmitted in autosomal dominant fashion, indicating that at least in these cases a genetic factor is involved. Cases of familial Alzheimer's disease (FAD) are indistinguishable from "classical" Alzheimer's disease except for a somewhat younger age of onset. The genetic locus for FAD has been mapped to the region 21q21-21q22.1 by virtue of its linkage to anonymous DNA probes whose location on chromosome 21 is known. The FAD locus is thought not to be included in the DS pathological region of the chromosome.

2. APP—The Precursor of the Beta-Amyloid Protein

The beta-amyloid protein is a 4,200-dalton polypeptide, which has been isolated from the amyloid cores of neuritic plaques that are one of the most prominent pathological findings in the brains of humans who have died of Alzheimer's disease. The beta protein is derived from a much larger protein, the amyloid precursor protein, (APP) which is expressed in many tissues, including the brain, of both healthy and sick subjects. Recently, a meaningful degree of similarity was found between APP and nexin II, a molecule that seems to inhibit proteases that cleave and thereby activate several types of growth factor molecules. It is plausible that the biological function of APP is to participate in the regulation of the activity of certain growth factors. The APP gene is in close proximity to the locus for FAD and was mapped to the proximal edge of the DS region. Reports of recombination occurring between these two loci indicate that they are distinct from one another with the FAD locus farther away from the pathological segment. The finding that both the APP and FAD loci map so close to the DS pathological region is intriguing, as DS patients who survive into their fourth decade invariably develop symptoms and pathology identical to those of Alzheimer's disease patients.

3. BCEI—The Gene for an Estrogen-Inducible mRNA that was Isolated from a Human Breast-Cancer Cell Line

The product of the gene is a small 84-amino-acid-long protein, which has been shown to be expressed by stomach mucosa cells. The exact function of this protein is unknown, although a similar porcine protein inhibits gastrointestinal motility and gastric secretion.

4. CBS—Cystathionine Beta Synthase

This is an enzyme of sulfur amino acid metabolism, which catalyzes the condensation of homocysteine and serine to form cystathionine, along the pathway converting methionine to cysteine. Comparison of the physical symptoms of patients of homocystinuria, an inborn error of metabolism that is most commonly caused by deficient CBS activity, with those of DS patients led to the conclusion that some of the physical symptoms characteristic of both diseases could be viewed as opposites of one another. This led to the suggestion by Jerome Lejeune in 1975, some 10 years before the assignment of the CBS gene to chromosome 21, that abnormalities in CBS activity could play a part in causing the symptoms of DS.

5. CD18—The Gene for the Beta Subunit of Lymphocyte Function–Associated Antigen (LFA-1) and Related Proteins

The LFA-1 molecule is intimately involved in mediating adhesive interactions between several species of cells involved in the immune reaction. Humans suffering from a deficiency in LFA-1 and its related antigens suffer from serious defects in the adhesive-dependent functions of B and T lymphocytes, monocytes, and granulocytes and from recurrent life-threatening infections.

6. COL6A1, COL6A2—Two Genes Coding for the Alpha-1 and the Alpha-2 Polypeptide Chains of Collagen Type 6

Collagens are a large family of trimeric, triple-helical proteins that form fibers in the extracellular matrix. Type 6 collagen is a ubiquitous structural protein, yet its precise functions are unknown. It is also a somewhat unique member of the collagen family, as more than two-thirds of each of its three constituent polypeptides are globular domains. [*See* COLLAGEN, STRUCTURE AND FUNCTION.]

7. CRYA1—The Gene for Alpha-1 Crystallin

Crystallins are major structural proteins found in lenses of vertebrates. Overexpression of this gene may be connected to the early cataracts that DS patients suffer from.

8. ETS2—The Gene for One of the Two Human Cellular Homologues of the Transforming Avian Erythroblastosis Retrovirus E26

Children affected with DS are prone to develop acute leukemia or to undergo a transient leukemia-like reaction. In acute myelogenous leukemia of subtype M2, a translocation of sequences from band 21q22 to chromosome 8 exists in many of the cases. However, although mapping studies have determined that apparently the ETS2 locus is contained in the segment that is translocated, the precise point of chromosome breakage is far removed, on a genetic scale, from the ETS2 gene. This gene is therefore not rearranged in this translocation.

9. ERG—The ETS-Related Gene

This gene codes for at least two proteins that share limited domains of similarity with the ETS-2-encoded protein. Like the ETS-2 gene, the ERG gene has been shown to be translocated to chromosome 8 in many patients with acute myelogenous leukemia of subtype M2. Although genetic linkage studies have mapped the ERG gene to a point closer to the precise point of chromosome breakage than the ETS-2 gene, it, too, has been shown not to be rearranged in this translocation.

10. HSPA3—A Possible Gene for the HSP70 Heat-Shock Response Protein

Cells in culture respond to heat shock and other stimuli by the induced synthesis of a small number of specific proteins and by repression of other genes that are normally active. The precise function of heat-shock proteins is unknown, but it is held that they are part of the mechanism that defends the cell in situations of stress. The human genome contains several copies of the Hsp70 gene. The assignment of the HspA3 gene to chromosome 21 is somewhat uncertain, and it seems that although chromosome 21 is involved in the expression of heat shock proteins, the manner by which this is done is unclear. [*See* HEAT SHOCK.]

11. IFNAR, IFNBR—The Receptor for Interferon Alpha and Beta, Respectively

Interferons are proteins that are produced by many different types of cells in response to various stimuli, including viruses and polynucleotides. Binding experiments suggest that interferons alpha and beta bind to the same receptor, but it has not been definitely proven that the IFNAR and IFNBR loci are one and the same. GDEs for the receptor have been observed in fibroblasts with varying copy numbers of chromosome 21. The cDNA for the interferon alpha/beta receptor has been cloned. Its expression in mouse cells conferred interferon alpha/beta sensitivity on them, although at levels lower than expected. It is therefore not inconceivable that another locus involved in response to these interferons is encoded on chromosome 21. [*See* INTERFERONS.]

12. IFNGT1—A Locus Found to Be Required for the Ability of Cells to Respond to Interferon Gamma

This locus is distinct from the interferon gamma receptor itself, which has been convincingly mapped to chromosome 6. Experiments using hamster–human hybrid cell lines show that the presence of chromosome 6 is sufficient for the binding of interferon gamma to cells. However, to confer a biologically measurable response to interferon gamma (the induction of the major histocompatibility complex antigens), the presence of chromosome 21 is also required. It has been therefore suggested that chromosome 21 encodes a component of the mechanism for cellular response to interferon gamma that is distinct from the receptor itself, possibly a transducer.

13. MX1—Myxovirus (Influenza) Resistance Locus

This locus encodes a protein that is induced by interferons of type alpha and beta. Its function is unknown, although it has been shown that in mice, a homologous interferon-induced protein is responsible for protection against influenza virus. [*See* INFLUENZA VIRUS INFECTION.]

14. PRGS, PGFT, and PAIS–The Genes for the Enzymes That Catalyze Three of the Reactions of the Purine Biosynthetic Pathway

Several lines of evidence suggest that all three enzymatic activities are encoded by one gene, probably located in the DS pathological segment of chromosome 21. A GDE has been reported for PRGS in human fibroblasts trisomic for chromosome 21, and DS patients have been known for a long time to suffer from impaired purine metabolism. However, as none of the above enzymatic activities is considered rate-limiting in purine biosynthesis, the linkage between their possible overproduction in DS and the disease's symptoms remains to be demonstrated.

15. PFKL—The Liver-Type Subunit of Phosphofructokinase

Phosphofructokinase catalyzes one of the major rate-limiting reactions in glycolysis, an essential pathway in the use of sugar energy in cells. GDEs for PFKL have been demonstrated in erythrocytes and in fibroblasts obtained from DS patients, and the composition of PFK from such erythrocytes shows an increased proportion of the liver-type subunit.

16. PNY1, PNY2, PNY3, PNY4, PNY5—Five Polypeptides of Known Size and Isoelectric Points but of Unknown Functions, Which Have Been Assigned to Chromosome 21

These are based on two-dimensional electrophoresis of extracts of a human–mouse hybrid cell line that contains chromosome 21 as its only human complement.

17. RNR4—A Locus for Ribosomal RNA Mapping to Band 21p12

This probably is not involved in causing DS.

18. S14—A Surface Antigen, The Identity or Function of Which Is Unknown

The precise subregion of the chromosome in which it resides is also unknown.

19. S100B—The Beta Subunit of the S-100 Protein, A Calcium-Binding Protein of Unknown Function

It is found mostly in the nervous system of vertebrates, predominantly in brain glial cells.

20. SOD1—Copper-Zinc Superoxide Dismutase

This enzyme catalyzes the reaction that transforms superoxide anion radicals into hydrogen peroxide as part of the organism's defense mechanisms against damage by free radicals. The hydrogen peroxide produced, which is in itself toxic, is disposed of by other enzymatic and nonenzymatic reactions. GDEs have been demonstrated for this enzyme in various types of blood cells, fibroblasts, and brain samples obtained from patients with DS. This gene was the first chromosome 21–encoded gene that was cloned and used in studies of GDEs in transfected cells and transgenic mice.

21. Several Other Genes Have Been Assigned to Chromosome 21

These, however, have not been regionally mapped.

Comparison of the findings obtained from physical maps to those of genetic maps of chromosome 21 reveals that as in other chromosomes, a large degree of recombination takes place near the distal part of the chromosome in band 21q22.3. Consequently, this region, which spans about 10% of the physical-cytogenetic length of the chromosome, takes up some 40% of its genetic length. It is of interest to note that the majority of chromosomal breaks in chromosome 21 that are associated with translocation of segments to other chromosomes, cluster in band 21q22, especially in subband 22.3.

The distal region of band 21q22.3 seems to contain more genes than expected. Some two-thirds of all unique probes assigned to and regionally localized on chromosome 21 map to band 21q22; of these, half map to band 22q22.3, as exemplified by the data in Table I. The clustering of genes to this region correlates well with other findings. Genes are generally thought to cluster in parts of the genome that are GC-rich (i.e., that contain a higher-than-average proportion of guanidine and cytidine residues); actively transcribed genes are believed to be found near regions of DNA that are rich in the dinucleotide CG and that are undermethylated. Band 21q22 fulfills both of these criteria, further attesting to its "gene-rich" character. It is therefore possible that the DS pathological segment contains more genes than can be estimated by assuming an even distribution of genes throughout the human genome.

As has been mentioned, there are several examples of DS patients whose karyotype seems normal. This result is attributed to chromosomal aberrations too fine to be detected by karyotyping and therefore limited in size to 2,000–3,000 kb of DNA. As this length is about a fifth of the size of band 21q22, a clear definition of the boundaries of such fine aberrations can serve to trim down the size of the pathological segment and make its analysis more manageable. Research into this question has so far suggested that triplication of the part of band 21q22 that is comprised of bands 21q22.2 and 21q22.3 is linked to the occurrence of DS, although the precise borders of this region are ill-defined at present. A major drawback to this approach is the precise clinical diagnosis of DS. The disease manifests itself in a large possible number of clinical and morphological symptoms, not all of which are expressed in each and every patient. In fact, the karyotypic finding of an extra copy of part of or all chromosome 21 is generally accepted as the ultimate proof that the patient in question does indeed suffer from DS. In the absence of this aid to the clinical diagnosis, it is feared that different patients with varying karyotypically undetectable triplications and somewhat different symptoms would all be similarly categorized as suffering from DS. Correcting for the variability in the disease's symptoms is extremely difficult but is necessary when interpreting the results of this type of studies.

IV. Model Systems for Investigation of the Molecular Biology of DS

Two main reasons motivated efforts to develop a suitable system for studying the molecular genetics of the syndrome. The first one stems from difficulties attendant in research on humans. Most of the pathological consequences of trisomy 21 are manifested during fetal development; research on human subjects, especially *in utero*, is ethically complicated and practically impossible. Therefore, attention was turned toward the development of model systems. A second reason has to do with the large number of genes residing in the pathological segment and the consequent need to identify and sort out the quintessence. It is not clear how many genes are involved in determining the characteristic DS phenotype and which one is doing what. Is overexpression of individual genes responsible for certain features associated with the syndrome (e.g., the high incidence of leukemia, the large protruding tongue, the low blood serotonin)? Or does imbalance in the expression of several genes act in a nonspecific fashion to produce the DS phenotype? A model system should be of help in answering these questions.

To date, two types of model systems have been developed: (1) a cellular system, consisting of cultured cells overexpressing candidate genes from chromosome 21; and (2) an animal model, employing either mice with trisomy 16 (animals in which many of the genes homologous to human genes from the region 21q22 are triplicated) or transgenic mice, carrying a gene from chromosome 21 in their genome and producing increased quantities of the gene product.

A. Cellular Model System: Transformed Cultured Cells Expressing Human Genes from the DS Locus

Recombinant DNA technology has made possible the expression of individual genes in established cell lines. This methodology offers a relatively simple approach for investigating GDEs at the cellular level and has already led to some understanding of how imbalanced expression of the genes may contribute directly or indirectly to the Down phenotype. [*See* RECOMBINANT DNA TECHNOLOGY IN DISEASE DIAGNOSIS.]

When a particular gene derived from the DS segment is introduced into cultured animal cells, the recipient cells resemble trisomy 21 cells except for one important difference: The imbalance is limited to one particular gene, rather than the whole chromosome. A cellular system of this type permits the study of the biochemical effects of the altered dosage of a particular gene in a defined background, irrespective of the overexpression of other genes from chromosome 21.

Although studies concerning the GDEs of a single gene have already been conducted on several candidate genes including ETS2, CBS, APP, and PFKL, the first and most detailed ongoing study is concerned with the gene encoding CuZnSOD, a key enzyme in the metabolism of oxygen free radicals. Overexpression of the CuZnSOD gene, because of gene dosage, may disturb the steady-state equilibrium of active oxygen species within the cell, resulting in oxidative damage to biologically important molecules. In particular, the polyunsaturated fatty acids of membranes may be affected in a process known as lipid peroxidation. Because brain function is highly dependent on membrane interactions and the brain might be particularly susceptible to lipoperoxidation damage, it was suggested that such a mechanism may in part be responsible for the mental retardation, hypotonia, and Alzheimer's disease pathology associated with the DS phenotype. Experimental studies were conducted by introducing a cloned human CuZnSOD gene into the rat PC12 cell line, which possesses characteristics of neuronal cells grown in culture. The exogenously introduced human gene was stably integrated into the host chromosome, giving rise to increased amounts of enzymatically active CuZnSOD. While outwardly maintaining their response to nerve growth factor and their typical appearance of cultured neurons, the cells expressing the extra gene had a greatly reduced capacity to take up neurotransmitters. Neurotransmitters transfer the signals from one neuron to the next at the junction between them—the synapse. Following detailed analysis of the phenomenon, it was discovered that in the transformant-CuZnSOD cells, the vacuoles (called chromaffin granules) responsible for accumulating neurotransmitters have a lesion in their membrane, possibly caused by lipid peroxidation, which prevents them from taking up the neurotransmitters at the normal rate. This deficiency could have important consequences for neurons in the brain that use a similar organelle (called the synaptic vesicle) for accumulation of neurotransmitters. If a released transmitter substance persists for an abnormally extended period, new signals cannot get through at the proper rate. This observation demonstrates that even at the cellular level, an imbalance in the expression of the CuZnSOD gene has a deleterious effect, which, if it occurs in the central nervous system, would produce alterations in neuron function, which would impair the transduction of signals and mimic the deficiencies apparent in DS.

A cellular model system of this type was also used in studies on the possible involvement of the APP gene in the neurodegenerative process characterizing the pathology of Alzheimer's disease, which constitutes a clinical symptom in DS patients older than the age of 40. In this case, PC12 cells were transfected with portions of the gene for the human APP, and colonies of cells containing the human DNA integrated into their genome were obtained. It was found that the PC12 cells expressing the APP gradually degenerated when induced to differentiate into neuronal cells by treatment with nerve growth factor. This observation indicates that a peptide derived from the APP gene product is neurotoxic, and therefore, overexpression of this gene in DS may result in the accumulation of a neurotoxic peptide in the brains of affected individuals.

B. Animal Models of DS

The examples described above illustrate the usefulness of the cellular model for studies of the biochemical events resulting from gene dosage. Exploitation of this system will further increase as more candidate genes assigned to the 21q22 segment are cloned. However, this type of model suffers from a serious drawback in that it does not address the more complex issues of development and morphogenesis (i.e., the physiological consequences of gene dosage at the level of the whole animal). The appropriate model for that purpose is obviously an animal model, preferably one that shares a large number of biological similarities with humans. For practical reasons the mouse has been the animal of choice.

1. Trisomy 16 and Chimeric Mice

Genetic homology exists between mouse chromosome 16 and human chromosome 21 (i.e., many of the genes residing at the DS region 21q22 of human chromosome 21 are found on the distal segment of mouse chromosome 16). Mice trisomic for this chromosome were therefore considered as an animal model system for studies of DS. Trisomy 16 mice are produced by mating normal females with males carrying two different Robertsonian translocations. Trisomy 16 mice exhibit a number of phenotypic characteristics observed in DS (e.g., the endocardial cushion defect, the aortic arch defect, and a shortened neck), and many informative studies are being performed on these animals. However, one of the major limitations of this model is that only few trisomic 16 fetuses survive to birth; they usually begin to die at day 14 of gestation, whereas those that are born alive survive but a few days. Mouse chromosome 16 constitutes a much larger portion of the genome than does human chromosome 21, thus the degree of genetic imbalance is considerably more extensive and results in fetus lethality. Therefore, developmental studies as well as attempts to identify putative ameliorative therapies are precluded, and the major research emphasis involving trisomy 16 mice has been placed in studies confined to the cellular level. Cell cultures of various types have been established from trisomic fetal or neonatal mice and studies on growth kinetics, life expectancy, and the sensitivity of various receptor systems have been carried out. In general, cells explanted from trisomy 16 tissues grow poorly in primary culture; trisomy 16 dorsal root ganglion cells degenerate and exhibit exaggerated electrical membrane properties.

To circumvent the early lethality of trisomy 16 fetuses and obtain both survival and postnatal development, an alternative procedure was developed in which chimaeric mice are produced having both trisomic and normal cells. Such chimeras are formed by fusing together an early embryo exhibiting trisomy 16 and a normal embryo. Chimeras of this type could be considered analogous to humans that are mosaic for trisomy 21, because they have both trisomic and normal cells within their tissues. In some chimeric mice, the proportion of trisomy 16 cells in the brain was as high as 50–70%. Studies on these animals revealed neurochemical abnormalities, as well as altered behavior (i.e., increased activity during the dark part of the light cycle with greater distance traveled, increased speed of movement, and excessive grooming activity).

Although the trisomy 16 and the chimeric 16 mice provide the opportunity for anatomical and physiological studies during gestation that cannot be performed in humans, this animal model leaves much to be desired, not only because of the early lethality mentioned above, but also because of the excess in genetic imbalance. Mouse chromosome 16 is considerably larger than human chromosome 21 and contains many genes, the human analogues of which reside on chromosomes other than 21. Mouse chromosome 16 represents 3.9% of the haploid autosomal complement of the mouse genome, whereas chromosome 21 constitutes only 1.9% of the human complement. In addition, several genes mapped at the DS region of chromosome 21 have been localized to chromosomes other than the mouse 16. Therefore genes that might be contributing to the etiology of DS are not implicated in the phenotype of the mouse trisomy 16. For these reasons, a better model for DS related studies would be trisomic mice, which contain only a portion of chromosome 16, the segment homologous to the DS region 21q22, or, even better, a triplication of one or several candidate genes from the DS locus. Trisomy mice of this kind will certainly have a better survival capacity and will also permit investigating the participation of individual genes in the DS phenotype. Such trisomy mice were recently obtained by introducing specific cloned genes into the mouse germ line.

2. Transgenic Mice Carrying Chromosome 21–Encoded Human Genes

Gene transfer into mice, leading to the creation of so-called transgenic mice, can be achieved by microinjecting a foreign gene into one of the pronuclei present in every fertilized egg. The exogenously introduced DNA becomes stably integrated into the mouse chromosome, and the resultant embryo develops into a mouse that carries an extra gene and transfers it to subsequent generations in a Mendelian fashion.

Advances in recombinant DNA techniques facilitated the development of transgenic mice as an *in vivo* model for genetic diseases. Transgenic mice harboring candidate genes from the DS region have an advantage over cultured cells with transfected genes in that they are closer to the natural situation; the inserted gene is present in every cell of the animal, and its influence is manifested throughout its entire developmental history. By overexpressing individual genes in transgenic mice, it might be possible to dissect the trisomy 21 phenotype, gene by gene. The first strains of transgenic mice harboring a candidate gene from the DS region were constructed in parallel with the CuZnSOD-cellular system, using as transgene the human CuZnSOD.

3. Transgenic-CuZnSOD Mice

Following the interesting observation of diminished uptake of neurotransmitters by cultured neuronal cells overexpressing the CuZnSOD gene, a cloned DNA segment containing this gene was microinjected into the male pronucleus of fertilized mouse eggs, and several strains of transgenic mice that carry the gene were obtained. These animals expressed the transgene in a manner similar to that of humans and showed an increased activity of the enzyme, from 1.6- to 6-fold in the brain and to an equal or lesser extent in several other tissues. Outwardly, the transgenic mice appeared normal, without any obvious deformities. This is not surprising, because there is no reason to expect that elevation of CuZnSOD activity alone will cause the major dysmorphic features of DS. Rather, we anticipate that overexpression of CuZnSOD will affect more subtle aspects of tissue function and integrity, particularly in those tissues that might be affected by altered metabolism of oxygen free radicals. Bearing in mind the effect observed in the CuZnSOD-cellular system, the concentration of the neurotransmitter serotonin was measured in the blood of transgenic-CuZnSOD mice and was found to be significantly lower than the corresponding value in nontransgenic littermate mice. This observation generated much interest, because reduced concentration of blood serotonin is a well-known clinical symptom in DS patients. When the deficiency was first noticed in the 1960s, it aroused considerable attention because of the possible relevance of serotonin uptake by blood platelets to neurotransmitter function in the central nervous system, and hence its involvement in the hypotonia and mental retardation of DS. At that time, attempts were made to raise the levels of blood serotonin in DS infants by administration of its precursor, 5-hydroxytryptophan; muscular tone, motor activity, and sleep abnormalities were reported to improve concomitantly with its administration. However, the development of infantile spasms, a severe seizure syndrome, in 17% of the patients receiving the drug brought these studies to a halt. Serotonin is an important neurotransmitter in the central nervous system, both in the embryonic state and in infants. It usually does not appear free in the blood circulation because of its efficient uptake by platelets, where it is accumulated and stored in the dense granules. Detailed analysis of platelets isolated from the transgenic mice bearing the extra CuZnSOD gene revealed that the uptake process in these granules is impaired, and this constitutes the cause for the reduced concentrations of blood serotonin in these mice. The dense granules of the platelets are in many respects similar to the vacuoles in the PC12 cells, which, as described in the previous section, were damaged by the elevated activity of CuZnSOD. It is intriguing that this same lesion appears both in the cellular-CuZnSOD system and the transgenic-CuZnSOD mice and that the consequent defect is a well-known deficiency diagnosed in DS. This observation is the first example in which a direct link between a clinical symptom of the syndrome and a GDE of an individual gene has been established. The transgenic-CuZnSOD mice have also abnormalities in the connections between nerve terminals and the muscles (the so-called neuromuscular junctions), which are similar to the defects observed in neuromuscular junctions of patients with DS. This is an additional indication of the connection between gene dosage of CuZnSOD and defects characterizing the syndrome.

4. Transgenic Mice and Alzheimer's Pathology

Since the construction of transgenic-CuZnSOD mice, additional strains of mice bearing other human chromosome 21 genes have been developed and studied. The relation between overexpression of the gene encoding APP and the Alzheimer's disease–type neuropathology found in adult DS patients is currently being investigated in mice carrying and overexpressing the gene for the amyloid precursor. Overproduction of APP was detected in neurons, particularly in the hippocampus, the deeper layer of the cortex, and in the Purkinje cells in the cerebellum. The effects of this elevated level of the amyloid precursor are presently being studied.

In summary, the etiology of DS is a complex process, involving many genetic factors that produce the profound disturbances of development and morphogenesis seen in DS. Progress toward understanding the molecular basis of the deficiencies is slowed down by our inability to conduct studies of GDEs in human patients. An animal model system consisting of transgenic mice, with only the DS genes triplicated, is therefore essential for meaningful exploration of this complex disease in the context of a living animal.

V. Future Prospects

To delineate, at the molecular level, the relative importance of different genes on chromosome 21 in determining the DS phenotype, the minimal chromosomal regions whose triplication produces the syndrome has first to be determined and the genes it contains identified. Despite rapid progress in the molecular analysis of the genetic structure of chromosome 21, the minimal size and precise localization of the DS region are still undefined. It is also not known whether the DS region is contiguous or exists in patches scattered over the 21q22 segment. Molecular and clinical investigation of DS individuals possessing unbalanced translocations of parts of chromosome 21 should permit the precise definition of the DS region. These clinicogenetic studies should be extensive, because some of the DS symptoms are variable and difficult to identify on the basis of an individual case.

The recent advances in human molecular genetics should provide the means to construct a detailed genetic and physical map of genes along chromosome 21. A physical map of the entire chromosome, consisting of ordered clones containing overlapping sets of DNA fragments in cosmid or yeast vectors, is also a prerequisite for determining the complete nucleotide sequence of chromosome 21. This undertaking is within the framework of the international effort to map and sequence the whole human genome. Knowing the nucleotide sequence will facilitate the identification of all the genes residing on this chromosome and eventually lead to the isolation of those mapped to the minimal DS region.

Availability of cloned genes will allow researchers to construct other cellular and animal model systems of the type described in this article and to investigate the biochemical and morphological consequences of imbalance of these genes. The ability to generate transgenic mice carrying several genes from defined regions of the DS segment will further facilitate the exploration of GDEs on development and morphogenesis. The powerful technique of gene transfer to mouse germ line, followed by a mating program between strains of transgenic mice, each carrying one candidate gene, permits the development of a mouse strain that overexpresses several transgenes. Eventually a battery of transgenic strains with a full complement of the DS region triplicated will become available. Such an animal model will not only lead to identification of the genes participating in the syndrome and enable detailed study of the Down abnormalities during the entire life span from embryogenesis to the fully developed animal, but it will also provide a test system for therapeutic or ameliorative procedures.

Although the application of therapy for DS is still a long way off, the animal models provide systems in which various therapeutic strategies that may prevent or ameliorate some of the pathologies associated with the syndrome can be tested. The molecular biological approach may include gene therapy, which, in the case of DS, will consist of genetic means to eliminate the overexpression of the genes that have been identified as causing clinical symptoms. This approach requires gene targeting (i.e., homologous recombination between DNA sequences residing in the chromosome and newly introduced DNA sequences). This technique, if sufficiently developed, may permit one day the selective inactivation of those extra genes that contribute to the Down phenotype. This technology is currently being used to introduce insertion mutations that will silence the human CuZnSOD gene carried by the transgenic-CuZnSOD mice.

Bibliography

Carritt, B., and Litt, M. (1989). Report of the committee on the genetic constitution of chromosomes 20 and 21. Human Gene Mapping 10 (1989); Tenth International Workshop on Human Gene Mapping. *Cytogenet. Cell Genet.* **51,** 351-371.

Cooper, D. N., and Hall, C. (1988). Down's syndrome and the molecular biology of chromosome 21. *Prog. Neurobiol.* **30,** 507–530.

Epstein, C. J. (1986). The consequences of chromosomal imbalance—Principles, mechanisms and models. *In* "Developmental and Cell Biology," vol. 18. (P. W. Barlow, P. B. Green, and C. C. Wylie, eds.). Cambridge University Press, Cambridge.

Epstein, C., and Patterson, D., eds. (1990). "21st Chromosome and Down Syndrome." Alan R. Liss, New York.

Stewart, G. D., Van Keuren, M. L., Galt, J., Kurachi, S., Buraczynska, M. J., and Kurnit, D. M. (1989). Molecular structure of human chromosome 21. *Annu. Rev. Genet.* **23,** 409–423.

Dreaming

ERNEST HARTMANN, *Tufts University School of Medicine; Sleep Research Laboratory, Lemuel Shattuck Hospital; Sleep Disorders Center, Newton Wellesley Hospital, Boston*

GLOSSARY

Delta activity Brain wave activity characterized by relatively low frequency of 0.5–4 cycles per second

Electroencephalogram Record of brain waves (electrical activity of the brain) obtained from electrodes placed on the scalp; the instrument used to make these recordings, consists of oscillographs and amplifiers and is known as an electroencephalograph

Neurotransmitter Chemical substance that transmits messages between neurons (nerve cells); the substance is usually released by one neuron, called presynaptic, into the space between cells called the synapse, where it has an excitatory or inhibitory effect on a second neuron, called the postsynaptic neuron

DREAMING IS MENTAL ACTIVITY that occurs during sleep. Aside from this, there is no universally agreed upon definition of dreaming; however, it is generally accepted that dreaming involves consciousness, but a different kind of consciousness from that of the waking state. Dreaming tends to be perceptual rather than conceptual, with a great deal of direct sensory experience and relatively little thinking. The sensory experience is most often visual, but 20–40% of dream reports mention auditory experience, and less percentages include touch, pain, smell, and taste, in that order. In a typical dream, the dreamer lacks the experience of free will, which is so characteristic of the waking state. In addition, dreams usually appear in isolated pieces, with sharp discontinuities between dreams and sometimes within parts of a single dream. Thus, our dream lives are not tied together by the threads of memory, which provide a sense of continuity in our waking lives. This difference in continuity allows us to distinguish our own waking states from our dreaming states, and lets us answer the ancient Chinese conundrum: How do I know I am a man who dreams (at night) he is a butterfly, and not a butterfly who dreams (by day) that it is a man?

I. Historical Importance of Dreaming

Almost all human cultures have given significance to dreaming, whether the dream is considered a voyage of the soul, a message from the gods, a prophecy of the future, a guide to the direction of one's life, or an aid in healing the body or mind.

A dream as a voyage taken by the soul or spirit is a very widespread idea and is not at all unreasonable, even though the words "soul" and "spirit" grate on the Western scientific ear. Obviously, in our dreams we are able to see and to converse with people who actually live far away, or even with people who are no longer living. If we wish to concretize and give a name to the part of us that does this seeing and conversing, it cannot be the body, peacefully sleeping in bed, nor the ordinary mind, which guides us in our daily routines, but something else. A word such as "soul" or "spirit" fills this place.

In fact, this aspect of dreaming can explain one of the most widespread beliefs about the soul: In numerous traditional cultures, it is believed that the soul of a person who has died remains on earth, among the living, for a number of months or years, visits its relatives, and so on. This view is not unknown even in modern Western culture, where mediums attempt to make contact with the souls of the recently departed. Surely this can be explained by the fact that relatives and friends are dreaming of the recently dead person, and, of course, this dreaming activity occurs most in the years immediately after the person's death, when he or she is well remembered. After 20–40 years, the chances are that the living are not dreaming about that particular person, whose soul is therefore believed to have finally departed.

A belief in prophetic dreams is found in many ancient cultures. The Chester Beatty Papyrus documents dream interpretations—frequently involving predictions of future events—from the 12th dynasty (1991–627 B.C.). The Old Testament contains many prophetic dream interpretations; perhaps the most famous is the Pharoah's dream of seven fat cows and seven lean cows, interpreted by Joseph as foretelling seven years of plentiful crops followed by seven lean years. In some cultures, skepticism is expressed about prophetic dreams: In the *Odyssey*, dreams are divided into false prophecies: ("passing through the gate of ivory") and true ("passing through the gate of horn").

In classical Greece and in many cultures of the Middle East, dreams also played a part in healing. Patients, or supplicants, seeking help with a physical or mental disease, or seeking for guidance in making plans for their lives, were treated at specially designated dream temples. Usually after various preparations and rituals, the patients went to sleep in the temple and carefully noted any dreams they experienced, which were then interpreted either by priests and priestesses or by the patients themselves in an effort to find a cure or solution. The cure or solution might involve a direct message or prescription from the gods but could also involve psychological insight in the dreamer. In the latter sense, this use of dreams continues to the present: dreams are used in psychoanalysis (see below) and a variety of psychotherapies to increase a patient's self-understanding. There has been resurgence of interest in using dream incubation (without a temple, but simply asking the dreamer to concentrate on a particular problem before sleep) as a means of solving personal problems.

Western science, however, has historically shown little interest in dreaming and, in fact, not much interest in sleep. By the nineteenth century, dreaming was considered a by-product of disordered mental functioning during sleep, of so little consequence that it was barely mentioned in textbooks of biology or psychology and was studied only by an occasional eccentric. Since 1900, scientific interest in dreaming has revived and flourished, nourished by two very different sources: Freud and psychoanalyis, starting around 1900, and Aserinsky and Kleitman's discoveries leading to the biology of dreaming, which began in the 1950s.

A. Freud and Psychoanalysis

Freud actually proposed not one but several interconnected theories about dreaming. First, he insisted that the dream is a meaningful mental product, created by the dreamer from recent waking events (called day residues) and from unconscious wishes. He felt that when analyzed in detail every dream represents the fulfillment of a wish; however, the wish is by no means always obvious. Freud felt that the direct expression of a wish or other drive material would be disruptive and awaken the sleeper; therefore, the "dreamwork" transforms and disguises the wish and other "latent dream thoughts," using mechanisms such as condensation (i.e., joining several thought elements to produce one dream image) and displacement (i.e., moving the emphasis and emotion from one element to another). [*See* PSYCHOANALYTIC THEORY.]

By asking the patient (the dreamer) to associate to each element of the dream, the analyst tries to interpret the dream by determining the wishes and other unconscious material that formed it. Freud considered the interpretation of dreams to be the most direct path to such unconsciousness material and, thus, called dreams the "royal road to the unconscious."

Attempts to test rigorously the various hypotheses embedded in Freud's theories have met with mixed results. For instance, evidence supports a very simplified version of wish fulfillment: Subjects dream more of drinking when they are thirsty (artificially deprived of water) than when they are not. However, it is certainly not established that every dream represents the fulfillment of a wish; nightmares do not easily fit, and Freud himself was not satisfied with his attempts to explain them. Also, laboratory evidence tends to contradict Freud's view that the dream functions to protect sleep.

Despite the lack of proof, Freud's views have been extremely influential. Most forms of psychoanalysis and psychotherapy practiced throughout the world are based directly or indirectly on Freud's work. Not all therapists subscribe to the totality of Freud's view on dreams, but insofar as dreams are used at all in therapy, they are seen as meaningful mental products and as the royal road, or at least one good road, to the patient's unconscious.

A number of Freud's followers accepted his main tenets but suggested significant amendments. Carl Gustav Jung, a prominent early psychoanalyst and ardent student of dreams, proposed that dreams are not only a road to the person's individual unconscious but also lead to the "collective unconscious," a postulated tendency of the entire species to dream in certain patterns and to dream of certain basic images or symbols, which Jung called archetypes. At first these views seemed far fetched and mystical; however, collective unconscious and archetypes may refer to aspects of the organization of the human cortex. These aspects impose constraints upon the individual's construction of his own dream, making dreams of different individuals similar in some basic ways—much as myths from very different cultures are strikingly similar.

Other researchers have emphasized that dreams often appear to be attempts at solving current problems of the dreamer. Many analysts and others have recently emphasized the importance of the manifest dream, insisting that the dream exactly as dreamt can provide important information and insight, without the need for deep interpretation.

In any case, dreams have been alive and well in the world of psychotherapy and self-understanding since 1900; however, they received scant attention from biologists and nonclinical psychologists until the mid-1950s.

B. Aserinksy and Kleitman: Rapid Eye Movement Sleep

Nathaniel Kleitman, a professor of physiology at the University of Chicago, had no particular interest in dreaming, but he was one of the few scientists to study human and animal sleep in great detail. He performed many research studies, often using himself as a subject; for instance, he spent weeks living underground in a cave to examine his sleep–wake patterns without the usual influence of dark and light. Eugene Aserinksy, his student, was interested in blinks and eye movements in children and decided to record these throughout a night of sleep. He found that on a number of occasions each night, the eyes moved in a conjugate pattern, similar to the movements involved in watching something while awake. Aserinksy and Kleitman hypothesized that the child was dreaming during these periods, which rapidly led to studies in adults demonstrating that awakenings during these periods of rapid eye movements (REMs) usually produced reports of dreams. These findings initiated a whole era of laboratory research on the biology of dreaming. The highlights of this reasearch will be summarized in the following sections.

II. A Night of Sleep in the Human Adult: What is Recalled When?

Many thousands of nights of sleep recorded in the laboratory demonstrated the following basic pattern: As a person falls asleep, the muscles gradually relax, pulse and respiratory rates decrease slightly, and the electroencephalogram (EEG) shows a decrease of waking alpha activity, which is replaced by random low-voltage activity without any clear rhythmic pattern. This is known as stage 1 sleep, sometimes called drowsiness or sleep onset rather than true sleep. In the next few minutes, the EEG begins to demonstrate specific sleep rhythms called sleep spindles—activity at 13–15 cycles per second in bursts 0.5–2 sec long. The onset of sleep spindles marks the onset of stage 2 sleep. The spindles continue, but the low-voltage background is gradually replaced by delta activity (0.5–4 cycles per second). Arbitrarily, a record in which 20% of each 30-sec epoch consists of delta activity is assigned to stage 3 sleep, and when 50% of each epoch consists of delta activity it is called stage 4 sleep. [*See* SLEEP.]

In a healthy young adult, the transition from stage 1 to stage 4 sleep occurs within 15–20 min. The next hour of sleep is typically stage 3 and stage 4 sleep. Then, rather suddenly, after 80–120 min, the EEG reverts to what looks like very light sleep—stage 1 sleep, which continues for 5–15 min. This episode of stage 1 is accompanied by rapid, conjugated eye movements and, in terms of awakening threshhold, the episode is not light but is approximately as deep as stage 4 sleep. This is the first REM period of the night. The remainder of the night consists of 90-min alternations between non-REM sleep (stages 2, 3, and 4) and REM sleep. Non-REM sleep gradually lightens, and mostly stage 2 sleep occurs toward the end of the night. The REM periods gradually lengthen, and the last one of the night typically lasts 20–30 min.

REM sleep is also known as D-sleep (desynchronized, or dreaming, sleep), whereas non-REM sleep is called S-sleep (synchronized sleep). Typically in a young adult, four or five REM periods last a total of 100–120 min, or 25% of total sleep time.

Numerous studies have examined mental activity at different times in a night of laboratory-recorded sleep. Awakenings during REM sleep result in a report identified as a dream by the sleeper, and by independent raters, 60–90% of the time. Awakenings within a few minutes of the onset of an REM period result in short dreams, and awakenings further into an REM period produce longer dreams. Awakenings during the later REM periods of the night result in more emotional dreams, and dreams containing more material from earlier in the dreamer's life. Awakenings from non-REM sleep do not necessarily produce reports of no content. About 50% of such awakenings result in a report of something going on—a thought, or image or fragment of a dream. Reports scored as dreams by raters occur in 10–40% of such awakenings in different studies. Reports scoreable as dreams also occur at one other time—during sleep onset, or stage 1 sleep; at these times, even though the person is barely asleep, awakenings result in very vivid, although usually short, dreamlike reports.

Overall, therefore, dreaming should not be equated simply with REM sleep. REM sleep is certainly the time when most typical dreams occur, and in many subjects every REM awakening produces a dream or at least a fragment, but the converse is not true—non-REM sleep cannot be considered a time when no dreaming occurs.

Different dreams during the same night generally show some relationship to one another. If a person is awakened in the laboratory during each REM period and asked to report a dream each time, the dream reports will not be identical, nor will they involve exactly the same dream setting or characters; however, they will usually show some similarity in theme or content, leading the experimenter in several studies to conclude that the dreams were dealing with the same issues or problems in the dreamer's life.

III. Brain Biology in REM Sleep

REM sleep occurs in almost all mammals (see Section III,B). Hundreds of studies, chiefly in the cat and the rat, have elucidated the brain physiology underlying REM sleep. In brief, REM sleep and the REM–non-REM cycle described above depend on centers in the pons, the midportion of the brainstem. An animal with a trans-section above the pons, or with the entire brain above the pons removed, continues to have cycles of REM and non-REM sleep.

Specifically, REM and non-REM sleep are regulated by the interaction of several small cell groups within the brainstem. A group of large cholinergic cells, sometimes referred to as the giant cell nucleus, appears to have a dominant role in turning REM sleep on, whereas two groups of cells—the serotonergic cells of the raphe nuclei and the noradrenergic cells of the locus coeruleus—act to turn REM sleep off and non-REM sleep on.

These three neurotransmitters—acetylcholine, serotonin, and norepinephrine—among others, play roles not only in regulating the states of waking and sleep, but their release in the forebrain helps to determine the characteristics of the states. The "dreamlike" qualities of mental experience during REM sleep, so different from the qualities of wakefulness, probably result from the almost complete lack of norepinephrine and serotonin activity in the forebrain during REM sleep.

A. Ontogeny of Dreaming: Dreams in Childhood

Dreams are recalled more or less throughout an individual's life, although with different frequencies in different individuals (see below). Do they begin at a particular time in childhood? And do they develop in a particular way?

REM sleep is definitely present in children. In fact, the amount of REM sleep decreases gradually during childhood; a newborn child spends 16 hr asleep, and approximately half that time is spent in REM sleep. Studies have also demonstrated the presence of REM sleep in the fetus. But what about actual dream reports?

Dreams have been recorded in laboratory settings from children as young as 4–5 yr old and in clinical settings from children as young as 2.5 yr. Reports of nightmares are quite frequent at 3–5 yr old. Apparently dreams, including nightmares, can be described by children as soon as their verbal skills are sufficient to describe any experience—usually around 2–3 yr old. However, some sort of dream probably occurs even before that time: In one case, the parents of a 1-yr-old child noticed certain movements during sleep in their son, who would then awaken crying, apparently upset. He continued to have these episodes occasionally for a number of months while learning to talk, and finally he was able to describe a frightening dream to his parents. Since the movements and upset crying were the same, the parents concluded that their child had probably had the same dream, or a similar dream, as early as 1 yr old.

Different ages showed a gradual development in terms of more characters, more humans, fewer animals, and more complex themes. Dreams appeared to develop gradually with age and maturity much as do other mental products—images, thoughts, stories—suggesting that dreams are indeed one sort of production of the human mind and that they can be examined and analyzed, like the others, by the methods of cognitive psychology.

B. Phylogeny of Dreaming: Do Animals Dream?

Phsyiological recordings of sleep have been performed in a wide variety of animal species. Both REM and non-REM sleep occur in the sleep of all mammals studied with one exception: the spiny anteater, or echidna. Birds have non-REM sleep and occasional short episodes of REM sleep; the latter are usually seen only in young (immature) birds and disappear or almost disappear in the adult. Reptiles definitely show non-REM sleep, and a few appear to have a rudimentary form of REM sleep. Amphibians and fish do not appear to have REM sleep, but they do show electrical changes suggesting a form of non-REM sleep. Invertebrates have been less methodically studied so far. Many invertebrate species demonstrate regular rest and activity cycles, with periods of inactivity that can be called behavioral sleep, but without all the electrophysiological characteristics of sleep in the vertebrates.

However, even in mammals who definitely have REM sleep, does this mean that they experience dreams? Leaving aside the philosophical question of whether or not we can ever be sure that another human being dreams, or thinks, we generally accept the report of another person who tells us he or she was dreaming. We are unable to ask animals whether or not they were dreaming, and they are unable to answer us in a language we understand. So perhaps the question cannot be answered; however, there are at least suggestive hints. First, we have all seen pets, especially young cats and dogs, running or playing in their sleep. We usually conclude that they must be dreaming. Lucretius, in his great work *De Rerum Naturae*, written 2,000 years ago, noted horses running in their sleep, and concluded they were dreaming. This is suggestive, but not proof.

Second, physiological recordings indicate that the pattern of brain activity in other mammals during REM sleep closely resembles that found in humans—for instance, there is intense activation in the visual cortex in many mammals, similar to that found during active waking, with search activity for instance. This certainly suggests, although it does not prove, that the animal is seeing something.

Third, in one study monkeys, trained for other reasons to press a bar whenever any picture was projected onto a screen in front of them, were found to be making bar-pressing movements at times during their sleep. Recordings were not made to determine times of REM sleep, and there were some other problems, but this sort of study is perhaps the closest we can come to an answer at present. Perhaps the researchers teaching speech to chimpanzees will at some time be able to ask them directly about dreams. Meanwhile, indirect evidence suggests a strong probability that animals other than humans do experience dreams.

IV. Nightmares

When someone says "I have terrible nightmares," she or he usually means one of two things, which have only recently been studied and clearly distinguished in sleep laboratory studies. The first phenomenon is the true nightmare: a long, vivid, frightening dream with increasingly scary content that awakens the sleeper. True nightmares occur during REM sleep, like most dreams, usually from very long REM periods in the second half of the night. The second phenomenon consists of waking in absolute terror, with a scream but without a dream; the sleeper either recalls no content or a single frightening image: "something was sitting on me" or "I was being crushed." This second phenomenon occurs during an arousal from deep non-REM sleep during the first hours of sleep. The second phenomenon is best called "night terror"; it is physiologically as well as psychologically different from the nightmare.

Nightmares are not rare. Questionnaire surveys suggest that the average adult experiences one to two definite nightmares per year. Although almost half of adults claim to have no nightmares, they usually mean no recent nightmares; on careful questioning, they often will remember one or two nightmares in childhood. Nightmares appear to be most common in children 3–6 yr old; at least 75% of children report at least some nightmares at this age. Most surveys suggest that nightmares are somewhat more common in women than in men, but careful studies suggest that women may be more willing to admit to having nightmares, rather than actually experiencing more.

A nightmare almost always consists of something harmful or dangerous happening to the dreamer; she or he is being chased, threatened, wounded, tortured, or killed. We ourselves are the victims in our nightmares. The only consistent exception occurs in mothers of young children who often dream that something is happening to their child rather than to themselves.

Nightmares can sometimes be initiated by acute trauma. In this case, the nightmare involves a repetitive playback of the traumatic event with small variations. Without trauma, nightmares tend to occur at stressful times, especially if the stress reminds the person of childhood vulnerability.

Some people tend to have nightmares all the time—every few days for many years, often since childhood. These people have been studied intensively. They are not necessarily very anxious people, nor do they necessarily have mental illness or serious problems; rather, they are people who are unusually open, appear defenseless or vulnerable, and are often highly creative (many artists, including some of the most famous, suffered from frequent nightmares). They are described as having "thin boundaries" in many different senses.

V. Other Special Dreams

The lucid dream is a special kind of dream that has recently aroused a great deal of interest. This is a dream in which one becomes aware that one is dreaming, within the dream. If one does not awaken, one can then direct the dream and make changes in it. Lucid dreams sound like a state in between dreaming and waking, but laboratory studies have demonstrated that lucid dreams occur in definite REM sleep. Lucid dreamers have been able to signal to the experimenter, using prearranged eye-movement signals, to indicate that they know they are dreaming. The EEG record shows the continuation of REM sleep before and after the signal.

Both dreams-within-dreams and false awakenings refer to a situation in which the dreamer dreams of having awakened from a dream but still finds himself within a dream from which he later truly awakens. Little is known about these dreams, except that they appear to occur most frequently in persons who also report nightmares or lucid dreams.

VI. Remembering and Forgetting Dreams: Who Remembers Dreams?

One of the most intriguing, and annoying, features of dreams is how easily they are forgotten. One recalls a dream on awakening and finds that after breakfast it is totally gone. This is probably related to the nature of the chemical environment at the forebrain during REM sleep, which is very different from the environment during active waking (see above).

Some people remember dreams very well, whereas others remember nothing. We all have four or five REM periods per night, each of which usually yields a dream if interrupted by an awakening. Theoretically, we should remember 30 dreams per week from our REM periods and a few others from sleep onset or non-REM sleep. When people are asked on questionnaires how many dreams per week they recall, the answers range from 0 to 21; the mean is about 2.5/wk. What accounts for these huge interpersonal differences? First, physiological factors appear to be involved. Several studies show that good dream recallers have slightly more REM sleep than poor recallers and also tend to awaken more often during the night, thus having a greater chance of awakening during an REM period. One study examined complicated measures of EEG frequency in both hemispheres during REM sleep and during waking and demonstrated that good dream recallers showed less difference on these measures between REM and waking than did poor dream recallers. In other words, good dream recallers "didn't have as far to go" when they switched from dreaming to waking.

Psychological factors are involved as well. Personality studies demonstrate that good dream recallers become more absorbed in daydreams and fantasies, have greater "tolerance for ambiguity," and have thinner boundaries (as discussed above), meaning they are more open, flexible, artistic, and vulnerable. Indeed, the nightmare sufferers and the lucid dreamers turn out to have a very high rate of dream recall. The dimension "thin versus thick boundaries" may actually encompass the physiological differences as well. Persons with relatively thick boundaries keep things in separate compartments and maintain very separate states, and their REM sleep state differs greatly from their waking state. This may account for their remembering few dreams.

VII. Does Dreaming Have a Function?

REM sleep, taking up a sizable fraction of mammalian sleep, almost certainly has a biological function, but there is no agreement as yet on the exact nature of that function. Evidence from recent studies of sleep deprivation and REM deprivation suggests a role in thermoregulation. Studies demonstrating increases in REM time after some kinds of learning suggest a role in learning and memory. Evidence also supports a role for REM sleep in restoring the functioning—perhaps the sensitivity—of forebrain systems using certain neurotransmitters, norepinephrine and perhaps serotonin; these systems are turned off (for repairs?) during REM sleep. These are brain systems that play a part in thermoregulation as well as in memory and learning, so that this proposed function is compatible with the others.

Whatever the exact biological function of REM sleep, it most likely fulfills this function whether or not any dreams are recalled; in fact, it is presumably fulfilling its function in the great majority of mammals whom we have no way of asking about dreams.

Does dreaming itself have a function then, aside from the function of REM-sleep? As mentioned, because a large number of apparently normal humans never recall a dream, it seems impropable that we will find a biological function for dreaming in this sense; however, nothing can stop us from making use of dreams in a variety of suprabiological or cultural ways. In this sense, for those who want to use them, dreams certainly can have a function in psychotherapy, problem-solving, and advancing our knowledge of ourselves.

Bibliography

Aserinksy, E., and Kleitman, N. (1953). Regularly occurring periods of eye motility and concomitant phenomena during sleep. *Science* **118,** 273–74.

Foulkes, D. (1985). "Dreaming: A Cognitive Psychology Approach." Erlbaum, Hillsdale, New Jersey.

Freud, S. (1900). "Die Traumdeutung." (English translation: "The Interpretation of Dreams," 1953.) Hogarth Press, London.

Hartmann, E. (1973). "The Functions of Sleep." Yale University Press, New Haven.

Hartmann, E. (1984). "The Nightmare." Basic Books, New York.

Hobson, J. (1988). "The Dreaming Brain." Basic Books, New York.

Hunt, H. (1989). "The Multiplicity of Dreams." Yale University Press, New Haven.

Kleitman, N. (1963). "Sleep and Wakefulnes." University of Chicago Press, Chicago.

Drug Analysis

DAVID B. JACK, *Fidia Research Laboratories*

Glossary

Chromatography Any system that produces a separation by partition between a stationary phase and a mobile one

Drug Any medicinal compound of animal, vegetable, mineral, or synthetic nature

Retention factor (R_f) In thin-layer chromatography, the distance traveled by the compound relative to the mobile phase

Retention time (t_R) Time taken for a compound to pass through a chromatographic system from injection to detection

DRUG ANALYSIS includes all methods that are generally used to determine the concentration of a drug in a matrix. This matrix may be relatively simple (e.g., a tablet) or complex (e.g., urine). This covers a wide range of techniques from the old and inexpensive to the new and costly.

I. Introduction

Drugs are substances used to treat or prevent disease, and they may be used alone or in combination. Drugs are administered in a variety of ways, depending on the nature of the drug, the disease, or the speed with which it is to act. Most drugs are given by mouth and rapidly enter the bloodstream; other routes include intravenous, intramuscular, and rectal administration as well as through the skin or by inhalation. They may only need be taken for a short time to cure a headache, for 1 or 2 weeks to treat an acute infection, or for life to control a chronic disease like high blood pressure.

To use drugs as safely as possible, it is important that the amount of active agent in each formulation (tablet, capsule, ointment, etc.) is carefully controlled and that the way it is given produces the required action with a minimum of side effects. Drugs are not just measured for therapeutic reasons: drug screening in sport, employment, and overdose is now common and likely to remain so. A wide variety of techniques are available, and the one chosen largely depends on the end in view. The techniques can be classified as either physical, chemical, or biological, and Table I lists the most important. One method may be used in preference to another because of the nature of the material containing the drug (e.g., tablet, ointment, blood, urine) or because of the amount of drug likely to be present (1 part in 10, 1 in 10^3, 1 in 10^6, and so on).

Drugs may often be converted in the body to more polar, water-soluble metabolites, which may exert no pharmacological effect or may exhibit a different effect, so it is important that any method measuring drugs in body fluids such as blood or urine can distinguish between the unchanged drug and its metabolites or, at least, can determine the unchanged drug without interference from metabolites.

II. Sample Preparation

Every analytical technique requires some form of sample preparation. This may range from the simple dilution of a cough mixture or plasma sample to a complex series of extractions. Some methods require little in the way of pretreatment of the sample,

TABLE I Techniques Used for Drug Analysis

Physical	Chemical	Biological
Radioisotope counting	Volumetric analysis	Immunoassay
Gravimetry (including thermal gravimetry)	Chromatography	Radioimmunoassay
Polarography	Gas–liquid	Enzyme immunoassay
Ion-selective electrodes	High-performance liquid	Fluoroimmunoassay
Spectroscopy (UV, VIS, IR, FLUOR)	Thin-layer	Luminescence immunoassay
Atomic absorption	Hyphenated techniques (GC-MS, etc.)	Microbiological
Spectrography		
Nuclear magnetic resonance		

especially if the drug is present in high concentration in a relatively simple matrix (e.g., an active ingredient in a tablet formulation). However, much drug analysis is carried out using complex biological fluids such as plasma or urine where there will be large amounts of interfering substances and where the drug may be present in only nanogram quantities. Techniques requiring little in the way of sample preparation are radioisotope counting, atomic absorption, volumetric analysis, some electrochemical methods, infrared (IR), ultraviolet (UV), and visible (VIS) spectroscopy, fluorescence, and many immunoassays. Techniques requiring considerable preparation are gravimetry and most types of chromatography although column switching in liquid chromatography can sometimes allow analysis without pretreatment. When drugs are present in relatively low concentration in complex media (e.g., plasma or urine), some means of separating the drug from interfering compounds and concentrating it is needed. The most frequently used methods are liquid–liquid and liquid–solid extraction.

A. Liquid–Liquid Extraction

This is based on the principle that the drug will selectively distribute between one of two immiscible liquids under a particular set of experimental conditions. By altering the pH, the ionization of the drug can be suppressed and its solubility in organic solvents increased. As a general rule, the least polar solvent possible should be chosen to extract the drug because this will ensure that much of the polar material is left behind in the aqueous phase. Some common solvents arranged in order of polarity are given in Table II. If acidic, neutral, and basic drugs are present, together these can all be separated by a sequential change in pH as is shown in Fig. 1. Generally this type of extraction procedure works well for drugs that are very lipid-soluble, and high recoveries (>90%) can often be obtained. However, some drugs may pose particular problems. For example, some are amphoteric, having both acidic and basic groups, and here the relative strengths of each group will determine the pH chosen for extraction. A number of drugs are ionized at all pH values and are too polar to be extracted. However, they can often be paired with a suitable ion of opposite charge to given an ion pair with an overall charge of zero and hence soluble in organic solvents. Careful choice of this counter-ion can lead to a very selective type of extraction indeed; the counter-ion can also be chosen because it possesses a particularly favorable property (e.g., fluorescence), and in this way the original drug can be measured at low concentrations even if it has no fluorescence itself. Liquid–liquid extraction is widely used, but it does have some disadvantages: it is costly in terms of organic solvent used, large quantities of solvent constitute a potential fire hazard, large volumes of waste solvent are generated and disposal can be expensive, and finally, this type of extraction is difficult to automate.

B. Liquid–Solid Extraction

This approach was developed to solve the above problems and involves passing the material contain-

TABLE II Common Organic Solvents Arranged in Increasing Polarity

Heptane	Dioxane
Hexane	Ethyl acetate
Carbon tetrachloride	Acetonitrile
Toluene	Pyridine
Diethyl ether	Ethanol
Chloroform	Methanol
Dichloromethane	Acetic acid
Acetone	(Water)

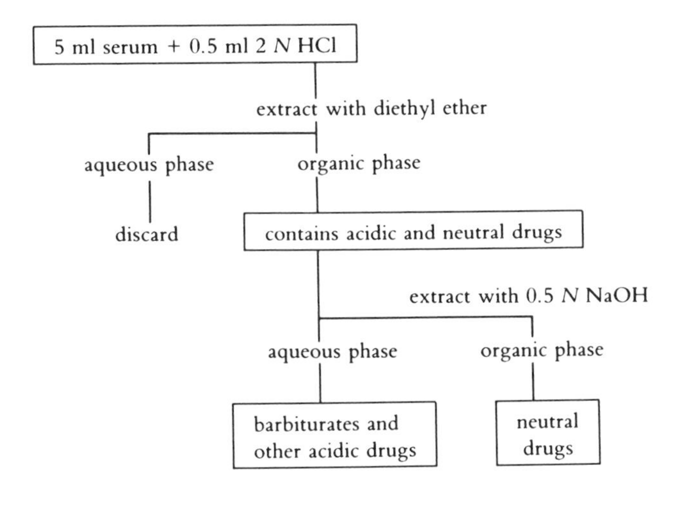

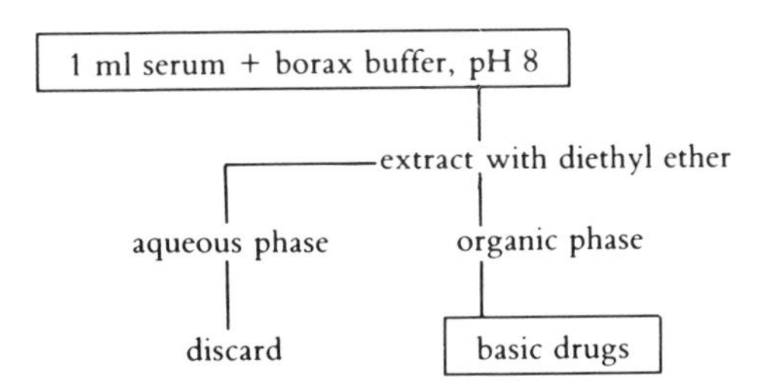

FIGURE 1 Separation of acidic, neutral, and basic drugs. [From Jack, D. (1984). "Drug Analysis by Gas Chromatography." Academic Press, Orlando.]

ing the drug over a matrix that will adsorb it while letting most of the other material through. The adsorbed drug can then be washed with a suitable buffer and then eluted with an organic solvent. This procedure is simple, economical, and much easier to automate. A range of solid phase cartridges is commercially available, and their properties and use are outlined in Table III. No matter which type of extraction is chosen, care should be taken to ensure that all glassware is scrupulously clean, and it should be borne in mind that some drugs, especially those with tertiary amine groups, bind strongly to glass in their un-ionized forms.

III. Analytical Techniques

As we have already seen, the choice of technique is determined by the amount of drug likely to be present and the nature of the matrix. The physical techniques will be considered first, followed by chemical, and finally, biological.

A. Physical Techniques

1. Radioisotope Counting

Administering a drug in a radioactively labeled form to an animal is an excellent way of gaining information on how a drug distributes itself in the body tissues. This approach obviously is not applicable to studies in humans, and unless some means of separating the unchanged drug from any metabolites is used, this approach will only give information about the "total" drug and will be of limited value in pharmacokinetic studies. Another use of the technique is to estimate recovery from complex extractions by adding labeled drug in a known amount at the beginning and counting what remains at the end.

2. Gravimetry

This method has been in use for hundreds of years, ever since accurate chemical balances were developed. It can be used when drugs are present in high concentration (e.g., the major component of a dosage form). Some simple initial extraction is usually applied to separate the drug, and it is then weighed.

3. Thermogravimetry

The change in weight of a sample on heating in a carefully controlled way is measured. It is widely applied to bulk drugs to investigate water content and stability.

4. Polarography

This can be applied to metals and many drugs and, in certain circumstances, can be applied directly to body fluids. The current is measured with changing potential between two electrodes. A drop-

TABLE III Comparison of Normal and Reverse-Phase Cartridges for Liquid–Solid Extraction

Cartridge	Silica	C 18
Polarity	Polar	Nonpolar
Sample applied in	Low-polarity solvent (e.g., hexane)	High-polarity solvent (e.g., water)
Wash solvent	Low polarity	High polarity
Eluting solvent	Increased polarity (e.g., buffer)	Decreased polarity (e.g., methanol, acetonitrile)
Elution order	Nonpolar first, most polar last	Most polar first, least polar last

ping mercury electrode is the cathode (negative electrode) and constantly renews itself, whereas a pool of mercury is the positive anode. A characteristic half-wave potential is produced for the compound or group undergoing reduction. A mathematical relation links current, mercury flow, and concentration.

5. Ion-Selective Electrodes

These can be used with solutions containing drugs and require little or no preparation. The basic design consists of an electrical sensing device separated from the sample by a selective barrier of conducting material. New liquid ion-exchange membranes, enzyme electrodes, and membranes containing macrocyclic compounds have extended the use of this technique and a number of drugs (e.g., antibiotics, alkaloids, and nonsteroidal anti-inflammatory agents) have been measured in the range of 10–40 μM.

6. Spectroscopic Techniques

These can be used when the drug is present in relatively high concentration. They are usually applied after some form of extraction procedure although, in the case of many dosage forms, this may not always be necessary. It uses light in the UV, VIS, and IR spectrum. The relatively high energy of the UV and VIS raises the electrons forming bonds in drug molecules to excited states, whereas lower energy absorption (IR) increases the vibrational energy of the bonds. These energy changes can be measured using appropriate equipment.

a. Ultraviolet and Visible Spectroscopy

These techniques are widely used for the quantitative analysis of compounds known to be present in formulations and body fluids. Measurement depends on the absorption of energy when the radiation traverses the sample. By preparing a series of standards containing known amounts of drug and by comparing their absorption with that of the sample, the amount of drug present can be calculated. The use of the technique has been extended by employing difference spectroscopy, in which different ionized forms of the drug are measured simultaneously, or derivative spectroscopy, in which slight changes in slope can be more easily detected or interference suppressed.

b. Infrared Spectroscopy When drugs absorb in this region, they show changes in their vibrational and rotational bond energies, which means that IR is useful as a tool to study structure. However, it can also be used quantitatively if the drug possesses any strongly absorbing groups (e.g., double or triple bonds). In principle it is more versatile than either the UV or VIS range, in which drugs will often only display one or two absorption peaks. Sample preparation is simple because the technique is generally applied to dosage forms. A number of tablets are ground and a sample compressed into a disc with a halide such as KBr (potassium bromide) or made into a mull with liquid paraffin. Standards are prepared in the normal way, and the absorption of the chosen band allows calculation of the amount of drug present.

7. Fluorimetry

This is more sensitive than either UV or VIS and is more selective because the sample can be irradiated at one wavelength and emit light at a higher wavelength. Fluorescence is sensitive to a number of factors (e.g., pH, temperature, and solvent). Glassware should be scrupulously clean and all materials used checked for interfering fluorophores. The linear range is narrower than with either UV or VIS, but its greater sensitivity has allowed it to be used successfully for a number of common drugs including catecholamines, phenothiazines, beta-adrenoceptor antagonists, and some antibiotics. After suitable chemical derivatization, it can also be used for drugs that do not possess any native fluorescence.

8. Miscellaneous Techniques

Atomic absorption is a sensitive and selective means of measuring metals and can be applied to biological fluids after only a simple dilution of the sample. This is because the energy sources (lamps) used are specific for the element in question. It is used frequently to monitor lithium concentrations in patients receiving lithium carbonate for manic depression. It can also be used for the trace analysis of zinc, copper, and aluminum. Spectrography can also be used in drug analysis, usually to determine trace contaminants, and X-ray diffraction has been used to provide structural information, but its use is hindered by the difficulty in obtaining high-quality crystals. Nuclear magnetic resonance is a powerful technique that has been used for quantitative and qualitative analysis. It is theoretically possible to measure several different drugs simultaneously, but the expense of the instrument and its demanding operating conditions have restricted its use.

B. Chemical Techniques

1. Volumetric Analysis

This is an old technique still used today. Here a reagent is chosen that will react chemically with the drug to be determined. Both are dissolved in a solvent and one titrated against the other using the appropriate indicator of the end point. The volume of reagent consumed is used to calculate the amount of drug present. In its simplest form, this is a cheap technique, but its sensitivity is limited and it is restricted to the analysis of drugs in dosage forms. Its range and sensitivity can, however, be extended by using electrochemical methods to detect the end point, and titrations can also be carried out in nonaqueous media.

2. Chromatography

Chromatography has expanded rapidly, particularly during the past 40 years, and it is now the most important technique used for drug analysis. An extensive literature is available on all aspects of its use. Essentially, the technique separates complex mixtures by allowing the individual components to partition between a stationary and a mobile phase. The stationary phase is usually contained within a column, and the mobile phase can be a gas (gas-liquid chromatography, GLC), a liquid (high-performance liquid chromatography, HPLC) or a supercritical fluid—a liquefied gas such as CO_2 (supercritical fluid chromatography, SFC). If the stationary phase is spread two-dimensionally, the process is called thin-layer chromatography (TLC), and when the particle size of this phase is carefully graded, it is called high-performance thin-layer chromatography (HPTLC).

The most important advantage of chromatography is that it can be made selective by (1) choosing the phases from a wide range of polarities and (2) using a number of sensitive and selective detectors. Under a given set of instrumental conditions, the time taken for a compound to move through the system from the start (injection) to the finish (detector) will be constant. This is called the *retention time* (t_R) and is measured usually in minutes. In HPLC, in which the mobile phase is a liquid, it may be measured either in minutes or volume (e.g., milliliters). In TLC the distance moved by the compound to be measured (e.g., drug) relative to the solvent front is called the *retention factor* (R_f). Column efficiency is defined by the number of theoretical plates, and this terminology derives from distillation theory. The larger the number of theoretical plates, the more efficient the column. The theory of chromatography is well-documented in a number of standard works.

Most modern analytical laboratories will possess both GLC and HPLC equipment, and their use is largely complimentary although HPLC is now probably more popular than GLC. A brief comparison is given in Table IV.

a. Gas Chromatography No matter whether GLC or HPLC is chosen, the choice of column is critical, because an unsuitable column will give poor results even with the most sophisticated equipment. Two types of column are used in GLC, packed and capillary. The former are relatively short (1–4 m) and wide bore (2–4 mm), whereas the latter are much longer (20–50 m) with a narrower bore (0.2–0.7 mm). The gas used as the mobile phase depends on the detector system but can be nitrogen, helium, or mixtures (e.g., argon: methane. [*See* GAS CHROMATOGRAPHY, ANALYTICAL.]

i. Packed Columns The column packing consists of a stationary phase coated on a relatively inert support to give a large surface area. The general efficiency of such columns is relatively low, about 400–2,000 theoretical plates; this may not be a disadvantage where the components of relatively simple extracts have to be separated [e.g., the active ingredients of a cough mixture or an extract of plasma in which the drug is present in high concentration (μg/ml)].

For analytical drug work, the percentage coating of the stationary phase on the support is low (1–3%).

ii. Support This is usually made from diatomaceous earth, which is composed almost entirely of silica in a pure form, which is carefully graded to give a narrow particle size distribution: this is important for an efficient column.

iii. Stationary Phase A good phase should be thermally stable and manufacturers' catalogues will indicate the recommended operating range for each. More than 200 phases of a wide range of polarity are available, but most drug analysis can be carried out with about a dozen. To achieve an efficient separation, it is important that the components of the mixture partition between the stationary phase and the gaseous mobile phase (note that no separation takes place in the mobile phase itself).

TABLE IV Comparison of GLC and HPLC

	GLC	HPLC
Stationary phase	range of polarity	range of polarity
Mobile phase	gas	liquid (organic/aqueous)
Operating temperature	subambient—350°C	subambient—50°C
Programming available	yes, temperature	yes, gradient elution
Sample volume (μl)	small <5	large <500
Range of detectors	wide, FID, N/PD, ECD	wide, UV, FLUOR, EC
Detector sensitivity	high, especially after derivatizing	high, with EC or derivatizing
Detector selectivity	good, N/PD and ECD	poor–moderate
Suitable for high-MW drugs	no	yes
Suitable for polar drugs	not without derivatizing	yes
Suitable for heat-labile drugs	no	yes
Automation available	yes	yes

The simple principle that "like dissolves like" can be applied in chosing a phase. For example, if a relatively nonpolar mixture of drugs is to be separated, a nonpolar phase should be chosen so that the components will truly dissolve and take a finite time to pass through the column. If, instead, a polar phase were chosen, the compounds would pass through quickly and emerge from the column without being separated.

In the early days of chromatography, columns were classified as being simply nonpolar, polar, or very polar, but columns can now be compared on quantitative terms. This is done by chromatographing a test mixture of a series of compounds containing a range of chemical groups on different stationary phases and comparing their retention times with those obtained on squalane, the least polar of the stationary phases. This gives a series of retention index differences that characterize each phase. The higher the retention index difference the more polar the phase. Tables of these *McReynolds constants,* as they are called, can be found in some suppliers' catalogues and are useful.

iv. Capillary Columns These are much longer and narrower than packed columns. The stationary phase can be bonded directly to the etched inner wall of the column, and an inert support is unnecessary. When this is done, the column is referred to as wall-coated open tubular (WCOT).

Supports such as microcrystalline sodium chloride can also be used, and these are first chemically bonded to the inner wall before coating the stationary phase to give support-coated open tubular (SCOT) columns. Most laboratories buy their capillary columns directly from commercial suppliers, although some still prefer to prepare their own. The thickness of the stationary phase can have an important effect on the separation because thin films (0.1–0.5 μm) give the most efficient separations with the shortest analysis times. However, greater amounts of sample can be handled by thicker films (1–1.5 μm). Column efficiency is high, and 10,000–50,000 theoretical plates can be obtained. The polarity of the stationary phase is less important with capillary columns, and much drug analysis is carried out on nonpolar phases such as OV 1. A great deal of packed column work is carried out at constant temperature. This is not always satisfactory, especially when mixtures containing high and low boiling components are present. In such cases, temperature programming is used. This option is now available on most modern instruments and is essential for capillary operation.

v. GLC Injection Systems Generally drug extracts from formulations or body fluids are dissolved in a volatile organic solvent such as hexane or toluene, and for injection on to packed columns, a suitable volume is taken up in a capillary syringe (1–10 μl) and injected through a silicone septum onto the column. The injection port temperature is always kept at a temperature higher than the column so that the injected material is efficiently swept onto the column. Because of the narrow bore, injection onto capillary columns is more complicated. Syringes with fine silica needles are available, and injection directly onto the column is possible. Such on-column injection is efficient. Variations such as split injection are available for concentrated solutions, in which only a preselected portion of the extract reaches the column while splitless injection is available for very dilute solutions, in which the entire injected volume enters the column; modern instruments usually allow both systems to be used. Headspace analysis is useful for volatile materials (e.g., the determination of ethyl alcohol in blood).

vi. Detectors A range of detectors are available, and flame ionization or electron capture are most frequently used for routine analytical work.

(*a*) *Flame Ionization Detection* Here the detection at the end of the column consists of a hydrogen flame burning in air. A potential difference is maintained between the flame tip and a collector electrode above. When an organic compound emerges from the column, it burns in the flame, giving an increased flow of electrons detected as a peak. This detector is robust, with a large linear range, and is relatively sensitive (down to about 0.1 μg injected) for organic compounds that do not contain large numbers of oxygen or halogen atoms.

Nitrogen/phosphorus detection (N/PD) is a modification of flame ionization detection (FID), which can be tuned to give a selective response to compounds containing nitrogen (most drugs) or phosphorus (many insecticides). This is particularly useful because when drugs are being measured in biological fluids, even after a complex extraction, many other substances will also be present. It is possible to measure some nitrogen-containing drugs down to 1 ng/ml plasma or less with capillary column instruments equipped with this detector.

(*b*) *Electron-Capture Detection* In this system a radioactive source (usually Ni^{63}) produces a stream of slow electrons by interacting with the carrier gas. Any drug containing halogen atoms, nitrogroups, or a highly conjugated system of double bonds will efficiently capture electrons. In practice this does not include many drugs, but chemical derivatization can easily produce electron-capturing compounds (see Section III,vii). The linear range is less than with FID, but subnanogram quantities of suitable drugs can be detected while reduced responses are produced by material that does not capture electrons. Thermal conductivity, photoionization, and flame photometric detection are also available but for a number of reasons they are not as useful as the ones discussed above. For some purposes, FID and electron-capture detection (ECD) can be used together to gain important structural information about drugs and their metabolites.

vii. Derivatization for GLC Although many drugs are volatile enough to be chromatographed directly, a number are too polar and have first to be converted to a less polar derivative. The conversion of carboxylic acids and amines to esters and amides is a good example. Silylation reactions are also widely used to give volatile trimethylsilyl (TMS) ethers. These latter types of derivative are generally more sensitive to water and are less stable than their nonsilylated counterparts. Derivatization can also be carried out to make the drug more electron-capturing. There is a wealth of chemical literature on chemical derivatization techniques.

b. High-Performance Liquid Chromatography This is now the most widely used technique for the determination of drugs in dosage forms and body fluids. This is due to its more general applicability (e.g., it can be applied to polar and nonpolar drugs and high molecular weight compounds without derivatization, and it is less demanding in terms of gases, injection systems, and detectors). The different forms of HPLC are normal phase, reversed phase, ion-pair, and size exclusion chromatography, and Table V gives an indication of the differences and how the right technique can be chosen. (*See* HIGH PERFORMANCE LIQUID CHROMATOGRAPHY.]

i. Columns These are generally made from stainless steel and are 10–25 cm in length and $\frac{1}{8}$–$\frac{1}{4}$ inch in internal diameter. Good separations can be obtained with normal phase columns such as silica gel, particularly when closely related compounds (e.g., drugs and their metabolites) are being investigated. Reversed-phase chromatography involves a nonpolar column with a more polar mobile phase, which usually contains water. Selective separations can be obtained because the degree of polarity can be carefully chosen, the mobile phase buffered at a selected pH, and the ionic strength controlled. Ion-exchange chromatography is now much less frequently used because it has largely been replaced by reversed-phase operation, which is more versatile. Size exclusion (gel permeation) chromatography depends on the different molecular sizes of the components of a mixture and is usually only used for compounds of MW >2,000 and, hence, is not applicable to most drugs. It does, however, play an important role in the separation of peptides and polymers.

TABLE V Choosing an HPLC Column

Drug	Chromatography	Column	Mobile phase	Example
Water-insoluble				
Nonpolar	Partition, reversed-phase	Bonded C_2–C_{18}	Polar	Antibiotics
Weakly polar	Adsorption, solid phase	Silica, alumina	Weakly polar	Alkaloids
Polar	Partition, normal phase	Bonded —CN, —NO_2 —NH_2	Nonpolar	Alkaloids
Water-soluble				
Anionic	Ion exchange	Anion exchanger	Polar	Sulfonamides
Both	Ion pairing	Bonded C_{18}	Polar	Many drugs
Cationic	Ion exchange	Cation exchangers	Polar	Catecholamines
Molecular weight >2000	Size exclusion	LiChrospher Microgel	Polar	Proteins, polymers

ii. Injection Systems These are much simpler than for GLC and usually involve a loop system where the extract, often dissolved in mobile phase, is injected into a loop isolated from the column. Rotation of a valve allows the extract to be efficiently swept onto the column for separation. Loops are usually of a fixed volume (20–500 μl), and hence larger volumes can be handled than by GLC. Again it is much easier to automate such a system.

iii. Detectors A range of different detectors are available some now approaching the sensitivity of GLC detectors.

(*a*) *UV Detection* This is the nearest we can get to a universal detection system for drug analysis with HPLC because most drugs exhibit some UV absorption even if only at low wavelengths. Detectors can be variable or fixed wavelength, the latter being obviously more versatile if a range of different drugs is to be measured. Under suitable circumstances, these detectors can measure drugs in blood down to about 50 ng/ml. Diode array systems are a recent development and allow the recording of a complete UV spectrum of a compound as it emerges from the column. This is particularly useful where identification is important (e.g., in drug metabolism or screening). It can also be used to monitor the column effluent at different wavelengths simultaneously and is useful to confirm whether a peak consists of a single component.

(*b*) *Fluorescence* (*FLUOR*) *Detection* This is more sensitive than UV, but fewer drugs possess a natural fluorescence. However, chemical derivatization can allow conversion to strongly fluorescent compounds (see Section III,iv). The ability to vary the excitation and emission wavelengths also makes the technique more selective, and low drug concentrations in plasma can be measured (5 ng/ml or less).

(*c*) *Electrochemical* (*EC*) *Detection* This can offer a sensitive detection system, and it has become increasingly popular in the past 10 years or so. Most detectors are electrolytic and depend on the oxidation or reduction of the drug to generate a current. Because of the electrochemical nature of the system, mobile phases have to be capable of dissolving suitable electrolytes, and hence their use is largely restricted to aqueous systems and reversed-phase operation.

iv. HPLC Derivatization In HPLC, volatility is not important, and here derivatization is aimed at coupling strongly UV-absorbing or fluorescent groups to drug molecules to allow them to be determined at low concentration in body fluids. Reaction conditions of temperature, time, solvent, and concentration have all to be carefully optimized. Reactions may be carried out before chromatography or as the compound emerges from the column. The latter approach is called *post–column derivatization*, and although this reduces the chances of decomposition, it restricts the choice of derivative.

c. Supercritical Fluid Chromatography This is a relatively recent development that is claimed to combine some of the best features of GLC and HPLC. The main advantage is that a supercritical fluid possesses the solvating properties of a true liquid but is much less viscous, and more efficient separations can be obtained using open tubular columns or conventional HPLC columns. The polarity of the mobile phase can be altered simply by changing the pressure, and the technique is highly compatible with FID. At present it works better with nonpolar than polar compounds because of the limited choice of mobile phases (e.g., carbon dioxide and nitrous oxide).

d. Thin-Layer Chromatography This is a versatile technique and, even in its most inexpensive form, can provide a great deal of useful information regarding the presence of contaminants in dosage forms and the structure of metabolites and as a simple means of drug screening. TLC can be regarded as two-dimensional HPLC, and TLC systems are often used in the early stages to develop conditions for a new HPLC separation.

The plates themselves are made of glass, plastic, or aluminum and can be normal phase (silica gel), cellulose, or reversed-phase using hydrocarbons 8–18 carbon atoms long. Plates can be obtained with a fluorophore added, which allows UV-absorbing spots to be readily detected. A wide range of spray reagents are available to allow different chemical groups to be detected (e.g., phenols, amines, and carboxylic acids).

TLC has the advantage that spots can be scraped off the plate and the compounds eluted for study by some other technique.

e. High-Performance Thin-Layer Chromatography This can be a useful means of drug analysis. For the best results, the sample "spot" size must be as small as possible, and this can rarely be achieved by hand. Automated sample application is not expensive, and volumes of as much as 50 μl can be applied.

Development is carried out usually in a glass tank, and the choice of stationary and mobile phase is carried out as in HPLC. The volume of the vapor phase can be reduced by using specially designed tanks, and in overpressure layer chromatography (OPC), linear development takes place with the minimum of mobile phase. Circular development (CTLC) is also possible, and gradient elution has also been used. Once the plate has been developed, the drugs can usually be quantitated by scanning densitometry in the UV or fluorescence mode.

f. Combination Techniques A number of so-called hyphenated techniques are used to provide powerful analytical instruments such as GC-MS (mass spectrometry), GC-IR, GC-IR-MS, LC-MS, and LC-FID. GC-MS is used mainly to obtain structural information on a drug or its metabolites or as a selective analytical instrument when set to detect a single mass ion. The most useful combination with HPLC is designed to allow FID to be used. Not unexpectedly, the combination of a technique that uses a liquid mobile phase with a flame detector has proved difficult, but instruments are now available and, as we have seen, SFC provides another approach to the problem. The combination of HPLC with the most sensitive form of IR, Fourier Transform, is able to provide good structural information, but mobile phases containing water or other polar solvents cannot be used, and hence reversed-phase chromatography is impossible.

3. Separation of Optical Isomers

Many drugs contain at least one optically active center and can, therefore, give rise to compounds of the same composition (isomer) that rotate the plane of polarization of light either to the right or to the left. Most synthetic methods produce racemic mixtures containing equal amounts of both isomers. Usually only one of the isomers possessed the desired pharmacological activity, but little attempt is made to separate the two because the other isomer is usually regarded as inert. However, it is now appreciated that in some cases the "inert" isomer may contribute substantially to unwanted side effects, and its kinetics may be different from the other isomer. Resolution and analysis of optical isomers can be achieved by HPLC, TLC, and even GLC, and this is an active area of research.

C. Biological Techniques

1. Immunoassays

These are used for the rapid determination of drugs in biological fluids, and the principle behind their application is simple. Many foreign substances are antigenic and, when introduced into animals, will generate an antibody (Ab) response, and the antibody produced will bind the foreign material. In practice most drugs have molecular weights too low to generate a significant antibody response and first have to be coupled with larger molecules such as albumin, the albumin used being from a species other than that in which the antibody is to be raised. The way in which the drug is coupled to the albumin (i.e., the 3-dimensional stereochemistry of the protruding drug molecule) will determine the specificity of the antibody produced (i.e., whether closely related compounds will also bind to the antibody).

The size of the animal chosen will generally govern the amount of antibody produced, and rabbits and horses are frequently used. Once the antibody has been isolated and purified, a dilution curve is prepared to find the antibody dilution that will respond most sensitively to the drug concentrations likely to be present in the specimens to be assayed.

How immunoassays work can be seen from the following simple example. If it is desired to measure drug *A* in patients, a small amount of the drug is prepared in a modified form, *A**, which may be radiolabeled, fluorescence-labeled, etc. A known amount of *A** is added to the patient plasma sample, and after mixing, the antibody to *A* is added, when the following reaction takes place:

$$A + \underset{\text{free}}{A^*} + Ab \rightleftharpoons A\text{-}Ab + \underset{\text{bound}}{A^*\text{—}Ab}$$

in which the unmodified drug (*A*) competes with *A** for binding to the antibody. By separating the free from the bound fraction and estimating the amount of *A** free, the amount of *A* in the original sample can be calculated. If the drug is labeled with an enzyme, fluorophore, or chemiluminescent substrate and binding to the antibody produces a change in property that can be detected without separation of bound from free, the assay is called *homogeneous*. This is obviously useful because little sample treatment is needed. If the label is in a form that produces no change in property (e.g., radiolabeling), a separation of bound from free is necessary, and the assay is called *heterogeneous*. Separation can be carried out in various ways. [*See* IMMUNOASSAYS, NONRADIONUCLEOTIDE.]

a. Radioimmunoassay A good example of the usefulness of immunoassays can be illustrated by the cardio-active agent, digoxin, a powerful drug given in low doses. The clinical signs of under- and overdosing are not easy to distinguish, and therapeutic monitoring is necessary. Digoxin has a high molecular weight and is unsuitable for GLC. It also does not have a strong enough UV absorption for HPLC. It is routinely and rapidly monitored by radioimmunoassay. Three isotopes are commonly used for radiolabeling. Their properties are compared in Table VI. Chemical synthesis can supply ^{3}H and ^{14}C-labeling of most drugs, but few contain iodine. Techniques exist for iodination of such drugs. These techniques, however, may modify the properties of the drug because of the introduction of such a large atom. [*See* RADIOIMMUNOASSAYS.]

b. Optical Immunoassays In this approach the reagents used are stable, and most assays are homogeneous. In enzyme-linked immunoassays, the drug to be determined is linked to an enzyme such as peroxidase or glucose-6-phosphate dehydrogenase. The sample volume required is generally small (10–50 μl), and all the materials needed for the assay are supplied commercially in kit form. Fluoroimmunoassays are useful and their sensitivity can be controlled by the choice of fluorophore. Fluorescein is the most popular, although others have also been used. Luminescence requires the drug to be linked to a substance that emits light and, because light is not introduced into the system, measurement is carried out in a luminometer in the dark, and increased sensitivity can be obtained. Luminol is frequently used as the label, and because luminescence is generally short-lived (about 500 ns), rapid mixing and signal integration is essential.

c. Advantages and Disadvantages of Immunoassays Immunoassays are rapid, and many hundreds if not thousands of patient specimens can be processed daily. This is far in excess of what can be achieved by any of the chromatographic methods, even if fully automated. They require relatively little equipment, especially the optical immunoassays, and can be set up in outpatient clinics or adjacent to hospital wards to provide immediate results. Their main disadvantage is that they are designed to

TABLE VI Properties of Isotopes Used for Immunoassays

Isotope	Symbol	Half-life	Radiation	Energy (KeV)	Counter	Advantages	Disadvantages
Tritium	^{3}H	12.3 years	beta	18	Liquid scintillation	Long half-life, low hazard	Counting expensive
Carbon	^{14}C	5,730 years	beta	155	Liquid scintillation	Long half-life, low hazard	Counting expensive
Iodine	^{125}I	60 days	X- and gamma	27	Crystal scintillation	No extraction, high specific activity	Iodination may alter binding

provide information on only one drug or a single group of drugs, unlike the chromatographic or spectroscopic systems that can easily be modified to look at completely different compounds. Price is also important, and the kits are expensive but bulk buying can bring down unit costs. However, because immunoassays are generally supplied commercially, kits are only available for drugs for which there is a demand, and new or experimental compounds have to be determined by other, usually chromatographic means.

2. Microbiological Assay

This is a simple and useful method of determining antibiotics in body fluids. In essence, agar plates containing a selected microorganism are prepared, and diluted specimens of patient plasma containing the antibiotic to be determined are introduced into wells cut in the plates. After a suitable incubation period, the area of microbial growth inhibition around each well is measured and compared with the areas around wells on the same plate containing standard concentrations of antibiotic. Much of the early work on antibiotic pharmacokinetics was carried out using this technique, but its main disadvantage is that only microbial activity is measured. If the antibiotic is metabolized and any of the metabolites possess similar activity, falsely high results will be obtained. If patients are also receiving other antibiotics, as frequently happens, these may also produce an inhibition in growth of the chosen organism.

For these reasons, although the method is still used in certain circumstances, it has been superseded largely by HPLC.

IV. Quality Control

No matter which method is chosen, it is important to ensure that it functions properly, with satisfactory precision and accuracy, and gives the "correct" result. Every method involves the preparation of some type of calibration curve with standards of known concentration, and in each batch processed, several other samples can be incorporated containing known amounts of drug covering the anticipated concentration range. These are "quality control" specimens and are not used in the construction of the calibration curve. Monitoring the day-to-day results for these samples can give important information on how a method is performing. Samples may also be reanalyzed at frequent intervals to give repeat analysis checks. Important as it is, intralaboratory control is not enough. Interlaboratory comparison is more demanding but ultimately more rewarding in terms of raised standards and increased operator confidence. A number of national and international schemes exist for drugs that are monitored regularly for therapeutic purposes (e.g., digoxin, diphenylhydantoin, theophylline). At regular intervals, laboratories are sent freeze-dried plasma samples containing relevant drugs, and these are analyzed in the same way as normal patient samples. The results are collected by a central body, and each laboratory receives a document showing how it performed in relation to the other participating laboratories. These data are often supplied in histogram form, and the identities of the other laboratories are not revealed. This type of approach is useful as an independent check on performance, and it rapidly identifies laboratories using unsatisfactory methods. For example, the determination of many anticonvulsant drugs by GLC was clearly shown to be inferior to HPLC by external quality control.

Bibliography

Aszalos, A., ed. (1986). "Modern Analysis of Antibiotics." Dekker, New York.

Gennaro, A. R., ed. (1990). "Remington's Pharmaceutical Sciences." Mack Publishing, Easton, Pennsylvania.

Grob, R. L. (1985). "Modern Practice of Gas Chromatography." Wiley, Chichester, United Kingdom.

Jack, D. B. (1984). "Drug Analysis by Gas Chromatography." Academic Press, Orlando, Florida.

Jack, D. B. (1990). Chemical analysis. *In* "Comprehensive Medicinal Chemistry," vol. 5. (J. Taylor, ed.). Pergamon Press, Oxford.

McDowall, R. D., Pearce, J. C., and Murkitt, G. S. (1986). Liquid–solid sample preparation in drug analysis. *J. Pharm. Biomed. Anal.* **4,** 3–21.

Moffat, A. C., ed. (1986). "Clarke's Isolation and Identification of Drugs," 2nd ed. Pharmaceutical Press, London.

Munson, J. W., ed. (1981). "Pharmaceutical Analysis Modern Methods." Dekker, New York.

Poole, C. F., and Schuette, S. A. (1984). "Contemporary Practice of Chromatography." Elsevier, Amsterdam.

Souter, R. W. (1985). "Chromatographic Separation of Stereoisomers." CRC Press, Boca Raton, Florida.

Wong, S. H. Y., ed. (1985). "Therapeutic Drug Monitoring and Toxicology by Liquid Chromatography." Dekker, New York.

Drugs of Abuse and Alcohol Interactions

MONIQUE C. BRAUDE, *Director, Graduate Women in Science (previously with the National Institute on Drug Abuse)*

Glossary

Antagonism Mechanism by which a drug inhibits the action of another drug

Metabolism Drug metabolism is a general term applied to the chemical processes taking place in living tissues after the administration of a drug that alter its original chemical composition

Synergism Quality of two drugs whereby their combined effects are greater than the algebraic sum of their individual effects

INTERACTIONS between alcohol and other drugs can result in synergism or antagonism. A direct interaction is the result of the pharmacologic effects of the drugs. Indirect effects include pharmacokinetic interactions (in which a drug alters the absorption, distribution, metabolism, or excretion of another) and tolerance phenomena, in which a drug alters the response of the target tissue to itself or another drug. Tolerance to alcohol has been reported after repeated use. Cross tolerance between alcohol and other drugs occurs when the physiologic changes induced by alcohol, for instance, carry over to another drug (e.g., a barbiturate) and the observed effect of the second drug is diminished. Interactions between alcohol and other drugs can usually be understood in terms of one or more of these phenomena.

The important consequences of psychotropic drug interactions for an individual and society include the following:

1) Unexpected degree of impairment in performing daily tasks associated with a serious potential for injury (driving a car, operating machinery).
2) Accidental death from overdose.
3) Increase in behavioral toxicity (cognitive functions, memory impairment, etc.).

The concomitant oral ingestion of alcohol and other drugs could theoretically affect the absorption of these drugs, but information is mostly lacking in this area and the existing evidence does not point to clinically significant effects. Alcohol in the body is primarily metabolized by the liver alcohol dehydrogenase to yield acetaldehyde. Research in the past 10 years also indicates that acute and chronic administration of alcohol can inhibit the biotransformation of many drugs that are normally degraded by liver enzymes. The most prominent result of this alcohol-induced inhibition of enzymes *in vivo* appears to be a prolongation of the plasma life of drugs such as pentobarbital, meprobamate, and amphetamines—but not phencyclidine (PCP). The amount of alcohol use and the direction of its use during a long or short period are also important factors in drug interactions.

Whereas acute alcohol intoxication usually inhibits the biotransformation of drugs, the long-term ingestion of alcohol can lead to hepatic microsomal enzyme induction and can produce enhancement of the formation of intermediate metabolites that may be toxic to the liver, thus increasing the hepatotoxic effect of alcohol. Such appears to be the case with cocaine. Finally, alcohol can influence the elimination of other drugs indirectly by causing hepatic dysfunction and/or nutritional deficiencies.

I. Interactions of Alcohol With Central Nervous System (CNS) Depressants and Benzodiazepines

Interactive effects between alcohol and CNS depressants such as barbiturates, benzodiazepines (e.g., valium), and methaqualone are frequent. In terms of CNS depression, use of these drugs with alcohol is usually additive. With drugs that are metabolized by the liver enzymes (e.g., phenobarbital and pentobarbital), synergism has been reported, leading to increased toxicity and death in humans.

Benzodiazepines have now significantly replaced barbiturates as anxiolytics and sedative hypnotics.

A. Alcohol and Opiates

Concomitant alcohol and narcotic abuse is common. More than half of the ''overdose'' cases with heroin or methadone are, in fact, cases in which concomitant abuse of alcohol has played a prominent role.

Several studies have shown that alcohol may have a biphasic effect on the disposition of another opioid. During chronic use of large quantities of alcohol and at times when high blood levels of alcohol are present, alcohol may inhibit the metabolism of other drugs such as methadone. But when alcohol is no longer present in the body, drug metabolism may be accelerated because, in the chronic alcoholic, alcohol has produced an enhancement of the liver detoxifying enzyme system. The presence of chronic alcoholic-induced liver disease may result in significant alterations in methadone disposition and prevent achievement of the steady state during methadone treatment. An interesting possible interaction between the opiate antagonist, naloxone, and ethanol gave promise for using an opiate antagonist to alleviate the ethanol-induced performance impairment. However, some studies using the opiate antagonist Naltrexone to assess the efficacy of this compound in attenuating acute alcohol intoxication in humans failed to find any significant differences between reported subjective levels of intoxication and detection of alcohol effects or blood-alcohol levels after Naltrexone compared with placebo administration.

B. Alcohol and Cannabinoids

Marijuana and alcohol are ubiquitous. The combined use of alcohol and marijuana in variable amounts, frequencies, and settings has become a well-established fact. Studies in rats showed that simultaneous administration of tetrahydrocannabinol (THC) and alcohol resulted in an increased rate and magnitude of alcohol tolerance and physical dependence. Complete alcohol tolerance was established within 12–16 days in animals receiving both drugs, whereas only minimal alcohol tolerance or cross-tolerance was detected in alcohol- or THC-treated rats, respectively, at this time. [*See* MARIJUANA AND CANNABINOIDS.]

A few studies in humans have indicated that the acute administration of THC, the primary psychoactive component of marijuana, in combination with alcohol results in enhanced impairment of physical and mental performance, more marked subjective effects such as pupil size, and a greater increase in pulse rate and conjunctival congestion compared with ingestion of either drug alone. However, the magnitude of the enhancement of the drug effects was not great.

More recent studies, using a battery of cognitive, perceptual, and motor function tests, also found that both THC and alcohol produce significant decrements in the general performance factors, but there was no evidence of any interaction between THC and alcohol, and the effects of a combination of both compounds were no more than additive. The same was true for cannabidiol (CBD). CBD did not produce any demonstrable effects either alone or in combination with alcohol. However, as CBD is one of the major components of some marijuana plants, there is still the possibility that there may be a THC–CBD–alcohol interaction. The combination of alcohol with CBD in humans may result in significantly lower blood-alcohol levels compared with alcohol given alone. However, there are few differences between the pharmacological effects of alcohol given alone or given with CBD.

In conclusion, it seems that there is no more than an additive effect between the cannabinoids and alcohol (both acting as CNS depressants). However, this may be dangerous in pregnancy as it was shown recently that alcohol produces a 100% enhancement of marijuana-induced fetotoxicity in rodents. This is especially significant in view of a study on concordant alcohol and marijuana use in women, showing strong correlations between alcohol and marijuana use. The heavy marijuana smokers reported drinking significantly more alcohol than the light marijuana smokers. For pregnant women who combine the use of alcohol and marijuana, this suggests a potential danger that may be far greater than that associated with using either drug alone.

C. Benzodiazepines and Alcohol

The most extensively studied drug interactions have been those of benzodiazepines and alcohol. Alcohol and benzodiazepines interact both at the pharmacokinetic and pharmacodynamic levels in a predictable manner qualitatively or quantitatively by enhancing each other's effects. However, benzodiazepines do not seem to potentiate the effects of alcohol as much as other sedatives do. [*See* PHARMACOKINETICS.]

Kinetically, acute doses of alcohol impair the disposition of benzodiazepines that are metabolized by demethylation or hydroxylation. For instance, it increases the blood levels of orally administered diazepam but not of oxazepam, which undergoes glucuronide conjugation. Chronic alcohol ingestion, however, increases the clearance of benzodiazepines that are demethylated or hydroxylated.

In general, alcohol and benzodiazepines when studied alone produce sedation and impair motor coordination, reaction time, memory acquisition, retention, and recall in a dose-related manner. Many studies indicate that the acute combination of both drugs produces the same impairment at a lower dose than given separately or reveals an impairment not apparent with the control dose of the drug alone. In this enhancing effect, alcohol appears to be the dominant partner. Although alcohol–benzodiazepine interactions may be less important than those involving alcohol and other psychotropic drugs (e.g., cannabinoids, neuroleptics, stimulants, and antidepressants), Chan points out that this combination is associated with drug-induced deaths, drug overdoses, traffic accidents, and fatalities.

Interestingly, a major clinical use of benzodiazepines has been in the short- and long-term treatment of alcoholics as benzodiazepines were found effective in alleviating alcohol-withdrawal symptoms. This is now controversial, as some studies have pointed out that alcoholics may be more at risk of developing benzodiazepine or alcohol–benzodiazepine dependence than the general population. There is therefore a dire need for large-scale controlled studies concerning the efficacy of benzodiazepines in the long-term treatment of alcoholics.

II. Combined Abuse of Alcohol and Stimulants: Amphetamines and Cocaine

Surprisingly, few studies have dealt with the interaction of alcohol and amphetamines in humans despite the widespread use of both. Most of the studies that have investigated the effects of combined alcohol and amphetamine use have used either animal or volunteer subjects rather than members of the subculture directly involved in such abuse. Although synergism between alcohol and amphetamines was demonstrated in some preclinical studies, a large majority have reported antagonism between the two. The clinical evaluation of patients for treatment of alcohol abuse shows that their clinical histories indicate the existence of two types of abuse patterns, one primarily concerned with the effects of amphetamine and the other with the effects of alcohol; the latter type of patients use amphetamine only to help them maintain a wakeful state.

A naturally occurring alkaloid, cocaine is a powerful and rapid CNS stimulant of short duration, which has recently become a prominent and favored drug of abuse. Cocaine is rapidly degraded in the body by various enzymes. Pretreatment with alcohol for 5 days in rodents at first reduces the immediate mortality rate produced by a large dose of cocaine, probably through a stimulation of the enzymes that speed up the degradation of cocaine. But this process can also produce greater amounts of intermediate metabolites that are toxic to the liver, delaying mortality (1–7 days after cocaine ingestion). These observations suggest that the early toxicity of cocaine is due to the parent drug and that ethanol pretreatment provides protection by enhancing the elimination of cocaine. However, the more rapid degradation of cocaine produces greater concentrations of norcocaine, an *N*-demethylated metabolite that is more toxic than cocaine. [*See* COCAINE AND STIMULANT ADDICTION.]

III. Conclusion

In summary, most of the abused compounds interact with alcohol in a predictable manner. What is important to remember is that the pharmacokinetic and pharmacodynamic interactions of a given drug with alcohol is in part dependent on the length of the individual exposure to alcohol. The end response may be enhanced in the acutely intoxicated individual who does not have a history of alcohol abuse. Conversely, it may be diminished in the otherwise physically healthy but alcohol-dependent individual. As mentioned with marijuana, the combined use of alcohol and abused substances in pregnancy requires further attention as it may be hazardous to the fetus and the offspring.

In general, appreciation of alcohol–drug interactions should provide a useful basis for the management of alcohol patients and the prevention of accidental overdoses.

Bibliography

Abel, E. L. (1985). Alcohol enhancement of marijuana induced fetotoxicity. *Teratology* **31,** 35–40.

Belgrave, B. E., Bird, K. D., Schesher, J. B. (1979). The effect of delta-9-tetrahydrocannabinol, alone and in combination with ethanol on human performance. *Psychopharmacology* **62,** 53–60.

Braude, M. C. (1986). Interactions of alcohol and drugs of abuse. *Psychopharmacol. Bull.* **22**(3), 717–721.

Braude, M. C., and Ginzburg, H. M., eds. (1986). Strategies for research on the interactions of drugs of abuse. *Natl. Inst. Drug Abuse Res. Monogr. Ser.* **68.**

Chan, A. W. K. (1984). Effects of combined alcohol and benzodiazepines. A review. *Drug Alcohol Depend.* **13,** 315–341.

Kipperman, A., and Fine, E. W. (1974). The combined use of alcohol and amphetamines. *Am. J. Psychiat.* **131**(11), 1277–1280.

Kreek, M. J., and Stimmel, B. (1984). Dual addiction—Pharmacological issues in the treatment of concomitant alcoholism and drug abuse. *Adv. Alcohol Subst. Abuse* **3**(4), 1–6.

Lex, B. W., Griffin, M. L., and Mello, N. K. et al. (1986). Concordant alcohol and marijuana use in women. *Alcohol* **3**(3), 193–200.

Mello, N. K., Mendelson, J. H., and Kuehnle, J. C. (1978). Human polydrug use: Marijuana and alcohol. *J. Pharmacol. Exp. Ther.* **207**(3), 922–935.

Muhoberac, B. B., Roberts, R. K., and Hoyumpa, A. M. (1984). Mechanism(s) of ethanol–drug interactions. *Alcoholism: Clin. Exp. Res.* **8**(6), 583–593.

Sellers, E. M., and Busto, V. (1982). Benzodiazepines and ethanol: Assessment of the effects and consequences of psychotropic drug interactions. *J. Clin. Psychopharmacol.* **2**(4), 249–262.

Smith, A. C., Freeman, R. W., and Harbison, R. D. (1981). Ethanol enhancement of cocaine induced hepatotoxicity. *Biochem. Pharmacol.* **30**(5), 453–458.

Dyslexia–Dysgraphia

ANNELIESE A. PONTIUS, *Harvard Medical School*

Glossary

Agnosias Disorders of recognition, associated with inaccurate representation in memory and faulty processing of accurately perceived sense impressions

Alpha rhythm A type of electroencephalographic recording that reflects brain activity of fast rhythmical oscillations in the electrical potential at the rate of 8 to 12 cycles per second, at a voltage of about 40–50 microvolts on average. It occurs in the normally inattentive brain at rest, as in drowsiness, light or hypnagogic sleep, narcosis, or when the eyes are closed, or even when open, if there is no intention to perceive, as in the initial stages of some meditation

Dyslexia "A disorder in children, who, despite conventional classroom experience, fail to attain the language skills of reading, writing, and spelling commensurate with their intellectual abilities." (World Federation of Neurology)

Ecological dyslexia In addition to essential criteria of dyslexia, large-scale ecologically determined severe underexposure and underuse of certain precursors to written language are factors.

Grapheme The product of the recoded phoneme into a visual sign, shaped with regard to the spatial-relational properties of the elements of a word.

Kimura figures-instruction A task requiring one to remember a set of six abstract figures presented on index cards

Phoneme The distinct, discrete unit of speech sounds, composing the spoken word

Recognition Cognitive function, viewed as a second step following encoding in memory (representation)

Representation Cognitive function with high-level cross- and supra-modal integrative processing (including memory); distinct from the primary level of modality specific perception

Theta rhythm A type of electroencephalographic recording reflecting brain activity, that consists of slow waves (4 to 8 cycles per second) at about 60 to 100 microvolts. It is produced, e.g., during a progressed stage of meditation and dreaming (REM sleep).

THE COGNITIVE COMPLEXITY of written language, its late acquisition (not yet attained by one-third of the world's population), and the range of individual distinctiveness of its disorders have remained a challenge over a century. The here-employed classical categorization, mainly meant as a heuristic device, is based on anatomico-clinical findings and is selected out of many other classificatory systems.

The operational definition of dyslexia (see Glossary) excludes the following as possible causal factors in dyslexia: psycho-socioeducational influences, autism, mental retardation, and severe sensory loss. Thus, by strict neurological definition, dyslexics have normal visual and auditory acuity. Presently, about 3.5–15% of otherwise well-developed North American children are labeled "dyslexic." At least 10–15% of adults are said to be mildly or functionally dyslexic.

This overview aims to alert researchers, clinicians and educators to the wide range of potential deficits implicating specific brain system dysfunctioning (see Section II). This can be due to lesions, immaturity or severe ecologically determined lack of exposure and practice (see Sections II,C,3,b and III,B,7). Thus, the emphasis is not on statistics, addressing frequency and degree of dysfunction, nor on specificity within a given modality. Rather, the goal is to present a neuro-functionally based ordering system to differentiate among the large diversity of specific symptoms and their typical grouping in syndromes. The awareness of such association among symptoms is of heuristic value, inviting the search for specific, potentially linked additional dysfunctions. Such diagnostic differentiation is necessary for theoretical reasons, as well as for specific, finely-honed remediation.

I. Hypotheses on Etiological Factors in "Developmental" and/or Brain Damage-Related "Acquired" Dyslexia–Dysgraphia

A. Familial Predisposition and Sex Difference

Dyslexia tends to be familial and is claimed to affect males three to four times as often as females. There is also a higher incidence in left-handers. A hypothesized sex-linked recessive genetic transmission, however, is not supported by available evidence, as is also the case in other forms of genetic inheritance (polygenetic or autosomal). Thus far, no single genetic model fits the data.

In addition, hormonal hypotheses have remained unconfirmed, positing testosterone interference with the usual lateralization of language functions in the left hemisphere, thereby contributing to dyslexia. [*See* CEREBRAL SPECIALIZATION.]

B. Subtle Brain Dysfunction (Pre-, Peri-, and/or Postnatal Onset)

1. Anatomico-Biological Factors

Cytoarchitectonic (cell assembly) abnormalities were reported in the brains of five male dyslexics. In addition, in all five cases, the right planum temporale failed to show the usual reduction in size seen in approximately 65% of normal brains. This lack of asymmetry and the cytoarchitectonic changes were judged to be significant in the pathogenesis of the "reading disorder."

Strict right-handedness, however, may be difficult to determine and exists only in 30% of a population. Cytoarchitectonic abnormalities of the sort seen in the dyslexic brains are only rarely seen in normal brains. When they are present, they are few in number and their distribution is different from that in the dyslexic case.

2. Neurochemical Factors

Thus far, neurochemical factors, such as a suggested reduction in certain neurotransmitters (dopamine and serotonin), have not been found to be specific for dyslexia but can pertain to learning disabilities in general.

3. Electrophysiological Factors

a. Auditory Event-Related Potentials Some evidence indicates that dyslexics may utilize the left hemisphere for tasks that normal readers accomplish with their right hemisphere. Activation of the left hemisphere occurs in some "disabled readers" during ongoing, relatively low-level reading-related cognitive tasks that employ "visual" components: letter-sound and letter-shape processing. Such a different use of neuronal systems is speculated to either reflect a deficit in right (rather than left) hemispheric functioning or in right–left hemispheric integration.

b. Brain Electrical Activity Mapping Recently, interest has switched from studying right–left differences against within-hemispheric (anterior–posterior) differences. Using the brain electrical activity mapping (BEAM) technique, one finds in dyslexics (1) increased alpha activity in both frontal lobes (left somewhat more than right). (Behaviorally, such findings can be viewed as a lowering of the level of frontal lobe-mediated attention in dyslexia which in turn is frequently found linked with attention deficit disorder, also implicating frontal lobe system immaturity or other dysfunctioning. (2) Further, there is a slight difference in theta activity in two tests. (i) The Kimura figures instruction task showed difference in theta activity in the left antero-(lateral)-frontal regions. (Note, however, that this task may tap especially interobject spatial relations rather than the intraobject relations, preferentially used for written language processes.) (ii) In a silent reading test, a difference in theta activity was found in the left midtemporal region and bilateral medial frontal region (left side more than right side).

With eye closing, such a small difference was also evident in the right parietal area.

C. Maturational Lag

Children with a developmental delay in attaining any of Luria's following phases of reading and/or writing skills (e.g., those with attention deficit disorder) (see Section I,B,3,b) can be expected to outgrow their dyslexic difficulties more readily than children with a certain implied brain damage, or subject to severe and long-lasting underuse in extreme ecological context (see Section III, B).

Step 1: Voluntary organized activity requires the conscious analysis (and synthesis) of the component sounds to be produced in a smooth sequence. This process requires the analysis of the phonetic complex, to detect its components, the phonemes.

A maturational lag at this level is analogous to disturbances of the linguistic (''acoustic'' ''phonetic'') components (see Section II,B) of written language.

Step 2: Graphemes, the optic structures corresponding to phonemes, have visuo-spatial properties, defined by visuo-spatial representation. These are specific for each grapheme and are required for encoding (writing) and for decoding (reading). This visuo-spatial representation requires complex spatial analysis and synthesis, considering the components' orientation (up–down, right–left), proportions, and ratios. If such intrapattern spatial representation (including memory representation) is inaccurate (e.g., too globalized), mistakes occur: such as erroneously reading and/or writing as if p = q = d = b; was = saw; from = form; N =Z; e = l; O = Q. Developmental delay at this level is analogous to disturbances of mainly spatial (''visual'') components of written language (see Section II,C).

Step 3: Graphemes are recoded into a system of motor acts step by step, until the process becomes increasingly more automatic with improved writing skills. Delays at this phase are analogous to disturbances of the motor components at various levels of integration (see Section II,A), particularly those mediated by the prefrontal lobe system: There are difficulties in sentence sequencing and in flexibility of planning strategies and of applying the rules of syntax. (Given that the prefrontal system is a supramodel one, its functioning also overlaps with that of other such systems (e.g., with spatial and ''quasi-spatial'' processing by the angular gyrus) (see Section II,C,3).

II. Anatomico-Functional Components of Dyslexia and Dysgraphia

Cognitive functions can be disturbed in various ways. Presently, the focus will be on certain probable sites of brain lesions or dysfunctions (implying anatomical and/or electrochemical changes).

Neuropathological constraints (based on lesion-acquired dyslexia and dysphasia) are here utilized as a structuring device to implicate specific brain areas or systems potentially dysfunctional to a milder degree in developmental dyslexia and dysgraphia.

In such structuring, the following principles pertain: (1) A lesion may not result in a complete, but only partial, loss of function, and a (compensatory) adaptation of the rest of the brain may occur, especially in young persons. (2) Each site contributes a specific quality to the ensuing functional disturbance. (3) Typically, entire specific brain systems are involved, and dysfunctions are overlapping, forming (specific) groupings (''syndromes''). Principles 2 and 3 have a heuristic diagnostic value in focusing on the site(s) of brain lesion and, at times, suggest a certain etiology (vascular accident, trauma, tumor, metabolic, etc.). Frequently, a diagnosis becomes even more focused when additional specific nonlanguage-related symptoms of a known syndrome exist. (4) For each set of the main components contributing to a given ''higher'' (i.e., late-emerging) cognitive function, such as written language, three levels, or zones, of interrelating brain systems exist. (i) Primary (projection) cortex for sensory input and motor output. (Most language-related functions, however, require higher levels of mediation.) (ii) Secondary zones, or ''association cortices'' (including their connecting white matter fiber tracts). Here, cross-modal input is processed, and programs are prepared. (iii) Tertiary zones, characterized by their late anatomical maturation (not before the seventh year of life. Note that prefrontal and certain parieto-occipital brain systems require several decades). There is overlapping participation and supramodal integration of many cortical (and subcortical) areas, with an increasingly abstract, more general representation of what is perceived and mediated. Thus, language can be used by the blind and the deaf. Writing can occur with any medium (e.g., body parts other than the hand, sky-writing by air planes, skating, computers, making knots in strings, coded representations in various art forms). In the tertiary zones, information is coded and functionally organized. In the left hemisphere (language-dominant in right-handers), such organization occurs especially with the help of language, while the right hemisphere (nondominant, or minor, in right-handers) mediates especially non-language-related spatial aspects.

A. Motor Components and Related Dysfunctions

1. Primary Motor Cortex

Motor cortex (left lower third of the prerolandic gyrus), adjacent areas, and connections to the motor association cortex. (i) With lesions of the monosynaptic connections from the motor cortex to the brainstem motor neurons (cortico-bulbar tract), impaired speech movements ensue: dysarthria, with potential secondary spelling difficulties (compare Section II,A,1,v). (ii) Motor dysgraphia shows impaired elements of writing movements. (iii) Efferent (kinetic) dysgraphia (left inferior part of the premotor cortex) is characterized by disturbed phonetic analysis and synthesis of words, which may be combined with section II A,1,iv). (iv) Impaired kinesthetic tracing, a necessary feedback from the speech muscles is found in afferent motor dysphasia as well as concomitant disturbance of "silent writing" (left posterior part of the premotor cortex). (v) Imprecise articulation (left posterior sensorimotor area) can lead to secondary inaccurate spelling of consonants. (Note that such articulatory imprecision may impair spelling in conjunction with certain street dialects, or with certain Pidgin versions of English, such as "tok pisin" of Papua New Guinea). [*See* SPEECH AND LANGUAGE PATHOLOGY.]

2. Secondary Zone: Motor Association Cortex

a. Broca's area is in the inferior prerolandic region of the frontal operculum (with projections from the primary motor cortex and to the same area in the contralateral hemisphere), involving both speech and language.

b. Frontal operculum or its outflow, when lesioned: initially leads to nonfluent speech disturbance with a transient buccofacial apraxia and, at times, with a secondary mild writing disturbance.

c. Corpus callosum (and other transcortical tracts) connect the right-sided area homologous to the left motor association cortex. In turn, the motor association cortex receives a projection from the posterior sensory association cortex (also see Section II,B,2 and II,D,1,a).

d. Cerebellum: When lesioned, one finds (i) dysgraphia and/or (ii) dysarthria in the context of ataxia also involving other body parts (poor coordination of muscles, such as used for writing or speaking, respectively) (compare Section II,A,1,iv).

3. Tertiary Cortical Zone for Motor Integration

a. Prefrontal Lobe System mediates the "highest" (most integrated) level of the regulating speech and language. This level includes Luria's conception of "inner language," which is required for the organization of any mature action behavior. When lesioned, or relatively immature (see Section I,C) for a certain task one finds:

b. In general, (i) frontal dysgraphia in the context of abulia and overall disorganization of action; (ii) in left- or right-sided lesions: aspontaneity, "pathological inertia" (tendency not to apply oneself to a task) in the language-related realm, impulsive guessing when reading and inadequate verification of the results of one's actions); (iii) impairment of various reading strategies requiring prefrontal mediation for anticipation and flexibly appropriate switching between various strategies and rules of action, narrated in writing and/or regarding regular or irregular spelling; (iv) deficient "chunking" (phonological segmentation is necessary especially for the decoding of continuous written language and for noncontent (context-dependent) words that emphasize structure (prepositions, conjunctions, auxiliary verbs, articles, and the copula). (Such words, which are mastered last and are absent in certain near stone-age native languages within their ecological context pose the greatest difficulties for developmental as well as for acquired dyslexics); (v) poor search by eye movements for active analysis of most significant elements of complex sentences during reading; (vi) loss of order of the elements in spoken and written language; (vii) reduced retention of required sequences; and (viii) perseveration of concrete units or of the principle of activity.

c. Specifically, when the left prefrontal convexity is impaired, its cortico-subcortical connections or the most caudal (backward placed) part of the second frontal convolution, dysorthographia (spelling and graphic disturbances) ensue.

B. Linguistic (Verbal, Acoustic, Phonetic) Components

In general, disturbances of written language can occur secondary to those of spoken language. Therefore, only certain specific difficulties will be listed.

1. Primary Cortical Region for Auditory Input

This region connects to Heschl's gyrus in the left transverse temporal gyrus. Here, lesions are not primarily essential in dyslexia and/or dysgraphia (both dysfunctions by strict neurological definition (see Glossary) include no perceptual deficits, auditory, or otherwise).

2. Secondary Zone: Left Auditory Association Cortex

This zone includes posterior planum temporale (see Section I.B.1) and adjacent regions in the left superior temporal gyrus, and the caudal part of the left, first temporal convolution (Wernicke's area), projecting to the frontal motor association cortex.

a. In general, lesions involve the linguistic (acoustic, phonetic) components of reading and writing, typically constituting mild forms of the neuropathological syndrome of sensory dysphasia (type I Wernicke's dysphasia) with fluent but paraphasic speech: (i) poor phonetic synthesis and analysis with omission of sounds; (ii) substitution of sounds with closely related ones; which may have different phonetic properties. (Such disturbances become especially apparent when trying to write unfamiliar words on the basis of their phonetic analysis or when writing a series of words or phrases.)

b. Specifically, the following associations between lesions and deficits pertain: (iii) Wernicke's or adjacent areas: occasionally dissociation between auditory and written comprehension; (iv) posterior to Wernicke's area: dysphasia with relatively more impairment of reading; (v) posterior parts of the superior temporal gyrus (here, type I Wernicke's sensory dysphasia can be associated with relatively more severely disturbed auditory or reading comprehension, depending on more anterior or posterior lesion extension, respectively); (vi) left temporo-occipital junction: impairment of word retrieval and of short-term memory.(vii)

In mid-temporal lesions, the naming deficits involve audio-verbal "speech memory." (Compare deficits in corpus callosum lesions (see Sections II,A,2,c and D,1,a) isolating the right hemisphere from the speech area. Further deficits in left ventrolateral and pulvinar of the thalamus functioning can occur (see Section II,D,2,b).

3. Tertiary Cortical Zone for Linguistic (Verbal, Acoustic Phonetic) Integration

Inferior parietal lobe system comprises Brodmann areas 39 and 40; (same as Section II,C,3) and its surrounding structures, or parieto-(temporo)-occipital system (Brodmann areas 37 and 21; (same as Section II,C,3), especially angular gyrus (highest level of supramodal integration). Thus there exists an overlap with other tertiary zones and between the subregions, here outlined mainly for emphasis, not implying absolute distinction between them.

a. Left Parieto-Occipital System In lesions one finds amnestic dysphasia (compare Section II,C,3,h,v) with deficient nominative functions as well as impaired visuo-spatial representation of corresponding objects to be named upon confrontation. The following abilities may be deficient: classificatory semantic schemes, distinction of the characteristic features of a shape (e.g., letters, words), of a real object, or of one depicted in a stylized manner, and description of the details of the drawing of a shape or of its completion when the drawing was started by someone else. Naming with correct pronunciation is made immediately possible by prompting with the first sound or first syllable of the forgotten word. This attests to an intact audio-verbal memory with maintained acoustico-verbal traces (in contrast to lesions of the (mid)temporal region; see Section II,B,2,b,v,vii; and II,C,3,h,i).

b. Left Inferior Parietal Lobe or its Surrounding Structures When impaired, one finds transcortical sensory dysphasia (type II Wernicke's dysphasia). Lesions also interfere with the contact between the angular region (see Sections II,B,3,d and II,C,3,d) and the rest of the brain, isolating the speech area, and resulting in dyslexia with dysgraphia, an impairment of comprehension of oral and written language. By contrast, the Broca area, arcuate fasciculus, and Wernicke area proper are spared, enabling intact audio-phonatory transposition and repetition.

c. Left Inferior Parietal or Parieto-(Temporo)-Occipital System Impairment is linked with (i) poor assembling of consecutive series of sounds into a word to be read simultaneously aloud. (ii)

With loss of prepositions, there is impaired comprehension of *spoken* and written communication, which is reduced to that of events and of individual words as in ecological-evolutionary restrictions, reflected in certain near stone-age native languages (also see Section II,C,2,e,ii,3,g).

d. Left Angular Gyrus When partially lesioned, one finds type III Wernicke's dysphasia characterized by predominant disturbances in the expression and comprehension of written language (see Section II,C,3,d).

C. Spatial (Visual) Representational Components

1. Primary Cortical Zone for Visual Input

This zone (Brodmann area 17) connects to the calcarine cortex of the occipital lobes, that has no primarily essential role in dyslexia or dysgraphia, in which visual perception is intact, by strict neurological definition (see Glossary).

2. Secondary Cortical Zone: Visual Association Cortex

This zone (Brodmann areas 18 and 19) comprises the occipital lobes and the adjacent parieto-occipital-temporal junction. In this system, the significance of direct peripheral input decreases, and the contact with other related cortical regions increases. (It appears that here perception loses its importance at the expense of the more integrated process of "representation," see Glossary).

This cortex uses a method of specific stepwise feature discrimination, and at a higher level (compare Section II,C,3), there is high level discrimination by modality specific regions (e.g., concerning color and possibly specific spatial relations).

In visuo-spatial representation, a general, essential distinction is necessary: intraobject (within a pattern) versus interobject spatial relations. Thus, in preliterate populations with highly developed interobject spatial-relational abilities (required for rapid global assessment), there is typically underevolved intrapattern spatial relational discrimination, which only becomes essential for such skills as literacy. Supporting this distinction are experiments with monkeys which delineated two visual cortical subsystems, apparently analogous to those subserving humans' visual abilities.

a. Inferior Temporal Cortex This cortex is connected to the occipital visual cortex by the inferior longitudinal fasciculus (see Section II,D,1,b,iii). When lesioned in the posterior part of the inferior temporal cortex in monkeys or further posteriorly in humans, there is impairment of the intricate (intrapattern) visual discrimination found in a subtype of visuo-spatial dyslexia whereas damage to the anterior part interferes with visual memory.

By contrast, interobject spatial relations with distance discrimination are mediated by occipital-parietal mechanisms, including the superior longitudinal fasciculus (see Section II,D,1,b,ii). Typically, in dyslexia and/or dysgraphia interobject visuo-spatial discrimination is intact. Thus, when only determining this ability, the presence of intrapattern visuo-spatial dyslexia tends to be overlooked.

The following subsections add further specification, although more by emphasis than by strictly distinct localization, in that there is much overlap.

b. Parieto-Occipital System (Bilateral or Right Side) (i) Spatial (visual, optic) dyslexia and/or dysgraphia with poor representation of spatial relations, requiring accurate subtle interrelations, ratios and orientation of the features within a given configuration (e.g., letter, word, face); (ii) poor position analysis of letters or words; (iii) defective visual analysis and synthesis and disturbed general integration of graphemes and their visual(-spatial) distinctiveness; (iv) poor simultaneous representation of letters or words, impaired survey of the whole system of presented sounds and/or signs in a whole word or in a series of sounds or signs as a single letter or word normally to be "perceived" simultaneously (literal dyslexia or verbal dyslexia, respectively). (Thus, when a word is perceived only by its first one or two letters, reading proceeds letter by letter.) Specifically, the findings of Levine & Calvanio are as follows:

c. Left Temporo-Occipital lesions: Here performance depends on the span of letters: as letter naming is limited to single letters only, so is reading. Horizontal three-letter arrays cannot be read.

d. Occipital Pole (Mainly Right Hemisphere) and/or Posterior Splenium Hemi-dyslexia: horizontal (as opposed to vertical) arrays of letters can be read only on one side of the visual field, and writing is performed only in half of the space, all in the absence of visual field defects.

e. Temporo-Parietal-Occipital System (also see Section II,C,3,c) (i) Poor recoding of identified phonemes into graphemes with unstable optic-acoustic connections; (ii) poor assembling of consecutive series of sounds and/or written signs into a word to be read simultaneously (Section II,B,3,c,i).

3. Tertiary Cortical Zone of Visuo-Spatial Representation

Inferior parietal system (Brodmann areas 39 and 40, same as Section II,B,3) and/or parieto(temporo)occipital system (Brodmann areas 37 and 21; same as Section II,B,3), especially including angular gyrus. Such overlapping between these and other tertiary (at times also secondary) zones is due to supramodel integration. The following subparts also overlap and are listed only for emphasis, not as indicative of strict delineation.

In general, the parieto-occipital system mediates two kinds of spatial synthesis necessary to process written language: concrete spatial relations (pertaining to any object, shape, or depiction), as well as symbolic, quasispatial (Luria) relations (pertaining especially to language processes). In principle, one must consider that even in this tertiary zone of visuo-spatial processing, the left hemisphere (with the language area) is the most important one, because most visuo-spatial aspects of percepts in humans are actively influenced by language: Precise spatial organization of a shape (including letters and words) or of any complex object or its depiction makes it possible to code them into precise categories, not merely in a generic way, but as individual members of categories. Such a coding is necessary for later retrieval (recognition) of even ambiguous individual features.

a. Left (or Right) Parieto-Temporal-Occipital System *In general,* despite lesions, there are intact acoustic and articulary bases of writing, as well as phonetic analysis and synthesis. Impaired writing occurs in the recoding of the identified phonemes into graphemes. Note that despite left temporal lesions, there are no such writing disturbances with hieroglyphic writing systems, in which the characters tend to denote actual concepts, not requiring acoustic and literal word analysis. Thus, in Japanese Kanji the verbal-linguistic subtype of dyslexia–dysgraphia occurs very rarely, distinguishing it from the visuo-spatial subtype of dyslexia–dysgraphia which occurs in lesions of the occipital-parietal cortex for visuo-spatial analysis and synthesis.

Thus, there appears to be a difference in visuo-spatial representation between the processing of any object, shape, or depiction as against signs denoting language.

Specifically, in left (or right) parieto-temporal-occipital lesions, there is deficient mediation of concrete spatial relations in graphic representation: (i) constructional dyspraxia, deficient reproduction of a constructed shape, even when asymmetrical (e.g., letter, word) from its component elements, requiring an exact position in space (before–behind, right–left, up–down, as in b = d = p = q; Z = N); from becomes form and was becomes saw. (ii) "Optic or visual-spatial dysgraphia": mirror writing and/or disorganized copying of letters or drawings. Milder cases have difficulties in producing a shape from memory or to mentally rotate a shape. All these impairments are based on a disturbed ability to retain the required spatial relations, ratios and positions of the lines forming the letters, despite clear distinction between the sounds or phonemes to be written.

b. Left and/or Right Occipital-Parietal Area, Particularly Inferior Parietal Lobule When lesioned, one can find the Gerstmann syndrome, with several (though not necessarily all) of the following dysfunctions: finger agnosia, dyscalculia, spatial and particularly right–left disorientation, dyslexia with dysgraphia, autotopagnosia (nonrecognition of own body, despite normal perception). Also frequently present are constructional disturbances of match stick or block designs (e.g., Kohs or other block design test) and of drawing (e.g., of schematized faces and/or topographic amnesia, even when tested in a purely visual (nonverbal) manner. It is important to distinguish between the mostly intact interpattern spatial representation (e.g., orientation maps between objects in the environment) and a disturbed intraobject spatial representation (such as within written words or within depicted face patterns (Fig. 1).

In support of the specificity of several essential symptoms found in dysfunctioning systems due to brain pathology, it is of note that Pontius delineated *ecological syndromes.* These are analogous to syndromes of Gerstmann, of prosopagnosia (see Section II,C,3,e,h,iv, and Section III,B), and of color dysnomia, all with disturbances of written language functions in not brain-damaged, contemporary, near stone-age adult populations. Such findings corroborate a shared underlying brain "circuitry"

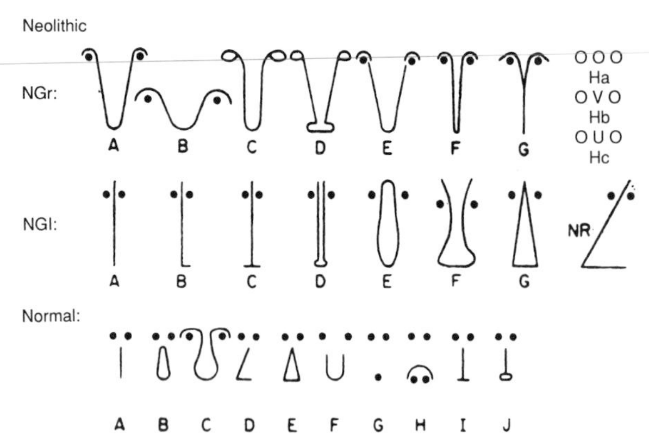

FIGURE 1 Frontal view of forehead, nose, eyes—sector of face patterns, measured with a ruler. "Neolithic face" patterns NGr and NGl: continuation of forehead into nose without any indication of a narrowing, indentation, or discontinuity at the bridge (root) of the nose, i.e., at the area between the eyes. Thus, a ruler placed at the upper and lower borders of the eyes measures the presence or absence of a Neolithic face configuration. Subtype NGr is present if at least at the lower border of the eyes (if not, even above it) the nose is as wide there as it is at its tip (NGr: A–Ha, b, c), discounting a bulbous enlargement of the cartilagenous parts around the tip of the nose (NGr: D). Note that patterns Ha, b, and c depict the same principle of obliteration of the bridge of the nose area: even at the *upper* border of the eyes, the nose is at least as wide here as it is at its "tip" area. Subtype NGl is present if the nose extends above the upper border of the eyes into the forehead (NGl: A–G). A half-profile type "NR" is not rated, as it is intermediate between Neolithic and normal faces. Normal face patterns (A–J) are characterized by a discontinuity between forehead and nose at the bridge of the nose area through (a) narrowing or (b) indentation, or (c) a beginning of the nose design only below the lower border of the eyes. In all face patterns, dots and circles can be interchanged without affecting the rating. Note that the DAPF is best performed automatically, without looking at a face, from memory, in a (playful) mode of relaxed attention. Thereby, face representation from memory is automatically depicted. The most telling results can be obtained from subjects naive in face drawing; otherwise, some false negative test results may ensue, masking a subtle spatial relational deficiency. The latter is likely to be severe, if a person trained in face drawing nonetheless depicts Neolithic face patterns. Note three neglected but essential aspects in the determination of subtle spatial representation, as necessary for reading and writing: (1) Face representation is not identical with face recognition (see glossary), (2) nor with visual perception, which requires focused attention. By definition, perception is intact in dyslexia. Thus, attention demanding tests, such as those using unfamiliar and/or complex geometrical figures or signs, are too inappropriate and/or insensitive to rule out visuo-spatial deficits in dyslexia. (3) Furthermore, there are two subtypes of spatial representation: interobject (used for maps) and intraobject representation of the spatial relations within the constraints of a pattern (face pattern or lexical items). [Reprinted with permission of publisher from: Pontius, A. A. "Links between literacy skills and accurate spatial relations in representations of the face: Comparisons of preschoolers, school children, dyslexics, and mentally retarded." *Perceptual and Motor Skills*, 1983, **57**, 659–666.]

given that similar linkage between specific dysfunctions pertains, whether in brain pathology or merely in situations of severe ecological underexposure and underuse (see Section III,B,7).

c. Parieto-Temporo-Occipital Junction and Angular Gyrus Small lesions lead to relatively isolated deficits in written language.

d. Angular Gyrus Lesions produce dyslexia with dysgraphia—loss of inner speech, analogous to illiteracy, as Geschwind referred to it (Section II,B,3,d).

e. Right Parieto-Occipital (Temporal) System When lesioned: in some cases the language system can remain intact, as can symbolic (including mathematical) operations and the understanding of complex logical grammatical structures. Instead, one finds disturbance of visual recognition and of the sense of familiarity. As active visual searching is maintained in an attempt to compensate for the visuo-spatial deficits in representation, there is guessing (e.g., during reading). In particular, while preserving generic recognition, there is disturbed recognition of the visuo-spatial representation of individual members of one or more categories with ambiguous shapes (e.g., letters, words, cars, chairs, cows or human faces). The latter kind of nonrecognition, prosopagnesia, is found mostly in bilateral occipital lesions and is associated with reading disturbances in half of the cases (see Section II,C,3,b).

f. Left Inferior Parieto-Occipital System In general, damage here brings about impairment of the representation of visuo-spatial components of writing while speech comprehension is relatively preserved.

g. Right Occipito-Parietal-(Temporal) System Lesions impair graphic representation (e.g., in writing and reading) of concrete spatial relations; in extreme cases, persons are unaware of the left half of the visual field (see Section II,C,3,III) during reading, writing, drawing or copying complex configurations (e.g., letters and words). Further, there is impairment of those aspects impinging on written language, which Luria termed quasispatial: symbolic, logico-grammatical, syntactic relationships (including prepositions, adverbial clauses, etc.). With loss of prepositions there is impaired comprehension of spoken and written communication, which becomes reduced to that of events and of individual words (see Section II,B,3,c,ii).

h. Right (Non-dominant) Parieto-Occipital System: Lesions are associated with impaired spatial aspects including these unrelated to language: (i) poor visual recognition of objects; (ii) impaired directionality with regard to the body and/or in external space: constructional apraxia, which can impinge on the writing process; (iii) unawareness of the left half of the visual field (see Section II,C,3,g) when reading, writing, or drawing or during spontaneous constructive activities (unilateral spatial agnosia); (iv) impaired sense of familiarity (e.g., of the face in prosop-agnosia (see Section II,C,3,b,e) (nonrecognition of the familiar face, linked with reading disturbances in half of the cases). (v) a specific kind of "amnestic dysphasia" (compare Section II,B,3,a), here associated with loss of sense of familiarity" in recognizing written communication. There is not merely an impairment of speech memory (see Section II,B,2,b,3,a), but especially of visuo-spatial representation and its classificatory spatial (and semantic) schemes, poor ability to distinguish the systems of characteristic features embodied in a shape (e.g., letter, word), to draw an object, to identify an object when depicted in a stylized manner, to complete such a drawing of an object started by someone else, or to describe its details. There is deficient visual representation of an object required to be named on confrontation, leading to a disturbed visuo-spatial basis of naming.

D. Contributory Subcortical Components

1. Fiber Tracts (white matter)

a. Interhemispheric connections are established by the *corpus callosum*. Its posterior part, the splenium (also see Section II,A,2,c), performs intercortical connections between homologous areas in left and right hemisphere. In lesions, Geschwind delineated various "disconnection syndromes" disrupting the functions of brain areas that are only unilaterally represented: (i) (quasi) pure word deafness disrupts the connection between the left and right primary auditory cortices on one hand, and Wernicke's area on the other, (ii) pure alexia (very rare) without (or only with minimal) agraphia and/or aphasia, also called agnosic alexia (see Section II,D,1,c): specifically, there is interruption between the splenium and the left geniculo-calcarine tract, disconnecting an otherwise intact left hemisphere from the visual input; (iii) in related disconnection symptomatology, it is controversial whether the result is a deficit to "see" letters, numbers, or colors (Luria) or to respond verbally to such stimuli (i.e., to name them on confrontation (see Section II,C,3,b) (Geschwind); or whether both kinds of deficits exist, especially in adults' left visual field. Apparently, isolation of the left hemisphere from direct and transcallosal visual input is associated with impairment of that kind of visual discrimination learned in conjunction with speech, whereas other visual discrimination is only mildly affected.

b. Intrahemispheric Fiber Tracts transfer impulses between specific association cortices: (i) In lesions of tracts linking motor and auditory association cortices of the left hemisphere, conduction dysphasia is claimed to ensue. This "associonistic" interpretation is, however, controversial, inasmuch as the ability to repeat and to read aloud can be preserved, although spontaneous language may include phonemic paraphasias (transformations, additions, and/or omissions). Relatively speaking, when repetition is disturbed, it is usually more severe than reading disturbance. The phonological code remains intact. (ii) In disconnection between left intrahemispheric occipito-frontal fiber tracts (also belonging to the superior longitudinal fasciculus, (see Section II,C,2,a) which links the specific visual and motor association cortices), poor oral reading results despite intact comprehension of written language. (In most cases, the lesion also involves the cortex of the supramarginal gyrus and affects the arcuate fasciculus in its most caudal part of the left parietal operculum). (iii) The specific tract of the inferior longitudinal fasciculus is the major intracerebral direct pathway for the transfer of "visual" impulses from the occipital visual cortex (areas 18 and 19; see Sections II,C,2 and II,C,3) to the memory mediation in the temporal lobe and limbic systems (see Sections II,E,1 and II,E,2). If interrupted, nonrecognition of perceived objects, such as written material, ensues in various forms of visual agnosias, (e.g., agnosic subtype of dyslexia (see Section II,D,1,a,ii).

2. Specific Subcortical Nuclei (Gray Matter)

It remains controversial whether such nuclei contribute directly to reading and writing or indirectly through the tracts (white matter) which lead to and from the following nuclei:

a. Basal Ganglia When lesioned, such as in Parkinsonism, poor articulation is claimed to result (potentially contributing to secondary spelling errors).

b. Thalamus The thalamusic nuclei maintain mutual axonal exchanges with the language zone. In addition lesions here are typically accompanied by a distinct kind of thalamic pain (without any other source) (compare Section III,A). [*See* THALAMUS.]

E. Contributory Limbic System Components

The limbic system (especially its hippocampal formation and amygdaloid nuclei), in conjunction with the *temporal lobes* (Section II,B,2) and the *inferior longitudinal fasciculus* (Section II,D,1,b,iii), mediates access to visual memory necessary for visual recognition. In bilateral lesions, stimuli (e.g., written material, faces) can be perceived but have lost their meaning: visual agnosias, including agnosic subtype of dyslexia (see Section II,D,1,b,iii). In unilateral lesions, dyslexia occurs only if the corpus callosum (Section II,D,1,a) is also dysfunctional.

1. Hippocampal Formation

This archicortex and its intralimbic connections are involved in the detection of novelty aspects and in the memory of learned, predictable regularities of various domains, including linguistic and individual visuo-spatial aspects of written signs (see Section III,A). When dysfunctional, memory for such detailed facts is more impaired than memory for procedures, or motor skills. [*See* HIPPOCAMPAL FORMATION.]

2. Amygdaloid Nuclei

These nuclei also contribute to the visual recognition of individual words and of other configurations. In general, contribution to visual recognition concerns more the identification of an object located directly in front rather than one in the periphery.

III. New Approaches, Suggesting Correlations between Written Language and Brain Systems

Some new approaches will be added to the already listed clinico-anatomical correlations (see Section II) and to the anatomico-biological and electrophysiological methods discussed in the context of etiological hypotheses (see Section I). The following approaches can address large populations of not brain-damaged, generally well-functioning populations in an attempt to suggest new aspects of the brain's mediation of reading and writing. Some clues about the roots and certain precursory components of these abilities might thereby be obtained.

A. Cerebral Blood Flow with Positron Emission Tomographic Studies

This new method holds great promise for the ultimate visualization (on a screen) of the neuronal substrate of cognitive functions. So far, it has been applied to the processing of a single common noun or of a syllablelike nonword (which follows gross rules of English spelling), and mostly by skilled readers at that. Experienced readers, however, typically process such overlearned stimuli as if they were processing a picture. Thus, such a holistic processing does not require semantic, orthographic, phonological or visuo-spatial analysis and synthesis of component parts, as is necessary for neophyte readers. Thus, a certain caveat applies to the interpretation of the results as if they were fully applicable to the complex function of reading (with the components suggested in Section II). Rather, thus far, aspects of the single-word tasks overlap with main aspects of picture recognition (well developed by 4 yr of age). Another factor to consider when determining neuronal activity (e.g., on the basis of measurements of increased blood flow) is that not only excitation requires energy, but so does inhibition.

Recent positron emission tomographic (PET) studies use water with oxygen-15 (half-life 123 sec) as the tracer for blood flow in the human brain during single-word processing. Measurement of regional cerebral blood flow is a marker of local neuronal activity. Seventeen normal subjects (intelligent and "highly skilled readers") performed several tasks. Each one took less than 1 min. During all tasks, the subjects fixated on a small crosshair presented on a color television monitor. There were four levels of complexity: (1) simple fixation; (2) passive observation (recognition) of a word presented either visually or auditorily; (3) verbal repetition of sounds (an output task, highlighting the pattern associated with verbal production. These single-word repetition tasks were found to bypass association cortex); (4) generation of a semantically appropriate verb in response to each presented noun (an association task). In addition, there were two monitoring tasks: a semantic one, in which members of a given semantic category (e.g., dangerous animals) had to be noted, and a rhyme-monitoring task, in which subjects had to judge whether visually presented pairs of words rhymed.

The PET studies are reported as delineating the localization of certain component mental operations, rather than the localization of an entire cognitive task such as reading.

With each task, few brain regions were activated. Auditory single-word representation activated primary auditory cortex, inferior anterior cingulate cortex, and superior temporal cortex. (Activation of the latter two regions occurred specifically upon word presentation.) In contrast, visual single-word presentation did not activate Wernicke's area (posterior temporal cortex). Visual presentation activated the primary visual cortex (the striate cortex in the occipital lobes), and the prestriate areas near the temporo-occipital border. The extrastriate areas and temporal cortex were specific to word presentation. An overlap occurred between activation by the auditory and visual word stimuli.

These PET studies suggested independence in the phonological input and output codes. Even though there was a common activation near the classical Broca area for word output on both visual and auditory single-word presentations, the Broca area was not activated by auditory single word input.

The results showed that no auditory reading task (a single word) activated the left temoporo-parietal cortex (Wernicke's area is traditionally implicated in phonological coding [although for more than one word or wordlike syllable]).

Among other findings, this led the researchers to a tentative interpretation in support of a "multiple-route" model in which visual input does not necessarily have to be phonologically recoded, as proposed by certain "single-route" models of cognitive functioning.

Whichever cognitive ability (reading or holistic single word or picturelike recognition) was actually assessed by such PET studies, both tasks can share certain aspects of memory and of attention: Various PET studies can distinguish between automatic activation and controlled processing by means of attention. Furthermore, there are two kinds of (interrelated) attention systems: (1) a more general system, largely mediated by the prefrontal system (such attention for action is required, for example, when attending to only one meaning of a word, while suppressing alternative meanings), and (2) the visual spatial attention system, mediated by the posterior parietal lobe, a portion of the thalamus (part of the pulvinar) (see Section II,D,2,b), and areas of the midbrain related to eye movements.

Memory is another aspect of language. Once an item has left current attention, "the hippocampus (see Section II,E,1) performs a computation needed for storage," which enables conscious retrieval, as the PET studies argue.

In general, the results from the PET studies discussed are congruent with those from brain lesion studies. Specific surprise findings in the PET studies include the greater tendency toward bilateral brain activation and the repetitive task bypassing the association cortex, which itself has access to the output system.

B. Neuroecological Correlation Method

A novel field of inquiry into dyslexia can implicate likely precursors to specific graphic representation of lexical signs: in general, mankind's early graphic representation in sacred art. Such art is often the only available record of virtually all human populations, cognitive functioning, dating back over millenia including the Neolithic period, by definition the last completely illiterate period. Of specific interest in the context of written language is the study of neolithic "artists" depicting the human face pattern (Fig. 1). Both configurations of face and of written signs are generally considered as sharing essential aspects, given that both can present an unlimited variety by subtle rearrangement among a few basic spatial features. A specific, largely overlooked subtype in testing visuo-spatial representation in both face and written signs concerns the spatial relations, orientation, and size ratios among the parts within their configurations. (This spatial ability must be distinguished from interpattern spatial relations, as used in mental maps [e.g., finding ones way among objects.]) It remains to be determined whether this common denominator of intrapattern variation, linking face and lexical signs, pertains exclusively to a correlational or even to a neurologically based causal relationship between face pattern and written signs (see Sections II,C,3,b, II,C,3,e, and II,C,3,h,iv).

A perusal of art of the neolithic (i.e., illiterate) period worldwide shows a prevalence (at least 3 of 4) of a specific misrepresentation within the face (Fig. 1, NGr, NGl) configuration, labeled "neolithic face" patterns. The same patterns prevailed in contemporary illiterates still living in near stone-age situations prior to acquiring literacy and rules of spelling. These patterns are characterized by an inaccurate continuity between forehead and nose whether in a roundish type (NGr) or in a longish type (NGl) (Fig. 1). This continuity obliterates the subtle spatial relations among the features of the upper part of the face around the bridge of the nose area. (The mouth is not important and is frequently omitted or is tiny and deep set in Neolithic art.)

Underlying neurophysiological factors contributing to this Neolithic inaccuracy were reasonable to postulate when it was discovered that such spatial inaccuracy begins to disappear in mankind's art as well as in contemporary peoples graphic representations with the advent of literacy (though it may resurface under the impact of alcohol or other drugs). About one third of dyslexics (as opposed to 6–11% of normal controls) produced the neolithic face patterns when asked to "Draw-A-Person-With-Face-In Front" (DAPF) [Fig. 1]. Two interrelated sets of factors must be considered: (1) A lack of exposure to specific external input, which promotes literacy. Probably, populations that live in severe ecological situations that neglect graphic-pictorial material, (e.g., of small patterns) have survival needs that promote different skills, unrelated to those tapped by literacy. Such an ecological context selects out the fostering of certain cognitive abilities (e.g., interobject, maplike spatial representation, and rapid global shape assessment of attacker/prey/food). By contrast other abilities are neglected (such as intraobject spatial representation required to process face- and lexical patterns). (2) Subtle brain dysfunction or underuse: The following brain areas were tentatively implicated (likely through lack of exposure and of usage): The mediation of intrapattern spatial relations implicates especially the parietal-occipital system (see Sections II,B,3 and II,C,3). (This suggestion gains further support by an apparently analogous link found in a neuropathological counterpart to the here proposed ecological syndrome: prosop-agnosia, linked with reading difficulties in half of the cases (see Section II,C,3,b, II,C,3,e, and II,c,3,h,iv). Such patients with occipital brain damage say they no longer "recognize" the familiar individual face, because they experience it as "flattened out, without any relief," and most of them also depict the face that way, which is of the "neolithic face" type (Fig. 1, NGr, NGl). Further possibly involved brain regions may be the hippocampus (see Section II,E,1, and Section III,A) concerning certain spatial representational globalization in memory, (implicated in chronic alcoholism), as well as certain prefrontal lobe mediation of flexibility (see Section II,A,3,a,iii).

A working hypothesis, initially based on such consistently repeated observations of Neolithic face depictions in diverse situations, was engendered by such inaccurate face representation being linked with illiteracy and later with the visual spatial subtype of dyslexia (discussed below). Inasmuch as these two reflections of apparently evolutionarily early or ecologically modified visuo-spatial cognitive functioning may share a common denominator, Neolithic face depiction may serve as an indicator of certain potentially poor literacy skills, meriting further scrutiny.

A fairly wide array of evidence has been gathered within the context of the neuroecological paradigm in support to this hypothesis.

1. *Contemporary near stone-age people* in remote areas of New Guinea, Indonesia, and South America, whose spontaneous sacred art (where produced) is characterized by neolithic face patterns, also preferred the same in their elicited face drawings. Surprisingly, the percentage of such globalized neolithic faces (in >1,000 subjects, collected over two decades) showed a significant positive correlation with the results from their direct reading and writing tests (UNESCO).
2. *Colonial North American stone masons,* whose rate of illiteracy (20.5%) is on record, and who left human face engravings on c. 1,500 tombstones, showed a positive correlation with their neolithic face depictions (19.6%).
3. *"Before and after" studies* with such near stone-age peoples showed a significant increase in accurate face depictions with the availability of literacy instruction, although not necessarily producing full literacy skills:
4. Schooling alone is apparently not sufficient for the emergence of literacy when ecological situations continue to lack specific stimuli with subtle intrapattern variety, (e.g., in small geometrical patterns). In contrast to 4% of their class mates of European descent, who drew neolithic faces, about a third of 269 *Australian Aboriginal school children* did so. All of these Aboriginals were at least of average intelligence, functioning within regular class settings. They fulfilled the requirements to be classified as dyslexics (see Glossary), or here as "ecological dyslexics" (distinguishing this subgroup from that listed under 6). The Aboriginals' overall bleak Northern Territory desert and artificial settlement surrounds were presumably not fostering the

development and utilization of those intra-object spatial skills necessary for reading.

5. In 407 *"Western" preschoolers,* the face depictions (collected by five different "blind" researchers in Europe and the United States) showed a consistant percentage of about one-quarter of the neolithic face pattern. It is likely that most of these children had already been exposed to graphic material seen on TV and small geometrical patterns depicted on fabrics and on paper.

6. *Out of 297 European dyslexics,* about one-third of them drew neolithic face configurations (Fig. 1, NGr, NGl) (as opposed to 6–11% of normal readers. These "Westerners" were naive as to face-drawing instruction, but otherwise grew up with exposure to TV and to other graphic pictorial materials, e.g. small geometric patterns. In distinction from "ecological dyslexics" [see 4] in general, in "Western" dyslexics internal factors are more likely to be implicated (e.g., subtle brain dysfunctioning) (see Section I,A, and I,B)].

7. Further support from ontogeny for the hypothesis is suggested by experiments with young *infants* (obviously illiterates): Their automatic, indiscriminate "smiling response" with wide-open mouth was elicited significantly faster ($p < 0.001$) by masks presenting neolithic faces (Fig. 1, NGr, NGl), (compared with control masks stressing the mouth, and even with the natural face) up to 18 wk of age, when the latter was preferred.

8. Finally, *prosop-agnosia* can occur in occipito (-parietal) brain damage (see Sections II,C,3,e, and II,C,3,h,iv). This neuro-pathological disturbance is characterized by experiencing (and drawing) the face configuration as "flattened out", analogous to typical "neolithic face" patterns (Fig. 1, NGr, NGl). Furthermore, half of such patients also have reading disturbances. Based on certain preliminary findings, it is hypothesized, that such brain damage may also occur in certain persons suffering from early phases of AIDS or Alzheimer presenility. If so, then "neolithic face" depiction in the DAPF test (Fig. 1) may constitute a sensitive early diagnostic indicator.

In summary, the neuroecological method implicates the following correlation in not brain-damaged illiterate (preliterate) populations worldwide: (1) When ecologically determined daily activities typically lack exposure to, and practice with subtle spatial relations within (small) geometrical configurations (as in graphic–pictorial material), such a lack is likely to be reflected in and (2) specific spatial-relational, globalized misrepresentation of the features within the face configuration, called "neolithic face" patterns (see Fig. 1, NGr, NGl).

Thus, when such "neolithic face" depictions are also found in drawings by "Western" dyslexics, such misrepresentation can be viewed as an indicator for the spatial (visual) subtype of dyslexia. Specifically, such a subtype of spatial (visual) dyslexia can be sub-classified as "ecological dyslexia" when such "neolithic face" patterns are depicted by populations in the process of emerging from preliteracy. On closer scrutiny, however, one finds that "ecological dyslexics" typically continue their daily activities within their original ecological setting despite receiving literacy instruction. Such instruction by itself, then, does not appear to be sufficient to attain literacy.

Bibliography

Arbib, M. A., Caplan, D., and Marshall, J. C. (eds.) (1982). "Neural Models of Language Processes." Academic Press, New York.

Duffy, F. H., and Geschwind, N. (eds.) (1985). "Dyslexia. A Neuroscientific Approach to Clinical Evaluation." Little Brown, Boston.

Galaburda, A. M. (1988). The pathogenesis of childhood dyslexia. *In* "Language Communication in the Brain" (F. Plum, ed.). Raven Press, New York.

Ingle, D. J., Mansfield, R. W., and Goodale, M. A. (eds.) (1982). "The Analysis of Visual Behavior." M.I.T. Press, Cambridge, Massachusetts.

Lecours, A. R., Lhermitte, F., and Bryans, B. (1983). "Aphasiology." Bailleire Tindall, London.

Luria, A. R. (1980). "Higher Cortical Functions in Man." Basic Books, New York.

Petersen, S. E., Fox, P. T., Posner, M. I., Mintun, M., and Raichle, M. E. (1988). Positron emission tomographic studies of the cortical anatomy of single-word processing. *Nature* **331 (6157),** 585.

Pirozzolo, F. S., and Wittrock, M. C. (1981). "Neuropsychology and Cognitive Processes in Reading." Academic Press, New York.

Pontius, A. A. (1980). Pre-literate people in New Guinea and Indonesia draw specifically distorted faces, as do "Western" dyslexics, using a paleo-visual-representational mode. *Experientia* **36,** 83.

Pontius, A. A. (1981). Geometric figure-rotation task and face representation in dyslexia: Role of spatial relations and orientation. *Percept. Mot. Skills* **53,** 607.

Pontius, A. A. (1989). Color and spatial error in block design in stone-age Auca Indians: Ecological underuse of occipital-parietal system in men and of frontal lobes in women. *Brain & Cognition* **10,** 54.

Seidenberg, M. S. (1988). Cognitive neuropsychology and language. The state of the art. *Cog. Neuropsychol.* **5,** 403.

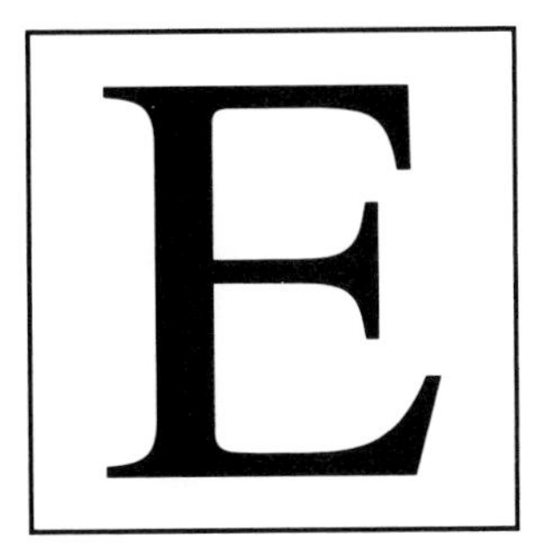

Ears and Hearing

JAMES O. PICKLES, *University of Queensland, Australia; University of Birmingham, United Kingdom*

Glossary

Decibel (dB) Relative measure of sound pressure, derived from the formula: number of decibels = 20 $\log_{10}$ (measured sound pressure/reference sound pressure). If the reference sound pressure is taken as 2×10^{-5} N/m^2, the measure becomes an absolute measure of sound pressure known as dB Sound Pressure Level (dB SPL)

Impedance In acoustics, the pressure needed to produce unit velocity of vibrations in the medium; the impedance for a plane wave in an infinite volume of the free medium (known as the specific impedance) is the pressure required to induce a unit velocity of vibration of the medium (units N sec/m^3)

Tuning curve (frequency-threshold curve) Curve of the stimulus intensity necessary to give a constant output of the system as the stimulus frequency is varied; it can be applied equally well, for example, to vibration of the basilar membrane (for constant amplitude of vibration), and responses of auditory neurons (for constant evoked firing rate); an advantage of the tuning curve is that it permits comparison of tuning of different stages of the auditory system

AUDITORY PERFORMANCE is determined in a number of important ways by the performance of the cochlea. Thus, for instance, both frequency resolution and the perception of distortion tones can be traced to the fundamental properties of cochlear mechanics and hair cells. Similarly, the absolute detection threshold can be traced to the cochlea, in conjunction with the outer and middle ears. Thus, understanding of the auditory system is closely allied to our understanding of cochlear physiology. The limitations to auditory performance imposed by the auditory central nervous system are much less clear.

I. The Outer Ear

The outer ear conveys the sound wave from the external field to the tympanic membrane (eardrum) and modifies the wave in two important ways. First, by virtue of its resonant cavities, the outer ear increases the sound pressure in the middle range of frequencies, enhancing sensitivity in this frequency range. Second, because the enhancement is directionally selective, the outer ear gives cues, which help in sound localization.

The outer ear has multiple resonant cavities. One important cavity is formed by the concha (Fig. 1), situated at the entrance to the external auditory meatus (ear canal). The other most important cavity is the meatus itself. Since the tympanic membrane offers a higher impedance to the passage of sound waves than does free air, the membrane reflects some of the sound energy that reaches the end of the canal. This sets up a quarter-wave resonance in the canal. The resonant wave has its node (low pressure) at the entrance to the canal, and its antinode (high pressure) at the tympanic membrane. In humans, this ear-canal resonance increases the pressure at the tympanic membrane by some 10–15 dB in a broad range around the resonant frequency of 2.5 kHz. The second most important resonance, in the concha, adds some 10 dB in the 5 kHz region.

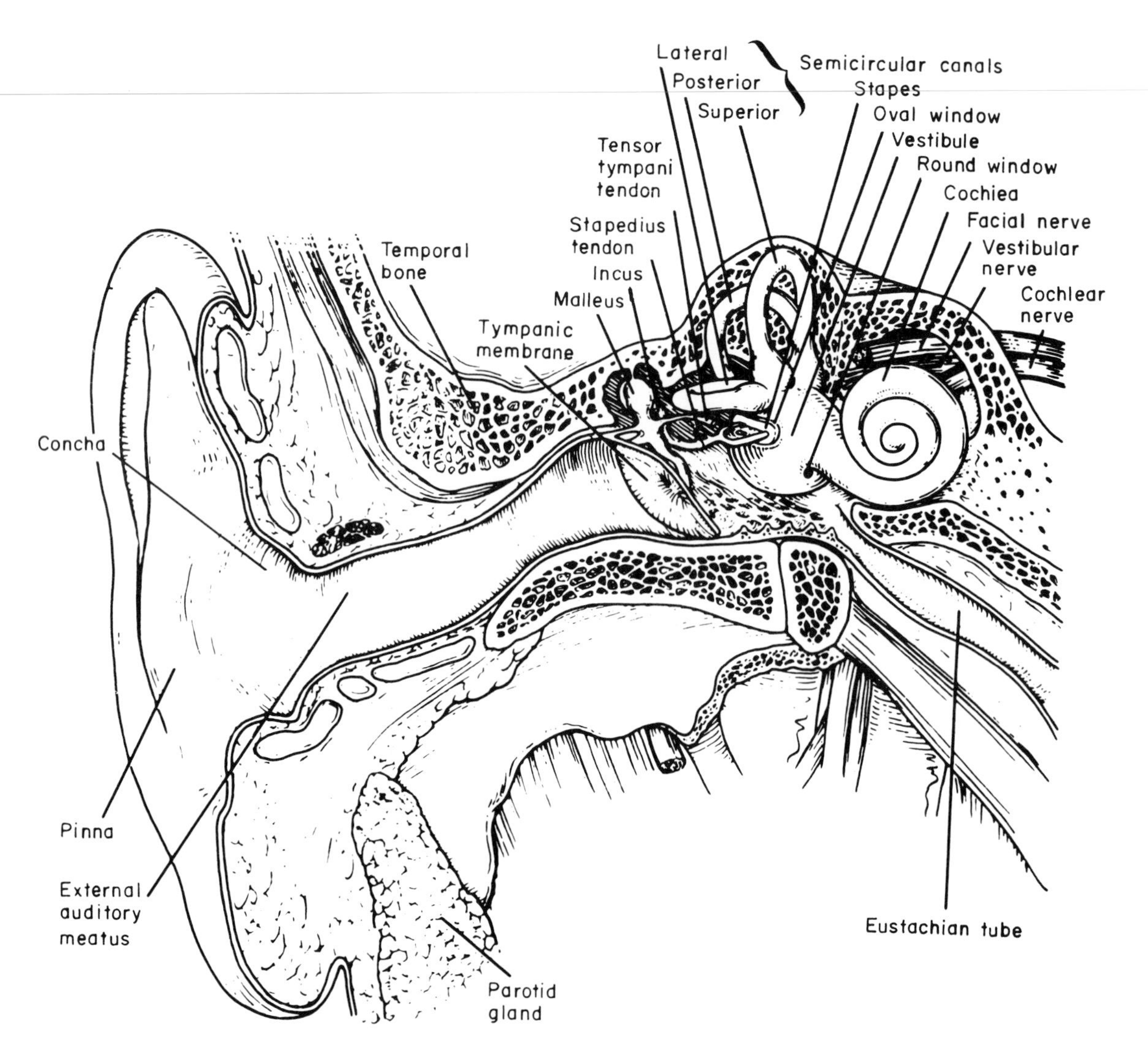

FIGURE 1 The outer, middle, and inner ears in humans. [Adapted, with permission, from R. G. Kessel and R. H. Kardon, 1979, "Tissues and Organs: A Text-Atlas of Scanning Electron Microscopy," W. H. Freeman and Company, New York.]

In humans, the sum of these two resonances, together with resonances in the pinna and reflections from the head, increases the sound pressure at the tympanic membrane by 10–20 dB above that in the free field, over the frequency range of 1.5–8 kHz.

The resonant amplification of sound pressure by the external ear depends on the direction of the sound source, in addition to the factors mentioned above. Although the most powerful cues for sound localization in humans are given by binaural (i.e., two-ear) listening (see below), the outer ear adds cues, which can be used for monaural localization and for distinguishing the elevation of the source. Sound waves are scattered from the raised rim at the edge of the pinna and from the rear wall of the concha. When a sound source is moved behind the ear, the scattered waves interfere destructively with the wave transmitted directly, reducing the input for sound frequencies >3 kHz. Because the convolutions of the pinna and concha are asymmetrical above and below the entrance of the ear canal, the interaction is also dependent on the elevation of the source. Thus, when a sound source is raised above the horizontal, the intensity at the ear canal increases in the 5–10 kHz range.

II. The Middle Ear

A. The Middle Ear as an Impedance Transformer

The acoustic vibrations are transferred from the tympanic membrane, via three small bones [the ossicles; namely the malleus, incus, and stapes (Fig.

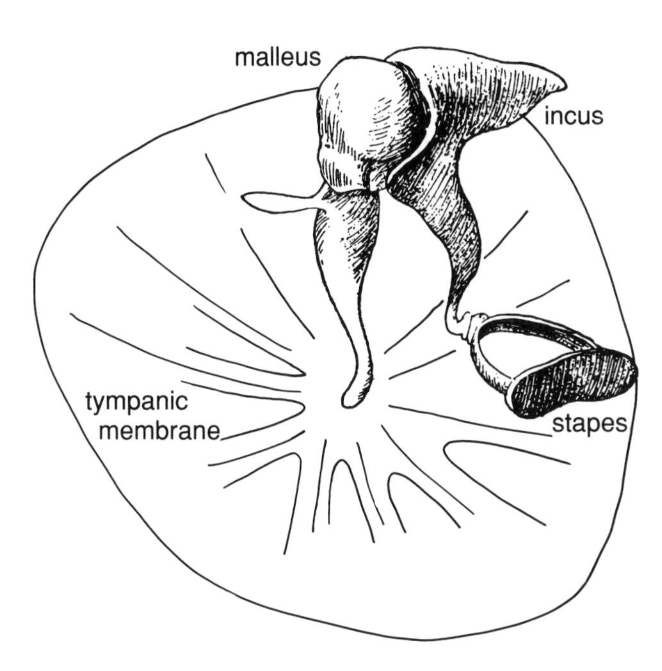

FIGURE 2 The tympanic membrane and middle ear ossicles in humans, as seen from within. The malleus and incus are suspended at their upper ends by ligaments (not shown).

2)], to the round window of the cochlea. Here, vibration of the stapes induces flow of the cochlear fluids. The cochlear fluids offer a much higher impedance to acoustic vibrations than does air. When sound waves meet a boundary with such an impedance jump, it can be calculated that much of the sound energy will be reflected, with only a small proportion transmitted. In the ear, we expect that only 1% of the incident energy to be transmitted to the cochlea if the sound waves met the oval window directly. The middle ear apparatus reduces this power loss substantially.

The middle ear reduces the power loss by acting as an impedance transformer. It changes the relatively large-amplitude, low-pressure vibrations of the tympanic membrane to low-amplitude, high-pressure vibrations at the round window. This is accomplished by three mechanisms: (1) the stapes footplate in the round window has a smaller effective area than the tympanic membrane; (2) the middle ear bones act as a lever, decreasing the amplitude of the movement at the round window but increasing its force; and (3) the tympanic membrane buckles as it moves, similarly decreasing the amplitude of the movement and increasing its force. At 1 kHz, the three factors together increase the ratio of force/displacement of the vibrations by a factor of 185 in the cat, the species for which we have the most information (by far the greatest contribution, a factor of 35 times, arises from the first mechanism). This produces a substantial reduction of the effective impedance of the cochlea as seen at the tympanic membrane. Nevertheless, the measured impedance at the tympanic membrane is rather higher than the impedance of air by a factor of about four. This is due to (1) the middle ear transformer ratio being smaller than the ideal value by a factor of two and (2) friction and other losses in the middle ear.

B. Transfer Function of the Middle Ear

Losses due to friction in the middle ear reduce transmission evenly over the whole frequency range. Further factors are more frequency-specific in their effects. The *stiffness* of the ligaments supporting the bones and membranes of the middle ear reduces transmission primarily at low frequencies (<1 kHz). At higher frequencies, the *mass* of the middle ear bones becomes dominant. In this frequency range, transmission also becomes less efficient because the stimulus is less effective at moving the tympanic membrane: The membrane vibrates as a number of separate zones, only some of which couple vibration to the malleus. Therefore, transmission is most efficient in the mid-range of frequencies (around 1 kHz), falling off at higher and lower frequencies. The combined effects of both the outer and middle ears in enhancing certain frequencies substantially accounts for the variation of our acoustic sensitivity with frequency, except at the very top and bottom of the frequency range, where cochlear factors are dominant.

C. Middle Ear Muscles

Transmission through the middle ear can be controlled by the middle ear muscles. The tensor tympani attaches to the malleus near the tympanic membrane, and the other—the stapedius muscle—attaches to the stapes. Contraction of the muscles increases the stiffness of the ossicular chain. As described above, an increase in stiffness reduces transmission at low frequencies (<1 kHz). In this range, middle ear muscles can reduce transmission by up to 20 dB. The middle ear muscles also seem to affect the vibration of the middle ear bones in further ways not understood, because the response at higher frequencies can be influenced, with certain irregularities in middle ear transmission being reduced.

Contraction of the middle ear muscles can be induced by loud sound (the acoustic reflex), by vocalizations and bodily movements, by tactile stimulation of the head, and, in some people, voluntarily. Above the intensity required to elicit the acoustic reflex (some 75 dB above the absolute threshold), the muscles help protect the ear from acoustic trauma, although the acoustically driven reflex is too slow to protect the ear from impulsive sounds such as gunshots. In addition, at high intensities, low-frequency sounds are particularly effective at masking high-frequency sounds. Contraction of the muscles, by selectively reducing the transmission of low-frequency sounds, may also be beneficial for the perception of complex sounds with low-frequency components, such as speech, at high intensities.

III. The Cochlea

A. General Anatomy

The cochlea is a spiral fluid-filled tunnel within the petrous (stony) part of the temporal bone (Fig. 1). It forms part of the labyrinth, the other division being the vestibular system, with which it is in continuity. The cochlear duct is divided into three compartments by two membranes known as the basilar membrane and Reissner's membrane (Fig. 3A). The compartments are called the scala vestibuli, scala media, and scala tympani. The cochlear duct with its three compartments spiral together around a central core called the modiolus, which contains the auditory nerve and the cochlear blood supply. In humans, the cochlear duct is about 35 mm long and is packed into a spiral of $3\frac{1}{2}$ turns in a space some 10 mm wide and 5 mm high. Lengths of duct in different mammals vary from 7 mm in the mouse to 57 mm in the elephant. The number of turns varies from half a turn in the relatively primitive monotremes to 4 turns in the guinea pig.

The organ of Corti (Fig. 3B), which contains the receptor apparatus, is situated on the basilar membrane. It is covered by a gelatinous and fibrous flap called the tectorial membrane. The position of the organ, on the interface between the scala media and the scala tympani with their fluids of very different composition, is considered critical for cochlear function.

B. Cochlear Fluids: Production and Electrochemistry

The scala media contains endolymph, which is unique for an extracellular fluid in that it has a high K^+ composition ($\approx$150 m*M*), and a low Na^+ concentration ($\approx$1 m*M*). It also has a high positive potential (+80 to +100 mV) with respect to the surrounding bone, the higher values being found in the more basal turns of the cochlea. In contrast, the perilymph, which is contained within the scala vestibuli and the scala tympani, has the ionic composition expected of most extracellular fluids and an electric potential similar to that of the surrounding bone.

The high positive potential and high K^+ concentration of the endolymph are directly dependent on active, ion-pumping processes within the stria vascularis, a structure situated on the wall of the scala media farthest away from the modiolus (Fig. 3A). As its name implies, the stria vascularis has a rich blood supply. The stria contains two types of cells: the more superficially located marginal cells and the more deeply located basal cells. The marginal cells have deep infoldings on the basolateral surface facing away from the scala media (i.e., on the surface facing the blood supply), the infoldings containing many mitochondria, indicative of high levels of energy-consuming activity. The marginal cells have high positive intracellular potentials, a few mV more positive than the endolymph, and it is suggested that they maintain the chemical and electrical composition of the endolymph. Although the details are still controversial, it is generally agreed that a Na^+–K^+-activated adenosine triphosphatase in the marginal cells is involved. Other processes, such as Na^+–K^+–Cl^- cotransport, also may play a part. [*See* COCHLEAR CHEMICAL NEUROTRANSMISSION.]

C. The Organ of Corti

Mechanical vibrations resulting from the incoming sound are transformed into neural activation in the organ of Corti (Fig. 3B). The hair cells form the mechanotransducing elements. The cells are situated with their apical ends in the reticular lamina on the top surface of the organ, with their sensory hairs, or stereocilia, projecting to touch or nearly

FIGURE 3 A. Cross section of a single turn of the cochlear duct. [Reproduced, with permission, from D. W. Fawcett, 1986, "A Textbook of Histology," W. B. Saunders, Philadelphia.] B. Cross section of the organ of Corti, as it appears in the basal turn. The modiolus is to the left of the figure. BV, blood vessel; IPC, inner phalangeal cell. [Slightly modified, with permission, from A. F. Ryan and P. Dallos, 1984, "Hearing Disorders" (J. L. Northern, ed.), pp. 253–266. Little, Brown and Company, Boston.]

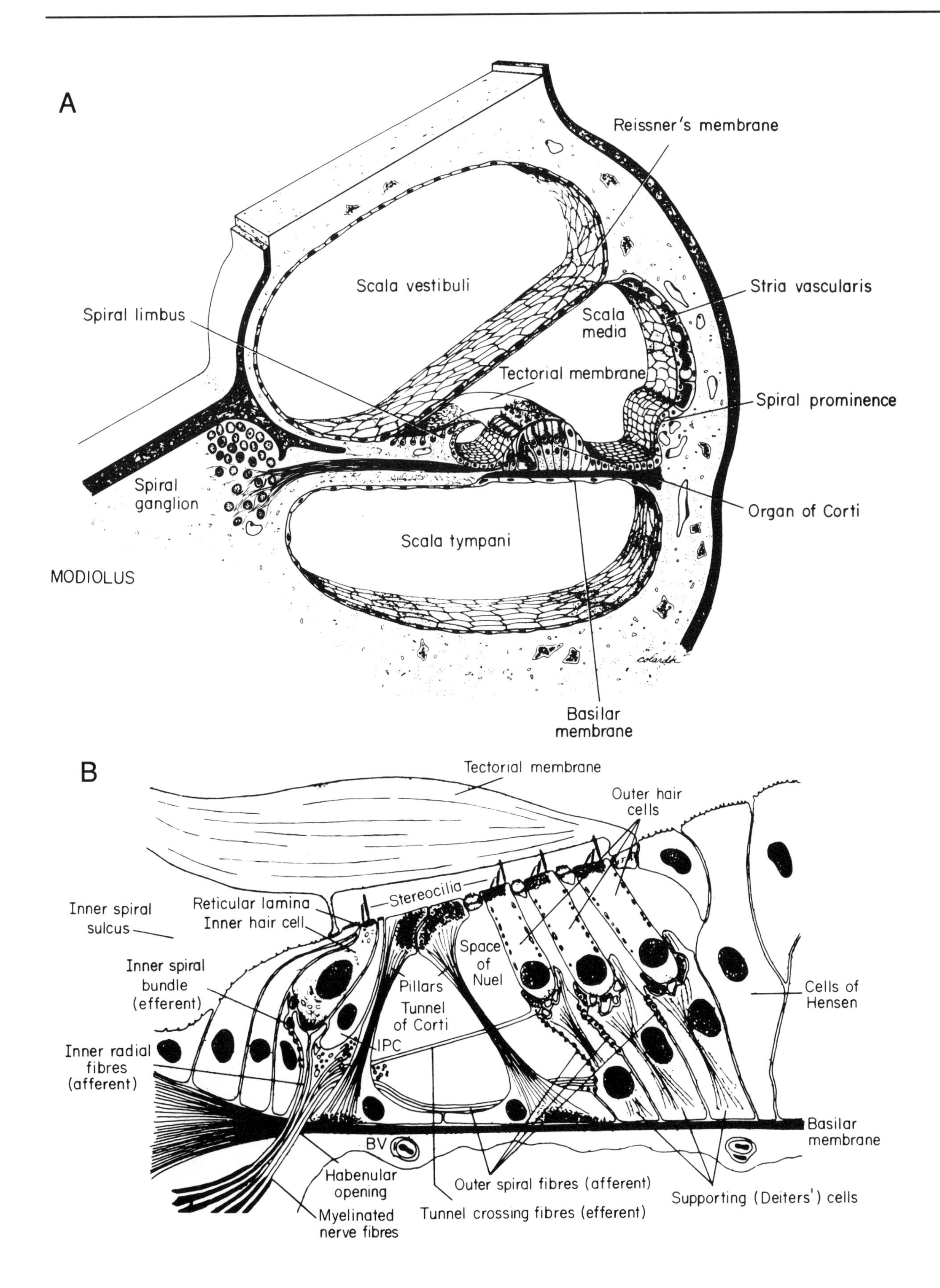
A
Reissner's membrane
Scala vestibuli
Stria vascularis
Spiral limbus
Scala media
Tectorial membrane
Spiral prominence
Spiral ganglion
Organ of Corti
Scala tympani
MODIOLUS
Basilar membrane
B
Tectorial membrane
Outer hair cells
Stereocilia
Inner spiral sulcus
Reticular lamina
Inner hair cell
Inner spiral bundle (efferent)
Space of Nuel
Pillars
Tunnel of Corti
Cells of Hensen
IPC
Inner radial fibres (afferent)
BV
Basilar membrane
Habenular opening
Outer spiral fibres (afferent)
Supporting (Deiters') cells
Myelinated nerve fibres
Tunnel crossing fibres (efferent)

touch the tectorial membrane. Figure 4A shows a schematic representation of the stereocilia and the other structures at the upper end of the hair cell. The hair cells are divided into two types, with complementary roles in cochlear function. The inner hair cells (Fig. 4B) are situated in a single row on the side of the organ of Corti nearest to the modiolus. They signal the vibrations of the cochlear partition to the central nervous system. The outer hair cells (Fig. 4C) are situated in three to five rows on the outer side of the organ of Corti. Their role is rather more problematic; however, they contribute to the sharp frequency selectivity and high degree of sensitivity of normal cochlear function. To analyze the mechanisms of hair cell function in the organ of Corti it has been necessary to bring together the results from a number of different types of experiments, including the basic biophysical analysis of the operations of single hair cells. Such experiments are rather difficult in the mammalian cochlea, and our most detailed information has come instead from the analysis of vestibular hair cells in other species.

D. Mechanotransduction in Cochlear Hair Cells

The structures in Fig. 4A, with minor variations, are common to all hair cells of the acousticolateral system, i.e., to hair cells of the cochlea and vestibular system and, in addition, to hair cells of the lateral line organ. The stereocilia are situated in several rows of graded height, with between 50 and 250 stereocilia per hair cell, depending on organ and species. The stereocilia are also richly interconnected by different sets of linkages. Linkages of one set connect the stereocilia laterally, so that all stereocilia move when some are deflected.

Mechanotransduction depends on deflection of the stereocilia. While it is known that transduction opens channels in the apical membrane of the hair cell, we have no concrete evidence as to the exact site and mechanisms of channel opening. However, it is suggested that the channels are near the tips of the stereocilia, because if the extracellular environment is explored with a microelectrode while the

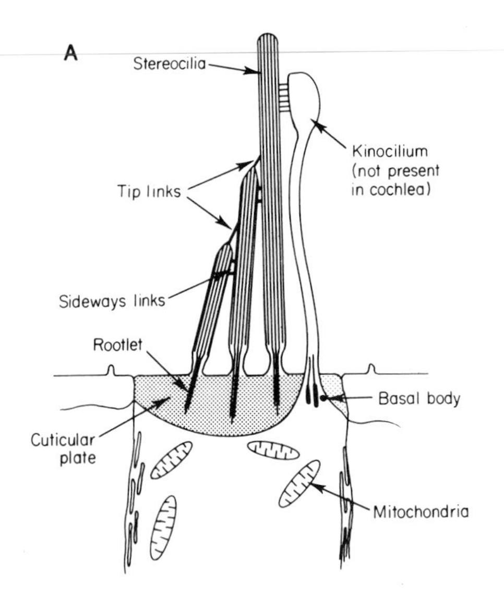

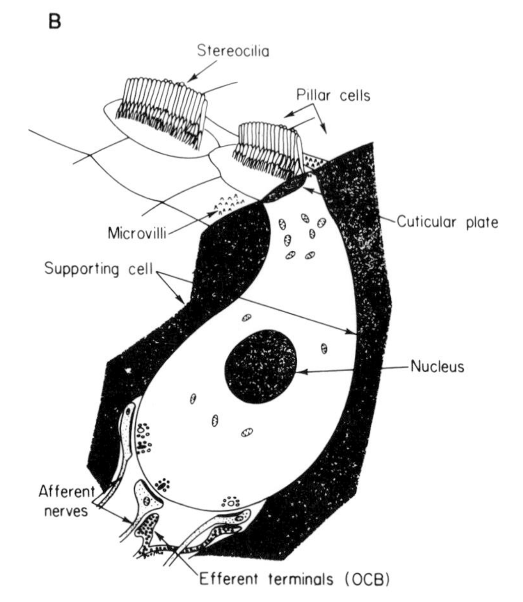

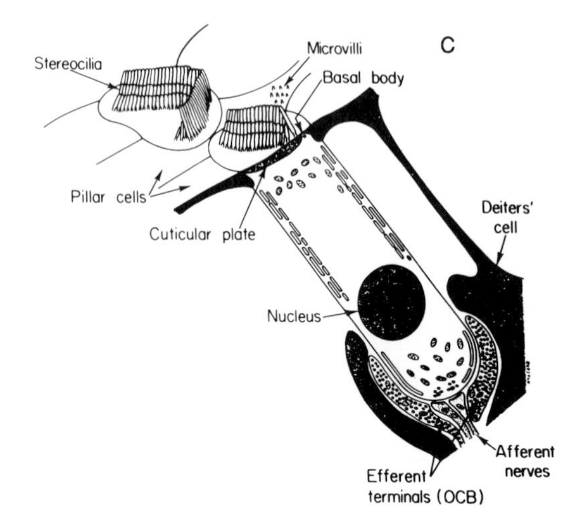

FIGURE 4 A. Schematic diagram of the common structures on the apical (transducing) pole of acousticolateral hair cells. In the cochlea, the kinocilium is present only in immature hair cells. B. Inner hair cells are shaped like a flask. C. Outer hair cells are shaped like a cylinder. In all parts of the figure, the modiolus would be situated to the left. [Copyright J. O. Pickles, 1987.]

stereocilia are manipulated, the greatest potential changes are found over the tips of the stereocilia. One theory is the following: Deflections of the bundle can be expected to produce a shear or sliding movement between the stereocilia of the different rows. It is suggested that the shear is detected by the tip links, which connect the stereocilia at their upper ends (Fig. 5). Stretch of the links would then open the channels by a direct mechanical pull. This is in agreement with the observation that deflection of the bundle in the direction of the tallest stereocilia is always excitatory, and deflection towards the shortest always inhibitory.

The relation between bundle deflection and intracellular voltage is an asymmetric and saturating sigmoid, with the changes in the depolarizing direction being larger than the changes in the hyperpolarizing direction (Fig. 6). In the formulation put forward by A. J. Hudspeth and his colleagues, the form of this function can be directly traced to the kinetics of the transducer channels and is determined by the Boltzmann distribution.

In the mammalian cochlea, it is expected that the stereocilia are stimulated by the vibration of the basilar membrane, which produces a shear movement between the tectorial membrane and the reticular lamina. Comparisons of the thresholds for activation of hair cells and auditory-nerve fibers suggest that depolarizations of 2 mV raise the firing rate of the auditory-nerve fibers to the hair cells by about 20 action potentials per second.

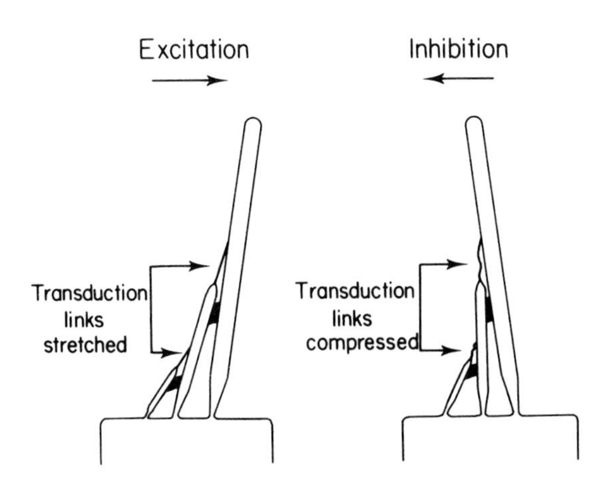

FIGURE 5 Hypothesis for transduction in hair cells, in which stretch of the tip links opens transducer channels at the tips of the stereocilia. [Reproduced, with permission, from J. O. Pickles, S. D. Comis, and M. P. Osborne, 1984, Cross-links between stereocilia in the guinea pig organ of Corti, and their possible relation to sensory transduction, *Hearing Res.* **15,** 103–112, Elsevier Science Publishers, Amsterdam.]

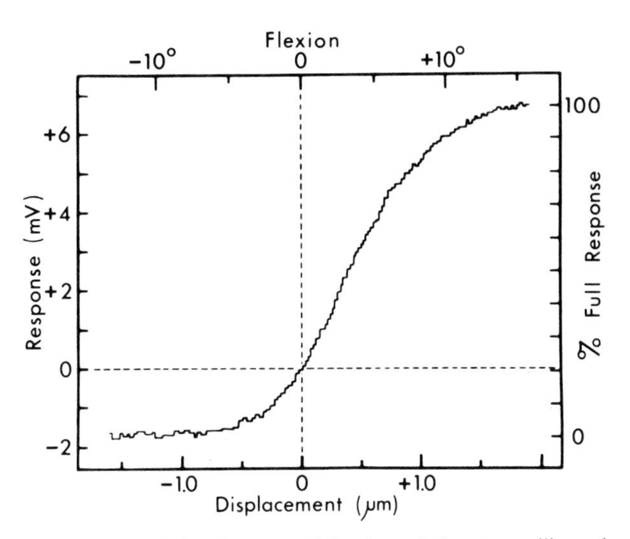

FIGURE 6 Relation between deflection of the stereocilia and intracellular voltage change in a hair cell of the bullfrog sacculus. [Reproduced, with permission, from A. J. Hudspeth and D. P. Corey, 1977, *Proc. Natl. Acad. Sci. U.S.A* **74,** 2407–2411.]

E. Cochlear Mechanics and the Frequency Selectivity of the Cochlea

In the mammal, the frequency selectivity of the hair cells, the frequency selectivity of the auditory central nervous system, and the frequency selectivity of much auditory performance are all a direct reflection of the selectivity of the mechanical vibrations of the basilar membrane. The early measurements of basilar membrane vibration by von Békésy showed the phenomenon of the traveling wave (Fig. 7A). This can be visualized as a series of ripples traveling up the cochlear duct from base to apex. The ripples grow in size as they travel apically, come to a peak, and then die away sharply. If the frequency of the input stimulus is constant, the envelope of the waveform is constant, and so the point of maximum vibration stays constant. The point of greatest response depends on the frequency of stimulation, low frequencies producing the greatest response at the apex of the cochlea, and high frequencies at the base.

The measurements of von Békésy were made in dead specimens; however, it has since been shown that the detailed pattern of the traveling wave is highly dependent on the cochlea being in a good physiological state. To ensure this, recent investigators have confined their measurements to single points along the cochlear duct, with the stimulus frequency being varied as the independent variable.

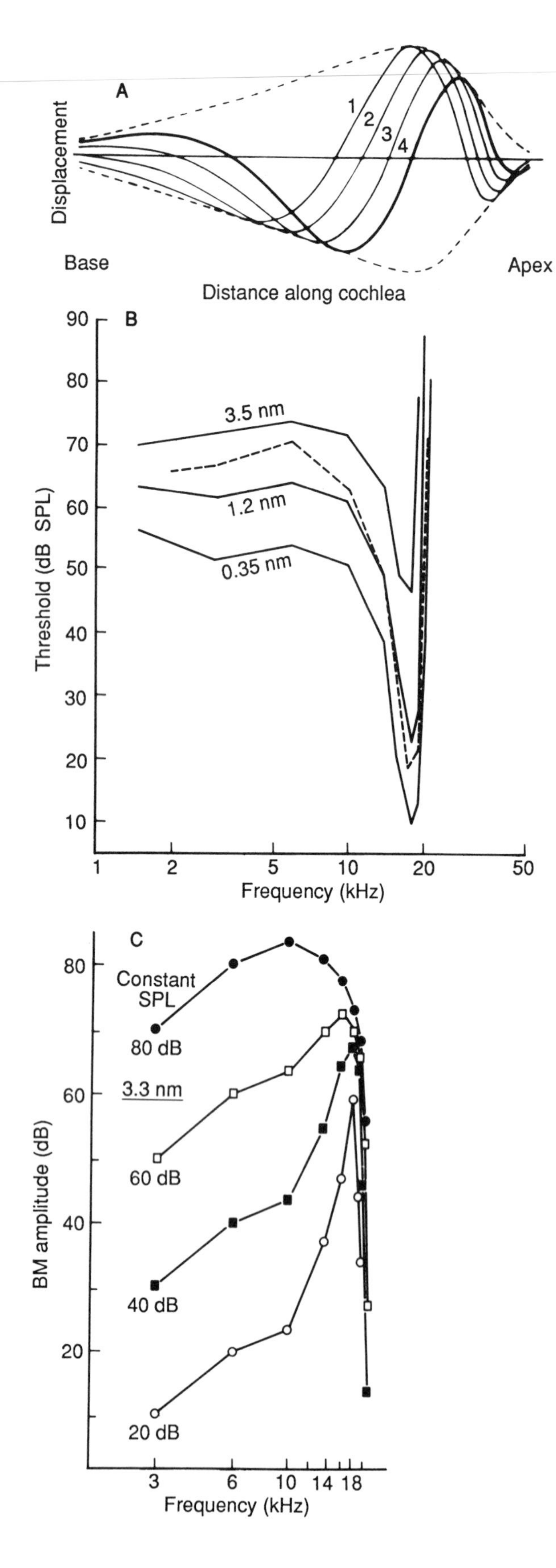

The recent results can be shown in two ways. One form is the tuning curve, which relates the sound intensity necessary to produce a certain, fixed level of response for different frequencies of stimulation (Fig. 7B). If the same data are instead used to plot the amplitude of response for stimuli of constant intensity as the stimulus frequency is varied, we have the amplitude functions of Figure 7C. The results differ from those of von Békésy in that the curves have steep slopes near the peak, indicating sharp frequency selectivity. In addition, the curve of Figure 7B also has a very low threshold at its minimum, and the curve of Figure 7C a relatively large amplitude at its peak, indicating great sensitivity in the responses.

The origin of the large amplitude and sharp frequency selectivity of the cochlear mechanics is controversial. Simple mathematical analyses of cochlear models show only low-sensitivity broadly tuned vibrations, similar to those found by von Békésy. The models that are most successful in showing vibrations comparable to those found experimentally are those in which the cochlear partition contains a source of mechanical energy. It is suggested that this is provided by the outer hair cells.

The suggestion, as yet unproven, is that outer hair cells actively produce movements when their stereocilia are deflected. The movements in turn feed back and augment the movement of the partition. In this way, the traveling wave is amplified as it moves up the cochlear duct. Since the increased

FIGURE 7 A. Traveling waves in the cochlea, according to the measurements of von Békésy. Frequency of stimulation: 200 Hz. [Reproduced with permission, from G. von Békésy, (1960). "Experiments in Hearing," McGraw-Hill, New York. B. Tuning curves for basilar membrane vibration at the 18-kHz point on the guinea pig cochlea. The numbers on the curves show the vibration amplitudes at which the curves were made. The dotted line shows a tuning curve for the firing of an auditory-nerve afferent, indicating similar tuning characteristics. [Reproduced, with permission, from P. M. Sellick, R. Patuzzi, and B. M. Johnstone, 1982, Measurement of basilar membrane motion in the guinea pig using the Massbauer technique, *J. Acoust. Soc. Am.* **72,** 131–141.] C. Amplitude functions for basilar membrane vibration at the 18-kHz point in the cochlea, measured at different intensities of stimulation. For a stimulus at 20 dB Sound Pressure Level (SPL), the vibration is sharply peaked, becoming more and more broadly peaked as the stimulus intensity is raised. [Reproduced, with permission, from B. M. Johnstone, R. Patuzzi, and G. K. Yates, 1986, Basilar membrane measurements and the travelling wave, *Hearing Res.* **22,** 147–153, Elsevier Science Publishers, Amsterdam.]

amplitude of vibration stimulates the outer hair cells still further, the vibration grows increasingly rapidly as it travels down the duct. But beyond a certain point in the duct, the mechanical parameters are such that a traveling wave is not possible for the particular frequency of stimulation, and the wave dies out abruptly. Thus, steep slopes are produced on either side of the peak of the traveling wave.

The mechanisms by which the outer hair cells generate motility is not known, although movements of various types (e.g., contraction or extension of the cell body, tilting of the cuticular plate, deflection of the stereocilia) have been shown in isolated hair cells stimulated *in vitro*. In addition, it is not yet known if these movements are of the right form to augment the vibration of the cochlear partition. However, unequivocal evidence indicates that some sort of movement-generating apparatus exists in the cochlea and the apparatus is closely allied to mechanotransduction. This evidence comes from the observation that under some circumstances the ear can emit sound.

To detect acoustic emissions, a small speaker and a sensitive microphone are sealed into the ear canal. If the ear is stimulated with a click, a long series of waves are found, following the direct pressure pulse produced by the stimulus. The waves can be easily seen in human subjects, the form of the waves being different for each subject. If the click is presented at a very low intensity (e.g., within some 10 dB of the subject's absolute threshold), then sometimes more energy is returned to the ear canal than was originally introduced. Indeed, in some subjects the stimulus triggers a long series of oscillations, which can continue for seconds, and in others continuous vibrations are detectable without any input. The latter observations show that the ear can generate movement rather than just passively reflect a proportion of the incident energy back into the ear canal. A number of experiments (e.g., the effect of toxic agents, stimulation of the olivocochlear bundle) further suggest that the motile mechanism is closely allied to the outer hair cells. Because the emissions need at least some normal hair cells (although they seem to reflect discontinuities in the properties of the cochlea along the duct), the emissions can be used as a basis for objective audiometry.

The degree of outer hair cell motility is probably under influence of the olivocochlear bundle, the pathway from the brainstem to the cochlea, some fibers of which give rise to synapses on the outer hair cells. Therefore, this forms a way in which the central nervous system can control the amplitude of vibration of the mechanical traveling wave. Evidence indicates that the bundle helps protect the ear from acoustic trauma, and it may be involved in selective auditory attention.

Since hair cells produce a nonlinear response to deflection of the stereocilia (Fig. 6), it can be expected that any motile response from the outer hair cells will also be nonlinear. In other words, there will be slow shifts in the position of the basilar membrane, together with harmonics of the stimulus frequency and, for multiple input frequencies, intermodulation products. Evidence indicates that such distortion components exist in the movement of the basilar membrane. The distortion components are produced at the point at which outer hair cell feedback is maximal (i.e., around the peak of the traveling wave). If the components are lower in frequency than the input frequency, they can give rise to traveling waves of their own, because they can then propagate away from the site of generation in the normal direction of the traveling wave (i.e., towards the apex of the cochlea). Two such distortion components are found at frequencies $f_2 - f_1$ and $2f_1 - f_2$ (where $f_2 > f_1$). These components can be detected both physiologically and psychophysically and may underlie some complex perceptual phenomena (e.g., the detection of musical consonance; auditory "streaming," i.e., the classification of sounds as being from a common source).

F. Gross Electrical Responses of the Cochlea

Correlates of hair-cell and auditory nerve-fiber activation can be measured by gross electrodes in and near the cochlea. Stimulation of the outer hair cells

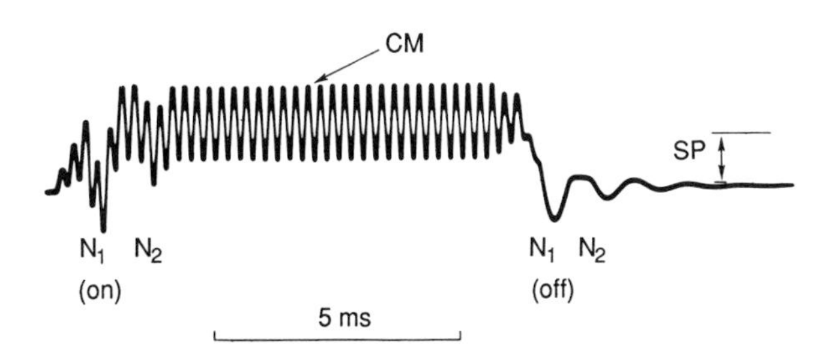

FIGURE 8 Diagram of the responses to a tone burst measured by an electrode on the round window. CM, cochlear microphonic; N_1 and N_2, compound action potentials; SP, summating potential (d.c. displacement of the microphonic). [Reproduced, with permission, from J. O. Pickles, 1988, "An Introduction to the Physiology of Hearing," 2nd ed., Academic Press, London.]

produces massed currents through the cells, giving rise to the cochlear microphonic, a potential which follows the waveform of the stimulus (Fig. 8). Nonlinearity in outer hair cell responses gives rise to d.c. components in the microphonic, producing the summating potential. Synchronized firing of auditory nerve fibers at the onset of a stimulus produces summed potentials called the compound action potential, or the N_1 and N_2 potentials. These potentials form important indicators of cochlear function in cases where it is not possible to measure the more detailed cochlear potentials.

IV. The Auditory Nerve

A. Anatomy

In the mammalian auditory nerve, 95% of fibers innervate the inner hair cells and the remainder innervate the outer hair cells. This indicates that the job of inner hair cells is to detect the movement of the cochlear partition and to transmit the information to the central nervous system. It must be presumed that the records that we have of auditory nerve-fiber activity arise from fibers innervating inner hair cells. In only one case do we have a proven recording from a fiber innervating an outer hair cell, and in this case the fiber showed no activity.

B. Tuning and Influence of Intensity

The firing of auditory nerve fibers as a function of frequency is primarily determined by the vibration of the cochlear partition. The tuning curves of auditory nerve fibers (Fig. 9) are similar to the tuning curves for vibrations of the basilar membrane (Fig. 7B). As far as current knowledge goes, any differences can probably be accounted for by the errors in measuring basilar membrane vibration.

The firing rate of auditory nerve fibers shows a sigmoidal relation with stimulus intensity, with the firing rate in most fibers reaching its maximum at a sound pressure of about 50 dB SPL (Sound Pressure Level), and staying constant for further increases in stimulus intensity. The intensity at which the firing rate reaches a just-noticeable increment above the background or spontaneous firing rate gives the threshold of the fiber: most of the fibers recorded in an animal have thresholds within 20 dB of the animal's absolute threshold. However, different fibers have different thresholds, those which show high spontaneous firing rates having lower thresholds than the others. Fibers of different

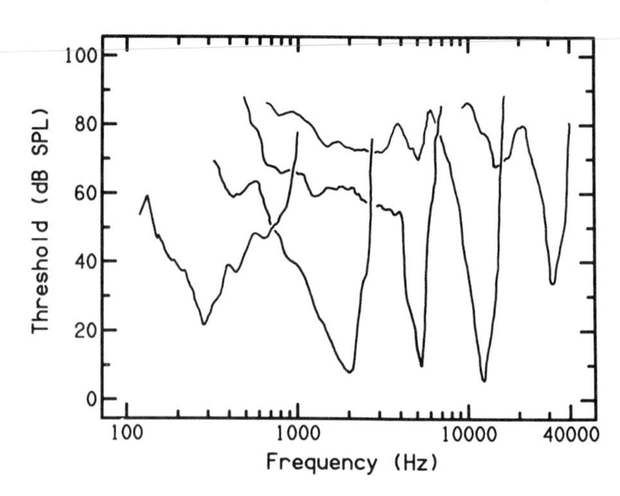

FIGURE 9 Tuning curves of five auditory fibers, all from the same cat. [Modified, with permission, from E. Javel, 1986, Basic response properties of auditory nerve fibers, *in* "Neurobiology of Hearing: The Cochlea" (R. A. Altschuler, R. P. Bobbin, and D. W. Hoffmann, eds.), pp. 213–245, Raven Press, New York.]

threshold and spontaneous rate innervate the same inner hair cells; therefore, it must be supposed that the variation in threshold results from differences in synaptic threshold.

C. Temporal Relations

The inner hair cell intracellular potential undergoes cyclic fluctuations in synchrony with the vibration of the basilar membrane. This evokes cyclic release of neurotransmitter (glutamate or aspartate are the favored candidates), producing phasic firing in auditory nerve fibers; however, such behavior is found only at low frequencies. As the stimulus frequency is raised above about 600 Hz, the number of phase-locked action potentials decreases. This occurs because at high frequencies the transducer current flows through the capacitance of the hair cell external membranes, decreasing the intracellular a.c. voltage response to the stimulus. How then do high-frequency stimuli evoke action potentials? The asymmetric relation between stereociliar deflection and intracellular depolarization (Fig. 6) means that for sinusoidal stimuli a net depolarizing potential is superimposed on the a.c. potential in the hair cell. The sustained depolarization contributes to the release of transmitter, but in a nonphase-locked manner. The balance between the two mechanisms changes with increases in frequency, so that above about 5 kHz no action potentials are phase-locked; therefore, temporal coding of auditory information is possible only below this frequency limit.

D. Response to Complex Stimuli

The cochlea functions nonlinearly, with the result that interactions are produced between the components of spectrally complex stimuli. One such interaction involves the generation of intermodulation products at frequencies $2f_1 - f_2$ (where $f_2 > f_1$) and $f_2 - f_1$, as described above. The products can be seen in auditory nerve activity in two ways. First, for frequencies of stimulation below the 5-kHz limit for phase-locking, the firing patterns of auditory nerve fibers contain intervals, which correspond to the periods of the distortion tones. Second, distortion products can activate auditory nerve fibers tuned to the frequency of the distortion products, even though the fibers may give no response to the primaries (i.e., to f_1 and f_2) when presented alone. This can be explained by a mechanism in which the distortion products are generated in the cochlea at the point at which the traveling waves to f_1 and f_2 overlap. The active mechanical process at this point then produces new traveling waves to the distortion tones, which peak at their own characteristic place.

A second nonlinear interaction can be seen in the operation of two-tone suppression. This is also thought, predominantly or in part, to arise from the nonlinear input–output functions of outer hair cells (Fig. 6). If one stimulus produces sufficient deflections to drive a cell to the flat part of the input–output function, a superimposed stimulus will not be able to produce any greater response. Therefore, the outer hair cells will not be able to contribute to the active amplification of the traveling wave to the second stimulus, and the peak of its traveling wave will not reach its normal size. Thus, the presence of one stimulus will reduce the response to the other, to the extent that the traveling wave to one overlaps the active process to the other. Two-tone suppression is particularly powerful when the suppressor lies just above the suppressed tone in frequency.

V. The Auditory Central Nervous System

A. Information flow

Auditory nerve fibers branch and innervate the three divisions of the cochlear nucleus (Fig. 10). The cells in the anteroventral cochlear nucleus have response properties similar to those of auditory nerve fibers. This nucleus therefore transmits auditory signals to the next stage, the superior olivary complex, with relatively little transformation. At

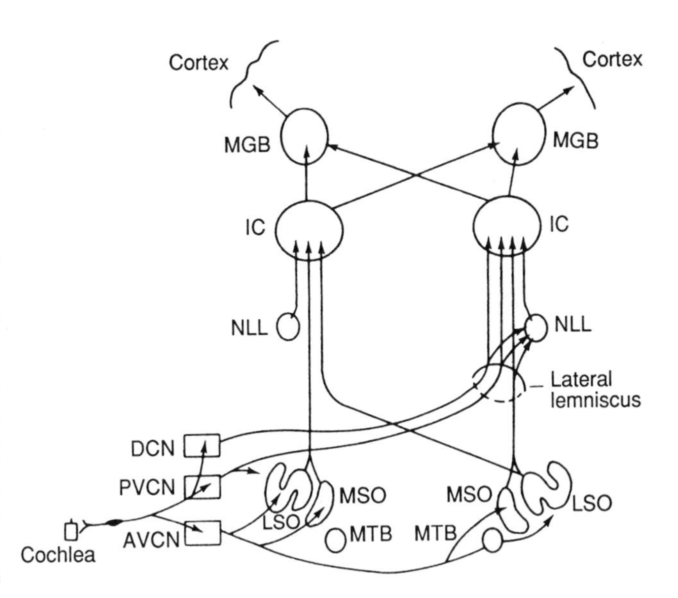

FIGURE 10 The main ascending pathways of the brainstem. Many minor tracts are not shown. AVCN, anteroventral cochlear nucleus; DCN, dorsal cochlear nucleus; IC, inferior colliculus; LSO, lateral superior olive; MGB, medial geniculate body; MSO, medial superior olive; MTB, medial nucleus of the trapezoid body; NLL, nucleus of the lateral lemniscus; PVCN, posteroventral cochlear nucleus. [Reproduced, with permission, from J. O. Pickles, 1988, "An Introduction to the Physiology of Hearing," 2nd ed., Academic Press, London.]

the other extreme, the dorsal cochlear nucleus has very complex neuronal responses, the cells showing tuning curves with many inhibitory areas, and complex temporal patterns of firing. The output of this nucleus bypasses the superior olivary complex to reach the inferior colliculus directly.

Significant binaural interaction in the auditory system first occurs in the superior olivary complex. The interaction is the basis of sound localization, the relevant cues being differences in the intensity and time of arrival of sounds at the two ears. Diffraction around the head produces significant interaural intensity differences for high-frequency stimuli. The lateral, or S-shaped, superior olivary nucleus responds to the intensity differences, sounds in one ear inhibiting the response to sounds in the other. The cells are also sensitive to differences in the time of arrival of transients in the envelopes of high-frequency stimuli. In contrast, the medial superior olive represents low-frequency stimuli. Such stimuli are in the range in which phase-locking is preserved, and here the sensitivity to differences in the times of arrival means that the cells are responsive to interaural phase.

Other subnuclei of the superior olivary complex relay information from one side of the brainstem to the other and also give rise to the fibers of the olivocochlear bundle (see above), the efferent pathway to the cochlea.

Activity from the superior olive is transmitted via the lateral lemniscus (some fibers synapsing in the nuclei of the lateral lemniscus) to the inferior colliculus, a substantial integrative and relay center of the brainstem. Here the location-specific information from the superior olive is combined with information from the complex pattern analyses of the dorsal cochlear nucleus. Thence, the information is transmitted via the specific thalamic relay for audition, the medial geniculate nucleus, to the auditory cortex.

Audition is represented in multiple areas in the cerebral cortex. In cats, the primarily auditory cortex is surrounded by fields called the secondary auditory (AII), anterior auditory, posterior ectosylvian, posterior, ventral, and ventral posterior fields. The secondary somatosensory area SII and the insulotemporal areas are also auditory fields. In primates, the auditory cortex is situated on the superior temporal plane in the lateral fissure, the primary field being surrounded by subsidiary cortical areas called the lateral, rostrolateral, and caudomedian fields, together with further auditory fields.

Cellular responses in the auditory cortex show tuning curves with single or multiple peaks. Some cells may respond only to complex auditory stimuli, such as frequency- or amplitude-modulated tones, or simulated speech sounds.

B. Principles of Organization

A number of principles of organization can be discerned in the central auditory system. First, the component nuclei are tonotopically organized; i.e., the cells are arranged in order of characteristic frequency. This facilitates interaction between neurons of adjacent characteristic frequency and may also represent economical packing of the afferent innervation. In the nuclei of the auditory brainstem and the thalamus, cells of the same characteristic frequency are situated in two-dimensional isofrequency planes across the nucleus. In the cortex, cells of the same characteristic frequency are situated in one-dimensional isofrequency strips across the cortical surface, separate tonotopic maps being found in several of the auditory cortical subareas.

Second, at and beyond the superior olivary complex, responses depend on the location of the sound source, cells in the higher nuclei being driven most strongly by stimuli on the contralateral side. Searches have also been made for location-specific information within each isofrequency plane or strip. Thus, in the primary auditory cortex cells of differing binaural interaction (e.g., cells excited by both ears, as against cells excited from one ear and inhibited from the other) are situated in discrete strips running at right angles to the isofrequency strips. Evidence also indicates that there is a crude map of auditory space within each isofrequency plane in the inferior colliculus. However, no correlate has been found in the mammal of the precise map of auditory space found in the owl, in its homologue of the inferior colliculus, the lateral dorsal mesencephalic nucleus.

Third, there is a general progression in the complexity of responses as one ascends the auditory system. For instance, tuning curves are exclusively simple in the auditory nerve and anteroventral cochlear nucleus, while both simple and complex tuning curves are found in the auditory cortex. However, the progression is far less clear than for instance in the visual system, and, except for sound localization, it has not been possible to be sure of the functional implications of any progression in complexity. Evidently, there are multiple parallel pathways for the extraction of much auditory information, and the hierarchical relation between the different stages and systems is not clear. Here it must not be forgotten that the auditory system also has a rich centrifugal innervation, with fibers running from the higher levels of the system to the lower. The olivocochlear bundle is one example of such a pathway. Therefore, the opportunity exists for the responses at the lower levels of the auditory system to reflect the complex processing of the highest.

VI. Biological Basis of Auditory Perception

A. Threshold

Measurement of the displacement of the basilar membrane suggests that the membrane vibrates by about 0.3 nm at an animal's measured behavioral threshold. If the organ of Corti and tectorial membrane move in a simple geometrical manner (by no means a reliable assumption), then at threshold the stereocilia would be deflected by about 10^{-2} degrees, the movement transmitted to the transducer channel being about 0.4 nm. This seems not an unreasonable value for the threshold movement at a channel. Because it is hypothesized that each of the hundred or so channels on a single hair cell is continually opening and closing under the influence of thermal energy, deflection merely causing a redistribution of the open and closed probabilities, the definition of threshold becomes a statistical one, and dependent on the amount of averaging employed. An animal's absolute threshold lies just below the best thresholds of the most sensitive of its auditory nerve fibers, probably a reflection of the averaging possible in an intact auditory nerve.

B. Frequency Resolution

Psychophysical frequency resolution is likely to be a direct reflection of the frequency resolution shown in basilar membrane vibration (Fig. 7B), as also reflected in the tuning of auditory nerve fibers (Fig. 9). While it is possible to measure similar tuning functions psychophysically in humans, care in interpretation is necessary if a detailed comparison is to be made. Functions in humans are plotted using more complex paradigms such as masking, where the threshold to one stimulus (the probe) is measured in the presence of another. Because the cochlea behaves nonlinearly, tuning to one stimulus may be affected by the presence of another, in experiments where the masker and probe are present simultaneously. Experiments where the masker and probe are present nonsimultaneously, such as forward masking, are thought to give more reliable measures of peripheral frequency resolution. Using such techniques, it is also possible to measure correlates of further phenomena found electrophysiologically in animals, such as two-tone suppression.

In experimental cases of cochlear pathology, the tuning curves of basilar membrane vibration rise in best threshold and increase in bandwidth. This most likely reflects loss of the active mechanism from the cochlear vibration. Psychophysical results in humans show similar changes, with an increase in threshold in the audiogram, and a loss of frequency resolution.

C. The Perception of Speech

Responses to speech sounds have been measured extensively in the auditory nerve and have shown how the different aspects of the stimulus are represented in the firing of auditory nerve fibers. The formants (spectral peaks) in vowels are for instance represented by increased levels of activity in the auditory nerve fibers tuned to the formant frequencies. At a higher level, the critical involvement of the auditory cortex in speech processing is demonstrated by disabilities in the production and understanding of speech after lesions of the speech areas, namely Wernicke's area in the dominant (normally left) cerebral hemisphere and Broca's area. The involvement of these areas is also shown through selective local increases in blood flow while listening to speech. It is also possible to use the asymmetry of the involvement of the two cerebral hemispheres in speech to tag anatomical specializations related to speech analysis: for instance, area 22, which contains Wernicke's area, is larger on the dominant side and contains certain neuronal specializations. However, caution is necessary, because not all hemispheric asymmetries in the auditory areas may be related to speech. In addition, recordings made in the auditory cortex of unanaesthetized human patients have shown that cells in the superior temporal gyrus respond efficiently to speech sounds and less effectively to other auditory stimuli. Wernicke's area was not explored in such operations, lest speech function be damaged. However, it is probably true to say that we have little or no information on the physiological mechanisms underlying the linguistic, as against the acoustic, aspects of speech.

Bibliography

Altschuler, R. A. , Bobbin, R. P., and Hoffman, D. W. (eds.) (1986). "Neurobiology of Hearing: The Cochlea." Raven Press, New York.

Hudspeth, A. J. (1989). How the ear's works work. *Nature* **341,** 397–404.

Irvine, D. R. F. (1987). The auditory brainstem. *Prog. Sens. Physiol.* **7,** 1–279.

Jahn, A. F., and Santos-Sacchi, J. (eds.) (1988). "Physiology of the Ear." Raven Press, New York.

Moore, B. C. J. (ed.) (1986). "Frequency Selectivity in Hearing." Academic Press, London.

Moore, B. C. J. (1989). "An Introduction to the Psychology of Hearing," Academic Press, London.

Pickles, J. O. (1988). "An Introduction to the Physiology of Hearing," 2nd ed. Academic Press, London.

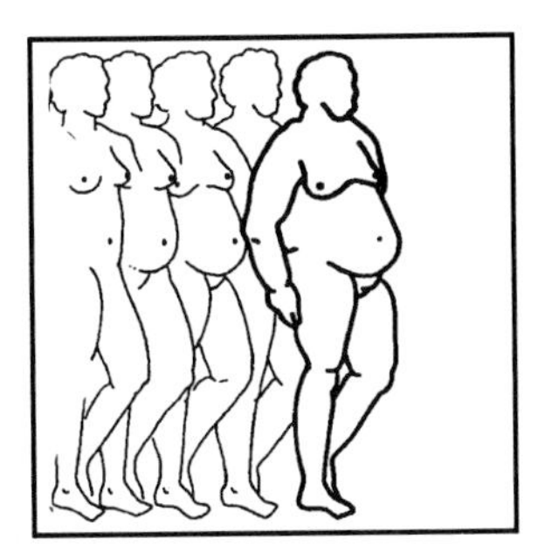

Eating Disorders

JANE MITCHELL REES, *University of Washington*

Glossary

Anorectic Pertaining to or affected with anorexia
Anorexia nervosa Nervous disorder in which a person eats very little food
Binge/gorge Episode of eating an unreasonably large amount of food
Bulimia nervosa Nervous disorder in which a person eats much more food than necessary and then voluntarily vomits or purges
Bulimic Pertaining to or affected with bulimia
Calorie Unit of measuring heat energy, commonly used to measure food energy
Developmental obesity Having an increase in weight beyond reasonable limits due to accumulation of fat throughout development
Endocrine Organs and structures of the body that produce hormones to regulate body processes
Hormones Chemical produced by an organ of the body that specifically regulates another organ and, therefore, body processes
Hyperphagia Eating larger than reasonable amounts of food
Purge To make the bowels empty out
Regurgitate To bring up undigested food
Rumination Disorder in which a person regularly brings up food to be chewed again after meals
Semistarvation Being nearly deprived of food for a long period

PERSONS WITH EATING disorders (anorexia nervosa, bulimia nervosa, rumination, and developmental obesity) fail to develop rhythms, or develop distorted rhythms, of eating and exercising. Their habits turn into obsessive abuses, taking food out of its normal function, namely that of maintaining life. The energy stored in their bodies as fat increases, decreases, or swings in wide variation, jeopardizing overall health. The disordered behaviors are linked to psychological inadequacies, as both causes and effects. People with eating disorders improve when the psychological basis of their habits are altered and the physical symptoms are treated by a team of health care specialists.

I. The Spectrum of Eating Disorders

When the physical symptoms of eating disorders are illustrated on a spectrum, as in Fig. 1, anorexia will be at one end and developmental obesity at the opposite end. In the middle of the spectrum, the figures represent individuals of normal or close to normal weight. Examples are those with bulimia nervosa and rumination, recently described disorders.

A. Anorexia Nervosa

In anorexia nervosa, excessive restriction of food energy (measured in calories) and strenuous exer-

Encyclopedia of Human Biology, Volume 3.

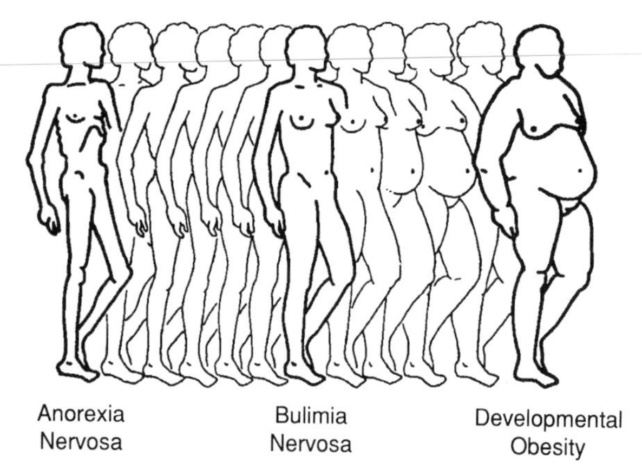

FIGURE 1 Spectrum of eating disorders. Underlying psychosocial characteristics are held in common while physical conditions vary across the spectrum. [Adapted, with permission, from J. M. Rees, 1984, Eating disorders, *in* "Nutrition in Adolescence (L. K. Mahan and J. M. Rees, eds.), p. 105. C. V. Mosby, St. Louis.]

cise depletes the energy pool of the body to the degree that the individual could die if not treated. The observable motivator of the affected person is a fear of fatness.

B. Bulimia Nervosa

In bulimia nervosa, the person who is trying to control body weight eats more food than he or she needs, or thinks is needed, and gets rid of it by vomiting and/or purging. Exaggerated feelings of inadequacy generally focus on the physique; the affected person is hypercritical of his or her anatomical form.

C. Rumination

The ruminator chews, swallows, and then regurgitates food and swallows it in a repetitive cycle after meals, apparently as a response to stress. Rumination is reportedly practiced most frequently by people who push themselves to be high achievers.

D. Developmental Obesity

The developmentally obese person grows up storing excess energy as fat, eating more than is necessary and exercising less than normal. This condition could be called hyperphagia nervosa to parallel the nomenclature of other eating disorders and distinguish it from overfatness not associated with psychological dependency on abnormal eating and exercise habits. There are sufficient other causes of storing excess energy as fat that the conditions are called "the obesities" by experts, even though the term obesity is still usually used to denote anyone weighing 20% more than normal body weight. [*See* OBESITY.]

II. Physical versus Psychological Problems

Most people focus on the physical characteristics of eating disorders because they are the most obvious. Average citizens can identify dancers who are skeletonlike rather than lithe and people who are so heavy they cannot move easily. Less commonly understood are the important underlying psychological problems associated with the disorders along the spectrum, problems that interfere with the ability of those affected to function normally and productively.

Most prominently, people with eating disorders use food inappropriately and obsessively in response to life stresses. Being preoccupied with weight and food, they have difficulty meeting developmental milestones and becoming independent adults. They have some degree of problem with experiencing bodily sensations arising within themselves as normal and valid, formulating their own value system instead of relying on social opinion, and realistically perceiving their body size as well as setting expectations for achievements and roles. Thinking flexibly and in the abstract, especially about themselves, is difficult. They do not conceive of food as nourishment, know the value of nourishing themselves, or the ways to do it.

III. Theories Regarding Etiology

Early psychoanalytic theories suggested eating disorders grew out of an inability to deal with sexual drives. In a later period, they were ascribed to abnormalities in the endocrine system. Recently, disturbances in the mother–child relationship and, more completely, interaction of the whole family are observed to foster the development of eating disorders and other psychosomatic diseases. A biopsychosocial explanation of the origins of disordered eating incorporates the previous theories, suggesting that a complex mix of characteristics leads to eating disorders. Thus, an individual born into a family whose interaction does not allow normal development of autonomy feels inadequate to cope with life and adopts abnormal eating and exercise behaviors. Patients may do this for the express purpose of changing their bodies or because they find gratification in the habits themselves. They may be fed inappropriately or taught aberrant eating and exercise behaviors by family members. Their bodies may also harbor a propensity for a biochemical imbalance.

However they originated, the abnormal behaviors and psychological disturbances often alter body function, including endocrine patterns. Thus, sexuality is indeed influenced through related changes in sex hormone production, an effect if not a cause of severe eating disorders, validating the original observations. The processes vary in each of the disorders, but family, individual psychological, and biochemical characteristics are all involved at different stages.

IV. Family Characteristics

The family that engenders eating disorders often tends to be overprotective and rigid rather than flexible. Conflicts are avoided and go unresolved because they are not faced openly. Children are enmeshed in the family, a characteristic that stands in the way of normal development. Because parents do not communicate well with each other, an affected child is pulled into their interactions as a kind of go-between, a situation termed triangulation. This role pressures the child, who is not free to be childlike and, in turn, is distracted from concentrating on his or her own development. While not all families follow the same patterns, disturbances leading to unreal expectations are usually a contributing factor.

V. Cultural Influences

Several cultural influences are partly responsible for the recent proliferation of eating disorders in the United States. Valuing a slim physique together with fat- and sugar-laden foods is a dichotomy that is played out in many lives, either casually, by restricting food for a while after eating rich meals, or in the extreme, as in eating disorders. Psychological stress levels and demands for achievement in modern life are also strong influences. Women are more vulnerable to eating disorders than in former eras because they are faced with greater role expectations combined with tightly circumscribed standards for physique. These cultural features exist to some degree in all affluent nations but have developed to the extreme in the United States.

Collectively, society does not have a reasonable concept of nourishment, the realization that human beings eat food to stay alive, and that materials they get in food are built into their bodies. They also do not realize that about three-fourths of the energy in the food they eat is used in the basic processes that keep them alive at rest, the other used in moving and working. The constant misuse of the word calorie encourages misunderstanding. To many people, calories are equated with mass or volume (chunks or lumps), or simply fat. When they realize the term is a unit of measure and energy is the thing being measured, they can understand more easily that it is necessary, positive, and dynamic and can be controlled and used. They begin to see food energy is like oxygen, a basic element of life; that it should not be restricted or overconsumed as influenced by fashion, or used frivolously; and, further, that their bodies are, themselves, needing to be cared for, not spoiled by overeating or punished with "dieting."

VI. Anorexia Nervosa

Because families of anorectics do not allow conflicts to surface, the symptoms seem suddenly to emerge in the children, usually female of model families. It is as if the patient, feeling inadequate to cope with the complex life of adolescence, regresses to the earliest ways of showing independence—regulating food—to feel she is in control of at least one facet of her life. The resistance of patients and their families to treatment stems from their desire at any cost to keep up the appearance of having no problems.

Quite a number of anorectics are overweight before starting to "diet." They take their expectations for slimness from models, dancers, and media stars, carrying them to extremes. A parent interested in food production or obsessive about exercise will have a strong influence. The affected child tends to be more passive than siblings and becomes enmeshed and affected. No predisposing biochemical vulnerability has been found so far, although some experts believe physical factors contributing to the onset of the disorder will be found. The body changes measurably throughout the course of the disease, in direct relationship to the amount and length of starvation, and recovers when nourishment is replenished.

According to surveys, about 1 in 100 affluent, highly educated and 1 in 200 of the general population of young women have anorexia nervosa. Only about 10% of patients in treatment are male. Throughout the anorexia and bulimia sections of this paper the female pronoun will be used because most anorectics and bulimics are female.

A. Physical Manifestation

The person with anorexia nervosa will initially try to reduce to an unreasonably low weight by eating less and less and exercising obsessively. She also may force herself to vomit or take laxatives to purge herself. At the same time, she isolates herself from peers and is argumentative with her family. The menstrual cycle ceases, so patients are infertile. The psychological disturbance may be responsible initially. When weight is sufficiently low for any reason, the menstrual cycle will stop due to decreased endocrine function, which is also responsible for the decrease in sex drive. Growth, sexual development, organ (including brain) makeup is interrupted during starvation.

Patients need comprehensive treatment with psychological, nutritional, and medical components. The focus will be the appropriate role of parents in setting limits while allowing the patient to develop without abnormal intrusion or pressures, rather than on the bizarre eating and exercise rituals alone. As the person with anorexia and the family respond to treatment, the anorectic will be able to accept new ideas and resume healthy eating and exercise behaviors as advised by nutrition and medical professionals. Progress toward physical health is monitored throughout treatment.

B. Crisis

If the anorectic, and her family, is untreated or does not respond well, her physical condition will often deteriorate to a state of outright starvation. With inadequate energy supply, the body fat stores are depleted, and muscles begin to be used up for immediate nourishment. Breathing and heart rate slow and the skin is dry and scaly; downy hair grows over the face, trunk, and appendages. Meanwhile, hair on the head is thin, dry, brittle, and falls out. The slow heart beat causes low blood pressure with fainting spells. Sleep, adjustment to temperature change, and bowel regularity are disturbed. Cracks may appear at the corner of the lips, fingernails become brittle, and tips of fingers, toes, and ear lobes turn blue. Body fluids and their contents are imbalanced and flow outside their usual compartments causing the body to swell abnormally.

Thoughts of the starving person are centered on food while coherent, creative thinking is impaired. The person is apathetic, dull, exhausted, uninterested in sex, and depressed.

A supportive but firm hospital program replenishes energy stores so that the body systems, including the brain, begin to function again. With weight gain, the life-threatening phase is over and psychotherapy can continue.

C. Long-Term Recovery

Full recovery requires a long period—*years*—of counseling and education. Leaving the hospital at near normal weight is the real beginning of treatment. Individual psychotherapy and family therapy is the key. The process is time-consuming because it is complex. The person will develop full potential as an autonomous adult after she and her family understand and can practice appropriate communication and interaction. Through information and changed attitudes and behavior gained in nutrition counseling, the anorectic will learn to control body weight at a reasonable level throughout the various stages and activity levels of the life cycle, without obsessive behaviors.

D. Outcome

Because of early recognition and modern treatment of anorexia nervosa, the death rate is down to 4%, or less, of treated patients as opposed to 10% in the 1970s. People can make a full physical, psychological, and social recovery if they have comprehensive treatment for a number of years. If patients and their families leave treatment after the physical crisis they continue to be obsessive and are not able to educate themselves, work at full potential, or to establish healthy families; they often have repeated periods of starvation, bulimia, or become obese.

VII. Bulimia Nervosa

Bulimia probably develops under the same set of circumstances as anorexia nervosa, although the goals and habits of many bulimics are not so self-destructive. Families are often more hostile and chaotic than those of anorectics and they develop symptoms a little later in life. They try to eat only small amounts of food to control body weight, but the body responds physically and psychologically to temporary starvation, triggering a binge. They vomit and/or purge to rid themselves of the food they have eaten.

Five to 10% of college females have bulimia nervosa while 20–25% practice some bulimic behaviors although they do not have the complete disorder. Less than 1% of college males have bulimia nervosa. Persons with bulimia are usually a little older (late adolescent to young adult) than those with anorexia nervosa.

A. Physical Symptoms

Bulimics may arise each morning thinking they will follow their ideal semistarvation eating plan and do so for most of the day. When they are alone and unoccupied, however, the gorging begins. It will end when they are overfilled and they force themselves to vomit. Vomiting gets easier with time, and they can do it without great effort. They may take laxatives and diuretics to purge themselves. Unless the habits are drastic, they can practice them for many years without serious health problems. Most prominently, the stomach acid they vomit will erode the teeth, irritate the throat, and inflame the esophagus. Salivary glands swell, lips crack, and blood vessels in the face may be broken by the pressure of vomiting. Their hands may be calloused from resting their upper teeth on them as they vomit. They usually are at near normal weight and are well enough nourished that they would remain fertile. The menstrual cycle is affected in as many as half, however, probably as result of psychological stress. They do not experience the interruptions in sexual function that anorectics do.

They need hospital treatment if they also are starving themselves or if they take many medications and vomit or purge so frequently that the body fluids and its contents are imbalanced, holes develop in the esophagus, or the kidneys are damaged.

B. Psychological Characteristics

Overall, bulimics feel secretive and guilty. They fear rejection, abandonment, and failure. An all or nothing attitude rules their expectations. They lack confidence related to judgements they make about their bodies. As part of the maladaptive concept of nourishment, they also harbor distorted beliefs about food and find it hard to accept their bodies and the way they function and look to others.

C. Treatment Goals

Psychological counseling is aimed at dispelling the negative feelings bulimic patients have about themselves and giving them the inner strength to cope with life as independent adults. Family counseling helps young persons still living at home to develop normally. With nutrition education and counseling, they nourish themselves adequately, set weight goals appropriately, and give up bulimic habits.

D. Outcome

Effective programs help people with bulimia to adjust psychosocially and improve their abnormal eating behaviors. Like anorectics, they need several *years* of therapy. Patients continue to improve while they are in treatment but unfortunately, also like anorectics, they often leave treatment before they have been fully rehabilitated.

VIII. Rumination

Some people ruminate as part of another eating disorder and others practice it as a sole disorder. The number and gender distribution of people who are affected is not well known and therapeutic programs are not common. It does not seem to cause nutritional problems or damaged teeth; one of the most common complaints of ruminators is that they have foul smelling breath. Helping people to deal with anxiety and monitor their regurgitation will most likely lead to success in giving up the habit.

IX. Obesity

Obesity is one of the biggest health problems in the United States, affecting as many as one-third of

Americans. The number of obese children alone rose by about 40% from the mid-1960s to the late 1970s. Debilitating conditions such as heart, blood vessel, and lung disease, diabetes, and some types of cancer are linked to obesity.

The prejudice of society, including health professionals in blaming the obese for causing their own problems by lack of will power, laziness, and gluttony, intensifies the dependency on simplistic treatments and hinders the search for reasonable solutions to obesity. In actuality, the obese need support and comprehensive long-term treatment to deal with difficult physical and psychological problems that arise from a multitude of factors, many outside their control. Many do not have psychological problems and as a group they do not demonstrate psychopathology. Treatment enables them to control the symptoms; there is no cure.

A. Categories of Obesity

1. Developmental Obesity

In a parallel to anorexia nervosa, disturbances in families and individuals may lead to developmental obesity as depicted in the illustration (Fig. 1). The family interaction stunts psychosocial development and encourages overeating and underexercising. Growing obesity, in turn, interferes with normal development by hampering actual movement and making the obese child feel negative about himself and his or her experiences with others.

2. Reactionary Obesity and Connections With Other Psychological Disturbances

Some developmentally obese are not affected all the time but revert to abuse of food, moving and interacting with others less as a reaction to loss or other trauma. Obesity is also associated with serious psychological disorders in some individuals.

3. Culturally Based and Behaviorally Fostered

The popularity of rich foods and emphasis on eating for gratification, celebration, and reward fosters obesity in some people who are not otherwise predisposed as well as intensifies the problems of those who are. Culture fosters obesity in families and individuals who do not know how to take precautions against influences such as advertising and constant availability of prepared food.

4. Physiological

Physically inherited body processes that store excess energy as body fat cause some obesities. However the obesity started, storage of excess energy activates other processes that work to sustain obesity much like a thermostat maintains heat. Both the inherited and acquired processes that conserve energy are survival traits, rooted in history when starvation was a great threat to humans.

B. Physical Factors

A person who is obese does not have to overeat to stay obese. If the person overeats, he or she will gain, and gain more easily, than a lean person. Compared with the lean body, the bodies of obese people save energy, for example, by producing less internal heat and pumping less sodium between cells. They store more fat because they have more energy to store and because the chemicals that allow fat into fat cells are more plentiful. Weight loss improves some, but not all, of these differences in the way energy is used and stored by the obese. The size of fat cells can be reduced, for example, but the number cannot. The reduced fat cells tend to fill again with ease. None of the processes can be treated with drugs at this point. Investigation of energy conservation processes has led to the set point theory: The cumulative effect is to maintain an excessive weight—the set point—which the body regains easily if weight is ever reduced.

Carrying excess weight stresses and causes disease in the heart, blood vessels, lungs, spine, and weight-bearing joints of the obese. A disordered endocrine system can interrupt the menstrual cycle in women, who may be infertile as result.

C. Psychological Factors

The developmentally obese have trouble separating from parents and tend to live passively and to feel they cannot accomplish what they set out to do. Secondly, a person who has become obese through purely physical factors as well as the developmentally obese suffers psychological damage through societal prejudices in education, employment, and social settings. Coping with isolation, name-calling, and scapegoating throughout childhood and adolescence affects personality development profoundly; they take on many of the psychological characteristics of the developmentally obese. A result is that many obese people feel guilty each time they eat rather than understanding that they need a certain amount of food to stay alive no matter what they weigh.

Finally, even the psychologically healthy individual faces a set of psychological problems in trying to control weight. He or she will have to learn to break tasks down into accomplishable segments with rewards, have the patience to wait for long-term results, and to enjoy the process. The effort is comparable to that required by an average person who for some reason would need to train for running a marathon or prepare to be a concert-level musician and must be conceptualized as being comparably demanding.

D. Cultural Factors

Overfeeding associated with psychological problems or even merely social customs in a family where the physical predisposition to obesity is genetically passed on is especially detrimental. In a society where using body energy is made obsolete by mechanization and energy-rich foods abound, the results are devastating. A significant number of people in the United States are two to three times heavier than normal people.

E. Treatment Objectives

The first goal of obesity treatment is to slow down the rate of gain, if the person is still gaining, and the second is to stop gaining. Because of individual variation in the physical potential to lose weight, focusing on weight loss as a short-term goal is unreasonable. Weight loss has proved nearly impossible to maintain if it is rapidly accomplished by severely restricting food. Semistarvation diets, which have been the mainstay of treatment for obesity, appear self-defeating. If a person eats much less than the amount of energy used, the systems that defend body weight are challenged, and the tendency to store excess fat increases in response to the starvation. Severe energy-restricting "diets" are the root of the dismal experiences the obese have in dealing with their problem. An even higher set point weight, bringing an even greater problem of control, may be avoided by concentrating on the knowledge, attitudes, and behaviors that control weight.

Learning to eat only the needed amount of food (especially fatty food), but not to restrict it; increasing body movement; and improving psychosocial adjustment helps patients stay at a weight as close to normal as their physical characteristics will allow. Any person trying to learn to control weight needs psychological support, individualized if the obesity is severe, and counseling if there is preexisting or acquired maladjustment.

Acknowledging that weight management is multifaceted and requires complex treatment increases the chances for controlling symptoms in any of the obesities. Most successful programs are carried out by teams of nutritionists and medical and psychosocial health specialists. As with the other eating disorders, a number of *years* are required while people learn and practice behaviors they will need to continue throughout their lives.

F. Outcome

Results of recorded studies show that about 95% of people fail to maintain a body weight they arrived at by a few months of drastic dietary treatment. These numbers do not include people who have learned to deal with obesity on their own or through individualized programs. Complete treatment strategies that facilitate more complete and sustainable long-term outcomes are beginning to be set up. This will increase the number of people who can control the symptoms of obesity in the near future.

X. Conclusion

A large portion of Americans are affected by eating disorders. Many others are overly concerned about weight and shape but do not have a complete disorder. Treatment helps disordered eaters put food and exercise in perspective and deal with psychosocial maladjustment. It is of primary importance that the complexity of the disorders be kept in mind by professionals and patients as they attempt to learn about, prevent, and deal with this spectrum of problems that impairs the function of so many in modern society.

Bibliography

Bennett, W. (1987). Dietary treatments of obesity. *Ann. N.Y. Acad. Sci.* **499,** 250–265.

Blinder, B. J. (1986). Rumination: A benign disorder? *Int. J. Eat Dis.* **5,** 385–386.

Bruch, H. (1973). "Eating Disorders." Basic Books, New York.

Comerci, G. (1990). Medical complications of anorexia nervosa and bulimia nervosa. *In* "Adolescent Medicine," Vol. 74 (J. A. Farrow, ed.). Medical Clinics of North America, pp. 1293–1310. W. B. Saunders, Philadelphia.

Garner, D. M., and Garfinkel, P. E. (eds.) (1985). "Handbook of Psychotherapy for Treatment of Anorexia Nervosa and Bulimia." Guilford Press, New York.

Keys, A., *et al.* (1950). "The Biology of Human Starvation," Vols. I and II. University of Minnesota Press, Minneapolis.

Minuchin, S., Rosman, B. L., and Baker, L. (1978). "Psychosomatic Families: Anorexia Nervosa in Context." Harvard University Press, Cambridge, Massachusetts.

Rees, J. M. (1984). Eating disorders. *In* "Nutrition in Adolescence" (L. K. Mahan and J. M. Rees, eds.), pp. 104–137. Times/Mirror Mosby, St. Louis.

Rees, J. M. (1990). Management of obesity in adolescents. *In* "Adolescent Medicine," Vol. 74 (J. A. Farrow, ed.). Medical Clinics of North America, pp. 1275–1292. W. B. Saunders, Philadelphia.

Stunkard, A. J. (ed.) (1980). "Obesity." W. B. Saunders, Co., Philadelphia.

Ecotoxicology

F. MORIARTY, *Consultant Ecotoxicologist*

Glossary

Bioaccumulation Increase of pollutant concentration in an aquatic organism from that in the ambient water, taking into account both direct intake from the water and intake from food

Bioconcentration Similar meaning to bioaccumulation, but indicates intake from water alone

Chemical speciation Occurrence of well-defined chemical entities such as ions, molecules, and complexes. Of particular relevance for heavy metals and some air pollutants

Ecosystem Community [an assemblage of species that forms a distinct system (e.g., the species in a lake)] and its habitat (e.g., the rocks, sediments, and water in a lake)

Environment All an organism's surroundings, both the habitat and the other plants and animals, including other members of its own species

Gene pool Collection of genes that is distributed among the individual members of a population

Habitat Inanimate, or abiotic, components of the environment, for an individual, population, or community

Pollutant Substance that occurs in the environment at least in part as a result of human activities and that has a deleterious effect on living organisms. Sometimes useful to distinguish from a contaminant, for which there is no reason, or no evidence, to suggest that there are deleterious biological effects

Population Group of individuals of the same species that are within a defined area, or, alternatively, that have the possibility of mating with each other, provided of course the needs are met for opposite sexes and maturity

ECOTOXICOLOGY deals with the effects of pollutants on ecosystems. Attention is often, and properly, focused on the consequences for individual organisms of exposure to pollutants. However, apart from our own species and for cultivated and domesticated species, the important effect is on populations of individual species, not on individual organisms. This shift of emphasis, from individual to population, makes a sharp distinction from environmental toxicology (*quod vide*). Thus the fact that a pollutant kills, say, half of the individuals in a species population may have little or no ecological significance, whereas a pollutant that kills no organisms but retards development may have a considerable ecological impact.

There are many potential pollutants. It has been estimated that, worldwide, about 63,000 chemicals are in common use, with 3,000 compounds accounting for almost 90% of the total weight of chemicals produced by industry. In addition, an estimated 200–1,000 new synthetic chemicals come onto the market each year.

A considerable amount of legislation and regulation has been developed in many countries during the past three decades, aimed at avoiding or controlling pollution, but the scientific basis for the development of effective control is often incomplete. Because of the potential economic and environmental costs that are often associated with any proposed controls, decisions can, therefore, be difficult to make, and these difficulties can be exacerbated by lack of agreement on what constitutes adequate proof. The degree of proof required can range from the balance of probabilities to absolute proof.

I. Types of Approach

Many sections of human society are concerned about the effects of pollution, approach the problems from different viewpoints, and consequently organize the subject in different ways. Three approaches are relevant here.

Administrators and regulators in governments tend to view the environment by habitat, when reference is made, for example, to air pollution, or of pollution in land, freshwater, or marine environments. This approach simply reflects the way in which administrators subdivide the world around them into sectors, but it is now becoming appreciated that sometimes efforts to reduce the degree of, say, air pollution can increase the degree of pollution in, for example, freshwater. There is therefore now a tendency to seek that combination of methods of control and of waste disposal that minimizes the total extent of environmental damage, rather than minimizing sectoral damage. The second approach occurs commonly in industry, where the immediate concern is for the risk of pollution from the chemicals produced by that particular industry, be they pesticides, heavy metals, detergents, or whatever.

Both these approaches, by habitat and by chemical class of pollutant, are perfectly valid, and most books and articles about pollution approach the subject in one or both of those ways. However, for a science of ecotoxicology, another approach is needed. It is more effective to ask rather different questions, which apply to all habitats and chemicals, questions such as the mode of action, exposure–dose relations, and interactions of one pollutant with other pollutants and with other environmental variables.

For an individual organism, a pollutant is just one of a large number of environmental variables that can affect it. The net effect on a population of an environmental change (e.g., exposure to a pollutant) can be measured by its impact on population size and on the population's gene pool. Stable populations can only change in size if at least one of four processes is affected: birth rate, death rate, gains from immigration, and losses from emigration. This is why sublethal effects can be more important ecologically than acute toxicity.

In practice these simple principles can be difficult to apply. Populations are often not stable—size and age distribution can change, sometimes rapidly—and rarely if ever are all environmental variables, bar the one of interest, constant. It follows that field observations of correlations between a change in the amount of pollutant present, in either the environment or the individual organisms, and change in population size or gene pool do not necessarily indicate an effect by the pollutant. Extrapolation from laboratory tests to field situations is also not straightforward, because the rest of the environment can profoundly modify an organism's response to pollutants.

II. Mode of Action

Pollutants can affect populations indirectly, by altering the habitat, or directly, by affecting individual organisms.

Alteration of the habitat can have complex consequences, whose precise details can be difficult to unravel. Eutrophication (the increase of plant nutrients in lakes, rivers, and coastal seas) is a classic example. It is a natural, slow process in many lakes, but human activities have accelerated the process enormously in many parts of the world. Nitrogen and phosphorus, commonly occurring as nitrate and phosphate, are usually the two major nutrients whose availability limits plant growth. Increased amounts of nutrient from, for example, nitrogenous fertilizers and phosphate in detergents can alter the abundance and species composition of both algae and higher plants, with consequent effects on the fauna. In extreme cases, sufficiently dense algal blooms develop to deplete the increased levels of nutrients; the algae then die and decompose fast enough to deoxygenate the water, when other species may also die. The important feature is that these changes in species composition and abundance do not result from any direct toxic effect by the pollutants, but result from differential effects on the abilities of different species to survive and reproduce themselves.

Direct toxic effects on individual organisms result from the combination of pollutant molecules with receptors. Thus the active forms of organophosphorus insecticides combine with, and inactivate, esterase enzymes, which are the receptors for these compounds. The critical biochemical lesion for acute poisoning from these insecticides is the inhibition of one of these enzymes, acetylcholinesterase (AChE), which breaks down acetylcholine at the synapses between nerves and at neuromuscular junctions. Inhibition of AChE enables acetylcholine to persist for much longer than usual, so that nerve function is grossly disturbed, which can then lead to death. We do not know the details of the mode of action for most toxic pollutants, only the symptoms of poisoning.

Many non-ionized organic compounds affect organisms, not by combining with specific receptors, but simply by being present within cells at sufficiently high concentrations. The theory is that like general anesthetics, they occlude space within the cells and so disrupt normal metabolic sequences.

III. Examples of Effects

Effective assessment and control of pollution requires a multidisciplinary approach. The complexities, and many of the relevant considerations, can best be illustrated by real problems caused by pollutants with different modes of action.

A. Impact on Habitat

Estuaries in general are vulnerable to pollution. Not only do they attract large towns or cities, but they can receive many pollutants via the freshwater rivers that feed into them, so that they often receive large quantities of human organic wastes and also agricultural and industrial wastes. Two general points emerge. First, human wastes break down eventually to harmless compounds such as carbon dioxide and water, so it is possible to release a certain amount of sewage effluent without causing widespread harm to the environment. For more persistent pollutants that do not break down readily or at all, greater efforts are desirable to avoid release, because once released into the environment they will be difficult to control. Second, we have to reconcile two conflicting attitudes, that environmental contamination should not be allowed to harm wildlife and that to reduce the amount of contamination released into the environment increases costs and may cause unemployment. A value judgment has to be made about the minimum acceptable degree of conservation or, conversely, the maximum acceptable degree of environmental damage.

The city of London, along the banks of the tidal River Thames, provides a well-documented example of estuarine pollution. Release of a sufficient quantity of organic human wastes into a waterway deoxygenates the water as bacteria metabolize the organic compounds, and eventually, when no oxygen remains, bacteria obtain oxygen from sulfates and so release sulfides, which are toxic to aquatic life. This stage developed in London during the years 1800–1850, when the human population doubled in size, human wastes, which had previously been deposited on land outside the city, were discharged into the river, and there was also additional contamination, in particular from gas and coke manufacture as one part of the Industrial Revolution. Successively better sewerage systems were developed during the next 130 years, which both released effluent at more suitable sites and produced effluent of higher quality (i.e., less damaging to the environment). In the 1970s, the management objectives became to support the fauna on the river bed, which is essential for sustaining sea fisheries, and to allow the passage of migrating fish at all states of the tide. The latter is taken, for operational purposes, to indicate that the average oxygen level in solution in the river in the third quarter of the year (when temperatures are highest) should not drop below 30% of the saturation value. Some considerable success has been achieved, symbolized by a salmon caught in 1974, the first recorded for 141 years.

It is becoming accepted in many parts of the world that human settlements and economic development should take care to sustain the yields of useful crops from the environment. Criteria other than purely economic are often involved, as in the above example of salmon migrating along the Thames estuary, which was achieved at considerable financial cost. The degree of care needed to protect human life and health can usually be agreed, but there is far less agreement about the appropriate degree of protection for wildlife.

B. Direct Toxicity

Widespread concern about the possible ecological effects of chemicals developed during the 1950s and 1960s when some agricultural pesticides were found to affect wildlife. Much attention focused on the peregrine falcon (*Falco peregrinus*) in both North America and Great Britain.

The size of the British breeding population appears to have been stable for several centuries, but it declined sharply and rapidly in the late 1950s.

Two insecticides were involved, *p,p'*-DDT and dieldrin.

From 1947 onward, peregrines started to lay eggs with thinner, weaker shells, and the incidence of broken eggs within clutches increased from 4% to 39%. This thinning was not caused directly by DDT, but by DDE, a persistent metabolite of DDT. Although it reduced breeding success, population size did not appear to have been affected. In North America, exposure to DDT alone was sufficient to reduce the size of the breeding population. In Great Britain the population did not decline until it also suffered acute toxicity from dieldrin.

Dieldrin was used extensively in the late 1950s as a seed dressing for cereal seed, and many incidents occurred in the Spring of dead and dying birds, including pigeons, that had been poisoned by eating dressed seed. Peregrines commonly feed on pigeons, and it was estimated that consumption of two or three pigeons heavily contaminated with dieldrin could be sufficient to kill a peregrine.

This episode exemplifies several features of general relevance:

1. The survey that established the decline of the British peregrine population was made in response to complaints by owners of racing pigeons that they were losing more birds than usual in races, because of increased predation by an increasing population of peregrines. Without those misplaced complaints, it is unlikely that we would have known so soon of the peregrine's decline in Great Britain, and this is a large bird of considerable interest to conservationists. The thinning of eggshells was only discovered 15 or so years after it had started, and only then because of detailed investigations. Effects on wildlife can easily pass unnoticed.
2. Conclusions from the survey of the peregrine's status were greatly facilitated by the existence of data about the population before synthetic insecticides were developed. Baseline data are invaluable for monitoring a population.
3. It was not easy to establish the causes of the peregrine's decline: Field data consisted of correlations, and specimens could not be taken for laboratory experiments both because the population was endangered and because this species is difficult to maintain in the laboratory. Moreover, many of the early laboratory experiments used standard laboratory species such as the chicken, pheasant, and quail, which appear to be immune to shell-thinning by DDE.
4. Shell-thinning was caused by a metabolite of DDT. For prediction of possible effects we need to consider not only the compound of interest but also possible metabolites and breakdown products.
5. The effect of DDE was completely unexpected.
6. It was a surprise that dieldrin should affect peregrines and other species of bird. To apply dieldrin as a seed-dressing is highly efficient, using minimum amounts of insecticide, placed exactly where it is needed to protect the crop. With hindsight, this environmental pathway (seed → seed-eating bird → predator) is not too surprising, but our ability to predict the detailed environmental distribution of pollutants is still limited.
7. It is a nice question what the effect of dieldrin on the British population would have been if breeding success had not already been impaired by DDE. For wildlife, we do need to consider the net effect of exposure to all pollutants.

C. Effects on the Gene Pool

Acute toxicity in successive generations from a pollutant may alter the gene pool of the population: It will increase the incidence of those genes that confer resistance and therefore the proportion of the population that is resistant.

Not all species contain the appropriate genetic variability, but many do to a wide range of pollutants. For example, a considerable number of plant species can develop resistance to one or more heavy metals, and some plant species can develop resistance to sulfur dioxide. The best-known aspect is the incidence of resistance to pesticides. By 1984, some populations of 447 species of insects and mites were known to be resistant to one or more insecticides, and similar problems have developed with herbicides, fungicides, and rodenticides.

Many different mechanisms can increase resistance, as suggested by Fig. 1. In general terms, there are at least six possibilities: changes in structure or, for animals, behavior that reduce exposure; reduced rate of intake; increased excretion rate; increased storage capacity at inert sites; reduced sensitivity to the pollutant at the site of action; and an increased rate of metabolism of the pollutant, although it should be noted that metabolism is not always a detoxifying mechanism. Sometimes resistance develops from more than one mechanism.

Pollutants that act by altering the habitat can also induce genetic changes. Melanic (dark-colored) specimens of the peppered moth (*Biston betularia*) occur rarely as chance mutations but can become the predominant form in areas of air pollution. These moths commonly rest during the day on tree trunks, and the less well the moths blend with their background the greater the risk of predation by

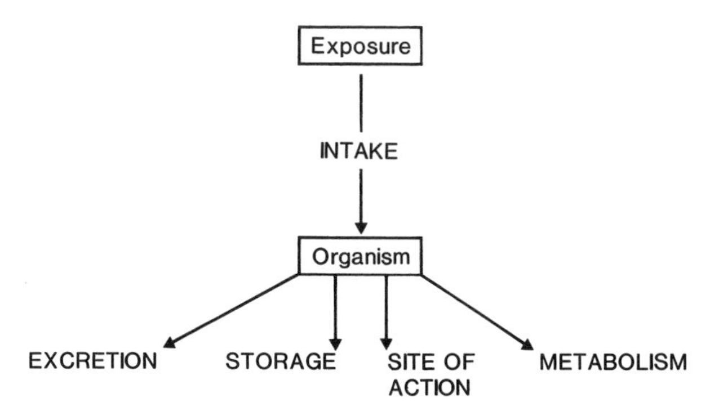

FIGURE 1 Scheme of the routes that molecules of pollutant can follow after intake.

birds. Air pollution tends to darken surfaces, particularly by deposits of soot, and to kill (particularly from sulfur dioxide) the lichens found on tree trunks, with the net result that melanic moths are then less conspicuous than the typical form. This differential predation by birds is thought to have caused the increased incidence of melanic moths around Manchester, England, where the first melanic specimen was only recorded in 1848, but by 1895, 98% of the individuals in that area were melanic.

IV. Prediction of Effects

We are now much more aware of the potential risks from pollutants to both our own and other species. In consequence, in many countries, the state now has to give permission before significant quantities of any new synthetic chemical can be marketed. The question also being considered is, which of the chemicals already in use need most urgently to be assessed for possible hazards? The underlying question is, how can we predict the ecological effects of a pollutant?

A. Exposure

The first, logical step is usually to attempt answers to a sequence of questions:

1. How much will be released into the environment?
2. How will the chemical be distributed within the environment?
3. How persistent will the chemical be?

Answers to these three questions indicate likely exposures, which is the first step for predicting direct effects (see Fig. 1). Release into the environment can be estimated from the amounts to be produced and the intended uses and methods of disposal. For volatile compounds (e.g., some organic solvents), we assume that all the amount produced will eventually reach the environment. In contrast, with polychlorinated biphenyls (PCBs) we have distinguished sharply between those uses such as plasticizers and lubricants, where release into the environment cannot be controlled, and those such as fluids in electrical transformers, where the compounds are in sealed units and their final fate is more easily controlled.

A pollutant's initial distribution in the environment is often envisaged as occurring between just three compartments (Fig. 2) and is assessed from physicochemical data, in particular vapor pressure and boiling point, solubility in water, and rate of hydrolysis.

Persistence is a key characteristic, already discussed for estuaries, although it can have several related meanings. For present purposes, it means for how long a chemical remains in the environment before it is converted either by organisms or in the physical environment into other compounds. Often more restricted meanings are used of how long a pollutant remains in one place, either in one part of the physical environment (e.g., the soil) or in an organism. Elemental pollutants such as heavy metals persist in the environment indefinitely, albeit in a range of chemical species, although their persistence within organisms may be much less.

Persistence in the environment is assessed from standard laboratory tests, which include degradation rates by microorganisms in both aerobic and

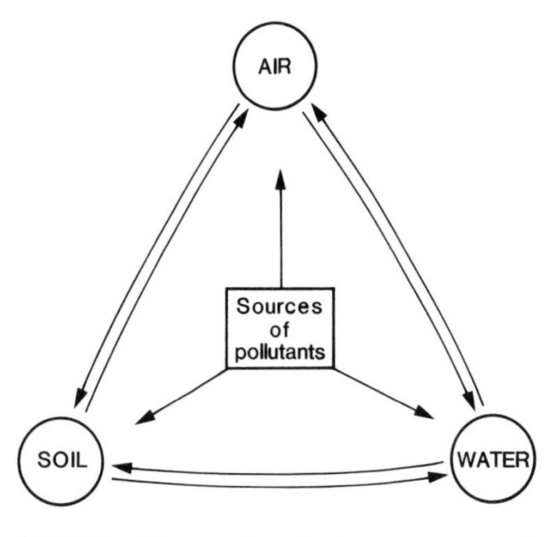

FIGURE 2 Scheme to describe the movement of pollutants from one or more sources into environmental compartments.

anaerobic conditions. It is a property not only of the particular pollutant, but also of the organism or habitat in which it occurs. The conventional measure of persistence is half-life, but this is usually a first approximation only: The rate at which a pollutant disappears (mass lost/unit mass/unit time) usually decreases with time. Mathematical models can then be used to predict amounts and concentrations of chemicals. The model used will depend on the particular circumstances. Suffice it to say that there are difficulties: The model usually entails a considerable simplification of the processes involved and is complicated by the fact that, in practice, concentrations of pollutant are rarely in a steady state.

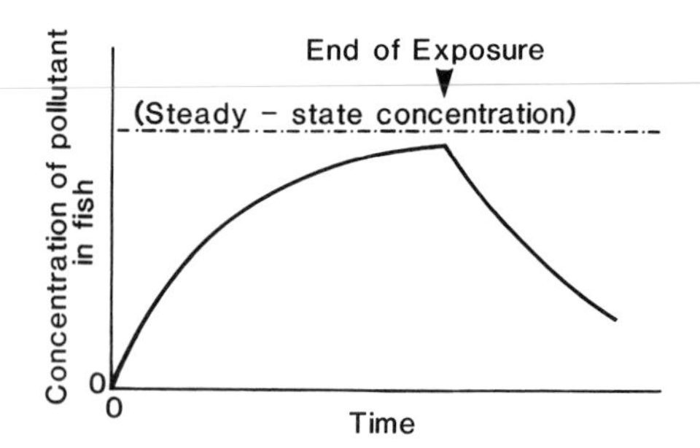

FIGURE 3 Idealized picture of the changes with time in the concentration of pollutant in an aquatic organism during exposure to a constant concentration of that pollutant in the water and after exposure has ended.

B. Dose

The term *dose*, taken from pharmacology, indicates the intake of pollutant by an organism from its exposure. It is usually measured as the concentration of pollutant in an organ, tissue, or whole body and is expressed in a variety of units (e.g., wet weight, dry weight, and lipid content). There is often the implicit and sometimes explicit assumption that the magnitude of the dose indicates the probability or degree of effect.

In the most general terms, there are two routes by which animals can acquire pollutants: by ingestion with their food, and from direct contact with their inanimate environment, across the body surfaces in general or the respiratory surfaces in particular. For most plants, direct contact is the only route—with the air or the soil. There are also two mechanisms by which organisms can lose pollutants (Fig. 1): by excretion, and by metabolism, when the pollutant is converted into one or more other compounds within the organism. With a constant exposure, the amount of pollutant within the organism tends to approach a steady state (Fig. 3), when the rate of intake equals the sum of the rates of excretion and metabolism.

Pollutants are not distributed uniformly within organisms. Fairly sophisticated compartmental models have been developed to describe amounts of pollutant within organisms during and after exposure, but although such models often fit equations to the data well, their theoretical basis is not well-established.

For aquatic animals exposed to organic compounds, a much simpler approach is commonly used. There is good evidence to suggest, at least for many pollutants, that food is not a significant route of intake. It has been found that for many organic compounds, the degree of bioconcentration increases with the *n*-octanol : water partition coefficient (commonly denoted as P or K_{ow}) (e.g., Fig. 4). The theory is that the degree of bioconcentration depends on the distribution ratio of the compound between the ambient water and the animal's fats, which is indicated by the partition coefficient. However, this use of P is indicative, not certain. Thus in Fig. 4, three compounds were "aberrant," and results for the other 27 compounds show that the predicted bioconcentration factors usually fall within one order of magnitude of the experimentally measured value.

Terrestrial animals will often acquire most of their pollutant burden from their food. The relation between concentrations of pollutant in predator and prey depends on many factors, but it is untrue that persistent pollutants inevitably increase in concentration along food chains, unless one defines persistent compounds as those that are only excreted and metabolized slowly.

Intake often depends on the minutiae of the exposure: the distribution and chemical form of the pollutant within the air, soil, water, or sediment. More attention to the pathways that pollutants follow in the environment could improve our ability to predict doses.

C. Effects

The common measures of dose indicate the amount of pollutant within the organism at one instant. Laboratory tests usually give organisms a single acute exposure or a chronic constant exposure. We usually do not know what influence fluctuations of dose with time and duration of dose have on the degree or likelihood of biological effect.

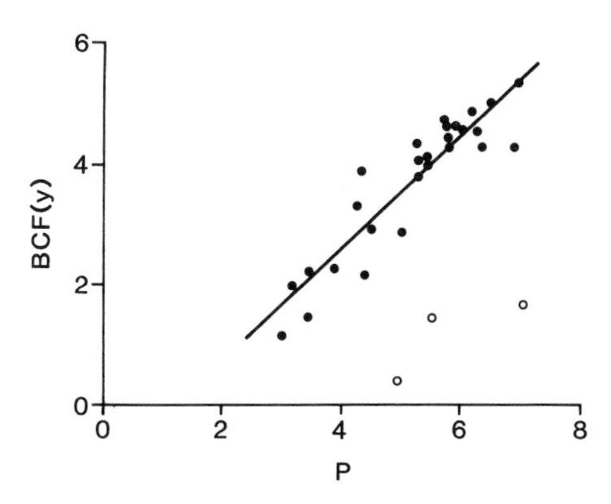

FIGURE 4 Linear regression of $\log_{10}$ of the bioconcentration factor (BCF) on $\log_{10}$ of the *n*-octanol : water partition coefficient (P). Results are for whole-body concentrations in fathead minnows (*Pimephales promelas*) exposed for 32 days in a continuous-flow system to one of 30 organic compounds and mixtures of compounds. Results for three compounds (○) are excluded from the regression, for which the residual deviation is 0.471. (Data from Veith et al., 1979). Veith, G. D., Defoe, D. L., and Bergstedt, B. V. (1979). Measuring and estimating the bioconcentration factors of chemicals in fish. *Journal of the Fisheries Research Board of Canada* **36,** 1040–1048.

There is currently much interest in quantitative structure–activity relations (QSARs). The presumption is that for a series of chemically related compounds, their relative degree of biological activity depends on speed of passage to the site of action, indicated by the partition coefficient and by the ease of attachment of molecules to the receptor. Equations are now being developed in which the measure of biological activity (the concentration of chemical required to produce a standard biological response) is a function of a series of terms for partition coefficient, electron density, and other aspects of molecular structure. These equations are developed empirically and are at an early stage.

Results of laboratory tests, no matter how comprehensive, cannot be applied directly to field situations. Ecosystems contain many species, and different species react differently to an input of pollutant for both genetic and environmental reasons. Moreover, the consequences of effects on individuals within one species population can be complex. The interactions between individuals within a population and between species within an ecosystem are often not well understood.

Studies are sometimes made on populations in the field or even on whole ecosystems, but most ecosystems are so complex that such studies inevitably entail a considerable amount of effort, and results can be difficult to interpret. Model ecosystems, or microcosms, are sometimes used in an attempt to gain the advantages of both relative simplicity and verisimilitude. They can range from fairly simple aquaria with sand, water, and a few selected plant and animal species to large marine enclosures with as much as 1,700 m^3 of seawater. The view is now developing that the chief value of these systems is not as simulations of nature but as experiments on selected components of the ecosystem. They can yield useful information on the pathways that pollutants follow in the habitat, but sometimes appear, for studies on populations, to suffer the worst of both worlds. They can be too artificial to represent natural ecosystems and too complicated for biological effects to be easily interpreted.

D. Assessment

The Organisation for Economic Co-operation and Development (OECD) has made considerable efforts to harmonize the tests required by different states when assessing new chemicals. The details can vary considerably, both between states and for different chemicals, but there is widespread agreement on the rationale, which is to estimate amounts released, environmental distribution, and concentration and to compare the predicted environmental concentrations with those needed to affect organisms. The tests of biological effects usually include tests of the exposure needed to kill a range of species, which usually include fish and crustacea, and can include, particularly for pesticides, birds, bees and other beneficial insects, and earthworms. Tests on bioaccumulation are commonly required, usually with a species of fish, but the details vary with intended uses of the chemical, quantities involved, and the results of earlier tests. Testing is performed by the manufacturer or supplier and continues until either a decision about permitted uses, if any, is made or the manufacturer withdraws its application.

This procedure takes little account of possible interactions that may occur between the chemical being considered and other pollutants that may also be present in the environment. There is some confusion about interactions, illustrated by the data in Table I. It is clear from that table that the degree to which photosynthesis is reduced in spinach plants exposed to both sulfur dioxide and hydrogen chloride is greater than the sum of the individual effects of exposure to either gas alone: The degree of effect by either pollutant depends on the presence or ab-

TABLE I Effects of Sulfur Dioxide and Hydrogen Chloride on the Rate of Photosynthesis, Measured as a Percentage of the Control Value, in Spinach Plants (*Spinacia oleracea*) after 53 Hours' Exposure

Exposure to HCl (mg/mm³ air)	Exposure to SO_2 (mg/mm³ air)	
$\frac{1}{M}$	0	0.9
0	100	89
1.35	82	50

From Moriarty, F. (1988) (adapted from Guderian 1977). "Ectoxicology. The Study of Pollutants in Ecosystems," 2nd ed. Academic, London.

sence of the other pollutant. More information is required before we can decide whether the two pollutants affect each other's biological activity. It may be that to double the exposure to either gas alone would also halve the rate of photosynthesis, in which case there is no interaction in the biological effect produced by exposure to these two gases. To determine whether two pollutants potentiate or antagonize each other's effects, one needs to determine the relation between the amounts of the two pollutants in equally toxic mixtures (Fig. 5).

It is generally recognized that these assessments of effects are crude, and a safety factor is therefore used, a factor by which the dose needed to produce an effect in the laboratory exceeds the predicted environmental concentration. Sometimes, in cases of doubt for pesticides, permission for limited use is given, when safety is monitored during use in the field.

Despite safety factors, there is still the risk of mistakes, of unacceptable ecological effects from new chemicals, but it is difficult to determine how great this risk is. The present system of assessment may seem more effective than it really is, because we do not know how many past or present ecological effects may have gone unnoticed. Moreover, apart from pesticides, which are released into the environment deliberately because of their biological activity, most new chemicals may not have serious ecological effects anyway. Both of these factors could make the system of assessment appear to be more effective than it really is.

V. Monitoring

Present abilities to understand and to predict the pathways that pollutants follow in the environment and the effects that they exert are inadequate. We need, therefore, to make repeated surveys (surveillance), or what is usually called *monitoring*. The latter term sometimes has a more restricted meaning of surveillance to check that the amounts of pollutants or the effects being studied conform to an already stated standard.

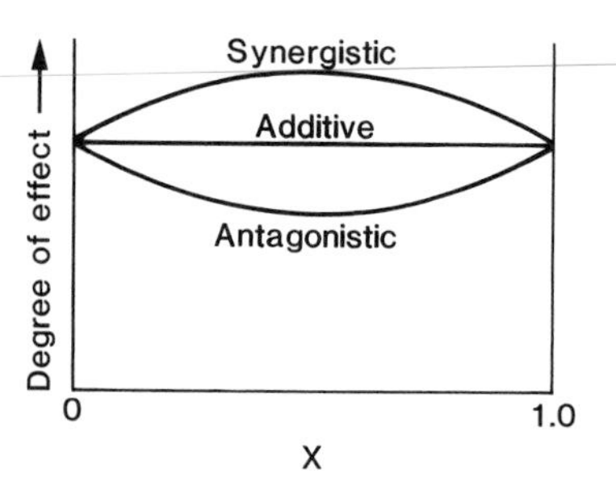

FIGURE 5 Scheme to show three possible types of response to a mixture of two chemicals when x and (1.0-x) are the proportions of two chemicals present in a mixture. Proportions are based on fixed amounts of the two chemicals. (Adapted from Loewe, 1928). Loewe, S. (1928). Die quantitativen Probleme der Pharmakologie. *Ergebnisse der Physiologie* **27,** 47–187.

Three types of information may be sought:

1. Rates of release of pollutants into the environment
2. Degree and changes of environmental contamination
3. Biological effects

Few monitoring schemes measure both amounts of pollutant in the environment and the biological effects, if any. Whatever the purpose of a monitoring scheme, it is essential that objectives are determined at the start, because the objectives dictate the details of the sampling program. It is also desirable that it should be known what specified degree of change can be detected with an appropriate degree of statistical significance. Many monitoring schemes do not meet these desiderata.

Measurements of amounts of pollutant, be they in nonliving samples or in organisms, can be related to standards (i.e., acceptable limits of contamination). Organisms can have advantages over nonliving samples: they acquire much higher concentrations of some pollutants, when chemical analysis can be easier. They also give a measure of the pollutant's availability, which is more relevant for the probability of biological effects than is a measure of the total amount in the environment. It is often also said that organisms integrate the amount of pollutant present in the environment during a period of time, but this is not strictly correct. Organisms do not retain pollutants indefinitely. They do give some sort of an average measure for fluctuating exposures, although the most recent exposures will contribute proportionately more than will earlier exposures.

A development of the past decade is to establish environmental specimen banks, collections taken from both living and inanimate components of our environment for indefinite storage at −190°C. These samples are being stored in anticipation of future, as yet unforeseen, needs in the expectation that this material will permit measurements of pollutants

1. As yet unknown or thought to be unimportant.
2. In sites or specimens that are at present of no concern, but which may become so in the future.
3. To give a historical record.
4. With new improved analytical techniques.

If these objectives can be fulfilled, these banks could be useful. Doubts stem from the difficulties of designing an effective sampling program when important details of the objectives are unknown. It must be accepted that although it is often easier to monitor amounts of pollutant rather than their biological effects, to rely on measurements of pollutant alone does ignore some potential problems:

1. It is sometimes impracticable to measure all contaminants.
2. Routine analytical techniques may be too insensitive.
3. The biological significance of pollutants, at the levels found, may not be fully appreciated.
4. Combinations of pollutants may interact.
5. Regular chemical measurements may miss occasional, significant, high values.

Conversely, it is often not a straightforward matter to relate biological changes, be they in functioning of individual organisms or of population size, to the degree of environmental contamination. Many variables can affect the dose–response relation, not all pollutants can readily be detected within organisms, and effects are sometimes indirect, mediated through other species or the abiotic environment. Moreover, one's concern is, at least sometimes, with the whole community. There are, in consequence, numerous indices that attempt to encapsulate in one number the abundance, presence, or absence of a range of species or the number and abundance of species present.

These indices do have many defects, of which perhaps the most fundamental is that it is often difficult to attach any biological meaning to the index. Monitoring of single named species is more incisive. Most species within a community are then likely to be ignored, but clear objectives can be stated from which practicable programs of monitoring can be devised, and any changes are relatively easy to interpret.

VI. Conclusions

We have to accept that for the foreseeable future, prediction and even detection of ecological effects by pollutants is often going to be difficult. Attitudes to pollution have changed, and it is beginning to be accepted that there is a need to consider the fate and effect of potential pollutants at all stages, from "cradle to grave." To a considerable extent, the political impetus for these new attitudes derives from public pressure, but what is often signally lacking is a consensus on the value judgments that underlie attempts to control effects by pollutants on wildlife.

Bibliography

Connell, D. W., and Miller, G. J. (1984). "Chemistry and Ecotoxicology of Pollution." Wiley, New York.

Kaiser, K. L. E., ed. (1987). "QSAR in Environmental Toxicology II." D. Reidel, Dordrecht, The Netherlands.

Moriarty, F. (1988). "Ecotoxicology. The Study of Pollutants in Ecosystems," 2nd ed. Academic, London.

OECD. (1981). "OECD Guidelines for Testing of Chemicals." Organisation for Economic Co-operation and Development, Paris.

Richardson, M., ed. (1986). "Toxic Hazard Assessment of Chemicals." The Royal Society of Chemistry, London.

Sheehan, P., Korte, F., Klein, W., and Bourdeau, P., eds. (1985). "Appraisal of Tests to Predict the Environmental Behaviour of Chemicals. SCOPE 25." Wiley, Chichester.

Wood, L. B. (1982). "The Restoration of the Tidal Thames." Adam Hilger, Bristol.

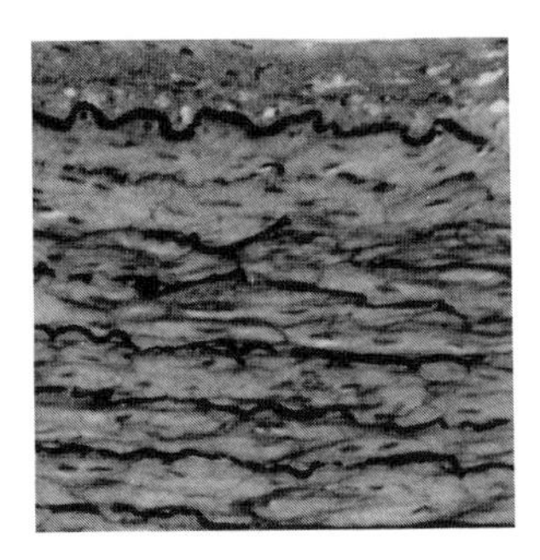

Elastin

ROBERT B. RUCKER and DONALD TINKER, *University of California, Davis*

Glossary

Alu sequences Alu Sequences of about 300 nucleotide pairs that are duplicated in human DNA so that they comprise 4–6% of the total DNA in the human genome

Coaccervation Causes temperature-dependent exclusion of hydrophobic polymers from polar environments; elastin coaccervate exists as a separate phase with the characteristics of an oil droplet

Ehlers–Danlos syndromes Syndromes representing a family of related connective tissue disorders, which are characterized by hyperextensible joints, hyperelasticity of skin, skin fragility, and tumorlike growths following trauma; both abnormal elastin and/or collagen fiber formation are components of the syndrome

Marfan's syndrome Major features are abnormalities of the heart and blood vessels as well as abnormal growth and development of bone and other connective tissues

Mesenchymal Mesenchymal tissue is the meshwork of connective tissue derived from the mesoderm, which is the middle layer of the three primary germ layers of the embryo; muscle, blood components, and epithelia are also derived from the mesoderm

Parenchyma Term that describes the functional elements of an organ as distinguished from its framework or stroma

Promoter Regulates RNA synthesis from given genes; in the promoter, some sites are characterized by specific nucleic acid sequences (e.g., so-called TATA, CAAT, or CG-rich boxes), and other sites bind specific transcriptional factors; TATA boxes are recognition sites that are important to RNA polymerase II activity; AP-1 and SP-1 sites are sequences known to bind to specific regulatory proteins in response to signals important to cell replication and growth cycles

Pseudoxanthoma elasticum Skin disease that is characterized by swelling and degeneration of elastic fibers

ELASTIN IS AN IMPORTANT structural protein found in the extracellular matrix of tissues that are required to be highly compliant. Elastin is composed of hydrophobic amino acid sequences that are separated by sequences enriched in alanine and lysine. Lysyl residues in the alanyl and lysyl sequences can be modified to facilitate the cross-linking of individual polypeptide chains of elastin. This results in an insoluble protein as the

Encyclopedia of Human Biology, Volume 3.

final product with a structure analogous to the organization of many natural and synthetic rubbers. The protein is regulated so that its synthesis coincides with specific periods in the development of given tissues and organs. Abnormal elastin metabolism can cause vascular, skin, and lung lesions, which are characterized by poor elasticity and laxity.

I. Introduction

Elastin is a unique protein that is important to the structure and development of many tissues. The synthesis of elastin is developmentally regulated by a gene that has unusual features. As a consequence, elastin is of interest to the developmental and molecular biologist, because biochemical changes in elastin can compromise and alter systematic growth and development. Elastin is also of interest to the protein chemist, because of its novel chemical and physical properties. Elastin is one of the few proteins in nature with the properties of a rubberlike elastomer. It possesses an unusual amino acid composition and undergoes complex posttranslational chemical modifications in the process of its assembly into mature fibers. Defects in elastin can also underlie important disease processes. Therefore, the protein is of interest to clinicians who deal with elastin-containing tissues such as skin, ligaments, arteries, and lung. Elastic fibers often function for exceedingly long periods. Elastin metabolic turnover in some tissues is best estimated in years. Consequently, genetic conditions or agents that alter elastin biosynthesis, processing, and deposition can have long-lasting effects.

II. The Elastic Fiber

Table I compares the collagen and elastin contents of selected human connective tissues. In such tissues, the arrangements of elastin fibers are diverse (e.g., extending from organized laminar sheets to structures that are interdispersed among other fibers). In tissues such as ligaments, elastin fibers are oriented in the direction of stress. In blood vessels, the fibers take on the laminar or sheetlike arrangements (see Fig. 1A). In skin, elastin fibers may be deposited as interlinked sheets of protein but with numerous fused and branched points. In lung parenchymal tissue, elastic fibers appear as laminar sheets that encapsulate individual alveoli.

TABLE I Collagen and Elastin Content of Selected Connective Tissues[a]

Tissue	Collagen	Elastin
Bone, mineral-free	85–93	
Achilles tendon	80–90	4–5
Skin	70–75	0.5–1.5
Cornea	65–75	
Cartilage	45–65	1–2
Ligament	15–20	75
Aorta	12–24	25–35
Liver	2–4	
Lung parenchyma	10–20	20–25

[a] g/100 g of dry weight.

Elastic fibers contain two components. Elastin is the predominant component (70–90%) and appears amorphous when examined at the electron micrograph level. In contrast, the second component is distinctly different and microfibrillar in appearance (see Fig. 1B). This component is composed of several glycoproteins that vary in size from about 25,000 to 340,000 daltons. The microfibrillar component may also include lysyl oxidase, the enzyme that catalyzes elastin cross-linking. Also, evidence indicates that proteins such as vitronectin and amyloid F may be absorbed onto aged elastic fibers. The microfibrillar proteins can exist in fibrillar arrangements or as aggregates independent of their association with elastin. Therefore, it is assumed that in elastic fibers they play roles in fiber organization. For example, it is speculated that a first step in elastin's assembly into fibers is by disulfide bond formation with one of the microfibrillar proteins. A chemical characteristic of the microfibrillar proteins is that they tend to be high in cysteine. The C-terminal amino acid sequence of elastin also contains several cysteine residues that may undergo disulfide bond cross-linking.

III. Chemical Composition and Physical Properties of Elastin and Elastic Fibers

Elastin is one of nature's most apolar proteins. It is also a very insoluble protein because of its extensive intermolecular (interpolypeptide chain) cross-linking. Elastin is resistant to solubilization by strong protein denaturants, autoclaving, treatment with hot alkali, or formic acid. Moreover, many of the novel amino acid sequences in the protein are

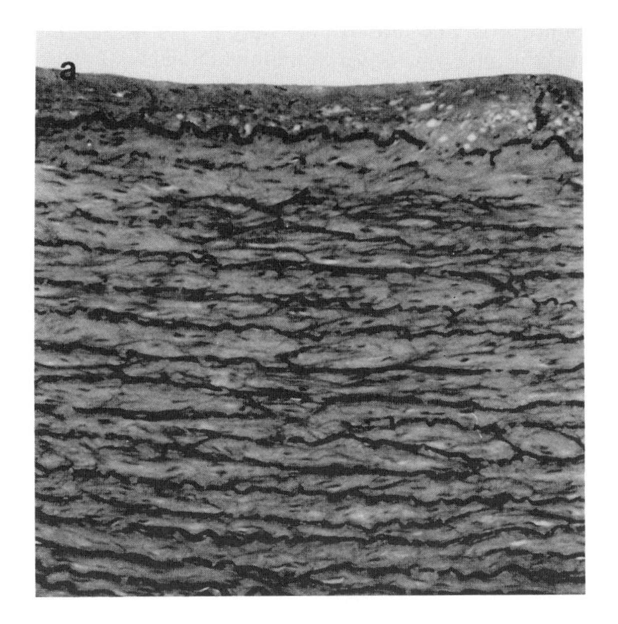

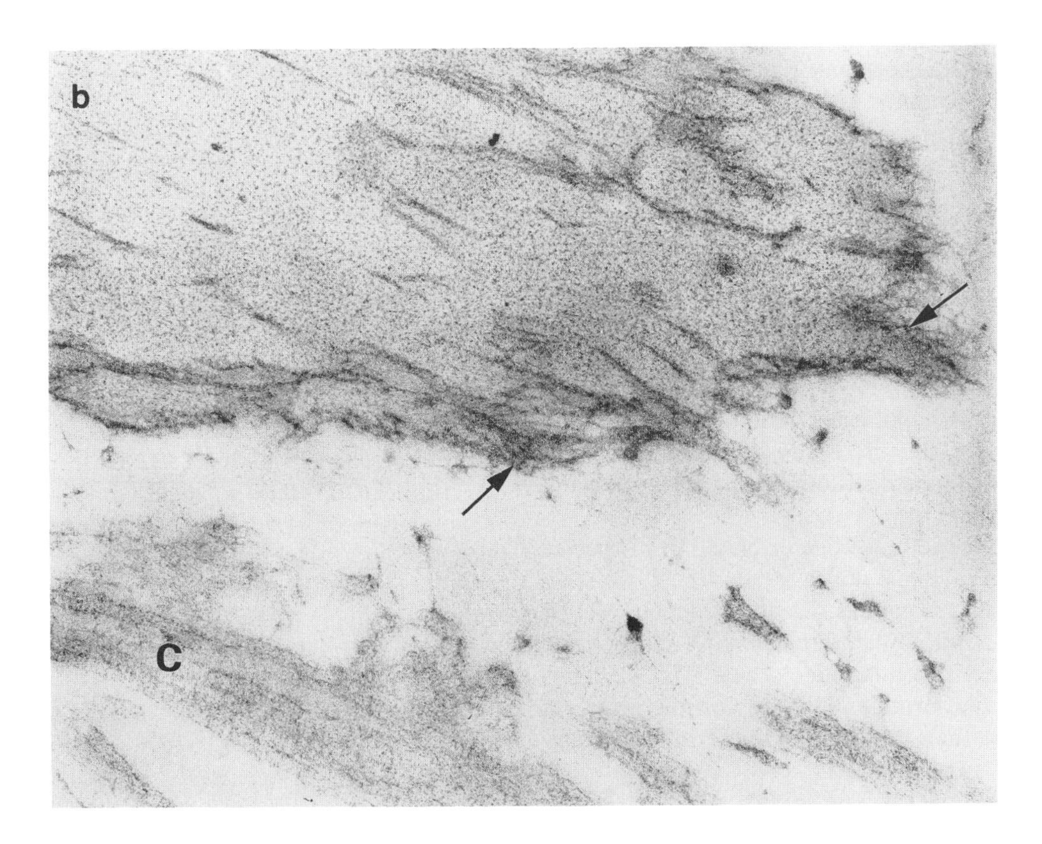

FIGURE 1 a, Light microscopic transverse section of human thoracic aorta from a healthy 27-yr-old male. The elastic fibers are darkly stained and are seen in a parallel arrangement. There is thickening of the intimal layer on the luminal surface within the thickened internal elastic lamella. Magnification ×350. b, Transverse section of an elastic fiber to show a peripheral rind of electron-dense microfibrils (arrows) surrounding the amorphous elastin. Collagen fibers (c) in transverse section are adjacent. Magnification ×70,000. [The light and electron micrographs were provided by Dr. E. G. Cleary and Dr. J. Kumaratilake, Department of Pathology, School of Medicine, University of Adelaide, Australia.]

resistant to mild hydrolytic conditions. The apolar sequences in elastin are enriched in glycine (Gly), valine (Val) and proline (Pro) (e.g., Gly-Gly-Val-Pro, Pro-Gly-Val-Gly-Val, or Pro-Gly-Val-Gly-Val-Ala. The Val-Pro imino bond is not cleaved as easily as typical peptide bonds. Elastin is also a basic protein. Many of the glutamyl and aspartyl residues in elastin are amidated; thus, cationic functions from lysyl and arginyl residues are in excess and not balanced by anionic functions, which normally arise from glutamyl or aspartyl residues. [*See* PROTEINS.]

However, long-range, reversible elasticity is the most important feature of elastin. The high degree of interchain cross-linking and the hydrophobic nature of the Gly-Val-Pro sequences result in a polymer that possesses properties analogous to rubber. The energy that results in elastic recoil comes about in part because of apolar–polar interactions between water and the hydrophobic portion of the elastin molecule. Interchain cross-linking also plays a role. The hydrophobic sequences in elastin are separated by short segments that are enriched in the amino acids alanine and lysine. Lysyl residues found in the alanyl- and lysyl-enriched sequences undergo specific chemical modifications that lead to interchain cross-links.

IV. Relationship between Physiological and Physical Properties

All animals with pulsating blood vessels contain some type of elastin. Many animals also have valves and ligamental structures that are elastic and contain elastin.

The elastins that have been isolated from species ranging from humans down to bony fishes contain lysine-derived cross-links. Elastins from chordates, however, do not contain cross-linking amino acids in easily measured quantities (Table II). Chordates have elasticlike proteins that are considerably less hydrophobic than the elastins found in higher animals.

In general, mammals and birds possess elastins that are highly cross-linked and hydrophobic. Reptiles and amphibians contain elastin with an intermediate percentage of hydrophobic amino acids. Bony fishes such as sharks contain lower amounts of hydrophobic amino acids. Such differences have evolved possibly because of the need to accommodate different fluid pressures as well as differences

TABLE II Lysine and Cross-Linking Amino Acid Content Commonly Found in Mature Elastin

Amino acid	Residues/1000 total
Desmosine[a]	1.5–2.0
Isodesmosine[a]	0.5–1.3
Lysinonorleucine[a]	0.8–1.5
Allysine[a]	0.5–2.0
Aldol condensation products[a]	0.8–1.5
Dehydrolysinonorleucine[a]	0.1–0.3
Others	2.0–10
Lysine	6–12

[a] Structures for desmosine, lysinonorleucine, allysine, aldol condensation products, and dehydrolysinonorleucine are given in Figures 3 and 4.

in body temperature. The hydrophobic portions of elastin amino acids are forced inward at lower temperatures, thus decreasing the potential for apolar–polar interactions. It may be predicted that, to accommodate high-fluid pressures or body temperatures, the more suitable forms of elastin should be relatively hydrophobic and highly cross-linked.

V. Tropoelastin and Cellular Sources of Elastic Fibers

Tropoelastin is the soluble precursor to elastin. This protein has been isolated from a number of different tissues and from various species of animals. The major source of elastin is the smooth muscle cell; however, cartilage cells, myofibroblasts, ligamentum fibroblasts, endothelial cells, and lung interstitial cell fibroblasts also secrete elastins.

Tropoelastin is not a single entity and occurs in several molecular forms. These forms arise from deletion and transposition of selected exons in the elastin gene. The single deletions and transpositions, however, do little to alter the overall amino acid composition of tropoelastin or its relative size (about 70,000 daltons). The deletions come about becasue of alternative processing of portions of tropoelastin mRNA molecules (Fig. 2). The deletions or transpositions may serve to produce insoluble elastin polymers that vary in their directional orientation or degree of branching.

VI. The Elastin Gene

Information is now available regarding structure of elastic genes, which has been useful in understand-

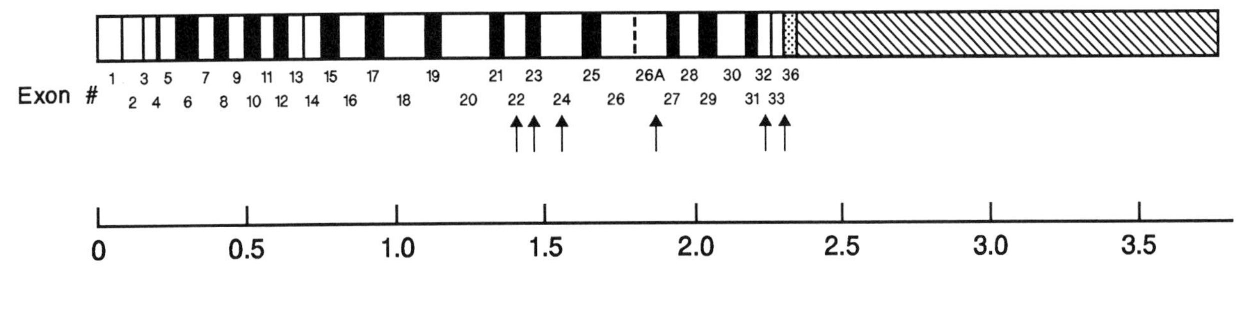

FIGURE 2 Diagram of human elastin cDNA. The cDNA is divided into exons, which are numbered and drawn to scale. Characterization of the cDNA and the complete human elastin gene has evolved primarily from the work of Joel Rosenbloom and his colleagues at the University of Pennsylvania. All exons are multiples of the three nucleotides, and exon–intron borders always split codons in the same way. This can permit "cassettelike" alternative splicing. The exons encoding cross-link domains are represented by solid bars, whereas the hydrophobic domains are open boxes. The arrows indicate known sites for alternative splicing. The dashed line that separates exon 26 and 26A indicates that in this particular exon, splicing occurs primarily in the 3′ portion. Exon 36 is novel in that it encodes for cysteine residues, which may be involved in tropoelastin attachment to microfibrillar proteins. In the region that corresponds to a nontranslated segment (stippled), there are two polyadenylation sites.

ing why there is variability in elastic fiber structure. The elastin gene has many unusual features compared with other known genes. In humans, there is a single gene, which contains nearly 40 kb of sequence that codes for an mRNA of about 3.5–3.7 kb. This gene is found on the long arms of chromosome 2. Chromosome 2 also contains genes for type II collagen, fibronectin, and type VI collagen.

The elastin gene has one of the highest intron–exon ratios (i.e., 19 to 1) of genes characterized to date. The exons that eventually code for the peptide sequences involved in cross-linking are interspersed among those exons that code for the hydrophobic regions. In addition to the human elastin gene, data are also available for the elastin genes that encode for bovine, chick, and rat elastins. The size of exons varies somewhat from one species to another. The spacing of introns to exons is also not well conserved among species. Alternative splicing is also a consistent and unusual feature that is associated with the processing of mRNA products from elastin genes. The splice borders are the same for each exon in elastin. Often a single cross-linking exon or hydrophobic exon is deleted or transposed. Alternative splicing gives rise to different mRNA molecules, which in turn translate into differing polypeptides. Consequently, the exons may be viewed as cassettes in which given segments of the molecule may be spliced out.

The human elastin gene is an Alu-enriched gene. At the 5′-promoter region, there are several CAT boxes but no TATA boxes in the promoter. The promoter also contains multiple initiation sites. Both ST-1 sites and AP-2-binding sites are present in the promoter. ST-1 and AP-2 are protein transcriptional factors. The AP-2 factor is responsive to changes in cellular cyclic adenosine monophosphate levels. Responsiveness to hormones and related physiological regulators (e.g., retinoids, hydroxylated sterols or dexamethasone) are also features of gene regulation. These responses, however, appear sensitive to the developmental phenotype of the cells making elastin. For example, with exposure to dexamethasone, increased elastin expression is observed when cells with a fetal phenotype are used in studies; when adult cells are used, typically no change in elastin expression is observed.

VII. Posttranslational Steps in Elastin Fiber Formation

Figure 3 outlines important posttranslational features of elastin fibrogenesis. First, about 3–8% of the total proline residues in elastin are hydroxylated. This step is analogous to the hydroxylation of prolyl residues in collagen and occurs intracellularly. The next phase in fibril formation appears to involve organization of elastin around the microfibrillar components. Lysyl residues in elastin are then modified to give rise to a number of unusual cross-linking amino acids. The cross-linking amino

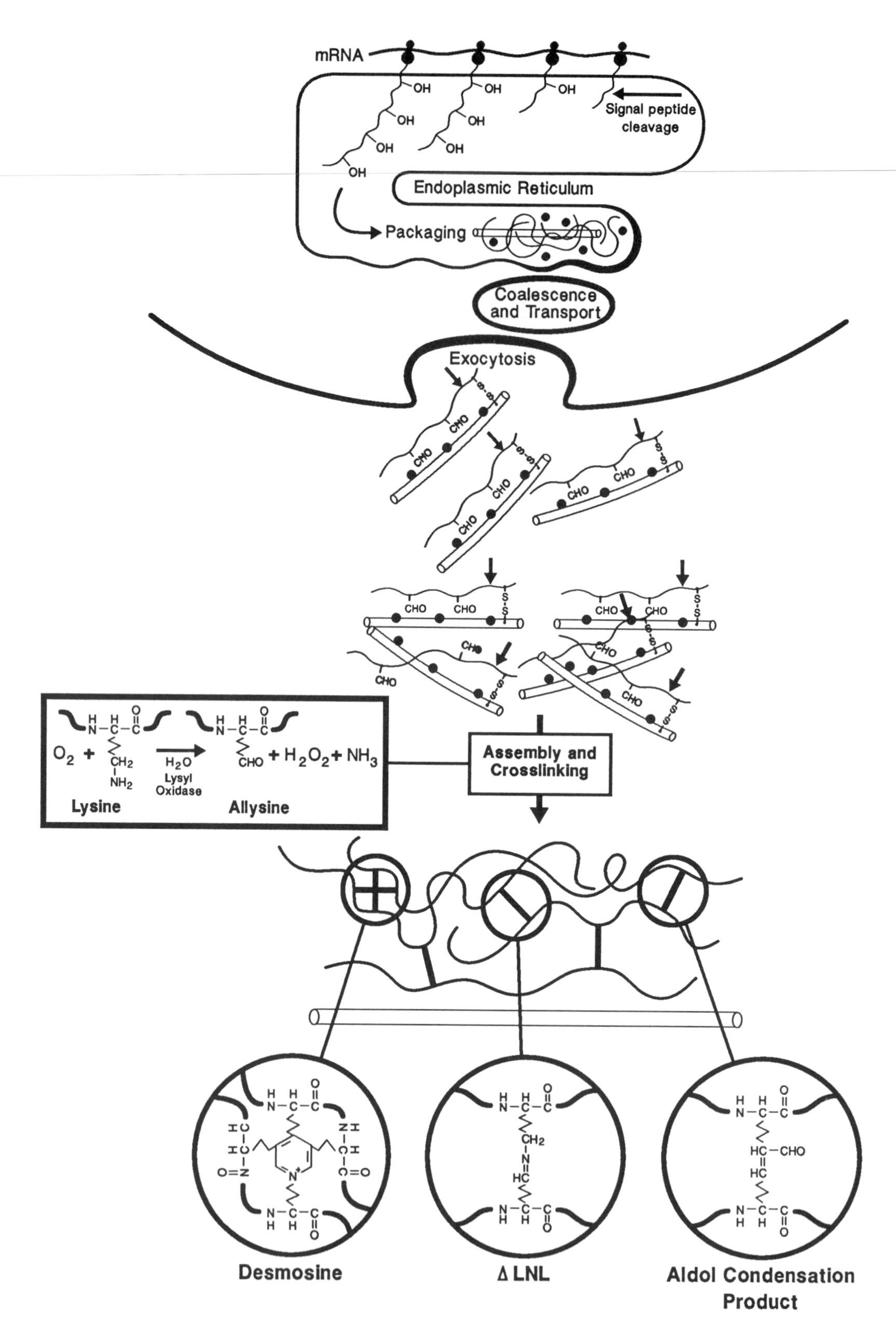

FIGURE 3 Steps in the posttranslational modification of elastin. The intracellular steps include signal peptide cleavage, prolyl hydroxylation, packaging, transport, and exocytosis by vesicles. The extracellular steps include association with microfibrillar protein components (rods) via sulfide linkages, conversion of selected lysyl residues to residues of allysine (the reaction defined in the box insert), cleavage of elastin associated with the microfibrils (arrows), and eventual coaccervation and cross-link formation. The major cross-links in elastin (shown in the magnified circles) are desmosine, derived from the condensation of lysyl residues with three allysyl residues; dehydrolysinonorleucine (ΔLNL), a Schiff-base product that can be reduced in elastin to form lysinonorleucine (see Fig. 4); and the aldol condensation product, derived from allysyl aldol condensation reagents.

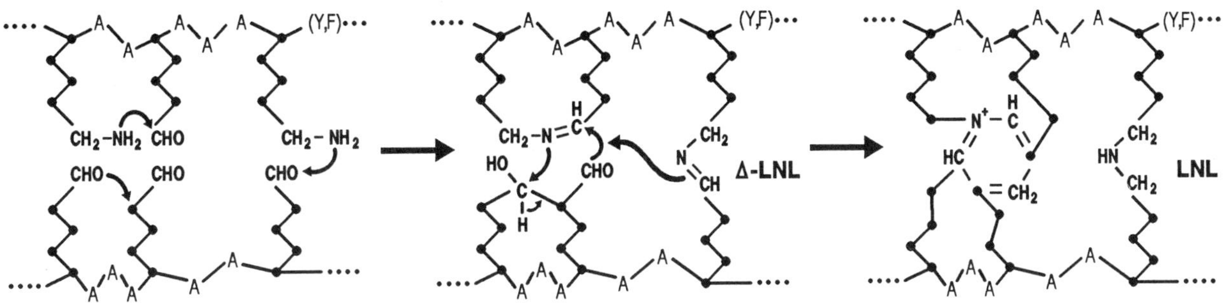

FIGURE 4 Oxidation of desmosine intermediates and reduction of dehydrolysinonorleucine (Δ-LNL) to form stable cross-links in elastin. The cross-linking regions in elastin arise from exons that encode predominantly alanyl and lysyl residues. It has also been observed that when an aromatic amino acid residue appears adjacent to a lysyl residue, the lysyl residue is often not oxidized. As a consequence, this favors dehydrolysinonorleucine formation or dictates whether desmosine or a related isomer forms. LNL, lysinonorleucine.

acids arise following the oxidative deamination of lysyl residues in tropoelastin. Once lysyl residues are oxidized (to peptidyl α-amino acid-δ-adipic-semialdehyde, allysine), cross-linking of tropoelastin occurs by a number of unusual condensation reactions. It is now well established that the cross-links in elastin are extensive and significantly influence both the biophysical and the biochemical properties of the protein. For the most part, the cross-links are necessary to constrain the elastic fiber so that upon stretching, the individual polypeptide chains are allowed to realign and restore fibril organization upon release of tension. When specific lysyl residues are modified and there is appropriate ordering and alignment of the lysyl-enriched portions of elastin, the condensation reactions occur spontaneously. Once the condensation products are formed, subsequent redox reactions result in stable and irreversible cross-links. A possible mechanism of condensation and, ultimately, stable cross-link formation is presented in Fig. 4. The mechanism suggests that the Schiff-base product of allysine and lysine, dehydrolysinonorleucine (Δ LNL), serves as the agent for the oxidation of hydroxydesmosine derivatives to form the tetrafunctional cross-links desmosine and its isomers. A pentafunctional variant of desmosine has also been isolated.

Much of the progress in our knowledge of the cross-linking amino acids in elastin and collagen resulted from studies that focused on the enzyme lysyl oxidase. Lysyl oxidase requires as cofactors copper and an orthoquinone component, currently speculated to be derived from peptidyl 6-hydroxy-dopa. Severe dietary copper deficiency will cause defective cross-linking. Agents that react with carbonyl functions, such as β-aminopropionitrile, various hydrazines, and semicarbazides, also promote defective cross-linking and aneurysms.

VIII. Role of Elastin in Organogenesis and Development

The deposition of elastin is correlated with specific physiological and developmental events. For example, in the mammalian lung, elastic fibers are concentrated around the opening of alveoli. Alveolar shape is thought to result in part from the molding or scaffolding influences of lung elastic fiber. Elastin synthesis and deposition are coordinated with alveolarization.

In developing arteries, elastin deposition appears to be coordinated with changes in arterial pressure and mechanical activity. The transduction mechanisms that link mechanical activity to elastin expression undoubtedly involve cell-surface receptors. Also, evidence indicates that cell movement in ligaments and arteries is related to the orientation of elastin. A current hypothesis is that elastin-synthesizing cells are attached to elastin through cell-surface receptors. Once attached, the synthesis of additional elastin and other matrix proteins and carbohydrates may be influenced by stretching or other factors that influence cellular shape.

One of the elastin receptors that has been studied most extensively is 67,000 daltons and has features that are similar, if not identical, to receptors that bind to laminin. Laminin and other proteins (e.g., fibronectin and entactin) facilitate cell–extracellular matrix communication by binding to both cell receptor proteins and extracellular matrix components. Very active investigations currently relate to how such attachments mediate biomechanical responses in connective tissues. The linkage is of obvious importance to understanding extracellular matrix formation during development or during tissue repair following injury. [*See* EXTRACELLULAR MATRIX; LAMININ IN NEURONAL DEVELOPMENT.]

IX. Other Functional Roles of Elastin

Elastin receptors are also found on the surface of cells that are not involved in extracellular matrix protein production. Phagocytic cells, such as the macrophage, appear to chemotax (migrate in response to a signal) to sites that are enriched in hydrophobic elastin peptides (e.g., VGVAPG) as a result of elastin degradation. Elastin is a long-lived protein that is normally not degraded, except following tissue injury. Consequently, it can be speculated that elastin may play an important role in chemotaxis. Elastin peptide levels are usually very low; therefore, any rise in elastin peptides could serve as a localized signal of mesenchymal tissue damage that may require the aid of cells with phagocytic functions.

A knowledge of the receptor–ligand association process is important to the understanding of certain diseases. Some tumor cells that metastasize to highly vascularized tissues attach to elastin by receptor-mediated processes. Once attached, the tumor cell secretes proteinases, thereby allowing penetration into the interstitial tissue.

X. Turnover of Elastin, Elastolytic Processes, and Elastin-Related Diseases

As noted, once elastin is synthesized and stabilized by cross-links, it does not undergo rates of turnover common to other proteins. In some tissues, measurable turnover is difficult to demonstrate, particularly when corrections are made for new tissue growth. Under normal situations, elastin peptides are released only in minute quantities as the result of degradation. Concentrations in human serum are normally in the nanogram per deciliter range. When urinary desmosine or elastin peptides are measured, the amounts correspond to no more than microgram quantities of elastin degraded per day. Because the body pool of elastin in humans is in gram amounts, the small quantities of elastin products that are excreted under normal situations clearly suggest that turnover is very slow. Likewise, when very young animals are injected with radioactively labeled lysine at those developmental periods corresponding to maximal expression of elastin, the labeled lysine that is incorporated into chemically stable cross-links, such as desmosine, may persist in elastic fibers for months to years; however, exceptions are evident, especially in tissues in which continual remodeling is required because of growth (e.g., the uterus).

For those diseases in which elastin destruction is of biological significance, the protein's "inertness" takes on considerable importance. Focal disruption of the elastic lamina of a blood vessel or alveoli need not be extensive to alter mechanical or biochemical properties. One of the best examples of how disruption of elastic fibers leads to pathology is pulmonary emphysema, a process that involves the proteolytic destruction of alveolar elastin. Emphysema appears to involve inappropriate elastolytic digestion by proteinases derived from neutrophils, macrophages, and/or even some mesenchymal cells. Furthermore, if elastin cross-linking becomes defective, elastin can serve as a substrate for many proteinases that normally do not degrade elastin (e.g., kallikrein or proteinases associated with blood coagulation).

Considerable effort and study have gone into understanding the process of elastolysis. Clearly, nature goes to some effort to modulate and counteract inappropriate proteolysis of extracellular proteins by the secretion of a wide variety of proteinase inhibitors. One such inhibitor is α_1-antiproteinase, an inhibitor made in the liver. Inactivation or genetic deficiencies of α_1-antiproteinase can promote obstructive pulmonary disease (e.g., emphysema) because of enhanced elastolysis.

Other diseases that involve elastin are disorders of skin. These diseases include pseudoxanthoma elasticum, Buschke-Ollendorf syndrome, cutis laxa, several Ehlers–Danlos syndromes, Menkes' disease, Marfan's syndrome, endocardial fibroelastosis, and elastoderma. These disorders are usually expressed genetically as either X-linked, autosomal recessive, or dominant.

The preceding description of the elastin gene suggests that several mechanisms may underlie the heritable diseases that involve elastin (e.g., mutation, insertion, or deletion of given exons, abnormal transcriptional or translational regulation, or mRNA instability). As a general rule, primary protein structural defects are often autosomal dominant and lethal. The disorders that are recessive in inheritance are usually less lethal and involve a processing enzyme or a gene regulatory element. For example, pseudoxanthoma elastictum is characterized by 35 times the normal RNA levels for elastin, whereas cutis laxa often results from a marked decrease in the levels of functional elastin mRNA. Menkes' disease, which is thought to be an X-linked recessive disorder, represents, in part, defects related to the transport of copper and lysyl oxidase function and, thus, affects elastin indirectly.

XI. Concluding Comments

Elastin is an important component of connective tissue in that it facilitates tissue extensibility. It imparts rubberlike properties to tissues such as lung, blood vessels, ligaments, skin, and the uterus. It is one of the most apolar proteins known to be synthesized by mammalian cells. In some tissues, highly cross-linked elastic fibers persist for months to years with little turnover. Slow protein turnover takes on considerable importance, because it can limit the possibility of appropriate repair following tissue injury.

Consequently, a knowledge of elastin's function is of obvious importance to human and veterinary medicine. Loss of elastin or elasticity contributes to a number of disease processes such as emphysema, atherosclerotic vessel wall disease, and cutaneous disorders.

Bibliography

Bashir, M. M., Indik, Z., Yeh, H., Ornstein-Goldstein, N., Rosenbloom, J., Abrams, W., Fazio, M., Vitto, J., and Rosenbloom, J. (1989). Characterization of the complete human elastin gene. *J. Biol. Chem.* **264,** 8887–8891.

Cleary, E. G. (1987). The microfibrillar component of elastic fibers. *In* "Connective Tissue Disease. Molecular Pathology of the Extracellular Matrix" (J. Vitto and A. J. Perejda, eds.), pp. 55–81. Marcel Dekker, New York.

Eyre, D. R., Paz, M. A., and Gallop, P. M. (1984). Crosslinking in collagen and elastin. *Annu. Rev. Biochem.* **53,** 717–748.

Franzblau, C., Pratt, C. A., Faris, B., Colannino, N. M., Offner, G. D., Mogayzel, P. J., and Troxler, R. F. (1989). Role of tropoelastin fragmentation in elastogenesis in rat smooth muscle cells. *J. Biol. Chem.* 264: 15115–15119.

Indik, Z., Yeh, H., Goldstein, N., Kucich, U., Abrams, W., Rosenbloom, J. C., and Rosenbloom, J. (1989). Structure of the elastin gene and alternative splicing of elastin mRNA: Implications for Human Disease *Am. J. Med. Genet.* **34,** 81–90.

Janes, S. M., Mu, D., Wemmer, D., Smith, A. J., Kaur, S., Maltby, D., Burlingame, A. L., and Klinman, J. P. (1990). A new redox cofactor in eukaryotic enzymes: 6-hydroxydopa at the active site of bovine serum amine oxidase. *Science* **248,** 981–987.

Kagan, H. M., Vaccaro, C. A., Bronson, E., Tang, S.-S., and Brody, J. S. (1986). Ultrastructural immunolocalization of lysyl oxidase in vascular connective tissue. *J. Cell Biol.* **103,** 1121-1128.

Mecham, R. P., Hinek, A., Griffin, G. L., Senior, R. M., and Liotta, L. A. (1989). The elastin receptor shows structural and functional similarities to the 67-kDa tumor cell lamin receptor. *J. Biol. Chem.* **264,** 16652–16657.

Parks, W. C., and Deak, S. B. (1990). Tropoelastin Heterogeneity: Implications for Protein Function and Disease. *Am. J. Respir. Cell Mol. Biol.* **2,** 399–406.

Tinker, D., and Rucker, R. B. (1985). Role of selected nutrients in the synthesis accumulation and chemical modification of connective tissue proteins. *Physiol. Rev.* **65,** 607–657.

Urry, D. W. (1978). Molecular perspectives of vascular wall structure and disease: The elastic component. *Perspect. Biol. Med.* **21,** 265-295.

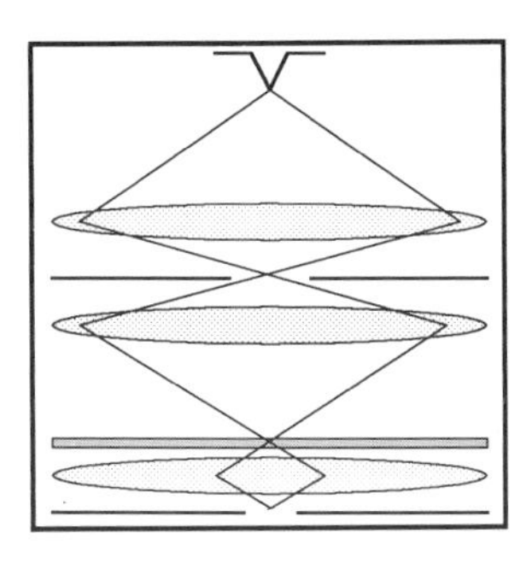

Electron Microscopy

PETER J. GOODHEW, *Liverpool University*

and

DAWN CHESCOE, University of Surrey

Glossary

Contrast Difference in intensity that permits a feature of interest to be distinguished from its (presumably less interesting) background. It is commonly quantified in terms of the local intensity I and the background intensity I_b as $(I - I_b)/I_b$

Resolution, or resolving power Closest spacing of two point objects at which they can be distinguished to be two objects rather than a single feature. This is not necessarily the same as the size of the smallest isolated object that could be seen

ELECTRON MICROSCOPY is a general term that embraces two major imaging techniques and a host of subsidiary activities such as the preparation of suitable samples and the use of the microscope for elemental analysis of small volumes. Scanning electron microscopy (SEM) is primarily used to reveal the external morphology of a small sample at magnifications typically between ×10 and ×10,000. Transmission electron microscopy (TEM) is used to reveal the internal microstructure (or ultrastructure) of a thin specimen at magnifications that are usually in the range of ×100 to ×1,000,000. Both microscopes can be used to perform elemental microanalysis using characteristic X-rays excited by the electron beam.

I. Introduction

The resolving power, d, of a light microscope is limited to some fraction (about one-third in the best case) of the wavelength of light, λ, via the relation

$$d = 0.6\lambda/\mu \sin \theta$$

where μ is the refractive index of the medium between the objective lens and the specimen and θ is the half-acceptance angle of the lens. Using green light of wavelength 500 nm, resolution is therefore limited to about 150 nm, and features closer than 150 nm apart in the specimen cannot be resolved as two objects but will appear as a single feature in the image. If electrons of wavelength 0.0037 nm are used, the attainable resolution would appear to be about 0.001 nm, or 1 pm, which is much smaller than a single atom. Although in practice such good resolution cannot actually be achieved, this logic has led to the development of a whole family of electron microscopes.

Because electrons carry a charge and are rather light, it is simple to construct electromagnetic lenses that act on a beam of electrons in the same way that a conventional thin glass lens acts on a beam of light. Using such lenses, the strength of which can be varied by varying the current through their windings, it is possible to make many different electron microscopes. Two configurations are in common use: (1) In SEM, a fine beam of electrons is used to probe the outer surface of a solid specimen revealing the external morphology; and (2) in TEM, a higher-energy electron beam is passed through a thin specimen, revealing the internal ultrastructure in a manner exactly analogous to the light microscope used in transmission.

Even high-energy electrons cannot travel far in a solid without being strongly scattered and eventu-

Encyclopedia of Human Biology, Volume 3.

ally stopped. Therefore SEM using 20-keV electrons will reveal the detail of the outermost few tens of nanometers of the sample, whereas TEM using 100-keV electrons can only provide useful images of specimens less than about 1 μm thick. Nevertheless, these capabilities have been found to be extremely useful, and many thousands of microscopes of both types are now in use around the world.

II. Scanning Electron Microscopy

An SEM works in a manner similar to a conventional TV camera. A beam of electrons is scanned across a small, usually rectangular, region of the specimen. A cathode ray tube (CRT) display is scanned in synchronism with the specimen, and its brightness is modulated according to the intensity of a signal emitted by the specimen. Most frequently this signal is related to the intensity of secondary electrons emitted from the surface, in which case the microscope is said to be operating in the secondary mode. However, many other signals [e.g., X-rays or light (cathodoluminescence, CL)] can be used, in which cases the SEM is said to be operating in X-ray or CL mode. [*See* SCANNING ELECTRON MICROSCOPY.]

Secondary mode SEM images are usually rather easy to interpret because the image contrast is similar to that experienced by the eye when looking at everyday objects. Thus, holes appear dark and hills usually have a bright side and a dark side in shadow.

III. Transmission Electron Microscopy

In a TEM the thin specimen is illuminated by a beam of energetic electrons. Electrons that are transmitted through the specimen with only a small loss of energy are used to form a projection image. The image is formed by a series of objective and projector lenses, which usually provide magnifications in the range of a few hundred to a million diameters.

Figure 1 is a ray diagram of a TEM. The essential components of any such microscope are an electron gun, two or more condenser lenses, an objective lens, several projector lenses, a viewing screen, and a camera. Additionally there are always three or more sets of apertures that permit the operator to limit the diameter of the electron beam at several points in the microscope column. Each of these components will be briefly considered.

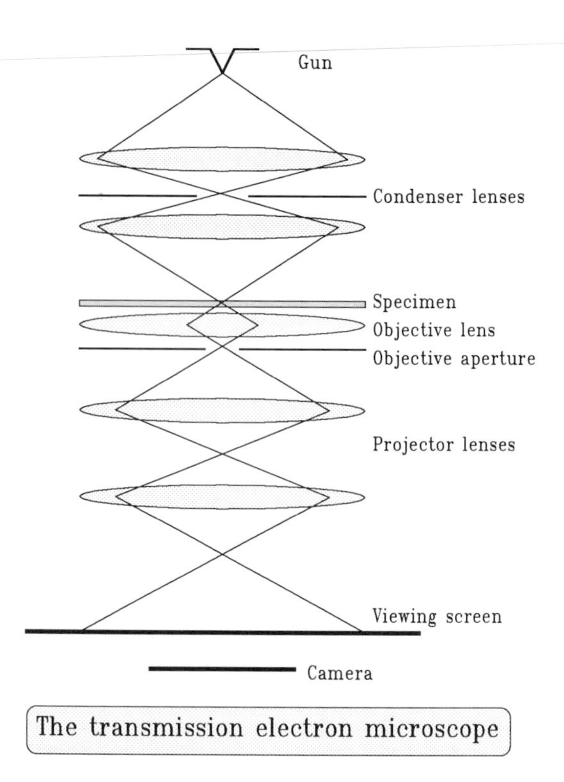

FIGURE 1 Outline ray diagram of a TEM illustrating the arrangement of the lenses and other major components with respect to the specimen.

The electron gun accelerates a beam of electrons through a potential usually in the range of 80–400 kV, giving each electron an energy of 80–400 keV. Higher energy electrons penetrate farther and can thus be used to study thicker samples. However, the cost of a microscope rises almost linearly with electron energy, so much biological microscopy is performed with microscopes capable of maximum electron energies of 100 or 120 keV.

All lenses in a TEM are electromagnetic and consist of a coil wound around a specially shaped magnetic core. Optically they can be considered to be "thin lenses," and ray diagrams for a TEM are similar to those that describe the operation of a projection light microscope. Electron microscope lenses differ from lenses for light in two important ways: (1) Their focal length can be changed easily by altering the current flowing through their coils; and (2) the image undergoes a rotation because the electrons travel a spiral path through each lens. The ability to vary the focal length of each lens makes it possible to use a TEM in many different ways and

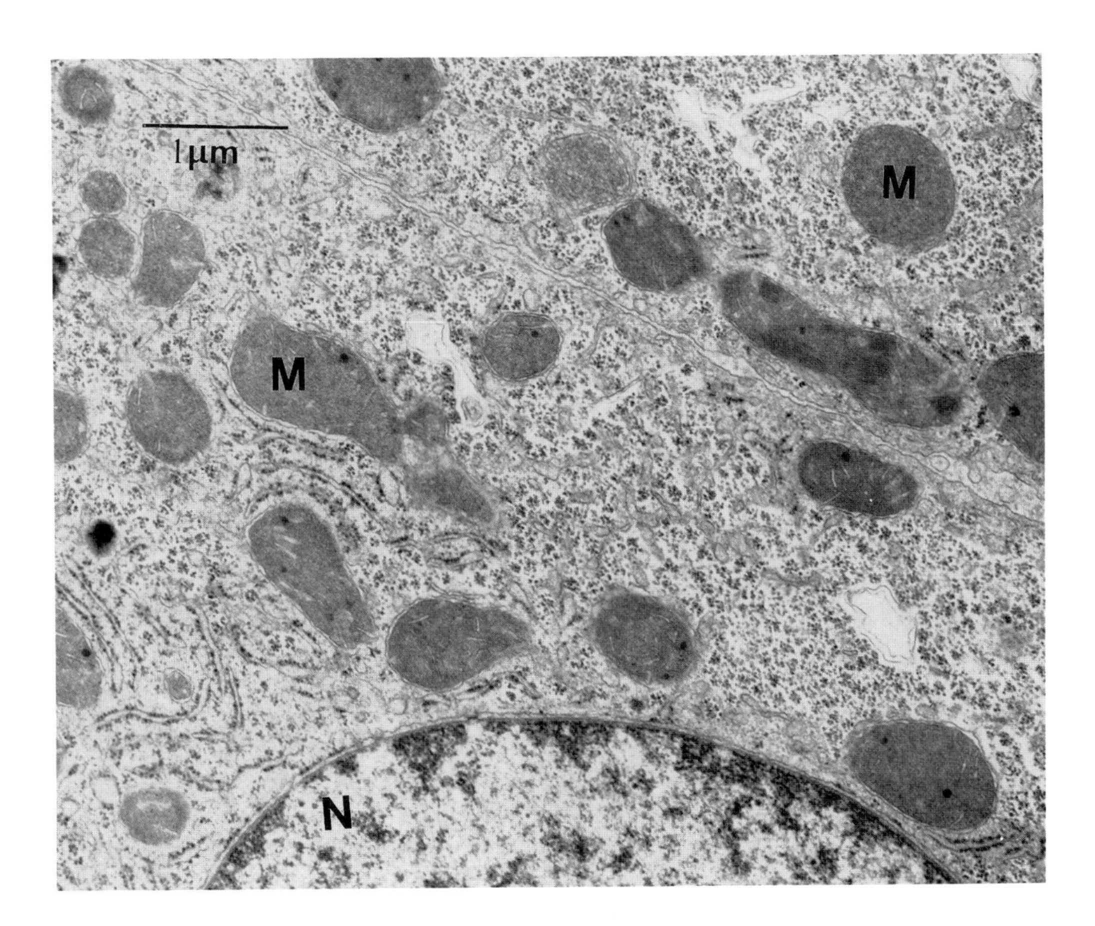

FIGURE 2 TEM micrograph of human liver showing a nucleus (N) and mitochondria (M).

to change focus, illumination, and magnification without moving either the specimen or the lenses. Image rotations, which may occur as lens strengths are altered, may confuse the novice but are not serious and are no longer present in many modern microscopes. Normal viewing of the image is via a window in the vacuum system, revealing the projection image on a green fluorescent screen. Micrographs are permanently recorded on photographic film loaded within the vacuum system directly below the viewing screen or directly into a computer via a TV camera in the same position.

Apertures are primarily used to control the diameter of (and current in) the illuminating beam and to enhance the image contrast (discussed below).

The micrograph of human liver shown in Fig. 2 illustrates the type of contrast to be seen in most biological material. Dark regions such as the nucleus and mitochondria in this figure are those that strongly scatter electrons.

Figure 3 shows how contrast is created in the TEM. Essentially all electrons incident on the specimen are transmitted. However, those that are scattered through an angle greater than α are stopped by the objective aperture and do not contribute to the image. Regions of the specimen that strongly scatter electrons (e.g., those that are much thicker or contain heavier elements) therefore lead to dark contrast in the image. Selection of a smaller objective aperture increases the number of scattered electrons that are stopped and therefore tends to increase contrast.

Because most tissue sections are intended to be parallel-sided (see Section VI), there should be no large changes in thickness across the specimen. Dark contrast therefore usually indicates the presence of a heavy element. Specimens are frequently stained with heavy elements to decorate and thus reveal particular features of the ultrastructure.

The resolution (resolving power) of a TEM is limited by aberrations in the electromagnetic lenses rather than by the electron wavelength. In a modern microscope it is possible to resolve two points that

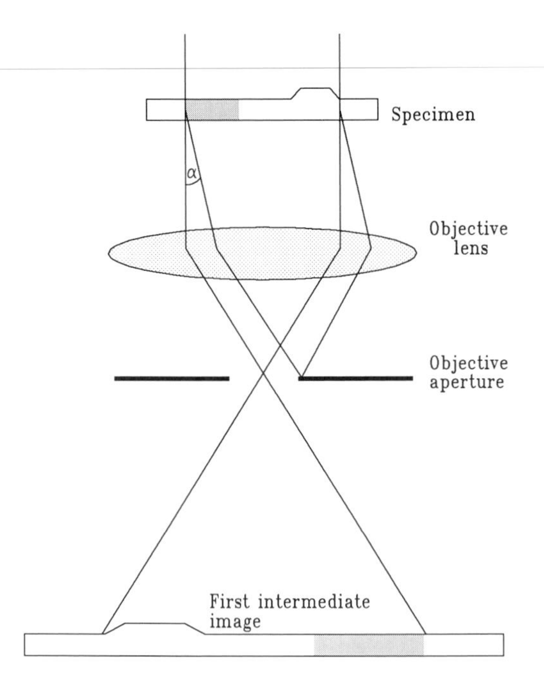

FIGURE 3 Ray diagram showing how the objective aperture is used to prevent electrons scattered through an angle greater than α contributing to the image. Without the objective aperture, there will be negligible contrast between different regions of the specimen.

are about 0.3 nm (3 Å) apart in an ideal specimen. However, biological specimens are rarely ideal for this purpose, and the resolution achievable on a particular specimen is more commonly determined by the specimen itself or by the degradation of the specimen under the influence of the beam. The high-energy electrons rapidly destroy the fine structure of the specimen, leaving a carbonaceous but chemically altered replica of the original. Stained features will be visible because of the heavy-metal atoms that they contain, which remain in the same place in the specimen. Many topographical features of the original specimen are preserved, and the image will probably look like the original specimen. However, the most finely resolved features are likely to be the heavy-metal deposits, which may now be rather larger than the ultrastructural features that they once decorated. In these circumstances, a more realistic resolution limit, imposed by the specimen, is 1–2 nm.

If high resolution is required, special experimental strategies are required, which are likely to involve "low-dose" microscopy of unstained material using short exposures to reduce the electron beam damage. It is then usually necessary to enlist the aid of computer image processing to extract the information from rather noisy images.

Figure 4 illustrates the different types of information available using the SEM and the TEM. Figure 4a shows a section of human gut cut on a microtome: The microvilli are clearly outlined by the stain, and it is evident that they have been sectioned at a variety of angles. Some microvilli are seen in almost circular cross section, whereas others have, by chance, been sectioned almost longitudinally. Figure 4b shows a similar gut specimen in the SEM: On the left at low magnification, the whole villus can be seen, whereas the magnified image of the boxed region on the right shows the microvilli end-on. It should be obvious that the TEM and SEM views give complementary information.

IV. Analytical Microscopy

The TEM is an excellent tool for performing local chemical analysis. Many of the electrons that are scattered as they pass through the specimen lose a significant amount of energy (from a few electron volts to several thousand electron volts) and excite one or more atoms to high-energy states. In a fraction of cases, the excited atoms will relax by emitting a characteristic X-ray. Chemical analysis can be performed either by measuring the energy of the X-ray or by measuring the amount of energy lost by the transmitted electron. These principles form the basis of the techniques commonly known as energy dispersive X-ray analysis (EDX) and electron energy loss spectroscopy (EELS).

Energy dispersive X-ray analysis is widely available on both TEM and SEM instruments. Characteristic X-rays excited within the specimen are detected by a semiconductor detector, which is typically sensitive to X-rays from elements 11 (sodium) to 92 (uranium) in the periodic table. Special "windowless" detectors can extend this range to the lighter elements with atomic number down to 5 (boron). Thus virtually all the elements of potential interest to biological microscopists are accessible to analysis. The volume of the specimen that can be analyzed can be as small as 10 nm × 10 nm × t. The thickness of the specimen, t, will typically lie in the

FIGURE 4 (a) TEM micrograph of sectioned human gut; (b) SEM micrograph of a similar gut sample. At the *right* is a magnified image of the region within the white box.

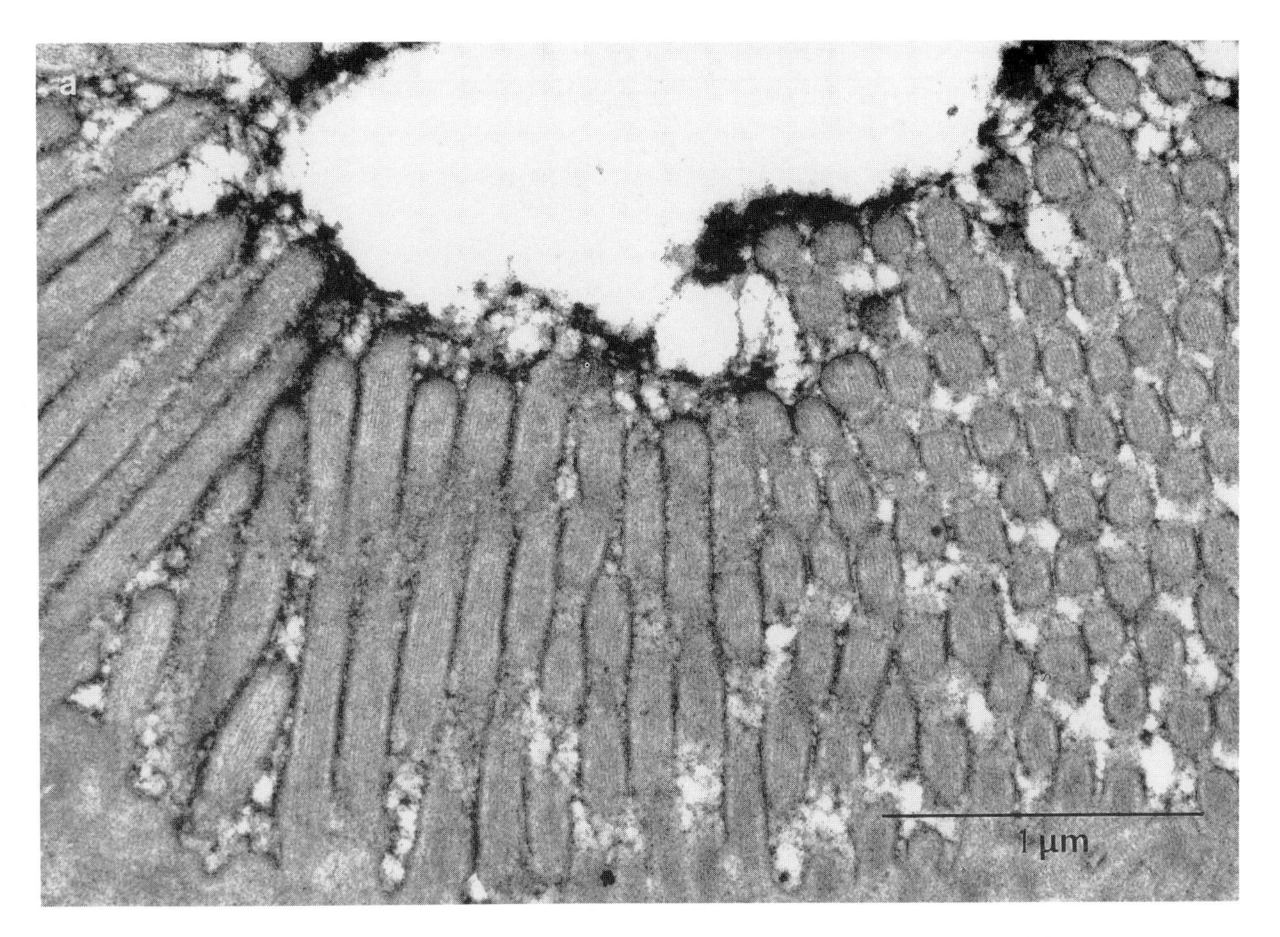
a
1 μm

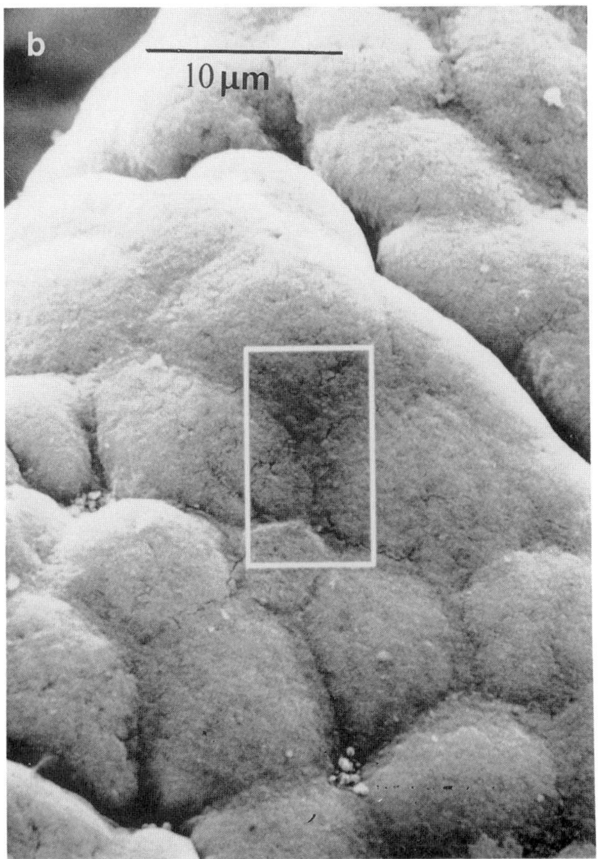
b
10 μm

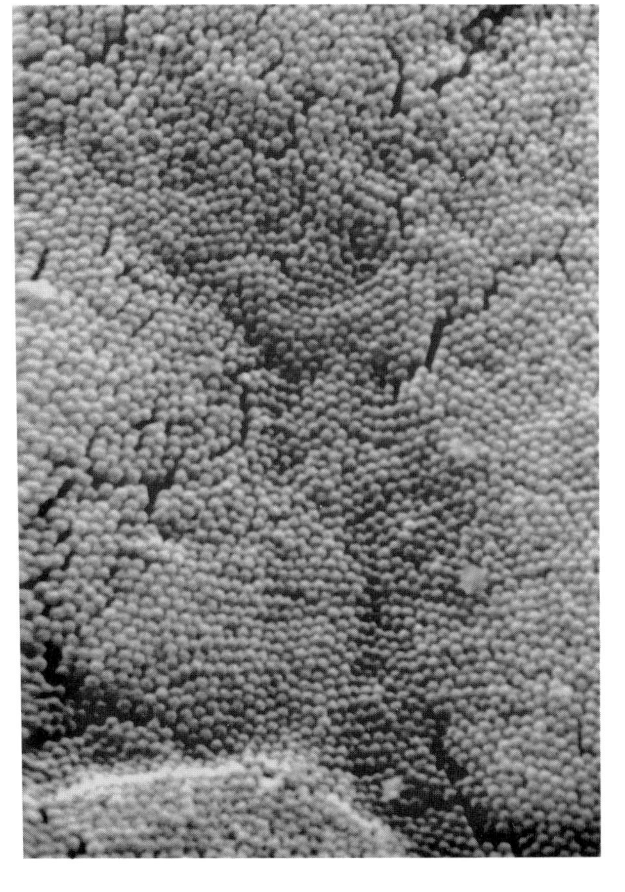

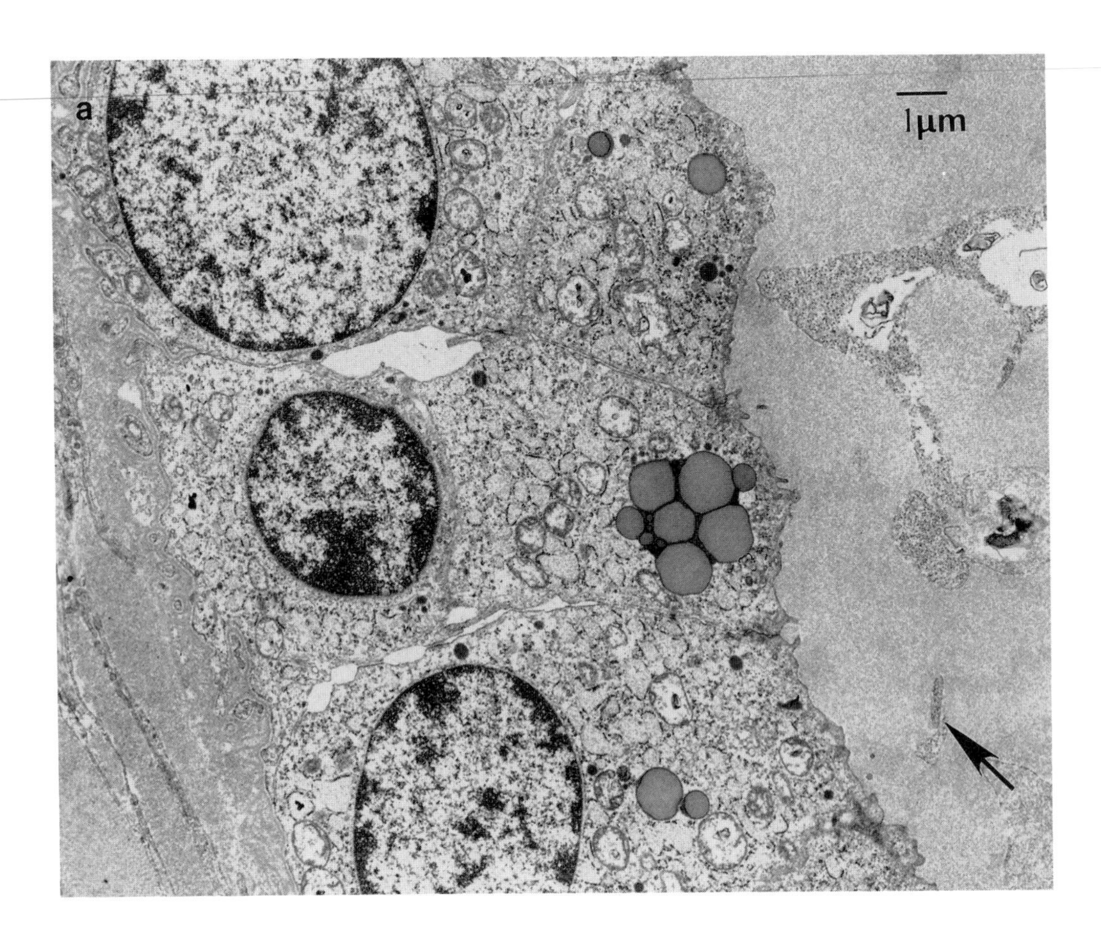

range of 20–500 nm, giving an analyzed volume of only about $10^{-23}m^3$, or about $10^{-17}g$. Because the signal (i.e., the number of emitted X-rays) is small, the analytical accuracy is not high, but the ability to localize the analysis to a known region provides an extremely powerful microanalytical tool. Figure 5 shows a section of human thyroid and the X-ray spectrum collected from the region of colloid labeled C. The small iodine signal that arises from the naturally occurring iodine within the colloid can be seen. The prominent osmium peak results from the fixation and staining procedures, whereas the chlorine peak comes from the embedding resin. The sulfur and calcium peaks are real features of the structure.

Electron energy loss spectroscopy (EELS) has been developed more recently than EDX and offers high sensitivity for light elements. Quantitative analysis is less well-developed using EELS than EDX, and the majority of analyses are carried out using the longer-established technique.

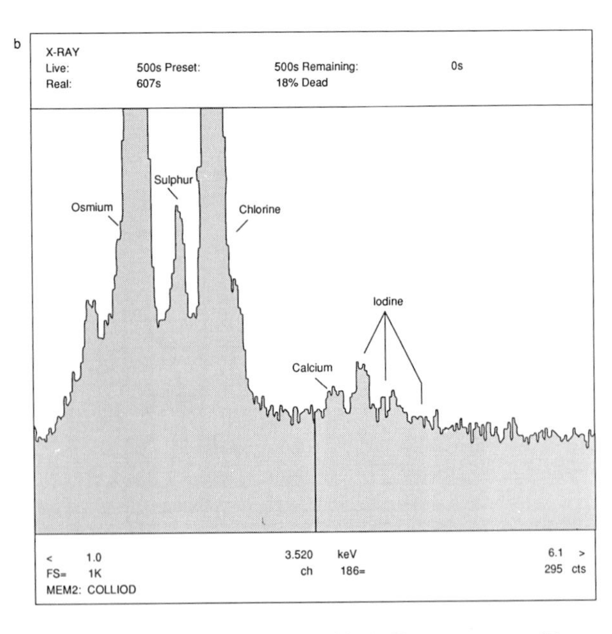

FIGURE 5 Section of human thyroid (a). X-ray spectrum (b) was collected from the region of colloid.

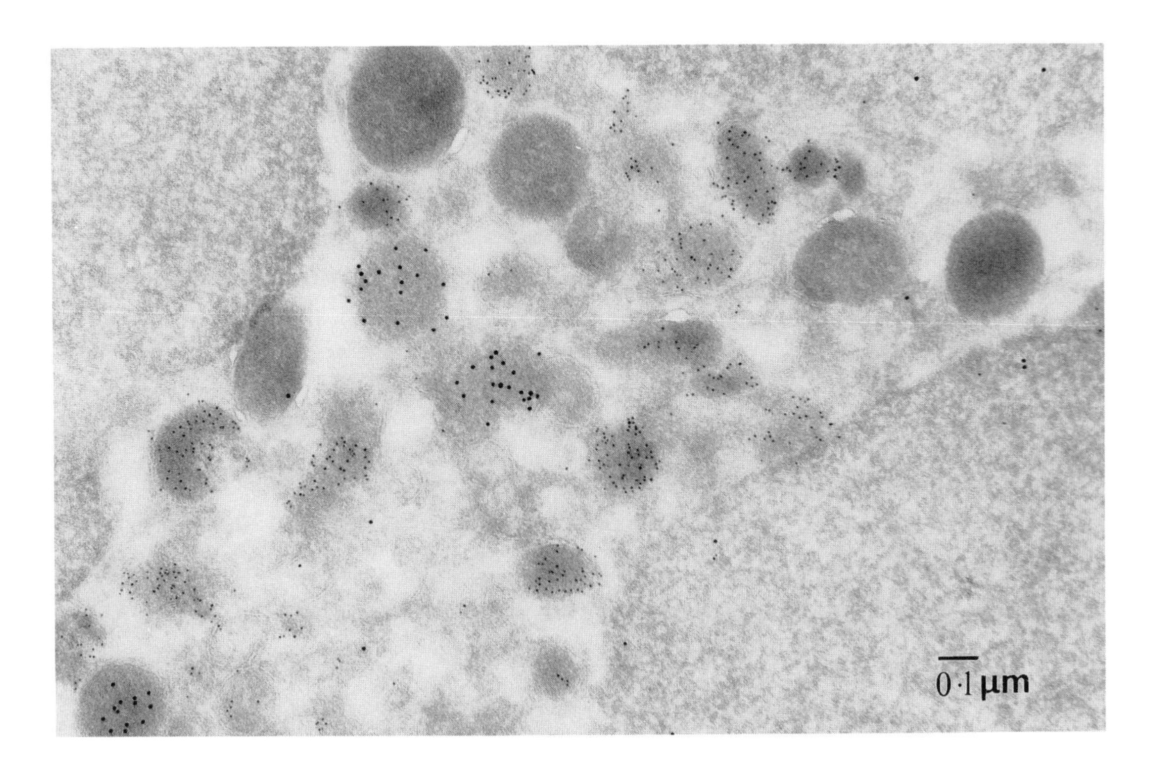

FIGURE 6 A gold-labeled section of human granulocytes. The section was cut while frozen and then postembedded. Two different antibodies (rabit antibody raised against lactoferrin and mouse monoclonal antibody raised against elastase) were labeled using two sizes of gold spheres (15 nm and 5 nm), which show as large and small black dots in the micrograph. The large spheres were coated with goat anti-mouse Ig, whereas the small spheres were coated with goat anti-rabbit Ig. It is clear that the different antibodies accumulate in separate granules. (Courtesy of Dr. J. E. Beesley.) Wellcome Research Lab, United Kingdom.

For analytical microscopy to be meaningful, it is obviously essential that the composition of the specimen is unchanged from that of the *in vivo* material. This means that special care has to be taken during specimen preparation (see Section VI). Clearly the electron beam can itself alter the chemical nature of the specimen while it is under observation. This is not of great concern if the changes involve carbon, hydrogen, and oxygen atoms, because it is unlikely that these would be of analytical interest. However, the microscopist should be aware that many ions of great interest (e.g., Na, K, Ca) are likely to be removed during specimen preparation and, even if they remain, can be mobile under the influence of the beam. Analyses involving these elements can therefore be erroneously interpreted.

V. Immunocytochemistry

The principle of this rapidly developing technique is to label a surface antigen site with a specific labeled antibody. Peroxidase can be used for labeling the antibody, because it is easily highlighted by subsequent osmium fixation and can therefore be made visible in the TEM. However, the use of antibodies labeled with colloidal gold is increasing and is already the dominant technique. Colloidal gold particles have the advantage that they can be seen in the image without further staining. The effectiveness of the technique depends on the specific antigen. It is possible with many specimens to fix them lightly with, for example, a dilute mixture of glutaraldehyde and formaldehyde. Depending on the antigen, labeling can take place before or after embedding. It is also possible in favorable circumstances to label frozen hydrated or freeze-substituted specimens (see Section VI). A gold-labeled section of human granulocytes is shown in Fig. 6.

VI. Preparation of Specimens

Specimen preparation is absolutely central to biological electron microscopy. The treatment of the sample at this stage will determine the quality of the image and analytical information that can be deduced by TEM. Preparing the specimen usually takes an order of magnitude longer time than performing the microscopy, and great importance therefore attaches to development of sound technique. The ideal TEM specimen would be thin, representative, and unaltered from its *in vivo* state. These ideals can rarely be attained, but there are three common approaches to preparing specimens that are at least interpretable. Briefly these techniques involve cutting thin slices (sections) from the embedded bulk specimen (ultramicrotomy), mounting small discrete entities (e.g., viruses) on a thin support film or replicating a carefully created internal surface (e.g., by freeze-fracture).

A. Section Cutting

The essential steps in the preparation of a thin section are fixation, staining, embedding, and ultramicrotomy. A small piece of the tissue material is typically fixed in a dilute 2–4% glutaraldehyde solution, washed in a buffer solution, and then impregnated with a 1–2% solution of osmium tetroxide. This acts both as a secondary fixative and as a heavy-metal stain. During fixation, efforts should be made to keep the pH, osmolarity, and ionic constitution of the material as nearly as possible unchanged.

At this stage it may be desirable to use a specific stain to render specific chemical (and thus biological) sites visible. However, for many samples the action of the osmium tetroxide in staining protein sites is adequate. There are many alternative staining methods, described, for instance, in volumes 3 and 5 of the Glauert series.

The specimen must now be dehydrated, by immersion in a series of water/ethanol mixtures with increasing ethanol content, finishing in a bath of pure ethanol. It may then be conditioned in propylene oxide before being embedded in resin. Once the resin is cured, the block (now about 8 mm diameter and 15 mm long) is gripped in the stage of an ultramicrotome, and sections of controlled thickness are shaved off it using a glass or diamond knife.

Thin sections can now be collected and transferred to the microscope, and they will usually be stained again, using, for example, lead citrate and uranyl acetate. Staining at this stage is more efficient than at an earlier stage because the specimen is so thin that penetration into the tissue is rapid.

Tissue sections prepared in this way are ideal for the purposes of comparative morphology, and a vast amount of microscopy is performed on sections prepared by this standard method. However, such sections are not well-suited to analytical microscopy because virtually all the sodium and potassium will have been removed during preparation, together with much of the calcium. Neither are these specimens ideal for immunocytochemistry because surface antigens are likely to have been altered by cross-linking and are thus unavailable for a labeled antibody.

B. Negative Staining

Small particulate materials, especially viruses, are difficult to handle and prepare by sectioning. They are usually dispersed, via a liquid medium, onto a carbon film, which can be supported on an electron microscope grid. They are then "negatively" stained: A solution of the chosen stain is flooded across the carbon film and collects around the outside of the virus particles. On drying, it leaves a deposit of heavy metal, defining the external mor-

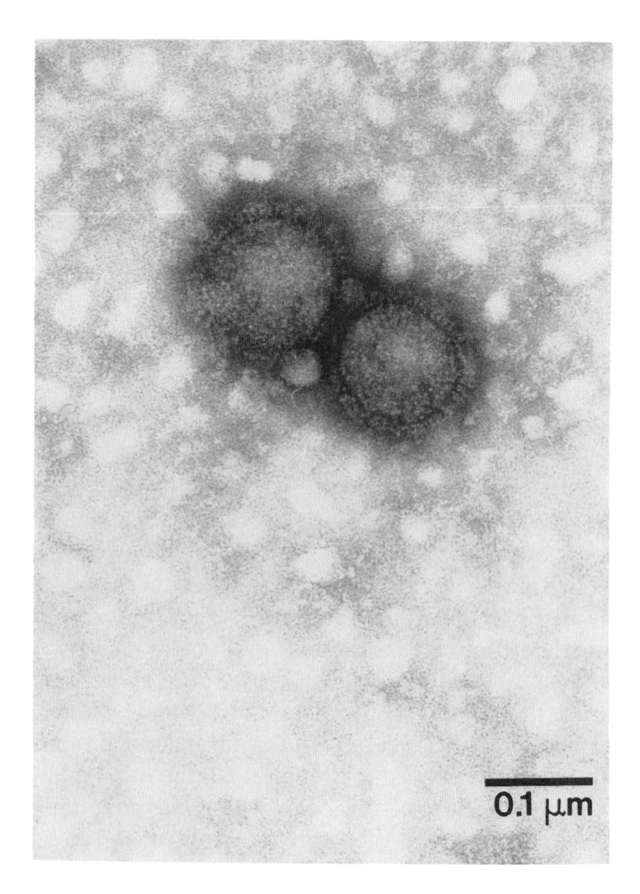

FIGURE 7 Pair of negatively stained influenza viruses.

phology of the particle. An example of a negatively stained influenza virus is shown in Fig. 7.

C. Cryo-Fixation

Rapid cooling techniques are becoming widely used as a means of preventing elemental migration and thus preparing useful specimens for analytical TEM. Two common methods of rapidly cooling the specimen and thus avoiding cell damage by ice formation are (1) plunging the specimen into liquid propane or (2) "slamming" the specimen onto a cold block. The cold specimen can then either be cut into sections while still frozen using a cryo-ultramicrotome or held at a higher temperature (still less than 0°C) until it has freeze-dried.

The advantage of specimens prepared in these ways is that there tends to be good preservation of the chemical structure and thus the distribution of elements. However, because the specimen is not stained, image contrast can be low and morphological information is not easy to extract.

In recent years several techniques have been developed that offer a compromise between frozen and room-temperature methods. For instance, it is possible to fix a sample with a very dilute (e.g., 0.5%) glutaraldehyde solution and/or stain it lightly before freezing. Alternatively, freeze-substitution can be attempted. After rapid freezing, the sample is held for several days in an organic solvent at about −90°C to allow substitution to take place and is then warmed to about −40°C and impregnated with a special resin that is still fluid at this temperature. The resin must be cured by irradiation with ultraviolet light while the sample is still cold, but the ultramicrotomy can then be performed at room temperature.

Bibliography

Chescoe, D., and Goodhew, P. J. (1989). "The Operation of Scanning and Transmission Electron Microscopes," Royal Microscopical Society Handbook no. 20. Oxford University Press, Oxford.

Elder, H. Y., and Goodhew, P. J., eds. (1989). "Proceedings of EMAG/MICRO 89," Institute of Physics Conference Series, vol. 98. Institute of Physics, Bristol.

Glauert, A. M., ed., (1973–present). "Practical Methods in Electron Microscopy" series, Elsevier, Amsterdam.

Goodhew, P. J. (1987). "The Transmission Electron Microscope:" a software package, Engineering Materials Software Series, Institute of Metals, London.

Goodhew, P. J., and Humphreys, F. J. (1988). "Electron Microscopy and Analysis." Taylor & Francis, London.

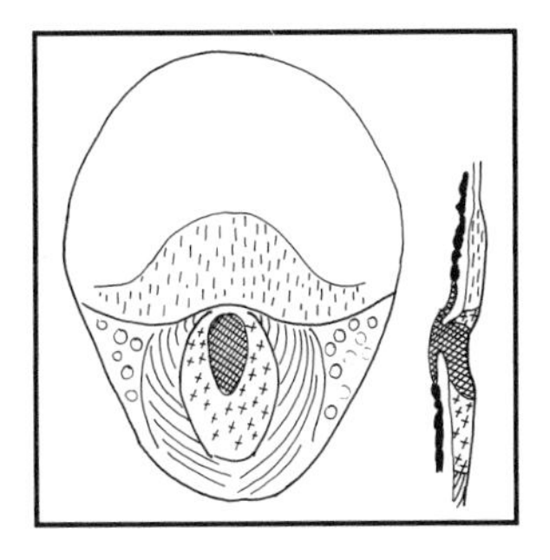

Embryo, Body Pattern Formation

JONATHAN COOKE, *National Institute for Medical Research, London*

Glossary

Commitment Setting of the course of development in a cell or cell group so that it can no longer be modified by exposure to normal stimuli that cause specification or respecification in its embryo of origin

Germ layers Three fundamental superimposed cell layers in development of the embryo; organs are characteristically contributed to by two adjacent germ layers

Histodifferentiation Functional specialization of different cell types, involving massive synthesis of special products and arrangement into special tissue structures, visualized in microscopy by histological techniques

Homology Relationship of structures in two organisms by common descent from a structure in a shared ancestral organism; can be a DNA sequence, the protein it encodes, an anatomical structure, or an episode of cellular activity in development

Models 1) Organism with an embryo type accessible to experimentation and presumed to show useful homology of mechanism with, ultimately, human development; 2) Quantitative (usually computer-run) simulation of a formally possible mechanism for pattern formation to assess performance characteristics

Morphogen Signal molecule whose localized place of synthesis, and interactions with others, enables control of a spatial pattern in which cells become diversely specified

Morphogenesis Chiefly, the aspect of development in which tissue reveals its pattern of diverse specifications and commitments by specific form-building movements and differentiations

Specification Cell state that will lead to development of a particular tissue type, or region from the normal body pattern, upon culture in isolation from any subsequent, respecifying stimuli such as may be experienced in the embryo or by experimental means (cp. commitment, above)

BODY PATTERN FORMATION is a distinct, early process of development in the embryos of higher multicellular animals that sets aside small regions in the tissue as precursors to particular parts of the future body. It occurs in a brief time frame in relation to the remainder of embryonic (fetal) development and, preceding the onset of differentiation in the various organs, is associated with the distinct set of mechanical activities, known as gastrulation, that reorganize the embryo's structure into layers. Normal features of newly set-up body patterns are the correct spatial arrangement, and rather close adherence to particular proportions, among the precursor regions. Positive mechanisms of regulation exist initially; i,e., proportions and completeness of the pattern of regionalization tend to attain normality despite earlier experimental manipulations, or natural variations in the overall amount of tissue available in individual embryos. Nevertheless, early biochemical and surgical interferences that leave a permanently disturbed body pattern, as well as direct visualization of the gene products involved, using molecular biological techniques, enable the study of the pattern formation mechanism using insects and lower vertebrates as ''model'' organisms. We might expect the molecular components of mechanisms to be highly conserved in evolution, as among different forms of embryo, but their strategy of deployment—the cell biology and anatomy of the process—to be much less so. Even the experimental approach on the model systems is difficult, so our real understanding of the beginning of mammalian (human) development is still extremely primitive, although it has just begun to accelerate. This article surveys theoretical requirements and possibilities for the pattern-forming mechanisms and summarizes our cell- and molecular-biological knowledge to date.

I. Body Pattern Formation as a Distinct Phase in the Higher Multicellular Life Cycle

Many people, including scientific specialists in other areas of biology, have surprisingly vague notions of the sequence of events in embryonic development. We are used to thinking of the nervous and endocrine systems as constituting the great integrative mechanisms of highly evolved multicellular life, which enable their possessors to live the distinctive lifestyle that would be unavailable to a heap of cells, a plant, or a primitive, spatially differentiated multicell such as a sponge. Yet these systems of organism-wide coordination have only reached their present pitch of complexity as aspects of the complexity and consistency of the anatomy achieved in the body of each species. Anatomy is a reflection of a precise spatial array of cellular activities in later morphogenesis, i.e., in form-building movements and in histodifferentiation to give specialized types. Evidence indicates that certain cell types of the body, though generated at many widely scattered locations within it, retain long-term ''memories'' of belonging to the particular geographical regions of their origin. They maintain the tissue characteristics proper to one region (e.g., the pattern of hair growth on skin, when transplanted elsewhere). The ill-understood but definite system of interactions that first regionalizes tissue in the early embryo, to give a properly proportioned and spatially ordered set of founder territories from which the anatomy is later constructed, is body pattern formation. It ranks as the third great integrative system of higher animal life forms.

As we shall finally see (Section V), body pattern formation probably uses the same categories of intercellular and intracellular signaling molecules that integrate bodily function later in life. The bulk of what we need to learn about developmental biology is simply cell biology, with the crucial agents of intercellular communication being relatively few, and some perhaps not yet identified. In at least one way, this early formation phase is unique; only then are cells being instructed as to which parts of the body, as well as to which functional types, their descendents will give rise. Thereafter, the intercellular signals are used to activate, or prevent activation, of preset options in differentiation or physiological function in target cells predesignated for each combination of signals.

The integrative systems of multicellularity seem to have evolved from relatively loosely, indefinitely controlled networks of interactions to more precise, specialized ones. Body pattern formation has been accompanied by compression in time, relegating the events to the earliest part of development, and the following of a precise sequence of cell-specification steps within that brief time. Initial evolution of multicellular bodies probably occurred by opportunistic harnessing of physiological or environmentally imposed differences arising among regions of a growing cell mass. These differences were co-opted as stimuli causing diversification of cell types, thus optimizing the survival ability of the whole. Subsequently, the stimuli for diversification and their spatial pattern became regularized, mainly from within the embryo by the cells themselves, and largely separated from a prolonged, later process of growth within the committed tissues. Thus, in most evolutionarily advanced animal forms, including mammals and the experimental models described later, pattern formation for the body as a whole is integrated, stereotyped, and compressed into 100 hours or less. This is preceded by a period of relatively uneventful cell multiplication (often, though perhaps not in mammals, a special, rapid kind) that provides enough tissue space, or a high enough cell number, for the process we are concerned with to take place in. Then, at the other end of the crucial, brief period, the embryo is still largely undifferentiated histologically but is in a very different state. It has become cryptically parceled into a sequence of regions, some of which may contain very few cells. If a region is subsequently removed or killed, or if the set of regions is incomplete or very wrongly proportioned due to physiological or genetic abnormality, the individual will be correspondingly incomplete or quantitatively imbalanced in anatomy. Each region has become committed, containing the unique precursor cells for a crucial component of the structure of a particular organ system or district of the body. The final pattern of commitments is achieved stepwise, with some regions going through transient states of specification where, on removal from the embryo and its signaling system, they will develop into definite structures other than those that would have been their normal fate in final commitment. The phenomenon of stepwise specification should help in unravelling the signals and cellular-response mechanisms involved.

Each region of commitment seems to be programmed for its future pattern of growth and for provision of particular sequences of terminally differentiated cell types or stem cells for such types. The program may be simple or complex (e.g., within the various regions of the brain or the blood-forming tissue). Further instructions *within* each initial region of commitment will, of course, occur to increase the complexity finally attained; but by "body pattern formation," we refer to the episode after which the primary parcelation into noninterchangeable regions, giving the large-scale plan of the body, has been completed. This is dramatically early within development as a whole and, especially in our own animal group, the vertebrates, is somehow associated with the integrated set of cell movements called gastrulation. These movements reorganize the sheet of cells, which will form the embryo, into the definitive three-layered structure (the germ layers) on which the body is based. Thus the crucial events occur, relatively, so early that in human development they begin around 15 days after conception but are essentially over 7 days after that. These are dates when few but western middle-class women make a point of being aware of pregnancy. A simple view would be that subsequent development is simply the playing out of a pattern of local schedules for growth and differentiation. Tissue could be surgically rearranged within the embryo with little effect on these schedules and, thus, with catastrophic effects on final anatomy. In experimental vertebrate embryos during or just after gastrulation, precursor tissue for the vertebrae or the brain can be reversed in polarity or exchanged between sites along the future body axis. This is followed by the final development of individuals with correspondingly misplaced or back-to-front structures embedded within an otherwise normal anatomy.

II. Evolutionary Considerations

A. The Embryo Body Itself

The transition to highly organized multicellular development, as opposed to that of various primitive and less tightly controlled body forms, may have occurred only once in order to give rise in evolution to all those forms that are now used as experimental models. In addition, despite proposals that rare, evolutionary steps might involve "hopeful monster" mutations that alter body patterns in massive but serendipitously adaptive ways, the more general correlation must surely be that the earlier in development the process is affected by a genetic change, the less the relative probability that that change will be advantageous and thus perpetuated. For these reasons, the assemblage of gene products providing the basis for body pattern formation might be expected to be largely homologous, as well as relatively highly conserved in structure, among such diverse animals as higher insects (fruit flies), annelid and round worms (leech and nematode), and lower and higher vertebrates. This expectation has been borne out by recent findings (see Section V). Thus, study of forms in which molecular genetic analysis is advanced must greatly increase our ability to make sense of vertebrate development where the list of known gene products grows more slowly, and the anatomical structure accompanying pattern formation is also more complex. However, complex problems arise when using the other model systems to arrive at a direct understanding of the mammalian (human) development.

Embryos themselves are subject to evolutionary pressures and, therefore, specialize in form and function, along with the bodies they give rise to. This means that, although the available menu of molecular signals is conserved, the selection made from that menu and the strategy of development at cell and anatomical levels are varied, often in quite a dramatic way, even among closely related forms. Within vertebrates, for instance, at the close of gastrulation (thus, pattern formation), the primitive anatomy of the body passes through a stage that is easily recognizable across all types. The anatomical arrangements in earlier development leading up to this, however, vary much more, and in ways that must be important for understanding mechanisms. Because our knowledge to date comes mainly from the fruit fly embryo, with significant additions from lower vertebrates (but little from mammals), certain idiosyncrasies of these most worked-upon model systems must be described at the outset. These idiosyncrasies, in particular, limit the direct extrapolation from those systems to mechanisms in our own body pattern formation.

In the specialized version of insect development, where the interplay of early-synthesized gene products has been best visualized and understood, the events occur in a syncytium, a layer of nuclei within cytoplasm that is not separated by true cell membranes until an advanced stage of pattern formation. A whole class of interactions, whereby a messenger RNA or protein product of gene activation in one nucleus can translocate in space to affect directly the gene activity of others, is thus possible and seems to occur in the fruit fly blastoderm, but it is ruled out for vertebrates, where pattern is formed in a truly cellularized tissue from the outset. Therefore, these first stages must involve largely different classes of gene product, as between vertebrate and fruit fly development.

A second limitation concerns the varied strategy among vertebrates themselves. The amphibian egg, which may represent the strategy ancestral to the group, employs a structural localization of some kind, which is set up by mechanical events immediately after fertilization. This, then, is a reference point, organizing the much later intercellular signaling events so that patterning and its orientation occur in relation to those cells that have "inherited" it during cleavage. In fish, birds, and mammals, the embryo body itself forms in a sheet of cells—the blastoderm or blastodisc—deriving from only part of the cell population of the original, cleaving egg. We know that the signaling events that initiate and orientate this pattern occur only after much cell division and in response to subtle cues extrinsic to the egg's structure (see Section V). This is important in relation to adequate theories of pattern formation (see Section IV). It means that pattern must be self-generating among a sheet of cells with initially homogeneous properties in embryos of our own type. Prelocalization of a gene product, or even of a particular physiologically maintained state, to certain cells is not required.

Nevertheless, once it is under way, the process of regionalization by spatial intercommunication among cells of sheetlike tissues probably employs the same molecular mechanisms among all vertebrates. Indeed, fruitfly development remains extremely relevant as a model, because evidence indicates homology of function of the gene products involved in this remaining, slightly later, aspect of pattern establishment as between it and vertebrates. These gene products mediate acquisition of permanent identities by cells as members of the regions that constitute the body pattern (see Section V).

B. Extra-Embryonic Structures in the Mammal

In addition to the delay of body pattern initiation until a multicellular blastodisc has formed, mammalian development is also complex in that considerable special commitment to differentiations is also occurring among the early cell populations that will never participate in the embryo body. These are connected to development of the special maintenance structures whereby the fetus will interface with the maternal blood supply and regulate its environment within the uterus. Their study is of great importance for understanding and potential control of mammalian reproductive biology, and they have been more amenable to detailed analysis so far than has true body pattern formation in mammals. The relative inaccessibility and minute size of the tissue in which the latter is occurring have been major limitations to direct knowledge of mammals. A significant part of this limitation has been removed, however, with recent technology for exactly visualizing the spatial patterns of expression in tissue of the products of particular, early-acting genes.

III. Structure of the Embryo During Pattern Formation

A. Gastrulation and Secondary Inductions

Figure 1 represents, in general form, the cell movements that occur while the embryo tissue is undergoing body pattern formation. At the beginning, the tissue is a sheet, one or a few cells thick, and breaks up into the obvious populations that will make the various structures only after gastrulation has produced the three fundamental germ layers. There is an arrangement whereby an orderly supply of cells leaves the original epiblast sheet at a rather re-

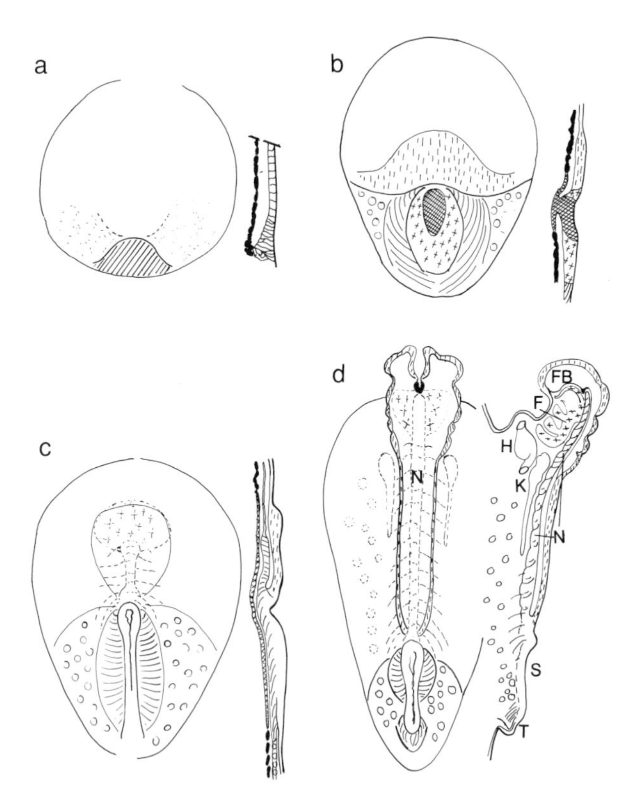

FIGURE 1 Sequence of morphological stages in body pattern formation of a generalized vertebrate. The figure is best studied after text-section III has been read. The sheet of tissue in which the embryo forms (blastoderm, or blastodisc) is shown in plan view from above (left) and sectional view from the future body's left side (at right) in each figure part. In mammalian development, this sheet is not, in fact, planar but a closed-ended cylinder, but representation of that fact obscures rather than aids understanding. In (a), the future dorsal axial midline, whence mesoderm is to arise in gastrulation, is represented as a thickened region of the upper cell layer where the posterior end of the axis will finally lie (obliquely hatched). Cells from within the stippled area may be in course of being specified as mesoderm. The lower layer of cells or hypoblast (black) may be important in inducing pattern formation but does not contribute to the embryo. In (b), the first cells of the future embryonic endoderm (inner cell layer) have begun to leave the upper cell layer in gastrulation, and the process of primary regionalization into future ectodermal, mesodermal, and endodermal derivatives is in train but has reached an unknown stage of refinement. Note that the embryonic endoderm (cross-hatched) is displacing the hypoblast, whereas subsequent regions will form a middle cell layer between ectodermal and endodermal layers. Codes for major regions in this and, where shown, in parts (c) and (d); head-mesoderm (crosses), dorsal axial mesoderm (notochord and somites) (oblique curved hatching), lateral-ventral mesoderm (circles), area destined for nervous system formation (not involuting) (vertical dashed). In (c), the region of active gastrulation has elongated to form the definitive, thickened primitive streak, or blastopore. The already gastrulated, anterior, dorsal axial mesoderm is also actively elongating and narrowing beneath the now morphologically distinct neuralized area, and its segmentation is beginning. It and the head mesoderm are shown by a dashed outline (i.e., below the surface) in plan view. Much posterior dorsal axial and ventral lateral mesoderm remains to gastrulate and probably does so through different but adjacent regions of the streak as it shortens (see part d). In (d), the body axis has taken shape and major secondary inductions (interactions between cell layers) have occurred, even though gastrulation is still incomplete through the shortened streak in the tail region. Major structures and organ systems labeled, in addition to previous regional coding. H = heart, K = primitive kidney, N = notochord (midline skeletal rod between two rows of somite segments), FB = forebrain region ahead of notochord underlain by the head mesoderm, F = face and mouth structure, contributed by head-mesoderm and -endoderm, with additions from the neural tube (see text). Endoderm as a whole (the gut lining) is omitted from (d) for clarity. S = remainder of primitive streak, T = tailbud.

stricted site to migrate and pass beneath that cell layer, coming to lie between it and a lower cell layer so that the three-layer structure is progressively formed. The rudiments of the body parts take shape and begin to differentiate in sequence, from the head (first-immigrating mesoderm and the parts of the other two layers that are finally associated with it) toward the trunk and tail (later-involuting mesoderm and associated cells of the other layers).

Immigrating mesoderm at the outset may be of two different types defined by cell behavior, in addition to revealing regionalizations within those two overall components. Mesoderm of the dorsal axial region, immediately upon immigration, forms a rather cohesive tissue that also adheres rather strongly to the overlying midline of the epiblast, while generating a forceful shape change (convergent extension) that turns the whole into an elongated axial formation. This version of gastrulation behavior is termed involution of mesoderm, and it has the most highly organized sequence, whereby the newly involuted cells at successive time points are destined for the front, middle, or back of the body. More laterally and ventrally fated mesoderm behaves more as individually immigrating cells, with less capacity to produce force by mutual adhesions and more capacity to spread into available space between the two layers.

The appearances of gastrulation in various vertebrates differ as to how the orderly supply of mesoderm cells from the epiblast is arranged to occur, from a reservoir of uninvoluted tissue, rather than in the structures formed after involution. The arrangement in mammals is the least understood at this level, but this need not dismay us unduly. Precise understanding of particular versions of their anatomical arrangement may not be relevant to understanding the underlying diversification of these cells, leading to their specification and commitment as members of different regions of body pattern.

The definitive endoderm layer, finally contributing the innermost of the three layers of the embryo, is probably derived from the first involuting cell population from the epiblast, preceding the "anteriormost" mesoderm. This displaces a previously underlying cell layer in the central, embryo-forming area of the blastoderm, or blastodisc, and is, in turn, migrated upon by the mesoderm sequence of tissue, when the latter inserts itself between the two other "germ layers" to give ectoderm (future nervous system and epidermis), mesoderm (all primary connective tissue and skeletal, muscular, excretory, and blood systems), and endoderm (lining epithelium of the gut and its derivatives) at each position in the body plan.

Sequences of local interactions (best defined as secondary inductions and briefly referred to in Section IV) exist whereby short-range instructive signals pass between cells of the different layers to coordinate their patterns of differentiation. The best studied of these are involved in establishing the regional axial pattern within the area of ectoderm, overlying the midline of the involuted mesoderm, that has been specified to give rise to the nervous system. These signals, emitted from that mesoderm, are thus an expression of the body pattern of regionalization already emerging within it. Generally, mesoderm is the controlling cell layer in most of the body, in the system of local inductions that ensures appropriate contributions to the organ systems from the other two layers. The situation may be different in the head, however, where a second population of mesodermlike cells, derived from the brain rudiment after its induction, enters the middle cell layer to collaborate in morphogenesis with the primary mesoderm. These cells may reimport the information as to which pieces of body structure are to be formed back into the middle cell layer, even though the primary mesoderm cells originally determined what brain regions the head neural ectoderm should form. This reversal of the normal flow of information may be one sign of a rather deep difference between an anterior and ventral domain of vertebrate body structure, devoted primitively to feeding apparatus, and the remaining, elongating axial domain that provides the locomotory (originally swimming) apparatus.

B. Primary Embryonic Induction

The above section shows that secondary inductions are essentially a means whereby the pattern of regionalization in one cell layer is imprinted onto superimposed ones. They then respond in a spatially appropriate manner in their own differentiation. In large-scale regionalization to make up the vertebrate body plan, the primary role is thereby given to pattern formation within the mesoderm itself, or at least within the cell population comprising anterior (pharyngeal) endoderm and the entire mesoderm. We need to understand a mechanism that sets aside, and specifies as (endo +) mesoderm, a subpopulation of the original epiblast cells. At the same time, this mechanism gives a regional diversity to that subpopulation so that it can show the spatial and temporal pattern of adhesive and locomotory behaviors that orders gastrulation and hence orders its final positions in the body plan. The evidence is that the regional characters of this tissue, which are expressed in an orderly way during gastrulation, lead directly to commitments as regions of the body, although the activities of gastrulation movements themselves perhaps refine and finalize the regionalization. Local inductive signals are, in normal development, given out during this process.

One, perhaps controversial, addition to our description of the above process of primary pattern inception (sometimes called primary embryonic induction) concerns the central nervous system. Before gastrulation really starts, the process may include the first specification of another area of epiblast, adjoining the future mesoderm, as neural. This area, left in the outer cell layer and significantly changing shape as the mesoderm migrates under it, is then subregionalized into the axial plan of the nervous system by the sequence of local inductive signals mentioned above. One confusing piece of terminology found in the literature of mammalian embryology should be addressed. The upper cell layer of the blastoderm, or blastodisc, from which the whole embryo is probably derived, is often called the "primitive ectoderm" before the start of gastrulation, even though, as we have seen, the endoderm and mesoderm, as well as both neural and epidermal parts of the pattern, are to come from it.

By the onset of the precise sequence of gastrulation movements, then, specification of neural, endodermal, and mesodermal territories, and regional diversification within at least the latter, are well advanced. Diversity within mesoderm exists at two levels: segregation into dorsal axial, convergently extending, vs. anteriormost and lateral sectors, with a different behavior, and the sequence of autonomous timing for cell behavior change *within* mesoderm of each sector, leading to the head-to-tail sequence of the body at gastrulation. In amphibian development, both these types of diversity occur via mechanisms significantly related to original positions of cells, during the induction process but before the times of involution. We do not yet know whether this is true in the mammalian version, or whether the diversification of mesoderm cells is by a mechanism not involving their relative positions at the time. In the following section (IV), we shall see that this is of importance for theory about mechanisms of patterning.

C. Segmentation

Even adult human anatomy shows a repeated unit of structure, with regional variations of architecture imposed upon its derivatives, in the axial skeleton and musculature and the peripheral nervous system below head-level. In more primitive aquatic vertebrates and especially in younger embryos or larval forms, this segmentation is much more prominent. It originates in segmentation of the dorsal axial mesoderm on either side of the midline rod (notochord) into a head-to-tail series of discrete units, the somites. Their cells contribute normally to provide the vertebrae, muscle blocks, and domains of dermal (skin) tissue and, secondarily, to organize the nervous system. They are of regular size at initial segregation, though this varies in different general regions of the axis and is later obscured by patterns of growth.

In the anterior part of the embryo head, the role of the mesodermal segmentation is less clear, because a repetitive pattern in the brain rudiment or even the endoderm (the gill bars) may control the pattern instead (see Section III,A). But behind this, somite pattern appears to be strongly controlled as to the constancy of segment numbers found in different individuals between the back of the skull and particular landmarks in the body, such as the articulation of the pelvic girdle with the vertebral column or the kidney rudiment. Regulation must be involved (see Sections I and IV), because total axial tissue available is considerably variable in different individual embryos during pattern formation, so that sizes of the individual segmented cell groups must be adapted in relation to this to ensure the constancy of numbers. All this might imply that the somites are a manifestation of fundamental, regular repeating domains of genetic activity, on which variation is played to compose the body pattern. This has been strikingly demonstrated in insect development, where the body is essentially made up of a fixed number of segments, as opposed to just containing one segmenting cell layer. These segments bear a fixed relationship to the early domains of gene activity that are known to give regional characters to the insect embryo blastoderm.

In fact, there is no evidence for and some evidence against the idea that mesodermal segmentation in vertebrates is so fundamental in the mechanism of constructing the body. Division of the muscle and skeleton into repeated units may initially have been simply an adaptation to locomotion in water, in fishlike vertebrate ancestors and their larvae, which has become deeply imprinted in the developing axial mesoderm. Under experimental and natural conditions, somite numbers can be made to vary slightly in relation to the regions of the body by extreme ambient temperatures, or transitions of physiological conditions, during development. It cannot be said, for instance, that somite number N will *always* form the last lumbar vertebrae, whereas N + 1 forms the first component of the sacrum (where N is on the order of 25). In insects, by contrast, such physiological perturbations, or genetic malfunctions, leave the unaltered species-typical number of body segments. Instead, they replace the normal characters of particular members of the series with those appropriate to other members, thus duplicating and deleting complementary segment types in the phenomenon of "homeosis."

IV. Theory of Biological Pattern Formation

A. The Primacy of the Mesoderm

Mechanisms of the relatively local and later-acting signals, termed secondary inductions between cell layers in Section III, are of great importance in coordinating their differentiation but will not be considered much further here. The signaling molecules are essentially unknown but are probably proteins. Intimate contact between inducing and responding cell layers, probably involving the special extracellular structure known as basement lamina, seems to be needed for the normal signals in some regions such as the posterior part of the neural induction by axial mesoderm. Elsewhere, more freely diffusible signal types seem to operate. An important principle may be that sequences of such signals are received by the responding tissue layer because the inducing mesoderm progressively translocates in relation to it during the migration of gastrulation.

B. The Generation of Diversity

The real heart of body-pattern formation is the process whereby the original cell sheet—the "primitive ectoderm" of mammalian embryologists—diversifies into a progressively regionalizing (endo +) mesoderm that will "drive" gastrulation, a neuralized area that will be further delimited and imprinted with regional specifications during gastrulation, and a remaining definitive ectoderm that will provide the body's epidermis. These events within one initial cell layer have the additional property of regulation; i.e., up to a certain stage, the *proportions* in which the available tissue is allocated to the parts of the future structure can be normalized in the face of natural or experimentally imposed variation in its total extent. Most of the slower-developing amphibian types used in experimental embryological research have considerable abilities of this sort, including replacement of an entire domain, such as the head-forming mesoderm, by respecification of surrounding cells if the former is surgically excised from its known normal location in the pregastrula or early gastrula embryo. The mammalian blastodisc almost certainly passes through such a profoundly regulative stage. What kind of mechanism could underlie such development?

Firstly, a mechanism for diversification of the states of the early cells is needed, which could initially be physiological, and lead only later to differential gene transcription, or else could result almost immediately in such diverse genetic activity. Secondly, a set of gene activities is required that "locks in" the specifications given by the generator of diversity, so that they become long-term memories as to which restricted regions of the future pattern can be contributed by particular cells and their descendants. The body pattern is characterized by (i) a particular orientation within the tissue, when, in mammals, no reference point is inherited structurally in the egg, and later by (ii) correct ordering and (iii) correct proportioning in the tissue for regions occupied by cells with the different specifications. The reference point, necessary to determine orientation and thus the axis of bilateral symmetry, must by definition involve an eccentric position being established as special within the blastodisc; however it is often assumed that features (ii) and (iii), because they are finally *expressed* as a spatial layout, are necessarily *generated* by the position-dependent specifications of cell diversity. But this is not a formally necessary assumption and may be largely a misleading one. Some formal possibilities for patterning mechanisms are represented in diagrammatic form in Fig. 2. [*See* DNA AND GENE TRANSCRIPTION.]

In the relatively well-understood fruitfly embryo, position-dependent specification *is* the dominant mechanism, involving interactions of gene products in space to form gradients and periodic patterns, which in turn cause precisely localized domains of activation for further genes. In the amphibian, it appears that cells' initial positions in the mesoderm-producing part of the embryo, and in particular their relative proximity to initial sources of mesoderm-inducing signals (see Section V), are at least part of the mechanism of their diversification. But even in the smaller, more rapidly developing types of amphibian egg, and especially in the minute tissue-space available during the early part of mammalian pattern specification, it is difficult to see how a morphogen gradient system containing sufficiently different levels of information to specify a diversity of cell states might stably exist. The property of regulation does seem to imply intercellular communication, whether this is by collaboration to form a gradient (Fig. 2a) or by emission of specific signals from cells in each of the states that the system is developing, which act as feedback to influence the commitment steps taken by cells elsewhere (Fig. 2b).

An extreme alternative possibility is the spontaneous capacity for generation of the diversely specified cells that are required, during cell division in the embryo precursor tissue, with the probabilities for entry into the various states such that the complete set of specificities is produced in proper proportions. On this model, regulation is automatic at early stages because random removal of cells from the tissue will ablate representative numbers from the intermixed cell types and leave the random gen-

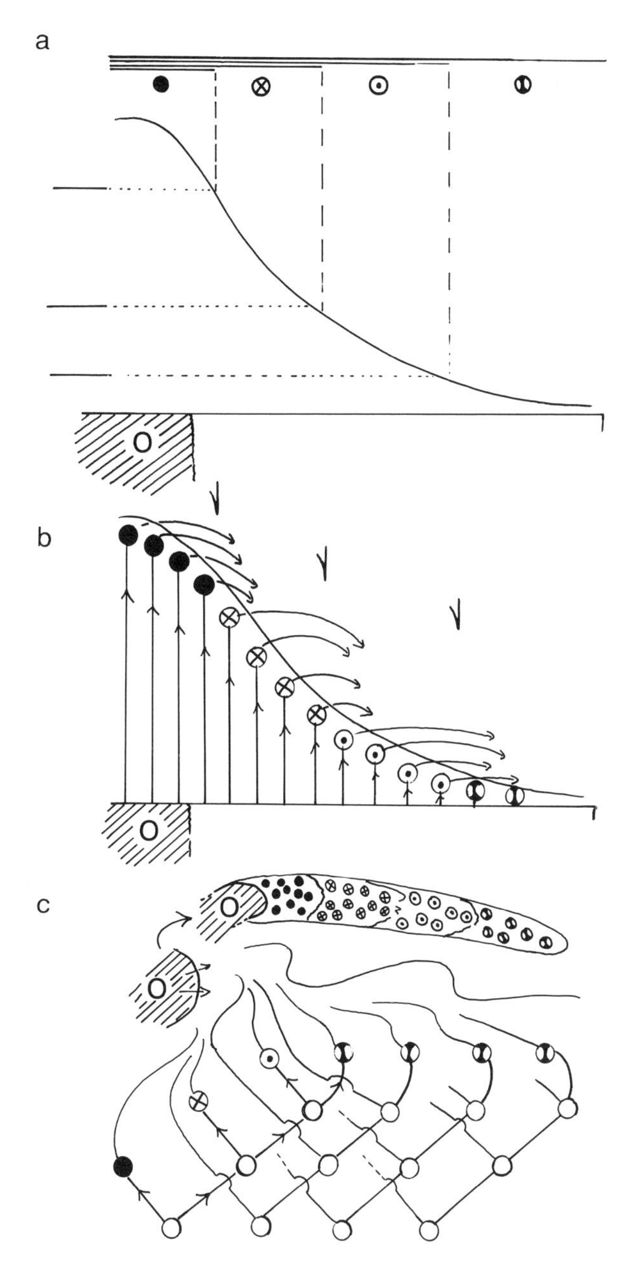

FIGURE 2 Three principles that could operate in the spatial diversification required for body pattern formation. All ideas start from the notion that a relatively restricted sector of the tissue (probably the future dorsal and anterior mesoderm/endoderm) becomes set aside as being in a special state and emits signals or "morphogen" molecules as the organizer (O), of body pattern (oblique hatched). Pattern is arbitrarily shown as consisting of four regions or cell states (represented by the symbols), which correspond to neither of the initial two states but are organized as a result of interactions between them. In (a) and (b), a gradient is set up by the "source" and "sink" activities of the organizer region and its surroundings. In (a), this is interpreted directly to give the cell states, which result from combinations of the active vs. inactive states of a small set of genes, each of which has its own threshold of morphogen signal level for switching between active and inactive states. Thus, the morphogen could be a repressor *or* an activator of the genes, but the cell type specified at one end of the system has all the genes turned off, whereas that at the other has them all activated. In (b), the same final arrangement is controlled through the gradient but in a less direct way. The gradient rank orders the *rate* of development of cells; cells near the organizer are determined first, and those far away last, in a race for occupancy of the hierarchy of available cell states. Cells entering each state in the hierarchy emit a specific diffusing signal that diverts less-developed cells toward the next preferred state and prohibits further access to their own state. Half-arrowheads mark positions at which buildup of each successive signal *diverts* the character of the wave front of cell determination to give the next state. In (c), the required cell types are generated initially without respect to position and are intermixed. An asymmetrical stem-cell lineage is shown, tied to the cell cycle, but any cell-autonomous generator of diversity may prove to be operating. The role of the organizer region is here to act as a focal point for cell migration through its emitted signals, with cells selectively positioning themselves through adhesive preferences, etc., during migration. None of these three mechanisms is likely to be a complete one, but the real mechanism may embody elements of all.

erator of diversity unaffected. To complete such a mechanism, because the cell diversity is finally manifested as a spatial pattern, a necessary capacity of specified cells is to migrate selectively relative to others during the movements of gastrulation, i.e., to sort out, in order to attain correct final position. The relatively small group of cells constituting the reference point for pattern orientation would emit signals orientating cell aggregation from elsewhere in the blastoderm, and the self assembly properties would do the rest (Fig. 2c). Whereas it is feasible to illustrate only the pure, alternative forms of mechanism in Fig. 2, it is almost certain that embryo formation in higher vertebrates occurs by a composite mechanism involving elements of both position-dependent specification and then refinement of pattern by selective aggregation of the diversified cells. In this case, the reference point for pattern and symmetry will exert its influence both as a focus for cell aggregation and as a high point in a morphogen gradient system. Such a relatively complex set of signaling interactions, involving elements of all three formalisms in Fig. 2, certainly underlies the pattern attained in one well-studied, very primitive multicellular model system, the cellular slime mold.

A fascinating recent idea, though a hard one to test, is as follows. Early embryo cells do seem to monitor elapsed "developmental time" in a striking way and may do this in relation to the progression of cell cycles while in an actively growing tissue. If, in a certain region of the blastodisc, conditions allowed cells to change their specification state systematically with time, but cell population pressure or aggregation toward the organizing center continuously removed cells from this special zone as development progressed, then a supply of tissue trapped into states of successively more posterior specification might be generated. The mammal embryo in particular is known to be laying down an axial pattern across a period of massive growth, and some evidence indicates a region specially active in cell division near the presumed site of origin of the axial tissues. Whatever the mechanism of the primary organizing and orientating center may be, its anatomical location (shifting with development) is most probably in the tissue that composes the first site of mesoderm immigration at the primitive streak (future head endoderm and mesoderm) and the hypoblast layer immediately underlying this. These cells or their homologues, in the different vertebrate embryo types where the experiment has been performed, organize extra body patterns when transferred elsewhere into host embryos.

C. Positional Memory in Embryo Cells

The major, second-acting part of any mechanism is the set of gene products that records the specifications or "position values" assigned to the cells as long-term commitments to body regions. It can be stated that these are not themselves gene products that trigger differentiation of the various histologically defined cell types. Each such cell type characteristically occurs throughout much of the body, but tail tissue is coded differently from head tissue before, say, cartilage or dermal cells have been committed to differentiation in either region. Constructing a body is not primarily a matter of assigning particular geographical patches to particular differentiated cell states. Direct visualization of the genetic strategy of pattern formation in the insect embryo using molecular technology, as well as the more limited results of such analysis in mammalian development, has clearly borne out this principle. These genes encode memories of cells' relative positions as such, and do this on a combinatorial basis, along the lines illustrated in Fig. 2a, although such a *direct* interpretation of a morphogen gradient may not be involved. Thus, each such gene is activated in an extensive, continuous domain along the future axis, suggesting a simple response to the diversifying mechanism (such as graded concentration of an intercellular signal, or cellular "age" at specification [see previous subsection]). An adequate number of uniquely encoded domains to specify a body plan is built up because these large domains, for the different genes, overlap extensively so that unique combinations of active genes characterize much more restricted regions. The "downstream" gene interactions, whereby such combinations of positional gene activity then dictate precise patterns of cell commitment to differentiations, are quite unknown. There is some recent evidence that, as might be expected from the demands of final structure, the sharpest and most fine-grained partitioning by "positional" gene activities in the vertebrate embryo occurs in the brain rudiment. Here, they may indeed be "segment-character" genes, as they are referred to in insect development, because they interact precisely with the segmental structure that seems to originate independently in the neural tissue and control aspects of the head pattern.

V. Current Candidates for Molecules Involved in Vertebrate Body Pattern Formation

A. Reference Point for Pattern Orientation and, Hence, Bilateral Symmetry

No vertebrate is thought to utilize structural prelocalization of a macromolecule as a source of axis formation from the egg stage. In amphibians however, symmetrization does occur by a physical reorganization of the massive egg cell to install a special state upon one narrow sector within it, during the one-cell stage after fertilization. The movements normally relate to gravity, but probably do not require a gravity vector. The mechanism seems to act by arranging that special inducing signals, giving rise to dorsal axial tissue specification, occur only within the narrow sector (see next subsection). In higher (mammalian) vertebrate embryos, establishment of a dorsal midline may indeed by accomplished as a similar patterning of inducing signals, i.e., one widespread around the blastodisc and one more restricted and eccentric in origin. But this cannot be due to special biochemical or physiological states inherited by cells from a position in the egg's structure. The reference point for axis formation and symmetry does not arise until multicellular tissue is present. An asymmetric relationship of the blastodisc to gravity or to local environmental variables (e.g., O_2 or CO_2 tensions) may trigger the breaking of symmetry, but the mechanism must be put in gear by a self-amplifying response that (a) "locks" the initially favored cells into a new state in which they produce dorsal axial type inducers and perhaps cell aggregation signals and (b) stabilizes the symmetry by suppressing (through intercellular signaling) the attainment of this organizing state by cells elsewhere.

B. The Intercellular Signals Generating Diversity for Body Pattern

There are currently no clues to molecular mechanisms for spontaneous, or cell-cycle–related, intracellular generation of diverse states of specification (Fig. 2c; Section IV). We have seen that, in vertebrates, position-dependent mechanisms of diversification must be by truly intercellular signals. Recently, exciting candidates for the first-acting members of such a signal cascade have been identified from amphibian development. These belong to families of small, secreted proteins whose first identified members were called "growth-factors" because of effects on the cell-cycle status of defined mammalian target cells in tissue culture. The changes of specification state that they evoke in responsive (i.e., essentially pregastrular aged) embryo cells correspond with the interactions embryologists have defined as primary inductions (see Section III,B); i.e., the diversification that precedes and accompanies gastrulation as opposed to the local interactions between the formed germ layers. Responses show some evidence of progressive change with the concentration of the molecules experienced by cells, as expected for a position-related (morphogen-like) patterning mechanism, and the active molecules fall into two functional classes, tending respectively to induce those structures made by the mesoderms of the two geographical sectors mentioned in Section III,A. Another "intercellular" molecule now implicated in patterning of the vertebrate body is retinoic acid or one of its relatives. In a concentration-dependent way, it can permit or prevent specification of particular sectors of axial body structure. The precise principle upon which such secreted molecules as these induce geographically patterned mesodermal and neural specification is currently unknown, but of particular interest is that among the first genes transcribed in response to them in development are members of the principal "homebox"-containing class to be mentioned in the next and final subsection.

C. Molecules Encoding Cell-Memories for the Regionality of the Body

In the fruit fly, which forms a pattern largely before advent of cell boundaries, the initial position-dependent regionalization, as well as the establishment of longer-term memories for regional identity, results from interactions among proteins that are sequence-specific DNA-binders, principally of two structural families. The function of such proteins as part of a morphogen mechanism of first regionalization is clearly impossible in fully cellularized development, where this role may be mediated by the smaller, secreted molecules mentioned above. The evidence is now overwhelming, however, that DNA binding proteins of these two functional classes, referred to popularly as the homebox-containing (helix-turn-helix) and "zinc-finger"-containing proteins, are central to intracellular recording of regional identity in vertebrates. Their genes may respond directly to morphogen concentrations, and also have a network of cross-interactions as controllers of one another's transcriptional activity within individual cell nuclei. Restricted axial domains of synthesis for these proteins, reminiscent of those in the fruit fly, are seen in relatively newly induced mesoderms and nervous systems of all vertebrate type embryos. Systematic interference with body pattern has resulted from experimental alteration of function of specific members. It is nevertheless premature to conclude that the logic of deployment of members of these gene families in vertebrate body pattern formation will closely correspond with that elucidated for insect development. [*See* DNA BINDING SITES.]

Bibliography

Slack, J. M. W. (1983). "From Egg to Embryo," 1st ed.; 2nd ed. expected 1991/2. Cambridge University Press.

Embryofetopathology

DAGMAR K. KALOUSEK, *University of British Columbia*

Glossary

Abortion or miscarriage Premature expulsion or removal of the conceptus from the uterus before it is able to sustain life on its own; eighteen developmental, or 20 gestational, weeks is considered the lower limit of viability; older fetuses can be delivered either as stillborn or premature, mature, and postmature newborn

Amnion Innermost membrane of the sac enclosing the embryo and fetus

Body stalk Early primitive umbilical cord connecting vascular systems of the embryo and developing placenta

Embryo Developing human from conception until the end of the eighth week, by which time all organ systems have been formed

Chorion Outermost of the two membranes that completely envelop the embryo and fetus

Chorionic sac Precursor of placenta

Conceptus or product of conception Includes all of the structures that develop from the zygote: the embryo–fetus and the placenta with its membranes

Decidua Maternal tissue derived from the mucosal lining of the uterus and that peels off at childbirth or abortion

Developmental or conceptional age Extends from the day of fertilization to the day of intrauterine death or expulsion of the live conceptus

Embryonic growth disorganization Represents generalized abnormal embryonic development, which can be divided into four categories varying from GD_1, most severe, with complete lack of embryonic development, to GD_4, with severely abnormal embryo being present

Gestational or menstrual age Extends from the first day of the last menstrual period to the expulsion or removal of the conceptus; therefore, gestational or menstrual age is 2 wk greater than developmental or conceptional age

Implantation Attachment of the blastocyst to the epithelial lining of the uterus occurring 6–7 days after fertilization of the ovum

Intrauterine retention period Refers to the time between the death of the embryo–fetus and its expulsion or removal

Placenta Vascular organ within the uterus connected to the fetus by the umbilical cord; serves as the structure through which the fetus receives nourishment from and eliminates waste into the circulatory system of the mother

Previable fetus Developing human from the beginning of the ninth week postconceptional until gestational viability reached (20 gestational wk)

Somite The paired, blocklike mass of mesoderm arranged segmentally along side the neural tube of the embryo, forming the vertebral column and segmental musculature

Uterus Hollow muscular organ of the mother in which the ovum is deposited and the conceptus develops; lay term is a womb

Yolk sac Extraembryonic structure that serves temporarily for the nourishment of the early embryo

EMBRYOFETOPATHOLOGY is a new discipline of human pathology dealing with the evaluation of the development of human embryos (up to 8 developmental wk) and previable fetuses (9–18 developmental wk). It has evolved in response to clinical interest in early monitoring of human pregnancy, intensive prenatal care, and increasing demands for

Encyclopedia of Human Biology, Volume 3.

genetic counseling. Embryofetopathologists investigate both spontaneously aborted embryos and fetuses and conceptions terminated due to prenatal diagnosis of an abnormality. The investigations consist of the evaluation of embryonic–fetal and placental development and detection of any developmental defect. The information is used both in determining the risk of repeated spontaneous abortion and in genetic counseling of the parents regarding risks of having normal or abnormal conceptus in future pregnancy and also for prenatal monitoring of future pregnancies.

I. Introduction

Pregnancy loss is surprisingly frequent. It has been estimated that about 50% of human conceptions spontaneously abort. The highest conceptus mortality exists in the early weeks of pregnancy usually before pregnancy is clinically recognized. Detailed hormonal studies at this very early stage suggest that many conceptions fail to implant. The rate of pregnancy loss after implantation and before clinical recognition is estimated to be around 15%. Another 15–20% of clinically recognized pregnancies spontaneously abort prior to 20 wk of gestation. There are numerous causes of pregnancy loss including chromosomal aberrations, developmental defects and inborn errors of metabolism in the conceptus, maternal illness, uterine structural anomalies, cervical incompetence, infection, and immunologic and hormonal imbalances. Depending on the particular stage of the pregnancy, when loss occurs, different etiologies are more common. The chromosomal defects are most common among early spontaneous abortions. On the other hand, pregnancy loss caused by infection is commonly detected among late abortions, especially after 16 wk of gestation. [*See* ABORTION, SPONTANEOUS; IMPLANTATION (EMBRYOLOGY).]

The early experience with the pathological examination of aborted human embryos has led to the description of basic types of embryonic maldevelopment and to the realization that embryonic developmental anomalies not only morphologically differ from defects at birth, but also are much more common than defects at birth. Many abnormal embryos are difficult to diagnose as having specific malformations and represent embryonic growth disorganization. The high incidence of abnormal embryos in abortion specimens has been confirmed by many investigators. In one of the largest morphological studies of early spontaneous abortion, among 1,226 specimens 84% of embryos were abnormal, with localized defects occurring in 5%, and generalized abnormal embryonic development in 79%. [*See* BIRTH DEFECTS; CHROMOSOME ANOMALIES.]

The findings among aborted fetuses are different from aborted embryos. The majority of aborted fetuses are morphologically normal, and their intrauterine demise or expulsion is caused mainly by infection, placenta-related problems, or cervical incompetence. Less than 20% of spontaneously aborted fetuses show developmental defects.

II. Types of Embryofetopathology Specimens

There are early abortion (corresponding to embryonic development) and late abortion (fetal development between 11 and 20 gestational wk) specimens.

A. Early Abortion Specimens

For practical purposes, early abortion specimens may be divided into complete and incomplete. A *complete specimen* of early abortion usually consists of a chorionic sac with an embryo, decidual tissue, and blood clots. The chorionic sac may be intact, ruptured, or fragmented. If the chorionic sac is not intact and an embryo is not found, the specimen is categorized as an *incomplete specimen*. Incomplete specimens of early abortion are common, often consisting of only fragments of chorionic sac, decidua, and blood clots.

An aborted intact chorionic sac is usually found embedded in maternal decidua, which resembles a cast of the uterine cavity (Fig. 1). Such specimens are carefully opened from the cervical end to release the intact chorionic sac (Fig. 2). The dimensions of the chorionic sac vary from 5 to 80 mm in diameter. The surface of a well-developed chorionic sac is completely covered by abundant chorionic villi, which are of uniform diameter with symmetrical branches and multiple buds along their length. Villi are best observed floating in saline solution under the dissecting microscope. Sparse villi with abnormal morphology are common in specimens of early abortion and usually indicate chromosomally abnormal conception.

The rupturing of the chorionic sac is generally associated with a rupture of the amniotic sac and

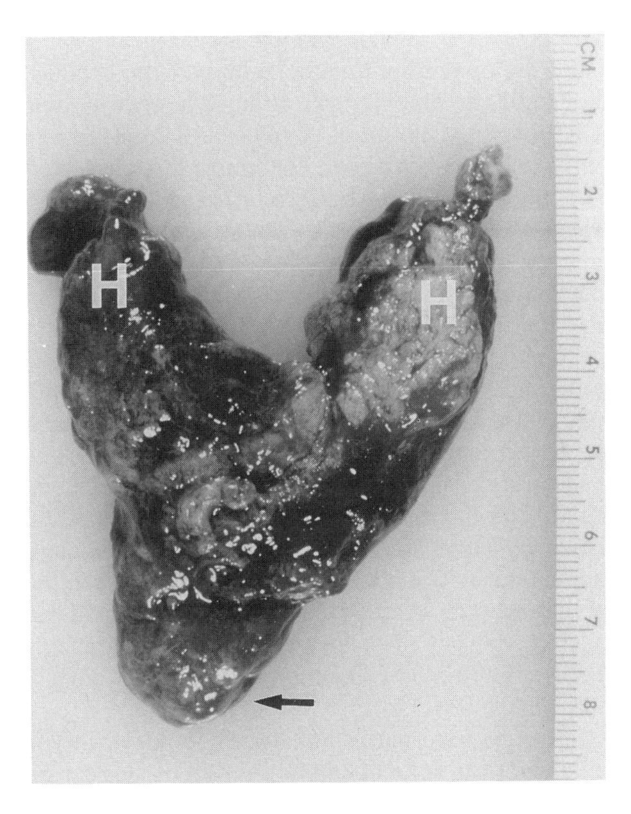

FIGURE 1 Decidual cast of uterine cavity with two labeled horns (H) and an arrow pointing to the cervical end.

FIGURE 2 Intact chorionic sac 1.3 cm in diameter covered with well-developed chorionic villi (arrows).

loss of amniotic fluid. The ruptured chorionic sac has a collapsed corrugated appearance. An embryo is frequently missing from such a sac. Even in the absence of the embryo, the evaluation of the amniotic sac, villi, body stalk, and yolk sac, both morphologically and histologically, allows categorization of such specimens as a normally or abnormally developing conceptus. A fragmented chorionic sac without a detectable embryo can yield information about embryonic development only if the villi are histologically examined.

B. Late Abortion Specimens

Simply due to their size, late abortion specimens are usually complete. They consist of an easily identifiable fetus (9–18 developmental wk or 11–20 gestational wk) and a placenta consisting of an umbilical cord, extra-embryonic membranes (amnion and chorion), and a disc made of villi. Both fetus and placenta should be examined to establish the cause of spontaneous abortion (Fig. 3).

III. Investigative Approach

A. Examination of Chorionic Sac

The size of the sac and the amount and the morphology of villi are evaluated initially. After opening of the intact chorionic sac, the presence or absence of amniotic sac, yolk sac, body stalk/cord, and embryo is noted. When the amniotic sac is present, its size and relationship to the chorionic sac are recorded. An abnormally large amniotic sac and its premature fusion with chorion are the most frequent abnormalities. The abnormal gross and histological morphology of the chorionic sac and its villi gives the embryopathologist insight into the sequence and timing of the insult during early embryogenesis.

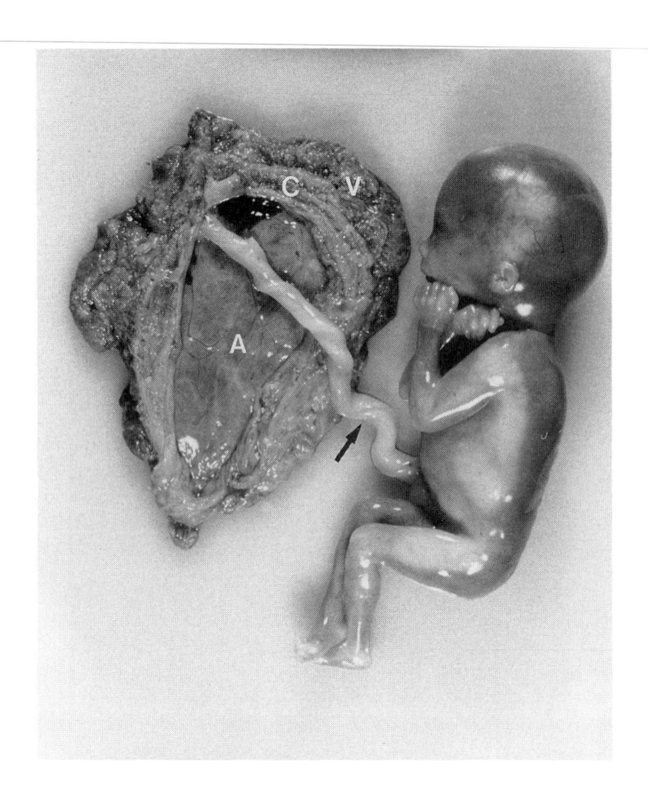

FIGURE 3 Fetus and its placenta at 16 wk of development. Note umbilical cord (arrow) connecting the vascular systems of the fetus and placenta. In the placenta, amnion (A), chorion (C), and placental villi (V) are indicated.

The histological abnormalities in the chorionic villi relate to the time at which embryogenesis was disturbed. If the villi show no embryonic vessels, swelling of its stroma and abnormal morphology of its epithelial lining embryonic death occurred before the establishment of the embryonic circulation in the placental villi (i.e., before the end of the third week). On the other hand, if the embryonic circulation was established, and then ceased because of embryonic death, the remnants of vascular channels can be observed.

After intrauterine embryonic death, the chorionic sac may remain for several weeks in the uterus, and significant changes such as collapse of villous vessels or an increase or decrease in villous fluid content occur within the chorionic villi. Calcifications in villi either focal or along basement membrane are a common finding. These changes must be distinguished from true abnormal development.

B. Examination of Embryos

The development of human embryos from fertilization until the embryo has attained a crown–rump (CR) length of 30 mm is divided into 23 stages, or horizons. In early stages, the embryo is characterized by the number of somites, whereas during later stages specific size of the embryo and its external features become important for its staging. The main external embryonic features of each stage are summarized in Table I.

Complete external examination of embryos is done under the dissecting microscope. The examination consists of detailed evaluation of characteristic facial, limb, and body features to provide the data for determination of embryonic developmental stage. The embryonic developmental age is established by correlating CR length and the developmental stage. The difference >14 days between the developmental age thus derived and gestational age calculated from the last menstrual period reflects the length of time of retention *in utero* after embryonic death.

Based on external examination, human embryos collected from spontaneously aborted tissues can be classified into four major categories:

1. Normal embryos, harmoniously developed for their developmental stage (Fig. 4)
2. Growth disorganized embryos, with highly abnormal embryonic development (Fig. 5)
3. Embryos with specific developmental defects (Fig. 6)
4. Macerated, damaged, unclassifiable embryos (Fig. 7)

Any discrepancy between the embryonic CR length and expected development for that stage suggests the existence of a developmental defect. For example, if a well-preserved fresh embryo measures 9 mm and shows an incomplete closure of the neural tube, this is diagnostic of an open neural tube defect, as a complete closure should take place before the embryos is 5 mm CR length. Accurate evaluation of normal embryonic development is best done by comparison of an embryo with photographs of normal human embryos from anterior, posterior, and lateral views as well as by comparison with actual preserved embryonic specimens illustrating the normal development of each embryonic stage (Fig. 8).

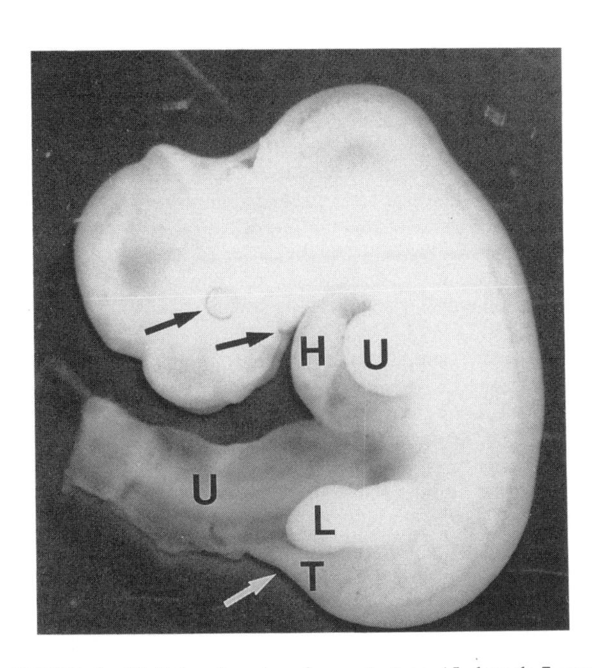

FIGURE 4 Well-developed embryo at stage 15, length 7 mm, showing developing eye (arrow), mouth (arrow), heart (H), upper limb (U), and lower limb (L). The presence of tail (T) is a normal finding at this stage. Umbilical cord (u) contains three vessels.

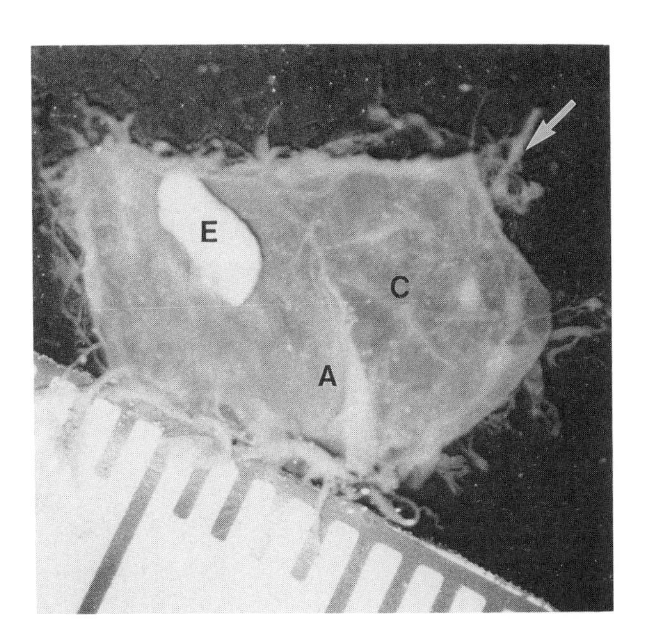

FIGURE 5 Opened abnormal chorionic sac with growth disorganized embryo (E). The embryo measures 3.5 mm in length but shows no evidence of differentiation. It appears as a cylinder of tissue with a barely recognizable head and tail end and a very short body stalk. In the chorionic sac, amnion (A), chorion (C), and villi (arrow) can be recognized. Very sparse and abnormal villous development is obvious.

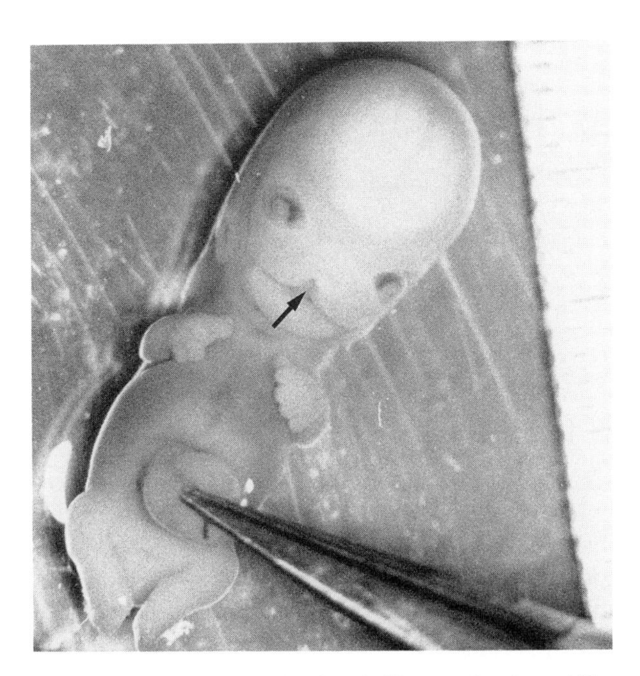

FIGURE 6 Embryo stage 21, length 23 mm, showing midline cleft of upper lip (arrow). No other developmental defects are present.

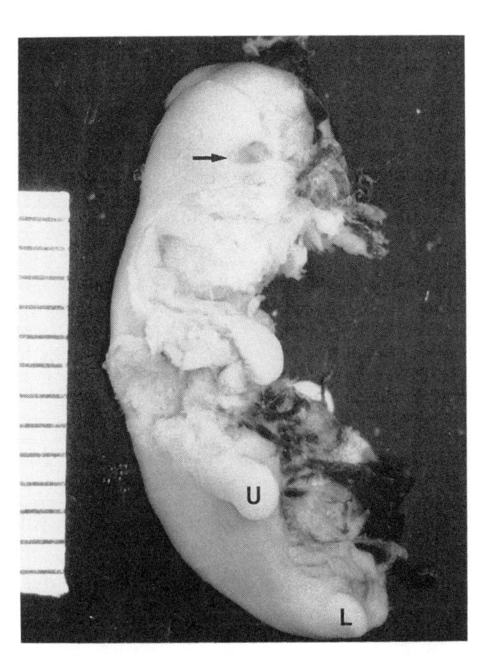

FIGURE 7 Embryo about stage 15 showing severe maceration and damage. Note complete loss of embryonic curvature and damage of face, neck, and abdomen. Eye is identified by an arrow. L, lower limb; U, upper limb.

TABLE I Developmental Stages in Human Embryos

Age (days)	Stage	No. of somites	Length (mm)	Main characteristics
20–21	9	1–3	1.5–3.0	Deep neural groove and first somites present. Head fold evident.
22–23	10	4–12	2.0–3.5	Embryo straight or slightly curved. Neural tube forming or formed opposite somites, but widely open at rostral and caudal neuropores. First and second pairs of branchial arches visible.
24–25	11	13–20	2.5–4.5	Embryo curved owing to head and tail folds. Rostral neuropore closing. Otic placodes present. Optic vesicles formed.
26–27	12	21–29	3.0–5.0	Upper limb buds appear. Caudal neuropore closing or closed. Three pairs of branchial arches visible. Heart prominence distinct. Otic pits present.
28–30	13	30–35	4.0–6.0	Embryo has C-shaped curve. Upper limb buds are flipperlike. Four pairs of branchial arches visible. Lower limb buds appear. Otic vesicles present. Lens placodes distinct. Attenuated tail present.
31–32	14	[a]	5.0–7.0	Upper limbs are paddle-shaped. Lens pits and nasal pits visible. Optic cups present.
33–36	15		7.0–9.0	Hand plates formed. Lens vesicles present. Nasal pits prominent. Lower limbs are paddle-shaped. Cervical sinus visible.
37–40	16		8.0–11.0	Foot plates formed. Pigment visible in retina. Auricular hillocks developing.
41–43	17		11.0–14.0	Digital, or finger, rays appear. Auricular hillocks outline future auricle of external ear. Trunk beginning to straighten. Cerebral vesicles prominent.
44–46	18		13.0–17.0	Digital, or toe, rays appearing. Elbow region visible. Eyelids forming. Notches between finger rays. Nipples visible.
47–48	19		16.0–18.0	Limbs extend ventrally. Trunk elongating and straightening. Midgut herniation prominent.
49–51	20		18.0–22.0	Upper limbs longer and bent at elbows. Fingers distinct but webbed. Notches between toe rays. Scalp vascular plexus appears.
52–53	21		22.0–24.0	Hands and feet approach each other. Fingers are free and longer. Toes distinct but webbed. Stubby tail present.
54–55	22		23.0–28.0	Toes free and longer. Eyelids and auricles of external ears are more developed.
56	23		27.0–31.0	Head more rounded and shows human characteristics. External genitalia still have sexless appearance. Distinct bulge caused by herniation of intestines still present in umbilical cord. Tail has disappeared.

[a] At this and subsequent stages, the number of somites is difficult to determine and therefore, is not a useful criterion.
Modified, with permission, from Moore K. L., 1982, Criteria for estimating developmental stages in human embryo, *in* "The Developing Human. Clinically Oriented Embryology," 4th ed. W. B. Saunders Co., Philadelphia.

Although the CR length is very important for staging normally developing embryos and embryos with focal defects, CR length cannot be used for embryos that are growth disorganized.

Damaged and severely macerated embryos also cannot be properly evaluated. Their developmental age can only be estimated, based on specific structure or organ development (e.g., hand, eye [Fig. 7]), but developmental defects cannot be diagnosed as maceration alone and can mimic focal defects such as open neural tube defect or cleft lip.

C. Examination of Fetus

Routine examination of the fetus consists of both external and internal examination, radiological examination, photographic documentation, and histological examination. Cytogenetic, biochemical, and microbiological studies are done selectively.

1. External Examination

The fetus is weighed and CR length and crown–heel (CH) length and head circumference are recorded. CR length is the main criterium used for

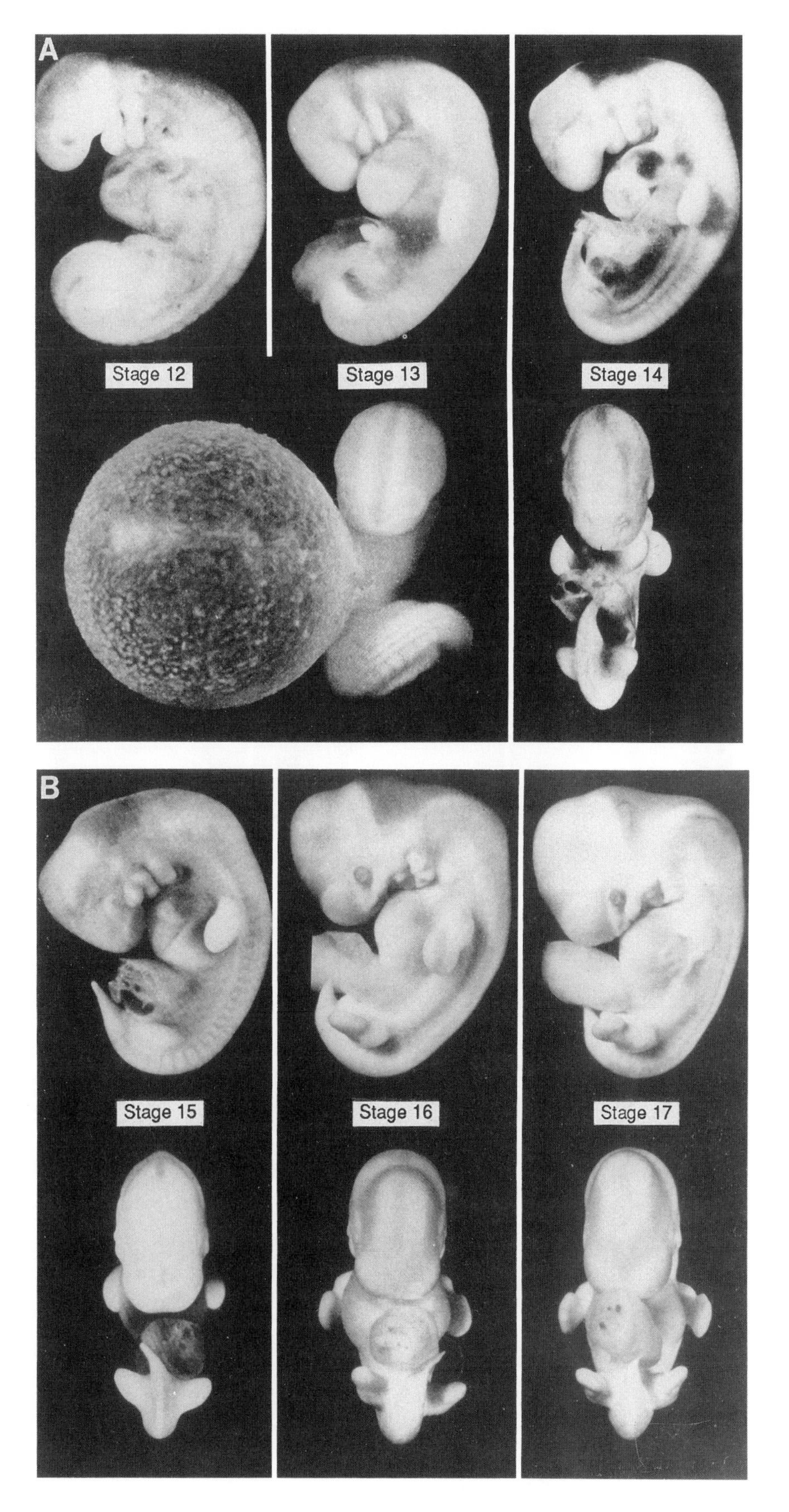

FIGURE 8 Illustration of embryonic stages 12–23. [Reproduced, with permission, from, N. Exalto, R. Rolland, T. K. A. B. Eskes, and G. P. Vooijs (1982). "De jonge swangerschap" Boehringer, Ingelheim.]

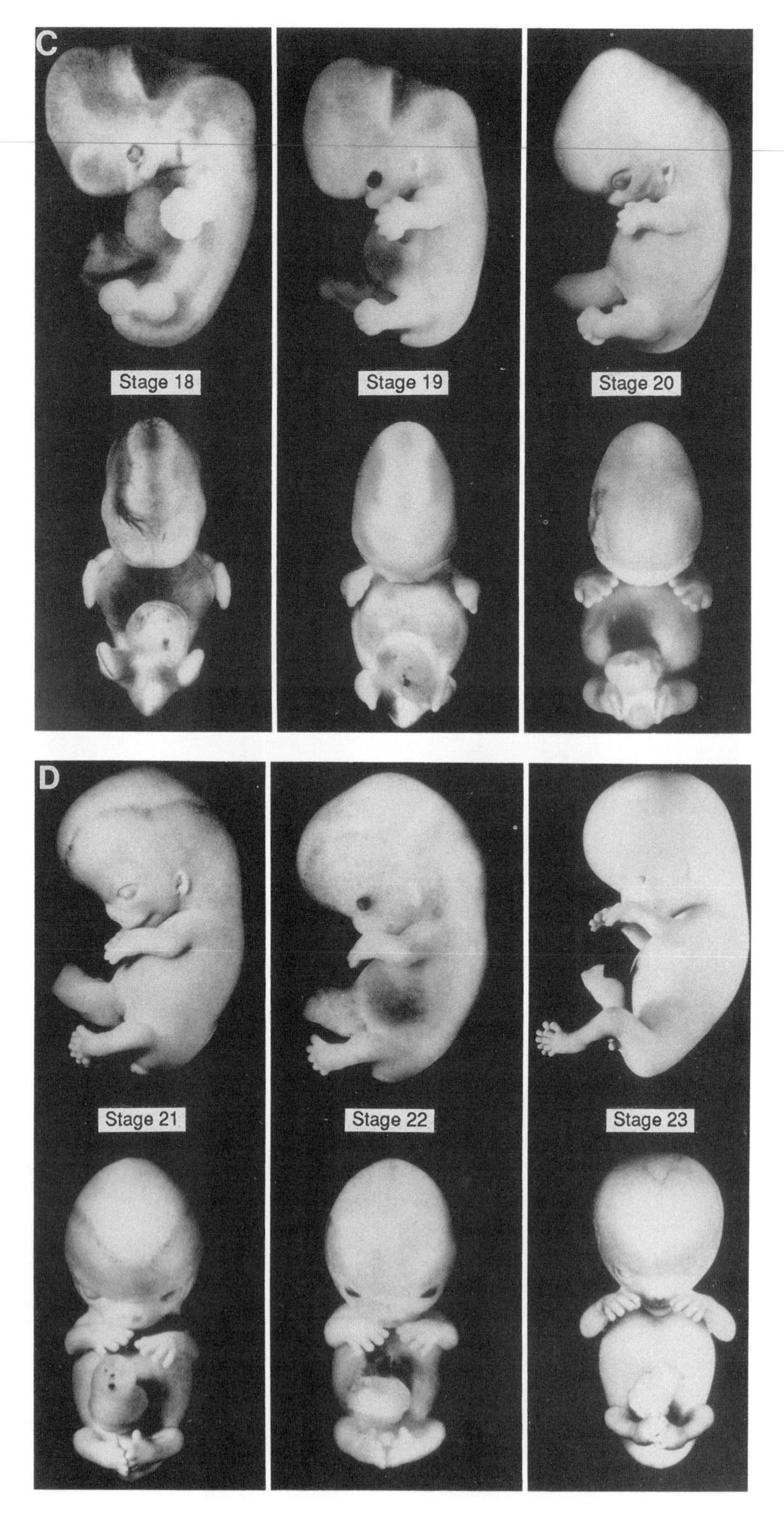

FIGURE 8 *(continued)*

establishing the fetal developmental age. Tables are also available for determining developmental age from the foot and hand lengths. These can be used instead of CR length when the specimen is incomplete or fragmented, or they can be used in addition to CR length for verification of the accuracy of CR measurement.

Detailed external examination of the fetus is an essential part of morphological studies. Shape of head, position of ears, presence or absence of scalp hair, and abnormalities of eyes, nose, mouth, palate, mandible, and ears are noted. The examination of limbs with their length recorded, the number and position of digits, and the presence or absence of flexion deformities as well as the evaluation of chest, abdomen, and external genitalia complete the external examination.

Macerated and damaged fetuses should not be disregarded even though they may appear to be grossly distorted (Fig. 9). In spite of molding, distortion and head collapse, the major external malformations, if present, such as neural tube defect, facial clefting, and missing or extra digits, can be diagnosed. The documentation of the absence of any developmental defects is as important as the diagnosis of an abnormality.

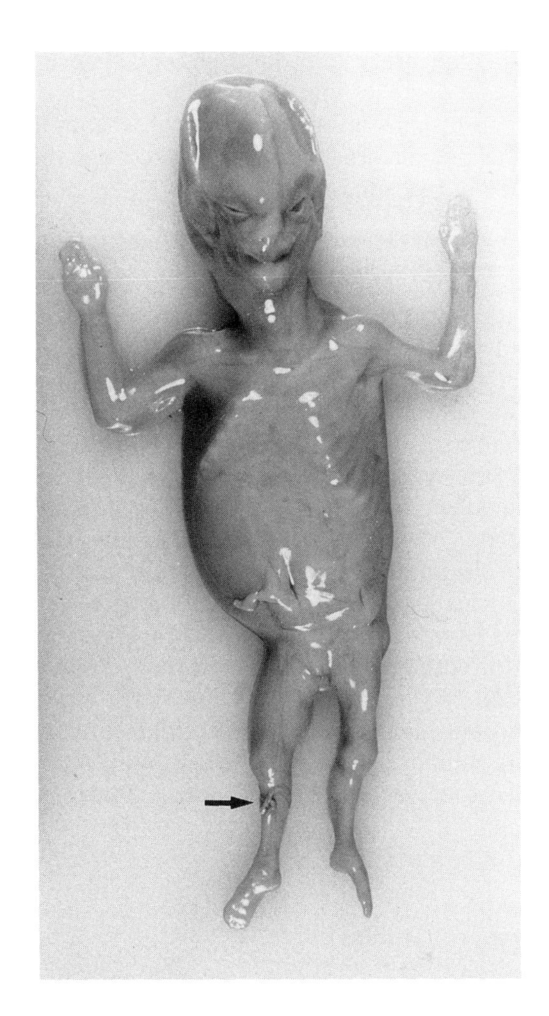

FIGURE 9 Macerated male fetus of 15 developmental wk showing the collapse of head, distortion of facial features, and peeling of skin (arrow).

2. Internal Examination

Dissection is usually done in the manner of a mini-autopsy with the help of the dissecting microscope. The internal examination is not directed only toward the diagnosis of the most obvious abnormalities such as the absence of an organ, its abnormal position, or incomplete formation. Dissection, using a magnifying lens or a dissecting microscope, identifies all internal malformations, including congenital heart defects. All internal organs are weighed, and the weight is compared with normal values for estimated developmental age.

The examination of brain is feasible only in nonmacerated fetuses, as it is usually completely liquefied in macerated fetuses.

3. Radiological Examination

Morphological assessment of the aborted fetus routinely includes radiographic examination for determination of bone developmental age and documentation of skeletal anomalies.

4. Photographic Documentation

Photographs and close-ups of all detected malformations enable identification of the range of abnormalities within a particular syndrome. Photographic documentation is also essential for comparisons, for re-evaluations of the case, and for allowing consultations by other syndromologists.

5. Histological Examination

Histological examination is necessary for evaluation of deviations from normal organ development and for identification of microscopic changes such as aspiration of infected amniotic fluid into fetal lungs.

D. Examination of Placenta

The morphology of the placenta of conceptions of 9–18 developmental wk is very similar to that of the placenta at term delivery. Essential steps in placental examination are establishing placental weight and shape, the site of insertion and the length of umbilical cord, number of cord vessels, evaluation of both fetal and maternal placental surface, and the appearance of placental membranes.

E. Cytogenetic Studies

Cytogenetic examination of abortion specimens consists of sampling the gestational sac or placenta, culturing the cells, and processing them for cytogenetic studies. Among early spontaneous abortions, the incidence of chromosomal defects is high. As many as 80% of specimens show a chromosomal numerical defect—chromosomal trisomy (one extra chromosome), sex chromosome monosomy (absence of a chromosome), triploidy (three sets of the germ cell chromosome number, instead of two), tetraploidy (four sets of the germ cell chromosome number), or chromosomal mosaicism (the presence of cells differing in chromosome number or structure). Chromosomal trisomies (one extra chromosome) are found in one-half of chromosomally abnormal specimens. Among late abortions, chromosomal defects are much less common, seen in <10% of specimens.

F. Biochemical Studies

Biochemical studies are done on selected specimens usually to confirm the prenatal diagnosis in a specimen of pregnancy termination. They involve DNA studies and enzyme studies.

G. Microbiology Studies

Microbiology studies are done selectively to confirm or rule out bacterial, viral, or parasitic infections.

IV. Biological and Clinical Significance of Embryofetopathology Investigations

A miscarriage is an upsetting event for a family because the pregnancy is usually well planned and frequently the mother is >30 yr of age, which means is close to the end of the reproductive period. Following abortion, the parents want to know:

1. if they did anything wrong during the pregnancy,
2. if there is any hereditary problem in the family predisposing them to defective offspring,
3. if it is going to happen again,
4. what did studies on the aborted tissues show, and
5. if there is prenatal diagnosis available in a future pregnancy to detect the "defect" seen in aborted tissues.

The embryofetopathologist, through identification of developmental defects and with good documentation of the development of both gestational sac–placenta and embryo–fetus, can address these questions (genetic counseling) on a sound basis. A meaningful report to the family physician and geneticist is the result of interpreting the findings on examination of the conceptus in conjunction with information from a comprehensive maternal reproductive, medical, and family history. [*See* Genetic Counseling.]

To illustrate the clinical benefit of detailed examination of products of conception, consider four examples. (1) In early spontaneous abortion specimens, the identification of both morphologically and cytogenetically normal embryos with no evidence of infection clearly points to a defect in implantation due to some uterine abnormalities or insufficient maternal hormonal support of the pregnancy. This finding will help the obstetrician in management of future pregnancy. (2) The diagnosis of growth disorganized embryos with chromosomal numerical abnormality indicates a lethal defect due to either abnormal gametes or due to an abnormal fertilization or cleavage. As all these events usually represent an accidental error, the parents can be reassured and encouraged to initiate a new pregnancy. (3) A different situation arises when chromosomal rearrangment is detected and is found to be inherited from one of the parents. Cytogenetic prenatal diagnosis is then offered in all future pregnancies. (4) The most significant finding from a genetic counseling point of view is the localized embryonic or fetal defect such as abnormal closure of spine and/or head (Fig. 10), cleft lip and palate (Fig. 11), or congenital heart defect. These are congenital anomalies, which can be carried to term and result in a newborn with developmental defect. The detection of such defects in spontaneously aborted embryo–fetus identifies the family at risk, as most of these defects have genetic cause or at least signifi-

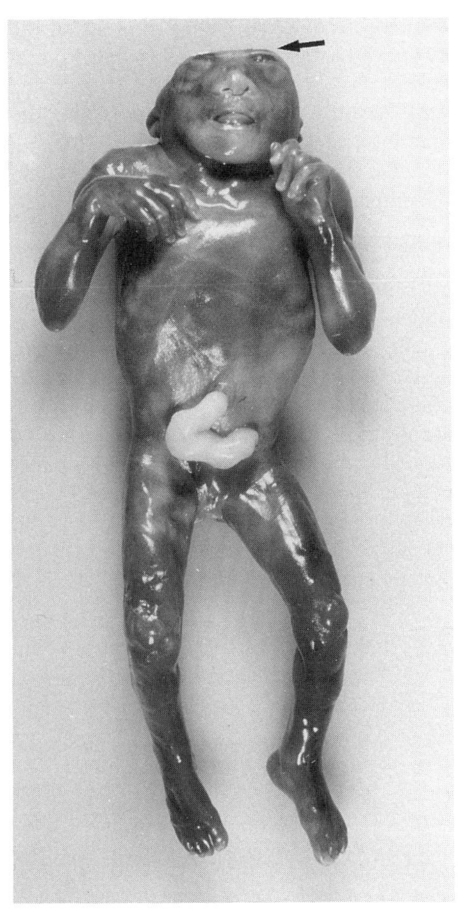

FIGURE 10 A 16-wk fetus with missing skull and brain (arrow). This condition is known as anencephaly.

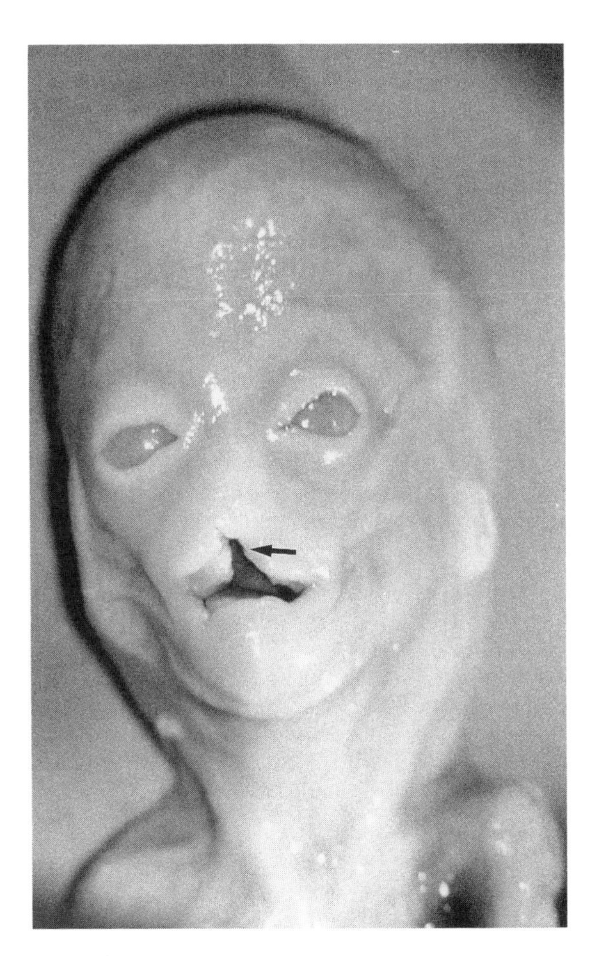

FIGURE 11 A macerated 11-wk fetus with midline cleft of both the upper lip and palate (arrow).

cant genetic contribution. The identification of the family allows follow up of any future pregnancy for the specific defect identified in spontaneously aborted conception.

The detailed routine examination of human aborted embryos and fetuses has advanced our understanding of intrauterine development and intrauterine diseases. It helps to establish that pregnancy loss is a common event and that in a majority of cases it represents nature's mechanism to remove defective conceptions. The combined morphological and cytogenetic studies revealed that >80% of early spontaneous abortions showed a chromosomal numerical defect mainly due to abnormal chromosome division in either maternal or paternal gametes. The studies of aborted embryos have confirmed that most of the developmental defects are established during the time of organ formation and tissue differentiation. Abortion studies indicate that the placenta plays an important role in maintenance of pregnancy, when placenta is normal, in spite of severe damage to embryo or fetus the pregnancy can survive until term and the abnormal infant may be born. This realization led to the development of intensive prenatal technology for the detection of a defective embryo–fetus. This includes amniocentesis (a procedure by which amniotic fluid is removed at 12–16 wk of gestation and then examined biochemically, cytogenetically and for DNA studies), chorionic villus sampling (an alternative procedure to amniocentesis involving removal of a small piece of the placenta for DNA, biochemical, or cytogenetic studies as early as 8 wk of gestation), ultrasound (gives very detailed images of various structures in the embryo and fetus—the embryonic heart beat can be seen during the first trimester, growth can be documented, the presence of twins can be observed, and the amount of amniotic fluid can be determined), amnioscopy (a fibro-optic scope inserted into the amniotic cavity for direct observation of specific fetal structures), or fetal blood sampling.

Following prenatal detection of fetal defect and pregnancy termination, detailed examination of products of conception by a specialized pathologist is indicated. The pathologist is expected not only to confirm the presence of the defect identified by prenatal diagnosis but also to identify any other associated abnormalities.

Bibliography

Kalousek, D. K. (1987). Anatomical and chromosomal anomalies in specimens of early spontaneous abortions: Seven year experience. *In* "Genetic Aspects of Developmental Pathology" (E. Gilbert and J. M. Opitz, eds.), pp. 153–168. Alan R. Liss, Inc., New York. (March of Dimes Birth Defects Foundation, Birth Defects: Original Article Series, Vol. 23, No. 1.)

Kalousek, D. K., and Poland, B. J. (1984). Embryonic and fetal pathology of abortion. *In* "Pathology of the Placenta" (E. D. V. K. Perrin, ed.). Churchill Livingston, New York.

Mall, F. P. (1908). A study of the causes underlying the origin of human monsters. *J. Morphol.* **19,** 3–368.

Mall, F. P. (1917). On the frequency of localized anomalies in human embryos and infants at birth. *Am. J. Anat.* **22,** 49–72.

Moore, K. L. (1982). "The Developing Human. Clinically Oriented Embryology," 3rd ed. W. B. Saunders Co., Philadelphia.

O'Rahilly, R., and Muller, F. (1987). "Development Stages in Human Embryos." Carnegie Institution of Washington, Publication 637.

Poland, B. J., Miller, J. R., Harris, M., and Livingston, J. (1981). Spontaneous abortion: A study of 1961 women and their conceptuses. *Acta Obstet. Gynecol. Scand. Suppl.* **102,** 5–32.

Rushton, D. I. (1984). The classification and mechanisms of spontaneous abortion. *Perspect. Pediatr. Pathol.* **8,** 269–286.

Winter, R. M., Knowles, S. A. S., Bieber, F. R., and Baraitser, M. (1988). "The Malformed Fetus and Stillbirth: A Diagnostic Approach." John Wiley & Sons Ltd., Chichester.

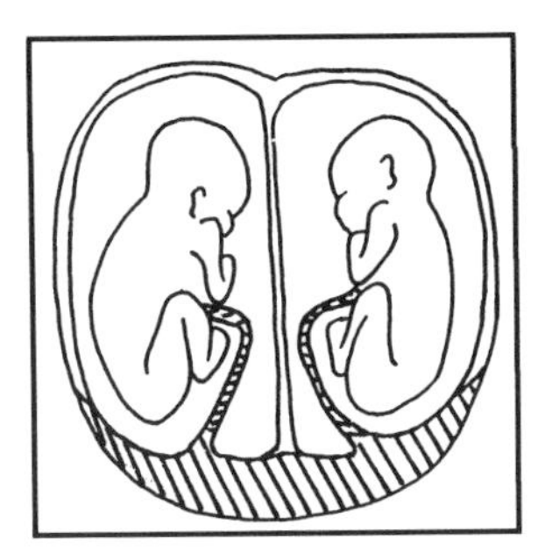

Embryology and Placentation of Twins

FERNAND LEROY, *Free University of Brussels and Saint Pierre Hospital*

Glossary

Blastocyst Preimplantation developmental stage consisting of a spherical and hollow structure formed by an outer layer of trophectoderm and an inner cell mass from which the embryo proper will be derived

Blastomeres Cells resulting from the first divisions of the fertilized egg

Gastrulation Developmental mechanism involving invagination of the upper embryonic layer (i.e., the ectoderm) and resulting in axial stratification of fundamental tissue components

Implantation Nestling of the blastocyst in the endometrium during the second week of development

Pituitary gonadotropin Peptide hormone controlling ovarian follicle growth and secretion, oocyte maturation, ovulation, and corpus luteum function

Oocyte Female germ cell undergoing maturation through two successive divisions, the first of which reduces the number of chromosomes by one-half (i.e., meiosis); both divisions are asymmetrical, producing the oocyte and a small "polar body"

Sex ratio Proportion of male infants among all births.

Trophectoderm Outer cellular layer of the blastocyst, destined to differentiate into the trophoblast (i.e., future placenta and chorion)

Zona pellucida Acellular envelope of the preimplantation conceptus composed of glycoproteins

Zygosity Embryological origin of twins (i.e., one egg–one sperm or two eggs–two sperm)

BEING UNUSUAL, the birth of twins has fascinated mankind since the earliest historic times. From the viewpoint of biological evolution, twinning is an obvious exception to the general rule, by which only one fetus at a time is carried by the human female. This is also true in most monkeys and anthropoid apes and seems to have derived from natural selection, favoring an arboreal mode of life; that is, it would indeed have been difficult for our simian ancestors to produce and raise a large litter up in a tree.

Some twins resemble each other very closely, and the confusion this can create has been a longstanding theme in fictional drama. Other twins, such as Jacob and Esau in the Bible, can be easily distinguished. The explanation is, of course, that some twins are derived from the division of a single fertilized ovum (i.e., a zygote) and others from the independent release and fertilization of two ova. Twins of the first type are genetically identical and hence of the same sex, while those of the second type are no more alike than ordinary brothers and sisters. The two types are therefore often called identical and fraternal, but in scientific usage they are termed monozygotic (MZ) and dizygotic (DZ).

Causal factors of DZ as well as MZ human twinning are still poorly understood. Both twin types, however, display a number of peculiarities, the study of which significantly broadens fundamental knowledge in genetics and in reproductive and developmental biology.

I. Different Types of Twins

The simplest scientific argument in favor of the existence of two types of twins results from examining the combination of sexes within pairs. Since the general sex ratio is close to 0.5, it might be assumed that oocytes have nearly equal chances of being fertilized by an X or a Y sperm. Therefore, there is an equal probability for twins resulting from independent fertilization by two sperms to be sexually different or identical. Throughout the world an excess of like-sexed twins versus male–female pairs has been observed, and this difference is best explained by postulating that it corresponds to a subpopulation of MZ twins. Frequencies of both types of twins can hence be calculated by the well-known Weinberg formula

$$MZ = \frac{L - U}{N} \text{ and } DZ = \frac{2U}{N}$$

where MZ and DZ are mono- and dizygotic frequencies, respectively; L and U are like- and unlike-sexed pairs, respectively and N is the total population.

There has been some debate about the validity of this method, namely on the basis that sexes among DZ pairs might not be determined independently. An excess of identically sexed pairs might, therefore, occur in this group. Conversely, the hypothesis of a preferential intrauterine loss of male twin fetuses, leading to the underestimation of like-sexed pairs, has also been raised against differential calculation. Be that as it may, it remains that versions of Weinberg's rule adapted to the prediction of zygosity distribution among triplets and quadruplets have given results that fitted remarkably well with frequencies observed. More recently, individual zygosity determination was carried out on about 2500 Belgian twin pairs, yielding figures in more than good agreement with those obtained by applying Weinberg's formula to the same material. Therefore, results of the numerous demographic and other studies using this method can be considered valid.

It has been hypothesized that a third type of twins can occur through fertilization by different sperm, of two cells arising from the same female germ cell (i.e., oocyte). Cases of dispermic mosaicism have been described in which the study of parental genotypes indicates that the two cellular types of which these individuals are composed differ not only in their paternal inheritance, but also partially in their maternal genes. In agreement with genetic exchange occurring between chromosomes at meiotic division (i.e., crossing over), it would appear that such anomalies are due to simultaneous fertilization of an egg and its second polar body by two different sperm. In view of the role of the zona pellucida, the occurrence of twins by this mechanism is difficult to vizualize, but cannot altogether be excluded.

Obviously, although not identical, twins belonging to such pairs would be genetically closer than those originating from two unrelated oocytes. Also, if sufficiently frequent, they would form a sizable part of the group of physically (and even sexually) different pairs. Therefore, among all phenotypically dissimilar twins, the observed rates of identity within pairs for genetic markers would be higher than their theoretical percentages of concordance if these latter are calculated on the basis of an all-DZ hypothesis (i.e., if all physically different twins arise from different oocytes). Such studies have been carried out on a series of blood group types, showing that the observed and theoretical rates of genetic identity for these markers are always close. Therefore, it can be concluded that in humans the third type of twins, if existing at all, is rare.

As shown by the examination of genetic markers such as blood groups and human leukocyte antigen (HLA), some twins can undoubtedly arise from different fathers who have had intercourse with the same woman within a short period of time. It is interesting that in many archaic societies it was believed that all twins were conceived by this mechanism, called superfecundation. Superfetation, which implies differential fertilization of two oocytes released at a several-week interval, has not yet been conclusively demonstrated in humans.

Several pairs have been described in which, despite an MZ origin proven through genetic analysis, twins were phenotypically different, sometimes even to the point of being of opposite sexes. Such pairs have been termed heterokaryotes. They differ in their chromosomes, one chromosome being inherited in excess by one cell while lacking in the other, at an early stage of development. Most heterokaryotes are composed of a normal male or female, while the other partner shows the physical traits and the genotype of Turner's (absence of one X or Y chromosome) or Down's syndrome (three copies of chromosome 21). In some of these cases, both twins are, in fact, double mosaics, in each of which a different cell type predominates. As in other instances of monosomy or trisomy (i.e., lack or excess, respectively of a chromosome), these anomalies are explained by an event of chromatid nondisjunction at the time of cell division. It is believed that in the case of heterokaryotic twins abnormal chromosome segregation would have occurred at early embryonic cleavage, rather than during gametogenesis, but in any case before splitting of the conceptus into the two twins. [*See* CHROMOSOME ANOMALIES; DOWN'S SYNDROME, MOLECULAR GENETICS.]

Individual zygosity determination has become of paramount importance in modern twin studies. Besides sex combination, morphology of the placenta and the fetal membranes might give a clue (see Section III). Many genetic markers (e.g., blood groups, placental isoenzymes, and HLA antigens) have been used successfully for defining twin zygosity. However, precise methods using recombinant DNA technology are bound to become master tools for this purpose.

II. Causality of Double Ovulation

Twinning rates show wide geographic variations. The highest values are found among black African populations, and the lowest are found in the Far East. Western European and North American countries, together with the Indian peninsula, exhibit intermediate frequencies. After breaking down into DZ and MZ figures, it appears that the MZ rate remains fairly constant everywhere (3.5–5.5 per 1000 births), whereas fluctuations are chiefly attributable to a strongly varying DZ frequency (from 2.5 in Japan to 45 in southwestern Nigeria, per 1000 births).

Factors that strongly influence the DZ rate are maternal age and pregnancy rank. Many studies have indicated that the chances of bearing fraternal twins rise steeply until about 40 years of age, after which ovaries seem to rapidly lose their capacity for double ovulation. It is clear, moreover, that at any given age the higher the number of previous pregnancies, the higher the DZ rate. In contrast, these factors have almost no influence on MZ frequency.

Nutrition also bears on DZ twinning. During World War II twinning rates decreased markedly in several countries submitted to food shortage. It has also been shown that tall heavy women are more prone to conceive DZ twins.

Other socioeconomic factors play an obvious, though ill-understood, role in the frequency of DZ twinning. Scandinavian unmarried mothers exhibit higher DZ twinning rates, even allowing for maternal age, and there is no reason to believe that they enjoy better nutrition than other women. DZ rates are also significantly elevated among pregnancies occurring within the first few months of wedlock or after protracted separation of the partners. On the whole there are several arguments in favor of the view that proneness to DZ twinning reflects high overall fecundability and fertility.

Ripening and ovulation of ovarian follicles are under the direct control of gonadotropins (i.e. sexual pituitary hormones), and it has been found that in mothers of twins the pituitary gland secretes higher amounts of these hormones, thereby favoring double ovulation. But why do these women show this disposition? The mere fact that there is a racial difference in DZ twinning rates suggests that genetic factors are involved in addition to the above-mentioned environmental influences. It is also popular knowledge that twinning tends to run in certain families. Also, mothers who have already delivered twins are prone to repeat this feat. From a host of investigations on the repeat frequency and familial incidence of twinning, it appears that proneness to double ovulation is genetically inherited. On the basis of these data, it has been proposed that the DZ twinning tendency is determined by a single pair of genes. Mothers carrying two recessive variants of this gene would be responsible for all spontaneous DZ pregnancies. These homozygous women represent 25% of the female population in Western countries. [*See* FOLLICLE GROWTH AND LUTEINIZATION; PITUITARY.]

The use of gonadotropic hormones and clomiphene citrate (an ovulatory drug acting at hypothalamic level) in the treatment of anovulation is responsible for a number of DZ twins and higher-rank multiple pregnancies. Also, treatment by *in vitro* fertilization followed by multiple embryo placement in the uterus is responsible for a 20–25% rate of twins and multiple pregnancies. It therefore could be anticipated that infertility treatment has a positive effect on overall DZ twinning rates. It has been suggested that *in vitro* fertilization as well as the use of ovulatory drugs might increase MZ frequency also, but factual arguments supporting these views remain scanty.

III. Embryology of Single-Ovum Twins

Of the two membranes surrounding the fetus, the chorion develops toward the end of the first week of pregnancy, while the amnion, which lies inside the chorion and immediately surrounds the fetus, is not differentiated until the second week. Theoretically, there are, therefore, three stages at which division of the conceptus might give rise to MZ twins (Fig. 1). If it occurs before the end of the first week, the twins develop separated amnia and choria, as is the case in all DZ pairs. If only the embryonic (i.e., inner cell) mass divides in two before formation of the amnion, the result will be a common chorion, but separate amnia. Finally, if the division takes place after differentiation of the amnion, twins not only share a common chorion, but are also contained in a single amniotic sac.

Three types of human MZ twins have indeed been described according to the morphology of fetal membranes and placenta, which might be dichorial–diamniotic (32%), monochorial–diamniotic (66%), or monochorial–monoamniotic (1–2%) (Fig. 2).

In practice, the chief difficulty lies in distinguishing between a secondarily fused dichorial placenta and a monochorial–diamniotic placenta. The components are, however, usually apparent from a naked-eye examination of membranes partitioning the gestational sacs. In a monochorial placenta the two layers of amnion appear translucent and peel away from each other, leaving nothing in between. In a dichorial placenta the septum is more opaque, and stripping the amnia leaves either a single fused layer or two separate layers of chorion attached to the fetal surface of the placenta. Histology of the dividing membranes is more time-consuming, but gives an almost infallible result.

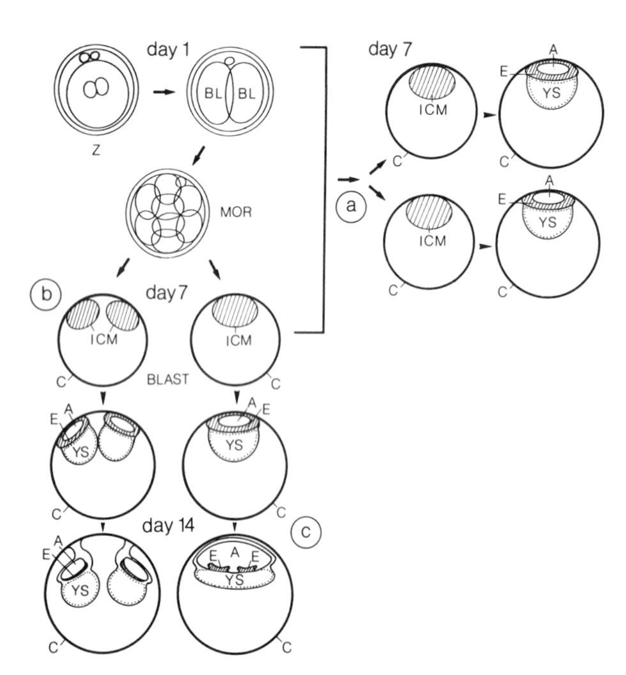

FIGURE 1 Early embryology of MZ twins. Modes of MZ twinning: (a) dichorial–diamniotic; (b) monochorial–diamniotic; (c) monochorial–monoamniotic. Z, Zygote (i.e., a fertilized egg); BL, blastomeres; MOR, morula; ICM, inner cell mass; C, chorion; BLAST, blastocyst; A, amnion; E, embryo; YS, yolk sac.

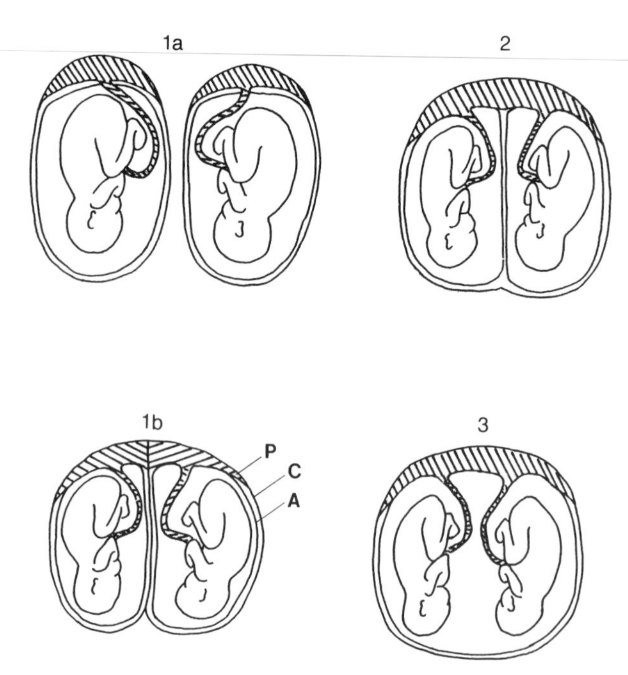

FIGURE 2 Placentation of twins: (1) diamniotic–dichorial, separated (a) or secondarily fused (b) (MZ or DZ); (2) diamniotic–monochorial (MZ); (3) monoamniotic (MZ, rare). P, Placenta; C, chorion; A, amnion.

It should thus be stressed that all monochorial placentas belong to MZ pairs, that all DZ twins have dichorial placentas, but that not all dichorial twins are DZ.

A. Dichorial MZ Twins

In most textbooks and monographs on human placentation, it appears as an inescapable dogma that dichorial MZ twins must arise through separation of the two cells deriving from the first division of the fertilized egg (i.e., the blastomeres), because *in vitro* they can each give rise to a complete embryo. Close scrutiny indicates that such a mechanism is unlikely to be involved in the natural production of MZ twins, in view of the presence of the zona pellucida, a thick membrane that surrounds the embryo. No one has described two distinctly separated morulae or blastocysts within the same zona, as should be present if the first two blastomeres produced two separate embryos. Their independent development is unlikely because, when kept or put together, blastomeres always stick to one another and organize between themselves the formation of a single embryo. This is true even when blastomeres of two different species (e.g., a rat and a mouse) are experimentally combined *in vitro* to produce a single chimeric blastocyst.

It is then possible that while traveling along the fallopian tube some eggs might escape from their zona, undergo separation of the early blastomeres, and give rise to MZ twins? Probably not, for it has been shown that naked mouse eggs at stages from one to eight cells, as well as single blastomeres isolated from such eggs, do not survive when transferred to the oviduct. Similar observations were reported in the rabbit, in which no blastocysts could be obtained from *in vivo* transfer of naked blastomeres originating from two- and four-cell eggs. Thus, culture *in vitro* appears to be the only possible way of achieving development of naked single blastomeres into blastocysts.

In addition to the possibility of a hostile effect of the tubal environment on unprotected eggs, the latter might also have some difficulty in achieving compaction (i.e., establishment of cohesive junctions between blastomeres when the embryo is composed of eight to 16 cells). It is known that eggs obtained from separated blastomeres compact easily when maintained under static *in vitro* conditions. However, it is possible that, *in vivo,* tubal wall motility prevents the establishment of tridimensional relationships, which are normally achieved through containment within the zona pellucida and are considered necessary for compaction.

A more likely explanation for the occurrence of dichorial MZ twins has been recently derived from the observation of *in vitro*-cultured cow eggs. Sometimes, when the embryo hatches out of the zona pellucida, instead of coming out through a large slit, the cow blastocyst progressively herniates through a small opening (Fig. 3). If both the inner cell mass and the trophectoderm participate in this process, an hourglass-shaped egg is formed, each part containing both cellular components. It is probable that the narrow connecting bridge will easily rupture, giving rise to separate twin blastocysts which can develop independently. In agreement with this interpretation, several births of dichorionic twin calves following transfer of a single day 7 blastocyst have been reported.

Experimental twinning has likewise been obtained in mice by simply nicking the zona pellucida at the blastocyst stage. Half-hatched human blastocysts have also been observed *in vitro*.

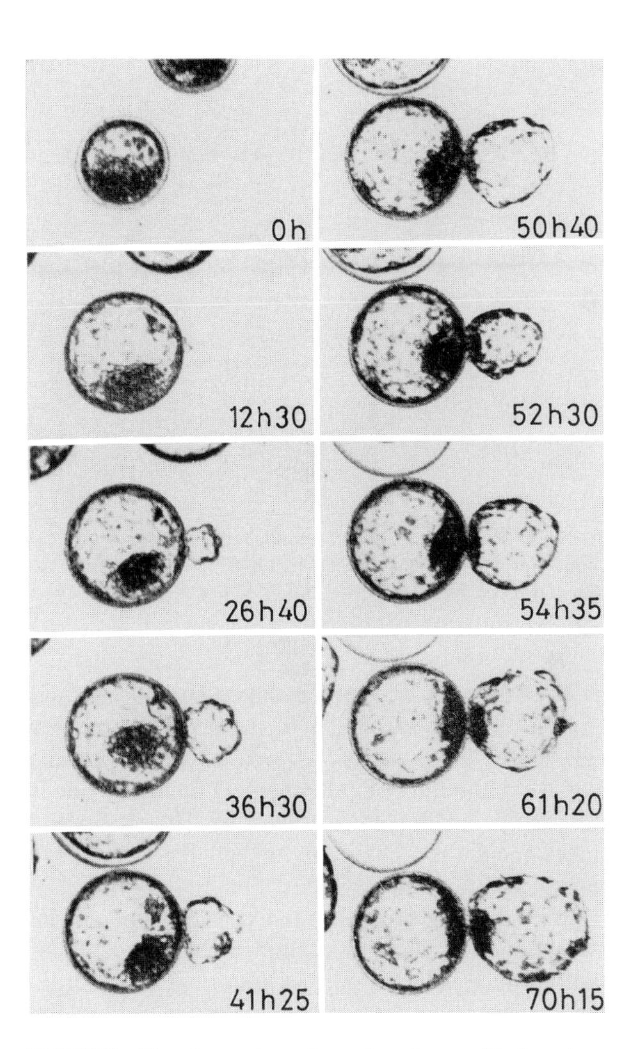

FIGURE 3 Time-lapse cinematography of a cow blastocyst undergoing atypical hatching, resulting in the separation of two complete twin blastocysts. [From A. Massip, P. Van der Zwalmen, J. Mulnard, and W. Zwijsen. *Vet. Rec.* **112,** 301 (1983), by permission of the authors.]

B. Monochorial Twins

The logical explanation for the formation of monochorial twins is that they arise at the blastocyst stage through division or duplication of the inner cell mass within a single trophectoderm. Such blastocysts have occasionally been found in species such as sheep or pigs, and the same type of duplication has also been obtained in mice by grafting a foreign inner cell mass into a host blastocyst.

In humans the earliest available specimen of monochorial twins belongs to the Carnegie Collection (No. 9009). It consists of twin didermic embryos (i.e., composed of two layers: ectoderm and

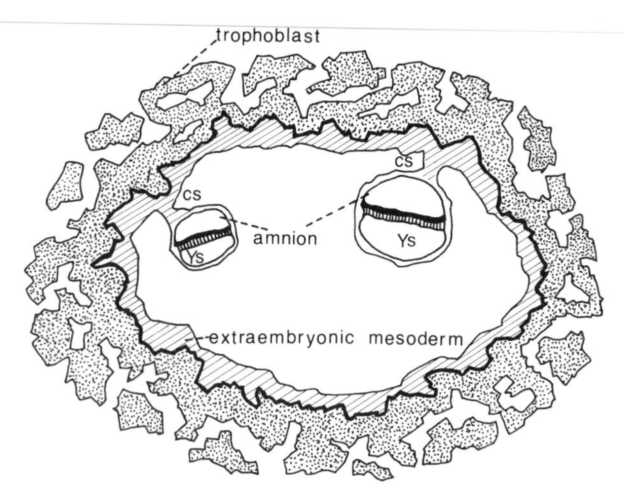

FIGURE 4 The earliest known monochorial twin embryos. cs, Connecting stalk; Ys, yolk sac. [After G. W. Corner, *Am. J. Obst. Gynecol.* **70,** 933 (1955).]

endoderm), each of which has its own yolk sac and amnion and is appended by a separate connecting stalk to the inner wall of a single chorionic vesicle (Fig. 4). Although both are about 17 days old, one is noticeably smaller than the other. In somewhat older specimens a common yolk sac has sometimes been found.

An interesting problem related to monochondrial twins concerns egg orientation at the time of implantation. It is admitted that in normal single pregnancies the attachment of the umbilical cord in the center of the placenta reflects the primary site of implantation. Since the human conceptus penetrates the endometrium by the region including the inner cell mass (i.e., the future embryo), further development of the connecting stalk through which the umbilical vessels will run is centered in the middle of the area where the placenta will differentiate.

Therefore, it is believed that marginal insertion of the cord occurs because of abnormal orientation of the inner cell mass at the time of implantation. In the case of monochorial twins, at least one of the two inner cell masses located at different poles of the same blastocyst will, of necessity, become orientated away from the endometrial surface. It follows that a higher frequency of marginal insertions of the cord is to be expected in the group of monochorial twins. Data compiled from the literature confirm that monochorial twin placentas show abnormal cord insertion much more frequently than do those of dichorial pairs. [*See* IMPLANTATION (EMBRYOLOGY).]

Placentation of most monochorial twins entails a remarkable anatomical feature: formation of vascular connections (i.e., anastomoses) between the two fetal circulations. The most frequent types are the isolated arterioarterial communication and the combination of arterioarterial with arteriovenous anastomoses. The latter are almost always located at the capillary level in a so-called common villous district. Such areas are centered on an umbilical artery of one fetus and a vein arising from its partner, which ramify and communicate into the depth of the placenta (Fig. 5). In contrast, arterioarterial and venovenous channels involve larger-caliber vessels running on the fetal surface of the placenta. Although the exchange of blood between the two fetuses raises a series of yet unsolved problems, it might be given a simplified interpretation based on correlations between anastomotic types and clinical observations. Broadly, several situations can occur. These are discussed below.

1. Hemodynamic Equilibrium

In placentas in which only arterioarterial and/or venovenous communications are present, blood can only circulate in these channels if pressures on each side are different. If such a difference occurs, blood pressure is readily equilibrated in both fetuses, since these vessels are of large caliber. Any hemodynamic disturbance occurring in one twin is immediately transmitted to his partner and, other things equal, their intrauterine growths will be parallel.

In most cases, however, large-caliber anastomoses are associated with arteriovenous communications. It follows that the differences in pressure between artery and vein are continuously compensated for by a reverse blood flux occurring in arte-

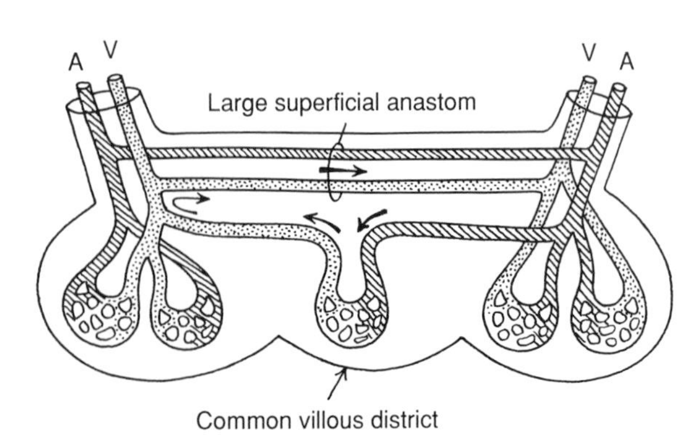

FIGURE 5 Vascular anastomoses in monochorial twin placenta. A, Umbilical artery; V, umbilical vein. Dotted and hatched areas represent venous and arterial vessels, respectively.

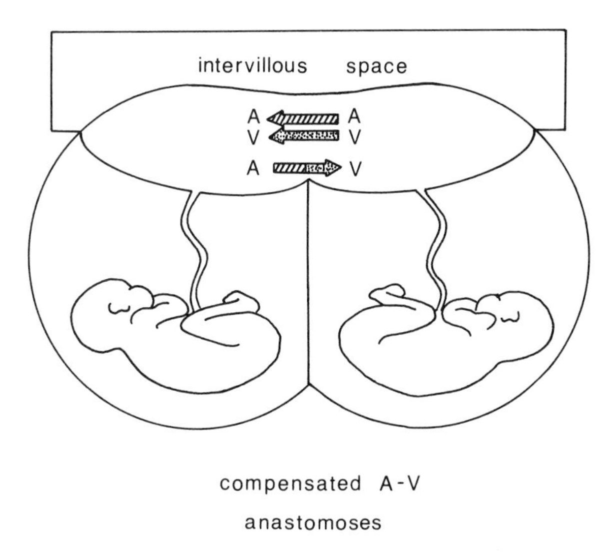

FIGURE 6 Third circulation of monochorial twins. Hemodynamic equilibrium. A, Arterial; V, venous. Dotted and hatched areas represent venous and arterial vessels, respectively.

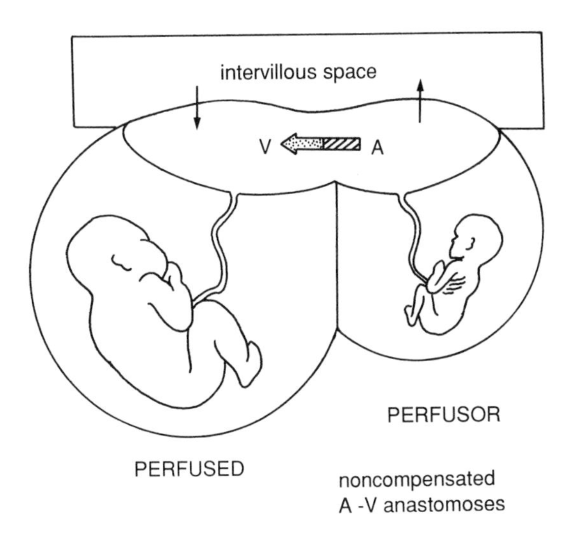

FIGURE 7 Third circulation of monochorial twins. Hemodynamic asymmetry. V, Venous; A, arterial. Dotted and hatched areas represent venous and arterial vessels, respectively.

rioarterial or venovenous channels (Fig. 6). This situation is also hemodynamically balanced, but a third circulation permanently takes place. It is likely that this blood shunt is often responsible for fetal growth impairment, since common villous areas might represent as much as one-fifth of the placental volume. This situation, however, is not the most detrimental.

2. Strong Hemodynamic Imbalance

In contrast, dramatic consequences arise from arteriovenous channels which are not compensated for (Fig. 7). Under such conditions one twin (i.e., the perfusor) will pump blood into the other (i.e., the perfused) until the perfusor's arterial pressure becomes equal to the perfused's venous pressure. Therefore, the transfused twin perishes from cardiac failure due to excessive plasmatic load, whereas the transfusor dies from hypoxia caused by blood depletion. For the same reasons excessive accumulation of amniotic fluid (hydramnios) develops on one side, while little of it forms on the other (oligoamnios). This is why, sometimes, the obstetrician comes across a case of very premature stillborn twins, one of which is hydropic, while the other appears underdeveloped.

3. Moderate Hemodynamic Imbalance

In most cases of the intertwin transfusion syndrome, the hemodynamic asymmetry is discrete, owing to imperfect compensation of arteriovenous communications. The transfusion from one fetus to the other is then very slow, and pregnancy is able to proceed closer to term and the mother can give birth to living babies. The perfusing twin will, nevertheless, be pale and small. Besides marked anemia, plasmatic proteins are depleted and the blood contains many immature red blood cells as a sign of bone marrow reaction to anemia. For reasons still unknown, the corresponding placental area is swollen because of villous edema. On the other hand, the transfused fetus will often be heavier than his or her partner. He or she will also show a deep purple complexion because of an excess of red blood cells, which entails high levels of blood bilirubin derived from their destruction. This increased destruction of red blood cells in turn causes liver and spleen enlargement, which is often found in such infants.

4. Fetus Papyraceus

Sometimes after undergoing intrauterine death, a twin becomes completely dried out and flattened by his normal partner's growth. This so-called fetus papyraceus is often thought to have arisen from intertwin transfusion. This is unlikely, however, because in one-half of such cases the placenta is dichorial, and it is known that in such placentas vascular anastomoses are rare. Therefore, it is believed that a fetus papyraceus can result from a diversity of causes entailing intrauterine death around midpregnancy, among which intertwin transfusion is only a special case.

5. Acardiac Fetus

Another feature relating to twins' third circulation is the rare anomaly known as acardiac fetus. Such fetuses have no heart and often lack the upper part of the body, as if cut at waist level. The growth of this hugely malformed fetus is possible only because of the existence of two large vascular anastomoses through which the malformed twin can be perfused by the heart of his normal partner. One of the linking channels is arteriorarterial; the other is venovenous. It is clear that circulation in the acardiac fetus is running opposite the normal direction. He receives poorly oxygenated blood through an umbilical artery from his cotwin, blood returning to the normal partner through the venous anastomosis. This malformed fetus is thus premanently lacking oxygen.

According to one theory, these abnormal hemodynamic conditions would be responsible for the absence of heart and cephalic development. However, it is difficult to visualize how vascular anastomoses could have been present and played a teratogenic role at early stages of development. Therefore, the view that this anomaly results from another yet unknown cause appears more plausible. Accordingly, the association with a normal twin would be fortuitous, but nevertheless mandatory, to allow the acardiac fetus to survive and grow.

So far, no vascular communications between DZ twins have been directly demonstrated. There is evidence, however, that in rare instances they must have existed somehow. The presence of a mixture of blood cells with characteristics of both twins (i.e., chimerism) has been described in a series of otherwise normal twins, often of opposite sexes. In most of these cases, each member of the pair was endowed with two red blood cell types carrying different group antigens. Since the mixture was the same in each partner, its existence is best explained by the exchange of erythropoietic cells during fetal life. Immunological tolerance between such twins has been demonstrated by showing that skin grafts from one to the other were not rejected even when the sexes were different.

C. Monoamniotic Twins

Twins contained in a single pouch are rare (1–2% of MZ pairs). This is fortunate, since not only are monoamniotic twins exposed to the hazards of intertwin transfusion and to a high rate of malformations, but often they perish through intrauterine entanglement of their umbilical cords.

Monoamniotic twinning is supposed to occur relatively late in development, since it results from duplication of the embryonic rudiment of the germ disk (i.e., the ectodermal plate). Normally, the signal for the formation of the amnion, which appears during the second week of development, is related to the presence of the embryo.

Therefore, to give rise to monoamniotic twins, splitting of the embryo has to occur after the seventh day following fertilization. At that time, however, the cells of the germ disk might not yet have differentiated into a single axial arrangement, which starts at about day 13 or 14, together with the appearance of the primitive streak. It is reasonable to assume that splitting occurring before axial differentiation is achieved will be incomplete, giving rise to conjoined twins. In technical terms of general embryology, the formation of monoamniotic twins corresponds to a process of double gastrulation (Fig. 8).

In some armadillos nature offers a unique example of monoamniotic multiple embryogenesis. In these animals the embryonic disk splits regularly into several parts. Each of these, while developing into an embryo, moves away from the others and draws its part of the amniotic chamber. The amnion finally becomes a branched structure, and each embryo lies in an individual secondary sac connected to the original site of the amnion through a narrow duct. Two species of armadillo, *Dasypus novemcinctus* and *Dasypus hybrida,* are known to produce four and six to nine fetuses, respectively, through this mechanism of polyembryony. It was suggested that because of delayed implantation in armadillos, this condition might be involved in the production of MZ multiples. However, many species undergo delayed implantation without showing any polyembryony.

Conjoined twins occur no more than about once in 500 MZ twin pregnancies (i.e., once or twice in 10^5 births). For still unknown reasons about two-thirds of these cases are girls. Seventy% of so-called Siamese twins are joined by some part of their anterior thorax, but almost any imaginable type and degree of fusion along the longitudinal body axis have been described. This confirms that such anomalies and also monoamniotic twins must be generated by either partial or total parallel duplication of the axial blueprint defined at gastrulation.

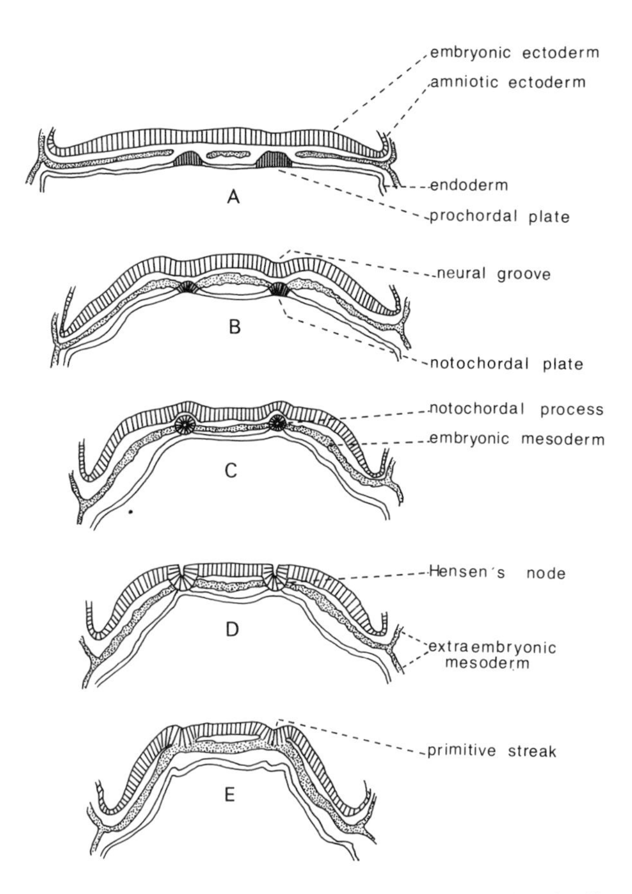

FIGURE 8 Putative mechanism of double gastrulation, leading to the formation of monoamniotic twins. (A–E) Successive transverse sections in the craniocaudal direction. (After Heuser's day 18 presomitic embryo. See W. J. Hamilton, J. D. Boyd, and H. W. Mossman. "Human Embryology," p. 55. Heffer, Cambridge, England, 1962.)

D. Congenital Anomalies

Besides the acardiac fetus and conjoined twins, which are clearly specific of MZ twinning, other malformations encountered in the general population are more frequent among twins. Since this increase is only found in the group of identically sexed pairs, it has been concluded that it pertains to MZ twins.

Galton's original hypothesis, that comparison between MZ and DZ twins might give a measure of environmental versus hereditary influences, has been applied to the analysis of birth defects. Such studies, however, are fraught with difficulties. A first obstacle relates to differential intrauterine conditions, to which most MZ twins are submitted. It is indeed difficult to evaluate which role such features as vascular anastomoses or abnormal implantation and cord insertion might have played in the occurrence of malformations and to differentiate their possible teratogenic effect on each twin. A low concordance rate for a given anomaly among MZ pairs does not, therefore, prove that genetic factors are unimportant determinants of the defect under study.

Another important limitation results from the rarity of both twinning and congenital malformations, which necessitates the analysis of large series of birth records. It has been estimated, for instance, that 40,000 births have to be monitored to detect just one twin with a cleft lip and/or palate. The lack of accurate individual zygosity determination has been another pitfall hampering the study of birth defects among twins.

It seems clear, nevertheless, that a series of malformative defects are more frequent among MZ twins than in corresponding populations of DZ pairs or singletons. These include neural defects (e.g., anencephaly and hydrocephalus), heart disease (namely, patent ductus arteriosus), cleft lip and/or palate, gut atresia, kidney malformation, syndactyly, and polydactyly. Except polydactyly, which shows a 100% concordance rate, all of these malformations appear to be discordant in more than one-half of the like-sexed pairs, with at least one twin carrying the defect. [*See* BIRTH DEFECTS.]

It has been suggested that anomalies occurring more frequently in MZ twins resemble those caused by mutations of homeotic genes in animals (i.e., genes controlling spatial and temporal patterns of embryonic development). Proteins specified by these genes regulate the genetic control of early development. They are synthesized in embryonic cells, which might become unequally distributed when fission of the conceptus takes place, causing a congenital defect in one twin, but not in the other. This hypothesis, however, remains to be demonstrated.

The excess of congenital malformations linked to monozygosity suggests that there might be a common factor in the causation of both. The fact that MZ twinning is itself a deviation from normal embryonic development reinforces this view. Unfortunately, the precise origin of most birth defects remains ill explained, and their study, therefore, has not provided a clue as to the causes of embryonic splitting.

IV. Possible Origins of Monozygotic Twins

The etiological factors giving rise to spontaneous MZ twinning in mammals remain unknown. Contrary to the case of DZ twinning, no clear genetic influence seems to be involved in the production of identical human twins.

In some invertebrates, as well as fish and birds, splitting of the embryo can be induced through cooling or through oxygen deprivation. No such results are available in mammals. It has been claimed that in humans the ratio of monochorial to dichorial placentation is particularly high among extrauterine twin pregnancies. Therefore, it was suggested that defective nutritional conditions might play a role in the formation of human MZ twins. However, this interpretation appears rather doubtful, because the mechanism of monochorial twinning is probably set in motion well before blastocyst attachment.

Another etiological hypothesis relates to preovulatory overripeness or retarded fertilization of the oocyte. Partial or complete axial duplication of embryos has been obtained through delayed insemination in some amphibians and even in mammals. As regards the human, it has been reported that mothers of MZ twins exhibit significantly more prolonged and irregular menstrual cycles than do those of DZ twins. Such abnormal ovulatory conditions might correspond to oocyte overmaturation.

Some authors have proposed that MZ twins are caused by a twinning impetus acting at random during early development. This would explain why the frequency distribution of the different types of placentation among MZ twins remains constant and in proportion to the duration of early developmental steps at which they are supposed to occur. However, as indicated, there is a fundamental difference in the embryological mechanism involved, namely, between the formation of mono- and dichorial MZ types. Therefore, a unitarian view on the causality of MZ twinning might not be realistic.

V. Intrauterine Growth of Twins

Ultrasound screening in early pregnancy has evidenced the phenomenon known as the vanishing twin. Not infrequently, one of the gestational sacs of a twin or multiple gestation progressively shrinks and disappears. Only rarely does examination of fetal adnexa of the remaining twin show some morphological trace of this event. Since authors vary extensively in their evaluation of the vanishing twin frequency, it is difficult to estimate to what extent it might affect MZ and DZ indices among the general population.

Low birth weight is a major problem befalling twins and other multiples. An important cause of this defect is premature birth. The delivery of twins occurs, on average at the 37th week of pregnancy, its duration being reduced to 35 and 34 weeks for triplets and quadruplets, respectively. Although the mechanism of labor onset in humans remains unclear, it might be assumed that uterine overstretching plays a role in the premature termination of multiple pregnancy. It is likewise possible that some humoral factor produced in greater amounts because of the existence of two fetoplacental units might reach a threshold level much sooner than in single pregnancies and induce early stimulation of uterine muscle activity.

However, mean birth weight of multiple-pregnancy babies remains lower than that of singletons at equivalent gestational age, since their intrauterine growth is significantly slowed during the last trimester of pregnancy. Twin babies are, therefore, not only premature, but also small for date. This has been confirmed by showing that their average biparietal diameter, as measured by ultrasound examination, lies somewhat below that of singletons at a similar intrauterine age.

The first explanation that comes to mind is that overcrowding of the uterus in multiple pregnancy prevents normal placental development. In twins, at all ages of gestation, placental weight and volume are lower than the double value of figures recorded in single pregnancies. However, at equivalent relative placental weight twins are still lighter than singletons, even when allowing for gestational age. In other words, the percentage of placental weight reduction in twins (i.e., 12%) is less than their relative body weight defect (20%). Therefore, twins' growth retardation cannot be totally accounted for by impaired placental development. There is evidence pointing to a role of insufficient blood supply to the uterus. It has been found that in multiple pregnancies uteroplacental circulation is much slower than normal and that twins have increased hemoglobin and red blood cell levels at birth. These data might be interpreted as a consequence of chronic intrauterine oxygen deprivation.

Curves of evolution using ultrasound assessment of biparietal diameters throughout pregnancy indicate that on average, MZ twins are more growth retarded than are DZ twins. This difference is due to the existence of vascular anastomoses in monochorial placentas, which are present in about two-thirds of MZ pairs. This feature also explains why largest body weight differences within a pair are found among MZ twins.

Bibliography

Bryan, E. M. (1983). "The Nature and Nurture of Twins." Baillière Tindall, London.

Bulmer, M. G. (1970). "The Biology of Twinning in Man." Oxford Univ. Press (Clarendon), Oxford, England.

Hill, A. V. S. (1985). Use of minisatellite DNA probes for determination of twin zygosity at birth. *Lancet* **2,** 1394.

Källen, B. (1986). Congenital malformations in twins: A population study. *Acta Genet. Med. Gemellol.* **35,** 167.

Leroy, F. (1985). Early embryology and placentation of human twins. *In* "Implantation of the Human Embryo" (R. G. Edwards, J. M. Purdy, and P. C. Steptoe, eds.), pp. 393–405. Academic Press, Orlando, Florida.

MacGillivray, I., Nylander, P. P. S., and Corney, G. (1975). "Human Multiple Reproduction." Saunders, London.

MacGillivray, I., Campbell, D. I., and Thompson, B. (1988). "Twinning and Twins." Wiley, Chichester, England.

Vlietinck, R., Derom, C., Derom, R., Van den Berghe, H., and Thiery, M. (1988). The validity of Weinberg's rule in the East Flanders prospective twin survey. *Acta Genet. Med. Gemellol.* **37,** 137.

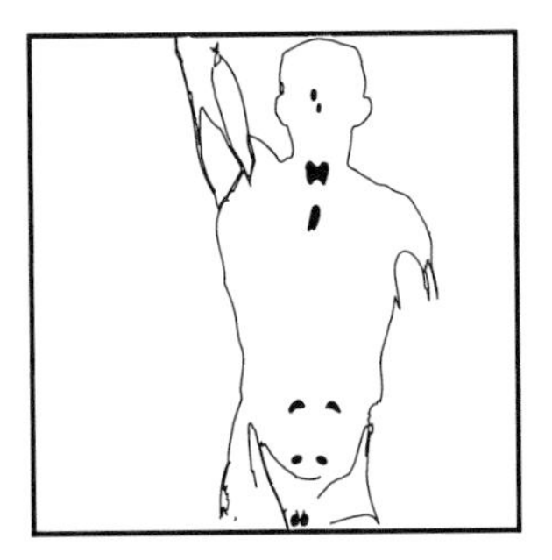

Endocrine System

HOWARD RASMUSSEN, *Yale University School of Medicine*

Glossary

Acinar Pertaining to a cell with a small saclike dilatation; can be found in various glands

Amino acids Any organic compound containing an amino ($-NH_2$) and a carboxyl ($-COOH$); building blocks for protein synthesis

Arachidonic acid Unsaturated fatty acid essential for human nutrition; precursor in the biosynthesis of leukotrienes, prostaglandins, and thromboxanes

Autonomic nervous system Portion of the nervous system concerned with the regulation of the activity of the cardiac muscle, smooth muscle, and glands

Chromosomes Structure in the nucleus containing a linear thread of DNA, which transmits genetic information and is associated with RNA and histones

Cytosol Liquid medium of the cytoplasm (the protoplasm of the cell)

Ectopic Located or produced away from its normal position

Enzymes Protein molecule that catalyzes chemical reactions of other substances without itself being destroyed or altered upon completion of the reaction

Fatty acids Any straight-chain single carboxyl group acid, especially those naturally occurring in fats, generally classified as saturated when they have no double bonds, monounsaturated when they have one double bond, and polyunsaturated when they have multiple double bonds.

Gene Segment of a DNA molecule that contains all the information required for synthesis of a product; biological unit of heredity

Glycerol Trihydric sugar alcohol (CH_2OH-$CHOH$, CH_2OH); alcoholic component of fats

Glycogen Chief carbohydrate storage material in animals, a long-chain polymer of glucose

Heterologous Different in either structure, position, or origin

Homologous Corresponding in structure, position, and origin

Ion Atom or radical having a charge (positive or negative) owing to the loss or gain of one or more electrons

Ketone Any of a large class of organic compounds containing the carbonyl group ($C{=}O$) whose carbon atom is joined to two other carbon atoms; ketone bodies include β-hydroxybutyric acid, acetoacetic acid, and acetone

Kinase Subclass of the transferase enzymes, which catalyze the transfer of a high-energy phosphate group from a donor compound (adenosine triphosphate or guanosine triphosphate) to an acceptor compound

Lactic acid End product of glycolysis involved in many biochemical processes

Mitochondria Small separate organelle found in the cytoplasm of cells, which is the principal site for the generation of energy (adenosine triphosphate)

Molar Measure of the concentration of solute, expressed as the number of moles of solute per liter

Multimeric Consisting of multiple subunits

Neurotransmitter Any of a group of substances that are released on excitation from the axon terminal of a presynaptic neuron of the central or peripheral nervous system and travel across the synaptic cleft to either excite or inhibit the target cell

Peroxidase Sub-subclass of enzymes of the oxidoreductase class that catalyze the oxidation of organic substrates by hydrogen peroxide, which is reduced to water

Phosphatases Any of a subclass of enzymes of the hydrolase class that catalyze the release of inorganic phosphate from phosphoric esters

Phosphodiesterases Any of a subclass of enzymes of the hydrolase class that catalyze the hydrolysis of one of the two ester linkages in a phosphodiester compound

Polymer Chemical compound consisting of repeating structural units

Proteolysis Splitting of proteins by hydrolysis of the peptide bonds with formation of smaller polypeptides

Pseudopodia Temporary cytoplasmic extrusion by means of which an ameba or other ameboid organism or cell moves about or engulfs food

Somatic Pertaining to or characteristic of the body

Transducer Translation of one form of signal to another

Very low-density lipoprotein Lipid–protein complex in which lipids are transported in the blood

TWO MAJOR SYSTEMS integrate cellular and tissue responses in multicellular organisms: the nervous and endocrine systems. Neural responses are characterized by their rapidity and their restricted nature; in contrast, endocrine responses are characterized by a wide range of temporal responses from relatively rapid to prolonged and by their diffuse nature. Even so, these systems are ultimately linked at the level of the hypothalamus, properly called a neuroendocrine organ. Also, the two systems possess important similarities. First, transmission of information from one cell to the other is via chemical messenger, and, second, the target cells possess specific receptors, or recognition molecules, for these chemical messengers. The chemical messengers in the nervous system are known as neurotransmitters, and those of the endocrine system as hormones. Initially, neurotransmitter receptors were thought to be linked to ion channels, and hormone receptors to membrane-associated enzymes or to chromosomes; however, evidence now indicates that these sharp distinctions are not as complete as once believed. A particular entity can serve as a neurotransmitter in one setting, and as a hormone in another. Furthermore, some hormones act on receptors linked to ion channels, and come neurotransmitters on membrane-associated enzymes. In addition, some neurotransmitters are released into the bloodstream and have a wide distribution in the body. In fact, the same chemical entity can serve to convey information from one cell to another by a variety of pathways. [*See* NEUROENDOCRINOLOGY.]

I. Hormone and Endocrine Tissues

A hormone was originally defined as an agent released into the bloodstream from cells in one organ and then transported to another organ, where it interacted with target cells to alter their function. During the first 50–60 years of this century, the term hormone came to mean an agent released into the bloodstream from specific endocrine organs. These classic endocrine organs were recognized as specific tissues whose sole function was to receive chemical signals from one or more bodily organs, process this information, and then release a specialized chemical mediator to alter the function of other organs. Hence, the notion of endocrine mediator and hormone were synonymous. However, in the past 20–25 years, it has become clear that the same chemical entity can act as either a neurotransmitter or as an autocrine, paracrine, neurocrine, or endocrine mediator (Fig. 1). A neurotransmitter is an agent released from a nerve ending (synapse) that acts on a second neuron. An autocrine agent is released into the cellular environment and acts on the same cell from which it was released; a paracrine agent acts on adjacent cells in the same tissue or organ (without entering the bloodstream); a neurocrine agent is released from nerve synapses and acts on one or more postsynaptic cells that are not neurons; and an endocrine agent acts on distant target cells. Thus, a more general definition of a hormone is an agent released from one cell that acts to alter the function of that cell, an adjacent cell, or a distant cell.

More recently, the definition of what constitutes an endocrine organ has also expanded. The original definition encompassed 10 highly specialized organs, each of whose function appeared to be solely that of secreting a particular hormone(s). It is now clear that practically every tissue in the body se-

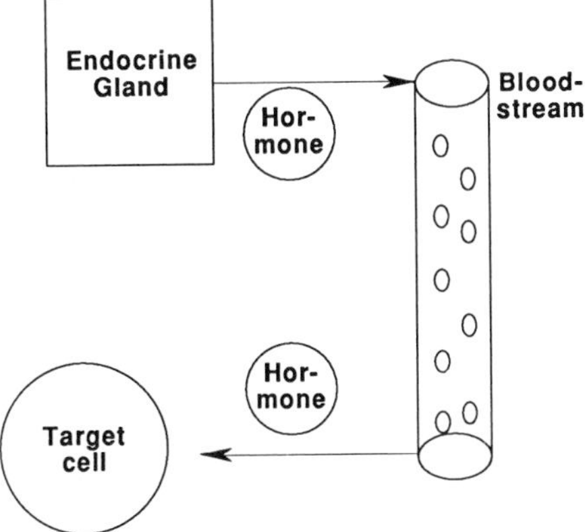

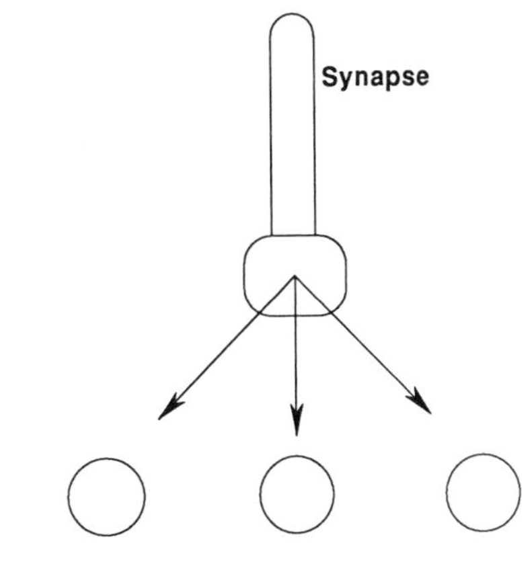

FIGURE 1 The various forms of hormone secretion: autocrine, in which the secreted product acts on its cell of origin; paracrine, in which the secreted product acts on an adjacent cell in the same tissue; neurocrine, in which a neuron secretes the product that acts on distant cells; and endocrine, in which the secreted product enters the bloodstream and is carried to a distant target cell.

cretes chemical messengers, which regulate either their own function or those of distant cells, tissues, or organs.

A. Classic Endocrine Tissues and Hormones

The classic endocrine organs (Fig. 2) are (1) hypothalamus, (2) anterior pituitary, (3) posterior pituitary, (4) thyroid, (5) parathyroid, (6) islets of Langerhans in the pancreas, (7) adrenal cortex, (8) adrenal medulla, (9) ovary, and (10) testes.

Three types of chemical entities have been identified as the products of these tissues (Tables I–III). The steroid hormones are listed in Table I. In this

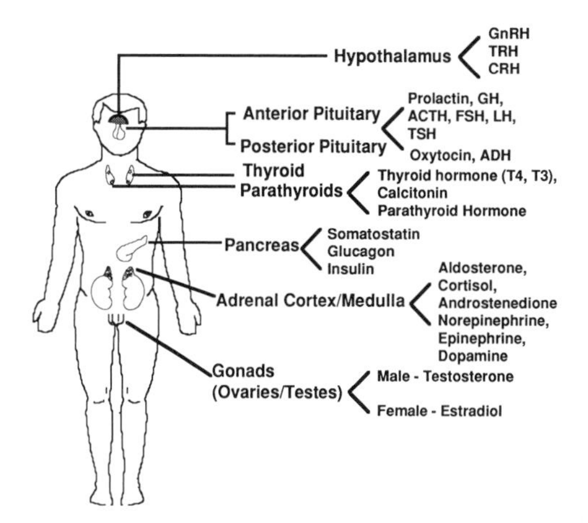

FIGURE 2 Classical endocrine glands and their hormones. ACTH, adrenocorticotrophic hormone; ADH, antidiuretic hormone; CRH, corticotropin-releasing hormone; FSH, follicle-stimulating hormone; GH, growth hormone; GnRH, gonadotropin-releasing hormone; LH, luteinizing hormone; TRH, thyrotropin-releasing hormone; TSH, thyroid-stimulating hormone.

TABLE I Steroidal Hormones

Hormone	Source	Site of action	Action
Estradiol	Ovary	Uterus	Growth of endometrium
		Breast	Growth of breast and other secondary sexual organs
		Bone	Growth of skeleton
Progesterone	Ovary	Uterus	Maturation of endometrium
Testosterone	Testes	Spermatogonia	Spermatogenesis
		Male sex organs	Growth and development
		Muscle and bone	Growth
Dihydrotestosterone	Several tissues convert testosterone to dihydrotestosterone	Male sex organs	Growth and development
Cortisol	Zona fasciculata of adrenal cortex	Many tissues—liver, connective tissue, bone, fat cells, lymphoid tissue	Multiple effects that are part of the response to stress
Aldosterone	Zona glomerulosa of adrenal cortex	Kidney Intestine	Increased retention of sodium and decreased retention of potassium
Androstenedione	Zona reticulosis of adrenal cortex	Secondary sex organs	Weak effects similar to testosterone
1,25-dihydroxyvitamin D_3	Kidney	Intestine	Increased absorption of calcium and phosphate
		Bone	Increased resorption of calcium and phosphate
24,25-dihydroxyvitamin D_3	Kidney and other tissues	Bone	Increased bone formation and mineralization

and the succeeding tables, the name, source, site of action, and principal actions are listed for each hormone. In the case of the steroid hormones, most have one or two principal actions; however, cortisol has a wide spectrum of effects, and the sex hormones have effects not only on the uterus and testes but also on all the secondary sex organs such as the breast and vagina in the female, and the penis, skeletal muscle, and hair growth in the male. The steroid hormones include the female sex hormones estradiol and progesterone from the ovary, the male hormones testosterone, and dihydrotestosterone from the testes and androstenedione from the adrenal cortex; and the adrenal cortical hormones aldosterone and cortisol. [*See* STEROIDS.]

The amine hormones (Table II) include epinephrine and norepinephrine from the adrenal medulla, serotonin and dopamine from the hypothalamus,

TABLE II Amine Hormones

Hormone	Source	Site of action	Action
Epinephrine vitamin D_3	Adrenal medulla	Liver, heart, blood vessels	
	Nerve endings	Fat cells, muscle, blood cells	Increases glucose production, breakdown of fat, heart rate, and many other actions
Norepinephrine	Adrenal medulla	Blood vessels	Contraction
	Nerve endings	Brain	Neurotransmitter
Dopamine	Nerve endings	Anterior pituitary	Inhibits prolactin secretion
		Brain	Neurotransmitter
Serotonin	Gut endocrine cells	Microvessels	Initiates inflammatory response
		Bronchial muscle	Contraction
Thyroxine (T_4)	Thyroid gland	Many different cells	Regulates basal metabolic rate
Triiodothyronine (T_3)	Thyroid and liver ($T_4 \rightarrow T_3$)	Many different cells	Regulates basal metabolic rate

TABLE III Classic Protein and Peptide Hormones

Hormone	Source	Site of action	Action
Insulin	Beta cells of endocrine pancreas	Liver, muscle, fat cells	Increases storage of glucose, fat, and protein
Glucagon	Alpha cells of endocrine pancreas	Liver	Increases glucose production
Glucagon	Gut	Liver	
Parathyroid hormone	Parathyroid gland	Kidney	Increases calcium retention, increases synthesis of 1,25-dihydroxyvitamin D_3, decreases phosphate retention
		Bone	Increases bone resorption
Calcitonin	C cells of thyroid gland	Bone	Inhibits bone resorption
Somatostatin	Hypothalamus, endocrine pancreas, gut endocrine cells	Anterior pituitary	Inhibits growth hormone secretion, inhibits insulin secretion, inhibits actions of several gut hormones
Angiotensin II	Bloodstream from a precursor	Blood vessels	Contraction
		Zona glomerulosa	Stimulates aldosterone secretion
		Brain	Neurotransmitter
		Kidney	Decreased sodium excretion
Atrial natriuretic peptide	Atria of heart	Kidney	Increased sodium and H_2O excretion
		Zona glomerulosa	Inhibits aldosterone secretion
		Blood vessels	Dilatation
Vasopressin	Hypothalamus Posterior pituitary	Kidney	Increased H_2O retention
Oxytocin	Hypothalamus	Breast	Milk secretion
	Posterior pituitary	Uterus	Contraction
Thyroid-stimulating hormone (TSH)	Anterior pituitary	Thyroid gland	Stimulates synthesis and secretion of thyroid hormone
Adrenocorticotrophic hormone (ACTH)	Anterior pituitary	Adrenal cortex	Stimulates synthesis and secretion of cortisol, androstenedione, and aldosterone
Follicle-stimulating hormone (FSH)	Anterior pituitary	Ovary	Increased estradiol synthesis and secretion
Luteinizing hormone (LH)	Anterior pituitary	Ovary	Increased synthesis and secretion of progesterone
		Testes	Increased synthesis and secretion of testosterone
Prolactin	Anterior pituitary	Breast	Increased milk products
Growth hormone	Anterior pituitary	Liver	Synthesis and release of somatomedin
Melanocyte-stimulating hormone	Intermediate lobe of the pituitary	Melanocytes	Skin darkening
Thyrotropin-releasing hormone	Hypothalamus	Anterior pituitary	Release of TSH
Corticotrophin-releasing hormone	Hypothalamus	Anterior pituitary	Release of ACTH
Gonadotropin-releasing hormone	Hypothalamus	Anterior pituitary	Release of FSH and LH
Growth hormone-releasing factor	Hypothalamus	Anterior pituitary	Release of GH
Somatomedin C or insulinlike growth factor	Liver	Muscle	Growth and proliferation

(*continued*)

TABLE III *(continued)*

Hormone	Source	Site of action	Action
		Bone and cartilage cells	Growth and proliferation
		Other cells	Growth and proliferation
Nerve growth factor	Several	Nerve cells	Growth and proliferation
Epidermal growth factor	Several	Epidermal cells and many others	Growth and proliferation
Large and increasing number of tissue-derived growth factors such as platelet-derived growth factors	Many	Large number of cells—several growth factors often work in combination	Growth and proliferation of a variety of different cell types
Cholecystokinin	Gut endocrine cells	Exocrine pancreas	Stimulates secretion of digestive enzymes
		Endocrine pancreas	Stimulates secretion of insulin
		Gallbladder	Stimulates contraction
Gastric inhibitory peptide	Gut endocrine cells	Stomach	Inhibits motility and acid secretion
		Endocrine pancreas	Stimulates insulin secretion
Secretin	Gut endocrine cells	Exocrine pancreas	Stimulates waste and salt excretion
Vasointestinal peptide	Gut endocrine cells	Intestine	Stimulates fluid and electrolyte secretion
		Smooth muscle	Stimulates contraction
Erythropoietin	Kidney	Bone marrow	Stimulates red cell production
Bombesin	Gut endocrine cells (others)	Many	Stimulates acid secretion
Gastrin	G cells in stomach	Parietal cells in stomach	Stimulates insulin secretion
Glucagonlike peptide-1	Gut endocrine cells	Endocrine pancreas	Stimulates insulin secretion
Endothelin	Endothelia cells	Smooth muscle	Stimulates contraction

histamine from mast cells, and thyroxine and triiodothyronine from the thyroid acinar cells. Again, only a few of the major effects of epinephrine, norepinephrine, and serotonin are listed. These substances are released from nerve terminals in the central nervous system in the autonomic nervous system, and in the enteric (intestinal) nervous system where they exert a variety of effects on secretory processes and on the state of contraction of smooth muscle cells in blood vessels, bronchi, and the intestinal wall.

The peptide and protein hormones are the largest and most diverse group (Table III), and each year additional members of this group continue to be identified. The list in Table III is incomplete in the sense that many of these peptides serve different functions in different locations. For example, thyrotropin-releasing hormone (TRH) acts on the pituitary to stimulate thyroid-stimulating hormone (TSH) secretion, but in the central nervous system TRH acts as a neurotransmitter. Likewise, a complete listing of the numerous peptide growth factors has not been given. Many of these stimulate, either alone or in combination with other growth factors, the growth and proliferation of cultured mammalian cells, but often the specific cell types they act on in an intact animal are unknown. Furthermore, amines, such as serotonin, or peptides, such as angiotensin, in addition to their classic actions, can also serve as one of the growth factors that stimulate the proliferation of certain classes of cells. It is particularly noteworthy that many intestinal peptides and amines that act as endocrine or neuroendocrine agents also are found in the central nervous system, where they are though to act on neurotransmitters. The classic peptide hormones include insulin and glucagon from the islets of Langerhans; calcitonin from the thyroid C cells; parathyroid hormone (PTH) from the parathyroid gland; TSH, follicle-stimulating hormone (FSH), luteinizing hormone, prolactin, adrenocorticotropin (ACTH), and growth hormone from specific cell types in the anterior pituitary; vasopressin and oxytocin from the posterior pituitary; and TRH, corticotropin-releasing hormone, growth hormone-releasing factor, gonadotrophin-releasing hormone, and somatostatin from specific loci in the hypothalamus. [*See* PEPTIDE HORMONES OF THE GUT.]

B. Nonclassic Endocrine Tissues and Hormones

Many tissues and organs that are not specific endocrine glands, nonetheless, manufacture and secrete hormones that are just as important to bodily function as are the classic hormones (Fig. 3). These nonclassic hormones fall into at least three classes: sterols, proteins and peptides, and eicosanoids (derivatives of the fatty acid arachidonic acid).

The only nonclassic steroidal hormones are derivatives of vitamin D_3 and include 24,25-dihydroxyvitamin D_3 and 1,25-dihydroxyvitamin D_3. The latter is manufactured in the kidney tubular cells.

By far the largest group of nonclassic hormones are peptides and proteins (Table III). Some of these are (1) atrial natriuretic peptide synthesized and secreted by modified heart muscle cells in the atria of the heart; (2) erythropoietin secreted by the kidney; (3) somatomedins (insulinlike growth factors), which are secreted by parenchymal cells of the liver; (4) a variety of peptide hormones secreted by cells in the intestinal tract including cholecystokinin, gastrin, gastric inhibitory peptide, glucagonlike peptide; vasoactive intestinal peptide, secretin, and bombesin; (5) a large and increasing number of tissue growth factors including nerve growth factor epidermal growth factor, and fibroblast growth factor, platelet-derived growth factor, and others; and (6) endothelin secreted by the endothelial cells lining the inner surface of blood vessels.

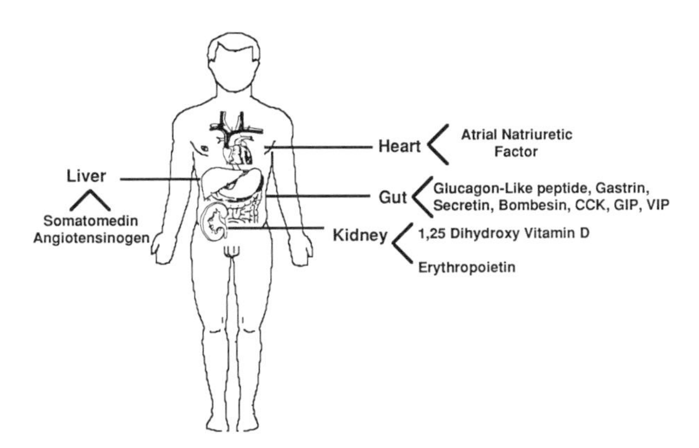

FIGURE 3 Some of the nonclassic endocrine organs and their hormonal products. CCK, cholecystokinin; GIP, gastric inhibitory peptide; VIP, vasoactive intestinal peptide.

Several other major peptide hormones are actually secreted as the inactive precursors hormogens into the bloodstream, where they undergo partial proteolysis to smaller active peptide hormones. Both bradykinin and angiotensin II are generated in this fashion.

Another large class of chemical messengers are the products of the further metabolism of arachidonic acid. The major classes of these compounds are prostaglandins, prostacyclin, thromboxanes, and leukotrienes. These are not thought to act as classic hormones but rather as paracrine and autocrine agents in the regulation of tissue function. Nonetheless, they interact with specific cell-surface receptors to initiate similar cellular responses, as do certain peptide hormones.

II. Organization of Endocrine System

There are two general features of endocrine systems: (1) they involve multiple hormones acting on a number of cell types, thus, achieving an integrated response; and (2) they display the property of feedback control.

If one oversimplifies somewhat, there is a direct correlation between the importance of the particular behavior to survival and the complexity of the endocrine inputs that regulate it. Thus, for example, it is essential to regulate plasma glucose concentration within certain limits. If the glucose concentration is too low, brain cell functions are impaired and coma occurs. If the plasma glucose is too high, glucose is lost in the urine, and high glucose also has toxic effects on the cells lining the walls of blood vessels.

When one eats a normal meal, the beta cell in the endocrine pancreas (the source of insulin) is informed by neural signals from both the head and intestine. These nerves release acetylcholine and cholecystokinin, respectively, and sensitize the beta cells so that when the glucose is absorbed from the intestine and causes an increase in blood glucose concentration, the beta cells respond with an increase in the insulin secretion. This effect of glucose is further enhanced by hormones such as cholecystokinin, gastric inhibitory peptide, and glucagonlike peptide-1, all of which act on the beta cell. Hence, a small increase in blood glucose concentration causes a large change in insulin secretory rate. Insulin acts on the three major organs to enhance the disposal of glucose. Insulin can be thought of as a storage hormone, which causes the cells in the liver, skeletal muscle, and adipose tissue to store the glucose in forms that can be utilized for energy production in times of fasting. In the liver, insulin causes some of the glucose to be converted into glycogen (a glucose polymer) and the rest into a form of fat (triglycerides), which is secreted into the bloodstream as very low-density lipoproteins (VLDLs). Insulin acts on the skeletal muscle to cause additional glucose to be stored in the muscle cell as glycogen. Insulin acts on the fat cell to convert glucose to α-glycerol phosphate and to break down the triglyceride (in VLDL) to glycerol and free fatty acid. These acids interact with α-glycerol to reform triglycerides in the fat cell. [*See* INSULIN AND GLUCAGON.]

During fasting, the blood glucose falls; however, it is essential that it not fall too low because the brain is nearly totally dependent on glucose for its energy metabolism, whereas most other tissues in the body can utilize free fatty acids or ketone bodies (partial breakdown productions of fatty acids) as their energy source. To maintain a minimal level of plasma glucose so that the brain continues to function, a variety of hormonal signals act in a coordinate way to (1) liberate free fatty acids from the tirglyceride in fatty cells; (2) instruct the liver to metabolize some of the free fatty acids to ketone bodies; (3) instruct the skeletal and heart muscle cells to preferentially utilize free fatty acids and ketone bodies for their energy needs; (4) cause the breakdown of liver glycogen and release the glucose (resulting from this breakdown) into the bloodstream to maintain a supply of glucose for the brain; (5) cause the breakdown of skeletal muscle glycogen to lactic acid, which is released into the bloodstream; (6) cause the breakdown of muscle proteins to amino acids, which are also released into the bloodstream; and (7) instruct the liver to synthesize glucose from lactic acid, glycerol, and amino acids (gluconeogenesis).

The signals that bring about a net release of fatty acids from adipose tissue are (1) a decrease in plasma insulin concentration and (2) an increase in plasma epinephrine and growth hormone. The major signal that instructs the liver to metabolize fatty acids to ketone bodies is a decrease in plasma insulin concentration. The preferential utilization of fatty acids and ketone bodies by skeletal muscle cells occurs because of lack of insulin and as a direct consequence of the increase in ketone body and free fatty acid concentrations in the blood. The breakdown of liver glycogen results from the direct actions of glucagon and epinephrine on the liver cell, and the breakdown of skeletal muscle glycogen results from the action of epinephrine on muscle cells. An increase in protein breakdown in the muscle is related to a rise in plasma cortisol concentration. The increased synthesis of glucose from lactate, glycerol, and amino acids results from the combined actions of a decreased plasma insulin concentration and an increased concentration of glucagon, epinephrine, and cortisol (the so-called counter-regulatory hormones). In addition, the increased supply of lactate and amino acids from muscle, and free fatty acids from adipose tissue, are necessary to sustain adequate rates of glucose production.

The relationship between the plasma Ca^{2+} concentration and the secretion of PTH can be utilized to illustrate a common property of endocrine systems: negative feedback control. When, for any reason, the concentration of Ca^{2+} in the blood plasma decreases slightly, this change is perceived by the cells of the parathyroid gland. These cells respond by increasing the secretion of PTH. As a result, the plasma PTH concentration rises. This rise is detected by PTH receptors on cells in kidney and bone. These cells respond in such a way that the plasma Ca^{2+} concentration rises. This rise is sensed by the parathyroid cells, which now reduce PTH secretion. [*See* PARATHYROID GLAND AND HORMONE.]

The control of endocrine responses are usually much more complicated. A large number are hierarchical. They involve specific cells in the hypothalamus secreting a releasing hormone, which acts, in turn, on a particular class of cells in the anterior pituitary, inducing them to release a tropic hormone that acts, in turn, on a peripheral endocrine gland. These systems also display feedback control, which characteristically involves the hormone product of the peripheral endocrine glands as the feedback modulator of the cells in both the hypothalamus and the pituitary (Fig. 4). For example, TRH secreted by cells in the median eminence of the hypothalamus act on thyrotrophs in the anterior pituitary to stimulate the secretion of TSH. The TSH acts, in turn, on the acinar cells of the thyroid gland to stimulate the release of the thyroid hormone thyroxine. The resulting increase in plasma thyroxine concen-

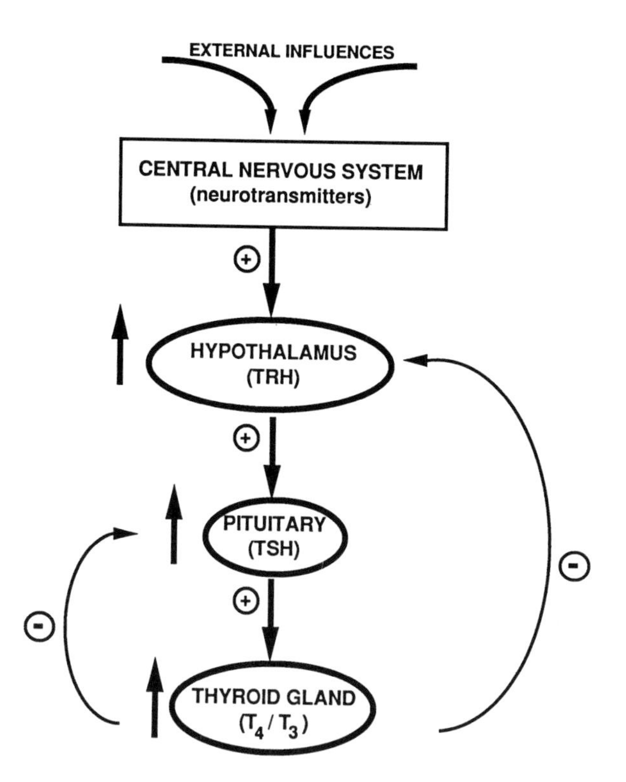

FIGURE 4 The principle of negative feedback control in an endocrine system as illustrated in the case of the interrelationship between thyrotropin-releasing hormone (TRH) from the hypothalamus, thyroid-stimulating hormone (TSH) from the pituitary, and thyroid hormones (T_4/T_3) from the thyroid. TRH stimulates TSH secretion, which in turn stimulates T_4 and T_3 secretion from the thyroid. T_4 and T_3 act as negative feedback controls at the level of the pituitary and the hypothalamus. In addition, neural inputs from the central nervous system affect TRH secretion in a positive way.

tration acts as a negative feedback signal on TRH-secreting cells in the hypothalamus and, more importantly, on the thyrotrophs in the anterior pituitary to inhibit TSH secretion. [*See* HYPOTHALAMUS; PITUITARY; THYROID GLAND AND ITS HORMONES.]

Similar but more complex feedback relationships exist in the system involved in the regulation of cortisol secretion from the adrenal zona fasciculata. Corticotropin-releasing hormone from the hypothalamus acts on corticotrophs in the pituitary to release ACTH, which in turn stimulates cortisol production from the adrenal cortex. Cortisol acts as a negative feedback regulator of ACTH production by interacting with cells in both the hypothalamus and the pituitary. However, superimposed on this feedback relationship is an additional signal from the central nervous system that dictates a diurnal variation in ACTH and cortisol secretion.

Even more complex patterns of control are evident in the control of other pituitary hormones. In particular, many of these hormones are secreted in a pulsatile fashion rather than in a continuous manner.

III. Characteristics of Hormones

Despite the diversity in their chemical structures, hormones share a number of common attributes. They are effective at very low concentrations (10^{-12}–10^{-8} M). Their actions depend on their combining with specific recognition proteins—receptors, which exist either on the surface or in the interior of the target cells. These receptors possess a high degree of specificity and affinity for a given hormone. The tissue and cellular distribution of these receptors determines which cells are targets for the action of the particular hormone. In some cases, the distribution of receptors for a specific hormone is quite limited (e.g., ACTH receptors are confined to adrenal cortical cells and some neurons in the central nervous system); in other cases, they are widely distributed (e.g., insulin receptors are present on a wide variety of cell types). The association of a hormone with its receptor leads to the activation of one or more transducing systems in the cells. As a general rule, specific cellular responses are regulated by multiple hormonal inputs, which are both stimulatory and inhibitory. Finally, the cellular response can be modulated by changing either the concentration of the particular hormone or the number, affinity, or cellular distribution of receptors.

Another common feature of hormones is their inactivation. If they are to serve as messengers, then both decreases as well as increases in their concentrations must be tightly regulated. In all cases, the enzymes responsible for their inactivation are different from the receptors for the given hormone. Also, the rate of change in plasma concentration is a reflection of the typical time course of action of the particular hormones. Many eicosanoids, amines, and peptide hormones have actions that are rapid in onset and in termination. Other peptides, thyroxine, and steroid and steroidal hormones act much more slowly and have prolonged effects. This difference is reflected in their biological half-lives in blood or bodily fluids. Eicosanoids and most amines have half-lives of seconds to minutes, peptide and protein hormones minutes, and thyroxine and steroid hormones hours to days.

Another class distinction is also apparent in terms of receptor location. Receptors for peptides and protein hormones, most amines, and eicosanoids are located on the cell surface, but receptors for thyroxine and steroidal hormones are located within the cell. [*See* CELL RECEPTORS.]

IV. Functions of Hormones

The endocrine system can be viewed as a messenger system that employs hormones to convey information from one cell, tissue, or organ to another. In the broadest sense, hormones coordinate and integrate the activities of the different organs and tissues of the body so that each serves the needs of the organism in a particular circumstance at a particular moment in time. From a physiological point of view, hormones regulate four major processes: reproduction, growth and development, maintenance of the internal environment, and regulation of nutrient utilization, energy storage, and energy metabolism.

A. Reproduction

In the case of reproduction, hormones not only regulate the functions of ovary and testes so that sperm and ova are produced at appropriate times and in appropriate numbers, but they act on a variety of other tissues to ensure the appropriate preparation of the uterus, for example, for implantation and successful survival of the fertilized ovum. Furthermore, in embryonic life the appropriate hormones appearing in the appropriate sequence are required for the development of the sexual organs.

B. Growth and Development

Endocrine control of somatic growth involves not only the production of growth hormone by the pituitary gland, but, as a consequence of growth hormone action, somatomedins (insulinlike growth factors) are produced in the liver. These growth factors act to stimulate the growth of bone, muscle, and other tissues. Appropriate concentrations of thyroid hormone, insulin, adrenal cortical hormones, and the sex hormones are also necessary for ordered skeletal growth and for the cessation of growth at an appropriate time. [*See* TISSUE REPAIR AND GROWTH FACTORS.]

In addition to this aspect of growth regulation, hormones are involved in the regulation of cell proliferation in many organs in the adult. Multiple, distinct tissue growth factors that regulate the proliferation of fibroblasts, epithelial cells, liver cells, and a variety of other cells have been identified. Commonly, for any one type of cell, several growth factors, acting on different intracellular signaling systems in the proper sequence, are necessary to induce a proliferative response. Abnormalities in these growth and proliferative responses can lead to the appearances of either benign or malignant tumors or, as in the case of blood vessels, to the altered proliferation of smooth muscle cells, which plays an important role in the pathogenesis of atherosclerosis. [*See* ATHEROSCLEROSIS.]

C. Maintenance of the Internal Environment

A general function of the endocrine system is that of maintaining the constancy of the internal environment. Our cells, whether in liver, muscle, kidney, nervous system, or bone, live in a highly stable but dynamic environment in which key constituents such as sodium, hydrogen, potassium, calcium, chloride, bicarbonate, phosphate, and magnesium ions must be maintained within very precise limits for these cells to function properly. These ionic constituents are constantly being lost in urine and/or skin and are being replenished from the diet. The organism must be able to coordinately regulate the function of intestine, bone, kidney, and other organs so as to efficiently store these dietary constituents at feeding time and to conserve them during fasting. The endocrine system orchestrates this coordinate behavior.

D. Nutrient Utilization and Energy Production

No function of hormones is of a more immediate and constant importance than that of regulating the efficient storage of ingested carbohydrates, fats, and amino acids into glycogen, fats, and proteins and of mobilizing these stored substances during periods of fasting or starvation. These substances are metabolized to tissue proteins, to other tissue constituents needed for cell growth and repair, or to CO_2 and H_2O with the generation of adenosine triphosphate (ATP), the common currency of cellular energy transactions. This ATP is employed to drive the work functions of the cell such as contraction, secretion, the synthesis of proteins and other complex molecules, and ion transport. [*See* ADENOSINE TRIPHOSPHATE (ATP).]

These different phases of energy metabolism are regulated by different hormones. Insulin, for example, can be considered the major energy storage hormone of the body. When its concentration increases, there is a more efficient uptake of glucose, amino acids, and fats into tissues such as liver, muscle, and adipose tissue. On the other hand, glucagon, epinephrine, and cortisol function to mobilize glucose directly from liver glycogen and indirectly from skeletal muscle glycogen, fatty acids from adipose tissue, and amino acids from muscle to supply adequate amounts of glucose and fatty acids for the use of other organs and tissues during periods of fasting or starvation.

V. Biosynthesis and Secretion of Hormones

Each chemically distinct class of hormones is synthesized by a different cellular route.

A. Peptides

The peptide and protein hormones are synthesized from amino acids via the classic DNA → RNA → protein pathway, and the finished product is packaged in secretory vesicles within the cell until an appropriate stimulus triggers its release (Fig. 5).

Release involves the fusion of the membranes of the vesicle with the surface (plasma) membrane of the cell, a process known as exocytosis (Fig. 5). A common feature of peptide hormone biosynthesis is that the initial protein product resulting from the translation of the specific messenger RNA is a larger molecule, a preprohormone, which then undergoes proteolytic modification to a prohormone and finally to the active hormone. For example, proinsulin is the initial product synthesized in the beta cell of the islets of Langerhans. After its synthesis, it is converted to proinsulin, which enters the immature secretory vesicle wherein it is converted by the action of specific proteases to insulin. In this active form, insulin is secreted into the bloodstream. In unusual cases, a secreted peptide is not biologically active until it is further processed in the bloodstream.

B. Steroids

In the case of the steroid hormones, the starting material is cholesterol. In the ovaries, testes, and adrenal, cholesterol is first converted to pregnenolone and then by successive modifications to the appropriate steroid hormone, estradiol or progesterone (ovary), testosterone (testes), and cortisol or aldosterone (adrenal). None of these are stored in secretory vesicles; rather, synthesis and secretion are concurrent (Fig. 6). [*See* CHOLESTEROL.]

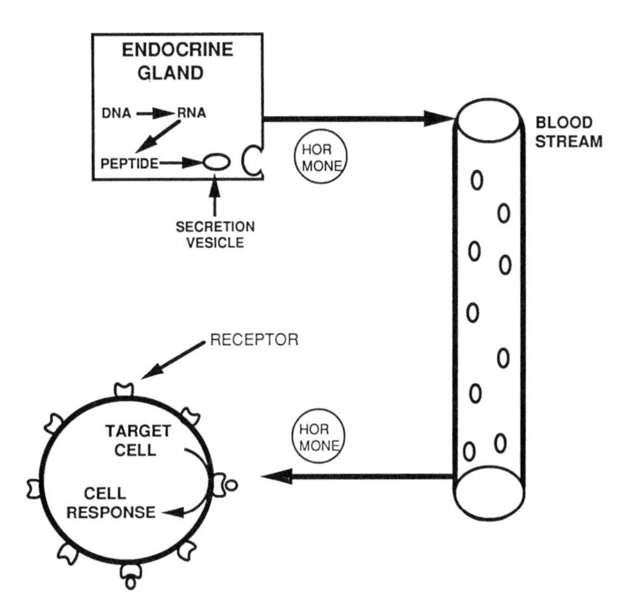

FIGURE 5 Secretion, transport, and cellular site of action of peptide hormones. The hormone is synthesized in the cells of the endocrine gland from amino acids. The product is stored in secretion vesicles, which on an appropriate stimulus fuse with the plasma membrane and thereby release their contents into the bloodstream. The hormone is transported in the blood, in the unbound state, to its target cell where it binds to specific receptors on the cell surface to initiate a cellular response.

The sterol hormone 1,25-dihydroxyvitamin D_3 is unique in that successive steps in its biosynthesis occur in different organs: cholesterol to 7-dehydrocortesterol to vitamin D_3 in the skin, vitamin D to 25-hydroxyvitamin D_3 in the liver, and 25-hydroxyvitamin D_3 to 1,25-dihydroxyvitamin D_3 in the kidney.

C. Amines

The synthesis of the amine hormones such as epinephrine involves the conversion of the amino acid tyrosine into the particular amines by a sequence of several steps. The finished product is stored in secretory vesicles; i.e., it is packaged and secreted like the peptide hormones.

The most unique pathway of biosynthesis and secretion is that found in the thyroid gland. This process takes place within acinar cells of the thyroid.

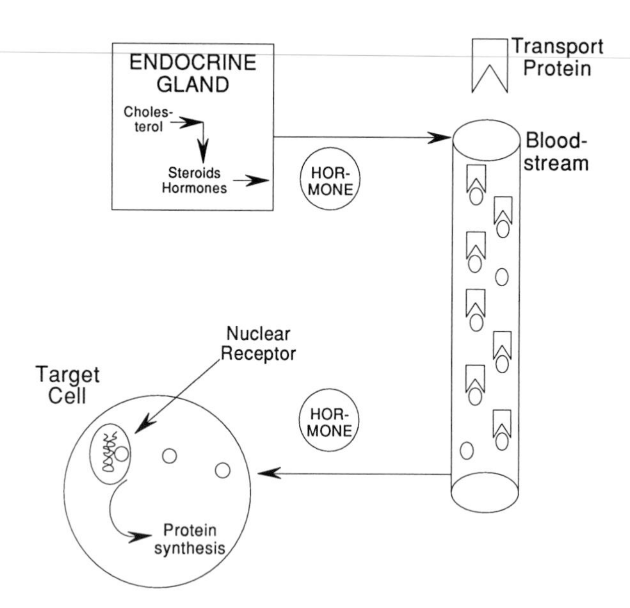

FIGURE 6 Secretion, transport, and site of action of steroid hormones. The hormone is synthesized from cholesterol and immediately secreted into the bloodstream where it combines to a specific transport protein. In this form, it is carried to the target cell where the hormone dissociates from the transport protein, enters the cell, and binds to an intracellular receptor, which in turn associates with sites on DNA within the nucleus.

These cells form into small spherical clusters within which an internal, extracellular space, the acinus, exists. The initial sequence of events involves the biosynthesis of a specific protein, thyroglobulin in the acinar cell. This is packaged in secretory vesicles and secreted into the acinar lumen. At the time of its secretion, the thyroglobulin is iodinated on a specific tyrosine residue by a specific peroxidase. Closely aligned iodinated tyrosine residues in the protein molecule are then coupled to form peptide-bound tri- and tetraiodothyronines; i.e., the hormone precursor is a large protein molecule in whose structure the small hormone (composed of two modified tyrosine residues) is contained. Upon stimulation by TSH, the acinar cells send out pseudopodia from their luminal surfaces, which engulf and internalize the iodinated thyroglobulin. This intracellular thyroglobulin undergoes breakdown to release the free thyroid hormones thyroxine and triiodothyronine from the other (basal) side of the cell into the bloodstream.

VI. Transport to Target Cells

The peptide and amine hormones are readily soluble in bodily fluids and, in general, are transported in the blood plasma in unbound forms (Fig. 5). An exception to this generalization are the somatomedins or insulinlike growth factors, which are transported, in large part, bound to specific carrier proteins.

The steroidal hormones and the thyronines are not readily soluble in water and, hence, are transported from site of synthesis to target cells by specific carrier proteins in the bloodstream (Fig. 6). These carrier proteins are of two general types: albumin and prealbumins, which have the ability to bind a large variety of small molecules with moderate affinity and low specificity, and specific carrier proteins (e.g., thyronine-binding globulin (TBG) or cortisol-binding globulin (CBG), which have high affinity and specificity for the particular hormone). These proteins bind their respective hormones tightly so that when a mixture of hormone and binding protein exists, as is the case in the blood plasma, a small amount of unbound (free) hormone and a large amount of bound hormone exist in equilibrium with each other.

In general, the binding capacity (a measure of the plasma content) of the carrier protein is considerably greater than the circulating concentration of the steroid hormone or thyronine. Hence only a very small amount of the hormone exists in the unbound state. It is this free (or unbound) form of the hormone to which the target cell responds. Furthermore, the content of a given carrier protein in the plasma can change independent of the content of the hormone that is transported.

VII. Measurement of Hormone Concentrations

Measurements of the concentrations of the various hormones in blood plasma or serum and/or in the urine is an extremely important means by which the endocrine physiologist studies the behavior of endocrine systems and by which the endocrine physician (endocrinologist) studies disorders of endocrine gland function.

In the case of catecholamines and steroid hormones, chemical methods for measuring their contents or those of their breakdown products in urine have been developed. However, such chemical methods have not been of great utility in the measurement of hormone concentrations in small samples of blood plasma or serum. This is particularly true of peptide hormones, because their low concentrations combined with their common chemical

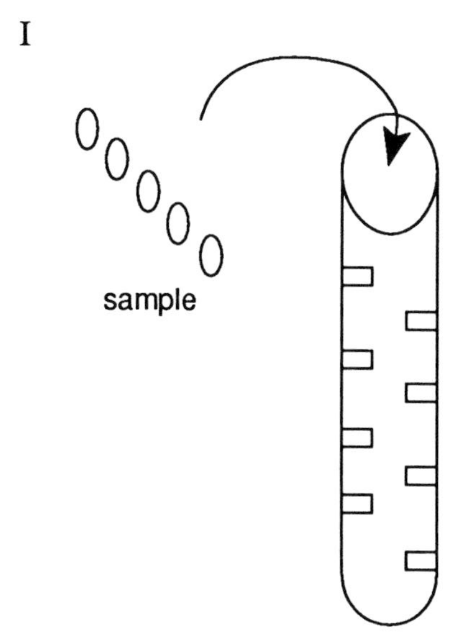

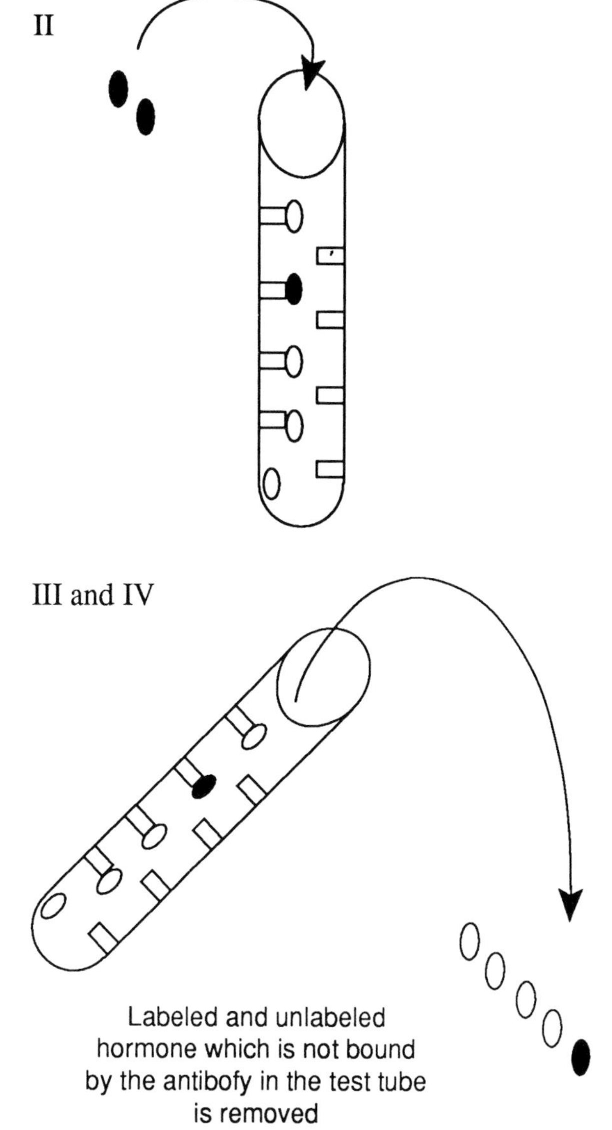

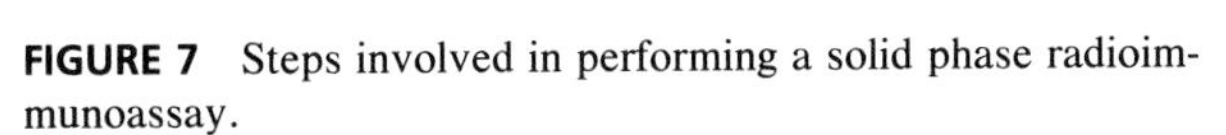

FIGURE 7 Steps involved in performing a solid phase radioimmunoassay.

I. Add blood sample with unknown amount of hormone to antibody coated tube.
II. Add small amount of radioactively labeled hormone to test tube containing sample.
III. Test tube decanted and amount of radioactivity left in the test tube is determined.
IV. Amount of radioactively labeled hormone left in the test tube is proportional to the amount of unlabeled hormone contained in the unknown sample. This is determined by comparing it to the displacement of known amounts of unlabeled hormone (Standard curve).

features has made a direct chemical approach nearly impossible. However, by using a different method known as radioimmunoassay (Fig. 7), it has been possible to measure, with a high degree of accuracy, the plasma concentrations of all the peptide hormones, the steroid and sterol hormones, the thyronines, and many other substances including a variety of commonly used drugs.

The method of radioimmunoassay depends on three reagents: (1) a specific, high-affinity antibody against the specific hormone or drug; (2) the production of a radioactively labeled hormone (in the case of peptide hormones, this is commonly achieved by iodinating specific amino acid residue [e.g., tyrosine]); and (3) after allowing antibody and labeled and unlabeled hormone to interact, a method for separating antibody-bound from free labeled hormone. The basic principle of the assay is quite simple: Standard amounts of labeled hormone and antibody are added to a series of tubes containing either standard amounts of unlabeled hormone or the hormone contained in a particular plasma or serum sample. The more unlabeled hormone (standard or unknown) the less labeled hormone becomes bound by the antibody, because the two compete for the limited number of binding sites on the antibody. By developing a standard curve from the analysis of standard amounts of unlabeled hormone, it is possible to define where on this curve the unknown sample falls and, therefore, the concentration of hormone in that sample. [*See* RADIOIMMUNOASSAYS.]

VIII. Cellular Effect of Hormones

At the cellular level, hormones, after combining with specific receptors, regulate the secretion of other hormones, control the transport of nutrients and ions across cell membranes, control the transcellular transport of nutrients and ions and across epithelial cells, regulate growth and proliferation, control metabolic rate, regulate the metabolism of sugar, fatty acids, and amino acids, and regulate the contraction and relaxation of smooth muscles.

In the case of peptide and amine hormone, hormone receptor interaction occurs at the cell surface. The interaction leads not only to an activation of some membrane transducer but also leads to the internalization of the hormone–receptor complex. Hence, the receptor population on the cell surface is constantly changing: they are lost by internalization and added by new synthesis or recycling. Chronic high extracellular concentrations of hormone lead to a steady state decrease in the number of receptors, a process known as down regulation. Conversely, chronic low concentrations of extracellular hormone lead to an increase in the number of surface receptors for the particular hormone, a phenomenon known as up regulation.

In addition to the effects of the homologous hormone on surface receptor number, there are many circumstances where exposure of cells to heterologous hormones, even of a different class, can alter the number of surface receptors for a particular hormone. Thus, an important way hormones exert their cellular effects is to change the expression of receptors for other hormones on their surface, thereby making them either more or less sensitive to the effects of these other hormones.

In any specific cell type, a given hormone usually exerts multiple effects, all of which are components of the integrated response of the cell. For example, when ACTH stimulates the synthesis of cortisol from cholesterol, the hormone stimulates cholesterol synthesis and the hydrolysis of stored cholesterol esters to ensure a sufficient supply of the precursor cholesterol; it also enhances the rate of production of ATP and nicotinamide adenine dinucleotide phosphate, components needed in the synthetic process, and it stimulates the transport of cholesterol from the cytosol into the mitochondria—the site of initial transformation of cholesterol occurs.

IX. Mechanisms of Action

In spite of the diversity of effects of hormones on target cells, there are a few common mechanisms by which hormones act. The first distinction is between steroidal hormones and thyronines on the one hand, and peptide and amine hormones on the other.

A. Steroids and Thyronines

Steroid hormones, 1,25-dihydroxyvitamin D_3, and thyronines exert their actions by combining with specific intracellular receptors in the cell (Fig. 8).

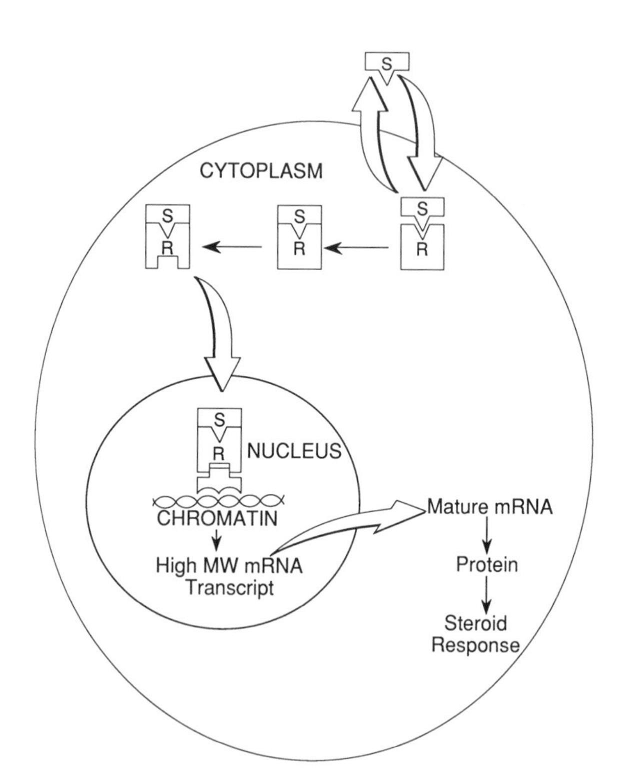

FIGURE 8 Mechanism of steroid hormone action. The steroid hormone (S) enters the cell by diffusion and binds with a cytoplasmic receptor (R). As a consequence, the receptor undergoes a conformational change, is taken up by the nucleus, and binds to specific DNA sequences, regulatory elements of specific genes. This results in transcription of the gene, and a high-molecular weight messenger RNA (MW mRNA) is produced, which is processed within the nucleus to a mature, lower-molecular weight mRNA. This is transported to the cytoplasm, where it is translated by ribosomes or polyribosomes into a protein. The increase in the content of the protein(s) leads to altered cell behavior.

These may be located either in the cytoplasm or the nucleus. When the hormone reaches the cell, it diffuses across the plasma membrane and binds to the receptor. The binding of the steroid (or thyronine) to the receptor leads to a change in its conformation so that it now binds to specific, high-affinity sites on the DNA in the nucleus. Within the nucleus, the steroid–receptor complex binds to specific regulatory elements with unique DNA sequences. Binding of the complex to the regulatory element initiates the synthesis of specific messenger RNAs, which in turn stimulate the synthesis of specific proteins. These proteins, in turn, bring about changes in cell function and/or cell growth.

B. Peptides and Amines

Peptide, protein, and amine hormones interact with receptors located on the cell surface. Interaction of hormone with receptors initiates events within the plasma membrane that lead to the activation of specific membrane transducers; i.e., they transduce the original extracellular message into an intracellular message or messages.

Five types of intracellular messengers have been recognized: cyclic nucleotides, Ca^{2+}, inositol phosphates, certain lipids, and phosphoproteins.

1. The Cyclic AMP Messenger System

Two major cyclic nucleotides have been identified: cyclic 3′,5′-adenosine monophosphate (cAMP) and cyclic 3′,5′-guanosine monophosphate (cGMP). Each is synthesized by a specific enzyme, adenylate and guanylate cyclase, respectively; each is degraded by one or more phosphodiesterases to their respective products, 5′AMP and 5′GMP; and each appears to act mainly by controlling the activity of a particular class of enzymes: cAMP-dependent and cGMP-dependent protein kinases. Both cAMP and cGMP have actions in multiple tissues, and their respective cyclases are activated by different hormonal receptors in different tissues. In addition, much more is known about the role of the cAMP in regulating cell function; its properties will be discussed in terms of the second messenger model of hormone action (Fig. 9).

Adenylate cyclase is associated with the plasma membrane. Interaction of the hormone with its receptor on the plasma membrane leads to an activation of the adenylate cyclase, which then catalyzes the formation of cAMP from ATP. The cAMP is released into the intracellular fluids where it serves as a second or intracellular messenger. The cAMP binds to a specific cAMP receptor protein, which is the regulatory subunit (R) of an enzyme, cAMP-dependent protein kinase, formed by this subunit bound to the catalytic (C) subunit. When cAMP associates with R, R and C dissociate. The released free C is active in catalyzing the combination of phosphates to a number of protein substrates on either serine or threonine residues, using ATP energy. The phosphoproteins thus formed are different in conformation and, hence, the catalytic efficiency from the original proteins. Classic examples of known proteins that become phosphorylated by cAMP-dependent protein kinase are phosphorylase kinase and glycogen synthase, key enzymes in the degradation and synthesis of glycogen, respectively. Phosphorylation of phosphorylase kinase increases its activity, whereas that of glycogen synthase decreases its activity. [*See* PROTEIN PHOSPHORYLATION.]

Many peptide and amine hormones act by either stimulating or inhibiting adenylate cyclase. Examples of stimulating hormones are epinephrine, glucagon, serotonin, PTH, calcitonin, TSH, ACTH, FSH, gastric inhibitory peptide, secretin, vasoactive intestinal peptide, and some growth factors. Examples of inhibitory hormones are epinephrine (acting via a different class of receptors), adenosine, angiotensin II, somatostatin, and dopamine.

The hormones that regulate adenylate cyclase act rapidly, and once the hormonal signal is terminated, the response usually ceases rapidly because a class of enzymes known as phosphodiesterases rapidly break down cAMP to 5′AMP, and another class, known as phosphoprotein phosphatases, dephosphorylate phosphoproteins. The reversible phosphorylation and dephosphorylation of cellular proteins is a nearly universal way in which peptide and amine hormones regulate cellular function.

2. Ca^{2+} Messenger System

Along with cAMP, Ca^{2+} ion is a nearly universal second messenger in hormone action. The generation of the Ca^{2+} messenger can occur by several different mechanisms, and its intracellular actions are more diverse. The concentration of Ca^{2+} in the extracellular fluid is 10^{-3} *M* and that within the cell cytosol is 10^{-7} *M*. In addition, Ca^{2+} is stored within the cell in a specialized organelle called the calcisome in nonmuscle cells and the sarcoplasmic reticulum in skeletal and cardiac muscle. An increase in Ca^{2+} concentration within the cell can occur as a

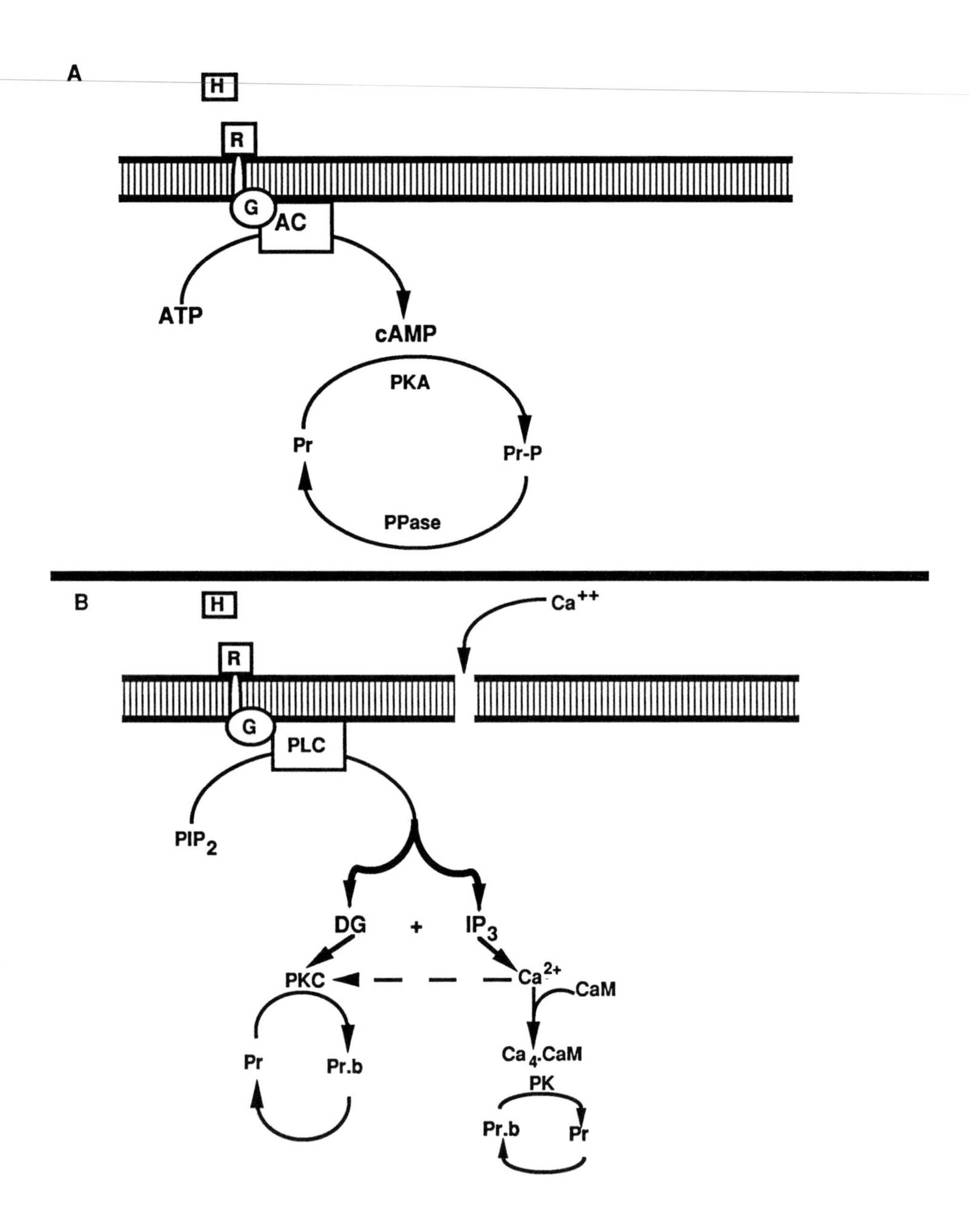

FIGURE 9 The cAMP and Ca^{2+} messenger systems. A. Interaction of hormone (H) with receptor (R) on the plasma membrane activates the enzyme adenylate cyclase (AC) via a specific type of membrane protein, the guanine nucleotide-binding protein (G). As a result, cAMP is produced from ATP. cAMP then binds to and activates a specific class of enzymes, the cAMP-dependent protein kinases (PKA), thereby stimulating the phosphorylation of a group of cellular proteins (Pr ⇒ Pr-P). Additionally, cAMP acts to inhibit another class of enzymes, the phosphoprotein phosphatases (PPases), which dephosphorylate these proteins. B. Interaction of hormone (H) with receptor (R) acting via a "G" protein stimulates the activity of phophoinositide-specific phopholipase C (PLC), causing the hydrolysis of phosphatidylinositol-4,5-bisphosphate (PIP_2) to give rise to diacylglycerol (DG) and inositol-1,4,5-trisphosphate (IP_3). The IP_3 causes the release of Ca^{2+} from an intracellular pool. In addition, hormone–receptor interaction leads to an increase in Ca^{2+} influx across the plasma membrane (hatched bars). These Ca^{2+} signals stimulate the activity of calmodulin (CaM)-dependent protein kinases, and along with DG the activity of protein kinase C (PKC) to catalyze the phosphorylation of cellular proteins. The extent of phosphorylation of these proteins is also determined by the activities of phosphoprotein phosphatases (PPases).

result of either an increase in Ca^{2+} entry from the extracellular fluids across the plasma membrane via specific Ca^{2+} channels in this membrane or by the release of Ca^{2+} from the calcisomes. In some cases, a hormone interacting with its surface receptor will cause only an increase in Ca^{2+} influx, but, more commonly, it causes both an increase in influx and a release of Ca^{2+} from the calcisome. In these latter cases (Fig. 9), hormone–receptor interaction almost always leads to the activation of the plasma membrane-associated enzyme phospholipase C, which catalyzes the hydrolysis of a specific membrane lipid, phosphatidylinositol-4,5-bisphosphate, resulting in the production of two messengers: water-soluble inositol-1,4,5-trisphosphate and membrane-soluble diacylglycerol. The inositol 1,4,5-trisphosphate acts as a signal to cause the release of Ca^{2+} from the calcisomes. The diacylglycerol causes the soluble, Ca^{2+}-insensitive form of a specific protein kinase, protein kinase C, to associate with the plasma membrane and in this way become Ca^{2+}-sensitive.

A change in intracellular Ca^{2+} concentration serves as a messenger by binding to specific Ca^{2+}-receptor proteins, or Ca^{2+}-activated enzymes. Binding of Ca^{2+} to a receptor protein leads to a change in the conformation of the protein, which then interacts with other proteins (enzymes), thereby altering their activity. In many cases, these enzymes are protein kinases. Hence, just as with cAMP, a common consequence of messenger Ca^{2+} is the activation of protein kinases.

Common Ca^{2+} receptor proteins are calmodulin and troponin C. Protein kinases activated by Ca^{2+}-calmodulin include myosin light-chain kinase, phosphorylase kinase, multifunctional protein kinase, and protein kinase C.

The specific receptors of many peptide and amine hormones, tissue growth factors, and ecosanoids are linked to the phosphoinositide-Ca^{2+} messenger system. The hormones include epinephrine, angiotensin II, bombesin, cholecystokinin, serotonin, certain prostaglandins, and thromboxanes.

3. Interactions between Ca^{2+} and cAMP

Although the two major messenger systems are presented as separate pathways for cell activation, they nearly always function together. They often regulate the same cellular response or components of a particular response. Hence, cAMP and Ca^{2+} serve as nearly universal, synarchic messengers in the activation of processes such as glycogen synthesis and breakdown, smooth muscle contraction and relaxation, aldosterone secretion, insulin secretion, the secretion of a wide variety of other peptides, amine, and steroid hormones, and in the growth and/or differentiation of many cell types; however, their relationship is not stereotyped. They can function in a coordinate, hierarchical, redundant, antagonistic, or a sequential relationship to modulate a specific cellular response.

4. Tyrosine Kinase-Linked Systems

A common consequence of peptide hormone–receptor interaction, for another class of peptide hormones, is the activation of a specific type of protein kinase, which adds a phosphate residue to tyrosine in proteins. In contrast, the cAMP- and Ca^{2+}-calmodulin-dependent protein kinases as well as protein kinase C catalyze the phosphorylation of either serine and/or threonine residues on their substrate proteins. The insulin receptor, and those of many other growth factors, possesses as an intrinsic part of its structure a protein kinase activity that catalyzes the phosphorylation of tyrosine residues in the protein substrate. In the case of many of these receptors, one of the substrates is the receptor itself; i.e., interaction of insulin with its receptor leads to the autophosphorylation of a tyrosine residue on the receptor protein. The unresolved issue is how this event gets transduced into insulin-mediated effects such as an increase in glucose transport across the cell membrane or an increase in the rate of protein synthesis. The best available evidence suggests that the insulin receptor is the first component of a protein kinase cascade that eventually leads to the activation of serine- and threonine-specific protein kinases and, hence, to the phosphorylation of a number of intracellular proteins. Thus, phosphoproteins themselves become the second messengers.

X. Disorders of Endocrine Function

Disorders of endocrine function result from several different abnormalities: underproduction or overproduction of normal hormones, the secretion of abnormal hormones, to resistance to the actions of hormones (Fig. 10). Because of the complex interrelationships between the functioning of the various components of the endocrine system, there is often overproduction of one hormone when another hormone is underproduced. Likewise, hormone overproduction may be a consequence of hormone resistance in some cases and the cause of hormone resistance in others.

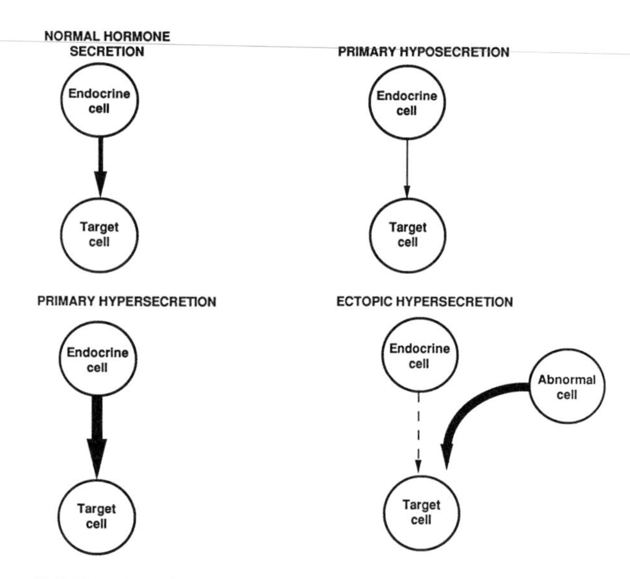

FIGURE 10 Disordered patterns of hormone secretion.

A. Abnormal Secretion

Hypersecretion of a hormone can be either primary or secondary. Primary hypersecretion is due to a disordered secretory system in the cells of origin in the hormone (e.g., primary hyperparathyroidism is a primary disorder of the synthesis and secretion of PTH from the parathyroid gland). Primary hyperparathyroidism can result from either an adenoma (benign tumor) or primary hyperplasia, and, in rare cases, from a carcinoma. In the case of the adrenal gland, primary overproduction can also result either from hyperplasia, adenoma, or carcinoma. In the majority of cases of primary hypersecretion, feedback control, although not completely normal, is still operative, but at an abnormal set point.

Secondary hypersecretion results either from an overstimulation by a normal tropic factor or by an antibody directed against and capable of activating the receptor for a tropic factor. Thus, secondary hyperparathyroidism occurs if there is a sustained decrease in the plasma calcium concentration, and secondary hyperthyroidism can result from an autoimmune production of antibodies against the surface receptor protein for TSH. These antibodies are capable of acting like TSH to stimulate the thyroid cells to produce thyroxine and triiodothyronine.

Primary and secondary hyposecretion are caused by underfunction of the primary gland or of a chronic state of understimulation by a tropic factor, respectively. Primary adrenal hyposecretion, or Addison's disease, is due to a failure of the adrenal cortical cells to secrete cortisol and aldosterone because of a disease of these cells, which may be due to infection, tumor, or autoimmune destruction of the tissue. Additionally, there can be either congenital absence of a particular gland or a normal-appearing gland in which one or more steps in the biosynthetic sequence is abnormal. Secondary hyposecretion results if there is a failure to secrete sufficient tropic hormone.

B. Ectopic Hormones

Ectopic hormone production (i.e., by cells that usually do not produce it) usually involves peptide or protein hormones and not steroid hormones, because in the case of the peptide, activation of a single gene is sufficient to cause the problem; however, in the case of steroid hormones, a coordinated activation of the multiple genes coding for the enzymes involved in steroid biosynthesis would be necessary. The only exception is the ectopic production of 1,25-dihydroxyvitamin D_3. This is possible because the conversion of 25-hydroxyvitamin D_3 to 1,25-dihydroxyvitamin D_3 in the kidney involves a single enzymatic step. Hence, activation of a single gene is sufficient for ectopic production and occurs in the macrophages in certain granulomatous diseases such as sarcoidosis and tuberculosis, as well as in cells of certain lymphomas.

Ectopic peptide hormones are produced by two mechanisms: either the activation of the gene for a particular hormone in the abnormal tissue or the activation of a gene that codes for a protein of similar but not identical structure.

An example of the first type of ectopic hormone production is ACTH, which is often produced by bronchogenic carcinoma cells. However, although the same gene is activated in the tumor cells as is the pituitary, the form of circulating hormone differs because the messenger RNA for the gene encodes a much larger ACTH itself, which has a different fate in the two cells. Normally, this long polypeptide undergoes considerable processing by which it is reduced to the ACTH secreted into the circulation. However, when the gene is activated ectopically, the processing is often incomplete so that a considerable portion of the circulating substance is of larger molecular weight. This so-called big ACTH is often less active biologically than ACTH, but even so may cause hyperactivity due to the large amount produced.

An example of the second type of ectopic hormone production is that of the secretion of a hormone, which bears with structural homology with PTH, but clearly produced by a different gene by squamous cell tumors, particularly of the head and neck. Owing to its structural homology with PTH, this gene product binds to activates the PTH receptor, producing a condition known as hypercalcemia of malignancy.

C. Resistance to Hormone Action

Resistance to the action of a hormone may involve either a deficiency or a structural abnormality in the receptor. In some cases, it is caused by alteration in the events that occur after hormone–receptor binding. Resistance can be caused by an inherited gene defect. This is the case for testosterone resistance in the condition known as testicular feminization, for resistance to the action of 1,25-dihydroxyvitamin D_3, or for some inherited conditions of resistance to insulin action. On the other hand, resistance is often acquired under conditions where constantly high concentrations of circulating hormone exist. An example is type II diabetes, in which a resistance to the actions of insulin exist because chronic high plasma concentrations of insulin induce a decrease in the number of insulin receptors on the cell surface of target tissues (downregulation).

Bibliography

Martin, C. R. (1985). "Endocrine Physiology." Oxford University Press, New York.

Rasmussen, H. (1981). "Calcium and cAMP as Synarchic Messengers." John Wiley and Sons, New York.

Tepperman, J., and Tepperman, H. M. (1987). "Metabolic and Endocrine Physiology." Yearbook Medical Publishers, Chicago.

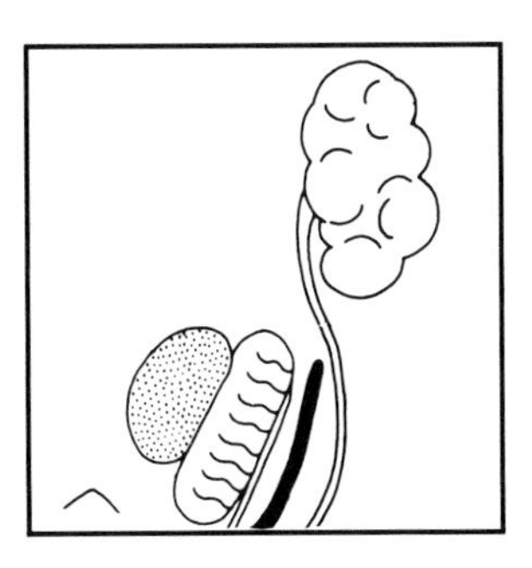

Endocrinology, Developmental

PIERRE C. SIZONENKO, *University of Geneva Medical School*

Glossary

Perinatal period First 28 days of life
Prepuberty From the second to the tenth year of life
Puberty Period during which the functions of reproduction are developing

ENDOCRINOLOGY in the pediatric age group differs from the adult, as it deals with a growing and maturing organism. The concept of growth should be associated with the maturation of the organs and, in particular, that of the endocrine glands. Each gland or organ has its own developmental patterns, scheduled to occur sharply at a preset time. Biological events are to be interpreted in view of embryology (morphogenesis and differentiation), chronology of events affecting growth, and maturation of organs and functions (progressive gene expression, synthesis of proteins, activity of enzymes). Growth and maturation depend on numerous hormones acting through endocrine mechanisms and growth factors, whose action is paracrine and/or autocrine.

Four developmental periods are described in the growing individual before adulthood is reached: fetal, perinatal, prepubertal, and pubertal.

I. Prenatal Endocrinology

The understanding of prenatal endocrinology in humans remains in a primitive stage compared with that of several animal models such as the rat, rabbit, sheep, and monkey.

Growth and maturation of the fetus depend on maternal, placental, and fetal hormonal factors. Frequently, the origin of some of these factors is impossible to trace precisely, because many hormones or factors can pass through the placenta or even undergo transformation within the placenta. In addition, the placenta secretes several hormones and factors. Therefore, the role of the placenta has rendered the studies of prenatal endocrinology particularly difficult in the human.

A. Hypothalamus and Posterior Pituitary

The differentiation of the diencephalon of the embryo occurs around 34 days postconception and the posterior pituitary primordium appears toward the 49th day. The capillaries of the hypothalamo-pituitary portal system develop between 60 and 100 days. At 21 wk gestation, the hypothalamo–pituitary portal system is complete and active. The transmission of neurosecretory material into the hypophyseal portal system with the concurrent development of the hypothalamus occurs between 14 and 18 wk gestation. At this time, the hypothalamic nuclei and fibers of the supraoptic tract appear with further differentiation of the pars tuberalis and the median eminence. Three types of neurohormones or factors have been identified very early in the fetal hypothalamus: (1) releasing or inhibiting factors or hormones, which are peptides; (2) aminergic neurotransmitters such as dopamine, norepinephrine, and serotonin; and (3) oxytocin and vasopressin nonapepides, which are synthesized in the paraventricular and supraoptic nuclei of the hypothalamus and transported by neurons to the posterior pituitary. [*See* HYPOTHALAMUS; PEPTIDES; PITUITARY.]

1. Neurotransmitters

Neurotransmitters as detected by monoamine fluorescence were observed in the hypothalamus as early as the 10th week of fetal life and in the median eminence at the 13th week. Nothing is presently known of the metabolism of monoamines in the human fetal hypothalamus.

2. Oxytocin and Vasopressin

Grains of secretion of oxytocin have been detected in the hypothalamus of 19-wk-old fetuses and in the posterior pituitary at 23 wk. At time of labor and delivery, the human fetus secretes important amounts of oxytocin. Oxytocin does not cross the placenta. Before birth, the high levels of oxytocin observed in the umbilical artery could play a role in the onset of labor. This is suggested also by the observation that anencephalic fetuses without hypothalamus usually cause prolonged pregnancy.

Vasopressin or antidiuretic hormone is found in the hypothalamus at 10–15 wk gestation and after 19–23 wk in the posterior pituitary.

B. The Anterior Pituitary and Target Organs

The capacity of the fetal anterior pituitary gland to synthesize and to store protein hormones is present during the first trimester of pregnancy.

1. Growth Hormone-Releasing Hormone, Growth Hormone-Release-Inhibiting Factor, Growth Hormone, and Growth Factors

Growth hormone secretion by the pituitary is under the control of two hypothalamic factors, the growth hormone-releasing hormone or factor and the growth hormone-release-inhibiting factor or hormone or somatostatin.

a. Growth Hormone-Releasing Hormone Growth hormone-releasing hormone neurones have been detected in the arcuate nucleus of the human fetus from the 28th week of development. Concentration of growth hormone-releasing hormone is observed in the median eminence neurons from the 31st week. These neurons retain an immature morphology until birth, suggesting that the control of the growth hormone secretion by growth hormone-releasing hormone is only effective at a late stage of gestation or during the perinatal period in humans.

b. Growth Hormone-Release-Inhibiting Hormone Growth hormone-release-inhibiting hormone has been found widely distributed throughout the central nervous system and in other organs such as the pancreas and stomach. Growth hormone-release-inhibiting hormone is observed in the hypothalamus at 20 wk pregnancy. There is a positive correlation between growth hormone-release-inhibiting hormone concentration in the hypothalamus and gestational age. Injection of growth hormone-release-inhibiting hormone in sheep fetuses *in utero* induces a decrease of plasma growth hormone at any time of gestation.

c. Growth Hormone and Growth Factors

Growth Hormone Pituitary acidophilic cells containing growth hormone have been detected by the 9th week of fetal life. Its content in growth hormone increases throughout pregnancy. In fetal serum, concentrations of growth hormone increase until the 20th–24th week. Serum growth hormone decreases thereafter until the end of gestation, but the levels observed are higher than those of the normal child and adult. Maternal or fetal growth hormone does not cross the placenta. Fetal growth hormone secretion can be stimulated *in vivo* by stress such as anoxia or acidosis at midgestation or at delivery time. The role of growth hormone in fetal growth has been challenged by experiments of nature such as anencephalic or apituitary fetuses and by children with idiopathic hypotuitarism who usually have normal birth length. Available evidence suggests that neither maternal nor fetal growth hormone is essential. Recently, a placental growth hormone has been isolated; its role remains unknown. The syncytiotrophoblastic cells of the placenta also secrete the human chorionic somatotrophic hormone, which has a considerable homology with pituitary growth hormone. This placental hormone, which circulates mainly in the maternal compartment, may play a major metabolic role in sparing energy in the mother, hence providing glucose and proteins necessary to the growth of the fetus.

Growth Factors Growth factors in the fetus are essentially *Insulin-like growth factor I* (IGF I), or *somatomedin C*, and *Insulin-like growth factor II* (IGF II). Possible fetal variants of IGF I and IGF II also have been described. Both factors stimulate multiplication of fetal cells and probably depend on the placental lactogenic hormone secreted by the trophoblastic cells, insulin, and nutritional factors such as glucose. Growth hormone probably plays a minor role. IGF II is mainly secreted in the human fetus and is thought to be involved in the growth of the brain. Insulin, which is structurally similar to IGF I and IGF II, has also been postulated as a growth factor in the fetus. The exact role of factors such as insulin or growth factors on fetal skeletal growth still remains uncertain. In addition, several other growth factors such as *epidermal growth factor, fibroblast growth factor*, and *nerve growth factor* have specific actions on fetal tissues either alone, in synergism (between them), or in association with thyroid hormones, androgens, and IGF I and IGF II. [*See* INSULIN-LIKE GROWTH FACTORS AND FETAL GROWTH; TISSUE REPAIR AND GROWTH FACTORS.]

2. Prolactin

Prolactin secretion is controlled by a hypothalamic inhibiting factor, the nature of which is still uncertain (the neurotransmitter dopamine or a gonadotropin-releasing hormone-associated protein named GAP). Presence of prolactin has been observed in the pituitary gland at 68 days gestation. Both prolactin pituitary content and serum concentration increase throughout gestation. The patterns of secretion of prolactin during pregnancy show a similar increase in the fetus and the mother, with a maximum occurring during late gestation and at term, in direct correlation with the increase of circulating estrogens during pregnancy. No placental transfer of prolactin has been reported. A role for fetal prolactin in water transport across the amniotic membranes to maintain a normal volume and composition of the amniotic fluid has been postulated.

3. The Hypothalamo–Pituitary–Thyroid Axis

The thyroid gland is controlled by thyrotropin, secreted by the pituitary. Thyrotropin itself mainly depends on a hypothalamic factor named thyrotropin-releasing hormone (TRH). [*See* THYROID GLAND AND ITS HORMONES.]

a. TRH TRH has been detected in human fetal brain 4–5 wk postconception. Its concentration increases progressively during gestation. Of the TRH present in the central nervous system, 20–30% is concentrated in the hypothalamus. Of the total brain TRH, 5% is found in the pituitary gland. Placenta and amniotic fluid contain important amounts of TRH. The role of placental TRH on the fetal pituitary–thyroid axis is yet unknown.

b. Thyrotropin Thyrotropin or thyrostimulating hormone has been observed in the pituitary and in the plasma of human fetuses as early as the 11th–12th wk, coincidental with the onset of iodine uptake by the fetal thyroid gland and the synthesis of iodothyronines. Serum thyrotropin concentration increases at midgestation, probably in relation to an increased secretion or an effect of thyrotropin. Thyrotropin does not cross the placenta.

c. The thyroid gland *Ontogenesis of the Thyroid Gland* Intracellular colloid formation with synthesis of thyroid hormones appears between 73 and 80 days gestation. Follicular structures are present from 80 days. Growth of the thyroid gland does not seem dependent on fetal thyrotropin. The synthesis of thyroglobulin occurs by the 29th day of rotropin. [*See* THYROID GLAND AND ITS HORMONES.]

Thyroid Hormone Secretion and Metabolism Thyroxine (T_4) was observed at 78 days in the human fetus plasma. Serum total and free thyroxine as well as total and free triiodothyronine (T_3) increase rapidly during gestation; however, the levels of total and free triiodothyronine are lower in late gestation than in normal children. There is a relative triiodothyronine deficiency in the human fetus with high concentrations of reverse triiodothyronine. Thyroxine, triiodothyronine, and reverse triiodothyronine are present in the amniotic fluid. There is no correlation between fetal serum thyroid hormones and amniotic fluid levels. Consequently, this observation does not allow for prenatal diagnosis of congenital hypothyroidism by amniotic fluid sampling.

By the 12th week of gestation, the presence of specific thyroxine-binding globulin and prealbumin is observed. Liver synthesis of thyroxine-binding globulin depends on maternal estrogens. Thyroxine binds preferentially to thyroxine-binding globulin rather than to prealbumin.

Placental Transfer of Thyroid Hormones In humans, placental transfer of thyroid hormones is limited. Less than 1% of a large dose of T_4 given to mothers in labor is transferred to the fetal circulation. Large doses of T_3 administered to women near term slightly decrease fetal serum T_4 concentration. No correlation exists between maternal and fetal serum concentrations of total and free T_3, T_4, or thyrotropin. There is a materno-to-fetal gradient of T_3 and free T_3 during pregnancy. A materno-to-fetal gradient of total and free T_4 is observed before 20 wk gestation. Near term, this gradient reverses and tends to favor the fetal compartment. Thus, because of the placental barrier, the fetal thyroid system develops and functions autonomously of the maternal system. The placenta is able to concentrate iodine in the fetus. Administration of iodine at high doses or of radioactive iodine to a pregnant woman is a danger for the very active fetal thyroid gland. Similarly, antithyroid drugs, which cross the placenta, can produce a goiter in the fetus. In maternal autoimmune thyrotoxicosis, the presence of thyrostimulating immunoglobulins, which can pass the placental barrier, may cause neonatal hyperthyroidism.

Role of Thyroid Hormones on Brain Development In spite of the numerous actions of thyroid hormones on metabolism and enzyme activities, they do not play a specific role in fetal somatic growth. However, they have a very important role in the developing nervous system of the fetus. Prenatal thyroidectomy in the monkey induces a decrease in the content of cerebral RNA and proteins. There is a decrease in the multiplication of neurons and glial cells both in the cerebrum and in the cerebellum. Thyroxine has a direct stimulatory effect on the maturation and the assembly of microtubules of the neuronal cells and on the development of synaptogenesis. Thyroid hormones increase the concentration of nerve growth factor in brain tissue.

4. Sexual Differentiation of the Fetus

Fetal sexual differentiation is an asymmetrical process, which consists in a series of events actively programmed at appropiate critical periods of fetal life and which leads to the sexual dimorphism observed at birth (Table I). Genetic factors and hormonal factors will alternate in this chain of programmed transformations of the gonads, the internal sex organs, and the external genitalia. *De facto*, the male genetic factors and the male hormones will orientate the fetus to maleness, femaleness resulting from the absence of any masculinizing genetic factor or hormone acting during the critical period of differentiation. Psychological sexual identity is acquired during postnatal life and is the result of psychosocial and hormonal imprinting.

a. Genetic Factors Sex genotype is determined at fertilization of the ovum by the spermatozoon. The sex chromosome Y will determine the genetic sex of the embryo, induce the male differentiation of the primordial gonad into a testis, and initiate a series of events that result in the development of the genital tract and the external genitalia. A small region of the Y chromosome called the sex-determining gene or testis-determining factor (Tdf) encodes a protein capable of activating the transcription of genes responsible for the differentiation of the testis. It has been hypothesized that the Y chromosome would also act through a glycoprotein named H-Y antigen. H-Y antigen is clearly a marker for the Y chromosome, but its absence does not necessarily imply the absence of Tdf as H-Y antigen maps on the long arm of the Y chromosome, a considerable distance from Tdf, which is on the short arm close to the pseudoautosomal region. The exact role for the H-Y antigen and its possible relationship to a spermatogenesis gene also located in this region remains unknown.

b. Gonadal Differentiation The gonadal primordium, which is common in both sexes, develops on the ventral surface of the mesonephros, where primordial germ cells migrate. Testicular tissue, in particular seminiferous tubes, is observed in the embryo at 43–49 days. Sertoli cells secrete a glycoprotein, the anti-Müllerian hormone, from the 7th week of fetal life. Leydig cells are found at 8 wk. Testosterone secretion by the Leydig cells starts at the same age and is maximal between 14 and 16 wk. This peak is concommitant with the peak of the placental secretion of human chorionic gonadotropin, suggesting that the secretion of testosterone is mainly under the control of human chorionic gonadotropin. Male anencephalic newborns who do not synthesize pituitary gonadotropins have normal male genitalia. The fetal testis migrates during gestation from the upper part of the abdomen to the inguinal canal. At birth, the testes are usually present in the scrotum.

Ovarian differentiation is a passive procedure. Ovarian organogenesis occurs at the 13th week and primordial follicles are observed from the 16th week. At 5 mo gestation, the ovary contains 7 million germinal cells. At birth, the number falls to 2

TABLE I Timing of Sexual Differentiation in the Human Fetus

Fetal age[a] (wk)	Crown–rump length (mm)	Sex-differentiating events
	blastocyst	Inactivation of one X chromosome
4	2–3	Development of Wolffian ducts
5	7	Migration of primordial germ cells in the indifferentiated gonad
6	10–15	Development of Müllerian ducts
7	13–20	Differentiation of seminiferous tubes
8	30	Regression of Müllerian ducts in the male fetus
8	32–35	Appearance of Leydig cells; first synthesis of testosterone
9	43	Total regression of Müllerian ducts; loss of sensitivity of Müllerian ducts in the female fetus
9	43	First meiotic prophase in ovogonia
10	43–45	Beginning of masculinization of external genitalia
10	50	Beginning of regression of Wolffian ducts in the female fetus
12	70	Fetal testis is in the internal inguinal ring
12–14	70–90	Male penile urethra is completed
14	90	Appearance of first spermatogonia
16	100	Appearance of first ovarian follicles
17	120	Numerous Leydig cells; peak of testosterone secretion
20	150	Regression of Leydig cells; diminished testosterone secretion
24	200	First multilayered ovarian follicles; canalization of the vagina
28	230	Cessation of ovogonia multiplication
28	230	Descent of testis

[a] Fetal age in weeks after the last menstrual period.

million. The fetal ovary is capable of secreting steroids from the 8th week.

c. Differentiation of the Genital Ducts Internal genital ducts are derived from the differentiation of the two pairs of ducts: the Wolffian ducts and the Müllerian ducts (Fig. 1). Experiments performed in gonadectomized male rabbit fetuses before the age of differentiation showed that Wolffian ducts degenerate and Müllerian ducts develop into tubes, the uterus, and the upper part of the vagina. In opposite experiments, locally implanted fetal testis induces in female fetuses regression of the Müllerian ducts and development of the Wolffian ducts. Local implants of testosterone induces development of the Wolffian ducts and no regression of the Müllerian ducts. These experiments lead to the concept of the anti-Müllerian hormone, which was later identified. Müllerian ducts are sensitive to anti-Müllerian hormone during a short period of fetal development, up to the 8th week. The Wolffian duct structures are also stabilized during a critical fetal period and apparently only by high local concentrations of testosterone.

d. Differentiation of the Urogenital Sinus and the External Genitalia In both sexes, the urogenital sinus and the external genitalia are similar up to the 9th week. Masculinization begins by a lengthening of the anogenital distance, followed by a fusion of the labioscrotal swellings in the midline, forming the scrotum, and of the rims of the urethral folds leading to the formation of the primordium of the male urethra (Fig. 2). Penile organogenesis is completed by 12–14 wk. The male differentiation of the external genitalia entirely depends on the secretion

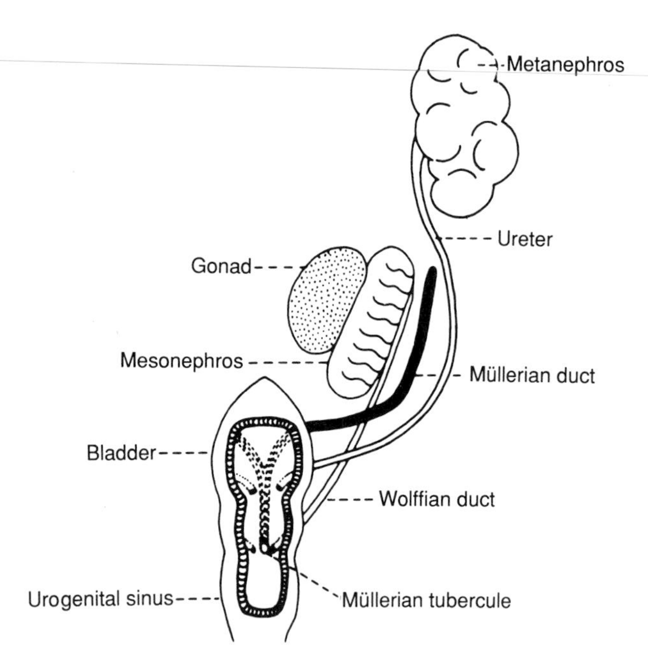

FIGURE 1 Undifferentiated stage of sex differentiation, with presence of fetal structures: Müllerian and Wolffian ducts and the urogenital sinus with the genital tubercle.

of testosterone by the fetal testis. In the female fetus exposed to androgens, different stages of the fusion of the labioscrotal swellings can be observed. Two main types of enzymatic abnormalities of testosterone secretion are known in the human: (1) excessive androgen secretion by the adrenal gland, which will cause virilization of female fetuses, as in congenital virilizing adrenal hyperplasia with female pseudohermaphroditism; and (2) enzymatic biosynthetic defects affecting either testosterone formation or the end-organ sensitivity to testosterone, which will cause a defect in the development of the male ducts and the external genitalia leading to male pseudohermaphroditism. Penile growth continues during pregnancy and normally depends on fetal testosterone. Testosterone acts directly on the differentiation of the epididymis, the vas deferens, and the seminal vesicle. Reduction of testosterone to dihydrotestosterone by 5α-reductase is necessary to obtain differentiation of the prostate, the prostatic utricule, the scrotum, and the penis (Fig. 3).

5. Hypothalamo–Pituitary–Gonadal Axis

As described above, secretions of the testis are necessary for the male sex differentiation and the hypothalamo-pituitary-gonadal axis matures during fetal life.

a. Gonadotropin-Releasing Hormone Gonadotropin-releasing hormone was observed in the brain of a 4.5-wk-old fetus and is mainly located in the hypothalamus. The concentration of gonadotropin-releasing hormone in the hypothalamus varies with age of gestation and sex.

b. Gonadotropins, Testosterone, and Estrogens Both follicle-stimulating hormone and luteinizing hormone are synthesized by the fetal pituitary under the stimulation of gonadotropin-releasing hormone. Both gonadotropins have been detected in the pituitary cells as early as 10 wk gestation. Only α-subunits of gonadotropins are found until the 10th week in the pituitary gland. Fetal pituitary follicle-stimulating hormone concentration increases between 150 and 210 days. In both sexes, fetal serum concentration of follicle-stimulating hormone peaks between 100 and 150 days, followed by a decline until term. Pituitary luteinizing hormone increases between 100 and 150 days gestation. In fetal serum, luteinizing hormone is also present but has been difficult to separate from human chorionic gonadotropin. Fetal human chorionic gonadotropin, which originates from the placenta, exhibits a peak by 90–120 days and then decreases. Interestingly, the pattern of change of human chorionic gonadotropin is related to that reported for serum testosterone in the male fetus. Testosterone peaks between 11 and 18 wk in the male fetus, corresponding to the time of the sex differentiation. A further decrease of testosterone is observed between 17 and 24 wk gestation.

In the female fetus, testosterone secretion is very low. Fetal circulating estrogens are mostly part of the placental production. The exact proportion of the estrogens originating from the fetal ovary is unknown.

6. The Hypothalamo–Pituitary Adrenal Axis, the Fetoplacental Unit

a. Corticotropin-Releasing Hormone Corticotropin-releasing hormone is present in the median eminence as early as the 16th week of pregnancy.

b. Adrenocorticotropin Hormone and Related Peptides A common precursor for adrenocorticotropin hormone (ACTH) and endorphins, named pro opiomelanocortin, has been isolated. [*See* chapter on pituitary gland hormones.] These ACTH-related peptides may play an important role

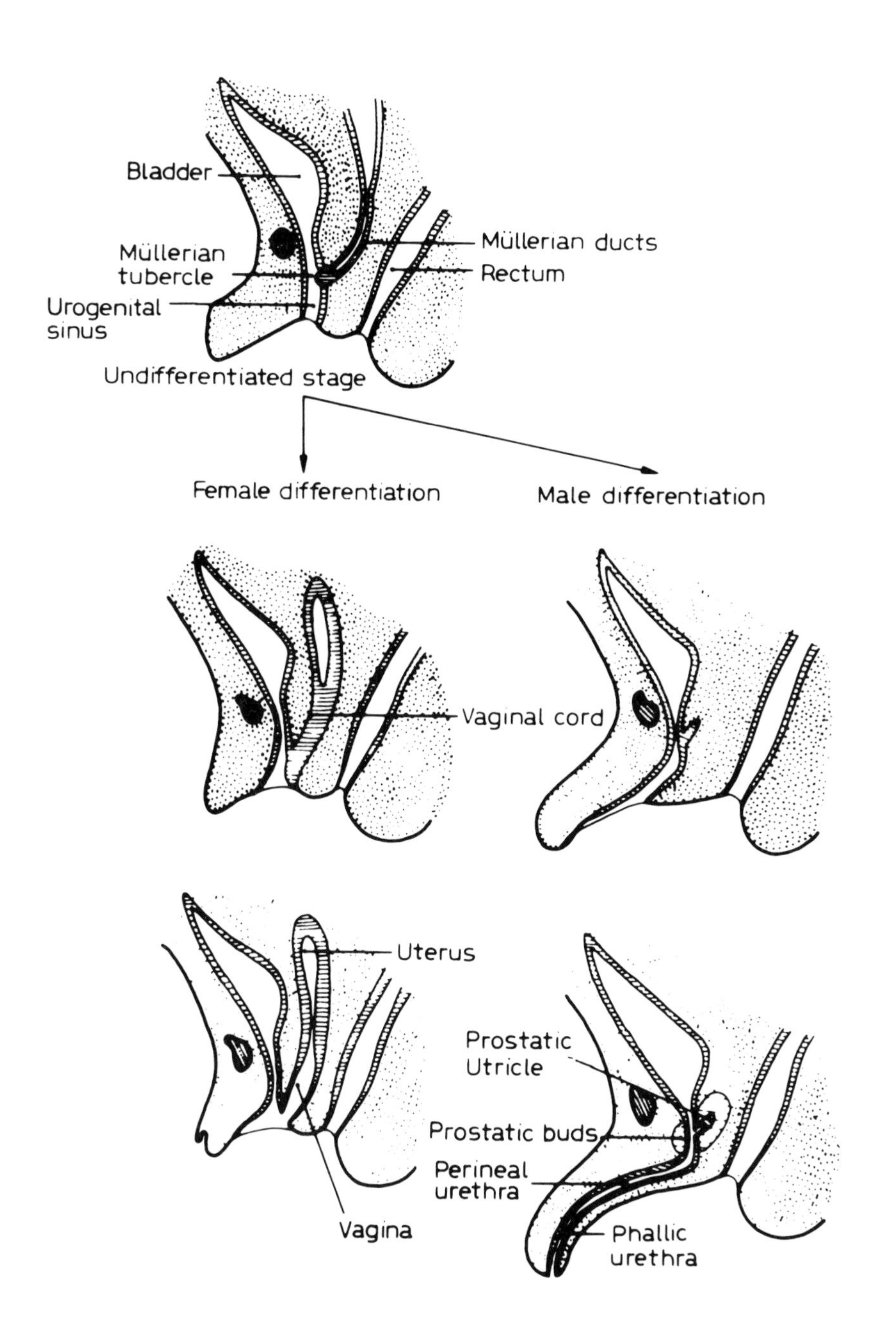

FIGURE 2 Differentiation of the urogenital sinus and the external genitalia. [Reproduced, with permission, from N. Josso (ed.), The intersex child, *in* "Pediatric and Adolescent Endocrinology," 1981, Vol. 8, p. 8, S. Karger AG, Basel.]

during fetal life. A change occurs in the pituitary content of ACTH-related peptides. At early gestation (12–18 wk), substantial amounts of ACTH, β-lipotropin (β-LPH), and β-endorphins are found in contrast to relatively low contents of α-melanocyte-stimulating hormone (α-MSH) and corticotropinlike intermediate lobe peptide (CLIP). From midgestation, α-MSH and CLIP become the predominant forms together with ACTH. Just before birth, the amount of the two cleavage products α-MSH and CLIP sharply increase.

ACTH is present in pituitaries of 14-wk-old fetuses. Transplacental passage of ACTH is very unlikely. High levels of serum ACTH are observed in 12-wk-old fetuses. The concentrations in serum decreased by 35–40 wk. Studies of human anencephalic fetuses with hypoplastic adrenals have given evidence for the critical role of ACTH in the normal growth of the adrenal cortex. A placental human

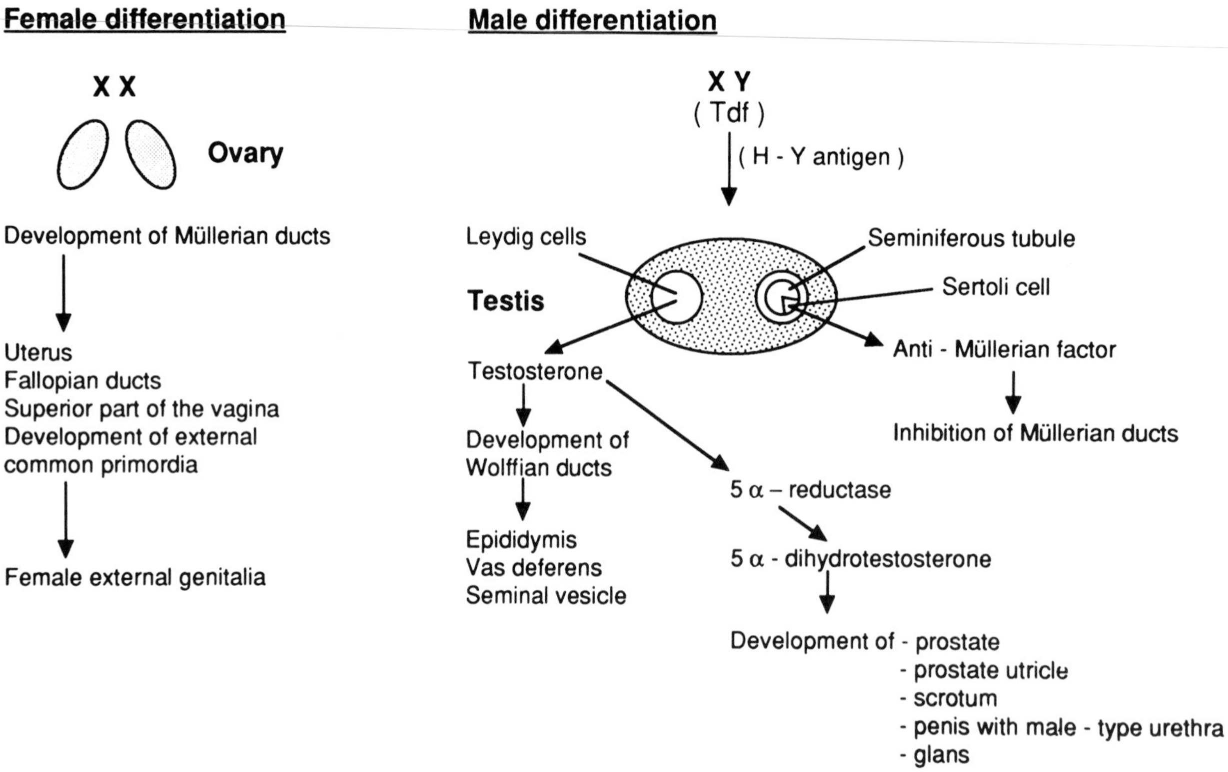

FIGURE 3 Genetic and hormonal factors acting on the differentiation of the gonads, internal sex organs, and external genitalia in female and male fetuses. Development of male sex structures depends on the presence of the anti-Müllerian factor, the secretion of testosterone, and its conversion to 5α-dihydrotestosterone. Tdf, testis-determining factor.

ACTH may stimulate the development of the fetal adrenal cortex. ACTH and β-endorphins are present in amniotic fluid, and the levels are increased in case of fetal distress.

c. Fetal Adrenal Cortex The fetal adrenal cortex plays a very important role in the fetus for its direct action and also for the supply of metabolites to the placenta. The maternal and fetal adrenal glands, the fetal liver, and the placenta constitute the *fetoplacental unit.*

Originating from the mesoderm, adrenocortical cells separate from the celomic epithelium and form two masses on either side of the aorta. Adjacent to the cortical cells are the medullary crest cells, which migrate from the neural crest; they will form the adrenal medulla glands. By 6–7 wk gestation, the fetal adrenal cortex is constituted of a thin outer layer and a large inner layer named the *fetal zone*, which represents 80% of the total fetal gland.

Inner Zone or Fetal Zone The principal characteristic of the fetal zone is the absence of the Δ^5-3β-hydroxysteroid deshydrogenase activity. Thus, the fetal zone is unable to synthesize Δ^4-3-ketosteroids from Δ^5-3β-hydrosteroids (Fig. 4). It has been suggested that the growth of the fetal zone during the first trimester of pregnancy depends on the human chorionic gonadotropin produced by the placenta. After the fourth or the fifth month, the maintenance of the adrenal cortex would depend on other trophic factors such as fetal ACTH and related peptides. In anencephalic fetuses, the adrenal glands develop normally during the first trimester of pregnancy and undergo involution later on, after the fifth month.

Outer Zone Cortisol secretion has been detected in the outer zone of the fetal adrenal cortex as early as 8–10 wk gestation. Aldosterone synthesis is present from the 15th week of gestation. A rise of fetal cortisol has been implicated in the onset of labor. Clinical experience has suggested that fetal adrenal insufficiency, as observed in anencephalic fetuses, may be the cause of delayed parturition. Shortened pregnancy has been described in cases of fetal adrenal hyperplasia secreting large amounts of cortisol.

Placental Transfer of Cortisol Placental transfer of cortisol from the mother to the fetus has been

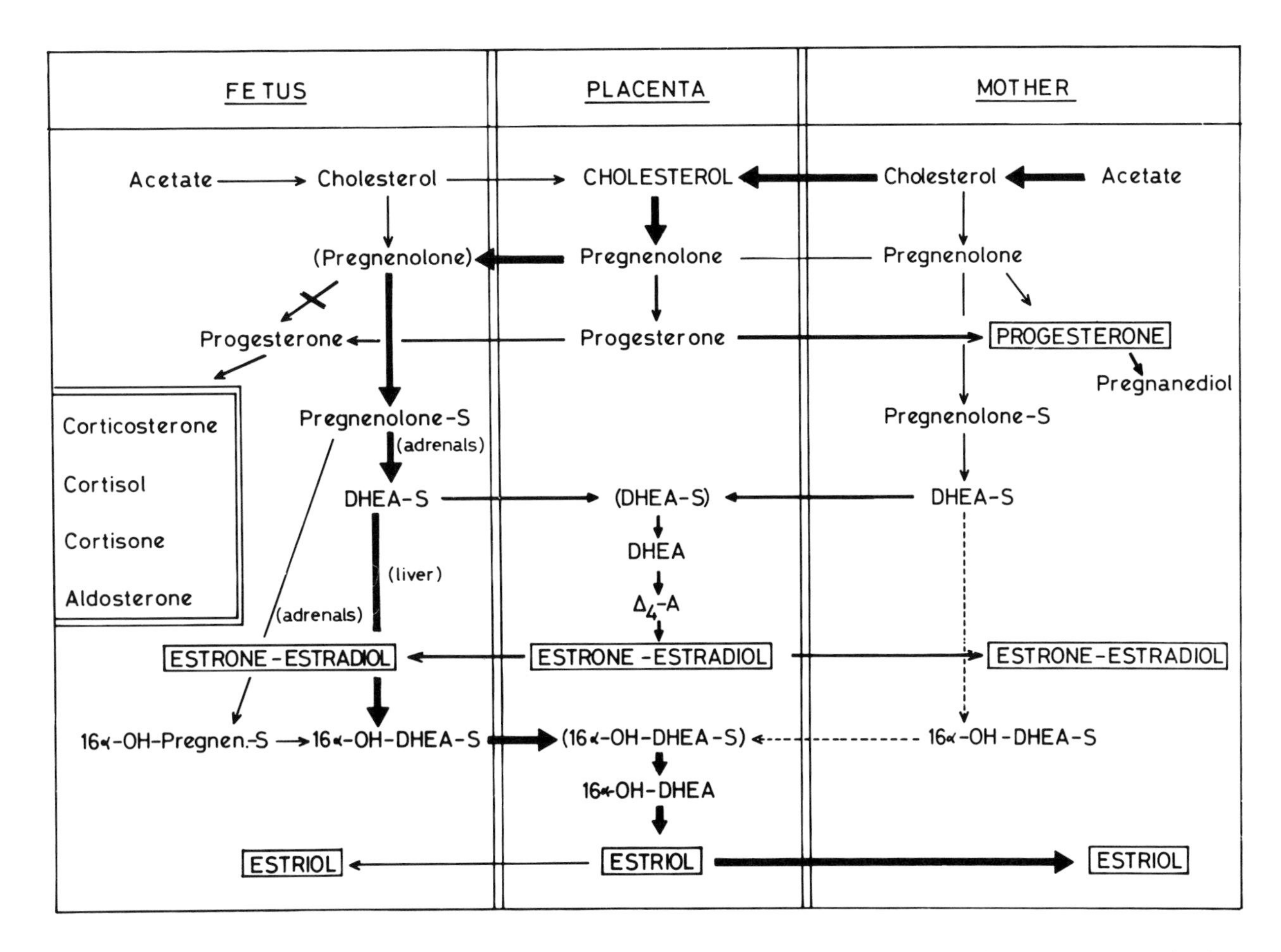

FIGURE 4 The fetoplacental unit: steroid biosynthetic pathways and interrelations of steroid metabolism among the mother, the placenta, and the fetus. [Reproduced, with permission, from P. C. Sizonenko and M. L. Aubert (1986). Pre- and perinatal endocrinology, *In* "Human Growth" Vol. 1 (F. Falkner and J. M. Tanner, eds.), Plenum Press, New York.]

demonstrated in humans. In cases of excessive maternal production of cortisol, adrenal insufficiency and low-plasma ACTH have been observed in the fetus. Fetal cortisol plays an important role in the development of surfactant factor and the maturation of pulmonary alveolar lining. Synthetic glucocorticoids acting like cortisol such as dexamethasone, methasone, or methylprednisone cross the placenta and are used as inducers of fetal surfactant factor. Thyroid hormones, estradiol, prolactin, thyrostimulating hormone-releasing factor, and β-adrenergic compounds would also play a role in the production of lung surfactant.

d. Fetoplacental Unit Because both the placenta and the fetus lack certain enzyme activities for complete steroidogenesis, the concept of a fetoplacental unit has arisen. Enzymes absent in the placenta apparently are present in the fetus and *vice versa*, and the integration of maternal, placental, and fetal functions can explain the production of the steroids made during the pregnancy.

The main enzymatic pathways are represented in Figure 4. Cholesterol produced by the mother from acetate is transformed into Δ^5-pregnenolone in the placenta. Δ^5-pregnolone is then transferred into the fetal adrenal glands to be converted into Δ^5-pregnenolone-sulfate and into dehydroepiandrosterone-sulfate (DHEA-S). DHEA-S is then hydroxylated into 16α-OH-DHEA-S in the fetal liver. 16α-OH-DHEA-S is transferred to the placenta, cleaved by sulfatases, and aromatized into estriol, which passes into the maternal compartment and is mainly excreted in the maternal urine. Hence, estriol synthesis is a complex process. Therefore, the assessment of fetal viability by urinary estriol will depend on functions of both fetal and maternal adrenal glands, fetal and maternal liver, sulfate cleavage, and aromatization by the placenta.

C. Calcium, Parathyroid Glands, Calcitonin, and Vitamin D

1. Fetal Calcium

The fetus has important needs for calcium, two-thirds of which has been acquired during the last trimester of gestation (26 mg/kg/day). In total, during pregnancy, the mother transfers 30 g of calcium to the fetus. Similarly, transplacental transfer of phosphorus during the last trimester is important (80 mg/kg/day). The plasma levels of 1,25-dihydroxycholecalciferol, the active metabolite of vitamin D, rise in the mother, permitting an increased intestinal absorption of calcium. Plasma total and ionized calcium levels in the fetus are higher than in the mother, suggesting an active materno-fetal gradient. Similar active transports of phosphorus and magnesium from the mother to the fetus have been suggested. Plasma levels of phosphorus and magnesium are higher in the fetus than in the mother.

2. Parathyroid Glands

The parathyroid glands derive from the endoderm of the 4th branchial pouch and the 3rd branchial cleft. Parathormone has been detected in parathyroid glands of fetuses at 14 wk gestation. Parathormone does not cross the placenta. Its role during fetal life is unknown. Newborns with congenital hypoparathyroidism have a normal skeleton and do not present hypocalcemic manifestations before 2 days of life. Parathormone secretion seems to depend on the maternal levels of plasma calcium as maternal chronic hypocalcemia stimulates hyperplasia of the fetal parathyroid glands, and maternal chronic hypercalcemia induces hypoplasia of the fetal glands. [*See* PARATHYROID GLAND AND HORMONE.]

3. Calcitonin

Calcitonin-secreting cells, of neuroectodermic origin, named C-cells, have been found very early in the fetal thyroid gland. Synthesis of calcitonin is present at 14 wk gestation. The role of calcitonin during fetal life is unknown.

4. Vitamin D and Its Metabolites

Vitamin D and its metabolites (25-hydroxycholecalciferol and 1,25-dihydroxycholecalciferol) are lower in the fetus than in the mother. Fetal plasma 25-hydroxycholecalciferol is highly correlated with the maternal levels and represents 60–70% of the maternal level. However, this gradient is low when the maternal level is low and elevated when the maternal concentration is high, suggesting a protective action of the placenta against vitamin D deficiency or vitamin D intoxication. 1,25-dihydroxycholecalciferol is present in fetal blood at 27 wk gestation. This synthesis is taking place in both the fetal kidney and the placenta. [*See* VITAMIN D.]

D. Pancreas

The fetal pancreas comes from an outgrowth of the duodenal endoderm. Differentiation of A-cells and B-cells occurs at 10 wk gestation. D-cells secreting somatostatin or pancreatic polypeptide are present at 17 wk.

1. Insulin

Insulin and C-peptide content of the human fetal pancreas is in direct correlation with the number of islet cells, and both peptides are present as early as the 12th week. Glucose is a poor stimulating agent of insulin release. Amino acids such as arginine are, on the contrary, a very potent factor for insulin secretion. These observations suggest different mechanisms for the release of insulin from the fetal B-cell. Only a very minimal fraction of plasma insulin passes the placenta. Insulin and C-peptide have been detected in the amniotic fluid. Amniotic fluid concentrations of both peptides originating mainly from the pancreas of the fetus are higher in diabetic pregnancies, supporting the observation that the pancreas of such fetuses are hyperplastic. [*See* INSULIN AND GLUCAGON.]

2. Glucagon

Glucagon has been detected as early as 10–12 wk gestation. Alanine stimulates glucagon secretion in the fetus. Glucagon does not cross the placenta in the human.

E. Adrenal Medulla

Adrenal medullary gland derives from the neuroectodermal tissue. The chromaffin cells have been observed very early, at 8 wk gestation. Some of them will differentiate as pheochromoblasts, invade the adrenal cortex, and give birth to the adrenal medulla. The fetal adrenal medulla develops mainly during the second half of gestation. However, the bulk of the chromaffin cells remain extra-adrenal and form the adrenergic neurons and ganglia along the aorta. This extra-adrenal tissue will involute later after birth. Catecholamines have been detected in the adrenal medulla at 15 wk gestation. The maternal catecholamines do not cross the placenta. They are present in the amniotic fluid and have been found to be higher in case of intrauterine growth retardation, particularly in the case of maternal smoking.

II. Perinatal Endocrinology

Birth is a *stressful event* in the sense that the newborn is separated from the mother and the placenta and must develop metabolic and hormonal mechanisms, which will permit the adaptation to extrauterine life (e.g., thermic control, food intake, day–night rhythm.) The newborn has a complete potential endocrine system, which will become fully operational only during the first weeks of life. [*See* ENDOCRINE SYSTEM.]

A. Hypothalamus and Posterior Pituitary

1. Oxytocin

At birth, umbilical arterial concentrations of oxytocin are higher than venous ones, suggesting that the fetus secretes oxytocin. Concentrations decrease after 30 min but remain higher than in adults. Oxytocin's role during the perinatal period remains unknown.

2. Vasopressin

At birth, the plasma levels of vasopressin are higher in infants born by vaginal delivery than those born by ceasarian section. Stress such as fetal hypoxia or diminished placental blood circulation stimulates vasopressin secretion. However, no correlation exists between plasma vasopressin and the usual criteria of fetal distress. A positive correlation between vasopressin and stages of cervical dilatation has been observed. The levels of vasopressin decrease rapidly after birth. The neonate is able to respond adequately to a water load or to a hypertonic infusion, providing evidence that the posterior pituitary and the osmoreceptor systems are functioning appropriately. Immature renal function explains more probably the decreased ability of the newborn infant to concentrate urine.

B. The Anterior Pituitary and Target Organs

1. Growth Hormone and Growth Factors

In the newborn, plasma concentrations of growth hormone are high compared with those of adults. They rise at 48 hr of life and then decrease progressively during the next 4 wk of life. In premature infants, the levels are higher than in normal babies. The newborn infant demonstrates a paradoxical response to glucose or to stress, but a normal response to hypoglycemia, amino acids, and growth hormone-releasing hormone. These observations suggest that the maturation of the secreting mechanisms for growth hormone secretion is not achieved. Sleep-associated secretion of growth hormone appears at 3 mo of life.

IGF I levels in cord blood are correlated to birth weight and length of newborns between 24 and 42 wk gestation. In infants with intrauterine growth retardation, IGF I levels are lower than those in normal infants. In some newborns with intrauterine growth retardation, a catch-up growth can be observed. Little is known of the mechanisms by which this postnatal catch-up occurs.

2. Prolactin

At birth, plasma prolactin is very high and decreases very quickly during the first 5 days of life but remain above the levels measured during childhood until the 6 wk of life. In anencephalic newborns, plasma concentrations of prolactin are similar to the normal newborn levels suggesting that the secretion of prolactin does not depend on a hypothalamic hormone stimulating the pituitary synthesis and the secretion of prolactin.

3. The Hypothalamo–Pituitary–Thyroid Axis

a. Thyrotropin-Releasing Hormone High levels of thyrotropin-releasing hormone are found in cord blood compared with maternal concentrations during the first 20–40 min after birth. These high concentrations are probably related to the rapid activation of the hypothalamo–pituitary–thyroid axis observed immediately after birth. In addition, the plasma of newborns contains a low amount of thyrotropin-releasing hormone degrading enzyme. This enzymatic activity appears in the newborn serum at 3 days of life.

b. Thyrotropin In the newborn, thyrotropin is higher than in maternal blood, rises further 10–30 min after birth, plateaus during the following 3–4 hr, and decreases at 48 hr of life. Fall in the body temperature of the newborn as well as the section of the cord are probably the triggering mechanisms. However, the prevention of the drop of the body temperature of the newborn does not suppress the thyrotropin peak.

c. Thyroid Hormones Free and total thyroxine concentrations in the cord blood are similar to the maternal levels; conversely, total and free triiodothyronine plasma levels are much lower than the maternal ones. The concentrations of reverse triiodothyronine are higher than those in maternal plasma. Following the rise of thyrotropin, thyroxine and triiodothyronine concentrations increase and remain elevated during the next 24–72 hr of life. The newborn switches from a biochemical triiodothyronine-deficient state to a condition of biochemical thyrotoxicosis with high triiodothyronine levels, which last 3–4 wk. The exact mechanism of this biological hyperthyroid state is not known. In premature and small-for-gestational-age babies, total and free thyroxine plasma concentrations are lower than in normal infants. Although there is a similar neonatal rise in thyrotropin as in normal neonates, this paradoxical state of transient hypothyroidism is probably due to a delayed maturation of the hypothalamo–pituitary–thyroid axis.

Because the hypothalamo–pituitary–thyroid axis is active at birth, screening programs for early diagnosis of congenital hypothyroidism based on the measurement of heel-prick blood thyrotropin at day 5 of life have been implanted. Some programs are based on measurement of thyroxine. Such screening permits early therapy of the congenital hypothyroidism, leading to the normal psychomotor development of the affected newborns.

4. The Hypothalamo–Pituitary Gonadal Axis

a. Gonadotropins In the newborn cord blood, plasma concentrations of human chorionic gonadotropin and α-subunits of gonadotropins are high and similar in the two sexes. Intact follicle-stimulating hormone and luteinizing hormone are low. In anencephalic newborns, the gonadotropin content of the pituitary is low, suggesting that gonadotropin-releasing hormone is necessary for the synthesis of intact molecules of gonadotropins.

After birth, follicle-stimulating hormone is higher in female infants than in males, and it remains so during the first 2 yr of life (Figs. 5 and 6). Levels of luteinizing hormone are low during the first week and increase later. During this period, gonadotropin concentrations in plasma are in the range seen during puberty, suggesting that the sensitivity of the negative feedback control regulating the hypothalamo–pituitary gonadal axis has not reached the degree of the childhood level. Gonadotropins are secreted during this period in a pulsatile fashion, as in the adult.

b. Gonadal Steroids In cord blood, testosterone concentration is slightly higher in the male than in the female infant. In peripheral blood, the difference of testosterone concentration is more marked. In male infants, plasma testosterone decreases on day 5, increases on day 10 with a peak value at 2 mo (Fig. 5). This testosterone rise is secondary to the elevation of luteinizing hormone. This rise is also observed in premature babies.

In the female newborn, the ovary contains numerous active follicles. In cord blood, high levels of

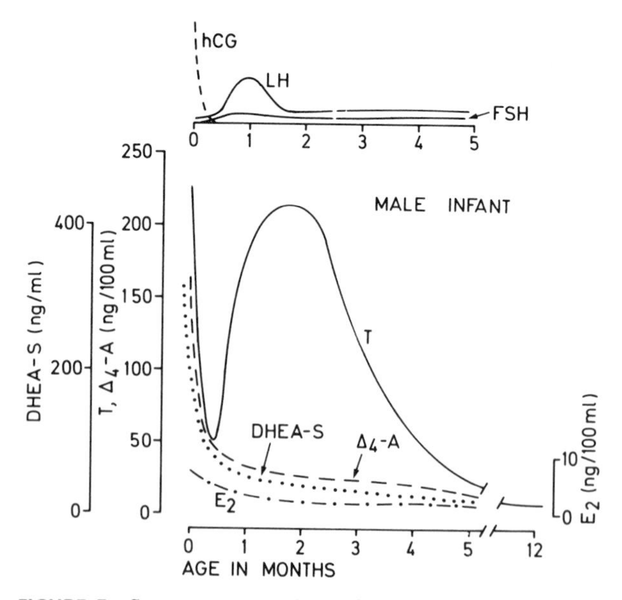

FIGURE 5 Serum concentrations of human chorionic gonadotropin (hCG), luteinizing hormone (LH), follicle-stimulating hormone (FSH), testosterone (T), Δ^4-androstenedione (Δ^4-A), dehydroepiandrosterone-sulfate (DHEA-S), and estradiol (E_2) in male infants during the first months of life. [Reproduced, with permission, from P. C. Sizonenko and M. L. Aubert (1986). Pre- and perinatal endocrinology, *In* "Human Growth" Vol. 1 (F. Falkner and J. M. Tanner, eds.), Plenum Press, New York.]

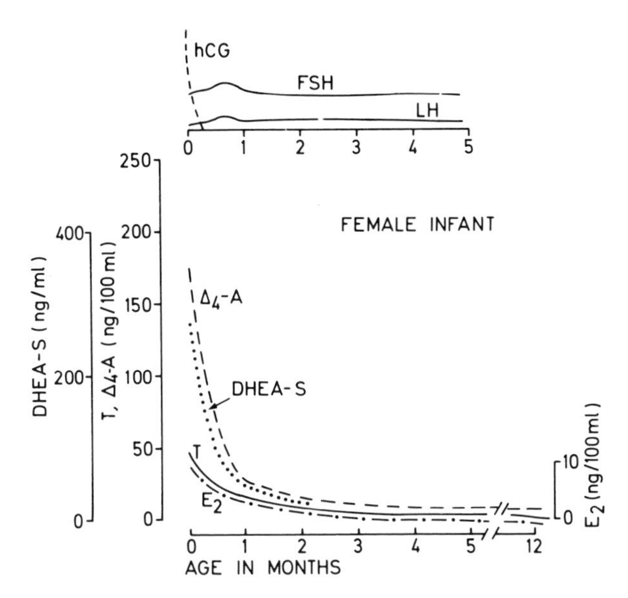

FIGURE 6 Serum concentrations of human chorionic gonadotropin (hCG), luteinizing hormone (LH), follicle-stimulating hormone (FSH), testosterone (T), Δ^4-androstenedione (Δ^4-A), dehydroepiandrosterone-sulfate (DHEA-S), and estradiol (E_2) in female infants during the first months of life. [Reproduced, with permission, from P. C. Sizonenko and M. L. Aubert (1986). Pre- and perinatal endocrinology. *In* "Human Growth" Vol. 1 (F. Falkner and J. M. Tanner, eds.), Plenum Press, New York.]

estradiol and estrone are present without sex difference. They decrease rapidly during the first 72 hr of life (Fig. 6).

5. The Hypothalamo–Pituitary Adrenal Axis

a. Perinatal Secretion of Cortisol Secretion of the steroids depending on the fetal zone, such as dehydroepiandrosterone, DHEA-S, 16α-OH-DHEA-S, and estriol decreases progressively during 8–10 wk postnatally (Figs. 5 and 6). Plasma cortisol levels are higher in newborns born by vaginal delivery than in infants born by ceasarian section, suggesting that the hypothalamo–pituitary adrenal axis is operative at birth. Peripheral plasma levels are high during the first weeks postnatally. Cortisol response to stress is present in the young infant. Secretion of cortisol is higher in infants and reaches adult value at the end of the first year. Nycthemereal rhythms of cortisol appear during the second month of life.

b. 17α-OH Progesterone 17α-OH-progesterone and progesterone decrease rapidly after birth, without difference between sexes. Measurement of 17α-OH-progesterone on day 5 of life has been proposed as a means to screen for congenital-virilizing adrenal hyperplasia due to 21-hydroxylase deficiency. In premature babies, levels of 17α-OH-progesterone are higher than in full-term newborns, rendering screening results often difficult to interpret.

c. Renin–Angiotensin–Aldosterone System Secretion of aldosterone in 1–8-day-old newborns is low as compared with that of older infants. The aldosterone concentrations rose in newborns whose mothers were put on a low-sodium diet and remained high during the first three days of life on a low-sodium diet. These results suggest that the renin-angiotensin-aldosterone system is fully active in the newborn. Plasma renin activity in newborns is inversely proportional to the sodium balance and the urinary excretion of sodium.

C. Calcium Metabolism

At birth in normal newborns, plasma total and ionized calcium levels, which are higher than those in the mother, decrease to a nadir on day 3 of postnatal life. Concomitantly, serum parathormone levels, which are low at birth, begin to rise, inducing a rise in plasma 1,25-dihydroxycholecalciferol. Plasma phosphorus also rises progressively, mainly depending on the milk formula. Breast milk maintains lower phosphorus levels than some milk formula rich in phosphorus. It has been suggested that transient neonatal hypocalcemia or tetany that can be observed during the first days of life, particularly in premature infants, can be caused by a functional hypoparathyroidism, sometimes exaggerated by a diet rich in phosphorus. Neonatal tetany can also be possibly induced by a low body calcium mass, as observed in premature babies, and/or low maternal levels of vitamin D. Supplementation of pregnant women with vitamin D has been suggested. Calcitonin levels are high at birth and decrease progressively until the first month of life. This hypersecretion of calcitonin may play a role in the regulation of plasma calcium during the neonatal period. In addition, the kidney tubule responsiveness to parathormone seems to be impaired in newborn infants. A tubular maturation with an increase of phosphorus clearance is observed during the first weeks of life.

D. Pancreatic Hormones

1. Insulin

In cord blood as well as in peripheral blood at birth, insulin levels are positively correlated with birth weight. Administration of glucose, arginine, or glucagon to newborn infants induces a sluggish response of insulin as compared with older infants, suggesting a progressive maturation of the mechanisms of secretion of pancreatic insulin: B-cells become increasingly sensitive to usual stimuli.

2. Glucagon

Blood glucose decreases 1 hr after birth. This fall, reaching a nadir within hours of birth, is associated with a significant increase in plasma glucagon concentration. The rise of glucagon, despite low levels of blood glucose, occurs only 24 hr after and is subsequently followed by the expected rise of glucose to normal values. Glucagon secretion is normally stimulated by arginine or alanine. Glucose, at this period of life, is a poor suppressor of glucagon.

E. Adrenal Medulla

After birth, most of the extra-adrenal chromaffin tissues undergo atrophy. In contrast, the adrenal medulla develops rapidly. Norepinephrine is the main catecholamine secreted during the neonatal period. During the first 3 yr after birth, epinephrine becomes the principal catecholamine secreted by the adrenal medulla, and urinary catecholamines excretion increases with age and body weight. The newborn is able to increase the secretion of norepinephrine and epinephrine in response to stress, hypoxia, or hypoglycemia.

III. Endocrinology of Prepuberty

This period of life from 2–10 yr is mainly characteristic by (1) a continuation of growth on a slow rate compared with the growth during the first 2 yr of life and the pubertal growth; (2) a period of quiescence of the hypothalamo–pituitary gonadal axis, after the perinatal activity described above and before the activation of the hypothalamo–pituitary gonadal axis, named the *gonadarche*, leading to the pubertal development; and (3) the occurrence at 7 yr of the maturation of the adrenal axis, named the *adrenarche*, before sexual maturation.

A. Growth Velocity and Bone Age

Growth has been divided into four periods characterized by the growth velocity. The first one, from birth to 2–3 yr, exhibits a fast growth velocity at birth, which declines progressively (Fig. 7). The second phase, from 2–3 yr to 10 yr is characterized by a slow decline of growth velocity. The third phase is the pubertal growth linked to sex hormones, which stimulate the bone growth and maturation. The fourth period consists of the rapid decline of the growth velocity, which arrests completely 3–4 yr after the pubertal growth velocity peak. Bone growth is responsible for the height of the child. Growth of the individual is linked to the

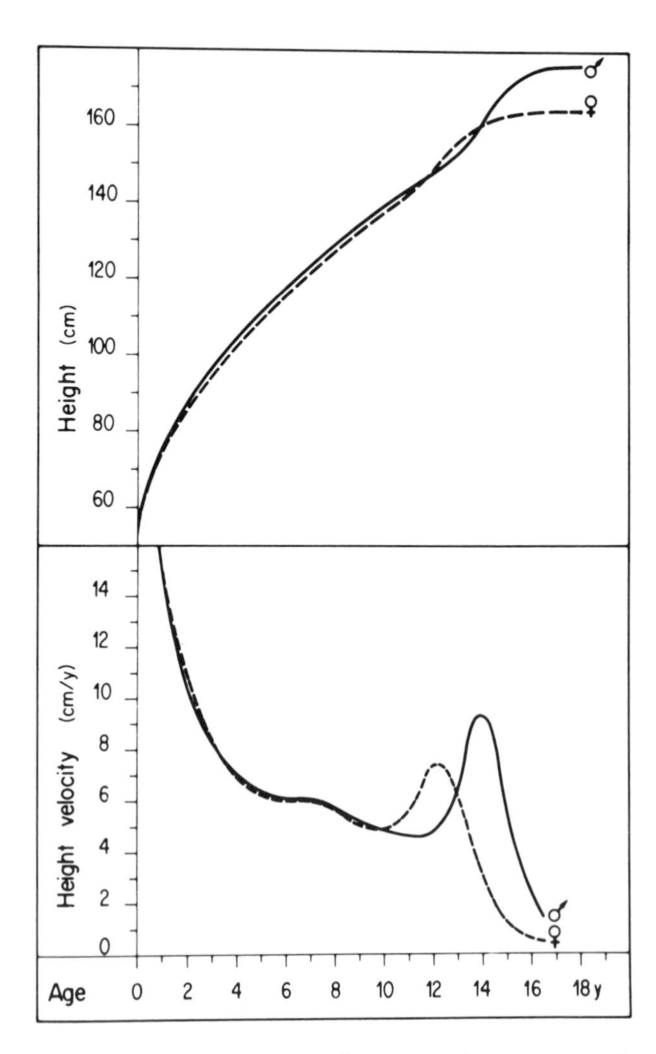

FIGURE 7 Normal growth: median values of height (cm) and height velocity for girls and boys. [Reproduced, with permission, from A. Prader (1990). Hormonal regulation of growth and the adolescent growth spurt. *In* "The Control of Onset of Puberty" p. 535. Williams & Wilkins Co., Baltimore.]

development of the epiphyseal cartilage, which is progressively calcified with age and incorporated to the metaphysis of the bone. Radiography of the bones, particularly of the hand and the wrist, permits the determination of the *bone age*. Based on the shape of the bones determined at each age, this concept of bone age has arisen. Usually, the bone age corresponds to the chronological age of the majority of individuals of the same sex. Prediction of the adult height can be calculated from the bone age. Several tables for such prediction are used, particularly in case of short or tall stature. [*See* GROWTH, ANATOMICAL.]

B. Hormonal Control of Growth

Growth hormone plays the major role in the postnatal growth of bones and the different organs. It acts either directly on the bone cartilage or indirectly through *growth factors* (Fig. 8). These growth factors are synthesized locally either in the target organs or in the liver. They are transported in the blood either in a free form or linked to binding proteins. Synthesis and secretion of growth factors are stimulated by growth hormone, insulin, and nutrition and are inhibited by glucocorticoids. *Insulin* has an important role on growth, although it is difficult to separate the direct effects of insulin (cell multiplication, energy metabolism, glucose, and amino acid transport, lipogenesis) from its indirect effects through the growth factors. *Thyroid hormones* are necessary for the bone growth and the bone maturation as well as the brain development (see above). They act synergistically with growth hormone. Hypothyroidism in children induces growth retardation and delay in bone maturation. In hyperthyroidism, thyroid hormones stimulate excessive growth and advance bone age. *Sex Hormones* (i.e., testosterone in males and estrogens, mainly estradiol, in females) stimulate growth. Excessive production of sex hormones as observed in sexual precocity or in congenital adrenal hyperplasia accelerates growth and, moreover, bone age. This advanced bone maturation leads to adult short stature. Delay in pubertal development is associated with short stature and bone age retardation. Sex hormones act directly on the cartilage but also increase growth hormone secretion, which has a synergistic action on growth. *Glucocorticoids* in excess have an inhibitory effect on growth and on bone maturation. They act directly on the cartilage by inhibiting the mineralization of the epiphysis, in-

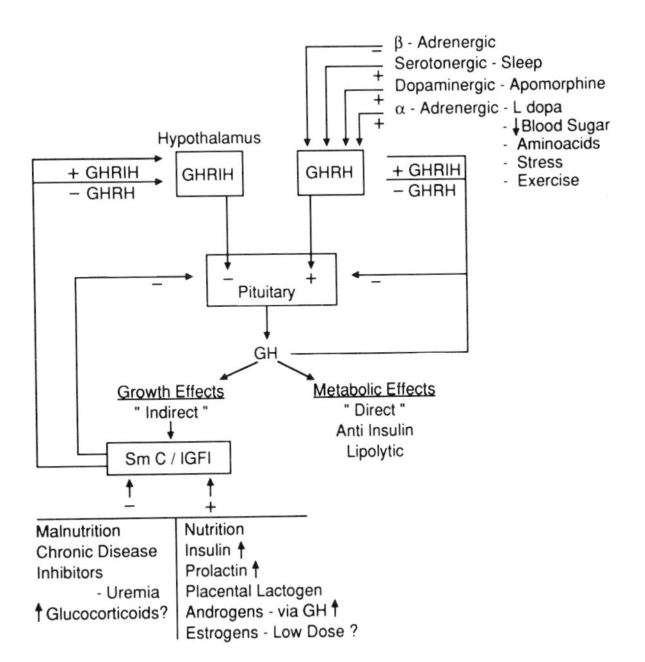

FIGURE 8 Hormonal regulation of growth. Pituitary growth hormone (GH) secretion is regulated by hypothalamic growth hormone-releasing hormone (GHRH) and growth hormone-release-inhibiting hormone (GHRIH). Neurotransmitters enhance or suppress GH secretion. GH mainly has an indirect effect on the cartilage through the growth factors called somatomedins (Sm C) or insulinlike growth factors (mainly IGFI). Recently, a direct action of GH on the cartilage has been postulated. GH and IGFI inhibit secretion of GH. GH also has a direct action on tissues like muscle and fat tissue and on carbohydrate metabolism. Sm C–IGFI production is enhanced by nutrition, insulin, prolactin, placental lactogen, and androgens. Sm C–IGFI production is diminished by malnutrition, chronic diseases, glucocorticoids, and high doses of estrogens. Inhibitors of Sm C–IGFI have been observed in malnutrition and renal insufficiency. [Reproduced, with permission, from M. H. MacGillivray (1987). Disorders of growth and development, *In* ''Endocrinology and Metabolism'' (P. Felig, J. D. Baxter, A. E. Broadus, and L. A. Froham, eds.), p. 1584. McGraw-Hill Book Company, New York.]

directly by decreasing the production of growth factors, by inhibiting the action of the growth factors on the cartilage, and finally by possibly decreasing the secretion of pituitary growth hormone. *Genetic factors* remain the main regulators of growth and cell division. Hormone actions depend on these genetic factors acting through gene regulation mechanisms. Absence of tissue receptors to growth hormone and deficiency of enzymes necessary for the normal synthesis of cartilage are examples of such abnormal regulation. Genetic factors are the most frequent causes of short or tall stature. *Nutrition* is also the main factor regulating growth of the or-

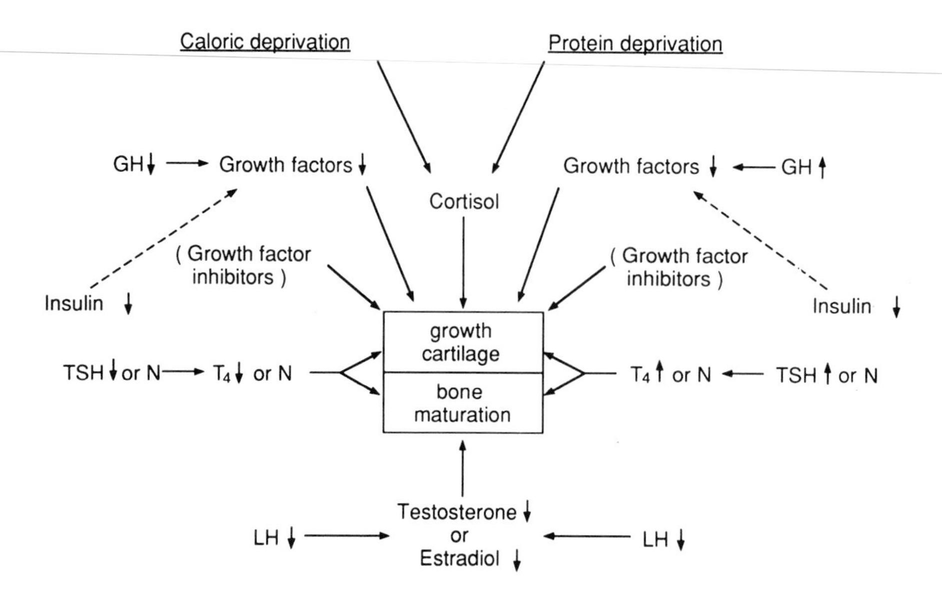

FIGURE 9 Hormonal adaptation to caloric or protein deprivation during childhood. Malnutrition decreases circulating growth factors and induces growth factor inhibitors. ↑, elevated levels; ↓, decreased levels; LH, luteinizing hormone; N, normal levels. [Adapted, with permission, from J. Bertrand, R. Rappaport, and P. C. Sizonenko (eds.) (1982). "Endocrinologie Pédiatrique" p. 195, Payot, Lausanne.]

ganism. Both caloric (marasmus) and protein (kwashiorkor) deprivation induces severe growth retardation (Fig. 9). *Psychosocial disorders*, such as maternal deprivation, named when severe, psychosocial dwarfism, can induce growth retardation.

C. Endocrine Disorders

Many of the disorders observed in children are also found during adulthood. Therefore, only disorders specifically related to pediatrics will be briefly described.

1. Growth Hormone Disorders

Growth hormone deficiency can be either isolated or associated with other deficiencies of the anterior and/or posterior pituitary. It can be familial by recessive or dominant autosomal transmission (by deletion of the gene of growth hormone) or, more commonly, sporadic. Etiology includes tumors of the hypothalamo–pituitary region. The main symptom is growth retardation with reduced growth velocity. Therapy with recombinant human growth hormone genetically synthesized induces catch-up growth with general normal or near-normal adult height. Hypersecretion of growth hormone is rare during childhood.

2. Thyroid Hormone Disorders

The main thyroid hormone disorder is congenital hypothyroidism due to either the absence of the thyroid gland (agenesis) or abnormal descent of the thyroid gland (ectopic gland). If not diagnosed during the neonatal period by screening (see above), symptoms appear slowly and constitute the classical aspect of myxoedema, with short stature and mental retardation. At this stage of the disease, therapy with thyroid hormones restores normal growth but does not prevent mental retardation. Hypothyroidism due to enzymatic defects of thyroid hormones synthesis is rare; the autoimmune disease of the thyroid gland, named thyroiditis (Hashimoto's disease), with hypothyroidism or hyperthyroidism, is more frequent. Hyperthyroidism (Graves–Basedow's disease), goiters, and cancer of the thyroid gland are also observed during this period of life.

3. Adrenal Gland Disorders

a. Adrenal Cortex Disorders Deficiencies of the adrenal cortex hormones (Addison's disease), hyperfunctions of the adrenal cortex (Cushing's syndrome, or disease), or tumors of the adrenal cortex are rare during childhood. Congenital adrenal hyperplasia is an autosomal recessive disorder affecting one of the enzyme of the metabolic pathway of cortisol. The most frequent disorder is 21-hydroxylase deficiency. Two forms have been observed: (1) one incomplete form with only excessive production of androgens, inducing in females sexual ambiguity and in both sexes accelerated growth and sexual precocity and (2) one complete form with, in addition to the cortisol deficiency and excessive secretion of androgens, a deficiency in the synthesis of aldosterone, provoking loss of sodium in the urine with possible severe dehydration of the infant. Hydrocortisone is administered as a substitute for cortisol. Mineralocorticoids should be added in the complete form. Cortisone suppresses the excessive ACTH secretion due to the absence of cortisol, reduces the abnormal production of the adrenal androgens, and, therefore, reduces the virilization process. Several other types of enzymatic deficiencies have been described. They can induce sexual ambiguity either by virilization of the female fetus or by nonvirilization of the male.

b. Adrenal Medulla Disorders Tumors of the adrenal medulla are infrequently observed during childhood. They constitute benign tumors of the chromaffin tissue named phaechromocytoma with arterial hypertension, excessive sweating, high temperature, and abnormal behavior. High levels of catecholamines and their metabolites are found in blood and urine.

4. Disorders of Calcium Metabolism

Calcium metabolism disorders are similar to these diseases observed during adulthood. In the case of parathormone deficiency, tetany is the main symptom and constitutes the disease named hypoparathyroidism. It can be linked with growth retardation and transmitted as a sex-linked recessive genetic defect or associated to aplasia of the thymus or to hypoplasia of the adrenal glands and/or moniliasis. Tetany can be observed also in the presence of high levels of plasma parathormone, suggesting a resistance to parathormone and constituting the pseudohypoparathyroidism.

5. Disorders of the Endocrine Pancreas

a. Diabetes Mellitus Diabetes mellitus in the child is due to insulin deficiency in relation to an autoimmune destruction of the B-cells; therefore, it is insulin-dependent. It is a lean diabetes in contrast to the diabetes observed in the obese patient (5% of the diabetic child).

b. Hypoglycemia Due to Hyperinsulin Many abnormalities of carbohydrate metabolism cause hypoglycemia in children. Among them, excessive secretion of insulin due to B-cell tumor or hyperplasia has been reported, particularly in young infants, presenting with ectopic B-cell islets within the pancreas, called nesidioblastosis.

IV. Endocrinology of Puberty

Puberty is defined as the period of life during which sexual maturation occurs (i.e., the growth of the gonads, the development of the external genitalia and the secondary sexual characteristics, and the establishment of the normal functions of reproduction). These changes are associated with increased growth, bone maturation with fusion of the epiphysis leading to the cessation of growth, and psychological, social, and behavioral modifications that are characteristic of adolescence. [*See* PUBERTY.]

A. Normal Pubertal Development

1. Age at Puberty

In females, normal puberty begins between the ages of 8.5 and 13.3 yr, with a median age of 10.9 yr; total pubertal growth is between 9.5 and 14.5 yr, with the mean pubertal growth spurt occurring around the median age of 12.2 yr at 9 cm/yr, from 6 to 11 cm; mean age of first menstrual bleeding, called menarche, is 12.9 yr, between 11.7 and 15.3 yr. In males, normal puberty starts between 9.2 and 14.2 yr, with a median age of 11.2 yr; testicular volumes increase >4 ml; total pubertal growth lasts between 10.5 and 17.5 yr, with a mean growth spurt observed at a median age of 13.9 yr at 10.5 cm/yr, from 7 to 15.5 cm. In both sexes, adolescents frequently follow an irregular pattern with some secondary sexual characteristics (e.g., axillary or pubic hair appearing prematurely or lagging behind).

2. Sexual Characteristics and Hormonal Changes

In females, the first change in the sexual characteristics is breast budding, which is followed by the appearance of axillary and pubic hair (Table II). Vulvae will subsequently mature with the development of the labia minora and majora, and the vaginal mucosa becomes pink. Menarche appears 2–3 yr after the first breast budding. Increase of the vol-

TABLE II Pubertal Stages and Growth Spurt in Girls[a]

		Chronological age[b]	Bone age
Breast development			
B1	Prepubertal		
B2	Budding of breast areola enlarged	8.5–**10.9**–13.3	8.5–**10.5**–13.2
B3	Enlargement of the breast with palpable glandular tissue	9.8–**12.2**–14.6	10.2–**12.0**–14.0
B4	Additional enlarged areola above tissular breast tissue	11.4–**13.2**–15.0	11.5–**13.5**–15.0
B5	Adult breast	11.6–**14.0**–16.4	12.5–**15.0**–16.0
Pubic hair			
PH1	Absent pubic hair		
PH2	Few, scanty hairs	8.0–**10.4**–12.8	8.5–**11.5**–13.0
PH3	Thick, wiry hair	9.8–**12.2**–14.6	10.5–**12.2**–14.5
PH4	Triangle-shaped hair	10.8–**13.0**–15.2	11.2–**13.2**–15.2
PH5	Adult female	11.6–**14.0**–16.6	
Growth spurt		10.2–**12.2**–14.2	10.0–**12.5**–14.5

[a] Breast and pubic hair development and growth spurt in relation to chronological and bone ages.
[b] Chronological and bone ages are given with the 95% confidence limits. Bold indicates average age. [Reproduced, with permission, from P. C. Sizonenko (1987) Normal sexual maturation, *Paediatrician* **14,** 191–207, S. Karger AG, Basel.]

ume of the ovaries and mainly of the uterus can be followed by ultrasonography of the pelvis. Plasma gonadotropins, prolactin, and estradiol increase progressively. Maturation of the ovary leads to menstrual bleedings and the appearance of ovulatory menstrual cycles, several months after the menarche. Plasma progesterone increases during the luteal phase of the ovulatory cycle. The first cycles after menarche are usually anovulatory.

In males, volume of the testes increase with appearance of axillary and pubic hair, growth of the penis, and development of the scrotum (Table III). Moustache, voice deepening due to the enlargement of the larynx, and acne will appear around 13 yr of age. Plasma gonadotropins and testosterone increase progressively. Spermatozoids can be found in the urine of boys at 14–15 yr. First ejaculations appear at the same age.

TABLE III Pubertal Stages and Growth Spurt in Boys[a]

		Chronological age[b]	Bone age
Genitalia stages			
G1	Prepubertal testes (TVI[c] < 4)		
G2	Pubertal testes (TVI between 4 and 5)	9.2–**11.2**–14.2	9.0–**11.5**–13.5
G3	TVI between 7 and 11	10.5–**12.9**–15.4	10.5–**13.2**–15.0
G4	TVI between 9 and 17	11.6–**13.8**–16.0	12.5–**14.5**–16.0
G5	Adult genitalia	12.5–**14.7**–16.9	
Pubic hair			
PH1	Absent pubic hair		
PH2	Few, scanty hairs	9.2–**12.2**–15.2	11.5–**13.5**–14.5
PH3	Thick, wiry hair	11.1–**13.5**–15.9	11.5–**14.2**–15.5
PH4	Triangle-shaped hair	12.0–**14.2**–16.4	12.5–**14.2**–16.5
PH5	Adult male	12.9–**14.9**–16.9	
Growth spurt		12.3–**13.9**–15.5	12.5–**14.5**–16.0

[a] Genitalia and pubic hair stages and growth spurt in relation to chronological and bone ages.
[b] Chronological and bone ages are given with the 95% confidence limits. Bold indicates average age.
[c] TVI, testicular volume index. [Reproduced with permission, from P. C. Sizonenko, 1987, Normal sexual maturation, *Paediatrician,* **14,** 191–207, S. Karger AG, Basel.]

In addition to the changes of circulating gonadotropins and sex steroids, sex steroid-binding globulins decrease. IGF I levels peak during the pubertal growth spurt, particularly in boys. Inhibin, which is secreted by the Sertoli cells of the testes and the granulosa cells of the follicle, increases in the serum as puberty develops.

B. Maturation of the Adrenal Cortex

The androgenic zone of the adrenal cortex (the zona reticularis) begins to mature as early as 7 yr in girls and 8 yr in boys. This is expressed by the rising levels of plasma DHEA and DHEA-S, followed by those of androstenedione. This early secretion of adrenal androgens represents a maturation of the adrenal cortex, possibly mediated by a yet poorly identified pituitary adrenal androgen-stimulating factor or by an intra-adrenal regulation of the secretion of the androgens by the adrenal cortex. Whether or not the adrenal androgens have any effect on the subsequent maturation of the hypothalamo–pituitary gonadal axis remains purely speculative, as many examples of nature show complete dissociation between the two maturational processes. Adrenal androgens have been made responsible for the development of axillary and pubic hair in adolescent girls.

C. Mechanisms of Puberty Onset

Numerous factors influence the age of puberty and its development: Some are genetic, like familial delay in puberty, or endogenous; others are exogenous, like nutrition or sport. Nutritional factors play an important role on the onset of puberty. Many studies have shown that menarche is related to a critical body composition, particularly fat content and a critical weight of 47.8 ± 0.5 kg for American girls. In anorexia nervosa, puberty is delayed. The theory of a critical body weight has been challenged by many experiments of nature: Tall girls have menarche at a greater weight than the critical weight but, in general, obese girls have menarche earlier. This critical weight theory probably reflects a temporal relationship rather than a real cause–effect relationship of the cerebral "appetite" center and an "onset of puberty" center.

Sexual maturation of the adolescent is under the endocrine control of the hypothalamus, which, through its gonadotropin-releasing hormone, regulates the secretion of the pituitary gonadotropins. In turn, the gonadotropins control the growth of the gonads and their functions (Figs. 10 and 11). Pituitary growth hormone and thyroid hormones are necessary for complete pubertal development. Night and day pulsatile secretion of growth hormone increases during this period. Growth factors such as IGF I increase considerably during sexual maturation. Adrenal glands through the adrenarche are also involved.

The central nervous system plays a key role in this endocrine mechanism. The nature of the impulses of the central nervous system, which controls the activity of the arcuate nucleus of the mediobasal hypothalamus, which in turn secretes gonadotropin-releasing hormone, is not yet understood. In animals, the pineal gland has been shown to exert an inhibitory action on the hypothalamus and the pubertal development by its secretion of melatonin. At the present time, whether or not this is also true in the human being is unknown. The hypothalamo–pituitary gonadal axis is already operative before puberty, and its maturation gradually achieves its adult functional level during puberty. Therefore, puberty likely represents the result of a slow integrated maturational process rather than the sudden awakening of an organ function that directs sexual maturation. Puberty begins spontaneously when a certain bone maturation has been achieved, irrespective of chronological age.

The hypothalamic–pituitary maturation consists of the rising production of gonadotropins, which are secreted in a pulsatile fashion. Amplitude of the secretory peaks of gonadotropins increases. In the human, whether or not the frequency of the secretory pulses changes is unknown. The results are the increased production of the gonadal steroids with a new resetting of the negative hypothalamo–pituitary negative feedback mechanism to the adult level (Fig. 12). The "gonadostat" theory, which suggests a low threshold of the hypothalamo–pituitary system before puberty and a decrease in the sensitivity at the onset of puberty as the primary mechanism for the onset of puberty, is no longer valid. This change in the sensitivity of the negative feedback mechanism is still present but is secondary to the main effect of the maturing hypothalamic–pituitary axis. In girls in late puberty, a positive feedback mechanism develops whereby increasing levels of estradiol triggers the pituitary secretion of luteinizing hormone, which induces rupture of the mature ovarian follicle and, subsequently, ovulation. This

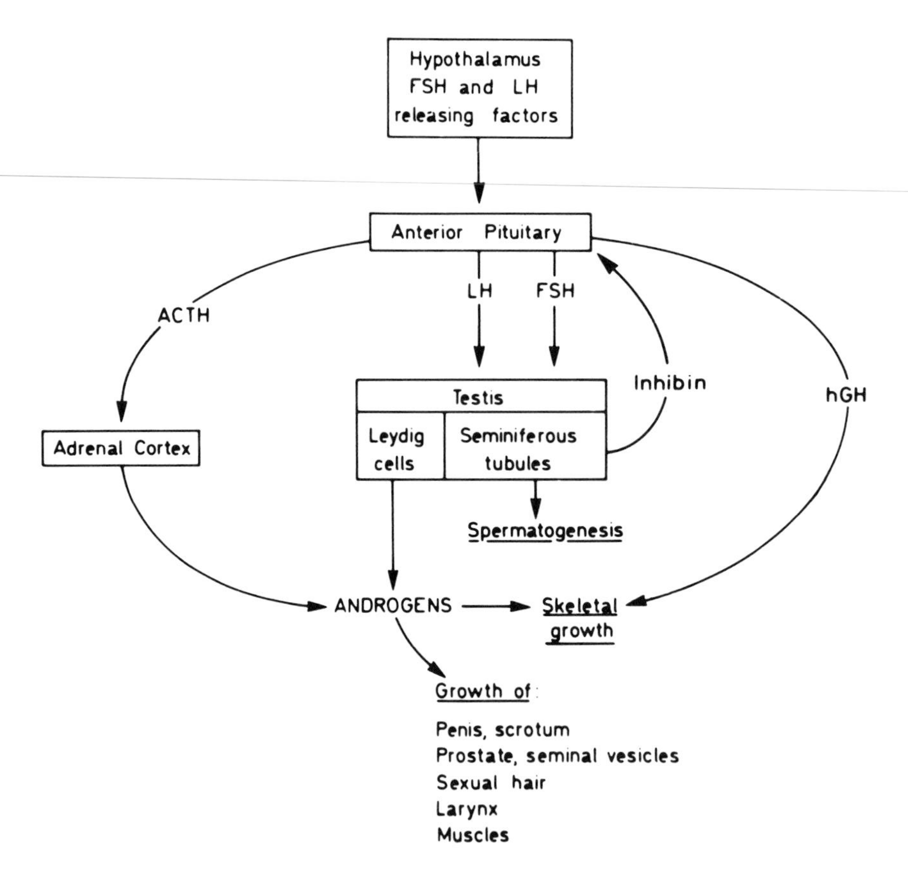

FIGURE 10 Endocrine control of sexual maturation in the male adolescent. [Reproduced, with permission, from P. C. Sizonenko, 1987, Normal sexual maturation, *Paediatrician* **14,** 191–201, S. Karger AG, Basel.]

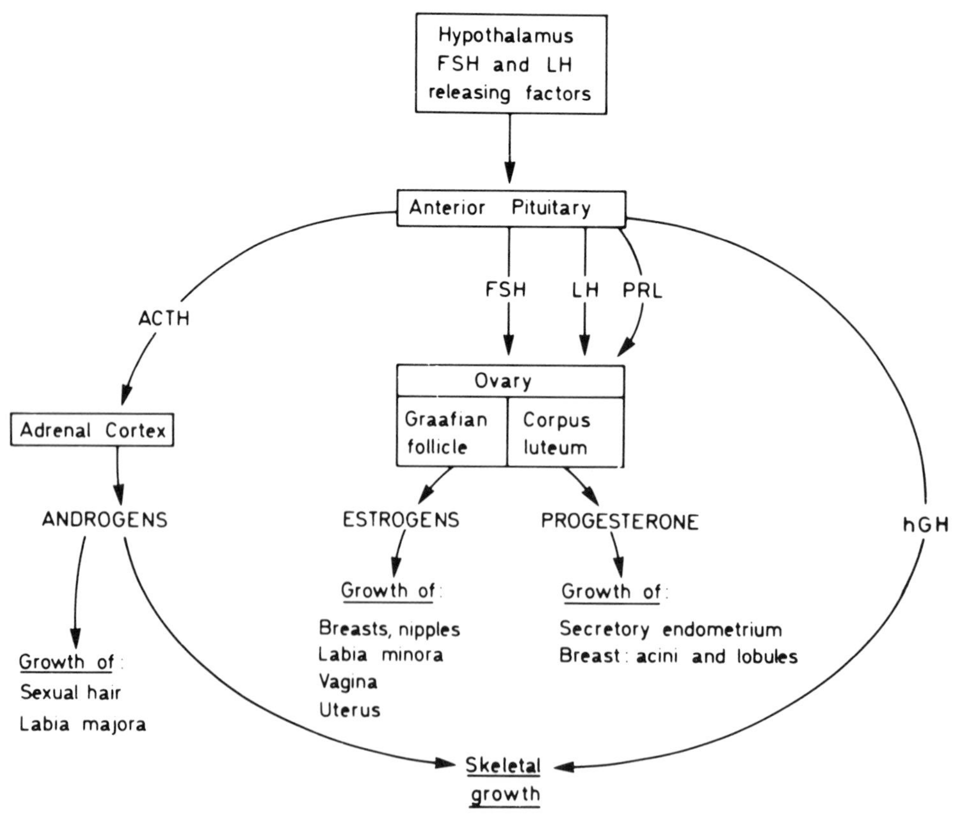

FIGURE 11 Endocrine control of sexual maturation in the female adolescent. [Reproduced with permission, from P. C. Sizonenko, 1987, Normal sexual maturation, *Paediatrician* **14,** 191–201, S. Karger AG, Basel.]

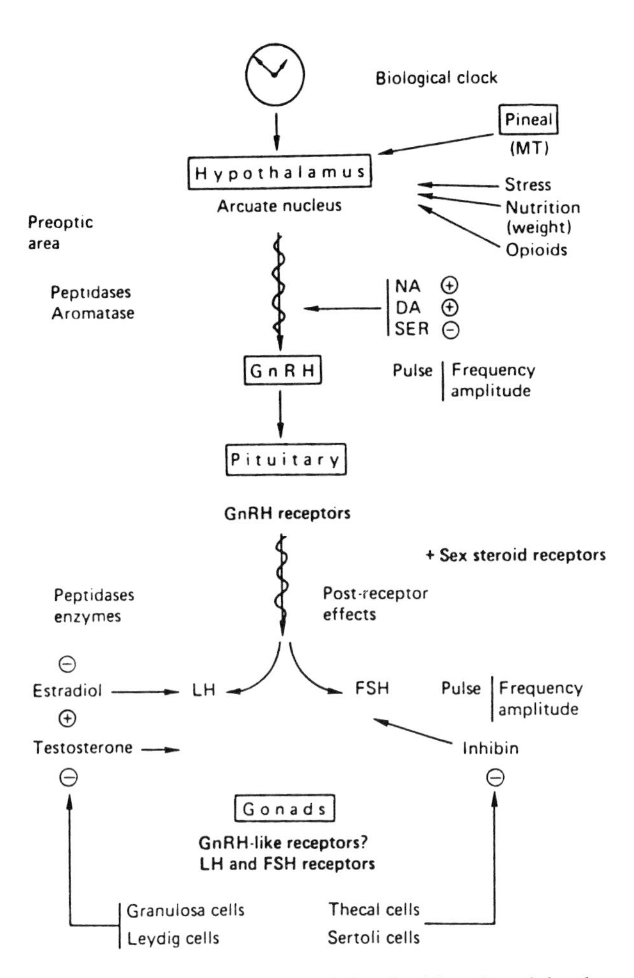

FIGURE 12 Neuroendocrine axis involved in pubertal development. Under the influence of a biological clock and the stimulating or inhibiting effect of cerebral factors such as the biogenic amines (norepinephrine [NA], dopamine [DA], and serotonin [SER]) or opiates, environmental factors such as stress, nutrition and the pineal gland (secreting, in particular, melatonin [MT]), the hypothalamus secretes gonadotropin-releasing hormone (GnRH) by pulses of higher amplitude and/or frequency. GnRH induces synthesis and stimulates secretion of pituitary gonadotropins (luteinizing hormone [LH] and follicle-stimulating hormone [FSH]). These hormones are also secreted in a pulsatile fashion. Acting on the ovary or the testis, they induce the ripening of the follicle or spermatogenesis and the secretion of estradiol or testosterone. In turn, estradiol and testosterone act on the hypothalamic–pituitary axis through a negative (in addition, in the girl, a positive) feedback mechanism. Inhibin, secreted by the testicular Sertoli cells and the follicle, also plays a role in the feedback mechanism. GnRH, LH, and FSH act on target cells through specific receptors. [Reproduced, with permission, from P. C. Sizonenko, 1987, Normal sexual maturation, *Paediatrician* **14,** 191–201, S. Karger AG, Basel.]

maturation leads to the cyclic pattern of gonadotropin secretion with monthly ovulation.

The pubertal development of the human is one step of the general evolution of the endocrine reproductive axis: During fetal life, the axis is very active at the time of sexual differentiation of the fetus, and during the perinatal period with the development of a postnatal genital "activation," the exact role of which is not known. These two steps are followed by a "quiescent" period, before the pubertal "awakening" of the reproductive axis (Fig. 13). The triggering mechanism for the awakening of the pubertal development is not yet known.

D. Disorders of Pubertal Development

1. Pubertal Physiological Discrepancies

Discordant manifestations, which may be present before or during normal puberty, may cause concerns to children, adolescents, and their parents. *Premature thelarche* consists of the isolated development of breasts in girls usually between 1 and 3 yr of age. This condition disappears spontaneously. *Premature adrenarche* is defined by the appearance of pubic and/or axillary hair before the age of 8 yr in girls and 9 yr in boys. This condition, which does not require any therapy, is followed by normal puberty at the normal age. Pubertal *gynecomastia* (i.e., development of some breast tissue) affects 40–60% of boys during pubertal development. Only a small percentage of the boys require ablation of the tissue. Usually, the breast tissue disappears spontaneously.

2. Precocious Puberty

Precocious puberty is defined as the development of sexual characteristics before the age of 8 yr in girls and 10 yr in boys. Two types of precocious puberty are described: (1) isosexual sexual precocity, or true precocious puberty, due to the premature activation of the hypothalamo–pituitary–gonadal axis (i.e., in the same direction as the genetic sex of the child, and (2) pseudoprecocious puberty, due to abnormal secretions of the adrenal gland or the gonad, particularly tumors of these two glands. In that case, the precocious puberty can be in the same direction as the sex of the child and the pseudoprecocious puberty is called isosexual pseudoprecocious puberty (estrogen-secreting tumors of the ovary in the girl, testosterone-secreting tumors of the testes in the boy). When in the opposite direction, it represents heterosexual pseudoprecocious puberty, such as androgen-secreting tumors of the adrenal gland or of the ovary, or congenital virilizing adrenal hyperplasia inducing masculinization in girls, or estrogen-secreting tumors of the adrenal gland or of the testis inducing

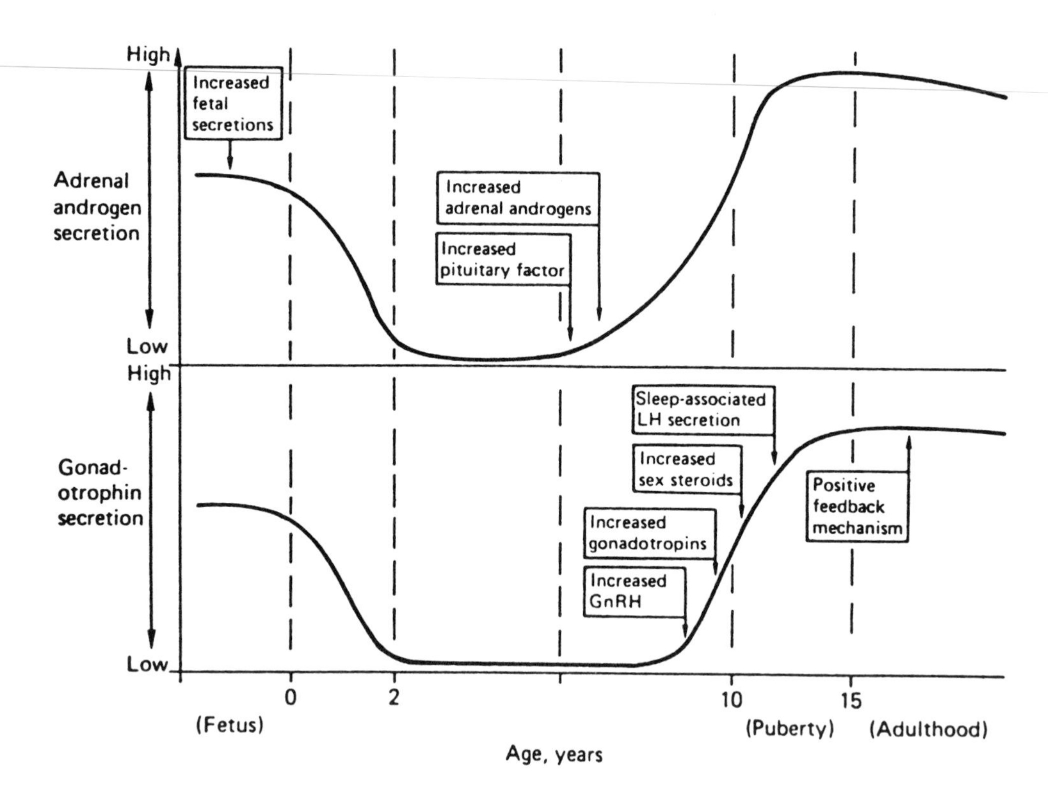

FIGURE 13 Current concepts of the maturation of the hypothalamic–pituitary–gonadal axis (gonadarche) and the secretion of the adrenal androgens (adrenarche) during prenatal life, infancy, childhood, and puberty. [Reproduced, with permission, from P. C. Sizonenko, 1987, Normal sexual maturation, *Paediatrician* **14,** 191–201, S. Karger AG, Basel.]

feminization in boys. Each type of precocious puberty has its own therapy. Central precocious puberty is presently treated by administration of agonists of gonadotropin-releasing hormone, which induces a desensitization of the pituitary receptors to gonadotropin-releasing hormone and a complete reduction of the secretion of the gonadotropins.

3. Delayed Puberty

Delayed puberty is defined as the absence of pubertal development (absence of breast development in girls, or increase of volume of the testes in boys) after the age of 13 yr in girls and 14 yr in boys. Delayed puberty can be a simple delay in adolescence (often familial) but could represent the manifestation of abnormalities called hypogonadism. The disease can affect either the hypothalamo–pituitary axis (this condition, with low levels of gonadotropins, is called hypogonadotropic hypogonadism) or the gonad itself (called hypergonadotropic hypogonadism with high levels of gonadotropins due to the absence of sex steroids acting on the negative feedback mechanism).

V. Conclusions

This chapter on pediatric endocrinology discusses the complexity of the mechanisms involved in the overall process of growth in children, from the fetal life to adulthood. It schematically presents the many hormonal factors that disturb or enhance growth. It also includes some of the factors involved in tissue generation, differentiation, and maturation (such as sexual differentiation), which are essential components of growth.

Bibliography

Delange, F., Fisher, D. A., and Malvaux, P. (eds.) (1985). Pediatric thyroidology. *In* "Pediatric and Adolescent Endocrinology," Vol. 14. Karger, New York and Basel.

Falkner, F., and Tanner, J. M. (eds.) (1986). "Human Growth," Vol. 1. "Developmental Biology and Prenatal Growth," 2nd ed. Plenum Press, New York and London.

Falkner, F., and Tanner, J. M. (eds.) (1986). "Human Growth," Vol. 3. "Methodology, Ecological, Genetic, and Nutritional Effects on Growth, 2nd ed. Plenum Press, New York and London.

Grumbach, M. M., Sizonenko, P. C., and Aubert, M. L. (eds.) (1990). "Control of Onset of Puberty." Williams & Wilkins, Baltimore.

Josso, N. (ed.) (1981). The intersex child. *In* "Pediatric and Adolescent Endocrinology," Vol. 8. Karger, New York and Basel.

MacGillivray, M. H. (1987). Disorders of growth and development. *In* "Endocrinology and Metabolism" (P. Felig, J. D. Baxter, A. E. Broadus, and L. A. Frohman, eds.). McGraw-Hill Book Company, New York.

Sizonenko, P. C., and Aubert, M. L. (1986). Pre- and perinatal endocrinology. *In* "Human Growth," Vol. 2. "Postnatal Growth and Neurobiology," 2nd ed. (F. Falkner and J. M. Tanner, eds.). Plenum Press, New York and London.

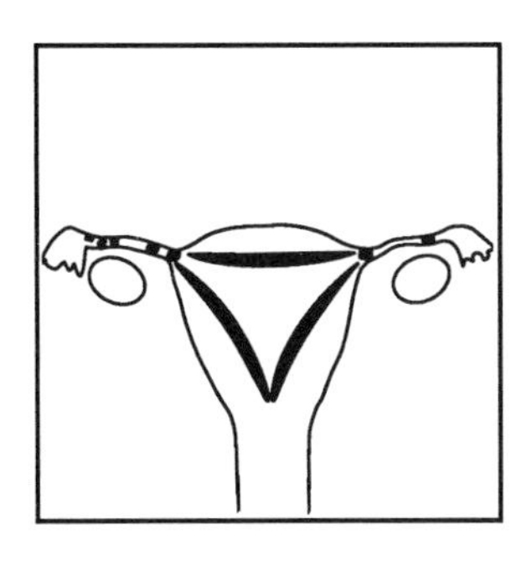

Endometriosis

ROBERT S. SCHENKEN, *The University of Texas Health Science Center at San Antonio*

Glossary

Adendomyosis Endometrial glands and stroma within the muscle wall of the uterus
Dysmenorrhea Painful menstruation
Dyspareunia Painful sexual intercourse
Endometrial glands and stroma Two major cellular components of endometrium
Endometrial implants Sites of endometriosis
Endometrial tissue Endometrial glands or stroma
Endometrioma Endometriosis that has formed a cystic structure
Endometriosis Presence of endometrial tissue outside the uterine cavity
Endometrium Tissue lining the uterine cavity
Episiotomy Surgical incision made in the vulva for delivery of the baby
Laparoscopy Surgical procedure to visualize the pelvic organs
Retrograde menstruation Reflux of menstrual blood and endometrial tissue through the fallopian tubes

ENDOMETRIOSIS IS A COMMON gynecologic condition characterized by the presence of endometrial tissue outside the endometrial cavity and uterine muscle. It is usually located in the pelvis but can occur nearly anywhere in the body. Endometriosis is rarely malignant; however, it is associated with many distressing and debilitating symptoms. Although the disease was first described over 100 years ago, considerable controversy remains regarding the incidence, histogenesis, pathogenesis, natural history, and optimal treatment of endometriosis. This controversy and our current knowledge will be discussed. The bibliography contains several references that provide more detailed information.

I. Epidemiology

The true prevalence of endometriosis is not known. Estimates of the prevalence has been based on visualization of the pelvic organs during surgery. The observed incidence varies according to the type of surgical procedure performed, the experience of the surgeon, and the indication for performing the operation. For example, the prevalence of pelvic endometriosis in women undergoing major surgery for all gynecologic indications is approximately 1%. When laparoscopy is performed to determine the cause of pelvic pain in reproductive-age women, endometriosis is found in 20%. Endometriosis can be found in 50% of teenagers undergoing laparoscopy for evaluation of dysmenorrhea (pain during menstruation) and in up to 33% of women undergoing laparoscopy during infertility evaluations. Thus, the prevalence of endometriosis ranges from about 1 to 50%.

The influence of socioeconomic status, race, and age on the prevalence of endometriosis has received considerable attention. Many researchers believe that endometriosis is more common in women of upper economic classes because they marry and have children later in life (delaying pregnancy is postulated to increase the risk of developing endometriosis). Whether this reflects a true increased incidence in upper socioeconomic groups or results from a better access to medical care remains unknown.

It was once thought that endometriosis was uncommon in Blacks. Recent evidence indicates that Blacks have a similar prevalence of endometriosis compared with Whites when the comparison is made in upper socioeconomic groups. However, the prevalence of endometriosis in black women undergoing gynecologic surgery in Nigeria is <1%. It is possible that race influences the chance of developing endometriosis, but the confounding variables including socioeconomic status obviate firm conclusions.

Endometriosis is most frequently diagnosed in the third and fourth decades of life. It has not been found in a prepubertal girl and is rarely diagnosed in postmenopausal women. These observations are likely related to the regulatory role of ovarian hormones on growth and development of endometriosis.

II. Pathogenesis

The origin of endometriosis (histogenesis) and the specific factors that cause the disease (etiology) have been extensively studied, but we still do not know why women develop endometriosis. To date, there are 11 theories on the histogenesis of endometriosis (Table I).

A. Histogenesis

The implantation theory proposes that endometrial tissue from the uterus is shed during menstruation and transported through the fallopian tubes where it gains access to and implants upon the pelvic structures. Endometrial tissue is, in fact, found in the fluid surrounding the pelvic organ of almost all women at the time of menstruation. Menstrual effluent also contains viable endometrial tissue that is capable of growth. This theory is the most likely explanation for the development of pelvic endometriosis.

TABLE I Theories of Histogenesis

Implantation theory
Mechanical transplantation theory
Lymphatic metastasis theory
Vascular metastasis theory
Direct extension theory
Uterotubal theory
Induction theory
Coelomic metaplasia theory
Müllerian cell rests theory
Wolffan cell rests theory
Composite theory

The mechanical transplantation theory is the probable explanation for endometriosis that develops in episiotomy, cesarean section, and other scars following surgery. Endometrial tissue from the uterus is directly transplanted to the incision site. Clearly, this cannot explain the development of endometriosis in women who have not had surgery.

Dissemination of endometrial cells or tissue through lymphatics and blood vessels explains the occurrence of endometriosis in locations outside the pelvic. Endometrial tissue can be found in pelvic lymph nodes and in the veins and lymphatics exiting the uterus. This tissue could access the systemic circulation through intracardiac defects and theoretically be transported to any site in the body.

The direct extension theory proposes that endometrium invades through the uterine musculature to implant and grow outside the uterus. Direct extension of endometrium to endometriortic implants has not been confirmed, but this theory does explain the occurrence of adenomyosis.

The uterotubal theory is a composite of the direct extension and implantation theories. Here, endometrial tissue extends directly from the uterus to the lumen of the fallopian tube and through the tubal wall where it could seed the pelvic cavity. This mechanism may be responsible for development of endometriosis in the fallopian tubes.

The coelomic metaplasia theory proposes that the lining of the abdominopelvic (the coelomic or peritoneal) cavity contains undifferentiated cells or cells capable of dedifferentiating into endometrial tissue. This theory is based on the embryologic studies demonstrating that all pelvic organs including the endometrium is derived from the same source—the cells lining the coelomic cavity.

The induction theory postulates shed endometrial cells refluxed through the fallopian tubes release specific substances that stimulate cells lining the coelomic cavity to form endometriosis. Thus, this theory is an extension of the coelomic metaplasia one, but neither theory has been conclusively proven.

The embryonic rest theory proposes that some of the embryonic cells that form the endometrium are retained outside the developing uterus and these cells can be activated to develop into endometriosis. Embryonic rest cells can be identified in the ovary and pelvis, but they are uncommon and thus unlikely to be responsible for most cases of endometriosis.

Finally, the composite theory was proposed to combine the implantation, vascular–lymphatic metastasis and direct extension theories. This theory is attractive in that it recognizes a multifaceted origin of endometriosis. The first two components of this theory are the most probable mechanisms for the development of endometriosis.

B. Etiology

Several factors may play a causative role in the development of endometriosis. They include factors predisposing to retrograde menstruation, altered immunity, and genetics.

Anatomic alterations of the pelvis that enhance tubal reflux of menstrual endometrium may increase a woman's chance of developing endometriosis. Teenage girls with obstruction of the vagina or cervix, conditions that prevent expulsion of menses into the vagina and increase the likelihood of tubal reflux, have a much higher incidence of endometriosis than expected. Uterine retroversion, a condition where the uterus is displaced toward the rectum, was once thought to predispose to retrograde menstruation and endometriosis. However, women with uterine retroversion do not have a higher incidence of retrograde menstruation or endometriosis. Recently, it was suggested that women with endometriosis have an anatomic or functional defect of the uterotubal junction, which permits easier passage of endometrial tissue from the uterus to the peritoneal cavity. This intriguing possibility has not been confirmed.

Overall, the incidence of retrograde menstruation appears similar in women with and without endometriosis. Thus, the development of endometriosis could depend on the quantity of endometrial tissue reaching the peritoneal cavity or on the frequency with which they are refluxed. Alternatively, the capacity of a woman's immune system to eliminate the refluxed menstrual debri may be the primary factor that determines why some women develop endometriosis.

Studies in women and monkeys with endometriosis have suggested that deficient cellular immunity results in an inability to recognize the presence of endometrial tissue in abnormal locations. Other studies have not consistently detected a difference in lymphocyte (cells involved in the immune response) cell populations in women with and without endometriosis. Antibodies to endometrial tissue and other proteins involved in the immune response have been found in sera and peritoneal fluid from women with endometriosis. This would suggest that some women have an immunologic response that protects against the development or progression of endometriosis. The immune system clearly has an important, albeit unclear, role in determining who will develop endometriosis. [*See* IMMUNE SYSTEM.]

The possibility of a familial tendency for endometriosis has been recognized for several decades. It is currently estimated that if a patient has endometriosis, there is a 7% likelihood of a first-degree relative being similarly affected. There are three possible explanations for the inheritance pattern of endometriosis: (1) polygenic–multifactorial, (2) a single mutant autosomal-dominant or -recessive gene, and (3) a single mutant gene occurring in a subset of patients with endometriosis. Although many diseases have been shown to be associated with HLA antigens, there is no HLA linkage for endometriosis.

III. Pathology

A. Locations

Endometriosis occurs most commonly in the pelvis. The primary sites, in decreasing order of frequency, are the ovaries, anterior and posterior cul-de-sacs, posterior broad ligaments, uterosacral ligaments, uterus, fallopian tubes, sigmoid colon and appendix, and round ligaments. Other sites less commonly involved include the vagina, cervix, rectovaginal septum, cecum, ileum, inguinal canals, abdominal or perineal scars, ureters, urinary bladder, and umbilicus. Exceptional cases of endometriosis are reported in the breast, pancreas, liver, gallbladder, kidney, urethra, extremities, vertebrae, bone, peripheral nerves, lung, diaphragm, and central nervous system. Although endometriosis may occur almost anywhere in the body, it is primarily a disease of the female pelvic organs.

B. Gross Appearance

In most patients, endometriosis is present in several organs or in multiple areas of the same organ. The appearance of endometriosis is quite variable. Im-

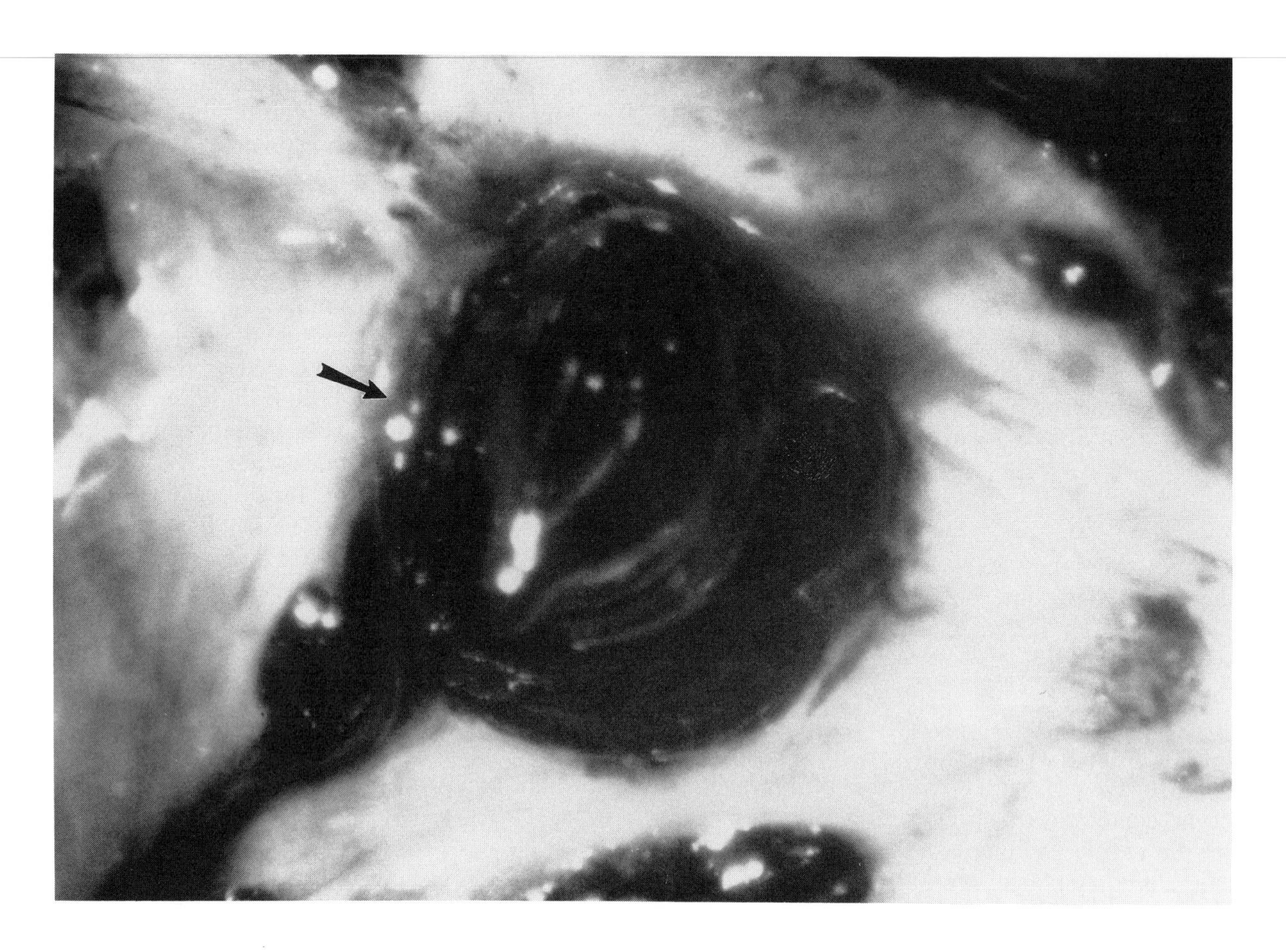

FIGURE 1 Raised reddish superficial nodule of endometriosis (arrow) on the surface of the peritoneum.

plants on the surface of organs may be superficial or invasive. Superficial implants appear as brownish discolorations (commonly known as ''powder burn'' lesions) or raised reddish to reddish blue nodules (Fig. 1). They vary from 1 mm to several centimeters in size and are irregularly shaped. Areas of endometriosis sometimes appear as whitish opacifications, translucent blebs, or raised flamelike patches. Invasive areas of endometriosis appear as reddish or reddish blue, irregularly shaped islands that may exceed several centimeters in size. They are frequently associated with a scarred or puckered peritoneal surface.

Endometriosis of the ovary may develop into cystlike structures known as endometriomas. These cysts result from accumulation of blood, fluid, and menstrual debris. The cyst content has a thick chocolatelike consistency.

Scarring or adhesions with apposition of adjacent pelvic structures is a common finding in women with endometriosis. This scarring may distort or obscure the normal pelvic anatomy and make identification of an endometriosis lesion difficult or impossible.

C. Microscopic Appearance

The microscopic appearance of endometriosis is similar to endometrium in the uterine cavity. The two major components of endometrium are the endometrial gland and stroma. Endometriosis contains these components, but unlike endometrium, fibrosis, hemorrhage, and cyst formation are often present in endometriotic tissue.

The endometrial glands may be regular, as in endometrium, or irregular. Collection of secretion and blood within the glands results in cystlike structures lined by glandular epithelium. Endometrial glands may show characteristic changes that occur in response to cyclic hormone changes during the menstrual cycle. However, the response to hormones in endometriotic tissue is quite variable, and some implants show complete lack of cyclic histologic changes. This is likely due to differences in the hormone receptors for estrogen and progesterone.

The stromal component of endometriosis is fairly characteristic of normal endometrial stroma but may occasionally resemble ovarian stroma. Endometriotic implants may contain only stroma when glandular elements have regressed.

Hemorrhage is a common feature of endometriotic implants. Accumulation of blood results in cyst formation and enlargement. Degradation products from blood (hemosiderim pigment) give a dark grainy appearance to implants. Long-standing endometriotic implants commonly show fibrosis resulting from chemical and mechanical irritation. In some cases, there is complete absence of glands and stroma with only fibrosis, histocytes and granulation tissue remaining.

IV. Symptoms

A variety of symptoms including pelvic pain, dysmenorrhea, dyspareunia (painful intercourse), abnormal menstrual bleeding, and infertility are associated with endometriosis. Many symptoms attributed to endometriosis also occur in association with other gynecologic disorders. A significant number of women with endometriosis are completely asymptomic, and the severity of endometriosis is not correlated with the presence or severity of symptoms.

The most common symptom of endometriosis is pelvic pain and dysmenorrhea. The pain may be chronic or acute. It typically begins several days prior to menses and lasts throughout menstruation. Pain is usually in the lower abdomen and pelvis, and it occasionally radiates to the back and thigh. Backache and a rectal pressure sensation are also common complaints.

Dyspareunia is usually encountered when endometriosis involves the uterosacral ligament, cul-de-sac or rectovaginal septum and when the uterus and ovaries are fixed in the cul-de-sac by adhesions. The dyspareunia is often described as deep, (i.e., when there is deep vaginal penetration).

Abnormal menstrual bleeding often takes the form of premenstrual spotting. This is more common in older patients with endometriosis. Other bleeding irregularities include prolonged menstrual flow and irregular cycle length, but these problems may be due to conditions other than endometriosis.

The association of endometriosis with infertility is well recognized. In the absence of anatomic distortion of the pelvic organs, the exact cause of infertility remains unknown. It is estimated that 20–50% of women with endometriosis are infertile.

V. Physical Findings

Physical findings in women with endometriosis are variable and depend on the location and severity of disease. Frequently, there are no obvious findings on examination. When findings are present, the most common is tenderness when palpating the cul-de-sac. Nodules of endometriosis on the uterosacral ligaments, enlarged ovaries due to endometriotic cysts, and a uterus fixed in the cul-de-sac by adhesions may also be detected by pelvic exam.

When endometriosis is suspected, it is helpful to perform the pelvic exam prior or during menstruation. Endometriotic implants are more tender and enlarged around the time of menstruation and, thus, are easier to detect.

Endometriosis of the umbilicus, vagina, and perineum and in surgical scars may appear as a firm indurated nodule ranging a few millimeters to several centimeters in size. These nodules are often tender and have a bluish or reddish blue appearance.

VI. Diagnosis

A diagnosis of endometriosis based on patient symptoms and physical exam findings is often errant. The primary conditions confused with endometriosis are pelvic inflammatory disease, benign or malignant ovarian tumors, hemorrhagic corpus luteum cysts, ectopic pregnancy, and pelvic adhesions. Additionally, many women suspected of having endometriosis on clinical grounds have no obvious pelvic abnormalities. The best way to diagnose endometriosis is to directly visualize the site of suspected involvement. Because endometriosis is located primarily on the pelvic organs, laparoscopy is the perferred technique to make an accurate diagnosis.

A. Techniques

Laparoscopy involves placement of a fiberoptic telescope through a 1-cm umbilical incision while the patient is under anesthesia. The pelvis and abdominal cavities can then be inspected for the presence of endometriotic implants. The accuracy of laparoscopy in diagnosing endometriosis depends on the skill of the surgeon and the presence of adhesion and other pelvic disorders that may obscure complete visualization of the pelvic organs. Biopsy and histologic study of suspicious areas is helpful when the diagnosis is questionable. Removal of adhesions with instruments placed through or guided by the laparoscope will usually permit adequate visualization of endometriosis, but occasionally a larger abdominal incision (laparotomy) and closer inspection of the suspected area is required.

The use of radiologic studies and blood tests to diagnose endometriosis are rarely helpful. Radiologic procedures including ultrasonography, standard X-rays, computerized tomography, and magnetic resonance imaging are not as sensitive or specific as laparoscopy. They may be helpful when one suspects endometriosis involving other areas such as bowel or urinary tract. Radioimmunoassay for a protein (called CA-125) found on endometriosis has been developed and tested in hopes of establishing a blood test for endometriosis. Unfortunately, the test is insufficiently sensitive, and patients having conditions other than endometriosis may test positive.

B. Staging

Several classification systems for endometriosis have been proposed. In most, taxonomy was established from the anatomic alterations detected during a surgical procedure to diagnose and treat endometriosis. The most widely used current classification system was introduced by the American Fertility Society in 1979 and revised in 1985. This system assigns a point score for the size and location of endometriosis and for the extent and location of associated adhesions. Based on the total additive score, the disease is catagorized as minimal, mild, moderate, or severe. This classification does not correlate with the severity of patient symptoms but roughly correlates with the chance for pregnancy after treatment of endometriosis.

VI. Treatment

Therapeutic management of endometriosis depends on the (1) severity of symptoms, (2) extent of disease, (3) location of disease, (4) desire for pregnancy, and (5) age of the patient. In many cases, no specific therapy is necessary. When treatment is required, either medication, surgery, or a combination of the two can be effective (Table II).

A. Expectant Management

Once the diagnosis of endometriosis is established, it is sometimes appropriate to avoid specific therapy. This is especially true when patients have minimal or no symptoms and have minimal or mild endometriosis. Young patients in this category can be managed expectantly, but some may benefit from cyclic oral contraceptives to retard progression of the disease and protect against unwanted pregnancy. Women wishing to conceive and having limited disease should also be observed without treatment for at least 1 yr. If pregnancy occurs, regression or complete resolution of the disease can be expected. If pregnancy does not occur, specific treatment of endometriosis is indicated. Women approaching menopause can often be managed expectantly, even when the disease stage is more advanced. The absence of ovarian hormone production after menopause will inhibit growth of endometriosis and eventually lead to its resolution.

B. Surgical Management

Surgery for endometriosis may be classified as "conservative" or "definitive," depending on whether or not the procedure preserves the possibility for pregnancy. Conservative surgery pre-

TABLE II Treatment Options

Expectant management
Medical therapy
Progestins
Danazol
Gonadotropin-releasing hormone analogues
Surgical therapy (laparoscopy or laparotomy)
Conservative—Retains uterus and ovarian tissue
Definitive—Removal of uterus and possibly ovaries
Combination therapy
Medical therapy before surgery
Medical therapy following surgery

serves the uterus and as much ovarian tissue as possible. Definitive surgery involves hysterectomy with or without removal of the fallopian tubes and ovaries.

Surgery is indicated when the symptoms are severe, incapacitating, or acute. Surgery is preferred over medial therapy when there is anatomic distortion of the pelvic organs, endometriotic cysts >2 cm, or obstruction of the bowel or urinary tract; when a woman is >35 yr old and desires pregnancy; and when symptoms have failed to resolve or they worsen under expectant or medical management.

Conservative surgery is accomplished either with laparoscopy or laparotomy. Complete treatment of endometriosis is usually possible at the time of laparoscopy to diagnose the condition. This approach offers the advantage of ablating the implants and adhesions while avoiding the possibility of disease or symptom progression and avoiding the expense and side effects of medical therapy. Potential disadvantages include inadvertent damage to adjacent organs such as bowel and bladder, infection, and mechanical trauma of pelvic structures, which may result in greater adhesion formation.

Conservative surgery usually includes excision, fulguration, or vaporization of endometriotic implants and removal of associated adhesion. The goal of surgery is to restore normal pelvic anatomy. Laparoscopy treatment offers potential advantages over laparotomy, including shorter hospital stay, anesthetic, and recuperation times. However, laparoscopic treatment may not be possible or advisable when dealing with extensive adhesions or invasive endometriosis located near structures such as uterine arteries, ureter, bladder, and bowel. In these cases, damage to the structure would result in significant morbidity; therefore, laparotomy with direct surgical treatment is preferred. Ancillary procedures performed at laparotomy include presacral neurectomy to interrupt sensory nerves innervating the pelvis and, thus, to relieve pain, and uterosacral plication and uterine suspension to avoid adhesion formation from the cul-de-sac to the posterior surface of the uterus, tube, and ovaries.

The efficacy of surgical treatment is judged by the completeness of implant and adhesion removal, degree of restoration of pelvic anatomy, prevention of recurrence, relief of pain, and enhancement fertility. The extent and location of the disease are the most important variables affecting the outcome of surgery. Overall, surgery is very effective in removing pathology and restoring normal anatomy. The risk of recurrence is estimated to be as much as 40% with 10 yr of follow-up. Pain relief is achieved in the majority of patients, and presacral neurectomy may provide additional pain relief. The chance for pregnancy following surgery depends on the stage of disease and presence of other infertility factors. Average pregnancy rates after surgery in patients with mild, moderate, and severe endometriosis are 61, 50 and 39%, respectively.

Definitive surgery for treatment of endometriosis is indicated when significant disease is present and future pregnancy is not desired, when incapacitating symptoms persist following medical therapy or conservative surgery, and when there is coexisting pelvic pathology that requires hysterectomy. Numerous factors affect the decision to perform a definitive procedure, the primary factor being the patients interest in maintaining child-bearing potential. Definitive surgery involves removal of the uterus with or without removal of the tubes and ovaries. The ovaries may be conserved in younger women to avoid the need for estrogen replacement therapy. Removal of both ovaries is appropriate when the ovaries are extensively damaged by endometriosis or when the woman is approaching menopause. Treatment with estrogen to prevent menopausal symptoms is indicated when the ovaries are removed, even when surgery has not removed all endometriotic implants. The chance for symptomatic recurrence in these cases is small except when endometriosis involves the bowel.

C. Medical Therapy

Clinical observation and recent research indicate that the growth and maintenance of endometriotic implants depends on ovarian steroids. This dependency has led to numerous attempts to "hormonally" simulate pregnancy or menopause—two physiologic states believed to inhibit or delay progression of endometriosis by interrupting cyclic ovarian hormone production. For example, progestins alone or in combination with estrogen hormonally mimic pregnancy, and gonadotropin-releasing hormone (FnRH) analogue induces a state of "pseudomenopause." The theoretical advantages of medical therapy include (1) avoidance of major surgery with the attendant risks of damaging pelvic organ and causing postoperative adhesion formation and (2) treatment of implants that are not visible at surgery. The major disadvantages of medical therapy are the associated side effects, recurrence rates following discontinuation of treatment, lack of effect on adhesions and endometrial cysts, and the inability to conceive because of the absence of ovulation during treatment. Consequently, medical therapy is not appropriate for women with advanced stages of endometriosis or older women desiring pregnancy.

The three medications most commonly used to treat endometriosis are progestins, danazol, and GnRH analogues. Each should be used only after a definitive diagnosis of endometriosis has been established by direct visualization of the implants.

Progestins inhibit endometriosis tissue growth by a direct effect on the implants causing initial decidualization and eventual atrophy. Progestins also inhibit pituitary gonadotropin secretion and ovarian hormone production. The most frequently used progestin is medoxyprogesterone acetate, and it is given either orally or as an injection. Side effects vary greatly among patients and include irregular menstrual bleeding, nausea, breast tenderness, fluid retention, and depression. The effectiveness of medroxyprogesterone in eliminating implants and the risk of recurrent endometriosis following treatment are not precisely known. Pain relief is uniformly excellent—>90% of women having good or complete relief. Pregnancy rates in patient with less severe stages of disease are equivalent to expectant management and conservative surgery.

Danazol is a synthetic steroid similar in structure to testosterone. Danazol's mechanisms of action are multiple: (1) inhibition of pituitary gonadotropin secretion, (2) direct inhibition of endometriotic implant growth, and (3) direct inhibition of ovarian enzymes responsible for estrogen production. Danazol is given orally in doses ranging from 100 to 800 mg daily for 3–6 mo. Side effects of treatment include weight gain, muscle cramps, decreased breast size, acne, hirsutism, and oily skin; other common complaints are hot flushes, mood changes, and depression. Although a majority of women taking danazol have side effects, only a small percentage of patients discontinue the drug because of unwanted effects. Danazol is effective in resolving implants when treating mild or moderate stages of disease. Large endometriotic cysts and adhesions do not respond well to danazol treatment. Most patients experience relief or improvement of pain symptoms within 2 mo of treatment. Pregnancy rates following treatment approximate 40% and appear to be independent of the disease severity. Danazol is no more effective than expectant management for treating infertility in patients with mild or moderate endometriosis.

The most recent advance in the medical treatment of endometriosis is the use of GnRH analogues. These compounds, administered over time, inhibit pituitary gonadotropin secretion by blocking the effect of endogenous GnRH, thus profoundly suppressing pituitary gonadotropin-stimulated ovarian estrogen secretion. These medications are administered by nasal spray or subcutaneous injection on a daily basis or by once-a-month intramuscular injections. Common side effects induced by the hypoestrogenic state include hot flushes, vaginal dryness, decreased libido, insomnia, breast tenderness, depression, headaches, and transient menstruation. A major concern of GnRH analogue treatment is the risk of osteoporosis; however, treatment with the medication for the recommended 6-mo periods causes little changes in bone density or total body calcium. Limited study to date indicates that GnRH analogues are very effective in reducing the size of endometriotic implants. There is no information on the risk of recurrence following treatment. GnRH analogues are as effective as other medical therapy in relieving pain symptoms and enhancing fertility.

Other medications including stilbesterol (a synthetic estrogen), methyltestosterone (a synthetic testosterone), and continuous oral contraceptives (so-called "pseudopregnancy") have been used to treat endometriosis. They are mentioned for historical interest only and play no role in contemporary medical treatment of endometriosis because of their side effects and limited efficacy. A final medication, gestrinone, is an antiprogestational steroid used extensively in Europe. It appears to be as effective as other forms of medical therapy, but it is not available for use in the United States.

D. Combination Medical–Surgical Therapy

A combination of medical and surgical therapy for endometriosis has been recommended. Medical therapy may be given prior to surgery in hopes of decreasing the size of endometriotic implants and the extent of surgery required. In cases where complete removal of implants is not possible or advisable, medical therapy may be used to eradicate residual disease. A variety of medications have been used in conjunction with laparotomy or laparoscopic conservative or definitive surgical treatment. Although some claim that preoperative medical therapy decreases the amount of surgical dissection required to remove implant, no evidence indicates that such therapy leads to improved resolution of pain symptoms, increased pregnancy rates, or decreased recurrence rates. Danazol following conservative surgery for severe stages of endometriosis may improve pregnancy rates, but this is controversial.

VIII. Unusual Manifestations

Endometriosis occasionally occurs in locations outside the pelvis. The most likely explanation for development of implant in areas outside the pelvis and abdomen is lymphatic or vascular transport of endometrial cells. Gastrointestinal or urinary tract endometriosis may occur by the above mechanism or by implantation of refluxed endometrial tissue within the peritoneal cavity. Endometriosis of the bowel or urinary tract may cause a variety of symptoms and occasionally obstruction. The diagnosis is often difficult because of the rarity of the condition. Diagnosis and treatment are usually accomplished surgically. Endometriosis of the lung and thoracic cavity does occur and may cause difficulty breathing and catamenial hemoptysis. The diagnosis is suggested by radiologic studies but is confirmed and treated by surgical excision of the implants. Abdominal wall, extremity, and peripheral nerve and muscle endometriosis usually presents as a bluish palpable nodule with local pain. Surgery is necessary for diagnosis and treatment. Spinal canal and brain endometriosis is exceptionally rare and usually presents with pain.

Two other unusual manifestations deserve mention. First, endometriosis has been reported in men being treated with high-dose estrogen for prostate cancer. In these cases, endometriosis is presumed to arise from estrogen stimulation of prostatic utricle, the embryologic remnant of the uterus. Second, endometriosis, on rare occasions, undergoes malignant transformation. In these cases, both benign and cancerous endometrial tissue is present on micropscopic examination.

IX. Conclusion

Endometriosis is a common gynecologic disorder that causes several distressing symptoms. The cause of endometriosis and its associated symptoms are poorly understood. Treatment of endometriosis is accomplished by medication or surgery. Relief of pain and enhancement of fertility can be achieved in most patients. In some cases, complete removal of the uterus, tubes, and ovaries is required. Endometriosis may occur outside the pelvis and in males, and in rare occasions the process becomes malignant.

Bibliography

Ballweg, M. L. (1987). "Overcoming Endometriosis: New Help from the Endometriosis Association." Gongnden and Week, Inc., New York.

Schenken, R. S. (1989). "Endometriosis: Contemporary Concept in Clinical Management." J. B. Lippincott Company, Philadelphia, Pennsylvania.

Speroff, L., Glass, R. H., and Kase, N. G. (eds.) (1989). Endometriosis and infertility. *In* "Clinical Gynecologic Endocrinology and Infertility," 4th ed. Williams and Wilkins, Baltimore, Maryland.

Endothelium-Derived Relaxing Factor

LOUIS J. IGNARRO, *UCLA School of Medicine*

Glossary

Cyclic GMP An intracellular chemical that acts as a secondary messenger in mediating the biological effects of hormones

Free radicals Chemically unstable and highly reactive molecules that can cause destruction of other chemicals and cells

Guanylate cyclase An enzyme that catalyzes the conversion of GTP to cyclic GMP in the intracellular compartment

Prostaglandins Polyunsaturated fatty acids formed in the body from arachidonic acid which function as autacoids

Receptors Molecules located outside or inside cells that interact selectively with chemicals or first messengers to trigger a cellular response

ENDOTHELIUM-DERIVED relaxing factor (EDRF) is an unstable chemical substance synthesized within and released from vascular endothelial cells, which line the inner, or intimal, surface of blood vessels. The two main biological actions of EDRF are the relaxation of vascular smooth muscle (i.e., vasodilation) and the inhibition of platelet clumping (i.e., aggregation) and adherence to the vascular endothelial surface. Various chemically diverse tissue hormones are capable of interacting with the endothelium and provoking the release of EDRF, which then diffuses into the underlying smooth muscle and nearby blood platelets. EDRF from artery, vein, and cultured vascular endothelial cells has been identified chemically and pharmacologically as nitric oxide (NO). NO is lipophilic (i.e., lipid soluble), rapidly diffuses through cell membranes, has a biological half-life of only several seconds, and undergoes spontaneous destruction in the presence of oxygen or certain oxygen-derived free radicals.

EDRF is a potent relaxant of vascular smooth muscle and an inhibitor of platelet function. NO is formed from the basic amino acid L-arginine and elicits its biological actions by activating the cytoplasmic enzyme guanylate cyclase, thereby stimulating the intracellular formation of cyclic GMP. In turn, cyclic GMP triggers subsequent steps down the cascade of biochemical pathways, leading to a decrease in the intracellular level of free calcium. Low intracellular free calcium levels are generally associated with smooth muscle relaxation and reduced platelet function, whereas high free calcium provokes smooth muscle contraction and platelet clumping. NO plays a physiological role in the local regulation of blood flow, blood pressure, and thrombus formation. Hypertension, myocardial infarction, and cerebral stroke are pathological states that may be attributed, at least in part, to a deficiency of NO formation resulting from vascular endothelial damage.

I. Discovery and Underlying Theory

A. Background

Although it had been appreciated for decades that the neurotransmitter acetylcholine causes vascular smooth muscle relaxation and a decrease in systemic blood pressure upon intravenous injection into mammals, isolated strips of arterial smooth muscle mounted in special bath chambers usually contracted upon the addition of acetylcholine. The reason for this difference in responsiveness was provided by R. F. Furchgott in 1980, who discovered that the relaxation of arterial smooth muscle by acetylcholine required the presence of an intact functioning endothelium on the arterial strips. Normal healthy blood vessels have a single layer of squamous epithelial cells (i.e., endothelium) lining their intimal surface (i.e., the inner surface in contact with blood), which functions to transport substances between smooth muscle and blood and to provide a vascular surface that retards platelet adherence and blood clotting. Acetylcholine interacts with endothelial cells and triggers the formation and/or release of EDRF. Vascular tissue with damaged or nonfunctional endothelium displays contractile instead of relaxant responses to acetylcholine. Thus, acetylcholine is an endothelium-dependent vasodilator, and one of the chemical substances released from the endothelium was termed endothelium-derived relaxing factor, or EDRF. [*See* SMOOTH MUSCLE.]

As investigators learned more about the chemical and pharmacological properties of EDRF, it became increasingly clear that there were many similarities between EDRF and NO. EDRF was proposed to be NO in 1986, conclusively demonstrated in 1987. Due to the lack of absolute specificity of current chemical procedures for NO, it has not yet been possible to distinguish unequivocally between authentic NO and unstable nitroso compounds that decompose spontaneously and liberate NO. Although it is possible that EDRF is a labile nitroso compound, it is clear that the principal actions of EDRF are attributed to NO itself.

B. Endothelium-Dependent Vasodilation

Numerous substances possessing diverse chemical properties cause endothelium-dependent vasodilation of arteries through the actions of NO. Some of these substances relax veins as well, by a similar mechanism. A partial list of endothelium-dependent vasodilators is presented in Table I. Although their chemistries differ, each chemical agent works in a similar fashion, by interacting with selective receptors on the endothelial cell surface, thereby triggering a cascade of biochemical reactions, culminating in the formation and release of NO. Many endothelium-independent vasodilators exist, some of which are categorized as nitrovasodilators and include several clinically important drugs, such as nitroglycerin and sodium nitroprusside. These drugs, used clinically to reduce high blood pressure, elicit their effects in smooth muscle cells after biotransformation to NO, a potent endothelium-independent vasodilator.

TABLE I List of Endothelium-Dependent Vasodilators

Amines and related agents
Acetylcholine
Histamine
Norepinephrine
Serotonin
Peptides
Angiotensin
Bradykinin
Platelet-activating factor
Substance P
Thrombin
Vasoactive intestinal polypeptide
Vasopressin
Polyunsaturated fatty acids
Arachidonic acid
Leukotrienes
Basic polyamino acids
Polyarginine
Polylysine
Polyornithine
Other chemicals
ATP
Calcium ionophore A23817
Melittin
Thimerosal

C. Inhibition of Platelet Function

Another important biological action of NO is inhibition of platelet clumping and adherence to the vascular endothelium. NO diffuses from the endothelial cells to the intimal surface in contact with the circulating blood and interacts with nearby or adherent platelets to prevent further platelet adherence and clumping. This action serves to retard local thrombus, or blood clot, formation. Endothelium-dependent vasodilators thus not only relax vascular smooth muscle, but also inhibit platelet function via the actions of NO.

D. Localized Actions

Many endothelium-dependent vasodilators are naturally occurring tissue hormones that elicit localized actions in the body. Since NO is chemically unstable, possessing a biological half-life of only several seconds, the actions of NO are confined to

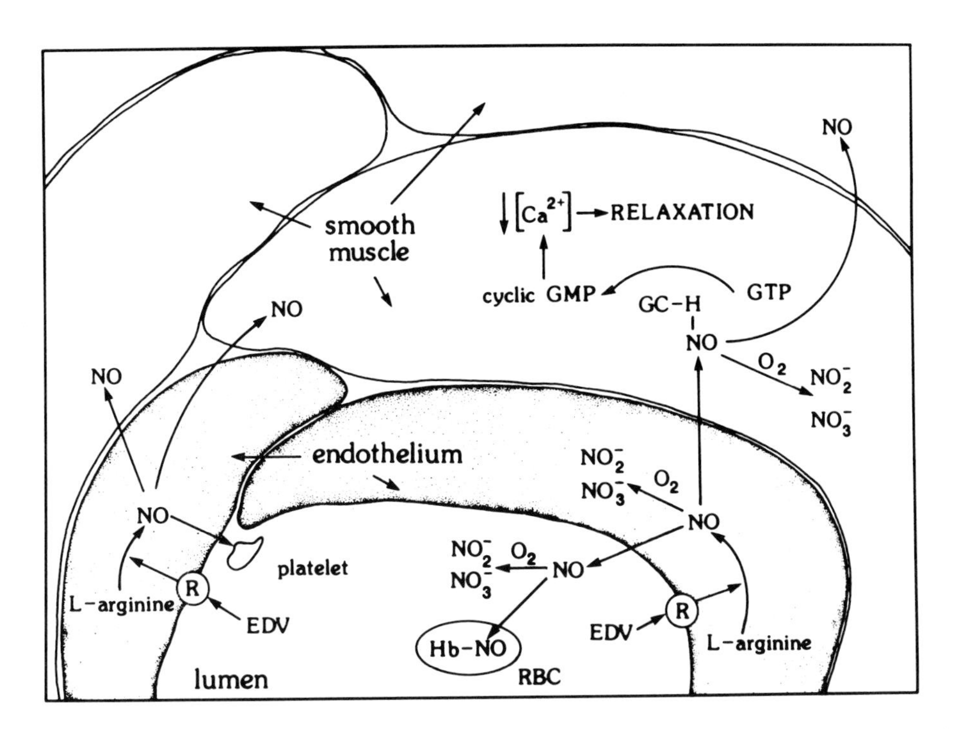

FIGURE 1 Bioassay cascade procedure used to study the release and properties of EDRF. Physiological salt solution is perfused by a roller pump through a segment of endothelium (E)-intact blood vessel or a column of isolated endothelial cells. The perfusate is superfused over three strips of E-denuded artery or vein that are precontracted by agents added to a separate superfusion line. Strips are mounted in a series (cascade), separated from one another by 3 seconds of flow time. Changes in smooth muscle tension are measured by transducers and recorded on chart paper. EDRF release from blood vessels or endothelial cells results in decremental relaxation responses down the cascade of strips due to the short half-life ($t_{1/2}$) of EDRF. (Inset) Recordings of typical relaxant responses, downward deflections signifying a loss of muscle tone or relaxation.

the immediate vicinity of its formation and release. Because of its local and fleeting action, NO may also be categorized as a tissue hormone. Thus, NO is important in the local regulation of blood flow, perhaps in small resistance vessels (i.e., arterioles) and capillary beds, and thrombus formation. Such localized actions of NO serve to maintain adequate tissue perfusion with blood and complement the body's more diffuse mechanisms of maintaining blood flow, especially in vital organs such as the heart and the brain.

II. Biological Actions

A. Demonstration of Release

The labile nature of NO initially made it difficult to demonstrate unequivocally its existence and release from vascular endothelial cells. Using bioassay techniques, in which an intact arterial segment is perfused through the lumen and the perfusate is allowed to bathe a nearby isolated and mounted strip or ring of endothelium-denuded artery, NO release from the perfused artery in the presence of added endothelium-dependent vasodilators can be easily demonstrated. Modified bioassay procedures, in which the perfusate is directed over a series of several arterial or venous strips arranged in a cascade, such that each strip is separated by about 3 seconds in flow time, reveal the short biological half-life of NO (Fig. 1). This technique has been used to demonstrate NO release from arteries, veins, and cultured arterial endothelial cells and has wide applications in biological research.

B. Chemical Properties

Even before the identification of the relaxing factor as NO, it was clear that it was a labile lipophilic chemical of low molecular weight capable of diffusing rapidly through several cell membrane barriers in order to reach its site of action in smooth muscle and platelets. In an environment rich in oxygen, NO is labile and undergoes spontaneous chemical inactivation. Inactivation is caused not only by oxygen, but also by certain oxygen-derived free radicals, such as superoxide anion, that are spontaneously generated in oxygen-containing solutions. In experimental situations, the addition of superoxide dismutase, an enzyme that catalyzes the destruction of superoxide anion, protects NO against rapid inactivation and thereby prolongs its duration of action. Hemoglobin and oxyhemoglobin have a high binding affinity for NO and attract and trap it, thereby preventing its entry into vascular smooth muscle or platelets. [*See* HEMOGLOBIN; SMOOTH MUSCLE.]

C. Pharmacological Properties

NO released from perfused blood vessels (i.e., arteries or veins) or from cultured vascular endothelial cells causes direct endothelium-independent relaxation of superfused isolated strips or rings of artery and vein in the same or different species. The relaxing factors from numerous sources possess the same chemical properties and also inhibit platelet clumping and adherence to vascular endothelial surface: they are all NO.

D. Mechanism of Action

NO elicits its two principal actions by the same mechanism. After diffusion into nearby vascular smooth muscle cells and locally circulating or adherent platelets, it activates the cytoplasmic enzyme guanylate cyclase, which catalyzes the conversion of GTP to cyclic GMP. In turn, cyclic GMP causes vascular smooth muscle relaxation and the inhibition of platelet function.

Like hemoglobin, cytoplasmic guanylate cyclase is a heme-containing protein from which the heme groups can be detached. In experiments designed to compare the properties of the heme-containing and heme-deficient forms of guanylate cyclase, it can be observed that NO activates guanylate cyclase by heme-dependent mechanisms. The activation of heme-containing guanylate cyclase by NO can be inhibited by the addition of excess heme or hemoglobin to enzyme reaction mixtures.

III. Pharmacology of Nitric Oxide

A. Activation of Guanylate Cyclase

NO is a colorless odorless gas that is soluble in aqueous and organic solvents and is a decomposition product of numerous unstable nitroso and nitro compounds. NO gas is rapidly oxidized to the brown pungent gas nitrogen dioxide (NO_2). In solution, NO is rapidly oxidized to NO_2^- (nitrite) and NO_3^- (nitrate). In the mid-1970s, NO gas was first shown to activate guanylate cyclase present in the soluble or cytoplasmic fraction of cells and to stimulate the intracellular accumulation of cyclic GMP in various tissues and species. Chemical agents that decompose to liberate NO also activate guanylate cyclase, as would be expected. Such chemical agents include the nitrovasodilators nitroglycerin, sodium nitroprusside, and amyl nitrite. The knowledge that nitrovasodilators liberate NO and stimulate cyclic GMP formation led to the discovery in 1979 that NO is a potent vasodilator that works through the actions of cyclic GMP. NO activates guanylate cyclase by heme-dependent mechanisms involving the binding of NO to the heme group of the enzyme, with consequent formation of the NO–heme–enzyme complex. This enzyme form represents the activated state of guanylate cyclase and catalyzes the conversion of GTP to cyclic GMP at a rate that is 50- to 200-fold greater than that of the unactivated enzyme. Guanylate cyclase activation by NO can be inhibited by methylene blue and related oxidizing agents.

B. Vascular Smooth Muscle Relaxation

The vascular smooth muscle relaxation induced by NO in the whole animal causes vasodilation, which can lead to a decrease in blood pressure. A more localized action of NO may improve regional blood flow without necessarily decreasing systemic blood pressure. Solutions of sodium nitroprusside spontaneously liberate NO, whereas nitroglycerin and amyl nitrite permeate smooth muscle cells and engage in biotransformation reactions, resulting in the formation of NO.

C. Inhibition of Platelet Function

The mechanism of action of NO on platelets is identical to that by which NO relaxes vascular smooth muscle: through cyclic GMP accumulation. In turn, cyclic GMP triggers other intracellular events that result in the inhibition of platelet function. Those nitrovasodilators that spontaneously liberate NO in solution (e.g., sodium nitroprusside) or that can be biotransformed to NO within platelets (e.g., isosorbide dinitrate and amyl nitrite) also inhibit platelet function.

D. Other Actions

In addition to vascular endothelium, leukocytes such as macrophages and neutrophils can synthesize NO or a closely related unstable nitroso compound. The function of NO in leukocytes is unclear, except that its formation is linked to the activated state of leukocytes. That is, leukocytes actively engaged in the phagocytosis (i.e., killing) of engulfed microorganisms also generate large quantities of NO, NO_2^-, and NO_3^-. The NO may function to destroy engulfed microorganisms, to further activate leukocyte function, and/or to dilate nearby blood vessels and inhibit local blood clotting. [*See* MACROPHAGES; NEUTROPHILS.]

IV. Secondary Messenger Role of Cyclic GMP

A. Secondary Messenger Concept

The secondary messenger concept pertains to the manner in which a signal is communicated from the extracellular to the intracellular compartment for the purpose of initiating a cellular response to an external stimulus. Often, the first messenger, such as a hormone, comes into contact with target sites or receptors on the cell surface. The hormone–receptor interaction triggers a response within the cell to generate a second signal, usually by increasing the concentration of another chemical, the secondary messenger, which then triggers the cellular response, thereby mediating the action of the first messenger. The secondary messenger may be cyclic GMP, cyclic AMP, calcium, or other chemical. NO acts in an even simpler way because it is lipophilic and diffuses into the cell to interact with an intracellular, rather than a membrane, receptor. The intracellular receptor for NO is the heme group bound to guanylate cyclase, and the secondary messenger is cyclic GMP.

B. Cyclic GMP

The information on cyclic GMP as a secondary messenger is relatively limited, essentially causing vascular smooth muscle relaxation and the inhibition of platelet function in response to NO. Cyclic GMP may also, however, play a role in regulating leukocyte function and cellular proliferation in general, but definitive conclusions cannot be drawn.

V. Formation, Release, and Metabolism

Unique enzymatic pathways are involved in the biosynthesis of endothelium- and macrophage-derived NO. The most plausible pathway is the enzymatic conversion of L-arginine to L-citrulline plus NO. Although the enzyme system has not yet been elucidated, experiments using radioisotopes have indicated that the basic (guanidino) amino nitrogen atom of L-arginine is oxidized and cleaved to yield NO and L-citrulline. The NO formed is labile and is rapidly oxidized to NO_2^- and NO_3^-, both of which can be detected in and around cells that form NO. The principal mammalian cells that can form these nitrogen oxides are vascular endothelial cells and activated macrophages.

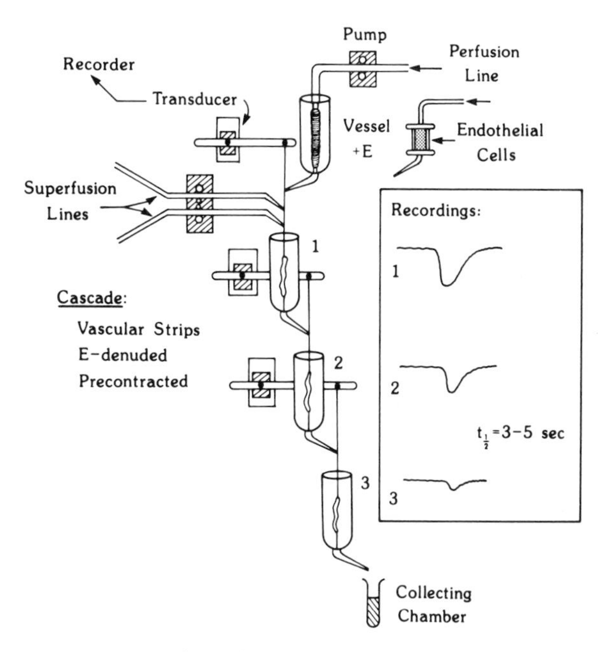

FIGURE 2 The formation, release, action, and metabolism of NO in response to circulating endothelium-dependent vasodilators (EDV). EDV interacts with receptors (R) on the luminal surface of endothelial cells, triggering the formation of NO from L-arginine. NO diffuses locally into the underlying smooth muscle to cause relaxation and into circulating or adherent platelets to cause the inhibition of platelet aggregation and adherence to the luminal surface of the endothelium. NO causes its effects in both smooth muscle and platelets by interacting with the heme (H) group of cytoplasmic guanylate cyclase (GC), thereby activating the enzyme and stimulating the formation of cyclic GMP from GTP. Cyclic GMP causes a lowering of intracellular free calcium levels, thereby resulting in vascular smooth muscle relaxation and the inhibition of platelet function. The actions of NO are rapidly terminated by spontaneous oxidation of NO_2^- and NO_3^- and by binding to hemoglobin (Hb) present in circulating red blood cells (RBC).

NO, being lipophilic and of low molecular weight, exits the endothelial cell by simple diffusion through the membrane. This mechanism requires no expenditure of cellular energy and allows the NO to exit the cell in all directions, thereby reaching the underlying smooth muscle and the nearby platelets on the opposite side as well. Although macrophage-derived NO serves primarily an intracellular function, it is highly likely that it diffuses out into the local environment, where it can act on nearby blood vessels and platelets.

NO is rapidly metabolized. In the presence of its natural environment of oxygen and oxygen-derived radicals, any NO formed will be oxidized rapidly to NO_2^- and NO_3^-, which are relatively inactive on blood vessels and platelets.

Figure 2 is a schematic illustration of the formation, release, action, and metabolism of endothelium-derived nitric oxide.

VI. Physiological Significance

A. Healthy State

Although the existence, identification, and pharmacological actions of the relaxing factor are clear, less than dogmatic conclusions can presently be drawn concerning its precise role in normal physiology. However, our knowledge of its chemical and biological properties, together with our current understanding of the actions of other regulatory molecules, allow some logical and informative speculation on its physiological significance.

NO plays a regulatory role in influencing local tissue blood flow and blood clot formation. A small quantity of NO is released constantly from the vascular endothelium, maintaining tissue perfusion by keeping arterioles open and free of blood clots. The release of NO is determined by stimuli originating in the body, such as the molecules listed in Table I, local increases in blood flow, and pressure, or shear, at the endothelium–blood interface. The platelets release ADP during clumping, and ADP can act locally to release NO. Thus, NO is likely to be involved in local, rather than more diffuse, or systemic, regulation of blood flow and platelet function. NO works in concert with a second endothelium-derived chemical, prostaglandin I_2 (i.e., prostacyclin), that, through the secondary messenger action of cyclic AMP, also causes vasodilation and inhibits platelet aggregation. When the two substances are present together, there is great enhancement of the inhibition of platelet aggregation, thereby ensuring minimal local blood clotting, if any.

B. Pathological or Disease States

In light of the biological actions of endothelium-derived NO, it is clear that cardiovascular disease states such as hypertension, myocardial infarction, and cerebral stroke may be attributed to its deficiency. Damage to vascular endothelial cells will result in a greatly diminished capacity to generate NO (as well as prostacyclin). Hypertension, atherosclerosis, and diabetes are often characterized morphologically by the abnormal appearance of the vascular endothelium and pathologically by elevated blood pressure and blood clot formation. Local vasoconstriction or vasospasm, with or without local thrombus formation, greatly curtails regional blood flow, with severe consequences when it occurs in the heart or the brain. [*See* ATHEROSCLEROSIS; HYPERTENSION; STROKE.]

Bibliography

Bassenge, E., and Busse, R. (1988). Endothelial modulation of coronary tone. *Prog. Cardiovasc. Dis.* **30,** 349.

Forstermann, U. (1986). Properties and mechanisms of production and action of endothelium-derived relaxing factor. *J. Cardiovasc. Pharmacol.* **8,** S45.

Furchgott, R. F. (1983). Role of endothelium in responses of vascular smooth muscle. *Circ. Res.* **53,** 557.

Ignarro, L. J. (1989). Endothelium-derived nitric oxide: Actions and properties. *FASEB J.* **3,** 31.

Ignarro, L. J. (1989). Biological actions and properties of endothelium-derived nitric oxide formed and released from artery and vein. *Circ. Res.* **64,** 1.

Marletta, M. A. (1988). Mammalian synthesis of nitrite, nitrate, nitric oxide, and N-nitrosating agents. *Chem. Res. Toxicol.* **1,** 249.

Moncada, S., Palmer, R. M. J., and Higgs, E. A. (1988). The discovery of nitric oxide as the endogenous nitrovasodilator. *Hypertension* **12,** 365.

Vanhoutte, P. M., Rubanyi, G. M., Miller, V. M., and Houston, D. S. (1986). Modulation of vascular smooth muscle contraction by the endothelium. *Annu. Rev. Physiol.* **48,** 307.

Waldman, S. A., and Murad, F. (1987). Cyclic GMP synthesis and function. *Pharmacol. Rev.* **39,** 163.

Energy Metabolism

BRITTON CHANCE, *University of Pennsylvania*

Glossary

Glycolytic activity Breakdown of sugars into simpler compounds such as pyruvate or lactate with the formation of small amounts of ATP

Metabolic controllers Inorganic phosphate (P_i); adenosine triphosphate (ATP), an "energy currency molecule"; adenosine diphosphate (ADP); phosphocreatine (PCr), an "energy currency molecule"

Metabolism Physical and chemical process common to all living organisms where energy is produced and transformed for survival purposes

Michaelis–Menten kinetics K_m, amount of a regulatory chemical to give half-maximal effect; V, velocity of adenosine triphosphate (ATP) synthesis necessary to equal that of ATPase; V_m, maximum velocity is defined as that of the maximal capability of the mitochondrial ATP synthesis as activated by the relevant control substrate (e.g., adenosine diphosphate [ADP]); ATPase, total breakdown of ATP in metabolic function

Mitochondrial respiratory chain Mitochondrion is an organelle that is the principal site of the generation of metabolic energy in the formation of ATP

Nicotine adenenine dinucleotide, reduced Formed from pyruvate and regulated by the activity of glycolysis

CONTROL OF ENERGY METABOLISM refers to the regulation of energy conservation to meet the needs of energy utilization.

I. Introduction

Years of work have gone into understanding the control of energy metabolism; it has been a literal "golden fleece" of most investigations on the biochemistry of human body organs. Energy metabolism is a key to the function of the replicative and metabolic systems. Enzymes, substrates, oxygen delivery, and energy conservation and degradation create a homeostatic state by which life can continue and multiply. Claude Bernard himself recognized *le milieu fixe* in 1878, and Walter B. Cannon recognized the physiological and some of the biochemical implications in 1939. A better understanding of feedback control in enzyme systems came with the quantitation of the exquisitely sensitive control of respiration by the adenosine diphosphate (ADP) level, with a K_m (the amount of a regulatory chemical to give half-maximal effect) of 10–20 μM establishing a "tight" negative feedback loop in which adenosine triphosphate (ATP) production can be precisely matched to ATP utilization. As indicated in Fig. 1, the control properties of mitochondria in which equal rates of output of ATP from the mitochondrial respiratory chain and the breakdown of ATP into ADP and inorganic phosphate (P_i) by functional activity identifies the steady state. ADP and P_i are fed back directly into the respiratory chain to resynthesize ATP and to ensure that no deficit occurs. Oxygen is reduced to water, and substrates are oxidized to carbon dioxide by flow in the respiratory chain. [ATP Synthesis by Oxidative Phosphorylation in Mitochondria; Adenosine Triphosphate (ATP).]

The important parameter of control of the mitochondrial respiratory chain is the characteristics for control by ADP and P_i (shown in the two bottom diagrams of Fig. 1); these diagrams are sometimes called "transfer characteristics". The rate of ATP synthesis is +dATP/dt. For excess P_i, this quantity

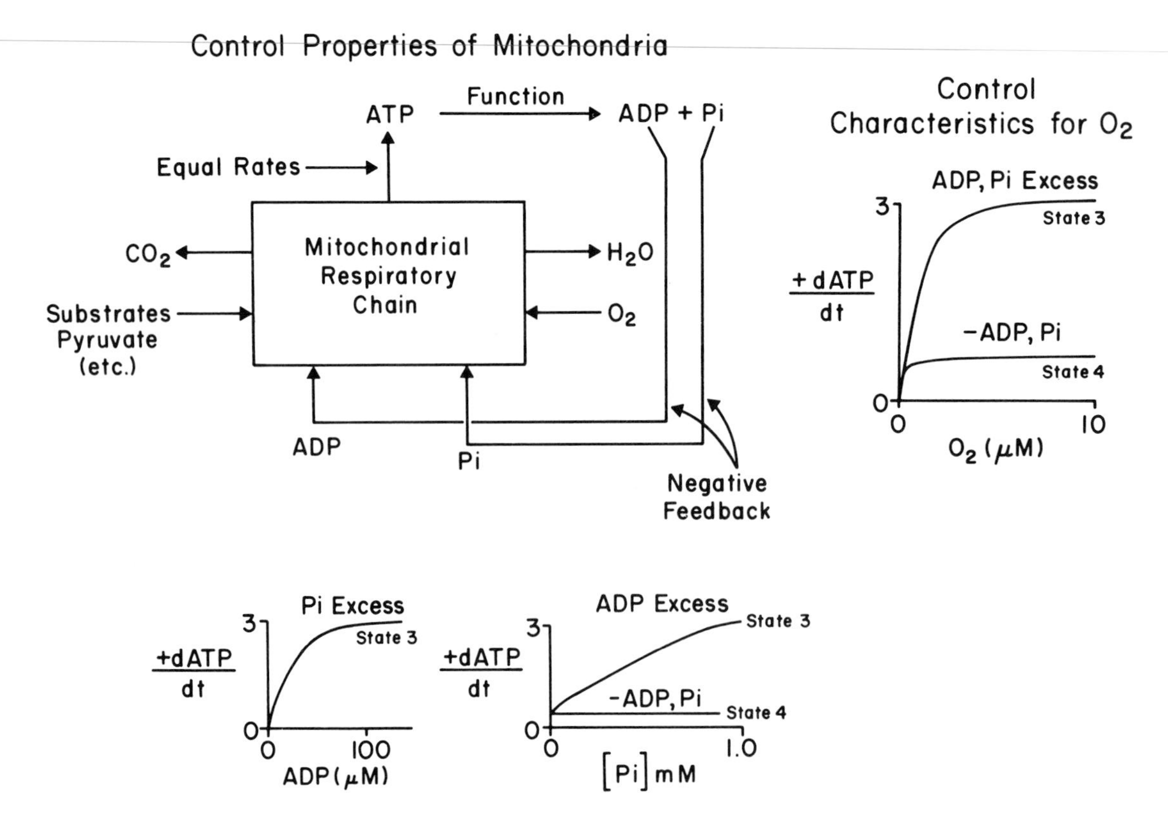

FIGURE 1 Feedback diagram for ADP (ATP?) control.

is hyperbolically related to the ADP concentration to give finally state 3, or maximal rate of ATP syntheses. Fig. 1 expresses this mathematically. Note that the scale for ADP control is 0–100 μM. The control characteristic for P_i control is on a millimolar scale, 0–1 mM, and for excess ADP has a K_m of >0.5 mM. However, if ADP is limiting, phosphate will have very little effect, as indicated by the state 4 transfer characteristic. Finally, we can have the control of respiration by the oxygen concentration. If ADP and P_i are in excess, the state 3 rate can be obtained under oxygen control; if ADP and P_i are absent, then the state 4 rate will be controlled with somewhat lower oxygen concentrations.

The controls of great interest are those where ADP is in control but oxygen may become limiting, under which condition ADP will rise to restore equality of ATP synthesis and ATP breakdown, providing biochemical homeostasis over a wider range of conditions than would be possible if ADP were not under control. One of the basic theorems of metabolic control is that the controller concentrations should be small compared with the affinity constants for optimal regulation (see below).

A striking and further recognition of the existence of feedback control in cells and tissues is provided by the experimental demonstration of biochemical oscillations; only systems that have feedback control operating beyond the bounds of stability can exhibit these oscillations.

Determination of the ADP level in tissues has been elusive because of the nearly inevitable breakdown of ATP in biochemical assays and the presence of ADP in different compartments within a single cell. The use of phosphorus magnetic resonance spectroscopy (MRS) as a basic research tool and as used in the clinical setting for the study of metabolizing systems in which the creatine kinase activity was sufficient to maintain creatine, ATP, phosphocreatine (PCr), and ADP near equilibrium, together with the MRS observation of PCr and P_i from the formation of P_i from PCr, revolutionized the study of metabolic control. In some organs of the body, tight feedback control is essential for organ function, as in heart and brain, and for the variation in demands for ATP synthesis, as in the rest–work transition for skeletal muscle.

II. Cell Compartmentation

The two main compartments of the eukaryotic cell, the highly organized cells of all higher organisms, are the cytoplasm and its mitochondrial spaces insofar as energy metabolism is concerned. Nuclei, endoplastic reticulum, and vacuolar spaces have their own roles and their particular contribution to cell function. But we need principally to concern ourselves with the cytoplasmic/mitochondrial relations, recognizing that the cytoplasm can contain the functional organelles, the myofibrils, etc. In the neuron, the synaptic junction, dendrites, axons, and transmembrane ion gradients constitute the corresponding "compartment" where functional activities demand ATP breakdown and its resynthesis. The mitochondrion is an autonomous organelle that has its own matrix space concentrations, transport mechanisms, and proton and electron gradients. For the purposes of this discussion, however, we will be able to explain metabolic control as a two-compartment system: in the cytoplasm, where ATP breakdown is employed to drive functional activity (i.e., in muscle contraction and ion transport), and in the mitochondrial space, where ATP is generated from oxygen reduction. In the cytoplasmic space, the glycolytic system under normoxic conditions is tightly regulated, so that the ADP and phosphate levels are only sufficient to provide adequate substrate for the mitochondria. Thus, the reduced nicotinamide⁻ adenenine dinucleotide (NADH) is available in adequate supplies so that ADP is the main regulator. Similarly, oxygen delivery is adequate under normoxic conditions, again resulting in ADP as the main regulator of oxidative metabolism. [*See* MITOCHONDRIAL RESPIRATORY CHAIN.]

III. Role of Chemical Controllers

Historically, phosphate was the main contender for regulation because, very obviously, it could be readily determined by the analytical biochemist and "made sense" as a regulator of metabolism. The idea that the affinity of ADP and phosphate for mitochondrial ATP synthesis would be different was pointed out, and, on the basis of simplistic kinetic arguments, namely the highest-affinity component would provide the most sensitive control, or ADP control. Researchers studied ascites tumor cells, yeast cells, skeletal muscle, and heart muscle, and sound evidence was produced in favor of ADP control. In the case of P_i, the remarkable discrepancy of the analytical chemical value and the actual changes in metabolic control was discovered in 1963.

The concept of perturbing the metabolic feedback loop by varying cell work and thus measuring the change of the metabolic control parameters, ADP and phosphate was realized only in the past decade with the use of phosphorus MRS (^{31}P MRS) of skeletal tissue where the work, ATP, P_i and PCr could be for the first time directly measured. It was immediately observed that P_i varied over very wide ranges initially it was equal to or greater than the *in vitro* K_m of mitochondria, and during work rose to 10–20 times the rest values. Thus, P_i is not regulating oxidative metabolism in the aerobic muscle. The K_{ADP} values calculated on the basis of the *in vitro* values for the creatine kinase equilibrium coincided almost exactly with the *in vitro* K_{ADP}. Thus, the ADP values required to maintain the appropriate ATP synthesis rate are correlated with that of the ATP breakdown rate due to muscle contraction. On this basis, the following simple Michaelis–Menten-type relationship was proposed in a new context:

$$V/V_m = \frac{1}{1 + (K_m/\text{ADP})}, \quad (1)$$

where the velocity V is defined as that of ATP synthesis, the maximum velocity V_m is defined as that of the maximal capability of the mitochondrial ATP synthesis, and the relevant control substrate is ADP. Equation 1 can be applied to the living system under conditions where the velocity of ATP breakdown is matched exactly by the velocity of ATP synthesis by adjusting the ADP level, phosphate being sufficiently high so that it is not a regulator of primary importance. For this relationship to hold, it is important to recognize that the measured quantity in skeletal muscle function is (external) mechanical work. However, the actual ATP breakdown is due to internal plus external work, and this work is provided by a complex system of muscle fibers of different types in different locations. Thus, the use of external work to measure internal ATPase can be precise only under the following conditions:

(1) Where the work load is such that the same fibers are used throughout the exercise; thus, recruitment of fibers and executing the work with movement of other parts of the body must be completely avoided.
(2) The pH should be no more than 6.9.

(3) The essential criterion of the steady state must be strictly observed, and is characteristic of slow twitch oxidative fibres capable of endurance performance. Fast twitch fibers are not capable of a steady-state function, the lack of adequate respiratory activity results in a system in which PCr and, indeed, ATP are continuously declining during exercise. Thus, slow twitch oxidative fibers must be recruited for the particular exercise. This means, in detail, that finger flexor muscles, which are often incapable of steady-state operation, can give results that would not be expected to meet the requirements of the steady state. Under certain conditions, the wrist flexor muscles have met the criterion of the steady state, whereas the leg muscles of the quadriceps group of the gastrocnemius muscle are most appropriate in humans, and correspondingly large, highly oxidative muscles in endurance performance animals are appropriate for study.

In other organs such as the liver, significant success in metabolic load perturbation has been achieved by stresses such as urogenesis, whereas in the brain epilepsy has been studied.

In the heart, the steady-state characteristics are more pronounced, and the requirements for steady-state function are more essential than in any other organ. Nature has proliferated control mechanisms so that not only is ADP maintained low, and thus available as a significant controller, but supplementary controls apparently involving NADH and possibly even oxygen delivery are superposed on ADP control, so that under most circumstances ADP itself is maintained in a homeostatic condition by regulations that are postulated to involve NADH delivery and oxygen delivery, as mentioned above. Thus, nature has followed the basic tenants of feedback control, namely (1) to employ a high gain system (i.e., a small change of concentration results in a large change of ATP synthesis [this favors ADP control over phosphate control]) and (2) to afford a multiplicity of high gain regulatory systems if one is to assure homeostatic conditions over the widest range of stresses (e.g., the heart may have three superposed feedback loops).

IV. Application to Diseases

The significant step forward from the concept of homeostasis as a physiological phenomenon to one of biochemical nature affords a much better coupling of feedback theory to health and disease. One of the striking examples has been the study of neonate brain, in which nearly 1,000 observations made in Philadelphia and in London have identified ADP values for which stability is obtained (measured as PCr/P_i of approximately 1 for which smaller values lead to brain atrophy and brain death). Thus, the margin between stability and life, and instability and death, is a close one for the neonate. The adult brain apparently operates at a greater margin of safety: PCr/P_i of 2–3 is usually observed. Genetic and metabolic diseases, which interrupt the delivery of substrate or delete an enzyme directly involved in ATP synthesis, cause loss of control and severe lactic acidosis in patients with deficiencies of the respiratory chain.

The most important concept emerging from these studies is that of stability and instability and criticality of metabolic control. The management of illnesses that involve oxygen or substrate delivery to the mitochondria are characterized by instability phenomena, which can be directly quantified by ^{31}P MRS of the particular organ and based on a single measurement; namely the PCr/P_i value of brain, heart, and skeletal tissues can indicate whether ADP is low enough to be in a stable region or has risen to values near the plateau of the rectangular hyperbola for metabolic control, thus providing the system with marginal stability.

V. Heterogeneity of Tissue Damage

Coupled to the concept of feedback stability and instability is the heterogeneity of metabolic demand (e.g., in brain, where some neurons may be extremely active as evoked by a particular physiological function and adjacent neurons may be highly inactive). Under these conditions, hypoxic stress could have a more dramatic effect on the neurons in which $\dot{V}$ (see Eq. 1) is high due to the necessary work level than on those in which $\dot{V}$ is low due to the lack of stimulation. Thus, hypoxic stress can inflict more damage on the highly functional neurons than on those that are substantially at rest.

Numerous other applications of these basic principles of metabolic control will emerge as greater localization of MRS is achieved and as noninvasive methods to measure oxygen delivery, particularly to heart and brain, become available for clinical study.

Bibliography

Chance, B., Leigh, J. S., Jr., Clark, B. J., Maris, J., Kent, J., and Smith, D. (1985). Control of oxidative metabolism and oxygen delivery in human skeletal muscle: A steady-state analysis of the work/energy cost transfer function. *Proc. Natl. Acad. Sci. USA* **82,** 8384–8388.

Chance, B., Leigh, J. S., Kent, J., McCully, K. Nioka, S., Clark, B. J., Maris, J. M., and Graham, T. (1986). Multiple controls of oxidative metabolism of living tissues as studied by ^{31}P MRS. *Proc. Natl. Acad. Sci. USA* **83,** 9458–9462.

Lehninger, A. L. (1965). "Bioenergetics." W. A. Benjamin Co., Menlo Park, California.

Stryer, L. (1975). "Biochemistry." W. H. Freeman, San Franciso.

Wright, B. (ed.) (1963). "Control Mechanisms in Respiration and Fermentation." The Roland Press Co., New York.

Environmental and Occupational Toxicology

BERNARD DAVID GOLDSTEIN, *UMDNJ-Robert Wood Johnson Medical School*

Glossary

Carcinogen Chemical or physical agent capable of causing cancer: sometimes discussed as direct or genotoxic agents, which have the ability, by themselves, to cause cancer and indirect or promoting agents, which cause the promotion or progression of the cancer process initiated by some other mechanism

Regulatory standard In environmental and occupational toxicology, usually a numerical statement of the amount of a chemical or physical agent that is permitted to be present in air, water, food, or soil or is allowed to be emitted from a specific pollutant source

Risk assessment Characterization of the risk of a given chemical or physical agent, or relevant mixtures, based on identification of its hazard, assessment of the extent of exposure, and understanding of the dose–response relation

Safety assessment Standardized testing of chemical or physical agents for toxicity performed in laboratory animals or *in vitro*

Teratogen Chemical or physical agent capable of producing a change resulting in a fetal abnormality

ENVIRONMENTAL AND occupational toxicology is concerned with the adverse effects of chemical agents in the workplace and the general environment. The science of toxicology is at the interface between chemistry and biology, dealing with the interaction between chemical or physical agents and biological systems. Growing evidence of the adverse effects to humans and to ecosystems of chemicals in the workplace and general environment has led to increased emphasis on the science of toxicology, including an understanding of the processes by which toxic agents produce their effects, and a focus on the ability to predict and prevent toxicity.

The traditional background of toxicology is the study of poisons. As a science, toxicology is usually traced back to Paracelsus, a 16th Century physician and alchemist who taught that the dose makes the poison. Simply stated, all chemicals can be conceived of as poisonous, there is only a question of the amount. Dose response remains a central feature of modern toxicology.

I. Introduction

In the late 17th century, Ramazzini, the father of occupational medicine, published a major treatise on diseases of workers. He clearly recognized that different diseases were associated with different work conditions, including problems caused by chemicals in workers such as chemists themselves. In the 19th century, with the growth of a chemical industry initially in Europe, the effects of chemicals on workers became even more apparent. Observations included aplastic anemia caused by benzene, methemoglobinemia, and bladder cancer in aniline dye factory workers, and the first deaths caused by compounds such as arsine. Toward the end of the 18th century, Sir Percival Pott noted an increased incidence of scrotal cancers among chimney sweeps caused by the presence of cancer-causing components of soot. In the next decades, there was a gradual recognition that good hygienic practices could prevent these tumors in chimney sweeps.

Today there are about 5 million known chemicals, approximately 75,000 of which are in commerce and therefore have the potential for human exposure at the workplace or in the general environment. A major focus of occupational health is the total prevention of exposure to all chemical agents. However, what may be called the first rule of public health is that accidents will happen and exposures will occur. Various national and international organizations have developed standards for occupational and environmental exposures to chemicals, creating guidelines for allowable exposures. These have greatly evolved through the years, with an increasing number of chemicals subject to more stringent limitations. There have been two driving forces for this phenomenon: (1) observations of the adverse impact of chemicals on workers and on the general population; and (2) a great increase in the extent of knowledge concerning the basic toxicology of chemicals.

Standards intended to protect the general public from chemical and physical agents have also been developed, with the breadth of chemicals covered and the stringency of the standards being greatly accelerated in the past three decades. With a rare exception, environmental standards are more stringent than occupational standards. This reflects the greater time of potential exposure (lifetime compared with 8 hours daily, 5 days weekly for 45 years) and the likelihood of increased susceptibility in the general population (children, the aged, and the infirm compared with individuals healthy enough to work).

Toxicology can be divided into two types of approaches for the purposes of this chapter: Safety assessment is the relatively formal approach in which chemicals are carefully tested in laboratory animals or *in vitro* for the likelihood that they will produce toxicity; toxicological science consists of approaches to understanding the mechanism by which chemicals produce adverse biological impacts. Safety assessment began with the simple counting of dead animals after exposure at different concentrations of the chemical being tested, a procedure known as the LD_{50}, the dose that kills 50% of the animals. As toxicological science has evolved, the insights obtained have been applied to safety assessment, leading to much more sophisticated animal tests of greater validity to the protection of humans. For noncarcinogens, the safety factors are now applied to the NOEL, the no observed effect level. For carcinogens, a risk assessment (extrapolation) process is used. Understanding of basic mechanisms of chemical action has also lead to development of highly useful *in vitro* tests applicable to safety assessment. However, such test tube approaches are currently only valid as supplements and, despite the wishes of everyone involved, cannot totally supplant the testing of chemicals in live animals.

II. Basic Concepts of Toxicology

There are three fundamental concepts of toxicology: each chemical or physical agent produces a specific pattern of toxic effects; the response to the agent is dependent on dose; and response in laboratory animals is useful in prediction of response in humans. Factors affecting the toxicity of chemicals can be considered in terms of physicochemical variables (e.g., solubility and vapor pressure) or in terms of physiological variables. Particularly prominent among the latter are factors affecting absorption, distribution, metabolism, and excretion. Of note is that metabolism does not necessarily result in a less toxic agent. There are many instances in which it is the metabolic product or process that is responsible for toxic effects, including cancer causation.

Toxicological processes can be described in many ways. Temporally, exposures are often divided into acute, subchronic, and chronic. Effects may be described as immediate or delayed and reversible or irreversible. Immediate effects from acute exposure are usually recognizable and readily attributed to the chemical insult. It is much more difficult to detect the insidious development of subtle irreversible effects leading over long periods of time to chronic disease resulting from chronic exposure.

A limited number of final common pathways lead to changes recognized as disease. Chemical and physical agents are capable of initiating each of these pathways: including alteration of energy use; disruption of the synthesis of macromolecules such as protein and DNA; membrane damage leading to cell death; somatic mutations leading to cancer; a variety of different alterations producing immune dysfunction; and fibrosis and other changes in the intercellular matrix leading to scarring and organ dysfunction.

In addition to these general pathways, a variety of toxicological processes is specific to individual organs. For example, the iodine concentrating powers of the thyroid make thyroid cancer the endpoint to be feared after exposure to radioactive iodine, and the metabolic capabilities of the liver render it particularly susceptible to compounds with reactive metabolite intermediates.

To the public cancer is undoubtedly the most feared endpoint of exposure to chemicals in the workplace or general environment. Approximately 85% of human cancer is due to environmental factors, but that is based on environment in the broadest sense, which includes anything that is not fully dependent on genetic factors. A much smaller percent of human cancer is due to exposure to chemicals in the workplace or general environment. The mechanism of chemical carcinogenesis is generally thought of in terms of a mutation of genetic material in an individual cell followed by progression to a multicellular cancer. The initial cancer cell and its progeny are not responsive to usual stimuli leading to cell maturation and death. Exogenous chemicals and physical agents may play a role not only in causing the DNA change that initiates cancer but also may exert their cancer-causing effects through as yet poorly understood genetic or nongenetic mechanisms that promote the biological progression and growth of the tumor.

A major concern in environmental and occupational toxicology is the determination of the causes of variations in human susceptibility. Advances in toxicological science have shown that there are four major reasons for enhanced susceptibility to a chemical. They can be thought of in terms of four mechanisms.

1. For a given ambient level, there can be an enhanced uptake of a chemical into the body. Such variations often occur because of lifestyle factors such as exercise leading to increased inhalation, or outdoors activities leading to greater exposure to sunlight.
2. For a given uptake into the body, there can be an increased delivery of a chemical or its metabolites to a target organ. This occurs most frequently with compounds for which there is a variation in the extent to which the metabolism can occur, for example, carcinogenic polycyclic aromatic hydrocarbons.
3. For the same target organ dose of the chemical or its metabolites, there can be an increase in effects in the target organ. Enhanced responsiveness can be seen with certain inherited disorders but may also reflect the simultaneous occurrence of exposure to another chemical or the presence of a disease.
4. For the same degree of organ damage, there can be an enhanced effect on the overall organism. Greater susceptibility of the individual to a given degree of organ dysfunction generally reflects impaired reserve. For example, a certain extent of loss of lung function might impact on the elderly and those with chronic obstructive pulmonary disease without being noticeable in a healthy teenager.

III. Routes of Exposure

An important consideration in toxicology is the route of exposure. Compounds may have markedly different effects dependent on whether they are inhaled, ingested, or absorbed through the skin. For example, rapidly reactive compounds such as formaldehyde will not get past the initial barrier of the upper airways if inhaled, the gut mucosa if ingested, or the outer layer of the skin. However, a lipophilic compound such as benzene, for which toxicity is dependent on metabolism, will readily penetrate lung or gut and to some extent will also be absorbed through the skin. Thus, for benzene the route of exposure is not as important as it is for relatively reactive compounds such as formaldehyde or ozone or for compounds for which physical properties limit distribution (e.g., asbestos inhalation).

There is a need to be concerned about appropriately duplicating actual human exposure routes when extrapolating from laboratory data. A number of "surprises" have occurred (e.g., plasticizer components being found insoluble when hot tea is added to a plastic cup, but readily extractable when lemon is present; or fire-resistant chemicals in children's pajamas not being absorbed when tested directly on skin but readily extractable and absorbed when urine is present). Similarly, the bioavailability of a component of food can be greatly altered by such factors as pH dependence, the presence of binding materials, etc., in the food. Uptake into the body can also be altered by the physiological status of the ingesting organ. Thus, the relatively water-soluble gas sulfur dioxide on quiet breathing does not penetrate the human respiratory tract beyond the nasal pharynx but on exercise with deep oral breathing can reach the lung; the pH of stomach contents will vary; and abraded or cut skin can be far more permeable than normal skin.

IV. Classes of Chemical and Physical Agents

Toxic agents are classified in many different ways, including chemical class (e.g., metals, hydrocarbons); physicochemical properties (e.g., gases, reductants); sources (air pollutants, food contaminants); and functions (e.g., solvents, plasticizers). To facilitate discussion of the toxicity of chemical and physical compounds, an overview of certain classes of compounds is provided below. This is by no means an exhaustive or comprehensive review of toxic agents. [*See* CARCINOGENIC CHEMICALS.]

A. Solvents

Human exposure to solvents at the workplace, the home, and the outdoor environment is common. A major class of solvents consists of those used to dissolve nonaqueous substances. In general, these are relatively hydrophobic agents that are required to have low vapor pressures for many uses (e.g., paints and glues). Many hydrocarbon solvents have anesthetic-like properties. In general, the higher the vapor pressure the more rapidly a compound will induce such central nervous system effects as dizziness, lack of coordination, euphoria, and giddiness, leading to narcosis and anesthesia. Chemical structure–activity relations for anesthetic-like endpoints can be readily discerned among solvent groups. Thus, among benzene and the alkyl benzenes, the lower molecular weight compounds more rapidly induce narcosis, whereas the higher weight compounds, once narcosis has been achieved, are more difficult to get out of the body. This represents the physicochemical characteristics of these compounds being translated into their ability to cross the blood-brain barrier and be dissolved in brain lipid. The limitations of using chemical structure–activity relation as the sole means to predict toxicity are also illustrated by this group of solvents. It is only benzene that produces severe bone marrow toxicity, all the others are without hematological effects. Other solvents also have specific toxicity related to unique structural elements or metabolism. Thus, both hexane and methyl-*n*-butylketone are potent producers of human peripheral neuropathies, whereas compounds that have slightly different structures (e.g., heptane and methyl ethylketone) are not. This is due to the formation of a common neurotoxic metabolite hexanedione.

B. Metals and Trace Elements

The toxicology of metals illustrates the diversity of human exposure to chemical and physical agents. All routes of entry into the body are possible: for example, lead can be taken up after ingestion in food, water, or soil; can be inhaled as a small particulate from automobile exhaust or smelter output; or can be absorbed up through the skin when in the tetraethyl form. Some metals are essential nutrients but may be toxic in inappropriate concentrations or locations. Toxicity may differ depending on the chemical form of the element, particularly if the metal is complexed to a carbon. Further, although the toxicology of organic arsenic differs somewhat from inorganic arsenic, both are markedly different from arsine, the hydrogenated gas that causes a dramatic destruction of circulating red blood cells, an effect that does not occur with the other forms. Valence also can be important: for example, hexavalent chromium is carcinogenic, whereas trivalent chromium is not. Solubility also plays a major role in toxicity, as does charge, which determines whether the metal crosses biological membranes.

Lead is a thoroughly studied metal. It affects many organs. Allowable occupational and environmental exposures have repetitively been lowered as new research has indicated yet additional effects of lead. As an environmental poison, the major concern has been central nervous system effects in children. The ability of lead to cross the blood-brain barrier appears to be greater in young children. At high levels of chronic or subchronic exposure to inorganic lead, encephalopathic effects occur, which are characterized by a wide variety of symptoms including restlessness, irritability, headaches, poor performance in school, and eventual progression to convulsions, coma, and death. An impressive series of studies has produced compelling evidence that at much lower body lead levels, perhaps equivalent to those currently present in the general population or even lower, lead produces a decrease in cognitive abilities of the developing brain. Far less conclusive evidence has led to a suggestion that low level body lead also causes an increase in blood pressure. At concentrations frequently present in the workplace, lead may produce peripheral neuropathies, renal and hematopoietic effects. Long-term renal toxicity and gouty nephropathy have been ascribed to lead. The hematopoietic effects, although less debilitating, are often used as biological markers for lead body burden and effect. At relatively low levels, lead interferes at different steps in heme synthesis leading to a buildup of delta aminolevulinic acid and free erythrocyte protoporphyrins (FEP), the latter being commonly used in screening tests. At higher levels, lead causes anemia both through a decrease in hemoglobin synthesis and by shortening red blood cell survival. Despite a decrease in blood lead levels because of the substantial removal of lead from gasoline in the United States, which is only slowly being followed in Europe and elsewhere, many environmental and workplace sources of lead remain.

Mercury is among the many metals that have an affinity for the kidney. In the inorganic Hg^{2+} form, mercury tends to accumulate in the kidney, leading to chronic renal disease. The affinity of organic mercurals for the kidney is primarily a function of the rapidity with which the alkyl mercury bond is dissociated *in vivo*. Longer chain organic mercurals tend to dissociate rapidly and have a renal affinity similar to that of inorganic mercury. However, methyl mercury, although predominately ending up in the kidney, has a relatively high concentration in the brain, and the major manifestation is central nervous system toxicity. Toxicity to the brain is also foremost when considering elemental mercury, which is able to penetrate to the brain because it is uncharged. Once there, it is oxidized to Hg^{2+} and becomes trapped in the brain. ''Mad as a hatter'' is a phrase derived from the neuropsychiatric manifestations of mercury vapor poisoning among hatters. The central nervous system effects of methyl mercury differ somewhat from elemental mercury, being focused more on the sensory system. Both forms produce tremor. Renal effects, predominantly of mercury salts, range from death caused by acute tubular damage leading to anemia, to chronic renal disease often characterized by proteinuria and to the nephrotic syndrome.

C. Gases and Aerosols

The toxicology of inhaled substances is dependent on the intrinsic chemical and physical properties of the agent and the physiology of the lungs and upper airways. The depth of penetration of inhaled gases is to a large extent dependent on their solubility in the highly moist respiratory tract and their inherent reactivity. For example, ozone is less soluble in water than is sulfur dioxide, so ozone is able to penetrate to the level of the terminal bronchiole rather than being removed in the upper airway. But ozone is so reactive that only perhaps 0.1% or less reaches the deep lung. The penetration of particles is to a large extent size-dependent, with larger particles being removed in the upper airway. Of those particles that penetrate deeply into the lung, the responsiveness of the airway will depend on a number of factors including the pH, solubility, and chemical composition of the particulate.

The short-term pulmonary response of concern is bronchoconstriction, the narrowing of airways. This is a nonspecific and to some extent protective response against the inhalation of respiratory irritants but is also the basis for difficulty in breathing among asthmatics and individuals with chronic obstructive pulmonary disease. Bronchoconstrictive responses were presumed to be the major cause of the approximately 4,000 deaths that occurred during the 1952 London smog episode. Occupational asthma covers a wide variety of acute bronchoconstrictive insults, mostly representing an allergic response to a workplace compound. Individual variation in susceptibility is common, but for some agents (e.g., toluene diisocyanate) a relatively large proportion of the population can be at risk. Typically, the acute response is greatest on return to work (Monday morning asthma) and ameliorates during weekends or vacations.

Two types of mechanisms of bronchoconstriction have been observed. One is a pharmacologic-type response occurring through direct chemical or vagus nerve mediation of airway smooth muscle tone such as occurs with allergies. This appears to be readily reversible and, in a situation such as sulfur dioxide exposure and certain allergens, modulates its effect on continued exposure. The second general mechanism of bronchoconstriction is associated with an overall inflammatory response (e.g., the response to phosgene and ozone), which worsens on further exposure and leaves more concern for eventual chronic effects.

Other acute responses to air pollutants include irritation of the throat and, particularly for photochemical air pollutants, the eyes. Of potentially greater health importance, a number of pollutants are able to potentiate bacterial respiratory tract infections, at least in laboratory animals. In the case of ozone and nitrogen dioxide, this appears to occur through interfering with the ability of alveolar macrophages to produce bactericidal free radicals after phagocytosis. [*See* TOXICOLOGY, PULMONARY.]

More difficult to determine, but of greatest consequences, are the potential long-term effects of inhaled compounds. Experience at the workplace has clearly demonstrated the relation of various dusts to chronic lung disease and to cancer. Most notable is asbestos exposure, which can lead to three diseases: asbestosis, a chronic fibrosing disease of the lung that causes death and debilitation because of replacement of gas exchange tissue with scar; mesothelioma, a usually fatal tumor of the outer lining of the lungs or abdomen; and bronchogenic lung cancer, a tumor usually associated with cigarette smoking. Of note is that there is a synergistic multiplicative interaction between cigarette smoking and asbestos exposure. A similar interaction is observed with exposure to radon in miners, which has led to further emphasis on the cessation of smoking for those who live in homes in high natural radon areas. Other occupational groups with a high risk of chronic lung disease include coal miners and textile workers. The causation or exacerbation of chronic lung disease by outdoor air pollutants has occurred in the past because of the gas–aerosol complex caused by the uncontrolled burning of fossil fuels and may now be occurring because of ozone and other photochemical oxidants stemming primarily from the use of automobiles and from acid particulates. [*See* ASBESTOS; TOBACCO SMOKING, IMPACT ON HEALTH.]

Air pollutants are not restricted to the outdoor environment. For many common pollutants, particularly solvents, exposure levels are likely to be higher indoors than out. Such problems as radon, offgassing of pollutants from plastics and construction materials, the presence of asbestos, and household solvents are of potential health importance.

V. Risk Assessment

Risk assessment as a formal process has become an important part of decisionmaking regarding control of environmental and occupational chemicals. In theory, the assessment of the extent to which a chemical presents a risk should be able to proceed independently of decisions concerning the management of the chemical; management decisions often involve other factors such as technical feasibility, economic cost, and a host of political considerations. Part of the reason for the recent emphasis on risk assessment has been a perception among the public that the political process has interfered with the unbiased assessment of the risk of chemical and physical agents. In the United States, a major recent impetus to risk assessment in the federal regulatory process was a report from the National Academy of Sciences in 1983 that described and defined the process.

Risk assessment consists of four components: hazard identification, dose response estimation, exposure assessment, and risk characterization.

A. Hazard Identification

Toxicity is an intrinsic property of a chemical or physical agent. Although all compounds are poisonous at a sufficient dose, the poison does not attack the same target organ in each case. For example, asbestos is capable of producing tumors in the lung but not in the brain. Identification of a human hazard usually occurs through high-dose human exposure in the workplace or through accidental or intentional poisoning. In most cases, hazard identification is reasonably straightforward, but in others, the lack of certainty greatly impinges on regulatory activities. For example, there are only approximately two dozen known human carcinogens defined as such on the basis of evidence in human beings. However, there are almost 200 compounds that have been identified to be carcinogenic in laboratory animals and thus have some degree of likelihood of also being carcinogenic in humans. Further, there are many other compounds for which there is concern about potential for carcinogenicity based on such findings as similar chemical structure to known carcinogens, positive short-term tests for mutagenicity, etc. In essence, there is a continuum of evidence ranging from great likelihood but no proof of human carcinogenicity, down to the most minuscule of evidence. The regulator is forced to draw a line through that continuum in deciding which compounds deserve to be regulated to protect human health. As reasonable scientists will differ as to the likelihood that the compound will turn out to be a human carcinogen, and as a straight line through such a continuum means that some compounds will be just above or below the line that is used for regulation, the hazard identification step for carcinogens will always lead to controversy.

B. Dose Response Estimation

The identification of hazard is a qualitative step. Quantitative risk assessment requires dose response data that can be obtained from animal studies or epidemiological studies of humans. In either case, extrapolation is necessary to determine the pertinence of the observation in the dose range and the population of concern. The problem of extrapolating from high dose to low dose has been central to questions concerning the appropriateness of risk assessment as a tool. A variety of models has been used to perform such extrapolations. Of first importance is to determine the expected shape of the dose response curve, which will differ depending on the endpoint and should be based on an understanding of the basic toxicology. Understanding the mechanism by which a given chemical produces cancer or other diseases can have a major impact on extrapolation of the data base to lower level human exposure. For example, much attention has been directed to the possibility that a different model is needed for genotoxic carcinogens, which are known to react directly with DNA, and nongenotoxic agents for which the effects on genetic material are not obvious. Among the latter, there may be a specific need for different models such as a receptor-based model for 2,3,7,8-tetrachlorodibenzodioxin, which seems to represent a new class of carcinogens acting through specific receptor interactions, in this case with a receptor related to those for estrogenic and other growth factors. A major recent advance in understanding dose has occurred through the use of physiologically based pharmokinetics. This approach provides better information on the actual dose to target tissue, thus permitting extrapolation from animals to humans in a more realistic fashion.

C. Exposure Assessment

The assessment of exposure is critical to risk assessment. No matter how hazardous a material, without exposure there is no risk. Although much attention has been focused on the dose response aspects of risk assessment, in many cases the scientific uncertainty associated with exposure is even greater and, as a rule, could be resolved much more readily. The methods for exposure assessment can be divided into three basic approaches. Using information about the sources and the location of the population at risk, we can use a variety of simple or sophisticated modeling techniques to determine the extent to which an individual might be exposed to the pollutant. The second approach is to determine analytically the ambient concentration of the compound of interest in air, water, soil, or food. Advances in personal monitoring techniques have led to improvement in the ability to typify the exposure of individuals. Third, biological monitoring involves taking biological samples from the population at risk. Thus the body burden of arsenic can be estimated through knowledge of its urinary excretion, or of lead from determination of blood lead level. New techniques capable of determining exposure through biological monitoring are being rapidly developed based on advances in toxicology and molecular biology. The ability to measure tiny amounts of organic chemicals adducted to macromolecules such as hemoglobin protein or blood cell DNA, and advances in analytical technique to measure urinary metabolites, has put many additional tools in the hands of those interested in exposure assessment.

D. Risk Characterization

After completing the various analytical steps, the risk must be characterized. It is in this risk assessment step that attempts to separate completely risk assessment from risk management often fail. One current issue concerns the extent to which the risk assessment process is appropriate for characterizing risk to the maximally exposed individual rather than simply the risk to a population. Another area of controversy in risk characterization concerns whether it is appropriate to give a single number, often as a plausible upper boundary of the risk estimate, or whether the risk should be quantified as a maximum likelihood estimate. In either case, there are those who advocate that the error boundary be placed around risk assessment figures in the hopes of improving communication of the uncertainties. Yet although the effort is to communicate to nonscientists, an error boundary can also mislead by ignoring the central tendency of the data.

The risk assessment process has a number of uncertainties associated with it. First and foremost, risk assessment must be understood as a process leading to information useful for regulation. Of great importance is that risk assessments be performed in an open, explicit, and logical way that has the confidence of the scientific and regulatory community. To enhance the process, various federal organizations, including the U.S. Environmental Protection Agency, have developed guidelines for the performance of risk assessment that describe the assumptions and processes. Deviation from the guidelines is possible but only if done openly and with a justification that will withstand rigorous scientific review.

The communication of risk to the public involves more than simply characterizing the risk. Individual perception of risk is conditioned by many factors, including the dread of the endpoint, the extent to which the individual feels a victim of external forces, the extent to which the individual participates or gains from the risk-taking activity, and, particularly in environmental issues, the sense of outrage reflecting the imposition of risk to individuals without their consent or involvement. These qualitative factors are subject to intense evaluation as the environmental regulatory community recognizes the need to obtain the informed consent of the public.

VI. Recent Trends in Environmental and Occupational Toxicology

A. Biological Markers

A biological marker is an indicator in a biological system of some event occurring elsewhere in that system. The use of biological markers is at least as old as Hippocrates, who diagnosed disease by inspecting the color of the urine. There has been renewed interest in the potential role for biological markers in environmental and occupational toxicology. This is due to recognition that advances in toxicology and in analytical sciences have led to the opportunity to use ethically obtained body samples from humans (e.g., blood, urine, saliva, and hair) as test organs to detect the presence or effect of relatively low levels of environmental agents.

Biological markers are divided into three types: markers of exposure, of effect, and of susceptibility. A continuum of events exists between exposure and disease so that certain markers can reflect both exposure and effect. It is ideal to be able to obtain information from one biological marker about both the extent of exposure and the degree to which an adverse effect is occurring. A reasonably good example is that of carboxyhemoglobin. Carbon monoxide combines with the oxygen-combining site of hemoglobin, thereby preventing the delivery of oxygen to the tissues. Measurement of carboxyhemoglobin provides an assay describing integrated levels of exposure to carbon monoxide in the past 8–12 hours as well as being an endpoint related to the basic mechanism of effect and thereby predictive of adverse consequences. An increased understanding of the mechanism of action of chemicals will lead to new biological markers capable of indicating both exposure and effect. In the case of genotoxic carcinogens, it will be essential to know the intermediate(s) responsible for the adverse effects of this agent and to be able to detect its presence bound to the nucleic acid of blood cells as a way of predicting its effect.

Biological markers of exposure and effect should be particularly powerful tools for risk assessment. The importance of biological markers to exposure assessment is described above. In addition, certain parts of the dose response curve are not now approachable by standard epidemiological or animal toxicology techniques. For example, at levels of benzene that are present in the general atmosphere, or in the usual instance of water contamination, it is necessary to extrapolate the risk from studies performed at much higher levels. It is inconceivable that a sufficient number of laboratory animals could be tested at one time to measure a cancer endpoint at realistic levels of benzene exposure, and it is beyond any reasonable possibility that standard epidemiological techniques will be able to assess such risk. However, in the future it might be possible to relate the presence of an adduct to DNA of a benzene metabolite with the extent to which benzene exposure had occurred. Determination of whether this occurs through a dose response pathway consistent with a one hit model (i.e., every molecule has a finite and equal chance of causing cancer) or through some other model would then be possible. Until that time, the prudence that characterizes public health approaches will preclude any change from the conservative assumption that any one molecule of a carcinogen has a finite risk of causing cancer.

Epidemiological studies will undoubtedly be greatly improved through use of biological markers of exposure. Much of the epidemiology relating chemicals to disease processes has relied on qualitative information (e.g., the presence of a worker in a specific location in the factory). The use of biological markers as indicators of the extent to which an individual has actually been exposed to a potential chemical hazard will greatly improve the strength of any observed epidemiological association and the power of the epidemiological study.

Biological markers of susceptibility may be of particular value in determining which individuals are at risk from exposure to chemicals. However, such markers will raise many ethical issues that should be addressed.

B. Multichemical Exposures

The focus in environmental and occupational toxicology has generally been on a single chemical. This is in keeping with the regulatory thrust, which tends to be one chemical in one medium. Unfortunately, we tend to be exposed to mixtures of chemicals either from the same source (e.g., in gasoline), from different sources in the same medium (e.g., in atmospheric ozone and sulfuric acid) or from different sources in different media (e.g., the apparent interaction between inhaled benzene and dietary ethanol). Study of the toxicology of such mixtures has been recognized as a major challenge that has been relatively untouched. Chemical interactions can be described under a number of different headings. When the effects of two chemicals are equivalent to the sum of the effects seen with either chemical acting alone, then the effects are said to be additive. If greater than the sum of the individual effects, it is known as synergism. Antagonism refers to a situation in which the effects are less than additive, and potentiation refers to a situation in which a chemical seemingly without any effect enhances the activity of another chemical. Although synergism is a major public concern. As the number of potential pollutant interactions is almost infinite, the only hope we have to deal intelligently with this problem is to understand the basic toxicological mechanisms by which these effects occur.

VII. Ecotoxicology

Ecotoxicology is a rapidly advancing field that is just beginning to get the attention it deserves. Although the chemicals that might affect human health are for the most part similar to the ones that affect ecosystems, this is not always the case. Furthermore, of central importance to understanding ecotoxicology is knowledge of the basic biology of the ecosystem.

There are four basic reasons to study the effects of environmental chemicals in ecological systems:

1. The canary principle—study the animal because it serves as sentinel for humans.
2. The ecosystem as a source of nutrition—study it because we eat it or we compete with it for food.
3. The need for propitiation—study it because of human needs other than nutrition (e.g., hunting, fishing, and the care of pets or lawns).
4. For its own sake—we recognize the importance, on this planet of limited size, of the protection of all living things.

Much attention has focused on the first three of these reasons, but it is likely to be the fourth that is most important in the long run.

Bibliography

Gallo, M., Gochfeld, M., and Goldstein, B. D. (1987). Biomedical aspects of environmental toxicology. *In* "Toxic Chemicals, Health and the Environment" (L. B. Lave and A. C. Upton, eds.), pp. 170–204. The John Hopkins University Press, Baltimore, Maryland.

Goldstein, B. D. (1988). Risk assessment/risk management is a three-step process: In defense of EPA's risk assessment guidelines. *J. Am. Coll. Toxicol.* **7,** 543–549.

Klaassen, C. D. (1986). Principles of toxicology. *In* "Casarett and Doull's Toxicology. The Basic Science of Poisons," 3rd ed. (C. D. Klaassen, M. O. Amdur, and J. Doull, eds.), pp. 11–32. Macmillan, New York.

National Research Council. (1989). "Biological Markers in Reproductive Toxicology." National Academy Press, Washington, D.C.

Wexler, P. (1988). "Information Resources In Toxicology," 2nd ed. Elsevier Science, New York.

Enzyme Inhibitors

MERTON SANDLER, *Royal Postgraduate Medical School, University of London*

H. JOHN SMITH, *University of Wales College of Cardiff*

GLOSSARY

Active site Area on the enzyme surface where the substrate is bound and catalysis occurs
Coenzyme Nonprotein component essential for the functioning of an enzyme
Enzyme Protein that, either alone or in combination with nonprotein components, catalyzes the reactions of its substrates
Substrate Substance that is modified by an enzyme

INHIBITORS ARE CHEMICAL AGENTS that decrease the ability of an enzyme to catalyze the reaction of its substrates. Enzyme inhibitors have been used extensively as tools to elucidate biochemical pathways in the body or, more importantly, as drugs to produce a useful clinical response by removal of the action of a particular enzyme (see Table I and examples in text). In the past, many drugs introduced into therapy as a result of microbiological or pharmacological screening tests have subsequently been shown to exert their action by inhibition of a specific enzyme in the body. On occasion, a drug introduced for one purpose has exhibited side effects due to its undesired inhibition of an enzyme in an unrelated biochemical pathway. Such findings have provided a "lead" compound as a starting point for the development of specific, potent inhibitors for the newly observed target enzyme. More recently, the impetus resulting from increased knowledge of the biochemical processes involved in the disease state has led to the rational design and synthesis of specific inhibitors toward identified target enzymes in such pathways.

I. Basic Concepts

A useful clinical response may be achieved by inhibiting a suitably selected enzyme, the blockade of which subsequently leads to a decrease in concentration of the product of the enzyme-catalyzed reaction or an increase in concentration of the substrate. Which effect is clinically important depends on the biochemical system concerned, as illustrated in Equation 1. In a biochemical chain (Eq. 1), consisting of its associated enzymes and substrates, where the final product (metabolite) has an action judged to be clinically undesirable, inhibition of the first enzyme in the chain, E_1, will reduce the metabolite concentration and alleviate the condition or disease (e.g., inhibition of xanthine oxidase to decrease conversion of purines to uric acid, which causes gout). Alternatively, where a metabolite is required for normal body functioning and is removed by a degradative enzyme, which is not part of the metabolites' synthetic chain (Eq. 2), inhibition of the degradative enzyme will lead to a desired clinical response as the concentration of the metabolite increases (e.g., inhibition of acetylcholinesterase to allow build up of acetylcholine at nerve endings in the treatment of glaucoma). This approach may also be used to preserve the action of a drug toward its target enzyme, where the drug is degraded rapidly either by the body or by a bacterial enzyme before it can exert its action (e.g., β-lactamase inhibitors). Here the enzyme inhibitor has the role of adjuvant drug.

TABLE I Some Other Target Enzymes for Drugs

Enzyme inhibited	Drug	Clinical use
Xanthine oxidase	Allopurinol	Gout
Carbonic anhydrase II	Acetazolamide, methazolamide, dichlorphenamide, ethoxzolamide	Glaucoma, anticonvulsant
Carbonic anhydrase	Sulthiame	Anticonvulsant (epilepsy)
Prostaglandin synthetase	Indomethacin, ibuprofen, naproxen, aspirin	Anti-inflammatory
Na^+,K^+-ATPase	Cardiac glycosides	Cardiac disorders
Riboxyl amidotransferase	6-Mercaptopurine, azathioprine	Anticancer therapy
L-dihydroorotate dehydrogenase	Biphenquinate	Anticancer agent
Aromatase	Aminoglutethimide, 4-hydroxyandrostenedione	Oestrogen-mediated breast cancer
H^+,K^+-ATPase	Omeprazole	Anti-ulcer agent
Trypsin and related enzymes	Gabexate mesylate, camostat mesylate	Pancreatitis and hyperproteolytic states
Plasmin	ε-aminocaproic acid (EACA), *p*-aminomethylbenzoic acid (*p*AMBA), tranexamic acid (AMCA)	Antifibrinolytic agent
Cholesterol synthesis enzyme	Meglutol	Hypolipidaemic
Aldhyde dehydrogenase	Nitrefazole	Alcoholism
Sterol 14α-demethylase of fungi	Miconazole, clotrimazole, ketoconazole, triconazole	Antimycotic
Aldose reductase	Sorbinil	Diabetes mellitus complications
Reverse transcriptase	Zidovudine	Acquired immunodeficiency syndrome
Succinic semialdehyde dehydrogenase	Sodium valproate	Epilepsy
Thymidine kinase and thymidylate kinase	Idoxuridine	Antiviral agent
DNA, RNA polymerases	Cytarabine (Ara-C)	Antiviral and anticancer agent
Aspartate transcarbamylase	Sparfosic acid (PALA)	Anticancer agent
Dihydrofolate reductase	Trimethoprim, methotrexate, pyrimethamine	Antibacterial, anticancer, and antiprotozoal agent
Transpeptidase	Penicillins, cephalosporins, cephamycins, carbapenems, monobactams	Antibiotics
Pyruvate dehydrogenase	Organo-arsenicals	Antiprotozoal agents
Alanine racemase	D-cycloserine	Antibiotic
Dihydropteroate synthetase	Sulphonamides	Antibacterial
Monoamine oxidase (MAO)	Iproniazid, phenelzine, isocarboxazid, tranylcypromine	Antidepressant
MAO A	Toloxatone, moclobemide, clorgyline	Antidepressant
MAO B	Selegiline [(−)−deprenyl]	Codrug with L-dopa for Parkinson's disease
Thymidylate synthetase	5-Fluorouracil	Anticancer agent
Aldehyde dehydrogenase	Disulfiram	Alcoholism
Bacterial urease	Acetohydroxamic acid	Chronic urinary infections
Formylglycinamide ribonucleotide aminotransferase	Azaserine	Anticancer
Peptidyl transferase	Chloramphenicol	Antibiotic
Ornithine decarboxylase	Eflornithine	Anticancer and antiprotozoal agent
Aromatic amino acid decarboxylase	Benserazide, carbidopa	Codrug with L-dopa for Parkinson's disease
Acetylcholinesterase	Neostigmine, physostigmine isoflurophate, ecothiophate	Glaucoma, myasthenia gravis

$$A \overset{E_1}{\underset{\uparrow}{\rightarrow}} B \overset{E_2}{\rightarrow} C \overset{E_3}{\rightarrow} D \text{ (metabolite)} \quad (1)$$

$$\underset{\text{(metabolite)}}{D} \overset{E}{\underset{\uparrow}{\rightarrow}} \text{inert product} \quad (2)$$

Occasionally, in bacterial infections, two drugs that act at different points in an essential biochemical chain of the microorganism (Eq. 3) may be administered. This approach is necessary when resistance to a single drug has built up due to prolonged use and resistant bacterial strains have emerged.

$$A \overset{E_1}{\underset{\uparrow}{\rightarrow}} B \overset{E_2}{\rightarrow} C \overset{E_3}{\underset{\uparrow}{\rightarrow}} D \quad (3)$$

Overall metabolite production by a biochemical chain may also be decreased by inhibiting an enzyme associated with production of a coenzyme essential for the functioning of an enzyme in the chain (Eq. 4).

$$A \overset{E_1}{\rightarrow} B \overset{E_2}{\rightarrow} C \overset{E_3}{\rightarrow} D$$
$$\text{coenzyme} \rightleftarrows \text{modified coenzyme} \quad (4)$$
$$E$$
$$\uparrow$$

II. Types of Inhibitors

A. Reversible Inhibitors

Reversible inhibitors bind to the enzyme by forming either an enzyme inhibitor (EI) or an enzyme-inhibitor substrate (EIS) complex through attractive forces, which do not involve covalent bond formation.

Competitive inhibitors compete with the substrate for the active site of the enzyme and, by forming an inactive EI complex, decrease the amount of enzyme available to combine with the substrate, as shown in Equation 5:

$$\begin{array}{l} E + S \overset{K_s}{\rightleftharpoons} ES \overset{k_2}{\rightarrow} E + P, \\ I \downarrow\uparrow K_i \\ EI \end{array} \quad (5)$$

where E is the enzyme, S is the substrate, I is the inhibitor, P is the product, K_s is the dissociation constant for the ES complex, k_2 is the rate constant for the breakdown of ES, K_i is the dissociation constant for the EI complex, and K_s is the substrate constant. The rate (v) of the enzyme-catalyzed reaction in the presence of a competitive inhibitor is modified in accordance with Equation 6:

$$v = \frac{V_{max}}{1 + \frac{K_m}{[S]}\left(1 + \frac{[I]}{K_i}\right)}. \quad (6)$$

where concentration terms are shown in square brackets and K_m is the Michaelis-Menten constant.

The extent that the rate of the reaction is slowed is determined by the inhibitor concentration and the value of K_i. A low value (10^{-6}–10^{-8} M) for K_i is characteristic for a potent inhibitor. The inhibition may be removed by increasing the substrate concentration (see Eq. 6), as would be expected for a competitive relationship between inhibitor and substrate.

The potency of an inhibitor is reflected in the K_i value, and this may be conveniently determined by the Lineweaver–Burk method. The rate of reaction (v) is determined for increasing substrate concentration in the absence and presence of a fixed concentration of inhibitor. Rearrangements of Equation 6 gives Equation 7. A plot of $1/v$ versus $1/[S]$ for the normal and inhibited reactions gives two lines that intersect on the $1/v$ axis at $1/V_{max}$ and on the $1/[S]$ axis at $-1/K_m$ (inhibitor absent) and $-1/K_m(1 + I/K_i)$, from which the value of K_i can be calculated.

$$\frac{1}{v} = \frac{1}{V_{max}} + \frac{K_m(1 + [I]/K_i)}{V_{max}[S]} \quad (7)$$

Noncompetitive inhibitors do not compete with the substrate for the active site but bind to the ES complex and prevent its breakdown to products (Eq. 8). The rate of the enzyme-catalyzed reaction (Eq. 9) is modified through the V_{max} parameter and, for a fixed inhibitor concentration, cannot be reversed by increased substrate concentration.

$$\begin{array}{l} E + S \rightleftharpoons ES \rightarrow E + P \\ I \downarrow\uparrow K_i \quad I \downarrow\uparrow K_i \\ EI + S \rightleftharpoons EIS \end{array} \quad (8)$$

$$v = \frac{V_{max}/(1 + I/K_i)}{1 + \frac{K_m}{S}} \quad (9)$$

A plot of $1/v$ versus $1/[S]$ in this instance gives converging lines for the normal and inhibited reactions, which intersect on the 1/[S] axis at $-1/K_m$, and at the $1/v$ axis at $1/V_{max}$ and $(1 + I/K_i)/V_{max}$, respectively. Most reversible enzyme inhibitors used as drugs are competitive so that noncompetitive inhibitors as well as other classes of reversible inhibitor (e.g., uncompetitive and mixed types) are not well known.

Examples

A key component of the renin-angiotensin system responsible for maintaining blood pressure is angiotensin-converting enzyme (ACE), which converts the inactive angiotensin I to the vascoconstrictor angiotensin II (Eq. 10).

$$\underset{\text{angiotensin I}}{\text{Asp-Arg-Val-Tyr-Ile-His-Pro-Phe} \updownarrow \text{His-Leu}} \xrightarrow{\text{ACE}} \underset{\text{angiotensin II}}{\text{Asp-Arg-Val-Tyr-Ile-His-Pro-Phe}} \quad (10)$$

Captopril (I) is a potent, reversible, tight-binding inhibitor of ACE. It is an orally active drug that lowers blood pressure in patients with hypertension. Further additions to this class of drugs are enalapril (II), which resembles captopril, and cilazapril (III), which was designed by computer-assisted graphic modelling to bind in a similar manner to captopril at the active site of ACE. Angiotensin I is considered to bind to the active site (Eq. 11) through an electrostatic bond formed between its terminal carboxylate ion and a protonated arginine residue on the protein, a hydrogen bond to the carbonyl of the amide and coordination with zinc of the carbonyl-oxygen atom of the amide bond to be broken. As illustrated in Equation 11, ACE inhibitors with similar stereochemistry have the same spatial arrangement of groups for binding to these points on the enzyme surface.

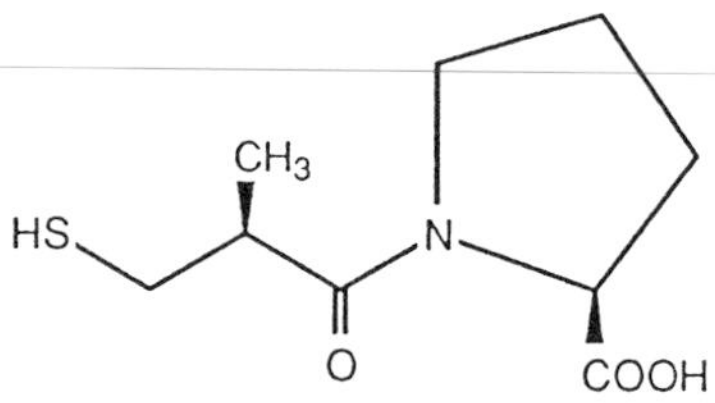

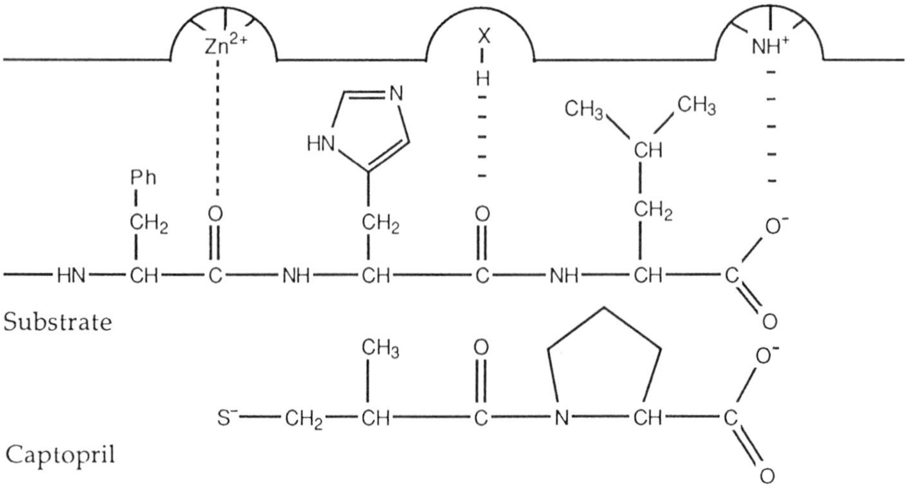

HS, CH_3, N, O, COOH

(I)

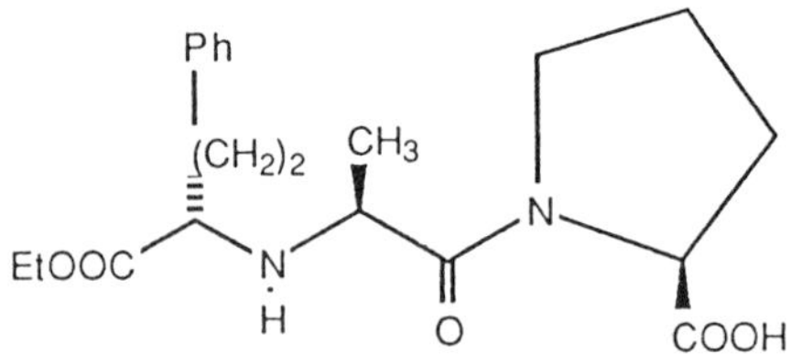

Ph, $(CH_2)_2$, CH_3, EtOOC, N, H, N, O, COOH

(II)

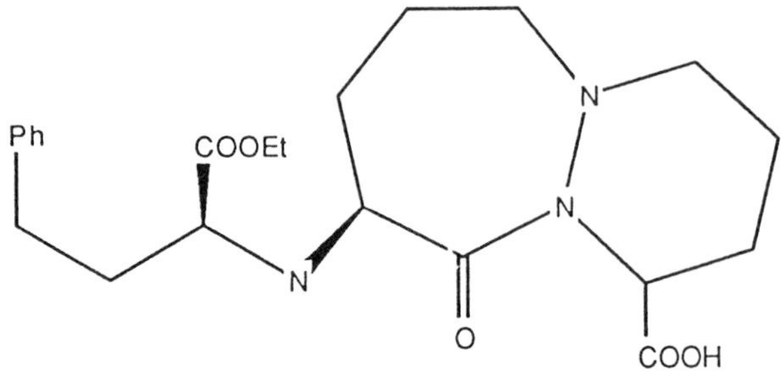

Ph, COOEt, N, N, N, O, COOH

(III)

Zn^{2+} X–H NH^+

Substrate: —HN—CH(CH_2Ph)—C(=O)—NH—CH(CH_2-imidazolyl)—C(=O)—NH—CH(CH_2CH(CH_3)$_2$)—C(=O)O^-

Captopril: S^-—CH_2—CH(CH_3)—C(=O)—N—CH—C(=O)O^- (proline ring)

(11)

A transition state analogue is a special type of competitive reversible inhibitor that binds more tightly to the enzyme than the substrate, as reflected in the K_i value (10^{-9}–10^{-11} M).

$$\begin{array}{ccccc} & & K_N\ddagger & & \\ E + S & \rightleftharpoons & E' + S\ddagger' & \rightarrow & E + P \\ K_s \downarrow\uparrow & & \downarrow\uparrow K_T & & \\ ES & \rightleftharpoons & ES\ddagger & \rightarrow & EP \\ & & K_E\ddagger & & \end{array} \qquad (12)$$

In an enzyme-catalyzed reaction, a high-energy activated complex, known as the transition state, is formed. The energy required for formation of the transition state is the activation energy and is the barrier to the reaction occurring naturally. A transition state analogue is a *stable* compound, which resembles the substrate portion of the unstable transition state.

The transition state for an enzyme-catalyzed reaction can be shown to be more strongly bound to the enzyme than the substrate(s), as follows. Equation 12 shows a substrate-enzyme reaction and the corresponding noncatalyzed reaction with the same bond making and breaking mechanism, where ES‡ and S‡′ represent the transition states, respectively, $K_E\ddagger$ and $K_N\ddagger$ the respective equilibrium constants, and K_s and K_T the respective association constants for formation of ES and the hypothetical binding of S‡′ to E. At equilibrium, $K_T K_N\ddagger = K_S K_E\ddagger$, which is modified to $K_T k_N = K_S k_E$, where k_E and k_N are the first-order rate constants for breakdown of the ES complex and nonenzymatic reaction, respectively. Rearrangement to $K_T = K_S k_E / k_N$ shows that because the ratio k_E/k_N is of the order 10^{10}, or more than the transition state, S‡′ must be bound more tightly to the enzyme than the substrate. Consequently, a transition state analogue, which bears some resemblance to the substrate part of the transition state for the enzyme-catalyzed reaction, will be a potent inhibitor of the enzyme.

Examples

Emphysema is a disease of the lung associated with loss of elastic recoil due to destruction of elastin by elastase. Human elastase is released by neutrophils in the lungs and, whereas in normal subjects it is removed by the plasma protein inhibitor, α_1-antitrypsin, an acquired (due to excessive smoking) or inherited deficiency of the inhibitor leads to the presence of excess elastase in the lung and lung degradation.

Elastase is a serine protease in which a serine hydroxyl group at the active site is activated by adjacent Asp-His residues to react with substrate in the ES complex to form (through a tetrahedral intermediate) an acyl enzyme with cleavage of a susceptible alanine or valine peptide bond (Eq. 13). The acyl enzyme is hydrolyzed through the Asp-His catalytic system, which activates water, on this occasion, to the carboxylic acid, with regeneration of the enzyme. The tetrahedral intermediate is stabilized by hydrogen bonding between its negatively charged oxygen and peptide NH groups (oxyanion hole).

The enzyme has an extended substrate-binding region of at least four or five subsites on either side of the bond to be hydrolyzed.

Certain aldehydes, such as elastatinal isolated from *Actinomycetes*, which contains an alaninal residue, and synthetic peptide aldehydes, which contain a valinal residue (e.g., $AdSO_2$-Lys-Pro-Val-H ($AdSO_2$, adamantanesulphonyl) are good inhibitors of the enzyme. The aldehyde inhibitors are considered to be transition-state analogues of the enzyme. They bind several orders of magnitude more strongly than the corresponding amide sub-

E—Ser·OH + R(C=O)NH–R₁ ⇌ ES Complex —k_2→ E—CH₂–O–C(R)(NH–R₁)(O⁻···H–X) ⟶ E—CH₂O—C(=O)R (Acyl enzyme) + R₁—NH₂ —k_3, HOH→ E + RCOOH

(13)

strates by forming a hemiacetal linkage between the aldehyde carbonyl group and the active site serine, so forming a tetrahedral intermediate reminiscent of that formed in the normal enzyme-catalyzed reaction (Eq. 14).

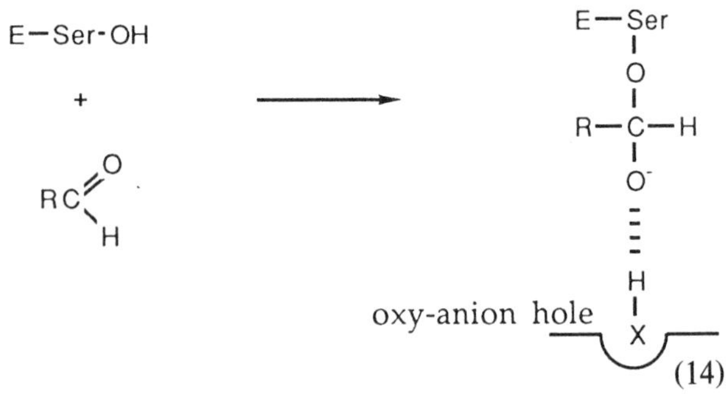

E—Ser-OH + RC(=O)H ⟶ E—Ser—O—C(R)(H)—O⁻ ⋯ H—X (oxy-anion hole) (14)

MeO-Suc-Ala-Ala-Pro-Boro-Val, which has the terminal structure -Pro-NH-CH($CH_2(CH_3)_2$)-$B(OH)_2$, is a trigonal boron compound, which is a transition-state analogue inhibitor of the enzyme. Trigonal boron compounds readily form compounds with OH^-, and these compounds form a tetrahedral boron compound with the serine alkoxide (Ser-O^-) group in a similar manner to that described for aldehydes, for example:

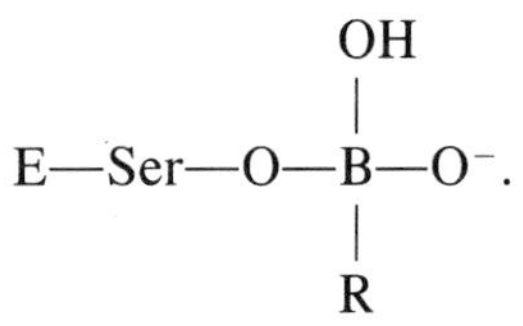

E—Ser—O—B(OH)(R)—O^-.

B. Irreversible Inhibition

1. Active Site-Directed Inhibitors

These resemble the substrate sufficiently closely to form an EI complex, resembling the ES complex, within which a reactive function on the inhibitor reacts with a functional group on an enzyme residue in or close to the active site, with formation of a covalent bond (Eq. 15). The inhibitor residue is firmly held at or near the active site of the enzyme and prevents access of substrate and subsequent catalysis.

$$\mathrm{E} + \mathrm{I} \underset{}{\overset{K_i}{\rightleftharpoons}} \underset{\text{complex}}{\mathrm{EI}} \xrightarrow{k_{+2}} \underset{\text{inhibited enzyme}}{\mathrm{E-I}} \quad (15)$$

A comparison of the relative potency of inhibitors toward an enzyme is achieved by comparison of their K_i and k_{+2} values, where k_{+2} is the first-order rate constant for breakdown of the complex to inhibited enzyme (Eq. 15). A more revealing comparison is given by the ratio k_2/K_i. These parameters may be determined as follows.

Integration of Equation 15 gives

$$k_1 t = \ln \mathrm{E} - \ln(\mathrm{E} - x), \quad (16)$$

where k_1 is the observed first-order rate constant, E is the initial enzyme content, and x is the inhibited enzyme content, and

$$k_1 = \frac{k_{+2}}{1 + \dfrac{K_i}{[\mathrm{I}]}}, \quad (17)$$

which, in the reciprocal form, is

$$\frac{1}{k_1} = \frac{K_i}{k_{+2}[\mathrm{I}]} + \frac{1}{k_{+2}}. \quad (18)$$

A plot of log(E − x) versus t (i.e., enzyme activity remaining with time), using a known concentration of inhibitor, gives a regression line with slope $-k_1/2.303$. A secondary plot of $1/k_1$ versus 1/[I] for different inhibitor concentrations gives a line where the intercept on the $1/k_1$ axis is $1/k_{+2}$ and that on the 1/[I] axis corresponds to $-1/K_i$.

Examples

Thrombin is a serine protease responsible for the clotting of blood by conversion of fibrinogen to fibrin. It is not present in the blood until released from prothrombin by a physiological stimulus. Effective chloromethyl ketone peptide or peptide aldehyde inhibitors of the enzyme possess a combination of the Pro-Arg present at the cleavage sites of factor XIII and prothrombin, together with D-Phe, rather than the Gly-Val-Arg of the natural substrate, fibrinogen. The potent irreversible inhibitor D-Phe-Pro-Arg-CH_2Cl is selective toward thrombin and at effective inhibitor concentrations, has little effect on other trypsinlike proteases such as factor Xa, plasmin, urokinase, and kallikrein, where an inhibitory action would upset normal body functions. Serine proteases are typically alkylated by peptide chloromethyl ketones on the nitrogen in the imidazole ring of the histidine residue, constituting the catalytic triad, Asp-His-Ser.

2. Mechanism-Based Enzyme Inactivators

These are also known as suicide substrates or k_{cat} inhibitors. These inhibitors act as substrates of the enzyme but are modified in such a way that *usually* a reactive function is formed, which covalently bonds with a reactive group present at the active site, so irreversibly inhibiting the enzyme. The reactive species (EI*) formed may degrade to innocuous products so that every turnover of the substrate may not lead to inhibition of the enzyme (Eq. 19). Partition between the inhibition event and product formation is measured by the partition ratio (k_{+4}/k_{+3}). Mechanism-based inactivators are not reactive species until armed by their specific target enzyme. Consequently, they show few toxic side effects due to this built-in specificity for their target.

$$E + I \underset{k_{-1}}{\overset{k_{+1}}{\rightleftharpoons}} \underset{\text{complex}}{EI} \xrightarrow{k_{+2}} EI^* \xrightarrow{k_{+3}} E - I \qquad EI^* \xrightarrow{k_{+4}} E + P \tag{19}$$

Kinetic analysis of Equation 19 follows that for Equations 16 and 17, where active site-directed inhibitors were considered, except that the kinetic parameters for the reaction are more complex. In the simplest case, where partitioning does not occur ($k_4 = 0$), then

$$k_1 = \frac{k_{+2}k_{+3}/(k_{+2} + k_{+3})}{1 + K_m{}^i/[I]} \tag{20}$$

and

$$K_m{}^i = \frac{k_{-1}k_{+3} + k_{+2}k_{+3}}{k_{+1}\,(k_{+2} + k_{+3})}, \tag{21}$$

where k_1 is the observed first-order rate constant and $K_m{}^i$ has replaced K_i in this more rigorous treatment.

Examples

Enzymes using pyridoxal phosphate as a coenzyme are concerned with amino acid modification and have provided a number of targets for the design of mechanism-based inactivators as potential clinical agents.

γ-aminobutyric acid (GABA) is the main inhibitory neurotransmitter in the brain, and its levels are regulated by GABA transaminase (GABA-T) (Eq. 22). Inhibitors of the enzyme allow a build-up of GABA, an effect that is useful in the treatment of epilepsy.

$$\underset{\text{GABA}}{H_2N{-}CH_2{-}CH_2{-}CH_2{\cdot}COOH} \xrightarrow{\text{GABA-T}} OHC{-}CH_2{-}CH_2{\cdot}COOH \tag{22}$$

Vigabatrin (IV; γ-vinyl GABA) is a time-dependent specific inhibitor of the enzyme and is used as an antiepileptic drug. One suggested mechanism for irreversible inhibition of the enzyme is shown in Equation 23.

Vigabatrin forms a Schiff base with pyridoxal phosphate present at the active site of the enzyme, followed by enzyme-catalyzed loss of a proton to form the vinylimine. Reaction between the terminal carbon of the vinylimine, which carries a positive charge, and a nucleophile (center of negative charge

(IV) + Co-Enzyme → Schiffs Base —H^+→ Vinylimine (: Nu-Enz) —H^+→ Inhibited enzyme (23)

or excess electrons [e.g., $-NH_2$]) on the enzyme surface leads to covalent bond formation and irreversible inhibition of the enzyme.

Certain bacteria produce β-lactamase enzymes and are resistant to the antimicrobial action of β-lactam antibiotics (e.g., penicillins [formula V] and cephalosporins. The enzyme hydrolyzes the β-lactam ring of the antibiotic to give ring-opened products that are inert as antibiotics.

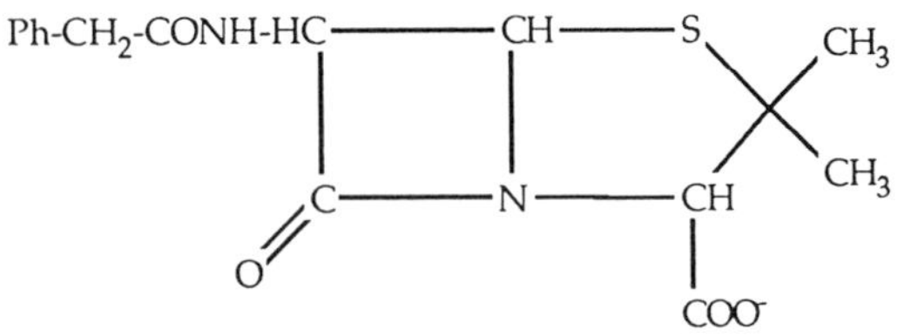

Ph-CH_2-CONH-HC—CH—S—C(CH_3)$_2$—CH(COO$^-$)—N—C(=O)

(v) Benzyl penicillin

The resistance of these bacteria to β-lactam antibiotics can be circumvented by the use of β-lactamase inhibitors as adjuvants in antibiotic therapy. β-lactamases are serine proteases and catalyze β-lactam ring hydrolysis, in the general manner previously described (Eq. 13), by initially forming an acyl enzyme (cf. Eq. 24).

Clavulanic acid, a mechanism-based inactivator of β-lactamase, is poor substrate of the enzyme. It forms an acyl enzyme, which is sufficiently stable to partition between the normal hydrolysis product and rearranged forms that are not hydrolyzed, leading to irreversible inhibition of the enzyme.

The rearrangements probably involve opening of the five-membered ring and removal of this species from equilibrium with the acyl enzyme by trapping of the oxygen anion as the more stable ketone form of the enolate anion. Clavulanic acid progressively inhibits the enzyme, and inhibition is complete after 150 turnovers. Here, chemical changes in the acyl enzyme, due to rearrangements, lead to irreversible inhibition, which is not the usual mechanism of action of this type of inactivator.

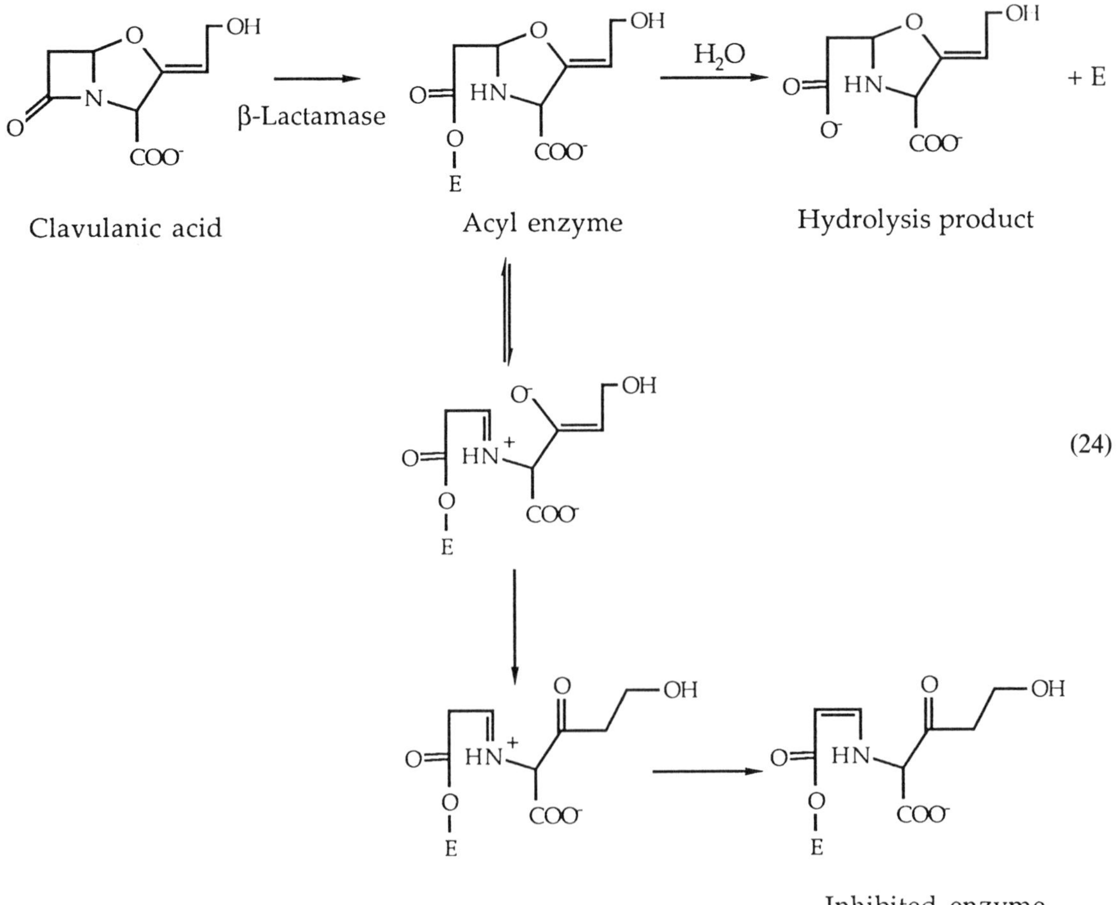

Clavulanic acid —(β-Lactamase)→ Acyl enzyme —(H_2O)→ Hydrolysis product + E

Acyl enzyme ⇌ (ring-opened enolate, O$^-$, HN$^+$, O–E) → (ketone form, HN$^+$, O–E) → Inhibited enzyme (24)

III. Selection of a Suitable Target-Enzyme Inhibitor Combination in Drug Design Strategy

The biochemical environment of a target enzyme may affect the selection of the type of inhibitor required satisfactorily to inhibit the enzyme.

In a biochemical chain of reactions in a steady state (Eq. 25), the concentration of the pool of the initial substrate (A) is under the influence of other chains and is reasonably constant. Reversible inhibition of E_1, which catalyzes the *first* step in the chain could be expected successfully to decrease production of the undesired metabolite (D). Generally, inhibition of an enzyme elsewhere in the chain would be expected to be less successful, due to build-up in concentration of the substrate concerned, which would overcome the block (see Eq. 6) because the throughput of (A) is reasonably constant. Rapid fluctuations in the level of (A), which could theoretically affect inhibition of E, have not been noted in practice.

$$A \xrightarrow{E_1} B \xrightarrow{E_2} C \xrightarrow{E_3} D \text{ (metabolite)} \qquad (25)$$

Irreversible inhibition progressively decreases the level of the target enzyme, wherever its position in the chain, and the biochemical environment of the enzyme is unimportant. However, the rate of synthesis of fresh enzyme by the body or pathogen to replace the inhibited enzyme, must be slower than the rate of inhibition of the enzyme, so as to maintain the desired effect. Consequently, the potency (k_2/K_i) of active site-directed inhibitors and the partition ratio for mechanism-based inactivators are important factors in their usefulness as potential drugs.

Specificity toward the target enzyme is an important consideration for potential clinical agents. Serious side effects could be apparent for all types of inhibitors where closely related enzymes with different biological roles (e.g., trypsinlike enzymes) are concerned. Active site-directed irreversible inhibitors, which could alkylate or acylate a range of essential tissue constituents containing amino or thiol groups (e.g., glutathione, proteins), could possess a more general toxicity.

The current emphasis on the design of mechanism-based inactivators as potential drugs revolves around their inherent lack of toxicity, specificity toward the target enzyme, and ability irreversibly to block a metabolic chain independent of the location of the target enzyme within the chain.

Bibliography

Barrett, A. J., and Salvesen, G. (eds.) (1986). "Proteinase Inhibitors." Elsevier, Amsterdam.

Sandler, M. (ed.) (1980). "Enzyme Inhibitors as Drugs." Macmillan, London.

Sandler, M., and Smith, H. J. (eds.) (1989). "Design of Enzyme Inhibitors as Drugs." Oxford University Press, Oxford.

Smith, H. J. (ed.) (1988). "Introduction to the Principles of Drug Design." 2nd ed. Wright, London.

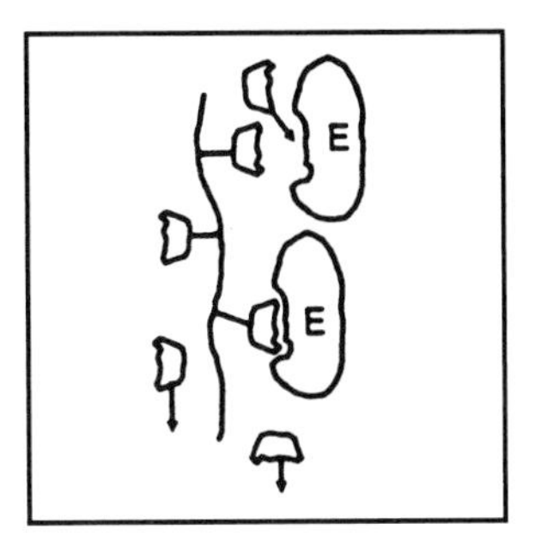

Enzyme Isolation

ROBERT K. SCOPES, *Center for Protein and Enzyme Technology, La Trobe University*

Glossary

Adsorbent Solid material that attracts otherwise soluble molecules to its surface

Denaturation Loss of structural integrity of a protein molecule, resulting in loss of its biological activity

Enzyme assay A test to determine the amount of enzyme activity present in a sample

Gel Filamentous molecules making a three-dimensional net, with large water-filled pores between the strands, through which large molecules, such as enzymes, can penetrate

Homogenize To break up a tissue in the presence of an extracting buffer so as to release cellular components into the soluble fraction

Hydrophobic Having the character of low solubility in water and high solubility in organic solvents

Ligand A molecule, generally small (i.e., a molecular weight of less than 1000), that interacts in a specific way with a macromolecule such as an enzyme; generally, a "biospecificity," representing a natural physiological process

Recombinant technique Use of DNA manipulation to enable genes to be expressed (i.e., to function) in a foreign "host" organism, giving rise to the synthesis of the protein encoded by the gene.

ENZYME ISOLATION is a phrase that encompasses many methodologies in protein chemistry. Enzymes are catalytic protein molecules responsible for all of the chemical reactions that occur in life. The isolation of enzymes is a necessary first stage in any detailed study; "isolation" here refers to a purification that may not completely isolate the specific enzyme molecules from all other material, but should go a long way toward obtaining a homogenous preparation of enzyme molecules suitable for studying the enzyme's structural, catalytic, and physiological roles. Techniques for protein isolation are varied, and the behavior of an enzyme studied for the first time is unpredictable. It is only *because* the properties of different protein molecules vary so much that we are able to isolate individual enzymes from the other proteins of the cell. An isolation scheme usually involves several steps and will most likely be unique for each enzyme.

I. Enzymes of the Human Body

A. Background

The whole of the life process is controlled by enzymes, molecules that catalyze the chemical reactions essential for all of the processes occurring in living organisms. The human body is no exception to this: Every move we make, every nutrient we digest, even every thought we have is ultimately controlled by enzymes and enzymelike interactions.

Every cell in our body contains thousands of different enzymes, each responsible for a particular function. The differing amounts and intracellular distribution of the enzymes cause one cell to differentiate from another, so that, for example, a liver cell does not behave the same way as a kidney cell. Each of these cells has a different complement of enzymes (though many not involved in the *specific* processes of the cell are common to both). A liver cell, arguably the most complex multifunctional type of cell in the body, has enzymes responsible for specific liver functions, such as regulating the glucose supply, detoxifying foreign substances, metabolizing fatty acids, and synthesizing serum proteins. The liver cell also has all of the enzymes necessary for the functioning of every cell in the body, for example, enzymes involved in protein and nucleic acid syntheses (and breakdown), cell growth and division, and in energy (ATP) generation.

One of the principal activities of biochemists in the 20th century has been the study of enzymes. Departments of biochemistry evolved in universities both from physiologists and from organic (i.e., natural product) chemists, and these origins are reflected in the dual nature of biochemists—those who are interested in the biological processes and those who wish to understand the chemistry behind them. The study of enzymes is the link between biology and chemical activity; in 19th-century terms, between the "life force" and basic chemistry. To study an enzyme, it is necessary to isolate it from the other constituents of the cell or tissue where it is found; little detailed work can be performed on unfractionated living material. Thus, over the years, methods have evolved for separating pure enzymes from the complex mixtures in which they exist naturally.

B. Enzymes from Other Animals

The study of an enzyme from humans may raise both ethical and practical problems. Whereas there is no problem in obtaining samples of plasma and, in most places, placental tissue, the source of other tissues must be cadavers or amputation, neither of which is satisfactory with a view toward investigating living, fresh, undiseased tissue. In addition, the possibility that technicians may contract diseases such as hepatitis or acquired immunodeficiency syndrome is a further obstacle. Consequently, most of what we know of human biochemistry has come from studying that of other mammals from baboons to rabbits, from horses to mice, on the assumption that the correspondence is close.

At the enzyme level, fortunately, the biochemistries are close; it is relatively rare for an enzyme isolated from another mammal to differ significantly from its human counterpart. At higher levels of cellular function, however, differences become more marked, for, after all, if all of the cells and organs of a pig behaved exactly as those of a human, we should not be able to distinguish a pig from a human! The distribution and amounts of enzymes, rather than their fundamental properties, cause differentiation in cellular behavior. Thus, we can be reasonably confident that studies of mammalian enzymes have direct relevance to human biology.

II. Purposes for Isolating Enzymes

A. Structural Studies

The isolation and study of an enzyme from a mammalian source are usually motivated by reasons beyond mere scientific curiosity. It is abundantly clear that any such study can impinge on medicine, as enzymes are responsible for most of our body's behavior, both normal and pathological. Those in the field of medicine aim to restore abnormal bodily behavior to normal and, wittingly or otherwise, frequently use drugs that attack abnormalities in the distribution (and sometimes behavior) of human enzymes. The detailed study of a misbehaving enzyme's structure and catalytic activity can lead to the design of a drug to counteract its misbehavior. Generally, such drugs will be inhibitors, slowing down or suppressing the activity of an enzyme that, for pathological reasons, is overactive in a given tissue. To design such a drug, a full study of the enzyme at the atomic level (e.g., X-ray crystallography) is necessary; thus, the pure enzyme must be isolated in reasonable quantity from a relevant comparable source.

B. Therapeutic Uses

In a few cases, and perhaps increasingly in the future, the lack of a normal enzyme function can be treated by infusion or injection of the enzyme itself. However, this can only presently be done for extracellular enzymes; ideally, the source should be human, to avoid immunological problems. Thus, factor VIII and other blood clotting factors are being used increasingly to treat hemophiliacs, and tissue plasminogen activator is being used to dissolve blood clots.

The problem of producing the proteins used in therapy is solved by DNA manipulation. It is possible to extract the human DNA (gene) responsible for the production of a particular protein or enzyme and introduce it to foreign cells in tissue culture, or to bacteria. These "host" cells then produce the gene product in safe conditions and in quantities that would be impossible using human tissues.

III. Isolation of Enzymes from Mammalian Tissues

A. Introduction

All enzymes are proteins.[1] Thus, the techniques for isolating enzymes can be the same as for isolating other proteins. However, because of their subtle structural design, many enzymes are more sensitive to inactivation (i.e., loss of enzymic activity) than other proteins with simpler functions. Consequently, techniques that have recently been widely used to isolate many proteins of interest (e.g., peptide hormones, viral antigens, and growth factors, which are mostly extracellular and are relatively sturdy molecules) may not be applicable to enzymes which are mostly intracellular and are being protected from the environment (i.e., less sturdy). The techniques used also depend on the reason for isolating the enzyme, in particular, the amount of end product required and whether it needs to be in the active state.

Useful work, directed toward gene isolation, can be done on a few micrograms of inactive enzyme, provided that it is homogeneously pure. On the other hand, the study of an enzyme's structure requires milligrams, if not hundreds of milligrams, of pure active product, so the scale of operations must be vastly greater. Generally, enzyme isolation involves one or more "classical" techniques for protein purification, perhaps followed by a more specifically designed step to select the chosen enzyme. These methods are described below. As every enzyme is different, there can be no generally applicable methodology, and the most successful processes frequently involve a step that may be unique to that enzyme or a small class of proteins.

B. Enzyme Extraction

The majority of enzymes exist in a soluble form, that is, in an aqueous environment: the cellular cytoplasm, the interior of organelles such as mitochondria, or extracellular fluids such as plasma or digestive juices. Other enzymes may be fixed in location, embedded in or attached to membranes or other solid material. To purify an enzyme, it is necessary to obtain an extract, an aqueous solution containing the dissolved enzyme. In most cases, all that is required (for intracellular enzymes) is to homogenize the tissues so that the cells are broken and the soluble enzymes are released into the suspending liquid. For membrane-bound enzymes, however, treatment with detergent (usually after isolating the organelles concerned) is generally needed, to separate the protein from the fatty component of membranes.

The first step is to identify the tissue that is the best, most convenient, source of the enzyme. It should contain the enzyme required in large amounts in a form easily released into the extraction buffer. Next, this tissue is broken down in a blender, using 2–5 vol of an appropriate buffer to one volume of tissue, yielding a "homogenate." The extract is prepared by centrifuging the homogenate to obtain a (fairly) clear aqueous solution containing all of the soluble components of the tissues. The enzyme could be less than one part in 1000 of all the proteins present.

Enzymes are protein molecules that are particularly vulnerable to a variety of stresses that cause them to lose their natural structure—they become "denatured." Such stresses must be avoided during the extraction and subsequent processing. Many things can cause enzyme denaturation and inactivation, including extremes of pH (acid or alkaline), high temperatures, organic solvents, many types of salts, and oxidizing agents. Generally these conditions are avoided unless the particular enzyme is particularly resistant to such stresses.

C. Enzyme Activity Measurement

It is axiomatic that an enzyme isolation cannot proceed without a method for measuring the enzyme's presence with reasonable precision. In virtually all cases, this is a measure of the enzyme's catalytic activity. Each enzyme requires a separate assay; hence, we can only generalize the methods used for detecting and quantitating activity. The assay method should be quick and easy, even if this means a sacrifice of accuracy. The amount of total protein should be known, but this is less important than knowing how much enzyme is present. During purification, as much enzyme activity is retained as possible, whereas the total protein decreases to a small amount.

Enzyme assay methods can be divided in many ways. One way is to group them into "continuous" methods, in which the progress of reaction can be

1. Exceptions are recently discovered pieces of RNA (ribozymes) with catalytic function that can be described as enzymes. This is not surprising, because RNAs can acquire three-dimensional folding similar to that found in proteins.

monitored continuously as it occurs, and "stopped" methods, in which the result of the enzyme's activity can only be determined after stopping the reaction and measuring how much product has been formed. These methods are not considered further here.

D. Fractionation Procedures

The most used methods for fractionating protein mixtures fall into one of three categories: precipitation (differential solubility), adsorption (batch or column chromatography), and methods in which the proteins remain soluble at all times.

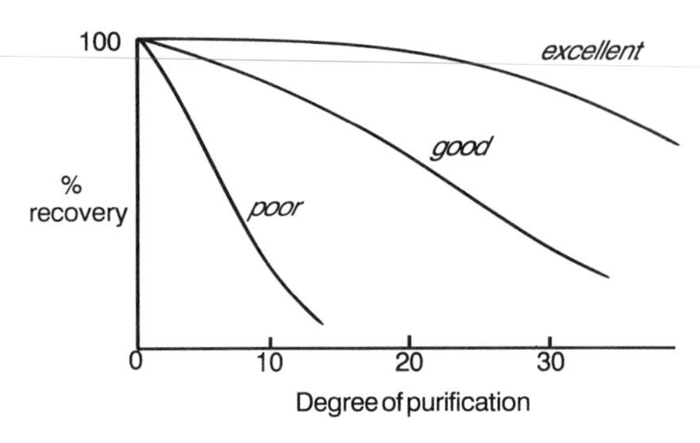

FIGURE 1 Relationship between overall recovery of an enzyme's activity and the purification factor (see text) achieved in the isolation step. Typical of "poor" steps are precipitation methods; nevertheless, these play an important part in an overall process.

1. Precipitation

Precipitation methods rely on the fact that in any mixture of proteins the solubility properties of individual proteins differ substantially, so that change in the properties of the solvent causes some proteins to aggregate, forming a precipitate, while others remain soluble.

Some of the ways in which solvent properties are changed include pH alterations (increased or decreased acidity), ionic strength (salt concentration), dielectric constant (organic solvent addition, which changes the strength of attraction between molecules), and water activity (addition of hydrophilic polymers to make less water available to the proteins).

A typical procedure is to take the extract containing all solubilized proteins and add a precipitant to a known concentration. For ionic strength increases (i.e., "salting out"), ammonium sulfate is normally used; up to 750 g of ammonium sulfate can be dissolved into 1 liter of extract. For changing dielectric constant, alcohol or acetone is added, up to 60% vol/vol, keeping the temperature close to 0° to minimize loss through denaturation.

After equilibration at a suitable concentration of precipitant, the aggregated proteins are removed by centrifugation and more precipitant is added to the supernatant. This may be repeated several times, until such a time when the enzyme of interest is found in the precipitate after it is redissolved in a suitable buffer. Such fractionation procedures typically result in only a modest amount of purification (i.e., a three- to 10-fold increase in specific activity, which is the activity per unit weight of total protein), with about 80% recovery of activity. They are most useful, however, as a first step, reducing a large volume of crude extract to a more easily handled volume of mainly proteinaceous material.

The aim of all fractionation schemes should be to maximize the recovery of active material, while minimizing the recovery of total protein (Fig. 1). The specific activity (i.e., the activity per milligram of protein) should increase at each step (the purification factor). The efficiency of each step is measured by the increase in specific activity, with due regard to maintaining a high percentage of recovery.

Precipitation steps in a fractionation procedure may be useful at any stage; they may be used at any step in purification simply to concentrate the protein mixture even if it does not contribute any significant purification.

2. Adsorption

The most valuable technique in enzyme purification is column chromatography. Selective adsorption of proteins to a suitable adsorbent (see below) due to some of the physicochemical properties of the proteins, followed by controlled elution procedures, can achieve purification factors of 10–100, with good recovery of activity. The basic equipment consists of a column (i.e., a glass tube filled with adsorbent), a magnetic stirrer to mix two salt solutions, thus creating a concentration gradient (see Fig. 2), and a means of collecting regular fractions emerging from the column A fraction collector, with continuous monitoring of column eluates for protein absorbance at 280 nm; in addition, a peristaltic pump to maintain a steady flow rate is also normally regarded as "basic" equipment (Fig. 2). More sophisticated automated equipment is

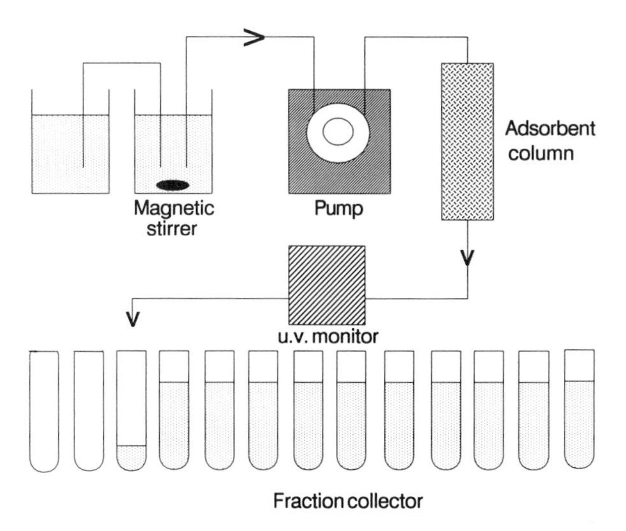

FIGURE 2 A column chromatographic step. A gradient of solvent is created with the aid of a magnetic stirrer, by adding graded amounts of high-concentration solute to a vessel in which the concentration is initially low, and is pumped through the column. Eluted proteins are monitored by ultraviolet absorption, typically at 280 nm, and pass into tubes in the fraction collector.

available and is widely used, especially for enzyme purifications routinely repeated.

The heart of the process is the column packing material and the mode in which it operates to adsorb proteins selectively. This may be by ion exchange (cation or anion), electrostatic interactions; hydrophobic adsorbents, interactions with hydrophobic amino acids on the protein surface; affinity adsorbent, biospecific interaction with enzyme at the site that binds the substrate, the ligand binding site; inorganic adsorbents such as hydroxyapatite, which interact by opposite charges and other modes not completely understood; dye adsorbents which are "multifunctional," interacting with protein surfaces in a variety of modes; metal chelates, forming coordinate metal–protein bonds between the column containing metal ions and the enzyme; and other specialized materials of unclear operational modes.

The column materials are mainly commercial products and consist of solid bead particles. The beads are sufficiently porous to allow protein molecules to diffuse through them. The functional adsorbent structure is chemically linked to the bead particle; as the main adsorption surface is internal, the effective surface area is large.

A protein mixture, typically obtained after a precipitation step, is applied to the column preequilibrated in an appropriate buffer. The enzyme required might not bind, in which case the column must adsorb most other proteins if the step is to be efficient. Normally, though, the desired enzyme binds, and, after a washing step, an elution procedure which separates the adsorbed proteins from each other is performed. Generally, a gradient of eluant (e.g., salt) is applied, that is, the column is perfused with liquid of progressively increasing salt concentrations; the more weakly bound proteins are eluted at low salt concentrations, and the strongly bound proteins are eluted only toward the end of the gradient, at higher salt concentrations (Fig. 3).

Adsorption chromatography is mainly a trial-and-error technique when beginning an enzyme purification that has not previously been attempted. The adsorption properties of the enzyme protein cannot be predicted from its known catalytic activity. However, "designer adsorbents," which are created specifically for the enzyme's catalytic properties, have been successfully used in many cases. These are the true affinity adsorbents, which interact with enzymes through the catalytic site. For this purpose, a substrate, an inhibitor of enzyme activity, or another effector known to bind to the enzyme is used as an immobilized ligand on the column; ideally, the only proteins that bind to this column are the enzymes that recognize and bind to the ligand.

After washing the column, the specifically bound enzymes are eluted either by using a buffer change (pH, ionic strength), which weakens the binding, or by displacing the enzyme with the buffer containing free ligand, which by combining the enzyme, frees it from the immobilized ligand (affinity elution; Fig. 4). Affinity chromatography is theoretically an ideal method for isolating an enzyme. In practice, complications such as difficult chemistry of immobiliza-

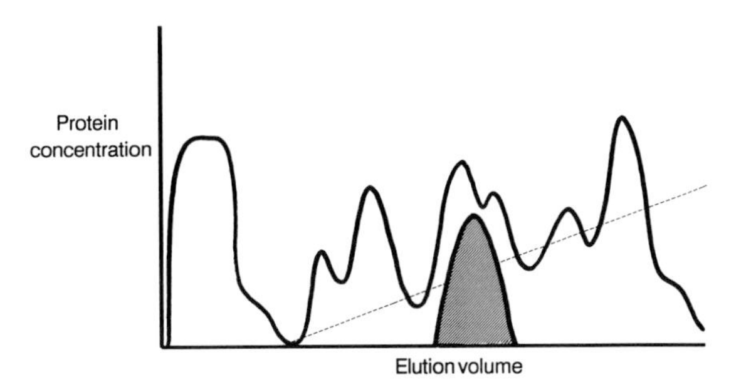

FIGURE 3 Idealized elution of proteins from an ion exchange column. Solid line, ultraviolet adsorption at 280 nm; dashed line, salt concentration gradient; shaded area, enzyme activity.

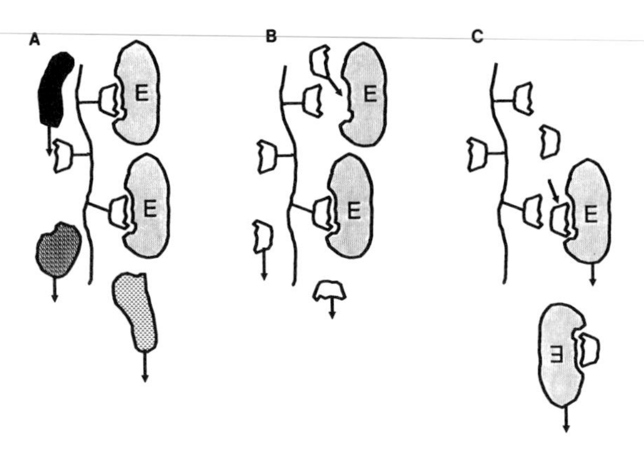

FIGURE 4 Principle of affinity elution from an affinity adsorbent. (A) The mixture of proteins is applied to the adsorbent, and the enzyme (E) binds to the immobilized ligand. (B) The column is then washed with buffer containing soluble ligand. (C) The enzyme exchanges from immobilized to soluble ligand, and so is specifically eluted from the column. In theory, the only enzymes eluted from the column are those that interact with the ligand being used.

tion and nonspecific adsorption mean that it is not suitable for all purposes.

A final example of adsorption chromatography is the ultimate specific column, an immunoadsorbent. Antibodies raised against a previously purified enzyme will generally bind to that enzyme and no other protein. The antibodies, either monoclonal or polyclonal, can be covalently attached to a matrix such as agarose and can pull out the desired enzyme from a complex mixture. Purification factors of 1000 and more are possible for enzymes existing in only trace amounts. The major problem with immunoadsorbents is how to displace the enzyme from the column, because the conditions required may be so harsh that the enzyme is inactivated. Relatively weakly binding monoclonal antibodies may be selected to overcome this problem. [*See* MONOCLONAL ANTIBODY TECHNOLOGY.]

3. Techniques in Which the Proteins Remain in Solution

Both precipitation and adsorption techniques can alter the structure of unstable enzymes; the least damaging methods are those that allow all the proteins to remain in aqueous solution at a neutral pH. Chief among these are electrophoretic methods and gel filtration.

Electrophoresis is the movement of a charged particle in an electric field. For nearly a century, attempts have been made to exploit electrophoresis for separating proteins according to their charges. Because of numerous technical problems, there is still no widely accepted electrophoretic technique, although there are many systems that have proven successful for individual cases. Nevertheless, electrophoresis, especially in laminar gels, has become the most widely used technique for analytical separations of protein. In an analytical system, only a few micrograms of sample is used, and there is no attempt to recover the separated proteins. For preparative work, several milligrams must be separated and the individual fractions must then be collected.

As a refinement of the electrophoresis system, a newer technique known as isoelectric focusing has been used successfully. Simple electrophoresis makes use of the fact that, at a fixed pH, a given protein has a certain charge that results in a certain speed of movement in an electric field. However, if the pH changes, the charge alters, so the protein's speed of migration changes also. At a particular pH, called the protein's isoelectric point, the charge on the protein is zero—all positives and negatives cancel out. As the pH approaches this value, the speed of migration in the electric field decreases to zero when it reaches the isoelectric point. Isoelectric focusing is a system that creates a gradient of pH such that each protein component moves until it reaches its particular isoelectric pH within the gradient. At this point, having no net charge, the protein molecule stops moving. The protein molecules "focus" to their respective isoelectric points. After the electric field is switched off, the protein can be eluted from the isoelectric strips (Fig. 5).

A much more extensively used technique is gel filtration. This is perhaps the most "gentle" of all methods; each molecule experiences only a diffusion through gel pores during the separation process. Porous beads are used, which allow small molecules, including the smallest proteins, to diffuse freely through them. The sizes of the pores in

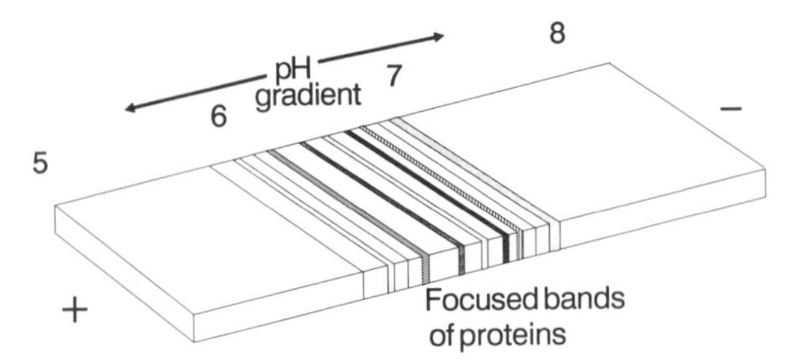

FIGURE 5 Separation of protein bands in a pH gradient generated by electrofocusing. If the separation medium is a porous gel material, it can be sliced and the separated bands can be eluted from the medium.

the beads vary, however, and some are too narrow to allow middle-sized proteins to pass. In some cases, all pores are too narrow to admit the largest protein molecules. Thus, large proteins pass down a column containing the beads, flowing mainly outside of the beads, and emerge quickly. The smallest molecules pass in and out of the beads freely, and so are retained in the column for longer times relative to the large proteins. With the optimum material, different sizes of protein molecules mixed together in the applied sample emerge from the column at different times and so are separated. In gel filtration, there is no adsorption, and the protein molecules remain in solution at all times.

IV. Recombinant Techniques for Expression and Isolation of Human Enzymes

The developments in molecular biology in the last decade have revolutionized our ability to produce large amounts of specific proteins and enzymes, including human products. Only a brief outline of the methods can be given here. In general, the procedure for isolating an enzyme making use of recombinant techniques involves the following:

1. The amino acid sequence of the enzyme must be analyzed. The enzyme must first be purified by conventional techniques described above, and a portion of polypeptide must be sequenced to define a stretch of synthetic DNA that can be expected to hydridize (i.e., interact through base pairing) with the gene that codes for the enzyme.
2. A "complementary DNA library" is prepared from the mRNA of a tissue that has a lot of the required enzyme in it, and this is screened with the synthetic DNA to isolate a clone containing the desired gene.
3. The gene is transferred to a host organism or tissue culture system that enables active expression of the enzyme from the recombinant gene.
4. The enzyme is purified from the culture, separating it from all other proteins present in the host system.

Taking these four steps in turn, the first is fairly routine, the requirements being a purified sample of the enzyme (not necessarily from a human source) and facilities for sequencing the protein and synthesizing the DNA (both can be done commercially).

The second step may be troublesome in that fresh human tissue is needed at this point; complementary DNA libraries are of variable quality, although techniques for creating them have greatly improved. [*See* CHROMOSOME-SPECIFIC HUMAN GENE LIBRARIES.]

The third step, expression in the host organism or tissue culture, is generally the most difficult. Human enzymes are often expressed in bacterial hosts as inactive protein, improperly folded. In eukaryotic cells such as yeast or animal tissue culture, the expression levels may be low. Often, there is also a problem of glycosylation; many proteins, mainly extracellular, have a carbohydrate addition that is not made in the host cell. Thus, the protein may be perfect as polypeptide chains, but not active in the correct way, owing to incorrect glycosylation.

Finally, purification can be troublesome, mainly because, if these enzymes are to be used clinically, they must be ultrapure and demonstrably free of any detectable proteins from the host organism. To achieve this result, it is possible to modify the gene so that the expressed protein has an unusual property that enables its easy isolation, but the unusual part must subsequently be clipped off.

Molecular biology techniques have allowed the production of many proteins, especially hormones, in previously unimaginable amounts. Modern molecular biology also makes it possible to modify proteins and enzymes so that they differ from the natural protein, a procedure that can be used to avoid many problems encountered in the use of enzymes for human clinical therapy.

Bibliography

Burgess, R. R., ed. (1987). "Proceedings of the UCLA Symposium on Protein Purification, Micro to Macro." Liss, New York.

Pharmacia Fine Chemicals (PFC). "Affinity Chromatography: Principles and Methods." PFC, Uppsala, Sweden.

Pharmacia Fine Chemicals (PFC). "Gel Filtration: Theory and Practice." PFC, Uppsala, Sweden.

Pharmacia Fine Chemicals (PFC). "Ion Exchange Chromatography: Principles and Methods." PFC, Uppsala, Sweden.

Scopes, R. K. (1987). "Protein Purification, Principles and Practice," 2nd ed. Springer-Verlag, New York.

Terkova, J., Chaiken, I. M., and Hearn, M. T. W., eds. (1986). Proceedings of the 6th International Symposium on Bioaffinity Chromatography and Related Techniques. *J. Chromatogr.* **376.**

Series: "Methods in Enzymology." Academic Press, San Diego, California. "The Enzymes" (P. D. Boyer, ed.), 3rd ed. Academic Press, San Diego, California.

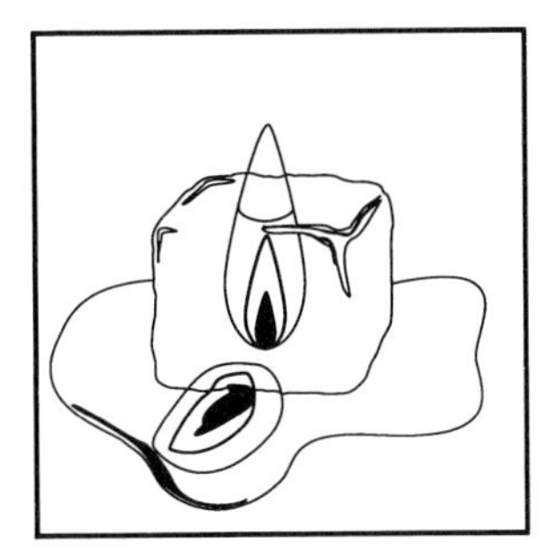

Enzyme Thermostability

TIM J. AHERN, *Genetics Institute, Inc.*

Glossary

Deamidation Hydrolysis of the amide ($CONH_2$) side chain of the amino acid asparagine and glutamine, resulting in the liberation of free ammonia

Enzyme conformation Three-dimensional structure of the polypeptide chain, amino acid side chains, and nonprotein adducts and cofactors of an enzyme

Enzyme thermoinactivation Loss of the biological activity of an enzyme that is accelerated by increasing temperature

THE STABILITY of an enzyme at high temperatures depends on many cooperative, intramolecular interactions maintaining its native structure. Enzymes can inactivate because of reversible processes that cause the protein to unfold or by irreversible processes that include aggregation, destruction of disulfide bonds, hydrolysis of peptide bonds, and deamidation of asparaginyl residues. These mechanisms occur both *in vitro* and *in vivo*. Enzymes exhibiting enhanced thermostability have been engineered as a result of our understanding how these processes inactivate enzymes.

I. Enzyme Conformation, Activity, and Stability

Within the structure of enzymes, the requirements for a certain degree of rigidity and integrity of composition are balanced against the need for flexibility and susceptibility to degradation. If enzymes had no rigidity, they would not be restricted to conformations suitable for binding specific molecules and catalyzing chemical reactions. Similarly, if enzymes were subject to rapid and random destruction of the covalent bonds comprising their primary structure, they quickly would be unable to maintain functional conformations. At the other extreme, if enzymes were *too* rigid, they also could not function, because interaction with other molecules requires that portions of enzymes be able to make limited, concerted motions. Finally, if enzymes were too resistant to degradation, their persistence and accumulation would pose critical problems to living systems, which depend on the recycling of molecules for new synthesis to meet evolving needs. For these reasons, enzymes are held together by forces that are just strong enough to maintain them and their unique functions within the biological milieu for a limited time, after which they undergo inactivation and degradation. It is the inherent weakness of the chemical bonds and noncovalent interactions maintaining the structure of enzymes that determines the degree of enzyme thermostability.

The thermostability of enzymes relies on (1) the overall strength of noncovalent interactions maintaining the native conformation(s), (2) the integrity of the amino acid residues of which polypeptide chains are composed, (3) the amide bonds linking them, (4), the disulfide bonds (if present) that form cross-links between cysteinyl residues, and (5) miscellaneous elements found in various proteins.

The noncovalent forces that maintain the higher levels of structural organization in proteins—hydrogen, ionic, and van der Waals (hydrophobic) interactions—are relatively weak. Furthermore, the stability of the native conformation apparently relies on the cooperative presence of all such interactions, because once unfolding begins, the process usually goes to completion.

With only a few exceptions, the common 20 amino acids comprising enzymes are so stable that they survive incubation in dilute hydrochloric acid at temperatures in excess of 100°C, conditions routinely used to hydrolyze proteins to form free amino acids in the course of composition analysis. However, the amide side chains of asparagine and glutamine, being similar to peptide bonds, also undergo hydrolysis not only under these but much less extreme conditions. The contribution of deamidation to enzyme thermoinactivation will be discussed below.

The peptide bond that unites amino acid residues of proteins in linear chains suits the requirements for limited stability well, because it can be synthesized and broken at a relatively low expenditure of energy, therefore permitting rapid turnover of enzymes. This bond also provides considerable freedom of motion of the polymerized amino acid residues, thus allowing a degree of flexibility. Disulfide bonds are also susceptible to irreversible chemical degradation. Such observations further underscore the tenuous nature of protein stability.

II. Thermal Inactivation

The behavior of enzymes at elevated temperatures illustrates the events leading to spontaneous enzyme destabilization, inactivation, and degradation. When an aqueous solution of an enzyme is heated, the following molecular events begin to take place. First, the enzyme molecules partly unfold because of a heat-induced disruption of noncovalent interactions maintaining the catalytically active conformation at room temperature. This process, which almost invariably leads to enzyme inactivation, is reversible for many enzymes; the native conformation and the enzymatic activity are completely recovered when the enzyme solution is promptly cooled. However, if heating persists, only a decreasing fraction of the enzymatic activity returns on cooling, signifying that other, *irreversible* processes take place.

III. Reversible Thermoinactivation

Reversible thermal unfolding (denaturation) of enzymes has been a subject of intensive investigation for several decades. The phenomenon is amenable to straightforward and exact thermodynamic analysis and as a consequence is conceptually well understood. If the activity of an enzyme lost because of exposure to elevated temperature can be regained by return to lower temperature, then as the name implies, the inactivation is "reversible." Perhaps the most marked reversible effect of elevated temperature on an enzyme is the increase in motion of its constituent parts to such a degree that what is known as the ordered, "native" conformation can be said to be lost, replaced by largely disordered, "denatured" conformations. This can be represented as

$$N \rightleftharpoons U \quad [1]$$

in which N is the native and U is the unfolded forms of an enzyme, respectively.

The native state is not a single, unique conformation because no enzyme is entirely rigid; even at subzero temperatures and in the crystalline state, the atoms comprising a protein undergo vibrations, rotations, and small translations of about 0.2–0.5 A. At intermediate temperatures, the range at which most enzymes exhibit their optimal activity (0–60°C), reversible translations of whole segments of protein structure, called "breathing," are observed. These and other concerted motions required for substrate binding, catalysis, and product release define the displacements referred to collectively as native conformation. Nevertheless, despite the existence of some freedom of movement, the predominant conformations are restricted by a complex balance of intramolecular, noncovalent interactions.

At elevated temperatures (in most instances, above approximately 50°C), extensive cooperative intramolecular motions may take place, which effectively denature an enzyme: At a given temperature that varies for each enzyme under specific environmental conditions of pH, ionic strength, etc., the native structure can no longer prevail against the drive toward increased entropy of the unfolded state expressed as rapid, random motion, and the protein loses most of its ordered secondary and tertiary structure as it undergoes what is known as denaturation. The unfolding of an enzyme in solution changes many easily measured physical characteristics such as viscosity, optical rotation, and ultraviolet absorbance (Fig. 1), which makes possible determination of the extent of unfolding.

In the course of denaturation, the amino acid residues comprising the active site of an enzyme are inevitably dispersed, and as a consequence, catalytic activity is lost. Loss of activity by unfolding can be regarded as the first step in nearly all enzyme thermoinactivation processes.

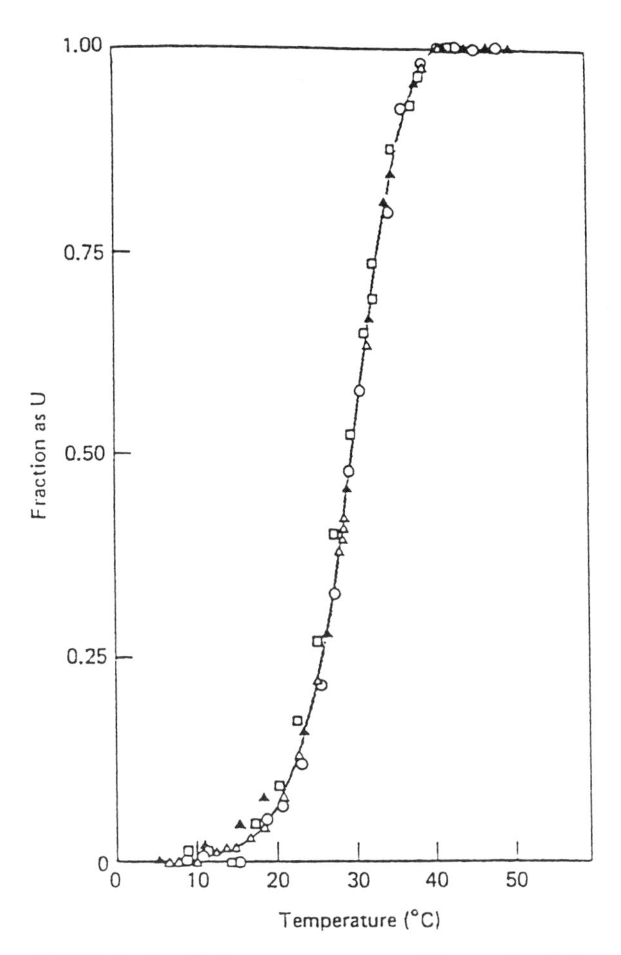

FIGURE 1 Equilibrium measurements of temperature-induced unfolding of bovine ribonuclease A in HCl-KCl at pH 2.1 and 0.019 ionic strength, measured by increase in viscosity (*squares*), decrease in optical rotation at 365 nm (*circles*), and decrease of UV absorbance at 287 nm (*open triangles*). *Filled triangles* show measurement of a second cooling from 41°C for 16 hr. [From Ginsberg, A., and Carroll, W. R. (1965). Biochemistry **4,** 2159–2174, with permission. Copyright by the American Chemical Society.]

This mechanism has been designated as a reversible process in Equation 1; an unfolded enzyme, once cooled below its characteristic transition temperature should, according to the "thermodynamic hypothesis," refold to form the active, native state. This model appears highly plausible because the native conformation is favored over all others during the *in vivo* synthesis of the disordered, nascent chain. It follows, then, that provided we choose the appropriate conditions for renaturation, even a randomly coiled protein should successfully refold to form the native state once again. This is generally true, at least for single-chained, monomeric enzymes, provided they have not undergone post-translational modifications (e.g., selective proteolysis and excision of portions of the polypeptide chain, as in the case of the protein insulin). For example, when an aqueous solution of lysozyme is heated to 100°C, well above the transition temperature of the enzyme, the catalytic activity is immediately lost. The reversibility of the thermoinactivation is illustrated by the recovery of 100% of the activity when the enzyme is promptly cooled to 25°C, well below the transition temperature. The reversible equilibrium between native and disordered conformations has been demonstrated in this way for other enzymes as well.

IV. Irreversible Thermoinactivation

All enzymes undergo conformational transitions at elevated temperatures, but prolonged incubation results in a loss of activity that is not readily reversible once the solution is cooled. Irreversible thermoinactivation of enzymes is the result of destruction of covalent bonds within the enzyme, as well as some conformational processes that are for all practical purposes "irreversible." Equation 2 represents the framework, distinguishing the various processes leading to thermal inactivation of enzymes.

Once unfolded, many enzymes become insoluble and form large particulates. Aggregation is a concentration-dependent process because of its polymolecular nature, and it can be explained by the amphiphilic nature of the surface of a disordered enzyme. Portions of the polypeptide chain that are normally buried tend to be much more hydrophobic than those exposed to solvent in the native structure. Once an enzyme is denatured, the exposed hydrophobic surfaces have a tendency to avoid interaction with the aqueous solvent because water

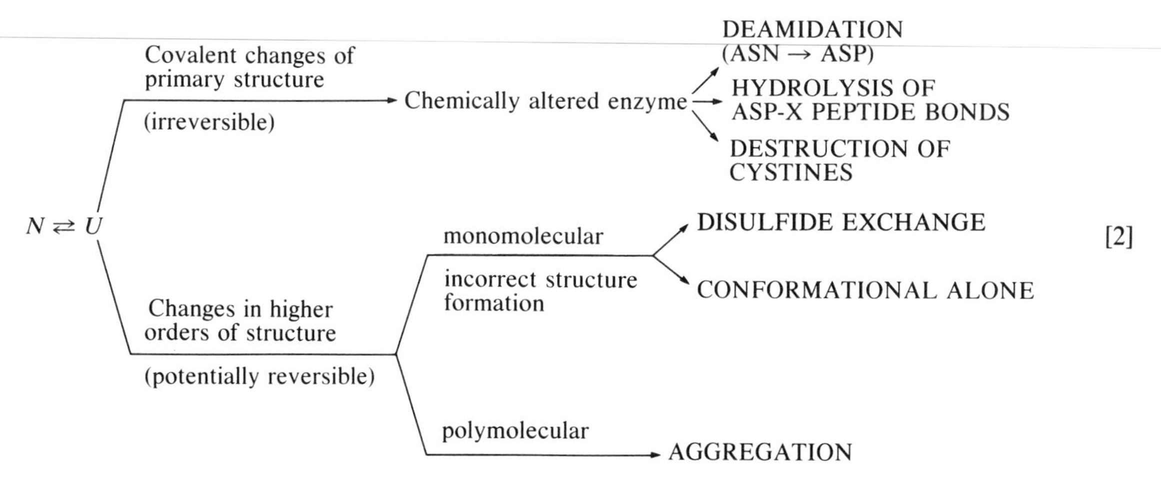

[2]

forms ordered clathrate structures around hydrophobic residues. The imposition of order on the solvent results in a decrease in entropy of the system as a whole. Thus, provided enzyme concentrations are high enough, such hydrophobic surfaces may form intermolecular interfaces via aggregation in an attempt to maximize the entropy of the solvent and thereby reduce the free energy of the system.

However, even the activity of dilute solutions of enzymes often cannot be recovered after prolonged heating followed by cooling. This irreversible thermal inactivation follows first-order kinetics, can be independent of the initial enzyme concentration (Fig. 2), and is not due to the formation of aggregates and can therefore be attributed to monomolecular processes (see Equation 2).

It is important to determine whether monomolecular, apparently irreversible thermoinactivation is caused by covalent changes of the primary structure or by changes in higher orders of structure, because the activity of conformationally altered enzymes that have undergone no irreversible deterioration has the potential to be regained. The existence of monomolecular, incorrect structure formation at high temperatures can be explained by the fact that there is more than one way to fold a protein: On denaturation, the tendency to bury hydrophobic groups, combined with the freedom of a protein to sample many conformational states, results in new, kinetically or thermodynamically stable structures that are catalytically inactive. Even after cooling, these incorrectly folded, "scrambled" structures may remain because a high kinetic barrier prevents spontaneous refolding to the native conformation. (Disulfide exchange, resulting in the mismatching of cysteinyl residues, can play a role in the formation of these scrambled structures. This mechanism is discussed separately below.)

Such processes can be distinguished from covalent mechanisms resulting in destruction of the polypeptide chain or chemical deterioration of the side chain residues. These latter processes are truly irreversible and define the basal rate of irreversible enzyme thermoinactivation.

Three techniques distinguish potentially reversible inactivation processes from covalent mechanisms affecting primary structure: (1) comparison of the rate of irreversible thermoinactivation in the presence and absence of strong denaturants such as

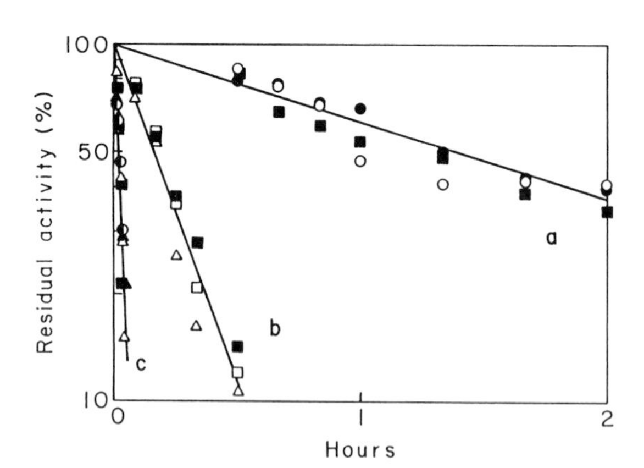

FIGURE 2 Time course of irreversible thermoinactivation of hen egg white lysozyme at 100°C as a function of pH and enzyme concentration. a, pH 4 (0.1 M Na acetate); b, pH 6 (0.01 M Na cacodylate); c, pH 8 (0.1 M Na phosphate). The concentrations of lysozyme were 1,000 (*open circles*), 100 (*filled circles*), 50 (*open squares*), 10 (*filled squares*), 5 (*open triangles*), 1 (*filled triangles*), and 0.5 μM (*half-filled circles*). After incubation for the time indicated, aliquots of enzyme solutions were cooled and assayed for residual catalytic activity. [From Ahern, T. J., and Klibanov, A. M. (1985). *Science* **228,** 1280–1284, with permission. Copyright by the AAAS.]

guanidine hydrochloride or acetamide. Denaturants disrupt noncovalent interactions in protein and maintain the enzyme molecules in a highly unfolded form; therefore, the enzyme is unlikely to assume inactive conformations separated from the native state by high activation energies. On subsequent cooling and dilution, the random coil should refold to the native structure. Conversely, denaturing agents are not expected to affect the rates of most covalent reactions. The above reasoning constitutes the first criterion: If addition of a denaturant stabilizes an enzyme against irreversible thermal inactivation, conformational processes are involved; if there is no effect, the rate of inactivation is probably due to covalent processes affecting the primary structure of the enzyme.

Potentially reversible mechanisms of thermoinactivation may be distinguished from destructive, covalent processes in enzymes by a second approach: (2) determine whether unfolding and refolding of the previously heated and cooled enzymes results in at least partial recovery of the enzymatic activity lost. Whereas the first approach attempts to *prevent* incorrect structures from forming, the second approach attempts to *recover* activity lost by incorrect structure formation. (If S-S bonds are present, the inactivated enzyme should be reduced and reoxidized during reactivation experiments.)

Finally (3), the magnitude of the rate of irreversible thermoinactivation of an enzyme can indicate whether the predominant mechanism is conformational or covalent. The half-lives of the covalent processes found to cause irreversible thermoinactivation are relatively large—of about 10 min to more than an hour at 100°C. Thus, if an enzyme inactivates irreversibly in less than 2 min at 100°C (e.g., lysozyme at pH 8) or inactivates rapidly at temperatures below 70°C at near-neutral pH, the inactivation is most likely predominantly due to conformational processes.

A. Potentially Reversible Mechanisms

In enzymes containing disulfide bridges, the formation of inactive monomolecular structures may be due to disulfide exchange, which results in incorrect pairing of cysteinyl residues. This reaction, which is known to occur in proteins at neutral and alkaline pH, requires the presence of catalytic amounts of thiols that promote the interchange by nucleophilic attack on the sulfur atoms of a disulfide. (How these thiols are produced in the course of heating an enzyme is described in Section IV,B.) The contribution of disulfide exchange in the formation of incorrect structures can be prevented if the enzyme is heated in the presence of thiol scavengers (e.g., *p*-mercuribenzoate and *N*-ethylmaleimide) or copper ion, which catalyzes the spontaneous air oxidation of thiols. The rate of heat-induced destruction of S-S bonds is almost entirely independent of the nature of the protein.

Enzymes that contain no disulfide bridges also undergo potentially reversible thermoinactivation. For example, alpha-amylases from *Bacillus amyloliquefaciens* and *B. stearothermophilus* contain 0 and 1 cysteinyl residue each, respectively, yet both are stabilized at least threefold against irreversible thermoinactivation at 90°C, pH 6.5, by the presence of 8 M acetamide. Furthermore, activity lost when the amylases are incubated in the absence of denaturants can be partially regained if the enzyme is briefly treated afterward with hot, concentrated denaturant. Therefore, it appears that not all potentially reversible thermoinactivation is due to disulfide exchange. The detailed nature of incorrect folding in such cases is an intriguing topic for future research that may help explain how enzymes from organisms that live at elevated temperature are more resistant to inactivation processes compared with enzymes from organisms that live at moderate temperature.

As mentioned earlier, similar procedures of renaturation have been successfully applied to enzymes that are observed to aggregate on heating. The noncovalent interactions (primarily hydrogen bonds and hydrophobic interactions) believed to be responsible for maintaining aggregates are apparently disrupted by the presence of a denaturant; once redissolved, the enzyme can refold to the native conformation when the denaturant is removed.

B. Irreversible Mechanisms

Conformational processes alone cannot account for the irreversible loss of activity of enzymes. Despite all measures to prevent potentially reversible inactivation, enzymes are nevertheless observed to inactivate irreversibly at high temperatures in aqueous solutions throughout the entire range of pH. The covalent mechanisms responsible are treated separately below.

Conceptually, covalent processes can affect enzyme structure in the following ways: cleavage of the polypeptide chain by hydrolysis, destruction of individual amino acid residues, destruction of disulfide bonds, and reactions involving metal ions, cofactors, and adducts caused by glycosylation, etc.

1. Peptide Chain Integrity

Hydrolysis of the polypeptide chain adjacent to aspartyl residues can account for significant irreversible thermoinactivation of enzymes at elevated temperatures under mildly alkaline conditions (e.g., pH 4). The Asp-X bond (where X is the amino acid residue bound to the carboxyl group of Asp) is the most labile peptide bond under those conditions. If heating of an enzyme continues, the peptide bonds on both sides of aspartyl residues are hydrolyzed, and free aspartic acid is released into the solution. Several pathways result in the release of aspartic acid from proteins (Fig. 3a).

The time course of hydrolysis of peptide bonds comprising the polypeptide chain(s) of an enzyme can be monitored after the appearance of the resulting fragments. The mechanism is deduced from the identities of the amino acids at the new carboxyl and amino termini must be created by the intrapeptide chain hydrolysis.

2. Amino Acid Destruction

Deamidation of asparaginyl residues within enzymes occurs at elevated temperatures under acidic, neutral, and alkaline conditions. Release of ammonia as a result of deamidation is thought to occur during the formation of an unstable intramolecular cyclic imide intermediate (Fig. 3b). When the cyclic ring hydrolyzes, a carboxyl side chain is left in place of the original, uncharged asparaginyl residue. Except under very acidic conditions, carboxyl groups are negatively charged; as a result, deamidated enzymes can be physically separated from their normal, nondeamidated forms by means of isoelectric focusing and thus quantified. The release of ammonia can be determined as well.

Use of the methods described above makes possible monitoring the extent of deamidation during thermoinactivation. For example, the time course of the initial evolution of ammonia during the heating of lysozyme occurs at a significant rate relative to enzyme inactivation (Fig. 4). The ammonia is ascribed to deamidation of asparagine residues because studies of model peptides have shown that the 14 asparagine residues of lysozyme are much more labile than its three glutamine residues.

In some cases, the loss of activity caused by deamidation can be measured by assay for enzymatic

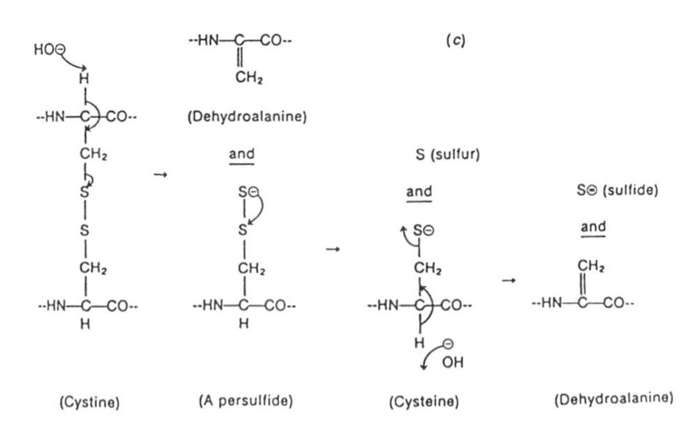

FIGURE 3 Proposed mechanisms of covalent reactions causing irreversible enzyme thermoinactivation. a, Hydrolysis of peptide bonds adjacent to Asp residues. [From Inglis, A. S. (1983). *Methods Enzymol.* **91,** 324–332, with permission. Copyright by Academic Press.] b, Deamidation of Asn residues, resulting in either L-aspartyl or L-isoaspartyl residues, depending on which amide linkage in the proposed succinimide intermediate is hydrolyzed. [From Clarke, S. (1985). *Annu. Rev. Biochem.* **54,** 479–506, with permission from Annual Reviews Inc.] c, Destruction of disulfide linkages via base-catalyzed O-elimination. [From Whitaker, J. R. (1980). *In* "Chemical Deterioration of Proteins." ACS Symposium Series, vol. 23, pp 320–326, with permission. Copyright by the American Chemical Society.]

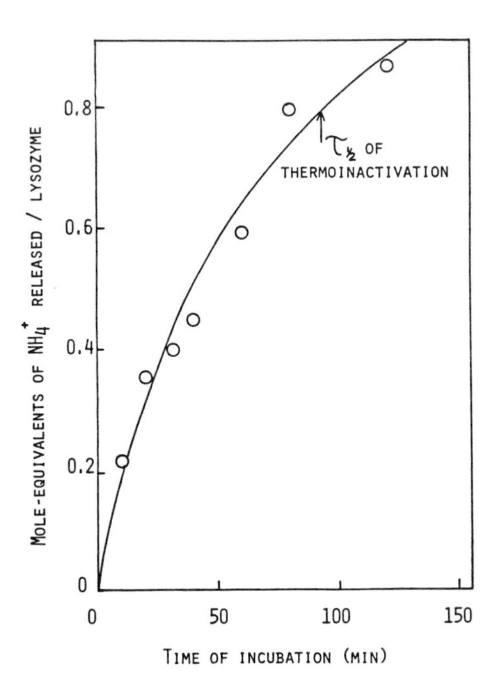

FIGURE 4 Release of ammonia during thermoinactivation of hen egg white lysozyme, 100°C, pH 4. Appearance of ammonia in solution was determined colorimetrically by a modified method of Forman, D. T. (1964). *Clin. Chem.* **10,** 497–508. Time at which only one half of initial catalytic activity could be recovered after cooling to 23°C is indicated.

TABLE I The Effect of Deamidation on the Relative Specific Activity of Enzymes

Enzyme	Relative specific activity
Hen egg white lysozyme	
native	1.00
monodeamidated	0.53
di- and trideamidated	0.21
Bovine pancreatic ribonuclease A	
native	1.00
monodeamidated	0.65
dideamidated	0.38
trideamidated	0.19
Cytochrome c	
native	1.00
monodeamidated	0.60
dideamidated	0.20
Yeast triosephosphate isomerase	
native	1.00
(Asn 78 → Asp 78)[a]	0.66

[a] The altered enzyme was produced by means of site-directed mutagenesis and expressed in *Escherichia coli*.

activity of deamidated forms of the thermoinactivated enzyme. The relative specific activities of deamidated species of various enzymes are depicted in Table I.

Deamidation may cause not only the conversion of asparagine residues to aspartic acid, but also the formation of peptide bonds incorporating the remains of the side chain. This can be explained by the proposed mechanism of deamidation, illustrated in Fig. 3b; the nature of the resulting polypeptide linkage depends on which bond in the succinimide intermediate is hydrolyzed. Deamidation resulting in enzyme inactivation occurs during incubation in aqueous solutions at whatever pH chosen.

3. Destruction of Disulfide Bridges

In addition to the described disulfide exchange, cystines also undergo irreversible destruction known as beta-elimination. The base-catalyzed abstraction of a proton from the alpha-carbon of a cysteine residue forming a disulfide bridge results in the cleavage of the cystine cross-link to form residues of dehydroalanine and thiocysteine (Fig. 3c). Dehydroalanine is a reactive species that can undergo an addition reaction with the epsilon-amino group of a lysine residue to form the novel intramolecular cross-link, lysinoalanine. The degradative product of the complementary Cys residue, thiocysteine, undergoes further deterioration to yield hydrosulfide ion (HS-) as one of many end products.

The formation of the degradative products described above accounts for the fate of approximately 90% of all the cystine lost during irreversible thermoinactivation of ribonuclease A. Furthermore, the contribution of beta-elimination to loss of enzymatic activity was demonstrated by the finding that the degree of stabilization of ribonuclease was proportional to the degree of protection against beta-elimination by reversible protection of the Cys residues by chemical modification.

4. Summary of General Mechanisms

Use of these techniques makes possible the calculation of the contribution of each inactivating process to the overall rate of irreversible thermoinactivation. Comparisons of the directly measured overall rates of enzyme thermoinactivation with the contributions of individual mechanisms are listed in Table II. Irreversible inactivation of lysozyme at 100°C is brought about (1) at pH 4 by a combination of deamidation and peptide hydrolysis; (2) at pH 6 by deamidation; and (3) at pH 8, by a combination of deamidation, destruction of disulfide bonds, and

TABLE II Rate Constants of Irreversible Thermoinactivation of Lysozyme and Ribonuclease—The Overall Process and Contributions of Individual Mechanisms to Thermoinactivation

Irreversible thermoinactivation		Rate constant (hr^{-1}) pH 4	pH 6	pH 8
Hen egg white lysozyme, 100°C				
Directly measured overall process		0.49	4.1	50
Due to individual mechanisms	Deamidation of Asn residues	0.45	4.1	18
	Hydrolysis of Asp-X peptide bonds	0.12		
	Destruction of crystine residues			6
	Formation of incorrect structures			32
Bovine pancreatic ribonuclease A, 90°C				
Directly measured overall process		0.13	0.56	23.4
Due to individual mechanisms	Deamidation of Asn residues	0.02	0.15	0.8
	Hydrolysis of Asp-X peptide bonds	0.10		
	Destruction of cystine residues		0.05	2.8
	Formation of incorrect structures[a]		0.31	19.4

[a] Shown to be due to thiol-catalyzed disulfide interchange.

formation of incorrect structures. Analogous findings were reported for the irreversible inactivation of ribonuclease at 90°C, with the additional finding that incorrect structure formation, observed at both pH 6 and 8, was shown to be due to thiol-catalyzed exchange; also, beta-elimination of cystine residues could account directly for approximately 10% of the loss of activity at pH 6.

Although these mechanisms adequately account for the irreversible loss of activity of the enzymes studied, it is likely that secondary mechanisms, resulting in only a small fraction of the overall rate of inactivation, are also at work. These may include (1) hydrolysis of the peptide chain adjacent to amino acid residues other than aspartate (peptide bonds involving glutamic acid, glycine, alanine, serine, and threonine have been reported to be cleaved during prolonged digestion in dilute acid); (2) deamidation of glutamine—as opposed to asparagine—residues; and (3) racemization of amino acid residues.

5. Other Mechanisms

In addition to the processes causing irreversible thermoinactivation outlined above, many additional degradative processes specific to enzymes containing unique constitutive elements may result in thermoinactivation as well. In addition to the 20 common amino acid residues, more than 100 unusual amino acids also exist in proteins, and half of them are susceptible to chemical deterioration [e.g., hydrolytic scission of side chain groups bound to indole, phenoxy, thioether, amino, imidazole, and sulfhydryl residues and the derivatives of Ser and Thr (e.g., O-glycosyl and O-phosphoryl groups) and Gln and Asn (e.g., methylated and glycosylated)]. Of the nonamino acid moieties associated with enzymes, the carbon–nitrogen bonds in purines and pyridines, glycosidic bonds, and phosphodiester bonds undergo hydrolytic breakdown at rates comparable to the hydrolysis of peptide bonds. If present, reducing sugars and fatty acid degradative byproducts can undergo the Maillard reaction with amino groups in enzymes to produce Schiff bases on removal of water. Metal ions can accelerate hydrolytic cleavage of peptide bonds.

In addition to these covalent, deteriorative reactions, simple dissociation of noncovalently bound prosthetic groups during thermally induced denaturation can be irreversible. For example, once molybdenum has been dissociated from the active center of sulfite oxidase during incubation at high temperature, no reactivation appears possible by cooling the enzyme solution and addition of an excess of the metal ion. Nevertheless, it is possible that the activity of some enzymes that lose their cofactors during thermoinactivation may be regenerated. For example, enzymes requiring metal–sulfur compounds can be reactivated after loss of their cofactors by addition of metal salts together with thiols or organic sulfides.

V. Enzyme Thermoinactivation *In Vivo*

Folding and degradation of enzymes *in vivo* is affected by features unique to the intracellular environment. Most pools of intracellular proteins are in a dynamic state of continual synthesis and degradation. The half-lives of proteins in the cell range from minutes to days and can even vary in response to external conditions (e.g., starvation *in vivo*) or withdrawal of serum during cultivation of cells.

The folding of proteins in eukaryotic cells occurs during or immediately after translocation of the peptide chains to the internal (luminal) side of the endoplasmic reticulum (ER). Covalent attachment of carbohydrates and fatty acids to the polypeptide chain may also contribute to the protein folding pathway. Several proteins in the ER and Golgi assist in the proper folding of nascent polypeptide chains. The formation of disulfide bonds in proteins that are eventually secreted is catalyzed by disulfide isomerase. Binding protein (BiP; also known as GRP78) has been implicated in the formation of oligomeric proteins and has been found to bind and retain improperly folded proteins within the ER. [*See* PROTEINS.]

Ubiquitin, a small protein found in all eukaryotic cells, covalently attaches to abnormal and some short-lived proteins via the epsilon-amino groups of lysinyl residues. Ubiquitination apparently marks the proteins for rapid degradation by intracellular proteases, the multicatalytic endopeptidase complex (the "proteasome"), and perhaps by ubiquitin itself. Specialized vesicles known as liposomes are the site of degradation of many membrane-bound proteins and long-lived cytosolic proteins. Other degradative pathways may be restricted to mitochondria, the ER, and Golgi, which degrade a portion of secretory and membrane proteins before they reach the cytosol.

The thermostability of enzymes may contribute to their susceptibility to intracellular degradative processes. Cellular proteins having disrupted structures, as a result of either posttranslational oxidative damage or substitution of amino acid residues by analogues, are more rapidly conjugated to ubiquitin than are native proteins. Research with temperature-sensitive mutants of proteins (e.g., the oligomeric VSV G glycoprotein) has shown that normal maturation and export from the Golgi occurs only below the nonpermissive temperature for the mutant; if the temperature is raised, oligomerization does not occur, and the monomer is retained within the Golgi.

Many proteins, especially structural proteins, are long-lived. Proteins of the eye lens can avoid total degradation and recycling until the death of the organism. Therefore, the initial effects of degradation that would trigger rapid protein turnover in most tissues may accumulate in lens proteins. For example, triosephosphate isomerase undergoes spontaneous deamidation of specific asparaginyl residues during the aging of erythrocytes and eye lens. Amino acid composition analysis has also shown that the aspartic acid content of tissues such as rat brain, liver, and heart also increases with age as the asparaginyl content concomitantly decreases. Deamidation of asparaginyl residues has also been observed in crystallized proteins by means of neutron scattering studies. The rate of deamidation of folded proteins depends on the local conformation and bonding structure. If the asparaginyl residue is sterically prevented from forming the imide intermediate, the rate of deamidation is markedly lowered.

Once deamidation of asparagine occurs, hydrolysis of the cyclic intermediate can produce either an aspartyl residue in the place of the original asparagine, or a novel, *iso*aspartyl side chain from what had been the C′ carbon of the original asparagine, and insertion of the remains of the asparaginyl side chain into the polypeptide chain (Fig. 3b). There may exist a mechanism *in vivo* for eliminating such beta-carboxyl linkages in polypeptide chains. A mammalian enzyme, carboxyl methyltransferase, preferentially methylates such abnormal isoaspartyl groups; on demethylation, the cyclic intermediate is reformed; subsequent hydrolysis yields a mixture of both normal aspartyl and iso-aspartyl residues. By this means, the inclusion of abnormal beta-carboxyl linkages into polypeptide chains can be reversed. Enzymatic pathways leading to the reversal of spontaneous damage to proteins are also capable of reducing methionyl residues in proteins that have undergone oxidation. The physiological importance of such reactions in the *in vivo* repair of proteins remains to be determined.

VI. Enzyme Thermostabilization

A. Thermophilic Enzymes

When compared with the enzymes of organisms that normally live below 40°C, those from thermophilic bacteria are extremely resistant to thermodenaturation. Some remain active despite prolonged heating near 90°C. Although structural information concerning thermophilic enzymes is scanty, it appears increased thermostability is due to small stability-enhancing changes throughout the proteins. Because the overall stabilization free energy of the native states of proteins is low, relatively small increments in the strength of existing bonds can result in significant increases in resistance to thermal denaturation. Hydrophobic interactions are stabilized by increases in temperature, so it is not surprising that thermophilic proteins have increased internal—and decreased external—hydrophobicity. Similar increases in the stability of alpha-helices and beta-sheet structures in thermophilic proteins have also been noted.

B. Protein Engineering

1. Enhancement of Thermostability by Design

It is reasonable to assume that the overall stability of enzymes can be increased by replacing those amino acid residues providing only weak contributions to the conformational stability of the molecule with others providing stronger interactions.

The superposition of hydrogen bond dipoles in alpha-helices, resulting in opposite charges at the ends of helical segments, was originally believed to be a relatively small electrostatic contribution to the sum of energetic interactions maintaining the native structures of proteins. Recently, however, researchers demonstrated that amino acid substitutions designed to increase stabilization via the helix dipoles in T4 lysozyme increased the melting transition temperature by as much as 4°C, reflecting a stabilization of the protein conformation of approximately 1.6 kcal/mol. The mutations introduced charged aspartyl residues that interacted electrostatically with the positively charged amino termini of helices in the protein. Earlier work showed that decreasing the charge from +2 to −1 on the amino-terminal residue of a helix in analogues of the S-peptide of ribonuclease S resulted in an increase of the melting temperature of the reconstituted enzyme by as much as 6°C compared with the protein containing the native S-peptide.

Secreted proteins are further stabilized by the presence of disulfide bonds. Engineering novel disulfide bonds into dihydrofolate reductase, T4 lysozyme, and subtilisin BPN′ have stabilized the active enzymes with respect to reversible unfolding. In one instance, addition of three disulfide bonds to T4 lysozyme increased the melting transition temperature by as much as 23.4°C.

Another approach that enhances the stability of a protein is to replace flexible residues (e.g., glycine), which require greater free energy than other, sterically constrained residues to restrict its conformation. Provided the substitutions do not introduce undesirable steric interactions, the mutated protein will have a decreased entropy of unfolding, resulting in a higher temperature of denaturation. In accord with this theory, substitution of one glycine and one alanine residue in T4 lysozyme increased the melting transition temperature by as much as 2°C, reflecting an increase of approximately 1 kcal/mol to the free energy of folding. Similarly, substitutions of two glycines in λ repressor increased the melting temperature of the N-terminal domain by 3–6°C.

Discrepancies between the amino acid sequences of enzymes having high degrees of overall sequence similarity but widely varying thermostability can be "corrected" to increase the thermostability of the less stable enzymes within the family. This technique has yielded a mutant form of the neutral protease from *Bacillus stearothermophilus* of enhanced thermostability, although many mutants designed by this criterion exhibited decreased thermostability.

2. Enhancement of Thermostability by Selection

The examples cited above illustrate rational approaches to stabilizing the conformation of proteins; by means of structural and functional data, the researchers conceived of specific changes that would theoretically improve the stability of a given enzyme. A second approach selects for improvement in the stability of an enzyme without prior knowledge of its structure. Mutants of kanamycin nucleotidyl transferase having increased thermostability were discovered by cloning and expressing the enzyme from a mesophilic organism in a thermophilic bacteria and selecting for colonies containing the activity of the enzyme at temperatures higher than the normal melting transition temperature of the enzyme. Such screening techniques also have yielded variants of T4 lysozyme and subtilisin BPN′ having enhanced thermostability.

There exist random mutations in staphylococcal nuclease that stabilize the enzyme against unfolding by guanidine hydrochloride. For example, whereas ~0.85 M guanidine hydrochloride results in the denaturation of the wild-type enzyme, as much as 1.3 M denaturant is required to unfold one of the mutants. Furthermore, three such stabilizing mutations were found to "correct" other mutations that had resulted in colonies having greatly reduced nuclease activity. One explanation is that the stabilizing substitutions may confer a global stability on the enzyme. In a similar fashion, the activities of nonfunctional and presumably unstable mutants of cytochrome c were partially restored by a second-site mutation (i.e., asparagine-57 replaced by isoleucine). Subsequent construction of the Asn57Ile mutant by site-directed mutagenesis resulted in an extraordinary 17°C increase in the transition temperature, corresponding to a greater than twofold increase in the free energy change for thermal unfolding.

3. Preventing Inactivation of Proteins Caused by Covalent Mechanisms

Replacement of specific asparagine residues in triosephosphate isomerase resulted in a nearly twofold decrease in the rate of irreversible thermoinactivation of the enzyme. Exposure to oxidative environments can result in the degradation of methionine. Subtilisin was stabilized against such oxidation by replacing a single methionine important for catalysis with serine, alanine, or leucine.

Bibliography

Ahern, T. J., Casal, J. I., Petsko, G. A., and Klibanov, A. M. (1987). Control of oligomeric enzyme thermostability by protein engineering. *Proc. Natl. Acad. Sci. U.S.A.* **84,** 675–679.

Ahern, T. J., and Klibanov, A. M. (1985). The mechanism of irreversible enzyme inactivation at 100°C. *Science* **228,** 1280–1284.

Anfinsen, C. B. (1973). Principles that govern the folding of protein chains. *Science* **181,** 223–230.

Argos, P., Rossman, M. G., Grau, U. M., Zuber, H., Frank, K. G., and Tratschin, J. D. (1979). Thermal stability and protein structure. *Biochemistry* **18,** 5698–5703.

Clarke, S. (1985). Protein carboxyl methyltransferases: two distinct classes of enzymes. *Annu. Rev. Biochem.* **54,** 479–506.

Karplus, M., and McCammon, J. A. (1981). The internal dynamics of globular proteins. *CRC Crit. Rev. Biochem.* **9,** 293–349.

Klibanov, A. M. (1983). Stabilization of enzymes against thermal inactivation. *Adv. Appl. Microbiol.* **29,** 1–29.

Lapanje, S. (1978). "Physicochemical Aspects of Protein Denaturation." Wiley, New York.

Matsumura, M., Signor, G., and Mathews, B. W. (1989). Substantial increase of protein stability by multiple disulphide bonds. *Nature* **342,** 291–293.

Pfeil, W. (1981). The problem of the stability of globular proteins. *Mol. Cell. Biochem.* **40,** 3–28.

Rothman, J. E. (1989). Polypeptide chain binding proteins: Catalysts of protein folding and related processes in cells. *Cell* **59,** 591–601.

Whitaker, J. R. (1980). *In* "Chemical Deterioration of Proteins." ACS Symposium Series, vol. 23. (J. R. Whitaker and M. Fujimaki, eds.), pp. 145–164. American Chemical Society, Washington, D.C.

White, R. H. (1984). Hydrolytic stability of biomolecules at high temperatures and its implications for life at 250°C. *Nature* **310,** 430–432.

Enzymes, Coenzymes, and the Control of Cellular Chemical Reactions

STEPHEN P. J. BROOKS, *Carleton University*

Glossary

E_a Activation energy for a reaction
ΔG Gibbs free energy for a system
$\Delta G^{0\prime}$ Standard free energy at pH 7.0
h Plank's constant (6.626×10^{-34} J/sec)
K_a Binding constant for an activator
K_{eq} Equilibrium constant for a reaction
K_i Binding constant for an inhibitor
K_m Michaelis constant for substrate binding to an enzyme active site. Defined as the concentration of substrate that produces half-maximal enzyme activity
N Avogadro's number (6.023×10^{23} molecules/mol)
P Product of a reaction
R Gas constant for an ideal gas (8.3144 J/deg/mol)
S Substrate for a reaction
D*S* Change in entropy associated with a reaction
T Temperature in degrees Kelvin
V_{max} Maximal velocity of an enzyme measured at saturating substrate concentrations

ENZYMES are either protein or nucleic acid molecules that act as catalysts to speed up reactions. Enzymes work so effectively that with respect to the multitude of chemical reactions that could take place in the cell, only the enzyme-catalyzed reactions occur because noncatalyzed reactions occur at too slow a rate. As a general rule, each enzyme catalyzes only a single reaction, meaning that they effectively determine the types of reactions that occur in the cell; enzymes direct cellular metabolism. Because enzymes are so powerful they are also strictly regulated by other cellular compounds and, sometimes, by the reactants themselves. It is this combination of reaction specificity and regulation of activity that helps maintain the constancy of cellular processes (cellular homeostasis) under widely different metabolic conditions and permits cells to perform various tasks. In animal liver, for example, cellular and blood glucose concentrations are regulated by controlling the enzymes responsible for storing glucose (in the form of glycogen) or releasing glucose (break down glycogen), depending on whether blood glucose levels are high (during eating) or low (in between meals). The liver regulates these processes through control of the enzymes, which act specifically on glycogen.

In the sections below, we shall see specifically how enzymes are controlled (Section I) and how their physical structure relates to their function (Section II). Finally we shall review the cofactors required for some enzymes to function (Section III). Several references are provided at the end of this review and should be referred to both for background information not provided in this review and for a more complete treatment of the subjects presented below.

I. How Enzymes Function

All biological processes, from the use of glucose to the building of new proteins to the reactions driving muscle contraction, are chemical reactions, and as such they obey the fundamental rules governing chemical processes. These rules determine (1) how fast a reaction happens, and (2) how much starting material reacts to make the product. A fundamental understanding of what governs chemical processes is essential to the understanding of how an enzyme works because chemical processes and enzyme action are intimately linked. A short discussion of factors that are enzyme-independent (i.e., those that depend only on the chemical properties of the reactants) is, therefore, important in understanding enzyme action.

A. Thermodynamic Aspects of Kinetic Function

1. Processes Not under Enzyme Control

In any chemical reaction the degree to which a particular reaction proceeds (how much product is formed) is determined only by the nature of the starting material and products. Quantification of the degree to which a reaction proceeds is measured by the equilibrium constant, defined as

$$K_{eq} = [\text{Product}]/[\text{Starting material}] \quad (1)$$

where the square brackets ([ ]) denote concentration. The K_{eq} value is measured when product and starting material concentrations are constant (equilibrium condition), and is independent of the path of molecular transformation and the time required to reach equilibrium. Because how far a reaction proceeds is, in essence, a function of the energy of the starting material, the products, and the system, it is convenient to analyze reaction energetics in terms of the energy change of the system associated with a reaction. In 1878, Willard Gibbs derived a function relating the change in free energy of a system to the change in enthalpy (ΔH, related to the internal energy of the system), the change in entropy (ΔS, related to the degree of randomness of the system) and the temperature (T):

$$\Delta G = \Delta H - T\Delta S \quad (2)$$

The ΔG value is an important criterion for determining reaction characteristics: a reaction can occur spontaneously (requires no energy input) only if ΔG is negative, and will not proceed without an input of energy when $\Delta G \geq 0$. Equation 2 shows that a reaction can be driven by a decrease in the enthalpy of the system (a negative ΔH results when the products have a lower energy than the starting material) and/or by an increase in the randomness of the system (ΔS). An example of the first type of process is the burning of wood, which releases a large amount of energy. An example of the second type of process is the dilution of salt in water.

It is possible to relate the K_{eq} value to the change in the total energy of the system (ΔG) and the energy change measured under standard conditions ($\Delta G^{0\prime}$) by Equation 3:

$$\Delta G = \Delta G^{0\prime} + RT \cdot \ln(K_{eq}) \quad (3)$$

where R is the gas constant and T is the temperature in degrees Kelvin (absolute degrees). At equilibrium, $\Delta G = 0$ (no further change can occur) and Equation 3 gives Equation 4:

$$\Delta G^{0\prime} = -RT \cdot \ln(K_{eq}) \quad (4)$$

Equation 4 shows that $\Delta G^{0\prime}$ and the equilibrium constant for a reaction are related by a simple equation that is independent of any kinetic parameters. This shows that it is possible to determine $\Delta G^{0\prime}$ values for any reaction simply by measuring the concentrations of the reactants at equilibrium. Tables of $\Delta G^{0\prime}$ values for several reactions (under standard conditions) can be found in many standard biochemistry and chemistry texts. These values are useful for predicting equilibrium concentrations of reactants. It is possible to measure how far a reaction is from the equilibrium point by comparing predicted and experimental values.

2. Transition State Theory and Reaction Rates

The above thermodynamic analysis shows that a reaction is spontaneous (i.e., can occur without the input of energy to the system) if ΔG is negative; however, it says nothing of how fast that reaction may occur. The factors determining the rate of the reaction are best illustrated by considering the energy diagram for an idealized reaction scheme shown in Fig. 1. For starting material (S) to react to form product (P), additional energy is required to break the chemical bonds of the starting material and/or produce the required reactant configurations. Wood, for example, does not combust without first lighting it, even though the burning of wood has a large negative ΔG.

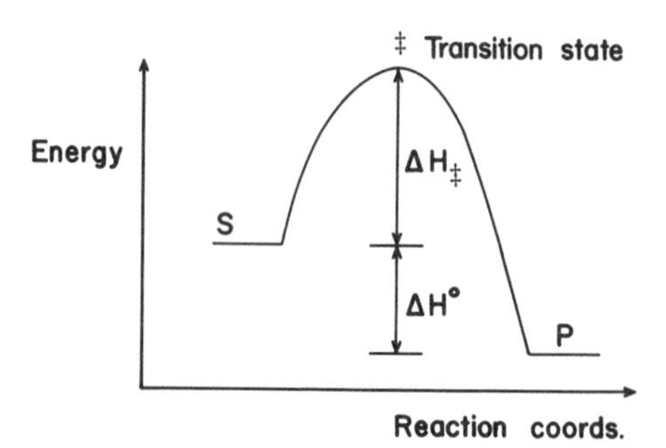

FIGURE 1 Idealized representation of a single reaction scheme. The energy level of the reactants is plotted versus hypothetical reaction coordinates (the reaction proceeds from left to right). S represents starting material and P products. The symbol ‡ represents the activated state, and ΔH‡ is the activation enthalpy for this particular reaction.

The way in which the activation energy influences the rate of a reaction is best described by transition state theory. The transition state is defined as a configuration of maximum potential energy through which the reactants must pass before continuing to form products. It is commonly denoted by a "‡" symbol (Fig. 1). The transition state is clearly distinguished from an intermediate, which is a meta-stable minimum on the reaction profile. In thermodynamic terms, the equilibrium constant for the formation of the transition state is given by Equation 5:

$$K^{\ddagger} = [A^{\ddagger}]/[A] \quad (5)$$

where $A^{\ddagger}$ represents the transition state. Using the thermodynamic principles developed above and quantum mechanics, it is possible to transform Equation 5 into Equation 6, an equation relating the enthalpy ($\Delta H^{\ddagger}$) and entropy ($\Delta S^{\ddagger}$) associated with transition state formation to the rate of the reaction:

$$k = (RT/Nh)\exp(\Delta S^{\ddagger}/R - \Delta H^{\ddagger}/RT) \quad (6)$$

where k is the reaction rate, R is the gas constant, N is Avogadro's number, and h is Plank's constant. Equation 6 explains much about reaction properties. Firstly, it shows that the reactions are temperature dependent, with rates increasing in direct proportion to the increase in temperature. Secondly, it shows that the reaction rate is directly influenced by the activation energy (E_a) of the reaction (in Equation 6, $E_a = \Delta H^{\ddagger} + RT$) with a larger E_a value, resulting in a smaller rate constant. Finally, it shows that the reaction rate is only dependent on the formation of the transition state. Once the transition state is formed, the reaction quickly proceeds to form products.

3. How Enzymes Influence Reaction Rates

The transition state theory outlined in Section I,A,2 also applies to enzymes because enzymes are simply catalysts that act to speed up reaction rates. Equation 6 shows that there are two different ways by which an enzyme may increase the rate of a reaction. Firstly, enzymes may lower the E_a by stabilizing a high-energy transition state, possibly by covalent bonding. Since a more stable transition state is easier to form ($K^{\ddagger}$ in Equation 5 is larger), the $\Delta H^{\ddagger}$ value in Equation 6 will be smaller (see Fig. 1) and a greater number of molecules will react to form product. This is illustrated in Equation 6, in which the magnitude of $\Delta H^{\ddagger}$ directly influences the rate constant; decreasing $\Delta H^{\ddagger}$ increases k. Coenzyme A, pyridoxyl phosphate, and thiamine phosphate are good examples of coenzymes that, when bound to enzymes, stabilize high-energy transition states (see Section III). Secondly, enzymes may increase the reaction rate by providing a favorable reaction entropy (decrease $\Delta S^{\ddagger}$). This can occur when reactive groups on both substrates and enzymes are oriented to optimize catalysis. Equation 6 shows that an increase in the transition state entropy term, $\Delta S^{\ddagger}$, will directly increase the reaction rate. Enzymes that hydrolyze sugar polymers are good examples of this type of catalysis (see Section II).

B. Kinetic Mechanisms for Enzyme Function

1. One-Substrate Reactions

As noted above, all enzymes are catalysts that increase reaction rates either by stabilizing high-energy transition states or by orienting reactive groups so that they are in close proximity. Exactly how enzymes participate in this process is the subject of this section. Consider the general reaction of Fig. 1 with S giving rise to P. The introduction of an enzyme catalyst to this process changes the details of the reaction mechanism because enzymes are physical entities that participate directly in the reaction mechanism. A general one-substrate reaction is shown in Equation 7:

$$S + E \underset{k_{-1}}{\overset{k_1}{\rightleftharpoons}} ES \underset{k_{-2}}{\overset{k_2}{\rightleftharpoons}} EP \underset{k_{-3}}{\overset{k_3}{\rightleftharpoons}} E + P \quad (7)$$

Equation 7 illustrates all the steps necessary to describe any one-substrate reaction, including the binding of substrate to enzyme (starting material in an enzyme reaction is usually referred to as substrate), the conversion of enzyme-bound substrate to enzyme-bound product, and the release of product from the enzyme. Note that Equation 7 does not formally include the formation of a transition state as a discrete step in the reaction mechanism. Its existence is, however, encompassed in the conversion of substrate into product (step 2). Equation 7 also demonstrates the temporal order for an enzyme-catalyzed reaction: (1) Substrate binds to the enzyme to form an enzyme-substrate complex (ES). (2) ES reacts to give the enzyme-product complex (EP), and (3) EP dissociates to free enzyme and product. This latter step regenerates the catalyst so that free enzyme is now capable of binding to new S to continue the reaction.

Two different enzyme-catalyzed reaction profiles are shown in Fig. 2A and B to illustrate the wide

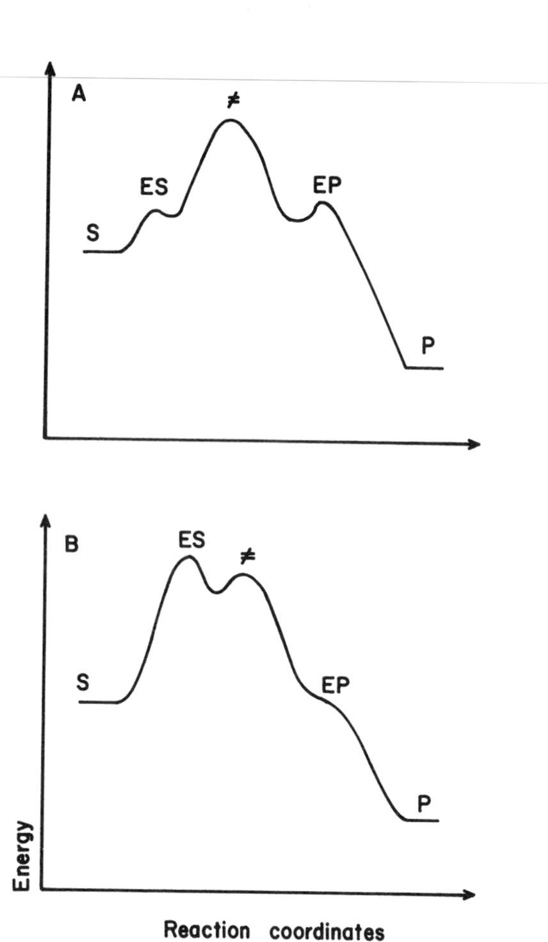

FIGURE 2 Energy diagram for idealized enzyme-catalyzed reactions. The energy level of the reactants and reacting species is plotted versus reaction coordinates. A: Reaction sequence in which the transition state is the most energetically unfavorable species. B: Formation of ES is the most energetically unfavorable step. S and ES, represent the free and enzyme bound substrate. P and EP, represent free and enzyme-bound product respectively.

variety of mechanisms that can account for a simple reaction such as that of Equation 7. In the reaction of Fig. 2A, formation of the transition state is the slowest step, as shown by the higher energy associated with its formation. In Fig. 2B, the formation of the *ES* complex is the slow step. It is important to identify the slow step because the overall reaction velocity will be determined only by the rate of this step; the slow step (and consequently the overall reaction rate) is not always related to the formation of the transition state.

In both reactions of Fig. 2, formation of *ES* and *EP* requires the input of energy, as shown by the higher energy of these species. This in normally true for any enzyme reaction because of the solvation effects. Any solute (e.g., a substrate) that is dissolved in water is surrounded by a layer of tightly bound water molecules. Before substrate can bind to the enzyme, this water layer must be stripped away to expose the reactive groups on the substrate (the same is true for the enzyme as we shall see in Section II). The amount of energy required for this process depends on the substrate and enzyme species and may vary considerably from reaction to reaction, as shown graphically in Fig. 2A and B.

The reaction of Equation 7 was drawn with arrows going in both directions to illustrate that all reactions are reversible; a finite flow of S to P and P to S always exists for any given reaction, even though at equilibrium the net flow is zero. Note also that each step of Equation 7 has a rate constant associated with it, which can be calculated from the individual concentrations and Equation 6 or, more commonly, is determined experimentally from enzyme studies. It is important at this point to distinguish between an overall reaction rate constant and the rate constants associated with the individual steps of Equation 7. The overall rate of any reaction is measured by following the formation of product over time, and it reflects the rate of the slowest step in the reaction mechanism. The microscopic rate constants (i.e., the individual rate constants associated with each step in the reaction mechanism) are often difficult to measure experimentally but correspond only to the step indicated in the reaction. These steps may or may not limit the rate of the overall reaction, depending on their magnitude relative to that of the other steps of the reaction sequence. Both overall and microscopic reactions have a transition state equivalent (i.e., point of maximum energy) and so may be adequately described by Equation 6. The microscopic rate constants are related to the K_{eq} values for formation of each species by Equation 8:

$$K_1 = k_1/k_{-1}, \qquad K_2 = k_2/k_{-2}, \qquad K_3 = k_3/k_{-3} \quad (8)$$

It is possible to reduce Equation 7 into a general scheme that describes enzyme behavior by assuming that the concentration of *EP* is zero:

$$S + E \underset{k_{-1}}{\overset{k_1}{\rightleftharpoons}} ES \xrightarrow{k_2} E + P \quad (9)$$

Experimentally it is relatively easy to generate this condition because most enzyme reactions are performed *in vitro* with an excess of substrate. If we measure the initial enzyme rate (measured within the first few minutes, before the product has a chance to accumulate), the concentration of EP is negligible and $[P] \approx 0$. This condition is referred to as the initial velocity assumption and greatly simplifies Equation 7 and the resulting initial velocity equations.

To reduce the scheme of Equation 9 into a usable equation describing an overall enzyme reaction, we must make an additional assumption about the specifics of the reaction mechanism. Depending on where the rate limiting step lies, it is possible to make one of two different assumptions (see Fig. 2). Both assumptions produce an equation of the general form

$$v = V_{max} \times [S]/(K_m + [S]) \quad (10)$$

where v is the observed rate of reaction (v = increase in product over time) and V_{max} is the maximal rate of the reaction measured at infinite substrate concentrations. Equation 10 is graphically shown in Fig. 3. It is easy to see why an enzyme-catalyzed reaction has a maximum rate, even though a noncatalyzed reaction does not. Because S binds E to form an ES complex with an equilibrium constant K_1 (= k_1/k_{-1}),

$$E + S \underset{k_{-1}}{\overset{k_1}{\rightleftharpoons}} ES \quad (11)$$

increasing the concentration of S while keeping E constant will eventually drive all the free E into the ES complex. At this point ES is the only enzyme species ($ES \approx$ [total enzyme]), so that there is no more free E for S to bind. The rate of the reaction is now independent of S and is given by the concentration of ES (= [total enzyme]) multiplied by the rate of its decomposition, k_2; $V_{max} = k_2 \times$ [total enzyme].

The K_m value (shown in Fig. 3) is defined as the concentration of substrate at which the velocity is equal to $V_{max}/2$. Practically, the K_m value is a measure of the affinity of enzyme for substrate, but the exact relation between the K_m value and the substrate affinity depends on what assumptions are made about the reaction mechanism. Michaelis and Menten assumed that the formation of the transition step was much slower than the rate of ES formation (Fig. 2A). In this case, the K_m value (the m stands for Michaelis) is equal to the dissociation constant for ES formation:

$$K_m = 1/K_1 = k_{-1}/k_1 \quad (12)$$

A more general assumption, first made by Briggs and Haldane, is that a steady state is reached in which the concentration of intermediate is constant: $d[ES]/dt = 0$. This leads to a kinetic definition for the K_m value:

$$K_m = (k_2 + k_{-1})/k_1 \quad (13)$$

Equation 13 applies when the binding reaction is the slow step (Fig. 2B) or when the rate of ES formation is not much different than the rate of transition state formation. *A priori* choice between these two definitions is impossible because a detailed kinetic study must be completed before one or the other can be ruled out. Fortunately, kinetic studies are not dependent on how the K_m is defined because Equation 10 applies equally well to both definitions.

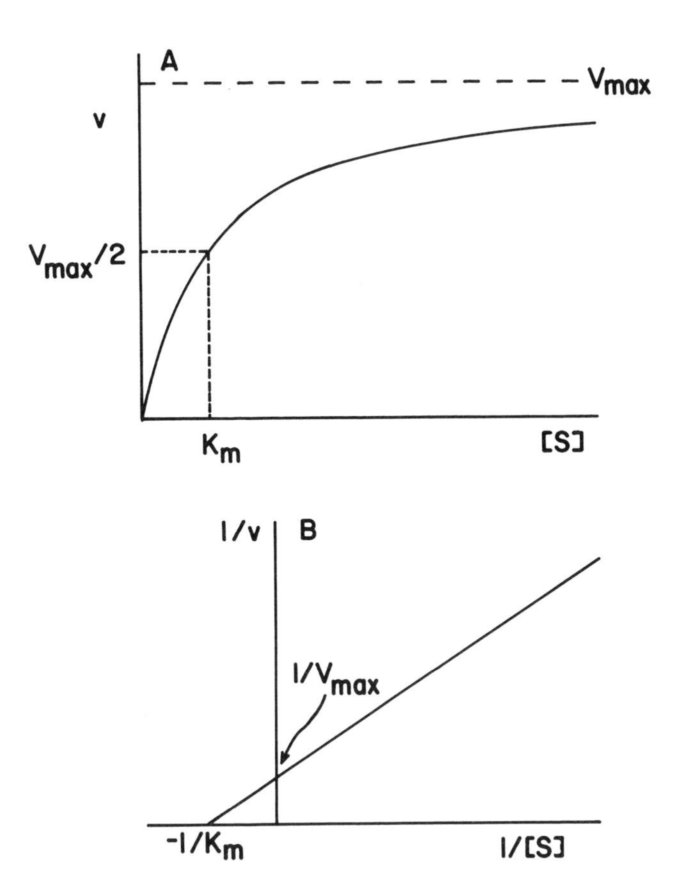

FIGURE 3 Reaction velocity of an enzyme-catalyzed reaction as a function of the substrate concentration. A: Normal plot of the velocity versus substrate. B: Lineweaver-Burk plot of $1/v$ versus $1/[S]$. The maximal velocity (V_{max}) is indicated, as well as the concentration of substrate that produces half-maximal saturation (K_m). Enzyme reaction kinetics are most commonly graphed using the Lineweaver-Burk plot, even though some authors dislike this plot because it magnifies errors at low substrate concentrations. It is used here for simplicity because the references at the end of this article use this graph exclusively.

Figure 3B demonstrates how K_m and V_{max} values are graphically determined. The method is that of Lineweaver and Burk, who rearranged Equation 10 into Equation 14:

$$1/v = K_m/V_{max} \cdot 1/[S] + 1/V_{max} \qquad (14)$$

The V_{max} and K_m parameters are obtained by extrapolating the line through the $1/[S]$ and 1/velocity data pairs to its intersection with the x and y axes. Thus, at infinite substrate concentrations $1/[S] = 0$, and $1/v = 1/V_{max}$; V_{max} is obtained from the reciprocal of the y-axis intercept. Equation 14 also shows that when $v = V_{max}/2$, $1/K_m = 1/[S]$, and the K_m value is obtained from the reciprocal of the x-axis intercept. We will use the Lineweaver-Burk plot to illustrate the effects of inhibitors later in this article.

2. Multisubstrate and Cooperative Reactions

The above analysis gave a simple equation relating substrate concentrations to observed reaction velocity at a fixed enzyme concentration. The result of this analysis, Equation 10, applies directly to systems in which one substrate reacts to give one product under initial velocity conditions. Numerous such equations (of varying complexity) exist that describe enzyme reactions involving two substrates and cooperative substrate binding. These equations are necessary because the majority of enzymes catalyze reactions involving more than one substrate–product pair. Many enzyme reactions also require coenzymes to complete their catalytic cycles (Section III,A). When these coenzymes are soluble (i.e., they can reversibly bind to enzyme active sites) they are treated kinetically as substrates. Thus, reactions between substrates and coenzymes are equivalent to two substrate reactions. Four of the most common mechanisms are presented in Table I for a reference. The first three kinetic patterns are distinguished from the fourth, the cooperative model, by the linearity of the Lineweaver-Burk plots when $1/v$ is plotted against the reciprocal of the first substrate concentration. Cooperative enzymes have nonlinear Lineweaver-Burk plots. Fig. 4 shows mechanisms for the three most common enzyme pathways with two substrates. The three reaction pathways are distinguished by the order that the substrates bind to the enzyme. In the random-equilibrium model (Fig. 4A) either substrate

Random

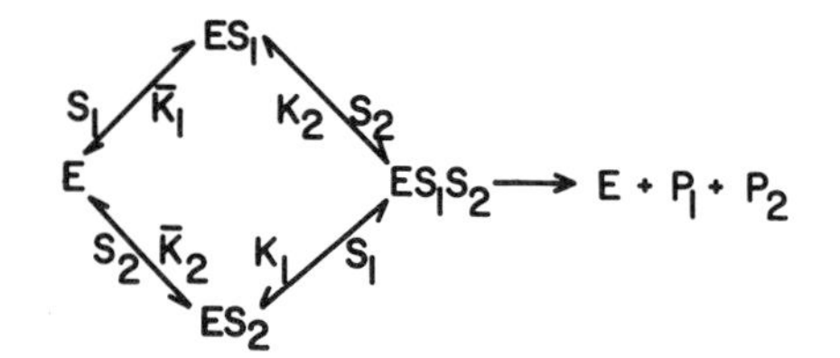

Ordered

$$E \underset{\bar{K}_1}{\overset{S_1}{\rightleftharpoons}} ES_1 \underset{K_2}{\overset{S_2}{\rightleftharpoons}} ES_1S_2 \longrightarrow E + P_1 + P_2$$

Ping Pong

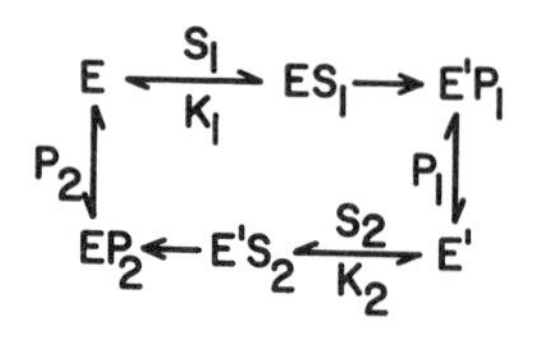

FIGURE 4 Common initial velocity of two-substrate mechanisms. S_1, first substrate; S_2 second substrate; E, free enzyme. (1) Random-equilibrium model. (2) Ordered-equilibrium model. (3) Substituted enzyme model (ping pong model). The $\bar{K}$ values represent true dissociation constants; the K values represent the dissociation constants in the presence of infinite amount of the opposite substrate. Mechanisms assume that the reactions are at equilibrium (Michaelis-Menten assumption).

TABLE I Common Multisubstrate and Cooperative Kinetic Equations[a]

Pattern	Variable $[S]$	V_{max}^{app}	K_m^{app}
Random, Ordered	$[S_1]$	$V_{max}[S_2]/(K_2 + [S_2])$	$(K_{12} + K_1[S_2])/(K_2 + [S_2])$
Substituted	$[S_1]$	$V_{max}[S_2]/(K_2 + [S_2])$	$K_1[S_2]/(K_2 + [S_2])$
Cooperative	$[S^h]$	V_{max}	K_m^h

[a] K_1 is the K_m value for S_1, K_2 is the K_m values for S_2, and $K_{12} = \bar{K}_1 \cdot K_2$ (see Fig. 4). The kinetic constants for the first three patterns were obtained by assuming that the binding reactions are at equilibrium (Michaelis-Menten assumption). The equation describing pattern 4 is an empirical relation derived by Hill to describe oxygen binding to hemoglobin. In cooperative enzyme kinetics, h represents the Hill coefficient, an arbitrary measure of the cooperativity of the system, and has no precise physical significance.

can bind to free enzyme. The ordered-equilibrium model (Fig. 4B) is observed when only the first substrate can bind to free enzyme; the second substrate can bind only the ES_1 complex. The substituted-enzyme model (or Ping-Pong mechanism, Fig. 4C) describes a situation in which one substrate binds free enzyme and reacts to give a covalently modified enzyme (E'). The covalently modified enzyme now reacts with the second substrate to give free enzyme and product. The latter mechanism is often observed with phosphate transfer mechanisms. Cooperative enzyme kinetics arise when an enzyme has more than one substrate binding site. Enzymes are often composed of more than one polypeptide chain (subunit), (see Section II,A). When this is true, a situation may arise where each subunit of a multisubunit enzyme has an active site. In simple cooperative models which can be described by the Hill equation: (1) all the subunits are identical so that the enzyme uses only one type of substrate (S_1 is the only substrate) and (2) the binding of substrate to one enzyme sites affects the binding of the second substrate molecule. Positive cooperativity (the most common form of cooperativity) thus describes a situation in which binding of one substrate increases the enzyme affinity for the second substrate.

Although the complete equations for the mechanisms shown in Fig. 4 are complex, they can be simplified into Equation 10 with V_{max}^{app} (apparent V_{max}) replacing V_{max}, and K_m^{app} (apparent K_m) replacing K_m. This means that the individual kinetic constants for these reactions can be obtained by manipulating substrate concentrations and plotting the resulting velocities in graphs similar to Fig. 3B. Table I shows the values of V_{max}^{app} and K_m^{app} for four common kinetic patterns. In the first three equations, V_{max}^{app} and K_m^{app} values are obtained from the intercepts when v is plotted against the corresponding substrate value shown in Table I. The exact kinetic equation describing the fourth mechanism (cooperative) is complex but may be adequately described by the Hill equation shown in Table I. Note that the Hill coefficient is simply an arbitrary measure of the increase in affinity affected by the binding of the first substrate, and has no precise physical meaning.

C. Enzyme Inhibition and Activation

In general, the cellular concentration of enzyme substrates and products does not change significantly even when the cell metabolic rate changes dramatically. This means that without the existence of other mechanisms for controlling enzyme rates, enzyme activities could not be regulated; enzyme velocity would always be equal to the value corresponding to a fixed substrate concentration (see Fig. 3). It is especially important for the cell to control enzymes found at the beginning of metabolic pathways. By regulating these enzymes, the cell can determine the activity of each pathway; the flux through the pathway is controlled by the amount of substrate provided by the initial enzyme. Enzyme inhibition and activation are important mechanisms for cellular control of individual reactions. In general, inhibitors (defined as substances that decrease enzyme activity) and activators (substances that increase enzyme activity) are compounds that serve as monitors of the energy state of the cell [e.g., adenosine triphosphate (ATP) or adenosine monophosphate (AMP)] or are products or substrates of the enzymes or pathways themselves. These compounds serve to link enzyme activity directly to the energy state of the cell or to the demand for substrate or product. In this section we will consider only reversible inhibitors and activators.

Three general classes of inhibitors exist, classified according to which enzyme form they bind (either E and/or ES in Equation 9) and what kinetic effect they have on the velocity versus substrate plots. (1) Competitive inhibitors compete directly with substrate for free enzyme and so affect the apparent substrate affinity for enzyme (K_m value increases with increasing inhibitor concentration). (2) Uncompetitive inhibitors bind to the ES complex and reduce the rate of product formation ($k_2' < k_2$ giving a decreased V_{max} value). (3) Mixed inhibitors bind both free and substrate-bound enzyme to give increased K_m values and decreased V_{max} values. The effects of inhibitors are best illustrated by considering a Botts-Moralis scheme (Equation 15). The Botts-Moralis scheme is simply Equation 9 expanded to include two new equilibria:

$$\begin{array}{ccccc} E & \underset{K_m}{\overset{S}{\rightleftharpoons}} & ES & \xrightarrow{k_2} & E + P \\ I \Big\Updownarrow K_i & & I \Big\Updownarrow K_i' & & \\ EI & \underset{K_m'}{\overset{S}{\rightleftharpoons}} & ESI & \xrightarrow{k_2'} & EI + P \end{array} \quad (15)$$

In Equation 15, K_i corresponds to the inhibitor dissociation constant for competitive inhibitors, and

K_i' to the inhibitor dissociation constant for uncompetitive inhibitors. The scheme of Equation 15 can be reduced to a generalized equation by assuming that the E, EI, ES, and ESI species are all in equilibrium. This is equivalent to the Michaelis and Menten assumption described above. As expected, solution of the equilibria in Equation 15 gives the generalized Equation 10 with $V_{max}{}^{app}$ replacing V_{max} and $K_m{}^{app}$ replacing K_m. Equation 16 shows how these new values relate to the individual binding constants of Equation 15.

$$V_{max}{}^{app} = [\text{total enzyme}] \cdot (k_2 + k_2' [I]/K_i')/(1 + [I]/K_i') \quad (16a)$$

$$K_m{}^{app} = K_m \cdot (1 + [I]/K_i)/(1 + [I]/K_i') \quad (16b)$$

The $V_{max}{}^{app}$ and $K_m{}^{app}$ values are the apparent V_{max} and K_m values measured at finite concentrations of inhibitor. They are obtained in the same way as V_{max} and K_m values by extrapolating $[S]$ to infinite concentrations (see Fig. 5). Table II lists some of the more commonly occurring inhibition patterns and illustrates, mathematically, how each inhibition pattern arises. Figure 5 shows how the Lineweaver-Burk plots behave for some of these inhibitors. For example, competitive inhibition, which occurs when an inhibitor binds only free E to compete directly with S for free enzyme, arises if ES has no affinity for inhibitor. This is expressed mathematically as $K_i' = K_m' = \infty$. Figure 5A shows the effect of competitive inhibition on the Lineweaver-Burk plots, and Fig. 5B shows how to obtain the kinetic constants from such an analysis. Other inhibition patterns are also presented in Table II and Fig. 5.

It is possible to use the Botts-Moralis scheme for analyzing reversible enzyme activation in the same way as enzyme inhibition, by replacing I in Equation 15 for A (for activator) and K_i for $1/K_m$ (K_m is the equilibrium constant for activator binding). Depending on the value of the individual constants,

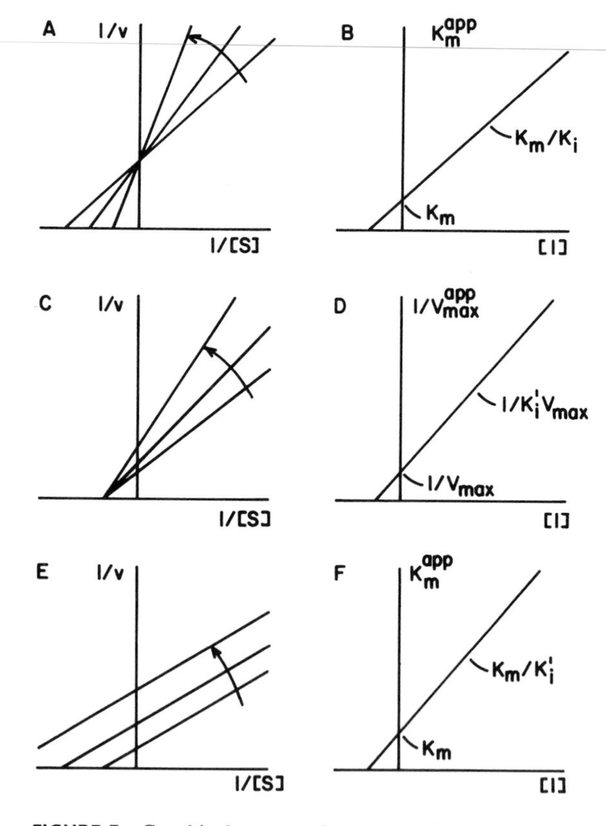

FIGURE 5 Graphical patterns for competitive, noncompetitive, and uncompetitive inhibition. A, C, and E represent primary plots of $1/v$ versus $1/[S]$ at increasing inhibitor concentrations (shown by *arrow*). B, D, and F represent secondary plots of indicated parameter as a function of increasing inhibitor concentration. Parameters are obtained as shown on the plots.

TABLE II Common Inhibition Patterns[a]

Pattern	Conditions	$V_{max}{}^{app}$	$K_m{}^{app}$
Competitive	$K_i' = K_m' = \infty$	V_{max}	$K_m \cdot (1 + [I]/K_i)$
Partially competitive	$K_m' > K_m$; $K_i' > K_i$; $k_2 = k_2'$	V_{max}	$K_m \cdot (1 + [I]/K_i)/(1 + [I]/K_i')$
Mixed (noncompetitive)	$K_m = K_m'$; $K_i = K_i'$; $k_2' = 0$	$V_{max}/(1 + [I]/K_i')$	K_m
Uncompetitive	$K_i = K_m' = \infty$; $k_2' = 0$	$V_{max}/(1 + [I]/K_i')$	$K_m/(1 + [I]/K_i')$
Partially noncompetitive	$K_i \neq K_i'$; $K_m \neq K_m'$; $k_2 \neq k_2'$	see Equation 16	

[a] Figure 4 lists the graphical patterns associated with these inhibition types.

kinetic patterns similar to those of Fig. 5 may be observed; however, activation is rarely as simple as inhibition, and nonlinear secondary plots often result. It is, perhaps, for this reason that well-defined common activation patterns (analogous to competitive, uncompetitive, and mixed inhibition patterns) have not been characterized. [*See* ENZYME INHIBITORS.]

II. Enzyme Function-Structure Relations

Enzyme kinetic studies provide a lot of information about the overall enzyme mechanism, but they provide no useful information about how an enzyme specifically catalyzes any given reaction. In this section we will examine the structure of an enzyme in some detail, with the aim of observing how enzymes perform their function.

A. Enzyme Structure

The old maxim "All enzymes are proteins but not all proteins are enzymes" was long regarded as a universal truth, but recently, with the discovery of auto-catalytic messenger RNA (mRNA) molecules (capable of splicing themselves), this old "truth" is no longer valid. It is unfortunate, however, that the paucity of information at this time on the structure-function relation of mRNA catalysts means that the present review will consider only protein-based enzymes.

Enzymologists have classically divided enzyme structure into four separate categories: primary, secondary, tertiary, and quaternary. The ordering of these four categories reflects the order of influence of one type of structure on the next. For example, the primary structure influences the degree of secondary structure but not the reverse.

Protein-based enzymes are complex molecules composed of a chain of amino acids linked together through peptide bonds. Each enzyme (often called a polypeptide) can be identified by its unique sequence of amino acids, and for this reason, the amino acid sequence is called an enzyme's *primary structure*. The primary structure is determined by the nucleotide bases on the DNA molecule, which code for the protein. The individual properties of the amino acids and the relative amount of each amino acid in the enzyme play an important role in determining the secondary, tertiary, and quaternary structure of an enzyme. Amino acids can be classified into four different classes depending on their R groups: hydrophobic, polar, negatively charged, and positively charged (see Table III). The properties of these R groups determine whether they will be highly solvated (polar, negative, and positive R groups tend to be on the outside of a protein, in direct contact with water) or will be hidden in the interior of the protein, away from water (hydrophobic).

The secondary structure refers to highly ordered substructures found within an enzyme. Two types of secondary structure have been identified: α-helices (Fig. 6B) and β-pleated sheets (Fig. 6C). The formation of these structures depends directly on the enzyme's amino acid sequence; amino acids with relatively small, uncharged polar R groups organize themselves into α-helices and amino acids with relatively small nonpolar R groups form β-pleated sheets. Both α-helices and β-pleated sheets may form a significant proportion of an enzyme's three-dimensional structure (see the picture of myoglobin, Fig. 6D). Proline is a special amino acid

TABLE III Types of Amino Acids

Class	Name	R group[a]
Nonpolar	alanine	$-CH_3$
	valine	$-CH(CH_3)_2$
	leucine	$-CH_2-CH(CH_3)_2$
	isoleucine	$-CH(CH_3)-CH_2-CH_3$
	proline	-OOC-(pyrrolidine ring, HN)
	phenylalanine	$-CH_2-\theta$
	tryptophan	$-CH_2$-indole
	methionine	$-CH_2-CH_2-S-CH_3$
Polar, uncharged	glycine	-H
	serine	$-CH_2-OH$
	threonine	$-CH(OH)-CH_3$
	cysteine	$-CH_2-SH$
	tyrosine	$-CH_2-\theta-OH$
	asparagine	$-CH_2-CO-NH_2$
	glutamine	$-CH_2-CH_2-CO-NH_2$
Polar, negative	aspartic acid	$-CH_2-COOH$
	glutamic acid	$-CH_2-CH_2-COOH$
Polar, positive	lysine	$-CH_2-CH_2-CH_2-CH_2-NH_2$
	arginine	$-CH_2-CH_2-CH_2-NH-C(NH_2^+)-NH_2$
	histidine	$-CH_2$-imidazole$^+$

[a] θ represents benzene. The basic amino acid structure is $R-CH(NH_2^+)-COO^-$. As indicated, amino acids are usually zwitterionic at pH 7. Charges on amino acid R groups are measured at pH 7. The complete structure of proline is shown.

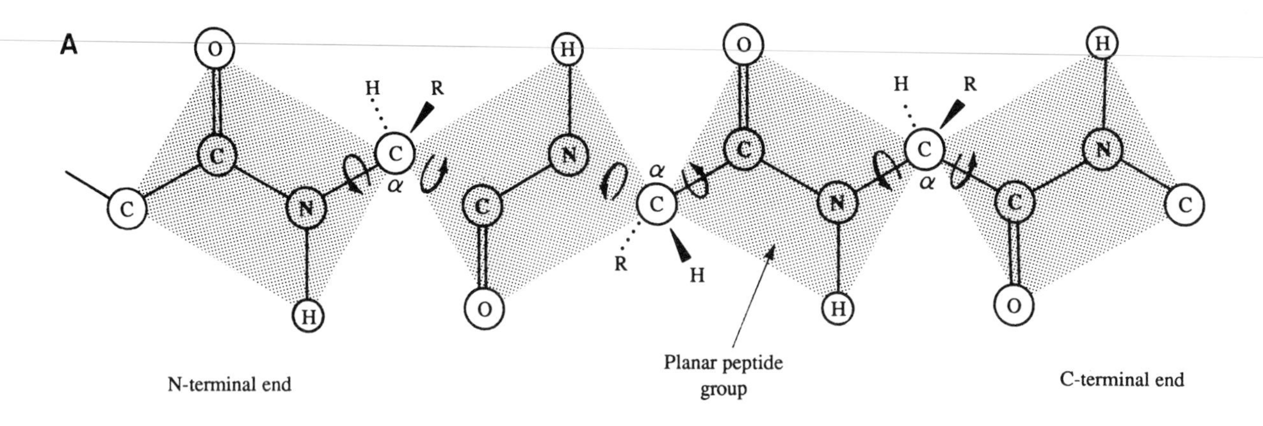
A
N-terminal end
Planar peptide group
C-terminal end

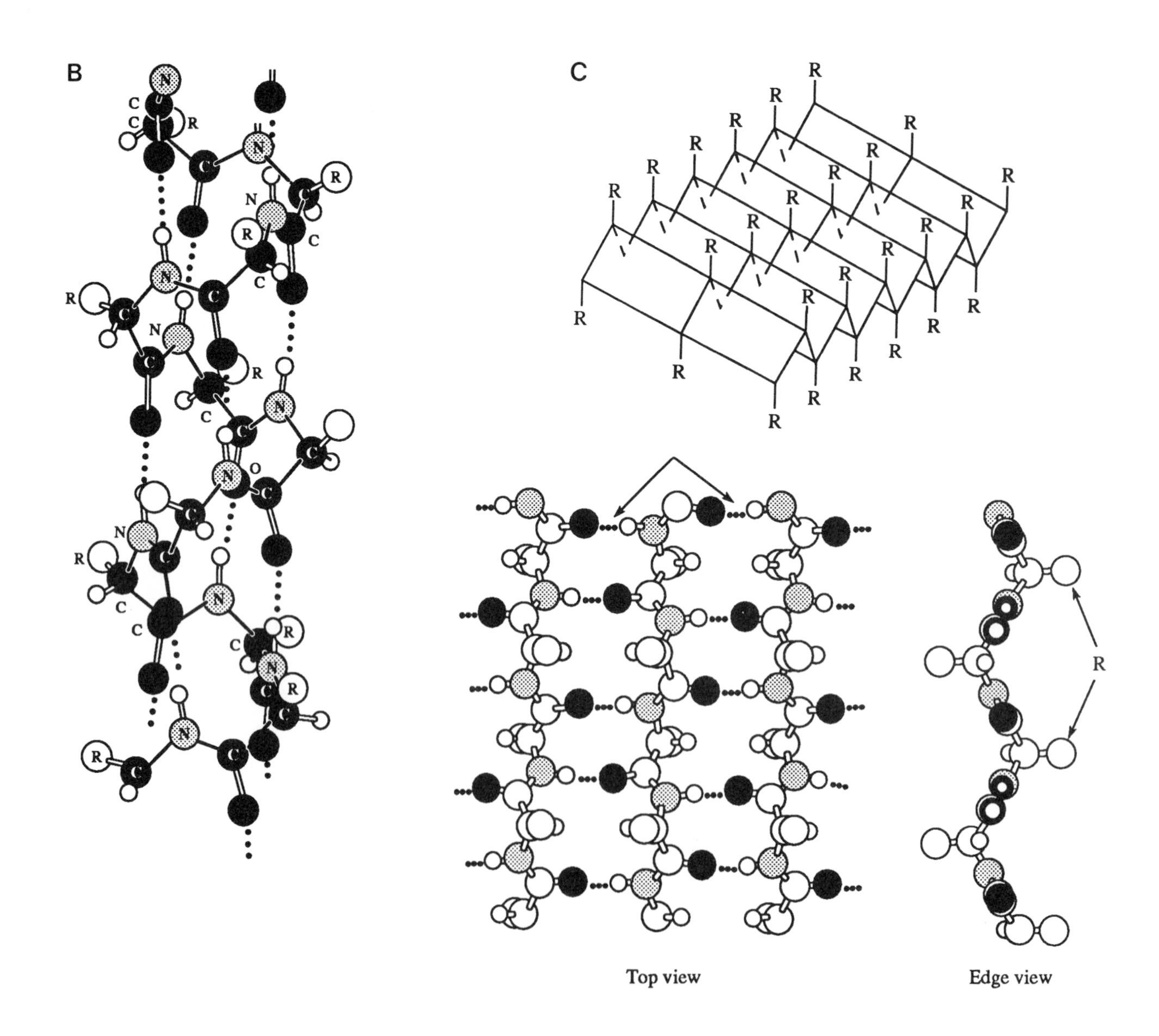
B
C
Top view
Edge view
R

because of its unique structure (Table III). Introduction of proline into an amino acid sequence introduces a permanent bend at this position. For this reason, the presence of proline in an α-helix or β-pleated sheet will disrupt the secondary structure at this point.

Tertiary enzyme structure refers to the final structure obtained when an amino acid chain is placed in water. Tertiary structure is largely determined by the interaction of R groups on the surface of the protein with the water and with other R groups on the protein surface. Polar R groups will move to the outside of the protein, nonpolar R groups will move toward the inside of the protein. The movement of certain parts of the polypeptide chain away from water will give a protein a characteristic three-dimensional shape. The final structure represents a minimum energy point and is a compromise between all the forces acting on it. Note that the reference point, in this discussion, is a water solution. It is possible to alter the tertiary structure of an enzyme by radically changing the solvating medium. For example, changing the solvating medium to an organic solvent or altering the pH, ionic strength, or temperature can radically change protein structure, often leading to irreversible denaturation. These changes cause proteins to lose structure either by solvating nonpolar R groups (dissolving the protein in an organic solvent) or by causing R groups on the surface of the protein to become nonsoluble so that proteins stick together (pH changes).

The fourth level of enzyme structural organization, quaternary structure, exists only in multisubunit enzymes. In fact, most enzymes are not single polypeptide chains but are either dimers, tetramers, or polymers of several identical subunits (each polypeptide chain is termed a subunit in a multiple subunit enzyme). In the case of multiple subunit enzymes, the R groups on each polypeptide chain may interact with each other, as well as with the solvating medium, to give a final enzyme structure that is different than that for the isolated subunit.

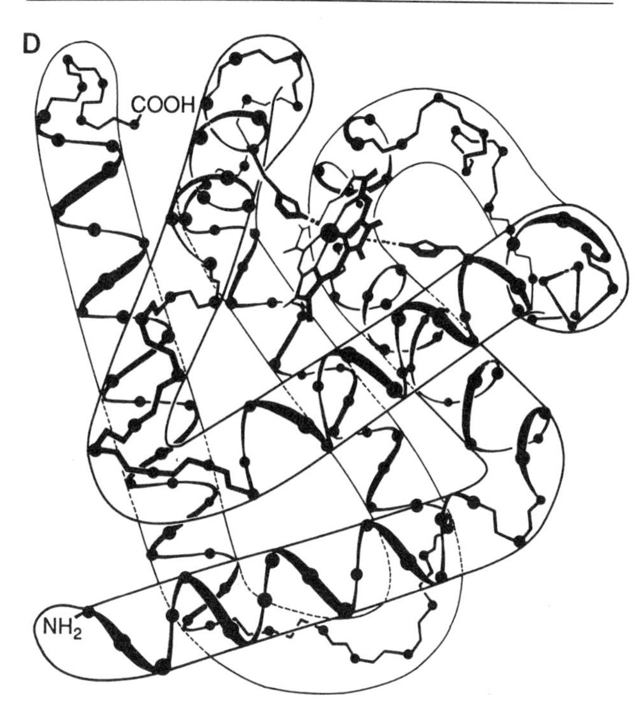

FIGURE 6 Structural elements of enzyme conformation. A: Protein primary structure. Amino acid components are commonly numbered starting with the N-terminal end. Peptide bonds (between two amino acids) are planar as indicated in the figure. [From Lehninger, A. L. (1975). "Biochemistry," 2nd ed. Worth, with permission.] B: Ball-and-stick model of α-helix. The α-helix shape is similar to a spring with the peptide bonds and α-carbon atoms forming the backbone of the structure. R groups stick out away from the α-helix. [From Haggis, G. H., Michie, D., Muir, A. R., Roberts, K. B., and Walker, P. H. B. (1964). "Introduction to Molecular Biology" John Wiley, with permission.] C: Top: Schematic representation of three parallel polypeptides in a β-sheet structure. [From Bennett, T. P. (1968). "Graphic Biochemistry." Macmillan, with permission.] Lower left: Ball-and-stick model of β-sheet showing the interchain hydrogen bonding between the carboxyl oxygens *(dark circles)* and the amide hydrogens *(white circles)*. Lower right: Ball-and-stick model of β-sheet showing R groups projecting out from the plane of the hydrogen bonds [From Springall, H. D. (1954). "The Structural Chemistry of Proteins." Academic Press, redrawn with permission.] D: Schematic diagram of the myoglobin backbone showing the high percentage of α-helical structure and heme prosthetic group. [From Neurath, H., ed. (1964). "The Proteins." Academic Press, with permission.]

B. How Enzyme-Catalyzed Reactions Occur: Lysozyme as a Model

When an enzyme has folded into its final structure, a specialized area or cleft on the protein surface exists, termed the *active site*. It is here that substrate binds, and it is here that the R groups responsible for binding substrate and for catalyzing the reaction are located. The active site of an enzyme is a highly conserved region of the protein, showing little change in different species, and is specifically configured to bind only one substrate (or one class of substrates). How well a particular substrate binds the active site depends on (1) which amino acids are present, (2) the positioning of the amino acids within the active site, and (3) other environmental factors such as the ionic strength, pH, and ion composition of the solvating medium. Although it has been convenient to think of the binding of substrate to active site as a "lock and key" fit, this is by no means a complete description of the interaction. Enzymes are not rigid structures but are somewhat flexible so that the conformation of enzyme and substrate are exactly complimentary only after the substrate binds. This dynamic binding mechanism is called *induced fit* and is illustrated by measurements of enzyme structure in the presence and absence of substrate. X-ray crystallography shows that enzyme structure is different when substrate is bound to the active site, a fact that is consistent with the proposed mechanism of enzyme action. Enzymes could not operate efficiently if substrate binding was optimized because a minimum in the energy graph, caused by a stable ES complex, would decrease the rate of catalysis; a large, negative $\Delta G^{0\prime}$ for ES formation would make $\Delta G^{0\prime}$ for the catalysis step more positive. It is the combination of active site geometry and enzyme conformational change that gives the enzyme-substrate complex its specificity. Even small changes in substrate chemical composition will dramatically alter its affinity.

The binding interaction between enzyme (or more specifically amino acid R groups at the enzyme active site) and substrate can also contribute to the catalytic efficiency of an enzyme. A good example of this effect is seen in the action of lysozyme, an enzyme found in mammalian tears and mucus, which lyses the sugars of bacterial cell walls; these walls are composed of alternating residues of *N*-acetylglycosamine (NAG) and *N*-acetylmuramic acid (NAM). Lysozyme binds six sugar residues at a time, three NAG and three NAM units. Catalytic action cleaves a single NAM–NAG linkage, NAM occupying the fourth spot in the active site. Cleavage is aided by physical strain on the NAM residue, forcing it to adopt an energetically unfavorable half-chair conformation (Fig. 7). This destabilization means that the ES complex is at a higher energy than the free S, effectively lowering the activation energy and leading to a rate enhancement.

Once substrate is bound to the active site, it is oriented so that amino acid R groups on the protein can participate directly in the catalytic mechanism. This is best illustrated in Fig. 7, which shows how aspartate 52 and glutamic acid 35 of lysozyme act to catalyze the NAG–NAM cleavage. Both groups are important to catalysis; glutamic acid provides the proton for acid catalysis, and aspartate stabilizes the resulting glycosyl-oxocarbonium ion. Lysozyme action thus shows that enzymes may increase the rate of catalysis by stabilizing the transition state (aspartate stabilizes the positive transition state ion). This mode of enzyme action is a common catalytic mechanism of other enzymes as well. For example, tight complexes between enzymes and synthetically manufactured transition state analogues are demonstrated by the fact that transition state analogues make highly specific and powerful enzyme inhibitors. Lysozyme action also demonstrates that enzymes catalyze reactions using simple chemical species (glutamate and aspartic acid in this case); the active site is not an isolated environment in which specialized chemistry can occur. Enzymes are, however, more efficient catalysts than simple chemical agents because (1) they orient the substrate so that optimal efficiency between the catalyst (e.g., protons for lysozyme) and the susceptible bond is achieved, and (2) they bind one substrate specifically so that other molecules are not acted on. This is in contrast to chemical catalysts that are frequently not substrate-specific nor stereospecific and, consequently, produce a significant proportion of undesirable products. Enzyme-catalyzed reactions are highly specific and so enzymes produce essentially 100% of a single product.

III. Coenzymes and Their Function

Lysozyme, considered in Section II above, is an example of an enzyme-catalyzed group transfer in which a glycosyl residue is transferred to water. Enzyme reactions, however, are not limited to simple acid-catalyzed water hydrolysis reactions. Several different types of enzyme reactions exist that can be loosely divided into four different classes. A short list of these are presented in Table IV, along with examples of reactions and enzymes listed beside each class. Many of the reactions shown in Table IV can be catalyzed using simple acid-base catalysis and so can be carried out with charged amino acid R groups (either positive or negative). The hydrolysis of NAM–NAG sugar residues is an example of such an acid-catalyzed reaction, which is catalyzed by aspartic acid 35 of lysozyme. However, several of the reactions in Table IV cannot be carried out with simple acid-base catalysts and require additional reactive groups not available on amino acid R groups. When this is true, coenzymes

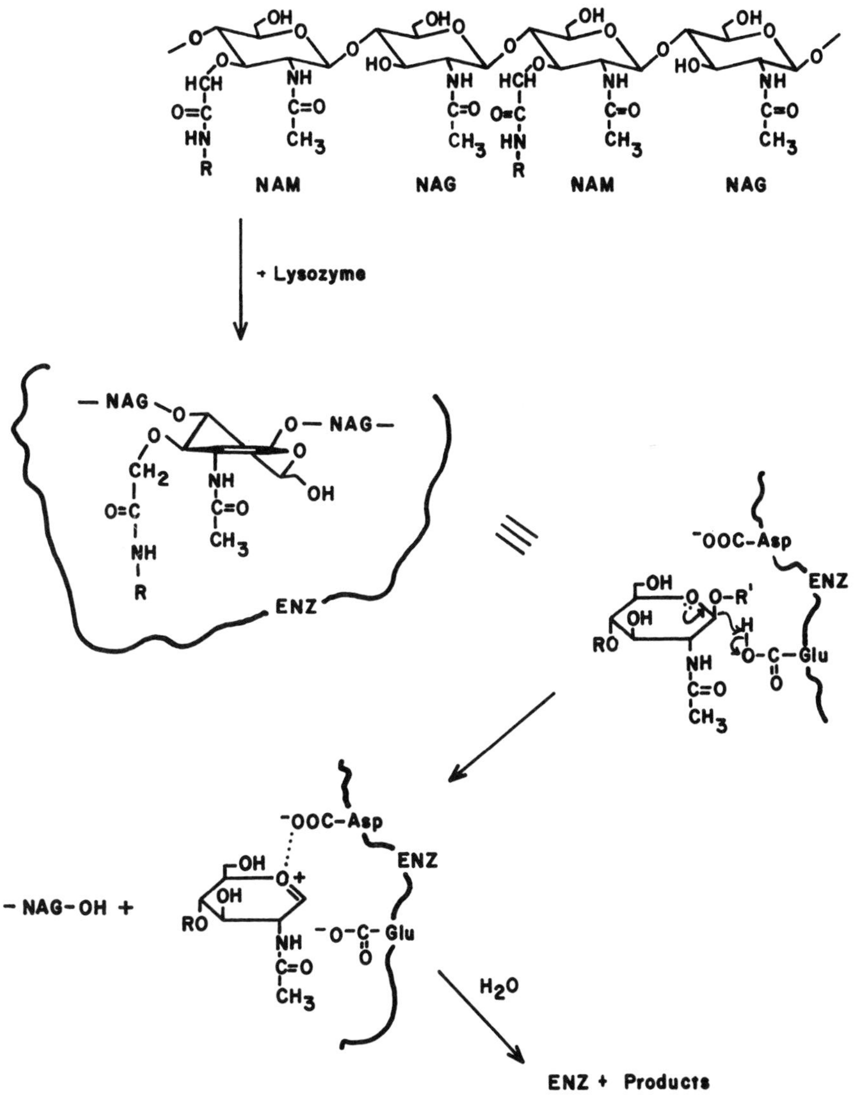

FIGURE 7 Lysozyme catalytic mechanism. NAM–NAG polymer binds to lysozyme to induce the half-chair conformation at position 4. Aspartate (Asp) and glutamate (Glu) participate directly in the reaction sequence. Glutamate donates a proton, and aspartate stabilizes the glycosyl-oxocarbonium ion (shown by *dots*). [From Walsh, C. (1979). "Enzymatic Reaction Mechanisms." W. H. Freeman, with permission.]

act as additional carriers that can accept and donate chemical groups, hydrogen atoms, or electrons (see Table V). Several of the reactions listed in Table IV indicate the requirement for coenzymes, demonstrating coenzyme importance in catalyzing cellular reactions. Coenzymes can be broadly classified into three groups: (1) compounds that transfer high-energy phosphate, (2) compounds that accept and donate electrons, and (3) compounds that activate substrates.

A. Coenzymes That Transfer High-Energy Phosphate

These compounds are highly phosphorylated derivatives of the nucleotide bases (e.g., adenosine and guanosine), which exist in either mono-, di-, and triphosphate forms. Usually, the diphosphate acts as a phosphate acceptor and the triphosphate acts as a phosphate donor. The monophosphate is made during biosynthetic reactions requiring a pyrophosphate intermediate: ATP $\rightarrow$ AMP + PP_i. All these compounds are extremely important to metabolic and biosynthetic reactions because they act as the

TABLE IV Classification of Enzyme Reactions[a]

Class	Types of reactions
Group transfers	Acyl transfer to water (proteases)
	γ-glutamyl and amino transfer (glutamine synthetase)
	Phosphoryl transfer I (phosphatases, ATPases, phosphodiesterases)
	Phosphoryl transfer II (kinases)
	Nucleotidyl and pyrophosporyl transfer
	Glycosyl transfer (glycogen synthetase, lysozyme)
Oxidations/reductions	NAD(P) requiring enzymes (alcohol dehydrogenase, lactate dehydrogenase)
	FAD requiring enzymes (succinate dehydrogenase)
	Monooxygenases I (FAD, FMN, and pterin-dependent reactions)
	Monooxygenases II (Copper-requiring enzymes)
	Metallo-flavoprotein oxidases
	Heme-containing oxidases (catalase, cytochrome *c* oxidase)
	Dioxygenases
Eliminations/isomerizations/rearrangements	Addition/elimination of water (aconitase, fumarase, enolase)
	Addition/elimination of ammonia (phenylalanine ammonia lyase)
	Isomerizations (phosphoglucose isomerase)
	Rearrangements (B_{12}-dependent enzymes)
C-C bond breaking	Decarboxylations (β-keto and β-hydroxyacids, α-amino acids)
	Carboxylations (biotin-dependent enzymes)
	Aldol reactions (aldolase)
	Pyridoxyl phosphate–requiring enzymes
	Tetrahydrofolate- and *S*-adenosyl methionine–requiring enzymes

[a] Reactions are listed by common enzyme mechanism and are grouped according to Walsh. [From Walsh, C. (1979). "Enzymatic Reaction Mechanisms." W. H. Freeman, with permission.]

TABLE V Coenzymes[a]

Classification	Coenzyme name (vitamin)	Abbreviation
High-energy phosphate transfer	Adenosine triphosphate	ATP
	Guanosine triphosphate	GTP
Electron acceptors	Nicotinamide adenine dinucleotide (niacin)	NAD
	Nicotinamide adenine dinucleotide phosphate (niacin)	NADP
	Flavin adenine dinucleotide (B_2)	FAD
	Riboflavin 5′-phosphate	FMN
	Lipoamide	—
	Coenzyme Q (ubiquinone)	CoQ
Substrate activators	Coenzyme A (pantothenic acid)	CoA
	Acyl carrier protein (phosphopantetheine)	ACP
	Biotin	—
	Thiamine pyrophosphate (B_1)	TPP
	Pyridoxyl phosphate (B_6)	PLP
	Coenzyme B_{12}	B_{12}
	Tetrahydrofolate (folic acid)	THF
	S-adenosyl methionine	SAM

[a] Biochemical names of the coenzymes are presented along with the vitamin they are derived from. Abbreviations are common but may not be universal.

energy currency of the cell. ATP is by far the most common cellular high-energy phosphate compound and occurs at concentrations approximately 100-fold higher than other high-energy phosphate intermediates. These compounds are considered as high-energy phosphates because hydrolysis of γ-phosphorus of ATP yields $\Delta G^{0\prime} = -30.6$ kJ/mole of energy corresponding to a K_{eq} value of approximately 10^8 in favor of adenosine diphosphate (ADP) and P_i formation (see Equation 4). This means that many reactions can be forced to completion by coupling them to ATP hydrolysis; a reaction with a positive $\Delta G^{0\prime}$ value can be made negative if the reaction is coupled to ATP hydrolysis (the overall $\Delta G^{0\prime}$ for a reaction is the sum of the individual $\Delta G^{0\prime}$ values). In accord with its function, ATP does not remain enzyme-bound in between catalytic cycles but diffuses from enzyme to enzyme to participate in other high-energy phosphate transfer reactions. For this reason ATP is usually thought of as a substrate because, like any other substrate, it must bind enzyme to form an *ES* complex. In this way the ATP generated by glycolysis and oxidative phosphorylation can be used in any number of energy-requiring processes such as biosynthetic pathways, membrane ion pumps, and protein synthesis. Because of its importance in cellular biochemistry, enzymes that use ATP are highly regulated because they control the cellular concentration of ATP through their action. These enzymes are usually found at the start of metabolic pathways and are often inhibited or activated by ATP, ADP, and

AMP, as well as by the end products of the metabolic pathway. [*See* ATP SYNTHESIS BY OXIDATIVE PHOSPHORYLATION IN MITOCHONDRIA; ADENOSINE TRIPHOSPHATE (ATP).]

B. Coenzymes That Accept and Donate Electrons

Coenzymes in this category participate in oxidation and reduction reactions, serving as two electron acceptors or two electron donators depending on the reaction. In accord with this function, several of these coenzymes have delocalized π-ring systems, which stabilize positive and negative charges. For example, extended π systems help stabilize positive charges present in the oxidized form of the coenzyme. During the normal oxidation-reduction cycle, these coenzymes can either reversibly bind enzyme active sites [NAD, NADP, and coenzyme Q (CoQ) diffuse from enzyme to enzyme] or remain closely associated with one enzyme (e.g., FAD, FMN). In the former case, NAD, NADP, and CoQ act as substrates to transfer either reducing or oxidizing equivalents from reaction to reaction. NAD and NADP participate in water-based reactions and CoQ participates in lipid (membrane) reactions. In the case of FMN and FAD coenzymes, the reduced riboflavin coenzyme transfers electrons to a soluble coenzyme such as NAD(P) or CoQ to complete its oxidation-reduction cycle. NAD and NADP can be considered as high-energy electron carriers, much like ATP is a high-energy phosphate carrier; these coenzymes are similar to ATP in their mode of action. Nicotinamide coenzymes thus transfer electrons from substrates to either (1) reduce compounds during biosynthetic reactions (e.g., during fatty acid synthesis), or (2) provide the electrochemical energy necessary for ATP synthesis in the mitochondrion. Like ATP using enzymes, nicotinamide coenzyme using enzymes are often highly regulated; however, unlike ATP using enzymes, they are not regulated by NAD or NADP concentrations but are sensitive to other substrate and product concentrations.

C. Coenzymes That Activate Substrates

Although this class includes several different types of reactions, all the coenzymes in this class react with substrate at the enzyme active site to produce a new, more reactive product. For example, coenzyme A (CoA) and acyl carrier protein (ACP) attach to acetyl residues (CH_3-CO-), forming acetyl-CoA and acetyl-ACP compounds. Both CoA and ACP are good chemical leaving groups, so that the reactivity of these acetyl derivatives is much higher than that for acetic acid ($\Delta G^{0\prime}$ for hydrolysis to acetic acid and CoA or ACP is large and negative). Formation of these compounds effectively creates a high-energy acetate derivative.

Biotin, a highly reactive double-ring system covalently attached to an enzyme by a long, flexible arm, serves as a carboxyl group carrier, adding -COOH groups to compounds by fixing the carbonate ion (HCO_3^-) dissolved in water. The reaction is energetically unfavorable and is coupled to the hydrolysis of ATP to drive the reaction to completion.

Thiamine pyrophosphate and pyridoxal phosphate catalyze several different types of reactions including α-condensation and α-cleavage reactions (thiamine pyrophosphate), transamination, decarboxylation, elimination, and aldol cleavage (pyridoxal phosphate). Both coenzymes can be considered as the reactive species in the enzyme active site, although neither is covalently attached to the enzyme. The thiamine pyrophosphate mechanism proceeds via a thiazolium dipolar ion, which can attack ketone or imide compounds. The newly formed adduct reacts readily to give products (carbon dioxide is often a product). Pyridoxal phosphate also forms an adduct with substrate (called a *Schiff's base*), which decomposes to give the products. These coenzymes increase the rate of catalysis by stabilizing the transition state to lower the overall activation energy. In both cases the substrate-coenzyme adduct is stabilized by the conjugated ring systems of the coenzyme, which act as either an electron sink or an electron source during the reaction.

Coenzyme B_{12} is another example of a compound that catalyzes a reaction by stabilizing the transition state. B_{12} is an unusual coenzyme because it contains a copper atom in the center of a corrin ring system coordinated to adenosine below and ribose above the plane of the corrin ring. The copper is the reactive center of the coenzyme and can easily accept and donate a single electron; coenzyme B_{12} catalyzes reactions using radical ion chemistry. It is this property that enables B_{12} to catalyze a wide range of reactions including group migrations, dehydrations, and oxidation-reduction reactions (often in conjunction with the oxidation-reduction coenzymes listed above).

Both *S*-adenosyl methionine (SAM) and tetrahydrofolate (THF) serve as methyl group carriers in much the same way that CoA and ACP serve as acetyl carriers. Because the chemistry of carbon-carbon bond breaking is complex, reactions involving THF and SAM often involve other coenzymes required for reduction or oxidation, as well as B_{12}. SAM and THF reactions are not common, but these coenzymes participate in two extremely important reactions: methylation of DNA to turn off gene transcription, and the methylation of homocysteine to form methionine, one of the 21 amino acids that make up proteins (see Table III).

Bibliography

Cornish-Bowden, A. (1979). "Fundamentals of Enzyme Kinetics." Butterworths, Toronto.

Lehninger, A. L. (1975). "Biochemistry." Worth, New York.

Metzler, D. E. (1977). "Biochemistry. The Chemical Reactions of Living Cells." Academic Press, New York.

Segal, G. (1975). "Enzyme Kinetics." John Wiley, New York.

Stryer, L. (1982). "Biochemistry," 3rd ed. W. H. Freeman, New York.

Sucking, K. E., and Sucking, C. J. (1980). "Biological Chemistry. The Molecular Approach to Biological Systems." Cambridge University Press, London.

Suelter, C. H. (1985). "A Practical Guide to Enzymology," Biochemistry, A series of monographs, vol. 3. John Wiley, New York.

Walsh, C. (1979). "Enzymatic Reaction Mechanisms." W. H. Freeman, San Francisco.

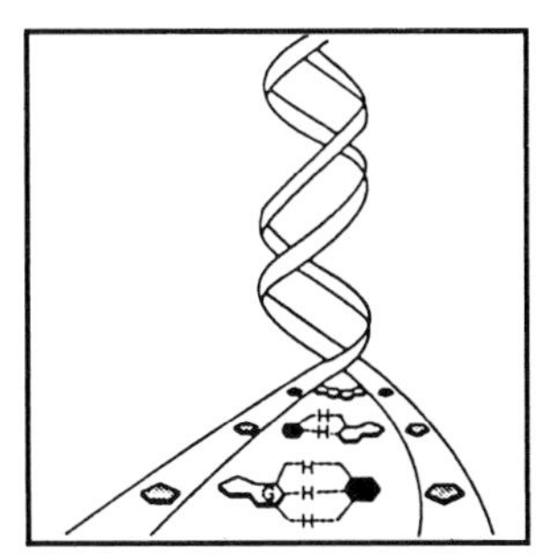

Enzymology of DNA Replication

ULRICH HÜBSCHER, *University of Zürich*

Glossary

Deoxyribonucleic acid (DNA) Carrier of the genetic information, which is encoded in the sequence of bases. It is present in chromosomes and chromosomal material of cell organelles (e.g., mitochondria and chloroplasts) and also in some viruses. DNA is a polymeric molecule made up of deoxyribonucleotide repeating units (composed of the sugar 2-deoxyribose, phosphate, and a purine or pyridine base) linked by the phosphate group joining the 3′ position of one sugar to 5′ position of the next

DNA polymerase Enzyme that catalyzes the addition of deoxyribonucleotide residues to one end of a primer hybridized to a template strand. The new nucleotide has to fit to the template according to the Watson-Crick base pairing rule, so that the template is faithfully copied

DNA replication Duplication of DNA before cell division. Both mother templates must be copied

Double helix Double-stranded DNA held together by hydrogen bonds. Because of the antiparallel arrangement of the two strands, the DNA usually forms a right-handed helix structure

Enzymes Catalytic proteins that mediate and promote the chemical processes of cell functions without being altered or destroyed

Okazaki fragments Short DNA fragments that are synthesized on one DNA strand during replication

Plasmids Extrachromosomal DNA in some bacteria. They may harbor antibiotic resistance genes and are important tools for recombinant DNA technology

Processivity Defines the numbers of monomers that are polymerized in one binding event by a polymerizing enzyme (e.g., by a DNA polymerase).

Semiconservative replication Duplication of DNA in a fashion in which the template strand is conserved and is used for an exact copy of a new complementary strand

Watson-Crick base pairing rule Double helix is formed by hydrogen bonds between the bases guanine and cytosine or adenine and thymine

ANY LIVING ORGANISM harbors genetic information coding for the protein molecules that are responsible for structure and function of a cell. Most organisms bear this information (comparable to the "software" of a computer) in the form of a long thread called *deoxyribonucleic acid* (DNA). The components of a DNA molecule are phosphoric acid, the pentose sugar deoxyribose, and the four bases adenine (A), guanine (G), cytosine (C), and thymine (T). It is the linear order of the four bases that encodes for the genetic information of the DNA. A human cell contains about 3×10^9 base pairs in its nucleus corresponding to a total length of 2 m. Before cell division, which generates two daughter cells from a single mother cell, the DNA must be duplicated.

Single–stranded DNA containing a 3'hydroxyl primer terminus + n [dATP, dGTP, dCTP, dTTP] —(Mg^{++}, DNA polymerase)→ Double–stranded DNA + nPP_i

FIGURE 1 DNA polymerase reaction.

I. DNA Replication Is a Vital, Universal Biological Task

DNA replication is the process that guarantees a faithful duplication of the genetic information. This vital event is crucial for the correct maintenance of the genetic information. Let us assume a cell nucleus corresponds to a tennis ball; then the DNA would be a thread of 20 km (13 miles). The process of DNA replication enables duplication of the 20-km thread in a tennis ball at an enormous speed and with an extremely high accuracy. Finally, the complete distribution of these two DNA threads to the two daughter nuclei during mitosis is another extremely challenging task of DNA replication.

The mechanism of DNA replication depends on the chemistry of DNA. The DNA backbone is built by a series of phosphodiester bridges between the carbon 3 of one deoxyribose and the carbon 5 of the next. Every deoxyribose itself is bound to one of the four bases via its carbon 1 residue. Millions of these phosphodiester bridges form a long molecule. At one (called the *3′ end*), the deoxyribose contains a free hydroxyl group at the carbon 3 and on the other (called the *5′ end*) a free hydroxyl group at the carbon 5 residue. The thermodynamically more stable form of the DNA is the double helix. The stability is guaranteed by the simple rule of base pairing between adenine and thymine or guanine and cytosine. In the helix the two strands have different free hydroxyl groups at their ends: at one end of the helix, one of the strands has a free carbon 3, the other strand has a free carbon 5. The arrangement of the two strands is therefore antiparallel and has, according to these free hydroxyl groups, on one strand a direction of $5' \rightarrow 3'$ and on the other of $3' \rightarrow 5'$. [*See* DNA and Gene Transcription.]

Basic mechanistic features of DNA replication emerged in the past three decades. Replication of double-stranded DNA is semiconservative. The two strands (called *mother strands*) serve each as templates for the synthesis of a new strand (called *daughter strand*). For this purpose, the DNA double helix has to be brought into single-stranded form before DNA synthesis. Because of the basic reaction mechanism of any DNA polymerase known (for details see Section III), the direction of synthesis of a DNA strand is always $5' \rightarrow 3'$. The difference derives from the antiparallel arrangement of the DNA and the fact that the overall process of replication proceeds in the same direction on both strands. Of the two mother strands, one is continuously copied as a whole and is called the *leading strand*, and the other is discontinously copied in small pieces and is called the *lagging strand*. During DNA synthesis, the discontinuously copied lagging strand consists of many pieces of DNA with still attached short ribonucleic acid (RNA) segments, which are called *Okazaki fragments*. These primers serve the DNA polymerases as starting points, as no DNA polymerase can start DNA synthesis *de novo* on a template, unless it is provided with a 3′ hydroxyl group of a previously synthesized primer with a complementary sequence to the template. Most primers in nature are short pieces of RNA. During replication the RNA primers are later removed and replaced by DNA; the Okazaki fragments are then joined one to another. As a result, both new strands must be synthesized with the same accuracy, copied entirely, and lead to two completely separated double strands (Fig. 1). [*See* DNA Synthesis.]

Considering these facts, it is not astonishing that DNA replication is a complex event. The logical consequence appears to be that a cell must possess many enzymes that can carry out all these steps in concerted action.

II. DNA Replication Requires Many Different Enzymes

A complex network of interacting proteins and enzymes is required for DNA replication. Much of our present understanding is derived from studies of the bacterium *Escherichia coli* and its viruses (called *bacteriophages*). The DNA replication system was also studied *in vitro* using genomes of bacteriophages (e.g., T7, T4, and ϕX174) or plasmids carrying the origin of bacterial replication as models. By genetical analysis, 25–30 proteins were isolated and their functional tasks are understood today at an elementary level. These proteins work together in coordinated order as multisubunit assemblies. These results served as a guideline for the search and the purification of similar proteins in eukaryotes. Here again, model systems for replication lead the way, especially with viral DNAs from simian virus 40, adenovirus, and herpesvirus.

For simplicity this article concentrates now exclusively on enzymes acting on a DNA in the process of replication (Table I). The structure that the DNA has at that time is designated as the replication fork. The classical model replication fork (see Color Plate 10) is an asymmetric structure with two distinct modes of DNA synthesis: continuous DNA synthesis on the leading strand, and discontinuous DNA synthesis on the lagging strand. A *DNA helicase* opens the DNA double helix by hydrolyzing two adenosine triphosphate (ATP) molecules per base pair to be melted. *One DNA polymerase* replicates the leading strand continuously. On the other strand, the lagging strand, a *DNA primase* synthesizes the RNA primers (4–10 nucleotides each) for the initiation of the DNA fragments (Okazaki fragments), which are then elongated by *another DNA polymerase* molecule (for details see Section III). The joining of these Okazaki fragments, whose average size in bacteria is about 1,000 and in eukaryotes about 150 nucleotides, is complicated by the presence of RNA primers at the 5′ hydroxyl end of each previously synthesized Okazaki fragment. They are removed by *ribnuclease H,* an enzyme that hydrolyzes RNA hybridized to DNA. Their removal leaves a gap, which has to be filled by a *third DNA polymerase,* and is then sealed by an enzyme called *DNA ligase*, which restores the continuity of the phosphodiester backbone.

TABLE I Some DNA Replication Enzymes and Their Roles

Enzyme	Role
DNA polymerase	Synthesis of DNA
DNA helicase	Unwinding of DNA
DNA primase	Synthesis of primers
Ribonuclease H	Removal of primers
DNA ligase	Ligation of DNA pieces
DNA topoisomerase	Conversion of topological DNA isomers
3′ → 5′ exonuclease	Removal of incorrectly incorporated bases (= proofreading for the DNA polymerase)

Because DNA polymerase can incorporate incorrect bases (Section III) an activity called *3′ → 5′ exonuclease* can remove bases that cannot form a proper base pair in the form of A to T and G to C. This activity is also called the *proofreading activity.* There exist DNA polymerases that contain a 3′ → 5′ exonuclease within the DNA polymerase polypeptides (e.g., DNA polymerase I of *Escherichia coli*, DNA polymerase from herpes simplex virus, DNA polymerase δ in general and DNA polymerase α from *Drosophila melanogaster*), whereas others do not (e.g., DNA polymerase III of *Escherichia coli* and DNA polymerase α).

When DNA is unwound by DNA helicase and newly synthesized by DNA polymerase, topological problems arise at the replication fork because the two strands are intertwined and cannot be separated as such. The entanglement is removed by enzymes called *DNA topoisomerases*, which cut the phosphodiester backbone of one (type I) or both (type II) strands. After the strands are disentangled, the phosphodiester bridge is resealed by the same enzyme. These nicking-closing-type enzymes are essential for DNA replication.

III. DNA Polymerases Are Key Enzymes for DNA Replication

DNA polymerases are the main actors for polymerization of deoxyribonucleoside triphosphates (dNTP) at the replication fork. These enzymes use the two strands of the mother DNA as templates to guide their DNA polymerization (Fig. 1) according to the Watson-Crick base pairing rule. In addition, to start DNA synthesis the enzyme needs a 3′ hydroxyl primer of short strand of nucleic acid (RNA or DNA) bound to the mother template strand. The four bases A, T, G, and C are required in the dNTP form. First, DNA polymerase binds to a template-primer junction. Then the corresponding dNTP binds to the active center of the enzyme, and the base is covalently attached as a deoxynucleoside monophosphate to the 3′ hydroxyl group of the primer releasing a pyrophosphate molecule. A divalent cation (usually magnesium) is required for enzymatic activities of all DNA polymerases known.

TABLE II Five Different Eukaryotic DNA Polymerases

Name	Likely biological function
α	Replication of nuclear DNA (lagging strand) Repair of nuclear DNA
β	Repair of nuclear DNA, recombination
γ	Replication of mitochondrial DNA
δ	Replication of nuclear DNA (leading strand)
ε	Repair of nuclear DNA Replication of nuclear DNA

The overall task of DNA polymerase to perform accurate and fast DNA synthesis is the reason for their complex structure. Five different DNA polymerases have been identified in eukaryotic cells (Table II). They are DNA polymerases α, β, γ, δ, and ε. DNA polymerases α, δ, and ε appear to possess important functional roles in nuclear DNA replication and are also involved in DNA repair mechanisms. DNA polymerase β is the main repair enzyme in the nucleus, and DNA polymerase γ is the replication enzyme for the mitochondrial DNA. Now consider functional aspects of DNA polymerases α and δ in relation to their likely roles at the replication fork.

A. DNA Polymerase α

The enzyme is a multipolypeptide complex that can be isolated as a four-subunit complex from yeast to human. The 160,000–200,000-Dalton subunit has DNA polymerase activity, the 40,000–50,000- and 50,000–60,000-Dalton subunits represent the DNA primase heterodimer, and the function of the 65,000–75,000-Dalton subunit is thought to be the protein that anchors the DNA polymerase/primase together. There is an almost complete conservation of this basic structure from yeast to human.

B. DNA Polymerase δ

This DNA polymerase was originally identified as an enzyme containing an intrinsic $3' \rightarrow 5'$ exonuclease proofreading activity (Table II). By this definition, the enzyme is biochemically different from DNA polymerase α, and, at least in yeast, is coded for by a different gene. There is evidence that two DNA polymerases containing a $3' \rightarrow 5'$ exonuclease exist. They can be distinguished by their response to proliferating cell nuclear antigen, a protein that is regulated through the cell cycle. DNA polymerase δ is dependent on proliferating cell nuclear antigen for processive activity and accuracy. DNA polymerase δ and proliferating cell nuclear antigen are strongly conserved in structure during evolution, supporting a crucial role in replication of the DNA. A second DNA polymerase, called δ_2 or II in yeast is processive in the absence of proliferating cell nuclear antigen and is now called DNA polymerase ε.

IV. DNA Replication Needs Two DNA Polymerases with a Complex Architecture

The function of a replication-specific DNA polymerase is to produce an accurate copy of the genetic material. The replicative complex, in which the DNA polymerase is active, must perform several functions other than that of the polymerization of dNTP. For example, it must take part in the unwinding of the two mother strands, so that these can be used as templates. On the leading strand the DNA must be synthesized in a highly processive way, whereas on the lagging strand, cooperation with DNA primases in the synthesis of oligomeric RNA is required to prime the discontinuous DNA synthesis. Furthermore, on the lagging strand, the DNA polymerase and its associated proteins must be frequently recycled. The fidelity of DNA replication is also achieved by enzymatic control (e.g., by proofreading with the aid of a $3' \rightarrow 5'$ exonuclease). Finally, the entire replication machinery must be able to cooperate with other enzymes, proteins, and factors involved in the control of cell cycle and DNA replication.

During DNA replication, both strands have to be replicated with the same speed and accuracy. The antiparallelity of the mother DNA strands raises the fundamental biological question of how nature has untangled the difficult problem of continuous DNA synthesis on one strand and a discontinuous one on the other. A model has been proposed in which two DNA polymerases act coordinately as a dimeric complex at the replication fork. One DNA polymerase replicates the leading strand continuously, whereas another replicates the lagging strand discontinuously. The lagging strand would bend back on itself to form a loop at the replication fork. This allows a dimeric DNA polymerase to move along both template strands in the "same" direction without violating the $5' \rightarrow 3'$ directionality rule. When the lagging strand DNA polymerase abuts on the

TABLE III Different Complexities of Replicative DNA Polymerases

	Polypeptides
Escherichia coli	
DNA polymerase III core	3
DNA polymerase III′	8
DNA polymerase III*	18
DNA polymerase III holoenzyme, asymmetric dimer	22
Eukaryotes including human	
DNA polymerase α core	1–2
DNA polymerase α/primase	4
DNA polymerase α holoenzyme	>10
DNA polymerase δ	2
DNA polymerase δ/proliferating cell nuclear antigen	3

Okazaki fragment synthesized during the previous round, the freshly synthesized DNA is threaded through the enzyme, allowing the DNA polymerase to recycle to a newly exposed single-stranded region more 3′ to the previous priming site for initiation of a new Okazaki fragment. In this way, a dimeric DNA polymerase would guarantee an efficient and coordinated progression of the replication fork (see Color Plate 11).

Studies on DNA polymerase III *Escherichia coli*, the main replicative enzyme, clearly indicated that forms of the multiprotein complex with several degress of complexity and functional capacity can be isolated (Table III). The simplest form is the DNA polymerase III core, which possesses three subunits, and the most complicated form is the DNA polymerase III holoenzyme dimer, which contains 22 polypeptides. The enzyme has an asymmetric form. The leading part is less complex than the lagging one. This is likely because the lagging part of the complex has to recycle frequently.

In eukaryotes it is also likely that DNA polymerases have similar complexity. In addition, it appears that the two different DNA polymerases δ and α share the task of continuous and discontinuous DNA replication. Both enzymes were identified as having different degrees of complexities (Table III). An attractive hypothesis is that these two DNA polymerases form a dimeric DNA polymerase complex, in which they perform different functions. The relevant properties of DNA polymerase δ and α, summarized in Table IV, propose DNA polymerase δ as the candidate for leading strand synthesis and DNA polymerase α for lagging strand synthesis, respectively. In particular, DNA polymerase δ is very processive and DNA polymerase α is only moderately processive and contains a DNA primase, obviously needed for frequent initiation at the lagging strand. In addition, DNA polymerase α is generally more complex than DNA polymerase δ, as we would expect from the enzyme that frequently has to recycle after completion of each Okazaki fragment.

V. Conclusive Summary

DNA replication is the process that guarantees the faithful duplication of the genetic information in advance of division of any living cell from bacteria to human. This universal event is crucial for the correct maintenance of the genetic information. Research progress in the past few years suggested that the mechanisms of DNA replication might be similar in most organisms tested. DNA synthesis at the replication fork requires the coordinated action of different types of enzymes. These appear to be organized in a higher order structure to ensure the

TABLE IV Properties of Eukaryotic DNA Polymerases δ and α for a Distinct Function at the Replication Fork

Property	DNA polymerase δ	DNA polymerase α
Preferred DNA to bind	double-stranded and single-stranded DNA	single-stranded DNA
Processivity	high	low
Associated DNA primase	no	yes
Associated 3′ → 5′ exonuclease	yes	in some preparations
Dependent on proliferating cell nuclear antigen	yes	no
Correlation with cell proliferation	yes	yes
Proposed for replication of	leading strand	lagging strand

precise and rapid duplication of the 3×10^9 base pairs present in the human genome.

DNA polymerases are the key enzymes for DNA replication. At least two different types of DNA polymerases with a complex architecture are necessary to replicate both strands of a mother DNA simultaneously. Dimerization of two DNA polymerase molecules would guarantee efficient, coordinated, and accurate replication of both DNA strands.

Bibliography

Alberts, B. M. (1987). Prokaryotic DNA replication mechanisms *Philos. Trans. R. Soc. Lond. B.* **317,** 395–420.

Baker, T. A., and Kornberg, A. (1990). "DNA Replication," 2nd ed. W. H. Freeman, San Francisco.

Burgers, P. M. J. (1989). Eukaryotic DNA polymerases α and δ: Conserved properties and interactions from yeast to mammalian cells. *Prog. Nucleic Acids Res. Mol. Biol.* **37,** 235–280.

Burgers, P. M. J., Bambara, R. A., Campbell, J. L., Chang, L. M. S., Downey, K. M., Hübscher, U. Lee, M. Y. W. T., Linn, S. M., So, A. G., and Spadari, S. (1990). Revised nomenclature for eukaryotic DNA polymerases. *Eur. J. Biochem.* **191,** 617–618.

"Enzymology of DNA Replication," vol. 951. (1988). *Biochim. Biophys. Acta*, special issue.

Fry, M., and Loeb, L. A. (1986). "Animal Cell DNA Polymerases." CRC Press, Boca Raton, Florida.

Hübscher, U. (1989). DNA polymerases δ and α: Possible coordinated actions at the DNA replication fork. *In* "Highlights in Modern Biochemistry" (A. Kotyk, J. Skoda, V. Paces, and V. Kostka, eds.), pp. 485–494. VSP International Science, Zeist, The Netherlands.

McHenry, C. S. (1988). DNA polymerase III holoenzyme of *Escherichia coli. Annu. Rev. Biochem.* **57,** 519–550.

Spadari, S., Montecucco, A., Pedrali-Noy, G., Ciarrocchi, G., Focher, F., and Hübscher, U. (1989). A double-loop model for the replication of eukaryotic DNA. *Mutat. Res.* **219,** 147–156.

Stillman, B., and Kelly, T., eds. (1988). "Eukaryotic DNA Replication. Cancer Cells," vol. 6. Cold Spring Harbor Laboratory, Cold Spring Harbor, New York.

Thömmes, P., and Hübscher, U. (1991). Eukaryotic DNA replication: Enzymes and proteins at the fork *Eur. J. Biochem.* (in press).

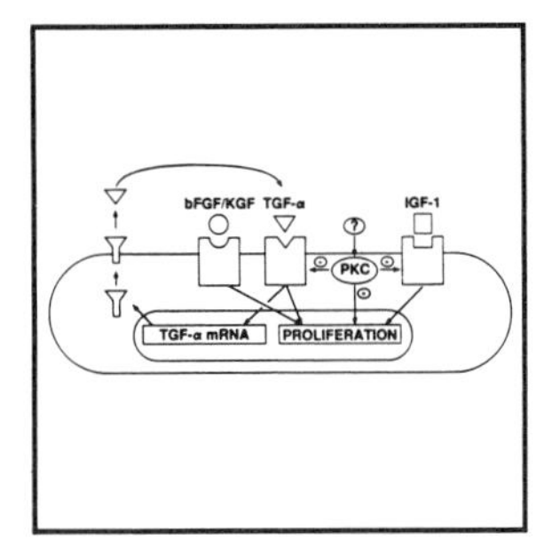

Epidermal Proliferation

JAMES T. ELDER, *University of Michigan*

Glossary

Autocrine In reference to cytokine–growth factor responsiveness, the cell responds to growth factors produced by the same cell; this response may take place intracellularly or on the cell surface

Basal cell Keratinocyte attached to the basement membrane of the dermoepidermal junction

Cytokine Any polypeptide that participates in intercellular signaling via secretion into the extracellular space and binding to receptors on target cells

Growth factor Any cytokine that induces cellular proliferation

Paracrine Responder cell is in the vicinity of the producer cell and may be a more or less differentiated cell of the same lineage, or a cell of a different lineage; no circulatory system is required to transport the factor from the producer to the responder cell.

Stratum corneum Uppermost layer of the epidermis, consisting of platelike squamous cells

THE EPIDERMIS is a specialized epithelium that forms the external portion of the largest organ of the body—the skin. The keratinocyte, a specialized epithelial cell, is the majority cell type of the epidermis. The keratinocyte must maintain a finely tuned balance between cell proliferation and loss through terminal differentiation in normal epidermal homeostasis and, yet, be able to greatly increase its proliferative rate in response to injury. A precise understanding of this regulation requires definition of the intercellular signals that regulate keratinocyte growth and differentiation. For each signal identified, it is necessary to define the cell type(s) responsible for its production, the target cell(s) upon which it acts, and the intracellular signaling pathways that mediate both its production and its ultimate effects on keratinocyte growth and differentiation. This chapter will review our available knowledge about the molecular signals that regulate epidermal proliferation. For this purpose, we will consider psoriasis an inflammatory and hyperplastic skin disease, as an ''experiment of nature'' which may be helpful in identifying the molecular signals that regulate epidermal homeostasis in normal epidermis.

I. Psoriasis, Wound Healing, and Epidermal Homeostasis

Psoriasis is a common skin disease characterized by markedly increased proliferation and altered differentiation of keratinocytes in the context of multiple biochemical, immunological, inflammatory, and vascular abnormalities. Despite its complexities, psoriasis has much to teach us about the balance between proliferation and differentiation in self-renewing tissues. Psoriasis has been likened to an unregulated process of cutaneous wound healing, because epidermal hyperplasia, inflammatory cell infiltrates, and vascular alterations characterize both processes. One reason for this similarity could be that both processes are mediated by the same set of intercellular signals, including growth factors, cytokines, and extracellular matrix components. Another reason is that each cell type present in the skin is constrained by its developmental history to display only a subset of potential responses to extracellular signals. In contrast, the clinical features of psoriasis and healing wounds differ markedly from those of various malignant neoplasms. These differences may offer important clues to the molecular etiology of psoriasis and the normal regulation of epidermal proliferation.

A. Psoriasis Is Not Cancer

While cancer is similar to psoriasis and healing wounds in that cellular proliferation is increased, cancer differs from psoriasis and healing wounds in several important ways. In normal individuals, cutaneous wound healing and psoriatic lesions are highly self-limited processes, which are confined to the skin. Psoriasis is often characterized by exacerbations and remissions, and lesions are frequently multifocal and often display considerable symmetry. In contrast, cancer usually arises in a single site and spreads locally or metastatically without remission or detectable symmetry. A substantial literature supports the concept that cancers arise due to somatic mutations affecting either the structure or the appropriate regulation of critical cellular genes (protooncogenes). The expression of several protooncogenes (including *c-myc, c-fos, c-jun, c-erbB, c-Ha-ras*) in psoriasis is not significantly different from those in normal skin. The clinical behavior of psoriasis and healing wounds is more consistent with altered regulation of normal intercellular cellular signaling processes, because the appearance of the skin can return to normal after wound healing is complete, after antipsoriatic therapy, or after spontaneous improvement of psoriasis. [*See* ONCOGENE AMPLIFICATION IN HUMAN CANCER.]

B. Local Regulation of Keratinocyte Proliferation

The behavior of wounded skin suggests that at least part of the proliferative signal inherent to the repair process must arise locally within the wound, because epidermal hyperplasia is not observed at distant sites. Overproduction of such a factor or factors required for keratinocyte growth could explain many features of the psoriatic phenotype, including its similarity to wound healing. Moreover, this local proliferative signal could result from either overproduction or enhanced responsiveness to factors that promote growth or to decreased production or responsiveness to factors that inhibit growth. [*See* KERATINOCYTE TRANSFORMATION, HUMAN.]

II. Positive Regulation of Epidermal Proliferation

A. Transforming Growth Factor-α: An Autocrine Keratinocyte Growth Factor

Transforming growth factor-α (TGF-α) is a potent growth factor for keratinocytes and other epithelia (see below). Whereas expression of protooncogene transcripts is comparable in normal skin and psoriatic lesions, TGF-α messenger RNA (mRNA) and protein levels are markedly increased in psoriatic lesions. These results identify TGF-α as a strong candidate for a locally produced growth factor hypothesized earlier.

TGF-α is the evolutionary homologue of epidermal growth factor (EGF) and binds to the EGF receptor with high affinity. While TGF-α was initially described as a growth factor for transformed cells and is found in a variety of tumors, TGF-α and related EGF receptor-binding peptides are also produced by a variety of normal embryonic and adult tissues. In contrast, EGF itself appears to be synthesized in only a limited subset of tissues. There is no detectable EGF mRNA in epidermis or cultured keratinocytes under conditions in which TGF-α mRNA is easily detectable. Thus, TGF-α appears to be the endogenous epidermal ligand for the EGF receptor. TGF-α is synthesized as a 160-amino acid precursor, which undergoes a variety of post-translational modifications, including phosphorylation, glycosylation, and proteolytic cleavage. A higher-molecular weight form of TGF-α is found in the plasma membrane of a variety of cell types (pro-TGF-α), from which TGF-α is released by proteolytic cleavage. TGF-α is mitogenic for many epithelial and fibroblastic cells, which express the EGF receptor. The membrane-associated form of pro-TGF-α has also been shown to have mitogenic activity, suggesting that cell-surface expression of TGF-α may be sufficient to stimulate the growth of adjacent cells. Moreover, several extracellular matrix (ECM) molecules, including thrombospondin, laminin, and tenascin, contain multiple EGF-like domains, suggesting that ECM molecules might directly stimulate the EGF receptor.

TGF-α induces its own expression in human and murine keratinocytes (autoinduction) and can also be induced in various epithelial tissues by other injurious stimuli such as ultraviolet light and partial hepatectomy. These observations suggest an important role for TGF-α in mediating epithelial proliferation in wound healing. The tumor-promoting phorbol ester tetradecanoyl phorbol acetate (TPA) is also a potent stimulus for expression of TGF-α, suggesting its possible role in the process of tumor promotion. Whether these various stimuli use common or divergent intracellular signal transduction pathways to induce TGF-α gene expression is not yet clear (see below). [*See* TRANSFORMING GROWTH FACTOR-α.]

B. EGF Receptor

TGF-α exerts its pleiotypic effects on target cells by binding to and activating the EGF receptor (Fig. 1). The EGF receptor is a 170-kDal phosphoglycoprotein containing an extracellular domain that binds EGF or TGF-α, a short, hydrophobic membrane-spanning domain, and a C-terminal domain, which displays ligand-dependent tyrosine kinase activity. *In vitro* mutagenesis experiments have demonstrated that the tyrosine kinase activity of the EGF receptor is essential for the transduction of pleiotypic responses to EGF, including cell growth. It is likely that TGF-α activates the receptor kinase in a very similar manner, although blocking experiments using monoclonal anti-EGF receptor antibodies have indicated that EGF and TGF-α may recognize the receptor in slightly different ways. Normal cells regulate the number of EGF receptors via a combination of endosomal degradation (downregulation) and transcriptional control. Transfection experiments in fibroblasts have shown that concurrent overexpression of TGF-α and EGF receptor leads to malignant transformation, whereas overexpression of either molecule alone is insufficient for transformation. In psoriasis, EGF receptor expression is comparable in normal and psoriatic skin. The lack of marked EGF receptor overexpression in psoriasis in the face of markedly elevated levels of TGF-α therefore distinguishes psoriasis and EGF receptor-mediated neoplastic transformation, is consistent with the clinical differences between psoriasis and cancer, and could help to explain the observation that the incidence of skin cancer is not higher in psoriatics than in the general population.

C. Paracrine Keratinocyte Growth Factors

When cultured in serum-free medium, epidermal keratinocytes require insulin in the culture medium for growth. This insulin requirement can be replaced by 100-fold lower concentrations of insulinlike growth factor 1 (IGF-1), suggesting that keratinocytes utilize the IGF-1 receptor rather than the insulin receptor. IGF-1 receptor mRNA is expressed by cultured keratinocytes and intact epidermis. However, neither cultured keratinocytes nor intact epidermis express detectable mRNA for insulin or IGF-1. Because circulating levels of insulin *in vivo* are much lower than those required for keratinocyte growth, it is likely that keratinocytes utilize IGF-1 as the ligand for the IGF receptor *in vivo*.

Keratinocyte growth is also stimulated by basic fibroblast growth factor (bFGF). Although some controversy exists at present, actively growing keratinocytes appear not to synthesize detectable mRNA for bFGF, acidic FGF, or a recently described homologue, keratinocyte growth factor (KGF). Cultured fibroblasts, however, express easily detectable IGF-1, bFGF, and KGF, and the size of each of these factors is such that they would be predicted to permeate the dermal–epidermal junction and gain access to keratinocytes. However, fibroblasts do not express detectable TGF-α mRNA. Therefore, whereas TGF-α appears to be an autocrine growth factor for keratinocytes, IGF-1, bFGF, and KGF appear to be paracrine keratinocyte growth factors (Fig. 1).

Studies involving transformed murine keratinocytes in serum-free medium have revealed that EGF acts synergistically with either bFGF or insulin to stimulate growth, whereas combinations of bFGF and insulin in the absence of EGF are inactive (Fig. 1). Since each of the receptors for these factors encodes a tyrosine kinase, these results suggest that phosphorylation of a certain combination of cellular substrates, which are differentially phosphorylated by the EGF, IGF-1, and bFGF-receptors, are required for progression through the cell cycle. It has long been established in fibroblasts that EGF acts to render the cells competent to enter the cell cycle, whereas insulin acts as a progression factor, allowing the cells to traverse the G1 phase of the cycle and enter into the S phase. Thus, it would

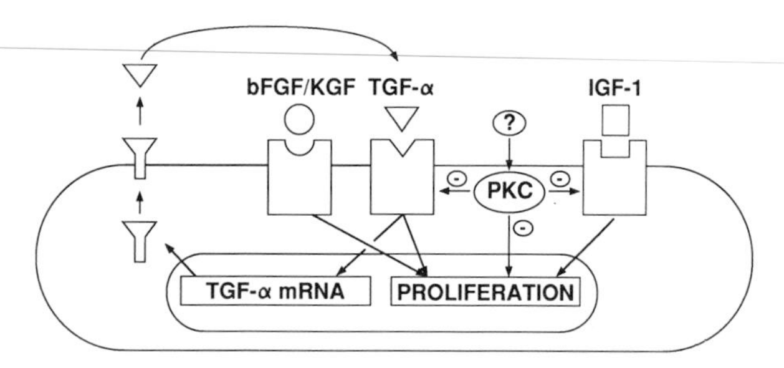

FIGURE 1 Signal transduction pathways mediating keratinocyte growth. The receptor depicted for TGF-α is the EGF receptor. The EGF receptor and IGF-1 receptors possess intrinsic tyrosine kinase activity. The receptor for bFGF/KGF has not been defined on the molecular level to date. Protein kinase C (PKC) negatively regulates the tyrosine kinase activities of the EGF and IGF-1 receptors. The extracellular mediator providing the physiologic stimulus for PKC activation is unknown. Of the three growth factors depicted, only TGF-α is synthesized constitutively by keratinocytes.

appear that a similar mechanism exists in keratinocytes, but that keratinocytes are an autocrine source of competence factors (TGF-α) and require a paracrine source of progression factors (IGF-1).

III. Negative Regulation of Epidermal Proliferation (Table I)

A. Transforming Growth Factor-β

The requirement for a steady-state balance between proliferation and terminal differentiation in the epidermis and other epithelia suggests that some form of communication takes place between the proliferative and differentiated cell compartments. Several different polypeptides have been proposed to fill this role by acting as negative mediators of epidermal growth. Transforming growth factor-β (TGF-β) was discovered by virtue of its ability to promote the growth of transformed cells in semisolid medium. However, it was later appreciated to be a potent antiproliferative agent for a variety of nontransformed and transformed epithelial cells, including epidermal keratinocytes. This effect is associated with a dramatic reduction in expression of the protooncogene *c-myc* mRNA in human and murine keratinocytes and appears to be mediated by the product of the tumor suppressor gene, RB. Many different cell types and tissues express TGF-β, which is synthesized as a latent homodimeric precursor containing two interchain disulfide bridges. Mature, biologically active TGF-β is generated from this complex by proteolytic cleavage and possibly by exposure to acidic environments *in vivo*. In many tissues, especially bone, TGF-β is an abundant component, but the bulk of the TGF-β appears to be in a biologically inactive state. It is now appreciated that there are multiple TGF-β species (TGF-β1 through TGF-β5 have been described), which are divergent members of a multigene family. In psoriasis, which is characterized by markedly increased epidermal proliferation, TGF-β1 mRNA levels are not significantly different in normal and lesional epidermis. Although the issue of how much biologically active TGF-β protein is present in psoriatic versus normal epidermis has not yet been adequately addressed, these data suggest that psoriatic epidermal hyperplasia is associated with overexpression of a growth stimulator (TGF-α) rather than loss of an antiproliferative stimulus (TGF-β). Moreover, while autoinduction of TGF-β has been observed in some cell types (e.g., fibroblasts), it has not been observed in keratinocytes. TGF-β causes a marked increase in keratinocyte spreading, and thus one of its normal biological roles in the epidermis may be to stimulate the flattening and spreading of keratinocytes, which occurs in the early phases of epidermal wound closure. Increased TGF-β immunoreactivity has also been observed in hair follicles, where it may play an important role in the regulation of the complex dermal–epidermal interactions involved in hair growth. [*See* TRANSFORMING GROWTH FACTOR-β.]

B. Interferon-γ

It is well-documented that interferon-γ (IFN-γ), while not a product of keratinocytes but rather of T cells (see below), also displays growth inhibitory properties toward keratinocytes *in vitro*. Paradoxically, IFN-γ induces a 2.5-fold increase in TGF-α gene expression after 24–48 hr in cultured keratinocytes, even as the cells reduce their proliferative rate. This paradox may be explained by the fact that IFN-γ treatment concurrently decreases the expression of the EGF receptor gene, which would render the cells refractory to the increased TGF-α present. *In vivo*, it is possible that increased TGF-α produced by more differentiated, EGF receptor-deficient cells stimulates the proliferation of less-differentiated, EGF receptor-rich basal and/or suprabasal cells. These results suggest that stimulation of autocrine growth factor expression in terminally differentiated cells may be an important mechanism by which the balance between proliferation and terminal differentiation is maintained in epidermal homeostasis. While speculative, these results suggest a plausible model linking psoriatic epidermal hyperplasia and the increased numbers of T cells present in psoriatic lesions (see below). Because cyclosporin A (CsA) inhibits the production of IFN-γ by T cells, this model could also explain the potent antipsoriatic effect of CsA. [*See* INTERFERONS.]

IV. Extracellular Matrix and Proteases as Regulators of Growth Factor Action

It was mentioned earlier that certain extracellular matrix (ECM) molecules have EGF-like domains, which could potentially stimulate the EGF receptor. In this regard, it is interesting that thrombospondin and fibronectin each promote migration of epidermal keratinocytes *in vitro*, because growth and migration of these cells appears to be closely linked. ECM molecules appear to have additional roles in modulating the effects of growth factors, principally by virtue of the ability of many growth factors and cytokines to bind to ECM molecules. Thus, both bFGF and granulocyte–macrophage colony-stimulating factor (GM-CSF) have been shown to bind to heparin-containing glycosaminoglycans, and heparin can potently potentiate or inhibit the effects of these growth factors *in vitro* depending on the experimental conditions employed. Finally, certain growth factors, including EGF and TGF-α and β, can potently stimulate the synthesis and secretion of ECM molecules.

The structure and composition of the epidermal basement membrane and many other ECM-containing structures are dramatically altered under conditions of injury and certain disease states. These alterations are mediated in large part by the actions of extracellular proteases and their inhibitors, such as plasminogen activator and plasminogen activator inhibitor-1, and these molecules are components of the ECM itself. How this multifactorial network of ECM molecules, proteases, and growth factors operates to regulate the proliferation of the epidermis and other organs continues to be a challenging area for experimental analysis. [*See* EXTRACELLULAR MATRIX.]

V. The Cutaneous Cytokine Network

A rapidly expanding literature is documenting the many cytokines that can be expressed by keratinocytes and other cell types present in the skin, especially under *in vitro* conditions. In many cases, however, it is not currently clear whether or not active forms of these cytokines are actually present in normal skin and in disease states, and whether or not they are activating the same cellular responses *in vivo* as they are capable of *in vitro*. While most cytokines have growth factor activity for certain immune and inflammatory cells, it is less clear in general that they have direct growth factor activity for epidermal keratinocytes.

Here we will briefly review the cell types present in normal and psoriatic skin and consider the expression of selected cytokines likely to be produced by them *in vivo* as they pertain to epidermal proliferation (see Table I). [*See* CYTOKINES IN THE IMMUNE RESPONSE.]

A. Keratinocytes

Keratinocytes express a variety of cytokines *in vitro*, including interleukins-1α and -β, granulocyte, macrophage, and granulocyte–macrophage colony-stimulating factors (IL-1α, IL-1β, G-CSF, M-CSF, and GM-CSF, respectively), tumor necrosis factor-α (TNF-α), IL-3, and IL-8. Since keratinocytes are the majority cell type of the epidermis, more data are currently available to compare the *in vivo* and *in vitro* expression of these cytokines than for other cell types present in the skin.

There is general agreement that normal epidermis contains large amounts of IL-1α immunoreactivity and biological activity, as much as 60 μg of IL-1α per individual. Paradoxically, however, several laboratories have reported that IL-1α mRNA is undetectable in epidermal RNA by RNA blot hybridization. This apparent contradiction remains to be resolved. Moreover, epidermis contains large amounts of an inhibitor of IL-1 bioactivity, whose relevance to the regulation of epidermal IL-1 effects *in vivo* is currently unclear. Evidence from our laboratories indicates that IL-1α bioactivity and immunoreactivity is reduced, while IL-1β immuno-

TABLE I Regulators of Epidermal Growth

Factor	Produced by keratinocytes?	Effect on keratinocyte growth *in vitro*
Transforming growth factor-α	Y	+
Epidermal growth factor	N	+
Insulinlike growth factor-1	N	+
Insulin	N	+
Basic fibroblast growth factor	N(?)[a]	+
Keratinocyte growth factor	N	+
Transforming growth factor-β_1	Y	−
Transforming growth factor-β_2	Y	−
Interferon-γ	N	−
Tumor necrosis factor-α	Y	+/−
Epidermal mitosis-inhibitory pentapeptide	Y	−
Epidermal G1-chalone	Y	−
Interleukin-1α and -β	Y	0 (human)/+ (mouse)
Interleukin-3	Y	+(?)
Granulocyte–macrophage colony-stimulating factor	Y	+(?)
Interleukin-6	N(?)	0(?)
Interleukin-8	Y	?

[a] (?) Indicates conflicting results.

reactivity and mRNA are increased in psoriatic lesions. Preliminary data from our laboratory indicates that mRNAs for GM-CSF and TNF-α are detectable in some samples from normal and psoriatic epidermis, while IL-8 mRNA is clearly increased in psoriatic lesions. In contrast to some recent reports, we find that IL-6 mRNA is undetectable in human epidermis. The roles of any of these cytokines in the stimulation of psoriatic epidermal hyperplasia remain speculative, although IL-3 and GM-CSF have been reported to stimulate keratinocyte proliferation in defined medium. It is noteworthy that at least two of these cytokines (IL-1 and TNF-α) stimulate the expression of the adhesion molecule ICAM-1 on keratinocytes. Expression of ICAM-1 allows binding of T lymphocytes to the epidermis, which may serve as a stimulus to epidermal proliferation (see below).

B. T Lymphocytes

As mentioned earlier, IFN-γ is expressed primarily by T cells and natural killer cells and exhibits growth inhibitory effects on keratinocytes. Since T cells are a prominent component of the infiltrating cells in psoriatic lesions, the effects of IFN-γ and other lymphokines on keratinocyte proliferation may be relevant to the pathophysiology of this disorder. It has been shown that keratinocytes from psoriatic patients are relatively refractory to the growth-inhibitory properties of IFN-γ, and it was suggested that this could be responsible for enhanced keratinocyte proliferation in psoriasis. However, the refractoriness to IFN-γ-induced growth inhibition relative to normal keratinocytes is relatively small (about 1.5-fold) and seems unlikely to be the sole cause of the marked (10-fold) increase in proliferation characteristic of psoriatic epidermis. At the present time, there is little convincing evidence for increased expression of IL-2, IL-3, IL-4, IFN-γ, or other T cell-derived cytokines in psoriatic lesions. Moreover, with the exception of IL-3, currently no evidence indicates that any of these lymphokines directly stimulate keratinocyte growth *in vitro*. [*See* LYMPHOCYTES.]

C. Antigen-Presenting Cells

In this category, monocyte–macrophages, Langerhans cells (epidermal dendritic cells with antigen-presenting capacity), and dermal dendritic cells (dermal dendrocytes) are grouped together. These cells are thought to stimulate proliferation of antigen-specific T cells by a combination of signals: antigen bound to class II major histocompatibility complex (MHC) molecules on the cell surface of the

APC (first signal) and elaboration of IL-1 (second signal). A dermal dendritic cell of bone marrow origin, distinct from Langerhans cells, was identified and shown to have antigen-presenting properties (see Fibroblasts, below). Increased numbers and functional alloantigen-presenting activity of APCs have been demonstrated in psoriatic lesions; however, the relevance of this observation to psoriatic epidermal hyperplasia has not been established. Cells of the monocyte–macrophage lineage produce a number of cytokines and growth factors upon activation by various stimuli; among these are IL-1, IL-3, IL-6, IL-8, TGF-α, and TGF-β. In evolving psoriatic lesions, it is possible that TGF-α elaborated by monocyte–macrophages or dendritic cells could trigger lesion development by stimulating keratinocyte expression of TGF-α. [*See* MACROPHAGES.]

D. Other Blood Elements and Mast Cells

Polymorphonuclear leukocytes (PMNs) are abundant in psoriatic lesions, forming microabscesses just below the stratum corneum of the epidermis. PMNs express IL-1α and -β in surprising amounts, as well as a number of other inflammatory mediators such as eicosanoids, proteases, and peroxidases. The platelet α-granule and mast cell granules contain a diverse array of growth factors, including a TGF-α-like peptide, TGF-β, and, as the name implies, platelet-derived growth factor, as well as numerous protease activities. Psoriatic lesions are characterized by gross exudation of plasma into the epidermal compartment, and stimuli capable of causing platelet degranulation may release large quantities of these factors into the epidermis in a localized fashion. The role of these factors as well as complement components such as C5a in the generation of psoriatic epidermal hyperplasia is poorly understood.

E. Melanocytes

While the primary role of melanocytes in the epidermis appears to be the synthesis of melanin and its transport to keratinocytes for purposes of photoprotection against ultraviolet light, its properties as a growth regulator for keratinocytes are largely unexplored. In part, this has been due to difficulties in establishing *in vitro* cultures of untransformed melanocytes. Recent advances in this area, however, have indicated that melanocytes display a proliferative response to bFGF and insulin/IGF-1. However, production of keratinocyte growth factors by melanocytes is only beginning to be explored *in vitro*.

F. Fibroblasts

The potential role of the fibroblast in the generation of paracrine keratinocyte growth factors was discussed previously. However, it is noteworthy that many of the spindle-shaped cells of the dermis previously called fibroblasts may actually be bone marrow-derived dermal dendrocytes of the monocyte–macrophage lineage; these dermal dendrocytes can be identified by surface staining for factor XIIIa and other bone marrow-derived cell markers and are found in increased abundance in psoriatic lesions. Fibroblastic cells derived from psoriatic lesions have been reported to stimulate the outgrowth of normal keratinocytes in collagen gels better than normal fibroblasts; however, these results have been difficult to duplicate under different culture conditions. Whether or not contamination of the fibroblastic cells with dermal dendrocytes is responsible for the conflicting results is, at present, unclear.

G. Endothelial Cells

Alterations in the cutaneous vasculature, specifically the superficial capillary plexus of the papillary dermis, occur very early in the development of psoriatic lesions and are the last morphologic changes to normalize following antipsoriatic therapy or spontaneous improvement of psoriasis. These changes consist of elongation, increased tortuosity, and dilatation of the capillary loop as a result of increased endothelial cell proliferation. Therefore, it is of interest that the epidermis is a more potent source of angiogenic factors than is the dermis in various *in vivo* assay systems, and that TGF-α is a potent angiogenic factor in the hamster cheek pouch model. However, a wide variety of cytokines possess angiogenic activity, and it would be premature to assign the psoriatic angiogenic activity sole to TGF-α at this time. Capillary permeability is increased in psoriasis, and this may allow increase egress of plasma growth factors into the epidermal milleu as well as provide a marked increase in total blood flow to the epidermal compartment. Recently, increased attention has also been given to the stimulation of binding to and diapedesis between endothelial cells by T cells and monocyte–

macrophages in response to IL-1, TNF-α, IFN-γ, and other cytokines, which may be abnormally active in psoriatic lesions. Clearly, these factors may be very important in the regulation of epidermal growth in psoriasis as well as other inflammatory disorders, many of which are characterized by increased epidermal proliferation.

H. Neurons

Free nerve fibers reach the dermoepidermal junction and penetrate the epidermis, and these fibers contain tachykinins such as substance P, calcitonin gene-related peptide, neuropeptide Y, and vasoactive intestinal peptide. These mediators have been shown to generate an efferent signal to the skin (the axon reflex) even though they are contained within sensory nerve fibers. It is the experience of every dermatologist that many dermatoses are exacerbated by psychic "stress." Elucidation of the molecular details of this phenomenon remains a major objective of dermatologic research.

VI. Signal Transduction Mechanisms and Growth Factor Responsiveness

The major signal transduction pathways thought to be active in keratinocytes are (1) the receptor-associated tyrosine kinases; (2) G proteins, which couple extracellular receptors to intracellular effectors including the adenylate cyclase–protein kinase A system and phospholipase C; (3) the phosphoinositide (PI) cycle, utilizing PI kinases and PI-specific phospholipase C to regulate release of calcium from intracellular stores; (4) calcium–phospholipid-dependent protein kinases (protein kinase C [PKC]); and (5) calcium–calmodulin-dependent protein kinases.

One example of how signal transduction pathways may interact to regulate growth factor responsiveness has potential therapeutic relevance to psoriasis and involves the interaction of PKC with the EGF receptor tyrosine kinase (Fig. 1). PKC-mediated phosphorylation of the threonine-654 residue of the EGF receptor reduces the tyrosine kinase activity of the receptor without reducing its affinity for ligand. In psoriasis, PKC levels are markedly reduced relative to normal epidermis. Thus, it is possible that the EGF receptor tyrosine kinase is more active in psoriatic epidermis. Because all responses to exogenous EGF (or TGF-α) appear to involve the tyrosine kinase activity of the EGF receptor, it is likely that the autoinductive response to TGF-α does so as well. Therefore, loss of PKC activity in psoriasis may release a constraint that is normally placed on autoinduction, resulting in increased TGF-α production and responsiveness, as well as increased epidermal proliferation. The explanation for reduced PKC levels in psoriasis is currently unclear but could involve downregulation of PKC due to an as-yet unidentified, chronic stimulus. Such a stimulus could be provided by infiltrating immune and/or inflammatory cells or nerves in the form of histamine, bradykinin, platelet-activating factor, or tachykinins such as substance P. This model has the advantage of integrating several biochemical and cellular abnormalities known to be present in psoriasis. Future studies in our laboratories will test the many hypotheses raised by this model at the molecular level.

Bibliography

Balkwill, F. R., and Burke, F. (1989). The cytokine network. *Immunology Today* **10(9),** 299.

Bishop, J. M. (1987). The molecular genetics of cancer. *Science* **235,** 305.

Braverman, I. M., and Sibley, J. (1982). Role of the microcirculation in the treatment and pathogenesis of psoriasis. *J. Invest. Dermatol.* **78,** 12.

Burgess, W. H., and Maciag, T. (1989). The heparin-binding (fibroblast) growth factor family of proteins. *Annu. Rev. Biochem.* **58,** 575.

Chen, W. S., Lazar, C. S., Poenie, M., Tsien, R. Y., Gill, G. N., and Rosenfeld, M. G. (1987). Requirement for intrinsic protein tyrosine kinase in the immediate and late actions of the EGF receptor. *Nature* **328,** 820.

Derynck, R. (1988). Transforming growth factor-α. *Cell* **54,** 593.

Elder, J. T., Fisher, G. J., Lindquist, P. B., Bennett, G. L., Pittelkow, M. R., Coffey, R. J., Ellingsworth, L., Derynck, R., and Voorhees, J. J. (1989). Overexpression of transforming growth factor α in psoriatic epidermis. *Science* **243,** 811.

Finch, P. W., Rubin, J. S., Miki, T., Ron, D., and Aaronson, S. A. (1989). Human KGF is FGF-related with properties of a paracrine effector of epithelial cell growth. *Science* **245,** 752.

Hanks, S. K., Quinn, A. M., and Hunter, T. (1988). The protein kinase family: Conserved features and deduced phylogeny of the catalytic domains. *Science* **241,** 42.

Kupper, T. S. (1989). The role of epidermal cytokines. *In* (J. Oppenheim and E. Shevach, eds.). "The Immunophysiology of Cells and Cytokines" Oxford University Press, New York.

Massague, J. (1987). The TGF-β family of growth and differentiation factors. *Cell* **49,** 437.

Pardee, A. B. (1987). Molecules involved in proliferation of normal and cancer cells: Presidential address. *Cancer Res.* **47,** 1488.

Sporn, M. B., and Roberts, A. B. (1988). Peptide growth factors are multifunctional. *Nature* **332,** 217.

Epilepsy

BRIAN B. GALLAGHER, *Medical College of Georgia*

Glossary

Agonist Substance that mimics the activity of the natural transmitter substance at a synapse

Antagonist Substance that binds to a receptor and blocks the activity of the natural transmitter substance (or its agonist) at a synapse

Electroencephalogram Recording of spontaneous electrical activity of the brain as seen by electrodes glued to the scalp

Receptor Specialized region of either presynaptic or postsynaptic neurons that binds a chemical substance released from the presynaptic neuron; activation of the receptor initiates a cascade of events related to excitation or inhibition

Re-uptake Transport of the chemical substance, released from the presynaptic region back into the presynaptic neuron

Seizure Clinical presentation of motor, sensory, autonomic, psychic, and cognitive events that result from a sustained paroxysmal abnormal discharge of central nervous system neurons

Synapse Specialized connection between neurons in which depolarization of the presynaptic neuron results in the release of at least one chemical substance, which binds to the postsynaptic site of another neuron

EPILEPSY IS A DISORDER of the central nervous system characterized by the occurrence of recurrent seizures. It is not a specific disease, rather it is a manifestation of abnormal neural activity caused by a wide variety of etiological factors. Seizures occur in an almost infinite multiformity of patterns. Nevertheless, most seizures can be classified on the basis of their clinical presentation and electroencephalographic (EEG) pattern or on the basis of the EEG pattern and a constellation of clinical factors that constitute a syndrome. Both methods will be described.

Epilepsy has a prevalence of about 1% of the population (excluding the syndrome of febrile seizures). The incidence of epilepsy is highest in childhood, with a peak occurrence during the first year of life. The incidence is relatively stable between the ages of 20 and 50 yr and then steadily increases during later years.

I. International Classification of Epileptic Seizures

The basis for this classification system is the clinical manifestation of the seizure and its EEG accompaniment (i.e., the ictal EEG pattern). Because it requires special effort to record an ictal EEG, the interictal EEG findings that are more frequently seen in the routine EEG will also be discussed. The International Classification System differentiates partial seizures from generalized seizures. Partial seizures have local spread from a focus of abnormal neurons. They may eventually become generalized seizures. Generalized seizures appear clinically and electroencephalographically to involve both hemispheres simultaneously. In actuality, most seizures probably arise from a focus of abnormal neurons, but the location of the focus in the brain and the directions of spread of the abnormal neuronal activity determine whether the seizure is classified as partial or generalized.

Partial seizures are divided into three categories: partial simple seizures, in which consciousness is not impaired; partial complex seizures, in which there is some degree of impaired consciousness; and partial seizures, which become secondarily generalized. The secondarily generalized portion of the seizure may be generalized tonic–clonic, generalized tonic, or generalized clonic.

Partial simple seizures are focal seizures that may involve motor systems, sensory systems, autonomic systems, or psychic phenomena, or a combination of these types, as long as consciousness is maintained throughout the duration of the seizure. When consciousness is impaired, the seizure is classified as a partial complex seizure. A partial complex seizure may begin as a partial simple seizure and evolve to a partial complex seizure, or consciousness may be impaired at the onset of the seizure. In contrast to partial simple seizures, the complex seizures usually have a rich array of clinical symptomatology, often involving automatisms, which may be relatively simple (lip smacking, picking at objects or clothing, etc.) or highly integrated acts of behavior (continuing to operate a machine or automobile during the seizure, walking out of the house during a seizure, etc.) Amnesia is frequently seen with partial complex seizures. The amnesia may extend retrograde to the onset of the seizure for variable durations of time, usually involves the actual duration of the seizure, and frequently extends for variable durations of time after the seizure (postictal amnesia). Clinically, amnesia is an important point because many patients are entirely unaware that a seizure has occurred unless a witness informs them of the event. Such patients will not be able to relate an accurate account of seizure frequency. For patients with partial complex seizures that secondarily generalize, determining whether or not there is a partial onset to the seizure is important because many such patients may be candidates for surgical treatment.

Six patterns of generalized seizures are recognized. Absence seizures almost always begin in childhood, are usually characterized by a brief interruption of consciousness with or without some accompanying motor manifestations, and have a characteristic EEG pattern of about 3-Hz spike-and-slow-wave complexes during the ictal event. Spike-and-slow-wave complexes <3–4 sec may occur in the EEG without obvious clinical manifestations. Atypical absence seizures deviate in clinical manifestations and EEG characteristics from typical absence seizures. It is important to distinguish these two patterns of absence seizures because the typical absences usually respond well to medical treatment, whereas atypical absences are often difficult to treat and are frequently associated with other generalized seizures.

Myoclonic seizures are brief jerks that may involve one or more extremity and/or truncal musculature. They may occur singly or as multiple successive events.

Clonic seizures involve repetitive contraction of bodily musculature, which is often but not necessarily bilaterally synchronous. Tonic seizures consist of contraction of bodily musculature. They are frequently associated with a loud cry at the onset as the larynx, diaphragm, and chest muscles contract, forcing pulmonary air through the larynx. Cyanosis occurs with these seizures. Tonic–clonic seizures begin with the tonic component and progress to the clonic phase. As the seizure continues, the clonic components gradually occur at longer intervals until the seizure ceases. There is almost always a pronounced autonomic component with micturition, sometimes defecation, profuse salivation, tachycardia, increased blood pressure, and often emesis at the termination of the seizure. The violence of these seizures is unpleasant for even the most experienced observer and is terrifying for the inexperienced. Even today, a significant number of people think of epilepsy as generalized tonic–clonic seizures and react to the word epilepsy with an inordinate amount of fear.

Atonic seizures involve a sudden loss of muscle tone. They principally occur in children and frequently cause severe injury because of the abrupt onset. If standing, the child will fall to the ground; if sitting, the child will usually pitch forward.

The International Classification of Epileptic Seizures provides a uniform description of most seizure types. Its value is in facilitating communication among individuals involved in the study and treatment of epilepsy; however, some seizures still do not fit well into the classification scheme. The older nomenclature, which used such terms as grand mal, petit mal, psychomotor, limbic, and minor motor, was quite ambiguous. The problem with the International Classification is that it does not describe epileptic syndromes that still have considerable clinical utility and prognostic value. [*See* SEIZURE GENERATION, SUBCORTICAL MECHANISMS.]

II. Epileptic Syndromes

A proposal for an International Classification of Epilepsies and Epileptic Syndromes is shown in Table I. More detailed descriptions than those provided in the following discussion can be found in the volume edited by J. Roger.

A. Febrile Convulsions

Some generalized seizures usually occur with a rising temperature during an acute febrile illness. They are age-specific, occurring between 6 mo and 5–6 yr of age. The seizures are brief in duration and are generalized tonic or tonic–clonic in character. The prevalence of febrile seizures is probably between 2 and 5%. If the seizure occurs during any infectious episode involving the central nervous system, it is not a febrile seizure.

It is frequently difficult to determine whether a seizure occurring during a fever is a febrile seizure or the beginning of recurrent seizures precipitated by the fever. Children with certain risk factors have an increased chance of developing recurrent seizures. In these children, the value of chronic antiepileptic treatment increases. These risk factors include (1) prolonged seizure (30 min or more), (2) seizure with focal features, (3) more than one seizure occurrence during the illness, or (4) seizure followed by postictal neurologic deficits. However, there are exceptions to each of these events in the syndrome of febrile seizures. A family history of febrile seizures may occur in about 10% of cases. The EEG is not helpful in distinguishing a febrile seizure from the onset of recurrent seizures. Diffuse or focal slowing is a common finding in the EEG, even up to 7–10 days following the episode. The presence of spike-and-slow-wave activity in the EEG does not have predictive value. It is important to determine as best as possible which seizures are febrile seizures because the prognosis for this syndrome is excellent. The risk for subsequent development of epilepsy in children with uncomplicated febrile seizures is about 1%; with one risk factor present, it rises to 2%, and with two more present, it is about 10%. In general, uncomplicated febrile seizures represent a benign condition, with a moderate (30–40%) chance of recurrence and a low risk of developing epilepsy (about 1%). Furthermore, no evidence suggests that prophylaxis prevents epilepsy.

TABLE I International Classification of Epilepsies and Epileptic Syndromes

1. Localization-related (focal, local, partial) epilepsies and syndromes
   - 1.1 Idiopathic with age-related onset
     - A. Benign childhood epilepsy with centrotemporal spikes
     - B. Childhood epilepsy with occipital paroxysms
   - 1.2. Symptomatic
     This category comprises syndromes of great individual variability, which are mainly based on anatomic localization, clinical features, seizure types, and etiologic factors (if known).
2. Generalized epilepsies and syndromes
   - 2.1 Idiopathic with age-related onset, in order of age
     - A. Benign neonatal familial convulsions
     - B. Benign neonatal convulsions
     - C. Benign myoclonic epilepsy in infancy
     - D. Childhood absence epilepsy (pyknolepsy)
     - E. Juvenile absence epilepsy
     - F. Juvenile myoclonic epilepsy (impulsive petit mal)
     - G. Epilepsy with grand mal seizures on awakening
       Other generalized idiopathic epilepsies, if they do not belong to one of the syndromes, can still be classified as generalized idiopathic epilepsies.
   - 2.2 Idiopathic and/or symptomatic, in order of age of appearance
     - A. West's syndrome (infantile spasms)
     - B. Lennox–Gastaut syndrome
     - C. Epilepsy with myoclonic-astatic seizures
     - D. Epilepsy with myoclonic absences
   - 2.3 Symptomatic
     - A. Nonspecific etiology
       - a. Early myoclonic encephalopathy
     - B. Specific etiology
       - a. Epileptic seizures may complicate many disease states
3. Epilepsies and syndromes undetermined as to whether they are focal or generalized
   - 3.1 With both generalized and focal seizures
     - A. Neonatal seizures
     - B. Severe myoclonic epilepsy in infancy
     - C. Epilepsy with continuous spike waves during slow-wave sleep
     - D. Acquired epileptic aphasia (Landau–Klefner syndrome)
   - 3.2 Without unequivocal generalized or focal features
     This heading covers all cases with generalized tonic–clonic seizures in which clinical and EEG findings do not permit classification as clearly generalized or localization-related, such as many cases of sleep grand mal.
4. Special syndromes
   - 4.1 Situation-related seizures
     - A. Febrile convulsions
     - B. Seizures related to other identifiable situations, such as stress, hormonal changes, drugs, alcohol, or sleep deprivation
   - 4.2 Isolated, apparently unprovoked epileptic events
   - 4.3 Epilepsies characterized by specific modes of seizure precipitation

B. Infantile Spasms (West's Syndrome)

West's syndrome has three features: seizures called infantile spasms, mental retardation, and a characteristic EEG pattern termed hypsarrhythmia or one of its variants. The incidence is estimated to be 1 out of 5,000, and the syndrome is age-related, occurring during the first 2 yr of life. Most cases begin during the first year, with a peak at 5 mo. Boys are affected at a slightly greater rate than girls.

The seizures or spasms consist of flexion, extension, or flexion–extension of various bodily musculature. The movements are usually bilateral and involve the neck, trunk, and/or limbs in either adduction or abduction. Mixed flexion–extension contractions occur slightly more frequently than flexion contractures, and extension contractures are the least frequent. The spasms are sudden in onset and brief in duration, usually lasting <15 sec. Most patients have a variety of spasm types, which occur while awake or asleep and frequently appear in clusters. It is not at all unusual for the spasms not to be recognized as seizures.

Mental retardation is present in 90–95% of patients and is more frequent in infants, in whom a cause for the syndrome, such as intrapartum hypoxia, can be identified. There are usually accompanying abnormalities in the neurologic examination.

The hypsarrhythmic EEG is a chaotic mixture of high-amplitude spikes, polyspikes, and high-amplitude slow waves occurring in a disorganized pattern. There are numerous variations in the EEG pattern.

The syndrome is usually divided into two classes: symptomatic, when an etiology can be established, and idiopathic, when no etiology is apparent. Outcome is poor in symptomatic patients, especially when neurologic or developmental abnormalities precede the spasms, other seizure types exist with or following the spasms, and/or the onset of the spasms is within the first 3 mo of life. A better prognosis exists for the few infants who are normal or nearly normal at the beginning of the spasms and in whom treatment is initiated rapidly.

C. Lennox–Gastaut Syndrome

Whether or not this is a syndrome in the sense of having some common underlying mechanism is debatable; however, it is clinically useful to consider it a syndrome because the prognosis is almost uniformly poor. It is characterized by three factors: multiple patterns of generalized seizures, including atypical absences, generalized tonic–clonic, tonic, atonic, and myoclonic; <3-Hz spike-and-slow-wave patterns in the EEG; and mental retardation. Although the real prevalence of the syndrome is not known, perhaps about 5% of patients with epilepsy are affected. Age of onset is usually before 8 or 9 yr, with a peak between the ages of 3 and 5 yr. A history of seizures may exist before the syndrome appears, and there is a significant incidence of prior infantile spasms. The syndrome may also begin with an episode of status epilepticus. Recurrent episodes of status epilepticus occur in as many as two-thirds of these patients. A cardinal feature of this syndrome is an unresponsiveness of the seizures to antiepileptic drugs. In general, as the child becomes older, the severity and frequency of the seizures lessen; however, the seizure disorder will usually still be severe, and the mental retardation remains permanent.

D. Benign Partial Epilepsies of Childhood

Three syndromes are recognized in this category: benign partial epilepsy with rolandic (centrotemporal) spikes, benign partial epilepsy with occipital spike waves, and benign partial epilepsy with affective symptoms. These patterns of benign epilepsy have some common features. The children are neurologically and mentally normal. The seizures usually begin during the first decade of life, rarely before 18 mo, with a peak between the ages of 4 and 8 yr. Frequently, there is a family history of seizure, which is often also benign. Background activity in the EEG is normal when the patient is awake and asleep. The seizures generally respond well to antiepileptic drugs and spontaneously cease during the second decade, usually by 16 yr of age.

In the syndrome of benign partial epilepsy with centrotemporal spikes, a characteristic EEG pattern is seen with high-voltage spikes that may be followed by a slow wave localized to the centrotemporal scalp regions. Sleep activates the spike discharges. Clinically, the seizure begins as a partial simple seizure, which may progress to a secondarily generalized seizure. At times, consciousness is impaired at the onset of the episode, resulting in a partial complex seizure. The typical seizure begins with sensory and motor involvement of the mouth, tongue, pharynx, and larynx; speech ceases and salivation occurs. The seizures may spread to involve the ipsilateral arm. Consciousness is maintained during this portion of the seizure. Most of the seizures occur nocturnally, but diurnal seizures may occur. The diurnal episodes are always partial simple seizures, whereas the nocturnal episodes are often secondarily generalized seizures. In the syndrome of benign partial epilepsy with occipital spike waves, the EEG is characterized by high-amplitude spike-wave discharges occurring over the occipital and posterior temporal regions of the scalp. These usually occur in bursts and are blocked by eye opening. Clinically, the seizures have visual and nonvisual symptoms. The visual ictal symptoms may include hemianopsia, loss of vision, flashing lights, visual hallucinations, or visual illusions. The nonvisual symptoms may involve a partial complex seizure with automatisms, a hemiclonic seizure, or, less frequently, a generalized tonic–clonic seizure. After the seizure, there may be a headache, which can resemble a migraine headache. Response to treatment with phenytoin, carbamazepine, or phenobarbital is usually good. The seizures spontaneously remit during the second decade.

In the syndrome of benign partial epilepsy with affective symptoms, the EEG is less specific than in the two preceding syndromes. There may be slow spike-wave discharges or rhythmic sharp-wave discharges involving frontal, temporal, or parietal regions. Clinically, the seizures are characterized by sudden fright or fear, vocalization, and running to someone nearby. There may be automatisms including chewing, swallowing, laughter, speech arrest, pallor, sweating, or abdominal pain. Consciousness is impaired but not completely lost. The seizures spontaneously remit during the second decade.

E. Generalized Tonic–Clonic Seizures on Awakening

It is well known that a relationship exists between generalized tonic–clonic seizures and the sleep–wake cycle. Although figures vary from study to study, a rough estimate is that about 40% of patients have nocturnal generalized tonic–clonic seizures, 40% have such seizures on awakening, and 20% or less have the seizure randomly during the cycle. In seizures that occur on awakening, at least 90% develop within 2 hr of awakening. About 75–85% of patients with awakening seizures have an onset of disease between the ages of 10 and 25 yr. There is often a family history of seizures and, frequently, the generalized tonic–clonic seizures are associated with myoclonic seizures or, less frequently, absence seizures. When mixed seizure patterns exist, a series of myoclonic jerks or absences preceding the generalized tonic–clonic seizures is typical. Some patients may have myoclonic jerks or absences without necessarily progressing to the generalized tonic–clonic seizure, whereas other patients may always progress to the major seizure.

An interesting psychologic observation about these patients is that they tend to be impulsive, unstable individuals who pursue contradictory goals without apparent awareness of the contradiction. A speculation concerning this trait is the possibility that there is a population of individuals with a genetic predisposition to these seizures. Persons with the psychologic characteristics described manifest the seizures because of their life-style, whereas others with more moderate life-styles do not manifest the genetic predisposition.

III. Genetics of Epilepsy

Certain inherited metabolic disorders are characterized by a high incidence of recurrent seizures (some amino acidurias, gangliosidoses, metachromatic leukodystrophy, etc.). A genetic factor(s) probably contributes to most if not all epilepsies. Inheritance is very apparent in febrile seizures, absence seizures, pure generalized tonic–clonic seizures, benign juvenile myoclonus, and rolandic epilepsy. Inheritance is also important in partial complex seizures but the genetic factors are more heterogeneous and the subclasses of partial complex seizures are, at present, less well delineated than some of the other epilepsy syndromes. Improving our knowledge, the genetics of epilepsy will have a major impact on the treatment of this disorder in the future.

IV. Transmitters and Epilepsy

A. γ-Aminobutyric Acid

A mechanism of neuronal control in the central nervous system involves tonic inhibition of neuronal circuits, which may be attenuated (disinhibition) or overcome by excitatory activity, to permit activation of the circuits. An important amino acid transmitter in inhibitory neuronal pathways is γ-aminobutyric acid (GABA). The receptor complex for GABA controls a chloride channel and activation of the receptor complex results in hyperpolarization and inhibition. Three binding sites in the receptor complex are recognized: (1) The GABA site—the pharmacological agent mucimol is an agonist, and bicuculline is an antagonist for this site. (2) The barbiturate-binding site—a variety of barbituric acid derivatives are agonists, and picrotoxin (a convulsant agent) is an antagonist for this site. (3) The benzodiazepine-binding site—agonist or antagonist activation of the barbiturate–benzodiazepine portions of the complex modulates the affinity of the GABA portion of the receptor for GABA.

The postsynaptic GABA receptor complex is thought to have an important role in epilepsy. Decreased effectiveness of the GABA receptor complex results in a reduction of inhibition and a consequent release of excitatory circuits. Such a mechanism seems to operate in the regulation of cortical and hippocampal pyramidal neurons, which are important areas for seizure foci in many of the epilepsies. GABA-mediated inhibition is also important in the cerebellum, which provides widespread inhibitory activity in the central nervous system. In addition, considerable experimental evidence indicates that GABA-mediated inhibition from efferent neurons of the pars reticularis of the substantia nigra has an important role in regulating seizure threshold.

B. L-Glutamic Acid

The amino acid L-glutamate appears to be a major excitatory transmitter in the central nervous system. Obtaining unequivocal proof of its function as a transmitter is complicated by its major role in intermediary metabolism in brain and spinal cord. Nevertheless, considerable data indicate that it is an excitatory transmitter with a major role in epilepsy. Glutamate receptors are found in high concentrations in cortex and hippocampus, which are important structures in the genesis of seizures. At present, four classes of glutamate receptors are known. They are classified according to the compounds that exhibit high-affinity binding to the receptors. The compounds are (1) *N*-methyl-D-aspartate (NMDA), (2) kainic acid, (3) quisqualate, and (4) L-2-amino-4-phosphonobutyrate (LAP4). [*See* CORTEX; HIPPOCAMPAL FORMATION.]

The NMDA receptors may mediate long-term neuronal events that are important in learning. The kainic acid and quisqualate receptors seem to be important for rapid excitatory synaptic transmission, which would make them important receptors in epilepsy. The LAP4 receptor may be a presynaptic receptor, at least in the hippocampus; therefore, it might have a role in regulating glutamate release. In addition to its role in intermediary metabolism and as a transmitter, glutamate also has neurotoxic properties in higher concentrations. These concentrations could be obtained in situations where there is excessive release of glutamate (e.g., a seizure) or where there is defective re-uptake of glutamate. This neurotoxic role of glutamate may be involved in the hippocampal neuronal loss that is frequently seen in partial complex seizures of temporal lobe origin. An improved understanding of glutamate involvement in epilepsy should result in new treatment strategies for epilepsy.

C. Biogenic Amines

Norepinephrine, dopamine, and serotonin are thought to be synaptic transmitters in the central nervous system. In experimental models of human epilepsy, manipulation of these amines indicate that they act to inhibit seizure activity. Their role in human epilepsy is not clear at present. In fact, pharmaceutical agents that apparently affect central nervous system amines have activity in psychiatric disorders and headaches but, thus far, not in epilepsy.

D. Neuropeptides

Even though the discovery of peptides with activity in the central nervous system is relatively recent, some of these compounds very likely have at least a modulatory activity on neuronal and synaptic events that are important for human epilepsy. High on the list of epilepsy-relevant peptides are β-endorphin, met-enkephalin, Leu-enkephalin, adrenocorticotrophin hormone, thyrotrophine-releasing hormone, vasopressin, and somatostatin. The biological properties of these compounds in the brain are not well understood at the present time. They are all components of nonsteady state systems, which begin with a large precursor protein that undergoes multiple proteolytic cleavages to release a cascade of active peptides with a wide variety of actions. It seems highly likely that a better understanding of these complex neuroendocrine systems will have a major impact on our comprehension of human epilepsy and normal brain function. [*See* NEUROENDOCRINOLOGY; PEPTIDES.]

Bibliography

Delgado-Escueta, A. B., Ward, A. A., Woodbury, D. M., and Porter, R. J. (eds.) (1986). "Advances in Neurology," Vol. 44. Raven Press, New York.

Commission on Classification and Terminology of the International League Against Epilepsy. (1985). Proposal for classification of epilepsies and epileptic syndromes. *Epilepsia* **26(3),** 268–78.

Flanigin, H. F., King, D. W., and Gallagher, B. B. (1983). Surgical treatment of epilepsy. *In* "Recent Advances in Epilepsy," Vol. II (B. S. Meldrum and T. A. Pedley, eds.), pp. 297–339. Churchill Livingstone, New York. 1985

Gallagher, B. B. (1988). "Seizures, Stupor and Comma" (P. Dove, ed.), Monograph 105. American Academy of Family Physicians.

Holmes, G. L. (1987). "Diagnosis and Management of Seizures in Children," Vol. 30. W. B. Saunders Company, Philadelphia.

Roger, J. (1985). "Epileptic Syndromes in Infancy, Childhood, and Adolescence. John Libbey Eurotext, London.

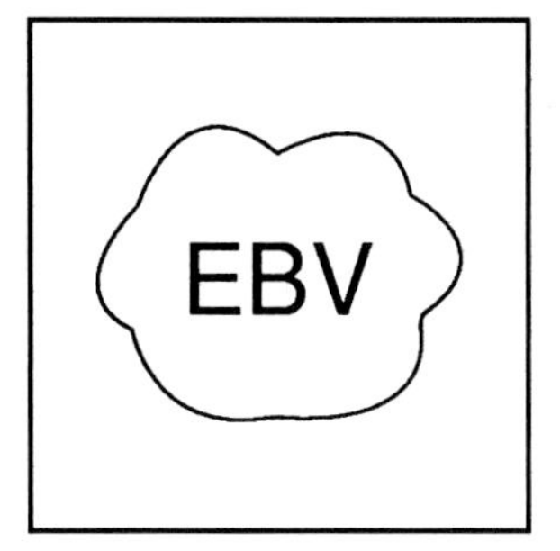

Epstein–Barr Virus

JOAKIM DILLNER, *Karolinska Institute*

Glossary

Episomal persistence Presence of the viral genome as covalently closed circular DNA (episomes), not directly associated with the cellular DNA

Immortalization Ability to confer an indefinite lifespan to infected cells; should not be confused with transformation, which in addition implies that infected cells are tumorigenic and have lost contact inhibition

Latency Mostly silent state of viral infection characterized by low expression of only a few viral genes and replication of viral DNA only concomitantly with the cellular DNA; there is no production of infectious virus and minimal or no cytopathic effects

Open reading frames Stretch of nucleotides in the viral DNA that is devoid of stop codons and, therefore, possibly could encode a protein

Viral lytic cycle Series of events that starts with activation of a latent infection, proceeds with expression of viral early genes, replication of viral DNA, expression of late genes, and finally lysis of the infected cell and release of infectious virus

EPSTEIN–BAR VIRUS (EBV) is a widespread human herpesvirus that is associated with four human diseases, all of which are characterized by an increased growth potential of the EBV-infected cell: infectious mononucleosis (IM), Burkitt's lymphoma (BL), nasopharyngeal carcinoma (NPC), and the EBV-carrying lymphoproliferative disorder in immunodeficient hosts. EBV replicates in the epithelial cells of the nasopharynx and the parotid gland. The virus is shed in the saliva and is frequently transmitted by kissing. From its replicative site in the oropharynx, the virus infects and immortalizes B-lymphocytes, whereafter viral latency is established in the B-lymphocytes for the remainder of the infected person's life. More than 90% of humans carry EBV in a latent form; therefore, it is assumed that interaction between the virus and the host's immune system will prevent disease for most infected people. In contrast to many other viruses, the virus does not become incorporated (integrate) into the cellular genome but persists in EBV-infected cells and in EBV-carrying tumors in a free ringlike molecular form (episomes). The association of EBV to two major human malignancies as well as its potent ability to immortalize B-lymphocytes has made EBV one of the most important examples for understanding the role of viruses in human cancer.

I. The EBV Infection

A. Viral Tropism

EBV is spread by oral contact, usually by the ingestion of saliva from an EBV-infected individual. The virus replicates in epithelial cells of the upper respiratory tract, particularly in the epithelial cells of the parotid gland. The virus persists in the infected individual throughout life, and infectious virus is intermittently released in the saliva of almost all healthy EBV-infected individuals. From the epithelial cells of the upper respiratory tract, the virus infects the B-lymphocytes that normally infiltrate them. The infectability by EBV is mainly restricted to B-lymphocytes, due to the fact that the EBV receptor, through which the virus gains entry into the cells, is the CR_2 receptor for the third component of complement (C3d). The same receptor may exist on the epithelial cells of the nasopharynx. [*See* LYMPHOCYTES.]

B. *In vitro* EBV Infection: The Lymphoblastoid Cell Line

The *in vitro* EBV-infected B-lymphocytes are immortalized into lymphoblastoid cell lines (LCLs). In contrast to the normal B-lymphocyte, the LCL has acquired the ability to produce its own B-cell growth factor (BCGF) and, consequently, possesses the ability to grow indefinitely in culture. Whereas the LCLs are dependent on BCGF for their growth, the BL-derived cell lines (see below) are not. The LCL is regarded as low-malignant, because it does not form tumors in nude mice and shows a low cloning frequency in soft agar. If EBV is inoculated into certain species of New World monkeys, such as cottontop marmosets, owl monkeys, and tamarins, polyclonal malignant lymphomas with phenotypic characteristics similar to the LCL arise. Similarly, B-lymphoblastoid tumors resembling the LCL may occur in EBV-infected patients with severe immunodeficiency syndromes.

II. EBV-Associated Diseases

A. BL

BL is a non-Hodgkin B-lymphocytic lymphoma of low differentiation. The disease accounts for nearly one-half of all cancers in children in the tropical regions of Africa. BL occurs with high frequency across tropical Africa and in Papua New Guinea, but only at altitudes <1,500 m (endemic form). BL also occurs outside these endemic areas but very rarely (sporadic form). The two forms differ in several aspects. The location of the endemic form of BL is usually in the jaws, whereas the sporadic form frequently is found in other locations, such as the long bones, kidneys, adrenals, thyroids, ovaries, or testes. The peak incidence of endemic BL is 6–10 yr of age, whereas the sporadic form usually occurs >15 yr of age. Some 98% of endemic Bls carry the EBV genome, often in multiple copies per cell, whereas only 20% of the sporadic form does. [*See* LYMPHOMA.]

In an EBV-carrying BL, 100% of the cells express the EBV-determined nuclear antigen (EBNA). Other EBV-determined antigens first found to be expressed in these tumors are the viral capsid antigen (VCA), early antigen (EA), and the membrane antigen (MA). Endemic BL patients consistently have elevated antibody titers against VCA. The geometric mean anti-VCA titer is eightfold higher than in a control group with other malignancies. An African child with a moderately elevated VCA titer has a risk of developing BL >30 times higher than the normal population. Moreover, these BL patients exhibit antibody titers to the restricted type of EA (EA-R) and to the MA, which are not normally seen among healthy EBV-seropositive donors. The EA-R and MA titers are correlated with the clinical activity of BL, high titers of anti-MA being observed in long-term survivors and dramatic rises in anti-EA-R levels correlating with disease onset or relapse.

The rare incidence of EBV-carrying sporadic cases of BL, which occur outside the endemic area, shows that some other factor must play a role in the etiology of BL as well. BL has been hypothesized to develop in three stages in the endemic regions. Phase 1 is the primary infection with EBV. While uninfected B-lymphocytes differentiate toward the plasma cell end stage, EBV-infected B-lymphocytes are arrested in an early differentiation stage and can continue to divide. These EBV-infected cells may risk chromosomal damage in direct proportion to the number of cell divisions. Phase 2 involves an environmental cofactor, most likely endemic malaria. Infection with *Plasmodium falciparum* may lead to an impairment of the ability of cytotoxic T cells to control the proliferation of the EBV-infected B cells, thereby further increasing the "pool" of EBV-infected B cells. Phase 3 involves the translocation of the distal part of chromosome 8 to chromosome 14 (or 2 or 22), which is a regular finding in BL, leading to the constitutive activation of the c-*myc* oncogene and subsequent monoclonal B-lymphocyte proliferation resulting in BL.

B. NPC

NPC is the most common tumor in the densely populated areas of Southern China. The tumor is also common in East and North Africa and in Arctic Eskimo populations. Irrespective of its geographic origin, 100% of NPC tumors carry multiple copies of EBV. In the tumor, both EBV DNA and EBNA can be detected in the epithelial cancer cells. As in

BL, antibody titers to VCA are elevated (approximately 10-fold). Antibodies against EA are regularly present in NPC patient sera; however, unlike in BL, they are mainly directed against the diffuse subspecificity (EA-D). Characteristically, these patients develop IgA antibodies to VCA, even before a detectable tumor mass has developed. Since an IgA response to VCA is not seen in other conditions, the IgA-VCA antibody test can be used routinely for early detection of NPC with high sensitivity and specificity. In the provinces Zang-Wu and Laucheng in Southern China, this VCA test has been employed on 185,000 inhabitants, resulting in the early diagnosis of 132 NPC cases.

The 100% correlation between EBV and NPC indicates that EBV has an etiological role in the development of this tumor, although the restricted geographical occurrence of the tumor indicates that genetic and/or environmental cofactors also contribute to its etiology. First-generation immigrants of Southern Chinese origin maintain a high frequency of NPC. Offspring of mixed marriages between Southern Chinese and non-Chinese groups show an intermediate frequency. Several reports also speak of familial aggregation of NPC. Thus, while in BL evidence indicates an environmental cofactor, the epidemiology of NPC speaks more in favor of genetic cofactors.

C. Acute primary EBV infection: IM

IM is a self-limiting disease in which many EBV-infected B cells are stimulated to proliferate (polyclonal proliferation), followed by the appearance of characteristic, atypical cytotoxic T cells. The disease has a peak incidence between 17 and 25 yr of age but can also occur in children and older adults. In lower socioeconomic groups, infection with EBV occurs during the early years of life and is usually not accompanied by clinical illness, but results in the appearance of virus-specific antibodies (seroconversion), which induce immunity to EBV for life. Clinical IM symptoms will develop in about 50% of previously uninfected adolescents exposed to EBV. IM is frequently transmitted by kissing, presumably as a result of ingestion of viral particles shed in the saliva. Prodromal symptoms of IM are headache, chills, and exhaustion. The typical IM syndrome lasts for about 3 wk and consists of fever, lymphadenopathy, skin rashes, pharyngitis, spleen enlargement, and some liver dysfunction. After 3–4 wk, the symptoms resolve without treatment and complete recovery almost always ensues. Recurrencies have not been documented.

During the early stages of IM the patients develop the characteristic heterophile antibodies, which agglutinate sheep erythrocytes and are used in tests for the diagnosis of IM (Paul–Bunnel–Davidsson test). A heterogeneous population of antibodies are induced by the EBV infection of B cells (polyclonal stimulation), and a similar phenomenon is seen during the early stages of IM. At this stage, up to 18% of circulating B cells may express EBNA, but this is diminished to some 0.1% already in the second week of illness. IgM antibodies to the VCA and the EA, mainly the diffuse subcomponent (EA-D), arise early in the course of IM. The IgM-VCA antibodies give way to IgG-VCA after the first weeks, whereas EA-D antibodies disappear. Antibodies to EBNA appear during convalescence, some 30–50 days after onset of disease. *Chronic mononucleosis* is a syndrome characterized by crippling exhaustion and recurrent opportunistic infections. At present, there is widespread confusion concerning this syndrome because it has been grossly overdiagnosed in recent years. The criteria for diagnosis should include strongly elevated antibody titers to VCA, high-antibody titers to EA, whereas antibodies to EBNA are low or absent owing to a decrease in the EBNA-1 antibodies. The syndrome should start as a normal IM, which does not disappear in the usual way. Patients presenting only with fatigue and marginally abnormal EBV titers should be classified as having a chronic fatigue syndrome. In an extreme case of chronic IM, the antiviral drug Acyclovir has brought about almost complete recovery, whereas it has no effect on chronic fatigue syndrome.

D. Lymphoproliferative Diseases Carrying the EBV Genome in Immunodeficient Hosts

1. Fatal IM

Although IM is characterized by B-cell proliferation and occasionally presents with severe clinical symptoms, lethal cases of IM are very rare. Typical features of fatal IM are depletion of T cells, abundance of plasma cells in the bone marrow, infiltration of brain, viscera by lymphocytes, and a widely disseminated proliferation of EBNA-positive B cells. The patients have heterophile antibodies. The syndrome frequently has a familial setting and may be linked to the X-linked lymphoproliferative syndrome (XLP), characterized by defective proliferation of B-lymphocytes and deficient immune response to EBV. Approximately 40% of XLP patients develop fatal IM, 40% develop malignant B-cell lymphoma, and 20% develop dysgammaglobulinemia, often associated with chronic EBV infection.

2. EBV-Carrying Lymphoproliferative Disorder

EBV-DNA-positive lymphoproliferative disorders of B-cell origin have been found in a number of other conditions associated with immune deficiencies, either genetically determined (ataxia telangiectasia) or medically induced (renal and cardiac transplant recipients). The lymphocyte proliferation is polyclonal at onset. Monoclonal lymphomas frequently arise in transplant recipients, but they do not carry specific karyotypic alterations. The lymphoproliferative disorders of transplant recipients may show complete regression once the immunosuppressive therapy is discontinued. [*See* IMMUNOBIOLOGY OF TRANSPLANTATION.]

E. Other Diseases

EBV has also been implicated to play a role in several other diseases, notably thymic and salivary gland carcinomas, the Guillain–Barré syndrome, Hodgkin's disease, and the Chediak–Higashi syndrome.

III. The EBV Genome

A. General Structure of EBV DNA

The EBV genome present in the virus particles is a linear double-stranded DNA molecule of 173,000 base pairs (bp) (Fig. 1). At each end of the molecule there are 4–12 copies of a 500-bp terminal repeat (TR). A recognized role of the TRs is to facilitate circularization of EBV DNA following infection. Inside the infected cell, viral DNA molecules become covalently linked, forming multiple covalently closed circular episomes. Multiple tandem internal repeats (IRs) of 3,071 bp separate the genome into two domains: a short unique region, U_S, of 15,000 bp and a long unique region, U_L of 150,000 bp (Fig. 1). The EBV DNA has been cloned as a set of overlapping fragments, and the complete nucleotide sequence has been determined. The possible protein-encoding regions, the open reading frames (ORFs), have been designated according to their position in the genome defined by fragments produced by the restriction enzyme BamHI. The system is best explained by an example: BKRF1 means BamHI K fragment Rightward open reading Frame number 1. The locations of the BamHI fragments in the EBV genome are depicted in Fig. 1. [*See* DNA AND GENE TRANSCRIPTION.]

B. Transcription of the EBV Genome in Immortalized Cells

In immortalized B cells that do not produce viruses at least five distinct regions are transcribed. The most abundantly transcribed region is in the BamHI C fragment, which encodes two small nonpolyadenylated RNAs transcribed by RNA polymerase III. These RNAs, designated as EBERs, for EBV-encoded RNAs, show structural similarities to the VA RNAs of adenovirus. They do not code for protein and exist in cells complexed with a host cell protein termed "La." It is highly likely that the EBERs, like adenovirus VA RNAs, facilitate translation of viral mRNAs.

The BamHI WYH region corresponds to the internal repeat and the adjacent 3′ region. The principal transcript from this region is a 3.0-kb mRNA, which has a 1.6-kb coding sequence toward its 3′ end, encoding a nuclear protein termed EBNA-2.

Several mRNAs from this region also contain a long coding region resulting from several small sequences in the Bam W and Y fragments spliced together. This ORF encodes another nuclear protein, termed EBNA-5. Transcription of mRNAs encoding EBNA-2 and EBNA-5 are initiated at a promoter in Bam W, which contains a lymphocyte-specific enhancer, probably involved in the control of transcription. The promoters for the EBNA-1, EBNA-3, EBNA-4, and EBNA-6 mRNAs are thought to be located further 5′, in the BamHI C fragment.

The BamHI WYH region is associated with EBV-transformation, because viral strains with a deletion in the region are incapable of transformation, and the transformability is restored by introducing that region into the virus. This association of the Bam WYH region and transformation was recently shown to be due to both EBNA-2 and EBNA-5.

The BAMHI E fragment contains three long ORFs, which encode part of the nuclear antigens EBNA-3, EBNA-4, and EBNA-6.

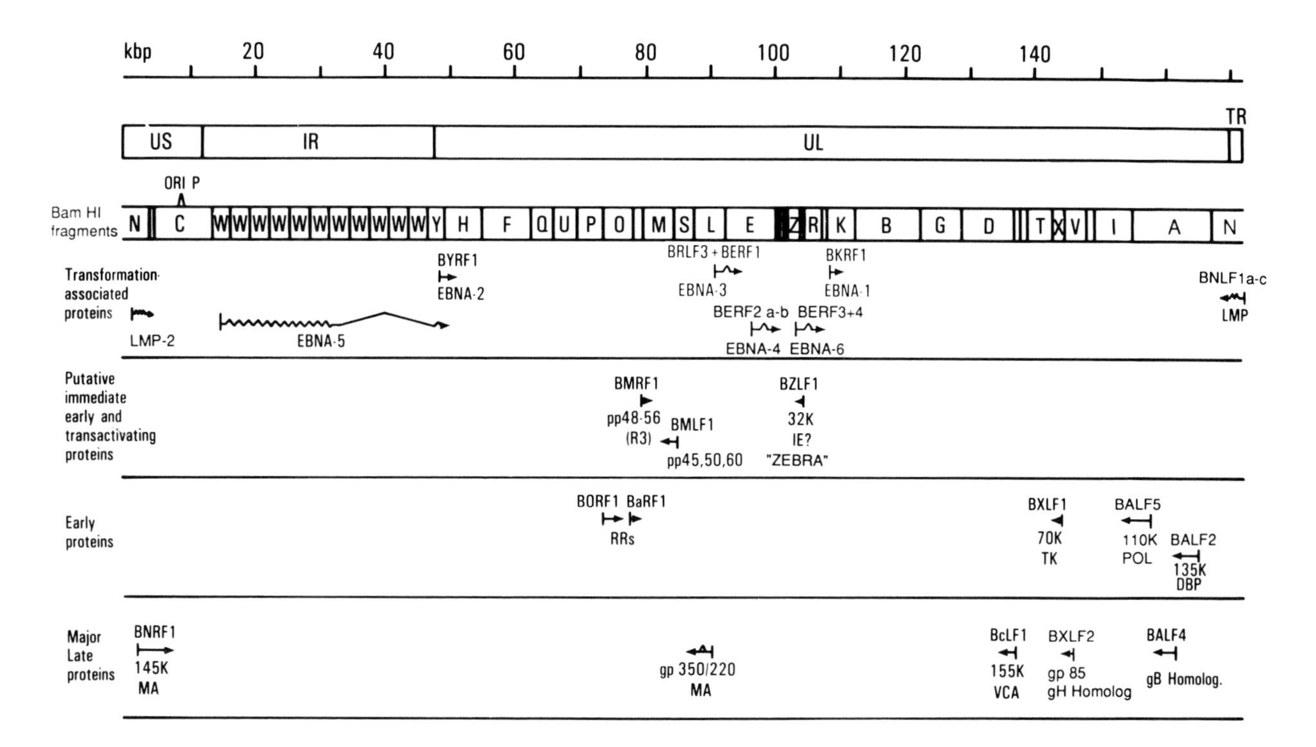

FIGURE 1 Genetic map of Epstein–Barr virus (EBV). Top: Relative size of the EBV genome in kilobase pairs (kbp). Upper bar: Organization of the genome in short unique sequence (US), internal repeat (IR), long unique sequence (UL), and terminal repeat (TR). Lower bar: Position of the BamHI restriction enzyme fragments. Gene map: Size, position, and direction of transcription of some major mRNA species are shown with arrows. The horizontal lines in the arrows denote coding sequences (exons). The names of open reading frames are given above each transcript (BYRF1, BamHI fragment Y Rightward open reading Frame number 1). DBP, major DNA-binding protein; EBNA, EBV-determined nuclear antigen; gB homologue, a protein showing homology to HSV glycoprotein B; gH homologue, a protein with homology to HSV glycoprotein H; IE, immediate early protein; K, kilodalton; LMP, latent membrane protein; MA, membrane antigen; POL, EBV DNA polymerase; pp, phosphoprotein; R3, a monoclonal antibody used to define these proteins, RRs, ribonucleotide reductases; TK, thymidine kinase; VCA, viral capsid antigen; ZEBRA, BamHI Z-encoded EB virus reactivating protein.

The BamHI K fragment hybridizes to a 3.7-kb mRNA with a 2.0-kb coding sequence, which encodes the nuclear antigen termed EBNA-1.

The BamHI N_{het} fragment gives rise to a 2.9-kb abundant mRNA, which encodes a membrane protein expressed during latent infection, termed LMP.

The BamHI C fragment encodes two very low-abundance spliced mRNAs with capacity to encode two membrane proteins, termed LMP-2.

IV. EBV-Encoded Proteins in EBV-Immortalized Cells: EBNAs 1–6 and LMP

A. EBNA

All latently EBV-infected cells contain the nuclear antigen EBNA, defined by anticomplement immunofluorescence (ACIF) with EBV immune sera. EBNA is an antigenic complex composed of at least six different proteins, designated EBNAs 1–6. The EBNA ACIF test primarily measures EBNA-1 and, in part, EBNA-2.

B. EBNA-1

EBNA-1 is a protein of a molecular weight between 65 and 92 kD, depending on viral isolates, which vary in the length of the third internal repeat array (IR3), included in the BKRF1 ORF. The IR3 is a simple array of only three triplet codons—GGG (glycine), GCA (alanine), and GGA (glycine)—repeated in an irregular fashion. Due to the IR3, EBNA-1 contains a 20–40-kDa copolymer containing only glycine and alanine. The major antigenic determinant of EBNA-1 is present within the glycine–alanine region and can be synthesized as a 20-amino acid peptide (referred to as p107). Antibodies to p107 are used in immunofluorescent staining as the routine method for specific detection of EBNA-1. p107 is also used clinically for detection of human anti-EBNA antibodies by the ELISA test: an IgM antibody response is diagnostic of recent EBV infection (IM), whereas an IgG response signals past EBV infection (healthy carrier state) and IgA antibodies to this peptide signals nasopharyngeal carcinoma.

Uninfected human cells contain proteins that react with the antibodies to p107. The anti-p107 IgM antibodies seen during acute infection react with these cellular proteins, whereas the anti-p107 IgG antibodies that develop during convalescence are specific for EBNA-1. The close relation between this major EBNA epitope and cellular proteins may explain why the IgG antibodies to EBNA do not appear until 1–2 mo after infection. It has also been speculated that this relationship may be a cause of EBV-associated autoimmunity.

The carboxy terminal part of the EBNA-1 is responsible for its DNA-binding activity. EBNA-1 has three specific binding sites on the EBV genome. Two of them are localized in the Bam C fragment, in a region termed oriP, which is necessary for EBV plasmid maintenance. The oriP has been reported also to act as a transcriptional enhancer, which is dependent on the binding of the EBNA-1 protein. Thus, one of the functions of EBNA-1 is to regulate plasmid maintenance as well as regulate its transcription. In these two respects, EBNA-1 is similar to the SV40 large T-antigen, a multifunctional protein that binds to and regulates the SV40 origin of replication and the SV40 enhancer. [*See* DNA BINDING SITES.]

Because the DNA-binding function of EBNA-1 is mediated by its carboxy-terminal sequence, the function of the glycine–alanine copolymer is puzzling. It may be involved in the unique feature of the EBNA-1 protein to bind diffusely to all the chromosomes. This is not the case for EBNA-2 or -3 or for other tumor antigens. A possible function of the EBNA-1 chromosome binding during cell division may be to secure equal distribution of episomes to the daughter cells. The EBV episomes are known to associate in a random pattern with the human chromosomes. Conceivably, the EBNA-1 protein might mediate the association by binding its C-terminal part to the EBV genome and its N-terminal part to a structural chromosomal protein.

C. EBNA-2

EBNA-2 is a phosphoprotein with unusual features: a long proline polymer (with an average of 26 prolines, but variable in different viral isolates), a 12-amino acid glycine–arginine repeat, and a highly charged, acidic carboxy terminus. It binds to both single- and double-stranded DNA and coprecipitates with a 32-kDa cellular protein.

Antibodies to EBNA-2 appear early during the course of IM, in contrast to the EBNA-1 antibodies, which do not appear until several months after onset of symptoms. An excess of antibody titer to EBNA-2 over that of EBNA-1 is indicative of primary EBV infection (IM) or of EBV reactivation (e.g., following immunosuppression).

All LCLs express EBNA-2, whereas biopsies of BL tumors and newly established BL-derived cell lines do not. This difference suggests that the protein is required only for the proliferation of low-malignant LCLs, which have an activated B-cell phenotype, but is not necessary for the growth of the highly malignant BL-derived lines, which have a B-cell phenotype resembling that of resting, ''memory,'' B cells. Furthermore, the Bam WYH region, which codes for EBNA-2, is deleted in four different BL-derived cell lines, indicating that its presence may even impose a selective disadvantage to the cells of the BL phenotype. When transfected into mouse fibroblasts, the EBNA-2 gene confers on the recipient cells the ability to grow in low serum concentration. When transfected into human EBV-negative BL-derived cells, the EBNA-2 gene induces the expression of the surface antigen CD-23, which is related to the receptor for the BCGF. This induction may be an important event in the EBV-induced immortalization.

D. Genetic Variation of EBNA-2: EBV Type A and Type B

DNA sequencing of the EBNA-2 coding region from several different viral isolates has shown that it can be present in two different types with only about 25% homology in the DNA sequence and 50% in the proteins. Based on the type of EBNA-2 gene, EBV strains are subtyped into type A and type B strains. The type A virus is the most common type of EBV in Europe and in the United States. The B-type virus has so far been mostly detected in isolates from the BL-endemic regions, where it is quite common also among healthy EBV carriers. *In vitro*, LCLs immortalized by type B viruses show a slower growth and grow more in large clumps than do LCLs immortalized by type A virus. The B-type EBNA-2 lacks most of the polyproline repeat region.

E. EBNA-3

EBNA-3 is a high-molecular weight (143–157 kDa) nuclear antigen that is encoded by a short ORF in the Bam L fragment (BLRF3) and by a long ORF in the Bam E fragment (BERF1), which varies in size depending on the EBV strain. It appears 2 days after the primary infection of tonsillar B cells and it is invariably expressed in LCLs. EBNA-3 binds to single-stranded DNA but has poor affinity to double-stranded DNA.

F. EBNA-4

EBNA-4 is a nuclear polypeptide with a molecular weight of about 180 kDa, variable among viral strains. The coding region is a part of the Bam E fragment that contains one short ORF (BERF2a) and one long ORF (BERF2b). It is also expressed 2 days after primary infection of B cells and appears to be expressed in most LCLs. EBNA-4 binds to both double- and single-stranded DNA, with preference for single-stranded DNA.

G. EBNA-5

EBNA-5 is an extraordinary complex nuclear antigen: as many as 28 different molecular weight forms, ranging in size from 20 to 130 kDa, have been described. Although each EBV-infected cell line usually has only one or a few major EBNA-5 molecular weight species, as many as 11 species have been described in the same cell. Between each molecular weights species there is a regular spacing of 6 kDa to the adjacent species, although sometimes smaller molecular weight differences of 2 and 4 kDa are seen. The genetic basis for the generation of these polypeptides is a fascinating, complex story.

In brief, the gene contains a series of repeats, each coding for a 66-amino acid peptide (about 6 kDa); messages of various length are made by different splicings, which include various numbers of repeats. Shorter repeats also exist, accounting for the 2 or 4 kDa differences.

EBNA-5 is invariably expressed in LCLs, whereas many BL-derived cell lines do not express this protein. A phenotypic drift of long-term established BL-derived cell lines toward a more LCL-like phenotype is accompanied by the appearance of EBNA-5 expression. EBNA-5 is a DNA-binding protein that binds more strongly to the nucleus than any other EBNA protein.

H. EBNA-6

EBNA-6 is a nuclear polypeptide of about 160 kDa that is encoded by a short ORF (BERF3) and a long ORF (BERF4), both situated in the BamHI E fragment, varying in size for different viral strains. It has been detected in several LCLs but is poorly characterized.

I. NAs Determined by Simian EBV-Like Viruses

EBV-like viruses have been isolated from four Old World primate species, namely gorillas, chimpanzees, orangutans, and baboons. They all affect B-lymphocytes, which then transform into LCLs. They all share 30–40% DNA sequence homology to EBV. They express VCA and EA that are highly cross-reactive among the different species. Like EBV, the simian EBV-like viruses also express multiple NAs. Serologic cross-reactivity is found mainly with EBNA-2, implying that the EBNA-2 protein has evolutionary conserved structures.

J. Latent Infection Membrane Protein

The latent infection membrane protein (LMP) (about 62 kDa molecular weight) contains six hydrophobic regions interspersed with short hydrophilic regions and, finally, a long hydrophilic carboxy terminus. It spans the plasma membrane six times and has both N- and C-termini at the cytoplasmic side. It is not glycosylated. The LMP-encoding mRNA from the BamHI N fragment is the most abundant EBV-specific mRNA in the immortalized cell.

By immunofluorescence, the protein is localized in "patches" at the cell membrane, in a pattern similar to that found for the cytoskeletal protein vimentin. The two proteins may be directly associated.

Transfection of rat fibroblasts or human keratinocytes with the LMP gene causes transformation.

K. LYDMA

LYDMA stands for "EBV-determined lymphocyte-detected membrane antigen" and designates the antigen(s) expressed on EBV-infected B-lymphocytes that serve as a target for EBV-specific lysis mediated by cytotoxic T cells. The antigen is defined with T cells derived from EBV-infected cultures, which also contain B cells, or with T cells stimulated by autologous LCLs. These MHC class I-restricted T cells often selectively lyse LCLs but not mitogen-induced B-lymphoblasts. LYDMA is probably a complex of several antigens, including LMP and EBNA-2.

L. Expression of Immortalization-Associated Proteins in EBV-Associated Diseases

This expression shows distinctively different patterns for the different diseases. The LCL and the EBV-carrying lymphoproliferative disease in immunodeficient hosts, which resembles the LCL phenotypically, express EBNAs (at least EBNAs 1, 2, 3, and 5) and the LMP. In contrast, BL and newly established BL-derived lines express only EBNA-1; however, upon cultivation *in vitro*, BL-derived cell lines can start to express some of these proteins. NPC cells invariably express EBNA-1; in about 50% of the cases, LMP. Primary EBV infection, IM, has not been analyzed, but primary EBV infection *in vitro* leads to expression of all the immortalization-associated proteins.

These expression patterns suggest that EBV-induced polyclonal proliferation of B cells during IM and EBV-carrying lymphoproliferative disease requires all seven immortalization-associated proteins. The attack of sensitized cytotoxic T cells may then induce disappearance (down-regulation) of EBNAs 2–6 and LMP, assuming that they constitute the LYDMA antigen, and the return of the B cells to their resting state with a latent, nonproliferative EBV infection. This resting B cell, expressing only EBNA-1, would then in rare cases be the origin of a BL following the typical chromosomal translocation event. The frequent expression of LMP in NPC indicates that this protein may be involved in the EBV-mediated transformation of epithelial cells. It should be noted, however, that very little is known of the expression of EBV in epithelial cells.

EBNA-1 stands out from the other EBV antigens, because EBNA-1 is always expressed in EBV-infected cells. Probably this is because it promotes plasmid maintenance and prevents integration. Presumably, EBNA-1 can not be detected by cytotoxic T cells, allowing it to be expressed in resting, latently infected cells. It is unlikely that EBNA-1 causes cell immortalization because the majority of humans, although infected with EBV, do not develop EBV-carrying tumors. More likely, candidates for immortalization are LMP and EBNA-2, which, at least in some *in vitro* systems, can have transforming functions.

V. Viral Proteins in Virus-Producing Cells

A. Detection, Classification, and Nomenclature of the Viral Lytic Cycle Proteins

In several EBV-carrying BL-derived cell lines and LCLs a small fraction of the cells (usually 1–5%) spontaneously express early and late viral antigens. The percentage of cells entering the viral lytic cycle can be greatly increased by treatment of infected cells with chemical inducers and other procedures. The proteins associated with the viral lytic cycle can then be detected by serological or biochemical methods.

Some 35 viral proteins, ranging in size from 350 to 18 kDa, have been identified. They are classified as either early or late depending on whether they are synthesized before or after the synthesis of viral DNA has begun. The majority of the proteins detected in virus-producing cells are early.

B. The EBV Gene Map

More than 80 major ORFs have been identified in the EBV DNA, 15 of which code for late proteins. The virus particle contains approximately 30 different proteins, suggesting that many late genes have not yet been identified. Up to 100 viral proteins have been mapped on the viral DNA. The gene map (Fig. 1) lists only a few of them. Note that early and late genes are interspersed throughout the genome. Most of the lytic cycle genes consist of a single ORF with the promoter region closely upstream. In contrast, the immortalization-associated genes have, in general, a very complex structure. [*See* GENETIC MAPS.]

C. Early Genes and Proteins

1. Viral Genes and Proteins Related to Activation of the Viral Lytic Cycle

Immediate early (IE) genes are operationally defined as a subset of early viral genes expressed in the absence of ongoing protein synthesis and, therefore, do not require virus-coded activation for their expression. To study them, protein synthesis is blocked by reversible inhibitors such as cycloheximide. After a period of time, cycloheximide is removed and the transcription-blocking drug actinomycin D is added. Under these conditions, IE gene products accumulate, whereas early and late genes are inactive.

As many as seven IE proteins, ranging in size from 120 to 48 kDa, have been described. The BamHI M fragment, ORF BMRF1, encodes proteins that are abundantly expressed in lytically infected cells, are highly antigenic, and constitute the main component of the EA-D antigen. The other ORF in the Bam M fragment, BMLF1, also encodes putative IE proteins.

Another approach to the identification of IE genes employs the defective EBV DNA produced by P3HR-1 cells, a long-time established BL cell line. When used for superinfection of cells carrying latent EBV genomes, this DNA activates replication of the latent EBV DNA. The reactivating ability of the DNA resides in the BamHI Z fragment, which presumably codes for the reactivating protein. This protein of 33 kDa has been designated "ZEBRA," for the BamHI Z-encoded EB virus reactivator. It is surprising, however, that in cycloheximide-treated cells it has not been possible to demonstrate expression of this protein or of mRNA mapping to BamHI Z. Thus, even though the definitive classification of the EBV IE genes is still controversial, the BamHI M genes and the BamHI Z gene seem to encode activators of transcription of viral genes functionally related in one way or another to activation of virus replication.

2. Viral Enzymes

The EBV DNA polymerase (110 kDa) is encoded by a single ORF—BALF5. Some regions of the EBV polymerase gene are homologous to the DNA polymerase gene of poxvirus, adenovirus, and cytomegalovirus. The EBV DNA polymerase is readily distinguished from those of its cellular counterparts by three properties: (1) stimulation by ammonium sulfate, (2) preferential utilization of synthetic templates, and (3) higher sensitivity to certain drugs such as phosphonoacetic acid and to acyclovir triphosphate. With respect to these three properties the EBV and the HSV DNA polymerases are very similar. A 45-kDa protein stimulates the activity of the EBV DNA polymerase.

The viral nuclease (an exonuclease) is associated with a 70-kDa molecule encoded by the Bam B fragment. It may provide nucleotide precursors for DNA synthesis by degrading cellular DNA, and it may also be involved in genetic recombination. Antibodies to the EBV exonuclease are preferentially found in sera of NPC patients.

Ribonucleotide reductase, which catalyzes the conversion of ribonucleoside diphosphates to deoxynucleoside diphosphates, is responsible for the synthesis of nucleotide precursors of DNA. It seems to contain two subunits responsible for catalysis and regulation, respectively, specified by two ORF (BORF1 and BaRF1) homologous to the two subunits of the HSV enzyme.

The viral thymidine kinase (TK) is encoded by the BXLF1 ORF. The EBV TK (70 kDa) has a good overall homology with the HSV-1 TK and also cross-reacts with monoclonal antibodies to the HSV TK. Replication of EBV is highly sensitive to drugs (e.g., Acyclovir) that require a viral TK for their phosphorylation activation.

The major DNA-binding protein (135 kDa) is the most abundant of the early proteins and is encoded entirely by the BALF2 ORF. It binds well to both single- and double-stranded DNA *in vitro*. It has a significant sequence homology with the corresponding HSV protein (ICP 8). It seems to be required for replication of the viral DNA, probably by changing its configuration. Its gene is deleted in the EBV genome carried by the BL cell line Raji. When these cells are induced to enter the EBV lytic cycle, there is production of early antigens, but no viral DNA synthesis and no late protein synthesis follows, presumably for lack of the 135-kDa protein. This phenomenon is also exploited in EBV serology: Induced Raji cells are the standard source of EA for immunofluorescence.

3. Major Immunogenic Early Proteins

Several of the early proteins are very abundant in virus-producing B cells. The 48–55-kDa BamHI M-encoded family of early polypeptides (discussed above) constitutes the major component of the diffuse subspecificity of the early antigen (EA-D). A 85-kDa early protein is detected by immunofluorescence as filamentous cytoplasmic structures and is the main target for EA-R antibodies. The gene encoding this polypeptide has not been identified. The 135-kDa DNA-binding protein is the the major early antigen detected by immunoprecipitation with human sera but is not detected by immunofluorescence.

D. Proteins of the Virus Particle

The EBV particle is eicosahedral, enveloped with typical herpesvirus morphology. A total of 33 polypeptides are associated with it. The major late antigen of virus-producing cells is a 155-kDa polypeptide. A 145-kDa virus envelope protein and the 155-kDa polypeptide constitute the VCA complex in virus-producing cells. The 155-kDa protein has been mapped to the BcLF1 ORF. Monoclonal antibodies to the 155-kDa major capsid protein are available.

The most prominent *virion envelope proteins* are two immunologically cross-reactive glycoproteins of 350 and 220 kDa, designated gp 350/220, both encoded by the Bam HI L fragment. The precursor polypeptides have molecular weights of 135 and 100 kDa, respectively; the higher molecular weights of the protein are due to glycosylation. They are the main targets of neutralizing antibodies and are also responsible for the binding of the virus to its cellular receptor; therefore, they are the main target for attempts to develop an EBV vaccine. An unrelated immunogenic glycoprotein of 85 kDa (gp85) is encoded by the BXLF2 ORF in the BamHI X fragment. gp85 has some limited homology to the glycoprotein H (gH) of HSV.

Another late protein, encoded by the BALF 4 ORF, is structurally homologous to the glycoprotein B (gB) of HSV 1, which is implicated in the internalization of HSV virions. The EBV BALF4 (110 kDa) protein may have a similar function. The envelope also contains the LMP protein.

An unglycosylated protein (145 kDa) is a major constituent of the virus particle. It appears to be encoded by the BNRF 1 ORF of the BamHI N and C fragments.

Bibliography

Baer, R., Bankier, A. T., Biggin, M. D., Deininger, P., Farrell, P. J., Gibson, T. J., Hatfull, G., Hudson, G. S., Satchwell, S. C., Seguin, C., Tuffnell, P. S., and Barrell, B. G. (1984). DNA sequence and expression of the B95-8 Epstein–Barr virus genome. *Nature* **310,** 207–211.

Dillner, J., and Kallin, B. (1988). The Epstein–Barr virus proteins. *Adv. Cancer Res.* **50,** 95–158.

Ernberg, I., and Kallin, B. (1984). Epstein–Barr virus and its association with human malignant diseases. *Cancer Surv.* **3,** 51–89.

Henle, W., and Henle, G. (1979). Seroepidemiology of the virus. *In* "The Epstein–Barr Virus" (M. A. Epstein and B. G. Achong, eds.), pp. 61–78. Springer Verlag, Berlin.

Kieff, E., Hennessy, K., Fennewald, S., Matsuo, T., Dambaugh, T., Heller, M., and Hummel, M. (1985). Biochemistry of latent EBV infection and associated growth transformation. *In* "Burkitt's Lymphoma: A Human Cancer Model" (G. M. Lenoir, G. O'Connor, and C. L. M. Olweny, eds.), pp. 323–329. Scientific publication no 60. International Agency for Research on Cancer, Lyon, France.

Knutson, J. C., and Sugden, B. (1989). Immortalization of B-lymphocytes by Epstein–Barr virus. *In* "Advances in Viral Oncology," Vol. 8, (G. Klein, ed.), pp. 151–712. Raven Press, New York.

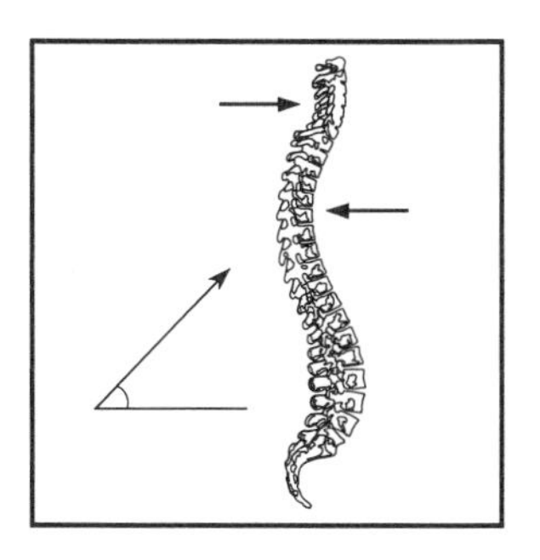

Ergonomics

K. H. E. KROEMER, *Virginia Polytechnic Institute and State University (VA TECH)*

Glossary

Anthropometry Measuring and describing the physical dimensions of the human body
Biomechanics Describing the physical behavior of the human body in mechanical terms
Ergonomics Discipline to study human characteristics for the appropriate design of the living and work environment; in the United States, often called human factors, or human factors engineering
Hawthorne effect Often subtle, certainly unplanned and unexpected, effect produced in persons taking part in an experiment by their knowledge that they are participating in the experiment
Humanization of work Goal of ergonomics to achieve ease and efficiency of technological systems involving humans
Industrial engineering Engineering branch concerned with the interactions among people, machinery, and energy
Industrial hygiene Controlling occupational health hazards that arise as a result of doing work
Work physiology (work psychology) Application of theoretical knowledge about the human body (and mind) to the evaluation of (industrial) work loads to derive recommendations for improving the work situation

ALL MAN-MADE TOOLS, devices, equipment, machines, and environments should serve, directly or indirectly, to further the safety, well-being, and performance of humans. Ergonomics (also called human factors in the United States) is the discipline to "study human characteristics for the appropriate design of the living and work environment."

Thus, ergonomics has two distinct aspects: (1) the study, research, and experiments used to determine the specific human traits and characteristics that one needs for engineering design and (2) the application and engineering in which tools, machines, shelter, environment, and work tasks and job procedures are designed to fit and accommodate the human. Of course, the actual performance of "human and equipment in the environment" is also a topic of study and research to assess the suitability of the design and to determine possible improvements.

I. History and Development

Fitting things to the human is as old as mankind. Primitive tools and weapons made of stone, bone, and wood were selected for their fit to the human hand and their suitability for the intended purpose. Purposeful shaping of these tools was the next step; creating and manufacturing followed. Fitting clothes and making shelters certainly were early and fundamental ergonomic activities.

With increasing complexity of the human society, organizational and management challenges developed. Purposeful training of workers and soldiers, for example, became necessary together with behavior formation and control. Roman soldiers underwent well-organized military conditioning until they could perform military exercises without sweat accumulating uselessly on their skin. "Drying the legions" relied, consciously or by experience, on the principle of training and adapting the physiological capabilities of the recruits to the physical requirements. For major projects, such as building the pyramids in old Egypt, assembling armies for warfare, and sheltering and supplying the inhabitants of ancient cities with food and water, sophisticated knowledge of human needs and desires was required, and careful planning and complex logistics needed to be mastered.

Artists, military officers, employers, and sports enthusiasts were interested in body build and physical performance. Medicine men treated illnesses and injuries. Anatomic and anthropological disciplines began to develop. About 400 B.C., Hippocrates described a scheme of four body types, which were supposedly determined by their fluids: the moist type was dominated by black gall; the dry type was influenced by yellow gall; the cold type was characterized by slime, and the warm type was governed by blood.

In the fifteenth, sixteenth, and seventeenth centuries, persons such as Leonardo da Vinci and Alfonso Borrelli mastered the complete knowledge of anatomy, physiology, and equipment design; they were artist, scientist, and engineer in one. In the eighteenth century, the sciences of anatomy and physiology diversified and accumulated specific and detailed knowledge; psychology began to develop as a separate science.

Well into the nineteenth century, the scientific disciplines tended to be theoretically oriented, solely devoted to understanding the complex human being. The stereotype is the scientist in a white coat leading a research-devoted life in the laboratory. But with increasing interest in industrial work and the old interest in military employment of the human, "applied" aspects of the pure sciences began to develop in the early 1800s. In France, Lamar, Lavoisier, Amar and Duchenne researched energy capabilities and strains of the working human body; Marey developed methods to describe human motions at work and to determine work payment systems. In England, the Industrial Fatigue Research Board considered theoretical and practical aspects of the human at work. In Italy, Mosso developed dynamometers and ergometers to research fatigue. In Scandanavia, Johannsson and Tigerstedt developed the scientific discipline of work physiology with a so-named institute in Germany founded by Rubner in 1913. In the United States, the Harvard Fatigue Laboratory was established in the 1920s.

In the first half of the twentieth century, work (or industrial) physiology and psychology were well-advanced and widely recognized, both in their theoretical work "to study human characteristics" and in the application of this knowledge "for the appropriate design of the living and work environment." Two distinct approaches to study human characteristics had developed: one was particularly concerned with physiological and physical traits of the human, and the other was mainly interested in psychological and social proprieties. Although these approaches greatly overlapped, the anatomical–physical and physiological aspects were studied mainly in Europe, and the psychological and social aspects in North America.

Based on a broad fundament of anatomical, anthropological, and physiological research, applied, or work, physiology had become of great concern, particularly during the hunger years associated with World War I in Europe. With marginal living conditions, the question of minimal or suitable nutrition to perform certain activities (e.g., the energy consumption while performing agricultural, industrial, military, and household tasks; the relationships between energy consumption and heart rate; the use of muscular capabilities; suitable body postures at work; design of equipment and work places to fit the human body) were researched for application purposes. Testing persons for their ability to perform physical and mental work, testing vigilance and attention, mental work load, driver behavior and road sign legibility, and related topics was known as psychotechnology in the 1920s in Europe.

In the United States, "most psychologists at this time [around 1900], were strictly scientific and deliberately avoided studying problems that strayed outside the boundaries of pure research." (Muchinsky, 1987, page 13). However, activities such as sending and receiving Morse code, perception and attention at work, using psychology in advertising, promoting industrial efficiency, and other basically psychological aspects were applied to problems in business and industry. A particularly important step was the development of intelligence testing to screen military recruits during World War I and later industrial workers, for assigning them to jobs suitable for their mental capabilities. Thus, the terms intelligence testing and industrial psychology appeared. (Muchinsky explains that the term "industrial psychology" first appeared as a typographical error, actually meant to read "individual" psychology.) Gould (1981) provides a partly amusing, partly disturbing account of the early years of intelligence testing.

Among the best-known results in industrial psychology are those of the experiments at the

Hawthorne Works in the mid-1920s. The study was designed to assess relationships between lighting and efficiency in work rooms where electrical equipment was produced. The bizarre finding was that the workers' productivity increased or remained at a satisfactory level whether or not the illumination was changed, obviously in response to the attention paid to the workers by the researchers. This became known as the Hawthorne effect.

Industrial psychology divided into several recognized branches, including personnel psychology, organizational behavior, industrial relations, and engineering psychology. During World War II, the "human factor" as part of a "man–machine system" became of major concern because, in many cases, the technological development had led to machines and operational systems that required more attention, more strength, and more endurance from individuals and teams than many humans could muster. For example, radar screens were to be observed over many hours with the intent of detecting some "blips" among others. High-performance aircraft made the pilot unable to operate hand controls properly under high g-loads, such as in sharp turns, or even made the pilot black out. Cockpits in tanks and aircraft were kept small and low, requiring that suitably small crew members be selected. Combat morale and performance were difficult to maintain under stressful conditions. Thus, military and related efforts at the home front generated the need to consider human physique and psychology, as an individual and in teams, purposefully and knowingly in the design of task, equipment, and environment.

Both in Europe and in North America, various names were used to describe the activities of psychologists, sociologists, physiologists, anthropologists, statisticians, and engineers who studied the human and used this information in design, selection, and training. On January 13 and 14, 1950, British researchers met in Cambridge, United Kingdom, and discussed, among other topics, the name of a society to be founded that would represent their activities. Among other terms, "ergonomics" (Grk. *ergon-*, work or effort; *-nomos*, law or surroundings) was proposed, a word invented by K. F. H. Murrell in late 1949. The new word ergonomics was neutral, not implying priority to physiology, psychology, functional anatomy, or engineering. It was easily remembered and recognized and could be used in any language. Ergonomics was formally accepted as the name of the new society at its council meeting on February 16, 1950 (Edholm and Murrell, 1974).

Several aspects are worthy of a brief note. The original proposal for the name vote included two alternative name suggestions. One was the Human Research Society, the other the Ergonomic Society. Note that there is no "s" at the end of "Ergonomic"; apparently, the "s" somehow slipped onto the voting ballot. The attachment of the "s" has made the derivation of an adjective or adverb relating to ergonomics rather difficult. (Incidentally, there are several claims that similar words had been coined and used earlier [e.g., in Poland].) The alternately proposed term, Human Research Society, bears some similarity to the term human engineering.

In the United States, similar discussions went on about a proper name to describe the activities of the new discipline. In 1956, a group of persons convened to establish a formal society. The name ergonomics was rejected, and instead the term human factors was selected to describe the professional activities. Often, the additional word engineering is used to indicate applications, such as in human (factors) engineering.

There has been, and still is, some discussion on whether or not human factors is different from ergonomics, which of the two relies more on psychology or physiology, or which is more theoretical or practical. Today, the two terms are usually considered synonymous, as exemplified by the Canadian Society: it uses human factors in its English name, and ergonomie in its French version. The term human factors is still mostly used in North America, while ergonomics is common around the globe.

II. Goals and Scope

"[Ergonomics] can be defined as the application of scientific principles, methods, and data drawn from a variety of disciplines to the development of engineering systems in which people play a significant role. Successful application is measured by improved productivity, efficiency, safety, and acceptance of the resultant system design. The disciplines that may be applied to a particular problem include psychology, cognitive science, physiology, biomechanics, applied physical anthropology, and industrial systems engineering. The systems range from the use of a simple tool by a consumer to a multiperson sociotechnical system. They typically include both technological and human components.''

''[Ergonomic] specialists from these and other disciplines are united by a singular perspective on the system design process: that design begins with an understanding of the user's role in overall system performance and that systems exist to serve their users, whether they are consumers, system operators, production workers, or maintenance crews. This user-oriented design philosophy acknowledges human variability as a design parameter. The resultant designs incorporate features that take advantage of unique human capabilities as well as build in safeguards to avoid or reduce the impact of unpredictable human error'' (National Research Council, 1983, pp. 2–3).

Ergonomics adapt the man-made world to the people involved. Ergonomics focus on the human as the most important component of our technological systems. Ergonomics span the range from basic research to engineering and managerial applications. Thus, the utmost goal of ergonomics is humanization of work. This goal may be symbolized by the ''EE'' of ease and efficiency for which all technological systems and their elements should be designed. This requires knowledge of the characteristics of the people involved, particularly of their dimensions, their capabilities, and their limitations.

There is a hierarchy of goals in ergonomics. The basic intent is to generate tolerable working conditions that do not pose known dangers to human life or health. This basic requirement assured, the next goal is to generate acceptable conditions upon which the people involved can voluntarily agree according to current scientific knowledge and under given sociological, technological, and organizational circumstances. Of course, the final goal is to generate optimal conditions, which are so well adapted to human characteristics, capabilities, and desires that physical, mental, and social well-being is achieved.

Ergonomics is neutral: It takes no sides, neither of employers nor of employees, neither for nor against technological progress. It is not a philosophy, but a scientific discipline.

III. Contributing Sciences

Ergonomics is a growing and changing science. That development stems from increasing and improving knowledge about the human and is driven by new applications and new technological developments.

As discussed earlier, a number of classic sciences provide the fundamental knowledge about the human. The anthropological basis consists of anatomy, describing the build of the human body; orthopedics, concerned with the skeletal system; physiology, dealing with the functions and activities of the living body, including the physical and chemical processes involved; medicine, concerned with illnesses and their prevention and healing; psychology, the science of mind and behavior; and sociology, concerned with the development, structure, interaction, and behavior of individuals or groups. Of course, physics, chemistry, mathematics, and statistics also supply knowledge, approaches, and techniques.

From these basic sciences, a group of more applied disciplines developed into the core of ergonomics. These include primarily anthropometry, the measuring and description of the physical dimensions of the human body; biomechanics, describing the physical behavior of the body in mechanical terms; work physiology, applying physiological knowledge and measuring techniques to the body at work; industrial hygiene, concerned with the control of occupational health hazards that arise as a result of doing work; and management, dealing with and coordinating the intentions of the employer and the employees. Of course, many other disciplinary areas have developed that also are part of ergonomics, contribute to it, or partly overlap it such as labor relations or safety engineering.

Several distinct application areas use ergonomics as basic components of their knowledge base or of their work procedures. Among these are industrial engineering, by definition concerned with the interactions among people, machinery, and energies; bioengineering, working to replace worn or damaged body parts; systems engineering in which the human is an important component of the overall work unit; safety engineering, which focuses on the well-being of the human; and military engineering, which relies on the human as soldier or operator. Naturally, other application areas are in urgent need of ergonomic information and data, such as computer-aided design in which information about the human must be provided in computerized form. Oceanographic, aeronautical, and astronautical engineering also rely intensively on ergonomic knowledge.

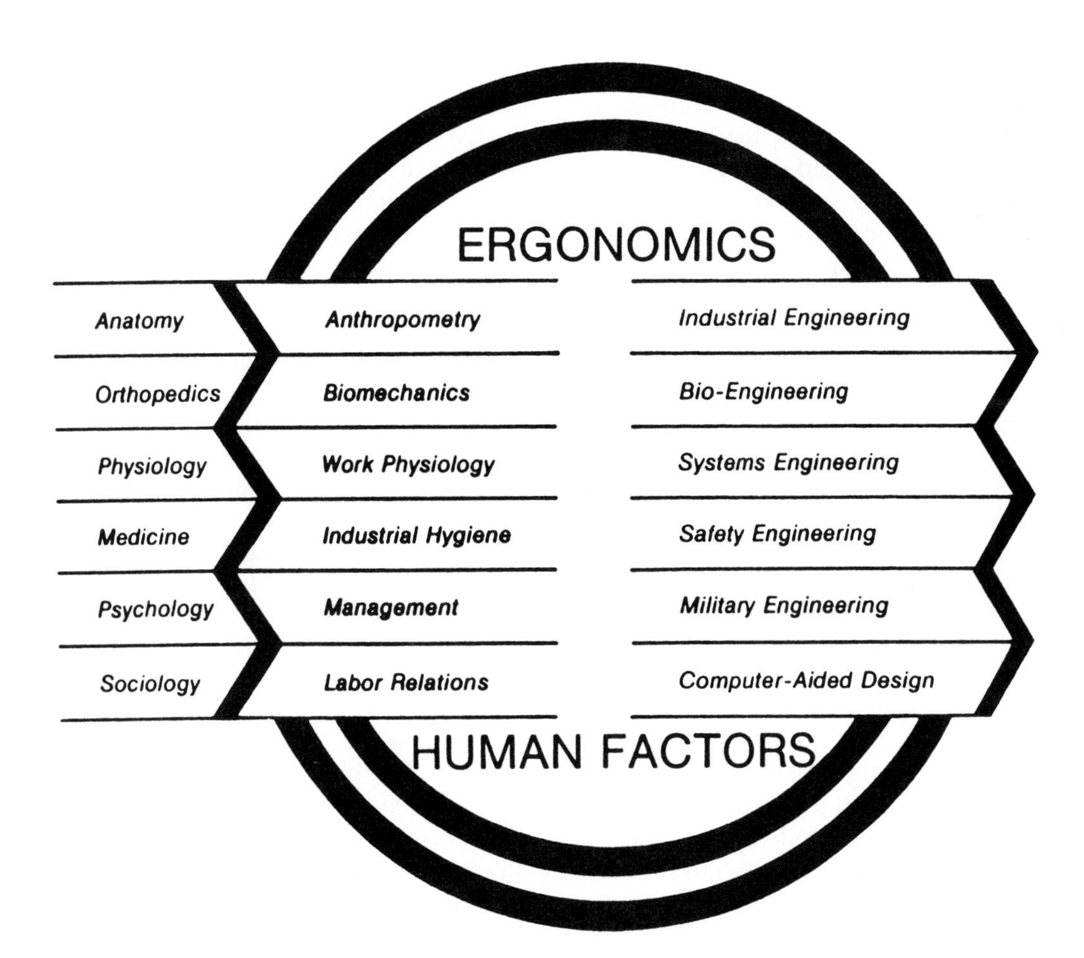

FIGURE 1 Ergonomics: contributing sciences, specialty areas, and primary users.

The development from the basic sciences to the applied ones in ergonomics, and the use of ergonomic knowledge in specific disciplines, is depicted schematically in Fig. 1. As more knowledge about the human becomes available, as new opportunities develop to make use of human capabilities in modern systems, and as needs for protecting the person from outside events arise, ergonomics change and develop.

IV. The Current Database in Ergonomics

The existing basis of ergonomic knowledge is deep, wide, and incomplete. It is deep because much of the wealth of information gathered in the "classic mother sciences" over decades and centuries contributes to ergonomic information. For example, anatomical and physiological knowledge is far-reaching and fairly complete.

The knowledge base of ergonomics is also wide, because many classic and new sciences contribute to ergonomic knowledge. For example, the knowledge about the functions of the body is paralleled with information about mental and social traits and needs.

Nevertheless, the ergonomic knowledge base is incomplete, particularly in rapidly developing areas such as sociology and biomechanics and even on "old" topics such as the elderly. Furthermore, it is incomplete because of new demands on the human in interaction with a new technical system; consider the exciting problems and challenges associated with long-duration space exploration by humans. Also, much of the information gathered in the traditional mother sciences was not developed with an ergonomic application in mind, nor did the research address new conditions such as in underwater habitat, modern warfare teams, or space exploration. However, one does not have to consider only "exotic" conditions of human–technology interactions. Even for rather mundane and everyday applications, information is not always available in appropriate quantity or quality. A 1985 evaluation of our knowledge base in anthropometry and biomechanics, for example, has shown that the information about the human body build and size is unsufficient to develop a complete or correct computerized model of the human body. This lack of suitable knowledge is even more pronounced in biomechanics, where not enough is known, for instance, about active muscular exertions (as opposed to reactive responses) of the human body to model its behavior in an automobile accident or in sports activities (Kroemer, Snook, Meadows, and Deutsch, 1988).

Attempting an overview about the currently available ergonomic database, one finds that a large body of written information exists. A typical listing of printed sources of information, available in the United States in 1989, includes about two dozen books, a variety of standards, and about half a dozen journals (see References).

V. Education in Ergonomics

Academic education in ergonomics/human factors in the United States is at the graduate level, i.e., after several years of studying disciplines such as engineering, psychology, physiology, physics, and chemistry, usually to the bachelor's level.

In the United States and Canada, the formal curriculum in human factors or ergonomics is often a specific option in colleges of engineering or psychology, with about half of the programs in either one. The trend seems to go toward academic education in engineering, almost without exception in industrial engineering departments. Here, one or two courses in ergonomics/human engineering are usually offered to undergraduates, while MS or Ph.D. formal specialization and intensive instruction is provided on the graduate level. The graduate degree is, generally, in industrial engineering or psychology, at the masters level. Very few schools in North America, or in the United Kingdom, convey in fact a degree specifically in ergonomics.

Many academic programs are housed in departments where some of the professors provide the core of ergonomic information, while additional instructors are called in from other departments to provide a wider perspective. Only a few universities have large enough departments in industrial engineering or psychology to rely exclusively or predominately on the faculty drawn from that department. The Human Factors Society provides detailed information on programs and courses in human factors of U.S. and Canadian universities. Similarly, the Ergonomics Society provides an overview of the programs in the United Kingdom and in many other countries.

The requirements for entering a graduate program in the ergonomics/human factors option vary from institution to institution, but typically, a bachelor's degree is required. If the applicant to the program does not have an undergraduate degree in the specific field (either psychology or engineering), the student may have to take additional courses to make up existing deficiences. The time from entering the program with a bachelor's degree to completing it with a master's degree is at least 2 years, but 4 years or longer to obtain a doctoral degree.

Employment of specialists in human factors in the United States used to be mostly in the military services, in occupational safety and health, industrial hygiene, or government agencies. However, with increasing importance placed on human health and performance in all occupations and professions, the demand for ergonomists/human factors engineers is now widespread and still growing. This growth is expected to continue because even traditional disciplines such as mechanical or architectural engineering realize that the proper "interfacing" of equipment with the human is an important aspect to assure performance, safety, and human well-being.

VI. Professional Organizations

The largest single (national) professional organization in ergonomics/human factors on earth is the Human Factors Society (P. O. Box 1369, Santa Monica, California 90406). The Human Factors Society had about 4,650 members in 1989, with various technical interest groups. The Human Factors Society publishes its own journal, *Human Factors,* and organizes an annual professional congress.

The Ergonomics Society is located in the Department of Human Sciences, University of Technology, Loughborough LE11 3TU, United Kingdom. Founded in 1950, it is the oldest professional organization for ergonomics. The Ergonomics Society had about 830 members in 1989, mostly in the United Kingdom. The Society supports two journals, *Ergonomics* and *Applied Ergonomics,* and organizes an annual professional congress.

More than two dozen national societies exist (Table I), most carrying the term ergonomics in their

a

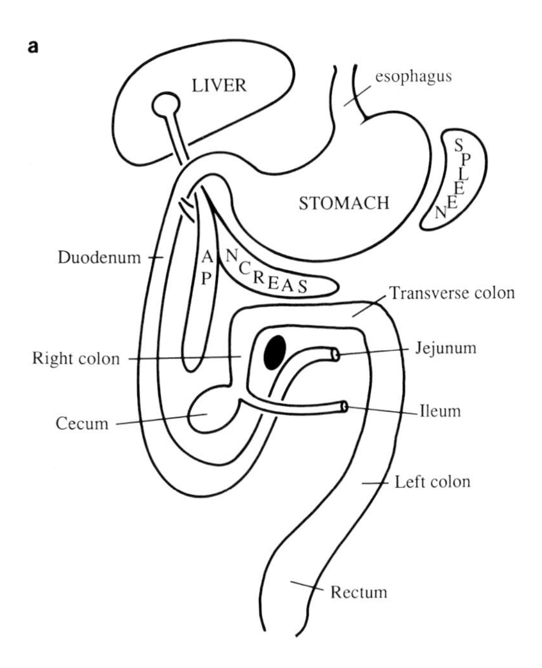

b

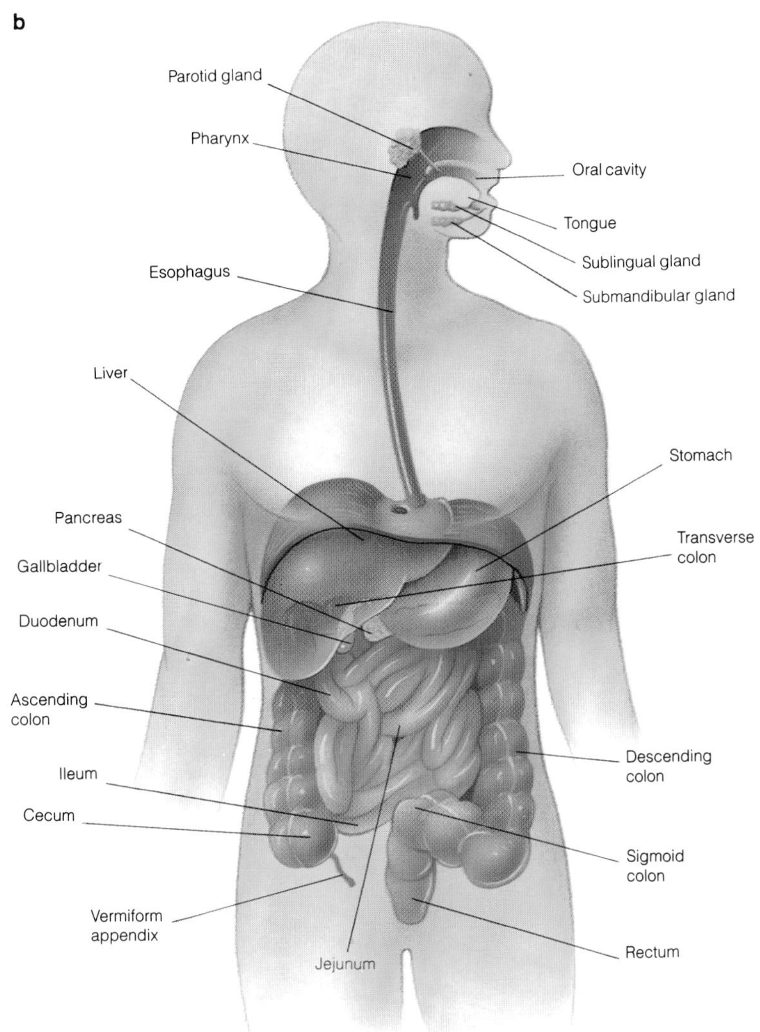

COLOR PLATE 1 Basic topography of (**a**) mammalian gastrointestinal tract and (**b**) layout in humans. [Source: Gaudin, A. J., and Jones, K. C. (1989). "Human Anatomy and Physiology." Harcourt Brace Jovanovich, San Diego. p. 441. Reproduced with permission.]

a

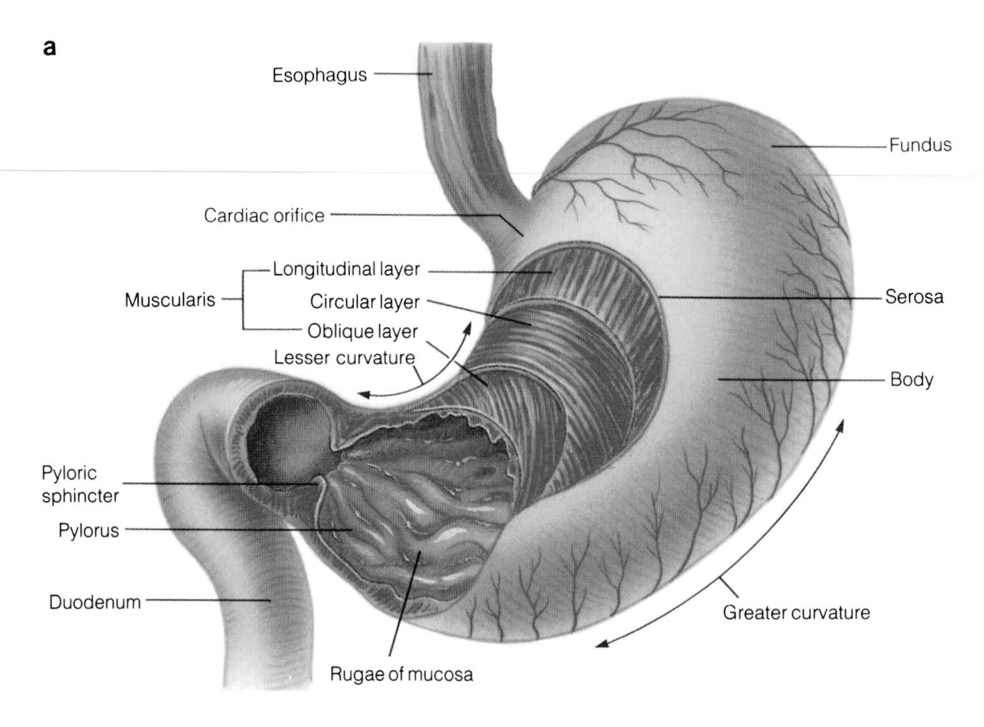

b

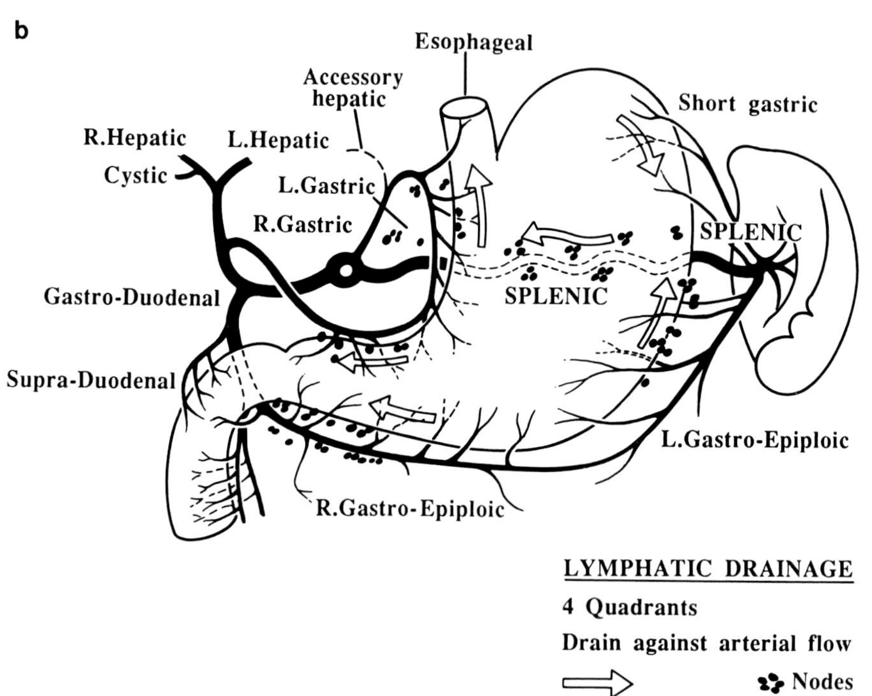

COLOR PLATE 2 Structure of stomach, showing (**a**) muscle layers and (**b**) blood supply and arterial drainage. [Source: Gaudin, A. J., and Jones, K. C. (1989). "Human Anatomy and Physiology." Harcourt Brace Jovanovich, San Diego. p. 451. Reproduced with permission.]

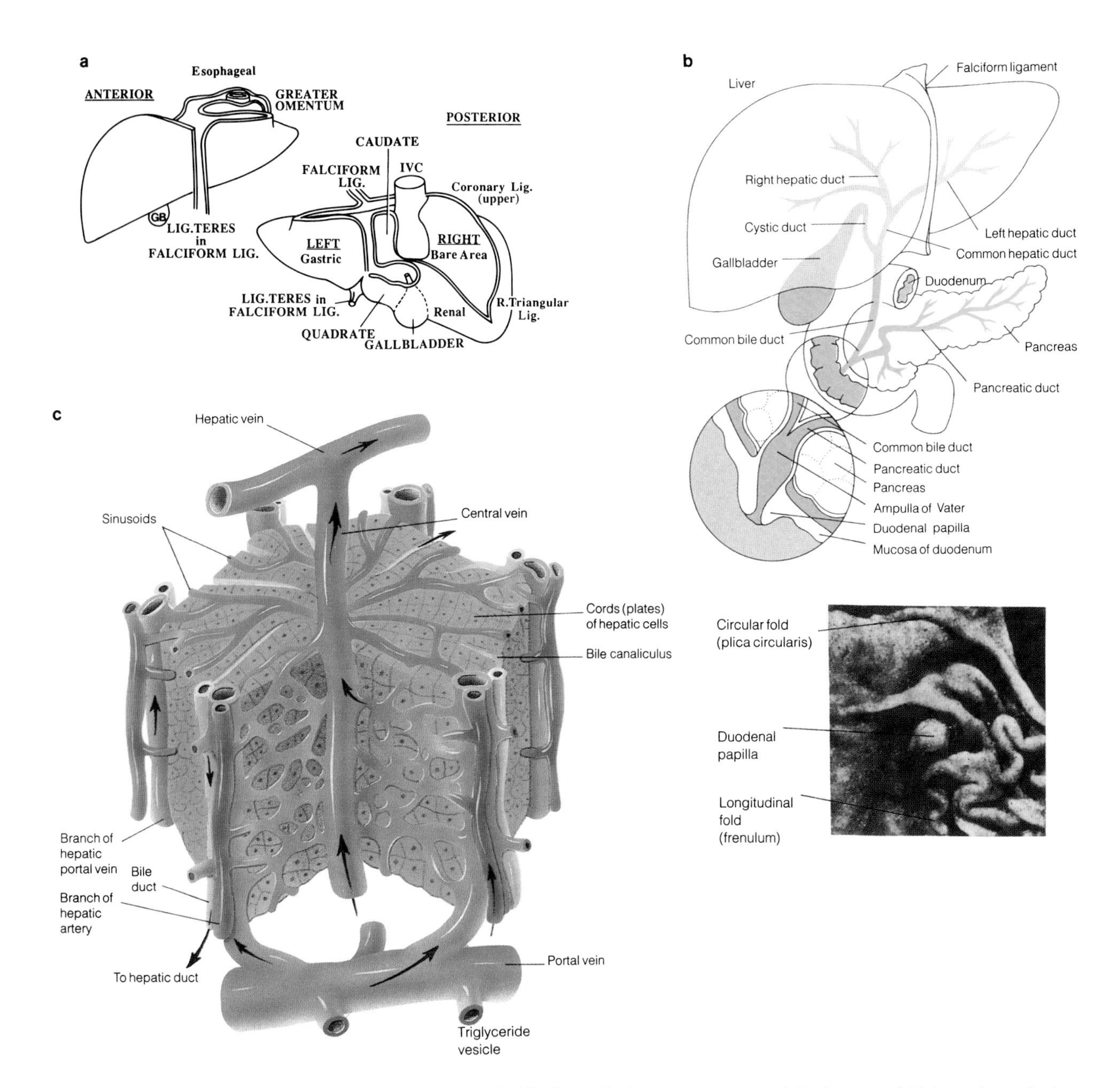

COLOR PLATE 3 Liver: (**a**) parietal and visceral surfaces, (**b**) bile duct and relations to pancreas and duodenum, and (**c**) internal organization of liver lobule. [Source: Gaudin, A. J., and Jones, K. C. (1989). "Human Anatomy and Physiology." Harcourt Brace Jovanovich, San Diego. p. 458, 469. Reproduced with permission.]

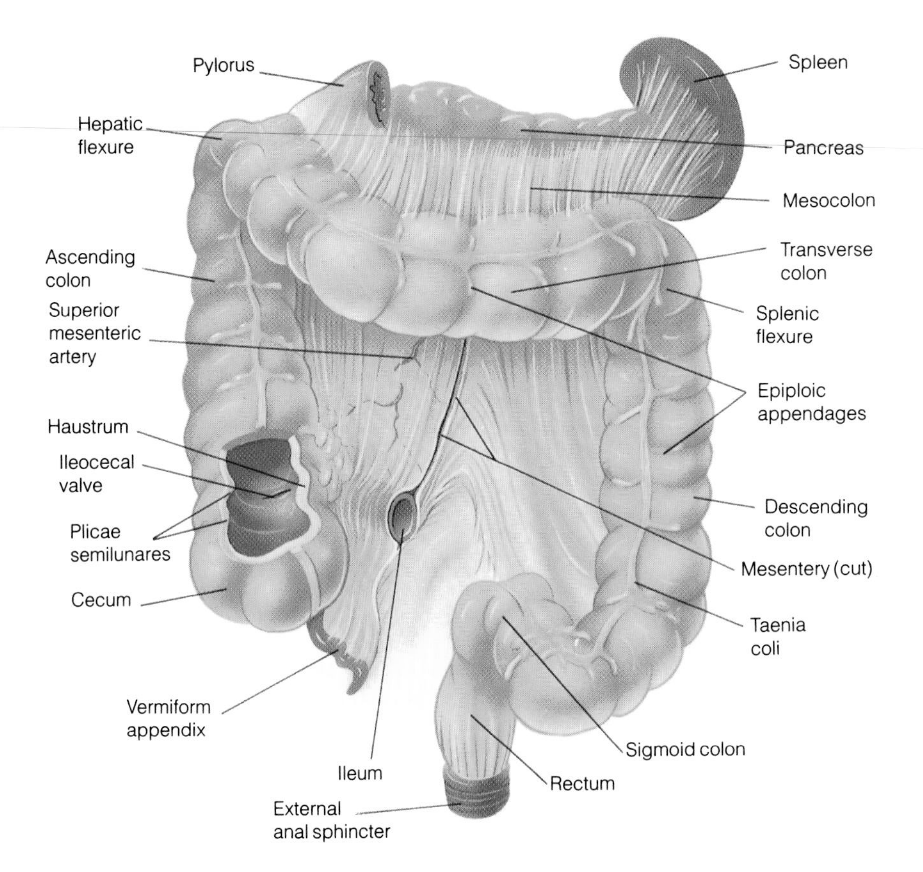

COLOR PLATE 4 Cecum and colon in relation to pancreas, spleen, and ileum. [Source: Gaudin, A. J., and Jones, K. C. (1989). "Human Anatomy and Physiology." Harcourt Brace Jovanovich, San Diego. p. 464. Reproduced with permission.]

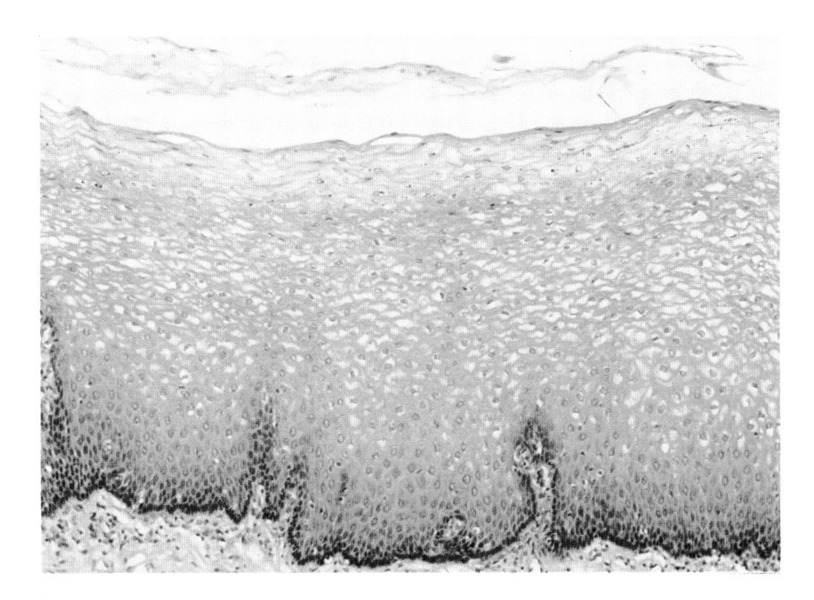

COLOR PLATE 5 Esophageal mucosa. The esophageal lining is shown with the luminal surface at the top. The stratified squamous epithelium has many scale-like layers of epithelium. The dark blue nuclei become dense near the base where cells are generated and then move upward toward the surface where they slough off. Similar epithelium lines the mouth and pharynx. [Photomicrograph courtesy of Dr. J. Kelly.]

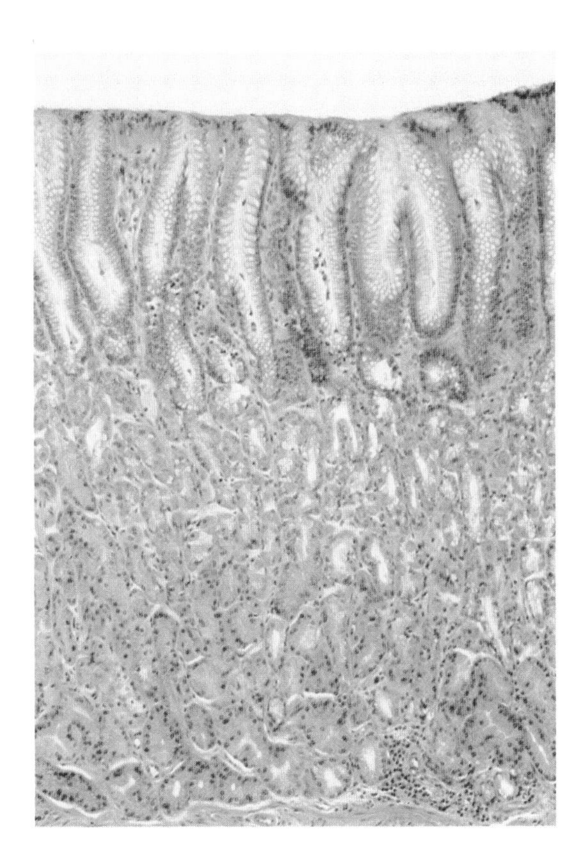

COLOR PLATE 6 Gastric mucosa with two layers. The upper third is composed of tubules lined by mucus-secreting columnar epithelium. The mucus appears paler and slightly foamy compared to the dark nucleus at the base of each cell. The lower two-thirds consists of glands which secrete acid and the enzyme, pepsinogen. The cells are arranged in circular configurations. As in Color Plate 5, the open space at the top represents the luminal cavity. [Photomicrograph courtesy of Dr. J. Kelly.]

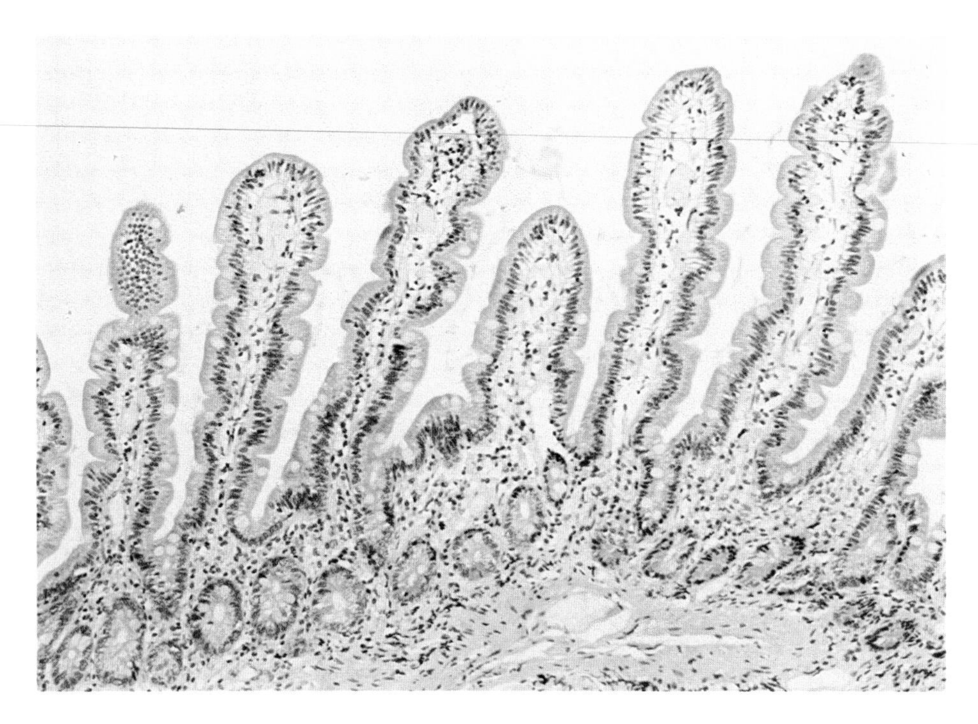

COLOR PLATE 7 The small intestine is lined by columnar cells. Multiple fingerlike villi protrude into the lumen and increase the surface area available for absorption. [Photomicrograph courtesy of Dr. J. Kelly.]

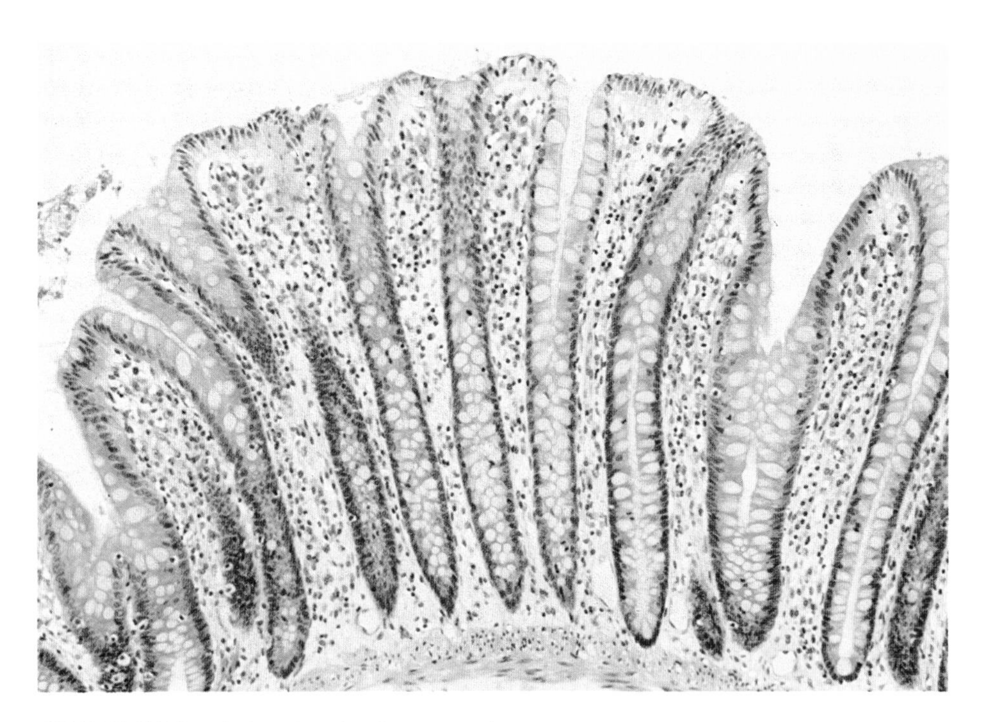

COLOR PLATE 8 The large bowel (colon) mucosa is composed of tubular glands lined by absorptive cells or mucus-secreting goblet cells. No villi are present.

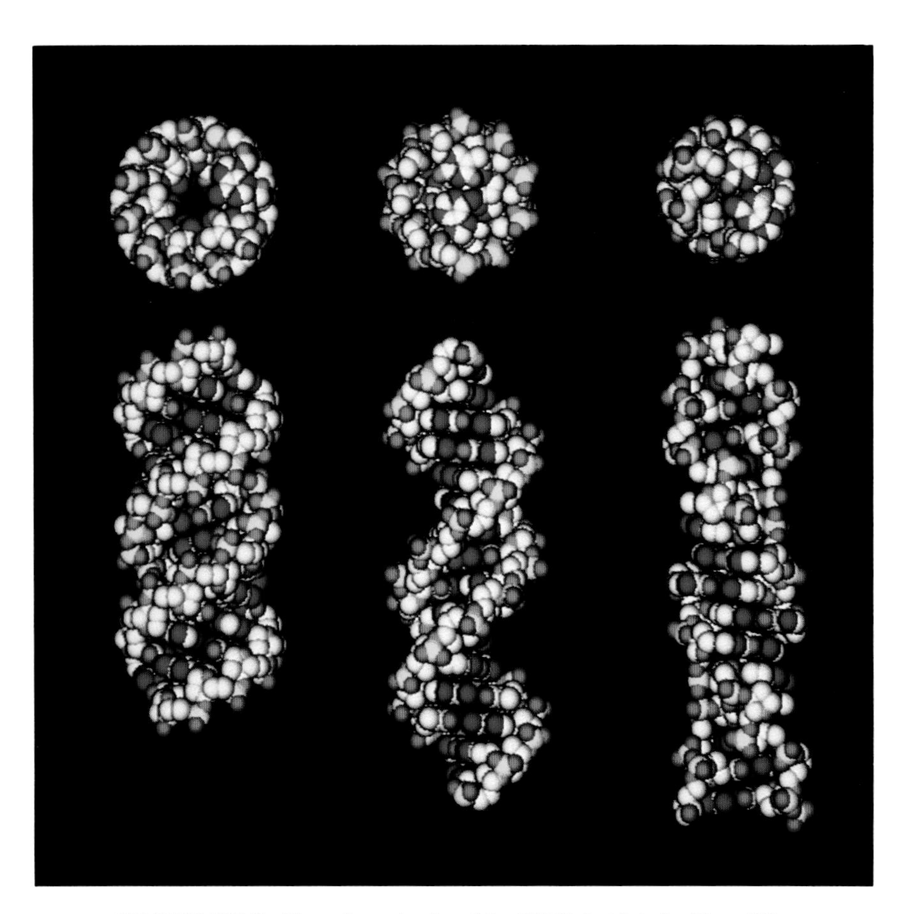

COLOR PLATE 9 Three-dimensional models of DNA double helix. Three different forms (A-, B-, and Z-forms) are shown. Their characteristics are summarized in Table I.

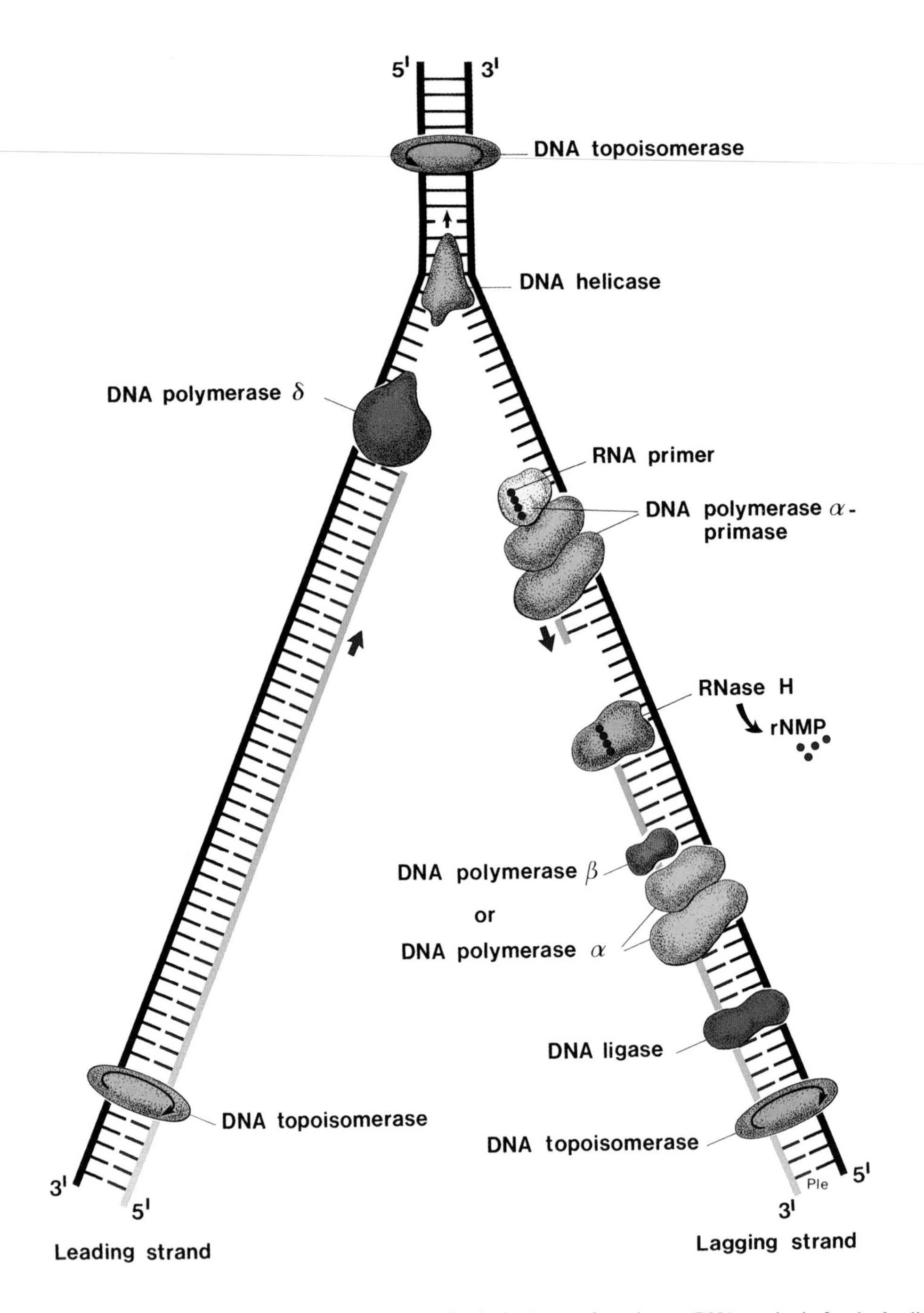

COLOR PLATE 10 DNA replication fork. The picture represents a classical scheme of continuous DNA synthesis for the leading strand (*left*) and discontinuous DNA synthesis for the lagging strand (*right*) at the replication fork. The arrows indicate the direction of DNA synthesis. [From S. Spadari, A. Montecucco, G. Pedrali-Noy, G. Ciarrocci, F. Focher, and U. Hübscher. (1989). *Mutat. Res.* **219**, 147–156.]

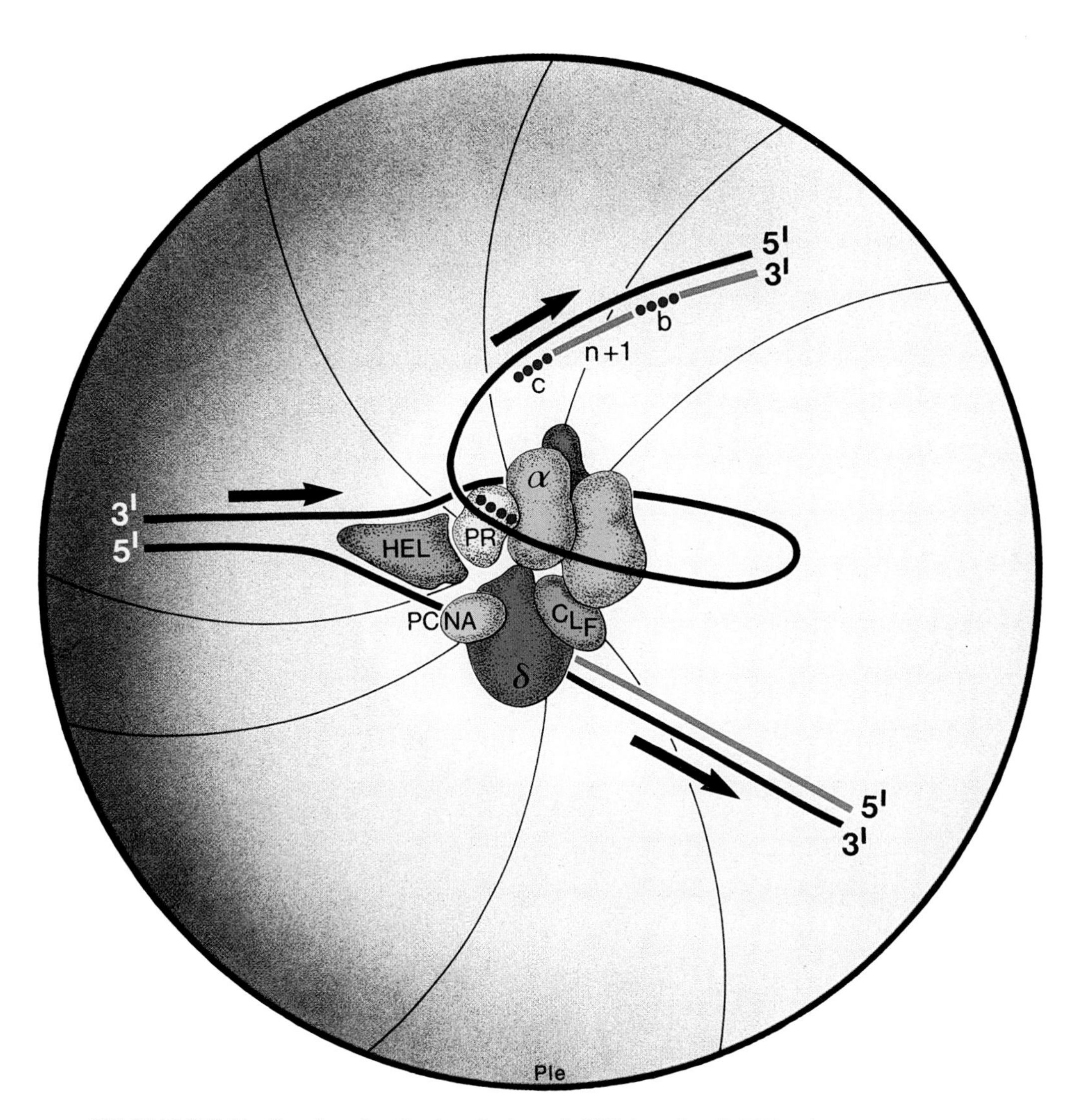

COLOR PLATE 11 Coordinated replication of eukaryotic DNA by a dimeric DNA polymerase. The model is based on an original suggestion by Bruce Alberts and co-workers for coordinated replication of bacteriophage T4 DNA. For explanations see text. α, DNA polymerase α with auxiliary proteins; δ, DNA polymerase δ; PR, DNA primase; HEL, DNA helicase; PCNA, proliferating cell nuclear antigen; CLF, clamp factor; ●●●●, RNA primers. Arrows indicate movement of DNA. [From S. Spadari, A. Montecucco, G. Pedrali-Noy, G. Ciarrocci, F. Focher, and U. Hübscher. (1989). *Mutat. Res.* **219**, 147–156.]

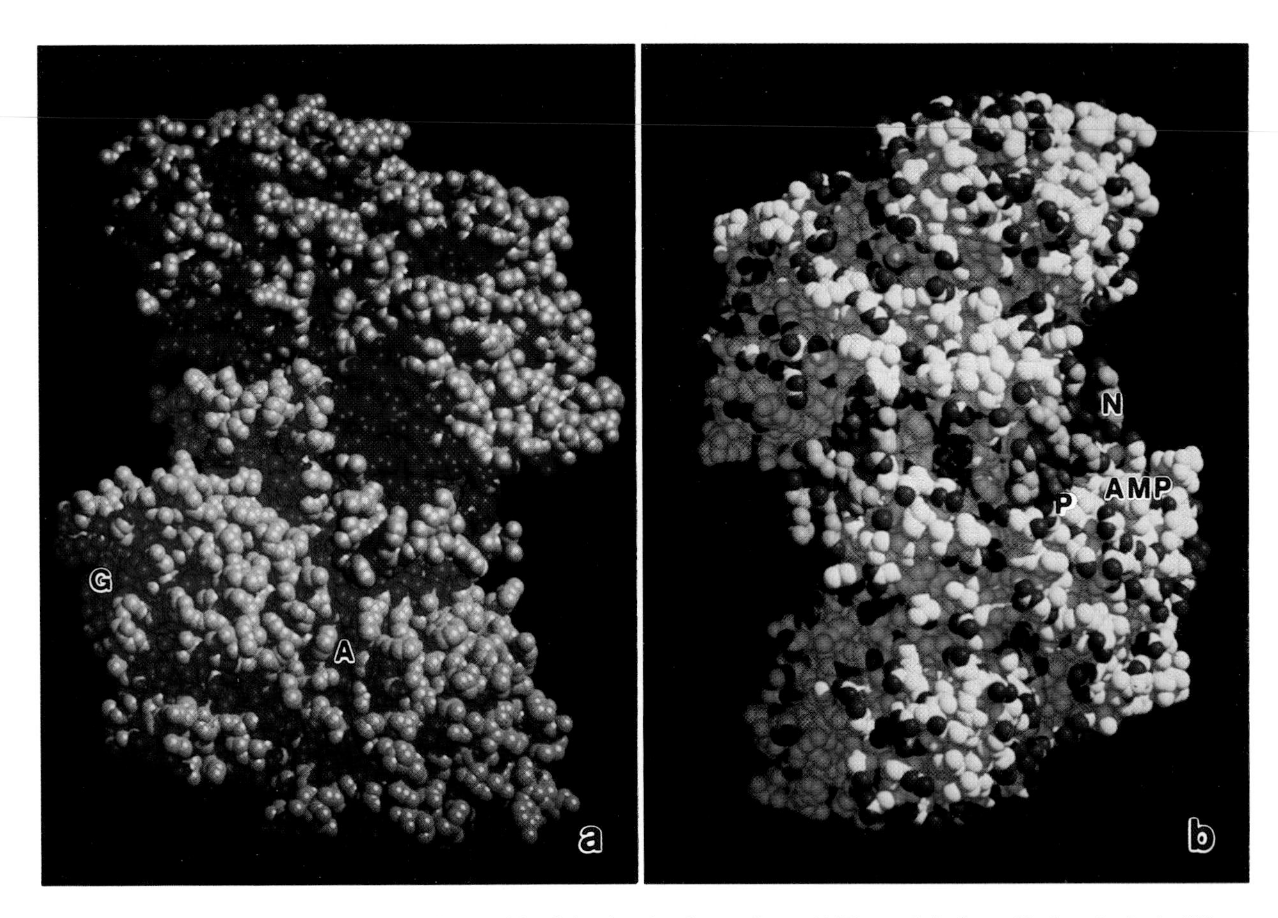

COLOR PLATE 12 Computer-drawn space-filling models of the phosphorylase *a* dimer. (a) The catalytic face with the subunits in different shades and A indicating entrance to active site, G the glycogen-binding site. (b) The catalytic face with N indicating the amino-terminal peptide, P the phosphate of serine 14, and AMP the binding site for AMP. [Drawings are courtesy of Dr. R. Read and Dr. A Muir, Department of Biochemistry, University of Alberta.]

TABLE I Ergonomic Societies and Recent Memberships

Country/region	Founding date	Recent membership Date	Recent membership Number	Population (million)	Members (per million)
Australia	1967	1988	665	15.8	42
Austria	1976	1986	40	7.5	5
Belgium	1986	1988	92	9.9	9
Brazil	1983	1988	244	143.3	2
Canada	1968	1988	295	25.6	12
Columbia	1987			30.7	
France	1963	1987	531	55.2	10
Germany (FR)	1958	1987	550	60.7	9
Hungary	1987	1988	90	4.6	20
India	1987	1988	35	768.0	—
Indonesia	1988	1988	120	176.8	1
Israel				4.3	
Italy	1965	1986	170	57.2	3
Japan	1964	1989	1,558	121.4	13
Korea (South)	1982	1988	250	43.9	6
New Zealand	1986	1988	86	3.3	26
Netherlands	1963	1986	576	14.5	40
Nordic (Denmark, Finland, Norway, Sweden)	1969	1986	550	22.6	24
Poland	1977	1988	416	37.5	11
Singapore	1988	1989	30	2.6	12
South Africa	1984	1988	107	34.3	3
Southeast Asia	1985				
United Kingdom	1949	1986	644	56.5	11
United States	1957	1988	4,710	242.2	19
Yugoslavia	1973	1989	50	23.2	2

From *Ergonomics* **32,** 8 (1989), p. 1030 and *Ergonomics* **33,** 3 (1990) p. 95.

title. Most are members of the International Ergonomics Association (founded in 1959), whose Secretary General is presently Dr. Hal W. Hendrick (Dean, College of Systems Sciences, University of Denver, Denver, Colorado 80208). In all, about 12,000 ergonomists are estimated to live mostly in North America, Europe, Japan, and Australia/New Zealand, with fairly large contingents also in Southeast Asia and South America. Ergonomists, often called "psychologists" with some modifying term, are found in Eastern Europe, the Soviet Union, and China.

Bibliography

Edholm, O. G., and Murrell, K. H. F. (1974). "The Ergonomics Research Society. A History 1949 to 1970." The Council of the Ergonomics Research Society (no location given).

Gould, S. J. (1981). "The Mismeasure of Man." Norton, New York.

Kroemer, K. H. E., Snook, S. H., Meadows, S. K., and Deutsch, S. (eds.) (1988). "Ergonomic Models of Anthropometry, Human Biomechanics, and Operator-Equipment Interfaces." National Academy Press, Washington, D.C.

Muchinsky, P. M. (1987). "Psychology Applied to Work," 2nd ed. Dorsey, Chicago.

National Research Council (ed.) (1983). "Research Needs for Human Factors." National Academic Press, Washington, D.C.

Books (*Textbook, **Engineering Design Guide)

*Astrand, P. O., and Rodahl, K. (1986). "Textbook of Work Physiology" 3rd ed. McGraw-Hill, New York.

**Bailey, R. W. (1982). "Human Performance Engineering: A Guide for Systems Designers" Englewood Cliffs, Prentice Hall, New Jersey.

**Boff, K. R., Kaufman, L., and Thomas, J. P. (eds.) (1986). ''Handbook of Perception and Human Performance'' Wiley, New York.
*Chaffin, D. B., and Andersson, G. B. J. (1984). ''Occupational Biomechanics'' Wiley, New York.
Chapanis, A. (ed.) (1975). ''Ethnic Variables in Human Factors Engineering'' The John Hopkins University Press, Baltimore, Maryland.
**Eastman-Kodak Company, (Vol. 1, 1983; Vol. 2, 1986). ''Ergonomic Design for People at Work'' Van Nostrand Reinhold, New York.
*Grandjean, E. (1988). ''Fitting the Task to the Man'' 5th ed., International Publications Service, New York, Taylor and Francis, London.
Helander, M. (ed.) (1981). ''Human Factors/Ergonomics for Building and Construction'' Wiley, New York.
Helander, M. (ed.) (1988). ''Handbook of Human-Computer Interaction'' Elsevier, Amsterdam and New York.
*Huchingson, R. D. (1981). ''New Horizons for Human Factors in Design'' McGraw-Hill, New York.
*Konz, S. (1990). ''Work Design: Industrial Ergonomics'' 3rd ed. Grid, Columbus.
*Kroemer, K. H. E., Kroemer, H. J., and Kroemer-Elbert, K. E. (1990) ''Engineering Physiology: Bases of Ergonomics/Human Factors'' 2nd ed. Van Nostrand Reinhold, New York.
**NASA/Webb (1978). ''Anthropometric Source Book'' (3 vols.). NASA Reference Publication 1024. Houston. NASA (NTIS, Springfield, VA 22161, order #79 11 734).
*Sanders, M. S. and McCormick, E. J. (1987). ''Human Factors in Engineering and Design'' 6th ed. McGraw-Hill, New York.
Oborne, D. J., and Gruneberg, M. M. (eds.) (1983). ''The Physical Environment at Work'' Wiley, New York.
Plog, B. A. (ed.) (1987). ''Fundamentals of Industrial Hygiene'' National Safety Council, Chicago.
**Salvendy, G. (ed.) (1987). ''Handbook of Human Factors'' Wiley, New York.

Standards

ANSI/HFS 100 (1988). American National Standard for Human Factors Engineering of Visul Display Terminal Workstations. Human Factors Society, Santa Monica, California.
ANSI/ASHRAE 55 (1981). Thermal Environment Conditions for Human Occupancy. American Society of Heating, Refrigerating, and Air-Conditioning Engineers, Atlanta, GA.
Department of Defense, Military Standard: Human Factors Engineering Design for Army Material, MIl-HDBK-759A. U.S. Government Printing Office, Washington, D.C. (Available from Naval Publications and Forms Center, 5801 Tabor Avenue, Philadelphia, PA 19120).
Department of Defense, Military Standard: Human Engineering Design Criteria for Military Systems, Equipment and Facilities, MIL-STD-1472C. U.S. Government Printing Office, Washington, D.C.*
Department of Defense, Military Handbook: Anthropometry of U.S. Military Personnel (metric). DOD-HDBK-743. U.S. Government Printing Office, Washington, D.C. (Available from Naval Publications and Forms Center, 5801 Tabor Avenue, Philadelphia, PA 19120).

Journals

Applied Ergonomics Butterworth Scientific, Guildford, England.
Biomechanics John Wiley, New York
Ergonomics Taylor and Francis, London.
Human Factors Human Factors Society, Santa Monica, California.
Journal of the American Industrial Hygiene Association AIHA, Akron, Ohio.
Spine Lippincott, Philadelphia.

Estrogen and Osteoporosis

SIJMEN DUURSMA, *University Hospital, Utrecht, The Netherlands*

Glossary

Estrogen Normal female hormone produced by the ovary during the whole period of the menstrual cycle

Progestogen Hormone produced by the ovary during the second half of the menstrual cycle

Osteoblasts Cells that produce new bone

Osteoclasts Cells that resorb bone

Osteoporosis Porotic bone, as a result of a negative balance between the activity of osteoblasts and osteoclasts; most common in postmenopausal women but also common in men with increasing age and in pathologic conditions; increases risk for fractures

ALBRIGHT NOTED a causal relationship between the postmenopausal state of women and the occurrence of vertebral and femoral fractures, which are observed in increasing numbers with increasing age. The skeleton consists of long and irregular shaped bones. The long bones, for example, a femur, have more compact (cortical) bone, which forms the solid outer shell of bones, giving them great strength. The irregular shaped bones, for example, the vertebrae, have more trabecular (or so called spongy or cancellous) bone, which is the interior meshwork of bones containing bone marrow. In healthy adults about 5% of bone is replaced yearly. This process starts with bone resorption by osteoclasts, cells that originate from stem cells that also produce the red and white blood cells. The bone defect is refilled with new bone, produced by osteoblasts, that originate from stem cells that also produce cells for the formation of cartilage, muscles, and fat tissue. Osteoblasts produce the organic matrix of bone tissue with collagen fibers, long protein chains. The local situation in the matrix permits calcium and phosphate ions to form crystals, hydroxyapatite, the major component of bone mineral. Osteoblasts also produce the enzyme alkaline phosphatase that plays a role in the mineralization process. The serum activity of this enzyme gives information about the osteoblastic bone forming activity. The amino-acid hydroxyproline is a degradation product of collagen and is produced by the osteoclastic bone resorption. The urinary excretion of this amino acid provides information about the process of bone resorption. To get information about the amount of bone in a person, techniques have been developed to measure bone mineral content in parts of the skeleton (e.g., the forearm or the lumbar spine, or in the total skeleton). In normal persons nearly all bone matrix has been mineralized and bone mineral content may be used as a measure for the amount of bone. However, in mineralization defects bone mineral content does not reflect the real bone mass. With these techniques the effect of therapeutical agents on bone cells and bone mass can be measured.

I. Estrogen and Bone

The development and activity of bone resorbing and producing cells is regulated by a complex interaction of systemic factors, supplied by the circulation, and locally produced factors. In addition, osteoclasts and osteoblasts produce coupling factors that adjust the activity of the two types of cells with opposite functions. The ratio between the number of bone cells and the amount of bone tissue is much higher in trabecular than in cortical bone. Changes in the systemic factors (e.g., estrogen) that regulate bone cell metabolism, have different effects on trabecular and cortical bone. A negative balance between the activity of osteoblasts and osteoclasts results in bone loss. Postmenopausal women yearly lose about 2–4% of their trabecular bone mass during the early postmenopausal years. About 10 yr after the menopause this bone loss results in vertebral fractures in about 25% of these women. Cortical bone loss needs some more time, the incidence of hip fractures increases sharply after the age of 65–70 yr.

The menopause is the moment of the last menstruation, a consequence of the decreasing estrogen production by the ovaries. Most common is the spontaneous loss of estrogen production in women of about 50 yr of age, but surgical oophorectomy, anorexia nervosa, and intensive physical training also lead to shortness of estrogen. The consequence of the relationship between the postmenopausal state and bone loss was estrogen replacement therapy. In postmenopausal women, treated with estrogen for 1–20 yr, the progress of osteoporosis was arrested, as judged by measurement of total height. In another study, oophorectomized women were treated with an estrogen preparation. They showed no decrease in height and bone mineral content, measured on the hand and forearm. It was concluded that estrogen prevents postmenopausal bone loss in both the axial and the peripheral skeleton. The effect of estrogen is dose-dependent. In a 25-yr study in women with severe postmenopausal osteoporosis, those receiving 0.625 mg conjugated estrogens had a reduction in vertebral fracture rate from 60 to 25 per 1,000 patient-yr, but those who received 1.25 mg conjugated estrogens had a fracture rate of only 3 per 1,000 patient-yr. A dose of 0.3 mg conjugated estrogens was not sufficient to prevent postmenopausal bone loss, but with the addition of 1,000 mg calcium no decrease of bone mineral content was observed.

The most common prescribed oral estrogen preparations are conjugated estrogens, a mixture of some steroids extracted from the urine of mares in gestation; estradiolvalerate, a synthetic estrogen with a structure nearly equal to that of human estrogen; and ethinylestradiol, another synthetic preparation. There is also a fair amount of use of oral micronized estradiol and a growing use of transdermal estradiol. The lowest effective dose for prevention of postmenopausal bone loss is 0.625 mg for conjugated estrogens, 2 mg for estradiolvalerate, and 25 μg for ethinylestradiol. The individual dose needed depends on the body weight. Estrogen is also produced by peripheral conversion of androstenedione in fat tissue; the more fat tissue a woman has the more estrogen she can produce. Measuring the serum concentration of estradiol can be a guideline to adjust the individual therapeutic dose. In a study with the use of estrogen administration via the skin with special plaster and oral conjugated estrogens, a relationship was observed between serum estradiol concentrations <120 pg/ml and the urinary calcium–creatinine and the hydroxyproline–creatinine ratios. These ratios provide information about bone resorption. Serum concentrations >120 pg/ml provided no additional bone-sparing effect.

The administration of estrogen causes growth of the endometrium and if continued over a long time, it enhances the risk for endometrial carcinoma. The cyclic combination with a progestogen preparation prevents the development of endometrial carcinoma by periodical bleeding. But does the addition of progestogen influence the effect of estrogen on bone metabolism? Progestogen causes a decline in the urinary calcium–creatinine ratio, which indicates a reduction in bone resorption. In a clinical study, a comparison was made among the effects of estradiolvalerate, a combination of estradiolvalerate with the progestogen preparation norgestrel, norgestrel alone, and another progestogen preparation, norethindrone. Norethindrone prevented bone loss as effectively as estradiolvalerate or estradiolvalerate in combination with norgestrel; however, norgestrel alone was ineffective in the prevention of bone loss. In other clinical studies, the positive effect of the combination of estrogen with cyclic progestogen prescription has been confirmed. Most postmenopausal women are not charmed by a preventive regime with continuation of premenopausal cyclic bleedings and menstrual discomfort. In a study of 55 women, 70 yr old, receiving estrogen–progestogen replacement therapy, 44% left the 1-yr study before its conclusion, mostly due to menstrual problems. However, in early postmenopausal women the daily combination of estradiolvalerate and cyproteronacetate prevented bone loss and caused only short periods with spot bleeding during the first 6 mo; later on no further bleeding occurred over the remaining 2-yr study period. No long-term studies are available about continuous combined therapeutic regimes.

II. How Does Estrogen Prevent Postmenopausal Bone Loss?

Because, until recently, it was not possible to demonstrate estrogen receptors in bone cells, an indirect effect of estrogen, mediated by a second factor, was assumed. It has been suggested that after menopause bone cells are more sensitive to parathyroid hormone, but no evidence supports this idea. Another suggestion was an increase in the synthesis of 1,25-dihydroxycholecalciferol after estrogen replacement therapy. 1,25-dihydroxycholecalciferol is the active metabolite of vitamin D, produced by the enzyme 1α-hydroxylase in the kidneys. It was found that a single injection of estradiol increased the 1α-hydroxylase activity in bird kidney homogenates. A rise in serum parathyroid hormone concentration, with secondary stimulation of the enzyme 1α-hydroxylase and the production of 1,25-dihydroxycholecalciferol, was also reported after estrogen treatment in patients with postmenopausal osteoporosis. In other studies, these observations could not be confirmed. It has generally been accepted that serum concentrations of calcium and phosphate, the activity of alkaline phosphatase, and the urinary excretion of calcium and hydroxyproline, are significantly higher in postmenopausal women compared with premenopausal women. These changes can not be explained by the supposed changes in the plasma concentrations of parathyroid hormone or 1,25-dihydroxycholecalciferol after the menopause, and seem to be secondary to the modified bone metabolism, rather than their cause.

III. Estrogen and Calcitonin

Calcitonin is a polypeptide hormone produced by the C-cells in the thyroid gland. The production and secretion is stimulated by an increase in serum calcium concentration. Calcitonin caused an increase in the number of osteoblasts and in bone length of mouse radius rudiments in a culture system. It also stimulated the multiplication of cultured fetal chicken osteoblasts. Whether or not this stimulation of mitogenesis in osteoblast precursors is relevant to human physiology is uncertain. Bone resorption, induced by parathyroid hormone in neonatal mouse vertebral bones, was inhibited by calcitonin. Calcitonin is the only hormone with a direct effect on osteoclasts that possess calcitonin receptors. The hormone inhibited isolated osteoclasts by eroding the surface of bone slices in culture.

Estradiol stimulates the secretion of calcitonin in cultured rat thyroid C-cells, but at concentrations higher than the serum estrogen level in postmenopausal women, using the normally advised dose to prevent postmenopausal bone loss. Administration of estrogen in postmenopausal women resulted in a significant rise in plasma calcitonin concentration at noon, but not at 9:00 A.M. In other clinical studies no changes in plasma calcitonin concentration could be observed during estrogen replacement therapy. Physiological levels of estradiol in postmenopausal women showed no relationship to serum calcitonin levels. However, basal levels of serum calcitonin were lower in postmenopausal women, when compared with premenopausal women and were strongly correlated with circulating estrone levels. From this study it is concluded that the calcitonin secretion capacity appears to be modulated by circulating estrone levels. Like high doses of estradiol, low doses of progesteron increased the secretion of calcitonin in rat thyroid C-cells. In estrogen–progestogen replacement regimes, usually with low doses of estrogen, the rise in serum calcitonin concentration may be, at least partly, the result of the progestogen preparation. An intermediate function for calcitonin between estrogen and bone metabolism remains to be possible, but it might be more plausible for progesteron. Further investigations about the interaction between calcitonin, estrogen, and progesteron are needed to clarify their role on bone metabolism.

IV. Estrogen and Prostaglandins

Nearly all mammalian cells have the capacity to produce prostaglandins by cleaving arachidonic acid from membrane phospholipid. The prostaglandins are rapidly metabolized by enzymes found in nearly all tissues. Their importance lies in their function as locally active substances. It has been suggested that estrogen might affect bone metabolism indirectly by inhibiting endogenous prostaglandin E_2 synthesis. Prostaglandin E_2 causes a stimulation of osteoblastic collagen synthesis, but is also a potent stimulator of bone resorption and calcium release from bone. Prostaglandin E_2 is reported to increase the number and activity of osteoclasts in bone cultures. Osteoblasts produce significant amounts of prostaglandin E_2, which may have a function in the coupling between the activity of osteoblasts and osteoclasts. Using advanced techniques, however, no direct effect of prostaglandin E_2 on bone resorption could be observed.

The effect of estrogen on prostaglandin E_2 synthesis was also studied in young rats. Oophorectomy prior to their sacrifice resulted in a twofold increase in prostaglandin E_2 release from parietal bones, and *in vivo* replacement therapy with estradiol inhibited the increased prostaglandin E_2 release. These results support the possibility that estrogen influences prostaglandin E_2 production, which can explain the development of osteoporosis and the preventive effect of estrogen replacement theory in postmenopausal women.

V. Estrogen and Insulinlike Growth Factors

Estrogen inhibits the production of insulinlike growth factor-I (IGF-I) in the liver, which results in an increase in plasma growth hormone concentration because of the negative feedback between these hormones. In a study with estrogen substitution over a 3-wk period, these effects on IGF-I and growth hormone could be confirmed in postmenopausal women. Further investigations on IGF-I, growth hormone, and parameters of bone metabolism supported the idea that IGF-I and growth hormone have an essential function in the regulation of bone metabolism and in the interaction between estrogen and bone cells. [*See* TISSUE REPAIR AND GROWTH FACTORS.]

For many years, it has been assumed that growth hormone needs the production of the intermediate substance IGF-I for its effects on tissues; however, during the last few years growth hormone has been demonstrated also to have a direct effect on some types of cells, including osteoblasts in cultures of fetal chicken calvaria. The effect of growth hormone might be caused by the local production of IGF-I, because, in physiological concentrations, IGF-I stimulates the synthesis of DNA and collagen in fetal rat calvaria. However, other studies demonstrated a direct effect of growth hormone on osteoblasts, without the intermediate production of IGF-I. Very fast effects of growth hormone on the process of osteoblast proliferation makes a direct effect plausible, because the IGF-I transcription needs much more time.

The decrease in IGF-I production and the rise in the serum growth hormone concentration in postmenopausal women using estrogen replacement therapy for 3 wk, was confirmed in other trials. Continuous estrogen–progestogen replacement therapy during a 2-yr period caused a stabile decrease in serum IGF-I by about 25% and a stabile increase in growth hormone concentration by about 200%. The concentrations of IGF-II did not change. This was expected because it is supposed that IGF-II plays a role in cell metabolism during fetal life and IGF-I in adults. The effect of estrogen replacement therapy on the serum concentrations of IGF-I and growth hormone in postmenopausal women is firmly established.

VI. Estrogen Receptors in Bone Cells

Estrogen receptors were not detected in several types of bones from humans and some animals during the 1970s. However, with new sensitive techniques, recently receptors could be demonstrated in human osteoblastlike osteosarcoma cells and in normal human osteoblastlike cells. On the other hand, an immunocytochemical study concluded that the number of estrogen receptors is low in human osteoblasts, and the function of these receptors in bone metabolism remains arguable. Recently, an increase in the mRNA for collagen could be demonstrated in human osteoblasts after addition of estradiol. This effect was mediated by the production of IGF-I in the osteoblasts, IGF-I antibodies prevented the mRNA production. The existence of estrogen receptors in bone cells has now been demonstrated unequivocally, but their function in osteoporosis and estrogen replacement therapy remains uncertain.

VII. Direct and Indirect Effect of Estrogen on Bone

In cell culture systems estrogen has a direct stimulatory effect on osteoblasts. However, estrogen replacement therapy in postmenopausal women causes a decrease in both bone formation and resorption. In cultured osteoblasts the effect of estrogen is mediated by an increased production of IGF-I, but in postmenopausal women a decrease in serum IGF-I concentration was observed after estrogen substitution. About 80% of the circulating IGF-I is produced by the liver, and the remaining 20% by peripheral cells like osteoblasts. Estrogen inhibits the IGF-I production in the liver and stimulates the production of this growth factor in osteoblasts. The inhibition in the liver cells is of more importance than the stimulating effect on bone cells, resulting in a fall in IGF-I available for osteoblasts and osteoclasts. The feedback between IGF-I and growth hormone causes a rise in plasma growth hormone concentration, which stimulates bone formation by the osteoblasts.

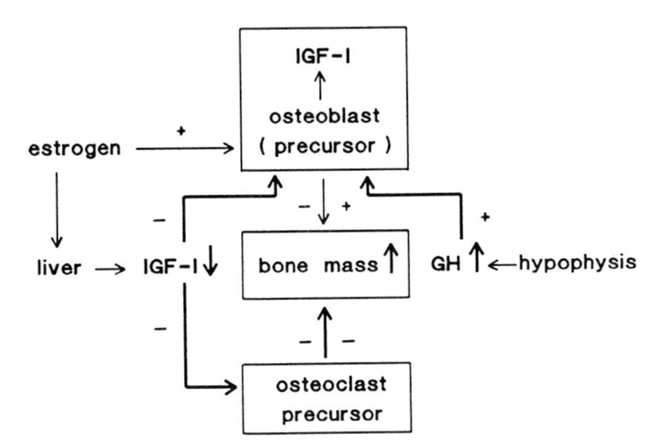

FIGURE 1 Schematic representation of the changes in insulin-like growth factor-I (IGF-I) and growth hormone (GH) after estrogen replacement therapy in postmenopausal women. Estrogen substitution reduces the IGF-I production in the liver and causes a decrease in plasma IGF-I concentration. IGF-I stimulates osteoblast and osteoclast precursors. A fall in plasma IGF-I concentration leads to inhibition of bone formation and resorption. The decrease in plasma IGF-I concentration causes an increase in plasma GH level. GH and estrogen induce local production of IGF-I in osteoblasts. A direct effect of GH on osteoblasts, independent of IGF-I, has also been demonstrated. After estrogen substitution the decrease in the systemic IGF-I concentration seems to be of more importance than the increase in the locally produced IGF-I by the osteoblasts. The net result is a decrease in bone formation and resorption with a change in their balance in a positive direction. However, proliferation and differentiation of bone cells depend also on other local and possibly systemic factors and factors that regulate the coupling between osteoblasts and osteoclasts. Interaction between these factors regulates the complex process of bone formation, bone resorption and the maintenance of bone mass during adult life.

In this model (Fig. 1) osteoporosis has been reduced to a disturbance between bone formation and resorption, as a result of a change in the balance between growth factors that regulate proliferation and differentiation of osteoblasts and osteoclasts. However, besides estrogen, calcitonin, prostaglandin E_2, IGF-1 and GH a number of locally produced factors participate in the complex regulation of cell proliferation and differentiation. The role of interleukin 1, produced by monocytes from bone marrow is unclear, although its production is inhibited by estrogen. Another growth factor is transforming growth factor β, that influences a number of cells, including bone cells. Knowledge about growth factors is rapidly increasing and in the near future new information will clarify the interaction between the effects of systemic and locally produced growth factors. [*See* TRANSFORMING GROWTH FACTOR-β.]

VIII. Summary

Osteoporosis is a common disturbance in bone metabolism resulting in a decrease in the amount of bone and an increase in fracture incidence, especially in postmenopausal women. Estrogen replacement therapy, mostly in combination with the prescription of progestogen, prevents postmenopausal bone loss. The mechanism by which estrogen influences bone metabolism is still under discussion. A direct effect of estrogen on bone cells is possible, but changes in mediator substances can also explain the effects. Calcitonin and prostaglandin E_2 are possible candidates for the role of mediator. Changes in growth factors, especially IGF-I in relationship to growth hormone, can also explain the alterations in bone metabolism after the menopause and after estrogen substitution. A combined effect of the mentioned and other factors is plausible. Proliferation and differentiation of bone cells is a complex process, with the interaction of systemic growth factors and locally produced factors. The details of this process are still unclear.

Bibliography

Duursma, S. A., Raymakers, J. A., Van Beresteyn, E. C. H., and Schaafsma, G. (1987). "Clinical aspects of osteoporosis." *World Rev. Nutr. Diet.* **50,** 92.

Duursma, S. A., Slootweg, M. C., and Bijlsma, J. W. J. (1988). How do estrogens prevent postmenopausal osteoporosis? *In* "Crossroads in aging" (M. Bergener, M. Ermini, and H. B. Stähelin, eds.). Academic Press, London.

Heersche, J. N. M. (1989). Bone cells and bone turnover—the basis for pathogenesis. *In* "Metabolic Bone Disease: Cellular and Tissue Mechanisms" (C. S. Tam, J. N. M. Heersche, and T. M. Murray, eds.). *CRC Press,* pp 1–19.

Jee, W. S. S. (1988). The skeletal tissues, in "Histology: Cell and Tissue Biology" (L. Weiss, ed.). Elsevier Biomedical, New York, Amsterdam.

Lobo, R. A., Whitehead, M. I., and Wadler, G. I. (eds.) (1989). Consensus Development Conference on Progestogens. *Int. Proceed. J.* **1,** 1.

Martin, T. J., and Raisz, L. G. (eds.) (1987). "Clinical Endocrinology of Calcium Metabolism." Marcel Dekker, Inc., New York.

Mundy, G. R., and Roodman, G. D. (1987). Osteoclast ontogeny and function. *Bone Min. Res.* **5,** 209.

Parsons, J. A. (ed.) (1982). "Endocrinology of calcium metabolism." Raven Press, New York.

Raisz, L. G., and Martin, T. J. (1983). Prostaglandins in bone and mineral metabolism. *Bone Min. Res.* **2,** 286.

Riggs, B. L., and Melton III, L. J. (1988). "Osteoporosis, Etiology, Diagnosis and Management." Raven Press, New York.

Sakamoto, S., and Sakamoto, M. (1986). Bone collagenase, osteoblasts and cell-mediated bone resorption. *Bone Min. Res.* **4,** 49.

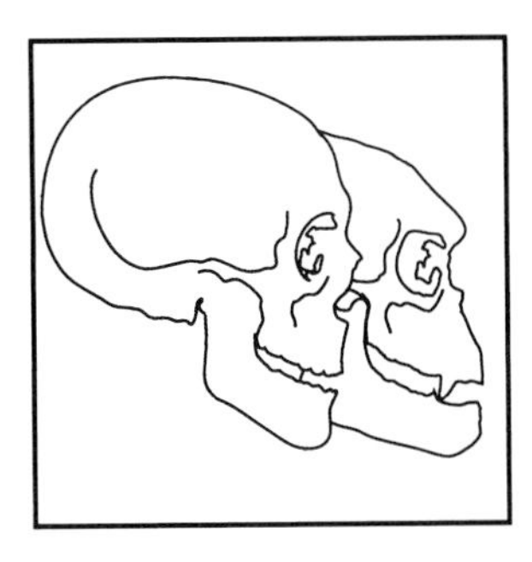

Evolution, Human

HENRY M. MCHENRY, *University of California, Davis*

Glossary

Australopithecus Extinct forms of humans who lived in Africa from about 5 to 1.3 million years ago. They were characterized by human-like bipedal posture, brain size about one third that of modern people, and relatively large cheek teeth. One species of this genus probably evolved into the genus *Homo*

Hominid Organisms belonging to the zoological family Hominidae which include both extinct and modern forms of humans such as *Australopithecus, Homo erectus,* and the Neanderthals

Hominoid Organisms belonging to the zoological superfamily Hominoidea including modern and fossil species of humans and apes

Molecular clock A method of computing the time of divergence between evolutionary lineages based on the genetic distances among living species

Pongid Organisms belonging to the zoological family Pongidae including modern great apes (chimpanzee, gorilla, and orangutan) and closely related fossil species

THE HUMAN EVOLUTIONARY lineage originated in Africa in the late Miocene, 8 to 5 million years ago, from an ancestor shared with the African great apes (chimpanzee and gorilla). Probably by 5 but certainly by 4 million years ago, our ancestors walked bipedally, had front teeth intermediate in shape between modern apes and people, and had a relative brain size about one-third that of *Homo sapiens*. By 2 million years, the teeth were basically human-like (except for the large size of the cheek teeth), and the relative brain size was about 50% larger than its predecessor. The species *Homo erectus* first appear 1.6 million years ago with a brain nearly twice the size of the first hominids. Not until about 1 million years ago did our family disperse out of Africa to colonize Asia and Europe, and not until about 300,000 years ago were brains expanded into the modern human range. Anatomically modern *H. sapiens* appears to have originated in Africa about 100,000 years ago. By about 35,000 years this form had occupied all the Old World including Australia. Although some people may have reached the Americas earlier, the first substantial population there dates to about 12,000 years ago.

This brief outline of current knowledge derives from the remarkably successful cooperation of molecular biologists, comparative morphologists, geologists, paleontologists, archaeologists, biological anthropologists, geneticists, ecologists, and individuals from many other fields. Such a sketch is, of course, tentative, pending further information and interpretation, but a great deal of hard evidence lies behind it.

I. Place in Nature

A. Taxonomy of *Homo sapiens*

Homo sapiens is one of about 180 living species within the order Primates, which is one of 18 mammalian orders. The order Primates contains prosimians (such as lemurs), monkeys, and apes. Our species is placed in the suborder Haplorhini along with tarsiers, monkeys, and apes. Within Haplorhini we are grouped with Old World monkeys, and apes in the infraorder Catarrhini, which in turn is divided into two superfamilies [i.e., Cercopithecoidea (Old World monkeys) and Hominoidea (apes and people)]. Traditionally the Hominoidea are divided into three families (i.e., Hylobatidae (gibbons and siamangs), Pongidae (orangutans, gorillas, and chimpanzees), and Hominidae (humans), but research since the 1960s has clearly demonstrated that this classification does not reflect the true phyletic relations. Several lines of evidence, especially molecular systematics, show that the African great apes (gorilla and chimpanzee) are more closely related to humans than to the Asian apes (gibbons, siamangs, and organutans). Some authors believe that the classification should reflect that fact by placing the African apes in the family Hominidae. By this scheme, the Hominidae is divided into subfamilies Paninae (gorilla and chimpanzee) and Homininae (humans). There are several other recently proposed revisions in the classification, but the most widely used is still the traditional division of Pongidae (great apes) and Hominidae (humans), with the understanding that this does not reflect the phyletic relations.

B. Genetic Relations

As early as 1904, G. H. F. Nuttall published "Blood Immunity and Blood Relationships," which explored the genetic relations among animals using immunological techniques. More than half a century lapsed before M. Goodman showed the power of molecular biology in clarifying our place in nature. Using the method of immunodiffusion he published in 1962 and 1963 what has become established fact: *H. sapiens* is much more closely related to African great apes than either is to the Asian great ape. Subsequent tests using a variety of methods (e.g., amino-acid sequencing, microcomplement fixation, DNA hybridizations, electrophoresis, and nucleotide sequencing) confirm the surprisingly close relations among humans, chimpanzees, and gorillas. The genetic similarity among these three is of a similar magnitude to the similarity between dog and fox, cat and lion, sheep and goat, and only slightly more than that between horse and donkey.

By the late 1960s V. Sarich and A. Wilson accumulated enough genetic comparisons among primates to show that immunologically detected differences in albumin (i.e., their ability to interact with certain antibodies) likely changed at a constant rate through time. By calibrating this change with a widely accepted date for the evolutionary divergence of two lineages, they derived a molecular clock for primate evolution that placed the origin of the hominid lineage at 4.2 million years ago. Subsequent studies by a variety of methods by numerous individuals, including the direct comparison of DNA, have shown that the molecular clock is generally correct, but irregularities in rate make it necessary to give significant error ranges. Recent estimates of the origin of the hominid lineage range from about 8 to 4 million years ago.

C. Traits Shared with Apes

Despite the obvious fact that *H. sapiens* is profoundly different from all other animals in many respects, we share a number of traits uniquely with apes that show our phylogenetic affinity with them. These shared and unique characteristics are most conspicuous in the forelimb and trunk anatomy. Apes and humans share exceptionally short lumbar regions of the spine, the lack of a tail, highly mobile shoulder and elbow joints, broad chests that are flattened from front to back, reduced ulnar olecranon processes, a vertebral column that protrudes into the chest cavity, long clavicles and acromial processes, a broad sternum, a diaphragm that is perpendicular to the spine, an obliquely positioned heart that adheres closely to the diaphragm, and abdominal organs that are closely attached to the posterior wall of the body cavity. These and other traits are probably related to the orthograde posture (upright) of the trunk. The molar teeth show detailed similarity in cusp number, fissure pattern, and overall form. A few traits are uniquely shared by humans and the African great apes, such as the presence of a frontal sinus that develops from the ethmoid bone and the fusion of the os centrale in the wrist. In addition, African apes and humans share certain molecular similarities such as a unique substitution of amino acids in the myoglobin chain at positions 23 and perhaps 110; two shared substitutions in the fibrinopeptide A and B chains; three transversions, eight transitions, and three deletions in the nuclear DNA sequence; and 16 substitutions in the mitochondrial DNA. [*See* COMPARATIVE ANATOMY.]

D. Unique Traits of Hominidae

As Lamarck, Huxley, Haeckel, Darwin, and other nineteenth century evolutionists pointed out, bipedalism was probably the primary change in the origin of our evolutionary lineage. Twentieth century discoveries of fossils confirm this. Habitually walking on the hindlegs with the forelimbs free for carrying and perhaps wielding weapons is the first fundamental change away from our common ancestors with the apes. This change required a major reorganization of the hindlimb and back away from the pattern common to most mammals. The most conspicuous anatomical differences are the shortened pelvic blade and the reorientation of the foot from a grasping organ to one in which the sole function is propulsion.

Modern human brain size is about three times greater than that of apes of the same body size. From the fossil record it is clear that brain size evolution occurred after the adoption of bipedalism. Most of the change occurred during the last 2 million years of evolution. Why the brain expanded is the subject of much speculation, but certainly the origin of language is related to this process.

II. Evolutionary History

A. The Stock from which Our Family Arose

The first substantial fossil evidence of the origin of the evolutionary lineage leading to modern catarrhines (Old World monkeys, apes, and humans) occurs in geological strata dating to 37 to 31 million years ago in the Fayum deposits of Egypt. Unfortunately, no fossil localities in Africa have produced sufficient mammalian remains before the rich and diversified primate fauna of the Fayum, so little is known about what came before. Primates are abundant in European and North American localities dated before 37 million years ago, but no catarrhines are present. A few mandibular fragments from Burma dating to about 40 million years ago hint at the possibility of catarrhines in Asia, but the evidence is tenuous so far.

The Fayum presents an extraordinary window on the early stages of the evolution of the primates between 37 and 31 million years ago. The climate was warm, the habitat forested, and a major river moved slowly through the site. Primates constituted a major component of the mammalian fauna. A tarsier-like form is represented (by a jaw fragment) as well as a loris-like creature (known only from one tooth). The most common primates are grouped under the family name of Parapithecidae (*Parapithecus* and *Apidium*), which are not catarrhines (i.e., they retain the primitive characteristic of having three premolar teeth). There are several species of catarrhines divided by their original describers into two genera, *Propliopithecus* and *Aegyptopithecus,* although some authorities recognize only one genus. *Propliopithecus* was small (2–3 kg), aboreal, quadrupedal, and sexually dimorphic. *Aegyptopithecus* was larger (about 6 kg), diurnal, and arboreal. Its postcrania show an adaptation for slow and deliberate branch climbing and quadrupedal walking. It had strong sexual dimorphism in body size, canines, and skull morphology. The teeth indicate fruit-eating. The best-preserved cranium shows a long, almost lemur-like snout, but other facial skeletons are much less prognathic. The brain is relatively small compared with modern catarrhines but catarrhine-like in having relatively smaller olfactory bulbs, larger visual cortex, and a more complex sulcal pattern.

By current paleontological evidence, catarrhines were confined to Africa until about 16 million years ago. Until this date Africa and Arabia were separated from Eurasia by the Tethys Sea. Unfortunately, there is a gap in the primate fossil record in Africa between 31 and 22 million years. By early Miocene times (21–17 million years ago) there was a rich sample of catarrhines in East Africa during a period when tropical forests extended much further east and north than they do now, although there are some indications that woodland or brushland grassland habitats existed as well.

The early Miocene primates of East Africa show the earliest evidence for the divergence between the two catarrhine superfamilies Cercopithecoidea (Old World monkeys) and Hominoidea (apes and people). Cercopithecoid fossils are relatively less abundant than hominoids. The hominoids are quite different from any living species and are best placed in one or more separate families. There are at least seven genera. The adaptive diversity among these hominoids is considerable. Body sizes range from 4 to 40 kg. Their diets were primarily of fruit, but some species were leaf eaters. Most were arboreal quadrupeds, but some may have been capable of forelimb suspensory behavior as well. The postcranium does not, however, reveal many (if any) of the distinctive traits that are shared by all the living hominoids. In fact, the early Miocene hominoids share very few derived traits with living hominoids, although two recently discovered genera, *Afropithecus* and *Turkanopithecus,* show greater resemblance to later hominoids.

Between 15 and 13 million years ago there was a major faunal change in the Old World because of the exchange of species between Africa and Eurasia. This event was apparently triggered by the breaking down of the Tethys Sea barrier, which had kept Afro-Arabian faunas isolated. Although forests were much more abundant than today, there is evidence for some open habitat and possibly grassland, particularly in the northern latitudes. There is evidence of climatic cooling at this time.

The first hominoids outside Africa occurred at this time (15–18 million years ago) in both Europe and Asia. The adaptive and phyletic diversity was much higher than among modern hominoids. Between 15 and 12 million years ago, at least four genera and nine species of large-bodied hominoids were in Eurasia, and possibly the African diversity was equally great, although fewer African fossil primates are known during this period. Most of these hominoids had thick molar enamel similar to that seen in modern orangutans and fossil hominids. They were apparently frugivorous, had marked sexual dimorphism, and looked somewhat like modern apes. Although the postcranial fossils are rarer and unassociated, they appear to share the suite of characteristics unique to modern hominoids. The divergence of the evolutionary lineage leading to the orangutan probably occurred by 12 million years ago, as shown by the remarkable discovery of a facial skeleton of *Sivapithecus* in Pakistan. This specimen shares numerous derived traits with the Asian great ape. There is a wealth of new material from Europe and especially China that shows the rich diversity of these ape-like forms in the Miocene compared with the impoverished diversity of modern ape species.

At about 10 or 11 million years ago, there is evidence for increasing open habitat in the Old World and another major faunal turnover. By about 7 million years, forest and woodland-adapted fauna were replaced by more open-country species in South Asia. In Africa the record is less well documented, but certainly the abundant forests by the early Miocene gave way to more and more patchy woodland and grassland habitats in the middle to late Miocene. The collection of middle to late African hominoids is rapidly expanding in recent years, but the picture is still far from clear. Yet no fossils are clearly linked uniquely to any of the living African great apes, although a palate from 8 million-year-old beds in Kenya shows some gorilla-like traits. There is a sprinkling of tantalizing bits of hominoid fossils in the late Miocene but nothing that can be linked specifically to living species until possibly 5 and certainly 4 million years ago when some fossils distinctly belong to the human evolutionary lineage.

B. First Bipeds

The molecular clock predicts the divergence of the human and African ape evolutionary lineages to be some time between 5 and 8 million years. During this period and later, the African habitats were drastically changing, with increasing seasonality of rainfall and spreading grassland. Areas of tropical rainforest were reduced, creating isolated pockets of forest in a sea of grass. Unfortunately, few fossil primates are known from this period: a *Sivapithecus*-like tooth at 9 million years ago, a palate with some gorilla-like features combined with unique traits at 8 million years, an ape-like molar crown at 7 million years, and a mandibular fragment with one molar tooth at 5–6 million years ago. The latter specimen from Lothagam in Kenya has some derived traits that appear to be shared with later hominoids (e.g., a relatively decreased molar length, entoconid size, and mandibular depth). It is associated with open-country fauna. By 4 million years, the presence of humans is more secure with fragments of a jaw, two arms, and a thigh. The thigh is reported to have an internal arrangement of trabeculae characteristic of later bipeds.

By 3.8–3.6 million years, the record of hominids becomes much richer by the discoveries at Laetoli in Tanzania. The site produced the remains of 23 hominids including jaws, associate dental rows, an infant skeleton, and three sets of footprint trails made by bipeds. The footprints are clearly hominid with distinctly convergent big toes and human-like proportions. The dental remains are said to be very similar to those found at Hadar in Ethiopia dating between 2.8 and 3.1 million years ago. The Hadar collection is wonderfully complete with one associated skeleton (A.L. 288-1, "Lucy"), the fragmentary remains of at least 13 individuals who apparently died together at one spot, and numerous other skeletal parts. The combined sample of Laetoli and Hadar yields an excellent picture of what the first human species, *Australopithecus afarensis,* looked like.

Australopithecus afarensis was fully bipedal as indicated by the Laetoli footprints and the Hadar postcranial skeletons. The pelvic blades are low, the sacrum wide, and the pelvic basin quite human-like in shape, although it is not identical to modern humans. The knees are characteristically human and not ape. The toes are relatively shorter than any ape but not reduced as much as they are in modern humans. The forelimbs are relatively quite small, and the wrists and hands show no adaptation for ape-like knuckle-walking. The skeleton retains many ape-like traits, but in most fundamental respects, it is adapted for bipedality. Its long, curved toes and fingers and many other ape-like features may imply that this first human species was a more adept tree-climber than later species of hominid. Although its pelvis and hindlimb are fundamentally human-like, basic differences imply a somewhat different form of bipedality from that seen in modern *H. sapiens*. The pelvic blades, for example, face more posteriorly; the thighs are relatively short; the knees appear to lack a human-like meniscus attachment; the ankle, in at least one specimen, slopes in an ape-like direction; the foot architecture has many ape-like traits; and the toes are relatively long and curved.

In other characteristics *A. Afarensis* possesses a mixture of hominid and pongid qualities. There was strong sexual dimorphism in body size, females weighing about 30 kg and males close to 45 kg. The brain size was about that of a modern chimpanzee (415 cc), which is about one-third the size of a modern human of the same body size. The skull is quite pongid-like, with a prognathic muzzle, an unflexed cranial base, and a strong development of the posterior fibers of the temporalis muscle. The canine is considerably reduced from the size seen in modern apes, but it is larger than that of modern humans. The lower first premolar is variable, but in many specimens it is quite ape-like in orientation and cusp number.

A similar form of hominid occurs at about 2.5 million years in South African cave deposits, *Australopithecus africanus*. In many postcranial parts it is remarkably similar to the Hadar hominids, although there are some differences, particularly in the hand. *A. africanus* is distinct from the Hadar and Laetoli hominids in having many cranial and dental traits more similar to later *Homo* (e.g., a reduced muzzle, a more flexed cranial base, and a biscuspid first premolar). Females appear to be as small as those from Hadar, but males are not quite as large. The average cranial capacity is slightly larger (442 cc), and there are no specimens as small as the smallest one from Hadar. The cheek teeth are relatively larger.

C. Extinct Cousins: The Robusts

From about 2.5 to 1.3 million years lived a variety of hominids referred to as "robust" australopithecines because of their hypertrophied masticatory system. At least two and probably four species are known so far. The earliest, referred to by some as *Australopithecus aethiopicus,* is known from East African sites dating to 2.5 million years ago. Its cheek teeth are enormous, as are all the supporting structures related to heavy chewing (i.e., massive jaws, strongly buttressed skull, enormous area of attachment for the muscles of mastication). In many ways the skull shares primitive characteristics with *A. afarensis* and pongids, (e.g., an unflexed cranial base, a prognathic muzzle, and a strong development of the posterior fibers of the temporalis muscle). In the later robust species, these traits are lost, and they appear more *Homo*-like. These species include *A. boisei,* found abundantly in East Africa between 2.3 and 1.3 million, and *A. robustus* of South Africa. Although body size is no greater than that in earlier hominid species, the brain size is about 100 cc larger. The postcrania are fragmentary and difficult to associate with these species, but what specimens there are indicate a remarkably human-like form. It is generally assumed that the massive development of the chewing apparatus was an adaptation to a vegetarian diet, perhaps consisting primarily of hard fruits and bulbs but not grass.

D. The Appearance of *Homo*

The first abundant evidence of what most investigators would refer to as *Homo* occurs in strata dated to about 2 million years ago, although fragmentary material is known slightly before that date. Between 1.9 and 1.6 million years ago, many specimens probably belong to the genus *Homo,* but the variability is higher than would be expected from a single species. Whether this variability indicates more than one species has not been resolved. The species name associated with this material is *habilis*. The *Homo*-like traits include an expanded brain (about 50% larger than *A. africanus* in relative size) and reduced cheek teeth. The smallest postcranial specimen was probably less than 1 m tall and may have weighed less than 25 kg. The largest probably weighed more than 60 kg. Relative to joint size, the hindlimbs were much less robust than those of *Australopithecus*. Stone tools first appeared in the archaeological record at 2.5 million years, and it is often inferred that *Homo* was responsible for them. Well-preserved living floors occur at least by 1.8 million years and show that these hominids were using tools for several activities including butchering, plant processing, and wood carving.

The species *H. erectus* first appeared at 1.6 million years ago in East and South Africa. Brain size in the larger specimens is twice that of *Australopithecus*. Relative cheek-tooth size was considerably reduced from the large size seen in earlier hominid species. Body size of the larger specimens probably exceeded 60 kg. It may be that body size sexual dimorphism was not as marked in early *H. erectus* as it apparently was in *H. habilis,* but as yet, no associated female skeletons are complete enough to be certain. At about the same time as the origin of *H. erectus* is the first appearance of the Acheulean material culture characterized by bifacially flaked large stone tools ("handaxes"), which are found throughout most of the Old World for the next 1.5 million years. There is a great deal of evidence supporting the hypothesis that the origin of *H. erectus* marked the beginning of major change in the adaptive strategy of Hominidae, although it remained a numerically minor part of the vertebrate fauna collected in Pleistocene beds despite strong bias of collectors.

E. Colonizing the Old World

By at least 1 million years ago, some populations of *H. erectus* had left their African homeland and colonized Eurasia. The best known of the early hominids outside of Africa are those from Java, dating to about 1 million years, although dating is much less precise than it is for earlier parts of the record. The imprecision is a result of the fact that the best-established method of establishing geological dates older than 1 million years is by the radioactive decay of potassium 40, but between that date and 40,000 years ago there are less precise methods of dating. One major clue is the shifting of the earth's magnetic polarity from north to south at specific times that can be detected in geological sediments. The last "reversal" (when the magnetic pole shifted back to the north) occurred 750,000 years ago.

The *H. erectus* of Java was found in 1893. The famous discovery of Eugene DuBois consisted of a skull cap with an estimated cranial capacity of 940 cc associated with a thigh that was well within the range of variation of *H. sapiens*. This association lead DuBois to name the creature *Pithecanthropus erectus* (erect ape-man). Unfortunately, few other specimens came to light until the 1930s. Between 1930 and the outbreak of the Asian part of World War II came a pulse of discovery in Java and China that established *H. erectus* as a well-documented species preceding our own. A wealth of new Javanese and Chinese fossils were discovered and later described, which gave a picture of a hominid of intermediate brain size (about 800–1000 cc, compared with the average for *H. sapiens* of 1,300 cc) with large brow ridges, a low cranial vault, teeth of intermediate size, and robust but modern-looking postcrania. Several fire layers are present in the best-known *H. erectus* site in China (Zhoukoudien), leading most investigators to infer the controlled use of fire. The first appearance of humans in the rest of Eurasia is less well documented. Some archaeological sites may date back to as early as 1 million years, but the dates are problematical. There are only a sprinkling of human remains in Western Europe before about 200,000 years, and they do not resemble *H. sapiens* very closely. Most if not all the European fossils before 35,000 years are best regarded as archaic *H. sapiens*.

F. Archaic *Homo sapiens*

The term "archaic *Homo sapiens*" refers to a heterogeneous collection of Old World hominids between about 400,000 and 35,000 years ago. The term is not precise, and there is a need for a formal taxonomic reappraisal. The contrast between what is now called archaic *H. sapiens* and anatomically modern *H. sapiens* is greater than that between archaic *H. sapiens* and *Homo erectus* except in one important characteristic, brain size. Brain size is usually within the modern human range of variation (1,000–1,700 cc), although there is one exceptionally small specimen (Sale of North Africa with 860 cc). They resemble *H. erectus* with their large faces, robust skeletons, brow ridges, and long, low skulls. Earlier specimens are associated with Acheulean culture, but by about 200,000 years ago stone tools became more sophisticated, particularly in their manufacture of prepared-core flake tools. The best-known variety of archaic *H. sapiens* is the Neanderthal.

Neanderthals were a relatively homogeneous group of archaic *H. sapiens* who occupied Europe and west Asia between about 200,000 and 35,000 years ago. Their facial morphology was unmistakable, with exaggerated midfacial prognathism leading to what must have been enormously protrusive noses. They were the first prehuman fossils known to science, and consequently they have played a major role in the interpretation of human evolution. Now that much more is known about prehistory, Neanderthals are seen in perspective as a relatively isolated extreme variant of archaic *H. sapiens* that held out against anatomically modern *H. sapiens* until 35,000 years ago when they disappeared.

G. Anatomically Modern *Homo sapiens*

Anatomically modern *H. sapiens* is distinctly different from the archaic form of the species in having small faces, high foreheads, a true chin, and longer and less robust limbs, especially the distal segments (forearm and shin). They are first known in Africa by about 100,000 years ago or slightly earlier. There is some evidence that they were in the Middle East by 90,000 years ago, but remains of archaic *H. sapiens* are much more common in that area until about 35,000 years ago when they disappear from the record. At this time the culture dramatically changes with the introduction of more finely worked stone blades, a wider variety of tools of all kind, and art including cave paintings and stone carvings. There is accumulating evidence that the morphological and behavioral changes between archaic and modern *H. sapiens* are profound enough to warrant changing the taxonomy to restrict the species *H. sapiens* to moderns only. Recent comparisons of nuclear and mitochondrial DNA of living humans appear to show that the ancestors of all modern humans derive from Africa 110,000 to 200,000 years ago. Current fossil and archaeological evidence supports an African origin.

H. Peopling of the Earth

There is no unanimity of opinion concerning the origin and spread of anatomically modern *H. sapiens*. Some scholars emphasize the apparent regional continuity between local forms of fossil hominids and living populations in the same area. For example, several traits characteristic of *H. erectus* in China, such as a high incidence of shovel-shaped incisors, resemble these of some populations of modern Chinese. However, all modern humans are more similar to one another than to archaic *H. sapiens* or *H. erectus;* this similarity may indicate a close genetic relation among modern humans, and a restricted geographical area of origin. Particularly telling is the fact that the maternally inherited mitochondrial DNA of all living people is best interpreted as having a single origin in Africa. Had local premoderns outside of Africa contributed to the modern gene pool, presumably that contribution would be detected in the maternal line. Much more work needs to be done in this area before any certain conclusions can be drawn.

Anatomically modern humans had reached Australia at least by 30,000 years, although some robust and rather archaic hominids are still found there by 10,000 years. Archaeological evidence shows that the colonization of the rest of the Pacific began several thousand years ago from the east, reaching the Marquesas Islands by about AD 300 and New Zealand by about AD 900. Abundant evidence shows that people arrived in the Americas from north Asia by at least 12,000 years ago. Archeological findings before that date are rare, and their authenticity is often challenged. It may be the case that small populations occupied America much earlier, but they remained at low population densities until 12,000 years ago.

Bibliography

Andrews, P., and Martin, L. B. (1987). Cladistic relationships of extant and fossil hominoids. *J. Human Evolution* **16,** 101–118.

Ciochon, R. L., and Corruccini, R. S., eds. (1983). "New Interpretations of Ape and Human Ancestry." Plenum Press, New York.

Day, M. H. (1986). "Guide to Fossil Man." University of Chicago Press, Chicago.

Delson, E., ed. (1985). "Ancestors: The Hard Evidence." A. R. Liss, New York.

Fleagle, J. G. (1988). "Primate Adaptation and Evolution." Academic Press, New York.

Harrison, G. A., Tanner, J. M., Pilbeam, D. R., and Baker, P. T. (1988). "Human Biology." Oxford University Press, Oxford.

Smith, F. H., and Spencer, F., eds. (1984). "The Origin of Modern Humans." A. R. Liss, New York.

Szalay, F. S., and Delson, E. (1979). "Evolutionary History of the Primates." Academic Press, New York.

Tattersall, I., Delson, E., and Van Couvering, J., eds. (1988). "Encyclopedia of Human Evolution and Prehistory." Garland Publishing, New York.

Tuttle, R. H. (1986). "Apes of the World." Noyes Publications, Park Ridge, New Jersey.

Evolving Hominid Strategies

PAUL GRAVES, *Southampton University*

Glossary

Encephalization Size of brain relative to body size
Home bases Fixed domicile at which offspring are cared for and to which subsistence materials are brought back from foraging
Hominid From the family name Hominidae, denotes those fossil species with whom we share common ancestry, exclusive of other primates
Social intellect Intelligence specifically adapted to the problems of social life
Taphonomy Study of the geological deposition and transformation of plant and animal remains

IN THE PAST 6 million years, the behavior and ecology of hominids have radically changed. Our earliest ancestors were very different in their anatomy and behavior from either living primates or modern humans. Traditional models of human evolution, which stressed the role of technology and the hunting of large mammals, are still to be found in current literature. But the rapidly expanding body of evidence from paleontology, archaeology, and primatology is fundamentally changing our understanding of the forces that shaped the biological and behavioral development of our species.

I. Hominids as Bipedal Primates

A. Descent from Apes

Since the publication of Darwin's *Origin,* the descent of humans from apes has been the subject of controversy and debate. Over the last century, a number of alternative models for human ancestry have been proposed. At one time it was believed that humans were directly and separately descended from early prosimians similar to the modern Tarsir. More recently, paleontological evidence led many to decide that the hominid lineage shared a last common ancestor with the orangutan rather than the African great apes. However, it is now virtually certain that the hominid lineage is most closely related to the chimpanzees (genus *Pan*). Immunological comparison and comparison of DNA sequences by hybridization have shown a very close genetic link between *Homo* and *Pan,* and the use of estimated rates of genetic divergence as a molecular clock has now enabled researchers to put a time scale on the whole hominoid (ape) radiation. The ancestors of African apes and orangutan diverged around 15–17 million years (Myr ago) ago. Subsequently, the *Pan*–hominid and gorilla lineages diverged at 10–8 Myr ago. A number of separate studies corroborate a divergence of *Pan* and the early hominids at 6–7 Myr ago. [*See* EVOLUTION, HUMAN.]

B. The Earliest Hominids

In recent years, our knowledge of the earliest hominids has been greatly enhanced by a number of fossil discoveries in the East African Rift Valley region. These hominids, of genus *Australopithecus,* exhibit an anatomy that shows strong affinity with other hominoid lineages. *Australopithecus* forelimbs exhibit a "retention" of features associated with a climbing, arboreal adaptation, whereas recent reanalysis of hominid body size suggests a considerable degree of sexual dimorphism (difference in body size between sexes). The earliest *Australopithecus,* species *afarensis,* has often been thought to be extremely small, at ca. 30 kg and ca. 4 ft (1.23 m) tall, but it now seems likely that some males weighed as much as 80 kg. This has considerable implications both for models of behavioral patterns and for estimates of relative brain size. The earliest hominids seem to have been less encephalized (had a lower brain–body ratio) than living apes such as the chimpanzee.

C. Bipedalism

The main distinguishing feature of all hominids is bipedalism. Despite debates concerning the efficiency of *afarensis* locomotion, archaeological evidence from fossil footprints and anatomical studies indicates that the earliest known hominids were effective bipeds. A number of explanatory models for the origin of this distinctive feature have been proposed.

1. Bipedalism may have freed the hand to use tools. The cognitive requirements of tool-making may, in turn, have set up a selective pressure for greater encephalization.
2. Bipeds would be able to carry food over some distance and, thus, efficiently exploit the carcasses of large animals. This would also allow hominids to occupy fixed home bases, because females would be provisioned by males.
3. Similarly, bipeds could carry their offspring over considerable distances. This would facilitate longer periods of infant dependency and, thus, greater encephalization.
4. An upright posture would have minimized the effects of mid-day solar radiation on body temperature, because an upright posture reduces the body surface area directly exposed to the sun.

However, the growing body of paleontological and archaeological evidence serves to cast doubt on several of these models.

D. Early Hominids as Bipedal Apes

The tendency to see hominid evolution as progressive has led many students of human evolution to expect their ancestors to run before they can walk! Tool use, although a feature common to hominids and apes, is not evident in the archaeological record until ca. 2.3 Myr ago (the earliest Oldowan pebble tools derive from Koobi Fora in Kenya). This would be some 3–4 Myr after the appearance of the first hominids. Moreover, we know that other primates use tools without being obligate bipeds. Some forms of tool use might be facilitated by bipedalism, such as the use of weapons, but there is no obvious causal connection between tool use and an upright posture. (For the relationship between tool-making and intelligence, see below.)

For the earliest hominids, similar problems exist in relating the origins of bipedalism to the carrying of food. No evidence indicates that the earliest hominids carried objects over any distance. It might be argued that *Australopithecenes* were carrying perishable objects such as plant materials or digging sticks; however, in the absence of archaeological remains, no positive inference can be made. Nevertheless, we know that, as in the case of tool use, other primates are capable of transporting food or other objects without bipedal locomotion. The claim by M. D. Leakey for the existence of fossil "living floors" at Olduvai Gorge (Tanzania) is controversial and relates to a later date (ca. 1–8 to 1.5 Myr ago). It seems unlikely that any hominids used fixed home bases until the Middle and Upper Paleolithic (100–20 thousand years [Kyr] ago). Even modern hunter–gatherers tend to use rather ephemeral camps as part of their settlement system, which might only be used for a matter of days or weeks. Such camps would be unlikely to be preserved in the archaeological record.

The carrying of infants presents similar problems. Other primates carry their infants in a variety of ways and often for considerable periods. Young chimpanzees may be carried by their mothers for up to 4 yr, with the main period of dependency in the first 2 yr of life. Infant-carrying hypotheses depend on the assumptions that (1) bipedalism led directly to greater encephalization and (2) the earliest hominids did not have the same body hair cover as living primates. It is now clear that hominids were bipedal for at least 2–3 Myr before they developed significantly larger brains. Moreover, because we cannot be sure when human body hair evolved to its present form, it is possible that *Australopithicene* or even early *Homo* infants were carried clinging to their mothers' fur, as is observed in other apes and monkeys.

Of all the progressively adaptive models proposed, only temperature regulation is at all convincing. Experimental studies suggest that temperature regulation would have been a significant factor for hominids living in open savannah, if they were active in the middle of the day when the tropical sun is at its zenith. Furthermore, it is clear that several features of human skin, such as the reduction of body hair and the development of sweat glands, may be associated with the requirements of keeping the body cool. Experiments have shown that while humans are capable of long periods of continued physical exertion such as running, other tropical mammals such as the large carnivores and herbivores can run only limited distances without succumbing to heat exhaustion.

However, the earliest hominids are likely to have inhabited more closed "mosaic" environments, consisting of open woodland and riverline gallery forest. Here they would not have been exposed to the same intensity of solar radiation, nor, indeed, would they have needed a capacity for sustained physical exertion. It seems more likely, then, that temperature regulation only became a significant selective factor when hominids colonized more open habitats (at ca. 2.5–2.3 Myr ago).

Thus, the combined evidence of anatomy, ecology, and archaeology does not support any particular progressive adaptive traits among the earliest (genus *Australopithecus*) hominids; rather, they should simply be considered as a bipedal form of ape. Bipedalism may be seen as an adaptation to the varied locomotor requirements of a mixed arboreal and terrestrial environment, under the constraints of an anatomy originally evolved for an arboreal existence. Hominids, like all apes, have a postcranial skeleton which reflects the requirements of brachiation, i.e., locomotion by swinging from the arms, as observed in contemporary gibbons. This has precluded a return to quadrupedal locomotion in all terrestrial apes, including the knuckle-walking gorilla and chimpanzee and the bipedal human. As bipedal apes, the earliest hominids exhibit an anatomy fairly typical of the great apes in terms of encephalization, sexual dimorphism, and forelimb specialization. Bipedalism might be seen as a preadaptation for later developments, but, given our current understanding of the chronology of hominid evolution, it was a preadaptation that took considerable time to become significant. This suggests that the action of other parameters was necessary for radical changes in intelligence and subsistence to occur.

II. Technology

A. Technological Determinism

Technology has generally been seen as a central causal factor in human evolution, mediating changes in subsistence, leading to increased intelligence and enhancing human physical capabilities. As such, Bergson's argument that man should be renamed *Homo faber* has continued to dominate analyses of hominid behavioral strategies. However, as in the bipedalism debate, the growth of the paleontological and archaeological record (and a more refined chronological framework) has cast some doubt on the causal efficacy of technological determination in human evolution.

B. Development of Technology

The earliest, Lower Paleolithic, tools date from ca. 2.3 Myr ago and are generally termed Oldowan (after the site of Olduvai Gorge). These tools are generally crude chopping or cleaving implements, made by striking four or five flakes from a cobble or pebble. Oldowan types persist until ca. 1.2 Myr ago, although similar crude implements are found in contexts as late as the Neolithic (ca. 3000 B.C.). By about 1.6 Myr ago, the first Acheulian assemblages appear, characterized by a tool type known as the handax, a large ovate or wedge-shaped tool formed by striking numerous parallel flakes along the axis of a cobble or core. Achuelian assemblages continue to dominate the archaeological record until ca. 3–200 Kyr ago. The handax itself is found in assemblages as late as 40 Kyr ago, albeit in a more refined form. However, the function of such objects is unknown; some are so large as to be practically useless for most subsistence tasks.

At around 200 Kyr ago, these "core tools" (formed on a large core of rock) are supplemented by new forms of flake-based implements. The earliest of these were produced using the Levallois technique, in which a circular, triangular, or oval flake was struck from a prepared core. Many of these flakes would have been retouched (shaped by further flaking) for use in cutting and scraping, although some appear to have been intended as the points of crude spears (although they would have

A

DATES Years Before Present	CLIMATE & VEGETATION	HOIMINIDS	AREAS COLONISED	TOOLS	OTHER INNOVATIONS
1,000 000	Stronger Glacial Cycles Begin	*Homo erectus*	S. Europe, S.W. Asia, N. Africa	Acheulian	
Beginning of Pleistocene 2,000 000	More open environments in E. and S. Africa	*Homo habilis* *A. boisei* *A. robustus*	China? S.E. Asia?	Oldowan	Fire? Scavenging?
Pliocene 4,000 000	First Glacials at Poles	*Australopithecus afarensis*	S. Africa E. Africa		
Miocene 6,000 000	Forest and Open Woodland throughout Eurasia and Africa	Divergence of Hominids and *Pan*			

FIGURE 1 Chronological table of human evolution. a. 6 Myr to 500 Kyr. b. 500 Kyr to the Holocene (present interglacial). (*Figure continues.*)

been too heavy for throwing spears). Levallois or Mousterian Middle Paleolithic techniques persist until ca. 40 Kyr ago, but already by this time many regions such as Southwest Asia and Africa yield evidence of a more refined, blade-based technique normally associated with the Upper Paleolithic (generally considered to be the period between 30 and 10 Kyr ago).

Blades are essentially long, parallel-sided flakes struck vertically around the circumference of a prepared cylindrical or conical core of flint, or some similar microcrystalline rock. A blade would generally be used as a "blank" for the preparation of more refined leaf-shaped spearheads or arrowheads or be broken into pieces to form smaller "bladelets," which would be mounted in a wooden or bone haft. Later Paleolithic stone technologies are also accompanied by the use of bone and ivory and by the appearance of art, both in cave paintings and in carvings of bone and ivory. Postglacial lithic technology is largely a refinement of Late Glacial Upper Paleolithic techniques.

C. Technology and Chronology

The chronological development of technology is indicative of its part in human evolution. The first stone tools appear in the archaeological record more or less simultaneously with the first members of the genus *Homo*. No *Australopithicene* has ever been found in exclusive and incontrovertible association with tools of any kind, although the lineage persists until ca. 1 Myr ago. This suggests the idea that higher intelligence is to be associated with the development of tools and associated subsistence strategies; however, the subsequent development of technology contradicts this.

The persistence of the Acheulian assemblage for around 1 Myr, with a few minor changes, does not correlate with other patterns of change and stasis. During this same million year period, the hominid

B

DATES Years Before Present	CLIMATE & VEGETATION	HOIMINIDS	AREAS COLONISED	TOOLS	OTHER INNOVATIONS
10,000	Holocene				
40,000	Last Glacial Maximum		N. America Melanesia		Art, Huts, Storage, Warm Clothing
			N. Eurasia	Upper Palaeolithic	Boats
Würm Glacial 100,000	Glaciers and Tundra in N. Eurasia during cold periods. Boreal and tropical forests	Modern *Homo sapiens*			Cave Sites with Fire
500,000	retreat. Areas of extreme aridity in N. Africa and Levant	*Homo sapiens Neanderthalensis* archaic *Homo sapiens*	*N.W.Europe S.USSR Central Asia*	Mousterian Middle Palaeolithic Levallois	Hunting

FIGURE 1 *(continued)*

brain and body developed to roughly their modern size, and the intermediate *Homo erectus* and early *Homo sapiens* chronospecies colonized the subtropical parts of the Old World and entered the temperate zone. While the occurrence of Levallois and Mousterian technologies represents something of a departure, the major changes in tool form and function do not occur until the Upper Paleolithic. Only here do we have evidence of effective projectile points, the use of the spear thrower and the bow, and the construction of substantial dwellings.

D. Function of Technology

Clearly, then, the chronology suggests that, for most of the past 6 Myr, technological development has not been a central causal agent in hominid evolution (Fig. 1). Ideas about the association of tools with hunting have been questioned (see below), and it now seems that technological development was in no way focal to the initial stages of the colonization of the globe. Indeed, in many areas of the Far East, there are no stone tools before ca. 50 Kyr ago, whereas hominids had been present in these areas for more than 1 Myr. If anything, we may say that technology has had a significant impact on human biology for the last 50–100 Kyr. Evidence for this is clear in the appearance of more gracile (less heavily built) 'modern' *Homo sapiens sapiens* some time between 100 and 40 Kyr ago. It is believed that these hominids were able to survive the conditions of glacial northern Eurasia because of superior technology, and that the use of such items as throwing spears, bows, and hafted tools of all kinds relieved the selective pressure that had produced their more robust ancestors such as *Homo erectus*.

III. Early Hominid Subsistence

A. Man the Hunter

Much as traditional accounts of human evolution have dwelled on the role of technology, so they have also focused on the role of hunting as a uniquely human adaptation among the primates. However, as with other theories, recent evidence contradicts this view, and new accounts of hominid subsistence are emerging. Over the last 30 yr, the concept of "man the hunter" has been in gradual retreat, as hominid capacity for hunting and uniqueness as a hunter have come to be doubted.

B. Hunters or Hunted?

Early ethnographic work led anthropologists to assume that hunting was, in some sense, the primitive state of humanity. Hence, when the first fossils of early *Homo sapiens,* and later of *Australopithecus,* were discovered, it was assumed that these creatures were hunters. Man, in Ardrey's words, was seen as the "killer ape"; his fossil remains surrounded by the stone tools he had used to kill his prey and the bones of those prey. The first doubts about these hypotheses arose from analysis of the faunal remains at early hominid sites and, in particular, their taphonomy (the conditions of deposition).

Work in South Africa and at East African sites has revealed that much of the fossil deposition at these sites was due to fluvial or carnivore activity. Not only were the bones of other animals the remains of carnivore kills; early hominids were also preyed upon by either the ancestors of the modern leopard or some form of the saber-toothed cat. Moreover, only limited evidence from "cut marks" on bone indicates that hominids were themselves butchering large animals. In many cases, the coincidence of large herbivore carcasses with lithic material is the work of fluvial activity in the kinds of river margin environments preferred by early hominids.

It is, in fact, hardly surprising that early hominids were not big game hunters; they were simply too small to tackle large game and were not equipped, as their descendants were, with effective weapons. Thus, most authorities would now argue that early hominids were scavengers; foraging for carnivore kills or preying upon large animals that had become trapped or disabled. This scavenged material might be supplemented by the hunting of small game, as is observed among other primates such as the chimpanzee and baboon. Indeed, it could be said that all primates have a basic hunting adaptation, given that they have traits such as binocular vision, which probably evolved to facilitate insectivory.

C. Seed Eaters

Most living primates are largely herbivorous, eating leaves or fruit, and, given that a hunting adaptation now seems less likely, the role of plant matter in the hominid diet has an increasing significance for the study of hominid strategies. Work conducted in the 1960s and 1970s has revealed that even modern hunters and gatherers are not as reliant on large game as was once thought. Hunting is a highly unpredictable business, and many societies rely on gathered material or the products of horticulture to sustain themselves when no meat is available. While no reliable figures are available, it seems that meat contributes only 25–30% of the diet of most hunter–gatherer societies (although exceptions exist), compared with estimates of perhaps 6% among the East African chimpanzee.

Unfortunately, plant remains are rarely preserved in the Paleolithic archaeological record, and much remains speculative. Analysis of early hominid dentition reveals that at least some early hominids such as *Australopithecus robustus* and *Australopithecus boisei* may have specialized in eating dry savannah plant foods such as seeds and corms. Moreover, it seems likely that all early hominids would have gathered a varied diet of plant materials in the mosaic environments of savannah river and lake margins. We can be fairly certain that meat eating was a part of hominid subsistence from about 2 Myr ago onward, but the relative importance of plant and animal matter continues to be in doubt.

D. Woman the Gatherer

Many, if not most, researchers in the domain of hominid gender roles, have assumed, as Lovejoy does, that early hominids were monogamous and pair-bonded. Indeed, this pair bond has been seen as the fundamental basis for more effective care of infants, hence creating the conditions for longer periods of parental dependence and, thus, larger brains. However, the biological and sociological evidence for the role of monogamy as a fundamental human trait is equivocal. It has been suggested that various primary and secondary sexual characteristics of humans, such as relatively small testicle size or large breasts, are to be associated with monogamous-pair bonding, but in fact, these traits cannot be causally linked to monogamy. Humans are virtually continually sexually active throughout adult life, whereas sexual activity among monogamous species tends to be infrequent. Moreover, it can be shown ethnographically that attitudes toward sexual partners and sexual practices reflect a tendency toward polygyny that is only suppressed, for social reasons, in modern societies.

Given this evidence, the fact that early hominids were not big game hunters has considerable implications for their socioecology. Traditionally, it has been assumed that male hunters or indeed scavengers would have provisioned their female partners, who would reside, for safety from predators, at a central home base; however, ethnography reveals that much is not most of the gathered material in hunter–gatherer diets is provided by women. In fact, it seems likely that the economic dependence of females on males is a fairly recent and context-specific phenomenon. One might suggest that some division of labor between sexes, or other groups, occurred quite early in the evolution of *Homo*. For each gender, we might expect differing life history tactics, reflecting differing reproductive strategies. Thus, for males, highly nutritious, but unpredictable, sources of meat from hunting and scavenging would be preferred, enhancing an individual's capacity to compete for social status and for mates. Females, on the other hand, would concentrate on lower value but more predictable sources of nutrition such as plant material. This would reflect the need for a more sustainable, long-term reproductive history, since the female reproductive capacity is limited by the time required for gestation. Social sharing of these materials, either with kin groups or potential mates, would constitute the basis for an ecology of social division of labor.

IV. Ice Age Hunters

A. Middle Paleolithic Adaptations

As noted above, stone tools from before the Middle Paleolithic are unlikely to have been used for hunting. These probably were used for the butchery of scavenged carcasses and the preparation of plant materials. It has been argued that even Middle Paleolithic populations were not hunters. In particular, faunal remains from sites such as Klasies River Mouth (in South Africa, dated ca. 150–90 Kyr ago) do not seem consistent with hunting. However, most authorities would agree that some form of large game hunting appeared during this period. The heavy Levallois Mousterian points were probably used in hunting large game by what is termed a confrontational method. Middle Paleolithic hominids such as the Neanderthals were extremely robust, while the projectile points suggest that they were used to form heavy thrusting spears. It is therefore suggested that large fauna were killed either by being mobbed by a large number of hunters or by being driven to their deaths (as at La Cotte in Jersey) over a convenient precipice. Such techniques would be highly dangerous, and, notably, Middle Paleolithic skeletons show a very high incidence of traumatic injuries, a surprising number of which were not actually fatal.

Most early evidence of hominid hunting derives from the last glacial cycle (the Wurm, or Wisconsin), which was the first period in which hominids continued to occupy temperate latitudes, despite deteriorating climatic conditions. In previous cycles, hominids probably only entered these latitudes in intermediate periods, between the cold glacial maxima and the warm interglacials when heavy boreal forest cover made most of northern Eurasia impassable. In this context, hunting may be seen as an adaptation to cold and, more generally, to arid climatic conditions or regions. Thus, in tropical regions we find that hunting populations occupied the more arid areas of Africa, the Middle East and Asia, while in northern latitudes hominid populations took advantage of the open glacial habitat to develop increasingly effective hunting strategies.

Traditionally, paleoanthropology has regarded glacial habitats as inhospitable, but in fact the reduction of tree cover in these periods led to a massive increase in populations of large fauna. In the early Wurm (125–30 Kyr ago) climatic conditions would have been mild enough to sustain the confrontational Middle Paleolithic hunters. Anatomical studies have shown that European *Homo sapiens neanderthalensis* had several specialized adaptations to cold conditions. Their relatively short limbs minimized body surface area and, thus, heat loss, while particularly large nasal cavities were probably adapted to the warming of cold air and the retention of moisture.

However, the more extreme conditions of the later Wurm (30–10 Kyr ago) probably required more sophisticated hunting strategies and social organization. In northern temperate latitudes, these more complex strategies are generally thought to be associated with the appearance of Upper Paleolithic technologies and the distinctive, anatomically "modern" *Homo sapiens sapiens*. This apparent replacement of indigeneous populations raises a number of difficult questions concerning the relationship among anatomical differences, behavior, and intelligence.

B. Modern Humans and the Upper Paleolithic

The so-called Middle-Upper Paleolithic transition is perhaps the most controversial area in hominid research. For at least half a century, paleoanthropology has debated the relationship between European and Near Eastern Neanderthal populations and the apparently immigrant populations of modern *H. s. sapiens*. The controversy has centered on whether indigenous populations evolved into "moderns" or they were replaced by modern humans who had evolved separately in Africa. Recent research in physical anthropology and microbiology is significantly resolving this debate, although a number of ecological and social questions remain open.

Research on human mitochondrial DNA has, through analysis of mutations and estimation of mutation rates, suggested that all living humans are descended from one ancestral population or even individual, originating in Africa at about 200 Kyr ago. This scenario is consistent with an origin of modern humans in Africa, as represented by fossils of early *H. s. sapiens* such as the Omo 1 and 2 skulls. It has been suggested that these modern *H. s. sapiens* evolved separately from the Eurasian *Homo sapiens neanderthalensis,* diverging from a common root represented by the Petralona, Steinheim, and Swanscombe skulls of early *H. sapiens*. The modern types first entered Southwest Asia around 100 Kyr ago (much earlier than had previously been thought) and colonized northern Eurasia and South Asia between 50 and 30 Kyr ago.

However, a number of questions remain open. Although, traditionally, modern types have been associated with the more advanced technology and subsistence techniques of the Upper Paleolithic, many of the earliest modern-type fossils are associated with Middle Paleolithic tools. Moreover, the blade-based lithics identified with the Upper Paleolithic are found in very early contexts in both North and South Africa. Therefore, it is difficult to decide on the nature of the link between anatomical and behavioral differences. Were the anatomical differences between moderns and Neanderthals actually a causal factor in behavioral differences? As noted above, most of the anatomical change in human evolution has taken place without concomitant change in technology. Given the fact that mitochondrial DNA is only transmitted through the female line, we may never know if living humans have inherited Neanderthal nuclear DNAs through the male line.

C. Hominid Mobility and Global Colonization

Certainly it would seem that the behavioral changes associated with the European Upper Paleolithic were directed toward a strategy for survival in cold arid climates. Growing evidence indicates very high levels of mobility among hominid populations in the last glacial, not least from the fact that all of northern Eurasia was colonized at this time, and, in the late glacial period (15–11 Kyr ago), hominids first crossed Beringia (now the Bering Strait) into Alaska. Moreover, it is clear that between 50 and 30 Kyr ago hominids were already successfully crossing open seas to colonize Australasia and Melanesia.

This high level of mobility, essential to survival in the severe climates of glacial northern Europe and Siberia, was associated with complex networks of social interaction. The distribution of raw materials, tool types and art objects across the western Soviet Union and eastern and western Europe demonstrates that artifacts were either carried or traded over great distances. These trading links may be associated with networks for exchange of mates and of information. Moreover, the existence of Upper Paleolithic sites at the very fringes of the northern ice caps suggests that annual or seasonal migrations may have involved journeys over many hundreds of kilometers. Perhaps, for example, from southwest France to Britain and Belgium and The Netherlands, or from the Amur river in southeast Siberia to the northeast Chukchi peninsula, although such journeys are hard to substantiate.

D. Art and Language

The coincidence of the appearance of art and of modern humans in the Upper Paleolithic is often thought to reflect the cognitive differences between *H. s. sapiens* and archaic *H. sapiens* such as the Neanderthals. The appearance of art perhaps implies the emergence of language; however, this may be doubted on archaeological, biological, and psychological grounds. Although the array of parietal (=cave) and mobilary (=carved) works of art in the Upper Paleolithic is impressive, many later archaeological periods are notable for the virtual absence of representational art. To claim that art is a necessary indicator of linguistic competence is, thus, problematic with respect to these later periods. No doubt the Upper Paleolithic origin of art does reflect considerable social change, but this need be no more significant that the appearance of cubism and nonrepresentational art in the twentieth century.

Biologically it has been argued that the laringeal tract of modern humans is different from that of early *H. sapiens*. Lieberman has modeled the throat and tongue of various hominids through inference from cranial and mandibular posture. He argues that Neanderthals would not have been able to form all of the vowels and consonants of modern human speech. However, most modern humans do not use >50% of possible phonetic sounds, suggesting that Neanderthals would have been perfectly capable of developing a language. Moreover, other physical differences such as in the structure of the hyoid bone have now been cast into doubt. Studies of endocranial casts suggest that lateral assymetry of the brain (generally associated with language in humans) begins early in the hominid line, if not before.

It is likely that the evolution of the hominid lineage would have seen progressive development of linguistic communication, rather than a unitary "origin" of language at a particular temporal locus. The development of language should reflect such factors as division of labor, mobility, and social organization, where explicit language would be needed to structure activity in time and space. It is likely, then, that the evolution of language would track the evolution of society. [*See* LANGUAGE, EVOLUTION.]

V. Social Evolution

A. Why Are Humans Intelligent?

Traditionally, explanations of human intelligence have concentrated on the cognitive requirements of tool-making and subsistence strategies involving hunting or scavenging from home bases, but as we have seen, these hypotheses do not totally accord with the evidence as it now exists. The development of technology does not keep pace with the development of the human brain to the extent that the brain of early *H. sapiens* had already reached its modern size well before the appearance of Upper Paleolithic complexities of technology. The alternative to a technodeterminist model for the origin of intelligence would be to consider the requirements of subsistence. In the early 1970s, Isaac suggested that social organization of parental care, or the sharing of resources, might explain the need for greater intelligence. These notions hint at the developments now taking place in psychology and primatology, which suggest that the requirements of the social axis of life may be more important than has previously been considered.

B. Social Intellect Hypothesis

Studies of primate behavior have revealed that monkeys and apes have limited capacities to deal with practical tasks; however, recent research has demonstrated that primates have much more subtle capacities when confronted with social tasks. Primates have very developed capacities for recognizing social status and, in the case of chimpanzees, or recognizing individual identities. In the wild, many species of monkeys are capable of complex communication concerning the nature and whereabouts of threats from predators. Moreover, they are also observed to use subtle techniques of deception to achieve social or sexual ends.

These observations and social psychological theories concerning the development of human intelligence have led to the formulation of what is now called the social intellect hypothesis. This suggests that the primary function of primate and human intellect is to deal with the complexities of social life. Most practical problems (except perhaps those encountered by modern humans) are relatively simple compared with the difficulties of achieving and maintaining social status or controlling the effects of agonistic (threatening or violent) encounters, which place stress on group cohesion. Practical problems are often confronted through the medium of social life, as in the case of avoiding predators or in finding subsistence material.

The social intellect hypothesis, therefore, suggests that the development of the human brain over the last 2 Myr may have had more to do with social causes than practicalities such as tool-making. This is hardly surprising when one considers the complex mechanisms required by social living. Humans have a highly developed ability to recognize and remember the identity of, and their relationship to, other members of their own species. This ability is absent in all other primates, except in a less-developed form in the chimpanzee and perhaps the gorilla. Moreover, of course, the large part of the human brain given over to the understanding and production of language is quintessentially a social function.

C. Social Intellect and Subsistence

The sharing of food may be a fundamental part of hominid adaptation. The unpredictability of diverse foraging strategies may be compensated by the sharing of resources, whereas large items such as carcasses may only be effectively exploited by groups. In fact, other primates do exhibit limited sharing behaviors, particularly with offspring or as "sexual favors" to potential or actual mates. However, the regular and habitual sharing of resources within a group requires recognition of complex individual relationships, if inequity and agonistic disputes are to be avoided. Concomitant analyses of social status and relationship are exactly what is predicted by the social intellect hypothesis.

Similarly, division of labor is accomplished by social means. The organization of society in time and space permits the efficient exploitation of resources within the home range. Where subsistence activities require more than individual foraging, task groups may be formed for specific purposes (e.g., to seek lithic raw material), and the products of such expeditions shared within the group. The collective appropriation of resources contrasts with most primate foraging, where individuals forage for themselves within a protective group, which must move around its home range *en masse*.

D. Social Organization and Colonization

The development of social organization may be inferred from the archaeological record. From the first appearance of genus *Homo* onward, the appropriation of raw materials and subsistence resources develops in geographical scale and temporal displacement. Even *H. erectus* is likely to have obtained lithic raw material from as far as 50 km from its home range. The exploitation of more arid, more inaccessible, and more seasonal environments would have required an increasingly logistic social organization of time and space. Similarly, the eventual appearance of large game hunting would have required considerable social organization in the production of effective weapons, the prosecution of the hunt, distribution of products, and the control of resources.

In the Middle and Upper Paleolithic, we find clear and considerable development of social organization. Raw materials are procured over increasingly long distances, and eventually distinctive regional cultures and practices appear. The burial of the dead, and care for the old or infirm are already apparent in some Neanderthal groups, suggesting that social relationships were permanent and not simply forgotten as soon as an individual no longer was productive or had died.

The Upper Paleolithic provides further ample evidence of the existence of social networks and of social structures extending over greater ranges of time and space. Colonization of landmasses separated by sea probably required the maintenance of social ties of great distances if colonists were not to be isolated and thus driven to extinction by the unpredictable nature of novel environments. Moreover, some of the caves of southwest Europe, such as Altamira, or the Upper Paleolithic villages of the Ukraine may represent centers of aggregation for populations, which would disperse in the spring and summer to exploit resources in more inhospitable northern latitudes. These aggregation centers were also focii of artistic activity, suggesting that the appearance of art is to be associated with large-scale complexity of Upper Paleolithic social organization.

All phases of hominid colonization, therefore, may be seen as the product of developing forms of social organization, where time and space are manipulated at the social level to maximize the exploitation of resources. The development of social complexity would have had wide ranging consequences, as in the development of language and the appearance of art. Increasing mobility of groups and individuals would lead to changes in the genetic structure of populations, with less likelihood of local isolation. It is probable that the encounter between modern *H. sapiens* and Neanderthals may also be understood in social terms. Differences in social organization and individual recognition might explain how, and indeed whether or not, the two populations interacted and intermingled.

Bibliography

Binford, L. R. (1983). "In Pursuit of the Past." Thames and Hudson, London.

Byrne, R., and Whiten, A. (1988). "Machiavellian Intelligence." Oxford University Press, London.

Foley, R. (1987). "Another Unique Species: Patterns in Human Evolutionary Ecology." Longman Scientific and Technical, London.

Gamble, C. S. (1987). "The Palaeolithic Settlement of Europe." Cambridge University Press, Cambridge.

Richards, G. (1987). "Human Evolution: An Introduction for the Behavioural Sciences." Routledge and Kegan Paul, London.

Exercise

FRANK BOOTH, *University of Texas Medical School at Houston*

Glossary

Adaptation Change in the chemical architecture of cells that minimizes disruption in the cellular or tissue biochemistry during a given intensity of physical exercise

Fitness Ability to undertake physical exercise or daily activities without undue fatigue. There are multiple types of fitness (e.g., aerobic fitness, strength fitness, coordination fitness, and flexibility fitness)

Maximal oxygen consumption The most oxygen that the body can use in aerobic exercise; synonymous with "maximal aerobic fitness"

Physical training Exercise bouts repeated for numerous days, leading to adaptations

PHYSICAL EXERCISE is the movement of bones caused by muscular contraction. The type of exercise and its resulting adaptation are dependent on the frequency, intensity, and duration of the physical work. A single bout of exercise can be identified as a daily workout. The frequency within the exercise bout would be the number of muscle contractions per minute. For example, the frequency of leg movements, and thus contractions of leg muscles, is greater in rapid running than in slow walking. The intensity of the exercise bout is the amount of work done per minute. For example, contracting arm muscles the same distance against 100 kg is of greater intensity than contracting them against 25 kg. Another example is that running is more intense than a slow walk because a greater distance, and thus more physical work, is done per unit of time when running. Duration is the length of the exercise bout.

An acute, or single, bout of exercise is distinguished from physical training, which is the summation of repeated daily bouts of exercise. A minimum frequency of two resistance training and three aerobic training bouts of exercise per week, repeated for many weeks, is necessary to produce training adaptations. Resistance training consists of contracting skeletal muscle against a near-maximal load (high intensity) for 30 contractions (low frequency) per day. As an adaptation the skeletal muscle enlarges after a few months. In contrast, aerobic exercises are low-intensity high-frequency activities that involve large masses of skeletal muscle (e.g., running, swimming, cycling, and rowing). Whereas resistance training requires near-30 maximal contractions per day for a training adaptation, submaximal aerobic exercise requires a minimum of 20–30 min/day to produce adaptations such as bradycardia, increased maximum cardiac output, and increased mitochondria density in skeletal muscle.

I. Aerobic Exercise

A. Quantification of Aerobic Exercise Intensity

In aerobic exercise oxygen is used by the muscles as they convert energy in glucose (sugar) and fatty acids (fats) to an energy form called ATP, which provides the direct source of energy for muscle fibers to shorten, thus causing limbs to move and exercise to occur. Measurement of the quantity of oxygen consumed during aerobic exercise indicates how many calories of sugar and fats were used and indirectly quantifies the caloric cost of the exercise. The greater the oxygen consumption during aerobic exercise, the more the calories used and the more the physical work done by the rhythmic activity (i.e., movement) of the limbs. For example, oxygen consumption increases each time the number of steps taken per minute increases. Normal walking uses less oxygen than a slow jog, which in turn requires less oxygen than a fast run. [*See* ADENOSINE TRIPHOSPHATE (ATP).]

B. Quantification of Fitness to Perform Aerobic Exercise

Knowledge of oxygen consumption during maximal aerobic exercise can also be used to classify the work fitness of the heart. The maximal number of steps that can be taken by the legs when running during a defined period depends on the fitness of the heart to do work. For example, an Olympic marathon runner will have a higher maximal oxygen consumption per minute of aerobic exercise than a sedentary person, who in turn will have a higher oxygen consumption than a bedridden individual. The marathon runner will have the highest heart fitness of the three, while the bedridden person will have the worst fitness of the heart to work. The more heart-fit a person is, the greater number of calories he can use per minute before fatigue. (''Aerobic fitness'' or ''aerobic work capacity'' is defined as the maximal amount of oxygen that can be used to make energy for work during physical exercise.)

C. What Limits Maximal Aerobic Fitness (Maximal Oxygen Consumption)?

Three aspects of maximal oxygen consumption (maximal caloric expenditure per minute) must be considered. They are: What limits it? What is its significance to an exercise task? How can it be altered? The factor limiting the maximal amount of oxygen that can be used per minute in converting sugar and fats to ATP during aerobic exercise in a normal young person at sea level is the maximal capacity of the heart to pump blood. Aerobic exercise requires the transfer of oxygen from air to muscle. Thus, oxygen in the air must be breathed into the lungs, where oxygen moves from alveoli (small air sacs in the lungs) to the pulmonary blood (blood leaves the right side of the heart and flows through the lungs, known as pulmonary blood flow), and following exit from the lungs, the blood is ''arterialized'' blood (i.e., it is rich in oxygen) and flows into the left heart, from which it is ejected by the left ventricle to flow to the rest of body (including the working muscles).

Oxygen is transferred to muscle cells through small blood vessels (i.e., capillaries), and oxygen then diffuses through the capillary membrane and the sarcolemma to mitochondria (the power plants of cells, where oxygen is used to convert foods to ATP). Thus, oxygen movement from the air to muscle mitochondria could be limited at numerous sites during its transfer. In healthy young humans at sea level, the limiting site for oxygen transfer in maximal aerobic exercise is the maximal ability of the heart to eject blood per minute.

D. What Is the Significance of Maximal Aerobic Fitness?

The second important aspect of the concept of maximal oxygen consumption in aerobic exercise is that its value determines how long a person can work before becoming fatigued (i.e., the person is unable to continue to exercise at that work magnitude). For example, if persons A and B both weighed 160 lb and walked at 3 mph, did light industrial work, or did housework, they would expend 5 cal of energy per minute and use 1 liter of oxygen per minute. If the maximal oxygen uptake for person A were 2 liters of oxygen per minute and for person B was 4 liters of oxygen per minute, then person A would be working at 50% of his maximal aerobic work capacity (1 liter of oxgyen used per 2 liters of oxygen consumption capacity) and person B would be working at 25% of his maximal aerobic potential (1 liter of oxygen used per 4 liters of oxygen consumption capacity) when they both walk at 3 mph.

The intensity of aerobic exercise for a given person is determined by the percentage of maximal aerobic work done, not by the absolute number of calories or oxygen used. Person A would fatigue sooner than person B if they were walking together at 3 mph, because person A would be working at a higher percentage of his capacity to use oxygen or calories. The reason for fatiguing sooner when the percentage of aerobic effort is higher is that the time one can exercise aerobically at a given oxygen consumption (in the above example, using 1 liter of oxygen per minute) is inversely related to the percentage of maximal oxygen that can be consumed per minute (Fig. 1).

Thus, person B can continue exercising at 25% of his maximal ability to use oxygen or calories when

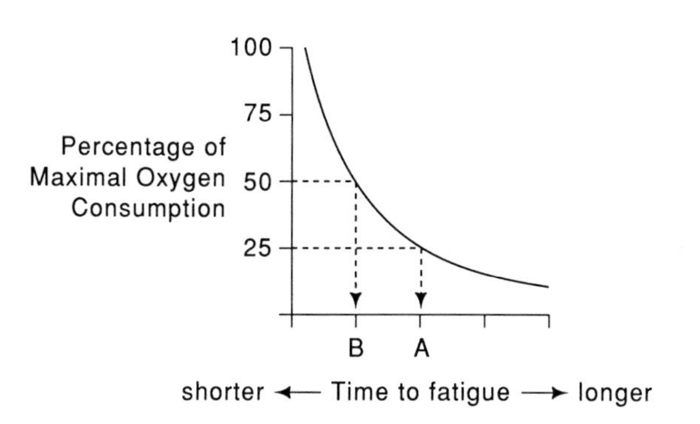

FIGURE 1 The time of work until the work cannot be continued any longer (due to fatigue) is inversely related to the percentage of maximal oxygen consumption invoked by the work being done. When person A has one-half the absolute value of maximal oxygen consumption of person B, and when persons A and B work at the same intensity or oxygen consumption, then person B will work a shorter time before fatigue, because person B is working at a higher percentage of his maximal oxygen consumption. [Reprinted, with permission, from R. H. Strauss (1984). p. 44. "Sports Medicine," Saunders, Philadelphia.

person A has fatigued, walking at the same speed. Thus, the maximal oxygen or calories that can be used per minute sets the comfort zone for work. A person with a high maximal oxygen consumption can either do the same workload for a longer period or undertake higher quantities of physical aerobic work for the same time than a person with a lower maximal oxygen consumption.

E. Can Aerobic Fitness for a Given Person Change?

Since maximal oxygen consumption per minute in a given person can be increased (by physical training) or decreased (by detraining, by sickness or disability, or by aging), a person's ability to work aerobically is determined by his state of physical training, health, and age. A healthy 20-year-old who experimentally undergoes a forced continuous period in bed for 20 days will experience physical deconditioning or detraining and a decrease in his potential to use oxygen or calories during aerobic exercise.

Likewise with aging from the age of 20 years, maximal oxygen consumption declines about 10% per decade. Thus, housekeeping chores will require a greater percentage of the maximal aerobic capacity of an 80-year-old than of a 20-year-old. The reason is the difference in their maximal oxygen consumptions. Housework requires the same aerobic cost of calories or oxygen at both ages, but the 80-year-old has a lower aerobic fitness, works at a higher percentage of his maximal oxygen consumption, and thus fatigues sooner than does a 20-year-old doing the same work.

F. Caloric Costs of Work

Work is defined as force times distance. In running, force is the weight of the body being lifted against gravity as a step is taken. Thus, work during running is the weight of the body times the distance moved. If small allowances for differences in body weight and speed of covering a distance are not made, approximately 100 kcal are used for each mile completed. The conversion of calories used to the amount of fat needed for this caloric expenditure follows. If a person were to add 1 mile more of distance to his usual daily exercise for 1 year, the caloric equivalent of 10 lb of body fat (36,500 kcal) would be used. Since only about 0.03 lb of stored fat is used for each additional mile on a single day, it is obvious that reduction of body fat by aerobic exercise is a long-term process (i.e., months to years) and is related to the additional distance covered, rather than to the speed at which the exercise is performed. The effect of the speed is small: Completing 1 mile by running uses only about 10 kcal more than walking.

Table I gives the caloric expenditures for certain

TABLE I Caloric Ladder for a 154-lb Person[a]

kcal/hr	Activity
1100	Running at 9 mph, walking upstairs, cross-country skiing at 8 mph
1000	—
900	Wrestling, rowing at 6 mph, running at 7 mph
800	Vigorous rope-skipping, boxing, soccer, swimming at 50 yd/min
700	Rapid ice-skating, bicycling at 13 mph, playing competitive squash
600	Running at 5 mph, playing paddleball, vigorous downhill skiing, cross-country skiing at 4 mph, playing basketball
500	Weight training; playing football, volleyball, or singles tennis; skating leisurely, bicycling at 11 mph
400	Swimming at 25 yd/min, walking downstairs, doing light calisthenics, playing doubles tennis
300	Walking at 3 mph, playing softball, bicycling at 8 mph, sailing, leisurely rope-skipping

[a] Caloric expenditure increases 10% for each additional 15 pounds in body weight.

sports. In brief, more calories are burned when larger numbers or masses of skeletal muscle are contracting (e.g., cross-country skiing uses more kilocalories than throwing a dart), when the exercise causes the body to be lifted more against gravity (e.g., walking upstairs uses almost three times as many kilocalories as walking downstairs), and when the body is moved over a longer distance (e.g., walking 5 miles uses almost five times as many kilocalories as running 1 mile).

G. Appetite and Aerobic Exercise

As discussed above, increased amounts of aerobic exercise use more calories. As a general rule, the increase in caloric intake because of increased appetite does not compensate for the increase in caloric expenditure, and some loss in body weight always occurs. Weight loss, however, is greater if caloric intake is maintained at the original level. [*See* APPETITE.]

II. Cardiovascular System

A. Effect of Aerobic Exercise on the Heart

Aerobic exercise continued daily for weeks improves the ability of the normal healthy heart to pump blood. For humans the minimum quantity of aerobic exercise needed to change the pumping capacity of the heart is 20–30 min/day, 3–4 days/week, at a heart rate corresponding to about 70% of the maximal rate. An approximate easy method for estimating the maximal heart rate is to subtract a person's age from 220. Thus, a 20-year-old has a maximal heart rate near 200 beats per minute and an 80-year-old has a maximal heart rate of about 140 beats per minute. However, 33% of people have a true maximal heart rate 10 beats either higher or lower than these estimates. A better way to obtain the maximal heart rate is to measure it during maximal exercise, but this is more difficult to do outside of a laboratory. The heart rate corresponding to 70% of the maximum is calculated by the formula:

$$70\% \text{ of maximal heart rate} = 0.7 \times (\text{maximal heart rate} - \text{resting heart rate}) + \text{resting heart rate}$$

For example, for a 20-year-old it would be:

$$0.7 \times (200 - 70) + 70 = 161 \text{ beats per minute at } 70\% \text{ of maximal exercise}$$

If the person in this example ran 3–4 days/week for 20–30 min/day at a heart rate of 161 beats per minute for 1 month (aerobic exercise training), the person's maximal heart rate would not change, but his heart rate at rest and at the same running intensity would be lower (the lower heart rate is termed bradycardia). Heart rate is one of two components determining the amount of blood ejected from the heart per minute (cardiac output). The formula for cardiac output is:

$$\text{Cardiac output} = \text{heart rate} \times \text{stroke volume}$$

The effects of aerobic exercise training on these three components are given in Table II.

TABLE II Magnitude of Increase in Heart Functions for Various Workloads after Aerobic Exercise Training

Heart function	At rest	Running at 5 mph	Running at maximal speed
Cardiac output (liters of blood per minute)	Same	Same[a]	Increase
Heart rate (beats per minute)	Less	Less[b]	Same
Stroke volume (milliliters of blood per beat)	More	More	More

[a] The increase in cardiac output when going from rest to running at 5 mph is the same before and after training.
[b] The increase in heart rate when going from rest to running at 5 mph is less after aerobic training than prior to it.

After aerobic exercise training stroke volume (i.e., the amount of blood ejected by a single heart beat) increases at rest, at 70% of maximal aerobic work, and at maximal heart rate. Other heart functions are also increased after 4–8 weeks of aerobic exercise training, as shown in Table III.

TABLE III Heart Function Values at Maximal Exercise Before and After Aerobic Exercise Training

Heart function	Before training	After training
Cardiac output (liters of blood per minute)	20	26
Heart rate (beats per minute)	200	200
Stroke volume (liters of blood per beat)	0.10	0.13

Maximal cardiac output (i.e., the heart's reserve to pump blood) was enhanced by 6 liters/minute after aerobic exercise training. Such a reserve could assist survival to an absolute challenge (e.g., if 10% of myocardial tissue stopped contracting from an illness); moreover, it permits the transfer of more oxygen from the lungs to the tissues at the maximal heart rate. The consequences of increasing maximal cardiac output are to make the caloric cost of housework a smaller percentage of the maximum number of calories that can be used per minute.

Information is becoming available to explain the mechanism by which aerobic training improves the fitness of the heart to work. One of the factors is a change in the activity of the sympathetic nervous innervation of the sinoatrial node, the heart pacemaker. The sympathetic system increases heart rate. During aerobic exercise its activity increases, causing an increase in heart rate. However, after aerobic training the sympathetic stimulation of the sinoatrial node is reduced, and the heart rate increases less when running at the same speed (Table IV).

It is not known how the body "senses" the percentage of maximal aerobic work effort and then modifies the activity of the sympathetic nervous system to adjust the heart rate. Aerobic training might improve the contractility of the heart (i.e., the amount of muscle tension at a given heart length) by increasing calcium influx during sarcolemmal depolarization. Higher intracellular free calcium would increase muscle tension, which would cause the improved stroke volume. [*See* EXERCISE AND CARDIOVASCULAR FUNCTION.]

B. Exercises Improving the Fitness of the Heart to Work

By definition, all exercises use more calories than the basal metabolic rate. As indicated in Section I,F, some types of exercise use more calories than others. Not all types of exercise produce aerobic or heart fitness (i.e., bradycardia and increased stroke volume). For example, if a healthy 20-year-old with a maximal oxygen consumption of 4 liters of oxygen per minute walks, he will use extra calories but will not develop heart fitness, because he will not be working at 70% of his maximal aerobic work capacity. On the other hand, if an 80-year-old with a maximal oxygen consumption of 2 liters of oxygen per minute walks, he would be working above 70% of his maximal aerobic capacity and will develop heart fitness. Thus, walking uses calories and produces heart fitness for an 80-year-old, whereas for a normal 20-year-old walking only uses calories but does not improve the aerobic work capacity.

Another exercise, weight-lifting, produces increased strength of those skeletal muscles that are trained, but does not improve the capacity of the heart to work. The reason is that weight-lifting and power training do not maintain an increased heart rate that is 70% of maximal heart rate for 30 consecutive minutes. Only circuit training, in which numerous strength exercises are performed sequentially without rest periods in between, produces some heart fitness. Weight training uses extra calories. On the other hand, aerobic exercises usually produce little or no increases in the strength of skeletal muscles.

The above examples indicate the specificity of the type of exercise on the adaptation of training. Other examples are given in Table V.

TABLE IV Degree of Lessening of the Sympathetic Discharge Related to the Percentage of Improvement in Maximal Oxygen Consumption[a]

Heart function	Before aerobic training	After aerobic training
Maximal oxygen consumption (liters of oxygen per minute)	3.0	4.0
Maximal heart rate	200	200
When running at 5 mph		
Oxygen consumption (liters of oxygen per minute)	2.1	2.1
Heart rate (beats per minute)	160	140
Percentage of maximal oxygen consumption	69%	54%
Percentage of maximal heart rate	70%	53%

[a] See Table V.

C. Partitioning of Cardiac Output to Tissues During Aerobic Exercise

The tissue that most needs the increased blood flow during exercise is the contracting skeletal muscle. Blood flow supplies the oxygen and metabolic fuels for the synthesis of ATP, the energy source required for muscle contraction. The normal blood flow to muscle is insufficient to supply the energy needs for vigorous contraction; hence, an increase in cardiac output is required in aerobic exercise. If all tissues were to share similar percentage increases in cardiac output during aerobic exercise, skeletal muscles would still not have sufficient arterial blood flow to perform maximal work. During exercise blood flow is shifted to the exercising skeletal muscle from other organs, by the following mechanisms. Sympathetic stimulation of arterioles of all tissues but the brain is increased, causing the

TABLE V Fitness Elements

Category of exercise training	Example of exercise	Uses calories?	Is heart fitness improved?	Is strength of skeletal muscle improved?	Is coordination improved?
Aerobic training	Jogging	Yes	Yes	No	Yes
Walking 20-year-old	Walking	Yes	No	No	Sometimes
Strength training	Weight-lifting	Yes	No	Yes	Yes
Anaerobic training	Sprinting	Yes	No	No	Minimal
Coordination training	Somersaulting	Yes	No	No	Yes

smooth muscle around the arterioles to contract, constricting the radius of the arteriolar lumen. This decrease in radius increases resistance to blood flow, so that less blood reaches the tissue. For example, in maximal aerobic exercise the amount of blood flowing to the liver, kidney, stomach, and intestines is reduced to 20% of nonexercise blood flow. [*See* SKELETAL MUSCLF.]

A similar process occurs in inactive skeletal muscle, whereas blood flow increases in heart and contracting skeletal muscle. In working muscle the metabolic byproducts of contraction (e.g., adenosine, carbon dioxide, and hydrogen and potassium ions) counteract the sympathetic system-induced constriction, resulting in dilation of the arterioles. The consequence of these events is to redistribute the flow resulting from the increase in cardiac output to the working skeletal muscle. When the requirement for blood flow to exercising skeletal muscle begins to exceed maximal cardiac output, a reflex from muscle activates the sympathetic nervous system so that the increase in muscle blood flow is lessened and arterial blood pressure is maintained in intense exercise. This reflex is attributed to protons (i.e., hydrogen ions) arising in part from an increased glycolytic rate in the working skeletal muscle.

D. Aerobic Exercise Is an Example of the "Fight or Flight" Syndrome

The increased sympathetic nervous system stimulation of arteriolar smooth muscle is a common survival event in the body. For example, it occurs in blood loss (i.e., hemorrhage), during which blood volume and arteriolar blood pressure decrease. If blood pressure falls too much, the force driving blood flow through tissues becomes insufficient, and tissues become hypoxic and die. A reflex survival mechanism released by the drop in blood pressure during hemorrhage signals the sympathetic nervous system to become more active, so that arterioles constrict and blood pressure increases.

During aerobic exercise the engagement of the sympathetic nervous system prevents a fall in arterial blood pressure. As opposed to hemorrhage, the signal to increase arterial blood pressure during exercise comes from the working muscle and some areas of the brain. However, as in hemorrhage, activation of the sympathetic nervous system by the survival mechanism of fight or flight prevents fainting due to hypotension during aerobic exercise (the flight).

III. Metabolic Responses

A. Communicative Roles of the Sympathetic Nervous System during Aerobic Exercise

Another example of how the sympathetic nervous system integrates all tissues to cooperate to maximize the ability to continue aerobic exercise is the regulation of sugar utilization. During aerobic exercise sympathetic nervous system discharges to the β cells of the pancreas cause a decrease in insulin secretion into the blood, with a consequent decrease in blood insulin levels.

As in juvenile diabetes, if insulin is low in the blood, sugar (glucose) will not enter skeletal mus-

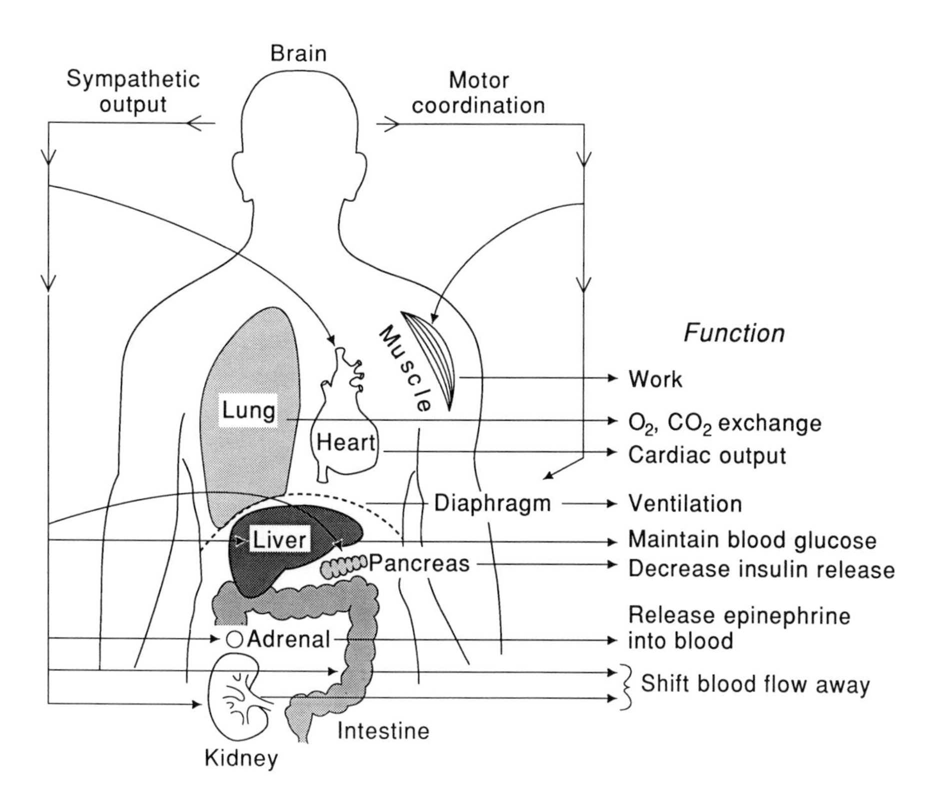

FIGURE 2 Integrative control of multiple organs during aerobic exercise. Both the nervous and hormonal systems communicate to all organs and tissues in the body during exercise and cause each organ system to undertake a specific function, so that the summation of all organ systems allows sufficient oxygen and caloric fuel for the contraction of large masses of skeletal muscle during exercise and allows for the elimination of metabolic byproducts, such as carbon dioxide (from the lungs), heat (from the skin), and hydrogen ions (from the kidneys).

cle. However, contraction of the muscle fiber causes the recruitment of glucose transporters from an internal storage site to the cell membrane. Increased numbers and activity of glucose transporters in the cell membrane override the effect of low blood insulin. Glucose uptake increases in the working, but not the resting, skeletal muscle. In this way blood sugar is redirected to muscles needing it. [*See* INSULIN AND GLUCAGON.]

This situation is analogous to the redistribution of blood flow toward exercising skeletal muscle. Low blood insulin also allows an increased breakdown and release of fatty acids from fat cells into the blood in a normal person during exercise (Fig. 2). When this occurs, a larger amount of fatty acids is used as fuel for muscle work, sparing the limited stores of glucose in the form of glycogen in the body until later during exercise. The net result of this regulatory role of the sympathetic nervous system is that the duration of aerobic exercise is longer before fatigue. Aerobic training accentuates this effect. When running at the same speed after training, less muscle glycogen is used as compared to before training. [*See* FATTY ACID UPTAKE BY CELLS.]

B. Heat as a Byproduct of Muscle Contraction

The formation of ATP from glucose and oxygen releases heat as well as other byproducts into the muscle, and this heat is carried from the muscle to the rest of the body by the blood flow, thus causing body temperature to increase during exercise. Some of this is lost to the environment, preventing body temperature from rising above 41°C, and consequently preventing heat stroke.

Both overheating and dehydration (i.e., loss of body water via sweat) can limit exercise. Thus, fluid replenishment during exercise and limitation of exercise at temperatures above 27°C are important. In addition, maximal aerobic capacity is reduced during exercise at high environmental temperature. For example, maximal oxygen consumption can decrease from 4 to 3 liters of oxygen per minute in the heat, while maximal cardiac output remains unchanged. The mechanism is the following. Exercising in the heat causes vasodilation of the skin, where heat from blood flowing from the core of the body can be lost to the external environment by convection (wind), by radiation (heat waves), by transfer from the skin to cooler external objects (conduction), or by the evaporation of sweat. The vasodilation of skin arterioles and small veins during exercise in the heat diverts some of the blood flow needed by the exercising skeletal muscles to the skin. Thus, the amount of blood supplied to skeletal muscle is reduced, and muscle work is limited by the lowered supply of oxygen.

IV. Adaptations

A. Plasticity or Adaptability of the Body to Aerobic Exercise Training

If aerobic exercise is performed 3–4 days/week for many weeks, two things are perceived by the person who is exercising. First, the performance of a certain amount of aerobic work (e.g., running at 5 mph) does not seem as stressful or as hard as before training. Second, performance of an absolute amount of aerobic work after training can be done for a longer duration before fatigue forces a stop to the exercise. Part of the reason for these perceptions is given in Section I,D and Table IV. Aerobic training increases maximal oxygen consumption with the consequent effect that an absolute workload requires a smaller percentage of the maximal oxygen that the body can use. Essentially, the body adapts to repeated daily bouts of exercise, so that there is less disruption of the internal environment surrounding and within cells (homeostasis). Biochemical and structural changes occur after training (adaptations), so that a given amount of exercise produces less stress on cells and on the organism. [*See* pH HOMEOSTASIS.]

Because the composition of cells and tissues is altered by the training, cells and tissues are said to exhibit plasticity. Two examples can be given. First, the increase in maximal cardiac output with aerobic training, which has been mentioned in Table III. Since cardiac output for a given submaximal aerobic exercise is unchanged by aerobic training, the heart works at a smaller percentage of its reserve after training. A second example is weightlifting training. Skeletal muscle enlarges so that a 100-lb load is distributed over a larger cross-sectional area of the muscle after training. Thus, the amount of the 100 lb supported by a 1-square-inch area is less after weight training of this muscle. In both of these examples, homeostasis for an absolute workload is less disrupted after cellular and organ adaptations to training.

B. Biochemical Basis of Adaptations to Aerobic Exercise

The tissue for which most is known about plasticity or adaptability to aerobic exercise is skeletal muscle. In response to aerobic training, the density of mitochondria and capillaries in skeletal muscle increases, but the muscular mass remains relatively unchanged. Adaptations to training are specific to the skeletal muscle that is trained. If a single leg is trained to bicycle, while the opposite leg does no aerobic training (on a specially constructed bike), mitochondria only increase in the muscle of the leg that trained. The opposite nonexercised leg has no change in mitochondria.

One of the major functions of the increased mitochondria is to spare the utilization of the body's stores of glucose during aerobic exercise. The mechanism is the following. The process of muscle shortening requires an energy source, ATP, to power the sliding of the muscle's contractile elements. Thus, during muscle contraction, ATP falls and ADP rises. After aerobic training, if the density of mitochondria in the trained skeletal muscle doubles, then the oxidative phosphorylation sites per unit volume of muscle doubled. ADP only rises one-half as much as its increase in nontrained muscle to obtain the same number of produced ATP molecules. The smaller increase in ADP in the aerobically trained skeletal muscle spares the utilization of glucose for ATP production. Glycolytic flux from glucose to pyruvate during exercise is increased less by the smaller ADP increase in skeletal muscle with higher mitochondrial density. Therefore, the trained skeletal muscle uses less glucose. [*See* ATP SYNTHESIS BY OXIDATIVE PHOSPHORYLATION IN MITOCHONDRIA.]

The trained muscle uses fat as an energy source to make ATP. Because the amount of glucose that is stored as glycogen is limited, the body usually begins to run low on glucose for fuel after running about 18 miles. When blood glucose concentrations begin to fall as glycogen stores are used up, the body is unable to continue exercising at the same speed, probably because nerves and red blood cells can only use glucose to make ATP. (Muscle can oxidize both glucose and fatty acids to make ATP.)

Thus, the training adaptation of glycogen sparing permits running longer at the same speed before fatiguing due to glycogen depletion. It is well known that eating sugar during running to exhaustion delays the onset of fatigue.

Present information suggests that exercise increases mitochondria by increasing the synthesis of mitochondrial proteins, owing to an increase in the concentration of mRNAs for mitochondrial proteins.

V. Summary

Muscle contraction permits animals and humans to move, and thus to survive. "Physical exercise" is a term employed to describe this activity. Physical exercise stresses the body and requires the integration of multiple organ systems to supply oxygen and fuel to the working skeletal muscles. Repeated bouts of exercise over a period of weeks results in adaptations that diminish the stress of exercise and permit more work prior to fatigue.

Bibliography

Brooks, G. A., and Fahey, T. D. (1987). "Fundamentals of Human Performance." Macmillan, New York.

Dempsey, J. A. (1988). Problems with the hyperventilatory response to exercise and hypoxia. *Adv. Exp. Med. Biol.* **227,** 277–291.

Gollnick, P. D., Riedy, M., Quintinskie, J. J., and Bertocci, L. A. (1985). Differences in metabolic potential of skeletal muscle fibres and their significance for metabolic control. *J. Exp. Biol.* **115,** 191–199.

Holloszy, J. O., and Coyle, E. F. (1984). Adaptations of skeletal muscle to endurance exercise and their metabolic consequences. *J. Appl. Physiol.* **56,** 831.

Lamb, D. R., and Murray, R. (1988). "Perspectives in Exercise Science and Sports Medicine. Volume I: Prolonged Exercise." Benchmark, Indianapolis, Indiana.

Nadel, E. R. (1985). Recent advances in temperature regulation during exercise in humans. *Fed. Proc., Fed. Am. Soc. Exp. Biol.* **44,** 2286–2292.

Rowell, L. B. (1988). Muscle blood flow in humans; How high can it go? *Med. Sci. Sports Exercise* **20,** S97–S103.

Saltin, B., and Gollnick, P. D. (1983). Skeletal muscle adaptability: Significance for metabolism and performance. *In* "Handbook of Physiology. Section 10: Skeletal Muscle" (L. D. Peachey, ed.). Williams & Wilkins, Baltimore, Maryland.

Strauss, R. H. (1984). "Sports Medicine." Saunders, Philadelphia, Pennsylvania.

Vollestad, N. K., and Sejersted, O. M. (1986). Exercise in human physiology. *Acta Physiol. Scand.* **128** (Suppl. **556**), 7–166.

Exercise and Cardiovascular Function

GEORGE E. TAFFET and CHARLOTTE A. TATE, *Baylor College of Medicine*

Glossary

Duration The length of time that the exercise intensity can be maintained per unit of time or distance

Dynamic exercise Isotonic; movement or physical activity in which work is performed; force is generated for external work

Exercise Force or tension developed by muscle; movement; physical activity above the basal, or resting, state

Intensity The amount of exercise that can be exerted per unit of effort; implies a quantity of effort

Static exercise Isometric; force is generated in muscle, but no external work is performed; resistance exercise

IN THE BASAL, or resting, metabolic state the cardiovascular system of a healthy adult pumps blood at only a small percentage of its potential capacity. The blood pumped from the heart (i.e., cardiac output) is primarily directed to the visceral organs, the brain, and the heart itself, with modest flow to the peripheral muscles. With the abrupt onset of physical activity (i.e., exercise), the increased energy demands of the working muscle require an augmentation of blood flow to the musculature for sustained physical activity. During exercise, therefore, a greater proportion of the cardiac output must be directed toward the working muscle for the delivery of oxygen and nutrients and for the removal of metabolic waste products and heat. Thus, the function of the cardiovascular system must be precisely regulated to meet the increased demands for blood flow by the working muscles. The adjustments and regulation of the cardiovascular system with exercise are considered here, and we describe the cardiovascular adjustments to dynamic exercise in which a large muscle mass is used. A brief description of the response to static exercise is given. The mechanisms underlying the physiological adjustments with exercise cannot be deeply analyzed here, however, and the reader is directed to the Bibliography for an in-depth treatment.

I. Cardiovascular Adjustments to Dynamic Exercise

A. Introduction

Whole-body oxygen consumption per unit of time ($\dot{V}O_2$) increases in the transition from rest to physical activity. The amount of oxygen consumed depends on the imposed absolute workload and can go from 2–3 ml/kg/min at rest to 60 ml/kg/min at maximal capacity in an active young adult male. The oxygen consumed for gardening, for example, is lower than that consumed for the lifting of heavy boxes. The intensity of work therefore determines in part the absolute oxygen consumption, which reflects the aerobic (i.e., oxygen-consuming) energy expenditure necessary to perform the exercise. The ability to sustain the increased workload with the new higher oxygen consumption is a measure of endurance, which depends in part on the relative workload (i.e., the percentage of one's maximal capacity to consume oxygen). The maximal capacity to consume oxygen (i.e., maximal oxygen consumption) is determined by progressively increasing the external workload until the amount of oxygen consumed no longer increases, despite an increase in the workload. The additional energy demands for the workloads higher than that required to elicit maximal oxygen consumption are met through anaerobic processes.

Maximal oxygen consumption is the best measure of cardiovascular fitness and provides a quantitative index of the maximal capacity of the cardiovascular system to deliver blood, with its oxygen and nutrients, to the working muscles and to remove metabolic waste products and heat. Maximal oxygen consumption is higher in an endurance-trained individual, although submaximal oxygen consumption (i.e., the oxygen consumed at a submaximal absolute workload) is equal in trained and untrained people of comparable age, body weight, and health status. The oxygen cost of doing submaximal tasks does not vary much among individuals of comparable body weight. This means that the same absolute submaximal workload taxes a greater percentage of the untrained person's maximal oxygen consumption. The importance of this concept lies in the quality of life for the endurance-trained person. The tasks of everyday living tax the maximal capacity of the trained person to a lesser extent, so that the task is relatively easier after an endurance training program.

B. Fick Equation

Oxygen consumption results from numerous processes, which are best represented by the Fick equation,

$$\dot{V}O_2 = \dot{Q} \times (a - \bar{v})O_2 \text{ difference}$$

where $\dot{V}O_2$ is oxygen consumption, $\dot{Q}$ is cardiac output, and $(a - \bar{v})O_2$ difference is the ability of the working muscles to extract and utilize oxygen from the perfusing blood. An increase in cardiac output and/or the $(a - \bar{v})O_2$ difference augments oxygen consumption. Most exercise physiologists agree that maximal oxygen consumption is limited by cardiac output (i.e., the primary determinant), whereas the ability to perform prolonged work at a submaximal workload might be limited by the $(a - \bar{v})O_2$ difference. The maximal capacity of the individual to perform aerobic work is limited by processes related to the cardiac output, rather than peripherally at the skeletal muscle. The components of the Fick equation are discussed below.

1. Cardiac Output

Cardiac output is the product of the heart rate times the stroke volume (i.e., the amount of blood pumped by the heart per beat). The systolic function of the heart is the actual pumping of the blood filling the ventricles via the contraction of the cardiac muscle cells. The diastolic function of the heart is the relaxation of the cardiac muscle cells following contraction, which allows the ventricles to fill with the incoming blood. Both systole and diastole, or contraction and relaxation, are important determinants of cardiac output. They are regulated by a number of factors which increase the chronotropic state of the heart (i.e., heart rate) and the positive inotropy (i.e., the increased rate of contraction and relaxation). Since cardiac output is so central to the adjustments of the cardiovascular system to dynamic exercise, much more emphasis is placed on the central regulating factors than on the peripheral adjustments. [*See* CARDIOVASCULAR SYSTEM, PHYSIOLOGY AND BIOCHEMISTRY.]

Cardiac output linearly increases as the work intensity increases. In a young adult male cardiac output increases from 5 liters/min at rest to 30 liters/min at maximal oxygen consumption. Females of comparable age, health, and degree of physical fitness typically have lower cardiac outputs at all intensities of work, primarily because of their typically smaller body size.

a. Heart Rate Although both heart rate and stroke volume increase with exercise to increase cardiac output, the augmentation of heart rate is the most important variable in most individuals. With the onset of dynamic exercise, there is a rapid increase in heart rate in proportion to the increase in $\dot{V}O_2$ and the external work. Multiple control mechanisms produce the increase in heart rate, including the parasympathetic and sympathetic nervous systems. Initially, the withdrawal of parasympathetic stimulation produces an increased rate, which is further augmented by sympathetic stimulation as exercise continues. Even at mild levels of dynamic exercise (i.e., 20–30% $\dot{V}O_2$ max), the sympathetic nervous system increases heart rate in proportion to demand. This effect is mediated by the β_1-adrenergic receptor, which responds to the neurotransmitter norepinephrine released at nerve terminals, because it can be blocked with β_1-selective antagonists such as atenolol. The circulating neurotransmitter, in contrast, does not appear to have much effect on heart rate. In fact, the heart rate of denervated or transplanted hearts, which have no sympathetic nervous input, cannot increase with exercise despite elevated circulating epinephrine and norepinephrine via endogenous overflow from the adrenal glands or from exogenous epinephrine infusion.

The other control of heart rate is intrinsic to the myocardium. The maximum heart rate appears to be approximately 220 beats per minute in young adult males, which is perhaps 10 beats per minute higher in females of comparable age. This limit is characteristic of the heart's internal pacemaker system, the sinoatrial node, and the conduction pathways. Electrical stimulation of the heart that circumvents the pacemaker system, either internal (e.g., arrhythmias) or external (e.g., pacemakers), can easily exceed the maximum heart rate, indicating that the intrinsic control of the maximum heart rate is at the sinoatrial node, which generates the signals, rather than the atrioventricular conduction pathway, which distributes the signal throughout the heart.

Exercise training does not alter the maximum heart rate; however, the submaximal heart rate at any given absolute workload and at rest is lower following an endurance training program. The resting bradycardia (i.e., low heart rate; 35–40 beats per minute in some athletes) and lower submaximal heart rate in a trained individual result primarily from a relative decrease in exercise-induced sympathetic tone, although an augmentation of parasympathetic tone has been implicated in some studies. This means that in an endurance-trained athlete, in whom the oxygen cost of performing a submaximal task is the same as in an untrained person, cardiac output can only be maintained by increasing the stroke volume. Furthermore, the increased maximal oxygen consumption in an endurance-trained person must be gained primarily by an increased cardiac output via an augmentation of stroke volume. Thus, stroke volume is critical to the oxygen consumption of an endurance-trained person.

b. Stroke Volume The stroke volume is the amount of blood ejected from the heart into the aorta with each contraction. The increase in stroke volume that occurs with dynamic exercise is quite small when compared to the 2- or 3-fold increase in heart rate. However, in a trained person whose submaximal heart rate is lower with an unchanged maximal heart rate, changes in stroke volume are the major way that cardiac output can increase with exercise, as stated above. Stroke volume during exercise is regulated by three primary factors: the Frank–Starling mechanism (preload), afterload or peripheral resistance, and contractility with (to a lesser extent) cardiac hypertrophy in highly trained athletes.

i. *Frank–Starling Mechanism* The force of contraction of the heart's ventricles varies as a function of the end-diastolic size, which is the absolute volume of the ventricle at the end of the filling period. This means that the larger the end-diastolic size, the greater is the force of contraction. This phenomenon is called the Frank–Starling mechanism, or Starling's law of the heart, and is the major way in which stroke volume is increased during exercise in which stroke volume is an important determinant of cardiac output, such as in an endurance-trained athlete. This mechanism is based on the relationship between length and active tension of the sarcomere, which is the basic contractile unit at the molecular level. Basically, this mechanism means that as the volume of the ventricle expands, the muscle is stretched passively, and this passive stretch increases the force of contraction. However, there is a point at which an increasing passive stretch actually decreases the force of contraction. This ordinarily does not happen in a normal heart.

For our purposes here the Frank–Starling curve is the integration of all of the responses of the heart to changes in its preload. Preload is a reflection of the filling of the heart, which is obviously dependent on venous return to the heart. The venous return is altered by the position of the body. For example, there is less filling (per unit time) in the upright position than in the supine position. When dynamic leg exercise is performed in the upright position, the blood pooled in the legs is forced back to the heart by the contraction of the leg muscles. Furthermore, the effort of ventilating the lungs produces negative pressure within the chest and pulls blood into the heart. This increased venous return results in an increased right-sided preload, and thus left-sided preload. The increase in left ventricular filling is greater for supine exercise than for upright exercise. The left ventricular end diastolic volume might be greater for supine exercise than for upright exercise, but similar peak left ventricular end-diastolic volumes occur in both postures at close to maximum levels of exercise intensity.

In an endurance-trained person, the total blood volume can increase as much as 10%. This expansion of total blood volume increases stroke volume via the Frank–Starling mechanism. The expansion of blood volume can also be achieved by the infusion of a physiologically compatible liquid (e.g., a dextran solution), which in turn augments stroke volume.

ii. *Afterload* Afterload is the tension or stress placed on the ventricular wall while the ventricle is ejecting blood. In part, it is determined by the systolic pressure in the ventricle; therefore, it is influenced by the systolic arterial blood pressure and the peripheral vascular resistance. One can envision afterload as a "wall of pressure" against which the heart must pump blood. When the wall is high, the heart must pump harder to overcome this resistance by pressure. By inference, the amount of blood pumped with each beat is lower when the wall is higher. Thus, mechanisms must be in place to insure that the pressures developed at the periphery are not excessive and can be reasonably overcome during dynamic exercise.

With the onset of dynamic exercise, systolic blood pressure increases from 120 mm Hg to 140–160 mm Hg, depending on the intensity of the exercise. This increase in systolic blood pressure results primarily from the increased blood flow within the first few minutes of physical activity, and it correlates positively with both the increased cardiac output and the increased relative intensity of exercise (i.e., the percentage of maximal oxygen consumption). Diastolic blood pressure, on the other hand, does not increase much during dynamic exercise. Therefore, there is a minor change in mean arterial blood pressure with dynamic exercise, at least when the intensity is less than 75% of one's maximal oxygen consumption. At the same time the peripheral resistance to blood flow, or vasomotor tone, is substantially reduced in the initial stages of all intensities of dynamic exercise. Overall, there is an exercise-induced reduction of peripheral resistance with a minor rise in arterial blood pressure, so that stroke volume is increased rapidly in the early stages of dynamic exercise. (The mechanisms underlying the decreased peripheral resistance with exercise are discussed in Section IV.)

iii. *Contractility* Contractility, the intrinsic state of the heart, is difficult to measure *in vivo*, primarily because there are numerous confounding variables. A true measure of contractility is only possible with an isolated heart in which these variables are controlled or with isolated heart muscle strips (e.g., papillary muscles). Experimental evidence supports the idea that contractility is increased to some degree in exercising animals, which is produced in part by β-adrenergic sympathetic stimulation. In experimental animals, in which preload and afterload can be manipulated independently, the maximum rate of pressure development ($+dP/dt$) is increased during exercise in exercise-trained hearts, indicating that exercise training results in an augmentation of intrinsic contractility. Also, when the heart of an exercising dog has the same preload and afterload as that of a resting animal, it ejects more blood, again suggesting an augmentation of contractility with increased physiological demand. The effect of exercise training on contractility in the unstressed state (i.e., at rest) is ill defined.

The speed of contraction ($+dP/dt$) has a subcellular basis at the level of the basic contractile apparatus (i.e., the sarcomere) in the cardiac muscle cell. The sarcomere is composed of two primary contractile proteins, actin and myosin, and two regulatory proteins, troponin and tropomyosin. Numerous studies have shown that the speed of contraction is related directly to the different forms of myosin (i.e., slow to fast forms), so that the faster myosin form is observed in the hearts of animals which have a faster $+dP/dt$. Relaxation (or $-dP/dt$) of the cardiac muscle cell after contraction is an important component of the contractile properties of cardiac muscle cells, especially for the optimal filling of the heart during diastole. However, relatively few studies have indicated any change in the relaxation of the cardiac muscle cells after exercise training, except under certain circumstances in which relaxation is impaired (e.g., old age).

iv. *"Physiological" Hypertrophy* Although the evidence is not concrete, another mechanism that can increase the stroke volume is physiological cardiac hypertrophy. ("Physiological" is used in the sense that the underlying mechanism is not caused by disease.) The left ventricular mass of highly trained athletes can be 50% higher than that of age-matched controls. Unlike the hypertrophy found in hypertension, an athlete's heart is enlarged primarily in end-diastolic cavity dimension, with lesser changes in end-systolic dimension and posterior wall and septal wall thickness. The hypertrophied heart has a 20–60% increase in stroke volume at rest and during exercise. With 3 months of vigorous training by either swimming or running, this hypertrophy can occur and then revert back to baseline values with as little as 3 weeks of detraining. This hypertrophy is an absolute increase in cardiac mass which does not result from an increase in the number of cardiac muscle cells (i.e., hyperplasia), but rather from the size of the individual cardiac muscle cells (i.e., hypertrophy).

2. Arteriovenous Oxygen Difference

Along with the increased blood flow to the working muscles, the ability of the muscle to extract oxygen and nutrients increases. This occurs through a variety of mechanisms, one of the most important physiological mechanisms being the increased surface area of the capillaries (via their opening) for the facilitation of this exchange. Since the $(a - \bar{v})O_2$ difference arises from both the arterial oxygen saturation and the mixed venous oxygen content, it is important to realize that the arterial oxygen saturation is nearly complete at 95% and does not change much with exercise. Therefore, the ability of the working muscle to extract and utilize the oxygen and nutrients coming to it from the heart is the primary determinant of the increased $(a - \bar{v})O_2$ difference observed during exercise.

II. Control of Peripheral Blood Flow during Dynamic Exercise

The regulation of blood flow during dynamic exercise is precisely matched to the energetic needs of the working muscles. With the onset of exercise in untrained people, the blood flow from the heart is shunted away from the viscera, where the majority of the blood flow is directed at rest, to the working muscles, including the heart (i.e., coronary blood flow). This means that the blood flow is directed away from the nonworking muscles as well. When exercise proceeds longer than 5 minutes, blood flow increases to the skin for the necessary dissipation of the heat produced during the muscular effort. Blood flow to the brain, however, is not compromised during exercise.

At rest most of the capillaries in skeletal muscle lie dormant. Thus, the opening of these capillaries increases blood flow and thereby increases the volume of blood perfusing the muscle without necessarily increasing the velocity of flow. The augmentation of vascularization by opening the capillaries magnifies the surface area necessary for exchange between the blood and the muscle fibers. Perfusion of the working muscles can increase by 20-fold or more. However, maximal vasodilatation or perfusion can only happen when the mass of the working muscle is small.

This opening of dormant capillaries to increase perfusion of the working muscles occurs primarily through two mechanisms: local control and neural factors. The local factors involved in peripheral vascular dilatation are metabolic byproducts (e.g., changes in temperature, pH, metabolites, and ions), which are generated by the muscle when it begins to work. This autoregulation of vascular dilatation is related directly to the metabolic needs of the working muscle and acts almost immediately at the beginning of exercise. The actual metabolites involved in the autoregulation are not known, but the increase in potassium and osmolality in the interstitial space caused by metabolite and ion release from the working muscle could be potent factors.

The neural control arises from the sympathetic stimulation with the onset of exercise. The sympathetic stimulation of blood flow to the working muscles comes from two subsystems of the sympathetic nervous system: the adrenergic constrictor system and the cholinergic dilator system. The adrenergic constrictor system works through the release of the neurotransmitter norepinephrine and results in vasoconstriction (thus, vasomotor tone), whereas the cholinergic dilator system is effected by the release of the neurotransmitter acetylcholine with subsequent vasodilatation. In the heart and the skeletal muscle the cholinergic nervous system acts to vasodilate the arterioles. Working in concert, these two systems act to vasodilate the working muscle and to vasoconstrict the viscera and the nonworking muscles. However, the local metabolic control probably can override any neural influence in the working muscle which might act to vasoconstrict.

Dynamic exercise produces an increased demand for myocardial blood flow. The coronary flow can increase 2- to 5-fold and is determined in part by the product of the arterial blood pressure times the heart rate. Since diastolic blood pressure is not increased much, the perfusion pressure in the coronary arteries cannot, by itself, improve supply. The major mechanism for increasing myocardial perfusion is the local autoregulation that the heart uses to dilate resistance vessels. Unlike skeletal muscle, the autoregulation can be controlled by adenosine and/or other purinergic receptor agonists.

With exercise training capillarization of both skeletal muscle and heart increases in many cases. When this is observed, the increase is in capillary density, rather than in the diameter of the vessels. This augmentation of capillarization allows for the greater perfusion of the trained muscle, thereby facilitating exchange between the blood and the muscle fibers.

III. Cardiovascular Adjustments to Static Exercise

Unlike dynamic exercise, in which the energy expenditure can be determined by the amount of oxygen consumed during work, the intensity of static exercise is determined by the amount of force or tension the muscle produces to perform the exercise. In practice this is determined by having a person resist a weight (e.g., lifting a barbell and holding it). (The lift, of course, is dynamic exercise.) Such exercise is also called resistance exercise. Another method in the laboratory setting is to use a hand dynamometer, which allows a person to squeeze to his or her maximum force (i.e., maximal voluntary contraction).

The most remarkable response to an acute resistance effort is seen in the response of blood pressure (pressor response), particularly the diastolic blood pressure, which does not change much during dynamic exercise. Both systolic and diastolic blood pressures increase in relation to the intensity of the voluntary contraction, as a percentage of the maximal voluntary contraction. Thus, mean arterial pressure increases. The heart rate increases as it does during dynamic exercise, but not to the same relative extent, unless the voluntary contraction exceeds 40–50% of the maximal voluntary contraction. For example, when a heavy weight is lifted, both heart rate and blood pressure increases, although not linearly, with an increase in active muscle mass. During a double leg-press in which a highly trained lifter is allowed to hold his breath, the heart rate can reach as high as 170 beats per minute (i.e., close to the maximum), with blood pressures of 320 mm Hg systolic and 250 mm Hg diastolic. The goal of the increased pressures is to overcome the high intramuscular pressures to maintain perfusion.

Components contributing to the elevated blood pressure include an increase in cardiac output, elevated intrathoracic and intraabdominal pressures (especially while holding one's breath), and a mild increase in peripheral resistance. During a static contraction, the heart rate and blood pressure increase as the contraction is held. Depending on the muscle, contractions that attain only 40% of the maximal voluntary contraction can reduce muscle blood flow through occlusion of the blood vessels by the static contraction of the force-generating muscles. Receptors in the muscle respond to the products of metabolism and stimulate the increases in systolic pressure through feedback mechanisms via unmyelinated (group IV) afferent nerves. In contrast, the receptors that stimulate the increase in heart rate appear to be tension receptors.

Functionally, the hearts of resistance-trained athletes are not different from sedentary controls at rest. Although the stroke volume of the athletes is larger than the controls in most studies, perhaps through cardiac hypertrophy, the difference is attributed to an increase in body size (i.e., lean body weight or body surface area) in the trained group. Systolic function might be slightly enhanced by resistance training at rest; however, the effect of training on systolic function during exercise is unknown, perhaps because of the difficulty in experimentally deriving functionally meaningful data. On the other hand, an impairment in diastolic function is characteristic of most types of pressure overload-induced cardiac hypertrophy. This has not been shown to be the case in noninvasive studies of resistance-trained athletes. Peak filling rates and other measures of diastolic function are greater than expected in the trained group; however, when normalized for the larger body size, there is no loss of diastolic function. This conclusion is made with some hesitancy. Only small numbers of subjects have been studied, and the tragic presence of anabolic steroids in these male subjects might be confounding the interpretation.

IV. Regulation of the Cardiovascular System during Exercise: Central versus Reflex Command

The regulation of the cardiovascular system during exercise is extremely complex and under precise control, so that under ordinary circumstances the cardiovascular response is in concert with the intensity of the exercise bout. Although the intensity of dynamic exercise is matched by an augmentation of cardiac output, and the intensity of static exercise is matched by increased arterial blood pressure, both types of exercise invoke an increased activity of the sympathetic nervous system and a decreased activity of the parasympathetic nervous system. The underlying neural mechanisms are unknown with any precision, despite considerable effort since the mid-1800s. The difficulty in deriving data which can be interpreted without ambiguity is that the response to exercise is a multifactorial process, involving numerous systems besides the cardiovascular system. The homeostatic mechanisms necessary for survival of the organism necessitate this complexity. However, two general theories have evolved in the past 100 years.

The first theory holds that the response of the cardiovascular system to exercise arises from the direct action of a central command descending from the motor centers of the cardiovascular control centers in the brain (i.e., the central neural command theory). The other theory hypothesizes that the response comes from reflexes emanating from afferent neural activity of receptors in the muscles and joints which are activated during exercise (i.e., the reflex neural command theory). Both mechanisms might have a final common pathway from the brain to the cardiovascular system. Both mechanisms can work separately or together, and it is probable that both mechanisms are operant during exercise to make the cardiovascular adjustments appropriate to the intensity of the exercise.

Bibliography

Armstrong, R. B. (1988). Distribution of blood flow in the muscles of conscious animals during exercise. *Am. J. Cardiol.* **62,** 9E–14E.

Astrand, P.-O., and Rodahl, K. (1977). "Textbook of Work Physiology. Physiological Bases of Exercise." McGraw-Hill, New York.

Braunwald, E., Ross, J. R., Jr., and Sonnenblick, E. H. (1976). "Mechanisms of Contraction of the Normal and Failing Heart." Little, Brown, Boston.

Hanson, P., and Nagle, F. (1987). Isometric exercise: Cardiovascular responses in normal and cardiac populations. *Cardiol. Clin.* **5,** 157–170.

McArdle, W. D., Katch, F. I., and Katch, V. L. (1986). "Exercise Physiology. Energy, Nutrition, and Human Performance." Lea & Febiger, Philadelphia, Pennsylvania.

Mitchell, J. H. (1985). Cardiovascular control during exercise: Central and reflex neural mechanisms. *Am. J. Cardiol.* **55,** 34D–41D.

Rowell, L. R. (1986). "Human Circulation Regulation during Physical Stress." Oxford Univ. Press, New York.

Saltin, B., and Rowell, L. B. (1980). Functional adaptations to physical activity and inactivity. *Fed. Proc., Fed. Am. Soc. Exp. Biol.* **39,** 1506–1513.

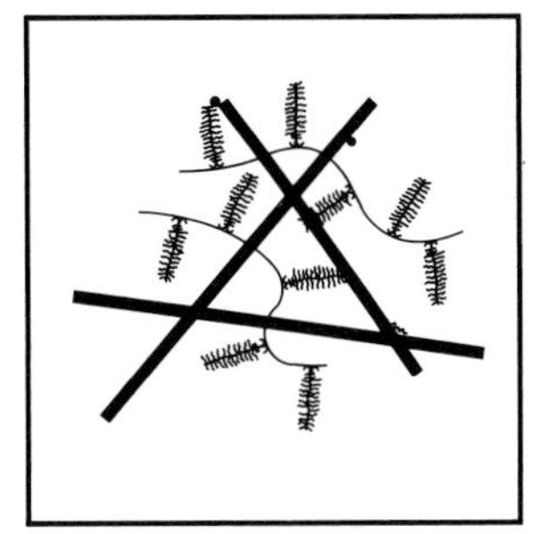

Extracellular Matrix

KENNETH M. YAMADA, *National Institutes of Health*

Glossary

Amino acid abbreviations used Ala, alanine; Arg, arginine; Asn, asparagine; Asp, aspartic acid; Cys, cysteine; Gln, glutamine; Glu, glutamic acid; Gly, glycine; His, histidine; Ile, isoleucine; Leu, leucine; Lys, lysine; Met, methionine; Phe, phenylalanine; Pro, proline; Ser, serine; Thr, threonine; Trp, tryptophan; Tyr, tyrosine; Val, valine

Basement membrane Thin, noncellular sheet of extracellular matrix underlying epithelia; the region closest to cells is termed the basal lamina, and the more loosely structured peripheral zone is termed the reticular layer

Cytoskeleton Cytoplasmic structural elements serving as an intracellular framework or "skeleton." For this article, the most relevant elements are actin microfilament bundles, comprised of F-actin polymers and associated proteins

Neural crest cells Migratory cells derived from the dorsolateral edge of the closing neural tube that form sensory and sympathetic ganglia and other structures

EXTRACELLULAR MATRIX is the noncellular material that provides the supporting, structural, externally multifunctional, regulatory environment of cells throughout the body. Its remarkably diverse constituents range from the structural collagens to adhesive glycoproteins, and from clear, jelly-like gels rich in hyaluronate to dense, rigid bone; it can vary in organization from locally homogeneous to elaborately structured. Major functions of the extracellular matrix include (1) providing structural support, tensile strength, or cushioning, (2) providing substrates and pathways for cell migration, and (3) regulating cellular differentiation and metabolic functions. Individual extracellular matrix molecules are often multifunctional, and their functions are now understood in terms of specific functional binding or recognition sites. Specific cellular receptors provide the mechanisms for cell surface interactions with these diverse and critically important structural and regulatory molecules.

I. Introduction

The extracellular matrix consists of a wide variety of specialized proteins, complex carbohydrates, and other molecules that surround, support, and organize cells and tissues. Without connective tissue, bone, and the unique local extracellular environments provided by the extracellular matrix, the human body would be little more than an amorphous mass of unorganized, poorly differentiated cells. Human growth and development, as well as tissue maintenance and wound repair, all require a host of essential extracellular molecules and their specific cellular receptors, which are regulated by complex patterns of gene expression. Table I presents a list of constituents of extracellular matrices and some binding functions of each.

II. Functions of the Extracellular Matrix

A. Functions in Structural Organization

As structural elements, extracellular matrix molecules provide support for cells and dictate the exter-

TABLE I Extracellular Matrix Molecules (partial listing)

Molecule	Molecular weight[a]		Binds to[b]
	Complex	Subunits	
Collagens	(Variety of types[c])		Cells, PGs, Fn, Ln, Vn, Nd
Fibronectin	~500,000[d]	~250,000[d]	Cells, Col, PGs, Fbn, amyloid P component
Laminin	~900,000	400,000, 210,000, 200,000	Cells, PGs, Col type IV, Nd, sulfatides
Proteoglycans	(Wide range of types[c])		Cells, PGs, Col, Fn, Ln
Elastin	Polymer	63,000[d]	Cells, microfibrils
Vitronectin (S-protein)		75,000	Cells, heparin, Col, (contains somatomedin B)
Thrombospondin	~450,000	~140,000	Cells, PGs, Col, Ln, Fbn, sulfatides, calcium
Tenascin	Hexamer	~230,000[d]	Cells, Fn
Entactin (nidogen)		150,000	Cells, Ln, Col type IV
Osteonectin (SPARC)		33,000	Calcium, other?
Anchorin CII		34,000	Cells, Col type II
Chondronectin	176,000	56,000	Cells, Col type II
Link protein		39,000	Cartilage PG, hyaluronate
Osteocalcin		5,800	Hydroxylapatite
Bone sialoprotein		34,000	Cells, hydroxylapatite
Osteopontin		33,000	Hydroxylapatite
Bullous pemphigoid antigen		230,000	(In epidermal hemidesmosomes)
Epinectin		70,000	Epithelial cells, heparin
Hyaluronectin		59,000	Hyaluronate, neurons
Amyloid P component	250,000	25,000	Fn, ? elastase

[a] If the molecules exist as regular complexes, the total size of the complex is indicated at the left; the right column indicates the size of each subunit or monomer. Molecular weights are generally derived from sequence data (lacking weight of carbohydrates) or from sodium dodecyl sulfate polyacrylamide gel electrophoresis.
[b] Many of these molecules are bound by cells, as well as binding to the indicated ligands. PG, proteoglycan; Fn, fibronectin; Ln, laminin, Vn, vitronectin; Nd, nidogen (entactin); Col, collagen(s); Fbn, fibrin/fibrinogen.
[c] Molecular weights vary substantially depending on the type.
[d] Molecular weights vary due to alternative splicing of precursor messenger RNA, which produces a family of closely related variants.

nal boundaries of most tissues. The epithelia of the body that cover skin, mucous membranes, and the interior of glands all rest on extracellular structures termed *basement membranes*. These thin sheets consist of an organized matrix of type IV collagen, laminin, and other molecules; one hypothesis about how basement membrane components are organized is shown in Fig. 1A. Basement membranes provide surfaces on which epithelial cells adhere and assume a polarized orientation. They also constitute a barrier separating epithelial cells from underlying connective tissue, and tumor cells that cross this boundary are generally considered malignant.

Other types of extracellular matrices are organized differently. Loose connective tissue produced by fibroblasts is comprised of extracellular matrix molecules such as type I collagen and proteoglycans. The sheets of collagenous material forming fascia, the cords of collagen forming ligaments and tendons, cartilage, bone, and even the clear vitreous humor of the eye all consist of different types of extracellular matrix secreted by specific cells as structural materials. The diversity of such matrices is particularly obvious when comparing the molecular organization of basement membrane with that of cartilage (Fig. 1). The specific molecules involved in such structural roles and the molecular interactions that organize them are considered further below. [*See* COLLAGEN, STRUCTURE AND FUNCTION; ELASTIN; PROTEOGLYCANS.]

B. Functions in Cell Migration

Extracellular matrix molecules such as fibronectin and laminin can provide pathways for cell migration. For example, studies of gastrulation (an early stage of embryonic development) in animals reveal that fibronectin provides a substrate for migration of cells as the basic body plan forms. Fibronectin and probably laminin are also important for the subsequent migration of embryonic neural crest cells throughout the body to form many of the structures of the face, the sympathetic and parasympathetic nervous systems, and skin pigment. To migrate, cells use specific receptors that bind to key sites in

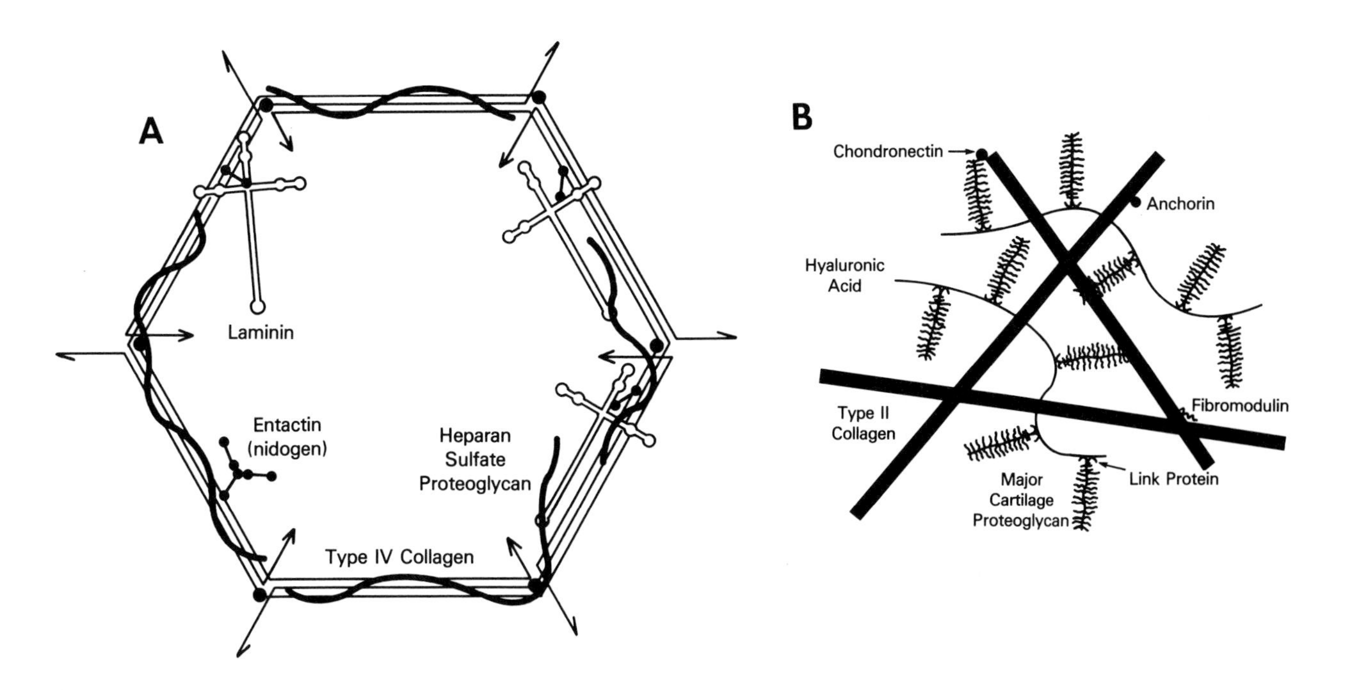

FIGURE 1 Comparison of the organization of two distinctive types of extracellular matrix. Current schematic models molecular organization of (A) basement membrane (upper basal lamina portion) and (B) cartilage matrix are shown as examples of complexity and diversity of extracellular matrices. Basement membrane consists of a scaffolding of type IV collagen, hypothesized by some to exist as sheets of hexagonal arrays (a single component hexagon is shown, which is connected to its neighbors by *single-sided arrows*); particularly functionally important 3-dimensional interconnections are formed by binding of tips of collagen molecules to one another (*regular arrows*). Major additional constituents are laminin, heparan sulfate proteoglycan, and entactin (nidogen). In contrast, cartilage (B) is more loosely organized, consisting of type II and other collagens, a major chondroitin sulfate-containing proteoglycan plus others, hyaluronic acid, and other, more minor protein constituents.

these proteins (see Section VI). Inhibiting the receptor-mediated interaction of cells with such extracellular substrates halts migration.

During wound healing, a meshwork of fibronectin linked to fibrin provides an excellent migratory substrate for epithelial cells as they close the wound, as well as a general scaffolding that migrating fibroblasts penetrate; these fibroblasts then secrete collagen and other molecules to form replacement connective tissue. The initial migratory/scaffolding proteins generally disappear after the wound is healed and are replaced by stable basement membrane and mature connective tissue. [*See* CONNECTIVE TISSUE.]

C. Functions in Cell Growth and Differentiation

Specific extracellular matrix molecules such as collagen, fibronectin, laminin, and heparan sulfate proteoglycan can each stimulate the growth or differentiation of different cell types. For example, collagen gels and artificially reconstituted basement membranes promote epithelial cell differentiation. In contrast, fibronectin blocks myoblast differentiation to form muscle while stimulating sympathetic neuron differentiation from neural crest cells. As yet, little is known about the actual mechanisms of action of these molecules nor how they are regulated and then coordinate with one another to help regulate cellular behavior.

III. Collagens and Elastin

Collagens comprise the most abundant class of protein in the human body. At least 13 different types of collagen have been identified to date, which fulfill a variety of different functions including providing crucial supportive and overall structural stability to tissues. Elastin is another major structural extracellular protein; collagens and elastin will be considered below only in the context of other matrix proteins and their receptors. A notable characteristic of these proteins is their ability to self-assemble into lengthy fibrils after secretion; this property simplifies the assembly of extracellular matrices.

IV. Fibronectin

In recent years, a number of noncollagenous glycoproteins have been identified in the extracellular matrix that can begin to account for its wide range of biological activities. Fibronectin is presently the best-studied of these glycoproteins, which can mediate or modulate cell adhesion, morphology, and migration. Although produced from only a single gene, fibronectin molecules can differ by the insertion or absence of three regions of protein termed ED-A (EDI), ED-B (EDII), or IIICS, which occurs by differences in processing (alternative splicing) of fibronectin precursor messenger RNA (Fig. 2). This diversity of fibronectins derived from a single gene provides a simple mechanism for generating a potential diversity of functions.

Fibronectin can mediate a host of adhesive and binding functions; for example, fibronectins can produce adhesion to surfaces of most cell types except erythrocytes. As shown in Fig. 2, each fibronectin molecule contains a linear array of modular binding units that bind to cells as well as to other key extracellular molecules including collagen, fibrin, and heparin (heparan or dermatan sulfate proteoglycans). Various pairings of these binding domains can account for the ability of fibronectin to form cross-links between different extracellular matrix proteins, as well as its ability to mediate cell adhesion to collagen or fibrin. It can even bind to a heparan sulfate proteoglycan in the plasma membrane to mediate binding to cell surfaces.

Most interactions of fibronectin with cells, however, appear to be mediated by specific receptors that can bind to certain key sequences in fibronectin as indicated in Fig. 2 and Table II. A key recognition unit in fibronectin and several other extracellular adhesion proteins needed for receptor binding is the amino acid sequence Arg-Gly-Asp (e.g., as seen within the fibronectin Gly-Arg-Gly-Asp-Ser sequence). Full binding specificity and strength, however, appears to require another region that functions in synergy with this basic adhesion sequence.

Cell-type specificity of recognition of fibronectin is mediated by yet another two sites in the IIICS region, present in only certain fibronectin molecules; they contain distinct amino acid recognition sequences (Table II). For example, although most cells adhere to the former Arg-Gly-Asp and synergy sequences, adhesive recognition of the latter two IIICS sites appears much more restricted (e.g., to cells derived from the neural crest, such as peripheral nervous system cells, and to lymphocytes). Besides these four adhesion regions, a fifth is involved in binding to heparin and to cells (Fig. 2, Table II), perhaps by binding to heparan sulfate proteoglycan on the cell surface.

Besides displaying striking adhesive activity, fibronectin can also promote the migration of many types of cells. It has been identified as a migratory substrate for embryonic gastrulating cells, neural crest cells and their derivatives, and primordial germ cells, plus mature corneal and epidermal cells and a variety of tumor cells.

The identification of crucial binding sequences in fibronectin has permitted the development of synthetic peptide inhibitors that competitively inhibit function. These nontoxic inhibitors can be used to probe the functions of such recognition sequences in intact organisms. Such studies have implicated the Arg-Gly-Asp sequence in embryonic cell migration as well as in tumor cell invasion and metastasis. For example, certain peptides from fibronectin can inhibit metastasis in animal model systems. [*See* METASTASIS.]

Finally, fibronectin can modify the morphology, growth, and differentiation of certain cells. Many of its morphological effects can be attributed to its adhesive activity, including cell flattening, decreased cell surface microvilli, cell–cell alignment, and organization of intracellular actin microfilament bundles. In addition, however, fibronectin can stimulate certain cells to proliferate, as well as inhibiting differentiation (e.g., blocking the fusion of myoblasts in muscle development and the synthesis of proteoglycans by cartilage cells); it can also promote the differentiation of certain neural crest cells to sympathetic nervous system cells. How fibronectin mediates its effects on growth and differentiation remains to be determined. To complicate matters, fibronectin and other extracellular molecules can also serve as carriers of growth factors such as transforming growth factor beta (TGF-β). Local accumulations of fibronectin, or its focal destruction by proteases, may lead to local release of such regulatory factors. [*See* TRANSFORMING GROWTH FACTOR-β.]

V. Laminin

At least three structurally distinct forms of the adhesive protein laminin have been identified from different cell sources (e.g., regular laminin, merosin,

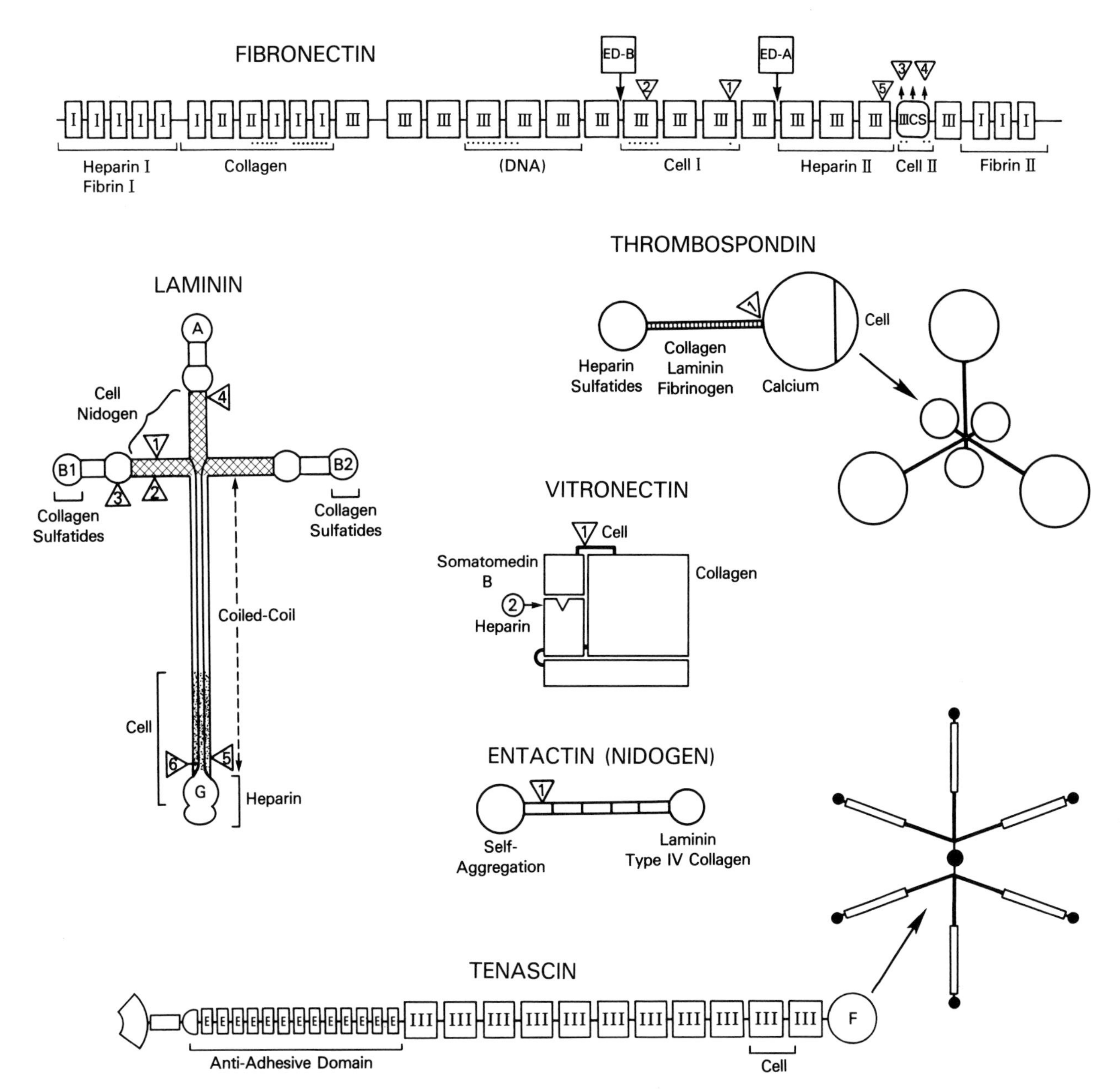

FIGURE 2 A variety of individual extracellular matrix molecules and their functional sites. These molecules often have multiple functional domains as indicated by labels (e.g., collagen- or heparin-binding domains), as well as recently identified specific amino acid sequences recognized by cellular receptors (*numbered triangles*). See Table I for characteristics of these molecules, and Table II for a listing of the specific recognition sequences identified to date. Fibronectin is composed of 3 types of repeating unit (i.e., types I, II, or III); it also contains a series of functional binding domains (e.g., for binding to heparin or to collagen). Particularly important binding sequences are indicated by *dots*. Two type III units termed ED-A and ED-B can be present or absent because of alternative splicing. The IIICS region can undergo more complex alternative splicing at 3 sites (*upward arrows*). Fibronectin forms dimers or multimers by disulfide bonding at the carboxyl terminus (*right end*). Laminin contains 3 noncovalently associated chains, termed A, B1, and B2. Its long arm is thought to consist of all 3 subunits entwined in a relatively rigid coiled-coil structure terminated by a globular or lobulated terminal domain (G). Thrombospondin assembles into a trimer with the larger globular domains outward. Vitronectin is globular and contains somatomedin B, which can be released by proteolysis. A novel cryptic heparin-binding domain (2) is exposed and active in only a small subpopulation of vitronectin molecules but is revealed by experimental denaturation (e.g., with urea). Entactin (nidogen) is a dumbbell-shaped molecule that can sometimes form clusters by self-aggregating. Tenascin is composed of a variety of domains including EGF-like repeats (E), fibronectin type III units, and a fibrinogen-like globular domain. Tenascin monomers often assembles into hexamers (hexabrachions).

TABLE II Specific Cell Adhesive Recognition Sequences in Matrix Proteins

Number[a] and name	Amino acid sequence recognized[b]	Cells binding the sequence
Fibronectin		
1. Cell-binding domain	Gly-*Arg-Gly-Asp*-Ser	Most cells tested
2. Synergistic sequence	Minimal sequence not yet defined	Most cells tested to date
3. IIICS: CS1 site	Asp-Glu-Leu-Pro-Gln-Leu-Val-Thr-Leu-Pro-His-Pro-Asn-Leu-His-Gly-Pro-Glu-Ile-*Leu-Asp-Val*-Pro-Ser	Most neural crest derivatives, lymphocytes
4. IIICS: REDV[c] site	Arg-Glu-Asp-Val	Melanoma cells, others?
5. Peptide I	Tyr-Glu-Lys-Pro-Gly-Ser-Pro-Pro-Arg-Glu-Val-Val-Pro-Arg-Pro-Arg-Pro-Gly-Val	Melanoma cells
Peptide II	Lys-Asn-Asn-Gln-Lys-Ser-Glu-Pro-Leu-Ile-Gly-Arg-Lys-Lys-Thr	Melanoma cells
Laminin		
1. YIGSR[c]	Tyr-Ile-Gly-Ser-Arg-	Many cell types
2. PDSGR[c]	Pro-Asp-Ser-Gly-Arg	Fibrosarcoma, melanoma, others
3. F9	Arg-Tyr-Val-Val-Leu-Pro-Arg-Pro-Val-Cys-Phe-Glu-Lys-Gly-Met-Asn-Tyr-Val-Arg	Fibrosarcoma, melanoma, glioma, endothelial cells
4. RGD[c] site	*Arg-Gly-Asp*-Asn	Endothelial, others?
5. p20	Arg-Asn-Ile-Ala-Glu-Ile-Ile-Lys-Asp-Ile	Neurons
6. IKVAV[c]	Ile-Lys-Val-Ala-Val	Many cell types
7. PA22-2	Ser-Arg-Ala-Arg-Lys-Gln-Ala-Ala-Ser-Ile-Lys-Val-Ala-Val-Ser-Ala-Asp-Arg	Neurons
8. LRE[c] site	Leu-Arg-Glu	Ciliary ganglion neurons
Entactin (nidogen)		
1. RDG[c] site	Ser-Ile-Gly-Phe-*Arg-Gly-Asp*-Gly-Gln-Thr-Cys	Mammary tumor cells
Vitronectin		
1. RGD[c] site	*Arg-Gly-Asp*-Val	Many cells

[a] Numbers correspond to numbering of open triangles in Fig. 2.
[b] See glossary for full name of amino acids; italics show putative minimal sequences, although in some cases the minimal sequence required for activity has not yet been determined.
[c] These sites are named according to the single-letter abbreviations of their crucial amino acids.

and S-laminin). The apparently most abundant form depicted in Fig. 2 is a prominent constituent of basement membranes, where nearly all laminin is found. It plays important roles in mediating adhesion of epithelial cells to flat surfaces. Laminin also contributes to the migration of certain embryonic cells near basement membranes (e.g., of certain neural crest cells and cardiac mesenchymal cells). Laminin can also appear transiently elsewhere for use in other functions (e.g., in directing neuronal outgrowth). Like fibronectin, laminin can mediate the adhesion and migration of cells, although it appears to be more specialized for epithelial and neuronal cells. Laminin also contains a series of amino acids sequences recognized by distinct cell surface receptors (Fig. 2, Table II). Besides a functional sequence that resembles the key fibronectin Arg-Gly-Asp sequence, others contain the sequence Tyr-Ile-Gly-Ser-Arg,Ile-Lys-Val-Ala-Val, and more as indicated in Fig. 2. At least two such synthetic peptides have been reported to inhibit tumor cell metastasis in animal studies modeled on those with fibronectin peptides. Unlike fibronectin, laminin also has the ability to adhere to cells by a novel mechanism involving binding to sulfated glycolipids in the plasma membrane. [*See* LAMININ IN NEURONAL DEVELOPMENT.]

Like fibronectin, laminin can bind to heparin and heparan sulfate proteoglycan; it also binds to type IV collagen and entactin (nidogen). These latter two binding activities are probably quite important in the assembly of basement membranes from these molecules (Fig. 1).

VI. Receptors for Extracellular Matrix Proteins

A. Overview

It has become increasingly apparent that extracellular matrix molecules interact with cells via specialized receptors in the plasma membrane (Fig. 3).

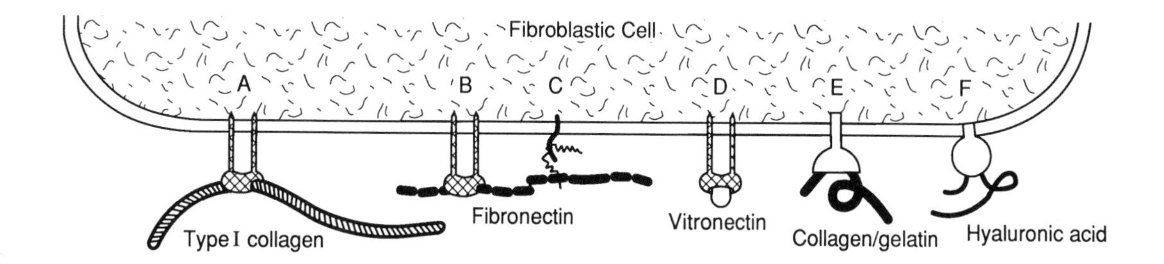

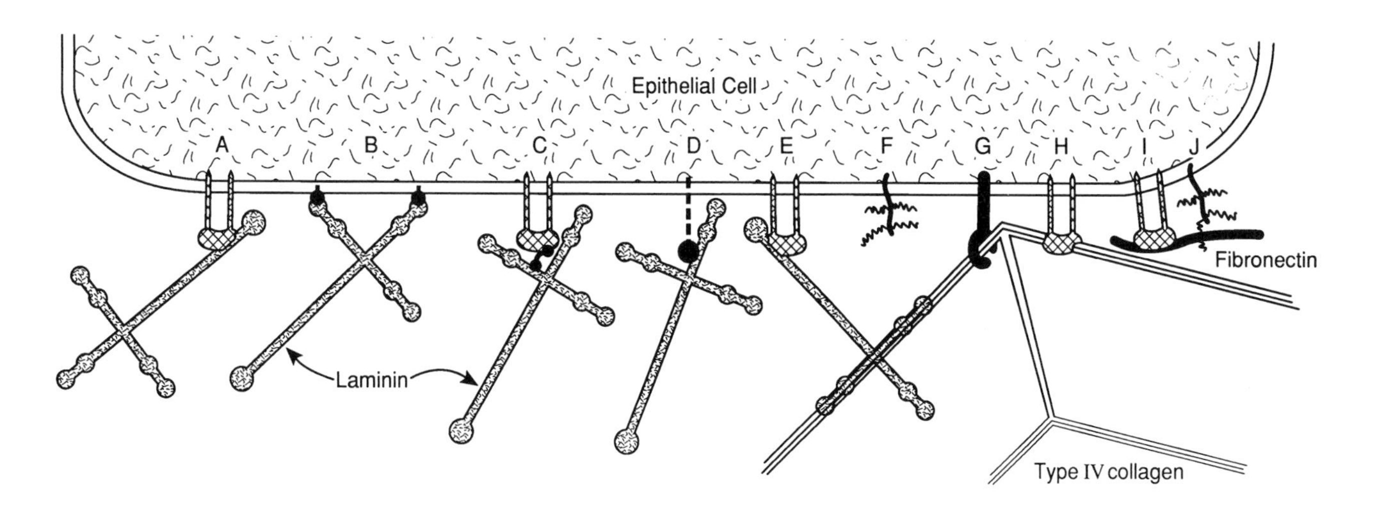

FIGURE 3 Cellular receptor systems for extracellular matrix molecules. Major receptors used by fibroblastic cells are compared with those used by epithelial cells. On fibroblastic and mesenchymal cells (*top*), at least 6 types of receptor can be found: (*A*) integrins binding to type I and other collagens (e.g., the $\alpha_2\beta_1$ and $\alpha_3\beta_1$ integrins); (*B*) integrins that bind fibronectin (e.g., $\alpha_5\beta_1$); (*C*) heparan sulfate proteoglycan(s) such as syndecan in mesenchymal cells that bind to fibronectin's heparin-binding site or collagen; (*D*) vitronectin receptors (e.g., $\alpha_v\beta_1$); (*E*) other collagen- and gelatin-binding receptor(s); and (*F*) a hyaluronic acid receptor. On various epithelial cells (*bottom*), major adhesive interactions can occur with laminin and type IV collagen: (*A*) integrins that bind laminin at a site just above the large globular domain (e.g., $\alpha_3\beta_1$ and $\alpha_6\beta_1$); (*B*) sulfated glycolipids (sulfatides) bind to the short arms of laminin; (*C*) the Arg-Gly-Asp site of entactin (nidogen) is bound by an unidentified integrin; the entactin then binds to laminin; (*D*) a laminin-binding protein of 67,000 binds to laminin and the plasma membrane as a peripheral membrane protein; its membrane-binding mechanism is unclear (*dashed line*); (*E*) type IV collagen is bound by laminin, which is in turn bound to the membrane by an integrin; (*F*) syndecan, a heparan sulfate-containing proteoglycan with unknown functions in mature epithelia; (*G*) other receptors for type IV collagen; (*H*) an integrin such as $\alpha_2\beta_1$ that directly binds type IV collagen; (*I,J*) in migrating epithelial cells separated from the basement membrane (e.g., during wound closure), integrins (*I*) and syndecan (*J*) can bind and use fibronectin as a migration substrate. Not all cells of each type possess each type of receptor, and specialized cells may also display other additional receptors not shown.

These receptors can differ in type and specificity depending on the type of cell examined, and various combinations of receptors can produce distinctive patterns of adhesion. The largest group of receptors belong to the integrin family (Table III), although other receptor mechanisms are also important. The existence of these receptor systems can account for cell specificity of interactions with different matrix proteins, and they provide straightforward mechanisms for mediating and regulating cell interactions with extracellular molecules. [*See* RECEPTORS, BIOCHEMISTRY.]

B. Integrin Receptor Family

There are at least 13 receptors in the human integrin family, with more members likely to be discovered (Table III). Each receptor has a distinctive pattern of expression during development and in different adult tissues, and each has its own specialized functions. In general, these receptors function in cell adhesion and migration. For example, adhesion of cells to collagens via the $\alpha_2\beta_1$ receptor may be important for overall tissue organization in embryonic development, because inhibition of its binding to

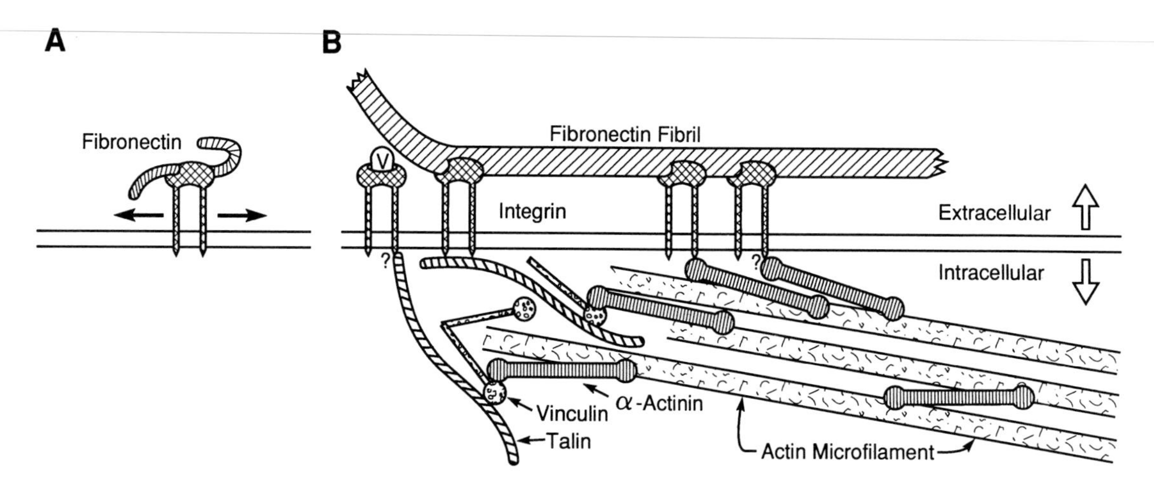

FIGURE 4 Transmembrane interactions of extracellular matrix receptors with intracellular molecules. (A), depicting a region of plasma membrane on a rapidly migrating cell, a single fibronectin molecule is bound by an integrin that is free to diffuse laterally and is not encumbered by complexes with intracellular molecules. In (B), depicting the membrane of a more stationary cell, integrin receptors bind to vitronectin (V) or to fibronectin fibrils on the extracellular face of the plasma membrane, as well as to intracellular cytoskeletal molecules such as talin or α-actinin as part of large complexes; the receptors are immobile when participating in these large transmembrane complexes. Question marks indicate interactions that have been suggested but that may require further proof.

collagen with monoclonal antibodies can block cellular migration within three-dimensional arrays of collagen. Even epithelial cells can adhere to collagen directly, which suggests that they can attach directly to the structural meshwork of type IV collagen in basement membranes by a mechanism distinct from adhesion mediated via laminin (Fig. 3).

Adhesion to fibronectin can occur through at least five different receptors, which may produce differing effects on cell behavior. For example, the $\alpha_5\beta_1$ receptor is required for migration of certain tumor cells, although it functions instead to form strong adhesions and an organized fibronectin matrix around certain normal fibroblasts. A number of

TABLE III Integrin Family of Cell Adhesion Receptors

Name	Alternative names	Molecular weight of α (reduced)[a,b]	Ligands bound[c]	Putative adhesion function
β_1 integrins (VLAs)				
$\alpha_1\beta_1$	VLA-1	210,000	Col, Ln	Cell-matrix
$\alpha_2\beta_1$	VLA-2, $GP_{Ia\text{-}IIa}$	165,000	Col	Cell-matrix
$\alpha_3\beta_1$	VLA-3	130,000 + 25,000	Fn, Ln, Col	Cell-matrix
$\alpha_4\beta_1$	VLA-4	150,000	Fn, other?	Cell-cell and -matrix
$\alpha_5\beta_1$	VLA-5, fibronectin receptor, $GP_{Ic\text{-}IIa}$	135,000 + 25,000	Fn	Cell-matrix
$\alpha_6\beta_1$	VLA-6	120,000 + 30,000	Ln	Cell-matrix
β_2 integrins (leukocyte adhesion receptors)				
$\alpha_L\beta_2$	LFA-1	180,000	I-CAM-1, I-CAM-2	Leukocyte cell-cell
$\alpha_M\beta_2$	Mac-1, CR-3	170,000	C3bi	Leukocyte cell-cell
$\alpha_X\beta_2$	p150,90	150,000	?	Leukocyte cell-cell
β_3 integrins (cytoadhesions)				
$\alpha_{IIb}\beta_3$	$GP_{IIb\text{-}IIIa}$	120,000 + 25,000	Fb, Fn, vWF, Vn, Tsp	Platelet cell-cell and cell-matrix
$\alpha_V\beta_3$	Vitronectin receptor	125,000 + 24,000	Vn, Fb, vWF, Tsp	Cell-matrix
Other integrins				
$\alpha_6\beta_4$	VLA-6_{alt}	120,000 + 30,000	Fn,Vn	Cell-matrix
$\alpha_V\beta_5$	Alternative vitronectin receptor	125,000 + 24,000	?	Cell-matrix

[a] After chemical reduction of disulfide bonds.
[b] Molecular weights of β subunits are: β_1 = 110,000; β_2 = 90,000; β_3 = 90,000; β_4 = 220,000; β_5 = 110,000.
[c] Col, collagen; Fn, fibronectin; Ln, laminin; Fb, fibrinogen; vWF, von Willebrand factor; Vn, vitronectin; Tsp, thrombospondin.

integrin receptors also mediate cell interactions with laminin. At least several of these receptors display transmembrane effects on organization of molecules in the cytoplasm of cells (Fig. 4). Specifically, several integrins including $\alpha_5\beta_1$, $\alpha_v\beta_3$, and $\alpha_{IIb}\beta_3$ (GPIIb-IIIa) are associated with the intracellular accumulation of organized arrays of actin-containing microfilaments; they may do so by forming receptor aggregates that bind linking proteins such as talin and α-actinin (Fig. 4). The functional consequence is that organized cytoskeletons form inside cells, affecting cell shape and function (e.g., promoting tenacious adhesion and halting cell migration).

C. Other Receptors and Binding Proteins

Other possible receptors besides integrins have been identified for collagens, laminin, elastin, and proteoglycans. The best characterized of these is a 67,000-dalton glycoprotein that is unusual in that it is reported to exist as a peripheral membrane protein rather than as one spanning the plasma membrane. This protein binds both laminin and elastin and is variously reported to function by binding to the peptide sequences Tyr-Ile-Gly-Ser-Arg, Leu-Gly-Thr-Ile-Pro-Gly, or galactoside sugars. This protein is increased on many tumor cells and is thought to contribute to the capacity of these cells to metastasize.

The wide variety of integrin and nonintegrin cell surface receptors can begin to account for the complexities of cell interactions with various extracellular matrices (Fig. 3). The regulation of these receptors and of the many known extracellular matrix molecules is under the control of a host of growth factors and cytokines such as TGF-β, as well as developmental regulatory systems. Such complexity of regulation might be expected, given the direct roles of extracellular molecules in a wide variety of cellular functions.

Bibliography

Bernfield, M., ed. (1989). Extracellular matrix. *Curr. Opin. Cell Biol.* **1,** 953.

Hay, E. D., ed. (1990). "Cell Biology of Extracellular Matrix," 2nd ed. Plenum Press, New York (*in press*).

Mayne, R., and Burgeson, R. E. (1987). "Structure and Function of Collagen Types." Academic Press, Orlando.

Mecham, R. P., and Wight, T. N. (1987). "Biology of Proteoglycans." Academic Press, Orlando, Florida.

Mosher, D. F. (1989). "Fibronectin." Academic Press, New York.

Piez, K. A., ed. (1984). "Extracellular Matrix Biochemistry." Elsevier, New York.

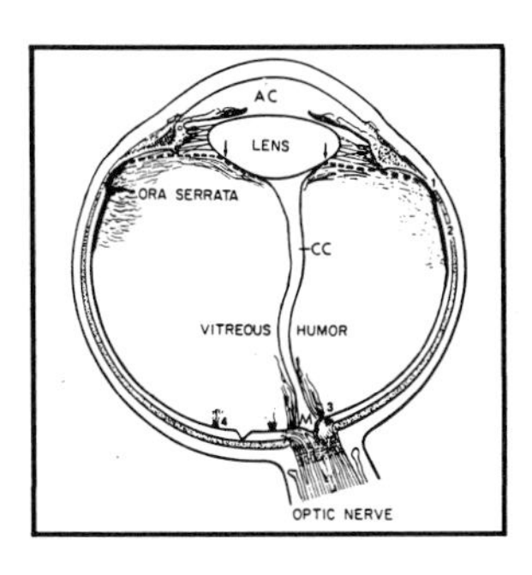

Eye, Anatomy

BRENDA J. TRIPATHI, RAMESH C. TRIPATHI,
SHARATH C. RAJA, *Visual Sciences Center,
The University of Chicago*

Glossary

Diopter Measure of the power of a lens, equal to the reciprocal of its focal length in meters; a converging lens is taken as positive, a diverging lens as negative

Dioptrics Science of the refraction of light by transmission

Emmetropia State of proper correlation between the refractive system of the eye and the axial length of the eyeball, with rays of light entering the eye parallel to the optic axis being brought to a focus exactly on the retina

Fornix conjunctivae Conjunctival cul-de-sac; retrotarsal fold; concave recess formed by the junction of the bulbar and palpebral portions of the conjunctiva, that of the upper lid being the f. conjunctivae superior, and that of the lower lid the f. conjunctivae inferior

Fovea centralis retinae Central fovea of retina, foveal centralis, fovea; conical central depression in the macula retinae where the retina is very thin so that rays of light have free passage to the layer of photoreceptors, mostly cones; this is the area of most distinct vision to which the visual axis is directed

Glaucoma Disease of the eye characterized by increased intraocular pressure due to restricted outflow of the aqueous through the trabecular meshwork and Schlemm's canal system, excavation and degeneration of the optic disk, and nerve fiber bundle damage producing characteristic defects in the field of vision

Melanocyte Cell bearing or capable of forming melanin; located in the skin and many structures of the eye

Myopia Shortsightedness; nearsightedness; near or short sight; a condition in which, in consequence of an error in refraction or of elongation of the globe of the eye, parallel rays are focused in front of the retina

Tonofilament Structural cytoplasmic protein, bundles of which together form a tonofibril; the protein of epidermal tonofilaments is keratin; the filaments are 8–10 nm in thickness

Zonular fiber (fibrae zonulares) Fine, elastic filaments that arise from the surface of the epithelium of the ciliary body as far back as the ora serrata and especially from the corona ciliaris and extend to the equatorial anterior and posterior surfaces of the lens of the eye; through their attachments, the ciliary muscle produces changes in the curvature of the lens during accommodation; collectively they form the zonula ciliaris or zonules of Zinn

VISUAL PERCEPTION of our environment involves two discrete processes: transduction of radiant energy into electrical impulses and the subsequent interpretation of these impulses by the brain. Our vision, defined as the translation of afferent neuronal impulses into meaningful symbols and patterns, can be shaped by our experience; however, we are born with sight, defined as the neuronal input of the retina to the brain.

The eye represents a higher degree of specialization than is found in any tissue in the body. It is the sensory apparatus for radiant energy of wavelengths between 400 and 750 nm, and it is responsible for some 40% of the total sensory input to the human brain. The main function of the eyeball is to provide an environment that is optimal for the function of its photoreceptive layer, the retina. All of the other complex structures of the eye, including the appendages, dioptric system, and protective and motor mechanisms, as well as the vascular and nerve supplies, facilitate the accurate and efficient functioning of the retina.

The anatomic structures mediating human photoreception can be classified under four main headings: the eyeball, its protective apparatus, its motor and supporting apparatus, and the visual pathway.

I. The Eyeball

The human eye, connected to the brain by the optic nerve, is recessed in a bony orbit. Although the eyeball is generally referred to as a globe, it is only approximately spherical, consisting of segments of two spheres placed one in front of the other (Fig. 1). The anterior one-sixth of the surface of the globe comprises the cornea; the posterior five-sixths is occupied by the sclera, which is less curved in the anterior region than it is behind the equator of the eyeball. The change in the curvature of the two spheres at the corneoscleral junction produces a shallow groove, the external scleral sulcus, which, *in vivo,* is filled largely by the conjunctiva and bulbar fascia. Originating from the sides of the pyramidal bony orbit and inserting on the sclera are the four rectus and two oblique extraocular muscles.

The geometric axis of the eye joins the central point of the cornea anteriorly and that of the scleral curvature posteriorly. However, the visual axis joins the fixation point and the fovea centralis of the retina (Fig. 1). The dimensions of the adult ametropic human eye are fairly constant; the average diameter is 24 mm sagitally (anterior–posterior), 23 mm vertically, and 23.5 mm horizontally. The sagittal diameter may vary the most, ranging from 21 to 26 mm in the normal eye, depending on its refractive state, but, with a high degree of axial myopia, this diameter may be as great as 29 mm. Overall, the male eye is about 0.5 mm larger than that of the female, and the eyes of Blacks are somewhat larger than those of Caucasians. The eye of the newborn is more nearly spherical than is the adult eye, with a sagittal diameter of 16–17 mm, but this increases rapidly to about 23 mm by 3 yr of age, and the adult size is attained by puberty. The average eyeball weighs about 7.5 g, and its volume is about 6.5 ml.

A. Surface Anatomy

Anteriorly, the white opaque sclera is demarcated from the transparent cornea by the bluish gray junctional zone of the limbus (2 mm wide vertically and 1.5 mm wide horizontally). In this zone, numerous small vessels emerge from the sclera (Fig. 2). An imaginary line, the spiral of Tillaux, connects the

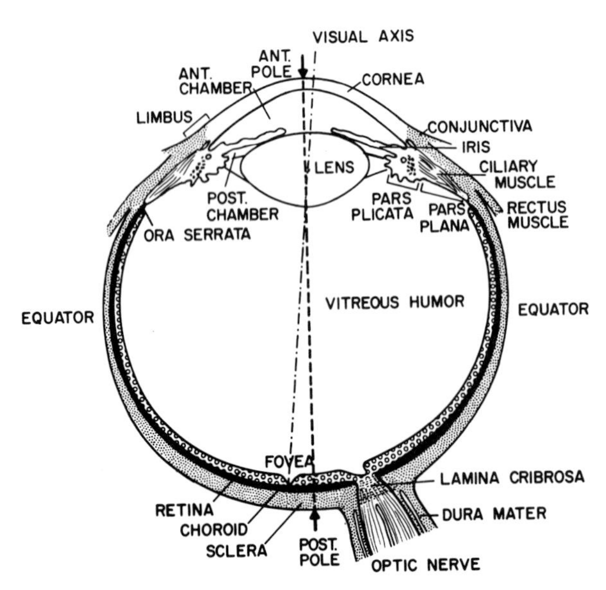

FIGURE 1 Diagrammatic representation of topography of a left human eye as seen in horizontal (meridional) section.

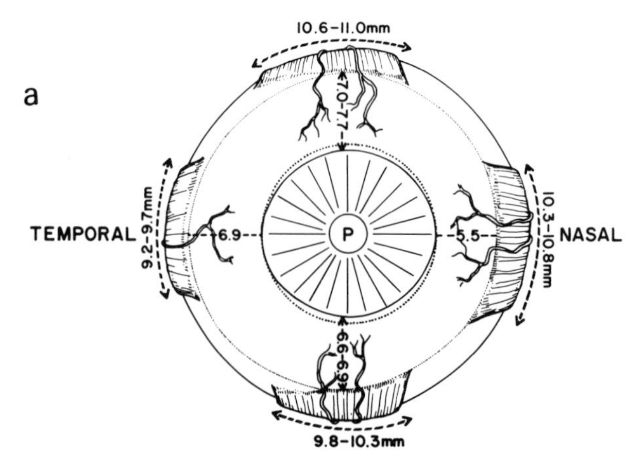

FIGURE 2a General topography of the eyeball seen from the front. The cornea is elliptical in shape, with the limbal zone (dotted line) being narrower horizontally than vertically. The spiral of Tillaux (dotted circle) forms an imaginary line by connecting the insertions of the four rectus muscles, corresponds approximately to the ora serrata. The width of the rectus muscle tendons (in mm) the distance of their insertions with respect to the corneal margin (in mm), and the vascular supply are also shown. P, pupil.

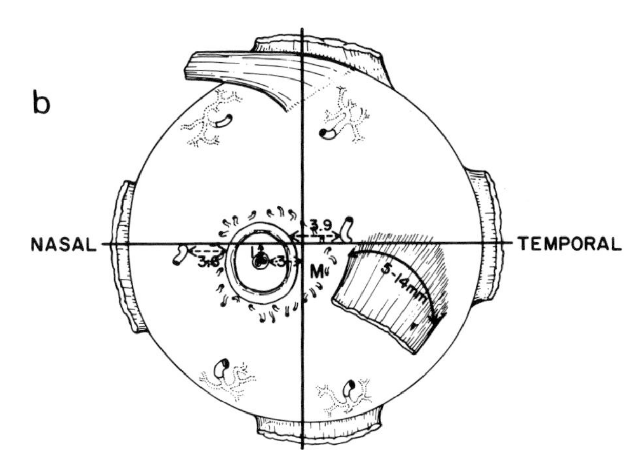

FIGURE 2b Diagrammatic representation showing general topography of the excised eyeball seen from the posterior aspect. The center of the optic nerve is located approximately 3 mm nasal and 1 mm inferior to the posterior pole. Some 20 short posterior ciliary arteries and 10 short posterior ciliary nerves pierce the sclera around the nerve. The two long ciliary arteries are located along the horizontal plane, 3.6 mm nasal and 3.9 mm temporal to the optic nerve. The macula (M) is located between the optic nerve and the curved insertion of the inferior oblique muscle tendon (IO), which may vary from 5 to 14 mm in width at its termination. The oblique scleral insertion (7–18 mm wide) of the superior oblique muscle tendon is temporal to the vertical meridian, mostly behind the equator. The approximate location of the two superior and the two inferior vortex veins is shown in relation to the vertical plane.

insertions of the four rectus muscles. In front of the insertions of the rectus muscles into the sclera, anterior ciliary arteries and veins pierce the sclera; two arteries accompany each rectus muscle, except the lateral rectus, which has only one associated artery (Fig. 2).

Posteriorly, the optic nerve emerges from the eyeball. The center of the optic nerve is displaced from the center of the geometric axis 3 mm nasally and 1 mm inferior to the posterior pole; this is responsible for the asymmetry of the three tunics (see below) of the eye. The optic nerve is covered by meningeal sheaths and is surrounded by 20 short posterior ciliary arteries and 10 short posterior ciliary nerves, all of which pierce the sclera in a ring shape (Fig. 2b). Two long ciliary nerves enter the sclera in the horizontal plane, one on the nasal and one on the temporal side, 3.6 mm and 3.9 mm, respectively, from the optic nerve. The four or more vortex veins, which drain the venous system of the uveal tract, emerge 3–6 mm behind the equator. In general, two superior and two inferior vortex veins pierce the sclera obliquely, above and below the optic nerve on each side, in the vertical plane. Each pair is separated from the other by the width of the superior or inferior rectus muscle, respectively. A posterior view of the globe also reveals the insertions of the oblique muscles. The muscular insertion of the inferior oblique is close to the fovea, whereas the tendinous insertion of the superior oblique is about 13 mm posterior to the limbus (Fig. 2b).

B. Tunics of the Eyeball

Bisection of the eyeball along a meridian discloses the fundamental architecture of the globe (Fig. 1). The three tunics of the eyeball, from the outside to the inside, are the fibrous coat of sclera and cornea, the uveal tract, and the retina.

1. The Fibrous Coat

a. The Sclera The sclera is the opaque, dull white, collagenous, viscoelastic outer coat of the eye. In addition to maintaining the intraocular pressure, the sclera keeps the ocular dimensions stable for optimum visual function. Enlargement of the eye, as occurs in congenital glaucoma and myopia, is caused by a gradual distention of the collagen bundles in the sclera beyond their elastic limit. Externally, the sclera is covered by the episclera and Tenon's capsule (fascia bulbi), which are connected by fine trabeculae. The anterior portion of the sclera is visible through the bulbar conjunctiva as the "white" of the eye. Broad collagen lamellae of variable diameter and of irregular distribution interweave with elastic fibers in an intricate fashion in the scleral stroma, thus increasing the tensile strength of the globe. The opacity of the sclera can be attributed to its high water content, the haphazard arrangement and variable diameter of its collagen fibrils, and a low concentration of ground substance. The sclera contains sparse fibroblasts, and the orientation of the collagen bundles varies in different regions. The innermost layer of the sclera, which separates it from the choroid and ciliary body, is called the lamina fusca, a layer rich in elastic tissue and pigmented cells. Its fine collagen bundles merge with the fibrous lamellae of the suprachoroidea and supraciliaris.

The sclera has two large openings, the anterior and posterior scleral foramina, and numerous smaller openings through which ocular nerves and blood vessels pass. The anterior foramen is not a

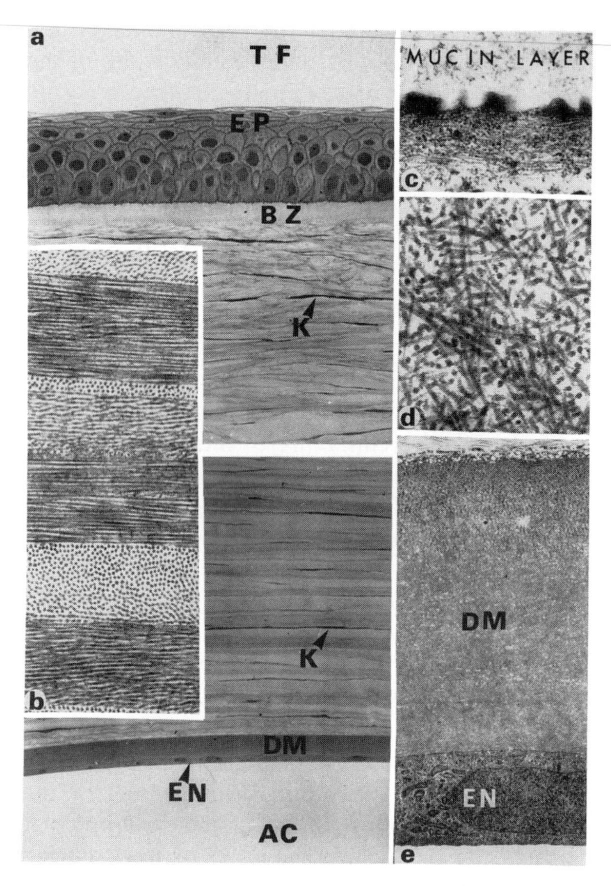

FIGURE 3 (a) Light micrograph of cornea showing its layered architecture. EP, multilayered squamous epithelium; BZ, Bowman's zone of stroma; K, keratocytes in anterior and posterior stroma; DM, Descemet's membrane; EN, endothelium. The cornea is bathed anteriorly by tear fluid (TF) and posteriorly by aqueous humor of the anterior chanber (AC) ×190. (b) Corneal stroma showing regular and parallel arrangement of collagen fibrils in individual lamellae that cross each other at right angles. ×12,440. (c) A superficial cell of corneal epithelium showing microvillous surface projections covered by mucinous layer of the tear film. ×15,890. (d) Bowman's zone of stroma constituted by a feltwork arrangement of collagen fibrils. ×19,140. (e) Descemet's membrane (DM) is the thickened basement membrane of the endothelium (EN). ×3,470. Figs. 3 b–e are electron micrographs.

true opening, because there is structural continuity between the cornea and sclera at the limbus. The posterior scleral foramen is formed by a sievelike apparatus, the lamina cribrosa, through which axons of the optic nerve pass (Fig. 1).

b. The Cornea The cornea consists of a clear, transparent, avascular, viscoelastic tissue with a smooth, convex external surface and a concave internal surface (Fig. 3). The main function of the cornea is optical; it forms the principal refracting surface of the dioptric system of the eye, accounting for 70% of total refractive power (45 diopters). When the eyelids are open, the cornea is separated from the air only by the precorneal tear film, a physiologic secretion that covers the external surface of the corneal and conjunctival epithelium. By filling minor surface irregularities, the tear film provides a smooth air–cornea interface for refraction. The precorneal tear layer is the main vehicle for the supply of nourishment to the cornea and for the maintenance of a nonkeratinized state of the corneal epithelium.

Structurally, the cornea consists of four layers: an epithelial layer with its basement membrane, the stroma with its anterior-modified Bowman's layer, Descemet's membrane (the thick basement membrane of the endothelium), and the corneal endothelium (Fig. 3).

The corneal epithelium is the most sensitive and probably the most highly organized stratified squamous epithelium in the body. It is composed of five to six cell layers. The superficial layers consist of flattened, polygonal cells, which disintegrate or are wiped away with the movement of the lids and by the precorneal tear film. Approximately 14% of these surface cells are exfoliated each day. The middle zone of the epithelium contains polygonal or "wing" cells; this zone is two or three cell layers thick and is derived from the mitotic activity of the basal cells. The basal zone, the deepest layer of the corneal epithelium, is responsible for its weekly turnover. The intercellular spaces of the corneal epithelium are filled with ground substance and adjacent cells are attached by numerous desmosomal junctions. The cytoplasm of the cells is rich in tonofilaments, which are oriented along the long axis of the cells and are condensed at the desmosomes. The tall columnar (20 μm × 10 μm) cells of the basal (germinal) layer are attached to their basement membrane by strong hemidesmosomal junctions. At the limbus, the corneal epithelium is continuous with that of the conjunctiva.

The corneal stroma is predominantly collagenous and constitutes approximately 85% of the corneal thickness (Fig. 3). Anteriorly, it is composed of a homogenous, acellular, collagenous latticework called Bowman's layer, or zone (8–14 μm thick), which is highly resilient and resists deformation and trauma as well as passage of bacterial organisms and foreign bodies. Bowman's zone is perforated in many places by unmyelinated nerves in transit to the corneal epithelium. The corneal stroma deeper to Bowman's zone is formed by parallel lamallae (numbering 200–250) of collagen fibrils. The overall arrangement of the collagen fibrils mimics a three-dimensional diffraction grating, with individual fibrils separated by less than half the wavelength of light, thus providing the structural basis for the transparency of the cornea. If this orderly arrangement of collagen fibrils is destroyed by pathologic processes, the architecture of the stromal fibrils is not renewed, and replacement occurs by the formation of scar tissue. The stromal ground substance consists primarily of two sulfated glycosaminoglycans, keratan sulfate and chondroitin-4-sulfate, which are secreted by native stromal keratocytes. The sparsely distributed keratocytes or stromal fibroblasts that constitute 3–5% of the stromal volume are flattened, thin, irregularly shaped cells.

Descemet's membrane, one of the thickest basement membranes in the body, forms the posterior boundary of the corneal stroma (Fig. 3). This layer, composed primarily of collagen, polysaccharides, and glycoproteins, is secreted by the corneal endothelium. The anterior (embryonic) zone is banded, but the posterior region is amorphous. The thickness of Descemet's membrane increases with age; it measures 3 μm at birth and 9 μm during adulthood.

The monolayer of corneal endothelium that lines the posterior surface of Descemet's membrane demarcates the anterior boundary of the aqueous cavity of the anterior chamber (Fig. 3). Approximately 500,000 closely fitted, hexagonal, flattened cells, which are 18–20 μm wide and 5–6 μm thick, comprise the corneal endothelium (Fig. 4). The cells contain a centrally placed, oval nucleus, as well as abundant mitochondria and other organelles, signifying the active metabolic state of the cells. Junctional complexes, consisting of zonulae occludentes and gap junctions, connect adjacent endothelial cells and preclude influx of aqueous humor into the corneal stroma, thus maintaining corneal turgescence and transparency. Because the endothelial cells normally do not undergo mitosis after birth, cell loss is generally compensated for by spreading and thinning of adjacent cells; this loss occurs with age and after trauma, inflammation, and degenerative insults. The structural and functional integrity of the corneal endothelium is vital to the maintenance of corneal transparency.

c. The Corneoscleral Limbus At the limbus, the corneal epithelium loses its regular structure and continues with the conjunctival epithelium. The compact transitional zone of corneoscleral tissue that forms the midlimbus is traversed by intrascleral vascular plexuses and nerves (Fig. 5). The limbal vessels provide nourishment for the peripheral, avascular cornea. The deep or inner limbus contains the canal of Schlemm and the trabecular meshwork, which provide the major pathway for drainage of aqueous humor. Dynamic and structural alterations in this region can lead to glaucoma.

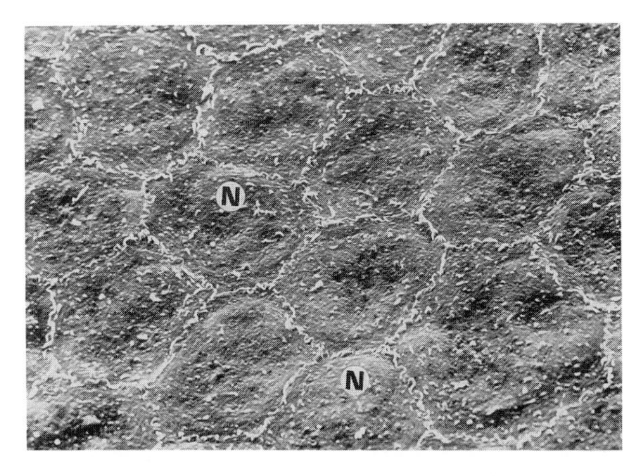

FIGURE 4 Scanning electron micrograph of normal human corneal endothelium. Individual cells are arranged regularly and are approximately hexagonal in shape, and contain an oval nucleus (N). Microvillous projections are present on the surface and especially along the cell borders. ×1,455.

The trabecular meshwork is composed of a lattice of superimposed, perforated sheets or beams that extend from the scleral spur and the anterior face of the ciliary muscle to the line of Schwalbe and deep corneal lamellae (Fig. 6). The core of each beam is formed by fibrous tissue that is covered by flattened trabecular cells. Changes in the tone of the ciliary muscle can alter the porosity of the meshwork. In aging and glaucomatous eyes, degenerative changes in the trabecular beams and increased accumulation of extracellular materials contribute to increased resistance to aqueous outflow.

The canal of Schlemm, located in the inner part of the limbus, is a toroidal structure that is supported on its inner aspect by the trabecular meshwork and on its outer aspect by the compact corneoscleral tissue. It is lined by a continuous single layer of endothelial cells, which contain giant vacuoles (Fig. 7). This unique structural configuration forms transcellular channels and is implicated in the bulk flow of aqueous humor into the canal. After draining into a deep scleral plexus, aqueous humor enters the

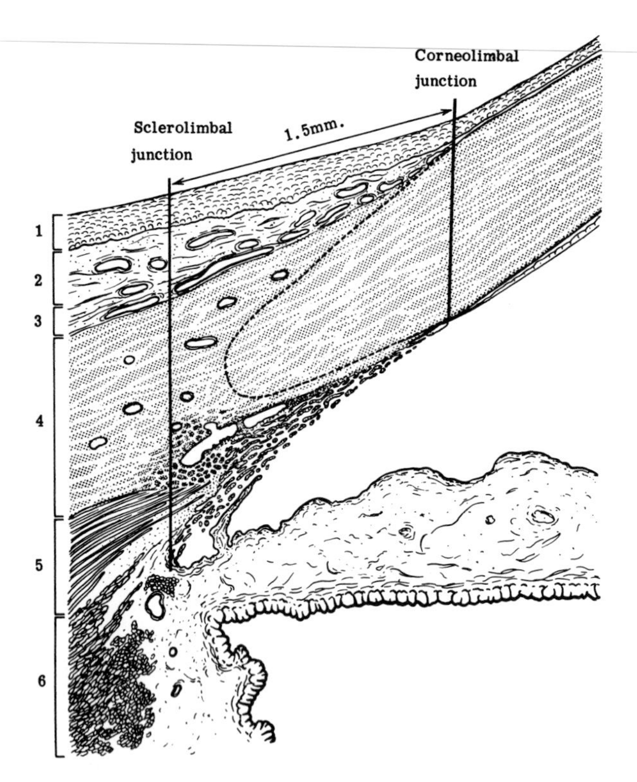

FIGURE 5 Diagrammatic representation of the limbus showing various structures from superficial to deep regions. 1, Conjunctiva; 2, conjunctival stroma; 3, Tenon's capsule and episclera; 4, limbal or corneoscleral stroma containing intrascleral plexus of veins and collector channels from Schlemm's canal; 5, meridional portion of ciliary muscle; 6, radial and circular portions of ciliary muscle. The so-called pathologist's limbus is about a 1.5 mm wide zone extending posteriorly from a line joining the peripheral termination of Bowman's zone and Descemet's membrane of the cornea. The histologist's limbus (dotted line) is represented by a cone-shaped termination of the corneal lamellae into the sclera.

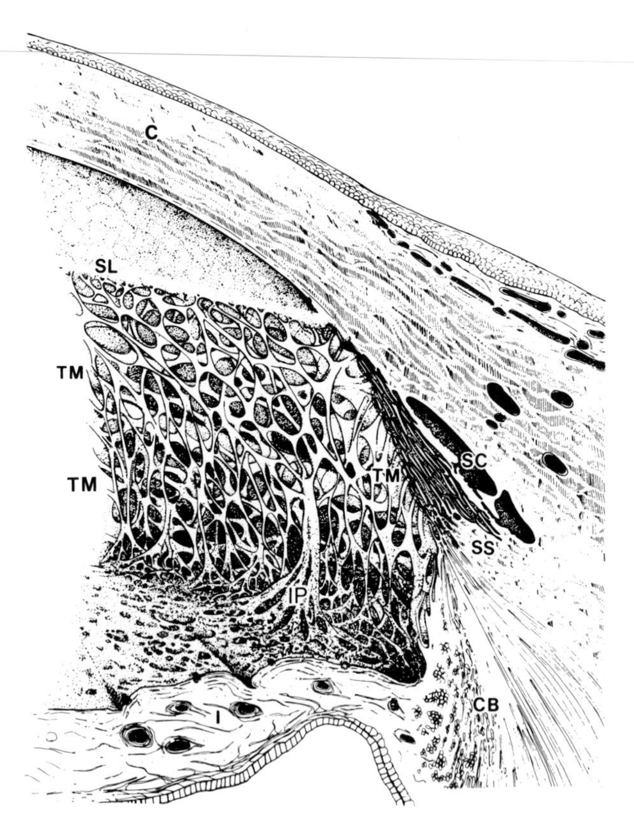

FIGURE 6 Semidiagrammatic representation of the angle of the anterior chamber of the human eye, as seen in a composite sectional view and three-dimensional gonioscopic view. The trabecular meshwork (TM) is made up of superimposed, circumferentially oriented fibrocellular sheets. Perforations in individual sheets (intratrabecular spaces) and spaces between adjacent sheets (intertrabecular spaces) provide the pathway for bulk flow of aqueous humor from the angle of the anterior chamber. The aqueous humor is drained finally into the canal of Schlemm (SC) and collector channels (CC). C, cornea; SL, line of Schwalbe; SS, scleral spur; CB, ciliary body; IP, iris process; I, iris.

venous circulation via the intra- and episcleral veins (Fig. 8). Several collector channels (aqueous veins of Ascher) bypass the deep scleral plexus and terminate directly in the subconjunctival veins after coursing through the sclera (Fig. 8).

2. The Uveal Tract

The uveal tract forms the pigmented vascularized tunic of the eye. It is divided into three discrete regions: the choroid, ciliary body, and iris (Fig. 1).

a. The Choroid The choroid has four main layers: the suprachoroidea, the stroma, the choriocapillaris, and Bruch's membrane. The suprachoroidea is a superficial layer of the choroid made up of collagen lamellae that blend into the sclera. The potential space between the sclera and choroid (suprachoroidal space) is traversed by the long and short ciliary arteries that supply the uveal tract. Before leaving the sclera, the existing veins converge to form ampullae that drain into the vortex veins. The choroidal stroma consists of an extensive network of vessels as well as loose collagenous and elastic tissue, fibroblasts, melanocytes, lymphocytes, and smooth-muscle "stars." The choriocapillaris, the capillary layer of the choroid, is anterior to the choroidal stroma. It consists of the largest vessels, which are flattened, closely packed, and highly permeable (fenestrated). These vessels supply nutrients and oxygen to the retinal pigment epithelium and the outer sensory retina and, exclusively, to the vessel-free area of the fovea. Bruch's membrane,

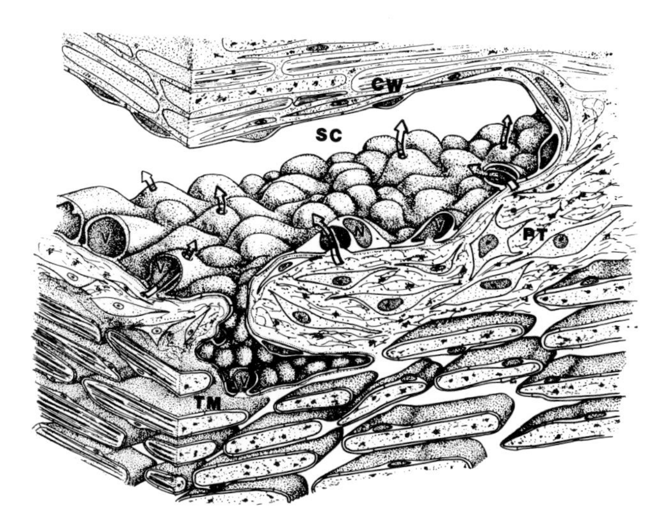

FIGURE 7 A composite three-dimensional schematic rendering of the walls of Schlemm's canal (SC) and adjacent trabecular meshwork (TM). The spindle-shaped endothelial cells lining the trabecular wall of Schlemm's canal are characterized by luminal bulges corresponding to unique macrovacuolar configurations (V) and nuclei (N). The macrovacuolar configurations are formed by surface invaginations on the basal aspect of individual cells which gradually enlarge to open eventually on the apical aspect of the cell surface, thus forming transcellular channels (arrows) for the bulk flow of aqueous humor down a pressure gradient. The endothelial lining of the trabecular wall is supported by a variable zone of cell-rich pericanalicular tissue (PT) below which lie the organized superimposed trabecular sheets having inter- and intratrabecular spaces that allow the flow of aqueous humor from the anterior chamber to the canal of Schlemm. The compact corneoscleral wall (CW) of Schlemm's canal is formed by the lamellar arrangement of collagen and elastic tissue.

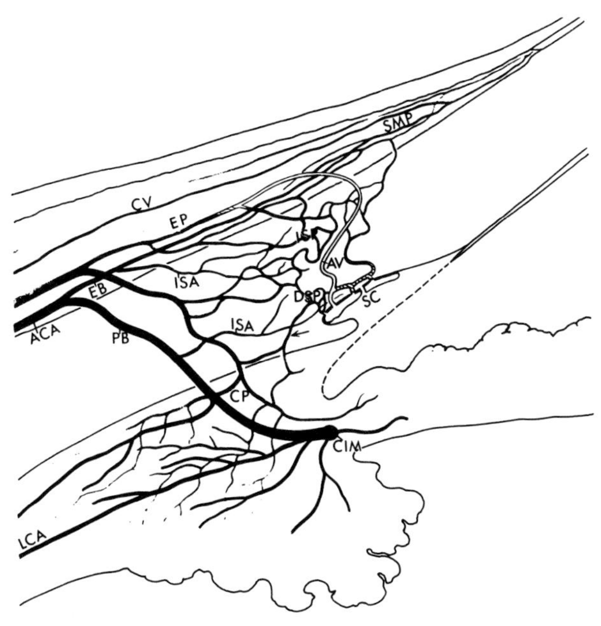

FIGURE 8 The limbal and ciliary vasculature shown diagrammatically. The deep scleral plexus (DSP) drains aqueous humor from the canal of Schlemm (SC) and is connected anteriorly to the intra- and episcleral plexuses of veins (ISP and EP). The aqueous veins (AV) arise directly from the canal of Schlemm or occasionally from the deep scleral plexus and join the episcleral veins. ISA, intrascleral arterial branches; PB, major perforating and EB, episcleral branches of anterior ciliary artery (ACA); CIM, circulus iridis major; SMP, superficial marginal plexus; CP, ciliary plexus; LCA, long posterior ciliary artery; CV, conjunctival vessel.

the innermost component of the choroid, has an anterior cuticular layer, which is derived from the retinal pigment epithelium, and a posterior layer, which functions as the basement membrane of the choriocapillaris. The core is formed by a network of collagen and elastic tissue.

b. The Ciliary Body The ciliary body is an anterior continuation of the choroid and retina and, as such, is divisible into uveal and neuroepithelial portions. The uveal portion includes the supraciliaris, the ciliary muscle, and the vessel layer.

The supraciliaris is a continuation of the suprachoroidea and is analogous to the suprachoroidal space in that it is only potentially present, coming into existence only in certain pathologic conditions. The supraciliaris has fewer collagen lamellae and less fibroelastic tissue than does the suprachoroidea. This space, along with the uveovortex veins, provides auxiliary drainage for aqueous humor that passes through the extracellular spaces of the anterior face of the ciliary body. The ciliary muscle is a ring of smooth muscle tissue that consists of longitudinal (meridional), radial, and circumferential fibers. This muscle plays an integral role in slackening the suspensory ligaments of the lens, which permits an increase in the convexity of the lens (accommodation). The vessel layer of the ciliary body supplies blood to the ciliary muscle and consists almost exclusively of fenestrated capillaries.

The neuroepithelial portion consists of the double-layered ciliary epithelium: A pigmented layer, which is continuous with the retinal pigment epithelium, and the cuboidal, or low columnar, nonpigmented epithelium, which is a direct continuation of the sensory retina. The apical surfaces of the two epithelial layers are juxtaposed. Aqueous humor is secreted by the epithelial cells of the ciliary processes, which are supplied by fenestrated capillaries. The ciliary epithelium is smooth posteriorly (this portion is called the pars plana), but becomes markedly convoluted anteriorly (pars plicata) and forms the finger-like projections of ciliary processes.

c. The Iris The iris is located in front of the lens and is a continuation of the ciliary body. It forms a delicate diaphragm between the anterior and posterior chambers of the eye (Fig. 9). The color of the iris varies among individuals, depending on the amount of pigment in the stromal cells. Stromal pigmentation increases rapidly during the first year of life, and hence many infants who were born ''blue-eyed'' gradually acquire darker pigmentation. The central aperture of the iris (the pupil) controls the amount of light entering the eye and provides an opening for the free flow of aqueous humor from the posterior chamber to the anterior chamber. The medial portion of the iris rests directly on the anterior surface of the lens.

The iris has two main components: the uveal portion (stroma) and the neuroepithelial portion. The stroma consists of a narrow anterior avascular border layer, with crypts of variable size, and a wider vascularized posterior stromal layer. The anterior border layer is formed by interlacing fibroblasts and pigmented melanocytes in an extracellular ground

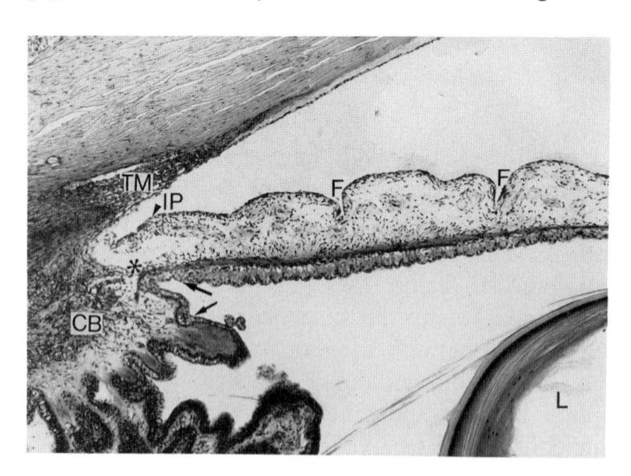

FIGURE 9 Transverse section through the mid and peripheral part of the iris. The dilatator muscle ends (asterisk) near the iris root which is the thinnest part of the iris. The posterior layer of the iris epithelium begins to lose its pigment and becomes continuous with the nonpigmented posterior epithelial layer of the pars plicata; the transitional zone is marked by arrows. Owing to partial dilation of the pupil, concentric furrows (F) are present anteriorly. The anterior border layer is partly continuous with the root of the iris process (IP). TM, trabecular meshwork, CB, ciliary body, L, lens. Partially bleached section. Photomicrograph ×39.

substance. In addition to these randomly distributed cells, clump cells, mast cells, macrophages, and lymphocytes are also present in the stroma. The collagen fibrils are arranged in a lattice pattern, which provides a scaffold during pupillary dilatation and constriction.

The posterior surface of the iris (neuroepithelial portion) is formed by a layer of pigmented cells, which are a continuation of the nonpigmented layer of the ciliary epithelium (Fig. 9). The myoepithelial layer of the iris, which constitutes the dilator pupillae muscle, is a continuation of the pigmented layer of the ciliary epithelium. The smooth muscle of the sphincter pupillae is located in the iris stroma around the pupil and controls the entry of light to the retina.

3. The Retina

The retina is derived from neuroectoderm and represents an extension of the brain in the eye. This thin, delicate layer of stratified nervous tissue is comprised of the retinal pigment epithelium and the sensory retina (Fig. 10). [*See* RETINA.]

a. Retinal Pigment Epithelium The retinal pigment epithelium is a single layer of hexagonal, cuboidal epithelial cells, 14–60 μm wide and 10–14 μm long, depending on the region of the retina. The cells are narrow, tall, and uniform in the foveal region, but they are wide and irregular toward the ora serrata. The retinal pigment epithelium extends from the margin of the optic nerve posteriorly to the ciliary epithelium anteriorly. The apices of the cells are joined by zonulae occludentes and, together with the cell borders, form the external blood–retinal barrier. The apical region of the cell surface is thrown into microvillous projections, which enclasp the outer segments of the photoreceptors. Granules of melanin, which are responsible for absorbing excess light, are concentrated mainly in the apical region of the cells. The primary functions of the cells include transport of nutrients from the choroidal vasculature to the photoreceptor layer, removal of metabolites, and active phagocytosis and recycling of photoreceptor disks that are shed daily from the outer segments of the photoreceptors. The cells secrete a basal lamina (the internal layer of Bruch's membrane) to which they are attached firmly. Because the sensory retina and the pigment epithelium are attached to each other only through the extracellular matrix, a separation of the two layers occurs in clinical conditions of retinal detachment.

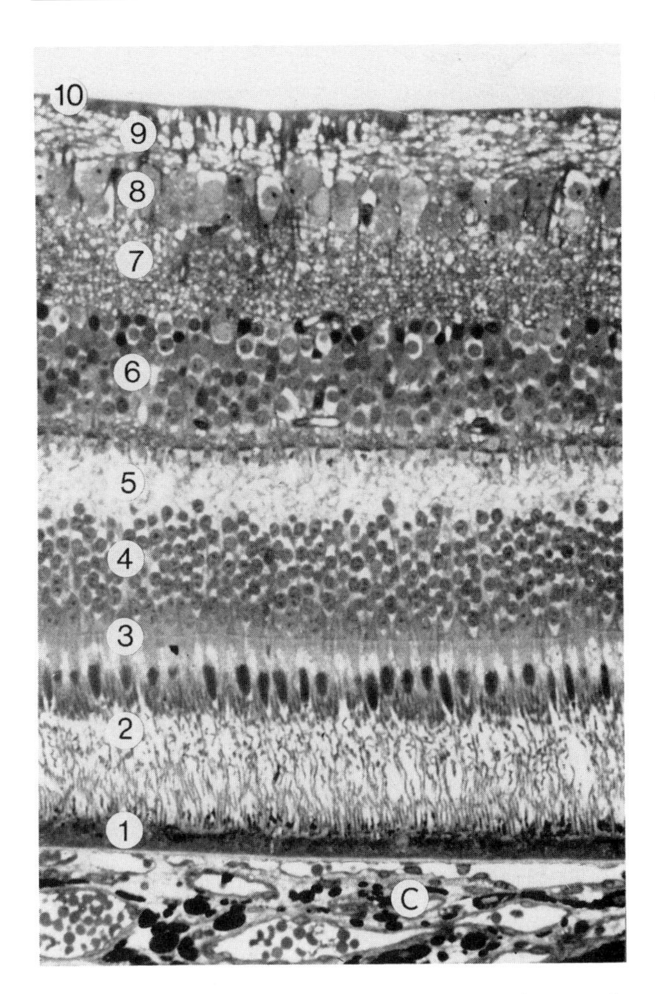

FIGURE 10 Transverse section of retina showing pigment epithelium (1) attached to choroid (C) through Bruch's membrane and sensory retina consisting of: 2, photoreceptor layer; 3, external limiting membrane; 4, outer nuclear layer; 5, outer plexiform layer; 6, inner nuclear layer; 7, inner plexiform layer; 8, ganglion cell layer; 9, nerve fiber layer; and 10, internal limiting membrane 280.

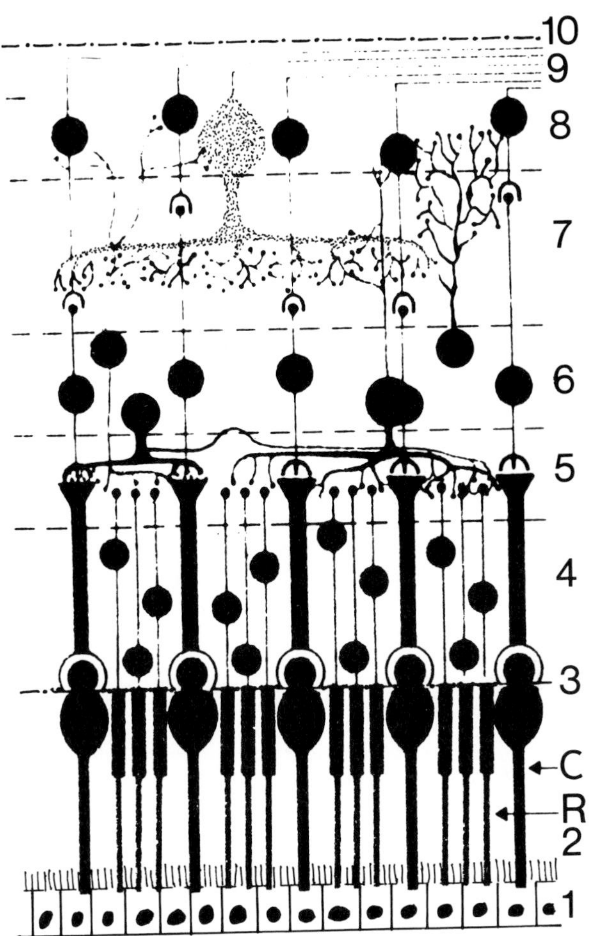

FIGURE 11 Diagrammatic representation of elements comprising the retina. 1, Pigment epithelium; 2, photoreceptor layer consisting of rods (R) and cones (C); 3, external limiting membrane; 4, outer nuclear layer; 5, outer plexiform layer; 6, inner nuclear layer; 7, inner plexiform layer; 8, ganglion cell layer; 9, nerve fiber layer; and 10, internal-limiting membrane.

b. Sensory Retina The sensory retina consists of a layer of photoreceptor cells, the axons of which synapse with intermediate cells for modulation of the photic response. Impulses modified by these modulator cells are relayed to an inner layer of ganglion cells, which transmit spike discharges (action potentials) through the optic nerve to the brain.

The general architecture of the retina is represented by 10 layers, including the retinal pigment epithelium (Figs. 10 and 11). The neuronal layers are not strictly distinct and should be considered merely as a theoretical paradigm of retinal organization:

1. The retinal pigment epithelium (described above).
2. The photoreceptor layer of the retina. These neuronal cells are distinguished functionally and structurally as rods and cones, each of these cells have three components: the end organ, consisting of a stack of membranous disks containing the visual pigment (the outer segments), the cell body with the nuclei, and the inner terminal fibers (Fig. 12). The 110–125 million rod photoreceptors are concentrated near the retinal periphery, whereas the cones (6.3–6.8 million), which are involved in photopic vision and color perception, have their maximum density at the fovea in the center of the macula (Fig. 13).
3. The external limiting membrane is formed by the attachment of photoreceptors to supporting elements called Muller cells.
4. The outer nuclear layer consists of densely packed cell bodies of photoreceptors with their nuclei and cytoplasm.

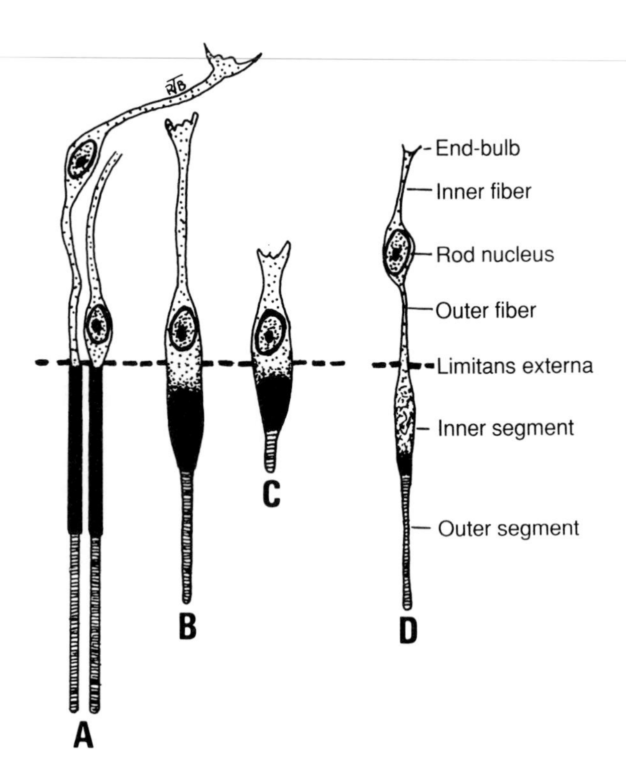

FIGURE 12 Schematic representation of photoreceptors of human retina. A, cones from the foveola; B, cone from midway between the ora serrata and the optic disc; C, cone from near the ora serrata; D, rod. Note the variations in the shape and size of these photoreceptor elements depending on their location in the retina.

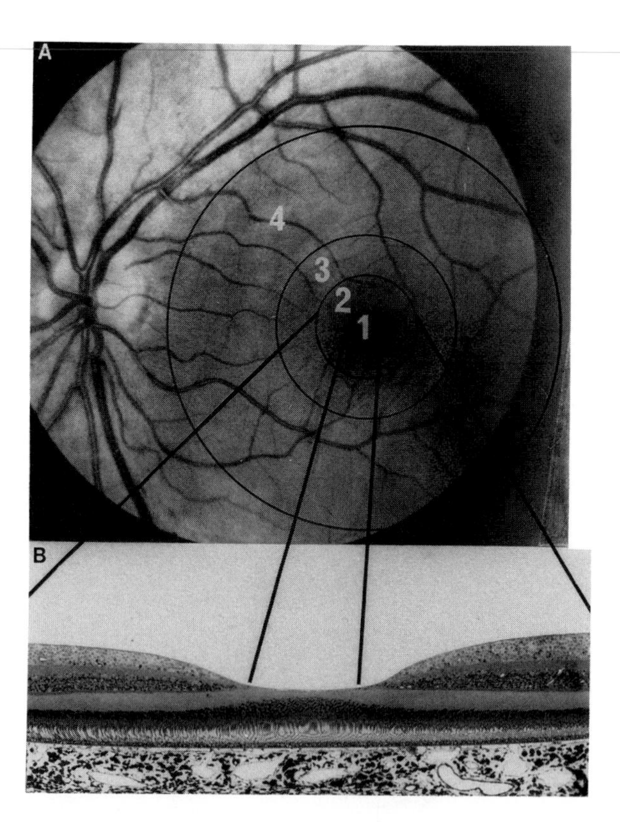

FIGURE 13 (a) Fundus photograph of left human eye showing topographic demarcation of the area centralis that measures 5.5–6 mm in diameter (outermost circle) and its subdivisions (inner circles). (From an anatomic standpoint, the zones demarcated are in fact horizontally elliptical rather than circular as shown diagrammatically here). The central area of the macular region is represented by the fovea centralis (2) approximately 1.85 mm in diameter, which has a central pit, the foveola (1), 0.35 mm in diameter. The anatomically distinguishable retinal belts surrounding the fovea centralis are the parafovea (3), 0.5 mm wide, and the perifovea (4), 1.5 mm wide. (b) Transverse section of foveal retina matched topographically to the fundus photograph shown in (a). Photomicrograph × 43.

5. The outer plexiform layer is defined by the synapses of first- and second-order retinal neurons. In this region, dendrites of the bipolar, horizontal cells and associated cells, which modulate and transmit stimuli from photoreceptors, form synapses with photoreceptors.

6. The inner nuclear layer consists of nuclei belonging to bipolar and associated cells; it also contains the deepest capillaries of the central retinal artery.

7. The inner plexiform layer is defined by synapses between the second (intermediate) and third (cerebral) retinal neurons located in this layer. The axons of bipolar and amacrine cells synapse with dendrites from ganglion cells.

8. The ganglion cell layer contains cell bodies (designated as W, X, and Y) of cerebral neurons and their supporting neuroglial cells.

9. The nerve fiber layer is composed of axons of the ganglion cells, which run parallel to the retinal surface and converge at the optic disk.

10. The internal limiting membrane is a basement membrane of both retinal and vitreal elements. Muller cells extend from the internal to the external limiting membranes and, together with astrocytes and accessory glia, provide skeletal support for the retina.

The structural organization of the retina shows variations that depend on the topography. The retina is thickest (about 0.5 mm) at the margin of the optic disc, and thinnest at the fovea (between 0.2 and 0.3 mm) and at the ora serrata (0.1 mm). The central retina, a specialized region approximately 6 mm in diameter (this includes 15 degrees of the visual field), has a central region, the fovea (1.5 mm in diameter, 5 degrees of the visual field). Because of its yellow tinge, this central area is called the macula lutea. The cell layers present at the fovea are

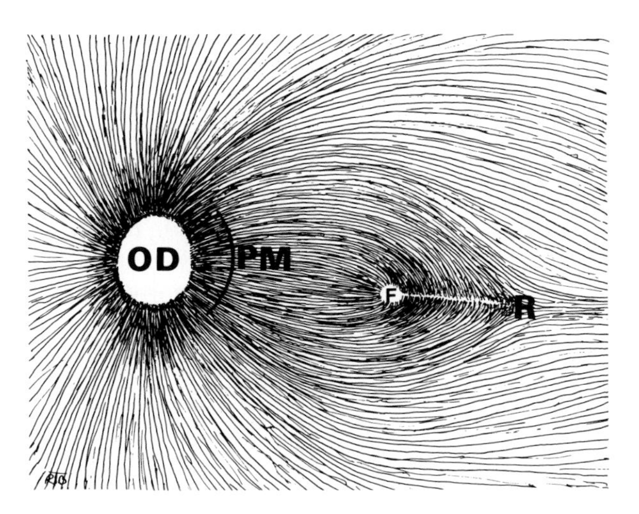

FIGURE 14 Schematic representation of the course and arrangement of the nerve fibers to the optic nerve. OD, optic disk; PM, papillo-macular bundle; F, foveola; R, raphe.

the pigment epithelium, the photoreceptors (exclusively cones, which are tall and slender), the outer nuclear layer, and the outer plexiform layer, which is designated as the layer of Henle (Fig. 13). The nerve fibers arising from the fovea are directed to the temporal region of the optic disc and constitute the papillomacular bundle (Fig. 14). At the edge of the fovea, the retina has full thickness (abundant ganglion cells are multilayered, Henle's layer is thick, and the cones are less slender and are intermingled with rods).

Except in the foveola (0.4 mm diameter), the retina is permeated by blood vessels to the level of the innermost part of the outer nuclear layer (Fig. 10). The structural and functional characteristics of these vessels are such that a competent blood–retinal barrier is normally maintained.

C. Chambers of the Eye

The eyeball contains three distinct chambers: the anterior chamber, posterior chamber, and vitreous cavity (Fig. 1).

1. Anterior Chamber

The anterior chamber is an ellipsoidal cavity bounded anteriorly by the corneal endothelium and peripherally by the trabecular meshwork (Figs. 1, 5, 6). Posteriorly, the anterior chamber is demarcated by the anterior surface of the iris and the pupillary portion of the lens. The apex of the chamber lies at the anterior face of the ciliary body. The average volume of aqueous humor in the anterior chamber is about 0.2 ml in the human eye.

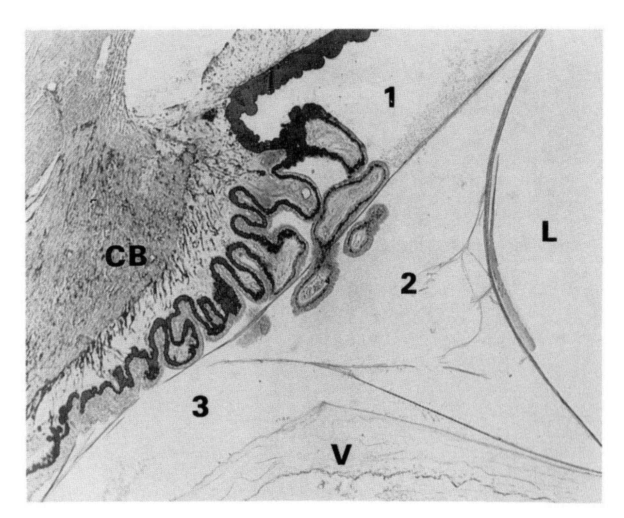

FIGURE 15 The three subdivisions of the posterior chamber in the human eye as seen in meridional section of the globe. 1, Posterior chamber proper or prezonular space; 2, zonular circumlental space or canal of Hanover; 3, retrozonular space or canal of Petit. Note the widening of the latter toward the ciliary body (CB) and narrowing toward the lens (L). V, condensed anterior vitreous. Photomicrograph $\times 30$.

2. Posterior Chamber

The posterior chamber is the space bounded anteriorly by the posterior surface of the iris, laterally by the ciliary processes, posteriorly by the anterior face of the vitreous, and medially by the lens equator (Fig. 15). It contains the freshly secreted aqueous humor (0.5 μl) formed by the ciliary processes.

3. Vitreous Cavity

The vitreous cavity of the eye occupies almost four-fifths of the total volume of the globe (5.2 ml). It is filled with a transparent, avascular, gellike substance of embryonic microcollagenous structure and is rich in hyaluronic acid. The vitreous is undoubtedly the most fragile connective tissue structure in the body. The boundaries of the vitreous cavity are demarcated by the internal surface of the retina and optic nerve posteriorly, by the pars plana laterally, and by the posterior surface of the lens anteriorly. The main sites of attachment for the vitreous fibrils are the ora serrata, the optic nerve, and the macular region (Fig. 16). Age-related degeneration, especially in myopic eyes, cause "floaters" to form in the vitreous.

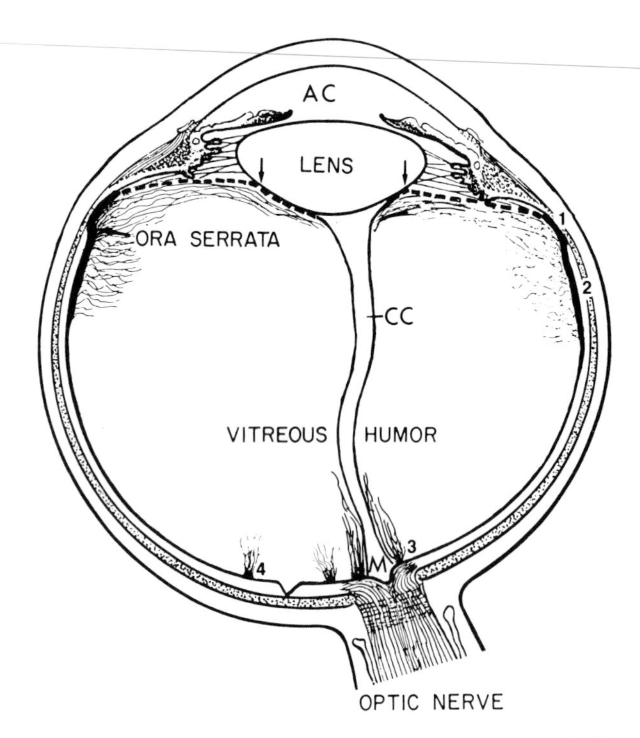

FIGURE 16 Anatomy of the vitreous depicted schematically. The vitreous body conforms to the shape of the cavity in which it is contained. Centrally, it is traversed by Cloquet's canal (CC) containing hyaloidean vitreous that extends in an S-shaped fashion from a tortuous funnel-shaped expansion on the posterior aspect of the lens to the optic nerve head with a smaller funnel-shaped expansion, the so-called area of Martegiani (M). Anteriorly, the vitreous face is in contact with the posterior lens surface separated only by the capillary space of Berger which extends peripherally into the retrozonular space of the posterior chamber, the so-called canal of Petit. In young eyes, the vitreous face is adherent to the posterior lens capsule in the region of the hyaloideocapsular ligament (arrows). The vitreous body is firmly attached (solid line) to about a 2 mm zone of the pars plana (1) and continues posteriorly to about a 4 mm span of the peripheral retina (2). The dotted line anteriorly represents the anterior vitreous face which, in places, is in contact with the posterior zonular fibers. Posteriorly, the vitreous face is attached to the edge of the optic disk (3) although not as firmly as at the vitreous base. A similar annular attachment, 3–4 mm in diameter, is seen in young eyes at the macular region (4). AC, anterior chamber.

D. The Lens

The human lens is a transparent, biconvex, semisolid, avascular structure (9 mm in diameter and 4 mm in central thickness), which is located behind the iris in a shallow depression of the anterior vitreous (Fig. 1). A system of zonular fibers that extend from the ciliary epithelium to the equatorial capsule holds the lens in place.

The lens is composed of three parts: the lens capsule, which exhibits regional variations in its thickness; the lens epithelium; and the lens substance, which consists of cortical and nuclear fibers (Fig. 17). The lens capsule (one of the thickest basal laminae in the body) completely encloses the lens substance. It is a thick, transparent, smooth, reflective, elastic, periodic acid Schiff-positive, acellular, collagenous basement membrane. Anteriorly beneath the capsule there is a single layer of cuboidal cells (the lens epithelium), which continue as far as the equator. The tapering and elongated lens cells or fibers (7–10 μm long, 8–10 μm wide, and 2–5 μm thick) comprise the lens substance; mature, anucleated lens fibers evolve from young, nucleated epithelial cells at the equator. The nuclei of the cells migrate to a location somewhat anterior to the equator to form the nuclear bow, and they eventually disintegrate. The superficial cortical cells are rich in microfilaments, especially actin, which have a major role in the accommodative process because of their inherent elasticity. With the superimposition of new cells, the older fibers progressively become displaced toward the center of the lens. Because the lens cells or fibers grow from the entire equator toward the anterior and posterior poles of the lens, they meet along radiating lines or sutures. Adjacent fibers interdigitate by forming complex ball-and-socket junctions.

Several zones or "nuclei" may be distinguished in the adult lens. The embryonic nucleus is formed by the posterior primordial lens cells, which elongate rapidly to fill the lens vesicle and lose their cell nuclei; hence, there is no epithelial cell layer beneath the posterior capsule as there is present anteriorly. The fetal nucleus has two distinctive Y-shaped sutures (erect anteriorly and inverted posteriorly) to which the lens fibers are joined. The sutures formed after birth become increasingly complex as the eye ages. The adult nucleus is formed by older lens fibers, which have an increased density. Histologically, the outlines of individual cells in the lens nucleus are indistinguishable. On slit-lamp biomicroscopy, a zone of external fibers shows a refractive index less than that of the nuclear zone and is distinguished as the lens cortex.

The zonular fibers (suspensory ligaments) form a circumferential suspensory apparatus, which maintains the lens in position (Fig. 15). Contraction of the ciliary muscle loosens the tension on the zonu-

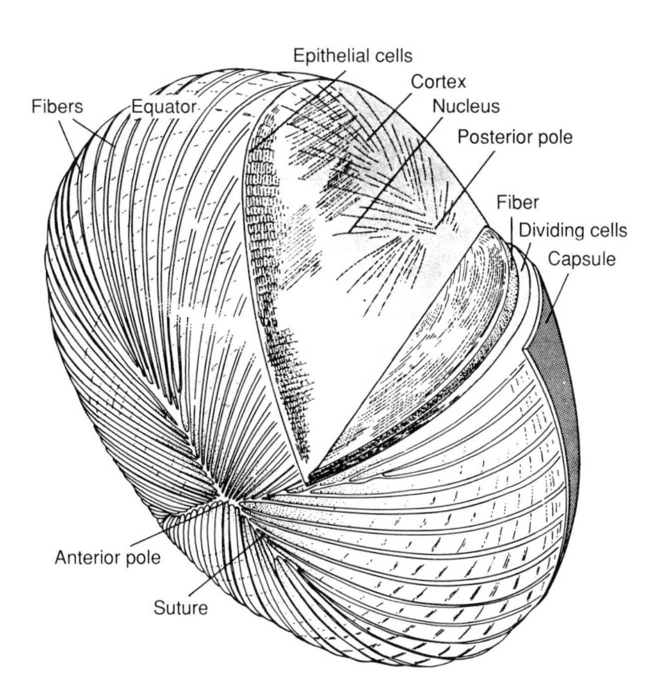

FIGURE 17 Schematic representation of the human lens. Growth of the lens occurs from the nuclear bow region near the equator. The new cells elongate and form lens fibers that wrap around the periphery of the lens and meet at the sutures where complex interdigitations occur. The outer fibers are nucleated and hexagonal in shape; as they are displaced inward by newer cortical fibers, and gradually they lose their nuclei and other intracellular organelles. [From R. Van Heyningen (1975). *Sci. Am.* **233,** 70–81. Copyright Scientific American, Inc. All rights reserved.]

lar fibers, which induces a greater curvature in the lens; this increases the accommodative power of the lens (by up to 17 diopters in the young eye).

II. The Protective Apparatus

A. The Orbit

The eyeball is housed and is freely mobile in the bony cavity, which is filled with a semispecialized fibrofatty tissue (Fig. 18). The orbit is formed by six bones: the maxilla and the palatine, frontal, sphenoid, zygomatic, and ethmoid bones (Fig. 19). The orbital cavity has a quadrilateral pyramid shape, with a triangular pyramid near its apex. The two principal posterior openings of the bony orbit, the optic foramen and the superior orbital fissure, allow passage of the optic nerve and blood vessels.

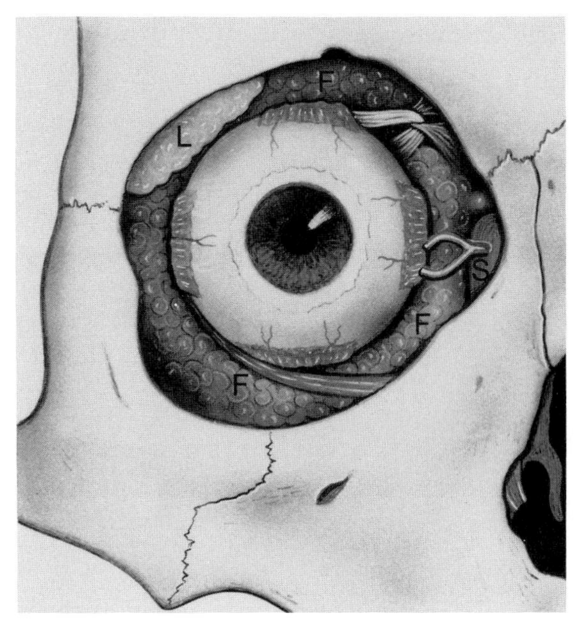

FIGURE 18 Globe in the orbit supported by orbital fat (F) and extraocular muscles. The lacrimal system, consisting of gland (L) and sac (S) are also shown. [From P. Henkind (1982). "The Eye: Anatomy and Measurements, Lippincott.]

B. The Eyelids

The eyelids consist of skin, muscles, and fibrous tissue (Fig. 20). They are lined internally by the mucous membrane of the conjunctiva and externally by thin skin without any subcutaneous fat. The muscles of the eyelids are the striated orbicularis oculi and the levator palpebrae superioris, as well as the small palpebral smooth muscle of Muller. Internally, a compact, fibrous tissue (the tarsal plate) contains modified sebaceous glands (the Meibomian glands), which open at the lid margin. The glands associated with the eyelashes are the glands of Moll (sudiferous) and of Zeis (sebaceous).

C. The Conjunctiva

The conjunctiva is a thin, translucent mucous membrane that lines the posterior surface of the eyelids (palpebral conjunctiva), is reflected on itself to form the fornices, and lines the anterior surface of the sclera (Figs. 20, 21). At the lid margin, the conjunctiva joins the skin (the mucocutaneous junction), and at the corneal margin, it is continuous with the corneal epithelium (Fig. 20). At the lacrimal punctum, the mucous membrane is continuous with the

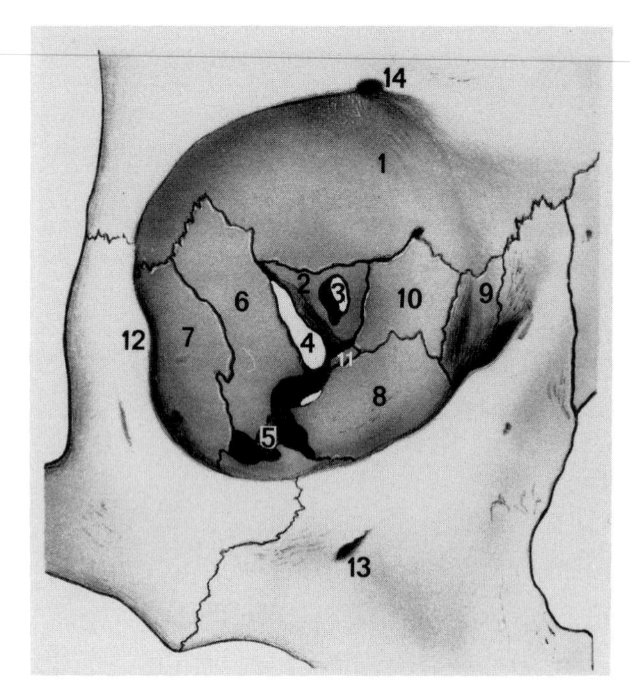

FIGURE 19 The right orbit viewed along its axis. 1, frontal bone; 2, lesser wing of sphenoid; 3, optic canal; 4, superior orbital fissure; 5, inferior orbital fissure; 6, greater wing of sphenoid; 7, zygomatic arch; 8, orbital plate of maxillary bone; 9, lacrimal bone and fossa; 10, ethmoid bone; 11, orbital process of palatine bone; 12, lateral orbital tubercle; 13, inferior orbital foramen; 14, supraorbital notch. [From P. Henkind (1982). The Eye: Anatomy and Measurements, Lippincott.]

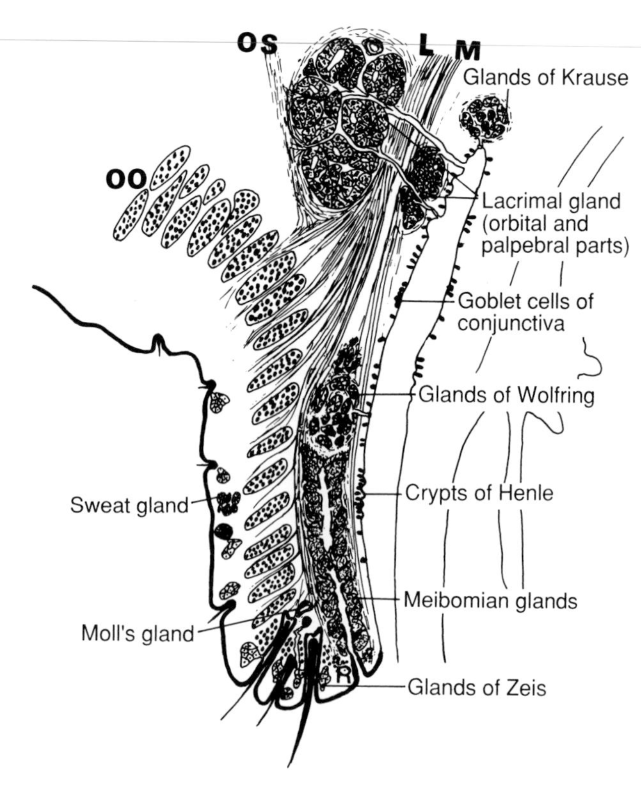

FIGURE 20 Diagrammatic representation of upper eyelid in vertical section. L, levator muscle; M, Muller muscle; OO, orbicularis oculi; R, muscle of Riolan; OS, orbital septum.

membrane lining the lacrimal canaliculi. On the nasal side of the globe and at the edge of the caruncle, the conjunctiva forms a crescent (plica semilunaris), which represents the third eyelid or nictating membrane of many animals. The bulbar conjunctiva is connected loosely to the underlying fascia bulbi, which, in turn, is attached loosely to the underlying episclera and sclera.

The epithelial lining of the conjunctiva and the cornea form the external protective covering of the open eye. Both are kept moist constantly by the tears, which prevent keratinization. The conjunctiva, like other mucous membranes in the body, consists of epithelium and substantia propria. The thickness of the epithelium (two to five cell layers) varies from region to region, being thinnest in the bulbar region and thickest at the mucocutaneous junction of the lid. Melanin granules are present in the cuboidal basal cells and occasionally in the middle layer of the epithelium. Mucus-secreting goblet cells are distributed irregularly in the conjunctiva but are most abundant in the fornices and sparse in the bulbar conjunctiva. The connective tissue of the stroma contains blood vessels, nerves, and glands; its compactness varies with the topography. The superficial layer is a fine, fibrous reticular network, which, after the age of 3 mo, contains nodules of lymphoid tissue. The vascularized deeper layers are made up of irregularly arranged bundles of collagen and some elastic tissue, as well as randomly distributed fibroblasts, melanocytes, mast cells, lymphocytes, and plasma cells. The fibrous tissue is continuous with the attached margin of the upper and lower tarsal plates and contains the smooth palpebral muscle of Muller.

D. The Lacrimal Apparatus

The main lacrimal gland (of the tubuloalveolar type) is located in the anterior lateral portion of the roof of the orbit (Fig. 18) and, together with its palpebral portions, opens into the superior fornix through several excretory ducts. The accessory lacrimal glands of Krause and Wolfring are distributed irregularly in the conjunctival stroma (Fig. 20). The blinking action of the lids distributes the tear layer

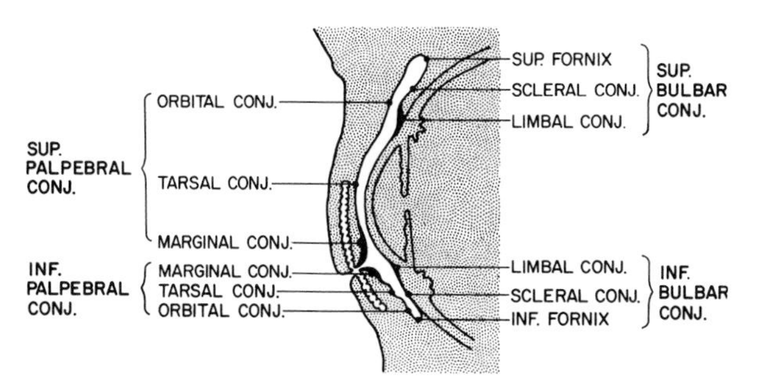

FIGURE 21 Diagrammatic representation of the conjunctival sac and its regional topography as seen in a vertical section of a closed eye.

over the keratoconjunctival surface and sweeps away foreign particles. The drainage system for the tears consists of the puncta, canaliculi, lacrimal sac, and the nasolacrimal ducts that open into the inferior nasal meatus (Fig. 22).

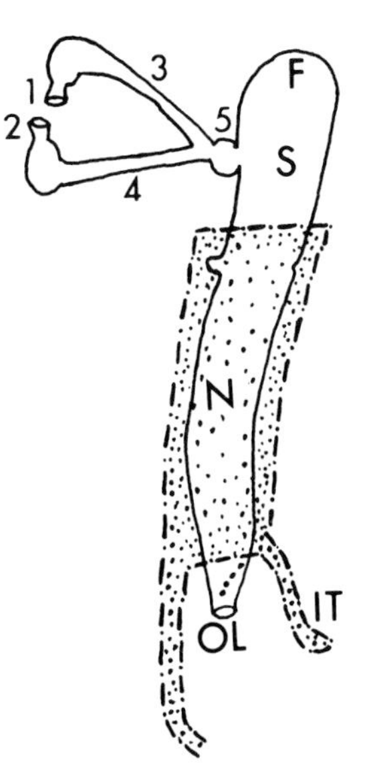

FIGURE 22 The lacrimal drainage system. 1, superior punctum; 2, inferior punctum; 3, superior canaliculus; 4, inferior canaliculus; 5, common canaliculus (sinus of Maier). F, fundus (5 mm); S, sac (8 mm); N, nasolacrimal duct (intraosseus part = 12–15 mm; intrameatal part = 5–6 mm); OL, osteum lacrimale opening in the inferior meatus; the constriction is formed by the plica lacrimalis or so-called valve of Hasner. IT, inferior turbinate bone.

III. Motor and Supporting Apparatus

A. The Extraocular Muscles

The motor apparatus of the eye is responsible for the accurate and rapid mobility of the globe. Six striated extraocular (extrinsic) muscles comprise the most important components of the motor apparatus (Figs. 2, 3, and 23). These muscles differ from other striated muscles in their unique two-fiber system: the fast-acting white fibers are involved in rapid ocular movements, and the slow-acting red muscles maintain a steady gaze. The extraocular muscles are also unusual in their high ratio of nerve to muscle fiber; this facilitates the precise coordination of muscle activity that is essential for binocular vision.

Four of the extrinsic muscles (medial, lateral, inferior, and superior rectus muscles; Figs. 2b, 23), are concerned primarily with inward, outward, downward, and upward eye movements, respectively. All of these muscles have a common origin in a ring-shaped fibrous structure that surrounds the optic nerve. The two oblique extrinsic muscles, the superior and inferior, are involved in intermediate movements of the globe. The superior oblique passes from the upper side of the posterior orbit through a fibro-cartilagenous pulley before inserting into the sclera. The inferior oblique muscle inserts directly at about the posterior pole of the eyeball. [*See* EYE MOVEMENTS.]

B. Vasculature of the Eye and Its Adenxa

The ophthalmic division of the internal carotid artery is the principal source of nutrients for the eye and the contents of the orbit. The eyelids and conjunctiva are well supplied with anastomotic branches of both the ophthalmic artery and divisions of the external carotid artery.

1. The Ophthalmic Artery

The ophthalmic artery is the first intracranial branch of the internal carotid artery, after it passes through the cavernous sinus. Branches of the ophthalmic artery include

1. The central retinal artery, which supplies the optic nerve and, after emerging from the optic disc, branches to nourish the retina as deep as the inner nuclear layer.
2. The posterior ciliary arteries, which consist of 6–20 branches. The majority (short ciliary arteries, Fig. 2b) enter the choroid to supply the choriocapillaris. Two branches (long posterior ciliary arteries) anastomose with branches of the anterior ciliary arteries to supply the ciliary body.
3. The recurrent meningeal artery, which anastomoses with the middle meningeal artery.
4. The terminal divisions of the lacrimal artery,

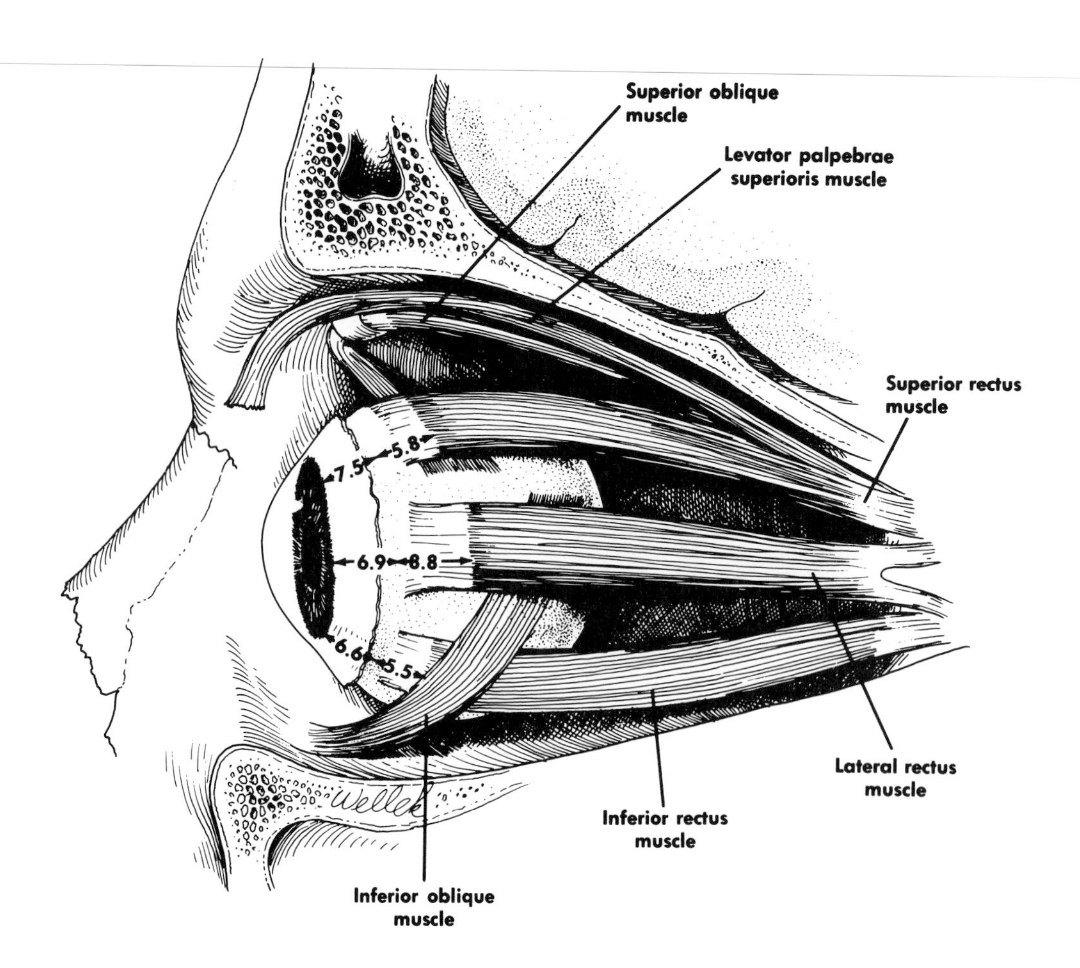

FIGURE 23 Extrinsic muscles of the eye. Both oblique muscle insert behind the equator of the globe. The inferior oblique muscle passes over the body of the inferior rectus muscle but beneath the lateral rectus muscle. The numbers indicate the distance of the insertion from the corneoscleral limbus and the length of the muscle tendon. (From Frank W. Newell (1982). Ophthalmology: Principles and concepts, ed. 4, St. Louis, C. V. Mosby Co.)

which supply the lacrimal gland, eyelid, and conjunctiva.

5. The anterior ciliary arteries, which are derived from arteries that supply the extraocular muscles.

6. The supraorbital artery, which supplies the upper eyelid, levator muscle, and periorbita.

7. The medial palpebral arteries.

8. The anterior and posterior ethmoid arteries.

9. The dorsalis nasal artery, which supplies the lacrimal sac.

10. The supratrochlear artery.

11. The episcleral and conjunctival arteries.

2. External Carotid Divisions to the Orbit and Adnexa

The blood supply to the eye and eyelids comes from the external facial artery, the superficial temporal artery, and the internal maxillary artery.

3. Venous Drainage of the Eye

The superior and inferior orbital veins are the principal pathways for the return of venous blood from the eye and its adnexa. The veins twist tortuously through the orbit before emptying into the intracranial cavernous sinus, an endothelium-lined venous cavity between the meningeal and periosteal layers of the dura mater. The internal carotid artery, together with the abducens, oculomotor, trochlear, and trigeminal nerves, passes through or adjacent to the cavernous sinus.

C. Innervation of the Eye

1. Motor Innervation of Intrinsic and Extraocular Muscles

The oculomotor (3rd cranial) nerve innervates the inferior oblique, superior, medial, and inferior rectus muscles, as well as the levator palpebrae superioris muscle. The superior oblique and lateral rectus muscles are innervated by the trochlear (4th cranial nerve) and abducens nerves (6th cranial nerve), respectively. Branches of the facial nerve are primarily motor components; they innervate the facial muscles as well as the submaxillary, sublingual, and lacrimal glands.

2. Sensory Innervation

The sensory innervation of all ocular tissues is provided by components of the ophthalmic division of the trigeminal nerve.

3. Autonomic Innervation

The sympathetic innervation of the eye and orbit is derived from preganglionic fibers of the superior cervical ganglion, which pass through the ciliary ganglion and then along branches of the ophthalmic artery (the external carotid artery) and along the nasociliary nerve (a branch of the ophthalmic division of the trigeminal nerve). Sympathetic fibers innervate sweat glands, muscles of the face and eyelids, uveal blood vessels, and Muller's muscle of the eyelid. The nerves for the dilator pupillae muscle of the iris do not pass through the ciliary ganglion.

The parasympathetic nerves, or efferent branches of the oculomotor nerve, synapse in the ciliary ganglion. Postganglionic efferent fibers innervate the ciliary muscle for the control of accommodation and of the sphincter pupillae muscle.

IV. The Visual Pathway

A. The Optic Nerve

The optic nerve is formed by an aggregation of approximately 1.3 million axons that arise from the retinal ganglion cells and extend to the lateral geniculate body in the brain, where they synapse. The optic nerve may be divided into three portions. The intraocular part, known as the optic disc or papilla, is about 1.5 mm in diameter and is formed by unmyelinated axons. This axonal aggregation is more compact on the temporal side (the papillomacular bundle) than it is on the nasal side (Fig. 14). Because there are no photoreceptor elements in the optic disc, this area represents the blind spot in the visual field. The central region of the disc has a physiologic pit through which the central retinal artery and vein emerge. In the intrascleral portion of the optic nerve, the axonal bundles acquire a myelin sheath and, together with the astroglia and oligodendroglia, pass through perforations in the sclera (the lamina cribrosa, Fig. 24). Within the orbit, the optic nerve (25–30 mm in length and about 3 mm in diameter) forms an S-shaped loop, which permits its extension during movement of the eyeball.

The optic nerve is covered by a pia mater, an arachnoid mater, and a dura mater similar to that of the brain (Fig. 24). The dura mater merges with the sclera, but the pia and arachnoid maters fuse at the lamina cribrosa, thus obliterating the flow of cerebrospinal fluid around the nerve. The orbital portion of the optic nerve continues posteriorly into the cranium through the optic foramen in the bony orbit.

B. Optic Radiations and Visual Cortex

In the brain, the optic nerve passes in a dorsal and medial direction to enter the formation of the optic chiasma (Fig. 25). The length of the cranial portion of the nerve varies (6–23 mm). The optic chiasma is a transversely oval structure that forms a bridgelike junction between the terminal portion of the two optic nerves. The nerves from the two eyes enter at the antero-lateral angles of the chiasma. The optic chiasma measures about 12 mm transversely, 8 mm sagitally, and 3–5 mm dorso-ventrally. Partial cross-over of the axons from the two optic nerves occurs in the optic chiasma, so that axons from the nasal portion of the retina on the contralateral (opposite) side and from the temporal retina on the ipsilateral (same) side combine to form the optic tracts, which emerge from the postero-lateral angles of the chiasma. Initially, the optic tracts are round, but they continue laterally and posteriorly as flattened bands. The nerve fibers from the retina end in two masses of gray matter: the lateral geniculate body and the superior colliculus.

The lateral geniculate body is subdivided into the dorsal and ventral nuclei. The ventral nucleus is homologous with the entire geniculate body of lower vertebrates. In humans, it is poorly developed and has no direct implications for vision. The dorsal nucleus is an oval structure with an elevation on the postero-lateral aspect. It serves as a relay

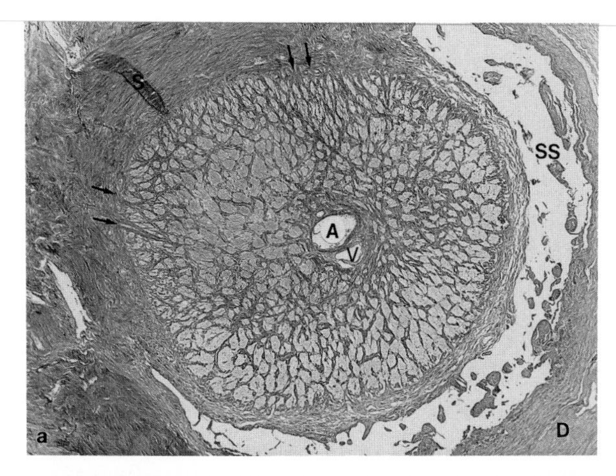

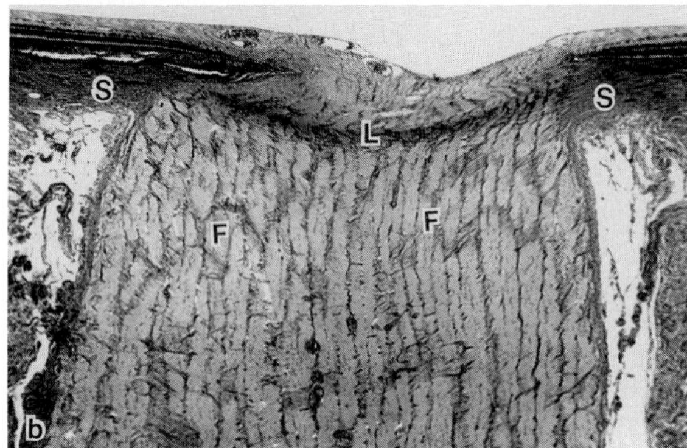

FIGURE 24 (a) Flat section of sclera passing through the lamina cribrosa. Note the connective tissue strands of the lamina cribrosa to be continuous with the sclera (S). Owing to a slightly tilted plane of section, the posterior continuation of the sclera with the dura mater (D) is seen separated from the optic nerve by arachnoid matter and subarachnoid space (SS). There is a dense connective tissue matrix around the central artery (A) and vein (V), which share a common adventitia. Wilder silver stain × 20. (b) Transverse section showing continuity of septal system of lamina cribrosa (L) with that of the sclera (S). Note the myelination of the nerve bundles (F) that occurs as the axons proceed posteriorly from the lamina cribrosa. Photomicrograph. × 30.

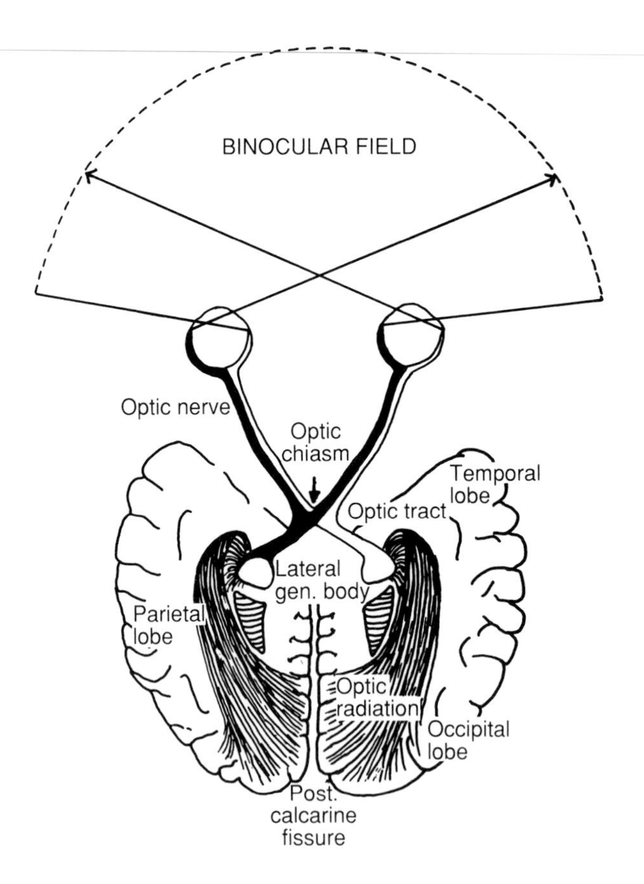

FIGURE 25 Diagrammatic representation of uniocular (145°) and binocular (120°) visual fields and the visual pathways in man.

station in the projection of the visual fibers to the occipital cortex. The geniculate body has a laminated structure, composed of alternate layers of white (medullated nerve fibers) and gray (synaptic junctions) matter. This lamination provides a sharp separation for the termination of the crossed and uncrossed fibers of the optic nerves, so that layers 1, 4, and 6 receive fibers from the contralateral eye and layers 2, 3, and 5 receive fibers from the ipsilateral eye. From the lateral geniculate body, the efferent nerve fibers pass to the occipital cortex as the geniculo-calcerine pathway or optic radiations (Fig. 25). The fibers terminate in the striate cortex and synapse primarily with the granular cells, which are situated in layer IV. These cells serve as the primary receptor elements, as well as interneurons through which visual impulses are transmitted to other layers, especially layers V and VI. Integration of visual stimuli occurs in higher-order visual associative areas, which are linked directly with the striate cortex. [*See* VISUAL SYSTEM.]

The remainder of the nerve fibers in the optic tract enter the hypothalamus (these fibers are implicated in circadean rhythm) and the superior colliculus, which is involved in coordinating reflex ocular movements.

Bibliography

Fine, B. S., and Yanoff, M. (1979). "Ocular Histology. A Textbook and Atlas," 2nd ed. Harper and Row, New York.

Hogan, M. J., Alvarado, J., and Weddell, J. E. (1971). "Histology of the Human Eye—An Atlas and Textbook." W. B. Saunders, Philadelphia.
Jacobiec, F. A. (1982). "Ocular Anatomy, Embryology, and Teratology." Harper and Row, Philadelphia.
Tripathi, R. C., and Tripathi, B. J. (1984). Anatomy of the human eye, orbit, and adnexa. *In* "The Eye," 2nd ed., Vol. 1 (H. Davson, ed.). Academic Press, London.
Warwick, R. (1976). "Wolff's Anatomy of the Eye and Orbit," 7th ed. W. B. Saunders, Philadelphia.

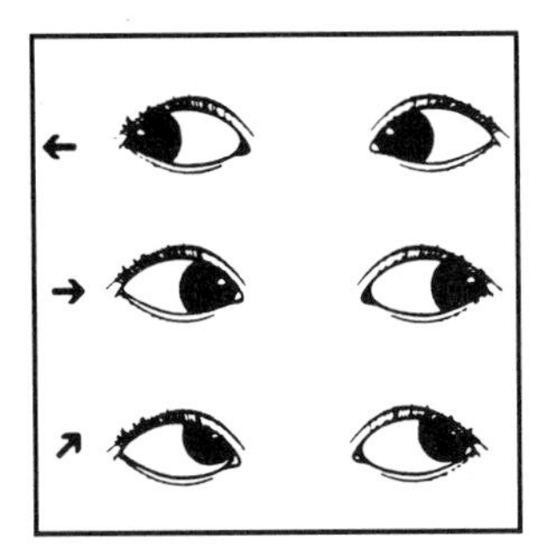

Eye Movements

BARBARA A. BROOKS, *The University of Texas Health Science Center at San Antonio*

Glossary

Brainstem Area at the base of the brain containing most of the cranial nerve nuclei, including the oculomotor nuclei

Conjugate movement Occurs when the lines of gaze in the two eyes move in the same direction, as during a saccade to the right or left

Oculomotor neurons Neurons with cell bodies in the oculomotor nuclei, whose axons constitute the oculomotor nerves and which synapse directly onto the eye muscle fibers; the "final common pathway" between the nervous and muscular systems

Premotor neurons Neurons in the brainstem that receive and process information from higher brain centers and feed directly onto the oculomotor neurons

Saccade Fast, ballistic eye movement that redirects the line of gaze

Smooth pursuit Following eye movement that enables continuous clear vision of a small moving object

Vergence movement Occurs when the lines of gaze in the two eyes move in opposite directions, as in changing focus from a far object to one nearby or vice versa

THE IMPORTANCE OF EYE movements in the maintenance of clear vision has long been appreciated, but only within the last century was the existence of more than one kind of human eye movement first recognized. In 1902, a differentiation was established between (1) very fast, short-duration eye movements, elicited by eccentric retinal stimuli and acting to place a visual target rapidly at the center of gaze, and (2) much slower movements, whose velocity varied with the velocity of a moving visual target and during which the foveated object was quite clear. The first classification is now known as rapid, or saccadic, eye movement, and the second as slow, or smooth pursuit, movement. There are variants on both themes, depending on the contribution of unconscious and reflexive mechanisms and on whether the two eyes move together (conjugate movements) or in different directions (disconjugate or vergence movements). The essential difference between the two basic classes has been confirmed by neurological observations and experimentation, which reveal separate brain mechanisms for their regulation. The common goal of all eye movements—whether reflexive or voluntary, slow or fast, conjugate or disconjugate—is to position the visual world properly upon the retinae of the eyes, thus ensuring clear, binocular vision.

I. The Globe and Its Extraocular Muscles

The six extraocular muscles include most fiber types known to exist in skeletal muscles and, like the skeletal muscles, are activated by the neurotransmitter acetylcholine at the neuromuscular junction. In primates, the majority of extraocular muscles are of the nonfatiguing, "fast twitch" type; metabolically slower muscle fibers are in the minority, and a few fibers combine metabolic properties of both fast and slow twitch types. The fibers show bursts of action potentials during the fast contractions of saccades, slowly changing rates of discharge during slow movements, and an exact relation between ongoing firing rate and position of the

Encyclopedia of Human Biology, Volume 3.

eye during steady gaze. The muscle fibers are driven very precisely by neural discharge in the "final common pathway" between nerve and muscle systems—the oculomotor neurons. The oculomotor neurons have cell bodies residing in three different nuclei of the brainstem—the oculomotor, abducens, and trochlear nuclei, collectively known as the oculomotor nuclei (see Table I). The oculomotor nuclei are under the influence of a variety of brain centers, including several classes of nearby cells that process information from higher control centers and provide direct input to the oculomotor neurons (the brainstem premotor neurons). [*See* EYE, ANATOMY.]

The oculomotor system can be contrasted with the skeletal motor system in several other important respects. The extraocular muscles are not required to pull against gravity; they operate against a fixed load determined by the viscoelastic properties of the globe and its contents. There is no stretch reflex in the eye muscles, although there are many centrally projecting muscle afferents. During movement, the muscles act in antagonistic pairs that rotate the eye around a fixed "joint," which is the geographical center of the globe; co-contraction of antagonists does not occur. The specific contribution of each muscle to an eye movement is a complicated function of the initial position of the eye in the orbit. Muscle tone is determined by a high rate of centrally controlled ongoing discharge in the oculomotor neurons. Contraction of a specific force in an eye muscle depends on an exact input from its motoneurons and produces a predictable and replicable angular displacement of the globe associated with reciprocal inhibition of the antagonist(s) on the same side. Synchronous movements of the two eyes require that appropriate inhibitory and excitatory neuronal signals cross the brainstem midline to coordinate the oculomotor nuclei for synergistic muscular action (see Fig. 1). Because of its relative simplicity, the study of the peripheral neural control of eye movement as well as measurement of the

TABLE I Simplified illustration of the main influence of each eye muscle on eye movement, when starting from direction of gaze straight ahead. Main action is shown for the *right* eye (next to the arrows).

Eye Muscle	Cranial Nerve # and Nucleus	Main Action on Eyeball	Illustration
Medial Rectus	III (Oculomotor)	Horizontal Adduction	→
Lateral Rectus	VI (Abducens)	Horizontal Abduction	←
Superior Oblique	IV (Trochlear)	Depression; Down and Outward	↙
Superior Rectus	III (Oculomotor)	Elevation; Up and Inward	↗
Inferior Oblique	III (Oculomotor)	Elevation; Up and Outward	↖
Inferior Rectus	III (Oculomotor)	Depression; Down and Inward	↘

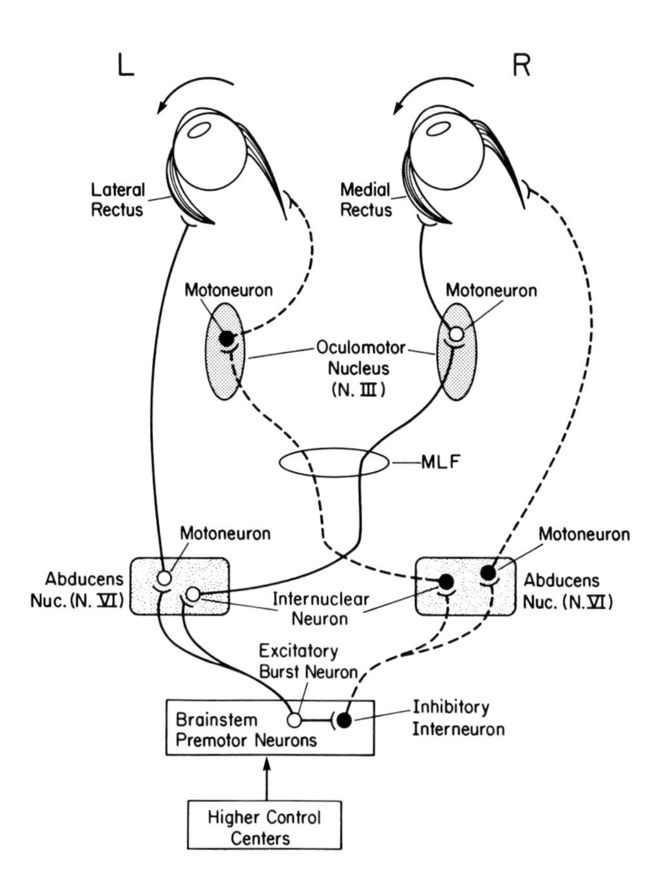

FIGURE 1 Neuronal activity required to produce a conjugate saccadic movement to the left. Pathways with unbroken lines and open neurons are excited, and pathways with broken lines and black neurons are inhibitory or silent during the movement. For this leftward movement, the lateral rectus muscle of the left eye (L) and the medial rectus of the right eye (R) must contract simultaneously. Brainstem premotor "burst" neurons discharge the excitatory chain, on command from a higher center. At the same time, the left medial rectus and the right lateral rectus must relax, so that the movement is unapposed. To accomplish this, an inhibitory interneuron in the brainstem is also activated by the burst neuron, which silences activity in the appropriate pathways for the duration of the movement. Excitatory and inhibitory information that must cross over the midline does so in a structure called the medial longitudinal fasciculus (MLF).

actual movements themselves are more amenable to quantitative analysis than the skeletal motor system, with its complicated trajectories and variable loads.

II. Types of Eye Movements

Eye movements can be sorted into types according to several descriptive criteria. The types described below occur continuously in alert, active mammals possessing binocular vision. Clear binocular vision requires that an object of interest affects corresponding areas of best visual acuity in each retina (the foveas). In other words, the direction of gaze in each eye must be aligned properly to avoid seeing double (diplopia); the focusing mechanisms (lens and cornea) ensure a clear image at the foveas. Animals lacking foveas such as rabbits, with laterally placed eyes and panoramic vision, have a very limited range of eye movements, which usually are an automatic accompaniment of body or head orientation.

A. Saccades

A saccade is a rapid eye movement that redirects the line of sight (see Fig. 2). Conjugate saccadic movements ensure that the fovea in each eye is stimulated by the same visual target. Saccades may be voluntary or involuntary and range from the primitive reflexes that stabilize the visual world during body and head motion (e.g., the "reset" phases of vestibular and optokinetic nystagmus) to the consciously initiated movements of active scanning with the head stationary. Scores of saccades are sequentially performed every minute during scanning of a busy visual environment and during close

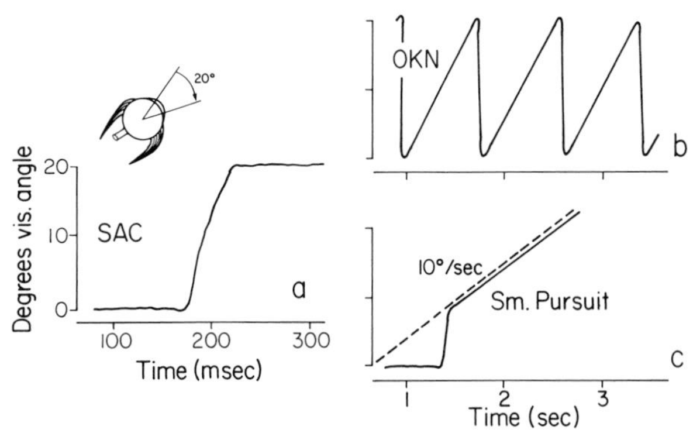

FIGURE 2 Waveforms of different eye movements. a. A saccadic movement moves the eye through 20° of visual angle. Entire movement requires 50 msec. b. Optokinetic nystagmus (OKN) generates a "sawtooth" waveform as the gaze follows vertical stripes moving smoothly in one horizontal direction. During the slow pursuit phase, a stripe is fixated until the eye deviates to its farthest extent; a saccadic fast phase then resets the globe back in the opposite direction, another stripe is fixated, and the pattern is repeated. A similar waveform is seen during vestibular nystagmus (see text). Note the time base. c. The broken line represents a target moving at 10°/sec across the visual field. The eye (unbroken line) makes a small "catch-up" saccade to fixate the target. Thereafter, the velocity of smooth pursuit eye movement matches the target as its image remains on the fovea. Sm., smooth.

activities such as reading. The fastest of all ocular movements, saccades attain maximum velocities of 700°/sec and amplitudes of 90° of visual angle (i.e., the arc described by the globe as it swivels around its orbital center). The trajectory of a saccade is stereotyped and smooth, with a rapid initial acceleration, a peak velocity halfway through the movement, and a gradual deceleration into a new orbital position (see Fig. 2). Velocity is not usually under voluntary control and can be affected by disease, inattention, fatigue, and drugs. As a general rule, larger saccades require more time (in msec) and are faster in velocity than small ones.

B. Smooth Pursuit

Smooth pursuit can be elicited by a small moving stimulus, which affects only the fovea and its immediate surrounding retina (see Fig. 3). The movement may occur in conjuction with head movement (e.g., as we watch players in a fast basketball game) or with the head stationary. Unlike saccades, smooth pursuit requires a moving visual target and cannot be voluntarily produced in the dark; its maximum velocity is around 100°/sec. It is best developed in animals such as the higher primates, who have binocular vision and well-developed foveas. Some bird and lizard species have excellent foveal vision but no pursuit system; afoveate animals, such as rabbits, cats, and goldfish, are essentially incapable of smooth pursuit movement. Because the object of smooth pursuit is to maintain clear vision of a moving target, the visual pathways from retina to cortex must be intact for it to occur. Motor signals must be generated to drive the muscles proportionately to the velocity of the target and to adjust eye position when the target "slips" off the foveal area (retinal error). The neural control pathways have not yet been completely worked out, but recent anatomical and neurophysiological evidence suggests that many cortical areas, the cerebellum, and possibly the basal ganglia, as well as brainstem premotor centers, participate in its generation and successful execution.

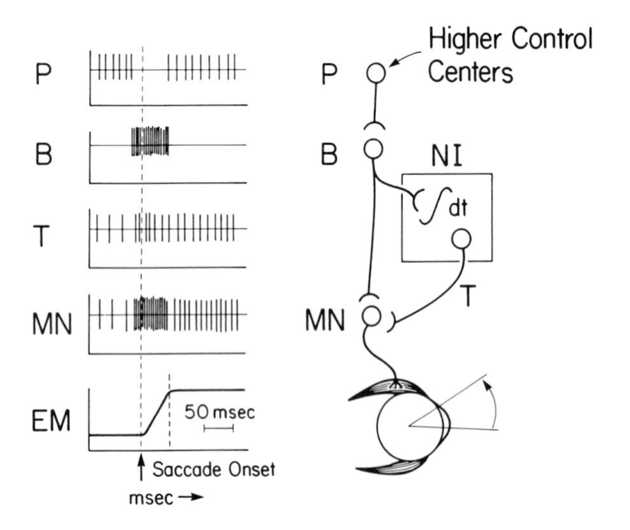

FIGURE 3 Simplified diagram relating activity in brainstem premotor neurons to the "pulse-step" discharge of motoneurons (MN). Higher control centers inhibit the pause cell (P) just before movement, permitting the burst cells (B) to generate the pulse. The pulse is integrated by the neural integrator (NI–dt) to provide the step change, which is observed in tonic neurons (T). The burst and tonic neurons converge on the motoneuron, resulting in saccadic eye movement (EM) followed by a new steady eye position.

C. Vestibulo-Ocular Reflex

Smooth, conjugate eye movements can be induced by vestibular sensory input. During active or passive head movements, which stimulate the semicircular canals and otoliths of the vestibular system, the eyes will deviate in the opposite direction to stabilize the eyeball in visual space. During prolonged, passive rotational stimulation, the slow compensatory phase is interrupted by rapid reset movements in a pattern called vestibular nystagmus (see Fig. 2). In the dark, the vestibular input to the oculomotor neurons is relatively direct through a three-neuron pathway from the semicircular canals. In lighted environments, visual information relayed through the cerebellum may modulate the direct vestibular circuit so that the vestibular reflex may be overriden or suppressed, as happens when a whirling dancer fixates the horizon. The slow phase of vestibular nystagmus survives the destruction of certain brainstem areas that regulate saccadic movement (the paramedian pontine reticular formation and rostral mesencephalon); the fast phase does not, indicating that nystagmic fast phases, at least partially, depend on the same brainstem locations that control voluntary saccades.

The vestibulo-ocular reflex (VOR) and visual input normally work together to keep visual images stable as we move our heads and bodies. Certain brainstem premotor neurons may integrate information from the two systems and may be the site of simple motor learning. For example, monkeys trained to make eye movements for reward can be fitted with special lenses that either slow down or speed up the image of the visual world during head turning. The normal VOR would be unable to compensate for this exaggerated image motion. After wearing the lenses for a few days, the VOR changes its output to match the new visual input. When the lenses are removed, the VOR reverts, slowly, to its former normal state. This type of experiment shows a remarkable plasticity of function, even for this relatively reflexive process.

D. Optokinetic Reflex

Optokinetic nystagmus can be elicited by purely visual stimulation with the head stationary, such as occurs when viewing the passing scene from a moving train (railway nystagmus; see Fig. 2). The reflex is elicited by wholesale movement of the entire visual environment and is comprised, like vestibular nystagmus, of a slow phase followed by a fast reset movement when the eye has reached maximum possible angular deviation. Since lesions of the cerebellar flocculus reduce the optokinetic response and floccular Purkinje neurons can code high velocity visual inputs, it appears that the cerebellum plays an important role in these visually guided compensatory movements. Other inputs from neck receptors and the visual system may also be integrated by the cerebellum to regulate eye–head coordination.

E. Vergence (or Disconjugate) Eye Movements

Vergence eye movements are visually mediated, are mainly horizontal, and occur when the eyes focus from a near object to a distant one (divergence), or from a far object to one close by (convergence). The net result is that the gaze axes of the two eyes move in opposite directions, as compared with conjugate (or versional) eye movements, which move the two axes in the same direction. The stimulation for vergence is either disparity of images affecting the two retinas (the same object affecting noncorresponding retinal points, as during voluntary crossing of the eyes), or blur. In ordinary conditions, both blur and disparity interact to generate vergence movements. Convergence requires simultaneous activation of the medial rectus of both eyes and inhibition of the lateral recti, divergence the opposite pattern. Most vergence movements are a combination of slow and fast components; they utilize the same motoneurons as the conjugate saccadic and smooth pursuit systems but may have special premotor command neurons exclusively devoted to either convergence or divergence, located near the oculomotor nuclei.

F. Miniature Eye Movements of Fixation

The eyes are never motionless. Even during steady fixation of an object, continuous, small, unconscious movements occur, rarely exceeding 15–20 min of visual angle. Slow ocular movements are called drifts and move the fovea over, and sometimes slightly away from, the target of interest. Tiny corrective saccades that return the fovea to the target are called flicks. Experiments have shown that drifts are essential for continuous clear vision and may be consciously controlled with considerable effort. Images that are totally motionless on the retina, such as those produced by the vasculature of the retina or while wearing special contact lenses with visual patterns, become invisible after a few seconds. Neural activity signalling motionless images shows a similar decline. Continuous motion of the retinal image, whether produced by eye movements or externally, is essential to clarity and continuity of visual experience.

III. Brainstem Control of Eye Movement

A. The Role of the Oculomotor Neurons

The two basic variants of conjugate volitional eye motion, slow pursuit and fast saccadic, require different patterns of discharge in the oculomotor neuron. Smooth pursuit and vestibular-optokinetic slow phases require a *velocity*-coded signal that is proportional to target velocity and retinal error during smooth pursuit and for complementing or cancelling head movement during eye–head coordination. During slow eye movements, which may variably accelerate and decelerate, the velocity signal for muscle contraction is coded in the firing rate of the motoneurons. The premotor organization of smooth pursuit movements is not as well documented as for the other eye movements. Some of the tonically firing cells in the brainstem modulate their discharge for ipsilateral eye velocity during smooth pursuit and vestibular slow phases and probably provide input to the motoneurons.

More is known regarding the saccadic control mechanism. Saccades require a signal proportional to the retinal position of the target relative to the center of gaze. Saccadic movements need (1) a powerful muscular contraction at movement onset to overcome the viscous drag of orbital tissues and (2) a steady muscular contraction to maintain the new eye position in the orbit against elastic restoring forces. The burst of motoneuron firing needed to initiate muscle contraction during saccades has been termed the pulse of innervation, while the tonic signal associated with new eye position following saccades is the step of innervation; therefore, a saccadic movement is controlled by a "pulse-step" of innervation. During steady fixation with the eyes looking straight forward, oculomotor neurons discharge at a relatively high but regular rate that varies from neuron to neuron. For a given motoneuron, this rate increases in a strong phasic burst about 5–8 msec before a saccade, if the neuron innervates an agonist muscle, which contracts during the movement (the "on," or pulling, direction). The velocity of the saccade is related to the burst frequency, whereas the duration of the burst determines its amplitude or size. The firing rate following the saccade is precisely matched to the new steady position and the amount of contraction required to hold the eye steady. The transform of eye *velocity* signal to eye *position* signal occurs at the premotor level in a neural integrator; (see Fig. 3). During saccades in the opposite, or "off," direction, the motoneuron is inhibited and then assumes a new lower tonic discharge corresponding to the new eye position (sufficient positional deviation in the off direction may tonically silence the motoneuron). The slope of the function relating eye position and discharge rate varies from neuron to neuron, as does the eye position threshold at which the neurons are recruited into firing.

B. Brainstem Premotor Mechanisms

The pons and mesencephalon contain premotor neurons that provide input to the oculomotor nuclei. Evidence from single cell recordings in alert monkeys trained to eye tracking shows essentially three types of cells: pause cells, burst cells, and tonic cells (see Fig. 3). Burst cells begin discharging at high frequency just prior to saccades; pause cells are tonically active during eye fixation and cease firing just before saccades; and tonic cells alter their steady discharge immediately after a saccade, assuming a rate proportional to the new eye position. The silence of burst cells during fixation is thought to result from tonic inhibition by pause cells; just before a saccade, a higher center inhibits the pause cell, thus disinhibiting the burst neuron, which begins to discharge 4–10 msec prior to onset of the saccade and continues for its duration. Burst cells provide the pulse in the pulse-step pattern delivered to the motoneuron. The discharge of the burst cells is integrated by specific circuits to change the rate of firing in tonic cells, thereby providing the step of the pulse-step command. During fixation, the pause neurons theoretically prevent burst neurons from extraneous firing and the resultant unintended ocular oscillations (opsoclonus and flutter).

Brainstem premotor neurons of all kinds—pause, burst, and tonic—have been described in anatomical locations that are clinically correlated with disturbances in saccadic eye movement. Horizontal saccades are controlled by the paramedian pontine reticular formation (PPRF), whereas vertical saccades depend on an intact rostral mesencephalon (rostral interstitial nucleus of the medial longitudinal fasciculus [riMLF]). Unilateral lesions of the PPRF eliminate saccades to the ipsilateral side; bilateral PPRF damage removes all fast horizontal eye movements. Bilateral lesions of the riMLF produce a loss of all vertical saccades; bilateral lesion of the medial and caudal PPRF disrupt fast movements in all directions, including nystagmic reset phases. None of these lesions appear to affect slow movements.

The brainstem regulation of smooth pursuit movements is not as thoroughly understood as it is for saccades. Structures containing neurons that encode signals important for smooth pursuit (e.g., eye velocity) include the dorsolateral pontine nucleus, the vestibular nuclear complex, the nucleus prepositus hypoglossi, parts of the mesencephalic reticular formation, and the PPRF.

Studies of brainstem neurotransmitters suggest that several excitatory and inhibitory amino acids (glutamate, aspartate, glycine) play a role in the premotor pathways.

IV. Higher Control Centers

There are a large number of higher control centers whose exact role in the programming of eye movements is not yet clear. Both smooth pursuit and saccadic eye movements can be affected by lesions in a variety of brain and brainstem areas, some of which they share in common.

A. Smooth Pursuit

Evidently, visual input is critical in the generation of accurate, voluntary smooth pursuit movements, so that an adequately functioning peripheral and central visual system is absolutely necessary (smooth pursuit movements cannot be made in the dark). Some areas in the frontal, temporal, and parietal lobes of the brain participate in the initiation of smooth pursuit, but the mechanisms are not well understood. The primary motor program is represented by neurons in the brainstem and cerebellum. All of these finely processed neural signals converge on the final common pathway, the oculomotor neuron.

B. Saccades

Input for visually guided saccades in primates is received primarily via the striate and peristriate visual cortex. Some cells in the posterior parietal cortex discharge in association with saccades to visual targets, provided that the target has meaning or significance to the animal (e.g., is remembered or recognized). Saccadic discharge in cells of the supplementary motor area of the frontal cortex less obviously depends on immediate visual stimulation or behavior and may precede eye movement by many hundreds of milliseconds. Single subcortical neurons with discharge patterns time-locked to saccades are found in the cerebellum, the superior colliculus, the substantia nigra of the basal ganglia, and the intralaminar nucleus of the thalamus. Direct electrical stimulation in many of these cortical and subcortical areas causes saccadic movement, directing the gaze to circumscribed locations in the visual field called the "movement field" of the stimulated cells; each cell has its own, differently located movement field. Two areas that provide apparently parallel pathways for the control of visually guided saccades are the superior colliculus and area 8 of the frontal cortex. Both structures project directly to the PPRF of the brainstem.

Combined lesions of the superior colliculus and area 8 result in a severe and permanent loss of saccades to visual targets, whereas lesions of either area alone produce only temporary deficits. Some evidence indicates that the superior colliculus can provide enough visual input to guide saccades in the absence of the visual (striate) cortex; patients with large, unilateral striate lesions can sometimes saccade with rough accuracy toward an object in the contralateral visual field, although they cannot conciously perceive its location or identity (blindsight). Clearly, the participation of multiple, bilateral structures in the regulation of eye movements explains why pathology of a single higher control center may only partially affect eye movements but not eliminate them.

We have briefly reviewed evidence based on the labors of many investigators applying the methods of single neuron recording, systems analysis, and the new anatomical connection tracing techniques. This combination of approaches has produced ingenious models of the oculomotor control systems, although many unanswered questions remain, including the exact relations between brainstem and subcortical and cortical components.

Bibliography

Carpenter, R. H. S. (1988). "Movements of the Eyes." Second Edition. Pion Press, London.

Eckmiller, R. (1987). Neural control of pursuit eye movements. *Physiol. Rev.* **67(3),** 797–857.

Freund, H. L., Büttner, U., Cohen, B., and North, J. (eds.) (1986). "The Oculomotor and Skeletalmotor Systems: Differences and Similarities." Progress in Brain Research, Vol. 64. Elsevier, Amsterdam.

Leigh, J. R., and Zee, D. S. (1983). "The Neurology of Eye Movement." Contemporary Neurology series 23. F. A. Davis, Philadelphia.

Robinson, D. A. (1964). The mechanics of human saccadic eye movement. *J. Physiol. (London)* **174,** 245–264.

Wurtz, R. H., and Goldberg, M. E. (eds.) (1989). "The Neurobiology of Saccadic Eye Movements." Reviews of Oculomotor Research, Vol. 3. Elsevier, Amsterdam.

F

Famine

JOANNE CSETE, *University of Wisconsin–Madison*

Glossary

Boom famine Famine caused by entitlement failure in spite of relatively great (theoretical) per-capita food availability (distinguished from *slump famine*)

Entitlement failure Inadequacy of a population group's access to sufficient food for survival; may take the form of *direct* entitlement failure, in which farm households producing food experience crop failure, or *exchange* entitlement failure, in which persons buying but not producing food are priced out of food markets

Famine early warning system (FEWS) Effort to collect, coordinate, analyze, and interpret data on a wide range of indicators—especially meteorologic, economic, agricultural, demographic, and nutritional—to predict and prevent famine or conditions predisposing to famine

Ketone bodies Products of incomplete oxidation of fatty acids whose production by the liver increases when fatty acid breakdown becomes rapid, as in prolonged fasting; their acid forms may serve as fuel for basic body functions in the absence of the ingestion of carbohydrates

FAMINE IS AN ACUTE CURTAILMENT of the food consumption and access to food of a given population that results in unusually high levels of death from severe undernutrition and related conditions. By its severity and the relative suddenness of the onset of its severe stages, it is distinct from the chronic undernutrition and high mortality associated with the long-term poverty suffered by many of the world's peoples. Famine should be defined not only in terms of its biomedical consequences—human mortality and physical suffering—but also by its profound social and economic impacts. Famines may destroy or alter the social organization on many levels, wreak havoc upon economic and political priorities, and thus greatly impede overall social and economic progress.

I. Factors Precipitating Famine

A. Immediate and Underlying Causes

Famines have made their mark upon societies throughout recorded history. They still threaten some parts of the world in spite of economic and agricultural progress in the twentieth century. The famine suffered by 21 nations of sub-Saharan Africa in 1984–85, in which tens of millions of Africans were at immediate risk of death by starvation, is a modern example. Much of Africa, parts of South Asia, and other localized areas of the world are chronically susceptible to some degree of famine in the modern era.

The immediate triggers of famines have generally been well documented. They include such calamitous natural occurrences as drought or other unfavorable weather, floods, earthquakes, cyclones, and human, animal, or crop disease. Drought has been an important cause of crop failure associated with famine in many parts of Africa and Asia in recent decades. Man-made factors that may be proximate causes of famines are war and civil disturbance, other disruption of markets or of market-related transportation and communication networks, and a range of political and strategic errors or misdeeds of governments that curtail the access to food of large numbers of people.

In many cases, it is rather more difficult to elucidate the longer-term underlying causes of famine, which determine the setting in which an immediate trigger may operate. Famines of the twentieth century have generally occurred in populations already characterized by poverty and dependent on seasonal harvests for their livelihood. Subsistence farmers and near-subsistence rural dwellers who rely on local markets have figured prominently among famine victims. Such populations generally have well-developed means of coping with threats to their access to food and thus of maintaining an ecologically balanced relationship between food availability and production and food needs. Where these coping mechanisms are threatened—by gradual deterioration (natural or man-made) of resource bases, long-term changes in climate, or chronic political or economic instability or repression, for example—the stage may be set for famine. The nature of underlying factors is such that there is considerable debate in the analysis of some famines as to the time of onset of famine per se, as opposed to the continuation of the chronically threatening, underlying situation.

The famine of 1984–1985 in Ethiopia (East Africa) illustrates the interplay of acute and chronic causal factors. Drought was the readily identifiable immediate cause of this famine which, as noted above, affected much of the African continent. In parts of Ethiopia, however, the death rate was much higher than in other countries due to what some observers have depicted as a complex of political, social, and economic factors and civil war upon which the drought was superimposed.

The government that took power in 1974 from the long-seated Emperor Haile Selassie (whose own downfall was precipitated by his mishandling of the Ethiopian famine of 1972–73) set upon an ambitious program of agricultural reform. Unfortunately, for a variety of reasons, the production cooperatives and state farms in which the government placed its hopes for increased food production and political solidarity did not achieve these goals. In addition, an official policy of resettlement of some agricultural communities apparently contributed to lowering their food productivity, among other problems. Few observers would blame the Ethiopian government's agricultural policies alone for the famine. It is widely thought, however, that some of these policies limited the capacity of peasant farmers to cope with the drought in traditional ways and reduced the food productivity of certain parts of the country that otherwise could have helped to compensate for the losses in more severely affected areas.

B. Sen's "Entitlement" Analysis

In the analysis of causes of famine, it is important to distinguish between the theoretical per-capita availability of food and the affected population's real ability to acquire food. The seminal work of economist Amartya Sen has clarified this distinction and greatly contributed to the analysis and classification of famines. By examining a number of famines throughout history, Sen identified some that took place in times of relatively abundant overall food supplies and concluded that a decline in the available food supply is not a necessary part of the famine scenario.

Sen introduced a system of classification of famines based on the notion of entitlement to food. In his formulation, the principal cause of famine is entitlement failure, which occurs when significant numbers of persons do not have the means by which to obtain sufficient food to remain alive. He described two types of entitlement failure: the *direct* entitlement failure suffered by farmers when crop failure cuts off their food supply and thus also their access to goods for which they would exchange food, and *exchange* entitlement failure suffered by food buyers who find themselves priced out of food markets because of sudden price jumps associated with increased demand or because their labor, skills, or products are no longer in demand. Famines that occur because of entitlement failure in spite of underlying food abundance are called "boom famines" by Sen (as opposed to "slump famines," which occur in the face of relative food shortage). He cites the widely studied Bengal famine of 1943 in India as an example of a boom famine.

Sen's entitlement framework has been criticized as difficult to verify empirically in some situations. He himself notes a number of limitations of the approach. The contribution of this framework to the understanding of famines is, however, undeniable. Entitlement analysis both underscores the error of defining famines in terms of overall per-capita food availability and portrays famines not only as food crises but as part of a social, economic, historical, and legal setting. Sen's contribution, moreover, suggests questions that should be asked by policymakers and donors who must formulate responses to famine situations in light of the reality that increasing the overall food supply alone may not address the most critical problems.

II. Consequences of Famine

A. Biomedical Consequences in Individuals

The death associated with severe and prolonged food deprivation is an agonizing one. In the absence of intake of adequate dietary energy, proteins essential for brain activity and other vital functions may be broken down for fuel. In prolonged food deprivation, the body calls upon metabolic and hormonal mechanisms to protect against this breakdown.

In brief and simple terms, the steps in the starvation process from the time of the last meal may be characterized as follows: (1) food is digested and absorbed; (2) adipose (fatty) tissues stores in the body are used to fuel liver functions leading to production of glucose, the body's main sugar fuel source, and to fuel muscle activity; (3) as adipose stores are reduced (over two or three days), the liver calls upon amino acids (the constituents of protein) from muscle proteins to produce glucose to fuel brain function; the liver also begins *ketogenesis,* a process that increases the efficiency of use of remaining fatty acids from adipose tissue through incomplete fatty acid breakdown, leading to the production of *ketone bodies,* which in turn help to fuel brain function and spare the muscles to some degree; (4) after about a week of fasting, the acid forms of ketones, or *ketoacids,* increase in the blood and help provide fuel to the brain; (5) as the second week of fasting nears, the brain is fueled preferentially by ketoacids rather than glucose, preventing the breakdown of remaining muscle protein, at least for a limited period. At this point and in the absence of other complications, the duration of survival depends on the quantity of remaining adipose stores.

Over an extended period, the brain appears not to function at its best when fueled by ketoacids rather than by glucose. Emotional changes, especially depression, are usually pronounced in persons suffering from starvation. As the body is focused on the protection of basic brain functions, so emotions and instincts become focused on simple survival, and social and kinship interactions decrease in value. Personal appearance also changes as muscle tone and body weight are rapidly lost. Edema or swelling of some body parts may occur. Many other signs of physiological and psychological deterioration have been documented in a range of famine situations.

In many famines, however, death can be attributed to infectious disease rather than to starvation itself. (It is often difficult to pinpoint the more important causal factor since undernutrition and infection exacerbate each other.) The calamitous events that may trigger famine are also likely to cause the breakdown of sanitation and basic health-care systems, increasing the likelihood of epidemics among persons in the famine region. In addition, undernutrition, even of a less severe degree than that experienced in famine, undermines the body's immune system that defends against infections. Diarrheal diseases gain an added foothold with deterioration of the gastrointestinal tract in the starved individual. In response to famine, moreover, affected populations may gather in overcrowded, undersupplied makeshift camps or feeding sites that are likely settings for transmission of infectious disease. Thus, deaths from dysentery, cholera and other diarrheal disease, diphtheria, louse-borne typhus, other vector-borne parasitic disease, and respiratory infections are commonly reported in famine situations. That famine often occurs in areas already suffering from high prevalence of infectious disease and low levels of preventive service delivery only increases the likelihood of high mortality from infectious disease. [*See* INFECTIOUS DISEASES; NUTRITION AND IMMUNITY.]

B. Consequences for the Society

It is frequently difficult to distinguish between the social effects of famine and those of the factors that triggered the famine, especially when famine is associated with underlying conditions of chronic poverty, food stress, and ecological imbalance between food needs and food production or distribution. Some observers, particularly social scientists, however, have suggested that the social consequences of famine itself are remarkable and worthy of more study than they have attracted in the past.

Many of the social consequences of famine are obvious. The search for food and survival by large numbers of debilitated and desperate persons cannot but wreak havoc upon social order and organization. When this search entails mass migration, societies are forced to reestablish, with minimal resources, their social, economic, and information systems in a new setting. The loss of life among family members and physical debilitation of those who survive contribute to the difficulty of the rehabilitation of society. The loss of food production and purchasing capacity of the famine-stricken population has serious consequences for neighboring regions and trading partners.

In a wider sense, the social consequences of famine may be determined by the reaction of the state and other institutions to the crisis. Famines on the Indian subcontinent in the 1950s and 1960s are thought by some observers to be the basis for the Indian government's subsequent support for newer high-yielding varieties of staple food crops (the so-called Green Revolution technologies), which have markedly increased food production potential in India. In some cases, governments or regimes may perceive the famine as a threat to their own survival and react with varying degrees of repression or generosity to placate or assist affected populations. Outside institutions that are drawn into famine situations as part of relief or development efforts may also leave their mark through the introduction of new technologies or new ways of approaching service to poverty-stricken populations.

III. Famine Relief and Rehabilitation

As might be expected, many aspects of famine are extremely understudied because of the difficulties of conducting research in famine situations. This is the case in the area of famine relief, where the same apparently impenetrable questions about the quantity and quality of relief and longer-term development efforts appropriate to a given situation seem to reemerge with each instance of famine. For logistical and ethical reasons, it is generally impossible to test different relief strategies as part of a controlled experiment in a famine situation. It has also been very difficult in recent famines in Africa, for example, to evaluate the effectiveness of relief measures carried out by different public and private institutions.

Famine relief is complicated by the fact that its planning and implementation take place on several levels and in many institutions and organizations. The government of a famine-stricken country may have national or subnational organizations or plans for famine management in place before the onset of famine. If there is the political will and the economic capacity to implement plans effectively, the effectiveness of relief operations should be improved by this preparedness. In practice, however, it is often the case that at the time of famine, governments have few resources upon which to call to implement their plans. Political concerns of the government may also impede the free flow of relief assistance. Some observers have again cited the Ethiopian famine of 1984–85 in this regard. From the governments's own reports, it was plainly possible to predict the famine, but political and other factors kept appropriate prevention and early relief measures from being put into place.

When a famine reaches the attention of persons other than those in the locality or country of its occurrence, a range of players other than the national government of the affected country may enter the scene. These include private voluntary agencies based in the wealthier nations, many of which have some capacity for famine relief operations; official development assistance agencies of the wealthier nations; and multilateral agencies such as those of the United Nations system. In principle, most of these agencies operate at the invitation and upon the orders of the government of the affected country. In practice, they may in some situations, officially or unofficially, enjoy a certain autonomy. In either case, the coordination of the activities of all of these bodies is an enormous task for a host government already burdened by many difficulties.

The first priority in short-term famine relief is usually getting food to the stricken population. An assessment of the needs of the stricken area is a logical first step in the relief of operation, but adequate time or data for a thorough evaluation are rarely available. Among the immediate decisions and questions about short-term relief that typically must be considered in a famine situation are the following:

- Are there local or nearby sources of food sufficient to meet the crisis? If not, what are the available routes for moving food into the area from other parts of the country and from outside donors? Who are the donors and what will their basic modes of operations be? What are the possible political and other constraints to easy movement of food to the stricken area?
- What distribution system is in place in the affected area once food arrives? How will equitable distribution be ensured?
- To what degree has the affected population already been starved? How will its current nutritional situation affect its ability to digest relief foods? What is

the estimated demographic composition of the affected population? What are the most nutritionally vulnerable groups in the affected area, and what are their special food needs? Are the available relief foods culturally acceptable to the recipients?

- How long is relief food likely to be necessary? How will continued supply flow be assured?
- How will the quantity and quality of food brought into the area and the distribution system affect longer-term development and resumption of food production and other income-generating activities in the population?

Indeed, consideration of the longer-term developmental impact of relief measures should be at the forefront of relief planning. Minimal disruption of the culture and social relationships of the affected population should be a primary objective. The types of longer-term development activities that may be undertaken after or in conjunction with short-term relief are too numerous to list, but the planning of longer-term rehabilitation of the population should go hand in hand with the relief process. Long-term development efforts and short-term relief activities should not have to compete with each other for resources and the attention of donors, as has appeared to be the case in African famines of the 1980s, but should be complementary. The successful execution of both depends heavily on an understanding of the immediate and underlying causes of the famine and on the political will to undertake appropriate measures.

IV. Famine Prediction and Prevention

Research efforts related to famine have in recent years focused on prediction and prevention of famines. The establishment of famine early warning systems (FEWS) has been of particular interest to researchers, donor agencies, and the governments of affected countries, particularly in Africa. A FEWS has as its goal the compilation, analysis and interpretation of time-series data on a number of indicators related to the potential for famine, all for the purpose of famine prevention. These indicators may be meteorologic (rainfall, temperature), economic (staple food prices, level of food stores, indicators of poverty), agricultural (harvest levels, prevalence of crop disease), demographic (unusual migrations, elevated mortality, unusual patterns of household formation or deterioration), and health-related or nutritional. Some experts also advocate the use of satellite photographs in time series as part of FEWS to examine changes in vegetation patterns.

One of the earliest FEWS activities was established in 1975 by the Food and Agricultural Organization of the United Nations in its Global Information and Early Warning System on Food and Agriculture (GIEWS). GIEWS uses secondary data collected from its sister agencies in the U.N. system, especially the World Meteorological Organization and the World Health Organization, as well as reports from the government departments and ministries in countries where it operates, and other agencies' reports. More recently, a number of bilateral donor agencies, including the U.S. Agency for International Development, have supported FEWS activities relying on primary data collection and prompt analysis in the country thought to be at risk of famine.

Data collection, analysis, and interpretation for the wide range of indicators that may be included in FEWS are difficult and costly under the best of circumstances, and countries most vulnerable to famine rarely provide the best of circumstances. Elucidating relationships between all of the indicators of interest and setting cut-off points to flag distress levels of various indicators have proven difficult in FEWS effort so far. Some observers, particularly social scientists, have criticized FEWS efforts for attempting a complex quantitative elaboration of underlying causes of famine that may be better captured by a basic understanding of the social relationships, ecological constraints, and perhaps political realities of a country or population group. Whether an array of indicators can be useful for prediction without a well-developed understanding or model of the social and economic framework of the area has been questioned. Others have pointed out that famine can appear in many guises, some of which may not be possible to detect, even with high-quality data collection and analysis for a range of indicators. In addition, it is certainly true that the best FEWS is of little value if there is no food or other resources, or political will, to act promptly in response to problems it identifies.

In spite of the many pitfalls that FEWS efforts may encounter, they have the potential to make an important contribution to the conceptualization of famine in some settings. In particular, if there is any level of understanding of the factors associated with high risk of famine, it is difficult for governments to treat famines as though they arise from nothing. The demonstration that famines are born of a complex of historical predisposing factors would be of great value to policymakers and planners who seek to prevent famine. The light that FEWS activities may shed on the nature of coping mechanisms in times of food-related stress short of famine may also justify their use.

Even in the absence of mechanisms for predicting famine, it may be possible to identify famine-prevention measures appropriate to a given setting. If seasonal fluctuation in food supply or households' access to food is a chronic problem, maintenance of food stores accessible to households or other measures to strengthen traditional food-related coping mechanisms may be indicated. Chronic or seasonal problems in the movement of food to non-food-producing regions may be addressed by improvements in transportation systems. Political and economic interventions to reduce poverty, which usually include increased economic and political participation of the poor in the society, may be the most effective famine-prevention measures.

V. Conclusion

Although some famines are triggered by natural disasters, most famines would not occur were it not for factors well under the control of humankind. Famine's enormous cost in terms of human life and social and economic decline justifies major efforts to predict and prevent famine. Though prediction of famine by early detection and other means is no small accomplishment, it is safe to say that the barriers to effective prediction and prevention of famine are more frequently political than technological. The elimination of the modern-day scourge of famine should be a high political and economic priority for all the world's peoples.

Bibliography

Mellor, J. W. (1986). Prediction and prevention of famine, *Federation Proceedings* **45,** no. 10, 2427.

Robson, J. R. K., ed. (1981). "Famine: Its Causes, Effects, and Management." Gordon and Breach, New York.

Scrimshaw, N. S. (1987). The phenomenon of famine, *Annual Review of Nutrition* **7,** 1.

Seavoy, R. E. (1986). "Famine in Peasant Societies." Greenwood Press, New York.

Sen, A. (1981). "Poverty and Famines: An Essay on Entitlement and Deprivation." Clarendon Press, Oxford.

Torry, W. I. (1984). Social science research on famine: A critical evaluation, *Human Ecology* **12,** no. 3, 227.

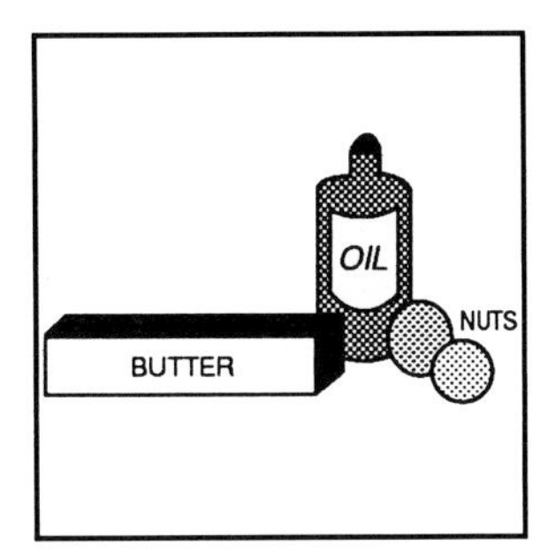

Fats and Oils (Nutrition)

PATRICIA V. JOHNSTON, *University of Illinois at Urbana–Champaign*

Glossary

Triglycerides Predominant chemical form of the constituents of fats and oils

P/S ratio Ratio of polyunsaturated fatty acids to saturated fatty acids (excluding the monounsaturated fatty acids)

Lipoproteins System of soluble proteins by which lipids are transported in the blood; they are combinations of lipids and proteins that are more readily transported in blood than lipids alone

Essential fatty acids Long-chain unsaturated fatty acids that cannot be synthesized in the body and must be provided in the diet for the maintenance of good health

Eicosanoids Biologically active 20-carbon compounds derived from the essential fatty acids

FATS AND OILS are part of a large group of mainly water-insoluble compounds known as lipids. They are composed of triglycerides in which three fatty acid chains are linked to a glycerol backbone. The difference between fats and oils is their physical state at room temperature: fats are solid and oils are liquid. Dietary fat is mainly in the form of triglycerides from fats and oils, with smaller amounts provided by phospholipids. Food sources of fat can be either animal or vegetable. Animal fats contain more saturated fatty acids than most vegetable sources. Many vegetable oils are rich in polyunsaturated fatty acids and are sources of the essential fatty acids. Fat is digested and absorbed in the small intestine and most of it enters the lymphatic circulation in the form of fatty particles known as chylomicrons. These are transformed in the liver into the water-soluble lipoproteins that transport lipids in the blood to the tissues. Fat serves as a concentrated form of energy, increases the satiety value of foods, and improves palatibility. Fat also serves as a carrier for the fat-soluble vitamins. In the body, fat serves as an energy reserve, as an insulating material, and (in the form of fat pads) as a protector of vital organs such as the kidneys. Many vegetable oils are rich in the essential fatty acids, linoleic and α-linolenic acids. These fatty acids are metabolized in the body to the precursors of the eicosanoids, the 20-carbon compounds that are potent regulators of many physiological functions.

I. Chemical Composition of Fats and Oils

A. Triglycerides and Phospholipids

Fats and oils are part of a broad group of compounds called *lipids,* a word derived from the Greek "lipos" meaning "fat." Lipids comprise a diverse group of compounds that are primarily insoluble in water but soluble in organic solvents such as hexane, diethyl ether, and chloroform. Fats and oils are among the simplest of lipids and are composed mainly of triacylglycerols or triglycerides as they are commonly known. The triglycerides consist of three fatty acid chains joined in an ester bond to a glycerol backbone as shown below:

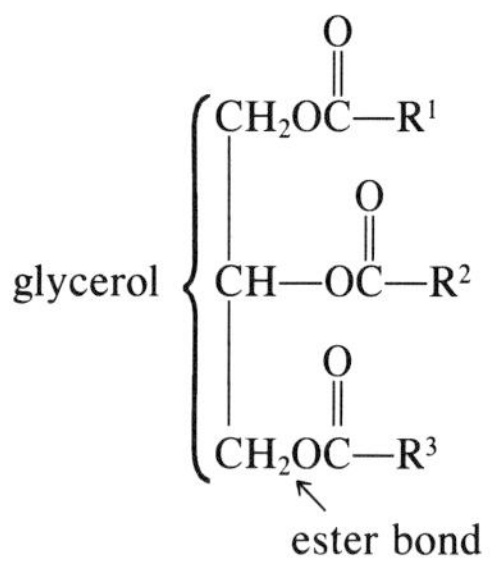

$$\text{glycerol}\begin{cases} CH_2OC(=O)—R^1 \\ CH—OC(=O)—R^2 \\ CH_2OC(=O)—R^3 \end{cases}$$

↖ ester bond

where R^1, R^2, and R^3 are fatty acyl chains that may be the same but are more often different. Fats and oils are mixtures of triglycerides; fats differ from oils in their physical states at room temperature: fats are solid and oils are liquid. It is usual when referring to dietary fats and oils to simply use the term *dietary fat* and it is assumed that this refers to the mixture of fats and oils derived from various food sources. The fat we consume therefore is primarily in the form of triglycerides, although small amounts of mono- and diglycerides having one and two fatty acyl chains only are usually present as additives. Sterols, such as cholesterol from animal fat sources and phytosterols from plants, are also consumed. Cholesterol is important as a constituent of cell membranes and as a precursor of some hormones. It is also synthesized by the liver. Depending upon the source, varying amounts of phospholipids will also be consumed. These lipids are a major constituent of all cell membranes. They consist of the glycerol backbone with only two fatty acids linked to it; the third is occupied by a phosphate group joined to an alcohol thus:

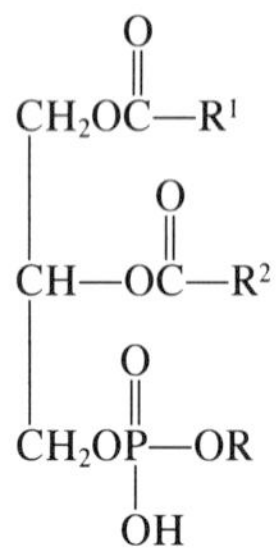

$$\begin{array}{l} CH_2OC(=O)—R^1 \\ | \\ CH—OC(=O)—R^2 \\ | \\ CH_2OP(=O)(OH)—OR \end{array}$$

where R^1 and R^2 are fatty acyl chains and R is choline, ethanolamine, serine, or the carbohydrate inositol. Traces of other lipids may be present but they contribute little to the nutritional value of the dietary fat. [*See* LIPIDS.]

B. Fatty Acid Constituents of Glycerides and Phospholipids

Fatty acids are chains of carbon atoms that have a methyl group (CH_3) at one end and a carboxylic acid group (COOH) at the other. Carbon atoms have a valence of four and therefore are attached to hydrogen atoms, another carbon atom, and a hydroxyl (OH) or oxygen. Thus a typical fatty acid has this structure:

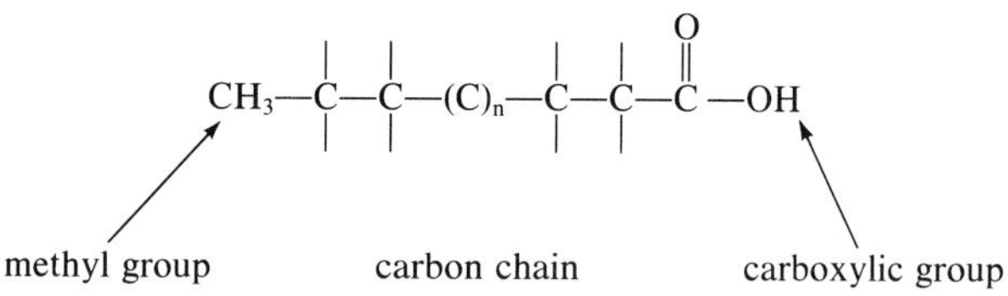

$$CH_3—C—C—(C)_n—C—C—C(=O)—OH$$

methyl group   carbon chain   carboxylic group

For all practical purposes fatty acids in foods have an even number of carbon atoms ranging from 4 to 22.

There are two main classes of fatty acids: saturated and unsaturated. In saturated fatty acids all four valence bonds of all carbons are satisfied. In unsaturated fatty acids some hydrogens along the chain are missing and carbons form bonds with each other, giving rise to olefinic or double bonds as shown below:

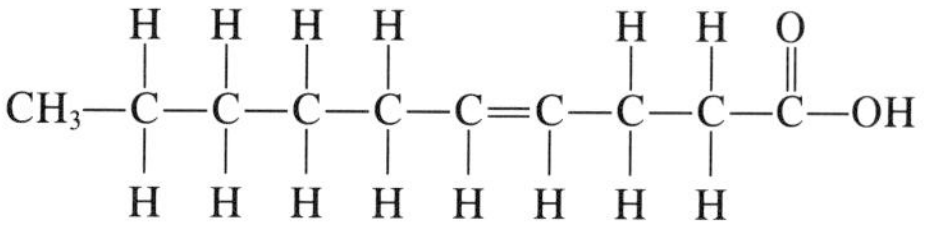

$$CH_3—CH_2—CH_2—CH_2—CH_2—CH{=}CH—CH_2—CH_2—C(=O)—OH$$

or more simply

$$CH_3(CH_2)_4—CH{=}CH—(CH_2)_2COOH.$$

1. Fatty Acid Nomenclature

Names of fatty acids are derived from the appropriate parent hydrocarbon. For example, if the fatty acid has eight carbon atoms the parent hydrocarbon is octane. Remove the terminal "e" and replace it with the suffix "oic" to give octanoic acid, $CH_3(CH_2)_6COOH$. If an acid has one double bond the parent hydrocarbon is octene and the fatty acid is octenoic acid. Fatty acids with one double bond are known as monenoic acids; those with two, dienoic acids. Similarly, we have trienoic, tetraenoic, pentaenoic, and hexaenoic for those acids with three, four, five, and six double bonds, respectively.

There are two ways in which the position of the double bond is indicated. In the first, the terminal carboxyl (COOH) group is termed carbon 1. Counting back towards the methyl (CH_3) end, oleic acid $CH_3(CH_2)_7CH{=}CH(CH_2)_7\overset{1}{C}OOH$ becomes 9,10-octadecenoic acid or simply 9-octadecenoic acid. Linoleic acid with two double bonds, CH_3-$(CH_2)_4CH{=}\overset{12}{C}H{-}CH_2CH{=}\overset{9}{C}H{-}(CH_2)_7\overset{1}{C}OOH$, is 9,12-octadecadienoic acid. The most common unsaturated fatty acids are shown in Table I.

Most naturally occurring fatty acids have conjugated double bonds, that is, the double bonds are separated by one or more single bonded carbon atoms. In most cases the double bonds are interrupted by a single methylene ($-CH_2-$) group. Conjugated double bonds, those adjacent to each other, do occur but they are rare. Because most naturally occurring fatty acids are unconjugated and their double bonds are interrupted by single methylene groups it is possible to use a shorthand system for naming fatty acids. In this system the terminal methyl group carbon is termed carbon 1 and given the designation omega (ω) or n-. Counting towards the carboxyl end, the first double bond from the ω is numbered and the series of fatty acids is named by the position of this first double bond. For example, if a fatty acid is described as 18 : 2ω6 (or n-6), we have all the information necessary to draw its structure. The number before the colon gives the number of carbons in the chain (18). The number after the colon gives the number of double bonds (2) and the number after the ω, the position of the first double bond from the methyl end (6). Since the double bonds are interrupted by a single methylene group, we can write the structure as follows:

$$\overset{\omega}{C}H_3(CH_2)_4\overset{6}{C}H{=}CHCH_2CH{=}CH{-}(CH_2)_7COOH$$

or named the other way, 9,12-octadecadienoic acid. Another example is 18 : 3ω3, which therefore has the following structure:

$$\overset{\omega}{C}H_3CH_2\overset{3}{C}H{=}CH{-}CH_2CH{=}CH{-}CH_2CH{=}CH{-}(CH_2)_7COOH$$

or 9,12,15-octadecatrienoic acid.

Most natural fatty acids have double bonds in the *cis* configuration. Compounds containing double bonds with atoms or groups on either side can have different spatial arrangements. If the atoms or groups on each side of the double bond are on the same side, the compound is said to be *cis*; if they are on the opposite side, the compound is *trans*. This is known as *geometric isomerism*. It is assumed unless stated otherwise that all double bonds are *cis*. *Trans* Double bonds arise on biohydrogenation or commercial hydrogenation. If a fatty acid has *trans* double bonds it should be stated. For example 18 : 2ω6 *cis, trans,* or 18 : 3ω3 all *trans*.

This shorthand system for naming fatty acids is not only convenient but also has biochemical significance. Animals cannot desaturate towards the methyl end. Elongation and desaturation can only take place towards the carboxyl end of the chain. Plants and microorganisms can, however, place double bonds towards the methyl end. It so happens that most natural fatty acids fall into one of four families when named using the ω system. Natural fatty acids fall into one of the series shown in Table II. Once a fatty acid is in a family any further elongations and desaturations in animal systems keep it in that family because of the lack of enzymes to desaturate towards the methyl end. Animals can synthesize the ω7 and ω9 series *de novo* thus:

$$16{:}0 \longrightarrow 16{:}1\omega7 \longrightarrow 18{:}1\omega7 \longrightarrow 18{:}2\omega7 \longrightarrow 18{:}3\omega7 \longrightarrow 20{:}3\omega7 \longrightarrow 20{:}4\omega7$$
$$16{:}1\omega7 \nearrow 16{:}2\omega7 \searrow 18{:}2\omega7$$

$$18{:}0 \longrightarrow 18{:}1\omega7 \longrightarrow 18{:}2\omega9 \longrightarrow 20{:}2\omega9 \longrightarrow 20{:}3\omega9$$

Animals lack the enzymes for synthesis of the ω6 and ω3 families and these must be ingested in the diet. Because of this the dietary precursors of these families are termed essential fatty acids (EFA). The EFA are very important nutritionally and they are discussed in detail in section IV. The precursors of the ω6 and ω3 families are 9, 12-octadecadienoic acid (18 : 2ω6 or linoleic acid) and 9, 12, 15-octade-

TABLE I Structures and Names of Common Unsaturated Fatty Acids

Structure	Systematic name	Trivial name
$CH_3(CH_2)_7CH{=}CH(CH_2)_7COOH$	9-octadecenoic acid	Oleic
$CH_3(CH_2)_4CH{=}CH{-}CH_2{-}CH{=}CH(CH_2)_7COOH$	9,12 octadecadienoic acid	Linoleic
$CH_3CH_2CH{=}CH{-}CH_2CH{=}CHCH_2CH{=}CH{-}(CH_2)_7COOH$	9,12,15-octadecatrienoic acid	α-Linolenic
$CH_3(CH_2)_4CH{=}CH{-}CH_2CH{=}CH{-}CH_2CH{=}CH{-}CH_2CH{=}CH(CH_2)_3COOH$	5,8,11,14-eicosatetraenoic acid	Arachidonic

TABLE II Families of Naturally Occurring Fatty Acids

Series	Methyl terminal of fatty acid
ω3	$CH_3CH_2CH{=}CH{-}$
ω6	$CH_3(CH_2)_4CH{=}CH{-}$
ω7	$CH_3(CH_2)_5CH{=}CH{-}$
ω9	$CH_3(CH_2)_7CH{=}CH{-}$

catrienoic acid (18 : 3ω3) or α-linolenic acid. When ingested by animals, including humans, they undergo elongation and desaturation as follows:

linoleic:

$$18{:}2\omega6 \rightarrow 18{:}3\omega6 \rightarrow 20{:}3\omega6 \rightarrow 20{:}4\omega6 \rightarrow 22{:}4\omega6 \swarrow 22{:}5\omega6$$

α-linolenic:

$$18{:}3\omega3 \rightarrow 18{:}4\omega3 \rightarrow 20{:}4\omega3 \rightarrow 20{:}5\omega3 \rightarrow 22{:}5\omega3 \swarrow 22{:}6\omega3$$

Because of the biochemical and nutritional significance of the omega naming system it will be the predominant system used throughout this text. Most fatty acids have trivial names and the use of these has persisted. Table III lists common fatty acids using their trivial names and omega nomenclature. Many other types of fatty acids occur in nature, including ones having hydroxy, keto, and cyclic groups, and branched chains. These are of no significance from a nutritional viewpoint.

TABLE III Some Common Fatty Acids

Trivial name	Omega nomenclature
Lauric	12 : 0
Myristic	14 : 0
Palmitic	16 : 0
Stearic	18 : 0
Palmitoleic	16 : 1ω7
Oleic	18 : 1ω9
Linoleic	18 : 2ω6
α-Linolenic	18 : 3ω3
γ-Linolenic	18 : 3ω6
Dihomo-γ-linolenic	20 : 3ω6
Arachidonic	20 : 4ω6
Timnodonic[a]	20 : 5ω3

[a] Also known as EPA (for eicosapentaenoic acid).

2. Distribution of Fatty Acid among Triglycerides

The simplest system for numbering a triglyceride molecule will be used here. In this system the carbons on the glycerol moiety are numbered 1, 2, and 3, and 1 and 3 are regarded as interchangeable. There is a stereospecific numbering system in which positions 1 and 3 are not regarded as interchangeable. Using the simple system a typical triglyceride will look as follows:

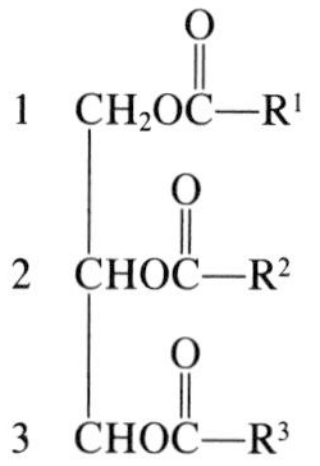

$$\begin{array}{ll} 1 & CH_2OC(=O){-}R^1 \\ & | \\ 2 & CHOC(=O){-}R^2 \\ & | \\ 3 & CHOC(=O){-}R^3 \end{array}$$

Fats and oils are mixtures of such triglycerides. Early in the development of lipid chemistry, researchers asked questions regarding the distribution of fatty acids among the triglycerides. The average fat or oil contains several different fatty acids. Was there some general rule by which one could predict the positions the various fatty acids would occupy? Several theories were proposed but technical difficulties hampered efforts to prove or disprove them. The most accepted theory, and one closest to determined values, is a modified version of the 1, 3, and 2 random distribution theory. This theory makes the 1 and 3 positions on the triglyceride identical and predicts that these will be occupied by identical kinds and proportions of acyl groups distributed at random. Similarly, in the 2-position acyl groups are distributed at random although a modification of the theory states that there is a preferential occupation of this position by 18-carbon unsaturated fatty acids. This theory is close to determined distribution patterns although some fats and oils are more random than others. The distribution pattern of the fatty acids confers particular properties on fats and oils that make them more or less suitable for particular food uses. Modification of natural fats and oils by hydrogenation, interesterification (or redistribution of fatty acid groups), and winterization to render them more useful for food use is common. For example, partial hydrogenation of vegetable oils for use in margarines and shortenings, and winterization of some vegetable oils to remove trisaturated species of triglycerides (by lowering the temperature to 5°C) for use as salad oils are usual in the edible oil industry. Winterization involves holding an oil at 5°C, allowing the trisaturated species of triglycerides to precipitate, and removing them by filtration.

The hydrogenation process leads to the transformation of some *cis* double bonds in fatty acids to the *trans* form. The amount of *trans* fatty acids formed depends upon the conditions of hydrogenation. Under present processing conditions and eating patterns only about 5–8% *trans* fatty acids are present in the diet. Some of this arises from nonprocessed sources such as milk and butter due to the biohydrogenation of fatty acids in the rumen of the cow. In the past there have been concerns regarding the possible health effects of *trans* fatty acids. Most health professionals now regard *trans* fatty acids as being similar to saturated fatty acids in that they have higher melting points than their *cis* counterparts. Hydrogenation conditions leading to high *trans* contents are much less common than several years ago. The winterization process merely removes more saturated triglyceride species from the salad oil and makes it remain liquid at refrigeration temperatures. Tempering and interesterification of fats are used to improve plasticity properties and have no apparent effect on nutritional value.

II. Food Sources of Fats and Oils

A. Trends in Fat Consumption

Throughout this century there has been an increase in the amount of fat consumed. For example, in 1987 the percentage of kilocalories from fat in the U.S. food supply was 43% whereas the desirable dietary goals suggest 30–35%. The amount of animal fat consumed has, however, decreased as the use of red meat, eggs, and dairy products has declined. Over half the available fat still comes from animal products. In the last decade however, both men and women in the United States have reduced their average daily fat intake by several grams. Moreover, there has been a shift to increased use of margarine, cooking oils, nondairy creamers, and skim milk rather than butter, shortening, and whole milk.

Some fats are readily recognized as visible fats; examples are butter, margarine, salad oils, and the fat surrounding meats. Such sources account for less than half of dietary fat, with the rest coming from invisible sources such as fat around meat fibers (*marbling*), homogenized milk, egg yolks, nuts, and whole-grain cereals. Total fat intake and types of dietary fat are influenced by many factors including economical, social, and cultural. More affluent societies tend to consume more fat. Many Asian countries have a lower fat intake than the United States. Some Mediterranean countries (such as Greece) consume considerable amounts of olive oil. While the efforts to reduce total fat and especially saturated fat are commendable, it should be remembered that intake of some fat is essential. Without fat in the diet we may lack essential fatty acids, calories, and sufficient fat-soluble vitamins.

B. Food Sources

The fat content and type of fat for some selected foods is shown in Table IV. Meats and foods of animal origin, such as many cheeses, are higher in fat than most foods of vegetable origin. Cheddar, edam, swiss, and colby cheeses contain as much as 8 grams of fat per ounce; feta, 6 grams; and cottage and skim milk ricotta, as little as 1 gram per ounce.

It was reported recently that, contrary to what was previously thought, not all saturated fat has an adverse effect on serum cholesterol and lipoproteins. Stearic acid (18 : 0), found in beef fat and hydrogenated vegetable oils, does not have the undesirable cholesterol-raising effect of other saturated fatty acids. Nevertheless, there are strong recommendations from health professionals' groups that people decrease their overall intake of saturated fat.

TABLE IV Fat Content and Type of Fat in Some Selected Foods

Foods	Total fat %	Fatty acid %	
		Saturated	Polyunsaturated[a]
Butter	81	62	4
Vegetable shortening	100	25–30	12–26
Poultry	5–20	30	20
Beef	10–40	50	4
Pork	20–35	39	11
Salmon	9	17	51
Tuna	5	41	30
Egg yolk	33	30	13
Peanuts	51	18	27
Avocado	18	19	12

[a] Mainly linoleic acid except in fish.

A useful guide to the saturation level of a fat is the P/S ratio, which is the ratio of polyunsaturated fatty acids to saturated fatty acids after exclusion of the monounsaturated fatty acids. There is now evidence that monounsaturated fatty acids have beneficial effects on serum lipids. It is probable that the P/S ratio will be replaced by the P–M/S ratio, in which the ratio of polyunsaturated plus monounsaturated fatty acids to saturated fatty acids is calculated. The higher the P/S and P–M/S ratios, the more desirable the fat. The ratios for some selected fats and oils and some foods are shown in Table V.

The potential usefulness of the P–M/S ratio as a health guide is particularly apparent in some cases. Note that the P/S ratio of peanut oil is relatively low (1.9) whereas its P–M/S ratio is 4.6. This is a reflection of the high level of the monounsaturated fatty acid, oleic acid (18 : 1ω9) in peanut oil. The same situation pertains to canola (low erucic acid rapeseed) and olive oils. Also note that selection of a margarine could be significant in that tub or soft-serve products made with a highly unsaturated oil like safflower have more desirable P/S and P–M/S ratios.

It will be noted that the two fish listed have relatively low P/S and P–M/S ratios. This should not detract from the importance of including and perhaps increasing the use of fish in the diet. With some exceptions, fish can contribute greatly to maintaining a lower total fat intake, particularly when the fish are baked or broiled with little or no added fat. Moreover, fish oils may have added health benefits in that they contain ω3 fatty acids—not in the form of the parent fatty acid of the family, 18 : 3ω3, but as the elongated desaturated products, eicosapentaenoic acid (20 : 5ω3 or EPA) and docosahexaenoic acid (22 : 6ω3 or DHA). More will be discussed about the ω3 family and these fatty acids in particular in Section IV.

TABLE V P/S and P–M/S Ratios for Selected Fats, Oils, and Foods

Foods	P/S ratio	P–M/S ratio
Fats and oils		
Corn	4.5	6.4
Soybean	4.1	5.8
Peanut	1.9	4.6
Canola	4.9	12.7
Olive	0.6	5.9
Coconut	0.02	0.16
Butter	0.05	0.51
Margarines		
Tub (safflower)	4.9	7.4
Stick (corn)	1.4	4.9
Animal fats		
Beef	0.06	0.78
Poultry	0.7	1.9
Fish		
Salmon	2.0	3.0
Tuna	0.5	1.0
Nuts		
Walnuts	7.0	9.5
Peanuts	1.9	5.6

III. Digestion, Absorption, and Transport of Fat

A. Digestion and Absorption

The absorption process to be described is the one taking place in the nonruminant animal. There are several differences in the process in ruminants.

Lipids are ingested largely in the form of triglycerides. Also ingested are plant and animal sterols, some phospholipids, and monoglycerides in the form of emulsifiers.

The absorption process is shown diagrammatically in Table VI. In the stomach the food is churned by the gastric motion, and a coarse oil-in-water emulsion of large (10,000 Å) particles is

TABLE VI The Absorption of Lipids

Section of gastrointestinal tract	Events
Stomach[a]	Churning to form oil-in-water emulsion of lipids.
Duodenum	Entry of bile salts and enzyme pancreatic lipase.
Upper jejunum	Initiation of absorption. Lipase starts to break down triglycerides. Formation of mixed micelles.
Lower jejunum	Penetration phase. Micelles are disrupted on intestinal mucosal cell surface.
Intestinal mucosal cell interior	Triglycerides are reformed from monoglycerides and chylomicrons leave cell to lymph (short-chain FA to portal blood).
Ileum	Resorption of bile salts.
Colon	Any undigested fat to stool.

[a] Lingual lipase, an enzyme at the base of the tongue initiates some hydrolysis of triglycerides to diglycerides.

formed. Here also the lipoproteins undergo proteolysis. The absorption process is initiated in the duodenum where bile salts and pancreatic lipase enter. The enzymic action on the triglycerides quickly forms diglycerides, slowly forms monoglycerides, and is very slow to give glycerol and free fatty acid (FA). Using a shorthand version of triglycerides without showing the components of glycerol backbone, this is the process:

$$\begin{array}{l} 1\ \text{OR} \\ 2\ \text{OR} \\ 3\ \text{OR} \end{array} \xrightarrow[\text{fast}]{\text{lipase}} \begin{array}{l} 1\ \text{OR} \\ 2\ \text{OR} \\ 3\ \text{OH} \end{array} + \text{FA} \rightarrow \begin{array}{l} \text{OH} \\ \text{OR} \\ \text{OH} \end{array} + \text{FA} \xrightarrow[\text{slow}]{\text{very}} \begin{array}{l} \text{OH} \\ \text{OH} \\ \text{OH} \end{array} + \text{FA}$$

Note that the monoglyceride formed is the 2-monoglyceride. This is important because the fatty acid in this position will largely persist throughout metabolism. The phosphoglycerides are also broken down by phospholipase A to give the lysophosphatides. These and the monoglycerides pass on down the intestine. [*See* Phospholipid Metabolism.]

The major area of absorption is the upper jejunum. Here mixed micelles of 50–100 Å are formed. This is facilitated by the detergent action of bile salts, lysophosphatides, and free fatty acids. Lysophosphatides are phosphoglycerides from which one fatty acid has been removed. Such compounds have a strong detergent action. The important molecular penetration phase then takes place. At the border of the intestinal epithelial cells the micelles are disrupted and the 2-monoglycerides, free fatty acids, and lysophosphatides enter the cell. This process is probably by diffusion as it is independent of energetic and enzyme effects.

The intracellular phase of absorption then follows. In this process triglycerides are resynthesized. There is then addition of proteins, and phospholipids to give lipid–protein particles known as *chylomicrons*. These pass out into the intercellular space into the lymph and via the thoracic duct to the blood. Thus most fatty acids ingested find their way to the blood via the lymph. Only short-chain (less than 10 carbons) fatty acids are directly absorbed into the portal blood.

A number of disorders can lead to malabsorption of fat. These include genetic enzyme defects, pancreatic disease, and inflammatory disorders such as colitis. When fat is not properly absorbed, light-colored, fatty stools will be excreted. This condition is known as *steatorrhea*. Such a condition also leads to loss of fat-soluble vitamins and can lead to a deficiency of linoleic acid, an essential fatty acid.

Cholesterol is absorbed free as a micelle with subsequent reesterification before transport into lymph in chylomicrons. [*See* CHOLESTEROL.]

B. Transport of Lipids in the Blood

Only chylomicron remnants reach the liver because under the influence of the plasma enzyme, lipoprotein lipase (clearing factor), they undergo partial loss of triglycerides. The lipoproteins then undergo progressive changes in the liver and extrahepatic sites.

Plasma lipoproteins are classified according to density and they are separated by differential centrifugation. This classification is purely one of convenience. In reality the plasma lipoproteins are a continuum. They are divided into two main classes, low and high density lipoproteins (LDL, HDL), which are further subdivided into very low density lipoproteins (VLDL), intermediate low density and low density lipoproteins (IDL, LDL), and various HDL fractions HDL_1, HDL_2 etc. This classification is shown in Table VII together with some physical and chemical characteristics of the fractions.

The VLDL are mainly formed in the liver but some are formed in the intestinal mucosal cell. The apoprotein B (apo B) formed in the liver is essential for the removal of the VLDL from the liver by exocytosis. The VLDL are then changed by the action of lipoprotein lipase in extrahepatic tissues with the release of free fatty acids. This gives IDL and then by an unclear mechanism yields LDL. Part of the IDL are changed to give nascent HDL, which exists in the form of a discoidal bilayer. This is believed to be the preferred shape for the enzyme lecithin-cholesterol acyl transferase (LCAT). Under the influence of LCAT, lysolecithin (phosphatidylcholine minus one fatty acid) is formed and the fatty acid released esterifies to cholesterol. This leads to a change in shape of the HDL to spherical. Note in Table VII that the amount of cholesterol increases as this process proceeds. The HDL is removed to the liver, where the cholesterol carried by it is transformed into bile salts.

HDL is of particular interest in risk of coronary heart disease. As noted HDL can transport cholesterol to the liver where it is excreted as bile salts. This is not the case with LDL. The cholesterol in HDL is often referred to as "good cholesterol" because it is so effectively removed from the blood-

TABLE VII Classification and Some Properties of Plasma Lipoproteins

	Low-density class				High-density class	
	Chylomicrons	VLDL	IDL	LDL	HDL 1	HDL 2
Hydrated density	0.93	0.94–1.01	1.01–1.05		1.09–1.14	
Average conc., mg/100 ml						
(male)	12 ± 2	129 ± 59	439 ± 99		300 ± 83	
(female)	13 ± 3	122 ± 63	389 ± 79		457 ± 115	
% protein	2	8	21		50	59
% lipid	98	92	79		50	41
Lipid composition (%)						
Cholesterol esters	3	13	47		40	25
Free cholesterol	1	8	10		4	5
Triglyceride	88	55	14		8	10
Phospholipids	8	20	28		48	56
Free fatty acids	1	2	1		—	3

stream and cannot contribute to atheromatous plaque formation in the intimal (inner) layer of the arteries, which is one of the features of atherosclerosis. LDL contains the highest level of cholesterol and its esters (Tab. VII). It is transported to many tissues where it is recognized by cells and taken up for degradation. Thus the cholesterol in LDL ends up mainly in the structural units of cell membranes. The infiltration theory of atherosclerosis states that normally the gaps between cells lining the arteries are too small for LDL to pass through. However, if the cells are damaged by trauma, LDL can infiltrate and build up at these sites. The smooth muscle cells of the arteries are stimulated, multiply, and readily take up LDL. Continued proliferation and uptake by the cells leads to the formation of fibrous plaques, which occlude the artery. LDL is, therefore, considered to be particularly atherogenic. HDL on the other hand is not recognized by receptors on arterial cells, so little of it is taken up and most of it goes to the liver where the cholesterol is metabolized to bile salts. High HDL levels are, therefore, considered to be a protective factor against atherosclerosis. [*See* Atherosclerosis.]

Note in Table VII that the HDL level of premenopausal females is higher than that of males. After the menopause it tends to decrease. This is consistent with the fact that premenopausal females have less heart disease than males in the same age group. Postmenopausally their risk increases as their HDL decreases.

Obesity, untreated diabetes, cigarette smoking, and hypertriglyceridemia are associated with lower HDL levels and increased incidence of atherosclerosis. Exercise and moderate daily alcohol consumption are associated with higher HDL levels.

As the lipoproteins enter the various cells of the body the fat portion is split off and is hydrolyzed again into glycerol and fatty acids. The fatty acids then undergo one of the following fates:

1. They are used as an immediate source of energy giving rise to carbon dioxide and water.
2. They are stored in cells, especially the specialized white adipocytes, in adipose tissue as a reserve form of energy.
3. They are incorporated into cell structures either in their original form or as elongated and desaturated products.
4. They are used in synthesis of essential metabolic compounds.

Depending upon the tissue, released glycerol may be used in resynthesis of triglycerides or to form glucose, thus giving rise to energy.

C. Control of Lipolysis in the White Adipocyte

We are concerned here with the white adipocyte (fat cell) as distinct from the brown fat cell, a specialized form of adipocyte occurring in embryonic life, the neonate, and hibernators. [*See* ADIPOSE CELL.]

Plasma clearing factor or lipoprotein lipase (LPL) is bound to endothelial cells of capillaries supplying blood to tissues. It liberates free fatty acids from triglycerides, chylomicrons, and lipoproteins arriving from the liver and via the lymph from intestinal absorption. Heparin in mast cells in the connective tissue assists in LPL release from capillaries. LPL is high in the fed state and low in fasting or starvation, when the mobilization of fat is required as an energy source. Insulin increases LPL activity and promotes lipogenesis. The hydrolysis of fat prior to its release from the adipocyte is a tightly controlled series of reactions. Because of this, all reactions leading to the release of stored lipid for formation of fatty acids and glycerol are separated from those pathways leading to lipogenesis and increased fat deposition. The reason for this tight control is that the specific function of the fat cell is to store fat in times of plenty and release it during periods of deprivation. It is important that these reactions related to storage and release are integrated with the needs of the whole organism.

The key factor is the hormone-sensitive lipase (HSL), which catalyzes triglyceride hydrolysis in adipose tissue. The HSL is high in fasting and low in fed animals. A brief exposure of the adipocyte to epinephrine and other hormones like glucagon and ACTH increases the activity of HSL. Exposure to insulin and prostaglandin E_1 (see section VI) decreases the activity. HSL is, therefore, the rate-determining enzyme in lipolysis. Its activity controls the overall rate of lipolysis. [*See* INSULIN AND GLUCAGON.]

D. Fatty Acid Oxidation, Ketosis, and Low Carbohydrate Reducing Diets

Fatty acid oxidation is the major pathway by which fatty acids are degraded to provide energy. The major mechanism is β-oxidation, that is, oxidation at the β carbon (position 2 from the carboxyl end) to yield a β-ketoacid, which is then cleaved. Two carbon units at a time are systematically removed this way until the whole molecule is broken down to yield molecules of acetyl coenzyme A (acetyl CoA) or a 2-carbon unit bound to the coenzyme A. Thus, palmitic acid (16 : 0) would yield eight molecules of acetyl CoA. The units of acetyl CoA then enter the tricarboxylic acid (TCA) cycle to be metabolized to carbon dioxide and water.

The liver has the capacity to divert some acetyl CoA from fatty acid or pyruvate oxidation to acetoacetate and β-hydroxy butyrate. Ultimately one can get the formation of free acetoacetic, β-hydroxybutric acids, and acetone. These are known collectively as *ketone bodies*. They can diffuse out of the liver into the bloodstream to peripheral tissues. Extrahepatic tissues like muscle can oxidize ketone bodies in the mitochondria via the TCA cycle. Normally the amount of ketones is low but in conditions of increased lipolysis (fasting and diabetes mellitus) massive production can occur.

Most acetoacetyl CoA arising in the liver is from head-to-tail condensation of two molecules of acetyl CoA from fatty acid oxidation.

$$2 \text{ acetyl CoA} \rightarrow CH_3\underset{\underset{O}{\|}}{C}\text{—}CH_2\text{—}\underset{\underset{O}{\|}}{C}\text{—SCoA}$$

acetoacetyl CoA

Further reactions can then occur to give β-hydroxy butyric acid and acetone the other ketone bodies.

Normally there are low blood and urine levels of ketone bodies. When the rate of formation in the liver exceeds the capacity of the peripheral tissues to utilize them a condition of ketosis results. The blood levels will rise (ketonemia) as will urine levels (ketonuria). The brain has the capacity to use ketone bodies for energy, especially in starvation.

High organic acid levels in the blood can be harmful and even fatal. Hydrogen ions from acetoacetate and β-hydroxy butyrate lead to a condition of acidosis in which blood pH decreases (becomes more acidic). Sodium and other electrolytes can be lost by chelation. If this is excessive there is water loss, dehydration, and coma—which if unchecked results in death. Uric acid has to compete with ketone bodies for excretion via the kidney. If this is excessive, a condition of gout will arise.

Conditions favoring ketosis are any in which fat breakdown (catabolism) is increased and in which carbohydrate catabolism is decreased. Juvenile onset diabetes is one such condition. Others include low carbohydrate/high protein diets for weight reduction. Such diets rely on putting the dieter into ketosis (which possibly has an anorexic effect). This obviously can be dangerous and such diets should be used only under medical supervision.

IV. Function of Fat in the Diet and in the Body

A. Role of Fat in the Diet

1. As a Source of Energy

Each gram of fat regardless of its food source provides 9 kcal compared to only 4 kcal for each gram of carbohydrate or protein. Even in lean cuts of meat with less than 30% fat, the fat will provide nearly 80% of the calories.

2. As a Carrier of Fat-Soluble Vitamins

There are four fat-soluble vitamins: A, D, E, and K. Dietary fat serves as a carrier for these vitamins and aids in their absorption. Much of our vitamin A supply comes from provitamins A in vegetable foods (such as carrots) that contain carotenoids. About 10% of the fat in the diet is required for efficient absorption of these vitamins. Elimination of fat in the diet would eventually lead to fat-soluble vitamin deficiencies. Any situation that leads to malabsorption of fat—inflammatory bowel disease, obstruction of the bile duct, or ingestion of peroxidized fat—can cause fat-soluble vitamin deficiency.

3. As a Satiety Value

The time taken for a meal to leave the stomach depends upon its fat content. The higher the fat content the longer it takes the stomach to empty. Fat is, therefore, an important contributor to the feeling of satiety after eating. This is of importance in planning weight reducing diets. It is easier to stick to a diet that is low in calories but allows fat in the form of salad oils or spreads on bread. The dieter feels more satisfied with such meals.

4. In the Palatability of Foods

Fat in food carries much of the flavor and aroma. Anyone who has tried to eat a very low fat diet becomes acutely aware of the role of fat in palatability. Very lean cuts of meat lack the flavor and tenderness of nicely marbled cuts. Much of chicken fat and cholesterol is in the skin or just below it. Removing it assists in lowering total fat and cholesterol but many individuals find that this greatly detracts from the taste.

B. Role of Fat in the Body

1. As an Energy Reserve

Body fat is the main store of energy in the body. As noted previously much of the fat for energy is stored in specialized cells called adipocytes and the release of this stored energy by lipolysis is a tightly controlled process under the regulation of hormone-sensitive lipase. It is considered normal and desirable for women to have 18–24% body fat and men 15–18%. People with reserves above these levels are considered overweight. Those with reserves 20–30% above these levels are considered obese. Excessive leaness, especially in females, can cause problems. Extremely thin females often have amenorrhea. Such females may also have difficulty in maintaining a pregnancy.

This bulk of storage fat is in white adipocytes or white fat cells. Brown fat cells, which have an extensive blood supply, are also sources of energy. Brown fat is more important in neonates, infants, and hibernating animals. Its purpose is to provide energy from oxidation of fat at a much faster rate than from white fat. An important triggering mechanism for brown fat oxidation is exposure to cold temperatures.

2. As an Insulator and Protector of Vital Organs

Deposits of fat beneath the skin serve to insulate the body and protect against changes in the environmental temperature and prevent excessive heat loss. Deposits around certain organs such as the kidneys, adrenals, and heart protect them from physical shock.

3. As a Source of the Essential Fatty Acids (EFA)

When Burr and Burr first described the essentiality of some fatty acids in 1929 they considered linoleic, α-linolenic and arachidonic acids to be essential for growth and the prevention of dermatitis in the rat. It was realized later that arachidonic acid ($20:4\omega6$) was formed from linoleic acid ($18:2\omega6$), so really only linoleic was truly essential. Later the essentiality of α-linolenic acid was questioned and its status wavered between essential and nonessential for many years. Here we shall use a broad concept of essentiality of a nutrient. An essential nutrient is one that the organism cannot synthesize, that appears to have a function, and that causes the organism to function less effectively without it. The absence of such essential nutrients in the diet may or may not manifest itself in overt clinical symptoms.

a. Deficiency and Function of Linoleic Acid, $18:2\omega6$ Linoleic acid deficiency has been demonstrated in all laboratory species examined as well as in humans. In the rat the symptoms are poor growth rate or failure to maintain weight and dermatitis in the later stages. There are also various metabolic effects such as cholesterol accumulation in some tissues. Other tissues (for example, the thymus) weigh less than normal. The deficiency is also associated with excessive water loss through the skin and excessive water intake (*polydipsia*). Males suffer testicular degeneration and in a severe deficiency reproduction is impossible. If deficient females are mated with normal males they can bear and suckle young if they are not too deficient. Extremely deficient animals will fail to lactate adequately to maintain viable litters. Poor wound healing is another symptom of the deficiency.

Most of the symptoms of linoleic acid deficiency can be reversed by feeding the fatty acid. The dermatitis in particular clears up quite readily. If α-linolenic acid (18 : 3ω3) is also absent from the diet there will be an increase in the so-called deficiency triene, 20 : 3ω9, in serum and in tissues. This is because the Δ6 desaturase enzyme involved in the further desaturation of 18 : 2ω6, 18 : 3ω3, and 18 : 1ω9 prefers the ω3 as a substrate followed by the ω6 and finally the ω9. When both 18 : 2ω6 and 18 : 3ω3 are absent the enzyme is left to desaturate the ω9 series, leading to a build-up of 20 : 3ω9, which normally is present in very small amounts. The desaturated, elongation product of 18 : 2ω6, arachidonic acid (20 : 4ω6), is normally a major tissue fatty acid. When both EFAs are absent there is a fall in 20 : 4ω6 levels and a rise in 20 : 3ω9, so the ratio 20 : 3ω9/20 : 4ω6 can be used as a measure of the severity of the deficiency. Ratios slightly above 0.4 indicate a marginal deficiency and ratios of 4.0 a severe deficiency.

Linoleic acid deficiency was not seen in humans until the advent of total parenteral (intravenous) feeding (TPN). Patients on TPN are often infants with intestinal problems and/or cancer patients. Both have poor stores of linoleic acid in their tissues. When the infusion has no lipid component to provide either EFA, the deficiency results. The ω6 fatty acid in serum will fall and the deficiency triene will increase, giving an elevated 20 : 3ω9/20 : 4ω6 ratio. There will be dermatitis, failure to gain weight, poor wound healing, and increased risk of infections. The condition is reversed by the addition of an EFA-containing emulsion to the infusion. A common one in the United States is a soybean oil, which therefore contains both linoleic and linolenic acids. Prior to the introduction of this emulsion several cases of TPN-induced EFA deficiency were reported. The FDA did not permit the use of lipids in TPN until several years ago because initial use of lipids had led to body temperature elevation and thrombi problems.

The function of linoleic acid is fairly well understood although a few of its deficiency symptoms have no clear explanation as yet. There appear to be two main functions of linoleic acid, one purely structural, the other as a precursor of a very important group of regulatory compounds, the prostaglandins and their related metabolites.

The structural roles are concerned with the function of long-chain, desaturated metabolites of linoleic acid in the maintenance of the integrity, fluidity, and permeability properties of cellular membranes. Of extreme importance is the role of fatty acids in membrane-based lipids in maintaining the optimum environment for enzymes. In EFA deficiency many enzyme activities are changed, including that of ubiquitous enzymes like Na^+-K^+-ATPase, and the changes are reversed on feeding EFA.

Much interest in linoleic acid now centers around its role as a dietary precursor of some of the most active prostaglandins. The desaturated, elongated products of linoleic acid are the direct precursors of these types of compounds. Dihomo-γ-linolenic acid (20 : 3ω6) gives the prostaglandin of the 1-series, arachidonic acid (20 : 4ω6) the prostaglandins of the 2-series, and timnodonic acid (20 : 5ω3) the 3-series (see Section IV,B,3,c, below).

b. α-Linolenic Acid, 18 : 3ω3 As noted earlier the original discoverers of EFA reported that α-linolenic acid was also essential. It was then noted that α-linolenic acid did not cure the dermatitis of rats on fat-free diets whereas linoleic acid did. Over the years some investigators have continued trying to prove that α-linolenic was or was not essential. Some confusion arose early in that EFA deficiency was usually brought about by feeding a fat-free or hydrogenated fat diet. Thus, such diets were devoid of both linoleic and α-linolenic acids. Later, as diets were refined, diets adequate in linoleic acid and lacking α-linolenic acid were fed. Such diets were fed through several generations of rats but no ill effects were observed on growth or reproductive performance even though tissues showed marked depletion of fatty acids of the ω3 series. There were some reports of neurological changes in primates on α-linolenic deficient diets and of learning disability in rats. The latter observation is probably explained by effects on the retina and is due therefore to lessened visual acuity. The retina is rich in ω3 fatty acids and the absence of α-linolenic acid in the diet interferes with cell repair in the retina. Nevertheless no really clear-cut syndrome recognizable as a nutrient deficiency was ever established. [*See* RETINA.]

Many nutritionists and food scientists dismissed α-linolenic acid as being nonessential. Others held the view that if a function can be defined for a nutrient that cannot be synthesized by the organism it should be regarded as essential. In 1982 it was reported that a young female placed on TPN using an emulsion of safflower oil developed neurological symptoms that were alleviated by changing to a soybean oil emulsion. Safflower oil is very high in linoleic acid (70+%) and very low in α-linolenic acid (ca. 0.5%). The Δ6 desaturase thus was flooded with ω6 fatty acids and had very little of the preferred ω3 acids. The consequence was a depletion of ω3 stores and an elevation of the ω6. At about the same time investigators began showing that at least one function of α-linolenic acid may be to modulate synthesis of the 1- and 2-series prostaglandins from linoleic acid. As noted earlier α-linolenic acid is the precursor of timnodonic acid, 20 : 5ω3, which is the substrate for production of prostaglandins of the 3-series.

Since the publication of the first case of apparent 18 : 3ω3 deficiency several other cases in humans have been described. Studies using monkeys have also shown neurological defects.

α-linolenic acid, therefore, appears to have two functions, one in the maintenance of membrane properties (especially in cells in which ω3 fatty acids are high), and the other in modulation of prostaglandin synthesis and the synthesis of other eicosanoids.

c. Prostaglandins and Related Metabolites: The Eicosanoids The prostaglandins (PGs) and their related metabolites are now such an extensive field that they could be the subject of many books. Here we shall consider enough about them to illustrate why and how they are becoming of interest to food scientists and nutritionists.

Prostaglandins are compounds based on the skeleton of prostanoic acid, a 20-carbon cyclic compound. They are essentially water-soluble, cyclic fatty acids. They were discovered over 50 years ago but they are produced in such small amounts they could not be studied for the first 30 or more years after their discovery. The technology was just not available. The PGs are produced by all cells studied except the erythrocyte. They are produced in response to some stimuli like hormones or trauma and they are usually destroyed at the site of their production. In other words cells usually have the enzymic machinery to both produce and catabolize the PGs. It is important to remember that PGs are not stored in tissues. One cannot speak of "PG content" but only of "capacity to synthesize PGs."

As we have seen the dietary precursors of PGs are the essential fatty acids linoleic and α-linolenic and the direct precursors arise from the elongation and desaturation of these:

linoleic	18 : 2ω6 →	18 : 3ω3 →	20 : 3ω6 →	20 : 4ω6
			↓	↓
			PGs of 1-series	PGs of 2-series
α-linoleic	18 : 3ω3 →	18 : 4ω3 →	20 : 4ω3 →	20 : 5ω3
				↓
				PGs of 3-series

The 1, 2, and 3 of the series name arises from the number of double bonds in the side chains of the PGs.

When the cyclooxygenase enzyme acts on 20 : 3ω6, 20 : 4ω6, or 20 : 5ω3 to produce PGs, the first products are very short-lived PGs called endoperoxides and given the names PGG and PGH. These endoperoxides of all three series undergo further reactions to produce PGs designated PGA through PGF and the very potent prostacyclin (PGI). Just as the three series of PGs differ only in the number of double bonds in their sidechains, these various PGs differ little in structure (constituents on five-membered rings for example) but they may differ markedly in biological action and may even have opposing effects in the same physiological system.

Prostaglandins are regulatory compounds that have many functions. The most important appears to belong to the 2-series and to a lesser extent the 1-series. (Less is known about the 3-series.) It is important to distinguish between physiological and pharmacological effects of prostaglandins. They may have one effect physiologically and another when given at drug levels. Some of the known involvements of prostaglandins are listed below:

- Control of intrarenal blood flow and blood pressure
- Regulation of microcirculation
- Control of vascular tone
- Regulation of neurotransmission
- Regulation of cell differentiation and division
- Regulation of the immune response
- Initiation of parturition
- Acceleration of luteolysis

- Facilitation of calcium binding to membranes
- Broncho-constriction and broncho-dilation
- Neuromuscular activity regulation
- Control of platelet aggregation and blood clotting
- Regulation of natriuresis

Many of the effects of prostaglandins are mediated via the cyclic nucleotides, cAMP and cGMP.

In addition to the classical PGs other metabolites of arachidonic acid were gradually discovered. These include the thromboxanes A and B (TXA, TXB). In addition to this pathway another catalyzed by lipoxygenase was discovered and this produces a series of hydroperoxy and hydroxy eicosatetraenoic acids (HPETEs and HETEs) and substances known as leukotrienes and lipoxins. Collectively, they are known as eicosanoids. Probably all the compounds produced from arachidonic acid can be produced for the corresponding 1- and 3-series from $20:3\omega6$ and $20:5\omega3$. It should be noted that different cells produce different products in various amounts and not all corresponding substances from different series are equally biologically active. Most PGs and the other eicosanoids are highly biologically active, hence they do not exist for long or they would be destructive to the cell and the whole organism. They are produced, do their work, and are destroyed at the site or on passage through the lung or kidney. They are "hormone-like" in their action but are not true hormones. Many of the lipoxygenase products are now known to have powerful physiological effects. Some of them inhibit the synthesis of the cyclooxygenase products, others are chemotactic. Some of the leukotrienes are the slow-reacting substances of anaphylaxis (sRSA). Anaphylaxis is an antigen-antibody reaction in which there is an unusual or exaggerated response to a foreign protein or other substance. sRSA is, therefore, of great interest in allergic responses. The picture of the role of all of these compounds is far from complete but it is clear that they exert profound regulatory effects on many body functions.

d. Modulation of Eicosanoids Synthesis by Dietary Fats and Oils Eicosanoid synthesis can be modified by drugs. For example aspirin and other nonsteroidal antiinflammatory drugs inhibit the cyclooxygenase and this, at least in part, explains their action against pain. Steroids inhibit the release of free arachidonic acid by the enzyme phospholipase A_2 and lead to less arachidonic acid entering the lipoxygenase pathway, explaining why they have therapeutic effects in diseases like rheumatoid arthritis and psoriasis. Since the precursors of the eicosanoids are the EFA it was obvious to investigate whether or not eicosanoid synthesis can be modulated by changing dietary fats and oils. There are now many examples of the success of this approach both in laboratory animals and humans. If the $18:2\omega6$ intake is lowered there is less $20:4\omega6$ in tissues and the production of PGs of the 2-series is decreased compared to that in subjects receiving higher intake of $18:2\omega6$. This is an extensive subject in itself. Just a few examples will suffice to illustrate the potential of this approach of intervening therapeutically via dietary fat.

A dietary intervention of popular interest in recent years has been the use of increased levels of $\omega3$ fatty acids for the prevention of risk factors for heart disease and stroke.

It has been known for sometime that Eskimos living a true Eskimo life-style have fewer heart attacks. They also have prolonged bleeding times and less sticky platelets than other population groups. Prolonged bleeding was noted by early Arctic explorers. An obvious connection is their high consumption of foods of marine origin. Marine oils are rich in $\omega3$ fatty acids, particularly in $20:5\omega3$, the precursor of the 3-series PGs.

There have now been many studies in this area. $\omega3$ fatty acids from both an $18:3\omega3$ source or a fish oil source ($20:5\omega3$) have been shown to lower serum cholesterol, although the $20:5\omega3$ is usually more effective than the $18:3\omega3$. Significantly it has been shown that $20:5\omega3$ from marine sources or $18:3\omega3$ transformed to $20:5\omega3$ in the body is effective in decreasing the stickiness of platelets, which can form white thrombi in blood vessels leading to heart attack or stroke. The mechanisms appears to be as follows. The endoperoxides from $20:4\omega6$ in platelets (PGG_2 and PGH_2) are highly aggregatory to the platelets, as is the thromboxane A_2, which also constricts blood vessels. Prostacyclin I_2 on the other hand disaggregates platelets and dilates blood vessels. There is, therefore, a balance between the actions of the PGG_2, PGH_2, and TXA_2, and PGI_2. If this balance is tipped in favor of the aggregatory eicosanoids, say by a damaged arterial wall where PGI_2 is produced, then thrombi may form, leading to heart attack or stroke. If the diet is high in $\omega3$ fatty acids this chance appears to be lessened for the following reasons. The $20:5\omega3$ leads to less $20:4\omega6$ in tissues. It also competes with $20:4\omega6$ for the cyclooxygenase enzyme. As a result synthesis of the 2-series PGs is decreased and some 3-series are synthesized. It so happens that the thromboxane from 20 : 5ω3 is not aggregatory to platelets but the PGI_3 is just as disaggregatory as, and a vasodilator like PGI_2. Thus, the balance is tipped towards less sticky platelets, a tendency to bleed, but less chance of heart attack.

Another example is the use of ω3 fatty acids in the treatment of psoriasis. There have been several reports that the ingestion of ω3 fatty acids, either 18 : 3ω3 or 20 : 5ω3 (the latter more often), alleviates the inflammation and skin proliferation in the disease. It has been known for some time that patients with psoriasis have excess free 20 : 4ω6 in their skin. Skin also has an active lipoxygenase enzyme that transforms 20 : 4ω6 to leukotriene B_4 (LTB_4). This leukotriene is highly chemotactic for neutrophils (it attracts white cells). These white cells enter the skin from capillaries below the surface and cause an inflammatory response. Additionally, LTB_4 is hyperproliferative to skin cells. Consequently, there is redness of the skin and accumulation of immature skin cells which pile up to form the psoriatic plaques. Leukotriene B_5 formed from 20 : 5ω3 is less chemotactic than LTB_4, which is produced in lesser amounts on an ω3 diet; as a result the psoriatic process is modified.

Thus we have seen that dietary fats and oils are not only important as sources of energy, enhancers of food flavor and aroma, and carriers of some vitamins but also have important functions in the body and can be used to modify health. Judicious choice of dietary fat and oils can lower serum cholesterol, a heart disease risk factor, and have potential for intervention in alleviating or preventing other diseases. A word of caution should be added. Dramatic changes in type of fat intake and especially the use of fish oil supplements should be done under medical supervision only. Fish-oil capsule use may lead to excessive bleeding on injury and may, depending on their source, contain oxidative polymers that could be deleterious to health.

Bibliography

Briggs, G. M., and Calloway, D. H. (1984). "Nutrition and Physical Fitness," 11th ed. Holt, Rinehart & Winston, New York..

Guthrie, H. A. (1989). "Introductory Nutrition." Times Mirror/Mosby College Publishing, St. Louis, Mo.

Kinsella, J. E. (1987). "Seafoods and Fish Oils in Human Health and Disease." Marcel Dekker, Inc., New York and Basel.

Simopoulos, A. P., Kifer, R. R., and Martin, R. E., eds. (1986). "Health Effects of Polyunsaturated Fatty Acids in Seafoods." Academic Press, New York.

Willis, A. L. (1984). Essential fatty acids, prostaglandins, and related eicosanoids. In "Present Knowledge in Nutrition." The Nutrition Foundation Inc., Washington, D.C.

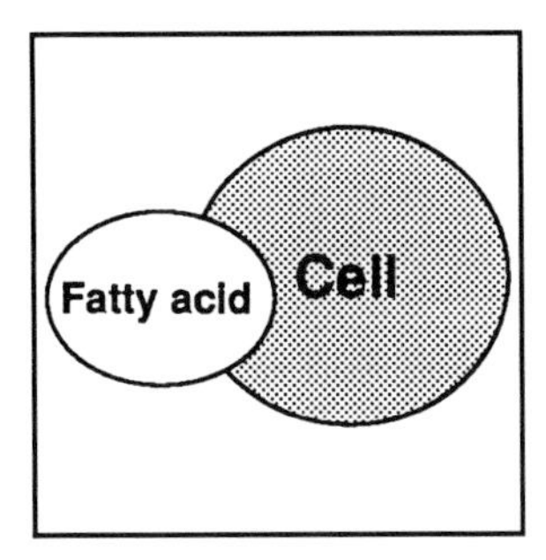

Fatty Acid Uptake by Cells

ROGER A. CLEGG, *Hannah Research Institute*

Glossary

Amphiphilic Having some characteristic of a hydrophilic substance and some of a hydrophobic one

Cyclic AMP Cyclic 3′5′ adenosine monophosphate; formed by the action of the plasma membrane enzyme adenylate cyclase on adenosine triphosphate

Facilitated diffusion Movement of a solute down a concentration gradient (therefore needing no net energy input) requiring the participation of a carrier or transporter system to enable passage across a diffusion barrier

Phospholipid bilayer Arrangement spontaneously adopted by phospholipids in aqueous medium into two single-layer leaflets, each with the polar headgroups orientated toward the water phase and the fatty acid "tails" interacting to form an apolar core to the membranous structure

THE ENTRY OF fatty acids into cells can take place by several distinct routes and mechanisms. Uptake of fatty acids provides cells with potential substrates for a variety of biochemical processes, different types of cell being specialized in respect of different processes. For any given metabolic and physiological state of an individual, disparate needs for exogenous fatty acid will exist among different cells and tissues, depending on their functional specialization. To ensure the appropriate partition of fatty acids, the pathways of uptake should be subject to regulation—possibly different ones from one tissue to another. In general, such metabolic coordination between tissues is accomplished by hormonal signaling. Although the number and chemical variety of hormones is great, individual cells that recognize them decode their signals into one (or sometimes more) biochemical "languages." The vocabularies of these languages comprise changes in the intracellular concentration of a few key information-transducer molecules—the second messengers of hormone action. The intracellular messenger whose mechanism of action is most thoroughly researches is the oldest member of this small family—cyclic AMP. In this article, the cellular uptake of fatty acids is considered in relation to their intracellular use; probable mechanisms whereby the uptake is accomplished and how these may be influenced by cyclic AMP are discussed.

I. Physical Chemistry of Fatty Acids

Fatty acids are a somewhat loosely defined group of chemically related mono-carboxylic acids having an acyl chain varying in length from a few carbons to a few tens of carbons. The overwhelming majority of fatty acids found in animal tissues and body fluids have unbranched acyl chains, usually with an even number of carbons. The exceptions are found among those branched and otherwise exotic (e.g., cyclic, hydroxy-derivatized) members of the series introduced by herbivores into the animal food chain from plant and algal lipids. The acyl chains may be either saturated or contain one or more double bonds, almost invariably in the *cis* conformation. Those fatty acids that are polyunsaturated possess their double bonds in unconjugated arrangements separated by single methylene groups. Some of the polyunsaturated fatty acids found in animals cannot be synthesized by humans and other mammals and thus comprise an essential constituent of their diets, derived mainly from plant lipids.

Those fatty acids having short acyl chains of one, two, and three carbons—acetate, proionate, and butyrate, respectively—are often considered separately from longer chain fatty acids on the basis of their markedly different physical properties. They are moderately volatile (boiling points: 118, 141, and 164°C) and freely soluble in water. Longer chain fatty acids become progressively more and more insoluble in aqueous media as chain length increases. Their physical properties are importantly dependent on temperature: Below a transition temperature, characteristic for each fatty acid, they exist in solid form. Above their transition temperature they melt to adopt a liquid crystalline form. The structural feature responsible for introducing the greatest variation in this transition temperature between fatty acids is their degree of unsaturation. Thus, for stearic acid (18 : 0—i.e., having 18 carbons with no double bonds), the transition temperature is 69.6°C; for oleic acid (18 : 1), 13.4°C; and for linoleic acid (18 : 2), −5°C. This transition temperature is of great importance in biological membranes where the proportion of various fatty acyl groups esterified into the component phospholipids of these membranes is a major determinant of the fluidity of the bilayer. This aspect of the biophysical properties of the phospholipid membrane bilayer itself exercises a profound influence over its biological/biochemical properties.

The carboxylic acid functionality endows fatty acids with the properties of weak acids (having pK_a values in the region of 4). This is, of course, well to the acid side of the pH values encountered in normal physiological situations both intracellularly (except within lysosomes) and extracellularly. Consequently, fatty acids will exist predominantly in their dissociated anionic (negatively charged) forms in physiological media. The dissociated carboxyl group is hydrophilic in character, whereas the acyl chain comprising the rest of the molecule is more or less hydrophobic, depending on chain length. Thus the entire molecule behaves as an amphiphile, with correspondingly detergent-like properties, forming monolayers at interfaces between polar and apolar phases and adopting micellar conformations in uniformly polar environments. When presented to one surface of a phospholipid bilayer (e.g., a cellular membrane), the acyl "tails" of fatty acid molecules will penetrate into the apolar core of the exposed leaflet of the bilayer, leaving the polar carboxylate "heads" interacting with the aqueous phase on the membrane surface. Many studies have shown that molecules (phospholipids, proteins) interdigitated between the phospholipid fatty acyl chains in this way can freely undergo rotational and lateral motion in the plane of the membrane. The rates of these motions are a function of membrane fluidity, which is itself influenced by temperature and phospholipid fatty acid composition as discussed above. In contrast to these types of motion, translocation of the carboxylate head of a fatty acid molecule through the apolar core of a biological membrane from one phospholipid leaflet of the bilayer to the other is not thermodynamically favored when this group is in its ionized, charged state. The considerable energy barriers opposing such "flip-flop" motion ensure that this process is exceedingly slow. Thus it is thermodynamically unfeasible that fatty acid anions should cross biological membranes by a simple passive diffusion process.

II. Fatty Acid Uptake in Relation to Cellular Economy

Many different types of cell engage in the synthesis of fatty acids (lipogenesis) from small molecule precursors, predominantly glucose in humans. The fatty acid products thus produced, together with the preformed fatty acids taken up by cells, are then processed through a number of intracellular metabolic pathways. In those animal models commonly employed in experimental studies aimed at understanding aspects of human nutrition, the processes of uptake and biosynthesis of fatty acids are both subject to regulation according to the dietary status of the subject. This is true for adipose tissue and muscle in the rat. Surprisingly, human adipose tissue supports only low rates of *de novo* lipogenesis; thus the fat content of individual human adipocytes and consequently the adiposity of the subject are primarily determined by coordinate regulation between the processes of fat breakdown and of fatty acid uptake and metabolic conversion to storage fat. [*See* ADIPOSE CELL.]

Before considering the processes of cellular fatty acid uptake, this section briefly summarizes the different metabolic fates that may be met by fatty acids when once inside the cell.

A. Fatty Acid Oxidation

A variety of cells and tissues use fatty acids oxidatively as energy-yielding substrates—of these, muscle (skeletal and cardiac muscle) and liver are quantitatively the most important in humans. The enzymic processes resulting in the complete beta-oxidation of fatty acids to carbon dioxide and water are located in the matrix compartment of mitochondria, enclosed by the inner mitochondrial membrane. However, in common with all other pathways for the metabolism of fatty acids, the first step in this process is the formation in the cytosol of coenzyme A esters of the fatty acids. The family of enzymes responsible for this esterification process have high affinities for free, nonesterified fatty acids, reflecting the situation that the free fatty acid concentration in cytosol is low (in the range of micromoles per liter). The translocation of the coenzyme A esters of fatty acids across first the outer then the inner mitochondrial membrane occurs via the mediation of a specific transport system (the "carnitine shuttle"), thereby ensuring the maintenance of functionally separate pools of coenzyme A in the cytosol and mitochondrial matrix. In the course of this shuttle process, fatty acyl groups are reversibly trans-esterified from coenzyme A to the nitrogenous base carnitine. Two apparently distinct membrane-bound enzymes, (carnitine-palmitoyl transferases) CPT 1 and CPT 2, catalyze the trans-esterifications at the outward-facing and inward-facing aspects of this shuttle. They are located on the outer and inner mitochondrial membranes, respectively.

B. Esterification

A family of enzymes located predominantly on membranes convert fatty acyl esters of coenzyme A to a variety of different esterified compounds. It is convenient to group these products according to their functional roles in the cell and organism.

The first group of esterified lipids are those that function as vital structural components of cellular membranes. Phospholipids predominate in this category, although they also have other more dynamic biochemical roles as discussed below. Phospholipid biosynthesis proceeds much more rapidly in those cells that are called on to sustain a high rate of membrane turnover and biosynthesis; examples of this type of cell include secretory and absorptive cells. At least one of the enzymes of phospholipid biosynthesis (cytidine triphosphate: phosphocholine cytidyltransferase) is under nutritional regulation in liver cells and probably also in other types of cell. Agents that increase the intracellular concentration of cyclic AMP (predominantly, glucagon in the case of liver cells) cause the translocation of this enzyme from a microsomal location where its full activity is expressed, to the cytosol where it is less active. Fatty acids have the reverse effect. [*See* PHOSPHOLIPID METABOLISM.]

The second group comprises storage compounds (triacylgylcerol and cholesterol ester). Adipose tissue is the primary site for accumulation of the former, whereas cholesterol ester stores are characteristic of steroidogenic tissues such as adrenal cortex and corpus luteum of the ovary. Close to half of the triacylglycerol stored in the body of a lean individual is found associated with skeletal muscle. This surprisingly large proportion occurs both in the form of fat droplets packed between myofibrils within the muscle cells and in adipocytes intruded among these cells. Both these major classes of esterified lipid are also synthesized in the liver, primarily for export to peripheral tissues as discussed below. Fatty acid esterification processes leading to triacylglycerol biosynthesis are regulated in the liver at the level of two enzymes [glycerophosphate : acyl transferase (GPAT) and phosphatidate phosphohydrolase (PAP)]. These enzymes catalyze early committed steps in the formation of triacylglycerol, and both are subject to reversible phosphorylation; PAP behaves as an ambiquitous enzyme, responding in concert with cytidyltransferase. Little detailed information is available concerning the mechanism of regulation of GPAT. Cholesterol esterification takes place by the transfer of fatty acyl groups from acyl-coenzyme A molecules under the direction of the enzyme acyl-coenzyme A: cholesterol acyl transferase (ACAT). Controversy currently surrounds the status of ACAT as a regulated enzyme. A direct role for cyclic AMP is not envisaged, although, as with the regulation of hydroxymethyl-glutaryl-coenzyme A reductase (HMG-CoA reductase) activity in the control of cholesterogenesis, its involvement more distally in a cascade of regulatory phosphorylations is virtually certain. [*See* CHOLESTEROL.]

The third group consists of fatty acyl compounds destined not for intracellular storage but for secretion from the cells elaborating them. In this group also, triacylglycerols and cholesterol esters predominate. In the liver, fatty acids derived from the diet (via the hepatic portal vein) and from lipolysis in the peripheral tissues (via the hepatic artery) are esterified into these two lipid classes when in surplus to the organ's needs for fatty acid as an energy source. They are then packaged into lipoprotein particles and secreted via a Golgi vesicle route into the circulation. Once exposed to extracellular aqueous media, these lipid droplets are maintained in the form of a stable emulsion by a surface monolayer of amphiphilic lipids (phospholipid and cholesterol) and associated proteins. Epithelial cells in the absorptive zones of the small intestine also elaborate and secrete lipoproteins; in this case, the lipid substrates for these biosyntheses are derived by uptake from the digesta within the gut lumen. Other specialized lipid-secreting cells are found in the sebaceous glands of the epidermis and in mammary tissue during lactation. Triacylglycerols are the predominant secretory lipid product of both these structures.

Fourth, there is a heterogeneous group of esterified fatty acids, contributing quantitatively in only a minor way to the overall complement of cellular lipid, that share in common a role in cellular signaling processes. Two of these are derived by hydrolytic breakdown of selected phospholipids of the plasma membrane. An important subclass of phosphatidyl inositols has phospholipid molecules in which the inositol group is polyphosphorylated. Such phosphatidyl inositol polyphosphates are specific substrates for a family of membrane-bound phospholipase C enzymes that when activated, hydrolyze them to diacylglycerols and inositol polyphosphates, which perform essential signaling functions within cells. Inositol phospholipids are also involved in the "tethering" of certain proteins to the surface of cells. These anchors are complex molecular assemblies of phosphatidyl inositol polyphosphate, ethanolamine, and glycan residues (abbreviated to PIG for phosphatidyl inositol glycan), covalently linked to the C terminus of the tethered protein (hence known as a PIG-tailed protein). Hydrolytic enzymes of the phospholipase C-type, specific for polyphospho-inositol phospholipids, can act on these complexes, leading to the release of the modified protein from the cell surface and the liberation of diacylglycerol in the plasma membrane. The structure of the PIG moiety closely resembles that of a family of minor inositol phospholipid glyco-conjugates found in cell surface membranes that can also be hydrolyzed by specialized phospholipase C enzymes to liberate molecules that may be involved in the intracellular signaling of insulin's actions. Phospholipids of the plasma membrane that have arachidonic acid esterified at the 2-position are also involved in cell signaling. In this case it is a phospholipase (or multiple phospholipases) of the A_2 type that hydrolyzes phospholipids of this family with the release of arachidonic acid. This fatty acid is the precursor for the synthesis of prostanoid and eicosaniod products—potent mediators of inflamatory responses between cells. Other minor phospholipid metabolites are known to have potent effects on a variety of cells. Of these, the best known is platelet-activating factor (PAF), which is derived from ether phospholipids of the plasma membrane by the two-step substitution of an acetyl for a long-chain fatty acyl group at position 2. Finally in this group of "minority" acylations, mention must be made of proteins that are fatty acylated. These fall into two categories: first, several proteins, including the p21 *ras* oncogene product and the catalytic subunit of cyclic AMP–dependent protein kinase, are myristoylated at their N termini. This modification is introduced while the protein is being synthesized and is metabolically rather stable. In contrast, the second category of acylated proteins, those that are palmitoylated (including at least one—p21 *ras*—that is also myristoylated), exhibit high rates of turnover of the acyl group. [*See* PLATELET-ACTIVATING FACTOR PAF-ACETHER.]

In summary, the role of fatty acids within living cells are diverse, ranging from generalized energy supply, through provision of structural and secretory multimolecular assemblies, to exquisitely sensitive modulation of cellular and enzymic function.

III. Sources of Fatty Acid Available to Cells

The presentation of fatty acids to cells in the aqueous medium that bathes them is predominantly in the form of esterified lipid. The levels of nonesterified fatty acid in peripheral plasma seldom rise above 1 mM; they are highest during fasting when the rate of fatty acid release from adipose tissue, resulting from the lipolytic mobilization of triacylglycerol stores, is greatest. Nonesterified fatty acids are essentially all bound to albumin during their transport in plasma. The amphphilic character of the fatty acid molecule endows it with detergent-like properties; the maintenance of a low concentration of free fatty acids is therefore a necessary strategy for the preservation of the structural integrity of cells and tissues interacting with plasma or capillary filtrate.

Esterified fatty acids are made available to cells and tissues as components of lipoprotein particles, which are are classified into three major classes. The triacylglycerol-rich lipoproteins (chylomicrons and VLDL) are the primary providers of fatty acids to peripheral tissues. VLDL particles are assembled more or less continuously in liver cells and small intestinal epithelial cells. During periods of postprandial absorption, another type of lipoprotein particle (i.e., chylomicrons) are elaborated in intestinal epithelial cells where triacylglycerol is reformed from the fatty acids and glycerol that are products of digestion (hydrolysis) of dietary fats and oils in the intestine. Hepatic lipoproteins are secreted directly into the circulation, leaving the liver via the hepatic vein. If this secretory process should become impaired, as occurs, for example, in cases of alcohol abuse, lipoprotein particles build up in the Golgi system of hepatocytes, leading to the well-known clinical syndrome of fatty liver. Intestinal lipoproteins pass first into the lymphatic vessels draining the gastrointestinal tract. Lymph is ultimately returned to the bloodstream at the point where major vessels of the vascular and lymphatic systems fuse at the thoracic duct. [*See* ATHEROSCLEROSIS.]

The rate of transfer across the capillary wall of nonesterified fatty acids from the capillary lumen to the extracellular fluid is 1,000-fold greater than the rate of albumin transfer. Thus it is unlikely that fatty acids remain bound to albumin during transfer, although microscopic studies of capillary blood vessels indicate the presence of "holes" between cells large enough to permit the egress of albumin, and indeed, albumin is a major protein constituent of extracellular fluid. An attractive hypothesis to account for these data is that fatty acids dissociate from albumin at the luminal surface of the capillary as the result of a specific high-affinity interaction with an endothelial cell surface component. Rapid transfer of the fatty acid to the tissue-facing surface of the capillary endothelial cell could then occur by the process of lateral diffusion discussed above. The rate of such motion in the plane of the membrane has been measured for several membrane proteins with attached ligands and found to be sufficient to account for the observed rates of fatty acid uptake by tissues from the circulation. Transfer of fatty acid back to albumin in the extracellular fluid would evidently require a reversal in the relative fatty acid binding affinities of albumin and the putative carrier protein on the endothelial cell surface. The mechanism of such a change is unknown at present.

Fatty acids from triacylglycerol molecules present in lipoproteins are also, like albumin-bound nonesterified fatty acids, released in the first instance within the capillary lumen and not directly at the surface of the cells that will ultimately use them. Access of VLDL and chylomicrons to these cells is obstructed by the size of the lipoprotein particles, which is greater than that of the pores in the capillaries. However, the user cells do participate indirectly in the release process in that they synthesize and secrete the enzyme (lipoprotein lipase), which ultimately effects the hydrolysis of lipoprotein triacylglycerol. This is achieved within the capillary lumen while lipoprotein lipase is attached to specialized glycosaminoglycan "anchors" on the surface of the capillary endothelial cell. Except under pathological circumstances, the concentration of lipoprotein triacylglycerol in serum is not sufficient to saturate lipoprotein lipase. Thus the rate of fatty acid uptake from this source by an individual tissue is determined both by the circulating concentration of lipoprotein triacylglycerol and by the amount of lipoprotein lipase within the capillary blood vessels supplying that tissue. As we shall see, cellular regulatory processes involving cyclic AMP operate in the control of both these variables.

Many *in vitro* studies with cultured mammalian cells have demonstrated the uptake of fatty acids from albumin complexes in proportion to the concentration at which they are supplied in the culture medium. Such uptake shows no evidence of saturability or a marked dependence of the rate of uptake on temperature. These properties suggest that uptake is via a passive diffusion process despite the theoretical unlikelihood of such a phenomenon (see Section I). Animal studies, however, suggest a different mechanism. When experimental animals were injected with radiolabeled fatty acids complexed to albumin, the pattern of differential fatty acid uptake between tissues was found to be distinctive. The labeled fatty acid was cleared from the plasma within a few minutes of administration. Brown fat and liver were tissues of high uptake;

moderate accumulation was seen in kidney cortex and gastric mucosa, whereas only low levels of fatty acid were found in brain and white adipose tissue. The degrees of uptake were not in proportion to the relative blood flow rates through these tissues—thus fatty acid uptake by tissues in the intact animal is not a simple function of blood flow rate and fatty acid concentration. These data support the suggested existence of tissue-specific fatty acid transporters. Selectivity among individual fatty acids is also a property of uptake by tissues. In general, polyunsaturated fatty acids are taken up more rapidly than saturated and mono-unsaturated members of the series.

In one specialized instance, fatty acid uptake by cells in a physiological context does almost certainly take place by passive nonfacilitated diffusion. The case in point is the uptake of short-chain fatty acids in the human large intestine. More than 90% of these volatile fatty acids, which are formed by bacterial fermentations in the lumen of the gut, are reabsorbed through the intestinal epithelial cells. The liver efficiently clears the absorbed acetate and butyrate from the portal blood; here, too, uptake is almost certainly by passive diffusion.

The availability of fatty acids of dietary origin is signaled to body tissues by the pancreatic hormones insulin and glucagon. In periods of fasting, insulin/glucagon ratios are low. This activates adenylate cyclase in the surface membrane of cells having glucagon receptors, with a consequent rise in the intracellular concentration of cyclic AMP. In adipose tissue this leads to an enhanced rate of lipolysis, providing nonesterified fatty acids to the circulation for uptake by the liver and conversion to lipoprotein. In the liver itself, increased levels of cyclic AMP lead to a decrease in the capacity for fatty acid esterification (see Section II), reflecting the diminution in fatty acid provision from the process of *de novo* synthesis. However, the decrease in esterification capacity is antagonized by elevated levels of circulating nonesterified fatty acids; thus, only in a prolonged fast when adipose tissue reserves become depleted will hepatic fatty acid esterification diminish. In cells presenting beta adrenoceptors on their outer surface, epinephrine and norepinephrine act similarly to glucagon, enhancing adenylate cyclase activity. By this means, these hormones cause a burst of lipolysis in white adipose tissue but are without effect on esterification in the liver because adult human liver cells have few beta receptors. [*See* INSULIN AND GLUCAGON.]

IV. Lipoprotein Lipase and Other Fatty Acid Hydrolases

A. Role in Fatty Acid Uptake from Lipoprotein Lipids

Hydrolysis of lipoprotein triacylglycerol by the enzyme lipoprotein lipase is an obligate stage in the uptake of fatty acids from this class of lipids. Studies with perfused tissues have indicated that not all the fatty acid liberated by lipoprotein lipase–mediated hydrolysis is taken up by the tissue in which hydrolysis occurs. That which is taken up has to traverse the capillary endothelium before reaching the "user cells" of the tissue in question. Elegant ultrastructural evidence has been presented to illustrate how this might occur by lateral diffusion in the outer leaflet of the phospholipid bilayers of the endothelial cell and adjacent user-cell surface membranes. This hypothesis requires that transient fusion of these monolayers should take place to allow the diffusing fatty acid to transfer from the surface of one cell onto that of an adjoining one. There is evidence from ultrastructural studies that such temporary fusions do occur in various cellular membranes, and transfer of lipophilic "reporter" molecules between cells *in vitro* can be induced by low concentrations of various fusiogenic agents including certain products of phospholipid hydrolysis. The hypothesis also proposes a topological relation between cell surface and intracellular membranes such that fatty acids may never need physically to cross a phospholipid bilayer to reach the membranous intracellular site of their further metabolism.

Cholesterol esters are also substrates for lipoprotein lipase, and one route for their uptake from lipoproteins by peripheral tissues involves their prior hydrolysis by this enzyme. However, receptor-mediated uptake of entire lipoprotein particles, with subsequent lysosomal hydrolysis of cholesterol esters, constitutes a much better-understood route for the uptake of this lipid from the lipoprotein particle most rich in it (i.e., LDL).

An extracellular lipolytic enzyme first characterized in liver (hence usually named *hepatic triacylglycerol lipase*—H-TGL) is now known to occur in a number of tissues that share in common the property of using cholesterol for the biosynthesis of other sterols. This lipase is distinguished from lipoprotein lipase both immunologically and on the basis of some of its catalytic properties (it is resistant to inhibition by salt, for instance, whereas lipoprotein lipase is salt-sensitive). There is evidence that H-TGL is involved in the uptake of cholesterol from cholesterol ester–enriched lipoproteins in steroidogenic tissues; it seems likely, parenthetically, that fatty acids derived from cholesterol ester hydrolysis will also be taken up by such tissues.

B. Regulation of Activity

In addition to its location on capillary endothelial cells, lipoprotein lipase is found on the surface of user cells in tissues capable of fatty acid use from lipoprotein triacylglycerol. The enzyme also occurs within these cells. *In vitro* studies with isolated cell preparations suggest that in its surface location, lipoprotein lipase may catalyze the vectorial hydrolysis of emulsified triacylglycerol. This is a process in which the fatty acid products are delivered into an intracellular compartment, as distinct from the extracellular one that contained the substrate, simultaneously with their liberation from that substrate. The physiological significance of this observation is uncertain because emulsified triacylglycerol is not normally presented to the surface of parenchymal cells within a tissue mass (see Section III). However, this behavior may give a clue as to how lipoprotein lipase functions on the surface of the capillary endothelial cell.

The attachment of lipoprotein lipase to the outer surface of cells such as adipocytes, myocytes, and fibroblasts is via the PIG-tail anchor discussed in Section III, and there is some evidence that its release from adipocyte-like cells is stimulated by insulin. This relates directly to the known characteristics of the nutritional regulation of lipoprotein lipase. Under fed conditions, activity is high in adipose tissue and low in muscle; during fasting, this differential is reversed, such that adipocytes no longer consume fatty acids from lipoprotein triacylglycerol, whereas these become available to muscle as an energy source in the absence of glucose from the diet. It is now known that these changes are brought about by a decrease in the circulating levels of insulin during fasting. The intracellular levels of lipoprotein lipase activity remain more or less constant regardless of dietary status: the insulin-provoked changes are expressed at the level of extracellular enzyme as illustrated in Table I. As discussed above, for a given tissue mass only part of this extracellular pool (that part located on the capillary endothelial cells) is physiologically involved in the uptake of fatty acids from lipoprotein triacylglycerol. Recent studies have shown that insulin increases the level of lipoprotein lipase mRNA in adipocytes. Newly translated lipoprotein lipase is catalytically inactive and unable to undergo further intracellular assortment until it is glycosylated in vesicles of the endoplasmic reticulum. Either before or after glycosylation, the enzyme may be directed toward intracellular degradation pathways rather than transported to the cell surface for ultimate export to the capillary endothelium. Controlled switching between these alternative pathways of degradation or further maturation evidently enables regulation of the levels of active lipoprotein lipase both in its intracellular and its extracellular forms. There is evidence that glucagon in muscle and epinephrine in adipose tissue, both acting via adenylate cyclase and raised intracellular concentrations of cyclic AMP, act as controllers of this switching. In muscle the partition of enzyme molecules is biased in favor of secretion, whereas in adipocytes, cyclic AMP mediates an enhancement in the rate of intracellular degradation. In both tissues, therefore, the effects of the cyclic AMP–mediated hormones are to antagonize those of insulin.

TABLE I Nutritional Regulation of Lipoprotein Lipase in Adipose and Mammary Tissue[a]

		Lipoprotein lipase activity (mU/g dry wt. of preparation)	
Tissue	Nutritional status	Intracellular	Extracellular[b]
Adipose	Fed	281	666
Adipose	Starved	214	49
Mammary	Fed	149	143
Mammary	Starved[c]	110	233

[a] Data from rat cell and tissue preparations.
[b] Includes ecto-cellular activity (that located on the outer surface of the cells considered).
[c] Mammary tissue responds analogously to muscle (see text), increasing its lipid-using capacity during starvation.

V. Facilitated Diffusion of Fatty Acids

Fatty acid binding proteins (FABPs) of similar molecular weight (about 14–15 kDa) have been known to exist in the cytosol of several types of mammalian (including human) cells for more than two decades. Three structurally distinct types of this protein have been distinguished, typified by liver, heart, and intestinal FABP. The primary structures of all three have been determined, revealing sufficient homology to support the proposal that they are the products of genes that have divergently evolved from a single ancestral gene. The precise function of these proteins remains to be elucidated: It has been suggested that they serve to protect the intracellular environment against the detergent effects of free fatty acids and that they may be involved in directing fatty acids to different routes of intracellular metabolism. The cytosolic concentration of these FABPs is adaptive and correlates with the rate of fatty acid uptake in various tissues. There is clear evidence of nutritional regulation mediated by insulin and counterregulatory hormones acting via cyclic AMP. All this circumstantially points to a role for FABP in the facilitation of fatty acid uptake. However, no information is currently available as to the molecular mechanisms of these actions, beyond the observation that FABP is not itself a physiological target for cyclic AMP–dependent protein kinase.

A FABP of molecular weight 40 kDa and pI 9.1 located in the plasma membrane has been extensively characterized in the recent past. Cells now known to express this protein include adipocytes, liver cells (on their basolateral membranes), intestinal cells, and cardiac muscle cells of the rat. An analogous protein is likely to occur in human cells. Fatty acid transport into these cells proceeds via a saturable, high-affinity (K_m ~0.3 μM for unbound oleate) uptake process, which is inhibited by antibodies against the 40-kDa FABP. Differences exist, however, in the detailed mechanism of this uptake between cell types; these indicate that other proteins may collaborate with the 40-kDa FABP to effect the overall uptake process. In the liver cell, for instance, uptake is sodium-dependent, whereas no such dependence is apparent in adipocytes. The properties of the 40-kDa FABP indicate that it is a peripheral membrane protein rather than a transmembrane protein inserted through the phospholipid bilayer. This would be consistent with the suggestion that it is an ecto-cellular protein capable of binding extracellular fatty acids and acting as the fatty acid binding component common to a family of multimolecular fatty acid uptake systems, differing in detailed composition and properties from one type of cell to another.

Adipocytes have also been the subject of an independent research program on transport of fatty acids in the context of their release from these cells during episodes of lipolysis. These studies have revealed the existence of a facilitated diffusion system that is markedly responsive to both insulin and catecholamines. The latter caused a 5–10-fold stimulation of fatty acid transport, while insulin antagonized this effect. The use of analogues of cyclic AMP enabled the conclusion that the catecholamine effect was mediated, as anticipated, by the activation of cyclic AMP–dependent protein kinase in response to enhanced intracellular concentrations of cyclic AMP. At present the relation between this hormone-responsive carrier and the 40-kDa FABP system is not documented. However, it is clear that the requirements of, for instance, the muscle cell for enhanced fatty acid uptake during starvation will be no less significant to the well-being of the whole organism than those of the adipocyte for enhanced fatty acid output. These considerations invite the speculation that future research will reveal the operation of systems in a variety of lipid-using cells, which regulate the activity of fatty acid transport into these cells in response to the hormonal stimuli that transduce the messages of physiological demand into the signals of cellular biochemistry.

Bibliography

Clegg, R. A. (1988). Regulation of fatty acid uptake and synthesis in mammary and adipose tissues: Contrasting roles for cyclic AMP. *In* "Current Topics in Cellular Regulation," vol. 29. (B. L. Horecker and E. R. Stadtman, eds.), pp. 77–128. Academic Press, New York.

Devaux, P. F. (1988). Phospholipid flippases. *FEBS Lett.* **234,** 8–12.

Glatz, J. F. C., Van der Vusse, G. J., and Veerkamp, J. H. (1988). Fatty acid–binding proteins and their physiological significance. *News Physiol. Sci.* **3,** 41–43.

Hoverstad, T. (1986). Studies of short-chain fatty acid absorption in man. *Scand. J. Gastroenterol.* **21,** 257–260.

Potter, B. J., Sorrentino, D., and Berk, P. D. (1989). Mechanisms of cellular uptake of free fatty acids. *Annu. Rev. Nutr.* **9,** 253–270.

Scow, R. O., and Blanchette-Mackie, E. J. (1985). Why fatty acids flow in cell membranes. *Prog. Lipid Res.* **24,** 197–241.

Sleight, R. G. (1987). Intracellular lipid transport in eukaryotes. *Annu. Rev. Physiol.* **41,** 193–208.

Feeding Behavior

ANDREW J. HILL and JOHN E. BLUNDELL, *Leeds University*

Glossary

Behavior Those activities of an individual that can be observed and recorded by another individual, including verbal reports made about subjective conscious experiences

Bingeing Rapid consumption of a large amount of food, accompanied by feelings of loss of control, and normally followed by a purge (e.g., vomiting)

Dietary restraint Individual's motivation to limit intake (or diet) to maintain or lose weight

Energy intake Amount of food consumed, expressed in terms of its potential supply of energy to the body to maintain physiological processes

Hunger Urge to eat, often accompanied by characteristic bodily sensations, normally indicating that eating is imminent

Nutrient selection Intake of basic nutrients such as protein, carbohydrate, and fat (macronutrients) and vitamins and minerals (micronutrients) from the choice of foods available

Satiety Inhibition over food consumption resulting from the consequences of ingestion

FEEDING behavior is a frequent and familiar human activity. Eating reflects a variety of physiological processes and is expressed within a psychological, social, and cultural environment. Human feeding behavior embraces the total amount of energy consumed and the distribution of intake over time, or the pattern of eating episodes such as meals and snacks. Human feeding behavior also involves food choices and selection from the products available. A variety of eating practices and habits exists which could be regarded as normal within their own context. In Western society abnormal eating (e.g., in bulimia nervosa or anorexia nervosa) has life-threatening medical consequences. One important area for study, therefore, is the distinction between normal and abnormal eating. In addition, the methods developed to assess the characteristics of human feeding behavior are an important part of this area. The detailed analysis of feeding behavior can be used as a tool to investigate abnormal conditions, as well as to disclose the operation of natural mechanisms. As the measurement of feeding behavior includes monitoring those events which lead up to, and follow, the actual intake of food, a variety of techniques have been developed for the use of the researcher. These range from the observation of the specific behaviors engaged in while eating to the description of long-term food consumption patterns.

I. Normal and Abnormal Eating

A. Overview

This article examines the scientific approach to the understanding of human feeding behavior as a motivated behavior, rather than describing the cultural patterns of cooking and eating that constitute the phenomenon we call cuisine. For humans eating is much more than simply the provision of fuel and nutrients in adequate quantities for our bodily processes. Eating assumes a place of great importance in our lives. It is typically a social activity and can form the core of a major celebration or cultural ritual. We usually spend 60 minutes or more per day engaged in eating or drinking, and a great deal of pleasure is derived from our contact with food. However, in considerations of human eating behavior, more attention is normally given to what is eaten than to the behavior itself, its antecedents, its consequences, and its expression.

In many areas of research, the assessment of behavior is reduced to a single number. For both theoretical and experimental convenience researchers often settle for a coarse level of measurement and end up measuring the consequences of behavior rather than the behavior itself. In feeding research this usually involves measuring the amount of food consumed, by weight or energy value. These variables, however, are not themselves behavior—they are the results of behavior. Naturally, for a complete understanding the consequences of behavior and the behavior itself both must be analyzed and interpreted. Consequences cannot properly be understood if we are ignorant of the behavior itself. [*See* BEHAVIOR MEASUREMENT IN PSYCHOBIOLOGICAL RESEARCH.]

Behavior has a form and a structure. It is composed of elements or acts arranged in particular sequences. The analysis of this structure can take place at various levels, ranging from the molecular to the molar. It is important to recognize that the structure of behavior is not arbitrary, but functional. In other words, behavior is expressed in such a way as to achieve certain objectives. For all organisms (i.e., animals and humans) behavior is central to the adaptive capacity of the organism. Most frequently, the function of behavior is to knit together physiological and environmental demands. Accordingly, the structure of behavior reflects the operation of these physiological and environmental influences and their interactions.

The concept of behavioral structure implies the existence of changes with time (i.e., temporal changes). Thus, structure cannot be assessed by a single instantaneous measurement, but only by successive or continuous measurements over time. The temporal dimension reveals the complexity and power of behavioral analysis. The sequence of elements which make up a behavior can change, and the rates of expression of these elements can also change. It follows that the way behavior is shaped in bringing about a change in intake is as important as the change itself. The behavioral adjustment can result from a physiological disturbance, a nutritional challenge, or a change in environmental demands. In each case the analysis of behavior can be used as an aid in diagnosing underlying causes. This principle is particularly relevant in clinical research, in which abnormalities in the physiological or psychological (i.e., environmental–social) domains are reflected in the altered structure of behavior as well as in the outcome of behavior.

The idea of using behavioral structure as a tool to disclose the operation of natural mechanisms, as well as to diagnose pathological conditions (or to assess the capacity of a pathological system), places demands on methodology. It is important that techniques and procedures can produce a valid representation of behavioral structure. It is for this reason that the approaches to the measurement of human eating behavior are of importance in their own right.

B. Expression

Human feeding behavior is a pattern of actions expressed to ensure our adequate supply of energy and specific nutrients for survival and adequate functioning. Feeding behavior can be characterized as a series of large (i.e., meals) and small (i.e., snacks) eating episodes, which leads to the ingestion of a certain amount of food (i.e., energy).

Feeding behavior also usually involves a selection of foods from an available range; this is often required in order to maintain an appropriate intake of essential macro- and micronutrients. This selection of foods can be broad or relatively restricted, depending on the adequacy of the supply and on individual food preferences. Given the familiar circumstance of a broad range of foods to choose from which are also affordable, what factors influence how people choose what to eat? A range of determinants arising from within the individual has been identified as potent forces guiding food selection. These include acquired preferences (acquired through learning mechanisms), conditioned aversions (i.e., powerful habitual rejections of particular foods), palatability (often assessed through an individual's preference for a food), and cravings (i.e., special preferences or yearnings for food). The mechanisms underlying these determinants of food choice are varied, and these factors are themselves worthy of more consideration than can be given here. [*See* ATTITUDES AS DETERMINANTS OF FOOD CONSUMPTION; FOOD ACCEPTANCE: SENSORY, PHYSIOLOGICAL, AND SOCIAL INFLUENCES.]

C. Normal Eating

What is meant by "normal" when we consider a behavior such as eating? Despite the universality of eating in the human behavioral repertoire, eating behavior is subject to major cultural variations in its expression. The most overwhelming cultural determinant is the adequacy of food supply. The developed and under-developed worlds differ in the quantity and diversity of foods available to their inhabitants. Symptoms of chronic malnutrition might reflect normal eating by many people, but only as the result of a desperately limited food supply. [*See* MALNUTRITION.]

Another, more subtle, cultural determinant of eating behavior is cuisine. The cuisine of a given culture not only refers to the style of food preparation, but places constraints on the acceptability of particular foods (e.g., pork, raw foods, insects, and spices) for its indiginates and on its mode of presentation and consumption. The concept of normal eating might therefore be of questionable validity as it embraces such a diversity of attitudes, practices, and provisions.

The developed, or Western, world is marked by its abundancy of food supply (apart from the most poor sections of society) and by the international nature of its cuisine. The more than adequate supply and variety available mean that weight gain might be the norm more often than weight maintenance or loss. However, the need to limit the amount of food eaten is felt by many people in technologically developed societies. Some of these people can be obviously overweight, or, in comparing themselves with society's advertised ideal body image, they might feel overweight or dissatisfied with certain parts of their body.

Dieting, or limiting the amount of food ingested, is therefore a major factor influencing the normal eating behavior of such people. For example, evidence suggests that the majority of women aged 18–25 at some time diet in an attempt to reduce or at least maintain (rather than increase) their body weight. These dieters do this by eliminating snacks or even meals (often breakfast) from their behavioral repertoire. In addition, they choose particular types of foods which are naturally low-energy foods or manufactured low-energy equivalents of normal foods and avoid other, often high-fat and highly desired, foods. Dieting, however, is not a naturally successful strategy even for weight maintenance. Episodes of food avoidance, which are seen as the most efficient way of losing weight and can last over days, rather than hours, are interspersed with periods of "overeating" to an extent which invalidates the entire process of abstinence. [*See* FOOD-CHOICE AND EATING-HABIT STRATEGIES OF DIETERS.]

The wish to restrict food intake or to diet has been called dietary restraint. This restraint is regarded as a personal disposition or trait typical of a certain type of person. Research on restraint shows that being motivated to diet confers a particular pattern on behavior. Thus, the dieter is characterized by a susceptibility to certain specific events which upset the imposed inhibition over eating. This is called disinhibition (i.e., inhibition of inhibition). The variety of these disinhibitors provides an insight into the reasons for the general lack of success of self-motivated dieting. Such disinhibitors can include being "tricked" into eating something, seeing someone overeating, drinking alcohol, and feeling anxious or depressed, among others. Clearly, while this can represent a frequently encountered form of eating behavior of many individuals, it is not necessarily a desirable pattern of behavior.

D. Abnormal Eating

Aberrations of human feeding behavior include overconsumption (i.e., hyperphagia), underconsumption (i.e., aphagia or anorexia), the altered intake of nutrients, and inappropriate food choices. Attention is paid briefly here to circumstances in which human feeding behavior can be considered to be abnormal in clinical terms. Strictly defined, obesity is a disorder of weight. Whether or not obesity is a disorder of feeding behavior, however, is a debated issue. One important point of consensus is that excess weight derives from an imbalance between energy input and energy expenditure. Obesity can result, therefore, from a surplus of incoming energy or a deficit in energy expenditure, or both. It is clear that, if sustained over a long period, a fairly small imbalance can lead to substantial fat deposition. [*See* EATING DISORDERS, OBESITY.]

Unfortunately, even if this is totally a product of excessive intake, it can be difficult to detect by monitoring food intake or eating behavior. Several investigations have followed the suggestion that the obese have defects in their perception of hunger or satiety, or that they display a characteristic obese eating style which lends itself to overconsumption. In general, the latter possibility has been hard to sustain. While they do not display a unique eating style, the obese do display some deficiencies in their expression of hunger and satiety. In addition, the obese seem to be especially sensitive to the palatability of food. However, the extent to which other variables (e.g., dieting) influence the behavior of the obese is unknown, as those mildly or moderately obese people who might frequently diet are rarely distinguished from nondieters or severely overweight individuals.

While some obese people claim to be compulsive eaters, the disturbance at a behavioral level in bulimia nervosa is part of the clinical definition. Episodic food binges, in which eating is typically out of control, are followed by some form of purgation, most commonly by vomiting or laxative abuse. These binges, which are often interspersed with periods of near-starvation, are fundamental, but not exclusive, diagnostic criteria for bulimia. Most bulimics are women who might binge several times a day. A week's ration might be consumed in a single sitting, followed by purging of what has been eaten in order to prevent any gain in weight. Indeed, bulimics are usually of normal body weight. Bingeing could become the dominant feature of the life style of a bulimic individual. Frequently, bingeing is a private activity, often going undetected by close family members.

Food and eating are also the dominant forces in the life of someone with anorexia nervosa, but they are manifest in a very different way from those of the bulimic. The restrictor form of anorexia nervosa is superficially the opposite of bulimia, as the anorexic imposes on him- or herself a near-starvation regimen. A lettuce leaf dressed with vinegar might constitute a meal, yet this does not imply a lack of concern with food and could be associated with considerable time spent cooking for other people. The sense of self-achievement and mastery derived from resisting hunger can reward this abstinence. Some of the physiological changes which take place in anorexia actually strengthen the resistance over eating, and this downward spiral can be difficult to break. Outwardly, the anorexic appears emaciated, a reflection of the much-reduced caloric intake. Also recognized now is the clinical picture of the anorexic who periodically binges. Here, the behavior incorporates both types of strategies, with long periods of starvation punctuated by episodes of bingeing and purging.

A substantially less life-threatening, but still abnormal, category of eating behavior, is pica. Pica is the consumption of a substance not normally regarded as appropriate for human consumption. Thus, it is not the *behavior* which is abnormal so much as the *target* of the behavior. It has been suggested that the consumption of substances such as ice, mineral clay, bone, washing starch, or lead can actually have some adaptive value. When there is a high nutritional need (e.g., during pregnancy, childhood, or malnourishment), people might derive some nutritive benefit from these apparently nonnutritive sources. This notion is attractive, although somewhat speculative. [*See* DIETARY CRAVINGS AND AVERSIONS DURING PREGNANCY.]

II. Determinants of Human Feeding

A. Biological Determinants

Eating is most often viewed as the behavioral expression of a biological demand for nutrition. Eating is part of the homeostatic process regulating our internal bodily state. Hunger is simply the subjective manifestation of the body's need for food. Therefore, a simple sequence is achieved: A physiological demand is experienced through hunger and is met by eating food. The elements that make up this closed system are varied, operate at different levels of physiological functioning, and interplay in a complex manner.

The basic physiological processes that operate when food is ingested and that serve to digest the material and make its constituent nutrients available to the body are well understood. Various physiological routines are involved in fuel utilization, and several routes are used to provide feedback about the state of the system. Physiological signals inform the brain so that behavior (i.e., eating or not eating) can be adjusted accordingly. The brain is the central integrator of information concerning the availability of glucose and other nutrients, stomach distension, and other signals. It is on the basis of this information that the subjective motivation to eat is believed to be activated. [*See* BRAIN.]

B. Biopsychological System

To assume that human feeding behavior is solely under the control of physiological processes ignores the power of psychological and environmental determinants of behavior. As already implied, environmental forces can shape and limit the expression of eating behavior. These ideas suggest that human feeding behavior is not the product of one particular mechanism, but is the output of a system—in this case a biopsychological system. The eating repertoire cannot be said to be the response of a single factor, but is the expression of a number of interacting factors. These factors exist in the biological or environmental domains. Moreover, the system has an integrity and acts in order to fulfill certain functions (e.g., the adequate supply of energy and nutrients or the preservation of the organism). A simplified systems view of feeding behavior is shown in Fig. 1. This conceptualization draws attention to the interrelationship between particular spheres of interest, including the external environment (cultural and physical), the behavioral act of eating (quantitative and qualitative components), the processes of ingestion and assimilation of foods, the storage and utilization of energy, and brain mechanisms implicated in the control system and mediating subjective states such as attributions and cognitions. Human feeding behavior therefore reflects an interaction between physiological events and environmental circumstances.

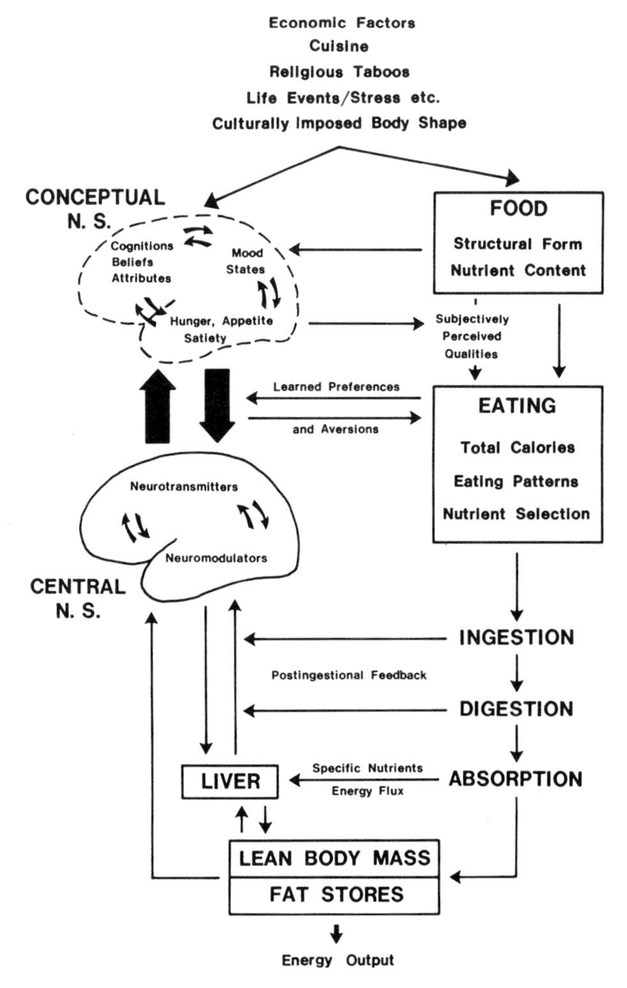

FIGURE 1 A conceptualization of certain significant aspects of the biopsychological system underlying the control of human feeding behavior. N. S., Nervous system.

The strength of a systems approach lies in the capacity to demonstrate links among different levels of the system. Central to this view is the proposal that alterations in any one domain provoke changes elsewhere. One example of this principle is dieting, and the change that ensues is unwanted and counterproductive. The strategy of reducing food intake does not simply deny energy to the body, but also leads to an adjustment in metabolism. Caloric restriction leads to a depression of metabolic rate, the change being severe when there is a marked and abrupt reduction in intake. This lowering of the metabolic rate is generally seen as an adaptive response to food restriction or a form of biological defense in the face of scarce commodities. The system is therefore integrative, pulling together biological events and psychological events, and demonstrates the intimate relationship in the interactions between these domains.

III. Approaches to Measurement

A. Overview

As human feeding behavior has a definable structure, its description and assessment can be carried out at different levels of analysis. The scientific study of human feeding behavior demands that detailed investigation requires the analysis of microevents in the behavioral repertoire as well as the study of large-scale long-term trends in consumption. Naturally, different types of procedures are required for these distinctive tasks.

The principle division is between direct and indirect techniques. The direct measurement of behavioral events is possible over short intervals, while long-term assessment usually depends on the indirect recording of food intake. Another distinction is between monitoring in a free natural environment, with attendant problems of control and precision, and measuring in artificial laboratory or clinic situations, in which accuracy is clearly easier to achieve. The particular limitations of data collected under these different circumstances present researchers with the dilemma of assigning importance to either the relatively inaccurate recording under natural circumstances or the very accurate recording in unnatural situations.

B. Dietary Studies of Individuals

Many of the methods for describing individual and group dietary practices are long established and have been extensively described and reviewed. Generally, individual dietary studies fall into two categories.

In the first category subjects are asked to record their present intake (e.g., by maintaining detailed records of all food eaten). These might be weighed records, in which all of the ingredients used in the preparation of food are precisely weighed, together with wastage at the end of the meal. Alternatively, they might be diaries of foods eaten, in which the quantities are described in terms of household measurements or estimation. The major drawback with these methods is a consequence of self-monitoring. It is possible, especially in certain groups, that constantly drawing attention to the process of selection and consumption of food leads some people to change their eating behavior.

The second category groups techniques used to aid the subject in remembering and describing their past intakes. The most common procedure elicits an inventory of the food eaten over the previous 24 hours. This is done either by asking subjects to note the foods on a checklist or by detailing the meals together with estimated amounts of food. This recall can be extended over a few days or longer.

Generally, recall techniques are quick and inexpensive and do not require specialist supervision. Their cost-effectiveness, however, might vary according to the required accuracy of the study. Many more 24-hour records are needed to accurately establish protein or fat intake than to describe energy intake.

Records of food intake, collected prospectively or retrospectively, are valuable sources of information concerning a person's eating behavior. Not only do they provide 24-hour summaries of nutrient intake, but they could provide a topography of eating during this period. Intake can be broken down into discrete episodes of eating, with average meal sizes and frequencies computed and distributions of snacks and meals plotted. However, it should also be recognized that the accuracy of data derived from food intake records is potentially compromised by the tables of food composition used to convert food records into quantities of nutrients, which represent the average composition of a particular food item.

C. Observational Studies

One apparently uncomplicated procedure for describing eating behavior is to observe and subsequently classify the entire sequence of behavior. In principle, because behavior is recorded in its totality, this strategy should be a powerful tool for describing human feeding.

Behavior can be monitored in naturalistic settings (e.g., cafes, restaurants, dining halls, or homes) or in the laboratory under controlled conditions. Analysis can be from live behavior or carried out from video recordings. Often, the observational method demands some form of sampling. Event sampling requires that every occurrence of a specified event during the course of the observational period is recorded. Time sampling means that whatever event is occurring at specified (brief) intervals during the observational period is monitored. For eating behavior, which normally spans a relatively short period (usually a meal), a number of significant events (e.g., taking a bite, pausing, and swallowing) can be continuously recorded by an observer. It is necessary to check the accuracy of the ratings of this observer by comparing them with an independent observer rating the same sequence (i.e., the coefficient of concordance). This is essential since even events such as taking a bite of food, which can be defined fairly unambiguously, can be recorded differently by two independent observers. This is particularly important when the start and end of an event are recorded as well as the overall frequency of events. Used carefully, observational procedures can be sensitive research devices, despite the intensive nature of the data collection and analysis.

D. Specialized Apparatus

During the last 20 years a number of specialized devices have been developed or adapted to improve the sensitivity, accuracy, or reliability of measuring food consumption. Most provide continuous monitoring of intake. Some are designed for liquid rather than solid food, and others allow a degree of food choice. Some demand a somewhat unnatural eating response, while others attempt to allow unhindered eating to take place. No device is perfect; they all have strengths and weaknesses.

Several researchers have devised forms of liquid food reservoirs, in which liquid food is either sucked or pumped at a steady rate during the depression of a button. The technique has been used in a variety of circumstances, including examining the consequences of oral and intragastric feeding on subsequent intake, the eating behavior of normal and obese subjects, and the effects of drugs on voluntary intake. This approach provides an objective method for studying certain parameters of ingestion. However, the dependence on liquid food, with the consequent restriction on variety of taste and texture, obviously limits the extent to which results can be generalized to more natural eating circumstances and situations. In summary, the technique scores high for internal validity, but somewhat lower for external validity.

A very different device is based on the recognition that human eating is composed of a sequence of contacts between the mouth and the eating utensil, and that the number of contacts is proportional to the amount consumed. This system of monitoring eating via eating utensils operates through specially constructed spoons and forks with handles that contain miniaturized telemetering equipment. Each contact between utensil and mouth creates a signal that is transmitted to a recording device. Studies with a prototype have shown the utensils to provide good records of two parameters: the number of bites (or spoonfuls) and the interbite interval. At the moment the full potential of this system is unevaluated, but it is clearly an unusual and original approach to measurement.

An alternative technique shifts attention away from the structure of eating behavior to the actual food being eaten. Thus, monitoring eating via the plate demands the accurate measurement of alterations in the weight of the food being eaten. This is achieved by the continuous weighing of the individual's plate with a concealed electronic balance on which the plate rests. The device is called the universal eating monitor, and it can be used with either solid food on plates or liquid foods (e.g., soups) in dishes. Periodic readings of the weight of the plate (e.g., every 3 seconds) are made by an on-line computer during the meal. From these readings a cumulative intake curve (i.e., the weight of the food removed from the plate over time) can be plotted. This curve is the major parameter of eating provided by this technique. Several investigations using this device have demonstrated its usefulness and sensitivity. The accurate readout of adjustments in the weight of the food consumed is similar to that obtained with the reservoir method. The great advantage of this technique, however, is that the food can be eaten normally from a plate instead of being sucked or pumped through a pipe.

Many of the techniques now used to study human eating are derived from strategies used to monitor feeding patterns in animals. The human equivalent of pellet dispensers and eatometers are food-dispensing machines. The solid-food dispenser is basically a commercial food vending machine modified to provide small food units (e.g., quarter-sandwiches) with weight, nutrient content, and caloric value accurately controlled. The removal of each food unit from the dispenser can be monitored automatically, and a cumulative record can be obtained of the individual's behavior (i.e., feeding profile) and the weight of the food consumed (i.e., energy intake). This device has the advantage that it uses common solid foods that are likely to be regularly consumed and does not put unusual demands on the eater. The resolving power of the device—its capacity to detect subtle or small adjustments in intake—is obviously restricted by the size of the individual food units. In its favor, however, is its capacity to monitor food selection. By stocking the machine with items varying in nutritional composition, for example, it becomes possible to measure individual preference for particular nutrients or tastes.

Returning once more to the behavior of the eater, one further approach has been to automate the recording of certain behavioral events which are available normally only through careful observation. Monitoring chewing and swallowing via a device called an edogram permits an objective insight into the microstructure of human eating. Swallowing is measured by changes in pressure in a small balloon resting on the Adam's apple and kept in place by an elastic collar. Chewing is simultaneously measured by a strain gauge that monitors jaw movements. Normally, a standardized test food is used with the apparatus (often a number of small open sandwiches). The output permits the calculation of microparameters such as the rate of chewing, duration of chewing between successive swallowing movements, and intrameal pauses without chewing. In addition, since the test food is composed of small consumable units of known weight, volume, and caloric value, the procedure allows continuous tracking of caloric intake. Consequently, the technique combines some of the best aspects of the utensil monitor and the plate monitor (i.e., the universal eating monitor).

E. Subjective Experience

The human capacity to communicate our internal experiences provides a further valuable source of information on eating behavior. Personal observations regarding the presence of certain bodily sensations or the intensity of hunger can be regarded as verbal behavior and, as such, are entirely available for quantification and measurement. As suggested earlier, subjective reports of hunger motivation are useful, as they might provide further insight into the processes guiding eating behavior. For this reason simultaneous recording of both subjective experience and overt eating behavior constitutes a powerful investigative strategy.

Subjective experiences concerning eating fall roughly into three categories: sensations, dispositions, and mood. Certain bodily sensations are well associated with times before or after eating. Commonplace experience associates gastric motility with hunger and gastric fullness with satiety. However, while sensations originating from the area of the stomach can be particularly prominent, they form only one component of a range of accompanying sensations. Discrete sensations such as lightheadedness, a bad taste in the mouth, dry throat, or a rumbling stomach, together with an overall feeling of bodily weakness, can occur if the individual has not eaten for several hours. At the same time the person might experience characteristic changes in affective state or mood (e.g., irritability and restlessness). Eating brings about a radical change in both sensation and affect, bodily sensations generally being more pleasant and accompanied by feelings of contentment and well-being.

The measurement strategies used to quantify these experiences are of varing degrees of complexity. The two most common methods are fixed-point scales and visual analogue scales. Fixed-point scales are quick and simple to use, and the data they provide are easy to analyze. However, the scales can vary greatly in complexity. In considering the appropriate number of points to be included in this type of scale, the freedom to make a wide range of possible responses (i.e., multipoint scales) must be balanced against the precision and reliability of the device (i.e., scales with few points). However, research seems to indicate that scales with an insufficient number of fixed points can be insensitive to certain changes in subjective experience. In addition, the fixed points themselves are important determinants of the way people use the scales and distribute their ratings.

One way of overcoming some of the failings of fixed-point scales is to abolish the points completely. Thus, visual analogue scales are horizontal lines (often 10 cm long), unbroken and unmarked except for word anchors at each end. The user of the scale is instructed to mark the line at a point that most accurately represents the intensity of the subjective feeling at that time. The researcher measures the distance to that mark in millimeters from one end, thus yielding a score of 0–100. By doing away with all of the verbal labels except the end definitions, visual analogue scales retain the advantages of fixed-point scales, while avoiding many of the problems with response distributions.

It should be recognized that scales asking people to rate their present level of hunger are very general questions. It is possible that individual subjects might use the term "hunger" differently from the manner intended by the experimenter. It might therefore be appropriate to present more than one scale which taps into different components of hunger motivation, such as "satiety," "desire to eat," "amount of food desired," and even "thirst." A highly specific measure of hunger motivation is to measure an individual's willingness to eat, or preference for, particular foods. A quantification of hunger motivation might be derived from a preference rating for a particular food or food category from the range of food preferred at that moment, the number of portions desired, or the number selected from a finite list.

F. Postscript—Changing Styles of Feeding Behavior

It is clear that certain cultures impose patterns on eating activities and modulate the expression of physiological mechanisms. Changes in cultural attitudes (e.g., about body shape) and in the nature of the food supply might also be expected to adjust the style of eating (i.e., feeding repertoire). In technologically advanced cultures (namely, Europe and North America) nutritional and catering developments are adjusting the pattern of overall consumption and the selection of nutrients. For example, instantly available (i.e., "fast") foods, often made palatable through taste and texture, can encourage an adjustment in the distribution of eating episodes. This could mean a shift from the meal–snack–meal pattern to a grazing style of ingestion, in which people ingest small amounts more or less continuously. If the attractive taste and texture are produced at the expense of macronutrient balance, then this grazing style can lead to an inappropriate intake of essential nutrients.

Human feeding behavior is not a rigid immutable form of human activity. It is adaptable and highly responsive to shifts in physiological, cultural, and nutritional conditions. It is therefore important to have reliable methods to monitor human feeding behavior and to be aware of alterations which could signal inappropriate physiological or cultural influences.

Bibliography

Barker, L. M. (1982). "The Psychobiology of Human Food Selection." AVI Publ., Westport, Connecticut.

Bellisle, F. (1979). Human feeding behaviour. *Neurosci. Biobehav. Rev.* **2,** 163.

de Castro, J. M., and Elmore, D. K. (1988). Subjective hunger relationships with meal patterns in the spontaneous feeding behaviour of humans: Evidence for a causal connection. *Physiol. Behav.* **43,** 159.

Polivy, J., and Herman, C. P. (1987). Diagnosis and treatment of normal eating. *J. Consult. Clin. Psychol.* **55,** 635.

Shepherd, R. (1989). "Handbook of the Psychophysiology of Human Eating." Wiley, Chichester, England.

Stunkard, A. J., and Stellar, E. (eds.) (1984). "Eating and Its Disorders." Raven, New York.

Walsh, B. T. (ed.) (1988). "Eating Behavior in Eating Disorders." Am. Psychiatr. Press, Washington, D.C.

Fertilization

W. RICHARD DUKELOW AND KYLE B. DUKELOW,
Michigan State University

Glossary

Acrosome Lysosome-like structure located on the tip of the sperm head, which contains enzymes that are released during the sperm penetration process

Acrosome reaction Phenomenon occurring on the sperm head whereby the sperm penetration enzymes are released

Capacitation Period of time of sperm incubation in the female reproductive tract (or under culture conditions) when sperm receptors are exposed as the sperm "achieves the capacity" to penetrate the egg

Egg Ovum or female gamete released from the follicle on the ovary, the sperm penetrate the egg to bring about fertilization

Embryo Product of sperm–egg fusion that develops into a new being, in the very early stages the embryo is called the zygote

Epididymis Strap-like organ attached to the testes in the male. The sperm mature and are stored in this organ until ejaculation

Fallopian tube Oviduct. A tube-like structure that allows transport of the egg from the ovary to the uterus, it consists of two parts, the ampulla (where fertilization occurs) and the isthmus proximal to the uterus distal

Fertilization General term that includes preparation of the sperm in both the male and female reproductive tracts, penetration of the egg by the sperm and approximation and disintegration of the pronuclear membranes within the egg to allow intermingling of the male and female chromosomal components

Spermatozoa Sperm, the male gamete that contains the male chromosomal material and carries this material to the female gamete or egg

Sperm plasma membrane Outermost membrane of the surface of the sperm, which plays an important role in capacitation, fusion, and release of the sperm penetrating enzymes

Zygote Fertilized egg or very early embryo

THE ROLE OF THE SPERM in creating a pregnancy was recognized from unique dog experimentation in the 1700s. By the middle of the nineteenth century the egg had been described and its role in the fertilization process recognized. In the late 1800s, using marine invertebrates (which are easy to study), scientists first observed the actual penetration of the egg by a sperm to create a zygote or potential embryo. Since then the process of fertilization has been studied extensively in a wide variety of animals. Because of the importance of reproduction and genetics to domestic animals, much of the early work in the reproductive field was carried out in these species. In the 1930s a great deal of interest was oriented to the fertilization process of our close relative, the monkey (nonhuman primate), and many pioneering efforts were made at that time.

An obvious aid to investigations of fertilization was the ability to observe fertilization and early embryonic development under in vitro ("culture dish") conditions. Studies on in vitro fertilization began in the 1890s but were not consistently successful until after the discovery of the capacitation phenomenon in 1951. The classic research of M. C. Chang resulted in the first production of mammals by *in vitro fertilization*. In 1969 the first successful *in vitro* fertilization of human eggs was reported by R. G. Edwards and in 1978 the first child was born by this same process. In the decade since that event over 15,000 children have been born by in vitro fertilization, and the technique has been applied to a large number of laboratory, domestic, and primate animals. These findings have allowed scientists to delve deeply into the biochemical and molecular aspects of fertilization so that today we have a sound understanding of the complicated process by which all animal life begins.

For the present description of the fertilization process we will first examine the development of the sperm in the male reproductive tract and the characteristics that the sperm acquire as they pass through the male and female reproductive tracts to reach the site of fertilization (the oviduct or fallopian tube) of the female.

I. Maturation of the Sperm in the Male Reproductive Tract

Sperm that are produced in the testes do not have the ability to move or to fertilize eggs. They pass, by fluid movement, into a strap-like organ called the *epididymis*. The epididymis, composed of convoluted tubules, has three distinct areas. The area closer to the testes is called the *head* (caput) of the epididymis. Distal to this is the *body* (corpus), and finally the *tail* (cauda) of the epididymis. It is within the epididymus that the sperm become vigorously motile and achieve the ability to fertilize. The maturation process varies in different species. Generally, however, the sperm first start making circular movements in the body or proximal tail portions of the epididymis. Sperm are then stored in the tail portion of the epididymis after they normally have achieved full motility and fertilizing ability. During the passage through the epididymis several biochemical events cause these changes. These changes are called *maturation* of the sperm; they involve mainly the sperm plasma membrane surrounding the acrosome and nucleus of the sperm. Some changes also occur in the nucleus or tail portion of the sperm, reflecting rearrangement of disulfide bonds. The epididymis is a very active organ, capable of absorbing and secreting fluids that alter the sperm. As the sperm pass through the epididymis they are sequentially exposed to a number of active macromolecules. As a result this passage changes the molecules that constitute the sperm membrane surface. The most evident changes are in the glycoproteins, i.e. proteins to which complex sugars are attached and that are brought about by the enzymes *galactosyltransferase* and *sialyltransferase* present in the epididymal fluid. The saccharide residues of the glycoproteins are altered, and the net negative surface charge of the sperm plasma membrane is changed. [*See* SPERM.]

Some membrane lipids also undergo change at this same time. The body of the epididymis has a high cholesterol synthesis rate, and cholesterol is one of the lipid molecules integrated into the sperm plasma membrane at this time. [*See* CHOLESTEROL.]

At ejaculation the sperm pass through the *vas deferens*, the *ejaculatory duct*, and then through the *urethra* for deposition into the female reproductive tract. In some species (for example, the Chinese hamster) the sperm are stored not only in the epididymis tail, but also in the vas deferens. During passage from the male reproductive tract the sperm are mixed with fluid secretions of the *seminal vesicles* (primarily fructose), the *prostate gland*, and the *bulbourethral* or *Cowper's gland*. These secretions neutralize the acidity of the urethra and provide nutrients and protection to the sperm until they reach the site of fertilization in the female. The combination of these fluids and the sperm represent the *semen* or ejaculate of the male.

II. Capacitation of the Sperm

In some species (rodents and the pig are examples) the semen is ejaculated directly into the uterus of the female. In most other species (including the human), the semen is deposited in the anterior portion of the vagina and then must traverse the *cervix* to reach the lumen of the *uterus* and thence to the *oviduct* or fallopian tube. For many years it was assumed that any delay from the time of sperm deposition until fertilization was caused by the transport of the sperm to the oviduct. Some classic experiments on the rabbit by Walter Heape in England at the turn of the century indicated that there was a 6-hour delay between sperm deposition and egg penetration. This delay was ascribed to sperm transport and this theory persisted in the literature for nearly 60 years. Subsequent research on cattle and guinea pigs showed that the time of transport was not measured in hours but rather 2.5–4 minutes! The discrepancy of these results remained unexplained until the discovery of the phenomenon that we now call *capacitation* by M. C. Chang in Massachusetts and C. R. Austin in Australia. This means that sperm, as they are produced by the male, are not capable of penetrating the eggs and in order to "achieve the capacity" (thence the name) they must first spend a period of residence within the female reproductive tract. It turns out that the time required for capacitation in the rabbit is 6 hours, in agreement with Walter Heape's finding.

The phenomenon of capacitation can be divided into two separate events. The first of these (for which the term capacitation is now used) appears to be unique to mammals, and occurs during the early passage of the sperm through the cervix, uterus, and oviduct. The second event is termed the *acrosome reaction,* and will be treated later.

Capacitation is a requirement of all mammals that have been studied, whether fertilization is natural (*in vivo*) or *in vitro*. Because the phenomenon is of short duration (3–4 hours in nonhuman primates and estimated at 7–11 hours in humans) it is difficult to study. Under natural conditions capacitation is normally achieved in the female reproductive tract, probably through initial incubation in the uterus and subsequent incubation in the isthmal portion of the oviduct, just prior to fertilization. Capacitation also occurs *in vitro*. This requirement gives rise to a delay between the time freshly ejaculated sperm are added to eggs and achievement of fertilization. Capacitation involves the biochemical activation of the sperm, probably through the production of cyclic nucleotides, which alter the metabolic processes of the sperm and allow the acrosome reaction to proceed. Indeed, addition of cyclic AMP to the fertilization medium reduces the time required for capacitation in both mice and nonhuman primates. Original descriptions of capacitation indicated that the process might involve the removal of a "coating" from the sperm plasma membrane, and modern molecular biological techniques suggest that this "coating" could represent removal of inhibitors from receptor sites on the sperm membrane, which would allow the acrosome reaction and attachment of the sperm to the egg to occur. Electron microscopic studies have failed to show distinct changes in the membrane surface due to the capacitation phenomenon alone, whereas the acrosome reaction is clearly evident later. A substance present in the seminal plasma of most species, termed *decapacitation factor* or *acrosome stabilizing factor* has the ability to reverse the capacitation phenomenon and render capacitated sperm "decapacitated." These sperm are unable to fertilize eggs until they have undergone a second capacitation period. This shows that capacitation is reversible.

The time and endocrine requirements for capacitation vary with different species. Much of the early work was done with the rabbit, which requires a fairly long (4–6 hour) period for capacitation. Four technical requirements must be fulfilled for studying capacitation: the ability to predict accurately the time of ovulation; a knowledge of the fertilizable life of the ovum; the ability to place sperm in the reproductive tract and recover them; and the ability to recover the egg for verification of fertilization. Ideally both an *in vitro* and *in vivo* fertilization system should be used to confirm capacitation. In most species one or more requirements cannot be fulfilled, which explains why the rabbit was used so extensively in the early days. Today it is possible to study other species as well. Among the studied species the required capacitation time appears to be: mouse, 1 hour; hamster, 3–4 hours; ferret, 3.5–11.5 hours; cat, 0.5–1 hour; sheep, 1.5 hours; pig, less than 2 hours; primates, 3–4 hours; and humans, 7–11 hours (estimated).

Capacitation can occur in the vagina, the uterus, or the oviduct, with different time requirements: for instance, 4–6 hours in the rabbit uterus, but 10 hours in the rabbit oviduct. *In vitro* capacitation occurs in high ionic strength media. There have been reports of the need for certain heat-stable, low-molecular weight substances present in follicular fluid and blood serum and very rich in the adrenal gland, but in general, research to identify specific capacitation factors has not been productive.

During natural fertilization *in vivo* the female must be under the influence of *estrogen* for capacitation to occur. In females in the follicular phase of the cycle capacitation will occur, but not in females receiving progesterone, or *pseudopregnant females*. Progesterone has an anticapacitation effect but only in the uterus, not in the oviduct. Some of the action of the "minipill" contraceptive agents may be against capacitation.

Many events occur in sperm during capacitation. As already indicated, adenyl cyclase and cAMP-dependent protein kinase play a role relative to the motility of capacitated sperm, which is markedly increased during capacitation. This effect is directly related to increased adenyl cyclase activity and increased *cAMP* availability. Activation of cAMP-dependent protein kinase appears to alter the sperm membranes through *phosphorylation of membrane proteins*. This may explain why capacitated sperm incorporate vital dyes (such as nigrosin-eosin live-dead stain) whereas freshly ejaculated sperm do not. Capacitated sperm also exhibit an increase in oxygen consumption and glycolytic activity after incubation in the female reproductive tract or in media containing tract fluids. [*See* PHOSPHORYLATION OF MICROTUBULE PROTEIN.]

There are also changes in intracellular ions during capacitation. Sperm normally have an ionic gradient across the sperm plasma membrane; the concentration of potassium is much higher inside than outside, and the reverse is true for sodium. This ionic gradient is maintained by an ATPase-mediated sodium/potassium ion exchange pump. In the guinea pig, there is a reduction in the intracellular potassium concentration after 2 hours of incubation when the sperm are ready for the acrosome reaction. There appears to be little change in the concentration of intracellular free calcium in rabbit sperm as a result of in vitro capacitation, in contrast to the acrosome reaction that involves a massive influx of extracellular calcium.

During capacitation a number of proteins are either attached to or released from the whole sperm surface. They are identified by immunological techniques, but their precise nature and their involvement in the capacitation process are unknown. Cholesterol in the sperm membrane may play an important role in capacitation. This substance has profound effects on membranes and affects active and passive ion permeability by regulating orientation, fluidity, and thickness of the membrane lipids. It has been postulated that during in vivo capacitation there is a gradual removal of cholesterol by albumin or some other component of the female genital tract, but firm evidence is lacking.

Many workers have attempted to capacitate sperm *in vitro* with various combinations of uterine, oviductal, or follicular fluid. At first, when *in vitro* fertilization of the human egg was announced (using freshly ejaculated sperm) many deduced that capacitation is not required for human sperm. In all cases, however, there is a delay from the time of sperm and egg mixing until sperm penetration occurs and this reflects the capacitation phenomenon. Capacitation could be attributed to the small amounts of follicular fluid with high concentrations of steroids that accompany the restricted eggs (with the associated cumulus oophorous cells), although many carefully defined media have now been used to culture sperm and eggs for *in vitro* fertilization. Nevertheless a number of interesting enzyme activities are present in the female reproductive tract at the time of capacitation, which may play a role.

A number of clinical applications of the capacitation process are important in humans. Human sperm maintain their fertility for at least 72 hours and perhaps longer, but the fertilizable life span of the human egg is shorter, probably from 6 to 12 hours. Under perfect conditions fertilization should therefore occur when sperm have just reached the capacitated state and when the egg has just ovulated. Such a timing should yield the highest conception rate with the lowest possibility for polyspermy, fertilization with aged gametes, or other factors that might yield congenital birth defects. The application of this principle to time coitus for low fertility couples or for intrauterine insemination suggests that sperm deposition should occur from 7 to 11 hours before the expected time of ovulation. Because there is suggestive evidence that estrogens have a beneficial effect on the fertilizable life of sperm and they enhance capacitation, such therapy could be useful.

The fact that once sperm have been capacitated their fertilizable life span in the oviduct is shortened is of interest from the standpoint of fertility enhancement. It would be of obvious value to provide a continuing supply of capacitated sperm at the site of fertilization in patients in whom the exact time of ovulation is unknown. Multiple insemination would thus be indicated. From the standpoint of contraception a naturally occurring antifertility factor in the oviduct could be developed artificially to block fertilization.

Sperm can be capacitated *in vitro* by removal of sperm from seminal plasma and incubation in a culture medium. Placement of the *in vitro* capacitated sperm in the reproductive tract to achieve conception has application for the treatment of some cases of infertility. Such manipulation is used where there is immunologically mediated sperm inactivation or when the uterus is incapable of capacitating sperm. The uterus and oviduct can be bypassed to bring about fertilization. Of course there is still the problem of implantation in a hostile uterine environment. Two treatment programs are in current use. The first, *intrauterine insemination,* is performed in women having hyperstimulated cycles. These women are closely monitored by ultrasound to determine the state of the ovarian follicles, and their estradiol serum levels are measured. Washed, incubated sperm are inserted into the uterine cavity when the egg is in the oviduct. Multiple inseminations with washed sperm significantly increase the pregnancy rate. In a second technique, *gamete intrafallopian tube transfer,* washed incubated sperm (capacitated) are placed together with the egg in the oviduct. With experimental studies this has been termed *xenogenous fertilization* since it often uses a foreign species for the fertilization. Finally, as discussed elsewhere, *in vitro* fertilization uses washed sperm to successfully fertilize human oocytes.

The use of antiestrogens and progestation agents has strong implications for capacitation. The antiovulatory properties of progesterone are well known. Lower levels (100–500 μg) of progestins prevent pregnancy but in many cases ovulation does occur. Three mechanisms have been postulated to account for the antipregnancy effect: an anticapacitation effect; rendering the cervical mucus spermicidal; and alteration of the mechanism of gamete transport through the reproductive tract. Evidence has been presented to substantiate all three effects and probably all act in concert to inhibit fertility.

III. The Acrosome Reaction and Activation of the Acrosomal Enzymes

Before discussing the *acrosome reaction* it is important to identify the specific organelles involved. The head of the sperm is composed of nuclear material and contains the genetic material necessary for the successful propagation of the species. Located on the tip of the sperm is a sac-like structure similar to a lysosome or to the zymogen granule of the pancreatic cells. This structure has been termed a "bag of enzymes" and has an *inner acrosomal membrane* next to the sperm head and an *outer acrosomal membrane* approximating the sperm plasma membrane that surrounds the complete sperm head. After capacitation and probably after attachment to the *zona pellucida* of the egg, the sperm undergoes the acrosome reaction. This is a vesiculation reaction between the outer acrosomal membrane and the plasma membrane of the sperm. This event results in the formation of pores on the surface of the sperm, which allow the release of the enzymatic contents of the acrosome. A number of enzymes have been isolated from the acrosome but the two that have received the most attention are *hyaluronidase* and *acrosin.* About one-half of the hyaluronidase is bound to the inner acrosomal membrane and the other half is free within the acrosome itself. The acrosome reaction occurs in a short period of time (from 2 to 15 minutes) and consists of an initial loss of free hyaluronidase and then a gradual and sequential loss or activation of enzymes bound to the inner acrosomal membrane. These events occur during the time that the sperm penetrates the outer vestments of the egg. At the present there is some confusion about whether the acrosomal reaction occurs after contact with the zona pellucida or during passage through the cumulus oophorous and corona radiata layers surrounding the egg. Most likely the former is the case but there may be species variations.

At this time we must briefly review the morphology of the egg as it is released from the follicle on the ovary. The egg cytoplasm is surrounded by the *egg plasma membrane* and has a nucleus containing the genetic material. The egg plasma membrane is surrounded by three different structures or layers. The closest to the plasma membrane is the *zona pellucida.* This glycoprotein layer serves to protect the egg and, in most species, incorporates a *block to polyspermy* in that once it is penetrated by a single sperm, other sperm are prevented from gaining entry. This mechanism prevents the incorporation of more than a single set of chromosomes and therefore the production of polyploid (multiple sets of chromosomes) individuals that would not survive. Around the zona pellucida is a small layer of cuboidal cells called the *corona radiata* and around this a layer of larger cells, more loosely packed and held together by a hyaluronic acid matrix, called the *cumulus oophorous.* For a sperm to penetrate to the cytoplasm of the egg it must first pass through these layers. In some species (such as the sheep, cow, monotremes, and marsupials) the cumulus oophorous layer is shed shortly after ovulation, commonly by changes in the bicarbonate content of the oviductal fluid. When this occurs, the corona radiata and the zona pellucida represent the only egg vestments that the sperm must pass before reaching the egg plasma membrane. In other species, including the human, the cumulus cells and the less conspicuous corona radiata must be penetrated before the zona pellucida is reached. This is accomplished through release of free hyaluronidase and that bound to the inner acrosomal membrane of the sperm. The cumulus cells from eggs incubated *in vitro* are readily dispersed by a hyaluronidase solution. Hyaluronidase is also released by dying sperm and this release must be differentiated from the "true" acrosome reaction. In some species, such as the guinea pig and the musk shrew, it is very easy to visualize the acrosome reaction. In other species—the golden and Chinese hamster—the acrosomes are smaller and phase contrast microscopy is required. The acrosomes of the rabbit, mouse, and human are thin and the acrosome reaction is more difficult to detect. The techniques used for studying the acrosome reaction in these species may result in death of the sperm; they include the acridine orange–UV method, the triple stain technique, the p-lectin method, the chlortetracycline-UV method and the naphthyl yellow/erythrosine B method.

There are species variations in the site from which the acrosome reaction initiates. In the rabbit it appears to be on the tip of the acrosome, whereas in the golden hamster, ram, and human, there is closer attachment at the equatorial segment of the acrosome cap region. In all cases, however, the release of enzymatic contents and the process of the acrosome reaction remain basically the same. Reference has already been made to the role of the zona pellucida and the cumulus oophorous in inducing the acrosome reaction. There is little doubt that the zona pellucida does have this ability in most species. The mouse has three zona pellucida proteins, one of which, ZP3, binds to the sperm plasma membrane over the acrosomal cap. The polypeptide chain in ZP3 molecules seems to serve as an initiator of the acrosome reaction. Whether the cumulus oophorous can induce the acrosome reaction is unknown.

The acrosome reaction may occur by two parallel pathways. One results independently of calcium ion concentration when the sperm plasma membrane interacts with the egg plasma membrane. There is then an increase in intracellular pH by sodium ion influx and hydrogen ion efflux. The second pathway is calcium dependent and represents membrane depolarization. Both of these methods open calcium channels, resulting in a large influx of extracellular calcium. The calcium goes through the plasma membrane and induces the fusion between the plasma membrane of the sperm and the outer acrosomal membrane resulting in an exocytosis (release) of acrosomal contents. A large number of molecules have been implicated in the sperm acrosome reactions including ions, sperm membrane proteins (including enzymes), cyclic nucleotides, calmodulin, and the acrosomal enzymes themselves.

Reference has already been made to the *hyperactivation* of sperm motility following capacitation. This hyperactivation plays an important role not only in the transport of the sperm over the short distance from the distal region of the isthmus to the ampulla, but also in the penetration of the outer vestments of the egg. The biochemical constituents of the medium will influence the hyperactivation, and the calcium uptake is essential for initiation of sperm hyperactivity in the guinea pig, hamster, and mouse. Bicarbonate seems to be necessary for hyperactivation of hamster sperm, and potassium, energy substrates, and albumins are all known to control hyperactivation in a number of other species. There is substantial evidence that the level of intracellular cAMP controls hyperactivation. The exact molecular mechanism is not known but obviously there are molecular activities that allow sperm to change from the rigid stiffness of tail-beating characterized by precapacitation sperm to the more active state. Intake of calcium may be a reflection of methylation of membrane phospholipids in the sperm tail membrane, which enhances the entrance of calcium into the sperm and may stimulate membrane-bound adenylate cyclase, thus increasing the intracellular concentration of cAMP.

IV. Sperm Penetration of the Egg

Reference has already been made to the role of hyaluronidase from the acrosome or perhaps also from the surface membrane of the sperm in penetration of the cumulus oophorous layer of the egg. There appears to be varying need for the hyaluronidase in various species. Thus in cattle, where the cumulus cells are normally dispersed within hours of ovulation, the role of hyaluronidase can be less than in other species. Originally it was believed that masses of sperm reached the oviduct and swam in heavy swarms about the egg. While this may be true for in vitro fertilization, in natural fertilization very few sperm actually reach the site of fertilization. Thus hyaluronidase is probably not released in a mass to disperse the cumulus oophorous cells but rather acts on the head of an individual sperm to ease penetration. Little is known of penetration through the corona radiata and many workers assume that penetration is by hyaluronidase. There is evidence, however, that a distinct *corona penetrating enzyme*

(CPE) exists, and is attached to the inner acrosomal membrane. This esterase-type enzyme facilitates penetration of the corona radiata.

The zona pellucida has a number of very important functions. One of these has already been mentioned, the *block to polyspermy* (sometimes called the *zona reaction*), which occurs soon after penetration by a single sperm, and is caused by a release of material from the cortical granules near the surface of the egg plasma membrane. The release modifies the composition of the zona pellucida. In the mouse this reaction is caused by the partial hydrolysis of the two zona pellucida proteins, ZP2 and ZP3. A second function of the zona pellucida is to prevent fertilization of eggs with sperm from a different species. This rarely occurs unless the zona pellucida is removed artificially from the eggs allowing sperm penetration. These findings suggest that through proteins on its head the sperm binds to species-specific receptor sites located in the zona pellucida of the unfertilized egg. Once fertilization has occurred these receptors are probably inactivated to prevent further penetration.

The zona pellucida of different species vary from 2 to 25 μm in thickness and contains from 2 to 35 ng of protein. It is a loose network of fibrillogranular strands and microvilli; processes extend from the egg and the surrounding follicle cells into it. The zona pellucida is also permeable to large macromolecules including immunoglobins and enzymes as well as small viruses.

Several proteins have been isolated from the zona pellucida of eggs of different species. In the pig the zona pellucida has four glycoproteins, whereas in the mouse and hamster it has three. The most extensive work on receptors found in the zona pellucida has been carried out in the mouse. The ZP3 molecule, which is a glycoprotein with an apparent molecular mass of 83, appears to serve both as a sperm receptor and as an acrosome reaction inducer. ZP3 consists of a 44-kDa polypeptide chain with 3–4 oligosaccharides covalently bound to asparagine residues (N-linked) and additional oligosaccharides attached to serine and threonine residues (O-linked). The ZP3 is associated with ZP2, another glycoprotein, to form the long filaments that constitute the zona pellucida. The third glycoprotein, ZP1 (200 kDa), is a dimer composed of identical polypeptide chains that crosslink the filaments of the other two proteins. Recent evidence suggests that the ZP3 O-linked oligosaccharides are important in sperm receptor function. Since there are many possible oligosaccharide structures in terms of composition, sequence, branching patterns, and conformation, they probably determine the species specificity of the sperm receptors. It is assumed that these oligosaccharides are altered after fertilization so that other sperm are not recognized or fail to bind to the modified receptors present. This modification is brought about by enzymes released from the cortical granules of the egg. Because the O-linked oligosaccharides in ZP3 appear to play the primary role the glycosidase of the cortical granules is the most likely enzyme.

The ZP3 O-linked oligosaccharides, however, do not induce the acrosome reaction in sperm in vitro. It has been suggested that after binding to sperm, the oligosaccharides must associate with a polypeptide chain for the acrosome reaction to occur. The ZP3 polypeptide chain itself may serve this function.

Several different sperm proteins may be involved in egg binding. These include lectins, glycosyltransferases, glycosidases, and proteinases. The first three of these specifically interact with sugars. The protein must be associated with the plasma membrane surrounding the sperm head. As an example, galactosyltransferase, an enzyme that normally stimulates the transfer of galactose from uridine-5′ diphosphate-galactose to terminal *N*-acetylglucosamine residues to form *N*-acetylgalactosamine, is a potential mediator of sperm binding to mouse eggs. The sperm galactosyltransferase would bind to *N*-acetylglucosamine on the egg. Under this scenario, sperm–zona pellucida binding would be similar to an enzyme-substrate reaction. Similarly, on the sperm, sialyltransferase and the zona pellucida sialic acid may interact to participate in the sperm–zona pellucida binding reaction. The concept that the sperm bind to the zona pellucida receptors through lectin-like proteins with saccharide-binding activity is interesting. Some proteins isolated from sperm bind to zona pellucida molecules but it is not clear to which groups—proteins or oligosaccharides.

Regardless of the mechanism all sperm must complete the acrosome reaction before passing through the zona pellucida. The sperm progress through the zona pellucida by vigorous movement of the sperm tail. Movement of the sperm through the zona pellucida leaves a thin hole, sometimes called the *penetration slit,* in the zona pellucida. It is assumed that enzymes bound to the inner acrosomal membrane "soften" the zona pellucida and allow penetration. This function has been attributed to a number of membrane-bound enzymes including acrosin, nonacrosin proteinase, hyaluronidase, arylsulfatase, glycosulfatase, and *N*-acetylhexosaminidase. Thus two mechanisms are proposed for zona pellucida penetration, one mechanical, by the power of the sperm tail, and the second enzymatic. Probably both mechanisms are involved. The passage of the sperm through the zona pellucida is comparatively rapid, from 7 to 30 minutes. Once past the zona pellucida the sperm crosses the perivitelline space and attaches itself to the vitelline membrane, in which it is gradually incorporated. This attachment involves the sperm plasma membrane (not the inner acrosomal membrane), and occurs above the equatorial segment. The surface of the egg has many microvilli except in the area above the metaphase spindle of the second meiotic division. The sperm head attaches itself in an area rich in microvilli and is gradually taken in.

V. Activation of the Egg

After incorporation of the sperm head (and in most species the midpiece and tail) into the egg cytoplasm, the metabolically quiescent egg initiates a series of events and this is referred to as *activation*.

The first step is the *exocytosis of the cortical granules and the resumption of meiosis*. The haploid nucleus of the egg then transforms into the *female pronucleus*. At the same time the sperm nucleus decondenses and becomes the *male pronucleus*. These two pronuclei come into close approximation near the center of the egg, their membranes disintegrate, and the chromosomes mix for the first mitotic division. This intermingling of chromosomes marks the end of the fertilization process and the beginning of embryonic development.

The biochemistry of egg activation is not well understood in mammalian species. There is evidence of a strong release of intracellular calcium in the hamster egg 10–30 seconds after attachment of the sperm to the egg plasma membrane. This release continues for over an hour, but its exact significance is unknown. Also, after sperm-egg fusion, the conversion of ATP to ADP increases with concomitant stimulation of the respiration and metabolism of the egg.

Upon penetration of the sperm into the egg cytoplasm and release of the contents of the egg cortical granules, the zona pellucida, as already mentioned, becomes refractory to sperm penetration (zona reaction). A secondary block occurs at the level of the egg plasma membrane, and is known as the *vitelline block*. The nature of this is poorly understood but it is very rapid. Both of the blocks to polyspermy at the zona pellucida and vitelline membrane levels effectively prevent penetration by more than one sperm. Human eggs seem to undergo a very strong zona reaction. Even when they are exposed to a large number of sperm, as during *in vitro* infertilization, many sperm can be bound to the surface of the zona pellucida, but few penetrate the inner regions of the zona. In humans, nevertheless, *polyspermic fertilization* does occasionally occur, perhaps due to a delayed cortical granule exocytosis and a subsequent delay in the zona reaction. A number of factors could cause this anomaly, including egg immaturity at the time of sperm penetration, excessive aging of the eggs in culture, or zona pellucida defects.

VI. Conclusion

During the past two decades a great deal has been learned about the basic mechanisms of sperm maturation, capacitation, the acrosome reaction, and penetration of the egg. The application of molecular techniques to ascertain the exact physiology and biochemistry of these events has been very useful. Many unanswered questions still remain, but new techniques are becoming available that will answer these questions. During the past year the gene encoding ZP3 has been cloned, suggesting that DNA sequences and cellular factors that regulate production of the sperm receptors might soon be identified. Knowing the amino acid sequence of the ZP3 polypeptide chain it should be possible to determine where in the ZP3 molecule, sperm receptor activity, induction of the acrosome reaction, and assembly of the zona pellucida filaments are regulated. The same degree of knowledge can be applied to all aspects of the fertilization process. These studies will not only yield a better understanding of the overall process but will provide important clues to the regulation of fertilization in humans and other animals. This knowledge will have important consequences with regard to the curing of human infertility and contraceptive applications.

Bibliography

Austin, C. R. (1985). Sperm maturation in the male and female genital tracts, *in* "Biology of Fertilization," vol. 2 (C. B. Metz and A. Monroy, eds.). pp. 121–155. Academic Press, New York.

Dukelow, W. R., and Williams, W. L. (1988). Capacitation of sperm, *in* "Progress in Infertility," 3d ed. (S. J. Behrman, R. W. Kistner, and G. W. Patton, Jr., eds.). pp. 673–687. Little, Brown, Boston.

Oliphant, G., Reynold, A. B., and Thomas, T. S. (1985). Sperm surface components involved in the control of the acrosome reaction, *American Journal of Anatomy* **174:** 269–283.

Wassarman, P. M. (1988). Eggs, sperm, and sugar: A recipe for fertilization, *News in Physiological Sciences* **3,** 120–124.

Wassarman, P. M. (1988). Fertilization in mammals, *Scientific American* (Dec.), 78–84.

Yanagimachi, R. (1988). Mammalian fertilization, *in* "The Physiology of Reproduction" (E. Knobil and J. Neill, *et. al.,* eds.) Raven Press, New York.

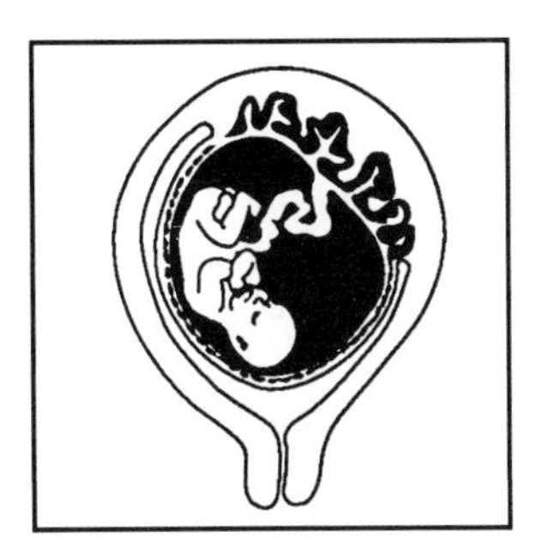

Fetus

FRANK D. ALLAN, *The George Washington University Medical Center*

Glossary

Amniotic fluid Fluid surrounding the embryo/fetus and within the amniotic cavity; produced by the lungs and the kidneys (in the form of urine)

Conceptus Product of conception (i.e., fertilization) and derivatives thereof, including the embryo and the extraembryonic membranes. The zygote, morula, blastocyst, and chorionic vesicle are early states of the developing conceptus

Differentiation Structural and functional elaboration or modification of cells, tissues, organs, or structures to perform a particular task

Embryo That portion of the conceptus which becomes the fetus or, at term, the newborn infant. Its development can be divided into stages designated the embryonic disk (bilaminar, then trilaminar) and the tubular, then definitive, embryo

Endometrium Lining (i.e., mucosa) of the uterus into which the conceptus implants. After implantation the endometrium becomes known as the decidua, a part of which forms the maternal component of the placenta

Extraembryonic membranes Nonembryonic tissues derived from the outer layer (i.e., the trophoblast) of cells of the blastocyst. Originally unilaminar, the trophoblast becomes the bilaminar (i.e., an outer syntrophoblast and an inner cytotrophoblast) and, finally, the trilaminar, chorion. A separate bilaminar delamination of trophoblast forms the amnion, which surrounds a cavity enclosing the embryo

Fetal circulation Pathway taken by blood within the fetus, the umbilical vessels, and the placenta

Fetal circulatory shunts Structures permitting fetal blood to pass to the placenta, then reenter the fetus (i.e., the umbilical arteries, vein, and ductus venosus); also, within the fetus, they bypass the lungs (i.e., the interatrial foramen and the ductus arteriosus). Closure of each shunt is a normal consequence of birth

Fetus Developing organism, from the onset of the ninth postfertilization week to birth

Meconium Product of the bowel, including epithelial debris, mucus, and bile

Oxygen saturation Relative content of oxygen carried in the blood, expressed as a percent

Placenta Disk-shaped specialized region of the chorion which interfaces with and includes the basal decidua of the endometrium; site of the transfer of substances between maternal and fetal bloodstreams

Primordium Earliest evidence or indication of an organ or structure constituting part of the embryo

Surfactant Lipid substance produced by cells lining the air spaces in the lung which reduces surface tension and makes inflation easier at the onset of respiration

DEVELOPMENT of the human organism and its associated membranes from conception (i.e., fertilization) to birth can be divided into three principal periods: preembryonic, embryonic, and fetal. The term ''conceptus'' designates all of the products of conception (the organism proper and the associated membranes).

During the preembryonic period the conceptus grows from a single cell (i.e., the zygote or fertilized ovum) to a multicellular sphere containing an embryonic disk. The sphere invades the lining of the uterus (i.e., implants), where it can develop, using nutrients which this lining is designed to provide.

Encyclopedia of Human Biology, Volume 3.

The preembryonic period requires most of the first 2 weeks of prenatal life.

The embryonic period encompasses development of the third through eighth weeks, during which growth and definition of both the embryo and the extraembryonic membranes are accomplished, including the establishment of the definitive placenta from the latter.

The fetal period encompasses the ninth week through the 38th week, or birth.

I. Introduction

A. Overview and Definitions

During the first week the conceptus undergoes repeated cell divisions, which form a hollow sphere of cells called the blastocyst. Its outer layer (i.e., the trophoblast) surrounds a cavity containing an internal mass of cells (i.e., the embryoblast) attached on one side. At 6 days the conceptus burrows into the endometrium, or lining of the uterus, where it continues development. The trophoblast gives rise to fetal membranes, including a major portion of the placenta, which isolate the embryo, yet provide maternally derived nutrients, oxygen, and other necessary substances.

Implantation and elaboration of the trophoblast during the second week are paralleled by growth and differentiation of the embryoblast into a bilaminar plate of cells. The upper layer of cells is designated the epiblast, the lower hypoblast. During the third week the plate or disk becomes trilaminar and elongated. The intermediate layer of the plate is called the mesoblast (i.e., mesoderm) and becomes vascularized (i.e., penetrated by vessels) at the end of the week. A primitive heart derived from the mesoblast initiates circulation. The plate then becomes tubular, with an expanded cranial end. In week 4 an infolding of surface cells (i.e., the epiblast) along the midline gives rise to the neural ectoderm, which forms primordia (i.e., precursors) of the brain and the spinal cord. The lower layer (i.e., hypoblast/endoderm) helps form a primitive gut within the tubular embryo as well as primordia of the internal organs (e.g., the lungs, liver, and pancreas). Bilateral swellings of the tubular embryo become the buds for upper and lower limbs. [*See* IMPLANTATION (EMBRYOLOGY).]

Thus, 1 lunar month after fertilization the conceptus is an implanted chorionic vesicle (derived from the trophoblast) containing a tubular embryo (derived from the embryoblast) with primordia of most body parts. Constituent cells are supplied with water, oxygen, minerals, and nutrients by an embryonic circulatory system linked to vessels within the chorion. The embryonic period continues to the end of the second lunar month, with elaboration of different tissues, organ primordia, systems, and other components. An embryo of 32 days is shown in Fig. 1. The embryo is clearly recognizable as human at the onset of the eighth week.

The embryo becomes a fetus at 57 days and measures about 32 mm (i.e., 1.25 inches) from crown to rump. Thereafter, the fetus grows about 1 cm/week throughout the remainder of prenatal life. At term, the end of 38 weeks, or 9.5 lunar months, the fetus, or neonate, measures about 32.5 cm (i.e., 13 inches) from crown to rump and 50 cm (i.e., 20 inches) from crown to heel.

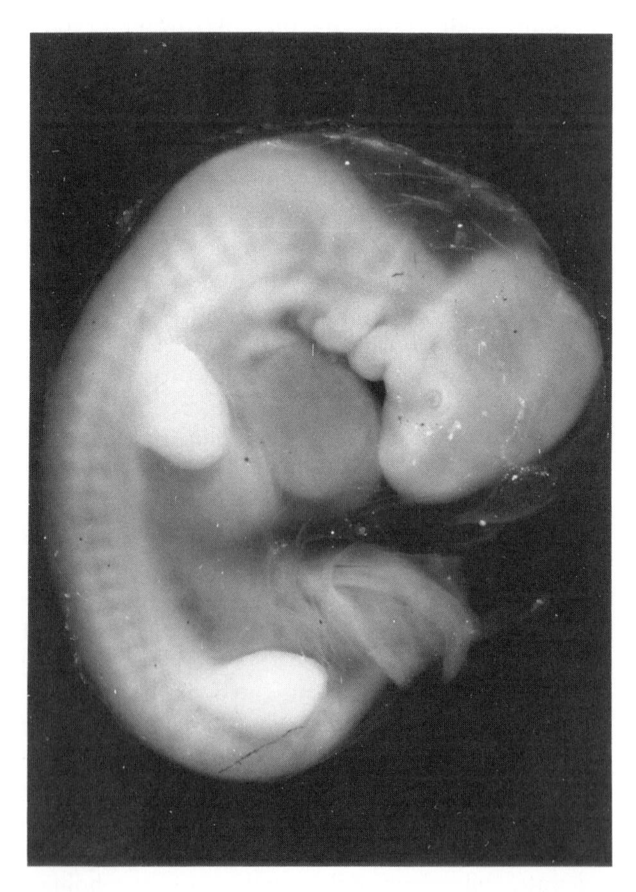

FIGURE 1 A 32-day embryo. Head at upper right overrides the heart and the liver (to the left and below), showing through the thin body wall. Upper and lower limb buds are paddlelike. The junction of the umbilical cord with the body wall is at lower center. © Frank D. Allan, Ph.D.

B. Early Fetal Period (Third Month)

At the onset of this period, the head is relatively large (nearly one-half of the embryo's crown–rump length), the segments of upper and lower extremities are discernible, and digits are separated. Eyelids are just about to close, fusing temporarily over the eye. Sex of the embryo is not obvious from the external genitalia; however, by 10 weeks this is possible. All internal structures are present. Many of these are tiny replicas of the organ or structure they are to become. The fetal period permits each existing body part (e.g., organ, gland, muscle, and bone) to enlarge by adding additional cells, increasing the size of existing cells, elaborating intercellular materials that add shape or form, and establishing functions in these parts. [*See* EMBRYO, BODY PATTERN FORMATION.]

At this time the eye shows all its basic components: the bulb with outer cornea and sclera, internal lens, retina, iris with pigment, and primitive muscle fibers. Extrinsic eye muscles have even developed microscopic characteristics (i.e., striations) that mark voluntary muscles in postnatal life.

For more than 1 month muscle cells in wall of the heart have been contracting, pumping blood through vessels permeating the embryonic tissues and, via the umbilical cord, into the placenta. There, vital substances carried by maternal blood surrounding the chorionic villi lie just two thin membranes away from the blood of the fetus. Division of the heart into chambers is complete, and a definitive vascular system carries blood to and from all body parts. As these parts grow, vessels within match this growth, hence providing a constant supply of substances supporting cell function, replication, migration, and differentiation.

Primordia of musculoskeletal structures of the extremities, head, and body proper (i.e., trunk) are present, although most of the skeleton is still cartilage. Even so, there are areas where cartilage is being converted to bone by the deposition of calcium. [*See* BONE, EMBRYONIC DEVELOPMENT.]

All components of the brain and the spinal cord are formed, and nerves link the stem of the brain and the spinal cord to all tissues and organs of the body. Nerves in the head and the neck link the skin and the primordia therein with the brain stem. Skin of the body wall, or trunk, and muscles and tissues within the trunk and of the extremities are linked by nerves to the spinal cord. Hollow organs within the body's cavities have nerves which link them with the brain or the spinal cord, and in some instances both. [*See* SPINAL CORD.]

The skin at this time is a delicate membrane, its epidermis two cells thick, supported by an embryonic connective tissue called mesenchyme, the precursor of the dermis. Invaginations from the epidermis into the dermis have established primitive hair follicles and glands. [*See* SKIN.]

Oral and nasal cavities lead from a primitive mouth (i.e., stomodeum) to the primitive pharynx, from which tubular respiratory and digestive tracts arise. Within the wall of each tract, layers comparable to those of the adult are present, but thin. The larynx, tracheobronchial tract, and lungs are formed as a diverticulum of the primitive pharynx. Similarly, the liver, biliary tract, and pancreas arise as diverticula of the gut (continuous with the pharynx) below the segment destined to become stomach. The elongating stomach and intestine outgrow the available space inside the body cavity at this time and herniate, temporarily, into an extension of the cavity in the umbilical cord. By 11 weeks the herniated portion of the bowel has returned to the body cavity and the extension is obliterated. A fetus of 11 weeks is shown in Fig. 2. [*See* DIGESTIVE SYSTEM, ANATOMY; LARYNX.]

Urinary and reproductive systems develop in intimate association and share certain primordia. By the fetal period the definitive kidneys are present and functioning, as are the ureters and the urinary bladder. The final segment of this system, the urethra, completes changes related to a definition of the external genitalia by the 11th week. Gonads have differentiated as testes or ovaries, and the primordia of both tracts (male and female) have appeared. Whether the male or female genital tract is to dominate and persist is dependent on genetic sex and is effected by the action of hormones produced in the testes or ovaries. [*See* REPRODUCTIVE SYSTEM, ANATOMY; URINARY SYSTEM, ANATOMY.]

At this time primordia of all endocrine organs have been established. The hypophysis, or pituitary gland, at the base of the brain, thyroid and parathyroid glands in the neck, and the suprarenal glands (intimately associated with the kidneys and the gonads) are well defined as highly vascularized clusters of epithelioid cells. By the onset of fetal life, differentiation of the initial cellular inhabitants of the embryo's germ layers has given rise to distinctive cell lineages which restrict the ultimate fate and purpose of component cells. Increasing numbers of cells representing each of the basic tissues (e.g.,

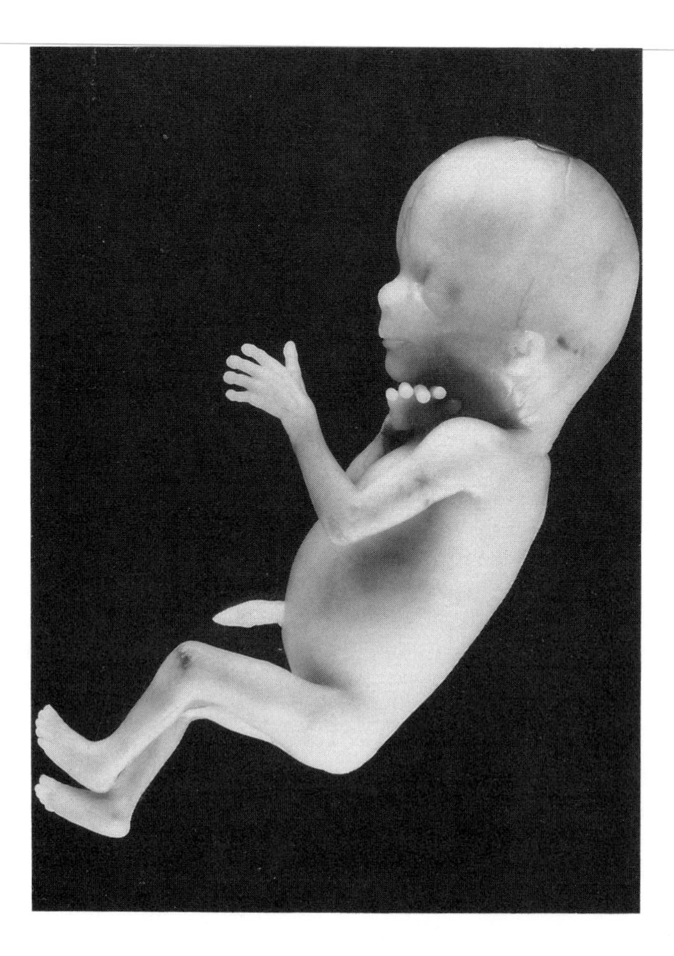

FIGURE 2 An 11-week fetus. Note the relative proportions and definition of the body parts. Note, too, the translucency of the skin. © Frank D. Allan, Ph.D.

epithelium, connective tissue, supporting tissues, blood, muscle, and nerve) have migrated to positions where they will best serve the organism.

II. Maternal–Fetal Relationships

Consideration of the physiology of the fetus requires an understanding of its anatomical and functional relationship to maternal tissues. As noted earlier, a portion of the conceptus, the trophoblast, gives rise to the so-called fetal membranes. After implantation the trophoblast forms the chorion, which serves as the fetal component of the interface. The endometrium (and blood from its intrinsic vessels) adjacent to the chorion serve as the maternal component.

A circumscribed portion of chorion and adjacent endometrium form the placenta, where maternal blood (from vessels in the endometrium opened during implantation) bathes fingerlike extensions of the chorion (e.g., chorionic villi). Fetal blood circulates through villous capillaries, with access through the villus wall to fluid and materials in maternal blood within the intervillous space (Fig. 3).

Within the chorionic vesicle the amnion, also derived from the trophoblast, forms a fluid-filled sac enveloping the embryo/fetus. The amnion is reflected from the sac onto the umbilical cord and is continuous at the umbilicus with the skin of the fetus. In its fluid bath the fetus is protected from fluctuations in temperature and effects of gravity (in large part), compression, or external mechanical stimuli.

A remarkable structure, the placenta serves a multitude of functions: (1) a lung or a respiratory organ, which accomplishes the transfer of gases; (2) a digestive tract, providing for absorption of fluids, minerals, and nutrients; (3) a liver for the storage of glucose; (4) a kidney, functioning to eliminate nitrogenous waste substances; (5) an endocrine gland, producing several hormones affecting maternal and

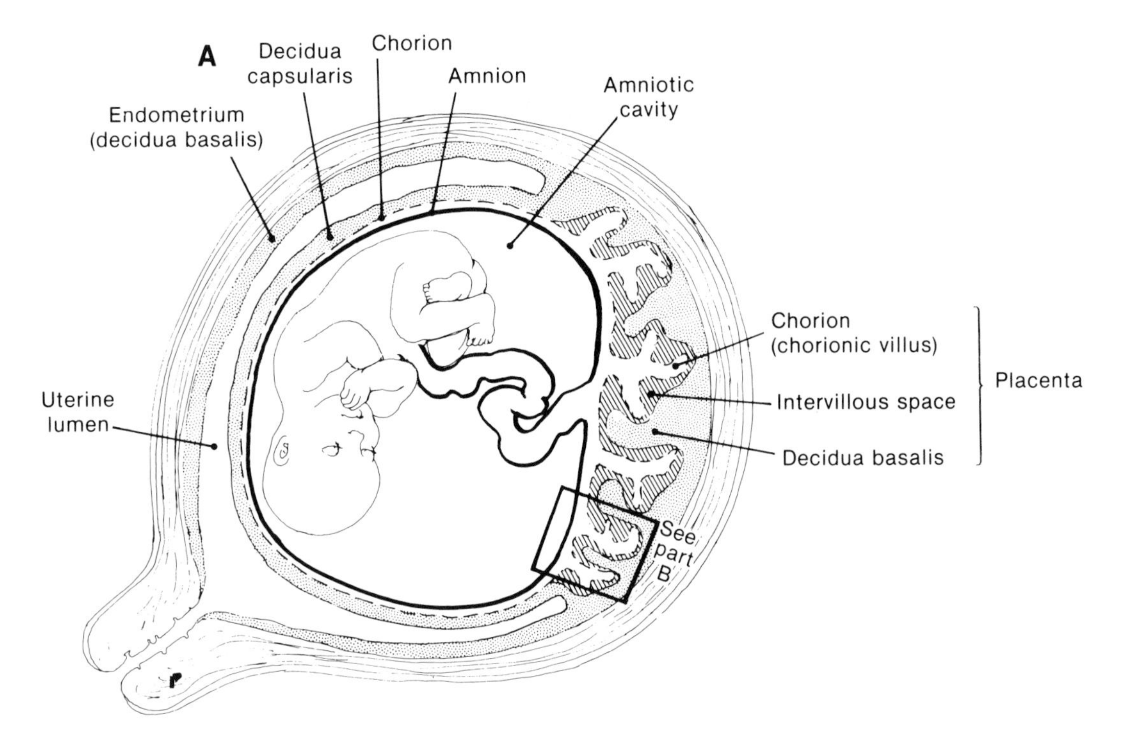

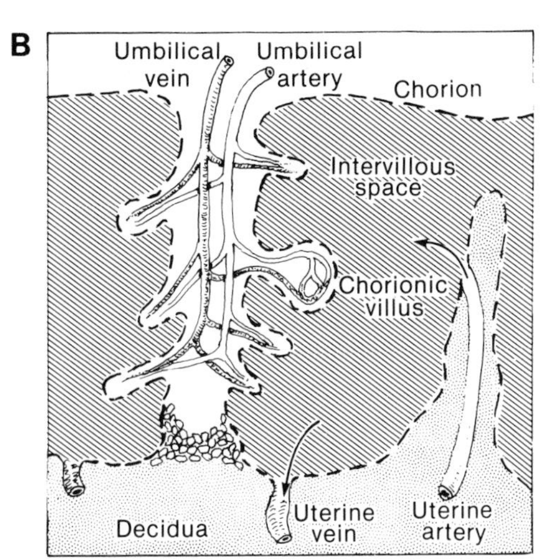

FIGURE 3 (A) Diagram of an uterus containing a fetus at the end of the third month, showing the extraembryonic tissues (i.e., chorion, amnion, and umbilical cord) within the uterine lining (i.e., decidua). Note the fetal and maternal components of the placenta. (B) Inset from lower right of (A) shows higher magnification of the chorionic villi and the intervillous space.

fetal tissues; and (6) a circulatory pump, since the contraction of smooth muscle in larger villi increases pressure within the intervillous space. In addition, the so-called placental barrier provides an important, although limited, protection, preventing, in variable degree, the entry of noxious substances into the fetal bloodstream. This is particularly true during the first trimester of intrauterine life.

III. Fetal Growth and General Features

Most fetal organs or structures at 4 months have established relationships and characteristics that will be carried into postnatal life (Figs. 4 and 5). Relative proportions, however, change appreciably. The head, with the brain and special organs,

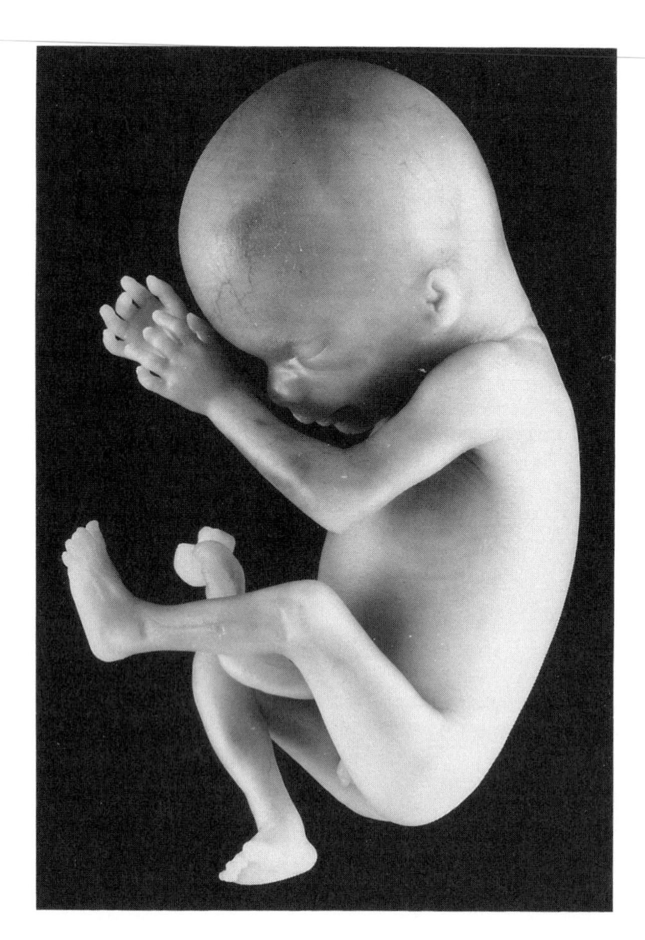

FIGURE 4 A 14-week fetus. Note flexion of the trunk and the lower extremities to fit the confines of the uterus. © Frank D. Allan, Ph.D.

usurps a large proportion of the total organism during prenatal life. Initially, the upper limbs and trunk are markedly larger and better developed than the lower limbs and trunk. Proportions still favor the head at birth (one-quarter of the body length compared to one-seventh in the adult). Within the thorax lungs occupy but a small portion of the available space compared to the heart, great vessels, and thymus gland. Organs in the abdominal cavity are relatively large, especially the liver and the spleen. Kidneys and adrenal glands are also relatively large structures and, with the stomach and the bowel, fill the cavity. In contrast, the pelvic cavity is small and the urinary bladder rises well above the pelvic brim. Genital organs are small and crowded into the limited space with a small terminal bowel and rectum.

Weeks 13–16 of prenatal life are marked by rapid growth of the fetus. Ossification of skeletal structures initiated earlier proceeds rapidly. During the following week (17–20) growth slows, but fetal movement increases. Hair on the head, eyebrows, and body (i.e., lanugo) appears. By week 20 the vernix caseosa, a cheesy substance which coats and protecs a thin still-translucent skin, is formed (Fig. 6). Near the end of the next 4-week period (i.e., week 24), fingernails have appeared, and, within the lung, cells elaborate surfactant, the presence of which marks the point at which the lungs can inflate and the fetus could survive premature delivery. During weeks 26–29 the respiratory apparatus matures and increases the likelihood of survival following premature delivery. The eyelids reopen and fat begins to fill and thicken the trunk. Skin becomes less translucent and toenails appear. Hair on the head and the amount of lanugo increase. The final weeks of life before birth (i.e., weeks 30–38) permit continued growth and an increased amount of fat in the skin. The fetus can grasp, and the breasts (both male and female) are well formed and often secrete colostrum at birth. Lanugo then decreases.

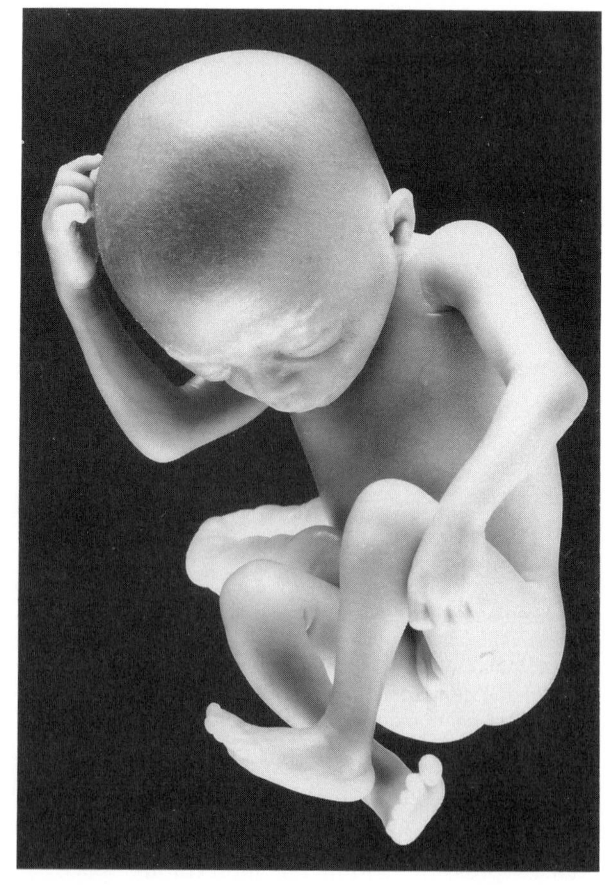

FIGURE 5 An 18-week female fetus. Note the beginning of eyebrows. © Frank D. Allan, Ph.D.

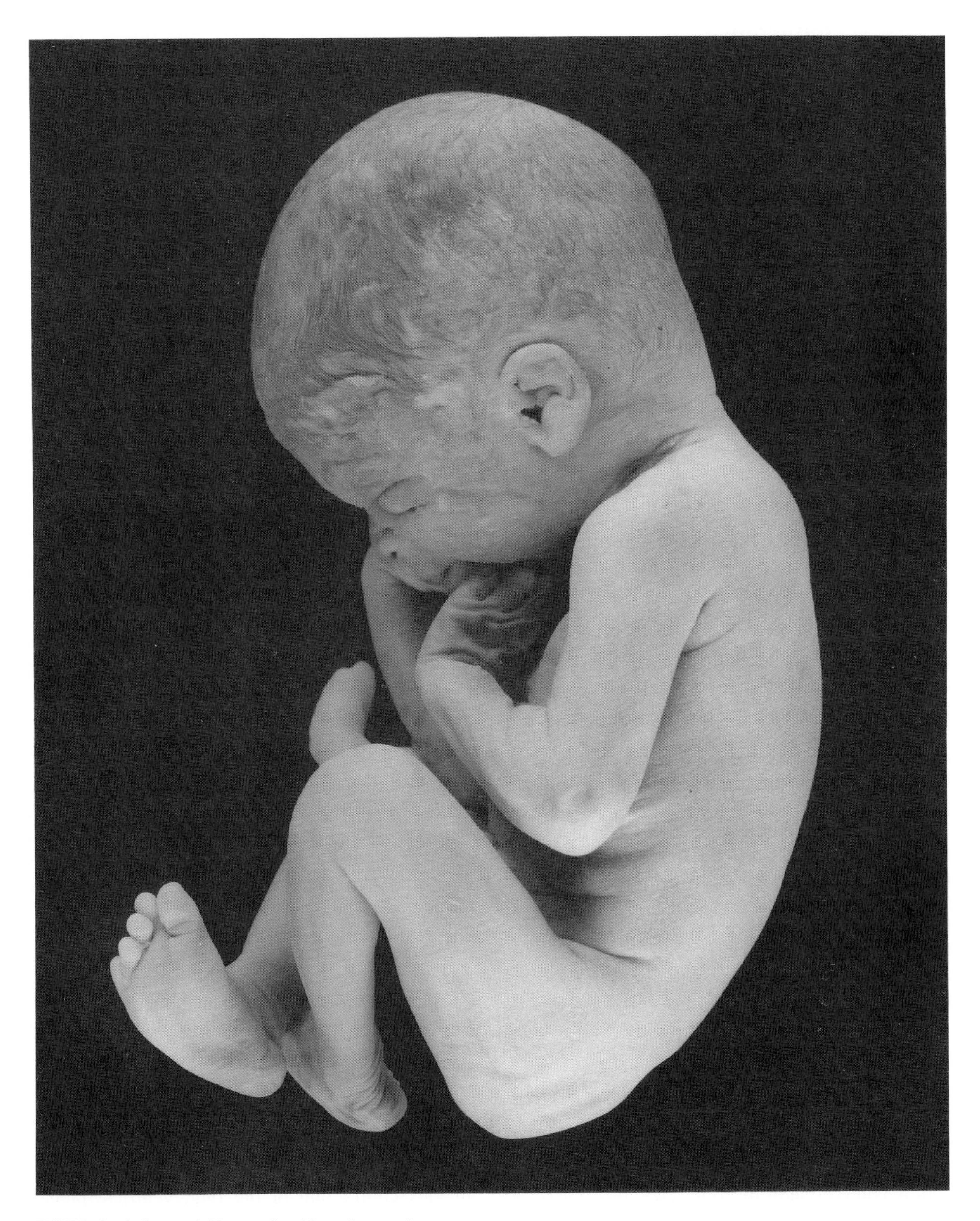

FIGURE 6 A fetus of 20+ weeks. Note the vernix caseosa smeared in the hair of the head. Also, note the opacity of the skin. © Frank D. Allan, Ph.D.

IV. Fetal Physiology

The basic function of cells and tissues of the developing organism is coexistent with growth and specialization of these cells and tissues before birth. Thus, all functions of cellular metabolism that affect growth, cell division, respiration, motility, and specialization of cells as tissue components are ongoing throughout embryonic and fetal life. Individual cells comprising the conceptus, embryo, or fetus must maintain their own well-being while performing specialized tasks benefiting the whole organism.

After cells specialize as one of the basic tissues and are incorporated into an organ, functions of the embryo/fetus can be considered at another level. It is not possible to deal with all such functions here. A few, however, are especially worthy of examination.

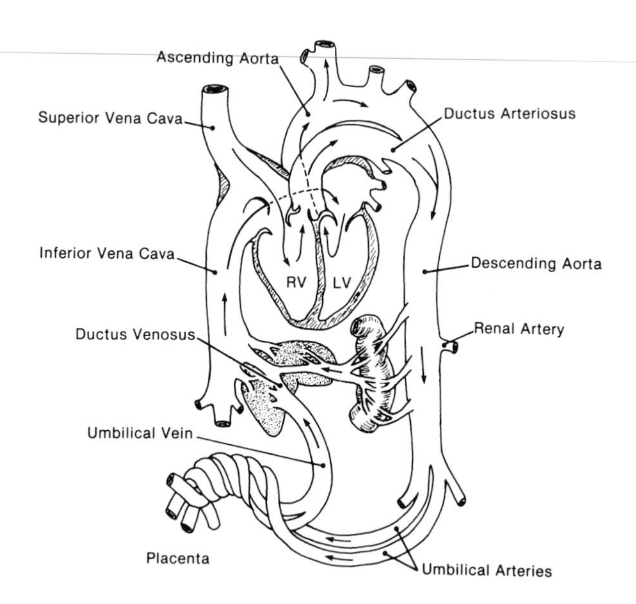

FIGURE 7 Fetal circulation. Only major vessels are indicated, with an attempt to represent smaller vessels which link with vascular plexuses in organs or tissues of the fetus and the associated placenta. RV, right ventricle; LV, left ventricle.

A. Early Fetal Function

In very early fetal life the function of cells, tissues, and the organs they constitute runs hand in hand with anatomical maturation. The fetus swallows amniotic fluid beginning with the third month. The fluid and some of its components pass through the lining of the gut and walls of adjacent capillaries, entering, thereby, the bloodstream. The absorbed fluid reaches the kidneys, which produce urine even at this early stage. Urine is expelled into the amniotic fluid, where it mixes with fluid produced by the lungs and sloughed epithelial cells from the amnion, or surface ectoderm.

During the first month of the fetal period, muscles twitch spontaneously and in response to reflex impulses initiated by occasional stimuli. Protected from the usual surface stimuli of postnatal life, the fetus' hands are, nevertheless, often held close to the face. Should finger or thumb touch the area around the mouth, a pursing of lips or even suckling movement might result.

B. Fetal Circulation

By the onset of the fetal period, the heart has four chambers: right atrium, right ventricle, left atrium, and left ventricle (Fig. 7). Blood returning to the heart enters its atria; blood leaving is expelled from and by contraction of the ventricles. Blood leaving the left ventricle enters the aorta and is distributed to all fetal tissues (except the lungs) and to the placenta by way of impaired umbilical arteries. The lungs are supplied by outflow from the right ventricle via the pulmonary arteries.

Having perfused fetal tissues, blood flows back to the right atrium using tributaries of superior and inferior venae cavae (draining the upper and lower halves of the body, respectively). Aortic blood flowing through umbilical arteries to chorionic vessels in the placenta is drained by tributaries of a single umbilical vein, returning it to the fetus via the umbilical cord. This flow is shunted by the ductus venosus through the liver to enter the inferior vena cava. Another shunt, the ductus arteriosus, transmits blood from the pulmonary trunk to the descending aorta.

Blood entering the right atrium from the superior vena cava (relatively unoxygenated) passes directly into the right ventricle, whereas blood entering from the inferior vena cava (including the oxygenated umbilical flow) is shunted through an opening (i.e., the interatrial foramen) into the left atrium. Mixing of the two streams in the right atrium is minimal. Blood within the left atrium is joined by blood returning from the lungs and passes into the left ventricle. This blood is pumped into the aorta to be distributed by branches arising therefrom. Simultaneous contraction of the right ventricle pumps blood into the pulmonary trunk, where a small portion passes to the nonfunctioning lungs, whereas the major portion passes through the ductus arteriosus into the aorta. And so the circuit is completed.

Fetal blood aerated within the placenta achieves 80% oxygen saturation (the most highly oxygenated) and returns to the heart through the umbilical vein and the ductus venosus. When this blood joins the inferior vena cava, carrying oxygenated depleted blood from the lower part of the body, the oxygen saturation decreases to about 70% saturation as it enters the right atrium. Blood entering the right atrium from the superior vena cava is only 30% saturated. Separation of the streams flowing through the right atrium from the two venae cavae is, therefore, significant. Blood passing through the foramen ovale, left atrium, and left ventricle reaches the aorta with an oxygen saturation of about 60%, whereas that passing into the right ventricle and into the pulmonary trunk is about 50% saturated. Beyond the junction of the aorta and the ductus arteriosus the oxygen saturation is approximately 55%. Since most of the terminal aortic blood flow enters the umbilical arteries, blood to the placenta has the same oxygen saturation (i.e., 55%). [*See* CARDIOVASCULAR SYSTEM, ANATOMY; CARDIOVASCULAR SYSTEM, PHYSIOLOGY AND BIOCHEMISTRY.]

C. Circulatory Changes at Birth

A remarkable series of changes in circulation takes place at birth, intimately related to the onset of respiration. Changes involve structure and function of the lungs and the circulatory pattern (Fig. 8).

During passage through the birth canal, umbilical blood flow is usually compromised to some degree. The fetus becomes hypoxic (i.e., receiving a diminished amount of oxygen) or, in severe cases, anoxic (i.e., deprived of oxygen). Obviously, when the umbilical cord is clamped or cut, the source of oxygen from maternal blood via the placenta is lost and carbon dioxide builds up in the fetal blood (i.e., hypercapnia). The fetal respiratory center in the brain is stimulated, and inspiratory efforts (i.e., contracton of the diaphragm and the intercostal muscles) result.

With inspiration, the air sacs (i.e., alveoli) within the lungs increase in size and number, and the capacity of the capillary network surrounding the sacs increases. Resistance in the pulmonary bed decreases; hence, blood flow through the bed increases. As a result the volume of blood returning to the left atrium from the lungs also increases.

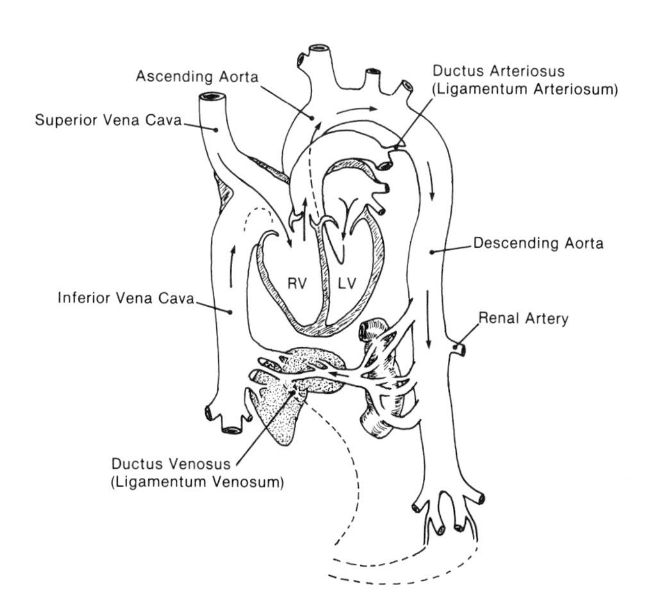

FIGURE 8 Postnatal circulation. Only major vessels are indicated. Note constriction of the fetal shunts (umbilical arteries, vein, and ductus venosus; ductus arteriosus) previously carrying blood to and from the placenta and those bypassing the lungs. Also, note the closure of the interatrial foramen (dashed lines in the right atrium). RV, right ventricle; LV, left ventricle.

With cessation of blood flow through the umbilical circuit, the umbilical vein and the ductus venosus collapse. Constriction of the umbilical arteries increases pressure within the aorta, and perfusion of all body tissues and organs is thereby increased. Blood pressure within the pulmonary system increases due to back-pressure from the aorta via the ductus arteriosus. The result is increased perfusion of the expanding lungs. Left atrial pressure also increases, and the valve of the interatrial foramen is functionally closed. Venous return to the right atrium increases as tributaries of the venae cavae receive the greater volume of blood perfusing the fetus. The right and left atrial pressures equalize, and the valve remains closed.

Continued respiration increases lung capacity, and oxygen saturation of the blood gradually increases. At an oxygen saturation of about 75% in aortic blood, the ductus arteriosus closes, and the entire output of the right heart passes to the lungs.

D. Pulmonary Apparatus

Remarkable as the changes in circulation are, they are no more remarkable than mechanisms which permit the lungs to change from a relatively cellular mass to one with increasing numbers of thin-walled air sacs, permitting an intimate relationship between blood-filled capillaries and the air sacs they embrace.

Structural changes and cellular maturation of the fetal lungs at 6 months lead to functional competency of this organ. Thus, the occasional air sacs found at the ends of the bronchial tree at that time increase in numbers. The expanded capillary network intimately related to the air sacs enlarges as more and more alveolar spaces are formed. Bronchial and bronchiolar passageways are lined with simple columnar or cuboidal ciliated epithelium. Similar cells constitute the end pieces of the bronchopulmonary tree and surround small fluid-filled sacs. Like any moist juxtaposed membranes, the cells lining the spaces are held together by surface tension. By the 24th week of intrauterine life, certain of the cells forming these end pieces elaborate a surfactant which minimizes surface tension and permits inflation of the potential spaces following repeated inspiratory efforts.

To survive delivery, the fetal lungs must be able to support respiration. At the end of the sixth lunar month, this ability, although marginal, is sometimes sufficient. Before this time few, if any, fetuses survive removal from the uterus. Obviously, those born later have increasing chances of survival. Effective respiration is directly related to the maturation of the alveolar–capillary interface and production of sufficient surfactant. Neural regulatory mechanisms for respiration are present much earlier (i.e., at the end of the first trimester). Respiratory movements are common *in utero* and increase in frequency as term approaches. It appears, nevertheless, that such mechanisms are suppressed until birth, when the stress on the fetus ends this suppression. [*See* RESPIRATORY SYSTEM, ANATOMY; RESPIRATORY SYSTEM, PHYSIOLOGY AND BIOCHEMISTRY.]

E. Urinary Apparatus

The kidneys, ureter, urinary bladder, and urethral passage are all functioning in simple fashion early in the fourth lunar month. Waste products of fetal metabolism are also being cleared from the blood within fetal capillaries of the placental chorionic villi. Nonetheless, functioning nephrons (i.e., functional units of the kidney) are present from this time on. Nephrons present in the fetus have small glomeruli (where arterial blood is filtered) and relatively short tubules (where the glomerular filtrate is processed). Research on animal fetuses suggests that the nephron of the definitive kidneys is functional by midgestation and awaits only the challenge of an increased load of metabolic waste following birth to fully mature.

F. Alimentary System

Cellular and tissue mechanisms for absorption and digestion are present by the end of the fourth month. Thus, microvilli of cells on intestinal villi (both villi and microvilli increase surface area of the lining of the intestine for absorption) and many intestinal digestive enzymes are present at this time. Gastric gland products appear in the fifth month. Swallowing of amniotic fluid occurs as early as the third month, and the frequency of this increases as the fetus ages. The swallowed fluid is absorbed in the gastrointestinal tract and cleared by the placenta or the kidneys. Peristalsis of the tract is present from the fourth month, although the muscular layers of the gut wall are relatively thin and its contractions are weak. Attempts to suckle have been seen *in utero* and in aborted fetuses of 3 months. Premature infants suckle effectively shortly after delivery and are able to ingest and process a relatively large volume of milk within the first few days following birth.

G. Regulatory Systems

Obviously, some regulation of organs, tissues, and cells is required to integrate various activities and responses during fetal life, even though this integration might not be as essential for survival as it is after birth. Regulation and integration are achieved by the nervous and endocrine systems.

Initial nervous activity begins in the third fetal month. Neuromuscular reflexes are more obvious in the upper part of the fetus, but soon spread to the trunk and the lower extremities. Such reflexes are responses (i.e., muscle contractions) to stimulation of tactile and pressure receptors within or near these muscles. Reflexes which develop during the third month involve the mouth and the head (e.g., sucking, swallowing, gagging, and moving the mouth and, therefore, the head toward a perioral stimulus). Electrical activity of the nervous system is discernible at the same time, but periods of electrical silence can occur up to mid-term. Movements of the fetus (i.e., quickening) are usually perceived by the mother at 4 months. By 7 months the electrical activity of the nervous system becomes continuous and rhythmical. Electric potentials appear following visual stimulation at the end of 6 months. Apparently, some visual function is present, and the response of the pupil to light can be elicited by this time.

Taste buds are functional at 6 months, and the modality for sweetness is well differentiated. Increased "drinking" of the amniotic fluid is effected when sweet substances are introduced.

Each of the endocrine glands is established early in the fetal period (i.e., the third month), and all are functioning, in variable degree, by midgestation. The pituitary gland, or hypophysis, has differentiated by the beginning of the second trimester and elaborates a number of hormones that have control over body tissues generally (growth) or specifically (trophic hormones) over other endocrine glands. The pars neuralis of the hypophysis produces its hormones (which cause the contraction of smooth muscle) at the same time. [*See* ENDOCRINE SYSTEM.]

The thyroid gland elaborates the thyroid hormones within the fourth month and has established a feedback system with the hypophysis by midgestation. Under hypophyseal control the thyroid hormone controls the general metabolism of cells. Parathyroid glands give evidence of functioning at the beginning of the fetal period and are active throughout pregnancy. Parathyroid hormone helps regulate the circulating level of calcium in the bloodstream. [*See* PARATHYROID GLAND AND HORMONE; THYROID GLAND AND ITS HORMONES.]

The cortex of the adrenal gland is formed and actively producing steroid hormones at the end of the fourth month. A special zone of fetal cortex is lost at birth, and definition of the permanent structure occurs postnatally. The adrenal medulla is an extension of neural tissue (embedded in the cortex) which forms neurohumors (epinephrine and norepinephrine) identical to those produced by the autonomic nervous system.

Other endocrine cells exist as islets within the pancreas and are distinguishable (i.e., component cells can be differentiated) at 3 months. They are elaborating hormones (e.g., glucagon and insulin) by the fourth month.

Bibliography

Barnes, A. C. (ed.) (1968). "Intrauterine Development." Lea & Febiger, Philadelphia, Pennsylvania.

England, M. A. (1983). "Color Atlas of Life before Birth." Yearbook Medical, Chicago, Illinois.

Moore, K. L. (1988). "The Developing Human," 4th ed. Saunders, Philadelphia, Pennsylvania.

Tanner, J. M., and Preece, M. A. (eds.) (1989). "The Physiology of Human Growth." Cambridge Univ. Press, Cambridge, England.

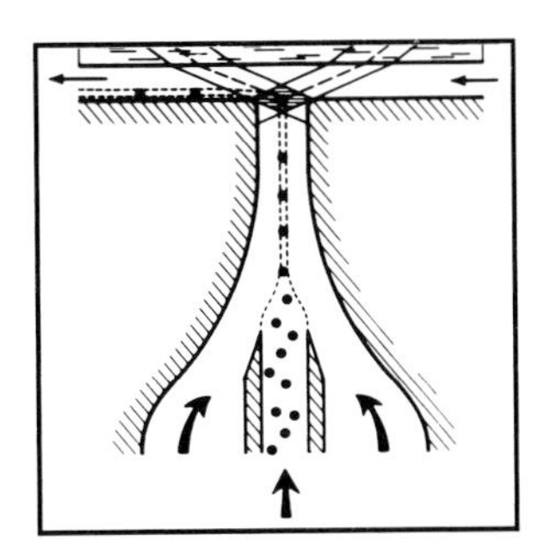

Flow Cytometry

OLE DIDRIK LAERUM, *University of Bergen*

Glossary

Analytical cytology Areas of cytology that deal with quantitative and qualitative analyses of cells
Aneuploidy Cellular DNA content different from a DNA index of 1 or 2
BrdU Bromdeoxyuridine; a synthetic analogue of a nucleotide base in DNA
DNA index Cellular DNA content as a quotient of DNA in diploid cells from same species
FITC Fluoroisothiocyanate; a fluorescent dye that emits green/yellow light after excitation with blue light
Fluorescence Emission of light caused by a light source of different, usually shorter, wavelength
Light scatter Spread of light caused by a cell as it passes a narrow laser beam

FLOW CYTOMETRY is a series of methods that enables automated quantitative and qualitative measurements on single cells in suspension. It is considered one of the major methodological breakthroughs in the field of analytical cytology during recent decades. Its methods have gained an important position in biological and medical research.

I. Principles

Flow cytometry is a series of methods where single cells in suspension pass through a light beam with high velocity by which automated quantitative and qualitative measurements can be performed. They include both the deviation of light as a measure of size and optical properties of the cells (light scattering), and the absorption of light and emission of fluorescence from cell components that have been stained with a fluorescent dye. Multiple cell parameters can be measured simultaneously, and with computer technology these parameters can be analyzed in relation to each other. The flow of cell suspension can be broken up into droplets that are electrically charged, and cells containing specific signals can be sorted out by passing a static electrical field. There are two main principles for flow cytometry. One is where cells pass a narrow laser beam; the other is by use of microscope optics where cells pass a light field provided by the microscope lamp, and fluorescence is measured through the optical system. These are shown in Fig. 1.

II. Different Types of Instruments in Current Use

Since the late 1960s, many laboratories have built their own machines based on the above-mentioned principles. In addition, several commercial machines became available in the early 1970s. These were built both as analyzers and as cell sorters using electrostatic deflection of charged droplets as a cell-sorting mechanism. Today, several laser-based analyzers are available. Some are highly automated,

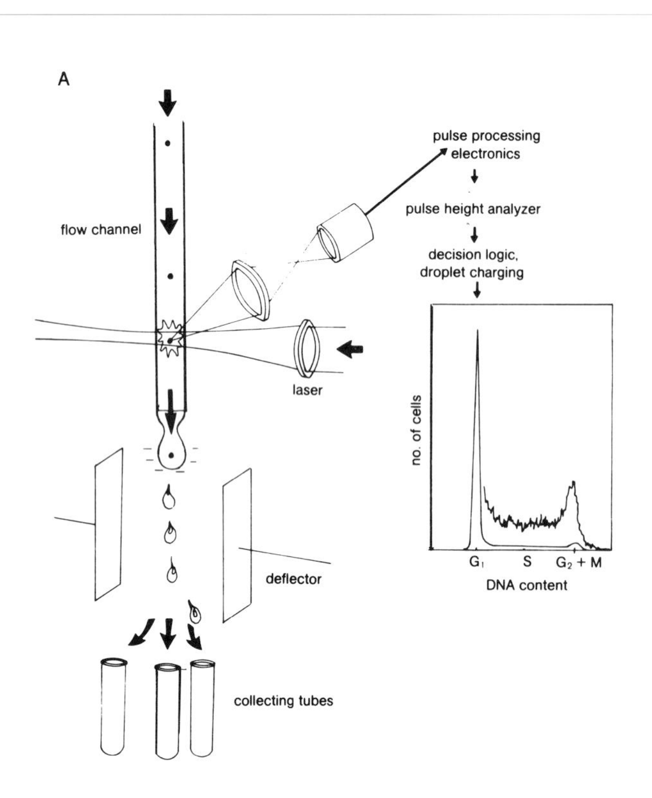

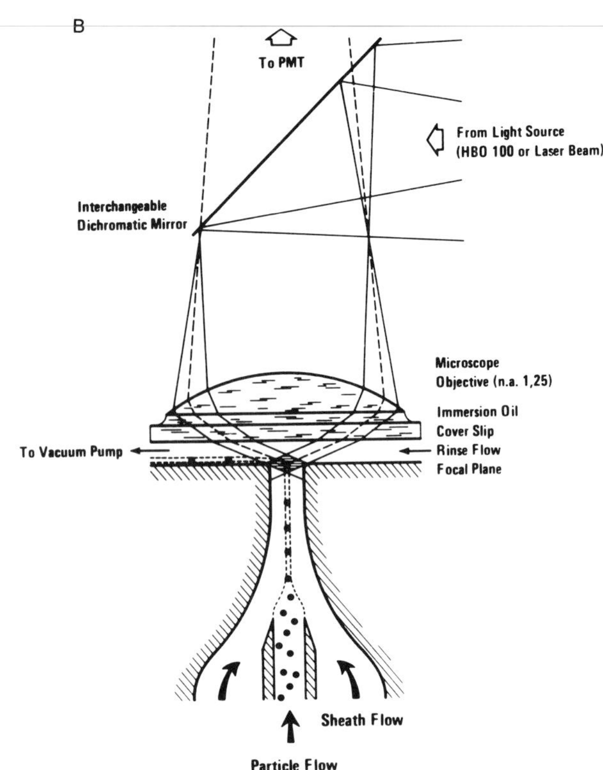

FIGURE 1 The two main principles of flow cytometry. a) Using a laser, where the measured cells can also be sorted in charged droplets passing a static electrical field, and b) by use of microscope optics, where excitation and emitted light go through the same optical system as described by Göhde (1973). [From Melamed *et al.*, 1979, "Flow Cytometry and Cell Sorting," John Wiley & Sons, New York.]

where cells, stained by simple methods, are automatically measured by use of multiple parameters. Most of the laser-based machines can also have cell-sorting devices. Their rate of measuring or sorting is usually between 1,000 and 5,000 single cells per second; however, for discriminating and sorting rare subpopulations, it takes a relatively long time to obtain a sufficient quantity of the desired cells. This can be overcome by performing flow analysis and sorting under high pressure, by which up to 20,000 single cells can be analyzed and sorted per second.

Several microscope-based systems are also on the market, either as closed-flow systems or as open-air systems where the cells pass over a coverslip in an illuminated field.

III. Cell Parameters That Can Be Measured

Although light scatter is not a reliable marker for cell size, it is useful for discriminating cells based on the combination of size and optical properties. If a small slit is placed between the recording light and the flow, the narrow light beam will scan each cell as it passes by. With a photomultiplier, the scattered light can be recorded and give a measure of the cell shape. The same can be done if mitotic cells are lysed so that single chromosome suspensions are obtained. Using this procedure, DNA content and shape and also the centromeric index of the chromosomes can be measured.

Light absorption was used earlier to some extent but has entirely been replaced by fluorescence measurements because fluorescence is a far more sensitive marker than light absorption.

IV. Common Preparation and Staining Techniques

Staining can be performed on fresh cells, on cells fixed in suspension by ethanol or formaldehyde, and by enzymatic resuspension of cells that have been fixed and embedded in paraffin.

To measure different cell components by fluorochrome staining, the dye must be specifically bound to certain cell components. If it also binds to other components, these can be removed beforehand by enzyme treatment. An example is the staining of DNA with ethidium bromide. This dye will also stain RNA and can be unspecifically bound to cell protein. Cells in suspension are therefore first treated with pepsin to remove the cytoplasm, and thereafter with RNAse to remove single-stranded RNA. This provides a specific staining of the remaining DNA.

Several fluorochromes can be used at the same time, provided that they bind specifically to different cell components and fluoresce at different wavelengths. One example is the staining of DNA with propidium iodide, which emits red fluorescence, and protein with FITC, which emits green/yellow fluorescence.

In order to obtain staining of the cells, their membranes have to be opened to give access to nonpenetrable dyes, either by fixation in suspension, e.g., by use of ethanol, or they can be lysed by use of detergents, leaving naked nuclei accessible for staining. The fact that some stains do not enter through an intact cell membrane can be used for viability testing, by which fluorescent cells are detected and scored as nonviable. Staining and measurement can also be performed on cell samples that have been frozen and stored at low temperatures for prolonged periods. For some studies, fixed cells are not a necessary requirement.

For different fluorochromes that possess the same excitation wavelength spectrum, one single laser can be used. If they have different excitation spectra, lasers with different wavelengths can be mounted serially, and the fluorescence emitted from the same cell can be measured with a slight time difference. For systems using a mercury lamp, the desired wavelengths can be picked up by using specific optical filters.

A new area was opened when monoclonal antibodies conjugated to a fluorochrome were used for quantitative measurements of different antigens on single cells by flow cytometry. With this method, direct as well as indirect immunofluorescence are used, provided that the emitted light is sufficient for discriminating the antigens from the background of light and electronic noise. If specific antibodies are available, the types of cell components that can be measured quantitatively are almost unlimited. [*See* MONOCLONAL ANTIBODY TECHNOLOGY.]

If cells take up a chromogenic substrate for a chemical or enzymatic reaction, the rate of this reaction can be measured quantitatively by use of the fluorescent product. In some cases, autofluorescence of naturally occurring substances can also be measured, although this is not as common as the other techniques.

V. Applications in Biology

Because there is almost no limitation of measurements that can be performed on single cells, including various cell components as well as the kinetics of different reactions, only main areas of applications will be mentioned.

A. Cell Kinetics

By specific staining of DNA and measurement of a cell population, its distribution around the cell cycle can be evaluated within a few minutes. Cells of the G_1 phase of the cell cycle show a fluorescence corresponding to two copies of each chromosome (DNA index = 1,0); cells in the DNA synthesis phase have an increasing amount of cellular DNA, whereas cells in the G_2 or mitotic phases double their DNA content (Figs. 2 and 3). Depending on the resolution of the technique, the fluorescence of the different phases show some overlap, but with computer programs using parametric or nonparametric statistical methods, the percent of cells in the different cell-cycle phases can be quantitated. Unless special preparation methods are used, G_1 and G_0 cells cannot be discriminated from each other, because they have the same DNA index = 1, neither can G_2 and mitotic cells. [*See* DNA SYNTHESIS.]

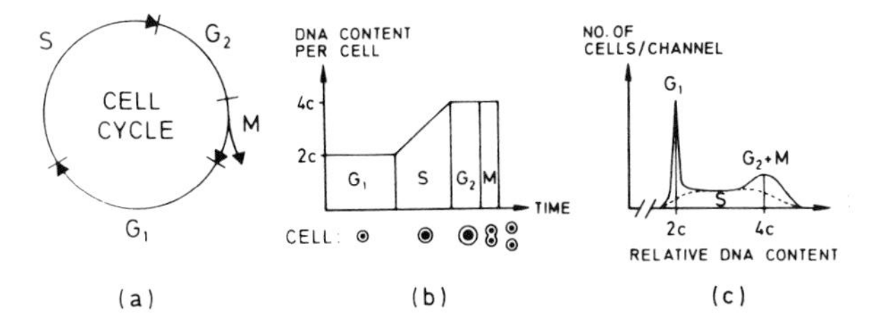

FIGURE 2 Schematic diagrams of a) cell cycle, b) cellular DNA content related to the cell cycle, and c) the resulting cellular DNA distribution curve, where the areas of the different cell-cycle phases are indicated.

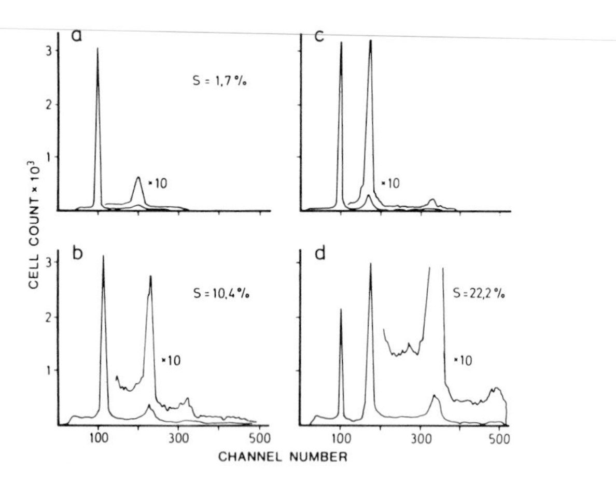

FIGURE 3 Example of DNA distribution curves in normal and malignant human cells obtained by aspiration from the uterus: a) normal endometrium at the sixteenth day of the cycle, b) a diploid endometrial carcinoma, c) ascitic fluid from a patient with an aneuploid tumor, and d) the primary aneuploid endometrial carcinoma (DNA index = 1.7). The proportions (%) of cells with S-phase DNA content are indicated for each curve. The areas of the curves corresponding to the S- and G_2-phases are also shown with an amplification of 10×. (From Iversen, O. E., 1986, *Am. J. Obstet. Gynecol.* **155,** 770–776.)

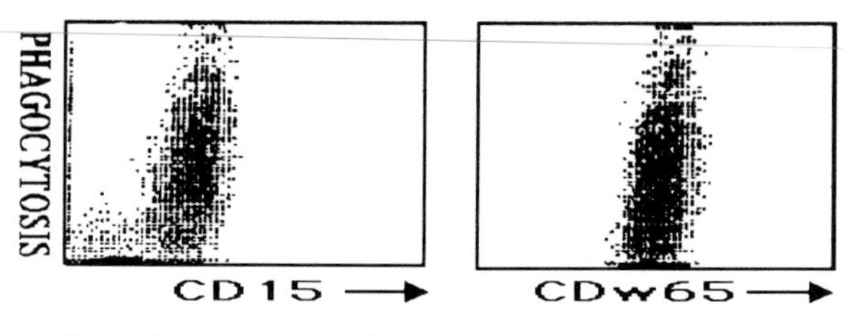

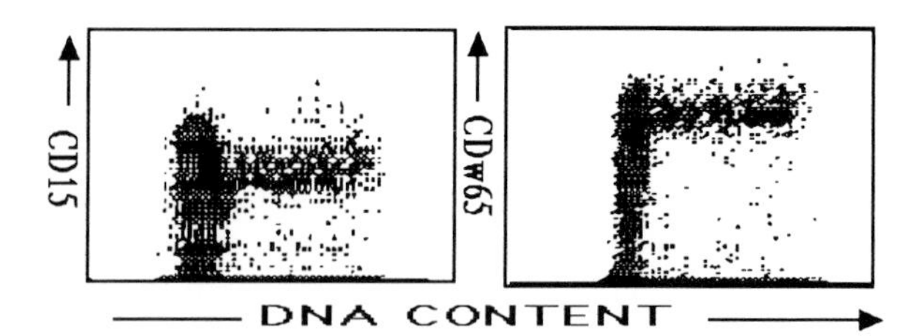

FIGURE 4 Example of cluster analysis of two simultaneously measured parameters in human bone marrow cells. The percentages of each subpopulation can be quantitated and further characterized by electronic gating. (Courtesy of F. Lund-Johansen.)

If cells synthesizing DNA are offered the nucleoside analogue BrdU, this will be incorporated into DNA instead of thymidine. Using fluorescent staining of the chromosomes or using FITC-labeled antibodies to BrdU, single-cell fluorescence will be dependent on how much BrdU it has taken up, i.e., its position in the cell cycle can be identified. Whereas an ordinary cellular DNA distribution curve only gives a static picture of the cell cycle, the BrdU technique gives the actual rate of cell cycling. Antibodies to different cell-cycle phases have also been used for their discrimination. Cell kinetics of, for example, human tumors are now being studied *in vivo* by using these labelling techniques and subsequent flow cytometry on biopsy materials.

B. Immunology and Leukocyte Functions

For many years, flow cytometry has been used for identification and quantitation of cells with different cell-surface antigens (Fig. 4). With monoclonal antibodies to such antigens, the numbers of different subclasses of lymphopoietic cells can rapidly be quantitated and separated in samples from both peripheral blood and lymphoid organs. The binding of complement and antibodies to immune cells can also be quantitated. In addition, the different aspects of phagocytosis can be identified and quantitated, either by use of fluorescent microorganisms or beads. By various combinations, adhesion, internalization, degradation, and exocytosis of bacteria exposed to human phagocytes can be measured sequentially and automatically. Defects in leukocyte functions can likewise be estimated. In addition, the killing of viable bacteria can be quantitated, as well as the degradation of both DNA and protein in bacteria exerted by phagocytes. Intracellular pH in granulocytes and macrophages degrading bacteria (phagosomal pH) can also be measured.

C. Hormone Research

By using fluorescent hormones, their binding and internalization can be studied and quantitated. The same applies to the binding of hormones to receptors and their internalization.

D. Measurements of Various Cell Components

Depending on the component that is stained, specific quantitative measurements can be performed of, for example, organelles, parts of the membrane, nuclear components, isolated chromosomes, RNA, lysosomes and components of the cytoskeleton.

E. Use in Cancer Research

So far flow cytometry in cancer research has concentrated on the identification of abnormal distribution of DNA in tumors, including aneuploidy, altered cell cycle parameters, and identification of abnormal chromosomes.

The slit-scan method and two-parameter measurements of, for example, cellular protein and DNA enables malignant cells to be discriminated from normal cells. In recent years, such parameters, and mainly DNA aberrations, have gained increasing interest for establishing prognosis in various tumor types.

F. Chromosome Analysis

Both normal and abnormal chromosomes can be measured and separated when chromosomes from mitotic cells are suspended. This applies to both their individual DNA content and their shape. So far, most success has been obtained by measuring cells from animals with few chromosomes (e.g., hamster, although human chromosome analysis is now performed routinely). Combined with different fluorescent DNA probes, the potential for identification of different genes is great.

G. Prokaryotes, Viruses, and Plant Cells

DNA, RNA, and protein can be measured in prokaryotic as in eukaryotic cells. This includes the distribution of DNA in growing bacterias, as well as their production of components with native fluorescence (e.g., chlorophyll). Also, viral DNA and RNA can be quantitated, and in some instances even single molecules can be detected with flow cytometry. Special devices are also available for the measurement of large cells, increasing the diameter of the sample flow. In addition, viral, bacterial, and plasmodial infections in cells can be identified and quantitated. Also, plant cells and algae can be analyzed with flow cytometric methods.

VI. Current Developments

Flow cytometry should be considered as a family of methods that combine high technology and automated measurements with advanced staining and preparation techniques for single cells. The flow cell and mechanical parts of the equipment have essentially remained the same during the last decade, although several practical improvements have been made. Developments have mainly been in preparation and staining techniques, by which almost all types of molecules and components can be identified and measured quantitatively in various cells. In addition, applications on different cell and organ types have rapidly expanded.

At present, one area of development is computer technology. Because large amounts of data are generated per second, the analysis is dependent on appropriate computer programs and large computers. This is especially the case when several different parameters are measured simultaneously and are analyzed in relation to each other. In recent years, great improvements have been achieved in multiparameter analysis and data handling. In this way multiple subpopulations within the same cell sample can be discriminated and measured for other parameters or sorted for further studies of biological or morphological parameters. This is facilitated by the use of a list mode computer function.

It is expected that in the near future simpler and less expensive machines will be available that will automatically measure multiple parameters on cells under highly standardized conditions.

Some areas of applications now under development are analyses of gene amplifications in mammalian cells, chromosome classification from both mammalian and plant cells, the measurement of chromatin in germ cells, identification of different gene products by in situ hybridization methods, and not least, clinical applications which lead to diagnostic and prognostic improvements in several human diseases.

Bibliography

Bjerknes, R., Bassoe, C.-F., Sjursen, H., Laerum, O. D., and Solberg, C. O. (1989). Flow cytometry for the study of phagocyte functions. *Rev. Infect. Dis.* **11,** 16.

Brown, S. C., and Bergounioux, C. (1988). Plant flow cytometry. *In* "Flow Cytometry: Advanced Research and Clinical Applications," Vol. II (A. Yen, ed.). CRC Press, Boca Raton, Florida.

Gray, J. W., ed. (1989). "Flow Cytogenetics." Academic Press, Ltd., London.

La Via, M. F., Hurtubise, P. E., Hudson, J. L., and Stites, D. P., eds. (1988). "Clinical Applications of Cytometry." *Cytometry,* Suppl., **3,** 1.

Laerum, O. D., and Farsund, T. (1981). Clinical applications of flow cytometry: A review. *Cytometry* **2,** 1.

Laerum, O. D., Lindmo, T., and Thorud, E., eds. (1980). "Flow Cytometry IV." Universitetsforlaget, Oslo.
Melamed, M. R. Mullaney, P. F., and Mendelsohn, M. L., eds. (1979). "Flow Cytometry and Cell Sorting." John Wiley & Sons, New York.
Shapiro, H. M., ed. (1988). "Practical Flow Cytometry," 2nd ed. Alan R. Liss, Inc., New York.
Van Dilla, M. A., Dean, P. N., Laerum, O. D., Melamed, M. R., eds. (1985). "Flow Cytometry: Instrumentation and Data Analysis." Academic Press, Ltd., London.
Yen, A., ed. (1989). "Flow Cytometry: Advanced Research and Clinical Applications," Vols. I and II. CRC Press, Boca Raton, Florida.

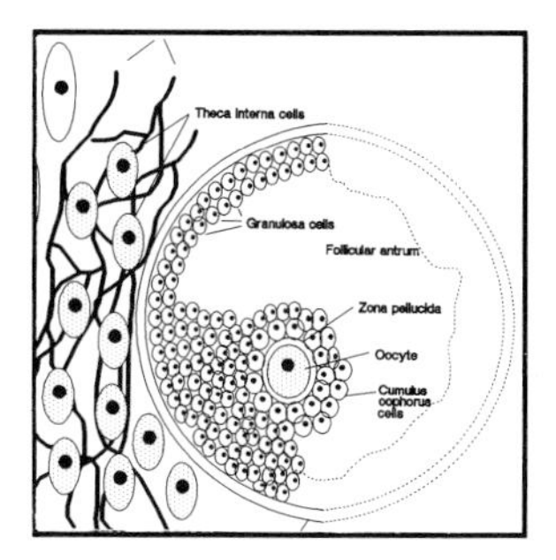

Follicle Growth and Luteinization

COLIN M. HOWLES, *Serono Laboratories (U.K.) Ltd.*

Glossary

Aromatization Enzymatic conversion of androgens to estrogens, which occurs mainly in granulosa cells; this aromatase enzyme system present in granulosa cells is stimulated by follicle-stimulating hormone

Atresia Degeneration of the follicle and its cells leading to death of the oocyte; occurs continuously at all stages of follicular development

Corpus luteum Develops after ovulation from the thecal and granulosa cells of the ruptured follicle; site of progesterone and estrogen synthesis; regresses if implantation of an embryo does not occur

Graafian follicle Well-developed antral follicle, which is destined to ovulate (named after the anatomist Reiner de Graaf, 1641–1673)

Granulosa cells Surround the oocyte throughout its development; possess aromatase enzymes that stimulate the conversion of androgens to estrogens; proliferate during follicular development

Luteinization Morphological and biochemical changes that occur in the follicle cells after ovulation

Menopause The last menses in women; marks cessation of cyclical ovarian activity

Menstrual cycle Ovary and female genital tract undergo cyclical changes over a 28-day period governed by biphasic variations in ovarian and gonadotrophin hormone secretion; the cycle is divided into two discrete parts, the follicular and luteal phases. Menses (bleeding associated with the loss of the endometrium) marks the beginning of the next menstrual cycle

Recruitment Occurs in the late luteal phase of the previous menstrual cycle when a group, or cohort, of small antral follicles are recruited into the final growth phase by raised blood levels of follicle-stimulating hormone

Selection One of the antral follicles (the most well developed in terms of granulosa cell number and aromatase action) is selected to continue development and become a mature Graafian follicle

Thecal cells Consist of two layers, the theca externa and interna cells (the latter possessing a rich blood supply); theca interna synthesize androgens under the influence of luteinizing hormone

THE FUNDAMENTAL UNIT of the ovary is the follicle that consists of the female germ cell (oocyte) surrounded by a series of specialized cell layers, the granulosa and the thecal cells (see Fig. 1). These cells are responsive to gonadotrophic hormone stimulation and cooperate to synthesize the most important ovarian steroid, estradiol-17β.

Every month during the woman's reproductive life, one oocyte is released (ovulation) from the single mature follicle that has completed development. If the oocyte is fertilized, the resultant pre-embryo is transported into the uterus where, if hormonal conditions are correct, it will implant and develop into a new and unique individual.

The optimum hormonal environment for implantation is coordinated by the corpus luteum, the remains of the follicle once the oocyte has been released. After ovulation, the follicle cells undergo luteinization. This is the collective term for a num-

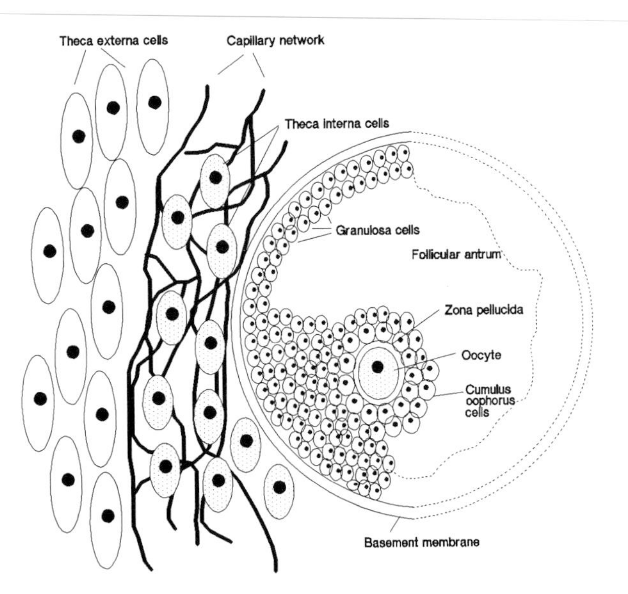

FIGURE 1 Diagram of a Graafian follicle showing the arrangement of follicle cells around the oocyte. The thecal cells consist of two layers: theca interna and theca externa. The theca interna cells possess a rich blood supply and are separated from the avascular granulosa cells by the basement membrane (follicle wall). (From Baird, 1984.)

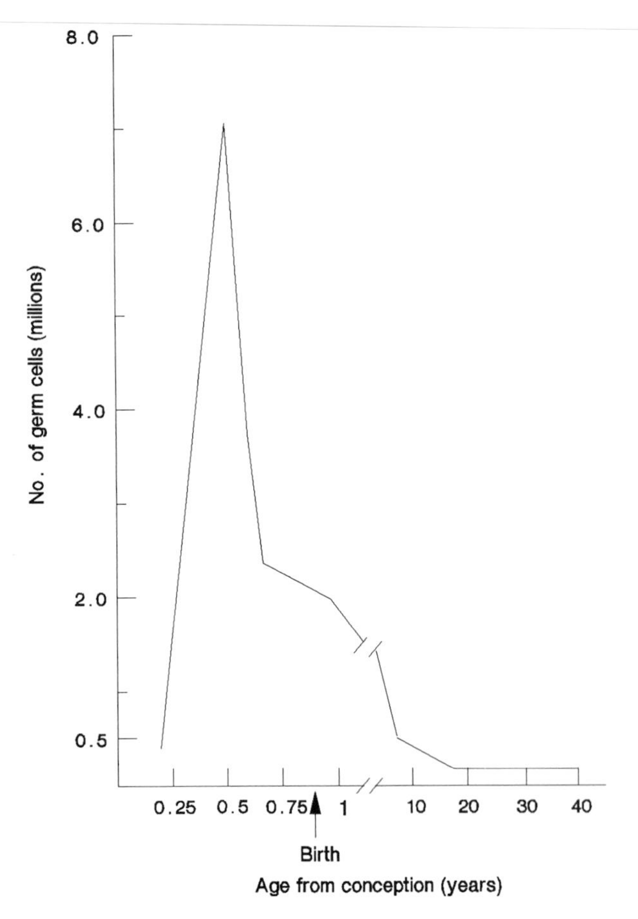

FIGURE 2 Total number of germ cells in the human female during life. Peak numbers occur before birth and then continuously decline until menopause. (From Baker, T. G., 1971, *Am. J. Obstet. Gynecol.* **110,** 746.)

ber of important biochemical and morphological changes that occur in the cells.

The physiological mechanisms that control the growth of the follicle, ovulation, and fertilization are certainly complex, but they have evolved to maximize the reproductive potential of the species.

I. Early Oogenesis

Unlike the male, the female germ cell (oogonia) undergoes mitotic proliferation in the ovary prior to the time of birth. All of these oogonia possess 46 chromosomes. From about the third month of gestation, increasing numbers of oogonia start to enter their first meiotic division, thereby becoming primary oocytes with a chromosome complement of 23. By the time of birth, or soon after, all female germ cells are primary oocytes. [*See* OOGENESIS.]

At all stages of follicle development, however, degeneration, or atresia, occurs. In the human female, about 99% of oocytes are lost in this way. The consequence of atresia is that the number of oocytes is gradually reduced from birth until the time of menopause in women, when very few oocytes can be detected in histological sections of the ovary. Changes in germ cell numbers during the life of a human female are illustrated in Fig. 2.

Soon after formation, the primary oocyte becomes surrounded by a single layer of flattened epithelial (granulosa) cells. This group of cells constitutes the primordial follicle. Further meiotic division of the primary oocyte is halted at the dictyotene stage of prophase and it may remain in this state for up to 50 years. It is not known why the oocytes are stored in this protracted meiotic state.

The control of the next stage of folliculogenesis is not well understood but seems to be independent of gonadotrophin support. The oocyte undergoes its major growth phase marked by massive synthetic activity and morphological changes. The granulosa cells become cuboid in shape and increase in number forming four or five layers around the oocyte, a preantral follicle.

However, during the next phase of follicular development, which occurs from puberty onward, growth is dependent upon continuous secretion of hormones from the anterior pituitary. At the beginning of each menstrual cycle, a group, or cohort, of follicles enters the gonadotrophin-dependent phase of growth. In the next section, the role of these hormones in the human female will be discussed.

II. Gonadotrophins and the Ovary

Gonadotrophin-releasing hormone (GnRH) is secreted episodically into the portal blood system linking the hypothalamus to the anterior pituitary. GnRH release is affected by steroid feedback from the ovary as well as external environmental cues. GnRH stimulates specialized cells in the pituitary (the gonadotrope cells) to secrete the gonadotrophins—luteinizing hormone (LH) and follicle-stimulating hormone (FSH). [*See* HYPOTHALAMUS; PITUITARY.]

The functioning of the adult ovary is dependent on the secretion of these gonadotrophic hormones. These interrelationships are diagrammatically represented in Fig. 3. The main properties of the two gonadotrophins, LH and FSH, and their action in the human female are shown in Table I.

The effect of LH and FSH on the ovary is to promote the synthesis of estradiol-17β, the predominant estrogen secreted by the ovary. In turn estradiol, acting on the hypothalamic pituitary axis, modifies LH and FSH secretion. During the luteal phase progesterone is the predominant ovarian steroid that modulates gonadotrophin secretion.

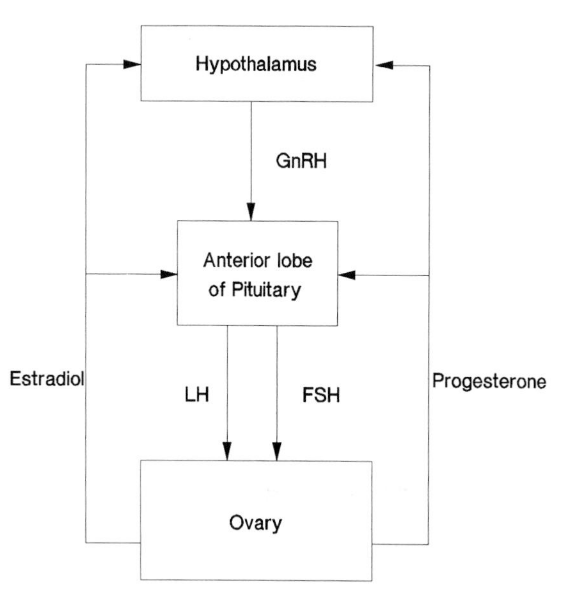

FIGURE 3 Hormone secretion between the hypothalamus, pituitary, and ovary. Estradiol and progesterone are the most important ovarian steroids that modulate gonadotrophin secretion in the follicular and luteal phases, respectively. Depending upon the stage of the menstrual cycle these steroids can have both positive and negative feedback effects upon gonadotrophin secretion (see text).

Before discussing in detail follicular growth, it is necessary to first consider how estradiol is synthesized by the ovary. Estradiol production is a coop-

TABLE I Properties and Action of the Gonadotrophic Hormones

	Gonadotrophins	
	LH	FSH
Secreted from:	Anterior pituitary under the influence of GnRH-release from the hypothalamus.	
Composition:	Glycoprotein made up of an alpha and beta chain with 16.4% carbohydrate residues.	Glycoprotein made up of an alpha (identical to that in LH) and beta chain with 25.9% carbohydrate residues.
Molecular weight:	34,000	32,600
Cells acted on:	Thecal cells of antral follicles. Granulosa cells of preovulatory follicles. Luteal cells of the corpus luteum.	Granulosa cells of preantral and antral follicles.

erative venture involving both the theca interna and the granulosa cells of the follicle.

III. The Two-Cell Theory of Follicular Steroidogenesis

The requirement for cell cooperation in the production of estradiol-17β was first demonstrated by Falck in 1959. In a series of classical experiments, he showed that estrogen formation by the rat follicle depends on the joint action of the theca interna and the granulosa cell layers.

Short in 1962, first proposed a two-cell theory to explain follicular steroidogenesis. This hypothesis was subsequently modified, as later experiments showed that, although thecal and granulosa cells can independently synthesize estrogens, the yield is greatly increased if both cell types are incubated together. This interaction of two cell types, independently stimulated by the two gonadotrophins, LH and FSH, led to the final realization of a "two-cell type, two gonadotrophin" theory for estradiol synthesis in the ovary, as illustrated in Fig. 4.

Although this theory requires the involvement of both gonadotrophins for normal follicular steroidogenesis to occur, some data suggest that normal folliculogenesis can occur independently of LH secretion. In these experiments, follicular development occurred normally, but estradiol production was suboptimal.

The principal androgens, androstenedione and testosterone, are produced by the thecal cells, which are stimulated by LH. The androgens pass into the blood or are transported across the basement membrane of the follicle into the granulosa cell where they are converted into estradiol by the aromatase enzyme stimulated by FSH. Estradiol is then secreted back into the blood or accumulates in the follicular fluid, keeping the intrafollicular environment highly estrogenic. For a mature follicle, the concentration of estradiol in the follicular fluid can be 1,000 times that circulating in the blood. This could further potentiate LH and FSH action, as estradiol has been shown to increase the sensitivity of the follicle to gonadotrophin stimulation.

The importance of this two-cell type–two gonadotrophin interaction for normal follicular development is aptly illustrated in the clinical condition known as polycystic ovarian disease (PCOD). Although the etiology of this condition is complex, it

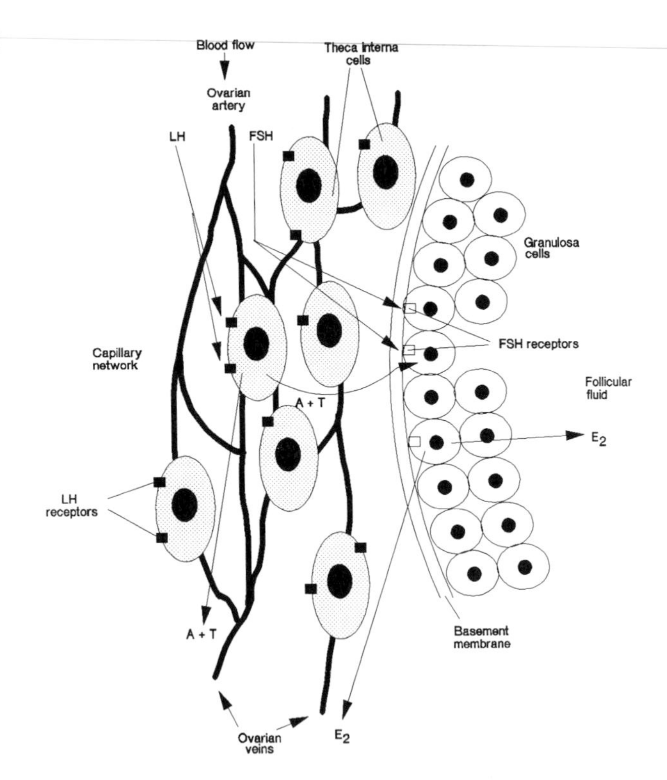

FIGURE 4 Schematic representation showing the action of the gonadotrophins on the follicle cells and the synthesis of estradiol. LH attaches to specific receptors (■) on theca interna cells. Under LH stimulation, these cells produce androgens (androstenedione, A; testosterone, T). These steroids are secreted into the blood or pass through the basement membrane of the follicle and enter the granulosa cells. FSH interacts with receptors (□) on the granulosa cells and activates the aromatase enzyme system, which converts androgens to estradiol (E_2). E_2 is either secreted out of the follicle and into the bloodstream or is concentrated in follicular fluid. (From Baird, 1984.)

is sufficient to say that in 80% of subjects LH secretion is higher than normal, which is a contributing factor to excess production of androgens by the thecal cells. Patients with PCOD can suffer with chronic anovulation and are more likely to be obese. The ovaries of these women contain numerous antral follicles that are androgenic rather than estrogenic. This can be rectified by the administration of FSH, which initiates normal follicular growth and estradiol production.

IV. Follicular Growth

It takes approximately 85 days, or three ovarian cycles, for the development of a preovulatory, or Graafian follicle, from an early preantral follicle. At

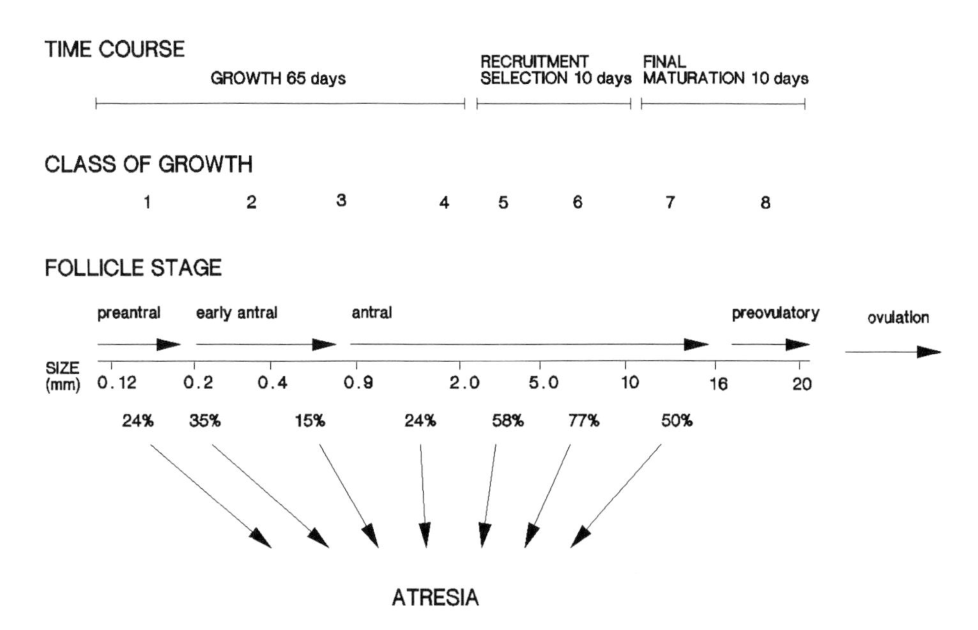

FIGURE 5 Time course of follicle growth. Class 1 is primary, and Classes 7 and 8 are Graafian follicles. At all stages of development, follicles degenerate (atresia). In the human female, only one follicle usually undergoes ovulation each month.

the onset of this development, a large number of primary follicles are capable of initiating growth. Those that do begin to grow are influenced by many factors, such as age of the woman as well as hormone activity and possibly nutritional status. Only one follicle will eventually become mature, ovulating and releasing the oocyte; the rest will undergo atresia.

The life cycle of a follicle during this 85-day period is shown in Fig. 5. There are eight classes of growing follicles, and at each stage many become atretic. Once the follicle reaches about 0.2 mm in diameter, fluid starts to collect between the granulosa cells. These pockets of fluid grow and merge together forming an antrum, hence the term antral follicle.

A. Follicular Recruitment

After about 65 days of growth, a final cohort of about 15–20 small antral follicles will enter the gonadotrophin-dependent phase of growth. Fig. 6 depicts the gonadotrophin and steroid hormone changes that occur during the follicular phase of the human menstrual cycle. The length of the follicular phase is primarily determined by the rate at which the principal antral follicle matures. This follicle exerts its control by secreting hormones that orchestrate the pattern of gonadotrophin release from the anterior pituitary. This process will be addressed in the next section.

If pregnancy does not occur, the corpus luteum regresses, leading to a decline in progesterone levels. Concomitant with this reduction in corpus luteum activity, FSH levels increase. Only those small antral follicles that have acquired gonadotrophin receptors, coincident with this intercycle rise in FSH, will be recruited into the next growth phase.

The most crucial event for the further development of an antral follicle is the activation of the aromatase system by FSH. It is now widely accepted that each small antral follicle (about 4 mm in diameter) has a threshold requirement for stimulation by FSH. This was first postulated by Jim Brown in 1978. Brown's hypothesis was formulated from the treatment of infertile patients with exogenously administered gonadotrophins of pituitary origin. It was found that the ovary could detect and respond to changes in blood levels of gonadotrophins in the region of 10–30%. Thus, a modest rise in FSH levels is all that is required to initiate the recruitment of antral follicles.

Because follicular development in the human female is highly asynchronous, at the time of the intercycle rise in FSH the ovaries contain a cohort of follicles with varying sensitivities to FSH. The follicle with the lowest FSH threshold will be the first to undergo activation of the aromatase system and begin estradiol production.

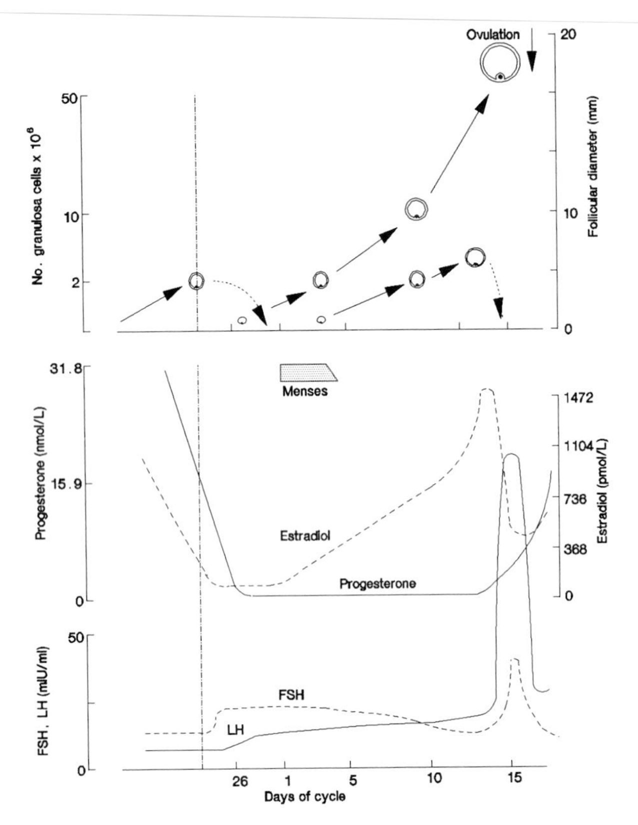

FIGURE 6 Top graph depicts the growth of antral follicles, and bottom graphs the plasma levels of gonadotrophins, estradiol, and progesterone during the follicular phase of a menstrual cycle. Only those follicles at the appropriate stage of development concomitant with the intercycle rise in FSH are recruited into the final growth stage. (From Baird, 1984.)

B. Hormone Production from the Follicle

The production of steroids and the increase in follicular size are intimately linked. The appearance of estrogen receptors in the granulosa cells stimulates cell proliferation and a further increase in the ability of the follicle to aromatize androgens to estradiol. (The hormone–cell interactions and steroid production of developing antral follicles are summarized in Table II.) This action is a classic example of a positive feedback loop, and it ensures an increasing capacity of the follicle to convert androgens to estradiol leading to a surge in estradiol production.

The increase in estradiol secretion from the ovary has been demonstrated to have a gonadotrophin-suppressive effect on the hypothalamic pituitary axis. In particular, FSH secretion begins to decline prior to the midfollicular phase of the menstrual cycle. However, it has been hypothesized for some years that the ovary also secretes a nonsteroidal factor (inhibin), which specifically inhibits FSH secretion. It was not until the early 1980s that experiments proved the existence of inhibin. These studies involved the use of follicular fluid that contains high concentrations of inhibin-like activity. When injected *in vivo,* FSH secretion was selectively inhibited.

More recently, inhibin has been purified and shown to be a glycoprotein of about 32,000 molecular weight. It is composed of two subunits linked by disulfide bridges. Inhibin is produced by granulosa cells in response to FSH and androgen stimulation, secreted into the circulation, or concentrated in follicular fluid. However, recent evidence suggests that inhibin does not play a role in the modulation of FSH secretion during follicular development. In the spontaneous menstrual cycle, changes in inhibin concentrations are not related to those in FSH. As we shall see later, inhibin becomes important during the luteal phase.

C. Selection of the Dominant Follicle

It seems, therefore, that FSH secretion in the follicular phase is inhibited by estradiol. The decline in blood FSH levels is in response to increased estradiol secretion from growing follicles. The result of this reduction in FSH during the spontaneous menstrual cycle is the "selection" of one antral follicle, which continues development and ultimately ovulates.

When the fall in FSH occurs, the selected follicle is less dependent on circulating levels of FSH. This is probably because it had the lowest FSH threshold at the onset of the intercycle FSH rise. The follicle will have had longer to activate its aromatase system, leading to higher estradiol production and greater granulosa cell proliferation than its rivals.

TABLE II Hormone–Cell Interaction and Subsequent Steroid Synthesis in Developing Antral Follicles

	Thecal cell	Granulosa cell
Hormone	LH	FSH (estradiol)
Steroid synthesis	Acetate/cholesterol to androgens	Androgens to estradiol

Once selected, the follicle is called a Graafian follicle. Selection occurs by about day 7 of the menstrual cycle. The rest of the cohort of follicles become atretic as FSH is suppressed below their own threshold level.

It is possible to overcome the follicle selection procedure by the administration of exogenous gonadotrophins. After gonadotrophin injections, FSH levels are elevated for a longer time period, thus allowing other follicles to continue development. This principle is practiced in patients undergoing *in vitro* fertilization treatment. Exogenous gonadotrophins can be given daily throughout the follicular phase of the cycle to promote "superovulation."

Another physiological process concerning selection may also contribute to the emergence of a dominant follicle. Rising levels of estradiol, in conjunction with FSH, induce the appearance of LH receptors on the outer layer of granulosa cells. Thus, there is a gradual change in distribution for gonadotrophin receptors, which may be critical for further follicular development. *In vitro* studies have shown that granulosa cells possessing both FSH and LH receptors responded identically to both hormones in terms of aromatase activity and steroid production. This may mean that the presence of both LH and FSH receptors on granulosa cells may further protect the emerging dominant follicle from declining FSH concentrations.

V. The LH Surge and Ovulation

Once LH receptors have been fully acquired by the granulosa cells, the Graafian follicle can enter the final or preovulatory phase of growth. This terminal growth phase is signaled by a surge in gonadotrophin output, the LH surge. As the large Graafian follicle reaches maturity estradiol output reaches a peak. In such a highly estrogenic environment, the pulse frequency of GnRH is more rapid and the sensitivity of the pituitary gonadotrope cells to GnRH is greatly enhanced. These events lead to a massive discharge of LH.

The effect of the LH surge is twofold: It causes profound changes to the structure and function of the follicle and, secondly, stimulates the resumption of meiosis in the oocyte.

The follicle undergoes a final rapid growth phase, mainly due to an increase in the volume of follicular fluid. Also, major changes occur in the endocrinological activity of the follicle cells.

Of major importance is that the granulosa cells can no longer produce estradiol by aromatization and, thus, lose their FSH and estradiol receptors. Instead, they start to synthesize progesterone through LH stimulation. These changes in hormone–cell interactions are summarized in Table III. This results in an increase in progesterone secretion from the follicle, concomitant with the rise in LH.

The production of progesterone may also be important in facilitating the positive feedback effects of estradiol on LH release. Recently, women undergoing an *in vitro* fertilization attempt have shown that a single injection of progesterone can elicit an LH surge. Thus, LH secretion from the pituitary is maximized, ensuring that final follicular maturation is completed.

While progesterone is being synthesized from the follicle, the chromosomes of the oocyte progress through the first meiotic division. Although the mechanisms are not fully understood, it has been proposed that rising levels of LH either inhibit the action or block the synthesis of an oocyte maturation inhibitor factor, thus allowing terminal maturation to occur. Recent studies have shown that if LH levels are raised prior to the LH surge then premature maturation of the oocyte can occur.

This phenomenon has been highlighted in patients undergoing superovulation for *in vitro* fertilization. In one study, high levels of LH were associated with failure of implantation and early pregnancy loss, whereas low levels of LH were associated with the establishment of ongoing pregnancy. Table IV summarizes the results of this study. This effect of LH on pregnancy outcome has been confirmed by other studies, including one in women not undergoing any infertility treatment. Further work is required to elucidate the mechanism of LH action on oocyte maturation.

TABLE III Hormone–Cell Interaction and Steroid Synthesis in a Mature Graafian Follicle

	Thecal cell	Granulosa cell
Hormone	LH	LH
Steroid synthesis	Acetate/cholesterol to androgens	Acetate/cholesterol to progesterone

TABLE IV LH Secretion in the Late Follicular Phase of Treatment and Outcome of *in vitro* Fertilization[a]

Outcome of treatment	Number of patients	Urinary LH secretion (IU/L/h ± sem)
Ongoing pregnancy	88	0.17 ± 0.01
Nonpregnant	92	0.20 ± 0.01
Miscarried	29	0.26 ± 0.03

[a] Refer to Howles and Macnamee, 1989.

Final meiotic division of the oocyte is peculiar in that the distribution of cellular material is grossly unequal. Only a very small amount of cytoplasm accompanies one half of the divided chromosomes; this forms what is called the first polar body, which is extruded to one side of the maturing oocyte.

Furthermore, the follicle wall undergoes dramatic changes. The rapid expansion of the follicle at this time stretches the follicle wall and, probably through the action of collagenase enzymes, particularly plasmin, and prostaglandins, the wall starts to break down. Where this occurs an outward bulge appears, the stigma, which eventually ruptures, releasing the follicle contents. By this time, the oocyte is only connected to the mass of granulosa cells by a very thin stalk of cells, which easily breaks allowing the oocyte to be extruded in the flow of follicular fluid.

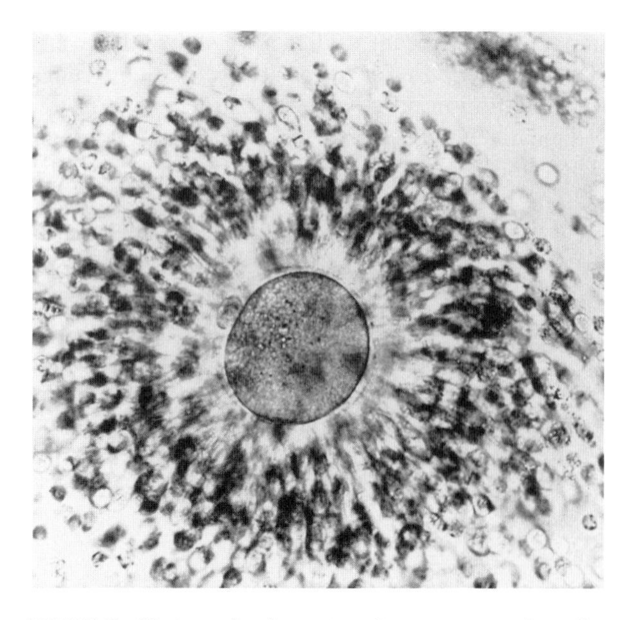

FIGURE 7 Photograph of a mature human oocyte, just after having been removed from the follicle for in vitro fertilization. The oocyte is surrounded by a mass of cumulus cells showing a dense corona radiata (sunburst appearance).

An oocyte, freshly harvested from a preovulatory follicle of a woman undergoing an *in vitro* fertilization treatment, is shown in Fig. 7. Note the sunburst arrangement of cumulus oophorus cells around the oocyte; this is highly characteristic of a mature human oocyte.

VI. Luteinization and the Formation of the Corpus Luteum

In the final stages leading up to ovulation, the follicle cells lose their LH receptors and become desensitized to further LH stimulation. However, about 2–3 days after ovulation, they recover their ability to respond to LH. In rodents, this recovery of receptors is stimulated by prolactin. Progesterone becomes the main secretory product of the collapsed follicle, which is now called the corpus luteum. The granulosa cells no longer divide but increase in size and undergo internal structural changes. These include the production of a carotenoid pigment called lutein, which gives the corpus luteum its yellowish appearance in many species. Such cells are said to be luteinized.

In addition to the secretion of progesterone, the corpus luteum of some higher primate species (including the human) produces estradiol. Both of these hormones are necessary to prepare the endometrium in the event of an embryo implanting.

During the luteal phase, gonadotrophin secretion is low, suppressed by the negative feedback effects of progesterone and estradiol. In addition to this negative feedback effect on gonadotrophin release, recent research has demonstrated that the corpus luteum also secretes large quantities of inhibin. Thus, inappropriate follicular development is completely inhibited during this phase of the cycle.

Normally, the corpus luteum starts to regress about 1 week after ovulation. However, if implantation occurs, regression of the corpus luteum and, hence, endometrial degeneration will not occur. It has been shown that the implanting embryo secretes a glycoprotein, human chorionic gonadotrophin, which extends the life of the corpus luteum. Human chorionic gonadotrophin has a similar action to LH and promotes hormone synthesis in the corpus luteum.

In the case of an embryo not implanting, the fall in progesterone and inhibin concentrations from the waning corpus luteum lead to a resumption of gonadotrophin secretion. Thus, the intercycle rise in FSH leads to a new phase of follicular recruitment and growth.

Bibliography

Baird, D. T. (1987). A model for follicular selection and ovulation: Lessons from superovulation. *J. Steroid Biochem.* **27,** 15.

Baird, D. T. (1984). *The ovary, in* "Reproduction in Mammals: Book 3" (C. R. Austin and R. V. Short, eds.). Cambridge University Press, Cambridge.

De Cherney, A. H., Tarlatzis, B. C., and Laufer, N. (1985). Follicular development: Lessons learned from human *in vitro* fertilization. *Am. J. Obstet. Gynecol.* **153,** 911.

Glasier, A. F., Baird, D. T., and Hillier, S. G. (1989). FSH and the control of follicular growth. *J. Steroid Biochem.* **32,** 167.

Hillier, S. G., Harlow, C. R., Shaw, H. J., Wickings, E. J., Dixson, A. F., and Hodges, J. K. (1988). Cellular aspects of pre-ovulatory folliculogenesis in primate ovaries. *Hum. Repro.* **3,** 507.

Howles, C. M. (1985). Follicle growth and luteinization. *In* "Implantation of the Human Embryo" (R. G. Edwards, J. M. Purdey, and P. C. Steptoe, eds.). Academic Press, London.

Howles, C. M., and Macnamee, M. C. (1989). The endocrinology of superovulation; the influence of luteinizing hormone. *Excerpta Medica Asia Ltd.,* Hong Kong.

Howles, C. M., Macnamee, M. C., and Edwards, R. G. (1987). The effect of progesterone supplementation prior to the induction of ovulation in women treated for *in vitro* fertilisation. *Human Repro.* **2,** 91.

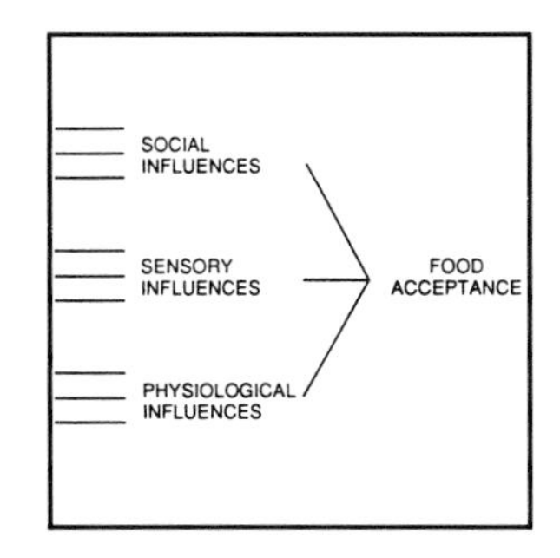

Food Acceptance: Sensory, Physiological, and Social Influences

D. A. BOOTH, *University of Birmingham, England*

Glossary

Cognition Objective thought, that is, mental processes capable of relating to some reality (complex information processing). This includes intention, emotion, and imagination as well as reasoning and perception. In current scientific usage cognitive processes do not have to be conscious (nor in a human mind).

Food Chemical mixture in a form that is regarded by an organism as suitable for ingestion. Alternatively, an actually nutritive material. (Ingestates consisting primarily of water are generally called drinks, and in that usage food is solid nutriment.) The totality of an organism's foods and drinks is its diet.

Physiological Refers to physical processes in any part of the body, including the gastrointestinal tract, liver, and brain, that (in this context) can have a normal causal effect on behavior toward food or drink. (Many other physiological processes in the central nervous system are also necessary for ingestion and all other behaviors or cognitions to occur and for sensory and social as well as physiological influences to be exerted.)

Sensory Perceptible aspect of the constitution of (in this context) a food or a drink. Sensory vocabulary is what can be used to describe physicochemical characteristics, but the use of a sensory descriptor score is not an objective measure of food constitution until shown to be such.

Social Pertaining either to one or more other people or to cultural, economic, or institutional processes operative in a society

THE ACCEPTANCE of food is a momentary act by which a person takes control of the ingestion of a food item, by purchasing it for eating, by preparing it for someone to eat, or by eating it. This implies that food acceptance behavior is physical intake of nutriment in response to its sensed characteristics, also depending on bodily needs and on socioeconomic context. The interactions of these influences on individual consumers' food acceptances might be more or less stable over time within an adult consumer, and more or less similar in structure and strengths across consumers within a culture. Nevertheless, sales, frequencies, or averages of responses from a sample of consumers, and average or individual intakes of a food, are the result of many diversely structured pieces of acceptance behavior and so cannot logically be assumed to provide accurate measures of that behavior; this requires the analysis of momentary situation-dependent influences on acceptance in representative individuals.

I. Cognitive Integration

A. Varieties of Food Acceptance Behavior

A variety of human actions can coherently and usefully be regarded as accepting a food.

1. Purchase Acceptance

Taking a product off the shelf and to the checkout counter of a food store is the act that is crucial for commerce. To explain and predict sales adequately,

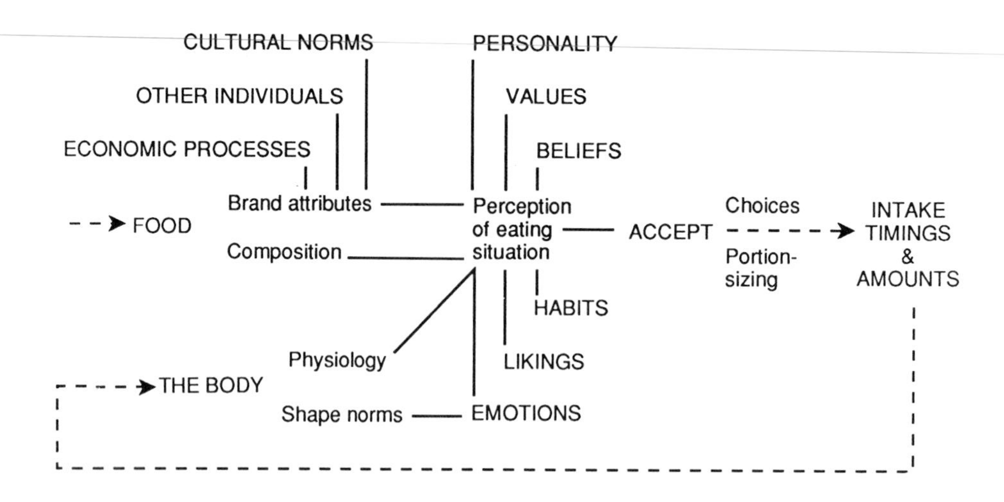

FIGURE 1 Immediate, mediating, and background sources of variation in people's food acceptance, with the classificatory or causal links between sources within and outside the individual mind.

we might need to understand more about the influences on food acceptance in the form of purchases.

Elaborate statistical modeling of preference data and/or sensory or other descriptive scores from consumer panels has increasingly been used in attempts to solve this problem. This is not, however, a mere data-analytic matter. The issues are the scientific ones of characterizing the causal processes within the behavior of consumers (Fig. 1). Research on the determinants of food acceptance therefore requires the science of cognitive psychology, which has been applied in other technologically complex areas (e.g., human–computer interaction and ergonomics) more broadly.

2. Eating Acceptance

Food is almost always bought to be eaten. People selecting for their own consumption from a store, buffet counter, or restaurant menu are effectively choosing what to eat. Furthermore, eating quality is likely to be a major influence on subsequent repurchasing behavior. Also, foods cannot affect health without being eaten.

Therefore, the preparation, serving, and ingestion of food already purchased (or grown domestically) are further examples of acceptance. Arguably, indeed, eating acceptance is the foundational criterion, or at least the central type, of food acceptance.

B. Scientific Concepts Related to Food Acceptance

Food acceptance, both at ingestion and during the "search" for food in shops and restaurants, is a phenomenon that has long been of major interest within the biological, medical, behavioral, and anthropological/sociological sciences.

1. Terminology in Major Disciplines

Eating habits and their social roles (e.g., food symbolism) are a major part of every human culture, both rural and industrialized. They have therefore been a focus of attention by many anthropologists and some sociologists.

The intake of materials resulting from food acceptance behavior is also a crucial influence on physiological processes (e.g., exchanges of water, sodium, energy, nitrogen, and other nutrients between the organism and the environment). However, studies of the intake of solids and fluids in regulatory physiology often lack the perceptual data to justify use of the behavioral concept of the sensory, physiological, or social control of food acceptance. As a result the physiologists and biopsychologists of food intake have often seriously distorted commonsense, but scientifically usable, concepts such as palatability, hunger, thirst, appetites, and satieties.

In the science of behavior itself, attention to food-oriented behavior has declined along with the dissolution of the specialism of "motivation" in psychology. Food intake has become a topic for neuroscientists. Food reinforcement is merely a tool for dissecting out learning mechanisms. Food attitudes only occasionally interest contemporary social psychologists. This lack of interest is paradoxical since eating and drinking are the most common cognitively rich human activities.

2. Terminology in Nutrition

Foods and beverages cannot affect health or disease unless they are in a diet (i.e., people eat and drink them). Dietary choices, sequences, and patterns, as perceived by the eater, are behavioral structures in the minds of individuals, equivalent to food acceptance. Dietary intakes are entirely the result of these cognitive–behavioral processes.

All intakes result from decisions made in terms of cultural constructs of foodstuffs and eating practices. There is concern over the effects of fermentables on the teeth, the timing of energy choices on the waistline, and preferences for foods high in saturates on the heart. Data on these dietary patterns are lost when weights of foods are totaled across the day. Clinical and community nutrition would advance by integrating soundly behavioral (and sociological) concepts equivalent to food acceptance into their research and practice.

3. Appetite: Organization in Food Acceptance Behavior

Appetite is the disposition to eat and drink, including the approaches to food in a shop, restaurant, or refrigerator.

This basic psychological sense of terms such as "appetite," "hunger," "thirst," and "satiety" must be distinguished from *ad libitum* dietary intake. Food intake is relevant to appetite, but only as a measure of the cumulative effect of expressions of appetite in the changing circumstances of a series of food acceptances. Thus, without full specification of the sensory, physiological, and social contexts influencing each momentary acceptance that contributed to the recorded total intake, the physical disappearance tells us nothing about the organization of the behavior.

It also must be appreciated that it is logically incoherent to take the word "appetite," even in its everyday usage, to refer to subjectively conscious mental processes alone. The appetite is in the directedness of what we actually do, and only secondarily in the private experiences that go with that observable performance. [*See* APPETITE.]

II. Empirical Unit of Acceptance Behavior

Food acceptance is a single momentary overt act of a person (or other organism) on a particular occasion. Thus, any acceptance of a food occurs in a particular context of alternative foods, bodily condition, physical and social environment, and other aspects of mental state. Also, any measure of behavior toward food that results from more than one act by one person toward one food in one situation might be confounded by variations over time in that person or by differences among people, foods, or situations.

A. Choice and Relative Acceptance

When two or more foods are present, the acceptance of one of them is an overt choice among the foods. In this context on this occasion, the chosen food has been preferred over the other food(s). Repetition of the choice by that person in similar contexts can establish that the choice was not random, or in some circumstances this can perhaps be assumed on other grounds.

Even if only one food is ever present at one time (i.e., monadic presentation), opportunities to accept can be provided in similar contexts for two or more foods in succession. Then, from the acceptance or refusal of the different foods, we can construct an observed relative acceptance.

This combination of acceptance and refusal on separate occasions can be represented by the same parameter as the combination of acceptances and refusals among the same set of foods presented simultaneously on repeated occasions (e.g., the proportion of acceptances of one food to the total acceptances from the set). Qualitative similarity of successive and simultaneous relative acceptances is evidence for the same underlying preference structure. Quantitative similarity is unlikely, however, since the strongest preference is freer to dominate behavior in the simultaneous test. The successive (i.e., monadic) test therefore has the advantage of sensitivity to less than the strongest disposition to accept a food. However, the real purchase or eating situation might be one of choice.

The issue that arises is always crucial: What are the objectives of the investigation? In particular, what situation of acceptance are we seeking to understand? In practice, using a familiar range of food products or variants of a food and monadic presentations within a fairly naturalistic situation provide sufficiently realistic and precise data to characterize and even to quantitate influences on either multi-item choice or single-item acceptance/rejection in the normal context.

B. Degrees of Acceptance and Graded Influences

This section has so far been couched in terms of acts of acceptance or rejection of particular foods in a particular situation. An acceptance is categorical—it occurs or it does not (except perhaps when two or more foods are put into the mouth or the shopping basket at once). Degrees of acceptance would be quantitatable as frequencies of acceptances (or refusals) in repeated tests.

Nevertheless, a single act can be used to generate an estimate of such frequencies over many yes/no tests. With people the simplest way to do this is to ask for a verbal judgment of the frequency with which they would accept a monadically presented item in some defined or assumed context. Alternatively, they can be asked to express a degree of liking or pleasantness; this appears to be more introspective, but that is exposed as an illusion of language when the hedonic rating is anchored on acts in specified situations. Then these ratings become the same as acceptance frequency ratings. In any case the qualitative validity and quantitative calibration of acceptance ratings of any sort are based on the actual frequencies of acceptance in the real-life decisions.

An influence on acceptance might also be inherently categorical (i.e., either present or absent), having no intermediate strengths. Brand name and food type are examples. However, these categories could be resolvable into sets of perceptible characteristics of the brand image or the food type's composition or uses; in other words, brand names or food types might scale onto several continua, whether sensory (e.g., sweet), physiological (e.g., filling), or social (e.g., breakfasttime or preschool age). The basis can be identified for either a nominal or an ordinal scale by finding the observable characteristic(s) of an influence on acceptance that regresses linearly and precisely onto degrees of acceptance by the individual.

C. Generalizing

The usual concept of the *acceptability* of a food is highly abstract and often rather unrealistic. Food acceptability, as commonly conceived, generalizes across consumers, eating or purchasing contexts, and sets of foods among which choices are being made. In consequence, sensory acceptability or palatability is commonly misconstrued as an inherent characteristic of the food. Even if the sensory influences on acceptance of a food did not vary among people or situations (as they commonly do), palatability would still be a characteristic of behavior toward the food, not of the food itself—of a causal relationship between a person and a food, not of one of the terms in that relationship.

At a moment in a meal, whether or not more of a food will be eaten depends on how boring and filling it has so far become and on anticipatory decisions when serving onto the platter or when purchasing the items for the meal. Whether another food now becomes accepted could depend on any of these factors for other available foods and on the next food's suitability to the upcoming stage of the meal. In Western cuisines the acceptabilities of meat pie and apple pie reverse between the first course and the dessert, although, of course, they do not compete when shopping for an entire meal.

Acceptance of a food is often also contingent on the time of day (cf. breakfast foods, between-meal hot drinks, and snack items), company at the table (e.g., spouse, children, and guests), season of the year (e.g., hot or cold weather), health concerns, specialness of the occasion, etc. Still further factors can be particularly influential at the point of purchase (e.g., price, advertising messages, and package design and information). Thus, a generalization about food acceptance must either specify the particular context assessed, in all its potentially influential aspects, or include the effect of an influence within the generalization.

Most challenging of all, people differ from one another in how they put all of these sensory, physiological, and social influences together into habitual acceptance decisions. Until recently, all approaches to influences on food acceptance have assumed that there are simply additive combinations among influences, and that therefore the analysis of data lumped across people would not be qualitatively, or even quantitatively, too misleading. In fact, many influences simply do not add up. Qualitatively different structures of interactions among influences also can be quite common. There has indeed long been evidence that the results of sensory tests with and without brand names sometimes differ greatly; that acceptabilities of food products after use often differ from, and can be even more diverse than, acceptabilities at first acquaintance and that the awareness parameter in a brand life-cycle model has nothing to do with the repurchase parameter.

D. Measurement of Influences within Acceptance Behavior

1. Disconfounding of Influences

Methodologically, the classes of influence on food acceptance must be distinguished and acceptances must be categorised according to the influence(s) on them.

a. Palatability and Satiety The palatability of a food can be assessed only by the difference that the purely sensory factors make to intake, or to some other measure of acceptance. The immediate sensory influences must not be confounded by other influences (e.g., secondary cognitive or physiological effects of palatability or the effects of the swallowed food).

Conversely, meal size cannot be a valid measure of postingestional effects unless both sensory effects and changes in reactions to sensory factors induced by the stimulation from them (e.g., "sensory satiety") have been excluded.

Even the fine details of the temporal pattern of eating movements or food intake are insufficient to diagnose influences within acceptance behavior. Dissociations (i.e., zero correlations) between parameters of the micropattern over different meals (or different eaters) would be evidence that there are distinct influences. What these influences are, though, can only be determined by varying putative causes and getting a closely related variation specifically in one of the dissociated parameters. To have valid measures, we must show how the ingestive movements in response to a pang do differ qualitatively from those in response to a craving.

b. Nutrient-Specific Effects Similarly, the choice of one nutrient preparation over another is not necessarily controlled by nutritional physiology. If the foods differing in nutrient composition also differ in sensory characteristics, with or without palatability differences, the relative acceptances of these foods might have nothing to do with nutrition.

For example, a wish to eat conventional snack foods, which are rich in carbohydrate and fat content and relatively low in protein content, should not be interpreted as a carbohydrate craving, or as being mediated by effects of low protein content on the brain, until an explanation in terms of a habit of eating this range of food types for their sensory characteristics has been excluded.

By the same criterion differences in food acceptance (whether as actual intakes or in appetite ratings), induced by foods that obviously differ in character as well as in nutrient composition, are not evidence for differences in the satiating or appetizing power of these nutrients, or indeed of the sensory differences (e.g., sweetness). People acquire strong expectations from experiences of recognizable foods. These learned responses can be sufficient to affect appetite hours after eating a particular type of menu. The acceptance data need not reflect any physiological action of the food.

2. Ascribed Differences versus Observed Differences

People can ascribe different strengths to influences on their food acceptance (e.g., "Taste is more important to me than texture," "I never get hunger pangs," or "I rate healthiness higher than palatability"). This sort of data has often been used to assess the relative importance of categories of influence. What people are aware of, especially in retrospect, can have little to do, however, with what is actually influencing them. It might reflect no more than their own positions on conventional theories and current topics of discussion.

That is to say, the wording of a rating cannot in itself make it a measure of what those words refer to, be it a subjective experience or an objective aspect of food composition, bodily state, or the context of the eating or purchasing occasion. So-called "direct" scaling is an illusion. Whether the rating uses a magnitude estimation procedure, visual analogs, or multiple categories, what the scores actually measure entirely depends on the precision of their relationship to an observed factor in the food, the body, or the context, or, indeed, to acceptance responses.

Furthermore, for a verbal test to identify influences on acceptance, its results would have to measure the relationship of differences in acceptance to differences in the putative influences, whether they be in the food's sensed characteristics, the physiological state affected, or the cultural role of the food.

3. Is Preference a Dimension?

This also means that it is not coherent to treat acceptance data (or preferences and likings) as quantitative values on a scale or dimension, as they are commonly treated in consumer behavior and economics. The real phenomenon, the causal process, or the theoretical entity is acceptance from within a perceived situation.

There is a psychological scale in preferences, but it is the relationship between the combination of the influences operative in a person and the acceptance, or hedonic response. This relationship is latent in these data by themselves, hidden behind or underneath the observations. It is not recoverable from the preference data alone unless the latent structure is simple enough to be guessed and we have enough data to discriminate this causal model from other models.

There are several psychological reasons that the structure of influences on the acceptance of food is usually much more complicated than allowed by the assumption of a preference dimension.

Some consumers might put the influences together in ways qualitatively different from others' preference structures. Also, what influences acceptance of a food might vary with the situation (e.g., whether other foods are present, expected, or recently experienced, or what the somatic or social context is).

Moreover, the preferred value of an influence sometimes is at an intermediate level. This is certainly the case for most of the salient sensory influences on acceptance. This creates a preference peak instead of the monotonic relationship between an influence and acceptance that standard modeling techniques assume. Different consumers have different peak values or ideal points for each influence. Most devastating of all to the grouped analysis of preference structures, people put together influences in idiosyncratic ways to generate overall acceptance, preference, and choice.

Therefore, we have only one general approach to the understanding and measurement of food acceptance behavior. This is to collect the individual's acceptance data together with data on the major influences relevant to the objectives of the investigation and to complete an analysis of each person's preference structure before attempting to generalize or aggregate.

4. Individual Measurement of Acceptance Structure

Each discrete influence on a person's food acceptance forms a subscale or psychophysical function on the acceptance responses. Even if this causal relationship is peaked, this peak is symmetrically linear whenever the consumer is allowed to express unbiased personal preference in a sufficiently familiar situation (Fig. 2). The basis of this theory is that people decide what they want on the basis of the differences of the alternatives available from the most preferred levels of the operative influences.

Then, in the probably not uncommon case of a food acceptance behavior that is explicable by a single integration point (Fig. 3), acceptance has a "cognitive algebra."

The personal acceptance structure can be expressed graphically as a response space. This maps the variation in acceptance by that individual against variations in all of the potential influences over their tolerated ranges. These personal response spaces can simply be summed across a representative sample of consumers, to provide highly precise, completely operational, and totally disaggregatable estimates of the response to particular propositions in any segment of the market.

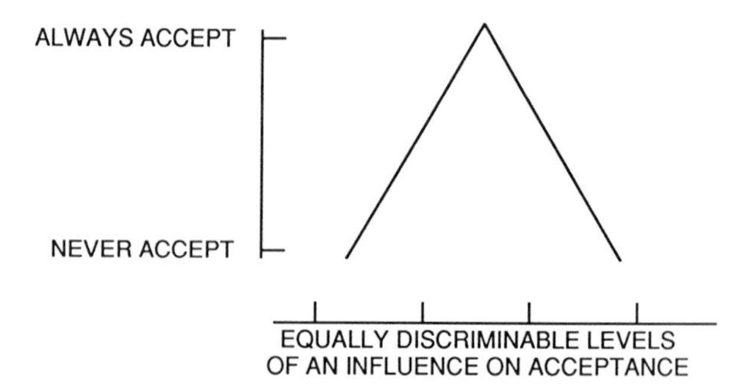

FIGURE 2 The acceptance triangle (also called the appetite triangle, the isosceles tolerance triangle, and the hedonic inverted V). It represents the cognitive mechanism by which the excitation of an individual consumer's disposition to accept food declines proportionally to the discriminable distance of the presented level of a discrete or integrated influence on acceptance from the personally acquired most facilitatory level of that influence.

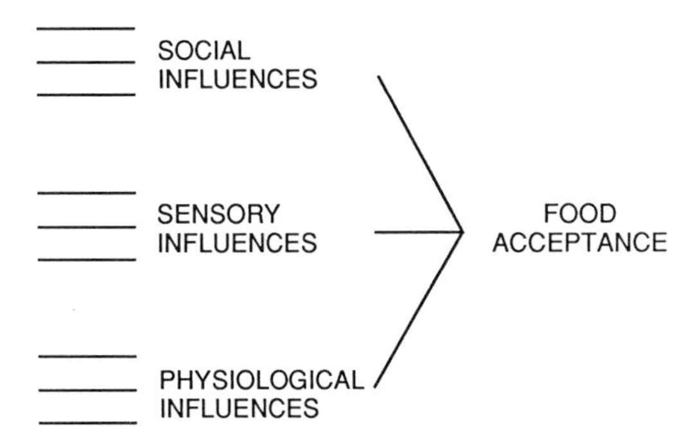

FIGURE 3 Different categories of immediate influence on an individual purchaser's or eater's acceptance of a food item, integrated through a single personal equation established by effects of past experience on inherited potentialities. This is the simplest type of causal model and should be refuted before multinodal attitudinal networks (such as in Fig. 1) are used in data analysis.

III. Sensory Influences

A. Limits on Sensory Evaluation

The above theoretical and methodological criteria for scientific understanding of influences on food acceptance have, until recently, appeared unachievably strict to those working on the sensory evaluation of foods, even when tackling relatively fundamental empirical or methodological issues. The lack of a psychologically scientific basis for sensory methods has not been noticed because the established atheoretical approaches to human data have helped food research and development personnel out of many problems that are beyond chemical or physical theory and measurement.

A technologically informed search for statistical differences or patterns in verbal responses to food samples has often sufficed to point to practical solutions to common problems, such as matching the current brand when the source of an ingredient changes, identifying and avoiding off-quality features (e.g., staling, spoiling, and taints), or working out the processes that yield the most popular version of a food for an acceptable level of expenditure by the manufacturer.

Moreover, the procedures for getting such pragmatic conclusions from descriptions of many characteristics of food samples by panels of expert, trained, or untrained assessors have been extended to more general issues. Careful use of quantitative descriptive analysis (i.e., "profiling") has illuminated fundamental problems in food perception, at least in cases in which a monotonic relationship exists in most or all assessors between the levels of each relevant physicochemical factor in the food and the rated intensity of a readily described sensation.

However, sensory interactions are more difficult to elucidate. This is because linear combinations of intensity ratings might not suffice, even if every panelist uses qualitatively the same rule for integrating the two or more physicochemical influences on the complex sensation. Furthermore, without a realistic formula for the interaction, quantitative estimates of the sensed effects of different formulations become unreliable. This could be of practical significance, as well as blocking the development of theory.

Most crucial of all for the extension of sensory evaluation to elucidate influences on food acceptance, there is typically a nonmonotonic relationship between the acceptance response and levels of a physicochemical factor or the relevant intensity scores. This means that the group average response might fail to contain appreciably more information than the distribution of most preferred levels across the panel, obtainable from a preference test without any sensory data. When more than one sensory factor is influencing acceptance, the interactions that enter into the response can be lost in the response space constructed from panel data.

Advanced statistical models have been developed to tackle these problems. These have sometimes revealed possibilities for product development not readily identified by conventional market research (or by experts or sensory panels). However, the power of such approaches to distinguish between qualitatively alternative consensus models of sensory integration into acceptance is usually low. The prevalence of the consensus is hard to estimate, and nonconsensus patterns (as well as their prevalences) are not identified.

The reliability of the estimates of the consensus optimum is low for all of these reasons and is much lower than those of a psychological approach at the same data-gathering cost. Even more serious, the validity of the estimates of response in the actual market is quite dubious. The polynomials or ideal points extracted are distorted by forcing everyone through the same set of samples without regard to personal ranges of tolerance. This also degrades the estimate of the individual's quantitative variant of the consensus preference structure. Worst of all from a pragmatic point of view, as well as for growth in scientific understanding, the conclusions of a pure data-fitting operation apply only to the data set. In contrast, if the data had been collected and analyzed in a way that could identify ways in which the tested sensory variations were influencing each consumer's acceptances in a normal situation of choice, then some generalizations to other stimulus sets could be made, as well as a rather precise aggregate response surface constructed for whatever subset of the panel was of interest.

B. Sensory Integration in Individuals' Acceptances

Most appearances, flavors, and textures are themselves based on two or more physicochemical factors in the foodstuff. Insofar as these integrated percepts enter consciously or unconsciously into acceptance, each individual operates according to an algebraic formula for combining the inputs.

1. Intramodal Integration

a. Visual and Tactile Pattern Psychological research on visual object recognition is, at the moment, grappling with highly abstract fundamental issues, often using verbal and pictorial materials. Visual integration phenomena directly relevant to food acceptance have yet to be examined.

The perception of mechanoreceptor patterns (i.e., tactile textures) has not been extensively studied. Some work on roughness and smoothness of touch to the skin of the arm should be extended to the feel of food in the mouth. Many of the textural sensations from homogeneous fluid foods are empirically related to various measures of viscosity, but the precise uses that the tongue and the palate make of viscous forces, and how they are combined with other physical actions of the fluid, remain to be elucidated. Solid and inhomogeneous foods are not understood psychophysically at all, and progress is slow because of the physical transformations of the stimulus during mastication.

b. Aroma Olfactory pattern recognition will probably be elucidated first by working "top-down" from the perceptual relationships among food aroma concepts, to complement the currently dominant "bottom-up" approach to odorant receptors and olfactory bulb neurophysiology. This would replace the traditional structure–activity approach of building statistical spaces out of supposed descriptions of subjective features corresponding to receptor types. Instead, personal summations and—more likely—interactions of molecular types in the recognition of familiar namable odor sources would be characterized by the discrimination analysis outlined above. [*See* OLFACTORY INFORMATION PROCESSING.]

c. Taste Gustatory integration is now gaining attention, although effective links to food acceptance have yet to be published.

Group analysis of raw ratings have been applied to mixtures of different sweeteners. The results have supported qualitative rules of combination; however, it has been pointed out that physical interactions could account for the observations and so these are not necessarily cognitive integration rules. Moreover, these designs are not strong enough to yield quantitative cognitive integration formulas; in any case this might require individual analysis.

There are many obvious hypotheses as to how tastant perceptions interact to contribute to well-known food flavors. Psychophysical data support the intuitive impression of generally subtractive relationships between sweetness and either sourness or bitterness. Individual combination of discriminable differences from ideal levels of sweetener and coffee solids in instant coffee provides further support, in the form of a sweetener–bittering ratio model of this integration.

The individualized discrimination approach to the tastants' contribution to likings for real foods could elucidate issues such as whether umami (the taste of, for example, monosodium glutamate in tomato or chicken) is a fifth "basic" taste or only a mixture of the four classic tastes, as it certainly can be described subjectively and classified neurophysiologically. [*See* TONGUE AND TASTE.]

2. Crossmodal Integration

An advantage of taking this cognitive approach to sensory influences on acceptance is that work can progress in parallel at different levels of interaction, or "chunking." The natural stimuli for food aromas, tastes, textures, and visual appearances can be used to study the higher level of integration into the entire food percept and on into its most preferred form for a certain purpose.

C. Sensory Influences on Later Food Acceptance

If the only perceptible differences between the foods in the test situation are inherent features (rather than verbal or pictorial labels or packaging, for example, or sight of the choices that others are making), then sensory influences must at least have contributed to the choice or relative acceptance of one food over the other(s) and indeed to any aftereffects of ingestion.

However, the perceived physical or chemical characteristics of a food can be heavily interpreted, and these attributions affect immediate or subsequent acceptance of foods as much as the "pure" sensations: for example, sweetness or creaminess might be sensually attractive, but feared for its usual health implications. The latter would be a clearly "cognitively" or conceptually mediated sensory influence.

Such effects are immediate, on the acceptance of the same food as is sensed, while it is being sensed. That is what is usually meant by palatability or sensory influences on acceptability. As well, however, the high acceptability of a food, or even the innately liked taste of sweetness itself, can elicit physiological responses that could affect subsequent acceptance of any foods. Proposed mediators of possible delayed effects of sweetness, for example, include neurally triggered changes in liver metabolism or the secretion of insulin from the pancreas or of endogenous opioids in the brain; there is, though, no satisfactory evidence for such effects in adults.

It is arbitrary to assume that a delayed sensory effect is mediated either conceptually or physiologically unless the alternative has been excluded by sensitive measurements. This cannot be done by a single verbal rating or just one biochemical assay, because a wide variety of mechanisms of conceptual and physiological mediation are possible, some quite complex. On the cognitive front, for example, sweetness from an artificial sweetener in a test food might be interpreted by someone who is trying to limit eating as indicating the consumption of a substantial amount of sugar calories, an idea which might so worry some people that they despair and lapse to unrestrained eating, instead of cutting back in compensation. We have evidence for this disparity between dieters and nondieters after eating soup with a "rich" aroma and texture. This might also happen with a low-calorie creamer.

IV. Physiological Influences

Naturally enough, it has mostly been biological psychologists and medical physiologists who have sought to characterize physiological influences on human appetite. As we have seen, however, this is not a biological issue in any sense that escapes thorough empirical analysis of the cognitive and psychosocial influences on the observed food acceptances (or on verbal data predictive of ingestion). On the other hand, progress on this set of issues cannot be made either by an exclusively psychological approach that lacks effective manipulations or measurements of the hypothesized processes under the skin. As yet, though, expertise in both human physiology and cognitive psychology, with access to the necessary facilities, is rare in one person or in an effective collaboration. As a result disappointingly little at present can be said that is definite about biological aspects of food acceptance in humans.

As was well-recognized 20 years ago, although it has often been neglected more recently, the main methodological problem is that physiological influences on food acceptance are normally thoroughly confounded with cognitive influences. Emptying of the stomach is rather precisely related to awareness of the lapse of time since the last meal. Filling of the stomach during a meal is even more closely associated with the perceived amount eaten. Different postingestional effects of the nutrient composition of foods cannot be separated from differences in the "image" of those foodstuffs and their sensed constituents.

Unfortunately, most of those experiments that have exposed postingestional influences of food eaten at a meal, by disguising its amount or composition, have recorded only the amounts or temporal patterns of subsequent food intake and have not attempted to measure or control postingestional effects or their timing. Differences in acceptance pattern among conditions or people fail to identify the physiological influences that are operative.

A sufficiently high dose of glucooligosaccharides produces a suppression of actual and rated food acceptance after 20–30 minutes. At that time most of this carbohydrate has left the stomach and glucose is being absorbed rapidly. This is consistent with the suppression of appetite by energy production from the glucose (thought, from animal experiments, to be in the liver). Action on receptors in the intestinal wall has not been excluded, though. This decrease in food acceptance is no greater than that related to guessing that caloric content of the meal is high; this shows the power of autosuggestion.

Infusion of fat into the upper intestine has been shown to delay the rise of rated hunger sensations, whereas this sort of effect is not produced by intravenous infusion. This points more conclusively to the importance of gut wall stimulation among the physiological influences on human food acceptance.

V. Social Influences

Consumer survey data and sales trends collected by the food business provide material for elaborate discussion of socioeconomic, interpersonal, and cultural influences on food acceptance. The psychological content would, however, be largely intuitive and speculative. This is because, once again, little research has been done that tests how each consumer makes purchasing decisions about foods, let alone chooses when, what, and how much to eat. Thus, the important task is to put social influences on acceptance on the same scientific footing as sensory influences.

Interpersonal and cultural influences are not inherently any less objective than sensed food qualities or perceived physiological processes. The difference between social factors and sensory and physiological factors is that many social influences are mediated linguistically; that is, they exist only because of communicable symbolic meaning. As a result they are not effectively describable in purely physicochemical terms. Nevertheless, the words or sentences applied to the food on the package or in the advertisement are entirely objective potential influences on speakers of the language who have been educated into the constructs being communicated. The same applies to nonverbal symbols in the pictorial "language" being used.

Some symbols have inherent potential for scaling on a dimension of influence (e.g., price, shelf-life date, and perhaps recommended storage temperature). These and essentially nominal labels both must be scaled, however. Scaling consists of analyzing out the relative strengths of influence of the labels on one or more underlying dimensions accounting for a descriptive or acceptance response. This is most realistically and hence effectively achieved by mapping the labels onto differences in response, in the same way that influences are identified from inherently graded physical influences (e.g., food composition).

A neat illustration is provided by the interaction of the labeled nature of a food constituent with the taste of that constituent, in a manner that is rational to the meaning of the label. The label for a sweetener might be "low calorie" or "sugar," for example. The influence of the label should multiply by the level of sweetener to generate an estimate of calorie content. This is indeed observed, so that the fattening potential of the food and its influence on food choice are rated to be higher for a strongly sweet-tasting food labeled "sugar-sweetened" than can be accounted for by the addition of the effects of the sugar label alone and sweetness alone.

Such interactions of the effect of prices with other salient attributes of the range of alternatives can provide realistically scaled price–purchase elasticity functions. In recent years such trade-offs between brand attributes have been characterized by nonmetric methods of group data analysis. For much the same data-gathering cost the individualized discrimination approach to cognitive model testing gives precise quantitative diagnoses of attribute interactions in each consumer sampled. The results not only imply practical recommendations for food design. They also tell us something theoretical about the mechanisms of acceptance in the tested situation.

Bibliography

Bennett, G. A. (1988). "Eating Matters. Why We Eat What We Eat." Heinemann Kingswood, London.

Boakes, R. A., Burton, M. J., and Popplewell, D. A. (eds.) (1987). "Eating Habits." Wiley, Chichester, England.

Booth, D. A. (1988). A simulation model of psychobiosocial theory of human food-intake controls. *Int. J. Vitam. Nutr. Res.* **58,** 55–69.

Booth, D. A. (1988). Practical measurement of the strengths of actual influences on what consumers do: Scientific brand design. *J. Market Res. Soc.* (*U.K.*) **30,** 127–146.

Diehl, J. M., and Leitzmann, C. (eds.) (1986). "Measurement and Determinants of Food Habits and Food Preferences." Dep. Hum. Nutr., Agric. Univ., Wageningen, The Netherlands.

Dobbing, J. (ed.) (1987). "Sweetness." Springer-Verlag, Berlin and New York.

Fieldhouse, P. (1986). "Food & Nutrition: Customs and Culture." Croom Helm, London.

Friedman, M. I., and Kare, M. R. (eds.) (1990). ''Chemical Senses: Appetite and Nutrition.'' Dekker, New York.

Manley, C. H., and Morse, R. E. (eds.) (1988). '' Healthy Eating—A Scientific Perspective.'' Allured, Wheaton, Illinois.

Ritson, C., Gofton, L., and McKenzie, J. (eds.) (1986). ''The Food Consumer.'' Wiley, Chichester, England.

Shepherd, R. (ed.) (1989). ''Handbook of the Psychophysiology of Human Eating.'' Wiley, Chichester, England.

Solms, J., and Hall, R. L. (eds.) (1981). ''Criteria of Food Acceptance: How Man Chooses What He Eats.'' Forster, Zurich, Switzerland.

Solms, J., Booth, D. A., Pangborn, R. M., and Raunhardt, O. (eds.) (1988). ''Food Acceptance and Nutrition.'' Academic Press, London, England.

Thomson, D. M. H. (ed.) (1989). ''Food Acceptability.'' Elsevier, London.

Thomson, D. M. H. (1989). Meeting report. The psychology of food. *Appetite* **13,** 229–232.

Food Groups

ELAINE B. FELDMAN and JANE M. GREENE, *Medical College of Georgia*

Glossary

Anthropometric Refers to measurements of individuals that reflect growth, development, and nutritional status. Common measurements are height, weight, skinfold thickness at various sites, and arm muscle circumference. Norms for healthy persons are published.

Collagen Characteristic protein of connective tissue and the most common protein in the body

Complex Carbohydrate Primarily from plant sources, they are made of polysaccharides (long chains of sugars), mainly starch and dextrins, that break down to yield intermediate polymers of sugars or simple sugars (mono- or disaccharides)

Cruciferous Vegetables of the mustard family—broccoli, cabbage, cauliflower, etc.—with a cross-shaped flower

Dietary fiber Nondigestible or partially digestible components of plants (fruits and vegetables) that may be complex carbohydrates including insoluble cellulose and lignin, and soluble pectin and gums

Fat Dietary fat, of animal or vegetable origin, includes triglycerides (fats and oils) that are esters of glycerol with three long-chain fatty acids. Fatty acids are saturated (hard), monounsaturated or polyunsaturated ($\omega 3$ and $\omega 6$ depending on the location of the terminal double bond), and of animal or vegetable origin. Cholesterol is a waxy steroid alcohol that is synthesized only by animal cells and therefore found only in animal foods, especially organ meats and egg yolk.

Nutrient Substances in the diet, contained in food, that are used by the body in varying amounts for growth, maintenance, and repair. They may be essential (the body cannot synthesize them) and include water, energy, carbohydrates, protein, fat, vitamins, and minerals. Micronutrients are required in the diet in amounts less than 1 g/day.

Protein Composed of a variety of amino acids, at least eight of which are essential. These nitrogen-containing compounds form the structure of the body (muscle) and perform vital functions (transport proteins, enzymes). The protein must be complete to sustain and maintain growth; that is, the proportion of essential amino acids must be at least 20%, and all individual essential amino acids must be present in appropriate proportions. The greater the percentage of nitrogen from essential amino acids, the greater the biologic value of the protein.

Vegetarian One who eats no animal products (vegan); or no animal products other than milk (lacto-vegetarian), eggs (ovovegetarian), or milk and eggs (lacto-ovo-vegetarian)

TO PROMOTE HEALTH and prevent disease, the daily diet must contain the 40+ essential nutrients. These are provided by a variety of foods, consumed as meals and snacks. The goal is to meet the nutrient needs for the vast majority of the healthy population, as determined by periodic review and recommendation by nutrition scientists (recommended dietary allowances). For convenience, foods are divided into groups so that a balanced diet, selected daily from a variety of foods from all the groups, in appropriate amounts, will provide all essential nutrients.

I. Sources of Population Nutrient Data

Information on the nutritional status of Americans is obtained from data on food production; imports and exports; marketing, distribution, and storage of food; patterns of food consumption by ethnic groups, families, and individuals; clinical nutrition surveys; studies of physical development; laboratory tests of nutrient levels; vital statistics on morbidity and mortality; and epidemiological information relating diet to disease. The American diet has changed markedly since 1900, with increased consumption of meat, poultry, fish, dairy products, sugar and other sweeteners, fats and oils, and processed fruits and vegetables and decreased consumption of grain products, potatoes, sweet potatoes, fresh fruit, vegetables and eggs. [*See* DIET.]

A. Ten-State Nutrition Survey and Health and Nutrition Examination Survey (HANES)

The Ten-State Nutrition Survey from 1968 to 1970 focused on low-income groups, evaluating the nutritional status and dietary practices of 40,000 persons. HANES, in 1971, evaluated a sample of 28,943 persons aged 1 to 74 from 65 locations in the 48 contiguous states. HANES provided information on dietary intake, clinical and biochemical findings, anthropometric data, hemoglobin, serum iron, transferrin saturation, and serum cholesterol levels. In 1974 a follow-up survey, HANES 2, was carried out. HANES 3 is scheduled over 6 years, from 1988 to 1994. Estimates will be developed after 2 years and these data will be made available by 1991.

B. Continuing Survey of Food Intakes by Individuals

In 1985 and 1986 the U.S. Department of Agriculture (USDA) conducted the yearly Continuing Survey of Food Intakes by Individuals (CSFII) using a 1-day dietary recall and 5 days of dietary data obtained by telephone. The CSFII is a major component of the National Nutrition Monitoring System, a set of federal activities that will provide regular information on the nutritional status of the U.S. population. The CSFII samples households of women 19–50 years old and their children 1–5 years old. Data were collected for 1500 women, 1100 men, and 500 children. In 1985, compared to 1977, men ate less meat (principally beef), whole milk, and eggs. They ate more fish, low-fat or skim milk, legumes, nuts and seeds, and carbonated soft drinks (regular and low calorie). The percent of calories decreased from fat and increased from carbohydrate.

In 1986 the survey reported that one-third of meals were consumed away from home and only half the people ate breakfast. Women consumed more skim and low-fat milk than whole milk. Children consumed equal amounts of these products and drank more carbonated beverages than fruit drinks or "ades." The vitamin intake of women was estimated to be less than recommended for vitamins E, B_6, and folacin, and calcium, magnesium, iron, and zinc. Children consumed less than the recommended daily allowance for iron and zinc. In 1986 the intake for women was reported as 37% of calories from fat (13% saturated, 14% monounsaturated, and 7% polyunsaturated), 46% from carbohydrate, and 17% from protein. For children the figures were 35% of calories from fat (14% saturated, 13% monounsaturated, and 6% polyunsaturated), 51% from carbohydrate, and 15% from protein.

II. Determinants of Food Intake

A. Food Choices

People choose foods—not nutrients—depending on cultural, social, personal, and situational factors, including ethnicity and family tradition. Fads affect food choices, and associations with rewards or punishment may explain some selections. The average American diet contains nearly 200 different foods. People avoid foods that cause unpleasant symptoms and select those that are well tolerated. Food choices may be restricted or influenced by poverty, lack of transportation, limited availability of foods in stores, poor food storage facilities, lack of cooking facilities or skills, or limited time for food preparation. Advertising and food labeling strongly influence choices.

In the United States, economic factors tend to limit the variety of food intake rather than directly determine an inadequate diet. Low-income populations may be poorly educated and thereby less understanding of the food group classifications; their choices are determined by their likes and dislikes, influenced by advertising, and the appearance of meal items.

B. Ethnic Preferences

Diets of ethnic minority groups in the United States may be influenced by the nature of the traditional diet, and the extent to which the diet has been adapted to the typical American diet. Dietary patterns of Black Americans resemble the traditional diet more than does the diet of other minorities. The black American diet shows a preference for "soul food" and contains yellow and dark green leafy vegetables, pork, fish, and poultry, which provide vitamins A and C, thiamin, and protein. The extensive use of frying, overcooking of vegetables, high consumption of sodium, and a low intake of milk and other dairy products are unhealthy aspects of the black diet.

Hispanic Americans eat diets high in carbohydrates in the form of tortillas and rice. Corn, onions, tomatoes, and sweet potatoes are the dominant vegetables, with few leafy green vegetables and fruits. Dried beans are the dominant protein source, but Americanization is causing an increase in animal protein. In general the fiber content is high, and the proportion of animal fat to total calories is less than in typical American diets. The diet may contain too much sodium and energy and may be deficient in calcium, iron, and vitamins A and C.

The primary source of energy in Asian/Pacific American diets is rice. Other carbohydrate foods such as wheat, noodles, and tubers are prominent. Compared to typical American diets these diets are higher in vegetables, fruits, fish and shellfish, but lower in meat and dairy foods. Native American Indians have varying dietary patterns depending on the tribe to which they belong and the geographic location. The traditional diets consist of a combination of traditional foods such as mutton, game, fish, tortillas, fried bread, fruits, roots, corn, and wild greens. The diet has been altered by the addition of processed foods such as bologna, potato chips, carbonated beverages and refined sugar products. The typical diet consumed today is high in carbohydrate, saturated fat, sodium, and sugar. Calcium intake is low.

III. Components of a Healthful Diet

Guidelines have been published that advise Americans on the components of a healthful diet. In 1977 the Senate Select Committee on Nutrition and Human Needs issued *Dietary Goals for the United States,* based on concerns that overnutrition may cause obesity, coronary heart disease, cancer, and stroke. This report set goals that Americans should consume less food, fat (especially saturated fat), cholesterol, refined sugar, and salt, and should increase consumption of fruits, vegetables, grain products, and unsaturated oils.

A. Nutrition Objectives for the Nation

National nutrition goals and objectives in relation to health promotion and disease prevention were first published in 1979 in the Surgeon General's Report on Health Promotion and Disease Prevention, *Healthy People*.

The 15 specific nutrition objectives to be attained by 1990 include reducing iron deficiency anemia in pregnant women; eliminating diet-induced growth retardation of infants and children; decreasing significant overweight by weight-loss regimens; reducing serum cholesterol and sodium ingestion; increasing breast feeding; educating the population on diet-related diseases and food composition; providing nutrition information in labels, via employee and school cafeterias, in state school systems, and in health contacts; and improving the national nutrition status monitoring system. Six of the 15 objectives seem attainable by the year 1990.

Nutrition objectives for the year 2000 will be measurable and addressed through surveys or surveillance systems, be feasible scientifically and technically, and include intervention strategies.

B. USDA/HEW Dietary Guidelines

In 1980, the USDA and the Department of Health, Education, and Welfare published *Dietary Guidelines* (updated in 1985). These recommend that we eat a variety of foods; maintain desirable weight; avoid too much fat, saturated fat, and cholesterol; eat foods with adequate starch and fiber; avoid too much sugar; avoid too much sodium; and drink alcoholic beverages in moderation.

C. Diet and Cancer Report

In 1982, the *Diet, Nutrition, and Cancer Report* of the National Academy of Sciences presented interim guidelines aimed at cancer prevention. The six items are similar to those listed above. Those specific for cancer prevention recommend that we minimize consumption of foods preserved by salt curing (including salt pickling) or smoking; minimize contamination of foods with carcinogens from any source; identify mutagens in food and remove or minimize their concentration unless the nutritive value of foods is jeopardized or other potential hazard introduced.

D. Surgeon General's *Report on Nutrition and Health,* 1988

Dietary changes are recommended in the Surgeon General's *Report on Nutrition and Health* in order to improve the health of Americans. The highest priority is to reduce intake of foods high in fats and increase intake of foods high in complex carbohydrates and fiber. The specific recommendations for most people (Table I) deal with fats and cholesterol, energy and weight control, complex carbohydrates and fiber, sodium and alcohol. Other issues for some people are fluoride, sugars, calcium, and iron.

IV. The Food Groups

A. The "Basic Four" Plan

In 1956 simple, specific guidelines enabling an American household to plan and consume meals that meet the recommended dietary allowances were established by the USDA in the form of four major *food groups*—the "basic four" plan. The plan recommends a specific number of daily servings from each group; milk and milk products, meat and meat substitutes, fruits and vegetables, and breads and cereals. A well-balanced diet for an adult should include two or more servings daily from the milk and meat groups, and four or more servings from the fruit and vegetable and bread and cereal groups (Fig. 1). The recommended number of milk servings increases for children and others with special needs.

Each group provides different nutrients (Table II). Individuals must select meals and snacks from each group in order to obtain all essential nutrients that are not distributed evenly within each group. Thus, servings of fruit and vegetables should include one vitamin C–rich food daily and one vitamin A–rich food every other day.

While adhering to the basic four plan *total nutrient intake* can vary widely. For example, depending

TABLE I Recommendations from the Surgeon General's Report on Nutrition and Health 1988[a]

Fats and cholesterol	Reduce consumption of fat (especially saturated fat) and cholesterol. Choose foods relatively low in these substances, such as vegetables, fruits, whole-grain foods, fish, poultry, lean meats, and low-fat dairy products. Use food preparation methods that add little or no fat.
Energy and weight control	Achieve and maintain a desirable body weight. To do so, choose a dietary pattern in which energy (caloric) intake is consistent with energy expenditure. To reduce energy intake, limit consumption of foods relatively high in calories, fats, and sugars, and minimize alcohol consumption. Increase energy expenditure through regular and sustained physical activity.
Complex carbohydrates and fiber	Increase consumption of whole-grain foods and cereal products, vegetables (including dried beans and peas), and fruits.
Sodium	Reduce intake of sodium by choosing foods relatively low in sodium and limiting the amount of salt added in food preparation and at the table.
Alcohol	To reduce the risk for chronic disease, take alcohol only in moderation (no more than two drinks a day), if at all. Avoid drinking any alcohol before or while driving, operating machinery, taking medications, or engaging in any other activity requiring judgment. Avoid drinking alcohol while pregnant.
Fluoride	Community water systems should contain fluoride at optimal levels for prevention of tooth decay. If available, use other appropriate sources of fluoride.
Sugars	Those who are particularly vulnerable to dental caries (cavities), especially children, should limit their consumption and frequency of use of foods high in sugars.
Calcium	Adolescent girls and adult women should increase consumption of foods high in calcium, including low-fat dairy products.
Iron	Children, adolescents, and women of childbearing age should be sure to consume foods that are good sources of iron, such as lean meats, fish, certain beans, and iron-enriched cereals and whole grain products. This issue is of special concern for low-income families.

[a] USDHHS, PHS Publication #88-50210.

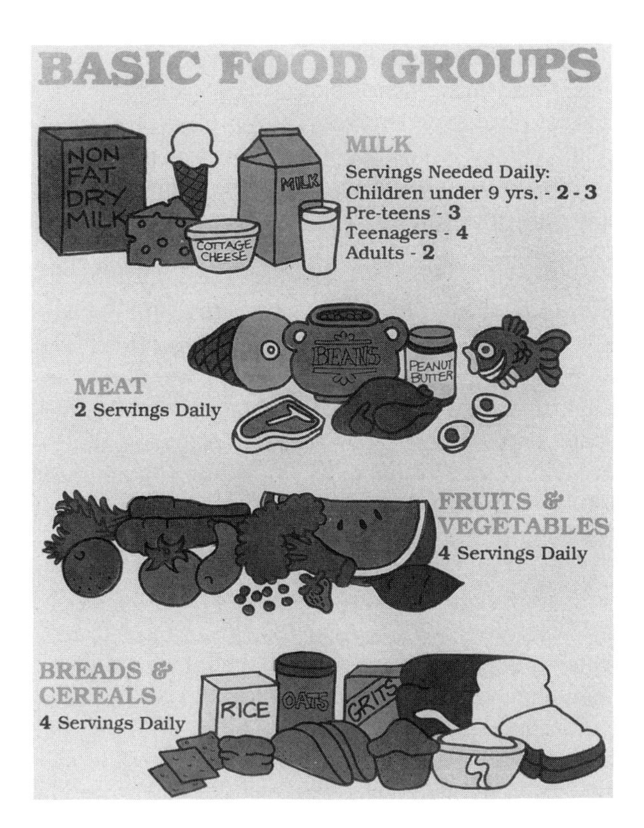

FIGURE 1 The four basic food groups and the servings needed daily. [From Feldman, E. B. (1988). "Essentials of Clinical Nutrition." F. A. Davis, Philadelphia, Figure 4–1, p. 88. Used with permission.]

TABLE III Equivalent Providers of Calcium in the Diet

Milk	1 cup
Yogurt	1 cup
Pudding	1 cup
Custard	1 cup
Non-fat dry milk	$\frac{1}{3}$ cup
Cottage cheese	$1\frac{1}{3}$ cup
Processed cheese	$1\frac{1}{3}$ oz
Cheddar cheese	$1\frac{1}{3}$ oz
Ice Cream	$1\frac{1}{2}$ cup

Source: From Feldman, E. B. (1988). "Essentials of Clinical Nutrition." F. A. Davis, Philadelphia, Table 4–6, p. 92. Used with permission.

on the specific foods chosen, vitamin E intake may range from 10 to 100 mg, and calories from 800 to 1800. Isocaloric servings may have dissimilar nutrient content, and sources of similar micronutrients may vary widely in calories. Roughly the same calcium content (300 mg) is found in milk, cheeses, and ice cream (Table III). Cheeses, eggs, peanut butter, and beans can be substituted for meat, while bread and cereal equivalents include biscuits, crackers, and pasta (Table IV). Eating only the recommended number of servings from each group may result in a diet deficient in calories, vitamins A and E, riboflavin, niacin, folacin, magnesium, iron, and zinc. Because the foods included provide a larger proportion of many nutrients than they do of calories, additional calories may be consumed by eating more servings of foods within the plan or by eating other unlisted food.

The basic four plan makes no recommendations for fat consumption and assumes that fat occurs in foods or is added in food preparation and seasoning. In practice more than enough fat is usually consumed in the American diet. Vegetable oils, lard, butter, margarine, and bacon are high in calories compared to the nutrients they contain (i.e., they are "calorie dense"). Sweets and alcohol are other calorie-dense foods. Sugar, honey, syrup, jelly, candy, soft drinks, wine, beer, and hard liquor provide relatively "empty" calories, i.e., they are foods that contribute calories with relatively little protein, vitamins, and minerals. It is important that people include beverages when reporting food consumption. [*See* FATS AND OILS, NUTRITION.]

TABLE II Food Groups and Some Major Nutrients

	Nutrient			Vitamins			Minerals	
Food group	Protein	CHO	Fiber	A	B	C	Ca	Fe
Milk	•			•	riboflavin		•	
Meat	•				•			•
Fruits and vegetables			•	•		•		•
Breads and cereals		•			•			•

Source: From Feldman, E. B. (1988). Essentials of Clinical Nutrition." F. A. Davis, Philadelphia, Table 4–4, p. 89. Used with permission.

TABLE IV Food Group Servings[a]

Protein source		Bread/cereal (grain) equivalents	
Beef, pork, poultry, lamb, fish	2 oz	Bread	1 slice
		Muffin	1
		Biscuit	1
Cheddar cheese	2 oz	Rolls	½
Cottage cheese	½ cup	Cornbread	1½ inch cube
Eggs	2	Crackers	4–6
Peanut butter	2 Tbs.	Cooked cereal, grits, pasta	½ cup
Dry beans, peas, cooked	1 cup		
		Breakfast cereal	¾ cup
		Tortilla	1
Fruits/vegetables ⅓–½ cup or ½–1 medium piece			

Source: From Feldman, E. B. (1988). "Essentials of Clinical Nutrition," F. A. Davis, Philadelphia, Table 4–7, p. 92. Used with permission.
[a] A serving is the approximate amount eaten of a given food in order to consume amounts of calories and other nutrients (i.e., protein, fat, carbohydrate) comparable to other foods in the group.

B. The Four Food Groups

1. Milk and Milk Products

Milk is a good source of many nutrients. Cow's milk protein is 80% casein. The whey includes lactalbumin and various immunoglobulins. Milk fat is easily digested and varies from 4% in whole milk to 2% in low-fat milk to less than 0.5% in skim milk. The carbohydrate in milk, lactose, is less sweet than sucrose and not readily digested in some ethnic groups and sick people. Calcium, present in large quantities, is generally absorbed more readily than the calcium in other foods. Milk contains minimal iron and is a useful source of riboflavin and nicotinic acid. Its low ascorbic acid content is destroyed by pasteurization. Vitamin D is generally added to milk.

The fat content of cream varies from 10% in half-and-half to 35% in whipping cream. The fat content of evaporated milk may vary, while condensed milk has added sugar. Skim milk contains the protein, calcium, and B vitamins in the original milk without fat and with less cholesterol. Yogurt is a nutritious and convenient food with variable fat content.

Most cheese is made from milk clotted using rennet, and contains the same protein and fat as milk and many of the other nutrients. Most cheeses contain 25–35% protein, 16–40 percent fat, and are rich in calcium, vitamin A, and riboflavin.

2. Meat, Poultry, Fish, Eggs, and Meat Substitutes

Meat is an excellent source of protein. Its digestibility relates to the amount of muscle protein vs. connective tissue, collagen, and fat. The collagen may vary from 2.5 to 23.6% and fat from 5 to 50%. Tenderness is associated with fat (marbling). Lean meat contains about 20% protein and 5–10% fat; the protein is of high biologic value. Pork and chicken have a higher protein-to-fat ratio than beef and lamb. Meats are usually rich in iron and zinc, contain little calcium, and are important sources of nicotinic acid and riboflavin. Muscle provides moderate amounts of vitamin B_{12}, but little vitamin A or ascorbic acid. Current recommendations limit meat to no more than six ounces a day.

Fish is an important source of protein of high biologic value. Lean fish, such as cod, haddock, and sole, contain less than 1% fat and about 10% protein. They are relatively low in calories and are easily digested. Fatty fish, such as herring, salmon, and sardines, contain 8–15% fish oil, doubling the calories. Halibut, mackerel, and trout have intermediate fat content. Fish roe contains 20–30% protein and 20% fat. Fish oils are rich sources of vitamins A and D, and long-chain, polyunsaturated ω-3 fatty acids. Iodine and fluoride are ample in marine fish, and small whole fish are high in calcium.

Shellfish have little fat and calories. The protein content of oysters, mussels, and other molluscs is about 15%. Oysters are the richest food source of zinc.

The average egg (60 g), contains 6 g protein, 6 g fat, and yields 80 calories. Egg proteins are mostly albumin, with the highest biologic value of all food proteins for human adults. The yolk is a fair source of vitamin A and contains significant amounts of B vitamins. The average large egg contains about 215 mg of cholesterol.

Textured vegetable protein derived from soybeans is flavored to resemble meat. The natural ingredients contain no vitamin B_{12}. Vegetable proteins have less of the amino acid methionine than do animal proteins (Table V) and may be lower in iron, thiamin, and riboflavin than meat.

Legumes are seeds of the family that includes peas, beans, peanuts, and lentils. Their high protein content (20 g per 100 g dry weight) qualifies legumes as meat substitutes. Their low content of sulfur-containing amino acids reduces the biologic value of the protein; they are rich in lysine, which is deficient in many cereals (Table V). A combination of

TABLE V Food Sources Providing Complementary Plant Proteins

Food	Amino acids deficient	Complementary protein
Grains	Isoleucine Lysine	Rice, corn, wheat, rice + legumes Wheat + peanut + milk Wheat + sesame + soybean Rice + sesame Rice + brewer's yeast
Legumes	Tryptophan Methionine	Legumes + rice Beans + wheat Beans + corn Soybeans + rice + wheat Soybeans + corn + milk Soybeans + wheat +sesame Soybeans + peanuts + sesame Soybeans + peanuts + wheat + rice
Nuts and seeds	Isoleucine Lysine	Peanuts + sesame + soybeans Sesame + beans Sesame + soybeans + wheat Peanuts + sunflower seeds
Vegetables	Isoleucine Methionine	Broccoli, Brussels sprouts, Cauliflower, Green peas, Lima beans } + sesame seeds Brazil nuts or mushrooms Greens + millet or converted rice

Source: From Feldman, E. B. (1988). "Essentials of Clinical Nutrition." F. A. Davis, Philadelphia, Table 4–11, p. 103. Used with permission.

legumes and cereal proteins may have a nutritive value as good as animal proteins and is an excellent source of fiber. Legumes are a good source of B vitamins, except riboflavin. Legumes lack ascorbic acid but sprouted legumes will prevent scurvy.

Soybeans are high in protein; the whole dry grain contains 40% protein and up to 20% fat. Soya also provides B vitamins. Peanuts contain about 20% fat. Other legumes include a variety of beans. Although their digestion and absorption is virtually complete, flatulence may be a by-product.

3. Fruits and Vegetables

Fruits have many pleasing flavors and can serve as desserts without excessive calories. Ascorbic acid is the only essential nutrient abundant in fruits. Fruits are sources of dietary fiber, and contain small quantities of carotene and B vitamins. Most fruits have little or no protein or fat, with 5–20% carbohydrate. Ripe fruits provide fructose and glucose as the major sugars. Bananas may serve as a useful energy source, but provide no protein. Fruits are high in potassium.

Vegetables include leaves, roots, flowers, stalks, and gourds. Their chief nutritional value is their carotene, ascorbic acid, folate content, and dietary fiber. Calcium and iron may be present in significant amounts but absorption is variable. Leafy vegetables may provide some B vitamins (riboflavin), but vegetables are poor sources of energy and protein.

Potatoes are the inexpensive food best capable of supporting life as the sole diet, with starch supplying most of the calories. The low protein content is of relatively high biologic value. Potatoes are high in fiber and a good source of potassium. They are easily digested and well absorbed.

The quality and nutritional value of canned and frozen vegetables compare favorably with fresh produce. A single serving of fruit or vegetable is one-half cup, or one medium-size piece. The National Research Council report on *Diet and Health* (1989) recommends five or more servings every day of vegetables and fruits, especially green and yellow vegetables and citrus fruits.

4. Breads and Cereals

Cereal grains are the most important single food in many countries, and are consumed as bread and in flour products. Corn, wheat, barley, oats, and rye are the principal cereals of North America.

Whole-grain cereals provide energy, good-quality protein, and appreciable amounts of calcium and iron. Cereals contain no ascorbic acid and practically no vitamin A; yellow corn contains significant amounts of carotene. Whole-grain cereals contain adequate amounts of B vitamins except for corn, in which the bound nicotinic acid is not biologically available. Milling and discarding of the outer portion of the seed diminishes the B vitamin content of wheat and rice.

Wheat may contain 10–20 g protein per 100 g, with lysine the limiting amino acid. In the average flour used to make white bread, protein provides about 13% of the energy. Whole-wheat flour contains three times as much dietary fiber as white flour. Whole wheat is also high in phytate, which binds minerals, making them unavailable. Thus, although whole-wheat flour contains appreciable calcium, iron, and zinc, absorption may be limited. In the United States, 100 g of white flour is enriched with up to 0.44 mg thiamin, 0.26 mg riboflavin, 3.5 mg nicotinamide, and 2.9 mg iron. In some states, calcium and vitamin D also may be added. Pasta is made from a hard variety of wheat but utilizes that portion relatively poor in B vitamins. It is frequently enriched.

Highly refined rice is almost devoid of vitamins. Parboiling fixes the vitamins so that they are not removed with milling and is the simplest preventive measure against beriberi (thiamin deficiency). Most rice contains 6.5–8.0 g of good quality protein per 100 g. The principal protein in corn is incomplete, lacking the amino acids lysine and tryptophan. The preparation of tortillas makes nicotinic acid biologically available from corn by heating the grains in lime water to soften them.

Oatmeal contains more protein (12 g per 100 g) and more oil (8.5 g per 100 g) than other common cereals and is rich in soluble fiber. Barley produces the malt for brewers and is the basis of the best beers and some whiskey. Bread made from rye flour is rich in B vitamins and also contains fiber.

The chief nutritive value of most dry breakfast cereals is derived from the addition of milk. Some cereals are fortified with B vitamins, iron, and, most recently, calcium.

5. Other Groups

Additional categories or groups of food include fats and oils, sugars and sweets, and beverages (alcoholic and nonalcoholic). Natural bottled water can include minerals and carbon dioxide. Seltzer can be low-calorie or be sweetened with sugar (sucrose), corn syrup, or other fruits or syrups (fructose).

V. Improving Meals as Sources of Nutrients

A. Nutritional Quality

Smaller portions of a wider variety of foods can improve the nutritional quality of meals without increasing the calories (Table VI). For example, 6 ounces of steak with a baked potato and a tomato contain fewer calories (480 kcal) than 10 ounces of steak and a tomato (655 kcal). Substituting a roll for a high-fat salad dressing transforms a 740 kcal chef's salad to a more healthful 590 kcal meal, with room for a fruit dessert. Decreasing the intake of fatty and protein-heavy foods and increasing the proportion of starchy foods and fruit will provide more satisfying meals that offer more food, fiber, vitamins, and minerals.

Simple changes in food preparation methods and seasoning (Table VII) lower calories without altering vitamins, minerals, or protein. More calories and nutrients can be obtained from healthful between-meal snacks such as milkshakes or peanut butter or other nut butter.

B. Determining Composition of Food

Food composition tables that provide average nutrient values based on quantitative analyses have been available for about 100 years. Most tables include data for five vitamins (vitamin A, thiamin, riboflavin, niacin, ascorbic acid), calcium, iron, energy, protein, carbohydrate, and fat. The USDA periodically publishes updated food composition tables (e.g., Handbook No. 8, *Composition of Foods*) in various sections. USDA Handbook No. 456 presents values for foods using household measures rather than the standard 100 g portions. [*See* NUTRITIONAL QUALITY OF FOODS AND DIETS.]

C. Labeling

The Food and Drug Administration (FDA) has set standards for food labeling to provide information on the nutritional content of food products. Labels must include the name and address of the manufacturer; the name of the product; the quantity of the

TABLE VI How to Improve Your Meals

	Common choice		Wiser choice	
	Menu	Kcal	Menu	Kcal
Breakfast	Orange juice, ½ cup		Orange juice, ½ cup	
	Black coffee		Skim milk, ½ cup	
	Fruit-flavored yogurt	320	Whole-grain cereal, ½ cup	
			Toast, 2 slices	
			Butter/margarine, 1 tsp.	325
Lunch	Tuna salad		Sliced turkey	
	Coleslaw, potato salad		Carrot sticks	
	Sliced tomato		Tomato	
	Crackers		Whole-wheat bread	
	Mineral water	930	Cantaloupe, ½	
			Milk, 1 cup	540
Dinner	Broiled chicken, ½		Broiled chicken leg	
	Tossed salad		Tossed salad	
	French dressing, 2 Tbs.		Dressing, 1 Tbs.	
	Green beans	410	Peas and onions, 1 cup	
			Roll	
			Grapes, small bunch	575
Snack	Omitted		Banana, popcorn (3 cups), 11 saltines, fruit or fruit juice and mineral water	150

Adapted from Feldman, E. B. (1988). "Essentials of Clinical Nutrition." F. A. Davis, Philadelphia, Table 4–8, p. 99. Used with permission.

contents; the ingredients listed in order of their predominance by weight; names of specific chemical preservatives used; a statement that artificial flavoring or coloring has been used; and serving size (reasonable for an adult male engaged in light physical activity) if the number of servings is stated.

Nutrition labels are required for foods that are advertised for their nutritional properties (e.g., low-calorie, low-fat) or enriched or fortified. The format includes: calories per serving; grams of protein; carbohydrate and fat per serving; and content of protein, calcium, iron, vitamin A, vitamin C, thiamin,

TABLE VII Suggestions to Decrease Calories

Suggestion	Examples	Amount	Calorie content
Use skim or low-fat dairy products	Whole milk	8 fl. oz.	160
	Skim milk	8 fl. oz.	90
Try seasonings or lemon juice instead of butter or margarine to bring out natural vegetable flavors.	Butter	1 tsp.	35
	Margarine	1 tsp.	35
	Lemon juice	1 tsp.	1
Broil, bake, stew, or roast trimmed meat instead of frying. Use a rack to hold roast out of drippings. Skim excess fat from stew and soup.	Vegetable oil	1 Tbs.	125
	Lard	1 Tbs.	115
	Vegetable shortening	1 Tbs.	110
Poach or boil eggs rather than frying or scrambling.	Poached egg	1	80
	Boiled egg	1	80
	Scrambled egg	1	110
	Fried egg	1	110

Source: From Feldman, E. B. (1988). "Essentials of Clinical Nutrition." F. A. Davis, Philadelphia, Table 4–9, p. 100. Used with permission.

riboflavin, and niacin, in percentages of the U.S. RDA (the recommended *daily* allowance based on the highest amount of each nutrient recommended for an adult by the National Academy of Science/National Research Council in 1968). Levels of other nutrients may be included but are not required. At present, listing the content of sodium, fat, cholesterol, simple sugars, etc. is voluntary although federal legislation mandating labeling has been proposed. The labels of foods containing saccharin must contain a warning. Table VIII provides a list of commonly used label terms.

VI. Special Dietary Needs

A. Vegetarian Diets

Some people adopt vegetarian diets, often out of philosophical or religious conviction. Forms of vegetarianism include: lacto-ovo-vegetarian, lacto-vegetarian, ovo-vegetarian, and strict or pure vegetarian (vegan) (see glossary). Vegetarians usually consume significantly less total fat, saturated fat and cholesterol, and more polyunsaturated fat, carbohydrate, and dietary fiber than in the usual American diet. Meeting protein needs and consuming all the essential amino acids is increasingly difficult in the more restricted forms of vegetarianism. The amino acids missing in any one grain or vegetable can be replaced by consuming several together in the same meal or adding dairy products (Table V). Vegan diets may be nutritionally inadequate if there is undue reliance on a single plant food source. Infants and children are particularly likely to develop symptomatic clinical nutritional deficiencies, with retarded growth and development. The four groups of plant foods relied on by vegans are legumes, cereal grains, fruits and vegetables, and nut and seeds.

TABLE VIII Commonly Used Label Terms

Term	Definition
Terms defined by Federal law[a]	
Low-calorie	No more than 40 kcal/serving, or 0.4 kcal/g
Reduced-calorie	At least $\frac{1}{3}$ fewer kcal than regular
Diet/dietetic	Low- or reduced-calorie, or one ingredient changed, substituted, or restricted
Sugarless, sugar-free, no sugar (sucrose)	May contain caloric sweeteners such as fructose, sorbitol, honey, or corn syrup
No added salt (salt-free)	No salt added in preparation Natural sodium remains, or soy sauce, or preservatives
Sodium-free	<5 mg Na/serving
Very low sodium	<35 mg Na/serving
Reduced sodium	<140 mg Na/serving
Lean	No more than 10% fat by weight
Extra lean	No more than 5% fat by weight
Leaner	At least 25% less fat by weight than original
Terms not defined by Federal law	
Lite or light	Can refer to color, taste, texture, calories, or weight
Natural	Meat or poultry without artificial flavors, colors, preservatives, or synthetic ingredients
Pure	One simple ingredient (and maybe water)
Organic	Implies no preservatives or pesticides used in growing, processing, or packaging (no guarantee)

[a] Other legal definitions include *enriched* (replacement of nutrients lost in the manufacturing process) and *fortified* (addition of nutrients to foods that did not originally contain them).

Other nutrients that may be deficient in vegan diets are vitamin B_{12}, vitamin D (for children not exposed to sunlight), riboflavin, calcium (especially for children and women), and iron (for women of childbearing age). Generous servings of green leafy vegetables, dried beans, sesame seeds, onions, and soybean milk will supply riboflavin and calcium, while beans, seeds, nuts, green leafy vegetables, dried fruits, and grains will supply iron.

B. The Athlete

Following the recommended dietary allowances will provide all necessary nutrients for a physical conditioning program. The diet that contributes to the best performance by the recreational or world-class athlete is that which is best for the nonathlete: a nutritionally balanced diet supplying appropriate quantities of water, energy, protein, fat, carbohydrate, vitamins, and minerals. Commercially promoted food supplements and drugs offer nothing to the healthy, well-nourished athlete and should be rejected.

VII. Modifications for the Life Cycle

A. Pregnancy and Lactation

The maternal diet must be increased in energy, protein, folic acid, calcium, and iron in order to meet the body's requirements in pregnancy. Food consumption should be increased to achieve a weight gain of 11–12.5 kg. Additional calories are usually provided by carbohydrate. The diet should include increased servings of the milk and meat group to provide increased protein. The calcium requirements are met by a higher intake of the milk group and foods rich in calcium. Iron supplementation is recommended, and many nutritionists recommend folate supplementation. The specifics of a daily diet for pregnancy are provided in Table IX.

The lactating woman's major increased needs are for calories, protein, and calcium, especially if lactation exceeds 3 months. These needs can be met by including 1 quart of whole milk per day in the diet.

B. Pediatric Needs

Infancy is the only time in life when a single food comprises the entire diet. While commercial infant formulas meet nutritional needs, breast-feeding is recommended to nourish the healthy term infant, particularly for the first 6 months. Formula may be replaced with up to 1 quart per day of cow's milk at about the first birthday. At that time the child will be consuming iron-rich foods such as fortified infant cereals or meats. Solid foods are added to the baby's diet during the first half of the first year (4–5 months), with the first food usually a single-grain cereal. Strained fruits and yellow vegetables are then added, followed by green vegetables, meats, egg yolks. Citrus, seafood, nut butters, chocolate, nitrate-containing vegetables, and egg whites are not introduced until the end of the first year. The older child should eat a variety of foods distributed

TABLE IX Daily Food Guide for Pregnant Women

Food group	Foods
Milk (3–4 glasses)	lowfat, buttermilk, skim, chocolate, canned, powdered
Lean Meat (2–3 servings)	Any fresh *lean* meat: beef, fish, chicken, pork, game liver; try to have liver at least one time a week. Substitutes for meat: eggs (2), cheese, peas, beans
Bread and Cereal (3–4 servings)	lightbread, cornbread, biscuits; or Cereals: grits, oatmeal, rice, dry cereal, cream of wheat
Vegetables and Fruits (4–5 servings)	1–2 Dark green leafy or dark yellow: turnip greens, collard greens, mustard greens, carrots, squash, pumpkin, sweet potatoes. 1–2 Fruits (or juice) or other vitamin C foods: Orange, Grapefruit, Cantaloupe, Tomato, Tomato Juice, Cabbage, raw, Watermelon. 1–2 Other fruits or vegetables: Potatoes, okra, green beans, corn, butter beans, peas, peaches, apples, grapes
Fat (No more than 1 tsp at each meal)	Use only *one* at each meal: margarine, butter, vegetable oil, shortening, fat meat, mayonnaise (salad dressing)

(From E. B. Feldman (1988). "Essentials of Clinical Nutrition." F. A. Davis, Philadelphia, Table 7-5, p. 168, by permission.)

among the four food groups. The adolescent requires increased calories, calcium and iron, not met (except for calories) by a diet of soft drinks, French fries, candy, and potato chips.

C. Geriatric Needs

Caloric needs are usually less, depending on the level of activity. Inactive elderly persons should consume foods of increased nutrient density, such as lean meat, fish, eggs, milk, and vegetables, and curtail fats and carbohydrates. Intakes of calcium, iron, zinc, vitamins A and D, and water-soluble vitamins may be inadequate or marginal, especially in low-income elderly. Good sources of these nutrients include dark-green vegetables, whole-grain and enriched cereals, pasta and bread, meat, fish, poultry, dried beans and peas, milk and milk products.

Bibliography

Agricultural Research Service (1975). "Composition of Foods," Agricultural Handbook no. 8. U.S. Department of Agriculture, Washington, D.C.

Feldman, E. B. (1988). "Essentials of Clinical Nutrition." F. A. Davis Co., Philadelphia.

National Academy Press (1989). "Diet and Health." Washington, D.C.

National Academy of Sciences (1989). "Recommended Dietary Allowances," 10th ed. Washington, D.C.

Pennington, Jean A. T. (1989). "Bowes and Church's Food Values of Portions Commonly Used," 15th ed. J. B. Lippincott Co., Philadelphia.

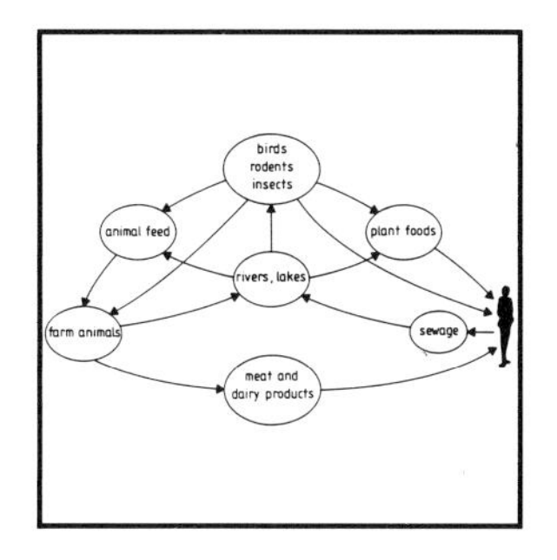

Food Microbiology and Hygiene

FRANK M. ROMBOUTS and ROBERT NOUT, *Wageningen Agricultural University*

Glossary

Chelate Strong binding by complex formation

Code of hygienic practice Describes good manufacturing practices to ensure microbiologically safe products of high quality

Food-borne pathogens Microbes that can cause food infection or food intoxication through their activity in the host or in the food

Food irradiation Treatment of food with ionizing radiation to extend shelf life and/or increase food safety

Hazard analysis critical control points Means of identifying and controlling the microbiologically important hazards in a food operation

Predictive food microbiology Prediction of shelf life or microbiological quality of foods by the use of a data base and mathematical models

Spoilage association Typical set of microbes that come to dominance when a food is allowed to spoil

FOOD MICROBIOLOGY AND HYGIENE belong to that branch of microbiology which deals with the control of microorganisms during the production, processing, storage, and distribution of food. Three categories of microorganisms can be distinguished. Two of these (i.e., spoilage and pathogenic organisms) are controlled through the prevention of contamination and proliferation. A third group, the fermentation organisms, is stimulated, or deliberately added as a starter culture to obtain fermented foods. Food microbiologists and hygienists are concerned with the microbial quality and safety of existing and new food products.

I. Contamination and Spoilage

Our food is derived from plants and animals, and these are naturally contaminated with microorganisms. Sound tissues of plants and animals are largely sterile, but large numbers of a wide variety of bacteria, yeasts, molds, and viruses occur on the surface of plants and on the coat, mucous membranes, and intestine of animals.

Other sources of contamination are harvesting, transport and processing equipment, and personnel handling raw materials and food products anywhere along the food chain.

Not all microorganisms entering food are of importance. The chemical nature of the product and a number of environmental factors determine which kinds of microorganisms will grow and cause spoilage. In this relationship the term "spoilage association" is frequently used and refers to the typical set of microorganisms which will come to dominance when a food product is allowed to spoil under certain environmental conditions. A few examples are given in Table I.

Three groups of factors are involved in food spoilage: (1) intrinsic factors (the physical and chemical properties of the food), (2) extrinsic factors (environmental factors, such as temperature, humidity, and the presence of oxygen), and (3) implicit factors (physiological and ecological features that enable microorganisms to respond to selective factors 1 and 2).

A. Intrinsic Factors

Both the speed and the type of spoilage are determined by a number of physical and chemical param-

TABLE I Spoilage Associations of Some Foods at Ambient Temperature

Food	Characteristics	Spoilage microorganisms
Fresh meat, fish	Moist, proteins, low in carbohydrates	*Pseudomonas, Acinetobacter, Alteromonas, Proteus, Enterobacter, Aeromonas*
Raw milk	Lactose, proteins, vitamins, minerals	*Lactococcus, Streptococcus, Lactobacillus*
Fruits, fruit juices	Fermentable sugars, low pH	*Saccharomyces, Candida, Hanseniaspora,* and other yeasts; *Penicillium, Botrytis,* and other molds; lactic acid bacteria
Bread, bakery products	Fairly dry, rich in carbohydrates	*Rhizopus, Penicillium, Geotrichum, Bacillus*
Bacon, ham	Fairly dry, proteins, low in carbohydrates	*Micrococcus, Lactobacillus, Streptococcus, Bacillus,* yeasts, and molds
Mineral waters	Extremely low in nutrients	*Flavobacterium, Cytophaga, Nocardia, Alcaligenes, Pseudomonas*

eters of food products. Most important are water activity, pH, the presence of oxygen and redox potential, the presence of nutrients, and natural antimicrobial systems.

1. Water Activity

Water activity (a_w) determines the availability of water to microorganisms for metabolism and growth. The a_w is equal to the ratio of the equilibrium vapor pressure of a food to that of pure water, and is slightly temperature dependent. In products with a low a_w, limitation of diffusional properties of nutrients can become an additional factor limiting growth.

Most perishable foods (e.g., fresh meat, fish, and milk) have an a_w close to 1 (i.e., >0.98) and spoil rapidly by bacteria requiring a high a_w (e.g., *Pseudomonas, Acinetobacter, Enterobacter, Proteus,* and *Bacillus* spp. Products with an a_w of 0.95–0.90 (e.g., dry ham, medium-age cheese, and many types of sausages) are likely to spoil by bacteria such as *Micrococcus, Lactobacillus, Bacillus,* yeasts (e.g., *Pichia, Candida,* and *Saccharomyces*), and fungi (e.g., *Rhizopus* and *Mucor*). Products with an a_w below 0.90 (e.g., crude smoked ham and salami-type sausage) spoil more slowly, and deterioration is usually caused by certain yeast and molds. Yeasts such as *Zygosaccharomyces rouxii,* which can grow at a high sugar concentration such as in honey or in fruit concentrates (a_w of 0.65–0.62), are termed osmophiles. Bacteria resisting high salt concentrations are called halophiles. They have an absolute requirement for higher concentrations of sodium chloride, as might be encountered in meat-curing brines.

2. pH

The pH value of most foods is in the range of 3.5–7.0. Most vegetables, cereal products, and fresh products of animal origin have a pH of around 6.0. If other factors (e.g., a_w) permit, these products spoil more or less rapidly through bacteria, which outgrow yeasts and molds at this pH. More acidic products (e.g., fruits, fruit juices, soft drinks, tomato products, beer, and wine) can spoil by yeasts, molds, and acid-tolerant bacteria.

Not only low pH, but also the presence of organic acids, plays a role in shelf life. Acetic acid is particularly effective against molds, yeasts and lactobacilli, and is a valuable preservative ingredient in acidic products such as dressings, mayonnaise, and mayonnaise-based salads.

3. Oxygen and Redox Potential

Products in which there is free access of air spoil primarily by aerobic organisms, which require oxygen for growth. This is true for the unprotected surfaces of many products, including meat, fish, and chicken, and for certain liquid foods (e.g., milk). Rapid growth and consumption of oxygen can lead to a depletion of dissolved oxygen and a decrease in redox potential. This can cause a shift in microflora toward facultative aerobic and strictly anaerobic microorganisms, which do not need oxygen, as is often the case in milk spoilage at ambient temperature, when lactic acid bacteria and occasionally butyric acid bacteria might finally dominate.

4. Chemical Composition

Spoilage is also determined by food as a substrate and a nutrient medium for microorganisms. Sugars, as a carbon source, are present in abundance in fruit juices and these are fermented to ethanol and other products, when yeast spoilage occurs. Ethanol can, in turn, serve as a carbon source for acetic acid bacteria to produce acetic acid, when sufficient oxygen is available. In tomato juice and vegetable juices, in which the pH is somewhat higher, lactic acid bacteria ferment the sugars to lactic acid. Fresh meat is, above all, a source of proteins, and the spoilage flora consists of gram-negative bacteria which are powerfully proteolytic. Proteolysis and metabolism of amino acids and peptides by these bacteria result in sliminess and off-odors caused by ammonia, amines, hydrogen sulfide, mercaptans, and other products. Foods with a high fat content (e.g., butter, margarine, and lard) can be spoiled by bacteria, molds, and yeasts which are lipolytic and cause rancidity through the liberation of fatty acids and accumulation of aldehydes and ketones.

5. Antimicrobial Systems

In both plant and animal products antimicrobial compounds occur, about which there is an increasing interest among food microbiologists. In plant products these are usually small molecules such as acids (e.g., salicylic acid in raspberries and benzoic acid in cranberries), phenolic compounds (e.g., tannin in walnuts and cocoa), alkaloids, and essential oils. These compounds work cooperatively with physical barriers such as peels, shells, skins, hulls, and rinds.

In the albumin of the avian egg, a whole series of proteins has antimicrobial properties, such as lysozyme (which causes lysis), ovotransferrin (which chelates ferric ions), ovidin (which chelates biotin), and ovoinhibitor (which inhibits several proteases).

Milk contains several antimicrobial systems, such as antibodies and a number of nonantibody proteins (e.g., lysozyme, lactoferrin, xanthine oxidase, and lactoperoxidase). With thiocyanate and hydrogen peroxide, both present in milk, lactoperoxidase forms the antimicrobial lactoperoxidase system. Lactoperoxidase catalyzes the oxydation of thiocyanate by hydrogen peroxide. The major oxidation product, hypothiocyanate, is an effective antimicrobial agent. It has recently been proposed to exploit the lactoperoxidase system in the preservation of raw milk by the addition of extremely small quantities of thiocyanate and hydrogen peroxide, in situations in which refrigeration is not possible because of a lack of electricity.

B. Extrinsic Factors

Extrinsic factors are temperature, relative humidity, and the gaseous environment of storage. Relative humidity and gaseous environment (e.g., oxygen and carbon dioxide) are at interplay with the concomitant intrinsic factors a_w and oxygen, described in Sections I,A,1 and I,A,3. The most important extrinsic factor is storage temperature. Although microbiological spoilage occurs over the wide range of temperatures of −5 to 60°C, low temperatures are effective in delaying spoilage. Microorganisms are killed at high temperatures, such as those used in pasteurization (<100°C) and sterilization (110–144°C) processes. With respect to the temperature requirement for growth, several groups of microorganisms exist (Table II). The storage temperature selects for those organisms with the highest growth rate at the temperature considered.

C. Implicit Factors

Specific growth characteristics of microorganisms (e.g., specific growth rate, growth factor requirements, and accumulation of end products) and mutual influences, antagonistic or synergistic, among organisms are commonly referred to as implicit factors of food spoilage.

The speed and the type of food spoilage, as well as microorganisms involved, are determined by the interplay of intrinsic, extrinsic, and implicit factors. Food spoilage is therefore an intricate ecological process.

II. Food-Borne Pathogens

Food can serve as a vehicle or growth medium for pathogens (e.g., bacteria, molds, viruses, amebae,

TABLE II Temperature Requirements of Microorganisms

Group	Minimum temperature (°C)	Optimum temperature (°C)	Maximum temperature (°C)
Psychrophiles	−10	10–15	18–20
Psychrotrophs	−5	25–30	35
Mesophiles	5–10	30–45	47
Thermotrophs	20–25	42–46	50–55
Thermophiles	40–45	55–75	60–90

and parasitic worms). A distinction can be made between food-borne infections and microbial food intoxications, depending on whether the pathogen itself or its toxic product (a microbial toxin, produced in the food) is the causal agent of disease. Table III presents the most important bacterial food-borne pathogens. Food infections caused by *Salmonella* and *Campylobacter* occur most frequently.

A. *Salmonella*

Salmonella is a food-borne pathogen causing gastroenteritis, an infection with diarrhea, vomiting, abdominal pains, and usually fever. The high incidence of this disease is a result of the existence of infection and contamination cycles, which have been intensively studied (Fig. 1). The primary habitat of *Salmonella* is the gut of humans and warm-blooded animals. Human patients and carriers can excrete large numbers of *Salmonella* with their feces, from which they are transmitted to surface waters (e.g., rivers and lakes). These function as sources of infection for farm animals, rodents, birds, and insects, which carry the infection either directly or indirectly, via contamination of animal feed, to the farm. During the transport and industrial slaughtering of pigs, poultry, and cattle, cross-contamination occurs, so that meat from *Salmonella*-free animals is also contaminated. The contamination level of broilers, for example, is often higher than 50%.

Contaminated meat and poultry are distributed to restaurants and catering facilities, and via retail outlets to homes. Through cross-contamination of equipment, surfaces, and hands and improper cleaning, other foods can become contaminated. Outgrowth to infectious dose occurs when heating, cooling, and cold storage are not adequate.

A rather small number of the 2200 serotypes of *Salmonella* are frequently encountered agents of gastroenteritis, among others: *S. typhimurium*, *S. virchov*, *S. brandenburg*, *S. infantis*, and *S. panama*. Of particular importance is the recent emergence, in both the United States and Europe, of a few phage types of *S. enteritidis* that cause severe gastroenteritis and that have been isolated from hens' eggs, a food product that was hitherto considered to be essentially free of pathogens. Evidence is accumulating that these phage types are capable of causing the transovarian contamination of eggs, and can therefore be spread by vertical transmission.

The incidence of *Salmonella* food infections is not accurately known, but is estimated by the World Health Organization (WHO) to be on the or-

TABLE III Bacterial Agents of Food-Borne Diseases

Organism	Pathogenicity	Incubation period (hours)	Duration of disease (days)
Salmonella	Infection	6–36	1–7
Shigella	Infection	6–12	2–3
Escherichia coli	Infection	12–72	1–7
Yersinia enterocolitica	Infection	24–36	3–5
Campylobacter jejuni	Infection	3–5 (days)	5–7
Listeria monocytogenes	Infection	Variable	—[a]
Vibrio parahaemolyticus	Infection	2–48	2–5
Aeromonas hydrophila	Infection	2–48	2–7
Staphylococcus aureus	Toxin in food	2–6	≤1
Clostridium botulinum	Toxin in food	12–96	1–8[b]
Clostridium perfringens	Toxin in intestine	8–22	1–2
Bacillus cereus[c]	Toxin in food	1–5	≤1
Bacillus cereus[d]	Toxin in intestine	8–16	>1

[a] Affects people with a predisposing factor; high mortality rate.
[b] High mortality rate; complete convalescence takes 6–8 months.
[c] Emetic type
[d] Diarrhoeal type

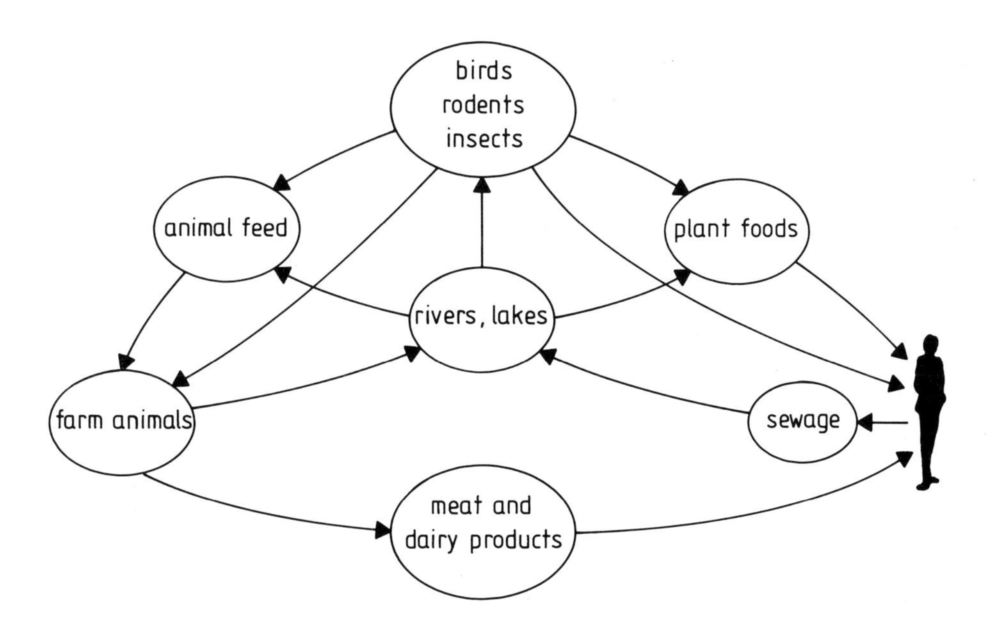

FIGURE 1 Contamination cycles of *Salmonella*.

der of 1 million, with several hundred deaths per year in Europe alone. The WHO strongly recommends an integrated approach to reduce the contamination levels of meat and poultry. Effective control should be possible by agricultural production free of infection, supplemented by hygiene efforts in food processing and preparation and by general preventive measures (e.g. heat, refrigeration, and chemical treatment or irradiation). [*See* SALMONELLA.]

B. *Campylobacter jejuni*

This spirally curved gram-negative bacterium appears to be responsible for 8–10% of all cases of acute enteritis diagnosed in The Netherlands. Campylobacteriosis can occur even more frequently than salmonellosis, which is diagnosed in about 5% of cases, and the symptoms of disease are at least equally serious. Because of lack of adequate isolation procedures this organism was only recognized around 1975 as an important agent of enteritis. Food of animal origin is often incriminated in campylobacteriosis (e.g., poultry in The Netherlands and raw milk in England).

It is not likely that the same contamination cycles exist for *Campylobacter* as for *Salmonella*. Important differences between the two pathogens are that *Campylobacter* does not grow below 25°C and is less widespread in the environment. However, it has been claimed that *Campylobacter* can persist in coccoid nonculturable forms which are still viable and can cause infections.

C. *Listeria monocytogenes*

This gram-positive rod-shaped bacterium is an environmental organism, capable of causing listeriosis. This disease typically affects people with a predisposing factor (e.g., immunosuppressed people, pregnant women, and neonates) and causes serious symptoms, such as meningitis, encephalitis, and abortion. Although much is unknown about its epidemiology, epidemics of listeriosis have been linked to food consumption in recent years. Food products incriminated in outbreaks in the United States and Canada were presumably recontaminated pasteurized milk, refrigerated coleslaw, and Mexican-style soft cheese. Although listeriosis does not occur frequently, foods are under intense scrutiny for *Listeria*, since mortality rates in the epidemics mentioned were extremely high (i.e., about 30%).

D. Other Food-Borne Pathogens

1. *Escherichia coli*

Escherichia coli is normally a harmless bacterium, with the intestine as its primary habitat. However, some strains (e.g., enteropathogenic, enteroinvasive, and enterotoxigenic) cause food infections, especially in babies and young children.

Apparently, these strains occur regularly in low numbers in food products and can incidentally grow to high numbers, setting off food infections. Recently enterohemorrhagic *E. coli* has caused severe outbreaks of infection due to meat products. Because of the high fatality rate (2–20%) it is now an organism of major concern.

2. *Staphylococcus aureus*

Staphylococcus aureus causes food poisoning. Contamination usually originates from personnel handling food. The skin and mucous membranes of humans and animals are its normal habitat. However, insufficiently cleaned and sanitized process equipment can also be an important source of contamination. The bacterium can grow in improperly cooled foods such as chicken, ham, sausages, and pastry products, and it produces a toxin in the food which remains active even after heating up to 100°C.

3. *Clostridium perfringens*

Clostridium perfringens, when ingested in large numbers, produces a toxin in the intestine. This food infection is often caused by consumption of meat and meat products that were not sufficiently heated to kill the fairly heat-resistant spores of this organism. Rapid cooling and cold storage of meat dishes that are not immediately consumed can prevent this food infection.

4. *Clostridium botulinum*

Clostridium botulinum, another anaerobic spore-forming bacterium, is a notorious food-poisoning organism. It produces a neurotoxin in foods with a pH above 4.5, and outbreaks of botulism are characterized by high mortality rates (i.e., 30%). Sterilized low-acid foods receive a ''botulinum cook'' to ensure elimination of the spores. Precooked chilled dishes are only heated to 75°C after packing and can present a hazard if not properly cooled during their entire shelf life.

5. *Bacillus cereus*

Bacillus cereus can give rise to two distinct types of food-borne disease, the emetic and the diarrhoeal syndromes. The emetic syndrome is believed to be associated with an emetic toxin preformed in the food. The symptoms are similar to those of *Staphylococcus aureus* intoxication. Cooked rice is often incriminated. The diarrhoeal type is caused by an enterotoxin and the symptoms are similar to those of *Clostridium perfringens* food poisoning. *Bacillus cereus* is a spore-forming organism, the spores being fairly heat resistant and their outgrowth is possible in a large variety of food products.

6. *Mycotoxigenic fungi*

Mycotoxigenic fungi are molds producing toxic substances when growing on food products. Notorious in this respect are *Aspergillus flavus* and *Aspergillus parasiticus*, producing the aflatoxins, secondary metabolites with carcinogenic properties. Molds used for food production are intensively studied for mycotoxin production, but so far no serious problems have been encountered. The occurrence of mycotoxins in plant products appears to be linked to extremes in weather conditions and an excessive occurrence of plague organisms in the growing season. The extreme dry growing seasons of 1988 and 1989 resulted in increased concentrations of aflatoxins in corn and peanuts in the United States.

E. Prevention of Food-Borne Disease

Factors that contribute to outbreaks of food-borne diseases are the consumption of contaminated raw foods (including seafood), inadequate cooling, inadequate heating, cross-contamination, and lapses of 12 hours or more between preparation and consumption. These are the principal factors for both food service establishments and homes. Major contributing factors associated with operations in food-processing plants are similar: contaminated raw ingredients, inadequate heat processing, improper cooling, contaminated persons handling foods, improper cleaning, and improper fermentation. There are many other factors that are a part of food protection schemes, but they appear to be of lesser importance.

III. Fermentation

The presence of high numbers of microorganisms in foods does not always indicate spoilage. Fermentation of foods involves the deliberate stimulation of part of the natural flora by providing conditions which favor its growth. These microorganisms might or might not still be present in the finished product, but the fermented food has usually acquired organoleptic (i.e., taste and small) properties which are widely different from the raw material.

Fermentation is one of the oldest methods of food processing. Some fermented products (e.g., cheese, yogurt, bread, beer, wine, and soy sauce) have experienced vast development and scale-up of production, including the use of intricate starter cultures. Others are still produced using age-old techniques under simple, even primitive, conditions.

For reasons of product integrity, taste and smell, and economics, most food fermentations cannot be performed profitably under strictly aseptic conditions. Fermented foods therefore can contain a variety of bacteria, yeasts, and molds originating from raw materials, process contamination, and inoculum. Not all of these organisms contribute positively to the desirable properties of the fermented product. In Table IV a number of fermented foods are listed, along with the microorganisms considered essential in their fermentation. In a number of examples (e.g., bread, beer, and some wines), the desirable fermentation can be achieved by a single microorganism. In others two or more species, with or without interactions, either simultaneously or in succession carry out a number of conversions, leading to the desirable properties of the final product.

TABLE IV Some Fermented Foods and the Important Microorganisms Involved

Raw material	Microorganism	Product
Cereal doughs	*Saccharomyces cerevisiae*	Bread
Barley malt	*Saccharomyces cerevisiae*	Beer
Milk	*Streptococcus thermophilus, Lactobacillus bulgaricus*	Yogurt
Soybeans, wheat flour	*Aspergillus oryzae, Aspergillus sojae, Zygosaccharomyces rouxii, Lactobacillus delbrueckii, Pediococcus halophilus*	*Shoyu (soy sauce)*
Soybeans	*Rhizopus oligosporus,* lactic acid bacteria	Tempe
Minced meat	*Lactobacillus* spp., *Pediococcus* spp., *Micrococcus* spp., *Penicillium chrysogenum,* various yeasts	Salami
Herring	*Micrococcus* spp., lactic acid bacteria	Tidbits

An example of synergism occurs in yogurt, prepared from milk by fermentation with *Lactobacillus bulgaricus* and *Streptococcus thermophilus. Lactobacillus bulgaricus* is proteolytic and provides *S. thermophilus* with peptides and essential amino acids, whereas *S. thermophilus* stimulates the growth of *L. bulgaricus* by the production of formate from pyruvate under anaerobic conditions. As a result the rate of lactic acid production is much higher than predictable from the acid production rates of the individual cultures.

Soy sauce, or shoyu, is an example of fermented food produced in a two-stage procedure. In the first (koji) stage cooked soybeans and roasted wheat flour are inoculated with *Aspergillus oryzae* and *Aspergillus sojae* and incubated for 3 days. A variety of lytic enzymes are produced, which give rise to fermentable sugars, amino acids, and peptides. In the second (moromi) stage 18% sodium chloride is added, and the mass is allowed to ferment for 6–8 months at ambient temperatures. Microorganisms essential in this stage are *Lactobacillus delbrueckii* and *Zygosaccharomyces rouxii.*

In wine fermentation *Saccharomyces cerevisiae* is the dominant and most desirable organism. Other yeasts and lactic acid bacteria are usually involved, as no stringent conditions of asepsis or pasteurization are applied. At least in high-quality red wines, a second fermentation should follow the alcoholic fermentation. This malolactic fermentation is provoked by *Leuconostoc oenos* and results in less sour wines of milder taste, due to the conversion of malic acid into lactic acid and carbon dioxide. No great Bordeaux wines would exist without malolactic fermentation.

Modern fermentation technology calls for starter cultures with new and improved fermentation capabilities, such as can be realized with recombinant DNA techniques. Examples of genetic traits to be incorporated in lactic starter cultures are the production of chymosine (a milk-clotting enzyme), bacteriophage resistance, and enhanced cheese-ripening properties in starters for cheese. Important properties to be incorporated into yeast starter cultures are the ability to convert poorly fermentable dextrins for the production of low-calory ''light'' beer and β-glucan-degrading ability to alleviate filtration problems in the brewing industry. However, at present no genetically modified microorganisms are legally tolerated in food production, and public discussions about their acceptability have taken on global proportions.

IV. Food Processing and Preservation

Most bulk crops are harvested in seasons and in geographical areas, far away from the large contingents of consumers. This implies that storage over the seasons and transport over large distances must be possible. In addition, consumers' demands have evolved into the direction of ready-prepared food products with built-in convenience. For these reasons agricultural produce must be processed and stored under adequate conditions. A large number of processing and preservation methods are available for this purpose. Some of these aim at killing microorganisms and preventing recontamination (e.g., bottling or canning and subsequent pasteurization, sterilization, or irradiation). In other preservation methods the prevention or retardation of growth of microorganisms is pursued (e.g., in drying, salting, acidification, freezing, and cooling). Most of these operations are now carried out according to strict rules and specifications of hygiene as laid down in codes of practice describing good manufacturing practices, to ensure microbiologically safe products of high quality.

There are a number of interesting developments in food processing and preservation that deserve attention. High-temperature short-time processing and aseptic packaging have taken on gigantic proportions in the last two decades. Liquid foods (e.g., milk, fruit juices, and soft drinks) can be easily heat-processed in heat exchangers, allowing high-temperature short-time treatments. With increasing temperature the rate of killing of vegetative microorganisms and spores increases more rapidly than the rate of chemical reactions, leading to a loss of nutrients and general quality deterioration. Aseptic packaging in multilayered cartons and coextruded laminates has, to a large extent, replaced glass and metal containers for these products. New developments go into the direction of high-temperature short-time treatment and aseptic packaging of liquid foods containing particulates (e.g., soups).

Consumer demands for convenience foods have spurred the development of precooked chilled foods. After cooking, assembling, and packaging these meals are merely pasteurized for a few minutes at 75°C, often by dielectric (i.e., microwave) heating. Since spores of nonproteolytic *Clostridium botulinum* can survive, one must rely on cold storage and distribution of these products below 4°C during the entire shelf life, to safeguard these products against growth and botulinum toxin production.

Modification of the gas atmosphere is increasingly used at various states of food manufacture and distribution, to the benefit of both producer and consumer. It partially provides an answer to the consumers' call for preservative-free foods. Carbon dioxide is the most effective gas for shelf-life extension, especially when combined with cooling. It extends the lag phase and decreases the growth rate of many, but not all, microorganisms. Lactic acid bacteria and *Brochothrix thermosphacta* are the dominant spoilage organisms in vacuum- and modified atmosphere-packed meats, stored under refrigeration.

Consumers' resistance toward preservatives such as benzoic acid and sorbic acid in products such as mayonnaise, dressings, and mayonnaise-based salads has led to the use of increased concentrations of food acidulants (e.g., acetic, lactic, and citric acids). However, because of sour taste, there are limitations to this practice, and a sufficiently long shelf life in cold storage can only be obtained by applying the most stringent rules of hygiene in the manufacture of these products.

Although emotionally rejected by the majority of the consumers, food irradiation is an effective process in reducing microorganisms in food products, extending shelf life and increasing food safety. In many countries food irradiation facilities have been installed. These usually contain a ^{60}Co source, emitting ionizing radiation, also called gamma radiation. The DNA in the bacterial cell is the primary target for radiation inactivation, while limited physical and chemical changes are induced in the food product.

The WHO has given approval for radiation dosages up to 10 kGy as being unconditionally safe. The dose of 1 Gy is equivalent to 1 J of energy absorbed per kilogram of product. Irradiation at levels of 2–5 kGy (i.e., radicidation) is effective in destroying salmonellae, campylobacters, and other non-spore-forming pathogens and could contribute much to reducing food infections from such products as raw meats and poultry. However, irradiated foods must be labeled as such and still find few buyers.

V. Microbiological Control and Monitoring

A. Control

Traditionally microbiological testing of foods has been focused on finished products. This was true for both the quality control laboratory of the manufacturers and the public health laboratories. The meaning of the data thus obtained depends largely on the sampling plan. However, due to the variable distribution of pathogens in foods, sampling plans are usually unsatisfactory. Consequently, it is often impossible to check for the absence of pathogens with reasonable confidence, no matter how rapid, accurate, and reproducible the microbiological analytical methods are. It has been recognized that massive microbiological routine testing is not a cost-effective approach to quality assurance.

The introduction in the 1970s of the hazard analysis critical control points concept has marked a change in philosophy toward the microbiological quality assurance of food. This concept provides a means for identifying the microbiologically important stages in a food operation and the means of their control. Introduction of this system starts with a detailed analysis of the hazards associated with the manufacture, distribution, and use of a food and leads to identification of the critical control points. Systematic and intensive monitoring and control are applied at these points. In applying these principles greater assurance of product safety is provided than would be possible with the traditional procedures.

Predictive food microbiology is an approach with respect to the prediction of shelf life and the microbiological safety of foods. Mathematical models and a data base are used to predict the microbiological status of a product of known composition and stored under a certain temperature regimen. The data base consists of data about growth and death of microorganisms, as measured in the laboratory under certain conditions of temperature, a_w, pH, etc. The conditions prevailing in the food are always different, but the models allow prediction of the likely outcome of conditions not specifically tested.

B. Monitoring

Routine microbiological testing used to be limited to enumeration of groups of microorganisms, followed by isolation and identification procedures for confirmation. These methods are time-consuming, rather tedious, and not fit for automation or on-line application. Also, they do not always provide us with information that is vital for proper judgment, such as the ability to produce toxin and the presence of toxin in food.

There have been rapid developments in techniques and instruments for the detection and enumeration of microorganisms that produce information faster than the traditional methods, as is required in modern quality assurance programs. The rapid techniques can be divided in those in which growth of microorganisms is required, usually taking 24 hours, and those in which cells are counted or metabolites are measured, directly in the product. Examples of the first group are impedimeters (Bactomatic, Princeton, New Jersey; Malthus, Stoke-on-Trent, United Kingdom), instruments which detect microbial activity by measuring changes in the electrical resistance of the growth medium. Other measuring principles applied in instruments of this category are nephelometry, radiometry, and microcalorimetry. In the second group direct counting can be done by direct epifluorescent filter techniques or by flow cytometry. In both techniques microorganisms are stained with a fluorescent dye or marker before microscopic counting. A successful direct method is the quantification of ATP of microbial origin by bioluminescence.

Other new approaches in the microbiological testing of foods are immunoassays and nucleic acid hybridizations. The application of enzyme-linked immunosorbent assays (ELISAs) is particularly successful in the detection of toxins (e.g., aflatoxins and staphyloenterotoxins) in food. ELISAs and latex agglutination kits for the detection of, for example, *Salmonella* spp. and *L. monocytogenes*, as well as various groups of fungi are now available. It is also possible to detect microorganisms or their virulence traits with DNA or RNA probes. As with the immunoassays, the DNA hybridization probes provide, in principle, any degree of specificity desired. *In vitro* DNA amplification with the aid of the polymerase chain reaction drastically increases the sensitivity of DNA probe detection. However, more research, especially on sample pretreatment, is required before DNA hybridization methods can be introduced into routine testing.

Food microbiologists and hygienists have the task of helping to reduce postharvest food losses and the incidence of food-borne disease. They must possess a thorough knowledge of the physiology and ecology of microorganisms in food and a good understanding of food science and technology in order to function optimally in the multidisciplinary team of food technologists that carries responsibility for providing the consumer with a continuous supply of wholesome and safe food.

Bibliography

Bourgeois, C. M., Mescle, J. F., and Zucca, J. (eds.) (1988). "Microbiologie Alimentaire. Volume 1: Aspect Microbiologique de la Sécurité et de la Qualité Alimentaires." Lavoisier, Paris.

Bryan, F. L. (1988). Risks of practices, procedures and processes that lead to outbreaks of foodborne diseases. *J. Food Prot.* **51,** 663.

Doyle, M. P. (ed.) (1989). "Foodborne Bacterial Pathogens." Dekker, New York.

Gould, G. W. (ed.) (1989). "Mechanisms of Action of Food Preservation Procedures." Elsevier, London.

Huis in't Veld, J., Hartog, B., and Hofstra, H. (1988). Changing perspectives in food microbiology: Implementation of rapid microbiological analyses in modern food processing. *Food Rev. Int.* **4,** 271.

Jay, J. M. (1986). "Modern Food Microbiology," 3rd ed. Van Nostrand-Reinhold, New York.

Wood, B. J. B. (ed.) (1985). "Microbiology of Fermented Foods," Vols. 1 and 2. Elsevier, London.

Food Toxicology

GEORGE S. BAILEY, RICHARD A. SCANLAN,
DANIEL P. SELIVONCHICK, AND DAVID E. WILLIAMS,
Oregon State University

Glossary

Ames assay Rapid screening test (developed by Bruce Ames) for genotoxic chemicals using mutant *Salmonella* strains

Amino acid pyrolysis product Any of a number of aromatic heterocyclic amines found in meats cooked at high temperatures

Cytochromes *P*-450 Family of enzymes, found predominantly in liver, which metabolize a wide variety of food-borne and other foreign chemicals

Electrophile Electron-deficient chemical capable of binding to electron-rich sites, such as are found in DNA

Generally recognized as safe (GRAS) Food additives that have been in use for a number of years with no known deleterious effects

High performance liquid chromatography Major technique used to separate and isolate food-borne toxic chemicals

Inhibitor Any chemical that reduces the probability that exposure to a carcinogen will result in tumor formation

Initiation Initial DNA-damaging event that can result in production of cancer

Mass spectrometry Extremely sensitive analytical technique used to determine the structural identity of a chemical

Monoclonal antibody Single antibody produced from a clone of cells that can be used in highly sensitive and specific immunoassays

Mycotoxins Carcinogenic secondary metabolites of various fungi that are unintentionally introduced into food

***N*-Nitrosamines** Carcinogenic chemicals having the form $R_2—N—N=O$ that are found in foods, tobacco, and other consumer products

Nuclear magnetic resonance Analytical technique used in structural analysis of chemicals

Promoter Any chemical or stimulus that enhances the probability that a DNA-damaging event (initiation) will lead to production of a tumor

FOOD TOXICOLOGY is the science that deals with the sources, nature, occurrence, and formation of toxic substances in foods. It is concerned with describing the harmful effects, the mechanisms and manifestations, and the limits of safety of these toxic substances in foods. In this sense, food includes all solid food (natural, synthetic, or manufactured) as well as liquid, i.e., alcoholic and other beverages.

I. Introduction and Historical Perspective

A. History

Modern food toxicology in the United States had its beginnings at the start of the twentieth century. National concern for food safety and recognition of the need for federal laws and regulations regarding the quality of the food supply began with the appointment of Dr. Harvey W. Wiley as chief chemist of the USDA in 1883. It was Dr. Wiley and his colleagues at the division of chemistry who began to analyze systematically the composition of foods using modern scientific methodologies. He was particularly concerned with the use of chemicals such as boric acid, formaldehyde, and salicylic acid as preservatives in foods. He dramatized the problem by his "poison squad" experiments, in which 12 USDA employees acted as human subjects to test the safety of these compounds. Each person ate food that contained measured amounts of these chemicals, and then were observed for any detectable evidence of harm. The results, published by Wiley during 1904–08, provoked widespread interest throughout the country. In 1906, the original Food and Drug Act was passed by Congress, after disclosures of the use of poisonous preservatives and dyes in foods, unsanitary conditions in meatpacking plants, and cure-all claims for worthless and dangerous patent medicines. The original Pure Food and Drug Act has been changed several times and now includes amendments for proof of safety for food additives (the Food Additive Amendment of 1958, which includes the famous Delaney Clause) and requirements that the use of synthetic color compounds be certified and limited in their use (the Color Additive Amendment of 1960). In 1954 the passage of the Miller Pesticides Amendment by Congress streamlined the procedures for setting safety limits for pesticide residues on raw agricultural commodities. These laws and amendments, the enforcement of which has been made possible by marked advances in analytical science, have greatly strengthened consumer protection.

B. Sources of Food Toxicants

Food toxicants can potentially be derived from three sources: naturally occurring toxicants, intentional food additives, and unintentional food additives. Each of these may be assigned to a toxicological category based on the concentration present in a specific food product: inactive or inert; active but nontoxic; and toxic. These properties are relative and can be influenced in one or more ways by several factors.

1. Naturally Occurring Toxicants

These toxicants are products of the metabolic processes of animals, plants, and microorganisms from which the food products are derived. In the course of evolution, man learned by trial and error to avoid those natural products that cause acute, easily recognizable poisoning or to develop processing methods that reduce or eliminate the toxicity. Many naturally occurring substances in food products are potent poisons that are capable under certain conditions of producing severe and sometimes fatal symptoms. Those products that produce immediate adverse effects can be avoided. Some of these rapidly acting toxic materials are easily removed; for example, protease inhibitors in lima beans can be inactivated by heating. By comparison, naturally occurring substances in food products that have a delayed toxicity are often difficult to recognize in food and a causal relationship between observed poisoning and the food can be slow in coming. [*See* TOXINS, NATURAL.]

2. Intentional Food Additives

These are chemicals that are added to foods to accomplish one or more objectives: improvement of nutritive value, maintenance of freshness, addition of some sensory property, or an aid in the processing. The number of food additives used has increased significantly in this century. The toxicological properties of many of these compounds remain unknown. Many additives have been used for a number of years without any apparent toxicity. These compounds are classified as generally recognized as safe (GRAS). At this time the GRAS list of food additives is under investigation on an item-by-item basis. The purpose of these investigations is to determine if any toxicological properties exist in view of new knowledge in this field.

3. Unintentional Food Additives

Food products may become contaminated inadvertently through several routes. In large part, these additions result from storage treatment, processing or other manipulation involved in food production and distribution, or enter foods through purely accidental circumstances. These substances range from sand to pesticides and radioactive fallout, with a corresponding range in toxicity from harmless to acutely toxic compounds. Through good manufacturing practices and food regulations an attempt is made to prevent the inclusion of harmful additives or at least to keep their inclusion below the toxic level.

C. Focus of This Article

Food toxicology may well have been the first prehistoric science. The need to recognize toxic substances in the environment was a matter of self-preservation from the time of humans' first appearance on earth. In time the idea was established that adverse responses to substances were often related to the ingestion of some common foods, often in a quantitative manner. This led to the concept of a *dose–response relationship*. That the adverse response was also a function of duration, or frequency of exposure, gave emphasis to the distinction between acute and chronic toxicity. *Acute toxic effects* are those that occur soon after exposure to a chemical agent. The acute effect is one that is observed within the first few hours or days after exposure and, in most cases within the first two weeks. *Chronic effects,* in contrast, are those that appear only after repetitive exposure to a substance, and often require months or years of continuous exposure. This article will focus only briefly on food-borne toxicities that result in acute responses, in particular neurological responses to marine animal toxins and allergic reactions associated with certain foods. The major focus will be on chronic toxicity, with a concentration on those food-borne substances involved with the initiation of cancer.

II. Acute Effects of Food-Borne Toxicants

A. Marine Animal Toxins; Sources and Neurological Responses

1. Shellfish

Consumption of contaminated shellfish results in symptoms of poisoning that include tingling or numbness in lips, face, and neck and/or dizziness, headache, and nausea in mild cases. In severe cases these symptoms may be accompanied by muscular paralysis, respiratory difficulties, and death.

The chemical responsible for this poisoning has been termed *saxitoxin*. Shellfish accumulate saxitoxin through consumption of dinoflagellates, a benthic phytoplankton. During certain times of the year and under appropriate environmental conditions, saxitoxin-producing organisms undergo a rapid period of reproduction and growth, resulting in blooms consisting of great numbers. These blooms have been termed "red tides" because the seawater takes on a red color. Shellfish feeding in these waters will accumulate saxitoxin. Control of paralytic shellfish poisoning is carried out by banning shellfish harvesting where either red tide or the toxin has been found.

2. Fish Poisoning

Ingestion of normally safe fish can occasionally lead to a toxic response, including gastrointestinal disorders and neurotoxicity that can lead to death. The illness associated with fish poisoning has been termed *ciguatera,* and no specific toxin has been isolated and identified. Ciguatoxic fish are tropical saltwater species that are bottom dwellers or feed on bottom dwellers. It has been suggested that these fish become ciguatoxic by consumption of blue-green algae that may contain the toxin(s), consumption of dinoflagellates, or the toxins are formed in the fish gut by bacteria in synergism with blue-green algae. Control of this poisoning has been mainly through education of the public as to what fish may be ciguatoxic and where they may be found.

3. Pufferfish Poisoning

Tetrodotoxin is an extremely toxic chemical that is found in certain organs of the pufferfish. This is a neurotoxin that is quick acting, with death resulting from respiratory paralysis. This fish is considered a delicacy in Japan and its health hazard is controlled by allowing only specially trained and licensed chefs to prepare pufferfish for human consumption, by separating the muscle cleanly from the liver and other viscera, which contain the toxin.

B. Food Allergies: Types and Responses

The term *food allergy* is used to identify a true immunologically based response to a food that affects the body's immune system. Examples are allergic reactions to common foods such as cow's milk, eggs, peanuts, shellfish, and fish. These adverse reactions to food are considered individualistic because they affect only a few people in the population; most individuals can eat the same foods with no ill effects. Allergic responses that are important with regard to food sensitivity are type I and IV allergies. *Type I allergies* represent allergic reactions to foreign proteins. The allergic response is usually immediate hypersensitivity occurring within a few minutes to several hours after consumption of the offending food. The observed symptoms of type I hypersensitivity result from the release of pharmacologically active substances such as histamine. This occurs from specific cells in the body known as *mast cells*, as a consequence of interaction between immunoglobulin E (IgE) and food substances that cause allergic reactions. *Type IV allergies* generally include delayed hypersensitivity-type allergic responses that occur over hours rather than minutes. This type of allergy involves the reaction of certain sensitive cells, usually lymphocytes, to the specific chemical substance that triggers the allergic reaction. Symptoms usually appear 6–24 hours after consumption of the particular food. [*See* ALLERGY.]

The symptoms of food allergies vary greatly with individuals and also depend on the type of sensitivity. Gastrointestinal symptoms are the most common; however, skin and respiratory systems may also be affected. A rare, but potentially life-threatening, result of food allergy is *anaphylactic shock* that can result in severe itching and hives, perspiration, constriction of the throat, breathing difficulties, lowered blood pressure, unconsciousness, and rapid death. Once diagnosed, allergic or hypersensitive people must alter their lifestyles to constantly monitor and avoid the offending foods. The food industry must continue to be alert to the needs of these particular individuals by providing accurate and complete lists of ingredients and by supplying additional information on specific ingredients when asked.

III. Carcinogens and Anticarcinogens in the Diet

A. Carcinogenesis

Cancer is initiated when strongly electrophilic chemicals, or physical agents such as ultraviolet light or irradiation, damage or react chemically with the DNA in somatic cells. The chemical structures of many covalent carcinogen-DNA adducts have now been described. A percentage of such adducts escape removal by DNA repair enzymes, and lead to DNA sequence changes (mutations) when damaged cells replicate. Recent research indicates that mutations that occur in specific genes, called *proto-oncogenes,* can lead to incorrect functioning of the gene products that regulate cell growth and differentiation. Once the cancer process has been initiated by formation of such mutations, a long latent period follows in which promotion of tumor development and progression to malignancy occurs in a fraction of initiated cells. Agents that enhance the postinitiation processes are called *tumor promoters,* whereas those that interfere with initiation or promotion are called *anticarcinogens.* [*See* ONCOGENE AMPLIFICATION IN HUMAN CANCER.]

Many scientists now believe that a majority of human cancer may arise through this initiation-promotion-progression pathway from carcinogenic compounds in the diet. The evidence for this belief comes from two principle sources. First, epidemiological studies suggest that certain cancers relate to diet rather than genetic background. For example, the incidence of stomach cancer is very high in Japan, but very low in various cultural groups in the United States, including second-generation Japanese-Americans. A reasonable inference is that causative factors in the diet underlie this difference. Second, controlled experiments have demonstrated unambiguously that a variety of chemicals present in our foods do cause cancer in laboratory animals when given at sufficient doses. These substances can be present in foods as contaminants, as natural components, or as a result of normal processing and cooking practices. Table I is a partial list of food-related chemicals that induce cancer in laboratory animals, and thus are of concern as potential human carcinogens.

A central issue in food safety is whether these compounds also would be carcinogenic to man and, if so, how much exposure would represent a cancer risk that society deems acceptable. The major difficulty lies in knowing how to extrapolate high-dose, chronic exposure animal studies to human populations, which may experience much lower average exposure and may differ in sensitivity compared to test animals. A great deal of attention is being given to the mathematical models used to extrapolate risk (expected tumor response) down to very low carcinogen doses relevant to human exposure. The procedure used to make these estimations is known as *quantitative risk assessment.* The uncertainty in these models remains a critical issue in risk assessment, since a difference between models of only 0.01% in estimated response constitutes 25,000 additional cases of cancer in the United States population.

Three classes of food-related compounds have received considerable attention as potential human carcinogens: the mycotoxins (especially aflatoxins) produced by certain molds; the protein pyrolysates produced by high-temperature broiling of meats; and the nitrosamines present in foods or formed in the body. The following sections will discuss these carcinogens in detail.

B. Mycotoxins

Mycotoxins are toxic secondary metabolites produced by certain pathogenic fungi, especially *As-*

TABLE I Carcinogenic Substances Found in Foods and Beverages

Carcinogen	Commodity
N-nitrosodimethylamine	Beer, bacon
N-nitrosopyrrolidine	Bacon
Ethanol	Beer, wine, spirits
Urethane	Wine, sake
Aflatoxin B1	Moldy grains, nuts, peanut butter
Aflatoxin M1	Milk (from aflatoxin B1 ingestion)
Saccharin	Diet soda
Hydrazines	Common mushrooms
Allyl isothiocyanate	Brown mustard
Pyrolysates	Broiled fish and meats
Benzo(*a*)pyrene	Charcoal broiled meats
Estragole	Basil
Formaldehyde	Shrimp, bread, cola, etc.
Nitropyrenes	Grilled chicken

pergillus, Penicillium, and *Fusarium* species. The role of these metabolites in fungal metabolism and survival is not well understood. They are introduced into the food supply as a result of fungal growth on food and feed crops during harvest, transport, and storage, and during the production of certain fermented and pickled foods. Examples of common fungal toxins in foods are given in Table II. [*See* MYCOTOXINS.]

Among these toxins, the *aflatoxins* (the structure of AFB_1 is shown in Fig. 1) have received the most attention because of the worldwide distribution of *Aspergillus* in warm climates and the potency of their toxins. Initial concern was raised from epizootics of "aflatoxicosis" (death by severe liver failure) in the early 1960s in domestic fowl and livestock fed moldy feeds, and the appearance of liver cancer in hatchery trout fed aflatoxin-contaminated cottonseed meal as part of their ration. While the sensitivity to acute effects varies among species and is difficult to extrapolate, human deaths from ingestion of moldy grains in famine-stricken India and Africa have been recorded.

Routine surveys of aflatoxin B1 (AFB1) in foods over the past two decades in Africa, Asia, and North America reveal variable contamination according to crop, location, and climatic conditions. *Aspergillus* infestation can occur in crops stressed from drought and insect damage, and becomes especially severe in nuts and grains stored under warm, moist conditions. AFB1 levels in such commodities as peanuts, maize, rice, wheat, brazil nuts, sorghum, and similar foods can range from undetectable (<1 part per billion) to over 300,000 parts per billion (ppb). The latter levels were associated with aflatoxicosis in pigs and would likely be lethal to humans if a modest amount were consumed.

From the preceding, it is clear that AFB1 is acutely toxic to humans when consumed at high levels. Perhaps more difficult to estimate is the cancer risk to man from chronic low-level dietary intake of aflatoxins. Studies in experimental animals indicate that AFB1 itself is harmless until metabolically activated to AFB1-8,9-epoxide by a specific cytochrome *P*-450 isozyme in the liver. Competing enzymes convert AFB1 to less toxic primary metabolites including aflatoxins M1, Q1, and P1, while secondary reactions produce polar metabolites (aflatoxicol glucuronides; AFB1-8,9-dihydrodiol; AFB1-glutathione conjugate) that are readily excreted and hence harmless. The susceptibility of any species to AFB1 toxicity and carcinogenicity depends in large part on the relative ratio of enzymes supporting activation and detoxication reactions. For example, rainbow trout are among the most susceptible species because *P*-450 activation is very efficient, glutathione detoxication is minimal, and the species is essentially deficient in the type of DNA repair needed to repair AFB1 damage. As little as 20 ppb dietary AFB1 for only two weeks leads to about 40% tumor response 9 months later in this

TABLE II Food-Borne Mycotoxins

Toxin	Fungal source	Common foods	Toxic effects in animals
Aflatoxin B1	*Aspergillus flavus*	Corn, rice, nuts	Liver cancer, necrosis
Aflatoxin M1	*A. flavus*	Milk (from contaminated food and feed)	Liver cancer
Ochratoxin A	*A. ocraceus*	Cereals, nuts	Liver, kidney necrosis
Citrinin	*Penicillium citrinin*	Yellow moldy rice	Liver cirrhosis, kidney necrosis
Patulin	*P. spp.*	Moldy grains, fruits	Paralysis, edema
Rubratoxins	*P. rubrum*	Corn	Liver necrosis, hemorrhage
Zearalenone	*Fusarium graminearum*	Corn, barley	Genital hypertrophy

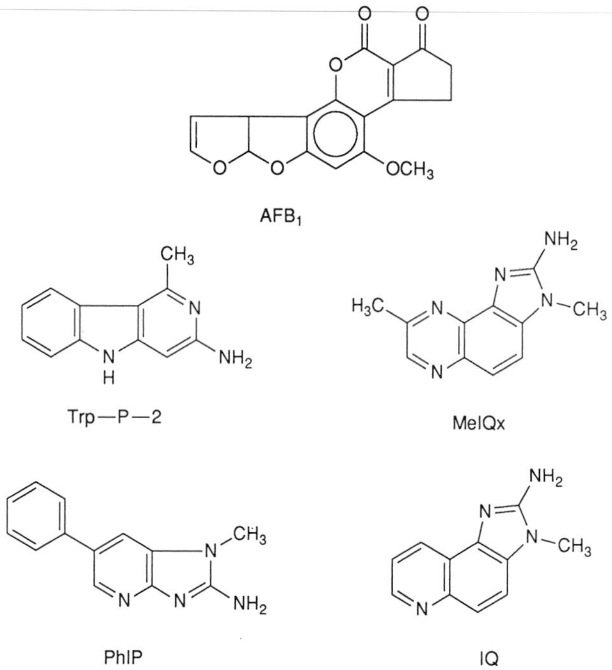

FIGURE 1 Chemical structures of AFB1 (aflatoxin B1) and the amino acid pyrolysis products Trp-P-2 (3-amino-1-methyl-5H-pyrido-[4,3-*b*]indole), MeIQx (2-amino-3,8-dimethylimidazo[4,5-*f*]quinoxaline), PhIP (2-amino-1 methyl-6-phenylimidazo[4,5-*b*]pyridine), and IQ (2-amino-3-methyl imidazo [4,5-*f*]quinoline).

species. By comparison, rats are more resistant, and mice almost totally resistant, because they are efficient at glutathione detoxication and repair of AFB1-DNA adducts in the liver. The few studies that have been conducted with human cells in culture suggest efficient detoxication and hence probable high resistance. This is significant because the FDA-allowable upper limit for AFB1 in foods for human consumption is 20 ppb, a highly carcinogenic dose in trout but not mice. The risk to humans from this exposure is probably low, but not precisely known.

Southern Guanxi, China, has among the highest incidence of human liver cancer in the world. Recent biochemical and epidemiological studies support earlier work in Africa and Asia in indicating a strong correlation between average annual AFB1 exposure and incidence of liver cancer among different subpopulations. The highest risk groups had annual AFB1 intake of ca. 50 mg AFB1/person/year, and a death rate of over 0.6% from liver cancer. Infection with hepatitis B virus also correlates with liver cancer and may be a synergistic factor in this disease in these countries. It seems clear that reasonable measures need to be taken to eliminate this potent toxin from the human diet as much as possible and practicable, especially in developing countries.

C. Amino Acid Pyrolysis Products

In the late 1970s, scientists in Japan discovered that, when cooked in a certain manner, protein-rich meats including beef, pork, lamb, fish, and chicken were sources of high levels of mutation-inducing chemicals, termed *amino acid pyrolysis products*. In the last 12 years, a number of laboratories have confirmed these findings, purified and characterized the compounds responsible, identified the probable precursors in the raw meat, established a number of cooking parameters that determine levels of these compounds, and outlined the mechanisms by which they produce mutagenesis. The finding that, to date, all of these amino acid pyrolysis products are carcinogenic in animals further implies a potential significant human health risk, especially for Western diets that are typically high in these protein-rich meats.

Amino acid pyrolysis products belong to the general chemical class known as *heterocyclic aromatic amines*. At least a dozen of these compounds have been isolated and their structures determined. The structures for four of the most common amino acid pyrolysis products are shown in Fig. 1. These compounds appear to be formed by reactions occurring at high temperature between creatinine, sugars, and amino acids including glycine, glutamic acid, tryptophan, and phenylalanine. The identity of these precursors has been confirmed in an experiment in which MeIQx was isolated from a test tube solution of creatinine, glycine, and glucose that had been subject to high temperatures. In addition, there is some correlation between tissue levels of creatine in raw meat and levels of amino acid pyrolysis products found after cooking. During cooking of meat creatine is converted to the cyclic creatinine, which is the source of the amino-imidazoazarene ring structure of these heterocyclic aromatic amines.

Which meats tend to produce high levels of these pyrolysis products? An example of levels found for three amino acid pyrolysis products in beef and fish is given in Table III. Findings to date indicate that all protein-rich foods from animals can be a source of these mutagens, with beef, chicken, pork, lamb, and fish having the highest levels. Eggs and egg

TABLE III Levels of Mutagenic Amino Acid Pyrolysis Products in Meats

Meat	Compound	Amount (ppb)
Beef	MeIQx	1–12.3
	PhIP	15
	IQ	0–1.9
Fish (walleye pollack)	MeIQx	6.4
	PhIP	69.2
	IQ	0.2

products only form mutagens if cooked at high temperatures for prolonged periods, and animal organs (such as liver and kidney) do not appear to produce these mutagens. Proteins from plant sources apparently form little amino acid pyrolysis products even when cooking times and temperatures are elevated.

The fact that heterocyclic aromatic amines are found in such low amounts (one millionth of a gram in 1000 g of cooked meat) has made the isolation and chemical characterization of these compounds a difficult and time-consuming task. Most researchers have used various extractions followed by high-performance liquid chromatography (HPLC) to purify the compounds and determined their structures by mass spectrometry and/or nuclear magnetic resonance. The relative amounts of the amino acid pyrolysis products are assayed throughout the purification procedure by taking advantage of their high potencies as mutagens in the Ames assay. These compounds typically produce 200,000–700,000 *Salmonella* revertants per gram, making them among the most potent compounds ever tested (for comparison benzo [*a*]pyrene, a polycyclic aromatic hydrocarbon known to cause cancer in mammals, is 300 times less potent). The time and expense involved in assaying levels of these compounds in meats makes routine assays impractical, and therefore, large-scale screening of consumer meat products is not currently possible. It has been estimated that 1 man-month is required just to isolate the compounds from 1 kg meat. Recent advances, such as using monoclonal antibodies to perform rapid screenings, may make monitoring possible in the near future.

A number of studies have demonstrated that the method of cooking can have a marked effect on the yield of mutagens from various meats. In most cases pan-frying or broiling produces the highest yields, whereas deep-frying or stewing results in much less mutagen in the cooked meat. Not surprisingly, cooking temperature and time also have pronounced effects. Cooking temperatures of 150°C, or final meat temperatures of 150°C or less (rare to medium-rare), do not appear to produce significant amounts of amino acid pyrolysis products. High temperatures, or cooking to a well-done state, produces the highest levels of these mutagens. As mentioned previously, there appears to be some positive correlation between the amount of mutagen produced and the initial level of creatine in the raw meat.

The amino acid pyrolysis products that have been isolated from various cooked meats are not mutagenic (and presumably not carcinogenic) by themselves but require biotransformation by oxidative enzymes in the liver to produce the toxic effect. The reaction involves hydroxylation of the exocyclic amino group to form the N-hydroxy heterocyclic aromatic amine, which can subsequently be acetylated. Both metabolites are chemically reactive and will bind to nucleophilic sites including the C-8 position of guanidine in DNA, the reaction thought to be responsible for the mutagenesis and carcinogenesis exhibited by this class of compounds.

The enzyme in liver responsible for the initial N-hydroxylation is cytochrome *P*-450. The cytochrome *P*-450 system is actually composed of a number of cytochromes *P*-450 (10–20 in most mammals), which are derived from distinct gene families. The expression of these various *P*-450s in an individual is under genetic and environmental control. In laboratory animals, the two *P*-450s responsible for the majority of bioactivation of amino acid pyrolysis products belong to the *P*-450 IA gene subfamily. These *P*-450s are normally present in low amounts but can be greatly induced by prior exposure to a number of environmental agents in the air, drinking water, and diet. The major chemical class of inducers of these *P*-450s are the polycyclic aromatic hydrocarbons, typified by benzo[*a*]pyrene. The *P*-450 IA subfamily is present in humans and liver fractions from humans and monkeys and effectively catalyzes N-oxidation of these heterocyclic aromatic amines in vitro, indicating that humans may be susceptible to the mutagenic and carcinogenic properties of these amino acid pyrolysis products. [*See* CYTOCHROME *P*-450.]

With respect to cancer, the few studies to date have been in rodent animal models. Feeding IQ to rats at levels of 0.03% in the diet for 2 years produced high levels of cancers of the liver and the intestine, and low levels of tumors in skin and oral cavity. Female rats dosed with IQ by a stomach tube developed primarily breast cancer. Mice fed 0.01 or 0.04% MeIQ for almost 2 years developed liver tumors, and mice injected subcutaneously as neonates developed liver and lung cancer. There have been few complete cancer studies with these compounds, probably due to the great difficulty and expense involved in obtaining enough compound to treat large numbers of animals. It is significant that essentially every one of the amino acid pyrolysis products tested to date has induced cancer in laboratory animals.

The cytochrome *P*-450 system is characterized by its response to environmental stimuli. The relative amounts of different *P*-450s can be modulated by dietary fat, plant flavonoids, and a host of foreign compounds. It may not be surprising, then, that the amount and type of dietary fat has been shown to influence the degree to which rats can metabolically activate amino acid pyrolysis products. Rats fed diets high in saturated (beef) or monosaturated (olive oil) fat have liver enzymes with a higher potential for pyrolysate activation than does the liver from animals fed diets rich in polyunsaturated fats (sunflower oil). Plant flavonoids, such as myricetin, quercetin, and anthraflavic acid have been demonstrated to inhibit the metabolic activation of amino pyrolysis products to mutagens *in vitro*. These findings are encouraging given that the human diet is rich in plant phenols that have been demonstrated to be effective inhibitors of cancer induced by other classes of chemicals. There is also at least one study in the literature that suggests certain dietary fibers could inhibit the production of mutagens in rats fed high levels of fried meat. Finally, the presence of other structurally similar heterocyclic aromatic amines (e.g., tryptamine), which are not precursors of amino acid pyrolysis products, appear to provide protection against both the formation and metabolic activation of these food mutagens.

What then is the health risk to humans from these compounds in our diet? Currently there is simply not enough information on this relatively new class of carcinogens to provide an accurate risk estimate. The estimated average daily intake is 0.1 mg per day for a 70 kg individual. For comparison, the TD_{50} value (the dose required to produce cancer in 50% of rats or mice tested) is 50 mg per day for a hypothetical 70 kg rat. Therefore, the estimated daily intake is 500 times less than the TD_{50}. One may question whether this is an adequate margin of safety, especially since essentially nothing is known about cancer incidences at lower doses. The widespread distribution of these compounds in our diet and their potent mutagenic and carcinogenic properties indicate their potential risk to human health and support the need for continuing research into the occurrence and carcinogenic properties of all of the amino acid pyrolysis products.

D. *N*-Nitrosamines

N-nitrosamines are formed in foods by chemical reaction of secondary or tertiary amines with a nitrosating agent. In foods, the nitrosating agent is usually nitrous anhydride. The most extensively studied nitrosamine is *N*-nitrosodimethylamine (NDMA), and its formation from dimethylamine and nitrous anhydride, a nitrosating agent, is shown in equation 1. Nitrous anhydride in turn is readily formed under acidic conditions from sodium nitrite, sometimes present in cured meats.

$$(CH_3)_2\text{—N—H} + N_2O_3 \rightarrow (CH_3)_2\text{—N—N=O} + HNO_2 \quad (1)$$

In 1956 two English scientists described the induction of liver tumors in rats by feeding NDMA. This was the first report that nitrosamines were carcinogenic.

Nitrosamines are a subgroup of a larger class of chemicals known as N-nitroso compounds, many of which produce carcinogenic, mutagenic, teratogenic, as well as acute toxic effects. Clearly, the occurrence of such compounds in the human food supply is cause for concern.

The suspicion that nitrosamines might form in foods resulted from studies carried out in Norway from 1960–64. Liver disorders in mink and several ruminants were related to the feeding of herring meal preserved with high levels of nitrite. In 1964 Norwegian scientists reported the isolation and identification of NDMA in toxic herring meal. Presumably dimethylamine, which occurs commonly in fish, reacted with the nitrite to form NDMA.

This important discovery in Norway initiated speculation as to whether nitrosamines might form in human foods, which regularly contain nitrosatable amines. This question has been a source of inquiry for food toxicologists around the world for the past 25 years.

Several generalizations apply to nitrosation reactions in acidic aqueous solution. The rate of nitrosation is pH dependent and is governed by the total amounts of nitrite and amine as shown in Equation 2.

$$\text{rate} = k_1[\text{amine}]\,[\text{nitrite}]^2 \qquad (2)$$

The pH optimum for nitrosation of most basic secondary amines is between 2.5 and 3.5, which is quite acidic.

This has several ramifications. Although foods are not as acidic, most of them are sufficiently acidic to allow nitrosation reactions to proceed at reduced rates. However, the 2.5–3.5 pH range is sufficiently close to the acidity of the human stomach to allow nitrosation to proceed in that medium. This, of course, has important implications for nitrosamine formation in the human body.

Nitrosation can be influenced by a wide range of accelerators and inhibitors. Examples of effective nitrosating agents are nitrosyl thiocyanate (NO—SCN) and nitrosyl chloride (NO—Cl). Undoubtedly foods contain a number of constituents, some of which accelerate or inhibit nitrosation.

Important in the inhibition of nitrosamine formation in foods are ascorbic acid (vitamin C), tocopherol (vitamin E), and sulfur dioxide. Erythorbic acid (isoascorbic acid), which is inactive as vitamin C, also inhibits nitrosation. Addition of ascorbic acid (or erythorbic acid) and tocopherol to cured meats, and of sulfur dioxide in making barley malt prevents nitrosamine formation. Presumably, ascorbic acid reduces nitrous anhydride to nitric oxide, a nonnitrosating species. Tocopherol acts like ascorbic acid; it is oxidized to tocoquinone while nitrous anhydride is reduced to nitric oxide. The combination of ascorbic acid and tocopherol is particularly effective in cured meats since the former is water soluble, and the latter is fat soluble.

Approximately 300 N-nitroso compounds, including many nitrosamines, have been assayed for carcinogenicity, and over 90% have produced cancer in experimental animals. *N*-nitrosodiethylamine (NDEA), the most widely tested nitrosamine, has been shown to be carcinogenic in 40 animal species.

Nitrosamines have some unique carcinogenic properties when compared to other N-nitroso compounds such as nitrosamides. To elicit a carcinogenic response, nitrosamines must first be activated by cytochrome *P*-450, whereas nitrosamides do not need activation: they are direct-acting carcinogens. Consequently nitrosamines often produce tumors at sites distant from the point of application while nitrosamides, which do not require metabolic activation, often produce tumors at the point of application. For a given species, nitrosamines often exhibit organ selectivity independent of the route of administration.

The literature dealing with analytical methodology for nitrosamines often refers to volatile and nonvolatile nitrosamines. Volatile nitrosamines are lower molecular weight, relatively nonpolar compounds such as NDMA, NDEA and *N*-nitrosopyrrolidine. Nonvolatile nitrosamines are higher molecular weight, more polar compounds like *N*-nitrosoproline and *N*-nitrosopeptides. In general, analytical methods are either limited or lacking for nonvolatile nitrosamines, whereas reliable methodology exists for detection of volatile nitrosamines in foods in the sub–parts per billion range.

Using these methods, foods from around the world have been analyzed for volatile nitrosamines. Table IV contains representative levels from investigations in numerous countries reported approximately ten years ago.

In cured meats, including bacon, nitrite is added during manufacture to protect against toxin formation by *Clostridium botulinum*. In addition, nitrite is a color fixatant and it prevents off flavors. Nitrite is the source of nitrosating agents, which in turn react with amines in the meat during storage and cooking to form nitrosamines. In the case of beer, direct-fired dried malt is the source of NDMA. Oxides of nitrogen, such as N_2O_3, form in the combustion gases and react with amines in the malt during drying to form NDMA.

The nitrosamine values in Table IV reflect the levels reported in the period 1977–82, and may be higher than levels in foods today. Through the use of nitrosation inhibitors such as ascorbic acid in cured meats, and by modifications of food processes, the food and beverage industries have markedly reduced nitrosamine formation in recent years. For instance, very recent reports indicate that the NDMA levels in beer today are 2–3% of levels reported a decade ago.

In addition to foods, nitrosamines have been found in a wide range of consumer products including tobacco products, cosmetics, rubber products, and packaging materials. Tobacco products and cigarette smoke contain relatively large amounts of tobacco-specific nitrosamines. Many scientists believe these nitrosamines play a major role in the carcinogenic effect of tobacco products.

In 1981 a committee of the National Academy of Sciences estimated relative exposures of U.S. residents to nitrosamines. By way of comparison, cigarette smoking contributed 17, beer contributed 0.97,

TABLE IV Volatile Nitrosamine Content of Foods[a]

Food	Nitrosamine[b]	Range[c] (μg/kg)	Occurrence
Fried bacon	NDMA, NPYR	1–50	Consistent
Cured meats	NDMA, NDEA NPYR	ND–10	Sporadic
Beer	NDMA	ND–5	Consistent
Cheese	NDMA	ND–10	Sporadic

[a] From reports 1977-1982, probably higher than current levels.
[b] NDMA, *N*-nitrosodimethylamine; NDEA, *N*-nitrosodiethylamine; NPYR, *N*-nitroso-pyrrolidine.
[c] ND, none detected.

and bacon contributed 0.17 μg of nitrosamines per person per day. As indicated previously, the contributions from beer and bacon are lower today.

Humans are exposed to nitrosamines formed in the body through physiological processes. They can be formed in the acidic stomach through reaction between amines and nitrosating agents formed from nitrite. Amines and nitrite can be derived from ingestion of food. Nitrite also can be formed from reduction of nitrate either contained in foods or synthesized in the body by normal physiological processes. Although this is an area of active research around the world, presently it is not possible to quantify exposure of humans to nitrosamines formed in the body.

Use of nitrite in cured meats, particularly bacon, results in the formation of low levels of nitrosamines, which at much higher levels have been shown to be carcinogenic in laboratory animals. Beer also has been shown to contain nitrosamines, principally NDMA. The significance, if any, to human health of this very low exposure to nitrosamines is unknown. Through use of chemical inhibitors and modifications in processing, the food industry has substantially reduced the volatile nitrosamine content of foods and beverages.

Improvements in analytical methodology will need to be forthcoming before we will know the nonvolatile nitrosamine content of our food supply. Likewise, additional research will be required before the significance of nitrosamine formation in the human body is understood.

E. Food-Borne Promoters and Inhibitors of Carcinogenesis

The human diet contains, in addition to carcinogenic substances, chemical compounds that can act as tumor promoters or inhibitors in experimental animals. Promoters include not only contaminants such as DDT and PCBs, but also basic nutritional components such as high dietary protein and excessive fats, which should be avoided. A great many compounds act as inhibitors or anticarcinogens in animal models. Anticarcinogens include diverse substances such as vitamins C and E, selenium, plant indoles and isothiocyanates, conjugated lipids and other antioxidants, beta-carotene, phenobarbital and other drugs, and allyl sulfide compounds from garlic. Human epidemiological studies indicate that high consumption of vegetables containing some of these components may be protective against certain cancers, and human intervention trials are now under way with carotenoid compounds. While there is interest that other dietary anticarcinogens may help reduce human cancer risk, there are as yet insufficient data to promote their use for deliberate chemoprevention.

Biochemical mechanisms of anticarcinogenesis are understood for animal models in a few specific instances. One class of anticarcinogen operates by inhibiting cancer initiation in animals. The inhibition results from induction of the enzymes that catalyze carcinogen detoxication pathways, from direct inhibition of the cytochrome *P*-450 enzymes involved in carcinogen activation, or from prevention of the formation of carcinogenic species from precursors *in vitro* or *in vivo*. An example is the compound indole-3-carbinol (I3C), a component from broccoli and Brussels sprouts that reduces liver DNA damage by AFB1 or NDEA by direct cytochrome *P*-450 inhibition in all animals, and by induction of detoxication enzymes in some. The magnitude of the protective effect increases with I3C dose over the range of possible human consumption, but the mode of action for this compound in humans is unknown. The second class of inhibitors operates through repression of postinitiation events in the promotion and progression of initiated cells to malignant neoplasms. Beta-carotene, some antioxidants, cyanates, and other anticarcinogens are among them. However, it is still unclear whether any of these compounds is protective in humans.

Of greater concern is that certain anticarcinogens, especially the anti-initiators, can protect in one animal experimental design but prove to be enhancing in another. Thus I3C protects if given before and during exposure to AFB1 or NDEA, but acts as a tumor promoter in trout and mice if given after carcinogen treatment. Similarly, the synthetic food antioxidant butylated hydroxytoluene can be protective, inhibitory, or ineffective, depending on the carcinogen, animal model, organ, and exposure protocol under investigation.

The relative potencies of these compounds for protective versus enhancing effects are not well understood; this must be known before their deliberate use for chemoprevention can be advocated. Another way of viewing the problem is that many of these potent modulating agents are essentially unavoidable in our daily diets. They probably influence our response to carcinogens, and we need to understand the nature of that influence. Variable consumption patterns almost certainly effect human cancer rates and complicate our ability to clearly identify the dietary and environmental chemicals responsible for *initiating* cancer in humans. The greatest hope for chemoprevention probably lies in the antipromotional agents such as beta-carotene and other retinoids, which appear to offer more general protection with less risk of enhancement.

IV. Future Directions

Research has expanded the knowledge base in food toxicology over the past quarter of a century. The majority of the focus has been on detection and understanding the mechanisms of action of carcinogens (such as mycotoxins, amino acid pyrolysis products, and nitrosamines), that find their way into our food. Despite some impressive advances, a number of fundamental questions remain unanswered. Can we develop simple, rapid, reliable analytical methods for detection of a wide range of deleterious compounds in foods? Can we accurately assess the effect on human health of trace amounts of these compounds in our food supply? Can we reliably predict the counteracting effects of beneficial compounds when ingested with or following carcinogen exposure? Hopefully these and other pertinent questions will be answered in the next quarter of a century.

Bibliography

Bailey, G., Selivonchick, D., and Hendricks, J. (1987). Initiation, promotion, and inhibition of carcinogenesis in rainbow trout, *Environmental Health Perspectives* **71,** 147.

Coon, J. M. (1988). "Food Toxicology. Principles and Concepts," Marcel Dekker, Inc., New York and Basel.

Hatch, R. T., Felton, J. S., and Knize, M. G. (1988). Mutagens Formed in Foods During Cooking, ISI Atlas of Science, 222.

Hotchkiss, Joseph H. (1987). A review of current literature on N-nitroso compounds in foods, *in* "Advances in Food Research" (C. O. Chichester, E. M. Mrak, and B. S. Schweigert, eds.). Academic Press, New York **31,** 54.

Institute of Food Technologists (1986). "Mycotoxins and Food Safety. The Scientific Status Summaries of the Institute of Food Technologists' Expert Panel on Food Safety & Nutrition." Institute of Food Technologists, Chicago.

Klaassen, C. D., Amdur, M. O., and Doull, J., eds. (1986). "Casarett and Doull's Toxicology. The Basic Science of Poisons," 3d ed. Macmillan, New York.

Scanlan, Richard A. (1983). Formation and occurrence of nitrosamines in foods, *Cancer Research* (suppl.) **43,** 2435_s.

Taylor, S. L., and Cumming, D. B. (1985). "Food Allergies and Sensitivities. A Scientific Status Summary by the Institute of Food Technologists' Expert Panel on Food Safety & Nutrition." Institute of Food Technologists, Chicago.

Williams, D. E., Dashwood, R., Hendricks, J. D., and Bailey, G. S. (1989). Anticarcinogens and tumor promoters in foods, *in* "Food Toxicology: Perspective on the Relative Risks" (S. L. Taylor and R. A. Scanlan, eds.). Marcel Dekker, New York.

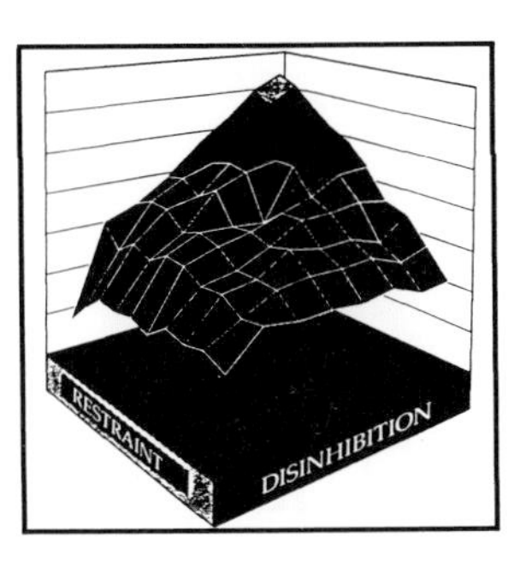

Food-Choice and Eating-Habit Strategies of Dieters

VOLKER PUDEL, NORBERT MAUS, *University of Göttingen*

Glossary

Behavior therapy Therapy consisting of the systematic application of learning principles and techniques to treat behavior disorders. Based on experimental studies of conditioning of both the classical and operant varieties

Concept of externality Model based on Schachter's "cognitive-physiological theory of emotions." He believes that the subjective interpretation of emotions is a function of the interaction between cognitive, or situational, variables in the environment and a state of physiological arousal.

DSM-III-R Revised version of the third edition of the "Diagnostic and Statistical Manual of Mental Disorders" by the American Psychiatric Association

Essential nutrients In describing the nutritional value of a food, it is necessary to describe both its calories and nutrients; important ingredients of food that are not used for producing energy, or not only for that purpose, are vitamins, linoleic acid, mineral elements, and amino acids

Relative body weight Comparison of a person's actual weight (W) and height (H), expressed as relative body weight (RBW), allows the classification of individuals as overweight and underweight; examples:

Broca-reference weight:	weight in kg = height in cm −100
Body mass index (BMI):	(W/H^2) = weight in kg/height in m^2

Wurtman hypothesis Carbohydrate-rich, protein-poor meal, by increasing brain serotonin synthesis, reduces the likelihood that the next meal will be of similiar composition; conversely, consumption of carbohydrate-poor, protein-rich meals diminishes brain serotonin levels, thus increasing the subjects's appetite for carbohydrates, often to the point of carbohydrate-craving

"SLENDERNESS REGARDLESS OF THE COSTS" still seems to be the norm in western societies—especially for the female population. The population still considers the "tool" for achieving this goal to be the diet, i.e., a consciously endured calorie restriction with frequent documentation of weight control. This behavior is described by the concept of "dieting behavior" or "restraint eating." Restrained eating may result in behavioral disturbances, such as a craving for sweet carbohydrates or ravenous appetite, and even bulimic symptoms.

I. Introduction

A very old dream of mankind has come true in western industrialized nations: sufficient food and food of good quality for everyone. But after only a short time, the dream of a land of milk and honey has turned into a nightmare of social medicine. Eating behavioral problems in an affluent society result in faulty nutrition and malnutrition as well as over- and undernutrition. In the Federal Republic of Germany alone, nutrition-related diseases consume DM 42 billion per year (DM 29 billion of direct costs and DM 13 billion of indirect costs). Understandably, medicine and behavioral science are looking for explanations as well as therapeutic and preventive procedures. Nevertheless, this research activity has not yet come up with clear and unequivocal answers. On the contrary, new questions arise from continuing research—as so often is the case in science—and many problems that had been considered solved now must be reconsidered.

In 1968, when psychiatrists began to measure human eating behavior, the approach was practically unknown in the FRG. By then, the regulation of food intake had already been declared as a subject of biochemical and neurophysiological research, but available findings were primarily from animal experiments. Frequently they were too vague and fragmentary to provide a full understanding of food intake.

II. Obesity

First of all, obesity—an obvious disorder of the regulation of food intake—served as a central model. Attempts at psychosocial explanations for obesity have focused primarily on mother–child interactions. They point out the importance of adequate learning to discriminate body awareness and emphasized, in terms of psychosomatic theory, the possibility of oral compensation through eating in stress situations. Ultimately, 30% of overweight people (women three times as often as men) react to stress situations (e.g., lack of success, irritation) with excessive eating. [*See* OBESITY.]

In the 1960s, knowledge about the cause of obesity could be simply summed up: People who eat too much get fat. For a therapy this conclusion was reversed: to avoid getting fat, it is necessary to eat less. However, as early as 1958, Albert Stunkard wrote "Most obese persons will not stay in treatment for obesity. Of those who stay in treatment, most will not lose weight, and of those who do lose weight most will regain it." His statement is still valid, even though in the last 10 years there was an initial optimism as a result of the dissemination of behavior therapy techniques and training programs against obesity. This optimism, however, had to yield to reality.

The reason for the behavioral research approach was epidemiological evidence that (1) classified, depending on the criteria employed, 25–50% of the German population as overweight and (2) proved obesity is a high risk factor, especially for cardiovascular diseases.

Research activities have centered around the sociological, psychosocial, and psychological determinants of food intake, the regulation of appetite,

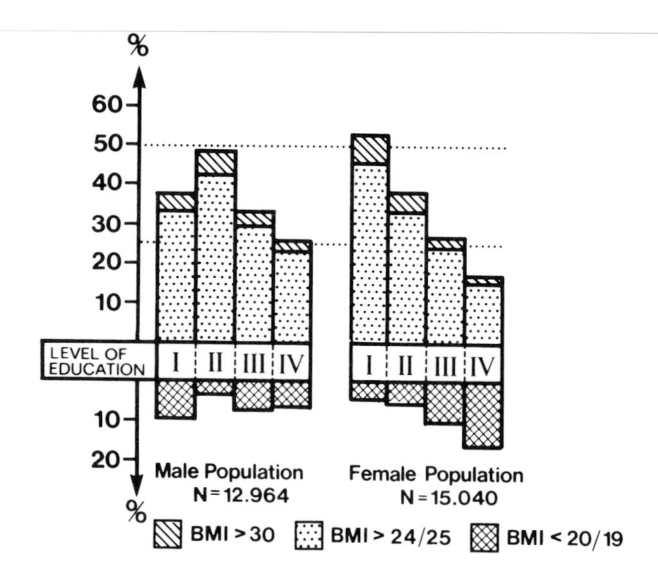

FIGURE 1 Relationship between body weight and level (I = low, IV = high) of education. BMI is body mass index.

and satiation. Their purpose was to determine why such a large part of the population generates a positive energy balance from eating.

In the industrialized nations, overweight status has repeatedly been found to be negatively correlated with *social class*. Figure 1 illustrates this relationship using several cross-sectional investigations based on representative samples of the FRG with a total of 28,000 people. This finding has always been interpreted as an indication that social determinants are important factors for obesity (i.e., choice of certain food or class-specific image of overweight people). But a similar relationship between obesity and social class could not be found for the children and juveniles in the FRG. In fact, the relationship found for adults proved to be a statistical artifact. If the correlative relationship is corrected for the different educational levels, the only remaining significant variable is age, with weight increasing with increasing age (see Fig. 2). Both biological and social factors can contribute to this weight increase.

Psychological approaches, aimed at characterizing an obese personality, have not been successful yet. However, it has sometimes been possible to demonstrate in laboratory experiments that, depending on the conditions, relatively large differences exist between the eating behaviors of normal and overweight people. The results can be summarized as follows: in their awareness of their bodily needs, such as appetite and satiation, overweight people tend to be more influenced by external stim-

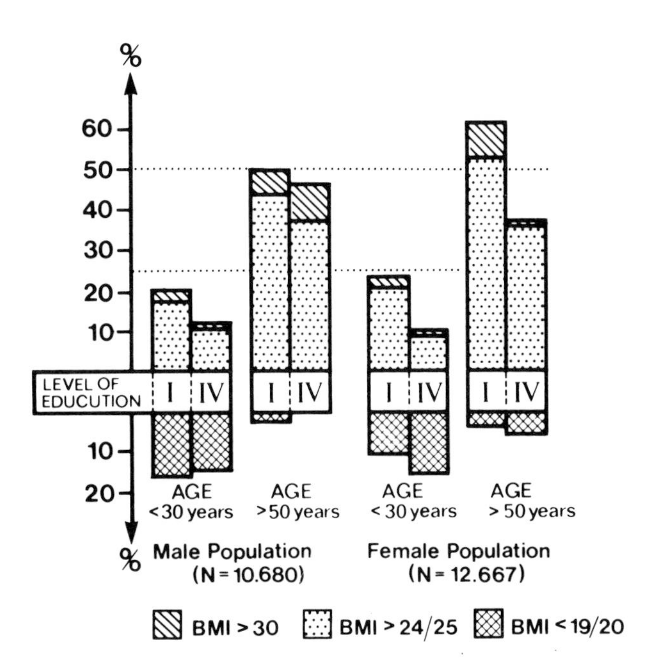

FIGURE 2 Relationship between body weight and level (I = low, IV = high) of education, referring for age.

uli—especially food cues—than by intrinsic signals. Their feeling of satiation during a meal (known as *psychic satiation*) seems delayed or weakened in its intensity.

According to this concept of externality, it seems reasonable that the disturbances of overweight people in their appetite and satiation regulation are conditional factors for obesity, produced by a hypercaloric diet in times of overabundance of food. In accordance with this *psychogenetic model,* normal weight subjects show an energy intake controlled adequately through internal, biochemical–neurophysiological regulatory systems, whereas in overweight subjects the energy intake is inadequately regulated by external (social and psychological) living conditions. On the basis of these considerations, behavioral, therapeutically oriented concepts and methods were rapidly developed for the treatment of overweight people under the aspect of social medicine (in addition to dietetic measures).

The idea of an enormous influence of situational and psychosocial conditions on the eating behavior of overweight people suggested that learning processes play an important role in the pathogenesis of these disorders. The behaviorally oriented training programs focus on the *relearning of eating behavior;* they attempt a deconditioning of the body's awareness of external stimuli, an increased sensibility for the body's awareness of itself. However, the expected benefit of behavioral therapy was not realized, perhaps because the assumption of exogenous induced disturbances in eating behavior amplified learning is not supportable as a general principle.

III. Restraint Eating As an Explanatory Factor

The expediency of a classification between normal weight and overweight, merely based on the body weight of subjects, soon had to be questioned, because results concerning the concept of externality could not be reproduced by a number of experiments. Although externality and other situational conditions showed a fundamental influence on eating behavior, not all of these findings could be related to obesity. In addition to problems with the methodological aspects of the experimental design, and to some extent the parameters used in the definition of experimental groups, a specific group of subjects moved increasingly into the center of attention.

This group consists of people with normal weight, who are always trying to control their eating behavior and their weight by using the most diverse methods. They maintain strict self-control, weigh themselves frequently, go on diets, and regulate their body weight quite deliberately. These normal weight people, who were classified as *latently obese*

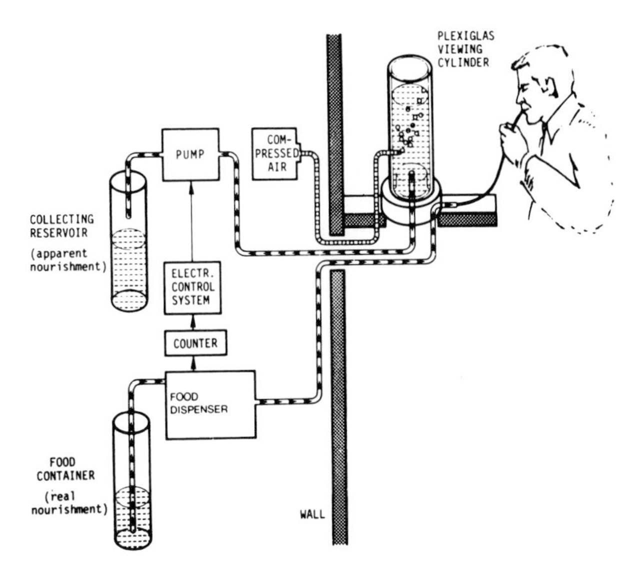

FIGURE 3 Schematic representation of the experimental setting. (From Pudel, Paul, and Maus, 1988.)

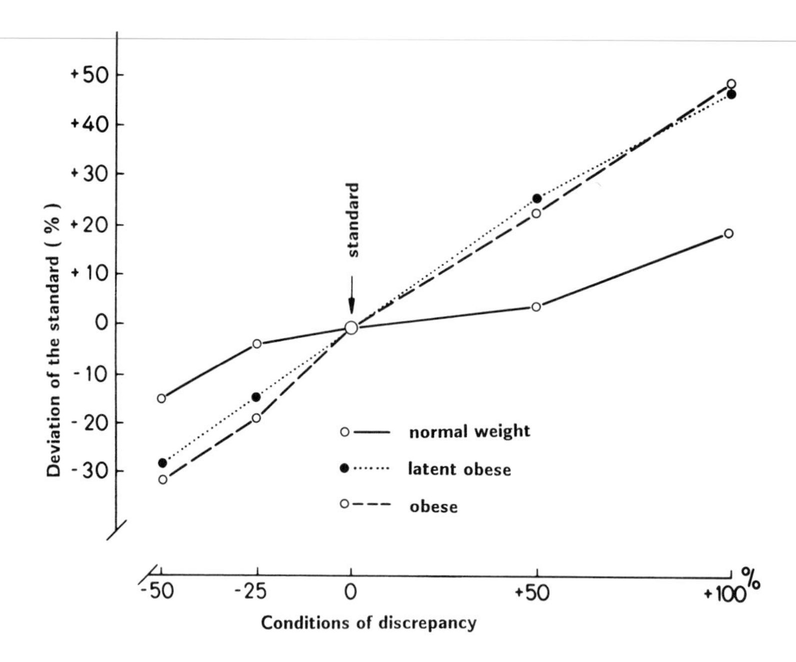

FIGURE 4 Results in normal weight subjects and latently obese and obese patients under the four conditions of discrepancy in comparison with the ingested amount in the standard setting. (From Pudel, Paul, and Maus, 1988.)

or as *restrained eaters*, demonstrated in experimental studies nearly identical behavior patterns to those of manifestly obese people; this shows that the experimental conditions are highly important, because they make a cognitive control of food intake more or less difficult.

In our experiments, a food dispenser was used to quantify the hypersensitivity to environmental cues and to make the postulated *external-stimulus–binding behavior* visible. This automatic recording instrument provides an optimal differentiation between internal and external cues. A schematic representation of the experimental setting is shown in Fig. 3.

In the experiment, the subjects take liquid food through a drinking tube, and in the transparent cylinder on the table in front of them they see the level decreasing. Subjects are instructed to drink until they reach a feeling of satiation. With the help of a hidden electronic control system, a pump, and an additional storage container, it is possible to lower the level in the viewing cylinder more quickly or more slowly. The subjects are unaware of this device. By disproportional pumping, it is possible to produce a discrepancy between the liquid food visible in the glass cylinder and the actual amount of liquid reaching the stomach, so it can be measured whether the stomach volume (*internal stimulus*) or the amount in the container (*external stimulus*) determines food intake.

The results show statistically significant differences between the three groups of subjects (see Fig. 4). All subjects reacted to the external stimulus (level of liquid food) in low discrepancy situations, but obese and latently obese subjects were much more strongly influenced by large discrepancies than normal weight subjects. They adjust their food intake to the level on the instrument and thereby take in highly different quantities.

IV. Satiation Disorder

If, for example, visual control over the quantity of food intake was prevented by covering the food container, latently and manifestly obese subjects were no longer able to regulate their consumption adequately. Variations in the caloric concentration of liquid food led to a proportional change of the volume only with normal weight subjects, but not with latently and manifestly obese subjects. The time–volume function of food intake for normal weight subjects follows a (biological) negatively accelerated satiation curve (see Fig. 5). This means that at the beginning of a meal these subjects consume large amounts of liquid food, progressively reducing their intake toward the end of the meal. In other words, the satiation process works and regulates the termination of food intake according to their stomach volume.

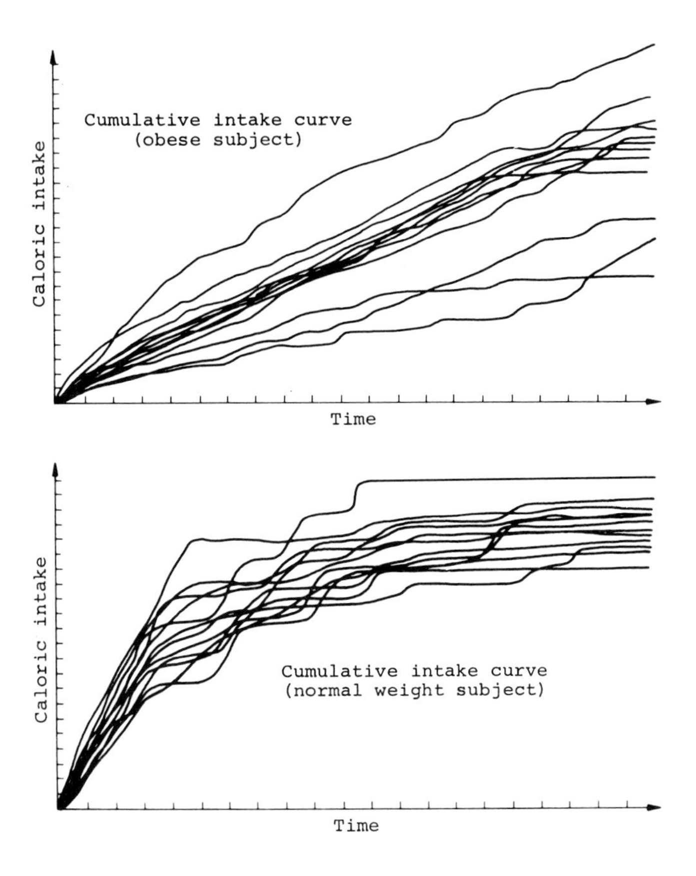

FIGURE 5 Caloric intake with time: the two different curves with and without slowing down at the end of the meal. Above, the linear intake curves, which are typical for latently and manifestly obese subjects; below, biological, negatively accelerated satiation curves of normal weight (unrestrained) subjects. (From Pudel, Paul, and Maus, 1988.)

For obese and latently obese subjects, a characteristically linear distribution of food intake was determined (see Fig. 5). This intake curve can be considered an indication of an altered satiation regulation. It shows no internal satiation signal for the latently and manifestly obese subjects, which becomes increasingly important and slows down consumption.

V. Slenderness Regardless of the Costs

After World War II, the prevalence of obesity increased rapidly in the Federal Republic of Germany. In view of the recognized risk factors, widespread information programs were initiated. Nearly all these campaigns of enlightenment were based on the quantitative aspects of increased caloric intake and a relative lack of exercise. The ideal weight, which at that time was 10–15% (sex-specific) below the Broca-reference weight, was proclaimed as a desirable and, even more important, a realistically attainable goal for every German citizen.

Parallel to these general information programs, numerous and sometimes fairly grotesque forms of calorie restriction were offered. These diets, presented as wonder or crash cures, guaranteed success that does not seem possible with just fewer calories. This tendency continued until the present time—the era of the Psycho, Hollywood, Manager, and Formula diets, slimming drinks, and reducing menus has not yet slackened.

This overabundance of slimming approaches supports the consumer's concept that it is really only a question of the proper diet and his or her own compliance to stick to the right diet. Any deviation from the suggested ideal weight (or that considered aesthetically pleasing to society) is thus a visible personal flaw, which testifies noncompliance. The formerly positive image of the overweight person (a period when the creditworthy belly was a social status symbol) has completely vanished. The general sympathy for these persons decreases dramatically in nearly one decade and the social discrimination

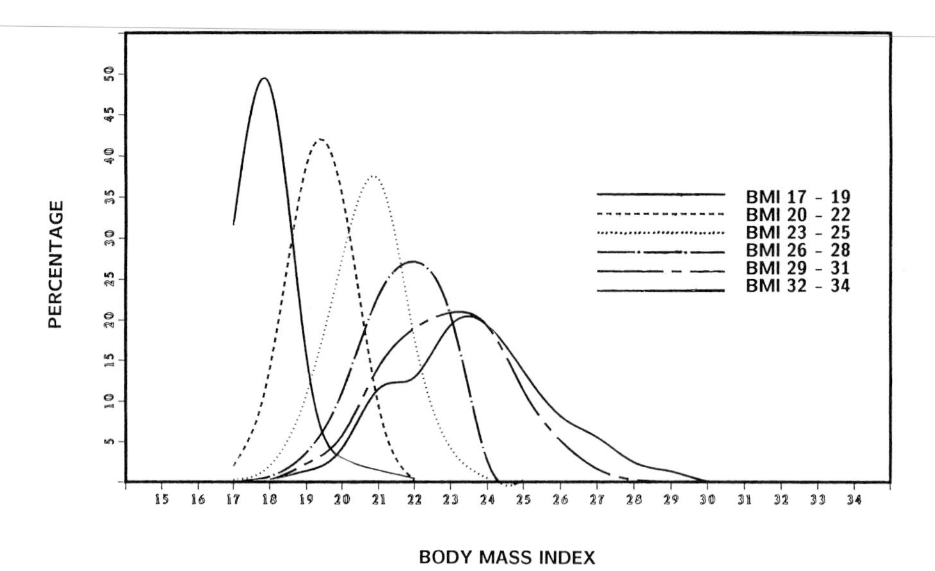

FIGURE 6 Aspired weight for different weight classes (women, n = 35,000).

of overweight people has been spreading for many years.

"Slenderness regardless of the costs" still seems to be the norm for a majority of the population—especially the female population. Epidemiological studies show that especially the younger members of the population are very weight- and calorie-conscious. A food-consumption survey involving 2,900 German families showed that girls between 13 and 18 years of age are consuming fewer calories than they did between the ages 10 and 12. Instead of taking in 200 calories more as they should, they are consuming nearly 200 calories less.

The results of biochemical analyses indicate that this group of relatively young women often fail to fulfill their minimum daily requirements with essential nutrients. Female dieters often endanger their metabolism and suffer from menstrual disorders and fertility problems.

VI. Dieting Behavior

In a comprehensive questionnaire, 35,000 readers of a large women's magazine were asked about their current body weight and their personal *target weight* (i.e., the weight which is realistically attainable). Additionally, they were asked what range of weight they personally desired. Although only 7.6% of the female readers definitely desired a weight loss (for medical reasons), the women showed a collective dissatisfaction with their actual body weight; 42% of them "dreamed" of a body weight between 18 and 20 BMI—a range that represents the inferior boundary of normal weight. The *desired weight* depends greatly on the actual weight; the lower the body weight, the lower the aspired weight (see Fig. 6).

This survey indicates that a subjective feeling of deviation from the norm is derived by many who are within the biological variation range but outside the socially defined, aesthetic niche. Not achieving the socially accepted norm is subjectively experienced as a flaw, which once again motivates these women to achieve it in an effort to obtain social reinforcement and appreciation, an important need developed in the socialization process of western industrial nations.

People still consider that to achieve this goal the tool is the diet. This behavior is described by the *concept of dieting behavior* or *restraint eating*.

Restrained eating scores measured with these tests show only low negative correlations with relative body weight, but are useful indicators for eating abnormalities that are evident in experimental situations. The observations lead to the hypothesis that the habit of dieting behavior, and not the current body weight, is more likely to be the decisive determinant, which is correlated with externality and the other characteristics of described eating behavior.

As a consequence of these results, the fundamental approach to behavior therapy of obesity must be changed because the behavioral patterns, which have been considered typical and possible causes of overweight, are now considered to be accompanying and perhaps conditioning manners for dieting behavior. These patterns no longer appear as a factor for gaining weight, but rather a likely consequence of efforts to lose weight by controlling and limiting food intake. This revised hypothesis is more compatible with the present results, clinical observations, epidemiological findings, and short-term effects of therapy.

More recent clinical findings about the prevalence and the psychopathology of *Bulimia nervosa* (overeating) suggest that the external conditions mentioned above provide the background for the nearly epidemiclike increase in the incidence of this serious eating disorder. Fear of gaining weight and the absolute striving for slenderness are the primary factors of *Bulimia nervosa*. Recurrent episodes of binge-eating and self-induced vomiting as the cardinal symptoms (DSM-II-R) are probably the results of slenderness forced with diet abuse, i.e., with intermittent fasting and severe calorie-restriction. Our investigations show that in test procedures for the classification of restrained eating, these female patients display an extremely high test value, higher than in other groups. [*See* EATING DISORDERS.]

We have, therefore, the justifiable hypothesis that in our society the individual weight of a person, based on its discrepancy from the subjectively desired weight, is a substantial factor determining choice of food and energy intake. The higher the subjectively felt distance, the greater the tendency to dieting behavior. With a comparatively insufficient knowledge of nutrition, dieting often leads to forms of deficient nutrition and/or the preference of seemingly healthy food (e.g., fruit, protein-based foods) and strict avoidance of sugar (or sweet foods). Consequently, 69% of all experienced dieters in the women's magazine study indicated that a hunger for sweets is their greatest problem while dieting. The extent of carbohydrates-craving (due to the comparatively large proportion of protein in many extreme diets), the basis of the Wurtman hypothesis, remains to be seen.

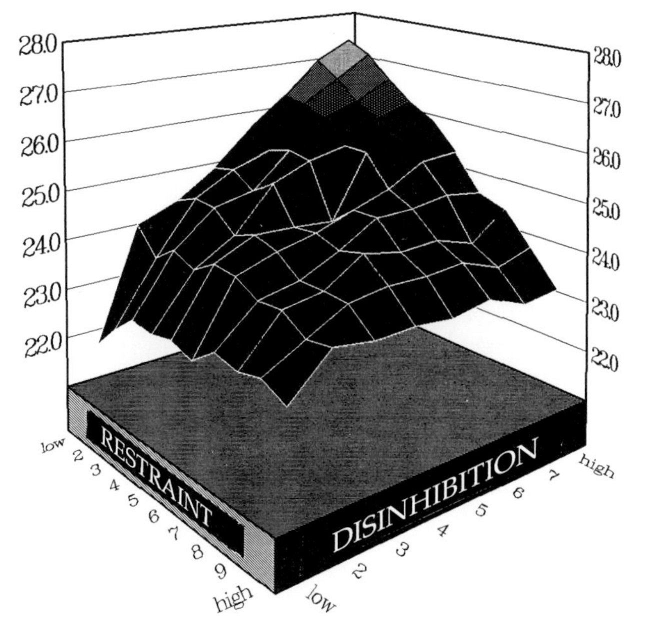

FIGURE 7 Intercorrelation between factors dietary restraint and disinhibition of control on the one hand and body weight (BMI) on the other.

The 35,000 readers of the women's magazine received an *Eating Inventory* test. Figure 7 shows the relationship between actual body weight, dietary restraint, and disinhibition of control. The sample was divided into three groups, each with low, medium, and high scores on both psychological factors. The factor dietary restraint reflects cognitive mechanisms for restricting food intake, whereas the factor disinhibition of control represents the lability in the efforts to control eating behavior and body weight (i.e., the disturbance of cognitive control).

The subgroups show a distinctively different distribution with respect to relative body weight. The higher the disinhibition of control, the higher the body weight. But in subjects scoring high dietary restraint, the correlation between disinhibition and body weight will be reduced, if the control of eating is enforced: High dietary restraint has a modifying effect in the direction of a lower body weight (independent of the degree of disinhibition). The reverse consideration illustrates that a state of low dietary restraint with increasing disinhibition of control is associated with gaining weight. Only the combination of both factors permits a clear control of body weight.

We now have to assume that for long-term weight control, the most promising method consists of a training program (i.e., cognitive behavior therapy) for improving mental control mechanisms. However, it should be considered that the disinhibition of control could increase with decreasing weight. The composition of energy-reduced diets in terms of both essential micronutrients and its macronutrient should be studied more carefully. "Slenderness regardless of the costs"—as witnessed in bulimic patients—but also slenderness with a slightly disturbed eating behavior should not be advocated, unless metabolic risks require the indication of a gradual and adequate loss of weight, which, in our opinion, can only be successful through continual cognitive control of the experienced bodily awareness. Only if control of eating overrides the disinhibition of control can the desired loss of weight occur by calorie-restriction. Whether this goal will soon become attainable through dietetic measures with modified behavior therapy remains to be seen.

Bibliography

Deutsche Gesellschaft für Ernährung, "Ernährungsbericht 1984," Druckerei Henrich, Frankfurt/Main. Deutsche Gesellschaft für Ernährung, "Ernährungsbericht 1988." Druckerei Henrich, Frankfurt/Main.

Herman, C. P., and Polivy, J. (1984). A boundary model for the regulation of eating. *In* "Eating and Its Disorders," Vol. 62, (A. J. Stunkard and E. Stellar, eds.). Association for Research in Nervous and Mental Disease, Raven Press, New York.

Maus, N., and Pudel, V. (1988). Psychological determinants of food intake. *In* "Food Acceptability" (D. M. H. Thomson, ed.). Elsevier Applied Science, London and New York.

Pudel, V., and Westenhöfer, J. (1989). "Der Fragebogen zum Eßverhalten—FEV." Hogrefe Verlag, Göttingen.

Pudel, V., Paul, Th., and Maus, N. (1988). Regulation of eating in obesity and bulimia nervosa. *In* "The Psychobiology of *Bulimia nervosa*" (K. M. Pirke, W. Vandereycken, and D. Ploog, eds.). Springer-Verlag, Berlin and New York.

Stunkard, A. J., and Messick, S. (1985). The three-factor eating questionnaire to measure dietary restraint, disinhibition and hunger. *Journal of Psychosomatic Research* **29/1,** 71.

Westenhöfer, J., Pudel, V., Maus, N., and Schlaf, G. (1987). Das kollektive Diätverhalten deutscher Frauen als Risikofaktor für Eßstörungen. *Aktuelle Ernährungsmedizin* **5/12,** 154.

Wurtman, R. J., and Wurtman, J. J. (1984). Nutrients, neurotransmitter synthesis, and the control of food intake. *In* "Eating and Its Disorders," Vol. 62 (A. J. Stunkard and E. Stellar, eds.). Association for Research in Nervous and Mental Disease, Raven Press, New York.

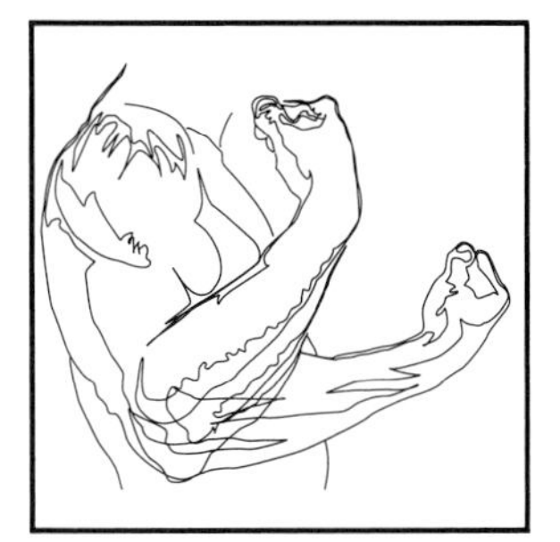

Forearm, Coordination of Movements

C. C. A. M. GIELEN, *University of Nijmegen*

Glossary

Biceps Muscle in the upper arm that contributes to flexion and supination of the arm
Electromyographic activity (EMG) Electrical activity of muscle related to activation
Force-velocity relation Relation that describes muscle force as a function of the rate of change in muscle length at a constant activation of muscle
Gastrocnemius Muscle in the lower leg that contributes to extension of the ankle
Intramuscular EMG EMG activity recorded with intramuscular wire or needle electrodes
Isometric contraction Contraction of a muscle when it is held at a constant length by keeping the limb at a fixed position
Isotonic contraction Contraction of muscle at constant force
Motor unit (MU) Ensemble of motoneuron in the spinal cord and its axon, and all muscle fibers innervated by that axon
Pronation Rotation of the hand along an axis through the forearm, counterclockwise to supination
Recruitment Phenomenon that more motor units become active with increase of muscle force in a fixed order
Recruitment threshold Muscle force at recruitment of a motor unit
Supination Rotation of the hand along an axis through the forearm such that the backside of the hand turns downward with the (right hand) thumb to the right
Triceps Muscle in the upper arm that contributes to elbow extension
Twitch Muscle force as a function of time after a pulse-like activation of muscle (fiber) or MU

THE COORDINATION of movements requires that a number of muscles are activated each with a precise intensity, duration, and timing between the activation of various muscles. Usually a distinction is made between coding of *amplitude* and *duration* of the movement and *direction* of the movement. Amplitude and duration of movements are varied by changing the intensity and duration of the muscle activation pattern in a specific way. Direction is coded by selection of a specific activation for (parts of) the relevant muscles. The muscle activation pattern is the result of contributions of the central nervous system and (afferent) signals reaching it from the periphery, depending on the type of movement and the phase of the movement.

I. Coordination of Movement Amplitude and Duration

In general, goal-directed movements have one accelerating phase and one decelerating phase if their trajectory is approximately along a straight line. This is not true any longer if an obstacle requires that the movement is curved or if the movement requires a specific curved trajectory. In these conditions the movement is segmented and consists of a sequence of trajectories, each with an accelerating and decelerating phase. [*See* MOVEMENT.]

Corresponding to the accelerating and decelerating phase, the activation of muscles has two components: one burst of EMG activity for the muscles that start the movement (agonist) and a burst of EMG activity in the muscles that brake the movement (antagonist). For movements made as fast as possible, a third burst of EMG activity may be observed in the agonist muscle. Although there are still some questions as to the functional role of this third burst, there is a general agreement that the third burst stabilizes the limb. The few data available suggest that the third burst in the thriphasic activation is not simply due to reflex components, but that it has, at least partially, a central origin. Its presence and amplitude are to some extent under voluntary control and can be changed independent of the amplitude and duration of the first and second burst. [*See* MUSCLE DYNAMICS.]

Remarkably, all goal-directed movements have a similar shape more or less independent of the amplitude and duration. Each velocity profile is bell-shaped; velocity profiles of movements with different amplitude and duration can be superimposed after proper scaling in time and/or amplitude. If the amplitude is varied, keeping the movement duration constant, the velocity profiles are scaled in amplitude proportionally to movement amplitude. After appropriate scaling all corresponding EMG bursts in the agonist muscles superimpose. This is not true for the EMG burst in the antagonist muscles, indicating a more complex activation of the antagonist muscles for braking movements as a function of movement duration and amplitude.

When movement duration is varied, keeping movement amplitude constant, the duration of the accelerating and decelerating phase of the movement is scaled by the same amount. However, the duration of the two phases varies with the inverse of the square of that scaling factor. This is not true for EMG in the antagonist muscle, presumably owing to muscle nonlinearities.

The observations reported above can be generalized to single- and multi-joint movements. This is surprising because in a multi-joint movement mono- and biarticular muscles are involved, each of which contributes with a different force at a different shortening or lengthening velocity. Moreover, because angular velocity is different for different joints, the intensity of activation of various muscles is different. All the available evidence suggests that movements are planned in a space-coordinate system and that the corresponding changes in joint angles are derived from the desired trajectory in space. A consequence is that the trajectory in space is approximately a straight line and that the activation of muscles is adjusted to achieve this goal.

The fact that the shape of the velocity profiles is more or less the same for single- and multi-joint movements despite the large variability in duration and amplitude has led to the idea of the *generalized* motor program. This idea implies that the activation pattern of the muscles is based on a general and abstract representation of a motor program, which acts as an elementary unit for any type of action and which can adjust movements for different task conditions by simply scaling a basic velocity profile. However, in movements like grasping and for movements aiming at targets of variable size, the velocity profiles were distinctly different from velocity profiles of fast movements to large targets. These data clearly indicate that planning and control of trajectories is task dependent and cannot always be reduced to a generalized motor program. A coherent model, which describes and explains the movement trajectories for different conditions, is lacking yet. [*See* MOTOR CONTROL.]

II. Coordination of Movement Direction

One of the main problems concerning the coordination of muscle activation for movements or torques in a particular direction is that for some motor tasks, the number of muscles acting across a particular joint is larger than strictly necessary. For example, there are at least seven muscles acting across the elbow joint, contributing to flexion/extension and supination/pronation (i.e., rotation along the longitudinal axis of the forearm). This gives rise to the possibility that a torque or movement in a particular direction can be realized by a large variety of different muscle activation patterns. Despite this redundancy, a unique activation pattern of muscles is observed in various subjects for each direction. Note that we do not imply that the motor system is redundant in general, because different muscles may have muscle fibers with different histochemical properties and with a different fiber architecture, which gives each muscle particular functional properties.

A. Muscle Activation in Isometric Contractions

To investigate the activation of muscles in detail, the activation of single motor units in human arm muscles has been measured. Muscle force is graded by two mechanisms: recruitment (i.e., the increase of the number of active motor units) and firing rate of motor units. The relative contribution of each of

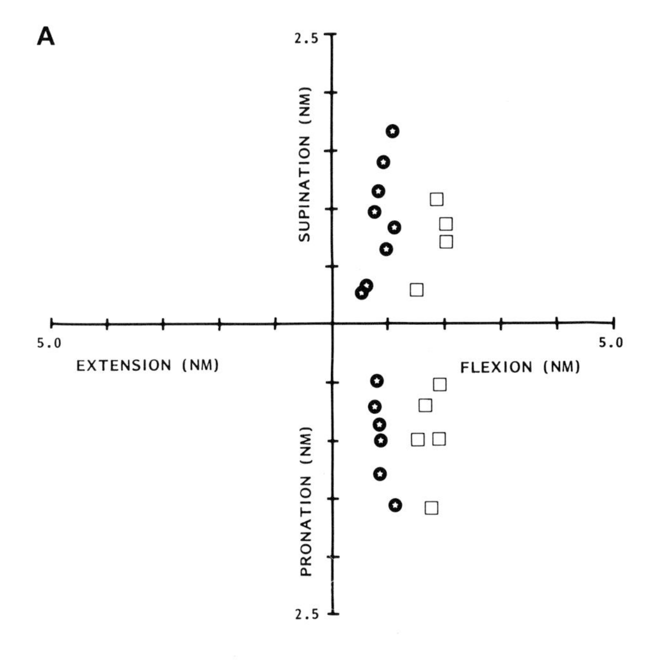

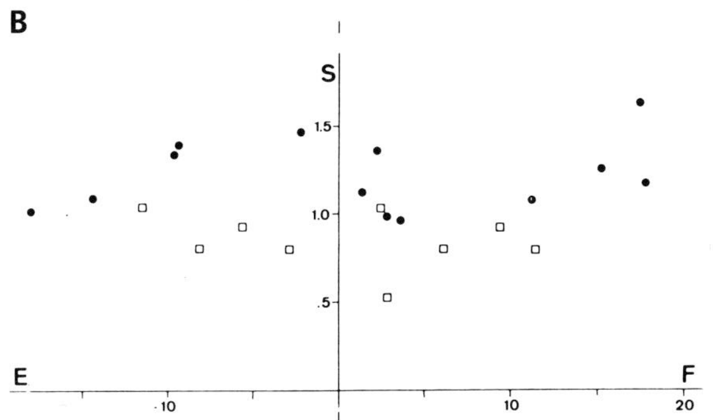

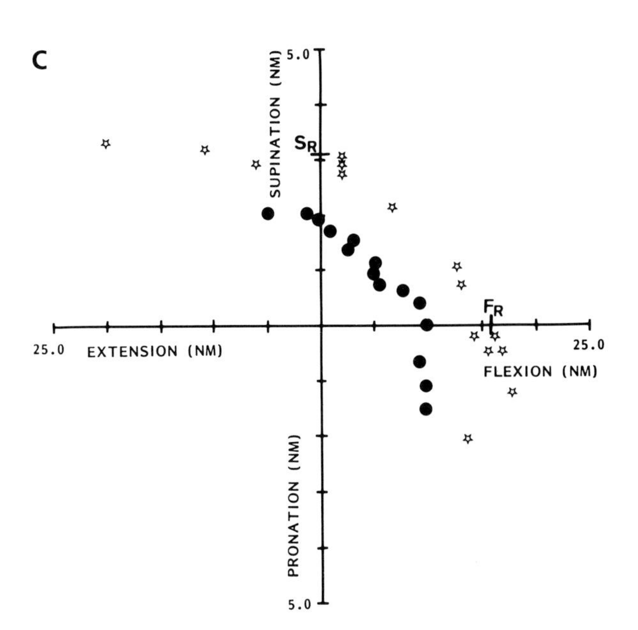

these mechanisms is different for proximal (i.e., closer to the center of the body) and distal (i.e., farther away from the center of the body) muscles, with a more prominent role for recruitment in proximal muscles and a more prominent role for rate modulation in distal muscles.

The motor units in a muscle all have a different recruitment threshold in the range from very small to very large forces. Because a motor unit is thought to be recruited every time the input to that motoneuron exceeds a particular level, the recruitment threshold of a motor unit may be considered as a condition in which the total synaptic input to that motor unit is constant. This phenomenon was used to study the input to motoneurons in different experimental conditions.

The human biceps brachii is frequently called a multifunctional muscle because m. biceps contributes to torques in flexion and supination direction. This multifunctional character is reflected in the activation of motor units. Figure 1 shows the recruitment threshold of several motor units in m. biceps brachii, caput longum. For the motor units in Fig. 1A, the recruitment threshold depends on the torque in flexion direction only. Whatever the torque in supination direction, the recruitment threshold for flexion remains the same. This indicates that motor units of this type are activated for flexion torques in the elbow.

Figure 1B shows the recruitment behavior of another type of motor units in m. biceps, caput longum. The recruitment threshold of motor units of this type depends on the torque in supination direction, whatever the torque in flexion direction.

The majority of motor units in m. biceps, caput longum, is recruited for torques in both flexion and supination direction. An example of this type of be-

FIGURE 1 Example of motor-unit behavior in m. biceps for three motor-unit subpopulations. With different symbols (*squares, circles,* and *stars*), recruitment behavior is illustrated for two units. Each symbol indicates the combination of torques at which a motor unit is recruited. Recruitment threshold is plotted in Newton meters (Nm) in flexion (F), extension (E), supination (S), and pronation. For one motor unit (C), the recruitment threshold for flexion F_r and supination S_r direction is indicated along the horizontal and vertical axis, respectively. A and B show recruitment behavior of motor units, the recruitment threshold of which depends only on flexion torque or supination torque, respectively. In C, data are shown for two motor units for which the recruitment threshold depends on torques in both flexion and supination direction. (Printed with permission from van Zuylen *et al.* 1988.)

havior is shown in Fig. 1C. All recruitment thresholds fall along a straight line in the flexion/supination plane, which can be described by the equation

$$\frac{F}{F_r} + \frac{S}{S_r} = 1$$

In this formula, F and S refer to the torque in flexion and supination direction, respectively, and F_r and S_r refer to the recruitment threshold in flexion and supination direction, respectively. For extension, the recruitment threshold of these motor units depends on the torque in supination direction only, and for pronation, the recruitment threshold depends on the torque in flexion direction only.

The combinations of torques in Fig. 1C, where the two motor units are recruited, fall along parallel lines. This is related to the fact that the ratio of recruitment thresholds in human biceps muscle is the same for all motor units of this type. This is illustrated in Fig. 2 where F_r and S_r are plotted for a representative sample of motor units in m. biceps brachii. The fact that all the data points in Fig. 2 fall approximately along a straight line indicates that the input to all motor units of this type is the same. Motor units may have a different recruitment threshold, but the relative input related to flexion and supination is the same for all motor units of the same type. The results in Fig. 1 are very consistent and reproducible over various subjects and indicate

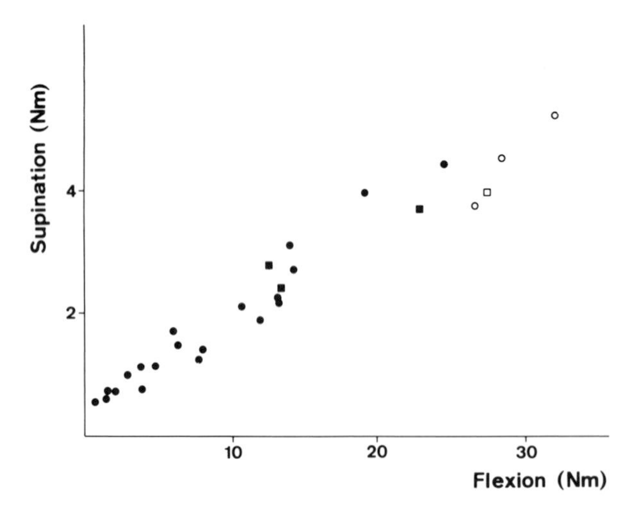

FIGURE 2 Recruitment threshold [in Newton meters (Nm)] for flexion and supination for motor units of the type shown in Fig. 1C. Circles and squares refer to data of motor units in the long head and short head, respectively, of m. biceps. The fact that all data points tend to fall along a straight line suggests that the ratio of recruitment thresholds in flexion and supination direction is constant. (Printed with permission from ter Haar Romeny *et al.* 1988.)

that the coordination of muscles for torque in a particular direction is unique except for a cocontraction term of antagonistic muscles.

The motor units of a particular subtype are not randomly distributed in the muscle but are clustered in a specific location of muscle. In fact, there is a specific relation between location of a motor unit in muscle and its recruitment behavior. This is illustrated in Fig. 3, which shows the intramuscular EMG activity recorded with intramuscular wire electrodes in the lateral and medial side of m. biceps brachii. When flexion torque is increased, activity at the lateral side increases, and when flexion torque decreases and supination torque increases, the center of activity shifts from the lateral to the medial side of m. biceps brachii.

These results indicate that the activation of the motoneuron population of m. biceps brachii is inhomogeneous. However, for each subpopulation the results are compatible with the notion of a homogeneous activation of a particular pool of motoneurons. The idea of homogeneous activation is related to the notion of the *size principle*. According to this principle, the same input is received by each of a set of motoneurons, and the recruitment threshold of each motoneuron is related to the size of the motoneuron cell body. This idea predicts that the recruitment order of motoneurons is the same in all conditions. Without further arguments, the size principle has been generalized in the literature to the whole motoneuron pool of a muscle, although this is not implied by the strict definition of the size principle. The data in Fig. 1 clearly reveal an inhomogeneous activation of the motoneuron pool of m. biceps, and similar data have been presented for almost all muscles. However, these data do not argue against the size principle. Rather the data in Fig. 1 suggest that the motoneuron pool of a muscle has several groups and that each group receives a distinct but homogeneous activation. Recent experiments in animal have confirmed the compartimentalization of motor-unit pool in a single muscle in several subpopulations and have shown that the motoneurons of motor units of the same subtype are clustered together in the spinal cord. This observation suggests that for movements or torques in different directions, different locations in the spinal cord may be activated.

The compartmentalization of motor units of a particular type in a specific part of muscle has important methodological implications because numerous studies have used surface EMG activity as a

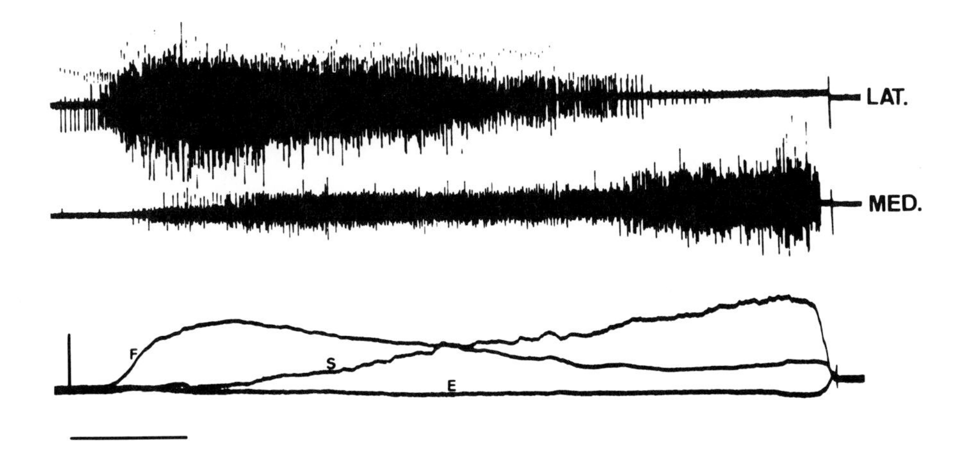

FIGURE 3 Intramuscular recordings of electromyographic activity obtained at the lateral side (*top trace*) and medial side (*middle trace*) in m. biceps caput longum when isometric flexion torque is initially increased and gradually decreased while supination torque is gradually increased. *Lower trace* shows torques in flexion (F), supination (S), and exorotation (E). Exorotation is a rotation in the shoulder joint along a longitudinal axis along the upper arm. (Printed with permission from ter Haar Romeny *et al.* 1988.)

measure of muscle activation. Because EMG activity recorded with surface electrodes reflects the weighted activity of some part of the muscle fibers, the EMG activity strongly depends on the position of the surface electrodes relative to the muscle. Therefore, EMG activity recorded with surface electrodes cannot provide a complete picture of muscle activation.

Because motor units that were active during flexion and supination were localized in different parts of m. biceps, the question came whether motor units in different subpopulations had a different mechanical advantage. The mechanical advantage of a muscle is the mechanical effect of muscle force to the force exerted by the limb. Evidently, this mechanical advantage is optimal when the pulling direction of muscle is orthogonal to the bone of insertion. To investigate this issue, the contribution of MU twitches in supination and flexion direction was determined. Because motor units may have a different twitch amplitude, the argument was that if all motor units contribute equally to flexion and supination torques, the ratio of twitches in flexion and supination direction should be the same for all motor units. Any systematic violation for motor units in different subpopulations would suggest a relation between recruitment behavior and mechanical advantage. It appeared that no relation whatsoever exists between the ratio of twitches in flexion and supination directions for motor units in human arm muscles. This indicates that the compartmentalization in m. biceps reflects a neural organization rather than a difference in mechanical advantage.

An inhomogeneous activation of the population motor units has been found in almost all muscles involved with flexion/extension and supination/pronation. It has also been found in muscles that were not considered multifunctional, such as m. triceps, which generates a torque in extension only. For a small sample of motor units in m. triceps, the recruitment threshold depends on torque in extension direction only. However, there is a larger group of motor units in m. triceps that is also activated for torques in supination and pronation direction. The activation for torques in supination direction can be understood from the fact that m. biceps is activated for torques in supination direction. Because m. biceps has a mechanical advantage with a component in flexion direction, a torque in supination direction can be obtained only if m. triceps compensates for the torque component in flexion direction by m. biceps. The activation of motor units in m. triceps for pronation can be understood from the contribution of m. pronator teres to torque in pronation and flexion direction.

The mechanical advantage of muscle changes as a function of joint angle. This is illustrated in Fig. 4, which shows the combinations of torque where a single motor unit in m. biceps is recruited at different flexion angles in the elbow joint. Clearly the ratio of recruitment thresholds in flexion and supination direction F_r/S_r changes as a function of joint angle. Similar changes in recruitment behavior have been observed in nearly all human arm mus-

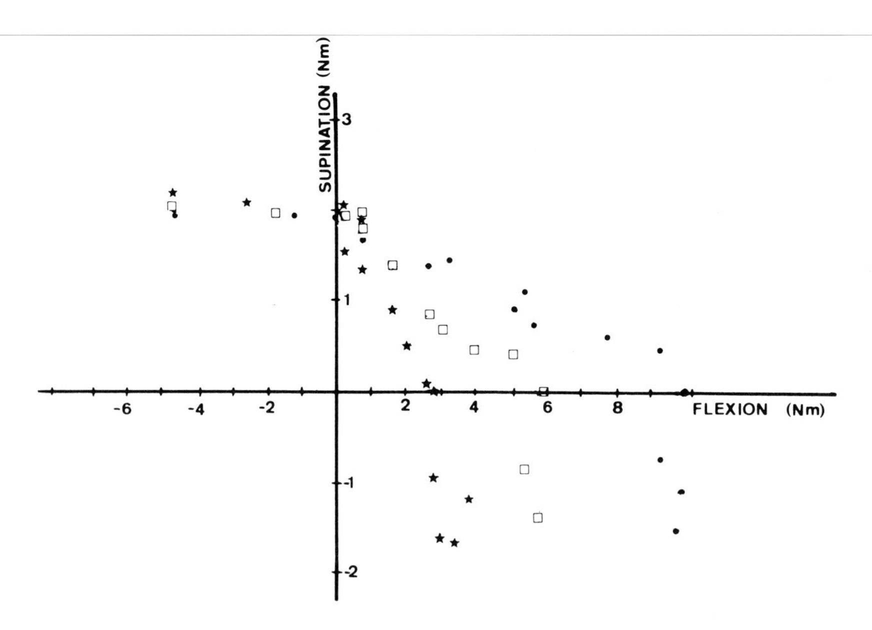

FIGURE 4 Recruitment threshold in Newton meters (Nm) for a single motor unit in m. biceps brachii (caput longum). Each data point indicates the recruitment threshold for a combination of forces in flexion/extension and supination/pronation. Different symbols refer to data obtained at different angles between forearm and upper arm: 100° flexion (*dots*), 145° flexion (*squares*), and 175° flexion (*stars*).

cles. These results indicate that the coordination of muscles and the precise distribution of muscle activations are functions of joint angle.

To provide an explanation for the inhomogeneous activation of the motor-unit pool of muscle, the concept of *task group* has been proposed, in which functional groups of alpha and gamma motoneurons and spindle afferents are programmed to achieve optimal control of movements. Alpha motoneurons are the motoneurons that activate muscle fibers and thereby produce force. The gamma-motoneurons activate the muscle spindles, thereby regulating their sensitivity. The problem with the notion of task group is that the criterion for "optimal" is not always properly defined. However, several suggestions have been made:

- Compartmentalization due to a different mechanical advantage. Some muscles have a broad attachment at the tendon such that different muscle fibers may have a different mechanical advantage (e.g., in cat sartorius muscle).
- Compartmentalization due to a different functional role. For example, some biarticular muscles are activated both during shortening and lengthening. In these two conditions, different motor-unit groups may be activated.
- Compartmentalization due to different mechanical properties of muscle fibers. For some muscles, e.g., gastrocnemius, muscle fibers in different muscle locations have different properties with respect to twitch contraction time, fatigability, twitch amplitude.
- Compartmentalization reflecting a neural organization, such as in the human m. biceps brachii.

Presumably, all these mechanisms may underly the compartmentalization of muscle in general.

B. Muscle Activation in Reflex-induced Contractions

In general it can be stated that the same recruitment order is found in isometric contractions and in contractions that result from the reflex-induced activation of muscle. However, for a proper description of the reflex-induced activation of motor units, a distinction has to be made for the different reflex components.

It is well known that reflex activity is segmented in several reflex components, each with a different functional role and presumably mediated by different neural pathways. The precise pathways involved in all reflex components are not yet known. For the first reflex component (also called tendon reflex, M-1 reflex, or short-latency reflex), it is generally accepted that it is predominantly caused by a short route involving a single neuron directly activated by stretch-sensitive muscle spindles (presumably I_a afferents); however, other pathways may also contribute to it. The main function of the short-

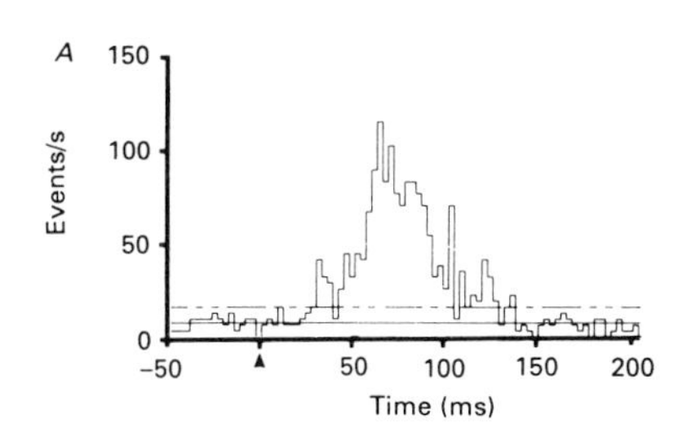

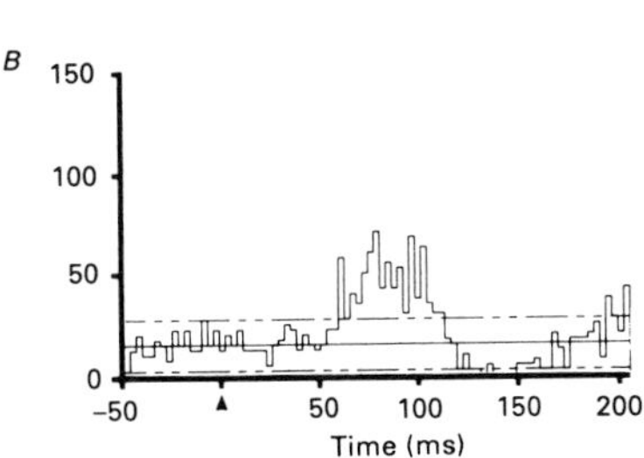

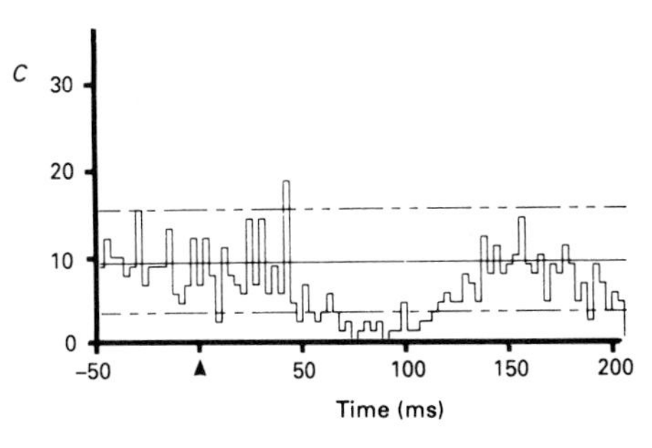

FIGURE 5 Motor-unit activity in m. triceps (A and B) and brachialis (C) elicited by torque perturbations in flexion direction (A) and pronation direction (B and C). Motor unit responses, plotted as the number of action potentials per unit of time, in m. triceps were tested for a preload in flexion direction. Motor units in brachialis were tested by torque perturbation superimposed on an extension preload. Note the excitation in m. triceps in the time interval between 50 and 100 msec and the decreased activity for the motor unit in m. brachialis in the same period. (Printed with permission from Gielen *et al.* 1988.)

latency reflex component is to compensate for the sudden decrease in muscle force caused by the break of actin-myosin bonds, which occurs when muscle fibers are stretched by more than 0.2% of their rest length.

The second reflex component is frequently called long-latency reflex component or M-2 reflex. Presumably more than a single type of sensor is involved. The experimental results so far have demonstrated a role for muscle spindle afferents (both I_a and type II endings) and cutaneous afferents. The pathways and mechanisms involved in the long-latency reflex are still unknown. Three main possibilities have been proposed:

1. The different components in the reflex EMG could be due to an afferent input in bursts
2. The long-latency reflex component could be mediated by more slowly conducting afferents
3. The long-latency reflex component may be mediated by pathways, involving many neurons in the spinal cord, or even a long-loop pathway involving the motor cortex

Some evidence supports each of the possibilities, and each may turn out to contribute to a greater or lesser extent.

The long-latency reflex component serves a role in the coordination of an adequate response. As explained before with regard to the activation of motor units in m. triceps, for torques in supination or pronation direction some muscles may be activated in isometric contractions even if the isometric torque is orthogonal to the torque that results from activation of that muscle. Similarly, m. triceps is activated in a long-latency reflex by torque perturbations in pronation direction, which do not stretch nor shorten the length of m. triceps. The time interval of activation of motor units in m. triceps after a torque perturbation in pronation direction corresponds to that of the long-latency reflex. This shows that muscles that are not stretched may reveal long-latency reflex activity if the adequate coordination of movements requires so. This reflex activity may be excitatory, as in the case of m. triceps in response to perturbations in pronation direction, or inhibitory such as observed for motor units in m. brachialis in response to perturbations in pronation direction (see Fig. 5).

The amplitude of short- and long-latency reflex components may be different in different muscles; however, within each reflex component, the recruitment order appeared to be the same, equal to that in isometric contractions.

C. Muscle Activation in Voluntary Movements

To study in detail the activation of motor units in voluntary contractions, single motor-unit activity has been studied in movements at constant velocity against a constant force acting in flexion or extension direction at the wrist. Because the relative activation of muscle changes as a function of joint angle, the recruitment behavior and firing rate were studied in a small range of positions centered around the same fixed elbow angle. This procedure also eliminates difficulties in interpretation due to the variable mechanical advantage of muscle at various joint angles. The constraint of constant velocity was imposed to compare recruitment behavior in various muscles in the same part of the force-velocity relation.

The activation of motor units in voluntary movements is quite different from that in isometric contractions. Already at very slow shortening velocities (as low as 2°/sec flexion in the elbow joint), the recruitment threshold of motor units in m. biceps is considerably decreased (about 30%) with respect to the isometric recruitment threshold. This decrease was found at the lowest contraction velocities that could be tested. For larger shortening velocities the recruitment threshold decreases more gradually, in quantitative agreement with the well-known force-velocity relation. The difference in recruitment threshold in isometric and isotonic conditions increases proportionally to the isometric recruitment threshold, and the ratio of recruitment thresholds in isometric and isotonic contractions at 2°/sec is approximately constant for all motor units in m. biceps.

In other muscles the recruitment behavior in isometric and isotonic contractions may differ from that observed in m. biceps. For example, in m. brachialis, the recruitment threshold in isometric and isotonic conditions was the same. If the angular velocity in the elbow joint is used to calculate the shortening velocity of muscle fibers in m. biceps and m. brachialis, the shortening velocity of fibers is approximately the same in the two muscles. This fact and the observation of a difference in recruitment behavior in isometric contractions and in movements in biceps and brachialis indicate that the relative activation of human arm muscles is different in isometric contractions and in movements.

Except for the difference in recruitment threshold of motor units, there is also a difference in firing rate at recruitment in isometric and isotonic contractions. For both muscles, firing rate at recruitment is lower for isotonic than for isometric contractions. Apparently, the larger fraction of motor units active at a given force during shortening velocities is compensated for by the lower firing rate of the motor units.

The observation that firing rate at recruitment may be different indicates that the idea that total synaptic drive on the motor-unit pool determines the behavior of the motor units of that muscle is not correct. Clearly two parameters are involved affecting both recruitment threshold and firing rate. This observation is in agreement with observations on motoneurons that have shown that at a constant intracellular input the response of the motoneuron may be affected by changes in membrane properties induced by neurotransmitters such as serotonin and noradrenalin.

At a higher level, the coordination of all muscles involved in a multi-joint movement reflects a rather stereotyped behavior. There is good evidence that the organization of limb movements involves a series of sensori-motor transformations between intrinsic and extrinsic coordinates. The central nervous system has at its disposal extra degrees of freedom in the performance of any motor task. Thus a given position of the hand in space, relative to the trunk, can be achieved with an infinite variety of postural orientations of the arm because the arm has seven degrees of freedom (three at the shoulder, one at the elbow, and three at the wrist) whereas only three parameters are required to describe hand location. Nevertheless, in the case of arm movements at the shoulder and elbow, it has been found that any given task is performed in a stereotyped manner with a low variability within and between subjects.

This observation is reflected in invariances in movements that hold true for entire classes of limb movements. For example, in drawing or handwriting, the magnitude of shoulder and elbow angular motions scales roughly with the size of the figure drawn. Furthermore, although shoulder and elbow movements are tightly coupled (constant phase relation), motions at distal joints are loosely coupled to those at the proximal joints (variable phase relation). Motion at distal joints increases the accuracy of the movement as indicated by the smaller variability of finger relative to wrist trajectories.

In general the relative activation of muscles in single- and multi-joint movements is very much constant. However, an exception has to be made for very fast movements, when muscles, that have predominantly "fast" muscle fibers may be activated preferably at the expense of muscles with predominantly "slow" muscle fibers. An example has been found in rapid movements in the cat ankle, where gastrocnemius seems to be activated almost exclusively with m. soleus almost silent.

III. Role of Central and Proprioceptive Inputs on Muscle Activation

Presumably because of the relatively large number of sensor types and their complex behavior, the precise role of muscle receptors and their contribution to movement control is unknown.

It is generally accepted that the initial accelerating part of fast goal-directed movements is made without the use of afferent feedback. Presumably the first effect of proprioceptive information on the motor program coming from the muscles themselves or the surrounding skin becomes evident about 115 msec after a perturbation or a deviation from the intended movement trajectory. Reflex mechanisms may affect the muscle activation pattern at shorter delay times, and they give rise to the same relative activation of muscles as observed in isometric contractions. Because it has been demonstrated that variability in the accelerating phase is initially high and that it decreases during the first 100 msec of the movement (i.e., before sensory information can become effective), this has been interpreted as evidence for the use of an internal feedback loop.

After the initial accelerating phase, sensory information may affect the activation pattern of muscles. On the basis of all available evidence, virtually everybody now agrees that muscle sensors contribute to position sense and to the control of movements. Recent experiments have demonstrated that the type of information used depends on the motor task. For example, it has been shown that an observer who is instructed to maintain his arm at a given target position while his biceps is vibrated by pulses with a repetition rate of about 100 Hz flexes the arm. (The subject cannot see the arm!) When he is instructed to move the arm fast to the target position, he will be surprised by the instruction because he thinks he is already at the target. However, when he is pushed to obey the instruction, he makes a movement and brings the arm accurately to the target. This suggests that certain motor acts employ a quite different map of limb position from that perceived. Electromyographic recordings ruled out a simple mass-spring strategy and demonstrated that subjects behaved as if the motor system really knew the correct position of the arm even though it was not perceived.

With regard to other types of sensors, the role of joint receptors in motor control is yet unknown. Most experimental observations indicate that joint receptors are only effectively excited when the limb is moved to one or other of its extremes. In the middle of the range, most, if not all, receptors are normally silent. However, there is evidence that joint receptors may modulate the reflex gain not only in the extremes of the physiological range, but throughout the full range of joint action.

Recently, more and more information suggests that cutaneous receptors do contribute to normal movements and to reflex activity. However, there is no consensus yet about their precise role. The same is true for the Golgi-tendon organs, although several hypotheses have been put forward as to their role, for example, for stiffness regulation.

Bibliography

Haar Romeny, B. M. ter, Denier van der Gon, J. J., and Gielen, C. C. A. M. (1984). Relation between location of a motor unit in the human biceps brachii and its critical firing levels for different tasks. *Exp. Neurol.* **85,** 631–650.

Jeannerod, M. (1988). The neural and behavioural organization of goal-directed movements. Clarendon Press, Oxford.

Sittig, A. C., Denier van der Gon, J. J., and Gielen, C. C. A. M. (1985). The attainment of target position during step-tracking movements despite a shift of initial position. *Exp. Brain Res.* **60,** 407–410.

Wadman, W. J., Denier van der Gon, J. J., and Derksen, R. (1980). Muscle activation patterns for fast goal-directed arm movements. *J. Human Movement Studies* **6,** 19–37.

Windhorst, U., Hamm, T. M., and Stuart, D. G. (1989). On the function of muscle and reflex partitioning. *Behavioral Brain Sci.* **12,** 629–681.

Zuylen, E. J. van, Gielen, C. C. A. M., and Denier van der Gon, J. J. (1988). Coordination and inhomogeneous activation of human arm muscles during isometric torques. *J. Neurophysiol.* **60,** 1523–1548.

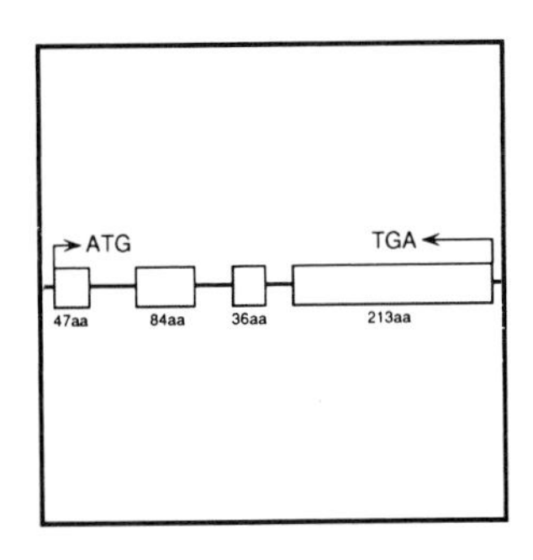

fos Gene, Human

INDER M. VERMA, *The Salk Institute*

Glossary

C-*jun*/*AP-1* Transcription factor, product of protooncogene *jun*
DNA motifs Sequences in genes needed for transcription factors to bind (e.g., TPA-responsive element and cAMP-responsive element)
Leucine "zipper" Hydrophobic region in proteins with heptads containing leucine residues
Plasmid Extrachromosomal DNA that can carry genetic information
Promoters Sequences in genes needed to initiate synthesis of mRNA
Protooncogenes Normal cellular genes capable of cellular transformation
Reporter gene Gene whose protein product can be used for functional assay
***trans*-Activation** Mechanism by which transcription of genes is activated
Transcription factors Proteins involved in transcribing RNA from genes
Transformation Conversion of normal cells to abnormal growth

THE REALIZATION that normal cells harbor genes that have the potential to induce neoplasia was a turning point in modern cancer research. Such genes, collectively referred to as oncogenes or protooncogenes, were first discovered through the agency of retroviruses, a group of viruses that cause tumors in experimental animals. Since human neoplasias generally do not display viral etiology [human acute T-cell leukemia and the acquired immunodeficiency syndrome (AIDS) being recent exceptions], the appearance of oncogenes on the tumor biology scene was not greeted with overt enthusiasm.

Through an entirely different avenue of experiments, it was shown that introduction of human tumor DNAs into established mouse cell lines (i.e., transfection) resulted in the appearance of foci of cells with altered growth properties, called transformed foci. Attempts to identify the active ingredient of the human DNA led to the surprising discovery that the nucleotide sequences mediating transformation were identical to previously identified retroviral oncogenes. The first celebrated example was that of the protooncogene RAS the altered form of which can be observed in at least 10–15% of all human tumor DNAs or tumor cell lines tested so far. The net for the search of oncogenes has been expanded by the successful approach of using DNA transfection methodology. To date, over two score of protooncogenes have been identified, and the list is growing.

Our laboratory has been studying protooncogene *fos*, the resident transforming gene of FBJ and FBR murine osteosarcoma viruses (FBJ- and FBR-MSV), which induces bone tumors *in vivo* and transforms fibroblasts *in vitro*. In this article I describe the structure of the human *fos* gene, the nature of its encoded product, its ability to induce cellular transformation, the regulation of its transcription, and its role in normal cells. [*See* ONCOGENE AMPLIFICATION IN HUMAN CANCER.]

I. Molecular Structure in the *fos* Gene

The complete sequences of FBJ- and FBR-MSV proviral DNAs and the cellular cognate of the viral *fos* gene in mouse and human cells have been reported previously. The nucleotide sequence of the human *fos* gene is shown in Fig. 1. The salient features are as follows:

1. From the presumptive 5′-cap nucleotide to the polyadenylation site, the human *fos* gene contains approximately 3400 nucleotides.
2. The *fos* gene has four exons (i.e., protein coding regions) that are interrupted by three introns (i.e., noncoding regions), predicting a transcript of about 2.1 kb, which is in good agreement with the 2.2-kb size of the fos mRNA after the poly(A) tail has been added.
3. Between the 5′ cap and ATG, the 5′-untranslated region is 157 nucleotides long. The coding region in the first exon is 141 nucleotides, capable of encoding 47 amino acids; the second exon is 252 nucleotides, encoding 84 amino acids; whereas the third exon is smaller, containing 108 nucleotides, capable of encoding 36 amino acids. The fourth and largest exon encodes 213 amino acids. Thus, the human fos protein is 380 amino acids long.
4. Between the chain terminator TGA (2598) where the coding region ends, and polyadenylation site (3403) there are 805 nucleotides, including the polyadenylation signal AATAAA (3381–3386). Additionally, the 3′-untranslated region contains an AT-rich region around positions 3220–3285 which has been implicated in reducing the stability of fos mRNA.
5. The sequence of over 700 nucleotides upstream of the 5′-cap nucleotide is shown. This region contains the regulatory signals for transcription. They include the TATA box, two AMP response elements (CRE), an AP-1 site, and the dyad symmetry element (DSE). Additionally, it has a number of direct repeats. These sequences are conserved between human and mouse *fos* genes (Fig. 1).
6. In the v-*fos* gene present in the virus, the homologous sequences are interrupted by four regions of nonhomology, three of which represent bona fide introns. The 104-nucleotide-long fourth region, which is present in both the mouse and human *fos* genes, has been deleted during the biogenesis of the v-*fos* gene. The loss of 104 nucleotides in the v-*fos* gene transcripts does not decrease the size of the fos protein, because of a switch to a different reading frame.
7. The 380-amino-acid human fos protein is remarkably similar in size to the 381-amino-acid v-fos protein. Figure 2 shows the amino acid sequence comparison of the human fos protein with the rat, mouse, chicken, *Xenopus,* and v-fos proteins. The mouse and human fos proteins share over 90% homology in the coding domain.
8. The introns of mouse and human *fos* genes share nearly 70% homology, but, remarkably, the 3′-untranslated region shows 90% sequence identities. In contrast, the 5′-untranslated region shows 54% sequence homology.

II. fos Protein

While the function of the fos protein remains unknown, considerable progress has been made in deciphering some of its biochemical properties. Many of the properties of the fos protein are also shared by other nuclear oncoproteins (e.g., *myc* and jun). Below, some of the properties of the human fos protein are listed:

1. The 380-amino-acid fos protein is localized in the nucleus.
2. The fos protein is posttranslationally modified by serine phosphoesterification. Most phosphorylation sites are located at the carboxy terminus.
3. It has a short half-life (i.e., ~70–90 minutes).
4. The fos protein belongs to the leucine "zipper" family of proteins, which includes other nucleoncoproteins such as jun, *myc,* and transcriptional factors such as CRE-binding protein and enhancer-binding protein.
5. fos and jun oncoproteins join together to form heterodimers. Although fos protein has been shown to associate with DNA, sequence-specific DNA binding is detected only when it forms a heterodimer with jun. The fos–jun heterodimer binds the DNA motif to the TPA-responsive element (TRE) to activate transcription.
6. The fos protein can autoregulate the transcription of the *fos* gene.

III. Transformation by Human fos Protein

FBJ-MSV was isolated from a spontaneous bone tumor in CFI strain of mice while FBR-MSV was isolated from bone tumors induced in experimental animals by radionuclides. Over 90% of the mice infected with these viruses develop tumors associated with bones and, occasionally, the peritoneum. Generally, metastases is not observed and complications of local tumor growth are responsible for morbidity and mortality. The tumors grow outward from the outer lining of the bone, the periosteum. In contrast, osteosarcomas in humans develop in the deep bony cortex, and metastases are prominent. Periosteal sarcomas that arise in humans are more benign and exhibit outgrowth patterns similar to that seen in FBJ-MSV-induced tumors.

FIGURE 1 Complete nucleotide sequence and deduced amino acid sequence of the human *fos* gene. The annotations include, from left to right: CRE, cAMP response element; SCM, *sis*-conditioned medium; DSE, dyad symmetry element also called serum response element); AP-1, the binding site for the c-jun protein; TATA, sequences required for the initiation of transcription; →, the 5′-cap region of the c-fos mRNA; $(A)_n$ signal, poly(A) addition signal; $(A)_n$, poly(A); The coding region and the start and finish sites of the intron are indicated, as well as the AT-rich destabilizing region (the region involved in mRNA destability). The composite figure indicating the salient features of the human *fos* gene is shown.

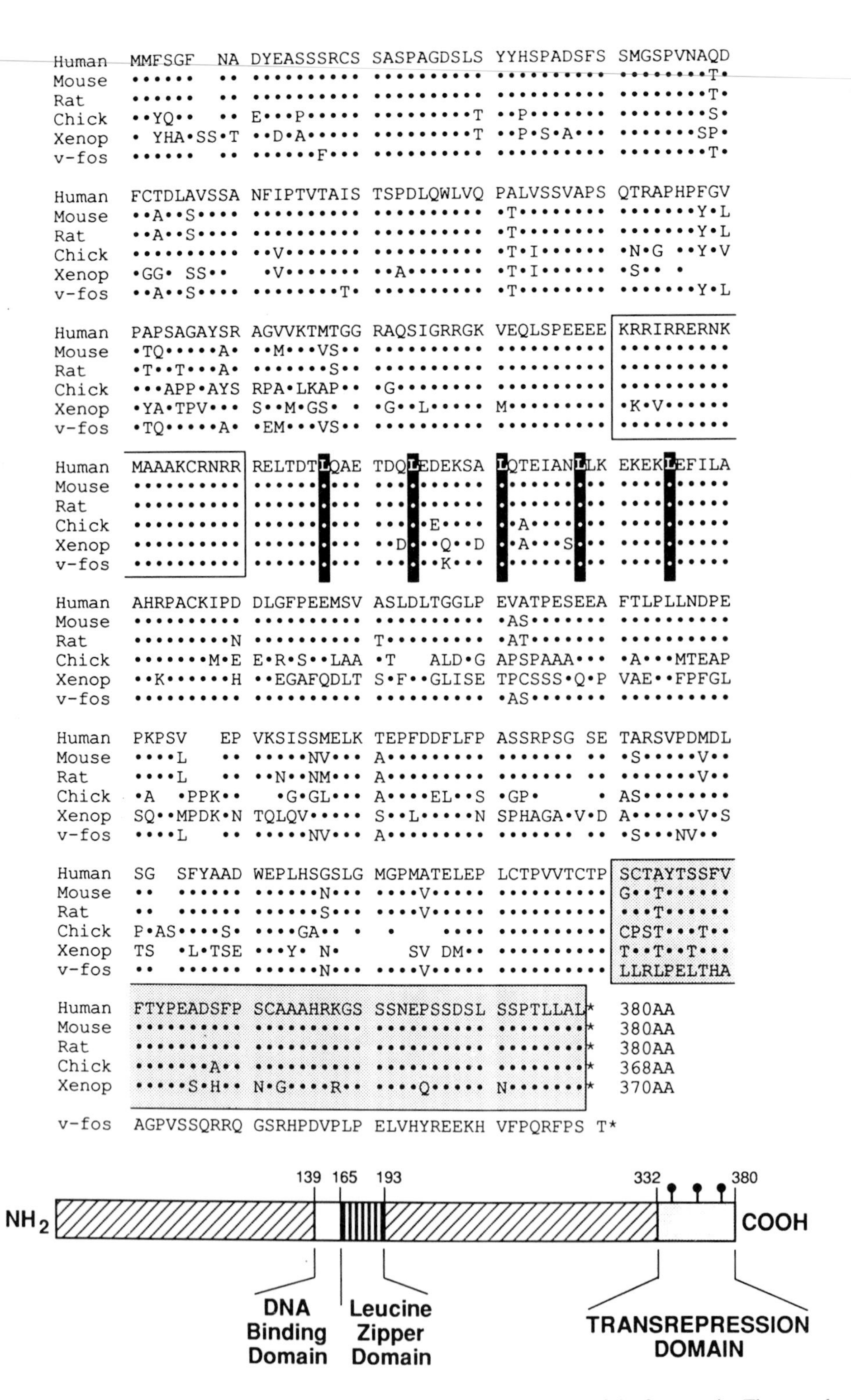

FIGURE 2 Features of the fos protein. The complete amino acid sequences of the human, mouse, rat, chicken, *Xenopus*, and v-fos proteins are shown. The DNA-binding domain (open box), the leucine zipper domain (leucines indicated in solid black), and the transrepression domain (shaded box) are shown. The salient features of the human fos protein are shown at the bottom.

Both FBJ- and FBR-MSV transform established fibroblast cell lines. Since viral oncogenesis is not the primary mode of induction of human tumors, it is particularly important to study the transforming potential of their cellular cognate, the protooncogene. Thus, it was intriguing to find that the cellular *fos* gene can also induce transformation, but requires at least two manipulations: (1) the addition of long-terminal-repeat (LTR) sequences, derived from retroviruses, presumably to increase transcription, and (2) the removal of sequences downstream from the coding domain. A number of recombinant constructs were generated *in vitro* that contained various portions of the viral (v) and cellular (c) *fos* genes. Briefly, the v-*fos* and c-*fos* genes were split into three parts: (1) the promoter region and the first 316 amino acids, originating either from the v-*fos* or the c-*fos* gene; (2) the carboxy terminus and 64 or 65 amino acids of the coding domain of either the v-*fos* or the c-*fos* gene; and (3) the 3′-noncoding domain [including the poly(A) addition signal], originating from either the v-*fos* or the c-*fos* gene. The results of transformation by various constructs demonstrate unequivocally that both the viral and cellular fos proteins can induce cellular transformation.

Constructs containing 3′-noncoding domain of human *fos* gene do not efficiently induce transformation. However, when the 3′-noncoding sequences are removed, transformation of cells is observed which is indistinguishable from the viral fos-transformed cells. The sequences in the noncoding domain involved in 3′ interactions consist of an AT-rich 67-bp region located some 600 nucleotides downstream from the end of the coding domain and about 120 nucleotides upstream of the poly(A) addition signal sequence. The precise influence of the 67-bp region in reducing the transforming potential of the *fos* gene is not understood, but it likely influences the stability of the fos mRNA.

The precise mechanism as to why *fos* induces bone tumors even though it is expressed in a wide variety of cell types remains obscure. Mice were made transgenic by injecting the human *fos* gene into the fertilized egg cell. The gene was under the control of a metallothionein promoter which can be activated by supplying the animal with metals (i.e., zinc) to cause expression of the gene. The gene also lacked the destabilizing sequences. These mice have bone deformities. Recently, however, it has been shown that deregulation of the *fos* gene in the thymus of transgenic mice also leads to hyperplasia. In this regard it is worth pointing out that the cell type specificity of a tumor caused by various oncogenic viruses remains a major challenge in oncology.

IV. Regulation of *fos* Gene

Expression of protooncogene *fos* is induced by a wide variety of agents, including growth factors, differentiation-specific agents, pharmacological agents, and stress. Induction is invariably rapid, and the expression is transient. Induction of the *fos* gene is likely carried out by the mediation of pathways involving the activation of either adenylate cyclase or protein kinase C. Expression of the *fos* gene by substances that activate adenylate cyclase is dependent on the CREs located at positions −60 and −350 (Fig. 1). The transcriptional induction of the *fos* gene by cAMP also requires the CRE-binding protein.

There is no direct proof that induction of the *fos* gene requires protein kinase C, but circumstantial evidence suggests that agents inducing *fos* gene expression are capable of phosphoinositide turnover, resulting in protein kinase C activation. The promoter region responsive to a wide variety of growth factors, mitogens, and differentiation agents contains a DSE (Fig. 1). A 67-kDa serum response factor that binds to the DSE element to induce *fos* gene expression has been identified and molecularly cloned. Recently, a 62-kDa protein has also been shown to bind to this element and forms a ternary complex with p67 serum response factor. The DSE or CArG-rich sequence has been identified in a number of other early-response genes inducible with serum. In most cases the fos DSE can compete to prevent induction, suggesting that similar cellular factors are used during serum induction.

A hallmark of the induction of the *fos* gene is its rapid and transient expression (Fig. 3). Expression of the *fos* gene is witnessed within minutes after addition of the inducer, which maximizes by 30–60 minutes and is essentially undetectable by 120 minutes. In the presence of inhibitors of protein synthesis, the fos mRNA is detectable for up to 4–6 hours, suggesting posttranscriptional stability. The mechanism of this stability is not well understood, but it might involve the AUUUA motif implicated in the instability of the fos mRNA. In the presence of inhibitors of protein synthesis such as cycloheximide or anisomycin, the *de novo* transcription of the *fos* gene continues longer (i.e., at least 180 minutes) and then assumes the basal level, suggesting the requirement of a labile protein to regulate its transcription. [*See* DNA AND GENE TRANSCRIPTION.]

Since the fos protein is one of the earliest proteins synthesized following induction, we hypothesized

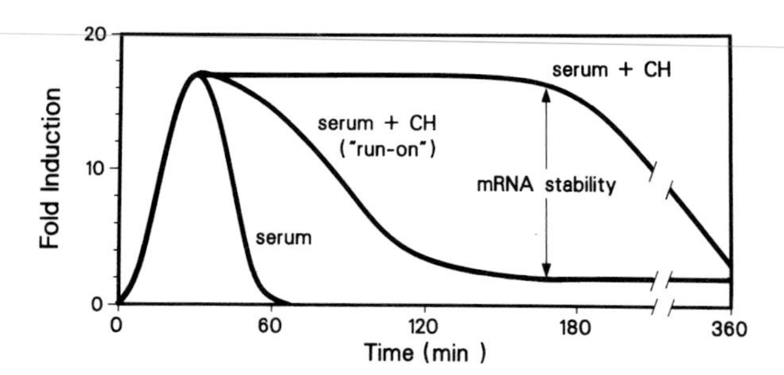

FIGURE 3 An idealized representation of the kinetics of induction of the *fos* gene in response to serum in the presence or absence of cycloheximide (CH). "Run-on," *De novo* transcription.

that the *de novo*-synthesized fos protein could play a role in regulating the transcription of the *fos* gene. To study the possible role of the fos protein in this regulation, plasmids containing upstream sequences (required for serum induction) of the *fos* gene linked to a reporter gene were cotransfected into NIH-3T3 fibroblasts together with *fos* gene included in a vector that permits its expression in the cells. Constructs containing 404 bp upstream from the *fos* transcription start site (Fig. 1) respond to induction with serum, but this induction is abolished if cotransfected with plasmids expressing the fos protein. Such repression is specific for serum induction, since little or no repression was observed when cells were serum starved.

In contrast to the c-fos protein, the v-fos protein is unable to suppress the transcription of the fos promoter. Since the v-fos and c-fos proteins differ at their carboxy termini, we propose that the carboxy terminus of the fos protein is essential for the transrepression of transcription from the fos promoter. Interestingly, the carboxy terminus of the fos protein is extensively phosphorylated, and mutations that prevent phosphorylation also diminish repression activity. Thus, modifications of the fos protein led to the modulation of its expression. The fos protein can also down-regulate the transcription of heat-shock promoter and has been implicated in the control of the adipocyte-specific gene *aP2*. The precise mechanism of repression by the fos protein remains obscure, but does require the DNA-binding and leucine zipper domains of the fos protein. Although the fos–jun complex can bind to the AP-1 site in the fos promoter, it is not essential for repression, since the transcription of only DSE-linked heterologous genes can be suppressed by the fos protein. We presume that the fos protein associates with another protein to form a complex which can then interact with DSE, or other factors binding to DSE, to turn off transcription.

V. Transactivation by fos Protein

The fos protein associates with a series of cellular proteins capable of binding to AP-1 sites in DNA. One AP-1 site is defined by a DNA sequence motif including the consensus TGACTCA, the TRE present in many TPA-inducible promoters. The product of the nuclear oncoprotein jun, which binds to the AP-1 site, was found to be the cellular homolog of transcription factor AP-1. It is now clear that both fos and jun belong to family of proteins which associate to form heterodimers and bind to the AP-1 site. The fos protein alone does not bind to TRE, whereas jun binds rather weakly. However, fos and jun together bind to TRE at least 10–100 times more efficiently. Fos and jun association is dependent on the leucine zipper domains in the two proteins. When individual leucine residues in either the fos or the jun leucine zipper domain (Fig. 2) are mutated, there is little effect on heterodimer formation, but if two consecutive leucine residues are changed to valine in the fos leucine zipper domain, no heterodimers are formed. If the leucine zipper domain is deleted from the jun protein, it can neither form homodimers nor associate with fos to form heterodimers. Thus, the leucine zipper domain contributes toward heterodimer formation.

Immediately preceding the leucine zipper domain is a region rich in basic residues which has been implicated in DNA binding. Deletion of this region (amino acid residues 133–160 in fos, 251–277 in jun) leads to the formation of a heterodimer, but the DNA binding ability is severely compromised. Mutation in the individual basic residues of the fos and jun proteins suggests: (1) Maximal loss in DNA binding activity is observed when reciprocal mutations are made in both proteins. (2) The binding of the fos–jun heterodimer is symmetrical, in agreement with the palindromic nature of the AP-1 binding site (TGACTCA/ACTGAGT). (3) Mutations in the DNA-binding domain of jun increased its affinity to associate with the fos protein, while reciprocal mutations in fos had no effect on the ability to associate with the jun protein. Because mutations in the DNA-binding domain of jun increase its binding affinity for fos, they function as transdominant negative mutants. These mutants abolish transcriptional transactivation by fos–jun heterodimers and thus can be exploited to study the role of fos and jun in normal and transformed cells.

VI. Function of fos in Normal Cells

fos is a pleiotropic protein which can exert its influence as both a negative and positive regulator of transcription. fos can suppress the transcription of its own promoter, of the heat-shock promoter, of the adipocyte differentiation-specific *aP2* gene, and perhaps the expression of other serum-inducible genes which contain CArG sequences. On the other hand, fos in collaboration with jun can activate the transcription of at least those genes which contain TREs. The negative regulatory domain of the fos protein has been localized to reside at its carboxy terminus, but the positively acting domain(s) of fos has not yet been deciphered. Since the fos protein is expressed during growth, differentiation, and development, one can envisage its role as follows: In order for differentiation to proceed, the transcription of many genes must be shut off.

For example, differentiation of monocytes to macrophages is invariably accompanied by a decline in the transcription of protooncogenes *myc* and *myb*. If c-*myc* is constitutively turned on in monocytes, they fail to differentiate to macrophages. Perhaps fos, which is rapidly induced during the differentiation of monocytes, acts as a repressor of transcription of these genes. On the other hand, when cells need to grow (e.g., during the transition from G_0 to G_1), they need the transcription of many genes to sustain growth. In such cases, fos acts as a positive regulator of transcription by associating with jun to turn on genes required for cell growth. At present only jun proteins can act as the partner of fos, but it is possible that there are other cellular transcription factors or proteins which associate with fos, thereby increasing its spectrum of function. With the availability of fos and jun expression vectors, jun transdominant negative mutants, and techniques of selectively inactivating *fos* gene in cells and animals by homologous recombination, it should be possible to delineate its function in the not too distant future.

Bibliography

Gentz, R., Rauscher, F. J., III, Abate, C., and Curran, T. (1989). Parallel association of fos and jun leucine zippers juxtaposes DNA binding domains. *Science* **243,** 1695.

Landschulz, W. H., Johnson, P. F., and McKnight, S. L. (1988). The leucine zipper: A hypothetical structure common to a new class of DNA binding proteins. *Science* **240,** 1759.

Norman, C., Runswick, M., Pollock, R., and Treisman, R. (1988). Isolation and properties of cDNA clones encoding SRF, a transcription factor that binds to the c-fos serum response element. *Cell* **55,** 989.

Sassone-Corsi, P., Lamph, W. W., Kamps, M., and Verma, I. M. (1988). fos-Associated cellular p39 is related to nuclear transcription factor AP-1. *Cell* **54,** 553.

Sassone-Corsi, P., Sisson, J. C., and Verma, I. M. (1988). Transcriptional autoregulation of the proto-oncogene *fos*. *Nature (London)* **334,** 314.

van Straaten, F., Muller, R., Curran, T., Van Beveren, C., and Verma, I. M. (1983). Nucleotide sequence of a human c-onc gene: Deduced amino acid sequence of human c-fos protein. *Proc. Natl. Acad. Sci. U.S.A.* **80,** 3183.

Verma, I. M. (1986). Proto-oncogene fos: A multifaceted gene. *Trends Genet.* **2,** 93.

Verma, I. M., and Graham, W. R. (1987). The fos oncogene. *Adv. Cancer Res.* **49,** 29.

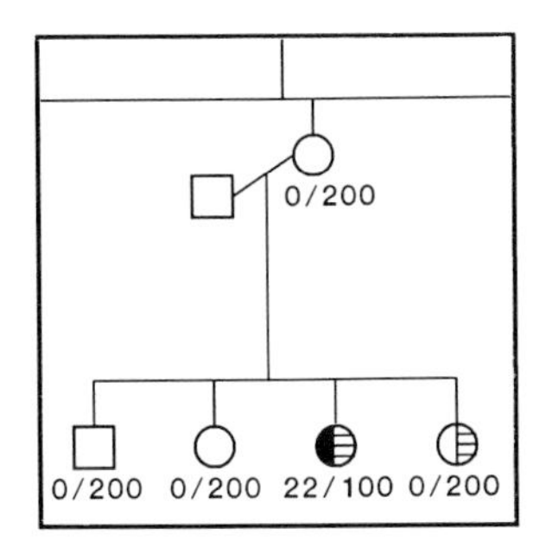

Fragile X Syndrome

RANDI J. HAGERMAN, *University of Colorado Health Sciences Center*

Glossary

Autism Syndrome characterized by impairment in reciprocal social interaction and in verbal and nonverbal communication, such that individuals appear isolated and in a world of their own

Heterozygote or carrier female Female who has one normal X chromosome and one fragile X chromosome; she may be unaffected or affected with the fragile X syndrome

Macroorchidism Large testicles; in the adult, a testicular volume of ≥30 ml represents macroorchidism

Mental retardation (MR) IQ less than 70

Nonpenetrant male Male who carries the fragile X mutation but is not affected by the syndrome and usually does not demonstrate the fragile X chromosome on cytogenetic testing

X chromosome One of the chromosomes that determines the sex of an individual; females have two X chromosomes and males have one X and one Y chromosome

X chromosome inactivation Process where by all females put one X chromosome to rest so that only one X chromosome is utilized for cell activities

X-linked MR Cause of MR that is characterized by an abnormal gene on the X chromosome; therefore, these disorders affect more males in a family tree than females and cannot be passed father to son because the father gives his Y chromosomes (not his X chromosome) to his sons

THE FRAGILE X SYNDROME is a genetic disorder associated with a fragile site or a partial break on the long end of the X chromosome at the q27.3 region. The abnormal gene associated with the fragile site causes a clinical picture of mental retardation often with autisticlike characteristics such as poor eye contact, hand-flapping, and hand-biting. Physical features are also associated with this syndrome including *macroorchidism*, or large testicles, a long narrow face, and large or prominent ears. Hyperextensibility of the finger joints, flat feet, and mitral valve prolapse are also commonly seen clinically. Males affected with fragile X syndrome are usually retarded, although the spectrum of involvement ranges from *learning disabilities* with a normal IQ to profound retardation. All males who demonstrate the fragile X chromosome are clinically affected by the syndrome, i.e., they usually demonstrate mental retardation and associated physical features. Approximately one-third of females who carry the fragile X chromosome are cognitively affected with a borderline IQ or with mental retardation. An additional one-third of carrier females are learning disabled with a normal IQ, and the last one-third are unaffected. Because this disorder is carried on the X chromosome, it is inherited in an X-linked fashion, i.e., it is passed on from mother to son and cannot be passed on from father to son. The overall incidence of individuals affected with this syndrome in the general population is approximately one in 1,000.

I. Epidemiology and Cytogenetic Aspects

The fragile X chromosome (Fig. 1) was first seen in 1969 in a family with four retarded males over three generations. This X chromosome is described as fragile because there is a partial break or disruption

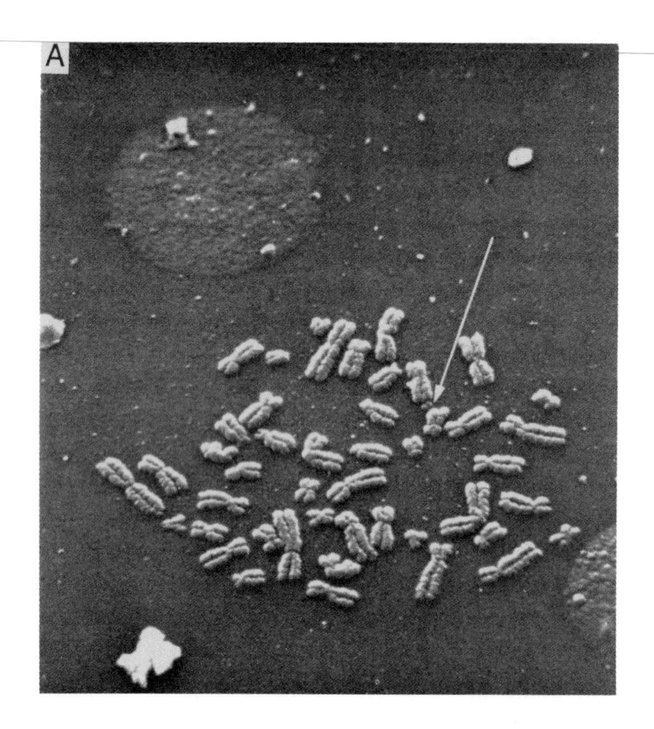

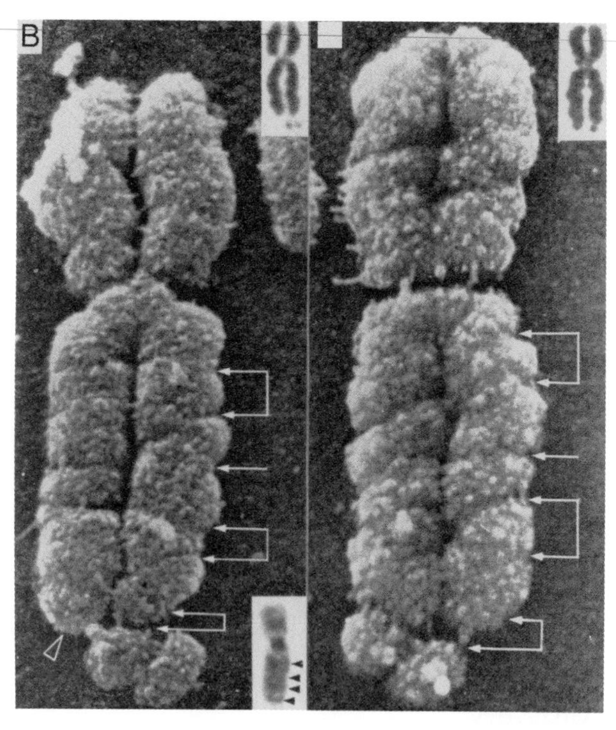

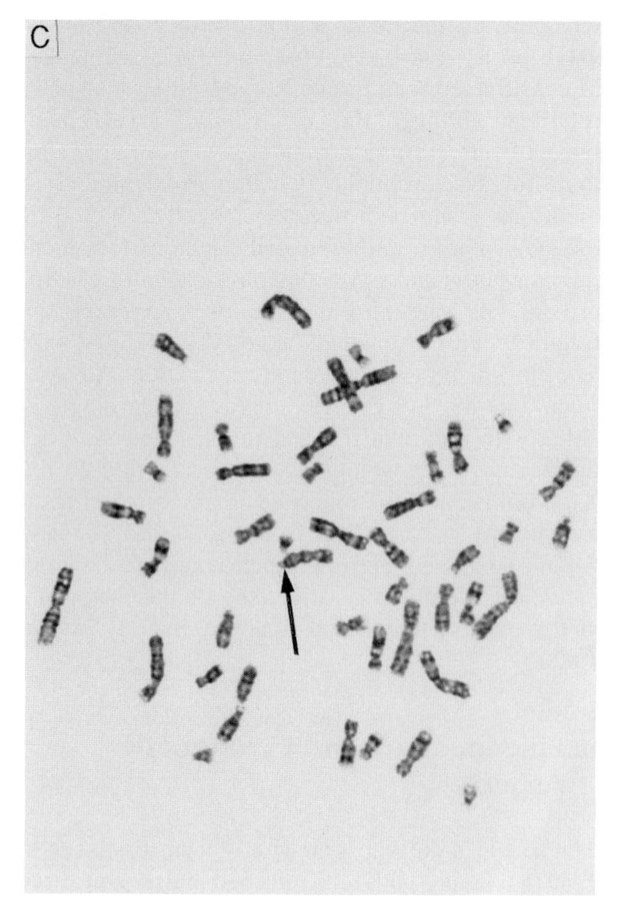

FIGURE 1 A, Scanning electron micrograph of chromosomes in metaphase with arrow pointing to the fragile site on the X chromosome. B, Close-up of two fragile X chromosomes with the lowest arrow pointing to the fragile site at Xq27.3. (Fig. A & B) (Reprinted, with permission, from Harrison, C. J., Jack, E. M., and Allen, T. D., *et al.*, 1983. The fragile X: A scanning electron microscope study. *Am. J. Med. Genet.* **20,** 280–285.) C, Metaphase spread viewed from light microscopy. The arrow is pointing to the fragile site at q27.3 on the X chromosome. (Courtesy of the Cytogenetics Laboratory at the Children's Hospital, Denver).

of the chromosome near the bottom. Work in Australia led to the association of the fragile X chromosome with the physical phenotype in the mid-1970s. However, individuals were rarely diagnosed with this disorder during the 1970s in the United States. This was because the demonstration of the fragile X chromosome is dependent on the use of a specific type of tissue culture media, which was first described by Sutherland in 1976 and more completely understood in 1979. The cytogenetic testing done on a peripheral blood sample requires that the cells be grown in tissue culture media that is deficient in folic acid and/or thymidine. Expression of the fragile X chromosome can also be induced by inhibitors of folic acid, such as methotrexate, or by inhibitors of thymidylate synthetase, such as 2-deoxy-5-fluorouridine (FUdR). Under these conditions, the fragile X chromosome will be seen in 1–50% of the

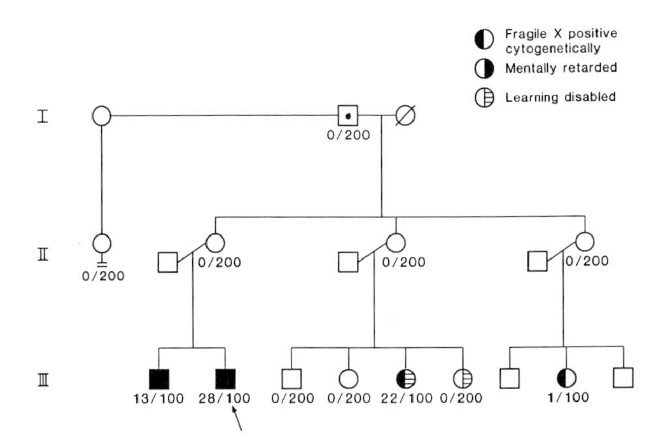

FIGURE 2 A fragile X pedigree. Cytogenetic results (number of fragile X positive lymphocytes/total number of lymphocytes evaluated) are noted below each patient. Generation I has a nonpenetrant male (grandfather of affected males), who has been documented to carry the fragile X gene by DNA studies.

cells analyzed. Although all of the cells carry a mutation at the fragile X site, the visualization of the fragile site at Xq27.3 is not seen in every cell for unknown reasons.

It is possible for an individual to carry the fragile X mutation and not demonstrate the fragile site in cytogenetic studies. This is seen in approximately 40% of carrier females. Carrier males, also called nonpenetrant males, have been identified in numerous pedigrees, and these individuals are usually the grandfathers of affected male children. Although the grandfather may carry the mutation, he usually does not demonstrate the fragile X chromosome on cytogenetic testing. However, he will pass on the mutation to all of his daughters and to none of his sons. His daughters are also unaffected female carriers, but approximately 40% of her sons will be retarded with fragile X syndrome. A typical fragile X pedigree can be seen in Fig. 2.

II. Physical Features

A. Males

Males affected by the fragile X syndrome usually demonstrate a triad of physical features, including macroorchidism or large testicles, large or prominent ears, and a long, narrow face.

Large or prominent ears are seen in approximately 80% of fragile X males at all ages. The ears are almost never malformed; however, they may demonstrate a cupping of the upper part of the pinna. In the prepubertal male, prominent ears are the most common feature that can be detected on physical examination (Fig. 3). Prepubertal individuals only occasionally demonstrate a long, narrow face, but this finding is more common in postpubertal males. Prognathism, or a prominence of the lower jaw, is also a frequent finding in older fragile X males. A high-arched palate is also commonly seen in males, and this may be associated with crowding of the teeth or malocclusion.

Macroorchidism is seen in approximately 80% of adult fragile X males, and it is quantified by a testicular volume of 30 ml or larger. Typically, adult fragile X males have a testicular volume of 50–60 ml, although individuals with a volume as large as 100 ml have been reported. Macroorchidism is much more difficult to diagnose in the prepubertal male, and only one-third of prepubertal individuals have a testicular volume that is larger than normal. The testicle increases most dramatically in size during the early pubertal years, perhaps related to gonadotropin stimulation.

Approximately 60% of prepubertal fragile X males demonstrate hyperextensibility of the finger joints. This is most easily detected by bending the fingers back at the knuckles. Hyperextensibility is defined by having the fingers bend to a 90° or greater angle extension at the knuckle (Fig. 4). Double-jointed thumbs and looseness of other joints, such as the elbow or ankle, are also commonly seen. Flat feet, often with pronation, are seen in over 50% of fragile X children and adults. Other features, which occur in 25–50% of fragile X males, include a pectus deformity of the chest or a pushed-in appearance to the chest, a weak eye muscle causing an in-turning or out-turning of one of the eyes, a single palmar or simian crease, and a large head circumference, particularly in early childhood. Computerized axial tomography (CAT) scans are usually normal in fragile X males; however, more recent studies of brain structure using magnetic resonance imaging (MRI) scans have documented atrophy of the cerebellar vermis, particularly the posterior region, in fragile X males. The correlation between this feature and clinical or behavioral findings in fragile X is not known at the present time.

Nonpenetrant or carrier fragile X males are usually cytogenetically negative, i.e., they do not demonstrate the fragile site on cytogenetic studies and usually do not demonstrate the physical features associated with the fragile X syndrome.

It has been postulated that a connective tissue

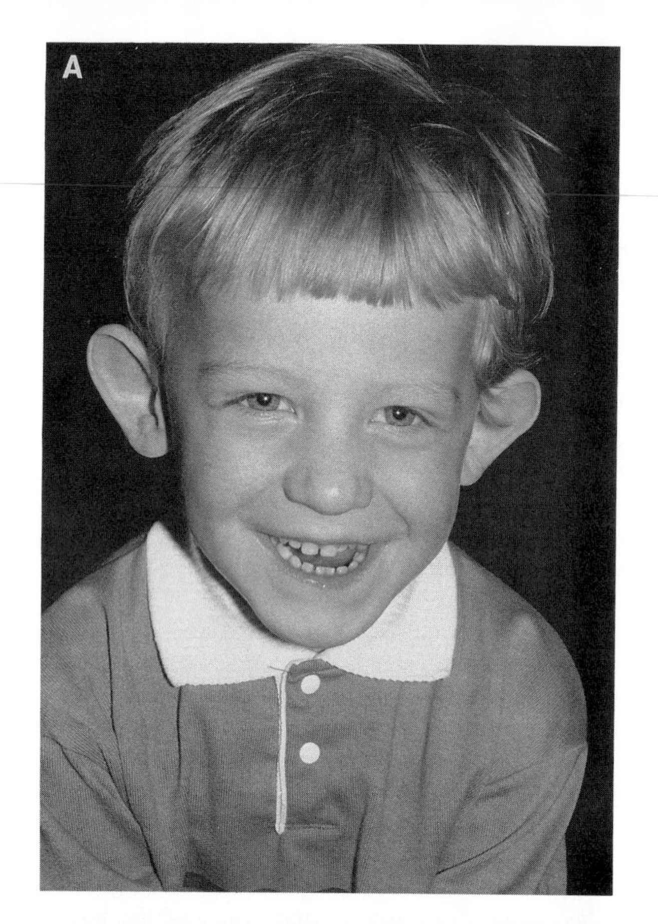
A

B
PANAMA
JACK

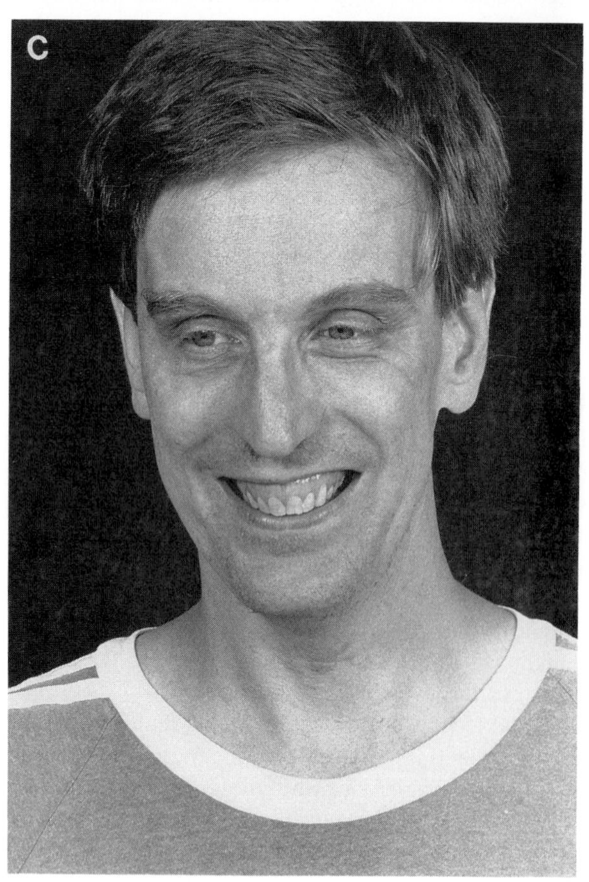
C

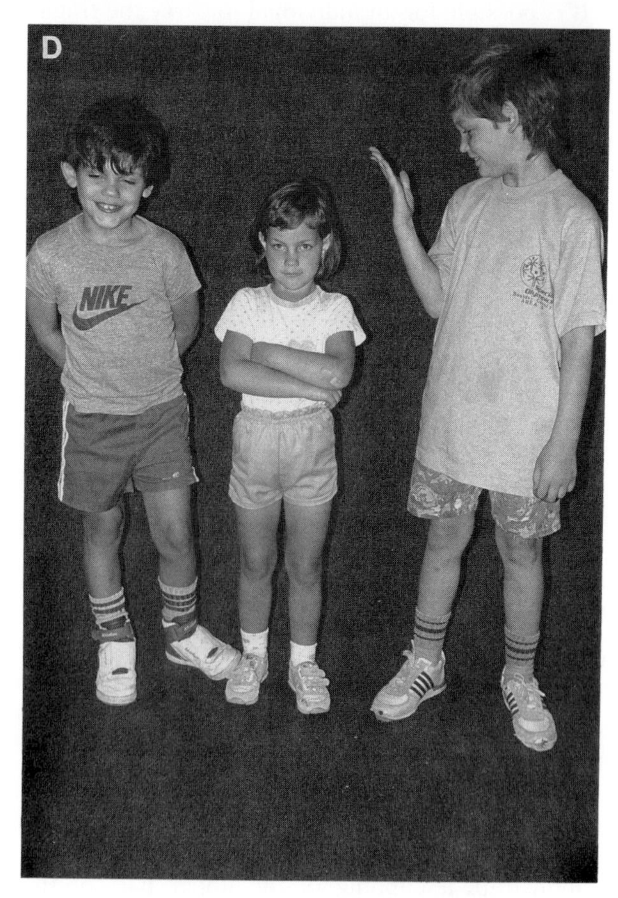
D
NIKE

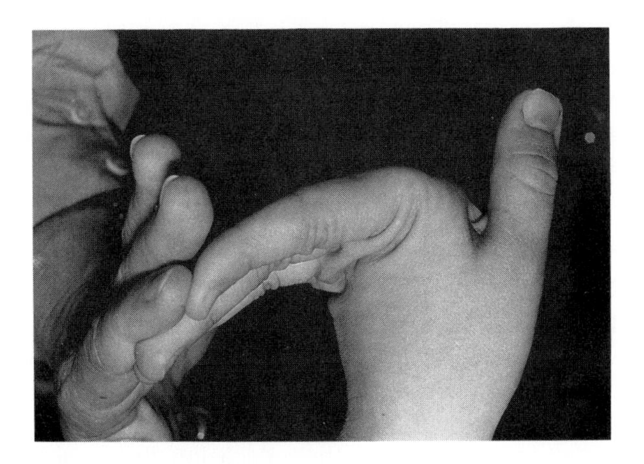

FIGURE 4 Hyperextensible finger joints. Note the fingers extend >90° from neutral position.

abnormality, or dysplasia, is associated with the fragile X syndrome because of the physical findings of flat feet, loose joints, and large or prominent ears. Other disorders of connective tissue, such as Marfan syndrome and Ehlers-Danlos syndrome, also demonstrate similar physical features. Because these other disorders commonly have cardiac problems associated with connective tissue abnormalities, fragile X patients have also been studied for cardiac abnormalities. Approximately 80% of adult fragile X males demonstrate mitral valve prolapse, which is a floppiness of the mitral valve that sometimes leads to regurgitation of blood through the mitral valve during systole. Mitral regurgitation, however, is only occasionally seen in males with fragile X syndrome. [*See* CONNECTIVE TISSUE.]

B. Females

Females who are intellectually impaired because of the fragile X syndrome are most likely to demonstrate the typical physical features. A long, narrow face and prominent ears are commonly seen in retarded or learning disabled females (Fig. 3D). Other findings, such as hyperextensible finger joints, a high-arched palate, and mitral valve prolapse, have also been documented in fragile X females. An occasional female is reported with large ovaries, but it is unclear whether this is a common finding. Unaffected carrier females rarely, if ever, demonstrate physical features of fragile X.

FIGURE 3 A, Young fragile X boy; note cupping of the ears. B, Young fragile X boy with only mildly prominent ears. C, Adult fragile X male; note long face. D, Two young fragile X boys and their sister who is also fragile X positive and learning disabled.

FIGURE 5 A family with three affected fragile X boys. Mother is an unaffected carrier female.

III. Behavioral Features

A. Hyperactivity

Behavior problems are commonly seen in fragile X males and affected females. The majority of young fragile X males demonstrate hyperactivity and a short attention span. They have difficulty in focusing their attention on a single task. Their attention will impulsively shift from one topic or object to another, and this is usually associated with an increased activity level. Their impulsivity can also be seen in their language, which is characterized by poor topic maintenance and a cluttering of thoughts and ideas, which can be communicated in a rapid and sometimes incomprehensible way. Often the diagnosis is made when the child is evaluated for hyperactivity or language delay, which is noted after 2 years of age. Younger fragile X boys may also demonstrate hypotonia, or poor muscle tone, in addition to their hyperextensible joints.

B. Autism

Other unusual features are commonly seen with hyperactivity and have been described as autisticlike. These features include poor eye contact, shyness or social interactional problems, hand-flapping when they are excited or angry, and other hand stereotypes or unusual hand movements. Obsessive and compulsive behavior, sometimes ritualistic behavior, can be seen and repetition or perserveration in both behavior and language are very common. Often a child will do an activity over and over again in a compulsive fashion or will speak in repetitive phrases. Other autistic features include tactile defensiveness or sensitivity or aversion to touch, overreactions to minimal stimuli, and frequent tantrums. Hand-biting is seen in the majority of young fragile X males and may persist into adulthood causing calluses on the finger where repetitive biting occurs.

Although the majority of young fragile X males demonstrate autistic features, a full diagnosis of infantile autism or autistic disorder is only seen in approximately 16% of fragile X males. This diagnosis is only made when a pervasive lack of relatedness and lack of interest in social interactions exist. The majority of fragile X boys are friendly and yet autisticlike features interfere with normal relatedness.

Populations of autistic males have been screened for fragile X syndrome, and it has been detected by cytogenetic studies in approximately 10% of autistic males. Fragile X syndrome, therefore, represents the most common inherited cause of autism that can be identified. The fragile X syndrome is also responsible for 30–50% of all forms of X-linked mental retardation. In institutional screening studies of retarded individuals, fragile X syndrome has been identified in 1–10% of various populations that have been studied. If one evaluates children or adults who are retarded for unknown reasons, fragile X can be detected in 2–10% of these individuals. Therefore, any child or adult who presents with mental retardation of unknown etiology or autism of unknown etiology should be screened for fragile X by performing cytogenetic studies on a peripheral blood sample using appropriate tissue culture conditions. Although the yield of fragile X positive individuals may be approximately 10%, this is a very significant yield in the evaluation of children or adults with developmental disabilities. [*See* AUTISM.]

C. Females

Females who are affected with fragile X syndrome with either learning disabilities or mild mental retardation often demonstrate attentional problems with distractibility but usually without significant hyperactivity. Although they occasionally demonstrate some autistic features, such as hand-flapping or hand-biting, they more commonly demonstrate shyness and social withdrawal. The shyness can be profound, particularly in adolescence, and feelings of social isolation leading to depression are quite common. Depression is also a finding that is frequently seen in mothers of fragile X children. It is unclear whether this finding is related to the environmental stresses of raising a difficult child with fragile X syndrome, or whether the fragile X mutation when present just in the carrier state can predispose an individual to depression. [*See* DEPRESSION.]

IV. Cognitive Features

A. Males

Intellectual abilities in fragile X males can range from severe learning disabilities in the normal IQ range to profound retardation. The majority of adult fragile X males are moderately retarded. Fragile X boys often present in the preschool years with IQ abilities in the low normal to borderline range. These higher cognitive abilities are usually a reflection of their areas of cognitive strength, which include single-word expressive vocabulary abilities, isolated visuoperceptual abilities, such as visual matching tasks, and imitation. Almost all fragile X males, however, demonstrate significant weaknesses in higher linguistic abilities, such as abstract reasoning skills, ability to generalize and solve novel problems, and memory deficits. These abilities, particularly the reasoning abilities, become more important in middle and later childhood and are reflected in the score on IQ testing. Many fragile X children will demonstrate a decrease in their IQ score through middle and later childhood and adolescence. Fragile X children do not lose cognitive abilities and, in fact, they continue to learn and increase their ability through time. However, the emphasis of cognitive testing in later childhood and adolescence stresses their weak areas, such as abstract reasoning. Math is a consistent weakness for almost all fragile X males, and it usually is their most profound deficit in the academic area. Many higher functioning fragile X males will learn to read, and they may even maintain reading and spelling abilities at their grade level.

B. Females

Approximately 30% of females who carry the fragile X gene are mentally impaired with a borderline or retarded IQ. Of females with a normal IQ, approximately 50% are learning disabled because of the fragile X gene. Females who are cognitively impaired with learning disabilities or mental retardation usually demonstrate their greatest deficit in mathematics. Cognitive testing using the Wechsler Intelligence Scale has demonstrated a profile of strengths and weaknesses in the subtest scores. A pattern of weakness can be seen in the subtests of Arithmetic, Digit Span, and Block Design. Areas of strength include the Vocabulary and Digit Symbol subtests. These profiles are seen in the analysis of group data but may be inconsistently present when looking at a particular individual. Cognitive profile testing cannot be used to identify carrier females.

Carrier female identification can be a difficult problem since many carriers are fragile X negative cytogenetically. At the present time, DNA probe studies are available as an indirect method when analyzing a family tree to predict the likelihood of a specific individual being a carrier of the fragile X gene. Within the next few years, the fragile site will likely be cloned and a specific DNA diagnosis will be available.

V. Inheritance Pattern

The inheritance pattern of the fragile X syndrome is unusual and unlike other disorders of X-linked mental retardation. The unusual characteristics include the presence of both carrier males and females who are completely unaffected by the syndrome. It is also unusual that the disorder has increasing involvement in subsequent generations (Fig. 2). That is, the generation that includes a nonpenetrant grandfather usually does not have other individuals who are retarded because of the fragile X syndrome. The daughters of the nonpenetrant male are generally also unaffected but then go on to have retarded children, both males and females. It has been postulated that a mutation takes place in the first generation and requires a subsequent mutation for full expression of the syndrome.

Laird (1987) has developed a theory to explain this unusual inheritance pattern that involves imprinting of the X chromosome. He postulates that the fragile X mutation does not per se cause the clinical syndrome but can be carried without any effects in both males and females. However, when the mutation passes through a carrier female, it may be normally inactivated by the process of Lyonization, in which one of the female X chromosomes is genetically turned off. When this X chromosome goes through the process of reactivation prior to oogenesis, Laird postulates that the fragile X mutation blocks reactivation at the fragile site. The fragile X mutation therefore causes an area of the X chromosome to be permanently inactivated or imprinted. When the imprinted fragile X chromosome is passed on to the subsequent generation, then the full syndrome is realized. Therefore, multiple genes at the fragile site may be responsible for the physical and cognitive features that have been described in this syndrome when these genes are collectively inactivated or imprinted. Although proof of this hypothesis has not yet been obtained, this theory appears to explain the unusual inheritance pattern seen in the fragile X syndrome.

VI. Treatment

A. Medical Follow-up

Medical follow-up of the fragile X syndrome is important for a variety of reasons. Medical problems are often associated with the connective tissue abnormalities in fragile X syndrome, and treatment may be required. Medical intervention or medication can be helpful for behavioral problems, particularly hyperactivity, and this will be addressed in depth. The physician can also be important in organizing and coordinating an overall treatment program that requires special education and individual therapies as described below. There is no cure for the fragile X syndrome, but an overall treatment plan can help fragile X children reach their maximal potential.

Fragile X children often suffer from frequent and recurrent middle ear infections in early childhood. This problem requires vigorous medical intervention, including antibiotics and often the placement of polyethylene tubes through the tympanic membrane to ventilate the middle ear space and to normalize hearing. If hearing is not normalized, the sequelae of language problems associated with a conductive hearing loss will further compound the language deficits that are associated with the fragile X syndrome.

The looseness of connective tissue, which is a frequent problem in fragile X, can cause significantly flat feet, which may require orthopedic intervention, and a high, narrow palate and mitral valve prolapse, which requires antibiotic prophylaxis to prevent subacute bacterial endocarditis. This prophylaxis is required when children undergo dental procedures or surgical procedures, which could contaminate the blood with bacteria.

Strabismus or a weak eye muscle may be seen in up to 56% of fragile X children and requires ophthalmological treatment, such as surgery or patching to strengthen the weak eye muscle. All fragile X males and cognitively impaired females should be evaluated by an ophthalmologist before 4 years of age to detect visual problems at an early stage.

A seizure disorder can be seen in up to 20% of fragile X children, and this requires anticonvulsant medication. EEG abnormalities, including a slowing of background activity and spike wave discharges similar to benign rolandic spikes, can be seen in approximately 50% of children with fragile X syndrome. When seizures are clinically apparent, they are usually infrequent and easily controlled with anticonvulsants. The seizures may be partial motor seizures, grand mal, petit mal, or temporal lobe seizures. The clinical seizures often disappear in adolescence, similar to benign, rolandic seizures. [*See* SEIZURE GENERATION, SUBCORTICAL MECHANISMS.]

Medical intervention for behavior problems, specifically hyperactivity, is often warranted. Central nervous system stimulant medication, such as methylphenidate, dextroamphetamine, or pemoline, can help young fragile X children to decrease hyperactivity and to improve their attention span. Medication, however, is only one aspect of a total treatment program, which should also include special education help, speech and language therapy, and occupational therapy.

B. Speech and Language Therapy

Often fragile X children present with very significant language delays, which are first identified between 2 and 3 years of age. Speech and language therapy can be helpful for the articulation deficits, which are commonly seen, but perhaps, most importantly, this therapy can enhance both expressive and receptive language abilities. The language therapist can work individually with children to improve auditory processing and attention and can also address early deficits in abstract reasoning skills and in the ability to generalize.

Pragmatic aspects of communication are also areas of deficit for fragile X children, perhaps because of the autistic features that are common. Group language therapy can enhance social communication abilities and can also address pragmatic aspects of communication.

C. Occupational Therapy

Occupational therapy is helpful for improving both fine and gross motor coordination even in very young fragile X children. Many of the behavioral difficulties associated with the fragile X syndrome have been ascribed to sensorimotor integration deficits. Fragile X children have difficulty in processing a variety of sensory input, including tactile, auditory, and visual input. They frequently become overwhelmed with stimuli, and this may precipitate behavior problems such as tantrums. Sensorimotor integration therapy is helpful in decreasing behavior difficulties and in helping a child feel comfortable and less anxious with a variety of stimuli. Hypotonia, motor incoordination, joint stability, motor planning, and calming techniques can all be addressed in occupational therapy.

D. Special Education

All children who are affected by the fragile X syndrome will require special education help. For higher functioning individuals, this may involve only pull-out remediation from a regular classroom setting. For more significantly retarded individuals, a self-contained placement is often necessary, particularly when severe hyperactivity and a short attention span are complicating features. Fragile X children often respond well to mainstreaming because they usually mimic the behavior of children who surround them. If these children have normal behavior, then the fragile X child is more likely to develop normal behavior patterns. If fragile X children are placed in a group of children with significantly deviant behavior, then fragile X children often mimic more deviant behavior.

Special education help is also essential for learning disabled individuals, particularly females and higher functioning males. Math deficits are the most common academic problem and individualized help is usually necessary. Many learning disabled individuals demonstrate high achievement in reading and spelling, although deficits can also be seen that require remediation. Attentional problems may also be present in normal IQ individuals, and stimulant medication, in conjunction with special education, is useful.

Counseling or psychotherapy can be helpful on an individual basis and in family work. Behavior modification techniques can be taught to parents to guide them in improving their fragile X child's behavior. Individual work can be useful, particularly for the adolescent or adult fragile X male, to help him deal with sexuality issues and aggression.

V. Genetic Counseling

Genetic counseling is probably the most important reason for identifying children and adults with the fragile X syndrome. Once a single individual has been identified, then the whole family pedigree must be evaluated so that carrier females can be identified and counseled appropriately concerning their risk of having subsequent children with the fragile X syndrome. Although the fragile X syndrome has a similar incidence to Down syndrome, the recurrence risk in fragile X is far higher than in Down syndrome. A carrier female has a 50% risk of passing the fragile X gene to her offspring. If the mother is an affected fragile X female, her risk of having a significantly affected child is increased. Because of the high recurrence risk, it is absolutely essential for medical professionals to identify this disorder which, at the present time, is significantly underdiagnosed. [*See* GENETIC COUNSELING.]

Bibliography

Brown, W. T., Jenkins, E. C., Cohen, I. L., Fisch, G. S., Wolf-Schein, E. G., Gross, A., Waterhouse, L., Fein, D., Mason-Brothers, A., Ritvo, E., Ruttenberg, B. A., Bentley, W., and Castells, S. (1986). Fragile X and autism: A multicenter survey. *Am. J. Med. Genet.* **23,** 341–352.

Chudley, A. E., and Hagerman, R. J. (1987). Fragile X syndrome. *J. Pediatr.* **110**(6), 821–831.

Davies, K., ed. (1989). "Fragile X Syndrome." Oxford University Press, London.

Hagerman, R. J., and McBogg, P. M., eds. (1983). "The Fragile X Syndrome: Diagnosis, Biochemistry and Intervention." Spectra Publishing Co., Inc., Dillon, Colorado.

Hagerman, R. J., Jackson, A. W., Levitas, A., Rimland, B., and Braden, M. (1986). An analysis of autism in 50 males with the fragile X syndrome. *Am. J. Med. Genet.* **23,** 375–380.

Hagerman, R. J., Cronister Silverman, A., eds. (1991). "Fragile X Syndrome: Diagnosis, Treatment and Research." Johns Hopkins University Press, Baltimore, Maryland.

Hagerman, R. J. (1987). Fragile X syndrome. *Curr. Prob. Pediatr.* **17,** 621–674.

Laird, C. D. (1987). Proposed mechanism of inheritance and expression of the human fragile X syndrome of mental retardation. *Genetics* **117,** 587–599.

Opitz, J. M., and Sutherland, G. R. (1984). Conference report, international workshop on the fragile X and X-linked mental retardation. *Am. J. Med. Genet.* **17,** 5–94.

Turner, G., Opitz, J. M., Brown, T. W., Davies, K. E., Jacobs, P. A., Jenkins, E. C., Mikkelsen, M., Partington, M. W., and Sutherland, G. R. (1986). Conference report: Second international workshop on the fragile X and on X-linked mental retardation. *Am. J. Med. Genet.* **23,** 11–67.

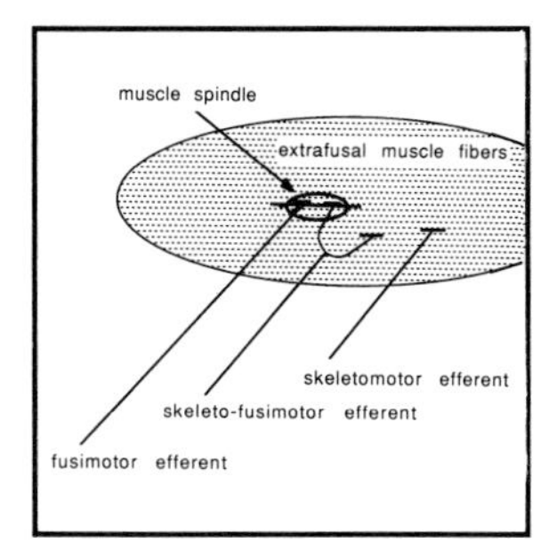

Fusimotor System

DAVID BURKE, *University of New South Wales*

Glossary

Alpha Skeletomotor
Beta Skeleto-fusimotor
Extrafusal fibers Force- and movement-generating muscle fibers, the bulk of the muscle, excluding the spindle
Fusimotor Motor to intrafusal muscle fibers in the muscle spindle
Gamma Fusimotor
Intrafusal fibers Muscle fibers in the muscle spindle
Skeletofusimotor Motor to both intrafusal and extrafusal muscle fibers
Skeletomotor Motor to extrafusal muscle fibers
Spinal monosynaptic pathway The simplest component of the stretch reflex, in which sensory fibers from muscle spindles directly excite motoneurons in the spinal cord
Stretch reflex Involuntary contraction of muscle, produced by activation of stretch-sensitive receptors

MUSCLE IS MORE than just a force-generating machine: It contains many receptors, some of which are encapsulated (e.g., the muscle spindle and tendon organ) but most of which are free nerve endings, responsive to injury and to metabolic, thermal, and mechanical stimuli within the muscle. The muscle spindle is a unique mechanoreceptor. It is innervated by motor nerve fibers through which the central nervous system can alter the response of spindle endings and thereby modify the messages it receives from the spindle endings. Muscle spindles are elongated sensory endorgans, lying in parallel with the contractile elements of muscle (Fig. 1). They are mechanoreceptors, primarily responsive to changes in length, be that produced by stretch or shortening of the whole muscle or by contraction of nearby muscle fibers. The Golgi tendon organ is the second major encapsulated mechanoreceptor in muscle, located at musculotendinous and musculofascial junctions (Fig. 1). Muscle fibers insert into the capsule of this receptor, which is therefore "in-series" with muscle fibers and responsive to their contraction.

The information from spindle endings is involved in a variety of reflex actions in the spinal cord and at higher levels in the central nervous system, including the classical monosynaptic pathway, and it constitutes one of the major sensory cues for kinaesthesia. This article will address the anatomy of the motor innervation of the muscle spindle and current concepts about how it functions in human subjects. [*See* PROPRIOCEPTORS AND PROPRIOCEPTION.]

I. Anatomical Considerations

A. General

The muscle spindle has a fusiform ("spindle"-like) appearance and contains a number of modified muscle fibers ("intrafusal" muscle fibers), which produce negligible force at the muscle tendon when they contract. The motor innervation of the spindle is directed to these intrafusal muscle fibers and is termed *fusimotor*. Because the contractile bulk of the muscle lies outside the spindle, the term *extrafusal* muscle is used when referring to the force-generating muscle fibers that constitute the bulk of muscle (Fig. 1). These muscle fibers are innervated by "skeletomotor" axons. Motor axons directed only to intrafusal muscle are of small diameter and

Encyclopedia of Human Biology, Volume 3.

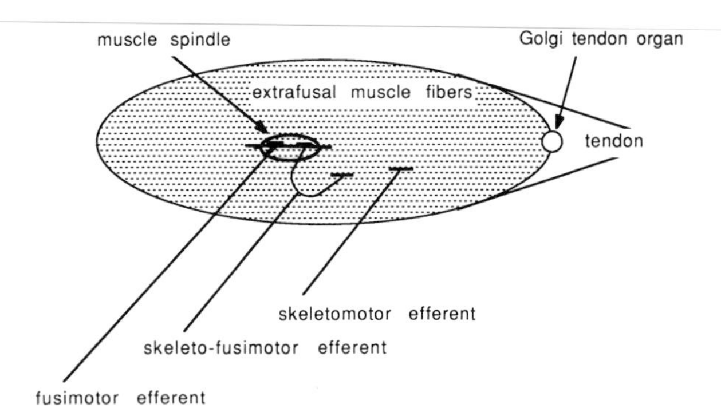

FIGURE 1 Simplified diagram of a muscle containing a single muscle spindle lying in parallel with "extrafusal" muscle fibers but considerably shorter than the muscle and a single Golgi tendon organ at the musculotendinous junction, in series with the extrafusal muscle. Motor innervation is by alpha motor axons (exclusively "skeletomotor"), gamma motor axons (exclusively "fusimotor"), or beta motor axons (directed to extra- and intrafusal muscle, hence "skeletofusimotor"). Afferent innervation is detailed in Fig. 2.

relatively slow conduction velocity and are often referred to as *gamma* efferent axons from "gamma" motoneurons. Similarly, skeletomotor axons are often called *alpha* and arise from "alpha" motoneurons. They are of larger diameter and have conduction velocities of some 30–60 m/sec. Some motor axons innervating extrafusal muscle send branches to nearby muscle spindles, thereby have a mixed function, and are termed *skeletofusimotor* or *beta*. [*See* MUSCLE DYNAMICS; MUSCLE, PHYSIOLOGY AND BIOCHEMISTRY.]

B. Muscle Spindles

Except for the facial and jaw-opening muscles, all skeletal muscles contain a number of muscle spindles, distributed throughout the bulk of the muscle, the absolute numbers varying from less than 50 for the intrinsic muscles of the hand to more than 1,000 for the thigh muscles. However, given the small size of the individual intrinsic muscles of the hand, the density (number of spindles per gram of muscle) is significantly higher for these muscles than for the much larger thigh muscles. It is generally believed that the greater density of spindles in the more distal muscles that operate the hand contributes to the precision with which the hand can be used, but attempts to demonstrate that proprioceptive sensations are more acute distally in the human upper limb have failed to do so. Presumably much of the precision with which the hand can be used resides not so much on the sensory side but on the motor side: in the greater access of corticospinal pathways transmitting the volitional drive from motor cortex to motoneurons innervating the more distal muscles. [*See* MOTOR CONTROL.]

Muscle spindles are commonly about 10 mm in length, although their size varies with the muscle in which they are located. Human spindles are slightly larger in absolute terms than in other animals (e.g., the cat), but they are relatively small when compared with the size of human muscles and, in particular, with the length of the extrafusal muscle fibers, with which they lie in parallel. As a result, human spindles commonly attach directly or indirectly to the adjacent extrafusal muscle fibers and are likely to be disturbed by the contraction of these muscle fibers. Hence, muscle spindles can be stretched or shortened not only by changes in overall muscle length, such as occur when a joint is rotated, but also by disturbances created when nearby extrafusal muscle fibers contract, even if that contraction does not produce joint movement and a change in overall muscle length.

Human spindles contain 2–14 intrafusal muscle fibers, which can be divided into three types on anatomical and physiological grounds. The more plentiful type is the "nuclear chain fiber" (commonly 3–10 per spindle), in which 20–50 nuclei lie side by side in a chain down the centre of the fiber (Fig. 2). There are two types of intrafusal muscle fiber in which the nuclei are congregated at the equator to form a swelling of the fiber, with some 5–6 nuclei abreast. These "nuclear bag fibers" can be differentiated anatomically into "bag_1" and "bag_2," the latter being thicker and longer and having a lot of surrounding elastic tissue. Each spindle always has one or more bag_1 fibers and usually one bag_2 fiber. For simplicity, the diagrammatic spindle in Fig. 2 contains only one nuclear bag fiber (a bag_1 fiber) innervated by a dynamic fusimotor axon. The visco-elastic properties of the different types of intrafusal fiber differ, the bag_1 fiber responding to stretch in a more viscous way, the bag_2 and chain fibers responding in a more elastic way. Thus, the deformation of the bag_1 fiber is maximal during the stretching movement, but its length tends to "creep" back when movement stops and the muscle is held in a stretched position. The resulting deformation of the sensory terminals on the bag_1 fiber thus produces a signal that is related to the speed of the stretch. However, the greater the extent of

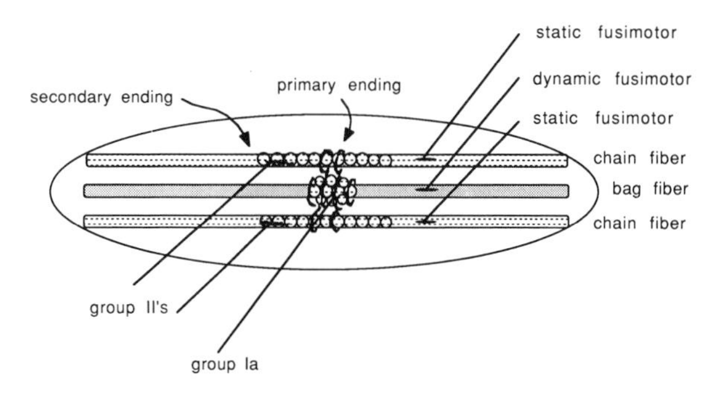

FIGURE 2 Simplified diagram of a muscle spindle containing a single nuclear bag (bag_1) and two nuclear chain fibers, with one primary ending (giving rise to a group Ia afferent) and two secondary endings (each giving rise to a group II afferent). Fusimotor innervation is directed to the bag_1 fiber (dynamic) and to the chain fibers (static).

stretch put on bag_2 and chain fibers, the greater will be the deformation of the sensory terminals on these fibers.

There are two types of sensory ending on the intrafusal fibers (Fig. 2). Each spindle has one primary ending, which commonly produces spirals around the equator of the intrafusal fibers. The primary ending gives rise to a single large myelinated afferent axon, the group Ia afferent, which in human subjects conducts at up to 60–70 m/sec in the upper limb and up to 50–60 m/sec in the lower limb. These velocities are much slower than the equivalent values in the cat (up to 120 m/sec). Given the greater length of nerve pathways in humans than in cats, sensory feedback from muscle will take much longer to reach the central nervous system in human subjects. It is likely then that the way in which the nervous system uses feedback from muscle spindles differs in different animal species. Each spindle may have a number of secondary endings, lying predominantly on the nuclear chain fiber, on one or both sides of the equator. There are one to five secondary endings per spindle, each giving rise to a small myelinated afferent axon (group II afferent). The conduction velocity for human group II afferents has not been determined, but on their size (and by analogy with other animals) it is likely to be significantly slower than the group Ia velocity. Because of the intrafusal muscle fibers on which their respective sensory endings lie, group Ia afferents transmit a signal proportional to the velocity *and* the extent of stretch, whereas the group II afferent encodes predominantly the extent of stretch.

C. Fusimotor Innervation

The poles of intrafusal muscle fibers are innervated by a number of small myelinated (gamma) efferent axons, which come from gamma motoneurons in the spinal cord. In addition, some spindles (perhaps 20–30%) receive branches of large myelinated efferent axons, which are directed to extrafusal muscles, and thereby innervate both extrafusal and intrafusal muscle fibers (Fig. 1). This skeletofusimotor (or beta) innervation is the sole motor innervation of spindles in amphibia but was thought to be uncommon in mammals and an evolutionary relic in humans. However, anatomical studies of primate spindles have shown it to be a significant form of spindle innervation, and physiological studies in the cat indicate that it may be an important modulator of spindle discharge in that species. Of necessity, beta innervation implies that any effect on the spindle is compulsorily tied to a muscle contraction. The evolutionary "advantage" of the gamma system is that it confers the potential for control of spindle output independent of the contractile state of the muscle.

Contraction of intrafusal muscle fibers caused by the activity of gamma or beta efferent axons can change both the resting deformation of the sensory terminals on those fibers and the additional deformation produced when the spindle is stretched (Fig. 2). Efferents innervating the bag_1 intrafusal fiber, be they gamma or beta, increase the sensitivity of the primary ending to the velocity of stretch and are therefore known as "dynamic" fusimotor (or skeletofusimotor) efferents. Efferents innervating the bag_2 and chain fibers have a number of actions including reducing the velocity sensitivity of the primary ending while increasing its overall discharge. They also increase the sensitivity of secondary endings to the extent of stretch. These efferents are known as "static" fusimotor (or skeletofusimotor) efferents. In the cat, it has been shown that stimulation within the central nervous system can affect dynamic fusimotor efferents independently from the static, and even the static efferents directed to chain fibers separately from those directed to bag_2 fibers. Comparable studies have not been performed in awake behaving animals, and methodological (and ethical) constraints have prevented such studies in human subjects. The extent to which the human nervous system ever accesses these subdivisions of the fusimotor system selectively and the conditions under which it normally does so have been the subject of speculation. Suffice it to say, the anatomical substrates for independent control exist in the human spindle.

In summary, there are three ways in which the discharge from the muscle spindle can be altered: by a change in length of the muscle; by the mechanical deformation produced by the contraction of nearby extrafusal muscle fibers; and by the activity of fusimotor efferents.

II. Theories of Fusimotor Function

With the gamma efferent system, attention has focused on its potential to adjust the discharge of spindles independently of the degree of contraction, with theories based on situations in which selective activation of gamma efferents might be functionally advantageous. Although there has been a lot of theorizing, the experimental evidence has been mainly negative. To date, there is only one circumstance in which selective activation of spindles endings of normal human subjects has been demonstrable. There is evidence that in standing human subjects, gamma efferents directed to spindles in tibialis anterior (a muscle that operates on the ankle) can be selectively activated by cutaneous sensory afferent, which convey information from receptors in the skin to the spinal cord from the sole of the foot. However, these afferents can also have reflex effects on the alpha motoneurons innervating tibialis anterior; the reflex effect on the gamma motoneurons is not really "selective," rather it occurs preferentially, with less intense afferent volleys than are required to activate alpha motoneurons. Discussion about how such reflexes contribute to the control of balance when standing and walking is beyond the scope of this article, but it is not immediately apparent what functional advantages, if any, are confered by this organization. A single instance in which a human spindle ending in the calf muscles has been activated selectively by extremely painful stimuli has also been documented: again the biological advantage of this particular finding is uncertain.

One purpose of the beta efferent system could be to adjust the sensitivity of spindle endings in a contracting muscle in relation to the contraction, perhaps to compensate for the unloading of the spindle when the contracting muscle shortens. However, those spindles not receiving beta innervation would still be unloaded unless their gamma innervation was simultaneously activated. There is some experimental evidence for this possibility, as discussed below.

A. Preparation for Movement

Changes in brain potentials precede any voluntary movement by hundreds of milliseconds, and reflex pathways are sensitized in advance of and in preparation for the movement. It was thought that this sensitization involved adjusting the responsiveness of muscle spindle endings so that their heightened discharge toned up reflex pathways. In addition, sensitized receptors would be better able to detect any disturbance that might compromise the intended movement. In reality, however, muscle spindle endings increase their discharge only after the muscle has started to contract. If the subject involuntarily tenses up in anticipation of the need to act, spindle endings will be activated in the tense muscles, but there is no evidence that motor preparation, even when given a warning, results in selective activation of the gamma efferent system. Attempts to train subjects to activate spindle endings using biofeedback techniques have been unsuccessful.

B. Initiation of Movement

The control of movement is often likened to a servo system as used by engineers. An influential theory in motor control postulated that some movements, particularly slow ones, were initiated indirectly using the gamma system to activate spindle endings, the afferents from which reflexly activated the alpha motoneurons of the desired muscle (using a pathway in the spinal cord). The final nail in the coffin of this theory was the demonstration that when a subject voluntarily contracted a muscle, the increase in spindle discharge occurred after rather than before the onset of the contraction. Perhaps the most popular current theory of the role of the fusimotor/muscle spindle system is that through its reflex pathways, the muscle spindle activity modulates contraction rather than initiating it. This possibility is considered further under the next heading. [*See* MOVEMENT.]

C. Maintenance of Spindle Discharge During Movement

If the purpose of the increase in fusimotor drive that accompanies a voluntary contraction is to maintain spindle feedback and thereby reflex assistance to the movement, one would expect spindle endings to maintain their discharge during voluntary movements that produce shortening of the active muscles. However, they do this only under specific circumstances. In free and relatively fast shortening movements, spindle endings fall silent. Their responsiveness to perturbations delivered during such movements is insufficient to produce much reflex assistance to the movement. It appears that free fast movements are performed "open-loop," without reflex assistance from muscle spindles. Indeed, given the delays that must occur in human subjects because of their longer pathways and slower conduction velocities (see Section I,B), powerful reflex compensation for a transient disturbance could occur too late to be of benefit and might instead disrupt subsequent stages of the movement. This is likely to be the situation in patients suffering from spasticity, with gross exaggeration of spinal reflex excitability.

If contractions are "isometric" (i.e., there is no external change in muscle length) or if movements proceed slowly against a load, the fusimotor activation that accompanies the contraction is capable of maintaining and even increasing the discharge of spindle endings and thereby of modulating the contraction. In such contractions, the reflex assistance improves the subject's ability to control the output of the active muscle. These are the types of contractions used when performing a delicate motor skill or when learning a totally new motor task. Movement is then slow and cautious, with the joint braced by simultaneous contraction of a muscle and its antagonist, movement allowed to proceed by adjusting this co-contraction. Under such circumstances the benefits of the fusimotor system can be readily appreciated.

D. Regulation of Intensity of Stretch Reflexes

The afferent input from muscle spindles can excite the motoneurons of their own muscle through a variety of pathways, the best known of which is the spinal monosynaptic pathway. The intensity of stretch reflexes can therefore be modified at three levels: by altering the muscle spindle input to the spinal cord; by altering the transmission of the afferent information to the motoneurons in the spinal cord; and by altering the excitability of the motoneurons. It cannot be denied that if other factors remain constant, a bigger afferent input into the spinal cord will evoke a larger reflex response, but it does not appear that a fusimotor mechanism is the major mechanism for altering the intensity of stretch reflexes. Powerful mechanisms exist within spinal circuitry for altering the excitability of all reflex pathways, even the simplest, the monosynaptic reflex. The fusimotor system does not appear to be called into action as a mechanism of altering reflex strength. For example, it is easy to demonstrate that tendon jerks in noncontracting muscles can be accentuated by the deliberate contraction of remote muscles. Potentiation of the knee and ankle jerks by fist clenching is commonly used by clinicians to determine whether a depressed reflex is absent or not. The process of potentiation (known as "reflex reinforcement") by the performance of this maneuver (known as the Jendrassik maneuver) was thought to be due to activation of the gamma efferent system, such that the spindle response to tendon percussion was enhanced. This view has proved erroneous: the reflex enhancement occurs centrally within spinal cord circuits, not peripherally at spindle level.

The view that the fusimotor system was not designed primarily to control reflex strength has implications for patients with neurological diseases that produce changes in the strength of reflexes. Based on the finding that excessive fusimotor drive occurred in and was necessary for the development of decerebrate rigidity in the cat after transection of the midbrain, the reflex abnormalities that occur in patients with spasticity and rigidity have been attributed to excessive fusimotor drive. Selective overactivity of dynamic fusimotor neurons was postulated to account for the exaggeration of tendon jerks and the phasic, velocity-dependent nature of muscle tone in spastic patients. Selective overactivity of static fusimotor neurons was postulated to account for the rigidity that occurs in Parkinson's disease. It is now known, however, that there are abnormalities of reflex transmission within the central nervous system in both disorders and that an excessive muscle spindle input is insufficient by itself to produce either condition. Whether overactivity of the fusimotor system ever occurs in these disorders is not known, but at most, it could only be a minor factor in the reflex disturbance.

III. Fusimotor System as a Motor System

When normal human subjects relax a muscle completely, the level of fusimotor drive directed to that muscle is low, insufficient to affect spindle discharge. This finding is perhaps not unexpected if the purpose of the fusimotor system is the control of movement. To be able to mobilize sensory feedback from muscle would be of greatest value during a motor act but would serve little purpose when the motor system was quiescent. To explain why a totally independent control of spindle discharge has evolved remains an elusive problem: Perhaps it lies in a variety of complex motor tasks not yet studied experimentally (e.g., in tasks involving inactive or relatively inactive synergists or antagonists to the active muscle group). In this respect, it may be relevant that cutaneous sensory inputs from the sole of the foot can preferentially affect gamma fusimotor neurons directed to tibialis anterior of standing human subjects. During normal quiet stance, balance is maintained by continuous activity in the calf muscle group, triceps surae, but there is little activity, if any, in the antagonist, tibialis anterior, unless the subject inadvertently sways backward. Under these circumstances, cutaneous inputs from the sole of the foot appear capable of adjusting the spindle feedback from the inactive tibialis anterior. Feedback from peripheral sources such as this is integrated with visual and vestibular inputs to produce the coordinated body movements that maintain balance.

Bibliography

Aniss, A. M., Diener, H.-C., Hore, J., Burke, D., and Gandevia, S. C. (1989). Reflex activation of muscle spindles in human pretibial muscles during standing. *J. Neurophysiol.* **64,** 671–679.

Boyd, I. A., and Gladden, M. H. eds. (1985). "The Muscle Spindle." Stockton, New York.

Burke, D. (1981). The activity of human muscle spindle endings during normal motor behavior. *In* "International Review of Physiology. Neurophysiology IV." (R. Porter, ed.), pp. 91–126. University Press, Baltimore.

Burke, D. (1983). Critical examination of the case for or against fusimotor involvement in disorders of muscle tone. *In* "Motor Control Mechanisms in Health and Disease, Advances in Neurology," vol. 39. (J. E. Desmedt, ed.), pp. 133–150. Raven Press, New York.

Burke, D., and Gandevia, S. C. (1989). The peripheral motor system. *In* "The Human Nervous System" (G. Paxinos, ed.), pp. 125–145. Academic Press, New York.

Hulliger, M. (1984). The mammalian muscle spindle and its central control. *Rev. Physiol. Biochem. Pharmacol.* **101,** 1–110.

Matthews, P. B. C. (1981). Evolving views on the internal operation and functional role of the muscle spindle. *J. Physiol. (Lond.)* **320,** 1–30.

Vallbo, A. B., Hagbarth, K.-E., Torebjork, H. E., and Wallin, B. G. (1979). Somatosensory, proprioceptive, and sympathetic activity in human peripheral nerves. *Physiol. Rev.* **59,** 919–957.

Gangliosides

A. SUZUKI and T. YAMAKAWA, *Tokyo Metropolitan Institute of Medical Science*

Glossary

Ceramide Hydrophobic portion of glycosphingolipids, consisting of sphingosine (a long-chain base; D-*erythro*-1,3-dihydroxy-2-amino-4,5-*trans*-octadecene and its analogs) and a fatty acid, which is attached to the sphingosine through an acid–amide linkage

Glycolipids Compounds consisting of a carbohydrate chain and a lipid. Glycosphingolipids and glycoglycerolipids differ in the presence of sphingosine and glycerol, respectively. Major glycolipids in mammalian cells are usually glycosphingolipids, and the term "glycolipids" is frequently used instead of "glycosphingolipids."

Hydrophobic interaction Interaction contributing to the aggregation of hydrophobic molecules or nonpolar portions of molecules in water in order to minimize their disruptive effect on the hydrogen-bonded network of water molecules. The aggregates are called micelles or liposomes.

Sialic acids Family of derivatives of neuraminic acid (5-amino-3,5-dideoxy-D-glycero-D-galactononulosonic acid), which is a nine-carbon sugar compound. Glycoconjugates containing sialic acid(s) are called sialoglycoproteins in the case of proteins and gangliosides in the case of glycolipids.

GANGLIOSIDES are a family of glycolipids composed of a ceramide and a carbohydrate chain containing sialic acid(s). Ceramide is the hydrophobic part, which anchors the carbohydrate chain to membranes. More than 100 gangliosides have been reported, which differ mainly in their carbohydrate structure. Because of these features, gangliosides can play roles in the recognition between cells or between cells and macromolecules. Recently, gangliosides have attracted attention as physiologically active substances, modulators for membrane receptors, receptors for toxins and viruses, and cancer-associated antigens.

I. Structural Characteristics

A. Discovery

The first indications of gangliosides were observed by K. Landsteiner and P. A. Levene of the Rockefeller Institute for Medical Research (New York) in 1925–1927 and by E. Walz of the University of Tübingen (Tübingen, Federal Republic of Germany) in 1927, in a chemical reaction involving color development for the detection of sialic acid in lipid fractions, but they did not realize gangliosides were pure chemical entities. In 1935–1938 E. Klenk of the University of Cologne (Cologne, Federal Republic of Germany) published a series of papers on the glycolipid that accumulated in the brains of patients with Tay–Sachs disease, and his work is regarded as being definitive as to the description and characterization of gangliosides.

Before reaching to the chemical structure of gangliosides, the structure of sialic acid, which is a component of gangliosides, was debated by Klenk and G. Blix, of the University of Uppsala (Uppsala, Sweden). Klenk isolated a new nitrogen-containing organic acid in a crystalline state from a ganglioside fraction obtained from bovine brain and named it *Neuraminsäure* in 1941, while Blix isolated the same acid from bovine submaxillary mucin, which is a glycoprotein, and he proposed the name "sialic acid" in 1952. They are the same substance, and the correct structure of neuraminic acid or sialic acid was finally proposed by A. Gottschalk of The Walter and Eliza Hall Institute (Melbourne, Australia) in 1955. S. Roseman established the skeleton of sialic acid in 1958, which is composed of *N*-acetylmannosamine and pyruvate. Thus, the structure of sialic acid is identified as shown in Fig. 1. The first report on the occurrence of a ganglioside in an extraneural tissue was made by T. Yamakawa in 1951, with the isolation of a sialyllactosylceramide or GM_3 from horse erythrocytes, which he named hematoside.

A variety of structural modifications of neuraminic acid have been reported so far, two major modifications being acylation of the amino group and acetylation of hydroxyl groups. Concerning the modification of the amino group, the gangliosides in normal human tissues only contain *N*-acetylneuraminic acid (NeuAc) and ones containing *N*-glycolylneuraminic acid (NeuGc) are only found in cancer tissues. Interestingly, mammalian species other than humans and chicken also contain NeuGc in normal tissues. Human tissues also contain some gangliosides, which are acetylated at the C-9 position of *N*-acetylneuraminic acid. We use "sialic acid" as the generic term for neuraminic acids with various modifications. Sialic acids of gangliosides include NeuAc, NeuGc, 4-*O*-acetyl-NeuGc, and 9-*O*-acetyl-NeuAc.

The structural elucidation of gangliosides was attempted by several laboratories, and finally the correct structures of a series of brain gangliosides were proposed in 1963 by R. Kuhn and H. Wiegandt of the Max–Planck Institute (Heidelberg, Federal Republic of Germany). Since then, the isolation and structural characterization of a variety of gangliosides have been reported. In Table I gangliosides, which are relatively abundant in human tissues, are listed with notes on their biological functions. In Fig. 2 the structure of one ganglioside, GM_1, is shown.

A nomenclature widely used for gangliosides is based on the system proposed by L. Svennerholm,

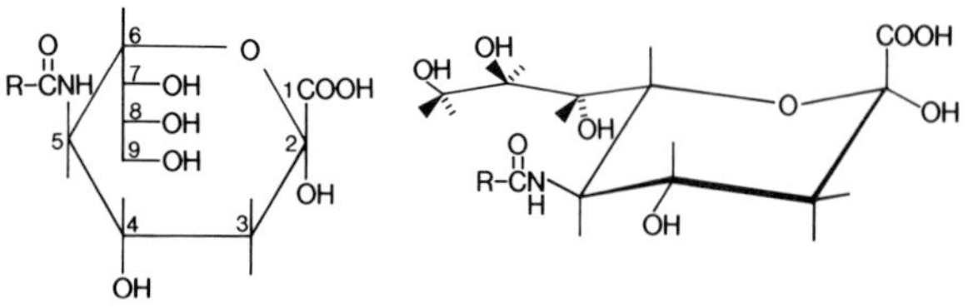

FIGURE 1 Structures of sialic acids. R: CH_3, N-acetylneuraminic acid; CH_2OH, N-glycolylneuraminic acid.

and another, used here, follows recommendations by the Nomenclature Committee of the International Union of Biochemistry. The latter nomenclature is based on that of neutral glycosphingolipids, which always comprise the backbone structures of gangliosides.

The structures of neutral glycosphingolipids are divided into five major types on the basis of their core oligosaccharide structures. The five groups are named gala-, ganglio-, lacto-, neolacto-, and globo-series glycolipids, according to the trivial names for their oligosaccharide structures. The core structures and their names are shown in Fig. 3. The gangliosides are therefore named sialo- or sialyl-R-osylceramide, where R is derived from the names of the neutral oligosaccharides (e.g., sialyllactosylceramide, sialylgangliotriaosylceramide, and sialylneolactotetraosylceramide). The position of the monosaccharide and the carbon atom of the monosaccharide to which sialic acid is attached should be determined because there are positional isomers. Thus, the position and the carbon number are indicated by a Roman numeral and a superscript Arabic numeral, respectively. The configuration of the ketosidic bond of sialic acids in gangliosides is also quite important and it was demonstrated to be α by Ledeen and Yu. Gangliosides reported so far exclusively contain sialic acids with α configuration. Table I lists the structures and names of gangliosides according to both nomenclatures.

B. Characteristics as Membrane Components

Gangliosides are glycosphingolipids which contain 1 mol or more of sialic acids in their molecules and thus share characteristics with glycosphingolipids, as amphipathic molecules, and with sialic acids, as acidic sugars. Glycosphingolipids contain both hydrophobic (or lipophilic) and hydrophilic structures. The hydrophobic structure is named ceramide, which is composed of a sphingosine (a long chain base) and a fatty acid. The name "sphingosine" was given by J. L. W. Thudichum to a component of the brain lipids cerebroside and sphingomyelin, the function of which was unclear, with the Sphinx in Greek mythology in mind. Ceramide thus can interact with other hydrophobic molecules (e.g., phospholipids and cholesterol) [*See* SPHINGOLIPID METABOLISM AND BIOLOGY.]

Lipid bilayers made of phospholipids and cholesterol are the basis of biological membranes, and gangliosides are able to act as membrane compo-

TABLE I Structures of Major Gangliosides Found in Human Tissue

Gala series			
GM$_4$[a]	I^{3}NeuAc-GalCer[b]	NeuAcα2–3Galβ1-Cer	A myelin marker
Ganglio series			
A series			
GM$_3$	II^{3}NeuAc-LacCer	NeuAcα2–3Galβ1–4Glcβ1-Cer	Modulator of epidermal or fibroblast growth factor receptors, causes the differentiation of HL-60 cells into macrophage-like cells, hematoside[c]
GM$_2$	II^{3}NeuAc-Gg$_3$Cer	GalNAcβ1–4(NeuAcα2–3)Galβ1–4Glcβ1-Cer	Accumulated in the brains of patients with Tay–Sachs disease, Tay–Sachs ganglioside[c]
GM$_1$	II^{3}NeuAc-Gg$_4$Cer	Galβ1–3GalNAcβ1–4(NeuAcα2–3)Galβ1–4Glcβ1-Cer	Receptor for cholera toxin, promotes neurite outgrowth
GD$_{1a}$	IV^{3}NeuAc-, II^{3}NeuAc-Gg$_4$Cer	NeuAcα2–3Galβ1–3GalNAcβ1–4(NeuAcα2–3)Galβ1–4Glcβ1-Cer	
GT$_{1a}$	IV^3(NeuAc)$_2$-, II^{3}NeuAc-Gg$_4$Cer[d]	NeuAcα2–8NeuAcα2–3Galβ1–3GalNAcβ1–4(NeuAcα2–3)Galβ1–4Glcβ1-Cer	
	IV^{2}αFuc-, II^{3}NeuAc-Gg$_4$Cer	Fucα1–2Galβ1–3GalNAcβ1–4(NeuAcα2–3)Galβ1–4Glcβ1-Cer	
	II^{3}NeuAc-Gg$_5$Cer	GalNAcβ1–4Galβ1–3GalNAcβ1–4(NeuAcα2–3)Galβ1–4Glcβ1-Cer	
B series			
GD$_3$	II^3(NeuAc)$_2$-LacCer	NeuAcα2–8NeuAcα2–3Galβ1–4Glcβ1-Cer	Melanoma antigen
GD$_2$	II^3(NeuAc)$_2$-Gg$_3$Cer	GalNAcβ1–4(NeuAcα2–8NeuAcα2–3)Galβ1–4Galβ1-Cer	Melanoma antigen
GD$_{1b}$	II^3(NeuAc)$_2$-Gg$_4$Cer	Galβ1–3GalNAcβ1–4(NeuAcα2–8NeuAcα2–3)Galβ1–4Glcβ1-Cer	Tetanus toxin receptor
GT$_{1b}$	IV^{3}NeuAc-, II^3(NeuAc)$_2$-Gg$_4$Cer	NeuAcα2–3Galβ1–3GalNAcβ1–4(NeuAcα2–8NeuAcα2–3)Galβ1–4Glcβ1-Cer	Tetanus and botulinus toxin receptor
GQ$_{1b}$	IV^3(NeuAc)$_2$-, II^3(NeuAc)$_2$-Gg$_4$Cer	NeuAcα2–8NeuAcα2–3Galβ1–3GalNAcβ1–4(NeuAcα2–8NeuAcα2–3)Galβ1–4Glcβ1-Cer	Neurite outgrowth activity, botulinus toxin receptor
C series			
GT$_3$	II^3(NeuAc)$_3$-LacCer	NeuAcα2–8NeuAcα2–8NeuAcα2–3Galβ1–4Glcβ1-Cer	
GT$_2$	II^3(NeuAc)$_3$-Gg$_3$Cer	GalNAcβ1–4(NeuAcα2–8NeuAcα2–8NeuAcα2–3)Galβ1–4Glcβ1-Cer	
GT$_{1c}$	II^3(NeuAc)$_3$-Gg$_4$Cer	Galβ1–3GalNAcβ1–4(NeuAcα2–8NeuAcα2–8NeuAcα2–3)Galβ1–4Glcβ1-Cer	
GQ$_{1c}$	IV^{3}NeuAc-, II^3(NeuAc)$_3$-Gg$_4$Cer	NeuAcα2–3Galβ1–3GalNAcβ1–4(NeuAcα2–8NeuAcα2–8NeuAcα2–3)Galβ1–4Glcβ1-Cer	
Ganglio series others			
GM$_{1b}$	IV^{3}NeuAc-Gg$_4$Cer	NeuAcα2–3Galβ1–3GalNAcβ1–4Galβ1–4Glcβ1-Cer	Erythrocytes
Lacto series			
IV^{3}NeuAc-Lc$_4$Cer		NeuAcα2–3Galβ1–3GlcNAcβ1–3Galβ1–4Glcβ1-Cer	
IV^{3}NeuAc-III^{4}αFuc-Lc$_4$Cer		NeuAcα2–3Galβ1–3(Fucα1–4)GlcNAcβ1–3Galβ1–4Glcβ1-Cer	Sialyl-*Le*^{a}
IV^3(NeuAc)$_2$-Lc$_4$Cer		NeuAcα2–8NeuAcα2–3Galβ1–3GlcNAcβ1–3Galβ1–4Glcβ1-Cer	
Neolacto series			
IV^{3}NeuAc-nLc$_4$Cer		NeuAcα2–3Galβ1–4GlcNAcβ1–3Galβ1–4Glcβ1-Cer	Sialylparagloboside,[c] erythrocytes
IV^{6}NeuAc-nLc$_4$Cer		NeuAcα2–6Galβ1–4GlcNAcβ1–3Galβ1–4Glcβ1-Cer	
IV^{3}NeuAc-, III^{3}αFuc-nLc$_4$Cer		NeuAcα2–3Galβ1–4(Fucα1–3)GlcNAcβ1–3Galβ1–4Glcβ1-Cer	Sialyl-*Le*^{x}
IV^3(NeuAc)$_2$-nLc$_4$Cer		NeuAcα2–8NeuAcα2–3Galβ1–4GlcNAcβ1–3Galβ1–4Glcβ1-Cer	
IV^{3}β(NeuAc-3Gal)-nLc$_4$Cer		NeuAcα2–3Galβ1–3Galβ1–4GlcNAcβ1–3Galβ1–4Glcβ1-Cer	
IV^{3}β(NeuAc-3GalNAc)-, II^{2}αFuc-nLc$_4$Cer		NeuAcα2–3GalNAcβ1–3Galβ1–4GlcNAcβ1–3(Fucα1–2)Galβ1–4Glcβ1-Cer	
VI^{3}NeuAc-nLc$_6$Cer		NeuAcα2–3Galβ1–4GlcNAcβ1–3Galβ1–4GlcNAcβ1–3Galβ1–4Glcβ1-Cer	
VI^{3}NeuAc-, V^{3}αFuc-nLc$_6$Cer		NeuAcα2–3Galβ1–4(Fucα1–3)GlcNAcβ1–3Galβ1–4GlcNAcβ1–3Galβ1–4Glcβ1-Cer	
VI^{3}NeuAc-, IV^{6}β(Galβ-4GlcNAc)-nLc$_6$Cer		NeuAcα2–3Galβ1–4GlcNAcβ1–3(Galβ1–4GlcNAcβ1–6)Galβ1–4GlcNAcβ1–3Galβ1–4Glcβ1-Cer	
VI^{3}NeuAc-, IV^{6}β(Fucα-2Galβ-4GlcNAc)-nLc$_6$Cer		NeuAcα2–3Galβ1–4GlcNAcβ1–3(Fucα1–2Galβ1–4GlcNAcβ1–6)Galβ1–4GlcNAcβ1–3Galβ1–4Glcβ1-Cer	
VI^{3}NeuAc-, IV^{6}β(NeuAc-3Galβ-4GlcNAc)-nLc$_6$Cer		NeuAcα2–3Galβ1–4GlcNAcβ1–3(NeuAcα2–3Galβ1–4GlcNAcβ1–6)Galβ1–4GlcNAcβ1–3Galβ1–4Glcβ1-Cer	
Globo series			
V^{3}NeuAc-Gb$_5$Cer		NeuAcα2–3Galβ1–3GalNAcβ1–3Galα1–4Galβ1–4Glcβ1-Cer	SSEA-4 in mouse, human teratocarcinoma
V^3(NeuAc)$_2$-Gb$_5$Cer		NeuAcα2–8NeuAcα2–3Galβ1–3GalNAcβ1–3Galα1–4Galβ1–4Glcβ1-Cer	
V^{3}NeuAc-, V^{6}NeuAc-Gb$_5$Cer		NeuAcα2–6(NeuAcα2–3)Galβ1–3GalNAcβ1–3Galα1–4Galβ1–4Glcβ1-Cer	Erythrocytes

[a] Nomenclature according to Svennerholm's system.

[b] Nomenclature recommended by the Nomenclature Committee of the International Union of Biochemistry.

[c] Trivial names.

[d] (NeuAc)$_n$: (NeuAcα2–8)$_{n-1}$NeuAcα2–3). The anomeric linkage of NeuAc- is of the α configuration. The β configuration is not found, and α is omitted.

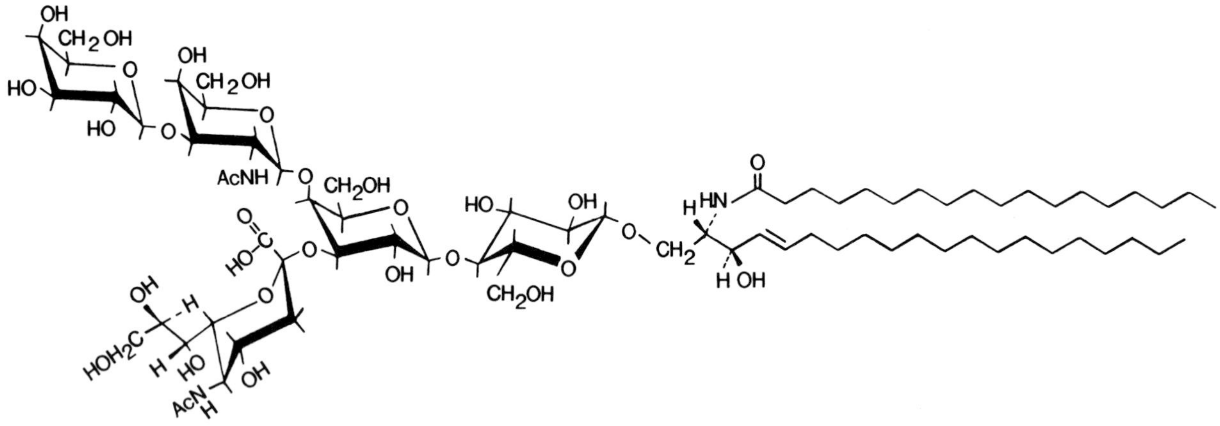

FIGURE 2 Structure of GM_1 ganglioside. The configuration of each monosaccharide is arbitrary, and the conformation of the molecule anchored into the cell membrane is a matter requiring further investigation.

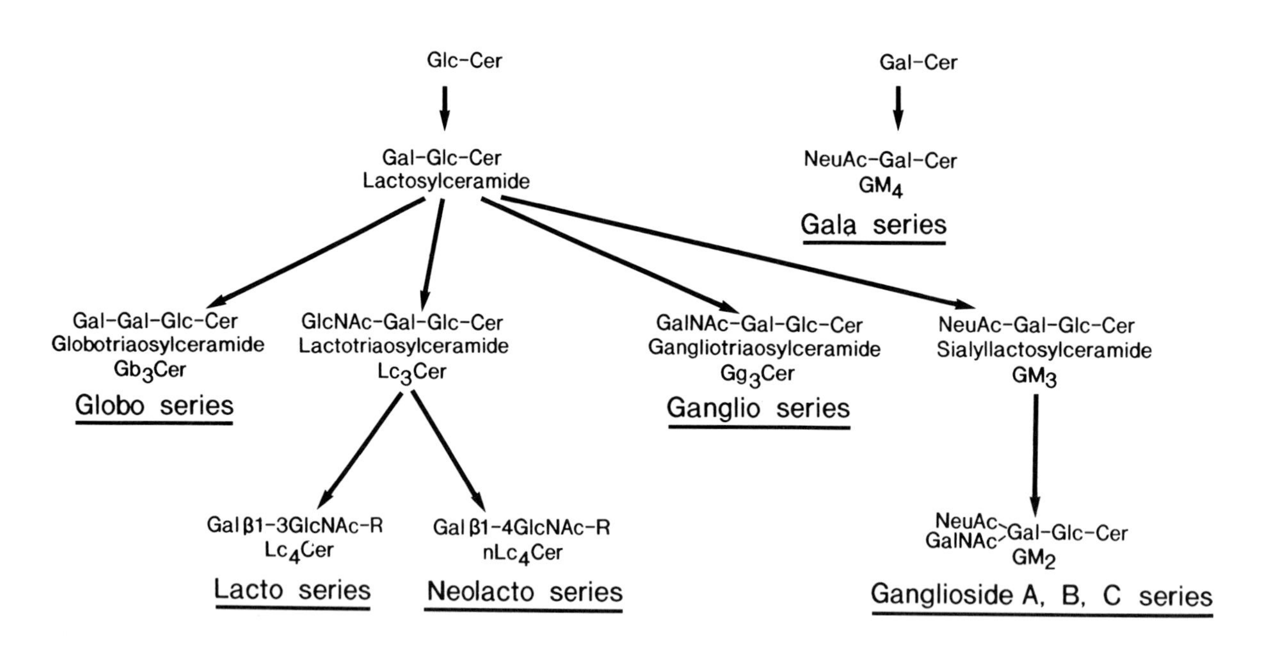

FIGURE 3 The five core structures of neutral glycosphingolipids for gangliosides.

nents because their ceramides can interact with these membrane components through hydrophobic interactions. A lipid bilayer or membrane is a sort of barrier, which separates the inside of the cell from the outside. Water and water-soluble substances occupy the spaces outside and inside the membrane in the case of mammalian cells, and, in these surroundings, gangliosides anchored by their ceramide portions to the membrane have biological functions. [*See* CHOLESTEROL; LIPIDS; PHOSPHOLIPID METABOLISM.]

In the plasma membrane of cells, however, ganglioside–carbohydrate chains are considered to be anchored to the outer leaflet of the lipid bilayer. This is due to the biosynthesis of their carbohydrate chains in cells. The carbohydrate chains of gangliosides are biosynthesized in the Golgi apparatus, which is a cellular organelle consisting of multilayers of membrane cisternae. Cells biosynthesize most carbohydrate chains of their constituents, including gangliosides, and of secretion products on the inner surface of Golgi cisternae, but not on the outer surface, which faces the inside of cells, or the cytosol. The inner leaflet of the Golgi membrane fuses with the outer leaflet of plasma membranes.

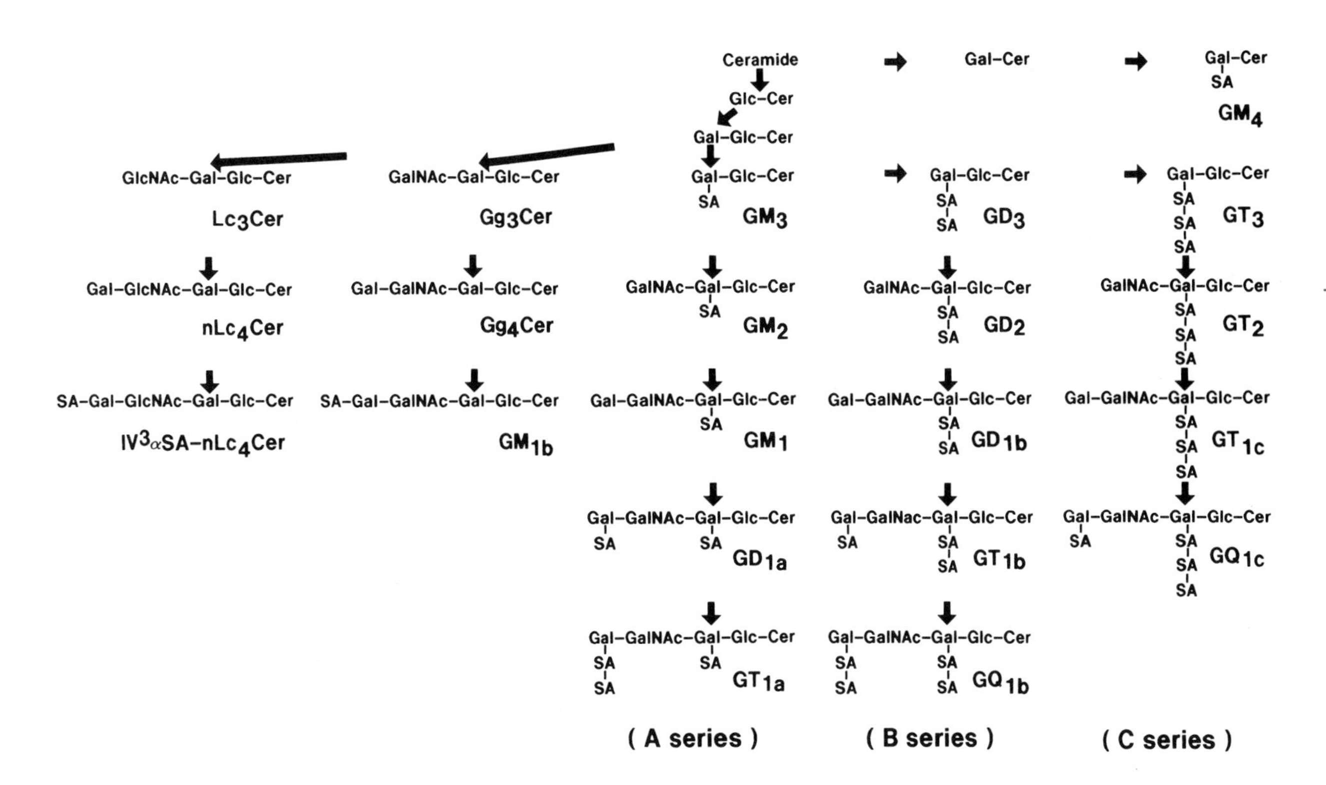

FIGURE 4 The major biosynthetic pathways in human tissues. The A-, B-, and C-series pathways and GM_4 synthesis occur in neural tissues. The neolacto-series pathway occurs in extraneural tissues. SA, Sialic acid.

Because of this mechanism of membrane biogenesis, the carbohydrate chains of gangliosides or other carbohydrate-containing molecules (e.g., membrane-bound glycoproteins) are extended into the space outside the cells. [*See* Golgi Apparatus.]

II. Regulation of Expression

A. Biosynthesis

Table I, which lists some gangliosides, indicates that gangliosides differ in their carbohydrate structures. The mechanisms producing the various kinds of carbohydrate structures are located in the Golgi apparatus, and glycosyltransferases are primarily responsible for the production of the various structures. Figure 4 shows the major biosynthetic pathways for gangliosides. Every step in the elongation of carbohydrate chains requires a glycosyltransferase, which sometimes has a broad, or sometimes a strict, substrate specificity. At present it is unknown how many glycosyltransferases in the Golgi apparatus are involved in ganglioside biosynthesis, but it is believed to be more than a few, at least several tens. Without this characteristic, variations in ganglioside structures cannot be produced in the tissues or cells of humans and other mammals.

Factors other than glycosyltransferases are also required for the production of ganglioside carbohydrates. These include sugar nucleotides as donors of sugars, which are transferred to precursor gangliosides or neutral glycolipids; enzymes involved in sugar nucleotide production; and translocators for the transport of sugar nucleotides into the internal space of the Golgi apparatus from the cytosol. Sugar nucleotides containing one or two high-energy phosphate bonds are synthesized in the cytosol, but CMP-NeuAc is exceptional, being synthesized in the nucleus and then transported to the cytosol.

It is also interesting that galactose (Gal), glucose (Glc), *N*-acetylgalactosamine (GalNAc), and *N*-acetylglucosamine (GlcNAc) form UDP derivatives, fucose (Fuc) and mannose (Man) GDP derivatives, and NeuAc, a CMP derivative. Two isomers are formed when a sugar attaches to any other molecules through the C1 carbon, an anomeric carbon atom. The isomers of all of these nucleotide sugars except CMP-NeuAc and GDP-Fuc are of the α configuration, but in gangliosides those of Gal and

GalNAc are either of the α or β configuration, those of Fuc and NeuAc are of the α configuration, and that of Glc is of the β configuration, indicating that the anomeric linkages in biosynthesized gangliosides are determined by the nature of the glycosyltransferases. It is also determined by transferases which of the four hydroxyl groups in hexoses and which of the three in hexosamines are used as the linkage position for a newly added sugar.

An interesting feature is that gangliosides found in various tissues of mammals, including humans, differ among tissues, but are constant for each tissue from the same species, matched as to age and sex, indicating that the mechanisms producing these characteristics are genetically determined. For example, in mammals the tissues that contain gangliosides at the highest concentration are neural tissues. The major biosynthetic pathways for A-, B-, and C-series gangliosides, and GM_4, as shown in Fig. 4, were deduced mainly from the data obtained on the isolation and structural determination of neural gangliosides and partially from the demonstration of biosynthetic activities *in vitro*. For the production of a large amount of gangliosides, neural tissues must activate a mechanism producing the gangliosides much more strongly than in the other tissues. At the same time, extraneural tissues in humans contain sialyllactosylceramide and neolacto-series gangliosides as the major gangliosides, indicating the activation of a different ganglioside-producing mechanism.

Another reason for the heterogeneity of gangliosides is species differences. Brain gangliosides in vertebrates are well conserved and are of the A, B, and C series, but gangliosides in visceral organs differ among species. In erythrocytes sialyllactosylceramide or GM_3 is the major ganglioside in horses, dogs, pigs, and humans, disialyllactosylceramide is found in cats, pandas, and bears, and A-series gangliosides are predominant in rats and mice. This heterogeneity is also considered to be genetically determined.

Individual differences in gangliosides were also reported for sialyl-Le^a (a derivative of *Lewis* antigens) in humans, sialic acid species of hematoside or GM_3 in dog erythrocytes, and A-series gangliosides in mouse liver. These three cases of heterogeneity (i.e., tissue-, species-, and individual-specific differences) should reflect how the biosynthetic mechanisms are activated. At present, however, we do not have much information on the activation or the mechanisms.

B. Changes in Ganglioside Expression

The tissue-specific composition of gangliosides is considered to be the result of changes in ganglioside expression during early embryogenesis, differentiation, and organogenesis. These changes are quite interesting because of the possibility that they play roles as signals for determination of the differentiation directions of cells in embryos and fetuses.

During early embryogenesis in mice, dramatic changes in carbohydrate expression have been demonstrated by immunostaining with monoclonal antibodies. Carbohydrate antigens are expressed at particular stages in mouse early embryogenesis, called stage-specific embryonic antigens (SSEAs). The epitopes of SSEA-1, -3, and -4 are Galβ1–4(Fucα1–3)GlcNAcβ, Galβ1–3GalNAcβ1–3Galα1–4Galβ, and NeuAcα2–3Galβ1–3GalNAcβ1–3Galα1–4Gal, respectively. Anti-SSEA-1 antibody and SSEA-1 oligosaccharide can arrest further embryogenesis after the morula stage. SSEA-4 is the only ganglioside determined so far, but other gangliosides might also be molecules involved in early mouse embryogenesis. In the human fetus the presence of SSEAs has also been demonstrated.

In neural tissues striking changes in ganglioside expression were demonstrated during development in the late stage of gestation and after birth. The main period of ganglioside accretion coincides with the period of accelerated neuronal membrane formation, just before and during the period of intensive synaptogenesis. The proliferative state of neuronal and glial precursor cells is characterized by the predominant expression of GD_3. The developmental period between neurogenesis and synaptogenesis seems to be characterized by the predominant synthesis of B-series gangliosides, mainly GQ_{1b}, the synthesis of A-series gangliosides being triggered last during development. An understanding of the regulation of these changes is quite important in relation to the development of neural functions.

Every mammal carries genetic information for the expression of gangliosides in its DNA sequences. Another important control mechanism for ganglioside expression is epigenetic control, which includes the following. Cells can change gangliosides when they are cultured with ground substances or different types of cells. Changes in ganglioside compositions are found in the brains of fish fed at low temperatures. Changes in gangliosides have also been reported in the brains of experimental animals as a result of learning. These changes in ganglioside contents and compositions are under epigenetic control and possibly are one of the biochemical bases for neuronal plasticity. [*See* PLASTICITY, NERVOUS SYSTEM.]

The tissue-specific expression of gangliosides, as described above, can be explained as the summation of cell type-specific expressions of gangliosides. It can be said, therefore, that the expression of gangliosides in individual cells is controlled or determined by genetic and epigenetic regulation. This concept can be extended to the idea that a particular ganglioside composition is a marker for a particular cell, depending on its particular stage of differentiation.

C. Three-Dimensional Structures

Gangliosides are transported onto the surfaces of cells, by biochemical mechanisms which are not well understood, with their carbohydrate portions extending out from the plasma membrane. The carbohydrates are recognized by lectins, toxins, antibodies, viruses, bacteria, and other recognition molecules or organisms, and this recognition process must be explained in terms of the three-dimensional structures of gangliosides anchored in membranes. A ganglioside has multiple binding sites, as shown using monoclonal antibodies. For one particular ganglioside, several monoclonal antibodies with different binding specificities can be produced. Three-dimensional structures of gangliosides are unknown, because it is difficult to obtain the crystals needed for X-ray crystallography. Recent advances in nuclear magnetic resonance (NMR) spectroscopy, however, give hope that it will be possible to analyze the structures of gangliosides anchored in membranes. The physiological functions of gangliosides are assumed to be mediated by structural changes in ganglioside-recognizing molecules, following the binding of gangliosides. An understanding, therefore, of the three-dimensional structures of gangliosides anchored in membranes will be the basis for an understanding of their biological functions.

The possibility was proposed that the carbohydrate chains of gangliosides are not straight on the cell-surface membrane, but are bent and thus lay on the membrane. One side of the surface of carbohydrate chains is rich in hydrophobic atoms or functional groups such as —CH, CH_3CONH—, and —CH_3, and the other side is rich in hydrophilic atoms such as —OH and —COO^-. The hydrophilic domain faces the membrane and might interact with polar groups of phospholipids. The hydrophobic face can be recognized by ganglioside–carbohydrate recognition molecules from the outside of the cells. This hypothesis will be verified by NMR-spectroscopic studies of gangliosides incorporated into micelles or the membranes themselves.

In the ceramide portion of gangliosides, there is heterogeneities in both sphingosines and fatty acids. As to human brain gangliosides, the major molecular species of sphingosines is 1,3-dihydroxy-2-amino-*trans*-4-octadecene (C18-sphingenine or C18 : 1-sphingosine). With aging or an increase in the molar composition of sialic acids, the proportion of C20-sphingenine (C20 : 1-sphingosine) to C18-sphingenine increases. The major fatty acid of brain gangliosides, other than GM_4, is stearic acid (C18 : 0 fatty acid), followed by palmitic acid. As to the gangliosides of visceral organs, the fatty acids are more heterogenous than those of the brain, palmitic acid and longer-chain fatty acids (C22 : 0, C24 : 0, and C24 : 1) being predominant. These heterogeneities of the ceramide portion were reported to affect the interaction between gangliosides and the molecules that recognize them. It is assumed that the three-dimensional structures of the carbohydrates of gangliosides in membranes are somehow modified when the ceramide portions are different.

D. Degradation

The degradation of gangliosides mainly occurs in lysosomes. A part of the plasma membrane is pinocytosed (i.e., engulfed into the cell), and the pinocytic vesicles fuse with primary lysosomes, which contain various carbohydrate hydrolases as well as other degradative enzymes. If one of the carbohydrate hydrolases is deficient, the cell cannot break down the carbohydrate structures which are susceptible to that enzyme's action, and carbohydrate-containing substances remain undegraded, thus accumulating in lysosomes. The damage to cells caused by such accumulation leads to dysfunctions or cell death, especially in tissues or cells exhibiting high activity as to the biosynthesis of carbohydrate-containing substances. Such metabolic disorders are the basis of a group of inherited diseases, and Tay–Sachs disease is a well-known ganglioside storage disease.

An understanding of the pathogenesis of Tay–Sachs disease was not reached until the ganglioside research by Klenk. He noticed and isolated a glycolipid, which is highly accumulated in patients' brain. He coined the term "ganglioside" because glycolipids exhibiting similar characteristics to the glycolipid of the patients' brains were abundant in gray matter and ganglions, which are masses of neural cells, in normal brains. The ganglioside accumulated in the brains of Tay–Sachs disease cases is GM_2 ganglioside, which contains a βGalNAc linkage at its terminus. This linkage is normally degraded by β-hexosaminidase in lysosomes. In the lysosomes of patients with Tay–Sachs disease, the activity of β-hexosaminidase is deficient. β-Hexosaminidases are dimers, consisting of α and β subunits, the dimers formed being α–α, α–β, and β–β; they require an activator protein for their activity. A defect of one of these three components causes Tay–Sachs disease. The clinical course of the disease is variable and dependent on which component is damaged. Recent molecular biological studies have revealed that various types of mutations occur in the genes of enzyme subunits and the activator protein and that the clinical manifestations differ slightly among mutations.

Another inherited disease concerning gangliosides is GM_1 gangliosidosis, in which a deficiency of β-galactosidase activity causes accumulation of the GM_1 ganglioside and oligosaccharides containing β-galactoside linkages at their termini. In this case mutations in the gene encoding β-galactosidase are considered to be responsible. Galactosialidosis is a disease in which the activities of β-galactosidase and α-sialidase are deficient. There is a protein which protects both enzymes from rapid degradation, and a deficiency of this protective protein is considered to be responsible for this disease.

The activities of some carbohydrate hydrolases have also been demonstrated in membranes or the cytosol, membrane-bound hydrolases having attracted particular attention because they can modify the carbohydrate structures of gangliosides in membranes *in situ*. The degraded components or partially degraded gangliosides are recycled to the Golgi apparatus for biosynthesis.

The linkages of sialic acids, which are modified at their hydroxyl groups through acetylation, are resistant to hydrolysis by sialidases; thus gangliosides having these sialic acids cannot be degraded further. Their degradation is controlled by a deacetylase.

E. Genes Controlling Ganglioside Expression

An example of a gene controlling glycoconjugates is the ABO blood group gene, which is located on human chromosome 9. On the ABO gene three alleles are present. The *A* allele encodes α-*N*-acetylgalactosaminyltransferase; the B allele, α-galactosyltransferase; and the *O* allele, no active transferases. This is the biochemical basis for the polymorphic expression of ABH blood group antigens. Genes controlling the expression of glycoconjugates, including gangliosides, have been demonstrated, and some have been mapped on human chromosomes (e.g., β-galactosyltransferase for the synthesis of Galβ1–4GlcNAc on chromosome 9; activator proteins for β-hexosaminidase and sulfatide sulfatase on chromosomes 5 and 10, respectively; and the β-hexosaminidase α and β subunits on chromosome 15 and 5, respectively). Four mouse genes were found to control the polymorphic expression of gangliosides, three of which were mapped on specific chromosomes. It is interesting, from the point of view of the evolution of genes involved in the regulation of glycoconjugate expression, that a gene for the expression of NeuGc-Gb_5Cer is closely linked to a gene for that of GlcNAc-Gb_5Cer, and both gene products are involved in biosynthetic reactions using Gb_5Cer as a common substrate.

Studies of gene controlling the expression of carbohydrates of glycoconjugates will establish a paradigm for a biological system using remarkably heterogenous structures, which have been produced during evolution.

III. Gangliosides in Cancer

The changes in the ganglioside composition in cancer cells were demonstrated by S. Hakomori and co-workers in 1968, using baby hamster kidney fibroblasts and their polyoma virus transformants. The changes in gangliosides in cancer tissues were summarized: (1) incomplete synthesis, with or without accumulation of precursor glycolipids; (2) induction of synthesis of new glycolipids; and (3) organizational changes of glycolipids in membranes.

Recent research involving the introduction of genes into cells has demonstrated that the introduction of oncogenes into normal fibroblasts induces changes in ganglioside expression. In several cases the changes were mediated by increases in glycosyltransferase activities. The precise mechanisms underlying the changes produced by malignant transformation, however, remain unknown. [*See* ONCOGENE AMPLIFICATION IN HUMAN CANCER.]

Extensive research on cancer-associated antigens and cancer immunity has been performed with the hope of clinical application to cancer diagnosis and therapy. The detection of gangliosides as cancer-associated antigens and immunotherapy with monoclonal antibodies against gangliosides are examples. The detection of sialyl-Le^a antigen in the sera of patients is applied to the diagnosis of pancreatic cancers, preferentially for the late stages. In mice tumors could be successfully treated by the injection of monoclonal antibodies against a glycolipid expressed in abundance on the cancer cells. Human and murine monoclonal antibodies against human melanoma cells recognizing the gangliosides GM_2, GD_2, and GD_3 were used in trials for the treatment of melanoma patients. Local and systemic injections of the monoclonal antibodies were successful in terms of regression of the tumor size or the suppression of metastatic invasions. At the same time antibodies against these gangliosides were detected in the sera of patients with melanomas, and a relationship between the prognosis of the cancer and the level of the antibodies was recognized. The possibility of active immunization for elevating the titer of these antibodies, therefore, is now being investigated.

As mentioned in Section II,B, the altered carbohydrate structures in cancer are known to mimic those in the stages of embryogenesis and the differentiation of fetuses. Antigens of this category are called oncofetal antigens. At present we do not know what biochemical events are the basis for the production of oncogenic and fetal antigens, and how to approach this problem is an important and fundamental subject.

IV. Biological Functions

A. Neural Functions

Among neurons of the brains of Tay–Sachs disease patients, some have swollen or giant neurites (i.e., meganeurites). These are packed with numerous membranous cytoplasmic bodies, composed of accumulating substances, including the GM_2 ganglioside. Many fine neurites originate from the meganeurites. In addition, the administration of brain gangliosides promotes the regeneration and reinnervation processes of both cholinergic and adrenergic nerve fibers in animals. These two observations suggest that brain gangliosides might induce neurite growth and axonal elongation. Many investigators have been interested in these effects and have tried to demonstrate them *in vitro*, using primary and established neural cell cultures. It was shown that mixtures of brain gangliosides—purified GM_1, GD_{1a}, and GQ_{1b}—increase the number of neurites and their length. A few-nanomolar concentration of GQ_{1b} produced these effects in human neuroblastoma cell lines. The carbohydrate structure of GQ_{1b} is strictly required for its effect: GD_{1a}, GD_{1b}, GT_{1a}, GT_{1b}, and GQ_{1c} are not active. On the basis of these results, it was proposed that neurons have receptors recognizing GQ_{1b}, which could play an important role in the formation of neuron networks.

Recently, the purification of a receptor recognizing GQ_{1b} was performed and reported. These results raise several questions. How are signals triggered and transmitted through cell-surface membranes after the binding of gangliosides to the receptors? How do the signals reach nuclei, and how is the biosynthesis of proteins and other components needed for neurite outgrowth and elongation triggered? Research to answer these questions is in progress.

Studies of subcellular distributions of gangliosides in neuronal cells have shown that they occur in synaptic membranes. Presynaptic terminals contain synaptic vesicles, which store chemical transmitters (e.g., acetylcholine). Neurotransmission takes place as the result of membrane fusion between synaptic vesicles and presynaptic membranes and the release of the chemical transmitter into the synaptic cleft, the narrow space between pre- and postsynaptic membranes. The efficiency of the vesicle fusion is affected by membrane components such as phospholipids, cholesterol, and gangliosides. Gangliosides can bind Ca^{2+}, which plays important roles in the release of neurotransmitters. It was proposed that the interaction between gangliosides in presynaptic membranes and Ca^{2+} is one of the events in this Ca^{2+} effect.

On the basis of these bioactive functions, the possibility of the treatment of diabetic neuropathy and senile dementia by the administration of gangliosides has been proposed, and a trial is now in progress. Studies on the roles of gangliosides in neural functions are one of major topics of ganglioside research. This is quite reasonable, because the brain contains the highest concentration of gangliosides among all human tissues or organs.

B. As Receptors for Toxins

A brain lipid fraction containing gangliosides can neutralize the toxic activity of tetanus toxin, which is produced and excreted by *Clostridium tetani*. The toxin produces characteristic tonic spasms of voluntary muscles in humans, called opisthotonos, suggesting that it acts in the spinal cord to suppress inhibitory neurons, the action of which relaxes extended muscles during movement caused by muscle contraction. In vitro, the toxin binds more strongly to GD_{1b} and GT_{1b} than to GM_2, GM_1, or GD_{1a}, suggesting that synapses with different functions could have different and specific ganglioside compositions.

In 1972, W. E. van Heyningen and collaborators reported that a mixture of gangliosides could abolish the toxic activity of cholera toxin. Extensive further research established that GM_1 is specifically recognized by and binds to the B subunit of the cholera toxin. A cholera toxin molecule consists of five B subunits and one A subunit. The B subunits exhibit binding activity for GM_1, and the binding of the B subunits to GM_1 on cell-surface membranes can cause insertion of the A subunit into the membranes through an unknown mechanism, possibly involving conformational changes of both the A and B subunits. The inserted A subunit can activate adenylate cyclase by ADP ribosylation. The elevated concentration of cAMP in the cells results in the hypersecretion of chloride and water and impaired absorption of sodium in the small intestine. However, the biochemical details of the processes from the elevation of cAMP to water loss are not known. The specific interaction between GM_1 and the B subunit can be used for the sensitive detection of GM_1.

The toxin produced by *Clostridium botulinum* is also neutralized by gangliosides, GT_{1b} and GQ_{1b} being the most effective.

Sialidase treatment of mammalian erythrocytes can destroy the receptor activity for influenza, Sendai, and Newcastle disease viruses. Sialylated compounds are the virus receptors on epithelial cells of the respiratory tract as well as erythrocytes. GT_{1a}, GQ_{1b}, and GP_{1c}, which have the terminal structure NeuAcα2–8NeuAcα2–3Galβ1–3GalNAc are possible receptors for the Sendai virus. The question of whether or not the terminal structure of glycoproteins or gangliosides is actually responsible for the virus infection awaits further research.

C. Regulation of Cell Growth and Differentiation

The exogenous addition of GM_3 inhibits the growth of epidermal cells cultured in chemically defined media containing epidermal growth factor (EGF). Other reports suggest that gangliosides might regulate the growth of cultured cells and that the effect might be produced through the modulation of growth factor receptor functions. The binding of growth factors to growth factor receptors and sequentially induced phosphorylation of the receptors are the initial events. It is thought that the exogenously added gangliosides become inserted into the plasma membranes and reach receptors, modifying their three-dimensional structures. In this way the gangliosides could modify the rate of receptor phosphorylation, which is triggered by the binding of growth factors to the receptors. The growth of fibroblasts appears to be controlled by the content of GM_3 in the plasma membrane, where fibroblast growth factor (FGF) interacts with the FGF receptors. A membrane-bound sialidase seems to control the content of GM_3 in the membranes, thus regulating cell growth. It is interesting that membrane-bound hydrolases, but not those in lysosomes, are involved in the turnover of membrane gangliosides. The modulation of cell growth by gangliosides seems to be a secondary control mechanism, which does not operate in an all-or-nothing fashion. This type of regulation or modulation would be physiologically important, but its mechanism is unknown.

HL-60 cells, a human myelogenous leukemia cell line, are induced to differentiate into macrophage-like cells by the exogenous addition of the GM_3 ganglioside, but they differentiate into granulocyte-like cells upon the addition of sialylparagloboside (i.e., IV^3NeuAc-nLc_4-Cer). These observations suggest that gangliosides could regulate or determine the direction of differentiation of immature cells. Little information is available on biochemical events triggered by the addition of gangliosides at present, but studies are in progress.

Extensive further studies on the role of gangliosides in physiological cellular function are required. For this purpose various approaches (e.g., enzymological, cell biological, and molecular biological) are needed. Given the limited space of this article, the authors have not been able to cover all aspects of ganglioside research. For further detailed information and discussion readers are referred to the excellent review articles listed in the Bibliography.

Bibliography

Hakomori, S.-I. (1981). Glycosphingolipids in cellular interaction, differentiation, and oncogenesis. *Annu. Rev. Biochem.* **50,** 733.

Kanfer, J. N., and Hakomori, S.-I. (1983). "Handbook of Lipid Research. Volume 3: Sphingolipid Biochemistry" (D. J. Hanahan, ed.). Plenum, New York.

Rahmann, H. (ed.) (1987). Gangliosides and Modulation of Neuronal Functions. *NATO Adv. Study Inst. Ser., Ser. H* **7.**

Sandhoff, K., Conzelmann, E., Neufeld, E. F., Kaback, M. M., and Suzuki, K. (1989). The GM_2 gangliosidoses. *In* "The Metabolic Basis of Inherited Disease" (C. R. Scriver, A. L. Beaudet, W. S. Sly, and D. Valle, eds.), 6th ed. McGraw-Hill, New York.

Schauer, R. (ed.) (1982). Sialic Acids: Chemistry, Metabolism and Function. *Cell Biol. Monogr.* **10.**

Wiegandt, H. (ed.) (1985). "Glycolipids. Volume 10: New Comprehensive Biochemistry". Elsevier, Amsterdam.

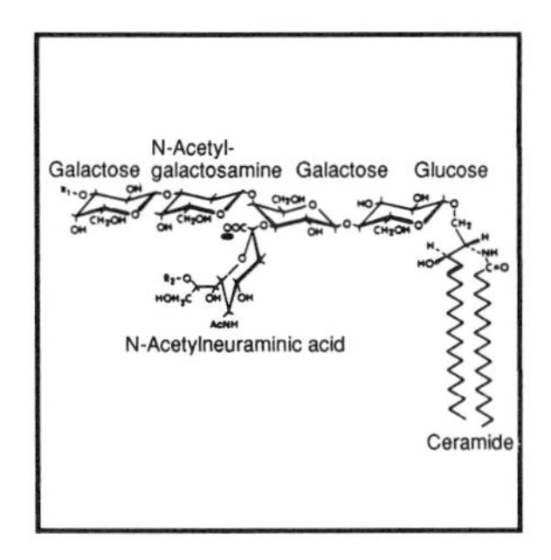

Gangliosides and Neuronal Differentiation

ROBERT W. LEDEEN, *Albert Einstein College of Medicine*

Glossary

Ganglio series Family of gangliosides based on GM1, most of which have the gangliotetraose (or closely related) structure

Ganglioside Glycosphingolipid that possesses at least one sialic acid on the oligosaccharide chain

Gangliotetraose structure GM1 ganglioside or an oligosialo derivative of it

GM1 ganglioside Galactosyl-*N*-acetylgalactosaminyl-(*N*-acetylneuraminyl)-galactosylglucosylceramide

Neuritogenic effect Promotion of neurite formation

Neuronotrophic effect General term for trophic influences on the neuron

THE HISTORICAL connection between gangliosides and neurons goes back to the 1930s, when Ernst Klenk first discovered gangliosides in the gray matter of normal and pathological brains and named them for their apparent localization in *Gangliozellen* (i.e., neurons). Although we now know them to be present in the plasma membrane of virtually all vertebrate cells, their unusually high concentration in the neuron, together with their distinctive structural and developmental patterns in these cells, has given rise to the idea of a special role(s) in neuronal development and functioning. This widely held view has been reinforced by many experiments over the past decade, demonstrating a significant trophic effect of gangliosides expressed *in vitro* upon the addition to neurons and neuroblastoma cells in culture and *in vivo* upon administration to animals suffering lesions to the central or peripheral nervous system. In many of these reports, it has been difficult to distinguish between neurite-promoting (i.e., neuritogenic) activity and a purely protective effect; the two might occur simultaneously in some systems. In this article the term "neuronotrophic" is used to denote both phenomena, as applied to the actions of exogenously administered gangliosides.

I. Ganglioside Changes in the Developing Nervous System

Gangliosides, defined as sialic acid-containing glycosphingolipids, contribute in a major way to the carbohydrate-rich glycocalyx on the surface of neurons. Structural complexity is one of the hallmarks of vertebrate gangliosides. Up to the present time over 90 molecular species possessing unique oligosaccharide structures have been identified in various vertebrate tissues. This number will undoubtedly continue to increase as improved isolation and analytical methods come into use. Thus far echinoderms are the only invertebrates found to contain sialoglycolipids, and these are structurally quite different from those of vertebrates. A significant portion of known vertebrate ganglioside structures are found in the brain, some variation being noted among different animal species. [*See* GANGLIOSIDES.]

Developing brains show dramatic changes in ganglioside pattern during the transition of neurons to the differentiated state. In addition to the marked increase in total ganglioside content, important qualitative changes occur, highlighted by the appearance of rising levels of the gangliotetraose family. These are the structures based on ganglioside GM1, shown in Fig. 1. The oligosialo members of this family are generated by the attachment of one or more sialic acids to this basic unit. In the rodent brain gangliosides of the hematoside family (i.e., those lacking hexosamine)—which are abundant in predifferentiated neurons—undergo rapid decline during the same period. The human brain was found to contain much less of this type.

Detailed study of the human brain has revealed a somewhat different developmental pattern among gangliotetraose species. Thus, GM1 increased 15-fold from the fifth gestational month in an almost linear fashion up to 5 years of age. GT1b was the major ganglioside during the first trimester and the first part of the second trimester, following which its concentration dropped to a minimum around birth and then increased gradually throughout life. The pattern for GD1b was similar. Ganglioside GD1a showed a sharp rise from the 10th gestational week to its maximum value soon after birth; this was followed by a plateau to around 5 years of age and then a steady decrease with aging. A correlation has been noted in humans and other species between developmental accretion of GD1a and process outgrowth and synaptogenesis. The unique pattern of GM1 could be attributed, at least in part, to the prominence of this species in myelin.

The ganglioside changes observed in the developing brain have been duplicated *in vitro* in primary neuronal cultures derived from cerebral hemispheres of chick embryo. The neurons in this system were observed to have two distinct phases with regard to ganglioside composition: an early phase corresponding to cell division, in which ganglioside content increased only slightly and included GD3 (hematoside) as the primary type, and a later phase corresponding to differentiation and synaptogenesis, in which the gangliotetraose species accumulated rapidly. *In vitro* systems of this kind thus provide excellent models for the further study of gangliosides' role in neuronal differentiation.

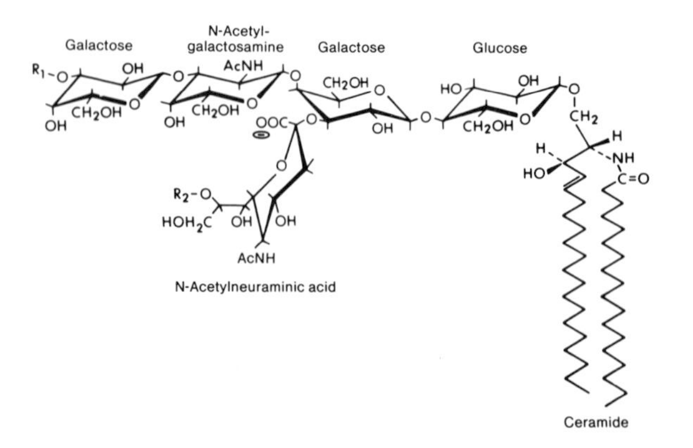

FIGURE 1 Structures of the five major gangliosides of the mammalian brain. GM1, $R_1 = R_2 = H$; GD1a, R_1 = NeuAc, $R_2 = H$; GD1b, $R_1 = H$, R_2 = NeuAc; GT1b, $R_1 = R_2$ = NeuAc; GQ1b, R_1 = NeuAc(2–8)NeuAc, R_2 = NeuAc. The zig zag lines of ceramide represent hydrocarbon chains. The nomenclature is that proposed by L. Svennerholm.

II. Neurite-Promoting Properties of Gangliosides

Many laboratories have now reported the striking neuritogenic effects of gangliosides added to primary neurons and neuroblastoma cell lines in culture. One of the most responsive cells is the mouse Neuro-2A neuroblastoma cell line, which shows prolific neurite outgrowth in the presence of 25–100 μM ganglioside (Fig. 2). Similar responses, with small qualitative and quantitative variations, were produced in these cells by a variety of gangliosides, ranging from the relatively simple sialosylgalactosyl ceramide (GM4) to the more complex tetrasialoganglioside GQ1b (Fig. 1). Further studies demonstrated the same kind of activity with synthetic sialoglycolipids that are not true gangliosides and are therefore unable to enter the metabolic pathways of the latter. Hence, it would appear that virtually any sialoglycolipid is capable of eliciting this kind of trophic response in a relatively nonspecific manner with Neuro-2A cells. This is to be contrasted with the marked specificity seen in the GOTO human neuroblastoma cell line, which responds only to GQ1b; however, few examples of this kind have come to light. Other neuroblastoma systems reported to respond somewhat like Neuro-2A include the B104, SB21B1, B50, and S20Y cell lines.

Several primary neuronal cultures have been shown to react similarly to gangliosides as neuritogenic agents. This was demonstrated first with dorsal root ganglia and later other peripheral nervous system neurons, including ciliary, sympathetic, and spinal root ganglia. In some cases gangliosides alone were not effective, but acted synergistically with other agents (e.g., nerve growth factor or ciliary neuronotrophic factor). This was also observed with the PC12 pheochromocytoma cell line. It has been postulated that the enhanced recovery of many nervous system lesions induced by gangliosides *in vivo* reflects the same neuritogenic properties observed *in vitro*, and studies with the sciatic nerve have, in fact, suggested that gangliosides aid regeneration through enhanced sprouting. However, in other cases the trophic effects of injected gangliosides appear to be more readily attributable to their functioning as survival factors (see Section II).

III. Protective Effects of Gangliosides

A biological property of exogenously administered gangliosides which is receiving considerable attention due to its biomedical implications is the protective effect toward neurons subjected to a variety of mechanical, toxic, and ischemic traumas. An example manifested *in vitro* is the enhanced survival afforded primary cultures of cerebellar and cortical neurons against glutamate toxicity. Prior exposure of the cells to approximately 100 μM ganglioside raised survival from approximately 20% in control cultures to around 80%. Similarly, neuroblastoma cells treated with the calcium ionophore A23187 suffered eventual cell death, which could be prevented by the same concentration of gangliosides.

The most striking illustration of this phenomenon is seen *in vivo* (e.g., in the facilitated recovery from mechanical lesions to the septohippocampal, nigrostriatal, and entorhinal pathways). Recent work has shown injected gangliosides to counteract retrograde degeneration of cholinergic neurons in the nucleus basalis magnocellularis following unilateral cortical lesion. Neurotoxic destruction of dopaminergic neurons in the mouse striatum caused by 1-methyl-4-phenyl-1,2,3,6-tetrahydropyridine (MPTP) was largely prevented by gangliosides, as were haloperidol-induced behavioral impairments following MPTP administration. Several other examples could be cited (Table I). In most of these studies involving central nervous system lesions, the ganglioside (frequently GM1) was administered intramuscularly or intraperitoneally, thereby requiring penetration of the blood–brain barrier. While the question is not entirely settled, the studies performed to date strongly suggest that sufficient GM1 does cross into the brain parenchyma to cause the observed effects.

TABLE I Systems Used to Demonstrate Ganglioside-Induced Neuronotrophic Effects *In Vivo*[a]

Peripheral nervous system
Mechanical lesions
Superior cervical ganglion–nictitating membrane (regeneration)
Sciatic nerve (regeneration, axonal transport)
Sciatic nerve—gastrocnemius muscle (muscle reinervation)
Superior gluteal nerve—gluteus maximus muscle (sprouting)
Newt limb buds (regeneration)
Diaphragm (electrogenic pump activity)
Neurotoxic lesions
Sciatic nerve: alloxan- or streptozotocin-induced diabetes (axonal transport)
Tibial nerve: acrylamide neuropathy (sprouting)
Sympathetic nerves: vinblastine sympathectomy (norepinephrine)
Central nervous system
Mechanical lesions
Optic nerve—goldfish (regeneration)
Entorhinal cortex (behavior)
Hippocampus—septal lesions (cholinergic parameters, behavior)
Nigrostriatum (histochemical and biochemical parameters, behavior)
Spinal cord (sprouting)
Nucleus basalis (retrograde degeneration)
Cerebral hemispheres: ischemia (membrane function, behavior)
Rubrospinal tract (RNA)
Neurotoxic lesions
Bulbospinal serotoninergic system: 5,7-dihydroxytryptamine (5-hydroxytryptamine and 5-hydroxyindoleacetic acid levels)
Cerebrocortical noradrenergic system: 6-hydroxydopamine (norepinephrine)
Nigrostriatal dopamine neurons: MPTP (neurochemical, histochemical, and behavioral parameters)
Nucleus basalis: ibotenic acid (anterograde degeneration)
Dentate gyrus: colchicine (behavior)
Intracerebroventricular injection of vincristine (behavior and neurochemical parameters)

[a] Listed are many of the *in vivo* animal models used to demonstrate ganglioside-facilitated recovery attributed either to enhanced neuronal regeneration or neuroprotective effects. In most cases the animals were rat. In addition to the nervous system component directly affected, the nature of the lesion is indicated, along with the parameter(s) measured (in parentheses). Specific references to some of these studies are given by R. W. Ledeen, *in* "Neurobiology of Glycoconjugates" (R. U. Margolis and R. K. Margolis, eds.), pp. 43–83. Plenum, New York, 1989 and by A. Consolazione and G. Toffano *in* "New Trends in Ganglioside Research" (R. W. Ledeen, E. L. Hogan, G. Tettamanti, A. J. Yates, and R. K. Yu, eds.) pp. 523–533. Tiviana, Padova, 1988.

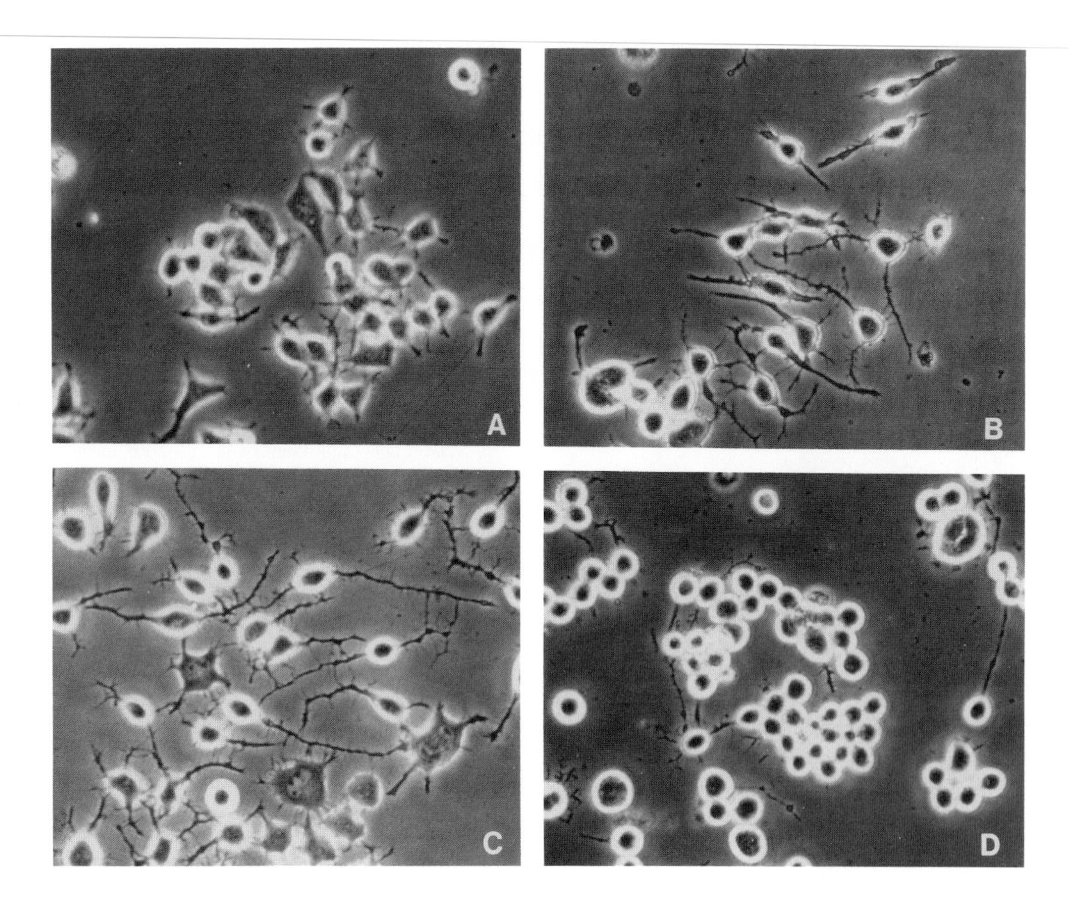

FIGURE 2 The neuronotrophic effect of exogenous gangliosides on murine Neuro-2A neuroblastoma cells. Growth conditions: Dulbecco's modified minimum essential medium, 5% fetal calf serum, 10 mg/100 mL gentamicin, 37°C, 5% CO_2/95% air. The following additions were made: (A) none (control); (B) bovine brain ganglioside mixture (0.1 m*M*); (C) GD1a (0.1 m*M*); (D) GD1a (0.1 m*M*) plus cholera toxin B subunit (5 μg/ml). Note the neuritogenic effect due to ganglioside mixture and GD1a alone and inhibition of this by cholera B.

IV. Mechanistic Considerations

In approaching the question of mechanism, it is necessary to distinguish between exogenous gangliosides affecting either *in vivo* or *in vitro* systems and those molecules which are endogenous or naturally occurring in the neuronal membrane. It is often assumed (although not yet proved) that these two aspects bear a fundamental relationship to each other. One reason for this assumption is the well-established fact that a small portion of exogenous ganglioside inserts spontaneously into the membrane and apparently behaves thereafter as endogenous ganglioside. However, the majority of exogenous ganglioside remains loosely attached to the cell surface, and at present it is not clear which of these two pools asserts the trophic effects. The broad spectrum of sialoglycolipid structures which show such activity (see Section II) would seem to argue against a mechanism in which the exogenous molecules are merely supplementing a suboptimal level of endogenous gangliosides through membrane insertion.

Another possibility now being explored is that exogenous gangliosides induce transient perturbations of the cell surface, perhaps through alteration of membrane fluidity, which cause direct or indirect modulation of certain plasma membrane components. These could include ion channels, receptors, and/or enzymes, several of which are known to be dramatically altered in the presence of gangliosides. Protein kinase C and various other kinases of neural membranes, for example, were found to be either activated or inhibited by these substances. Evidence has been presented for specific ganglioside-reactive kinases, including an ectoprotein kinase on the surface of certain neuroblastoma cells. Gangliosides have been found to affect cellular proliferation in some cases (see Section V).

The mechanisms by which endogenous gangliosides assert their functional role also remain to be elucidated, although it is likely that structural requirements in this case are considerably more stringent than for exogenous gangliosides. The programmed appearance of gangliotetraose species in the plasma membranes of differentiating neurons (see Section I) points to a primary role for such structures at this phase of development as well as at later stages, when high levels of these persist. Special significance is attached to GM1 ganglioside following several reports demonstrating blockage of differentiation and/or regeneration by anti-GM1 antibody or cholera B subunit, a highly specific GM1-binding ligand. This has been shown, for example, with the Neuro-2A cell line (Fig. 2). Comparison of the ganglioside patterns of several neuroblastoma cells revealed higher levels of GM1 in those cell lines most responsive to neuritogenic stimuli. The precise manner in which GM1 asserts its influence is not known, although some studies have suggested a cofactor role with various receptors and/or ion channels.

V. Effects on Nonneuronal Cells

It has become clear from recent work that exogenous gangliosides influence the behavior of other cell types besides neurons. Cultured astroglial cells of the central nervous system, for example, respond to concentrations of about 10^{-5} *M* (in the absence of serum) with increased proliferation. In addition, gangliosides have the property of blocking the conversion of such cells from a flat epithelioid form to a stellate morphology caused by the activation of cAMP-synthesizing systems. The HL-60 human promyelocytic cell line, in the presence of ganglioside GM3, is induced to differentiate toward the monocyte–macrophage phenotype, in contrast to the effect of dimethyl sulfoxide, which favors the granulocyte pathway. Swiss 3T3 cells and various other cell lines respond to exogenous gangliosides as regulators of cellular proliferation, possibly through modulation of growth factor receptors. It is not unreasonable to expect that additional cell types will be discovered that respond in some significant manner and that such phenomena will eventually help to elucidate the biological roles of gangliosides in the vertebrate membrane.

VI. Summary

The outstanding feature of gangliosides, in addition to their unusually high concentration in the mature neuronal membrane, is the striking developmental changes which they undergo during neurite outgrowth and synaptogenesis. The rapid accretion of gangliotetraose species during this period suggests a primary role for these structures, especially GM1, since anti-GM1 antibodies and cholera toxin B subunit inhibit this stage of differentiation. Exogenous gangliosides exhibit trophic properties toward a large variety of primary neurons and neuroblastoma cell lines in culture, acting either as neuritogenic agents or as survival factors. This phenomenon shows relatively little structural specificity, since a wide variety of sialoglycolipids cause essentially the same effects; however, sialic acid is required in most cases, since removal of this group from the glycolipid produces inactive molecules. Physicochemical perturbation of the membrane, leading to changes in the activity of enzymes, receptors, or other membrane components, has been proposed as the operative mechanism. Gangliosides are now recognized as effectors of interesting physiological changes in astroglia and a variety of other nonneuronal cells. Their ability to modulate growth factor receptors of such cells has become an important area of research. Finally, gangliosides have been observed to operate as neuronotrophic factors in a large number of animal models, providing a rationale for their current investigation as potential biomedical agents.

Bibliography

Ando, S. (1983). Gangliosides in the nervous system. *Neurochem. Int.* **5,** 507–537.

Bremer, E. G., Schlessinger, J., and Hakomori, S.-I. (1986). Ganglioside-mediated modulation of cell growth. *J. Biol. Chem.* **261,** 2434–2440.

Consolazione, A., and Toffano, G. (1988). Ganglioside role in functional recovery of damaged nervous system. *In* "Neurochemical and Neuroregenerative Aspects" (R. W. Ledeen, E. L. Hogan, G. Tettamanti, A. J. Yates, R. K. Yu, eds.), pp. 523–533, Liviana, Padova, Italy.

Dreyfus, H., Ferret, B., Harth, S., Gorio, A., Freysz, L., and Massarelli, R. (1984). Effect of exogenous gangliosides on the morphology and biochemistry of cultured neurons. *In* "Ganglioside Structure, Function, and Biomedical Potential" (R. W. Ledeen, R. K. Yu,

M. M. Rapport, and K. Suzuki, eds.), pp. 513–524. Plenum, New York.

Favaron, M., Manev, H., Alho, H., Bertolino, M., Ferret, B., Guidotti, A., and Costa, E. (1988). Gangliosides prevent glutamate and kainate neurotoxicity in primary neuronal cultures of neonatal rat cerebellum and cortex. *Proc. Natl. Acad. Sci. U.S.A.* **85,** 7351–7355.

Ledeen, R. W. (1984). Biology of gangliosides: Neuritogenic and neuronotrophic properties. *J. Neurosci. Res.* **12,** 147–159.

Ledeen, R. W. (1985). Gangliosides of the neuron. *Trends Neurosci. (Pers. Ed.)* **8,** 169–174.

Ledeen, R. W. (1989). Biosynthesis, metabolism, and biological effects of gangliosides. *In* "Neurobiology of Glycoconjugates" (R. U. Margolis and R. K. Margolis, eds.), pp. 43–83. Plenum, New York.

Leon, A. D., Benvegnu, D., Dal Toso, R., Presti, D., Facci, L., Giorgi, O., and Toffano, G. (1984). Dorsal root ganglia and nerve growth factor: A model for understanding the mechanism of GM1 effects on neuronal repair. *J. Neurosci. Res.* **12,** 277–287.

Mahadik, S. P., and Karpiak, S. K. (1988). Gangliosides in treatment of neural injury and disease. *Drug Dev. Res.* **15,** 337–360.

Nagai, Y., and Tsuji, S. (1988). Cell biological significance of gangliosides in neural differentiation and development: Critique and proposals. In "New Trends in Ganglioside Research: Neurochemical and Neuroregenerative Aspects" (R. W. Ledeen, E. L. Hogan, G. Tettamanti, A. J. Yates, and R. K. Yu, eds.), pp. 329–350, Liviana, Padova, Italy.

Seifert, W., and Fink, H.-J. (1984). *In-vitro* and *in-vivo* studies on gangliosides in the developing and regenerating hippocampus of the rat. In "Ganglioside Structure, Function, and Biomedical Potential" (R. W. Ledeen, R. K. Yu, M. M. Rapport, and K. Suzuki, eds.), pp. 535–545, Plenum, New York.

Svennerholm, L., Bostrom, K., Fredman, P., Mansson, J.-E., Rosengren, B., and Rynmark, B.-M. (1989). Human brain gangliosides: Developmental changes from early fetal stage to advanced age. *Biochim. Biophys. Acta* **1005,** 109–117.

Varon, S., Skaper, S. D., and Katoh-Semba, R. (1986). Neuritic responses to GM1 ganglioside in several in vitro systems. *In* "Gangliosides and Neuronal Plasticity" (G. Tettamanti, R. Ledeen, K. Sandhoff, Y. Nagai, and G. Toffano, eds.), pp. 215–230, Liviana, Padova, Italy.

Yu, R. K. (1988). Regulation of protein phosphorylation by gangliosides. *In* "New Trends in Ganglioside Research: Neurochemical and Neuroregenerative Aspects" (R. W. Ledeen, E. L. Hogan, G. Tettamanti, A. J. Yates, and R. K. Yu, eds.), pp. 461–471. Liviana, Padua, Italy.

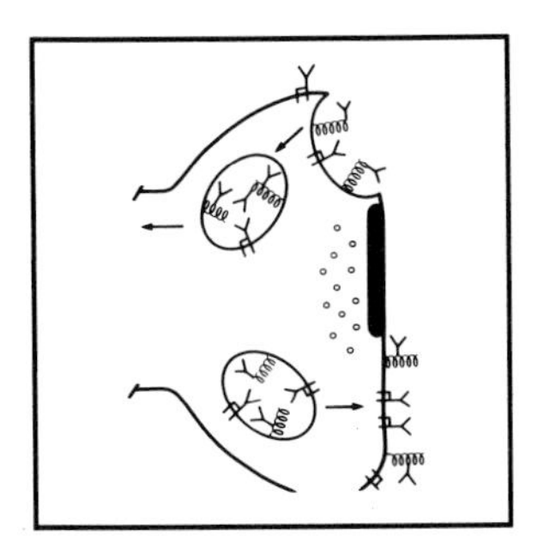

Ganglioside Transport

ROBERT W. LEDEEN, *Albert Einstein College of Medicine*

GLOSSARY

Anterograde axonal transport Transport of substances in the axon away from the cell body, toward the nerve ending

Ganglio-series Family of gangliosides based on GM1, most of which have the gangliotetraose (or a closely related) structure

Ganglioside Glycosphingolipid that possesses at least one sialic acid on the oligosaccharide chain

Gangliotetraose structure GM1 ganglioside or an oligosialo derivative of it

GM1 ganglioside Galactosyl-*N*-acetylgalactosaminyl (*N*-acetylneuraminyl)-galactosyl-glucosylceramide

Oligosialogangliosides Gangliosides having two to five sialic acids

Retrograde axonal transport Transport of substances in the axon back toward the cell body

GANGLIOSIDES form part of the carbohydrate-rich layer that surrounds mammalian cells and determines their surface properties. In neurons they comprise a much larger proportion of this membrane-associated coating than in most other cells. Gangliosides were once thought to be localized primarily at the synapse. Although their precise localization in the neuron is still somewhat controversial, increasing evidence suggests they are distributed over a large portion of the neuronal surface including cell body, synaptic region, and the extensive network of dendrites and axons. These latter, which are often several thousand times greater in length than the cell body where synthesis occurs, require efficient transport/transfer mechanisms to convey gangliosides to the far reaches of this complex cell. Research over many years has elucidated several aspects of this process as well as that of "reverse flow" involving return to the cell body. These studies quite naturally employed animal models, as in most other areas of ganglioside research, but it is reasonable to assume that processes that are common to a variety of vertebrates, from fish to rabbits, apply as well to the human. Axonal transport of gangliosides has been studied in both the central and peripheral nervous systems, with generally similar results. In the course of this work, several ancillary findings relating to ganglioside distribution and metabolic behavior have come to light. The recent discovery of glycolipid transfer proteins has expanded our general concept of ganglioside movement. The subject of dendritic transport unfortunately has not yet been investigated in relation to gangliosides, perhaps owing to the lack of a suitable anatomical model. Experimental ingenuity will be needed in the future to overcome this rather large gap in our understanding of ganglioside transport. [*See* GANGLIOSIDES.]

I. Axonal Transport in the CNS

The general procedure for axonal transport studies has been to inject a radiolabeled precursor into the tissue of a living animal in the vicinity of neuronal cell bodies, followed by detection of radiolabeled substances that have migrated a measured distance along the axon during a specified period of time. The optic system lends itself well to this approach because of the clear anatomical separation of cell bodies from axons (optic nerve/tract) and nerve endings (optic tectum, superior colliculus, lateral geniculate nucleus). This involves one cell type, the ganglion cells, whose cell bodies reside in the retina. Beginning with the goldfish, it was shown that intraocular injection of ganglioside precursors (e.g., glucosamine and *N*-acetylmannosamine) gives rise to radiolabeled gangliosides, which are translocated from the retina to the optic tectum. Because the optic system is generally crossed, axonal migration occurs to the tectum of the opposite side; the tectum of the same side is then used as control tissue to indicate background labeling that occurs systemically via the circulation. Subtraction of the latter from the former gives the amount of axonally transported radiolabel. Similar studies were subsequently carried out with the optic systems of chick, rat, rabbit, and other kinds of fish. [SEE RETINA; VISUAL SYSTEM.]

In such studies it is necessary to rule out the possibility that the precursor undergoes transport and is subsequently assembled in the axon or nerve terminal, but this possibility could be excluded. One might argue intuitively against such a mechanism on the basis of the large number of enzymes, substrates, and relevant organelles that would have to be transported. Direct experimental evidence against it often consists in the failure to observe movement of radiolabeled precursors into the soluble compartment of axons–nerve endings. In most of those cases where such movement was thought to occur, it subsequently turned out to be attributable to extraaxonal diffusion of administered radiolabel.

The velocity of anterograde ganglioside transport away from the cell body corresponds to the fast phase of protein transport, estimated at 70–100 mm/day in the goldfish optic system and approximately two to four times as fast in the rat or rabbit. This wave, now recognized as also characteristic of phospholipids, glycoproteins, and membrane components in general, stands in contrast to the slower phases involving cytoskeletal elements and other classes of proteins.

Because the gangliosides of the nervous system are a complex mixture of structural types, it was of some interest to determine the molecular pattern undergoing axonal transport. Using the rabbit optic system, it was shown that all detectable gangliosides migrate out of the retina at the same speed. That study further revealed that the types of ganglioside being transported in this single neuron class (retinal ganglion cells) are qualitatively similar to those of whole rabbit brain. From this it can be inferred that the complex molecular pattern of gangliosides in brain arises primarily from the multiplicity of structures in individual neurons, rather than summation of simple but variable patterns in different neuronal types. However, at higher sensitivity it might be possible to detect specificity in regard to one or more ganglioside structures, as shown for a group of minor gangliosides that appear to be unique to a particular class of neuron. There also may be differences in regard to the relative proportions of individual gangliosides, which undoubtedly do exist.

Some interesting conclusions have also been drawn from such studies in regard to ganglioside localization in the neuron. One is that axons do not simply serve as conduits for the flow of gangliosides to nerve endings, but themselves receive a fraction of those transported. Axonal and nerve ending membranes thus appear to function as a unit in the uptake and turnover of these molecules. This behavior has been contrasted with that of glycoproteins, which are targeted primarily for the nerve ending, a much smaller portion entering the axon. Another important difference is reflected in the fact that 30% or so of the gangliosides synthesized in the cell body region (e.g., retina) is committed to transport, compared with about 12% for glycoproteins and 1–2% for proteoglycans. These findings are consistent with other studies showing that although all classes of glycoconjugate are represented in membranes of the axon and nerve ending, gangliosides predominate. They lead to the prediction that analysis of the axon membrane will reveal an especially heavy preponderance of these substances.

The form in which gangliosides migrate down the axon has not been well-defined. There are indications they are part of prepackaged lipid-protein vesicles, of which there may be more than one type. The importance of protein was emphasized by the observation that injection of protein-synthesis inhibitors along with ganglioside precursors effectively blocked axonal transport of the newly synthesized gangliosides. Because gangliosides have been shown to occur in coated vesicles and some types of synaptic vesicles (organelles known to undergo fast transport away from the cell body), these would appear to constitute important transport vectors. Additional forms designed especially for membrane renewal may also exist. Axonal transport has been considered an example of endomembrane flow involving mechanisms of plasma membrane renewal common to all cells. In this scheme, a key role has

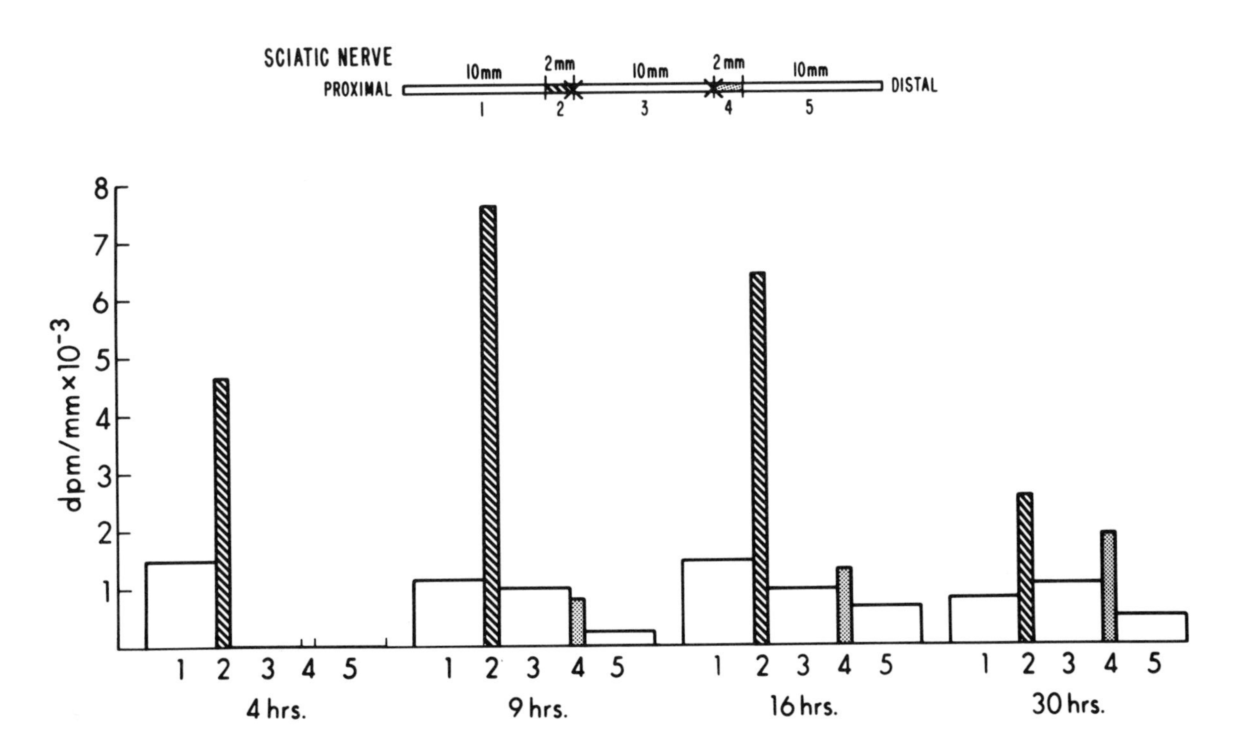

FIGURE 1 Profiles of ganglioside ^{3}H-radioactivity in sciatic nerve segments (*lower diagram*). Ligatures were tied (*upper diagram*) at the times indicated after [^{3}H]glucosamine injection into motoneurons, and accumulations progressed for 2–3 hr. Gangliosides were isolated in pure form from each segment and counted. Radioactivity accumulation at segment 2 represented flow away from cell bodies, and that at segment 4, flow in the opposite direction. The latter was first discernible at 16 hr after injection. [From Aquino, D. A., *et al.* (1985). *J. Neurochem.* **45,** 1262, with permission.]

been ascribed to the Golgi apparatus present in the cell body for the packaging of such vesicles as well as for biosynthesis of the gangliosides themselves.

Developmental questions have been addressed indirectly through analysis of neuronal growth cones, structures that constitute the neurite's leading edge during neuronal differentiation. Using newly developed methods for isolating these particles from embryonic rat brain, the plasma membranes obtained after osmotic disruption were shown to contain appreciable levels of ganglioside. This indicated that neuritic transport begins early in the life of a neuron, even before the process of differentiation has been completed. It is of some interest that glycolipids were shown by microscopic studies to be the predominant form of cell surface glycoconjugate at this early stage of development; although some glycoproteins are also present at that time, the majority of these appear to enter the neuronal membrane at a later stage. [*See* GANGLIOSIDES AND NEURONAL DIFFERENTIATION.]

II. Axonal Transport in the PNS

Ganglioside transport has been studied more recently in the peripheral nervous system, with results that are generally similar to those obtained with the optic system. Thus, using rat sciatic nerve, gangliosides were shown to undergo fast transport away from the cell body at a rate of approximately 300–400 mm/day. As with the CNS, all molecular species appeared to migrate simultaneously.

Use of the sciatic nerve has permitted comparison of transported gangliosides in two different neuronal populations, motor and sensory. In both cases, fast transport was observed with generally similar molecular patterns in which species of the gangliotetraose type predominated. However, sensory neurons also had structures characteristic of other ganglioside families, along with neutral glycosphingolipids that were lacking in motoneurons. Despite the differences, these and related observations suggest that a preponderance of ganglio-series structures is an intrinsic feature of neurons in general.

An interesting difference in the outflow patterns of gangliosides and glycoproteins, after labeling of both by glucosamine injection into the dorsal root ganglia, was that gangliosides did not show the well-defined crest of radioactivity characteristic of glycoproteins, but rather a pattern with an attenuated crest in a series of rather flat curves representing different times. This result can be interpreted as indicating extensive exchange of gangliosides between mobile and stationary axonal structures, in contrast to glycoproteins, which are targeted primarily to the nerve ending. This is similar to the results obtained with the optic system discussed above.

A major advantage in using the PNS is that longer nerves (e.g., the sciatic) facilitate study of transport in both directions. The method of choice has been the double ligation model, which permits observation of both flows simultaneously. The procedure is to inject radiolabeled precursor into the vicinity of the cell bodies (e.g., dorsal root ganglia, lumbosacral spinal cord) and then after variable periods of time tie two ligatures 1 cm apart on a "downstream" segment of the exposed sciatic nerve. Depending on the direction of flow, the radioactive molecules accumulate on one or the other side of each ligature. After a few hours, the tissue is dissected for analysis. By this means gangliosides were shown to undergo bidirectional transport in both motoneurons and sensory neurons of rat sciatic nerve. A typical experiment of this kind is shown in Fig. 1. Essentially similar results were obtained for glycoproteins. The velocity of transport toward the cell body could not be calculated directly, but the early return of labeled gangliosides appeared consistent with the relatively rapid velocities (equivalent to one-half anterograde flow) previously estimated for proteins.

Use of the double ligation model has facilitated comparison of the molecular species migrating in the two directions. Such experiments would help to answer long-standing questions concerning possible metabolic changes that might occur during the transport of gangliosides to and return from axonal and nerve ending membranes. It has been conjectured, for example, that because the enzyme neuraminidase is present in synaptic plasma membranes, a portion of the oligosialogangliosides residing there might be converted to GM1. Using the model depicted in Fig. 1, the procedure is to compare the gangliosides extracted from segment two, which is toward the cell body in respect to the first ligature (representing transport away from the cell body), with those extracted from segment four, which is on the opposite side in respect to the second ligature (representing transport toward the cell body). The general pattern distributions in relation to sialic acid number turned out to be similar (Fig. 2), indicating no apparent change in the relative proportions of mono-, di-, tri-, or tetrosialogangliosides; this suggested no major metabolic alterations due to neuraminidase in axons or nerve endings during the course of the round trip. However, the possibility of metabolic change(s) of a mi-

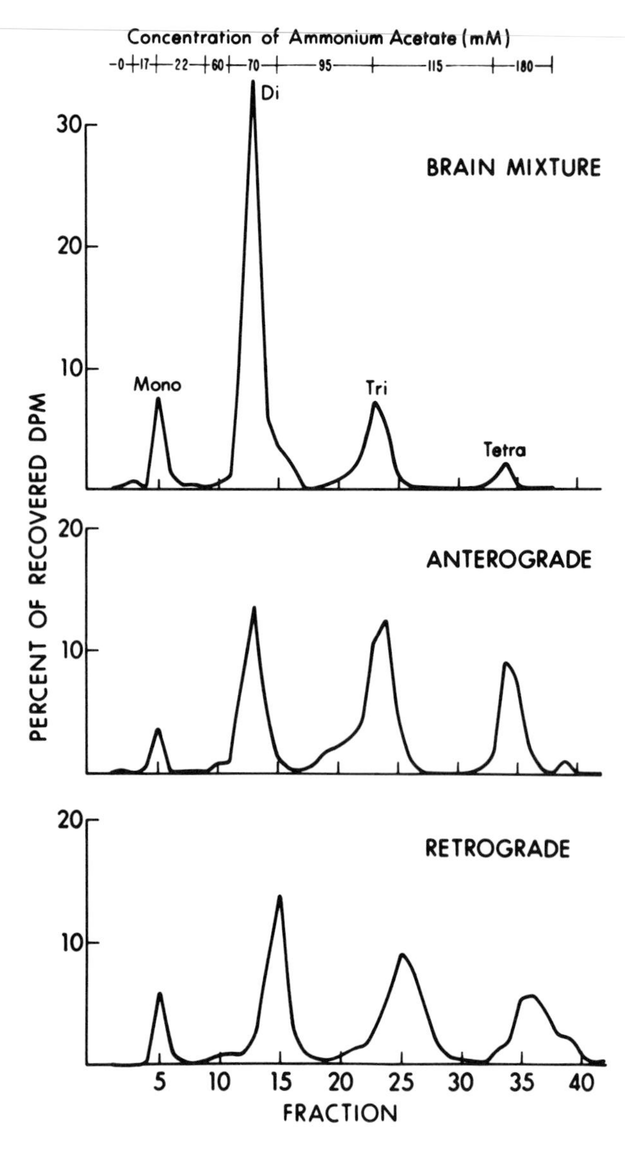

FIGURE 2 Fractionation of gangliosides on DEAE-Sephadex. A brain mixture is compared with gangliosides undergoing transport away from the cell body (anterograde) or toward the cell body (retrograde) transport in motoneurons of sciatic nerve. Gangliosides were labeled before transport by injection of [^{3}H]glucosamine into the lumbosacral spinal cord. [From Aquino, D. A., *et al.* (1985). *J. Neurochem.* **45**, 1262, with permission.]

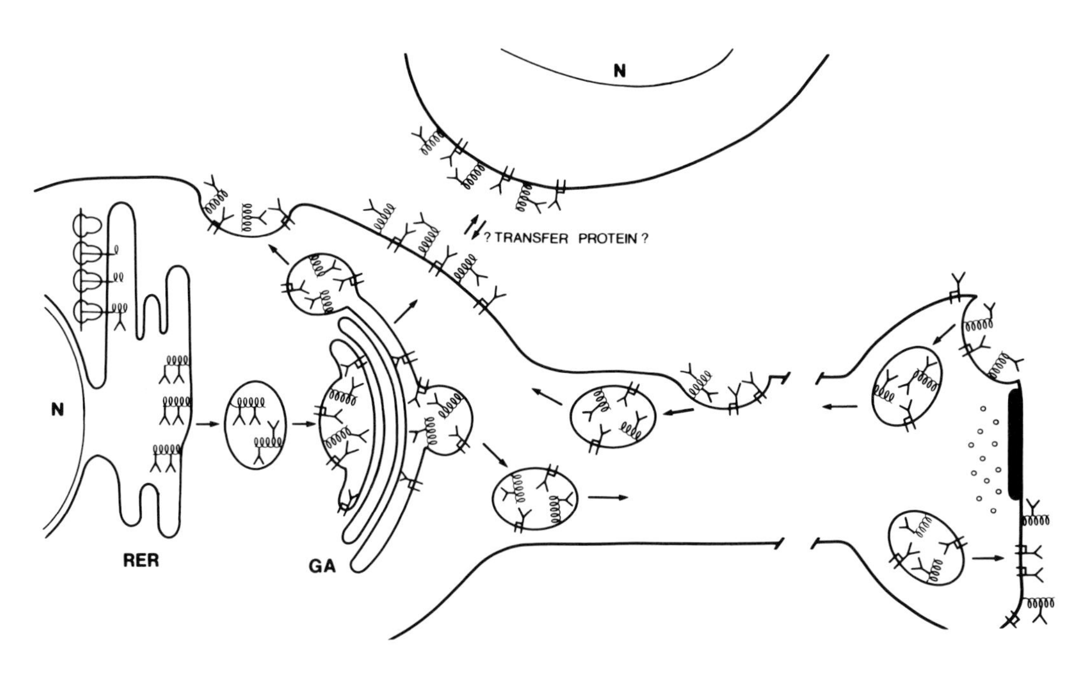

FIGURE 3 Aspects of ganglioside and glycoprotein movement in the neuron, as currently visualized. Asparagine-linked glycoprotein (coil symbol with the carbohydrate symbol attached) commences synthesis in the rough endoplasmic reticulum (RER) and passes thru the Golgi apparatus (GA). Gangliosides (symbol) are also synthesized at least partially in the GA, but the role of the RER is not known. From the GA, glycoconjugates migrate to the plasma membrane and to axonal and nerve-ending membranes via fast axonal transport. Return to the cell body is also depicted. Intracellular migration, including axonal transport, is thought to occur in vesicles, the carbohydrate portions being sequestered in the luminal compartments. Intercellular movement of gangliosides, catalized by ganglioside-transfer protein, is presented as a hypothetical process. N = nucleus. [From Ledeen, R. W., *et al.* (1987). *In* "Gangliosides and Modulation of Neuronal Functions" (H. Rahmann, ed.), pp. 259–274. Springer-Verlag, Berlin, Heidelberg, with permission.]

nor nature remains open until the compositions of these two pools can be examined in more detail.

Observation of retrograde axonal transport provides at least a partial answer to the question of the mechanism of ganglioside turnover. The phenomenon of bidirectional transport indicates that at least a portion of the gangliosides situated in axonal and nerve-ending membranes return to the cell body for catabolism and/or recycling. There are suggestions that other mechanisms may also operate because the amount returning is possibly less than the amount leaving the cell body. One such mechanism is shedding from the surface; this was previously shown to occur for glycoproteins of certain animal cells and more recently for gangliosides as well. It could account for at least part of the gangliosides detected in the soluble compartment of brain. Another possibility is *in situ* degradation within the axon and nerve ending. However, although selective protein degradation has been reported in these structures, there is as yet no direct evidence pertaining to gangliosides.

III. Ganglioside Transfer Proteins

Although anterograde and retrograde axonal transport can account for some of the movement of gangliosides within the neuron, it is likely that additional mechanisms are needed for distribution to and from their sites of use. Transfer of gangliosides between membranes, analogous to the well-characterized transfer of phospholipids, may be one such mechanism because proteins capable of catalyzing such transfer and/or exchange have been isolated from the soluble compartment of brain. This activity was displayed toward all the major brain gangliosides and some neutral glycolipids as well, but not toward phosphatidylcholine. Transfer proteins possess a molecular weight of approximately 20,000 Da, similar to activator proteins for certain glycohydrolase enzymes, which also possess glycolipid transfer activity; however, they are clearly different proteins.

The physiological function of these transfer proteins is still unknown, as is their cellular locus. If they are cytosolic proteins, the question arises as to how they would interact with intracellular gangliosides, most of which are believed to be sequestered within the luminal portion of organelles (Fig. 3).

Because transmembrane movement ("flip-flop") of glycolipids is considered unlikely, they would not readily become accessible to transfer. At this stage we cannot preclude the existence of vesicles or organelles with gangliosides on the cytoplasmic surface nor the possibility of a role within organelles (e.g., the Golgi matrix). Such possibilities can be better assessed when the localization of ganglioside transfer proteins has been determined. Should they turn out to be extracellular, a catalytic role in intercellular transfer would need to be considered (Fig. 3).

IV. Summary

The predominant—perhaps sole—site of ganglioside synthesis in the neuron is the endoplasmic reticulum/Golgi complex, from which they enter the transport machinery of the axon (and presumably dendrites). Fast transport away from the cell body is the primary form of movement into the axons and nerve endings of both the CNS and PNS. Axonal transport back to the cell body, revealed by the double ligation method applied to sciatic nerve, accounts for the turnover of at least some of the membrane-bound gangliosides of the axon and nerve ending. Similarities in the gross patterns of gangliosides moving in the two directions suggest there are few if any metabolic alterations of these molecules during transport and use. Comparison studies have revealed that transported gangliosides are targeted to both axonal and nerve ending membranes, in contrast to glycoproteins, which are destined primarily for the nerve ending. Ancillary findings were that virtually all gangliosides undergo transport simultaneously and that the molecular diversity detected in a single class of neurons (retinal ganglion cells) is qualitatively similar to that of whole brain. Hence the complexity of ganglioside structural types arises from the pattern within individual neurons. It would appear that ganglio-series structures are an intrinsic feature of neurons in general. Ganglioside transfer proteins, which catalyze exchange and/or net transfer of glycolipids between membranes, were shown to occur in the soluble fraction of brain, suggesting another mode of intra- or intercellular ganglioside movement. Current concepts of ganglioside transport after their synthesis in the cell body are summarized in Fig. 3.

Bibliography

Aquino, D. A., Bisby, M. A., and Ledeen, R. W. (1985). Retrograde axonal transport of gangliosides and glycoproteins in the motoneurons of rat sciatic nerve. *J. Neurochem.* **45,** 1262–1267.

Aquino, D. A., Bisby, M. A., and Ledeen, R. W. (1987). Bidirectional transport of gangliosides, glycoproteins and neutral glycosphingolipids in the sensory neurons of rat sciatic nerve. *Neuroscience* **20,** 1023–1029.

Gammon, C. M., Vaswani, K. K., and Ledeen, R. W. (1987). Isolation of two glycolipid transfer proteins from bovine brain: Reactivity toward gangliosides and neutral glycophingolipids. *Biochem.* **26,** 6239–6243.

Goodrum, J. F., Stone, G. C., and Morell, P. (1989). Axonal transport and intracellular sorting of glycoconjugates. *In* "Neurobiology of Glycoconjugates" (R. U. Margolis and R. K. Margolis, eds.), pp. 277–308. Plenum Press, New York.

Harry G. J., Goodrum, J. F., Toews, A. D., and Morell, P. (1987). Axonal transport characteristics of gangliosides in sensory axons of rat sciatic nerve. *J. Neurochem.* **48,** 1529–1536.

Landa, C. A., Defilpo, S. S., Maccioni, H. J. F., and Caputto, R. (1979). Deposition of gangliosides and sialoglycoproteins in neuronal membranes. *J. Neurochem.* **37,** 813–823.

Ledeen, R. W., Aquino, D. A., Sbaschnig-Agler, M., Gammon, C. M., and Vaswani, K. K. (1987). Fundamentals of neuronal transport of gangliosides. Functional implications. *In* "Gangliosides and Modulation of Neuronal Functions" (H. Rahmann, ed.), pp. 259–274. Nato ASI Series, vol. H7. Springer-Verlag, Berlin, Heidelberg.

Ledeen, R. W., Skrivanek, J. A., Nunez, J., Sclafani, J. R., Norton, W. T., and Farooq, M. (1981). Implications of the distribution and transport of gangliosides in the nervous system. *In* "Gangliosides in Neurological and Neuromuscular Function, Development, and Repair" (M. M. Rapport and A. Gorio, eds.), pp. 211–223. Raven Press, New York.

Rosner, H., Wiegandt, H., and Rahmann, H. (1973). Sialic acid incorporation into gangliosides and glycoproteins of the fish brain. *J. Neurochem.* **21,** 655–665.

Yamada, K., Abe, A., and Sasaki, T. (1986). Glycolipid transfer protein from pig brain transfers glycolipids with β-linked sugars but not with α-linked sugars at the sugar-lipid linkage. *Biochem. Biophys. Acta.* **879,** 345–349.

Yates, A. J., Tipnis, U. R., Hofteig, J. H., and Warner, J. K. (1984). Biosynthesis and transport of gangliosides in peripheral nerve. *In* "Ganglioside Structure, Function and Biomedical Potential" (R. Ledeen, R. K. Yu, M. M. Rapport, and K. Suzuki, eds.), pp. 155–168. Plenum Press, New York.

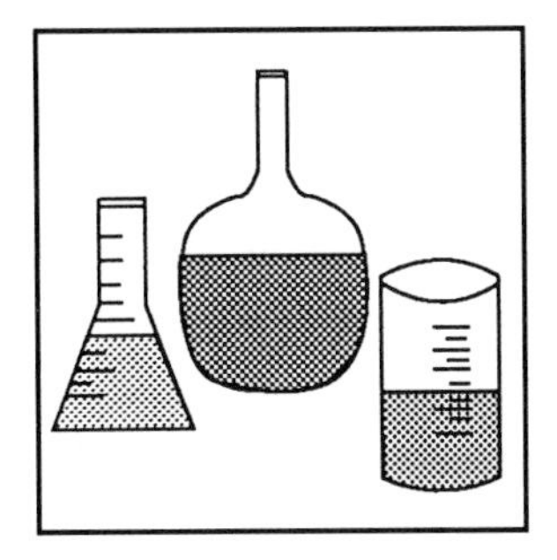

Gas Chromatography, Analytical

WALTER JENNINGS, *University of California–Davis*

Glossary

A Packing factor term of the van Deemter (packed column) equation

***α* (Greek "alpha")** Relative retention. The ratio of the adjusted retention times of any two solutes, measured under identical conditions. Because alpha is never less than 1.0, it is the function of the more retained solute relative to that same function of the less retained solute:

$$\alpha = [t'_{R(B)}]/[t'_{R(A)}] = [K_{D(B)}]/[K_{D(A)}] = [k_{(B)}]/[k_{(A)}]$$

This parameter is termed the "separation factor" by IUPAC

B Longitudinal diffusion term of the van Deemter (packed column) and Golay (open tubular column) equations

***β* (Greek "beta")** Column phase ratio. The volume of column occupied by mobile (gas) phase relative to the volume occupied by stationary (liquid) phase. In the open tubular column,

$$\beta = (r - 2d_f)/(2d_f) \approx 0.5r/d_f$$

The use of d instead of r and of different units for r and d_f are common student errors

c_M, c_S Solute concentrations in mobile and stationary phases, respectively

C Resistance to mass transport (mass transfer) term in the van Deemter equation; C_M and C_S denote mass transport from mobile to stationary, and from stationary to mobile phases, respectively

d Column diameter. Both mm and μm are commonly used; the latter, although consistent with the units used for d_f, implies three significant figures, which is rarely true

d_f Thickness of the stationary phase film, usually in μm

D Diffusivity; D_M and D_S represent diffusivities in mobile and stationary phases, respectively

ECD Electron capture detector

f_1, f_2 Gas compressibility (pressure drop) correction factors

F Volumetric flow of the mobile phase, usually cm^3/min. IUPAC uses this same symbol for "nominal linear flow" and F_c for volumetric flow; ASTM uses F_a for "gas flow rate from column"

FID Flame ionization detector

FPD Flame photometric detector

GC/FTIR Gas chromatography/Fourier transform infrared spectroscopy. Coupled instrumentation in which bands sequentially eluted from the gas chromatograph are subjected to vapor phase infrared spectroscopy

GC/MS Gas chromatography/mass spectrometry. Coupled instrumentation in which bands sequentially eluted from the GC are introduced to a mass spectrometer

h Length of the column equivalent to one theoretical plate. Less precisely termed "height equivalent to a theoretical plate" (which accounts for the ASTM symbol "HETP"), this value is obtained by dividing the column length by the theoretical plate number and is usually expressed in mm:

$$h = L/n$$

When measured at $u_{opt'}$ the result is termed "h_{min}"

I.D. Inner diameter of the column

k Partition ratio. The ratio of the amounts of a solute in stationary (liquid) phase and mobile (gas) phase, which is equivalent to the ratio of the times the solute spends in the two phases. Because all solutes spend t_M time in mobile phase

$$k = [t_R - t_M]/t_M = t'_R/t_M$$

ASTM prefers the term *capacity factor*; neither nomenclature uses the symbol k', which must be considered archaic

$\boldsymbol{K_D}$ Distribution constant. The ratio of the concentrations of a solute in stationary and mobile phases:

$$K_D = c_S/c_M$$

ASTM uses the symbol K and the term *partition coefficient* for this parameter

L Length of the column

n Theoretical plate number.

$$n = [t_R/\sigma]^2$$

where σ is the standard deviation of the peak

$\boldsymbol{n_{req}}$ Number of theoretical plates required to separate two solutes of a given alpha and given partition ratio to a given degree of resolution

NPD Nitrogen phosphorus detector. These are of two types: in the thermionic type, a bead of an alkali metal is heated by the lighted flame of a conventional FID; the plasma type employs an electrically heated bead in an unlighted detector

O.D. Outer diameter of the column

r Radius of the column

$\boldsymbol{R_s}$ Peak resolution. A measure of separation as evidenced by both the distance between the peak maxima and the peak widths. Both IUPAC and ASTM definitions are based on w_b measurements, which must be determined by extrapolation. If peaks are assumed to be Gaussian.

$$R_s = 1.18[(t_{R(B)} - t_{R(A)})/(w_{h(A)} + w_{h(B)}]$$

σ (Greek "sigma") Standard deviation of a Gaussian peak

$\boldsymbol{t_M}$ Gas hold-up time. The time (or chart distance) required for the elution of a nonretained substance (e.g., mobile phase). *Hold-up time* and *hold-up volume* are preferred to the terms *dead time* and *dead volume*

$\boldsymbol{t_R}$ Retention time. The time (or distance) from the point of injection to the point of the peak maximum

$\boldsymbol{t'_R}$ Adjusted retention time. Equivalent to the residence time in stationary phase, this is determined by subtracting the gas hold-up volume from the solute retention time:

$$t'_R = t_R - t_M$$

This should not be confused with the term *corrected retention time*, which is defined differently

u Average linear velocity of the mobile phase:

$$u = L(\text{cm})/t_M(\text{sec})$$

v Linear velocity, usually of the solute band

V Volume; V_M and V_S represent volume of mobile and stationary phases, respectively

$\boldsymbol{w_b}$ Peak width at base. This is determined by measuring the length of baseline defined by intercepts extrapolated from the points of inflection of the peak. Equivalent to four standard deviations (σ) in a Gaussian peak

$\boldsymbol{w_h}$ Peak width at half height. Measured across the peak halfway between the baseline and the peak maximum, this can be determined directly without extrapolation. In Gaussian peaks, equal to 2.35 standard deviations (σ)

ANY ANALYTICAL PROCEDURE, whether qualitative or quantitative, requires the separation or differentiation of the components of interest; chromatographic techniques, with their tremendous potential for achieving separations, repetitively partition the materials to be separated between two phases, one of which is usually stationary, the other mobile. In gas chromatography (GC), the mobile phase is a gas, and the stationary phase is most commonly a liquid, although solid absorbants are extremely useful for certain special separations. GC is the most powerful of the separation sciences, but as the sample components must spend some of their time in the gas phase, it demands vaporization of the sample components. Hence it may be unsuitable for materials that are altered by heat (including many biological materials). Nonvaporizing techniques such as liquid chromatography (LC) or supercritical fluid chromatography (SFC) would appear to be more attractive for such compounds than GC. But lower mobile phase diffusivities and detector limitations create special problems in both LC and SFC, and SFC has so far proved unsatisfactory for the separation of polar compounds. As compared with these other chromatographic methods, GC generally offers so many advantages (e.g., speed, sensitivity, separating power, convenience) that we have extended its use to these thermally fragile compounds through, for example, derivatization techniques that enhance the volatilities of higher boiling solutes.

In addition, GC differs from other analytical techniques in that its immense powers of separation, coupled with the apparent simplicity of the process, permit the generation of usable data by even unskilled analysts. Although this untrained user can generate useful information, much of the immense separating potential of this inherently powerful technique has usually been sacrificed to the limited knowledge of this casual analyst. More knowledgeable analysts can achieve better separations of that same mixture in shorter times and at higher sensitivities by (1) selecting proper design parameters (e.g., the instrument and the column), and (2) optimizing operational parameters (e.g., temperature conditions and gas flows).

Practicing chromatographers generally belong to one of two groups, which can be differentiated by the diversity of their analytical efforts. One group deals with the routine analysis of similar samples (e.g., the qualitative and quantitative analysis of prostaglandins, or the analysis of C-18 fatty acids); members of this group can optimize the GC system to the highest degree because they can begin by selecting optimal "design parameters." The most critical of these is usually the selection of the column (L, d, d_f, stationary phase)[1] that is ideal for a given analysis. The other group handles many different types of samples through the course of the day; these may range from the detection of drug metabolites in blood or urine to the quantitation of respiratory gases. Members of this group generally prefer to leave a "standard" column (30 m × 0.25 mm, d_f 0.25 μm) in position, because it is the most satisfactory for these varied applications. Both groups can benefit by adjusting "operational parameters" [primarily flow rate (or u), and temperature parameters] to improve a difficult separation on the one hand or to trade excessive resolution for shorter analysis times and higher sensitivities on the other.

I. Design Parameters

A. Column Selection

Gas chromatographers sometimes refer to the column as "the heart of the instrument." Certainly it is THE vital organ; all the other devices service this column. The injectors are designed to introduce a short plug of sample to the column, the detectors note the emergence of solutes from the column, and the chromatographic oven plays the vital role of controlling the column temperature (and hence, solute volatilities). Although it is possible to generate poor results even with a good column, it is impossible to generate good results with a poor column—the column is all important to successful analytical results in GC.

Two extreme column types are used: packed, and open tubular. The latter are more economical to use, in that they can achieve superior separations in shorter times at higher sensitivities, and their greater inertness lends increased credence to qualitative and quantitative aspects of the analysis. Most of our considerations, although pointed specifically at the open tubular column, are also at least partially applicable to the packed column (discussed later).

Those factors that influence column selection are apparent from the resolution equation:

$$n_{\mathrm{req}} = 16Rs^2[\alpha/(\alpha - 1)]^2[(k + 1)/k]^2 \qquad (1)$$

These various terms relate to

1. column length (affecting n)
2. column diameter[2] (affecting n and β)
3. stationary phase (affecting K_D, k, and α)
4. phase ratio of the column[2] (affecting k)
5. temperature of operation (affecting K_D, k)

The first four of these become fixed with the selection of the column; the last is under the control of the operator and is designated an "operational parameter." Other operational parameters that will affect results and that should be optimized (but that are not reflected in the resolution equation *per se*) are the linear velocity of the carrier gas (u) and the shortness of the band introduced to the column (injection efficiency).

The resolution equation can also take the form

$$Rs = 1/4\ n\ [(\alpha - 1)/\alpha][k/(k + 1)] \qquad (2)$$

As the resolution requirement increases (larger values of Rs), larger values of n, α, and/or k are necessitated; resolution (Rs) of 1.5 will usually achieve baseline separation and results in a resolution multiplier in Equation 1 of 36. Rs 0.75 may be sufficient for qualitative analysis (i.e., in establishing that there are two compounds at that point); this first multiplication factor then becomes 9 instead of 36, and the analysis requires only one-fourth as many theoretical plates. Although the resolution equation deals with only two compounds, the approach is

1. See Glossary for symbols and nomenclature.

2. Because $\beta = r/2d_f$, either β, or r and d_f should be specified.

applicable to more complicated mixtures through selection of the most difficult pair. If those components are separated, everything is separated.

There are two routes for manipulating n (i.e., increasing column length and decreasing column diameter), but either of these necessitates higher column inlet pressures and generates other difficulties. Solute partition ratios (k) should be neither too small nor too large; values that are too small demand more theoretical plates to achieve separation (the factor $[(k + 1)/k]^2$ in Equation 1 becomes large), and those that are too large lead to long analysis times and lower sensitivity. Ideally, k should fall in the range of about 2–7 and is manipulated through the choice of stationary phase, column temperature, and column phase ratio. Relative retentions (alpha) are affected by temperature, but those effects are not always predictable; the most useful route for manipulating alpha is the choice of stationary phase. Recent developments in low thermal mass connectors that simplify column coupling have greatly extended opportunities for precise manipulation of alpha.

B. Operational Parameters

Even the proper column can give poor results; attention must also be directed to proper column installation, to the choice of carrier gas, to the velocity (u) of that carrier gas,[3] to the sample introduction technique, and to the temperature profile to which the column is then subjected.

II. Selecting the Column

A. Packed Versus Open Tubular

Packed columns are filled with a granular material coated with a film of the stationary (liquid) phase and are still widely used despite the fact that compared with the open tubular column, they deliver inferior separations in longer times at lower sensitivities, and results are quantitatively and qualitatively less reliable. Although the initial purchase price of the packed column is significantly lower than that of the open tubular column, the lower cost per analysis of the latter soon erodes what would initially appear to be a cost advantage. The continued use of the packed column can be ascribed to (1) user inertia, and (2) their "user friendliness," which is largely attributable to their higher flows of carrier gas. Large diameter open tubular columns have done much to displace these outmoded packed columns, because they too will tolerate large gas flows, can be used with packed column injectors, and in short, display this same "user friendliness." The modern "Megabore™" column uses state-of-the-art technology, much of which is proprietary, to chemically bond the cross-linked stationary phase to the inner periphery of fused silica tubing, 0.53 mm I.D. and 10–30 m in length. Although less efficient than true capillary columns (0.18–0.35 mm I.D.), they are capable of developing approximately 2,000 theoretical plates per meter; hence a 30-m Megabore delivers at best about 60,000 theoretical plates. Because few packed columns can deliver more than 5,000 theoretical plates, even a short Megabore operating under nonoptimal conditions will out-perform the packed column.

3. Because u varies with T, complex separations can be improved by optimizing pressure/flow conditions to provide the optimum u for the average k of the most critical region when it is about halfway through the column.

B. Length and Diameter

Ideally, the column should be long enough to deliver the required number of theoretical plates, and no longer. Longer columns suffer two disadvantages: analysis times become long (solutes must traverse more column at a lower velocity per unit length). Longer columns also require higher inlet pressures, and for theoretical reasons that are beyond the scope of this offering, this may imperil the overall separation of complex mixtures. The requisite length is largely determined by the column diameter. For more difficult separations, smaller diameter columns may be required; as the column diameter decreases, many of the advantages of the open tubular system (i.e., improved separation, shorter analysis times, higher sensitivities) can all be realized to an even greater degree, but the demands for excellence in instrument design and operator technique both increase.

III. Selecting the Stationary Phase

The factors that should influence the choice of stationary phase have been discussed in detail elsewhere. Because the lower polarity methyl silicone columns are somewhat more tolerant of abuse and usually exhibit longer lifetimes, they are the column of choice in most cases. Where separation is difficult because solutes have similar relative retentions (see Equations 1,2), a more selective stationary phase may be required.

1. Stationary Phase Film Thickness, d_f

Thin-film columns (d_f 0.1 μm) are useful for the analysis of higher boiling solutes but have limited capacity; "standard" columns (d_f 0.25–0.35 μm) offer the most useful compromise between loadability and resolution. Thick-film columns are useful for the analysis of highly volatile materials but possess several disadvantages for more normal solutes. Most reputable column manufacturers maintain technical support personnel whose considerable experience in a variety of separations can be of immense benefit to customers in column selection.

IV. Special Problems Created by Biological Samples

Biological samples cover the range from respired gases to thermally labile high boiling solutes dispersed through complex matrices. Hence one area for consideration becomes that of sample preparation, whereas another becomes that of the analysis *per se*. Sample preparation assumes special importance, because nonvolatile sample residues are a primary cause of premature column failure, and biological samples obtained from sera and other tissues are frequently contaminated with nonvolatile materials. Even a slight accumulation of such substances can lead to increased detector signal, which is often perceived as "column bleed." Where the mass spectrometer is used as a detector (GC/MS systems), spectra of the molecules comprising the "bleed" show that they are degradation products of the stationary phase. It is now recognized that silicone stationary phases are degraded by oxygen or by sample residues, and the degradative reactions are temperature-dependent. At higher temperatures, silicone fragmentation results when nonvolatile residues attack the silicone. Troubleshooting these cases is facilitated by subtracting all peaks that can be attributed to silicone and concentrating on what is left [i.e., the remaining (nonsilicone) ions are often indicative of the causative agent].

A major advantage of columns with cross-linked surface-bonded stationary phases is that they can be rinsed with a variety of solvents, and sample residues can often be removed without damage to the column. It is preferable, of course, to employ cleaner samples, and several methods, ranging from liquid–liquid extractions to solid phase extractions, are available for the cleanup of biological samples.

For the analysis of lower boiling compounds, nonvolatile sample residues can be avoided by injecting the vapor over the sample (headspace), but such samples are usually extremely dilute and may require a preconcentration procedure. Larger volumes of these gaseous samples can be passed through a small quantity of a solid that adsorbs the entrained volatiles. These are then desorbed to the chromatographic column in a much reduced volume of carrier gas (purge-and-trap). Headspace samples are normally dominated by the more volatile (i.e., low k) solutes. From the relation $K_D = \beta k$, which can also be expanded to the form $K_D = 0.5r/d_f\,k$, it becomes evident that solute partition ratios can be increased through the use of more retentive stationary phases or adsorptive-type columns, operated at lower temperatures (both of these lead to larger values for K_D), and by smaller diameter columns and/or thicker films of stationary phase (lower β). Although each of these approaches has limitations, they can usually be judiciously manipulated to accomplish our goals.

These same parameters must be manipulated in the opposite direction in the case of more strongly retained (and higher molecular weight) compounds. To decrease the partition ratios of these large k solutes, we resort to larger diameter columns coated with thinner films of less retentive stationary phases (usually the methyl silicones), operated at higher temperatures. In this latter case, we have one other variable we can employ. Because $k = t'_R/t_M$ and $u = L/t_M$, we can see that $t'_R = K_D[2d_f\,L]/[ru]$, and it becomes evident that elution times can be shortened by increasing the carrier gas velocity.

V. Component Identification

Gas chromatography is often used as a criterion for compound identification. It is not always recognized that GC retentions establish not what a solute is, but what it cannot be; it cannot be any number of other compounds whose behavior can be shown to be different on that column under those conditions. High-resolution (capillary) columns merely narrow the possibilities through the exclusion of many more compounds. If the sample is known to consist of only methyl esters of fatty acids and if the column employed is capable of separating all such esters under the conditions employed, retention behavior becomes a reasonable criterion of identification. The agreement of GC retentions between a known compound and a given solute on two columns of different selectivity strengthens an assignment to some degree, but assignments on both columns are still based on solute vapor pressures. For assigning identifications to the components of more complex or unknown mixtures, it is preferable to follow the GC separation with a second mode of solute characterization that is based on entirely different solute properties. GC/MS, in which solutes are first differentiated on the basis of their volatilities and second on the basis of their mass spectral fragmentation patterns, is extremely useful in this regard. Similarly, GC/FTIR, in which the solutes are first differentiated on the basis of volatilities and second on their infrared absorption patterns, lends increased credence to identifications. The possibility that two different solutes will exhibit precisely the same GC retention AND the same MS fragmentation pattern (or absorb at the same infrared frequencies) is extremely remote.

Bibliography

Jennings, W. (1987). "Analytical Gas Chromatography." Academic Press, New York, London, San Francisco.

Jennings, W., and Rapp, A. (1983). "Sample Preparation for Gas Chromatographic Analysis." Huethig Publishing, Heidelberg.

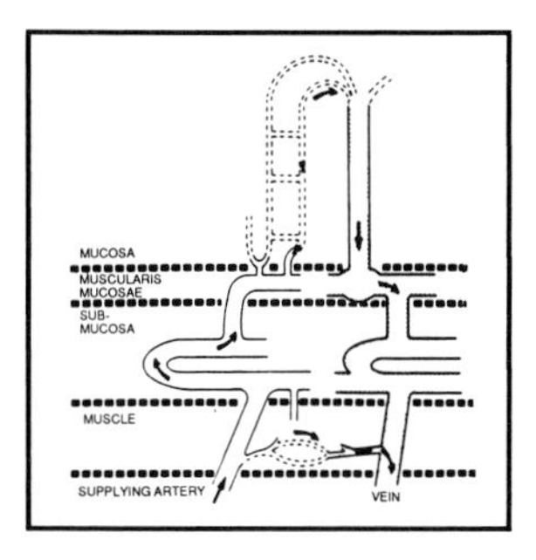

Gastric Circulation

EDWARD H. LIVINGSTON AND PAUL H. GUTH, *VAMC, UCLA, and Center for Ulcer Research and Education, Los Angeles*

Glossary

Acid secretion Metabolically active process of the stomach that secretes strong acid into the lumen, assisting in the digestion of food

Artery Vessel supplying oxygenated blood from the heart to a tissue

Capillary Small thin-walled vessel located between the arterial and venous sides of the circulation; where the exchange of oxygen and nutrients occurs

Erosion Damage confined to the mucosa and not extending through the muscularis mucosa

Mast cells Cells within the gastric mucosa that contain histamine. When histamine is released, it stimulates gastric acid secretion

Microcirculation Network of small arterioles, venules, and capillaries that are closely interfaced with the body cells. It is at this level of the circulation that nutrients and oxygen are exchanged and blood flow is controlled.

Mucosa Mucous membrane consisting of a specialized epithelial layer on the luminal side of the gastric wall

Muscularis externa Layer of smooth muscle on the outer portion of the stomach wall which is invested with a thin layer of epithelium known as the serosa

Muscularis mucosa Thin layer of smooth muscle at the base of the mucosa (between the mucosa and the submucosa)

Portal vein Vein draining blood from the stomach and the intestines into the liver. From the liver the blood enters the vena cava and then travels to the heart.

Sphincter Circularly oriented groups of smooth muscles surrounding various portions of the gastrointestinal tract (e.g., the pylorus) which contract or relax to regulate the diameter of the structure at that location. These serve to regulate the passage of contents at their locations.

Submucosa Loose connective tissue just beneath the mucosa

Ulceration Local area of tissue destruction in the stomach extending through the entire mucosa, including the muscularis mucosa. When perforation occurs, the damage extends through the entire wall of the stomach.

Vein Vessel draining blood from a tissue and connected to increasingly larger veins that ultimately enter the heart

GASTRIC circulation is the system of arteries, veins, and capillaries that provides for the flow of blood to, through, and out of the stomach. The stomach is a major digestive organ in the gastrointestinal tract. It lies between the esophagus and the duodenum in the abdominal cavity. The proximal stomach serves two major functions. The fundus is the temporary storage portion and stretches to accommodate incoming food. The relaxation of this portion with feeding is known as receptive relaxation and is mediated by the vagus nerve. As the stomach empties, the fundus gradually contracts to its original size. Digestion of food into particles of a size that can be absorbed by the intestine first begins with the act of chewing, breaking up the largest of particles. Food particles entering the stomach are large and must be broken down further before they enter the small bowel. The fundus and the midportion of the stomach, known as the corpus, secrete large amounts of highly concentrated (i.e., up to 0.15 N HCl) acid from its parietal cells. Another population of cells, the chief cells, secrete a proteolytic enzyme, pepsin, which, in combination with the acid, breaks down the food particles chemically. The antrum is the muscular distal portion of the stomach, which is capable of generating powerful contractions that break up the food particles by mechanical means. When the particles are small enough they are "pumped" by the antral contractions from the stomach into the duodenum.

With all of these functions to perform, the stomach necessarily has a high metabolic demand and therefore is richly vascularized. Blood supplies the necessary oxygen and nutrients to support the metabolic activity required for the secretory and contractile functions. Acid secretion requires the delivery of chloride and carbon dioxide to generate the hydrochloric acid that is secreted. Blood also carries away the metabolites accumulated from these processes. On average the stomach secretes about 1 liter of fluid per day, all of which is derived from the blood. This secretion is needed to facilitate the partial digestion and passage of ingested food distally into the gastrointestinal tract.

I. Anatomy

Soon after the esophagus traverses the diaphragmatic hiatus, it enters the abdomen and terminates at the cardia of the stomach (Fig. 1). The stomach lies in the left upper quadrant of the abdomen and extends from left to right. The lower side of the stomach is called the greater curvature and the upper side, the lesser curvature. It fans out to the corpus, then tapers to the antrum and the pylorus. The pylorus is fixed in position on the right side of the abdomen. The fundus, corpus, and antrum do not have attachments anteriorly or posteriorly and are therefore relatively free to move as the stomach fills with food. [*See* DIGESTIVE SYSTEM, ANATOMY.]

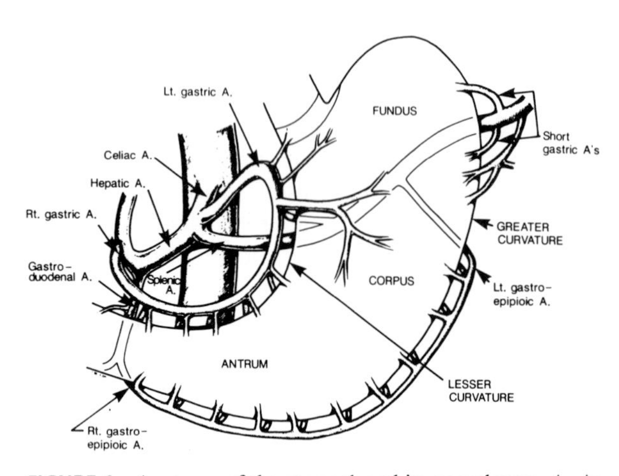

FIGURE 1 Anatomy of the stomach and its vasculature. A, Artery.

The first major branch of the aorta after it enters the abdominal cavity is the celiac axis (Fig. 1). Branches of this artery provide the major blood supply to the stomach. The first branch of the celiac axis is the left gastric artery, which courses from the cardia to the pylorus along the lesser curvature. The left gastric artery supplies the distal esophagus, cardia, and the superior aspect of the corpus, antrum, and pylorus. The right gastric artery originates from the common hepatic artery, a branch of the celiac axis. This artery is found on the superior surface of the pylorus, where it anastomoses with the left gastric artery and supplies the superior pylorus. Both left and right gastric arteries send branches to the anterior and posterior walls of the stomach. The inferior portion of the stomach is supplied by the gastroepiploic arteries. The left gastroepiploic artery branches from the splenic artery, the second branch of the celial artery. This artery follows the greater curvature, anastomosing with the right gastroepiploic artery, which courses along the right side of the greater curvature, originating from the gastroduodenal artery, a branch of the common hepatic artery. The fundus is supplied by the short gastric arteries which branch from the splenic artery in the region of the splenic hilum.

The vascular supply of the stomach has a large number of interconnections, assuring that occlusion of one or several arteries will not lead to gastric ischemia. The right and left gastric arteries anastomose at the antropyloric area, the right and left epiploic arteries anastomose along the greater curvature, and the branches of the gastric and gastroepiploic arteries anastomose with each other along the anterior and posterior aspects of the stomach. Within the stomach all of these vessels terminate in the submucosal arteriolar plexus, which connects all of the supplying vessels together. The gastroepiploic arteries anastomose with branches of the superior mesenteric artery, assuring vascular supply even if the celiac axis blood flow were to become compromised.

The veins of the stomach follow the course of the arteries, but terminate in various branches of the portal vein. The proximal stomach drains into the splenic vein, and the distal stomach empties into the superior mesenteric vein.

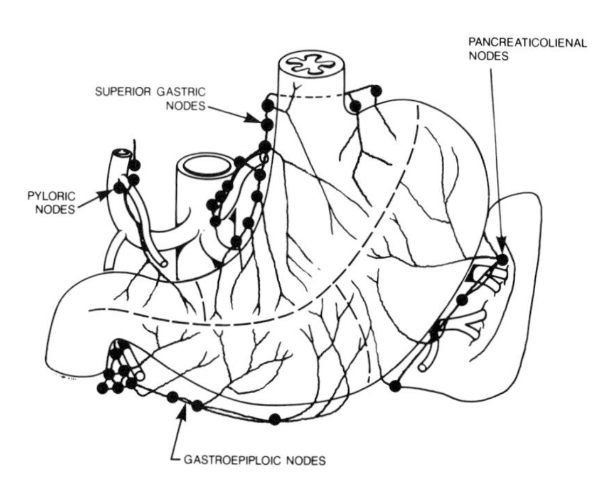

FIGURE 2 Lymph drainage of the stomach.

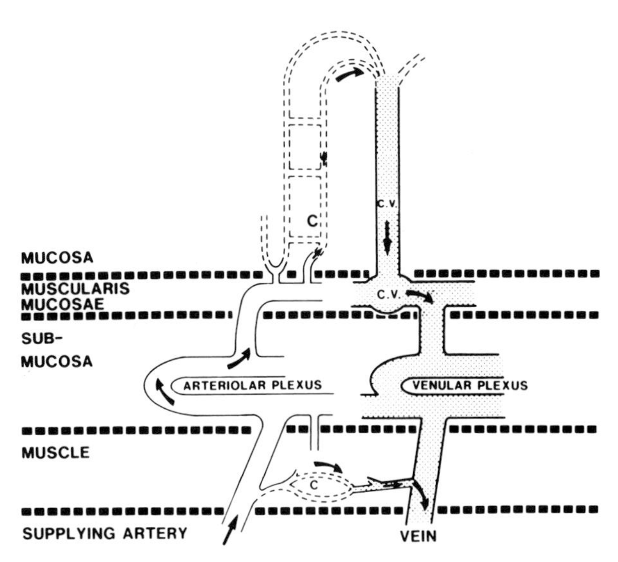

FIGURE 3 Gastric microcirculation. C, Capillaries; C.V., collecting vein.

The interstitial spaces are drained by the lymphatics, which follow the course of the arteries, coalesce, and enter the lymph nodes (Fig. 2). The majority of the stomach is drained by lymphatics, which cluster about the left gastric artery, known as the superior gastric lymph nodes. The second largest area is drained by the right gastroepiploic nodes and the pyloric nodes. The fundic region drains into the left gastroepiploic nodes and the pancreaticosplenic lymph nodes. A small portion of the pylorus and the antrum is drained by the right gastric lymph nodes.

II. Microvascular Anatomy

The arteries supplying the stomach course along the external surface of the stomach and eventually branch into small vessels which perforate the muscularis externa (Fig. 3). As these small arteries travel through the muscle layer, they branch into smaller arteries, which supply the muscularis externa. The arteries then terminate in the submucosal arteriolar plexus. The submucosal plexus of arterioles (i.e., very small arteries) is a network of highly interconnected arcades of vessels. Vessels branch from this network, enter the mucosa, and terminate at the base of the mucosa in capillaries which traverse the mucosal layer and enter collecting venules in the luminal one-third of the mucosa. These venules traverse the mucosa and enter a plexus of venules (i.e., very small veins), which architecturally resemble the arterial plexus. Veins from the plexus join draining veins from the muscle layer, traverse the muscularis externa, and coalesce to form the large veins of the stomach.

The capillaries of the corpus mucosa originate from arterioles at the base of the mucosa and extend to the luminal portion of the mucosa, entering draining venules. The capillaries form a network that surrounds the acid- and pepsin-secreting glands. Characteristic of the gastric mucosal capillaries is that they originate from the submucosal arterioles, enter the mucosa, travel the entire thickness of the mucosa, and then turn at the mucosal surface to enter the collecting venules in the upper one-third of the mucosa.

The blood supply to the stomach is highly redundant, not only because of the multiple interconnecting arteries supplying it, but also because of the rich network of anastomosing arterioles in the submucosa. This ensures adequate blood supply for the stomach even if one or more vessels should become damaged. One can envision an evolutionary advantage for the redundant blood supply: Blood supply is critically important for mucosal resistance against the corrosive effects of luminal acid. Should the blood supply be interfered with, severe mucosal damage and even life-threatening hemorrhage could occur. The highly redundant blood supply guards against this possibility.

III. Physiology

A. Methods of Study

To assess the contribution of blood flow to various physiological and pathological conditions, it is necessary to accurately measure it. In experimental animals the amount of blood passing through some unit volume of tissue per unit time can be determined by measuring all of the venous drainage. This would give the total volume flow for all layers of the tissue. Flow in different tissue compartments (i.e., the mucosa, submucosa, or the muscularis externa) differs, depending on their individual metabolic demands. The majority of flow is distributed to the mucosa, which has the highest metabolic activity. Flow alterations in the mucosa might lead to pathological conditions such as ulcers. The available methods applicable to blood flow measurement in the human gastric mucosa are based on the clearance or extraction of a tracer, determination of flow velocity through capillaries, or measurement of the amount of hemoglobin in the tissue's vessels.

1. Clearance Techniques

Clearance techniques are based on the clearance of a tracer from the blood into the tissue or vice versa, whose concentration in the tissue can be measured. For a freely diffusible tracer (i.e., one that is soluble in blood and diffuses out of the blood into the tissue through the capillaries such that it rapidly achieves equilibrium between tissue and blood) the amount of tracer extracted from the blood by one pass through the capillaries is proportional to the flow (volume of blood per unit time) and the concentration difference

$$Q_t = F(C_a - C_v)dt$$

where Q_t is the quantity, in moles, of tracer in the tissue that changes over time; F is the flow, in milliliters per second; C_a is the concentration of tracer in the supplying arterial blood, in moles per liter; C_v is the concentration of tracer in the venous blood; and dt is the time interval, in seconds, over which the experiment is performed. The quantity of a material per unit volume in which it is dissolved is its concentration:

$$C_t = Q_t/V$$

where C_t is the concentration of the tracer in the tissue, in moles per liter; Q_t is the total quantity of substance, in moles, in the volume of tissue V, in liters. Combining the above two equations yields

$$C_tV = F(C_a - C_v)dt$$

and

$$C_t = F(C_a - C_v)dt/V$$

At equilibrium the ratio of concentration of the tracer in venous blood and in tissue is defined by the partition coefficient λ:

$$\lambda = C_v/C_t$$

Substituting

$$C_t = F(C_a - C_t\lambda)dt/V$$

for the clearance of a tracer from previously saturated tissue (after the gas is removed from the inspired air), the C_a falls to 0:

$$C_t = F(-C_t\lambda)dt/V$$

The solution of this differential equation is

$$C_t = C_0e^{-\kappa t}$$

where C_o is the tissue concentration of tracer at complete tissue saturation, and κ is $\lambda F/V$.

Thus, by measuring the rate constant for tissue clearance of the tracer (or uptake), the blood flow per unit volume can be determined. For determination of the constant, the tracer must be measured continuously over time. The presence of a tracer can be determined by radioactive decay or electrochemical reaction.

^{85}Kr has been used for the measurement of human gastric blood flow by injecting the tracer intraarterially until the stomach tissue was saturated with tracer. Upon reaching tissue saturation, the injection was stopped and the disappearance of tracer was determined by observation of radioactive decay.

Hydrogen gas concentration can be measured by the electrolytic oxidation of molecular hydrogen at the surface of a platinum electrode. The hydrogen gas clearance technique is a polarographic technique in which current (due to electrons released as the hydrogen is oxidized) passing through an electrode held at constant voltage is proportional to the concentration of the hydrogen being oxidized at the electrode surface. This technique measures flow in a small area of the mucosa around the electrode surface.

2. Partition Techniques

Partition techniques are based on the principle that weak bases (i.e., with pKa's of 5–10) remain uncharged in the pH 7.4 environment of the blood. The uncharged molecules diffuse freely across the plasma membranes of the gastric cells. Upon reaching the acid environment of the lumen, they ionize (i.e., become charged) and lose their lipid solubility.

Consequently, they partition in the gastric lumen. The rate of luminal accumulation in the face of a constant plasma concentration of the tracer is proportional to the gastric blood flow. Aminopyrine, a weak base measurable by fluorescence or labeling with ^{14}C, was used extensively for many years for gastric blood flow determination. However, it was found to be actively secreted by parietal cells, leading to overestimation of blood flow in the presence of stimulated acid secretion. Partition techniques have been largely replaced by inert gas clearance techniques.

3. Laser Doppler Technique

The laser Doppler technique is based on the Doppler principle. If the tissue is illuminated with light of a uniform frequency, the reflected light will have a shift in frequency that is proportional to the velocity of cells moving in the field of observation. The velocity can be determined from the frequency (i.e., color) shift, and if the volume of tissue under study is known, the blood flow can be derived. This technique has the advantage that the incident and reflected light can be carried within fiber-optic light guides. The fiber-optic cable can be narrow and passed through an endoscope to measure gastric mucosal blood flow in humans. The measurement can be continuous over time. The major difficulties of the technique are failure to maintain adequate optical coupling between the tissue and the probe, resulting in motion artifact and, once coupling is established, the possibility of excessive pressure, causing mucosal ischemia (i.e., stoppage of blood flow). In addition, the resulting blood flow measurement is in volts, and, as yet, there is no valid way to convert this into blood flow units (i.e., milliliters per minute per gram of tissue).

4. Reflectance Spectrophotometry

In this technique white light illuminates the tissue under study, and the reflected light is analyzed spectrophotometrically. By measuring the intensity of the reflected light at the wave lengths characteristic for hemoglobin and oxygenated hemoglobin, the relative concentration of hemoglobin and the fraction of it that is oxygenated can be measured. These measurements are not absolute and only provide indices of the tissue's hemoglobin content and hemoglobin oxygen saturation. However, the hemoglobin concentration is reduced with diminished blood flow and increases when blood flow increases. Similarly, the proportion of hemoglobin that is oxygenated is reduced with reduced blood flow. As with the laser Doppler technique, the incident and reflected light can be carried along fiber-optic light guides and used for measurement of human gastric mucosal blood flow through the endoscope. The same limitations apply: The optical coupling is difficult to maintain (although motion artifacts are not so great as with the laser Doppler), and if too much pressure is applied to the end of the fiber-optic cable, blood flow in the tissue is stopped.

B. Intrinsic Regulation

Blood flow to the stomach must supply needed oxygen and nutrients to the cells as well as carry away metabolic waste and back-diffusing acid. Therefore, systems regulating the amount of blood flow entering the tissue must be sensitive to the metabolic needs of the tissue or to the presence of acids or waste. Intrinsic regulation of blood flow involves the mechanisms by which changes in tissue perfusion are triggered by locally occurring events. This is different from extrinsic regulation, which is the control of blood flow by mechanisms which function primarily outside the stomach.

The fundamental mechanism for the regulation of blood flow entering a capillary bed is the control of arteriolar diameter. It is at the arteriole that the major resistance to flow is encountered. Arterioles have the capability of completely closing or dilating widely. These vessels are sensitive to the effects of the vasoactive agents discussed Section III,C and IV. The importance of these vessels in the regulation of flow entering the capillary bed can be inferred from Poiseuille's law:

$$Q = \frac{\pi \Delta P r^4}{8 \nu l}$$

which states that the flow through a vessel (Q, in milliliters per second) is equal to the constant π times the change in pressure (ΔP) across the vessel times the radius of the vessel raised to the fourth power (r^4), all divided by the product of eight times the viscosity constant for blood (η) times the length of the vessel (l). From this equation it is readily apparent that the amount of blood flow through a vessel is exquisitely sensitive to the diameter of the vessel. Thus, the arterioles, which are capable of large diameter changes, are the single most important regulator of the amount of blood entering the tissue.

A second potential mechanism for the control of

blood flow is change in the number of capillaries receiving blood flow. In most tissues this is accomplished by opening or closing of the precapillary sphincters. At any given time a certain fraction of the capillaries is perfused. To increase the delivery of oxygen and nutrients, more sphincters are opened, perfusing a greater number of capillaries, thereby increasing the surface area available for capillary exchange. However, no evidence for this type of mechanism exists for gastric microcirculation. In the stomach submucosal arteriolar diameter changes have the most significant regulatory influence on blood flow.

The blood flow into the gastric tissue, and therefore arteriolar diameter, must be sensitive to a variety of factors. An important role of blood flow is to provide oxygen to the tissues. It has been hypothesized that one mechanism for the control of blood flow is the effect of oxygen on the arterioles. Contraction of the arterioles is an active metabolic phenomenon dependent on the presence of oxygen. If tissue levels of oxygen decrease, less would be available to support the arteriolar contraction. In this setting the arterioles would dilate, increasing the perfusion of the tissue. This sort of mechanism probably does not exist in the stomach, because of the anatomic arrangement of the vessels. In the stomach the arterioles that regulate mucosal blood flow are in the most basal portion of the mucosa, up to 1 mm away from the superficial mucosa. The likelihood that precapillary arterioles can detect the oxygen demands of tissue that far away is small. Similarly, no evidence exists that deficits of specific nutrients required for tissue metabolism can trigger increases in blood flow. Most likely, the response to gastric mucosal needs for blood flow results from the accumulation of products, rather than the substrates of metabolism.

One of the more likely regulators of blood flow is adenosine. Adenosine is a nucleotide that transports energy within cells by carrying phosphate groups. Utilization of energy occurs by phosphorylated adenosine's releasing its phosphate groups. As a consequence, when cells are metabolically active, adenosine accumulates inside and outside the cell. Most studies of the relationship between adenosine and blood flow regulation have been performed in the heart. It appears that within the cardiac circulation adenosine serves as the dominant regulator of blood flow. During conditions of low oxygenation or increased cardiac work, adenosine accumulates in the interstitial space, resulting in vasodilatation and increased blood flow.

Adenosine appears to be a potent arteriolar dilator in all tissues studied. Studies of gastric circulation confirm that adenosine probably contributes to the regulation of gastric mucosal blood flow. Topical application of adenosine to gastric microvessels results in gastric mucosal hyperemia (i.e., an increased amount of blood). Increases in acid secretion lead to large increases in metabolic demand, concomitant with large increases in gastric mucosal blood flow. Treatment with the specific adenosine inhibitor, 8-phenyltheophylline, inhibits the blood flow response to stimulated acid secretion.

Histamine might also participate in the regulation of gastric mucosal blood flow. In the presence of injurious agents, histamine is released by mucosal mast cells. Histamine applied locally to the stomach results in increased mucosal blood flow and increases capillary permeability. It has been hypothesized that when the mucosa is damaged, histamine is released and signals the need for increased blood flow. The increased blood flow is needed to neutralize back-diffusing acid and carry away toxic substances.

The eicosanoids are a group of chemicals derived from arachidonic acid. They are divided into two major classes: the prostaglandins, which are synthesized by cyclooxygenase, and the leukotrienes, synthesized by lipoxygenase. Prostaglandins and leukotrienes mediate a variety of physiological responses and are released by nearly all tissues. The different prostaglandins and leukotrienes differ in their side-chain substitutions and have widely varying properties. Some are vasodilatory; others produce vasoconstriction. Capillary permeability can be altered by these compounds. Both prostaglandins and leukotrienes are released by cells within the gastric mucosa and contribute to the regulation of mucosal blood flow.

Diameter changes of the submucosal arterioles control mucosal blood flow. How the mucosa signals the submucosal arterioles to change their diameter in response to mucosal blood flow demands is not clear. A number of substances found within the mucosa are released either as a result of metabolic activity or to initiate processes such as acid secretion which result in augmented metabolic activity. Many of these agents, (e.g., histamine, prostaglandins, and some gut peptides) are vasoactive. Theoretically, they could diffuse from the mucosa to the submucosa and, in turn, regulate mucosal flow by their action on the submucosal vessels. However, the probability that any of these agents could reach the submucosa is low. The interstitial space contains a number of enzymes that deactivate the mediators before they can travel far from where they are released. By rapidly deactivating these highly potent substances, distant unwanted effects of these agents are prevented. One hypothesis concerning the regulation of mucosal blood flow is based on the observation that nerves traverse the mucosa and could provide pathways for reflex control of the submucosal vessels. In this way the mediators found within the mucosa could stimulate mucosal sensory nerves which, in turn, transmit a signal to other nerves that control submucosal arteriolar diameter. There is evidence that this sort of pathway exists for the hyperemic response to injurious agents penetrating into the mucosa.

Aside from regulating the amount of blood flow entering the tissue in response to metabolic needs, the local circulatory system must maintain adequate perfusion in the face of changes that might occur in the systemic circulation. Two intrinsic regulatory phenomena exist to perform this task: autoregulation and reactive hyperemia.

Autoregulation is the ability of the tissue to maintain nearly constant levels of perfusion despite wide variations in the pressure of the supplying arteries. In most tissues the blood flow changes little, despite wide variations in perfusion pressure. It is thought that the smooth muscle of arterioles has the ability to contract in proportion to the pressure exerted against it. Thus, when the perfusion pressure increases, the arteriolar smooth muscle contracts to hold constant the blood flow to the tissue. Another possible mechanism is that concentrations of oxygen or metabolites (e.g., adenosine or carbon dioxide) adjacent to the arterioles regulate their diameter. If perfusion pressure increased and flow through the tissue increased, these metabolites would be washed away at higher rates than during basal conditions, reducing their concentrations and leading to vasoconstriction and a return of flow to basal levels. Conversely, a decrease in perfusion pressure would lead to lower flow through the capillary bed, a rise in the concentrations of the metabolites, and a consequent vasodilatation and return of flow to basal levels.

In the stomach autoregulation has been difficult to establish. Rather than autoregulate, blood flow tends to parallel perfusion pressure in the stomach in intact animals. With sympathetic denervation, however, autoregulation of the stomach approximates that of other tissues. Study of the oxygen extraction of the tissue demonstrates that this parameter is held nearly constant, despite wide variations in perfusion pressure or blood flow. Thus, autoregulation of the stomach occurs not so much as has been described in other tissues with constancy of blood flow in the face of changes in perfusion pressure, but rather as constancy of oxygen delivery to the tissue, despite variations in perfusion pressure.

Reactive hyperemia is the rebound increase in blood flow observed following a period of ischemia. Ischemia can be defined as a deficit of blood supply in relation to the metabolic needs of the tissue. When the restriction of blood supply is removed, there is a short-lived increase in blood flow above basal values. During the period of ischemia, tissue oxygen levels fall and metabolites such as carbon dioxide and adenosine accumulate. These effects serve to dilate the arterioles so that when the restriction of blood flow is removed, the flow is high. As the metabolites are washed away and tissue oxygen content is restored, the arterioles again constrict and flow returns to the basal level.

The relationship between gastric blood flow and acid secretion has been extensively studied. Studies using techniques not affected by the process of acid secretion (e.g., the hydrogen gas clearance technique) have indicated that there is most likely a linear relationship between gastric mucosal blood flow and acid secretion.

One difficulty that arises in the study of the acid secretion–blood flow relationship is that some of the agents used to control acid secretion have direct vasoactive effects. For example, histamine is a potent vasodilator, in addition to its effect on acid secretion. Histamine can potentially overestimate the positive relationship between blood flow increases resulting from stimulated acid secretion, because in itself it increases blood flow by producing vasodilation. When used in too high a dose, histamine obscures the relationship by causing vasodilatation throughout the body, resulting in systemic hypotension, reducing gastric blood flow, while acid secretion remains elevated. Thus, in studying the relationship it is important to account for potential primary vasoactive effects of the agents being studied.

C. Extrinsic Regulation

The distribution of blood flow to different organs is regulated by the nervous system. This is important for the maintenance of homeostasis of the entire organism. The brain requires a constant supply of well-oxygenated blood for normal function. The other organs of the body can tolerate temporary fluctuations in their blood supply. Nervous system control of the blood supply coordinates the blood supply to all of the organs with regard to the overall needs of the entire organism.

A variety of nuclei in the brain and the brain stem serve to control gastric blood flow. These receive inputs from sensory nerves distributed throughout the body and send out fibers that reach the vessels of the stomach. The most important nuclei are those contained within the hypothalamus, which mediate sympathetic nervous system function, and those of the medulla, which regulate the parasympathetic nervous system.

The sympathetic nerve supply to the stomach is derived from the sixth to the 10th spinal nerves and terminates in the celiac ganglion. From the ganglion postganglionic fibers reach the stomach as discrete nerves, as nerves mixed with vagal fibers, or as nerves accompanying the arteries. The arterioles of the submucosal plexus are richly innervated with terminals from these nerves. There is some evidence that the capillaries of the mucosa are also innervated with sympathetic fibers. The sympathetic nervous system is activated to enhance an animal's response in a "fight-or-flight" situation. Therefore, pulse rate and blood pressure rise, blood flow to the brain and muscles increases, and blood flow to the viscera decreases. It is thought that the reduction in visceral blood flow is to provide more blood supply to organs, necessary to respond to environmental threats.

Sympathetic nerves release norepinephrine as their transmitter. Stimulation of the sympathetic nervous system also activates the adrenal medulla, causing it to release epinephrine. These neurotransmittors act on receptors located on the arteriolar smooth muscle. There are two major classes of receptors: α-adrenergic receptors are vasoconstrictive, and β receptors are vasodilatory. Norepinephrine reduces gastric mucosal blood flow by activating the α receptors of the arterioles of the submucosal plexus. Epinephrine can produce either vasodilatation, by activating the β receptors, or vasoconstriction, by its action on α receptors. The activity of these neurotransmitters on the gastric circulation is complex, and study is complicated by the systemic cardiovascular effects of both agents. To reliably identify their actions on the gastric circulation, they must be administered locally by close intraarterial infusion. Studies of both epinephrine and norepinephrine intraarterial infusions have revealed that they initially decrease gastric mucosal blood flow, but then, after several minutes, mucosal blood flow increases. This phenomenon is known as autoregulatory escape. When the sympathetic nerves supplying the stomach are stimulated, the autoregulatory escape is not complete, and the overall effect is the reduction of gastric mucosal blood flow. Teleologically, one can imagine that the purpose of reduced gastric blood flow in the fight-or-flight state is to make the blood available to the brain when the animal is threatened.

The parasympathetic nerve supply to the stomach is derived from the vagus nerve. Stimulation of the vagus releases acetylcholine, which serves to increase gastric motility, acid secretion, and mucosal blood flow. Similar to the adrenergic receptors, cholinergic receptors are located on the smooth muscle cells of the submucosal arterioles. Activation of the receptors with acetylcholine results in vasodilatation. Stimulation of the vagus results in increased gastric acid secretion, which could result secondarily in increased blood flow. However, studies in which the acid secretory response has been blocked have revealed that blood flow increases still occur, indicating a direct neurally mediated vasodilatory response. Furthermore, stimulation of the vagus nerve results in rapid vasodilatation of the submucosal arterioles, too rapid to be secondary to the increased metabolic demand of acid secretion.

In the vagus there are many populations of fibers. Although many of the actions of vagal nerve stimulation can be explained by the action of acetylcholine, only a small percentage of the vagal fibers are cholinergic. Many are adrenergic and others contain peptide neurotransmitters. Stimulation with low-frequency electrical impulses activate low-threshold (i.e., activated with impulse durations of less than 0.5 msec) noncholinergic fibers, and this type of stimulation results in gastric vasodilatation not mediated by acetylcholine. The cholinergic high-threshold fibers, when stimulated, increase gastric mucosal blood flow by a combination of direct neurally mediated vasodilatation as well as by increasing gastric acid secretion.

The visceral afferent nerves are a group of small fibers found in the gastric mucosa that are sensory. They detect changes in the mucosal environment and transmit information to the central nervous system. These fibers contain neuropeptides such as calcitonin gene related peptide which are known vasodilators. Recently, evidence has been obtained indicating that when these fibers are stimulated, they release these transmitters into the tissue, resulting in increased mucosal blood flow.

IV. Pathophysiology

The stomach is continuously exposed to the harshest environment of all of the tissues of the body. The mechanisms involved in the protection against injury appear to hinge on adequate blood flow. Most laboratory studies of gastric injury use agents known to produce gastric mucosal injury in humans (i.e., aspirin, ethanol, and bile acids). When the stomach is exposed to these agents in the presence of hypotension and, therefore, reduced gastric mucosal blood flow, injury is much worse than in animals with normal blood flow. Conversely, if blood flow is increased, the injury is less. Blood flow is needed to supply the necessary oxygen and metabolites used in mounting a protective response.

One important factor in mucosal damage is acid. Agents that damage the mucosa produce more injury in the presence of gastric acid than when administered alone. Acid alone can injure the tissue. Therefore, one important component of mucosal defense is acid neutralization. Blood contains bicarbonate; thus, one function of blood flow in gastric mucosal protection is the delivery of bicarbonate to neutralize and carry away the acid.

It has been hypothesized that the way alcohol, aspirin, or bile salts damage the tissue is by breaking the gastric–mucosal barrier to acid back-diffusion. The barrier is the surface mucosal epithelium, which prevents the entry of luminal substances into the gastric tissue. Once the barrier is broken, acid from the gastric lumen enters the underlying tissues, damaging them. When this occurs, mucosal blood flow increases to neutralize and carry away the incoming acid and thus reduce the amount of damage. Recent studies have indicated that the hyperemia observed in response to barrier-breaking is mediated by sensory nerves. Inhibition of the function of these nerves eliminates the mucosal hyperemic response to barrier-breaking and greatly exacerbates the injury. Thus, adequate blood flow is necessary not only for maintenance of tissue integrity under normal circumstances, but also for protecting the mucosa against exogenous injurious agents.

It has been hypothesized for over 100 years that a blood flow deficit might be the cause of gastric ulceration. However, while vascular disease resulting in reduced gastric blood flow in humans is rare, gastric ulceration is common. Furthermore, as pointed out in Section III, the gastric vascular supply is highly redundant and a reduction in blood flow on a structural basis would require the simultaneous loss of many supplying vessels. If gastric ulceration results from a blood flow deficit, the deficit probably results from a disordered physiological response to the need for increased blood flow. The abnormality probably occurs in a local area, and it is in this area where the ulceration begins.

Bibliography

Gannon, B., Browning, J., O'Brien, P., and Rogers, P. (1986). Mucosal microvascular architecture of the fundus and body of human stomach. *Gastroenterology* **86,** 866–875.

Granger, D. N., and Bulkley, G. (ed.) (1981). "Measurement of Blood Flow: Applications to the Splanchnic Circulation." Williams & Wilkins, Baltimore, Maryland.

Granger, D. N., and Kvietys, P. R. (1981). The splanchnic circulation: Intrinsic regulation. *Annu. Rev. Physiol.* **43,** 409–418.

Guth, P. H., Leung, F. W., and Kauffman, G. L. (1989). Physiology of gastric circulation. *In* "Handbook of Physiology—The Gastrointestinal System I" (J. Wood, ed.). Am. Physiol. Soc., Washington, D.C.

Miller, T. A. (1988). Gastroduodenal mucosal defense: Factors responsible for the ability of the stomach and duodenum to resist injury. *Surgery* **103,** 389–397.

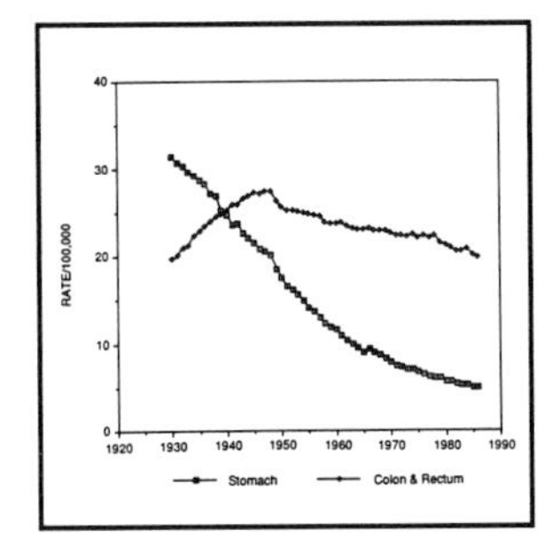

Gastrointestinal Cancer

PELAYO CORREA, *Louisiana State University Medical Center*

Glossary

Dietary fiber Undigestible food residue

Dysplasia Histopathologic pattern of abnormal cell proliferation characterized by loss of polarity and nuclear abnormalities, not invasive but considered precancerous

Foveolar cells Mucus secreting epithelial cells lining the surface and the pits (foveolae) of the gastric mucosa

Goblet cells Specialized epithelial cells secreting acid mucins, which accumulate in the cytoplasm as a globus, which is later discharged

Hamartoma Overgrowth of mature tissues and cells that normally occur in the affected organs

Leiomyoma Benign tumor of smooth muscle

Leiomyosarcoma Malignant tumor of smooth muscle

Oxyntic Specialized epithelial cells, which secrete hydrochloric acid and are located in the corpus and fundus of the gastric mucosa

Peristalsis Wave contractions of the smooth muscle layers of the digestive organs

THIS ARTICLE provides a narrative account of the salient features of the biology of gastrointestinal cancer in humans. It emphasizes the precancerous process as well as the epidemiology and etiology of cancer in the context of the interaction between genetic and environmental influences. It points out human ecology as the main determinant of risk and elaborates on the role of secular changes that have taken place in the human environment as a consequence of urbanization. The histologic features of the digestive system that are relevant to gastrointestinal cancer are summarized. The natural history and the clinical features of the tumors are briefly described. Finally, the available biologic information on the cancerous and precancerous process is used to provide guidelines of primary and secondary prevention.

I. Gastrointestinal Cancer: A Reflection of Human Ecology

Approximately one-third of all cancers in humans originate in the gastrointestinal mucosa. This proportion is similar in most populations, but there are marked interpopulation differences in the specific location and type of tumors. Marked shifts in this frequency with time have been observed in several populations, the most prominent of which is depicted in Fig. 1, which shows the time trends of mortality rates from cancers of the stomach and large bowel in U.S. populations. The marked shifts in the frequency of digestive cancer in the relatively short lapse of a few decades are a reflection of the rather marked changes in human ecology, particularly evident in new dietary habits in response to changes in life-style.

Although the reasons for such shifts are still somewhat controversial, there is general agreement that they mostly represent changes in dietary practices, most probably driven by the complex phenomenon of "urbanization." The change in gastrointestinal cancer frequency observed so

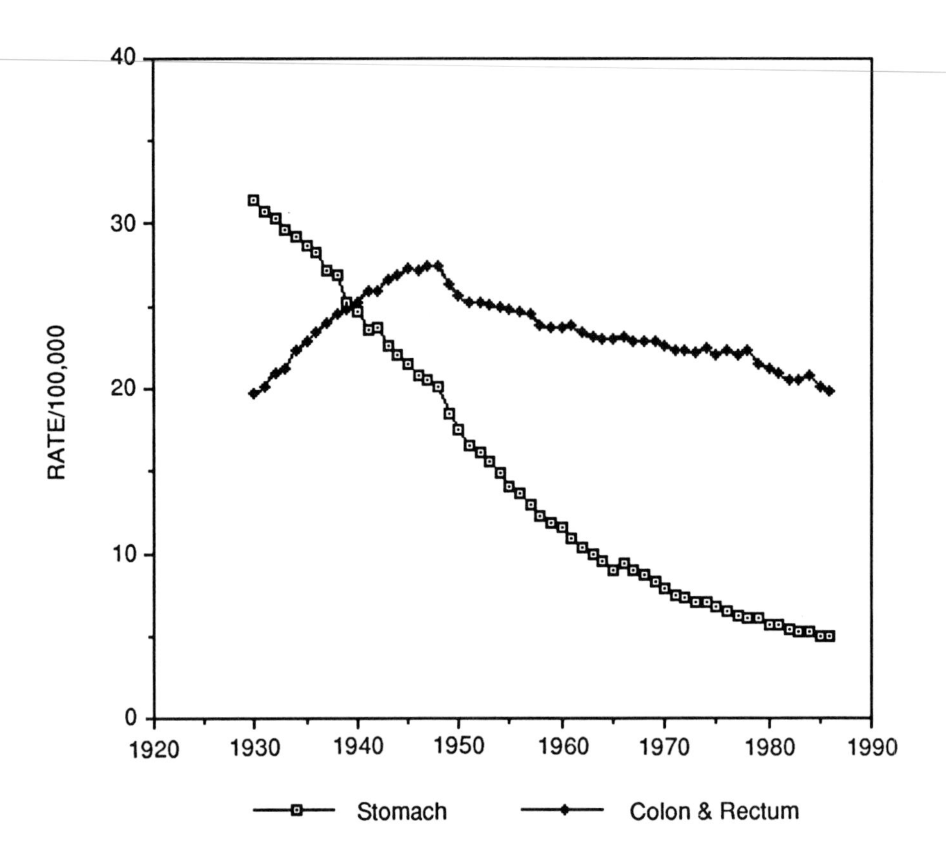

FIGURE 1 Trends of mortality rates for gastric and colorectal cancer. (Data compiled by E. Silverberg, American Cancer Society.)

dramatically in the United States has also occurred in other countries, although it seems to have started somewhat later and so far reached less contrasting proportions. Many populations of the world still display the 1930 U.S. pattern of high gastric and low colorectal cancer rates. This is especially true of the Andean and mountainous parts of Central and South America and central Europe as well as some populations that have kept a more "primitive" diet (e.g., the American Indians). The diet of such populations has several common characteristics. It consists mostly of locally grown agricultural products rich in starchy grains and roots and of other vegetable foods rich in complex carbohydrates. Such food items are rarely consumed fresh (unprocessed). They are usually ingredients of complex soups and other items extensively cooked at high temperatures. Extensive cooking may have resulted from empirical observations that diarrheic diseases followed the consumption of raw vegetables, owing to contamination with human excreta containing infective bacteria and parasites (e.g., cysts of *Entamoeba histolytica*). The practice of high-temperature cooking prevents infectious and parasitic diseases but also deprives the consumer of "protective" factors, which are present in fresh fruits and vegetables.

High-gastric/low-colorectal cancer risk diets are usually low in animal foods rich in fats and good quality proteins, which in present-day primitive societies are expensive. These items (i.e., beef, eggs, chicken, and milk) have become more affordable in societies that have "benefited" from technological advances in agro-industry.

Another frequent habit in societies with high gastric/low colorectal cancer rates is excessive salt intake. Most of the world consumes amounts of salt that are at least one order or magnitude greater than the presumed dietary requirement. The reasons for this human preference for excessive salt intake are not clear. Newborn infants show no preference for salty fluids versus water, but by the age of 2–3 years they prefer salty food over its unsalted counterpart. The primate ancestors of humans were likely to have existed on a vegetarian diet poor in salt, which makes animals search out for sources of salt in their environment. Primitive human diets are predominately vegetarian and need to be supplemented with salt intake. Salt was historically a highly valued commodity, a marked contrast with today's industrialized societies in which salt is cheap and plentiful. The taste for salty foods is largely acquired and can be modified: Subjects accustomed to high-salt intake prefer salty foods but, when they change to a low-salt diet, dislike salty foods. The salt consumption differs greatly between populations: from higher than 25 g/day/subject to lower than 4 g in some isolated Indian tribes. The etiologic role of excessive salt in cardiovascular diseases and in cancer has been widely recognized. The Japanese government has sponsored an extensive education campaign, which in two decades has brought the average per capita daily intake of salt from approximately 24 g to approximately 11 g. The rate of gastric cancer in Japan, the highest in the world, did not change for about 10 years after the salt intake reduction, but at the present time is falling rapidly. [*See* SALT PREFERENCE IN HUMANS.]

Most primitive societies have a bulky diet with abundant undigestible residues of cellulose and similar compounds (so-called dietary "fiber"). This habit requires frequent defecation, which becomes a problem in urbanized societies. Dietary changes from primitive to modern and from rural to urban societies have brought marked reduction in undigestible dietary residues, which require less frequent defecation and increase the bowel transit time. This change has been accomplished gradually over the centuries, mostly by increasing the consumption of animal foods (low in residue) and reducing the undigestible residues from vegetables by decreasing their intake. For that purpose the technology in "refining" vegetables has been used, as in the case of highly refined wheat flour, which has much less residue than the original whole wheat flour. These changes have no apparent effect on gastric cancer frequency but do seem mainly responsible for the increase in colorectal cancer frequency. Societies that are heavily dependent on animal meats (e.g., Scotland, Australia, and Argentina) have the highest rates of colon cancer. When the consumption of animal products emphasizes dairy products such as eggs, milk, and cheese, the colorectal cancer rates are reduced but the rates of coronary heart disease are increased. [*See* DIETARY FIBER, CHEMISTRY AND PROPERTIES.]

The rates of colorectal cancer in the United States probably started to climb before 1930, as suggested by Fig. 1. Vital statistics before 1930 are unreliable and preclude inferences concerning the actual time of initiation of this "epidemic." Most primitive societies have rates that are lower than 10 deaths per 100,000 inhabitants per year.

Not all countries belong to one of the two patterns depicted in the United States in 1930 versus 1980. Some have excesses of both gastric and colorectal cancer, as in the case of France and Uruguay, whereas others have low rates of both tumors, as in the case of most Africa and in northern Brazil. Such populations offer opportunities to investigate specific dietary practices linked to specific tumors. It appears that excessive residue in the diet does not necessarily contribute to gastric cancer rates if irritants such as excessive salt are avoided and protectors such as fresh fruits and vegetables are provided. Cancer protectors may also be found in fish, with a high content of Omega 3 fatty acids (a special kind of unsaturated acids), which may exert a protective role similar to that of fresh fruits and vegetables.

Not all gastric and colorectal cancers are so markedly influenced by dietary factors. Infections and other nondietary related factors play a prominent role in some cases. Genetic factors determine if specific individuals are susceptible or resistant to the carcinogenic influences in the environment.

II. Digestive Organs: Tissues and Cells

The gastrointestinal tract is a series of interconnected hollow organs specialized in digestive functions responsible for the breakdown of food, the absorption of its components, and the mobilization of the undigestible residue. The breakdown and absorption of food is accomplished by the inner glandular layer of these tubular organs, called the *mucosa*. The mobilization of residues is the function of a series of juxtaposed layers of smooth muscle whose zonal contractions and relaxation (peristalsis) moves the food bolus forward. The muscular and mucosal layers are separated by the submucosa, a layer made of loose connective tissue that allows sliding movements between the other layers. The external surface of these tubular organs is covered by a thin layer of connective tissue lined by mesothelial (peritoneum) cells, which allow minimal friction from the peristaltic movements. A rich network of nerve filaments and nerve cells is present and regulates the digestive and peristaltic functions. Such regulation has also a hormonal component represented by endocrine cells that secrete specialized polypeptide hormones. The specialized cells and tissues described intermix with "support" cells and tissues, mostly the connective tissue composed of fibroblasts and their products (reticullum and collagen fibers), the vascular tissue through which blood and lymph flows, and the lymphoid tissue whose cells take care of the immunologic functions. [*See* DIGESTIVE SYSTEM, PHYSIOLOGY AND BIOCHEMISTRY; PEPTIDE HORMONES OF THE GUT.]

All the above cells can give rise to neoplasia, but the great majority of them (adenomas and adenocarcinomas) originate in the mucosal glands. Somewhat less frequent are tumors originating in the smooth muscle cells (leiomyomas and leiomyosarcomas) and the lymphoid organs (lymphomas). Tumors originating in endocrine cells are infrequent, and tumors originating in other cells (nerve cells, mesothelial cells) are even more rare. Our discussion will concentrate on glandular tumors because of their frequency and because more is known about their etiology.

The gastric mucosa has two main components with specialized functions: the antrum (pyloric) and the corpus-fundus (oxyntic). Both are lined by foveolar cells secreting specialized neutral mucin, which serves as a lubricant and probably as a protector against acid secretions, apparently with help from bicarbonate, which is also secreted by foveolar cells. Blood group substances are also secreted along with the mucin. The glands of the oxyntic mucosa contain three different cell types: (1) the parietal cells (also called *oxyntic cells*) secrete hydrochloric acid and intrinsic factor; (2) the chief cells secrete pepsinogens (I and II), which on release into the acid gastric secretions are activated to form the proteolytic enzyme pepsin (this enzyme starts the process of digestion of ingested proteins); and (3) the endocrine cells, which in the oxyntic mucosa secrete some known polypeptide hormones such as serotonin and somatostatin as well as some hormones whose classification and function is still unknown.

The cells of the antral glands secrete mostly mucins and pepsinogen II. The endocrine cells of the antrum secrete gastrin, a stimulator of acid-pepsin secretion. Somatostatin, bombesin, and serotonin are also secreted by antral cells. The so-called gland necks are composed of special "neck" cells, which are the only ones capable of replication in the normal stomach. The loss of cells at the surface or in deeper glands leads to replication of neck cells, which force migration and specialization of the displaced cells. Neck cells secrete both mucins and pepsinogens, but they in actuality represent "stem" cells, which contain the genetic information to synthetize (after migration) the more specialized secretions of the glands and surface epithelial cells.

The small intestinal mucosa is formed by villi that increase the absorptive capacity of the organ. Such villi are taller and presumably more efficient in individuals living in temperate zones than in tropical climates. Migration from tropical to temperal zones results in improvement of overall bowel morphology. At the base of the villi are the crypts, where replication of stem cells takes place. Such cells migrate to the surface and are shed at a rapid rate: The small intestinal epithelium renovates itself faster than any other tissue. This may be one of the explanations for the rarety of epithelial tumors in this organ: Cells with abnormal mitosis or other DNA abnormalities are discarded before they have a chance to form new abnormal viable clones. The villi are mostly covered by enterocytes, columnar cells specialized in absorption and digestion. They contain α-1 antitripsin, apolipoprotein, and lyzozyme. Their surface is covered by tall and crowded microvilli (identified as a "brush border" in light microscopy) rich in disaccharidases and peptidases. Alternating with enterocytes are goblet cells, which secrete sialomucins. The endocrine cells of the small intestine contain several polypeptide hormones including cholecystokinin, glucagon, gastrin, motilin, secretin, and somatostatin. Such cells give rise to carcinoids, also called *APUD tumors*.

The large intestinal mucosa (cecum, ascending, transverse, descending, sigmoid colon, and rectum) is lined by crypts arranged perpendicular to the surface, lined predominantly by mucus-secreting columnar and goblet cells. They secrete a mixture of two acid mucins: Sialomucins predominate in the proximal colon, and sulfomucins predominate in the distal colon and rectum. The base of the crypts of the small and large intestine contain Paneth cells with large eosinophilic granules of unknown function. The granules are rich in lysozyme and immunoglobulins A and G, suggesting a role in the control of the luminal bacterial flora.

III. Heterogeneity of Gastrointestinal Cancer

The great majority of gastrointestinal tumors originate from the glandular epithelium and are therefore appropriately called *adenomas* (if benign) and *adenocarcinomas* (if malignant). Beyond this general affinity, subtypes of adenocarcinomas are determined by the organ of origin, the type of epithelial cells they mimic, and their biologic behavior.

Most of the gastric malignant tumors belong to one of two types, determined mostly by the cohesion of their cells. If the tumor cells adhere to each other by desmosomes and tight junctions, they form neoplastic glandular structures that resemble intestinal glands. This characteristic has given these tumors their name: they are called *intestinal-type* or *expansive tumors*. If the intercellular junctions are absent, the tumor cells diffusely infiltrate the gastric wall and do not form well-defined structures. For such reason they are called *diffuse-type* or *infiltrating carcinomas*. As a rule, the intestinal type of gastric carcinomas grow less rapidly, are more frequent in males and older persons, and more frequently are ulcerated than their diffuse counterparts.

Despite such obvious differences in biologic behavior, no clear differences in the characteristic of the cells forming intestinal or diffuse types of tumors have been documented. They both produce mucins in an irregular and anarchic pattern: gastric-type (neutral) mucins, small intestinal–type (sialic) mucins, and colorectal (sulfated) mucins have been reported in both types of tumors. Abnormal pepsinogens have also been reported. Both types of gastric adenocarcinomas may display patterns that resemble more primitive cells (anaplasia).

Adenocarcinomas of the small intestine are rather rare. The ampulla of Vater, where bile meets the intestinal content, is the most frequent site of origin. Other carcinomas originate in previously existing "benign" polyps such as those seen in Peutz-Jeghers syndrome or in familial polyposis.

Adenocarcinomas of the large intestine are usually mucin-producing tumors that vary in their degree of differentiation. They are especially frequent in the sigmoid and rectosigmoid portions. Most are believed to originate from preexisting adenomas.

Endocrine tumors are somewhat less frequent in the gastrointestinal tract. In some organs such as the vermiform appendix and the recto-anal junction, they tend to remain small and limited to their site of origin. Those that invade and metastasize originate more frequently in the small intestine and less frequently in the stomach and colon. Those tumors secreting 5-hydroxytryptamine (5HT) are frequently called *carcinoids*, and those secreting one or more of the polypeptide hormones are named after their main secretion (i.e., glucagonoma). 5HT is metabolized to 5-hydroxyindole acetic acid (5-HIAA) by monoamine oxidase in the liver; 5-HIAA is excreted in urine and is useful in the diagnosis of the carcinoid syndrome. The latter is characterized by episodes of facial flushing (which can be induced with alcohol), cardiac lesions, and episodic diarrhea. The cardiac lesions consist of right-sided fibrosis, which may lead to stenosis of the tricuspid and pulmonary valves. Although several biologically active substances may be involved in the pathogenesis of the syndrome, serotonin is probably the major contributor because of its abundance in midgut carcinoids with which the syndrome is almost exclusively associated.

Gastrointestinal lymphomas as a rule originate in lymphoid cell populations that specifically home to these digestive organs. Contrary to other lymphomas, they tend to remain localized to their site of origin for a long time and disseminate to other organs only rarely. In Western countries, these tumors are seen predominantly in the stomach, followed by the small intestine, and less frequently the large intestine. They have received the names of MALT or GALT because of their tendency to remain localized to the site of origin (mucosa-associated lymphoid tissue and gut-associated lymphoid tissue). The tumors apparently originate from perifollicular B cells. They respond well to locally directed therapy. In the Middle East, a special type called the *Mediterranean lymphoma* predominates; it is principally localized in the small intestine. This tumor is part of a disease complex called immunoproliferative small intestinal disease (IPSID) first described in the Tropics and studied in detail in the Middle East. IPSID is clinically manifested as a malabsorption syndrome and predominants in young individuals. In its initial stages it behaves as an infectious disease and can be cured with broad spectrum antibiotics. In more advanced stages it behaves as a frank lymphoma, involving extensively the small intestine. Histologically the intestinal villi are expanded by a dense infiltrate of mononuclear leukocytes that in advanced stages extends to the muscular layer. Lymph node involvement is a late and rare event. The tumor cells are B lymphocytes and plasma cells that synthetize an abnormal α_1 heavy-chain protein (it is sometimes called *α_1 heavy-chain disease*). A nonspecific antigenic drive originated by luminal bacteria is suspected in its pathogenesis. In north Africa and the Middle East, another peculiar lymphoma is observed in the gastrointestinal tract of children—Burkitt's lymphoma. These are rapidly spreading tumors made of dense agglomerations of primitive stem cells and occasional phagocytic macrophages, giving a microscopic pattern that has been compared with a starry sky. Another variant of gastrointestinal lymphoma is made of T cells, mostly associated with gluten intolerance and celiac disease. The tumor is derived from the intraepithelial T lymphocytes. [*See* LYMPHOMA.]

The stromal tissues of the gastrointestinal tract give rise to tumors that have most frequently been classified as leiomyomas (benign) or leoimoyosarcomas (malignant), even though distinct features of smooth muscle differentiation are often lacking. Small nodular proliferations of smooth muscle, so-called seedling leiomyomas, are frequent autopsy findings, especially in the esophagus, stomach, and small intestine; they have no clinical significance. Malignant tumors form spherical masses that compress and ulcerate the mucosa and may produce blood-borne metastasis.

IV. Precancerous State

Most malignant gastrointestinal tumors are preceded by mucosal lesions that can be identified by morphologic techniques. The better-known cancer precursors have been described in the stomach and in the large bowel.

The intestinal (or expansive) type of gastric cancer is preceded by the following apparently sequential lesions: superficial gastritis (SG), chronic atrophic gastritis (CAG), intestinal metaplasia (IM), and dysplasia. The first two lesions involve cell damage and repair processes, whereas the last two probably represent cell mutational events. Cell products are gradually altered as the process advances, particularly the mucins, digestive enzymes, and the so-called fetal antigens. The normal neutral gastric mucin is altered in amount but not in quality during SG and CAG. When the atrophic foci express abnormal intestinal phenotypes (IM), neutral mucins are gradually replaced by acid mucins. These are mostly sialomucins in mature metaplasia, also called *type I-II, complete,* or *small intestinal–type*. More advanced metaplastic changes (type III, incomplete, or colonic) express sulfomucins characteristic of the distal large bowel; they are frequently observed in the vicinity of dysplastic and early neoplastic foci.

The normal gastric pepsinogen I (precursor of pepsin, the only gastric digestive enzyme) gradually decreases as atrophy advances, and the normal chief cells are replaced by intestinal cells. Pepsinogen I can be measured in the blood, and its low level is a good indicator of advanced gastric atrophy (loss of chief cells) and indirectly of high gastric cancer risk. Intestinal digestive enzymes make their abnormal appearance in the gastric mucosa when it becomes intestinalized (IM). A full set of intestinal enzymes (sucrase, trehalase, leucine aminopeptidase, and alkaline phosphatase) are present in small intestinal metaplasia, which for that reason has been labeled *complete metaplasia*. Such enzymes gradually disappear when the metaplasia becomes "colonic" or "incomplete."

Gastric dysplasia is characterized by proliferation of abnormal closely packed tubular glands, which secrete little or no mucin and are lined by cells with large, hyperchromatic, and crowded nuclei. The cells' phenotype is neoplastic, but they lack the invasive capacity and remain within the bounds of the basal membrane that surrounds each gland. Gastric dysplastic lesions are mostly flat or slightly depressed, but occasionally the proliferating glands form a small mass that protrudes into the lumen and is called *adenomatous polyp*. Not all dysplastic foci progress to invasive cancer, but they are indicators of high risk and frequently coincide with small foci of invasive carcinoma present in other areas of the mucosa.

In the large intestine, a process of abnormal (benign) gland proliferation leads to the formation of tubular adenomas (also called *adenomatous polyps*), which are dragged distally by the fecal stream forming a pedicle that separates the polyp proper from the mucosa from which it originated. Colonic adenomas increase in size and prevalence with age and are present in more than 50% of subjects older than 60 years in populations with high colon cancer rates (e.g., the United States, western Europe, Australia, and Argentina). In populations with low colon cancer rates (e.g., Africa), colonic adenomas are practically nonexistent. The high prevalence of colonic adenomas in Western societies obviously indicates that most polyps never progress to malignancy: It has been estimated that less than 5% of them do. Most colon carcinomas, however, are thought to originate from adenomatous polyps.

Several genetic alterations were detected in different stages of the colorectal precancerous process discussed in this Section. In the hyperproliferative stage that precedes polyp formation, gene mutations or losses were observed in chromosome 5q. Clonal expansion of one of the hyperproliferating cells may lead to the formation of small adenomas in which the genome is hypomethylated. Hypomethylation may lead to aneuploidy and the loss of suppressor gene alleles. K-*ras*, which appears to occur in one cell of the small adenomas and throughout clonal expansion, results in larger and more dysplastic tumors. A key event is the loss of a large portion of chromosome 17p, associated with the progression of adenomas to carcinomas. Allelic losses in chromosomes 17p lead to the removal of the wild type p53 gene which functions as a tumor suppressor. Losses of tumor suppressor genes in other chromosomes (18q) may also occur. The accumulated losses of suppressor genes correlates with tumor aggressiveness and capacity to metastasize. Although the order in which these genetic alterations appear may be of some importance, it seems that the accumulation of the alterations is the major determinant of progression of precancerous changes to ultimate invasion.

V. Causes of Gastrointestinal Cancer

Most malignant neoplasms in humans are the end result of the interplay of forces that induce or inhibit anarchic cell replication. Such interplay is especially prominent and complex in the gastrointestinal organs, in which it usually operates for many years before a final cancer outcome. Schematic models of the carcinogenic process have been proposed and usually described in epidemiologic terms as a "web of causation." The etiologic forces fall into two main categories: genetic, and environmental. The latter term taken in its broadest sense refers to all circumstances surrounding the individual, including factors under his or her control, known as the voluntary environment (i.e., smoking, diet) as well the involuntary environment (i.e., industrial pollution). Within each category of etiologic forces are factors that favor and factors that impede the progress of carcinogenesis. In the genetic categories, such forces confer either susceptibility or resistance to the host. In the environmental category, such forces are usually known as carcinogens (initiators and promoters) or inhibitors. The entire process is the modulation of the carcinogenic process.

Causal associations in humans are usually established with epidemiologic methodology, which has been applied more extensively to adenocarcinomas of the stomach and the large bowel as summarized below. With some exceptions, the etiology of other gastrointestinal tumors is largely unknown.

The genetic influences in gastrointestinal cancer are best exemplified by the familial polyposis syndromes, the most frequent of which is familial polyposis coli (FPC). This condition observed in approximately one of every 7,000 births is characterized by the development of increasing numbers of adenomas in the large bowel mucosa until it is literally covered with such tumors. In resected specimens, the large bowel contains more than 1,000 grossly recognizable adenomas, but more are identified histologically, making them literally innumerable. The disease is inherited in autosomal dominant fashion, and therefore the children of an affected parent have a 50% chance of developing the disease. As a rule the polyps can be seen in the third decade of life and become symptomatic in the fourth. If not resected, one or more of the polyps will inevitably develop into carcinoma; the mean age of diagnosis of carcinoma is 39 years. The susceptibility to carcinogenesis is more evident in the large bowel epithelial cells, but it is also evident in other gastrointestinal organs; even skin fibroblasts have an increased susceptibility to malignant transformation by oncogenic viruses and chemical carcinogens. Although the genetic influence in this disease is overriding, environmental factors are suspected to play a role. The clearest example of such influences is the occurrence of gastric adenomas in most FPC patients living in areas of high gastric cancer risk such as Japan but less frequent occurrence in FPC patients in Western populations with low gastric cancer risk. Other familial polyposis syndromes resulting in multiple colon adenomas are the Gardner's syndrome and the Turcot syndrome, whose patients develop additional tumors of other organs, especially the bones and the central nervous system. Polyps somewhat different from the adenomas, in which tubular glands predominate, are the hamartomas or hamartomatous polyp, in which other tissues are prominent (e.g., smooth muscle, fibroblasts, and blood vessels). Hamartomatous polyps are seen in Peutz-Jeghers syndrome

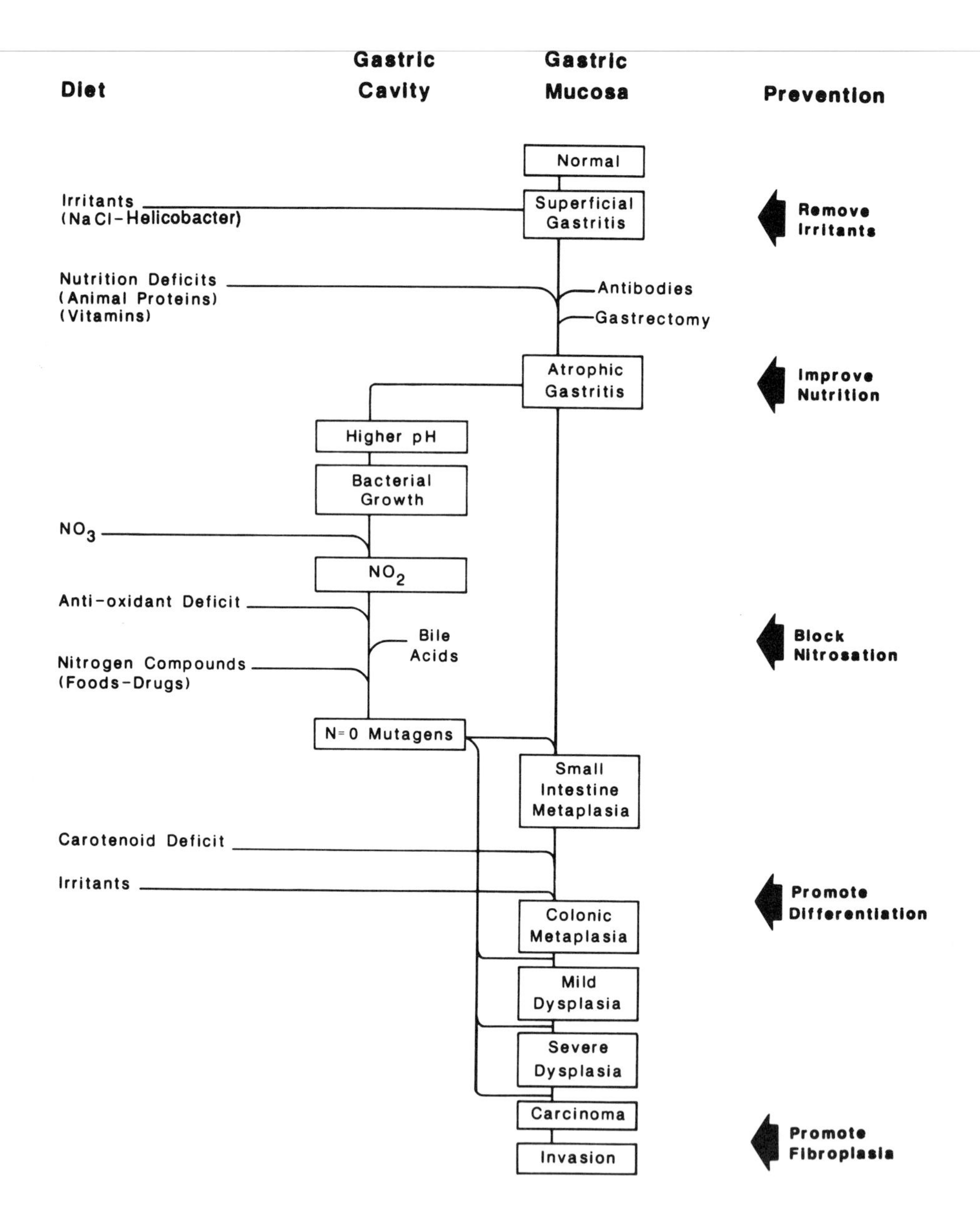

FIGURE 2 Etiologic model of gastric carcinogenesis in humans.

and in the juvenile polyposis syndrome. The former displays polyps mostly in the small intestine and less frequently in the colon and stomach; it is associated with patchy pigmentations around the lips. The malignant transformation of hamartomatous polyps is much less frequent than that observed in FPC.

Strong genetic influences are suspected in the pathogenesis of diffuse gastric carcinoma because of its association with blood group A, its limited interpopulation variability, and its stability in migrant populations. Our knowledge of the etiology and pathogenesis of diffuse gastric carcinoma is too limited to draw meaningful conclusions.

The intestinal type of gastric carcinoma, however, has been the subject of extensive interdisciplinary research, which has led to the postulation of an etiologic model outlined in Fig. 2. The key event in the process is a transformation of the genotype of normal cells into that of neoplastic cells whose replication is no longer under the control of physiologic forces. Such sine qua non event is probably a mutation or a series of sequential mutations, which call for a genotoxic etiologic factor. This factor has not been identified, but it has been postulated to be an

N-nitroso compound synthetized in the gastric cavity by the action of nitrite on nitrogen-containing substances. The nitrite is produced in the gastric lumen by anaerobic bacteria containing reductases, which transform ingested nitrate. The anaerobic bacteria colonize the stomach only when its normally acid pH (around 2.0) is elevated to levels around 5.0 or higher. These come about by focal loss of parietal cells as a result of multifocal chronic atrophic gastritis, which is frequently complicated by intestinal metaplasia as explained above. The genotoxic event is therefore dependent on a long chain of previous (not genotoxic) events having to do with loss of cells and their inadequate replacement. Although the forces leading to cell loss and epithelial hyperplasia are not entirely understood, at least one factor [i.e., excessive salt (NaCl) intake] has been identified in epidemiologic and experimental studies. Other factors are under scientific scrutiny (e.g., infection with *Helicobacter pylori*, a prevalent human infection that is especially severe in populations at high gastric cancer risk). The proposed genotoxic event (i.e., the intragastric synthesis of *N*-nitroso compounds) can be blocked by antioxidants such as vitamin C, whose ingestion may be less than optimal in some high-risk populations. Even if the genotoxic event takes place, its effects can be inhibited or can be retarded by "protective" forces such as antioxidant micronutrients in the diet. Prominent among these protectors is beta-carotene, which has been found deficient in subjects with gastric dysplasia and has shown an inhibitor role in experimental gastric carcinogenesis. Ascorbic acid further plays a role in retarding the progression of cancer by increasing the fibroplastic barrier, which stands in the way of invasion. The complexity of the gastric cancer etiologic model contrasts with the simplicity of experiments in which large doses of carcinogens are administered to inbred strains of susceptible experimental animals. The proposed model is probably closer to human reality than the experiments, but such reality may be even more complex, as suggested by human genetic studies. One such study is the segregation analysis of CAG, a cancer precursor, the results of which are expressed in the model depicted in Fig. 3. This model proposes the existence of an autosomally transmitted major recessive susceptibility gene that is prevalent in the population. The expression of such a gene results in CAG, and its penetrance is modulated by age and by having an affected mother, which probably represent

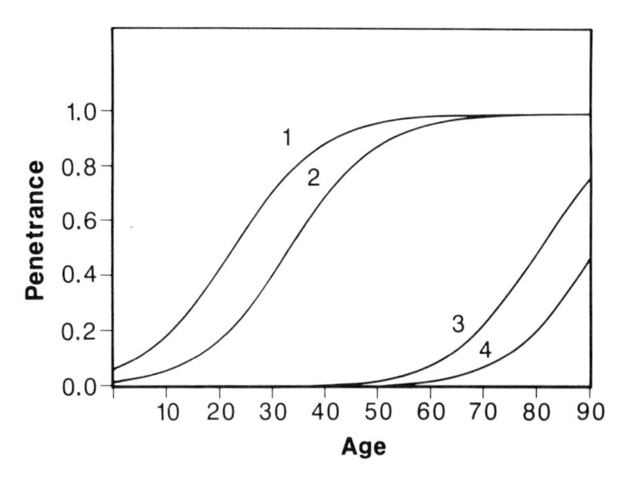

FIGURE 3 Theoretical model of the genetic etiology of chronic atrophic gastritis derived from segregation analysis of family clusters. Estimated penetrance fractions (CAG prevalence): (*1*) homozygous recessive (AA), mother affected; (*2*) homozygous recessive (AA), mother unaffected; (*3*) carrier (AB) or noncarrier (BB), mother affected; (*4*) carrier (AB) or noncarrier (BB), mother unaffected. [From Bonney G. E., et al. (1986). *Genet. Epidemiol.* **3,** 213–224, with permission.]

environmental forces. The mother's influence may be related to preparation of food for the family. The prevalence of CAG reaches approximately 80% at age 40 in homozygous individuals whose mother is affected by CAG; homozygous subjects whose mother is not affected reach such a high prevalence of CAG years later. Carriers of the gene and subjects presumably without it may develop gastritis much later because of excessive environmental influences, but they never reach high prevalence levels because competing disease risks decimate the population. The proposed model takes into account genetic and environmental interactions and may be useful in exploring neoplastic diseases other than gastric cancer.

The etiology of colorectal cancer offers considerable contrasts with stomach cancer; the demographic distributions approximate mirror images of each other. Colorectal cancer incidence rates are low in Africa and high in Western societies known for their high consumption of meat. After living in England and then in Africa for many years, Burkitt postulated an etiologic model outlined in Fig. 4. The model basically proposes that the high-residue diet of primitive populations prevents colon cancer by inducing rapid transit time and dilution of the bile acids, which otherwise may be carcinogenic. However, the mechanism by which prevention is accomplished is still unknown. Recent research on the

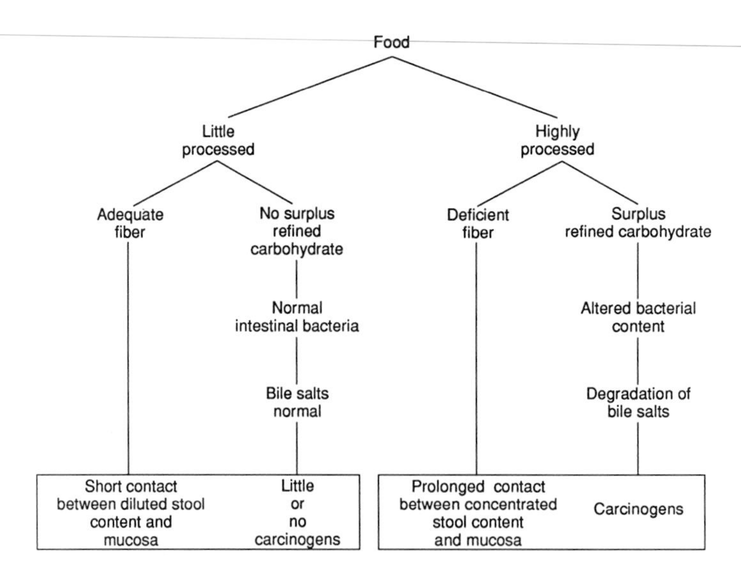

FIGURE 4 Diagramatic representation of the possible relation between diet and cancer of the large bowel. [From Burkitt, D. (1971). *Cancer* **28,** 3–13, with permission.]

causes of colon cancer has centered on the etiologic factors discussed below.

The search for genotoxic agents using mutagenesis tests with bacteria *in vitro* has centered on compounds found in fried foods and in feces. Several compounds formed by pyrolisis during food preparations at high temperature are highly mutagenic, and induce nuclear aberrations in colonic cells. Although these compounds have not been identified in the human fecal stream, one of them [i.e., 2-amino-3-methylimid-azo (4-5f) quinoline (IQ)] has been found carcinogenic to the liver in subhuman primates. Whether any of the pyrolitic compounds plays a role in human colon carcinogenesis is still undetermined. Fecal mutagens are more abundant in Americans on a Western diet than in African primitive societies. A sizable portion of that mutagenicity is due to fecapentenes formed by fecal bacteria; considerable epidemiologic and experimental work, however, has failed to support fecapentenes as human carcinogenes. Another group of compounds suspected to play a genotoxic role are products of cholesterol metabolism, especially 3-keto-steroids, which induce chromosomal aberrations in colonic cells and which could be formed from cholesterol either in the colon or in the preparation or storage of food, and induce chromosomal aberration in colonic cells in culture. Proof of their relevance to humans is still missing. [*See* CHOLESTEROL; FOOD MICROBIOLOGY AND HYGIENE.]

Other dietary factors of possible relevance in human colorectal carcinogenesis are calcium deficiency and fecal pH. Calcium salts significantly reduce the toxic effects of bile acids, which are well-established cancer promoters on the colon. High-fat diets both increase the level of bile acids in the fecal stream and create a relative deficiency of calcium in the intestine. Evidence implicating suboptimal calcium intake in colon carcinogenesis in humans is, however, insufficient. The importance of fecal pH has recently been emphasized, because it is several units lower (more acidic) in populations at low colon cancer risks than in their high-risk counterparts. At high pH (about 9) most fecal bile acids are in solution. As the pH decreases, more of the acids are in their protonated form and drop out of solution, so that at pH 6, their concentration is very low. Diets low in fat and high in fiber and starch may prevent cancer at least in part by lowering fecal pH. [*See* BILE ACIDS.]

VI. Natural History

As described above, the precancerous process for most gastrointestinal tumors is so prolonged that it takes many years to reach the stage of invasive neoplasia. After it reaches such a stage, two phases have been observed. In the first one, the malignant tumor remains small and localized for years, but when the localized stage is surpassed, the tumor cells invade neighboring structure and metastasize to other organs at an accelerated pace. [*See* METASTASIS.]

In the stomach, the localized stage is characterized by confinement of the tumor cells to the mucosa and later to the submucosa. For them the term *early cancer* has been coined in Japan. Cell kinetics studies conducted in Japan have estimated that the localized stage lasts longer than 15 years on average. These tumors are classified according to their architecture: Type I is elevated, type II is approximately flat, and type III is ulcerated. They all have an extensive free surface through which malignant cells are exfoliated. When localized ("early") carcinomas are resected, the cure rate is high: more than 95% 5-years survival in Japan and somewhat lower rates in other countries. In Japan, screening campaigns using mobile X-ray units monitor apparently normal individuals. There are some indications that Japanese have a more favorable prognosis than other races after a diagnosis of cancer in the gastrointestinal as well as in other organs. When the localized phase is surpassed and the tumor invades the muscularis, the prognosis is irreversibly bad: less than 10% survival at 5 years. The tumor cells invade the perigastric lymph nodes and then spread through lymphatic and blood vessels to produce metastasis in distant organs, especially the liver, the peritoneum, and the cervical lymph nodes.

The localized stage in the large bowel is characterized by the presence of microcarcinomas in adenomatous polyps. Such foci are practically absent in small polyps (<1 cm in diameter) but increase in prevalence as the size of polyp increases. The occurrence of lymph node metastasis from resected polyps containing small superficial carcinoma is low: 1–9%. Factors that increase the probability of metastasis are poor differentiation of the carcinoma, extension to the surgical margin, and short stalk. Adenocarcinomas most frequently overcome the localized stage by invading the submucosa and penetrating the lymphatic and blood vessels. The metastasis to the mesenteric lymph nodes then extends to the peritoneal organs, the liver, and other distant organs. The local invasion of the intestinal wall is of considerable prognostic importance. It is the basis of the Duke's classification, which considers three stages: (A) when the invasion is limited to the mucosa, submucosa, and inner part of the muscularis; (B) when it reaches the serosa; (C) when regional lymph nodes have metastasis; and (D) distant metastasis. The 5-year survival rates after treatment are 99% for Duke's (A); 85% for (B); 67% for (C); and 14% for (D). These rates vary in different series and are considerably lower in blacks and lower socioeconomic groups.

VII. Prevention

Present knowledge about etiology as well as about precancerous and early cancerous lesions makes it possible to plan prevention strategies that cover two fields: primary prevention to avoid the development of the disease, and secondary prevention to avoid death from the disease.

Primary prevention of stomach cancer apparently has been taking place in most Western communities. This is evidenced by declining incidence rates. It is not clear which of the many recent changes in our society is responsible for the decline, but the invention and widespread use of refrigerators appears to have been a major factor. It has allowed fresh fruits and vegetables to be available yearround and decreased our dependency on food preservations methods based on excessive use of salt and nitrites. Most communities with above average consumption of salt and below average consumption of fresh fruits and vegetables have high gastric cancer rates and should benefit from changes in such dietary practices.

Secondary prevention of gastric cancer depends on the identification of high-risk populations and individuals who carry precancerous or early cancerous lesions. This can be accomplished with markers of gastric atrophy (e.g., low pepsinogen I blood levels). Recent advances in endoscopic techniques have made it possible to screen high-risk individuals and take multiple gastric biopsies with a minimum of risk. If a small (localized or early) carcinoma is found, gastrectomy offers a high probability of cure. If dysplastic changes are found, close surveillance with frequently repeated gastroscopies is indicated. Research is being conducted on induction of regression of premalignant lesions with antioxidants such as retinoids and carotenoids. Retinoids accomplish such regression in experimental systems but are toxic to humans. Carotenoids are practically nontoxic but have been less explored. It remains to be determined if chemoprevention can be of practical value in humans. The prevalence of intestinal metaplasia of the gastric mucosa is high in many communities, and dietary or chemical prevention of its progression to cancer is an important interesting challenge.

Primary prevention of colorectal cancer is a more difficult matter because there is no general consensus as to which of the etiologic factors is relevant and amenable to intervention. Increase in the intake of undigestible components of food (''roughage'' or ''fiber''), as well as decrease in fat intake, is generally recommended but has not been tested in large populations.

Secondary prevention of colorectal cancer again depends on the identification of high-risk groups. In two groups of individuals, the risk is so high (close to 100%) that total colectomy is indicated. These are patients with the familial colon polyposis syndrome (FPC) and individuals with chronic (more than 10 years) active ulcerative colitis whose colonic mucosa shows unequivocal signs of dysplasia.

Otherwise, secondary prevention addresses the management of polyps of the large bowel, a common condition. One way to identify colonic polyps or early carcinoma is based on the search for occult blood in the stools. This is accomplished with commercially available kits for home use based on guaiac or similar substances. These kits are useful and may discover carcinomas with high probability of curable resection. However, both false-positive and false-negative results are frequent. The second stage in the identification of premalignant or early malignant lesions is endoscopy and biopsy of suspected lesions. The above methods are generally available but are not as yet being used with a frequency that makes a real impact in the incidence and mortality rates for colorectal cancer.

Bibliography

Beauchamp, G. (1987). The human preference for excess salt. *Am. Scientist* **75,** 27–33.

Bruce, W. R. (1987). Recent hypothesis for the origin of colon cancer. *Cancer Res.* **47,** 4237–4242.

Correa, P. (1988). A human model of gastric carcinogenesis. *Cancer Res.* **48,** 3554–3560.

Fearon, E. R., and Vogelstein, B. (1990). A genetic model of colon tumorigenesis. *Cell* **61,** 759–767.

Joossens, J. V., Hill, M. J., and Geboers, J. (1985). ''Diet and Human Carcinogenesis.'' Excerpta Medica, Amsterdam.

National Research Council. (1982). ''Diet, Nutrition and Cancer.'' National Academy Press, Washington, D.C.

Schottenfeld, D., and Fraumeni, J. (1982). ''Cancer Epidemiology and Prevention.'' W. B. Saunders, Philadelphia (*new edition in press*).

Whitehead, R. (1989). ''Gastrointestinal and Oesophageal Pathology.'' Churchill Livingston, New York.

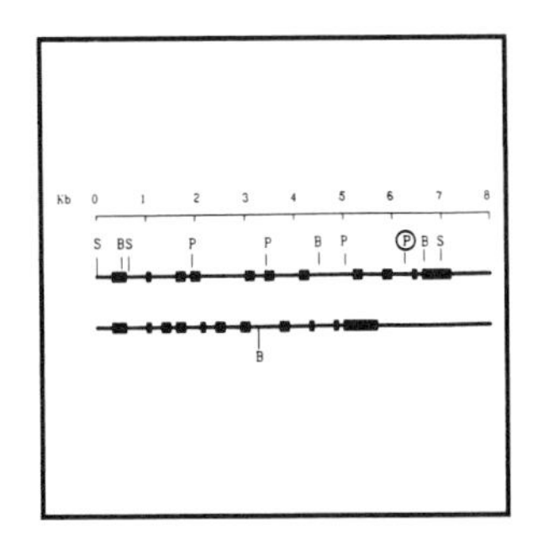

Gaucher Disease, Molecular Genetics

ARI ZIMRAN, *Shaare Zedek Medical Center, Jerusalem*

JOSEPH SORGE, *Stratagene*

ERNEST BEUTLER, *Research Institute of Scripps Clinic*

Glossary

Alu sequences Repeating sequences found in the human genome, flanked, characteristically, by inverted repeats. Alu sequences are believed to have the capability of moving from one part of the genome to another.

Gaucher disease Disease characterized by storage of the glycolipid glucocerebroside in macrophages, particularly in liver, spleen, and bone marrow.

Glucocerebrosidase A β-glucosidase that is required for the degradation of glucocerebroside. It cleaves the β-glucosidic bond by which glucose is attached to sphingosine.

Leader sequence Hydrophobic amino acid sequence that helps to direct a newly synthesized protein through the endoplasmic reticulum. It is then cleaved from the rest of the protein.

Pv1.1$^+$ and Pv1.1$^-$ Designation of a polymorphism affecting the glucocerebroside genes based on the presence or absence of PvuII restriction site.

GAUCHER DISEASE is probably the most common fat storage disorder. It is characterized by an accumulation of the sphingoglycolipid glucocerebroside, mainly in macrophages, and it is caused by an inherited deficiency of the lysosomal enzyme glucocerebrosidase. Type I ("adult") disease is by far the most common form, with involvement largely limited to the liver, spleen, and bone marrow; while several other organ systems may be affected, the central nervous system is spared. This form is especially prevalent among the Ashkenazi Jewish population, although it does occur in all ethnic groups. There is a great variability in the severity of the clinical manifestations, ranging from patients who are totally asymptomatic to patients with severe debilitating disease. Most of the symptoms are related to enlargement of the spleen with a lowered platelet count and bone pains, occasionally with fractures.

Types II and III (neuronopathic forms) are extremely rare. Type II ("infantile") disease is characterized by extensive brain abnormalities in addition to the abdominal manifestation, with onset of symptoms during the first 6 months of life and death before 2 years of age. This form does not have any ethnic predilection. Type III ("juvenile") disease is similar to type I in regard to the systemic symptoms and signs, but the nervous system manifestations are similar to those in the "infantile" type and develop after age 5 or during adolescence, with death usually occurring in early adulthood. One form of this variant is relatively prevalent in the northern Swedish provinces of Norbotten and Vsterbotten.

I. History

Gaucher disease was first described in a 32-year-old woman in 1882 by the French pathologist Phillippe Gaucher, who actually assumed the disorder was a primary neoplasm of the spleen. Twenty-five years later Marchand noticed the storage of foreign material by the reticuloendothelial cells. This storage material was defined as the cerebroside, a group of lipids, by Epstein and Lieb in 1924. The etiology of the disease was defined in 1965 by R. O. Brady and by A. D. Patrick as an enzymatic deficiency of the glucocerebrosidase. The first practical biochemical diagnostic test was developed in 1970 by E. Beutler and W. Kuhl, and the cDNA was cloned and sequenced by J. Sorge *et al.* in 1985.

II. Cloning and Characterization of Glucocerebrosidase cDNA

The cloning of the full-length cDNA for human glucocerebrosidase provided important insights into the biology of the enzyme at the levels of the protein, mRNA and DNA, and made it possible to isolate, characterize, and identify lesions causing Gaucher disease. The finding of 1545 coding nucleotides corresponds to a precursor protein of 515 amino acids. An interesting finding in the glucocerebrosidase cDNA was the existence of two ATG initiating codons (Fig. 1). The upstream, inframe ATG would initiate the synthesis of a protein containing a 39 amino acid leader sequence. The downstream inframe ATG would initiate a 19 amino acid leader sequence. This NH_2-terminal amino acid leader sequence is presumably required for transport across the membranes of the rough endoplasmic reticulum and is cleaved from the mature protein as it is secreted into the lysosome. Experiments using oligonucleotide mutagenesis to remove one or the other ATG have demonstrated that both upstream and downstream ATGs can function independently to produce active glucocerebrosidase enzyme in cultured fibroblasts. The significance of the presence of both ATGs is still unclear. It has been speculated that different human tissues preferentially utilize one or the other ATG or that different signal peptides are processed differently in various human tissues.

```
TTC TCT TCA TCT AAT GAC CCT GAG GGG ATG GAG TTT TCA AGT CCT
Phe Ser Ser Ser Asn Asp Pro Glu Gly Met Glu Phe Ser Ser Pro

TCC AGA GAG GAA TGT CCC AAG CCT TTG AGT AGG GTA AGC ATC ATG
Ser Arg Glu Glu Cys Pro Lys Pro Leu Ser Arg Val Ser Ile Met

GCT GGC AGC CTC ACA GGT TTG CTT CTA CTT CAG GCA GTG TCG TGG
Ala Gly Ser Leu Thr Gly Leu Leu Leu Leu Gln Ala Val Ser Trp

GCA TCA GGT|GCC CGC CCC TGC ATC CCT
Ala Ser Gly|Ala Arg Pro Cys Ile Pro
```

FIGURE 1 Sequence of 5′ portion of coding region of glucocerebrosidase cDNA including the initiator ATGs (underlined). The vertical line represents the known cleavage site for the mature placental enzyme. (From J. A. Sorge, C. West, W. Kuhl, L. Treger, and E. Beutler. 1987. The human glucocerebrosidase gene has two functional ATG initiator codons. *Am. J. Hum. Genet.* **41**:1016–1024. Reprinted with permission of the University of Chicago Press.)

III. Structure and Function of the Human Glucocerebrosidase Gene and Pseudogene

Approximately 16 Kb downstream from the glucocerebrosidase gene is a highly homologous pseudogene. Both the active gene and the pseudogene have recently been cloned and sequenced in their entirety. The active gene contains 11 exons extending from base pair 355 to base pair 7232 in the 7604 nucleotides that have been sequenced. The pseudogene is shorter than the active gene; its length is 5769 base pairs, and it is approximately 96% homologous to the functional gene. However, there are large "deletions" in the pseudogene introns 2,4,6, and 7, consisting of "Alu" sequences 313, 626, 320, and 277 bp in length, respectively. There is also a 55 bp deletion from a part of exon 9 flanked by a short inverted repeat, as well as base pair changes scattered throughout the gene. It has been suggested that the ancestral glucocerebrosidase gene lacked the Alu sequences and that they have actually been inserted into the introns of the functioning gene. A schematic representation of the glucocerebrosidase gene and pseudogene with its restriction map is shown in Fig. 2.

The gene promoter has also been identified and its functionality tested. It contains two TATA and two CAT-like boxes upstream of the major RNA transcriptional initiation site. The functionality of the promoter has been confirmed by coupling it to a bacterial gene coding for the enzyme chloramphenicol acetyltransferase and assaying its activity in transfected cells. Unlike the active gene's promoter, the enzyme activity directed by the pseudogene promoter was found to be very low, however, pseudogene mRNA has been detected.

The availability of the complete sequences of

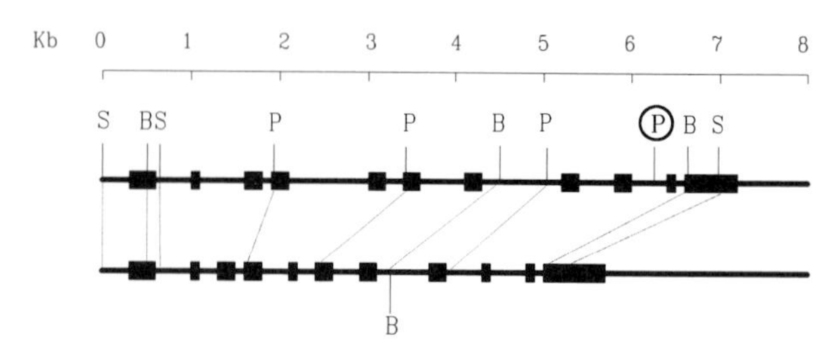

FIGURE 2 Restriction map of the glucocerebrosidase gene (top) and pseudogene (bottom) based on a published sequence. B, BamH1; P, PstI; S, SacI. The PstI site that is circled is unique to the functional gene and is one of the features that has been found to be useful in separating functional from pseudogene for the purposes of diagnosis.

both glucocerebrosidase genes (active and pseudogene) have allowed the development of strategies useful for rapid diagnosis of Gaucher disease on the molecular level.

IV. Mutations in the Glucocerebrosidase Gene

Several point mutations that are associated with Gaucher disease have been defined recently. These mutations are summarized in Table I by their position on the cDNA and on the genomic sequence. Most are point mutations. Another type of mutation is the result of unequal crossover between the glucocerebrosidase gene and its pseudogene.

A. Mutation 1448: C → T (444)

This point mutation was originally identified in a genomic clone from a patient with type II disease, and this base pair substitution has been identified subsequently in all three forms of Gaucher disease. While in the neuronopathic forms (types II and III) it has been detected in both the heterozygous and homozygous state, in the adult type it occurs only as a heterozygous mutation. Patients with adult type Gaucher disease, who are compound heterozygotes for this mutation, usually have moderate to severe disease with extensive skeletal involvement. Most of these patients are not Jewish. Overall, it is the second most common mutation affecting adult type Gaucher patients and the most frequent mutation in patients with type II and type III Gaucher disease.

The normal sequence of the pseudogene matches this sequence. This raises the possibility that its occurrence in Gaucher patients represents the result of some recombination events between the active gene and its pseudogene, such as gene conversion or unequal crossing over. The latter phenomenon has been documented. It is also relevant for molecular diagnosis, as it creates a need for separating the active gene from the pseudogene.

The 1448 mutation creates a new restriction site. It can be detected easily by amplification of pseudogene-free DNA fragments containing the mutation site and digesting them with the NciI endonuclease or by using the technique of allele-specific oligonucleotide hybridization.

B. Mutation 1226: A→ (370)

This point mutation was found exclusively in patients with type I disease. It is especially prevalent among Jewish patients and accounts for more than

TABLE I Point Mutations in Gaucher Disease

Position						
cDNA	Genomic	Base pair substitution	Exon	Amino-acid change normal → mutant mature protein number	Restriction site created (+) or removed (−)	Disease type
Common Mutations						
1226	5841	A → G	9	Asn → Ser (370)	+ CviJI[a]	I
1448	6433	T → C	10	Leu → Pro (444)	+ NciI	I,II,III
?						
Uncommon Mutations						
476	3060	G → A	5	Arg → Gln (119)	+ BstNI, EcoRII	I
535[b]	3119	G → C	5	Asp → His (140)	+ BspHI, NlaII	I
1093[b]	5309	G → A	8	Glu → Lys (326)	+ BbvII, MboII	I
					− Uba26	
580	3164	A → C	5	Lys → Gly (155)		
1361	5976	C → G	9	Pro → Arg (415)	+ HhaI	II
764	4113	T → A	7	Phe → Tyr (216)	+ KpnI	I
1297	5912	G → T	9	Val → Leu (394)	− HgiEII[a], TaqII[a]	I
1342	5957	G → C	9	Asp → His (409)	− StyI	I
1343	5958	A → T	9	Asp → Val (409)	− AflIII	I
1090	5306	G → A	8	Gly → Arg (325)	+ Bsu36I-AvrII	II
1141	5357	T → G	8	Cys → Gly (342)	− HaeI[a]	II

[a] Commercially unavailable at present.
[b] Both mutations have been detected on the same allele.

75% of the Gaucher alleles, compared with 36% among non-Jewish patients.

Studies correlating the mutations detected and the clinical severity of Gaucher disease have shown an association of the 1226 mutation with a mild phenotypic expression of the disorder. Thus, patients who are homozygous for this point mutation usually have a mild course of Gaucher disease; some are totally asymptomatic. The remainder usually develop symptoms related to enlargement of the spleen and decrease of the platelet count, such as easy bruising, nosebleeds, and abdominal discomfort. Patients who are compound heterozygotes for the 1226 mutation will have a clinical course of disease influenced by the severity of the mutation on the other allele. When combined with the 1448 mutation, the disease course tends to be more severe and is characterized by extensive bone involvement; yet the 1226 mutation protects these patients from the development of central nervous system involvement.

C. Uncommon Point Mutations (Single Reports)

A growing number of various point mutations are being detected in Gaucher disease. Unlike the 1226 and 1448 mutations, which together represent more than 80% of the Gaucher alleles and therefore are detected in all the centers studying the disease, the following point mutations have been identified only in single cases and so far have been reported by a single research group.

1. Mutation 476: G→−A (470)

Detected by sequencing of a cDNA from a Jewish patient with adult type Gaucher disease, this mutation has also been identified in family members of the patient and was shown on expression studies to be the cause for the glucocerebrosidase deficiency. It has not been detected in 70 other unrelated Gaucher patients.

2. Mutation 1361: C→G (415)

This mutation was also identified after sequencing of a full-length cDNA, made from a fibroblast cell line (GM0877) from a patient with type II Gaucher disease, and was shown to cause the enzyme deficiency by transfection expression studies. The genotype of the patient was 1361/1448. This mutation has been found only in that single case and could not be identified among more than a hundred other DNA samples from Gaucher patients tested.

3. Other Point Mutations

Six other point mutations appear in Table I. These point mutations are based on as yet unpublished reports. Mutations 530 and 1093 appeared on the same allele from a non-Jewish patient with type I disease whose other allele had the 580 mutation. While these three point mutations were identified in other family members of the propositus case, they could not be found in more than 55 other DNA samples from unrelated Gaucher patients. Mutation 1053 may be more common; it has been reported in 2 of 12 unrelated adult type Gaucher patients. However, it could not be detected in 60 other samples tested in another laboratory.

D. Glucocerebrosidase Fusion Gene

A different kind of mutation has been identified in a number of patients with Gaucher disease. cDNA has been found to represent a transcript from a fusion gene, in which the 5′ end is the active gene and the 3′ end is the pseudogene. Analyses of genomic DNA have shown that this fusion gene was created as a result of a molecular recombination event of unequal crossing over rather than from a gene conversion. Although rare, this fusion gene is an experiment of nature that provides data regarding the molecular anatomy of the glucocerebrosidase genes and has significant implications in respect to the molecular pathogenesis and diagnosis of Gaucher disease. It indicates that the functional gene is located 5′ to the pseudogene and provides a means for calculating the intergenic distance between the two glucocerebrosidase genes, which is approximately 16 Kb (Fig. 3).

V. DNA Polymorphism in Gaucher Disease

Although few single base-pair changes have been reported in sequences of different DNA clones or between the genomic clones, only one frequent DNA restriction polymorphism of the glucocerebrosidase gene complex, the PvuII polymorphism, has been found. It provided the first evidence at the molecular level for the existence of genetic heterogeneity in Gaucher disease, previously suspected on the basis of clinical observations and family studies. Such heterogeneity exists not just within each of the disease forms, but also among individuals of

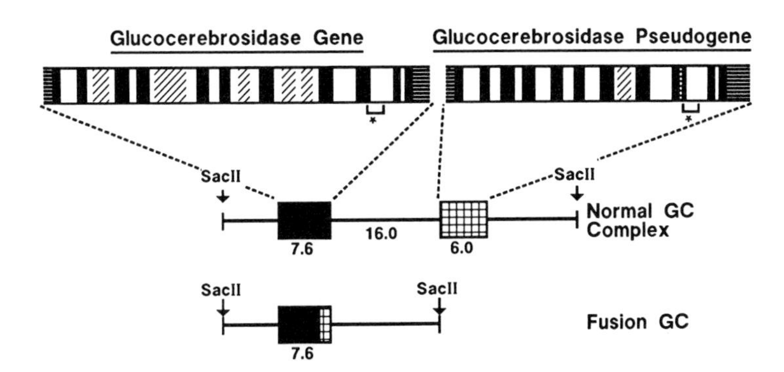

FIGURE 3 The formation of a fusion gene from the glucocerebrosidase (GC) gene and pseudogene. The portion of the genes in which the crossing over occurred are shown (*). SacII restriction analysis of the glucocerebrosidase gene–pseudogene complex shows that the fragment containing the normal glucocerebrosidase gene–pseudogene complex is approximately 49 Kb long, and the one containing the fusion gene is approximately 27 Kb long. The fusion gene is the same length as the functional GC gene (7.6 Kb), because all of the missing segments of the pseudogenes (which made it approximately 1.6 Kb shorter) are 5′ to the point of crossing over. Most of these segments are Alu sequences and are represented as diagonal dashed areas on the GC gene diagram. The 55 base pair deletion in exon 9 of the pseudogene is represented by white dots. The horizontal lines in exons 1 and 11 are the noncoding regions. Since the length of both the GC gene and pseudogene are known, as are the lengths of the SacII fragments, the intergenic distance can be calculated as being about 16 Kb. (From A. Zimran, J. A. Sorge, E. Gross, M. Kubitz, C. West, and E. Beutler. (1990). Glucocerebrosidase fusion gene in Gaucher disease—implications for the molecular anatomy, pathogenesis and diagnosis of this disorder. *J. Clin. Invest.* 85, **219,** 1990. Reprinted with permission from the Rockfeller University Press.).

the same ancestry, such as the Jewish population. The finding of PvuII polymorphism in diverse racial groups further suggests its ancient origin.

The PvuII polymorphism has been characterized by the presence or absence of a 1.1 Kb fragment (see Fig. 4); thus, genes that generated the 1.1 Kb fragment after digestion with PvuII were designated $Pv1.1^+$, and those from which it was absent were designated $Pv1.1^-$.

Having the complete sequence of both $Pv1.1^+$ and $Pv1.1^-$ human glucocerebrosidase genes, the site of the PvuII polymorphism could be localized to intron 6, consisting of a G → A single base substitution at position 3931 on the genomic sequence, and methods have been developed for rapid detection of this polymorphism using the polymerase chain reaction. Analysis of more than 50 Gaucher patients with various genotypes has shed light on the genetic evolution of Gaucher mutations (see Table II). The first observation is the complete linkage between the $Pv1.1^-$ genotype and the common "Jewish" mutation 1226. This suggests that the 1226 mutation arose only once in a $Pv1.1^-$ allele. In contrast, it seems likely that the 1448 mutation arose more than once since one of the patients homozygous for the 1448 mutation was also homozygous for $Pv1.1^-$ and the other two homozygous for $Pv1.1^+$. Finally, the fact that Jewish patients heterozygous for the 1226 allele and one unidentified allele (the 1226/?genotype) were predominantly $Pv1.1^+/Pv1.1^-$ suggests that one of the unknown alleles may be relatively common and linked to the $Pv1.1^+$ genotype.

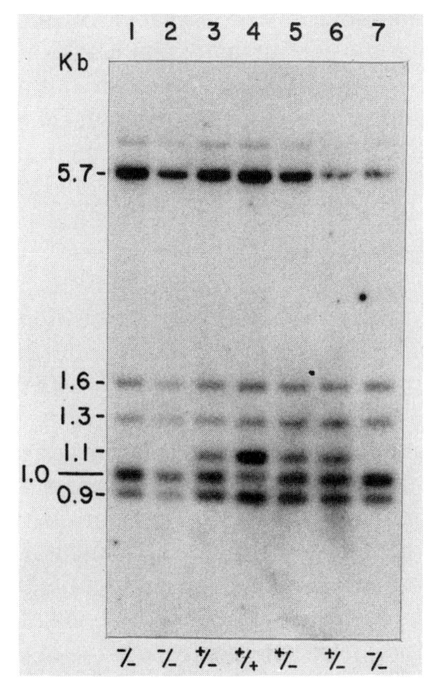

FIGURE 4 Southern blots of DNA digested with Pvu II. When developed with a glucocerebrosidase cDNA probe, it is apparent that the DNA from some individuals has a 1.1 Kb fragment, which is missing from others. The genotype is given at the bottom of each lane. (From E. Beutler. 1988. Gaucher disease: New developments. *Curr. Hematol. Oncol.* **6:**1–26. Reprinted with permission from Yearbook Medical Publishers, Inc.).

TABLE II PvuII Polymorphism and Genotype of 64 Gaucher Disease Patients

Gaucher genotype	Number of patients	PvuII genotype +/+	+/−	−/−
1226/1226	27	—	—	27
1226/1448	6	—	3	3
1226/?	21	—	16	5
1226/Xovr	1	—	—	1
1448/1448	3	2	—	1[a]
1448/1361	1	—	—	1[b]
1448/?	2	—	1	1
?/?	2	2	—	—
764/?	1	—	1	—

[a] Cell line GM0877
[b] Cell line GM1260

VI. Diagnosis of Gaucher Disease

Classically, the diagnosis of Gaucher disease has been established by histological examination of the bone marrow, spleen, or liver and identification of Gaucher cells. The development of a reliable enzymatic assay for the β-glucosidase of peripheral blood lymphocytes or, if available, of cultured skin fibroblasts not only obviates the need for performing an invasive diagnostic procedure but is actually more accurate, since a variety of clinical disorders, including certain infections, leukemias, and lymphomas, can cause the appearance of Gaucher-like cells. On the other hand, low β-glucosidase activity is pathognomonic for Gaucher disease. Enzymatic assay also provides the opportunity to perform a prenatal diagnosis of Gaucher disease by measuring the β-glucosidase activity of amniotic fluid cells. Yet the enzymatic assay itself is not without limitation. The detection of heterozygotes is not sufficiently accurate to allow for precise calculation of the gene frequency for epidemiologic purposes or for genetic counseling. It does not correlate with the severity of the disease among patients with the adult form and cannot even differentiate between this form of the disease and the neuronopathic forms. Thus, prenatal diagnosis of the disease usually leads to termination of the pregnancy even though the potential severity of the disease may not justify it.

With the recent progress in molecular biologic techniques, diagnosis of genetic disorders can be performed at the DNA level by identification of the mutations causing the disease. Gaucher disease is not an exception. However, unlike sickle cell anemia, which is caused by a single point mutation, there are multiple mutations associated with Gaucher disease, and therefore not all Gaucher cases can be diagnosed by DNA analysis. Mutation analysis can establish the complete and accurate diagnosis when the mutations on both alleles are identified, or it can provide important information even when only one of the Gaucher alleles is identified. Finding the 1226/1226 genotype can be encouraging since it suggests a mild clinical course. At the other extreme, finding 1448/1448 or 1448/1361 genotypes predicts a neuronopathic form of Gaucher disease. Detection of 1226 mutation on one allele with 1448 (or an as yet unidentified mutation) on the other allele may not be sufficient for an accurate prognostication of the severity of the disease; yet it can definitely exclude the fatal neuronopathic forms.

When more mutations are detected, DNA analysis will become increasingly useful in prenatal diagnosis, genetic counseling, and population genetic studies. [*See* DNA Markers as Diagnostic Tools.]

All methods used today for the analysis of known mutations of Gaucher disease include the use of the polymerase chain reaction (PCR). Separation of the active glucocerebrosidase gene from the pseudogene is accomplished by different approaches, including allele specific oligonucleotide amplification, allele-specific oligonucleotide hybridization, and analysis of color PCR with an automatic DNA sequencer.

VII. Gene Therapy of Gaucher Disease

Currently available treatment for Gaucher disease is purely symptomatic in nature. Splenectomy relieves symptoms caused by the enlarged organ or by low blood counts. Orthopedic procedures, including joint replacement, can improve joint mobility and control pain. Experimental treatment with enzyme replacement has recently been found to be effective. In a few cases, cure of the disease has been achieved by bone marrow transplantation, but this procedure, even when a HLA-matched donor is available, carries a high risk, and the patient may succumb to its complications. [*See* Bone Marrow Transplantation.]

In the future, cure of the disease may be achieved

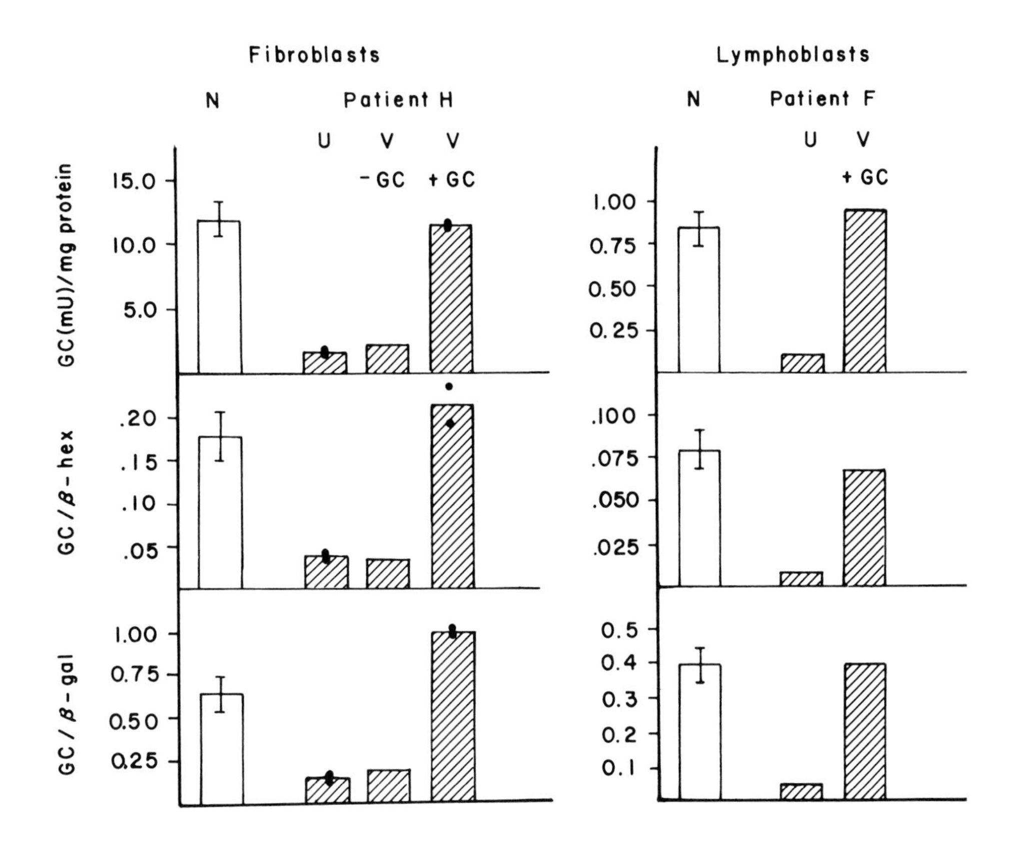

FIGURE 5 The restoration of glucocerebrosidase activity in fibroblasts of two type I Gaucher disease patients. N represents normal control cells. Enzyme assays were carried out on uninfected cells (U), cells infected with a retrovirus containing the neo gene but not the glucocerebrosidase cDNA (V − GC), and a retrovirus containing both the neo gene and the glucocerebrosidase cDNA (V + GC). (From J. Sorge, W. Kuhl, C. West, *et al.* 1987. Complete correction of the enzymatic defect of type I Gaucher disease fibroblasts by retroviral-mediated gene transfer. *Proc. Natl. Acad. Sci. U.S.A.* **84**:906–909.)

by gene replacement therapy. This mode of treatment for many different genetic disorders has become the topic of intensive research because of the availability of cloned genes and of the methodology required to introduce them into mammalian cells in a functional state. In the case of Gaucher disease, glucocerebrosidase cDNA has been successfully inserted into cultured fibroblasts and lymphoblasts from Gaucher patients, and the enzymatic defect has been corrected (see Fig. 5). Gaucher disease is suitable for this treatment modality since the cells expressing the enzymatic deficiency are progeny of bone marrow precursor cells, which are easily accessible and manipulated *in vitro*. Infection of such hematopoietic stem cells from Gaucher patients by glucocerebrosidase cDNA in a recombinant retrovirus and autotransplantation of the transformed cells could cure the disease without exposing the patients to the high risks of allogeneic bone marrow transplantation. Currently, there are two major obstacles to the implementation of gene therapy. First, the efficiency of transformation of bone marrow cells is too low. Second, the expression of the retroviral genes in primitive cells, unlike mature cells such as cultured skin fibroblasts, is very limited and effectively suppressed. Once these barriers are overcome, gene replacement therapy may become a reality.

VIII. Conclusions

Gaucher disease is one of a small number of human genetic disorders that have been studied at the molecular level in considerable detail. Such investigations have already led to improved diagnosis and the ability to predict disease outcome. Gaucher disease is an outstanding target for the implementation of gene transfer therapy, and it is to be hoped that it will be possible to offer this type of treatment to patients with this disease in the not-too-distant future.

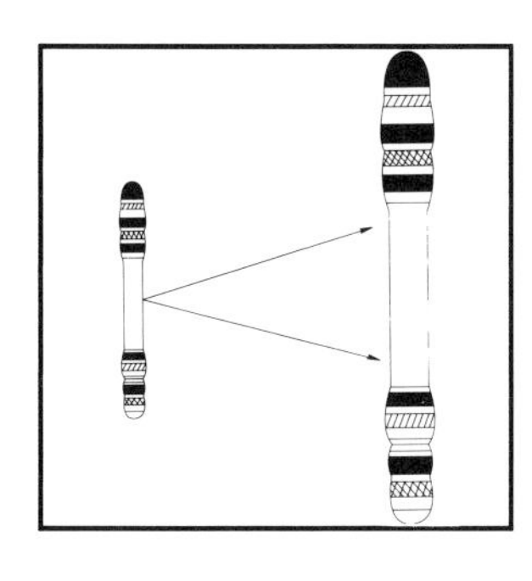

Gene Amplification

SHYAM DUBE, *University of Maryland*

JOSEPH R. BERTINO, *Memorial Sloan–Kettering Cancer Center*

Glossary

Amplicon Unit-repeated sequence or amplification unit presents in multiple copies and arranged in linear tandem arrays in amplified DNA

Amplification control elements *cis*-Acting DNA sequence element identified in *Drosophila* chorion clusters and shown to be absolutely essential for the amplification of chorion genes. Although their identification has been best documented in *Drosophila*, it is likely that such elements are involved in other instances of gene amplification

Amplification enhancer regions DNA sequence elements identified in *Drosophila* chorion clusters and shown to have an enhancing function in the amplification of chorion DNA. While amplification can occur in the absence of these regions, it is markedly increased in their presence. Although their identification has been best documented in *Drosophila*, it is likely that such elements are involved in other instances of gene amplification

Double-minute chromosomes Self-replicating extrachromosomal elements observed in tumors, tumor cell lines, and cells carrying amplified genes. They lack centromeres, which precludes their attachment to the mitotic spindle, resulting in their erratic segregation and frequent loss from the daughter cells

Gene amplification Overreplication of a small region of a cell's genome to produce multiple extra copies of a specific DNA sequence

Homogeneously staining region Chromosomal region that lacks the characteristic irregularly spaced dark bands observed in G-banded preparations of most chromosomes. They stain uniformly lightly, uniformly darkly, or exhibit fine bands at regular intervals against a background of lighter uniform staining

A FUNDAMENTAL trait of living systems is their ability to preserve the gross structure of their genome across cell generations while retaining the flexibility to accommodate alterations in the genome in response to developmental needs or environmental pressure. They are thus able to ensure the maintenance of their identity while improving the chances of their survival. One of the means by which alterations in the genome are accomplished is gene amplification. By this process a small region of a cell's genome is overreplicated, producing multiple extra copies of a specific DNA sequence. A cell is thus able to cope with sudden needs for an unusually large amount of a specific gene product. It is also likely that mechanisms of this general nature have played a key role in the generation of multigene families in genome evolution for the development of complex genomes of higher organisms.

I. Gene Amplification and Development

Developmentally regulated gene amplification has been best documented in two types of cells that need to make an unusually large amount of one gene product relatively quickly. One is the developing egg (i.e., oocyte) of many animals, which must make and store huge numbers of ribosomes. The other is the follicle cell of insects. In young *Xenopus* oocytes ribosomal RNA genes are amplified about 1000-fold, producing many free ribosomal DNA molecules that cluster together to form about 1000 extra nucleoli. The follicle cells of insects synthesize and secrete the proteins that form the hard coat, or chorion, of insect eggs. In *Drosophila*, just before the chorion proteins are needed, the DNA sequences coding for them are amplified 20-fold in the sex-linked chorion cluster and 60- to 80-fold in the chorion cluster of the third chromosome. While in *Drosophila* the problem of rapid synthesis of chorion genes is solved in developmental time, silk moths provide an interesting instance in which this has been done in evolutionary time, such that the present-day silk moths come endowed with approximately 25 copies of these genes in their germ line for developmentally regulated expression during choriogenesis. [*See* GENES.]

II. Gene Amplification and Environmental Pressure

While developmentally specific gene amplification occurs in the absence of selection, in a vast majority of other instances amplification is revealed only by selection. For example, tumors or tissue culture cells treated with a variety of drugs become drug resistant by amplifying the gene encoding the enzyme inhibited by the particular drug. This phenomenon has been observed in every system investigated, including bacteria, fungi, plants, insects, and other prokaryotes or eukaryotes. Indeed, inhibitors that kill cells or arrest their growth are frequently used for selecting cells with amplified genes in them.

An interesting example of gene amplification revealed under environmental pressure is provided by the flax plant. In response to any one of a variety of unusual conditions (e.g., too much phosphate or too little nitrogen in the soil), flax plants can produce seeds that generate daughter plants with growth properties that differ from those of the parent plant. Major changes in the DNA are found to occur in this sudden transition. One type of daughter plant contains 10% more DNA in each cell than does the normal parent, because of the differential amplification of many different DNA sequences, although the nature of these sequences has not been elucidated.

III. Structure of Amplified DNA

Amplified DNA forms a variety of structures. It is found in tandem repeats of a chromosomal segment, including the selected gene and variable amounts of flanking DNA. This repeat structure exists in the form of small acentric extrachromosomal elements called double-minutes, or in an integrated form in chromosomes forming expanded or homogeneously staining chromosomal regions (Fig. 1). Amplified ribosomal DNA in *Xenopus* oocytes is in the form of extrachromosomal circles, while amplified chorion genes in *Drosphila* are found in a multiforked structure generated by multiple rounds of initiation within the chromosome. In drug-resistant *Leishmania* amplified DNA forms extrachromosomal supercoiled circles.

IV. Regulation of Gene Amplification

The best-documented evidence on how gene amplification might be regulated comes from studies of the amplification of chorion genes in *Drosophila*, which

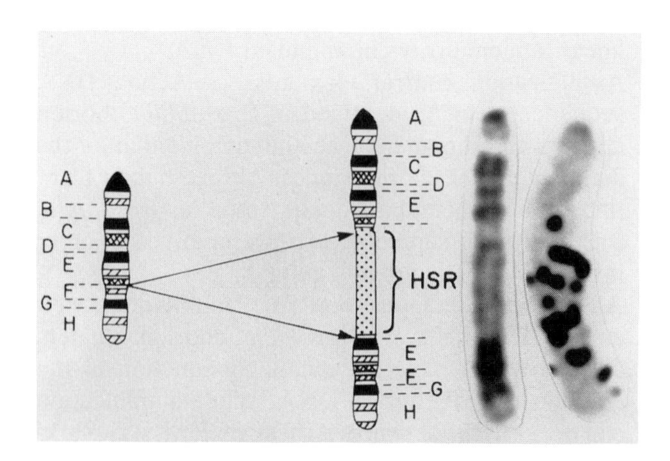

FIGURE 1 A normal mouse chromosome 2 from a methotrexate-resistant subline containing amplified (200-fold) dihydrofolate reductase genes. The homogeneously staining region (HSR) appears to be located within a reduplicated interstitial segment of chromosome 2 (region E). A subtle pattern of repeated gray bands is seen within the HSR. Enlargement of the marker chromosome from an *in situ* hybridization experiment demonstrates the clustering of silver grains in the region corresponding to the HSR. [From *Cytogenet. Cell Genet.* **29,** 143–152 (1981). Reprinted by permission of S. Karger AG, Basel.]

have revealed the involvement of both *cis* elements and *trans*-acting factors. Two types of *cis* elements have been identified: an amplification control element (ACE) and multiple amplification enhancer regions (AERs). It was found that while small ACE-containing fragments could autonomously amplify with correct developmental specificity, they could not sustain many rounds of amplification without AERs, thus defining the distinguishing features of the two *cis* elements.

Although the genetic functions needed in *trans* for normal amplification have not been well characterized, their requirement has been established by demonstrating that mutations in certain genes unlinked to the two chorion clusters markedly reduce amplification without changing the chorion transcription per gene copy. One or more *trans*-acting factors could be related to DNA replication functions, to which amplification is especially sensitive.

V. Induction of Gene Amplification

Although DNA amplification is a spontaneous process, its frequency can be increased by the exposure of cells to certain drugs (e.g., hydroxyurea) which interfere with DNA synthesis or to such agents as ultraviolet radiation, X-rays, and carcinogens, which damage DNA and lead to the activation of repair mechanisms. In addition, arsenate and transient hypoxia have also been shown to increase amplification frequencies. Upon restoration of DNA synthesis after any of the above treatments, it is believed that a subset of cells with a DNA content greater than 4C is generated, and it is from this subset that cells with increased frequencies of gene amplification emerge. It is not certain whether the generation of cells with greater DNA content per cell is achieved by multiple initiations of DNA replication within a single cell cycle or by a mechanism involving an alteration in the normally ordered progression from the S phase into mitosis, such that mitosis is delayed and occurs during the subsequent S phase.

VI. Amplification and Drug Resistance

Since the first demonstration that resistance to methotrexate, a specific powerful inhibitor of its target enzyme, dihydrofolate reductase, was due to amplification of the gene coding for this protein,

TABLE I Amplified Copies of Specific Genes in Drug-Resistant Cell Lines[a]

Selective drug	Amplified gene
Methotrexate	Dihydrofolate reductase
5-Fluorodeoxyuridine	Thymidylate synthase
PALA	CAD
Hydroxyurea	Ribonucleotide reductase
Deoxycoformycin	Adenosine deaminase
Coformycin and adenine	AMP deaminase
Difluoromethyl ornithine	Ornithine decarboxylase
Doxorubicin, colchicine, vinca alkaloids	p170 Glycoprotein
Albizzin	Asparagine synthetase
Hypoxanthine–aminopterin–thymidine	HGPRTase (mutant)
Compactin	Hydroxymethylglutaryl-coenzyme A reductase
6-Azauridine, pyrazofurin	UMP synthetase
Methionine sulfoximine	Glutamine synthetase
Ouabain	Na^+/K^+-transporting ATPase
Histidinol	Histidyl-tRNA synthetase
Tunicamycin	*n*-Acetylglucosaminyltransferase
Berrelidin	Threonyl-tRNA synthetase
Cadmium	Metallothionein

[a] PALA, *N*-Phosphoacetyl-L-aspartate; CAD, carbamoyl-phosphate synthetase-aspartate carbamoyltransferase-dihydroorotase; HGPRTase, hypoxanthine–guanine phosphoribosyltransferase.

almost 20 other examples of resistance to specific inhibitors as a result of gene amplification have been described (Table I). Except for amplification of the p170 glycoprotein, associated with multidrug resistance (MDR), the amplified gene code for an enzyme inhibited by the inhibitor. In the case of multidrug resistance, the protein overexpressed is a membrane protein, believed to be important in the energy-dependent efflux of several antitumor agents, including doxorubicin, colchichine, vinca alkaloids, and actinomycin D. Selection with any of these drugs leads to resistance to others in this class.

While alteration of transcriptional regulation might also result in the overexpression of proteins without gene amplification, except for a few examples (e.g., multidrug resistance in patient tumors and arginosuccinate synthetase increase in canavanine resistant fibroblast cells), gene amplification is a major mechanism of drug resistance in mammalian cell lines. Other mechanisms of resistance might also occur and are not mutually exclusive (e.g., a highly resistant methotrexate cell line could contain amplified and altered dihydrofolate reductase genes).

In general, the amplification of drug-resistant genes occurs in stepwise fashion, and the gene copy number increases as the selection pressure is increased. Single-step selection with high levels of a drug is more likely to produce resistant cells with other resistant mechanisms. As mentioned, amplified genes can reside on one or more chromosomes or on extrachromosomal elements called double-minute chromosomes. In the latter circumstance the resistance can be unstable. During recent years examples of gene amplification in tumors from patients resistant to methotrexate and 5-fluorodeoxyuridine have been reported, but the incidence and importance of this phenomenon in the clinic are not well established.

VII. Oncogene Amplification in Tumors

Amplification of cellular oncogenes which occur in homogeneously staining regions or in double-minute chromosomes have been reported with increasing frequency in fresh tumor specimens from patients. Table II lists some of the oncogenes that have been found to be amplified in various human tumors. These amplified oncogenes appear to be related to the degree of aggressiveness of the patient's tumor. For example, high levels of amplification of N-*myc* and C-*myc* are associated with a poor prognosis in patients with neuroblastoma and small-cell lung cancer, respectively. Analogous to drug resistance, alterations in the coding sequence and structure of the oncogene product could also result in a phenotype with a growth advantage, as noted for the *ras* oncogene. [*See* ONCOGENE AMPLIFICATION IN HUMAN CANCER.]

TABLE II Amplified Oncogenes in Human Cancers

Oncogene	Cancer
C-*myc*	Lung, breast, Burkitt's lymphoma
N-*myc*	Neuroblastoma
L-*myc*	Lung
Ki-*ras*	Lung, ovarian, leukemia
HER-2/Neu	Breast, ovarian
Epidermal growth factor receptor (*erb-B*)	Squamous cell carcinoma, glioma
C-*myb*	Acute lymphocytic leukemia

VIII. Coamplification—Inadvertent and Advertent

In most cases of gene amplification associated with drug resistance, the size of the amplicon, or repeating unit, is much larger than the size of the gene coding for the target protein. In some instances other genes have been found to be coamplified, as "piggyback" genes that presumably are located adjacent to the primary gene. For example, in some, but not all, cells that have amplified *MDR* genes, a gene coding for a calcium-binding protein of unknown function (sorcein) can also be amplified. Our laboratory has also found that some methotrexate-resistant human tumor lines that have high levels of dihydrofolate reductase gene amplification also contain an increase in copies of hydroxymethylglutaryl-coenzyme A reductase genes. Recently, a neuroblastoma cell line containing an amplified oncogene, N-*myc*, was shown also to contain amplified sequences homologous to ornithine decarboxylase. Thus, amplification of any gene can be associated with coamplification of an adjacent gene and could impart phenotypic changes to the cell not expected unless the coamplified gene is detected.

In the above circumstances the coamplified process is inadvertent (i.e., not planned or expected). In other situations introduction of a drug-resistant gene into a cell, together with another gene, whose product is desired, could be a useful technique to obtain overexpression of the latter gene, after stepwise selection with the appropriate drug. For example, cotransfection of cells with mammalian expression vectors coding for dihydrofolate reductase and human tissue-type plasminogen activator (t-PA) and subsequent selection in methotrexate have provided a useful cell line for the production of t-PA.

IX. Mechanisms of Gene Amplification

The mechanisms involved in the generation of multiple copies of an amplified gene in mammalian cells are not well understood. However, there are two favored models, not necessarily mutually exclusive, for the formation of linear amplified structures observed in mammalian cells.

The overreplication/resolution model proposes that a chromosomal domain is overreplicated, as in the case of chorion genes in *Drosophila*, in which the bidirectional migration of the replication fork and the generation of a multiforked structure are well documented. In contrast to *Drosophila*, however, in which the multiforked structure is not resolved, resolution of this structure in mammalian cells is proposed to occur by extensive recombination, resulting in the generation of an unbranched transmissible DNA molecule, either free double-minutes or integrated into a preexisting chromosome (i.e., homogeneously staining region). This model allows large increases in the copy number per cell generation.

The deletion-plus-episome model proposes that amplification starts with deletion of a region of DNA encompassing the selected gene and *cis*-acting sequences required for autonomous replication. The autonomously replicating elements, called episomes, segregate unequally and increase progressively over time, due to the selective advantage they impart. These molecules replicate once per cell cycle, and in this model, therefore, amplification to high-copy levels is achieved gradually.

Both models have attractive features, but also some problems. In the first model it is not clear how a multiforked domain produced by multiple initiations within a cell cycle would generate a particular novel joint more than once, unless a recombination "hot spot" was present, for which there is no evidence at present. Since many amplified genes have been found in exact copies of the same unit, joined head to tail or as a tandem array of inverted repeats, it is unlikely that such structures are derived by random recombinations from a multiforked domain. Alternative overreplication models invoking only one initiation have been proposed to explain the generation of exact copies, but there is no direct evidence for such models. The difficulty that the second model faces concerns the question of what causes the first circle to generate episomes. While the model proposes that recombination within replication intermediates could explain the genesis of these structures, evidence for this is lacking.

Bibliography

Delidakis, C., Swimmer, C., and Kafatos, F. C. (1989). Gene amplification, an example of genome rearrangement. *Curr. Opin. Cell Biol.* **1,** 488–496.

Hamlin, J. L., Milbrant, J. D., Heinz, N. H., and Azizkahn, J. C. (1984). Sequence amplification in mammalian cells. *Int. Rev. Cytol.* **90,** 31–82.

Schimke, R. T. (1984). Gene amplification in cultured animal cells. *Cell* **37,** 705–713.

Schimke, R. T. (1988). Gene amplification in cultured cells. *J. Biol. Chem.* **263,** 5989–5992.

Stark, G. R. (1986). DNA amplification in drug resistant cells and tumors. *Cancer Surv.* **5,** 1–23.

Stark, G. R., Debatisse, M., Giulotto, E., and Wahl, G. M. (1989). Recent progress in understanding mechanisms of mammalian DNA amplification. *Cell* **57,** 901–908.

Wahl, G. M. (1989). The importance of circular DNA in mammalian gene amplification. *Cancer Res.* **49,** 1333–1340.

Warr, J. R., and Atkinson, G. F. (1988). Genetic aspects of resistance to anticancer drugs. *Physiol. Rev.* **68,** 1–26.

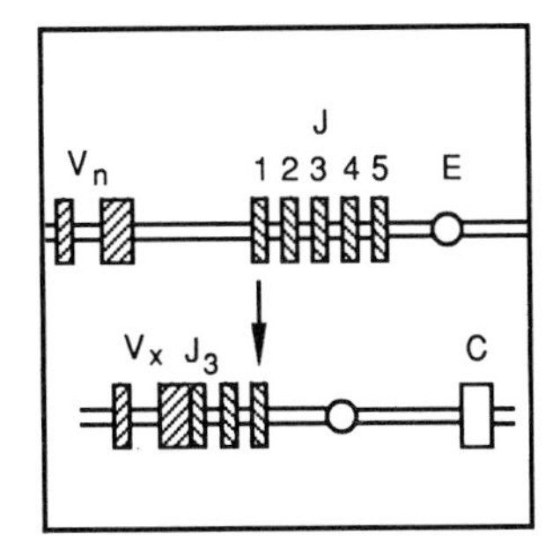

Genes

H. ELDON SUTTON, *The University of Texas at Austin*

Glossary

Allele Alternate form of a gene
Chromosome Structure found in the nuclei of eukaryotes that contains DNA and proteins; the means by which a cell distributes genes during cell division
Complementation Formation of a normal phenotype when two "alleles" are on different members of a pair of homologous chromosomes; ordinarily this indicates that the mutations are in different loci
DNA Deoxyribonucleic acid; the molecule that serves for the primary storage of genetic information; a double-stranded structure composed of nucleotides
Eukaryotic Having cells with nuclei; prokaryotes, such as bacteria, do not have nuclei
Exon Segments of a gene that appear in messenger RNA after processing of the primary RNA transcript
Gene Basic unit of heredity
Intron Segments of a gene that are removed during processing of the primary RNA transcript
Mendelian Heredity characterized by segregation of either of two alleles into gametes and recombination in zygotes
Mutation Any abrupt heritable change in the genetic material
Protein Large linear molecules composed of amino acid subunits; the sequence of amino acids is coded by genes
Transcription Process by which genetic information encoded in DNA is copied into RNA

GENES ARE THE BASIC units of biological heredity. They are the elements of the blueprint, the genotype, received by a cell from its parents from which it builds a complete organism. The characteristics of that organism, the phenotype, is a product of the interaction of the genotype with the environment. Genetic information is encoded in the nucleotide sequences of DNA in a linear, quaternary code. This information eventually is translated into the kinds and amounts of proteins that an organism can synthesize. Variations in the coded information account for the inherited variations that exist among individuals in a population, among populations within a species, and among different species. Since present-day genes in all species evolved from a few primordial genes by duplication and diversification, the evolutionary relationships among genes and among species can often be noted in the similarities of gene structure.

I. Definitions of a Gene

A. Genes as Units of Variation

The observations of Mendel and the early Mendelians were concerned with inherited variation. The physical structures responsible for the variations were, of course, unknown. Only by observing variation could one deduce the existence of a unit that, in its various forms (allelomorphs or, more commonly, alleles), accounted for the inheritance of different phenotypes. This unit was given the name gene in 1909.

Genes were observed to have various properties. They could mutate to form a different allele, a fact that became incorporated into the concept of a gene. The gene as a unit of mutation is not a generally useful concept, however, since we now know that mutation ignores gene boundaries. Genes, or rather specific combinations of alleles, could be separated by crossing-over, giving rise to the idea of a gene as a bead on a string. Again, the gene as a unit of recombination failed with the discovery that crossing-over occurs within genes as well as between.

B. Genes as Units of Function

A major shift in the concept of a gene occurred with studies in the 1930s and 1940s on gene function. Each mutant allele was associated with loss of function of a single specific enzyme. The normal ("wild-type") allele was therefore responsible for production of functional enzyme. These results were anticipated by studies on metabolic diseases in humans, but such studies had little impact because they represented too great a jump from established knowledge and, because the observations were on humans, experimental studies were limited.

Enzymes are proteins, the only important exceptions being some RNA molecules that have recently been found to catalyze certain reactions. The connection between genes and proteins was established in 1956 with the demonstration that the amino acid sequence of human sickle-cell hemoglobin is altered as compared with the sequence in normal hemoglobin. Therefore, one function of genes is to specify the amino acid sequence (primary structure) of proteins. Other studies showed that mutations in the tryptophan synthetase gene of *Escherichia coli* can be arranged by genetic analysis into a linear sequence and that the corresponding amino acid changes in the protein product have the same linear sequence; i.e., the two structures are colinear. Therefore, one could consider a gene as a DNA sequence that codes for a corresponding amino acid sequence in a protein.

C. Genes as Units of Transcription

As knowledge of the molecular events of DNA function and protein synthesis have expanded, the definition of a gene has also evolved. It is now accepted that the function of a gene is largely described by the amount and structure of the RNA that is transcribed. Furthermore, some genes, such as ribosomal RNA or transfer RNA genes, have RNA as their final product rather than protein. Therefore, a gene can be more generally defined as a unit of transcription. Because the amount of transcribed RNA is regulated by adjacent DNA nucleotide sequences, these must also be included within the boundaries of the gene, even though they are not transcribed. These contiguous regulatory sequences are to be distinguished from more remote DNA segments (enhancer regions) that also influence transcription of particular genes. [*See* DNA AND GENE TRANSCRIPTION.]

D. Pseudogenes

DNA probes have allowed the detection of specific nucleotide sequences and, hence, of any gene or other DNA region that contains sequences complementary to the probe. Although many probes detect a single nucleotide sequence in the genome, others detect more. In some instances where several complementary copies exist, only one is ever transcribed; the remainder are pseudogenes, i.e., genes whose structures resemble those of functional genes but are not functional. Pseudogenes are thought to have arisen primarily by gene duplication with subsequent changes that caused loss of function. Some appear to have resulted from insertion of cDNA; i.e., a DNA copy is made from transcribed and processed RNA and is then inserted back into the DNA but without the essential regulatory regions.

E. Alleles

An allele is defined as an alternate form of a gene (alternate being understood in the modern sense as a variant DNA sequence). Until it became possible to analyze gene structure and function at the molecular level, the demonstration that two genetic variants were modifications of the same gene was formally impossible; rather, arguments were based on the absence of crossing-over (which produces recombinant products) when the two variants were in heterozygous combination in the same diploid cells. Supporting evidence was provided by the inability of the two variants to complement each other, suggesting that the defects were in the same functional group. We now know that crossing-over can occur within genes, albeit extremely rare, providing false evidence of nonallelism. Also, there are interpretations of complementation other than nonallelism.

Nonallelism of distant loci can readily be established by the frequency of recombination between two variants; however, if two variants map very close to each other (i.e., there is a very low frequency of recombination between them), convinc-

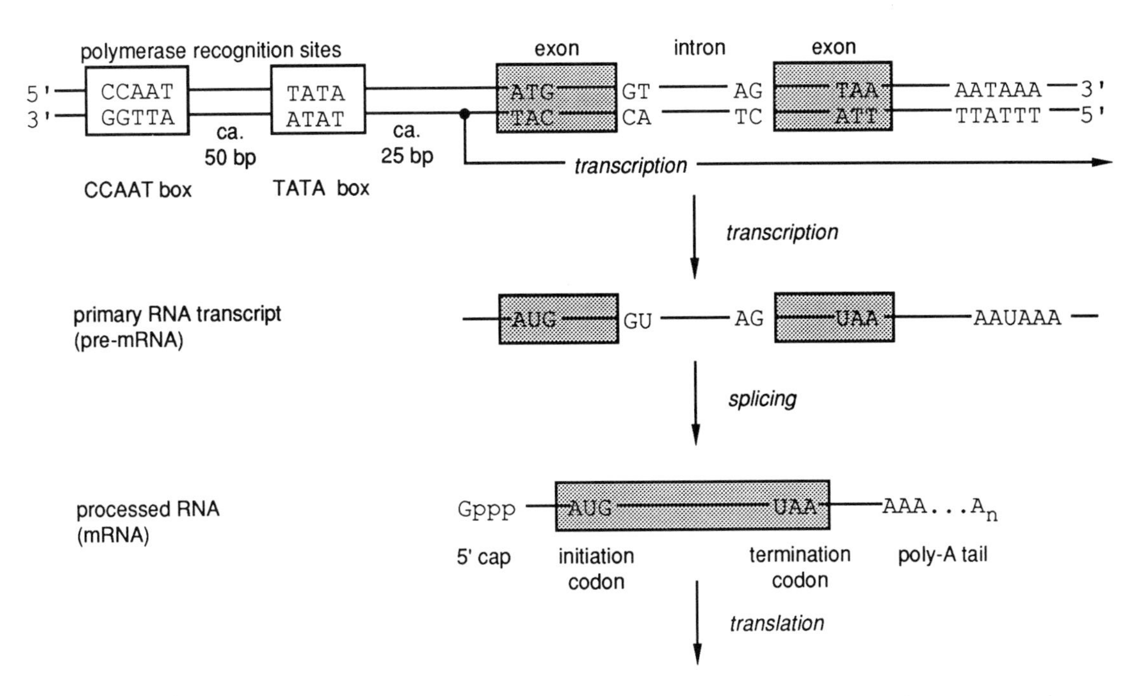

FIGURE 1 Diagram of a typical eukaryotic gene. The sense strand is at the top, and the antisense strand, which serves as the template for transcription, is on the bottom. Only one intron is shown, but the coding regions (exons) may be interrupted by dozens of introns. The primary RNA transcript is processed to remove introns (splicing) and the 5′ cap and poly-A tail are added to form messenger RNA (mRNA). The information coded in mRNA is used in translation to determine the sequence of amino acids in the corresponding protein. [Adapted, with permission, from H. E. Sutton, (1988). "An Introduction to Human Genetics" 4th ed. p. 228. Harcourt Brace Jovanovich, San Diego.]

ing evidence of allelism or nonallelism can only be provided by knowledge of the DNA changes associated with the variants.

Because any change in nucleotide sequence can be transmitted as a stable variant, the number of possible alleles at any locus is very large. Many will be lethal, but many others, perhaps most, are likely to be viable, at least in heterozygous combination with a "normal" allele.

II. Structure and Function of Eukaryotic Genes

A. DNA Structure of Genes

Typical eukaryotic genes that code for proteins have the structure shown at the top of Fig. 1. The top strand is the sense strand of a DNA double helix, the 5′ to 3′ direction conventionally written from left to right. The complementary 3′ to 5′ strand, the antisense strand, serves as the template for RNA transcription. The 5′ to 3′ direction of the sense strand and of the transcribed RNA corresponds to the N-terminal to C-terminal direction of the corresponding polypeptide chain.

The protein-coding regions, present in exons, are separated by regions that do not code. These introns or intervening sequences are transcribed into RNA but must be removed before protein synthesis. The number of exons varies greatly. For example, the globin genes have three; other genes are known to have several dozen. The relative sizes of the exons and introns are quite variable, with many introns being larger than the entire coding region of a gene.

B. Transcription of Genes

RNA transcription refers to copying the nucleotide sequence of a segment of DNA into a complementary sequence of RNA (Fig. 1). Transcription is initiated some dozens of nucleotides in the 5′ direction (upstream) from the coding region and continues in the 3′ direction (downstream) past the coding region. The exact point of initiation is determined by the location of the sites at which the RNA polymerase that catalyzes transcription is bound. The point of termination of transcription is not well defined. The nucleotide sequence AATAAA occurs at some distance downstream from the coding region, and transcription extends beyond that, presumably until some nucleotide sequence signals the termination.

The amount of transcription, indeed, whether or not transcription occurs at all, is dependent on regulatory elements in the 5′ direction from the point at which transcription is initiated. Two important nucleotide sequences are recognized: TATA or some close variation (the TATA box) occurs some 25 nucleotides 5′ from the initiation point, and CAAT (the CAAT box) some 50 nucleotides from the initiation point. These sequences are the binding sites for RNA polymerase, and transcription does not occur if they are absent. Mutations in other upstream regions of specific genes may alter the rate of transcription, although attempts to develop general rules that relate DNA structure to transcription have been only partially successful.

A number of mutations downstream from the coding regions are known to reduce protein synthesis. These are thought to affect translation rather than transcription.

The coding region and its contiguous regulatory regions were noted earlier as constituting a gene when defined as a transcription unit. Enhancer regions introduce some ambiguity into this definition. An enhancer is defined as a DNA segment that enhances transcription of a specific gene that is located on the same chromosome but that is ordinarily noncontiguous. The enhancer may be in either the 5′ or 3′ direction from the coding region and may be thousands of nucleotides distant. Yet, it affects only the gene on the same chromosome and not the same gene on a homologous chromosome. Therefore, there must be a physical interaction between the enhancer region and the target gene and not interaction by means of a diffusible gene product.

Enhancer regions have been detected for relatively few genes, but their demonstration is difficult. Mutations in an enhancer region can lead to loss of product of the target gene, just as mutations can within the coding or regulatory regions of the target gene. Only by analysis of the DNA structure of a mutant or by fine structure mapping can the distinction occasionally be made.

Transcription is also regulated by the protein products of other genes. Such transregulation is apparently responsible for some of the events of differentiation and for regulation of physiological functions. While it is useful to recognize the function of certain genes as regulatory, the structures of regulatory genes themselves are not known to be different from other genes.

C. Comparison of Eukaryote Genes with Other Genes

The basic patterns of genetic coding, transcription, and translation are remarkably similar throughout the biological world. Nevertheless, there are variations in detail, some of which are instructive about the evolutionary relationships among different major biological groups. The gene structure of eukaryotes (i.e., organisms whose cells have nuclei and chromosomes) have been described above. These patterns apply to such diverse eukaryotes as humans, plants, and yeast.

The genes of prokaryotes (organisms whose cells lack nuclei and chromosomes) have some characteristic differences from eukaryotes. One difference is the absence of introns. Prokaryote genes code without interruption from the initiation codon to the termination codon. Another difference is the organization of some prokaryote genes into operons (i.e., gene structures in which DNA segments that code for separate polypeptides are part of the same transcription unit under the control of a single 5′ regulatory region).

Mitochondrial genes are similar to prokaryote genes rather than eukaryote nuclear genes in some respects, supporting the idea that mitochondria arose as prokaryote symbionts. For example, mitochondrial genes have few introns. Several of the mitochondrial codons vary slightly from those of eukaryote nuclear genes. Because all mitochondrial genes are functional at all times, mitochondrial genes do not need the more complex regulatory systems of nuclear genes. But in most respects, mitochondrial genes conform to the general structure of genes found in all biological systems.

D. Gene Rearrangement

All the cells of multicellular organisms typically contain the same complement of genes and chromosomes that were present in the fertilized egg. An obvious exception are the products of meiosis—eggs and sperm—with half the chromosome complement (and half the genes). Another exception in mammals is red blood cells. As red cells mature, their nuclei disintegrate, producing a cell that is anuclear. Other specialized cells may be polyploid (liver) or multinuclear (muscle). In each of these variations, the structure of individual genes is unchanged. [*See* MEIOSIS.]

There are now several well-studied examples in eukaryotes in which DNA rearrangement occurs as a part of normal differentiation. The only example in mammals involves certain genes necessary for immunity. The great variety of potential antibody responses to antigenic stimulation are possible largely because of the ability of the antibody genes to rearrange during the development of the lymphocytes. An example of this rearrangement is shown in Fig. 2. Each gene is actually a complex of gene parts. In the case of antibodies produced by B cells, the genes that code for light chains (L chains) consist of three components: V regions (which code for the N-terminal "variable" part of the polypeptide chain), J regions (which code for a joining segment), and C regions (which code for the C-terminal "constant" segment). Heavy-chain genes have an additional component—the D region—that is located between the J and C regions and adds diversity to the final product. During development of a stem cell destined to become a B cell, stretches of DNA are eliminated so that the mature B cell has a contiguous array of a V segment and a J segment (Fig. 2).

This new array serves as the transcription unit for that gene complex. Since the joining of DNA segments is random and varies from one chromosome to another, not every rearrangement produces a functional transcription unit. If the first attempt is successful, no rearrangement occurs on the homologous chromosome. If the first attempt is not successful, rearrangement on the homologous chromosome occurs. That also may or may not be successful. In the case of the L chains, there are two such complexes, on chromosomes 8 and 22. The H chain is represented by a single complex on chromosome 14. A cell that successfully rearranges to produce an L chain and an H chain is retained in the immune repertory to produce that particular antibody. Many cells are not successful and are lost.

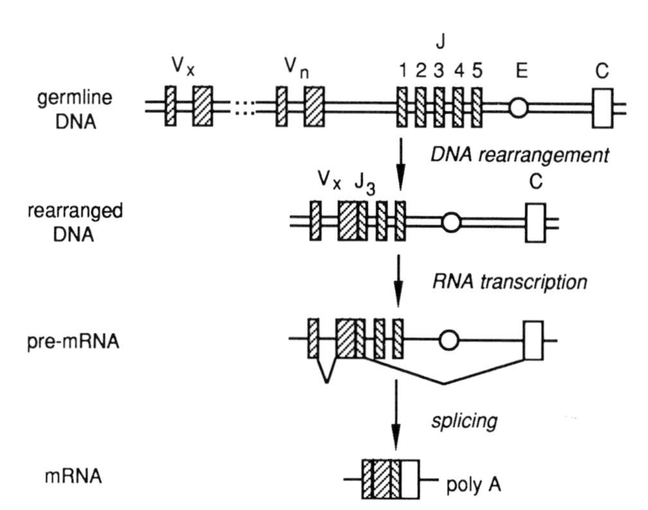

FIGURE 2 Genetic structure of the κ chain immunoglobulin complex. During maturation of a B-lymphocyte, a V region is joined directly to a J region by deletion of the intervening DNA. [Reproduced, with permission, from H. E. Sutton, (1988). "An Introduction to Human Genetics" 4th ed. p. 405. Harcourt Brace Jovanovich, San Diego.]

The random rearrangement of the DNA has the potential to produce an enormous number of different successful combinations, thus generating the required antibody diversity. When an antibody-producing cell is stimulated to grow by presence of the corresponding antigen (or rarely by malignant transformation), the rearranged DNA is faithfully replicated as any other DNA would be, expanding the number of cells that code for that particular antibody. Because the rearrangements occur only in the cells of the immune system, all other somatic cells continue to have the original DNA, sometimes referred to as germline DNA.

Rearrangement of the genes that code for T-cell receptors also occurs in a similar manner. The antibody genes and the T-cell receptor genes are evolutionarily related and are part of a gene family (see below). Not all members of the family can rearrange, and rearrangement is not known to occur in any other mammalian gene family.

E. Imprinting

According to traditional Mendelian rules, the activity of a gene is independent of the parent from which it was received. While this is true of the vast majority of genes studied, there are now a few well-documented examples of differential activity of an allele, depending on whether it was received from the father or the mother. The basis for imprinting is not certain, but the most likely explanation is transmission of the state of methylation of the DNA. Genes that are highly methylated are typically not very active. Traditional models of DNA replication do not take the methylation pattern into consideration, and, indeed, for most genes it must not be important, but for some, it may be. Just as the inactive X chromosome remains inactive through numerous mitotic replications, some genes may be transmitted in an inactive form, even through meiosis. Activation would appear to be possible in some instances.

III. Organization of Human Genes

A. Number of Genes

The number of nucleotide pairs in the DNA of a haploid set of human or other mammalian chromosomes is approximately 3 billion. If genes were closely packed and a typical gene required 3,000 nucleotides for the coding and regulatory regions and for intron sequences, there would be room for 1.5 million genes.

Both genetic arguments and experimental observations indicate this number of genes to be much too high. The genetic arguments hinge on the mutation rate of genes. If detrimental mutations occurred at a rate of one per million loci per gamete, a figure that is lower than the observed rates, each zygote would have on the average three new mutations plus all the mutations that were transmitted from prior generations. Various models suggest that such a number of mutations would overwhelm the ability of natural selection to eliminate them.

Two kinds of observations also support a lower number of genes. One is the direct analysis of nucleotide sequences. There are substantial stretches of noncoding DNA between functional genes, DNA for which there is no obvious or demonstrated function. Another consideration is the great variation in the total amount of DNA among various species. Mammals are relatively similar to each other in having some 3 billion nucleotide pairs per haploid genome; however, such closely related vertebrates as toads (*Bufo bufo*) have twice as much DNA as humans, and the amphibian *Amphiuma* has 76 billion nucleotides. *Tradescantia*, a plant, has 53 billion nucleotides. We humans are reluctant to concede that amphibians and plants are genetically that much more complex than we are. The alternate hypothesis is that most of the DNA in those species does not code for genes, and the same may be true for *Homo sapiens*.

The conventional estimate of the number of human genes is 50,000–100,000. This estimate is not based on secure analytical arguments but rather on speculation that we need at least 50,000 loci to code for all the required mammalian functions, and much more than 100,000 would create problems with the mutation rate.

B. Genes and Chromosomes

Each chromosome appears to have a single molecule of DNA stretching from one end of the chromosome to the other. Therefore, the thousands of genes on a single chromosome must be arrayed along this molecule. How the genes are organized with respect to each other is largely unknown. There are a number of clusters of genes with related functions and structures. Such clusters could have functional significance. Or they could be a reflection of evolutionary origins by tandem duplication from a single ancestral gene, with chance not yet having separated them through chromosome rearrangement. Probable examples of both are known.

The location of genes on particular chromosomes appears to be largely a matter of evolutionary accident. The chromosomes of mammals have undergone many rearrangements, with new, highly functional combinations of genes being generated many times. In addition, many healthy persons are known who have inversions or balanced translocations. So long as the genomic complement is complete, there are many arrangements that are functional.

This does not mean that all genes are oblivious to their neighbors. There appear to be clusters of genes that are regulated as a unit, and presumably any breaking up the cluster by chromosome rearrangement would be detrimental. Evidence is lacking at present on the number of such clusters and their typical sizes. Nor do we have any idea how tolerant genes or gene clusters are of their location; we can only observe the successful rearrangements and not the unsuccessful.

If one divides the number of nucleotides by the approximate number of genes in a mammalian genome, the nucleotides per gene is very much larger than would be accounted for by the known requirements of structural genes and their regulatory regions. This excess amount of DNA appears to be distributed along the chromosomes between coding regions, and some also occurs as introns. Much of this excess DNA is unique sequence DNA; i.e., the particular sequence of nucleotides occurs only once in the genome. Coding regions of genes are other examples of unique sequence DNA.

Some 30% of the total DNA in mammals is repetitive DNA. In the case of highly repetitive DNA, there are millions of copies of short nucleotide sequences. Middle repetitive DNA consists of thousands to hundreds of thousands of relatively longer sequences. The *Alu* sequence in humans, an example of middle repetitive DNA, is approximately 300 nucleotides in length, with some 500,000 copies dispersed throughout the genome. The function of repetitive DNA is unknown. It may have some as yet undescribed function, or it may be "junk" DNA—DNA that has no function other than to replicate itself.

Chromosomal regions are classified as euchromatin and heterochromatin on the basis of their staining behavior, the heterochromatin being more easily stained by nuclear stains. In general, euchromatin consists of unique sequence DNA and is the location of functional genes. Heterochromatin consists of repetitive DNA and lacks functional genes. As more is learned about the fine structure of DNA and chromosomes, there are likely to be exceptions to these general statements, but they are well documented as rules in many species. Heterochromatic regions may occur in any part of a chromosome, but the regions around the centromere are likely to be heterochromatic.

IV. Evolution of Genes

A. Universality of the Genetic Code

The fact that all genes in all species use virtually the same genetic code and have most of the same organizational features can most readily be interpreted as indicating that all genes are descendants of one primordial gene, or at least one form of primordial life in which there was free exchange of nucleic acid. In addition to this common heritage, however, similarities in gene structure among groups of genes suggest a more recent divergence from each other and inform us about the evolutionary relationships not only of the genes but also of various taxa.

B. Mutations in Genes

A particular locus may evolve through adaptive or neutral mutations, processes that are very slow but that nevertheless account for the great diversity of homologous loci among species. Mutations may be nucleotide substitutions, additions, or deletions. The consequences of such small changes—small in the physical sense—may be inconsequential for gene function or they may be very large, particularly if they occur in critical regions, such as the DNA sequences that regulate transcription. Most mutations are neutral or detrimental and will ultimately be eliminated by natural selection, but occasionally these random changes in the structure of a gene may result in a more functional locus that will be favored by selection.

Mutation may also result in duplication or deletion of entire genes. Deletion of a critical gene is likely to be detrimental to the organism, and the mutant form will be eliminated. Duplication may be tolerated and allows mutations to occur and new gene functions to evolve without losing the original gene function. Such repeated gene duplication presumably accounts for the large number of loci in the repertory of higher organisms.

In addition to duplication and divergence of genes, other rearrangements among genes may be adaptive. For example, the selective advantage of introns within the coding regions of genes has not been proved. One attractive hypothesis is that each exon codes for a separate protein domain. Indeed, there are many illustrations of this principle, but other genes appear to violate it. In situations in which each exon corresponds to a functional domain, one could imagine that entire domains might be translocated to other genes, thereby providing an additional function. Such seems to be the case, for example, in several transmembrane proteins, in which the exons that code for the transmembrane domain are more closely related structurally than would be expected by parallel evolution. It is thought that the different domains may have evolved separately and then been joined by chromosome rearrangements to produce new functional units.

Such "exon shuffling" must be understood in the traditional concepts of evolutionary changes. There is no reason to suppose that exons are traded about other than by rare, chance chromosome rearrangements. On the other hand, the occasional finding of repetitive sequences such as *Alu* within introns might be expected to promote mispairing of chromosomes. If this were followed by crossing-over or translocation, new genes would be generated.

C. Gene Families

A number of groups of proteins—and their corresponding genes—share structural features and functions. Such groups are referred to as gene families. One example would be the globins that form hemoglobins and myoglobins. Each human haploid genome has several α-globinlike loci on chromosome 16, several β-globinlike loci on chromosome 11, and a myoglobin locus on chromosome 22; however, examination of the nucleotide sequences of these genes, the organization of introns and exons, the amino acid sequences, the three-dimensional structure of the polypeptide, and the functions provides convincing evidence that all evolved from a single, ancient globinlike gene. The detection of globinlike DNA sequences in such remote relatives as yeast suggests that the globin family may be quite old and quite extensive.

Other gene families that have been identified include the serine proteases and the immunoglobins. Serine proteases are enzymes that catalyze the breakdown of peptide bonds, with serine at the active site of the enzyme. Examples are trypsin, chymotrypsin, and pepsin. Other members of the family include haptoglobin, a plasma protein not known to have proteolytic activity. Even though the serine proteases have diverged greatly in the types of peptide bonds attacked and the conditions under which maximal activity occurs, their shared structures support a common origin.

The genes that code for immunoglobulins (Ig) and T-cell receptors (TCR) are part of an especially interesting superfamily of genes. Among the properties shared by Ig and TCR genes and no other known mammalian genes is the rearrangement of DNA during maturation of lymphocytes. Such rearrangement in somatic cells must be extremely rare, although DNA rearrangement also occurs in germline genes that code for trypanosome antigens. Other members of the Ig–TCR superfamily do not undergo rearrangement, nor do the Ig and TCR genes in any tissue other than lymphocytes. Thus, programmed DNA rearrangement can occur but is exceedingly rare. [*See* LYMPHOCYTES.]

D. Phylogenetic Comparisons of Genes

As the ultimate repository of biological heritage, DNA has the potential to tell us how closely related all existing species are. From homologous nucleotide sequences, it is possible to construct phylogenetic trees that account for the evolutionary origins of all species. For the most part, existing information is based on amino acid sequences of proteins, which are, of course, the products of genes and, hence, reflect the nucleotide sequences of coding regions; however, data on DNA sequences in various species is rapidly accumulating.

As yet, no remarkable revisions of phylogeny have been required by the DNA sequence data. On the other hand, the consistency of DNA structure and fossil or other phylogenetic evidence support the value of DNA as a means of reconstructing phylogeny. DNA has proved valuable in understanding the relationships among closely related species, subspecies, or even local populations.

V. Genetic Variation in Humans

A. Mutation and Genetic Variation

Mutation is a fundamental property of genes. The changes in genetic information, which we call mutation, may have little or no effect on the function of a gene, or they may interfere profoundly with function. A single nucleotide substitution may have no effect on the gene product or it may make the gene unable to produce any product at all. Therefore, many mutant alleles are normal variants, whereas others interfere with normal development.

Mutations are rare events. The measurement of mutation rates is difficult in humans but is on the order of one per 100,000 gametes per locus for many mutations that cause dominant disease. The rate of production of normal variant alleles, essentially nucleotide substitutions, is probably the same order of magnitude. The most likely fate of a new mutation is elimination, either by natural selection or by chance, even for favorable mutations. A few will attain substantial frequencies, and rarely one may displace the original wild-type allele, again either by natural selection or by chance. [*See* MUTATION RATES.]

B. Normal Genetic Variation in Humans

The existence of alternate alleles in a population allows each member of the population to be genetically unique. With the exception of identical twins, each person possesses a combination of alleles that has never existed before, nor will it exist again. This is assured by the variety of choices available at tens of thousands of loci.

In discussing genetic variability, distinguishing between polymorphic loci and monomorphic loci is useful. Monomorphic loci are those at which there is only one common allele. Polymorphic loci are those with two or more common alleles. An arbitrary definition of "common" is 1% or greater. Monomorphic loci may have rare variant alleles, not one of which has a frequency as large as 1%. In the case of polymorphic loci such as ABO and Rh blood groups two or more alleles have frequencies greater than 1%. The genes that account for most of the individual variation are polymorphic.

There are various estimates of the proportion of loci that are polymorphic. Such estimates are based on variations in the protein products of genes, usually detected by electrical charge variations that lead to variations in electrophoretic mobility. Studies of enzymes in *Drosophila* and in humans indicate at least one-third of loci to be polymorphic. On the other hand, studies that include structural proteins give a lower frequency, on the order of 6% in humans. In either case, there is ample variation to account for the many genetic differences that distinguish each person. No population-based studies have yet been reported on DNA variations that might influence the function of loci without altering the structure of the gene product. We may assume that additional polymorphisms occur at this level. [*See* POLYMORPHISM, GENES.]

In addition to polymorphic alleles, there are rare normal variants at many loci. Because of their rarity, few have been studied in detail. The fact that they are found in normal persons does not mean that there is no effect on the phenotype. Usually we have no way to relate the individuality of the phenotype to variation at particular loci. We suspect, however, that the phenotypic differences that we observe among individuals are often due to the cumulative effects of variations at many loci. As information has accrued on the extent of genetic variation, the view has changed from one in which occasional variation occurs, rigidly limited by natural selection, to one in which variation is very widespread.

C. Genetic Variation and Disease

We assume that mutations occur at all 50,000–100,000 human genes, and many of the mutant alleles cannot function normally. Some will be eliminated immediately as dominant lethals, and we may never observe the mutant form. Others may cause disease. The number of loci at which mutant alleles are known to cause disease exceeds several hundred, and the true number is probably in the thousands. Some inherited diseases such as cystic fibrosis, Tay-Sachs disease, and sickle-cell anemia are relatively common; others may be very rare, affecting one person in tens of thousands.

The impact of genetic variation on health and health care has been estimated in various ways, none very satisfactorily. Some 1% of newborns have some genetic defect, often trivial but sometimes fatal. Even diseases more characteristic of older ages such as heart disease and cancer often have a large genetic component. Understanding the full extent of the genetic contribution to disease and the specific genes involved should enable us to reduce the burden somewhat, though perhaps never to zero.

Bibliography

Alberts, B., Bray, D., Lewis, J., Raff, M., Roberts, K., and Watson, J. D. (1989). "Molecular Biology of the Cell," 2nd ed. Garland Publishing, New York.

Darnell, J., Lodish, H., and Baltimore, D. (1986). "Molecular Cell Biology." Scientific American Books, New York.

Sutton, H. E. (1988). "An Introduction to Human Genetics," 4th ed. Harcourt Brace Jovanovich, San Diego.

Watson, J. D., Hopkins, N. H., Roberts, J. W., Steitz, J. A., and Weiner, A. M. (1987). "Molecular Biology of the Gene," Vol. 1, 4th ed. Benjamin/Cummings, Menlo Park, California.

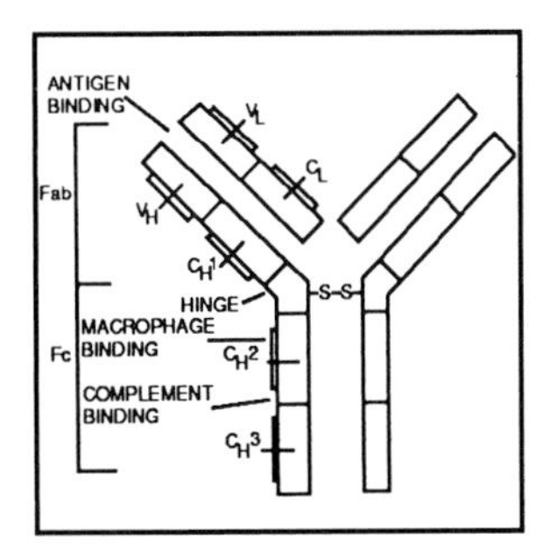

Genetically Engineered Antibody Molecules

SHERIE L. MORRISON, *University of California, Los Angeles*

Glossary

Antibody-dependent cellular cytotoxicity (ADCC) Cell killing reaction in which Fc receptor bearing killer cells recognize target cells via specific antibodies

Chimeric Assembled from diverse sources not normally found associated

Complement Group of serum proteins important for the lysis of foreign cells and pathogens; they also play an important role in phagocytosis

Complement-dependent cytotoxicity Cell killing mediated using complement

Constant region Portion of the antibody molecule exhibiting little variation and determining the isotype of the antibody

Drug resistance Ability to grow in the presence of drugs that are normally toxic

Fc Portion of the antibody responsible for binding the antibody receptors on cells and activating complement

Glycosylation Attachment of carbohydrate (sugar) residues; proteins containing carbohydrate residues are glycoproteins

Hybridoma Cell derived by fusion of a normal cell, usually a lymphocyte, with a tumor cell

Isotype Class of the antibody

Monoclonal antibody Homogenous antibody derived from a single clone of antibody-producing cells

Plasmid Circular DNA segment that replicates extrachromosomally in bacteria

Spheroplasts Bacteria whose cell walls have been removed by treatment with lysozyme

Transfection Introduction of foreign DNA into a cell; the foreign DNA may be transiently expressed or integrated into the chromosome and stably replicated and expressed

Variable region Variable portion of the antibody molecule that is responsible for antigen binding

Vector Piece of DNA, usually from a plasmid or virus, used to deliver genetic information to a cell by transfection

ANTIBODIES have long been appreciated for their exquisite specificity in binding the antigenic determinant they recognize. Because of this specificity, antibodies seemed the ideal "magic bullet" for targeting diagnosis or therapy to specific cells. However, even the monoclonal antibodies produced by hybridoma cell lines have some properties that make them less than ideal for this purpose: (1) It has proven difficult to produce monoclonal human antibodies; and (2) the isotype of the resulting antibodies often are inappropriate for the desired biologic properties. Genetically engineered antibodies produced by expressing cloned genes in the appropriate cell type provide an approach for producing antibodies with superior properties. It is now possible to produce chimeric antibodies with gene segments derived from diverse sources. Variable regions can be expressed joined to constant regions from either the same or different species; constant regions with improved biologic properties can be produced. In addition, molecules that bind antigen but are never found in nature (e.g., fusions of antibodies with nonantibody proteins) can be manufactured.

I. Properties of Antibody Molecules

Antibodies are among the most versatile of proteins. They are designed to achieve specificity of binding so that they can distinguish foreign substances from the naturally occurring components of the body. They also are capable of inducing biologic activities that can destroy and eliminate undesirable substances such as bacteria.

All antibodies have a similar structural organization and are formed by polypeptide chains held together by noncovalent forces and disulfide bridges. In the basic structure, two pairs of identical heavy (H) and light (L) chains form a bilaterally symmetric structure (Fig. 1). The polypeptide chains fold into globular domains separated by short peptide segments. The H chain has four or five domains depending on the isotype. The L chain has two domains. The N-terminal domains of each chain constitute the variable region that carries the antigen combining site and determines the specificity of the antibody. The remainder of the antibody constitutes the constant region, which is responsible for the effect of the antibody in the body.

Antibodies of the same specificity can have different H-chain constant regions and, therefore, exhibit different effects. In humans the different constant regions produce antibodies of different isotypes: IgM, IgD, IgG1–4, IgA1, IgA2, and IgE. IgM, IgA1, and IgA2 differ from the other isotypes in that they are constituted of multiples of the basic H_2L_2 structure, and they contain an additional polypeptide chain, the J chain. There are also two different isotypes of light chains, κ and λ. The L-chain isotypes do not appear to influence the effect of the antibody molecule. All antibodies are glycoproteins with the carbohydrate content varying among different isotypes. The carbohydrate present in the constant region of IgG antibodies has been shown to be essential for many of its effects.

Antibodies are unusual proteins in that the genes that encode them must be assembled after birth to produce a functional protein. Hundreds of different H- and L-chain variable region genes are present in the genome. For a variable region to be expressed, it must be positioned next to a J segment for the L chain or a DJ segment for H chain by a DNA rearrangement. The variable regions of the chains are expressed first with an IgM constant region, but as the immune response matures, further DNA rearrangement occur at the H-chain locus, and the variable regions are expressed with different constant regions (Fig. 2). In the case of the L chains, the variable region of the κ chain is rearranged first. As soon as a functional κ-chain protein is produced, rearrangement stops. If both κ-chain alleles have rearranged but failed to produce a functional protein, rearrangement at the λ locus occurs. Each antibody-producing cell produces only one functional H chain and one functional L chain, a phenomenon called *allelic exclusion*. It should be noted that the domain structure seen in the antibody molecule is reflected in the structure of the genes that encode it (Fig. 1).

During the normal immune response, a wide variety of antibodies are produced. These include antibodies with different variable regions, which recognize the same antigen and antibodies with the same variable region associated with different constant regions. Different individuals will make different immune responses. This heterogeneity in the immune response has made it difficult to use antisera for many applications. [*See* IMMUNE SURVEILLANCE.]

A significant breakthrough was made when it became possible to produce *hybridoma antibodies*. Hybridoma antibodies are *monoclonal* (i.e., they are the product of a single antibody-producing cell) and are, therefore homogeneous, with a single variable region associated with only one constant region. Although these homogeneous antibodies have many advantages, they still have some inherent limitations. Although it has become routine to produce rat and mouse hybridoma antibodies with antigen binding specificities appropriate for use in basic research and *in vitro* clinical diagnosis, it has proven

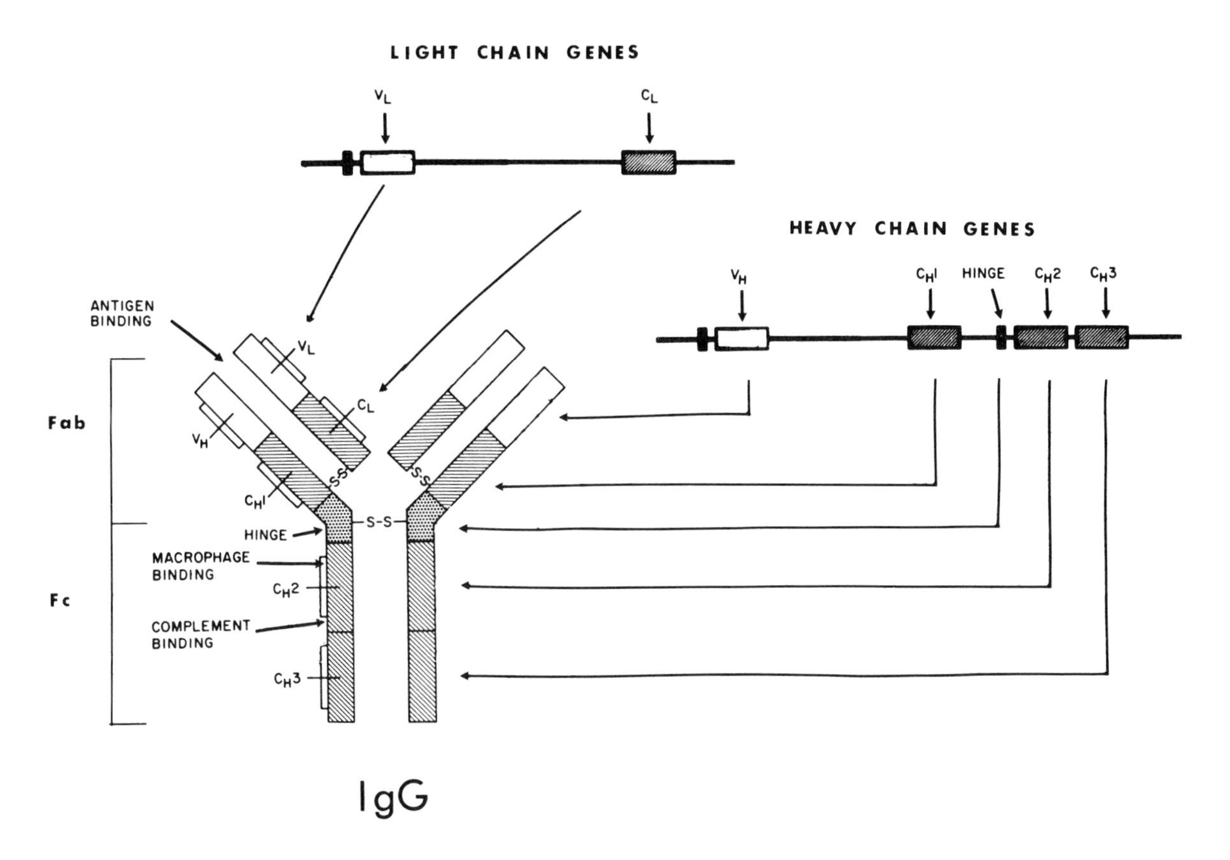

FIGURE 1 Diagram of an antibody molecule and of the genes that encode the heavy and light chains. The antibody molecule is divided into discrete functional domains: two domains (V_L and C_L) constitute the light chains, whereas five domains (V_H, $C_H{}^1$, hinge, $C_H{}^2$, and $C_H{}^3$ make up the heavy chain. The variable region domains (V_L and V_H) make the antibody-combining sites and are designated the Fv region. The effector functions of the molecule (e.g., the ability to activate complement or bind to cellular receptors) are properties of the constant region domains. The hinge provides flexibility in the antibody molecule, facilitating antigen binding and some effector functions. In the genes, each domain is encoded by a discrete exon (indicated by *boxes*), separated by intervening sequences (introns) indicated by the *line*; the intervening sequences are present in the primary transcript but are removed from the mature mRNA by splicing. The heavy and light chains both contain hydrophobic leader sequences (indicated by the *black exon*) necessary for their secretion. This leader sequence is present in the newly synthesized heavy and light chains but is cleaved from them after they enter the endoplasmic reticulum and therefore is not present in the mature antibody molecule. The enzyme papain cleaves the molecule into an Fab fragment containing the antibody binding site and an Fc fragment. [Modified from S. L. Morrison. *Science* **229,** 1202–1207, with permission. (1985).]

difficult to produce human hybridoma antibodies useful for *in vivo* immunotherapeutic applications. In addition, hybridoma antibodies are homogeneous not only with respect to their variable regions but also with respect to their constant regions and hence might not have the desired effects. In addition, they may not possess the exact desired binding specificity or affinity. [*See* MONOCLONAL ANTIBODY TECHNOLOGY.]

One approach to producing improved antibody molecules is to use recombinant DNA techniques to produce antibodies with improved antigen binding specificities and effects. An advantage of this approach is that we are not limited to producing antibodies as they exist in nature. Instead, we can produce antibody molecules with improved properties such as binding specificities, pharmacokinetics, and effector functions such as complement activation and Fc receptor binding. Additionally, we can introduce into the antibody molecule novel functions not normally found there. The availability of these antibody molecules promises to revolutionize our ability to use antibody molecules for diverse applications.

II. Chimeric Antibody Molecules

In Greek mythology, the chimera is a she-monster with a lion's head, a goat's body, and a dragon's tail. Chimera has come to characterize molecules

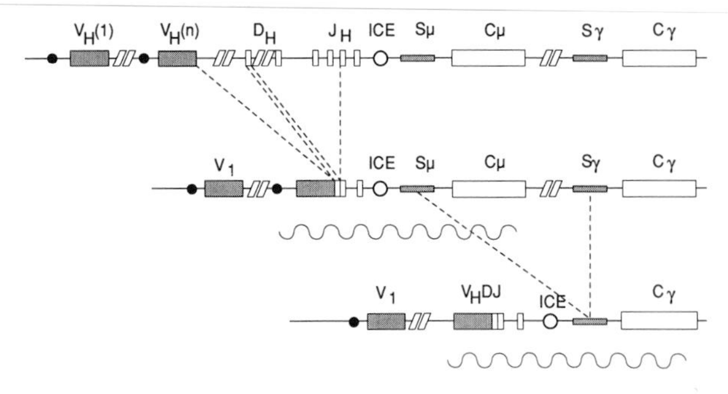

FIGURE 2 Structure of the heavy-chain locus before expression (shown at the *top*), during expression of an IgM protein (shown at the *middle*), and after switching to production an IgG antibody (shown at the *bottom*). The locus has been simplified to show only one IgG gene, whereas there are several. ICE designates the enhancer that is found in the intervening sequence. Before expression the V_H, D_H, and J_H exist as discrete segments of DNA, shown at the *top* of the figure. To produce a functional heavy-chain gene, a V must be joined to a D and J_H segment of DNA; the DNA segments between the joined elements are usually deleted from the genome. The first heavy chain produced during B-cell development is of the IgM isotype shown at the *middle* of the figure. As B-cell development proceeds, class switching occurs by DNA, rearrangement, and antibodies are produced with a different constant region joined to the same variable region. In the example, shown at the *bottom* of the figure, an antibody of the IgG isotype is produced.

put together from diverse gene segments not normally found associated. Chimeric antibody genes are assembled from diverse gene segments not normally found associated. Chimeric antibodies can be intraspecies, where variable and constant regions are from the same species, or interspecies, where, for example, the variable region is of mouse origin and the constant region is of human origin. The term *chimeric antibody* also refers to hybrid gene segments derived from different antibody genes or in which a segment is assembled from more than one source. Lastly, chimeric antibody can include antigen binding specificities joined to nonantibody protein structures.

Among the most useful chimeric antibodies are *mouse/human chimerics,* in which the variable region is derived from a murine hybridoma cell line and the constant regions are human. By using the variable regions from the murine antibodies, a wide range of antigen binding specificities can be used. The resulting chimeric antibodies are superior to completely murine antibodies for use in humans because they have reduced immunogenicity and are better able to interact with the human immune system. Mouse/human chimeric antibodies also have advantages over the available human antibodies, which have a limited range of antigen binding specificities, and are often of an inappropriate isotype.

Genetically engineered antibody molecules may possess properties not found in the naturally occurring antibody molecules. Thus genetic engineering has been used to enhance some effector functions and to decrease or eliminate others. The antibody combining site also can be altered. In addition, antibody combining sites can be genetically coupled to nonantibody protein molecules such as toxin, enzymes, growth factors, or hormones. In the resulting antibody-fusion proteins, the specificity of the antibody can be used to target the activity of the associated protein.

Genetically engineered antibody molecules can be produced in bacteria, yeast, and mammalian cells. To produce a functional antibody, proper assembly and proper glycosylation must take place. These requirements and the need for the antibody to be secreted determine the choice of the cell type used to produce the antibody molecule.

III. Antibody Production in Bacteria and Yeast

Expression of gene products in bacteria has a great deal of appeal. Bacteria can be grown in large quantities at relatively little cost and can be used to produce large quantities of proteins efficiently. However, antibodies are glycoproteins, and bacteria do not glycosylate proteins, so any function of the antibody molecule that depends on glycosylation will not be exhibited by antibodies synthesized in bacteria. In addition, production of a functional antibody molecule requires correct assembly and disulfide bond formation between the H and L chains, which is frequently difficult to obtain in bacteria.

Attempts to generate intact functional antibodies by expression in bacteria have met with limited success. The majority of the H and L chains produced in bacteria end up as insoluble material that accumulates in inclusion bodies. Attempts to renature these insoluble products resulted in antibodies exhibiting minimal function.

Bacterial expression of antibody fragments has been more successful. ''Fc-like'' fragments from IgE antibodies also were made as insoluble protein in inclusion bodies. However, these could be solubilized and renatured to form functional fragments that exhibited the functional properties of IgE (e.g., the ability to trigger the cells involved in allergic reaction). Because proteins synthesized in *Escherichia coli* do not contain carbohydrate, it is clear that the carbohydrate is not required for this biologic property.

The Fab (Fig. 1) fragment from an antihuman carcinoma antibody and the Fv fragment, which contains only the variable region of the phosphorylcholine binding myeloma McPC603, have been expressed in bacteria, demonstrating that it is possible to get assembly of different polypeptide chains in bacteria. When the two chains were made in the same cell, they assembled into functional antibody that showed no difference in binding affinities from the naturally occurring antibodies. Clearly, antibody fragments produced in bacteria cannot at this time be used for applications that require an intact antibody molecule or any of the functional properties associated with its glycosylation.

Antibodies and antibody fragments have also been produced in yeast, which glycosylate their proteins. However, the added carbohydrate differs in structure from the carbohydrate added by mammalian cells. Antibody-dependent cellular cytotoxicity (ADCC) and complement-dependent cytotoxicity (CDC) are two effector functions that are not exhibited by antibodies lacking carbohydrate. Antibodies possessing the yeast carbohydrate were able to mediate ADCC but were unable to mediate CDC. Therefore, antibodies require carbohydrates similar in structure to those added by mammalian cells to exhibit their full range of biologic functions.

IV. Gene Transfection of Mammalian Cells

Expression of genes in mammalian cells such as myeloma and hybridoma cell lines, which efficiently express, glycosylate, assemble, and secrete functional antibody molecules, provides an alternative to expression in bacteria and yeast. The introduced genes, however, are not usually expressed at high level.

Several methods are available for introducing foreign DNA into mammalian cells. $CaPO_4$-precipitated DNA is effective in fibroblasts but not in lymphoid cells. In contrast, both electroporation and protoplast fusion are effective methods for lymphoid cells. For electroporation, the cells are suspended in a solution of DNA containing the genes to be expressed and are then subjected to an electrical pulse, which makes pores in the cells and enables them to take up and express the DNA. For protoplast fusion, the DNA containing the genes is propagated as a plasmid in bacteria. The bacteria are then converted to spheroplasts using lysozyme and fused to the cell to be transfected. With both electroporation and protoplast fusion, transfection frequencies between 10^{-3} and 10^{-6} can be routinely achieved, and are sufficient to isolate the desired transfectant cell line. Myeloma cell lines that do not produce antibodies are routinely used.

V. Vectors Used for Transfection

Even under optimal conditions, gene transfection into eukaryotic cells is an inefficient procedure. Vectors containing selectable markers are therefore used to select the rare stably transfected cell lines. The most commonly used vectors to date, the pSV2 vectors shown in Fig. 3, contain three essential ele-

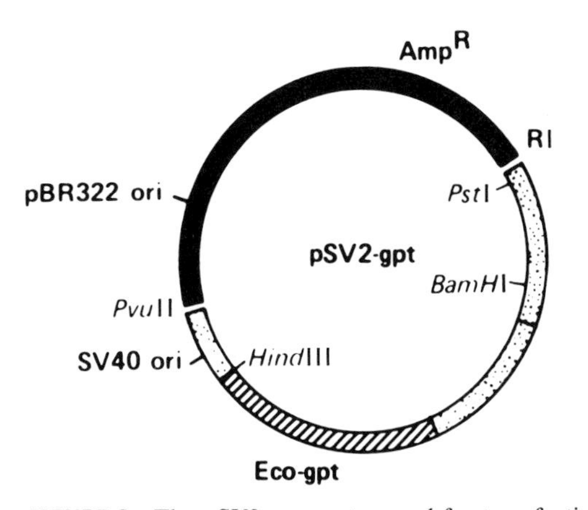

FIGURE 3 The pSV2-*gpt* vector used for transfecting eukaryotic cells. The vector is a plasmid, a closed circle of double-stranded DNA. Contained within the vector are sequences, indicated by the *black region*, derived from the plasmid pBR322. These consist of the origin of replication and the β-lactamase gene, which provides resistance to the antibiotic ampicillin. The gene providing a marker selectable in eukaryotic cells, in this case, *gpt*, the expression that enables eukaryotic cells to grow in mycophenolic acid, is indicated by the *hatched area* and was derived from the bacterium *E. coli*. DNA segments derived from the virus SV40 are indicated by *stippling*. These include the SV40 ori, which contain the early promoter used for expressing the *gpt* gene and splice and poly A addition sites located 3′ of the *gpt* gene. The Eco R1, Pst I, and Bam HI restriction endonuclease cleavage sites are located in regions of DNA not necessary for function in prokaryotic or eukaryotic cells and provide convenient sites for insertion of genes of interest. [From S. L. Morrison and V. T. Oi. *Annu. Rev. Immunol.* **2,** 239–256, with permission. (1984).]

ments: a plasmid origin of replication, a gene encoding a biochemically selectable phenotype in bacteria, and a gene encoding a biochemically selectable marker in eukaryotic cells. When the objective is to produce proteins for subsequent analysis and use, stable transfectants are produced, which contain the vectors stably integrated into the chromosome of the cell.

It is essential that the vectors can be propagated as plasmids in bacteria to obtain DNA in sufficient quantities for genetic manipulation. To achieve this objective, some plasmids contain the origin of replication and β-lactamase gene from the plasmid pBR322, which confers on the bacteria propagating these plasmids resistance to ampicillin. Other vectors contain the chloramphenicol acetyl transferase gene, which confers resistance to chloramphenicol. With both vectors it is easy to obtain large quantities of DNA for *in vitro* manipulation.

Because only rare animal cells (10^{-3}–10^{-6}) incorporate and stably express exogenous DNA, it is also necessary to include biochemical markers that can be used to select this minor cell population. The most commonly used markers are drug-resistance markers, connected to a viral (SV40) promoter, splice junction, and poly A addition signal so that they can be expressed in eukaryotic cells. Two selectable bacterial genes commonly used are (1) the phosphotransferase gene from Tn5 transposon (designated *neo*) and (2) the xanthine-guanine phosphoribosyl transferase gene (*xgprt* or *gpt*). The product of the *neo* gene inactivates the antibiotic G418, an inhibitor of protein synthesis in eukaryotic cells. Biochemical selection with *gpt* expression is based on the fact that the enzyme encoded by this gene uses xanthine as a substrate for purine nucleotide biosynthesis, whereas the homologous eukaryotic enzyme uses only hypoxanthine. Thus, when the conversion of inosine monophosphate to xanthine monophosphate is blocked by mycophenolic acid, cells provided with xanthine can survive only if they express the bacterial *gpt* gene. Biochemical selection with G418 or mycophenolic acid depends on two entirely different mechanisms; therefore, they can be used in vectors to select independently for the expression of exogenous recombinant DNA gene segments.

To create functional antibody molecules, the genes encoding H and L chains must be transfected into the same cell, and both polypeptides must be synthesized and assembled. Both the H- and L-chain genes can be inserted into a single vector and then transfected; this approach generates large, cumbersome vectors that are difficult to manipulate. A second approach is to transfect sequentially the H- and L-chain genes, using different drug-resistance markers to select for the expression of the different vectors. Alternatively, both genes can be introduced simultaneously on different DNA fragments using either electroporation or protoplast fusion. The latter approach is usually the most efficient in creating complete antibody molecules.

VI. Obtaining Variable Regions for Expression

The variable regions needed to produce the desired antibody can be obtained as cDNA or genomic clones from the appropriate antibody-producing cell line. By obtaining variable regions from human antibody-producing cell lines, it is possible to produce totally human antibodies. This approach is appropriate for rescuing low-producing human cell lines and for obtaining human immunoglobulins with the desired isotype.

For a gene to be expressed in mammalian cells, it must be provided with an appropriate promoter and enhancer. The promoter is the region in which transcription of the gene is initiated; the enhancer is a regulatory element that increases the level of expression. Both the promoter and enhancer may be obtained from either viruses or eukaryotic genes. Both promoters and enhancers can be tissue-specific (i.e., they will function only in certain cell types). Therefore, the choice of promoter and enhancer must be matched with the cell type in which the genes will be expressed.

The organization of the antibody H- and L-chain genes facilitates the isolation of the desired expressed variable regions as genomic clones. Hundreds of different H- and L-chain variable regions are present in the genome. However, as described above, for a variable region to be expressed it must be positioned next to a J segment (for the L chain) or a DJ segment (for H chain) (Fig. 2). Therefore, in an antibody-producing cell, the expressed variable regions can be distinguished from the hundreds of nonexpressed variable regions because of the proximity to a J segment. Molecular probes are available for the J segments, and the expressed variable region can be identified using a J-region probe without any prior information as to the sequence of the expressed variable region. An advantage in using genomic clones is that the variable region is obtained with its own promoter.

Expression of variable regions derived by cloning cDNA (i.e., DNA complementary to messenger RNAs) provides an alternative approach. The cDNAs are cloned into a plasmid and introduced into bacteria; a J-region probe is used to identify bacteria containing antibody genes. A recent variation on this approach has been devised using the polymerase chain reaction (PCR) technique, which allows the amplification of only the desired variable regions.

Two approaches can be used to construct vectors to express immunoglobulin cDNA clones. In one approach, the variable regions obtained from H-chain cDNAs are substituted for the corresponding variable region in a genomic clone. This approach depends on the presence of the appropriate restriction endonuclease cleavage sites to permit the DNA exchange; these sites may either occur naturally or be engineered into the vectors and cDNA clones.

In a second, more general approach, a heterologous transcription unit is created. An entire immunoglobulin gene is created by joining a cDNA variable region with a cDNA constant region. The immunoglobulin genes are then inserted into a transcription unit containing the necessary signals. The choice of promoter and enhancer elements will depend on the cell type in which the gene is to be expressed; frequently, viral sequences are used. Using this approach, variable regions from mouse hybridomas have been expressed with both human H- and L-chain constant regions.

Oligonucleotide synthesis is becoming a more efficient and less costly process. The technology already exists for synthesizing complete immunoglobulin domains. It is conceivable that synthesis will replace cloning as the preferred method for acquiring variable regions of the appropriate sequence and combining specificity.

VII. Production of Chimeric Antibodies

The procedure for the production of chimeric antibodies in mouse myeloma and hybridoma cells is outlined in Fig. 4.

Genomic clones of immunoglobulin variable and constant region genes are easy to manipulate using genetic engineering techniques because of their di-

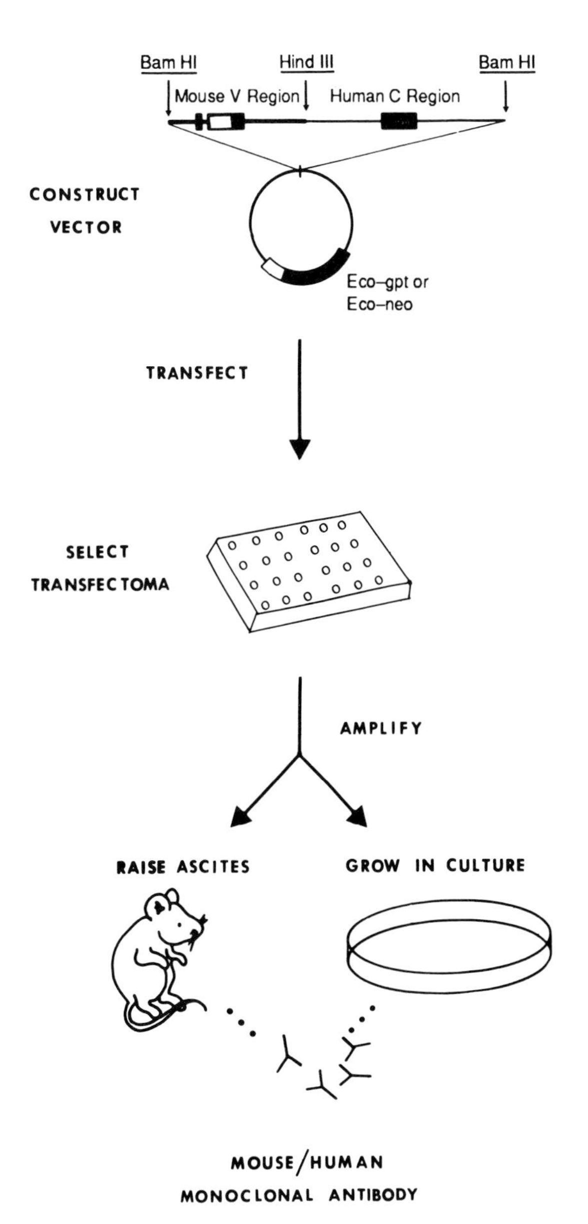

FIGURE 4 Steps in the production of a genetically engineered antibody molecule. The first step is to construct the gene of interest in the appropriate expession vector. The gene segments can be obtained from genomic or cDNA clones. In this example, the variable region from a murine hybridoma is joined to a human constant region; both were obtained by genomic cloning. The vectors are then transfected into an appropriate recipient cell; if a complete antibody molecule is to be produced, both heavy- and light-chain genes must be expressed. The transfected cells must be treated with the appropriate drug selection so that the rare cells in which the transfected genes are expressed can survive and nonselected cells will perish. The transfected cells producing the antibody of interest are then expanded and used to either raise ascites in mice or grown in tissue culture; the genetically engineered antibodies can be isolated from the ascitic fluid or culture supernatant. [From S. L. Morrison. *Science* **229,** 1202–1207, with permission. (1985).]

vision into separate coding regions, the exons. The distinct domains of an immunoglobulin polypeptide are each encoded by a discrete exon. The intervening DNA sequences (introns) separating each domain provide ideal sites for manipulating the antibody gene shown in Fig. 1; because the intervening sequences are removed by splicing when mRNA is made, alterations within them will not affect the structure of the protein. In addition, the RNA splice junction between variable and constant region exons in both H- and L-chain genes is similar, making it is easy to manipulate the Ig structure by deleting, exchanging, or altering the order of exons. The exon structure of the immunoglobulin gene can also be exploited to construct antibody-cassette expression vectors, which are then assembled in various ways.

The first mouse/human chimeric antibodies created used variable regions derived from mouse myeloma cell lines producing antibodies specific for small molecular weight substances and constant regions from human cells. When the H- and L-chain genes were transfected into the same recipient mouse cell, they directed the synthesis of H and L chains that assembled into functional chimeric antibody molecules. These were glycosylated and secreted and retained their antigen binding specificity. Comparable results were obtained when the H- and L-chains variable regions from hybridomas specific for cell surface antigens or tumor-specific antigens were used. Therefore, it seems to be a general rule that the specificity of binding of a murine variable region is unchanged when it is expressed associated with a human constant region. It would also seem likely that human variable regions can be expressed associated with different constant regions without changing their combining specificity.

VIII. Use of Chimeric Antibodies to Study Structural Correlates of Antibody Effector Functions

Functions such as complement fixation, ADCC, and Fc receptor binding are mediated by the H-chain constant region, different isotypes of which display different biologic properties as shown in Table I. Although some structures important for these activities have been defined, the exact molecular correlates of many antibody effector functions remain unknown. With the use of recombinant chimeric antibodies, the study of structure–function relations within the antibody molecule can now be approached systematically. It is now possible to study chimeric antibodies in which the same variable region is joined to different human constant regions. They can be used to determine the amino acid sequences responsible for specific effector functions. The following is an example.

The human IgG subclasses share extensive sequence homology throughout their constant regions with the exception of the so-called hinge region, which shows considerable variation in length and amino acid composition. One question of interest is the extent to which the hinge region modifies effector functions. It has been suggested by several groups that hinge length (i.e., the number of amino acids residues in the hinge region) determines the degree of segmental flexibility and affects complement and Fc receptor binding by IgG molecules. IgG3 has an extended hinge encoded by four exons, the last three of which are identical; it is flexible and fixes complement well. In contrast, human IgG4 has a short hinge encoded by a single exon; it is a rigid molecule and does not fix complement.

To assess the role played by the hinge, a panel of IgG3 molecules of differing hinge length was constructed and the hinge regions exchanged between IgG3 and IgG4. The experiments demonstrated that the presence of at least one of the hinge segments is essential for complement activation, consistent with the idea that an intact H_2L_2 molecule is required for maintaining complement binding activity. However, a single hinge exon is sufficient to endow IgG3 with its full capacity to activate complement, whereas lengthening the hinge to seven exons results in a molecule with impaired ability to activate complement. Providing IgG4 with the hinge of IgG3 does not result in an antibody that can activate complement, indicating that additional structural aspects of the IgG4 molecule interfere with its ability to activate complement.

Chimeric antibodies were used to study the role of glycosylation. All IgG molecules contain a consensus glycosylation sequence (Asn-*X*-Thr-, where *X* represents any amino acid) in the C_H2 domain. The presence of this carbohydrate is important for some of the effector functions of Ig, including Fc receptor binding and complement activation. When *in vitro* site directed mutagenesis was used to remove that glycosylation site from C_H2 of chimeric mouse/human antibodies, the resulting carbohydrate-deficient antibodies, although properly as-

TABLE I Some Properties of Human Immunoglobulins

	IgG				IgA				
	IgG1	IgG2	IgG3	IgG4	IgA1	IgA2	IgM	IgD	IgE
Average concentration in normal serum (mg/ml)	8	4	1	0.4	3.5	0.4	1	0.03	0.0001
Molecular weight (kDa)	150	150	180	150	150–600	150–600	900	180	190
Half-life in serum (days)	23	23	8	23	6	6	5	3	2.5
Active in complement fixation-classical pathway	+++	±	+++	–	–	–	+++	–	–
Binds to high-affinity Fc receptor	+++	0	+++	+	–	–	–	–	–
Sensitizes human mast cells for anaphylaxsis	–	–	–	–	–	–	–	–	+
Present in mucosal secretions	–	–	–	–	++	++	±	–	–

sembled, secreted, and capable of binding antigen, were more sensitive to proteases than their corresponding wild-type IgGs, and some had a shorter serum half-life in mice. Thus, absence of carbohydrate profoundly changes some properties of the antibody molecule while leaving others intact. The precise influence of carbohydrate structure on antibody function remains to be determined.

A property of antibody molecules critical for their *in vivo* function is the ability to bind to cellular Fc receptor. Chimeric IgG1 and IgG3 bind the high-affinity Fc receptor (FcγR1) with a high affinity; chimeric IgG4 binds with approximately 10-fold lower affinity, and binding by chimeric IgG2 is not detectable. Using site-directed mutagenesis, residue 235 of mouse IgG2b has been pinpointed as interacting with this Fc receptor.

IX. Genetically Engineering the Antibody Variable Region

A potential limitation to the *in vivo* use of chimeric mouse/human antibodies is the immune response that may arise to the murine variable regions. A way to minimize this response would be to make the mouse variable regions more human. When the three-dimensional structure of antibody variable regions is compared, they are found to be similar. They all contain framework regions of similar structure that determine the position of complementary-determining residues (CDRs); these in turn form the antibody-combining site. This conservation of structure makes it feasible to transfer the CDRs from one variable region to another. The CDRs from mouse monoclonal antibodies can be substituted into the framework of a human antibody, with the resulting chimeric antibodies continuing to bind their specific antigen. The possible applications of this "variable region grafting" remain to be evaluated.

X. Antibody Fusion Proteins

Novel proteins that possess the binding specificity of antibody molecules can be created by replacing the constant region DNA sequence of an antibody molecule with a sequence derived from another molecule such as an enzyme, toxin, growth factor, or biological response modifier. Such molecules have potential use in immunoassays, in diagnostic imaging, and in immunotherapy.

Several enzymes have been joined to antibody-combining sites, including the nuclease from *Staphylococcus aureus* and the Klenow fragment of *E. coli* DNA polymerase I. In both cases the antibodies continued to bind antigen, and the enzymes were catalytically active, albeit with reduced specific activity. In other studies, the constant region of an antibody was replaced by the sequence of insulin-like growth factor 1. The resulting antibodies continued to bind antigen and bound to the IGF1 receptor present on cells, but with reduced affinity. Binding to growth factor receptors provides an additional means of targeting antibodies to specific cell populations.

Antibody specificity and a new effector function

can also be combined to produce an antibody-targeted pharmacological reagent. It seems likely that plasminogen activators with enhanced fibrin specificity can be made by joining a portion of tissue-type plasminogen activator (t-PA) to a fibrin-specific antibody. To achieve this aim, the DNA sequences encoding a portion of the t-PA were used to replace some of the constant region exons of a mouse $\gamma 2b$ antibody specific for fibrin. When the fusion protein was expressed with its specific L chain, the resulting molecule continued to bind fibrin and contained the peptidolytic activity of native t-PA. Molecules of this nature may be used for targeting activities to specific locations in the body.

XI. Chimeric Antitumor Antibodies

Murine mAbs that recognize tumor-associated antigens are potentially useful for recognizing tumors for diagnostic imaging and for therapy via ADCC, CDC, or antibody-conjugated anticancer agents. However, the use of these murine antibodies has been limited by the development of a human antimouse immunoglobulin antibody response and/or failure of the antibody to possess the desired biologic properties (e.g., serum clearance pharmacokinetics and ability to interact optimally with the human immune system). Mouse/human chimeric antibodies comprised of variable regions derived from murine hybridomas and human constant regions should be superior for administration to humans because of their reduced immunogenicity, altered pharmacokinetics, and ability to function more effectively with human effector cells.

Variable regions from a number of murine mAbs reactive with tumor-associated antigens have been cloned, and chimeric antibodies have been prepared using human constant region genes. In all cases, the resulting chimeric proteins retained the binding specificity of the murine mAbs. Chimeric antitumor antibodies have been shown to exhibit the biological properties appropriate for their respective human immunoglobulin subclasses. Several human monoclonal and antitumor mAbs were developed to date, but most are of the IgM isotype, limiting their usefulness. These could be changed to one of the human IgG isotypes, using the approach outlined above.

In some instances the substitution of human constant regions has resulted in enhanced antitumor activity or has imparted biological activity to a nonfunctional murine antibody possessing desirable binding specificity. For example, conversion of antitumor antibodies to chimeric antibodies with a human γ_1 constant region results in a chimeric antibody that mediated ADCC in the presence of human effector cells at a concentration 100 times lower than that required for the murine antibody. Similarly, a murine mAb that recognizes the CD20 antigen expressed in both normal and malignant B cells lacks cytolytic activity; conversion of the antibody to chimeric mouse/human IgG1 results in an antibody possessing the antigen binding specificities of the parental antibody as well as the capability to mediate ADCC with human effector cells and CDC with human complement.

In the management of cancer, the most immediate advantage of chimeric mAbs may be the reduction of the human antimurine antibody response, which is mostly directed toward the murine Fc region of the Ig molecule. The human antimurine antibody response creates two problems: the potential of an allergic response with anaphylaxis, and the rapid clearance of the administered antibody, which would prevent the vast majority of the mAb from reaching the tumor site. Thus, in virtually all the previously reported human therapy trials using multiple administrations of murine mAbs, only the first and/or perhaps the second mAb administrations efficiently reached the tumor site. The use of chimeric antibody diminishes this problem.

One of the major advantages in the use of genetically engineered mAbs is the ability to modify the Ig molecule to alter pharmacokinetics. We may wish to slow down plasma clearance of a mAb so that it will have the opportunity to mediate ADCC or complement-mediated cytolysis or to speed up the plasma clearance of an mAb conjugated to a radioisotope, in which the circulating conjugate may cause damage to normal cells. Genetically engineered Igs provide several ways to alter pharmacokinetic properties. These include (1) large alterations in size by the addition or deletion of domains and more subtle alterations in size using smaller deletions, (2) construction of Fv molecules, (3) alterations in glycosylation, and (4) mutations in sequences controlling clearance rates. [*See* PHARMACOKINETICS.]

Finally, antitumor mAbs may be modified so that they can act as more efficient vehicles for the delivery of antitumor drugs, toxins, radionuclides, or biological response modifiers. This can be achieved by either directly ligating the antitumor agent into

TABLE II Some Potential Applications of Genetically Engineered Chimeric Antibodies

Genetic Modification	Potential Uses
Variable regions attached to new constant region derived either from same or different species	Produce isotype variants within a species; can be used to make the available IgM human antibodies more useful by converting them to IgG antibodies
	Can be used to make mouse/human chimerics with the variable region from mouse and the constant region from humans. The resulting chimeric antibodies should have the following properties: reduced immunogenicity when used *in vivo* in humans; increased ability to interact with the human immune system; improved pharmacokinetics
Variable regions modification before joining to a constant region	Humanize mouse variable regions by positioning mouse binding regions (complementarily determining regions) within human framework regions
	In vitro site-specific mutagenesis to produce variable regions with altered binding constants
Modification of constant regions by domain shuffling or deletion or *in vitro* mutagenesis	Improved biologic properties
	Improved pharmacokinetics and tissue distribution
Antibody sequences joined to nonantibody sequences	Joining to drugs or toxins for their more efficient delivery
	Join to enzymes for either *in vivo* application or to make better reagents for *in vitro* applications
	Insert biologic response modifiers to increase their effectiveness or to improve targeting
	Insert sequences to facilitate labeling with radioisotopes

the Ig molecule or by the ligation of efficient linkers for drugs, radioisotopes, or bioresponse modifiers into the Ig molecule.

XII. Additional Applications and Future Prospects

Many potential applications of chimeric or genetically engineered antibodies exist in addition to use as anticancer agents (Table II). Monoclonal and chimeric antibodies have been produced to cell surface antigens, which are important for immune interactions. These antibodies have great potential as immune response modifiers to facilitate organ transplantation and address problems of autoimmunity. It is also possible that antibodies may be developed to approach the problem of immune nonreactivity such as is seen in AIDS. Chimeric antibodies also have potential applications in treatment of infectious diseases. They should be far superior to antisera derived from heterologous sources (e.g., horses) and should address the heterogeneity and limited availability of antisera derived from human sources. Chimeric antibodies also have great potential for use as vaccines devised to exploit the anti-idiotypic network. [*See* IDIOTYPES AND IMMUNE NETWORKS.]

Bibliography

Better, M., Chang, C., Robinison, R., and Horwitz, A. H. (1988). *Escherichia coli* secretion of an active chimeric antibody. *Science* **240,** 1041.

Boulianne, G. L., Hozumi, N., and Shulman, M. J. (1984). Production of functional chimeric mouse–human antibody. *Nature (London)* **312,** 643.

Brüggemann, M., William, G. T., Bendon, C. J., Clark, M. R., Walker, M. R., and Jefferis, R. (1987). Comparison of the effector functions of human immunoglobulins using a matched set of chimeric antibodies. *J. Exp. Med.* **166,** 1351.

Haber, E., Quertermous, T., Matsueda, G. R., and Runge, M. S. 1989. Innovative approaches to plasminogen activator therapy. *Science* **243,** 51.

Jones, P. T., Dear, P. H., Foote, J., Neuberger, M. S., and Winter, G. (1986). Replacing the complementarity determining regions in a human antibody with those from a mouse. *Nature (London)* **321,** 522.

Köhler, G., and Milstein, C. (1975). Continuous cultures of fused cells secreting antibody of predefined specificity. *Nature (London)* **256,** 495.

Lui, A. Y., Robinson, R. R., Hellstrom, K. E., Murray, E. D., Chang, C. P., Hellstrom, I. (1987). Chimeric mouse-human IgG1 antibody that can mediate lysis of cancer cells. *Proc. Natl. Acad. Sci. U.S.A.* **84,** 3439.

Morrison, S. L., Johnson, M. J., Herzenberg, L. A., and Oi, V. T. (1984). Chimeric human antibody molecules: mouse antigen binding domains with human constant region domains. *Proc. Natl. Acad. Sci. U.S.A.* **81,** 6851.

Mulligan, R. C., and Berg, P. (1981). Selection for animal cells that express the *Escherichia coli* gene coding for xanthine-guanine phosphoribosyltransferase. *Proc. Natl. Acad. Sci. U.S.A.* **78,** 2072.

Neuberger, M. S., Williams, G. T., and Fox, R. O. (1984). Recombinant antibodies possessing novel effector functions. *Nature (London)* **312,** 604.

Skerra, A., and Plückthun, A. (1988). Assembly of a functional immunoglobulin Fv fragment in *Escherichia coli. Science* **240,** 1038.

Southern, P. J., and Berg, P. (1982). Transformation of mammalian cells to antibiotic resistance with a bacterial gene under control of the SV40 early region promoter. *J. Mol. Appl. Genet.* **1,** 327.

Sun, L. K., Curtis, P., Raksowicz-Szulczynska, E., Ghrayeb, J., Chang, N., Morrison, S. L., and Koprowski, H. (1987). Chimeric antibody with human constant regions and mouse variable regions directed against a carcinoma-associated 17-1A antigen. *Proc. Natl. Acad. Sci. U.S.A.* **84,** 214.

Tao, M.-H., and Morrison, S. L. (1989). Studies of aglycosylated chimeric mouse-human IgG. Role of carbohydrate in the structure and effector functions mediated by human IgG constant regions. *J. Immunol.* **143,** 2595.

Tan, L. K., Shopes, R., Oi, V. T., and Morrison, S. L. (1990). The hinge region: Influence on complement activation, C1q binding and segmental flexibility in chimeric human Igs. *Proc. Natl. Acad. Sci.* **87,** 162.

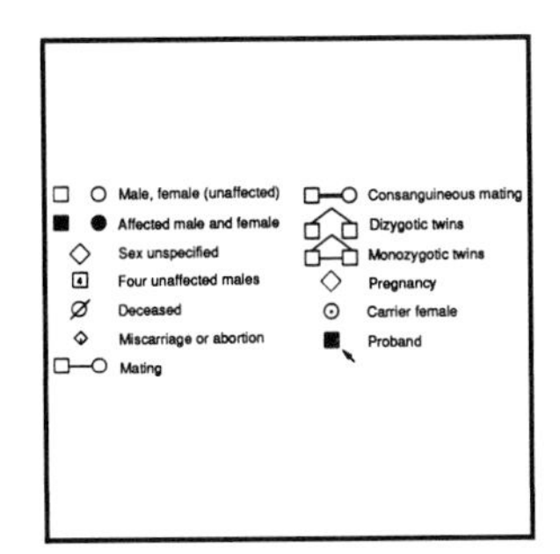

Genetic Counseling

VIRGINIA L. CORSON, *Johns Hopkins University School of Medicine*

Glossary

Carrier Individual who possesses the gene that determines an inherited disorder and who is usually healthy at the time of study

Chromosome Tightly coiled structures in each cell that contain the genetic blueprint coded by individual genes

Gene Units of information that determine physical characteristics and biochemical properties of the body

Multifactorial Determined by multiple genetic and environmental factors

Recessive Mode of inheritance in which a nonworking gene is inherited from each carrier parent

GENETIC COUNSELING is a communication process that addresses the human problems associated with the occurrence, or the risk of occurrence, of an inherited disorder or a birth defect in the family. Medical facts including the diagnosis, prognosis, and available management are discussed, as well as the role of heredity and the risk of recurrence. Reproductive options are investigated as the family is encouraged to pursue a course of action that seems appropriate to their goals and values. The counselor strives to facilitate the family's adjustment to a disorder or their risk of a genetic problem while offering psychological and emotional support. Genetic counseling services have become an integral part of health care as a result of an increasing awareness of the clinical significance of the principles of human genetics.

I. Indications for Counseling

A. Affected Family Member

Approximately 3% of newborns have some birth defect (e.g., a cleft lip or heart defect) that will require medical intervention. One baby in 200 is born with a chromosome abnormality (e.g., Down's syndrome). The parents of these children with mental retardation, physical abnormalities, or a suspected genetic disorder will seek genetic counseling to obtain further information about the etiology and implications of their child's problem. Some genetic conditions do not have manifestations until later in life, and an evaluation will be suggested during the teenage or adult years as symptoms arise. In addition, relatives in the extended family will have questions regarding the implications of a disorder and may seek counseling after a diagnosis has been made in one family member. [*See* BIRTH DEFECTS; CHROMOSOME ANOMALIES.]

Although traditional genetic counseling is not widely sought for common disorders of adulthood (e.g., diabetes, heart disease, cancer, or emotional illness), the understanding of genetic factors contributing to their occurrence continues to grow, and counseling will expand for these indications. Through early diagnosis of clinical symptoms and the development of predictive genetic markers, relatives of affected individuals can seek appropriate medical care or modify pertinent environmental influences.

B. Prospective Parents

Increasing numbers of couples contact a genetics center with concerns about risks to the fetus or their

TABLE I Estimated Risk of All Chromosome Abnormalities in Live Births for Specific Maternal Ages

Maternal age	Frequency of chromosome abnormalities
25	1/476
27	1/455
29	1/417
31	1/385
33	1/286
35	1/179
37	1/123
39	1/81
41	1/49
43	1/31

future offspring. Many are referred to discuss the mother's age-related risks for chromosome abnormalities as seen in Table I. Pregnant women in their mid-30s and older are routinely offered prenatal testing to rule out Down syndrome and other chromosome abnormalities. Other couples may have a previously affected child or some other family history concern for which prenatal diagnosis is available. Pregnant women exposed to potentially harmful drugs or infections seek information about possible risks to the fetus. Some couples are referred for counseling and additional testing in follow-up of abnormal prenatal test results obtained by their obstetrician. Finally, some couples have experienced multiple pregnancy losses and are seen to evaluate a possible genetic basis for their reproductive difficulties.

C. Ethnic Background

Individuals of particular ancestries are at increased risk to be carriers of certain recessive genetic disorders. Screening tests and prenatal diagnosis are available to these persons. Eastern European (Ashkenazi) Jews and French-Canadians are at increased risk to be carriers for Tay-Sachs disease, a degenerative, neurological disorder that is usually lethal by 3–5 years of life. Through measurement of the enzyme hexosaminidase A in a blood sample, carriers can be identified and counseled. Blacks are at greater risk to be carriers for sickle cell anemia and thalassemia, two serious blood disorders. Thalassemia carriers are also more frequent in persons of Greek, Italian, Middle Eastern, Indian, and Southeast Asian backgrounds.

II. Health Care Setting and Providers

As the demands for counseling services increased in the 1970s, genetics clinics were established in university medical centers where specialists trained in human genetics were able to see patients and pursue research into understanding the basis of inherited problems. Satellite clinics affiliated with these medical centers were begun in more rural communities or in areas without easy access to a university center. Additional facilities arose in private community hospitals or physicians' offices, largely in response to the growing need for prenatal diagnostic services.

A team of trained individuals is best suited to address the broad range of problems encountered in human genetics and to provide the laboratory support needed for diagnostic testing. Physicians trained in medical genetics are responsible for the medical evaluation and treatment plans of affected patients. Genetic counselors, health professionals with specialized graduate degrees, usually coordinate the clinic activities, participate in the communication process, and assume most of the prenatal genetic counseling responsibilities. Other members of the team may include personnel from the genetics laboratories, nurses, and medical social workers. Specialists from other disciplines such as ophthalmology, cardiology, orthopedics, and radiology are consulted for diagnostic testing and treatment of some patients.

The other professional important to the genetic counseling process is the primary care physician who refers patients and manages their medical follow-up. These pediatricians, obstetricians, and internists must consider the possibility of a genetic diagnosis in a significant fraction of their patient population.

III. Counseling Process

A. Obtain Family History

One of the cheapest tools of the geneticist is the family history, or pedigree. Valuable clues to a diagnosis can be revealed through obtaining information about other family members. Multiple miscarriages or early infant deaths can be indicators of chromosomal or metabolic conditions. Ethnic background can be important, as some problems are more common in certain populations, and consan-

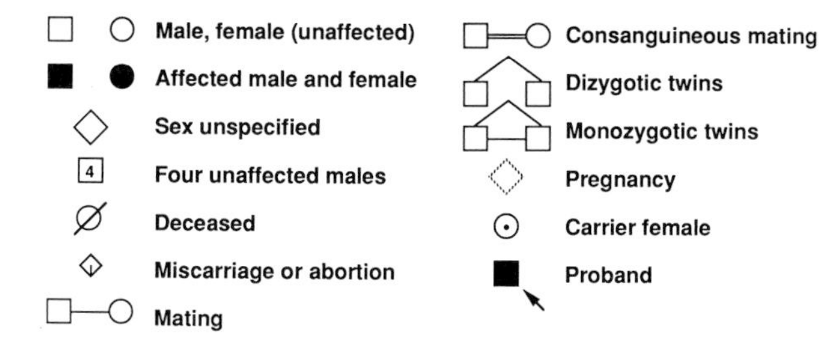

FIGURE 1 Symbols used in constructing a pedigree.

guinity (mating within the family) points suspicion toward a recessive inheritance pattern. The conventions seen in Fig. 1 have arisen for pedigree construction and are appropriate for any clinical setting.

B. Review Medical Records and Clinical History

Appropriate medical records regarding the individual under evaluation as well as other pertinent family members should be reviewed by the genetics team. If indicated, a physical examination is performed, and additional factors such as developmental milestones and the mother's pregnancy history are discussed.

C. Evaluate for Diagnosis and Interpret Laboratory Data

Based on the family history and clinical presentation, possible diagnoses will be considered for the individual under evaluation. Laboratory studies or specialist consultations may be indicated to substantiate further or to rule out a particular diagnosis. For prospective parents, the indication for referral will be addressed and options for evaluation through parental testing or prenatal diagnosis presented.

D. Discuss Prognosis, Management, and Inheritance

After a diagnosis is made, the discussion will focus on the significance of this disorder for the family. Expectations about future capabilities and limitations of the affected individual are addressed, as well as plans for medical interventions and therapies. Genetic implications are communicated as they apply to different family members, and carrier detection is offered to appropriate individuals. Issues such as the variation in severity of any diagnosis must be addressed also.

E. Provide Psychosocial Support

The occurrence, or even the risk of occurrence, of a genetic disorder or birth defect can be devastating for the family. Emotional sequelae must be acknowledged and addressed as families deal with their anger, denial, guilt, and depression. Marital problems can be exacerbated, and difficulties with the other children may arise. The counselor's awareness of this process will facilitate dialogue with the parents through reassurance that their emotions are normal reactions and part of a healthy adjustment. Appropriate resources (e.g., support groups and literature references) can be helpful during this period of time. Other economic and community services will be offered as needed.

F. Examine Reproductive Options

Couples at an increased risk for a genetic disorder face a number of reproductive alternatives. For many genetic conditions, prenatal diagnosis is available and will offer an acceptable alternative. Amniocentesis, the withdrawal of fluid from the sac surrounding the fetus, is most commonly performed at 15–18 weeks gestation with a risk of miscarriage estimated to be 0.3%. Chorionic villus sampling (CVS), an aspiration of early placental tissue, is a newer procedure performed at 9–12 weeks gestation and is thought to have a slightly greater risk of miscarriage than amniocentesis. Other options for prenatal testing include ultrasound for evaluation of the fetal anatomy and maternal serum α-fetoprotein (AFP) screening for open spine defects and Down syndrome. Through these techniques, a number of chromosomal, single-gene or multifactorial disorders can be diagnosed. If an abnormality is detected through prenatal diagnosis, the options of pregnancy termination or improved obstetric management at delivery can be offered.

Some couples will decline future childbearing after weighing the risks and the burden that the affected child and/or future children could place on their life-style. Adoption may be considered by some of these families. For reasons related or unrelated to the genetic diagnosis, a percentage of couples will separate and change marital partners, thus lowering their genetic risks. Under special circumstances, donor sperm or eggs may be used to lower a genetic risk when prenatal diagnosis is not available or unacceptable.

IV. Principles of Effective Counseling

A. Counseling Guidelines

When a counseling session begins, it may be beneficial to determine the counselee's preconceptions or knowledge about the disorder in question. Misconceptions about the etiology, inheritance, or prognosis can interfere with the communication process. Both parents should be encouraged to attend so that each has the benefit of hearing the information presented and the opportunity to ask questions. Facts should be explained biologically with the aid of diagrams whenever possible and at a level appropriate for the family's comprehension. Information should be presented when facts are available and when the parents are ready for additional implications of a diagnosis.

The counselor needs to address the feelings of parental guilt that often accompany the birth of a handicapped child. Parents can often be reassured that specific environmental events that occurred during the pregnancy were not responsible for an abnormality. For couples at risk for a specific recessive disorder, the knowledge that *all* individuals are carriers for 5–10 deleterious genes can be helpful in alleviating the feeling that they are different from others.

Much information will be shared during an initial counseling session, and in many cases, some form of follow-up contact with the family will be essential to their long-term comprehension. A written summary, a return visit, or a telephone contact with the genetic counselor can be effective mechanisms to reinforce the primary discussion.

B. Risk and Burden

The impact of a diagnosis of a genetic disorder on family units varies according to their perceptions of the burden. Emotional and financial implications of the disease contribute to the stress felt by parents and siblings. For many families, the long-term care requirements of a progressively degenerating disorder carry demanding economic and psychological strains. In contrast, the sudden, unexpected loss of a congenitally malformed child shortly after birth may be accompanied by intense pain, which will diminish over time. The loss is as real, but the emotional and financial demands not as great.

As with burden, perception of the risk magnitude is open to individual interpretation. Some persons will view a 1% risk as "high," whereas others will view that as "low." Many have difficulty in comprehending risk factors in general, and it can be helpful to use an odds likelihood as well as a percentage. Thus, a 3% recurrence risk can be described as one chance in 33, or a 25% risk is one chance in four. The alternative risk that a problem will *not* occur is another important perspective to offer, so that families can put their likelihood for a favorable outcome in perspective. For single-gene disorders, the recurrence risk is the same for each pregnancy.

Reproductive decisions will be made on the basis of this perceived burden and the recurrence risk of the disorder. Genetic counseling can aid in the decision process through identification of these two factors and an exploration of the couple's perceptions of risk and burden.

V. Goals

Through this communication with genetic counselors and medical geneticists, individuals can achieve an increased understanding of the specific disorder of concern in their family. Genetic implications of the diagnosis are addressed, and risks to various family members outlined. Personal decision-making is supported by the genetics team as families face questions of reproductive planning, carrier testing, and treatment for affected individuals. Options for childbearing (e.g., adoption or prenatal diagnosis) are discussed openly for the family's consideration. Ultimately, the goal of genetic counseling is to facilitate increased coping with the realities, or the prospects, of genetic disease and its impact on the family.

Bibliography

Filkins, K., and Russo, J. F., eds. (1990). "Human Prenatal Diagnosis," 2nd ed. Marcel Dekker, New York.
Harper, P. S. (1988). "Practical Genetic Counselling," 3rd ed. Butterworth, London.

Marks, J. H., Heimler, A., Reich, E., et al., eds. (1989). "Genetic Counseling Principles in Action: A Casebook." Birth Defects Original Article Series 25, vol. 5. Alan R. Liss, New York.

Thompson, J. S., and Thompson, M. W. (1986). "Genetics in Medicine," 4th ed. W. B. Saunders, Philadelphia.

Weaver, D. D. (1989). "Catalog of Prenatally Diagnosed Conditions." Johns Hopkins University Press, Baltimore, Maryland.

Genetic Diseases

MAX LEVITAN, *Mount Sinai School of Medicine, City University of New York*

Glossary

Autosomes Kinds of chromosomes normally present doubly in both males and females; in humans there are 22 pairs of autosomes
Chromosomal aberration Change in the number or position of the genes or chromosomes
Dominant Trait or gene that is expressed even if there is only one gene for it at the pertinent locus on a pair of homologous chromosomes
Genetic diseases Pathological conditions caused or influenced by abnormalities in the genetic material
Holandric inheritance Determined by genes on the Y chromosome
Locus Position of a gene on its chromosome
Maternal inheritance Heredity of traits determined by DNA on cytoplasmic organelles; being passed on almost exclusively via the ovum and not dependent on meiosis, their transmission does not follow Mendelian principles
Mutant Gene or individual that manifests a changed characteristic as a result of mutation
Mutation Change in one or more nucleotides of DNA; in its broadest usage it also includes chromosomal aberrations
Polygenic inheritance Transmission of traits that involve the additive contributions of many loci
Recessive Trait or gene that is not expressed unless both loci of a pair of homologous chromosomes are alike
Simple inheritance Characteristics are attributable to the effects of a single locus
X chromosome Chromosome normally present doubly in the female, but singly in the male, in humans and many other organisms
X linked Genes, or characteristics determined by them, whose loci are on the X chromosome
Y chromosome Chromosome type normally present only in males

GENETIC DISEASES are those pathological conditions that are caused or influenced by an abnormality in the genetic material. If the root cause is the change in one or more nucleotides in a single gene, it is usually referred to as a mutation—the gene, and often the individual manifesting it, being referred to as a mutant. When the cause stems from a disturbance in the number or position of the genes or chromosomes, it is usually referred to as a chromosomal aberration. The position of a gene on its chromosome is its locus.

A mutated gene that is part of one of the 22 pairs of chromosomes normally present in both males and females and the associated genetic disease are referred to as autosomal. However, if it is part of the X chromosome, the one that is normally present doubly in the female but singly in the male, it is referred to as X linked (formerly called "sex linked"). A few genes are known on the Y chromosome, normally present only in males (i.e., holandric inheritance), and on the mitochondrial cytoplasmic organelles; the latter and associated diseases are maternally inherited, being transmitted via the ovum.

I. Simply Inherited Diseases

When the genetic disease is attributable to a change in a single gene, it is said to be simply inherited. A simply inherited disease is autosomal recessive if the affected person has abnormal autosomal genes for the same locus on the chromosomes received from both parents, autosomal dominant if the affected individual must have only one abnormal gene at the pertinent autosomal locus, or X linked if the gene in question is on the X chromosome. [*See* GENES; GENETICS, HUMAN.]

Simply inherited autosomal recessive diseases generally involve genes that normally control the production of enzymes, organic catalysts that enable cells to carry on metabolic processes rapidly at the relatively low temperatures of the human body. When only one such abnormal gene is present, the amount of the enzyme is usually decreased, but enough enzyme is produced via the normal gene present to enable the metabolic process to go on normally.

McKusick's catalog lists 626 well-established autosomal recessive diseases. Most are rare, occurring in one per 50,000 births or fewer, but some are more common, particularly in certain ethnic groups. In persons of northern European descent, for example, cystic fibrosis occurs in about one of 2500 births. By contrast, its incidence in Oriental and black populations is about one per 250,000 births. Similarly, the frequency of Tay–Sachs disease (type I gangliosidosis) is about one per 4000 births in Ashkenazic Jews, about 10,000 times its frequency in Sephardic Jews or Gentiles. [*See* CYSTIC FIBROSIS, MOLECULAR GENETICS.]

Known enzyme deficiencies lend themselves to prenatal diagnosis by growing embryonic cells obtained from pregnant woman by various techniques and determining whether they can produce the pertinent enzyme. Molecular genetics has extended these procedures to detect with increasing accuracy the presence or absence of recessive genes (e.g., the ones involved in cystic fibrosis). Prenatal diagnosis is especially useful when a previous child has been born with the recessive defect. Since each subsequent child has one chance in four of being similarly affected, the parents often would hesitate to risk having further children. Prenatal diagnosis demonstrating that the fetus is probably normal can allay these anxieties. The contrary finding, that the fetus is probably affected, poses the psychological and ethical dilemma between selective abortion and the birth of a defective child.

The genes responsible for autosomal dominants generally affect the structural proteins of cells (e.g., hemoglobin, collagen, and myosin). Usually, a person producing less than the normal amount of such a protein, because one of his genes for it is abnormal, manifests a detectable abnormality.

Although McKusick's catalog lists 1443 well-established autosomal dominants, only about 500 can be considered diseases, for many of the dominants determine normal variants of serum proteins and cell-surface molecules that act as receptors for circulating substances or provide antigenic specificities to the cells. Some of these variants, however, can have pathological consequences. Hemolytic disease of the newborn, for example, can result from certain combinations of Rh, ABO, Duffy, Kidd, and other blood groups in parents and child. Another large group causes physical abnormalities, such as various forms of polydactyly (extra fingers or toes) or brachydactyly (short fingers), that can be noted by a physician (or dentist, if they affect the teeth), but generally have no pathological sequelae.

Sometimes autosomal dominants cause relatively mild diseases in single dose, but rather severe ones when present doubly. The more serious diseases are therefore, in effect, recessive. Prominent examples are the gene for S hemoglobin (the SS person being affected with sickle cell anemia) and the genes that result in the reduced production of various globins (the persons with double doses being affected with thalassemias, many of them so severe as to be lethal early in life). [*See* SICKLE CELL HEMOGLOBIN.]

Similarly, familial hypercholesterolemia results from mutations at the locus for production of the cells' receptors for low-density lipoprotein. Cells with one normal and one abnormal gene are less able to capture cholesterol from the blood for its normal uses to build cell membranes, secrete steroid hormones, and produce bile salts. This results in a greater tendency for abnormal accumulations of low-density lipoprotein and its deposition in the linings of critical vessels (e.g., the arteries of the heart), and this increases the likelihood of myocardial infarction (i.e., heart attack). About one in 500 people have single doses of these defective genes, making familial hypercholesterolemia perhaps the most common of all simple Mendelian disorders. The rarer persons with double dose almost invariably develop high cholesterol levels in the blood and fatal cardiac effects at an early age. [*See* ATHEROSCLEROSIS; CHOLESTEROL.]

Of the 139 well-established X-linked conditions, most qualify as diseases, although many (e.g., red–green color blindness, testicular feminization, and several types of deafness) are not seriously debilitating. Included, however, are serious clotting defects (i.e., hemophilias), muscular dystrophies, and immunodeficiencies.

II. Polygenic Inheritance

The role of genes in such common diseases as diabetes, hypertension, peptic ulcer, bipolar (e.g., manic–depressive) mental disorders, and schizophrenia is currently not well understood, partially because conditions with different etiologies are often grouped under the same clinical rubric, and many involve complex interactions of genetics and environment. Although reports appear from time to time that single important genes governing these traits have been discovered, most geneticists believe that the underlying physiology of these conditions depends on the additive contribution of many genes. These produce a wide spectrum of variation, with individuals having more than a threshold number labeled as affected. Such a multiple-factor, or polygenic, hypothesis probably also accounts for the genetic aspect of such birth defects as spina bifida, cleft palate (with or without cleft lip), congenital dislocation of the hip, pyloric stenosis, clubfoot, Hirschsprung's disease, and some congenital heart defects.

III. Chromosomal Aberrations

Multiple congenital defects, some mimicking those determined by polygenes, are characteristic of most disorders that result from chromosomal aberrations. These can be divided into two classes: (1) aberrations of number, the affected individual having more than or fewer chromosomes per cell than the normal 23 pairs, and (2) aberrations of chromosome structure, situations in which one or more genes are missing, are present in more than their usual number per cell, or are present in unusual locations. Aberrations of number are usually due to errors in cell division, whereas errors of structure usually stem from unrepaired chromosomal breakage. However, many aberrations of structure have pathological effects only because they lead to offspring with deficient or excessive numbers or parts of chromosomes. That a deficiency of genes could be harmful is understandable, but the basis for the ill effects of an excess is not well understood. Apparently, normal well-being depends on an evolutionarily developed "balance" among the genetic materials that can be disrupted by an excess as well as by a deficiency. [*See* CHROMOSOME ANOMALIES.]

Aberrations of number are most viable if they involve the X or Y chromosomes. Indeed, females with an extra X chromosome (47,XXX) and males with an extra Y (47,XYY) show few, if any, abnormalities. 47,XYYs do tend to be quite tall, and this can be responsible for their greater than random tendency to be involved in criminal activities. (This does not mean that tall men are usually XYYs, nor does it mean that XYYs are usually criminals.) Males with extra X chromosomes (e.g., 47,XXY, 48,XXXY, and 49,XXXXY) exhibit Klinefelter's syndrome; they tend to be infertile and to exhibit various degrees of mental deficiency.

Females with more than one extra X chromosome (e.g., 48,XXXX, and 49,XXXXX) are likely to be mentally defective and to have other symptoms. Females with only a single X chromosome (i.e., 45,XO) tend to have Turner's syndrome. Its prominent features are short stature, absence of secondary sex characteristics, and ovaries consisting of streaks of connective tissue. There is some evidence that the viable XOs lost the other sex chromosome after fertilization, so that they are or were mosaics of normal and abnormal cells, whereas nonmosaic 45,XOs (the other sex chromosome having been absent from the egg or sperm) usually die *in utero*.

Deficiency of a whole autosome is invariably lethal. Spontaneous abortuses sometimes contain all of the chromosomes in triplicate (i.e., triploids) or quadruplicate (i.e., tetraploids). Individuals with a single extra autosome (i.e., trisomy) appear more frequently. Those that come to term invariably have multiple congenital anomalies and, unless they are mosaics who have many normal cells as well, usually die within a few weeks or months after birth. The exceptions are trisomies for the smallest autosome, labeled number 21. Trisomy 21 results in Down syndrome, as the English physician J. L. H. Down first described it clearly; in the literature it is also (unfortunately) called mongolism, or mongolian idiocy. It appears in about one of 700 births, more often from older mothers than from younger ones. Although also generally beset with congenital anomalies, including varying degrees of mental deficiency, 21-trisomics who survive the perinatal period have a good chance of viability and, in the case of females, can even be fertile, with the potential for normal offspring as well as trisomics. [*See* DOWN'S SYNDROME, MOLECULAR GENETICS.]

Many partial trisomies and deficiencies have been described. As their effects vary considerably, depending on the nature or amount of the chromosome involved, few clear syndromes have been delineated to date. It is known, however, that defects formerly attributed to gene mutations are often due to deletions of whole genes or their critical parts. This is particularly true of many of the thalassemias.

Partial trisomies (sometimes also called duplications) and deficiencies can result from the misalignment of normal chromosomes during germ cell production. Deficiencies can also be produced when a chromosome breaks in two places on the same side of its spindle attachment point (i.e., centromere) and reheals without incorporating the substance between the breaks. If the two breaks are on different sides of the centromere, the two terminal pieces can be lost, and fusion of the two broken ends results in a ring chromosome. Many partial trisomies and deficiencies are, however, the byproducts of sperm or ovum production in carriers of inversions or translocations. Such carriers are normal, but they can have abnormal offspring. An inversion is produced when a chromosome breaks in two places and the substance between the breaks is reincorporated in reverse order. Reciprocal translocations, found in about two per 1000 births, occur when two chromosomes suffer one break each, and the chromosomes exchange the broken-off pieces. In the rarer insertional translocation or transposition the chromosome fragment resulting from two breaks in one chromosome is inserted at a third break. The latter break can be in the same chromosome, in which case it is often called a shift, or it can be in another chromosome. Some Down syndromes result from carriers of reciprocal translocations involving chromosome 21.

IV. Genetics of Tumors and Cancers

Susceptibility to tumors or cancers has long been recognized as a genetic disease, but the exact role of heredity has not been clear. A number of syndromes characterized by the development of neoplasia seem to be simply inherited. One of the best understood is xeroderma pigmentosum, a skin disorder resulting from mutants of loci that normally produce enzymes used to repair DNA that has been damaged by ultraviolet light. Persons with single doses of these genes often exhibit a large degree of freckling, but do not develop the malignancies. Several other recessive disorders predispose to leukemias or lymphomas, but they can result in tumors at other sites as well. Simply inherited dominant forms most often involve tumors of the endocrine glands, skin, or mucosa. Being dominants, the family history can often alert physicians to relatives at risk and, by early detection of cancerous or precancerous lesions, prevent more serious effects. This has been particularly successful in several forms of multiple intestinal polyposis. Similar considerations apply to apparent "cancer families," whose members seem unusually susceptible to a wide variety of neoplasms. Some of the most frequent cancers (e.g., breast, uterus, stomach, and colon) do not fit models of simple inheritance. Prevailing opinion is that neoplastic development depends on the simultaneous presence of two genetic defects, the so-called "two-hit" theory: (1) probably mutation or absence of both members of a pair of genes that normally act as tumor suppressors, and (2) mutation of a class of growth-regulating genes known as oncogenes, because, like oncogenic viral genes with similar structures, they are capable of causing the transformation of certain tissue cultures from regular monolayer growth to growth in multilayered jumbles, resembling the way cancer cells grow in culture. [*See* CHROMOSOME PATTERNS IN HUMAN CANCER AND LEUKEMIA.]

Bibliography

Levitan, M. (1988). "Textbook of Human Genetics," 3rd ed. Oxford Univ. Press, New York.

McKusick, V. A. (1988). "Mendelian Inheritance in Man: Catalogs of Autosomal Dominant, Autosomal Recessive, and X-linked Phenotypes," 6th ed. Johns Hopkins Univ. Press, Baltimore, Maryland.

Milunsky, A. (ed.) (1986). "Genetic Disorders and the Fetus: Diagnosis, Prevention, and Treatment," 2nd ed. Plenum, New York.

Vogel, F., and Motulsky, A. G. (1986). "Human Genetics: Problems and Approaches," 2nd ed. Springer-Verlag, Berlin and New York.

Genetic Maps

MICHAEL DEAN, *Frederick Cancer Research and Development Center*

Glossary

Allele One of the forms of a gene
Centimorgan (cM) Measure of genetic distance equal to 1% recombination. In humans one cM, is on the average, 1 million base pairs
Dominant Allele that shows its phenotypic effect, even the presence of another (recessive) allele. Codominant alleles are expressed simultaneously
Eukaryote Organism composed of a cell or cells with membrane-bound nuclei and organelles and multiple chromosomes; includes all plants and animals
Genome Full complement of genetic material of an organism. The DNA sequences contained in all the chromosomes
Genotype Genetic makeup of an individual, including silent alleles. The genotype determines the phenotype, the expressed characteristics of an individual
Heterozygote Individual who has two different alleles for the same gene. Homozygotes have identical alleles
Interference Effect of recombination at one position on the recombination frequency at an adjacent position. The value can be positive or negative
Linkage Proclivity for two genes on the same chromosome to be inherited together. The distance is measured in percent recombination or centimorgans (cM)
Meiosis Division of the nucleus that results in the production of haploid germ cells (sperm and eggs)
Molecular cloning Isolation and insertion of a DNA segment into a vector that can be propagated in large quantity. Typically fragments from complex organisms are cloned into bacteria
Operon Group of adjacent genes that perform a coordinated process, usually sequential reactions in a metabolic pathway
Recombination (crossing-over) Shuffling of genetic material that accompanies sexual reproduction and results in new combinations of genes in the progeny
Restriction enzyme Bacterial protein that cleaves DNA at specific positions, allowing physical maps to be constructed. The workhorse of recombinant DNA and genetic engineering
Somatic cell hybrid Mixed cell line formed by the fusion of cells from two different species, typically containing the full chromosome complement of one species and a partial set of the other

IN THE 1860s Gregor Mendel described the first principles of genetics, and in the early 1900s analysis of chromosomes from the fruit fly *Drosophila melanogaster* led to the concept of genetic mapping. A genetic map is the blueprint of an organism. It is the linear arrangement of the genes of that species, constructed by measuring the distance between individual genes or genetic markers. The making of a genetic map is one of the first steps toward the complete characterization of the genetic material of an organism, the determination of the total nucleotide sequence of the species. Mapping of genes that cause disease or unusual phenotypes is a crucial step toward the isolation of each gene. Genetic maps have also proved to be valuable in the study of species biology. Today, genetic maps have begun to be pieced together for dozens of species from bacteria to humans.

I. Mapping in Experimental Organisms

A. Bacteria

Most of the gene mapping to date has been performed on experimental organisms. Many of the organisms studied have fairly simple genetic structures, which allow the ready performance of genetic experiments. The most complete genetic map of a free-living species is that of the bacterium *Escherichia coli*. This organism has been studied in great detail by both biologists and geneticists. Many of the genes of *E. coli* have been identified. Each gene encodes a single protein with a specific function, such as the ability to utilize a certain sugar molecule or synthesize an amino acid. Because this bacterium grows rapidly, and its genetic material is relatively small and can be easily manipulated, maps could be quickly constructed.

The genetic map of *E. coli* is displayed in Fig. 1. A crucial discovery in the development of genetics in *E. coli* was the finding that bacteria can mate or exchange genetic material. Mating occurs when a "male" bacterium binds to another recipient bacterium. Male bacteria contain a extrachromosmal segment of DNA, the F factor, which encodes a structure known as a pilus present at the bacterial surface, through which genetic material can be transferred. In this process, known as conjugation, the F factor itself is transferred. Bacterial strains were identified in which the F factor was incorporated into the bacterial genome. When these bacteria conjugate, parts of the donor's DNA are transferred with the F factor. It was found that for a given strain this process always starts at the same place in the genome (the point of integration of the F factor) and always proceeds in the same direction at a constant pace. By interrupting the conjugation process at different times (it takes 90 min to transfer the whole genome), the order of genes can be determined. That is, the genes closest to the F factor are always transferred first, whereas distal genes take longer.

Additional methods for mapping bacterial genomes also involve the transfer of segments of naked DNA, a process termed *transformation,* or the use of viruses to carry bits of genetic material, termed *transduction*. In either case the principle is the same. Genetic markers that are close to each other are transferred together more often than markers that are far apart. By determining the frequency in which pairs of markers travel together, a map of the genes can be built. Recently the entire genome of *E. coli* has been molecularly cloned and the fragments mapped by restriction enzymes. This is a step toward the completion of the ultimate map, the complete nucleotide sequence. To date, only the genomes of certain viruses have been completely sequenced.

One important outcome of the mapping of genes in *E. coli* was the discovery that many genes whose proteins are made at the same time and work together in a biochemical pathway are near each other in the genome. Blocks of genes that are coordinately regulated are called *operons*.

B. Mapping in Eukaryotic Organisms

Genetic maps of higher organisms are constructed by following the inheritance of genetic markers from parent to offspring. Many of the concepts of genetics were developed by analyzing the inheritance of visible traits (for instance, the color of the eyes, the length of the bristles) in the fruit fly *Drosophila melanogaster*. T. H. Morgan determined in 1911 that certain traits were inherited together more often than expected by chance, suggesting that they were physically linked. A. H. Sturtevant used this approach to build the first map of a eukaryotic chromosome. Figure 2 shows the genetic location of several traits found on a portion of the *Drosophila* second chromosome, along with a depiction of these traits. The genetic position of several mutations that cause visible traits in the fly are shown along with a depiction of those traits.

We now know that the DNA of all eukaryotes is organized into chromosomes. Most individuals are diploid [i.e., they carry two copies of each chromosome (except the sex chromosomes), one inherited from each parent]. During meiosis, when sperm and egg cells are generated, there is the opportunity for genetic exchange between the chromosome pairs. This process is known as *recombination* or *crossing-over* and leads to the reassortment of alleles (different forms of a gene). Markers that are close together on a chromosome rarely recombine, and those farther apart recombine more frequently. This feature allows the order of genes to be determined by measuring the recombination rate of pairs of markers. Thus a genetic map of a sexual organism is generated by determining the recombination frequency of a set of genes on a chromosome. [*See* CHROMOSOMES; MEIOSIS.]

Recombination, however, is not a simple pro-

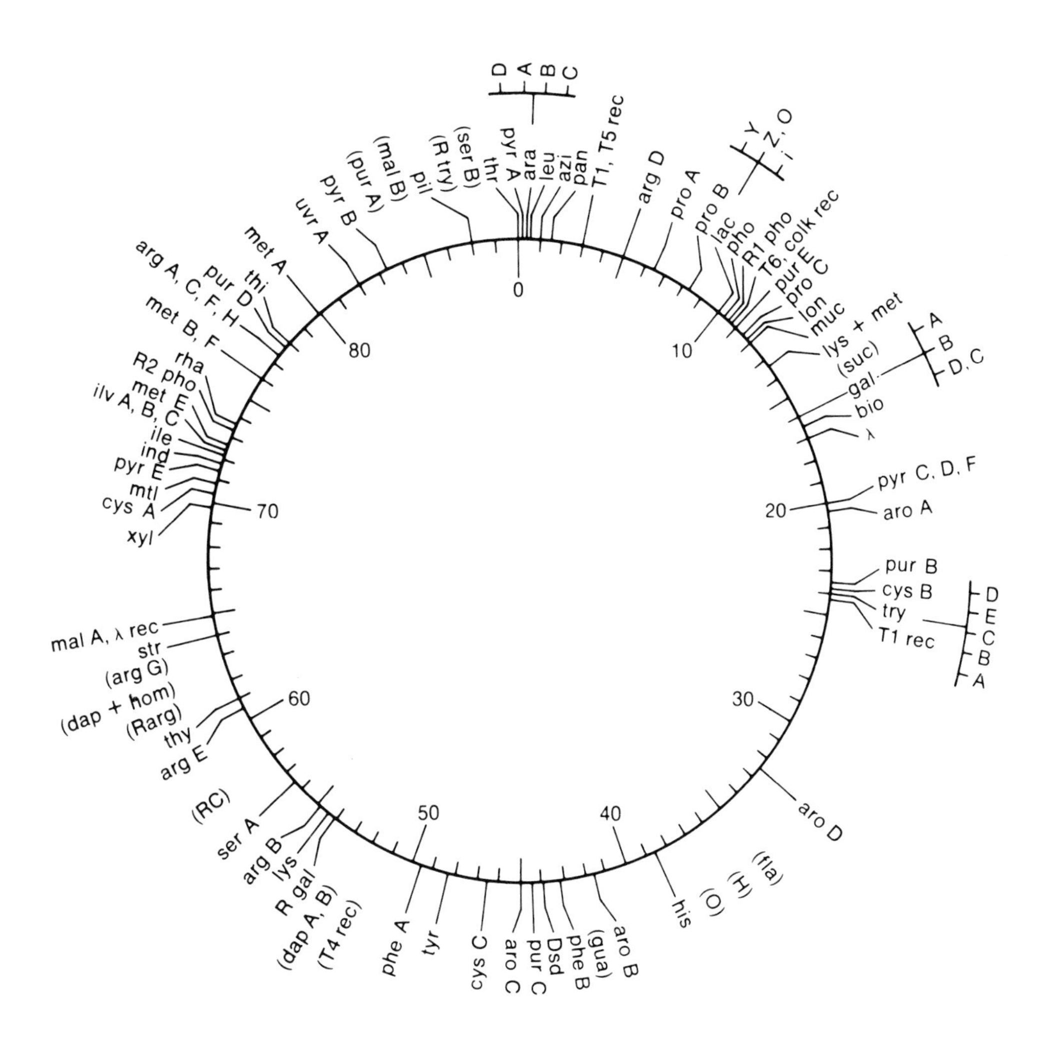

FIGURE 1 A genetic map of *E. coli*. The map is displayed in units of minutes. The position of several operons is shown outside of the central map. [From Taylor, A. and Thoman, M. (1964) Genetics **50,** 667, with permission.]

cess. Although it does give definitive evidence of the order of markers, it does not occur evenly over a chromosome, and it occurs at different rates in the two sexes. *Drosophila* offers an extreme example, in which all recombination occurs in females and none in males. A cross-over in one part of a chromosome can also affect recombination further down the chromosome, a phenomenon known as *interference*. An example would be three genes, A, B, and C, in which there is 20% recombination between A and B and 30% recombination between B and C. We would expect that a double recombinant with recombination between A and B and also between B and C would occur at a frequency of 6% ($0.20 \times 0.30 = 0.06$). However, the frequency of the double recombinant may be only 2%. In other words, when a recombination happens between A and B, crossing-over between B and C is interfered with or vice versa. This may be a merely physical phenomena; there is probably only so much you can bend the DNA that carries the genes, but there may be more to it. The degree of interference varies from species to species and region to region.

The assembly of the maps of more than 100 organisms has begun, including organisms of all complexities, from viruses to mammals, including humans and plants. Well-studied creatures include yeast, the simplest and best understood eukaryotic cell; corn, in which many of the principles of segregation were discovered as well as the presence of mobile genetic elements similar to the bacterial F factor, *Caeorhabditis elegans,* a small transparent worm whose every cell can be followed visually from birth to adulthood; and the domestic cow, an important food resource for much of the world. The field of genetics is diverse and touches all the other biological sciences.

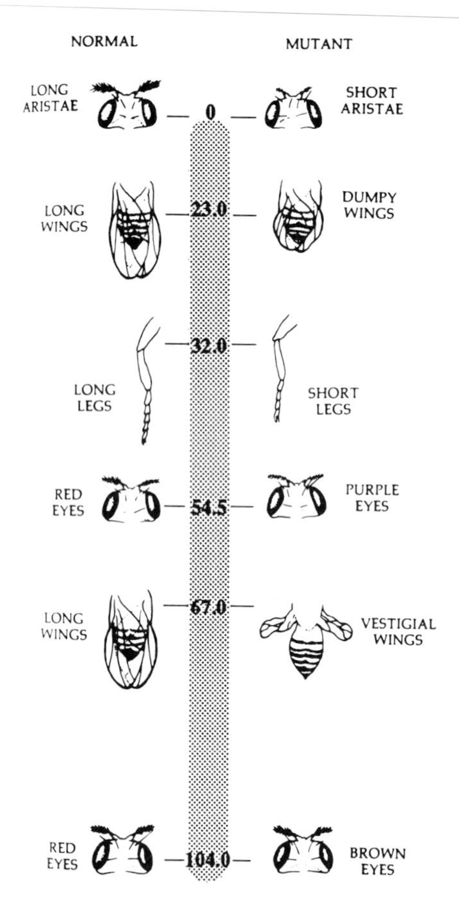

FIGURE 2 Map of several genes on *Drosophila melanogaster* chromosome 2. The position of several genes is shown along with the genetic distance between them, calculated by the recombination frequency. [From Curtis, H., and Barnes, N. S. (1989). "Biology." Worth Publishers, New York, with permission.]

C. Mapping the Lab Mouse

Although the studies on bacteria, flies, molds, and plants laid the groundwork for the discipline of genetics, researchers sought an organism that would provide a more useful model for human biology. The clearly superior choice has been the laboratory mouse, *Mus musculus*. Although we might not think we have a great deal in common with a rodent, consider that we are both mammals. Both species give birth to live young, lactate, and grow hair. Mice and humans both have complex brains, intricate behavior patterns, and the ability to learn. We both have much the same organ systems, with quite similar metabolic pathways. Mice, like humans, develop cancer and suffer from viral infections, neurological disorders, and birth defects.

The key to the establishment of mouse genetics was the discovery of pure (inbred) strains with visible genetic traits, whose inheritance could be followed. For centuries the Japanese, mostly as a hobby, bred and collected mice with interesting colors of fur. These lines formed stocks for many of the present strains of experimental animals. By creating purebred strains and crossing them, the genes that determine coat color have been mapped to linkage groups (i.e., chromosomes). During the past several decades, similar mapping has been achieved for a host of traits. Careful observation of traits arising by mutation and the maintenance of such strains, mainly at the Jackson Laboratory in Bar Harbor, Maine, have led to the collection of hundreds of mutant animals. Mouse models for human diseases such as muscular dystrophy, blindness, albinism, obesity, and diabetes have been developed. In addition, there are strains with high susceptibility to cancer and resistance to various viruses. For the most part, these phenotypes are controlled by single genes, which have been located on the genetic map of the mouse. One by one these interesting genes are being isolated and characterized, advancing our understanding of mammalian biology.

Mouse genetics has profited enormously from the development of molecular biology. By using cloned DNA fragments to probe specific parts of the genome, any gene or DNA segment can be placed on the mouse genetic map. The principle is illustrated in Fig. 3. Restriction enzymes can be employed to break mouse DNA into fragments of characteristic lengths, which can then be detected with radioactively labeled cloned DNA segments. Sometimes the length of the fragments detected are different between different strains of mice, allowing the inheritance of that particular DNA segment to be followed. This difference is referred to as a restriction fragment length polymorphism (RFLP). RFLPs have been employed to great advantage in the interspecies backcross. This type of experiment starts with two different subspecies of mice that can breed and form fertile offspring. Animals from the first (F1) generation of mating are then bred (backcrossed) to animals from one of the parent species, and the offspring of this mating are analyzed. In the example shown in Fig. 3, spretus mice are mated to the black/6 (Bl/6) inbred strain and the female F1 animals backcrossed to Bl/6 mice. Because of the relatively large genetic distance between spretus and Bl/6, differences in the size of restriction fragments are frequently detected. The offspring inherit a Bl/6 fragment from their father, because he is pure

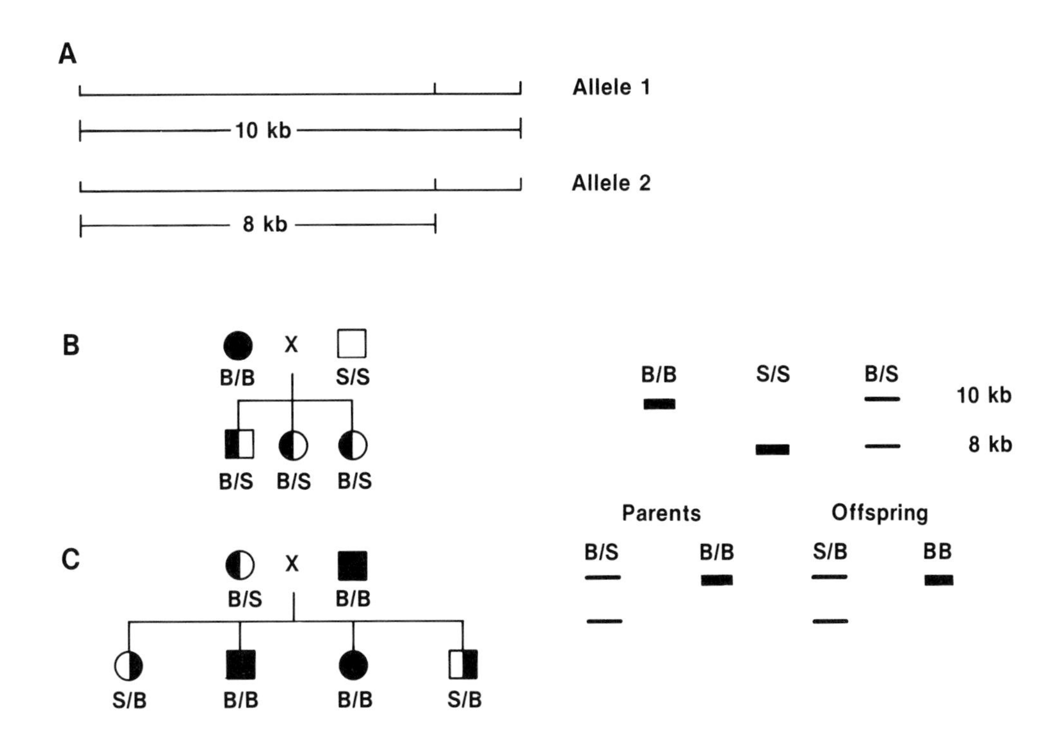

FIGURE 3 Restriction fragment length polymorphism (RFLP) and the interspecies backcross. **A:** An illustration of an RFLP detected using a specific probe and the restriction enzyme X. Two alleles are detected; allele 1 is 10 kb and found in Bl/6; allele 2 is 8 kb and is specific to spretus. **B:** Generation of the F1 animals for a backcross. Female (*round symbols*) inbred Bl/6 mice are mated to male (*square symbols*) spretus mice (a closely related species of mice). The resulting offspring inherit one chromosome from each of their parents and are all genetically identical. Also shown is the pattern of an RFLP in these animals in which the 10-kb allele is present in Black/6 and the 8-kb allele in spretus. **C:** The backcross. Female F1 animals are mated back to male Black/6 animals. All animals inherit a Black/6 chromosome from the father and either a Black/6, spretus, or a recombined chromosome from the mother.

Bl/6. They will inherit either a spretus or Bl/6 band from the mother, in approximately a 1 : 1 ratio. If the inheritance of several markers is determined in the same set of animals, we should find either that they segregate randomly (are inherited together only about half the time) or that they display linkage (are inherited together most of the time). By determining the proportion of animals that display recombination between two markers, the genetic distance between them can be computed. An example is shown in Table I. Ten backcross animals were typed with probes for two genes, A and B. The data are expressed as S or B to designate whether the animal inherited the spretus or the Bl/6 form of the gene from the heterozygous parent. For eight of the 10 animals, the same form of the gene was inherited. This suggests that the two genes are linked and the recombination distance between them is 20%, two recombinations of 10 meioses. By studying several markers from the same chromosome, a map can be constructed.

Many mammalian genes have been identified based on their ability to code for specific proteins. The DNA that encodes a number of these genes have been isolated. In some cases, alterations in the DNA sequence of the gene (mutations) that cause disease has been identified. Because physical mapping with DNA fragments does not rely on the identification of a visible trait, a large number of markers can be followed in the same set of animals. This technique also allows any gene to be mapped, whether or not mutations in the gene cause a discernible phenotype. By combining the mapping of mutants with the physical mapping of DNA seg-

TABLE I Backcross Mapping of Two Genes[a]

	Backcross animal									
	1	2	3	4	5	6	7	8	9	10
Gene A	S	S	S	B	S	B	B	B	B	S
Gene B	S	S	S	B	S	S	B	S	B	S

[a] RFLPs for two genes were used to type 10 spretus(S)/Bl/6(B) backcross animals. Two recombinations are observed between genes A and B.

ments, the position of the genes that cause mutations can be accurately placed, an important step toward the isolation of the actual gene. The number of genes placed on the mouse genetic map is increasing at an ever faster pace, as are the number of exciting discoveries from this effort.

II. Human Genetic Maps

A. Overview

The genetics of humans is most complex and difficult to study. It cannot follow the principle of mouse genetics, which depends on the generation of pure strains of individuals, crosses between selected individuals, and the generation and selection of useful mutants. Human genetic analysis must depend on the use of the experiments that nature provides. Human geneticists have made use of genetic diseases and other heritable traits in large families and closed societies with some degree of inbreeding. The reward has been the characterization of hundreds of genetic diseases in humans, the identification and molecular cloning of many of the genes involved in disease, and the ability to make accurate predictions and diagnoses for affected families with several disorders. [*See* GENETIC DISEASES.]

The earliest human genetic markers used were proteins present in multiple forms in the population. Probably the most familiar are the proteins of the ABO blood group, which are found on the surface of blood cells. There are three alleles in this system (A, B, and O) and four possible phenotypes (A, B, AB, and O) (Table II). O is actually the absence of this protein on the cell (a null allele), and A and B are dominant over O. That is, if we inherit the A allele from our mother and the O allele from our father, the red blood cells will express the A protein on the surface and will be blood type A. A and B are codominant, so if we inherit both alleles, both proteins are expressed and blood type is AB. From the genotypes of the parents, we can predict the possible genotypes of their offspring and follow the inheritance of the alleles in each child.

The pool of human markers expanded with the identification of isozymes, enzymes present in multiple allelic forms. However, the number of isozymes is rather small, and the size of the human genome is rather large (23 chromosome pairs and 3 billion nucleotides). Although the principle of using protein markers to detect linkage with human genetic diseases was proposed by J. B. S. Haldane as early as 1937, human genetic mapping did not start in earnest until the 1970s, not through genetics but by using somatic cell hybrids. Somatic cell hybrids are constructed by fusing the cells of two different species, in this case of humans and a rodent, typically mouse or hamster. The fused cell contains the chromosomes of both species but tends to gradually lose those of one of the species. In human-rodent hybrids, the human chromosomes are lost in a fairly random fashion. By isolating a series of cell lines after this loss, a panel of hybrids can be generated, each containing a few or a single human chromosome. Each cell line can then be characterized by karyotyping (visually identifying the chromosomes) to determine which chromosomes it has retained. By detecting the presence or absence of a gene in a panel of hybrids, that gene can be assigned to a chromosome. For instance, the enzyme lactate dehydrogenase A was found to be present in all hybrids that contained human chromosome 11 and in none of the hybrids that did not. Somatic cell hybrids have been used to map genes in many mammalian species, and the technique is still used as a first step in the mapping of genes. However, somatic cell hybrids do not allow the accurate determination of the position of the gene on the chromosome or the ordering of genes. An alternative approach is to use X-rays to fragment the DNA before forming the hybrid. This breaks up the DNA into smaller segments, and these radiation hybrids can be used to make physical maps.

TABLE II ABO Blood Groups

Blood group	Genotype
O	O/O
A	A/A, A/O
B	B/B, B/O
AB	A/B

B. Linkage Analysis

The ordering of genes takes advantage of the segregation of markers within families. Many disease genes have been mapped to the X chromosome by observing the segregation of the disease. X-linked recessive disorders are inherited mostly by males. A female, who has two X chromosomes, is a carrier when she has one disease gene allele and one normal allele. Half of her sons will inherit her X chromosome, which carries the mutation, and because they do not receive another X chromosome with the normal allele to compensate, they will suffer the disease. Her daughters will receive one of her X chromosomes, and a normal X chromosome from the father. The result is a disease that affects mostly males and is carried by females. Diseases such as hemophilia A, Duchenne's muscular dystrophy, and color blindness are examples of the genetic price males pay for having only one X chromosome. [*See* HEMOPHILIA, MOLECULAR GENETICS; MUSCULAR DYSTROPHY, MOLECULAR GENETICS.]

The first linkage of human genes was described by J. Bell and J. B. S. Haldane in 1937. By analyzing the inheritance of two X-linked diseases (hemophilia and color blindness) within a set of pedigrees, they were able to determine the genetic distance between them. The major difficulties in mapping human genes arise because the individuals for study are selected from an outbred population. This means that no assumptions can be made about the genetic makeup of a particular person. One specific problem is that of determining the phase or arrangement of a group of markers on a chromosome. In the example in Fig. 4A, two parents and an offspring are shown, along with the genotypes for three genes (A, B, and C), each with two alleles (A, a; B, b; C, c). The goal is to determine which alleles the child inherited from each parent, and the arrangement of the alleles in the parental chromosomes (i.e., the phase). Consider first genes A and B. Both the mother and the child are homozygous for A and heterozygous for B. They must each have a chromosome that is A-b and another that is A-B. This is designated by the line drawn between the markers. The father, however, is heterozygous for both A and B, and we cannot tell the phase of his alleles. We also cannot tell which alleles the child inherited and from which parent, except that she must have inherited the father's A allele.

The situation gets worse when we add gene C, in which each person is heterozygous. We get no information about the phase of the C alleles. Even though there is information, it cannot be used to tell which chromosomes were passed on. However, the phase can be sorted out by the addition of just one grandparent (Fig. 4B). Because the grandfather is homozygous for all three genes, he must have two copies of an A-B-C chromosome. His son must have inherited one of these, and therefore, his other chromosome must be a-b-c. The daughter clearly

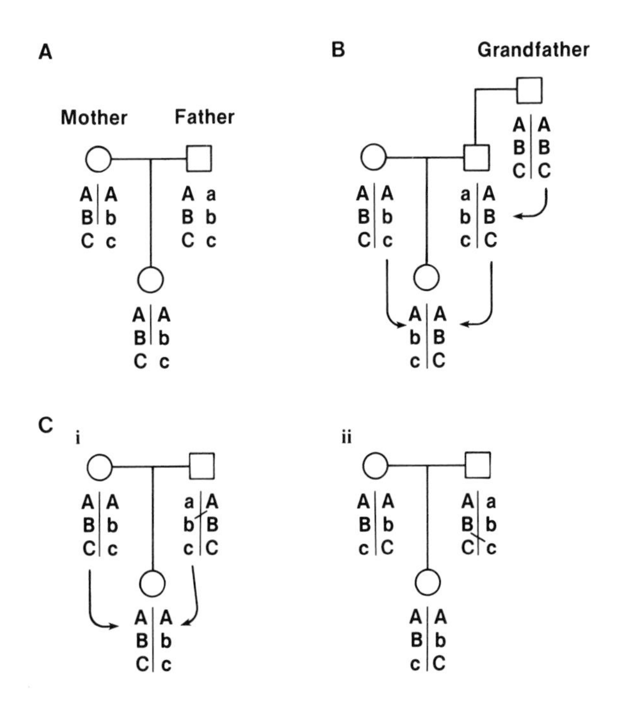

FIGURE 4 Human linkage analysis. **A:** The genotypes for three genes (A, B, C) each with two alleles (A, a; B, b; C, c) in a nuclear family. Females are depicted by *circles* and males by *squares; horizontal lines* between individuals designate mates, and *vertical lines* point to offspring. Vertical lines between alleles separate alleles that lie on different homologous chromosomes. **B:** Information provided by the addition of information from a grandparent. *Arrows* show the path of inheritance of chromosomes. **C:** Potential recombination events. Shown are a few of the possible recombinations that could have occurred and yielded the same data. (i) A recombination in the father's chromosomes between genes A and B. (ii) Cross-over between genes B and C.

received an A-B-C chromosome from her father and an A-b-c chromosome from the mother. Thus by including, whenever possible, information from grandparents, the phase of the markers can often be determined.

Our conclusion above was based on the assumption that no recombination events have occurred in any of the chromosomes passed on to the daughter. However, Fig. 4C shows that several single recombinations could have occurred and still produced the same data. In (i), a recombination in the father between A and B produces an A-b-c chromosome, and the child could then have received the mother's A-B-C chromosome, quite a different result than what we predicted in Fig. 4B. Panel (ii) shows another possibility, and many more involve multiple recombination events. The point is that with such data we cannot be completely sure of what happened. How can we make a map when the inheritance of each chromosome cannot be unambiguously followed?

Making genetic maps of human chromosomes depends on the use of certain mathematical and statistical treatments that take into account the uncertainty associated with the data. One of these is the calculation of the log of the odds (LOD) score. Basically these manipulations allow us to calculate the probability that two genes are linked and that a group of linked genes are arranged in a certain order (a genetic map). This can never be absolutely certain, but when the odds become great enough, even the most skeptical become convinced. Usually odds of 1,000 to 1 are sufficient to convince the geneticist.

The LOD score is the logarithm of the odds that two genes are linked. Thus a LOD score of 3 means that it is 1,000 times more likely that two genes are linked than that they are unlinked. In practice, the LOD score is calculated by following the coinheritance in a set of families of a genetic marker and a genetic disease. Each time the marker and the disease gene are inherited together, the LOD score goes up, and for each apparent recombination, the score decreases. When all the data are added up, if the score is at or below zero, the two can be considered unlinked. The process is repeated with additional markers until a LOD score above 3 is reached, which is taken as proof that the marker and disease gene are linked. If the chromosome location of the marker is known, the disease gene has been mapped to a chromosome. LOD scores less than −2 are considered as evidence that two loci are unlinked.

C. Huntington's Disease

Huntington's disease is an autosomal dominant disorder of the nervous system, which illustrates both the power of human genetics and the difficult social issues raised by such work. It has been known that the disease is inherited since it was described by George Huntington in 1872. Those who inherit this genetic defect suffer from a progressive deterioration of the nervous system, which typically begins when they are in their 40s.

Huntington's disease patients have a 50% chance of transmitting the disease to each of their children, although most patients do not know that they have the disease until they have passed their childbearing years. Research into the biochemical basis of the disease has been frustrated by the complexity of the brain and the lack of a suitable animal model. In the early 1980s, several groups began to pursue a genetic approach, to search and find the gene using linkage analysis. With the gene in hand, the protein involved could be identified and studied. This process has come to be known as "reverse genetics," to distinguish it from the classical approach of proceeding from protein to gene to mutation. In 1983, linkage was detected between a DNA marker on the short arm of chromosome 4 and Huntington's disease. A key to this success was the use of a large, extended Venezuelan family suffering from this disease. This was the first major success of the use of linkage to locate the disease gene without any other clues as to its location.

Intensive research has focused on identifying additional markers on the short arm of chromosome 4 to further pinpoint the location of the disease gene and to develop genetic diagnosis. The discovery of closely linked markers allows predictions to be made in HD families as to who does or does not carry the defective gene. Figure 5 illustrates a hypothetical family in which a woman with the disease

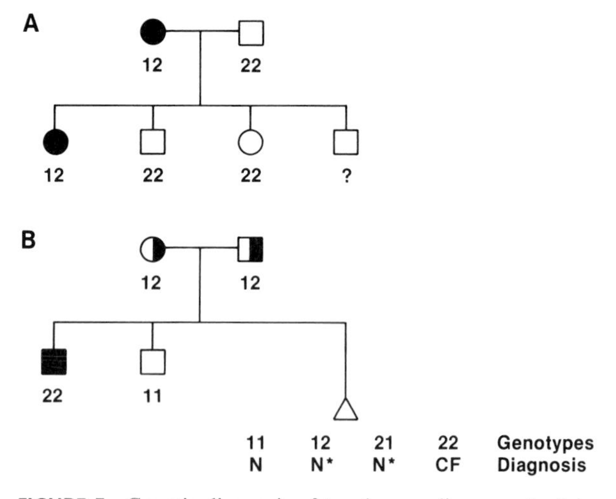

FIGURE 5 Genetic diagnosis of two human diseases. **A:** A hypothetical pedigree of a family with Huntington's disease, a dominant disease. The affected individuals are shown in *filled symbols*, with the genotypes for a closely linked marker shown underneath. 1,2 are the two alleles of a closely linked marker. The disease was passed from the mother to the oldest daughter. The youngest son is the individual requesting diagnosis. **B:** Pedigree of a cystic fibrosis family. Because this is a recessive disease, both parents must be carriers (*half-filled symbols*). The oldest son is affected, and his brother is predicted to be homozygous normal. The fetus is shown by the *triangle*, along with the four possible genotypes and their predicted outcomes. N*, a healthy carrier.

has had four children. She is heterozygous for a closely linked marker (1,2), whereas her husband is homozygous for (2,2). Her oldest daughter is also affected with the disease and has inherited the 1 allele of the marker from her mother. Assuming that no recombinations have occurred, the 1 allele must be on the same chromosome as the HD gene. At this point, the other children can be typed for the marker, and a prediction can be made as to their carrier state. (Remember this can only be a prediction because of the chance of recombination. The accuracy of the prediction will be determined by the genetic distance between the marker and the disease gene.)

It is easy to see that this new ability to predict the carrier state of people early in life creates a complex dilemma. A negative result predicts that the individual will be free of disease and can have children without fear of passing the gene to them. A positive result foretells a life that will involve progressive suffering and loss of contact with the outside world. Should we want to know the answer? The testing is always accompanied by extensive counseling to help patients cope with this information.

At this writing the location of the HD gene has been narrowed to a small region on the tip of chromosome 4. When the gene itself is identified, research on this disorder can proceed, providing insight into its mechanism and potential therapies. Hopefully, this research will also provide powerful insight into the workings of the nervous system. [*See* HUNTINGTON'S DISEASE.]

D. Cystic Fibrosis

Cystic fibrosis (CF) is the most common fatal genetic disease in the Western world, affecting about one in every 2,000 children born to Caucasian parents. It is 10 times rarer in Blacks and almost unheard of in Orientals. Like Huntington's disease, CF is a complex disorder, and its molecular basis has eluded investigators for decades. However, the past 5 years have seen a flurry of discoveries on CF, which have led to a much clearer understanding of the disorder. Clinically the disease involves two major organs (the lung and the pancreas). In most CF patients, the pancreas fails to secrete digestive enzymes, leading to an inability to absorb fats and deficiencies in fat-soluble vitamins. The lungs of CF patients become increasingly congested with a thick mucus, leading to bacterial infections and eventually death. The most consistent clinical symptom for diagnosis is an abnormally high concentration of sodium and chloride ions in the sweat of CF individuals. The levels are usually three times higher than in normal individuals. In the lungs, pancreas and sweat ducts there are secretory cells, and the process of secretion is largely controlled by the transport of ions into and out of the cell. In CF patients, the regulation of ion transport is defective, leading to obstructions in the pancreas, congestion in the lung, and an ion imbalance in the sweat.

Despite the detailed knowledge gained by studying cells from CF patients, the defective protein was not identified by this approach. Several groups collected samples from families affected by CF and began using linkage analysis to try to locate the mutated gene. Progress was slow until 1985, when four successive reports of linkage were announced. The two closest markers were each found to lie on opposite sides of the gene, at a recombination distance of 1%. This allowed, for the first time, the ability to perform genetic testing and prenatal diagnosis on CF families. An example is shown in Fig. 5B. A couple has a child with CF. Because the disease is recessive, the child must have inherited a defective gene from each parent; they are both carriers. Each parent is heterozygous for a closely linked marker, and the CF child is homozygous for allele 2 of the marker. In this situation, allele 2 is predicted to lie on the CF chromosome in both parents. The patient's brother is clinically normal. However, he could be a CF carrier, and carriers have no clinically detectable traits. The genetic analysis shows that he inherited the 1 allele from both parents and is therefore predicted to be homozygous for the normal chromosome. This analysis can be extended to any subsequent pregnancies in this family. Fetal DNA can be obtained by chorionic villus sampling at 8–10 weeks of gestation or by amniocentesis at 16–18 weeks. The four possible genotypes are shown in Fig. 5B. The fetus has a 1:4 chance of being homozygous normal, a 2:4 chance to be a carrier, and a 1:4 chance of having CF. Again, these are predictions, with the accuracy dependent on the distance between the marker and the mutation. However, with close markers on both sides of the gene, the accuracy is greater than 99%, because they allow the detection of any recombinant in the interval between them.

This discovery was a tremendous breakthrough for some families with a history of CF. Because the disease is recessive, however, most carriers do not know that they carry a defective gene, and thus most CF children are born to completely unsuspecting parents. Remember that the marker does not test for the mutation itself; it can only be used to trace the inheritance of a mutation that is known to be present in the family. To test for carriers in the general population and to learn more about the disease, the gene and the mutation(s) had to be found. This breakthrough was accomplished in September 1989.

By examination of the inheritance of RFLP markers in hundreds of families, it was determined that the CF gene was between the MET gene and the marker D7S8 (Fig. 6A). This distance between MET and D7S8 was determined to be approximately 1,000 kilobases. By successively cloning sequences toward the gene from both sides, eventually the gene was reached and identified. Several pieces of evidence support the idea that the correct gene has been identified. The gene is in the location predicted by all the available genetic data, the predicted protein has a structure similar to a family of transport proteins localized in the cell membrane, and the gene is mutated in CF patients. Figure 6C shows the nature of the mutation present on the majority of CF chromosomes. The defect is due to the deletion of three nucleotides within the coding region of the protein. The mutuation occurs in a region of the molecule that appears to be involved in the interaction of the protein with ATP. This binding is thought to provide the energy needed for the protein to perform its transport functions. Thus the mutation is rather subtle, leading to the production of a slightly smaller protein that is partially disabled in function. The situation is similar to the case of sickle cell anemia, which involves a substitution of a single amino acid in a globin gene, which carries oxygen. The substitution leads to a reduction in oxygen binding and to structural abnormalities in the cell. It is presumed in CF and known for globins that a more drastic mutation or complete loss of function of the gene would be lethal.

Figure 6 displays three types of genetic maps. C shows the nucleotide sequence of a portion of the gene. The nucleotide sequence is the most detailed type of map, as it displays the complete information encoded by the genome. B shows a restriction enzyme map of a large portion of the gene. Restriction maps are generated by digesting DNA with restriction enzymes, resolving the fragments by gel electrophoresis, determining the size of the fragments, and determining the order of the fragments in the original segment of DNA. This type of map is useful for elucidating the size and structure of genes. A shows the cleavage position of several restriction enzymes that cleave infrequently in mammalian DNA, every 100,000 base pairs or more. These large fragments can be separated by a technique known as pulsed-field gel electrophoresis, which uses pulsing fields of current to increase the resolution of large DNA molecules. In the CF locus, this method was used to link together distant genetic markers, and it is useful for determining the distance between genes. [*See* CYSTIC FIBROSIS, MOLECULAR GENETICS.]

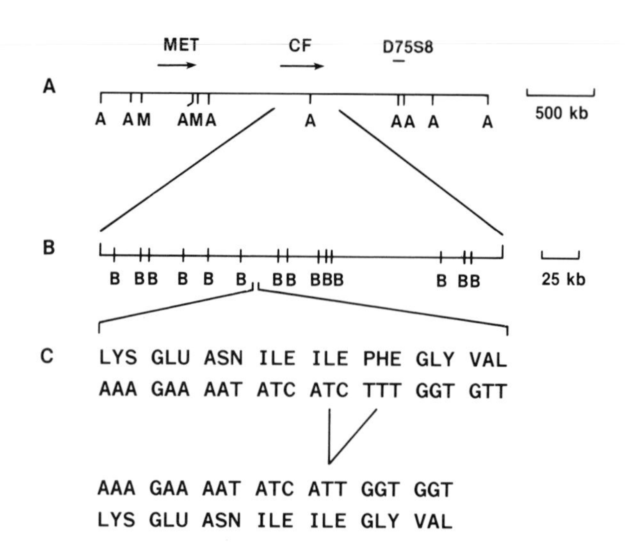

FIGURE 6 Physical maps of the cystic fibrosis locus. **A:** A long-range restriction map generated by pulsed-field gel electrophoresis. MET, an oncogene; D7S8, a DNA fragment detecting an RFLP. A, M, cutting sites of restriction enzymes NaeI, MluI. **B:** Restriction map of the CF gene. B, Bam HI restriction enzyme site. **C:** Sequence of the portion of the CF gene containing the most common mutation. The nucleotide and amino acid sequence of a portion of the CF gene is shown along with the three-nucleotide deletion that causes the most common form of the disease.

E. Future Prospects

There now exists genetic maps for each human chromosome with a resolution of 5–10%. Congress has approved funding for the Human Genome Initiative, a large-scale cooperative program that has the goal of producing a 1-cM linkage map, the cloning and ordering of restriction fragments covering the entire genome (physical map), and eventually the determination of the complete nucleotide sequence of our species. The project has been compared with the effort to place a human on the moon, although much international cooperation is expected. This effort will permit a rapid advance in our understanding of human biology. [*See* GENOME, HUMAN.]

In the meantime, rapid progress is being made on the study of many disease genes. The genes for the common forms of muscular dystrophy as well as retinoblastoma, neurofibromatosis and Wilm's tumor (three inherited cancers) have been isolated by molecular and genetic techniques. These discoveries have provided tremendous insights into the workings of these disorders. Linkage has been detected between RFLPs for several disorders, including another inherited cancer, von Hippell-Lindau syndrome; a form of epilepsy; a heritable form of asthma; and tentatively to certain types of schizophrenia. Although the process leading to potential cures is slow, genetic analysis provides the crucial first step toward identifying the causes of these complex diseases. [*See* DNA MARKERS AS DIAGNOSTIC TOOLS.]

Bibliography

Dean, M. (1988). Molecular genetics of cystic fibrosis. *Genomics* **3,** 93–99.

Gilliam, T. C., Tanzi, R. E., and Haines, J. L. (1987). Localization of the Huntington's disease gene to a small segment of chromosome 4 flanked by D4S10 and the telomere. *Cell* **50,** 565–571.

Lalouel, J. -M., and White, R. (1987). Chromosome mapping with DNA markers. *Sci. Am.* **258,** 40–48.

O'Brien, S. J., ed. (1987). "Genetic Maps," vol 4. Cold Spring Harbor Laboratory, Cold Spring Harbor, New York.

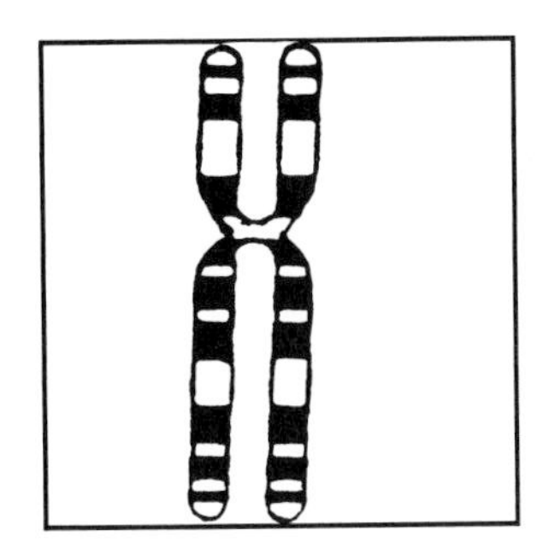

Genetics and Abnormal Behavior

STEVEN G. VANDENBERG and DAVID L. DILALLA, *University of Colorado*

Glossary

Aneuploidy Incorrect number of chromosomes, or loss or addition of part of a chromosome

Haplotype Particular combination of a series of closely linked genes

Linkage Co-occurrence of loci in close proximity on same chromosome, leading to nonindependent assortment

Lyonization Functional inactivation of all but one X chromosome when more than one X chromosome is present

Monosomy Presence of only one of an expected pair of chromosomes (e.g., Turner's syndrome)

Path analysis Statistical analysis of effect of each variable on a second variable, independent of other variables; paths between variables are presented in visual diagrams

Penetrance Proportion of individuals of a given genotype who manifest a specific trait

Pleiotropy Effect of a single gene on two or more different traits

Quantitative genetics Study of inheritance of characteristics of individuals, which differ in degree and not kind, quantitative rather than qualitative differences

Restriction fragment length polymorphism (RFLP) Genetic differences between individuals in the length of small segments of DNA visible after DNA is cut by a restriction enzyme

Trisomy Presence of three chromosomes instead of a pair, (e.g., trisomy 21)

OUR UNDERSTANDING of hereditary factors contributing to abnormal behavior is uneven, partly because abnormal behaviors differ greatly in form, outcome, and etiology. The term *genetic* has come to mean hereditary determination rather than the earlier usage, which meant "developmental." Most persons have come to accept the new meaning, although the old use can still be found in some quarters, including a journal titled *Genetic Psychology* and in the writings of Jean Piaget.

I. What Is Meant by Genetic Effects on Behavior?

Psychology and related fields have oft been warmed by the heat of the nature–nurture debate. Is human behavior best viewed as "hard-wired" so that environmental forces must be exceedingly strong to modify predetermined patterns, or is the newborn human infant a "tabula-rasa" whose behavior and developing personality are almost completely molded by the effects of the environment? Clearly, neither of the above positions can do justice to the complexity of human behavior. Rather, recent evidence points to an intricate pattern of interrelations between inherited tendencies and environmental forces that shape human development.

The role of some genetic factors in many types of abnormal behavior is supported by a variety of techniques. The number of fully established genetic conditions is still rather small, however, and in some of these the hereditary control may not be complete. In most of the disorders to be discussed here, the role of environmental factors is important, even though the precise pathways of environmental influence are still unclear. It should be kept in mind that environment in this context is a broad term that includes prenatal insults and intrauterine biochemical and hormonal conditions, as well as postnatal social and nutritional influences.

Most neurological disorders are to a large degree genetic in origin. Rather narrowly defined disorders such as muscular dystrophy, Huntington's, Alzheimer's, Tourette's syndrome, and Parkinson's disease have been described. Some of these disorders have their onset later in life and cause those affected great mental anguish because it is not known which of their children may be at risk for developing the disease. Other neurological disorders (e.g., Tay Sach disease) are present at an early age. We will not discuss these types of neurological abnormalities, but rather will be concerned here with conditions generally considered to be in the sphere of mental disorders.

II. Methods of Investigation

Research in behavioral genetics focuses on human as well as animal populations. Research impossible to undertake with humans can be conducted using animals and subsequently applied to the human case. For example, "rotating" mice have been selectively bred to produce strains that show a preference for circling either to the right or left. These animals have been used to test the effects of various drugs on the symptoms of Parkinson's disease. Mice also have been bred for seizure susceptibility and preference for alcohol and nicotine to develop an analogue for the study of human problems of addiction to alcohol, cigarettes, and other drugs.

Human studies in behavioral genetics typically employ one of several basic research designs. In family studies (also called *pedigree studies*), the prevalence of a particular illness in relatives of the "proband" (the original clinical case) is studied. A family pedigree is constructed and single gene hypotheses derived from Mendelian models can be tested to determine whether dominant or recessive, autosomal or sex-linked inheritance can be accepted. For most behavioral abnormalities, it is now thought that such simple single gene models are not likely to be correct. Polygenic models (two or more genes influencing the disorder or trait) are more likely to apply. An assumption of this approach is that the liability for the disease is normally distributed in the population. When an individual's liability exceeds some threshold, the disease becomes

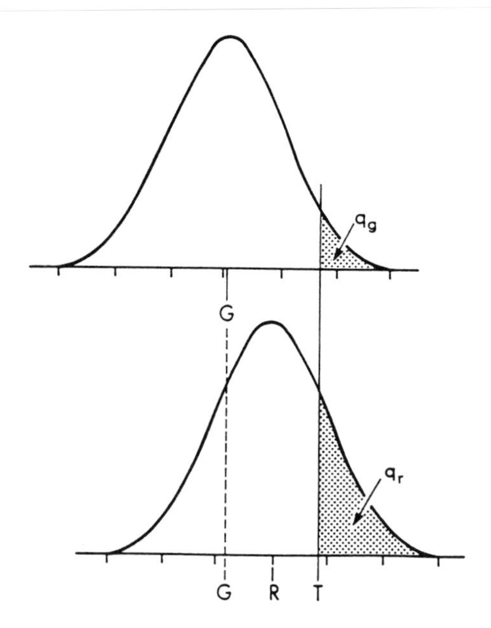

FIGURE 1 Single threshold model of inheritance of liability toward illness. T is the disease threshold. Top distribution represents general population. G, mean liability; q_g refers to affected individuals. *Lower curve* represents liability for relatives of affected individuals with mean liability of R and q_r affected relatives. [Modified from D. S. Falconer, (1981). "Introduction to Quantitative Genetics" (2nd Edition). Longman, New York.]

apparent. Whether the threshold is reached is due partly to genetic and partly to environmental factors. Figure 1 illustrates a single threshold model for onset of disease.

Genetic and environmental effects on quantitative characteristics of individuals can also be investigated using the family method. By assessing similarities and differences among family members who differ with respect to their degree of genetic relatedness, estimates of the importance of genetic and environmental influences on behavior can be made. Most of these statistical techniques are based on correlations between individuals (e.g., parent and offspring).

Linkage studies also take advantage of the pedigree approach. Recent linkage studies are frequently based on the identification of restriction fragment length polymorphisms (RFLPs), which can be used to map the genome. RFLPs are small pieces of DNA that are evident after DNA is cut by an enzyme known as a *restriction endonuclease*. Differences in lengths of the restriction fragments are inherited according to Mendelian patterns. If a single gene for a disorder is embedded within (or is extremely close to) a particular RFLP, the transmission of the disease gene can be tracked in families. Linkage studies usually exploit large multigenerational families in which there are a large number of cases of the disorder of interest. The analysis involves sampling DNA of family members to determine whether there is a high probability that a specific RFLP on a particular chromosome is linked with the presence of the disorder. [*See* DNA Markers as Diagnostic Tools.]

Also in the methodological arsenal of the behavioral genetic researcher is the twin method. Although proposed by Galton as a means of testing the relative influence of nature and nurture, Galton did not realize that there are identical and fraternal twins. It was not until much later that the first papers were published in which the similarities of monozygotic (MZ) and dizygotic (DZ) twins were compared.

The simplest method of twin analysis consists of a direct comparison of concordances of MZ and DZ twin pairs, realizing that on average, DZ twins share half their genes, whereas MZ twins are genetically identical. The basic premise is that any difference between MZ twins must be due to environmental effects. Similarly, to the extent that MZ twins are more similar than DZ twins on some measure (assuming that MZ and DZ twin pairs experience the same types of environments, the so-called equal environments hypothesis) the greater MZ similarity must be due to genetic factors. A broad estimate of the degree of heritability can be obtained by doubling the difference between the MZ and DZ correlations for a given measure. Heritability gives an indication of the percentage of variability among individuals that is caused by genetic factors. Analogously, the percentage of variance caused by environmental factors is sometimes referred to as environmentality.

A number of other complex statistical techniques based on path analysis have been developed for analyzing twin as well as other types of behavioral genetic data. These techniques allow us to focus simultaneously on a number of variables that may influence a trait or behavior. This is particularly useful in the case of comorbidity, the co-occurrence of two disorders in the same individual. These methods have also been extended to include repeated measures of target variables to produce information on longitudinal problems including estimates of the ways in which genetic and environmental effects may be modified during the course of development. A detailed discussion of these techniques is beyond our scope here, but the interested reader is referred to C. C. Li's introductory text.

Another mainstay in behavioral genetic research is the study of adoptees and their biological and adoptive relatives. This method has at its core the natural "experiment" wherein children are removed from the care of their biological parents early in life and are reared by persons to whom they are presumably genetically unrelated. Genetic predisposition and environmental influence are thus neatly separated. Any similarities between adopted children and their adoptive parents are due to environmental effects or cultural transmission, whereas similarities between adopted children and their biological parents represent the effect of shared genes. This assumes "random" placement of children into adoptive homes so that adoptive parents are uncorrelated with biological parents. To the extent that adoptive parents are systematically chosen so that they are similar to their adopted children's biological parents, selective placement has occurred. The effects of selective placement can be assessed by including in the statistical models variables indicative of selective placement such as race, social status, or correlates of these and testing whether their inclusion increases the fit between the model and the observed data.

A complication to all the methods we have discussed is assortative mating or marriage. In the polygenic models that have been derived from basic Mendelian laws, the basic assumption is that every combination of alleles is equally likely to occur. Because it has become obvious that husbands and wives are often more similar than would be expected from this rule for many important characters such as intelligence and some personality traits, techniques have been devised to take this assortment into account. The details of these procedures are too laborious to discuss here, but Falconer's introductory book on quantitative genetics provides an entry into this literature.

III. Classification of Abnormal Behavior

One of the pervasive difficulties encountered in the study of the genetics of aberrant behavior is satisfactorily defining the phenotype of interest. Unlike some genetic disorders such as Down's syndrome, which have relatively clear boundaries and can be precisely identified, many behavioral abnormalities have fuzzy, overlapping boundaries. Even more confusing, the symptoms of various disorders can sometimes mimic each other, so-called phenocopies.

Cultural norms, which may change with time, often influence what is deemed to constitute mental illness. In most societies, for example, a distinction is made between criminals and mentally ill persons. However, it is difficult, if not impossible, to draw a clear line between these groups. To facilitate diagnosis, treatment, and research, a number of diagnostic conventions for mental disorders have been created. In the United States, the most widely used diagnostic standard is the *Diagnostic and Statistical Manual of Mental Disorders*, third edition revised (DSM III-R) of the American Psychiatric Association. The DSM III-R is, in the main, consistent in terminology with *International Classification of Diseases*, ninth edition (ICD-9) published by the World Health Organization. Both the DSM and the ICD are scheduled for revision in 1993.

Since the beginning of modern psychiatry, there has been a curious interchange of meaning between the words *neurosis* and *psychosis*. Originally, neurosis was thought to be a biological defect of the nerves, whereas psychosis was a problem of emotional adjustment. Today psychosis refers to serious psychological difficulty including symptoms such as hallucinations and thought disorder (e.g., delusional thinking including paranoia, disorganized thought, loose associations between thoughts), which make it difficult or impossible for affected individuals to function in society. These symptoms are now often thought to have biochemical and perhaps genetic origins.

The term *neurosis* now is reserved for patients with problems of living that are less threatening to others, although they may be painful or disturbing to the affected individual and his or her family. Neurotic problems include chronic worry and anxiety as well as other "conflicts" experienced in everyday life; such problems rarely require extended hospitalization, although many individuals seek extended outpatient psychotherapy. In the sections that follow, we will present the current evidence related to genetic influences on some of the major mental disorders and behavioral abnormalities.

IV. Discussion of Specific Abnormalities

A. Mental Retardation

Because mental retardation is rather easily recognized and in more severe cases is due to biochemical damage to the brain, many mental retardation conditions are known that are caused by a single gene. Mental retardation is often due to aneuploidy: affected individuals may have an extra chromosome or part of a chromosome, or may be missing part of a chromosome. In the special case of Turner syndrome, one of the sex chromosomes is missing altogether (XO). We also may consider all those genes in their unmutated state to be responsible for the normal mental state of the individual, although it is probably true that there are some genes that are more particularly involved in the "normal" distribution of individual differences in intelligence quotient (IQ).

Binet's initial ideas for determining the relative standing of a child by assessing his/her mental age were widely accepted in the early 1900s. Around 1912, Stern modified Binet's approach by calculating IQ derived from the ratio of mental age to chronological age and suggesting, perhaps unintentionally, a stable attribute. Still later, intelligence quotients were based on the degree to which an individual's score deviated from the population mean and were standardized so that IQ scores were comparable at different ages.

Research on the nature of the intellectual deficits of retarded children has suffered from lack of a commonly accepted test for primary abilities. The Wechsler Intelligence Scale for Children-Revised (WISC-R) is often too difficult for retarded individuals who may even fail the easiest items, so that no estimate of their ability is possible on some of the 10 subtests. The Stanford Binet is somewhat more useful for assessing individuals with limited abilities, but it provides only one score based on a heterogeneous set of items and does not measure several abilities. The Pacific Multifactor test was reported to measure six relatively independent abiilties at age 3 in normal children and at mental age 2 to 3 in retarded children. However, the test was abandoned shortly after the death of its prime author. Reasonable hope is voiced by some that there may be characteristic patterns of deficits and relative strengths in various types of mental retardation. However, it is good to remember Luria's admonition that the brain, even of a seriously impaired child, is so versatile that each individual will have his or her own distinct pattern of deficits and strengths. The existence of savants can be used to bolster either position. Because mental retardation is a rather broad term, we will consider several conditions that often lead to subaverage mental ability.

B. PKU

One of the early success stories of human genetics was the discovery of phenylketonuria (PKU), which, if untreated, leads to severe mental retardation. The disorder is characterized by the absence in newborn infants of phenylalanine hydroxylase (PAH), an enzyme that metabolizes phenylalanine. This leads to abnormally high blood levels of phenylalanine and interferes with the myelinization of neurons in the developing brain. Soon after the discovery of this chain of events, it was shown that a diet low in phenylalanine, if instituted early enough, could easily prevent the outcome of mental retardation. After the child has reached the age of about 7, the diet may be discontinued, under medical supervision, although this may cause other problems to develop. Later in life pregnant women may have to resume a low phenylalanine diet to protect the embryo.

It has been established that the PAH gene is located on chromosome 12 at 12q22-q24.1, a region 90 kilobases long of 13 exons, which codes for a protein of 451 amino acids. Eight RFLPs have been detected with 1,536 possible combinations of codons, haplotypes, of which only 12 have been found in a Danish sample. In contrast, Polynesians have several different haplotypes, shared to a remarkable extent by individuals from other parts of southeast Asia. These populations seem to lack those PAH haplotypes given the arbitrary labels 2 and 3, which are found in Caucasians with PKU. Other mutations may exist in other parts of the world. In northern Africa and in some families from southern France, a partial deficiency is due to a single mutation, which replaces a glutamic acid with lysin in the enzyme; these individuals are not mentally retarded. These details give an idea of the complexities often encountered in genetic systems controlling human traits. Still more complications are expected for traits that are controlled by multiple genes and those in which environmental factors come into play. [*See* PHENYLKETONURIA, MOLECULAR GENETICS.]

C. Trisomy 21/Down's Syndrome

Another success story was the discovery in 1959 of an extra chromosome in persons with Down's syndrome. When staining methods became available, which permitted identifying individual chromosomes as well as small structural changes in them, it became clear that some individuals had excess genetic material in the form of an extra copy (trisomy) of chromosome 21 or a part of it. This mechanism explained why phenotypically normal individuals

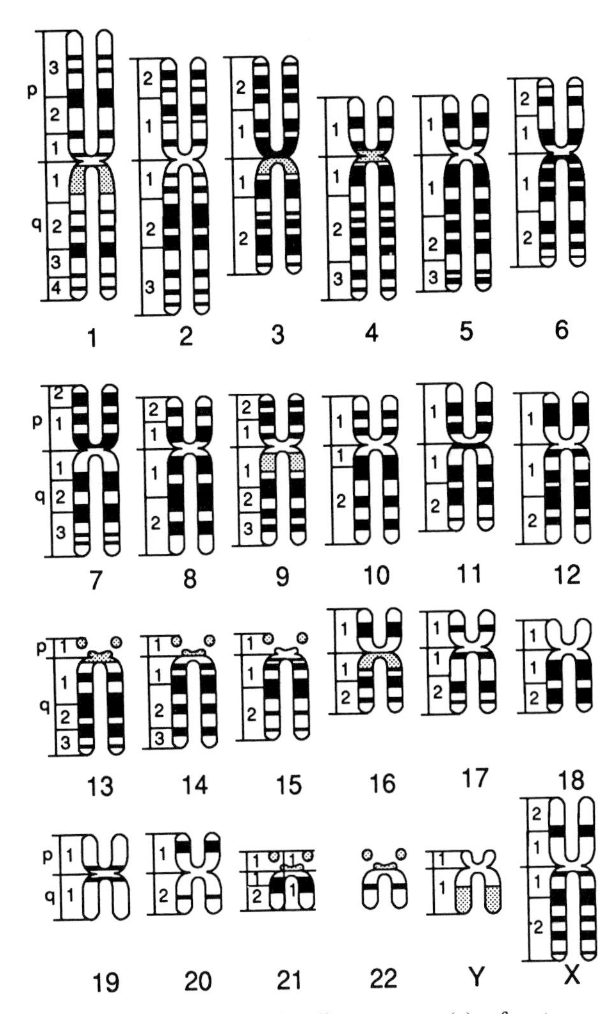

FIGURE 2 Chromosome banding patterns: (p) refers to upper two arms of each chromosome, (q) to lower two arms. In each chromosome, the left chromatid shows G banding observed in mid-metaphase, and the right chromatid shows that observed in late metaphase. Reprinted by permission from J. J. Yunis (1976). High resolution of human chromosomes. *Sci.*, **191,** 1268–1270. The American Academy for the Advancement of Science.

can have a high risk of producing offspring with Down syndrome: in most cases, the abnormality arises from a defect during the process of meiosis. Figure 2 shows the normal complement of human chromosomes with bands resulting from two different stains. [*See* DOWN'S SYNDROME, MOLECULAR GENETICS.]

Down's syndrome is estimated to occur in one of every 700 births, making it a fairly common genetic abnormality. The risk of having a child with Down's syndrome increases with maternal age, from one in 300 between ages 30 to 35 to one in 50 for women older than age 44. This increased risk is presumably because of increased likelihood in older women of nondisjunction (the two chromosomes of a pair remaining together) of individual chromosomes during meiosis (when they should separate). Most of these lead to miscarriage, but nondisjunction of chromosome 21, which the smallest may not disturb the developmental process to the same degree. Individuals with Down's syndrome tend to have a wide range of clinical features including upward slanting eyelids with folding of the tissue around the eyelids, hearing problems, skeletal problems, and severe mental retardation. It is estimated that the average IQ of institutionalized persons with Down's syndrome is below 50; the average IQ in the general population is 100 and the usually accepted cutoff for mental retardation is 70.

We will not discuss aneuploidies further except to point out that the number of sex chromosomes has been found to vary more than any other chromosome without leading to extreme abnormalities, even though there is a negative correlation between number of excess chromosomes and IQ level summarized in Fig. 3.

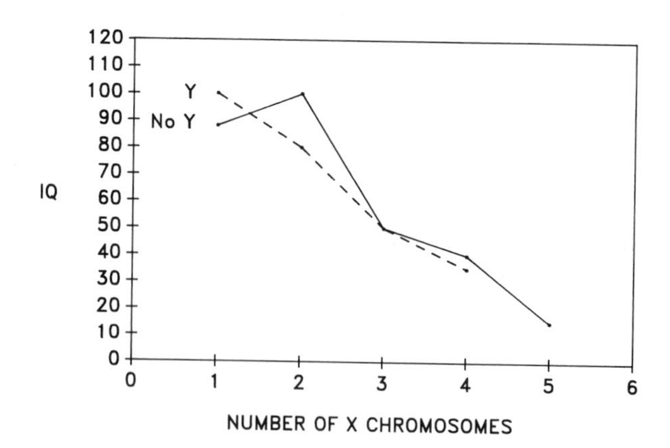

FIGURE 3 Relation between number of X chromosomes and IQ. *Dashed line* represents individuals with one Y chromosome; *solid line* refers to individuals with no Y chromosome [Modified from L. Moor (1967). Niveau intellectuel et polygonosomie: Confrontation du caryotype et du niveau mental de 374 malades dont le caryotype comporte un exces de chromosomes x ou y [Intellectual level and polyploidy: A comparison of karyotype and intelligence of 374 patients with extra X or Y chromosomes]. *Rev. Neuropsych. Infant.* **15,** 325–348.]

D. Fragile X

One other cause of mental retardation introduces a number of other complications. Before the methods for staining chromosomes to produce bands were introduced, Lubs reported an association between what he called a fragile site on the X chromosome and reading problems. He used the term *fragile* because he believed that the chromosome was about to break. This is not actually likely, and some authors prefer the term *marker X*. However, the abbreviation fraX is used to label the locus for the gene. The paper was not given much attention, and no replication was published until considerably later when it was shown that a particular medium in which cells are cultured brings out fragile sites. Since then there has been an explosion of research, and fragile sites have been identified on all chromosomes, although the only one with clear clinical consequences so far is the one at the end of the X chromosome. The fragile X condition is estimated to have a frequency of one per 2,000 in males and a smaller but still considerable prevalence in females. In many cases, physical anomalies exist including atypical facial appearance with large ears, and large testes in males (macroorchidism).

The pattern of inheritance of fragile X is not simply classical sex linkage, such as is found, for instance, in anomalous color vision. The lyonization process in which one of the two Xs is turned off (inactivated) has been invoked to explain why some women with most normal Xs turned off were affected. Further complications include reports of phenotypically normal males who had affected children and negative correlations between age and IQ in women who had affected children and, thus, were "obligate carriers." There are indications in linkage studies that there is genetic heterogeneity. Until the production of RFLPs for this region of the X chromosome, no clear understanding of patterns of inheritance is likely. Figure 4 depicts the fragile X site. [*See* FRAGILE X SYNDROME.]

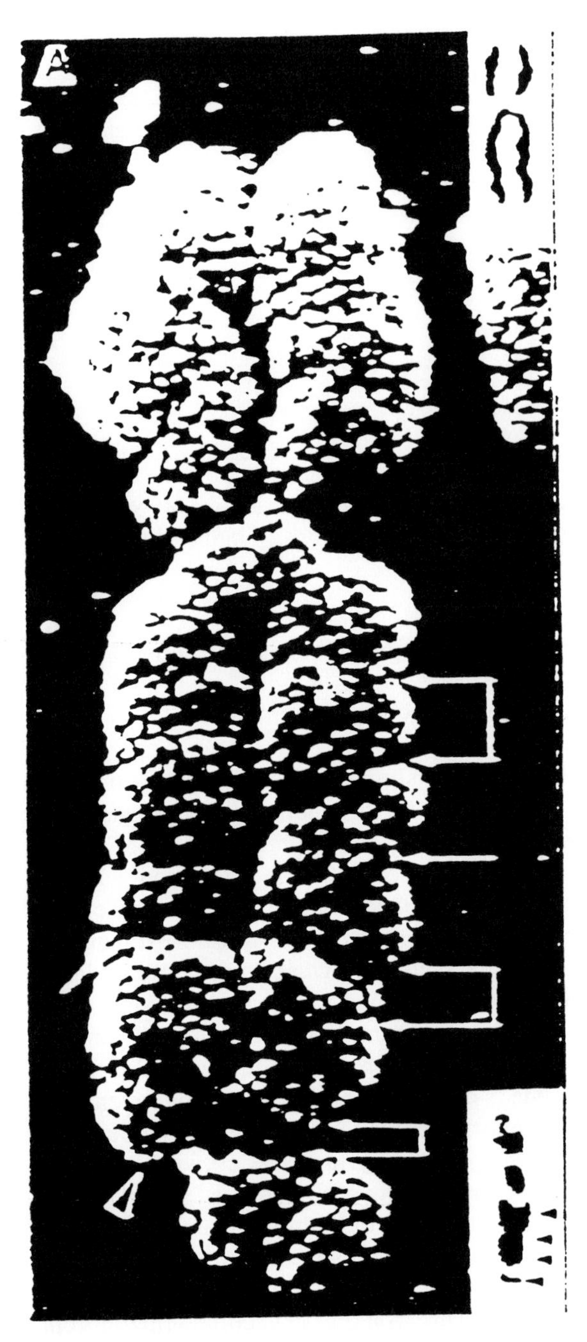

FIGURE 4 Electron scanning photo of the fragile X site, showing that the quaternary coiling is loosened at the end of the long arm. Reproduced, with permission from C. J. Harrison, E. M. Jack, T. D. Allen, and R. Harris (1983). The fragile X; a scanning electron microscope study. *J. Med. Genet.* **20,** 280–285.

E. Schizophrenia

This is perhaps the most puzzling and devastating mental illness. As codified in the DSM III-R, schizophrenia represents a pervasive disturbance in personality, a generally characteristic disordering of thought processes that includes delusions (often of paranoia and of being controlled by an outside force), bizarre sensory perceptions (most often in the form of auditory hallucinations), and inappropriate emotional responses. Onset of symptoms usually occurs during late adolescence to early adulthood, with a smaller proportion of cases having a late adulthood onset. Epidemiological studies have shown the lifetime prevalence of schizophrenia to be approximately 1%. [*See* SCHIZOPHRENIC DISORDERS.]

Debate continues over where the affective psychoses fit relative to schizophrenia. In the United States, use of the concept of schizo-affective disorder testifies to the difficulty in drawing a line between the two conditions, especially in a first psychiatric interview. Even when depression can be ruled out, the clinical presentation of schizophrenia varies so widely that some persons despair of finding a common genetic/biological cause. It is probably best to speak of the "schizophrenias" in recognition of the likelihood that there is a spectrum of disorders related to schizophrenia.

Genetic heterogeneity is one potential determinant of a schizophrenia spectrum. The concept of genetic reaction range, formulated by Gottesman, is also relevant to a spectrum disorder. Reaction range refers to the limits of potential phenotypes being fixed by the genotype, whereas the actual phenotype is dependent on environmental influences. An important implication is that the same genotypes can lead to different phenotypes (as is the case in MZ twins discordant for schizophrenia) and different (but closely related) genotypes can lead to the same phenotype.

Research supporting a genetic component in the etiology of schizophrenia has come from a variety of sources. Table I presents a summary of family studies on the risk of schizophrenia in the relatives of schizophrenic probands. These results highlight the familial nature of schizophrenia and show that risk to relatives drops off sharply as we move from first to second and third degree relatives. It is also worth noting that these data show that the vast majority of relatives of a schizophrenic proband are not schizophrenic.

1. Adoption Studies

In Heston's pioneering study of children adopted away from schizophrenic mothers (see Table II), a significant proportion of those children became

TABLE I Risk for Schizophrenia in Relatives of Schizophrenic Probands[a]

Relationship to affected individual	Morbid risk (%)	No. of studies
First degree relatives		
Parents	5.6	14
Siblings	10.1	13
Children	12.8	7
Children of two schizophrenics	46.3	5
Second degree relatives		
Half siblings	4.2	5
Uncles/aunts	2.4	3
Nephews/nieces	3.0	6
Grandchildren	3.7	5
Third degree relatives		
First cousins	2.4	3
Genetically unrelated		
Spouses	2.3	4

[a] [Modified, with permission, from I. I. Gottesman, and J. Shields, (1982). "Schizophrenia: The Epigenetic Puzzle." Cambridge University Press, Cambridge.]

schizophrenic despite having been reared in homes separate from their biological mothers. In another study the adopted-away offspring of male and female schizophrenics were followed up. "Cross fostering" studies have also been conducted, investigating children adopted from schizophrenic biological parents to nonschizophrenic adoptive parents, as well as children unlucky enough to have been adopted from nonschizophrenic parents and placed with adoptive parents one of whom would later become schizophrenic. All these studies converge on the conclusion that sharing a rearing environment with a schizophrenic parent does not explain the observed familial aggregation of schizophrenia. In all the studies, having a biological parent with schizophrenia conferred significantly higher risk on adopted-away offspring than would be expected among members of the general population.

TABLE II Risk of Schizophrenia in Heston's Study of Children Adopted Away from Schizophrenic Mothers and from Nonpsychiatric Controls[a]

	Mother schizophrenic	Mother nonpsychiatric
Number of children	47	50
Mean age at follow-up	35.8	36.8
Morbid risk %	16.6	0

[a] [Modified, with permission, from I. I. Gottesman, and J. Shields, (1982). "Schizophrenia: The Epigenetic Puzzle." Cambridge University Press, Cambridge.]

TABLE III Concordance Rates for Schizophrenia[a] in Recent Twin Studies[b]

	MZ Pairs		Same Sex DZ Pairs	
Investigator	Total	Concordance (%)	Total	Concordance (%)
Tienari	17	35	20	13
Kringlen	55	45	90	15
Fischer	21	56	41	27
Pollin et al.	95	43	125	9
Gottesman and Shields	22	58	33	12
Weighted Average Concordance		46		14

[a] Schizophrenia or schizophreniform psychosis.
[b] [Modified, with permission, from I. I. Gottesman, and J. Shields, (1982). "Schizophrenia: The Epigenetic Puzzle." Cambridge University Press, Cambridge.]

2. Twin Studies

The first twin studies of schizophrenia provided strong evidence for a genetic influence, given the substantially higher concordance rate for MZ (average concordance of 65%) as compared with DZ twins (average concordance of 12%). These older twin studies have stood the test of time remarkably well, despite methodological advancements in the twin method as applied to psychopathology. Table III summarizes concordance rates for MZ and DZ twins from several recent studies. The effect of genotype is clear, but the important role of environment cannot be denied when we note that less than 50% of the genetically identical cotwins of schizophrenic twins become schizophrenic. Finally, a dramatic demonstration of the role of genetic factors in schizophrenia is the finding that children of the unaffected member of MZ twin pairs discordant for schizophrenia are as likely to become schizophrenic as children of the ill twin. (see Table IV).

Critics of the twin method have noted that the higher concordance for schizophrenia in MZ twins may be due to MZ twins being more likely to share a pathological environment than DZ twins. Were this the case, we would expect an overrepresentation of MZ twins in samples of schizophrenics. However, reviews of major studies of schizophrenia, including studies of twins reared together and apart, have found no such overrepresentation of MZ twins or of twins in general in schizophrenic samples.

TABLE IV Risk of Schizophrenia and Schizophreniform Psychosis in Offspring of Schizophrenic and Normal Twins[a]

	Total offspring	Affected offspring	Age corrected morbid risk (%)
MZ Twin parents			
Schizophrenic (n = 11)	47	6	16.8 ± 6.6
Normal cotwins (n = 6)	24	4	17.4 ± 7.7
DZ Twin parents			
Schizophrenic (n = 10)	27	4	17.4 ± 7.7
Normal cotwins (n = 20)	52	1	2.1 ± 2.1

[a] [Modified, with permission, from I. I. Gottesman and A. Bertelsen, (1989). Confirming unexpressed genotypes for schizophrenia. Arch. Gen. Psychiatry, **46,** 10, p. 867.]

3. Linkage Studies

Two studies that attempted to locate major schizophrenia-related loci via linkage analyses were recently reported. One provided evidence for linkage to chromosome 5, but another study did not. These conflicting results highlight the difficulty that will be encountered in attempting to "solve" the genetics of schizophrenia. It is possible that the discrepant findings of linkage studies reflect the influences of different single genes on various "forms" of schizophrenia. However, it seems more likely that looking for single gene solutions will fail to solve the schizophrenia puzzle.

Prevalence rates for relatives of schizophrenic individuals argue against a simple Mendelian single gene dominant or recessive pattern; schizophrenia is more likely to be polygenically influenced, for various reasons: (1) range of disability is from mild to severe including "subthreshold" effects, which can be detected in nonaffected individuals; (2) there are proportionately more affected relatives of severely ill individuals as compared with mildly ill individuals; (3) as the number of affected family members increases, risk to other members of the family increases; (4) there is a sharp decline in risk as degree of genetic relatedness to the affected individual decreases (e.g., brother versus first cousin); and (5) there is generally a distribution of affected cases on both paternal and maternal sides of the family. Additionally, studies of the relatively small number of schizophrenia-by-schizophrenia matings support polygenic control, as prevalence patterns in offspring of such matings do not conform to Mendelian expectations for single gene dominant or recessive transmission. In a polygenic system, each gene is assumed to exert a small additive influence on the phenotype. The effects of each gene of the system are also assumed to be mediated by environmental influences. Thus, even with a relatively simple polygenic system, an extremely large number of factors can influence expression of the phenotype.

4. Affective Disorders

The predominant clinical feature of the affective disorders is disturbance of mood either in the form of depression or mania. For bipolar affective disorder (sometimes referred to as *manic depressive illness*), an individual must have experienced a full-blown manic episode characterized by such symptoms as uncharacteristically euphoric mood, increased activity, impulsivity, grandiosity, and distortions in perceptual processes. Experience of depressed mood is not required, although affected individuals often experience cyclic periods of depression and mania. Individuals with unipolar forms of depression experience periods of depressed mood, disturbance in attention and concentration, feelings of helplessness and hopelessness, disturbance in appetite, loss of interest in usual activities, and psychomotor retardation. In some cases, distortions of perceptual processes are also present. Results from the Epidemiological Catchment Area (ECA) program indicated lifetime prevalence rates ranging from 6.1% to 9.5% for any affective disorder. [*See* Mood Disorders.]

It has been long recognized that depression often "runs" in families and that first degree relatives of affected individuals tend to have higher than expected rates of affective disorder. This speaks to the familiality of depression but does not illuminate potential genetic influences, as it is plausible that environmental effects within families could lead to depressive symptomatology. Studies of twins and adoptees have convincingly shown that there is a genetic component in the etiology of unipolar and bipolar forms of depression. Concordance rates from twin studies of affective illness are presented in Table V. The ratio of MZ to DZ twin pair concordances is approximately 4 : 1. Adoption studies provide additional support for the influence of genetic factors on depression. For example, a study conducted by Mendlewicz and Rainier on the adoptive and biological parents of depressed adoptees found the prevalence of depression to be 31% for biological parents versus 12% for adoptive parents. [*See* Depression.]

Recently there has been great interest in tracking

TABLE V Concordance Rates for Affective Illness in Monozygotic (MZ) and Dizygotic (DZ) Twins[a]

Study	MZ Twins (concordant pairs/total pairs) (%)	DZ Twins (concordant pairs/total pairs) (%)
Luxenberger (1930)	3/ 4 (75.0)	0/ 13 (0.0)
Rosanoff et al. (1935)	16/ 23 (69.6)	11/ 67 (16.4)
Slater (1953)	4/ 7 (57.1)	4/ 17 (23.5)
Kallman (1954)	25/ 27 (92.6)	13/ 55 (23.6)
Harvald and Hauge (1965)	10/ 15 (66.7)	2/ 40 (5.0)
Allen et al. (1974)	5/ 15 (33.3)	0/ 34 (0.0)
Bertelsen (1979)	32/ 55 (58.3)	9/ 52 (17.3)
Totals	95/156 (65.0)	39/278 (14.0)

[a] [Modified, with permission, from N. Andreason (in press). The Genetics of Depression. *In* "Colorado Symposium on Clinical Psychology: Depression." (B. Bloom and K. Schlesinger eds.) Erlbaum and Associates, Hillsdale, New Jersey.]

down the depression gene or genes. However, a number of difficulties are inherent in this pursuit. The results of twin studies cited above testify to the incomplete penetrance of any genes for depression. Were penetrance complete and environmental influence negligible, MZ twins would have full concordance for the disorder. Another difficulty is that the relation between unipolar and bipolar forms of depression is unclear. The clinical presentation of persons diagnosed with depression often is varied, and it seems likely that there are environmentally based "phenocopies"—forms of depression that meet diagnostic criteria but which are caused by environmental stressors. It also seems probable that at least some forms of depression are influenced by polygenes, sets of genes that together predispose an individual toward depression. Finally, recent evidence supports the notion that depression may be heterogeneous at the genetic level.

In summary, there is compelling evidence in support of the role of genetic factors in the development of depression. Environmental factors for depression may lie in the "psychological" sphere or could be related to nonpsychological events such as perinatal injury.

5. Delinquency/Criminality

The conventional wisdom regarding delinquent and criminal behavior has been that they are societal problems principally determined by environmental influences. Many persons have found the idea of genetic influences on delinquent and criminal behavior to be distasteful, and some investigators have rejected the idea of genetic effects for ideological reasons. To be sure, some of the early studies of genetics of criminality suffered from major methodological shortcomings. However, recent additions to the literature provide evidence that cannot be simply ignored. [*See* CRIME, DELINQUENCY AND PSYCHOPATHY.]

In twin and adoption studies of delinquency and criminality, as in other areas, definition of the phenotype of interest is a major stumbling block to good research. An added difficulty is the high base rate of delinquent behavior in adolescence, which makes it difficult to detect genetic predisposition toward antisocial behavior. Finally, there may be etiological heterogeneity leading to various types of offenders. For example, individuals may be delinquent in adolescence and continue criminal behavior in adulthood; adolescents may exhibit delinquency but not become criminal as adults; and there may be adult criminals who did not engage in delinquent behavior during adolescence.

Genetic studies of delinquency per se do not support the role of genetic etiological factors. For example, average concordance rates for delinquency across twin studies are in the range of 0.87 for MZ twins and 0.72 for DZ twins. The high concordance rates for both types of twins speak to the high base rate for delinquency, and the degree of concordance for DZ twins in particular points to the importance of the family environment. However, some studies have found heritabilities as high as 70% for specific forms of antisocial behavior (e.g., aggressive behavior) as opposed to broadly defined delinquency. One explanation for these differences is that for many adolescents, delinquency represents a developmental phase strongly influenced by the social environment, whereas some delinquents are influenced, at least in part, by genetic factors and continue to engage in criminal behavior during adulthood.

Studies related to genetic effects on adult criminality are much stronger than those for delinquency, including the results of methodologically sound twin and adoption studies. Twin studies show substantially higher criminality concordance rates for MZ (approximately 0.5) as compared with DZ (approximately 0.2) twins, leading to heritability estimates of roughly 55%. The effect of shared environment also has been shown to be substantial, in the range of 20%.

Adoption studies show that having a biological

TABLE VI Percentage of Male Adoptees Who Have a Criminal Record[a]

Adoptive father	Biological father	
	Criminal	Noncriminal
Criminal	36.2	11.5
Noncriminal	21.5	10.5

[a] [Modified, with permission, from B. Hutchings and S. A. Mednick (1975). Registered criminality in the adoptive and biological parents of male criminal adoptees. *In* "Genet. Res. Psychiatry," Proc. Annu. Meet. Am. Psychopathol. Assoc. R. R. Fieve, D. Rosenthal and H. Brill (eds.). pp. 105–116. Johns Hopkins University Press, Baltimore.]

parent who was criminal places adoptive offspring at higher risk for exhibiting criminal behavior during adulthood. At greatest risk are those individuals for whom both biological and adoptive parents are criminal. The adoption studies also point to the substantial role of environment in shaping criminal behavior. For example, adopted-away children whose biological parents had no criminal history were still found to have higher rates of criminality than the population average (Table VI).

To sum up this rather controversial topic, delinquency per se does not appear to be heritable, although aggressive behavior in adolescence does show heritable patterns. For criminality, numerous well-designed studies point toward a definite genetic etiological influence. These studies also highlight clear environmental influences on delinquent and criminal behavior; it is a small percentage of adopted-away children of criminal parents who run afoul of the law.

V. Concluding Remarks

There is growing consensus among behavioral geneticists that the influence of shared family environment on behavioral traits generally is small, whereas the influence of genetic factors and individual experiences of children of the same parents is often considerable. Our understanding of environmental factors is still too vague to be directly applicable to individual prediction, but it provides good starting places for research. Research, however, will be difficult because it will have to be based on after-the-fact observation and deduction rather than experimental manipulation, except for rare cases such as the removal of a child from a bad environment with attempts made to provide a better one. Perhaps with time, a number of such cases can be accumulated. Another topic on which there is some consensus is that many behavioral abnormalities are probably heterogeneous in origin, even though they resemble one another superficially.

There is a possibility that we have overestimated the effects of single "deviant" genes on the mental status of an individual and underestimated the effects of "normal" genes. This makes it all the more important to study the genetic factors influencing normal variation in personality and behavior. Except for what might be called physical defects, most of the variation in normal behavior—and most of the behavioral abnormalities—are due to the complex interaction of environmental and genetic factors.

Bibliography

American Psychiatric Association. (1987). "Diagnostic and Statistical Manual of Mental Disorders," 3rd ed., revised. American Psychiatric Association, Washington, D.C.

Ehrman, L., and Parsons, P. A. (1976). "The Genetics of Behavior." Sinauer Associates, Sunderland, Massachusetts.

Falconer, D. S. (1981). "Introduction to Quantitative Genetics," 2nd ed. Longman Scientific and Technical, New York.

Fuller, J. L., and Thompson, W. R. (1978). "Foundations of Behavior Genetics." Mosby, St. Louis.

Gottesman, I. I. (1990). "Schizophrenia Genesis: Origins of Madness." W. H. Freeman, New York.

Li, C. C. "Path Analysis—A Primer." (1975). Boxwood Press, Pacific Grove, California.

Plomin, R., Defries, J. C., and McClearn, G. E. (1980). "Behavioral Genetics: A Primer." W. H. Freeman, San Francisco.

Scarr, S. W., and Carter-Saltzman, L. (1982). Genetics and intelligence. *In* "Handbook of Intelligence" (R. J. Sternberg, ed.). Cambridge University Press, Cambridge.

Storfer, M. (1990). "Intelligence and Giftedness: Heredity and Early Environment." Jossey-Bass, San Francisco.

Vandenberg, S. G., Singer, S. M., and Pauls, D. L. (1986). "The Heredity of Behavior Disorders in Adults and Children." Plenum, New York.

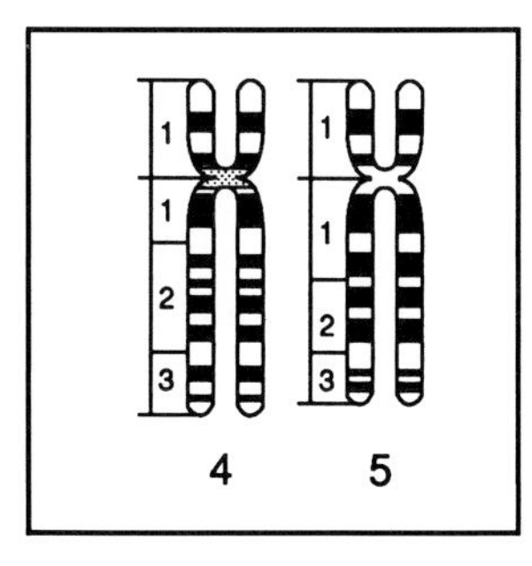

Genetics, Human

H. ELDON SUTTON, *The University of Texas at Austin*

I. Mendelian Heredity
II. Complex Inheritance
III. Mitochondrial Inheritance

Glossary

Allele Alternate form of a gene
Autosome Any chromosome other than a sex chromosome. Humans have 22 pairs of autosomes.
Dominant Allele that is expressed when only one copy is present
Genotype Specific set of genes present in an organism, whether or not they are expressed
Heritability That part of the variation of a trait in a population that is due to a variation in genotypes
Linkage Occurrence of genes on the same chromosome. Genes that are close together have reduced recombination in meiosis.
Meiosis Type of cell division in germ cells that reduces the number of chromosomes by one-half, producing gametes that have only one of each kind of chromosome
Mitosis Type of cell division that occurs in somatic cells and in earlier divisions of the germ cell line. Each daughter cell receives an exact copy of the parental set of chromosomes.
Phenotype Observable organism, resulting from interaction of the genotype and the environment
Recessive Allele that alters the phenotype only when both chromosomes have the allele
Segregation Separation into different daughter cells of genes that are together in a parent cell
Sex chromosome Chromosome whose variation in number is associated with sex. In humans the X and Y chromosomes are sex chromosomes.

GREGOR MENDEL'S studies in *Pisum sativum* (i.e., flowering peas) led to the conclusion that inherited variations result from combinations of discrete stable determinants (i.e., genes). For most traits these occur in pairs, except in gametes, which receive only one of each pair during meiosis. Fertilization of an egg creates new pairs. The similarity of transmission of genes and of chromosomes in cell division led to the recognition that genes are located on chromosomes.

Genes can occur in many alternate forms (i.e., alleles). Dominant alleles are expressed in the phenotype when a single copy is present. Recessive alleles are expressed only when both members of the pair are similar. These rules describe many of the human pedigrees of autosomal dominant and recessive transmission of traits. Genes on the X and Y sex chromosomes have somewhat different patterns of transmission, because females have two X chromosomes and males have one X and one Y chromosome. Modern studies of the molecular nature of genes and gene action have led to understanding of these events and the occasional apparent deviations that occur. [*See* GENES.]

The inheritance of many traits is more complex. These traits can reflect variation at multiple gene loci as well as in the environment. The influence of genetic variation nevertheless can be detected by comparing similarities among close relatives, who share many genes, with similarities among unrelated persons.

Mitochondria are cytoplasmic organelles that have DNA, which codes for some of the proteins and for RNA. Variation in mitochondrial genes is transmitted only by females, who supply the cytoplasm for ova. Such transmission is known as maternal, or cytoplasmic, inheritance.

I. Mendelian Heredity

Interest in the transmission of inherited traits precedes modern science by hundreds of years. Indeed, the X-linked pattern of transmission for human traits such as hemophilia and color blindness was known, although the rules were not based on understanding the mechanics or stochastic nature of heredity. It remained for Mendel in 1865 to determine experimentally rules that apply to all eukaryotic organisms for the genetic transmission of traits.

A. Mendelian Principles

1. Genetic Segregation

Mendel's experiments with *P. sativum* were successful in large part because he selected clearly alternative traits to observe. We now know that such qualitative variations are much more likely to result from variation of a single gene than are quantitative traits. His initial crosses of two true-breeding stocks assured that only one kind of genetic determinant—the gene in modern terminology—was present at most loci of the parental stocks.

The basic observation made by Mendel was that in a cross of two parental stocks differing for a trait (e.g., round versus wrinkled peas), only one trait was observed in the F_1 (i.e., first filial) generation. This can be expressed symbolically:

Parents	Round × wrinkled
F_1	Round

Avoiding another error of many of his predecessors, Mendel carried out another cycle of breeding and found that the determinants of wrinkled peas were neither modified nor lost in the F_1 generation:

F_1	Round × round
F_2	3 Round : 1 wrinkled

Further breeding showed that the round peas were of two types. One-third produced only round progeny when bred among themselves, whereas two-thirds continued to produce both round and wrinkled peas. Wrinkled peas from the F_2 generation produced only wrinkled progeny when bred among themselves.

From these observations Mendel concluded that heredity depended on the transmission of determinants that occurred in various combinations in individual plants, that the expression of the associated characteristics depended on the combinations of determinants, and that determinants were recovered from combinations without having been altered.

The number of such determinants could be deduced by the ratio of peas in the F_2 generation. Based on the fact that the same results were obtained irrespective of which parent possessed which characteristic, Mendel concluded that the contribution of each parent was identical. The minimum number of determinants in a gamete (i.e., pollen or ovule) would be one; therefore, the minimum number in a plant must be two. The crosses could then be symbolized:

Parents	RR × WW
Gametes	R × W
F_1	RW
Gametes	R,W × R,W
F_2	1RR + 2RW + 1WW

The ratio of round to wrinkled peas in F_2 would be a simple statistical combination, producing three round to one wrinkled if the combination RW produced round peas. This was the ratio observed experimentally. Proposing a greater number of determinants would generate different ratios. Hence, Mendel concluded that the number of determinants in each gamete is one. The above observations comprise the law of genetic segregation.

In modern terminology the determinants are genes, and the alternate forms are alleles. The trait that is expressed in F_1 is dominant, and the unexpressed trait is recessive. If the same allele is received from both parents, the organism is homozygous for that locus. If two different alleles are present, the organism is heterozygous. "Genotype" refers to the particular combination of alleles of an individual. The expression of these genes produces the phenotype, the result of the genotype acting in a particular environment. A cross such as that described in which a single trait is considered is called a monohybrid cross.

The expected genetic ratios in the F_2 generation provide a powerful test of the validity of a particular genetic hypothesis. The genotypic ratios are those that refer to the actual genotypes: In the example above 1:2:1 is the ratio of RR:RW:WW. The phenotypic ratios refer to the observable differences—3 round : 1 wrinkled in the example. For allelic combinations at some loci, heterozygotes can be distinct from either homozygote, in which case genotypic and phenotypic ratios are the same.

2. Chromosomal Basis of Heredity

In 1902 W. S. Sutton and T. Boveri independently recognized that the distribution of chromosomes in cell division and in gamete formation are the same as for mendelian genes. They proposed that genetic information is stored on chromosomes and is replicated and distributed in this form, a theory that has been proved amply.

Humans have 46 chromosomes, consisting of 23 pairs. Twenty-two of the pairs are similar in males and females and are called autosomes. The remaining pair are sex chromosomes, consisting of two X chromosomes in females and one X and one Y chromosome in males. The chromosomes that compose each autosomal pair are homologous (i.e., they have the same array of genes). The X and Y chromosomes have a small region of homology at the tip of the short arms, but they are not otherwise homologous. [*See* CHROMOSOMES.]

During division of somatic cells, each daughter cell receives an exact duplicate of the chromosomes of the parental cell in a type of division known as mitosis (Fig. 1). In meiosis, the cell divisions that produce germ cells, each daughter cell receives only one of each pair of parental chromosomes (Fig. 2). Homologous chromosomes pair physically in meiosis. During this period crossing-over can occur. This involves a physical exchange of chromosome segments and can occur once or several times in each pair of chromosomes, depending on the size. Crossing-over allows alleles that were originally on different chromosomes to be on the same chromosome; the reciprocal cross-over product would separate alleles that were originally coupled on the same chromosome. [*See* MEIOSIS; MITOSIS.]

3. Independent Assortment

Mendel's experiments were based on seven different traits (e.g., round versus wrinkled peas and yellow versus green pods). Identical results were obtained for each of the seven traits. Furthermore, in dihybrid crosses, in which simultaneous segregation of two traits was considered, the results from each trait were independent of each other. That is, in the F_2 generation the likelihood of a round or a wrinkled pea was independent of the likelihood of a yellow or a green pod. Thus, the original parental combinations of alleles were not preserved in the gametes produced by the F_1 plants. This is known as the law of independent assortment.

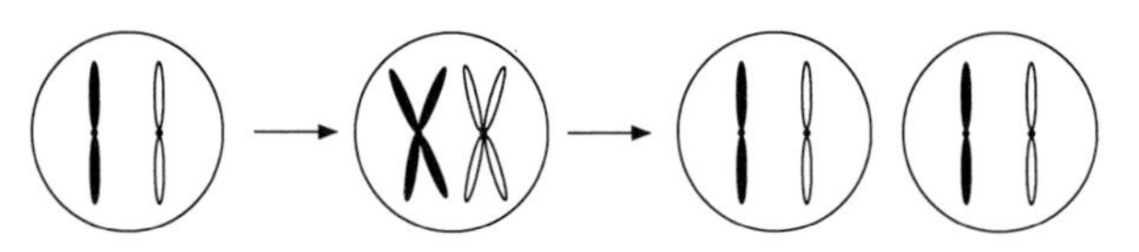

FIGURE 1 Transmission of chromosomes in mitosis. Each daughter cell receives an exact copy of the parental set of chromosomes.

4. Linkage

We now know that a large number of genes (i.e., 50,000–100,000 in mammals) are located on a small number of chromosomes (i.e., 23 pairs in humans). Since the chromosomes remain intact in various parts of the cell cycle, not all pairs of genes should assort independently. Those on different chromosomes do. Those that are widely separated on the same chromosome also assort independently as a result of crossing-over that might occur between them (Fig. 3). However, if two such loci are close together, the likelihood of crossing-over between them is low, and the parental combinations of alleles are largely preserved in meiosis. Such loci are described as linked. Two or more linked loci constitute a linkage group.

A gamete with new combinations of alleles is described as recombinant, whether the new combination arises from the independent assortment of chromosomes or from crossing-over between loci on the same chromosome. Since the frequency of crossing-over between two loci is approximately proportional to the distance between them, one can express their proximity to each other by the percentage of recombination. For example, if an F_1 individual were formed from the parental gametes *AB* and *ab*, where *A* and *a* are alleles at one locus and *B* and *b* are alleles at another, then the nonrecombinant (parental) gametes produced by F_1 would be *AB* and *ab*; the recombinant gametes would be *Ab* and *aB*. If the recombinants composed 10% of the total, the loci would be 10 centimorgans (i.e., map units) apart.

By comparing pairs of loci, one can construct a linear map showing the order of loci and the distances between adjacent loci for each chromosome. Such a map is called a genetic map, based entirely on genetic recombination. It is similar to a physical map in that loci are in the same order, but the rate of crossing-over need not be uniform along a chromo-

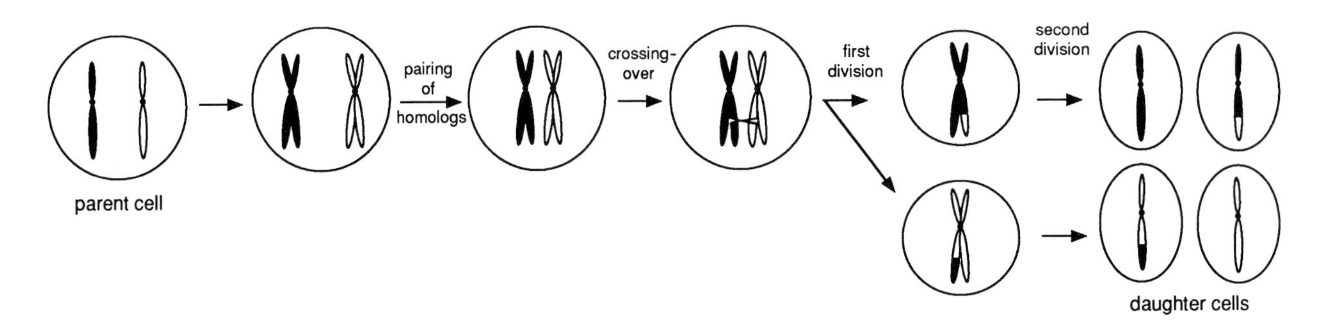

FIGURE 2 Distribution of chromosomes in meiosis. Early in meiosis each chromosome pairs with its homolog. Crossing-over occurs during this period. The chromosomes then separate, and the members of a homologous pair go into different daughter cells. The chromosomes then complete division, and a second cell division occurs. The end result is four potential gametes, each of which has one copy of each of the original set of chromosomes.

some. As a result the genetic map can show distortions of scale compared to the physical map. [*See* GENETIC MAPS.]

B. Human Pedigrees

The rediscovery of mendelism in 1900 promptly led to searches for mendelian traits in other species. The generality of mendelian heredity was soon established for both plants and animals, including *Homo sapiens*. Indeed, over 3000 mendelian traits have now been identified in humans, and the rate of discovery continues to increase. A compilation describing all known mendelian traits is published at regular intervals under the supervision of V. A. McKusick and is accessible by computer.

In analyzing transmission patterns in mammals, it is necessary to distinguish between genetic loci located on autosomes and those on the sex chromosomes. It is often useful to summarize both genotypic and phenotypic information in the form of a pedigree. Some of the conventions used in drawing pedigrees are shown in Fig. 4. The types of pedigrees commonly encountered depend on the relative frequencies of the normal "wild-type" and variant alleles. For the most part interest is in rare traits that are associated with inherited disease, and discussion here is focused on these situations. [*See* GENETIC DISEASES.]

1. Autosomal Dominant Inheritance

The first mendelian trait described in humans was reported in 1905 by W. C. Farabee. Brachydactyly is characterized by short broad fingers. It is a rare relatively benign trait that is easily recognized. It also has high penetrance (i.e., all persons with the appropriate genotype express the trait). A pedigree of the original family described by Farabee is shown in Fig. 5.

Analysis of pedigrees is particularly useful for rare dominant traits. Because persons homozygous for the rare allele would be exceedingly rare, there are essentially only two genotypes observed, *Aa* and *aa*, where *A* represents the dominant allele and *a*, the recessive. Each genotype thus corresponds to a different phenotype.

Several characteristics of autosomal dominant inheritance are useful in analyzing human pedigrees. (1) Every affected person should have an affected parent. The only exception would be instances in which a new mutant gene appears, but mutation is rare and should not be encountered often. (2) The dominant allele should be transmitted equally to sons and daughters by either parent. Hence, the number of affected males and females should be equal. (3) There should not be preferential transmission from parents of one sex to children of the same or opposite sex.

Expression of a trait can, of course, vary between males and females or can be influenced by other

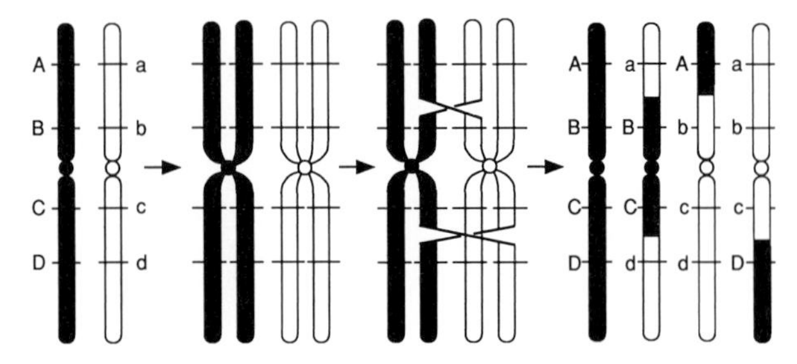

FIGURE 3 Crossing-over during meiosis. Crossing-over permits the recombination of alleles on a particular chromosome. Thus, a chromosome in a gamete is a composite of the original homologous pair of chromosomes from which it is derived.

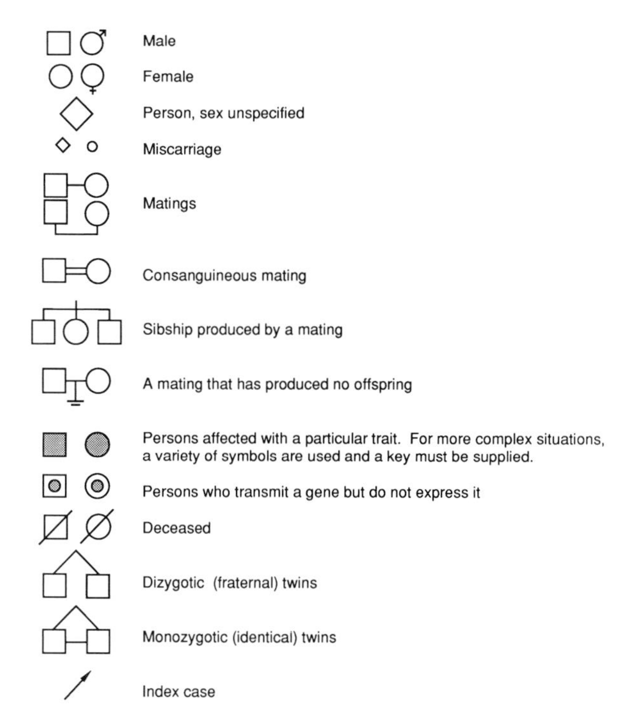

FIGURE 4 Symbols commonly used in depicting pedigrees of inherited traits. Roman numerals are often used to designate generations, I being the oldest. Arabic numerals are used to designate individuals, beginning with 1. Thus, the fourth person in the second generation is II-4. (Reproduced by permission from H. E. Sutton, "Introduction to Human Genetics," 4th ed. Harcourt Brace Jovanovich, San Diego, California, 1988.)

factors. Penetrance can also be less than 100%. It is therefore necessary to rule out other explanations when deviations from the ideal transmission pattern occur. The demonstration that a trait is dominantly transmitted by the above criteria requires that it not confer sterility or impose a health burden that is incompatible with reproduction. Were it to do so, the only affected persons would be those who received a newly mutant allele, which they could not transmit.

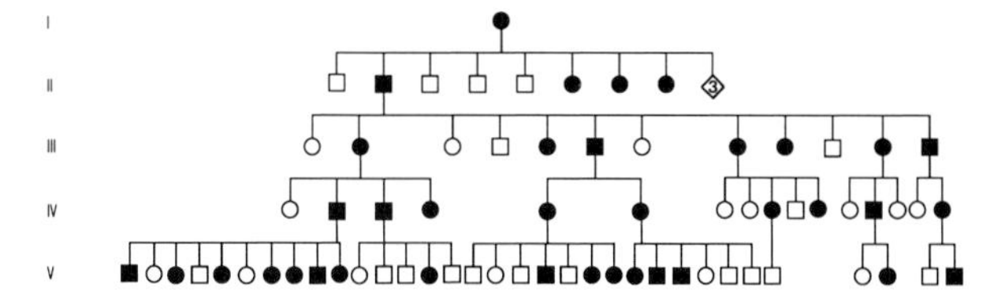

FIGURE 5 Pedigree of brachydactyly, an autosomal dominant trait. This was the first example of mendelian heredity reported in humans. (From W. C. Farabee (1905), "Inheritance of Digital Malformations in Man" Papers of The Peabody Museum of American Archaeology and Ethnology, Harvard University, Vol. 3, pp. 69–77, Cambridge, Mass.)

There are many examples of autosomal dominant inheritance in humans, of which Huntington disease, achondroplasia, and some forms of osteogenesis imperfecta (i.e., brittle bones) are especially well known. [*See* HUNTINGTON'S DISEASE.]

2. Autosomal Recessive Inheritance

The expression of recessive traits depends on the presence of the recessive allele on both chromosomes. This requires that both parents transmit the allele. Males and females should be equally often affected (i.e., homozygous), although a detrimental trait can be expressed more severely in one sex or the other. Since both parents of a child with a recessive trait must be heterozygous, the typical pedigree for a recessive trait contains a single sibship with affected persons.

A common finding for rare recessive traits is an increase in consanguinity in the parents of affected persons. In such matings both parents are much more likely to be heterozygous for the same recessive allele received from one of the common ancestors than would two unrelated persons. Each child from such a mating would therefore have a 25% chance of being homozygous for the allele. The increase in consanguinity of parents of children with rare traits is often the most compelling argument that the traits are recessively inherited.

Because of the typical pattern of isolated affected sibships, pedigree analysis is generally of little value in proving that a particular trait is inherited as a rare autosomal recessive. Such familial clustering can result from environmental as well as genetic causes. Many deleterious conditions are inherited as autosomal recessive traits. These include phenylketonuria, Tay–Sachs disease, and cystic fibrosis. [*See* CYSTIC FIBROSIS, MOLECULAR GENETICS; PHENYLKETONURIA, MOLECULAR GENETICS.]

3. Pseudoautosomal Inheritance

There is a small region of homology between the human X and Y chromosomes, in both cases near the ends of the short arms (Fig. 6). Crossing-over regularly occurs in meiosis in this region, leading to the exchange of terminal segments between the X and Y chromosomes. Genes in this region show the same pattern of transmission as autosomal genes, hence the term pseudoautosomal. The older literature uses the term "partial sex linkage" for these genes. So far the only functional gene placed in this

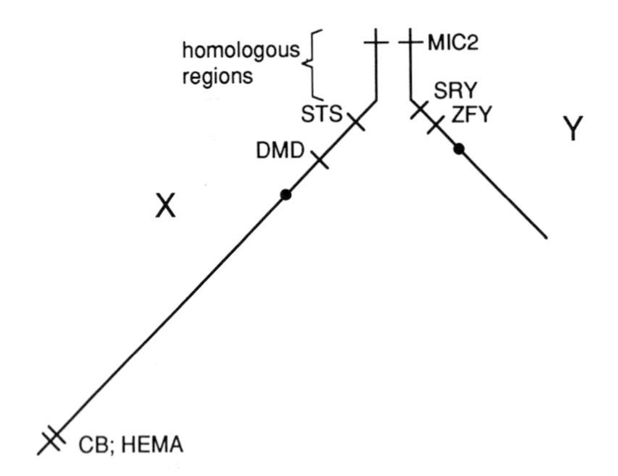

FIGURE 6 Regions of homology and nonhomology of the human X and Y chromosomes. Gene symbols are as follows: *MIC2*, a cell-surface antigen; *SRY*, sex-determining region Y; *ZFY*, a gene possibly involved in sex determination; *STS*, steroid sulfatase; *DMD*, Duchenne muscular dystrophy; *CB*, deutan color blindness; *HEMA*, hemophilia A.

region in humans is *MIC2*, responsible for a cell surface antigen.

4. X-Linked Dominant Inheritance

The terms "X linked" and "sex linked" are synonymous. This pattern of transmission differs from the autosomal pattern in that a male transmits his one X chromosome to all of his daughters but none of his sons, whereas a female transmits one of her two X chromosomes to each child, irrespective of the sex of the child. Therefore, a male with a rare X-linked dominant trait transmits it to all of his daughters but none of his sons. A heterozygous female transmits it to one-half of her daughters and one-half of her sons. Since females have twice as many X chromosomes as males, the frequency of an X-linked dominant trait is twice as great among females than among males. Good examples of X-linked dominant transmission, other than of normal traits, are rare.

5. X-Linked Recessive Inheritance

The distinctive pattern of transmission in families was responsible for X-linked recessive traits being the first recognized. The inheritance of hemophilia is noted in the *Talmud*, and the rules of transmission of hemophilia were established in 1820 by C. F. Nasse and became known as Nasse's law.

In order for a female to express an X-linked recessive trait, she must be homozygous for the variant allele. Since males have only one X chromosome, they express the trait when that X chromosome has the variant allele (i.e., they are hemizygous for the trait). An affected male cannot transmit the trait to his sons, since they receive his Y chromosome. He transmits the recessive allele to all of his daughters, who will be heterozygous carriers unless their mother also transmits a recessive allele. Therefore, affected females can occur only if their fathers are also affected. A consequence of this pattern of transmission is the fact that all affected males in a pedigree are related to each other through heterozygous females. Thus, an affected male can have affected maternal uncles or an affected maternal grandfather, but not both. Hemophilia, red–green color blindness, and Duchenne muscular dystrophy are common examples of X-linked recessive inheritance. A pedigree of hemophilia among the royal families of Europe is shown in Fig. 7. [*See* HEMOPHILIA, MOLECULAR GENETICS; MUSCULAR DYSTROPHY, MOLECULAR GENETICS.]

6. Y-Linked Inheritance

Since chromosomes have genes, a logical supposition is that the Y chromosome has genes. Except for the small portion of the Y chromosome that is homologous to the X chromosome, there appear to be few, if any, genes except that (or those) necessary for sex determination. This is reasonable, considering that no very important genes could be put on a chromosome that is absent in females.

There have been reports of genes that appear to be transmitted according to rules expected for Y linkage. These rules are that a trait should occur only in males and, when present in a male, must be transmitted to all of his sons but none of his daughters. The only such trait that has withstood rigorous examination is maleness itself. The gene for male determination in humans is located on the short arm of the Y chromosome proximal to the region of X chromosome homology (Fig. 6). A candidate gene has been identified, but how sex is determined at the molecular level in humans is incompletely known.

C. Genetic Segregation in Human Families

The power of mendelian analysis in experimental crosses lies in the strong predictions that it makes of genetic ratios. These predictions depend on some knowledge of parental genotypes, as in a monohybrid cross in which the parents are known to be homozygous. Indeed, if a trait is determined by a single pair of segregating alleles, then the genetic

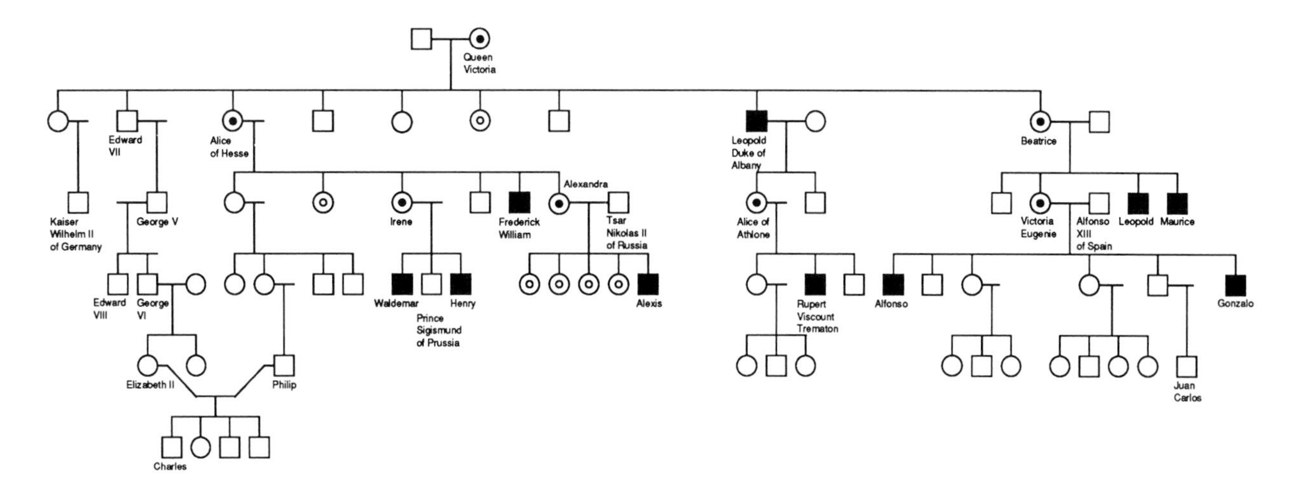

FIGURE 7 Pedigree of hemophilia A in the royal families of Europe. □, Normal male; ○, normal female, presumed not heterozygous; ■, hemophilic male; ⊙, normal female, proved heterozygous; ◎, normal female, at risk, status not established. (Reproduced by permission from H. E. Sutton and R. P. Wagner, "Genetics: A Human Concern." Macmillan, New York, 1985.)

ratio in the F_2 generation must be 1 : 2 : 1. This prediction can be tested by observing the phenotypes in the F_2 generation.

Genetic ratios can also be tested in human families. However, great care must be exercised not to violate the assumptions that underlie their use.

1. Problem of Small Family Size

Genetic ratios refer to stochastic (i.e., random), rather than deterministic, events. Therefore, the reliability of observations depends on the number of offspring produced from a particular cross. Too few observations are associated with large error and might fail to invalidate a false hypothesis. In experimental genetics the offspring from crosses of known genotype can generally be pooled to produce sufficient data to achieve any desired statistical validity. In individual human families the number of offspring is much too small for reliable comparisons with mendelian predictions. Families can be pooled, but at the risk of mixing different genotypes that are phenotypically similar.

Another problem is the bias introduced when families are ascertained (located) through affected offspring. The family that happens to produce two affected offspring is twice as likely to be selected in a sample as a similar family that produced only one affected offspring. Had the same family, by chance, produced only nonaffected offspring, it would not have been selected at all.

Nevertheless, it is useful to be able to test segregation ratios in humans. The segregation ratio is important evidence in support of autosomal recessive inheritance, and there are occasions when deviations from the theoretical segregation ratios can disclose significant biological events. Because the nature of the ascertainment biases are often known, corrections can be applied that allow the calculation of correct ratios.

2. Simple Sib Method of Correction

A correction for ascertainment, the simple sib method, is useful in situations in which a rare trait is proposed to be inherited as an autosomal recessive trait. Typically, affected persons occur in isolated sibships, which can contain one or more affected people as well as nonaffected people. According to the hypothesis to be tested, all such sibships are produced from matings of the type $Aa \times Aa$, where A is the normal dominant allele and a, the recessive allele. Offspring should be in the proportion 3 $A/-$: 1 aa.

Location of the affected sibships is by means of an affected member. The procedure for correction assumes that the likelihood of including a sibship in the sample is directly proportional to the number of affected persons. Thus, sibships that happen to include no affected members would be missed. Furthermore, a sibship with two affected members is exactly twice as likely to be located as a sibship with one affected member. These assumptions apply when only a small sample of the possible sibships in the base population is chosen. Another assumption is that sibships with more affected members are not more likely to occur in the population base from which the sample is drawn than are sibships with fewer affected members.

If these conditions are met, the ascertainment bias can be removed by subtracting the index case from the total. Thus, for an autosomal recessive trait a ratio of 75% nonaffected : 25% affected should be found among the sibs of the index cases. The actual results can be tested against these theoretical expectations, using appropriate statistical methods.

Other methods of correction make other assumptions about the methods of ascertainment. With the use of computers, it is possible to develop complex models of segregation to test against observations, which incorporate such variables as penetrance, modifying genes, aberrant segregation, and unequal survival.

D. Factors that Modify Mendelian Expectations

Although the validity of mendelian transmission is well documented, a number of discoveries alter and extend our understanding of how genes act as well as more complex genetic situations.

1. Dominance and Recessiveness

Dominance was earlier defined as the allele that is expressed in the F_1 generation in a monohybrid cross. Mendel himself found that in some experimental crosses the F_1 phenotype is intermediate between those of the two homozygotes.

As additional systems have been studied, the concepts of dominance and recessiveness have changed. A dominant allele is one that is expressed in the phenotype when only one copy is present. There is no implication that homozygous and heterozygous genotypes have the same phenotype. Often, they do not. Nor is there the suggestion that phenotypic expression of the second allele is suppressed in heterozygotes. Often, the two alleles are expressed independently of each other; the term "codominant" applies to such pairs of alleles. A recessive allele is one that affects the phenotype only when the allele is present on both chromosomes.

At the molecular level dominance and recessiveness are readily understood. One functional allele can be adequate for normal development and metabolism. A defective allele would then be recessive, since heterozygous combination of that allele with a normal allele would produce a normal phenotype. On the other hand, if one functional allele cannot supply normal cell requirements for the gene product, a nonfunctional allele would be dominant. For some loci a variant allele might produce a product that interferes with normal function, an effect that is likely to be dominant.

The terms "dominant" and "recessive" are useful in describing transmission patterns in families or experimental crosses. They should always be used with the understanding that an isolated allele is neither dominant nor recessive. Rather, the terms have meaning only in the context of the interaction of alleles in a biological system.

2. Allelic Diversity

Genetics is often taught as if each locus had only two alleles. Such pedagogical practice obscures the fact that there might be many alleles. Consideration of the many nucleotides that compose a gene makes it obvious that the number of nucleotide substitutions, deletions, and additions within the coding regions as well as changes in regulatory regions should make the potential number of allelic variations enormous.

This is indeed the case. To be sure, many of the variations would be eliminated rapidly by natural selection after a transient existence. However, new alleles are continuously being produced by mutation, many of them detrimental, some normal variants, and perhaps rarely one that is functionally more adaptive than the parental allele from which it arose.

The most extensively investigated gene systems in human populations in terms of allelic variations are the globin genes and the gene for glucose-6-phosphate dehydrogenase. Although only a few alleles at these loci achieve substantial frequencies, over 100 alleles have been identified as rare variants at each of the loci. Not every locus is so tolerant of variation, but there is no reason to believe that these loci are unusual.

3. Genetic Heterogeneity

Molecular analysis of many mutant alleles associated with disease has verified great allelic heterogeneity at many loci. The mutant alleles responsible for Duchenne muscular dystrophy result from many independent mutations and are of many different types. Diseases inherited as recessive traits often involve two different nonfunctional alleles in an affected person, described as a compound heterozygote. In some instances a particular mutant allele might be prevalent, even though there are also many other variant alleles with the same effect. In the case of cystic fibrosis (CF), there is one common *CF* allele that has a one-codon deletion, but other alleles have the same phenotypic effect.

The products of different alleles can, of course, be associated with different phenotypes. One example is found in deficiencies of hypoxanthine phosphoribosyltransferase, an X-linked recessive trait. A modest deficiency produces a rare form of gout. Allelic variants that produce no active enzyme cause Lesch–Nyhan syndrome, a condition that involves mental deficiency, compulsive biting of fingers and lips, and other defects. Mutations at a single locus therefore might be responsible for more than one disorder.

The great heterogeneity of mutations suggests that some of the variation in phenotype of inherited traits in outbred populations is due to different combinations of alleles. In small inbred populations allelic variation is likely to be less, and persons who express an inherited trait are more likely to be genetically homogeneous.

4. X Chromosome Inactivation

A puzzle that concerned geneticists for many years was the mechanism of dosage compensation of X-linked loci. Most of the loci known to be on the X chromosome have no direct relationship to sex. Females with two X chromosomes have twice the number of copies of most of these genes than do males. However, no differences could be found in function of the genes in males and females.

The solution was found in X chromosome inactivation, a hypothesis first clearly formulated by M. Lyon and demonstrated to apply to mammals, including humans. In each XX zygote (i.e., fertilized egg) both X chromosomes are active. However, early in embryogenesis one of the X chromosomes in each cell becomes largely inactive, leaving only one active X chromosome, as in males. Through subsequent cell divisions, the same X chromosome remains inactive, replicating somewhat later than the active X chromosome.

At the time of the original inactivation, the choice of which chromosome is to be inactivated is an independent random event in each cell. Some cells have one chromosome active; some will have the other. Females are therefore a mosaic of cells and tissues that differ with respect to which X chromosome is active. Because inactivation occurs in the early embryo, any particular cell at that time might serve as the progenitor for a large number of cells in the developed individual. On average approximately one-half of the cells have a specific active X chromosome. But chance might favor one or the other X chromosome in a particular tissue. For this reason many females who are heterozygous for an X-linked recessive trait show some expression of the trait. At the cellular level each cell is found definitely to express one or the other allele, but not both.

II. Complex Inheritance

Analysis of mendelian traits has been highly successful in humans. Mendelian traits are those discrete variants shown to be due to variations at single loci. There are many other traits of great interest that are more complex in their transmission. Indeed, prior to the rediscovery of Mendel's work, most efforts were directed to the study of these complex traits. Preeminent among the scientists working on complex traits in humans was Francis Galton, who invented several statistical techniques to solve problems in human genetics, techniques that are still in use in many fields.

A. Basis of Complex Traits

With the demonstration that mendelian heredity applies generally to plants and animals, an obvious question was how one could explain such traits as height, behavior, or crop yields in mendelian terms. These traits do not occur in simple alternative forms. Rather, they are distributed as continuous variables. Other traits (e.g., cleft lip in humans or the number of toes in guinea pigs) might occur as discrete inherited variables, but still are not due to variation at a single locus.

An answer lay in the hypothesis that these are multigenic traits. These can also be called multifactorial traits, a term that admits the role of environmental, as well as genetic, variation. Combinations of alleles at several loci constitute the genotype that produces a certain phenotype. Different combinations of alleles might alter the phenotype by small increments. Furthermore, a particular phenotype can result from several different combinations of alleles, making it difficult to infer the genotype from the phenotype.

B. Analysis of Complex Traits

Even though complex traits cannot be tested with segregation analysis and cannot be partitioned directly into variation at specific loci, the existence of a genetic component of variation can be tested. The underlying premise is that the alleles that contribute to a multigenic trait are found in greater number in close relatives than in more distantly related or unrelated persons.

1. Intrafamilial Correlations

The method of study introduced by Galton was the comparison of family members for similarity. Galton's studies were performed from approximately 1865 to 1900, between Mendel's original report and its rediscovery in 1900. Thus, Galton knew nothing of mendelian theory. His analyses were based only on the idea that a child receives hereditary determinants equally from each parent. From this it follows that sibs have one-half of their heredity in common, grandparents and grandchildren have one-fourth in common, half-sibs have one-fourth, first cousins have one-eighth, etc.

To compare the observed similarity among related persons with theory, Galton devised the correlation coefficient, which standardizes the variation on a dimensionless scale and expresses the overall similarity between pairs of persons (e.g., parent and child) for the trait. The correlation coefficient can vary from +1, if the two measures are always the same, to −1, if they consistently vary in opposite directions. If the variation is independent, the correlation coefficient is 0. With the eventual understanding of mendelian heredity, it was possible to calculate the expected correlations for various degrees of relationship, as shown in Table I, on the assumption that all variation is genetic.

An example of a trait whose variation is virtually entirely due to heredity is given in Table I. Fingerprints are determined in the embryo and do not change after that. The pattern intensity is measured by the number of ridges crossed if a line is drawn from the center of the pattern on each finger to the triradius in the case of a loop, or to both triradii in the case of a whorl. Arches have no triradii and are scored as 0. The sum of the values for the 10 fingers is the overall measure for the person. This measure turns out to be complex in its genetic determination, with no indication of environmental influences. This does not mean, of course, that environment is not important in the development of fingerprint ridges. Rather, it means that variation in environment is not important in variation of ridge counts, whereas genetic variation is.

TABLE I Correlations Between Relatives for Total Dermal Ridge Counts[a]

Relationship	No. of pairs	Observed correlation coefficient	Theoretical correlation coefficient
Parent-child	810	0.48	0.50
Mother-child	405	0.48 ± 0.04	0.50
Father-child	405	0.49 ± 0.04	0.50
Father-mother	200	0.05 ± 0.07	0.00
Midparent-child[b]	405	0.66 ± 0.03	0.71
Sib-sib	642	0.50 ± 0.04	0.50
Monozygotic twins	80	0.95 ± 0.01	1.00
Dizygotic twins	92	0.49 ± 0.08	0.50

[a] From S. B. Holt (1961) *Brit. Med. Bull.*, **17**, 247.
[b] Midparent is the average of the values of the two parents.

2. Twin Studies

Galton realized that twins might be especially useful in the analysis of inherited traits. This is because some twins are identical genetically, whereas fraternal twins are two sibs that happen to be conceived at the same time because of double ovulation. These are referred to as monozygotic and dizygotic twins, respectively. By comparing the variation within pairs of monozygotic twins and dizygotic twins of the same sex, one can avoid the effects of different ages and environments that characterize comparisons of single-born sibs.

Twin studies are widely used to assess the magnitude of genetic variation of complex traits. They have other special uses as well. For example, since monozygotic twins are genetically identical, any variation within a pair must be nongenetic. The discordance that occurs for many traits within monozygotic pairs provides an opportunity to identify environmental factors that differ for the members of a pair that can contribute to the appearance of the trait.

3. Adoption Studies

For some traits, particularly behavioral traits, the underlying assumptions of the twin method can be questioned. Monozygotic twins might, in fact, have more similar environments than do dizygotic pairs, because monozygotic twins are perceived as identical. An alternative approach is to compare related persons, one of whom is adopted into a different setting at birth. Typically, adopted children are compared with their biological parents versus their foster parents. To the extent that the adopted child resembles the biological parent, the underlying cause is assumed to be shared genes. A powerful

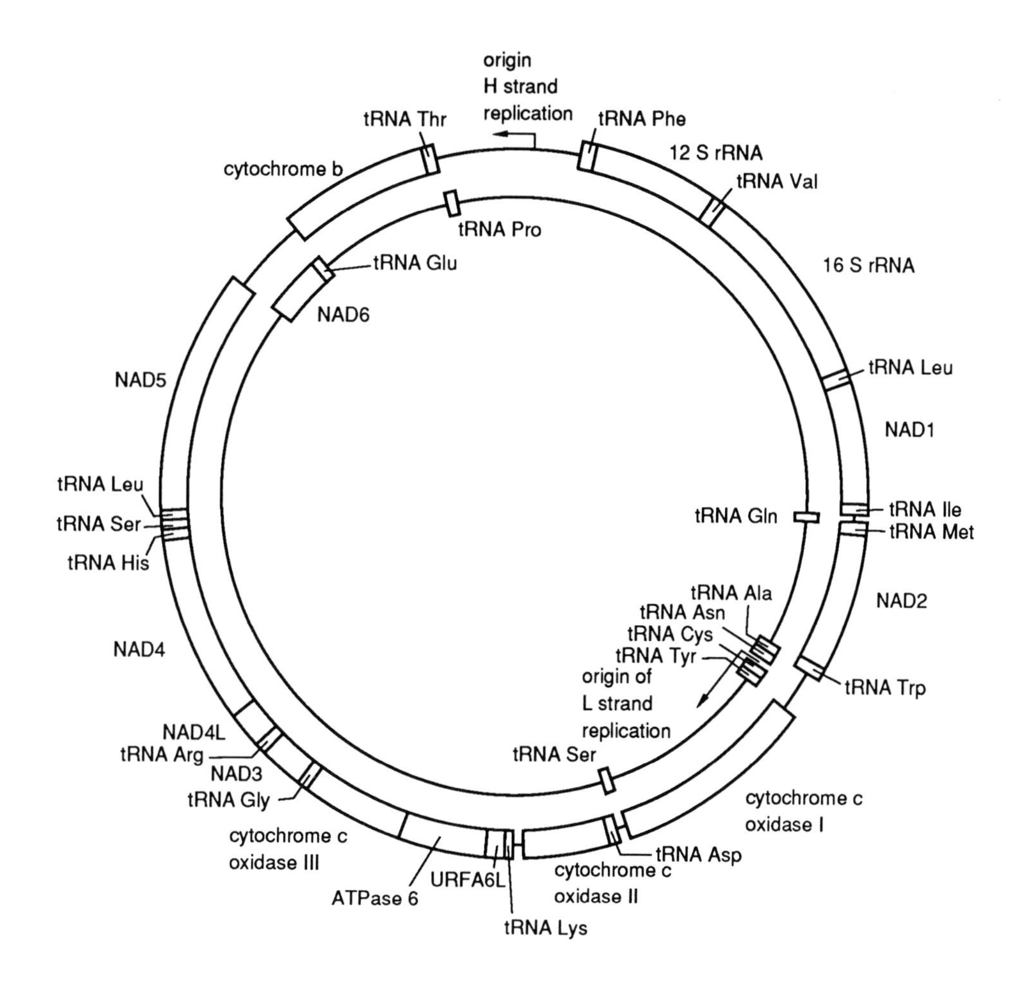

FIGURE 8 Genetic structure of human mitochondria. The various genes are concerned primarily with protein synthesis and coding for cytochrome and related proteins. [Reproduced by permission from H. E. Sutton, (1988). "Introduction to Human Genetics," 4th ed. Harcourt Brace Jovanovich, San Diego.]

comparison would be monozygotic twins separated at birth, but these are too few in number for most studies.

C. Nature and Nurture—The Meaning of Heritability

Geneticists often wish to express the portion of the variability in a complex trait that can be attributed to genetic variability. This term, "heritability," can be formulated in various ways. Typically, the denominator is the total variation of a trait in the population. The numerator is the genetic variation, which, together with the variation attributed to environment, should equal the total variation.

Every phenotype is the result of a genetic blueprint executed in a specific environment. In this context it is meaningless to talk about whether heredity or environment is more important, but it is meaningful to attribute the variation in a trait to heredity or environment. If a population is genetically uniform for the alleles that determine a trait, then the only source of variation is environment. The trait is not less genetic, but the variation is. Conversely, increasing the genetic heterogeneity might increase the total variation but decrease the fraction due to environment.

There have been many acrimonious debates on the relative contributions of heredity and environment to a particular trait. Most often, these debates are based on misunderstanding of heritability: It is a measure of how much genetic and environmental variation occur in the population, not whether the trait is genetically determined

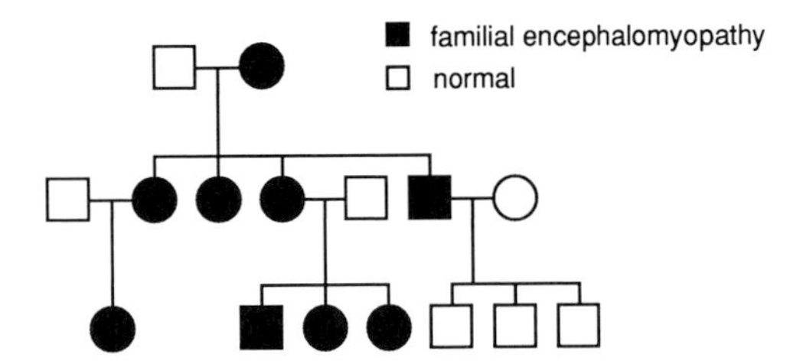

FIGURE 9 Pedigree of familial mitochondrial encephalomyopathy showing maternal transmission of the disorder. All children of affected females are affected regardless of sex. No child of an affected male is affected. [Based on a report by D. C. Wallace, X. Zheng, M. T. Lott, J. M. Shoffner, J. A. Hodge, R. I. Kelley, C. M. Epstein, and L. C. Hopkins (1988). "Cell," Vol. **55**, pp. 601–610.]

III. Mitochondrial Inheritance

Mitochondria are the only cell organelles other than nuclei in animals that have DNA. In plants chloroplasts also have DNA, and the principles of mitochondrial inheritance also apply.

A. Genetic Structure of Mitochondria

The amount of DNA in mitochondria is small compared to nuclear DNA: 16,569 bp versus 3 billion in humans. The number of genes coded in mitochondrial DNA is small, but these include all of the genes necessary for RNA transcription and protein translation, including a complete set of transfer RNA genes (Fig. 8). In addition, there are genes that code for key electron transport proteins that function in the use of oxygen for the production of ATP.

Not all proteins found in mitochondria are coded in mitochondrial DNA. Indeed, most are coded in the nucleus and find their way into mitochondria after synthesis. Genetic variation in mitochondrial proteins therefore could be transmitted in typical mendelian patterns. Those proteins coded by mitochondrial DNA have very different transmissions.

B. Maternal Inheritance

There are thousands of mitochondria in the cytoplasm of an ovum. By contrast, a sperm has a small number in the midpiece. When an ovum is fertilized, few, if any, of the paternal mitochondria enter the zygote. Therefore, the mitochondria in the developing embryo are maternal in origin.

These biological facts determine the transmission of traits coded in mitochondrial DNA, known as maternal, or cytoplasmic, inheritance. Whatever variations are present in the mother are transmitted to every child; fathers make no contribution of mitochondrial DNA. These principles are amply borne out by studies of variants of mitochondrial DNA. There are few mitochondrial variations known that alter the phenotype. A pedigree of one such disorder is shown in Fig. 9.

Bibliography

McKusick, V. A. (1990). "Mendelian Inheritance in Man," 9th ed. Johns Hopkins Univ. Press, Baltimore, Maryland.

Sutton, H. E. (1988). "Introduction to Human Genetics," 4th ed. Harcourt Brace Jovanovich, San Diego, California.

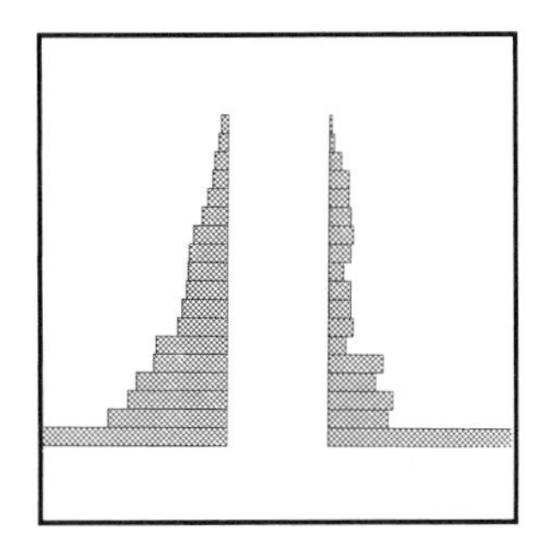

Genital Cancer and Papillomaviruses

SUSANNE KRÜGER KJAER, *Danish Cancer Registry*

AND HERBERT PFISTER, *University Erlangen-Nürnberg*

Glossary

Hybridization Double-strand (duplex) formation between complementary single-stranded molecules of DNA or RNA. Stringent hybridization conditions discriminate between highly and imperfectly matched molecules; "relaxed" conditions allow duplex formation between imperfectly matched sequences

Proto-oncogenes Cellular genes that fullfil physiologic functions in the regulation of cell proliferation and differentiation but may become oncogenic if their expression is changed qualitatively and/or quantitatively

GENITAL CANCER refers to cervical, vaginal, vulvar, and penile carcinomas. Clinical studies and especially data from molecular biology have provided evidence that certain types of human papillomavirus (HPV) may be of etiological importance for genital cancer. Papillomaviruses are small DNA viruses that induce proliferative squamous epithelial lesions (papillomas) in many higher vertebrates. At least 60 HPV types are differentiated because they show less than 50% cross-hybridization of their DNAs.

I. Epidemiological Aspects of Genital Cancer

A. Geographic Differences in Incidence

Cancer of the cervix uteri is a problem of global concern, and it has been estimated that some 465,000 new cases occur worldwide every year. The age-standardized incidence rates vary at least 20-fold between different parts of the world (Fig. 1). In developing countries, cervical cancer is the most frequent type of female cancer, and of all new cases diagnosed in the world, approximately 80% occur in women from the economically underdeveloped areas.

There are also substantial geographic differences in the incidence of penile cancer (Fig. 1). In the industrialized countries, it is relatively rare (only about 0.5% of all male cancers). By contrast, it seems to be frequent in parts of Africa and Asia, where it accounts for as much as 15% of all cancers in men. The indication of a correlation between incidence rates of cervical and penile cancer (Fig. 1) may suggest that both carcinoma types are related to the same factors.

B. Risk Factors

For several decades, epidemiological studies have suggested that a female sexual behavior characterized by multiple sexual partners and early age at first intercourse is associated with an increased risk of cervical cancer. Furthermore, a promiscuous behavior of the male partner (i.e., many sexual partners, sexual contact with prostitutes) may also increase the cervical cancer risk in otherwise low-risk women.

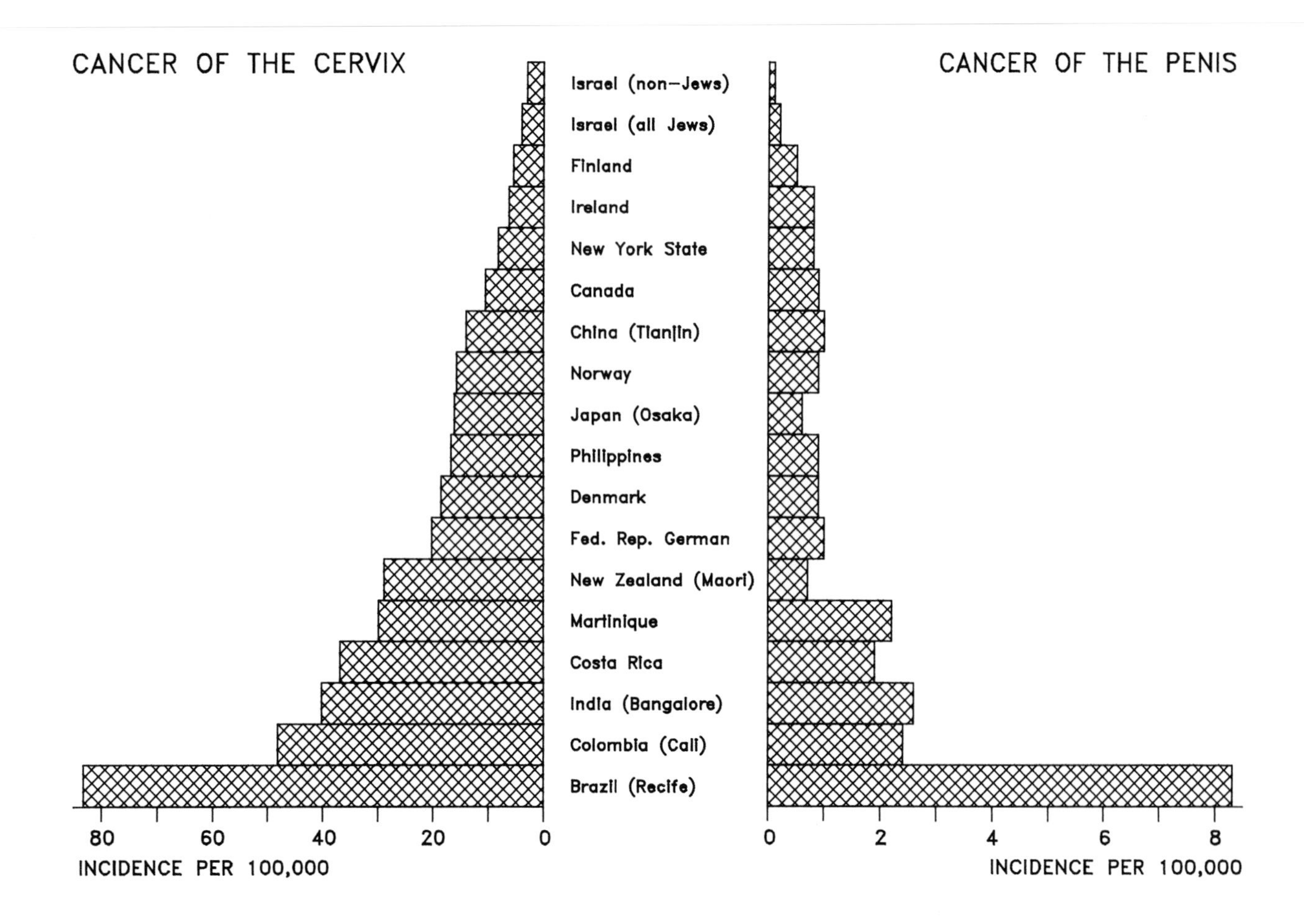

FIGURE 1 Age-standardized incidence rates of cervical and penile cancer in selected countries. [Based on data from Muir, C., et al. (1987). "Cancer Incidence in Five Continents," Vol. V. International Agency for Research on Cancer, Lyon.]

The possible risk determinants of penile cancer have been less intensively studied, but personal hygiene and the event of circumcision are often mentioned as important factors.

On this epidemiological background, a sexually transmitted factor has been suggested to play an etiological role in both cervical and penile carcinogenesis. The existence of such a common factor affecting both partners is supported by several studies showing that cervical cancer is significantly more frequent in marital partners of men with cancer of the penis than in a group of control women.

II. Human Papillomaviruses in Genital Tumors

A. Virus Types

Human papillomaviruses affect cutaneous and mucosal squamous epithelium. Twenty-two HPV types reveal a clear-cut preference for the genital tract (Table I) and appear associated with genital warts (condylomata acuminata) and precancerous lesions (intraepithelial neoplasias), which represent a spectrum from cauliflower-like growths to flat, inconspicuous tumors. The genetic relation (Table I) of the genital HPVs partly correlates with their tropism, but some types (40, 42, 44, 51, 57) are more closely related to HPVs from nongenital skin warts or oral lesions than to other isolates from the genital mucosa.

The DNA of 12 different HPV types has been detected in genital cancers. Sequences related to DNA of HPV10, which is usually found in flat skin warts, were encountered in a few vulvar and cervical cancers (Table I).

TABLE I Human Papillomaviruses Detected in Genital Tumors

HPV type	Related types	Condyloma acuminatum	Intraepithelial neoplasia	Genital cancer
6	11, 13, 55	++[a]	++	+
10	3			(+)
11	6, 13, 55	++	++	+
16	31		++	+++
18	45		++	++
30	53		+	
31	16		++	+
33	52, 58		+	+
34			+	
35			+	+
39			+	+
40	7		+	
42	32		+	
43			+	
44	13, 55	+	+	
45	18		+	+
51	26		+	+
52	33		+	++
54		+		
55	6, 11, 13, 44	+		
56			+	+
57	2, 27		+	
58	33		+	
59			+	

[a] Number of + reflects the prevalence.

B. Virus Replication

The basal keratinocytes of the epidermis and all malignant cells are not permissive for HPV (i.e., they do not support virus propagation) but allow only the expression of so-called early viral genes (labeled E to indicate their activity *e*arly in the viral life cycle). The differentiating, suprabasilar keratinocytes of HPV-associated benign and precancerous lesions are permissive, so that the synthesis of infectious virus particles occurs in these cells. Virus-specific cytopathic effects like koilocytosis and dyskeratosis are most prominent in the upper epidermal layers. Koilocytes are characterized by cytoplasmic vacuolization, atypical nuclei, and often binucleation. They represent a reliable indicator of HPV infection. Their absence does not exclude a productive HPV infection, however, because HPV16, for example, usually induces only few or even no such cells. The dyskeratocyte is also a mature squamous cell but with a dense cytoplasm. The nuclei are similar to those of the koilocytes. [*See* KERATINOCYTE TRANSFORMATION, HUMAN.]

III. Frequency of HPV Infections and HPV Types

A. Detection of HPV Infection

Different approaches have been used in the diagnosis of HPV infection (e.g., clinical examination, cytology, colposcopy, and DNA hybridization methods). Condylomata acuminata are the classic, papillary genital lesions, which most often occur on the vulva, penis, vagina, the perianal region, and less frequently, the cervix. They are macroscopically visible and thus clinically detectable.

In contrast, most cervical HPV lesions are nonpapillomatous and macroscopically invisible. These flat condylomas appear with varying degrees of cellular atypia. The present general concept is that flat HPV lesions are the earliest stage of cervical intraepithelial neoplasia (CIN), which is graded into I, II, and III in accord with increasing severity. Flat lesions also occur in the vagina (VAIN, vaginal intraepithelial neoplasia). At external sites, they exist both in the vulva region (VIN), and in the penile region (PIN).

The cytomorphologic basis for the diagnosis of cervical HPV infection by means of Papanicolaou's smear is the presence of either of two types of cells: the koilocyte and the dyskeratocyte. Koilocytosis has received most attention, whereas dyskeratosis has often been ignored, even though it may be the only cytologic sign of HPV infection in some cases. The cytological diagnosis of HPV infection is a fast, noninvasive, and inexpensive method. However, the sensitivity of the test, with DNA hybridization as the "gold standard," has been reported to be as low as 5–35%. Moreover, studies of agreement between pathologists indicate the existence of a substantial inter- and intraobserver variability.

Examinations with the colposcope, a low power microscope with which the organs are examined, are used in the diagnosis of internal as well as external lesions. Although the diagnosis of condylomata acuminata is usually evident, colposcopic examination may be of help in the detection of smaller lesions and early lesions. To make lesions more prominent, the epithelia can be treated with 5% acetic acid, which leads to "acetowhite" plaques and to the visualization of pathological vessels, which are both characteristic but not pathognomonic of HPV infection. Colposcopy is particularly important for the examination of the external skin of vulva and penis, where smears do not provide enough cells from deeper epithelial layers for proper cytologic evaluation. Finally, another important purpose of colposcopy is to direct the biopsy to areas of most significant disease.

Latent HPV infections, which induce neither clinical symptoms nor histologic or cytologic changes, can only be detected by nucleic acid hybridization techniques. Specific probes labeled with radioactive isotopes or other markers are prepared from cloned HPV DNAs and are detected by autoradiography or immunochemical reactions. Various hybridization methods have been used, which differ in sensitivity (likelihood of an infected individual being test-positive) and specificity (likelihood of a noninfected individual being test-negative).

Southern blot hybridization starts with purified DNA, which is digested with specific endonucleases that cut DNA at short, well-defined sequences. After separation of the DNA fragments by gel electrophoresis, the DNA is denatured into single strands, transferred to a filter, and hybridized with the probe under stringent or relaxed conditions. Southern blotting is very sensitive and is considered the most accurate method for HPV diagnosis. HPV type-specific DNA fragment patterns support the classification, which becomes increasingly difficult on the basis of hybridization alone because of significant cross-reactivity between DNAs of many HPV types. Even with Southern blotting, however, interlaboratory variability in detection and identification of HPV types has been noted, which could in part account for the reported variation in HPV prevalence.

Extracted DNA may be directly immobilized in single-stranded forms as a dot on a filter and hybridized locally (dot blot hybridization). Because of background problems, the test is less sensitive and gives a higher number of false-positive reactions. The same drawbacks apply to filter *in situ* hybridization, in which sample cells are directly filtered onto nitrocellulose and lysed by alkali before hybridization.

The enzymatic amplification of HPV DNA before hybridization, the so-called polymerase chain reaction (PCR), provides the most sensitive method for diagnosis of HPV infection. It allows the detection of a few copies of HPV DNA in biopsy samples.

The only method that permits a precise correlation of hybridization signals with pathology is *in situ* hybridization within cytological preparations or tissue sections. This technique is less sensitive than Southern blotting.

B. Incidence of Condylomas

In the United Kingdom, data collected from sexually transmitted diseases (STD) clinics indicate a 2.3-fold increase in the incidence of reported genital warts from 1971 (29.8 per 100,000 persons) to 1982 (71.3 per 100,000 persons), making HPV infection the fourth most common STD in that country. The number of consultations to private physicians in the United States for condylomata increased from 169,000 in 1966 to 946,000 in 1981. The majority of infected persons was between 15 and 29 years of age, with the age-group 20–24 years being at greatest risk.

All reports on condylomata acuminata from 1950 to 1978 were collected for a study in a well-defined population in Rochester, Minnesota. The annual incidence of genital warts increased from 13 per 100,000 persons in 1950–1954 to 106 per 100,000 in 1975–1978. Also in this population, the infection rate was highest in persons 20–24 years of age.

It is important to note that the rates in the abovementioned areas should be considered as minimal incidence rates as the figures are based only on reported cases. In conclusion, it seems that the incidence of genital warts has increased in the past 20 years. [*See* SEXUALLY TRANSMITTED DISEASES (PUBLIC HEALTH).]

C. Prevalence of Koilocytosis

In general, the prevalence of HPV infection as detected by means of Papanicolaou's smear ranges from 0.4% to 2.2% among women attending gynecological departments or outpatient clinics for routine screening. Similar results have been obtained in investigations based on mass screening programs.

In line with HPV infection being sexually transmitted, the prevalence found in smears from women attending STD clinics tends to be much higher (8–13%) than in women from, for example, family planning clinics (0.2%).

It has been suggested that the prevalence of HPV infection diagnosed by means of cytology increases over time. However, most formal studies indicated that the rise in the HPV infection rate could largely be explained by an increased awareness of HPV infection in recent years.

D. Prevalence of HPV Types

Using different hybridization methods, the prevalence of HPV types has been studied in different geographical regions in women with invasive cervical cancer or precancerous lesions, in women with normal cervix, in women with cancer of the vulva, and in men either with penile cancer or without any symptomatic penile lesions.

Cervix (only studies of more than 20 cases are included)

Squamous Cell Carcinoma In studies of this type of carcinoma as many as 67% of cases have been reported to harbor HPV16 (range, 36–67%); the prevalence of HPV18 ranges from 0 to 25%, 0–11% contain HPV6 or 11, and in 1–5%, HPV31 has been diagnosed. Other relevant types listed in Table I have also been found in a few cases of squamous cell carcinoma each.

Adenocarcinoma Prevalence studies are hampered by case numbers being small. Most results are based on less than 10 cases. The prevalence of HPV16 and 18 has been found in the range of 0–33%.

Cervical Intraepithelial Neoplasia (CIN) It has been found that 14–75% of CIN lesions contain HPV16, 0–13% HPV18, 5–30% type 6 or 11, and in 20%, HPV31 was detected. These results concern all grades of CIN (I–III). In general, the prevalence of HPV16 or 18 is higher in more advanced lesions.

Normal Cervix The prevalence of HPV16 has been reported to vary from 3–30% and that of HPV16/18 ranges from 0–43%. An infection rate of 0–14% has been detected for HPV6 or 11, and type 31 has been found in 0–1%.

Vulva

The presence of HPV DNA has also been studied in carcinomas of the vulva. The findings, which are based on reports with 7–9 cases, indicate that HPV16 is the most frequent type, the prevalence being 29–57%. HPV18 was diagnosed in 0–33%, and type 6 was found in 22–33%.

Penis

Similarily, squamous cell carcinomas of the penis have been shown to contain DNA from HPV. The prevalence of HPV16 has ranged from 0 to 50%, and 9–39% of the tumors harbored type 18. HPV6 or 11 was found in 6% of the cases. In a study of 530 healthy men without signs of penile HPV infection, some 6% were HPV DNA positive (type 6/11, 16/18, or both).

In general, it is not known to what extent this variation in the prevalence of HPV within each group of lesions reflects a real difference between populations. Several aspects should be taken into consideration in interpretation of results and comparisons between studies. Except for a few comprehensive studies, most of the results are based on relatively small case-series, and differences reported may in part be due to chance. Moreover, demographic information as social background and age distribution is in general not available; as it has been suggested that the rate of HPV infection increases with age, some of the discrepancies may partially be explained by different age distributions. Another drawback in making comparisons between investigations is the frequent lack of information on the referral source of the study population.

Although the true incidence and prevalence of HPV infection is not known, it is, first, obvious that HPVs are extremely widespread in most if not all populations. Preliminary studies with PCR indicate that the prevalence may considerably exceed 40%. The frequencies of HPV6, 11, 16, and 18 infections seem to be within the same order of magnitude. Second, the prevalence of HPV types 16 and 18 consistently increases with increasing grades of CIN, and HPV16 is the most common type found in invasive cervical cancer. In contrast, HPV6 and 11 have been found to be more frequent in condylomata acuminata and lower grades of CIN than in normal cervical tissue. This led to the concept of HPVs with higher and lower oncogenic potential.

IV. Risk of Individual HPV Infections

Malignant conversion of condylomata acuminata has been documented but represents an extremely rare event, occurring mainly after a long persistence. Giant condylomata (Buschke-Löwenstein tumors) are invasively growing, rarely metastasizing, verrucous carcinomas. Like normal genital warts, they contain HPV6 or 11 DNA.

Human papillomavirus-associated flat lesions in the cervix were shown to progress in 5–20% of the cases to a more severe degree of CIN within a few years, and it is generally accepted that they would progress to carcinomas if left untreated. The clinical course of flat lesions at external genital sites of both sexes seems to be more benign.

Even though results from prospective cohort studies vary considerably, they generally indicate that lesions containing HPV16 or 18 have a higher probability of progression to more advanced CIN or even to invasive cancer than lesions containing HPV6 or 11. In contrast, the fate of normal cervical tissue with and without HPV infection is unknown.

The association between cervical neoplasia and HPV infection with types 6, 11, 16, and 18 has also been evaluated in case-control studies (comparison of women with cancer in respect to normal women for HPV infection). In the most recent and comprehensive study, infection with HPV16 and 18 was closely related to the risk of cervical cancer. The relative risk increased from about 2 to 9.1 (95% CI: 6.1–13.6) with increasing strength of the hybridization reactions (filter *in situ* hybridization). HPV6 or 11 was a somewhat weaker predictor of cervical cancer risk. The highest risk was observed in women positive for both 6/11 and 16/18.

Only one population-based, cross-sectional study of the prevalence of HPV DNA has been published so far. It was conducted in two geographical areas with a 5.7-fold difference in the incidence of cervical cancer. Surprisingly, the prevalence of HPV16 or 18 was 1.5 times higher in the low-incidence than in the high-incidence area for cervical cancer (see also Section VII,A).

V. Role of HPV in Tumor Progression

There can be little doubt that HPVs are the cause of the benign and precancerous tumors they are associated with: (1) Condylomata have been transmitted from person to person using cell-free filtrates of the warts; (2) histologically similar lesions occur in sexual partners and contain the same type of HPV; (3) normal human foreskin or cervical tissue injected *in vitro* with a HPV11 isolate and implanted beneath the renal capsule of athymic mice developed histological characteristics of condylomas and mild intraepithelial neoplasias; (4) transfection of HPV16 or 18 DNA into primary human foreskin keratinocytes or cervical epithelial cells results in immortalization, increased cellular proliferation, and altered differentiation. The viral genes E6 and E7 are required to trigger these effects. The *in vitro* changes may correlate with an increased cell supply from the basal layer *in vivo* in combination with delayed terminal differentiation. Both effects lead to a thickening of the epidermal prickle cell layer, the most consistent feature of HPV-associated lesions. (5) HPV16-transfected keratinocytes quickly change to an aneuploid karyotype, which is in line with frequently occurring abnormal mitoses in HPV16-positive lesions. (6) Transfection of HPV16 DNA in stratifying keratinocyte cultures induces histological abnormalities that resemble those seen in intraepithelial neoplasia *in vivo*. At higher passage level, abnormal cells appear throughout the full thickness of the *in vitro*–generated epithelium.

The role of HPV in malignant conversion is less clear. The persistence of HPV DNA in genital cancers and the continued expression of the transforming proteins E6 and E7 suggest that viral functions are basically required for carcinoma development and for the maintenance of the malignant state, although these observations are certainly no final proof. Three explanations may be offered for the small number of squamous cell cancers that appear to be negative for HPV DNA: (1) DNA of unknown HPV types persists at low levels but cannot be detected because of limited cross-hybridization with available HPV DNA probes; (2) HPV DNA was lost because it was no longer required at an advanced tumor state; and (3) HPV was not involved in the genesis of this minority of cervical cancers.

In attempting to understand the role of HPV in tumor progression, differences have been analyzed in the physical state of the viral DNA, in viral gene expression between benign and malignant tumors, and in more or less oncogenic HPVs.

The DNA of HPV16 and 18 usually persists extrachromosomally in premalignant lesions, but integration into the host cell genome may occur even in mild dysplasias, and most of or all the viral DNA appears integrated in many cancers. More than 30% of HPV16-positive cervical cancers and one cancer-derived cell line gave no evidence for integration, however, which, therefore may not be absolutely required for tumor progression. Other virus types such as HPV33 or 52 seem to be even less prone to integration than HPV16.

The insertion of viral DNA theoretically allows special mechanisms of carcinogenesis. Transcription control signals of the integrated viral genome could activate cellular oncogenes; the disruption of the circular viral genome may imply qualitative and quantitative changes in viral gene expression. It is noteworthy that the viral genes E1 and E2, which encode proteins involved in the positive and negative regulation of viral transcription, are destroyed in the majority of the cancers. This can be viewed as a consequence of selection in the course of tumor progression. In one case, flanking cellular sequences downstream from the HPV16 genes E6 and E7 were shown to enhance transcription, thus leading to transforming activity of the virus-cell DNA hybrid when tested with NIH3T3 mouse fibroblasts.

Integration of viral DNA results in the production of viral-cellular fusion transcripts. The cell-derived termini may affect the half-life and thus the levels of E6 and E7 transcripts, which could account for increased amounts of E6 and E7 proteins, which are involved in cell transformation.

On the cellular side, integration is not site-specific. It may occur on different chromosomes such as 3, 8, 12, 13, or 20 in the vicinity of fragile sites, oncogenes, and chromosome breakpoints that are also characteristic of hematologic malignancies and solid tumors. In the cervical cancer-derived cell lines HeLa and C4-1, the integration sites were mapped in the general vicinity of the cellular proto-oncogene *myc,* and elevated levels of *myc* mRNA were noted in both cases relative to other cervical carcinoma cell lines. The more generally observed increase in *myc* expression in cervical cancers, however, is probably due to indirect mechanisms.

Extrachromosomally persisting HPV6 DNA from vulvar carcinomas revealed duplications, insertions, or deletions of DNA sequences preferentially in the transcription control region. Some of these changes were shown to affect the viral enhancer or the transformation capacity. This may represent another way to stimulate the expression of oncogenic viral functions. [*See* DNA AND GENE TRANSCRIPTION.]

Viral transcription patterns in cancers and premalignant lesions differ by the lack of E4-, L1-, and L2-specific RNAs in carcinomas. The levels of transcripts from the E6-E7 region are similar, but final conclusions have to await a detailed analysis of *in situ* hybridizations. It will be particularly important to differentiate the overlapping messages for E6 and E7 because E7 seems to play the major role in oncogenic transformation and represents the most abundant viral protein in cancer-derived cell lines.

Oncogenic and less oncogenic genital HPVs, interestingly, generate the E7 mRNA by different mechanisms. In HPV16 and 18, its transcription is obtained by splicing out an intron from the original E6 mRNA, whereas a separate promoter is used for the E7 mRNA in HPV6 and HPV11, which lack equivalent splice signals. The splice event in HPV16 and 18 furthermore leads to the translation of carboxy terminally truncated E6* proteins. The meaning of these differences for pathogenicity, if any, is not clear.

VI. Synergism of Additional Risk Factors

The high proportion of asymptomatic cervical HPV infection, which has been found in women with otherwise normal cervices, clearly indicates that HPV is not sufficient for the development of cervical cancer. Several observations point to a synergism between HPV infection and certain cofactors (e.g., smoking, herpes simplex virus infection, use of oral contraception) in the production of cervical cancer. It has been suggested that such factors modify cellular genes with the consequence of a disturbed intracellular surveillance, in turn leading to lack of control of persisting viral genes by the host cell.

Carcinogens may affect cellular proto-oncogenes or tumor suppressor genes. An implification of one or both proto-oncogenes called *myc* and *ras* can be observed in advanced stages of cervical cancers, with the number of positive cases varying considerably between different studies. A 4–20-fold overexpression of the c-*myc* gene was noted with 35% of stage I or II tumors (still confined to the cervix or extending to the vagina and/or the parametrium). It is of interest in this context that *in vitro* transformation experiments have shown a cooperative effect of HPV16 and the oncogene *ras*. [*See* ONCOGENE AMPLIFICATION IN HUMAN CANCER.]

The inactivation or loss of recessive growth control genes in cancer cells may be more important than activation of a dominant oncogene because tumorigenicity is recessive with respect to the normal phenotype. Hybrids between a cervical cancer cell (HeLa) and human keratinocytes are nontumorogenic as long as the normal chromosome 11 is present. This chromosome is therefore likely to encode a tumor suppressor gene, which is no longer active in HeLa cells. Under specific culture conditions, the transcription of HPV18 appears suppressed in nontumorigenic hybrids and released in tumorigenic segregants, showing an intriguing correlation between the loss of growth control and the derepression of HPV gene expression.

VII. Medical Aspects

A. Risk Factors for HPV Transmission

As HPV infection and cervical neoplasia appear closely related, there is understandable concern about virus transmission. In studies of women with a cytological or histological diagnosis of condyloma or women who themselves gave a history of condylomata acuminata, multiple sexual partners have been found to be a significant determinant of HPV infection; a history of other sexually transmitted diseases has also been suggested to be associated with the risk of HPV infection. Examination of sexual partners of patients with condylomas or intraepithelial neoplasias showed that 30–60% suffered from comparable lesions. These findings are in line with a venereal mode of transmission, which might be at least reduced by the use of barrier contraception.

By contrast, studies that compared the exposures of infected and noninfected women, using DNA hybridization as the diagnostic method, surprisingly showed that women with "multiple" sexual partners were not at higher risk of being infected with HPV than women with "few" partners.

This result could be due to problems inherent in the biological system and/or to the diagnostic test: Repeated HPV infection in individuals who are highly active sexually might stimulate the immune response, which would in turn eliminate HPV replicating cells, thus affecting HPV detectability. An alternative explanation may be that factors exist especially in women with multiple sexual partners (e.g., other cervical infections), which interfere with the ability to detect HPV by certain hybridization methods.

If there are no pitfalls with the detection tests, the logical explanation for the above data would be that HPV is transmitted also by other routes than through sexual contact. This may be supported by preliminary data on HPV16-specific DNA and antibodies in children. If nonsexual transmission is important, infection will tend not to be associated with sexual activity. Given these uncertainties, it is not possible to provide clear recommendations on the prevention of genital HPV infections.

We are presently left with the discrepancy between clinically manifest HPV infection showing the epidemiology of a venereal disease and latent infection not being correlated with sexual activity. This paradox might be explained if factors associated with high sexual activity played an important role in the activation of the widespread, preexisting, latent HPV infections. In addition, the virus load transmitted from florid lesions by sexual intercourse is likely to be higher than the infectious dose obtained by other transmission routes, thus increasing the probability of a clinically apparent course.

If it holds true that a concordance between risk factors for inapparent HPV infection and cervical neoplasia cannot be confirmed, we have to conclude that the activation of latent infections represents the first essential step in tumor development.

B. Therapy of Papillomavirus-Associated Lesions

Treatment is undertaken as prophylaxis against malignancy and for disease control. Most procedures are only aimed at the physical destruction or the surgical removal of the tumor and do not interfere directly with the viral infection. Freezing by liquid nitrogen or application of caustic podophyllin are frequently used for genital warts. Surgical techniques include excision, electrocautery, and carbon dioxide laser vaporization, which is rapidly becoming the method of choice for treating intraepithelial neoplasia. Persisting HPV DNA may be demonstrated in the surrounding normal tissue after removal of lesions as we would expect, so that the patients may not be cured from HPV infection even after clinical cure. Treatment with α-interferon appears more promising in this regard because interferon displays antiviral activities in addition to its antiproliferative and immunomodulatory properties. It is highly effective in inducing papilloma regression but has not yet been shown to affect the persistence of HPV DNA either *in vivo* or *in vitro*. Hence, there is no approved antiviral therapy so far.

C. HPV Diagnosis and Cancer Prevention

The early detection of HPV-induced genital lesions, particularly those at the cervix uteri, may be warranted medically because these lesions are regarded as precursors of genital cancer. Routine screening is based on the cytologic examination of exfoliated cells collected from the cervix and stained according to Papanicolaou. This test was designed to detect precancerous lesions long before their association with HPV was established.

Human papillomavirus infection can be confirmed by the demonstration of group-specific viral antigens using immunological methods and by the detection of virus-specific nucleic acid. In both cases a negative test result does not exclude an HPV etiology. Only 30–80% of *bona fide* HPV-induced lesions are positive for the group-specific antigen, and the limited homology among different HPV DNAs prevents the regular demonstration of some viral DNAs with the presently available, limited number of test probes.

In view of the ubiquity of HPV infections, the demonstration of viral footprints *per se* does not provide clinically relevant information regarding prognosis on tumor development. There is presently no way to predict the behavior of latent infections and no therapy to eliminate the persisting virus. We should therefore refrain from a large-scale screening for HPV infection.

The tight association between HPV and genital cancer raises the possibility that HPV DNA might be used as tumor cell marker. For example, it is possible to take advantage of the high sensitivity of DNA hybridization techniques to screen lymph nodes of patients with HPV-positive cervical cancers for early metastases.

Circumstantial evidence that the type of HPV largely determines the risk of malignant conversion and the aggressive behavior of a cancer raised interest in diagnostic virus classification. Routine application has to await prospective epidemiologic studies, however, and therapy concepts should not presently be modified according to virus type.

D. Prospects for Vaccination

The possibility of HPV playing a necessary role in genital cancer development raises the prospect of tumor prevention by vaccination. The naturally occurring, spontaneous regression of HPV-induced lesions can be ascribed to immune mechanisms, indicating that the host is basically able to control HPV infection. The defense is usually weak, however, and sometimes not detectable at all. This is probably due to low levels of viral antigens, so that it appears promising to stimulate the immune system by vaccination. To treat an established infection, a vaccine should be able to induce immunity directed against HPV-infected tumor cells, which would at the same time prevent the growth of new lesions. This can only be achieved by early viral antigens, which are expressed in the proliferating cells of the tumor. The potential efficacy of immunity directed against various viral proteins is presently under study in animal model systems.

It is unlikely that HPV infection itself can be prevented by the induction of neutralizing antibodies because HPVs are easily able to enter their target cell in the epidermis without antibody contact. The cancer risk of latently infected persons possessing an antipapilloma cell immunity is not known, but the clinical observation that genital cancers usually are preceded by precursor lesions suggest that latently infected cells are at low risk.

Bibliography

Brinton, L. A., and Fraumeni, J. F., Jr. (1987). Epidemiology of uterine cervical cancer. *J. Chronic Dis.* **39,** 1051–1065.

Muir, C., Waterhouse, J., Mack, T., Powell, J., and Whelau, S., eds. (1987). "Cancer Incidence in Five Continents," vol. V. (IARC Scientific Publications No.88) International Agency for Research on Cancer, Lyon.

Munoz, N., Bosch, X., and Jensen, O. M. eds. (1989). "Human papillomavirus and cervical cancer", (IARC Scientific Publications No. 94) International Agency for Research on Cancer, Lyon.

Pfister, H., ed. (1990). "Papillomaviruses and Human Cancer." CRC Press, Boca Raton, Florida.

Syrjänen, K., Gissmann, L., and Koss, L. G., eds. (1987). "Papillomaviruses and Human Disease." Springer, New York.

Zur Hausen, H. (1989). Papillomaviruses in Anogenital Cancer as a Model to understand the Role of Viruses in Human Cancers. *Cancer Res.* **49,** 4677–4681.

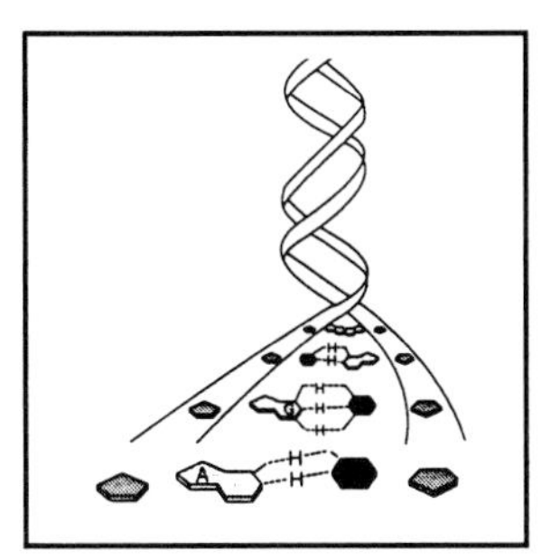

Genome, Human

PHILIP GREEN, *Washington University School of Medicine*

Glossary

Chromosome Threadlike structure in the cell nucleus consisting of a single DNA molecule (up to several hundred million nucleotides in length), together with associated proteins necessary for maintaining its structural integrity (human cells contain 23 pairs of chromosomes)

Chromosome map Schematic indication of the relative locations of particular DNA sites in a chromosome. In linkage maps, distances between sites are expressed in terms of the rate at which crossing over between the sites occurs during meiosis; in physical maps, distance is expressed as the number of nucleotides between the sites

Deoxyribonucleic acid (DNA) Molecule that carries genetic information. It consists of two complementary helical strands, each of which is a polymerized sequence of nucleotide subunits.

Gene Segment of the DNA molecule encoding the information necessary for the cell to construct a protein or RNA molecule. Genes are the "functional units" of the genome

Nucleotide A molecule consisting of a phosphate group, a sugar group (which together form part of the DNA strand backbone), and one of the bases adenine, cytosine, guanine, or thymine

THE GENOME contains the hereditary information necessary to specify the human organism. This information is encoded in the sequence of some 6 billion nucleotide subunits composing the polymer DNA in the nuclei of human cells. Over 99% of this sequence is identical in essentially all humans and distinguishes us as a species from other organisms; differences at various points in the other 1% ensure that no two individuals, except identical twins, have exactly the same genetic complement and account for the extensive inherited variation seen in the human population.

The genome includes on the order of 100,000 genes, segments of the DNA containing the necessary instructions to manufacture protein molecules in the appropriate cells. Since essentially all biological structures and processes involve proteins, and the structures of all proteins are specified by genes, knowledge of the genome provides a focus for organizing our understanding of human biology. Identifying all genes has accordingly become a major priority of current biological research.

I. Genetic Information

A. DNA and Chromosomes

Every human develops from a single cell. One of the central preoccupations of 20th-century biology has been to understand how the genetic material in this cell encodes the information necessary to specify a complete individual.

The structure of the molecule which carries this information, DNA, was found by James Watson and Francis Crick in 1953 (Fig. 1). It is a double helix in which each of the two helical strands consists of a sequence of chemical subunits, called nucleotides. These are of four different types denoted A, C, G, and T after the names of their component bases adenine, cytosine, guanine, and thymine. Each base in one strand is paired with a unique complementary base in the other strand via hydro-

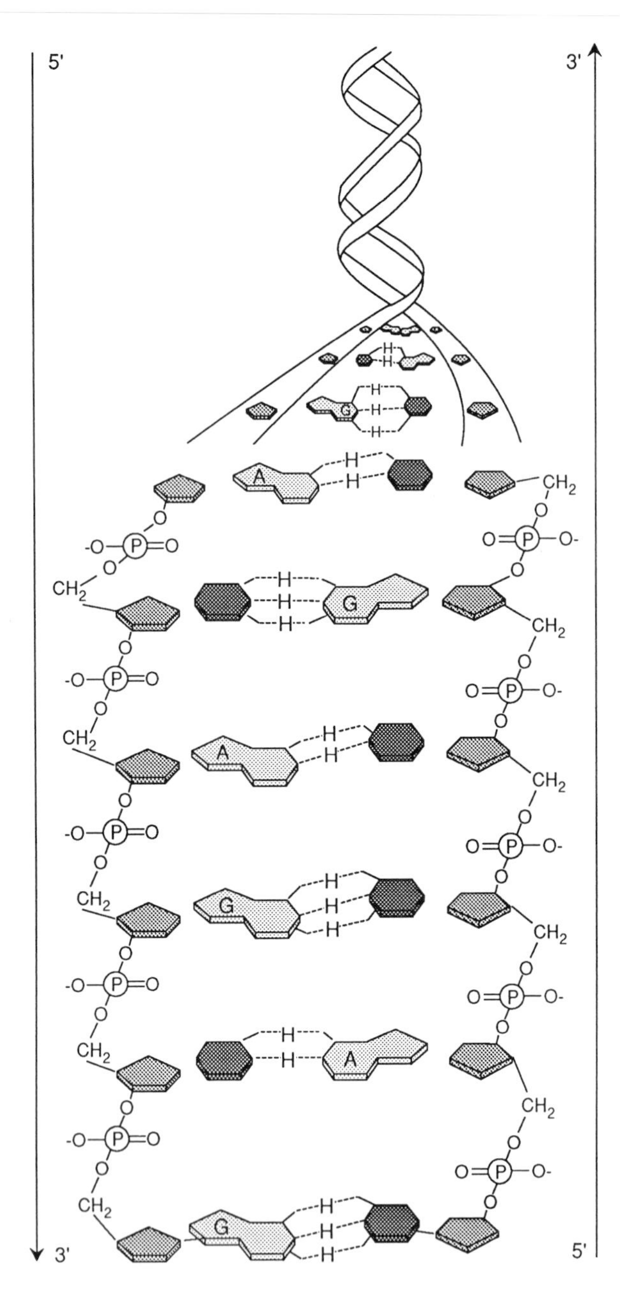

FIGURE 1 Structure of DNA. Note that the two sugar-phosphate chain runs in opposite directions. This orientation permits the complementary bases to pair. From, P. W. Davis and E. P. Solomon (1986). "The World of Biology." 3rd ed. pp. 208. Saunders College Publishing, Philadelphia.

gen bonds: T is always paired with A, and C with G. The bases are covalently attached to a strand backbone formed from alternating sugar and phosphate groups (every nucleotide contributing one of each). Each strand has a direction: any nucleotide has a 5′, or upstream, neighboring nucleotide, as well as a 3′, or downstream, neighbor (the numerical terminology derives from the sites in the sugar group to which the adjacent phosphates are attached). The two strands run in opposite directions. Any particular DNA molecule thus could be described by giving the sequence of letters representing the bases (in 5′-to-3′ order) in one of the strands. It is no exaggeration to say that biology has been revolutionized by the insight that the genetic information is contained in the specific nucleotide sequence of the DNA molecule. [*See* DNA AND GENE TRANSCRIPTION.]

The structure of DNA also immediately suggests the means for its replication: Since the base sequence of either strand determines the other strand, two faithful replicas of the molecule can be created by separating the two strands and synthesizing new strands complementary to each. This synthesis is carried out in the cell by the enzyme DNA polymerase, which appends the complementary nucleotides one by one in the 5′-to-3′ direction. DNA replication occurs prior to cell division, one replica then being segregated to each daughter cell. As a result nearly all of the 10 trillion cells in the mature human organism have DNA complements identical to that in the single-celled zygote from which it started its complex development.

The genome is packaged in the form of chromosomes, each of which consists of a single long DNA molecule together with proteins that maintain its structural integrity. There are 24 different human chromosomes (Fig. 2): the sex chromosomes X and Y and 22 autosomes, composing a total of approximately 3 billion base pairs (bp) of DNA. Most human cells are diploid, containing two copies of each autosome and either two Xs (in females) or an X and a Y (in males). Genetic information is transmitted to the offspring via haploid cells (formed from germ-line diploid cells during meiosis) that contain one chromosome from each of the 23 pairs. These haploid cells, the sperm and the egg, each contribute one-half of the genetic material of the zygote formed by their union. [*See* CHROMOSOMES.]

B. Genes

Most information in the genome is in the form of genes, segments of the DNA molecule which encode the instructions necessary for the cell to construct two kinds of macromolecules: proteins and RNA. Proteins, polymers whose chemical subunits are 20 different kinds of amino acids, play crucial functional roles in nearly all cellular processes and are essential components of most cellular struc-

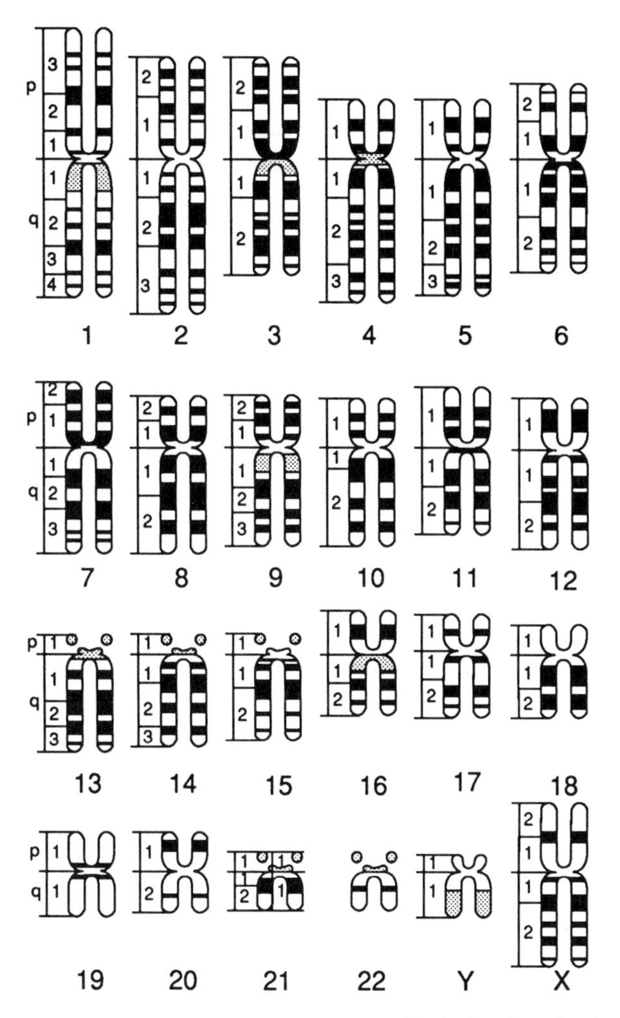

FIGURE 2 The human chromosomes, with designations for the Giemsa-stained metaphase bands (see Section IV,B,2). Modified from J. J. Yunis (1976). High resolution of human chromosomes. *Science* **191**.

tures. The vast majority of the estimated 100,000 human genes encode proteins. RNA is a polymer with four types of nucleotide units similar to those of DNA (but with slightly different backbone sugar groups and the base uracil instead of thymine). It is usually single stranded, but can base-pair with a DNA or RNA strand of complementary sequence in the same way that two DNA strands do. RNA molecules also play a number of important functional and structural roles in the cell, particularly in various stages of the process by which protein-encoding genes are expressed (i.e., the nucleotide sequence information is converted into the protein molecule.) [*See* GENES.]

The structural features of protein-encoding genes are best understood in terms of the stages of the expression process. In the first stage, transcription, the enzyme RNA polymerase synthesizes (in the 5′-to-3′ direction) an RNA molecule, the transcript, complementary to a segment of one of the two DNA strands. The genetic information to control this process is contained in the promoter, a part of the gene which typically includes several short sequences recognized by nuclear proteins that interact with the polymerase. One promoter component is usually a short sequence rich in As and Ts, the TATA box, necessary to accurately position the polymerase to start transcription. Other promoter elements, which are usually located within a region of 100 or so nucleotides 5′ to the TATA box, determine the rate at which transcripts are initiated (and thus, indirectly, the number of protein molecules produced in the cell). Some of these elements mediate changes in transcription rate in response to certain signals, such as those conveyed by hormones or growth factors; others determine the types of cells in which transcription occurs.

The transcript starts at a point 25–30 bases downstream from the TATA box and usually extends for several thousand nucleotides further downstream. A curious feature of most human genes is that the coding region, the part of the gene which encodes the protein structure, is interrupted by one or more noncoding segments called introns. [The coding segments (called exons) often correspond to structural domains in the protein molecule, which suggests that genes might have been assembled over the course of evolution from exon "modules." Known genes vary widely in their number of exons, from one to more than 50.] These intronic sequences are spliced out of the transcript to leave an RNA molecule with a single contiguous coding region, typically consisting of 1000 or so nucleotides. Introns often vastly exceed the coding region in length, totalling more than 1 million nucleotides in some genes. They contain short nucleotide sequences to permit their recognition by the splicing machinery, but are otherwise not believed, in general, to contain significant information of functional importance in gene expression.

The processed RNA molecule (now referred to as mRNA) is then transported out of the cell nucleus to a ribosome, where it is translated into the protein molecule: Starting with a trinucleotide ATG (which encodes the amino acid methionine) near the beginning of the mRNA molecule, each successive nucleotide triplet, or codon, in the mRNA molecule is "read," and the amino acid which corresponds to it (via the genetic code) is attached to the nascent peptide chain. Translation stops, and the completed protein is released from the ribosome, when a termination, or "stop," codon (i.e., TAA, TAG, or TGA) is reached.

In addition to the genes that encode proteins, there are others which are transcribed to yield functionally important RNA molecules not translated into proteins. These "RNA genes" are transcribed by different RNA polymerases from the one which transcribes protein genes, and many of them have promoter sequences contained wholly or partially within the transcribed sequence, rather than 5′ to it.

C. Chromosome Maintenance

The genome also contains sequences necessary for maintaining chromosome integrity and ensuring proper chromosomal replication and segregation to the daughter cells at cell division. The telomeres are specialized structures at the chromosome ends to ensure the proper replication of the end and protect it from cellular enzymes which would otherwise degrade it. Human telomeres include several hundred to several thousand copies of the six-nucleotide sequence TTAGGG, repeated in tandem, and might include other functionally important sequences. The centromere is the chromosomal structure to which the spindle is attached during cell division, to ensure segregation of the chromosomes to the daughter cells. The DNA sequences essential for centromeric function have not been identified, but may include the so-called α-satellite sequences, which consist of several hundred thousand tandem repeats of simple-sequence DNA found in the centromeric region. [*See* TELOMERES, HUMAN.]

Chromosomes also have (as yet unidentified) specific sequences to signal the sites at which DNA replication is to start and other sequences at which various protein components of the chromosomes bind the DNA.

D. Augmenting the Genome's Information: Methylation

There are several ways that the information in the genome's nucleotide sequence can be modified or augmented to affect gene expression. Rearrangements of the DNA in some genes (e.g., antibody genes) occur in cells of the immune system to generate a large number of different protein molecules of closely related structure. A more widely used modification is covalent attachment of a methyl group to certain C bases having a 3′ guanine neighbor. Not all Cs occurring in such CG dinucleotides are methylated; and a particular C might be methylated in some cells, yet unmethylated in others. Certain such Cs occurring near genes are generally unmethylated in those tissues in which the genes are expressed, and methylated in tissues in which the gene is not expressed, suggesting that the methylation is involved in gene expression. Also possibly involved in expression are regions known as "HTF (Hpa II tiny fragment) islands," several hundred base pairs or more in length which are rich in unmethylated Cs and are often found near genes.

Methylation might also play a role in X inactivation. Early in female embryogenesis one of the two X chromosomes in each cell assumes a more condensed (i.e., heterochromatic) state, which is stably inherited by the replicas of this chromosome through subsequent cell divisions. The condensed X is known to be more heavily methylated than the other X, and most of its genes are not expressed. The purpose of X inactivation is probably to ensure that X chromosome genes, most of which are irrelevant to sex determination, have the same number of active copies (i.e., one) in female cells as in male cells.

Methylation also appears to be involved in imprinting, a phenomenon in which the expression of certain genes depends on the parent from whom the gene is inherited. In at least some cases imprinting is associated with methylation patterns which are stably maintained in the development of somatic (non-germ-line) tissues, but undergo sex-specific alterations in the germ line. Imprinting might underlie the observation that the severity of certain genetic diseases depends on the parental origin of the disease gene.

E. Changes in the Genetic Information

The central role of the genome makes it essential that it be passed down virtually unchanged, both from cell to cell within the organism and from generation to generation. As a result mechanisms have evolved to ensure that DNA replication is extraordinarily accurate and that damage to the DNA is detected and repaired efficiently. Nevertheless, the action of chemical agents, errors in the DNA replication process, or radiation do occasionally alter (i.e., mutate) the DNA sequence in a particular cell, such that the changed sequence is inherited by the descendants of that cell. Mutations occurring in germ-line cells might be passed on to the offspring; in contrast, a somatic mutation affects only tissues derived from the mutated cell within the individual in whom the mutation occurs.

Mutations are of several types, including substitution of one nucleotide for another, deletion or insertion of one or a few nucleotides, and, more rarely, deletions, duplications, or rearrangements of more extensive regions (up to several million nucleotides or more in length). A relatively frequent type of nucleotide substitution occurs when a methylated C loses an amine group, which changes it to a T. Thus, CG dinucleotides tend to mutate to TG or (when the mutation occurs on the complementary strand) to CA. This type of mutation has occurred frequently enough during human evolution that CG dinucleotides are found much less commonly in the genome than would be expected from the frequencies of Cs and Gs.

Germ-line mutations are important as the raw material for evolutionary change. Any mutation initially occurs in a single individual; it becomes fixed in the species only if, by some subsequent generation, every member of the species is descended from that individual. This can occur if the mutation confers a selective advantage on the individual carrying it; however, statistical considerations show that it also, by chance, occurs occasionally with neutral mutations that have no significant effect on the organism. It is, in fact, likely that the great majority of mutations that have been incorporated into the genome during human evolution are of the latter type.

A mutation occurring in a gene might or might not have a significant effect on the structure or function of the gene product, or its expression; when it does, it is likely to be deleterious (because most genes have already undergone millions of years of evolutionary fine tuning, and most improvements are likely to have been implemented already). In humans these deleterious mutations (if they are not lethal) can cause genetic diseases, over 3000 of which are known. Although most of these are relatively rare, significant genetic components are increasingly being identified for common diseases such as diabetes, some psychiatric disorders, heart disease, and many cancers. Genetic diseases are an important resource for the study of human biology, because they provide the only examples of the effects on the whole organism of altering the structure or expression of a gene. [*See* GENETIC DISEASES.]

Mutations provide genetic diversity. Their efficient utilization for species evolution depends on the ability to mix the genetic material from different individuals, so that ultimately several advantageous mutations might be combined within a single individual. In part this is ensured by sexual reproduction and independent segregation of chromosomes at meiosis, but an important additional mechanism, crossing-over, is necessary to allow mutations occurring at different positions (i.e., loci) on different copies of the same chromosome to be combined on the same chromosome. Crossing-over occurs during meiosis, when the two homologous chromosomes in each pair are aligned next to each other. Roughly speaking, both chromosomes break at the same position, and each piece is rejoined to the other chromosome so as to create two new copies of the chromosome, which are mosaics of the previous copies. The breaks can occur anywhere along the chromosome, but more frequently in some regions (particularly near the ends of the chromosomes) than others. In male meiosis the X is paired with Y, but cross-overs between these chromosomes occur only in the pseudoautosomal region, consisting of about 2 million bp near one telomere of both the X and the Y.

F. "Junk" DNA

Genes and chromosome maintenance information are thought to account for no more than 5–10% of the DNA content of the genome. The function(s), if any, of the remaining 90% (in which we include the introns as well as the intergenic regions) is unknown. Although there are undoubtedly additional types of functionally important sequences remaining to be discovered, the information content in the intergenic regions is not likely to be as dense as in the genes: Comparisons of human intergenic sequences with the corresponding sequences in evolutionarily closely related mammals indicate that mutations are accumulating there at a substantially higher rate than in coding regions, almost certainly an indication that they contain less information for selective forces to maintain. Much of this sequence presumably reflects the continual flux (over millions of generations) of random rearrangements, duplications, and other accumulated mutations, which could lead occasionally to new useful genes, but for the most part are neutral. Presence of a sequence in the genome does not entail functional significance.

There are some recognizable features in this noncoding DNA. Perhaps 50% of it consists of several types of repeated sequences, which are found in numerous copies in the genome. The most prevalent of these is the Alu element, a sequence of some 300 bp found in over 500,000 copies dispersed throughout the genome and accounting for between 5% and 10% of all human DNA. The Alu element is closely related in sequence to the gene for a small RNA molecule (7SL RNA), a component of the cellular machinery to target proteins to particular compartments, and is probably derived from it evolutionarily, but is not believed to serve any similar function within the cell. The Alu element is transcribed and, like small RNA genes, has a promoter wholly included within the transcribed sequence. This could account for its evolutionary spread through the genome: one notion is that RNA molecules are occasionally "reverse-transcribed" into DNA by an enzyme, reverse transcriptase, which synthesizes a DNA strand complementary to the RNA sequence. Following synthesis of the complementary DNA strand, the DNA molecule could then be inserted at the site of a chromosome break. If this process occurs in germ-line cells, it might be incorporated into the genome of subsequent generations. (Reverse transcription and insertion into the genome have also occurred occasionally with the mRNA from protein-encoding genes, but since the promoters of these genes are not contained within the transcript, these "processed pseudogenes" lack promoters and generally are not expressed.)

Alu (and other repeated elements) could thus simply be examples of "selfish," or parasitic, DNA, sequences which have accidentally evolved the capacity to take advantage of the cell's machinery to spread themselves through the genome. On the other hand, these repeated elements could play an important role in facilitating evolutionary change, either by altering the expression of nearby genes or by mediating occasional genome rearrangements (by causing misalignment of the two homologous chromosomes during meiosis).

II. Genome Organization

Many genes are found in gene clusters, or families, whose members are located near each other in the same chromosomal region and have closely related nucleotide sequences. Often, the genes in a family encode proteins with similar, but not identical, functions or differ in their patterns of expression (e.g., the tissues or developmental stages at which they are transcribed). Each cluster has likely arisen during evolution from a single progenitor gene by several rounds of duplication (perhaps caused by misalignment of the two chromosomal regions during meiosis, accompanied by crossing-over within the misaligned region); the duplicate copies then form the raw material for further mutational changes, which can lead to altered expression or function. Many clusters include nonexpressed pseudogene members, the result of duplication events that failed to evolve advantageously and instead acquired inactivating mutations.

In some cases proximity of the genes in a family could play a functionally important role in their expression. An example is the genes encoding the RNA molecules used as structural components of ribosomes: These ribosomal RNA genes occur in large numbers of adjacent virtually identical tandemly repeated units near the ends of the acrocentric chromosomes (i.e., those in which the centromere is near one end of the chromosome) 13, 14, 15, 21, and 22 and are coordinately expressed in association with the nucleolus during the construction of ribosomes.

A fact of unknown significance is that the genome is apparently a mosaic of subregions which differ widely in C/G content, the percentage of their nucleotides which are Cs or Gs. Each subregion, or isochore, is at least several hundred thousand base pairs in length and is relatively homogeneous in C/G content. The C/G content can be one of a small number of different values ranging from about 40% to about 55%. Genes tend to be located in the most C/G-rich isochores (although some are found in the other isochores).

III. Genome Variation: Polymorphism

Genes or other genomic DNA segments that occur in two or more reasonably common variant forms (i.e., alleles) in the human population are said to be polymorphic. The variants represent old mutations which have been inherited by an appreciable fraction of the current population. Most of these mutations are probably neutral, although some gene polymorphisms reflect competing selective advantages of the different alleles. Studies of several gene regions from different individuals suggest that roughly one in 100 to one in 1000 nucleotide sites is polymorphic, with any two copies of the same region differing at about one in every 1000 nucleotides (although it is unclear whether these figures are characteristic of the genome as a whole); by comparison the genomic sequences of a human and a chimpanzee (the most closely related primate species) in such regions differ at 1–2% of the sites.

Polymorphisms in gene coding regions account for much of the inherited observed variation in the human population, underlying such diverse traits as blood groups, metabolic differences, and (perhaps) aspects of personality. These polymorphisms are nearly always nucleotide substitutions, since most deletions or insertions disrupt the reading frame, leading to an entirely different (and usually nonfunctional) protein product.

An important use of polymorphisms is in distinguishing different copies of the same chromosome. For example, when an individual is heterozygous (i.e., has two different alleles) at a polymorphic site, it becomes possible to track the inheritance of that chromosomal region in the children of that individual. This provides a powerful tool for localizing disease genes through linkage analysis (see Section IV,B,1). Polymorphisms are also useful for studying the origins of human population groups, which often differ in the frequencies of particular alleles.

A disproportionate number of nucleotide substitution polymorphisms occur at CG dinucleotides, reflecting the propensity of methylated C to mutate to T. Another important class of polymorphic loci are the minisatellite, or variable number of tandem repeat, loci. These sites (whose functional significance, if any, is unknown) consist of up to several hundred copies of a short (i.e., typically less than 20-base) sequence, repeated in tandem, the alleles differing in the numbers of copies of the repeated element. One recently discovered class of minisatellite loci consists of tandem repeats of the dinucleotide GT. These are often found near genes and could play some role in their expression.

Some minisatellite loci are highly polymorphic, having multiple alleles which are each relatively rare. These sites provide a powerful tool for distinguishing the DNA of different individuals, and accordingly have important medical and forensic applications. Although minisatellite loci are found throughout the genome, the highly polymorphic ones often tend to be located near the ends of chromosomes, presumably reflecting a higher mutation rate in the telomeric region. [*See* POLYMORPHISM, GENES.]

IV. Methods for Studying the Genome

Most current knowledge about the genome has been obtained using a variety of powerful tools developed over the last 20 years to detect and manipulate DNA, RNA, and protein molecules. These techniques have made it possible to explore the organization of the genome, to find a number of genes within it, to determine the nucleotide sequences of these genes, and in many cases to analyze their various functional components by testing the effects of changes in promoter elements and coding sequences.

A. Tools for Manipulating DNA *in Vitro*

Methods have been developed to detect, amplify, synthesize, separate, alter, and determine the nucleotide sequence of DNA molecules in the laboratory. Many of these rely on enzymes purified from bacteria, the byproducts of extensive biochemical studies of the mechanisms by which bacterial cells process DNA. Among the more important techniques are:

1. Chemical synthesis of oligonucleotides, short single-stranded DNA molecules (usually less than 100 nucleotides in length) of any desired sequence.

2. The polymerase chain reaction (PCR), a method for amplifying a particular sequence from genomic DNA. Two oligonucleotides (primers) are synthesized corresponding to sequences flanking the region to be amplified, one of them upstream of this region on one strand, the other upstream of it on the other strand. These are then added to a mixture containing the genomic DNA, a thermostable bacterial DNA polymerase, and nucleotides. The mixture is heated to separate the DNA molecules into single strands (by breaking the hydrogen bonds between complementary bases). Upon cooling, the oligonucleotides anneal to each strand; the polymerase then synthesizes complementary strands starting at these oligonucleotides and continuing downstream through the desired region. Each repeated cycle of heating, cooling, and DNA synthesis doubles the number of copies of the sequence between the oligonucleotides, so that in a short time it is possible to amplify this sequence 1 million-fold or more. Since the amount of starting material can be

extremely small—the DNA from a single human cell has been used—this provides an extremely powerful method both for detecting whether a particular sequence is present and for making sufficient amounts of it for further analysis or manipulation. Its main limitations are that the nucleotide sequence of the region to be amplified must be at least partially known, and at present it is difficult to amplify regions more than a few thousand base pairs in length.

3. Cleavage of DNA molecules with restriction enzymes, which recognize specific nucleotide sequences (usually 4–6 bases in length) within a DNA molecule and cut it there. For example, the enzyme *Eco*RI recognizes the nucleotide sequence GAATTC and cleaves between the G and the first A. Since this sequence also occurs on the opposite strand, *Eco*RI "digestion" of DNA produces a set of double-stranded fragments with the single-stranded four-nucleotide overhang AATT at both ends. The overhang on one fragment is complementary to the overhang on any other fragment and can anneal to it; the enzyme DNA ligase can then be used to covalently link two such restriction fragments together or circularize a single linear fragment. Restriction digestion in conjunction with ligation provides a powerful general method for making recombinant DNA molecules composed of pieces from several different sources. Several hundred restriction enzymes of varying sequence specificities have been purified from various bacterial species.

4. Size fractionation of mixtures of DNA molecules (e.g., restriction digestion fragments) by gel electrophoresis. The DNA molecules (which are negatively charged at normal pH) are impelled through a gel by an electrical field; separation according to size results from the fact that large molecules are retarded more by the gel matrix than are small ones. Agarose gels can resolve molecules differing by a few percent in length and ranging in size from several hundred to 50,000 bp. Acrylamide gels can resolve single- or double-stranded DNA molecules of up to several hundred bases in length which differ in length by as little as a single nucleotide.

Large DNA fragments are poorly resolved by ordinary electrophoresis, because they orient themselves for efficient passage through the gel such that retardation by the matrix is no longer proportional to length. Pulsed field gel electrophoresis circumvents this phenomenon by periodically changing the field direction to reorient the molecules and thereby restore proportional retardation of large fragments. This permits resolution of molecules up to several million base pairs in length. [*See* PULSED FIELD GEL ELECTROPHORESIS.]

5. Cloning, a powerful method both for fractionating a complex DNA source into individual pieces and for making as many copies as desired of any particular piece. Cloning is usually a prerequisite to nucleotide sequence determination and to functional studies of a particular DNA segment. The DNA molecules to be cloned are attached (typically by restriction digestion and ligation) to a vector DNA molecule, and the ligated molecules are introduced into microorganism host cells. The vector includes sequences that permit replication of the ligated molecule within the host, and usually also a marker gene to permit selective growth or identification of only those cells which have successfully incorporated the DNA. *Escherichia coli* is commonly used as a bacterial host, and, as vectors, either plasmids, which replicate as circular DNA molecules and include antibiotic resistance genes as markers, or bacteriophage, viruses which infect bacteria, are used. DNA fragments of up to about 45,000 bp can be cloned in this way. Larger DNA fragments (of several hundred thousand base pairs in length) can be cloned as artificial chromosomes in yeast. Traditional cloning of small DNA fragments is being supplanted for many purposes by PCR, which can be regarded as *in vitro* cloning.

Clone libraries are collections of clones of fragments from a particular organism's DNA. Libraries fractionate a complex DNA source into manageable smaller pieces, which can be screened (e.g., by hybridization, described below) with a probe for a particular DNA sequence to obtain a clone containing that sequence. Bacteriophage, plasmid, and yeast artificial chromosome (YAC) libraries of genomic DNA are all used in this manner. YAC libraries in particular are proving to be an important tool for analysis of the human genome due to the large size of the cloned fragments. [*See* CHROMOSOME-SPECIFIC HUMAN GENE LIBRARIES.]

The importance of gene coding regions and their low density within the genome make it useful to efficiently identify these sequences. Libraries of cloned coding regions can be constructed by using the enzyme reverse transcriptase to synthesize single-stranded DNA complementary in sequence to the mRNA molecules purified from a particular tissue type; these cDNA molecules are then converted to double-stranded DNA molecules using DNA polymerase, ligated to plasmid or bacteriophage vector sequences, and introduced into the bacterial host. Expression cDNA libraries use specialized vector sequences which allow the bacterial host to transcribe and translate the cloned coding region. These libraries can then be screened using methods (e.g., labeled antibodies) which directly detect the protein, rather than the DNA.

6. Hybridization, a powerful method for detecting a sequence of interest within a complex mixture. It is based on the tendency for two single-stranded DNA molecules of complementary sequence to seek each other out and anneal, or hybridize. A cloned, or synthetic, DNA segment (the probe) is radioactively or fluorescently labeled, and its strands are separated and allowed to anneal with target DNA whose strands have also been separated. Usually, the target DNA has been fractionated (either electrophoretically or in the form of

a clone library) and transferred to a solid support, typically a nylon membrane. (Transfers from electrophoretic gels are known as Southern blots.) The location on the membrane at which the probe anneals indicates the fragment size or clone containing the probe sequence.

7. Determining the nucleotide sequence of a small cloned or PCR-amplified DNA segment. This depends on the ability to generate, for each of the four types of nucleotides, a family of single-stranded DNA fragments which are subsequences of the cloned segment, and have the same 5′ end but terminate at variable 3′ positions corresponding to the various occurrences of that particular nucleotide in the segment. These fragments can be created either (1) by partial digestion with chemicals which cleave DNA strands at occurrences of the nucleotide (Maxam–Gilbert sequencing) or (2) by DNA polymerase-mediated synthesis of fragments complementary to one of the original segment strands, using a reaction mix which contains, in addition to the usual nucleotides, a modified form of one nucleotide that, when incorporated by the polymerase, terminates the nascent strand at that nucleotide (Sanger sequencing). The relative lengths of these fragments (to single-base resolution) are then determined by acrylamide gel electrophoresis; correlating the terminating nucleotide of each fragment with its length yields the nucleotide sequence. In a single gel run it is possible to determine the sequence of a DNA molecule of several hundred nucleotides.

B. Mapping

Mapping (i.e., determining the locations of genes or other DNA segments within the genome) provides useful information for cloning genes, for inferring evolutionary relationships between genes and organisms, and for organizing our knowledge of the genome. There are several mapping techniques in current use. [*See* GENETIC MAPS.]

1. Linkage maps (historically the first type of genomic map to be constructed, using experimental organisms) delineate the order of polymorphic sites along a chromosome. The map distance between two sites is defined as the average number of cross-overs occurring between them per meiosis, the unit of measurement usually being the centimorgan (corresponding to an average of one cross-over per 100 meioses; on average 1 cM in the human genome corresponds to a physical distance of about 1 million bp, but there is substantial local variation in this value). Because rates of crossing-over differ in male and female meioses, map distances must be specified separately for each sex.

Linkage maps are constructed from studies of the inheritance of alleles at polymorphic loci in a panel of families. When an individual is heterozygous at each of two different loci, it is often possible to determine whether or not particular alleles at the two loci are coinherited by his or her children. For loci near each other on a chromosome, an allele at one locus is coinherited with that allele at the other locus which happens to lie on the same copy of the chromosome in the parent, except when a cross-over occurs between them. Loci on different chromosomes show no allele coinheritance. Thus, inheritance studies allow one to identify when polymorphic loci are on the same chromosome and to determine the relative distances and order for loci on the same chromosome. In comparison to studies of experimental organisms, the analysis of human linkage data is complicated by the facts that the allele phase (i.e., the particular copy of the parental chromosome on which a given allele lies) is often not known, any particular individual is likely to be heterozygous for only a subset of the loci being studied, and it is impractical to study large numbers of families. This has necessitated the development of specialized mathematical and statistical techniques (i.e., multilocus linkage analysis) to extract the maximum amount of information from the data.

Convenient for mapping purposes are restriction fragment-length polymorphisms (RFLPs), DNA sequence variations, which affect the length of particular genomic DNA restriction fragments, either by nucleotide substitution within a restriction enzyme recognition site or insertion or deletion between sites (e.g., minisatellites). RFLPs are scored by hybridizing a probe which detects the variable fragment to Southern blots of genomic DNA digested with the appropriate restriction enzyme. RFLP linkage maps with an average spacing of 5–10 cM between sites have been constructed for each human chromosome. These maps have revealed several interesting facts: Crossing-over occurs about twice as frequently in female meiosis as in male meiosis, with the exception of certain regions (found mostly near the ends of chromosomes) where males have a higher rate. Polymorphisms are unevenly distributed in the genome, the most highly polymorphic loci more often found near the telomeres than in other parts of the chromosomes, and on the autosomes than on the X.

2. Cytogenetic maps are based on the discovery that preparations of metaphase human chromosomes stained with certain dyes (e.g., Giemsa) reveal a characteristic pattern of alternating light and dark bands for each chromosome, a total of about 300 bands in the genome as a whole (Fig. 2). The physical basis for this differential staining is unclear, although Giemsa light (R) bands are known to be relatively more C/G rich than the Giemsa dark (G) bands. Medical genetics has benefited enormously from the use of cytological staining to identify and classify mutations due to gross chromosome rearrangements, such as translocations (in which a part of one chromosome is joined to another), deletions, and

aberrations in chromosome numbers (such as trisomy 21, the basis of Down syndrome).

This ability to identify individual chromosomes and portions of chromosomes has led to several useful mapping techniques. Somatic cell hybrids, fusions of a human cell with a rodent cell, preferentially lose human chromosomes during replication, so that after several cell divisions only one or a few human chromosomes are retained with the rodent chromosomes. The chromosome containing a particular human DNA segment can often be identified by hybridizing a probe for that segment to DNA prepared from a panel of such hybrids retaining different human chromosome complements. In an important extension of this technique, radiation hybrid mapping, the human chromosomes are fragmented by radiation prior to the cell fusion; hybrid panels constructed in this manner allow the mapping of DNA segments relative to each other within a chromosome.

In situ hybridization of a radioactively labeled human DNA probe to metaphase chromosomes can be used to localize the probe sequence to a specific chromosomal band (on average, a region of about 10 million bases). Recent fluorescent labeling techniques permit the simultaneous visualization of several nearby probes, which allows higher-resolution ordering of probes within a band.

3. Restriction maps indicate the order and distances [usually measured in kilobases (kb), or thousands of nucleotides] between restriction enzyme sites in a particular genomic region. Restriction enzymes with recognition sequences of 4–6 bp generally produce genomic fragments averaging a few hundred to a few thousand base pairs in length and are used to construct high-resolution restriction maps of regions up to 30 kb or so in length. These maps are usually acquired primarily from analysis of restriction digests of small overlapping clones spanning the region in question.

In contrast, long-range restriction maps spanning several million base pairs rely on the use of a small number of restriction enzymes (having recognition sequences of eight or more nucleotides in length and/or containing the rare dinucleotide CG) which produce much larger fragments, with a size of several hundred kilobases or larger. These maps are usually constructed by hybridizing probes from the region to Southern blots of pulsed-field gels to identify the sizes of fragments containing the probe sequence.

4. Overlapping clone maps, sets of DNA clones which cover particular genomic regions, are particularly useful because they provide a replica of the genome in pieces of manageable size. They are usually constructed in either of two ways, both involving the use of clone libraries: fingerprinting, which involves generating characteristic sequence or restriction site data (e.g., the fragment sizes for a particular restriction enzyme) for each clone in the library and identifying overlaps via shared portions of the fingerprint patterns; or library screening, using DNA probes from the region, with overlaps now implied by the presence of shared probe sequences. The latter method includes as a special case chromosome walking. In that method, as clones are added to the map, new probes are generated from the ends of the cloned segment and used to rescreen the library for the next neighboring clones.

Overlapping YAC clone maps are presently being constructed using both methods for a number of genomic regions.

C. How Genes Are Found

Once a protein has been isolated, cloning the corresponding gene is relatively straightforward. If part of its amino acid sequence can be determined, oligonucleotides corresponding to part of the gene sequence can be synthesized (using the genetic code) and used to probe a genomic or cDNA library. Alternatively, if the protein can be detected *in vitro* (e.g., via an antibody which binds the protein), it can often be found in a cDNA expression library without having to determine the amino acid sequence first.

When the protein is not directly detectable—for example, if it is known only through the existence of a genetic disease—finding its gene is much more difficult. The approximate location for disease genes can often be identified using the techniques of linkage mapping to find a polymorphic DNA locus which tends to be coinherited with the disease gene in families with the disease. Having identified the chromosomal location of the polymorphic locus, it is then necessary to search through cloned DNA from this region to find the gene. Most successful such searches to date have depended on the existence of more specific clues to location, such as chromosomal rearrangements (i.e., translocations or deletions) disrupting the gene in some individuals with the disease. In the absence of these, it becomes necessary to identify candidate genes within the region of interest (by screening cDNA libraries for transcribed sequences or by identifying sequences which are evolutionarily conserved among several organisms), sequence them, and look for differences between the sequences of affected and normal individuals. This is rendered more difficult by the necessity to distinguish polymorphisms from the disease mutation, but has been successful in a few instances. The availability of overlapping clone maps covering most of the genome should greatly aid such searches.

V. Human Genome Project

In recent years it has become apparent that a systematic approach to finding all human genes would have many benefits. Present cloning approaches, despite impressive successes, have yielded only a few percent of the estimated 100,000 genes. Availability of the entire genome sequence would greatly accelerate progress: Once the sequence of a small part of any particular gene were determined, the rest would be immediately accessible. Furthermore, it is likely that there are many genes or other functionally important regions of the genome whose presence is totally unsuspected, and which therefore might not be accessible to current cloning efforts.

For these reasons determination of the complete structure of the genome has become the subject of a major initiative which is both international and interdisciplinary in scope, involving input from mathematicians, computer scientists, engineers, chemists, and physicists as well as biologists. As currently envisaged, this genome project will have several goals:

1. A high-resolution (i.e., average spacing of 2–5 cM) linkage map of the genome, to facilitate the mapping of disease genes and provide landmark sites of known chromosomal order for use in constructing physical maps. This will be facilitated by the use of a common panel of families for inheritance studies, which permits researchers from different laboratories to combine the data for different polymorphic sites into a single map. The DNA from such a panel of 60 large three-generation families is currently being distributed by the Centre d'Étude du Polymorphisme Humain.

2. Physical maps of several types, including radiation hybrid maps, long-range restriction maps, *in situ* hybridization maps, and overlapping clone maps. These disparate maps will be merged into a single common map giving the positions of sequence tagged sites (STSs), short regions consisting of a few hundred bases of known sequence, for which PCR assays have been developed. A set of STSs of known order, spaced at approximately 100-kb intervals along each chromosome, will make any genomic region immediately accessible (by providing PCR probes for screening genomic libraries). The STS concept provides a common language for physical mapping, much as the use of a shared set of reference families provides a common language for linkage mapping.

3. Determination of the 3 billion nucleotide genomic sequence. This will most likely depend heavily on sets of overlapping clones constructed in Goal 2 and will require substantial prior technological research to increase the speed and decrease the cost of current sequencing methods. Since polymorphic variation implies that there is no unique human sequence, it would also be desirable to have at least a partial catalog of the polymorphic sites.

4. It is important to recognize that the sequence is not an end in itself, but is only useful insofar as we can identify the genes and find out what the corresponding proteins do in the cell. Gene identification directly from sequence data is difficult due to the interrupted structures of most genes and the fact that a relatively small proportion of the genome is in coding regions. Sequencing the genomes of several "model" organisms (including the yeast *Saccharomyces cerevisiae*, the nematode *Caenorhabditis elegans*, the mouse *Mus musculus*, and the fly *Drosophila melanogaster*) will provide a useful resource for identifying the functionally important regions (many of which will be evolutionarily conserved and thus have similar sequences in several species) and will allow experimental manipulations to test hypotheses about gene function. These and other approaches to determining gene function can be expected to occupy human biologists for many years after the entire human genome sequence is known.

Bibliography

Committee on Mapping and Sequencing the Human Genome (B. Alberts, Chairman) (1988). "Mapping and Sequencing the Human Genome." Nat. Acad. Press, Washington, D.C.

Donis-Keller, H., Green, P., Helms, C., Cartinhour, S., Weiffenbach, B., Stephens, K., Keith, T., Bowden, D., Smith, D., Lander, E., Botstein, D., Akots, G., Rediker, K., Gravius, T., Brown, V., Rising, M., Parker, C., Powers, J., Watt, D., Kauffman, E., Bricker, A., Phipps, P., Muller-Kahle, H., Fulton, T., Ng, S., Schumm, J., Braman, J., Knowlton, R., Barker, D., Crooks, S., Lincoln, S., Daly, M., and Abrahamson, J. (1987). A genetic linkage map of the human genome. *Cell* **51,** 319–337.

Lewin, B. (1987). "Genes III." Wiley, New York.

Sambrook, J., Fritsch, E. F., and Maniatis, T. (1989). "Molecular Cloning: A Laboratory Manual," 2nd ed. Cold Spring Harbor Lab., Cold Spring Harbor, New York.

Vogel, F., and Motulsky, A. G. (1986). "Human Genetics: Problems and Approaches," 2nd ed. Springer-Verlag, Berlin and New York.

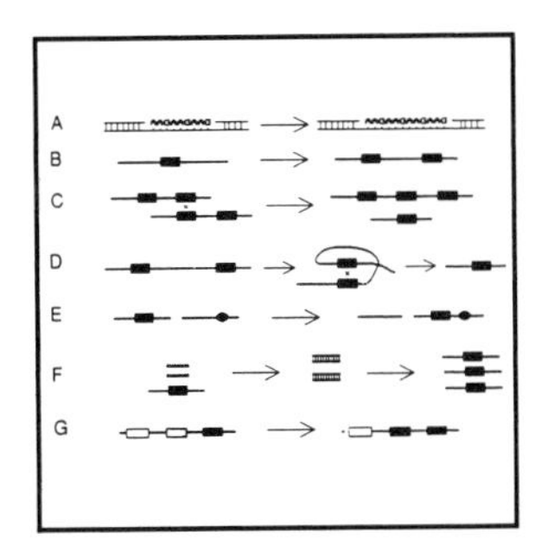

Genomes in the Study of Primate Relations

JONATHAN MARKS, *Yale University*

Glossary

Evolution Descent and divergence with modification

Homology Relationship between parts of two different organisms, due to their descent from a common ancestor

Homoplasy Similarity between parts of two different species, not attributable to common ancestry, but to parallel evolution (e.g., adaptation to a similar environment or a similar mutation)

Natural selection Differential survival and reproduction of organisms with particular characters. Selection for the mean value or "normal" character is called stabilizing or purifying selection; selection away from the mean or norm is called directional selection.

Phylogeny Historical relationships among a group of organisms

STUDIES comparing the genetic structure of species have been of great use in the study of physical (or biological) anthropology, the study of our place in nature. Genetic studies provide an independent test of phylogenetic hypotheses derived from anatomical studies. These are most often useful when anatomical studies have been ambiguous. Nevertheless, it is not easy to interpret comparisons among genomes, since their evolutionary rates and modes are diverse.

I. Introduction

For most of the 20th century, there had been a considerable diversity of opinion as to the relationships of humans to the great apes (chimpanzee, gorilla, and orangutan). Genetic studies based on the intensity of immunological cross-reactions of blood serum, which inferentially measured the similarity of antigenic proteins among species, which was a reflection of genetic similarity, were performed in the early 1960s by Morris Goodman. These showed that the closest genetic similarity existed among the human, chimpanzee, and gorilla—and that the category "great apes" was therefore phylogenetically artificial, because the chimpanzee, gorilla, and orangutan were not genetically most similar to one another, and therefore probably not one another's closest relatives. Thus, a study of molecular relationships was able to resolve a phylogenetic question derived from anatomical ambiguities.

According to the modern theory of evolution, evolutionary change is the genetic transformation of populations of organisms through time and across space. Anatomies (i.e., phenotypes) are the external products of genomes, to a considerable extent. Whereas anatomies are highly responsive to environmental conditions and can be molded by natural selection in similar ways in different lineages, genomes, for the most, part are not. Thus, the phylogenetic "noise" of parallel evolution should be reduced if one compares genomes, rather than phenotypes. Genetic phylogenies are therefore sometimes thought to represent truer histories than anatomically based phylogenies. This is quite naive, however, as the evolutionary processes operating within the genome are now known to be quite complex, poorly understood, and often yield patterns that are difficult to interpret.

Evolutionary studies of the genome transcend the difficulty of parallel evolution (i.e., homoplasy)

caused by similar responses to the environment in different lineages, yet have several difficulties of their own. First, only four nucleotides are possible at any given position (a deletion could be considered a fifth), so the universe of possibilities is limited—a mutation from A to G at any given position is likely to occur many times in different lineages. Second, the genome is filled with redundancy, which can complicate genetic comparisons among species, especially in gross comparisons such as DNA hybridization (see below). Any DNA segment, therefore, has many homologs in another species. In anatomical evolution this is called serial homology; in genomic evolution it is called paralogy. [*See* GENOME, HUMAN.]

The concept of homology is critical in evolutionary studies. Two structures are homologous if they are descended from a common ancestor. The adult α_1-globin gene in humans (Fig. 1) is homologous to the adult α_1-globin gene in gorillas; that is, they are both descended from a common ancestor: the α_1-globin gene of a Miocene ape such as *Proconsul africanus*. The human α_1 gene, however, is also homologous to the α_2 gene alongside it; that is, they are both descended from a common ancestor, an ancestral α-globin gene present in an Oligocene catarrhine such as *Aegyptopithecus zeuxis*. In the evolution of the monkeys and apes, a duplication of the α-globin gene occurred. As shown in Fig. 1, several duplications had already preceded that one. These duplicates are said to be paralogous DNA segments, or paralogs; the particular α-globin genes across species are said to be orthologous DNA segments, or orthologs.

Homologies within species far outnumber those across species in the comparison of any pair of genomes. There are, for example, at least 14 known homologies of the human blood globin genes. In humans there are also hundreds of thousands of homologies of the *Alu* repeats, short (i.e., about 300-bp) DNA sequences interspersed in the DNA of primates.

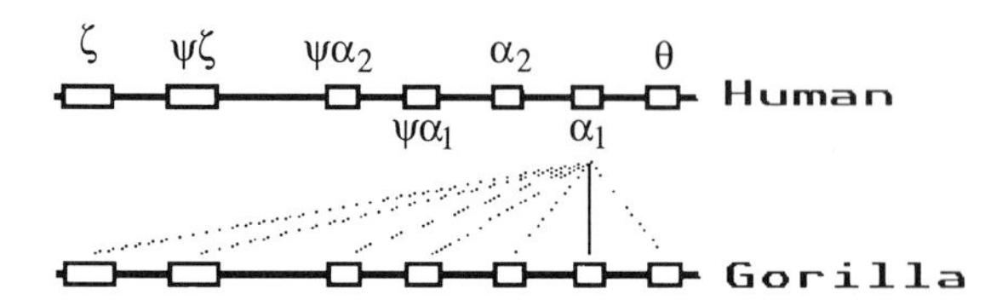

FIGURE 1 Patterns of homology in the genome. In the α-globin gene cluster of the human, each gene is paralogous to every other. The human α_1-globin gene is orthologous to one gene in the gorilla (solid vertical line) and paralogous to several others (dotted lines).

Another difficulty in the study of phylogeny from genomes is the problem of typology (i.e., judging the characteristics of a population by a single individual). Anatomists often have knowledge of the variation in the characters they study, in order to gauge whether or not the character in question actually differentiates two populations from one another, or is merely a reflection of the individuality of the specimen being examined. Genomic studies, particularly DNA sequencing, are sufficiently labor intensive that one is often forced to generalize about the genome of a species, despite the knowledge that the information has only been obtained from a single individual and there are likely to be polymorphisms in the population.

II. Genomic Evolutionary Processes

A number of diverse processes are now known to be operating within the genome, creating diversity within and across species. These modes of mutation can analytically be divided into three kinds: point mutations, genome mutations, and chromosomal mutations. Generally, point mutations tend to accumulate at a fairly regular tempo over time and are the classical mutations familiar to those who have studied genetics in the 20th century. Genomic mutations are responsible for the pervasive paralogy in the genome, and both these and chromosomal mutations seem to accumulate irregularly in a lineage.

A. Point Mutations

Point mutations are changes which can occur to a single nucleotide or to adjacent nucleotides in the DNA molecule. Most often, these are the substitution of one nucleotide for another, but they might also be the deletion of nucleotide or the insertion of a nucleotide. These could have a major phenotypic effect, or none at all, depending on where the mutation occurs. A point mutation can occur within a gene or between genes (i.e., in intergenic DNA). Within a gene the mutation can occur within coding or noncoding regions. Within coding regions the mutation can result in the coding of a different amino acid or can be a silent mutation, with the same amino acid inserted in the protein.

Natural selection operates on organismal phenotypes and allows favorable phenotypes to proliferate at the expense of alternative phenotypes. Because all organisms are the products of eons of natural selection, most random changes to an organismal phenotype are injurious and are therefore ''weeded out'' by natural selection. When we compare DNA sequences among species, therefore, we find that coding sequences are always the most similar, as they are most strongly subject to natural selection. Genic noncoding DNA is next most similar across species, and intergenic DNA is the most divergent, presumably virtually invisible to natural selection, as it produces no obvious phenotype.

This highlights an important difference between evolution at the anatomical and genomic levels. In anatomical evolution one expects to find *similarity* among related species (since heredity is fundamentally conservative), and one studies their differences, which are explained by directional natural selection, or adaptations to diverse environments. On the other hand, since so little of the genome interacts with the environment, the accumulation of DNA divergence over time is guided principally by genetic drift, a statistical property of heredity in populations. One therefore expects to find *differences* among DNA sequences of related species, and is instead struck by similarity, which is then attributed to purifying natural selection, indicative of a coding sequence with an important function. Indeed, the homeo box, which may be crucial in the developmental genetics of complex organisms, was recognized for precisely this reason: a high degree of conservatism between flies and humans, which indicated great functional significance for the region.

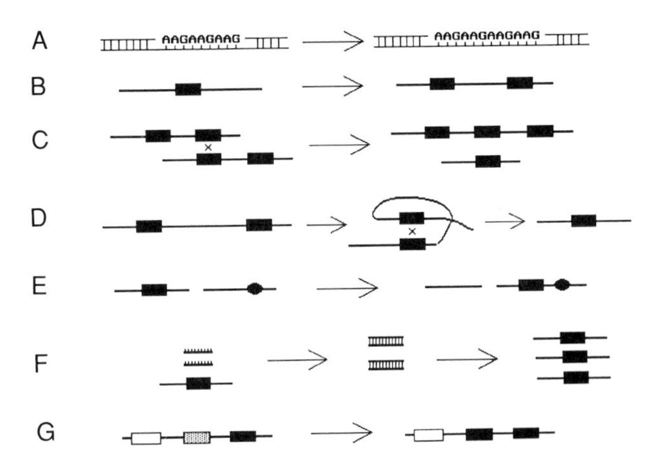

FIGURE 2 Modes of genomic mutation. (A) Strand slippage. (B) Tandem duplication. (C) Unequal crossing-over. (D) Intrachromosomal recombination. (E) Transposition. (F) Retrotransposition. (G) Gene correction.

B. Genomic Mutations

While point mutations formed the bedrock of evolutionary genetic theory throughout most of the 20th century, molecular genetics has uncovered a series of other mutational modes which act to differentiate the genomes of individuals, and ultimately the genetic compositions of populations. These are illustrated in Fig. 2 and discussed below.

Strand slippage involves a mistake on the part of the DNA polymerase molecule during DNA replication, causing the insertion or deletion of a short series of nucleotides. This occurs most often during a run of identical nucleotides, or short stretches of tandem repeats, and results in the deletion or insertion of a block of DNA—one or a few tandem repeats, or nucleotides.

Tandem duplication involves the insertion of a copy of a DNA segment adjacent to the original. This ''rubber stamp'' process is the ultimate origin of multigene families. There are three evolutionary fates for a gene duplicate. First, it can continue to make the same gene product as the original, augmenting the function of the first gene. Second, it can accumulate mutations that alter the gene product and its regulation, so long as the original remains intact. In this way the copy can come to take on a specialized function in a particular tissue or at a particular developmental stage. Third, the copy can accumulate mutations that render it inactive; in this case it is called a pseudogene.

Unequal crossing-over is a meiotic consequence of repeated DNA segments clustered together. While crossing-over is part of the normal process of meiosis, recombining paternal and maternal alleles on a single chromosome, this ordinarily occurs between orthologous DNA segments. If there are tandemly duplicated blocks of DNA, the meiotic machinery can temporarily mispair the chromosomes long enough for a crossing-over to occur between paralogous DNA segments. The evolutionary consequence is a duplication and reciprocal deletion in different gametes.

Intrachromosomal recombination can occur as a consequence of tandemly duplicated segments. A loop formed by a single DNA strand folding back on itself can permit the crossing-over process. This would delete the DNA in the loop from the chromosome.

Transposition is the insertion of a piece of DNA into a new location. This is a common property of the relationship of viral DNA to eukaryotic genomes. Here, the genetic element itself is movable, or transposable.

Retrotransposition involves the transcription of an RNA molecule from a DNA template, reverse transcription of the RNA into DNA, and insertion of the DNA copy elsewhere in the genome. The genome can thus be flooded by such RNA-derived copies, and this is the apparent source of the *Alu* repeats. Additionally, some processed pseudogenes appear to be formed in this manner.

Gene correction is a poorly understood process by which duplicate copies of a DNA segment are kept identical to one another. Usually, these duplicates are adjacent, as in the α_2- and α_1-globin genes, but they can also lie on different chromosomes (e.g., the genes for ribosomal RNA).

C. Chromosomal Mutations

The third mode of change to be considered is that which can occur to entire chromosomes, or large segments, and is thereby visible by the light microscope. Balanced changes do not affect the overall quantity of genetic material, but merely rearrange it; these occur and are readily visible among the phylogenetic relatives of humans. Most significantly, they do not appear to have an effect of the organismal phenotype. Balanced chromosomal changes seem to reorder large blocks of genetic material, without generally affecting the function of the genes involved. Unbalanced changes, on the other hand, add or delete genetic material and usually results in pathological phenotypes, unless the DNA in question is genetically inert. These chromosomal changes are of four principal kinds (a fifth, polyploidy, the duplication of entire chromosome sets, does not play a role in mammalian evolution, but is significant in the evolution of other groups, and probably occurred in the very remote ancestry of our lineage). The four types are discussed below. [*See* CHROMOSOME ANOMALIES.]

Inversions result from the breakage of a chromosome in two places, a reversal of polarity of the segment, and subsequent fusion of the breaks. These can often be inferred when comparing the chromosomes of the apes and humans.

Translocations result from the breakage of two different chromosomes and the subsequent improper rejoining of the segments. These are the most frequent chromosomal aberrations detectable in human populations, but are rarely detectable across primate species, since they usually have negative effects on the fertility of the bearer.

Fusions and fissions reduce or enlarge the characteristic number of chromosomes per cell. In the human lineage a fusion of two small chromosomes has occurred, creating a large chromosome (i.e., chromosome 2) and reducing the human complement of chromosomes from 24 pairs to 23 pairs.

Heterochromatin amplification or reduction is the expansion or contraction, respectively, of darkly staining material, usually associated with highly redundant satellite DNA. In humans these tend to vary in size, but not location, whereas across species they can be quite labile. In humans this material exists at the centromere of all chromosomes, as well as below the centromere of chromosomes 1, 9, and 16 and at the distal end of the Y chromosome. Chimps and gorillas, by contrast, have this heterochromatic material at the tips of most of their chromosomes. Because this probably represents nonfunctional DNA, it has an evolutionary role in spite of being an unbalanced change.

III. Human Evolution at the Genome Level

Comparisons of the genome have yielded some fascinating, and sometimes counterintuitive, inferences about the recent genetic evolution of humans. In the 1960s and 1970s a number of genetic studies (generally based on protein similarity, as a reflection of gene function and structure) showed that, contrary to the phenotypes, in which humans are clearly different from chimpanzees and gorillas, the genomes of humans were not clearly different. Indeed, the genetic distances that could be measured invariably showed that the distances among the human, chimpanzee, and gorilla were both very small and sufficiently similar to one another as to preclude a clear phylogenetic inference. The split among the three species thus came to be regarded as a genetic "trichotomy," the orangutan lying outside this.

A simple deduction from these observations is that when a small portion of the genome is examined, the sampling error is probably great enough to render any apparent "breaking" of the trichotomy dubious. Indeed, some genetic features can be invoked to link the human and the chimpanzee (e.g., the form of human chromosome 9 and its homologs), the human and the gorilla (e.g., the form of the Y chromosome), or the chimpanzee and the gorilla (e.g., heterochromatin at the tips of the tips of the chromosomes). The few existing DNA sequences of these species have found some nucleotide substitutions that appear to link each pair of species. The DNA sequence from the β-globin cluster overall appears to favor a human–chimpanzee link, while the DNA sequence for involucrin (a protein made by a class of skin cells) appears to favor a chimpanzee–gorilla link, as do the phenotypic data.

More significant than the branching sequence, however, is the recognition that all of the human uniquenesses—habitual bipedalism, brain expansion and language, culture, canine tooth reduction, loss of fur, concealment of ovulation in females, elaboration of the precision hand grip (e.g., holding a pencil), etc.—were accompanied by very little in the way of genetic change. In fact, all of these phenotypic ways in which humans differ from chimpanzees were accompanied by the same amount of genetic change that accompanies the phenotypic ways *gorillas* differ from chimpanzees; and it is difficult to name many of these, aside from body size.

There is no question that genetic change causes evolutionary change, so the human uniquenesses are caused by genetic changes, but these genetic changes are relatively quite few in number. Yet not one was probably a macromutation: a pathological hairless child with a big head and language, born to an australopithecine mother. Rather, the changes in the phenotype per generation were almost certainly very minor. Since the linkage among genotypes, development, and phenotypes—the question of how one gets a phenotype from a genotype—is still one of the major unsolved problems in human biology, this inference is not threatening to the theoretical structure of evolutionary biology. Phenotypic changes, especially large ones, generally occur over the span of many generations, and involve changes in the action of several genes.

Because the structure of the functioning gene products appear to differ so little among humans, chimpanzees, and gorillas, it is inferred that the causes of the phenotypic differences lie in the regulation of the genes. Thus, the human uniquenesses are presumed to be attributable to largely the same gene products, but turned on and off at slightly different times, affecting the relative growth rates of body parts, and ultimately resulting in the classic human phenotype.

Another point worthy of consideration is that if the genetic difference between a human and a chimpanzee is small, and much smaller than the difference between the two species phenotypically, then the genetic differences between any two human groups must be much smaller still, and probably also much less striking than any phenotypic differences between these groups.

We can make some generalizations about the modes of evolution in the primate genome. First, the size of the mutation is not correlated with the magnitude of its phenotypic effect. Thus, a single nucleotide difference between two genomes might have no effect at all or might have a significant and perceptible effect. This is easily deducible from the study of genetic pathologies within the human species. When we extend this observation across species, it helps us account for the paradoxical observation that chimpanzees and gorillas are so extraordinarily similar to us genetically, but are very different from us phenotypically.

Second, the genomes of species can change from those of their relatives in quite significant ways in short periods. The functional significance of such changes is unclear, as they involve principally nongenic DNA. Thus, despite the extraordinary similarity of the functional genic DNA of the chimpanzee to the human already noted, chimpanzee's have been measured to have 10% more DNA per cell than humans (see Table I). The significance of this DNA, if any, is unknown.

Third, some of these genomic processes can be effective in creating reproductive barriers between populations, and therefore can be more important in speciation than in adaptation. A major role has been hypothesized for tandem duplications of repetitive

TABLE I Genome Size of Old World Monkeys and Apes, Relative to Humans

Group	Percentage of human genome size
Human (*Homo*)	100
Chimpanzee (*Pan*)	110
Gorilla (*Gorilla*)	102
Orangutan (*Pongo*)	117
Gibbons (Hylobatidae)	79.2–85.2
Cercopithecidae	
Colobinae	103–123
Cercopithecinae	
Cercopithecini	100–149
Papionini	98–107

DNA in speciation, a process which has been called "molecular drive."

Fourth, these modes of genetic evolution are qualitatively different, and are therefore difficult to compare quantitatively. Are two genomes that differ by four chromosomal inversions more similar than two genomes that differ by two translocations and the emergence of a novel repetitive sequence?

Indeed, attempts to quantify genomic differences can be misleading. For example, if we count the haploid chromosome number as a measure of chromosomal similarity, we would find that the human ($n = 23$) and the chimpanzee ($n = 24$) are as different as the black gibbon (*Hylobates concolor*; $n = 26$) and the siamang (*Hylobates syndactylus*; $n = 25$). However, this would mask the fact that all of the human chromosomes can be almost perfectly aligned against homologs from the chimpanzee, while few of the chromosomes from the two *Hylobates* species can be matched to one another. However, by other standards of genetic distance, the two gibbon species are no more divergent from one another than is the human from the chimpanzee. [*See* HUMAN GENOME AND ITS EVOLUTIONARY ORIGIN.]

IV. Rates of Molecular Evolution

The case of the two gibbon species having highly divergent chromosomes while being nevertheless closely related is an apparent paradox. Should not closely related species be similar in their genetic makeup?

In general, of course, they are. Nevertheless, there are numerous cases of discordant evolutionary rates for some of these genomic evolutionary processes, especially for chromosomal changes. Probably most extreme is a situation from the artiodactyls, in which two species of a deer known as the muntjak, similar enough to form viable hybrids, nevertheless have a diploid number of $2n = 46$ in one species, and $2n = 7$ in the male and $2n = 6$ in the female of the other species.

The rate of evolution of chromosomes is thus not well related to the rate of evolution of phenotypes. Two different but closely related species can have karyotypes that are virtually identical (as in the case of many of the baboons and the macaques), radically different (as among the gibbons and the guenons), or only slightly different (as in the great apes and the humans).

The reason for this variation in evolutionary rates is probably that changes in chromosome morphology affect the pairing of chromosomes in meiosis, and hence fertility. These changes are therefore likely to play a part in speciation (i.e., the formation of reproductively isolated communities) in some groups, while the phenotypes are directly concerned with adaptation to the environment. In groups that form new species via a chromosomal mechanism, chromosomal diversity between closely related species is relatively high, whereas among groups that form new species by other genetic means, the chromosomes are more similar.

Nucleotide substitutions are more useful for quantifying evolutionary divergence, since one can (in theory) simply count the number of differences among homologous parts of the genome. Here, most detectable nucleotide substitutions are involved in neither adaptation nor speciation; they are neutral, or nearly so. The rate of accumulation of neutral mutations within a given lineage should be roughly constant over time. Thus, given the amount of genetic difference between two species and an estimate of the rate at which neutral substitutions accumulate, one should be able to judge how long the two species have been accumulating mutations separately. That is, the number of nucleotide differences discernible between homologous DNA regions of two species should be indicative of the number of neutral nucleotide substitutions that have occurred over that region, which should be indicative of the time the two species have been diverging. This is the concept of the "molecular clock."

Like most generalizations in science, the molecular clock is valid only to a first approximation. One immediate difficulty is that to implement it in the comparison of two genomes from different species, one must ignore other evolutionary modes (e.g., strand slippage). Often in the comparison of several species' genomes, however, one encounters fairly clear-cut examples (e.g., a string of five adenines in a row in one species, homologous to six adenines in another species and eight adenines in yet another species). A second difficulty is that gene sequences are the least neutral parts of the genome, and thus have selection operating on them to a greater extent than do intergenic DNA sequences. Thus, one should examine intergenic DNA preferentially for these purposes. A third difficulty is that the more divergent two species are, the greater the likelihood that a single nucleotide difference reflects more than one nucleotide substitution. This is the problem of "multiple hits" at the same nucleotide site, which must be statistically accounted for.

For all of these caveats, the molecular clock appears to work, however crudely. Taking the above factors into consideration, a study of nearly 11 kb from a pseudogene in the β-globin gene cluster shows that the human DNA sequence is about 1.7% different from those of a chimpanzee and a gorilla; the same regions from these three species are about 3.5% different from the homologous region of an orangutan; the region from these four species is about 7.9% different from that of an Old World monkey, the rhesus macaque (a cercopithecid); and the region from these five species is about 11.9% different from that of a New World primate, the spider monkey. Although the rates of molecular change have varied somewhat, there is, nevertheless, a strong association between the amount of detectable genetic divergence and the recency of common ancestry.

V. Evolutionary Analysis of Genomic Data

An important component in using the amount of difference to infer phylogenetic relationships is to show that the rate of change is similar in the various lineages. This is accomplished by means of the relative rate test, shown in Fig. 3.

Consider three species: A, B, and C. The amount of difference between A and B is 2, and the amount of difference between B and C and between A and C is 4 (Fig. 3A). We would be tempted to infer that A and B are more closely related to one another than either is to C (Fig. 3B). However, this is valid *only* if the rates of evolution are similar in all of the lineages. If, on the other hand, C were evolving rapidly, then the amount of difference between C and anything else would be very high, regardless of the branching sequence. Thus, the hypothetical data just given are compatible with A and C being closest relatives, if C is accumulating differences much more rapidly than either A or B (Fig. 3C).

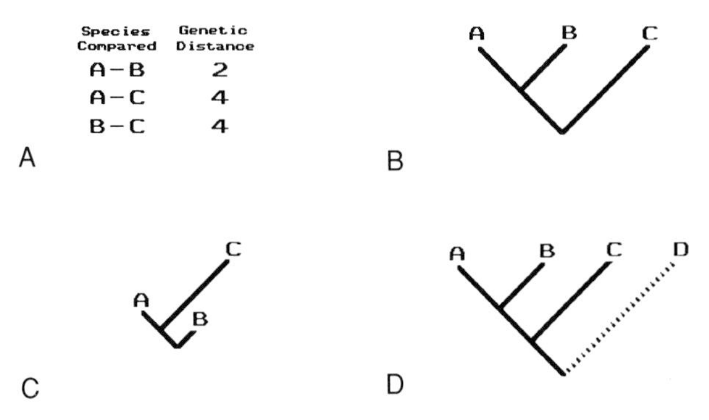

FIGURE 3 The relative rate test. (A) A set of genetic distances measured between pairs of species. (B) Their probable relationships, assuming equal evolutionary rates. (C) Their possible relationships, if the character being measured is evolving more rapidly in species C than in A and B. (D) Comparison of A, B, and C to an outgroup, D, can detect disparities in evolutionary rates across lineages.

The relative rate test invokes a fourth species outside of those being compared (call it D) and asks how different A, B, and C are from D. If C is evolving rapidly, then it should be more different from D than A or B is. If the genetic distances between A and D, B and D, and C and D are all very similar, then we can infer that there is no relative difference in the evolutionary rates among A, B, and C.

An alternative strategy to inferring phylogenetic relationships on the basis of similarity is to count the specific DNA changes themselves. Then, assuming that a nucleotide substitution is a fairly rare event, a tree is constructed that invokes the fewest nucleotide substitutions to account for the data. This is known as a parsimony method (i.e., the tree with the fewest substitutions is the most parsimonious), in contrast to a phenetic method, such as the one just described, in which overall similarity or difference dictates relationship.

Parsimony methods have their limitations as well. There is no guarantee that the most parsimonious tree is the correct one, for there is no guarantee that any particular nucleotide substitution is sufficiently rare as to preclude its having occurred independently in different lineages. Indeed, mutational "hot spots" do occur, and there appears to be a small but significant amount of parallel evolution (i.e., homoplasy) that confuses the fine-scale determination of branching sequences from DNA sequence data.

Since there is no guarantee that the most parsimonious DNA tree is the correct one, for fine-scale branchings one must often judge among the few most nearly parsimonious trees to find the one that is most concordant with other data sets. In other words, the closer the split among three species, the more difficult it is to tell which two are the most closely related, by any method.

What is quite extraordinary through all of these handicaps is that, for the most part, using a fairly small genetic sequence (e.g., hemoglobin) one can reconstruct with a high degree of accuracy the same phylogenetic relationships inferred on the basis of phenotypes. Where the phenotypic and genomic trees conflict in their fine structures, it is often difficult to tell which is wrong. Is the phenotypic tree mistakenly using a similar adaptation in distantly related species to infer phylogeny incorrectly? Or is the DNA segment under study not evolving in the most parsimonious manner, or perhaps not representative of the genome as a whole?

VI. DNA Hybridization

Another method for inferring phylogenetic relationships, based on a larger sample of the genome, uses the thermal stability of imperfectly paired DNA strands as a guide to how different the two strands are. If the two DNA strands come from different species, the DNA molecule is said to be a heteroduplex. Since mutations have accumulated along the lineages since the species diverged from one another, a heteroduplex molecule is held together with fewer hydrogen bonds than a homoduplex DNA molecule (i.e., one whose two strands are derived from the same species). Therefore, it should require less energy to break the bonds holding the heteroduplex together, and the strands should dissociate from one another at a lower temperature than the two well-paired strands of homoduplex DNA.

Hybrids can be prepared by isolating the least redundant half of the genome, shearing it to fragments of about 500 bp in length, labeling it radioactively, boiling it (to dissociate the DNA strands), then incubating it for several days with a large excess of single-stranded DNA from a different species. Labeled DNA segments are unlikely to pair with one another by sheer numbers, and therefore the only labeled double-stranded DNA to emerge from the incubation should be heteroduplex. The behavior of the heteroduplex DNA can be monitored by tracking the radioactivity.

As the hybrid DNA is heated, its strands dissociate (i.e., the DNA is said to be denatured, or melted). Plotting the amount of single-stranded DNA versus the temperature enables the calculation of a melting temperature characteristic of the DNA sample. The difference in the melting temperatures of heteroduplex and homoduplex DNA could be a measure of the genetic change that has accumulated since the two species diverged.

In practice, this method has not been as reliable as hoped, and small spurious differences—differences in DNA melting temperatures attributable not so much to a shift in the location of the melting curve as to statistical artifacts—are easily confused for real evolutionary differences. These artifacts are caused in part by the mispairing of DNA sequences due to the extensive repeats within the genome. Thus, in the α-globin cluster shown in Fig. 1, any of the paralogous genes could be pairing, however poorly, with any other paralog in the other species. Since the paralogous matches will be more divergent than the orthologous matches, a hybridization experiment with a lot of paralogous pairing can have an artifactually deflated melting temperature.

DNA hybridization has certainly supported the close relationship among humans, chimpanzees, and gorillas, inferred on the basis of narrower genetic similarities. Indeed, from the available data, it appears that the DNA analyzed from these three species are no more than 1–2% different from one another.

VII. Mitochondrial DNA and Human Origins

There is, in addition to the DNA located in the cell's nucleus, a minute amount of DNA in the mitochondria (i.e., cytoplasmic organelles involved in generating ATP for metabolic energy). In humans the mitochondrial DNA (mtDNA) is a compact molecule of about 16,500 bp in length, with none of the redundancy and excess characteristic of the nuclear genome.

mtDNA has several interesting properties. First, it does not undergo recombination; there is no union of maternal and paternal alleles, as in nuclear chromosomes. Second, mtDNA is inherited clonally through the maternal line. At the time of fertilization the zygote inherits mtDNA only from the egg, not from the sperm. Therefore, a child is a mitochondrial clone of its mother and unrelated to its father, in contrast to the situation in the nuclear genome, in which it is equally closely related to both. Third, it is relatively easily isolated, and is therefore readily amenable to study. Fourth, it evolves at a rapid pace—in humans about 10-fold faster than nuclear DNA—which makes it amenable to the study of relatively short spans of evolutionary time.

Applied to the trichotomy, mtDNA has been as ambiguous as the other genetic method. However, this rapidly evolving DNA was used to study a more recent evolutionary event: the divergence of human groups from one another.

Using the cut sites of various restriction endonucleases, the DNA sequences of mtDNA genomes from nearly 150 people from various populations were sampled. The most parsimonious tree found separated nearly all of the people in the sample, derived from all of the continents, including Africa, from seven African mtDNAs. It can be concluded that Africans have far greater diversity in their mtDNA genomes than do populations from other continents, which in turn implies that mtDNAs have been evolving longer in Africa than elsewhere. Thus, the root of the family tree of the human species should be placed in Africa. The deepest branch of that mtDNA tree was dated at about 200,000 years ago, on the basis of the amount of difference and the clock rate at which mtDNA diversity appears to be accumulating.

These conclusions were in considerable harmony with an interpretation of the fossil material advocated by several paleontologists. This interpretation sees modern humans as having evolved 100,000–200,000 years ago in Africa, then emigrating and supplanting anatomically archaic populations of Asia and Europe. Other interpretations of the fossil record have humans on each continent evolving largely *in situ,* effectively having diverged from one another hundreds of thousands of years earlier. The mtDNA data provide an independent test of these two paleontological hypotheses and support the former.

Bibliography

Cann, R. L., Stoneking, M., and Wilson, A. C. (1987). Mitochondrial DNA and human evolution. *Nature (London)* **325,** 31.

Cavalier-Smith, T. (ed.) (1985). "The Evolution of Genome Size." Wiley, New York.

Djian, P., and Green, H. (1989). Vectorial expansion of the involucrin gene and the relatedness of the hominoids. *Proc. Natl. Acad. Sci. U.S.A.* **86,** 8447.

Dover, G. A. (1982). Molecular drive: A cohesive model of species evolution. *Nature (London)* **299,** 111.

Koop, B. F., Tagle, D. A., Goodman, M., and Slightom, J. L. (1989). A molecular view of primate phylogeny and important systematic and evolutionary questions. *Mol. Biol. Evol.* **6,** 580.

MacIntyre, R. J. (ed.) (1985). "Molecular Evolutionary Genetics." Plenum, New York.

Marks, J. (1983). Hominoid cytogenetics and evolution. *Yearb. Phys. Anthropol.* **25,** 125.

Marks, J. (1989). Molecular micro- and macro-evolution in the primate alpha-globin gene family. *Am. J. Hum. Biol.* **1,** 555.

Pellicciari, C., Formenti, D., Redi, C. A., and Manfredi Romanini, M. G. (1982). DNA content variability in primates. *J. Hum. Evol.* **11,** 131.

Sarich, V. M., and Wilson, A. C. (1967). Rates of albumin evolution in primates. *Proc. Natl. Acad. Sci. U.S.A.* **58,** 142.

Sarich, V. M., Schmid, C. W., and Marks, J. (1989). DNA hybridization as a guide to phylogeny: A critical appraisal. *Cladistics* **5,** 3.

Stringer, C. B., and Andrews, P. (1988). Genetic and fossil evidence for the origin of modern humans. *Science* **236,** 1263.

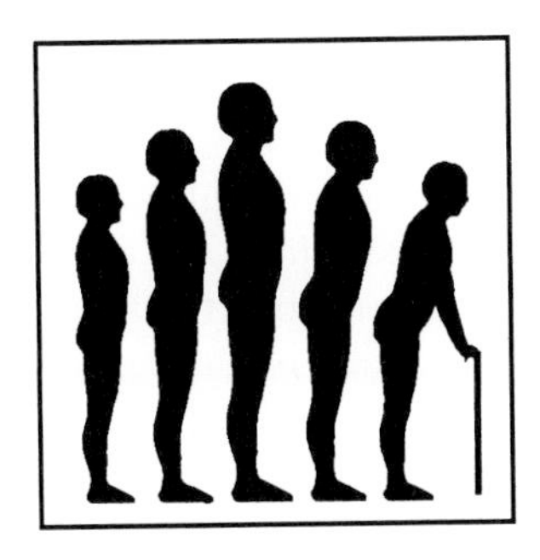

Gerontology

ROBERT M. BUTLER and BARBARA KENT, *Mount Sinai Medical Center*

Glossary

Age-dependent disease Disease that appears to be directly related to chronological aging (e.g., coronary artery disease, osteoporosis, Alzheimer's disease, and Parkinson's disease)

Age-related disease Disease related to aging without a direct time relationship (e.g., cancer: 50% of all cancers arise after age 65, 80% after 50, but there are suggestions of a decreased incidence after 80)

Aging Time-dependent biological changes occurring throughout the life span and leading ultimately to death. "Chronological age" refers to time since conception or birth. Physiological age is a ranking determined by individual performance based on normative group performance over the life span

Average life expectancy Years of life from birth to death, given current mortality rates from disease and accident; length of life of 50% of any birth group

Biomarker Biological property that changes with age and can be used to predict life span. A biomarker of aging is a biological measurement that changes with age

Life span Number of years from birth to death; average length of life for longest-lived individuals; species specific

Longevity Length of life

Senescence Postreproductive period

"GERONTOLOGY" is a 20th-century term applied to the study of the biology of aging by Elie Metchnikoff in 1907. Throughout human history how and why we live as long as we do have been questions of great fascination. The concepts of aging and longevity and the cessation of life (i.e., death) have shaped our conception of social organization, economics, religion, and philosophy. The biological tenets to explain aging, however, have remained elusive. Recently, the techniques of recombinant DNA, hybridoma, and transgenic animals promise application of the scientific method to the study of the basic mechanisms of aging. Today gerontology encompasses the study of aging from biological, social, psychological, economic, and, indeed, all perspectives. The biology of human aging is emphasized here.

Human aging is determined by the interplay of intrinsic aspects of aging and longevity with environmental factors. Appreciation of environmental influences on the kinds and rates of biological change in humans comes from systematic longitudinal studies of diverse human populations. Indeed, we know that the antecedents of health and disease include the environment, broadly defined, including human behavior as well as the physical and cultural environment, genetic factors, and aging. Of these three, research on aging has received the least attention, yet aging profoundly affects the character of health and disease with the passage of time. It is important in gerontology to separate, at least conceptually, aging and disease. Many diseases become more prevalent with age, and in the past some disease-related changes were erroneously attributed to aging. Challenges in studying normal human aging are proper screening and the elimination of disease from normative groups.

I. Biological Theories of Aging

A. General

Evidence of biological aging can be found in almost all forms of life, from the simplest to the most complex. Aging changes are well documented on a descriptive level: function *x* decreases, substance *y* accumulates, or structure *z* changes appearance in a characteristic way. Interestingly, time-related changes found in single-celled organisms seem to be preserved in evolution and are exhibited during aging in higher forms of life, including humans. The accumulation of the aging pigment, lipofuscin, is an example. Not only is the presence of lipofuscin a function of age, but the rate of accumulation in cellular vacuoles is proportional to the aging rate of the species. Thus, neurons from a 3-year-old (i.e., close to the end of the normal life span) rat are found to be 95% filled with the aging pigment, while humans must age many more years, to 70 or 80, before such quantities of lipofuscin accumulate in the neurons.

There are many other similarities in aging among species, including changes in cell membrane fluidity, cross-linking of macromolecules, mitochondrial size changes, and loss of fecundity. The underlying mechanisms of aging remain elusive, however. There is no unified single theory of aging, and it remains impossible to separate intrinsic aging processes from external (e.g., environmental and toxic) events. A number of theories of aging are currently under scrutiny. These theories can be roughly divided into two types: determinant and stochastic, or random.

B. Determinant

The determinant theories hold that aging is an orderly programmed process, orchestrated by genetic information, which is played out at a predetermined time to govern the steps in the aging process and, ultimately, the time of death. The evidence that genetics plays a role in aging is overwhelming. An obvious example from nature is the species-specific life span. By definition species are kept reproductively distinct by virtue of the uniqueness of their genetic pool. The variance in life span between individuals within a species is small compared to the variance in life span among species. Within a group of mammals, for example, there is great diversity in life span, the shortest-lived species (the shrew) living about 1.5 years and the longest-lived (the elephant and the human) living to over 80 years.

This observation can be brought into sharper focus using the inbred rodent strains which have strain-specific life expectancies and life spans. The individuals of an inbred strain are genetically identical and might differ from another strain by only a few alleles, or in some cases only one allele. Nevertheless, these minute genetic differences are reflected in significant changes in the mortality statistics and life spans among strains. Along the same line hybrid vigor is a genetic phenomenon that commonly affects the aging process. That is, the F_1 generation between two inbred strains usually has a longer life span than either of the parent strains. Selective breeding techniques have led to strains of aging-accelerated mice. The problem in studying accelerated aging in these mice is the danger that the life-shortening genetic change is really a pathology-producing change in disguise.

Many other lines of evidence support the involvement of genes in the aging process. The fact that, in most species, the female lives longer than the male suggests either protection by a double dose of the X chromosome or lack of exposure to some deleterious effect of the Y chromosome. In addition, it is commonly found that ancestors of people living to 90–100 years had extended longevity, suggesting a heritable component to human aging. Likewise, in a fashion analogous to the inbred mice, identical twins are more similar in life span than fraternal twins.

These observations not withstanding, there is no particular group of people bound by culture, environment, or race which is longer lived than another. The reports of the mid-1960s and early 1970s of genetically isolated groups both in the Georgian area of Russia and in the Andies in Equador reaching very long life spans have not proved to be correct. Rather, it was found that age is venerated in these societies and that overestimating one's age is a common occurrence.

There are broad questions such as whether genetic mechanisms really apply in aging. Sir Peter Medawar observed that there is no evolutionary significance to aging. He considered aging to begin after the reproductive period and therefore concludes that there is no selective pressure for aging. It could be argued, however, that in many of the longer-lived species and certainly in humans, the time involved in the rearing of the offspring creates a selective advantage for longer life. An offspring whose parents live long enough to nurture and protect it has a better opportunity of living to adulthood and continuing the gene pool than one orphaned because of the lack of longevous genes. Also contrary to Medawar's observation is the idea of the inertia of good health programmed by the genes. According to this hypothesis, the more successful the genes are in regulating the developmental phase of life leading up to the point of perpetuation of the species, the longer will be postreproductive period. In other words, genes conferring robustness and good health during development could have a selective advantage, allowing individuals with these genes to start the postdevelopment period in better health, so that the senescent period takes longer. [*See* GENES.]

Further credence for a determinant theory of aging comes from experiments using tissue culture, carried out by Leonard Hayflick in the 1960s. Events of the cell cycle are under exquisite genetic control. In order for cells to grow in tissue culture, they must continually divide by repetition of the cell cycle. Given the proper balance of nutrients and factors in the medium, fibroblasts can be grown in culture until they reach confluence. At this point they can be divided in half and replated on fresh culture vessels, where they again divide until confluence is reached. This process, called doubling, can be repeated until no more doublings occur and cells stop cycling and die. Cells are then said to have reached their replication, or Hayflick, limit.

The Hayflick limit has interesting gerontological implications because of the correlation between the maximum number of potential doublings of a group of fibroblasts and the life span of the donors. There is a direct proportionality between fibroblast doubling potential and life span for animals whose maximum life span ranges from 4 years (mice fibroblasts, with a low doubling potential) to 80–110 years (humans, with twice the number of doublings) to 175 years (the longest-lived species, the Galápagos turtle, with the greatest number of doublings). Of even more relevance is the observation that fibroblasts from young members of a species have a greater doubling potential than older individuals of the same species, suggesting the winding down of a clock, accumulation of a toxic substance, or depletion of an essential factor. Cells which normally divide in the body demonstrate a Hayflick limit. Cytosurgical techniques in which the nucleus of a young cell is put in an enucleated old cell and vice versa have led to the localization of factors controlling replication to the nucleus. Indeed, when cell cycle events go awry, as in cells transformed by oncogenes, the Hayflick limit is never reached (i.e., the cells become immortal).

The biology of cancer and aging raises significant and interesting questions. There are 30 known oncogenes, some of which are, in a sense, "antiaging" genes, or gerontogenes. The alteration in senescent human fibroblasts seems to be repression of specific protooncogenes along with multiple changes in gene expression. These observations support the view of a determinant role in the process of terminal differentiation and cell senescence. Cancer, then, might be considered "failed aging."

C. Stochastic

The stochastic theories of aging espouse the idea that aging is caused by a gradual accumulation of randomly occurring deleterious events. Although no one can deny the role of the genetic determinants in biological aging, proponents of the stochastic theories ascribe greater importance to the random effects of environment and random molecular events in the cell in causing aging. Generally, aging is thought of as a "wear-and-tear" process, leading ultimately to death. The stochastic theories can be subdivided into several interrelated aging theories, depending on the importance prescribed to the damaging element or to the part of the organism effected.

Denham Harman formulated the free-radical theory of aging in the mid-1950s. Free radicals are molecules to which an extra electron is attached. They are unstable and can react with almost any biomolecule to perpetuate free-radical reactions and cause damage. According to this theory, free radicals, generated during the course of normal metabolism or acquired from ultraviolet radiation or xenobiotics, cause random damage to cells and biomolecules, and this damage accumulates with time, leading to senescence and eventual death of the organism.

Several features of this theory make it appealing. First, environmental sources of free radicals are plentiful, and studies of lower organisms show a strong inverse relationship between free-radical exposure and life span. Of more interest from a biological point of view is the observation that 5% of the oxygen taken in to support aerobic metabolism is converted to free radicals (i.e., superoxide radicals). A corollary to this hypothesis is that animals that use large quantities of oxygen per gram of tissue have shorter life spans than those that use less. Indeed, an inverse relationship between specific metabolic rate of a range of mammals and maximum life span potential is found. Thus, the hyperactive little shrew whose heart rate can approach 780 beats per minute uses a relatively large amount of oxygen per gram of tissue and has a life span of around 1.5 years. The long-lived elephant has a metabolic rate about one-thirtieth that of the shrew.

Free radicals are too reactive to be measured directly. Their presence can be implied by measuring markers, or "footprints," of free-radical damage. Lipofuscin is thought to be a free-radical footprint generated from free-radical reaction with phospholipids through the intermediate malondialdehyde. Indeed, ethane and pentane gas in expired air indicate lipid peroxidation secondary to free-radical damage. DNA is also a target molecule for free-radical damage. Because DNA repair is a high priority to the cell, the best indication to date of free-radical damage to DNA must be measured indirectly by the accumulation of thymidine dimers in the urine. There is an inverse relationship between these urinary markers of free-radical damage and life span.

Free radicals are generated in cells by a number of intracellular processes in addition to respiration. Prostaglandin synthesis, cytochrome *P*-450, and the inflammatory reaction of macrophages are a few of the sources of intracellular free radicals. Both endogenous and exogenous antioxidants limit damage from free radicals. The genome of animal cells encodes a protein, superoxide dismutase, that scavenges superoxide free radicals and dismutes them to hydrogen peroxide. Hydrogen peroxide is also quite reactive and is probably most responsible for cellular lipid peroxidation. An additional genetically encoded protein, catalase, causes the degradation of hydrogen peroxide to water. When hydrogen peroxide is allowed to come into contact with iron, hydroxyl radicals are formed. They are an extremely reactive species with no known scavenger. Other free-radical scavengers include vitamin C, vitamin E, β-carotene, and urate.

The quantity of superoxide dismutase in cells from many different species varies directly with life span, suggesting that defense against free radicals impedes the aging process. This relationship does not hold up for other intracellular defense enzymes. However, autooxidation occurs much more rapidly in tissues from short-lived animals than from ones with longer life spans, indicating that overall protection from free radicals correlates with longer life. Experiments were performed in which groups of mice were fed substances thought to act as free-radical scavengers. In one of these studies, vitamin E was found to increase the life expectancy of a group of mice, but did not affect the maximum life span. It was felt that vitamin E might forestall age-dependent diseases, but does not have the hoped-for ability to extend life.

Most age-dependent diseases have a free-radical component in their etiology. Emphysema is a good example. Here, the compliance changes of the small airways seen in patients with genetic- or cigarette smoking-induced emphysema are similar to those found in older individuals. Presumably, the free radicals produced as a result of both smoke irritation with subsequent phagocytic activity in the lung and the direct inhibition of antiproteases from free radicals in smoke act in concert to speed up the aging process in the lung; that is, the higher dose of free radicals to which smokers expose themselves increases the rate of progressive accumulation of damage. That cigarette smoking exacerbates so many of the diseases of aging is, in itself, evidence in support of the free-radical theory. Other age-dependent diseases in which free radicals play a primary role include atherosclerosis, cataractogenesis, arthritis, Parkinson's disease, Alzheimer's disease, and osteoporosis.

The only reproducible method for extending life span in mammals is by caloric restriction. Known as the McKay effect, this has been demonstrated only for rodents. It might apply to longer-lived species, but experiments have not been performed to prove this. When mice or rats have limited access to calories, but are not malnourished or vitamin deficient, they live longer than *ad libitum*-fed controls of the same strain. Caloric restriction can significantly increase life span and therefore is thought to fundamentally affect the aging process.

Interestingly, decreased caloric intake also delays many of the physiological changes with age, as well as the age-dependent diseases thought to be related to free-radical damage. Among other salutary effects, caloric restriction slows lipofuscin accumulation, delays the time of onset of immune and reproductive losses, increases hepatic catalase activity, increases cell membrane fluidity, and delays the time of onset of cataracts and cancer. A tempting explanation for these observations is that caloric restriction retards aging and age-dependent diseases by decreasing free-radical generation or by increasing their dismutation. The hypothesis that caloric restriction extends life span by decreasing the specific metabolic rate has been tested and does not seem to hold up. In fact, there are no differences in metabolic rate between calorie-restricted and *ad libitum*-fed rats. Current work at the molecular level indicates an increase in endogenous free-radical scavengers in calorie-restricted rodents.

Other stochastic theories of aging are also based on progressive damage at the cellular level, but this might or might not require free radicals in the initiation event. L. Orgel's error catastrophe theory, proposed in 1963, has been, for the most part, disclaimed, but its heuristic value remains important. This theory holds that as cells age, errors in DNA, from cross-linking or free-radical damage, accumulate and contribute to an ever-growing repertoire of modified proteins until a fatal buildup of defects in cellular machinery leads to an error catastrophe and cell death. In testing this theory, biochemists have discovered that errors in DNA are so faithfully repaired that there is rarely any change in transcription translation through the life of the cell. The amino acid sequences in proteins from young and old cells are identical. With modern techniques to study conformational changes in proteins, however, it is becoming clear that proteins in older cells fold differently from their younger counterparts. This difference in conformation is postulated to be the result of longer transient times in the older cell.

Experiments using radioactively labeled protein markers in red blood cells showed a change in the way proteins are catabolized in older cells. Lysosomes of older cells are not as efficient at picking up and degrading proteins, thereby allowing them to accumulate in the cell. Not only does this allow time for proteins to assume new shapes, it also might overwhelm the cytosolic ubiquitin-dependent pathway for the degradation of smaller proteins. Therefore, while young and old cells synthesize proteins equally well, the ability of the cell to rid itself of proteins (the disposal system) could be more problematic in older cells. The altered shape of proteins, arising from longer cell transit times, could have further implications in the autoimmune diseases accompanying aging.

II. Biological Theories of Longevity

Biological theories of longevity emphasize the effect of coping paradigms on ultimate life span. For example, in clones of one-celled paramecium bombarded by ultraviolet light and allowed to recover under blacklight, treatment with sunlight induced an enzyme to repair the DNA damage. This gave rise to the idea of longevity assurance. A plethora of studies in many species ensued, showing a positive correlation between life span and the ability of an organism to repair damaged DNA. Likewise, cells from older animals have decreased repair abilities compared to cells from younger animals. In tissue culture cells nearing their Hayflick limit are unable to repair ultraviolet DNA damage as well as cells in earlier doublings. These studies support the importance of the repair DNA in the maintenance of life. [*See* REPAIR OF DAMAGED DNA.]

A single gene capable of increasing life span has been identified in the nematode, *Caenorhabditis elegans*. This tractable worm has long been a favorite model for studying the genetics of aging because of its two chromosomes mapped in exquisite detail and its similarity to other aging models; that is, the McKay effect holds true in *C. elegans* and there is timed somatic cell death during the life course of the animal, similar to the thymic involution seen in higher animals. By screening mutants for longer life, a colony was found that outlived controls by 70% in both life expectancy and maximum life span. Called the age-I mutant, the genetic locus has been mapped and offers exciting possibilities for the use of molecular biological techniques in discerning how the gene slows the aging process and prolongs life.

III. Human Aging

Common-sense observations have made clear the inevitability of decline and death in humans. It has only been since the 1960s, however, that the various origins of decline have been systematically differentiated. It is now known that much that had been attributed to human aging per se is due, instead, to a variety of other conditions, such as social adversity, disease, personality, poor conditioning, and toxins (e.g., cigarette smoke and alcohol). When criteria to exclude disease are carefully applied in the selection of a group of young and old people (i.e., a cross-sectional study) to be studied physiologically, declines previously thought to be unavoidably linked to age (e.g., the ability of the heart to increase cardiac output during exercise) are no longer demonstrable. Such studies, along with longitudinal studies now in progress, are leading to a revision of many stereotypes attributed to aging. Instead of a linear decline in function with age, a new rectangularized relationship is emerging in which individuals remain healthy and robust until close to the end of the life span. The length of the human "health span" as this century comes to a close is approaching the human life span.

The 20th century has seen an extraordinary increase in life expectancy: some 25 years in the industrialized world. This is nearly equal to what had been attained in the preceding 5000 years of human history. This new longevity is obviously a social achievement, not a function of biological evolution. The need to better understand aging because of the rising increase in the elderly population due to this longevity revolution was one of the driving forces for the 1974 Research on Aging Act, which created the National Institute on Aging. Established in 1975, the National Institute on Aging has led to a markedly increased investment in aging research in the United States.

Gerontological studies have pointed the way to interventions to extend the health span. Exercise is touted to mitigate many of the deleterious effects of aging. For example, exercise and calcium are used in the maintenance of bone density to postpone the onset of osteoporosis. Indeed, gerontology, in its applications in medicine and geriatrics, is poised to intervene more dramatically in human aging. The purpose of research into the biology of aging is to control, prevent, or reverse the factors which contribute to aging. Some age-related factors go beyond the standard definitions of disease and include overall functional status and factors that predispose to disease (e.g., osteopenia, age-dependent decreased bone density, which can result in osteoporosis or disease). This illustrates a problem in distinguishing the borderline between age and disease. The constant rise in systolic blood pressure with age at some point becomes a pathological entity called hypertension. Likewise, impaired glucose tolerance that seems to be a normal concomitant of age might cross the line to maturity-onset diabetes. [*See* BONE DENSITY AND FRAGILITY, AGE-RELATED CHANGES; HYPERTENSION.]

Normal aging is difficult to define for the human population. Even when measurable diseases are excluded as carefully as possible, there is a wide diversity in function with increasing age. In fact, the one constant in most clinical studies of functional aging changes is the increase in variance in the measured parameter with age. Whether it be oxygen consumption, renal clearance, glucose tolerance, or systolic blood pressure, to name a few that are measured cross-sectionally as a function of age, the overall pattern of the average can be in a deleterious direction, but invariably elderly individuals are found whose functions are no different from those of 20-year-olds. Did these elderly people not age as fast as their contemporaries, or did they start at a more advantageous level in youth for the measured function? Longitudinal studies are needed to answer this question.

Because aging is a ubiquitous biological phenomenon, many gerontologists have looked for explanations at the level of the cell to build arguments from single-celled organisms that apply to more complicated ones. The fundamental aging process in more complex organisms might not occur at the cellular level, but might be the manifestation of reaching limits in critical systems, which determines aging and life span.

One critical aspect of human aging is the decline in immune function. Starting with thymic involution beginning at puberty, there is a decrease in T cell function with age and a concomitant decrease in the control of antibody production. As the immune system declines with age to perhaps 20% of its peak function in adolescence, there is the prospect for intervention, should means of restoring or maintaining immune function be established. On the other hand, the decreased ability of the elderly to respond to antigen might be protective. Autoimmune diseases such as rheumatoid arthritis might be diminished by senescence of the immune system. Clearly, the pros and cons of interfering with immune senescence must be carefully evaluated. [*See* AUTOIMMUNE DISEASE; POLYMORPHISM OF THE AGING IMMUNE SYSTEM.]

The glucocorticoid cascade hypothesis of aging suggests that the emergence of hyperadrenal corticoidism, largely caused by stress as encountered during life, might be responsible for a host of age-associated problems. In initial observations using a rodent model, the recovery of circulating levels of corticosterone after a stress in older rats was found to be dramatically delayed from the recovery in younger animals. Older animals were also found to have higher basal levels of adrenocorticotropic hormone (ACTH), with an accompanying decreased sensitivity of the adrenal gland to ACTH stimulation. Insensitivity to negative feedback control and resultant hypersecretion occurs throughout the adrenocortical axis with age. This can be explained by the observation that specific groups of neurons in the hippocampus of aged rat brains are found to be deficient in corticosteroid receptors. It seems that during life, stresses accumulate until at some point, the down-regulation of corticosteroid receptors is great enough to interfere with hippocampal feedback inhibition, resulting in the hypersecretion of corticosteroids. Additional support for the glucocorticoid cascade hypothesis is found in an examination of the age-related diseases that have a significant component of hyperadrenocorticism (e.g., immunosuppression, muscle atrophy, arteriosclerosis, osteoporosis, and steroid diabetes).

The neuroendocrine system has been studied in relation to aging from the perspective of female reproductive capacity, developing a model of a negative feedback defect in some ways similar to the glucocorticoid hypothesis. Using the decrease in ovarian follicular number and its correlation with acyclicity as reproducible biomarkers of aging in rodents, the unexpected result is that transplantation of young ovaries into old mice ovariectomized in youth caused the initiation of estrous cycles, whereas young ovaries transplanted into old female mice who had aged with their ovaries intact did nothing to reinstitute the estrous cycle. It is postulated that the ovarian secretion, estradiol, is necessary for aging of the hypothalamopituitary axis necessary for the negative feedback control of the estrous cycle. [*See* NEUROENDOCRINOLOGY.]

Supporting this postulate is the observation that when young ovariectomized mice are given a short (i.e., 3-month) elevation of estradiol via their drinking water, cycling is maintained with difficulty after reimplantation of young ovaries, whereas young ovariectomized mice not treated with estradiol resume normal cycling when reimplanted with young ovaries. Further, several markers of reproductive aging, including glial hyperactivity in the arcuate nucleus, are accelerated by chronic exposure of young animals to estradiol and delayed in chronically ovariectomized animals. Glial hyperactivity is usually associated with damaged neurons, so this could indicate a cumulative effect of estradiol to slowly terminate hypothalamic negative feedback control over the estrous cycle. The deterioration of negative feedback homeostatic control systems due to accumulated damage from circulating steroids in other models of aging is similar.

Some of the theories of aging are more general, while some are more specific, for human aging. None is sufficient to explain aging adequately. The theories are not mutually exclusive; in fact, they are closely interrelated. Modern gerontology, then, has as its challenge the necessity of dissecting aging from disease and testing the theories of aging and longevity until a coherent scientifically tenable explanation for biological aging emerges. The payload for understanding human aging, of course, will be the enormous impact on health, health expectations, and our ability to help each individual reach maximum potential in terms of ability to contribute to society and to have meaning in life regardless of age. To conclude, pioneer American gerontologist Nathan Shock once said, "As we learn more and more about diseases and come to be able to control them, there won't be much left to old age except aging."

Bibliography

Finch, C. E., and Hayflick, L. (eds.) (1985). "Handbook of the Biology of Aging." Van Nostrand-Reinhold, New York.

Hall, D. A. (1984). "The Biomedical Basis of Gerontology." Stonebridge Press, Bristol, England.

Kent, B., and Butler, R. N. (eds.) (1988). "Human Aging Research: Concepts and Techniques." Raven, New York.

Warner, H. R., Butler, R. N., Sprott, R. L., and Schneider, E. L. (eds.) (1987). "Modern Biological Theories of Aging." Raven, New York.

Gerontology, Physiology

BERTIL STEEN, *Gothenburg University, Sweden*

Glossary

Aging Normal, physiological processes regarding structure and function with advancing age from the stage maturity is reached

Biological age Age of an individual in relation to his genetically determined life span

Biological aging Aging processes related to biology as opposed to psychological and social aging

Chronological age Duration of time passed since the individual was born

Functional age Age of an individual in relation to individuals of the same chronological age

Gerontology Multidisciplinary scientific field devoted to the study of biological, psychological, and social aging

THIS ARTICLE DEALS WITH structural and functional changes inevitably occurring with advancing age from the stage maturity is reached until the death of the individual.

Biological aging is conceptually a normal process and belongs to the field of physiology. Thus, this presentation is part of the gerontological arena, not of geriatric medicine.

However, the manifestations of aging are influenced by environmental factors such as diet, physical activity, smoking and abuse of alcohol, and disease, and the extent of such manifestations as well as the speed with which they appear are, thus, to a certain degree influenced by public health preventive activities.

Symptoms and signs seen in advanced age constitute a mixture of the manifestations of normal aging and those of disease, which makes it difficult to differentiate between health and disease in the higher age groups. Knowledge of gerontology is, therefore, a prerequisite to study and practice geriatric medicine.

I. Physiological Aging in General

Most functions of the human body show different phases throughout life. After conception and during early life there are phases best characterized as growth, maturity, and improvement. After these comes a phase of aging that goes on until the death of the individual.

Based upon mainly cross-sectional population studies this phase of aging has been described as continuous and accompanied by inevitably and constantly decreasing functional levels, this decline beginning shortly after maturity is reached.

Some functions clearly reflect this. For example, tissue elasticity in structures such as the ocular lens deteriorates very early in life. Other examples are perceptual speed (manifesting itself in for example increasing reaction time) and rate of oxygen consumption at rest; both these functions show an almost linear decline throughout life beginning as early as age 20–25.

However, recent longitudinal studies have shown a much more complicated picture. Some functions seem to be quite uninfluenced by age, such as the oxygen carrying capacity of the blood. Others do not change markedly until about 70–75 years of age, when the decline begins to be more obvious. This point where the functional curve will start to decline more rapidly is very different for different individuals and for different functions. Furthermore, the age-related changes may be very small in healthy individuals even at higher age groups, whereas deterioration seems to be much more marked in individuals with certain diseases. This is especially obvious regarding mental aging, in which memory and verbal and logical ability have been shown to be rather unchanged up to age 80 in healthy individuals, whereas patients with cardiovascular disease show marked deterioration. [*See* AGING AND LANGUAGE.]

The variability regarding different functions around the chronological age-related average values is very marked, and seems to increase throughout life. In other words, chronological age can be used with more accuracy in childhood and early adult life than in later life, where functional age is much more important. At the chronological age of 70 some functions show distributions of up to 20 years upwards and downwards in healthy individuals. This means, for example, that some healthy 70-year-olds have functional levels that are as high as those of some healthy 30-year-olds. Part of this variability certainly has genetic causes, but physical and mental training also play roles in this respect.

At least in developed countries, normal aging has fewer and less serious consequences to the average elderly person in spite of the fact that the prevalence rates of most diseases, somatic as well as mental, show a very marked increase from age 60 and upwards. However, some kinds of functional decline have practical consequences for most elderly people, such as the aging processes relating to visual acuity.

In general, those functions that presuppose some stress (such as mental functions demanding speed and physical functions demanding maximal performance) suffer more from age than the corresponding functions at rest. Furthermore, functions requiring coordination (such as neuromuscular precision) tend to be more dependent on age.

II. Aging at the Cellular Level

Tissue and organ aging are preceded by basal aging processes at the molecular and cellular levels.

Many studies in this field relate to the division potential of cells, where it has been said that the limited number of generations of normal cells is an expression of cellular aging. The conclusion from such studies is that the genome reorganization occurring during the division cycle creates a shift in cell functions, most markedly seen in postmitotic cells such as neurons and myocardial cells.

There is a decrease of the amount of functionally important DNA fractions in aging cells, while the total amount of DNA seems to be unchanged. Furthermore, damage occurs continuously in the DNA molecules, triggered in part by free radicals and peroxides. The efficacy of DNA repair processes in the cell is related to the life span of the species, and has also been shown in vitro to be lower with advancing age. A decreasing efficacy of the genetic programs can result in an increased production of defective proteins. [*See* REPAIR OF DAMAGED DNA.]

Cellular aging cannot be described as a general decrease of cellular activity, since the production of some substances increases while that of others decreases. Some authors claim that cellular aging is related more to lack of capacity to coordinate between synthesis and degradation of cell products than to changed production per se.

Intracellular deposition of waste products is characteristic in postmitotic cells, such as neurons and myocardial cells. One example is lipofuscin in secondary lysosomes, which can occupy a large proportion of the volume of the cells. Others are lipids, hyaline granulae, and vacuoles.

Free radicals, normally produced in aerobic metabolism, are toxic and might contribute to changes occurring with advancing age. There are, however, protective mechanisms, such as enzymes (for example, selenium-containing glutathione peroxidase). It is still unclear whether an increased administration of vitamin E, selenium, or other antioxidative substances influences the aging process in humans.

The mechanisms of cellular aging are still unknown, but the bulk of evidence points towards a combination of different processes.

III. Aging in Organ Systems

A. Central Nervous System

The distinction between normal aging of the brain and cerebral disease such as the age dementia syndromes—among them Alzheimer's disease—is not clear-cut, and the diagnosis of age-related cerebral disease is therefore difficult. Structural changes occur in both normal aging and disease and are often not *qualitatively* but only *quantatively* different.

[*See* ALZHEIMER'S DISEASE; DEMENTIA IN THE ELDERLY.]

With increasing age brain weight decreases by about 10% between age 20 and 80 in healthy individuals. However, variation is considerable. The atrophy is predominantly cortical, and is accompanied by a compensatory enlargement of the ventricles. Other structural age-related changes are loss of neurons, compensatory gliosis, accumulation of lipofuscin, senile plaques, neurofibrillary tangles, and granulovacuolar degeneration. The brain content of lipids, carbohydrate, protein, and minerals remains the same or declines only slightly, and the water content increases with age. Such changes, which to some degree correlate with the degree of neuronal loss and damage at the synaptic level, may make the brain increasingly susceptible to toxic substances and disease. A marked reduction in radioimmunoassayable peptide levels has been observed with advancing age. Among the substances particularly vulnerable to aging and influenced during disease are the glycolytic enzymes hexokinase and phosphofructokinase, and the different neurotransmitters. It has also been suggested that sympathetic hyperactivity in old age may interfere with cognitive function. [*See* BRAIN.]

Some deterioration of extrapyramidal function seems to occur with advancing age, which may be the explanation for the motor changes of gait and gestures sometimes resembling those of parkinsonian patients. Deterioration of most sensory systems is well known in the literature; so too are sleep disturbances, which often involve changes of the biological rhythm.

The psychological effects of normal cerebral aging are influenced by environmental processes. They can be influenced by mental training, and, furthermore, a large proportion of the psychological differences that have been found between younger and higher age groups can be partly explained by cohort changes.

Perceptual speed decreases with advancing age, and reaction time grows longer. This depends almost entirely on an increase in time required for central processing. Cognitive function very often shows no or very slight decrease with advancing age, and if so then late in life. However, the importance of general health to cognitive function is marked, and a "terminal decline" has been described during the two years before death. Most studies show a decreasing creativity during senescence. Short-term memory does not seem to show any obvious deterioration with advancing age, while short-term learning shows some impairment with age due in part to longer central processing time. [*See* LEARNING AND MEMORY.]

B. Cardiovascular System

Structural age-related changes in the heart include an increase of the amount of collagen and fat, especially in the septal and atrial regions, and sclerosis of valves. Lipofuscin accumulates with aging, leading to "brown atrophy." The sinus node shows a reduced number of cells and a relative increase of connective tissue and fat. Heart volume increases physiologically with age, at least up to age 75. In the vascular system deteriorating elastin function and an increase of the amount of collagen lead to increasing vascular stiffness.

The influence by age on resting heart rate is very little. There is, however, an obvious decrease in maximal heart rate response to physical exercise caused by a decreased automaticity of the sinus node. The disposition toward atrial arrhythmias in the elderly is increased. Cardiac output during exercise is lower than earlier in life both in the supine and sitting positions, while resting cardiac output is influenced in the supine position only. The most important explanation is a reduction of stroke volume at rest and of heart rate in exercise. An effect of these structural and functional changes is a decrease of oxygen consumption capacity during physical activity. Contributing factors have been suggested such as the known resistance to β-adrenergic mediated responses to exercise, and the increased vascular input impedance. [*See* EXERCISE AND CARDIOVASCULAR FUNCTION.]

The structural changes in the vascular system give rise to an increase of total peripheral resistance. Systolic and diastolic blood pressure will increase with advancing age as well as the pulse pressure. These blood pressure changes are seen at rest, but increase with physical exercise. Arterial blood pressure is also influenced by changes of water and electrolyte balance as well as of humoral factors.

Mean arterial blood pressure, thus, increases with age in most people despite the fact that the decrease in cardiac output and the increase in luminal volume of the great arteries work in the opposite direction.

C. Renal System

As in most organs there is also a reduction of cell mass in the kidneys. The loss of renal mass is particularly obvious in the renal cortex, and there is a loss of glomeruli. These changes are paralleled by a reduction of renal blood flow both in absolute figures and relative to cardiac output.

The glomerular filtration rate decreases in the order of 1% annually from age 50. Since skeletal muscle mass also decreases throughout life there are no significantly higher levels of creatinine in the serum despite the reduction of glomerular filtration.

Maximal tubular water reabsorption capacity also decreases with age. The ability of the kidneys to handle a water or an acid load is lower in high age than earlier in life. Furthermore, the reactions to such challenges grow more sluggish with advancing age.

D. Respiratory System

Structural changes in the lungs with advancing age include deterioration of elastic properties, for example, in the alveolar walls, and decrease of alveolar volume and surface area. Furthermore, changes in surrounding structures such as calcified costal cartilage and the narrowing of intravertebral space influence the breathing process.

Total lung capacity remains rather unchanged throughout life, whereas the residual volume after maximal expiration increases, leaving a smaller vital capacity for active gas exchange. A major reason for this is the increasing tendency for expiratory small duct and alveolar collapse as a result of changed elastic properties. An additional factor is the increasing stiffness of the thoracic structures.

In spite of these structural and functional changes alveolar ventilation seems adequate at higher ages, as mirrored by an unchanged carbon dioxide pressure. However, the partial pressure of oxygen decreases significantly with age because of the expiratory collapse of alveoli, and as a consequence of that an uneven distribution of ventilation relative to circulation—again an effect of deteriorating elastic properties. A minor explanation may be decreased alveolar diffusion capacity from air to blood.

The age-related respiratory changes play a minor role to the average elderly person. However, they may add risks in pulmonary disease, such as pneumonia. It must also be added that smoking may dramatically change the physiological properties of the respiratory system. [*See* TOBACCO SMOKING, IMPACT ON HEALTH.]

E. Gastrointestinal System

Salivary secretion shows a small decrease with age. Tasting ability declines and a shift between different taste sensations may occur.

Age related esophageal dysfunction includes inability to relax, especially in the lower esophageal sphincter, and the appearance of nonperistaltic muscle waves. These derangements might give rise to dysphagia and nutritional difficulties.

An age-related mucosal atrophy is common throughout the gastrointestinal tract. The consequences are moderate with the exception of the fact that in about 25% of 60-year-olds gastric acid production is virtually absent. Also, the production of intrinsic factor and pepsin decreases with age.

Structural and functional changes with age have been described in the liver, where the organ weight somewhat decreases. However, the metabolic capacity is affected very little.

A decrease of the absorption capacity regarding fat, protein, and calcium, and a reduced ability of the calcium absorption system to adapt to low dietary calcium intake with advancing age, have been reported. The age-related changes in the colon are few and very moderate.

In general, the physiological aging of the gastrointestinal system yields very slight effects, especially when the large reserve capacity is taken into consideration.

F. Endocrine System

Although some endocrine derangements in disease may resemble aging processes, no bulk of evidence attaches aging per se to endocrine deficiencies. The prevalence of endocrine disorders certainly increases with age, but age does not seem to have a causative bearing on such changes.

Hyperplasia is seen in most endocrine organs with advancing age, with the exception of parathyroid glands.

There is no support for a changed hypophyseal function with age. Thus, the production of gonadotropins has the ability to increase in the feedback system from the gonads.

Pathological derangements of thyroid function

are quite common in the elderly. Normal aging processes include fibrotic changes in the thyroid gland with an increase of collagen and lymphocytes. These structural changes seem, however, not to give rise to any marked thyroid dysfunction. Also regulation of the thyroid function seems to be well preserved with advancing age. It may be concluded that an observed significantly deranged thyroid function in the elderly most certainly depends on disease rather than aging.

Production of cortisol seems to decrease with advancing age, while the cortisol concentration in blood remains unchanged.

Regarding endocrine pancreas function, fasting blood sugar levels increase and glucose tolerance decreases with age.

Gonadal function deteriorates in females and males; males show decreasing levels of testosterone. At menopause the production of estrogens from the ovaries is discontinued. However, steroids—for example, androstenedione—are still produced. An obvious consequence of the hormonal changes in females is atrophy of genitalia, which may produce irritation and infection.

G. Immune System

Aging is accompanied by a decreasing ability to cope with environmental stress. There are many causes relating to this incompetence. An increasingly immunoincompetent system plays a role in this context, since the immune system protects the body against foreign invasion by viruses, bacteria, and fungi, and the growth of other invasive antigens such as cancer cells.

Macrophages seem not to be adversely affected by aging in their handling of antigens. There is strong evidence suggesting that aging may affect thymus-derived T cells, which modulate antibody response. An altered function of T lymphocytes, leading to a deterioration of both cell-mediated and antibody-mediated immunological defense, has been described.

Such changes seem, however, not to exist in all individuals, and there are longitudinal population studies showing very slight differences by age in these respects.

In general, fewer changes occur in humoral than in cellular immunity. Most evidence suggests that the immunological decline with age is primarily related to changes in the T-cell component of the immune system.

H. Skin

The alteration of the appearance of the skin in elderly people is due to a combination of aging processes per se and environmental damage due, for example, to sun exposure.

The basis for the dryness, wrinkling, laxity, and uneven pigmentation seen in most of the elderly comprises changes in the epidermis and dermis. A decreasing ability to proliferate has been described in epidermal cells, and there is a flattening of the dermoepidermal junction. The epidermal turnover rate diminishes in the order of 50% from early to later life. The ability of DNA repair in skin fibroblasts seems to decrease with age, and the skin repair rate (relating to situations such as wound healing and regeneration of blister roofs) declines with age.

The number of active exocrine sweat glands and sweat production decrease. Sebum production decreases as well, paralleling the decrease of gonadal androgen, despite the fact that there is no atrophy of sebum glands.

The number of melanocytes decreases, and the ability to produce vitamin D after ultraviolet light exposure seems to deteriorate with advancing age. Loss of dermal thickness is characteristic of the aging skin.

The aging skin is vulnerable to pressure sores and trauma. Prurigo is common due to dryness of the skin, even in the absence of skin disease. Thermoregulation is often impaired due to deranged vascular response to heat or cold, decreased sweat production, and loss of subcutaneous fat.

I. Muscular System

Total muscle mass decreases with advancing age due to a decreasing number of muscle fibers, while there are very small changes in fiber size. Some observations indicate that there is a marked decrease of number of motor units but more fibers in existing motor units, pointing to an ability for dynamic reorganization of fibers.

There seem to be very insignificant changes with age regarding the proportion of "slow" to "fast" muscle fibers, although some studies indicate a reduction in "fast" fiber size. A reduced size of type 2 fibers is associated with denervation and inactive muscles.

As a result of the structural changes, both isometric and dynamic muscle strength decline; coordination and flexibility also decrease with age in the same order as the loss of muscle mass. Some data show that especially isometric muscle performance is impaired to a great extent. This is of practical importance since correcting movements where isometric muscle activity is involved is essential to avoid falls.

Physical training improves both muscular strength and coordination even in the higher age groups, and maximal aerobic power has been shown to increase with training at age 70–75. This is of special geriatric importance since muscle function plays an important protective role regarding the risk for falls and hip fractures.

J. Bone

Peak bone mass is achieved by age 25–30; thereafter an age-related endosteal bone loss starts; this begins earlier in females than in males. There is great variation in the rate of decrease between individuals. Males seem to lose 0.3–0.5% of bone mass per year, trabecular bone loss being slightly more marked than that of cortical bone. The corresponding figure for females is higher and is close to 1% per year, and young women have about 30% higher mineral content in the skeleton compared to elderly women, while the corresponding difference for men is only about 15%.

The amount of bone mineral is related to the strength of the skeleton and, therefore, to the risk of getting fractures. However, other protective factors are involved, such as muscle strength. [*See* Bone Density and Fragility, Age-Related Changes.]

The border between normal aging of the skeleton and osteoporosis is not clear-cut. Osteoporosis and hip fractures currently show an increase in many developed countries. Many factors may contribute to accelerated bone aging and osteoporosis. Such primary risk factors include, apart from age, sex and race (Black people being less susceptible), a sedentary lifestyle, smoking, alcohol abuse, and environmental conditions favoring falls. Adequate intake of dietary calcium is also essential, although the level of recommended intake is still a matter of dispute.

K. Metabolism and Body Composition

Energy metabolism is lower in advanced age than earlier in life. The major explanation for this is probably the decreasing number of cells in the organs, loss of metabolizing tissue, and reduced physical activity, although some intracellular age-induced changes on the enzyme level might also be responsible. Energy expenditure is, thus, lower in old age than earlier in life. This seems to be related less to a decrease of basal metabolic rate than to a decrease of energy expenditure relating to physical activity, and diet-induced and temperature-induced thermogenesis. [*See* Energy Metabolism.]

Decreasing number of cells in the organs and increasing disuse of skeletal muscle tissue with age result in a decreasing body cell mass. At age 70 skeletal muscle has been shown to have lost about 40% of its maximal weight in early adult life as compared to about 20% for the liver, and about 10% for the kidneys and lungs.

Most data have shown an increasing amount of body fat with age, especially when expressed as a proportion of body weight. In the eighth decade of life, however, body fat seems to decrease, earlier in females than in males. Subcutaneous fat is deposited more on the trunk than on the extremities in old compared to young individuals, and there is an increase of deep adipose tissue relative to subcutaneous fat with age.

Body water decreases with advancing age in most studies. However, this trend might be less significant in the highest age groups since it might be counteracted by the fact that a change in cardiac and renal function might increase the amount of body water. However, the most important reason for the well-known decrease of body weight during the eighth decade of life has shown to be a decreasing amount of body water, especially extracellular water.

IV. Functional Aging and Functional Capacity

As mentioned above the functional age of an elderly individual may differ very much from chronological age. Therefore, the importance of assessment of functional capacity of elderly people is obvious. Functional capacity comprises levels and profiles of different functions in an individual in relation to chronological age and the average functional capacity of people of the same chronological age. Such functions should include not only biological and medical parameters but also psychosocial capacity.

Attempts to measure functional age and functional capacity have been done using wide batteries of tests. Certain sensory, psychomotor, and motor functions have been claimed to be particularly useful in this context. Also the assessment of functioning in activities of daily living has shown to be useful.

Since the combination of many different age-related functional deteriorations is obviously of more importance to the average elderly person than effects by age regarding single functions, the attempts to follow functional capacity with age seem to be an urgent field of investigation. Attention has also been paid to the manifestations of disease in the overall judgment of the functional capacity of the aged individual.

Interesting observations in ongoing longitudinal and sequential studies of representative elderly populations seem to indicate a slightly higher degree of performance regarding some physical and mental functions in "young elderly" (age 65–75) now, compared to people of the same age just one or two decades ago. These cohort differences are, however, not a universal phenomenon for all aging functions, and this trend has to be confirmed in further studies.

Bibliography

Bergener, M., ed. (1987). "Psychogeriatrics: An International Handbook." Springer Publishing Company, New York.

Brocklehurst, J. C., ed. (1978). "Textbook of Geriatric Medicine and Gerontology," 2d ed. Churchill Livingstone, Edinburgh, London, and New York.

Cape, R. D. T., Coe, R. M., and Rossman I., eds. (1985). "Fundamentals of Geriatric Medicine." Raven Press, New York.

Chernoff, R., and Lipschitz, D. A., eds. (1988). "Health Promotion and Disease Prevention in the Elderly." Raven Press, New York.

Horwitz, A., Macfadyen, D. M., Munroe, H. N., Scrimshaw, N. S., Steen, B., and Williams, T. F., eds. (1989). "Nutrition in the Elderly." Oxford University Press, Oxford.

Libow, L. S., and Sherman, F. T., eds. (1981). "The Core of Geriatric Medicine." C. V. Mosby Company, St. Louis.

Masoro, E. J. (1986). Physiology of aging, *in* "Geriatric Dentistry" (P. Holm-Pedersen and H. Löe, eds.). Munksgaard, Vojens.

Pathy, M. S. J., ed. (1985). "Principles and Practice of Geriatric Medicine," 1st ed. Wiley, Bath.

Terry, R. D., ed. (1988). "Aging and the Brain." Raven Press, New York.

Glutathione

OWEN W. GRIFFITH, *Cornell University Medical College*

Glossary

Electrophile Molecule having an electron-deficient atom that can react with certain electron-rich atoms (i.e., nucleophiles). Reaction of electrophiles with nucleophilic atoms of cellular proteins and nucleic acids generally destroys or prevents the normal functioning of the protein, nucleic acid, or other molecule attacked.

Free radical Molecule having a single unpaired electron. Such molecules are often nonspecifically reactive with a variety of cellular constituents. Unsaturated membrane lipids are particularly susceptible to free-radical damage.

K_m Michaelis constant. In characterizing the effect of substrate concentration on the rate of an enzyme-catalyzed reaction, the K_m represents the substrate concentration necessary for the reaction to proceed at one-half of its maximal rate (assuming all other substrates are at saturating concentrations). K_m has units of concentration, and each substrate has a characteristic K_m. Enzymes typically have a high affinity for substrates with low K_m's.

Nucleophile Molecule having an atom with nonbonded electrons able to react with electrophiles. Nucleic acids, proteins, and other cell constituents contain numerous nucleophilic atoms often essential to their structure or function.

Oxidation–reduction Oxidation is a chemical process in which an atom loses electron density or entire electrons. Reduction is the opposite: a gain in electron density or electrons. Biological oxidation–reduction reactions always occur simultaneously; one reactant is oxidized and the other is reduced. In the aerobic environment of mammalian cells, oxidative reactions are driven directly or indirectly by the energetically favorable reduction of oxygen to water. Reduction of biological molecules, on the other hand, often requires the input of additional metabolic energy.

Peroxide Molecule containing two oxygens joined by a single bond [e.g., hydrogen peroxide (H—O—O—H) or organic peroxides (R—O—O—H, where R— is an organic molecule)]. Peroxides are moderately strong oxidizing agents.

Thiol/disulfide Thiols (i.e., sulfhydryls) are reduced sulfur compounds containing an —SH group. Thiols are readily oxidized to disulfides, which contain the —S—S— structure (i.e., 2 R—SH + H_2O_2 → R—S—S—R + 2 H_2O). Disulfides can be symmetrical (e.g., R—S—S—R) or mixed (e.g., R—S—S—R′, where R ≠ R′). Although disulfides resemble peroxides in some respects, they are more stable and thus less reactive.

GLUTATHIONE (GSH) is a tripeptide (γ-glutamylcysteinylglycine) (Fig. 1) present in all human tissues, generally in high (i.e., millimolar) concentration. Two aspects of GSH structure are of particular importance: (1) Whereas glutamate in proteins and most peptides is bonded through its α-carboxyl, the glutamic acid residue of GSH is joined to cysteine through its γ-carboxyl; this unusual γ-glutamyl bond prevents the hydrolysis of GSH by ordinary proteases and peptidases. (2) The sulfhydryl (i.e., —SH) group of the cysteinyl residue of GSH accounts for about 90% of the total low-molecular-weight thiols in most tissues; GSH is thus the most prevalent and one of the most reactive nucleophiles in tissues.

GSH is synthesized in the cytosol of all cells from its constituent amino acids by the sequential action of two ATP-dependent enzymes. Within cells GSH plays an important role in the detoxification of a variety of compounds, including free radicals, organic peroxides, hydrogen peroxide, and reactive electrophiles. GSH is not degraded within cells, but is transported intact across the plasma membrane to the extracellular space. Following transport GSH turnover is initiated on the external surface of certain cells by the γ-glutamyl transpeptidase-mediated cleavage of its γ-glutamylcysteine bond. Cysteinylglycine, a dipeptide product of this reaction, is subsequently hydrolyzed to cysteine and glycine by any of several peptidases.

Tissues vary greatly in the amount of γ-glutamyl transpeptidase present. Liver, a tissue richly supplied with cysteine by both the diet and *de novo* synthesis, makes and secretes substantial amounts of GSH into the plasma, but is deficient in γ-glutamyl transpeptidase; it degrades little of the GSH it releases. Instead, GSH released from the liver is taken up by the blood and transported throughout the organism. Peripheral tissues with significant amounts of γ-glutamyl transpeptidase (e.g., kidney, lung, pancreas, and lymphocytes) degrade not only the GSH released by their own cells, but also that delivered by the blood. Tissues able to degrade GSH take up the amino acids and peptides released (i.e., glutamate, cysteine, glycine, and certain dipeptides). In this fashion the hepatic synthesis and peripheral tissue degradation of GSH provide for the interorgan transport of cysteine, an amino acid often limiting in cellular nutrition.

Inherited disorders affecting most of the enzymes involved in GSH metabolism have been reported. Although all of the disorders are rare, studies of affected patients have significantly elucidated the metabolism of GSH in humans. Pharmacological manipulations of GSH levels and metabolism are also possible. Therapies designed to increase GSH levels offer the possibility of augmenting the protective functions of GSH. Therapies that decrease GSH levels are of interest in the sensitization of tumor cells to radiation and chemotherapy and in the treatment of certain parasitic disorders.

Since mammalian GSH metabolism has been examined most extensively in mice and rats, much of the specific information presented is based on studies in these animals. To the extent that studies of human enzymes or metabolism have been reported, the findings are similar to those in rodents. Where available, data from human studies are presented and are identified as such.

I. Chemistry

A. Structure and Physical Properties

GSH is a tripeptide composed of the protein amino acids L-glutamate (Glu), L-cysteine (Cys), and glycine (Gly) (the structures of these amino acids, GSH, and γ-glutamylcysteine, an intermediate in the synthesis of GSH, are shown in Fig. 1). At physiological pH both of the carboxyl groups and the α-amino group of GSH are ionized; the tripeptide is thus an anion with a net charge of -1 (the pKa values of GSH, determined by nuclear magnetic resonance spectroscopy, are 2.05, 3.40, 8.72, and 9.49 for the glutamyl α-carboxyl, glycyl carboxyl, amino, and thiol groups, respectively). GSSG, the disulfide formed from GSH, is a dianion at physiological pH. Because of their small size, absence of hydrophobic groups, and presence of multiple ionic groups, both GSH and GSSG dissolve rapidly and extensively in water; solutions of greater than 1 *M* are easily prepared at neutral pH. In solution GSH and GSSG show considerable conformational mobility. Bonds rotate freely, and bond lengths and angles are normal.

High water solubility, permitting significant intracellular or extracellular accumulations without precipitation, is a biologically important property of GSH and GSSG. Although cysteine is also very soluble, it is readily oxidized in aerobic environments to its disulfide, cystine, which is insoluble. When accumulated to concentrations causing precipitation (e.g., >0.6 m*M*), cystine is toxic (cf. the inherited disorder cystinosis, in which intralysosomal cystine accumulation results in cystine crystallization, lysosomal destruction, and, in severe cases, death). It is probable that GSH has been selected by evolution, in part, to provide a safe means for the intracellular storage of cysteine; mechanisms for re-

FIGURE 1 Reactions of GSH synthesis, turnover, and metabolism. The circled numbers correspond to the following enzymes or metabolic processes: 1, γ-glutamylcysteine synthetase; 2, glutathione synthetase; 3, γ-glutamyl transpeptidase; 4, γ-glutamylcyclotransferase; 5, 5-oxoprolinase; 6, cysteinylglycinase; 7, glutathione *S*-transferase; 8, free-radical inactivation by GSH; 9, enzymatic and nonenzymatic oxidation of GSH by oxygen; 10, enzymatic and nonenzymatic GSH-dependent transhydrogenations; 11, glutathione peroxidase; 12, glutathione reductase; 13, various reactions requiring GSH as a cofactor; 14, cysteine conjugate *N*-acetyltransferase. P_i, Inorganic phosphate. [Modified by permission from O. W. Griffith and H. F. Friedman, *in* "Synergism and Antagonism in Chemotherapy" (T.-C. Chou and D. Rideout, eds.). Academic Press, San Diego 1990.]

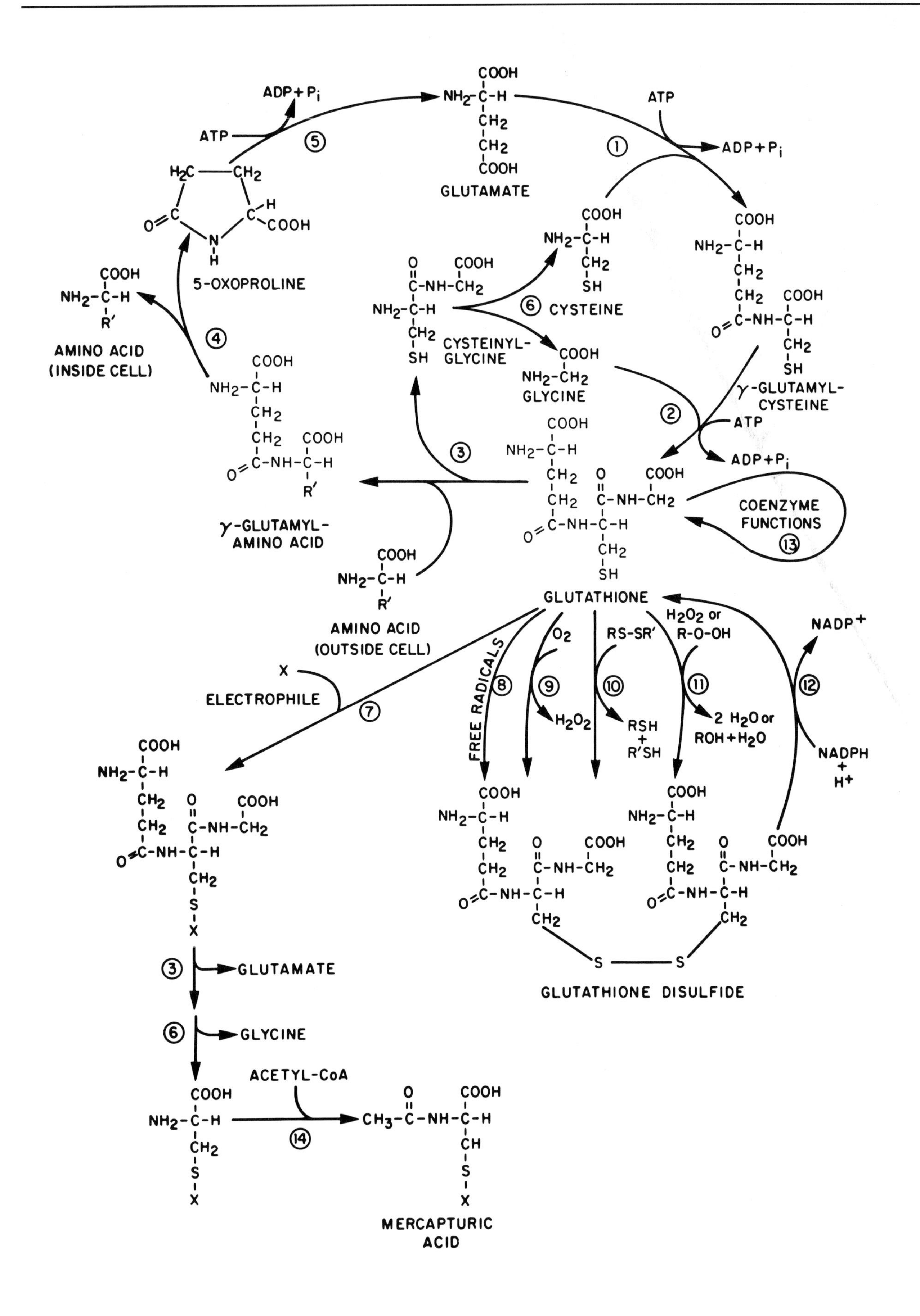
ADP + Pi
ATP
5
COOH
NH2-C-H
CH2
CH2
COOH
GLUTAMATE
ATP
ADP + Pi
1
5-OXOPROLINE
AMINO ACID (INSIDE CELL)
4
CYSTEINE
6
CYSTEINYL-GLYCINE
GLYCINE
γ-GLUTAMYL-CYSTEINE
2
ATP
ADP + Pi
3
γ-GLUTAMYL-AMINO ACID
AMINO ACID (OUTSIDE CELL)
COENZYME FUNCTIONS
13
GLUTATHIONE
X
ELECTROPHILE
7
FREE RADICALS
8
O2
9
H2O2
RS-SR′
10
RSH + R′SH
H2O2 or R-O-OH
11
2 H2O or ROH + H2O
NADP+
12
NADPH + H+
GLUTATHIONE DISULFIDE
3
GLUTAMATE
6
GLYCINE
ACETYL-CoA
14
MERCAPTURIC ACID

leasing cysteine from GSH are discussed in Section II.

B. Sulfhydryl Chemistry

The protective functions of GSH are dependent on the chemistry of its sulfhydryl group. As indicated in Fig. 1, GSH can be reversibly oxidized to GSSG. Both free radicals and peroxides are detoxified by reactions that result in GSH oxidation. Since GSSG is rapidly reduced back to GSH by an NADPH-dependent enzyme (see Section III), GSH plays an essentially catalytic role in the detoxification of free radicals and peroxides. In this respect it should be noted that GSH, in contrast to many other thiols (e.g., cysteine), is only slowly oxidized by molecular oxygen in the absence of specific catalysts (Reaction 10, Fig. 1). The chemistry of the GSH sulfhydryl thus facilitates its role as a protectant from cytotoxic oxidatants, while minimizing its spontaneous reaction with molecular oxygen, a ubiquitous and essential constituent of mammalian cells.

The sulfur atom of sulfhydryls is electron rich. Since the electrons are in large easily distorted orbitals (i.e., are polarizable), sulfhydryls are among the strongest and most reactive nucleophiles found in biological systems. The sulfhydryl of GSH is typically reactive and readily attacks a variety of potentially toxic electrophiles. These reactions, catalyzed by glutathione *S*-transferases (see Section III), result in the chemical joining of GSH to the electrophile (Reaction 7, Fig. 1); the products are chemically stable, thus nontoxic, sulfides. In contrast to the reductive detoxification of free radicals and peroxides, detoxification of electrophiles irreversibly consumes GSH. The sulfide product is metabolized to release glutamate and glycine, but the nutritionally valuable cysteine residue is not recovered.

II. Biosynthesis and Turnover

A. Enzymology of GSH Synthesis

GSH is synthesized in all mammalian cells by the sequential action of two cytoplasmic enzymes. The first, γ-glutamylcysteine synthetase, catalyzes the ATP/Mg^{2+}-dependent formation of L-γ-glutamyl-L-cysteine from L-glutamate and L-cysteine (Reaction 1, Fig. 1); cleavage of ATP (K_m = 0.2 m*M*) to ADP and inorganic phosphate (P_i) drives the reaction to completion. Under physiological conditions the enzyme is highly specific for L-glutamate (K_m = 1.6 m*M*) and is relatively specific for L-cysteine (K_m = 0.3 m*M*). L-α-Aminobutyrate, a minor metabolite of L-methionine, is present *in vivo* and is a moderately active substrate of the enzyme (K_m = 1.0 m*M*); when L-α-aminobutyrate reacts in place of L-cysteine in the GSH biosynthetic pathway, the ultimate product is L-γ-glutamyl-L-α-aminobutyrylglycine, known as ophthalmic acid, due to its initial discovery in the lens. Although less extensively studied, the structural and catalytic properties of γ-glutamylcysteine synthetase isolated from the human erythrocyte appear to be identical to those described for the rat kidney enzyme.

GSH synthesis is completed by glutathione synthetase, an enzyme catalyzing the ATP/Mg^{2+}-dependent synthesis of GSH from L-γ-glutamyl-L-cysteine ($K_m < 50$ μM) and glycine (K_m = 33 μM) (Reaction 2, Fig. 1). ATP (K_m = 0.76 m*M*) is again cleaved to ADP and P_i, driving the reaction nearly to completion. Under physiological conditions the enzyme is highly specific for the natural substrates, although, as noted, L-γ-glutamyl-L-α-aminobutyrate is converted to ophthalmic acid. *In vitro*, the enzyme is relatively nonspecific with respect to the L-γ-glutamyl moiety of synthetic γ-glutamylcysteine analogs, but does not accommodate many structural modifications to the L-cysteinyl moiety.

B. GSH Levels and Their Regulation

Intracellular GSH levels range from about 0.5 to 10 μmol/g in a wide range of mouse and rat tissues. In healthy animals each tissue maintains a characteristic GSH level, although some tissues, particularly liver, show a diurnal fluctuation in response to feeding. In many tissues GSH levels change rapidly postmortem, and few human tissues have been examined under conditions likely to preserve *in vivo* concentrations. Nevertheless, the data available suggest that human GSH levels are comparable to or a little lower than those in rodents (Table I). In all species extracellular GSH levels are much lower than intracellular levels. Plasma, cerebrospinal fluid, and urine concentrations are all in the micromolar range (Table I).

Intracellular GSH levels are controlled and determined by the tissue-specific relative rates of synthesis and utilization. Since essentially all γ-glutamylcysteine formed is quickly converted to GSH, the rate of GSH synthesis is normally regulated by the rate of the γ-glutamylcysteine synthetase reac-

TABLE I Glutathione Levels

Tissues, cells, or physiological fluids	GSH concentration Mouse	Rat	Human
Tissues (μmol/g of tissue)			
Brain	2.08 ± 0.15		
Heart	1.35 ± 0.10		
Lung	1.52 ± 0.13		
Spleen	3.43 ± 0.35		
Liver	7.68 ± 1.22	4.51 ± 0.27	3.55 ± 0.89
Pancreas	1.78 ± 0.31	1.66 ± 0.07	
Kidney	4.13 ± 0.15	2.56 ± 0.09	
Small intestine mucosa	2.94 ± 0.16		
Colon mucosa	2.11 ± 0.19		
Skeletal muscle	0.78 ± 0.05	0.75 ± 0.08	
Cells			
Erythrocytes (μmol/ml of cells)	2.0 ± 0.2		1.9 ± 0.3
T lymphocytes (nmol/10^6 cells)			0.6–1.0
Mixed peripheral blood leukocytes (nmol/10^6 cells)			1.4–2.2
Physiological fluids (μM)			
Plasma	37.0 ± 2.5	22–27	4.98 ± 0.65
Urine	2–5		Negligible
Cerebrospinal fluid	5.05 ± 0.25		Negligible

tion. This reaction is, in turn, controlled by the amount of enzyme present, by the amount of GSH already in the cell (GSH is a nonallosteric feedback inhibitor; $K_i = 2.3$ mM), and by cysteine availability. Many diets are relatively deficient in the sulfur-containing amino acids cyst(e)ine (i.e., cysteine and cystine) and methionine. Since cysteine biosynthesis depends on sulfur derived from methionine, diets deficient in cyst(e)ine, methionine, or both ultimately limit the supply of cysteine available for GSH synthesis. Correspondingly, several studies in animals and humans indicate that tissue GSH levels can be increased modestly by giving L-cysteine or L-cysteine precursors parenterally or in the diet (see Section IV,B). In human infants, in whom L-cysteine synthesis from L-methionine is limited by the immaturity of some liver enzymes, increased dietary intake of L-cysteine might significantly increase tissue GSH levels.

In the absence of pathological stresses, GSH utilization is controlled primarily by the rate at which GSH is transported out of the cell. Rates of transport vary greatly among tissues (e.g., high in liver, kidney, lung, and leukocytes, but very low in erythrocytes). In all tissues the rate of GSH transport out of the cell is balanced with the synthetic rate in order to maintain intracellular GSH levels at the concentration typical of the tissue. It is noteworthy that the flow of GSH out of a cell represents transport down a sizable concentration gradient; GSH transport to the extracellular space is thus thermodynamically favored, whereas uptake of intact GSH by cells is not. In fact, most studies reported to date indicate that intact GSH is not significantly taken up into any cell type, even if the concentration gradient is made favorable by increasing the extracellular concentration. A consequence of the unidirectionality of GSH transport is that extracellular GSH must be degraded if its constituent amino acids are to be recovered.

Studies with erythrocytes and liver suggest that the transport of GSH and GSSG out of the cell are mediated and controlled by separate mechanisms. Since GSSG is normally present in only trace amounts intracellularly, its rate of transport usually does not significantly affect cellular GSH homeostasis. However, under conditions of oxidative stress, the intracellular concentration of GSSG increases, and, in at least some tissues, its rate of transport out of the cell increases proportionately. Under this circumstance the rate of GSH synthesis and the total rate of glutathione transport out of the cell might not be in balance; intracellular total glutathione levels (i.e., GSH plus 2× GSSG) might decrease substantially. Although loss of cellular glutathione can have serious consequences, it has been suggested that the rapid transport of GSSG out of cells effectively protects the intracellular environment from this product of oxidative stress.

C. Extracellular Turnover

As noted in the definition paragraph, the unusual γ-glutamyl bond in GSH renders the tripeptide resistance to most proteases and peptidases. *In vivo*, GSH degradation is initiated only by γ-glutamyl transpeptidase, an enzyme bound to the external surface of the plasma membrane of many, but not all, cells. Particularly high levels of γ-glutamyl transpeptidase are found in certain epithelial tissues, including the brush border membrane of the kidney proximal tubule, the microvilli of the intestinal lumen, the choroid plexus, and the ciliary body; at each of these locations, it is important that GSH be degraded so that its constituent amino acids can be recovered by the whole organism or specific tissue. γ-Glutamyl transpeptidase catalyzes both transpeptidation and hydrolytic reactions involving the γ-glutamyl bond of GSH (Eqs. 1 and 2, respectively; Eq. 1 is shown as Reaction 3 in Fig. 1).

$$\text{GSH} + \text{amino acid} \rightarrow \gamma\text{-glutamylamino acid} + \text{cysteinylglycine} \quad (1)$$

$$\text{GSH} + H_2O \rightarrow \text{glutamate} + \text{cysteinylglycine} \quad (2)$$

The amino acid indicated in Eq. 1 can be any of a wide variety of L-amino acids; cystine, glutamine, and methionine are particularly active. Some small peptides can also act as acceptors in place of the amino acid. Glycylglycine, for example, reacts at a high rate and is used to assay the enzyme *in vitro*. Even a second molecule of GSH can react in place of the amino acid; in this case the product is γ-glutamyl-GSH. This reaction has been shown to occur *in vivo* in both humans and rodents. Since γ-glutamyl transpeptidase catalyzes hydrolysis and transpeptidation reactions involving the γ-glutamyl bond of GSH, GSSG, γ-glutamylamino acids, γ-glutamyl-GSH, and glutamine (the last is "γ-glutamyl-ammonia" an amino acid present in high concentration in plasma), the extracellular metabolism of GSH and related γ-glutamyl compounds is considerably more complicated than is indicated in Fig. 1 or Eqs. 1 and 2; nevertheless, the products shown in Eqs. 1 and 2 are quantitatively the most important.

Cysteinylglycine and cystinyl-*bis*-glycine, its corresponding disulfide (see below), are hydrolyzed to their constituent amino acids (i.e., cysteine, cystine, and glycine) by both extracellular and intracellular peptidases; the extracellular enzymes cysteinylglycinase and aminopeptidase M are particularly active toward cystinyl-*bis*-glycine and cysteinylglycine, respectively. Hydrolysis of cysteinylglycine is shown as Reaction 6 in Fig. 1.

As noted in Section I, GSH is relatively resistant to spontaneous oxidation to GSSG. In contrast, cysteine and cysteinylglycine are easily oxidized. Such oxidation, catalyzed nonenzymatically by trace metals and catalyzed enzymatically by thiol oxidases, occurs in the extracellular environment to form cystine, cystinyl-*bis*-glycine, and various mixed disulfides. These disulfide products can, in turn, oxidize GSH to GSSG (Eq. 3). There is also specific oxygen-dependent enzymatic oxidation of GSH itself within the kidney tubule and of some other sites. At present the physiological role, if any, of extracellular thiol oxidation is unknown. One consequence of such oxidation, however, is that the extracellular thiol/disulfide ratio (typically 2–0.5, depending on the thiol and disulfide considered) is much lower than the intracellular ratio (GSH/GSSG >100 in the cytosol of most cells). Since the breaking and forming of disulfide bonds within and between proteins are affected by the thiol/disulfide ratio, it is possible that the relatively low extracellular thiol/disulfide ratio facilitates the formation of specific protein–protein or protein–ligand disulfide bonds.

$$2\ \text{GSH} + \text{cystinyl-}bis\text{-glycine} \rightarrow \text{GSSG} + 2\ \text{cysteinylglycine} \quad (3)$$

The several products of the extracellular γ-glutamyl transpeptidase, dipeptidase, and thiol oxidation reactions are taken up by cells of the surrounding tissue. Separate transport systems mediate the uptake of γ-glutamylamino acids, glutamate, cysteine, cystine, cysteinylglycine, and glycine. Of the several products mentioned, only γ-glutamyl-GSH is thought not to be taken up; it is either excreted in the urine or is further metabolized by γ-glutamyl transpeptidase to yield products that can be taken up. Disulfides taken up by the cell (e.g., cystine and cystinyl-*bis*-glycine) are reduced to the free thiols by intracellular GSH (Reaction 10, Fig. 1; see Section III). Cysteinylglycine is intracellularly hydrolyzed, presumably by a nonspecific peptidase.

D. Intracellular Metabolism of γ-Glutamylamino Acids: The γ-Glutamyl Cycle

Mammalian tissues vary markedly in their ability to take up γ-glutamylamino acids; tissues with high levels of γ-glutamyl transpeptidase (e.g., kidney and pancreas) tend to be most effective in γ-glutamylamino acid uptake. Within the cytosol γ-glutamylamino acids are metabolized by γ-glutamylcyclotransferase to form 5-oxo-L-proline (i.e., pyroglutamate) and free amino acid (Reaction 4, Fig. 1). The free amino acid formed is incorporated directly into the intracellular amino acid pool and is thus available for a variety of metabolic purposes. 5-Oxo-L-proline, on the other hand, is metabolized only by 5-oxoprolinase (Reaction 5, Fig. 1), an ATP/Mg^{2+}-dependent enzyme that hydrolyzes 5-oxo-L-proline to L-glutamate; ADP and P_i are coproducts.

In 1970 A. Meister and M. Orlowski realized that the reactions of GSH synthesis and turnover constitute a metabolic cycle, the γ-glutamyl cycle, which could function as a mechanism for transporting amino acids into the cell. Thus, GSH synthesized within the cell is transported out of the cell and reacts there with an extracellular amino acid in a γ-glutamyl transpeptidase-catalyzed reaction to form the corresponding γ-glutamylamino acid; uptake and intracellular hydrolysis of the γ-glutamylamino acid effectively transport the amino acid into the cell. Hydrolysis of 5-oxoproline and resynthesis of GSH, along with recovery of cysteinylglycine (or cysteine plus glycine), complete the cycle. Subsequent studies using specific inhibitors of various reactions of the cycle have clearly established that amino acids are transported by the γ-glutamyl cycle in γ-glutamyl transpeptidase-rich tissues (e.g., kidney).

Although the cycle might be important in the transport of specific amino acids (e.g., cystine) that are excellent γ-glutamyl transpeptidase substrates, it is noted that the γ-glutamyl cycle is neither the only nor the quantitatively most important mechanism of amino acid transport in any tissue examined to date. The γ-glutamyl cycle, defined so as to account for hydrolysis as well as transpeptidation of GSH by γ-glutamyl transpeptidase, does fully account for GSH turnover in γ-glutamyl transpeptidase-rich tissues. For γ-glutamyl transpeptidase-poor tissues (e.g., liver) GSH turnover occurs by an interorgan γ-glutamyl cycle (e.g., liver and kidney).

III. Biological Functions

A. Glutathione Reductase and Maintenance of Intracellular Thiols

Although several physiological and pathological processes oxidize GSH to GSSG, the intracellular GSH/GSSG ratio is, as noted, typically greater than 100. The enzyme responsible for the rapid and nearly complete reduction of GSSG is glutathione reductase, a ubiquitous flavoprotein present in both the cytosol and the mitochondrial matrix (Reaction 12, Fig. 1). As indicated in Eq. 4, GSSG reduction requires the stoichiometric oxidation of NADPH to $NADP^+$; $NADP^+$ is then reduced back to NADPH by any of several cellular dehydrogenases. In most cells glucose-6-phosphate dehydrogenase and 6-phosphogluconate dehydrogenase, enzymes involved in the pentose phosphate pathway of carbohydrate metabolism, are believed to be the most important sources of the NADPH needed to maintain a reduced GSH pool. Inherited partial deficiency of glucose-6-phosphate dehydrogenase is a moderately common human disorder that typically affects the erythrocytes in particular. Under conditions of oxidative stress (e.g., as induced by the antimalarial drug chloroquine), the rate of GSSG formation is high, and the erythrocytes of affected patients cannot maintain a high GSH/GSSG ratio. Failure to maintain a high GSH/GSSG ratio compromises the membrane-protective functions of GSH and leads to erythrocyte destruction; hemolytic anemia is observed clinically.

$$\text{GSSG} + \text{NADPH} + \text{H}^+ \rightarrow 2\ \text{GSH} + \text{NADP}^+ \quad (4)$$

Human tissues contain a number of thiols and disulfides, in addition to GSH and GSSG. Cysteine, cystine, cysteinylglycine, cystinyl-*bis*-glycine, and γ-glutamylcysteine have been mentioned; there are, in addition, a variety of thiol-containing enzymes and other proteins as well as the coenzymes lipoic acid and coenzyme A. These compounds are maintained predominantly in the reduced form, but are not subject to direct reduction by NADPH. Instead, disulfides such as cystine are reduced by GSH, as shown in Reaction 10, Fig. 1. Equations 5a and 5b show in detail the mechanism of cystine reduction by GSH (CyS—SG is the mixed disulfide between cysteine and GSH). Note that the overall process, catalyzed by thiol transhydrogenases, reduces cystine to two molecules of cysteine, while oxidizing two molecules of GSH to GSSG. Since GSSG is readily and completely reduced back to GSH by glutathione reductase, the equilibrium positions of Eqs. 5a and 5b are well to the right *in vivo*; the cysteine/cystine ratio, like the GSH/GSSG ratio, is also high intracellularly.

$$\text{Cystine} + \text{GSH} \rightarrow \text{cysteine} + \text{CyS-SG} \quad (5a)$$

$$\text{Cys-SG} + \text{GSH} \rightarrow \text{cysteine} + \text{GSSG} \quad (5b)$$

It is important to emphasize that neither the GSH/GSSG ratio nor the cysteine/cystine ratio is infinite or invariant. Drugs or other conditions producing oxidative stress can significantly decrease the ratios. Even under normal conditions GSSG and cystine are present intracellularly, and the activity of some enzymes is regulated by the GSH/GSSG ratio. Among those enzymes affected by GSH/GSSG ratios within normal physiological limits are hydroxymethylglutaryl-coenzyme A reductase and phosphofructokinase, enzymes catalyzing rate-limiting reactions in cholesterol biosynthesis and glycolysis, respectively. It is apparent that, for a limited number of enzymes, alterations in the GSH/GSSG ratio represent an important mechanism of metabolic control.

B. GSH and Detoxification of Peroxides

Hydrogen peroxide (i.e., H—O—O—H) is a coproduct of several intracellular enzymes (e.g., superoxide dismutase, D-amino acid oxidase, and monoamine oxidase) and is produced during mitochondrial respiration by the incomplete reduction of oxygen (oxygen is normally reduced by a four-electron process to water, but is reduced by a two-electron process to H_2O_2 up to 5% of the time). Organic peroxides (R—O—O—H) are formed mainly from unsaturated fatty acids, usually by reaction of a fatty acid free radical with molecular oxygen. Both hydrogen peroxide and organic peroxides are subject to spontaneous homolytic cleavage to form highly destructive free radicals (e.g., H—O—O—H → 2 HO$^{\cdot}$, where HO$^{\cdot}$ is hydroxide radical). Furthermore, in the presence of certain metals, particularly iron, hydrogen peroxide and superoxide (i.e., $O_2^{\overline{\cdot}}$, also a normal cellular metabolite) react to form hydroxide radical and singlet oxygen, the latter also a very destructive species. While peroxide formation is apparently an inescapable consequence of aerobic life, it is obviously important that the cellular levels of peroxides and superoxide be kept at a minimum to prevent their conversion into highly damaging species.

Several defenses have evolved. Superoxide is destroyed by superoxide dismutase with the formation of ordinary oxygen and hydrogen peroxide (Eq. 6). Catalase, a peroxisomal enzyme, catalyzes the dismutation of hydrogen peroxide (Eq. 7), but is inactive toward organic peroxides. Both hydrogen peroxide and organic peroxides are reduced by glutathione peroxidase, a selenium-dependent enzyme present in both the cytosol and the mitochondrial matrix (Reaction 11, Fig. 1). In addition, some isoforms of glutathione *S*-transferase (see Section III,D) also catalyzes the GSH-dependent reduction of organic peroxides; hydrogen peroxide is not reduced. It is apparent that the reduction of peroxides by either glutathione peroxidase or glutathione *S*-transferase causes the concomitant oxidation of GSH to GSSG; as discussed previously, GSSG is reduced back to GSH by glutathione reductase. Since glutathione peroxidase is quantitatively more important than catalase in reducing physiological levels of hydrogen peroxide, cellular defenses against both hydrogen peroxide and organic peroxides depend heavily on the intracellular GSH, glutathione reductase, and a continuing supply of NADPH. At pathologically high levels of hydrogen peroxide, the role of catalase becomes proportionately more important, but does not supplant that of glutathione peroxidase.

$$2\ O_2^{\overline{\cdot}} + H^+ \rightarrow O_2 + H_2O_2 \quad (6)$$

$$2\ H_2O_2 \rightarrow O_2 + H_2O \quad (7)$$

C. GSH and Prevention and Repair of Free-Radical Damage

Free radicals are intermediates in several biological reactions essential for life. As already mentioned, other free radicals (e.g., OH$^{\cdot}$) are highly reactive and tissue-damaging species capable of causing cell death. Ionizing radiation, either from the environment or administered as a treatment of certain cancers, splits a variety of chemical bonds to yield free radicals. Since cells contain high concentrations of water, a common effect of ionizing radiation is the homolytic cleavage of H_2O to HO$^{\cdot}$ and H$^{\cdot}$. Cytotoxic free radicals are also formed metabolically. Carbon tetrachloride is metabolized to form $Cl_3C^{\cdot}$, for example. In general, free radicals in which the unpaired electron is on carbon or oxygen are highly reactive and cytotoxic species. Such species might add to the carbon–carbon double bonds of membrane lipids, producing lipid radicals and ultimately lipid peroxides. Free radicals such as $HO^{\cdot}$ and $Cl_3C^{\cdot}$ might abstract $H^{\cdot}$ from important biomolecules such as nucleic acids and proteins, leaving those molecules are unstable reactive free radicals.

GSH plays a role in both the prevention and repair of free-radical damage. Since sulfur-centered free radicals are less reactive than most carbon- and oxygen-centered radicals, direct reaction of GSH with, for example, $HO^{\cdot}$ yields HOH and $GS^{\cdot}$, a much less cytotoxic species. Combination of two $GS^{\cdot}$ species yields GSSG, which is reduced to two GSH, as described in this Section. Mammalian cells might also contain GSH-dependent enzymes capable of detoxifying free radicals. The role of GSH and glutathione peroxidase in reducing the lipid peroxides formed by the reaction of lipid free radicals with molecular oxygen was described in Section III,B.

D. GSH Reaction with Electrophiles: Mercapturic Acid Synthesis

Reactive electrophiles are formed in the course of normal metabolism and during the metabolism of various drugs and other foreign molecules. Many such electrophiles are reactive and add nonenzymatically to any of a wide variety of cellular nucleophiles, including essential nucleic acids and proteins. Since such reactions generally destroy the functionality of the nucleic acid or protein affected, the cell has evolved enzymes that specifically facilitate the reaction of such electrophiles with GSH. Since GSH is an expendable strongly nucleophilic molecule present in high concentrations, reaction with GSH spares more valuable nucleic acids and proteins from electrophilic damage. Reaction of GSH with electrophiles occurs spontaneously and is also catalyzed by a family of related enzymes, the glutathione *S*-transferases (Reaction 7, Fig. 1). The enzyme-catalyzed reaction is quantitatively important for most electrophiles. The product is no longer reactive and is transported out of the cell. Extracellularly, it is metabolized similarly to GSH; the γ-glutamyl moiety is removed by γ-glutamyl transpeptidase, and the bond between the substituted cysteine residue and glycine is cleaved by a dipeptidase. The resulting *S*-substituted cysteine (i.e., CyS-electrophile) is then taken up by various tissues (e.g., kidneys), and the amino group of the cysteine residue is acetylated in an acetyl-coenzyme A-dependent reaction. The resulting product, an *S*-substituted-*N*-acetyl-L-cysteine, is referred to as a mercapturic acid and is excreted in the urine (Fig. 1; pathway indicated by Reactions 7, 3, 6, and 14).

It is noteworthy that GSH, or at least its essential cysteine residue, is irreversibly lost through the formation of a mercapturic acid; in this important respect mercapturic acid formation differs from all of the other protective functions mentioned. Among the common drugs metabolized to mercapturic acids is acetaminophen (e.g., Tylenol), which is activated in the liver to an electrophile which then reacts with GSH. Although the hepatic supply of GSH is more than adequate for normal doses of acetaminophen, massive doses of the drug produce the reactive metabolite in amounts that deplete hepatic GSH. In the absence of GSH, reaction of the activated drug with other cellular nucleophiles occurs and leads to hepatic damage and, in severe cases, death.

E. GSH as a Coenzyme

GSH acts as a coenzyme in several enzymatic reactions (grouped as Reaction 13, Fig. 1). In the glyoxylase reaction glyoxylase I first catalyzes an internal oxidation–reduction reaction of the hemimercaptal formed nonenzymatically between methylglyoxal and GSH to produce *S*-lactoylglutathione (i.e., $CH_3—CO—CHOH—SG \rightarrow CH_3—CHOH—CO—SG$). Glyoxylase II then catalyzes the hydrolysis of this product to D-lactate and GSH. The *cis–trans* isomerization of maleylacetoacetate to fumarylacetoacetate, a reaction involved in the catabolism of the amino acid tyrosine, similarly requires GSH as a coenzyme. Formaldehyde dehydrogenase, an enzyme reversibly interconverting formaldehyde and formic acid, and two enzymes involved in the formation of certain prostaglandins also exhibit a specific requirement for GSH. Houseflies require GSH to detoxify the insecticide 1,1′-(2,2,2-trichloroethylidene)bis[4-chlorobenzene] (DDT); the latter reaction apparently does not occur in humans. [*See* ENZYMES, COENZYMES, AND THE CONTROL OF CELLULAR CHEMICAL REACTIONS.]

IV. Medical Aspects

A. Inherited Disorders of GSH Metabolism

Inherited deficiencies affecting several of the enzymes of the γ-glutamyl cycle have been reported; none is common. In 1970 systemic deficiency of glutathione synthetase was the first such disorder reported. The affected patient was a 19-year-old mentally retarded boy of normal height and weight, exhibiting signs of organic cerebral damage with spastic quadraparesis and cerebellar disturbances. The biochemical defect was discovered during attempts to elucidate a severe acidosis that developed following a surgical procedure. Although the acidosis was effectively controlled with bicarbonate, further studies established that the patient excreted 24–35 g of 5-oxo-L-proline per day in his urine. Once it was determined that glutathione synthetase was severely deficient in the patient, it became apparent that the near absence of GSH released γ-glutamylcysteine synthetase from feedback control and permitted a substantial overproduction of γ-glutamylcysteine. Since this product could not be converted to GSH, it was metabolized intracellularly to 5-oxoproline and cysteine by γ-glutamylcyclotransferase. Production of 5-oxoproline exceeded the capacity of 5-oxoprolinase, and the excess spilled into the plasma, causing acidosis. 5-Oxoproline was eventually excreted in the urine, leading to the clinical characterization of the disorder as 5-oxoprolinuria. To date 16 patients with systemic glutathione synthetase deficiency have been identified; a smaller number are affected by a deficiency involving only the erythrocytes, a relatively benign condition.

Two patients exhibiting nearly complete deficiency of γ-glutamylcysteine synthetase have been reported. These patients reportedly exhibit hemolytic anemia, spinocerebellar degeneration, peripheral neuropathy, myopathy, and amino aciduria. Erythrocyte GSH levels are less than 3% of normal levels, but other tissues exhibit significantly higher GSH levels (e.g., 25% of normal levels in skeletal muscle). The enzyme deficiency is thus not complete.

Two patients with apparently generalized deficiency of γ-glutamyl transpeptidase have also been described. As expected, these patients cannot degrade GSH released by various tissues (mainly the liver); such GSH accumulates in the plasma, is filtered at the kidney glomerulus, and is excreted in the urine. One patient excreted 850 mg of GSH per day. Further studies established that in neither patient is the enzyme deficiency complete; residual γ-glutamyl transpeptidase catalyzes the extracellular synthesis of γ-glutamylcystine (from GSH and cystine) and γ-glutamyl-GSH (from two molecules of GSH). These products are also found in the urine. Although both patients are moderately retarded, the disorder has been sought mainly among patients of mental institutions, and it is not yet established that γ-glutamyl transpeptidase deficiency accounts for the mental deficit.

Three individuals with 5-oxoprolinase deficiency have been reported. Although the enzyme defect is apparently not complete, the affected individuals excrete up to 9 g of 5-oxoproline per day; this value, equivalent to 70 mmol/day, represents a minimum estimate of the daily γ-glutamyl cycle flux in humans. In contrast to patients with glutathione synthetase deficiency, the individuals with 5-oxoprolinase deficiency have normal tissue GSH levels. The condition appears to be benign.

Since glutathione reductase is a flavin-dependent enzyme, acquired deficiency of glutathione reductase is possible when diets deficient in riboflavin are consumed. Moderate glutathione reductase deficiencies due to inadequate flavin availability are, in fact, common, but appear to be without clinical effect. Severe but incomplete deficiency of the glutathione reductase protein has been documented in the erythrocytes of only three patients, all siblings. One patient suffered a hemolytic crisis after eating fava beans (a source of oxidative stress); two suffered cataracts, possibly due to failure to maintain an adequately high GSH/GSSG ratio. Other signs and symptoms were not reported.

Complete deficiency of glutathione peroxidase has not yet been reported, although there are certain Jewish and Mediterranean populations in which tissue levels are about one-half of normal. Since selenium is necessary for the synthesis of active enzyme, acquired deficiency among individuals consuming a selenium-poor diet is common in some parts of the world. No pathological consequences clearly attributable to these incomplete inherited or acquired deficiencies have been reported. [*See* SELENIUM IN NUTRITION.]

B. Pharmacological Control of GSH Metabolism: Therapeutic Implications

The multiplicity of enzymatic, transport, and protective functions attributed to GSH has stimulated efforts to develop agents useful in the deliberate pharmacological manipulation of GSH levels and GSH-dependent metabolisms. As noted, various drugs (e.g., acetaminophen) and toxic agents (e.g., many aromatic chemicals) cause tissue damage when their detoxification requires GSH in quantities greater than those present in the tissue. High intracellular levels of GSH also minimize radiation damage and could facilitate the repair of radiation-damaged DNA.

With these facts in mind, several approaches to increasing tissue levels of GSH have been developed. Since GSH synthesis is often limited by the availability of cysteine, GSH levels can be moderately increased by direct administration of L-cysteine or L-cysteine precursors [e.g., *R*-thiazolidine-4-carboxylate (i.e., L-thioproline), *N*-acetyl-L-cysteine, and L-2-oxothiazolidine-4-carboxylate]; the latter two compounds are enzymatically converted by L-cysteine intracellularly by *N*-acetylase and 5-oxoprolinase, respectively. More recently, Meister and colleagues have established that, in contrast to GSH, glutathione ethyl ester (i.e., glycyl carboxylate esterified) and related compounds are transported into many mammalian cells. Esterases within cells hydrolyze the carboxylate ester to produce free GSH within the cytosol.

Since tissue GSH levels can be markedly increased in animals given GSH esters, human trials of these compounds are likely in the near future. In this context it should again be noted that GSH itself is not effectively taken up by intact cells. Parenteral or oral administration of GSH does not, therefore, directly increase cellular GSH levels. GSH is, however, broken down extracellularly (see Section II,C) to provide cysteine, which can be taken up and used for GSH synthesis. In this respect GSH is similar to cysteine and other cysteine precursors.

Whereas the maintenance of normal to supranormal tissue GSH levels is normally of benefit to the organism, the ability to selectively decrease tissue GSH levels can facilitate certain therapeutic interventions. In particular, a variety of *in vitro* studies have established that the radiation therapy and certain chemotherapies of tumors is more effective if tumor GSH is first depleted. Several regimens designed to effect such depletion have been developed. One of the least toxic and most effective approaches relies on the administration of buthionine sulfoximine (BSO), a specific and potent inhibitor of γ-glutamylcysteine synthetase. Following administration of BSO, GSH synthesis is blocked, whereas GSH utilization, primarily transport of GSH out of the cell, continues unabated; in tissues with moderately high rates of GSH turnover, GSH depletion occurs within a few hours. Human tumors grown as xenografts in athymic nude mice have been shown to be sensitized to both radiation and various chemotherapies following administration of BSO.

It is also noteworthy that drug resistance in tumors (i.e., the clinical observation that doses of anticancer drugs initially causing tumor remission are not effective in recurrent tumors) has, in some cases, been correlated with increased levels of GSH in the recurrent tumor. Preliminary studies in human ovarian carcinoma and other tumors indicate that administration of BSO abolishes the increased GSH levels and restores drug sensitivity. Finally, studies with mice infected with *Trypanosoma brucei brucei*, an animal model of the human parasitic disease African sleeping sickness (i.e., trypanosomiasis), indicate that treatment with BSO depletes trypanosomal as well as host GSH; treated mice exhibit extended survival and are, in some cases, cured. Further studies have suggested that GSH-depleted trypanosomes were unable to control the oxidative stresses inherent in their normal metabolic reactions; GSH-depleted cells of the host (i.e., the mouse) survived because they had alternative defenses against oxidative stress and depended on metabolic reactions that generated lower levels of cytotoxic oxidants.

Acknowledgment

Studies performed in the author's laboratory were supported in part by National Institutes of Health grant DK26912. I thank Ernest B. Campbell and Michael A. Hayward for expert technical assistance in these studies.

Bibliography

Griffith, O. W. (1985). Glutathione and cell survival. *In* "Cellular Regulation and Malignant Growth" (S. Ebashi, ed.), pp. 292–300. Jpn. Sci. Soc. Press, Tokyo.

Griffith, O. W., and Friedman, H. F. (1990). Inhibition of metabolic drug inactivation: Modulation of drug activity and toxicity by perturbation of glutathione metabolism. *In* "Synergism and Antagonism in Chemotherapy" (T.-C. Chou and D. Rideout, eds.). Academic Press, San Diego, California. In press.

Meister, A. (1989). Metabolism and function of glutathione. *In* "Glutathione: Chemical, Biochemical and

Medical Aspects, Part A'' (D. Dolphin, O. Avramović, and R. Poulson, eds.), pp. 367-474. Wiley (Interscience), New York.

Meister, A., and Anderson, M. E. (1983). Glutathione. *Annu. Rev. Biochem.* **52,** 711–760.

Meister, A., and Larsson, A. (1989). Glutathione synthetase deficiency and other disorders of the γ-glutamyl cycle. *In* ''The Metabolic Basis of Inherited Disease'' (C. R. Scriver, A. L. Beaudet, W. S. Sly, and D. Valle, eds.), 6th ed., pp. 855–868. McGraw-Hill, New York.

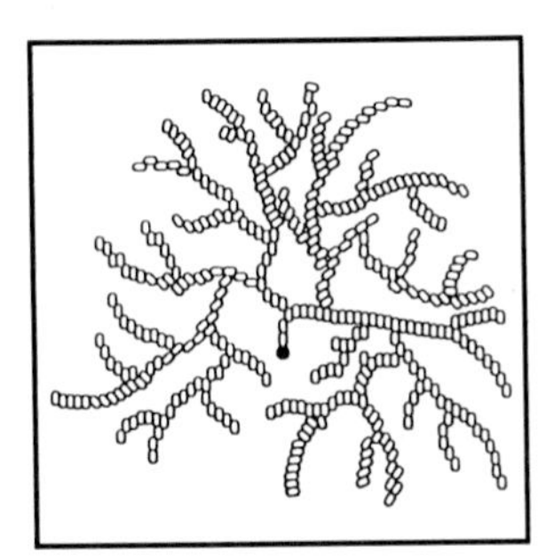

Glycogen

NEIL B. MADSEN, *University of Alberta*

Glossary

Active site or catalytic site Location on an enzyme protein where the substrate(s) bind and are catalytically transformed into product(s)

Allostery Control mechanism in enzyme activity in which a site other than the active site is occupied by an allosteric effector molecule, which may be an activator or an inhibitor

Amino terminal Beginning of a polypeptide chain is occupied by an amino acid with a free amino group, its carboxyl group joined to the amino group of the next residue

Coenzyme Small molecule that binds at the active site of an enzyme and participates in the catalytic process

Conformation One of many shapes or structures that a protein molecule may assume

Feedback inhibitor Compound that may be the last in a metabolic pathway and that inhibits an early enzymatic step in the pathway, usually via allostery

Metabolite Any small molecule that undergoes enzymatic transformation in the cell

Polypeptide chain Amino acids combine by the removal of water between the carboxyl group of one and the amino group of the next (forming peptide bonds), yielding long polypeptide chains

Subunit Most enzymes, which are proteins, are dimers or tetramers of polypeptide chains, each folded into a globular shape termed a *subunit*, which may or may not be identical

T and R states T (taut) state is considered the enzymatically inactive conformation of an allosteric enzyme, whereas the R (relaxed) state is the active conformation

GLYCOGEN is the major storage form of glucose in animal organisms. Glucose molecules are joined together to form long chains with many branch points, resulting in the buildup of huge spherical molecules that are found in large amounts in muscle and liver tissues. In muscle the breakdown of glycogen provides anaerobic (i.e., in the absence of oxygen supply) energy for short, intense bursts of muscular activity (e.g., sprints) and contributes aerobic (i.e., with oxygen available) energy for long sustained activity (e.g., marathon running). Liver glycogen is a reserve source of blood glucose during short periods of fasting (e.g., overnight). During the synthesis of glycogen, several enzymes are employed to transform free glucose to phosphorylated and activated forms, and finally the enzyme glycogen synthase adds the glucose units to the growing chains, whereas another enzyme inserts the branch points. When glycogen breakdown is required, the enzyme phosphorylase cleaves the glucose chains, using phosphate, while a debranching enzyme removes the branch points. Glycogen phosphorylase and synthase are under reciprocal and coordinated controls so that glycogen synthesis and degradation may not occur simultaneously. Thus low blood sugar levels cause hormonal changes that activate, via several intermediate steps, liver phosphorylase and inactivate the glycogen synthase. In muscle the breakdown of glycogen is coordinated with contraction because both processes are stimulated by calcium, which is released in response to nerve impulses. The structure of phosphorylase has been determined in great detail, and it is possible to explain the molecular mechanisms for the various physiological controls that regulate this key enzyme of glycogen metabolism.

I. The Glycogen Molecule

A. Discovery and Chemistry

Claude Bernard, the great French physiologist, discovered glycogen in 1857 during the course of his studies on sugar metabolism and showed that the liver synthesized the substance and used it to maintain the correct level of blood sugar, enunciating also the principle of the maintenance of equilibria for such parameters as blood sugar. Figure 1 is an electron micrograph of a single glycogen particle or molecule isolated by gentle means from rabbit liver. Such rosette-like structures are typical of liver glycogen and have a molecular weight of as much as 500 million, whereas the glycogen molecule from muscle is the size of one of the small spheres making up the liver particle and has a molecular weight of as much as 10 million. All the enzymes needed for the synthesis and degradation of glycogen, as well as those concerned with regulation, are found bound to the glycogen particles. Glycogen is exclusively composed of the sugar glucose and a small amount of a protein known as glycogenin on which the formation of glycogen begins but which we will not consider further. Because each glucose residue in glycogen has a molecular weight of 162, we can see that there can be from several thousand to more than a million glucose residues in one glycogen molecule. As noted below, some tissues store large amounts of sugar in the form of glycogen, but so much sugar as individual glucose molecules would cause the cells to burst from excessive osmotic pressure, whereas a single glycogen molecule exerts no more osmotic pressure than a single glucose molecule.

Glycogen is a large branched polymer of glucose. As shown in Fig. 2, long chains of glucose are formed by the elimination of water between carbon 1 of one glucose and carbon 4 of another, forming α-1,4 links. Every four to five glucose molecules in the chain, a branch is formed by an α-1,6 link. Figure 2 shows two outer branches of the molecule with the 4-hydroxy groups available for further elongation. In principle, there would be one reducing 1-hydroxy group per glycogen molecule, but in fact this is coupled to a tyrosine residue of the initiator protein, glycogenin. Continued lengthening of the glucose chains with the introduction of branch points builds up a large tree-like structure, which is illustrated schematically as ''mature glycogen molecule'' in Fig. 3.

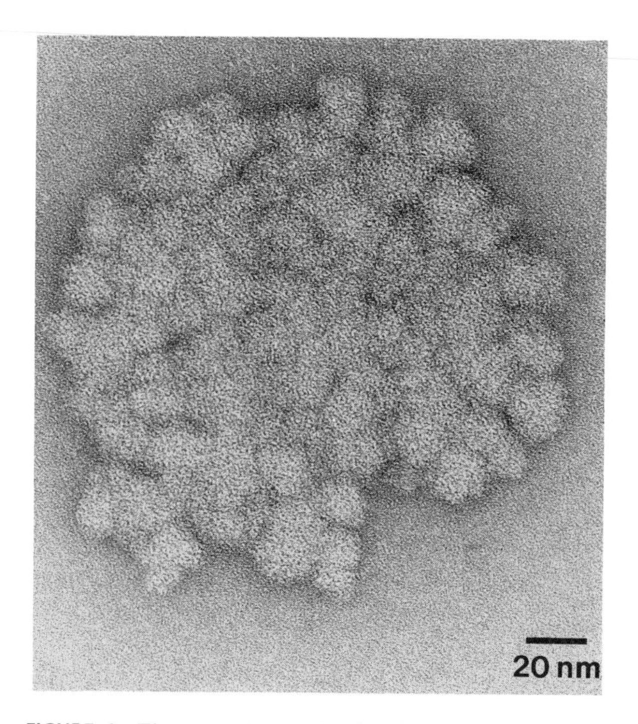

FIGURE 1 Electron micrograph of a single glycogen particle from liver, stained with 2% uranyl acetate. (Courtesy of Dr. D. G. Scraba and Mr. R. D. Bradley. Univ. of Alberta.)

B. Occurrence in Human Tissues and Biological Roles

Tissues containing glycogen include liver, skeletal muscle, heart, kidney, and brain, with most others having at least some. The liver of a well-fed man is approximately 5% glycogen, the skeletal muscle 1.3%, and the brain only 0.1%, so that liver and muscle glycogen are the most significant physiologically. Although the content of glycogen in the liver is higher than in muscle, because of the different tissue masses, the liver can store only 80–100 g of glycogen compared with about 450 g in the total muscle.

In skeletal muscle, the role of glycogen is to provide a source of phosphorylated glucose units (especially 6-phosphate), which can enter the glycolytic pathway and provide energy for muscular contraction. For example, during a sprint, there is

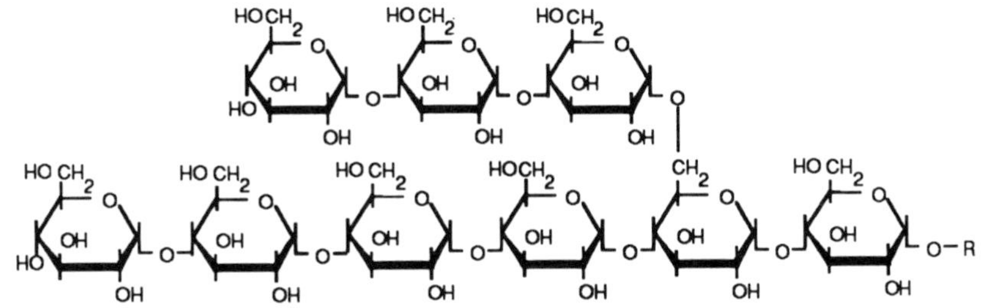

FIGURE 2 Two outer branches of a glycogen molecule, showing how the glucose units are joined by α-1,4 links (*horizontal rows*), with a branch point consisting of an α-1,6 link connecting the two branches. The two branches terminate in nonreducing ends, whereas R represents the rest of the molecule.

sufficient glycogen in the muscles to provide energy for only 20 sec of maximal activity, and most of the energy is provided by the breakdown of glycogen and the anaerobic production of adenosine triphosphate (ATP) required for muscle contraction, because the extreme energy demands do not permit time for external sources (e.g., blood glucose and fatty acids), which produce ATP through a more elaborate pathway, to play a role. Marathon runners, however, employ an aerobic metabolism and in the later stages derive 60% of their energy from oxidation of fatty acids as well as blood sugar derived from the breakdown of liver glycogen. Muscle glycogen stores are depleted slowly in long-distance running, but exhaustion is associated with its complete utilization. The regulation of glycogen utilization is obviously under exquisitely sensitive control to meet a variety of demands, as discussed in Section III. [*See* ADENOSINE TRIPHOSPHATE (ATP).]

As first suggested by Claude Bernard, the liver has the chief responsibility for the maintenance of blood sugar levels at a safe minimum and also contributes to the reduction of high levels. When blood sugar levels rise after a meal, the liver rebuilds its glycogen stores to maximal levels. During short fasting periods (e.g., overnight) these stores are depleted to replenish the blood glucose used by other tissues such as the brain. For example, a 24-hr fast lowers the glycogen content from 5% to 0.7%. During longer fasts, some tissues begin to metabolize fatty acids and proteins for energy, but the liver then has to make glucose from protein to supply the brain. The glycogen content of other tissues (e.g., the heart) can provide an emergency supply of energy during brief periods of oxygen deprivation.

II. Metabolism

A. Enzymes Involved in the Formation of Glycogen

1. Phosphoglucomutase is the enzyme that links the glycolytic pathway to glycogen metabolism. Glucose 6-phosphate, formed in the glycolytic pathway by hexokinase, is diverted from the degradative pathway by its conversion to glucose 1-phosphate by this first enzyme. The reaction is shown in Fig. 3, and it is freely reversible under physiological conditions, because when the enzyme operates long enough on either of the glucose phosphates, it reaches a chemical equilibrium in which there is 95% of the 6-phosphate and 5% as the 1-phosphate. Phosphoglucomutase can operate in both the formation and the degradation of glycogen.

2. UDP-glucose pyrophosphorylase takes the glucose 1-phosphate formed in the previous reaction and reacts it with uridine triphosphate (UTP) to form pyrophosphate (PP_i) and UDP-glucose (uridine diphosphate glucose), as shown in Fig. 3. Although this reaction is freely reversible, it is rendered irreversible under physiological conditions by the hydrolysis of one of the products, PP_i, carried out by the pyrophosphatase ubiquitous in cells. The other product, UDP-glucose, is an activated form of glucose, which is well-suited as the donor of glucose for the synthesis of glycogen and other large sugar molecules.

3. Glycogen synthase adds glucose units from UDP-glucose to the 4-hydroxy groups of the glucosyl termini of preexisting glycogen molecules known as primers, causing the formation of long chains of glucose molecules joined in α-1,4-glycosidic linkages, as shown in Fig. 3. The original primers are short glucose chains joined to the tyrosine residue of the initiator protein. Because glycogen synthase catalyses the first unique, committed reaction leading to glycogen formation, it is under rigorous metabolic and hormonal control, as we will discuss in Section III.

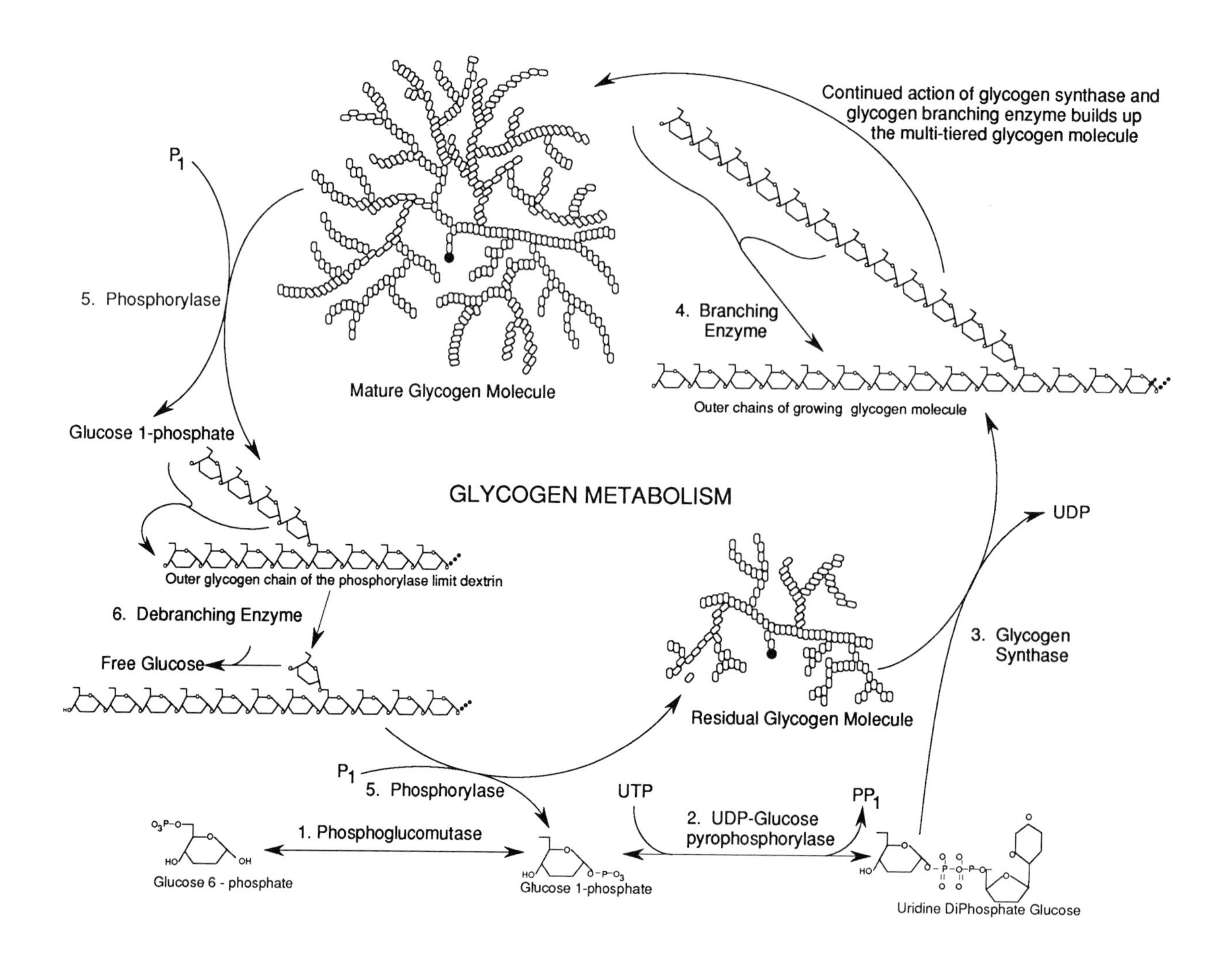

FIGURE 3 Scheme showing the actions of enzymes that synthesize and degrade glycogen.

4. Branching enzyme forms branches in the long linear glucose chains resulting from the previous reaction. Long linear chains become insoluble, but with the introduction of branches, a compact, highly soluble spherical molecule comes into being. The number of terminal residues is also increased enormously, so that both the formative and degradative enzymes have more sites on which to work, resulting in faster rates. Branching enzyme exhibits a precise specificity, taking a block of six to seven glucose units from a growing chain, breaking the α-1,4 link, and transferring the block to the 6 position of a glucose residue in an adjacent chain to form a new α-1,6-linked branch, as illustrated in Fig. 3. The new branch is from four to six residues from the next interior branch. It is interesting to note from thermodynamic data that the formation of branches is favored over the reverse reaction, and indeed, this enzyme does not play a role in the degradation of glycogen, the removal of the α-1,6 links being effected by a separate enzyme with a different mechanism.

B. Enzymes Involved in the Degradation of Glycogen

1. Phosphorylase was discovered in 1936 when Carl and Gerty Cori demonstrated that the addition of inorganic phosphate to muscle extracts resulted in the conversion of glycogen to a new hexose monophosphate, which they proved to be glucose 1-phosphate. They also showed that the latter compound could be converted into glucose 6-phosphate by muscle and that the phosphorylase reaction was stimulated by adenosine monophosphate (AMP), perhaps the earliest demonstration of allostery. By 1940, they and their colleagues had crystallized phosphorylase (placing it among the first crystalline enzymes) and elucidated the chemistry of the phosphorolysis reaction that it catalyses. As illustrated in Fig. 4, the enzyme carries out the cleavage of the terminal glycosidic bond of the glucose chains in glycogen, using inorganic phosphate (phosphorolysis) instead of water, which is used in many cleavage reactions, (hydrolysis), thus incorporating the phosphate at the C-1 position of the released glucose unit. Although the equilibrium constant of the phosphorolysis reaction is only 0.28 (meaning that only 28% of added phosphate would be converted to glucose 1-phosphate), under physiological conditions the reaction is driven in the direction of glycogen degradation because the ratio of inorganic phosphate to glucose 1-phosphate is 100 or more, so that a high glycogen concentration could not be maintained without the stringent control of phosphorylase, as discussed in Section III. Furthermore, because of the design of its active site, phosphorylase cannot cleave the final four glucose units from an α-1,6 branch, as shown in Fig. 3. Because these branch points are symmetrically arranged in concentric tiers in the spherical glycogen molecule, the action of phosphorylase results in a glycogen with short outer chains, known as a *limit dextrin*.

Phosphorylase contains in its active site a tightly bound coenzyme that is essential for its function. This is pyridoxal phosphate, a derivative of vitamin B_6, and because there is so much phosphorylase in muscle, more than 80% of the vitamin B_6 stores in muscle are bound to this enzyme. Nevertheless, a deficiency of this vitamin shows up as disorders of protein metabolism, in which it plays major roles, rather than of glycogen metabolism. The manner in which pyridoxal phosphate functions in phosphorylase is unique for this coenzyme, involving the interaction of its phosphate with that of the substrate, whereas, in other enzymes, different parts are involved.

2. Debranching enzyme acts on the phosphorylase limit dextrin in the manner depicted in Fig. 3. It transfers a block of three glucose units from one outer branch (the one joined to another branch by an α-1,6 link) to the outer branch containing the "side-chain." It is thus an α-1,4 $\rightarrow$ α-1,4-glycosyl transferase. This action leaves a single glucose unit joined by an α-1,6 link to the "main-chain." The debranching enzyme then hydrolyzes this link, releasing a free glucose molecule, and thus exhibiting α-1,6-glucosidase activity. It may be noted that both the transferase and glucosidase activities of the debranching enzyme are carried out by a single monomeric polypeptide chain of 160,000 molecular weight, the first example of a "double-headed" enzyme in higher organisms. As shown in Fig. 3, the action of the debranching enzyme on the phosphorylase limit dextrin results in a smaller glycogen molecule with long outer branches on which phosphorylase may act further. Successive alternating actions of phosphorylase and debranching enzyme will obviously reduce the initially large glycogen molecule to the small size found in liver after a fast or in muscle after prolonged and exhausting exercise.

3. Glucose 6-phosphatase is the final enzyme in the pathway for complete degradation of glycogen

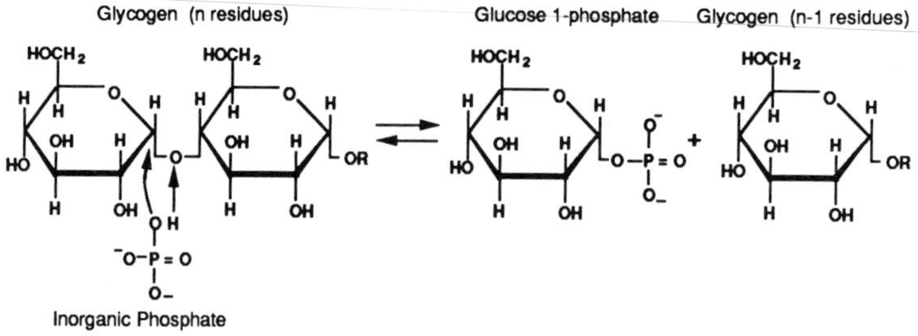

FIGURE 4 The reaction catalyzed by glycogen phosphorylase. The carbon numbering scheme used for glucose is indicated for glucose 1-phosphate.

to free glucose, and removes the phosphate group after phosphoglucomutase has changed the glucose 1-phosphate, arising from glycogen, into glucose 6-phosphate. It is especially important in liver where its hydrolysis of glucose 6-phosphate permits the export of glucose to the blood, thus fulfilling the liver's role in maintaining blood sugar during exercise and fasting. It is also found in the kidney and intestines but not in muscle and brain, because the latter tissues are the major users of glucose.

C. Glycogen Storage Diseases

A number of genetically determined diseases characterized by accumulations of glycogen in liver, muscle, or other organs have been traced to inherited deficiencies of the various enzymes of glycogen metabolism. The first and most famous of these, characterized by the massive amount of normal glycogen in the liver, was described by E. von Gierke in 1929. It causes low blood sugar and failure to thrive in infants unless carefully controlled by dietary measures. In 1952, Carl and Gerty Cori showed that glucose 6-phosphatase was missing from the livers of these patients, allowing no release of glucose from the degradation of glycogen. In another disease, the debranching enzyme is missing, and normal glycogen with short outer branches accumulates in liver and muscle. Absence of the branching enzyme results in a normal amount of glycogen in liver and spleen, but it has long outer branches and results in liver failure during infancy. A final example concerns the absence of phosphorylase from muscle. Patients with this problem are unable to perform vigorous exercise because of painful cramps. It is obvious that there would be problems because the vital energy derived from glycogen degradation is not available, but the immediate cause of the cramps appears to be the accumulation of the breakdown product of ATP (the immediate energy source for contraction) [i.e., adenosine diphosphate (ADP)], because the latter cannot be recycled back to ATP.

III. Regulation of Glycogen Metabolism

A. Metabolically Interconvertible Enzymes

1. Covalent attachment of a phosphate by phosphorylase kinase to a specific serine residue on phosphorylase makes this enzyme active under physiological conditions. The phosphorylase that was discovered in 1936 required AMP for activity and is now termed *phosphorylase b*. In 1942 a new form was isolated and crystallized, designated *phosphorylase a,* which does not require AMP for activity, although it binds it even more tightly. Binding of AMP improves activity of the enzyme at low substrate concentration. Eventually it was discovered that both the *a* and *b* forms are dimers of a 97,400–molecular weight subunit containing 842 amino acids in a single polypeptide chain. Phosphorylase *a* contains a phosphorylated serine at position 14, (counted from the amino terminal), in each of its subunits. This is the only chemical difference between the *a* and *b* forms, but there are changes in the three-dimensional structure of the protein, which will be discussed in Section IV,B. The phosphoserine is close to the binding site for the AMP and may be regarded as a covalently bound activator. Both phosphoserine and AMP are on the opposite side of the dimer from the catalytic active site, which can be defined by the presence of the coenzyme, pyridoxal 5′-phosphate, bound to lysine 680. Phosphorylase *b* is subject to allosteric inhibition by glucose 6-phosphate, which may be regarded as a feedback inhibition by the chief product of glycogen breakdown in muscle (and the first

metabolite in glycolysis). Energy control of phosphorylase *b* is exerted via the allosteric inhibition by ATP (which binds to the same site as does AMP) and the allosteric activation by AMP, as well as the substrate activation by inorganic phosphate. The latter two compounds indicate low energy levels, whereas ATP indicates a high energy level. Phosphorylase *a* has "escaped" allosteric control because it no longer requires AMP nor is it inhibited significantly by ATP or glucose 6-phosphate. Phosphorylase activity can thus be turned on or off by hormonal or nervous controls regardless of the levels of metabolites, permitting a more sophisticated whole organism regulation to be imposed on the cellular level.

2. Phosphorylase kinase is a complicated enzyme consisting of four copies of each of four different types of subunits, termed α, β, γ, and δ. The entire molecule can thus be denoted $\alpha_4\beta_4\gamma_4\delta_4$ and has a total molecular weight of 1.3×10^6. The function of each subunit has been defined by many years of experimental work, and the catalytic activity has been assigned to the γ subunit. The δ subunit was found to be the calcium binding protein calmodulin, and its function explains the obligate requirement of the kinase for calcium ion, which confers partial activity on the enzyme. Full activity is only achieved when phosphate groups are incorporated into specific serine residues in both the β and α subunits, so that these two subunits are concerned with regulating the activity via intersubunit-transmitted conformational changes. The enzyme that phosphorylates the β and α subunits is the cyclic AMP (cAMP)-dependent protein kinase.

3. cAMP-dependent protein kinase is activated, as implied, by cAMP, which is formed in response to hormonal signals, as described below. The protein kinase is composed of two regulatory subunits and two catalytic subunits, forming an inactive tetramer, which can be denoted R_2C_2. When cAMP binds to the regulatory subunits, the catalytic subunits dissociate as free active monomers, which can phosphorylate the phosphorylase kinase. In addition, the cAMP-dependent protein kinase also phosphorylates specific serine residues on glycogen synthase *a*, forming synthase *b*, but in this case, unlike the phosphorylase case, the phosphorylated form of the synthase is physiologically inactive. We will return to this enzyme in Section III,B.

4. cAMP is known as a "second messenger" because it is formed in response to hormones (the first messengers) binding to specific receptors on the outside of the cell membrane. The hormones do not enter the cell but cause conformational changes to be transmitted across the membrane (via proteins that span the membrane) to the enzyme adenylate cyclase on the inside of the cell membrane, causing it to carry out the transformation of ATP into cAMP, as diagrammed in Fig. 5. Among the better known hormones, adrenalin is secreted in response to emotional states such as fright or anger and it binds to specific receptors on muscle cells where it initiates the chain of events that leads to the breakdown of glycogen, as detailed in Section III,B. Glucagon, however, is secreted in response to low blood sugar, and it binds to specific receptors on liver cells, which again leads to the breakdown of glycogen into the form of glucose that is released into the bloodstream.

B. Glycogen Cascade

1. The glycogen cascade is diagrammed in Fig. 6. The term *cascade* is used because a series of events cascade one into the next to connect the primary hormonal or nervous signals with the ultimate effect on the breakdown or synthesis of glycogen. We have described most of the components of the glycogen cascade, and in this section we will discuss its operation and the principles involved.

2. Hormonal stimulation of glycogen breakdown is initiated by the binding of hormones to their specific receptors. Glucagon binding to its receptors on the liver cell, or epinephrine to the muscle cells, activates adenylate cyclase and causes its conversion to cAMP. A separate enzyme, known as a diesterase, removes the cAMP when the hormonal stimulus is removed. The newly formed second messenger, cAMP, binds to the regulatory subunits of the protein kinase, releasing the active catalytic subunits, which can now cause the phosphorylation of the β and α subunits (in that order) of phosphorylase kinase. It should be noted that the phosphorylated phosphorylase kinase is not active unless sufficient calcium ion is present. In the liver, the calcium level is probably raised by hormonal action, and also the phosphorylated kinase is more sensitive to calcium and can act with the normal low cellular level. In muscle it is unlikely that there is much phosphorylase kinase activity until calcium is released from intracellular stores by the action of the same nerve impulse that caused muscular contraction. Thus we see a vital link between contrac-

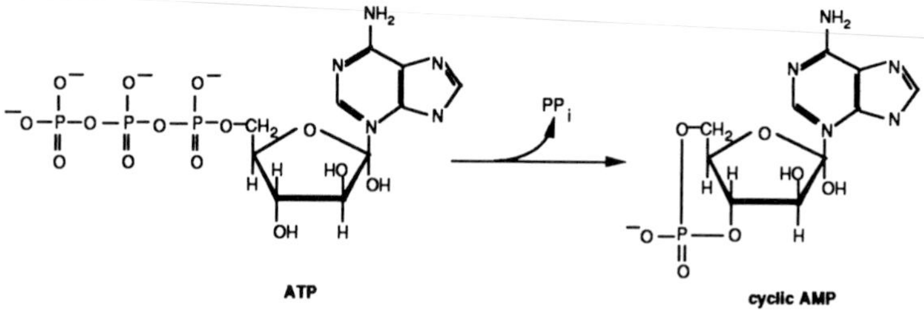

FIGURE 5 The reaction catalysed by adenylate cyclase.

tion and the breakdown of glycogen, which provides energy for it. Even in the absence of hormonal stimulation, calcium can cause sufficient stimulation of phosphorylase kinase to promote glycogen breakdown, a good provision because epinephrine release is not normally associated with many forms of exercise (e.g., walking up stairs), but a large amount of energy is required. On Fig. 6 we attempt to show that nervous stimulation of muscles causes not only the contraction but also the breakdown of glycogen to provide energy for the contraction, and the unifying link between the two processes is the release of calcium ion.

The next step in the glycogen cascade is the action of phosphorylase kinase to transform phosphorylase *b* to *a* by phosphorylation of serine 14 of the phosphorylase. We have already discussed the fact that this causes the physiological activation of phosphorylase, so that it can now carry out the conversion of glycogen into glucose 1-phosphate, leading to glucose 6-phosphate in muscle and free glucose in liver.

3. Amplification of signal is one of the most interesting features of the glycogen cascade. Four different enzymes are interposed between the hormonal signal and the phosphorolysis of glycogen, and the first three of these act, directly or indirectly, to activate the next enzyme in the sequence. Because an enzyme is a biological catalyst and each molecule acts on many molecules of its substrate per second, each time an enzyme in the cascade is "turned on," it causes an amplification of the signal, much as turning on a radio tube or transistor results in a current far more powerful than the activating current. The threefold repetition of this type of amplification in the glycogen cascade allows a small concentration of hormone in the blood, less than 1×10^{-8} M, to activate 2×10^{-7} M protein kinase, thence 4×10^{-6} M phosphorylase kinase and finally 1×10^{-4} M glycogen phosphorylase (concentrations in muscle tissue). In response, the rate of glycogen breakdown caused by strenuous muscular exercise (e.g., sprinting) increases by about 1,000-fold, from 0.05 μmole/min/g of tissue in resting muscle to 60 μmole/min. A concomitant feature of the glycogen cascade is that it takes a little time for the maximal rate of glycogen breakdown to be achieved, perhaps 2–3 sec, because the signal has to be passed through several steps.

4. Reciprocal effects on the breakdown and synthesis of glycogen are built into the regulation of glycogen metabolism and are illustrated to some extent in Fig. 6. The first and most important of these is that cAMP-dependent protein kinase phosphorylates not only phosphorylase kinase but also glycogen synthase, resulting in the inactivation of the latter under physiological conditions. This feature makes obvious sense because it prevents the resynthesis of glycogen from glucose phosphates when the latter are needed elsewhere. Phosphorylase kinase can also act on glycogen synthase to ensure further that it is switched off when glycogen breakdown is required.

5. Insulin also has a profound effect on glycogen metabolism, an effect usually opposite to the two hormones epinephrine and glucagon, which turn on glycogen breakdown. In contrast to glucagon, insulin is released from the pancreas in response to high blood sugar, such as occurs after a meal. In some tissues (e.g., muscle), it facilitates the transport of glucose across the cell membrane, as illustrated in Fig. 6. In liver it induces the formation via protein synthesis of a tissue-specific form of hexokinase, leading to the rapid phosphorylation of glucose to glucose 6-phosphate. Finally, it activates the protein phosphatase, which acts on both phosphorylase and glycogen synthase, resulting in the inactivation of the former and the activation of the latter. It may be seen that all these effects of insulin are consistent with the increase in glycogen synthesis, which has long been known to accompany insulin secretion. It may be added that although insulin leads to activation of the protein phosphatases that hydrolyze phosphate off the serine residues of pro-

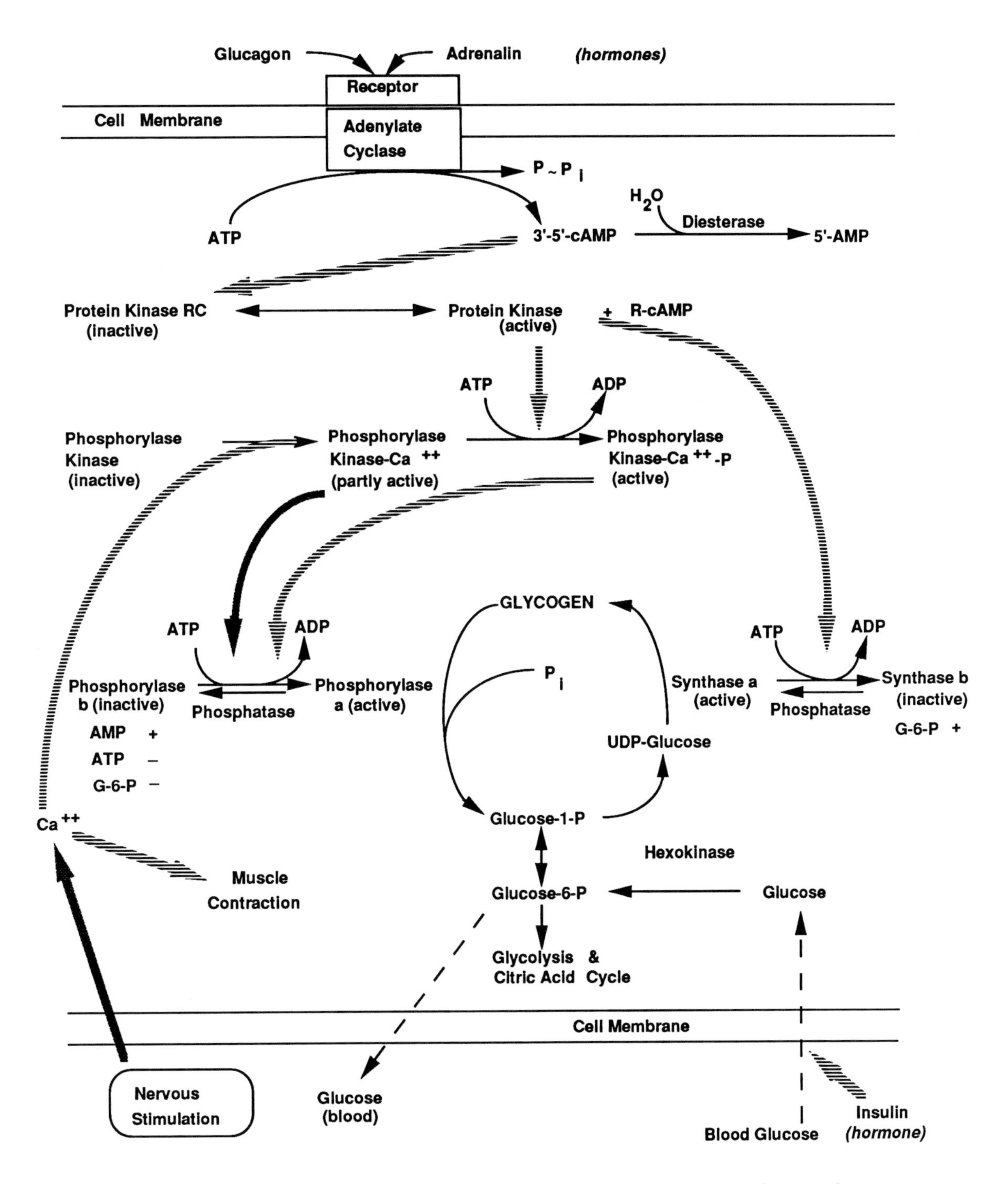

FIGURE 6 Scheme to illustrate the control of glycogen metabolism, including the glycogen cascade (turn on of glycogen breakdown in response to glucagon, or epinephrine, and calcium) and the effect of insulin.

teins, hormones that raise cAMP levels cause the indirect inhibition of these phosphatases, another example of the antagonism between the two types of hormones. [*See* INSULIN AND GLUCAGON.]

IV. Structure–Function Relations in Phosphorylase

A. Structure of the Molecule in the Crystal

The powerful technique of X-ray crystallography permits the determination of protein structure to atomic resolution, 2 Å or less, so that the position of the atoms can be determined in three-dimensional space. Phosphorylase is the only one of the enzymes concerned with glycogen metabolism to have had its structure determined, and this was a formidable task because there are 842 amino acid residues in each subunit, plus the coenzyme and a phosphate, meaning that there are almost 7,000 nonhydrogen atoms (determining the position of hydrogens in proteins is beyond our capability at this time). The phosphorylase structure has considerable general interest because it is under so many different kinds of control and binds so many different types of molecules, and we can gain insights into these features by the technique of X-ray crystallography.

Color Plate 12 shows a computer graphics representation of the normal dimeric form of the phosphorylase *a* molecule obtained by X-ray crystallography, with each sphere representing one nonhydrogen atom but slightly larger than usual to represent the hydrogens we know are present. In this representation of the dimer, one subunit is shaded lightly and its identical neighbor is dark. The two subunits are related by a twofold axis of symmetry such that if we take any feature of one and rotate it by 180° about the center, the identical feature is generated on the other subunit. Note how intimately the two subunits interact at what we call the subunit interface. It is hard to visualize how we could alter the shape of one subunit without forcing the other to also change shape, thus conserving symmetry. In fact, this intuitively obvious physical fact is the basis of allosteric cooperativity. When a small molecule binds to a specific site on one subunit and forces a change in conformation, the other subunit must also change shape, making it easier for a second molecule to bind to its specific site.

It is possible to determine the location of the binding sites for small molecules and proteins, as well as details of their interactions, by X-ray crystallography. For example, the binding of AMP to phosphorylase *a* was determined by soaking a crystal in a solution of AMP, thus allowing the nucleotide to diffuse through the solvent channels between the protein molecules and bind to its specific sites, one per subunit. When the crystal was analyzed by X-rays and an electron density map calculated, the extra density caused by the AMP was found to be superimposed on that because of the native protein. By this means, the binding sites of the substrates, activators, and inhibitors of phosphorylase could all be located. On Color Plate 12 we indicate the entrance to the active site, the actual site where the glucose 1-phosphate or phosphate substrates bind, close to the phosphate of the pyridoxal 5′-phosphate, located deep within the subunit. Incidentally, this location of the active site explains why phosphorylase can only remove glucose units from glycogen to within four from a branch point, because it takes four units to reach from the outside to the site of action. Glucose also binds within this active site, occupying much the same position as the glucose part of glucose 1-phosphate, and so it inhibits by competing with the substrate. Also indicated on Color Plate 12 is the binding site for glycogen, determined by the binding of short glucose polymers. This is the site by which phosphorylase binds tightly to some of the terminal branches of the large glycogen molecule, but it acts on other terminal branches that can go in and out of the active site while the enzyme remains bound to the glycogen. We know from kinetic studies, not crystallographic, that glycogen synthase also has binding sites for glycogen separate from its catalytic sites.

We call the side of the phosphorylase dimer shown in Color Plate 12a the catalytic face of phosphorylase because it contains the active sites and also it is this side that binds to the substrate glycogen. If we turn the molecule around by 180°, we look at the opposite side, which we term the *control face* because on this side is found the binding site for AMP as well as the serine 14 that becomes phosphorylated. The control face of the dimer is depicted in Color Plate 12b, in a different molecular graphics coding than that used in Color Plate 12a. On this model we indicate the site where AMP binds when present, the phosphate of serine 14 in phosphorylase *a*, and the amino-terminal peptide that contains the serine 14.

B. Regulation of Phosphorylase Studied at Atomic Resolution

We have mentioned that glucose inhibits by competing with glucose 1-phosphate or inorganic phosphate at the active site. The crystals grown in the presence of glucose contain an inactive form of phosphorylase *a*, not only because of the presence of glucose but also because glucose stabilizes a loop of the polypeptide chain, which must move to allow the phosphate substrates to bind. The conformation of phosphorylase *a* induced by glucose is the inactive T state. Substrates and AMP counteract glucose and stabilize the active R state. Phosphorylase phosphatase binds to the R state but cannot act on the serine 14 phosphate because the latter is tucked into a fold in the protein. In the glucose-inhibited T state, the phosphate of serine 14 is pushed outward enough by the binding of glucose to allow phosphorylase phosphatase to act on it. In the liver, when the glucose level rises after a meal, this change provides a mechanism to inactivate phosphorylase *a* by converting it to phosphorylase *b*, ensuring that the buildup of glycogen in response to increased glucose will not be hindered. Furthermore, the phosphorylase phosphatase, after inactivating the phosphorylase, can then activate glycogen synthase. These metabolite controls on liver glycogen metabolism are in addition to the hormonal controls discussed earlier, and the exact relation between the two is still being worked out. However, via X-ray crystallography, we can see the changes in the molecule induced by the competing inhibitors, activators, and substrates, and thus provide a molecular explanation for the regulation of the phosphatase. It is tempting to speculate that the inhibition by glucose may have been a primitive feedback regulation that has evolved into a regulation of the interconversion of active and inactive phosphorylases.

It has been mentioned before that glucose 6-phosphate, which can be considered the metabolic end product of glycogen breakdown in the muscle, inhibits phosphorylase *b* by competing for the same site as does AMP. Phosphorylase *b* normally exists in the inactive T state, but AMP causes a conformational change leading to the active R state, which binds substrate much better, whereas substrate improves the binding of AMP. By X-ray crystallography we can see the communication between the AMP and catalytic sites, some 30 Å apart, as evidenced by changes in the connecting protein structure. ATP also inhibits phosphorylase *b* by binding to the AMP site and preventing the latter from binding and activating. Thus we have a molecular mechanism for the control of phosphorylase by energy demands, as discussed in Section III,A.

The most sophisticated regulation of phosphorylase is the conversion between the *b* and *a* forms by covalent phosphorylation and dephosphorylation, respectively. Because both forms of phosphorylase have had their structures determined independently to high resolution by scientists in two different laboratories, who then compared their results, we can assess the effects of phosphorylating serine 14 on the structure and hence interpret the effects on function. In phosphorylase *b*, the first 16 residues from the N terminal cannot be "seen" crystallographically, and this, with other evidence from protein chemistry, suggests that this segment of polypeptide chain is disordered and waving about, free from the main body of the protein. In phosphorylase *a*, residues 5–16 are ordered and lie across the interface between the two subunits, as pointed out in Color Plate 14, with the negative phosphate of serine 14 making charge interactions with positive arginine residues from both subunits. This extra contact between the subunits accounts for their lesser tendency to dissociate, and their tighter coupling of symmetry-related changes. In addition, the binding site for AMP is better formed, so that it binds AMP 100 times more strongly, and there are changes at the active site, which lead to tighter binding of substrates. The proportion of phosphorylase *a* in the active R conformation is 100 times more than for phosphorylase *b*, accounting for its "escape" from allosteric controls, discussed above, and the energy for this conformational shift is derived from the neutralization of positive charges by the phosphate on serine 14, allowing the amino-terminal peptide to bind the subunits more tightly in a more favorable conformation. The X-ray crystallographic studies on phosphorylase are beginning to provide answers on the mechanisms for regulation of enzymes via both covalent modification and allostery, and we can look forward to more detailed explanations in the future.

Bibliography

Boyer, P. D., and Krebs, E. G., ed. (1986). "Control by Phosphorylation. The Enzymes," 3rd ed., vol. XVII. (Chaps. 1, 3, 8–12). Academic Press, New York.

Madsen, N. B. (1990). Glycogen phosphorylase and glycogen synthetase. *In* "A Study of Enzymes," vol. II. (S. A. Kuby, ed.). CRC Press, Boca Raton, Florida.

Newsholme, E. A., and Leech, A. R. (1983). "Biochemistry for the Medical Sciences." (Chaps. 5, 7, 9, 11, 16). John Wiley, New York.

Sprang, S. R., Acharya, K. R., Goldsmith, E. J., et al. (1988). Structural changes in glycogen phosphorylase induced by phosphorylation. *Nature* **331,** 215–221.

Stryer, L. (1988). "Biochemistry," 3rd ed. (Chap. 19). W. H. Freeman, New York.

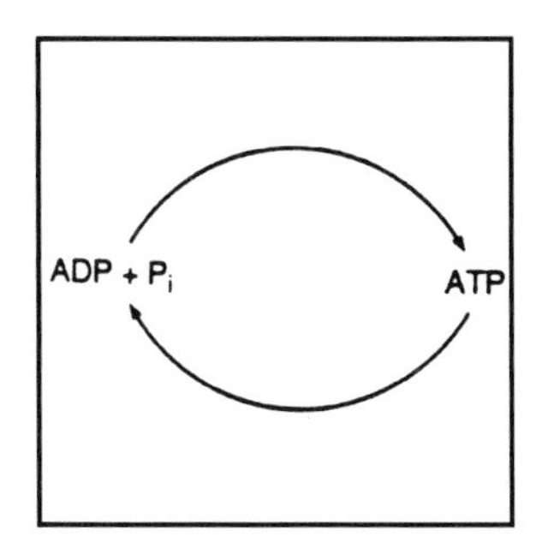

Glycolysis

EFRAIM RACKER, *Division of Biological Sciences, Cornell University–Ithaca*

Glossary

Allostery A control mechanism in enzyme activity (or other protein functions) by occupation of an allosteric effector at a site in the protein other than the catalytic site

Fermentation Any anaerobic process that breaks down organic compounds, resulting in the generation of ATP

High-energy phosphate compounds Conventional term for phosphorylated compounds (e.g., acetylphosphate or phosphoenolpyruvate) that have a large heat of hydrolysis, ΔH (8–12 kcal/mol), and are thermodynamically suited for a transfer of the phosphoryl group to ADP by phosphotransferases

Homeostasis Mechanism in which the steady-state concentration of a biological compound (e.g., blood sugar) is kept constant, usually with the expenditure of energy

Nature's economy is clever, simple and beautiful. ATP is made only when it is needed. Man's economy is crazy. Products that are not needed, are advertised on TV and given to me as Christmas presents (e.g., electric backscratchers).

Oxidative or photophosphorylation Process catalyzed by mitochondria, bacteria, or chloroplasts in which the transport of electrons either from food substrates or from H_2O (using light energy) generates a proton-motive force (proton gradient plus membrane potential) which is used to synthesize ATP from ATP and P_i

Pasteur effect An important control mechanism of feedback inhibition in which ATP, the bioenergetic product of glycolysis, inhibits the utilization of sugar

Product inhibition Control mechanism in which the product of an enzyme-catalyzed reaction inhibits the enzyme

Rate-limiting step Reaction in a pathway which acts as a pacemaker (e.g., an enzyme which increases the rate of metabolism when its activity is increased). There can be more than one pacemaker in a pathway.

"Tight coupling" Expression used to describe the compulsory linkage of oxidative processes during glycolysis or during electron transport to the synthesis of ATP; represents the basic economic principle in bioenergetics

GLYCOLYSIS IS A cellular pathway that catalyzes the conversion of glucose to lactic acid. The function of this pathway is to convert the energy stored in glucose into ATP, which is the most universal and important energy currency in living cells. Each molecule of glucose converted to two molecules of lactic acid yields two molecules of ATP. Heating glucose in a test tube in the presence of sodium hydroxide will also yield lactic acid, but the energy is released as heat and cannot be used by cells to synthesize proteins, nucleic acids, and other cellular ingredients. Moreover, glycolysis yields only lactic acid, whereas heat yields a mixture of compounds. Glycolysis is a complex pathway involving about a dozen enzymes and several cofactors. Students of evolution, who propose that glycolysis was the primary mode of energy generation when life on earth emerged, should be reminded of Renee Dubos, who once pointed out that if this were true, glycolysis must have sprung from the forehead of Zeus, like Minerva, fully armed. More likely, gradients of sodium or proton ions were used as primary sources of energy when we first started. In the course of evolution, however, glycolysis has emerged as the major pathway in the generation of ATP. In the absence of air, glycolysis is not a very efficient process, but in the presence of air, pyruvate, one of its end products, instead of being reduced to lactate, is burned into CO_2 and water via the Krebs cycle by the very efficient pathway of oxidative phosphorylation in mitochondria, specialized organelles present in all eukaryotic cells. By generating a proton gradient (i.e., a difference in hydrogen ion concentration across a membrane) during oxidation, glycolysis allows the production of 32 molecules of ATP per molecule of glucose metabolized.

Glucose itself is an intermediate in the flow of energy in the living world. The energy of the sun, transmitted as light, is utilized by plants and other photosynthetic cells in other organelles, the chloroplasts, or in bacteria to cleave water and to generate via an electron transport chain, similar to that found in mitochondria, a proton motive force. This proton-motive force, consisting of a proton gradient and an electrical potential across the membrane, is utilized by an ATP-driven proton pump acting in reverse to synthesize ATP from ADP and inorganic phosphate (P_i). This ATP then catalyzes energy-requiring processes in the cell. Thus, there are in living cells three major pathways for the synthesis of ATP: photophosphorylation, oxidative phosphorylation, and glycolysis or fermentation. Fermentation is a broader term for the breakdown of a variety of sugars and other organic molecules, leading to the production of ATP without the direct participation of oxygen. The fragments generated by fermentation contain less energy than the initial (e.g., carbohydrate) molecule, and this energy difference is utilized to synthesize ATP from ADP and P_i. The products of fermentation vary among living cells, of which yeast is perhaps the most popular, fermenting glucose or sucrose to ethyl alcohol and living happily ever after.

Space limitations do not permit me to dwell on the history of the elucidation of the glycolytic pathway. It is a fascinating account, though, of the glories and failings of the human mind, from which students can derive many lessons. It was believed in the early decades of this century that methylglyoxal was a key intermediate in the formation of lactic acid from glucose. This widely accepted hypothesis which served to block further progress was eliminated by showing that the removal of glutathione, a necessary cofactor for the conversion of methylglyoxal to lactic acid, by dialysis from a crude yeast extract, failed to affect glycolysis. The elucidation of the individual steps of the glycolytic pathway served as a role model for subsequent analyses of multistep pathways, including that of the oxidation of pyruvate in mitochondria (formulated by Sir Hans Krebs). Mitochondrial oxidative phosphorylation is quantitatively the most important mechanism in animal cells for ATP synthesis. By oxidizing the carbons of pyruvate stepwise one by one to water and CO_2 via the many intermediates of the Krebs cycle, nature invented a machine more efficient than most made by humans.

Before discussing the pathway of glycolysis, with its intricate catalysts, described in minute detail in textbooks of biochemistry, I would like to discuss the Pasteur effect. This is a very important feature of glycolysis regulation which is studiously deleted from modern textbooks. In 1861, Pasteur published the historical discovery that, per gram of glucose utilized, more yeast is formed in the presence of air than in its absence. These experiments demonstrated that, as in New York City, "life without air" is possible, but expensive. It was the first demonstration that aerobic metabolism (via oxidative phosphorylation) is more efficient than anaerobic glycolysis. Pasteur also observed that, in the absence of air, glucose was more rapidly used per gram of yeast, than in air, thereby compensating for the inefficient fermentation process. The inhabition of glucose utilization by oxygen was called the "Pasteur effect," and subsequently was found to be a major regulatory mechanism that serves for the preservation of food throughout the living world. I return to the mechanism of the Pasteur effect at the end of this article. A brief recapitulation of the Pasteur effect is shown below.

1. Glucose is used up more slowly aerobically than anaerobically.
2. However, yeast grows faster aerobically than anaerobically.
3. From this, we must conclude that the aerobic cell converts glucose into biological energy (ATP) more efficiently than does the anaerobic cell.
4. Cells must therefore possess a mechanism by which glucose utilization is suppressed under aerobic conditions.

I. Glycolytic Pathway: An Introduction

Some bacteria, such as the pneumococcus that causes pneumonia in humans, use glycolysis as their only source of energy. Colin MacLeod, an outstanding American biologist, who participated in the revolutionary discovery of nucleic acids as the transmitters of genetic information in pneumococci, used to marvel that his beloved pneumococcus utilized glucose by the same enzymatic pathway as the brain of Plato. But instead of oxidizing the end product of glycolysis, the less inspired fermenting bacteria must get rid of the waste products by devising a specific transporter that excretes lactate together with a proton. They must also dissipate the heat generated by this inefficient mechanism. Most bacteria are better off, though, because they also have oxidative or photosynthetic pathways. Some have additional mechanisms of ATP formation, for example, by the phosphoroclastic cleavage (in which P_i cleaves a carbon–carbon bond) of pyruvate or of fructose-6-phosphate yielding acetylphosphate, which is used as the donor in the formation of ATP from ADP. [*See* ADENOSINE TRIPHOSPHATE (ATP).]

For didactic purposes, I divide glycolysis into five stages, discussed in the following sections.

II. Stage 1: The Overture

We see from Table I that the first stage involves the expenditure of two ATP molecules, which transform glucose into fructose 1,6-bisphosphate. It may seem a funny way to conduct the business of making ATP by first spending it. Quite a bit of ATP energy is wasted in heat during these two steps. The hydrolysis of the sugar phosphate esters ("low-energy phosphates") liberate much less energy than does the hydrolysis of ATP (a "high-energy phosphate compound"). In other words, the "energy market value" of simple phosphate esters is much lower than that of the phosphate anhydride in ATP. Why have such wasteful steps survived during the evolution from the pneumococcus to Plato? What are the fringe benefits derived from these steps to justify the waste?

As it turns out, these ATP expenditures are good investments and represent good economy. First, as we shall see, the investment loss is not as bad as it may seem, because the phosphorylation steps induce changes in the sugar molecules that permit later, in a single pathway, the recovery of ATP by clever oxidation steps.

A second major benefit to the cells is that by phosphorylation the sugar becomes a carrier of the negatively charged phosphate groups. Bernard Davis, a well-known Harvard biochemist, once wrote an article entitled "On the Importance of Being Ionized." Like Oscar Wilde, Davis is a man of simple tastes; he is always satisfied with the best. And for a cell the best is to make sure that the food that is injested is not allowed to leave. Charged ions do not readily traverse the phospholipid bilayers of the plasma membrane without a passport—a specifically designed transporter, such as the previously mentioned proton–lactate channel or the phosphate transporter. In contrast, uncharged molecules such as glycerol or ethyl alcohol are readily lost from the cells if they are not rapidly metabolized. This was taken advantage of during World War I by diverting yeast to make glycerol from sugar and by using the excreted glycerol for the production of explosives.

The first step in stage 1 is the conversion of glucose to glucose-6-phosphate by an enzyme called hexokinase. The discovery of this enzyme by Otto Meyerhof is, I believe, the first reconstitution experiment in history. Meyerhof added to an extract from muscle that was capable of degrading glycogen, but not glucose, to lactate, a protein fraction from yeast, incapable of fermentation. Together, the two preparations glycolyzed vigorously. Thus, the muscle extract, which does not need hexokinase for generating glucose-6-phosphate from glycogen, served as an assay in the purification hexokinase present in the yeast fraction. Where there is an assay and a will, there is a way to purification. Years later, simpler assays were devised and crystalline hexokinase was isolated.

The second step in stage 1 of glycolysis is the conversion of glucose-6-phosphate to fructose-6-phosphate by an enzyme called glucose-phosphate isomerase. The third step is the phosphorylation of fructose-6-phosphate to fructose 1,6-bisphosphate

TABLE I Flow of Glycolysis[a]

Stage 1: From Glucose to Fructose 1,6-Bisphosphate[b]

Glucose + ATP ⇌ (Hexokinase) Glucose-6-phosphate + ADP + H^+

Glucose-6-phosphate ⇌ (Glucosephosphate isomerase) Fructose-6-phosphate

Fructose-6-phosphate + ATP → (Phosphofructokinase) Fructose 1,6-bisphosphate + ADP + H^+

Stage 2: From Fructose 1,6-Bisphosphate to Glyceraldehyde-3-phosphate[c]

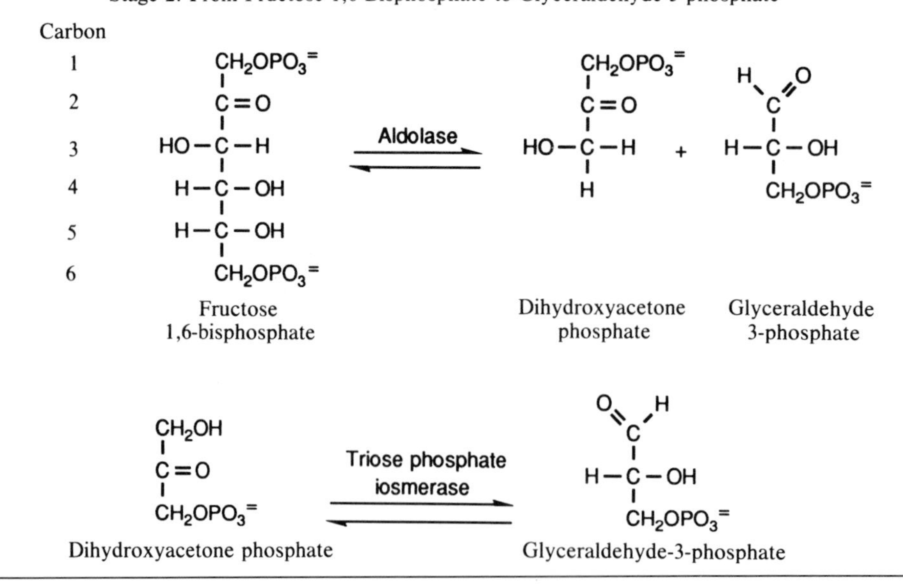

TABLE I *(continued)*

Stage 3: From Glyceraldehyde-3-phosphate to Phosphoglycerate[d]

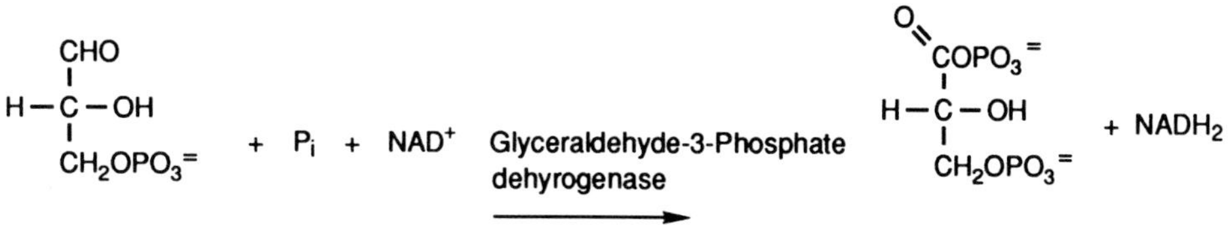

$$CHO{-}CH(OH){-}CH_2OPO_3^{=} + P_i + NAD^+ \xrightarrow{\text{Glyceraldehyde-3-Phosphate dehyrogenase}} O{=}C(OPO_3^{=}){-}CH(OH){-}CH_2OPO_3^{=} + NADH_2$$

Glyceraldehyde-3-phosphate → Diphosphoglycerate

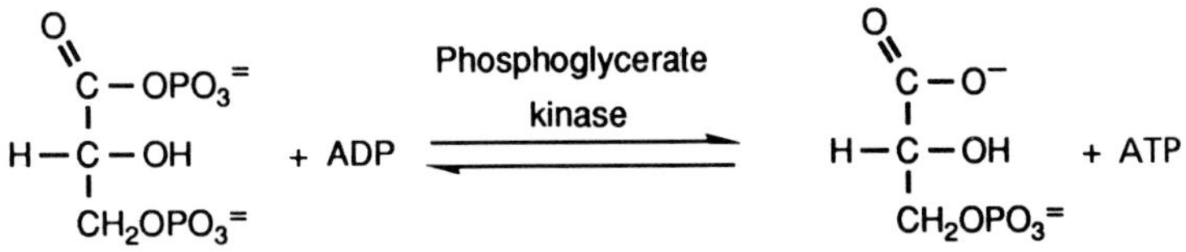

$$O{=}C(OPO_3^{=}){-}CH(OH){-}CH_2OPO_3^{=} + ADP \underset{}{\overset{\text{Phosphoglycerate kinase}}{\rightleftharpoons}} O{=}C(O^-){-}CH(OH){-}CH_2OPO_3^{=} + ATP$$

1,3-Diphosphoglycerate ⇌ 3-Phosphoglycerate

Stage 4: From Phosphoglycerate to Pyruvate[e]

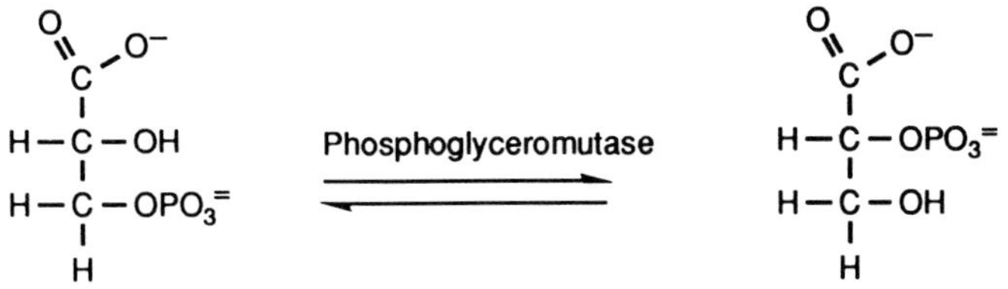

$$O{=}C(O^-){-}CH(OH){-}CH(H){-}OPO_3^{=} \overset{\text{Phosphoglyceromutase}}{\rightleftharpoons} O{=}C(O^-){-}CH(OPO_3^{=}){-}CH(H){-}OH$$

3-Phosphoglycerate ⇌ 2-Phosphoglycerate

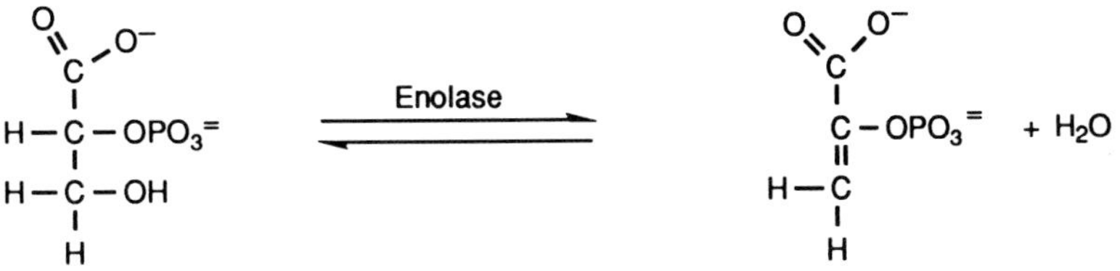

$$O{=}C(O^-){-}CH(OPO_3^{=}){-}CH(H){-}OH \overset{\text{Enolase}}{\rightleftharpoons} O{=}C(O^-){-}C(OPO_3^{=}){=}CH_2 + H_2O$$

2-Phosphoglycerate ⇌ Phosphoenolpyruvate

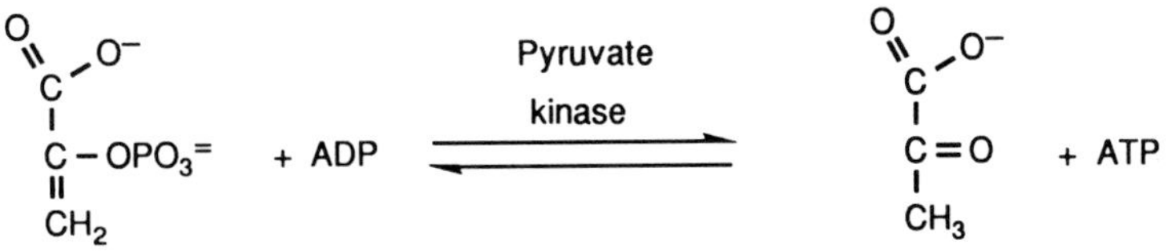

$$O{=}C(O^-){-}C(OPO_3^{=}){=}CH_2 + ADP \overset{\text{Pyruvate kinase}}{\rightleftharpoons} O{=}C(O^-){-}C{=}O{-}CH_3 + ATP$$

Phosphoenolpyruvate ⇌ Pyruvate

Stage 5: Reduction of Pyruvate to Lactate[f]

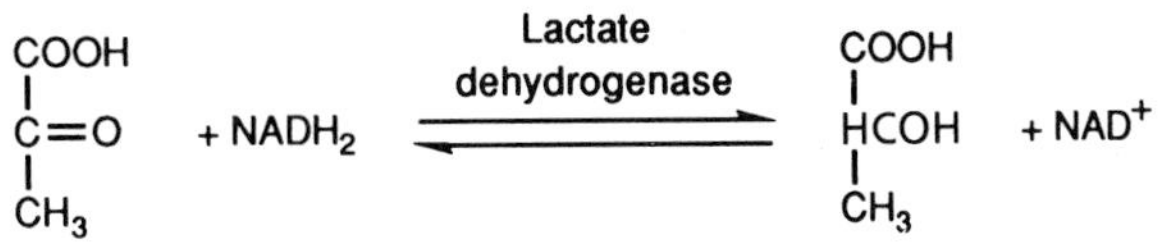

$$COOH{-}C{=}O{-}CH_3 + NADH_2 \overset{\text{Lactate dehydrogenase}}{\rightleftharpoons} COOH{-}HCOH{-}CH_3 + NAD^+$$

---

[a] The sugar structues are presented in the ring form as well as in the open chain form. In nature, we find a mixture of both forms.

[b] Stage 1. The overture: Conversion of glucose to fructose 1,6-bisphosphate.

[c] Stage 2. Cleavage of the sugar molecule: Cleavage of fructose 1,6-diphosphate, which contains six carbon atoms, to dihydroxyacetone phosphate and glyceraldehyde-3-phosphate, each having three carbon atoms. This step is followed by the conversion of dihydroxyacetone phosphate to glyceraldehyde-3-phosphate.

[d] Stage 3. Energy harvest I: Oxidation of glyceraldehyde-3-phosphate to 3-phosphoglycerate and formation of the first ATP.

[e] Stage 4. Energy harvest II: Conversion of 3-phosphoglycerate to pyruvate and formation of the second ATP.

[f] Stage 5. Epilogue: Reduction of pyruvate to lactate or to other fermentation products in microorganisms such as ethyl alcohol.

The enzyme that achieves the phosphorylation of fructose-6-phosphate into fructose 1,6-bisphosphate, with the expenditure of yet another ATP, is called 6-phosphofructokinase. This protein contains two special regions: One acts as a catalyst, that is, it brings the two substrates—fructose-6-phosphate and ATP—close together and allows them to interreact to yield ADP and fructose 1,6-bisphosphate. The second site binds a variety of cellular molecules called effectors, such as ATP, ADP, fructose 2,6-bisphosphate, phosphocreatine, citrate, or inorganic phosphate. As a result of this binding, the shape of the protein molecule changes and the activity of the catalytic site may be either decreased or increased. In other words, the second (noncatalytic) site is a controlling or allosteric site. Phosphofructokinase is one of the most important allosteric enzymes in glycolysis, because an inhibition of its action diminishes the utilization of sugar and thereby allows the preservation of energy reserves. By lowering the activity of 6-phosphofructokinase, fructose-6-phosphate and glucose-6-phosphate accumulate. The latter compound, in turn, inhibits the activity of hexokinase, a phenomenon called product inhibition, another simple widely used mechanism in metabolic regulations. Thus, the cells break down only the amount of sugar that they need for making the amount of ATP energy required for living. The most important compound acting as a negative allosteric effector of phosphofructokinase is ATP. When too much ATP is generated and not enough P_i and ADP are available, phosphofructokinase is inhibited. The enzyme also responds to positive allosteric effectors, compounds that turn the enzyme on again. P_i, ADP, AMP, and cAMP are such positive effectors. When the cell needs a lot of energy and breaks down ATP, it generates ADP and P_i, which in turn activate the machinery of glycolysis that replenishes the ATP level.

III. Stage 2: Cleavage of Fructose 1,6-Bisphosphate to Dihydroxyacetone Phosphate and Glyceraldehyde-3-phosphate

The formation of fructose-1,6-bisphosphate in the previous stage represents the clever synthesis of an almost symmetrical molecule which is cleaved by the enzyme aldolase into two interconvertible triosephosphates. I call this process clever because it allows the cell to deal eventually with a single substrate, glyceraldehyde-3-phosphate. If the second expenditure of ATP were not invested, the cleavage of the hexose monophosphate would yield one triose and one triosephosphate and would require two separate sets of enzymes to deal with the two different substrates.

Note that carbons 3 and 4 of the fructose 1,6-bisphosphate molecule become carbon 1 of glyceraldehyde-3-phosphate, carbons 1 and 6 become carbon 3, and carbons 2 and 5 become carbon 2. This feature is important for investigations of the fate of the individual carbons by using glucose labeled at a single carbon site with radioactive carbon 14.

IV. Energy Harvest I—Stage 3: Oxidation of Glyceraldehyde-3-phosphate to 3-Phosphoglycerate and Formation of ATP

In organic chemistry, the oxidation of an aldehyde to an acid is usually obtained by the intervention of some oxidizing agent, such as bromine. The energy resulting from the oxidation step, however, is lost as heat.

The enzyme glyceraldehyde-3-phosphate dehydrogenase is more ingenious. The simplest formulation of its mode of action is as follows (see Table II): The enzyme (E) which contains a sulfhydryl group (SH) interacts with the aldehyde group of glyceraldehyde-3-phosphate, yielding a product shown in Table II that is oxidized to an enzyme-bound thiolester. In the oxidation step (step 2), NAD^+ serves as the hydrogen acceptor to form $NADH_2$ ($NADH + H^+$). This is the key reaction of glycolysis. The energy of oxidation, instead of dissipating as heat, is preserved by formation of the energy-rich thiolester substrate–enzyme intermediate. A thiolester is a "high-energy" compound because it releases about 10 kcal of heat when hydrolyzed by water. The cell preserves the energy of oxidation of the aldehyde to the acid by allowing P_i to enter, instead of water (step 3). This process, called phosphorolysis (instead of hydrolysis), results in the formation of a new high-energy intermediate, 1,3-bisphosphoglycerate. With the help of an enzyme called phosphoglycerate kinase, the phosphate attached via an oxygen at carbon 1 is transferred to

TABLE II Oxidation of Glyceraldehyde-3-phosphate to 1,3-Diphosphoglycerate[a]

$R = -CHOH-CH_2OPO_3^{=}$

Step 1.	$E-SH + HC(=O)-R$ Enzyme + glyceraldehyde-3-phosphate	$\longrightarrow$	$E-S-CH(OH)-R$ Enzyme–substrate product
Step 2.[b]	$E-S-CH(OH)-R + NAD^+$ Enzyme–substrate product	$\longrightarrow$	$E-S-C(=O)-R + NADH_2$ Enzyme thiolester
Step 3.[c]	$E-S-C(=O)-R + P_i$ Enzyme thiolester	$\longrightarrow$	$E-SH + R-C(=O)OPO_3^{=}$ Enzyme + 1,3-diphosphoglycerate
Step 4.[d]	$R-C(=O)OPO_3^{=} + ADP$ 1,3-diphosphoglycerate	$\longrightarrow$	$R-C(=O)OH + ATP$ 3-phosphoglycerate

[a] E, Enzyme; SH, sulfhydryl group.
[b] Oxidation.
[c] Phosphorolysis.
[d] Phosphotransfer.

ADP to form ATP (phosphotransfer). The steps of oxidation, phosphorolysis, and phosphotransfer catalyzed by glyceraldehyde-3-phosphate dehydrogenase and phosphoglycerate kinase (see Table I, stage 3), are "tightly coupled." Only when the phosphorylated product is removed by phosphoglycerate kinase can another molecule of aldehyde be oxidized. This control mechanism makes certain that virtually no oxidation takes place, unless both P_i and ADP are present to form ATP. We see here how simply and ingeniously nature regulates its energy budget and how the Pasteur effect is achieved. When there is an excess of ATP, but little ADP and P_i, there is no oxidation and the glycolytic machinery comes to a standstill. When ATP is utilized (e.g., during muscular contraction), for biosynthesis or for transport processes, P_i and ADP are liberated and gear the glycolytic machinery that regenerates ATP into motion.

V. Energy Harvest II—Stage 4: Conversion of 3-Phosphoglycerate to Pyruvate and Formation of ATP

The product of the oxidation and ATP-forming steps is phosphoglycerate. In a series of consecutive steps listed in Table I, the phosphate is moved from carbon 3 in 3-phosphoglycerate to carbon 2 in phosphenolpyruvate, and H_2O is pushed out of the molecule. These interconversions catalyzed by the enzymes phosphoglyceromutase and enolase, yield phosphoenolpyruvate, another high-energy compound. This is, in turn, used by the enzyme pyruvate kinase to transfer phosphate to ADP to form ATP, with formation of pyruvate. The principle of the reaction leading to ATP formation is basically the same as in stage 3, in which an oxidation of an aldehyde and a reduction of NAD^+ takes place. In stage 4, the oxidoreduction occurs within the three-carbon molecule. Carbon 2 of 2-phosphoglycerate becomes more oxidized by losing hydrogen, while carbon 3 is reduced to a methyl group. The pyruvate kinase reaction is another key point in glycolytic control, because it is dependent on the availability of ADP. It is allosterically inhibited by ATP and stimulated by fructose 1,6-bisphosphate. In some cells, the enzyme is phosphorylated and regulated by protein kinases that inhibit enzyme activity. Names for enzymes such as pyruvate kinase and phosphoglycerate kinase are confusing because, during glycolysis, the reactions they catalyze occur in the forward reaction, that is, from 1,3-diphosphoglycerate to phosphoglycerate with transfer of phosphate to ADP, whereas most kinases (e.g., hexokinase) transfer the phosphate from ATP to the substrate. But the kinase term is simpler and may serve as a reminder that the reactions are reversible and in fact can operate in reverse when glycogen is formed from pyruvate.

VI. Epilogue—Stage 5: Reduction of Pyruvate to Lactate: Formation of Other Fermentation Products in Microorganisms

Pyruvate, the product of stage 4, is reduced in the presence of lactate dehydrogenase by $NADH_2$, which was generated during the oxidation of glyceraldehyde-3-phosphate and now returns to NAD^+ and lactate is formed. Thus, the hydrogen cycle through NAD^+ closes and oxidoreductions are balanced. Two molecules of lactic acid contain the same number of oxygens and hydrogens as one molecule of glucose. The oxidoreductions take place internally, without need for molecular oxygen. In yeast, no lactate is formed. Instead, carbon 1 of pyruvate is released as CO_2, while the remaining hydrogen moves to carbon 2 to give rise to acetaldehyde containing two carbons (CH_3CHO). Acetaldehyde is then reduced by $NADH_2$ to yield ethyl alcohol (CH_3CH_2OH), the end product of alcoholic fermentation, and once more oxidation reduction balance is achieved. The oxidized form of the coenzyme (i.e., NAD^+) is regenerated by forming lactate in one case, alcohol in the other.

When things "get sour" in the winery, for example, when the fermentation vat gets contaminated with certain bacteria, acetaldehyde is oxidized to acetic acid (also known as vinegar). Pasteur cured this wine disease by eliminating the infectious bacteria. Some bacteria make formic acid, a compound also produced by ants. Some microorganisms can produce compounds such as glycerol and acetone, used in the manufacture of explosives. As mentioned earlier, during World War I, Karl Neuberg in Germany found a way to force yeasts to accumulate glycerol (CH_2OH $CHOH$ CH_2OH) instead of ethanol by chemically trapping acetaldehyde and forcing dihydroxyacetone to serve as the hydrogen acceptor for $NADH_2$. Chaim Weizmann in England developed another fermentation process in which acetone (CH_3COCH_3) was produced and was also used during World War I for the production of explosives. As in World War II, scientific discoveries can be used to enhance the viciousness of war. Some bacteria accumulate acetoin (CH_3-$CHOHCOCH_3$), which has been used as a starting material for the manufacture of synthetic rubber. Pharmaceutical companies use bacterial fermentation processes to manufacture drugs when it is either too difficult or too expensive to use chemical synthesis. Just as atomic energy, fermentation can be used for good and evil purposes.

Let us now recapitulate what happens to the energy balance during glycolysis (Table III). For each molecule of glucose, two molecules of lactate are formed. Forming fructose 1,6-bisphosphate from glucose requires expenditure of two ATP molecules. Each of the two triosephosphates, which are made by splitting fructose 1,6-bisphosphate, yield one ATP during the oxidation of glyceraldehyde-3-phosphate in the presence of ADP and P_i. The energy balance is now even; the "wasted" ATP is recovered. In stage 4, from two molecules of phosphoglycerate, two molecules of phosphoenolpyruvate are made, generating two molecules of ATP in the presence of ADP. We have now accumulated two ATPs for each glucose fermented. Because glycolysis is a tightly coupled process, these two extra ATP molecules must be reconverted into ADP and P_i if we want glycolysis to continue. ADP and P_i are cofactors of glycolysis present in a cell in relatively

TABLE III Balance Sheet[a]

Expenditure		Earning
	Glucose	
2 ATP →	↓	
	Fructose-1,6-bisphosphate	
	↓	
	2 Triosephosphate	
	↓ + 2 P_i + 2 ADP →	2 ATP
	2 Phosphoglycerate	
	↓	
	2 Phosphoenolpyruvate	
	↓ + 2 ADP →	2 ATP
	2 Pyruvate	
	↓	
	2 Lactate	

[a] Overall: glucose + 2 P_i + 2 ADP → 2 lactate + 2 ATP + 2 H_2O.

small amounts and must continuously be regenerated (Fig. 1). Various work processes (e.g., transport, muscle contraction, and syntheses) contribute to this "ATPase" activity and thereby activate the glycolytic pathway.

From a thermodynamic point of view, the yield of two ATPs per glucose represents an overall efficiency of about 50%, which is not bad compared with man-made energy transformers. Moreover, many fringe benefits are derived from the glycolytic pathways. It provides, by diversion of intermediates, the glycerol moiety for fats, ribose and deoxyribose for nucleic acids, and some amino acids for protein snythesis. No wonder evolution has conserved this pathway and the human brain shares it with pneumococcus and with many other microorganisms and plants.

Although the basic reactions of the glycolytic pathway are indeed identical in diverse cells, there are important differences in the distribution of the individual enzymes that participate in the pathway. By varying the proportions of the catalysts, nature altered these pathways in such a manner that it appears as if different qualitative mechanisms exist.

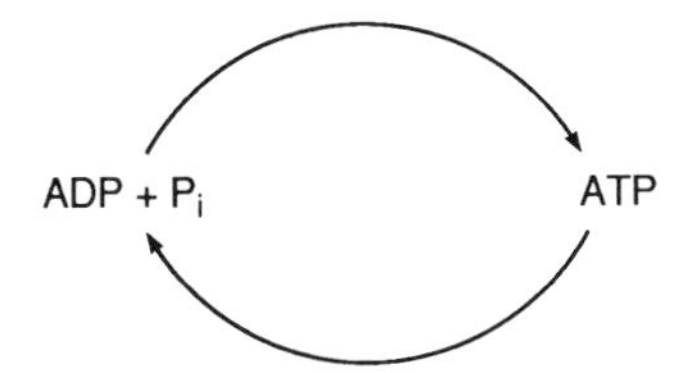

FIGURE 1 The P_i + ADP ↔ ATP cycle in glycolysis and oxidative phosphorylation.

For example, sodium fluoride, which is added to toothpaste to prevent tooth decay, is a poison when too much is ingested. We know the exact location of action of fluoride. By forming a complex with the enzyme enolase and Mg^{2+}, a cofactor required for enzyme activity, it inhibits the conversion of phosphoglycerate into phosphoenolpyruvate. Some bacteria contain so little enolase that the entire glycolytic pathway is controlled by the activity of this enzyme. We say that in this case enolase is one of the "rate-limiting" steps or that it is a "pacemaker" of the pathway. Other cells, for example, some tumors, contain a huge excess of enolase, over 10 times that required. We can add enough fluoride to such cells to poison 90% of enolase activity without interfering with the rate of glycolysis. The same amount of fluoride would wipe out glycolysis in a cell that has enolase as a pacemaker.

This basic principle of varying the relative concentration of the glycolytic enzymes is being used by nature to vary the pattern and the pacemakers of glycolysis even in different tissues of the same organism. For example, the brain of mammals contains a huge amount of hexokinase, which is needed to assure a steady utilization of glucose, the principal nutrient of brain cells. Brain cells work all the time and therefore need a steady supply of energy. On the other hand, muscle cells have different requirements. When muscles work, they need a lot of energy, but they need much less during rest. As a consequence, our muscles store a reservoir of energy in the form of glycogen, a network of interconnected glucose molecules. Muscles therefore contain a large amount of an enzyme which, in the presence of P_i liberated by ATP hydrolysis during muscular contractions, breaks down glycogen to glucose-1-phosphate. The latter is converted into glucose-6-phosphate and thus enters the glycolytic pathway. Hexokinase activity in muscle is quite low, but there is enough to catalyze the slow but continuous conversion of glucose to glycogen during rest periods. [*See* GLYCOGEN; MUSCLE, PHYSIOLOGY AND BIOCHEMISTRY.]

Sometimes the pattern of glycolysis is altered by inserting an additional enzyme which diverts an intermediate in another direction. For example, the liver contains an enzyme called glucose-6-phosphatase which diverts glucose-6-phosphate by hydrolyzing it to glucose and inorganic phosphate, with liberation of heat. This seems an awful waste of energy, yet the overall metabolic story makes good sense. One of the functions of the liver is the homeostasis of the glucose level of blood. This is just a fancy way of saying that the organ acts as a regulator, keeping the blood sugar level constant. For this purpose, the liver stores glycogen and has a very active glycogen-degrading enzyme, just like muscle; but, in addition, it has glucose-6-phosphatase, which is not present in muscle. When hormones such as adrenaline, which is secreted by the adrenal glands, reaches the liver cell, the glycogen-degrading enzymes are activated by a remarkable complex series of events involving a secondary messenger called cAMP. As a result, glucose-1-phosphate is formed, converted to glucose-6-phosphate, and finally glucose is released into the blood.

VII. Oxidative Phosphorylation

Evolution did not stop at the glycolytic pathway; it created oxidative phosphorylation, an even more complex and much more efficient machinery for energy generation. The respiratory chain of the mitochondria oxidizes the products such as pyruvate, fatty acids, and amino acids that arise from metabolism of the carbohydrates, fats, and proteins in our food. These oxidation processes generate the electrons that are transported via the respiratory chain of mitochondria or of bacteria and are tightly coupled to phosphorylation and to the generation of ATP from ADP and P_i. As in the case of glycolysis, without these two cofactors oxidations stop, a phenomenon called respiratory control. Since both glycolysis and oxidative phosphorylation are tightly coupled and require ADP and P_i, there is a fierce competition for these rate-limiting cofactors. When respiration wins out, glycolysis is inhibited and we observe the Pasteur effect; when glycolysis is overwhelming, as in some cancer cells, we observe the Crabtree effect, an inhibition of respiration. But the Pasteur effect is more complex than inhibition of the oxidation of glyceraldehyde-3-phosphate, because of ADP–P_i limitation. This limits the oxidation of glyceraldehyde-3-phosphate and the formation of lactate, but how is the control of glucose utilization achieved? There is a remarkable set of allosteric control mechanisms superimposed at the sites of hexokinase and phosphofructokinase action. The latter is probably among the most regulated enzymes known in biochemistry. The checks and balances are more complex than the controls exerted by Congress and the judiciary branch on the executive branch of our government. The first set of allosteric controls was mentioned earlier: ATP inhibits; P_i, ADP, and AMP stimulate. Other negative controls include phosphocreatine, citrate, and fructose 1,6-bisphosphate. A more recently discovered allosteric effector of phosphofructokinase is fructose 2,6-bisphosphate, which is a potent activator of the enzyme that has reawakened the interest of biochemists in "old hat" glycolysis. The control of the steady-state concentration of fructose 2,6-bisphosphate is in itself staggering, involving several enzymes and cAMP. It is likely that in some cancer cells fructose 2,6-bisphosphate plays a role in the high rate of aerobic glycolysis. The inhibition of phosphofructokinases triggered by high ATP and low P_i and ADP results in the accumulation of fructose-6-phosphate and glucose-6-phosphate. The latter is a powerful feedback inhibitor of hexokinase, thereby affecting the utilization of glucose. This inhibition of hexokinase is perhaps also responsible for a diminished influx of glucose via the glucose transporter. This speculation is based on somewhat thin evidence provided by experiments on a phenomenon called catabolite repression, glucose utilization via glycolysis represses other catabolic pathways (e.g., the utilization of galactose). In yeast, this repression is shown to involve one form of hexokinase.

Without belittling the importance of these control mechanisms operating at the hexose levels and several other controls catalyzed by protein kinases and phosphatases (see the Bibliography), the bottom line of the Pasteur effect is the competition of glycolysis and oxidative phosphorylation for ADP and P_i. The classical demonstration over five decades ago that dinitrophenol, a proton ionophore and uncoupler of oxidative phosphorylation, releases the inhibition imposed on glycolysis by respiration, implicates ADP and P_i as the key regulators in the Pasteur effect. Dinitrophenol not only uncouples respiration from phosphorylation, but also stimulates the mitochondrial ATPase by releasing it from its inhibition by electron transport, to which it is normally tightly coupled.

Glycolysis is altered in diseases. There is a pathology of glycolysis in diseases of liver, muscle, or red blood cells, in which the utilization of glucose or glycogen is impaired because of a hereditary alteration of an enzyme. In most cancer cells, aerobic glycolysis is greatly enhanced, but the mechanism is still uncertain. Now, with the tools of modern molecular biology, it can be shown that transfection of a normal cell with a single oncogene (e.g., *ras*) increases the rate of glycolysis four- to fivefold. Transfections with single oncogenes should allow us to focus on a more specific field of inquiry, namely, the pathway of signal transduction induced by an oncogene. The study of such pathological situations may enhance our knowledge of the normal events in the areas of energy control and expenditures.

We know a great deal about the energy budget of glycolysis and oxidative phosphorylation, but our knowledge of the expenditure budget is pathetic. There are two major ways of generating ATP in animal cells, but thousands of ways of spending it. These expenditures, resulting in the formation of ADP and P_i, represent the sum of the "ATPase" reactions that take place in the cell and that keep glycolysis and respiration going. As noted earlier, these processes are required not only for ATP generation, but for the production of intermediates for many synthetic pathways operating in living cells. From a bioenergetic point of view, our ignorance of expenditures is tragic, but not serious, because we are rich. Our potential of ATP generation by far exceeds our needs as long as nutrients are supplied and the bioenergetic machineries are not damaged. We need to know more about our expenditures in order to understand the various ATPase activities that contribute to the rate of glycolysis and are likely to be involved in the high aerobic glycolysis characteristic for most cancer cells.

Bibliography

Pilkis, S. J., Raafad El-Maghrabi, M., and Claus, T. H. (1988). *Annu. Rev. Biochem.* **57,** 755–783.

Racker, E. (1976). "A New Look at Mechanisms in Bioenergetics." Academic Press, New York.

Racker, E. (1985). "Reconstitutions of Transporters, Receptors, and Pathological States." Academic Press, Orlando, Florida.

Stryer, L. (1988). "Biochemistry," 3rd ed., Ch. 15. Freeman, New York.

Van Schaftingen, E. (1987). *Adv. Enzymol.* **59,** 315–395.